Review of
Medical Microbiology & Immunology

A Guide to Clinical Infectious Diseases

Seventeenth Edition

Senior Author
Warren Levinson, MD, PhD

Authors
Peter Chin-Hong, MD
Elizabeth A. Joyce, PhD
Jesse Nussbaum, MD
Brian Schwartz, MD

New York   Chicago   San Francisco   Athens   London   Madrid   Mexico City
Milan   New Delhi   Singapore   Sydney   Toronto

This book was set in Minion Pro by KnowledgeWorks Global Ltd.
The editors were Michael Weitz and Christina M. Thomas.
The production supervisor was Richard Ruzycka.
Project management was provided by Deepanshu Manral, KnowledgeWorks Global Ltd.
Cover photo: *Candida auris*, an emerging drug-resistant fungus. 3D illustration by Kateryna Kon.
Photo credit: Shutterstock/Kateryna Kon.
This book is printed on acid-free paper.

Faculty who adopt the textbook are entitled to Power Point slides of **All Images** from the text, simply email: customersuccess@mheducation.com

Contents

PART I

BASIC BACTERIOLOGY   1

PART II

CLINICAL BACTERIOLOGY   101

PART III

BASIC VIROLOGY   217

PART IV

CLINICAL VIROLOGY   283

Authors

Senior Author

Warren Levinson, MD, PhD
Professor of Microbiology
Department of Microbiology & Immunology
University of California, San Francisco
San Francisco, California

Authors

Peter Chin-Hong, MD
Professor of Medicine
Department of Medicine
Division of Infectious Diseases
University of California, San Francisco
San Francisco, California

Elizabeth A. Joyce, PhD
Professor of Microbiology
Department of Microbiology & Immunology
University of California, San Francisco
San Francisco, California

Jesse Nussbaum, MD
Assistant Adjunct Professor of Medicine
Department of Medicine
Division of Infectious Diseases
University of California, San Francisco
San Francisco, California

Brian Schwartz, MD
Professor of Medicine
Department of Medicine
Division of Infectious Diseases
University of California, San Francisco
San Francisco, California

Preface

This book is a concise review of the medically important aspects of microbiology and immunology. It covers both the basic and clinical aspects of bacteriology, virology, mycology, parasitology, and immunology. It also discusses important infectious diseases, using an organ system approach.

Its two major aims are (1) to assist those who are preparing for the USMLE (National Boards) and (2) to provide students who are currently taking medical microbiology courses with a brief and up-to-date source of information. The goal is to provide the reader with an accurate source of clinically relevant information at a level appropriate for those beginning their medical education.

This new edition presents current, medically important information in the rapidly changing fields of microbiology and immunology. It contains many color micrographs of stained microorganisms as well as images of important laboratory tests. It also includes many images of clinical lesions and highlights current information on antimicrobial drugs and vaccines.

These aims are achieved by using several different formats, which should make the book useful to students with varying study objectives and learning styles:

1. A narrative text for complete information.
2. A separate section containing summaries of important microorganisms for rapid review of the high-yield essentials.
3. Sample questions in the USMLE (National Board) style, with answers provided after each group of questions.
4. A USMLE (National Board) practice examination consisting of 80 microbiology and immunology questions. These questions are written in a clinical case format and simulate the computer-based examination. Answers are provided at the end of each block of 40 questions.
5. Self-assessment questions at the end of the chapters so you can evaluate whether the important information has been mastered. Answers are provided.
6. Clinical case vignettes to provide both clinical information and practice for the USMLE.
7. A section titled "Pearls for the USMLE" describing important epidemiologic information helpful in answering questions on the USMLE.

The following features are included to promote a successful learning experience for students using this book:

1. The information is presented succinctly, with stress on making it clear, interesting, and up to date.
2. There is strong emphasis in the text on the clinical application of microbiology and immunology to infectious diseases.

3. In the clinical bacteriology and virology sections, the organisms are separated into major and minor pathogens. This allows the student to focus on the most important clinically relevant microorganisms.
4. Key information is summarized in useful review tables. Important concepts are illustrated by figures using color.
5. Important facts called "Pearls" are listed at the end of each basic science chapter.
6. Self-assessment questions with answers are included at the end of the chapters.
7. The 654 USMLE (National Board) practice questions cover the important aspects of each of the subdisciplines on the USMLE: Bacteriology, Virology, Mycology, Parasitology, and Immunology. A separate section containing *extended* matching questions is included. In view of the emphasis placed on clinical relevance in the USMLE, another section provides questions set in a clinical case context.
8. Brief summaries of medically important microorganisms are presented together in a separate section to facilitate rapid access to the information and to encourage comparison of one organism with another.
9. Fifty clinical cases are presented as unknowns for the reader to analyze in a brief, problem-solving format. These cases illustrate the importance of basic science information in clinical diagnosis.
10. Color images depicting clinically important findings, such as infectious disease lesions, Gram stains of bacteria, electron micrographs of viruses, and microscopic images of fungi, protozoa, and worms, are included in the text.
11. There are 11 chapters on infectious diseases from an organ system perspective. They are written concisely and are appropriate for a medical student's introduction to this subject. These chapters include Bone & Joint Infections, Cardiac Infections, Central Nervous System Infections, Gastrointestinal Tract Infections, Pelvic Infections, Upper Respiratory Tract Infections, Lower Respiratory Tract Infections, Infections of the Skin & Skin Structures, Urinary Tract Infections, Sepsis & Septic Shock, and Eye Infections.

After teaching both medical microbiology and clinical infectious disease for many years, I believe that students appreciate a book that presents the essential information in a readable, interesting, and varied format. I hope you find that this book meets those criteria.

Warren Levinson, MD, PhD
San Francisco, California
January 2022

Acknowledgments

In this 17th edition, the senior author, Warren Levinson, would like to express great appreciation for the ongoing valuable, informative writings of the four coauthors: Elizabeth A. Joyce, Jesse Nussbaum, Brian S. Schwartz, and Peter Chin-Hong.

Elizabeth A. Joyce, PhD, is a Professor of Microbiology and Immunology in the School of Medicine, University of California, San Francisco. Elizabeth teaches in and directs courses in the microbiology curriculum in the Schools of Medicine, Dentistry, and Pharmacy.

Jesse Nussbaum, MD, is an Assistant Adjunct Professor of Medicine in the Division of Infectious Diseases in the School of Medicine, University of California, San Francisco (UCSF). Jesse is an infectious disease specialist who cares for patients at UCSF HIV/AIDS clinics. In his research, he studies how the innate immune system interacts with T lymphocytes.

Brian S. Schwartz, MD, is a Professor of Clinical Medicine in the School of Medicine, University of California, San Francisco, specializing in infectious diseases. His clinical and research interests include the diagnosis, treatment, and prevention of infections in immunocompromised patients. He and Dr. Peter Chin-Hong share the leadership of the Microbiology course for medical students.

Peter Chin-Hong, MD, is a Professor of Clinical Medicine in the School of Medicine, University of California, San Francisco, specializing in infectious diseases. He directs the immunocompromised host infectious diseases program at UCSF. His research focuses on human papillomavirus, Chagas disease, and other donor-derived infections in transplant recipients. He and Dr. Brian S. Schwartz share the leadership of the Microbiology course for medical students.

The senior author thanks his cousin, Ralph Levinson, MD, for his expert review of Chapter 80 on Eye Infections. Ralph Levinson is a retired Health Sciences Professor of Ophthalmology in the Jules Stein Eye Institute at the University of California, Los Angeles.

The senior author is indebted to the editor of the first five editions, Yvonne Strong; to the editor of the sixth edition, Cara Lyn Coffey; to the editor of the seventh and ninth editions, Jennifer Bernstein; to the editor of the eighth edition, Linda Conheady; to the editor of the tenth and eleventh editions, Sunita Dogra; to the editor of the twelfth edition, Rebecca Kerins; to the editor of the thirteenth and fifteenth editions, Caroline Define; to the editor of the fourteenth edition, Nupur Mehra; to the editor of the sixteenth edition, Dr Anchal Kaushik; and to the editor of the seventeenth edition, Sailina Saini; all of whom ensured that the highest standards of grammar, spelling, and style were met.

The invaluable assistance of the senior author's wife, Barbara, in making this book a reality is also gratefully acknowledged.

The senior author dedicates this book to his father and mother, who instilled a love of scholarship, the joy of teaching, and the value of being organized.

How to Use This Book

1. **CHAPTER CONTENTS:** The main headings in each chapter are listed so the reader can determine, at a glance, the topics discussed in the chapter.

2. **TEXT:** A concise, complete description of medically important information for the professional student. Includes basic and clinical bacteriology (pages 1–216), basic and clinical virology (pages 217–398), mycology (fungi) (pages 399–426), parasitology (pages 427–492), immunology (pages 493-594), and ectoparasites (pages 595–602).

    The text also includes 11 chapters on infectious diseases. These chapters include Bone & Joint Infections (pages 603–607), Cardiac Infections (pages 608–613), Central Nervous System Infections (pages 614–622), Gastrointestinal Tract Infections (pages 623–630), Pelvic Infections (pages 631–638), Upper Respiratory Tract Infections (pages 639–644), Lower Respiratory Tract Infections (pages 645–650), Infections of the Skin & Skin Structures (pages 651–658), Urinary Tract Infections (pages 659–662), Sepsis & Septic Shock (pages 663–666), and Eye Infections (pages 667–674).

3. **SUMMARIES OF ORGANISMS:** A quick review for examinations describing the important characteristics of the organisms (pages 675–714).

4. **SELF-ASSESSMENT QUESTIONS:** USMLE-style questions with answers are included at the end of the chapters.

5. **PEARLS FOR THE USMLE:** Eleven tables containing important clinical and epidemiologic information that will be useful for answering questions on the USMLE (pages 725–732).

6. **USMLE-TYPE QUESTIONS:** 654 practice questions that can be used to review for the USMLE and class examinations (pages 733–774). Of the 654 questions, 60 questions are presented in a clinical-case format starting on page 767.

7. **USMLE PRACTICE EXAM:** Two 40-question practice examinations in USMLE format (pages 775–784).

8. **PEARLS:** Summary points at the end of each basic science chapter.

9. **CLINICAL CASES:** 50 cases describing important infectious diseases with emphasis on diagnostic information (pages 715–724).

PART I BASIC BACTERIOLOGY

C H A P T E R

1

Bacteria Compared With Other Microorganisms

CHAPTER CONTENTS

Microbes That Cause Infectious Diseases

Important Features of Microbes

Properties of Eukaryotes & Bacteria

Terminology

Pearls

Self-Assessment Questions

Practice Questions: USMLE & Course Examinations

MICROBES THAT CAUSE INFECTIOUS DISEASES

The agents of human infectious diseases belong to five major groups of organisms: bacteria, fungi, protozoa, helminths, and viruses. Bacteria belong to the Bacteria domain, whereas fungi (yeasts and molds), protozoa, and helminths (worms) are classified in the Eukarya domain. Bacteria, fungi, and protozoa are unicellular or relatively simple multicellular organisms. In contrast, helminths are complex multicellular organisms. Viruses are noncellular and therefore are quite distinct from the other organisms.

IMPORTANT FEATURES OF MICROBES

Many of the essential characteristics of these organisms are described in Table 1–1.

(1) **Structure.** Cells have a nucleus or nucleoid (see below), which contains DNA; this is surrounded by cytoplasm, where proteins are synthesized and energy is generated. Viruses have an inner core of genetic material (either DNA or RNA) but no cytoplasm, and so they depend on host cells to provide the machinery for protein synthesis and energy generation.

(2) **Method of replication.** Cells replicate either by binary fission or by mitosis, during which one parent cell divides to make two progeny cells while retaining its cellular structure. Bacteria replicate by binary fission, whereas eukaryotic cells replicate by mitosis. In contrast, viruses disassemble, produce many copies of their nucleic acid and protein, and then reassemble into multiple progeny viruses.

Furthermore, viruses must replicate within host cells because, as mentioned previously, they lack protein-synthesizing and energy-generating systems. With the exception of rickettsiae and chlamydiae, which also require living host cells for growth, bacteria can replicate extracellularly.

(3) **Nature of the nucleic acid.** Cells contain both DNA and RNA, whereas viruses contain either DNA or RNA, but not both.

PROPERTIES OF EUKARYOTES & BACTERIA

Eukaryotes (fungi, protozoa, and helminths) can be distinguished from bacteria based on their structure and the complexity of their organization.

1

TABLE 1–1 **Comparison of Medically Important Organisms**

Characteristic	Viruses	Bacteria	Fungi	Parasites
Cell wall	No	Yes	Yes	+/−[1]
Approximate diameter (μm)[2]	0.02–0.2	1–5	3–10 (yeasts)	15–25 (trophozoites)
Nucleic acid	Either DNA or RNA	Both DNA and RNA	Both DNA and RNA	Both DNA and RNA
Type of nucleus	None	No distinct nuclear compartment	Membrane-bound nucleus	Membrane-bound nucleus
Ribosomes	Absent	70S	80S	80S
Mitochondria	Absent	Absent	Present	Present
Nature of outer surface	Protein capsid and lipoprotein envelope	Rigid wall containing peptidoglycan	Rigid wall containing chitin	Flexible membrane
Motility	None	Some	None	Most
Method of replication	Not binary fission	Binary fission	Budding or mitosis[3]	Mitosis[4]

[1]The cyst forms of parasites have cell walls but the trophozoite forms do not.

[2]For comparison, a human red blood cell has a diameter of 7 μm.

[3]Yeasts divide by budding, whereas molds divide by mitosis.

[4]Helminth cells divide by mitosis, but the organism reproduces itself by complex, sexual life cycles.

(1) Fungi, protozoa, and helminths have a **true nucleus** with multiple chromosomes surrounded by a nuclear membrane and use a mitotic apparatus to ensure equal allocation of the chromosomes to progeny cells.

(2) The **nucleoid** of a bacterial cell typically consists of a single circular molecule of DNA and lacks a nuclear membrane and mitotic apparatus.

The characteristics of prokaryotic bacterial cells and eukaryotic human cells are compared in Table 1–2.

In addition to the different types of nuclei, the two classes of cells are distinguished by several other characteristics:

(1) Eukaryotic cells contain **organelles**, such as mitochondria and lysosomes, and larger (80S) ribosomes, whereas bacteria contain no organelles and smaller (70S) ribosomes.

(2) Most bacteria have a unique rigid external cell wall that contains **peptidoglycan**, a polymer of amino acids and sugars. Eukaryotes do not contain peptidoglycan. Either they are bound

TABLE 1–2 **Characteristics of Prokaryotic Bacterial Cells and Eukaryotic Human Cells**

Characteristic	Prokaryotic Bacterial Cells	Eukaryotic Human Cells
DNA within a nuclear membrane	No	Yes
Mitotic division	No	Yes
Chromosome number	Usually 1	More than 1
Membrane-bound organelles, such as mitochondria and lysosomes	No	Yes
Size of ribosome	70S	80S
Cell wall containing peptidoglycan	Yes	No

by a flexible cell membrane, or, in the case of fungi, they have a rigid cell wall with chitin, a homopolymer of N-acetylglucosamine, typically forming the framework.

(3) The eukaryotic cell membrane contains **sterols**, whereas no prokaryote, except the wall-less *Mycoplasma*, has sterols in its membranes.

Motility is another characteristic by which these organisms can be distinguished. Most protozoa and some bacteria are motile, whereas fungi and viruses are nonmotile. The protozoa are a heterogeneous group that possesses three different organs of locomotion: flagella, cilia, and pseudopods. The motile bacteria move only by means of flagella.

TERMINOLOGY

Bacteria, fungi, protozoa, and helminths are named according to the binomial Linnaean system that uses genus and species. For example, regarding the name of the well-known bacteria *Escherichia coli*, *Escherichia* is the genus and *coli* is the species name. Similarly, the name of the yeast *Candida albicans* consists of *Candida* as the genus and *albicans* as the species. Viruses typically have a single name, such as poliovirus, measles virus, or rabies virus. Some viruses have names with two words, such as herpes simplex virus, but those do not represent genus and species.

PEARLS

- The agents of human infectious diseases are **bacteria, fungi (yeasts and molds), protozoa, helminths (worms), and viruses**.
- Bacteria, fungi, protozoa, and helminths are **composed of cells**, whereas viruses are **noncellular**.

- Bacterial cells do not have a membrane-bound nucleus, whereas human, fungal, protozoan, and helminths cells do have a membrane-bound nucleus.
- All cells contain **both DNA and RNA**, whereas viruses contain **either DNA or RNA, but not both**.
- In addition to a flexible membrane, bacterial and fungal cells are surrounded by a **rigid cell wall**, whereas human, protozoan, and helminth cells are bounded only by a **flexible cell membrane**.
- Most bacteria have cell walls that contain **peptidoglycan**, whereas fungal cell walls contain chitin.

SELF-ASSESSMENT QUESTIONS

1. Bacteria and viruses are both capable of growing within a human host and causing disease. Which one of the following statements most accurately describes the features of bacteria and viruses?
   (A) Bacteria have mitochondria allowing them to generate energy, whereas viruses are reliant on the host cell for energy.
   (B) Because viruses lack a nucleolus, they can replicate independently of their hosts, whereas bacteria rely on host proteins to replicate their genomes.
   (C) Bacterial ribosomes are composed of the same RNA and proteins as viral ribosomes.
   (D) Viruses can only replicate within a cell, whereas bacteria can replicate independently of host cells.

2. Bacteria, fungi (yeasts and molds), viruses, and protozoa are important causes of human disease. Which one of the following microbes contains either DNA or RNA but not both?
   (A) Bacteria
   (B) Molds
   (C) Viruses
   (D) Protozoa
   (E) Yeasts

3. Which one of the following contains DNA that is not surrounded by a nuclear membrane?
   (A) Bacteria
   (B) Molds
   (C) Protozoa
   (D) Yeasts

ANSWERS

(1) **(D)**
(2) **(C)**
(3) **(A)**

PRACTICE QUESTIONS: USMLE & COURSE EXAMINATIONS

Questions on the topics discussed in this chapter can be found in the Basic Bacteriology section of Part XIII: USMLE (National Board) Practice Questions starting on page 733. Also see Part XIV: USMLE (National Board) Practice Examination starting on page 775.

Structure of Bacterial Cells

SHAPE & SIZE OF BACTERIA

Bacteria are classified by shape into three basic groups: **cocci (round), bacilli (rods),** and **spirochetes (spiral shaped)** (Figure 2–1). Some bacteria are variable in shape and are said to be **pleomorphic** (heterogeneous shape). The shape of a bacterium is determined by its rigid cell wall. The microscopic appearance of a bacterium is one of the most important criteria used in its identification.

In addition to their characteristic shapes, the arrangement of bacteria is important. For example, certain cocci occur in pairs (**diplococci**), some in chains (**streptococci**), and others in grapelike clusters (**staphylococci**). The arrangement of rods and spirochetes is medically less important.

Bacteria range in size from about 0.2 to 5 μm (Figure 2–2). The smallest bacteria (*Mycoplasma*) are about the same size as the largest viruses (poxviruses) and are the smallest organisms capable of existing outside a host. The longest bacteria are the size of some yeasts and human red blood cells (7 μm).

STRUCTURE OF BACTERIA

The structure of a typical bacterium is illustrated in Figure 2–3, and the important features of each component are presented in Table 2–1.

Cell Wall

The cell wall is the outermost component common to all bacteria (except *Mycoplasma* species, which are bounded by a cell membrane, not a cell wall). Some bacteria have surface features external to the cell wall, such as capsule, flagella, and pili, which are less common components and are discussed next.

The cell wall is located external to the cytoplasmic membrane and is composed of **peptidoglycan** (see page 6). The peptidoglycan provides structural support and maintains the characteristic shape of the cell.

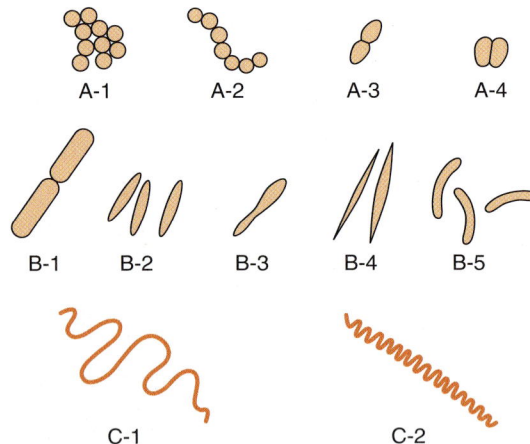

FIGURE 2–1 Bacterial morphology. **A:** Cocci in clusters (e.g., *Staphylococcus;* A-1); in chains (e.g., *Streptococcus;* A-2); in pairs with pointed ends (e.g., *Streptococcus pneumoniae;* A-3); in pairs with kidney bean shape (e.g., *Neisseria;* A-4). **B:** Rods (bacilli): with square ends (e.g., *Bacillus;* B-1); with rounded ends (e.g., *Salmonella;* B-2); club-shaped (e.g., *Corynebacterium;* B-3); fusiform (e.g., *Fusobacterium;* B-4); comma-shaped (e.g., *Vibrio;* B-5). **C:** Spirochetes: spiral shape with relaxed coils (e.g., *Borrelia;* C-1); spiral shape with tight coils (e.g., *Treponema;* C-2). (Reproduced with permission from Joklik WK, Willett HP, Amos DB: *Zinsser Microbiology,* 20th ed. New York, NY: McGraw Hill; 1992.)

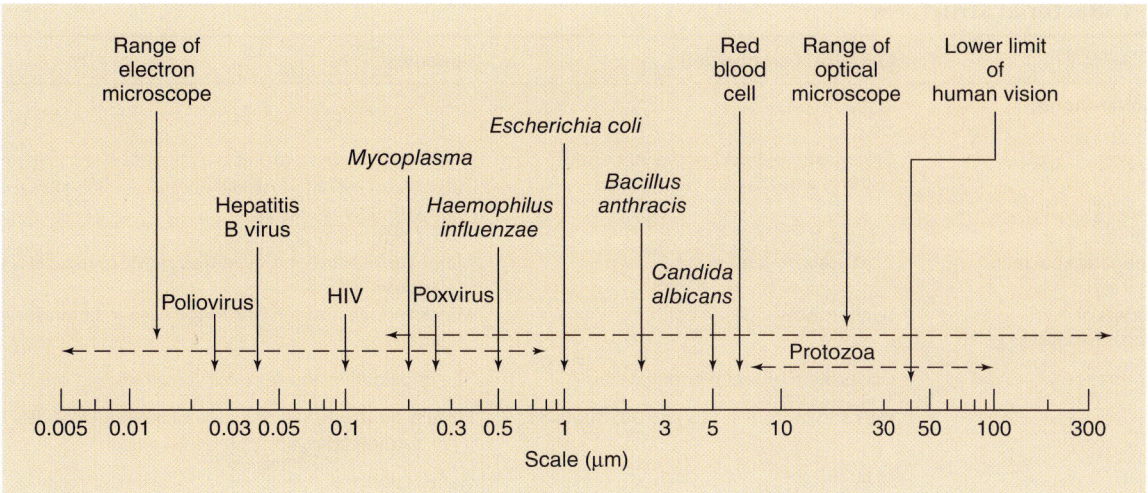

FIGURE 2–2 Sizes of representative bacteria, viruses, yeasts, protozoa, and human red cells. The bacteria range in size from *Mycoplasma,* the smallest, to *Bacillus anthracis,* one of the largest. The viruses range from poliovirus, one of the smallest, to poxviruses. Yeasts, such as *Candida albicans,* are generally larger than bacteria. Protozoa have many different forms and a broad size range. HIV, human immunodeficiency virus. (Reproduced with permission from Joklik WK, Willett HP, Amos DB: *Zinsser Microbiology,* 20th ed. New York, NY: McGraw Hill; 1992.)

Cell Walls of Gram-Positive and Gram-Negative Bacteria

The structure, chemical composition, and thickness of the cell wall differ in gram-positive and gram-negative bacteria (Table 2–2, Figure 2–4A, and "Gram Stain" box).

(1) The peptidoglycan layer is much thicker in gram-positive than in gram-negative bacteria. Many gram-positive bacteria also have fibers of teichoic acid that protrude outside the peptidoglycan, whereas gram-negative bacteria do not have teichoic acids.

(2) In contrast, the gram-negative bacteria have a complex outer layer consisting of lipopolysaccharide (LPS), lipoprotein, and phospholipid. Together with the cell wall, this gram-negative architecture is referred to as the "envelope." Lying between the outer-membrane layer and the cytoplasmic membrane in gram-negative bacteria is the **periplasmic space**, which is the site, in some species, of enzymes called β-lactamases that degrade penicillins and other β-lactam drugs.

The cell wall in gram-positive organisms or cell envelope in gram-negative organisms has several other important properties:

(1) In gram-negative bacteria, the envelope contains **endotoxin**, an LPS (see pages 9 and 42).

(2) Both gram-positive and gram-negative bacteria contain polysaccharides and proteins on their surface that are antigens useful in laboratory identification.

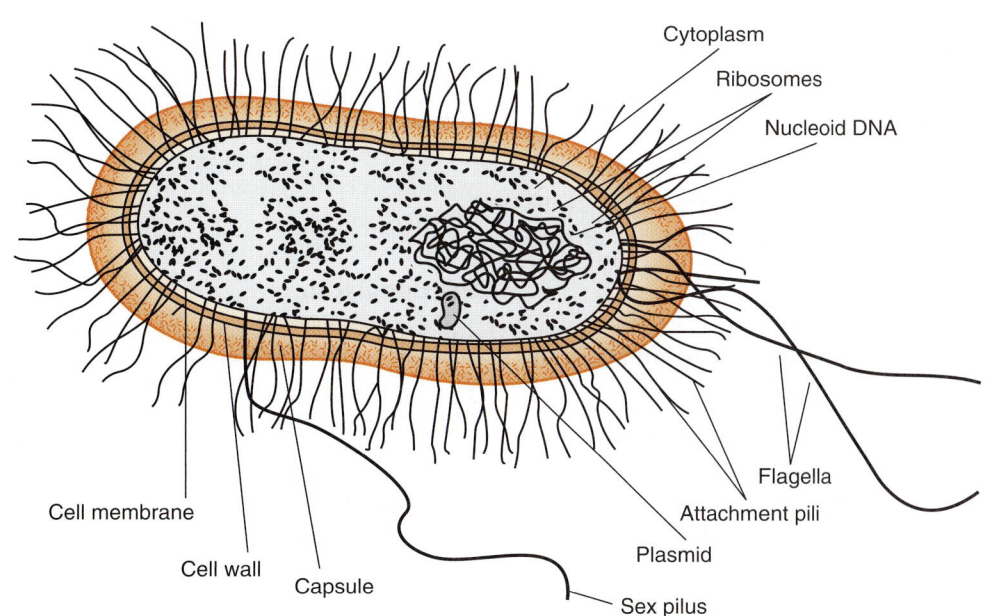

FIGURE 2–3 Bacterial structure. (Reproduced with permission from Ryan K: *Sherris Medical Microbiology,* 4th ed. New York, NY: McGraw Hill; 2004.)

TABLE 2–1 Bacterial Structures

Structure	Chemical Composition	Function
Essential components		
Cell wall		
Peptidoglycan	Glycan (sugar) backbone with peptide side chains that are cross-linked	Gives rigid support, protects against osmotic pressure, is the site of action of penicillins and cephalosporins, and is degraded by lysozyme
Outer membrane of gram-negative bacteria	1. Lipid A 2. Polysaccharide	Toxic component of endotoxin Major surface antigen used frequently in laboratory diagnosis
Surface fibers of gram-positive bacteria	Teichoic acid	Major surface antigen but rarely used in laboratory diagnosis
Plasma membrane	Lipoprotein bilayer without sterols	Site of oxidative and transport enzymes
Ribosome	RNA and protein in 50S and 30S subunits	Protein synthesis; site of action of aminoglycosides, erythromycin, tetracyclines, and chloramphenicol
Nucleoid	DNA	Genetic material
Mesosome	Invagination of plasma membrane	Participates in cell division and secretion
Periplasm	Space between plasma membrane and outer membrane	Contains many hydrolytic enzymes, including β-lactamases
Nonessential components		
Capsule	Typically polysaccharide	Protects against phagocytosis
Pilus or fimbria	Glycoprotein	Two types: (1) mediates attachment to cell surfaces; (2) sex pilus mediates attachment of two bacteria during conjugation
Flagellum	Protein	Motility
Spore	Keratin-like coat, dipicolinic acid	Provides resistance to dehydration, heat, and chemicals
Plasmid	DNA	Contains a variety of genes for antibiotic resistance and toxins
Granule	Glycogen, lipids, polyphosphates	Site of nutrients in cytoplasm
Glycocalyx	Polysaccharide	Mediates adherence to surfaces

(3) **Porin** proteins play a role in facilitating the passage of small, hydrophilic molecules into the cell. Several types of porin proteins can be found in the outer membrane of gram-negative bacteria, allowing the entry of essential substances such as sugars, amino acids, vitamins, and metals as well as many antimicrobial drugs such as penicillins. Porins have also been identified in gram-positive bacteria, where they are anchored to the cell wall.

Cell Walls of Acid-Fast Bacteria

Mycobacteria (e.g., *Mycobacterium tuberculosis*) have an unusual cell wall, resulting in their inability to be Gram-stained (Figure 2–4B). These bacteria are said to be **acid-fast** because they resist decolorization with acid–alcohol after being stained with carbolfuchsin.

TABLE 2–2 Comparison of Cell Walls of Gram-Positive and Gram-Negative Bacteria

Component	Gram-Positive Cells	Gram-Negative Cells
Peptidoglycan	Thicker; multilayer	Thinner; few layers
Teichoic acids	Yes	No
Lipopolysaccharide (endotoxin)	No	Yes

This property is related to the high concentration of lipids, called **mycolic acids**, in the cell wall of mycobacteria.

Note that *Nocardia asteroides*, a bacterium that can cause lung and brain infections in immunocompromised individuals, is **weakly acid-fast**. The meaning of the term "weakly" is that if the acid-fast staining process uses a weaker solution of hydrochloric acid to decolorize than that used in the stain for mycobacteria, then *N. asteroides* will *not* decolorize. However, if the regular-strength hydrochloric acid is used, *N. asteroides* will decolorize.

In view of their importance, three components of the cell wall (i.e., peptidoglycan, LPS, and teichoic acid) are discussed in detail here.

Peptidoglycan

Peptidoglycan is a complex, interwoven network that surrounds the entire cell and is composed of a single covalently linked macromolecule. It is found *only* in bacterial cell walls. It provides rigid support for the cell, is important in maintaining the characteristic shape of the cell, and allows the cell to withstand low osmotic pressure. A representative segment of the peptidoglycan layer is shown in Figure 2–5. The term **peptidoglycan** is derived from the peptides and the sugars (glycan) that make up the molecule. Synonyms for peptidoglycan are **murein** and **mucopeptide**.

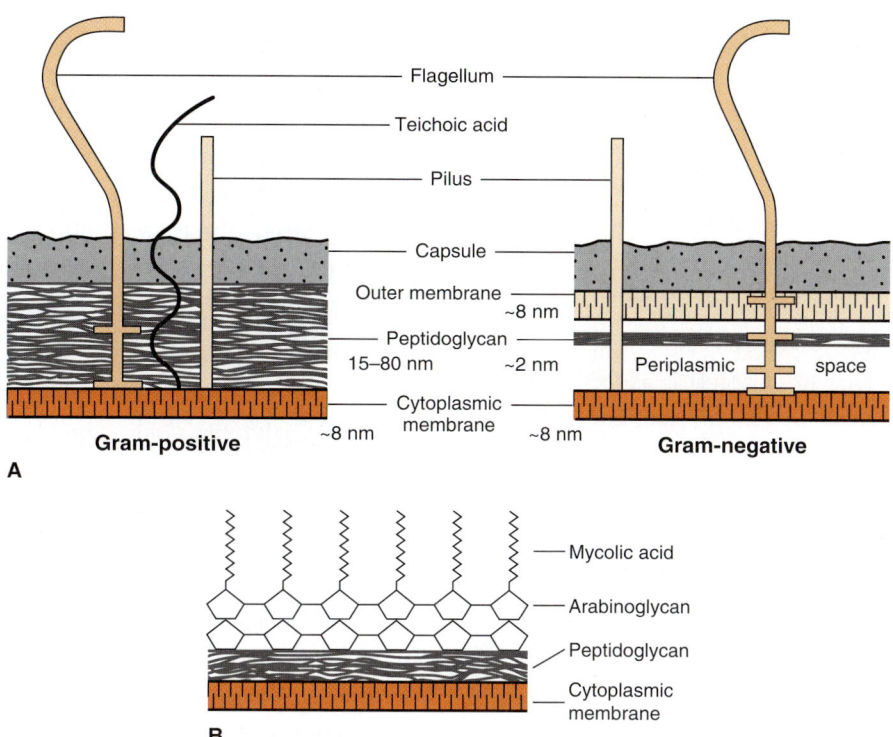

FIGURE 2–4 Bacterial cell wall structure. **A:** Cell walls of gram-positive and gram-negative bacteria. Note that the peptidoglycan in gram-positive bacteria is much thicker than in gram-negative bacteria. Note also that only gram-negative bacteria have an outer membrane containing endotoxin (lipopolysaccharide [LPS]) and thus have a periplasmic space where β-lactamases are found. Several important gram-positive bacteria, such as staphylococci and streptococci, have teichoic acids. **B:** Cell wall of *Mycobacterium tuberculosis:* Note the layers of mycolic acid and arabinoglycan that are present in members of the genus *Mycobacterium* but not in most other genera of bacteria. (A, Reproduced with permission from Ingraham JL, Maaløe O, Neidhardt FC: *Growth of the Bacterial Cell.* Sunderland, MA: Sinauer Associates; 1983.)

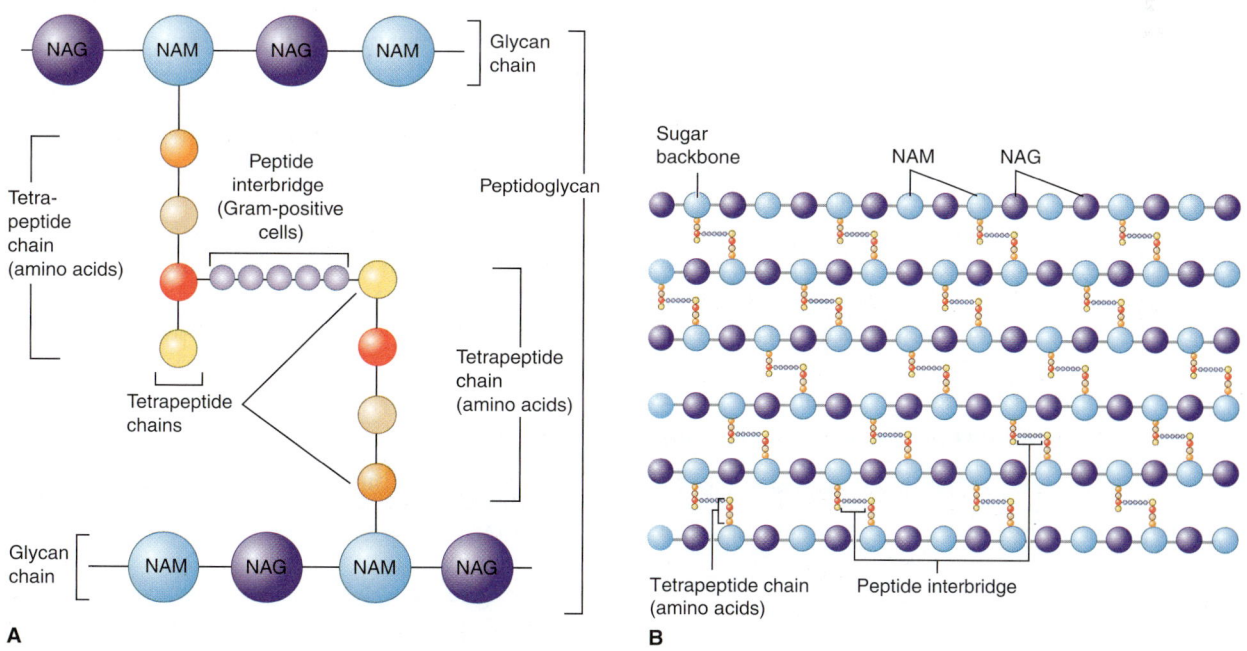

FIGURE 2–5 Peptidoglycan structure. **A:** Peptidoglycan is composed of a glycan chain (NAM and NAG), a tetrapeptide chain, and a cross-link (peptide interbridge). **B:** In the cell wall, the peptidoglycan forms a multilayered, three-dimensional structure. NAG, *N*-acetylglucosamine; NAM, *N*-acetylmuramic acid. (Reproduced with permission from Nester EW, Anderson D, Roberts CE, et al: *Microbiology: A Human Perspective,* 6th ed. New York, NY: McGraw Hill; 2009.)

GRAM STAIN

This staining procedure, developed in 1884 by the Danish physician Christian Gram, is the most important staining procedure in microbiology. It separates most bacteria into two groups: the gram-positive bacteria, which stain purple, and the gram-negative bacteria, which stain red. The Gram stain involves the following four-step procedure:

(1) The crystal violet dye stains all cells purple.

(2) The iodine solution (a mordant) is added to form a crystal violet–iodine complex; all cells continue to appear purple.

(3) The organic solvent, such as acetone or ethanol, extracts the purple dye/iodine complex from the lipid-rich, thin-walled, gram-negative bacteria to a greater degree than from the lipid-poor, thick-walled, gram-positive bacteria. The gram-negative organisms appear colorless; the gram-positive bacteria remain purple.

(4) The red dye safranin stains the decolorized gram-negative cells red/pink; the gram-positive bacteria remain purple.

The Gram stain is useful in two ways:

(1) In the identification of many bacteria.

(2) In influencing the choice of antibiotic because, in general, gram-positive bacteria are more susceptible to penicillin G than are gram-negative bacteria.

However, not all bacteria can be seen in the Gram stain. Table 2–3 lists the medically important bacteria that cannot be seen and describes the reason why. The alternative microscopic approach to the Gram stain is also described.

Note that it takes approximately 100,000 bacteria/mL to see 1 bacterium per microscopic field using the oil immersion (100×) lens. So the sensitivity of the Gram stain procedure is low. This explains why a patient's blood is rarely stained immediately but rather is incubated in blood cultures overnight to allow the bacteria to multiply. One important exception to this is meningococcemia in which very high concentrations of *Neisseria meningitidis* can occur in the blood.

Figure 2–5 illustrates the carbohydrate backbone, which is composed of alternating *N*-acetylmuramic acid and *N*-acetylglucosamine molecules. Attached to each of the muramic acid molecules is a tetrapeptide consisting of both D- and L-amino acids, the precise composition of which differs from one bacterium to another. Two of these amino acids are worthy of special mention: diaminopimelic acid, which is unique to bacterial cell walls, and D-alanine, which is involved in the cross-links between the tetrapeptides and in the action of penicillin. Note that this tetrapeptide contains the rare D-isomers of amino acids; most proteins contain the L-isomer. The other important component in this network is the peptide cross-link between the two tetrapeptides. The cross-links vary among species; in *Staphylococcus*

aureus, for example, five glycines link the terminal D-alanine to the penultimate L-lysine.Because peptidoglycan is present in bacteria but not in human cells, it is a good target for antibacterial drugs. Several of these drugs, such as penicillins, cephalosporins, and vancomycin, inhibit the synthesis of peptidoglycan by inhibiting the transpeptidase that makes the cross-links between the two adjacent tetrapeptides (see Chapter 10 for a description of these antibiotics).

Lysozyme, an enzyme present in human tears, mucus, and saliva, can cleave the peptidoglycan backbone by breaking its glycosyl bonds, thereby contributing to the natural resistance of the host to microbial infection. Lysozyme-treated bacteria may swell and rupture as a result of the entry of water into the cells, which have a high internal osmotic pressure.

TABLE 2–3 Medically Important Bacteria That Cannot Be Seen in the Gram Stain

Name	Reason	Alternative Microscopic Approach
Mycobacteria, including *M. tuberculosis*	Too much lipid in cell wall so dye cannot penetrate	Acid-fast stain
Treponema pallidum	Too thin to see	Dark-field microscopy or fluorescent antibody
Mycoplasma pneumoniae	No cell wall; very small	None
Legionella pneumophila	Poor uptake of red counterstain	Prolong time of counterstain
Chlamydiae, including *C. trachomatis*	Intracellular; very small	Inclusion bodies in cytoplasm
Rickettsiae	Intracellular; very small	Giemsa or other tissue stains

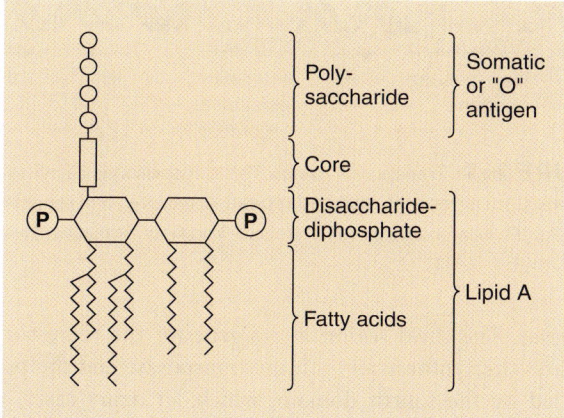

FIGURE 2–6 Endotoxin (lipopolysaccharide [LPS]) structure. The O-antigen polysaccharide is exposed on the exterior of the cell, whereas the lipid A faces the interior. (Reproduced with permission from Brooks GF, Jawetz E: *Medical Microbiology*, 19th ed. New York, NY: McGraw Hill; 1991.)

Lipopolysaccharide

The LPS of the outer membrane of the cell wall of gram-negative bacteria is **endotoxin**. It is responsible for many of the features of the disease, such as fever and shock (especially hypotension), caused by these organisms (see page 42). It is called endotoxin because it is an integral part of the cell envelope, in contrast to exotoxins, which are actively secreted from the bacteria. The constellation of symptoms caused by the endotoxin of one gram-negative bacterium is similar to another, but the severity of the symptoms can differ greatly. In contrast, the symptoms caused by exotoxins of different bacteria are usually quite different.

The LPS is composed of three distinct units (Figure 2–6):

(1) A phospholipid called lipid A, which is responsible for the toxic effects.

(2) A core polysaccharide of five sugars linked to lipid A.

(3) An outer polysaccharide consisting of up to 25 repeating units of three to five sugars. This outer polymer is the important somatic, or O, antigen of several gram-negative bacteria that is used to identify certain organisms in the clinical laboratory. Some bacteria, notably members of the genus *Neisseria*, have an outer lipooligosaccharide (LOS) containing very few repeating units of sugars.

Teichoic Acid

Teichoic acids and lipoteichoic acids are fibers anchored to the cell wall or cell membrane, respectively, that extend from the outer layer of the gram-positive cell wall. The medical importance of teichoic acids lies in their ability to **induce inflammation and septic shock when caused by certain gram-positive bacteria**; that is, they activate the same pathways as does endotoxin (LPS) in gram-negative bacteria. Teichoic acids also mediate the attachment of staphylococci to mucosal cells. Gram-negative bacteria do not have teichoic acids.

Cytoplasmic Membrane

Just inside the peptidoglycan layer of the cell wall lies the cytoplasmic membrane, which is composed of a phospholipid bilayer similar in microscopic appearance to that in eukaryotic cells. They are chemically similar, but eukaryotic membranes contain sterols, whereas prokaryotes generally do not. The only prokaryotes that have sterols in their membranes are members of the genus *Mycoplasma*. The membrane has four important functions: (1) active transport of molecules into the cell, (2) energy generation by oxidative phosphorylation, (3) synthesis of precursors of the cell wall, and (4) secretion of enzymes and toxins.

Cytoplasm

The cytoplasm has two distinct areas when seen in the electron microscope:

(1) An amorphous matrix that contains ribosomes, nutrient granules, metabolites, and plasmids.

(2) An inner, nucleoid region composed of DNA.

Ribosomes

Bacterial ribosomes are the site of protein synthesis as in eukaryotic cells, but they differ from eukaryotic ribosomes in size and chemical composition. Bacterial ribosomes are 70S in size, with 50S and 30S subunits, whereas eukaryotic ribosomes are 80S in size, with 60S and 40S subunits. The differences in both the ribosomal RNAs and proteins constitute the basis of the selective action of several antibiotics that inhibit bacterial, but not human, protein synthesis (see Chapter 10 for a description of these antibiotics).

Nucleoid

The nucleoid is the area of the cytoplasm in which DNA is located. The DNA of most prokaryotes is a single, circular molecule; however, there are important exceptions. For instance, the genome of *Vibrio cholerae*, the causative agent of cholera, is composed of two circular chromosomes. *Borrelia burgdorferi*, the spirochete that causes Lyme disease, is composed of a linear chromosome and multiple circular and linear plasmids (see below). The size of bacterial genomes varies widely, with the smallest genome containing just over 130 genes and the largest containing approximately 11,600 genes. By contrast, human DNA has approximately 25,000 genes.

Because the bacterial nucleoid contains no nuclear membrane, no nucleolus, no mitotic spindle, and no histones, there is little resemblance to the eukaryotic nucleus. One major difference between bacterial DNA and eukaryotic DNA is that bacterial DNA has no introns, whereas eukaryotic DNA does.

Plasmids

Plasmids are double-stranded, circular DNA molecules that are capable of replicating independently of the bacterial chromosome. Although plasmids are usually extrachromosomal, they can be integrated into the bacterial chromosome. Plasmids occur in both gram-positive and gram-negative bacteria, and several different types of plasmids can exist in one cell:

(1) **Transmissible** plasmids can be transferred from cell to cell by conjugation (see Chapter 4 for a discussion of conjugation). They are large (molecular weight [MW] 40–100 million) since they contain about a dozen genes responsible for synthesis of the sex pilus and for the enzymes required for transfer. They are usually present in a few (1–3) copies per cell.

(2) **Nontransmissible** plasmids are small (MW 3–20 million) since they do not contain the transfer genes; they are frequently present in many (10–60) copies per cell.

Plasmids carry the genes for the following functions and structures of medical importance:

(1) Antibiotic resistance, which is mediated by a variety of enzymes, such as the β-lactamase of *S. aureus*, *Escherichia coli*, and *Klebsiella pneumoniae*.

(2) Exotoxins, such as the enterotoxins of *E. coli*, anthrax toxin of *Bacillus anthracis*, exfoliative toxin of *S. aureus*, and tetanus toxin of *Clostridium tetani*.

(3) Pili (fimbriae), which mediate the adherence of bacteria to epithelial cells.

(4) Resistance to heavy metals, such as mercury, the active component of some antiseptics (e.g., Merthiolate and Mercurochrome), and silver, which is mediated by a reductase enzyme.

(5) Resistance to ultraviolet light, which is mediated by DNA repair enzymes.

(6) Bacteriocins, which are toxic proteins produced by certain bacteria that are lethal for other bacteria. Two common mechanisms of action of bacteriocins are (i) degradation of bacterial cell membranes by producing pores in the membrane and (ii) degradation of bacterial DNA by deoxyribonuclease (DNase). Examples of bacteriocins produced by medically important bacteria are colicins made by *E. coli*, pyocins made by *Pseudomonas aeruginosa* and lysostaphins made by *S. aureus*. Bacteria that produce bacteriocins have a selective advantage in the competition for food sources over those that do not. The medical importance of bacteriocins is that they may be useful in treating infections caused by antibiotic-resistant bacteria.

Transposons

Transposons are pieces of DNA that move readily from one site to another either within or between the DNAs of bacteria, plasmids, and bacteriophages. Because of their unusual ability to move, they are nicknamed "jumping genes." Transposons can code for drug-resistant enzymes, toxins, or a variety of metabolic enzymes and can either cause mutations in the gene into which they insert or alter the expression of nearby genes.

Transposons typically have four identifiable domains. On each end is a short DNA **sequence of inverted repeats**, which are involved in the integration of the transposon into the recipient DNA. The second domain is the gene for the transposase, which is the enzyme that mediates the excision and integration

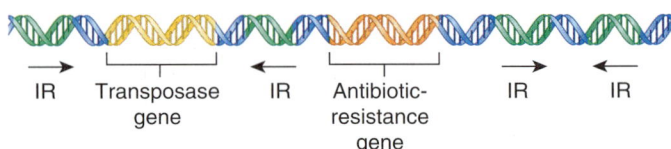

FIGURE 2–7 Transposon genes. This transposon is carrying a drug-resistance gene. IR, inverted repeat. (Reproduced with permission from Willey JM, Sherwood L, Woolverton: *Prescott's Principles of Microbiology*. New York, NY: McGraw Hill; 2009.)

processes. The third region is the gene for the repressor that regulates the synthesis of both the transposase and the protein encoded by the fourth domain, which, in many cases, is an enzyme mediating antibiotic resistance (Figure 2–7). Note that for simplicity, the repressor gene is not shown in Figure 2–7.

Antibiotic resistance genes are transferred from one bacterium to another primarily by **conjugation** (see Chapter 4). This transfer is mediated primarily by plasmids, but some transposons, called **conjugative transposons**, are capable of transferring antibiotic resistance as well.

In contrast to plasmids or bacterial viruses, transposons are not capable of independent replication; they replicate as part of the DNA in which they are integrated. More than one transposon can be located in the DNA; for example, a plasmid can contain several transposons carrying drug-resistant genes. **Insertion sequences** are a type of transposon that have fewer bases (800–1500 base pairs), since they do not code for their own integration enzymes. They can cause mutations at their site of integration and can be found in multiple copies at the ends of larger transposon units.

Structures Outside the Cell Wall

Capsule

The capsule is a gelatinous layer covering the entire bacterium. It is typically composed of polysaccharide. The sugar components of the polysaccharide vary from one species of bacteria to another and frequently determine the serologic type (serotype) within a species. For example, there are approximately 95 different serotypes of *Streptococcus pneumoniae*, which are distinguished by the antigenic differences of the sugars in the polysaccharide capsule.

The capsule is important for four reasons:

(1) It is a determinant of virulence of many bacteria since it limits the ability of phagocytes to engulf the bacteria. Variants of encapsulated bacteria that have lost the ability to produce a capsule are usually nonpathogenic.

(2) Specific identification of an organism can be made by using antiserum against the capsular polysaccharide. In the presence of the homologous antibody, the capsule will swell greatly. This swelling phenomenon, which is used in the clinical laboratory to identify certain organisms, is called the **Quellung reaction**.

(3) Capsular polysaccharides are used as antigens in certain vaccines because they are capable of eliciting protective antibodies.

For example, the capsular polysaccharide of *S. pneumoniae*, *Neisseria meningitidis,* and *Haemophilus influenzae* is the immunogen in the current vaccine against these bacteria.

(4) The capsule may play a role in the adherence of bacteria to human tissues, which is an important initial step in causing infection.

Flagella

Flagella are long, whip-like appendages that move the bacteria toward nutrients and other attractants, a process called **chemotaxis**. The long filament, which acts as a propeller, is composed of many subunits of a single protein, flagellin, arranged in several intertwined chains.

Flagellated bacteria have a characteristic number and location of flagella: some bacteria have one, and others have many; in some, the flagella are located at one end, and in others, they are all over the outer surface. Only certain bacteria have flagella. Many rods do, but most cocci do not and are therefore nonmotile. Spirochetes move by using a flagellum-like structure called the **axial filament**, which wraps around the spiral-shaped cell to produce an undulating motion.

Flagella are medically important for two reasons:

(1) Some species of motile bacteria (e.g., *E. coli* and *Proteus* species) are common causes of urinary tract infections. Flagella may play a role in pathogenesis by propelling the bacteria up the urethra into the bladder.

(2) Some species of bacteria (e.g., *Salmonella* species) are identified in the clinical laboratory by the use of specific antibodies against flagellar proteins.

Pili (Fimbriae)

Pili are hairlike filaments that extend from the cell surface. They are shorter and straighter than flagella and are composed of subunits of pilin, a protein arranged in helical strands. They are found mainly on gram-negative organisms.

Pili have two important roles:

(1) They mediate the **attachment** of bacteria to specific receptors on the human cell surface, which is a necessary step in the initiation of infection for some organisms. Mutants of *Neisseria gonorrhoeae* that do not form pili are nonpathogens.

(2) A specialized kind of pilus, the sex pilus, forms the attachment between the donor and the recipient bacteria during conjugation (see Chapter 4).

Glycocalyx (Slime Layer)

The glycocalyx is a polysaccharide coating that is secreted by many bacteria. It covers surfaces like a film and allows the bacteria to **adhere firmly** to various structures (e.g., skin, heart valves, prosthetic joints, and catheters). The glycocalyx is an important component of biofilms (see page 32). The medical importance of the glycocalyx is illustrated by the finding that it is the glycocalyx-producing strains of *P. aeruginosa* that cause respiratory tract infections in cystic fibrosis patients, and it is the glycocalyx-producing strains of *Staphylococcus epidermidis* and viridans streptococci that cause endocarditis. The glycocalyx also mediates adherence of certain bacteria to the surface of teeth. This plays an important role in the formation of plaque.

Bacterial Spores

These highly resistant structures are formed in response to adverse conditions by two genera of medically important gram-positive rods: the genus *Bacillus*, which includes the agent of anthrax, and the genus *Clostridium*, which includes the agents of tetanus and botulism. Spore formation (sporulation) occurs when nutrients, such as sources of carbon and nitrogen, are depleted (Figure 2–8). The spore forms inside the cell and contains bacterial DNA, a small amount of cytoplasm, cell membrane, peptidoglycan, very little water, and most importantly, a thick, keratin-like coat that is responsible for the remarkable resistance of the spore to heat, dehydration, radiation, and chemicals. This resistance may be mediated by **dipicolinic acid**, a calcium ion chelator found only in spores.

Once formed, the spore has no metabolic activity and can remain dormant for many years. Upon exposure to water and the appropriate nutrients, specific enzymes degrade the coat, water and nutrients enter, and germination into a potentially pathogenic bacterial cell occurs. Note that this differentiation process is *not* a means of reproduction since one cell produces one spore that germinates into one cell.

The medical importance of spores lies in their **extraordinary resistance to heat** and chemicals. As a result of their resistance to heat, sterilization cannot be achieved by boiling. Steam heating under pressure (autoclaving) at 121°C, for at least 15 minutes, is required to ensure the sterility of products for medical use. Spores are often not seen in clinical specimens recovered from patients infected by spore-forming organisms because the supply of nutrients is adequate.

Table 2–4 describes the medically important features of bacterial spores.

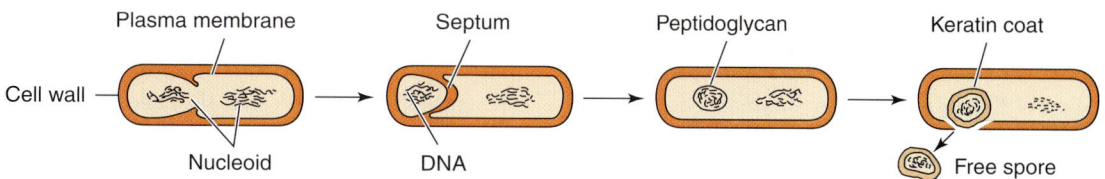

FIGURE 2–8 Bacterial spores. The spore contains the entire DNA genome of the bacterium surrounded by a thick, resistant coat.

TABLE 2–4 Important Features of Spores and Their Medical Implications

Important Features of Spores	Medical Implications
Highly resistant to heating; spores are not killed by boiling (100°C), but are killed at 121°C.	Medical supplies must be heated to 121°C for at least 15 minutes to be sterilized.
Highly resistant to many chemicals, including most disinfectants, due to the thick, keratin-like coat of the spore.	Only solutions designated as sporicidal will kill spores.
They can survive for many years, especially in the soil.	Wounds contaminated with soil can be infected with spores and cause diseases such as tetanus (*C. tetani*) and gas gangrene (*C. perfringens*).
They exhibit no measurable metabolic activity.	Antibiotics are ineffective against spores because antibiotics act by inhibiting certain metabolic pathways of bacteria. Also, spore coat is impermeable to antibiotics.
Spores form when nutrients are insufficient but then germinate to form bacteria when nutrients become available.	Spores are not often found at the site of infections because nutrients are not limiting. Bacteria rather than spores are usually seen in Gram-stained smears.
Spores are produced by members of only two genera of bacteria of medical importance, *Bacillus* and *Clostridium*, both of which are gram-positive rods.	Infections transmitted by spores are caused by species of either *Bacillus or Clostridium*.

PEARLS

Shape & Size

- Bacteria have three shapes: **cocci** (spheres), **bacilli** (rods), and **spirochetes** (spirals).
- Cocci are arranged in three patterns: pairs (diplococci), chains (streptococci), and clusters (staphylococci).
- The size of most bacteria ranges from 1 to 3 μm. *Mycoplasma*, the smallest bacteria (and therefore the **smallest cells**), are 0.2 μm. Some bacteria, such as *Borrelia*, are as long as 10 μm; that is, they are longer than a human red blood cell, which is 7 μm in diameter.

Bacterial Cell Wall

- All bacteria have a cell wall composed of **peptidoglycan** except *Mycoplasma*, which are surrounded *only* by a cell membrane.
- Gram-negative bacteria have a **thin** peptidoglycan covered by an outer lipid-containing membrane, whereas gram-positive bacteria have a **thick** peptidoglycan and no outer membrane.
- The outer membrane of gram-negative bacteria contains **endotoxin (lipopolysaccharide, LPS)**, the main inducer of septic shock. Endotoxin consists of **lipid A**, which causes the fever and hypotension seen in septic shock, and a polysaccharide called **O antigen**, which is useful in laboratory identification.
- Between the inner cell membrane and the outer membrane of gram-negative bacteria lies the **periplasmic space**, which is the location of **β-lactamases**—the enzymes that degrade β-lactam antibiotics, such as penicillins and cephalosporins.

- Peptidoglycan is found *only* in bacterial cells. It is a network that covers the entire bacterium and gives the organism its shape. It is composed of a sugar backbone (**glycan**) and peptide side chains (**peptido**). The side chains are cross-linked by **transpeptidase**—the enzyme that is inhibited by penicillins and cephalosporins.
- The cell wall of mycobacteria (e.g., *Mycobacterium tuberculosis*) has **more lipid** than either the gram-positive or gram-negative bacteria. As a result, the dyes used in the Gram stain do not penetrate into (do not stain) mycobacteria. The **acid-fast stain** does stain mycobacteria, and these bacteria are often called acid-fast bacilli (acid-fast rods).
- **Lysozymes** kill bacteria by cleaving the glycan backbone of peptidoglycan.
- The cytoplasmic membrane of bacteria consists of a phospholipid bilayer (without sterols) located just inside the peptidoglycan. It regulates active transport of nutrients into the cell and the secretion of toxins out of the cell.

Gram Stain

- **Gram stain** is the most important staining procedure. Gram-positive bacteria stain *purple*, whereas gram-negative bacteria stain *pink*. This difference is due to the ability of gram-positive bacteria to *retain the crystal violet–iodine complex in the presence of a lipid solvent*, usually acetone–alcohol. Gram-negative bacteria, because they have an outer lipid-containing membrane and thin peptidoglycan, lose the purple dye when treated with acetone–alcohol. They become colorless and then stain pink when exposed to a red dye such as safranin.

- Not all bacteria can be visualized using Gram stain. Some important human pathogens, such as the bacteria that cause tuberculosis and syphilis, cannot be seen using this stain.

Bacterial DNA

- The bacterial genome typically consists of a **single chromosome of circular DNA** located in the nucleoid.
- **Plasmids** are extrachromosomal pieces of circular DNA that encode both exotoxins and many enzymes that cause antibiotic resistance.
- **Transposons** are small pieces of DNA that frequently move between chromosomal DNA and plasmid DNA. They carry antibiotic-resistant genes.

Structures External to the Cell Wall

- **Capsules** are antiphagocytic; that is, they limit the ability of neutrophils to engulf the bacteria. Almost all capsules are composed of *polysaccharide*; the polypeptide capsule of anthrax bacillus is the only exception. Capsules are also the antigens in several vaccines, such as the pneumococcal vaccine. Antibodies against the capsule neutralize the antiphagocytic effect and allow the bacteria to be engulfed by neutrophils.

Opsonization is the process by which antibodies enhance the phagocytosis of bacteria.

- **Pili** are filaments of protein that extend from the bacterial surface and mediate **attachment** of bacteria to the surface of human cells. A different kind of pilus, the sex pilus, functions in conjugation (see Chapter 4).
- The **glycocalyx** is a polysaccharide "slime layer" secreted by certain bacteria. It **attaches bacteria firmly** to the surface of human cells and to the surface of catheters, prosthetic heart valves, and prosthetic hip joints.

Bacterial Spores

- **Spores** are medically important because they are **highly heat-resistant** and are not killed by many disinfectants. Boiling will *not* kill spores. They are formed by certain gram-positive rods, especially *Bacillus* and *Clostridium* species.
- Spores have a thick, keratin-like coat that allows them to survive for many years, especially in the soil. Spores are formed when nutrients are in short supply, but when nutrients are restored, spores germinate to form bacteria that can cause disease. Spores are *metabolically inactive* but contain DNA, ribosomes, and other essential components.

SELF-ASSESSMENT QUESTIONS

1. The initial step in the process of many bacterial infections is the adherence of the organism to mucous membranes. The bacterial component that mediates adherence is the:

   (A) lipid A
   (B) nucleoid
   (C) peptidoglycan
   (D) pilus
   (E) plasmid

2. In the Gram stain procedure, bacteria are exposed to 95% alcohol or to an acetone/alcohol mixture. The purpose of this step is:

   (A) to adhere the cells to the slide
   (B) to retain the purple dye within all the bacteria
   (C) to disrupt the outer cell membrane so the purple dye can leave the bacteria
   (D) to facilitate the entry of the purple dye into the gram-negative cells
   (E) to form a complex with the iodine solution

3. In the process of studying how bacteria cause disease, it was found that a rare mutant of a pathogenic strain failed to form a capsule. Which one of the following statements is the most accurate in regard to this unencapsulated mutant strain?

   (A) It was nonpathogenic primarily because it was easily phagocytized.
   (B) It was nonpathogenic primarily because it could not invade tissue.
   (C) It was nonpathogenic primarily because it could only grow anaerobically.

   (D) It was highly pathogenic because it could secrete larger amounts of exotoxin.
   (E) It was highly pathogenic because it could secrete larger amounts of endotoxin.

4. *Mycobacterium tuberculosis* stains well with the acid-fast stain, but not with the Gram stain. Which one of the following is the most likely reason for this observation?

   (A) It has a large number of pili that absorb the purple dye.
   (B) It has a large amount of lipid that prevents entry of the purple dye.
   (C) It has a very thin cell wall that does not retain the purple dye.
   (D) It is too thin to be seen in the Gram stain.
   (E) It has histones that are highly negatively charged.

5. Of the following bacterial components, which one exhibits the most antigenic variation?

   (A) Capsule
   (B) Lipid A of endotoxin
   (C) Peptidoglycan
   (D) Ribosome
   (E) Spore

6. β-Lactamases are an important cause of antibiotic resistance. Which one of the following is the most common site where β-lactamases are located?

   (A) Attached to DNA in the nucleoid
   (B) Attached to pili on the bacterial surface
   (C) Free in the cytoplasm
   (D) Within the capsule
   (E) Within the periplasmic space

7. Which one of the following is the most accurate description of the structural differences between gram-positive bacteria and gram-negative bacteria?

    (A) Gram-positive bacteria have a thick peptidoglycan layer, whereas gram-negative bacteria have a thin layer.

    (B) Gram-positive bacteria have an outer lipid-rich membrane, whereas gram-negative bacteria do not.

    (C) Gram-positive bacteria form a sex pilus that mediates conjugation, whereas gram-negative bacteria do not.

    (D) Gram-positive bacteria have plasmids, whereas gram-negative bacteria do not.

    (E) Gram-positive bacteria have capsules, whereas gram-negative bacteria do not.

8. Bacteria that cause nosocomial (hospital-acquired) infections often produce extracellular substances that allow them to stick firmly to medical devices, such as intravenous catheters. Which one of the following is the name of this extracellular substance?

    (A) Axial filament

    (B) Endotoxin

    (C) Flagella

    (D) Glycocalyx

    (E) Porin

9. Lysozyme in tears is an effective mechanism for preventing bacterial conjunctivitis. Which one of the following bacterial structures does lysozyme degrade?

    (A) Endotoxin

    (B) Nucleoid DNA

    (C) Peptidoglycan

    (D) Pilus

    (E) Plasmid DNA

10. Several bacteria that form spores are important human pathogens. Which one of the following is the most accurate statement about bacterial spores?

    (A) They are killed by boiling for 15 minutes.

    (B) They are produced primarily by gram-negative cocci.

    (C) They are formed primarily when the bacterium is exposed to antibiotics.

    (D) They are produced by anaerobes only in the presence of oxygen.

    (E) They are metabolically inactive, yet can survive for years in that inactive state.

ANSWERS

    (1) **(D)**
    (2) **(C)**
    (3) **(A)**
    (4) **(B)**
    (5) **(A)**
    (6) **(E)**
    (7) **(A)**
    (8) **(D)**
    (9) **(C)**
    (10) **(E)**

PRACTICE QUESTIONS: USMLE & COURSE EXAMINATIONS

Questions on the topics discussed in this chapter can be found in the Basic Bacteriology section of Part XIII: USMLE (National Board) Practice Questions starting on page 733. Also see Part XIV: USMLE (National Board) Practice Examination starting on page 775.

Growth

GROWTH CYCLE

When we think of growth, this is usually in reference to changes in weight (**how many ounces** has a baby gained?) or height (**how many inches** has a child grown?). However, thinking about bacterial growth usually refers to **the number** of bacteria. Bacteria reproduce by **binary fission**, a process by which one cell divides into two cells, two divides into four, four divides into 8, etc. This is referred to as exponential growth (logarithmic growth), which is illustrated by the following relationship:

Number of cells	1	2	4	8	16
Exponential	2^0	2^1	2^2	2^3	2^4

Thus, 1 bacterium will produce 16 bacteria after 4 generations.

The doubling (generation) time of bacteria ranges from as little as 20 minutes for *Escherichia coli* to as long as 18 to 24 hours for *Mycobacterium tuberculosis*. The concept of exponential growth (and the short doubling time of some organisms) explains how very large numbers of bacteria can occur in short periods of time. For example, 1 *E. coli* can produce over 1 million in about 7 hours. Doubling time varies not only with the species, but also with the amount of nutrients, the temperature, the pH, and other environmental factors.

The growth cycle of bacteria has four phases. Figure 3–1 illustrates the typical phases of a standard growth curve when a small number of bacteria are inoculated into a liquid medium and the progeny are counted at frequent intervals.

(1) **Lag phase:** Metabolic activity occurs but cells do not divide. This can last for a few minutes up to many hours.

(2) **Log (logarithmic or exponential) phase:** Rapid cell division occurs. Many antibiotics, such as penicillin, are most efficacious during this phase because they act by disrupting biosynthetic processes carried out by the bacterial cell when they are actively dividing.

(3) **Stationary phase:** Nutrient depletion or toxic products cause growth to slow until the number of new cells produced balances the number of cells that die (also called steady state).

(4) **Death phase:** A decline in the number of viable bacteria.

OBLIGATE INTRACELLULAR GROWTH

Most human bacterial pathogens can be cultivated on artificial media in the laboratory. The medium is typically composed of purified chemicals such as sugars, amino acids, and essential salts, and often contains sheep's blood, which supplies other essential nutrients.

However, certain human bacterial pathogens, notably *Chlamydia* and *Rickettsia* (see Chapters 25 and 26, respectively) and *Ehrlichia* and *Anaplasma* (see Chapter 26), can *only* grow within living cells and are referred to as **obligate intracellular pathogens**. The main reason for this is that they lack the ability to produce sufficient adenosine triphosphate (ATP) and must use ATP produced by the host cells.

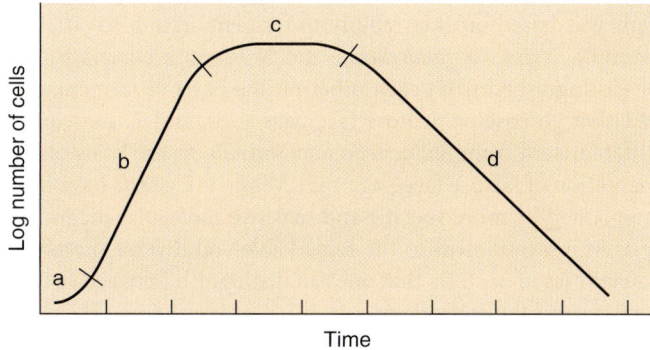

FIGURE 3–1 Growth curve of bacteria: a, lag phase; b, log phase; c, stationary phase; d, death phase. (Reproduced with permission from Joklik WK, Willett HP, Amos DB: *Zinsser Microbiology*, 20th ed. New York, NY: McGraw Hill; 1992.)

AEROBIC & ANAEROBIC GROWTH

For most organisms, an adequate supply of oxygen enhances metabolism and growth. The oxygen acts as the hydrogen acceptor in the final steps of energy production catalyzed by the flavoproteins and cytochromes. However, this generates toxic molecules, such as hydrogen peroxide (H_2O_2) and superoxide (O_2^-). Some bacteria possess two enzymes, which work in concert to detoxify these: **Superoxide dismutase** O_2^- to H_2O_2, followed by **catalase**, which reduces H_2O_2 to harmless water and oxygen molecules. These enzymes are important because bacteria lacking them are unable to grow in the presence of oxygen (and are thus considered to be **anaerobic**; see below). In addition, the catalase test serves as an important diagnostic test in the clinical laboratory.

Growth in the presence of oxygen is often used to characterize bacteria into one of 3 groups:

(1) Some bacteria, such as *M. tuberculosis*, are **obligate aerobes**; that is, they require oxygen to grow because their ATP-generating system is dependent on oxygen as the hydrogen acceptor.

(2) Other bacteria, such as *E. coli*, are **facultative anaerobes**; they use oxygen, if it is present, to generate energy by respiration, but they can use the fermentation pathway to synthesize ATP in the absence of sufficient oxygen.

(3) The third group of bacteria consists of the **obligate anaerobes**, such as *Clostridium tetani*, which cannot grow in the presence of oxygen because they lack either superoxide dismutase or catalase, or both. Obligate anaerobes vary in their response to oxygen exposure; some can survive but are not able to grow, whereas others are killed rapidly.

Note, however, that for medical purposes, the aerobes and the facultative bacteria are often just called aerobes to distinguish them from the anaerobes. This is useful because the anaerobic bacteria are handled differently from the aerobes in the laboratory and are often treated with different antibiotics.

FERMENTATION OF SUGARS

Historically, identification of many important human pathogens was based on their ability to ferment certain sugars. For example, *Neisseria gonorrhoeae* and *Neisseria meningitidis* can be distinguished from each other on the basis of fermentation of either glucose or maltose (see page 126), and *E. coli* can be differentiated from *Salmonella* and *Shigella* on the basis of fermentation of lactose (see page 145). While these tests have been supplanted by more specific and sensitive molecular diagnostic tests, it is a testament to the remarkable and diverse metabolic capabilities of bacteria that one can distinguish between species by their metabolic profiles.

The term **fermentation** refers to the breakdown of a sugar (such as glucose or maltose or lactose) to pyruvic acid and then, usually, to lactic acid. This is the process by which facultative bacteria generate ATP and can occur in the absence of oxygen.

When oxygen is present, the pyruvate produced by glycolysis enters the Krebs cycle (oxidation cycle, tricarboxylic acid cycle) and is metabolized to two final products, CO_2 and H_2O. The Krebs cycle generates much more ATP than the glycolytic cycle; therefore, facultative bacteria grow faster in the presence of oxygen. Facultative and anaerobic bacteria ferment, but aerobes, which can grow only in the presence of oxygen, do not. Aerobes, such as *Pseudomonas aeruginosa*, produce metabolites that enter the Krebs cycle by processes other than glycolysis, such as the deamination of amino acids.

During fermentation acidic end products, like lactate, are generated, which can be detected by an indicator that changes color according to pH. For example, if a sugar is fermented in the presence of phenol red (an indicator), the pH becomes acidic and the medium turns yellow. If, however, the sugar is not fermented, no acid is produced and the phenol red remains red. This simple test can be used to differentiate between closely related organisms with different fermentation profiles.

IRON METABOLISM

Iron, in the form of ferric ion, is required for the growth of bacteria because it is an essential component of cytochromes and other enzymes. The amount of iron available for pathogenic bacteria in the human body is very low because the iron is sequestered in iron-binding proteins such as transferrin. To obtain iron for their growth, bacteria produce iron-binding compounds called **siderophores**. Siderophores, such as enterobactin produced by *E. coli*, are secreted by the bacteria, capture iron by chelating it, then attach to specific receptors on the bacterial surface, and are actively transported into the cell where the iron becomes available for use. The fact that bacteria have such a complex and specific mechanism for obtaining iron testifies to its importance in the growth and metabolism of bacteria.

PEARLS

- Bacteria reproduce by **binary fission**, whereas eukaryotic cells reproduce by mitosis.

- The bacterial growth cycle consists of four phases: the **lag** phase, during which nutrients are incorporated; the **log** phase, during which rapid cell division occurs; the **stationary** phase, during which as many cells are dying as are being formed; and the **death** phase, during which most of the cells are dying because nutrients have been exhausted.

- Some bacteria can grow in the presence of oxygen (**aerobes** and **facultatives**), but others die in the presence of oxygen (**anaerobes**). The use of oxygen by bacteria generates toxic products such as **superoxide** and **hydrogen peroxide**. Aerobes and facultatives have enzymes, such as **superoxide dismutase** and **catalase**, that detoxify these products, but anaerobes do not and are killed in the presence of oxygen.

SELF-ASSESSMENT QUESTIONS

1. Figure 3–1 depicts a bacterial growth curve divided into phases a, b, c, and d. In which one of the phases are antibiotics such as penicillin most likely to kill bacteria?

    (A) Phase a
    (B) Phase b
    (C) Phase c
    (D) Phase d

2. Some bacteria are obligate anaerobes. Which of the following statements best explains this phenomenon?

    (A) They can produce energy both by fermentation (i.e., glycolysis) and by respiration using the Krebs cycle and cytochromes.
    (B) They cannot produce their own ATP.
    (C) They do not form spores.
    (D) They lack superoxide dismutase and catalase.
    (E) They do not have a capsule.

ANSWERS

(1) **(B)**
(2) **(D)**

PRACTICE QUESTIONS: USMLE & COURSE EXAMINATIONS

Questions on the topics discussed in this chapter can be found in the Basic Bacteriology section of Part XIII: USMLE (National Board) Practice Questions starting on page 733. Also see Part XIV: USMLE (National Board) Practice Examination starting on page 775.

Genetics

INTRODUCTION

There are several unique aspects of microbial genetics that largely account for the great genotypic and phenotypic diversity, the ability to cause disease, and the propensity to develop resistance to virtually any antibiotic used to treat bacterial infections. Bacteria have a simple genetic organization relative to eukaryotic organisms. They are haploid, usually possessing a single chromosome and therefore a single copy of each gene. This is in contrast to eukaryotic cells (such as human cells), which are **diploid**, meaning they have a pair of each chromosome and therefore have two copies of each gene. In diploid cells, one copy of a gene (allele) may be expressed as a protein (i.e., be dominant), whereas another allele may not be expressed (i.e., be recessive). In haploid cells, any gene that has acquired a mutation will result in a cell synthesizing either a mutant protein or no protein at all depending on the type of mutation.

MUTATIONS

A mutation is a change in the base sequence of DNA that can result in the insertion of a different amino acid or stop codon into a protein and the appearance of an altered phenotype. Mutations result from three types of molecular changes:

(1) **Base substitution.** This occurs when one base is inserted in place of another. It takes place at the time of DNA replication, either because the DNA polymerase makes an error or because a mutagen alters the hydrogen bonding of the base being used as a template in such a manner that the wrong base is inserted. When the base substitution results in a codon that simply causes a different amino acid to be inserted, the mutation is called a

missense mutation; when the base substitution generates a termination codon that stops protein synthesis prematurely, the mutation is called a **nonsense mutation**. Nonsense mutations almost always destroy protein function.

(2) **Frameshift mutation.** This occurs when one or more base pairs are added or deleted, which shifts the reading frame on the ribosome and results in incorporation of the wrong amino acids "downstream" from the mutation and in the production of an inactive protein.

(3) **Transposons** and **insertion sequences.** Mutations can occur when pieces of DNA, such as transposons and insertion sequences, are inserted into the genome. This can cause profound changes in both genes that are disrupted by the insertion, as well as in adjacent genes, whose transcription can be affected.

Mutations can be caused by chemicals, radiation, or viruses. Chemicals act in several different ways.

(1) Some, such as nitrous acid and alkylating agents, alter the existing base so that it forms a hydrogen bond preferentially with the wrong base (e.g., adenine would no longer pair with thymine but with cytosine).

(2) Some chemicals, such as 5-bromouracil, are base analogs, since they resemble normal bases. Because the bromine atom has an atomic radius similar to that of a methyl group, 5-bromouracil can be inserted in place of thymine (5-methyluracil). However, 5-bromouracil has less hydrogen-bonding fidelity than does thymine, and so it binds to guanine with greater frequency. This results in a transition from an A-T base pair to a G-C base pair, thereby producing a mutation.

(3) Some chemicals, such as benzpyrene, which is found in tobacco smoke, bind to the existing DNA bases and cause

frameshift mutations. These chemicals, which are frequently carcinogens as well as mutagens, intercalate between the adjacent bases, thereby distorting and offsetting the DNA sequence.

X-rays and ultraviolet light can also cause mutations.

(1) X-rays have high energy and can damage DNA by: (a) breaking the covalent bonds that hold the ribose phosphate chain together, (b) producing free radicals that can attack the bases, and (c) altering the electrons in the bases and thus changing their hydrogen bonding.

(2) Ultraviolet radiation, which has lower energy than X-rays, causes the cross-linking of the adjacent pyrimidine bases to form dimers. This cross-linking (e.g., of adjacent thymines to form a thymine dimer) results in inability of the DNA to replicate properly.

Conditional lethal mutations are of medical interest because they may be useful in vaccines (e.g., influenza vaccine). The word *conditional* indicates that the mutation is expressed only under certain conditions. The most important conditional lethal mutations are the temperature-sensitive ones. Temperature-sensitive organisms can replicate at a relatively low, permissive temperature (e.g., 32°C) but cannot grow at a higher, restrictive temperature (e.g., 37°C). This behavior is due to a mutation that causes an amino acid change in an essential protein, allowing it to function normally at 32°C but not at 37°C because of an altered conformation at the higher temperature. An example of a conditional lethal mutant of medical importance is a strain of influenza virus currently approved for vaccine use. This vaccine contains a virus that cannot grow at 37°C and hence cannot infect the lungs and cause pneumonia, but it can grow at 32°C in the nose, where it can replicate and induce immunity.

TRANSFER OF DNA WITHIN BACTERIAL CELLS

Transposons transfer DNA from one site within the bacterial genome to another site. They do so by synthesizing a copy of their DNA and inserting the copy at another site in the bacterial chromosome or a plasmid. The structure and function of transposons are described in Chapter 2, and their role in antimicrobial drug resistance is described in Chapter 11. The transfer of a transposon to a plasmid and the subsequent transfer of the plasmid to another bacterium by conjugation (see below) contribute significantly to the spread of antibiotic resistance.

Transfer of DNA within bacteria also occurs by **programmed rearrangements** (Figure 4–1). These gene rearrangements account for many of the antigenic changes seen in *Neisseria gonorrhoeae* and *Borrelia recurrentis*, the cause of relapsing fever. (They also occur in trypanosomes, which are discussed in Chapter 52.) A programmed rearrangement consists of the movement of a gene from a site where the gene is not expressed (silent sites) to a different site where transcription and translation occur (active sites). There are many silent genes that encode variants of a variety of surface antigens, and the insertion of a new gene into the active site in a sequential, repeated programmed manner is the source of the consistent antigenic variation. These gene movements allow the organism to evade the existing host immune response.

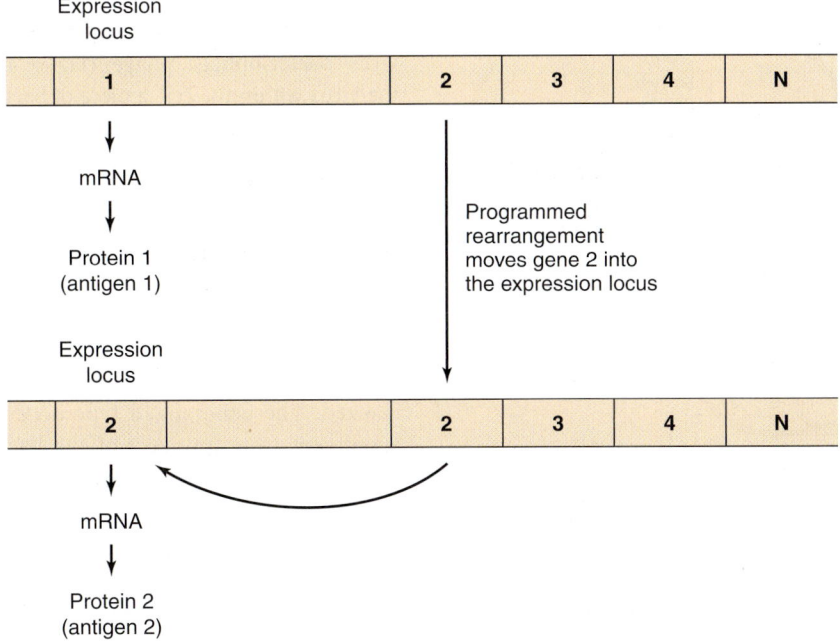

FIGURE 4–1 Programmed rearrangements. In the top part of the figure, the gene for protein 1 is in the expression locus, and the mRNA for protein 1 is synthesized. At a later time, a copy of gene 2 is made and inserted into the expression locus. By moving only the copy of the gene, the cell always keeps the original DNA for use in the future. When the DNA of gene 2 is inserted, the DNA of gene 1 is excised and degraded.

TABLE 4–1 Comparison of Conjugation, Transduction, and Transformation

Transfer Procedure	Process	Type of Cells Involved	Nature of DNA Transferred
Conjugation	DNA transferred from one bacterium to another	Prokaryotic	Chromosomal or plasmid
Transduction	DNA transferred by a virus from one cell to another	Prokaryotic	Any gene in generalized transduction; only certain genes in specialized transduction
Transformation	Naked DNA in the immediate environment taken up by a cell	Prokaryotic	Any DNA

TRANSFER OF DNA BETWEEN BACTERIAL CELLS

The transfer of genetic information from one cell to another can occur by three methods: conjugation, transduction, and transformation (Table 4–1). From a medical viewpoint, the two most important consequences of DNA transfer are (1) **that antibiotic resistance genes are spread from one bacterium to another primarily by conjugation** and (2) **that several important exotoxins are encoded by bacteriophage genes and are transferred by transduction.**

1. Conjugation

Conjugation is the mating of two bacterial cells, during which DNA is transferred from the donor to the recipient cell (Figure 4–2). The mating process is controlled by an **F (fertility) plasmid** (F factor), which carries the genes for the proteins required for conjugation. One of the most important proteins is pilin, which forms the **sex pilus** (conjugation tube). Mating begins when the pilus of the donor bacterium carrying the

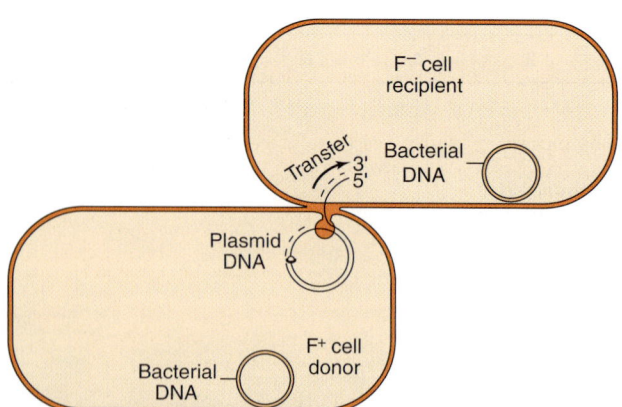

FIGURE 4–2 Conjugation. An F plasmid is being transferred from an F$^+$ donor bacterium to an F$^-$ recipient. The transfer is at the contact site made by the sex pilus. The new plasmid in the recipient bacterium is composed of one parental strand (solid line) and one newly synthesized strand (dashed line). The previously existing plasmid in the donor bacterium now consists of one parental strand (solid line) and one newly synthesized strand (dashed line). Both plasmids are drawn with only a short region of newly synthesized DNA (dashed lines), but at the end of DNA synthesis, both the donor and the recipient contain a complete copy of the plasmid DNA.

F factor (F$^+$) attaches to a receptor on the surface of a recipient bacterium, which does not contain an F factor (F$^-$), resulting in a direct connection between the cytoplasm of the donor and recipient cells. After an enzymatic cleavage of the F factor DNA, one strand is transferred across the conjugal bridge (mating bridge) into the recipient cell. The process is completed by synthesis of the complementary strand to form a double-stranded F factor plasmid in both the donor and recipient cells. The recipient is now an F$^+$ cell that is capable of transmitting the plasmid further.

Resistance plasmids (R plasmids) can also be transferred by conjugation. R plasmids can carry one or more genes for a variety of enzymes that can degrade antibiotics and modify membrane transport systems. For example, R plasmids encode the β-lactamases of *Staphylococcus aureus*, *Escherichia coli*, and *Klebsiella pneumoniae*. In addition, they encode the proteins of the transport system that actively export sulfonamides out of the bacterial cell. Note that R plasmids can be transferred not only to cells of the same species but also to other species and genera. (See Chapter 11 for more information about R plasmids.)

2. Transduction

Transduction is the transfer of cell DNA by means of a bacterial virus (**bacteriophage, phage**) (Figure 4–3). During the growth of the virus within the cell, a piece of bacterial DNA is incorporated into the virus particle and is carried into a new recipient cell at the time of infection. Within the recipient cell, the phage DNA can integrate into the cell DNA and the cell can acquire a new trait—a process called **lysogenic conversion** (see the end of Chapter 29). This process can change a nonpathogenic organism into a pathogenic one. Diphtheria toxin, botulinum toxin, cholera toxin, Shiga toxin of *E. coli*, and erythrogenic toxin (*Streptococcus pyogenes*) are encoded by bacteriophages and can be transferred by transduction.

There are two types of transduction: generalized and specialized. The **generalized** type occurs when the virus carries a segment from any part of the bacterial chromosome. This occurs because the cell DNA is fragmented after phage infection and pieces of cell DNA the same size as the viral DNA are incorporated into the virus particle at a frequency of about 1 in every 1000 virus particles. The **specialized** type occurs when the bacterial virus DNA that has integrated into the cell DNA is excised and carries with it an adjacent part of the cell DNA. Since most lysogenic (temperate) phages integrate at specific sites in the bacterial DNA, the adjacent cellular genes that are transduced are usually specific to that virus.

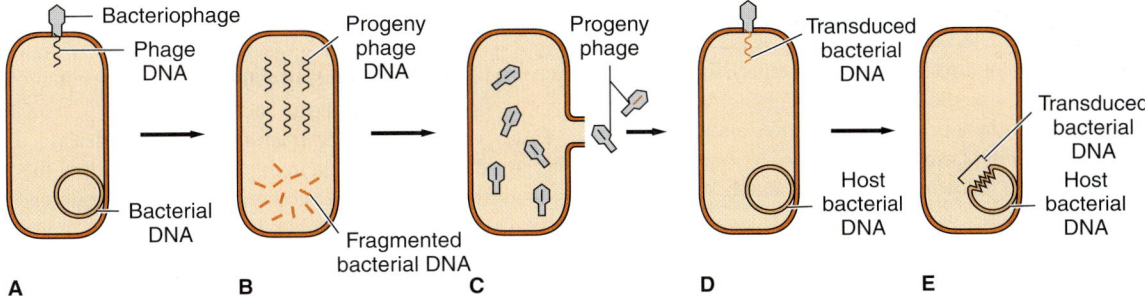

FIGURE 4–3 **Transduction. A:** A bacteriophage infects a bacterium, and phage DNA enters the cell. **B:** The phage DNA replicates, and the bacterial DNA fragments. **C:** The progeny phages assemble and are released; most contain phage DNA, and a few contain bacterial DNA. **D:** Another bacterium is infected by a phage-containing bacterial DNA. **E:** The transduced bacterial DNA integrates into host DNA, and the host acquires a new trait. This host bacterium survives because no viral DNA is transduced; therefore, no viral replication can occur. (Another type of transduction mechanism is depicted in Figure 29–10.)

3. Transformation

Transformation is the transfer of DNA itself from one cell to another. This occurs by either of the two following methods. First, in nature, dying bacteria may release their DNA, which may be taken up by recipient cells. Certain bacteria, such as *Neisseria*, *Haemophilus*, and *Streptococci*, synthesize receptors on the cell surface that play a role in the uptake of DNA from the environment.

Second, in the laboratory, an investigator may extract DNA from one type of bacteria and introduce it into genetically different bacteria. The experimental use of transformation has revealed important information about DNA. In 1944, it was shown that DNA extracted from encapsulated smooth pneumococci could transform nonencapsulated rough pneumococci into encapsulated smooth organisms. This demonstration that the transforming principle was DNA marked the first evidence that DNA was the genetic material.

RECOMBINATION

Once the DNA is transferred from the donor to the recipient cell by one of the three processes just described, it can integrate into the host cell chromosome by recombination. There are two types of recombination:

(1) **Homologous recombination**, in which two pieces of DNA that have extensive homologous regions pair up and exchange pieces by the processes of breakage and reunion.

(2) **Nonhomologous recombination**, in which little, if any, homology is necessary.

Different genetic loci govern these two types, and so it is presumed that different enzymes are involved. Although it is known that a variety of endonucleases and ligases are involved, the precise sequence of events is unknown.

PEARLS

- Bacteria have only one copy of their genome DNA (i.e., they are **haploid**). In contrast, eukaryotic cells have two copies of their genome DNA (i.e., they are **diploid**). Bacterial DNA is typically circular; human nuclear DNA is linear.

- The transfer of DNA within bacterial cells occurs by two processes: movement of transposons and programmed rearrangements. **Transposons** are small pieces of DNA that move readily from one site on the bacterial chromosome to another or from the bacterial chromosome to a plasmid. Medically, transposons are important because they commonly **carry antibiotic resistance genes**. The transfer of transposons on plasmids to other bacteria by conjugation contributes significantly to antibiotic resistance.

- **Programmed rearrangements** are the movement of genes from inactive sites into active sites, where they are expressed as new proteins. Medically, this is important because bacteria can acquire new proteins (antigens) on their surface and evade the immune system. Two important organisms in which this occurs are *Neisseria gonorrhoeae*, the cause of gonorrhea, and *Trypanosoma brucei*, a protozoan that causes African sleeping sickness.

- The transfer of DNA between bacterial cells occurs mainly by two processes: conjugation and transduction. **Conjugation** is the process by which DNA, either plasmid or chromosomal, is transferred directly from one bacterium to another. For conjugation to occur, the donor bacterium must have a "fertility" plasmid (F plasmid) that encodes the proteins that mediate this process, the most important of which are the proteins that form the **sex pilus**. The DNA transferred by conjugation to the recipient bacterium is a new copy that allows the donor to keep a copy of the DNA. Plasmids carrying antibiotic resistance genes are commonly transferred by conjugation.

- **Transduction** is the process by which DNA, either plasmid or chromosomal, is transferred from one bacterium to another by a **virus**. The transferred DNA integrates into the chromosomal DNA of the recipient, and new proteins, such as exotoxins, are made—a process called **lysogenic conversion**.

- **Transformation** is the process by which DNA itself, either DNA released from dying cells or DNA purified in the laboratory, enters a recipient bacterium.

SELF-ASSESSMENT QUESTIONS

1. The emergence of antibiotic-resistant bacteria, especially in enteric gram-negative rods, is a medically important phenomenon. This most commonly occurs by a process that involves a sex pilus and the subsequent transfer of plasmids carrying one or more transposons. Which one of the following is the name that best describes this process?

    (A) Conjugation
    (B) Transduction
    (C) Transformation
    (D) Translocation
    (E) Transposition

2. Several important pathogenic bacteria have the ability to translocate pieces of their DNA in a process called *programmed rearrangements*. Which one of the following is the most important known consequence of this ability?

    (A) The number of plasmids increases significantly, which greatly enhances antibiotic resistance.
    (B) The amount of endotoxin increases significantly, which greatly enhances the ability to cause septic shock.
    (C) The surface antigens of the bacteria vary significantly, which greatly enhances the ability to avoid opsonization by antibody.
    (D) The ability of the bacterium to be lysogenized is significantly increased, which greatly enhances the ability to produce increased amounts of exotoxins.
    (E) The ability of the bacterium to survive intracellularly is greatly increased.

3. Which statement is the most accurate regarding transposons?

    (A) They encode enzymes that degrade the ends of the bacterial chromosome.
    (B) They are short sequences of DNA that often encode enzymes that mediate antibiotic resistance.
    (C) They are short sequences of RNA that silence specific regulatory genes.
    (D) They are a family of transfer RNAs that enhance mutations at "hot spots" in the bacterial genome.

4. *Corynebacterium diphtheriae* causes the disease diphtheria by producing diphtheria toxin. The gene encoding the toxin is integrated into bacterial genome during lysogenic conversion. The toxin gene was acquired by which process?

    (A) Conjugation
    (B) Transduction
    (C) Transformation
    (D) Translocation
    (E) Transposition

ANSWERS

(1) **(A)**
(2) **(C)**
(3) **(B)**
(4) **(B)**

PRACTICE QUESTIONS: USMLE & COURSE EXAMINATIONS

Questions on the topics discussed in this chapter can be found in the Basic Bacteriology section of Part XIII: USMLE (National Board) Practice Questions starting on page 733. Also see Part XIV: USMLE (National Board) Practice Examination starting on page 775.

Classification of Medically Important Bacteria

PRINCIPLES OF CLASSIFICATION

The current classification of bacteria is based primarily on morphologic and biochemical characteristics. A scheme that divides the medically important organisms by genus is shown in Table 5–1. For pedagogic purposes, this classification scheme deviates from those derived from strict taxonomic principles in two ways:

(1) Only organisms that are described in this book in the section on medically important bacteria are included.

(2) Because there are so many gram-negative rods, they are divided into three categories: respiratory organisms, zoonotic organisms, and enteric and related organisms.

The initial criterion used in the classification is the nature of the cell wall (i.e., is it rigid, flexible, or absent?). Bacteria with rigid, thick walls can be subdivided into free-living bacteria, which are capable of growing on laboratory medium in the absence of human or other animal cells, and non–free-living bacteria, which are obligate intracellular parasites and therefore can grow only within human or other animal cells. The free-living organisms are further subdivided according to shape and staining reaction into a variety of gram-positive and gram-negative cocci and rods with different oxygen requirements and spore-forming abilities. Bacteria with flexible, thin walls (the spirochetes) and those without cell walls (the mycoplasmas) form separate units.

Using these criteria, along with various biochemical reactions, many bacteria can be readily classified into separate genera and species. However, there have been several examples of these criteria placing bacteria into the same genus when DNA sequencing of their genome reveals they are significantly different and should be classified in a new or different genus. For example, an organism formerly known as *Pseudomonas cepacia* has been reclassified as *Burkholderia cepacia* because the base sequence of its DNA was found to be significantly different from the DNA of the members of the genus *Pseudomonas*.

PEARLS

- The classification of bacteria is based on various criteria, such as the nature of the cell wall, staining characteristics, ability to grow in the presence or absence of oxygen, and ability to form spores.
- The criterion currently used is the base sequence of the genome DNA. Several bacteria have been reclassified based on this information.

PRACTICE QUESTIONS: USMLE & COURSE EXAMINATIONS

Questions on the topics discussed in this chapter can be found in the Basic Bacteriology section of Part XIII: USMLE (National Board) Practice Questions starting on page 733. Also see Part XIV: USMLE (National Board) Practice Examination starting on page 775.

TABLE 5–1 Classification of Medically Important Bacteria

Characteristics	Genus	Representative Diseases
I. Rigid, thick-walled cells		
A. Free-living (extracellular bacteria)		
1. Gram positive		
a. Cocci	*Streptococcus*	Pneumonia, pharyngitis, cellulitis
	Staphylococcus	Abscess of skin and other organs
b. Spore-forming rods		
(1) Aerobic	*Bacillus*	Anthrax
(2) Anaerobic	*Clostridium*	Tetanus, gas gangrene, botulism
c. Non–spore-forming rods		
(1) Nonfilamentous	*Corynebacterium*	Diphtheria
	Listeria	Meningitis
(2) Filamentous	*Actinomyces*	Actinomycosis
	Nocardia	Nocardiosis
2. Gram negative		
a. Cocci	*Neisseria*	Gonorrhea, meningitis
b. Rods		
(1) Facultative		
(a) Straight		
(i) Respiratory organisms	*Haemophilus*	Meningitis
	Bordetella	Whooping cough
	Legionella	Pneumonia
(ii) Zoonotic organisms	*Brucella*	Brucellosis
	Francisella	Tularemia
	Pasteurella	Cellulitis
	Yersinia	Plague
(iii) Enteric and related organisms	*Escherichia*	Urinary tract infection, diarrhea
	Enterobacter	Urinary tract infection
	Serratia	Pneumonia
	Klebsiella	Pneumonia, urinary tract infection
	Salmonella	Enterocolitis, typhoid fever
	Shigella	Enterocolitis
	Proteus	Urinary tract infection
(b) Curved	*Campylobacter*	Enterocolitis
	Helicobacter	Gastritis, peptic ulcer
	Vibrio	Cholera
(2) Aerobic	*Pseudomonas*	Pneumonia, urinary tract infection
(3) Anaerobic	*Bacteroides*	Peritonitis
3. Acid-fast	*Mycobacterium*	Tuberculosis, leprosy
B. Non–free-living (obligate intracellular parasites)	*Rickettsia*	Rocky Mountain spotted fever, typhus, Q fever
	Chlamydia	Urethritis, trachoma, psittacosis
II. Flexible, thin-walled cells (spirochetes)	*Treponema*	Syphilis
	Borrelia	Lyme disease
	Leptospira	Leptospirosis
III. Wall-less cells	*Mycoplasma*	Pneumonia

6

The Human Microbiome

INTRODUCTION

The **human microbiome** is the term used to describe the distinct microbial communities that inhabit different host environments on the body's skin and mucosal surfaces. Historically, microbiologists referred to microbial populations routinely found on and in the body as **normal flora**. The term *microbiome* also encompasses all of the genetic material associated with these normal constituents. As you will read below, the *genetic* capabilities of any given normal flora organism can have profound impacts on the interactions that the microbe has with the host. The establishment of the human microbiome is initiated immediately after birth and is a necessary and normal part of human development.

Until relatively recently, our understanding of the organisms that compose the human microbiome relied on cultivation to isolate organisms in pure culture. This approach is limited in its usefulness for several reasons. First, the vast majority of microbes associated with humans cannot be cultivated ex vivo. Second, the ability to culture a microbe does not yield any information on the relative abundance of that organism in the niche under investigation. Finally, growing an organism out of its environment in pure culture gives little, if any, information on the complexity and interdependence of the microbial communities in that niche.

The development of sophisticated molecular techniques over the past decade (see Chapter 9 for more detail) has revealed enormous numbers of bacteria, yeasts, and protozoa that are associated with the human microbiome, many of which were previously unknown. Current estimates suggest that there is an approximately equal number of prokaryotic cells on and in the human body as there are human cells, most of which are gut-associated. This remarkable statistic is even more notable considering that the average adult gut is home to ~1000 bacterial species, each of which contains ~2000 genes, cataloging a total of 2,000,000 gut-associated microbial genes. This is 100 times more than the ~20,000 genes encoded in the entire human genome.

Variation in the abundance and complexity of the microbiome constituents is observed within an individual over time and also between individuals. However, longitudinal characterization of the human gut microbiome has shown that within the first few years of life, our microbial communities become relatively stable and unique to each individual unless perturbed, such as by antibiotic treatment.

Once established, members of the microbiome are considered **permanent residents** of the associated body sites, such as the skin, oropharynx, colon, and vagina (Tables 6–1 and 6–2). These microbes are often referred to as **commensals**, which are organisms that derive benefit from another host but do not damage that host.

Members of the microbiota vary in both abundance and type from one body site to another. Internal organs usually are sterile, although the central nervous system, blood, lower bronchi and alveoli, liver, spleen, kidneys, and bladder experience occasional transient microbial intrusions, often introduced after modest trauma, such as flossing of teeth, or abrasions on the skin.

A distinction should be made between the established members of the microbiome and the **carrier state**. The term *carrier* implies that an individual has become **colonized** with a potential pathogen and therefore can be a source of infection of others. It is most frequently used in reference to a person with an asymptomatic infection or to someone who has recovered from a disease but continues to carry the organism and can serve as a reservoir of infection for others.

It is important to be aware that members of the normal flora can cause disease when they gain access to other body sites. Examples of this include *Escherichia coli* and *Bacteroides*

TABLE 6–1   Summary of the Members of Normal Flora and Their Anatomic Locations

Members of the Normal Flora[1]	Anatomic Location
Bacteroides species	Colon, throat, vagina
Candida albicans	Mouth, colon, vagina
Clostridium species	Colon
Corynebacterium species (diphtheroids)	Nasopharynx, skin, vagina
Enterococcus faecalis	Colon
Escherichia coli and other coliforms	Colon, vagina, outer urethra
Gardnerella vaginalis	Vagina
Haemophilus species	Nasopharynx
Lactobacillus species	Mouth, colon, vagina
Neisseria species	Mouth, nasopharynx
Propionibacterium acnes	Skin
Pseudomonas aeruginosa	Colon, skin
Staphylococcus aureus	Nose, skin
Staphylococcus epidermidis	Skin, nose, mouth, vagina, urethra
Viridans streptococci	Mouth, nasopharynx

[1]In alphabetical order.

fragilis, both are normal flora organisms of the intestinal tract, which cause urinary tract infections (*E. coli*) and peritonitis (often a combination of *E. coli* and *B. fragilis*).

The human microbiome can confer resistance to pathogen colonization. The nonpathogenic resident bacteria occupy attachment sites on the mucosa that can interfere with colonization by pathogenic bacteria. The ability of members of the normal flora to limit the growth of pathogens is called **colonization resistance**. If the composition of normal flora is altered (e.g., by diet) or suppressed by antibiotics, pathogens may grow and cause disease. For example, antibiotic use can reduce the normal colonic flora, allowing the growth of *Clostridium difficile*, which can lead to pseudomembranous colitis.

The human microbiome harbors a diverse reservoir of antibiotic resistance genes. There is a substantial repertoire of resistance genes in the bacteria that comprise the gut microbiome. Since antibiotic resistance determinants are readily exchanged between bacteria through horizontal gene transfer, these genes can serve as a reservoir of resistance that is accessible to pathogens.

MICROBIOME OF THE INTESTINAL TRACT

In normal fasting people, the stomach contains few organisms, primarily because of its low pH. The small intestine usually contains small numbers of streptococci, lactobacilli, and yeasts, particularly *Candida albicans*. Larger numbers of these organisms are found in the terminal ileum.

The largest and most complex microbial population in humans resides in the colon. Roughly 20% of feces consists of primarily anaerobic bacteria at approximately 10^{11} organisms/g. In most people, within the colon, the two most abundant phyla of bacteria are the Firmicutes followed by the Bacteroidetes, although there is a great inter-individual variation in the population. The Firmicutes are gram-positive rods and members of the genera *Clostridium* and *Faecalibacterium* are prominent

TABLE 6–2   Medically Important Members of the Normal Flora

Location	Important Organisms[1]	Less Important Organisms[2]
Skin	*Staphylococcus epidermidis*	*Staphylococcus aureus, Corynebacterium* (diphtheroids), various streptococci, *Pseudomonas aeruginosa*, anaerobes (e.g., *Propionibacterium*), yeasts (e.g., *Candida albicans*)
Nose	*S. aureus*[3]	*S. epidermidis, Corynebacterium* (diphtheroids), various streptococci
Mouth	Viridans streptococci	Various streptococci, *Eikenella corrodens*
Dental plaque	*Streptococcus mutans*	*Prevotella intermedia, Porphyromonas gingivalis*
Gingival crevices	Various anaerobes (e.g., *Bacteroides, Fusobacterium*, streptococci, *Actinomyces*)	
Throat	V. streptococci	Various streptococci (including *Streptococcus pyogenes* and *Streptococcus pneumoniae*), *Neisseria* species, *Haemophilus influenzae, S. epidermidis*
Colon	*Bacteroides fragilis, Escherichia coli*	*Bifidobacterium, Eubacterium, Fusobacterium, Lactobacillus*, various aerobic gram-negative rods, *Enterococcus faecalis* and other streptococci, *Clostridium*
Vagina	*Lactobacillus, E. coli*,[3] group B streptococci[3]	Various streptococci, various gram-negative rods. *B. fragilis, Corynebacterium* (diphtheroids), *C. albicans*
Urethra		*S. epidermidis, Corynebacterium* (diphtheroids), various streptococci, various gram-negative rods (e.g., *E. coli*)[3]

[1]Organisms that are medically significant or present in large numbers.

[2]Organisms that are less medically significant or present in smaller numbers.

[3]These organisms are not part of the normal flora in this location but are important colonizers.

TABLE 6–3 Major Bacteria Found in the Colon

Bacterium[1]	Number/g of Feces	Important Pathogen
Bacteroides, especially *B. fragilis*	10^{10}–10^{11}	Yes
Bifidobacterium	10^{10}	No
Eubacterium	10^{10}	No
Coliforms	10^7–10^8	Yes
Enterococcus, especially *E. faecalis*	10^7–10^8	Yes
Lactobacillus	10^7	No
Clostridium, especially *C. perfringens*	10^6	Yes

[1]*Bacteroides*, *Bifidobacterium*, and *Eubacterium* (which make up more than 90% of the fecal flora) are anaerobes. Coliforms (*Escherichia coli*, *Enterobacter* species, and other gram-negative organisms) are the predominant facultative anaerobes.

organisms. The Bacteroidetes are gram-negative rods and the genera *Bacteroides* and *Prevotella* are important members. Species of Proteobacteria (gram-negative rods such as *Escherichia* and *Salmonella*) and Actinobacteria (gram-positive rods such as *Actinomyces*) make up the bulk of the remainder. The major bacteria found in the colon are listed in Table 6–3.

Colonization of the neonatal gut begins immediately after birth, predominated by microbes from the mother. The infant gut microbiota is low in diversity, fluctuating in response to diet. The microbiomes of breast-fed infants are distinct from those that are formula-fed. Weaning has a pronounced impact on the infant microbiome. Once solid food is introduced, the types and quantities of complex carbohydrates and dietary fiber impact the diversity of microbiome composition. Microbial diversity continues to increase and begins to stabilize and take on a more adult-like configuration starting around 3 years of age.

There is mounting evidence that the microbiome composition plays important roles in several disease states, such as weight control (obesity), and several inflammatory diseases, such as the two main inflammatory bowel diseases (IBDs)—Crohn's disease and ulcerative colitis. The effect on obesity is revealed by studies involving the transfer of fecal bacteria between strains of inbred mice. For example, fecal bacteria from obese mice transplanted into germ-free strains of nonobese mice resulted in the nonobese mice becoming obese. It appears that the fecal bacteria metabolize more of the input food, making more calories available to the mice. In other experiments, fecal transplants from identical (monozygotic) human twins, one obese and the other not obese, were transplanted into germ-free mice. The mice that received the fecal transplant from the obese twin gained significantly more weight than the mice that received the fecal transplant from the nonobese twin.

IBD is characterized by dysbiosis of the microbiome. Several studies have suggested that the microbiomes of patients with IBD have significantly lower abundances of putative beneficial microorganisms, particularly from the phyla Bacteroidetes and Firmicutes, and more from the phyla Actinobacteria and Proteobacteria, than healthy subjects. In addition, the composition of the microbiomes from patients with IBD fluctuated considerably more than those of healthy individuals.

The gut microbiome provides instructions to the developing immune system in the intestinal tract. Research in germ-free animal models reveals that the gut microbiota plays a critical immunomodulatory role in the development of gut-associated lymphoid tissues (GALT). Germ-free animals show low serum levels of antibodies and do not produce CD8 intraepithelial lymphocytes. In addition, variation in microbiome composition influences the proportion of Th1, Th2, and Th17 T cells (see Chapter 60). These observations all suggest that an intact, healthy microbiota impacts the development of adaptive immune responses.

The gut microbiome contributes to nutrition and synthesis of several vitamins. Gut bacteria aid digestion by breaking down otherwise indigestible plant fibers into short-chain fatty acids, such as propionate and butyrate, that intestinal cells can access. They also synthesize a variety of micronutrients including several of the B vitamins and vitamin K and have a major impact on the absorption of key minerals, such as iron.

The gut microbiome plays an important role in the metabolism of some drugs. For example, certain gut bacteria produce an enzyme that alters dioxin to a less readily absorbed form. Tacrolimus, an immunosuppressive drug, can be inactivated by certain gut bacteria. On the other hand, some drugs are activated by gut bacteria. Lovastatin, the cholesterol-lowering drug, is administered as a prodrug that is metabolized by gut bacteria into the active form.

The gut microbiome may play an important role in mental health. There is evidence that the gut microbiome may influence susceptibility to depression, schizophrenia, and other mental disorders. A **"Gut-Brain Axis"** is postulated to explain various findings in experimental animals. Research to support the existence of a gut-brain axis involves severing the vagus nerve and observing various changes in the mental health of the test animals.

MICROBIOME OF THE SKIN

The skin has a microbiome that is less complex than that of the intestinal tract. The predominant organism on the skin is *Staphylococcus epidermidis*, which in this location is a nonpathogen but can cause disease when it reaches certain sites, such as artificial heart valves and prosthetic joints. It is found on the skin much more frequently than its pathogenic relative *Staphylococcus aureus* (see Table 6–2). There are about 10^3 to 10^4 organisms/cm^2 of skin. Most of them are located superficially in the stratum corneum, but some are found in the hair follicles and act as a reservoir to replenish the superficial flora after hand washing. Anaerobic organisms, such as *Propionibacterium* and *Peptococcus*, are situated in the deeper follicles in the dermis, where oxygen tension is low. *Propionibacterium acnes* is a common skin anaerobe that is implicated in the pathogenesis of acne.

The yeast *C. albicans* is also a member of the normal flora of the skin. It can enter a person's bloodstream when needles pierce the skin (e.g., in patients with intravenous catheters or in those who use intravenous drugs). It is an important cause of systemic infections in patients with reduced cell-mediated immunity.

MICROBIOME OF THE RESPIRATORY TRACT

The microbiome of the nose, throat, and mouth contains a wide spectrum of organisms but the lower bronchi and alveoli typically contain few, if any, organisms. The nose is colonized by a variety of streptococcal and staphylococcal species and some people are long-term carriers of the important pathogen *S. aureus*. Occasional outbreaks of disease caused by *S. aureus*, particularly in newborns, can be traced to nasal, skin, or perianal carriage by health care personnel.

The throat contains a mixture of viridans streptococci, *Neisseria* species, and *S. epidermidis* (see Table 6–2). These nonpathogens occupy attachment sites on the pharyngeal mucosa and inhibit the growth of the pathogens *Streptococcus pyogenes*, *Neisseria meningitidis*, and *S. aureus*, respectively.

In the mouth, viridans streptococci make up about half of the bacteria and are found on a variety of oral surfaces, including the teeth. Plaque that builds up on the enamel surface of teeth is composed of salivary proteins that deposit on the enamel as well as gelatinous, high-molecular-weight glucans secreted by colonizing streptococcal bacteria, which form a structure for an ordered succession of different organisms to colonize. *Streptococcus mutans*, a member of the viridans group, is of special interest since it is found in large numbers (10^{10}/g) in the dental plaque of patients with dental caries. The *S. mutans* established in the plaque produces a large amount of acid, which demineralizes the enamel and initiates caries. Other viridans streptococci found in the oral cavity, such as *Streptococcus sanguinis*, are also among the leading causes of subacute bacterial (infective) endocarditis. These organisms can enter the bloodstream and attach to damaged heart valves.

Anaerobic bacteria, such as species of *Bacteroides*, *Prevotella*, *Fusobacterium*, *Clostridium*, and *Peptostreptococcus*, are found in the gingival crevices, where the oxygen concentration is very low. If aspirated, these organisms can cause lung abscesses, especially in debilitated patients with poor dental hygiene. In addition, the gingival crevices are the natural habitat of *Actinomyces israelii*—an anaerobic actinomycete that can cause abscesses of the jaw, lungs, or abdomen.

MICROBIOME OF THE GENITOURINARY TRACT

The vaginal flora of adult women consists primarily of *Lactobacillus* species (see Table 6–2). Lactobacilli are responsible for producing the acid that keeps the pH of the adult woman's vagina low. Lactobacilli appear to prevent the growth of potential pathogens, since their suppression by antibiotics can lead to overgrowth by *C. albicans*. Overgrowth of this yeast can result in *Candida* vaginitis.

The vagina is located close to the anus and can be colonized by members of the fecal flora. For example, women who are prone to recurrent urinary tract infections harbor organisms such as *E. coli* and *Enterobacter* in the opening of the vaginal cavity. About 15% to 20% of women of childbearing age carry group B streptococci in the vagina. This organism is an important cause of sepsis and meningitis in the newborn and is acquired during passage through the birth canal. The vagina is colonized by *S. aureus* in approximately 5% of women, which predisposes them to toxic shock syndrome.

Urine in the bladder is sterile in the healthy person, but during passage through the outermost portions of the urethra, it often becomes contaminated with *S. epidermidis*, coliforms, diphtheroids, and nonhemolytic streptococci. The area around the urethra of women and uncircumcised men contains secretions that carry *Mycobacterium smegmatis*, an acid-fast organism. The skin surrounding the genitourinary tract is the site of *Staphylococcus saprophyticus*, a cause of urinary tract infections in women.

PEARLS

- **Normal flora** are those microorganisms that are the **permanent residents** of the body in all humans. The **microbiome** refers to normal flora organisms and additionally includes the genetic composition and capabilities of these organisms. Some people can be transiently **colonized**, either for short or long periods, with certain organisms, but those are not considered members of the normal flora. **Carriers** (also called chronic carriers) are individuals in whom pathogenic organisms are present in significant numbers and therefore are a source of infection for others.

- Normal flora organisms inhabit the body surfaces exposed to the environment, such as the **skin, oropharynx, intestinal tract,** and **vagina.** Members of the normal flora differ in number and kind at various anatomic sites.

- Members of the normal flora are **low-virulence** organisms. In their usual anatomic site, they are nonpathogenic. However, if they leave their usual anatomic site, especially in an immunocompromised individual, they can cause disease. Normal flora organisms can also horizontally acquire genes from other members of the microflora, which can impact their virulence.

- **Colonization resistance** occurs when members of the normal flora occupy receptor sites on the skin and mucosal surfaces, thereby preventing pathogens from binding to those receptors.

Important Members of the Normal Flora

- **Skin.** The predominant member of the normal flora of the skin is *Staphylococcus epidermidis*. It is an important cause of infections of prosthetic heart valves and prosthetic joints. *Candida albicans*, **a yeast** also found on the skin, can enter the bloodstream and cause disseminated infections, such as endocarditis in intravenous drug users. *Staphylococcus aureus* is also present on the skin, but its **main site is in the nose**. It causes abscesses in the skin and in many other organs.

- **Oropharynx.** The main members of the normal flora of the mouth and throat are the **viridans streptococci**, such as *Streptococcus sanguinis* and *Streptococcus mutans*. Viridans streptococci are the most common cause of subacute endocarditis.

- **Gastrointestinal tract.** The stomach contains very few organisms because of the low pH. The colon contains the **largest** **number of normal flora** and the most diverse species, including both anaerobic and facultative bacteria. There are both gram-positive and gram-negative rods and cocci. The members of the colonic normal flora are an important cause of disease outside of the colon. The two most important members of the colonic flora that cause disease are the anaerobe *Bacteroides fragilis* and the facultative *Escherichia coli*. *Enterococcus faecalis*, a facultative bacterium, is also an important pathogen.

- **Vagina. Lactobacilli** are the predominant normal flora organisms in the vagina. They keep the pH of the vagina low, which inhibits the growth of organisms such as *C. albicans*, an important cause of vaginitis.

- **Urethra.** The outer third of the urethra contains a mixture of bacteria, primarily *S. epidermidis*. The female urethra can become colonized with fecal flora such as *E. coli*, which predisposes to urinary tract infections.

SELF-ASSESSMENT QUESTIONS

1. The colon is the site of the largest number of normal flora bacteria. Which of the following bacteria is found in the greatest number in the colon?

    (A) *Bacteroides fragilis*
    (B) *Clostridium perfringens*
    (C) *Enterococcus faecalis*
    (D) *Escherichia coli*
    (E) *Lactobacillus* species

2. A 76-year-old woman with a prosthetic (artificial) hip comes to you complaining of fever and pain in that joint. You are concerned about an infection by *Staphylococcus epidermidis*. Using your knowledge of normal flora, what is the most likely source of this organism?

    (A) Dental plaque
    (B) Mouth
    (C) Skin
    (D) Stomach
    (E) Vagina

3. Your patient is a 30-year-old woman with a previous history of rheumatic fever who has had fever for the past 2 weeks. On examination, you find a new heart murmur. You suspect endocarditis and do a blood culture, which grows a viridans group *Streptococcus* later identified as *S. sanguinis*. Using your knowledge of normal flora, what is the most likely source of this organism?

    (A) Duodenum
    (B) Skin
    (C) Throat
    (D) Urethra
    (E) Vagina

4. An outbreak of postsurgical wound infections caused by *S. aureus* has occurred in the hospital. The infection control team was asked to determine whether the organism could be carried by one of the operating room personnel. Using your knowledge of normal flora, which of the following body sites is the most likely location for this organism?

    (A) Colon
    (B) Gingival crevice
    (C) Mouth
    (D) Nose
    (E) Throat

ANSWERS

(1) **(A)**
(2) **(C)**
(3) **(C)**
(4) **(D)**

PRACTICE QUESTIONS: USMLE & COURSE EXAMINATIONS

Questions on the topics discussed in this chapter can be found in the Basic Bacteriology section of Part XIII: USMLE (National Board) Practice Questions starting on page 733. Also see Part XIV: USMLE (National Board) Practice Examination starting on page 775.

7

Pathogenesis

PRINCIPLES OF PATHOGENESIS

A microorganism is a **pathogen** if it is capable of causing disease; however, some organisms are highly pathogenic (i.e., they often cause disease), whereas others cause disease rarely. **Opportunistic** pathogens are those that rarely, if ever, cause disease in immunocompetent people but can cause serious infection in patients with reduced host defenses (immunocompromised).

Virulence is a quantitative measure of pathogenicity and is measured by the number of organisms required to cause disease. The 50% lethal dose (LD_{50}) is the number of organisms needed to kill half of the hosts that are exposed to the pathogen, while the 50% infectious dose (ID_{50}) is the number needed to cause infection in half of the exposed hosts. Organisms with a *lower* LD_{50} (or ID_{50}) are said to be *more* virulent than those with a higher LD_{50} (or ID_{50}) because fewer organisms are needed to cause death or disease.

The **infectious dose** of an organism required to cause disease varies greatly among the pathogenic bacteria. For example, *Shigella* and *Salmonella* both cause diarrhea by infecting the gastrointestinal tract, but the infectious dose of *Shigella* is less than 100 organisms, whereas the infectious dose of *Salmonella* is on the order of 100,000 organisms. The infectious dose of bacteria depends primarily on their **virulence factors** (e.g., whether their pili allow them to adhere well to mucous membranes, or they produce exotoxins or endotoxins, or possess a capsule to protect them from phagocytosis, or if they can survive various nonspecific host defenses such as acid in the stomach).

Some pathogenic bacteria are referred to as **parasites** because their presence is detrimental to the host. For example, *Chlamydia* and *Rickettsia* are called **obligate intracellular parasites** because they can grow only within host cells. Most bacteria are facultative parasites because they can grow within cells, outside cells, or on bacteriologic media. The other use of the term *parasite* refers to the protozoa and the helminths, which are discussed in Part VI of this book.

WHY DO PEOPLE GET INFECTIOUS DISEASES?

People get infectious diseases when microorganisms overpower our host defenses (i.e., when the balance between the organism and the host shifts in favor of the organism). The organism or its toxins are present in sufficient amount to induce various symptoms, such as fever and inflammation, which we interpret as those of an infectious disease.

From the organism's perspective, the two critical determinants in overpowering the host are the **number of organisms** to which a person is exposed and the **virulence** of these organisms. Clearly, the greater the number of organisms, the greater is the likelihood of infection. It is important to realize, however, that a small number of highly virulent organisms can cause disease just as a large number of less virulent organisms can. The virulence of an organism is determined by its ability to produce various **virulence factors**, several of which were described previously.

The production of specific virulence factors also determines what disease the bacteria cause. For example, a strain of *Escherichia coli* that produces one type of exotoxin causes watery (nonbloody) diarrhea, whereas a different strain of *E. coli* that produces another type of exotoxin causes bloody diarrhea. This chapter describes several important examples of specific diseases related to the production of various virulence factors.

From the host's perspective, the two main arms of our host defenses are innate immunity and acquired immunity, the latter of which includes both antibody-mediated and cell-mediated immunity. A reduction in the functioning of any component of our host defenses shifts the balance in favor of the organism and increases the chance that an infectious disease will occur. Some important causes of a reduction in our host defenses include genetic immunodeficiencies such as agammaglobulinemia and acquired immunodeficiencies such as acquired immunodeficiency syndrome (AIDS), drug-induced immunosuppression in patients with organ transplants, and cancer patients who are receiving chemotherapy. Patients with diabetes and autoimmune diseases also may have reduced host defenses. An overview of our host defenses is presented in Chapters 8 and 57.

In many instances, a person can acquire an organism without subsequent development of disease because the host defenses were successful. Such **asymptomatic infections** are common and are typically recognized by detecting antibody against the organism in the patient's serum.

TYPES OF BACTERIAL INFECTIONS

The term **infection** has more than one meaning. One meaning is that an organism has infected the person (i.e., it has entered the body of that person). For example, a person can be infected with an organism of low pathogenicity and not develop symptoms of disease. Another meaning of the term *infection* is to describe an infectious disease, such as when a person says, "I have an infection." In this instance, infection and disease are being used interchangeably, but it is important to realize that the word infection does not have to be equated with disease. Usually, the meaning will be apparent from the context.

Bacteria cause disease by two major mechanisms: (1) **toxin production** and (2) **invasion** and **inflammation**. Toxins fall into two general categories: **exotoxins** and **endotoxins**. Exotoxins are secreted proteins, whereas endotoxins are lipopolysaccharides (LPS) that form an integral part of the cell wall of gram-negative bacteria. While not actively released from the cell, endotoxins can cause fever, shock, and other generalized symptoms. Both exotoxins and endotoxins by themselves can cause symptoms; the presence of the bacteria in the host is not required. Invasive bacteria, on the other hand, grow to large numbers locally and induce an inflammatory response consisting of erythema, edema, warmth, and pain. Invasion and inflammation are discussed later in the section entitled "Determinants of Bacterial Pathogenesis."

Many, but not all, infections are **communicable** or **contagious** (i.e., they are spread from host to host). For example, tuberculosis is communicable (i.e., it is spread from person to person via airborne droplets produced by coughing), but botulism is not, because the exotoxin produced by the organism in the contaminated food affects only those eating that food.

Epidemics are infections that occur much more frequently than usual, and are referred to as **pandemics** if there is worldwide distribution. An **endemic** infection is constantly present at a low level in a specific population. In addition to infections that result in overt symptoms, many are **inapparent** or **subclinical** and can be detected only by demonstrating a rise in antibody titer or by isolating the organism. Note that a person with an asymptomatic infection does not have symptoms but nevertheless can be the source of organisms that can cause symptomatic disease in others.

Some infections result in a **latent** state, after which reactivation of the growth of the organism and recurrence of symptoms may occur. Certain other infections lead to a **chronic carrier** state, in which the organisms continuously grow with or without producing symptoms in the host. Chronic carriers (e.g., "Typhoid Mary") are an important source of infection of others and hence are a public health hazard.

The determination of whether an organism recovered from a patient is actually the cause of the disease involves an awareness of two phenomena: normal flora and colonization. Members of the **normal flora** are permanent residents of the body and vary in type according to anatomic site. When an organism is obtained from a patient's specimen, the question of whether it is a member of the normal flora is important in interpreting the finding. **Colonization** refers to the presence of a new organism that is neither a member of the normal flora nor the cause of symptoms. It can be a difficult clinical dilemma to distinguish between a pathogen and a colonizer, especially in specimens obtained from the respiratory tract, such as throat cultures and sputum cultures.

STAGES OF BACTERIAL PATHOGENESIS

Most bacterial infections are acquired from an external source. However, some are caused by members of the normal flora and, as such, are not transmitted directly prior to the onset of infection.

A generalized sequence of the stages of infection is as follows:

(1) Transmission from an external source into the portal of entry.

(2) Evasion of primary host defenses such as skin or stomach acid.

(3) Adherence to mucous membranes, usually by bacterial pili.

(4) Colonization by growth of the bacteria at the site of adherence.

(5) Disease symptoms caused by toxin production or invasion accompanied by inflammation.

(6) Host responses, including both innate and adaptive immunity during steps 3, 4, and 5.

(7) Progression or resolution of the disease.

DETERMINANTS OF BACTERIAL PATHOGENESIS

1. Transmission

Understanding the mode of transmission of infectious agents is extremely important from a public health perspective, because interrupting the **chain of transmission** is an excellent way to prevent infectious diseases. The mode of transmission of many infectious diseases is "human-to-human," but infectious diseases are also transmitted from nonhuman sources such as soil, water, and animals. **Fomites** are inanimate objects, such as towels, that serve as a source of microorganisms that can cause infectious diseases. Table 7–1 describes some important examples of these modes of transmission.

Although some infections are caused by members of the normal flora, most are acquired by transmission from external sources. Pathogens exit the infected patient most frequently from the respiratory and gastrointestinal tracts; hence, transmission to the new host usually occurs via airborne respiratory droplets or fecal contamination of food and water. Organisms can also be transmitted by sexual contact, urine, skin contact, blood transfusions, contaminated needles, or biting insects. The transfer of blood, either by transfusion or by sharing needles during intravenous drug use, can transmit various bacterial and viral pathogens. The screening of donated blood for *Treponema pallidum* (the cause of syphilis), human immunodeficiency virus (HIV), human T-cell lymphotropic virus, hepatitis B virus, hepatitis C virus, and West Nile virus has greatly reduced the risk of infection by these organisms.

The major bacterial diseases **transmitted by ticks** in the United States are Lyme disease, Rocky Mountain spotted fever, ehrlichiosis, relapsing fever, and tularemia. Of these five diseases, Lyme disease is by far the most common. Ticks of the genus *Ixodes* (deer tick) transmit three infectious diseases: Lyme disease, ehrlichiosis, and babesiosis, a protozoan disease. *Dermacentor* ticks (dog tick) transmit several diseases: Rocky Mountain spotted fever, tularemia, ehrlichiosis, anaplasmosis, and tick paralysis.

Bacteria, viruses, and other microbes can also be transmitted from mother to offspring, a process called **vertical transmission**. The three modes by which organisms are transmitted vertically are across the placenta, within the birth canal during birth, and via breast milk. Table 7–2 describes some medically important organisms that are transmitted vertically. (**Horizontal transmission**, by contrast, is person-to-person transmission that is not from mother to offspring.)

There are four important portals of entry: respiratory tract, gastrointestinal tract, genital tract, and skin (Table 7–3). Important microorganisms and diseases transmitted by water are described in Table 7–4.

The important bacterial diseases transmitted by foods are listed in Table 7–5, and those transmitted by insects are listed in Table 7–6. The specific mode of transmission of each organism is described in the subsequent section devoted to that organism.

Animals are also an important source of organisms that infect humans. They can be either the source (**reservoir**) or the mode of transmission (**vector**) of certain organisms. Diseases for which animals are the reservoirs are called **zoonoses**. The important zoonotic diseases caused by bacteria are listed in Table 7–7.

2. Adherence to Cell Surfaces

Certain bacteria have specialized structures (e.g., **pili**) or produce substances (e.g., **capsules** or **glycocalyces**) that allow them to adhere to the surface of human cells, thereby enhancing their ability to cause disease. These adherence mechanisms are essential for organisms that attach to mucous membranes; mutants that lack these mechanisms are often nonpathogenic. For

TABLE 7–1 Important Modes of Transmission

Mode of Transmission	Clinical Example	Comment
I. Human-to-human		
A. Direct contact	Gonorrhea	Intimate contact (e.g., sexual or passage through birth canal)
B. No direct contact	Dysentery	Fecal–oral (e.g., excreted in human feces, then ingested in food or water)
C. Transplacental	Congenital syphilis	Bacteria cross the placenta and infect the fetus
D. Bloodborne	Hepatitis B	Transfused blood or intravenous drug use can transmit bacteria and viruses; screening of blood for transfusions has greatly reduced this risk
II. Nonhuman-to-human		
A. Soil source	Tetanus	Spores in soil enter wound in skin
B. Water source	Legionnaire's disease	Bacteria in water aerosol are inhaled into lungs
C. Animal source		
1. Directly	Cat-scratch fever	Bacteria enter in cat scratch
2. Via insect vector	Lyme disease	Bacteria enter in tick bite
3. Via animal excreta	Hemolytic-uremic syndrome caused by *E. coli* O157	Bacteria in cattle feces are ingested in undercooked hamburger
D. Fomite source	Staphylococcal skin infection	Bacteria on an object (e.g., a towel) are transferred onto the skin

TABLE 7–2 Vertical Transmission of Some Important Pathogens

Mode of Transmission	Pathogen	Type of Organism[1]	Disease in Fetus or Neonate
Transplacental	*Treponema pallidum*	B	Congenital syphilis
	Listeria monocytogenes[2]	B	Neonatal sepsis and meningitis
	Cytomegalovirus	V	Congenital abnormalities
	Parvovirus B19	V	Hydrops fetalis
	Toxoplasma gondii	P	Toxoplasmosis
Within birth canal/at the time of birth	*Streptococcus agalactiae* (group B *Streptococcus*)	B	Neonatal sepsis and meningitis
	Escherichia coli	B	Neonatal sepsis and meningitis
	Chlamydia trachomatis	B	Conjunctivitis or pneumonia
	Neisseria gonorrhoeae	B	Conjunctivitis
	Herpes simplex type 2	V	Skin, CNS, or disseminated infection (sepsis)
	Hepatitis B virus	V	Hepatitis B
	Human immunodeficiency virus[3]	V	Asymptomatic infection
	Candida albicans	F	Thrush
Breast milk	*Staphylococcus aureus*	B	Oral or skin infections
	Cytomegalovirus	V	Asymptomatic infection
	Human T-cell leukemia virus	V	Asymptomatic infection

CNS = central nervous system.

[1]B, bacterium; V, virus; F, fungus; P, protozoa.

[2]*L. monocytogenes* can also be transmitted at the time of birth.

[3]HIV is transmitted primarily at the time of birth but is also transmitted across the placenta and in breast milk.

TABLE 7–3 Portals of Entry of Some Common Pathogens

Portal of Entry	Pathogen	Type of Organism[1]	Disease
Respiratory tract	*Streptococcus pneumoniae*	B	Pneumonia
	Neisseria meningitides	B	Meningitis
	Haemophilus influenzae	B	Meningitis
	Mycobacterium tuberculosis	B	Tuberculosis
	Influenza virus	V	Influenza
	Rhinovirus	V	Common cold
	Epstein-Barr virus	V	Infectious mononucleosis
	Coccidioides immitis	F	Coccidioidomycosis
	Histoplasma capsulatum	F	Histoplasmosis
Gastrointestinal tract	*Shigella dysenteriae*	B	Dysentery
	Salmonella typhi	B	Typhoid fever
	Vibrio cholerae	B	Cholera
	Norovirus	V	Gastroenteritis
	Rotavirus	V	Gastroenteritis
	Hepatitis A virus	V	Hepatitis A
	Poliovirus	V	Poliomyelitis
	Trichinella spiralis	H	Trichinosis
Skin	*Staphylococcus aureus*	B	Impetigo, boils, cellulitis, folliculitis
	Clostridium tetani	B	Tetanus
	Rickettsia rickettsii	B	Rocky Mountain spotted fever
	Rabies virus	V	Rabies
	Trichophyton rubrum	F	Tinea pedis (athlete's foot)
	Plasmodium vivax	P	Malaria
Genital tract	*Neisseria gonorrhoeae*	B	Gonorrhea
	Treponema pallidum	B	Syphilis
	Chlamydia trachomatis	B	Urethritis
	Human papillomavirus	V	Genital warts
	Herpes simplex virus 2	V	Genital herpes
	Candida albicans	F	Vaginitis

[1]B, bacterium; V, virus; F, fungus; P, protozoa; H, helminth.

TABLE 7–4   Transmission of Important Waterborne Diseases

Portal of Entry	Pathogen	Type of Organism[1]	Disease
Gastrointestinal tract			
1. Ingestion of drinking water	*Salmonella* species	B	Diarrhea
	Shigella species	B	Diarrhea
	Campylobacter jejuni	B	Diarrhea
	Norovirus	V	Diarrhea
	Giardia lamblia	P	Diarrhea
	Cryptosporidium parvum	P	Diarrhea
2. Ingestion of water while swimming[2]	*Leptospira interrogans*	B	Leptospirosis
Respiratory tract			
Inhalation of water aerosol	*Legionella pneumophila*	B	Pneumonia (Legionnaire's disease)
Skin			
Penetration through skin	*Pseudomonas aeruginosa*	B	Hot-tub folliculitis
	Schistosoma mansoni	H	Schistosomiasis
Nose			
Penetration through cribriform plate into meninges and brain	*Naegleria fowleri*	P	Meningoencephalitis

[1]B, bacterium; V, virus; P, protozoa; H, helminth.
[2]All of the organisms that cause diarrhea by ingestion of drinking water also cause diarrhea by ingestion of water while swimming.

TABLE 7–5   Bacterial Diseases Transmitted by Foods

Bacterium	Typical Food	Main Reservoir	Disease
I. Diarrheal diseases			
Gram-positive cocci			
Staphylococcus aureus	Custard-filled pastries; potato, egg, or tuna fish salad	Humans	Food poisoning, especially vomiting
Gram-positive rods			
Bacillus cereus	Reheated rice	Soil	Diarrhea
Clostridium perfringens	Cooked meat, stew, and gravy	Soil, animals, or humans	Diarrhea
Listeria monocytogenes	Unpasteurized milk products	Soil, animals, or plants	Diarrhea, neonatal sepsis
Gram-negative rods			
Escherichia coli	Various foods and water	Humans	Diarrhea
E. coli O157:H7 strain	Undercooked meat	Cattle	Hemorrhagic colitis, hemolytic-uremic syndrome (HUS)
Salmonella enteritidis	Poultry, meats, and eggs	Domestic animals, especially poultry	Diarrhea
Salmonella typhi	Various foods	Humans	Typhoid fever
Shigella species	Various foods and water	Humans	Diarrhea (dysentery)
Vibrio cholerae	Various foods (e.g., seafood) and water	Humans	Diarrhea
Vibrio parahaemolyticus	Seafood	Warm salt water	Diarrhea
Campylobacter jejuni	Various foods	Domestic animals	Diarrhea
Yersinia enterocolitica	Various foods	Domestic animals	Diarrhea
II. Nondiarrheal diseases			
Gram-positive rods			
Clostridium botulinum	Improperly canned vegetables and smoked fish	Soil	Botulism
Listeria monocytogenes	Unpasteurized milk products	Cows	Sepsis in neonate or mother
Gram-negative rods			
Vibrio vulnificus	Seafood	Warm salt water	Sepsis
Brucella species	Meat and milk	Domestic animals	Brucellosis
Francisella tularensis	Meat	Rabbits	Tularemia
Mycobacteria			
Mycobacterium bovis	Milk	Cows	Intestinal tuberculosis

TABLE 7–6 Bacterial Diseases Transmitted by Insects

Bacterium	Insect	Reservoir	Disease
Gram-negative rods			
Yersinia pestis	Rat fleas	Rodents (e.g., rats and prairie dogs)	Plague
Francisella tularensis	Ticks (Dermacentor)	Many animals (e.g., rabbits)	Tularemia
Spirochetes			
Borrelia burgdorferi	Ticks (Ixodes)	Mice	Lyme disease
Borrelia recurrentis	Lice	Humans	Relapsing fever
Rickettsiae			
Rickettsia rickettsii	Ticks (Dermacentor)	Dogs, rodents, and ticks (Dermacentor)	Rocky Mountain spotted fever
Rickettsia prowazekii	Lice	Humans	Epidemic typhus
Ehrlichia chaffeensis	Ticks (Dermacentor, Ixodes)	Dogs	Ehrlichiosis
Anaplasma phagocytophilum	Ticks (Ixodes)	Dogs, rodents	Anaplasmosis

example, the **pili** of *Neisseria gonorrhoeae* and *E. coli* mediate the attachment of the organisms to the urinary tract epithelium, and the **glycocalyx** of *Staphylococcus epidermidis* and certain viridans streptococci allows the organisms to adhere strongly to the endothelium of heart valves. The various molecules that mediate adherence to cell surfaces are called **adhesins**.

After bacteria attach, they often form a protective matrix called a **biofilm** consisting of various polysaccharides and proteins. Biofilms form especially on foreign bodies such as prosthetic joints, prosthetic heart valves, and intravenous catheters, but they also form on native structures such as heart valves. Biofilms protect bacteria from both antibiotics and host immune defenses such as antibodies and neutrophils. They also retard wound healing, resulting in chronic wound infections, especially in diabetics. Biofilms play an important role in the persistence of *Pseudomonas* in the lungs of cystic fibrosis patients and in the formation of dental plaque, the precursor of dental caries and periodontal disease.

The production of biofilms by bacteria such as *Pseudomonas* is controlled by the process of **quorum sensing**, which

TABLE 7–7 Zoonotic Diseases Caused by Bacteria

Bacterium	Main Reservoir	Mode of Transmission	Disease
Gram-positive rods			
Bacillus anthracis	Domestic animals	Direct contact	Anthrax
Listeria monocytogenes	Domestic animals	Ingestion of unpasteurized milk products	Sepsis in neonate or mother
Erysipelothrix rhusiopathiae	Fish	Direct contact	Erysipeloid
Gram-negative rods			
Bartonella henselae	Cats	Skin scratch	Cat-scratch disease
Brucella species	Domestic animals	Ingestion of unpasteurized milk products; contact with animal tissues	Brucellosis
Campylobacter jejuni	Domestic animals	Ingestion of contaminated meat	Diarrhea
Escherichia coli O157:H7	Cattle	Fecal–oral	Hemorrhagic colitis
Francisella tularensis	Many animals, especially rabbits	Tick bite and direct contact	Tularemia
Pasteurella multocida	Cats	Cat bite	Cellulitis
Salmonella enteritidis	Poultry, eggs, and cattle	Fecal–oral	Diarrhea
Yersinia enterocolitica	Domestic animals	Fecal–oral	Diarrhea
Yersinia pestis	Rodents, especially rats and prairie dogs	Rat flea bite	Sepsis
Mycobacteria			
Mycobacterium bovis	Cows	Ingestion of unpasteurized milk products	Intestinal tuberculosis
Spirochetes			
Borrelia burgdorferi	Mice	Tick bite (Ixodes)	Lyme disease
Leptospira interrogans	Rats and dogs	Urine	Leptospirosis
Chlamydiae			
Chlamydia psittaci	Psittacine birds, especially parrots and parakeets	Inhalation of aerosols	Psittacosis
Rickettsiae			
Rickettsia rickettsia	Rats and dogs	Tick bite (Dermacentor)	Rocky Mountain spotted fever
Coxiella burnetiid	Sheep	Inhalation of aerosols of amniotic fluid	Q fever
Ehrlichia chaffeensis	Dogs	Tick bite (Dermacentor)	Ehrlichiosis
Anaplasma phagocytophilum	Dogs, rodents	Tick bite (Ixodes)	Anaplasmosis

allows bacteria to coordinate the synthesis of particular proteins according to the density of the bacterial population. When the concentration of bacteria is low, these proteins are not expressed; but once the population reaches a critical high cell density, the individual members sense this and begin to synthesize these proteins, resulting in phenotypic changes that benefit the population as a whole. Examples of behaviors that are controlled by quorum sensing include biofilm formation, expression of virulence factors, and antibiotic resistance, all of which can contribute to pathogenesis.

Foreign bodies, such as artificial heart valves and artificial joints, predispose to infections. Bacteria can adhere to these surfaces, but phagocytes adhere poorly owing to the absence of selectins and other binding proteins on the artificial surface (see Chapter 8).

3. Invasion, Inflammation, & Intracellular Survival

One of the main mechanisms by which bacteria cause disease is **invasion** of tissue followed by **inflammation**. (The inflammatory response is described in Chapter 8.) The other main mechanism, **toxin production**, and a third mechanism, **immunopathogenesis**, are described later in this chapter.

Several enzymes secreted by invasive bacteria play a role in pathogenesis. Among the most prominent are the following:

(1) **Collagenase** and **hyaluronidase**, which degrade collagen and hyaluronic acid, respectively, thereby allowing the bacteria to spread through subcutaneous tissue; they are especially important in cellulitis caused by *Streptococcus pyogenes*.

(2) **Coagulase**, which is produced by *Staphylococcus aureus* and accelerates the formation of a fibrin clot from its precursor, fibrinogen (this clot may protect the bacteria from phagocytosis by walling off the infected area and by coating the organisms with a layer of fibrin). Coagulase is also produced by *Yersinia pestis*, the cause of bubonic plague, which is discussed in Chapter 20.

(3) **Immunoglobulin (Ig) proteases**. There are several examples of organisms that produce enzymes that degrade IgA and IgG. *N. gonorrhoeae*, *Haemophilus influenzae*, and *Streptococcus pneumoniae* produce IgA proteases, which inactivate IgA at the mucosal surface. This leads to better adherence of these organisms to mucous membranes. *S. pyogenes* produces an enzyme that specifically cleaves IgG heavy chains, which reduces opsonization and complement activation, enhancing the virulence of this organism.

In addition to these enzymes, several virulence factors contribute to invasiveness by limiting the ability of the host defense mechanisms, especially phagocytosis, to operate effectively.

(1) The most important of these antiphagocytic factors is the **capsule** external to the cell wall of several important pathogens such as *S. pneumoniae*, *Neisseria meningitidis*, and *H. influenzae*. The polysaccharide capsule prevents the phagocyte from adhering to the bacteria. Note that anticapsular antibodies allow more effective phagocytosis to occur (a process called **opsonization**) (see Chapter 8). The vaccines against *S. pneumoniae*, *H. influenzae*, and *N. meningitidis* contain the capsular polysaccharide of the organism as the antigen that induces protective anticapsular antibodies.

(2) A second group of antiphagocytic factors are the cell wall proteins of the gram-positive cocci, such as the M protein of the group A streptococci (*S. pyogenes*) and protein A of *S. aureus*. The M protein is antiphagocytic, and protein A binds to the Fc portion of IgG and prevents the activation of complement. These virulence factors are summarized in Table 7–8.

(3) Leukocidins are pore-forming toxins that degrade the cell membrane of neutrophils and macrophages. The Panton-Valentine leukocidin produced by methicillin-resistant strains of *S. aureus* (MRSA) is a good example.

Bacteria can cause two types of inflammation: **pyogenic** and **granulomatous**. In pyogenic (pus-producing) inflammation, **neutrophils** are the predominant cells. Some of the most

TABLE 7–8 Surface Virulence Factors Important for Bacterial Pathogenesis

Organism	Virulence Factor	Used in Vaccine	Comments
Gram-positive cocci			
Streptococcus pneumoniae	Polysaccharide capsule	Yes	Determines serotype
Streptococcus pyogenes	M protein	No	Determines serotype[1]
Staphylococcus aureus	Protein A	No	Inhibits phagocytosis by binding to Fc region of IgG, which prevents activation of complement
Gram-negative cocci			
Neisseria meningitides	Polysaccharide capsule	Yes	Determines serotype
Gram-positive rods			
Bacillus anthracis	Polypeptide capsule	No	
Gram-negative rods			
Haemophilus influenzae	Polysaccharide capsule	Yes	Determines serotype
Klebsiella pneumoniae	Polysaccharide capsule	No	
Escherichia coli	Protein pili	No	Causes adherence
Salmonella typhi	Polysaccharide capsule	Yes	Not important for other salmonellae
Yersinia pestis	V and W proteins	No	

[1]Do not confuse the serotype with the grouping of streptococci, which is determined by the polysaccharide in the cell wall.

important pyogenic bacteria are the gram-positive and gram-negative cocci listed in Table 7–8. In granulomatous inflammation, **macrophages and helper T cells** predominate. The most important organism in this category is *Mycobacterium tuberculosis*. No bacterial enzymes or toxins that induce granulomas have been identified. Rather, it appears that bacterial antigens stimulate the cell-mediated immune system, resulting in sensitized T-lymphocyte and macrophage activity. Phagocytosis by macrophages kills most of the bacteria, but some survive and grow within the macrophages in the granuloma.

Two important diseases, diphtheria and pseudomembranous colitis, are characterized by inflammatory lesions called **pseudomembranes**. These lesions are thick, adherent, grayish or yellowish exudates on the mucosal surfaces of the throat in diphtheria and on the colon in pseudomembranous colitis. The term *pseudo* refers to the abnormal nature of these membranes in contrast to the normal anatomic membranes of the body, such as the tympanic membrane and the placental membranes.

Intracellular survival is an important attribute of certain bacteria that enhances their ability to cause disease. These bacteria are called "intracellular" pathogens and commonly cause granulomatous lesions. The best-known of these bacteria belong to the genera *Mycobacterium, Legionella, Brucella,* and *Listeria*. The best-known intracellular fungus is *Histoplasma*, which is often found within macrophages. These organisms can be cultured on microbiologic media in the laboratory and therefore are *not* obligate intracellular parasites, which distinguishes them from *Chlamydia* and *Rickettsia*. The intracellular location provides a protective niche from antibody and neutrophils that function extracellularly.

Intracellular bacteria use several different mechanisms to allow them to survive and grow inside cells. These include (1) inhibition of the fusion of the phagosome with the lysosome, which allows the organisms to avoid the degradative enzymes in the lysosome; (2) inhibition of acidification of the phagosome, which reduces the activity of the lysosomal degradative enzymes; and (3) escape from the phagosome into the cytoplasm, where there are no degradative enzymes. Members of the genera *Mycobacterium* and *Legionella* are known to use the first and second mechanisms, whereas *Listeria* species use the third.

The invasion of cells by bacteria is dependent on the interaction of specific bacterial surface proteins called **invasins** and specific cellular receptors belonging to the integrin family of transmembrane adhesion proteins. The movement of bacteria into the cell is a function of actin microfilaments. Once inside the cell, these bacteria typically reside within cell vacuoles such as phagosomes. Some remain there, others migrate into the cytoplasm, and some move from the cytoplasm into adjacent cells. Infection of the surrounding cells in this manner allows the bacteria to evade host defenses. For example, *Listeria monocytogenes* aggregates actin filaments on its surface and is propelled in a "sling-shot" fashion, called **actin rockets**, from one host cell to another.

The "Yops" (*Yersinia* outer-membrane proteins) produced by several *Yersinia* species are important examples of bacterial virulence factors that act primarily after invasion of human cells by the organism. The most important effects of the Yops are to inhibit phagocytosis by neutrophils and macrophages and to inhibit cytokine production (e.g., tumor necrosis factor [TNF] production) by macrophages. For example, one of the Yops of *Y. pestis* (Yop J) is a protease that cleaves signal transduction proteins required for the induction of TNF synthesis. This inhibits the activation of our host defenses and contributes to the ability of the organism to cause bubonic plague.

The genes that encode many virulence factors in bacteria are clustered in **pathogenicity islands** located on the bacterial chromosome or plasmids. For example, in many bacteria, the genes encoding adhesins, invasins, and exotoxins are adjacent to each other on these islands. Nonpathogenic variants of these bacteria do not have these pathogenicity islands. It appears that these large regions of the bacterial genome were transferred as a block via conjugation or transduction. Pathogenicity islands are found in many gram-negative rods, such as *E. coli, Salmonella, Shigella, Pseudomonas,* and *Vibrio cholerae,* and in gram-positive cocci, such as *S. pneumoniae*.

After bacteria have colonized and multiplied at the portal of entry, they may invade the bloodstream and spread to other parts of the body. Receptors for the bacteria on the surface of cells determine, in large part, the organs affected. For example, certain bacteria or viruses infect the brain because receptors for these microbes are located on the surface of brain neurons. The *blood–brain barrier*, which limits the ability of certain drugs to penetrate the brain, is not thought to be a determinant of microbial infection of the brain. The concept of a blood–brain barrier primarily refers to the inability of hydrophilic (charged, ionized) drugs to enter the lipid-rich brain parenchyma, whereas lipophilic (lipid-soluble) drugs enter well.

4. Toxin Production

The second major mechanism by which bacteria cause disease is the production of toxins. A comparison of the main features of **exotoxins** and **endotoxins** is shown in Table 7–9.

Exotoxins

Exotoxins are produced by several gram-positive and gram-negative bacteria, in contrast to endotoxins, which are present only in gram-negative bacteria. The essential characteristic of exotoxins is that they are **secreted** by the bacteria, whereas endotoxin is a component of the cell wall. Exotoxins are polypeptides whose genes are frequently located on plasmids or lysogenic bacterial viruses (bacteriophages). Some important exotoxins encoded by bacteriophage DNA are diphtheria toxin, cholera toxin, and botulinum toxin.

Exotoxins are among the **most toxic** substances known. For example, the fatal dose of tetanus toxin for a human is estimated to be less than 1 μg. Because some purified exotoxins can reproduce all aspects of the disease, we can conclude that certain bacteria play no other role in pathogenesis than to synthesize the exotoxin. Exotoxin polypeptides are good antigens and induce the synthesis of protective antibodies called antitoxins, some of

TABLE 7–9 **Main Features of Exotoxins and Endotoxins**

Property	Exotoxin	Endotoxin
	Comparison of Properties	
Source	Certain species of gram-positive and gram-negative bacteria	Cell wall of gram-negative bacteria
Secreted from cell	Yes	No
Chemistry	Polypeptide	Lipopolysaccharide
Location of genes	Plasmid or bacteriophage	Bacterial chromosome
Toxicity	High (fatal dose on the order of 1 μg)	Low (fatal dose on the order of hundreds of micrograms)
Clinical effects	Various effects (see text)	Fever, shock
Mode of action	Various modes (see text)	Includes TNF and interleukin-1
Antigenicity	Induces high-titer antibodies called antitoxins	Poorly antigenic
Vaccines	Toxoids used as vaccines	No toxoids formed and no vaccine available
Heat stability	Destroyed rapidly at 60°C (except staphylococcal enterotoxin)	Stable at 100°C for 1 hour
Typical diseases	Tetanus, botulism, diphtheria	Meningococcemia, sepsis by gram-negative rods

TNF = tumor necrosis factor.

which are useful in the prevention or treatment of diseases such as botulism and tetanus. When treated with formaldehyde (or acid or heat), the exotoxin polypeptides are converted into **toxoids**, which are used in protective vaccines because they retain their antigenicity but have lost their toxicity.

Many exotoxins have an **A–B subunit** structure; the A (or active) subunit possesses the toxic activity, and the B (or binding) subunit is responsible for binding the exotoxin to specific receptors on the membrane of the human cell. The binding of the B subunit determines the specific site of the action of the exotoxin. For example, botulinum toxin acts at the neuromuscular junction because the B subunit binds to specific receptors on the surface of the motor neuron at the junction. Important exotoxins that have an A–B subunit structure include diphtheria toxin, tetanus toxin, botulinum toxin, cholera toxin, and the enterotoxin of *E. coli* (Figure 7–1).

The A subunit of several important exotoxins acts by catalyzing the addition of adenosine diphosphate ribose (ADP-ribose) to the target protein in the human cell (**ADP-ribosylation**). The modification of target proteins with ADP-ribose often

inactivates it but can also hyperactivate it, either of which can cause the symptoms of disease. For example, diphtheria toxin and *Pseudomonas* exotoxin A ADP-ribosylate elongation factor-2 (EF-2), an essential factor required for eukaryotic protein synthesis. This modification inactivates EF-2, freezing the translocation complex, and results in the inhibition of protein synthesis.

On the other hand, cholera toxin and *E. coli* toxin ADP-ribosylate G_s protein, thereby activating it. This causes an increase in adenylate cyclase activity, a consequent increase in the amount of cyclic adenosine monophosphate (AMP), and the production of watery diarrhea. Pertussis toxin is an interesting variation on the theme. It ADP-ribosylates G_i protein and inactivates it. Inactivation of the inhibitory G proteins turns on adenylate cyclase, causing an increase in the amount of cyclic AMP, which plays a role in causing the symptoms of whooping cough.

Exotoxins are released from bacteria by specialized structures called **secretion systems**. Some secretion systems transport the exotoxins into the extracellular space, but others transport the exotoxins directly into the mammalian cell. Those that transport the exotoxins directly into the mammalian cell are especially effective because the exotoxin is not exposed to antibodies in the extracellular space.

Several classes of bacterial secretion systems (at least nine!) have been identified, but the **type III secretion system** (also called an injectosome) is particularly important in virulence. This secretion system is mediated by a needlelike projection (sometimes called a "molecular syringe") and by transport pumps in the bacterial cell membrane. The importance of the type III secretion system is illustrated by the finding that the strains of *Pseudomonas aeruginosa* that have this secretion system are significantly more virulent than those that do not. Other medically important gram-negative rods that utilize similar systems include *Shigella* species, *Salmonella* species, *E. coli*, and *Y. pestis*.

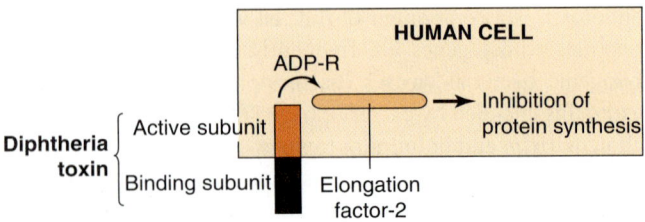

FIGURE 7–1 Mode of action of diphtheria toxin. The toxin binds to the cell surface via its binding subunit, and the active subunit enters the cell. The active subunit is an enzyme that catalyzes the addition of ADP-ribose (ADP-R) to elongation factor-2 (EF-2). This inactivates EF-2, and protein synthesis is inhibited.

TABLE 7–10 Important Bacterial Exotoxins

Bacterium	Disease	Mode of Action	Toxoid Vaccine
Gram-positive rods			
Corynebacterium diphtheriae	Diphtheria	Inactivates EF-2 by ADP-ribosylation	Yes
Clostridium tetani	Tetanus	Blocks release of the inhibitory neurotransmitter glycine by proteolytic cleavage of releasing proteins	Yes
Clostridium botulinum	Botulism	Blocks release of acetylcholine by proteolytic cleavage of releasing proteins	Yes[1]
Clostridium difficile	Pseudomembranous colitis	Exotoxins A and B inactivate GTPases by glucosylation	No
Clostridium perfringens	Gas gangrene	Alpha toxin is a lecithinase; enterotoxin is a superantigen	No
Bacillus anthracis	Anthrax	Edema factor is an adenylate cyclase; lethal factor is a protease that cleaves MAP kinase, which is required for cell division	No
Gram-positive cocci			
Staphylococcus aureus	1. Toxic shock syndrome	Is a superantigen; binds to class II MHC protein and T-cell receptor; induces IL-1 and IL-2	No
	2. Food poisoning	Is a superantigen acting locally in the gastrointestinal tract	No
	3. Scalded skin syndrome	Is a protease that cleaves desmoglein in the skin	No
Streptococcus pyogenes	Scarlet fever	Is a superantigen; action similar to toxic shock syndrome toxin of *S. aureus*	No
	Necrotizing fasciitis	Exotoxin B is a protease that cleaves E-cadherin in skin	
Gram-negative rods			
Escherichia coli	1. Watery diarrhea	Labile toxin stimulates adenylate cyclase by ADP-ribosylation; stable toxin stimulates guanylate cyclase	No
	2. Bloody diarrhea	Shiga toxin inhibits protein synthesis in enterocytes by removing adenine from 28S ribosomal RNA	No
Shigella dysenteriae	Bloody diarrhea	Shiga toxin inhibits protein synthesis in enterocytes by removing adenine from 28S ribosomal RNA	No
Vibrio cholerae	Cholera	Stimulates adenylate cyclase by ADP-ribosylation	No
Bordetella pertussis	Whooping cough	Stimulates adenylate cyclase by ADP-ribosylation; inhibits chemokine receptor	Yes[2]

[1]For high-risk individuals only.
[2]The acellular vaccine contains pertussis toxoid and four other proteins.

The mechanisms of action of the important exotoxins produced by toxigenic bacteria are described in the following discussion and summarized in Tables 7–10 to 7–12. The main location of symptoms of disease caused by bacterial exotoxins is described in Table 7–13.

Gram-Positive Bacteria

The exotoxins produced by gram-positive bacteria have several different mechanisms of action and produce different clinical effects. Some important exotoxins include diphtheria toxin, which inhibits protein synthesis by inactivating EF-2; tetanus toxin and botulinum toxin, which are neurotoxins that prevent the release of neurotransmitters; and toxic shock syndrome toxin (TSST), which acts as a superantigen causing the release of large amounts of cytokines from helper T cells and macrophages. The mechanisms of action and the clinical effects of exotoxins produced by gram-positive bacteria are described next.

TABLE 7–11 Important Mechanisms of Action of Bacterial Exotoxins

Mechanism of Action	Exotoxin
ADP-ribosylation	Diphtheria toxin, cholera toxin, *Escherichia coli* heat-labile toxin, and pertussis toxin
Superantigen	Toxic shock syndrome toxin, staphylococcal enterotoxin, and erythrogenic toxin
Protease	Tetanus toxin, botulinum toxin, lethal factor of anthrax toxin, and scalded skin toxin
Lecithinase	*Clostridium perfringens* alpha toxin

TABLE 7–12 Exotoxins That Increase Intracellular Cyclic AMP

Bacterium	Exotoxin	Mode of Action
Vibrio cholerae	Cholera toxin	ADP-ribosylates G_s factor, which activates it, thereby stimulating adenylate cyclase
Escherichia coli	Labile toxin	Same as cholera toxin
Bordetella pertussis	Pertussis toxin	ADP-ribosylates G_i factor, which inactivates it, thereby stimulating adenylate cyclase
Bacillus anthracis	Edema factor of anthrax toxin	Is an adenylate cyclase

TABLE 7–13 Main Location of Symptoms of Disease Caused by Bacterial Exotoxins

Main Location of Symptoms	Organism	Mode of Action Exotoxin
Gastrointestinal tract		
1. Gram-positive cocci	*Staphylococcus aureus*	Enterotoxin is a superantigen
2. Gram-positive rods	*Clostridium difficile*	Inactivates GTPases in enterocytes
	Clostridium perfringens	Superantigen
	Bacillus cereus	Superantigen
3. Gram-negative rods	*Vibrio cholera*	Stimulates adenylate cyclase
	Toxigenic *Escherichia coli*	Stimulates adenylate cyclase
	E. coli O157	Inactivates protein synthesis
Nervous system		
1. Gram-positive rods	*Clostridium tetani*	Inhibits glycine release
	Clostridium botulinum	Inhibits acetylcholine release
Respiratory tract		
1. Gram-positive rods	*Corynebacterium diphtheriae*	Inactivates protein synthesis
2. Gram-negative rods	*Bordetella pertussis*	Stimulates adenylate cyclase; inhibits chemokine receptor
Skin, soft tissue, or muscle		
1. Gram-positive cocci	*S. aureus* (scalded skin syndrome)	Protease cleaves desmosome in skin
	S. aureus (MRSA strains)	PV leukocidin is a pore-forming toxin that disrupts cell membrane
	Streptococcus pyogenes (scarlet fever)	Erythrogenic toxin is a superantigen
2. Gram-positive rods	*C. perfringens*	Lecithinase cleaves cell membranes
	Bacillus anthracis	Edema factor is an adenylate cyclase; lethal factor is a protease
Systemic		
1. Gram-positive cocci	*S. aureus*	Toxic shock syndrome toxin is a superantigen

MRSA = methicillin-resistant *Staphylococcus aureus*; PV = Panton-Valentine.

(1) Diphtheria toxin, produced by *Corynebacterium diphtheriae*, inhibits protein synthesis by ADP-ribosylation of EF-2 (see Figure 7–1). (Note that exotoxin A of Pseudomonas aeruginosa, a gram-negative rod, has the same mode of action.)

The resulting death of the affected cells leads to two prominent symptoms of diphtheria: pseudomembrane formation in the throat and myocarditis.

The exotoxin activity depends on two functions mediated by different domains of the molecule. The toxin is synthesized as a single polypeptide that is nontoxic because the active site of the enzyme is masked. This molecule is cleaved and modified to yield two active polypeptides. Fragment A, derived from the amino-terminal end of the exotoxin, yields an enzyme that catalyzes the transfer of ADP-ribose from nicotinamide adenine dinucleotide (NAD) to EF-2, inhibiting protein synthesis. Fragment B, derived from the carboxy-terminal end, binds to receptors on the outer membrane of eukaryotic cells and mediates transport of fragment A into the cells.

As the bacteria synthesize and secrete the full-length exotoxin, the carboxy-terminal end binds to host cell membrane receptors. The toxin is transported across the cell membrane, triggering cleavage and modification that result in active fragment A, which then targets and inactivates EF-2. The specificity for this ADP-ribosylating enzyme is due to a unique amino acid, a modified histidine called diphthamide, that is present only on eukayotic EF-2. Since all eukaryotic cells carry out protein synthesis, there is no tissue or organ specificity. Prokaryotic and mitochondrial protein synthesis are not affected because

a different, nonsusceptible elongation factor is involved. The enzyme activity is remarkably potent; a single molecule of fragment A will kill a cell within a few hours. Other organisms whose exotoxins act by ADP-ribosylation are *E. coli*, *V. cholerae*, and *Bordetella pertussis*.

The *tox* gene, which codes for this exotoxin, is carried by a lysogenic bacteriophage called beta phage. As a result, only *C. diphtheriae* strains lysogenized by this phage cause diphtheria. (Nonlysogenized *C. diphtheriae* can be found in the throat of some healthy people.) This is an important example of *lysogenic conversion*, the process by which bacteria acquire new traits when lysogenized by a bacteriophage (see Chapter 4). Regulation of exotoxin synthesis is controlled by the interaction of iron in the medium with a *tox* gene repressor synthesized by the bacterium. As the concentration of iron increases, the iron-repressor complex inhibits the transcription of the *tox* gene.

(2) Tetanus toxin, produced by *Clostridium tetani*, is a **neurotoxin** that prevents release of an inhibitory neurotransmitter involved in muscle relaxation. When the inhibitory neurons are nonfunctional, the excitatory neurons are unopposed, leading to muscle spasms and a spastic paralysis. Tetanus toxin (tetanospasmin) is composed of two polypeptide subunits encoded by plasmid DNA. The heavy chain of the polypeptide binds to gangliosides in the membrane of the neuron; the light chain is a protease that degrades the protein(s) responsible for the release of the inhibitory neurotransmitters (γ-aminobutyric acid [GABA] and glycine). The toxin released at the site of the peripheral wound may

travel either by retrograde axonal transport or in the bloodstream to the internuncial neurons of the spinal cord. Inhibiting the release of the GABA and glycine leads to convulsive contractions of the voluntary muscles, best exemplified by spasm of the jaw and neck muscles ("lockjaw").

(3) Botulinum toxin, produced by *Clostridium botulinum*, is a **neurotoxin** that blocks the release of a different neurotransmitter, acetylcholine, at the synapse of the neuromuscular junction, producing a flaccid paralysis. Approximately 1 μg is lethal for humans; it is one of the most toxic compounds known. The toxin is composed of two polypeptide subunits held together by disulfide bonds. One of the subunits binds to a receptor on the neuron; the other subunit is a protease that degrades the protein(s) responsible for the release of acetylcholine. There are six serotypes of botulinum toxin (A–F), with toxins A, B, E, and F being the most important for human disease. Some serotypes are encoded on a plasmid, some on a temperate bacteriophage, and some on the bacterial chromosome.

(4) Two exotoxins are produced by *Clostridium difficile*, both of which are involved in the pathogenesis of pseudomembranous colitis. Exotoxin A is an enterotoxin that causes watery diarrhea. Exotoxin B is a **cytotoxin** that damages the colonic mucosa and causes pseudomembranes to form. Exotoxins A and B are glucosyltransferases that modify target signal transduction proteins (Rho GTPases), which interferes with their signal transduction function. Glucosylation by exotoxin B causes disaggregation of actin filaments in the cytoskeleton, leading to apoptosis and cell death.

(5) Multiple toxins and tissue degrading enzymes are produced by *Clostridium perfringens* and other species of clostridia that cause gas gangrene, although no single species of *Clostridium* makes all of the toxins. The best characterized is the **alpha toxin**, which is a **lecithinase** that hydrolyzes lecithin in the cell membrane, resulting in destruction of the membrane and widespread cell death. Other enzymes produced are collagenase, protease, hyaluronidase, and deoxyribonuclease (DNase). Several other clostridial toxins with hemolytic and necrotizing activity have also been described. Certain strains of *C. perfringens* produce an enterotoxin that causes watery diarrhea. This enterotoxin acts as a superantigen similar to the enterotoxin of *S. aureus* (described below).

(6) Three exotoxins are produced by *Bacillus anthracis*, the agent of anthrax: edema factor, lethal factor, and protective antigen. The three exotoxins associate with each other, but each component has a distinct function. **Edema factor** is an adenylate cyclase that raises the cyclic AMP concentration within the cell, resulting in loss of chloride ions and water and consequent edema formation in the tissue (see Table 7–12). **Lethal factor** is a protease that cleaves a phosphokinase (MAP kinase) required for the signal transduction pathway that controls cell growth. Loss of the phosphokinase results in the failure of cell growth and consequent cell death. **Protective antigen** binds to a cell surface receptor and forms pores in the human cell membrane that allow edema factor and lethal factor to enter the cell. The name *protective antigen* is based on the finding that antibody

against this protein protects against disease. The antibody blocks the binding of protective antigen, preventing edema factor and lethal factor from entering the cell.

(7) TSST is a **superantigen** produced primarily by certain strains of *S. aureus* and *S. pyogenes*. TSST binds directly to class II major histocompatibility (MHC) proteins on the surface of antigen-presenting cells (macrophages) without intracellular processing. This complex interacts with the T-cell receptor of many helper T cells, resulting in nonspecific T cell activation (see the discussion of superantigens in Chapter 58). This causes the release of large amounts of interleukins (ILs), especially IL-1, IL-2, and TNF. These cytokines produce many of the signs and symptoms of toxic shock.

(8) Staphylococcal enterotoxin is also a superantigen, but because it is ingested, it acts locally on the lymphoid cells lining the small intestine. The enterotoxin is produced by *S. aureus* in the contaminated food and causes food poisoning, usually within 1 to 6 hours after ingestion. The main symptoms are vomiting and watery diarrhea. The prominent vomiting seen in food poisoning is caused by serotonin (5-hydroxytryptamine) produced by mast cells stimulating the enteric nervous system, which activates the vomiting center in the brain.

(9) Exfoliatin is a protease produced by *S. aureus* that causes scalded skin syndrome. Exfoliatin cleaves desmoglein, a protein in the desmosomes of the skin, resulting in the detachment of the superficial layers of the skin. Exfoliatin is also called epidermolytic toxin.

(10) Panton-Valentine (PV) leukocidin is a pore-forming exotoxin produced by methicillin-resistant *Staphylococcus aureus* (MRSA). It destroys white blood cells, skin, and subcutaneous tissue. The two subunits of the toxin assemble in the cell membrane to form a pore through which cell contents exit into the extracellular space.

(11) Erythrogenic toxin, produced by *S. pyogenes*, causes the rash characteristic of scarlet fever. It is also a superantigen encoded by a lysogenic bacteriophage. Nonlysogenic bacteria do not cause scarlet fever, although they can cause pharyngitis.

(12) Exotoxin B is a protease produced by strains of *S. pyogenes* that cause necrotizing fasciitis. These strains are called "flesh-eating" streptococci. Exotoxin B cleaves E-cadherin in the skin, immunoglobulins, and complement proteins.

Gram-Negative Bacteria

The exotoxins produced by gram-negative bacteria also have several different mechanisms of action and produce different clinical effects. Two very important exotoxins are the enterotoxins of *E. coli* and *V. cholerae* (cholera toxin), which induce an increase in the amount of cyclic AMP within the enterocyte, resulting in watery diarrhea (see Table 7–12).

(1) The **heat-labile enterotoxin** produced by *E. coli* causes **watery, nonbloody diarrhea** by stimulating adenylate cyclase activity in cells in the small intestine (Figure 7–2). The resulting increase in the concentration of cyclic AMP causes excretion of the chloride ion, inhibition of sodium ion absorption, and

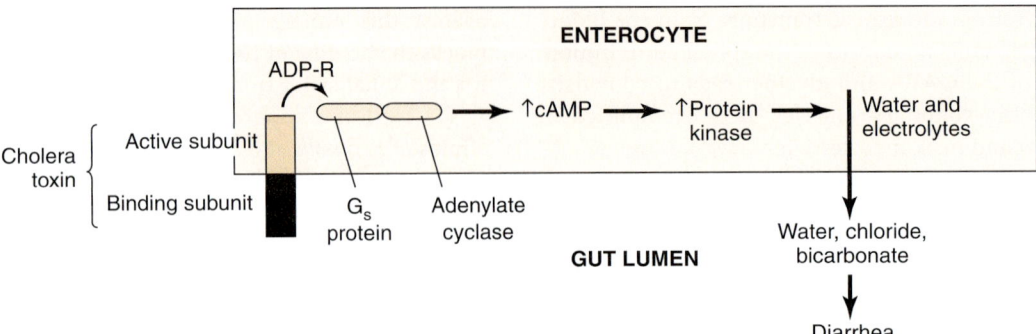

FIGURE 7–2 Mode of action of *Escherichia coli* and *Vibrio cholerae* enterotoxins. The enterotoxin (e.g., cholera toxin) binds to the surface of the enterocyte via its binding subunit. The active subunit then enters the enterocyte. The active subunit is an enzyme that catalyzes the addition of ADP-ribose (ADP-R) to the G_s regulatory protein. This activates adenylate cyclase to overproduce cyclic adenosine monophosphate (AMP). As a consequence, cyclic AMP–dependent protein kinase activity increases, and water and electrolytes leave the enterocyte, causing watery diarrhea.

significant fluid and electrolyte loss into the lumen of the gut. The heat-labile toxin, which is inactivated at 65°C for 30 minutes, is an A-B toxin. The B subunit confers specificity to the enterocytes in the small intestine by binding to a ganglioside receptor in the cell membrane. This enables the A subunit to enter the cell where it ADP-ribosylates its target G_s protein. This locks the G_s protein in the "on" position, which stimulates adenylate cyclase to synthesize cyclic AMP. This, in turn, activates cyclic AMP–dependent protein kinase, an enzyme that phosphorylates ion transporters in the cell membrane, resulting in the loss of water and ions from the cell. Most of the genes for the heat-labile toxin and for the heat-stable toxin (described next) are carried on plasmids.

In addition to the labile toxin, there is a **heat-stable toxin**, which is a polypeptide that is not inactivated by boiling for 30 minutes. The heat-stable toxin affects cyclic guanosine monophosphate (GMP) rather than cyclic AMP. It stimulates guanylate cyclase and thus increases the concentration of cyclic GMP, which inhibits the reabsorption of sodium ions and causes diarrhea.

(2) **Shiga toxin** is an exotoxin produced primarily by strains of *E. coli* with the O157:H7 serotype. These enterohemorrhagic strains cause **bloody diarrhea** and are the cause of outbreaks associated with eating undercooked meat, especially hamburger in fast-food restaurants. The toxin is named for a very similar toxin produced by *Shigella dysenteriae*. The toxin is a glycosidase that inactivates protein synthesis by removing adenine from a specific site on the 28S rRNA in the large subunit of the human ribosome.

Shiga toxin is encoded by a temperate (lysogenic) bacteriophage. When *E. coli* Shiga toxin enters the bloodstream, it can cause **hemolytic-uremic syndrome** (HUS). Shiga toxin binds to receptors on the glomerulus of the kidney and on the endothelium of small blood vessels. Inhibition of protein synthesis results in death of vascular epithelial cells, leading to renal failure and microangiopathic hemolytic anemia. Certain antibiotics, such as ciprofloxacin, can increase the amounts of Shiga toxin produced by *E. coli* O157, which predisposes to HUS.

(3) The enterotoxins produced by *V. cholerae*, the agent of cholera (see Chapter 18), and *Bacillus cereus*, a cause of diarrhea, act in a manner similar to that of the heat-labile toxin of *E. coli* (see Figure 7–2).

(4) Pertussis toxin, produced by *B. pertussis*, the cause of whooping cough, is an exotoxin that catalyzes the transfer of ADP-ribose from NAD to an inhibitory G protein. Inactivation of this inhibitory regulator has two effects: one is to stimulate adenylate cyclase activity, leading to an increase in cyclic AMP concentration within the affected cells (see Table 7–12). This results in edema and other changes in the respiratory tract, leading to the cough of whooping cough. It also inhibits the signal transduction pathway used by chemokine receptors, causing the marked **lymphocytosis** seen in patients with pertussis. The toxin inhibits signal transduction by all chemokine receptors, resulting in an inability of lymphocytes to migrate to and enter lymphoid tissue (spleen, lymph nodes). Because they do not enter tissue, there is an increase in their number in the blood (see the discussion of chemokines in Chapter 58).

Endotoxins

Endotoxins are an integral part of the cell wall of gram-negative bacteria, in contrast to exotoxins, which are actively released from the cell (see Table 7–9). In addition, endotoxins are lipopolysaccharides, whereas exotoxins are polypeptides. The enzymes that produce the lipopolysaccharides are encoded by genes on the bacterial chromosome, whereas exotoxins are usually encoded by plasmid or bacteriophage DNA. The toxicity of endotoxins is low in comparison with that of exotoxins. All endotoxins produce the same generalized effects of **fever** and **hypotension (shock)**, although the endotoxins of some organisms are more effective than those of others (Figure 7–3). Endotoxins are weakly antigenic; they induce protective antibodies so poorly that multiple episodes of toxicity can occur. No toxoids have been produced from endotoxins, and endotoxins are not used as antigens in any available vaccine.

A major site of action of endotoxin is the **macrophage**. Endotoxins (LPS) are released from the surface of gram-negative

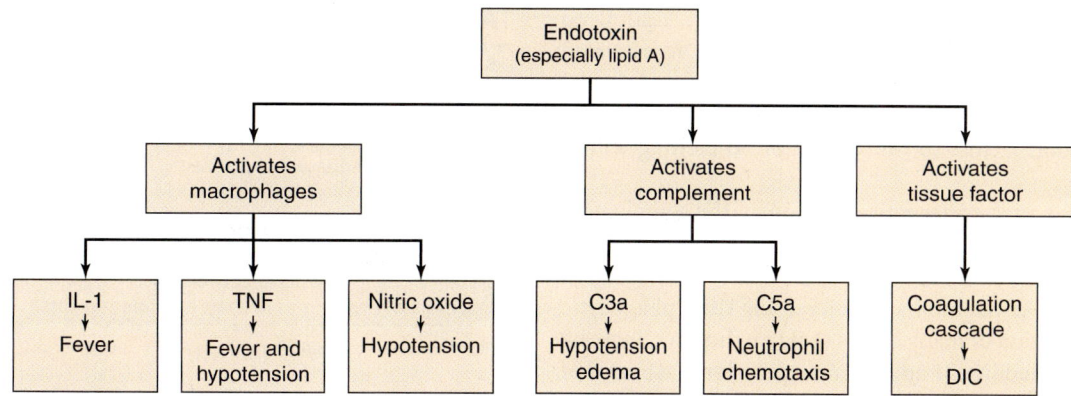

FIGURE 7–3 Mode of action of endotoxin. Endotoxin is the most important cause of septic shock, which is characterized primarily by fever, hypotension, and disseminated intravascular coagulation (DIC). Endotoxin causes these effects by activating three critical processes: (1) activating macrophages to produce interleukin-1 (IL-1), tumor necrosis factor (TNF), and nitric oxide; (2) activating the alternative pathway of complement to produce C3a and C5a; and (3) activating tissue factor, an early component of the coagulation cascade.

bacteria in small pieces of outer membrane that bind to LPS-binding protein in the plasma. This complex binds to a receptor on the surface of macrophages called CD14, which activates toll-like receptor-4 (TLR-4). A signal cascade within the macrophage is then activated, resulting in the synthesis of cytokines such as IL-1, TNF, and nitric oxide (see later and Figure 7–3).

The findings of fever and hypotension are salient features of **septic shock**. Additional features include tachycardia, tachypnea, and leukocytosis (increased white blood cells, especially neutrophils, in the blood). Septic shock is one of the leading causes of death in intensive care units and has an estimated mortality rate of 30% to 50%. The endotoxins of gram-negative bacteria are the best-established causes of septic shock, but surface molecules of gram-positive bacteria (which do not have endotoxins) can also cause septic shock.

Two features of septic shock are interesting:

(1) Septic shock is different from toxic shock. In septic shock, the bacteria are in the bloodstream, whereas in toxic shock, it is the toxin that is circulating in the blood. The clinical importance of this observation is that in septic shock, blood cultures are usually positive, whereas in staphylococcal toxic shock, they are usually negative.

(2) Septic shock can cause the death of a patient even though antibiotics have killed the bacteria in the patient's blood (i.e., the blood cultures have become negative). This occurs because septic shock is mediated by cytokines, such as TNF and IL-1, which continue to act even though the bacteria that induced the cytokines are no longer viable.

The structure of the LPS is shown in Figure 2–6. The toxic portion of the molecule is **lipid A**, which contains several fatty acids. β-Hydroxymyristic acid is always one of the fatty acids and is found only in lipid A. The other fatty acids differ according to species. The polysaccharide core in the middle of the molecule protrudes from the surface of the bacteria and has the same chemical composition within members of a genus.

The long-chain polysaccharide on the exterior portion of endotoxin is called the somatic (O) antigen. It is an important antigen of many gram-negative bacteria and is composed of 3, 4, or 5 sugars repeated up to 25 times. Because the number of permutations of this array is very large, many antigenic types exist. For example, more than 1500 antigenic types have been identified for *Salmonella* based on different sugars in the O antigen. There are many types of *E. coli* based on their O antigens, the most important of which is *E. coli* O-157, the cause of hemolytic-uremic syndrome Some bacteria, especially *N. meningitidis* and *N. gonorrhoeae*, have lipooligosaccharide (LOS) containing very few repeating sugar subunits in the O antigen.

The biologic effects of endotoxin (Table 7–14) include the following:

(1) **Fever** due to the release by macrophages of IL-1 (endogenous pyrogen) and IL-6, which act on the hypothalamic temperature-regulatory center.

(2) **Hypotension**, shock, and impaired perfusion of essential organs due to nitric oxide–induced vasodilation, TNF-induced increased capillary permeability, bradykinin-induced vasodilation, and increased capillary permeability.

(3) **Disseminated intravascular coagulation (DIC)** due to activation of the coagulation cascade, resulting in thrombosis,

TABLE 7–14 Effects of Endotoxin

Clinical Findings[1]	Mediator or Mechanism
Fever	Interleukin-1 and interleukin-6
Hypotension (shock)	Tumor necrosis factor, nitric oxide, and bradykinin
Inflammation	C5a produced via alternative pathway of complement attracts neutrophils
Coagulation (DIC)	Activation of tissue factor

DIC = disseminated intravascular coagulation.

[1]Tumor necrosis factor triggers many of these reactions.

a petechial or purpuric rash, and tissue ischemia, leading to failure of vital organs. The coagulation cascade is activated when tissue factor is released from the surface of endothelial cells damaged by infection. Tissue factor interacts with circulating coagulation factors to cause widespread clotting within capillaries.

(4) Activation of the alternative pathway of the complement cascade, resulting in recruitment of neutrophils, inflammation, and tissue damage.

(5) Activation of macrophages, increasing their phagocytic ability, and activation of many clones of B lymphocytes, increasing antibody production. (Endotoxin is a polyclonal activator of B cells, but not T cells.)

The end result of the above five processes is called the **systemic inflammatory response syndrome**, or SIRS. The most common clinical signs of SIRS are fever, hypotension, tachycardia, tachypnea, and leukocytosis.

Damage to the vascular endothelium plays a major role in both the hypotension and DIC seen in septic shock. Damage to the endothelium allows the leakage of plasma and red cells into the tissue, resulting in the loss of blood volume and consequent hypotension. Damaged endothelium also serves as a site of platelet aggregation and activation that leads to the thousands of endovascular clots manifesting as DIC.

The evidence that endotoxin causes these effects comes from the following two findings: (1) purified LPS, free of the organism, reproduces the effects, and (2) antiserum against endotoxin can mitigate or block these effects.

Clinically, the presence of DIC in the patient can be assessed by the D-dimer laboratory test. D-dimers are cleavage products of fibrin (fibrin split products) that are detected in the blood of patients with DIC.

Endotoxins do not cause these effects directly. Rather, they elicit the production of cytokines such as **IL-1** and **TNF** from macrophages.[1] TNF is the central mediator because purified recombinant TNF reproduces the effects of endotoxin and antibody against TNF blocks the effects of the endotoxin. Endotoxin also induces macrophage migration inhibitory factor, which also plays a role in the induction of septic shock.

Note that TNF in small amounts has beneficial effects (e.g., causing an inflammatory response to the presence of a microbe), but in large amounts, it has detrimental effects (e.g., causing septic shock and DIC). It is interesting that the activation of platelets, which results in clot formation and the walling off of infections, is the same process that, when magnified, causes DIC and the necrosis of tumors. It is the ability of TNF to activate platelets that causes intravascular clotting and the consequent infarction and death of the tumor tissue. The symptoms of certain autoimmune diseases such as rheumatoid

TABLE 7–15 Beneficial and Harmful Effects of TNF

Beneficial effects of small amounts of TNF
Inflammation (e.g., vasodilation), increased vascular permeability
Adhesion of neutrophils to endothelium
Enhanced microbicidal activity of neutrophils
Activation and adhesion of platelets
Increased expression of class I and II MHC proteins
Harmful effects of large amounts of TNF
Septic shock (e.g., hypotension and high fever)
Disseminated intravascular coagulation
Inflammatory symptoms of some autoimmune diseases

MHC = major histocompatibility complex; TNF = tumor necrosis factor.

arthritis are also mediated by TNF; however, these symptoms are not induced by endotoxin but by other mechanisms, which are described in Chapter 66. Some of the important beneficial and harmful effects of TNF are listed in Table 7–15.

Endotoxins can cause fever in the patient if they are present in intravenous fluids. In the past, intravenous fluids were sterilized by autoclaving, which killed any organisms present but resulted in the release of endotoxins that were not heat-inactivated. For this reason, these fluids are now sterilized by filtration, which physically removes organisms without releasing endotoxin.

Endotoxin-like pathophysiologic effects can occur in **gram-positive** bacteremic infections (e.g., *S. aureus* and *S. pyogenes* infections) as well. Since endotoxin is absent in these organisms, a different cell wall component—namely, lipoteichoic acid—causes the release of TNF and IL-1 from macrophages.

Endotoxin-mediated septic shock is a leading cause of death, especially in hospitals. The results of attempts to treat septic shock with hemoperfusion to adsorb endotoxin or by administering antibodies specific to lipid A and TNF have been mixed.

5. Immunopathogenesis

In certain diseases, such as rheumatic fever and acute glomerulonephritis, it is not the organism itself that causes the symptoms of disease but the immune response to the presence of the organism. For example, in rheumatic fever, antibodies are formed against the M protein of *S. pyogenes*, which cross-react with joint, heart, and brain tissue. Inflammation occurs, resulting in arthritis, carditis, and chorea that are the characteristic findings in this disease.

BACTERIAL INFECTIONS ASSOCIATED WITH CANCER

The fact that certain viruses can cause cancer is well established, but the observation that some bacterial infections are associated with cancers is just emerging. Several documented examples include (1) the association of *Helicobacter pylori* infection with gastric carcinoma and gastric mucosal-associated lymphoid tissue (MALT) lymphoma, and (2) the association of

---

[1]Endotoxin (LPS) induces these factors by first binding to LPS-binding protein in the serum. This complex then binds to CD14, a receptor on the surface of the macrophage. CD14 interacts with a transmembrane protein called *toll-like receptor*, which activates an intracellular signaling cascade, leading to the activation of genes that encode various cytokines such as IL-1, TNF, and other factors.

Campylobacter jejuni infection with MALT lymphoma of the small intestine (also known as alpha-chain disease). Support for the idea that these cancers are caused by bacteria comes from the observation that antibiotics can cause these cancers to regress if treated during an early stage.

DIFFERENT STRAINS OF THE SAME BACTERIA CAN PRODUCE DIFFERENT DISEASES

S. aureus causes inflammatory, pyogenic diseases such as endocarditis, osteomyelitis, and septic arthritis, as well as nonpyogenic, exotoxin-mediated diseases such as toxic shock syndrome, scalded skin syndrome, and food poisoning. How do bacteria that belong to the same genus and species cause such widely divergent diseases? The answer is that individual bacteria produce different virulence factors that endow those bacteria with the capability to cause different diseases.

The different virulence factors are encoded on plasmids, on transposons, on the genome of temperate (lysogenic) phages, and on pathogenicity islands. These transferable genetic elements may or may not be present in any single bacterium, which accounts for the ability to cause different diseases. Table 7–16 describes the different virulence factors for three of the most important bacterial pathogens: *S. aureus*, *S. pyogenes*, and *E. coli*. Figure 7–4 describes the importance of pathogenicity islands in determining the types of diseases caused by *E. coli*.

TYPICAL STAGES OF AN INFECTIOUS DISEASE

A typical acute infectious disease has four main stages (Figure 7–5):

(1) The **incubation period**, which is the time between the acquisition of the organism (or toxin) and the beginning of symptoms (this time varies from hours to days to weeks, depending on the organism).

(2) The **prodrome period**, during which nonspecific symptoms such as fever, malaise, and loss of appetite occur.

(3) The **specific-disease period**, during which the overt characteristic signs and symptoms of the disease occur.

(4) The **recovery period**, also known as the **convalescence period**, during which the illness abates and the patient returns to the healthy state. IgG and IgA antibodies protect the recovered patient from future encounters by the same organism.

After the recovery period, some individuals become **chronic carriers** of the organisms and may shed them while remaining clinically well. Others may develop a **latent infection**, which can recur either in the same form as the primary infection or manifesting different signs and symptoms. Although many infections cause symptoms, many others are **subclinical** (i.e., the individual remains asymptomatic although infected with the organism). In subclinical infections and after the recovery period is over, the presence of antibodies is often used to determine that an infection has occurred.

TABLE 7–16 Different Strains of Bacteria Can Cause Different Diseases

Bacteria	Diseases	Virulence Factors	Mode of Action
Staphylococcus aureus			
1. Exotoxin mediated	Toxic shock syndrome	Toxic shock syndrome toxin	Superantigen
	Food poisoning (gastroenteritis)	Enterotoxin	Superantigen
	Scalded skin syndrome	Exfoliatin	Protease cleaves desmoglein
2. Pyogenic	Skin abscess, osteomyelitis, and endocarditis	Enzymes causing inflammation and necrosis	Coagulase, hyaluronidase, leukocidin, lipase, and nuclease
Streptococcus pyogenes			
1. Exotoxin mediated	Scarlet fever	Erythrogenic toxin	Superantigen
	Streptococcal toxic shock syndrome	Toxic shock syndrome toxin	Superantigen
2. Pyogenic (suppurative)	Pharyngitis, cellulitis, and necrotizing fasciitis	Enzymes causing inflammation and necrosis	Hyaluronidase (spreading factor)
3. Nonsuppurative (immunopathogenic)	Rheumatic fever	Certain M proteins on pilus	Antibody to M protein cross-reacts with cardiac, joint, and brain tissue
	Acute glomerulonephritis	Certain M proteins on pilus	Immune complexes deposit on glomeruli
Escherichia coli			
1. Exotoxin mediated	Watery, nonbloody diarrhea (traveler's diarrhea)	Labile toxin	Activation of adenylate cyclase increases cyclic AMP; no cell death
	Bloody diarrhea (associated with undercooked hamburger); O157:H7 strain	Shiga-like toxin (verotoxin)	Cytotoxin inhibits protein synthesis; cell death occurs
2. Pyogenic	Urinary tract infection	Uropathic pili	Pili attach to Gal–Gal receptors on bladder epithelium
	Neonatal meningitis	K-1 capsule	Antiphagocytic

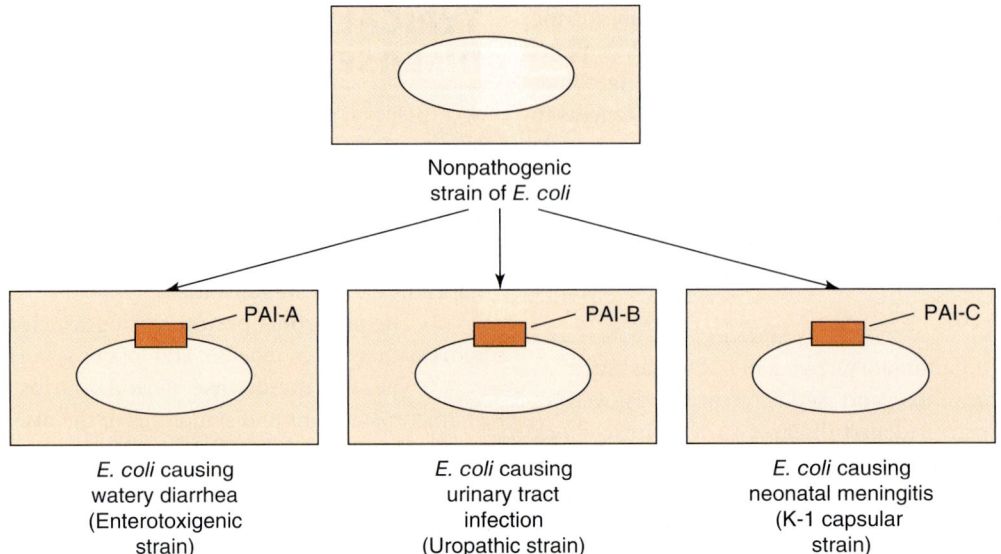

Nonpathogenic
strain of *E. coli*

PAI-A

PAI-B

PAI-C

E. coli causing
watery diarrhea
(Enterotoxigenic
strain)

E. coli causing
urinary tract
infection
(Uropathic strain)

E. coli causing
neonatal meningitis
(K-1 capsular
strain)

FIGURE 7–4 Pathogenicity islands encode virulence factors that determine the type of infection. The top of the figure depicts a nonpathogenic strain of *Escherichia coli* that does not contain a pathogenicity island (PAI) in the genome DNA. The black oval line within the *E. coli* cell is the genome DNA. PAIs can be transferred, by either conjugation or transduction, from another enteric gram-negative rod into the nonpathogenic strain of *E. coli*. Acquisition of a PAI that encodes virulence factors endows the nonpathogenic *E. coli* with the ability to cause specific diseases. In this figure, PAI-A encodes an enterotoxin, PAI-B encodes the pili that bind to urinary tract epithelium, and PAI-C encodes the enzymes that synthesize the K-1 capsular polysaccharide. This results in three different strains of *E. coli* capable of causing three different infections.

DID THE ORGANISM ISOLATED FROM THE PATIENT ACTUALLY CAUSE THE DISEASE?

Because people harbor microorganisms as members of the permanent normal flora and as transient passengers, this can be an interesting and sometimes confounding question. The answer depends on the situation. One type of situation relates to the problems of a disease for which no agent has been identified and a candidate organism has been isolated. This is the problem that Robert Koch faced in 1877 when he was among the first to try to determine the cause of an infectious disease, namely, anthrax in cattle and tuberculosis in humans. His approach led to the formulation of **Koch's postulates**, which are criteria that he proposed must be satisfied to confirm the causal role of an organism. These criteria are as follows:

(1) The organism must be isolated from every patient with the disease.

(2) The organism must be isolated free from all other organisms and grown in pure culture in vitro.

(3) The pure organism must cause the disease in a healthy, susceptible animal.

(4) The organism must be recovered from the inoculated animal.

At the time, these principles provided a general but rigorous guideline for collecting evidence in support of identifying etiological agents of disease. We now recognize that there are infectious agents that do not fulfill all of Koch's postulates, and proving disease causation rests on a constellation of observations. Exceptions to Koch's postulates include the following:

(1) Pathogens isolated from patients who are not manifesting symptoms (i.e., asymptomatic carriers) such as *Salmonella typhi*.

(2) Pathogens that cannot be cultivated on existing media, such as *T. pallidum* (the causative agent of syphilis) or prions (infectious proteins that cause Creutzfeldt-Jakob disease).

(3) Exposures to pathogens, for example, *M. tuberculosis*, that do not always result in disease in all hosts.

A second consideration for determining whether an isolated organism is actually responsible for a disease involves the

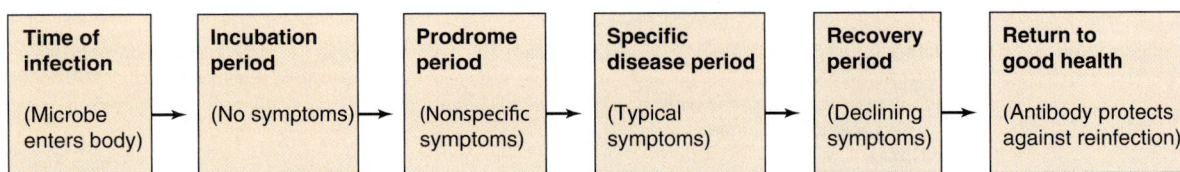

Time of infection	Incubation period	Prodrome period	Specific disease period	Recovery period	Return to good health
(Microbe enters body)	(No symptoms)	(Nonspecific symptoms)	(Typical symptoms)	(Declining symptoms)	(Antibody protects against reinfection)

FIGURE 7–5 Typical stages of an infectious disease. After infection, the patient progresses through four main stages: incubation period, prodrome period, specific disease period, and recovery period. The patient then typically returns to good health and has antibody that protects against reinfection and disease.

diagnosis of a patient's illness. In this instance, the signs and symptoms of the illness can often suggest a constellation of possible causative agents. The recovery of an agent in *sufficient numbers* from the *appropriate specimen* is usually sufficient for an etiologic diagnosis. This approach can be illustrated with two examples: (1) in a patient with a sore throat, the presence of a few β-hemolytic streptococci is insufficient for a microbiologic diagnosis, whereas the presence of many would be sufficient, and (2) in a patient with fever, α-hemolytic streptococci in the throat are considered part of the normal flora, whereas the same organisms in the blood are likely to be the cause of bacterial endocarditis.

In some infections, no organism is isolated from the patient, and the diagnosis is made by detecting a rise in antibody titer to an organism. For this purpose, the titer (amount) of antibody in the second or late serum sample should be **at least four times** the titer (amount) of antibody in the first or early serum sample.

PEARLS

- The term **pathogen** refers to those microbes capable of causing disease, especially if they cause disease in immunocompetent people. The term **opportunistic pathogen** refers to microbes that are capable of causing disease only in immunocompromised people.
- **Virulence** is a measure of a microbe's ability to cause disease (i.e., a highly virulent microbe requires fewer organisms to cause disease than a less virulent one). The ID_{50} is the number of organisms required to cause disease in 50% of the population. A low ID_{50} indicates a highly virulent organism.
- The virulence of a microbe is determined by **virulence factors**, such as capsules, exotoxins, or endotoxins.
- Whether a person gets an infectious disease or not is determined by the balance between the number and virulence of the microbes and the competency of that person's host defenses.
- Many infections are **asymptomatic** or **inapparent** because our host defenses have eliminated the microorganism before it could multiply to sufficient numbers to cause the symptoms of disease.
- The term **infection** has two meanings: (1) the **presence of microbes** in the body and (2) the **symptoms of disease**. The presence of microbes in the body does not always result in symptoms of disease (see the previous bullet).
- Bacteria cause the symptoms of disease primarily by two main mechanisms: **production of toxins** (both exotoxins and endotoxins) and **induction of inflammation**. A third mechanism, immunopathogenesis (e.g., rheumatic fever), occurs less commonly.
- Most bacterial infections are **communicable** (i.e., capable of spreading from person to person), but some are not (e.g., botulism and *Legionella* pneumonia).
- Three epidemiologic terms are often used to describe infections: **endemic** infections are those that occur at a persistent, usually low level in a certain geographic area, **epidemics** are those infections that occur at a much higher rate than usual, and **pandemics** are those infections that spread rapidly over large areas of the globe.

Determinants of Bacterial Pathogenesis

Transmission

- The modes of transmission of microbes include both **human-to-human** and **nonhuman-to-human** processes. Nonhuman sources include animals, soil, water, and food.
- Human-to-human transmission can occur either by **direct contact**, such as airborne droplets or fecal-oral or indirectly via a **vector** such as an insect, notably ticks or mosquitoes, or via a fomite such as a towel. Animal-to-human transmission can also occur either by direct contact with the animal or indirectly via a vector.
- The main "portals of entry" into the body are the **respiratory tract, gastrointestinal tract, skin,** and **genital tract**.
- Human diseases for which animals are the reservoir are called **zoonoses**.

Adherence to Cell Surfaces

- **Pili** are the main mechanism by which bacteria adhere to human cells. They are fibers that extend from the surface of bacteria that **mediate attachment** to specific receptors on cells.
- **Glycocalyx** is a polysaccharide "slime layer" secreted by some strains of bacteria that **mediates strong adherence** to certain structures such as heart valves, prosthetic implants, and catheters.

Invasion, Inflammation, & Intracellular Survival

- Invasion of tissue is enhanced by enzymes secreted by bacteria. For example, **hyaluronidase** produced by *Streptococcus pyogenes* degrades hyaluronic acid in the subcutaneous tissue, allowing the organism to spread rapidly.
- **IgA protease** degrades secretory IgA, allowing bacteria to attach to mucous membranes.
- The **capsule** surrounding bacteria is **antiphagocytic** (i.e., it retards the phagocyte from ingesting the organism). Mutant strains of many pathogens that do not produce capsules are nonpathogenic.

- **Inflammation** is an important host defense induced by the presence of bacteria in the body. There are two types of inflammation, **pyogenic** and **granulomatous**, and bacteria typically elicit one type or the other. **Pyogenic inflammation**, the host defense against pyogenic (pus-producing) bacteria such as *S. pyogenes*, consists of neutrophils (and antibody and complement). **Granulomatous inflammation**, the host defense against intracellular, granuloma-producing bacteria, such as *Mycobacterium tuberculosis*, consists of macrophages and CD4-positive T cells. The type of inflammatory lesion is an important diagnostic criterion.

- **Intracellular bacterial pathogens** can live within cells and are protected from various cellular (macrophages and neutrophils) and humoral (antibodies) host defenses. Note that many of these bacteria (e.g., *M. tuberculosis*) are not obligate intracellular parasites (which can grow only within cells), but rather have the ability to enter and survive inside cells.

Exotoxins

- **Exotoxins** are **polypeptides secreted** by certain bacteria that alter specific cell functions resulting in the symptoms of disease. They are produced by both gram-positive and gram-negative bacteria, whereas endotoxin is found only in gram-negative bacteria.

- Exotoxins are **antigenic** and induce antibodies called **antitoxins**. Exotoxins can be modified to form **toxoids**, which are antigenic but not toxic. Toxoids, such as tetanus toxoid and diphtheria toxoid, are used to immunize against disease.

- Many exotoxins have an **A–B subunit** structure in which the A subunit is the **active** (toxic) one and the B subunit is the one that **binds** to the cell membrane and mediates the entry of the A subunit into the cell.

- Exotoxins have different mechanisms of action and different targets within the cell and therefore cause a variety of diseases with characteristic symptoms (see Tables 7–9 and 7–10).

Several exotoxins are enzymes that attach ADP-ribose to a cell component (**ADP-ribosylation**). Some exotoxins act by **proteolytic cleavage** of a cell component, whereas others act as **superantigens**, causing the overproduction of cytokines.

Endotoxins

- **Endotoxins** are **lipopolysaccharides (LPS)** located in the outer membrane only of gram-negative bacteria. They are not secreted by bacteria.

- **Lipid A** is the toxic component of LPS. It induces the **overproduction of cytokines**, such as tumor necrosis factor, interleukin-1, and nitric oxide, from macrophages, which causes the symptoms of septic shock, such as fever and hypotension. In addition, LPS activates the **complement cascade** (alternate pathway), resulting in increased vascular permeability, and the **coagulation cascade**, resulting in increased vascular permeability and **disseminated intravascular coagulation**.

- Endotoxins are poorly antigenic, do not induce antitoxins, and do not form toxoids.

Typical Stages of an Infectious Disease

- There are often four discrete stages. **The incubation period** is the time between the moment the person is exposed to the microbe (or toxin) and the appearance of symptoms. The **prodrome period** is the time during which nonspecific symptoms occur. The **specific-illness period** is the time during which the characteristic features of the disease occur. The **recovery period** is the time during which symptoms resolve and health is restored.

- After the recovery period, some people become **chronic carriers** of the organism, and in others, **latent** infections develop.

- Some people have **subclinical** infections during which they remain asymptomatic but can serve as a source of infection for others. The presence of antibodies reveals that a prior infection has occurred.

SELF-ASSESSMENT QUESTIONS

1. Handwashing is an important means of interrupting the chain of transmission from one person to another. Infection by which of the following bacteria is most likely to be interrupted by handwashing?

   (A) *Borrelia burgdorferi*
   (B) *Legionella pneumophila*
   (C) *Staphylococcus aureus*
   (D) *Streptococcus agalactiae (group B Streptococcus)*
   (E) *Treponema pallidum*

2. Vertical transmission is the transmission of organisms from mother to fetus or newborn child. Infection by which of the following bacteria is most likely to be transmitted vertically?

   (A) *Chlamydia trachomatis*
   (B) *Clostridium tetani*
   (C) *Haemophilus influenzae*

   (D) *Shigella dysenteriae*
   (E) *Streptococcus pneumoniae*

3. The cells involved with pyogenic inflammation are mainly neutrophils, whereas the cells involved with granulomatous inflammation are mainly macrophages and helper T cells. Infection by which of the following bacteria is most likely to elicit granulomatous inflammation?

   (A) *Escherichia coli*
   (B) *Mycobacterium tuberculosis*
   (C) *Neisseria gonorrhoeae*
   (D) *Streptococcus pyogenes*
   (E) *Staphylococcus aureus*

4. Which of the following sets of properties of exotoxins and endotoxins is correctly matched?

   (A) Exotoxins—polypeptides; endotoxins—lipopolysaccharide
   (B) Exotoxins—weakly antigenic; endotoxins—highly antigenic

(C) Exotoxins—produced only by gram-negative bacteria; endotoxins—produced only by gram-positive bacteria

(D) Exotoxins—weakly toxic per microgram; endotoxins—highly toxic per microgram

(E) Exotoxins—toxoid vaccines are ineffective; endotoxins—toxoid vaccines are effective

5. Which of the following sets consists of bacteria **both** of which produce exotoxins that increase cyclic AMP within human cells?

(A) *Vibrio cholerae* and *Corynebacterium diphtheriae*

(B) *Clostridium perfringens* and *Streptococcus pyogenes*

(C) *Escherichia coli* and *Bordetella pertussis*

(D) *Corynebacterium diphtheriae* and *Staphylococcus aureus*

(E) *Bacillus anthracis* and *Staphylococcus epidermidis*

6. Which of the following sets of bacteria produces exotoxins that act by ADP-ribosylation?

(A) *Corynebacterium diphtheriae* and *Escherichia coli*

(B) *Clostridium perfringens* and *Staphylococcus aureus*

(C) *Clostridium tetani* and *Bacillus anthracis*

(D) *Enterococcus faecalis* and *Mycobacterium tuberculosis*

(E) *Escherichia coli* and *Streptococcus pyogenes*

7. Which of the following bacteria produces an exotoxin that inhibits the release of acetylcholine at the neuromuscular junction?

(A) *Bacillus anthracis*

(B) *Bordetella pertussis*

(C) *Clostridium botulinum*

(D) *Corynebacterium diphtheriae*

(E) *Escherichia coli*

8. A 25-year-old man with abdominal pain was diagnosed with acute appendicitis. He then had a sudden rise in temperature to 39°C and a sudden fall in blood pressure. Which of the following is the most likely cause of the fever and hypotension?

(A) An exotoxin that ADP-ribosylates elongation factor-2.

(B) An exotoxin that stimulates production of large amounts of cyclic AMP.

(C) An endotoxin that causes release of tumor necrosis factor.

(D) An endotoxin that binds to a class I MHC protein.

(E) An exoenzyme that cleaves hyaluronic acid.

9. Several biotech companies have sponsored clinical trials of a drug consisting of monoclonal antibody to lipid A. Sepsis caused by which of the following sets of bacteria is most likely to be improved following administration of this antibody?

(A) *Bordetella pertussis* and *Clostridium perfringens*

(B) *Escherichia coli* and *Neisseria meningitidis*

(C) *Pseudomonas aeruginosa* and *Bacillus anthracis*

(D) *Staphylococcus epidermidis* and *Staphylococcus aureus*

(E) *Streptococcus pneumoniae* and *Staphylococcus aureus*

10. Regarding endotoxin, which of the following is the **MOST** accurate?

(A) Endotoxin is a polypeptide, the toxic portion of which consists of two D-alanines.

(B) Endotoxin is produced by both gram-positive cocci and gram-negative cocci.

(C) Endotoxin acts by binding to class II MHC proteins and the variable portion of the beta chain of the T-cell receptor.

(D) Endotoxin causes fever and hypotension by inducing the release of interleukins such as interleukin-1 and tumor necrosis factor.

(E) The antigenicity of endotoxin resides in its fatty acid side chains.

ANSWERS

(1) **(C)**
(2) **(A)**
(3) **(B)**
(4) **(A)**
(5) **(C)**
(6) **(A)**
(7) **(C)**
(8) **(C)**
(9) **(B)**
(10) **(D)**

PRACTICE QUESTIONS: USMLE & COURSE EXAMINATIONS

Questions on the topics discussed in this chapter can be found in the Basic Bacteriology section of Part XIII: USMLE (National Board) Practice Questions starting on page 733. Also see Part XIV: USMLE (National Board) Practice Examination starting on page 775.

Host Defenses

PRINCIPLES OF HOST DEFENSES

Host defenses are composed of two complementary, frequently interacting systems: (1) **innate (nonspecific)** defenses, which protect against microorganisms in general; and (2) **adaptive (specific)** immunity, which protects against a particular microorganism.

Innate defenses can be classified into three major categories: (1) physical barriers, such as intact skin and mucous membranes; (2) phagocytic cells, such as neutrophils, macrophages, and natural killer cells; and (3) proteins, such as complement, lysozyme, and interferon. Figure 8–1 shows the role of several components of the nonspecific defenses in the early response to bacterial infection. Adaptive immunity is mediated by antibodies and T lymphocytes. Chapter 57 describes these host defenses in more detail.

There are two main types of host defenses against bacteria: the **pyogenic** response and the **granulomatous** response. Certain bacteria, such as *Staphylococcus aureus* and *Streptococcus pyogenes*, are defended against by the pyogenic (pus-producing) response, which consists of antibody, complement, and neutrophils. These pyogenic bacteria are often called *extracellular pathogens* because they do not invade cells. Other bacteria, such as *Mycobacterium tuberculosis* and *Listeria monocytogenes*, are defended against by the granulomatous response, which consists of macrophages and CD4-positive (helper) T cells. These bacteria are often called *intracellular pathogens* because they can invade and survive within cells.

INNATE (NONSPECIFIC) IMMUNITY

Skin & Mucous Membranes

Intact skin is the first line of defense against many organisms. In addition to the physical barrier presented by skin, the fatty acids secreted by sebaceous glands in the skin have antibacterial and antifungal activity. The increased fatty acid production that occurs at puberty is thought to explain the increased resistance to ringworm fungal infections, which occurs at that time. The low pH of the skin (between 3 and 5), which is due to these fatty acids, also has an antimicrobial effect. Although many organisms live on or in the skin as members of the normal flora, they are harmless as long as they do not enter the body.

Another important defense is the mucous membrane of the respiratory tract, which is lined with cilia and covered with mucus. The coordinated beating of the cilia drives the mucus up to the nose and mouth, where the trapped bacteria can be expelled. This mucociliary apparatus, the **ciliary elevator**, can be damaged by alcohol, cigarette smoke, and viruses; the damage predisposes the host to bacterial infections. Other protective mechanisms of the respiratory tract involve alveolar macrophages, lysozyme in tears and mucus, hairs in the nose, and the cough reflex, which prevents aspiration into the lungs.

Loss of the physical barrier provided by the skin and mucous membranes predisposes to infection. Table 8–1 describes the organisms that commonly cause infections associated with the loss of these protective barriers.

The nonspecific protection in the gastrointestinal tract includes hydrolytic enzymes in saliva, acid in the stomach, and various degradative enzymes and macrophages in the small intestine. The vagina of adult women is protected by the low pH generated by lactobacilli that are part of the normal flora.

Additional protection in the gastrointestinal tract and in the lower respiratory tract is provided by **defensins**, a large family of antimicrobial peptides expressed predominantly by leukocytes and epithelial cells. These small, positively charged (cationic) proteins have broad-spectrum microbicidal activity

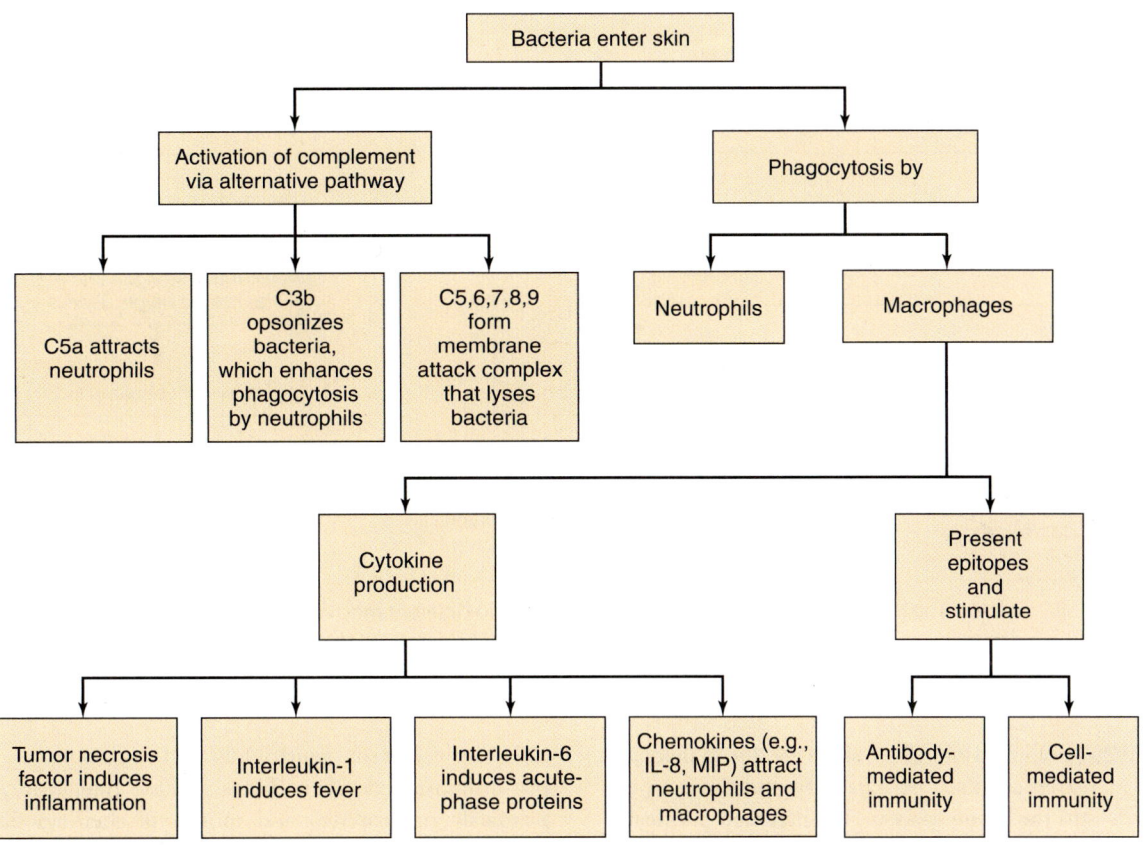

FIGURE 8–1 Early host responses to bacterial infection. IL-8 = interleukin-8; MIP = macrophage inflammatory protein.

against both gram-positive and gram-negative bacteria. They kill bacteria using a number of different mechanisms.

The bacteria of the normal flora of the skin, nasopharynx, colon, and vagina occupy these ecologic niches, preventing

pathogens from multiplying in these sites, a process called **colonization resistance**. The importance of the normal flora is appreciated in the occasional case when antimicrobial therapy suppresses these beneficial organisms, thereby allowing organisms such as *Clostridium difficile* and *Candida albicans* to cause diseases such as pseudomembranous colitis and vaginitis, respectively.

The Blood–Brain Barrier

The **blood–brain barrier** retards the entry of microbes and drugs from the capillaries into the brain. It is formed by the strong tight junctions between the endothelial cells. Microbes, such as bacteria and viruses, have difficulty passing through uninflamed endothelium but inflammation weakens the tight junctions and allows passage of microbes into the brain. Similarly, drugs, for example, penicillins, pass more easily through inflamed capillaries than through uninflamed vasculature. In addition, lipophilic drugs pass through the blood–brain barrier more easily than do charged hydrophilic drugs. Molecules required for normal brain function, such as glucose, have specific transporters that allow passage into the brain.

Inflammatory Response & Phagocytosis

The presence of bacteria within the body provokes a protective inflammatory response (Figure 8–2). This response is characterized by the clinical findings of redness, swelling, warmth,

TABLE 8–1 Damage to Skin and Mucous Membranes Predisposes to Infection Caused by Certain Bacteria

Predisposing Factor	Site of Infection	Bacteria Commonly Causing Infection Associated with Predisposing Factor
Intravenous catheters	Skin	*Staphylococcus epidermidis*, *Staphylococcus aureus*
Diabetes	Skin	*S. aureus*
Burns	Skin	*Pseudomonas aeruginosa*
Cystic fibrosis	Respiratory tract	*P. aeruginosa*[1]
Trauma to jaw	Gingival crevice	*Actinomyces israelii*
Dental extraction	Oropharynx	Viridans streptococci[2]
Oral mucositis secondary to cancer chemotherapy	Mouth but also entire gastrointestinal tract	Viridans streptococci, *Capnocytophaga gingivalis*

[1]Bacteria less commonly involved include *Burkholderia cepacia* and *Stenotrophomonas maltophilia*.

[2]Viridans streptococci do not cause local infection after dental extraction but can enter the bloodstream and cause endocarditis.

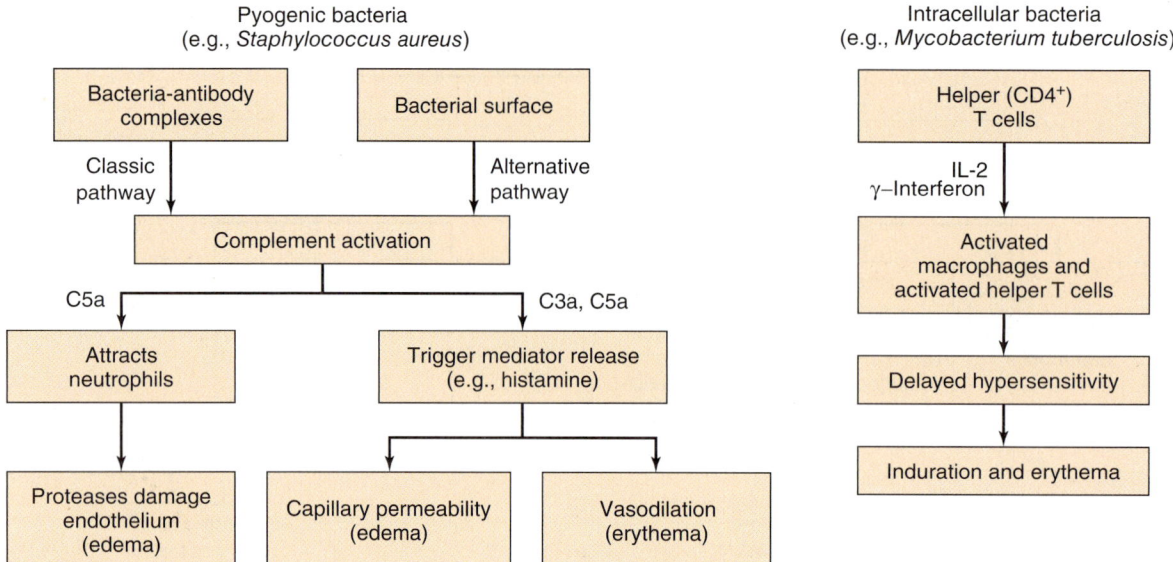

FIGURE 8–2 Inflammation. The inflammatory response can be caused by two different mechanisms. **Left:** Pyogenic bacteria (e.g., *Staphylococcus aureus*) cause inflammation via antibody- and complement-mediated mechanisms. **Right:** Intracellular bacteria (e.g., *Mycobacterium tuberculosis*) cause inflammation via cell-mediated mechanisms. IL-2 = interleukin-2.

and pain at the site of infection. These signs are due to increased blood flow, increased capillary permeability, and the escape of fluid and cells into the tissue spaces. The increased permeability is due to several chemical mediators, of which **histamine, prostaglandins, bradykinin**, and **leukotrienes** are the most important. Complement components, C3a and C5a, also contribute to increased vascular permeability. **Bradykinin** is also an important mediator of pain.

Neutrophils and **macrophages**, both of which are phagocytes, are an important part of the inflammatory response. Neutrophils predominate in acute pyogenic infections, whereas macrophages are more prevalent in chronic or granulomatous infections.

Macrophages are phagocytic and produce two important "proinflammatory" cytokines: **tumor necrosis factor (TNF) and interleukin-1 (IL-1)**. The synthesis of IL-1 from its inactive precursor is mediated by proteolytic enzymes (caspases) in a cytoplasmic structure called an **inflammasome**.

Certain proteins, known collectively as the **acute-phase response**, are produced early in inflammation by the liver. The best known of these are **C-reactive protein which opsonizes bacteria** and **mannose-binding protein**, which activates of complement (see Chapter 58).

Lipopolysaccharide (endotoxin)-binding protein is another important acute-phase protein produced in response to gram-negative bacteria. **Interleukin-6 (IL-6)** is a proinflammatory cytokine that is the main inducer of the acute-phase response. Macrophages are the principal source of IL-6, but other types of cells produce it as well. **Gamma interferon**, which activates macrophages and enhances their microbicidal action, is produced by activated helper T cells.

Neutrophils and macrophages are attracted to the site of infection by **chemokines** (*chemo*tactic cyto*kines*), which are produced by tissue cells in the infected area, local endothelial

cells, and resident neutrophils and macrophages. Important pro-inflammatory chemokines include Interleukin-8 (attracts primarily neutrophils) and monocyte chemotactic protein 1 (MCP-1) and macrophage inflammatory protein (MIP) (attract macrophages and monocytes) (see Chapter 58).

As part of the inflammatory response, bacteria are engulfed (phagocytized) by polymorphonuclear neutrophils (PMNs) and macrophages. PMNs make up approximately 60% of the leukocytes in the blood, and their numbers increase significantly during infection (leukocytosis) due to the production of granulocyte-stimulating factors (granulocyte colony-stimulating factor [G-CSF] and granulocyte-macrophage colony-stimulating factor [GM-CSF]; see Chapter 58) by macrophages soon after infection.

The process of phagocytosis can be divided into three steps: **migration, ingestion**, and **killing**. Chemokines (like IL-8 and complement component C5a) induce **migration of PMNs** to the infection site. Circulating PMNs are signaled to adhere to the endothelium through interactions with surface-localized proteins expressed on the endothelium (intracellular adhesion molecule [ICAM]) and PMN surface proteins (**selectins** and **integrins**).

ICAM proteins on the endothelium are induced by inflammatory mediators, such as IL-1 and TNF (see Chapter 58), produced by macrophages in response to bacterial infection. The increase in surface-exposed ICAM proteins ensures that PMNs selectively adhere to the site of infection. Increased permeability of capillaries as a result of histamine, kinins, and prostaglandins allows PMNs to migrate through the capillary wall to reach the focus of infection. This migration is called **diapedesis** and takes several minutes to occur.

The **bacteria are phagocytosed** by PMNs into a membrane-bound vacuole (**phagosome**). This engulfment is enhanced when immunoglobulin G (IgG) antibodies or the C3b component of

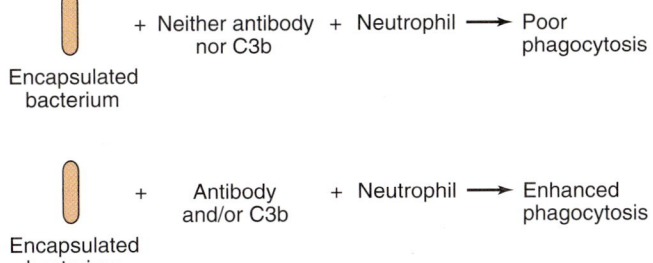

FIGURE 8–3 Opsonization. **Top:** An encapsulated bacterium is poorly phagocytized by a neutrophil in the absence of either immunoglobulin G (IgG) antibody or C3b. **Bottom:** In the presence of either IgG antibody or C3b or both, the bacterium is opsonized (i.e., it is made more easily phagocytized by the neutrophil).

complement (both **opsonins**) is bound to the surface of the bacteria in a process called **opsonization** (Figure 8–3). (The outer cell membranes of both PMNs and macrophages have receptors both for the Fc portion of IgG and for C3b.)

At the time of engulfment, a pathway known as the **respiratory burst** is triggered, which results in the production of a superoxide radical and hydrogen peroxide. These highly reactive compounds (often called *reactive oxygen intermediates*) are microbicidal. **NADPH oxidase** is the enzyme that synthesizes superoxide and **superoxide dismutase** is the enzyme that synthesizes hydrogen peroxide.

Nitric oxide (NO) is another important microbicidal agent. It is a *reactive nitrogen intermediate* that is synthesized by an inducible enzyme called nitric oxide synthase in response to stimulators such as endotoxin. NO participates in oxidative killing of ingested microbes phagocytosed by neutrophils and macrophages. Overproduction of NO contributes to the hypotension seen in septic shock because it causes vasodilation of peripheral blood vessels.

The **killing of the organism** within the phagosome is a two-step process that consists of degranulation followed by production of **hypochlorite**, which is probably the most important microbicidal agent. In degranulation, membrane-bound lysosomal granules fuse with the phagosome, emptying cytolytic enzymes, such as nucleases and proteases, into the vesicle. This converts the phagosome into a structure called the phagolysosome, where the actual killing of microorganisms occurs. The most important bactericidal mechanism is the production of **hypochlorite ion**, which damages bacterial cell membrane. **Myeloperoxidase** is the enzyme that synthesizes hypochlorite.

Additional bactericidal mechanisms involve lactoferrin which chelates iron from the bacteria, lysozyme which degrades peptidoglycan in the bacterial cell wall, and cationic proteins which damage bacterial membranes.

Macrophages also migrate, engulf, and kill bacteria by using essentially the same processes as PMNs do, but there are several differences:

(1) Macrophages do not make hypochlorite ion; however, they do produce hydrogen peroxide and superoxide by respiratory burst.

(2) Certain organisms such as the agents of tuberculosis, brucellosis, and toxoplasmosis are preferentially ingested by macrophages rather than PMNs and may remain viable and multiply within these cells; granulomas formed during these infections contain many of these macrophages.

Reduced Phagocytosis Predisposes to Bacterial Infections

Phagocytosis is a crucial host defense mechanism. Mutations or treatments resulting in reduced phagocytic function predisposes the host to bacterial infections, especially infections caused by certain organisms (Table 8–2):

(1) Children with genetic defects in phagocytic processes have an increased number and severity of infections. Two examples of these defects are **chronic granulomatous disease**, where the phagocyte cannot kill the ingested bacteria due to a defect in NADPH oxidase, which results in a failure to generate H_2O_2; **Chédiak–Higashi syndrome**, involving abnormal lysosomal granules that don't fuse with the phagosome, and are subsequently unable to kill bacteria.

(2) Frequent infections occur in **neutropenic** patients, especially when the PMN count drops below 500/μL as a result of immunosuppressive drugs or irradiation. These infections are frequently caused by opportunistic organisms (i.e., organisms that rarely cause disease in people with normal immune systems).

(3) **Splenectomy** removes an important source of both phagocytes and immunoglobulins, and predisposes to sepsis caused by three encapsulated pyogenic bacteria: *Streptococcus pneumoniae*, *Neisseria meningitidis*, and *Haemophilus influenzae*. *S. pneumoniae* causes approximately 50% of episodes of sepsis in splenectomized patients. Patients with sickle cell anemia and other hereditary anemias can autoinfarct their spleen, resulting in a loss of splenic function and a predisposition to sepsis caused by these bacteria.

TABLE 8–2 Reduced Phagocytosis Predisposes to Infection Caused by Certain Bacteria

Type of Reduction	Cause of Reduction	Bacteria Commonly Causing Infection Associated With the Type of Reduction
Decreased number of neutrophils	Cancer chemotherapy, total-body irradiation	*Staphylococcus aureus*, *Pseudomonas aeruginosa*
Decreased function of neutrophils	Chronic granulomatous disease	*S. aureus*
	Diabetes	*S. aureus*
Decreased function of spleen	Splenectomy, sickle cell anemia	*Streptococcus pneumoniae*, *Neisseria meningitidis*, *Haemophilus influenzae*

(4) People who have **diabetes mellitus**, especially those who have poor glucose control or episodes of ketoacidosis, have an increased number of, and more severe, infections compared with people who do not have diabetes. The main host defense defect in these patients is reduced neutrophil function, especially when hyperglycemia and acidosis occur.

Two specific diseases highly associated with diabetes are **malignant otitis externa** caused by *Pseudomonas aeruginosa* and **mucormycosis** caused by molds belonging to the genera *Mucor* and *Rhizopus*. In addition, there is an increased incidence and severity of community-acquired pneumonia caused by bacteria such as *S. pneumoniae* and *S. aureus* and of urinary tract infections caused by organisms such as *Escherichia coli* and *C. albicans*. Candidal vulvovaginitis is also more common in diabetic patients. Diabetic patients also have many foot infections because atherosclerosis compromises the blood supply and necrosis of tissue occurs. Skin infections, such as ulcers and cellulitis, and soft tissue infections, such as necrotizing fasciitis, are common and can extend to the underlying bone, causing osteomyelitis. *S. aureus* and mixed facultative anaerobic bacteria are the most common causes.

Fever

Infection causes a rise in the body temperature that is attributed to **endogenous pyrogen** (IL-1) released from macrophages. Fever may be a protective response because a variety of bacteria and viruses grow more slowly at elevated temperatures.

ADAPTIVE (SPECIFIC) IMMUNITY

Adaptive immunity results either from exposure to the organism (active immunity) or from receipt of preformed antibody made in another host (passive immunity).

Passive adaptive immunity provides temporary protection against an organism and is acquired by receiving preformed antibodies from another person or animal. Passive immunization also occurs normally in the form of immunoglobulins passed through the placenta (IgG) or breast milk (IgA) from mother to child. This protection is very important during the early days of life when the child has a reduced capacity to mount an active response. For example, immunizing mothers with the TDaP vaccine, which protects against tetanus, diphtheria, and pertussis, provides immunity not only to the mother but also to the infant due to the passage of IgG from the mother to the infant.

Passive immunity has the important advantage that its protective abilities are present immediately, whereas active immunity has a delay (a few days to a few weeks) depending on whether it is a primary or secondary response. However, passive immunity has the important disadvantage that the antibody concentration decreases fairly rapidly as the proteins are degraded, and so the protection usually lasts for only a month or two. The administration of preformed antibodies can be lifesaving in certain diseases that are caused by powerful exotoxins, such as botulism and tetanus. Serum globulins, given intravenously, are a prophylactic measure in patients with hypogammaglobulinemia or bone marrow transplants. In addition, they can mitigate the symptoms of certain diseases such as hepatitis caused by hepatitis A virus.

Active adaptive immunity is protection based on exposure to the organism in the form of overt disease, subclinical infection (i.e., an infection without symptoms), or a vaccine. This protection has a slower onset but longer duration than passive immunity. The **primary response** usually takes 7–10 days for the antibody to become detectable. An important advantage of active immunity is that an **anamnestic (secondary)** response occurs (i.e., there is a rapid response [approximately 3 days] of large amounts of antibody to an antigen that the immune system has previously encountered). Active immunity is mediated by both antibodies (immunoglobulins) and T cells:

(1) Antibodies protect against organisms by a variety of mechanisms—neutralization of toxins, lysis of bacteria in the presence of complement, opsonization of bacteria to facilitate phagocytosis, and interference with adherence of bacteria and viruses to cell surfaces. If the level of IgG drops below 400 mg/dL (normal = 1000–1500 mg/dL), the risk of pyogenic infections caused by bacteria such as staphylococci increases.

Because antibodies, especially IgG, rise to a protective level slowly (~7–10 days after infection), they do not play a major role in combating primary infection but rather protect against a second infection by that organism or against hematogenous dissemination of the organism to distant sites in the body at a future time.

(2) T cells mediate a variety of reactions, including cytotoxic destruction of virus-infected cells, activation of macrophages, and delayed hypersensitivity. T cells, especially Th-1 cells (see Chapter 58), and macrophages are the main host defense against mycobacteria such as *M. tuberculosis* and systemic fungi such as *Histoplasma* and *Coccidioides*. T cells also help B cells to produce antibody against many antigens.

Table 8–3 describes the essential host defense mechanisms against bacteria, which include both humoral immunity against

TABLE 8–3 Essential Host Defense Mechanisms Against Bacteria

Essential Host Defense Mechanism	Type of Bacteria or Toxin	Important Examples
Antibody-mediated opsonization	Encapsulated pyogenic bacteria	*Streptococcus pneumoniae, Streptococcus pyogenes, Staphylococcus aureus, Neisseria meningitidis, Haemophilus influenzae, Pseudomonas aeruginosa*
Antibody-mediated neutralization	Exotoxins	*Corynebacterium diphtheriae, Clostridium tetani, Clostridium botulinum*
Cell-mediated	Intracellular bacteria	*Mycobacterium tuberculosis,* atypical mycobacteria, *Legionella pneumophila, Listeria monocytogenes*

pyogenic bacteria and exotoxins, and cell-mediated immunity against several intracellular bacteria.

FAILURE OF HOST DEFENSES PREDISPOSES TO INFECTIONS

The frequency or severity of infections is increased when certain predisposing conditions exist, which fall into two main categories: immunocompromised patients or patients that have foreign bodies (indwelling catheters or prosthetic devices). Foreign bodies predispose because host defenses do not operate efficiently in their presence. Table 8–4 describes the predisposing conditions and the most common organisms causing infections when these predisposing conditions are present.

Certain diseases and anatomic abnormalities also predispose to infections. For example, patients with diabetes often have *S. aureus* infections for a few reasons: these patients have extensive atherosclerosis, which causes relative anoxia to tissue, and they have a defect in neutrophil function. Patients with sickle cell anemia often have *Salmonella* osteomyelitis, probably because the abnormally shaped cells occlude the small capillaries in the bone. This traps the *Salmonella* within the bone, increasing the risk of osteomyelitis.

Patients with certain congenital cardiac defects or rheumatic valvular damage are predisposed to endocarditis caused by viridans streptococci. Neutrophils have difficulty in penetrating the vegetations formed on the valves in endocarditis. Patients with an aortic aneurysm are prone to vascular infections caused by *Salmonella* species.

Patients with reduced host defenses often have a muted response to infection (e.g., a low-grade [or no] fever and a mild [or absent] inflammatory response). For this reason, a high index of suspicion regarding the presence of infection must be present when evaluating patients who are immunocompromised, especially those who are intentionally immunosuppressed, such as transplant recipients.

TABLE 8–4 Conditions That Predispose to Infections and the Organisms That Commonly Cause These Infections

Predisposing Condition	Organisms Commonly Causing Infection
Immunocompromised state	
Low antibody	Pyogenic bacteria (e.g., *Staphylococcus aureus, Streptococcus pneumoniae*)
Low complement (C3b)	Pyogenic bacteria (e.g., *S. aureus, S. pneumoniae*)
Low complement (C6,7,8,9)	*Neisseria meningitidis*
Low neutrophil number	Pyogenic bacteria (e.g., *S. aureus, S. pneumoniae*)
Low neutrophil function as in CGD	*S. aureus* and *Aspergillus fumigatus*
Low CD4 cells as in AIDS	Various bacteria (e.g., mycobacteria), various viruses (e.g., CMV), and various fungi (e.g., *Candida*)
Presence of foreign bodies	
Urinary catheters	*Escherichia coli*
Intravenous catheters	*Staphylococcus epidermidis, Candida albicans*
Prosthetic heart valves	*S. epidermidis, C. albicans*
Prosthetic joints	*S. epidermidis*
Vascular grafts	*S. epidermidis, S. aureus, Salmonella enterica*

AIDS = acquired immunodeficiency syndrome; CGD = chronic granulomatous disease; CMV = cytomegalovirus.

- The **acute-phase response** consists of proteins (e.g., C-reactive protein, mannose-binding protein, and LPS-binding protein) that enhance the host response to bacteria. Interleukin-6 is the main inducer of this response.

- Neutrophils and macrophages are attracted to the site of infection by **chemokines**, which are small polypeptides produced by cells at the infected site. **Interleukin-8** and **C5a** are important chemokines for neutrophils.

- In response to most bacterial infections, there is an **increase in the number of neutrophils** in the blood. This increase is caused by the production of **granulocyte-stimulating factors** by macrophages.

- Both neutrophils and macrophages **phagocytose** bacteria, but macrophages (and similar cells called dendritic cells) also **present antigen** to CD4-positive (helper) T cells, whereas neutrophils do not. **Dendritic cells are the most important antigen-presenting cells**.

- After neutrophils are attracted to the infected site by chemokines, they attach to the endothelium first using **selectins** on the endothelium, then by the interaction of **integrins** on the neutrophils with intracellular adhesion molecule (ICAM) proteins on the endothelium. The concentration of ICAM proteins is increased by cytokines released by activated macrophages, which results in neutrophils being attracted to the infected site.

- Neutrophils then migrate through the endothelium (**diapedesis**) and ingest the bacteria. **IgG and C3b are opsonins**, which enhance ingestion of the bacteria. There are receptors for the heavy chain of IgG and for C3b on the surface of the neutrophils.

- Killing of the bacteria within the neutrophil is caused by **hypochlorite, hydrogen peroxide**, and **superoxides. Lysosomes** contain various degradative enzymes and fuse with the phagosome to form a **phagolysosome** within which the killing occurs.

- Severe, recurrent **pyogenic infections** occur in those who have **inadequate neutrophils**. For example, people with defective neutrophils, people with fewer than 500 neutrophils/μL, and those who have had a splenectomy or who have diabetes mellitus are at increased risk for pyogenic infections.

Adaptive Immunity

- **Passive immunity** refers to protection based on the transfer of preformed antibody from one person (or animal) to another person. Passive immunity provides **immediate but short-lived protection** (lasting a few months). Examples of passive immunity include administration of antitoxin, passage of IgG from mother to fetus across the placenta, and passage of IgA from mother to newborn through breast milk.

- **Active immunity** refers to protection based on the formation of both **antibodies and cell-mediated immunity after exposure** either to the microbe itself (with or without disease) or to the antigens of the microbe in a vaccine. Active immunity provides **long-term protection but is not effective** for days after exposure to the microbe. In the **primary response**, antibody is detectable in 7–10 days, whereas in the **secondary response**, antibody is detectable in approximately 3 days.

- The main **functions of antibodies** are to **neutralize bacterial toxins and viruses, opsonize bacteria, activate complement** to form a membrane attack complex that can kill bacteria, and **interfere with attachment to mucosal surfaces**. IgG is the main opsonizing antibody, IgG and IgM activate complement, and IgA interferes with attachment to the mucosa.

- The main functions of **cell-mediated immunity are to protect against intracellular bacteria** and to **kill virus-infected cells**. Helper T cells (and macrophages) protect against intracellular bacteria, whereas cytotoxic T cells kill virus-infected cells.

Reduced Host Defenses

- **Reduced host defenses** result in an increase in the frequency and severity of infections. The main causes include various genetic immunodeficiencies, the presence of foreign bodies, and the presence of certain chronic diseases, such as diabetes mellitus and renal failure.

SELF-ASSESSMENT QUESTIONS

1. Which one of the following host defense processes is the **MOST** important in preventing the action of exotoxins?
   (A) Binding of cytokines to exotoxin-specific receptors inhibits the attachment of exotoxins.
   (B) Degradation of exotoxins by the membrane attack complex of complement.
   (C) Lysis of exotoxins by perforins produced by cytotoxic T cells.
   (D) Neutralization of exotoxins by antibody prevents binding to target cell membrane.
   (E) Phagocytosis of exotoxins by neutrophils and subsequent destruction by hypochlorite.

2. An inflammatory response in the skin is characterized by erythema (redness). Which of the following is the most important cause of this erythema?
   (A) C3b component of complement
   (B) Gamma interferon
   (C) Histamine
   (D) Hypochlorite
   (E) Superoxide

3. A 1-year-old child with repeated infections was diagnosed with chronic granulomatous disease (CGD). A defect in which of the following is the cause of CGD?
   (A) Gamma interferon receptor
   (B) LFA-integrins
   (C) Mannose-binding protein

(D) NADPH oxidase

(E) Nitric oxide

4. Opsonization is the process by which:

(A) Bacteria are made more easily phagocytized.

(B) Chemokines attract neutrophils to the site of infection.

(C) Neutrophils migrate from the blood through the endothelium to reach the site of infection.

(D) The acute-phase response is induced.

(E) The alternate pathway of complement is activated.

ANSWERS

(1) **(D)**

(2) **(C)**

(3) **(D)**

(4) **(A)**

PRACTICE QUESTIONS: USMLE & COURSE EXAMINATIONS

Questions on the topics discussed in this chapter can be found in the Basic Bacteriology section of Part XIII: USMLE (National Board) Practice Questions starting on page 733. Also see Part XIV: USMLE (National Board) Practice Examination starting on page 775.

Laboratory Diagnosis

APPROACH TO LABORATORY DIAGNOSIS

Clinical microbiology laboratories play a fundamental and indispensable role in providing reliable and timely information regarding the identification of infectious disease agents. Clinicians use this information not only to make or confirm a diagnosis but also to guide clinical decisions and treatment options, so this information needs to be definitive, significant, and relevant to the case under consideration. Obtaining accurate results that can be interpreted with high confidence is dependent on the laboratory receiving high-quality patient specimens.

Since specimen selection and collection are typically the responsibility of the medical staff, clinicians should (1) understand the pathogenesis of the infection and ensure proper collection of an adequate quantity of specimen from the body site that is most likely to yield growth of the infecting organism, while avoiding contamination from the normal flora; (2) ensure that the integrity of the specimen is not compromised during transport and that the specimen is handled in such a way to preserve the viability of any anaerobes or fastidious organisms; and (3) provide ancillary information to guide the laboratory personnel who will process and analyze the specimen.

As diagnostic microbiology testing becomes more complex, clear communication and strong partnerships between microbiology lab professionals and clinicians will remain a top priority. Clinicians communicate crucial clinical information about a patient to microbiology lab staff that will then allow lab personnel to direct clinicians toward context-appropriate tests and optimized sample collection methods that have the most diagnostic value, which will ultimately lead to better outcomes for patients.

Traditionally, diagnosis has relied on **culture, microscopic and phenotypic characterization** of an organism (Table 9–1), and **serologic testing**, in which the organism is identified by the detection of organism-specific antibodies in the patient's serum. More recently, advances in the fields of **molecular biology** and **genomics** have resulted in more accurate and rapid identification of causative organisms. There are now several US Food and Drug Administration (FDA)-approved nucleic acid and proteomic-based assays for identifying infectious agents (discussed below), and their routine implementation in the clinical laboratory setting has resulted in improved patient care, better antibiotic stewardship resulting in decreased antimicrobial resistance rates, and increased efficiency of the laboratory and healthcare facility in the processing and analysis of clinical samples.

TABLE 9–1 General Approach to the Diagnosis of a Bacterial Infection

1. Obtain a specimen from the infected site.
2. Stain the specimen using the appropriate procedure (e.g., Gram stain or acid-fast stain). For any bacteria seen, the gram reaction (positive or negative), shape (e.g., cocci or rods), size, and arrangement (e.g., chains or clusters), should be noted, as well as whether only one or more than one type of bacteria is present. The microscopic appearance is *not* sufficient to speciate an organism, but it often allows an educated guess to be made regarding the genus of the organism and thereby guides empiric therapy.
3. Culture the specimen on the appropriate media (e.g., blood agar plates). In most instances, plates should be streaked to obtain isolated colonies (i.e., a "pure culture"). The plates should be incubated in the presence or absence of oxygen as appropriate.
4. Identify the organism using the appropriate tests (e.g., sugar fermentation, DNA probes, antibody-based tests such as agglutination, or immunofluorescence). Note special features such as hemolysis and pigment formation.
5. Perform antibiotic susceptibility tests.

METHODS OF LABORATORY DIAGNOSIS

Microscopic Examination

If a specimen is collected from a "sterile" body site that does not harbor a "background" of normal flora (e.g., sterile tissues, cerebral spinal fluid, joint fluid, or urine), a sample of the specimen can be prepared for microscopic examination using an appropriate staining method, like the Gram stain or acid-fast stain. If bacteria are seen in the specimen, their shape (e.g., cocci or rods), size, and arrangement (e.g., chains or clusters) and whether they are gram-positive, gram-negative, or acid-fast should be noted and can be useful in their identification. It is also important to determine whether only one or more than one type of bacteria is present. The microscopic appearance is typically not sufficient to definitively identify an organism, but often allows an educated guess to be made regarding the taxonomic classification (genus) of the organism and thus guides empiric therapy that can be initiated without waiting for growth of the organism.

Culture-Based Methods

Several methods for diagnosing bacterial or fungal infections require the suspected pathogen to be isolated in pure culture from a properly obtained clinical specimen. Historically this has been accomplished by using an agar-based medium, for instance, blood agar plates, and streaking the specimen over the agar surface in a manner to obtain well-isolated colonies. The agar plates are then incubated under atmospheric conditions that will support the growth of a variety of different microorganisms.

Media can be "selective," containing compounds that only allow certain bacteria to grow (e.g., antibiotics, salts, or dyes), and/or "differential," because they contain other compounds that allow one type of bacteria to be distinguished from another based on a biochemical reaction (e.g., detecting hemolysis on blood agar plates or pigment formation). Table 9–2 contains a list of various bacteriologic agars commonly used in the diagnostic laboratory and the function of these agars. Once pure, well-isolated colonies are obtained, further phenotypic characterization (e.g., analyzing biochemical and enzymatic activities) and antibiotic susceptibility testing (see Chapter 11) can be performed.

In modern Clinical Microbiology Laboratories, the process of cultivation has been improved upon substantially with the development and introduction of novel nylon flocked swabs (Eswabs) and accompanying transport containers. The Eswabs have significantly improved pathogen recovery rates and allow for the elution of >90% of sample organisms into liquid transport media as opposed to agar-base transport media.

There are several advantages to liquid-based cultivation. Having high recovery of a homogenous distribution of organisms in the transport medium facilitates the use of automated plating instruments, allows for many more plates to be inoculated from the same clinical specimen, and has shifted the practice of microbiology from labor-intensive, tubed biochemical identification and susceptibility testing methods to a more accurate and efficient automated sample processing pipeline using plates and assays that can be incubated, monitored, and read automatically.

TABLE 9–2 Commonly Used Bacteriologic Agars and Their Function

Name of Agar[1]	Bacteria Isolated on the Agar	Function or Properties of the Agar
Blood	Various bacteria	Detect hemolysis
Bordet-Gengou	*Bordetella pertussis*	Increased concentration of blood allows growth
Charcoal-yeast extract	*Legionella pneumophila*	Increased concentration of iron and cysteine allows growth
Chocolate	*Neisseria meningitidis* and *Neisseria gonorrhoeae* from sterile sites	Heating the blood inactivates inhibitors of growth
Chocolate agar plus X and V factors	*Haemophilus influenzae*	X and V factors are required for growth
Egg yolk	*Clostridium perfringens*	Lecithinase produced by the organism degrades egg yolk to produce insoluble precipitate
Eosin-methylene blue	Various enteric gram-negative rods	Selects against gram-positive bacteria and differentiates between lactose fermenters and nonfermenters
Löwenstein-Jensen	*Mycobacterium tuberculosis*	Selects against gram-positive bacteria in respiratory tract flora and contains lipids required for growth
MacConkey	Various enteric gram-negative rods	Selects against gram-positive bacteria and differentiates between lactose fermenters and nonfermenters
Tellurite	*Corynebacterium diphtheriae*	Causes tellurite to become tellurium, which has black color
Thayer-Martin	*N. gonorrhoeae* from nonsterile sites	Chocolate agar with antibiotics to inhibit growth of normal flora
Triple sugar iron (TSI)	Various enteric gram-negative rods	Distinguishes lactose fermenters from nonfermenters and H_2S producers from nonproducers

[1]Names are listed in alphabetical order.

A substantial body of research has shown that automation of specimen processing and incubation has decreased institutional costs, technician errors, and time-to-specimen identification, all of which should improve patient care and safety.

Since specimen selection and collection are of paramount importance in yielding high-quality laboratory results, common specimen collection sites and methods are discussed below.

Blood Cultures

Blood cultures are performed most often when sepsis, endocarditis, osteomyelitis, meningitis, or pneumonia is suspected. The bacteria most frequently isolated from blood cultures are two gram-positive cocci, *Staphylococcus aureus* and *Streptococcus pneumoniae*, and three gram-negative rods, *Escherichia coli*, *Klebsiella pneumoniae*, and *Pseudomonas aeruginosa*. Certain pathogenic fungi including yeast (*Candida* species and *Cryptococcus neoformans*) and molds can also be isolated from blood cultures.

For blood cultures, the site for venipuncture is cleaned with an antiseptic to prevent contamination by members of the skin flora, usually *Staphylococcus epidermidis*, and decrease the risk of infection-related complications. Blood obtained is added to a rich growth medium in a bottle that contains an indicator for carbon dioxide (CO_2) production. Standard practice is to inoculate 10 mL of blood into each of two bottles per culture set, with one bottle incubated aerobically and one anaerobically. Production of CO_2 within the bottle indicates organism metabolism and growth. Gram stain, subculture, and antibiotic sensitivity tests can then be performed. In some hospitals, molecular methods are used to identify the organism (see later in this chapter).

Throat Cultures

Throat cultures inoculated onto blood agar plates are used to detect the presence of group A β-hemolytic streptococci (*Streptococcus pyogenes* or GAS), an important, treatable cause of pharyngitis. They are also used when diphtheria, gonococcal pharyngitis, or thrush (*Candida*) is suspected.

In the past several years, the FDA has approved point of care (POC) devices that test for GAS antigens (Rapid Antigen Detection Tests [RADT]) and, most recently, a molecular diagnostic test based on detection of GAS-specific DNA sequences (see **Molecular Diagnostic Methods** section below). This *n*ucleic *a*cid *a*mplification *t*est (NAAT) is much more sensitive and specific than the RADT, and can detect fewer than 50 bacterial cells/mL of GAS from a throat swab in as little as 15 minutes, which is significantly faster than the 24 hours required for plate cultivation followed by bacitracin susceptibility testing to confirm GAS infection.

Since accurate diagnosis and appropriate antimicrobial therapy is recommended to prevent post-GAS infection complications (rheumatic fever or acute glomerulonephritis), POC testing is preferable to traditional plate culture. The rapid definitive diagnosis of infection facilitates initiation of definitive antimicrobial therapy, which aligns with institutional efforts on antimicrobial stewardship to improve quality of care.

Whether the specimen is being obtained for plate culture or POC testing, obtaining a high-quality specimen is important. The collection swab should not only touch the posterior pharynx but also both tonsils and tonsillar fossae, and be streaked onto a blood agar plate to obtain single colonies, or can be used to inoculate the POC devices. GAS will form β-hemolytic colonies on blood agar plates after 24 hours of incubation at 35°C. Further testing looking at bacitracin susceptibility (traditional) or MALDI-TOF analysis (described below) can be used to determine whether the organism is likely to be a group A *Streptococcus*.

Gram stain is typically *not* done on a throat swab because it is impossible to distinguish between the appearance of the normal flora streptococci and *S. pyogenes*.

Sputum Cultures

Sputum cultures can be performed to determine infectious etiologies of pneumonia or to test for active pulmonary tuberculosis. The most frequent bacterial cause of community-acquired pneumonia is *S. pneumoniae*, whereas *S. aureus* and gram-negative rods, such as *K. pneumoniae* and *P. aeruginosa*, are common causes of hospital-acquired pneumonias.

It is important that the specimen for culture really be sputum and not saliva or nasopharyngeal secretions from the upper airway. A reliable specimen has more than 25 leukocytes and fewer than 10 epithelial cells per 100× field. An unreliable sample can be misleading and should be rejected by the laboratory. If the patient cannot cough and the need for a microbiologic diagnosis is strong, tracheal aspirate, bronchoalveolar lavage, or lung biopsy may be necessary. Because these procedures bypass the normal flora of the upper airway, they are more likely to provide an accurate microbiologic diagnosis. A preliminary assessment of the cause of the pneumonia can be made by Gram stain if large numbers of typical organisms are seen.

Culture of the sputum on blood agar can reveal the presence of colonies, with identification established using various serologic or biochemical tests or by matrix-assisted laser desorption/ionization time-of-flight (MALDI-TOF) mass spectrometry (see below). Cultures of *Mycoplasma* are infrequently done; diagnosis is usually confirmed by a rise in antibody titer. If *Legionella* pneumonia is suspected, the organism can be cultured on charcoal-yeast agar, which contains the high concentrations of iron, sulfur, and cysteine required for growth.

If active tuberculosis is suspected, an acid-fast stain can be done to look for the organism in the sputum; however, this microscopic method is not very sensitive. The sputum can also be cultured on special media but requires at least 6 weeks of incubation to cultivate *Mycobacterium tuberculosis*. There is now a NAAT (Xpert MTB/RIF assay) available to detect *M. tuberculosis* infection. This test can be completed in less than 2 hours and confirms the presence of *M. tuberculosis* and can also detect if the infecting strain is resistant to rifampin.

Aspiration pneumonia and lung abscesses represent other types of infections affecting the respiratory tract. If these are suspected, culturing for anaerobic bacteria is important.

Cerebrospinal Fluid Cultures

Cerebrospinal fluid (CSF) cultures are performed primarily when a neurologic infection such as meningitis, meningoencephalitis, or transverse myelitis is suspected. CSF specimens from tissue-centric cases, including encephalitis, brain abscess, and subdural empyema, may show negative cultures. The most important causes of acute bacterial meningitis are three encapsulated organisms: *Neisseria meningitidis, S. pneumoniae,* and *Haemophilus influenzae.*

Because acute meningitis is a medical emergency, the specimen should be taken immediately to the laboratory. The Gram-stained smear of the sediment of the centrifuged sample guides the immediate empirical treatment. If meningitis caused by acid-fast bacteria such as *M. tuberculosis* is suspected, an acid-fast stain and culture of CSF should be performed, although for the reasons discussed above, nucleic acid amplification methods should also be used for rapid identification.

The fungus *C. neoformans*, a cause of meningitis, particularly in human immunodeficiency virus–infected patients, can also be cultured from CSF. Most laboratories test the CSF with a specific and sensitive latex agglutination test for *C. neoformans* (cryptococcal antigen).

Stool Cultures

Most cases of acute diarrhea are self-limiting and do not require empiric antimicrobial therapy or stool culture. However, testing will be performed for patients experiencing severe, persistent, or bloody diarrhea, patients who are immunocompromised, outbreak-associated diarrhea, and healthcare-associated diarrhea. While Norovirus is the most common overall cause of diarrhea in the United States, the most common bacterial pathogens causing diarrhea are *Salmonella, Shigella, Campylobacter, and E. coli* O157 strains (STEC). *Clostridium difficile* should be suspected for patients who develop nosocomial diarrhea, particularly after antibiotic treatment. The patient's stool can be tested for the presence of the *C. difficile* toxins using an enzyme immunoassay or for the presence of a toxigenic *C. difficile* strain using a NAAT.

When bacterial culture is recommended, feces collected during the acute phase of symptoms is the specimen of choice. Specimens should be processed by the clinical lab within 2 hours of collection to maximize detection of the organisms. The selection of primary plating media used for routine culture varies across laboratories, but typically includes (1) MacConkey agar; (2) a selective/differential medium (e.g., eosin-methylene blue agar [EMB]) to maximize recovery of *Salmonella* and *Shigella*; (3) a medium to recover *Campylobacter* (e.g., Campy-CVA), which must be incubated in a microaerobic atmosphere at 42°C; and (4) a medium to recover *E. coli* O157, like MacConkey-sorbitol medium. Antigen detection assays to test for Shiga toxin I and II or *C. difficile* toxin A/B should also be performed.

MacConkey and EMB agars are both selective and differential. They are selective because they allow gram-negative rods to grow but inhibit many gram-positive organisms. Their differential properties are based on the fact that *Salmonella* and *Shigella* do not ferment lactose, whereas many other enteric gram-negative rods do. On EMB agar, colonies of *E. coli*, a lactose fermenter, appear **purple** and have a **green sheen**. In contrast, colonies of nonlactose fermenters, such as *Salmonella* and *Shigella*, appear **colorless**.

More and more clinical labs have incorporated MALDI-TOF technology (discussed in the **Molecular Diagnostics** section) into their diagnostic tool repertoire, which is the preferred culture-based diagnostic method both from a cost and time-effective perspective. This method compares mass spectral libraries representing thousands of reference strains with a patient isolate to make sensitive and reliable identification of the unknown bacterial or fungal pathogen within minutes of sample preparation.

This technology cannot distinguish between all important pathogens. Currently, MALDI-TOF cannot accurately differentiate among all species within certain groups of organisms, such as the *Enterobacter cloacae* complex, *Burkholderia cepacia* complex, and *Streptococcus bovis* species group. In addition, *Shigella* and *E. coli* cannot be reliably differentiated by MALDI-TOF. Since *E. coli* is one of the most frequent organisms encountered in clinical microbiology laboratories, alternative strategies such as employing TSI (Triple Sugar Iron) slants or EMB agar, or (slide agglutination tests (discussed below) to distinguish *E. coli* from *Shigella* species.

There are also several culture-independent diagnostic tests (CIDTs) now available for suspected GI pathogens, which are based on immune assays to detect toxins, or nucleic acid amplification to detect pathogens. These tests are more expensive than culture-based testing, but are highly sensitive and specific, and typically increase detection of all bacterial GI pathogens, especially those that are harder to cultivate (i.e., *Campylobacter* and *Shigella*). The main advantages of CIDTs are the speed at which a diagnosis can be obtained (hours versus days for cultivation) and their reliability. CIDTs should be confirmed by culture especially if susceptibility testing is indicated or if there is a suspected outbreak that would require a Public Health investigation.

Urine Cultures

Urine cultures are performed primarily when pyelonephritis or cystitis is suspected. By far the most frequent cause of urinary tract infections is *E. coli*. Other common agents are *Enterobacter, Proteus,* and *Enterococcus faecalis.*

Urine in the bladder of a healthy person is sterile, but it acquires organisms of the normal flora as it passes through the distal portion of the urethra. To avoid these organisms, a midstream specimen, voided after washing the external orifice, is used for urine cultures. In special situations, suprapubic aspiration or catheterization may be required to obtain a specimen. Because urine is a good culture medium, any organisms present in the specimen can multiply, leading to erroneous results regarding type and number of organisms present at the time of collection. Thus, it is essential that the cultures be done within 1 hour after collection or stored in a refrigerator at 4°C for not more than 18 hours.

It is commonly accepted that a bacterial count of at least 100,000/mL must be found to conclude that significant bacteriuria is present (in *asymptomatic* persons). There is evidence that a bacterial count as low as 1000/mL is significant in *symptomatic* patients. For this determination to be made, quantitative or semiquantitative cultures are performed. There are several techniques: (1) a calibrated loop that holds 0.001 mL of urine can be used to streak the culture; (2) serial tenfold dilutions can be made and samples from the dilutions streaked; and (3) a screening procedure suitable for the physician's office involves an agar-covered "paddle" that is dipped into the urine—after the paddle is incubated, the density of the colonies is compared with standard charts to obtain an estimate of the concentration of bacteria.

Genital Tract Cultures

Genital tract cultures can be performed on specimens from individuals with an abnormal discharge or on specimens from asymptomatic contacts of a person with a sexually transmitted disease. One of the most important genital tract pathogens is *Neisseria gonorrhoeae*. Laboratory diagnosis of gonorrhea can be made by microscopic examination of a Gram-stained smear and culture of the organism, but is now typically done with nucleic acid techniques. Culture may still be important to determine antimicrobial susceptibility in cases of treatment failure.

Specimens are obtained by swabbing the urethral canal (for men), the cervix (for women), or the anal canal (for men and women). A urethral discharge from the penis is frequently used. Because *N. gonorrhoeae* is very delicate, the specimen should be inoculated quickly onto medium such as a Thayer–Martin chocolate agar plate.

Gram-negative diplococci found *intracellularly* within neutrophils on a smear of a urethral discharge from a man have over 90% probability of being *N. gonorrhoeae*. Because smears are less specific when made from swabs of the endocervix and anal canal, cultures or nucleic acid-based testing are necessary. The finding of only *extracellular* diplococci suggests that these Neisseriae may be members of the normal flora and that the patient may have nongonococcal urethritis.

Nongonococcal urethritis and cervicitis are also extremely common infections. The most frequent cause is *Chlamydia trachomatis*, an obligate intracellular pathogen that cannot grow on artificial medium. Nucleic acid testing is typically used to diagnose this organism.

Because *Treponema pallidum*, the agent of syphilis, cannot be cultured, diagnosis is made primarily by serology and sometimes by microscopy if dark-field microscopy is available. The presence of motile spirochetes with typical morphologic features seen by dark-field microscopy of the fluid from a painless genital lesion is sufficient for the diagnosis. The serologic tests fall into two groups: (1) the nontreponemal antibody tests such as the Venereal Disease Research Laboratory (VDRL) or rapid plasma reagin (RPR) test and (2) the treponemal antibody tests such as the fluorescent treponemal antibody-absorption (FTA-ABS) test. These tests are described in Chapter 24.

Wound & Abscess Cultures

A variety of different organisms have been described in association with wound and abscess infections, many of which are polymicrobial. The bacteria most frequently isolated differ according to anatomic site and predisposing factors. Abscesses of the brain, lungs, and abdomen are frequently caused by anaerobes such as *Bacteroides fragilis* and gram-positive cocci such as *S. aureus* and *S. pyogenes*. Members of the soil flora such as *Clostridium perfringens* are important causes of traumatic open-wound infections, while surgical-wound infections are commonly associated with skin flora including various staphylococci, streptococci, and *Propionibacterium acnes*. Infections of dog or cat bites are often due to *Pasteurella* species (~50% of cases), whereas human bites usually involve viridans streptococci, especially *Streptococcus anginosus*, and mouth anaerobes, such as *Prevotella* and *Fusobacterium*.

Anaerobes are frequently involved in these types of infection, so it is important to place the specimen in anaerobic collection tubes and transport it promptly to the laboratory. Because many of these infections are due to multiple organisms, including mixtures of anaerobes and nonanaerobes, the specimen should be cultured on several different media under different atmospheric conditions. The Gram stain can provide valuable information regarding the range of organisms under consideration.

Sometimes an organism is not recovered by culturing, either because it is nonculturable on bacteriologic media or it is intermittently present or only present in limited numbers, and other techniques must be used. Table 9–3 describes some approaches to making a diagnosis when the cultures are negative, which include immunologic and molecular methods discussed below.

Serologic Methods

These methods are described in more detail in Chapter 64. However, it is of interest here to present information on how serologic reactions aid the microbiologic diagnosis. There are essentially two basic approaches: (1) using known antibody to

TABLE 9–3 How to Diagnose a Bacterial Infection When the Culture Is Negative

1. Detect antibody in the patient's serum. Detecting IgM antibody indicates a current infection. A fourfold or greater rise in antibody titer between the acute serum sample and the convalescent serum sample also indicates a current infection. (A major drawback with the use of acute and convalescent serum samples is that the convalescent sample is usually taken 10–14 days after the acute sample. By this time, the patient has often recovered and the diagnosis becomes a retrospective one.) A single IgG antibody titer is difficult to interpret because it is unclear whether it represents a current or a previous infection. In certain diseases, a single titer of sufficient magnitude can be used as presumptive evidence of a current infection.
2. Detect antigen in the patient's specimen. Use known antibody to detect presence of antigens of the organisms (e.g., fluorescent antibody to detect antigens in tissue, latex agglutination to detect capsular polysaccharide antigens in spinal fluid).
3. Detect nucleic acids in the patient's specimen. Use polymerase chain reaction (PCR) and DNA probes to detect the DNA or RNA of the organism.

identify the microorganism, and (2) using known antigens to detect antibodies in the patient's serum.

Identification of an Organism with Known Antiserum

Slide Agglutination Test—Antisera can be used to identify *Salmonella* and *Shigella* by causing agglutination (clumping) of the unknown organism. Antisera directed against the cell wall O antigens of *Salmonella* and *Shigella* are commonly used in hospital laboratories. Antisera against the flagellar H antigens and the capsular Vi antigen of *Salmonella* are used in public health laboratories for epidemiologic purposes.

Latex Agglutination Test—Latex beads coated with specific antibody are agglutinated in the presence of the homologous bacteria or antigen. This test is used to determine the presence of the capsular antigen of the yeast *C. neoformans*.

Enzyme-Linked Immunosorbent Assay—In this test, a specific antibody that is linked to an easily assayed enzyme is used to detect the presence of the homologous antigen. This test is useful in detecting a wide variety of bacterial, viral, and fungal infections and is discussed in greater detail in Chapter 64.

Fluorescent Antibody Tests—A variety of bacteria can be identified by exposure to known antibody labeled with fluorescent dye, which is detected visually in the ultraviolet microscope. Various methods can be used, such as the direct and indirect techniques (see Chapter 64).

Identification of Serum Antibodies with Known Antigens

Slide or Tube Agglutination Test—In this test, serial two-fold dilutions of a sample of the patient's serum are mixed with standard bacterial antigen suspensions. The highest dilution of serum capable of agglutination is the titer of the antibody. As with most tests of a patient's antibody, at least a fourfold rise in titer between the early and late samples must be demonstrated for a diagnosis to be made. This test is used primarily to aid in the diagnosis of typhoid fever, brucellosis, tularemia, plague, leptospirosis, and rickettsial diseases.

Serologic Tests for Syphilis—The detection of antibody in the patient's serum is frequently used to diagnose syphilis, because *T. pallidum* does not grow on laboratory media. There are two kinds of tests.

(1) The nontreponemal tests use a cardiolipin–lecithin–cholesterol mixture as the antigen, not an antigen of the organism. Cardiolipin is a lipid extracted from normal beef heart. Flocculation (clumping) of the cardiolipin occurs in the presence of antibody induced by infection by *T. pallidum*. The VDRL and RPR tests are nontreponemal tests commonly used as screening procedures. They are not specific for syphilis but are inexpensive and easy to perform.

(2) The treponemal tests use *T. pallidum* as the antigen. The two most widely used treponemal tests are the FTA-ABS and the *T. pallidum* particle agglutination (TPPA) tests. In

the FTA-ABS test, the patient's serum sample, which has been absorbed with treponemes other than *T. pallidum* to remove nonspecific antibodies, is reacted with nonviable *T. pallidum* on a slide. Fluorescein-labeled antibody against human immunoglobulin G (IgG) is then used to determine whether IgG antibody against *T. pallidum* is bound to the organism. In the TPPA test, the patient's serum sample is mixed with gelatin particles that have been sensitized with *T. pallidum* whole cell antigens. Patient serum that contains antibodies to *T. pallidum* will react with the gel particle, resulting in agglutination that will appear like a uniformly distributed smooth mat of particles in the microtiter plate. A negative result appears as a compact button at the bottom of the microtiter plate.

Cold Agglutinin Test—Patients with *Mycoplasma pneumoniae* infections develop autoimmune antibodies that agglutinate human red blood cells in the cold (4°C) but not at 37°C. These antibodies occur in certain diseases other than *Mycoplasma* infections; thus, false-positive results can occur. NAATs are preferred for the diagnosis of pneumonia caused by *M. pneumoniae*.

Molecular Diagnostic Methods

The field of molecular diagnostics is dynamic and rapidly evolving. The methods described below, and in some cases mentioned above, have been adopted by clinical microbiology labs not only due to their increased sensitivity and specificity and reduced turnaround time, as compared with more traditional diagnostics described above, but also because early and accurate diagnosis has a significant and positive impact on patient care. It is not the intent here to discuss all assays that are currently available or under development. Rather, it is important for clinicians to be cognizant of the pace of development of molecular techniques in the realm of clinical microbiology and to consult with the clinical microbiology lab when considering assays that are most appropriate for any given patient.

Genomic Tests

Molecular diagnostic tests can be broadly categorized into those that evaluate nucleic acids (DNA or RNA) and those that assay for proteins or enzymatic activity. There are three types of nucleic acid–based tests used in the diagnosis of bacterial diseases: NAATs, nucleic acid probes, and nucleic acid sequence analysis; many of these tests have become a routine part of clinical microbiology diagnostics.

Nucleic acid–based tests can be performed rapidly, are highly specific and quite sensitive (especially the amplification tests), and can often be performed directly from the clinical specimen, mitigating the need to wait for culture results. They are therefore especially useful for those bacteria that are difficult to culture, such as *Chlamydia* and *Mycobacterium* species.

NAATs use polymerase chain reaction (PCR) or other amplifying process to increase the number of bacteria-specific DNA or RNA molecules so the sensitivity of the test is significantly higher than that of unamplified tests. Contemporary assays can target a single pathogen or use multiplexed panels containing multiple targets to identify pathogens associated with particular clinical syndromes (e.g., pneumonia, endocarditis, or meningitis).

As mentioned above, examples of FDA-approved NAAT tests include *C. trachomatis* and *N. gonorrhoeae* in urine samples in sexually transmitted diseases, the Luminex respiratory virus panel for detection of seven different respiratory viruses, and the BioFire Film Array meningitis/encephalitis (ME) panel, which simultaneously detects 14 common infectious agents in CSF.

Tests that use nucleic acid probes are designed to detect bacterial DNA or RNA directly (without amplification) using a labeled DNA or RNA probe that will hybridize specifically to the bacterial nucleic acid from a cultured organism. These tests are simpler to perform than the amplification tests but are less sensitive.

Nucleic acid sequence analysis of ribosomal RNA (rRNA) can be used to identify bacteria or fungi. This is considered a universal approach because it is based on amplification of genes that are highly conserved within a given organism type, such as the 16S and 23S rRNA genes in bacteria and the 28S rRNA and internal transcribed spacer (ITS) genes in fungi. A bacterium that had never been previously cultured, *Tropheryma whipplei*, was identified using this approach.

An emerging technique in diagnostic microbiology is metagenomic sequencing analysis. This approach aims to comprehensively identify infectious agents by randomly amplifying and sequencing all of the DNA and RNA in clinical samples. The typically small fraction of "nonhost" (nonhuman) sequences corresponding to potential pathogens is then mapped to large reference databases, such as the National Center for Biotechnology Information (NCBI) GenBank, to identify sequences from any virus, bacterium, fungus, or parasite that is present. Currently, this test is only available on a limited basis from a few select laboratories, such as the University of California, San Francisco Clinical Microbiology Laboratory, although tests based on this approach are rapidly being developed.

Many of these assays are often performed in concert with other culture-dependent methods; these approaches only confirm the presence of a nucleic acid target and do not prove the presence of a viable organism. In addition, the exquisite sensitivity of nucleic acid amplification–based methods makes it challenging to ensure that results are due to the actual presence of the target organism, and not postcollection contamination.

Proteomic Tests

Molecular proteomic platforms have also been developed, and one that is now being used in many clinical laboratories is MALDI-TOF mass spectrometry. MALDI-TOF technology measures particles based on their mass-to-charge ratio. In this technique, organisms that have been cultured and purified from clinical samples (bacteria and some types of fungi have been successfully analyzed) are embedded in a matrix material that, when excited by a laser, transfers charge from the matrix to the microbial macromolecules (proteins and nucleic acids) causing desorption of the newly ionized particles. These charged particles are then separated by their mass-to-charge ratio, yielding a mass spectral signature that is unique to a specific genus and often to the species level. Using bioinformatics, these MALDI-TOF spectra can be compared to standardized databases, and those that are highly aligned are identified as a match with a stated level of confidence. This assay, which takes less than a minute to complete once the sample is loaded into the machine, has been shown to be highly accurate, efficient, and cost-effective.

Despite the emergence of a diverse array of molecular and biochemical diagnostic tools, culture-based approaches to diagnosing infectious disease remain a mainstay of the clinical microbiology lab. A combination of new methodologies and classic techniques is central to the successful and accurate identification of microorganisms encountered in the clinical setting.

PEARLS

- The **laboratory diagnosis** of infectious diseases includes microscopic, **culture-based, immunologic (serologic)**, and **molecular (nucleic acid– and protein-based) tests**.

Microscopic Tests

- Bacteriologic tests typically begin with **staining** the patient's specimen and **observing** the organism in the microscope. The Gram stain and the acid-fast stain are two important examples.

Culture-Based Tests

- Microscopic examination is followed by **culturing** the organism, typically on blood agar, and then **performing various tests** to identify the causative organism. Obtaining a **pure culture** of the bacteria is essential to accurate diagnosis.

- **Blood cultures** are useful in cases of **sepsis** and other diseases in which the organism is often found in the bloodstream, such as endocarditis, meningitis, pneumonia, and osteomyelitis.

- **Throat cultures** are most useful to diagnose **pharyngitis** caused by *Streptococcus pyogenes* ("strep throat"), but they are also used to diagnose diphtheria, gonococcal pharyngitis, and thrush caused by the yeast *Candida albicans*.

- **Sputum cultures** are used primarily to diagnose the cause of **pneumonia** but also are used in suspected cases of tuberculosis.

- **Spinal fluid cultures** are most useful in suspected cases of **meningitis**. These cultures are often negative in encephalitis, brain abscess, and subdural empyema.

- **Stool cultures** are useful primarily when the complaint is **bloody diarrhea** (dysentery, enterocolitis) rather than watery diarrhea, which is often caused by either enterotoxins or viruses.

- **Urine cultures** are used to determine the cause of either **pyelonephritis** or **cystitis**.
- **Genital tract cultures** are most often used to diagnose **gonorrhea**. *Chlamydia trachomatis* is difficult to grow, so nonbacteriologic methods such as nucleic acid amplification tests (NAATs) are now used more often than culture. The agent of syphilis cannot yet be cultured on bacteriologic medium, so the diagnosis is made serologically.
- **Wounds and abscesses** can be caused by a large variety of organisms. Cultures should be incubated both in the presence and in the absence of oxygen because **anaerobes** are often involved.

Serologic Tests

- Serologic tests can determine whether **antibodies are present in the patient's serum** as well as detect the **antigens of the organism in tissues or body fluids**.
- In these tests, the antigens of the causative organism can be detected by using specific antibody often labeled with a dye such as fluorescein (fluorescent antibody tests). The presence of antibody in the patient's serum can be detected using antigens derived from the organism. In some tests, the patient's serum contains antibodies that react with an antigen that is not derived from the causative organism, such as the VDRL test, in which beef heart cardiolipin reacts with antibodies in the serum of patients with syphilis.

- In many tests in which antibodies are detected in the patient's serum, an acute and convalescent serum sample is obtained, and at least a **fourfold increase in titer** between the acute and convalescent samples must be found for a diagnosis to be made. The reason these criteria are used is that the presence of antibodies in a single sample could be from a prior infection, so a significant (fourfold or greater) increase in titer is used to indicate that this is a current infection. **IgM antibody** can also be used as an indicator of current infection.

Molecular Diagnostics

- Molecular tests can detect the presence of bacterial DNA, RNA, or protein in patient specimens. These tests are both sensitive and specific, and results are available within a clinically useful time frame. They have become the diagnostic "gold standard" for many infections.
- NAATs use the polymerase chain reaction (PCR) to detect *C. trachomatis* and *Neisseria gonorrhoeae* in urine samples in sexually transmitted disease clinics. These tests are also used to identify *Mycobacterium tuberculosis* in sputum samples.
- The specificity of these tests resides in the ability of the DNA or RNA probe to bind to DNA or RNA present only in the bacteria to be identified.
- Emerging molecular tests in diagnostic microbiology, such as MALDI-TOF, enable rapid and specific identification of bacteria and many fungi.

SELF-ASSESSMENT QUESTIONS

1. If the venipuncture site is inadequately disinfected, blood cultures are most often contaminated with which of the following bacteria?

    (A) *Escherichia coli*
    (B) *Haemophilus influenzae*
    (C) *Pseudomonas aeruginosa*
    (D) *Staphylococcus epidermidis*
    (E) *Streptococcus pneumoniae*

2. The main purpose of performing a throat culture is to detect the presence of which of the following bacteria?

    (A) *Neisseria meningitidis*
    (B) *Staphylococcus aureus*
    (C) *Staphylococcus epidermidis*
    (D) *Streptococcus pneumoniae*
    (E) *Streptococcus pyogenes*

3. A sputum culture will be rejected (i.e., it will not be stained or cultured) by the clinical laboratory if:

    (A) it is streaked with blood.
    (B) it contains IgA antibody.
    (C) it contains many more epithelial cells than neutrophils.
    (D) it contains pus.
    (E) it contains sulfur granules.

4. The identification of *Salmonella* and *Shigella* in stool cultures using eosin-methylene blue (EMB) media is dependent on which of the following properties?

    (A) *Salmonella* and *Shigella* produce a blue colony in the presence of methylene blue.
    (B) *Salmonella* and *Shigella* produce a colorless colony because they do not ferment lactose.
    (C) *Salmonella* and *Shigella* produce a green colony because they use the bile in the media.
    (D) *Salmonella* and *Shigella* produce a red colony in the presence of eosin.
    (E) *Salmonella* and *Shigella* produce a yellow colony because they ferment glucose.

ANSWERS

(1) **(D)**
(2) **(E)**
(3) **(C)**
(4) **(B)**

PRACTICE QUESTIONS: USMLE & COURSE EXAMINATIONS

Questions on the topics discussed in this chapter can be found in the Basic Bacteriology section of Part XIII: USMLE (National Board) Practice Questions starting on page 733. Also see Part XIV: USMLE (National Board) Practice Examination starting on page 775.

Antibacterial Drugs: Mechanism of Action

ANTIMICROBIAL DRUG STEWARDSHIP

The discovery of antimicrobials is one of the great advances in medicine, and their use has substantially reduced morbidity and mortality worldwide. Unfortunately, with widespread antibiotic use, we have witnessed the emergence of multidrug-resistant pathogens and reduced efficacy of many of our most powerful antimicrobials. In addition, we have also recognized many adverse effects of antimicrobials, most notably the rising rates of *Clostridioides difficile* colitis. Further, the cost of medical care is greatly increased due to overuse of antibiotics and treating infections caused by resistant organisms. It is critical that health professionals understand the key concepts behind antimicrobial stewardship at the time they learn about microbial pathogens and antimicrobials.

The worldwide problem of antibiotic resistance makes the need for antimicrobial stewardship evident. The Centers for Disease Control and Prevention estimates that in the United States over 2 million infections with multidrug-resistant pathogens occur each year, leading to more than 35,000 deaths. These pathogens include methicillin-resistant *Staphylococcus aureus* (MRSA) and extended-spectrum β-lactamase–producing gram-negative rods (e.g., *Escherichia coli* and *Klebsiella pneumoniae*). Hospital-associated infections, many of which are caused by antibiotic-resistant bacteria, are estimated to cost billions of dollars each year.

The basic principles of good stewardship are threefold: (1) reduce inappropriate use of antibiotics, (2) encourage targeted treatment with narrow-spectrum drugs, and (3) limit adverse effects (Table 10–1). Inappropriate antibiotic use can occur for many reasons, including the providers' desire to adhere to patient's wishes, even when it is not medically appropriate. The most common example of inappropriate antibiotic use is the prescribing of antibiotics for a viral respiratory tract infection. It is estimated that half of all prescriptions given for upper respiratory infections (pharyngitis, sinusitis) are inappropriate.

The concept of targeted treatment refers to making a microbiologic diagnosis promptly and using the most specific antibiotic that has the best safety profile for the patient. If multiple broad-spectrum antibiotics are used as empiric therapy early in infection, then switching to a narrow-spectrum antibiotic as soon as possible should be done. Cultures should be sent prior to starting antibiotics so that the drugs will not reduce the likelihood of isolating the causative organism. Further, switching from intravenous antibiotics to oral dosing reduces the risk of catheter-associated infections. The net result of targeted treatment is to reduce the ability of antibiotics to select for resistant mutants present in the bacterial population.

Limiting the occurrence of adverse effects caused by antibiotics is another major goal of antimicrobial stewardship. Minimizing the duration of antibiotic use to only as long as clinically indicated is a key intervention since duration of exposure correlates closely with the risk of many types of adverse effects. Patients with reduced renal function should have the dose of some antibiotics adjusted based on their estimated glomerular filtration rate.

Antibiotic allergies should be identified and explored in detail. Although reaction to antibiotics may be commonly reported, they may not always be of significance such as

TABLE 10–1 Basic Principles of Antimicrobial Drug Stewardship

Current Problems in the Use of Antibiotics	Role of Antimicrobial Drug Stewardship in Mitigating These Problems
Inappropriate use of antibiotics	1. Use antibiotics only when a microbiologic diagnosis indicates effectiveness 2. Empiric therapy should be tailored to the most likely pathogen(s) 3. Send appropriate cultures before starting antibiotics
Overuse of broad-spectrum antibiotics	1. Use narrow-spectrum antibiotics whenever possible 2. Require approval for the use of advanced-generation broad-spectrum antibiotics
High rate of adverse effects	1. Stop antibiotics as soon as appropriate to reduce adverse effects, such as antibiotic-associated colitis caused by *Clostridioides difficile* 2. Be aware of the effect of the patient's renal function on the dose of antibiotic prescribed 3. Be aware of the patient's hypersensitivity to specific antibiotics 4. Determine whether the patient's declared hypersensitivity is correct and clinically significant 5. Warn patients regarding certain idiosyncratic drug reactions, such as photosensitization

some alleged hypersensitivity reactions. If optimal treatment requires a drug such as penicillin to which the patient says he or she is allergic, then skin testing can be employed to determine the accuracy of that claim. In addition, patients should be warned regarding the possibility that certain drugs may cause adverse effects. For example, certain photosensitizing antibiotics may cause a rash when the patient is exposed to sunlight.

The reasons that inappropriate use of antibiotics occurs are varied. Probably, the most important is the lack of knowledge or awareness of the physician. Risk avoidance on the part of the physician is also common. Inadequate microbiologic information plays a role as well. Patient expectation and direct demands for antibiotics contribute to the problem.

In summary, antimicrobial stewardship refers to the effort to improve the treatment of infectious diseases by the appropriate use of antibiotics. This is critical in this era of rising rates of multidrug-resistant pathogens. Targeted therapy with the most appropriate single antibiotic will hopefully improve clinical outcomes and reduce the cost of care.

PRINCIPLES OF ANTIMICROBIAL THERAPY

The most important concept underlying antimicrobial therapy is **selective toxicity** (i.e., selective inhibition of the growth of the microorganism without damage to the host). Selective toxicity is achieved by exploiting the differences between the metabolism and structure of the microorganism and the corresponding features of human cells. For example, penicillins and cephalosporins are effective antibacterial agents because they prevent the synthesis of peptidoglycan that is found in bacteria but not human cells. Antimicrobial drugs target four major sites in the bacterial cell: cell wall, ribosomes, nucleic acids, and cell membrane (Table 10–2 and Figure 10–1).

There are far more antibacterial drugs than antiviral drugs. This is a consequence of the difficulty of designing a drug that will selectively inhibit viral replication. Because viruses use many of the normal cellular functions of the host in their growth, it is not easy to develop a drug that specifically inhibits viral functions and does not damage the host cell.

Broad-spectrum antibiotics are active against several types of microorganisms (e.g., tetracyclines are active against many gram-negative rods, chlamydiae, mycoplasmas, and rickettsiae). **Narrow-spectrum** antibiotics are active against one or very few types (e.g., vancomycin is primarily used against certain gram-positive cocci, namely, staphylococci and enterococci).

BACTERICIDAL & BACTERIOSTATIC ACTIVITY

In some clinical situations, it is essential to use a bactericidal drug rather than a bacteriostatic one. A **bactericidal drug kills bacteria**, whereas a **bacteriostatic drug inhibits their growth but does not kill them** (Figure 10–2). The salient features of the behavior of bacteriostatic drugs are that (1) the bacteria can grow again when the drug is withdrawn, and (2) host defense mechanisms, such as phagocytosis, are required to kill the bacteria. Bactericidal drugs are particularly useful in certain infections (e.g., those that are immediately life-threatening; those in patients whose polymorphonuclear leukocyte count is below 500/μL; and endocarditis, in which phagocytosis is limited by the fibrinous network of the vegetations and bacteriostatic drugs do not effect a cure).

TABLE 10–2 Mechanism of Action of Important Antibacterial Drugs

Mechanism of Action	Drugs
Inhibition of cell wall synthesis	
Inhibits cross-linking (transpeptidation) of peptidoglycan	Penicillins, cephalosporins, imipenem, aztreonam, vancomycin
Inhibits other steps in peptidoglycan synthesis	Cycloserine, bacitracin
Inhibition of protein synthesis	
Acts on 50S ribosomal subunit	Chloramphenicol, erythromycin, clindamycin, linezolid
Acts on 30S ribosomal subunit	Tetracyclines and aminoglycosides
Acts on t-RNA synthetase	Mupirocin
Inhibition of nucleic acid synthesis	
Inhibits nucleotide synthesis	Sulfonamides, trimethoprim
Inhibits DNA synthesis	Quinolones (e.g., ciprofloxacin)
Inhibits mRNA synthesis	Rifampin
Alteration of cell membrane function	
Disrupts membranes	Polymyxin, daptomycin
Other mechanisms of action	
Inhibits mycolic acid synthesis	Isoniazid
Acts as electron sink and damages DNA	Metronidazole
Inhibits arabinogalactan synthesis	Ethambutol
May inhibit fatty acid synthesis	Pyrazinamide

MECHANISMS OF ACTION

INHIBITION OF CELL WALL SYNTHESIS

1. Inhibition of Bacterial Cell Wall Synthesis

Penicillins

Penicillins (and cephalosporins) act by inhibiting **transpepti-dases**, the enzymes that catalyze the final cross-linking step in the synthesis of peptidoglycan (see Figure 2–5). For example, in *S. aureus*, transpeptidation occurs between the amino group on the end of the pentaglycine cross-link and the terminal carboxyl group of the D-alanine on the tetrapeptide side chain. Because the stereochemistry of penicillin is similar to that of a dipeptide, D-alanyl-D-alanine, penicillin can bind to the active site of the transpeptidase and inhibit its activity.

Two additional factors are involved in the action of penicillin:

(1) Penicillin binds to a variety of proteins in the bacterial cell membrane and cell wall, called **penicillin-binding proteins (PBPs)**. Changes in PBPs are in part responsible for an organism's becoming resistant to penicillin.

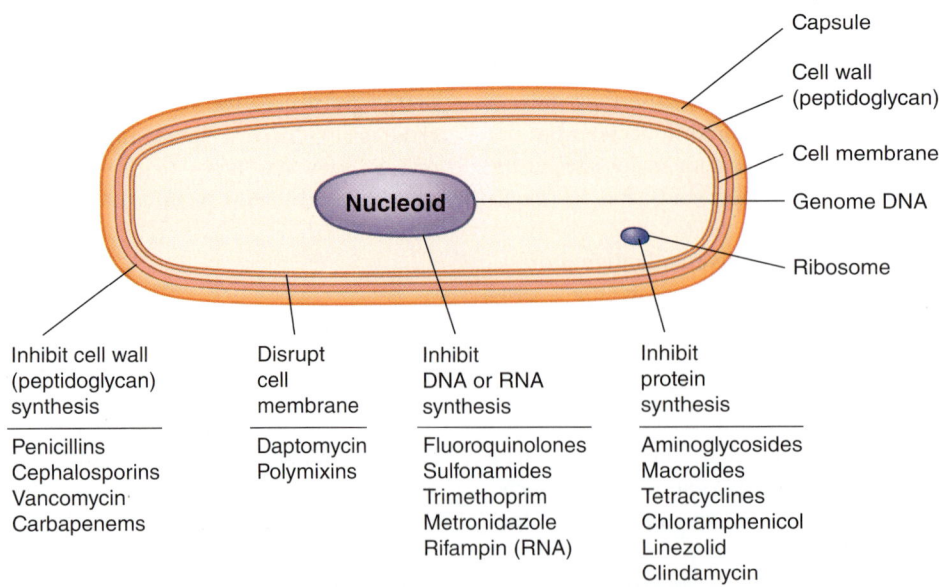

FIGURE 10–1 Model of typical bacterial cell showing sites of action of important antibacterial drugs.

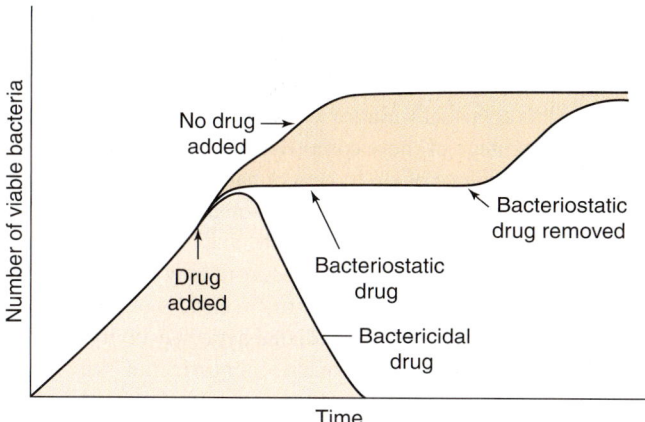

FIGURE 10–2 Bactericidal and bacteriostatic activity of antimicrobial drugs. Either a bactericidal or a bacteriostatic drug is added to the growing bacterial culture at the time indicated by the arrow. After a brief lag time during which the drug enters the bacteria, the bactericidal drug kills the bacteria, and a decrease in the number of viable bacteria occurs. The bacteriostatic drug causes the bacteria to stop growing, but if the bacteriostatic drug is removed from the culture, the bacteria resume growing.

(2) **Autolytic enzymes** called murein hydrolases (murein is a synonym for peptidoglycan) are activated in penicillin-treated cells and degrade the peptidoglycan. Some bacteria (e.g., strains of *S. aureus*) are **tolerant** to the action of penicillin because these autolytic enzymes are not activated. A tolerant organism is one that is inhibited but not killed by a drug that is usually bactericidal, such as penicillin. Penicillin-treated cells die by

rupture as a result of the influx of water into the high-osmotic-pressure interior of the bacterial cell.

Penicillin is bactericidal, but it **kills cells only when they are growing**. As cells grow, new peptidoglycan is synthesized, and transpeptidation occurs. However, in nongrowing cells, no new cross-linkages are required, and penicillin is inactive. Penicillins are therefore **more active during the log phase** of bacterial cell growth than during the stationary phase (see Chapter 3 for the bacterial cell growth cycle).

Penicillins (and cephalosporins) are called β-lactam drugs because of the importance of the β-lactam ring (Figure 10–3). An intact ring structure is essential for antibacterial activity; cleavage of the ring by penicillinases (**β-lactamases**) inactivates the drug. The most important naturally occurring compound is benzylpenicillin (penicillin G), which is composed of the 6-aminopenicillanic acid nucleus that all penicillins have, plus a benzyl side chain (see Figure 10–3). Penicillin G is available in three main forms:

(1) Aqueous penicillin G, which is metabolized most rapidly.

(2) Procaine penicillin G, in which penicillin G is conjugated to procaine. This form is metabolized more slowly and is less painful when injected intramuscularly because the procaine acts as an anesthetic.

(3) Benzathine penicillin G, in which penicillin G is conjugated to benzathine. This form is metabolized very slowly and is often called a "depot" preparation.

Benzylpenicillin is one of the most widely used and effective antibiotics. However, it has four disadvantages, the first three of

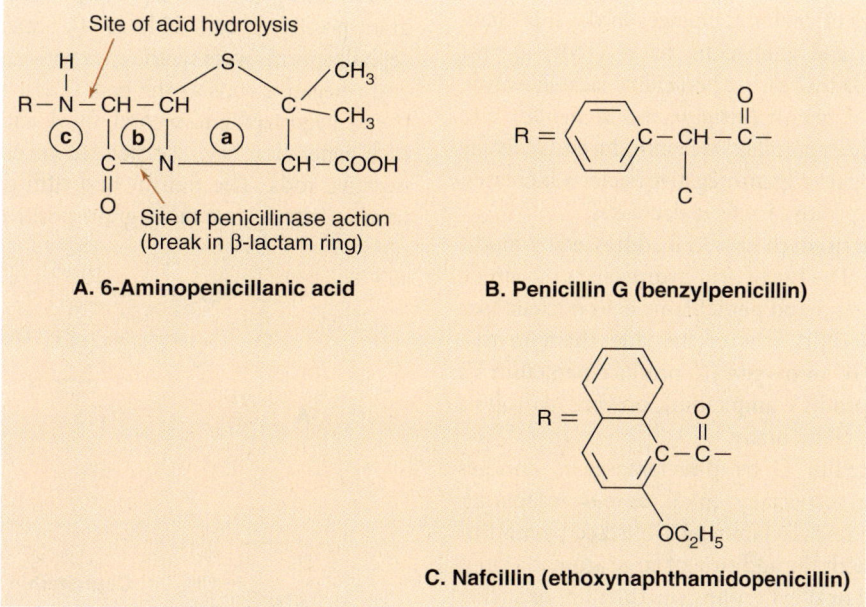

FIGURE 10–3 Penicillins. **A:** The 6-aminopenicillanic acid nucleus is composed of a thiazolidine ring (a), a β-lactam ring (b), and an amino group (c). The sites of inactivation by stomach acid and by penicillinase are indicated. **B:** The benzyl group, which forms benzylpenicillin (penicillin G) when attached at R. **C:** The large aromatic ring substituent that forms nafcillin, a β-lactamase–resistant penicillin, when attached at R. The large ring blocks the access of β-lactamase to the β-lactam ring.

TABLE 10–3   Activity of Selected Penicillins

Drug	Clinically Useful Activity[1]
Penicillin G	Gram-positive cocci, gram-positive rods, *Neisseria*, spirochetes such as *Treponema pallidum*, and many anaerobes (except *Bacteroides fragilis*) but none of the gram-negative rods listed below
Ampicillin or amoxicillin	Certain gram-negative rods, such as *Haemophilus influenzae, Escherichia coli, Proteus, Salmonella,* and *Shigella* but not *Pseudomonas aeruginosa* or *Klebsiella pneumoniae*
Ticarcillin	*P. aeruginosa*, especially when used in synergistic combination with an aminoglycoside
Piperacillin	Similar to ticarcillin but with greater activity against *P. aeruginosa* and *K. pneumoniae*
Nafcillin or dicloxacillin	Penicillinase-producing *Staphylococcus aureus*

[1]The spectrum of activity is intentionally incomplete. It is simplified for the beginning student to illustrate the expanded coverage of gram-negative organisms with successive generations and does not cover all possible clinical uses.

which have been successfully overcome by chemical modification of the side chain. The four disadvantages are

(1) **limited effectiveness against many gram-negative rods** due to the inability of the drug to penetrate the outer membrane of the organism.

(2) **hydrolysis by gastric acids**, so that it cannot be taken orally.

(3) **inactivation by β-lactamases**.

(4) **hypersensitivity reactions**, especially anaphylaxis, in some recipients of the drug. This disadvantage has not been overcome by chemical modification.

The effectiveness of penicillins against gram-negative rods has been increased by a series of chemical changes in the side chain (Table 10–3). Ampicillin and amoxicillin have activity against several gram-negative rods that earlier penicillins lack. However, these drugs are not useful against *Pseudomonas aeruginosa* and *K. pneumoniae*. Hence, other penicillins were introduced. Generally speaking, as the activity against gram-negative bacteria increases, the activity against gram-positive bacteria decreases.

Acid hydrolysis in the stomach has been addressed by modification of the side chain. The site of acid hydrolysis is the amide bond between the side chain and penicillanic acid nucleus (see Figure 10–3). Minor modifications of the side chain in that region, such as addition of an oxygen (to produce penicillin V) or an amino group (to produce ampicillin), prevent hydrolysis and allow the drug to be taken orally.

Inactivation of penicillin G by β-lactamases is another important disadvantage, especially in *S. aureus* infections. Access of the enzyme to the β-lactam ring is blocked by modification of the side chain with the addition of large aromatic rings containing bulky methyl or ethyl groups (methicillin, oxacillin, nafcillin, etc.; see Figure 10–3). β-Lactamases inhibitors, such as clavulanic acid, tazobactam, sulbactam, and avibactam, are structural analogues of penicillin that have little antibacterial activity but bind strongly to β-lactamases. Thus, prescribing

these in combination protects the penicillin. Combinations, such as amoxicillin and clavulanic acid (Augmentin) and piperacillin plus tazobactam (Zosyn), are in clinical use.

Penicillins are usually nontoxic at clinically effective levels. The major disadvantage of these compounds is hypersensitivity, with a reported prevalence of 1% to 10% of patients. The most serious of the hypersensitivity reactions is IgE–mediated, which can give rise to bronchospasm, urticarial rash, and anaphylactic shock (see Chapter 65). Fortunately, this occurs in only 0.5% of patients. Death as a result of anaphylaxis occurs in 0.002% of patients (1 in 50,000 patients). IgG and cell-mediated hypersensitivity reactions are more common and can include nonurticarial skin rashes, hemolytic anemia, nephritis, and drug fever. A maculopapular drug-induced rash is quite common. While these manifestations are not considered to be "true" allergy and are not life threatening, they are adverse reactions, and an alternative antibiotic should be considered for patients with a history of these symptoms.

To determine whether the patient's allergy is clinically significant, a skin test using penicilloyl-polylysine as the test reagent can be performed. A wheal and flare reaction occurs at the site of injection in allergic individuals. If the patient's disease requires treatment with penicillin, the patient can be desensitized under the supervision of a trained allergist.

Cephalosporins

Cephalosporins are β-lactam drugs that, like penicillins, also inhibit the cross-linking of peptidoglycan. The structures, however, are different. Cephalosporins have a six-membered ring adjacent to the β-lactam ring and are substituted in two places on the 7-aminocephalosporanic acid nucleus (Figure 10–4), whereas penicillins have a five-membered ring and are substituted in only one place.

The first-generation cephalosporins are active primarily against gram-positive cocci (Table 10–4). Similar to the penicillins, new cephalosporins were synthesized with expansion of activity against gram-negative rods as the goal. These new cephalosporins have been categorized into second, third, and fourth generations, with each generation having expanded coverage against certain gram-negative rods. The fourth- and fifth-generation cephalosporins have activity against many gram-positive cocci as well.

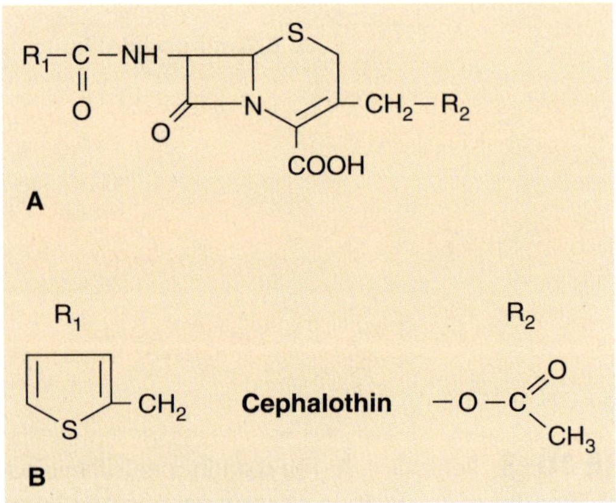

FIGURE 10–4 Cephalosporins. **A:** The 7-aminocephalosporanic acid nucleus. **B:** The two R groups in the drug cephalothin.

TABLE 10–4   Activity of Selected Cephalosporins[1]

Generation of Cephalosporin	Drug	Clinically Useful Activity
First	Cefazolin, cephalexin	Gram-positive cocci such as staphylococci and streptococci except enterococci and MRSA
Second	Cefuroxime Cefoxitin	*Haemophilus influenzae* *Bacteroides fragilis*
Third	Ceftriaxone Ceftazidime	Enteric gram-negative rods such as *Escherichia coli*, *Klebsiella*, and *Proteus*. Also *Neisseria gonorrhoeae* *Pseudomonas aeruginosa* and other enteric gram-negative rods
Fourth	Cefepime	Enteric gram-negative rods that produce extended-spectrum β-lactamases; *Staphylococcus aureus* (not MRSA) and penicillin-resistant *Streptococcus pneumoniae*
	Cefiderocol	Enteric gram-negative rods that produce extended-spectrum β-lactamases
Fifth	Ceftaroline	Gram-positive cocci and gram-negative rods that cause bacterial pneumonia and gram-positive cocci that cause skin infections including MRSA
	Ceftolozane	Enteric gram-negative rods that produce extended-spectrum β-lactamases; *P. aeruginosa*; used in combination with tazobactam

MRSA = methicillin-resistant *Staphylococcus aureus*.

[1]The spectrum of activity is intentionally incomplete. It is simplified for the beginning student to illustrate the expanded coverage of gram-negative organisms with successive generations and does not cover all possible clinical uses.

Cephalosporins are effective against a broad range of organisms, are generally well tolerated, and produce fewer hypersensitivity reactions than do the penicillins. Despite the structural similarity, a patient allergic to penicillin has only about a 10% chance of being hypersensitive to cephalosporins also. Most cephalosporins are the products of molds of the genus *Cephalosporium*; a few, such as cefoxitin, are made by the actinomycete *Streptomyces*.

The inactivation of cephalosporins by β-lactamases (cephalosporinases) is an important clinical problem. β-Lactamase inhibitors such as tazobactam and avibactam are combined with certain cephalosporins to prevent inactivation of the cephalosporin. For example, the US Food and Drug Administration (FDA) has approved the combination of ceftazidime/avibactam (Avycaz) and ceftolozane/tazobactam (Zerbaxa) for the treatment of intra-abdominal infections and complicated urinary tract infections (UTIs) caused by antibiotic-resistant gram-negative rods.

Carbapenems

Carbapenems are β-lactam drugs that are structurally different from penicillins and cephalosporins. For example, imipenem (*N*-formimidoylthienamycin), a commonly used carbapenem, has a methylene group in the ring in place of the sulfur (Figure 10–5). Imipenem has one of the widest spectrums of activity of the β-lactam drugs. It has excellent bactericidal activity against many gram-positive, gram-negative, and anaerobic bacteria

(Table 10–5). It is effective against most gram-positive cocci (e.g., staphylococci and streptococci), most gram-negative cocci (e.g., *Neisseria meningitidis*), many gram-negative rods (e.g., *Pseudomonas*, *Haemophilus*, and members of the family Enterobacteriaceae such as *E. coli*), and various anaerobes (e.g., *Bacteroides* and *Clostridium*).

Imipenem is especially useful in treating infections caused by gram-negative rods that produce extended-spectrum β-lactamases that make them resistant to all penicillins and cephalosporins. Carbapenems are often the "drugs of last resort" against bacteria resistant to multiple antibiotics and are thus reserved for hospital settings.

Imipenem is prescribed in combination with cilastatin, which is an inhibitor of dehydropeptidase, a kidney enzyme that inactivates imipenem. Other carbapenems, such as ertapenem and meropenem, are not inactivated by dehydropeptidase and are not prescribed in combination with cilastatin.

Imipenem is not inactivated by most β-lactamases; however, carbapenemases produced by *K. pneumoniae* that degrade imipenem and other carbapenems have emerged. To address the problem of carbapenemase-producing gram-negative rods, the FDA has approved a combination of meropenem–vaborbactam (Vabomere) for the treatment of complicated UTIs caused by *E. coli*, *K. pneumoniae*, and others. Vaborbactam is a carbapenemase/β-lactamase inhibitor.

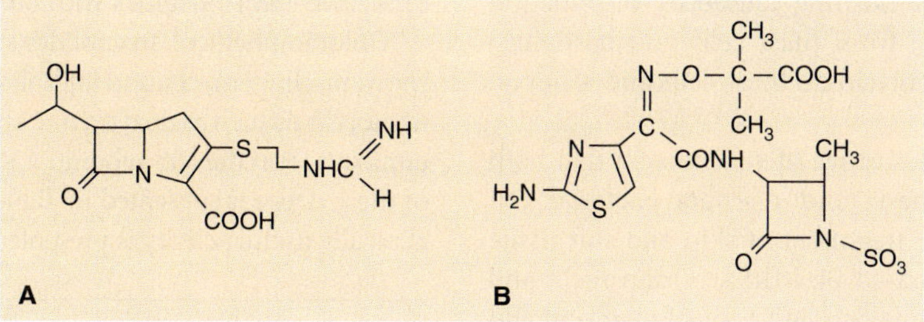

FIGURE 10–5   A: Imipenem. **B:** Aztreonam.

TABLE 10–5 Clinical Activity of Selected Carbapenems

Drug	Clinical Activity	Comment
Imipenem	1. Gram-positive cocci such as *Staphylococcus aureus* but not MRSA[1] 2. Gram-negative rods including *Pseudomonas aeruginosa* 3. Treatment of mixed aerobic-anaerobic infections 4. Treatment of febrile neutropenic patient	Given in combination with cilastatin, a renal dehydropeptidase inhibitor
Meropenem	Similar to imipenem but slightly more activity against gram-negative rods	Not degraded by renal dehydropeptidase
Doripenem	Similar to imipenem but slightly more activity against gram-negative rods	Not degraded by renal dehydropeptidase
Ertapenem	Similar to imipenem but *not* effective against *P. aeruginosa*	Not degraded by renal dehydropeptidase

[1]MRSA = methicillin-resistant *S. aureus*.

Monobactams

Monobactams are also β-lactam drugs that are structurally different from penicillins and cephalosporins. Monobactams are characterized by a β-lactam ring without an adjacent sulfur-containing ring structure (i.e., they are monocyclic) (see Figure 10–5). Aztreonam, currently the most useful monobactam, has excellent activity against many gram-negative rods, such as Enterobacteriaceae and *Pseudomonas*, but is inactive against gram-positive and anaerobic bacteria. It is resistant to most β-lactamases. It is very useful in patients who are hypersensitive to penicillin because there is no cross-reactivity.

Vancomycin

Vancomycin is a glycopeptide that **inhibits cell wall peptidoglycan synthesis by blocking transpeptidation** but by a mechanism different from that of the β-lactam drugs. Vancomycin binds directly to the D-alanyl-D-alanine portion of the pentapeptide, which blocks the transpeptidase from binding, whereas the β-lactam drugs bind to and inhibit the activity of the transpeptidase itself.

Vancomycin is a bactericidal agent **effective against certain gram-positive bacteria**. Its most important use is in the treatment of infections caused by *S. aureus* strains that are resistant to the penicillinase-resistant penicillins such as nafcillin and methicillin (e.g., MRSA). Note that vancomycin is not a β-lactam drug and, therefore, is not degraded by β-lactamase.

Vancomycin is also used in the treatment of infections caused by *Staphylococcus epidermidis*, penicillin-resistant *Streptococcus pneumoniae*, and enterococci. Strains of *S. aureus*, *S. epidermidis*, and enterococci with partial or complete resistance to vancomycin have been recovered from patients.

A well-known adverse effect of vancomycin is "red man" syndrome. "Red" refers to the flushing caused by vasodilation induced by histamine release from mast cells and basophils. This is a direct effect of vancomycin on these cells and is not an IgE-mediated response.

Telavancin is a synthetic derivative of vancomycin that both inhibits peptidoglycan synthesis and disrupts bacterial cell membranes. It is used for the treatment of skin and soft tissue infections, especially those caused by MRSA. Oritavancin and dalbavancin are lipoglycopeptide derivatives of vancomycin and teicoplanin, respectively. These drugs inhibit the transpeptidases and transglycosylases required to synthesize the peptidoglycan of gram-positive bacteria. They are effective in the treatment of infections caused by *S. aureus*, including MRSA, and *Enterococcus*, including vancomycin-resistant enterococci (VRE).

Cycloserine & Bacitracin

Cycloserine is a structural analogue of D-alanine that inhibits the synthesis of the cell wall dipeptide D-alanyl-D-alanine. It is used as a second-line drug in the treatment of multidrug resistant (MDR) tuberculosis. Bacitracin is a cyclic polypeptide antibiotic that prevents the dephosphorylation of the phospholipid that carries the peptidoglycan subunit across the cell membrane. This blocks the regeneration of the lipid carrier and inhibits cell wall synthesis. Bacitracin is a bactericidal drug useful in the treatment of superficial skin infections but nephrotoxicity limits its systemic use.

Fosfomycin

Fosfomycin inhibits peptidoglycan synthesis by blocking an enzyme required for the production of *N*-acetyl muramic acid, a major component of the peptidoglycan backbone. Fosfomycin is approved for use in the treatment of uncomplicated UTIs in women caused by gram-negative rods such as *E. coli* and *Proteus* species. It is not a β-lactam drug so is useful in treating UTI caused by extended spectrum β-lactamase-producing (ESBL) *E. coli*.

INHIBITION OF PROTEIN SYNTHESIS

Several drugs inhibit protein synthesis in bacteria without significantly interfering with protein synthesis in human cells. This selectivity is due to the differences between bacterial and human ribosomal proteins, RNAs, and associated enzymes. Bacteria have 70S[1] ribosomes with 50S and 30S subunits, whereas human cells have 80S ribosomes with 60S and 40S subunits.

Chloramphenicol, macrolides such as azithromycin and erythromycin, clindamycin, and linezolid act on the 50S subunit, whereas tetracyclines such as doxycycline and aminoglycosides such as gentamicin act on the 30S subunit. A summary of the modes of action of these drugs is presented in Table 10–6, and a summary of their clinically useful activity is presented in Table 10–7.

---

[1]"S" stands for Svedberg units, a measure of sedimentation rate in a density gradient. The rate of sedimentation is proportionate to the mass of the particle.

TABLE 10–6 Mode of Action of Antibiotics That Inhibit Protein Synthesis

Antibiotic	Ribosomal Subunit	Mode of Action	Bactericidal or Bacteriostatic
Aminoglycosides	30S	Blocks functioning of initiation complex and causes misreading of mRNA	Bactericidal
Tetracyclines	30S	Blocks tRNA binding to ribosome	Bacteriostatic
Chloramphenicol	50S	Blocks peptidyl transferase	Both[1]
Macrolides	50S	Blocks translocation	Primarily bacteriostatic
Clindamycin	50S	Blocks peptide bond formation	Primarily bacteriostatic
Linezolid	50S	Blocks early step in ribosome formation	Both[1]
Telithromycin	50S	Same as other macrolides (e.g., erythromycin)	Both[1]
Streptogramins	50S	Causes premature release of peptide chain	Both[1]

[1]Can be either bactericidal or bacteriostatic, depending on the organism.

TABLE 10–7 Spectrum of Activity of Antibiotics That Inhibit Protein Synthesis[1]

Antibiotic	Clinically Useful Activity	Comments
Aminoglycosides		
Streptomycin	Tuberculosis, tularemia, plague, brucellosis	Ototoxic and nephrotoxic
Gentamicin and tobramycin	Many gram-negative rod infections including *Pseudomonas aeruginosa*	Most widely used aminoglycosides
Amikacin	Same as gentamicin and tobramycin	Effective against some organisms resistant to gentamicin and tobramycin, especially Pseudomonas
Neomycin	Preoperative bowel preparation	Too toxic to be used systemically; use orally since not absorbed
Plazomicin	Complicated urinary tract infections	Active against gram-negative rods resistant to other aminoglycosides
Tetracyclines		
Doxycycline	Rickettsial and chlamydial infections, *Mycoplasma pneumoniae*	Not given during pregnancy or to young children
Tigecycline	Skin infections caused by various gram-positive cocci and intra-abdominal infections caused by various facultative and anaerobic bacteria (see text)	Adverse effects similar to tetracyclines
Chloramphenicol	*Haemophilus influenzae* meningitis, typhoid fever, anaerobic infections (especially *Bacteroides fragilis*)	Bone marrow toxicity limits use to severe infections
Macrolides		
Azithromycin	Genital tract infections caused by *Chlamydia trachomatis* and pneumonia caused by *Mycoplasma, Legionella,* and *Chlamydophila pneumoniae*	Generally well tolerated but some diarrhea; long half-life (68 hours); good oral availability
Erythromycin	Similar to Azithromycin	2 hours half-life
Clarithromycin	Used mainly in the treatment of *Helicobacter* and *Mycobacterium avium-intracellulare* infections	6 hours half-life
Clindamycin	Anaerobes such as *Clostridium perfringens* and *B. fragilis*	Pseudomembranous colitis is a major side effect
Linezolid	Vancomycin-resistant enterococci, methicillin-resistant *Staphylococcus aureus* and *Staphylococcus epidermidis,* and penicillin-resistant pneumococci	Generally well tolerated
Telithromycin	Community-acquired pneumonia caused by various bacteria, including multidrug-resistant *Streptococcus pneumoniae*	Many bacteria that are resistant to other macrolides are susceptible to telithromycin
Streptogramins	Bacteremia caused by vancomycin-resistant *Enterococcus faecium*	No cross-resistance between streptogramins and other drugs that inhibit protein synthesis
Retapamulin	Skin infections caused by *Streptococcus pyogenes* and methicillin-sensitive *S. aureus*	Topical application only
Mupirocin	Skin infections caused by *S. pyogenes* and *S. aureus*	

[1]The spectrum of activity is intentionally incomplete. It is simplified for the beginning student to illustrate the expanded coverage of gram-negative organisms with successive generations and does not cover all possible clinical uses.

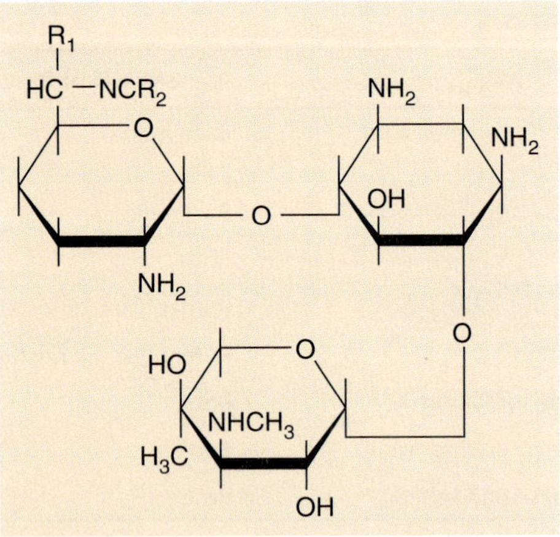

FIGURE 10–6 Aminoglycosides. Aminoglycosides consist of amino sugars joined by a glycosidic linkage. The structure of gentamicin is shown.

1. Drugs That Act on the 30S Subunit

Aminoglycosides

The aminoglycosides in common clinical use are gentamicin, tobramycin, and amikacin. Aminoglycosides are bactericidal drugs especially useful against many gram-negative rods. Certain aminoglycosides are used against other organisms (e.g., streptomycin is used in the therapy of MDR tuberculosis, and gentamicin is used in combination with penicillin G against enterococci). Aminoglycosides are named for the amino sugar component of the molecule, which is connected by a glycosidic linkage to other sugar derivatives (Figure 10–6).

In 2018, the FDA approved a new aminoglycoside, plazomicin, for the treatment of complicated UTIs. It is active against many aerobic gram-negative rods, such as *E. coli*, *Klebsiella*, *Proteus*, and *Pseudomonas*, including those resistant to other aminoglycosides.

The two important modes of action of aminoglycosides have been documented best for streptomycin; other aminoglycosides probably act similarly. Both **inhibition of the initiation complex** and **misreading of messenger RNA** (mRNA) occur; the former likely more important for the bactericidal activity of the drug. An initiation complex composed of a streptomycin-treated 30S subunit, a 50S subunit, and mRNA will not function; that is, no peptide bonds are formed, no polysomes are made, and a frozen "streptomycin monosome" results. Misreading of the triplet codon of mRNA so that the wrong amino acid is inserted into the protein also occurs in streptomycin-treated bacteria. The site of action on the 30S subunit includes both a ribosomal protein and the ribosomal RNA (rRNA). As a result of inhibition of initiation and misreading, membrane damage occurs and the bacterium dies. Another mode of action namely, that aminoglycosides inhibit ribozyme-mediated self-splicing of rRNA may also contribute to its inhibitory activity.

Aminoglycosides have certain limitations in their use: (1) They have a toxic effect both on the kidneys and on the auditory and vestibular portions of the eighth cranial nerve. To avoid toxicity, serum levels of the drug, blood urea nitrogen, and creatinine should be measured. (2) They are poorly absorbed from the gastrointestinal tract and cannot be given orally. (3) They penetrate the spinal fluid poorly and must be given intrathecally in the treatment of meningitis. (4) They are ineffective against anaerobes, because their transport into the bacterial cell requires oxygen.

Tetracyclines

Tetracyclines are a family of antibiotics with bacteriostatic activity against a variety of gram-positive and gram-negative bacteria, Mycoplasma, Chlamydiae, and Rickettsiae. They inhibit protein synthesis by **blocking the aminoacyl transfer RNA (tRNA) from entering the acceptor site** on the 30S ribosomal subunit. However, the selective action of tetracycline on bacteria is not at the level of the ribosome, because tetracycline in vitro will inhibit protein synthesis equally well in purified ribosomes from both bacterial and human cells. Its selectivity is based on its greatly *increased uptake* into susceptible bacterial cells compared with human cells.

Tetracyclines, as the name indicates, have four cyclic rings with different substituents at the three R groups (Figure 10–7). The various tetracyclines (e.g., doxycycline, minocycline, and oxytetracycline) have similar antimicrobial activity but different pharmacologic properties. In general, tetracyclines have low toxicity but are associated with some important side effects. One is suppression of the normal flora of the intestinal tract, which can lead to diarrhea and overgrowth by drug-resistant bacteria and fungi. Second is that suppression of *Lactobacillus* in the vaginal normal flora results in a rise in pH, which allows *Candida albicans* to grow and cause vaginitis. Third is brown staining of the teeth of fetuses and young children as a result of deposition of the drug in developing teeth; tetracyclines are avid calcium chelators. For this reason, tetracyclines are contraindicated for use in pregnant women and in children younger than 8 years of age. Tetracyclines also chelate iron, and so products containing iron, such as iron-containing vitamins, should not be taken during therapy with tetracyclines. Photosensitivity (rash upon exposure to sunlight) can also occur during tetracycline therapy.

Tigecycline (Tygacil) is the first clinically available member of the glycylcycline class of antibiotics. They have a structure similar to tetracyclines and have the same mechanism of action as tetracyclines and have a similar range of adverse effects.

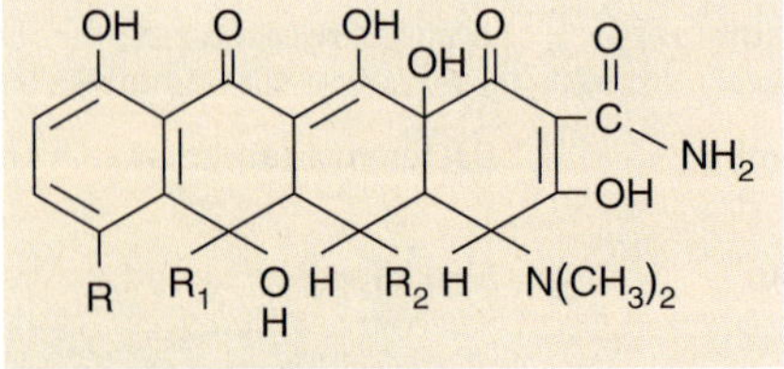

FIGURE 10–7 Tetracycline structure. The four-ring structure is depicted with its three R sites. Chlortetracycline, for example, has R = Cl, R_1 = CH$_3$, and R_2 = H.

Tigecycline is used to treat skin and skin structure infections caused by *S. aureus*, group A and group B streptococci, VRE, *E. coli*, and *Bacteroides fragilis*. It is also used to treat complicated intraabdominal infections caused by a variety of facultative and anaerobic bacteria. In 2018, the FDA approved **eravacycline**, a drug closely related to tigecycline, for the treatment of complicated intraabdominal infections.

Also in 2018, the FDA approved **omadacycline** for the treatment of acute bacterial skin and soft tissue infections and community-acquired bacterial pneumonia. Omadacycline is a new-generation tetracycline that is effective against gram-positive cocci, for example, *S. aureus*, enterococci, as well as *S. pneumoniae*. It is also effective against *Haemophilus influenzae*, *E. coli*, *Legionella*, and *Mycoplasma*. It is less prone to various bacterial resistance mechanisms than many other tetracyclines.

2. Drugs That Act on the 50S Subunit

Chloramphenicol

Chloramphenicol is active against a broad range of organisms, including gram-positive and gram-negative bacteria (including anaerobes). It is bacteriostatic against certain organisms, such as *Salmonella typhi*, but has bactericidal activity against the three important encapsulated organisms that cause meningitis: *H. influenzae*, *S. pneumoniae*, and *N. meningitidis*.

Chloramphenicol inhibits protein synthesis by binding to the 50S ribosomal subunit and **blocking the action of peptidyl transferase**; this prevents the synthesis of new peptide bonds. It inhibits bacterial protein synthesis selectively because it binds to the catalytic site of the transferase in the 50S bacterial ribosomal subunit but not to the transferase in the 60S human ribosomal subunit. Chloramphenicol inhibits protein synthesis in the mitochondria of human cells to some extent, since mitochondria have a 50S subunit (mitochondria are thought to have evolved from bacteria). This inhibition may be the cause of the dose-dependent toxicity of chloramphenicol to bone marrow.

Chloramphenicol is a comparatively simple molecule with a nitrobenzene nucleus (Figure 10–8). Nitrobenzene is a bone marrow depressant and is likely to be involved in the hematologic problems reported with this drug. The most important side effect of chloramphenicol is bone marrow toxicity, of which there are two types. One is a dose-dependent suppression, which is more likely to occur in patients receiving high doses for long periods and is reversible when administration of the drug is stopped. The other is aplastic anemia caused by an idiosyncratic reaction to the drug. This reaction is not dose-dependent, can occur weeks after administration of the drug has

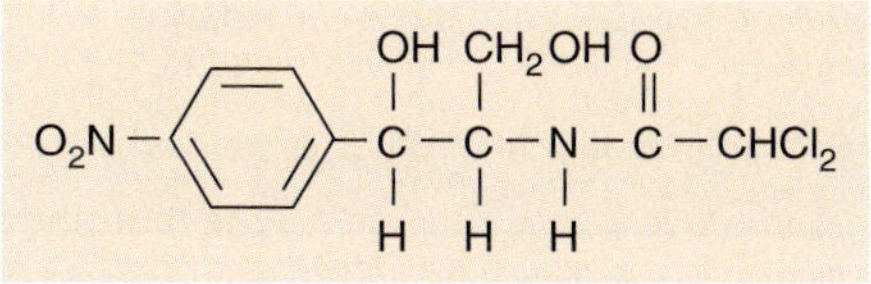

FIGURE 10–8 Chloramphenicol.

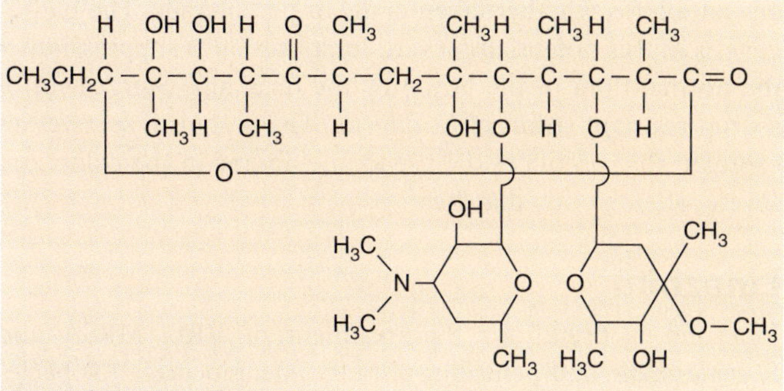

been stopped, and is not reversible. Fortunately, this reaction is rare, occurring in about 1 in 30,000 patients.

One specific toxic manifestation of chloramphenicol is "gray baby" syndrome, in which the infant's skin appears gray and vomiting and shock occur. This is due to reduced glucuronyl transferase activity in infants, resulting in a toxic concentration of chloramphenicol. Glucuronyl transferase is the enzyme responsible for detoxification of chloramphenicol.

Macrolides

Macrolides are a group of bacteriostatic drugs with a wide spectrum of activity. The name *macrolide* refers to their large (13–16 carbon) ring structure (Figure 10–9). Azithromycin, erythromycin, and clarithromycin are the main macrolides in clinical use. Azithromycin is used to treat genital tract infections caused by *Chlamydia trachomatis* and respiratory tract infections caused by *Legionella*, *Mycoplasma*, *Chlamydia pneumoniae*, and *S. pneumoniae*. Erythromycin has a similar spectrum of activity but has a shorter half-life and so must be taken more frequently and has more adverse effects, especially on the gastrointestinal tract. Clarithromycin is used primarily in the treatment of *Helicobacter* infections and in the treatment and prevention of *Mycobacterium avium-intracellulare* infections. An important adverse effect of clarithromycin is prolongation of the QT interval, which may increase the risk of cardiac death.

Macrolides inhibit bacterial protein synthesis by binding to the 50S ribosomal subunit and blocking translocation. They prevent the release of the uncharged tRNA after it has transferred its amino acid to the growing peptide chain. The donor site remains occupied, a new tRNA cannot attach, and protein synthesis stops.

Clindamycin

The most useful clinical activity of this bacteriostatic drug is against anaerobes, both gram-positive bacteria such as *Clostridium perfringens* and gram-negative bacteria such as *B. fragilis*.

Clindamycin binds to the 50S subunit and blocks peptide bond formation by an undetermined mechanism. Its specificity for bacteria arises from its inability to bind to the 60S subunit of human ribosomes.

The most important side effect of clindamycin is pseudomembranous colitis, which, in fact, can occur with virtually

FIGURE 10–9 Erythromycin.

any antibiotic, whether taken orally or parenterally. The pathogenesis of this potentially severe complication is suppression of the normal flora of the bowel by the drug and overgrowth of a drug-resistant strain of *C. difficile*. The organism secretes an exotoxin that produces the pseudomembrane in the colon and severe, often bloody diarrhea.

Linezolid

Linezolid is useful for the treatment of VRE, MRSA and *S. epidermidis*, and penicillin-resistant pneumococci. It is bacteriostatic against enterococci and staphylococci but bactericidal against pneumococci.

Linezolid binds to the 23S rRNA in the 50S subunit and inhibits protein synthesis, but the precise mechanism is unknown. It appears to block some early step (initiation) in ribosome formation. Tedizolid is a second-generation drug in the same class as linezolid but is approximately 10 times more effective. It is used for the treatment of skin and soft tissue infections caused by a similar range of bacteria as linezolid and has a similar mechanism of action.

Telithromycin

Telithromycin (Ketek) is the first clinically useful member of the ketolide group of antibiotics. It is similar to the macrolides in general structure and mode of action but is sufficiently different chemically such that organisms resistant to macrolides may be sensitive to telithromycin. It has a wide spectrum of activity against a variety of gram-positive and gram-negative bacteria (including macrolide-resistant pneumococci) and is used in the treatment of community-acquired pneumonia, bronchitis, and sinusitis.

Streptogramins

A combination of two streptogramins, quinupristin and dalfopristin (Synercid), is used for the treatment of bloodstream infections caused by vancomycin-resistant *Enterococcus faecium* (but not vancomycin-resistant *Enterococcus faecalis*). It is also approved for use in infections caused by *Streptococcus pyogenes*, penicillin-resistant *S. pneumoniae*, MRSA, and methicillin-resistant *S. epidermidis*.

Streptogramins cause premature release of the growing peptide chain from the 50S ribosomal subunit. The structure and mode of action of streptogramins are different from all other drugs that inhibit protein synthesis, and there is no cross-resistance between streptogramins and these other drugs.

Retapamulin

Retapamulin (Altabax) is the first clinically available member of a new class of antibiotics called pleuromutilins. These drugs inhibit bacterial protein synthesis by binding to the 23S RNA of the 50S subunit and blocking attachment of the donor tRNA. Retapamulin is a topical antibiotic used in the treatment of skin infections, such as impetigo, caused by *S. pyogenes* and methicillin-sensitive *S. aureus*. In 2019, lefamulin (Xenleta), a second pleuromulin, was approved for the treatment of community-acquired bacterial pneumonia in adults.

3. Drugs That Inhibit tRNA synthetase

Mupirocin is effective against gram-positive cocci, especially *S. aureus*, (both MRSA and MSSA), by inhibiting tRNA synthetase. It is used topically in treating minor skin infections such as impetigo. It is also used to reduce nasal carriage of *S. aureus* in hospital personnel.

INHIBITION OF NUCLEIC ACID SYNTHESIS

The mode of action and clinically useful activity of the important drugs that act by inhibiting nucleic acid synthesis are summarized in Table 10–8.

1. Inhibition of Precursor Synthesis

Sulfonamides

Either alone or in combination with trimethoprim, sulfonamides are used to treat a variety of bacterial diseases such as

TABLE 10–8 Mode of Action and Activity of Selected Nucleic Acid Inhibitors[1]

Drug	Mode of Action	Clinically Useful Activity
Sulfonamides (e.g., sulfamethoxazole)	Inhibit folic acid synthesis; act as a competitive inhibitor of PABA	Used in combination with trimethoprim for UTI caused by *Escherichia coli*; otitis media and sinusitis caused by *Haemophilus influenzae*; MRSA; *Pneumocystis* pneumonia
Trimethoprim	Inhibits folic acid synthesis by inhibiting DHFR	Used in combination with sulfonamides for the uses described above
Fluoroquinolones (e.g., ciprofloxacin, levofloxacin)	Inhibit DNA synthesis by inhibiting DNA gyrase	Ciprofloxacin is used to treat GI tract infections caused by *Shigella* and *Salmonella*, and UTI caused by enteric gram-negative rods. Levofloxacin is used to treat respiratory tract infections, especially those caused by penicillin-resistant *Streptococcus pneumoniae*
Rifampin	Inhibits mRNA synthesis by inhibiting RNA polymerase	Used in combination with isoniazid and other drugs to treat tuberculosis

DHFR = dihydrofolate reductase; GI = gastrointestinal; MRSA = methicillin-resistant *Staphylococcus aureus*; PABA = para-aminobenzoic acid; UTI = urinary tract infection.
[1]The spectrum of activity is intentionally incomplete. It is simplified for the beginning student to emphasize the most common uses.

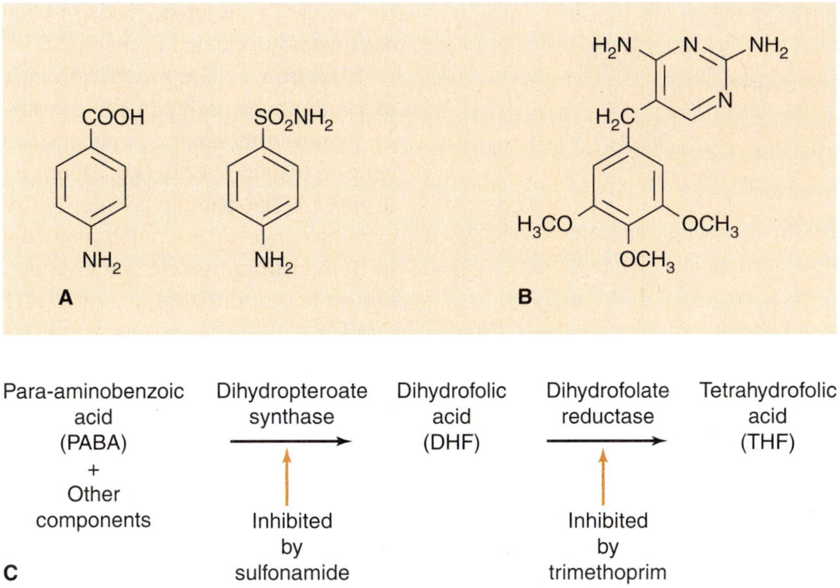

FIGURE 10–10 Mechanism of action of sulfonamides and trimethoprim. **A:** Comparison of the structures of *p*-aminobenzoic acid (PABA) and sulfanilamide. Note that the only difference is that PABA has a carboxyl (COOH) group, whereas sulfanilamide has sulfonamide (SO_2NH_2) group. **B:** Structure of trimethoprim. **C:** Inhibition of the folic acid pathway by sulfonamide and trimethoprim. Sulfonamides inhibit the synthesis of dihydrofolic acid (DHF) from its precursor PABA. Trimethoprim inhibits the synthesis of tetrahydrofolic acid (THF) from its precursor DHF. Loss of THF inhibits DNA synthesis because THF is required to transfer a methyl group onto uracil to produce thymidine, an essential component of DNA. (Data from Corcoran JW, Hahn FE. *Mechanism of Action of Antimicrobial and Antitumor Agents.* New York, NY: Springer-Verlag; 1975.)

UTIs caused by *E. coli*, otitis media caused by *S. pneumoniae* or *H. influenzae* in children, shigellosis, nocardiosis, and chancroid. In combination, they are also the drugs of choice for two additional diseases, toxoplasmosis and *Pneumocystis* pneumonia. The sulfonamides are a large family of bacteriostatic drugs that are produced by chemical synthesis. In 1935, the parent compound, sulfanilamide, became the first clinically effective antimicrobial agent.

The mode of action of sulfonamides is to block the synthesis of tetrahydrofolic acid, which is required as a methyl donor in the synthesis of the nucleic acid precursors adenine, guanine, and thymine. Sulfonamides are **structural analogues of *p*-aminobenzoic acid** (PABA). PABA condenses with a pteridine compound to form dihydropteroic acid, a precursor of tetrahydrofolic acid (Figure 10–10). Sulfonamides compete with PABA for the active site of the enzyme dihydropteroate synthetase.

The basis of the selective action of sulfonamides on bacteria is that many bacteria synthesize their folic acid from PABA-containing precursors, whereas human cells require preformed folic acid as an exogenous nutrient because they lack the enzymes to synthesize it. Human cells, therefore, bypass the step at which sulfonamides act. Bacteria that can use preformed folic acid are similarly resistant to sulfonamides.

The *p*-amino group on the sulfonamide is essential for its activity. Modifications are therefore made on the sulfonic acid side chain. Sulfonamides are inexpensive and infrequently cause side effects. However, drug-related fever, rashes, photosensitivity (rash upon exposure to sunlight), and bone marrow suppression

can occur. They are the most common group of drugs that cause erythema multiforme and its more severe forms, Stevens-Johnson syndrome and toxic epidermal necrolysis.

Trimethoprim

Trimethoprim inhibits the production of tetrahydrofolic acid by a different mechanism from that of the sulfonamides, namely it inhibits the enzyme **dihydrofolate reductase** (see Figure 10–10). Its specificity for bacteria is based on its much greater affinity for bacterial reductase than for the human enzyme.

Trimethoprim is used most frequently together with sulfamethoxazole. Note that both drugs act on the same pathway but at different sites to inhibit the synthesis of tetrahydrofolate. The advantages of the combination are that (1) bacterial mutants resistant to one drug will be inhibited by the other and (2) the two drugs can act **synergistically** (i.e., when used together, they cause significantly greater inhibition than the sum of the inhibition caused by each drug separately).

Trimethoprim-sulfamethoxazole is clinically useful in the treatment of UTIs, *Pneumocystis* pneumonia, and shigellosis. It is also used for prophylaxis in neutropenic patients to prevent opportunistic infections.

2. Inhibition of DNA Synthesis

Fluoroquinolones

Fluoroquinolones are bactericidal drugs that block bacterial DNA synthesis by inhibiting **DNA gyrase (topoisomerase)**. Fluoroquinolones, such as ciprofloxacin (Figure 10–11),

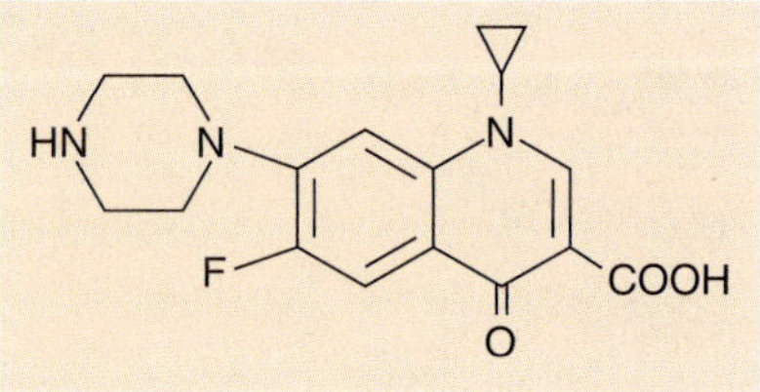

FIGURE 10–11 Ciprofloxacin. The triangle indicates a cyclopropyl group.

levofloxacin, norfloxacin, ofloxacin, and others, are active against a broad range of organisms that cause infections of the lower respiratory tract, intestinal tract, urinary tract, and skeletal and soft tissues. Nalidixic acid, which is a quinolone but not a fluoroquinolone, is much less active and is used only for the treatment of UTIs. Fluoroquinolones should not be given to pregnant women and children under the age of 18 years because they damage growing bone and cartilage.

In 2018, the FDA approved **delafloxacin** for the treatment of acute bacterial skin and soft tissue infections. Delafloxacin is a fluoroquinolone that is effective against gram-positive cocci, like *S. aureus*, *S. pyogenes*, and *E. faecalis*. It is effective against gram-negative bacteria, for example, *E. coli*, *Klebsiella*, *Pseudomonas*, and Legionella. It is also effective against *Mycoplasma* and anaerobes such as *Bacteroides*. It is the only fluoroquinolone approved for the treatment of MRSA.

The FDA issued a warning regarding the possibility of Achilles tendonitis and tendon rupture associated with fluoroquinolone use, especially in those over 60 years of age and in patients receiving corticosteroids, such as prednisone. The FDA recommends that fluoroquinolones *not* be used in the treatment of acute sinusitis and uncomplicated UTIs. Another important adverse effect of fluoroquinolones is peripheral neuropathy, the symptoms of which include pain, burning, numbness, or tingling in the arms or legs. **Another rare but serious adverse effect is rupture of the aorta.**

3. Inhibition of mRNA Synthesis

Rifampin is used primarily for the treatment of tuberculosis in combination with other drugs and for prophylaxis in close contacts of patients with meningitis caused by either *N. meningitidis* or *H. influenzae*. It is also used in combination with other drugs in the treatment of prosthetic-valve endocarditis caused by *S. epidermidis*. In 2018, the FDA approved the use of oral rifampin for the treatment of traveler's diarrhea caused by *E. coli*. With the exception of the short-term prophylaxis of meningitis and treatment of traveler's diarrhea, rifampin is given in combination with other drugs because **resistant mutants appear at a high rate** when it is used alone.

The selective mode of action of rifampin is based on **blocking mRNA synthesis** by bacterial RNA polymerase without affecting the RNA polymerase of human cells. Rifampin is red, and the urine, saliva, and sweat of patients taking rifampin often turn orange; this is disturbing but harmless. Rifampin is excreted in high concentration in saliva, which accounts for

its success in the prophylaxis of bacterial meningitis since the organisms are carried in the throat.

Rifabutin, a rifampin derivative with the same mode of action as rifampin, is useful in the prevention of disease caused by *M. avium-intracellulare* in patients with severely reduced numbers of helper T cells (e.g., acquired immunodeficiency syndrome [AIDS] patients). Note that rifabutin does not increase cytochrome P450 as much as rifampin, so rifabutin is used in human immunodeficiency virus (HIV)/AIDS patients taking protease inhibitors or non-reverse transcriptase inhibitors (NRTI).

Fidaxomicin (Dificid) inhibits the RNA polymerase of *C. difficile*. It is used in the treatment of pseudomembranous colitis and in preventing relapses of this disease. It specifically inhibits *C. difficile* and does not affect the gram-negative normal flora of the colon.

ALTERATION OF CELL MEMBRANE FUNCTION

There are few antimicrobial compounds that act on the cell membrane because the structural and chemical similarities of bacterial and human cell membranes make it difficult to provide sufficient selective toxicity.

Polymyxins are a family of polypeptide antibiotics of which the clinically most useful compound is polymyxin E (colistin). Colistin is active against gram-negative rods, especially *P. aeruginosa*, *Acinetobacter baumannii*, and carbapenemase-producing Enterobacteriaceae. Most strains of these highly antibiotic-resistant bacteria are sensitive to colistin, although rare isolates from patients are resistant.

Polymyxins are cyclic peptides composed of 10 amino acids, six of which are diaminobutyric acid. The positively charged free amino groups act like a cationic detergent to disrupt the phospholipid structure of the cell membrane.

Daptomycin is a cyclic lipopeptide that disrupts the cell membranes of gram-positive cocci. It is bactericidal for organisms such as *S. aureus*, *S. epidermidis*, *S. pyogenes*, *E. faecalis*, and *E. faecium*, including methicillin-resistant strains of *S. aureus* and *S. epidermidis*, vancomycin-resistant strains of *S. aureus*, and vancomycin-resistant strains of *E. faecalis* and *E. faecium*. It is approved for use in complicated skin and soft tissue infections caused by these bacteria.

ADDITIONAL DRUG MECHANISMS

Isoniazid, or isonicotinic acid hydrazide (INH), is a bactericidal drug highly specific for *Mycobacterium tuberculosis* and other mycobacteria. It is used in combination with other drugs to treat tuberculosis and by itself to prevent tuberculosis in exposed persons. Because it penetrates into human cells well, it is effective against the organisms residing within macrophages. The structure of isoniazid is shown in Figure 10–12.

INH **inhibits mycolic acid synthesis**, which explains why it is specific for mycobacteria and relatively nontoxic for humans.

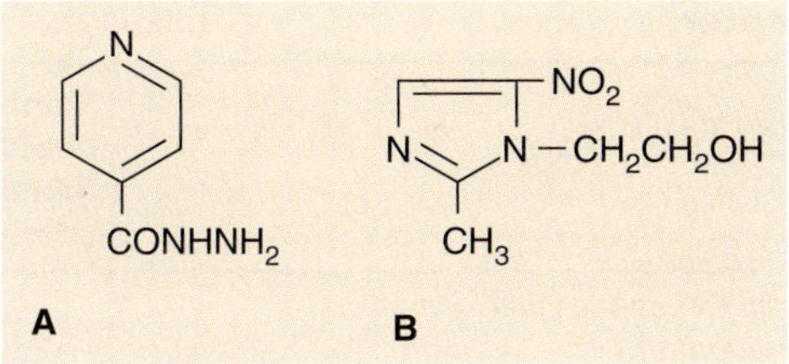

FIGURE 10–12 **A:** Isoniazid. **B:** Metronidazole.

The drug inhibits a reductase required for the synthesis of the long-chain fatty acids called mycolic acids that are an essential constituent of mycobacterial cell walls. The active drug is a metabolite of INH formed by the action of catalase peroxidase because deletion of the gene for these enzymes results in resistance to the drug. Its main side effect is liver toxicity. It is given with pyridoxine to prevent neurologic complications.

Metronidazole (Flagyl) is bactericidal against anaerobic bacteria such as *B. fragilis*. It is used to treat bacterial vaginosis and can be used for nonserious colitis caused by *C. difficile*. It is also effective against infections caused by certain protozoa such as *Giardia* and *Trichomonas*. Metronidazole is a prodrug that is activated to the active compound within anaerobic bacteria by ferredoxin-mediated reduction of its nitro group.

This drug has two possible mechanisms of action, and it is unclear which is more important. The first, which explains its specificity for anaerobes, is its ability to act as an **electron sink**. By accepting electrons, the drug deprives the organism of required reducing power. In addition, when electrons are acquired, the drug ring is cleaved and a toxic intermediate is

formed that damages DNA. The precise nature of the intermediate and its action is unknown. The structure of metronidazole is shown in Figure 10–12.

The second mode of action of metronidazole relates to its ability to inhibit DNA synthesis. The drug binds to DNA and causes strand breakage, which prevents its proper functioning as a template for DNA polymerase.

Nitrofurantoin is a urinary tract antiseptic that is useful in the treatment of uncomplicated lower UTIs. It is concentrated in the urine to reach bactericidal levels but does not reach cidal levels systemically so is not useful for infections outside the urinary tract.

Nitrofurantoin acts by binding to DNA. Its selective toxicity for bacteria is dependent upon the ability of bacteria to form larger amounts of the highly reactive reduced form of the drug compared to the amount formed in human cells.

Ethambutol is a bacteriostatic drug used in the treatment of infections caused by *M. tuberculosis* and many of the atypical mycobacteria. It acts by inhibiting the synthesis of arabinogalactan, which functions as a link between the mycolic acids and the peptidoglycan of the organism.

Pyrazinamide (PZA) is a bactericidal drug used in the treatment of tuberculosis but not in the treatment of most atypical mycobacterial infections. PZA is particularly effective against semi-dormant organisms in the lesion, which are not affected by INH or rifampin. The mechanism of action of PZA is uncertain, but there is evidence that it acts by inhibiting a fatty acid synthetase that prevents the synthesis of mycolic acid. It is converted to the active intermediate, pyrazinoic acid, by an amidase in the mycobacteria.

CHEMOPROPHYLAXIS

In most instances, the antimicrobial agents described in this chapter are used for the *treatment* of infectious diseases. However, there are times when they are used to *prevent* diseases from occurring, a process called **chemoprophylaxis**.

Chemoprophylaxis is used in three circumstances: prior to surgery, in immunocompromised patients, and in people with normal immunity who have been exposed to certain pathogens. Table 10–9 describes the drugs and the situations in which they are used. For more information, see the chapters on the individual organisms.

Of particular importance is the **prevention of endocarditis in high-risk patients undergoing dental surgery** by using

amoxicillin perioperatively. High-risk patients include those who have unrepaired damage to their heart valve, have a prosthetic heart valve, or have previously had infective endocarditis. Prophylaxis to prevent endocarditis in patients undergoing gastrointestinal or genitourinary tract surgery is not recommended.

Cefazolin is often used to prevent staphylococcal infections in patients undergoing orthopedic surgery, including prosthetic joint implants, and in vascular graft surgery. Chemoprophylaxis is unnecessary in those with an implanted dialysis catheter, a cardiac pacemaker, or a ventriculoperitoneal shunt.

PROBIOTICS

In contrast to the chemical antibiotics previously described in this chapter, probiotics are well-characterized strains of living microorganisms that, when administered in adequate amounts, confer a scientifically demonstrated health benefit on the host. The suggested basis for the possible beneficial effect lies either in providing colonization resistance by which the nonpathogen

excludes the pathogen from binding sites on the mucosa, in enhancing the immune response against the pathogen, or in reducing the inflammatory response against the pathogen. For example, the oral administration of live *Lactobacillus rhamnosus* strain GG significantly reduces the number of cases of nosocomial diarrhea in young children. Also, the yeast *Saccharomyces*

TABLE 10–9 Chemoprophylactic Use of Drugs Described in This Chapter

Drug	Use	Number of Chapter(s) for Additional Information
Penicillin	1. Prevent recurrent pharyngitis in high-risk patients who have had rheumatic fever 2. Prevent syphilis in high-risk patients exposed to *Treponema pallidum* 3. Prevent pneumococcal sepsis in splenectomized young children	15 24 15
Ampicillin	Prevent neonatal sepsis and meningitis in children born of mothers carrying group B streptococci	15
Amoxicillin	Prevent endocarditis caused by viridans streptococci in high-risk patients with damaged heart valves undergoing dental surgery	15
Cefazolin	Prevent staphylococcal surgical wound infections	15
Ceftriaxone	Prevent gonorrhea in high-risk patients exposed to *Neisseria gonorrhoeae*	16
Ciprofloxacin	1. Prevent meningitis in high-risk patients exposed to *Neisseria meningitidis* 2. Prevent anthrax in high-risk patients exposed to *Bacillus anthracis* 3. Prevent infection in neutropenic patients	16 17 68
Rifampin	Prevent meningitis in high-risk patients exposed to *N. meningitidis* and *Haemophilus influenzae*	16, 19
Isoniazid	Prevent progression of *Mycobacterium tuberculosis* in high-risk patients recently infected who are asymptomatic[1]	21
Erythromycin	1. Prevent pertussis in high-risk patients exposed to *Bordetella pertussis* 2. Prevent gonococcal and chlamydial conjunctivitis in newborns	19 16, 25
Tetracycline	Prevent plague in high-risk patients exposed to *Yersinia pestis*	20
Trimethoprim-sulfamethoxazole	Prevent recurrent urinary tract infections	18
Mupirocin	Reduce colonization of *S. aureus* in nose of hospital personnel	15

[1]Chemoprophylaxis with isoniazid is also viewed as treatment of asymptomatic individuals (see Chapter 21).

boulardii reduces the risk of antibiotic-associated diarrhea caused by *C. difficile*. Adverse effects are few; however, serious complications have been reported in highly immunosuppressed patients and in patients with indwelling vascular catheters. It is important to note that there is very little regulatory oversight of the probiotic industry. Typically, probiotics are regulated as food supplements and are thus not subject to stringent standards that address efficacy, safety, and quality. Thus, neither consumers nor healthcare providers can easily evaluate most of the claims of vaguely described "health benefits" of these products. In fact, there are currently no probiotics that satisfy the regulatory requirements to be labeled as a therapy that can cure, mitigate, or prevent disease.

PEARLS

- For an antibiotic to be clinically useful, it must exhibit **selective toxicity** (i.e., it must inhibit bacterial processes significantly more than it inhibits human cell processes).
- There are four main targets of antibacterial drugs: **cell wall, ribosomes, cell membrane**, and **nucleic acids**. Human cells are not affected by these drugs because our cells do not have a cell wall, and our cells have different ribosomes, nucleic acid enzymes, and sterols in the membranes.
- **Bactericidal** drugs kill bacteria, whereas **bacteriostatic** drugs inhibit the growth of the bacteria but do not kill. Bacteriostatic drugs depend on the phagocytes of the patient to kill the organism. If a patient has too few neutrophils, then bactericidal drugs should be used.

Inhibition of Cell Wall Synthesis

- **Penicillins** and **cephalosporins** act by inhibiting **transpeptidases**, the enzymes that cross-link peptidoglycan. Transpeptidases are also referred to as **penicillin-binding proteins (PBPs)**. Several medically important bacteria (e.g., *Streptococcus pneumoniae*) manifest resistance to penicillins based on mutations in the genes encoding PBPs.
- Exposure to penicillins activates **autolytic enzymes** that degrade the bacteria. If these autolytic enzymes are not activated (e.g., in certain strains of *Staphylococcus aureus*), the bacteria are not killed and the strain is said to be **tolerant**.

- Penicillins kill bacteria when they are growing (i.e., when they are synthesizing new peptidoglycan). Penicillins are therefore **more active during the log phase** of bacterial growth than during the lag phase or the stationary phase.
- Penicillins and cephalosporins are **β-lactam drugs** (i.e., an intact **β-lactam ring** is required for activity). **β-Lactamases** (e.g., penicillinases and cephalosporinases) cleave the β-lactam ring and inactivate the drug.
- **Modification of the side chain** adjacent to the β-lactam ring endows these drugs with **new properties**, such as expanded activity against gram-negative rods, ability to be taken orally, and protection against degradation by β-lactamases. For example, the original penicillin (benzyl penicillin, penicillin G) cannot be taken orally because stomach acid hydrolyzes the bond between the β-lactam ring and the side chain. But ampicillin and amoxicillin can be taken orally because they have a different side chain.
- **Hypersensitivity** to penicillins, especially **IgE-mediated anaphylaxis**, remains a significant concern.
- **Cephalosporins** are structurally similar to penicillins: both have a β-lactam ring. The first-generation cephalosporins are active primarily against gram-positive cocci, and the second, third, and fourth generations have expanded coverage against gram-negative rods.
- **Carbapenems, such as imipenem**, and monobactams, such as aztreonam, are also β-lactam drugs but are structurally different from penicillins and cephalosporins.
- **Vancomycin** is a **glycopeptide** (i.e., it is not a β-lactam drug), but its mode of action is very similar to that of penicillins and cephalosporins (i.e., it **inhibits transpeptidases**).

Inhibition of Protein Synthesis

- **Aminoglycosides** and **tetracyclines** act at the level of the 30S ribosomal subunit, whereas **chloramphenicol, erythromycins**, and **clindamycin** act at the level of the 50S ribosomal subunit.
- **Aminoglycosides** inhibit bacterial protein synthesis by binding to the 30S subunit, which **blocks the initiation complex**. No peptide bonds are formed, and no polysomes are made. Aminoglycosides are a family of drugs that includes gentamicin, tobramycin, and streptomycin.
- **Tetracyclines** inhibit bacterial protein synthesis by **blocking the binding of aminoacyl tRNA** to the 30S ribosomal subunit. The tetracyclines are a family of drugs; doxycycline is used most often.
- **Chloramphenicol** inhibits bacterial protein synthesis by **blocking peptidyl transferase**, the enzyme that adds the new amino acid to the growing polypeptide. Chloramphenicol can cause bone marrow suppression.
- **Erythromycin** inhibits bacterial protein synthesis by **blocking the release of the tRNA** after it has delivered its amino acid to the growing polypeptide. Erythromycin is a member of the macrolide family of drugs that includes azithromycin and clarithromycin.
- **Clindamycin** binds to the same site on the ribosome as does erythromycin and is thought to act in the same manner. It is effective against many anaerobic bacteria. Clindamycin is one

of the antibiotics that predisposes to pseudomembranous colitis caused by *Clostridioides difficile* and is infrequently used.

Inhibition of Nucleic Acid Synthesis

- **Sulfonamides** and **trimethoprim** inhibit **nucleotide** synthesis, **quinolones** inhibit **DNA** synthesis, and **rifampin** inhibits **RNA** synthesis.
- **Sulfonamides** and **trimethoprim** inhibit the **synthesis of tetrahydrofolic acid**—the main donor of the methyl groups that are required to synthesize adenine, guanine, and thymine. **Sulfonamides** are structural analogues of *p*-aminobenzoic acid, which is a component of folic acid. **Trimethoprim** inhibits **dihydrofolate reductase**—the enzyme that reduces dihydrofolic acid to tetrahydrofolic acid. A combination of sulfamethoxazole and trimethoprim is often used because bacteria resistant to one drug will be inhibited by the other.
- **Quinolones** inhibit DNA synthesis in bacteria by **blocking DNA gyrase** (topoisomerase)—the enzyme that unwinds DNA strands so that they can be replicated. Quinolones are a family of drugs that includes ciprofloxacin, ofloxacin, and levofloxacin.
- **Rifampin** inhibits RNA synthesis in bacteria by **blocking the RNA polymerase** that synthesizes mRNA. Rifampin is typically used in combination with other drugs because there is a **high rate of mutation of the RNA polymerase gene**, which results in rapid resistance to the drug.

Alteration of Cell Membrane Function

- **Polymyxins**, such as colistin, are positively charged cyclic peptides that disrupt bacterial cell membranes. They are active against enteric gram-negative rods.
- **Daptomycin** is a cyclic lipopeptide that disrupts bacteria cell membranes. It is active against gram-positive cocci such as *Staphylococcus* and *Enterococcus*.

Additional Drug Mechanisms

- **Isoniazid inhibits the synthesis of mycolic acid**—a long-chain fatty acid found in the cell wall of mycobacteria. Isoniazid is a **prodrug** that requires a bacterial **peroxidase (catalase) to activate isoniazid** to the metabolite that inhibits mycolic acid synthesis. Isoniazid is the most important drug used in the treatment of tuberculosis and other mycobacterial diseases.
- **Metronidazole** is **effective against anaerobic bacteria and certain protozoa** because it **acts as an electron sink**, taking away the electrons that the organisms need to survive. It also forms toxic intermediates that damage DNA.

Chemoprophylaxis

- Antimicrobial drugs are used to prevent infectious diseases as well as to treat them. Chemoprophylactic drugs are given primarily in three circumstances: to prevent surgical wound infections, to prevent opportunistic infections in immunocompromised patients, and to prevent infections in those known to be exposed to pathogens that cause serious infectious diseases.

SELF-ASSESSMENT QUESTIONS

1. Cefazolin is often given prior to surgery to prevent postsurgical wound infections. Which of the following best describes the mode of action of cefazolin?

   (A) It acts as an electron sink depriving the bacteria of reducing power.

   (B) It binds to the 30S ribosome and inhibits bacterial protein synthesis.

   (C) It inhibits transcription of bacterial mRNA.

   (D) It inhibits transpeptidases needed to synthesize peptidoglycan.

   (E) It inhibits folic acid synthesis needed to act as a methyl donor.

2. Which of the following drugs inhibits bacterial nucleic acid synthesis by blocking the production of tetrahydrofolic acid?

   (A) Ceftriaxone

   (B) Erythromycin

   (C) Metronidazole

   (D) Rifampin

   (E) Trimethoprim

3. Regarding both penicillins and aminoglycosides, which of the following is the most accurate?

   (A) Both act at the level of the cell wall.

   (B) Both are bactericidal drugs.

   (C) Both require an intact β-lactam ring for their activity.

   (D) Both should not be given to children under the age of 8 years because damage to cartilage can occur.

   (E) They should not be given together because they are antagonistic.

4. Listed below are drug combinations that are used to treat certain infections. Which of the following is a combination in which **both** drugs act to inhibit the **same** metabolic pathway?

   (A) Ceftriaxone and azithromycin

   (B) Isoniazid and rifampin

   (C) Penicillin G and gentamicin

   (D) Sulfonamide and trimethoprim

5. Regarding penicillins and cephalosporins, which of the following is the most accurate?

   (A) Cleavage of the β-lactam ring will inactivate penicillins but not cephalosporins.

   (B) Penicillins act by inhibiting transpeptidases but cephalosporins do not.

   (C) Penicillins and cephalosporins are both bactericidal drugs.

   (D) Penicillins and cephalosporins are active against gram-positive cocci but not against gram-negative rods.

   (E) Renal tubule damage is an important adverse effect caused by both penicillins and cephalosporins.

6. Regarding antimicrobial drugs that act by inhibiting nucleic acid synthesis in bacteria, which of the following is the most accurate?

   (A) Ciprofloxacin inhibits RNA polymerase by acting as a nucleic acid analogue.

   (B) Rifampin inhibits the synthesis of messenger RNA.

   (C) Sulfonamides inhibit DNA synthesis by chain termination of the elongating strand.

   (D) Trimethoprim inhibits DNA polymerase by preventing the unwinding of double-stranded DNA.

7. Regarding aminoglycosides and tetracyclines, which of the following is the most accurate?

   (A) Both classes of drugs are bactericidal.

   (B) Both classes of drugs inhibit protein synthesis by binding to the 30S ribosomal subunit.

   (C) Both classes of drugs inhibit peptidyl transferase, the enzyme that synthesizes the peptide bond.

   (D) Both classes of drugs must be acetylated within human cells to form the active antibacterial compound.

   (E) Both classes of drugs cause brown staining of teeth when administered to young children.

8. The next three questions ask about the adverse effects of antibiotics, which are an important consideration when deciding which antibiotic to prescribe. Which antibiotic causes significant neurotoxicity and must be taken in conjunction with pyridoxine (vitamin B_6) to prevent these neurologic complications?

   (A) Amoxicillin

   (B) Ceftriaxone

   (C) Isoniazid

   (D) Rifampin

   (E) Vancomycin

9. Of the following antibiotics, which causes the most phototoxicity (rash when exposed to sunlight)?

   (A) Nafcillin

   (B) Ciprofloxacin

   (C) Gentamicin

   (D) Metronidazole

   (E) Sulfamethoxazole

10. Which of the following antibiotics causes "red man" syndrome?

   (A) Azithromycin

   (B) Doxycycline

   (C) Gentamicin

   (D) Sulfamethoxazole

   (E) Vancomycin

ANSWERS

(1) **(D)**
(2) **(E)**
(3) **(B)**
(4) **(D)**
(5) **(C)**
(6) **(B)**
(7) **(B)**
(8) **(C)**
(9) **(E)**
(10) **(E)**

PRACTICE QUESTIONS: USMLE & COURSE EXAMINATIONS

Questions on the topics discussed in this chapter can be found in the Basic Bacteriology section of Part XIII: USMLE (National Board) Practice Questions starting on page 733. Also see Part XIV: USMLE (National Board) Practice Examination starting on page 775.

Antibacterial Drugs: Resistance

C H A P T E R

11

PRINCIPLES OF ANTIBIOTIC RESISTANCE

There are four major mechanisms that mediate bacterial resistance to drugs (Table 11–1). (1) Bacteria produce enzymes that **inactivate drugs**. (2) Bacteria **synthesize modified targets** against which the drug has a reduced effect. (3) Bacteria **reduce permeability** to the drug such that an effective intracellular concentration of the drug is not achieved. (4) Bacteria **actively export drugs** using a "**multidrug-resistance pump**." The multidrug-resistant (MDR) pump imports protons and, in an exchange-type reaction, exports a variety of foreign molecules including certain antibiotics, such as tetracyclines.

Most drug resistance is due to a genetic change in the organism, either a chromosomal **mutation** or the acquisition of a **plasmid** or **transposon**.

The term **high-level** resistance refers to resistance that cannot be overcome by increasing the dose of the antibiotic. A different antibiotic, usually from another class of drugs, is used. Resistance mediated by enzymes such as β-lactamases often results in high-level resistance, as all the drug is destroyed. **Low-level** resistance refers to resistance that can be overcome by increasing the dose of the antibiotic. Resistance mediated by mutations in the gene encoding a drug target is often low level, as the altered target can still bind some of the drug but with reduced strength.

TABLE 11–1 Mechanisms of Drug Resistance

Mechanism	Important Example	Drugs Commonly Affected
Inactivate drug	Cleavage by β-lactamase	β-Lactam drugs such as penicillins, cephalosporins, and carbapenems
Modify drug target in bacteria	1. Mutation in penicillin-binding proteins 2. Mutation in protein in 30S ribosomal subunit 3. Replace alanine with lactate in peptidoglycan 4. Mutation in DNA gyrase 5. Mutation in RNA polymerase 6. Mutation in catalase–peroxidase	Penicillins Aminoglycosides such as streptomycin Vancomycin Quinolones Rifampin Isoniazid
Reduce permeability of drug	Mutation in porin proteins	Penicillins, aminoglycosides, and others
Export of drug from bacteria	Multidrug-resistance pump	Tetracyclines, sulfonamides, quinolones

TABLE 11–2 Medically Important Bacteria That Exhibit Significant Drug Resistance

Type of Bacteria	Clinically Significant Drug Resistance
Gram-positive cocci	
Staphylococcus aureus	Penicillin G, methicillin/nafcillin
Streptococcus pneumoniae	Penicillin G
Enterococcus faecalis, *Enterococcus faecium*	Penicillin G, aminoglycosides, vancomycin
Gram-negative cocci	
Neisseria gonorrhoeae	Penicillin G, fluoroquinolones
Gram-positive rods	
None	
Gram-negative rods	
Haemophilus influenzae	Ampicillin
Pseudomonas aeruginosa	β-Lactams,[1] aminoglycosides
Enterobacteriaceae[2]	β-Lactams,[1] aminoglycosides, carbapenems
Acinetobacter baumannii[3]	β-Lactams,[1] aminoglycosides
Mycobacteria	
Mycobacterium tuberculosis[4]	Isoniazid, rifampin
Mycobacterium avium-intracellulare	Isoniazid, rifampin, and many others

[1]β-Lactams are penicillins, cephalosporins, and carbapenems.

[2]The family Enterobacteriaceae includes bacteria such as *Escherichia coli, Enterobacter cloacae, Klebsiella pneumoniae,* and *Serratia marcescens.* Many strains produce extended-spectrum β-lactamases (ESBLs) and some are multidrug resistant (MDR).

[3]Acinetobacter is resistant to many antibiotics, including carbapenems such as imipenem.

[4]Some strains of *M. tuberculosis* are resistant to more than two drugs.

Hospital-acquired infections are significantly more likely to be caused by antibiotic-resistant organisms than are community-acquired infections. This is especially true for hospital infections caused by *Staphylococcus aureus* and enteric gram-negative rods such as *Escherichia coli* and *Pseudomonas aeruginosa.* Antibiotic-resistant organisms are common in the hospital setting because widespread antibiotic use in hospitals selects for these organisms. Furthermore, hospital strains are often resistant to multiple antibiotics. This resistance is usually due to the acquisition of plasmids carrying several genes that encode the enzymes that mediate resistance.

Two types of resistant gram-negative rods are especially important in hospital-acquired infection multidrug-resistant (**MDR**) gram-negative rods and **extended spectrum beta-lactamase** (**ESBL**) producing gram-negative rods. MDR gram-negative rods produce beta-lactamases that inactivate penicillins, cephalosporins, and carbapenems. ESBL gram-negative rods produce beta-lactamases that inactivate second, third, and fourth generation cephalosporins. Table 11–2 describes certain medically important bacteria and the main drugs to which they are resistant.

GENETIC BASIS OF RESISTANCE

Chromosome-Mediated Resistance

Chromosomal resistance is due to a mutation in the gene that codes for either the target of the drug or the transport system in the membrane that controls the uptake of the drug.

The frequency of spontaneous chromosomal mutations usually ranges from 10^{-7} to 10^{-9}, which is much lower than the frequency of acquisition of resistance plasmids.

The treatment of certain infections with two or more drugs is based on the following principle. If the frequency that a bacterium mutates to become resistant to antibiotic A is 10^{-7} (1 in 10 million) and the frequency that the same bacterium mutates to become resistant to antibiotic B (operating under a different mechanism) is 10^{-8} (1 in 100 million), then the chance that the bacterium will become resistant to both antibiotics is the product of the two probabilities, or 10^{-15}. It is therefore highly unlikely that the bacterium will become resistant to *both* antibiotics. Stated another way, although an organism may become resistant to one antibiotic, it is likely that it will be effectively treated by the other antibiotic.

Plasmid-Mediated Resistance

Plasmid-mediated resistance is very important from a clinical point of view for three reasons:

(1) It occurs in many different species, especially gram-negative rods.

(2) Plasmids frequently mediate resistance to multiple drugs.

(3) Plasmids have a high rate of transfer from one cell to another, usually by conjugation.

Resistance plasmids (resistance factors, R factors) are extrachromosomal, circular, double-stranded DNA molecules that carry genes for a variety of enzymes that degrade antibiotics and modify membrane transport systems (Figure 11–1). Table 11–3 describes the most important mechanisms of resistance for several important drugs.

R factors can carry one or more antibiotic resistance genes. The medical implication of this is twofold: the first is that a

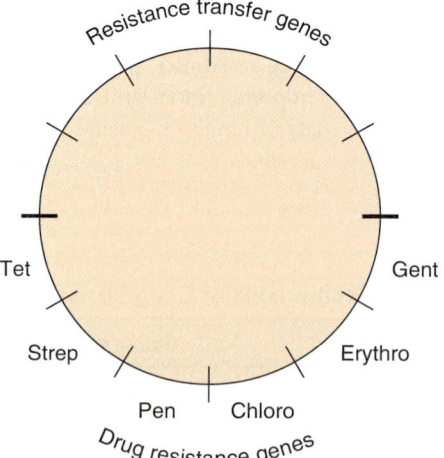

FIGURE 11–1 Resistance plasmid (R plasmid, R factor). Most resistance plasmids have two sets of genes: (1) resistance transfer genes that encode the sex pilus and other proteins that mediate transfer of the plasmid DNA during conjugation and (2) drug resistance genes that encode the proteins that mediate drug resistance. The bottom half of the figure depicts (from left to right) the genes that encode resistance to tetracycline, streptomycin, penicillin (β-lactamase), chloramphenicol, erythromycin, and gentamicin.

TABLE 11–3 R-Factor–Mediated Resistance Mechanisms

Drug	Mechanism of Resistance
Penicillins and cephalosporins	β-Lactamase cleavage of β-lactam ring
Aminoglycosides	Modification by acetylation, adenylation, or phosphorylation
Chloramphenicol	Modification by acetylation
Erythromycin	Change in receptor by methylation of rRNA
Tetracycline	Reduced uptake or increased export
Sulfonamides	Active export out of the cell and reduced affinity of enzyme

bacterium containing such a plasmid can be resistant to more than one class of antibiotics (e.g., penicillins and aminoglycosides), and second, the use of an antibiotic that selects for an organism resistant to one antibiotic will select for an organism that is resistant to any antibiotics whose resistance genes are carried by the plasmid. For example, if an organism has the R plasmid depicted in Figure 11–1, then the use of penicillin will select for an organism resistant not only to penicillin but also to tetracyclines, aminoglycosides (e.g., streptomycin and gentamicin), chloramphenicol, and erythromycin.

In addition to producing drug resistance, R factors have two important properties: (1) they can replicate independently of the bacterial chromosome, and as such a cell can contain many copies; and (2) they can be transferred not only to cells of the same species but also to other species and genera.

In addition to conveying antibiotic resistance, R factors can confer resistance to metal ions and resistance to certain bacterial viruses by coding for restriction endonucleases that degrade the DNA of the infecting bacteriophages.

Transposon-Mediated Resistance

Transposons are genes that are transferred either within or between larger pieces of DNA such as the bacterial chromosome and plasmids. A typical drug resistance transposon is composed of three genes flanked on both sides by shorter DNA sequences (see Figure 2–7). The three genes code for (1) transposase, the enzyme that catalyzes excision and reintegration of the transposon; (2) a repressor that regulates synthesis of the transposase; and (3) the drug resistance gene.

SPECIFIC MECHANISMS OF RESISTANCE

Penicillins & Cephalosporins—Cleavage by **β-lactamases** (penicillinases and cephalosporinases) is by far the most important mechanism of resistance (see Figure 10–3). β-Lactamases have different properties. For example, staphylococcal penicillinase is inducible by penicillin and is secreted outside of the bacterium. In contrast, some β-lactamases produced by several gram-negative rods are constitutively produced, are located in the periplasmic space near the peptidoglycan, and are not secreted outside of the bacterium.

The β-lactamases produced by various gram-negative rods have different specificities: some are more active against cephalosporins, others against penicillins. Clavulanic acid, tazobactam, sulbactam, and avibactam are penicillin analogues that bind strongly to β-lactamases and inactivate them. Combinations of these inhibitors and penicillins (e.g., clavulanic acid plus amoxicillin [Augmentin] and piperacillin plus tazobactam [Zosyn]) can overcome resistance mediated by many but not all β-lactamases.

ESBLs inactivate extended-spectrum cephalosporins (second- and third-generation cephalosporins), such as ceftriaxone, cefotaxime, and ceftazidime, as well as penicillins and first-generation cephalosporins. They are produced by several enteric bacteria, notably E. coli, Klebsiella, Enterobacter, and Proteus. ESBLs endow the bacteria with resistance to all penicillins, cephalosporins, and monobactams, such as aztreonam. Carbapenems, such as imipenem, are the drug of choice to treat infections caused by ESBL-producing bacteria. However, some ESBL-producing bacteria have acquired resistance to carbapenems and can be treated only with colistin, a polypeptide antibiotic that does not have a β-lactam ring.

In 2009, a new strain of highly resistant Klebsiella was isolated in India carrying a plasmid that encoded **New Delhi metallo-β-lactamase** (NDM-1). This plasmid confers high-level resistance to many antibiotics including carbapenems and has spread from Klebsiella to other members of the Enterobacteriaceae. Resistant Enterobacteriaceae carrying NDM-1 have emerged in many countries, including the United States.

Resistance to penicillins is also caused by changes in the **penicillin-binding proteins (PBPs)** in the bacterial cell membrane. These changes account for both the low- and high-level resistance exhibited by Streptococcus pneumoniae to penicillin G and for the resistance of S. aureus to nafcillin and other β-lactamase–resistant penicillins. The resistance of methicillin-resistant S. aureus (MRSA) to almost all β-lactams is attributed to the presence of PBP2a, which is found particularly in MRSA. Note that PBP2a can perform its transpeptidase function but does not bind penicillins. The relative resistance of Enterococcus faecalis to penicillins may be due to altered PBPs.

Resistance to penicillin is also caused by **poor permeability** of the drug, as in the case of low-level resistance of Neisseria gonorrhoeae. Note that high-level resistance to penicillin is due to the presence of a plasmid coding for a penicillinase.

Some isolates of S. aureus demonstrate yet another form of resistance, called **tolerance**, in which growth of the organism is inhibited by penicillin but the organism is not killed. This is attributed to a failure of activation of the autolytic enzymes, murein hydrolases, which degrade the peptidoglycan.

Carbapenems—Resistance to carbapenems, such as imipenem, is caused by carbapenemases that degrade the β-lactam ring. This enzyme endows the organism with resistance to penicillins and cephalosporins as well. Carbapenemases are produced by many enteric gram-negative rods, especially Klebsiella, Escherichia, and Pseudomonas. In 2018, a carbapenem-resistant

Klebsiella pneumoniae was isolated from a patient with a urinary tract infection that contained the genes for three different carbapenemases.

Carbapenem-resistant strains of *K. pneumoniae* are an important cause of hospital-acquired infections and are resistant to almost all known antibiotics. Note that the FDA has approved a combination of meropenem–vaborbactam for the treatment of complicated urinary tract infections caused by *E. coli*, *K. pneumoniae*, and others. Vaborbactam is an inhibitor of carbapenemases and other beta-lactamases.

Vancomycin—Resistance to vancomycin is caused by a change in the peptide component of peptidoglycan from D-alanyl-D-alanine to D-alanine-D-lactate, to which vancomycin does not bind. Of the four gene loci mediating vancomycin resistance, VanA is the most important. It is carried by a transposon on a plasmid and provides high-level resistance to both vancomycin. The VanA locus encodes those enzymes that synthesize D-alanine–D-lactate as well as several regulatory proteins.

Aminoglycosides—Resistance to aminoglycosides occurs by three mechanisms: (1) modification of the drugs by plasmid-encoded phosphorylating, adenylylating, and acetylating enzymes (the most important mechanism); (2) chromosomal mutation in the gene that codes for the target protein in the 30S subunit; and (3) decreased permeability of the bacterium to the drug.

Tetracyclines—Resistance to tetracyclines is primarily the result of failure of the intracellular inhibitory concentration of the drug to be reached inside the bacteria. This is due to plasmid-encoded processes that either reduce the uptake of the drug or **enhance its transport** out of the cell.

Chloramphenicol—Chloramphenicol resistance occurs when the plasmid-encoded acetyltransferase modifies and inactivates the drug.

Erythromycin—Resistance is primarily due to a plasmid-encoded methylase that modifies the 23S rRNA, thereby blocking binding of the drug. An efflux pump that reduces the concentration of erythromycin within the bacterium causes low-level resistance to the drug. Esterase-producing enteric gram-negative rods cleave the macrolide ring and inactivates the drug.

Sulfonamides—Resistance to sulfonamides is mediated primarily by two mechanisms: (1) a plasmid-encoded transport system that *actively exports* the drug out of the cell, and (2) a chromosomal mutation in the gene coding for the target enzyme dihydropteroate synthetase, which reduces the binding affinity of the drug.

Trimethoprim—Resistance to trimethoprim is due primarily to mutations in the chromosomal gene that encodes dihydrofolate reductase, the enzyme that reduces dihydrofolate to tetrahydrofolate.

Quinolones—Resistance to quinolones is due primarily to chromosomal mutations that modify the bacterial DNA gyrase.

Rifampin—Resistance to rifampin is due to a chromosomal mutation in the gene encoding the bacterial RNA polymerase, resulting in ineffective binding of the drug. Because resistance occurs at high frequency (10^{-5}), rifampin is not prescribed alone for the *treatment* of infections. It is used alone for the *prevention*

of certain infections because it is administered for only a short time (see Table 10–8).

Isoniazid—Resistance of *Mycobacterium tuberculosis* to isoniazid is due to mutations in the organism's catalase–peroxidase gene. Catalase or peroxidase enzyme activity is required to synthesize the metabolite of isoniazid that actually inhibits the growth of *M. tuberculosis*.

Ethambutol—Resistance of *M. tuberculosis* to ethambutol is due to mutations in the gene that encodes arabinosyl transferase, the enzyme that synthesizes the arabinogalactan in the organism's cell wall.

Pyrazinamide—Resistance of *M. tuberculosis* to pyrazinamide (PZA) is due to mutations in a gene encoding bacterial amidase that converts PZA to its active drug, pyrazinoic acid.

NONGENETIC BASIS OF RESISTANCE

There are several nongenetic reasons for the failure of drugs to inhibit the growth of bacteria:

(1) Bacterial abscesses can be challenging for cells (neutrophils) and molecules (antibiotics) to penetrate effectively. Surgical drainage is often necessary adjunct to chemotherapy.

(2) Bacteria can be in a resting state (i.e., not growing); they are therefore insensitive to cell wall inhibitors such as penicillins and cephalosporins. Similarly, *M. tuberculosis* can remain dormant in tissues for years, during which time it is less sensitive to drugs. If host defenses are lowered and the bacteria begin to multiply, they become more susceptible to the drugs, indicating that a genetic change did not occur.

(3) The presence of foreign bodies, such as surgical implants and catheters or other penetrating material (splinters and shrapnel), makes successful antibiotic treatment more difficult. This is in part due to the fact that bacteria can form biofilms on the foreign bodies (see Chapter 7). Biofilm-associated bacteria can be highly tolerant of antibiotic concentrations that would kill the same bacteria existing outside of a biofilm.

(4) If the patient fails to take the drug as prescribed (noncompliance, nonadherence) or if the drug is not effective (wrong prescription or the isolate is resistant) also account for drug failures.

SELECTION OF RESISTANT BACTERIA BY OVERUSE & MISUSE OF ANTIBIOTICS

There are several reasons why overuse or misuse of antibiotics enhance the selection of resistant mutants:

(1) Prescribing multiple antibiotics when one would be sufficient, prescribing unnecessarily long courses of therapy, using antibiotics in self-limited infections, and unnecessary prescribing of antibiotics for prophylaxis before and/or after surgery.

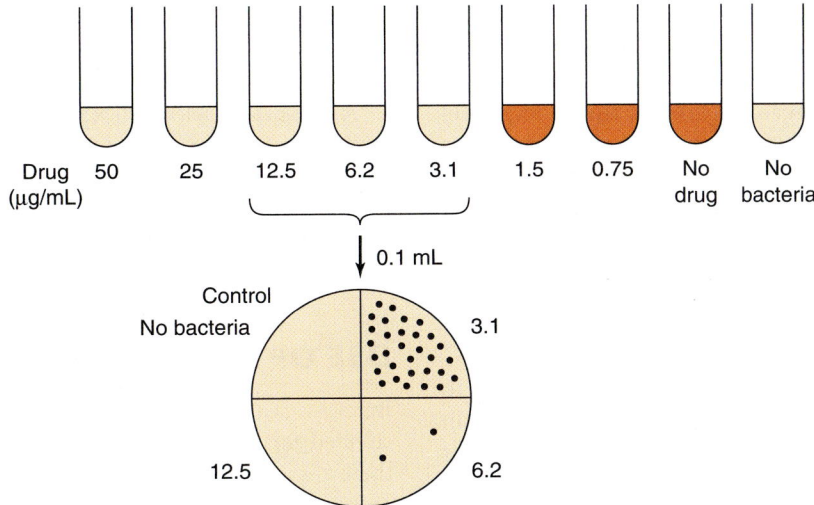

FIGURE 11–2 Determination of minimal inhibitory concentration (MIC) and minimal bactericidal concentration (MBC). **Top:** The patient's organism is added to tubes containing decreasing amounts of the antibiotic. After incubation overnight, growth of the bacteria is observed visually. The lowest concentration of drug that inhibits growth (i.e., 3.1 μg/mL) is the MIC. However, at this point, it is not known whether the bacteria have been killed or whether the drug has only inhibited their growth. **Bottom:** To determine whether that concentration of drug is bactericidal (i.e., to determine its MBC), an aliquot (0.1 mL) from the tubes is plated on an agar plate that does not contain any drug. The concentration of drug that inhibits at least 99.9% of the bacterial colonies (i.e., 6.2 μg/mL) is the MBC.

(2) In many countries, antibiotics are sold over the counter to the general public, which encourages inappropriate and indiscriminate use of the drugs.

(3) Antibiotics are used in animal feed to prevent infections and promote growth. This selects for resistant organisms in the animals and may contribute to the pool of resistant organisms in humans.

ANTIBIOTIC SENSITIVITY TESTING

Antibiogram

An antibiogram is the term used to describe the results of antibiotic susceptibility tests performed on the bacteria isolated from the patient. These results are the most important factor in determining the choice of antibiotic with which to treat the patient. Other factors such as the patient's renal function and hypersensitivity profile must also be considered in choosing the antibiotic.

There are two types of tests used to determine the antibiogram: (1) the tube dilution test that determines the minimal inhibitory concentration (MIC) and (2) the disk diffusion (Kirby-Bauer) test that determines the diameter of the zone of inhibition (see following discussion and Figures 11–2 and 11–3).

Minimal Inhibitory Concentration

For many infections, the results of sensitivity testing are important in the choice of antibiotic. These results are commonly reported as **MIC**, which is defined as the lowest concentration of drug that inhibits the growth of the organism. The MIC is determined by inoculating the purified patient isolate into a series of tubes containing twofold dilutions of the drug (see Figure 11–2). After incubation at 35°C for 18 hours, the lowest

concentration of drug that prevents visible growth of the organism is the MIC. This provides the physician with a precise concentration of drug to guide the choice of both the drug and the dose.

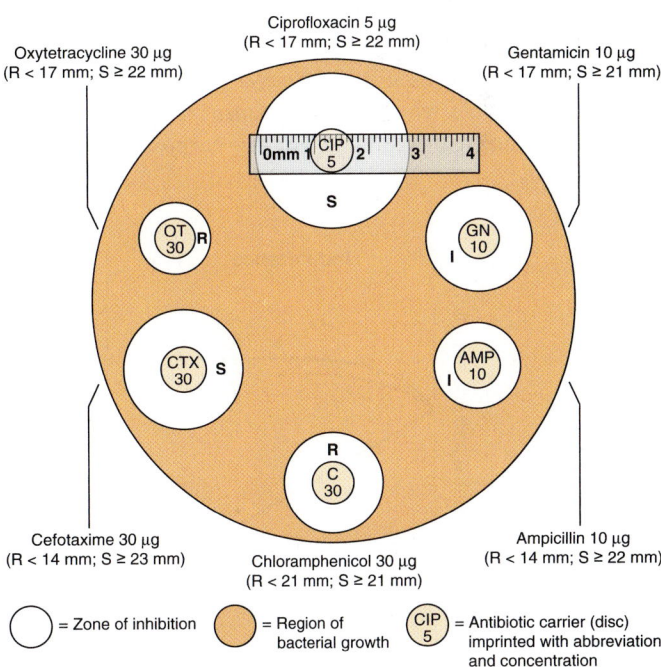

FIGURE 11–3 Antibiotic sensitivity testing. A zone of inhibition surrounds several antibiotic-containing disks. A zone of certain diameter or greater indicates that the organism is sensitive. Some resistant organisms will grow all the way up to the disk. R, Resistant; I, Intermediate; S, Sensitive. (Reproduced with permission from Cowan MK, Talero KP. *Microbiology: A Systems Approach*, 2nd ed. New York, NY: McGraw Hill; 2009.)

A second method of determining antibiotic sensitivity is the disk diffusion method, in which disks impregnated with various antibiotics are placed on the surface of an agar plate that has been inoculated with the organism isolated from the patient (see Figure 11–3). After incubation at 35°C for 18 hours, during which time the antibiotic diffuses outward from the disk, the diameter of the zone of inhibition is determined. The size of the zone of inhibition is compared with standards to determine the sensitivity of the organism to the drug.

Minimal Bactericidal Concentration

For certain infections, such as endocarditis, it is important to know the concentration of drug that actually kills the organism rather than the concentration that merely inhibits growth. This concentration, called the **minimal bactericidal concentration (MBC)**, is determined by taking a small sample (0.01 or 0.1 mL) from the tubes used for the MIC assay and spreading it over the surface of a drug-free blood agar plate (see Figure 11–2). Any organisms that were inhibited but not killed now have a chance to grow because the drug has been diluted significantly. After incubation at 35°C for 48 hours, the lowest concentration that has reduced the number of colonies by 99.9%, compared with the drug-free control, is the MBC. Bactericidal drugs usually have an MBC equal or very similar to the MIC, whereas bacteriostatic drugs usually have an MBC significantly higher than the MIC.

β-Lactamase Production

For severe infections caused by certain organisms, such as *S. aureus* and *Haemophilus influenzae*, it is important to know as soon as possible whether the organism isolated from the patient is producing β-lactamase. For this purpose, rapid assays

testing for the enzyme can yield an answer in a few minutes, as opposed to an MIC test or a disk diffusion test, both of which take 18 hours.

A commonly used procedure is the chromogenic β-lactam method, in which a colored β-lactam drug is added to a suspension of the organisms. If β-lactamase is made, hydrolysis of the β-lactam ring causes the drug to turn a different color in 2 to 10 minutes. Disks impregnated with a chromogenic β-lactam can also be used.

USE OF ANTIBIOTIC COMBINATIONS

In most cases, the single best antimicrobial agent should be selected for use because this minimizes side effects. However, there are several instances in which two or more drugs are commonly given:

(1) To treat serious infections before the identity of the organism is known.

(2) To achieve a synergistic inhibitory effect against certain organisms.

(3) To prevent the emergence of resistant organisms. (If bacteria become resistant to one drug, the second drug will kill them, thereby preventing the emergence of resistant strains.)

Two drugs can interact in one of several ways (Figure 11–4). They are usually indifferent to each other (i.e., additive only). Sometimes there is a **synergistic** interaction, in which the effect of the two drugs together is significantly greater than the sum of the effects of the two drugs acting separately.

A synergistic effect can result from a variety of mechanisms. For example, the combination of a penicillin and an aminoglycoside such as gentamicin has a synergistic action against enterococci (*E. faecalis*) because penicillin damages the cell

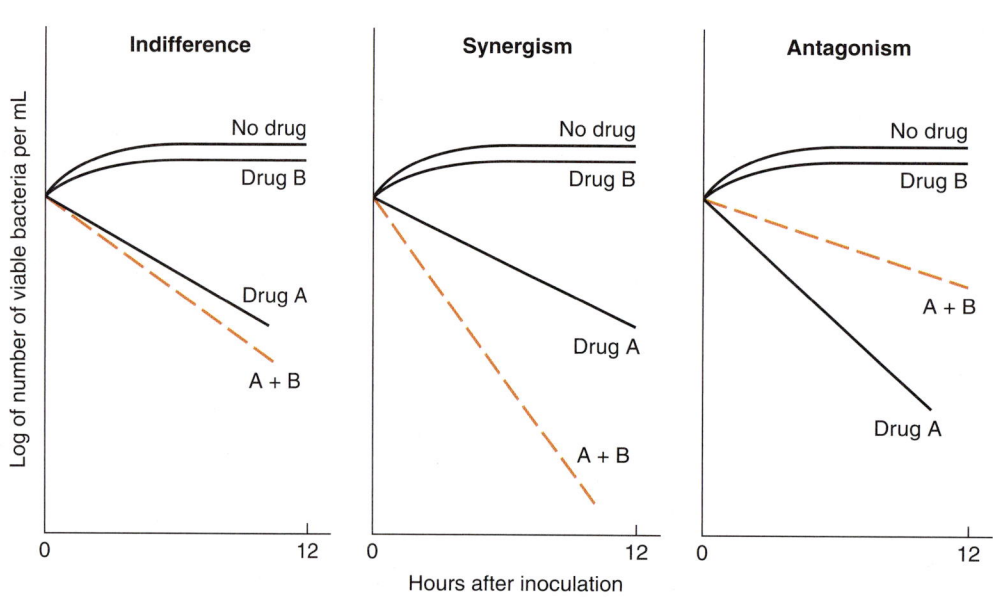

FIGURE 11–4 Drug interaction. The solid lines represent the response of bacteria to drug A alone, drug B alone, or no drug. The dotted lines represent the response to drug A and drug B together.

wall sufficiently to enhance the entry of aminoglycoside, which inhibits protein synthesis. When given alone, neither drug is effective. A second example is the combination of a sulfonamide with trimethoprim. In this instance, the two drugs act on the same metabolic pathway, such that if one drug does not inhibit folic acid synthesis sufficiently, the second drug provides effective inhibition by blocking a subsequent step in the pathway.

Although antagonism between two antibiotics is rare, one example is clinically important. This involves the use of penicillin G combined with the bacteriostatic drug tetracycline in the treatment of meningitis caused by *S. pneumoniae*. Antagonism occurs because the tetracycline inhibits the growth of the organism, thereby preventing the bactericidal effect of penicillin G, which kills only growing organisms.

PEARLS

- The four main mechanisms of antibiotic resistance are (1) **enzymatic degradation** of the drug, (2) **modification of the drug's target**, (3) **reduced permeability** of the drug, and (4) **active export** of the drug.
- Most drug resistance is the result of a genetic change in the organism, caused either by a chromosomal mutation or by the acquisition of a plasmid or transposon.

Genetic Basis of Resistance

- **Chromosomal mutations** typically either **change the target of the drug** so that the drug does not bind or **change the membrane** so that the drug does not penetrate well into the cell. Chromosomal mutations occur at a low frequency (perhaps 1 in 10 million organisms) and often affect only one drug or one family of drugs.
- **Plasmids cause drug resistance by encoding enzymes** that degrade or modify drugs. Plasmid-mediated resistance occurs at a **higher frequency** than chromosomal mutations, often affecting **multiple drugs** or families of drugs.
- **Resistance plasmids** (R plasmids, R factors) usually carry two sets of genes. One set encodes the enzymes that degrade or modify drugs, and the other encodes the proteins that **mediate conjugation**, the main process by which resistance genes are transferred from one bacterium to another.
- **Transposons** are small pieces of DNA that **move from one site on the bacterial chromosome to another** or from the bacterial chromosome to plasmid DNA. **Transposons often carry drug resistance genes**. Many R plasmids carry one or more transposons.

Specific Mechanisms of Resistance

- **Resistance to penicillins and cephalosporins** is mediated by three main mechanisms: (1) degradation by β-lactamases, (2) mutations in the genes for penicillin-binding proteins, and (3) reduced permeability. **Degradation by β-lactamases is the most important**.
- Resistance to vancomycin is caused by a change in the D-alanyl-D-alanine part of the peptide in peptidoglycan to D-alanine-D-lactate, resulting in an inability of vancomycin to bind.

- Resistance to aminoglycosides is mediated by three main mechanisms: (1) modification of the drug by **phosphorylating, adenylylating, and acetylating enzymes**; (2) mutations in the genes encoding one of the 30S ribosomal proteins; and (3) reduced permeability.
- Resistance to **tetracyclines** is often caused by either reduced permeability or **active export** of the drug from the bacterium.
- Resistance to erythromycins is primarily caused by a plasmid-encoded enzyme that **methylates the 23S ribosomal RNA**, thereby blocking binding of the drug.
- Resistance to sulfonamides is due primarily to plasmid-encoded enzymes that actively export the drug from the bacterium.
- Resistance to quinolones is primarily caused by **mutations** in the gene encoding the bacterial DNA gyrase.
- Resistance to rifampin is primarily caused by **mutations** in the gene encoding the bacterial RNA polymerase.
- Resistance to isoniazid is due primarily to the **loss of the bacterial peroxidase (catalase)** that activates isoniazid to the metabolite that inhibits mycolic acid synthesis.

Nongenetic Basis of Resistance

- Nongenetic reasons why bacteria may not be inhibited by antibiotics are that drugs may not reach bacteria located in the center of an abscess and that certain drugs, such as penicillins, will not affect bacteria that are not growing. Also, the presence of foreign bodies makes successful antibiotic treatment more difficult.

Antibiotic Sensitivity Testing

- The **minimal inhibitory concentration (MIC)** is the lowest concentration of drug that **inhibits the growth** of the bacteria isolated from the patient. In this test, it is not known whether the inhibited bacteria have been killed or just have stopped growing.
- The **minimal bactericidal concentration (MBC)** is the lowest concentration of drug that **kills** the bacteria isolated from the patient. In certain diseases, such as endocarditis, it is often necessary to use a concentration of drug that is bactericidal.

Use of Antibiotic Combinations

- Two or more antibiotics are used under certain circumstances, such as to treat life-threatening infections before the cause has been identified, to prevent the emergence of resistant bacteria during prolonged treatment regimens, and to achieve a synergistic (augmented) effect.

- A **synergistic effect** is one in which the effect of two drugs given together is much greater than the sum of the effect of the two drugs given individually. The best example of synergy is the marked killing effect of the combination of a penicillin and an aminoglycoside on enterococci compared with the minor effect of either drug given alone.

SELF-ASSESSMENT QUESTIONS

1. The spread of antibiotic resistance from one bacterium to another is a well-recognized and clinically important phenomenon. Which one of the following mechanisms is most likely to be involved with the spread of resistance?

   (A) Acetylation
   (B) Conjugation
   (C) Programmed rearrangement
   (D) Protoplast mobility
   (E) Translation

2. Regarding the specific mechanisms by which bacteria become resistant to antimicrobial drugs, which one of the following is the most accurate?

   (A) Some bacteria contain an enzyme that cleaves the ring of aminoglycosides.
   (B) Some bacteria contain clavulanic acid, which binds to penicillin G and inactivates it.
   (C) Some bacteria contain a mutated gene encoding an altered transpeptidase, which makes them resistant to doxycycline.
   (D) Some bacteria contain a mutated gene that encodes an altered RNA polymerase, which makes them resistant to rifampin.
   (E) Some bacteria contain an altered ribosomal protein, which makes them resistant to isoniazid.

3. The susceptibility of bacteria to an antibiotic is often determined by using the MIC assay. Regarding the MIC assay, which one of the following is the most accurate?

   (A) MIC is the lowest concentration of the bacteria isolated from the patient that inhibits the activity of a standard dose of antibiotic.
   (B) MIC is the lowest concentration of antibiotic that inhibits the growth of the bacteria isolated from the patient.
   (C) MIC is the lowest concentration of antibiotic that kills the bacteria isolated from the patient.
   (D) MIC is the lowest concentration of antibiotic in the patient's serum that inhibits the activity of a standard dose of antibiotic.

4. The MIC of the patient's organism to penicillin is 1 μg/mL and the MIC to gentamicin is 8 μg/mL. However, the MIC to a combination of penicillin and gentamicin is 0.01 μg/mL. Which one of the following terms is the most accurate to describe this effect?

   (A) Activation
   (B) Antagonism
   (C) Reassortment
   (D) Recombination
   (E) Synergism

5. Regarding the mechanisms of resistance to specific drugs, which one of the following is the most accurate?

   (A) Certain strains of *Enterococcus faecalis* produce D-lactate rather than D-alanine, which causes them to be resistant to vancomycin.
   (B) Certain strains of *Escherichia coli* produce ergosterol, which causes them to be resistant to gentamicin.
   (C) Certain strains of *Neisseria gonorrhoeae* produce a mutant peptidyl transferase, which causes them to be resistant to tetracycline.
   (D) Certain strains of *Streptococcus pyogenes* produce a β-lactamase, which causes them to be resistant to erythromycin.

ANSWERS

(1) **(B)**
(2) **(D)**
(3) **(B)**
(4) **(E)**
(5) **(A)**

PRACTICE QUESTIONS: USMLE & COURSE EXAMINATIONS

Questions on the topics discussed in this chapter can be found in the Basic Bacteriology section of Part XIII: USMLE (National Board) Practice Questions starting on page 733. Also see Part XIV: USMLE (National Board) Practice Examination starting on page 775.

Bacterial Vaccines

PRINCIPLES OF BACTERIAL VACCINES

Several bacterial diseases can be prevented by using immunizations that induce either active or passive immunity. **Active** immunity is induced by vaccines prepared from bacteria or their products. This chapter presents a summary of the types of vaccines (Table 12–1); detailed information regarding each vaccine is located in the chapters on the specific organisms. **Passive** immunity is provided by the administration of preformed antibody in preparations called immune globulins. The immune globulins useful against bacterial diseases are described later.

Passive–active immunity involves giving both immune globulins to provide immediate protection and a vaccine to provide long-term protection. This approach is described later in the section on tetanus antitoxin.

ACTIVE IMMUNITY

Bacterial vaccines are composed of capsular polysaccharides, inactivated protein exotoxins (toxoids), killed bacteria, or live, attenuated bacteria. The available bacterial vaccines and their indications are described next. Table 12–2 lists the bacterial

TABLE 12–1 Current Bacterial Vaccines

Usage	Bacterium	Disease	Antigen
Common usage	Corynebacterium diphtheriae	Diphtheria	Toxoid
	Clostridium tetani	Tetanus	Toxoid
	Bordetella pertussis	Pertussis (whooping cough)	Acellular (purified proteins) or killed organisms
	Haemophilus influenzae	Meningitis	Capsular polysaccharide conjugated to carrier protein
	Streptococcus pneumoniae	Pneumonia	Capsular polysaccharide or capsular polysaccharide conjugated to carrier protein
	Neisseria meningitidis	Meningitis	Capsular polysaccharide or capsular polysaccharide conjugated to a carrier protein; factor H binding protein for group B meningococci
Special situations	Salmonella typhi	Typhoid fever	Live organisms or capsular polysaccharide
	Vibrio cholerae	Cholera	Live organisms in the United States Killed organisms in countries where cholera is endemic
	Yersinia pestis	Plague	Killed organisms
	Bacillus anthracis	Anthrax	Partially purified proteins
	Mycobacterium bovis (BCG)	Tuberculosis	Live organisms
	Francisella tularensis	Tularemia	Live organisms
	Rickettsia prowazekii	Typhus	Killed organisms
	Coxiella burnetii	Q fever	Killed organisms

TABLE 12–2 Vaccines Recommended for Children Age 0–6 Years[1]

Bacterial Vaccines	Viral Vaccines
Diphtheria toxoid, tetanus toxoid, acellular pertussis (DTaP)	Hepatitis A
Haemophilus influenzae type b (Hib)	Hepatitis B
Meningococcal (only for children with high-risk conditions)[2]	Influenza
Pneumococcal	Measles, mumps, rubella (MMR) Poliovirus, inactivated Rotavirus Varicella

[1]Vaccines are listed in alphabetical order. A complete description of the vaccine schedule is available on the Centers for Disease Control and Prevention website, www.cdc.gov.

[2]High-risk conditions such as asplenia and HIV infection. Meningococcal vaccine for children without high-risk conditions should be given at age 11 years.

(and viral) vaccines recommended for children from 0 to 6 years of age as of 2021. Table 12–3 lists some important bacterial (and viral) vaccines recommended for travelers. Location and duration of travel are important factors when determining whether a vaccine should be recommended. Additional information regarding vaccines for travelers can be found at the website for the Centers for Disease Control and Prevention: www.cdc.gov/travel.

Capsular Polysaccharide Vaccines

(1) Both versions of the vaccine against *Streptococcus pneumoniae* contain the capsular polysaccharide of the bacteria as the immunogen. One version contains the capsular polysaccharide of the 23 most prevalent serotypes. It is recommended for persons older than 60 years of age and adult patients of any age with such chronic diseases as diabetes and cirrhosis or with compromised spleen function or splenectomy. A second version containing the capsular polysaccharide of 13 pneumococcal serotypes coupled to a carrier protein (diphtheria toxoid) is available for the protection of young children who do not respond well to the unconjugated vaccine. The function of the carrier protein is explained in Figure 57–3.

A potential problem regarding the use of the pneumococcal vaccine (or a vaccine against any organism with multiple serotypes) is that of **serotype replacement**. Will the vaccine reduce the incidence of disease caused by the serotypes in the vaccine

TABLE 12–3 Some Important Vaccines Recommended for Travelers

Bacterial Vaccines	Viral Vaccines
Meningococcal Cholera Typhoid	Hepatitis A Hepatitis B Polio Yellow fever Rabies Japanese encephalitis

but not the overall incidence of disease because other serotypes that are not in the vaccine will now cause disease? In fact, this occurred. An increase in invasive pneumococcal disease caused by serotype 19A, a serotype not in the previous vaccine, was observed. In view of this, serotype 19A is included in the current 13 serotype vaccine.

(2) There are more than a dozen serogroups of *Neisseria meningitidis* that are classified according to the capsular polysaccharide. Serogroups A, B, C, W, and Y are the primary causes of meningococcal disease throughout the world. In the United States, two quadrivalent meningococcal polysaccharide vaccines conjugated to either diphtheria toxoid or tetanus toxoid are available that protect against serogroups A, C, W, and Y. Immunization with conjugate vaccines elicits a much more robust antibody response, particularly in children, than vaccine formulations containing only polysaccharide as the main antigen.

Until recently, efforts to develop a vaccine against serogroup B strains were unsuccessful because the capsule of that serogroup is composed of a polysaccharide that is also found on the surface of many host tissues. This makes it a poorly immunogenic antigen, and any antibodies raised against the polysaccharide could potentially elicit an autoimmune response by cross-reacting with host cell surfaces. Nonpolysaccharide vaccines against serogroup B were licensed beginning in 2014 (see below).

(3) *Haemophilus influenzae* vaccine contains the type b polysaccharide conjugated to diphtheria toxoid or other carrier protein. It is given to children between the ages of 2 and 15 months to prevent meningitis. The capsular polysaccharide alone is a poor immunogen in young children, but coupling it to a carrier protein greatly enhances its immunogenicity. A combined vaccine consisting of this vaccine plus the diphtheria, tetanus, and pertussis (DTP, DTaP) vaccine is available.

(4) One of the vaccines against typhoid fever contains the capsular polysaccharide of *Salmonella typhi*. It is indicated for persons living or traveling in areas where there is a high risk of typhoid fever and for persons in close contact with either infected patients or chronic carriers.

Toxoid Vaccines

(1) *Corynebacterium diphtheriae* vaccine contains the toxoid (formaldehyde-treated exotoxin). Immunization against diphtheria is indicated for every child and is given in three doses at 2, 4, and 6 months of age, with boosters given 1 year later and at intervals thereafter.

(2) *Clostridium tetani* vaccine contains tetanus toxoid (formaldehyde-treated exotoxin) and is given to everyone both early in life and later as boosters for protection against tetanus.

(3) *Bordetella pertussis* vaccine contains pertussis toxoid but includes other proteins as well. Therefore, it is described in the next section.

Purified Protein Vaccines

(1) There are two types of *B. pertussis* vaccines: an acellular vaccine containing purified proteins and a vaccine containing

whole killed bacteria. The acellular vaccine is now recommended in the United States. The principal antigen in the acellular vaccine is inactivated pertussis toxin (pertussis toxoid), but other proteins, such as filamentous hemagglutinin and pertactin, are also required for full protection. Pertussis toxin for the vaccine is inactivated *genetically* by introducing two amino acid changes that eliminate its toxic (ADP-ribosylating) activity but retain its antigenicity. It is the first vaccine to contain a genetically inactivated toxoid. The vaccine is indicated for every child as a protection against whooping cough. It is usually given in combination with diphtheria and tetanus toxoids (DTP or DTaP vaccine).

(2) *Bacillus anthracis* vaccine contains "protective antigen" purified from the organism. It is given to persons whose occupations place them at risk of exposure to the organism.

(3) As mentioned above, challenges developing a polysaccharide-based vaccine for *N. meningitidis* serogroup B were overcome with the introduction of vaccines containing one or more surface-exposed proteins expressed by *N. meningitidis* serotype B. Factor H binding protein is main antigen in these vaccines against serogroup B meningococci. These vaccines are discussed in depth in Chapter 16.

Live, Attenuated Bacterial Vaccines

(1) The vaccine against tuberculosis contains a live, attenuated strain of *Mycobacterium bovis* called bacillus Calmette-Guérin (BCG) and, in some countries (but not the United States), is recommended for children at high risk for exposure to active tuberculosis.

(2) One of the vaccines against typhoid fever contains live, attenuated *S. typhi*. It is indicated for persons living or traveling in areas where there is a high risk of typhoid fever and for persons in close contact with either infected patients or chronic carriers.

(3) The vaccine against tularemia contains live, attenuated *Francisella tularensis* organisms and is used primarily in people who are exposed in their occupation, such as laboratory personnel, veterinarians, and hunters.

(4) An oral, live attenuated cholera vaccine (Vaxchora) is used in the United States for travelers to areas where cholera caused by serogroup O1 is endemic.

Killed Bacterial Vaccines

(1) Another *Vibrio cholerae* vaccine contains killed organisms. It is not available in the United States but is used in many other countries where cholera is endemic.

(2) *Yersinia pestis* vaccine contains killed organisms and is indicated for persons at high risk for contracting plague.

(3) The vaccine against typhus contains killed *Rickettsia rickettsiae* organisms and is used primarily to immunize members of the armed forces.

(4) The vaccine against Q fever contains killed *Coxiella burnetii* organisms and is used to immunize those who are at high risk for being exposed to animals infected with the organism.

PASSIVE IMMUNITY

Antitoxins (immune globulins) can be used for either the treatment or prevention of certain bacterial diseases. The following preparations are available:

(1) **Tetanus** antitoxin is used in the treatment of tetanus and in its prevention (prophylaxis). Antitoxin neutralizes unbound toxin to prevent disease. Administering antitoxin is effective as a treatment if it is given promptly after exposure to limit toxin activity. Antitoxin can also be delivered prophylactically as a preventative therapy to inadequately immunized persons with contaminated ("dirty") wounds. The antitoxin is made in humans to avoid hypersensitivity reactions. In addition to the antitoxin, these people should receive tetanus toxoid. This is an example of **passive–active** immunity. The toxoid and the antitoxin should be given at different sites in the body to prevent the antitoxin from neutralizing the toxoid.

(2) **Botulinum** antitoxin is used in the treatment of botulism. Because the antitoxin can neutralize unbound toxin to prevent the disease from progressing, it should be given promptly. It contains antibodies against botulinum toxins A, B, and E, the most commonly occurring types. The antitoxin is made in horses, so hypersensitivity may be a problem.

(3) **Diphtheria** antitoxin is used in the treatment of diphtheria. The antitoxin can neutralize unbound toxin to prevent the disease from progressing; therefore, the antitoxin should be given promptly. The antitoxin is made in horses, so hypersensitivity may be a problem.

(4) Bezlotoxumab, a monoclonal antibody against exotoxin B of *Clostridium difficile*, is effective in preventing relapses of pseudomembranous colitis.

WANING IMMUNITY TO VACCINES

The duration of protection provided by vaccines can decline (wane) with time. It also varies greatly from one vaccine to another. Evidence that waning immunity is a medical problem is based on the observation that some cases of vaccine-preventable diseases occur in those who have been immunized. For example, in the 2019 outbreak of measles in the United States, about 3% of adults who contracted measles received the MMR (measles, mumps, and rubella) vaccine.

Some vaccines provide adequate protection for months and others for many years. For example, the protection induced by the acellular pertussis vaccine wanes rapidly during the first few years, but the protection induced by the tetanus and diphtheria vaccine lasts for at least a decade.

One suggested explanation for waning immunity is that the ability of memory cells to survive and to retain function declines with time. The response of memory cells may depend upon the precise nature of the antigen used in the vaccine. Determining the criteria required to induce long-lived immunity will be important in developing more effective vaccines.

PEARLS

- Immunity to certain bacterial diseases can be induced either by immunization with bacterial antigens (**active immunity**) or by administration of preformed antibodies (**passive immunity**).

Active Immunity

- Active immunity can be achieved by vaccines consisting of (1) **bacterial capsular polysaccharides, toxoids, whole bacteria** (either killed or live, attenuated), or (2) **purified proteins** isolated from bacteria.

- **Vaccines containing capsular polysaccharide** as the immunogen are directed against *Streptococcus pneumoniae, Haemophilus influenzae, Neisseria meningitidis,* and *Salmonella typhi*. The capsular polysaccharide in the pneumococcal vaccine, the meningococcal vaccine, and the *H. influenzae* vaccine is conjugated to a carrier protein to enhance the antibody response.

- Two vaccines contain **toxoids** as the immunogen, the vaccines against **diphtheria** and **tetanus**. A **toxoid is an inactivated toxin** that has lost its ability to cause disease but has retained its immunogenicity. (The pertussis vaccine also contains toxoid but contains other bacterial proteins as well.)

- Three vaccines contain purified bacterial proteins as the immunogen. The most commonly used is the **acellular pertussis vaccine**, which in combination with diphtheria and tetanus toxoids is recommended for all children. The group B meningococcal vaccine contains factor H binding protein as the main immunogen. The **vaccine against anthrax** also contains purified proteins but is recommended only for individuals who are likely to be exposed to the organism.

- The **BCG vaccine** against tuberculosis **contains live, attenuated** *Mycobacterium bovis* and is used in countries where the disease is endemic. One of the vaccines against typhoid fever contains live, attenuated *S. typhi*. The cholera vaccine used in the United States contains live, attenuated *Vibrio cholerae*.

- The vaccines against plague, typhus, and Q fever contain killed bacteria. The cholera vaccine used in many countries where cholera is endemic which contains killed *V. cholerae*. These vaccines are used only to protect those likely to be exposed.

Passive Immunity

- Passive immunity in the form of **antitoxins** is available for the prevention and treatment of **tetanus, botulism**, and **diphtheria**. In addition, a monoclonal antibody against exotoxin B of *Clostridium difficile* (bezlotoxumab) prevents relapses of pseudomembranous colitis. These four diseases are caused by exotoxins. Antitoxins (antibodies against the exotoxin) bind to exotoxins and prevent their toxic effects (i.e., they **neutralize** the toxins).

Passive–Active Immunity

- This involves providing both immediate (but short-term) protection in the form of antibodies and long-term protection in the form of active immunization. An excellent example of the use of passive–active immunity is the prevention of tetanus in an unimmunized person who has sustained a contaminated wound. Both tetanus antitoxin and tetanus toxoid should be given. They should be given at different sites so that the antibodies in the antitoxin do not neutralize the toxoid.

SELF-ASSESSMENT QUESTIONS

1. Which one of the following is the immunogen in the vaccine against *Streptococcus pneumoniae*?

   (A) Capsular polysaccharide
   (B) Endotoxin
   (C) Formaldehyde-killed organisms
   (D) Pilus protein
   (E) Toxoid
   (F) Factor H binding protein

2. Disease caused by which one of the following bacteria is prevented by a toxoid vaccine?

   (A) *Bacteroides fragilis*
   (B) *Corynebacterium diphtheriae*
   (C) *Neisseria meningitidis*
   (D) *Salmonella typhi*
   (E) *Vibrio cholerae*

3. Disease caused by which one of the following bacteria is prevented by a vaccine in which the immunogen is covalently bound to a carrier protein (conjugate vaccine)?

   (A) *Bacillus anthracis*
   (B) *Clostridium tetani*
   (C) *Haemophilus influenzae*
   (D) *Mycobacterium tuberculosis*
   (E) *Streptococcus pyogenes*

4. Which one of the following is the immunogen in the vaccine against group B *Neisseria meningitidis*?

   (A) Capsular polysaccharide
   (B) Endotoxin
   (C) Formaldehyde-killed organisms
   (D) Pilus protein
   (E) Toxoid
   (F) Factor H binding protein

5. Passive immunity is used to prevent or treat disease caused by which one of the following sets of bacteria?

   (A) *Clostridium tetani* and *Clostridium botulinum*
   (B) *Escherichia coli* and *Staphylococcus aureus*
   (C) *Neisseria meningitidis* and *Bacillus anthracis*
   (D) *Streptococcus pneumoniae* and *Haemophilus influenzae*
   (E) *Streptococcus pyogenes* and *Salmonella typhi*

ANSWERS

(1) **(A)**
(2) **(B)**
(3) **(C)**
(4) **(F)**
(5) **(A)**

PRACTICE QUESTIONS: USMLE & COURSE EXAMINATIONS

Questions on the topics discussed in this chapter can be found in the Basic Bacteriology section of Part XIII: USMLE (National Board) Practice Questions starting on page 733. Also see Part XIV: USMLE (National Board) Practice Examination starting on page 775.

Sterilization & Disinfection

STERILIZATION, DISINFECTION, & STANDARD PRECAUTIONS

The purpose of sterilization and disinfection procedures is to prevent transmission of microbes to patients. In addition to sterilization and disinfection, other important measures to prevent transmission are included in the protocol of "**standard precautions**" (previously known as Universal Precautions). Standard precautions should be used in interaction with *all* patients, as any particular patient can be a reservoir of transmissible bacteria, viruses, or other microbes. These precautions include (1) hand hygiene, (2) respiratory hygiene and cough etiquette, (3) safe injection practices, and (4) proper disposal of needles and scalpels. Further, if exposure to body fluids or aerosols is likely, personal protective equipment (PPE) such as masks or face shields, gloves, gowns, and protective eyewear should be used. The precautions taken should be specific for the task rather than for the particular patient.

There are also **transmission-based precautions** that supplement the standard precautions and should be employed when the patient is infected (or suspected to be infected) with a highly transmissible organism. The three categories of transmission-based precautions are contact, droplet, and airborne and are described in Table 13–1. For additional information, please consult the Centers for Disease Control and Prevention (CDC) website (http://www.cdc.gov/hai/), where healthcare-associated infections (HAI) are discussed.

PRINCIPLES OF STERILIZATION & DISINFECTION

Sterilization is the killing or removal of *all* microorganisms, including bacterial spores, which are highly resistant to heat and disinfectants. Sterilization is usually carried out by autoclaving, which consists of exposure to steam at 121°C under a pressure of 15 lb/in^2 for 15 minutes. Surgical instruments that can be damaged by moist heat are usually sterilized by exposure to ethylene oxide gas, and most intravenous solutions are sterilized by filtration.

Disinfection is the killing of many, but not all, microorganisms. For adequate disinfection, pathogens must be killed, but some organisms and bacterial spores may survive. Disinfectants vary in their tissue-damaging properties from the corrosive phenol-containing compounds, which should be used only on inanimate objects, to less toxic materials such as ethanol and iodine, which can be used on skin surfaces. Chemicals used to kill microorganisms on the surface of skin and mucous membranes are called **antiseptics**.

Table 13–2 describes the clinical uses of common disinfectants and modes of sterilization.

TABLE 13–1 Infection Control Precautions and Practices

Type of Precaution	Example of Type of Patient or Type of Infection	Important Precaution Practice Employed
Standard	All patients	1. Hand hygiene 2. Respiratory hygiene and cough etiquette 3. Safe injection practices 4. Proper disposal of needles and scalpels
Standard	If exposure to blood, secretions, or body fluids is likely to occur	Personal protective equipment (PPE) such as mask, face shield, goggles, gloves, or gown
Contact	1. Stool incontinence, e.g., *Clostridium difficile*, norovirus 2. Generalized rash, e.g., varicella (chickenpox) 3. Draining wounds	1. Wear gloves and gown 2. Disinfect room
Droplet	1. Respiratory viruses, e.g., influenza 2. *Bordetella pertussis* 3. Early infection with *Neisseria meningitidis*	1. Face mask or face shield for patient and provider 2. Disinfect room
Airborne	1. Tuberculosis 2. Measles 3. Varicella (chickenpox) when patient is coughing	1. Isolation room; negative pressure room 2. Face mask or face shield for patient and provider; N-95 respirator, if available 3. Disinfect room

TABLE 13–2 Clinical Use of Disinfection and Sterilization

Clinical Use	Commonly Used Disinfectant or Method of Sterilization
Disinfect surgeon's hands prior to surgery	Chlorhexidine
Disinfect surgical site prior to surgery	Iodophor
Disinfect skin prior to venipuncture or immunization	70% ethanol
Disinfect skin prior to blood culture or inserting vascular catheter	Tincture of iodine followed by 70% ethanol, or iodophor, or chlorhexidine
Disinfect air in operating room (when not in use)	Ultraviolet light
Disinfect floor of operating room	Benzalkonium chloride (Lysol)
Disinfect stethoscope	70% ethanol
Cleanse wounds	Thimerosal, chlorhexidine, hydrogen peroxide
Cleanse burn wounds	Silver sulfadiazine
Cleanup of blood spill from a patient with hepatitis B or C (disinfect area)	Hypochlorite (bleach, Clorox)
Sterilize surgical instruments and heat-sensitive materials (e.g., endoscopes, respiratory therapy equipment)	Ethylene oxide or glutaraldehyde
Sterilize non–heat-sensitive materials (e.g., surgical gowns, drapes)	Autoclave
Sterilize intravenous solutions	Filtration
Preservative in vaccines	Thimerosal (not used in childhood vaccinations) or phenol

RATE OF KILLING OF MICROORGANISMS

Death of microorganisms occurs at a certain rate dependent primarily on two variables: the concentration of the killing agent and the length of time the agent is applied. The number of microorganisms killed is directly proportional to the time and concentration of the chemical agent that the organisms are exposed to. The relationship is usually stated in terms of survivors because they are easily measured by colony formation.

CHEMICAL AGENTS

Chemicals vary greatly in their ability to kill microorganisms. They act primarily by one of the three mechanisms: (1) disruption of the lipid-containing cell membrane, (2) modification of proteins, or (3) modification of DNA. Each of the following chemical agents has been classified into one of the three categories, but some of the chemicals act by more than one mechanism.

DISRUPTION OF CELL MEMBRANES

Alcohol

Ethanol is widely used to clean the skin before immunization or venipuncture. It acts mainly by disorganizing the lipid structure in membranes, but it denatures proteins as well. Ethanol requires the presence of water for maximal activity (i.e., it is far more effective at 70% than at 100%). Ethanol will not kill bacterial spores and therefore cannot be used for sterilization.

Detergents

Detergents are "surface-active" agents composed of a long-chain, lipid-soluble, hydrophobic portion, and a polar hydrophilic group. These surfactants interact with the lipid in the cell membrane and disrupt the membrane. **Quaternary ammonium compounds (e.g., benzalkonium chloride)** are cationic detergents widely used for skin antisepsis.

Phenols

Phenol was the first disinfectant used in the operating room (by Lister in the 1860s), but it is rarely used as a disinfectant today because it is too caustic, damaging membranes and denaturing proteins. **Chlorhexidine** is a chlorinated phenol that is widely used as a hand disinfectant prior to surgery ("surgical scrub") and in the cleansing of wounds.

MODIFICATION OF PROTEINS

Chlorine

Chlorine is used as a disinfectant to purify the water supply and to treat swimming pools. It is also the active component of **hypochlorite (bleach, Clorox)**, which is used as a disinfectant. Chlorine is a powerful oxidizing agent that kills by cross-linking essential sulfhydryl groups in enzymes to form the inactive disulfide.

Iodine

Iodine is the most effective skin antiseptic used in medical practice and should be used prior to obtaining a blood culture and installing intravenous catheters because contamination with skin flora such as *Staphylococcus epidermidis* can be a problem. Iodine, like chlorine, is an oxidant that inactivates sulfhydryl-containing enzymes.

Iodine is supplied in two forms:

(1) **Tincture of iodine** (2% solution of iodine and potassium iodide in ethanol) is used to prepare the skin prior to blood culture. Because tincture of iodine can be irritating to the skin, it should be removed with alcohol.

(2) **Iodophors** are complexes of iodine with detergents that are frequently used to prepare the skin prior to surgery because they are less irritating than tincture of iodine. They are also used to disinfect the skin of healthcare personnel. **Povidone-iodine (Betadine)** is an iodophor commonly used as an antiseptic.

Heavy Metals

Mercury and silver have the greatest antibacterial activity of the heavy metals and are the most widely used in medicine. They act by binding to sulfhydryl groups, thereby blocking enzymatic activity. **Thimerosal** (Merthiolate) and **merbromin** (Mercurochrome), which contain mercury, are used as skin antiseptics. **Silver nitrate** drops are effective in preventing gonococcal neonatal conjunctivitis (ophthalmia neonatorum). Silver sulfadiazine is used to prevent infection of burn wounds.

Hydrogen Peroxide

Hydrogen peroxide is used as an antiseptic to clean wounds. Hydrogen peroxide is an oxidizing agent that attacks sulfhydryl groups, thereby inhibiting enzymatic activity. Its effectiveness is limited by the organism's ability to produce catalase, an enzyme that degrades H_2O_2.

Formaldehyde & Glutaraldehyde

Formaldehyde, which is available as a 37% solution in water (formalin), denatures proteins and nucleic acids. Both proteins and nucleic acids contain essential $-NH_2$ and $-OH$ groups, which are the main sites of alkylation by the hydroxymethyl group of formaldehyde. **Glutaraldehyde**, which has two reactive aldehyde groups, is 10 times more effective than formaldehyde and is less toxic. In hospitals, it is used to sterilize respiratory therapy equipment, endoscopes, and hemodialysis equipment.

Ethylene Oxide

Ethylene oxide gas is used extensively in hospitals for the sterilization of heat-sensitive materials (surgical instruments and plastics). It kills by alkylating proteins and nucleic acids, and is classified as a mutagen and a carcinogen.

Acids & Alkalis

Strong acids and alkalis kill by denaturing proteins. Although most bacteria are susceptible, it is important to note that *Mycobacterium tuberculosis* and other mycobacteria are relatively resistant to 2% NaOH, which is used in the clinical laboratory to liquefy sputum prior to culturing the organism. Weak acids, such as **benzoic, propionic, and citric acids**, are frequently used as food preservatives because they are bacteriostatic.

MODIFICATION OF NUCLEIC ACIDS

A variety of dyes not only stain microorganisms but also inhibit their growth. One of these is crystal violet (gentian violet), an antiseptic used to treat fungal infections of the skin. Its action is based on binding of the positively charged dye molecule to the negatively charged phosphate groups of the nucleic acids. Malachite green, a triphenylamine dye like crystal violet, is a component of Löwenstein–Jensen medium, which is used to grow *M. tuberculosis*. The dye inhibits the growth of unwanted organisms in the sputum during the 6-week incubation period.

PHYSICAL AGENTS

The physical agents act either by imparting energy in the form of heat or radiation or by removing organisms through filtration.

HEAT

Heat energy can be applied in the form of moist heat (either boiling or autoclaving), or dry heat, or by pasteurization. In general, heat kills by denaturing proteins, but membrane damage may also be involved. Moist heat sterilizes at a lower temperature than dry heat because water aids in the disruption of noncovalent bonds (e.g., hydrogen bonds), which hold protein chains together in their secondary and tertiary structures.

Moist heat sterilization (**autoclaving**) is the most frequently used method of sterilization. Because bacterial **spores are resistant to boiling** (100°C at sea level), they require a higher temperature to inactivate them. This cannot be achieved unless the pressure is increased. For this purpose, an autoclave chamber, in which steam, at a pressure of 15 lb/in^2, reaches a temperature of 121°C and is held at that temperature for 15 to 20 minutes is used to kill organisms. To test the effectiveness of the autoclaving process, spore-forming organisms, such as members of the genus *Clostridium*, are used.

Sterilization by dry heat, on the other hand, requires temperatures in the range of 180°C for 2 hours. This process is used primarily for glassware and is used less frequently than autoclaving.

Pasteurization, which is used primarily for milk, consists of heating the milk to 62°C for 30 minutes followed by rapid cooling. ("Flash" pasteurization at 72°C for 15 seconds is often used.) This is sufficient to kill the vegetative cells of the milk-borne pathogens (e.g., *Mycobacterium bovis*, *Salmonella*, *Streptococcus*, *Listeria*, and *Brucella*), but not to sterilize the milk.

RADIATION

The two types of radiation used to kill microorganisms are **ultraviolet (UV) light and X-rays**. The greatest antimicrobial activity of UV light occurs at 250–260 nm, which is the wavelength region of maximum absorption by the purine and pyrimidine bases of DNA. The most significant lesion caused by UV irradiation is the formation of thymine dimers. As a result, DNA replication is inhibited and the organism cannot grow. Cells have repair mechanisms against UV-induced damage that involve either cleavage of dimers in the presence of visible light (photoreactivation) or excision of damaged bases, which is not dependent on visible light (dark repair). Because UV radiation can damage the cornea and skin, the use of UV irradiation in medicine is limited. However, it is used in hospitals to kill airborne organisms, especially in operating rooms when they are not in use. Bacterial spores are quite resistant and require a dose up to 10 times greater than do the vegetative bacteria.

X-rays have higher energy and penetrating power than UV radiation and kill mainly by the production of free radicals (e.g., production of hydroxyl radicals by the hydrolysis of water). These highly reactive radicals can break covalent bonds in DNA, thereby killing the organism.

X-rays kill vegetative cells readily, but spores are remarkably resistant. X-rays are used in medicine for sterilization of heat-sensitive items, such as sutures and surgical gloves, and plastic items, such as syringes.

FILTRATION

Filtration is the preferred method of **sterilizing certain solutions** (e.g., those with heat-sensitive components). In the past, solutions for intravenous use were autoclaved, but heat-resistant endotoxin in the cell walls of the dead gram-negative bacteria caused fever in recipients of the solutions. Therefore, solutions are now filtered to make them **pyrogen-free** prior to autoclaving.

The most commonly used filter is composed of nitrocellulose and has a pore size of 0.22 micrometer. This size will retain all bacteria and spores. Filters work by physically trapping particles larger than the pore size and by retaining somewhat smaller particles via electrostatic attraction of the particles to the filters.

PEARLS

- Sterilization is the **killing of all** forms of microbial life, including bacterial spores. **Spores** are **resistant to boiling**, so sterilization of medical equipment is typically achieved at 121°C for 15 minutes in an autoclave. Sterilization of heat-sensitive materials is achieved by exposure to ethylene oxide, and liquids can be sterilized by filtration.

- **Disinfection** is **reducing the number of bacteria** to a level low enough that disease is unlikely to occur. Spores and some bacteria will survive. For example, disinfection of the water supply is achieved by treatment with chlorine. Disinfection of the skin prior to venipuncture is achieved by treatment with 70% ethanol. Disinfectants that are mild enough to use on skin and other tissues, such as 70% ethanol, are called **antiseptics**.

- The killing of microbes by either chemicals or radiation is proportional to the **dose**, which is defined as the product of the concentration multiplied by the time of exposure.

- Chemical agents kill bacteria by one of three actions: disruption of lipid in cell membranes, modification of proteins, or modification of DNA.

- Physical agents kill (or remove) bacteria by one of three processes: heat, radiation, or filtration.
- Heat is usually applied at temperatures above boiling (121°C) to kill spores, but heat-sensitive materials such as milk are exposed to temperatures below boiling (**pasteurization**) that kill the pathogens in milk but do not sterilize it.

- Radiation, such as **ultraviolet light** and X-radiation, is often used to sterilize heat-sensitive items. Ultraviolet light and X-radiation **kill by damaging DNA**.
- Filtration can sterilize liquids if the pore size of the filter is small enough to retain all bacteria and spores. Heat-sensitive liquids (e.g., intravenous fluids) are often sterilized by filtration.

SELF-ASSESSMENT QUESTIONS

1. Regarding sterilization and disinfection, which one of the following is the most accurate statement?
   (A) Seventy percent alcohol is a better antiseptic than iodine, so 70% alcohol should be used to disinfect the skin prior to drawing a blood culture rather than iodine.
   (B) Disinfectants kill both bacterial cells and bacterial spores.
   (C) During sterilization by autoclaving, the temperature must be raised above boiling in order to kill bacterial spores.
   (D) Transmission of milk-borne diseases can be prevented by pasteurization, which kills both bacterial cells and spores.
   (E) Ultraviolet light used in the operating room to disinfect the room kills bacteria primarily by causing oxidation of lipids in the cell membrane.

2. Which one of the following chemicals is used to sterilize heat-sensitive materials, such as surgical instruments, in the hospital?
   (A) Benzalkonium chloride
   (B) Phenol
   (C) Ethylene oxide
   (D) Thimerosal
   (E) Tincture of iodine

ANSWERS

(1) **(C)**
(2) **(C)**

PRACTICE QUESTIONS: USMLE & COURSE EXAMINATIONS

Questions on the topics discussed in this chapter can be found in the Basic Bacteriology section of Part XIII: USMLE (National Board) Practice Questions starting on page 733. Also see Part XIV: USMLE (National Board) Practice Examination starting on page 775.

Overview of the Major Pathogens & Introduction to Anaerobic Bacteria

CHAPTER CONTENTS

OVERVIEW OF THE MAJOR PATHOGENS

The major bacterial pathogens are presented in Table 14–1 and described in Chapters 15 through 26. So that the reader may concentrate on the important pathogens, the bacteria that are less medically important are described in a separate chapter (see Chapter 27).

Table 14–1 is divided into organisms that are readily Gram stained and those that are not. The readily stained organisms fall into four categories: gram-positive cocci, gram-negative cocci, gram-positive rods, and gram-negative rods. Because there are so many kinds of gram-negative rods, they have been divided into three groups:

(1) Organisms associated with the enteric tract
(2) Organisms associated with the respiratory tract
(3) Organisms from animal sources (zoonotic bacteria)

For ease of understanding, the organisms associated with the enteric tract are further subdivided into three groups: (1) pathogens both inside and outside the enteric tract, (2) pathogens inside the enteric tract, and (3) pathogens outside the enteric tract.

As is true of any classification dealing with biologic entities, this one is not entirely precise. For example, *Campylobacter* causes enteric tract disease but frequently has an animal source. Nevertheless, despite some uncertainties, subdivision of the large number of gram-negative rods into these functional categories should be helpful to the reader.

The organisms that are not readily Gram stained fall into six major categories: *Mycobacterium* species, which are acid-fast rods; *Mycoplasma* species, which have no cell wall and so do not stain with Gram stain; *Treponema* and *Leptospira* species, which are spirochetes, too thin to be seen when stained with Gram stain; and *Chlamydia* and *Rickettsia* species, which are very small, intracellular bacteria and are difficult to visualize within the cytoplasm of the cell.

INTRODUCTION TO ANAEROBIC BACTERIA

Important Properties

Anaerobes are characterized by their ability to grow only in an atmosphere containing less than 20% oxygen (i.e., they grow poorly if at all in room air). They are a heterogeneous group

TABLE 14–1 **Major Bacterial Pathogens**

Type of Organism	Genus
Readily Gram stained	
Gram-positive cocci	Staphylococcus, Streptococcus, Enterococcus
Gram-negative cocci	Neisseria
Gram-positive rods	Corynebacterium, Listeria, Bacillus, Clostridium, Actinomyces, Nocardia
Gram-negative rods	
Enteric tract organisms	
Pathogenic inside and outside tract	Escherichia, Salmonella
Pathogenic primarily inside tract	Shigella, Vibrio, Campylobacter, Helicobacter
Pathogenic outside tract	Klebsiella–Enterobacter–Serratia group, Pseudomonas, Proteus–Providencia–Morganella group, Bacteroides
Respiratory tract organisms	Haemophilus, Legionella, Bordetella
Organisms from animal sources	Brucella, Francisella, Pasteurella, Yersinia
Not readily Gram stained	
Not obligate intracellular bacteria	Mycobacterium, Mycoplasma, Treponema, Leptospira
Obligate intracellular bacteria	Chlamydia, Rickettsia

composed of a variety of bacteria, from those that can barely grow in 20% oxygen to those that can grow only in less than 0.02% oxygen. Table 14–2 describes the optimal oxygen requirements for several representative groups of organisms. The obligate aerobes, such as *Pseudomonas aeruginosa*, grow best in the 20% oxygen of room air and not at all under anaerobic conditions. Facultative anaerobes such as *Escherichia coli* can grow well under either circumstance. Aerotolerant organisms such as *Clostridium histolyticum* can grow to some extent in air but multiply much more rapidly in a lower oxygen concentration. Microaerophilic organisms such as *Campylobacter jejuni* require a reduced oxygen concentration (approximately 5%) to grow optimally. The obligate anaerobes such as *Bacteroides fragilis* and *Clostridium perfringens* require an almost total absence

of oxygen. Many anaerobes use nitrogen rather than oxygen as the terminal electron acceptor.

The main reason why the growth of anaerobes is inhibited by oxygen is the reduced amount (or absence) of catalase and superoxide dismutase (SOD) in anaerobes. Catalase and SOD eliminate the toxic compounds hydrogen peroxide and superoxide, which are formed during production of energy by the organism (see Chapter 3). Another reason is the oxidation of essential sulfhydryl groups in enzymes without sufficient reducing power to regenerate them.

In addition to oxygen concentration, the oxidation–reduction potential (E_h) of a tissue is an important determinant of the growth of anaerobes. Areas with low E_h, such as the periodontal pocket, dental plaque, and colon, support the growth of anaerobes well. Crushing injuries that result in devitalized tissue caused by impaired blood supply produce a low E_h, allowing anaerobes to grow and cause disease.

Anaerobes of Medical Interest

The anaerobes of medical interest are presented in Table 14–3. It can be seen that they include both rods and cocci and both gram-positive and gram-negative organisms. The rods are divided into the spore formers (e.g., *Clostridium*) and the nonspore formers (e.g., *Bacteroides*). In this book, three genera of anaerobes are described as major bacterial pathogens, namely, *Clostridium, Actinomyces,* and *Bacteroides. Streptococcus* is a genus of major pathogens consisting of both anaerobic and facultative organisms. The remaining anaerobes are less important and are discussed in Chapter 27.

Clinical Infections

Many of the medically important anaerobes are part of the normal human flora. As such, they are nonpathogens in their normal habitat and cause disease only when they leave those sites. The two prominent exceptions to this are *Clostridium botulinum* and *Clostridium tetani*, the agents of botulism and tetanus, respectively, which are soil organisms. *C. perfringens*, another important human pathogen, is found in the colon and in the soil.

TABLE 14–2 **Optimal Oxygen Requirements of Representative Bacteria**

Bacterial Type	Representative Organism	Growth Under Following Conditions	
		Aerobic	Anaerobic
Obligate aerobes	Pseudomonas aeruginosa	3+	0
Facultative anaerobes	Escherichia coli	4+	3+
Aerotolerant organisms	Clostridium histolyticum	1+	4+
Microaerophiles	Campylobacter jejuni	0	1+[1]
Obligate anaerobes	Bacteroides fragilis	0	4+

[1] *C. jejuni* grows best (3+) in 5% O_2 plus 10% CO_2. It is also called **capnophilic** in view of its need for CO_2 for optimal growth.

TABLE 14–3 Anaerobic Bacteria of Medical Interest

Morphology	Gram Stain	Genus
Spore-forming rods	+	*Clostridium*
	–	None
Non–spore-forming rods	+	*Actinomyces, Bifidobacterium, Eubacterium, Lactobacillus, Propionibacterium*
	–	*Bacteroides, Fusobacterium*
Non–spore-forming cocci	+	*Peptococcus, Peptostreptococcus, Streptococcus*
	–	*Veillonella*

Diseases caused by members of the anaerobic normal flora are characterized by abscesses, which are most frequently located in the brain, lungs, female genital tract, biliary tract, and other intra-abdominal sites. Most abscesses contain more than one organism, either multiple anaerobes or a mixture of anaerobes plus facultative anaerobes. It is thought that the facultative anaerobes consume sufficient oxygen to allow the anaerobes to flourish.

Three important findings on physical examination that arouse suspicion of an anaerobic infection are a foul-smelling discharge, gas in the tissue, and necrotic tissue. In addition, infections in the setting of pulmonary aspiration, bowel surgery, abortion, cancer, or human and animal bites frequently involve anaerobes.

Laboratory Diagnosis

Two aspects of microbiologic diagnosis of an anaerobic infection are important even before the specimen is cultured: (1) obtaining the appropriate specimen and (2) rapidly transporting the specimen under anaerobic conditions to the laboratory. An appropriate specimen is one that does not contain members of the normal flora to confuse the interpretation. For example, such specimens as blood, pleural fluid, pus, and transtracheal aspirates are appropriate, but sputum and feces are not.

In the laboratory, the cultures are handled and incubated under anaerobic conditions. In addition to the usual diagnostic criteria of Gram stain, morphology, and biochemical reactions, the special technique of gas chromatography is important. In this procedure, organic acids such as formic, acetic, and propionic acids are measured.

Treatment

In general, surgical drainage of the abscess and administration of antimicrobial drugs are indicated. Drugs commonly used to treat anaerobic infections are penicillin G, cefoxitin, chloramphenicol, clindamycin, and metronidazole. Note, however, that many isolates of the important pathogen *B. fragilis* produce β-lactamase and are thus resistant to penicillin. Note also that aminoglycosides such as gentamicin are not effective against anaerobes because they require an oxygen-dependent process for uptake into the bacterial cell.

SELF-ASSESSMENT QUESTIONS

1. The main reason why some bacteria are anaerobes (i.e., they cannot grow in the presence of oxygen) is because:
   (A) they do not have sufficient catalase and superoxide dismutase.
   (B) they have too much ferrous ion that is oxidized to ferric ion in the presence of oxygen.
   (C) they have unusual mitochondria that cannot function in the presence of oxygen.
   (D) transcription of the gene for the pilus protein is repressed in the presence of oxygen.

2. Which one of the following sets consists of bacteria that are both anaerobes?
   (A) *Actinomyces israelii* and *Serratia marcescens*
   (B) *Campylobacter jejuni* and *Vibrio cholerae*
   (C) *Clostridium perfringens* and *Bacteroides fragilis*
   (D) *Mycobacterium tuberculosis* and *Pseudomonas aeruginosa*
   (E) *Mycoplasma pneumoniae* and *Corynebacterium diphtheriae*

ANSWERS

(1) **(A)**
(2) **(C)**

PRACTICE QUESTIONS: USMLE & COURSE EXAMINATIONS

Questions on the topics discussed in this chapter can be found in the Clinical Bacteriology section of Part XIII: USMLE (National Board) Practice Questions starting on page 737. Also see Part XIV: USMLE (National Board) Practice Examination starting on page 775.

Gram-Positive Cocci

INTRODUCTION

There are two medically important genera of gram-positive cocci: *Staphylococcus* and *Streptococcus*. Two of the most important human pathogens, *Staphylococcus aureus* and *Streptococcus pyogenes*, are described in this chapter. Staphylococci and streptococci are nonmotile and do not form spores.

Both staphylococci and streptococci are gram-positive cocci, but they are differentiated by two main criteria:

(1) Microscopically, staphylococci appear in grapelike clusters, whereas streptococci are in chains.

(2) Biochemically, staphylococci produce catalase (i.e., they degrade hydrogen peroxide), whereas streptococci do not.

Additional information regarding the clinical aspects of infections caused by the organisms in this chapter is provided in Part IX, entitled Infectious Diseases beginning on page 603.

STAPHYLOCOCCUS

1. Staphylococcus aureus

Diseases

S. aureus causes abscesses (Figure 15–1), various pyogenic infections (e.g., endocarditis, septic arthritis, and osteomyelitis), food poisoning, scalded skin syndrome (Figure 15–2), and toxic shock syndrome. It is one of the most common causes of hospital-acquired pneumonia, septicemia, and surgical-wound infections, including the site of insertion of cardiac pacemakers. It is an important cause of skin and soft tissue infections (SSTIs), such as folliculitis (Figure 15–3), cellulitis, and impetigo (Figure 15–4). It is a common cause of bacterial conjunctivitis.

Methicillin-resistant *S. aureus* (MRSA) is the most common cause of skin abscesses in the United States. MRSA is also an important cause of pneumonia, necrotizing fasciitis, and sepsis in immunocompetent patients. *S. aureus*, especially MRSA, is the most common cause of infections in users of intravenous drugs.

Kawasaki syndrome is a disease of unknown etiology that may be caused by certain strains of *S. aureus*.

Important Properties

Staphylococci are spherical gram-positive cocci arranged in irregular grapelike clusters (Figure 15–5). All staphylococci produce **catalase**, whereas no streptococci do (catalase degrades H_2O_2 into O_2 and H_2O). Catalase is an important virulence

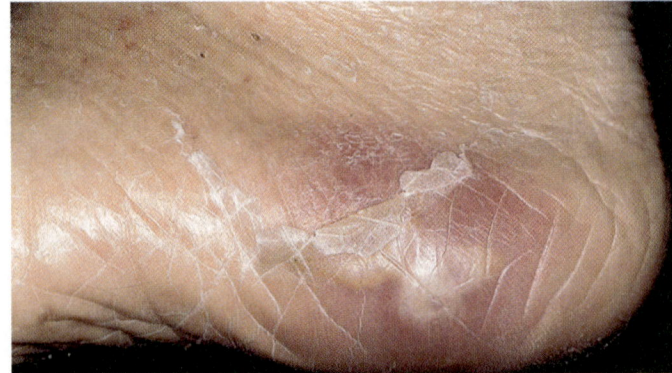

FIGURE 15–1 Abscess on foot. Note central raised area of whitish pus surrounded by erythema. An abscess is the classic lesion caused by *Staphylococcus aureus*. (Reproduced with permission from Wolff K, Johnson R: *Fitzpatrick's Color Atlas & Synopsis of Clinical Dermatology*, 6th ed. New York, NY: McGraw Hill; 2009.)

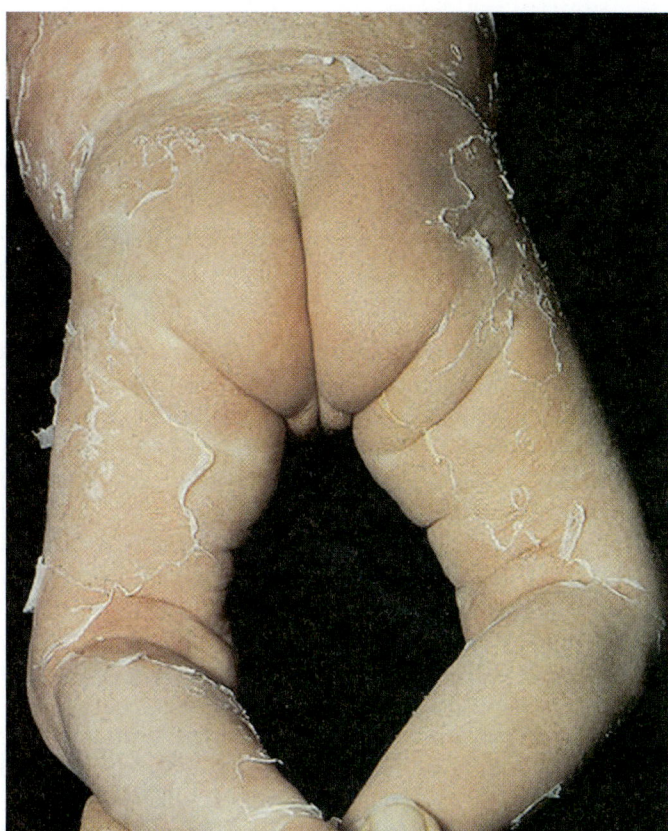

FIGURE 15–2 Scalded skin syndrome. Note widespread areas of "rolled up" desquamated skin in infant. Caused by an exotoxin produced by *Staphylococcus aureus*. (Reproduced with permission from Wolff K, Johnson R: *Fitzpatrick's Color Atlas & Synopsis of Clinical Dermatology*, 6th ed. New York, NY: McGraw Hill; 2009.)

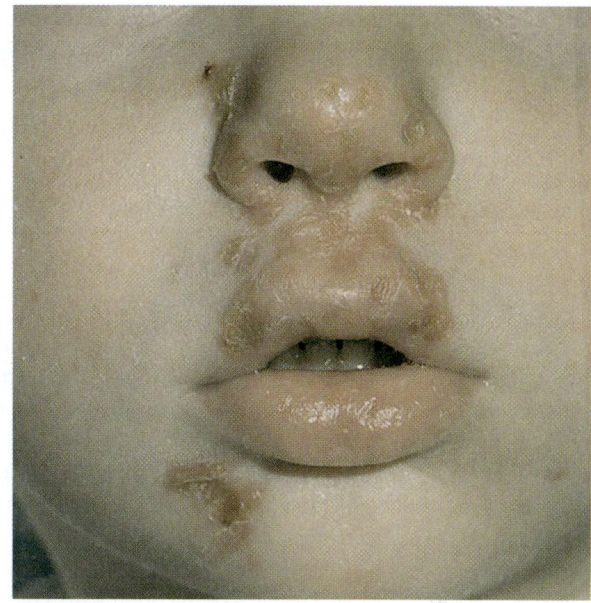

FIGURE 15–4 Impetigo. Lesions of impetigo are crops of vesicles with a "honey-colored" crust. Impetigo is caused by either *Staphylococcus aureus* or *Streptococcus pyogenes*. (Reproduced with permission from Wolff K, Johnson R: *Fitzpatrick's Color Atlas & Synopsis of Clinical Dermatology*, 6th ed. New York, NY: McGraw Hill; 2009.)

factor. Bacteria that make catalase can survive the killing effect of H_2O_2 within neutrophils.

Three species of staphylococci are important human pathogens: *S. aureus*, *Staphylococcus epidermidis*, and *Staphylococcus saprophyticus* (Table 15–1). Of these three, *S. aureus* is by far the most common and causes the most serious infections. *S. aureus* is distinguished from the others primarily by **coagulase** production (Figure 15–6). **Coagulase** is an enzyme that causes plasma to clot by activating prothrombin to form thrombin. Thrombin

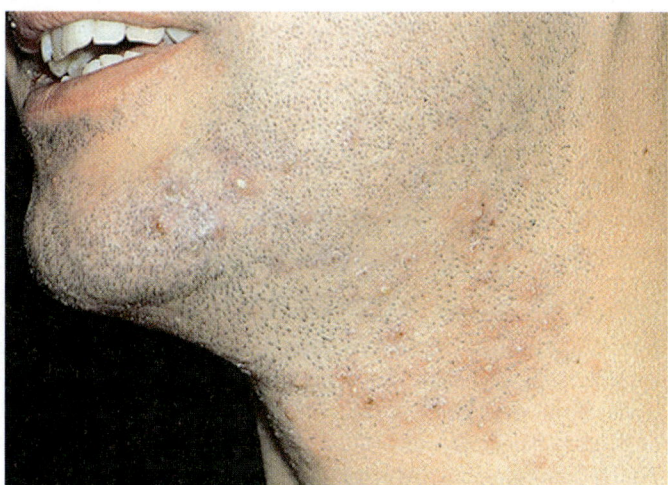

FIGURE 15–3 Folliculitis. Note the multiple, small pustules on the chin and neck. *Staphylococcus aureus* is the most common cause of folliculitis. (Reproduced with permission from Wolff K, Goldsmith LA, Katz SI, et al: *Fitzpatrick's Dermatology in General Medicine*, 7th ed. New York, NY: McGraw Hill; 2008.)

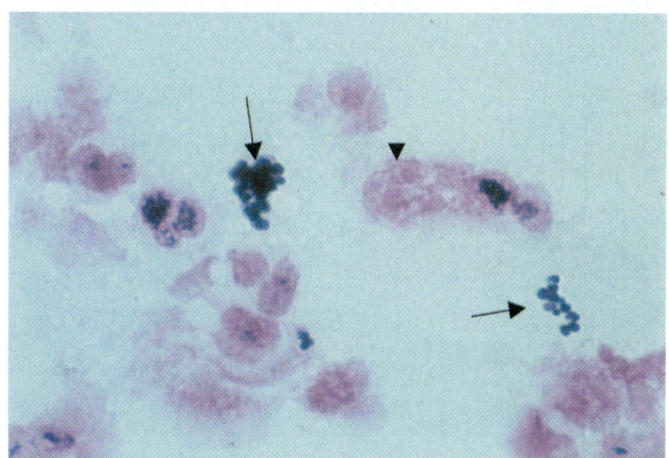

FIGURE 15–5 *Staphylococcus aureus*—Gram stain. Arrows point to two "grapelike" clusters of gram-positive cocci. Arrowhead points to neutrophil with pink segmented nuclei. (Used with permission from Professor Shirley Lowe, University of California, San Francisco School of Medicine.)

TABLE 15–1 Staphylococci of Medical Importance

Species	Coagulase Production	Typical Hemolysis	Important Features[1]	Typical Disease
S. aureus	+	β	Protein A on surface	Abscess, food poisoning, toxic shock syndrome
S. epidermidis	–	None	Sensitive to novobiocin	Infection of prosthetic heart valves and hips; common member of skin flora
S. saprophyticus	–	None	Resistant to novobiocin	Urinary tract

[1]All staphylococci are catalase-positive.

then catalyzes the activation of fibrinogen to form the fibrin clot. *S. epidermidis* and *S. saprophyticus* are often referred to as coagulase-negative staphylococci and are described in the section below.

Two other characteristics further distinguish these species, namely, *S. aureus* usually ferments mannitol and hemolyzes red blood cells, whereas *S. epidermidis* and *S. saprophyticus* do not. Hemolysis of red cells by hemolysins produced by *S. aureus* is the source of iron required for growth of the organism. The iron in hemoglobin is recovered by the bacteria and utilized in the synthesis of cytochrome enzymes used to produce energy.

More than 90% of *S. aureus* strains contain plasmids that encode **β-lactamase**, the enzyme that degrades many, but not all, penicillins. β-Lactamase–resistant penicillins such as nafcillin are used to treat infections caused by those strains of *S. aureus*.

Note, however, that many strains of *S. aureus* are resistant to the β-lactamase–resistant penicillins, such as methicillin and nafcillin, by virtue of changes in the **penicillin-binding proteins** (PBPs) in their cell membrane. Genes on the bacterial chromosome called *mecA* genes encode these altered PBPs.

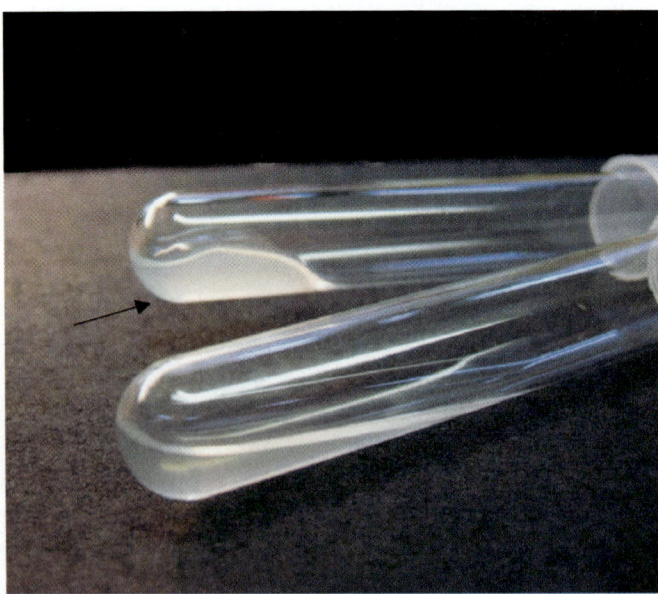

FIGURE 15–6 Coagulase test. Upper tube inoculated with *Staphylococcus aureus;* lower tube inoculated with *Staphylococcus epidermidis.* Arrow points to clotted plasma formed by coagulase produced by *S. aureus.* (Used with permission from Professor Shirley Lowe, University of California, San Francisco School of Medicine.)

The most important of these PBPs is PBP2a that can perform its transpeptidase function but does not bind penicillins.

These strains are commonly known as MRSA or nafcillin-resistant *S. aureus* (NRSA). MRSA causes both healthcare-acquired (HCA-MRSA) and community-acquired (CA-MRSA) infections. MRSA currently accounts for more than 50% of *S. aureus* strains isolated from hospital patients in the United States. CA-MRSA is a very common cause of community-acquired staphylococcal infections. Almost all strains of CA-MRSA produce P-V leukocidin (see later), whereas relatively few strains of HCA-MRSA do so. The most common strain of MRSA in the United States is the USA300 strain.

Strains of *S. aureus* with intermediate resistance to vancomycin (VISA) and with full resistance to vancomycin (VRSA) have also been detected. The cassette of genes that encodes vancomycin resistance in *S. aureus* is the same as the cassette that provides vancomycin resistance in enterococci. These genes are located in a transposon on a plasmid and encode the enzymes that **substitute D-lactate for D-alanine** in the peptidoglycan.

S. aureus has several important cell wall components and antigens:

(1) **Protein A** is the major protein in the cell wall. It is an important virulence factor because it binds to the Fc portion of immunoglobulin (Ig) G. The Fc part is occupied and is not free to bind to the Fc receptor on neutrophils and macrophages, so phagocytosis of *S. aureus* by those cells does not occur. The coagulase-negative staphylococci do not produce protein A.

(2) **Teichoic acids** are polymers of ribitol phosphate. They mediate adherence of the staphylococci to mucosal cells. **Lipoteichoic acids** play a role in the induction of septic shock by inducing cytokines such as interleukin-1 (IL-1) and tumor necrosis factor (TNF) from macrophages (see the discussion of septic shock in the Endotoxin section of Chapter 7).

(3) Polysaccharide capsule is also an important virulence factor. There are 11 serotypes based on the antigenicity of the capsular polysaccharide, but types 5 and 8 cause 85% of infections. Some strains of *S. aureus* are coated with a small amount of polysaccharide capsule, called a microcapsule. The capsule is poorly immunogenic, which has made producing an effective vaccine difficult.

(4) Surface receptors for specific staphylococcal bacteriophages permit the "phage typing" of strains for epidemiologic purposes. Teichoic acids make up part of these receptors.

(5) The peptidoglycan of *S. aureus* has endotoxin-like properties (i.e., it can stimulate macrophages to produce cytokines and can activate the complement and coagulation cascades). This explains the ability of *S. aureus* to cause the clinical findings of septic shock yet not possess endotoxin.

Transmission

Humans are the reservoir for staphylococci. The nose is the main site of colonization of *S. aureus*, and approximately 30% of people are colonized at any one time. People who are chronic carriers of *S. aureus* in their nose have an increased risk of skin infections caused by *S. aureus*.

The skin, especially of hospital personnel and patients, is also a common site of *S. aureus* colonization. Hand contact is an important mode of transmission, and handwashing decreases transmission.

S. aureus is also found in the vagina of approximately 5% of women, which predisposes them to toxic shock syndrome. Additional sources of staphylococcal infection are shedding from human lesions and fomites such as towels and clothing contaminated by these lesions.

Disease caused by *S. aureus* is favored by a heavily contaminated environment (e.g., family members with boils) and a compromised immune system. Reduced humoral immunity, including low levels of antibody, complement, or neutrophils, especially predisposes to staphylococcal infections.

Diabetes is a major predisposing factor to infections by *S. aureus*. *S. aureus*, especially MRSA, is the most common cause of infections in users of intravenous drugs. Patients with chronic granulomatous disease (CGD), a disease characterized by a defect in the ability of neutrophils to kill bacteria, are especially prone to *S. aureus* infections (see Chapter 68).

Pathogenesis

S. aureus causes disease both by producing toxins and by inducing pyogenic (pus-producing) inflammation. The typical lesion of *S. aureus* pyogenic infection is an **abscess**. Abscesses undergo central necrosis and usually drain pus to the outside (e.g., furuncles and boils), but organisms may disseminate via the bloodstream as well. **Foreign bodies**, such as sutures and intravenous catheters, are important predisposing factors to infection by *S. aureus*.

Coagulase is an important virulence factor in the formation of an abscess. It causes formation of a fibrin clot that walls off the bacteria and prevents access of neutrophils to the site of infection.

Several important toxins and enzymes are produced by *S. aureus*. The three clinically important exotoxins are enterotoxin, toxic shock syndrome toxin, and exfoliatin.

(1) **Enterotoxin** causes **food poisoning** characterized by prominent vomiting and watery, nonbloody diarrhea. It acts as a **superantigen** within the gastrointestinal tract to stimulate the release of large amounts of IL-1 and IL-2 from macrophages and helper T cells, respectively. The prominent vomiting is caused by serotonin (5-hydroxytryptamine) released from mast cells, which stimulates the enteric nervous system via the vagus nerve to activate the vomiting center in the brain. Enterotoxin is fairly heat-resistant and is therefore usually not inactivated by brief cooking. It is resistant to stomach acid and to enzymes in the stomach and jejunum. There are six immunologic types of enterotoxin, types A–F.

(2) **Toxic shock syndrome toxin** (TSST) causes **toxic shock**, especially in tampon-using menstruating women or in individuals with wound infections. Toxic shock also occurs in patients with nasal packing used to stop bleeding from the nose. TSST is produced locally by *S. aureus* in the vagina, nose, or other infected site. The toxin enters the bloodstream, causing a toxemia. Blood cultures typically do not grow *S. aureus*.

TSST is a **superantigen** and causes toxic shock by stimulating the release of large amounts of IL-1, IL-2, and TNF (see the discussions of exotoxins in Chapter 7 and superantigens in Chapter 60). Approximately 5% to 25% of isolates of *S. aureus* carry the gene for TSST. Toxic shock occurs in people who do not have antibody against TSST.

(3) **Exfoliatin** causes **"scalded skin" syndrome** in young children. It is "epidermolytic" and acts as a **protease that cleaves desmoglein** in desmosomes, leading to the separation of the epidermis at the granular cell layer. Localized production of exfoliatin by *S. aureus* results in **bullous impetigo**.

(4) Several exotoxins can kill leukocytes (leukocidins) and cause necrosis of tissues in vivo. Of these, the two most important are alpha toxin and P-V leukocidin. **Alpha toxin** causes marked necrosis of the skin and hemolysis. The cytotoxic effect of alpha toxin is attributed to the formation of holes in the cell membrane and the consequent loss of low-molecular-weight substances from the damaged cell.

P-V leukocidin is a **pore-forming toxin** that kills cells, especially white blood cells, by damaging cell membranes. The two subunits of the toxin assemble in the cell membrane to form a pore through which cell contents leak out. The gene-encoding P-V leukocidin is located on a lysogenic phage. P-V leukocidin is an important virulence factor for CA-MRSA and plays a role in the severe SSTI caused by this organism. A severe necrotizing pneumonia is also caused by strains of *S. aureus* that produce P-V leukocidin. Approximately 2% of clinical isolates of *S. aureus* produce P-V leukocidin.

(5) The enzymes include **coagulase**, fibrinolysin, hyaluronidase, proteases, nucleases, and lipases. Coagulase, by clotting plasma, serves to wall off the infected site, thereby retarding the migration of neutrophils into the site. Staphylokinase is a fibrinolysin that can lyse thrombi.

Unlike *S. aureus*, coagulase-negative staphylococci do not produce exotoxins. Thus, they do not cause food poisoning or toxic shock syndrome. They do, however, cause pyogenic infections (see later).

Clinical Findings

The important clinical manifestations caused by *S. aureus* can be divided into two groups: pyogenic (pus-producing) and

TABLE 15–2 Important Features of Pathogenesis by Staphylococci

Organism	Type of Pathogenesis	Typical Disease	Predisposing Factor	Mode of Prevention
S. aureus	1. Toxigenic (superantigen)	Toxic shock syndrome Food poisoning	Vaginal or nasal tampons Improper food storage	Reduce time of tampon use Refrigerate food
	2. Pyogenic (abscess) a. Local b. Disseminated	Skin infection (e.g., impetigo, surgical-wound infections) Sepsis, endocarditis[1]	Poor skin hygiene; failure to follow aseptic procedures IV drug use	Cleanliness; handwashing; reduce nasal carriage Reduce IV drug use
S. epidermidis	Pyogenic	Infections of intravenous catheter sites and prosthetic devices	Failure to follow aseptic procedures or remove IV catheters promptly	Handwashing; remove IV catheters promptly
S. saprophyticus	Pyogenic	Urinary tract infection	Sexual activity	

IV = intravenous.

[1]For simplicity, many forms of disseminated diseases caused by *S. aureus* (e.g., osteomyelitis, arthritis) were not included in the table.

toxin-mediated (Table 15–2). *S. aureus* is a major cause of skin, soft tissue, bone, joint, lung, heart, and kidney pyogenic infections. Pyogenic diseases are the first group described, and toxin-mediated diseases are the second group.

Staphylococcus aureus: Pyogenic Diseases

(1) SSTIs are very common. These include abscess (see Figure 15–1), impetigo (see Figure 15–4), furuncles, carbuncles (Figure 15–7), paronychia, cellulitis, folliculitis (see Figure 15–3), necrotizing fasciitis (Figure 77–10), hidradenitis suppurativa, conjunctivitis, eyelid infections (blepharitis and hordeolum), and postpartum breast infections (mastitis). Lymphangitis can occur, especially on the forearm associated with an infection on the hand.

Severe necrotizing SSTIs are caused by MRSA strains that produce P-V leukocidin. These infections are typically community-acquired rather than hospital-acquired. In the United States, CA-MRSA strains are the most common cause of SSTIs. These CA-MRSA strains are an especially common cause of infection among the homeless and intravenous drug users. Athletes who engage in close personal contact such as wrestlers and football players are also at risk.

Note that MRSA are also an important cause of SSTI in the hospital. HCA-MRSA cause approximately 50% of all nosocomial *S. aureus* infections. Molecular analysis reveals that the CA-MRSA strains are different from the HCA-MRSA strains.

(2) Septicemia (sepsis) can originate from any localized lesion, especially wound infection, or as a result of intravenous drug abuse. Sepsis caused by *S. aureus* has clinical features similar to those of sepsis caused by certain gram-negative bacteria, such as *Neisseria meningitidis* (see Chapter 16).

(3) Endocarditis may occur on normal or prosthetic heart valves, especially right-sided endocarditis (tricuspid valve) in intravenous drug users. (Prosthetic valve endocarditis is often caused by *S. epidermidis.*)

(4) Osteomyelitis and septic arthritis may arise either by hematogenous spread from a distant infected focus or be introduced locally at a wound site. *S. aureus* is a very common cause of these diseases, especially in children.

(5) *S. aureus* is the most common cause of postsurgical wound infections, which are an important cause of morbidity and mortality in hospitals. For example, *S. aureus* and *S. epidermidis* are the most common causes of infections at the site where cardiac pacemakers are installed.

(6) Pneumonia can occur in postoperative patients or following viral respiratory infection, especially influenza. Staphylococcal pneumonia often leads to empyema or lung abscess.

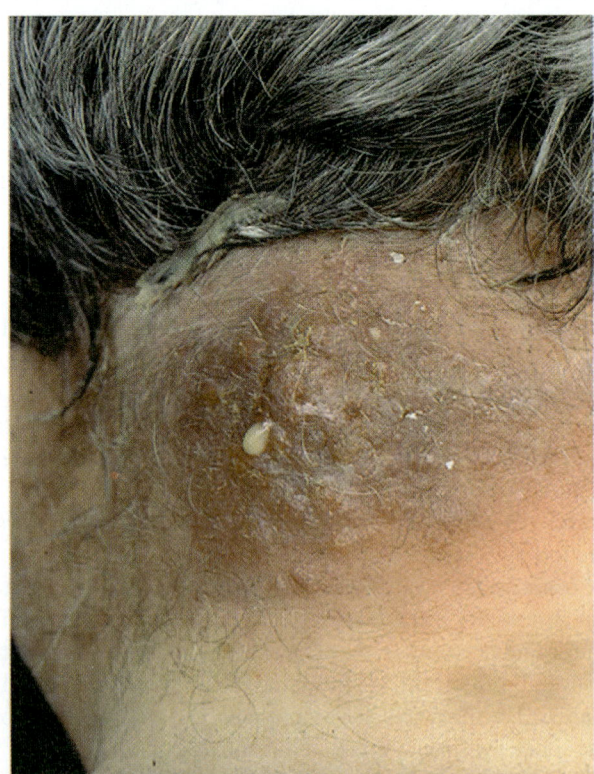

FIGURE 15–7 Carbuncle. A carbuncle is a multiheaded abscess often located on the back of the neck. Note drop of yellowish pus near the center of the lesion. Carbuncles are caused by *Staphylococcus aureus*. (Reproduced with permission from Wolff K, Johnson R: *Fitzpatrick's Color Atlas & Synopsis of Clinical Dermatology*, 6th ed. New York, NY: McGraw Hill; 2009.)

In many hospitals, it is the most common cause of nosocomial pneumonia in general and especially of ventilator-associated pneumonia in intensive care units. CA-MRSA causes a severe necrotizing pneumonia.

(7) Conjunctivitis typically presents with unilateral burning eye pain, hyperemia of the conjunctiva, and a purulent discharge. The organism is transmitted to the eye by contaminated fingers. *S. aureus* is the most common cause overall, but *Streptococcus pneumoniae* and *Haemophilus influenzae* are more common in children. Gonococcal and nongonococcal (caused by *Chlamydia trachomatis*) conjunctivitis is acquired by infants during passage through the birth canal.

(8) Abscesses can occur in any organ when *S. aureus* circulates in the bloodstream (bacteremia). These abscesses are often called "metastatic abscesses" because they occur by the spread of bacteria from the original site of infection, often in the skin.

Staphylococcus aureus: Toxin-Mediated Diseases

(1) Food poisoning (gastroenteritis) is caused by ingestion of enterotoxin, which is preformed in foods and hence has a short incubation period (1–8 hours). In staphylococcal food poisoning, vomiting is typically more prominent than diarrhea.

(2) Toxic shock syndrome is characterized by fever; hypotension; a diffuse, macular, sunburn-like rash that goes on to desquamate; and involvement of three or more of the following organs: liver, kidney, gastrointestinal tract, central nervous system, muscle, or blood.

(3) Scalded skin syndrome is characterized by fever, large bullae, and an erythematous macular rash. The rash looks like a burn, hence the name "scalded-skin." Large areas of skin slough, serous fluid exudes, and electrolyte imbalance can occur. Hair and nails can be lost. Recovery usually occurs within 7 to 10 days. This syndrome occurs most often in young children.

(4) Bullous impetigo is characterized by vesicles containing clear fluid that coalesce to form bullae (Figure 15–8). Bullous impetigo is caused by localized production of exfoliatin by *S. aureus*.

Staphylococcus aureus: Kawasaki Disease

Kawasaki disease (KD) is a disease of unknown etiology that is discussed here because several of its features resemble toxic shock syndrome caused by the superantigens of *S. aureus* (and *S. pyogenes*). KD is a vasculitis involving small- and medium-size arteries, especially the coronary arteries. It is the most common cause of acquired heart disease in children in the United States.

Clinically, KD is characterized by a high fever of at least 5 days in duration; bilateral nonpurulent conjunctivitis; lesions of the lips and oral mucosa (e.g., strawberry tongue, edema of the lips, and erythema of the oropharynx); cervical lymphadenopathy; a diffuse erythematous, maculopapular rash; and erythema and edema of the hands and feet that often ends with desquamation.

The most characteristic clinical finding of KD is cardiac involvement, especially myocarditis, arrhythmias, and

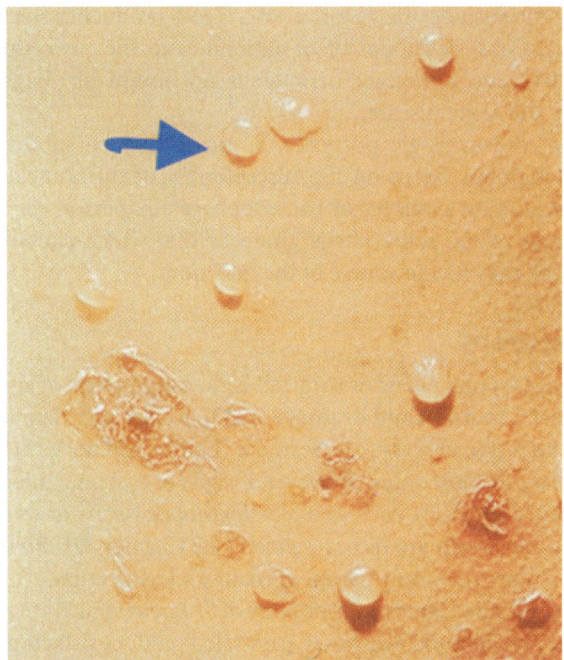

FIGURE 15–8 Bullous impetigo. Blue arrow points to one of several vesicles that will coalesce to form bullae, the characteristic lesion of bullous impetigo. (Reproduced with permission from Kang S, Amagai M, Bruckner AL, et al: *Fitzpatrick's Dermatology in General Medicine*, 9th ed. New York, NY: McGraw Hill; 2019.)

regurgitation involving the mitral or aortic valves. The main cause of morbidity and mortality in KD is **aneurysm of the coronary arteries**.

KD is much more common in children of Asian ancestry, leading to speculation that certain major histocompatibility complex (MHC) alleles may predispose to the disease. It is a disease of children younger than 5 years, often occurring in mini-outbreaks. It occurs worldwide but is much more common in Japan.

There is no definitive diagnostic laboratory test for KD. Effective therapy consists of high-dose intravenous immune globulins (IVIG) plus high-dose aspirin, which promptly reduce the fever and other symptoms and, most importantly, significantly reduce the occurrence of aneurysms.

Laboratory Diagnosis

Smears from staphylococcal lesions reveal gram-positive cocci in grapelike clusters (see Figure 15–5). Cultures of *S. aureus* typically yield golden-yellow colonies that are usually β-hemolytic. *S. aureus* is **coagulase-positive** (see Figure 15–6). Mannitol-salt agar is a commonly used screening device for *S. aureus*. *S. aureus* ferments mannitol, which lowers the pH, causing the agar to turn yellow, whereas *S. epidermidis* does not ferment mannitol and the agar remains pink.

In toxic shock syndrome, isolation of *S. aureus* is not required to make a diagnosis as long as the clinical criteria are met. Laboratory findings that support a diagnosis of toxic shock syndrome include the isolation of a TSST-producing strain of *S. aureus* and development of antibodies to the toxin during convalescence, although the latter is not useful for diagnosis during the acute disease.

For epidemiologic purposes, *S. aureus* can be subdivided into subgroups based on the susceptibility of the clinical isolate to lysis by a variety of bacteriophages. A person carrying *S. aureus* of the same phage group as that which caused the outbreak may be the source of the infections.

Treatment

Drainage (spontaneous or surgical) is the cornerstone of abscess treatment. **Incision and drainage (I&D)** is often sufficient treatment for a skin abscess (e.g., furuncle [boil]). Antibiotics are not necessary in most cases. However, if signs of systemic infection are seen, antibiotics, such as oral trimethoprim-sulfa or intravenous vancomycin are recommended (see Chapter 77). Previous infection provides only partial immunity to reinfection.

In the United States, 90% or more of *S. aureus* strains are resistant to penicillin G. Most of these strains produce **β-lactamase**. Such organisms can be treated with β-lactamase–resistant penicillins (e.g., nafcillin or cloxacillin), some cephalosporins, or vancomycin. Treatment with a combination of a β-lactamase–sensitive penicillin (e.g., amoxicillin) and a β-lactamase inhibitor (e.g., clavulanic acid) is also useful.

Approximately 20% of *S. aureus* strains are **methicillin-resistant** or nafcillin-resistant by virtue of altered PBPs. These resistant strains of *S. aureus* are often abbreviated **MRSA** or NRSA, respectively. Such organisms can produce sizable outbreaks of disease, especially in hospitals.

The drug of choice for these staphylococci is vancomycin, to which gentamicin is sometimes added. Daptomycin is also useful. Trimethoprim-sulfamethoxazole or clindamycin can be used to treat non–life-threatening infections caused by these organisms. Note that MRSA strains are resistant to almost all β-lactam drugs, including both penicillins and cephalosporins. Ceftaroline fosamil is the first β-lactam drug useful for the treatment of MRSA infections.

Strains of *S. aureus* with intermediate resistance to vancomycin (VISA strains) and with complete resistance to vancomycin (VRSA strains) have been isolated from patients. These strains are typically methicillin-/nafcillin-resistant as well, which makes them very difficult to treat. Daptomycin (Cubicin) can be used to treat infections by these organisms. Quinupristin-dalfopristin (Synercid) is another useful choice.

The treatment of toxic shock syndrome involves correction of the shock by using fluids, pressor drugs, and inotropic drugs; administration of a β-lactamase–resistant penicillin such as nafcillin; and removal of the tampon or debridement of the infected site as needed. Pooled serum globulins, which contain antibodies against TSST, may be useful.

Mupirocin is very effective as a topical antibiotic in skin infections caused by *S. aureus*. It has also been used to reduce nasal carriage of the organism in hospital personnel and in patients with recurrent staphylococcal infections. A topical skin antiseptic, such as chlorhexidine, can be added to mupirocin.

Some strains of staphylococci exhibit **tolerance** (i.e., they can be inhibited by antibiotics but are not killed). (That is, the ratio of minimum bactericidal concentration [MBC] to minimum inhibitory concentration [MIC] is very high.) Tolerance may result from failure of the drugs to inactivate inhibitors of autolytic enzymes that degrade the organism. Tolerant organisms should be treated with drug combinations (see Chapter 10).

Prevention

There is no vaccine against staphylococci. Cleanliness, frequent handwashing, and aseptic management of lesions help to control spread of *S. aureus*. Persistent colonization of the nose by *S. aureus* can be reduced by intranasal mupirocin or by oral antibiotics, such as ciprofloxacin or trimethoprim-sulfamethoxazole, but is difficult to eliminate completely. Shedders may have to be removed from high-risk areas (e.g., operating rooms and newborn nurseries). Cefazolin is often used perioperatively to prevent staphylococcal surgical-wound infections.

2. Coagulase-negative staphylococci (*Staphylococcus epidermidis* & *Staphylococcus saprophyticus*)

Diseases

S. epidermidis causes prosthetic valve endocarditis and prosthetic joint infections. It is the most common cause of central nervous system shunt infections and an important cause of sepsis in newborns. *S. saprophyticus* causes urinary tract infections, especially cystitis.

Important Properties

Staphylococci are spherical gram-positive cocci arranged in irregular grapelike clusters (see Figure 15–5). All staphylococci produce **catalase**, whereas no streptococci do (catalase degrades H_2O_2 into O_2 and H_2O). Catalase is an important virulence factor. Bacteria that make catalase can survive the killing effect of H_2O_2 within neutrophils.

S. epidermidis and *S. saprophyticus* are often referred to as coagulase-negative staphylococci. **Coagulase** is an enzyme that causes plasma to clot by activating prothrombin to form thrombin. Thrombin then catalyzes the activation of fibrinogen to form the fibrin clot (see Figure 15–6).

Two other characteristics further distinguish these species, namely, *S. aureus* usually ferments mannitol and hemolyzes red blood cells, whereas *S. epidermidis* and *S. saprophyticus* do not.

Transmission

Humans are the reservoir for staphylococci. *S. epidermidis* is found primarily on the human skin and can enter the bloodstream at the site of intravenous catheters that penetrate through the skin. *S. saprophyticus* is found primarily on the mucosa of the genital tract in young women and from that site can ascend into the urinary bladder to cause urinary tract infections.

S. epidermidis infections are almost always hospital-acquired, whereas *S. saprophyticus* infections are almost always community-acquired.

Pathogenesis

The virulence of *S. epidermidis* and *S. saprophyticus* is significantly less than that of *S. aureus*, a property that is, in part, attributed to their failure to produce protein A, an important virulence factor made by *S. aureus*.

Unlike *S. aureus*, the two coagulase-negative staphylococci do not produce exotoxins. Thus, they do not cause food poisoning or toxic shock syndrome. They do, however, cause pyogenic infections. For example, *S. epidermidis* is a prominent cause of pyogenic infections on prosthetic implants such as heart valves and hip joints, and *S. saprophyticus* causes pyogenic urinary tract infections, especially cystitis.

Strains of *S. epidermidis* that produce a glycocalyx are more likely to adhere to prosthetic implant materials and therefore are more likely to infect these implants than strains that do not produce a glycocalyx. Hospital personnel are a major reservoir for antibiotic-resistant strains of *S. epidermidis*.

Clinical Findings

S. epidermidis is part of the normal human flora on the skin and mucous membranes but can enter the bloodstream (bacteremia) and cause metastatic infections, especially at the site of implants. It commonly infects intravenous catheters and prosthetic implants (e.g., prosthetic heart valves [endocarditis], vascular grafts, and prosthetic joints [arthritis or osteomyelitis]) (see Table 15–2). *S. epidermidis* is also a major cause of sepsis in neonates and of peritonitis in patients with renal failure who are undergoing peritoneal dialysis through an indwelling catheter. It is the most common bacterium to cause cerebrospinal fluid shunt infections.

S. saprophyticus causes urinary tract infections, particularly in sexually active young women. Most women with this infection have had sexual intercourse within the previous 24 hours. This organism is second to *Escherichia coli* as a cause of community-acquired urinary tract infections in young women.

Staphylococcus lugdunensis is a relatively uncommon coagulase-negative staphylococcus that causes prosthetic valve endocarditis and skin infections.

Laboratory Diagnosis

Smears from staphylococcal lesions reveal gram-positive cocci in grapelike clusters (see Figure 15–5). Cultures of coagulase-negative staphylococci typically yield white colonies that are nonhemolytic. The two coagulase-negative staphylococci are differentiated by their reaction to the antibiotic novobiocin: *S. epidermidis* is sensitive, whereas *S. saprophyticus* is resistant.

There are no serologic or skin tests used for the diagnosis of any staphylococcal infection.

Treatment & Prevention

S. epidermidis is highly antibiotic-resistant. Most strains produce β-lactamase but are sensitive to β-lactamase–resistant drugs such as nafcillin. These are called methicillin-sensitive strains (MSSE). Some strains are methicillin/nafcillin-resistant (MRSE) due to altered PBPs. The drug of choice for MRSE is vancomycin, to which either rifampin or an aminoglycoside can be added. Removal of the catheter or other device is often necessary. *S. saprophyticus* urinary tract infections can be treated with trimethoprim-sulfamethoxazole or a quinolone, such as ciprofloxacin.

There is no vaccine against the coagulase-negative staphylococci. Prompt removal of intravenous catheters reduces the risk of infections caused by *S. epidermidis*.

STREPTOCOCCUS

Streptococci of medical importance are listed in Table 15–3. All but one of these streptococci are discussed in this section; *S. pneumoniae* is discussed separately at the end of this chapter because of its importance.

Diseases

Streptococci cause a wide variety of infections. *S. pyogenes* (group A *Streptococcus*) is the leading bacterial cause of pharyngitis (Figure 15–9) and cellulitis (Figure 15–10). It is an important cause of impetigo (see Figure 15–4), erysipelas, necrotizing fasciitis, scarlet fever, and streptococcal toxic shock syndrome. It is also the inciting factor of two important immunologic diseases, namely, rheumatic fever and acute glomerulonephritis. *Streptococcus agalactiae* (group B *Streptococcus*) is the leading cause of neonatal sepsis and meningitis. *Enterococcus faecalis* is an important cause of hospital-acquired urinary tract infections and endocarditis. Viridans group streptococci are the most common cause of endocarditis (Figure 15–11). *Streptococcus bovis* (also known as *Streptococcus gallolyticus*) is an uncommon cause of endocarditis.

Important Properties

Streptococci are spherical gram-positive cocci arranged in chains or pairs (Figure 15–12). All streptococci are **catalase-negative**, whereas staphylococci are catalase-positive (see Table 15–3).

One of the most important characteristics for identification of streptococci is the type of hemolysis (Figure 15–13).

(1) **α-Hemolytic** streptococci form a green zone around their colonies as a result of incomplete lysis of red blood cells in the agar. The green color is formed when hydrogen peroxide produced by the bacteria oxidizes hemoglobin (red color) to biliverdin (green color).

TABLE 15-3 Streptococci of Medical Importance

Species	Lancefield Group	Typical Hemolysis	Diagnostic Features[1]
S. pyogenes	A	β	Bacitracin-sensitive
S. agalactiae	B	β	Bacitracin-resistant; hippurate hydrolyzed
E. faecalis	D	α or β or none	Growth in 6.5% NaCl[2]
S. bovis[3]	D	α or none	No growth in 6.5% NaCl
S. pneumoniae	NA[4]	α	Bile-soluble; inhibited by optochin
Viridans group[5]	NA	α	Not bile-soluble; not inhibited by optochin

[1]All streptococci are catalase-negative.

[2]Both *E. faecalis* and *S. bovis* grow on bile-esculin agar, whereas other streptococci do not. They hydrolyze the esculin, and this results in a characteristic black discoloration of the agar.

[3]*Streptococcus bovis* is a nonenterococcal group D organism.

[4]NA, not applicable.

[5]Viridans group streptococci include several species, such as *S. sanguinis, S. mutans, S. mitis, S. gordonii, S. salivarius, S. anginosus, S. milleri,* and *S. intermedius.*

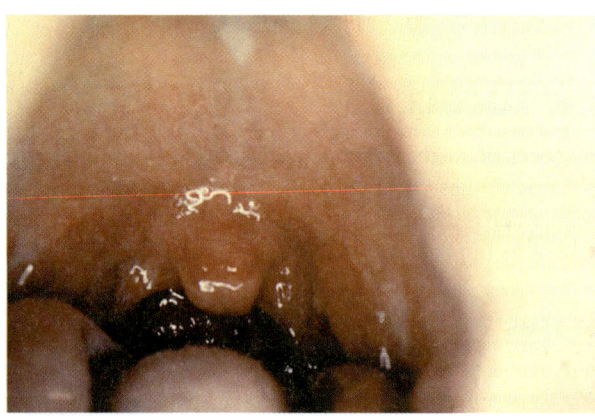

FIGURE 15-9 Pharyngitis. Note erythema of soft palate, uvula, and posterior pharynx and swelling of the uvula. The most common bacterial cause of pharyngitis is *Streptococcus pyogenes*. Note: The curved white lines on the uvula and the palate are artifacts of photography. (Reproduced with permission from Centers for Disease Control and Prevention. CDC #6323.)

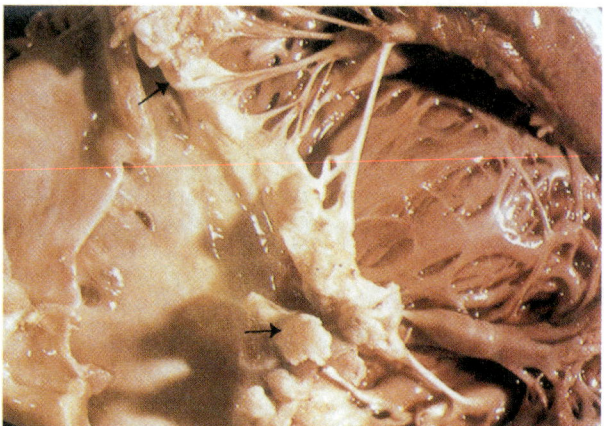

FIGURE 15-11 Endocarditis. Note vegetations (black arrows) on mitral valve. Viridans streptococci are the most common cause of subacute bacterial endocarditis. (Reproduced with permission from Longo DL, Fauci AS, Kasper DL, et al: *Harrison's Principles of Internal Medicine*, 18th ed. New York, NY: McGraw Hill; 2012.)

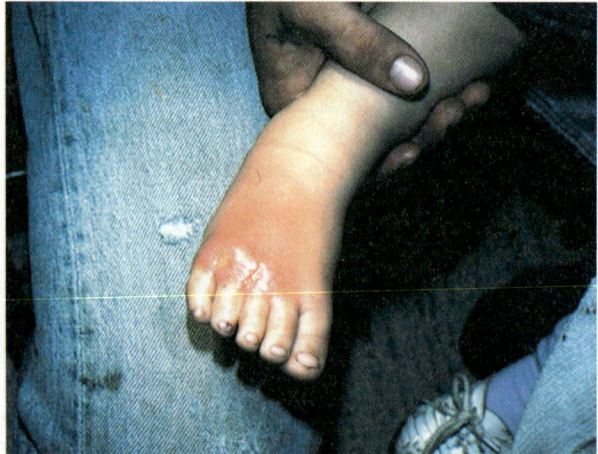

FIGURE 15-10 Cellulitis. Note erythema and swelling of the dorsum of the foot. *Streptococcus pyogenes* is the most common cause of cellulitis. (Reproduced with permission from Usatine RP, Smith MA, Mayeaux EJ Jr, et al: *The Color Atlas of Family Medicine*. New York, NY: McGraw Hill; 2009. Photo contributor: Richard P. Usatine, MD.)

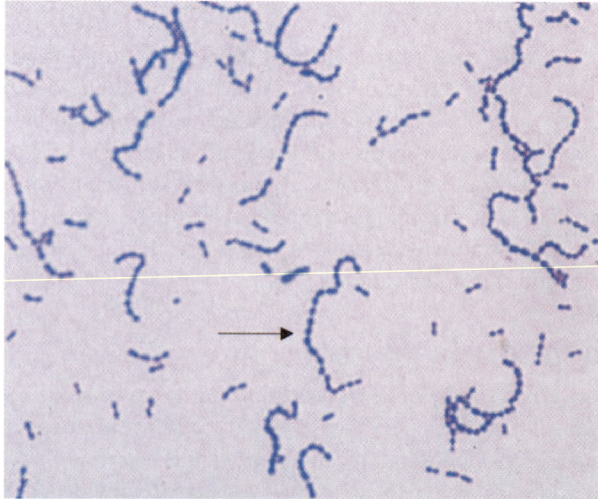

FIGURE 15-12 *Streptococcus pyogenes*—Gram stain. Arrow points to a long chain of gram-positive cocci. (Used with permission from Professor Shirley Lowe, University of California, San Francisco School of Medicine.)

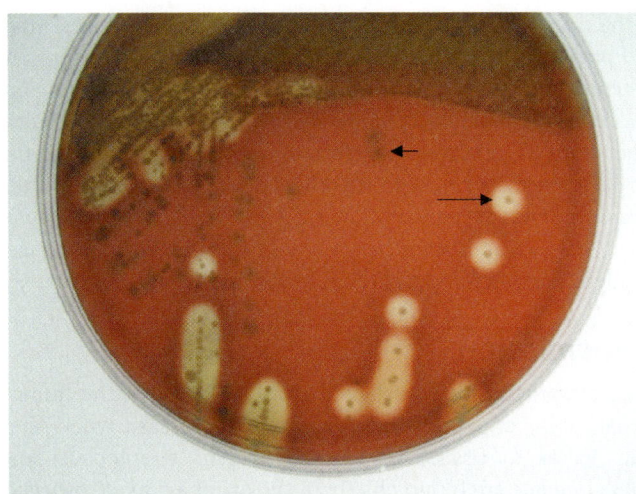

FIGURE 15–13 α-Hemolysis and β-hemolysis on blood agar. Short arrow points to an α-hemolytic colony, probably a viridans group *Streptococcus*. Long arrow points to a β-hemolytic colony, probably *Streptococcus pyogenes*. The specimen was a throat swab taken from a person with a sore throat. (Used with permission from Professor Shirley Lowe, University of California, San Francisco School of Medicine.)

(2) **β-Hemolytic** streptococci form a clear zone around their colonies because complete lysis of the red cells occurs. β-Hemolysis is due to the production of enzymes (hemolysins) called streptolysin O and streptolysin S (see "Pathogenesis" later).

(3) Some streptococci are nonhemolytic (γ-hemolysis).

There are two important antigens of β-hemolytic streptococci:

(1) **C carbohydrate** determines the *group* of β-hemolytic streptococci. It is located in the cell wall, and its specificity is determined by an amino sugar. For example, group A β-hemolytic streptococci (*S. pyogenes*) are distinguished from group B β-hemolytic streptococci (*S. agalactiae*) because they have a different C carbohydrate.

(2) **M protein** is the most important virulence factor of *S. pyogenes*. It protrudes from the outer surface of the cell and blocks phagocytosis (i.e., it is **antiphagocytic**). It inactivates C3b, a component of complement that opsonizes the bacteria prior to phagocytosis (see Chapter 63). Strains of *S. pyogenes* that do *not* produce M protein are nonpathogenic.

M protein also determines the *type* of group A β-hemolytic streptococci. There are approximately 100 serotypes based on the M protein, which explains why multiple infections with *S. pyogenes* can occur. Antibody to M protein provides type-specific immunity.

Strains of *S. pyogenes* that produce certain M protein types are **rheumatogenic** (i.e., cause primarily rheumatic fever), whereas strains of *S. pyogenes* that produce other M protein types are **nephritogenic** (i.e., cause primarily acute glomerulonephritis). Although M protein is the main antiphagocytic component of *S. pyogenes*, the organism also has a polysaccharide capsule that plays a role in retarding phagocytosis.

Classification of Streptococci

β-Hemolytic Streptococci

These are arranged into groups A–U (known as Lancefield groups) on the basis of antigenic differences in C carbohydrate. In the clinical laboratory, the group is determined by precipitin tests with specific antisera or by immunofluorescence.

Group A streptococci (*S. pyogenes*) are one of the most important human pathogens. They are the most frequent bacterial cause of pharyngitis and a very common cause of skin infections. They adhere to pharyngeal epithelium via pili composed of lipoteichoic acid and M protein. Many strains have a hyaluronic acid capsule that is antiphagocytic. The growth of *S. pyogenes* on agar plates in the laboratory is inhibited by the antibiotic bacitracin, an important diagnostic criterion (Figure 15–14).

Group B streptococci (*S. agalactiae*) colonize the genital tract of some women and can cause neonatal meningitis and sepsis. They are usually bacitracin-resistant. They hydrolyze (break down) hippurate, an important diagnostic criterion.

Group D streptococci include enterococci (e.g., *E. faecalis* and *Enterococcus faecium*) and nonenterococci (e.g., *S. bovis*). Enterococci are members of the normal flora of the colon and are noted for their ability to cause urinary, biliary, and cardiovascular infections. They are very hardy organisms; they can grow in hypertonic (6.5%) saline or in bile and are not killed by penicillin G. As a result, a synergistic combination of penicillin and an aminoglycoside (e.g., gentamicin) is required to kill enterococci. Vancomycin can also be used, but vancomycin-resistant enterococci (VRE) have emerged and become an important and much feared cause of life-threatening nosocomial infections.

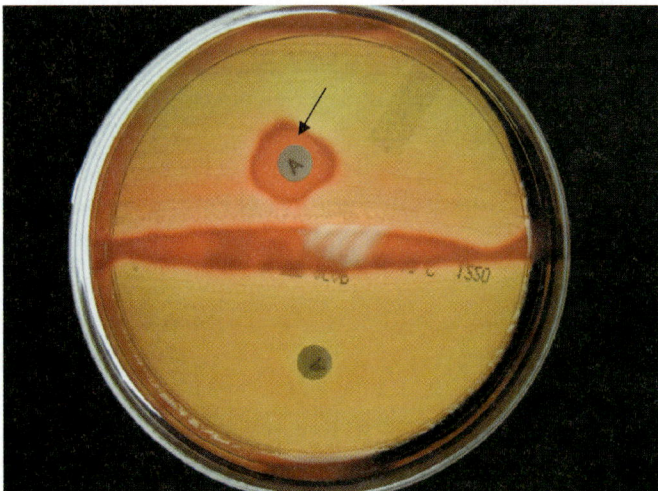

FIGURE 15–14 Bacitracin test. Arrow points to zone of inhibition of growth of group A streptococci (*Streptococcus pyogenes*) caused by bacitracin that has diffused from the disk labeled A. Upper half of blood agar plate shows β-hemolysis caused by group A streptococci, except in the region around the bacitracin disk. Lower half of blood agar plate shows β-hemolysis caused by group B streptococci (*Streptococcus agalactiae*), and there is no zone of inhibition around the bacitracin disk. (Used with permission from Professor Shirley Lowe, University of California, San Francisco School of Medicine.)

More strains of *E. faecium* are vancomycin-resistant than are strains of *E. faecalis*.

Nonenterococcal group D streptococci, such as *S. bovis*, can cause similar infections but are much less hardy organisms (e.g., they are inhibited by 6.5% NaCl and killed by penicillin G). Note that the hemolytic reaction of group D streptococci is variable: most are α-hemolytic, but some are β-hemolytic, and others are nonhemolytic.

Groups C, E, F, G, H, and K–U streptococci infrequently cause human disease.

Non–β-Hemolytic Streptococci

Some streptococci produce no hemolysis; others produce α-hemolysis. The principal α-hemolytic organisms are *S. pneumoniae* (pneumococci) and the viridans group of streptococci (e.g., *Streptococcus mitis, Streptococcus sanguinis,* and *Streptococcus mutans*). Pneumococci and viridans streptococci are distinguished in the clinical laboratory by two main criteria: (1) the growth of pneumococci is inhibited by optochin, whereas the growth of viridans streptococci is not inhibited; and (2) colonies of pneumococci dissolve when exposed to bile (bile-soluble), whereas colonies of viridans streptococci do not dissolve.

Viridans streptococci are part of the normal flora of the human pharynx. They may intermittently enter the bloodstream and cause infective endocarditis. *S. mutans* synthesizes polysaccharides (dextrans) that are found in dental plaque and lead to dental caries. *Streptococcus intermedius* and *Streptococcus anginosus* (also known as the *S. anginosus-milleri* group) are usually α-hemolytic or nonhemolytic, but some isolates are β-hemolytic. They are found primarily in the mouth and colon and cause abscesses of the brain, lung, and liver.

Peptostreptococci

These grow under anaerobic or microaerophilic conditions and produce variable hemolysis. Peptostreptococci are members of the normal flora of the gut, mouth, and female genital tract and participate in mixed anaerobic infections. The term *mixed anaerobic infections* refers to the fact that these infections are caused by multiple bacteria, some of which are anaerobes and others are facultatives. For example, peptostreptococci and viridans streptococci, both members of the oral flora, are often found in brain abscesses following dental surgery. *Peptostreptococcus magnus* and *Peptostreptococcus anaerobius* are the species frequently isolated from clinical specimens.

Transmission

Most streptococci are part of the normal flora of the human throat, skin, and intestines but produce disease when they gain access to tissues or blood. Viridans streptococci and *S. pneumoniae* are found chiefly in the **oropharynx**; *S. pyogenes* is found on the **skin** and in the oropharynx in small numbers; *S. agalactiae* occurs in the **vagina** and colon; and both the enterococci and anaerobic streptococci are located in the **colon**.

Pathogenesis

Group A streptococci (*S. pyogenes*) cause disease by three mechanisms: (1) **pyogenic inflammation**, which is induced locally at the site of the organisms in tissue; (2) **exotoxin production**, which can cause widespread systemic symptoms in areas of the body where there are no organisms; and (3) **immunologic**, which occurs when antibody against a component of the organism cross-reacts with normal tissue or forms immune complexes that damage normal tissue (see the section on poststreptococcal diseases later in the chapter). The immunologic reactions cause inflammation (e.g., the inflamed joints of rheumatic fever), but there are no organisms in the lesions (Table 15–4).

The **M protein** of *S. pyogenes* is its most important antiphagocytic factor, but its capsule, composed of hyaluronic acid, is also antiphagocytic. Antibodies are not formed against the

TABLE 15–4 Important Features of Pathogenesis by Streptococci

Organism	Type of Pathogenesis	Typical Disease	Main Site of Disease (D), Colonization (C), or Normal Flora (NF)
S. pyogenes (group A)	1. Pyogenic		
	a. Local	Impetigo, cellulitis	Skin (D)
		Pharyngitis	Throat (D)
	b. Disseminated	Sepsis	Bloodstream (D)
	2. Toxigenic	Scarlet fever	Skin (D)
		Toxic shock	Many organs (D)
		Necrotizing fasciitis	Skin, fascia, and muscle (D)
	3. Immune-mediated (poststreptococcal, nonsuppurative)	Rheumatic fever	Heart, joints (D)
		Acute glomerulonephritis	Kidney (D)
S. agalactiae (group B)	Pyogenic	Neonatal sepsis and meningitis	Vagina (C)
E. faecalis (group D)	Pyogenic	Urinary tract infection, endocarditis	Colon (NF)
S. bovis (group D)	Pyogenic	Endocarditis	Colon (NF)
S. pneumoniae	Pyogenic	Pneumonia, otitis media, meningitis	Oropharynx (C)
Viridans streptococci	Pyogenic	Endocarditis	Oropharynx (NF)

capsule because hyaluronic acid is a normal component of the body and humans are tolerant to it.

Group A streptococci produce four important **enzymes related to pathogenesis:**

(1) **Hyaluronidase** degrades hyaluronic acid, which is the ground substance of subcutaneous tissue. Hyaluronidase is known as **spreading factor** because it facilitates the rapid spread of *S. pyogenes* in skin infections (cellulitis).

(2) **Streptokinase** (fibrinolysin) activates plasminogen to form plasmin, which dissolves fibrin in clots, thrombi, and emboli. It can be used to lyse thrombi in the coronary arteries of heart attack patients.

(3) **DNase** (streptodornase) degrades DNA in exudates or necrotic tissue. Antibody to DNase B develops during pyoderma; this can be used for diagnostic purposes. Streptokinase–streptodornase mixtures applied as a skin test give a positive reaction in most adults, indicating normal cell-mediated immunity.

(4) **IgG degrading enzyme** is a protease that specifically cleaves IgG heavy chains. This prevents opsonization and complement activation thereby enhancing the virulence of the organism.

In addition, group A streptococci produce five important **toxins and hemolysins:**

(1) **Pyrogenic exotoxins** are a group of toxins that include **pyrogenic exotoxin A**, the cause of streptococcal toxic shock syndrome, and **erythrogenic toxin**, the cause of scarlet fever. These toxins are superantigens and act by the same mechanism as the toxic shock syndrome toxin of *S. aureus* (see earlier in this chapter). Both of these toxins are encoded by bacteriophages that lysogenize *S. pyogenes*.

The term "pyrogenic" means to produce fever. These superantigens induce fever and other symptoms by stimulating production of IL-1 and other proinflammatory cytokines by T cells and macrophages.

(2) **Exotoxin B** is a protease that rapidly destroys tissue and degrades Igs. It is produced in large amounts by certain strains of *S. pyogenes,* the so-called **"flesh-eating" streptococci that cause necrotizing fasciitis.**

(3) **Streptolysin O** and **streptolysin S** are pore-forming toxins which disrupt cell membranes resulting in damage to tissue and death of phagocytes. They cause the β-hemolysis seen surrounding colonies of *S. pyogenes* on blood agar plates in the clinical laboratory. Streptolysin O is inactivated by oxygen whereas streptolysin S is stable to oxygen. It is streptolysin S that causes the β-hemolysis when colonies grow on the surface of the agar.

Antibody to streptolysin O develops after infection by *S. pyogenes*. Presence of this antibody is used as evidence of prior infection by *S. pyogenes* in the diagnosis of rheumatic fever. Antibody to streptolysin S does not develop because it resembles a human protein and we are tolerant to it.

Pathogenesis by group B streptococci (*S. agalactiae*) is based on the ability of the organism to induce an inflammatory response. However, unlike *S. pyogenes*, no cytotoxic enzymes or exotoxins have been described, and there is no evidence for any immunologically induced disease. Group B streptococci have a polysaccharide capsule that is antiphagocytic, and anticapsular antibody is protective.

Pathogenesis by *S. pneumoniae* and the viridans streptococci is uncertain, as no exotoxins or tissue-destructive enzymes have been demonstrated. The main virulence factor of *S. pneumoniae* is its antiphagocytic polysaccharide capsule. Many of the strains of viridans streptococci that cause endocarditis produce a glycocalyx that enables the organism to adhere to the heart valve.

Clinical Findings

S. pyogenes causes three types of diseases: (1) **pyogenic** diseases such as pharyngitis and cellulitis, (2) **toxigenic** diseases such as scarlet fever and toxic shock syndrome, and (3) **immunologic** diseases such as rheumatic fever and acute glomerulonephritis (AGN). (See next section on poststreptococcal diseases.)

S. pyogenes (group A *Streptococcus*) is the most common bacterial cause of **pharyngitis** (sore throat). Streptococcal pharyngitis (strep throat) is characterized by throat pain and fever. On examination, an inflamed throat and tonsils, often with a yellowish exudate, are found, accompanied by tender cervical lymph nodes. If untreated, spontaneous recovery often occurs in 10 days, but rheumatic fever may occur (see next section on poststreptococcal diseases). Untreated pharyngitis may extend to the middle ear (otitis media), the sinuses (sinusitis), the mastoids (mastoiditis), or the meninges (meningitis). Continuing inability to swallow may indicate a peritonsillar or retropharyngeal abscess.

If the infecting streptococci produce erythrogenic toxin and the host lacks antitoxin, scarlet fever may result. A "strawberry" tongue is a characteristic lesion seen in scarlet fever. *S. pyogenes* also causes another toxin-mediated disease, streptococcal toxic shock syndrome, which has clinical findings similar to those of staphylococcal toxic shock syndrome (see page 109). However, streptococcal toxic shock syndrome typically has a recognizable site of pyogenic inflammation and blood cultures are often positive, whereas staphylococcal toxic shock syndrome typically has neither a site of pyogenic inflammation nor positive blood cultures.

Group A streptococci cause **SSTIs**, such as cellulitis, erysipelas (Figure 15–15), necrotizing fasciitis (streptococcal gangrene), and impetigo (see Figure 15–4). They also cause endometritis (puerperal fever), a serious infection of pregnant women, and sepsis.

Necrotizing fasciitis is often called the "**flesh-eating**" disease. In addition to *S. pyogenes*, *Clostridium perfringens* and MRSA are important causes. The clinical aspects of necrotizing fasciitis are described in Chapter 77.

Impetigo, a form of pyoderma, is a superficial skin infection characterized by "honey-colored" crusted lesions. Lymphangitis can occur, especially on the forearm associated with an infection on the hand.

Group B streptococci cause **neonatal sepsis** and **meningitis**. The main predisposing factor is prolonged (longer than 18 hours) rupture of the membranes in women who are colonized with

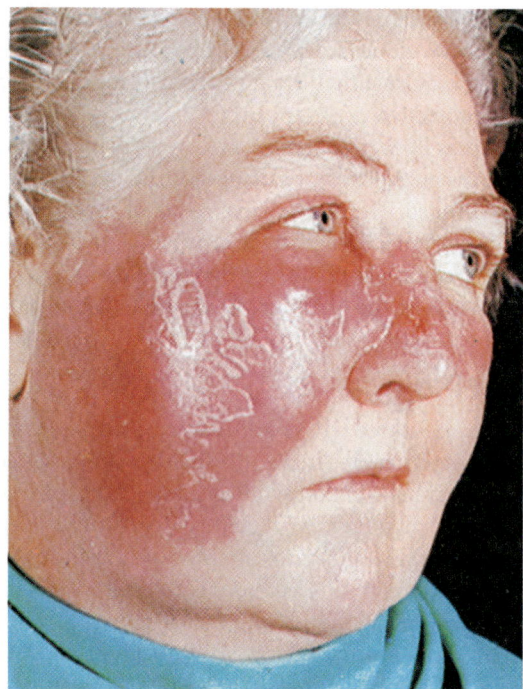

FIGURE 15–15 Erysipelas. Note well-demarcated border of the inflamed area. *Streptococcus pyogenes* is the most common cause of erysipelas. (Reproduced with permission from Longo DL, Fauci AS, Kasper DL, et al: *Harrison's Principles of Internal Medicine*, 18th ed. New York, NY: McGraw Hill; 2012.)

the organism. Children born prior to 37 weeks of gestation have a greatly increased risk of disease. Also, children whose mothers lack antibody to group B streptococci and who consequently are born without transplacentally acquired IgG have a high rate of neonatal sepsis caused by this organism. Group B streptococci are an important cause of neonatal pneumonia as well.

Although most group B streptococcal infections are in neonates, this organism also causes infections such as pneumonia, endocarditis, arthritis, cellulitis, and osteomyelitis in adults. Postpartum endometritis also occurs. Diabetes is the main predisposing factor for adult group B streptococcal infections.

Viridans streptococci (e.g., *S. mutans, S. sanguinis, S. salivarius,* and *S. mitis*) are the most common cause of infective **endocarditis**. They enter the bloodstream (bacteremia) from the oropharynx, typically after **dental surgery**. Signs of endocarditis are fever, heart murmur, anemia, and embolic events such as splinter hemorrhages, subconjunctival petechial hemorrhages, and Janeway lesions. The heart murmur is caused by vegetations on the heart valve (see Figure 15–11). It is 100% fatal unless effectively treated with antimicrobial agents. About 10% of endocarditis cases are caused by enterococci, but any organism causing bacteremia may settle on deformed valves. At least three blood cultures are necessary to ensure recovery of the organism in more than 90% of cases.

Viridans streptococci, especially *S. anginosus, S. milleri,* and *S. intermedius,* also cause brain abscesses, often in combination with mouth anaerobes (a mixed aerobic–anaerobic infection). Dental surgery is an important predisposing factor to brain abscess because it provides a portal for the viridans streptococci and the anaerobes in the mouth to enter the bloodstream

(bacteremia) and spread to the brain. Viridans streptococci are involved in mixed aerobic–anaerobic infections in other areas of the body as well (e.g., lung abscesses and abdominal abscesses, including liver abscesses).

Enterococci (*E. faecalis, E, faecium*) cause **urinary tract infections**, especially in hospitalized patients. Indwelling urinary catheters and urinary tract instrumentation are important predisposing factors. Indwelling intravenous catheters including central lines, predispose to bloodstream infections with Enterococci. Enterococci also cause endocarditis, particularly in patients who have undergone gastrointestinal or urinary tract surgery or instrumentation. They also cause intra-abdominal and pelvic infections, typically in combination with anaerobes.

S. bovis, a nonenterococcal group D *Streptococcus,* causes **endocarditis**, especially in patients with carcinoma of the colon. This association is so strong that patients with *S. bovis,* bacteremia, or endocarditis should be investigated for the presence of colonic carcinoma.

Peptostreptococci are one of the most common bacteria found in brain, lung, abdominal, and pelvic abscesses.

Poststreptococcal (Nonsuppurative) Diseases

These are disorders in which a local infection with group A streptococci is followed weeks later by inflammation in an organ that was *not* infected by the streptococci. The inflammation is caused by an **immunologic (antibody)** response to streptococcal M proteins that cross-react with human tissues. Some strains of *S. pyogenes* bearing certain M proteins are nephritogenic and cause AGN, and other strains bearing different M proteins are rheumatogenic and cause acute rheumatic fever. Note that these diseases appear several weeks after the actual infection because that is the length of time it takes to produce sufficient antibodies.

Acute Glomerulonephritis

AGN typically occurs 2 to 3 weeks after skin infection by certain group A streptococcal types in children (e.g., M protein type 49 causes AGN most frequently). AGN is more frequent after skin infections than after pharyngitis. The most striking clinical features are hypertension, edema of the face (especially periorbital edema) and ankles, and "smoky" urine (due to red cells in the urine). Most patients recover completely. Reinfection with streptococci rarely leads to recurrence of glomerulonephritis.

The disease is initiated by **antigen–antibody complexes on the glomerular basement membrane**. Complement is activated and C5a attracts neutrophils that secrete enzymes that damage the endothelium of the glomerular capillaries. There is no antibiotic therapy for AGN. Administration of penicillin after the onset of symptoms is not effective.

Acute Rheumatic Fever

Approximately 2 weeks after a group A streptococcal infection—usually pharyngitis—rheumatic fever, characterized by fever, migratory polyarthritis, and carditis, may develop. The carditis damages myocardial and endocardial tissue, especially the mitral and aortic valves, resulting in vegetations on the valves, raising the risk of endocarditis. Chorea, manifesting as uncontrollable, spasmodic movements of the limbs or face may also occur.

The presence of antibody to streptolysin O (ASO) demonstrates that a prior infection with *S. pyogenes* has occurred. An elevated erythrocyte sedimentation rate is often seen. As there is no specific laboratory test for rheumatic fever, a set of findings called he Jones Criteria are used to make the diagnosis (see Chapter 70).

Rheumatic fever is due to an **immunologic cross-reaction** between antibodies formed against M proteins of *S. pyogenes* and proteins on the surface of joint, heart, and brain tissue. It is an autoimmune disease greatly exacerbated by recurrent streptococcal infections. If streptococcal infections are treated within 8 days of onset, rheumatic fever is usually prevented. After a heart-damaging attack of rheumatic fever, reinfection must be prevented by long-term prophylaxis.

Note that group A streptococcal *skin* infections do not cause rheumatic fever. Most cases of pharyngitis caused by group A streptococci occur in children between ages 5 and 15 years, and hence rheumatic fever occurs in that age group. In the United States, less than 0.5% of group A streptococcal infections lead to rheumatic fever, but in developing tropical countries, the rate is higher than 5%. Rheumatic heart disease remains a significant global health burden.

Laboratory Diagnosis

Microbiologic

Gram-stained smears are useless in streptococcal pharyngitis because viridans streptococci are members of the normal flora and cannot be visually distinguished from the pathogenic *S. pyogenes*. However, stained smears from skin lesions or wounds that reveal streptococci are diagnostic. Cultures of swabs from the pharynx or lesion on blood agar plates show small, translucent β-hemolytic colonies in 18 to 48 hours. If **inhibited by bacitracin** disk, they are likely to be group A streptococci (see Figure 15–14).

Group B streptococci are characterized by their ability to **hydrolyze hippurate** and by the production of a protein that causes enhanced hemolysis on sheep blood agar when combined with β-hemolysin of *S. aureus* (CAMP test). Group D streptococci **hydrolyze esculin in the presence of bile** (i.e., they produce a black pigment on bile-esculin agar). The group D organisms are further subdivided: the enterococci **grow in hypertonic (6.5%) NaCl**, whereas the nonenterococci do not.

Although cultures remain the gold standard for the diagnosis of streptococcal pharyngitis, a problem exists because the results of culturing are not available for at least 18 hours, and it is beneficial to know while the patient is in the office whether antibiotics should be prescribed. For this reason, rapid tests that provide a diagnosis in approximately 10 minutes were developed.

The rapid test detects bacterial antigens in a throat swab specimen. In the test, specific antigens from the group A streptococci are extracted from the throat swab with certain enzymes and are reacted with antibody to these antigens bound to latex particles. Agglutination of the colored latex particles occurs if group A streptococci are present in the throat swab. The specificity of these

tests is high, but the sensitivity is low (i.e., false-negative results can occur). If the test result is negative but the clinical suspicion of streptococcal pharyngitis is high, a culture should be done.

A rapid test is also available for the detection of group B streptococci in vaginal and rectal samples. This polymerase chain reaction (PCR)-based assay detects the DNA of the organism, and results can be obtained in approximately 1 hour.

Viridans group streptococci form α-hemolytic colonies on blood agar and must be distinguished from *S. pneumoniae* (pneumococci), which is also α-hemolytic. Viridans group streptococci are resistant to lysis by bile and will grow in the presence of optochin, whereas pneumococci will not. The various viridans group streptococci are classified into species by using a variety of biochemical tests.

Serologic

ASO titers are high soon after group A streptococcal infections. In patients suspected of having rheumatic fever, an **elevated ASO titer** is typically used as evidence of previous infection because throat culture results are often negative at the time the patient presents with rheumatic fever. Titers of anti–DNase B are high in group A streptococcal skin infections and serve as an indicator of previous streptococcal infection in patients suspected of having AGN.

Treatment

Group A streptococcal infections can be treated with either penicillin G or amoxicillin, but neither rheumatic fever nor AGN patients benefit from penicillin treatment *after* the onset of the two diseases. In mild group A streptococcal infections, oral penicillin V can be used. In penicillin-allergic patients, erythromycin or one of its long-acting derivatives (e.g., azithromycin) can be used. However, erythromycin-resistant strains of *S. pyogenes* have emerged that may limit the effectiveness of the macrolide class of drugs in the treatment of streptococcal pharyngitis. Clindamycin can also be used in penicillin-allergic patients. *Streptococcal pyogenes* is not resistant to penicillins.

Invasive group A streptococcal infections such as necrotizing fasciitis and streptococcal toxic shock syndrome can be treated with a combination of clindamycin and intravenous gamma globulins.

Endocarditis caused by most viridans streptococci is curable using prolonged penicillin treatment. However, enterococcal endocarditis can be eradicated only by a penicillin or vancomycin combined with an aminoglycoside.

Strains of enterococci (especially *E. faecium*), resistant to multiple drugs (e.g., penicillins, aminoglycosides, and vancomycin) have emerged, particularly in the hospital setting. Resistance to vancomycin in enterococci is mediated by a cassette of genes that encode the enzymes that substitute D-lactate for D-alanine in the peptidoglycan. The same set of genes encodes vancomycin resistance in *S. aureus*.

VRE are now an important cause of nosocomial infections. There is no reliable antibiotic therapy for these organisms. At present, linezolid (Zyvox) and daptomycin (Cubicin) are typically used to treat infections caused by VRE.

Nonenterococcal group D streptococci (e.g., *S. bovis*) are not highly resistant and can be treated with penicillin G.

The drug of choice for group B streptococcal infections is either penicillin G or ampicillin. Some strains may require higher doses of penicillin G or a combination of penicillin G and an aminoglycoside to eradicate the organism. Peptostreptococci can be treated with penicillin G.

Prevention

Rheumatic fever can be prevented by prompt treatment of group A streptococcal pharyngitis with penicillin G or oral penicillin V. Prevention of streptococcal infections (usually with benzathine penicillin once each month for several years) in persons who have had rheumatic fever is important to prevent recurrence of the disease. There is no evidence that patients who have had AGN require a similar penicillin prophylaxis.

In patients with damaged heart valves who undergo invasive dental procedures, endocarditis caused by viridans streptococci can be prevented by using amoxicillin perioperatively. To avoid unnecessary use of antibiotics, it is recommended to give amoxicillin prophylaxis only to patients who have the highest risk of severe consequences from endocarditis (e.g., those with prosthetic heart valves or with previous infective endocarditis) and who are undergoing high-risk dental procedures, such as manipulation of gingival tissue. It is no longer recommended that patients undergoing gastrointestinal or genitourinary tract procedures receive prophylaxis.

The incidence of neonatal sepsis caused by group B streptococci can be reduced by a two-pronged approach: (1) All pregnant women at 35 to 37 weeks of gestation should be screened by doing vaginal and rectal cultures. If cultures are positive, then penicillin G (or ampicillin) should be administered intravenously at the time of delivery. (2) If the patient has not had cultures done, then penicillin G (or ampicillin) should be administered intravenously at the time of delivery to women who have not delivered within 18 hours after rupture of membranes, or whose labor begins before 37 weeks of gestation, or who have a fever at the time of labor. If the patient is allergic to penicillin, either cefazolin or vancomycin can be used.

Oral ampicillin given to women who are vaginal carriers of group B streptococci does not eradicate the organism. Rapid screening tests for group B streptococcal antigens in vaginal specimens can be insensitive, and neonates born of antigen-negative women have, nevertheless, been diagnosed with neonatal sepsis. Note, also, that as group B streptococcal infections have declined as a result of these prophylactic measures, neonatal infections caused by *E. coli* have increased.

There are no vaccines available against any of the streptococci except *S. pneumoniae* (see following section).

STREPTOCOCCUS PNEUMONIAE

Diseases

S. pneumoniae causes pneumonia, bacteremia, meningitis, and infections of the upper respiratory tract such as otitis media, mastoiditis, and sinusitis. Pneumococci are the most common cause of

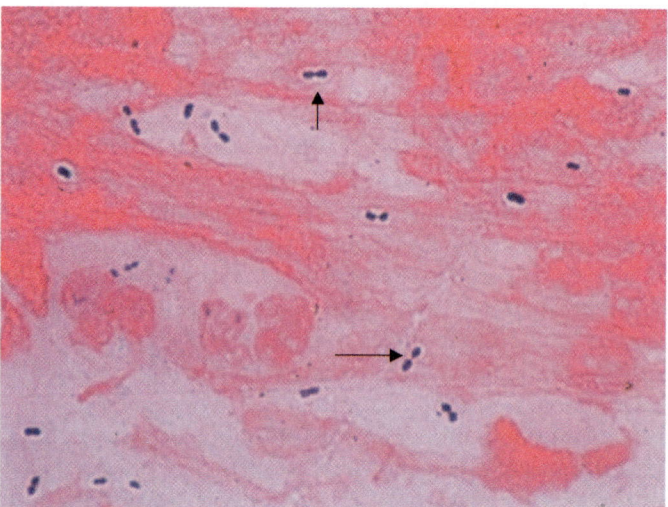

FIGURE 15–16 *Streptococcus pneumoniae*—Gram stain. Arrows point to typical gram-positive diplococci. Note that the clear area around the organism is the capsule. (Used with permission from Professor Shirley Lowe, University of California, San Francisco School of Medicine.)

community-acquired pneumonia, meningitis, sepsis in splenectomized individuals, otitis media, and sinusitis. They are a common cause of conjunctivitis, especially in children. Note that *S. pneumoniae* is also known as the pneumococcus (plural, pneumococci).

Important Properties

Pneumococci are gram-positive lancet-shaped cocci arranged in pairs (**diplococci**) or short chains (Figure 15–16). (The term *lancet-shaped* means that the diplococci are oval with somewhat pointed ends rather than being round.) On blood agar, they produce α-hemolysis. In contrast to viridans streptococci, they are lysed by bile or deoxycholate, and their growth is inhibited by optochin (Figure 15–17).

Pneumococci possess **polysaccharide capsules** that have 91 antigenically distinct types (serotypes) based on the different sugars in the polysaccharide. With type-specific antiserum, capsules swell (**quellung reaction**), and this can be used to identify the type.

Capsules are important virulence factors (i.e., they interfere with phagocytosis). Specific antibody to the capsule opsonizes the organism, thereby facilitating phagocytosis. Such antibody develops in humans as a result either of infection (asymptomatic or clinical) or of administration of polysaccharide vaccine. Capsular polysaccharide elicits primarily a β-cell (i.e., T-independent) response.

Another important surface component of *S. pneumoniae* is a teichoic acid in the cell wall called **C-substance** (also known as **C-polysaccharide**). It is medically important not for itself, but because it reacts with a normal serum protein made by the liver called **C-reactive protein** (CRP). CRP is an "acute-phase" protein that is elevated as much as 1000-fold in acute inflammation. CRP is not an antibody (which are γ-globulins) but rather a β-globulin. (Plasma contains α-, β-, and γ-globulins.) Note that CRP is a nonspecific indicator of inflammation and is elevated in response to the presence of many organisms, not just *S. pneumoniae*.

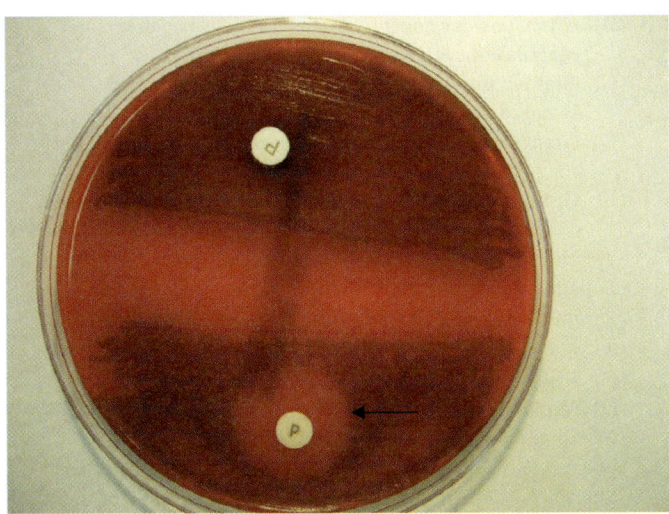

FIGURE 15–17 Optochin test. Arrow points to zone of inhibition of growth of *Streptococcus pneumoniae* caused by optochin that has diffused from the disk labeled P. In the lower half of the blood agar plate, there is α-hemolysis caused by *S. pneumoniae*, except in the region around the optochin disk. The arrow points to the outer limit of the zone of inhibition. Upper half of blood agar plate shows α-hemolysis caused by a viridans streptococcus, and there is no zone of inhibition around the optochin disk. (Used with permission from Professor Shirley Lowe, University of California, San Francisco School of Medicine.)

Clinically, CRP in human serum is measured in the laboratory by its reaction with the carbohydrate of *S. pneumoniae*. Another medical importance of CRP is that an elevated CRP appears to be a better predictor of heart attack risk than an elevated cholesterol level.

Transmission

Humans are the natural hosts for pneumococci; there is no animal reservoir. Because a proportion (5%–50%) of the healthy population harbors virulent organisms in the oropharynx, pneumococcal infections are not considered to be communicable. Resistance is high in healthy young people, and disease results most often when predisposing factors (see following discussion) are present.

Pathogenesis

The most important virulence factor is the capsular polysaccharide, and anticapsular antibody is protective. Lipoteichoic acid, which activates complement and induces inflammatory cytokine production, contributes to the inflammatory response and to the septic shock syndrome that occurs in some immunocompromised patients. Pneumolysin, the hemolysin that causes α-hemolysis, may also contribute to pathogenesis.

Pneumococci produce **IgA protease** that enhances the organism's ability to colonize the mucosa of the upper respiratory tract by cleaving IgA. Pneumococci multiply in tissues and cause inflammation. When they reach alveoli, there is outpouring of fluid and red and white blood cells, resulting in consolidation of the lung. During recovery, pneumococci are phagocytized, mononuclear cells ingest debris, and the consolidation resolves.

Factors that lower resistance and predispose persons to pneumococcal infection include (1) alcohol or drug intoxication or other cerebral impairment that can depress the cough reflex and increase aspiration of secretions; (2) abnormality of the respiratory tract (e.g., viral infections), pooling of mucus, bronchial obstruction, and respiratory tract injury caused by irritants (which disturb the integrity and movement of the mucociliary blanket); (3) abnormal circulatory dynamics (e.g., pulmonary congestion and heart failure); (4) **splenectomy**; and (5) certain chronic diseases such as sickle cell anemia and nephrosis. Patients with sickle cell anemia auto-infarct their spleen, become functionally asplenic, and are predisposed to pneumococcal sepsis. Trauma to the head that causes **leakage of spinal fluid** through the nose predisposes to pneumococcal meningitis.

Clinical Findings

Pneumonia often begins with a sudden chill, fever, cough, and pleuritic pain. Sputum is a red or brown "rusty" color. Bacteremia occurs in 15% to 25% of cases. Spontaneous recovery may begin in 5 to 10 days and is accompanied by development of anticapsular antibodies. Pneumococci are a prominent cause of otitis media, sinusitis, mastoiditis, conjunctivitis, purulent bronchitis, pericarditis, bacterial meningitis, and sepsis. Pneumococci are the leading cause of sepsis in patients without a functional spleen.

Laboratory Diagnosis

In sputum, pneumococci are seen as lancet-shaped gram-positive diplococci in Gram-stained smears (see Figure 15–15). They can also be detected by using the quellung reaction with multitype antiserum. On blood agar, pneumococci form small **α-hemolytic** colonies. The colonies are **bile-soluble** (i.e., are lysed by bile), and growth is **inhibited by optochin** (see Figure 15–16).

Blood cultures are positive in 15% to 25% of pneumococcal infections. Culture of cerebrospinal fluid is usually positive in meningitis. Rapid diagnosis of pneumococcal meningitis can be made by detecting its capsular polysaccharide in spinal fluid using the latex agglutination test. A rapid test that detects urinary antigen is also available for the diagnosis of pneumococcal pneumonia and bacteremia. The urinary antigen is the C polysaccharide (also known as the C substance), *not* the capsular polysaccharide. Because of the increasing numbers of strains resistant to penicillin, antibiotic sensitivity tests must be done on organisms isolated from serious infections.

Treatment

Most pneumococci are susceptible to penicillins and erythromycin, although a significant resistance to penicillins has emerged (see next paragraph). In severe pneumococcal infections, penicillin G is the drug of choice, whereas in mild pneumococcal infections, oral penicillin V can be used. A fluoroquinolone with good antipneumococcal activity, such as levofloxacin, can also be used. In penicillin-allergic patients, erythromycin or one of its long-acting derivatives (e.g., azithromycin) can be used.

In the United States, about 25% of isolates exhibit low-level resistance to penicillin, primarily as a result of changes in PBPs.

An increasing percentage of isolates, ranging from 15% to 35% depending on location, show **high-level resistance**, which is attributed to multiple **changes in PBPs**. They do *not* produce β-lactamase. Vancomycin is the drug of choice for the penicillin-resistant pneumococci, especially for severely ill patients. Ceftriaxone or levofloxacin can be used for less severely ill patients. However, strains of pneumococci tolerant to vancomycin have emerged. (Tolerance to antibiotics is described on pages 69 and 85.) Strains of pneumococci resistant to multiple drugs, especially azithromycin, have also emerged.

Prevention

Two pneumococcal vaccines are available: the 13-valent conjugate vaccine (Prevnar13) and the 23-valent polysaccharide vaccine (Pneumovax 23).

Despite the efficacy of antimicrobial drug treatment, the mortality rate of pneumococcal infections is high in immunocompromised (especially splenectomized) patients and children under the age of 5 years. Such persons should be immunized with the **13-valent pneumococcal conjugate vaccine** (Prevnar 13). The immunogen in this vaccine is the pneumococcal polysaccharide of the 13 most prevalent serotypes conjugated (coupled) to a carrier protein (diphtheria toxoid). The unconjugated 23-valent pneumococcal vaccine (Pneumovax 23) should be given to healthy individuals at age 65 years or older.

These vaccines are safe and effective and provide long-lasting (at least 5 years) protection. Immunization of *children* reduces the incidence of pneumococcal disease in *adults* because children are the main source of the organism for adults and immunization reduces the carrier rate in children.

A booster dose of Pneumovax is recommended for (1) people older than 65 years and (2) people between the ages of 19 and 64 years who are immunocompromised, for example, those who are asplenic, infected with human immunodeficiency virus (HIV), receiving cancer chemotherapy, or receiving immunosuppressive drugs to prevent transplant rejection.

A potential problem regarding the use of the pneumococcal vaccine is that of **serotype replacement**. Will the vaccine reduce the incidence of disease caused by the serotypes in the vaccine but not the overall incidence of pneumococcal disease because other serotypes that are not in the vaccine will now cause disease? In fact, an increase in invasive pneumococcal disease caused by serotype 19A, which was not in the previously used 7-valent vaccine, occurred. This led to the production of the current conjugate vaccine containing 13 serotypes, including 19A.

SELF-ASSESSMENT QUESTIONS

1. You are in the clinical laboratory looking at a Gram stain when the laboratory technician comes up to you and says, "I think your patient has Staph epi (short for *S. epidermidis*) bacteremia." Which one of the following sets of results did the technician find with the organism recovered from the blood culture?

   (A) Gram-positive cocci in chains, catalase-positive, coagulase-positive

   (B) Gram-positive cocci in chains, catalase-negative, coagulase-negative
   (C) Gram-positive cocci in clusters, catalase-positive, coagulase-negative
   (D) Gram-positive cocci in clusters, catalase-negative, coagulase-positive
   (E) Gram-positive diplococci, catalase-negative, coagulase-positive

2. Superantigen production by *Staphylococcus aureus* is involved in the pathogenesis of which one of the following diseases?
   (A) Impetigo
   (B) Osteomyelitis
   (C) Scalded skin syndrome
   (D) Septicemia
   (E) Toxic shock syndrome

3. Which one of the following is the virulence factor produced by *Staphylococcus aureus* that prevents the activation of complement and thereby reduces opsonization by C3b?
   (A) Catalase
   (B) Coagulase
   (C) Endotoxin
   (D) Protein A
   (E) Teichoic acid

4. The main reason why methicillin-resistant *Staphylococcus aureus* (MRSA) strains are resistant to methicillin and nafcillin is:
   (A) they produce β-lactamase that degrades the antibiotics.
   (B) they have altered penicillin-binding proteins that have reduced binding of the antibiotics.
   (C) they have mutant porin proteins that prevent the antibiotics from entering the bacteria.
   (D) they have plasmid-encoded export proteins that remove the drug from the bacteria.

5. A pore-forming exotoxin produced by *Staphylococcus aureus* that kills cells and is important in the severe, rapidly spreading necrotizing lesions caused by MRSA strains is:
   (A) coagulase.
   (B) enterotoxin.
   (C) exfoliatin.
   (D) P-V leukocidin.
   (E) staphyloxanthin.

6. Of the following antibiotics, which one is the most appropriate to treat a severe necrotizing skin infection caused by an MRSA strain of *Staphylococcus aureus*?
   (A) Amoxicillin
   (B) Ceftriaxone
   (C) Ciprofloxacin
   (D) Gentamicin
   (E) Vancomycin

7. An outbreak of serious pneumococcal pneumonia and sepsis among inmates in an overcrowded prison has occurred. Laboratory analysis determined that one serotype was involved. The prison physician said that the pneumococcal vaccine might have limited the outbreak. Which one of the following structures of the pneumococcus is responsible for determining the serotype and is also the immunogen in the vaccine?
   (A) Capsule
   (B) Flagellar protein
   (C) O antigen
   (D) Peptidoglycan
   (E) Pilus protein

8. Which one of the following best describes the pathogenesis of rheumatic fever?

   (A) An exotoxin produced by *Streptococcus pyogenes* that acts as a superantigen damages cardiac muscle.

   (B) An exotoxin produced by *Streptococcus pyogenes* that ADP-ribosylates a G protein damages joint tissue.

   (C) Antibody to the capsular polysaccharide of *Streptococcus pyogenes* cross-reacts with joint tissue and damages it.

   (D) Antibody to the M protein of *Streptococcus pyogenes* cross-reacts with cardiac muscle and damages it.

   (E) Endotoxin produced by *Streptococcus pyogenes* activates macrophages to release cytokines that damage cardiac muscle.

9. Which one of following laboratory tests is the most appropriate to distinguish *Streptococcus pyogenes* from other β-hemolytic streptococci?

   (A) Ability to grow in 6.5% NaCl

   (B) Activation of C-reactive protein

   (C) Hydrolysis of esculin in the presence of bile

   (D) Inhibition by bacitracin

   (E) Inhibition by optochin

10. Infections by which one of the following bacteria are typically treated with penicillins such as amoxicillin, because they exhibit neither low-level resistance nor high-level resistance and synergy with an aminoglycoside is not required in order for penicillins to be effective?

    (A) *Enterococcus faecalis*

    (B) *Staphylococcus aureus*

    (C) *Staphylococcus epidermidis*

    (D) *Streptococcus pneumoniae*

    (E) *Streptococcus pyogenes*

11. Your patient in the emergency room has a 5-cm ulcer on her leg that is surrounded by a red, warm, and tender area of inflammation. You do a Gram stain on pus from the ulcer and see gram-positive cocci in chains. Culture of the pus grows small β-hemolytic colonies that are catalase-negative and are inhibited by bacitracin. These results indicate that the organism causing her lesion is most likely:

    (A) *Enterococcus faecalis*.

    (B) *Staphylococcus aureus*.

    (C) *Streptococcus agalactiae*.

    (D) *Streptococcus pneumoniae*.

    (E) *Streptococcus pyogenes*.

12. The Jones family of four had a delicious picnic lunch last Sunday. It was a warm day, and the food was in the sun for several hours. Alas, 3 hours later, everyone came down with vomiting and non-bloody diarrhea. In the emergency room, it was found that Mrs. Jones, who prepared the food, had a paronychia on her thumb. Which one of the following is the most likely causative organism?

    (A) *Enterococcus faecalis*

    (B) *Staphylococcus aureus*

    (C) *Staphylococcus epidermidis*

    (D) *Streptococcus agalactiae*

    (E) *Streptococcus pyogenes*

13. A 20-year-old sexually active woman reports dysuria and other symptoms of a urinary tract infection. Gram stain of the urine reveals gram-positive cocci. Which one of the following sets of bacteria is most likely to cause this infection?

    (A) *Staphylococcus aureus* and *Streptococcus pyogenes*

    (B) *Staphylococcus saprophyticus* and *Enterococcus faecalis*

    (C) *Streptococcus agalactiae* and *Staphylococcus epidermidis*

    (D) *Streptococcus pneumoniae* and *Enterococcus faecalis*

    (E) *Streptococcus pyogenes* and *Streptococcus pneumoniae*

14. Your patient is a 2-week-old infant who was well until 2 days ago, when she stopped feeding and became irritable. She now has a fever to 38°C, has developed a petechial rash all over her body, and is very difficult to arouse. In the emergency room, a blood culture and a spinal tap were done. Gram stain of the spinal fluid showed gram-positive cocci in chains. Culture of the spinal fluid on blood agar revealed β-hemolytic colonies that grew in the presence of bacitracin and hydrolyzed hippurate. Which one of the following is the most likely causative organism?

    (A) *Staphylococcus aureus*

    (B) *Streptococcus agalactiae*

    (C) *Streptococcus mutans*

    (D) *Streptococcus pneumoniae*

    (E) *Streptococcus pyogenes*

15. Your patient is a 50-year-old woman who has a community-acquired pneumonia caused by *Streptococcus pneumoniae*. Antibiotic susceptibility tests reveal an MIC of less than 0.1 mg/mL to penicillin G. Which one of the following is the best antibiotic to treat the infection?

    (A) Clindamycin

    (B) Gentamicin

    (C) Metronidazole or doxycycline

    (D) Penicillin G or levofloxacin

    (E) Vancomycin

16. Your patient is a 70-year-old man with endocarditis caused by *Enterococcus faecalis*. Which one of the following is the best combination of antibiotics to treat the infection?

    (A) Azithromycin and trimethoprim-sulfamethoxazole

    (B) Chloramphenicol and rifampin

    (C) Doxycycline and levofloxacin

    (D) Metronidazole and clindamycin

    (E) Penicillin G and gentamicin

ANSWERS

(1) **(C)**

(2) **(E)**

(3) **(D)**

(4) **(B)**

(5) **(D)**

(6) **(E)**

(7) **(A)**

(8) **(D)**

(9) **(D)**

(10) **(E)**

(11) **(E)**

(12) **(B)**

(13) **(B)**

(14) **(B)**

(15) **(D)**

(16) **(E)**

SUMMARIES OF ORGANISMS

Brief summaries of the organisms described in this chapter begin on page 675. Please consult these summaries for a rapid review of the essential material.

PRACTICE QUESTIONS: USMLE & COURSE EXAMINATIONS

Questions on the topics discussed in this chapter can be found in the Clinical Bacteriology section of Part XIII: USMLE (National Board) Practice Questions starting on page 737. Also see Part XIV: USMLE (National Board) Practice Examination starting on page 775.

Gram-Negative Cocci

NEISSERIA

Diseases

The genus *Neisseria* contains two important human pathogens: *Neisseria meningitidis* and *Neisseria gonorrhoeae*. *N. meningitidis* mainly causes meningitis and meningococcemia (Figure 16–1). In the United States, it is the leading cause of death from infection in children. *N. gonorrhoeae* causes gonorrhea (Figure 16–2), the second most common notifiable bacterial disease in the United States (Tables 16–1 and 16–2). It also causes neonatal conjunctivitis (ophthalmia neonatorum) (Figure 16–3) and pelvic inflammatory disease (PID). Note that *N. meningitidis* is also known as the meningococcus (plural, meningococci), and *N. gonorrhoeae* is also known as the gonococcus (plural, gonococci).

Additional information regarding the clinical aspects of infections caused by the organisms in this chapter is provided in Part IX, entitled Infectious Diseases beginning on page 603.

Important Properties

Neisseriae are gram-negative cocci that resemble paired kidney beans (Figure 16–4).

(1) *N. meningitidis* (meningococcus) has a prominent **polysaccharide capsule** that enhances virulence by its antiphagocytic action. The capsule also is the immunogen in the vaccine that induces protective antibodies (Table 16–3). Meningococci are divided into at least 13 serologic groups on the basis of the antigenicity of their capsular polysaccharides. Five serotypes cause most cases of meningitis and meningococcemia: A, B, C, Y, and W-135.

Serotype A is the leading cause of epidemic meningitis worldwide. Serotype B causes most of the disease in the United States. This is because the group B polysaccharide is not immunogenic in humans and therefore is not part of the

vaccines that contain the capsular polysaccharide of the other four groups. In 2014, a vaccine against group B meningococci containing factor H binding protein as the immunogen was approved.

(2) *N. gonorrhoeae* (gonococcus) has no polysaccharide capsule but has more than 100 serotypes based on the antigenicity of its pilus protein. There is **marked antigenic variation** in the

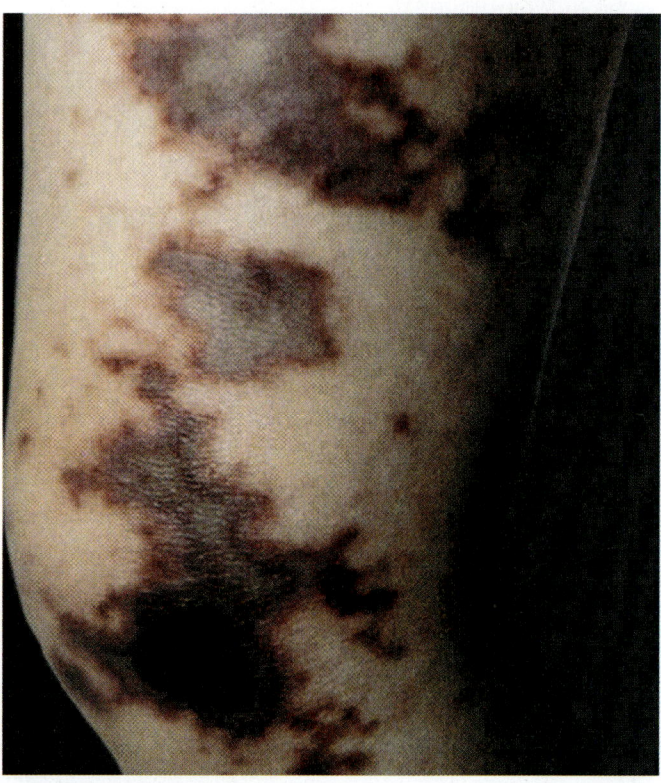

FIGURE 16–1 Meningococcemia. Note purpuric lesions on leg caused by endotoxin-mediated disseminated intravascular coagulation (DIC). (Reproduced with permission from Wolff K, Johnson R: *Fitzpatrick's Color Atlas & Synopsis of Clinical Dermatology*, 6th ed. New York, NY: McGraw Hill; 2009.)

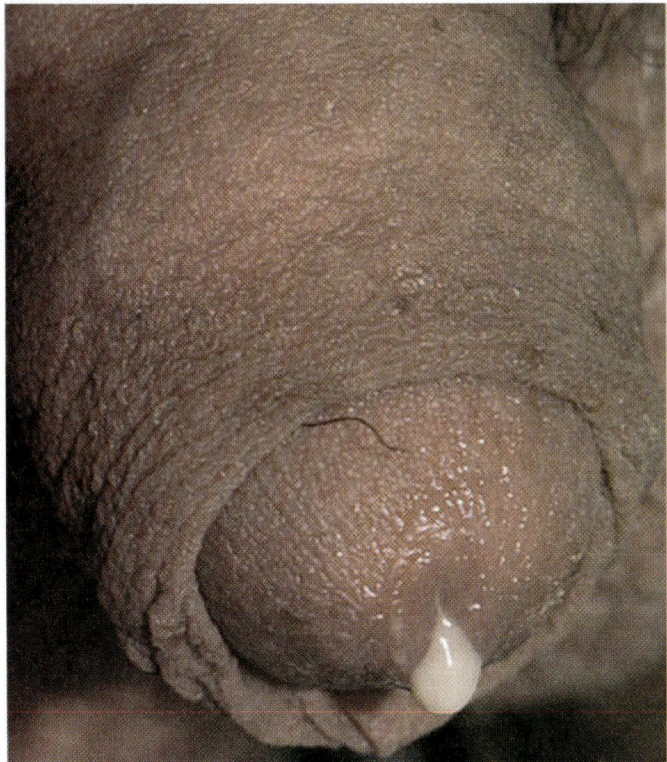

FIGURE 16–2 Gonorrhea. Note purulent urethral discharge caused by *Neisseria gonorrhoeae*. (Reproduced with permission from Wolff K, Johnson R: *Fitzpatrick's Color Atlas & Synopsis of Clinical Dermatology*, 6th ed. New York, NY: McGraw Hill; 2009.)

gonococcal pili as a result of programmed chromosomal rearrangement (see Chapter 4 and Figure 4–1). The large number of serotypes allows repeated gonococcal infections to occur. Gonococci have three outer membrane proteins (proteins I, II, and III). Protein II plays a role in attachment of the organism to cells and varies antigenically as well.

Neisseriae are gram-negative bacteria and contain endotoxin in their outer membrane. Note that the endotoxin of Neisseriae consist of lipo**oligo**saccharide (LOS), in contrast to the lipo**poly**saccharide (LPS) found in enteric gram-negative rods. Both LPS and LOS contain lipid A, but the oligosaccharide part of LOS contains few sugars, whereas the polysaccharide part of LPS contains a long repeating sugar side chain.

The growth of both organisms is inhibited by toxic trace metals and fatty acids found in certain culture media (e.g., blood agar plates). They are therefore cultured on "chocolate" agar containing blood heated to 80°C, which inactivates the inhibitors. Neisseriae are **oxidase-positive** (Figure 16–5) (i.e., they possess the enzyme cytochrome *c*). This is an important laboratory diagnostic test in which colonies exposed to phenylenediamine turn purple or black as a result of oxidation of the reagent by the enzyme (see Figure 16–2).

The genus *Neisseria* is one of several in the family Neisseriaceae. A separate genus contains the organism *Moraxella catarrhalis*, which is part of the normal throat flora but can cause such respiratory tract infections as sinusitis, otitis media, bronchitis, and pneumonia. *M. catarrhalis* is described in Chapter 27.

1. Neisseria meningitidis

Pathogenesis & Epidemiology

Humans are the only natural hosts for meningococci. The organisms are transmitted by **airborne droplets**; they colonize the membranes of the nasopharynx and become part of the transient flora of the upper respiratory tract. Carriers are usually asymptomatic. From the nasopharynx, the organism can enter the bloodstream and spread to specific sites, such as the meninges or joints, or be disseminated throughout the body (meningococcemia).

About 5% of people become chronic carriers and serve as a source of infection for others. The carriage rate can be as high as 35% in people who live in close quarters (e.g., military recruits); this explains the high frequency of outbreaks of meningitis in the armed forces prior to the use of the vaccine. The carriage rate is also high in close (family) contacts of patients. Outbreaks of meningococcal disease also have occurred in college students living in dormitories.

Two organisms cause more than 80% of cases of bacterial meningitis in children older than 2 months of age: *Streptococcus pneumoniae* and *N. meningitidis*. Of these organisms, meningococci, especially those in group A, are most likely to cause **epidemics of meningitis**.

Group B meningococci cause many cases of meningitis in developed countries because they are not present in the capsular polysaccharide vaccine. They are not included in the vaccine because they are not immunogenic in humans. Overall, *N. meningitidis* ranks second to *S. pneumoniae* as a cause of meningitis but is the most common cause in persons between the ages of 2 and 18 years.

Meningococci have four important virulence factors:

(1) A **polysaccharide capsule** that enables the organism to resist phagocytosis by polymorphonuclear leukocytes (PMNs). The capsule is the immunogen in several commonly used vaccines against meningococci.

TABLE 16–1 Neisseriae of Medical Importance[1]

Species	Portal of Entry	Polysaccharide Capsule	Maltose Fermentation	β-Lactamase Production	Available Vaccine
N. meningitidis (meningococcus)	Respiratory tract	+	+	None	+
N. gonorrhoeae (gonococcus)	Genital tract	–	–	Some	–

[1]All Neisseriae are oxidase-positive.

TABLE 16-2 Important Clinical Features of Neisseriae

Organism	Type of Pathogenesis	Typical Disease	Treatment
Neisseria meningitidis	Pyogenic	Meningitis, meningococcemia	Penicillin G
Neisseria gonorrhoeae	Pyogenic		
	1. Local	Gonorrhea (e.g., urethritis, cervicitis)	Ceftriaxone[1] plus doxycycline[2]
	2. Ascending	Pelvic inflammatory disease	Cefoxitin plus doxycycline[1,2]
	3. Disseminated	Disseminated gonococcal infection	Ceftriaxone[1]
	4. Neonatal	Conjunctivitis (ophthalmia neonatorum)	Ceftriaxone[3]

[1]Other drugs can also be used. See treatment guidelines published by the Centers for Disease Control and Prevention.

[2]Add doxycycline for possible coinfection with *Chlamydia trachomatis*.

[3]For prevention, use erythromycin ointment or silver nitrate drops.

(2) **Endotoxin**, which causes fever, shock, and other pathophysiologic changes (in purified form, endotoxin can reproduce many of the clinical manifestations of meningococcemia).

(3) An **immunoglobulin A (IgA) protease** that helps the bacteria attach to the membranes of the upper respiratory tract by cleaving secretory IgA.

(4) **Factor H binding protein (FHbp)** on meningococci, which binds Factor H, an inhibitor of complement factor C3b. The presence of Factor H on the surface of meningococci reduces the opsonizing activity of C3b and reduces the amount of membrane attack complex produced (see complement action in Chapter 63). FHbp is the immunogen in the vaccine against group B meningococci.

Resistance to disease correlates with the presence of antibody to the capsular polysaccharide. Most carriers develop protective antibody titers within 2 weeks of colonization. Because immunity is group-specific, it is possible to have protective antibodies to one group of organisms yet be susceptible to infection by organisms of the other groups.

Complement is an important feature of the host defenses, because people with complement deficiencies, particularly in the **late-acting complement components** (C6–C9), have an increased incidence of meningococcal bacteremia. Patients receiving eculizumab, a terminal complement inhibitor used in the treatment of paroxysmal nocturnal hemoglobinuria, have a 1000-fold increase in meningococcal disease.

Clinical Findings

The two most important manifestations of disease are **meningococcemia** (see Figure 16–1) and **meningitis**. Meningococcemia typically presents as septic shock, namely, high fever, hypotension, and a petechial or purpuric rash. The most severe form of meningococcemia is the life-threatening **Waterhouse–Friderichsen syndrome**, which is characterized by high fever, shock, widespread purpura, disseminated intravascular coagulation, thrombocytopenia, and adrenal insufficiency. Adrenal insufficency is caused by infarction of the adrenals.

Meningococcemia can result in the seeding of many organs, especially the meninges. The symptoms of meningococcal meningitis are those of typical bacterial meningitis, namely, fever, headache, stiff neck, and an increased level of PMNs in the spinal fluid.

Laboratory Diagnosis

The principal laboratory procedures are smear and culture of blood and spinal fluid samples. A presumptive diagnosis of

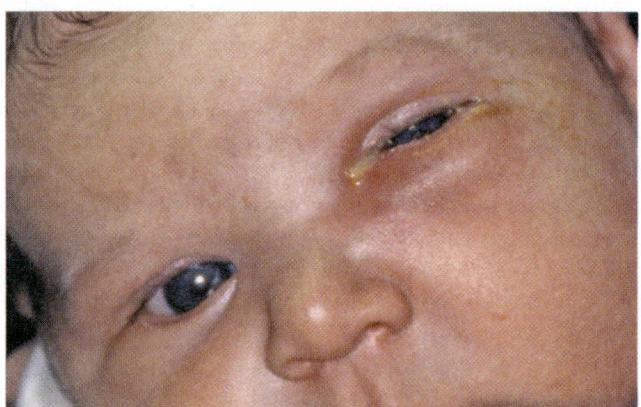

FIGURE 16–3 Neonatal conjunctivitis (ophthalmia neonatorum) caused by *Neisseria gonorrhoeae*. Note purulent exudate at the corner of the medial side of child's left eye. The other common cause of neonatal conjunctivitis is *Chlamydia trachomatis*. Photo contributor: David Effron, M.D. (Reproduced with permission from Knoop KJ, Stack LB, Storrow AB, et al: *The Atlas of Emergency Medicine*, 4th ed. New York, NY: McGraw Hill; 2016. Photo contributor: David Effron, M.D.)

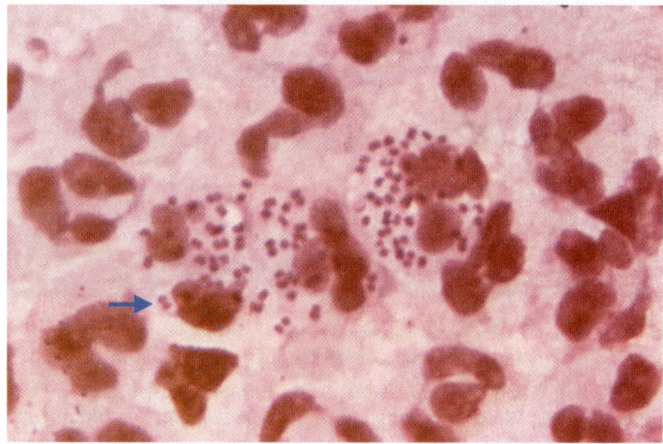

FIGURE 16–4 *Neisseria gonorrhoeae*—Gram stain. Blue arrow points to a typical gram-negative diplococcus within a neutrophil. Many other gram-negative dipocci can be seen within the neutrophils. (Used with permission from Dr. Bill Schwartz, Centers for Disease Control and Prevention.)

TABLE 16–3 Properties of the Polysaccharide Capsule of the Meningococcus[1]

1. Enhances virulence by its antiphagocytic action
2. Is the antigen that defines the serologic groups
3. Is the antigen detected in the spinal fluid of patients with meningitis
4. Is the antigen in the vaccine

[1]The same four features apply to the capsule of the pneumococcus and *Haemophilus influenzae*.

meningococcal meningitis can be made if gram-negative cocci are seen in a smear of spinal fluid (see Figure 16–4). The organism grows best on chocolate agar incubated at 37°C in a 5% CO_2 atmosphere. A presumptive diagnosis of *Neisseria* can be made if oxidase-positive colonies of gram-negative diplococci are found (see Figure 16–5).

The differentiation between *N. meningitidis* and *N. gonorrhoeae* is made on the basis of sugar metabolism tests: meningococci metabolize maltose, whereas gonococci do not (both organisms metabolize glucose). Immunofluorescence can also be used to identify these species. Tests for serum antibodies are not useful for clinical diagnosis. However, a procedure that can assist in the rapid diagnosis of meningococcal meningitis is the latex agglutination test, which detects capsular polysaccharide in the spinal fluid.

Polymerase chain reaction (PCR) tests are also available. They provide rapid and sensitive results but the organism must still be isolated by using culture methods to provide antibiotic susceptibility information.

Treatment

Either ceftriaxone or penicillin G is the drug of choice for meningococcal infections. Strains resistant to penicillin have

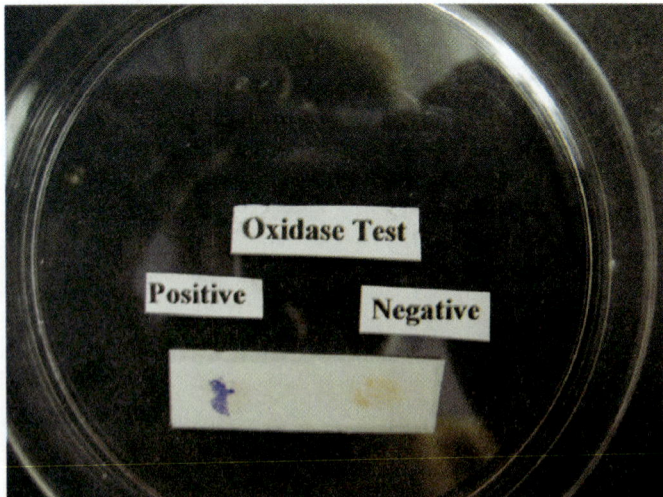

FIGURE 16–5 Oxidase test. A drop of the oxidase reagent was placed on the left and right side of the filter paper. Bacteria from a colony of *Neisseria gonorrhoeae* were rubbed on the drop on the left, and the purple color indicates a positive test (i.e., the organism is oxidase-positive). Bacteria from a colony of *Escherichia coli* were rubbed on the drop on the right, and the absence of a purple color indicates a negative test (i.e., the organism is oxidase-negative). (Used with permission from Professor Shirley Lowe, University of California, San Francisco School of Medicine.)

rarely emerged, but sulfonamide resistance is common. In 2019, strains of meningococci resistant to penicillin and ciprofloxacin were recoverd from patients in several states.

Prevention

Chemoprophylaxis and immunization are both used to prevent meningococcal disease. Either rifampin or ciprofloxacin can be used for prophylaxis in people who have had close contact with the index case. These drugs are preferred because they are efficiently secreted into the saliva, in contrast to penicillin G.

The meningococcal vaccines are described in Table 16–4. The vaccines against groups A, C, Y, and W-135 meningococci contain the polysaccharide capsule as the immunogen. The vaccine against group B meningococci contains FHBP as the main immunogen.

In the United States, the vaccines against groups A, C, Y, and W-135 meningococci are **conjugate vaccines**, that is, the capsular polysaccharide is conjugated to a carrier protein.

There are four forms of the polysaccharide vaccine for use in the United States: (1) Menactra contains the four polysaccharides conjugated to diphtheria toxoid as the carrier protein; (2) Menveo contains the four polysaccharides conjugated to a nontoxic mutant of diphtheria toxin as the carrier protein; (3) Menquadfi contains the four polysaccharides conjugated to tetanus toxoid as the carrier protein; and (4) MenHibrix contains two polysaccharides (C and Y) plus the capsular polysaccharide of *Haemophilus influenzae*, all conjugated to tetanus toxoid.

Menomune, the unconjugated vaccine, contains only the four polysaccharides (not conjugated to a carrier protein). It is not available in the United States but is used in other countries. Another vaccine created for use in the meningitis belt of Africa, called MenAfriVac, is a conjugate vaccine that contains only the group A polysaccharide.

The conjugate vaccines induce higher titers of antibodies in children than does the unconjugated vaccine. The vaccines induce similar antibody titers in adults. Note that none of these vaccines contain the group B polysaccharide because it is not immunogenic in humans. The conjugate vaccine is recommended for children at the age of 11 to 12 years, which will reduce the incidence of meningococcal disease in teenagers and young adults. The conjugate vaccine is also recommended for children younger than 11 years with high-risk conditions, such as asplenia and HIV infection. Travelers to areas where an epidemic of meningococcal disease is occurring should receive the conjugate vaccine. College students living in dormitories are encouraged to receive the conjugate vaccine.

The vaccine against group B meningococci contains FHbp as the immunogen. It induces antibody against the binding protein, thereby inhibiting the ability of the bacteria to bind Factor H on its surface. This enhances the action of complement, an important host defense, because Factor H blocks complement component C3b from binding to the bacterial surface. Stated another way, if Factor H cannot bind to the surface of the bacteria, that allows C3b to bind. C3b is an important opsonizer that enhances the immune response against the meningococci.

TABLE 16–4 Meningococcal Vaccines

Serogroups Covered	Immunogen	Carrier Protein	Where Available	Name of Vaccine
A, C, Y, W-135	Capsular polysaccharide	Diphtheria toxoid	United States	Menactra, Menveo
A, C, Y, W-135	Capsular polysaccharide	Tetanus toxin	United States	Menquadfi
A, C, Y, W-135	Capsular polysaccharide	None	Countries other than the United States	Menomune
A	Capsular polysaccharide	Diphtheria toxoid	Africa's meningitis belt	MenAfriVac
C, Y plus *Haemophilus influenzae*	Capsular polysaccharide	Tetanus toxoid	United States	MenHibrix
B	Factor H binding protein	None	United States	Trumenba
B	Factor H binding protein, NadA, NHBA, OMV	None	United States	Bexsero

NadA = neisserial adhesin A; NHBA = neisserial heparin binding antigen; OMV = outer membrane vesicles (which contain PorA as the antigen).

The FHbp used in the vaccine is made by recombinant DNA techniques in *Escherichia coli*. The vaccine is approved for use in people between ages 10 and 25 years. In 2015, a second vaccine against group B meningococci containing four surface proteins (FHbp, NadA [neisserial adhesin A], NHBA [neisserial heparin binding antigen], and OMV [outer membrane vesicle; PorA]) was approved.

2. Neisseria gonorrhoeae

Pathogenesis & Epidemiology

Gonococci, like meningococci, cause disease only in humans. The organism is usually transmitted **sexually**. Newborns can be infected during birth. Because gonococci are quite sensitive to dehydration and cool conditions, sexual transmission favors its survival. Gonorrhea is usually symptomatic in men but often asymptomatic in women. Genital tract infections are the most common source of the organism, but anorectal and pharyngeal infections are important sources as well.

Pili constitute one of the most important virulence factors because they mediate attachment to mucosal cell surfaces and are antiphagocytic. Piliated gonococci are usually virulent, whereas nonpiliated strains are avirulent. Two virulence factors in the cell wall are **endotoxin (lipooligosaccharide, LOS)** and the **outer membrane proteins**. The organism's **IgA protease** can hydrolyze secretory IgA, which could otherwise block attachment to the mucosa. Gonococci have no capsules.

The main host defenses against gonococci are antibodies (IgA and IgG), complement, and neutrophils. Antibody-mediated opsonization and killing within phagocytes occur, but repeated gonococcal infections are common, primarily as a result of antigenic changes of pili and the outer membrane proteins.

Gonococci infect primarily the mucosal surfaces (e.g., the urethra and vagina). From the mucosal site, bacteria can enter the bloodstream and cause disseminated disease. Certain strains of gonococci cause disseminated infections more frequently than others. The most important feature of these strains is their resistance to being killed by antibodies and complement.

The occurrence of a disseminated infection is a function not only of the strain of gonococcus but also of the effectiveness of the host defenses. Persons with a deficiency of the late-acting complement components (C6–C9) are at risk for disseminated infections, as are women during menses and pregnancy. Disseminated infections usually arise from asymptomatic infections, indicating that local inflammation may deter dissemination.

Clinical Findings

Gonococci cause both localized infections, usually in the genital tract, and disseminated infections with seeding of various organs. Gonococci reach these organs via the bloodstream (gonococcal bacteremia).

Gonorrhea in men is characterized primarily by urethritis accompanied by dysuria and a purulent discharge (see Figure 16–2). Epididymitis can occur.

In women, infection is located primarily in the endocervix, causing a purulent vaginal discharge and intermenstrual bleeding (cervicitis). The most frequent complication in women is an ascending infection of the uterine tubes (**salpingitis, PID**), which can result in **sterility** or ectopic pregnancy as a result of scarring of the tubes. Women with PID may also have a rare perihepatitis characterized by right upper quadrant pain. This is called Fitz-Hugh-Curtis syndrome and is characterized by "violin-string" adhesions. This syndrome is also caused by *Chlamydia trachomatis*.

Disseminated gonococcal infections (DGIs) commonly manifest as arthritis, tenosynovitis, or pustules in the skin. Disseminated infection is the most common cause of septic arthritis in sexually active adults. The clinical diagnosis of DGI is often difficult to confirm using laboratory tests because the organism is not cultured in more than 50% of cases.

Other infected sites include the anorectal area, throat, and eyes. Anorectal infections occur chiefly in women and homosexual men. They are frequently asymptomatic, but a bloody or purulent discharge (proctitis) can occur. In the throat, pharyngitis occurs, but many patients are asymptomatic.

In newborn infants, purulent conjunctivitis (ophthalmia neonatorum) (see Figure 16–3) is the result of gonococcal infection acquired from the mother during passage through the birth canal. The incidence of gonococcal ophthalmia has declined greatly in recent years because of the widespread use of prophylactic erythromycin eye ointment (or silver nitrate) applied shortly after birth. Gonococcal conjunctivitis also occurs in adults as a result of the transfer of organisms from the genitals to the eye.

Other sexually transmitted infections (e.g., syphilis and nongonococcal urethritis caused by *C. trachomatis*) can coexist with gonorrhea; therefore, appropriate diagnostic and therapeutic measures must be taken.

Laboratory Diagnosis

The diagnosis of urogenital infections depends on Gram staining and culture of the discharge (see Figure 16–4). However, nucleic acid amplification tests are widely used as screening tests (see later).

In **men**, the finding of gram-negative diplococci **within PMNs** in a urethral discharge specimen is sufficient for diagnosis (see Figure 16–4). In **women**, the use of the Gram stain alone can be difficult to interpret; therefore, cultures should be done. Gram stains on cervical specimens can be falsely positive because of the presence of gram-negative diplococci in the normal flora and can be falsely negative because of the inability to see small numbers of gonococci when using the oil immersion lens. Cultures must also be used in diagnosing suspected pharyngitis or anorectal infections.

Specimens from mucosal sites, such as the urethra and cervix, are cultured on **Thayer-Martin medium**, which is a chocolate agar containing antibiotics (vancomycin, colistin, trimethoprim, and nystatin) to suppress the normal flora. The finding of an oxidase-positive colony (see Figure 16–5) composed of gram-negative diplococci is sufficient to identify the isolate as a member of the genus *Neisseria*. Specific identification of the gonococcus can be made either by its fermentation of glucose (but not maltose) or by fluorescent antibody staining. Note that specimens from sterile sites, such as blood or joint fluid, can be cultured on chocolate agar without antibiotics because there is no competing normal flora.

Nucleic acid amplification tests, **often abbreviated NAAT**, detect the presence of gonococcal nucleic acids in patient specimens. These tests are widely used for screening purposes, produce results rapidly, and are highly sensitive and specific. They can be used on urine samples, obviating the need for more invasive collection techniques. Note that serologic tests to determine the presence of antibody to gonococci in the patient's serum are *not* useful for diagnosis.

Treatment

A single dose of ceftriaxone is the drug of choice for uncomplicated urogenital, rectal, or pharyngeal gonorrhea. Doxycycline should be added to treat possible chlamydial infection. Azithromycin can be used if patient is pregnant. If the patient is allergic to penicillins or cephalosporins, a regimen such as gentamicin plus azithromycin can be used. Partner therapy with oral cefixime plus doxycycline can be used.

Because reinfection rates are high, patients who have been treated for gonorrhea should be retested 3 months after treatment. A test-of-cure follow-up culture or NAAT is not necessary following treatment for genital or rectal gonorrhea but should be performed if the patient had pharyngeal gonorrhea.

Treatment of complicated gonococcal infections, such as PID, typically requires hospitalization. Treatment regimens are complex and beyond the scope of this book.

Prior to the mid-1950s, all gonococci were highly sensitive to penicillin. Subsequently, isolates emerged with low-level resistance to penicillin and to other antibiotics such as tetracycline and chloramphenicol. This type of resistance is encoded by the bacterial chromosome and is due to reduced uptake of the drug or to altered binding sites rather than to enzymatic degradation of the drug.

Then, in 1976, **penicillinase-producing (PPNG)** strains that exhibited high-level resistance were isolated from patients. Penicillinase is plasmid-encoded. PPNG strains are now common in many areas of the world, including several urban areas in the United States, where approximately 10% of isolates are resistant. Isolates resistant to fluoroquinolones, such as ciprofloxacin, have become a significant problem, and fluoroquinolones are not recommended as treatment. Isolates resistant to sulfonamides and tetracyclines have also been recovered. In 2017, the World Health Organization (WHO) reported that several strains of gonococci resistant to all known antibiotics have been isolated.

Prevention

The prevention of gonorrhea involves the use of condoms and the prompt treatment of symptomatic patients and their sex partners. Cases of gonorrhea must be reported to the public health department to ensure proper follow-up and contact tracing. A major problem is the detection of asymptomatic carriers. Gonococcal conjunctivitis in newborns is prevented most often by the use of erythromycin ointment. Silver nitrate drops are used in some countries. No vaccine is available.

SELF-ASSESSMENT QUESTIONS

1. Regarding the differences between *Neisseria meningitidis* (meningococci) and *Neisseria gonorrhoeae* (gonococci), which one of the following is the most accurate statement?

   (A) Meningococci are oxidase-positive, whereas gonococci are not.

   (B) Meningococci have a thick polysaccharide capsule, whereas gonococci do not.

   (C) Meningococci have lipid A, whereas gonococci do not.

   (D) Meningococci produce penicillinase, whereas gonococci do not.

   (E) Meningococci synthesize IgA protease, whereas gonococci do not.

2. Your patient is a 14-year-old girl who was sent home from school because she had a fever of 102°C and a severe headache and was falling asleep in class. When her fever rose to 104°C, her mother took her to the emergency room, where a blood pressure of 60/20 mm Hg and several petechial hemorrhages were found. Gram-negative diplococci were seen in a Gram stain of the spinal fluid. Which one of the following is most likely to cause the fever, hypotension, and petechial hemorrhages?

   (A) Endotoxin
   (B) IgA protease
   (C) Oxidase
   (D) Pilus protein
   (E) Superantigen

3. Regarding the patient in Question 2, which one of the following is the best antibiotic to treat the infection?

   (A) Azithromycin
   (B) Doxycycline
   (C) Penicillin G
   (D) Rifampin
   (E) Trimethoprim-sulfamethoxazole

4. Regarding the differences between *N. meningitidis* (meningococci) and *N. gonorrhoeae* (gonococci), which one of the following is the most accurate statement?

   (A) Humans are the reservoir for both organisms.
   (B) Many clinical isolates of meningococci produce β-lactamase, but clinical isolates of gonococci do not.
   (C) Meningococci have multiple antigenic types, but gonococci have only one antigenic type.
   (D) The conjugate vaccine against gonorrhea contains seven types of the pilus protein as the immunogen.
   (E) The main mode of transmission for both organisms is respiratory droplets.

5. Your patient is a 20-year-old man with a urethral exudate. You do a Gram stain of the pus and see gram-negative diplococci with neutrophils. Which one of the following is the best antibiotic to treat the infection?

   (A) Ceftriaxone
   (B) Gentamicin
   (C) Penicillin G
   (D) Trimethoprim-sulfamethoxazole
   (E) Vancomycin

ANSWERS

(1) **(B)**
(2) **(A)**
(3) **(C)**
(4) **(A)**
(5) **(A)**

SUMMARIES OF ORGANISMS

Brief summaries of the organisms described in this chapter begin on page 675. Please consult these summaries for a rapid review of the essential material.

PRACTICE QUESTIONS: USMLE & COURSE EXAMINATIONS

Questions on the topics discussed in this chapter can be found in the Clinical Bacteriology section of Part XIII: USMLE (National Board) Practice Questions starting on page 737. Also see Part XIV: USMLE (National Board) Practice Examination starting on page 775.

Gram-Positive Rods

INTRODUCTION

There are five medically important genera of gram-positive rods: *Bacillus, Clostridium, Corynebacterium, Listeria,* and *Gardnerella. Bacillus* and *Clostridium* form spores, whereas *Corynebacterium, Listeria,* and *Gardnerella* do not. Members of the genus *Bacillus* are aerobic, whereas those of the genus *Clostridium* are anaerobic (Table 17–1).

These gram-positive rods can also be distinguished based on their appearance on Gram stain. *Bacillus* and *Clostridium* species are longer and more deeply staining than *Corynebacterium* and *Listeria* species. *Corynebacterium* species are club-shaped (i.e., they are thinner on one end than the other). *Corynebacterium* and *Listeria* species characteristically appear as V- or L-shaped rods. *Gardnerella vaginalis* is a short gram-variable rod.

Additional information regarding the clinical aspects of infections caused by the organisms in this chapter is provided in Part IX, entitled Infectious Diseases beginning on page 603.

SPORE-FORMING GRAM-POSITIVE RODS

BACILLUS

There are two medically important *Bacillus* species: *Bacillus anthracis* and *Bacillus cereus.* Important features of pathogenesis by these two *Bacillus* species are described in Table 17–2.

TABLE 17–1 Gram-Positive Rods of Medical Importance

Growth	Anaerobic Growth	Spore Formation	Exotoxins Important in Pathogenesis
Bacillus	–	+	+
Clostridium	+	+	+
Corynebacterium	–	–	+
Listeria	–	–	–
Gardnerella	–	–	–

1. Bacillus anthracis

Disease

B. anthracis causes anthrax (Figure 17–1), which is common in animals but rare in humans. Human disease occurs in three main forms: cutaneous, pulmonary (inhalation), and gastrointestinal. In 2001, an outbreak of both inhalation and cutaneous anthrax occurred in the United States. The outbreak was caused by sending spores of the organism through the mail. There were 18 cases, causing five deaths in this outbreak.

Important Properties

B. anthracis is a large gram-positive rod with square ends, frequently found in chains (Figure 17–2). Its antiphagocytic capsule is composed of D-glutamate. (This is unique—capsules of other bacteria are polysaccharides.) It is nonmotile, whereas other members of the genus are motile. Anthrax toxin is encoded on

TABLE 17–2 Important Features of Pathogenesis by *Bacillus* Species

Organism	Disease	Transmission/Predisposing Factor	Action of Toxin	Prevention
B. anthracis	Anthrax	1. Cutaneous anthrax: spores in soil enter wound 2. Pulmonary anthrax: spores are inhaled into lung	Exotoxin has three components: protective antigen binds to cells; edema factor is an adenylate cyclase; lethal factor is a protease that inhibits cell growth resulting in cell death (necrosis)	Vaccine contains protective antigen as the immunogen
B. cereus	Food poisoning	Spores germinate in reheated rice, then bacteria produce exotoxins, which are ingested	Two exotoxins (enterotoxins): 1. Similar to cholera toxin, it increases cyclic AMP 2. Similar to staphylococcal enterotoxin, it is a superantigen	No vaccine

one plasmid, and the enzymes that synthesize the polyglutamate capsule are encoded on a different plasmid.

Transmission

Spores of the organism persist in soil for years. Humans are most often infected cutaneously at the time of trauma to the skin, which allows the **spores on animal products**, such as hides, bristles, and wool, to enter. Spores can also be inhaled into the respiratory tract. Pulmonary (inhalation) anthrax occurs when spores are inhaled into the lungs. Gastrointestinal anthrax occurs when contaminated meat is ingested.

Inhalation anthrax is not communicable from person to person, despite the severity of the infection. After being inhaled into the lung, the organism moves rapidly to the mediastinal lymph nodes, where it causes hemorrhagic mediastinitis. Because it leaves the lung so rapidly, it is not transmitted by the respiratory route to others.

Pathogenesis

Pathogenesis is based primarily on the production of two exotoxins, collectively known as anthrax toxin. The two exotoxins,

edema factor and **lethal factor**, each consists of two proteins in an A–B subunit configuration. The B, or binding, subunit in each of the two exotoxins is **protective antigen**. The A, or active, subunit has enzymatic activity.

Edema factor, an exotoxin, is an **adenylate cyclase** that causes an increase in the intracellular concentration of cyclic adenosine monophosphate (AMP). This causes an outpouring of fluid from the cell into the extracellular space, which manifests as edema. (Note the similarity of action to that of cholera toxin.)

Lethal factor is a protease that cleaves the phosphokinase that activates the mitogen-activated protein kinase (MAPK) signal transduction pathway. This pathway controls the growth of human cells, and cleavage of the phosphokinase inhibits cell growth, leading to apoptosis, cell death, and the necrotic skin lesion (black eschar). Protective antigen forms pores in the human cell membrane that allows edema factor and lethal factor to enter the cell. The name *protective antigen* refers to the fact that antibody against this protein protects against disease.

Clinical Findings

The typical lesion of cutaneous anthrax is a painless ulcer with a black eschar (crust, scab) (see Figure 17–1). Local edema is striking. The lesion is called a **malignant pustule**. Untreated cases progress to bacteremia and death.

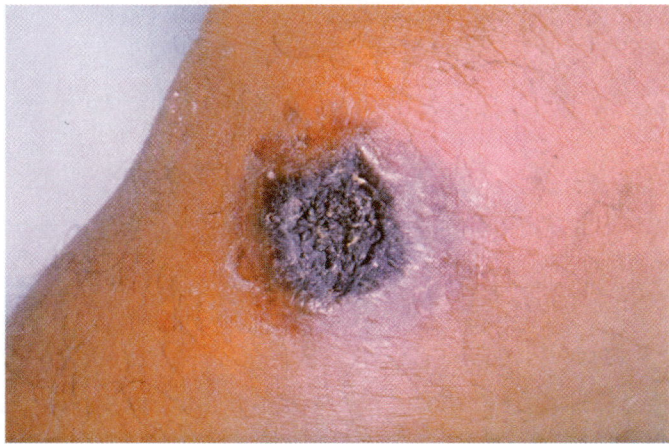

FIGURE 17–1 Skin lesion of anthrax. Note the *black eschar*, a necrotic lesion covered by a crust, caused by lethal factor, an exotoxin produced by *Bacillus anthracis*. Note the area of edema surrounding the eschar, which is caused by another exotoxin called *edema factor*. (Used with permission from Dr. James H. Steele, Centers for Disease Control and Prevention. CDC #2033.)

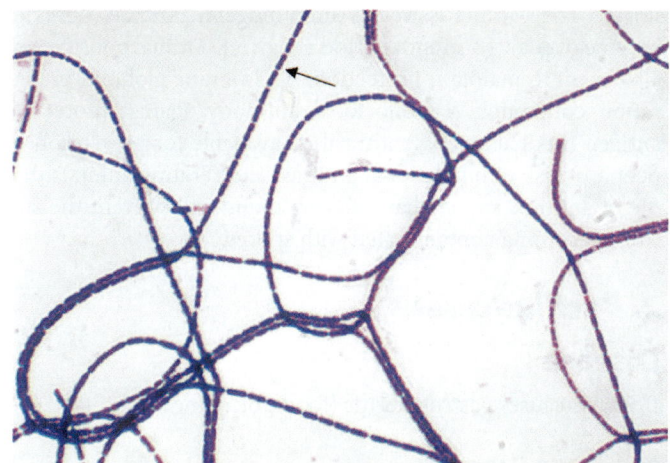

FIGURE 17–2 *Bacillus anthracis*—Gram stain. Arrow points to one large "box car–like" gram-positive rod within a long chain. (Reproduced with permission from Public Health Image Library, Centers for Disease Control and Prevention.)

Pulmonary (inhalation) anthrax, also known as "wool-sorter's disease," begins with nonspecific respiratory tract symptoms resembling influenza, especially a dry cough and substernal pressure. This rapidly progresses to hemorrhagic mediastinitis, bloody pleural effusions, septic shock, and death. Although the lungs are infected, the classic features and X-ray picture of pneumonia are not present. Mediastinal widening seen on chest X-ray is an important diagnostic criterion. Hemorrhagic mediastinitis and hemorrhagic meningitis are severe life-threatening complications. The symptoms of gastrointestinal anthrax include vomiting, abdominal pain, and bloody diarrhea.

Laboratory Diagnosis

Smears show large, gram-positive rods in chains (see Figure 17–2). Spores are usually not seen in smears of exudate because spores form when nutrients are insufficient, and nutrients are plentiful in infected tissue. Nonhemolytic colonies form on blood agar aerobically. Colonies on blood agar typically have a characteristic flared "comet's tail" appearance.

In case of a bioterror attack, rapid diagnosis can be performed in special laboratories using polymerase chain reaction (PCR)-based assays. Another rapid diagnostic procedure is the direct fluorescent antibody test that detects antigens of the organism in the lesion. Serologic tests, such as an enzyme-linked immunosorbent assay (ELISA) test for antibodies, require acute and convalescent serum samples and can only be used to make a diagnosis retrospectively.

Treatment

Ciprofloxacin is the drug of choice. Doxycycline is an alternative drug. No resistant strains have been isolated clinically.

Prevention

Ciprofloxacin or doxycycline was used as prophylaxis in those exposed during the outbreak in the United States in 2001. People at high risk can be immunized with cell-free vaccine (BioThrax) containing purified protective antigen as immunogen. The vaccine is weakly immunogenic, and six doses of vaccine over an 18-month period are given. Annual boosters are also given to maintain protection. An immune globulin preparation containing a monoclonal antibody against protective antigen (raxibacumab, Anthrasil) is available for prevention in people at risk of inhalational anthrax. Incinerating animals that die of anthrax, rather than burying them, will prevent the soil from becoming contaminated with spores.

2. Bacillus cereus

Disease

B. cereus causes gastroenteritis (food poisoning).

Transmission

Spores on grains such as rice survive steaming and rapid frying. The spores germinate when rice is kept warm for many hours (e.g., **reheated fried rice**). The portal of entry is the gastrointestinal tract.

Pathogenesis

B. cereus produces two enterotoxins. The mode of action of one of the enterotoxins is the same as that of cholera toxin (i.e., it adds adenosine diphosphate [ADP] ribose, a process called ADP-ribosylation, to a G protein, which stimulates adenylate cyclase and leads to an increased concentration of cyclic AMP within the enterocyte). The mode of action of the other enterotoxin resembles that of staphylococcal enterotoxin (i.e., it is a superantigen).

Clinical Findings

There are two syndromes. (1) One syndrome has a short incubation period (4 hours) and consists primarily of nausea and vomiting, similar to staphylococcal food poisoning. (2) The other has a long incubation period (18 hours) and features watery, nonbloody diarrhea, resembling clostridial gastroenteritis.

Laboratory Diagnosis

This is not usually done.

Treatment & Prevention

Only symptomatic treatment is given. There is no specific means of prevention. Cooked rice should not be kept warm for long periods.

CLOSTRIDIUM

There are four medically important species: *Clostridium tetani, Clostridium botulinum, Clostridium perfringens* (which causes either gas gangrene or food poisoning), and *Clostridium difficile*. All clostridia are anaerobic, spore-forming, gram-positive rods (Figure 17–3). Important features of pathogenesis and prevention are described in Table 17–3.

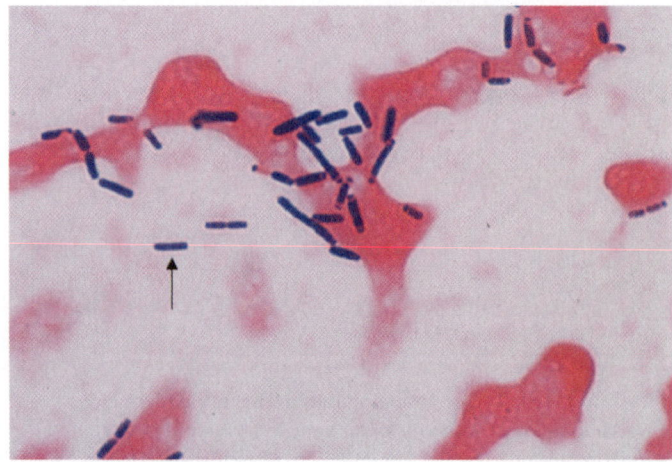

FIGURE 17–3 *Clostridium perfringens*—Gram stain. Arrow points to a large gram-positive rod. (Used with permission from Professor Shirley Lowe, University of California, San Francisco School of Medicine.)

TABLE 17–3 Important Features of Pathogenesis by *Clostridium* Species

Organism	Disease	Transmission/Predisposing Factor	Action of Toxin	Prevention
C. tetani	Tetanus	Spores in soil enter wound	Blocks release of inhibitory transmitters (e.g., glycine)	Toxoid vaccine
C. botulinum	Botulism	Exotoxin in food is ingested	Blocks release of acetylcholine	Proper canning; cook food
C. perfringens	1. Gas gangrene 2. Food poisoning	Spores in soil enter wound Exotoxin in food is ingested	Lecithinase Superantigen	Debride wounds Cook food
C. difficile	Pseudomembranous colitis	Antibiotics suppress normal flora	Cytotoxin damages colon mucosa	Appropriate use of antibiotics

1. Clostridium tetani

Disease

C. tetani causes tetanus (Figure 17–4).

Transmission

Spores are widespread in soil. The portal of entry is usually a **wound** site (e.g., where a nail penetrates the foot), but the spores can also be introduced during "skin-popping," a technique used by drug addicts to inject drugs into the skin. Germination of spores is favored by necrotic tissue and poor blood supply in the wound. Neonatal tetanus, in which the organism enters through a contaminated umbilicus or circumcision wound, is a major problem in some developing countries.

Pathogenesis

Tetanus toxin (tetanospasmin) is an exotoxin produced by vegetative cells at the wound site, *not* by the spores. This polypeptide toxin is carried intra-axonally (retrograde) to the central nervous system, where it binds to ganglioside receptors and blocks release of inhibitory mediators (e.g., glycine

and γ-aminobutyric acid [GABA]) at spinal synapses. Tetanus toxin is encoded by a plasmid, unlike botulinum toxin which is encoded by a lysogenic bacteriophage.

Tetanus toxin and botulinum toxin (see later) are among the most toxic substances known. They are both proteases that cleave the proteins involved in mediator release from the neurons.

Tetanus toxin has one antigenic type, unlike botulinum toxin, which has eight. There is therefore only one antigenic type of tetanus toxoid in the vaccine against tetanus.

Clinical Findings

Tetanus is characterized by strong muscle spasms (spastic paralysis, tetany). Specific clinical features include **lockjaw** (trismus) due to rigid contraction of the jaw muscles, which prevents the mouth from opening; a characteristic grimace known as **risus sardonicus**; and exaggerated reflexes. **Opisthotonos**, a pronounced arching of the back due to spasm of the strong extensor muscles of the back, is often seen (see Figure 17–4). Respiratory failure ensues. A high-mortality rate is associated with this disease. Note that in tetanus, **spastic paralysis** (strong muscle contractions) occurs, whereas in botulism, **flaccid paralysis** (weak or absent muscle contractions) occurs.

Laboratory Diagnosis

There is no microbiologic or serologic diagnosis. Organisms are rarely isolated from the wound site. *C. tetani* produces a **terminal spore** (i.e., a spore at the end of the rod). This gives the organism the characteristic appearance of a "tennis racket."

Treatment

Tetanus immune globulin (tetanus antitoxin) is used to neutralize the toxin. The role of antibiotics is uncertain. If antibiotics are used, either metronidazole or penicillin G can be given. An adequate airway must be maintained and respiratory support given. Benzodiazepines (e.g., diazepam [Valium]) should be given to prevent spasms.

Prevention

Tetanus is prevented by immunization with tetanus **toxoid** (formaldehyde-treated toxin) in childhood and every 10 years thereafter.

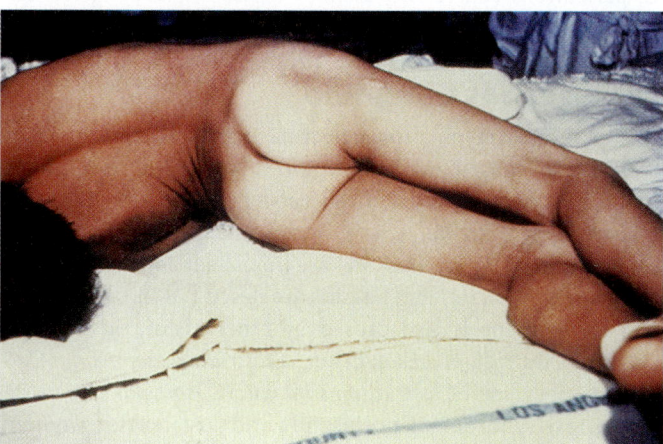

FIGURE 17–4 Tetanus. Note the marked hyperextension of the back, a position called *opisthotonos*, caused by tetanus toxin, an exotoxin that inhibits the release of mediators of the inhibitory neurons in the spinal cord. (Reproduced with permission from Centers for Disease Control and Prevention. CDC #6373.)

Tetanus toxoid is usually given to children in combination with diphtheria toxoid and the acellular pertussis vaccine (DTaP).

When trauma occurs, the wound should be cleaned and debrided, and tetanus toxoid booster should be given. If the wound is grossly contaminated, **tetanus immune globulin**, as well as the toxoid booster, should be given and penicillin administered. Half of the immune globulins should be infiltrated into the wound and the other half given intramuscularly at a site separate from the tetanus toxoid.

Tetanus immune globulin (tetanus antitoxin) is made in humans to avoid serum sickness reactions that occur when antitoxin made in horses is used. The administration of both immune globulins and tetanus toxoid (at different sites in the body) is an example of **passive–active immunity**.

2. Clostridium botulinum

Disease

C. botulinum causes botulism.

Transmission

Spores, widespread in soil, contaminate vegetables, and meats. When these foods are canned or vacuum-packed without adequate sterilization, spores survive and germinate in the anaerobic environment. Toxin is produced within the canned food and **ingested preformed**. The highest-risk foods are (1) alkaline vegetables such as green beans, peppers, and mushrooms and (2) smoked fish. The toxin is relatively heat-labile; it is inactivated by boiling for several minutes. Thus, disease can be prevented by sufficient cooking.

Pathogenesis

Botulinum toxin is absorbed from the gut and carried via the blood to peripheral nerve synapses, where it **blocks release of acetylcholine**. It is a protease that cleaves the proteins involved in acetylcholine release. The toxin is a polypeptide encoded by a lysogenic phage. Along with tetanus toxin, it is among the most toxic substances known. There are eight immunologic types of toxin; types A, B, and E are the most common in human illness. Botox is a commercial preparation of exotoxin A used to remove wrinkles on the face. Minute amounts of the toxin are effective in the treatment of certain spasmodic muscle disorders such as torticollis, "writer's cramp," and blepharospasm.

Clinical Findings

Descending weakness and paralysis of cranial nerves, including diplopia, dysphagia, ptosis, and respiratory muscle failure, are seen. No fever is present. In contrast, Guillain-Barré syndrome is an ascending paralysis (see Chapter 66).

Two special clinical forms occur: (1) wound botulism, in which spores contaminate a wound, germinate, and produce toxin at the site and (2) infant botulism, in which the organisms grow in the gut and produce the toxin there. Ingestion

of honey containing the organism is implicated in transmission of infant botulism. Affected infants develop weakness or paralysis and may need respiratory support but usually recover spontaneously. In the United States, infant botulism accounts for about half of the cases of botulism, and wound botulism is associated with drug abuse, especially skin-popping with black tar heroin.

Laboratory Diagnosis

The organism is usually not cultured. Botulinum toxin is demonstrable in uneaten food and the patient's serum by mouse protection tests. Mice are inoculated with a sample of the clinical specimen and will die unless protected by antitoxin. Enzyme-linked immunoassay (EIA) tests are also used to detect the toxin, and PCR tests are used to detect the DNA encoding the toxin.

Treatment

Treatment with antiserum should *not* be delayed while waiting for laboratory results. The heptavalent antitoxin containing all seven types (A–G) is preferred to the trivalent antitoxin containing types A, B, and E. Respiratory support is provided as well. The antitoxin is made in horses, and serum sickness may occur. A bivalent antitoxin (types A and B) purified from the plasma of humans immunized with botulinum toxoid is available for the treatment of infant botulism.

Prevention

Proper sterilization of all canned and vacuum-packed foods is essential. Food must be adequately cooked to inactivate the toxin. Swollen cans must be discarded (clostridial proteolytic enzymes form gas, which swells cans).

3. Clostridium perfringens

C. perfringens causes two distinct diseases, gas gangrene and food poisoning, depending on the route of entry into the body.

Disease: Gas Gangrene

Gas gangrene (myonecrosis, necrotizing fasciitis) is one of the two diseases caused by *C. perfringens* (Figure 17–5). Necrotizing fasciitis is often called the "**flesh-eating**" disease. In addition to *C. perfringens*, *Streptococcus pyogenes* and methicillin-resistant *Staphylococcus aureus* (MRSA) are important causes. The clinical aspects of necrotizing fasciitis are described in Chapter 77.

Gas gangrene is also caused by other histotoxic clostridia such as *Clostridium histolyticum*, *Clostridium septicum*, *Clostridium novyi*, and *Clostridium sordellii*. (*C. sordellii* also causes toxic shock syndrome in postpartum and postabortion women.)

Transmission

Spores are located in the soil; vegetative cells are members of the **normal flora of the colon and vagina**. Gas gangrene is

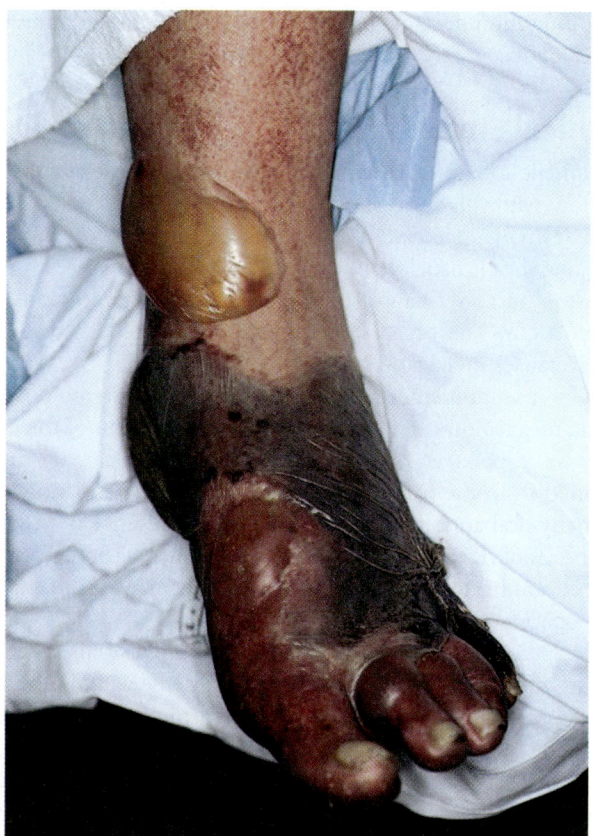

FIGURE 17–5 Gas gangrene. Note large area of necrosis on lateral aspect of foot. Necrosis is mainly caused by lecithinase produced by *Clostridium perfringens*. Gas in tissue is a feature of gangrene produced by these anaerobic bacteria. A large gas- and fluid-filled bulla is seen near the ankle. (Used with permission from David Kaplan, MD.)

associated with war wounds, automobile and motorcycle accidents, and septic abortions (endometritis).

Pathogenesis

Organisms grow in traumatized tissue (especially muscle) and produce a variety of toxins. The most important is **alpha toxin** (lecithinase), which damages cell membranes, including those of erythrocytes, resulting in hemolysis. Degradative enzymes produce gas in tissues.

Clinical Findings

Pain, edema, cellulitis, and gangrene (necrosis) occur in the wound area (see Figure 17–5). If crepitus is palpated in the affected tissue, it indicates gas in the tissue. This gas is typically hydrogen produced by the anaerobic bacteria. Hemolysis and jaundice are common, as are blood-tinged exudates. A foul-smelling, bloody vaginal discharge can occur in endometritis. Shock and death can ensue. Mortality rates are high.

Laboratory Diagnosis

Smears of tissue and exudate samples show large gram-positive rods. Spores are not usually seen because they are formed primarily under nutritionally deficient conditions. The organisms are cultured anaerobically and then identified by sugar fermentation reactions and organic acid production. *C. perfringens* colonies exhibit a double zone of hemolysis on blood agar. The colonies also produce a precipitate in egg yolk agar caused by the action of its lecithinase. Serologic tests are not useful.

Treatment & Prevention

Penicillin G plus clindamycin is the treatment of choice. Wounds should be cleaned and debrided. Incision and drainage of focal lesions should be done. Penicillin may be given for prophylaxis. There is no vaccine.

Disease: Food Poisoning

Food poisoning is the second disease caused by *C. perfringens*.

Transmission

Spores are located in **soil** and can contaminate **food**. The heat-resistant spores survive cooking and germinate. The organisms grow to large numbers in reheated foods, especially meat dishes.

Pathogenesis

C. perfringens is a member of the normal flora in the colon but not in the small bowel, where the enterotoxin acts to cause diarrhea. The mode of action of the enterotoxin is the same as that of the enterotoxin of *S. aureus* (i.e., it acts as a superantigen).

Clinical Findings

The disease has an 8- to 16-hour incubation period and is characterized by watery diarrhea with cramps and little vomiting. It resolves in 24 hours.

Laboratory Diagnosis

This is not usually done. There is no assay for the toxin. Large numbers of the organisms can be isolated from uneaten food.

Treatment & Prevention

Symptomatic treatment is given; no antimicrobial drugs are administered. There are no specific preventive measures. Food should be adequately cooked to kill the organism.

4. Clostridium difficile (Clostridioides difficile)

Disease

C. difficile causes antibiotic-associated pseudomembranous colitis (Figure 17–6). *C. difficile* is the most common nosocomial (hospital-acquired) cause of diarrhea. It is the leading

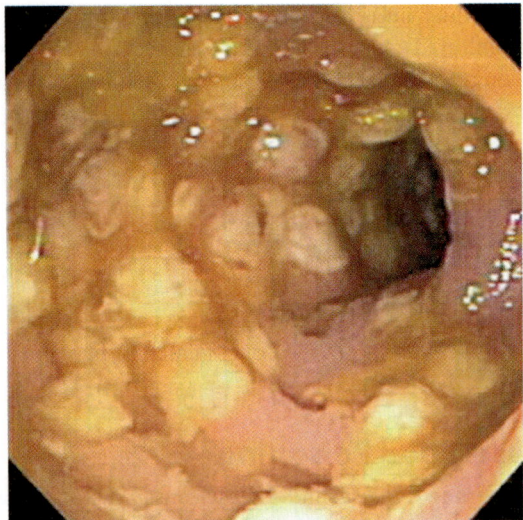

FIGURE 17–6 Pseudomembranous colitis. Note yellowish plaquelike lesions in colon. Caused by an exotoxin produced by *Clostridium difficile* that inhibits a signal transduction protein, leading to death of enterocytes. (Reproduced with permission from Usatine RP, Smith MA, Mayeaux EJ Jr, et al: *The Color Atlas of Family Medicine*, New York, NY: McGraw Hill; 2009. Photo contributor: E.J. Mayeaux, Jr., MD.)

infectious cause of gastrointestinal-associated deaths in the United States. The name *C. difficile* has recently been changed to *Clostridioides difficile* but for this edition the previous name will be used.

Transmission

The organism colonizes the **large intestine** of approximately 3% of the general population and up to 30% of hospitalized patients. Note that most people are not colonized, which explains why most people who take antibiotics do not get pseudomembranous colitis. *C. difficile* is transmitted by the fecal–oral route. Either the spores or the bacterial organism itself can be transmitted.

The majority of cases occur in hospitalized patients, but about one-third of cases are community-acquired. The hands of hospital personnel are important intermediaries.

Pathogenesis

Antibiotics suppress drug-sensitive members of the normal flora of the colon, allowing *C. difficile* to multiply and produce large amounts of **exotoxins A and B**. Both exotoxin A and exotoxin B are glucosyltransferases (i.e., enzymes that glucosylate [add glucose to] a G protein called Rho GTPase). The main effect of these exotoxins is to cause depolymerization of actin, resulting in a loss of cytoskeletal integrity, apoptosis, and death of the enterocytes. Exotoxin B is thought to play the leading role in producing the signs and symptoms of human disease.

Clindamycin was the first antibiotic to be shown to predispose to pseudomembranous colitis, but many antibiotics are known to predispose to this disease. At present, third-generation cephalosporins are the most common because they are so frequently used. Ampicillin and fluoroquinolones are also commonly implicated. In addition to antibiotics, **cancer chemotherapy** and

proton pump inhibitors predispose to pseudomembranous colitis. *C. difficile* rarely invades the intestinal mucosa.

Clinical Findings

C. difficile causes diarrhea associated with **pseudomembranes** (yellow-white plaques) on the colonic mucosa (see Figure 17–6). (The term *pseudomembrane* is defined in Chapter 7 on page 37). The diarrhea is usually not bloody, and neutrophils are found in the stool in about half of the cases. Fever and abdominal pain often occur. The organism rarely enters the bloodstream and rarely causes metastatic infection.

The pseudomembranes are visualized by sigmoidoscopy. Toxic megacolon can occur, and surgical resection of the colon may be necessary. Pseudomembranous colitis can be distinguished from the transient diarrhea that occurs as a side effect of many oral antibiotics by testing for the presence of the toxin in the stool. Even with adequate treatment, the organism may not be eradicated from the colon, and recurrences occur at a rate of approximately 15% to 20%.

In 2005, a new, more virulent strain of *C. difficile* emerged. This *hypervirulent* strain causes more severe disease, is more likely to cause recurrences, and responds less well to metronidazole than the previous strain. The strain is also characterized by resistance to quinolones. It is thought that the widespread use of quinolones for diarrheal disease may have selected for this new strain.

Laboratory Diagnosis

The presence of exotoxins in the filtrate of a patient's stool specimen is the basis of the laboratory diagnosis. It is insufficient to culture the stool for the presence of *C. difficile* because people can be colonized by the organism and not have disease.

A useful screening test is the enzyme immunoassay for the glutamine dehydrogenase (GDH) of *C. diificile*. GDH is produced by all isolates of *C. diificile* so detection of GDH as an antigen supports the diagnosis. Testing for the presence of the toxin or its genes is then warranted.

There are two types of tests used to make the laboratory diagnosis. One detects the exotoxin itself, and the other detects the genes that encode the exotoxin. To detect the *exotoxin itself*, an ELISA test employing antibody to the exotoxin is used. To detect the *genes that encode the exotoxin*, a PCR assay to determine the presence of the *toxin gene DNA* is used. The DNA-based test has greater sensitivity and specificity than the ELISA test. However, these nucleic acid amplification tests (NAATs) should be interpreted with caution because a person may only be colonized by *C. difficile* and be recorded as positive when, in fact, *C. difficile* is not the cause of the patient's disease.

Treatment

The causative antibiotic should be withdrawn. Either fidaxomicin or vancomycin, administered orally, is the drug of choice for the initial episode and for recurrences. Metronidazole is no longer recommended but can be used for the initial episode of nonsevere cases, if neither fidaxomicin or vancomycin can be used.

Vancomycin can be given orally or per rectum or, in very severe cases, both ways. In life-threatening cases, vancomycin plus metronidazole should be used but surgical removal of the colon may be required.

In many patients, treatment does not eradicate the carrier state, and recurrent episodes of colitis can occur. Fidaxomicin is used both in the treatment of pseudomembranous colitis and in preventing relapses of this disease. It is effective in life-threatening cases. Bezlotoxumab, a monoclonal antibody against exotoxin B of *C. difficile*, is effective in preventing relapses.

Fecal transplantation is another therapeutic approach. It involves administering bowel flora from a normal individual either by enema or by nasoduodenal tube to the patient with pseudomembranous colitis. This approach is based on the concept of bacterial interference (i.e., to replace the *C. difficile* with normal bowel flora). Very high cure rates are claimed for this technique, but aesthetic considerations have limited its acceptance.

Prevention

There are no preventive vaccines or drugs. Because antibiotics are an important predisposing factor for pseudomembranous colitis, they should be prescribed only when necessary. Fidaxomicin is useful in preventing relapses of this disease. In the hospital, strict infection control procedures, including rigorous handwashing, are important. Probiotics, such as the yeast *Saccharomyces*, may be useful to prevent pseudomembranous colitis.

NON–SPORE-FORMING GRAM-POSITIVE RODS

There are three important pathogens in this group: *Corynebacterium diphtheriae*, *Listeria monocytogenes*, and *G. vaginalis*. Important features of pathogenesis and prevention of *C. diphtheriae* and *L. monocytogenes* are described in Table 17–4.

CORYNEBACTERIUM DIPHTHERIAE

Disease

C. diphtheriae causes diphtheria (Figure 17–7). Other *Corynebacterium* species (diphtheroids) are implicated in opportunistic infections.

Important Properties

Corynebacteria are gram-positive rods that appear **club-shaped** (wider at one end) and are arranged in palisades or in V- or L-shaped formations (Figure 17–8). The rods have a beaded appearance. The beads consist of granules of highly polymerized polyphosphate—a storage mechanism for high-energy phosphate bonds. The granules stain **metachromatically** (i.e., a dye that stains the rest of the cell blue will stain the granules red).

Transmission

Humans are the only natural host of *C. diphtheriae*. Both toxigenic and nontoxigenic organisms reside in the upper respiratory tract and are transmitted by **airborne droplets**. The organism can also infect the skin at the site of a preexisting skin lesion. This occurs primarily in the tropics but can occur worldwide in indigent persons with poor skin hygiene.

Pathogenesis

Although exotoxin production is essential for pathogenesis, invasiveness is also necessary because the organism must first establish and maintain itself in the throat. Diphtheria toxin inhibits protein synthesis by **ADP-ribosylation of elongation factor-2** (EF-2). The toxin affects all eukaryotic cells regardless of tissue type but has no effect on the EF-2 in prokaryotic cells.

The toxin is a single polypeptide with two functional domains. The binding (**B**) domain mediates binding of the toxin to glycoprotein receptors on the cell membrane. The active (**A**) domain possesses enzymatic activity that cleaves nicotinamide from nicotinamide adenine dinucleotide (NAD) and transfers the remaining ADP-ribose to EF-2, thereby inactivating it. Other organisms whose exotoxins act by ADP-ribosylation are described in Tables 7–10 and 7–11.

The DNA that codes for diphtheria toxin is part of the DNA of a temperate bacteriophage called beta phage. During the lysogenic phase of viral growth, the DNA of this virus integrates into the bacterial chromosome and the toxin is synthesized. *C. diphtheriae* cells that are not lysogenized by this phage do not produce exotoxin and are nonpathogenic.

The host response to *C. diphtheriae* consists of the following:

(1) A local inflammation in the throat, with a fibrinous exudate that forms the tough, adherent, gray **pseudomembrane** characteristic of the disease.

(2) Antibody that can neutralize exotoxin activity by blocking the interaction of the binding domain with the receptors, thereby preventing entry into the cell. The immune status of a person can be assessed by Schick's test. The test is performed by

TABLE 17–4 Important Features of Pathogenesis by *Corynebacterium diphtheriae* and *Listeria monocytogenes*

Organism	Type of Pathogenesis	Typical Disease	Predisposing Factor	Mode of Prevention
C. diphtheriae	Toxigenic	Diphtheria	Failure to immunize	Toxoid vaccine
L. monocytogenes	Pyogenic	Meningitis; sepsis	Neonate; immunosuppression	No vaccine; pasteurize milk products

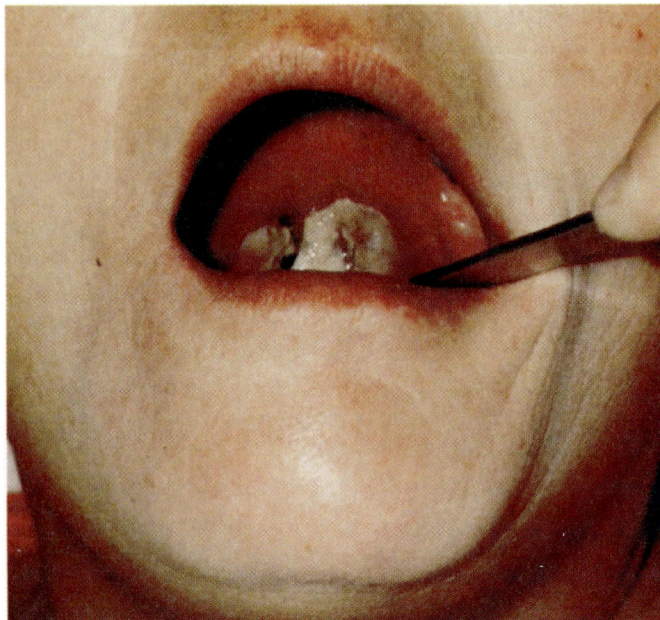

FIGURE 17–7   Diphtheria. Note whitish-gray pseudomembrane covering posterior pharynx and marked inflammation of palate and pharynx. Caused by diphtheria toxin, an exotoxin that inhibits protein synthesis by inhibiting elongation factor-2. (Used with permission from Dr. Peter Strebel.)

intradermal injection of 0.1 mL of purified standardized toxin. If the patient has no antitoxin, the toxin will cause inflammation at the site 4 to 7 days later. If no inflammation occurs, antitoxin is present and the patient is immune. The test is rarely performed in the United States except under special epidemiologic circumstances.

Clinical Findings

Although diphtheria is rare in the United States, physicians should be aware of its most prominent sign, the thick, gray,

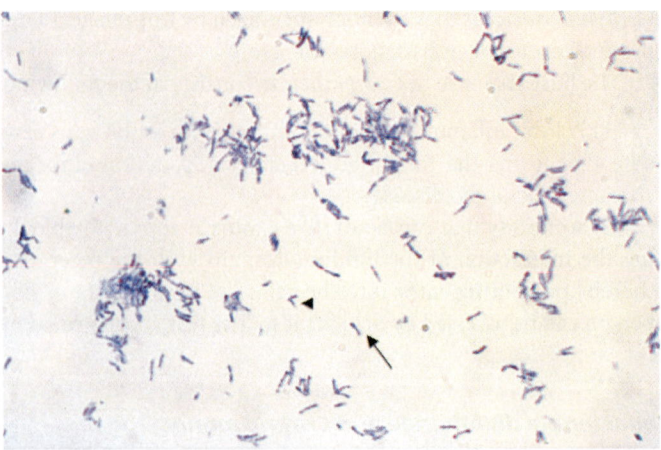

FIGURE 17–8   *Corynebacterium diphtheriae*—Gram stain. Arrow points to a "club-shaped" gram-positive rod. Arrowhead points to typical V- or L-shaped corynebacteria. (Reproduced with permission from Public Health Image Library, Centers for Disease Control and Prevention.)

adherent **pseudomembrane** over the tonsils and throat (see Figure 17–7). (The term *pseudomembrane* is defined in Chapter 7 on page 37.) The other aspects are nonspecific: fever, sore throat, and cervical adenopathy. There are three prominent complications:

(1) Extension of the membrane into the larynx and trachea, causing airway obstruction.

(2) Myocarditis accompanied by arrhythmias and circulatory collapse.

(3) Nerve weakness or paralysis, especially of the cranial nerves. Paralysis of the muscles of the soft palate and pharynx can lead to regurgitation of fluids through the nose. Peripheral neuritis affecting the muscles of the extremities also occurs.

Cutaneous diphtheria causes ulcerating skin lesions covered by a gray membrane. These lesions are often indolent and often do not invade surrounding tissue. Systemic symptoms rarely occur. In the United States, cutaneous diphtheria occurs primarily in the indigent.

Laboratory Diagnosis

Laboratory diagnosis involves both isolating the organism and demonstrating toxin production. It should be emphasized that the decision to treat with antitoxin is a clinical one and cannot wait for the laboratory results. A throat swab should be cultured on Loeffler's medium, a **tellurite plate**, and a blood agar plate. The tellurite plate contains a tellurium salt that is reduced to elemental tellurium within the organism. The typical gray-black color of tellurium in the colony is a telltale diagnostic criterion. If *C. diphtheriae* is recovered from the cultures, either animal inoculation or an antibody-based gel diffusion precipitin test is performed to document toxin production. A PCR assay for the presence of the toxin gene in the organism isolated from the patient can also be used.

Smears of the throat swab should be stained with both Gram stain and methylene blue. Although the diagnosis of diphtheria cannot be made by examination of the smear, the finding of many tapered, pleomorphic gram-positive rods can be suggestive. The methylene blue stain is excellent for revealing the typical metachromatic granules.

Treatment

The treatment of choice is **antitoxin**, which should be given immediately on the basis of clinical impression because there is a delay in laboratory diagnostic procedures. The toxin binds rapidly and irreversibly to cells and, once bound, cannot be neutralized by antitoxin. The function of antitoxin is therefore to neutralize unbound toxin in the blood. Because the antiserum is made in horses, the patient must be tested for hypersensitivity, and medications for the treatment of anaphylaxis must be available. Serum sickness (see Chapter 65) may occur after administration of antiserum made in horses.

Treatment with penicillin G or erythromycin is also recommended, but neither is a substitute for antitoxin. Antibiotics inhibit growth of the organism, reduce toxin production, and decrease the incidence of chronic carriers.

Prevention

Diphtheria is very rare in the United States because children are immunized with **diphtheria toxoid** (usually given as a combination of diphtheria toxoid, tetanus toxoid, and acellular pertussis vaccine, often abbreviated as DTaP).

Diphtheria toxoid is prepared by treating the exotoxin with formaldehyde. This treatment inactivates the toxic effect but leaves the antigenicity intact. Immunization consists of three doses given at 2, 4, and 6 months of age, with boosters at 1 and 6 years of age. Because immunity wanes, a booster every 10 years is recommended. Immunization does not prevent nasopharyngeal carriage of the organism.

LISTERIA MONOCYTOGENES

Diseases

L. monocytogenes causes meningitis and sepsis in newborns, pregnant women, and immunosuppressed adults. It also causes outbreaks of febrile gastroenteritis. It is a major cause of concern for the food industry.

Important Properties

L. monocytogenes is a small gram-positive rod arranged in V- or L-shaped formations similar to corynebacteria. The organism exhibits an unusual **tumbling** motility that distinguishes it from the corynebacteria, which are nonmotile. Colonies on a blood agar plate produce a narrow zone of β-hemolysis that resembles the hemolysis of some streptococci.

Listeria grows well at cold temperatures, so storage of contaminated food in the refrigerator can increase the risk of gastroenteritis. This paradoxical growth in the cold is called "cold enhancement."

Pathogenesis

Listeria infections occur primarily in two clinical settings: (1) in the fetus or in a newborn as a result of transmission **across the placenta** or **during delivery** and (2) in pregnant women and immunosuppressed adults, especially renal transplant patients. (Note that pregnant women have reduced cell-mediated immunity during the third trimester.)

The organism is distributed worldwide in animals, plants, and soil. From these reservoirs, it is transmitted to humans primarily by ingestion of unpasteurized milk products, undercooked meat, and raw vegetables. Contact with domestic farm animals and their feces is also an important source.

In the United States, listeriosis is primarily a foodborne disease associated with eating unpasteurized cheese and delicatessen meats. Following ingestion, the bacteria appear in the colon and then can colonize the female genital tract. From this location, they can infect the fetus if membranes rupture or infect the neonate during passage through the birth canal.

The pathogenesis of *Listeria* depends on the organism's ability to invade and survive within cells. Invasion of cells is mediated by internalin made by *Listeria* and E-cadherin on the surface of human cells. The ability of *Listeria* to pass the placenta, enter the meninges, and invade the gastrointestinal tract depends on the interaction of internalin and E-cadherin on those tissues.

Upon entering the cell, the organism produces **listeriolysin**, which allows it to escape from the phagosome into the cytoplasm, thereby escaping destruction in the phagosome. Because *Listeria* preferentially grows intracellularly, cell-mediated immunity is a more important host defense than humoral immunity. Suppression of **cell-mediated immunity** predisposes to *Listeria* infections.

L. monocytogenes can move from cell to cell by means of **actin rockets**—filaments of actin polymerize—and propel the bacteria through the membrane of one human cell and into another.

Clinical Findings

Infection during pregnancy can cause abortion, premature delivery, or sepsis during the peripartum period. Newborns infected at the time of delivery can have acute meningitis 1 to 4 weeks later. The bacteria reach the meninges via the bloodstream (bacteremia). The infected mother is either asymptomatic or has an influenzalike illness. *L. monocytogenes* infections in immunocompromised adults can be either sepsis or meningitis.

Gastroenteritis caused by *L. monocytogenes* is characterized by watery diarrhea, fever, headache, myalgias, and abdominal cramps but little vomiting. Outbreaks are usually caused by contaminated dairy products, but undercooked meats such as chicken and hot dogs and ready-to-eat foods such as coleslaw have also been involved.

Laboratory Diagnosis

Laboratory diagnosis is made primarily by Gram stain and culture. The appearance of gram-positive rods resembling **diphtheroids** and the formation of small, gray colonies with a narrow zone of β-hemolysis on a blood agar plate suggest the presence of *Listeria*. The isolation of *Listeria* is confirmed by the presence of motile organisms, which differentiate them from the nonmotile corynebacteria. Identification of the organism as *L. monocytogenes* is made by sugar fermentation tests. A PCR assay is available.

Treatment

Treatment of invasive disease, such as meningitis and sepsis, consists of ampicillin with or without gentamicin. Trimethoprim-sulfamethoxazole can also be used. Resistant strains are rare. *Listeria* gastroenteritis typically does not require treatment.

Prevention

Prevention is difficult because there is no immunization. Limiting the exposure of pregnant women and immunosuppressed patients to potential sources such as farm animals, unpasteurized milk products, and raw vegetables is recommended. Trimethoprim-sulfamethoxazole given to immunocompromised patients to prevent *Pneumocystis* pneumonia can also prevent listeriosis.

GARDNERELLA VAGINALIS

Disease

G. vaginalis is the main organism associated with bacterial vaginosis. This disease is the most common vaginal infection of sexually active women.

Important Properties

G. vaginalis is a small, facultative **gram-variable rod**. The term "gram-variable" refers to the observation that some organisms are purple while others are pink in a Gram-stained specimen. Structurally, it has a gram-positive cell wall, but the wall is thin and older organisms tend to lose the purple color.

Pathogenesis

The pathogenesis of bacterial vaginosis is uncertain. *G. vaginalis* is often found in association with anaerobes such as *Mobiluncus* and *Prevotella* and nonanaerobes such as *Mycoplasma hominis* and *Ureaplasma urealyticum*. Together they cause the symptoms of bacterial vaginosis.

Bacterial vaginosis is not thought to be transmitted by sexual activity. Rather, it is considered to be a dysbiosis in which the *Lactobacillus* that are found as normal flora in the vagina are replaced by these other organisms.

Clinical Findings

Bacterial vaginosis is characterized by a malodorous, white or gray-colored vaginal discharge. The discharge has a characteristic "fishy" odor. Inflammatory changes are typically absent, which is why it is called a "vaginosis" rather than a "vaginitis." Mild itching may occur. Women with bacterial vaginosis have a higher incidence of preterm deliveries, and consequently, a higher incidence of morbidity and mortality occurs in their newborn children.

Laboratory Diagnosis

Clue cells, which are vaginal epithelial cells covered with bacteria, are an important laboratory finding seen in a microscopic examination of the vaginal discharge (Figure 17–9). In addition, the **"whiff" test**, which consists of treating the vaginal discharge with 10% KOH and smelling a pungent, "fishy" odor, is often positive. However, trichomoniasis, which can also cause a positive whiff test, must be ruled out before a diagnosis of bacterial vaginosis can be made. A pH of greater than 4.5 of the vaginal discharge supports the diagnosis of bacterial vaginosis.

Treatment and Prevention

The drug of choice is metronidazole. Treatment of sexual partners is not recommended as it is not considered to be transmissible. There is no vaccine.

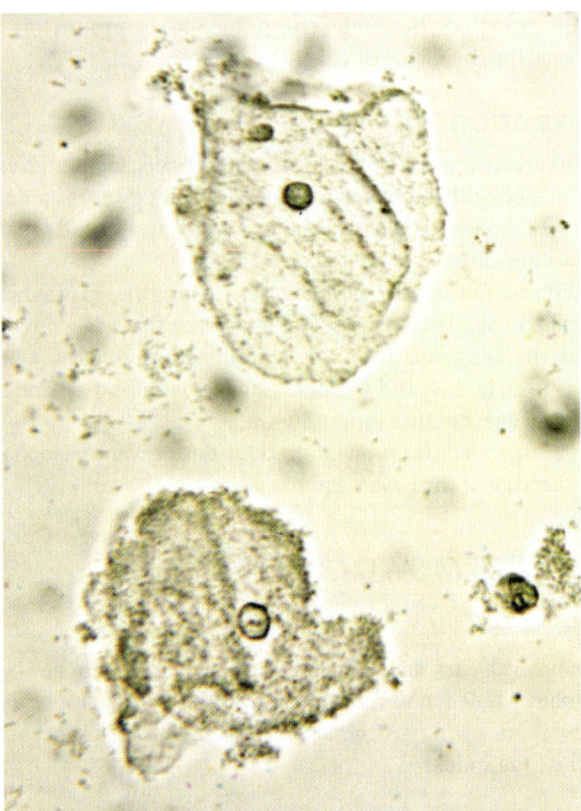

FIGURE 17–9 Clue cells in bacterial vaginosis. Note that the lower epithelial cell is a "clue cell" because its surface is covered with bacteria. The upper epithelial cell is *not* a "clue cell" because its surface has few bacteria. (Reproduced with permission from Usatine RP, Smith MA, Mayeaux EJ Jr, et al: *The Color Atlas of Family Medicine*, New York, NY: McGraw Hill; 2009. Photo contributor: Mayeaux EJ, Jr, MD.)

SELF-ASSESSMENT QUESTIONS

1. Which one of the following is a club-shaped, gram-positive rod that causes disease by producing an exotoxin that kills cells by inhibiting elongation factor-2, resulting in the inhibition of protein synthesis?

   (A) *Bacillus anthracis*
   (B) *Bacillus cereus*
   (C) *Clostridium perfringens*
   (D) *Corynebacterium diphtheriae*
   (E) *Listeria monocytogenes*

2. Which one of the following is a large gram-positive rod that causes necrosis of tissue by producing an exotoxin that degrades lecithin, resulting in the lysis of cell membranes?

   (A) *Bacillus anthracis*
   (B) *Bacillus cereus*
   (C) *Clostridium perfringens*
   (D) *Corynebacterium diphtheriae*
   (E) *Listeria monocytogenes*

3. Which one of the following sets of bacteria causes disease characterized by a pseudomembrane?

   (A) *Bacillus anthracis* and *Listeria monocytogenes*
   (B) *Bacillus cereus* and *Clostridium perfringens*
   (C) *Bacillus cereus* and *Clostridium tetani*
   (D) *Corynebacterium diphtheriae* and *Clostridium difficile*
   (E) *Corynebacterium diphtheriae* and *Listeria monocytogenes*

4. Disease caused by which one of the following sets of bacteria can be prevented by a toxoid vaccine?

   (A) *Bacillus anthracis* and *Clostridium botulinum*
   (B) *Bacillus anthracis* and *Clostridium perfringens*
   (C) *Bacillus cereus* and *Clostridium tetani*
   (D) *Corynebacterium diphtheriae* and *Clostridium tetani*
   (E) *Corynebacterium diphtheriae* and *Listeria monocytogenes*

5. Your patient in the pediatric intensive care unit is a 2-week-old boy with a high fever and the signs of meningitis. Gram stain of the spinal fluid reveals small gram-positive rods. Colonies on blood agar show a narrow zone of β-hemolysis. Which one of the following is the most likely cause of his neonatal meningitis?

   (A) *Bacillus anthracis*
   (B) *Bacillus cereus*
   (C) *Clostridium perfringens*
   (D) *Corynebacterium diphtheriae*
   (E) *Listeria monocytogenes*

6. Regarding the patient in Question 5, which one of the following is the best antibiotic to treat the infection?

   (A) Doxycycline
   (B) Gentamicin
   (C) Metronidazole
   (D) Trimethoprim-sulfamethoxazole
   (E) Vancomycin

7. Your patient is a 40-year-old woman with diplopia and other signs of cranial nerve weakness. History reveals she grows her own vegetables and likes to preserve them in jars that she prepares at home. She is fond of her preserved string beans, which is what she ate uncooked in a salad for dinner last night. Which one of the following is the most likely cause of this clinical picture?

   (A) *Bacillus anthracis*
   (B) *Clostridium botulinum*
   (C) *Clostridium perfringens*
   (D) *Clostridium tetani*
   (E) *Listeria monocytogenes*

8. Your patient is a 30-year-old man with a 2-cm lesion on his arm. It began as a painless papule that enlarged and, within a few days, ulcerated and formed a black crust (eschar). He works in an abattoir where his job is removing the hide from the cattle. A Gram stain of fluid from the lesion reveals large gram-positive rods. Which one of the following bacteria is likely to be the cause?

   (A) *Bacillus anthracis*
   (B) *Clostridium botulinum*
   (C) *Clostridium perfringens*
   (D) *Clostridium tetani*
   (E) *Listeria monocytogenes*

9. Your patient is a 30-year-old man who was brought to the emergency room following a motorcycle accident in which he sustained a compound fracture of his leg. He now has a high fever and a rapidly spreading cellulitis with crepitus in the area of the fracture. Large gram-positive rods are seen on the exudate. Necrotic tissue was debrided. Which one of the following is the best antibiotic to treat the infection?

   (A) Azithromycin
   (B) Ciprofloxacin
   (C) Gentamicin
   (D) Penicillin G
   (E) Vancomycin

10. Your patient is a 65-year-old woman who is several days post-op following removal of her carcinoma of the colon. She now spikes a fever and has a cough, and chest X-ray shows pneumonia. While being treated with the appropriate antibiotics, she develops severe diarrhea. You suspect she may have pseudomembranous colitis. Which one of the following is the best antibiotic to treat the infection?

    (A) Ceftriaxone
    (B) Doxycycline
    (C) Gentamicin
    (D) Metronidazole
    (E) Trimethoprim-sulfamethoxazole

ANSWERS

(1) **(D)**
(2) **(C)**
(3) **(D)**
(4) **(D)**
(5) **(E)**
(6) **(D)**
(7) **(B)**
(8) **(A)**
(9) **(D)**
(10) **(D)**

SUMMARIES OF ORGANISMS

Brief summaries of the organisms described in this chapter begin on page 675. Please consult these summaries for a rapid review of the essential material.

PRACTICE QUESTIONS: USMLE & COURSE EXAMINATIONS

Questions on the topics discussed in this chapter can be found in the Clinical Bacteriology section of Part XIII: USMLE (National Board) Practice Questions starting on page 737. Also see Part XIV: USMLE (National Board) Practice Examination starting on page 775.

Gram-Negative Rods Related to the Enteric Tract

INTRODUCTION

Gram-negative rods are a large group of diverse organisms (Figures 18–1, 18–2, and 19–1). In this book, these bacteria are subdivided into three clinically relevant categories, each in a separate chapter, according to whether the organism is related primarily to the enteric or the respiratory tract or to animal sources (Table 18–1). Although this approach leads to some overlap, it should be helpful because it allows general concepts to be emphasized.

Gram-negative rods related to the enteric tract include a large number of genera. These genera have therefore been divided into three groups depending on the major anatomic location of disease, namely, (1) pathogens both within and outside the enteric tract, (2) pathogens primarily within the enteric tract, and (3) pathogens outside the enteric tract (see Table 18–1).

The frequency with which the organisms related to the enteric tract cause disease in the United States is shown in Table 18–2. *Salmonella*, *Shigella*, and *Campylobacter* are frequent pathogens in the gastrointestinal tract, whereas *Escherichia*, *Vibrio*, and *Yersinia* are less so. Enterotoxigenic strains of *Escherichia coli* are a common cause of diarrhea in developing countries but are less common in the United States. The medically important gram-negative rods that cause diarrhea are described in Table 18–3. Urinary tract infections are caused primarily by *E. coli*; the other organisms occur less commonly. The medically important gram-negative rods that cause urinary tract infections are described in Table 18–4.

Additional information regarding the clinical aspects of infections caused by the organisms in this chapter is provided in Part IX, entitled Infectious Diseases beginning on page 603.

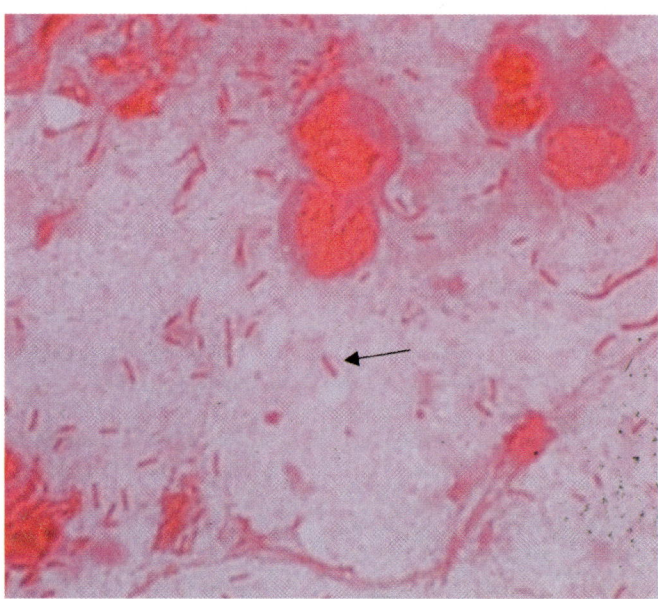

FIGURE 18–1 *Escherichia coli*—Gram stain. Arrow points to a gram-negative rod. (Used with permission from Professor Shirley Lowe, University of California, San Francisco School of Medicine.)

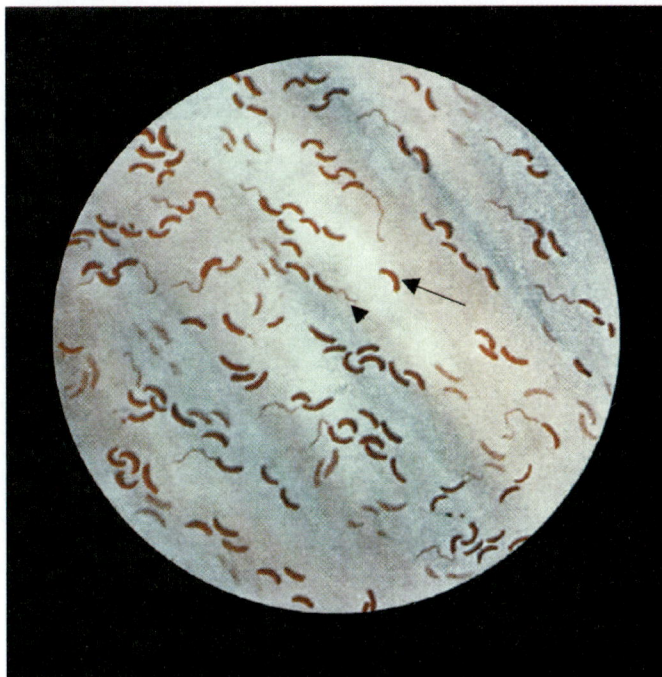

FIGURE 18–2 *Vibrio cholerae*—Gram stain. Long arrow points to a curved gram-negative rod. Arrowhead points to a flagellum at one end of a curved gram-negative rod. (Reproduced with permission from Public Health Image Library, Centers for Disease Control and Prevention.)

TABLE 18–1   Categories of Gram-Negative Rods

Chapter	Source of Site of Infection	Genus
18	Enteric tract 1. Both within and outside	*Escherichia, Salmonella*
	2. Primarily within	*Shigella, Vibrio, Campylobacter, Helicobacter*
	3. Outside only	*Klebsiella–Enterobacter–Serratia* group, *Proteus–Providencia–Morganella* group, *Pseudomonas, Bacteroides, Prevotella, Fusobacterium*
19	Respiratory tract	*Haemophilus, Legionella, Bordetella*
20	Animal sources	*Brucella, Francisella, Pasteurella, Yersinia*

TABLE 18–2   Frequency of Diseases Caused in the United States by Gram-Negative Rods Related to the Enteric Tract

Site of Infection	Frequent Pathogens	Less Frequent Pathogens
Enteric tract	*Salmonella, Shigella, Campylobacter*	*Escherichia, Vibrio, Yersinia*
Urinary tract	*Escherichia*	*Enterobacter, Klebsiella, Proteus, Pseudomonas*

Patients infected with certain enteric pathogens such as *Shigella, Salmonella, Campylobacter,* and *Yersinia* have a high incidence of certain autoimmune diseases such as reactive arthritis and Reiter syndrome (see Chapter 66). In addition, infection with *Campylobacter jejuni* predisposes to Guillain-Barré syndrome.

Before describing the specific organisms, it is appropriate to describe the family Enterobacteriaceae, to which many of these gram-negative rods belong.

ENTEROBACTERIACEAE & RELATED ORGANISMS

The Enterobacteriaceae is a large family of gram-negative rods found primarily in the colon of humans and other animals, many as part of the normal flora. These organisms are the major facultative anaerobes in the large intestine but are present in relatively small numbers compared with anaerobes such as *Bacteroides*. Although the members of the Enterobacteriaceae are classified together taxonomically, they cause a variety of diseases with different pathogenetic mechanisms. The organisms and some of the diseases they cause are listed in Table 18–5.

Features common to all members of this heterogeneous family are their anatomic location and the following four metabolic processes: (1) they are all facultative anaerobes; (2) they all ferment glucose (fermentation of other sugars varies); (3) none have cytochrome oxidase (i.e., they are oxidase-negative); and (4) they reduce nitrates to nitrites as part of their energy-generating processes.

These four reactions can be used to distinguish the Enterobacteriaceae from another medically significant group of organisms—the nonfermenting gram-negative rods, the most important of which is *Pseudomonas aeruginosa*.[1]

P. aeruginosa, a significant cause of urinary tract infection and sepsis in hospitalized patients, does not ferment glucose or reduce nitrates and is oxidase-positive. In contrast to the Enterobacteriaceae, it is a strict aerobe and derives its energy from oxidation, not fermentation.

Pathogenesis

All members of the Enterobacteriaceae contain endotoxin in their cell walls. In addition, several exotoxins are produced (e.g., *E. coli* and *Vibrio cholerae* secrete exotoxins, called *enterotoxins*, that activate adenylate cyclase within the cells of the small intestine, causing diarrhea) (see Chapter 7). In addition, *E. coli* O157 produces Shiga toxin that causes hemolytic-uremic syndrome.

---

[1]The other less frequently isolated organisms in this group are members of the following genera: *Achromobacter, Alcaligenes, Eikenella, Flavobacterium, Kingella,* and *Moraxella* (see Chapter 27) and *Acinetobacter* (see Chapter 19).

TABLE 18–3 Gram-Negative Rods Causing Diarrhea

Species	Fever	Leukocytes in Stool	Infective Dose	Typical Bacteriologic or Epidemiologic Findings
Enterotoxin-mediated				
1. *Escherichia coli*	–	–	?	Ferments lactose
2. *Vibrio cholerae*	–	–	10^7	Comma-shaped bacteria
Invasive-inflammatory				
1. *Salmonella* (e.g., *S. enterica*)	+	+	10^5	Does not ferment lactose
2. *Shigella* (e.g., *S. dysenteriae*)	+	+	10^2	Does not ferment lactose
3. *Campylobacter jejuni*	+	+	10^4	Comma- or S-shaped bacteria; growth at 42°C
4. *E. coli* (enteropathic strains)	+	+	?	
5. *E. coli* O157:H7	+	+/–	?	Transmitted by undercooked hamburger; causes hemolytic-uremic syndrome
Mechanism uncertain				
1. *Vibrio parahaemolyticus*[1]	+	+	?	Transmitted by seafood
2. *Yersinia enterocolitica*[1]	+	+	10^8	Usually transmitted from pets (e.g., puppies)

[1]Some strains produce enterotoxin, but its pathogenic role is not clear.

TABLE 18–4 Gram-Negative Rods Causing Urinary Tract Infection[1] or Sepsis[2]

Species	Lactose Fermented	Features of the Organism
Escherichia coli	+	Colonies show green sheen on EMB agars
Enterobacter cloacae	+	Causes nosocomial infections and often drug-resistant
Klebsiella pneumoniae	+	Has large mucoid capsule and hence viscous colonies
Serratia marcescens	–	Red pigment produced; causes nosocomial infections and often drug resistant
Proteus mirabilis	–	Motility causes "swarming" on agar; produces urease
Pseudomonas aeruginosa	–	Blue-green pigment and fruity odor produced; causes nosocomial infections and often drug-resistant

EMB = eosin-methylene blue.

[1]Diagnosed by quantitative culture of urine.

[2]Diagnosed by culture of blood or pus.

TABLE 18–5 Diseases Caused by Members of the Enterobacteriaceae

Major Pathogen	Representative Diseases	Minor Related Genera
Escherichia	Urinary tract infection, traveler's diarrhea, neonatal meningitis	
Shigella	Dysentery	
Salmonella	Typhoid fever, enterocolitis	*Arizona, Citrobacter, Edwardsiella*
Klebsiella	Pneumonia, urinary tract infection	
Enterobacter	Pneumonia, urinary tract infection	*Hafnia*
Serratia	Pneumonia, urinary tract infection	
Proteus	Urinary tract infection	*Providencia, Morganella*
Yersinia	Plague, enterocolitis, mesenteric adenitis	

Antigens

The antigens of several members of the Enterobacteriaceae, especially *Salmonella* and *Shigella*, are important; they are used for identification purposes both in the clinical laboratory and in epidemiologic investigations. The three surface antigens are as follows:

(1) The cell wall antigen (also known as the somatic, or O, antigen) is the outer polysaccharide portion of the lipopolysaccharide (see Figure 2–6). The O antigen, which is composed of repeating oligosaccharides consisting of three or four sugars repeated 15 or 20 times, is the basis for the serologic typing of many enteric rods. The number of different O antigens is very large (e.g., there are approximately 1500 types of *Salmonella* and 150 types of *E. coli*).

(2) The H antigen is on the flagellar protein. Only flagellated organisms, such as *Escherichia* and *Salmonella*, have H antigens, whereas the nonmotile ones, such as *Klebsiella* and *Shigella*, do not. The H antigens of certain *Salmonella* species are unusual because the organisms can reversibly alternate between two types of H antigens called phase 1 and phase 2. The organisms may use this change in antigenicity to evade the immune response.

(3) The capsular or K polysaccharide antigen is particularly prominent in heavily encapsulated organisms such as *Klebsiella*. The K antigen is identified by the quellung (capsular swelling) reaction in the presence of specific antisera and is used to serotype *E. coli* and *Salmonella typhi* for epidemiologic purposes. In *S. typhi*, the cause of typhoid fever, it is called the Vi (or virulence) antigen.

Laboratory Diagnosis

Specimens suspected of containing members of the Enterobacteriaceae and related organisms are usually inoculated onto two media, a blood agar plate and a selective differential medium such as MacConkey's agar or eosin-methylene blue (EMB) agar. The *differential* ability of these latter media is based on **lactose fermentation**, which is the most important metabolic criterion used in the identification of these organisms (Table 18–6). On these media, the nonlactose fermenters (e.g., *Salmonella* and *Shigella*) form colorless colonies, whereas the lactose fermenters (e.g., *E. coli*) form colored colonies. On EMB agar, *E. coli* colonies have a characteristic **green sheen**. The *selective* effect of the media in suppressing unwanted gram-positive organisms is exerted by bile salts or bacteriostatic dyes in the agar.

TABLE 18–6 Lactose Fermentation by Members of the Enterobacteriaceae and Related Organisms

Lactose Fermentation	Organisms
Occurs	*Escherichia, Klebsiella, Enterobacter*
Does not occur	*Shigella, Salmonella, Proteus, Pseudomonas*
Occurs slowly	*Serratia, Vibrio*

TABLE 18–7 Triple Sugar Iron (TSI) Agar Reactions

	Reactions[1]			
Slant	Butt	Gas	H₂S	Representative Genera
Acid	Acid	+	–	*Escherichia, Enterobacter, Klebsiella*
Alkaline	Acid	–	–	*Shigella, Serratia*
Alkaline	Acid	+	+	*Salmonella, Proteus*
Alkaline	Alkaline	–	–	*Pseudomonas*[2]

[1]Acid production causes the phenol red indicator to turn yellow; the indicator is red under alkaline conditions. The presence of black FeS in the butt indicates H_2S production. Not every species within the various genera will give the above appearance on TSI agar. For example, some *Serratia* strains can ferment lactose slowly and give an acid reaction on the slant.

[2]*Pseudomonas*, although not a member of the Enterobacteriaceae, is included in this table because its reaction on TSI agar is a useful diagnostic criterion.

An additional set of screening tests, consisting of triple sugar iron (TSI) agar and urea agar, is performed prior to the definitive identification procedures. The rationale for the use of these media and the reactions of several important organisms are presented in the box titled "Agar Media for Enteric Gram-Negative Rods" and in Table 18–7. The results of the screening process are often sufficient to identify the genus of an organism; however, an array of 20 or more biochemical tests is required to identify the species.

Another valuable piece of information used to identify some of these organisms is their motility, which is dependent on the presence of flagella. *Proteus* species are very motile and characteristically **swarm** over the blood agar plate, obscuring the colonies of other organisms. Motility is also an important diagnostic criterion in the differentiation of *Enterobacter cloacae*, which is motile, from *Klebsiella pneumoniae*, which is nonmotile.

If the results of the screening tests suggest the presence of a *Salmonella* or *Shigella* strain, an agglutination test can be used to identify the genus of the organism and to determine whether it is a member of group A, B, C, or D.

Coliforms & Public Health

Contamination of the public water supply system by sewage is detected by the presence of coliforms in the water. In a general sense, the term *coliform* includes not only *E. coli* but also other inhabitants of the colon such as *Enterobacter* and *Klebsiella*. However, because only *E. coli* is exclusively a large intestine organism, whereas the others are found in the environment also, it is used as the indicator of fecal contamination.

In water quality testing, *E. coli* is identified by its ability to ferment lactose with the production of acid and gas, its ability to grow at 44.5°C, and its characteristic colony type on EMB agar. An *E. coli* colony count above 4/dL in municipal drinking water is indicative of unacceptable fecal contamination. Because *E. coli* and the enteric pathogens are killed by chlorination of the drinking water, there is rarely a problem with meeting this standard. Disinfection of the public water supply is one of the most important advances of public health in the twentieth century.

AGAR MEDIA FOR ENTERIC GRAM-NEGATIVE RODS

Triple Sugar Iron (TSI) Agar

The important components of this medium are ferrous sulfate and the three sugars namely glucose, lactose, and sucrose. Glucose is present in one-tenth the concentration of the other two sugars. The medium in the tube has a solid, poorly oxygenated area on the bottom, called the butt, and an angled, well-oxygenated area on top, called the slant. The organism is inoculated into the butt and across the surface of the slant.

The interpretation of the test results is as follows: (1) If lactose (or sucrose) is fermented, a large amount of acid is produced, which turns the phenol red indicator yellow both in the butt and on the slant. Some organisms generate gases, which produce bubbles in the butt. (2) If lactose is not fermented but the small amount of glucose is, the oxygen-deficient butt will be yellow, but on the slant, the acid will be oxidized to CO_2 and H_2O by the organism, and the slant will be red (neutral or alkaline). (3) If neither lactose nor glucose is fermented, both the butt and the slant will be red. The slant can become a deeper red-purple (more alkaline) as a result of the production of ammonia from the oxidative deamination of amino acids. (4) If H_2S is produced, the black color of ferrous sulfide is seen.

The reactions of some of the important organisms are presented in Table 18–7. Because several organisms can give the same reaction, TSI agar is only a screening device.

Urea Agar

The important components of this medium are urea and the pH indicator phenol red. If the organism produces urease, the urea is hydrolyzed to NH_3 and CO_2. Ammonia turns the medium alkaline, and the color of the phenol red changes from light orange to reddish purple. The important organisms that are urease-positive are *Proteus* species and *Klebsiella pneumoniae*.

Antibiotic Therapy

The appropriate treatment for infections caused by members of the Enterobacteriaceae and related organisms must be individually tailored to the antibiotic sensitivity of the organism. Generally speaking, a wide range of antimicrobial agents are potentially effective (e.g., some penicillins and cephalosporins, aminoglycosides, chloramphenicol, tetracyclines, quinolones, and sulfonamides). The specific choice usually depends on the results of antibiotic sensitivity tests.

Note that many isolates of these enteric gram-negative rods are **highly antibiotic-resistant** because of the production of β-lactamases, including extended-spectrum β-lactamases (ESBL; see Chapter 11) and other drug-modifying enzymes. These organisms undergo conjugation frequently, at which time they acquire plasmids (R factors) that mediate multiple drug resistance. For example, plasmid-encoded New Delhi metallo-β-lactamase causes resistance to penicillins, cephalosporins, monobactams, and carbapenems.

PATHOGENS BOTH WITHIN & OUTSIDE THE ENTERIC TRACT

ESCHERICHIA

Diseases

E. coli is the most common cause of urinary tract infection and gram-negative rod sepsis. It is one of the two important causes of neonatal meningitis and is the bacterium most frequently associated with "traveler's diarrhea," a watery diarrhea. Some strains of *E. coli*, especially the O157:H7 strain, cause bloody diarrhea and hemolytic-uremic syndrome.

Important Properties

E. coli is a straight gram-negative rod (see Figure 18–1), in contrast to the curved gram-negative rods of the genera *Vibrio*, *Campylobacter*, and *Helicobacter*.

E. coli is the most abundant facultative anaerobe in the colon and feces. It is, however, greatly outnumbered by the obligate anaerobes such as *Bacteroides*.

E. coli **ferments lactose**, a property that distinguishes it from the two major intestinal pathogens, *Shigella* and *Salmonella*.

It has three antigens that are used to identify the organism in epidemiologic investigations: the O, or cell wall, antigen; the H, or flagellar, antigen; and the K, or capsular, antigen. Because there are more than 150 O, 50 H, and 90 K antigens, the various combinations result in more than 1000 antigenic types of *E. coli*. Specific serotypes are associated with certain diseases (e.g., O55 and O111 cause outbreaks of neonatal diarrhea).

Pathogenesis

The reservoir of *E. coli* includes both humans and animals. The source of the *E. coli* that causes urinary tract infections is the patient's own colonic flora that colonizes the urogenital area. The source of the *E. coli* that causes neonatal meningitis is the mother's birth canal; the infection is acquired during birth. In contrast, the *E. coli* that causes traveler's diarrhea is acquired by ingestion of food or water contaminated with human feces. Note that the main reservoir of enterohemorrhagic *E. coli* (EHEC) O157 is cattle and the organism is acquired in undercooked beef, for example, hamburgers.

E. coli has several clearly identified components that contribute to its ability to cause disease: pili, a capsule, endotoxin, and three exotoxins (enterotoxins), two that cause watery diarrhea and one that causes bloody diarrhea and hemolytic-uremic syndrome.

Intestinal Tract Infection

The first step is the adherence of the organism to the cells of the jejunum and ileum by means of **pili** that protrude from the bacterial surface. Once attached, the bacteria synthesize **enterotoxins** (exotoxins that act in the enteric tract), which act on the cells of the jejunum and ileum to cause diarrhea. The toxins are strikingly cell-specific; the cells of the colon are not susceptible, probably because they lack receptors for the toxin. Enterotoxigenic strains of *E. coli* (ETEC) can produce either or both of two enterotoxins.

(1) The heat-labile toxin (LT) acts by stimulating **adenylate cyclase**. Both LT and cholera toxin act by catalyzing the addition of adenosine diphosphate-ribose (a process called ADP-ribosylation) to the G protein that stimulates the cyclase. This irreversibly activates the cyclase. The resultant increase in intracellular cyclic adenosine monophosphate (AMP) concentration stimulates cyclic AMP-dependent protein kinase, which phosphorylates ion transporters in the membrane. The transporters export ions, which cause an outpouring of fluid, potassium, and chloride from the enterocytes into the lumen of the gut, resulting in watery diarrhea. Note that cholera toxin has the same mode of action.

(2) The other enterotoxin is a low-molecular-weight, heat-stable toxin (ST), which stimulates guanylate cyclase.

The enterotoxin-producing strains **do not cause inflammation**, do not invade the intestinal mucosa, and cause a watery, nonbloody diarrhea. However, certain strains of *E. coli* are enteropathic (enteroinvasive) and cause disease not by enterotoxin formation but by invasion of the epithelium of the large intestine, causing bloody diarrhea (dysentery) accompanied by inflammatory cells (neutrophils) in the stool.

Certain enterohemorrhagic strains of *E. coli* (i.e., those with the O157:H7 serotype) (Shiga toxin-producing *E. coli* [STEC]) also cause bloody diarrhea by producing an exotoxin called **Shiga toxin**, so called because it is very similar to that produced by *Shigella* species. Shiga toxin is a glycosidase that acts by removing an adenine from the large (28S) ribosomal RNA, thereby stopping protein synthesis. Shiga toxin is encoded by temperate (lysogenic) bacteriophages. Shiga toxin is also called verotoxin because it has a cytopathic effect on Vero (monkey) cells in culture.

These O157:H7 strains are associated with outbreaks of bloody diarrhea following ingestion of undercooked hamburger, often at fast-food restaurants. The bacteria on the surface of the hamburger are killed by the cooking, but those in the interior, which is undercooked, survive. Also, direct contact with animals (e.g., visits to farms and petting zoos) has resulted in bloody diarrhea caused by O157:H7 strains. *E. coli* O157 has a low ID_{50} of approximately 100 organisms.

Hemolytic-Uremic Syndrome

Some patients with bloody diarrhea caused by O157:H7 strains also have a life-threatening complication called **hemolytic-uremic syndrome (HUS)**, which occurs when Shiga toxin enters the bloodstream and binds to small blood vessels, for example, capillaries. This syndrome consists of hemolytic anemia, thrombocytopenia, and acute renal failure. The hemolytic anemia and renal failure occur because there are receptors for Shiga toxin on the surface of the endothelium of small blood vessels, especially on the surface of kidney glomeruli. Death of the endothelial cells of small blood vessels results in a microangiopathic hemolytic anemia in which the red cells passing through the damaged area become grossly distorted and then lyse. These distorted cells are called **schistocytes**. Thrombocytopenia occurs because platelets adhere to the damaged endothelial surface. Death of the kidney epithelial cells leads to renal failure. Treatment of diarrhea caused by O157:H7 strains with antibiotics, such as ciprofloxacin, increases the risk of developing HUS by increasing the amount of Shiga toxin released by the dying bacteria.

Urinary Tract Infections

Certain O serotypes of *E. coli* preferentially cause urinary tract infections. These **uropathic** strains are characterized by pili with adhesin proteins that bind to specific receptors on the urinary tract epithelium. The binding site on these receptors consists of dimers of galactose (**Gal-Gal dimers**). These pili are also called P fimbria or pyelonephritis-associated pili (PAP).

Cranberry juice contains flavonoids that inhibit the binding of pili to receptors and may be useful in the prevention of recurrent urinary tract infections. The motility of *E. coli* may aid its ability to ascend the urethra into the bladder and ascend the ureter into the kidney.

Systemic Infection

The other two structural components, the **capsule** and the **endotoxin**, play a more prominent role in the pathogenesis of systemic, rather than intestinal tract, disease. The capsular polysaccharide interferes with phagocytosis, thereby enhancing the organism's ability to cause infections in various organs. For example, *E. coli* strains that cause neonatal meningitis usually have a specific capsular type called the K1 antigen. The endotoxin of *E. coli* is the cell wall lipopolysaccharide, which causes several features of gram-negative sepsis such as fever, hypotension, and disseminated intravascular coagulation.

Th-17 helper T cells that produce interleukin-17 are an important host defense against sepsis caused by enteric bacteria such as *E. coli* and *Klebsiella*. Patients infected with human immunodeficiency virus (HIV) experience a loss of Th-17 cells and are predisposed to sepsis caused by *E. coli* and *Klebsiella*.

Clinical Findings

E. coli causes a variety of diseases both within and outside the intestinal tract. The main clinical findings, the major pathogenetic factors, and the main laboratory results are described in Table 18–8.

TABLE 18–8 Clinical Aspects of *Escherichia coli*

Clinical Finding/Disease	Major Pathogenetic Factor	Main Laboratory Result
Findings within the intestinal tract		
Watery, nonbloody diarrhea (traveler's diarrhea)	Enterotoxin that increases cyclic AMP	No RBC or WBC in stool
Bloody diarrhea caused by *E. coli* O-157; hemolytic-uremic syndrome (HUS)	Shiga toxin (verotoxin) inhibits protein synthesis	RBC in stool; schistocytes in blood smear
Findings outside of intestinal tract		
Urinary tract infection	Gal-gal pili bind to bladder mucosa	WBC in urine, positive urine culture
Neonatal meningitis	K-1 capsular polysaccharide is antiphagocytic	WBC in spinal fluid, positive CSF culture
Sepsis, especially in hospital	Endotoxin induces fever, hypotension, and DIC	Leukocytosis, positive blood culture

AMP = adenosine monophosphate; CSF = cerebrospinal fluid; DIC = disseminated intravascular coagulation; RBC = red blood cell; WBC = white blood cell.

(1) **Clinical findings within the intestinal tract:**

Diarrhea caused by **enterotoxigenic *E. coli*** (ETEC) is usually **watery**, nonbloody, self-limited, and of short duration (1–3 days). It is frequently associated with travel (traveler's diarrhea, or "turista").[2]

Infection with **EHEC**, on the other hand, results in a dysentery-like syndrome characterized by **bloody diarrhea**, abdominal cramping, and fever similar to that caused by *Shigella*.

The O157:H7 strains of *E. coli* (STEC) also cause bloody diarrhea, which can be complicated by **HUS**. This syndrome is characterized by kidney failure, hemolytic anemia, and thrombocytopenia. The hemolytic anemia is caused by exotoxin-induced capillary damage, which results in damage to the red cells as they pass through the capillaries. These distorted, fragmented red cells called **schistocytes** can be seen on blood smear and are characteristic of a microangiopathic hemolytic anemia. Strains of *E. coli* other than O157:H7, such as O154:H4, can also cause HUS.

HUS occurs particularly in children who have been treated with fluoroquinolones or other antibiotics for their diarrhea. For this reason, antibiotics should not be used to treat diarrhea caused by EHEC.

(2) **Clinical findings outside of the intestinal tract:**

E. coli is the leading cause of community-acquired **urinary tract infections**. These infections occur primarily in women; this finding is attributed to three features that facilitate ascending infection into the bladder, namely, a short urethra, the proximity of the urethra to the anus, and colonization of the vagina by members of the fecal flora. It is also the most frequent cause of nosocomial (hospital-acquired) urinary tract infections, which occur equally frequently in both men and women and are associated with the use of indwelling urinary catheters. Urinary tract infections can be limited to the bladder or extend up the collecting system to the kidneys. If only the bladder is involved, the disease is called *cystitis*, whereas infection of the kidney is called *pyelonephritis*. The most prominent symptoms of cystitis

are pain (dysuria) and frequency of urination; patients are usually afebrile. Pyelonephritis is characterized by fever, flank pain, and costovertebral angle tenderness; dysuria and frequency may or may not occur.

E. coli is also a major cause, along with the group B streptococci, of **meningitis** and sepsis in neonates. Exposure of the newborn to *E. coli* and group B streptococci occurs during birth as a result of colonization of the vagina by these organisms in approximately 25% of pregnant women.

E. coli is the organism isolated most frequently from patients with hospital-acquired sepsis, which arises primarily from urinary, biliary, or peritoneal infections. Peritonitis is usually a mixed infection caused by *E. coli* or other facultative enteric gram-negative rod plus anaerobic members of the colonic flora such as *Bacteroides* and *Fusobacterium*.

Laboratory Diagnosis

Specimens suspected of containing enteric gram-negative rods, such as *E. coli*, are grown initially on a blood agar plate and on a differential medium, such as EMB agar or MacConkey's agar. *E. coli*, which ferments lactose, forms pink colonies, whereas lactose-negative organisms are colorless. On EMB agar, *E. coli* colonies have a characteristic **green sheen**. Some of the important features that help distinguish *E. coli* from other lactose-fermenting gram-negative rods are as follows: (1) it produces indole from tryptophan, (2) it decarboxylates lysine, (3) it uses acetate as its only source of carbon, and (4) it is motile.

E. coli O157:H7 does not ferment sorbitol, which serves as an important criterion that distinguishes it from other strains of *E. coli*. The isolation of enterotoxigenic or enteropathogenic *E. coli* from patients with diarrhea is not a routine diagnostic procedure. The diagnosis of bloody diarrhea caused by *E. coli* O-157 (STEC) can be made by using an enzyme immunoassay for Shiga toxin in the stool.

Treatment

Treatment of *E. coli* infections depends on the site of disease and the resistance pattern of the specific isolate. For example, an uncomplicated lower urinary tract infection (cystitis) can be treated using oral trimethoprim-sulfamethoxazole or nitrofurantoin. Pyelonephritis can be treated with ciprofloxacin or ceftriaxone.

---

[2]Enterotoxigenic *E. coli* is the most common cause of traveler's diarrhea, but other bacteria (e.g., *Salmonella*, *Shigella*, *Campylobacter*, and *Vibrio* species), viruses such as norovirus, and protozoa such as *Giardia* and *Cryptosporidium* species are also involved.

However, *E. coli* sepsis requires treatment with parenteral antibiotics (e.g., a third-generation cephalosporin, such as cefotaxime, with or without an aminoglycoside, such as gentamicin). For the treatment of neonatal meningitis, a combination of ampicillin and cefotaxime is usually given.

Antibiotic therapy is usually *not* indicated in mild traveler's diarrhea. Administration of loperamide (Imodium) may shorten the duration of symptoms. Rehydration is typically all that is necessary in this self-limited disease. Azithromycin is the drug of choice for moderate or severe traveler's diarrhea. In 2018, rifamycin was approved for the treatment of travelers' diarrhea.

However, diarrhea caused by *E. coli* O157 should *not* be treated with antimotility drugs as their use increases the risk of HUS. Giving antibiotics to patients with *E. coli* O157 diarrhea is controversial as it may increase the risk of HUS. Treatment of diarrhea caused by *E. coli* O157 is typically supportive.

Treatment of HUS is typically supportive but dialysis may be necessary if renal failure occurs. If severe anemia or thrombocytopenia occurs, transfusions may be necessary. Plasma exchange does not alter the course of HUS.

Prevention

There is no specific prevention for *E. coli* infections, such as active or passive immunization. However, various general measures can be taken to prevent certain infections caused by *E. coli* and other organisms. For example, the incidence of urinary tract infections can be lowered by the judicious use and prompt withdrawal of catheters and, in recurrent infections, by prolonged prophylaxis with urinary antiseptic drugs (e.g., nitrofurantoin or trimethoprim-sulfamethoxazole). The use of cranberry juice to prevent recurrent urinary tract infections appears to be based on the ability of flavonoids in the juice to inhibit the binding of the pili of the uropathic strains of *E. coli* to the bladder epithelium.

Some cases of sepsis can be prevented by prompt removal of or switching the site of intravenous lines. Traveler's diarrhea can sometimes be prevented by the prophylactic use of doxycycline, ciprofloxacin, trimethoprim-sulfamethoxazole, or Pepto-Bismol. Ingestion of uncooked foods and unpurified water should be avoided while traveling in certain countries.

SALMONELLA

Diseases

Salmonella species cause enterocolitis, enteric fevers such as typhoid fever, and septicemia with metastatic infections such as osteomyelitis. They are one of the most common causes of bacterial enterocolitis in the United States. *S. typhi*, *S. paratyphi* A and *S. paratyphi* B are common causes of enteric fevers.

Important Properties

Salmonellae are gram-negative rods that **do not ferment lactose** but do produce H_2S—features that are used in their laboratory identification. Their antigens—cell wall O, flagellar H, and capsular Vi (virulence)—are important for taxonomic and epidemiologic purposes. The O antigens, which are the outer polysaccharides of the cell wall, are used to subdivide the salmonellae into groups lettered A through I. There are two forms of the H antigens, phases 1 and 2. Only one of the two H proteins is synthesized at any one time, depending on which gene sequence is in the correct alignment for transcription into mRNA. The Vi antigens (capsular polysaccharides) are antiphagocytic and are an important virulence factor for *S. typhi*, the agent of typhoid fever. The Vi antigens are also used for the serotyping of *S. typhi* in the clinical laboratory.

There are three methods for naming the salmonellae. Ewing divides the genus into three species: *S. typhi*, *Salmonella choleraesuis*, and *Salmonella enteritidis*. In this scheme, there is one serotype in each of the first two species and 1500 serotypes in the third. Kaufman and White assign different species names to each serotype; there are roughly 1500 different species, usually named for the city in which they were isolated. *Salmonella dublin* according to Kaufman and White would be *S. enteritidis* serotype *dublin* according to Ewing. The third approach to naming the salmonellae is based on relatedness determined by DNA hybridization analysis. In this scheme, *S. typhi* is not a distinct species but is classified as *Salmonella enterica* serotype (or serovar) *typhi*. All three of these naming systems are in current use.

Clinically, the *Salmonella* species are often thought of in two distinct categories, namely, the typhoidal species (i.e., those that cause typhoid fever) and the nontyphoidal species (i.e., those that cause diarrhea [enterocolitis] and metastatic infections, such as osteomyelitis). The typhoidal species are *S. typhi* and *S. paratyphi* A and B. The nontyphoidal species are the many serotypes of *S. enterica*. Of the serotypes, *S. enterica* serotype *choleraesuis* is the species most often involved in metastatic infections.

Pathogenesis & Epidemiology

The three types of *Salmonella* infections (enterocolitis, enteric fevers, and septicemia) have different pathogenic features.

(1) **Enterocolitis** is characterized by an invasion of the epithelial and subepithelial tissue of the small and large intestines. Strains that do not invade do not cause disease. The organisms penetrate both through and between the mucosal cells into the lamina propria, with resulting inflammation and diarrhea. Neutrophils limit the infection to the gut and the adjacent mesenteric lymph nodes; bacteremia is infrequent in enterocolitis. In contrast to *Shigella* enterocolitis, in which the infectious dose is very small (on the order of 100 organisms), the dose of *Salmonella* required is much higher, at least 100,000 organisms. Various properties of salmonellae and shigellae are compared in Table 18–9. Gastric acid is an important host defense; gastrectomy or use of antacids lowers the infectious dose significantly.

(2) In **typhoid** and other enteric fevers, infection begins in the small intestine, but few gastrointestinal symptoms occur. The organisms enter, multiply in the mononuclear phagocytes

TABLE 18–9 Comparison of Important Features of Salmonella and Shigella

Feature	Shigella	Salmonella Except Salmonella typhi	Salmonella typhi
Reservoir	Humans	Animals, especially poultry and eggs	Humans
Infectious dose (ID$_{50}$)	Low[1]	High	High
Diarrhea as a prominent feature	Yes	Yes	No
Invasion of bloodstream	No	Yes	Yes
Chronic carrier state	No	Infrequent	Yes
Lactose fermentation	No	No	No
H$_2$S production	No	Yes	Yes
Vaccine available	No	No	Yes

[1]An organism with a low ID$_{50}$ requires very few bacteria to cause disease.

of Peyer's patches, and then spread to the phagocytes of the liver, gallbladder, and spleen. This leads to bacteremia, which is associated with the onset of fever and other symptoms, probably caused by endotoxin. Survival and growth of the organism within phagosomes in phagocytic cells is a striking feature of this disease, as is the predilection for invasion of the gallbladder, which can result in establishment of the **carrier state** and excretion of the bacteria in the feces for long periods.

(3) **Septicemia** accounts for only about 5% to 10% of *Salmonella* infections and occurs in one of two settings: a patient with an underlying chronic disease, such as **sickle cell anemia** or cancer, or a child with enterocolitis. The septic course is more indolent than that seen with many other gram-negative rods. Bacteremia results in the seeding of many organs, with **osteomyelitis**, pneumonia, and meningitis as the most common sequelae. **Osteomyelitis in a child with sickle cell anemia** is an important example of this type of salmonella infection. Previously damaged tissues, such as infarcts and **aneurysms**, especially aortic aneurysms, are the most frequent sites of metastatic abscesses. *Salmonella* are also an important cause of vascular graft infections.

The epidemiology of *Salmonella* infections is related to the ingestion of food and water contaminated by human and animal wastes. *S. typhi*, the cause of typhoid fever, is **transmitted only by humans**, but all other species have a significant animal as well as human reservoir. Human sources are either persons who temporarily excrete the organism during or shortly after an attack of enterocolitis or chronic carriers who excrete the organism for years. The most frequent **animal source is poultry and eggs**, but meat products that are inadequately cooked have been implicated as well. Dogs and other pets, including turtles, snakes, lizards, and iguanas, are additional sources.

Clinical Findings

After an incubation period of 12 to 48 hours, enterocolitis begins with nausea and vomiting and then progresses to abdominal pain and diarrhea, which can vary from mild to severe, with or without blood. Usually the disease lasts a few days, is self-limited, causes nonbloody diarrhea, and does not require medical care except in the very young and very old. HIV-infected individuals, especially those with a low-CD4 count, have a much greater number of *Salmonella* infections, including more severe diarrhea and more serious metastatic infections than those who are not infected with HIV. *Salmonella typhimurium* is the most common species of *Salmonella* to cause enterocolitis in the United States, but almost every species has been involved.

In typhoid fever, caused by *S. typhi*, and in enteric fever, caused by organisms such as *S. paratyphi* A, B, and C (*S. paratyphi* B and C are also known as *Salmonella schottmuelleri* and *Salmonella hirschfeldii*, respectively), the onset of illness is slow, with fever and constipation rather than vomiting and diarrhea predominating. Diarrhea may occur early but usually disappears by the time the fever and bacteremia occur. After the first week, as the bacteremia becomes sustained, high fever, delirium, tender abdomen, and enlarged spleen occur. **Rose spots** (i.e., rose-colored macules on the abdomen) are associated with typhoid fever but occur only rarely. Leukopenia and anemia are often seen. Liver function tests are often abnormal, indicating hepatic involvement.

The disease begins to resolve by the third week, but severe complications such as intestinal hemorrhage or perforation can occur. About 3% of typhoid fever patients become chronic carriers. The carrier rate is higher among women, especially those with previous gallbladder disease and gallstones.

Septicemia is most often caused by *S. choleraesuis*. The symptoms begin with fever but little or no enterocolitis and then proceed to focal symptoms associated with the affected organ, frequently bone, lung, or meninges.

Laboratory Diagnosis

In enterocolitis, the organism is most easily isolated from a stool sample. However, in the enteric fevers, a blood culture is the procedure most likely to reveal the organism during the first 2 weeks of illness. Bone marrow cultures are often positive. Stool cultures may also be positive, especially in chronic carriers in whom the organism is secreted in the bile into the intestinal tract.

Salmonellae form non–lactose-fermenting (colorless) colonies on MacConkey's or EMB agar. On TSI agar, an alkaline slant and an acid butt, frequently with both gas and H$_2$S (black color in the butt), are produced. *S. typhi* is the major exception; it does not form gas and produces only a small amount of H$_2$S. If the organism is urease-negative (*Proteus* organisms, which can produce a similar reaction on TSI agar, are urease-positive), the *Salmonella* isolate can be identified and grouped by the slide agglutination test into serogroup A, B, C, D, or E based on its O antigen. Definitive serotyping of the O, H, and Vi antigens is performed by special public health laboratories for epidemiologic purposes.

Salmonellosis is a notifiable disease, and an investigation to determine its source should be undertaken. In certain cases of enteric fever and sepsis, when the organism is difficult to recover, the diagnosis can be made serologically by detecting a rise in antibody titer in the patient's serum (Widal test).

Treatment

Enterocolitis caused by *Salmonella* is usually a self-limited disease that resolves without treatment. Fluid and electrolyte replacement may be required. Antibiotic treatment does not shorten the illness or reduce the symptoms; in fact, it may prolong excretion of the organisms, increase the frequency of the carrier state, and select mutants resistant to the antibiotic. Antimicrobial agents are indicated only for neonates or persons with chronic diseases who are at risk for septicemia and disseminated abscesses. Plasmid-mediated antibiotic resistance is common, and antibiotic sensitivity tests should be done. Drugs that retard intestinal motility (i.e., that reduce diarrhea) appear to prolong the duration of symptoms and the fecal excretion of the organisms.

The treatment of choice for enteric fevers such as typhoid fever and septicemia with metastatic infection is either ceftriaxone or ciprofloxacin. Note that a multidrug-resistant strain of *S. typhi* has emerged. It is resistant to many antibiotics but remains sensitive to azithromycin. However, in some areas of South Asia, strains resistant to azithromycin have emerged.

Ampicillin or ciprofloxacin should be used in patients who are chronic carriers of *S. typhi*. Cholecystectomy may be necessary to abolish the chronic carrier state. Focal abscesses should be drained surgically when feasible.

Prevention

Salmonella infections are prevented mainly by public health and personal hygiene measures. Proper sewage treatment, a chlorinated water supply that is monitored for contamination by coliform bacteria, cultures of stool samples from food handlers to detect carriers, handwashing prior to food handling, pasteurization of milk, and proper cooking of poultry, eggs, and meat are all important.

Two vaccines against typhoid fever are available in the United States, One contains the Vi capsular polysaccharide of *S. typhi* (given intramuscularly), and the other contains a live, attenuated strain (Ty21a) of *S. typhi* (given orally). Both vaccines are about 50% to 80% effective. The vaccine is recommended for those who will travel or reside in high-risk areas and for those whose occupation brings them in contact with the organism.

A conjugate vaccine against typhoid fever containing the capsular polysaccharide (Vi) antigen coupled to a carrier protein (tetanus toxoid) is safe and immunogenic in young children but is not available in the United States at this time.

PATHOGENS PRIMARILY WITHIN THE ENTERIC TRACT

SHIGELLA

Disease

Shigella species cause enterocolitis. Enterocolitis caused by *Shigella* is often called shigellosis or bacillary dysentery. The term *dysentery* refers to bloody diarrhea.

Important Properties

Shigellae are **non–lactose-fermenting**, gram-negative rods that can be distinguished from salmonellae by three criteria: they produce no gas from the fermentation of glucose, they **do not produce H2S**, and they are **nonmotile**. All shigellae have O antigens (polysaccharide) in their cell walls, and these antigens are used to divide the genus into four groups: A, B, C, and D.

Pathogenesis & Epidemiology

Shigellae are the most effective pathogens among the enteric bacteria. They have a **very low ID$_{50}$** (see page 30). Ingestion of as few as 100 organisms causes disease, whereas at least 10^5 *V. cholerae* or *Salmonella* organisms are required to produce symptoms. Various properties of shigellae and salmonellae are compared in Table 18–9.

Shigellosis is only a **human disease** (i.e., there is no animal reservoir). The organism is transmitted by the fecal–oral route. The four Fs—fingers, flies, food, and feces—are the principal factors in transmission. Foodborne outbreaks outnumber waterborne outbreaks by 2 to 1. Outbreaks occur in day care nurseries and in mental hospitals, where **fecal–oral** transmission is likely to occur. Children younger than 10 years account for approximately half of *Shigella*-positive stool cultures. There is no prolonged carrier state with *Shigella* infections, unlike that seen with *S. typhi* infections.

Shigellae cause disease almost exclusively in the gastrointestinal tract. They produce bloody diarrhea (dysentery) by invading the cells of the mucosa of the distal ileum and colon. Local inflammation accompanied by ulceration occurs, but the organisms rarely penetrate through the wall or enter the bloodstream, unlike salmonellae. Although some strains produce an enterotoxin (called *Shiga toxin*), invasion is the critical factor in pathogenesis. The evidence for this is that mutants that fail to produce enterotoxin but are invasive can still cause disease, whereas noninvasive mutants are nonpathogenic.

Shiga toxins are encoded by lysogenic bacteriophages. Shiga toxins very similar to those produced by *Shigella* are produced by EHEC O157:H7 strains that cause enterocolitis and HUS.

Clinical Findings

After an incubation period of 1 to 4 days, symptoms begin with fever and abdominal cramps, followed by diarrhea, which may be watery at first but later contains blood and mucus. The disease varies from mild to severe depending on two major factors: the species of *Shigella* and the age of the patient, with young children and elderly people being the most severely affected. *Shigella dysenteriae*, which causes the most severe disease, is usually seen in the United States only in travelers returning from abroad. *Shigella sonnei*, which causes mild disease, is isolated from approximately 75% of all individuals with shigellosis in the United States. The diarrhea frequently resolves in 2 or 3 days; in severe cases, antibiotics can shorten the course. Serum agglutinins appear after recovery but are not protective because the organism does not enter the blood. The role of intestinal IgA in protection is uncertain.

Laboratory Diagnosis

Shigellae form non–lactose-fermenting (colorless) colonies on MacConkey's or EMB agar. On TSI agar, they cause an alkaline slant and an acid butt, with no gas and no H_2S. Confirmation of the organism as *Shigella* and determination of its group are done by slide agglutination.

One important adjunct to laboratory diagnosis is a methylene blue stain of a fecal sample to determine whether neutrophils are present. If they are found, an invasive organism such as *Shigella*, *Salmonella*, or *Campylobacter* is involved rather than a toxin-producing organism such as *V. cholerae*, enterotoxigenic *E. coli* (ETEC), or *Clostridium perfringens*. (Certain viruses also cause diarrhea without neutrophils in the stool.)

Treatment

The main treatment for shigellosis is fluid and electrolyte replacement. In mild cases, no antibiotics are indicated. In severe cases, a fluoroquinolone (e.g., ciprofloxacin) is the drug of choice, but strains resistant to fluoroquinolones have emerged and antibiotic sensitivity tests must be performed. Trimethoprim-sulfamethoxazole is an alternative choice. Antiperistaltic drugs are contraindicated in shigellosis, because they prolong the fever, diarrhea, and excretion of the organism.

Multidrug-resistant strains of *S. sonnei* have caused outbreaks of shigellosis in the United States. These strains are resistant to ampicillin, ceftriaxone, and trimethoprim-sulfa.

Prevention

Prevention of shigellosis is dependent on interruption of fecal–oral transmission by proper sewage disposal, chlorination of water, and personal hygiene (handwashing by food handlers). There is no vaccine, and prophylactic antibiotics are not recommended.

VIBRIO

Diseases

V. cholerae, the major pathogen in this genus, is the cause of cholera. *Vibrio parahaemolyticus* causes diarrhea associated with eating raw or improperly cooked seafood. *Vibrio vulnificus* causes cellulitis and sepsis. Important features of pathogenesis by *V. cholerae*, *C. jejuni*, and *Helicobacter pylori* are described in Table 18–10.

Important Properties

Vibrios are curved, **comma-shaped**, gram-negative rods (see Figure 18–2). *V. cholerae* is divided into serogroups according to the nature of its O cell wall antigen. Members of the O1 and O139 serogroups cause epidemic disease, whereas non-O1 organisms either cause sporadic disease or are nonpathogens. The O1 organisms have two biotypes, called classic and El Tor, and three serotypes, called Ogawa, Inaba, and Hikojima. (Biotypes are based on differences in biochemical reactions, whereas serotypes are based on antigenic differences.) These features are used to characterize isolates in epidemiologic investigations. Serogroup O139 organisms, which caused a major epidemic in 1992, are identified by their reaction to antisera to the O139 polysaccharide antigen (O antigen).

Note that only the O1 and O139 organisms cause cholera because only they produce cholera toxin. They produce cholera toxin because they are lysogenized by a bacteriophage that carries the gene for the toxin (see below). The non-O1 strains can cause milder outbreaks of diarrhea but not cholera.

V. parahaemolyticus and *V. vulnificus* are **marine organisms**; they live primarily in the ocean, especially in warm salt water. They are **halophilic** (i.e., they require a high NaCl concentration to grow).

1. *Vibrio cholerae*

Pathogenesis & Epidemiology

V. cholerae is transmitted by **fecal contamination** of water and food, primarily from human sources. Human carriers are frequently asymptomatic and include individuals who are either in the incubation period or convalescing. The main animal reservoirs are marine shellfish, such as shrimp and oysters. Ingestion of these without adequate cooking can transmit the disease.

TABLE 18–10 **Important Features of Pathogenesis by Curved Gram-Negative Rods Affecting the Gastrointestinal Tract**

Organism	Type of Pathogenesis	Typical Disease	Site of Infection	Main Approach to Therapy
Vibrio cholerae	Toxigenic	Watery diarrhea	Small intestine	Fluid replacement
Campylobacter jejuni	Inflammatory	Bloody diarrhea	Colon	Antibiotics[1]
Helicobacter pylori	Inflammatory	Gastritis; peptic ulcer	Stomach; duodenum	Antibiotics[1]

[1]See text for specific antibiotics.

A major epidemic of cholera, spanning the 1960s and 1970s, began in Southeast Asia and had spread over three continental areas of Africa, Europe, and Asia. In 1991, another wave of cholera began in Peru and had spread to many countries in Central and South America. The organism isolated most frequently was the El Tor biotype of O1 *V. cholerae*, usually of the Ogawa serotype.

The factors that predispose to epidemics are poor sanitation, malnutrition, overcrowding, and inadequate medical services. Quarantine measures failed to prevent the spread of the disease because there were many asymptomatic carriers. In 1992, *V. cholerae* serogroup O139 emerged and caused a widespread epidemic of cholera in India and Bangladesh.

The pathogenesis of cholera is dependent on colonization of the small intestine by the organism and secretion of cholera toxin. For colonization to occur, large numbers of bacteria must be ingested because the organism is particularly sensitive to stomach acid. Persons with little or no stomach acid, such as those taking antacids or those who have had gastrectomy, are much more susceptible. Adherence to the cells of the brush border of the gut, which is a requirement for colonization, is related to secretion of the bacterial enzyme mucinase, which dissolves the protective glycoprotein coating over the intestinal cells.

After adhering, the organism multiplies and secretes an **enterotoxin** called choleragen (cholera toxin). This exotoxin can reproduce the symptoms of cholera even in the absence of the *Vibrio* organisms. The mode of action of cholera toxin is described in the next paragraph and in Figure 7–3 in the chapter on bacterial pathogenesis.

Choleragen consists of an A (active) subunit and a B (binding) subunit. The B subunit, which is a pentamer composed of five identical proteins, binds to a ganglioside receptor on the surface of the enterocyte. The A subunit is inserted into the cytosol, where it catalyzes the addition of ADP-ribose to the G_s protein (G_s is the stimulatory G protein). This locks the G_s protein in the "on" position, which causes the persistent stimulation of **adenylate cyclase**. The resulting overproduction of cyclic AMP activates cyclic AMP-dependent protein kinase, an enzyme that phosphorylates an ion transporter (namely, the cystic fibrosis conductance transporter) in the cell membrane. This results in the loss of water and ions, such as Na^+, Cl^-, K^+, and HCO_3 from the cell. The watery efflux enters the lumen of the gut, resulting in a massive watery diarrhea that contains neither neutrophils nor red blood cells. Morbidity and death are due to **dehydration** and **electrolyte imbalance**. However, if treatment is instituted promptly, the disease runs a self-limited course in up to 7 days.

The genes for cholera toxin and other virulence factors are carried on a single-stranded DNA bacteriophage called CTX. Lysogenic conversion of non–toxin-producing strains to toxin-producing ones can occur when the CTX phage transduces these genes. The pili that attach the organism to the gut mucosa are the receptors for the phage.

Non-O1 *V. cholerae* is an occasional cause of diarrhea associated with eating shellfish obtained from the coastal waters of the United States.

Clinical Findings

Watery diarrhea in large volumes is the hallmark of cholera. There are no red blood cells or white blood cells in the stool. **Rice-water stool** is the term often applied to the nonbloody effluent. There is no abdominal pain, and subsequent symptoms are referable to the marked dehydration. The loss of fluid and electrolytes leads to cardiac and renal failure. Acidosis and hypokalemia also occur as a result of loss of bicarbonate and potassium in the stool. The mortality rate without treatment is 40%.

Laboratory Diagnosis

The approach to laboratory diagnosis depends on the situation. During an epidemic, a clinical judgment is made and there is little need for the laboratory. In an area where the disease is endemic or for the detection of carriers, a variety of selective media[3] that are not in common use in the United States are used in the laboratory.

For diagnosis of sporadic cases in this country, a culture of the diarrhea stool containing *V. cholerae* will show colorless colonies on MacConkey's agar because lactose is fermented slowly. The organism is oxidase-positive, which distinguishes it from members of the Enterobacteriaceae. On TSI agar, an acid slant and an acid butt without gas or H_2S are seen because the organism ferments sucrose. A presumptive diagnosis of *V. cholerae* can be confirmed by agglutination of the organism by polyvalent O1 or non-O1 antiserum. A retrospective diagnosis can be made serologically by detecting a rise in antibody titer in acute- and convalescent-phase sera.

Treatment

Treatment consists of prompt, adequate replacement of water and electrolytes, either orally or intravenously. Glucose is added to the solution to enhance the uptake of water and electrolytes. Antibiotics such as azithromycin are not necessary, but they do shorten the duration of symptoms and reduce the time of excretion of the organisms.

Prevention

Prevention is achieved mainly by public health measures that ensure a clean water and food supply. An oral, live attenuated vaccine called Vaxchora is available in the United States for travelers to areas where cholera caused by serogroup O1 is endemic. Other oral vaccines containing killed organisms or the B subunit of cholera toxin are available in countries where cholera epidemics occur.

The use of tetracycline for prevention is effective in close contacts but does not prevent the spread of a major epidemic. Prompt detection of carriers is important in limiting outbreaks.

---

[3]Media such as thiosulfate-citrate-bile salts agar or tellurite-taurocholate-gelatin are used.

2. *Vibrio parahaemolyticus*

V. parahaemolyticus is a marine organism transmitted by **ingestion of raw or undercooked seafood**, especially shellfish such as oysters. It is a major cause of diarrhea in Japan, where raw fish is eaten in large quantities, but is an infrequent pathogen in the United States, although several outbreaks have occurred aboard cruise ships in the Caribbean. Little is known about its pathogenesis, except that an enterotoxin similar to choleragen is secreted and limited invasion sometimes occurs.

The clinical picture caused by *V. parahaemolyticus* varies from mild to quite severe watery diarrhea, nausea and vomiting, abdominal cramps, and fever. The illness is self-limited, lasting about 3 days. *V. parahaemolyticus* is distinguished from *V. cholerae* mainly on the basis of growth in NaCl: *V. parahaemolyticus* grows in 8% NaCl solution (as befits a marine organism), whereas *V. cholerae* does not. No specific treatment is indicated, because the disease is relatively mild and self-limited. Disease can be prevented by proper refrigeration and cooking of seafood.

3. *Vibrio vulnificus*

V. vulnificus is also a marine organism (i.e., it is found in warm salt waters such as the Caribbean Sea and along the Eastern coast of the United States). It causes severe skin and soft tissue infections (**cellulitis, necrotizing fasciitis**), **especially in shellfish handlers**, who often sustain skin wounds. It can also cause a rapidly fatal **septicemia in immunocompromised people who have eaten raw shellfish** containing the organism. Hemorrhagic bullae in the skin often occur in patients with sepsis caused by *V. vulnificus*. Chronic liver disease (e.g., cirrhosis) predisposes to severe infections. The recommended treatment is doxycycline.

CAMPYLOBACTER

Diseases

C. jejuni is a frequent cause of enterocolitis, especially in children. *C. jejuni* infection is a common antecedent to Guillain-Barré syndrome. Other *Campylobacter* species, such as *C. intestinalis*, are rare causes of systemic infection, particularly bacteremia.

Important Properties

Campylobacters are curved, gram-negative rods that appear either **comma-** or **S-shaped**. They are **microaerophilic**, growing best in 5% oxygen rather than in the 20% present in the atmosphere. *C. jejuni* grows well at 42°C, whereas *Campylobacter intestinalis*[4] does not—an observation that is useful in microbiologic diagnosis.

Pathogenesis & Epidemiology

Domestic animals such as cattle, chickens, and dogs serve as a source of the organisms for humans. Transmission is usually **fecal–oral**. Food and water contaminated with animal feces are the major sources of human infection. Foods, such as poultry, meat, and unpasteurized milk, are commonly involved. Puppies with diarrhea are a common source for children. Human-to-human transmission occurs but is less frequent than animal-to-human transmission.

C. jejuni is a major cause of diarrhea in the United States; it was recovered in 4.6% of patients with diarrhea, compared with 2.3% and 1% for *Salmonella* and *Shigella*, respectively. *C. jejuni* is the leading cause of diarrhea associated with consumption of unpasteurized milk.

Features of pathogenesis by *Campylobacter* are described in Table 18–10. Inflammation of the intestinal mucosa often occurs, accompanied by blood in stools. Systemic infections (e.g., bacteremia) occur most often in neonates or debilitated adults.

Clinical Findings

Enterocolitis, caused primarily by *C. jejuni*, begins as watery, foul-smelling diarrhea followed by bloody stools accompanied by fever and severe abdominal pain. Systemic infections, most commonly bacteremia, are caused more often by *C. intestinalis*. The symptoms of bacteremia (e.g., fever and malaise) are associated with no specific physical findings.

Gastrointestinal infection with *C. jejuni* is associated with Guillain-Barré syndrome, the most common cause of acute neuromuscular paralysis. Guillain-Barré syndrome is an autoimmune disease attributed to the formation of antibodies against *C. jejuni* that cross-react with antigens on neurons (see Chapter 66). Infection with *Campylobacter* is also associated with two other autoimmune diseases: reactive arthritis and Reiter syndrome. These are also described in Chapter 66.

Laboratory Diagnosis

If the patient has diarrhea, a stool specimen is cultured on a blood agar plate containing antibiotics[5] that inhibit most other fecal flora.

The plate is incubated at 42°C in a microaerophilic atmosphere containing 5% oxygen and 10% carbon dioxide, which favors the growth of *C. jejuni*. It is identified by failure to grow at 25°C, oxidase positivity, and sensitivity to nalidixic acid. Unlike *Shigella* and *Salmonella*, lactose fermentation is not used as a distinguishing feature. If bacteremia is suspected, a blood culture incubated under standard temperature and atmospheric conditions will reveal the growth of the characteristically comma- or S-shaped, motile, gram-negative rods. Identification of the organism as *C. intestinalis* is confirmed by its failure to grow at 42°C, its ability to grow at 25°C, and its resistance to nalidixic acid.

Treatment & Prevention

Erythromycin or ciprofloxacin is used successfully in *C. jejuni* enterocolitis. The treatment of choice for *C. intestinalis* bacteremia is an aminoglycoside.

---

[4]Also known as Campylobacter fetus subsp. fetus.

[5]For example, Skirrow's medium contains vancomycin, trimethoprim, cephalothin, polymyxin, and amphotericin B.

There is no vaccine or other specific preventive measure. Proper sewage disposal and personal hygiene (handwashing) are important.

HELICOBACTER

Diseases

H. pylori causes gastritis and peptic ulcers. Infection with *H. pylori* is a risk factor for gastric carcinoma and is linked to mucosal-associated lymphoid tissue (MALT) lymphomas.

Important Properties

Helicobacters are curved gram-negative rods similar in appearance to campylobacters, but because they differ sufficiently in certain biochemical and flagellar characteristics, they are classified as a separate genus. In particular, helicobacters are strongly urease-positive, whereas campylobacters are urease-negative.

Pathogenesis & Epidemiology

H. pylori attaches to the mucus-secreting cells of the gastric mucosa. The production of large amounts of ammonia from urea by the organism's urease, coupled with an inflammatory response, leads to damage of the mucosa. Loss of the protective mucus coating predisposes to gastritis and peptic ulcer (see Table 18–10). The ammonia also neutralizes stomach acid, allowing the organism to survive. Epidemiologically, most patients with these diseases show *H. pylori* in biopsy specimens of the gastric epithelium.

The natural habitat of *H. pylori* is the human stomach, and it is probably acquired by ingestion. However, it has not been isolated from stool, food, water, or animals. Person-to-person transmission probably occurs because there is clustering of infection within families. The rate of infection with *H. pylori* in developing countries is very high—a finding that is in accord with the high rate of gastric carcinoma in those countries.

MALT lymphomas are B-cell tumors located typically in the stomach, but they occur elsewhere in the gastrointestinal tract as well. *H. pylori* is often found in the MALT lesion, and the chronic inflammation induced by the organism is thought to stimulate B-cell proliferation and eventually a B-cell lymphoma. Antibiotic treatment directed against the organism often causes the tumor to regress.

Clinical Findings

Gastritis and peptic ulcer are characterized by recurrent pain in the upper abdomen, frequently accompanied by bleeding into the gastrointestinal tract. No bacteremia or disseminated disease occurs.

Laboratory Diagnosis

The organism can be seen on Gram-stained smears of biopsy specimens of the gastric mucosa. It can be cultured on the same media as campylobacters. In contrast to *C. jejuni*, *H. pylori* is urease-positive. Urease production is the basis for a noninvasive diagnostic test called the "urea breath" test. In this test, radiolabeled urea is ingested. If the organism is present, urease will cleave the ingested urea, radiolabeled CO_2 is evolved, and the radioactivity is detected in the breath.

A test for *Helicobacter* antigen in the stool can be used for diagnosis and for confirmation that treatment has eliminated the organism. The presence of immunoglobulin (Ig) G antibodies in the patient's serum can also be used as evidence of infection.

Treatment & Prevention

The concept that underlies the choice of drugs to treat gastritis and peptic ulcer is to use antibiotics to eliminate *Helicobacter* plus a proton-pump inhibitor to reduce gastric acidity. Quadruple-drug therapy is used because resistance, especially to metronidazole, has emerged. One regimen consists of a combination of tetracycline, bismuth, metronidazole (or tinidazole), and a proton pump inhibitor (PPI) such as omeprazole. Another regimen consists of amoxicillin, clarithromycin, metronidazole (or tinidazole), and a PPI.

There is evidence that treating *Helicobacter* gastritis with antibiotics can prevent gastric carcinoma. There is no vaccine against *Helicobacter*.

PATHOGENS OUTSIDE THE ENTERIC TRACT

KLEBSIELLA–ENTEROBACTER–SERRATIA GROUP

Diseases

These organisms are usually opportunistic pathogens that cause nosocomial infections, especially pneumonia and urinary tract infections. *K. pneumoniae* is an important respiratory tract pathogen outside hospitals as well.

Important Properties

K. pneumoniae, *Enterobacter cloacae*, and *Serratia marcescens* are the species most often involved in human infections. They are frequently found in the **large intestine** but are also present in soil and water. These organisms have very similar properties and are usually distinguished on the basis of several biochemical reactions and motility. *K. pneumoniae* has a **very large polysaccharide capsule**, which gives its colonies a striking mucoid appearance. *S. marcescens* produces **red-pigmented colonies** (Figure 18–3).

Pathogenesis & Epidemiology

Of the three organisms, *K. pneumoniae* is most likely to be a primary, nonopportunistic pathogen; this property is related to its antiphagocytic capsule. Although this organism is a primary

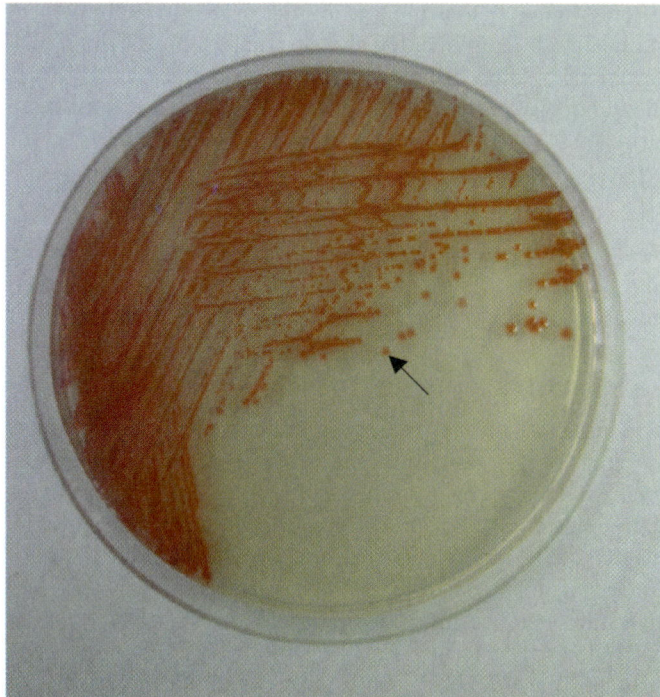

FIGURE 18–3 *Serratia marcescens*—red-pigmented colonies. Arrow points to a red-pigmented colony of *S. marcescens*. (Used with permission from Professor Shirley Lowe, University of California, San Francisco School of Medicine.)

pathogen, patients with *K. pneumoniae* infections frequently have predisposing conditions such as advanced age, chronic respiratory disease, diabetes, or alcoholism. The organism is carried in the respiratory tract of about 10% of healthy people, who are prone to pneumonia if host defenses are lowered.

Enterobacter and *Serratia* infections are clearly related to hospitalization, especially to invasive procedures such as intravenous catheterization, respiratory intubation, and urinary tract manipulations. In addition, outbreaks of *Serratia* pneumonia have been associated with contamination of the water in respiratory therapy devices. Prior to the extensive use of these procedures, *S. marcescens* was a harmless organism most frequently isolated from environmental sources such as water.

Serratia also causes endocarditis in users of injection drugs. As with many other gram-negative rods, the pathogenesis of septic shock caused by these organisms is related to the endotoxins in their cell walls.

Clinical Findings

Urinary tract infections and pneumonia are the usual clinical entities associated with these three bacteria, but bacteremia and secondary spread to other areas such as the meninges and liver occur. It is difficult to distinguish infections caused by these organisms on clinical grounds, with the exception of pneumonia caused by *Klebsiella*, which produces a thick, mucoid, bloody sputum (**currant-jelly sputum**) and can progress to necrosis and abscess formation.

There are two other species of *Klebsiella* that cause unusual human infections rarely seen in the United States. *Klebsiella ozaenae* is associated with atrophic rhinitis, and *Klebsiella rhinoscleromatis* causes a destructive granuloma of the nose and pharynx.

Laboratory Diagnosis

Organisms of this group produce lactose-fermenting (colored) colonies on differential agar such as MacConkey's or EMB, although *Serratia*, which is a late (slow) lactose fermenter, can produce a negative reaction. These organisms are differentiated by the use of biochemical tests.

Treatment & Prevention

Because the antibiotic resistance of these organisms can vary greatly, the choice of drug depends on the results of sensitivity testing. Isolates from hospital-acquired infections are frequently resistant to multiple antibiotics. Strains of *K. pneumoniae* that produce extended-spectrum β-lactamases (ESBL) are an important cause of hospital-acquired infections and are resistant to almost all known antibiotics. An aminoglycoside (e.g., gentamicin) and a cephalosporin (e.g., cefotaxime) are used empirically until the results of testing are known. In severe *Enterobacter* infections, a combination of imipenem and gentamicin is often used.

Some hospital-acquired infections caused by gram-negative rods can be prevented by such general measures as changing the site of intravenous catheters, removing urinary catheters when they are no longer needed, and taking proper care of respiratory therapy devices. There is no vaccine.

PROTEUS–PROVIDENCIA–MORGANELLA GROUP

Diseases

These organisms primarily cause urinary tract infections, both community- and hospital-acquired.

Important Properties

These gram-negative rods are distinguished from other members of the Enterobacteriaceae by their ability to produce the enzyme phenylalanine deaminase. In addition, they produce the enzyme **urease**, which cleaves urea to form NH_3 and CO_2. Certain species are very motile and produce a striking **swarming** effect on blood agar, characterized by expanding rings (waves) of organisms over the surface of the agar (Figure 18–4).

The cell wall O antigens of certain strains of *Proteus*, such as OX-2, OX-19, and OX-K, cross-react with antigens of several species of rickettsiae. These *Proteus* antigens can be used in laboratory tests to detect the presence of antibodies against certain rickettsiae in patients' serum. This test, called the Weil-Felix reaction after its originators, is being used less frequently as more specific procedures are developed.

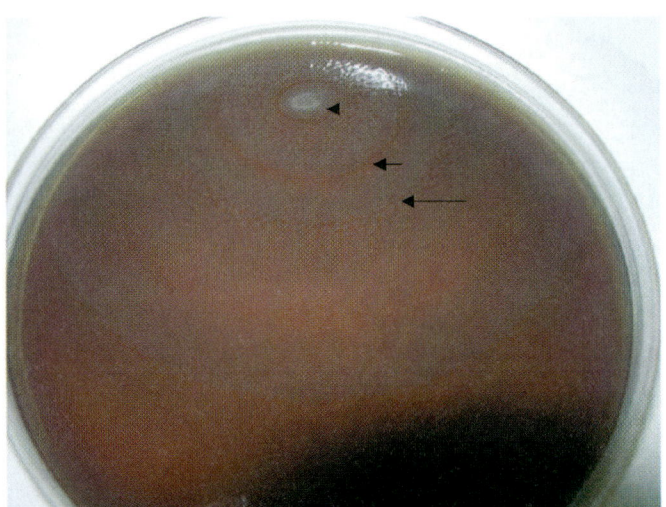

FIGURE 18-4 *Proteus* species—swarming motility on blood agar. Arrowhead points to the site where *Proteus* bacteria were placed on the blood agar. Short arrow points to the edge of the first ring of swarming motility. Long arrow points to the edge of the second ring of swarming motility. (Used with permission from Professor Shirley Lowe, University of California, San Francisco School of Medicine.)

In the past, there were four medically important species of *Proteus*. However, molecular studies of DNA relatedness showed that two of the four were significantly different. These species have been renamed: *Proteus morganii* is now *Morganella morganii*, and *Proteus rettgeri* is now *Providencia rettgeri*. In the clinical laboratory, these organisms are distinguished from *Proteus vulgaris* and *Proteus mirabilis* on the basis of several biochemical tests.

Pathogenesis & Epidemiology

The organisms are present in the human colon as well as in soil and water. Their tendency to cause urinary tract infections is probably due to their presence in the colon and to colonization of the urethra, especially in women. The vigorous motility of *Proteus* organisms may contribute to their ability to invade the urinary tract.

Production of the enzyme urease is an important feature of the pathogenesis of urinary tract infections by this group. Urease hydrolyzes the urea in urine to form ammonia, which raises the pH, producing an alkaline urine. This encourages the formation of stones (calculi) called **struvite** composed of magnesium ammonium phosphate. Struvite stones often manifest as staghorn calculi in the renal pelvis. They obstruct urine flow, damage urinary epithelium, and serve as a nidus for recurrent infection by trapping bacteria within the stone. Because alkaline urine also favors growth of the organisms and more extensive renal damage, treatment involves keeping the urine at a low pH.

Clinical Findings

The signs and symptoms of urinary tract infections caused by these organisms cannot be distinguished from those caused by *E. coli* or other members of the Enterobacteriaceae. *Proteus* species can also cause pneumonia, wound infections, and septicemia. *P. mirabilis* is the species of *Proteus* that causes most community- and hospital-acquired infections, but *P. rettgeri* is emerging as an important agent of nosocomial infections.

Laboratory Diagnosis

These organisms usually are highly motile and produce a **swarming** overgrowth on blood agar, which can frustrate efforts to recover pure cultures of other organisms. Growth on blood agar containing phenylethyl alcohol inhibits swarming, thus allowing isolated colonies of *Proteus* and other organisms to be obtained. They produce non–lactose-fermenting (colorless) colonies on MacConkey's or EMB agar. *P. vulgaris* and *P. mirabilis* produce H_2S, which blackens the butt of TSI agar, whereas neither *M. morganii* nor *P. rettgeri* does. *P. mirabilis* is indole-negative, whereas the other three species are indole-positive—a distinction that can be used clinically to guide the choice of antibiotics. These four medically important species are urease-positive. Identification of these organisms in the clinical laboratory is based on a variety of biochemical reactions.

Treatment & Prevention

Most strains are sensitive to aminoglycosides and trimethoprim-sulfamethoxazole, but because individual isolates can vary, antibiotic sensitivity tests should be performed. *P. mirabilis* is the species most frequently sensitive to ampicillin. The indole-positive species (*P. vulgaris*, *M. morganii*, and *P. rettgeri*) are more resistant to antibiotics than is *P. mirabilis*, which is indole-negative. The treatment of choice for the indole-positive species is a cephalosporin (e.g., cefotaxime). *P. rettgeri* is frequently resistant to multiple antibiotics.

There are no specific preventive measures, but many hospital-acquired urinary tract infections can be prevented by prompt removal of urinary catheters.

PSEUDOMONAS

Diseases

P. aeruginosa causes infections (e.g., sepsis, pneumonia, and urinary tract infections) primarily in patients with lowered host defenses. It also causes chronic lower respiratory tract infections in patients with cystic fibrosis, wound infections (cellulitis) in burn patients (Figure 18–5), and malignant otitis externa in diabetic patients. It is the most common cause of ventilator-associated pneumonia.

(*P. aeruginosa* is also known as *Burkholderia aeruginosa*.) *Pseudomonas cepacia* (renamed *Burkholderia cepacia*) and *Pseudomonas maltophilia* (renamed *Xanthomonas maltophilia* and now called *Stenotrophomonas maltophilia*) also cause these infections, but much less frequently. *Pseudomonas pseudomallei* (also known as *Burkholderia pseudomallei*), the cause of melioidosis, is described in Chapter 27.

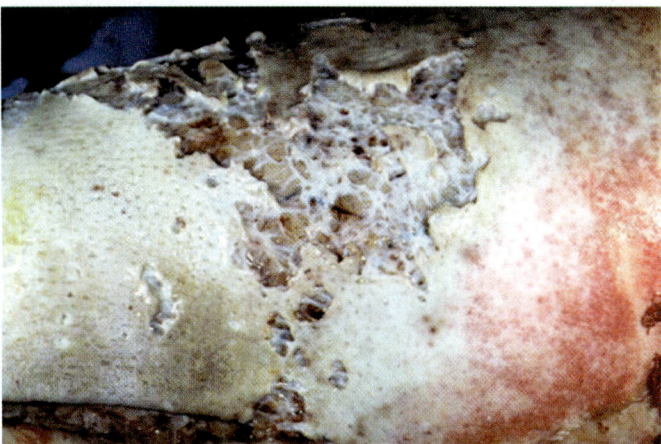

FIGURE 18–5 Cellulitis caused by *Pseudomonas aeruginosa*. Note the blue-green color of the pus in the burn wound infection. (Used with permission from Dr. Robert L. Sheridan.)

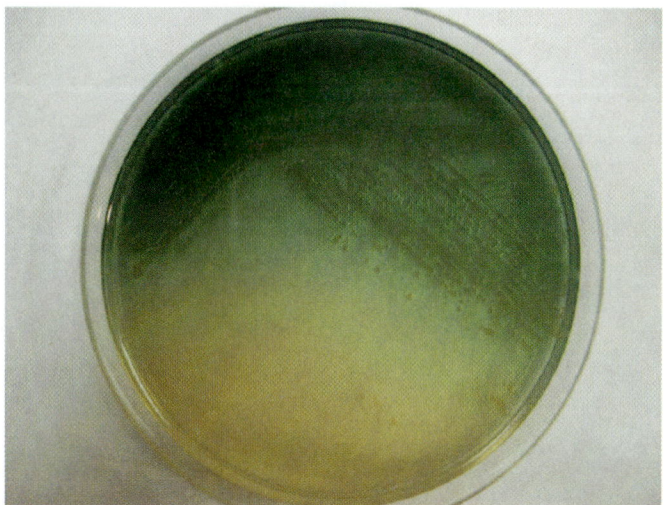

FIGURE 18–6 *Pseudomonas aeruginosa*—blue-green pigment. Blue-green pigment (pyocyanin) produced by *P. aeruginosa* diffuses into the agar. (Used with permission from Professor Shirley Lowe, University of California, San Francisco School of Medicine.)

Important Properties

Pseudomonads are gram-negative rods that resemble the members of the Enterobacteriaceae but differ in that they are strict aerobes (i.e., they derive their energy only by oxidation of sugars rather than by fermentation). Because they do not ferment glucose, they are called **nonfermenters**, in contrast to the members of the Enterobacteriaceae, which do ferment glucose. Oxidation involves electron transport by cytochrome c (i.e., they are **oxidase-positive**).

Pseudomonads are able to grow in **water** containing only traces of nutrients (e.g., tap water), and this favors their persistence in the hospital environment. *P. aeruginosa* and *B. cepacia* have a remarkable ability to withstand disinfectants; this accounts in part for their role in hospital-acquired infections. They have been found growing in hexachlorophene-containing soap solutions, in antiseptics, and in detergents.

P. aeruginosa produces two pigments useful in clinical and laboratory diagnosis: (1) **pyocyanin**, which can **color the pus in a wound blue** and (2) pyoverdin (fluorescein), a yellow-green pigment that fluoresces under ultraviolet light, a property that can be used in the early detection of skin infection in burn patients. In the laboratory, these **pigments diffuse into the agar, imparting a blue-green color** that is useful in identification. *P aeruginosa* is the only species of *Pseudomonas* that synthesizes pyocyanin (Figure 18–6).

Strains of *P. aeruginosa* isolated from cystic fibrosis patients have a prominent slime layer (glycocalyx), which gives their colonies a very mucoid appearance. The slime layer mediates adherence of the organism to mucous membranes of the respiratory tract and prevents antibody from binding to the organism.

Pathogenesis & Epidemiology

P. aeruginosa is found chiefly in soil and water, although approximately 10% of people carry it in the normal flora of the colon. It is found on the skin in moist areas and can colonize the upper respiratory tract of hospitalized patients. Its ability to grow in simple aqueous solutions has resulted in contamination of respiratory therapy and anesthesia equipment, intravenous fluids, and even distilled water.

P. aeruginosa is primarily an opportunistic pathogen that causes infections in hospitalized patients (e.g., those with extensive burns), in whom the skin host defenses are destroyed; in those with chronic respiratory disease (e.g., cystic fibrosis), in whom the normal clearance mechanisms are impaired; in those who are immunosuppressed; in those with neutrophil counts of less than 500/μL; and in those with indwelling catheters. It causes 10% to 20% of hospital-acquired infections and, in many hospitals, is the most common cause of gram-negative nosocomial pneumonia, especially ventilator-associated pneumonia.

Pathogenesis is based on multiple virulence factors: endotoxin, exotoxins, and enzymes. Its endotoxin, like that of other gram-negative bacteria, causes the symptoms of sepsis and septic shock. The best known of the exotoxins is exotoxin A, which causes tissue necrosis. It inhibits eukaryotic protein synthesis by the same mechanism as diphtheria exotoxin, namely, ADP-ribosylation of elongation factor-2. It also produces enzymes, such as elastase and proteases, that are histotoxic and facilitate invasion of the organism into the bloodstream. Pyocyanin damages the cilia and mucosal cells of the respiratory tract.

Strains of *P. aeruginosa* that have a "type III secretion system" are significantly more virulent than those that do not. This secretion system transfers the exotoxin from the bacterium directly into the adjacent human cell, which allows the toxin to avoid neutralizing antibody. Type III secretion systems are mediated by transport pumps in the bacterial cell membrane. Of the four exoenzymes known to be transported by this secretion system, Exo S is the one most clearly associated with virulence. Exo S has several modes of action, the most important of which is ADP-ribosylation of a Ras protein, leading to damage to the cytoskeleton.

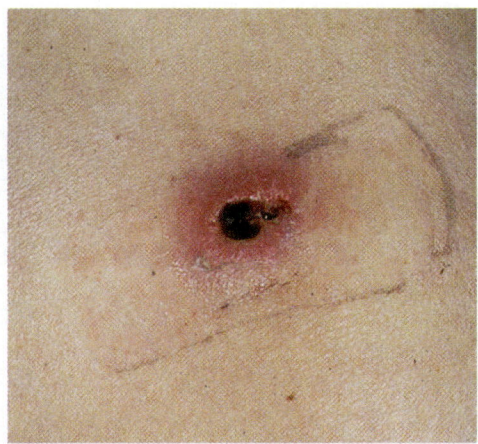

FIGURE 18–7 Ecthyma gangrenosum. Necrotic skin lesion caused by *Pseudomonas aeruginosa*. (Reproduced with permission from Wolff K, Johnson R, Saavedra A: *Fitzpatrick's Color Atlas & Synopsis of Clinical Dermatology*, 8th ed. New York, NY: McGraw Hill; 2017.)

Clinical Findings

P. aeruginosa can cause infections virtually anywhere in the body, but urinary tract infections, pneumonia (especially in **cystic fibrosis** patients), and wound infections (especially burns) (see Figure 18–5) predominate. It is an important cause of hospital-acquired pneumonia, especially in those undergoing mechanical ventilation (ventilator-associated pneumonia). From these sites, the organism can enter the blood, causing sepsis. The bacteria can spread to the skin, where they cause black, necrotic lesions called **ecthyma gangrenosum** (Figure 18–7). Patients with *P. aeruginosa* sepsis have a mortality rate of greater than 50%. It is an important cause of endocarditis in intravenous drug users.

Severe external otitis (malignant otitis externa) and other skin lesions (e.g., folliculitis) occur in users of swimming pools and hot tubs (hot tub folliculitis) in which the chlorination is inadequate. *P. aeruginosa* is the most common cause of osteomyelitis of the foot in those who sustain puncture wounds through the soles of gym shoes. Corneal infections caused by *P. aeruginosa* are seen in contact lens users.

In addition to *P. aeruginosa*, *Stenotrophomonas* and *Burkholderia* also cause chronic lung infections in patients with cystic fibrosis.

Laboratory Diagnosis

P. aeruginosa grows as non–lactose-fermenting (colorless) colonies on MacConkey's or EMB agar. It is **oxidase-positive**. A typical metallic sheen of the growth on TSI agar, coupled with the blue-green pigment on ordinary nutrient agar (see Figure 18–6), and a fruity aroma are sufficient to make a presumptive diagnosis. The diagnosis is confirmed by biochemical reactions. Identification for epidemiologic purposes is done by bacteriophage or pyocin[6] typing.

Treatment

Because *P. aeruginosa* is **resistant to many antibiotics**, treatment must be tailored to the sensitivity of each isolate and monitored frequently; resistant strains can emerge during therapy. The treatment of choice is an antipseudomonal penicillin (e.g., piperacillin/tazobactam or ticarcillin/clavulanate) plus an aminoglycoside (e.g., gentamicin or amikacin). Ceftazidime is also effective. For infections caused by highly resistant strains, colistin (polymyxin E) is useful. The drug of choice for urinary tract infections is ciprofloxacin. The drug of choice for infections caused by *B. cepacia* and *S. maltophilia* is trimethoprim-sulfamethoxazole.

Prevention

Prevention of *P. aeruginosa* infections involves keeping neutrophil counts above 500/μL, removing indwelling catheters promptly, taking special care of burned skin, and taking other similar measures to limit infection in patients with reduced host defenses.

BACTEROIDES & PREVOTELLA

Diseases

Members of the genus *Bacteroides* are the most common cause of serious anaerobic infections (e.g., sepsis, peritonitis, and abscesses). *Bacteroides fragilis* is the most frequent pathogen. *Prevotella melaninogenica* is also an important pathogen. It was formerly known as *Bacteroides melaninogenicus*, and both names are still encountered.

Important Properties

Bacteroides and *Prevotella* organisms are anaerobic, non–spore-forming, gram-negative rods. Of the many species of *Bacteroides*, two are human pathogens: *B. fragilis*[7] and *Bacteroides corrodens*.

Members of the *B. fragilis* group are the predominant organisms in the human colon, numbering approximately 10^{11}/g of feces and are found in the vagina of approximately 60% of women. *P. melaninogenica* and *B. corrodens* occur primarily in the mouth.

Pathogenesis & Epidemiology

Because *Bacteroides* and *Prevotella* species are part of the normal flora, **infections** are endogenous, usually arising from a break in a mucosal surface, and are not communicable. These organisms cause a variety of infections, such as local abscesses at the site of a mucosal break, metastatic abscesses by hematogenous spread to distant organs, or lung abscesses by aspiration of oral flora.

---

[6]A pyocin is a type of bacteriocin produced by *P. aeruginosa*. Different strains produce various pyocins, which can serve to distinguish the organisms.

[7]*B. fragilis* is divided into five subspecies, the most important of which is *B. fragilis* subsp. fragilis. The other four subspecies are *B. fragilis* subsp. *distasonis, ovatus, thetaiotaomicron*, and *vulgatus*. It is proper, therefore, to speak of the *B. fragilis* group rather than simply *B. fragilis*.

Predisposing factors such as surgery, trauma, and chronic disease play an important role in pathogenesis. Local tissue necrosis, impaired blood supply, and growth of facultative anaerobes at the site contribute to anaerobic infections. The facultative anaerobes, such as *E. coli*, utilize the oxygen, thereby reducing it to a level that allows the anaerobic *Bacteroides* and *Prevotella* strains to grow. As a result, many anaerobic infections contain a mixed facultative and anaerobic flora. This has important implications for therapy; both the facultative anaerobes and the anaerobes should be treated.

The polysaccharide capsule of *B. fragilis* is an important virulence factor. The host response to the capsule plays a major role in abscess formation. Note also that the endotoxin of *B. fragilis* contains a variant lipid A that is missing one of the fatty acids and consequently is 1000-fold less active than the typical endotoxin of bacteria such as *Neisseria meningitidis*.

Enzymes such as hyaluronidase, collagenase, and phospholipase are produced and contribute to tissue damage. Enterotoxin-producing strain of *B. fragilis* can cause diarrhea in both children and adults.

Clinical Findings

The *B. fragilis* group of organisms is most frequently associated with intra-abdominal infections, either peritonitis or localized abscesses. Pelvic abscesses, necrotizing fasciitis, and bacteremia occur as well. Abscesses of the mouth, pharynx, brain, and lung are more commonly caused by *P. melaninogenica*, a member of the normal oral flora, but *B. fragilis* is found in about 25% of lung abscesses. In general, *B. fragilis* causes disease below the diaphragm, whereas *P. melaninogenica* causes disease above the diaphragm. *Prevotella intermedia* is an important cause of gingivitis, periodontitis, and dental abscess.

Laboratory Diagnosis

Bacteroides species can be isolated anaerobically on blood agar plates containing kanamycin and vancomycin to inhibit unwanted organisms. They are identified by biochemical reactions (e.g., sugar fermentations) and by production of certain organic acids (e.g., formic, acetic, and propionic acids), which are detected by gas chromatography. *P. melaninogenica* produces characteristic black colonies (Figure 18–8).

Treatment & Prevention

Members of the *B. fragilis* group are resistant to penicillins, first-generation cephalosporins, and aminoglycosides, making them among the most antibiotic-resistant of the anaerobic bacteria. Penicillin resistance is the result of β-lactamase production. Metronidazole is the drug of choice, with cefoxitin, clindamycin, and chloramphenicol as alternatives. Aminoglycosides are frequently combined to treat the facultative gram-negative rods in mixed infections. The drug of choice for *P. melaninogenica* infections is either metronidazole or clindamycin. β-Lactamase-producing strains of *P. melaninogenica* have been isolated from patients. Surgical drainage of abscesses usually

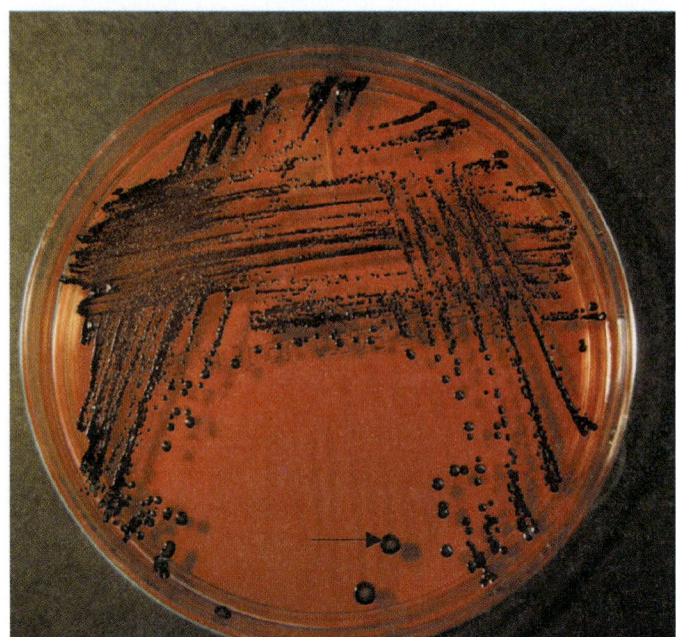

FIGURE 18–8 *Prevotella melaninogenica*—black pigmented colonies. Arrow points to a black pigmented colony of *P. melaninogenica*. (Used with permission from Professor Shirley Lowe, University of California, San Francisco School of Medicine.)

accompanies antibiotic therapy, but lung abscesses often heal without drainage.

Prevention of *Bacteroides* and *Prevotella* infections centers on perioperative administration of a *cephalosporin*, frequently cefoxitin, for abdominal or pelvic surgery. There is no vaccine.

FUSOBACTERIUM

Fusobacterium species are long, anaerobic, gram-negative rods with pointed ends (Figure 18–9). They are part of the human normal flora of the mouth, colon, and female genital tract and are isolated from brain, pulmonary, intra-abdominal, and

FIGURE 18–9 *Fusobacterium nucleatum*—Gram stain. Note the long, thin gram-negative rods with pointed ends. (Used with permission from Dr. Dowell VR, Jr. Public Health Image Library, Centers for Disease Control and Prevention.)

pelvic abscesses. They are frequently found in mixed infections with other anaerobes and facultative anaerobes.

Fusobacterium nucleatum occurs, along with various spirochetes, in cases of Vincent's angina (trench mouth), which is characterized by a necrotizing ulcerative gingivitis. *Fusobacterium necrophorum* causes Lemierre's disease, which is an anaerobic infection of the posterior pharyngeal space accompanied by thrombophlebitis of the internal jugular vein and metastatic infectious emboli to the lung.

The laboratory diagnosis is made by culturing the organism anaerobically. The drug of choice for *Fusobacterium* infections can be penicillin G, clindamycin, or metronidazole. There is no vaccine.

SELF-ASSESSMENT QUESTIONS

1. Your patient is a 75-year-old man with an indwelling urinary catheter following prostatectomy for prostate cancer. He now has the sudden onset of fever to 40°C, blood pressure of 70/40 mm Hg, and a pulse of 140 bpm. You draw several blood cultures, and the laboratory reports that all are positive for a gram-negative rod that forms red-pigmented colonies. Which one of the following bacteria is the most likely cause of this infection?

    (A) *Escherichia coli*
    (B) *Klebsiella pneumoniae*
    (C) *Proteus mirabilis*
    (D) *Pseudomonas aeruginosa*
    (E) *Serratia marcescens*

2. You're a public health epidemiologist who is called to investigate an outbreak of bloody diarrhea in 16 people. You find that it is associated with eating rare hamburgers in a particular fast-food restaurant. A culture of the remaining uncooked hamburger grows a gram-negative rod that produces a dark purple colony on EMB agar, which is evidence that it ferments lactose. Which one of the following bacteria is the most likely cause of this outbreak?

    (A) *Escherichia coli*
    (B) *Salmonella enterica*
    (C) *Salmonella typhi*
    (D) *Shigella dysenteriae*
    (E) *Vibrio cholerae*

3. Your patient has third-degree burns over most of his body. He was doing well until 2 days ago, when he spiked a fever, and his dressings revealed pus that had a blue-green color. Gram stain of the pus revealed a gram-negative rod that formed colorless colonies on EMB agar. Which one of the following bacteria is the most likely cause of this infection?

    (A) *Campylobacter jejuni*
    (B) *Escherichia coli*
    (C) *Haemophilus influenzae*
    (D) *Pseudomonas aeruginosa*
    (E) *Salmonella enterica*

4. Regarding the patient in Question 3, which one of the following is the best combination of antibiotics to treat the infection?

    (A) Azithromycin plus gentamicin
    (B) Doxycycline plus gentamicin
    (C) Metronidazole plus gentamicin
    (D) Piperacillin/tazobactam plus gentamicin
    (E) Vancomycin plus gentamicin

5. Regarding the members of the family Enterobacteriaceae, which one of the following statements is the most accurate?

    (A) All members of the family are anaerobic, which means they must be cultured in the absence of oxygen.
    (B) All members of the family ferment lactose, which is an important diagnostic criterion in the clinical laboratory.
    (C) All members of the family have endotoxin, an important pathogenetic factor.
    (D) All members of the family produce an enterotoxin, which ADP-ribosylates a G protein in human enterocytes.

6. You're on a summer program working in a clinic in a small village in Ecuador. There is an outbreak of cholera, and your patient has massive diarrhea and a blood pressure of 70/40 mm Hg. Which one of the following would be the most appropriate action to take?

    (A) Administer antimotility drugs to diminish the diarrhea.
    (B) Administer intravenous saline to replenish volume.
    (C) Administer tetracycline to kill the organism.
    (D) Perform stool cultures and fecal leukocyte tests to make an accurate diagnosis.

7. Your patient is a 20-year-old woman with diarrhea. She has just returned to the United States from a 3-week trip to Peru, where she ate some raw shellfish at the farewell party. She now has severe watery diarrhea, perhaps 20 bowel movements a day, and is feeling quite weak and dizzy. Her stool is guaiac-negative, a test that determines whether there is blood in the stool. A Gram stain of the stool reveals curved gram-negative rods. Culture of the stool on MacConkey's agar shows colorless colonies. Which one of the following bacteria is the most likely cause of this infection?

    (A) *Escherichia coli*
    (B) *Helicobacter pylori*
    (C) *Proteus mirabilis*
    (D) *Pseudomonas aeruginosa*
    (E) *Vibrio cholerae*

8. Your patient is a 6-year-old boy with bloody diarrhea for the past 2 days accompanied by fever to 40°C and vomiting. He has a pet corn snake. Blood culture and stool culture from the boy and stool culture from the snake (taken very carefully!) revealed the same organism. The cultures grew a gram-negative rod that formed colorless colonies on EMB agar. Which one of the following bacteria is the most likely cause of this infection?

    (A) *Helicobacter pylori*
    (B) *Proteus mirabilis*
    (C) *Salmonella enterica*
    (D) *Shigella dysenteriae*
    (E) *Vibrio cholerae*

9. Your patient is a 25-year-old woman with pain on urination and cloudy urine but no fever or flank pain. She has not been hospitalized. You think she probably has cystitis, an infection of the urinary bladder. A Gram stain of the urine reveals gram-negative rods. Culture of the urine on EMB agar shows colorless colonies, and a urease test was positive. Swarming motility was noted on the blood agar plate. Which one of the following bacteria is the most likely cause of this infection?

    (A) *Escherichia coli*
    (B) *Helicobacter pylori*
    (C) *Proteus mirabilis*
    (D) *Pseudomonas aeruginosa*
    (E) *Serratia marcescens*

10. Your patient has abdominal pain, and a mass is discovered in the left lower quadrant. Upon laparotomy (surgical opening of the abdomen), an abscess is found. Culture of the pus reveals *Bacteroides fragilis*. Regarding this organism, which one of the following is the most accurate?

    (A) A stage in the life cycle of *Bacteroides fragilis* involves forming spores in the soil.

    (B) *Bacteroides fragilis* is an anaerobic gram-negative rod whose natural habitat is the human colon.

    (C) *Bacteroides fragilis* produces black colonies when grown on blood agar.

    (D) Pathogenesis by *Bacteroides fragilis* involves an exotoxin that increases cyclic AMP by ADP-ribosylation of a G protein.

    (E) The toxoid vaccine should be administered to prevent disease caused by *Bacteroides fragilis*.

11. Regarding the patient in Question 10, which one of the following is the best antibiotic to treat the infection?

    (A) Doxycycline

    (B) Gentamicin

    (C) Metronidazole

    (D) Penicillin G

    (E) Rifampin

12. Your patient in the gastrointestinal clinic is a 50-year-old insurance salesman with what he describes as a "sour stomach" for several months. Antacids relieve the symptoms. After taking a complete history and doing a physical examination, you discuss the case with your resident, who suggests doing a urea breath test, which tests for the presence of urease. Which one of the following bacteria does the resident think is the most likely cause of the patient's disease?

    (A) *Helicobacter pylori*

    (B) *Proteus mirabilis*

    (C) *Salmonella enterica*

    (D) *Serratia marcescens*

    (E) *Shigella dysenteriae*

13. Your patient is a 35-year-old woman with epilepsy who had a grand-mal seizure about 2 months ago. She comes to see you now because she has been coughing up foul-smelling sputum for the past week. Chest X-ray reveals a cavity with an air-fluid level. Gram stain of the sputum reveals gram-negative rods, and culture reveals black colonies that grow on blood agar only in the absence of air. Which one of the following bacteria is the most likely cause of this infection?

    (A) *Bacteroides fragilis*

    (B) *Campylobacter jejuni*

    (C) *Klebsiella pneumoniae*

    (D) *Prevotella melaninogenica*

    (E) *Proteus mirabilis*

14. Regarding *Fusobacterium nucleatum*, which one of the following is most accurate?

    (A) Its natural habitat is the soil.

    (B) It is an anaerobic gram-negative rod with pointed ends.

    (C) The drug of choice for infections caused by *F. nucleatum* is azithromycin.

    (D) Laboratory diagnosis is based on detecting the ability of the exotoxin to kill cells in tissue culture.

ANSWERS

(1) **(E)**
(2) **(A)**
(3) **(D)**
(4) **(D)**
(5) **(C)**
(6) **(B)**
(7) **(E)**
(8) **(C)**
(9) **(C)**
(10) **(B)**
(11) **(C)**
(12) **(A)**
(13) **(D)**
(14) **(B)**

SUMMARIES OF ORGANISMS

Brief summaries of the organisms described in this chapter begin on page 675. Please consult these summaries for a rapid review of the essential material.

PRACTICE QUESTIONS: USMLE & COURSE EXAMINATIONS

Questions on the topics discussed in this chapter can be found in the Clinical Bacteriology section of Part XIII: USMLE (National Board) Practice Questions starting on page 737. Also see Part XIV: USMLE (National Board) Practice Examination starting on page 775.

Gram-Negative Rods Related to the Respiratory Tract

INTRODUCTION

There are four medically important gram-negative rods typically associated with the respiratory tract, namely, *Haemophilus influenzae*, *Bordetella pertussis*, *Legionella pneumophila*, and *Acinetobacter baumannii* (Table 19–1). *H. influenzae* and *B. pertussis* are found only in humans, whereas *L. pneumophila* is found primarily in environmental water sources. *A. baumannii* is found in environmental water sources but also colonizes the skin and upper respiratory tract.

Additional information regarding the clinical aspects of infections caused by the organisms in this chapter is provided in Part IX, entitled Infectious Diseases beginning on page 603.

HAEMOPHILUS

Diseases

H. influenzae used to be the leading cause of meningitis in young children, but the use of the highly effective "conjugate" vaccine has greatly reduced the incidence of meningitis caused by this organism. It is still an important cause of upper respiratory tract infections (otitis media, sinusitis, conjunctivitis, and epiglottitis) and sepsis in children. It also causes pneumonia in adults, particularly in those with chronic obstructive lung disease. *Haemophilus ducreyi*, the agent of chancroid, is discussed in Chapter 27.

Important Properties

H. influenzae is a small, pleomorphic gram-negative rod (coccobacillary rod) with a polysaccharide capsule (Figure 19–1). It is one of the three important **encapsulated pyogens**, along with the pneumococcus and the meningococcus. Serologic typing is based on the antigenicity of the capsular polysaccharide. Of the six serotypes (a–f), **type b** is the most important. Type b used to cause most of the severe, invasive diseases, such as meningitis and sepsis, but the widespread use of the vaccine containing the type b capsular polysaccharide as the immunogen has greatly

TABLE 19–1 Gram-Negative Rods Associated with the Respiratory Tract

Species	Major Diseases	Laboratory Diagnosis	Factors X and V Required for Growth	Vaccine Available	Prophylaxis for Contacts
Haemophilus influenzae	Meningitis[1]; otitis media, sinusitis, pneumonia, epiglottitis	Culture; capsular polysaccharide in serum or spinal fluid	+	+	Rifampin
Bordetella pertussis	Whooping cough (pertussis)	Fluorescent antibody on secretions; culture	–	+	Azithromycin
Legionella pneumophila	Pneumonia	Serology; urinary antigen; culture	–	–	None
Acinetobacter baumannii	Ventilator-associated pneumonia	Culture	–	–	None

[1]In countries where the *H. influenzae* b conjugate vaccine has been deployed, the vaccine has greatly reduced the incidence of meningitis caused by this organism.

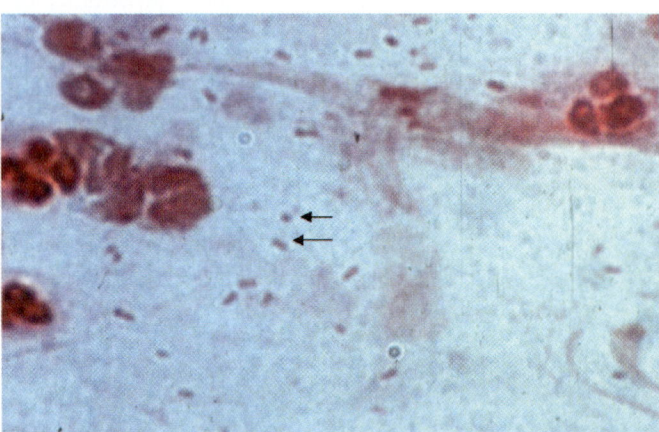

FIGURE 19–1 *Haemophilus influenzae*—Gram stain. Arrows point to two small "coccobacillary" gram-negative rods. (Used with permission from Professor Shirley Lowe, University of California, San Francisco School of Medicine.)

reduced the incidence of invasive disease caused by this type. The type b capsule is composed of polyribitol phosphate.

Unencapsulated strains can also cause disease, especially mucosal diseases of the upper respiratory tract such as sinusitis and otitis media, but are usually noninvasive. Growth of the organism on laboratory media requires the addition of two components, **heme (factor X)** and **NAD (factor V)**, for adequate energy production.

Pathogenesis & Epidemiology

H. influenzae infects only humans; there is no animal reservoir. It enters the body by the inhalation of airborne droplets into the **respiratory tract**, resulting in either asymptomatic colonization or infections such as otitis media, sinusitis, or pneumonia. The organism produces an IgA protease that degrades secretory IgA, thus facilitating attachment to the respiratory mucosa.

After becoming established in the upper respiratory tract, the organism can enter the bloodstream (bacteremia) and spread to the meninges. Meningitis is caused primarily by the encapsulated strains, but nonencapsulated strains are frequently involved in otitis media, sinusitis, and pneumonia. Note that the incidence of meningitis caused by capsular type b has been greatly reduced because the vaccine contains the type b polysaccharide as the immunogen. Pathogenesis of *H. influenzae* involves its antiphagocytic capsule and endotoxin; no exotoxin is produced.

Most infections occur in children between the ages of 6 months and 6 years, with a peak in the age group from 6 months to 1 year. This age distribution is attributed to a decline in maternal IgG in the child coupled with the inability of the child to generate sufficient antibody against the polysaccharide capsular antigen until the age of approximately 2 years.

Clinical Findings

Meningitis caused by *H. influenzae* cannot be distinguished on clinical grounds from that caused by other bacterial pathogens (e.g., pneumococci or meningococci). The rapid onset of fever,

headache, and stiff neck, along with drowsiness, is typical. Sinusitis and otitis media cause pain in the affected area, opacification of the infected sinus, and redness with bulging of the tympanic membrane. *H. influenzae* is second only to the pneumococcus as a cause of these two infections.

Other serious infections caused by this organism include septic arthritis, cellulitis, and sepsis, the latter occurring especially in splenectomized patients. Rarely, **epiglottitis**, which can obstruct the airway, occurs. A swollen "cherry-red" epiglottis is seen. This life-threatening disease of young children is caused almost exclusively by *H. influenzae*. Pneumonia in elderly adults, especially those with chronic respiratory disease, can be caused by untypeable strains of *H. influenzae*.

Laboratory Diagnosis

Laboratory diagnosis depends on isolation of the organism on heated-blood ("chocolate") agar enriched with two growth factors required for bacterial respiration, namely, factor X (a heme compound) and factor V (NAD). The blood used in chocolate agar is heated to inactivate nonspecific inhibitors of *H. influenzae* growth.

An organism that grows only in the presence of both growth factors is presumptively identified as *H. influenzae*; other species of *Haemophilus*, such as *Haemophilus parainfluenzae*, do not require both factors. Definitive identification can be made with either biochemical tests or the capsular swelling (quellung) reaction. Additional means of identifying encapsulated strains include fluorescent antibody staining of the organism and counterimmunoelectrophoresis or latex agglutination tests, which detect the capsular polysaccharide.

Treatment

The treatment of choice for meningitis or other serious systemic infections caused by *H. influenzae* is ceftriaxone. From 20% to 30% of *H. influenzae* type b isolates produce a β-lactamase that degrades penicillinase-sensitive β-lactams such as ampicillin but not ceftriaxone. It is important to institute antibiotic treatment promptly because the incidence of neurologic sequelae (e.g., subdural empyema) is high. Untreated *H. influenzae* meningitis has a fatality rate of approximately 90%. *H. influenzae* upper respiratory tract infections, such as otitis media and sinusitis, are treated with either amoxicillin-clavulanate or trimethoprim-sulfamethoxazole.

Prevention

The vaccine contains the capsular polysaccharide of *H. influenzae* type b **conjugated to diphtheria toxoid** or other carrier protein. Depending on the carrier protein, it is given some time between the ages of 2 and 15 months. This vaccine is **much more effective** in young children than the unconjugated vaccine and has reduced the incidence of meningitis caused by this organism by approximately 90% in immunized children. Meningitis in close contacts of the patient can be prevented by rifampin. Rifampin is used because it is secreted in the saliva to a greater extent than ampicillin. Rifampin decreases respiratory carriage of the organism, thereby reducing transmission.

BORDETELLA

Disease

B. pertussis causes whooping cough (pertussis).

Important Properties

B. pertussis is a small, coccobacillary, encapsulated gram-negative rod.

Pathogenesis & Epidemiology

B. pertussis, a pathogen **only for humans**, is transmitted by **airborne droplets** produced during the severe coughing episodes. The organisms attach to the ciliated epithelium of the upper respiratory tract but do not invade the underlying tissue. Decreased cilia activity and subsequent death of the ciliated epithelial cells are important aspects of pathogenesis.

Pertussis is a highly contagious disease that occurs primarily in infants and young children and has a worldwide distribution. The number of cases has declined in the United States because use of the vaccine is widespread. However, outbreaks of pertussis during the years 2005, 2010, and 2012 have led to concern about waning immunity to the vaccine and to the recommendation that an additional booster immunization be given (see "Prevention").

The following several factors play a role in the pathogenesis:

(1) Attachment of the organism to the cilia of the epithelial cells is mediated by a protein on the pili called filamentous hemagglutinin. Antibody against the filamentous hemagglutinin inhibits attachment and protects against disease.

(2) **Pertussis toxin** stimulates adenylate cyclase by catalyzing the addition of adenosine diphosphate ribose—a process called ADP-ribosylation—to the inhibitory subunit of the G protein complex (G_i protein). This results in prolonged stimulation of adenylate cyclase and a consequent rise in cyclic adenosine monophosphate (AMP) and in cyclic AMP-dependent protein kinase activity. This results in edema of the respiratory mucosa that contributes to the severe cough of pertussis. The toxin also has a domain that mediates its binding to receptors on the surface of respiratory tract epithelial cells. It is an A-B subunit toxin.

Pertussis toxin also causes a striking **lymphocytosis** in the blood of patients with pertussis. The toxin inhibits signal transduction by chemokine receptors, resulting in a failure of lymphocytes to enter lymphoid tissue such as the spleen and lymph nodes. Because the lymphocytes do not enter lymphoid tissue, there is an increase in their number in the blood (see the discussion of chemokines in Chapter 58). The inhibition of signal transduction by chemokine receptors is also caused by ADP-ribosylation of the G_i protein.

(3) The organisms also synthesize and export adenylate cyclase. This enzyme, when taken up by phagocytic cells (e.g.,

neutrophils), can inhibit their bactericidal activity. Bacterial mutants that lack cyclase activity are avirulent.

(4) Tracheal cytotoxin is a fragment of the bacterial peptidoglycan that damages ciliated cells of the respiratory tract. Tracheal cytotoxin appears to act in concert with endotoxin to induce nitric oxide, which kills the ciliated epithelial cells.

Clinical Findings

Whooping cough is an acute tracheobronchitis that begins with mild upper respiratory tract symptoms followed by a severe paroxysmal cough, which lasts from 1 to 4 weeks. The paroxysmal pattern is characterized by a series of hacking coughs, accompanied by production of copious amounts of mucus, that end with an inspiratory "whoop" as air rushes past the narrowed glottis. Despite the severity of the symptoms, the organism is restricted to the respiratory tract and blood cultures are negative. A pronounced leukocytosis with up to 70% lymphocytes is seen. Although central nervous system anoxia and exhaustion can occur as a result of the severe coughing, death is mainly due to pneumonia.

The classic picture of whooping cough described above occurs primarily in young children. In adults, B. pertussis infection often manifests as a paroxysmal cough of varying severity lasting weeks. The characteristic whoop is often absent, leading to difficulty in recognizing the cough as caused by this organism. In the correct clinical setting, adults with a cough lasting several weeks (often called the 100-day cough) should be evaluated for infection with B. pertussis.

Laboratory Diagnosis

The organism can be isolated from nasopharyngeal swabs taken during the paroxysmal stage. Bordet-Gengou[1] medium used for this purpose contains a high percentage of blood (20%–30%) to inactivate inhibitors in the agar. Identification of the isolated organism can be made by agglutination with specific antiserum or by fluorescent antibody staining. However, the organism grows very slowly in culture, so direct fluorescent antibody staining of the nasopharyngeal specimens can be used for diagnosis. Polymerase chain reaction-based tests are rapid, specific, and highly sensitive and should be used if available.

Isolation of the organism in patients with a prolonged cough is often difficult. Serologic tests that detect antibody in the patient's serum can be used for diagnosis in those patients.

Treatment

Azithromycin is the drug of choice. Note that azithromycin reduces the number of organisms in the throat and decreases the risk of secondary complications but has little effect on the course of the disease at the "prolonged cough" stage because the toxins have already damaged the respiratory mucosa. Supportive care (e.g., oxygen therapy and suction of mucus) during the paroxysmal stage is important, especially in infants.

---

[1]The French scientists who first isolated the organism in 1906.

Prevention

There are two types of vaccines: an acellular vaccine containing purified proteins from the organism and a killed vaccine containing inactivated *B. pertussis* organisms. The **acellular vaccine** contains five antigens purified from the organism. It is the vaccine currently used in the United States.

The main immunogen in this vaccine is inactivated pertussis toxin (pertussis toxoid). The toxoid in the vaccine is pertussis toxin that has been inactivated genetically by introducing two amino acid changes, which eliminates its ADP-ribosylating activity but retains its antigenicity. It is the first vaccine to contain a genetically inactivated toxoid. The other pertussis antigens in the vaccine are filamentous hemagglutinin, pertactin, and fimbriae types 2 and 3. The acellular vaccine has fewer side effects than the killed vaccine but has a shorter duration of immunity. Outbreaks of pertussis are attributed to "waning immunity," a manifestation of the short duration of vaccine-induced immunity.

The pertussis vaccine is usually given combined with diphtheria and tetanus toxoids (DTaP) in three doses beginning at 2 months of age. A booster at 12 to 15 months of age and another at the time of entering school are recommended. Because outbreaks of pertussis have occurred among teenagers, a booster for those between 10 and 18 years old is recommended. A pertussis booster dose is recommended for adults as well. To protect newborns, pregnant women should receive pertussis vaccine. Antipertussis IgG will pass the placenta and protect the newborn.

The killed vaccine is no longer used in the United States because it is suspected of causing various side effects, including postvaccine encephalopathy at a rate of about one case per million doses administered. The killed vaccine is in use in many countries other than the United States.

Azithromycin is useful in prevention of disease in exposed, unimmunized individuals. It should also be given to immunized children younger than 4 years who have been exposed because vaccine-induced immunity is not completely protective.

LEGIONELLA

Disease

L. pneumophila (and other legionellae) causes pneumonia, both in the community and in hospitalized immunocompromised patients. The genus is named after the famous outbreak of pneumonia among people attending the American Legion convention in Philadelphia in 1976 (Legionnaires' disease).

Important Properties

Legionellae are gram-negative rods that **stain faintly with the standard Gram stain**. They do, however, have a gram-negative type of cell wall, and increasing the time of the safranin counterstain enhances visibility. Legionellae in lung biopsy sections do not stain by the standard hematoxylin-and-eosin (H&E)

procedure; therefore, special methods, such as the Dieterle silver impregnation stain, are used to visualize the organisms.

L. pneumophila causes approximately 90% of pneumonia attributed to legionellae. There are 16 serogroups of *L. pneumophila*, with most cases caused by serogroup 1 organisms. There are about 30 other *Legionella* species that cause pneumonia, but most of the remaining 10% of cases are caused by two species, *Legionella micdadei* and *Legionella bozemanii*.

Pathogenesis & Epidemiology

Legionellae are associated chiefly with **environmental water sources** such as air conditioners, hot tubs, and water-cooling towers. Outbreaks of pneumonia in hospitals have been attributed to the presence of the organism in water taps, sinks, and showers.

Legionellae can replicate to large numbers in **free-living amebas** in these water sources. The amebas also enhance the survival of legionellae. Under adverse environmental conditions, the amebas encyst, ensuring both their own survival and the survival of the intracellular legionellae as well.

The portal of entry is the respiratory tract, and pathologic changes occur primarily in the lung. However, in severe cases, bacteremia occurs, accompanied by damage to the vascular endothelium in multiple organs, especially the brain and kidneys. The major virulence factor of the organism is lipopolysaccharide (endotoxin). No exotoxins are produced.

The typical candidate for Legionnaires' disease is an older man who smokes and consumes substantial amounts of alcohol. Patients with acquired immunodeficiency syndrome (AIDS), cancer, or transplants (especially renal transplants) or patients being treated with corticosteroids are predisposed to *Legionella* pneumonia, which indicates that **cell-mediated immunity** is the most important defense mechanism. Despite airborne transmission of the organism, person-to-person spread does *not* occur, as shown by the failure of secondary cases to occur in close contacts of patients.

Clinical Findings

The clinical picture can vary from a mild influenza-like illness to a severe pneumonia accompanied by mental confusion, nonbloody diarrhea, proteinuria, and microscopic hematuria. Although cough is a prominent symptom, sputum is frequently scanty and nonpurulent. Hyponatremia (serum sodium ≤130 mEq/L) is an important laboratory finding that occurs more often in *Legionella* pneumonia than in pneumonia caused by other bacteria. Most cases resolve spontaneously in 7 to 10 days, but in older or immunocompromised patients, the infection can be fatal.

Legionellosis is an **atypical pneumonia**[2] and must be distinguished from other similar pneumonias such as *Mycoplasma* pneumonia, viral pneumonia, psittacosis, and Q fever.

---

[2]A pneumonia is atypical when its causative agent cannot be isolated on ordinary laboratory media or when its clinical picture does not resemble that of typical pneumococcal pneumonia.

Pontiac fever is a mild, flulike form of *Legionella* infection that does not result in pneumonia. The name "Pontiac" is derived from the city in Michigan that was the site of an outbreak in 1968.

Laboratory Diagnosis

Sputum Gram stains reveal many neutrophils but no bacteria. The organism **fails to grow on ordinary media** in a culture of sputum or blood, but it will grow on charcoal-yeast agar, a special medium supplemented with iron and cysteine. Diagnosis usually depends on a significant increase in antibody titer in convalescent-phase serum by the indirect immunofluorescence assay.

Detection of *L. pneumophila* antigens in the urine is a rapid means of making a diagnosis. The urinary antigen test is available only for serogroup 1 organisms. If tissue is available, it is possible to demonstrate *Legionella* antigens in infected lung tissue by using fluorescent antibody staining. The cold-agglutinin titer does not rise in *Legionella* pneumonia, in contrast to pneumonia caused by *Mycoplasma*.

Treatment & Prevention

Azithromycin or erythromycin (with or without rifampin) is the treatment of choice. Certain fluoroquinolones, such as levofloxacin and trovafloxacin, are also drugs of choice. These drugs are effective not only against *L. pneumophila* but also against *Mycoplasma pneumoniae* and *Streptococcus pneumoniae*. The organism frequently produces β-lactamase, and so penicillins and cephalosporins are less effective.

Prevention involves reducing cigarette and alcohol consumption, eliminating aerosols from water sources, and reducing the incidence of *Legionella* in hospital water supplies by using high temperatures and hyperchlorination. There is no vaccine.

ACINETOBACTER

Acinetobacter species are small coccobacillary gram-negative rods found commonly in soil and water, but they also colonize the skin and upper respiratory tract. They are opportunistic pathogens that readily colonize patients with compromised host defenses.

A. baumannii, the species usually involved in human infection, causes disease chiefly in a hospital setting usually associated with respiratory therapy equipment (ventilator-associated pneumonia) and indwelling catheters. Pneumonia and urinary tract infections are the most frequent manifestations. Outbreaks of carbapenem-resistant *A. baumannii* infections have occurred in COVID-19 patients in several hospitals in the United States. Laboratory diagnosis is made by culturing the organism.

A. baumannii is remarkably antibiotic resistant, and some isolates are resistant to all known antibiotics. Imipenem is the drug of choice for infections caused by susceptible strains. Colistin is useful in carbapenem-resistant strains. There is no vaccine. Previous genus names for this organism include *Herellea* and *Mima*.

SELF-ASSESSMENT QUESTIONS

1. Your patient is a 75-year-old man who has smoked cigarettes (two packs a day for more than 50 years) and consumed alcoholic drinks (a six-pack of beer each day) for most of his adult life. He now has the signs and symptoms of pneumonia. Gram stain of the sputum reveals neutrophils but no bacteria. Colonies appear on buffered charcoal yeast (BYCE) agar but not on blood agar. Which one of the following bacteria is most likely to be the cause of his pneumonia?

   (A) *Bordetella pertussis*
   (B) *Haemophilus influenzae*
   (C) *Klebsiella pneumoniae*
   (D) *Legionella pneumophila*
   (E) *Pseudomonas aeruginosa*

2. Regarding the patient in Question 1, which one of the following is the best antibiotic to treat the infection?

   (A) Azithromycin
   (B) Ceftriaxone
   (C) Gentamicin
   (D) Metronidazole
   (E) Piperacillin/tazobactam

3. Your patient is a 6-year-old boy who is complaining that his ear hurts. His mother says this began yesterday and that he has a fever of 103°F. On physical examination, you see a perforated eardrum that is exuding a small amount of pus. Using a swab, you obtain a sample of the pus and do a Gram stain and culture. The Gram stain reveals small coccobacillary rods. There is no growth on a blood agar plate, but a chocolate agar plate supplemented with X and V factors grows small gray colonies. Which one of the following bacteria is the most likely cause of his otitis media?

   (A) *Bordetella pertussis*
   (B) *Haemophilus influenzae*
   (C) *Klebsiella pneumoniae*
   (D) *Legionella pneumophila*
   (E) *Pseudomonas aeruginosa*

4. It's time to play "What's my name?" I am a small gram-negative rod that causes an important respiratory tract disease. I produce an exotoxin that ADP-ribosylates a G protein. One remarkable feature of my disease is a great increase in lymphocytes. I don't cause disease commonly in the United States now because of the widespread use of the vaccine that induces antibodies against five of my proteins, one of which is the exotoxin. The identity of the mystery organism is mostly likely which one of the following?

   (A) *Bordetella pertussis*
   (B) *Haemophilus influenzae*
   (C) *Klebsiella pneumoniae*
   (D) *Legionella pneumophila*
   (E) *Pseudomonas aeruginosa*

5. Your patient is a 75-year-old woman with a 110-pack-year history of cigarette smoking who now has a fever of 39°C and a cough productive of yellowish sputum. Gram stain of the sputum shows small gram-negative rods. There is no growth on blood agar, but colonies do grow on chocolate agar supplemented with hemin and NAD. Which one of the following bacteria is the most likely cause of her pneumonia?

   (A) *Bordetella pertussis*
   (B) *Haemophilus influenzae*
   (C) *Klebsiella pneumoniae*
   (D) *Legionella pneumophila*
   (E) *Pseudomonas aeruginosa*

6. Your patient is a 5-year-old boy with a high fever and signs of respiratory tract obstruction. Visualization of the epiglottis shows inflammation characterized by marked swelling and "cherry-red" appearance. Which one of the following is the best antibiotic to treat the infection?

   (A) Ampicillin
   (B) Ceftriaxone
   (C) Doxycycline
   (D) Gentamicin
   (E) Metronidazole

ANSWERS

(1) **(D)**
(2) **(A)**
(3) **(B)**
(4) **(A)**
(5) **(B)**
(6) **(B)**

SUMMARIES OF ORGANISMS

Brief summaries of the organisms described in this chapter begin on page 675. Please consult these summaries for a rapid review of the essential material.

PRACTICE QUESTIONS: USMLE & COURSE EXAMINATIONS

Questions on the topics discussed in this chapter can be found in the Clinical Bacteriology section of Part XIII: USMLE (National Board) Practice Questions starting on page 737. Also see Part XIV: USMLE (National Board) Practice Examination starting on page 775.

Gram-Negative Rods Related to Animal Sources (Zoonotic Organisms)

INTRODUCTION

Zoonoses are human diseases caused by organisms that are acquired from animals. There are bacterial, viral, fungal, and parasitic zoonoses. Some zoonotic organisms are acquired directly from the animal reservoir, whereas others are transmitted by vectors, such as mosquitoes, fleas, or ticks.

There are four medically important gram-negative rods that have significant animal reservoirs: *Brucella* species, *Francisella tularensis*, *Yersinia pestis*, and *Pasteurella multocida* (Table 20–1).

Additional information regarding the clinical aspects of infections caused by the organisms in this chapter is provided in Part IX, entitled Infectious Diseases beginning on page 603.

BRUCELLA

Disease

Brucella species cause brucellosis (undulant fever).

Important Properties

Brucellae are small gram-negative rods without a capsule. The three major human pathogens and their animal reservoirs are *Brucella melitensis* (goats and sheep), *Brucella abortus* (cattle), and *Brucella suis* (pigs).

Pathogenesis & Epidemiology

The organisms enter the body either by ingestion of **contaminated milk products** especially queso fresco, or **through the skin** by direct contact in an occupational setting such as an abattoir. They localize in the **reticuloendothelial system**, namely, the lymph nodes, liver, spleen, and bone marrow. Many organisms are killed by macrophages, but some survive within these cells, where they are protected from antibody. The host response is granulomatous, with lymphocytes and epithelioid giant cells, which can progress to form focal abscesses. The mechanism of pathogenesis of these organisms is not well defined, except that endotoxin is involved. No exotoxins are produced.

TABLE 20–1 Gram-Negative Rods Associated with Animal Sources

Species	Disease	Source of Human Infection	Mode of Transmission from Animal to Human	Diagnosis
Brucella species	Brucellosis	Pigs, cattle, goats, sheep	Dairy products; contact with animal tissues	Serology or culture
Francisella tularensis	Tularemia	Rabbits, deer, ticks	Contact with animal tissues; ticks	Serology
Yersinia pestis	Plague	Rodents	Flea bite	Immunofluorescence or culture
Pasteurella multocida	Cellulitis	Cats, dogs	Cat or dog bite	Wound culture
Bartonella henselae	Cat-scratch disease and bacillary angiomatosis	Cats	Cat scratch or bite; bite of cat flea	Serology or Warthin-Starry silver stain of tissue

Imported cheese made from unpasteurized goats' milk produced in either Mexico or the Mediterranean region has been a source of *B. melitensis* infection in the United States. The disease occurs worldwide but is rare in the United States because pasteurization of milk kills the organism.

Clinical Findings

After an incubation period of 1 to 3 weeks, nonspecific symptoms such as fever, chills, fatigue, malaise, anorexia, and weight loss occur. The onset can be acute or gradual. The undulating (rising-and-falling) fever pattern that gives the disease its name occurs in a minority of patients. Enlarged lymph nodes, liver, and spleen are frequently found. Pancytopenia occurs. *B. melitensis* infections tend to be more severe and prolonged, whereas those caused by *B. abortus* are more self-limited. Osteomyelitis is the most frequent complication. Secondary spread from person to person is rare.

Laboratory Diagnosis

Blood cultures or bone marrow cultures typically yield the organism. Recovery of the organism requires the use of enriched culture media and incubation in 10% CO_2. The organisms can be presumptively identified by using a slide agglutination test with *Brucella* antiserum, and the species can be identified by biochemical tests. If organisms are not isolated, analysis of a serum sample from the patient for a rise in antibody titer to *Brucella* can be used to make a diagnosis. In the absence of an acute-phase serum specimen, a titer of at least 1:160 in the convalescent-phase serum sample is diagnostic. A PCR test is available.

Treatment & Prevention

The treatment of choice is doxycycline to which either streptomycin or rifampin is added. There is no significant resistance to these drugs. Prevention of brucellosis involves pasteurization of milk, immunization of animals, and slaughtering of infected animals. There is no human vaccine.

FRANCISELLA

Disease

F. tularensis causes tularemia.

Important Properties

F. tularensis is a small, pleomorphic, gram-negative rod. It has a single serologic type. There are two biotypes, A and B, which are distinguished primarily on their virulence and epidemiology. Type A is more virulent and found primarily in the United States, whereas type B is less virulent and found primarily in Europe. Type A is associated with rabbits, whereas type B is more commonly found in rodents and water sources.

Pathogenesis & Epidemiology

F. tularensis is remarkable in the wide variety of animals that it infects and in the breadth of its distribution in the United States. It is enzootic (endemic in animals) in every state, but most human cases occur in the rural areas of Arkansas and Missouri. It has been isolated from more than 100 different species of **wild animals**, the most important of which are rabbits, deer, and a variety of rodents.

The bacteria are transmitted among these animals by vectors such as **biting flies** and **ticks**, especially the *Dermacentor* ticks that feed on the blood of wild rabbits. The tick maintains the chain of transmission by passing the bacteria to its offspring by the transovarian route. In this process, the bacteria are passed through ovum, larva, and nymph stages to adult ticks capable of transmitting the infection.

Humans are accidental "dead-end" hosts who acquire the infection most often by being bitten by the vector or by having skin contact with the animal during removal of the hide. Rarely, the organism is ingested in infected meat, causing gastrointestinal tularemia, or is inhaled, causing pneumonia. Rarely, the person is infected by a penetrating skin lesion in a water environment, such as a fish-hook injury. There is no person-to-person spread. The main type of tularemia in the United States is tick-borne tularemia from a rabbit reservoir.

The organism enters through the skin, forming an ulcer at the site in most cases. It then localizes to the cells of the reticuloendothelial system, and granulomas are formed. Caseation necrosis and abscesses can also occur. Symptoms are caused primarily by endotoxin. No exotoxins have been identified.

Clinical Findings

Presentation can vary from sudden onset of an influenza-like syndrome to prolonged onset of a low-grade fever and adenopathy. Approximately 75% of cases are the "ulceroglandular" type, in which the site of entry ulcerates and the regional lymph nodes are swollen and painful. Other, less frequent forms of tularemia include glandular, oculoglandular, typhoidal, gastrointestinal, and pulmonary. Disease usually confers lifelong immunity.

Laboratory Diagnosis

Attempts to culture the organism in the laboratory are rarely undertaken, because there is a high risk to laboratory workers of infection by inhalation, and the special cysteine-containing medium required for growth is not usually available. The most frequently used diagnostic method is the agglutination test with acute- and convalescent-phase serum samples. Fluorescent antibody staining of infected tissue can be used if available.

Treatment & Prevention

Streptomycin is the drug of choice. There is no significant antibiotic resistance. Prevention involves avoiding both being bitten by ticks and handling wild animals. There is a live, attenuated bacterial vaccine that is given only to persons, such as fur trappers, whose occupation brings them into close contact with wild animals. The vaccine is experimental and not available commercially but can be obtained from the U.S. Army Medical Research Command, Fort Detrick, Maryland. This and the bacillus of Calmette-Guérin (BCG) vaccine for tuberculosis are the only two live bacterial vaccines for human use.

YERSINIA

Disease

Y. pestis is the cause of plague, also known as the black death, the scourge of the Middle Ages. It is also a contemporary disease, occurring in the western United States and in many other countries around the world. Two less important species, *Yersinia enterocolitica* and *Yersinia pseudotuberculosis*, are described in Chapter 27.

Important Properties

Y. pestis is a small gram-negative rod that exhibits bipolar staining (i.e., it **resembles a safety pin**, with a central clear area). Freshly isolated organisms possess a capsule composed of a polysaccharide–protein complex. The capsule can be lost with passage in the laboratory; loss of the capsule is accompanied by a loss of virulence. It is one of the **most virulent** bacteria known and has a strikingly low ID_{50} (i.e., 1–10 organisms are capable of causing disease).

Pathogenesis & Epidemiology

The plague bacillus has been endemic in the wild rodents of Europe and Asia for thousands of years but entered North America in the early 1900s, probably carried by a rat that jumped ship at a California port. It is now endemic in the wild rodents in the western United States, although 99% of cases of plague occur in Southeast Asia.

The enzootic (sylvatic) cycle consists of transmission among **wild rodents by fleas**. In the United States, prairie dogs are the main reservoir. Rodents are relatively resistant to disease; most are asymptomatic. Humans are accidental hosts, and cases of plague in this country occur as a result of being bitten by a flea that is part of the sylvatic cycle.

The urban cycle, which does not occur in the United States, consists of transmission of the bacteria among urban rats (the reservoir), with the **rat flea** as vector. This cycle predominates during times of poor sanitation (e.g., wartime), when rats proliferate and come in contact with the fleas in the sylvatic cycle.

The events within the flea are fascinating as well as essential. The flea ingests the bacteria while taking a blood meal from a bacteremic rodent. A thick biofilm containing many organisms forms in the upper gastrointestinal tract that prevents any food from proceeding down the gastrointestinal tract of the flea. This "blocked flea" then regurgitates the organisms into the bloodstream of the next animal or human it bites.

The organisms inoculated at the time of the bite spread to the regional lymph nodes, which become swollen and tender. These swollen lymph nodes are the **buboes** that have led to the name **bubonic plague**. The organisms can reach high concentrations in the blood (bacteremia) and disseminate to form abscesses in many organs. The **endotoxin-related symptoms**, including disseminated intravascular coagulation and cutaneous hemorrhages, probably were the genesis of the term **black death**.

In addition to the sylvatic and urban cycles of transmission, respiratory droplet transmission of the organism from patients with pneumonic plague can occur.

The organism has several factors that contribute to its virulence: (1) the envelope capsular glycoprotein, called F-1, which protects against phagocytosis; (2) endotoxin; (3) an exotoxin; and (4) two proteins known as V antigen and W antigen that are essential for virulence. The V antigen regulates the secretion of various proteins from the organism into the human cell and suppresses the induction of tumor necrosis factor (TNF) and interferon-gamma, two important components of the host defense response. The action of the exotoxin is unknown.

Other factors that contribute to the extraordinary pathogenicity of *Y. pestis* are a group of virulence factors collectively called **Yops (*Yersinia* outer proteins)**. These are injected into the human cell via type III secretion systems and inhibit phagocytosis and cytokine production by macrophages and neutrophils. For example, one of the Yops proteins (YopJ) is a protease that cleaves two signal transduction pathway proteins required for the induction of TNF synthesis. This inhibits the activation of our host defenses and contributes to the ability of the organism to replicate rapidly within the infected individual.

Clinical Findings

Bubonic plague, which is the most frequent form, begins with pain and swelling of the lymph nodes draining the site of the flea bite and systemic symptoms such as high fever, myalgias, and prostration. The affected nodes enlarge and become exquisitely tender. These buboes are an early characteristic finding. Septic shock and pneumonia are the main life-threatening subsequent events. Pneumonic plague can arise either from inhalation of an aerosol or from septic emboli that reach the lungs. Untreated bubonic plague is fatal in approximately half of the cases, and untreated pneumonic plague is invariably fatal.

Laboratory Diagnosis

Smear and culture of blood or pus from the bubo is the best diagnostic procedure. Great care must be taken by the physician during aspiration of the pus and by laboratory workers doing the culture not to create an aerosol that might transmit the infection. Giemsa or Wayson stain reveals the typical safety-pin appearance of the organism better than does Gram stain. Fluorescent antibody staining can be used to identify the organism in tissues. A rise in antibody titer to the envelope antigen can be useful retrospectively.

Treatment

The treatment of choice is a combination of streptomycin and a tetracycline such as doxycycline, although streptomycin alone can be used. Levofloxacin can also be used. There is no significant antibiotic resistance. In view of the rapid progression of the disease, treatment should not wait for the results of the bacteriologic culture. Incision and drainage of the buboes are not usually necessary.

Prevention

Prevention of plague involves controlling the spread of rats in urban areas, preventing rats from entering the country by ship or airplane, and avoiding both flea bites and contact with dead wild rodents. A patient with plague must be placed in strict isolation (quarantine) for 72 hours after antibiotic therapy is started. Only close contacts need to receive prophylactic tetracycline, but all contacts should be observed for fever. Reporting a case of plague to the public health authorities is mandatory.

A vaccine consisting of formalin-killed organisms provides partial protection against bubonic but not pneumonic plague. This vaccine was used in the armed forces during the Vietnam War but is not recommended for tourists traveling to Southeast Asia.

PASTEURELLA

Disease

P. multocida causes wound infections associated with cat and dog bites.

Important Properties

P. multocida is a short, encapsulated gram-negative rod that exhibits bipolar staining.

Pathogenesis & Epidemiology

The organism is part of the normal flora in the mouths of many animals, particularly **domestic cats and dogs**, and is transmitted by **biting**. About 25% of animal bites become infected with the organism, with sutures acting as a predisposing factor to infection. Most bite infections are polymicrobial, with a variety of facultative anaerobes, especially *Streptococcus* species, and anaerobic organisms present in addition to *P. multocida*. Pathogenesis is not well understood, except that the capsule is a virulence factor and endotoxin is present in the cell wall. No exotoxins are made.

Clinical Findings

A rapidly spreading cellulitis at the site of an animal bite is indicative of *P. multocida* infection. The incubation period is

brief, usually less than 24 hours. Osteomyelitis can complicate cat bites in particular, because cats' sharp, pointed teeth can implant the organism under the periosteum.

Laboratory Diagnosis

The diagnosis is made by finding the organism in a culture of a sample from the wound site.

Treatment & Prevention

Penicillin G is the treatment of choice. There is no significant antibiotic resistance. People who have been bitten by a cat should be given ampicillin to prevent *P. multocida* infection. Animal bites, especially cat bites, should not be sutured.

BARTONELLA

Disease

Bartonella henselae is the cause of cat-scratch disease (CSD) and bacillary angiomatosis (BA). CSD is one of the most common zoonotic diseases in the United States.

Important Properties

B. henselae is a small, pleomorphic gram-negative rod. It is a fastidious organism and will not grow on routine blood agar. It can be cultured on specialized media in the clinical laboratory.

Pathogenesis & Epidemiology

Cat scratches or bites, especially from kittens, are the main mode of transmission of *B. henselae* to humans. The organism is a member of the oral flora of many cats. There is evidence that it is transmitted from cats to humans by the bite of cat fleas. Exposure to cat urine or feces does not pose a risk of transmission. Person-to-person transmission of *B. henselae* does not play a significant role in infection. *B. henselae* is a low-virulence organism, and disease is self-limited in immunocompetent individuals.

The pathogenesis of angiomas that occur in *Bartonella* infections in immunocompromised individuals is uncertain. One current explanation is that infection of endothelial cells by *Bartonella* induces the synthesis of angiogenesis factor that causes endothelial cells to proliferate.

Clinical Findings

In immunocompetent people, *B. henselae* causes **CSD**. This disease is characterized by fever and tender, enlarged lymph nodes, typically on the same side as the scratch (Figure 20–1). A papule at the site of the scratch may precede the lymphadenopathy. CSD has a prolonged course but eventually resolves, even without antibiotics. A small percentage of those infected develop systemic disease, such as endocarditis or encephalitis.

In immunocompromised individuals, especially patients with acquired immunodeficiency syndrome (AIDS), *B. henselae*

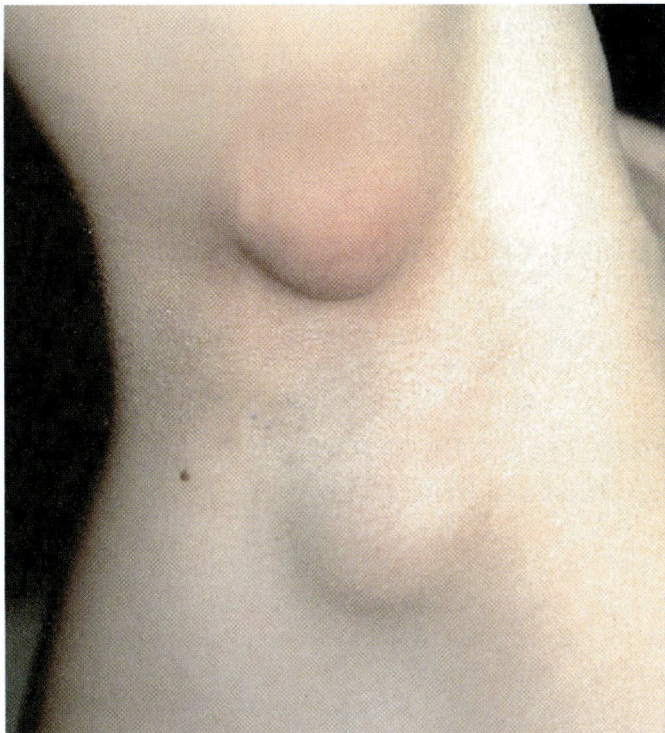

FIGURE 20–1 Cat-scratch disease. Note the two enlarged, inflamed axillary lymph nodes in a patient with cat-scratch disease. (Reproduced with permission from Wolff K, Johnson R: *Fitzpatrick's Color Atlas & Synopsis of Clinical Dermatology*, 6th ed. New York, NY: McGraw Hill; 2009.)

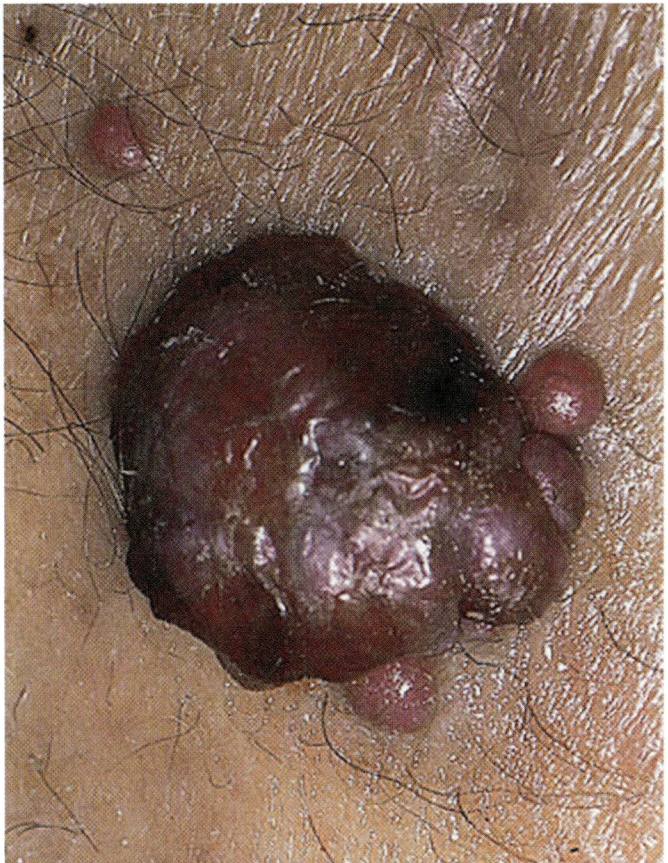

FIGURE 20–2 Bacillary angiomatosis. Note the cherry-red hemangioma-like skin lesion. (Reproduced with permission from Wolff K, Johnson R: *Fitzpatrick's Color Atlas & Synopsis of Clinical Dermatology*, 6th ed. New York, NY: McGraw Hill; 2009.)

causes **BA**. BA is characterized by raised, cherry-red vascular lesions in the skin and visceral organs (Figure 20–2). The lesions appear papular or nodular. Bacillary peliosis (peliosis hepatis) is similar to BA except that in peliosis, the lesions occur primarily in the liver and spleen.

Laboratory Diagnosis

The diagnosis of CSD is usually made serologically. Antibodies against *B. henselae* antigens can be detected in a patient's serum by a variety of immunologic tests. The organism can be cultured on artificial media but takes 5 days or longer to grow and so is not usually done. The diagnosis of BA is often made by finding pleomorphic rods in biopsy tissue using the Warthin-Starry silver stain. Pathologic examination of tissue from the lesion will distinguish BA from Kaposi's sarcoma. A PCR test for *B. hensalae* is available.

Treatment & Prevention

No antibiotic therapy is typically recommended for CSD. If the patient has severe lymphadenitis, azithromycin is the drug of choice. Treatment of BA with doxycycline or erythromycin is effective. There is no significant antibiotic resistance. Antibiotics are not recommended for people who have sustained a cat scratch. There is no vaccine.

SELF-ASSESSMENT QUESTIONS

1. Your patient is a 10-year-old boy who has a high fever and swollen, painful axillary lymph nodes on the left side. His mother says that he brought home a dead rat a few days ago. You suspect he may have bubonic plague. Regarding the causative organism, which one of the following is most accurate?

   (A) It has a very low ID_{50}.
   (B) It is transmitted from rodents to humans by ticks.
   (C) It is endemic primarily in the states along the East Coast of the United States.
   (D) Its main virulence factor is an exotoxin that induces interleukin-2 (IL-2) production by CD4-positive helper T cells.
   (E) Infection should be treated with high doses of penicillin G intravenously.

2. Your patient is a 20-year-old man who was bitten on the hand when he tried to break up a fight between two cats yesterday. He now has a red, hot, tender, swollen lesion at the bite site that has spread rapidly across his hand. Which one of the following bacteria is the most likely cause of his cellulitis?

   (A) *Brucella melitensis*
   (B) *Francisella tularensis*
   (C) *Pasteurella multocida*
   (D) *Yersinia pestis*

3. Your patient is a 30-year-old woman who reports that she has had intermittent fever of 102°F, sweating, and fatigue for the past 1 month or so. She has lost her appetite and has lost about 10 pounds in that period. She enjoys eating unpasteurized goat cheese. On examination, hepatosplenomegaly is detected. A blood count reveals pancytopenia. Which one of the following bacteria is the most likely cause of this infection?

   (A) *Brucella melitensis*
   (B) *Francisella tularensis*
   (C) *Pasteurella multocida*
   (D) *Yersinia pestis*

4. Regarding *Bartonella henselae*, which of the following is most accurate?

   (A) *Bartonella henselae* is an anaerobic, spore-forming, gram-positive rod.
   (B) The natural habitat of *Bartonella henselae* is the cat's mouth.
   (C) *Bartonella henselae* causes cellulitis in immunocompromised patients such as AIDS patients.
   (D) Diagnosis in the clinical laboratory depends on detecting antibodies in the patient's serum that will agglutinate cardiolipin.
   (E) The drug of choice for *Bartonella henselae* infections is metronidazole.

ANSWERS

(1) **(A)**
(2) **(C)**
(3) **(A)**
(4) **(B)**

SUMMARIES OF ORGANISMS

Brief summaries of the organisms described in this chapter begin on page 675. Please consult these summaries for a rapid review of the essential material.

PRACTICE QUESTIONS: USMLE & COURSE EXAMINATIONS

Questions on the topics discussed in this chapter can be found in the Clinical Bacteriology section of Part XIII: USMLE (National Board) Practice Questions starting on page 737. Also see Part XIV: USMLE (National Board) Practice Examination starting on page 775.

Mycobacteria

C H A P T E R

21

CHAPTER CONTENTS

Introduction
Mycobacterium tuberculosis
Atypical Mycobacteria
Mycobacterium leprae

Self-Assessment Questions
Summaries of Organisms
Practice Questions: USMLE & Course Examinations

INTRODUCTION

Mycobacteria are aerobic, **acid-fast** bacilli (rods) (Figure 21–1). They are neither gram positive nor gram negative (i.e., they are stained poorly by the dyes used in Gram stain). They are virtually the only bacteria that are acid-fast. (One exception is *Nocardia asteroides*, the major cause of nocardiosis, which is also acid-fast.)

The term *acid-fast* refers to an organism's ability to retain the carbol fuchsin stain despite subsequent treatment with an ethanol–hydrochloric acid mixture. The high lipid content (approximately 60%) of their cell wall makes mycobacteria acid-fast.

The major pathogens are *Mycobacterium tuberculosis*, the cause of tuberculosis, and *Mycobacterium leprae*, the cause of leprosy. Atypical mycobacteria, such as *Mycobacterium*

avium-intracellulare complex (MAI, MAC) and *Mycobacterium kansasii*, can cause tuberculosis-like disease but are less frequent pathogens. Rapidly growing mycobacteria, such as *Mycobacterium chelonae*, occasionally cause human disease in immunocompromised patients or those in whom prosthetic devices have been implanted (Table 21–1). The clinical features of three important mycobacteria are described in Table 21–2.

Additional information regarding the clinical aspects of infections caused by the organisms in this chapter is provided in Part IX, entitled Infectious Diseases beginning on page 603.

MYCOBACTERIUM TUBERCULOSIS

Disease

This organism causes tuberculosis. Worldwide, *M. tuberculosis* causes more deaths than any other single microbial agent. Approximately one-third of the world's population is infected with this organism. Each year, it is estimated that 1.7 million people die of tuberculosis and that 9 million new cases occur. An estimated 500,000 people are infected with a multidrug-resistant strain of *M. tuberculosis*.

Important Properties

M. tuberculosis **grows slowly** (i.e., it has a doubling time of 18 hours, in contrast to most bacteria, which can double in number in 1 hour or less). Because growth is so slow, cultures of clinical specimens must be held for 6 to 8 weeks before being recorded as negative. *M. tuberculosis* can be cultured on bacteriologic media, whereas *M. leprae* cannot. Media used for its growth (e.g., Löwenstein-Jensen medium) contain complex nutrients (e.g., egg yolk) and dyes (e.g., malachite green). The dyes inhibit the unwanted normal flora present in sputum samples.

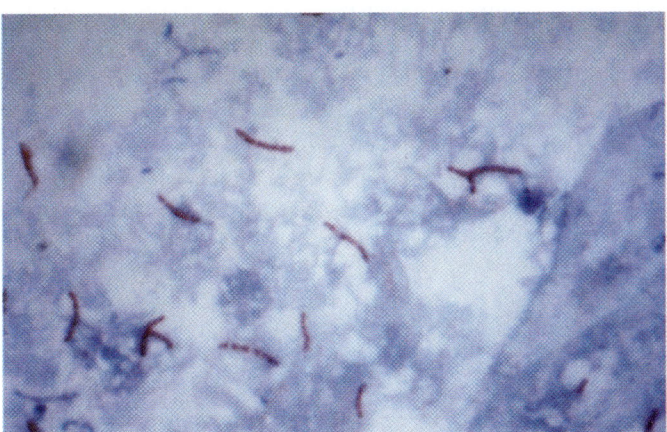

FIGURE 21–1 *Mycobacterium tuberculosis*—acid-fast stain. Long red rods of *M. tuberculosis* are seen on a blue background. (Used with permission from Dr. George Kubica, Public Health Image Library, Centers for Disease Control and Prevention.)

TABLE 21–1 Medically Important Mycobacteria

Species	Growth on Bacteriologic Media	Preferred Temperature In Vivo (°C)	Source or Mode of Transmission
M. tuberculosis	Slow (weeks)	37	Respiratory droplets
M. bovis	Slow (weeks)	37	Milk from infected animals
M. leprae	None	32	Prolonged close contact
Atypical mycobacteria[1]			
M. kansasii	Slow (weeks)	37	Soil and water
M. marinum	Slow (weeks)	32	Water
M. avium-intracellulare complex	Slow (weeks)	37	Soil and water
M. fortuitum-chelonae complex	Rapid (days)	37	Soil and water

[1]Only representative examples are given.

M. tuberculosis is an **obligate aerobe**; this explains its predilection for causing disease in highly oxygenated tissues such as the upper lobe of the lung and the kidney. The acid-fast property of *M. tuberculosis* (and other mycobacteria) is attributed to long-chain (**C78–C90**) fatty acids called **mycolic acids** in the cell wall.

Cord factor (trehalose dimycolate) is correlated with virulence of the organism. Virulent strains grow in a characteristic "serpentine" cordlike pattern, whereas avirulent strains do not. The organism also contains several proteins, which, when combined with waxes, elicit delayed hypersensitivity. These proteins are the antigens in the **purified protein derivative (PPD)** skin test (also known as the tuberculin skin test). A lipid located in the bacterial cell wall called phthiocerol dimycocerosate is required for pathogenesis in the lung.

M. tuberculosis is relatively resistant to acids and alkalis. NaOH is used to concentrate clinical specimens; it destroys unwanted bacteria, human cells, and mucus but not the organism. *M. tuberculosis* is resistant to dehydration and therefore survives in dried expectorated sputum; this property may be important in its transmission by aerosol.

Strains of *M. tuberculosis* resistant to the main antimycobacterial drug, isoniazid (**isonicotinic acid hydrazide, INH**), as well as strains resistant to multiple antibiotics (called **multidrug-resistant** or **MDR** strains), have become a worldwide problem. This resistance is attributed to one or more chromosomal mutations, because no plasmids have been found in this organism. One of these mutations is in a gene for mycolic acid synthesis, and another is in a gene for catalase-peroxidase, an enzyme required to activate INH within the bacterium. Resistance to rifampin is caused by mutations in the gene encoding the RNA polymerase of the organism.

Transmission & Epidemiology

M. tuberculosis is transmitted from person to person by respiratory aerosols produced by coughing. The source of the organism is a cavity in the lung that has eroded into a bronchus. The portal of entry is the respiratory tract, and the initial site of infection is the lung. In tissue, it resides chiefly within reticuloendothelial cells (e.g., **macrophages**). Macrophages kill most, but not all, of the infecting organisms. The ones that survive can continue to infect other adjacent cells or can disseminate to other organs.

Humans are the natural reservoir of *M. tuberculosis*. Although some animals, such as cattle, can be infected, they are not the main reservoir for human infection. Most transmission occurs by aerosols generated by the coughing of "smear-positive" people (i.e., those whose sputum contains detectable bacilli in the acid-fast stain). However, about 20% of people are infected by aerosols produced by the coughing of "smear-negative" people.

In the United States, tuberculosis is almost exclusively a human disease. In developing countries, *Mycobacterium bovis* also causes tuberculosis in humans. *M. bovis* is found in cow's milk, which, unless pasteurized, can cause gastrointestinal tuberculosis in humans.

The disease tuberculosis occurs in only a small number of infected individuals. In the United States, most cases of tuberculosis are associated with reactivation in elderly, malnourished men. The risk of infection and disease is highest among socioeconomically disadvantaged people, who have poor housing and poor nutrition. These factors, rather than genetic ones, probably account for the high rate of infection among Native Americans, African Americans, and Native Alaskans.

TABLE 21–2 Clinical Features of Important Mycobacteria

Organism	Main Site of Infection	Skin Test in Common Use	Multiple-Drug Therapy Used	Vaccine Available
M. tuberculosis	Lungs	Yes	Yes	Yes
M. avium-intracellulare	Lungs	No	Yes	No
M. leprae	Skin, nerves	No	Yes	No

TABLE 21-3 Risk Factors for Infection and Reactivation

Risk Factors for Infection	Risk Factors for Reactivation
Foreign born/residence in a country with high rate of tuberculosis	HIV/AIDS
Close contact with a person with active disease	TNF-α blocker drugs such as infliximab
Homeless or reside in homeless shelter	Drugs to prevent transplant rejection
Incarceration in jail or prison	Corticosteroids
Intravenous drug use	Diabetes
Health-care worker	Smoking

AIDS = acquired immunodeficiency syndrome; HIV = human immunodeficiency virus; TNF-α = tumor necrosis factor-α.

In the United States, there are approximately 15 million people with latent tuberculosis and 10,000 cases of active disease. Most cases of active disease in the United States are caused by reactivation of latent infection. The risk factors for infection and reactivation (progression) to disease are listed in Table 21–3.

Pathogenesis

An overall scheme of pathogenesis by *M. tuberculosis* is shown in Figure 21–2. It describes primary tuberculosis, which typically results in a Ghon focus in the lower lung. Primary tuberculosis can heal by fibrosis, can lead to progressive lung disease, can cause bacteremia and miliary tuberculosis, or can cause hematogenous dissemination resulting in no immediate disease but with the risk of reactivation in later life.

If the primary infection heals without causing disease, it is called a **latent infection**. Of those exposed to *M. tuberculosis*, approximately 90% develop latent infection and approximately 10% develop disease. Of those who have latent infection, approximately 10% progress to active disease (reactivation) at a later time, whereas 90% remain latent.

Figure 21–2 also describes secondary tuberculosis with a cavity in the upper lobes. This can cause disease directly or result in reactivation disease in later life with central nervous system lesions, vertebral osteomyelitis (Pott's disease), or involvement of other organs.

M. tuberculosis produces no well-recognized exotoxins and does not contain endotoxin in its cell wall. However, *M. tuberculosis* produces two proteins that appear to play a role in pathogenesis. One is tuberculosis necrotizing toxin (TNT), which cleaves nicotinamide adenine dinucleotide (NAD) within macrophages resulting in death of the infected macrophage. The other is early secreted antigen-6 (ESAT-6), a protein that reduces the innate immune response by reducing gamma interferon production, thereby enhancing the virulence of the organism. The precise role of these proteins in pathogenesis remains to be determined.

The organism preferentially infects **macrophages** and other reticuloendothelial cells. *M. tuberculosis* survives and multiplies within a cellular vacuole called a phagosome. The organism produces a protein called exported repetitive protein that prevents the phagosome from fusing with the lysosome, thereby allowing the organism to escape the degradative enzymes in the lysosome.

Lesions are dependent on the presence of the organism and the host response. There are two types of lesions:

(1) **Exudative lesions**, which consist of an acute inflammatory response and occur chiefly in the lungs at the initial site of infection.

(2) **Granulomatous lesions**, which consist of a central area of giant cells containing tubercle bacilli surrounded by a zone of epithelioid cells. These giant cells, called **Langhans' giant cells**, are an important pathologic finding in tuberculous lesions. A **tubercle** is a granuloma surrounded by fibrous tissue that has undergone central **caseation necrosis**. Tubercles heal by fibrosis and calcification.

The primary lesion of tuberculosis usually occurs in the lungs. The parenchymal exudative lesion and the draining lymph nodes together are called a **Ghon complex**. Primary lesions usually occur in the **lower lobes** because airflow to those areas is greatest. In contrast, reactivation lesions usually occur in the **apices** because oxygenation is greatest there and lymph flow is low allowing the organism to remain at the site. Reactivation lesions also occur in other well-oxygenated sites such as the kidneys, brain, and bone. Reactivation is seen primarily in immunocompromised or debilitated patients.

Spread of the organism within the body occurs by two mechanisms:

(1) A tubercle can erode into a bronchus, empty its caseous contents, and thereby spread the organism to other parts of the lungs, to the gastrointestinal tract if swallowed, and to other persons if expectorated.

(2) It can disseminate via the bloodstream to many internal organs. Dissemination can occur at an early stage if cell-mediated immunity fails to contain the initial infection or at a late stage if a person becomes immunocompromised.

Immunity & Hypersensitivity

After recovery from the primary infection, resistance to the organism is mediated by **cellular immunity** (i.e., by CD4-positive T cells and macrophages). The CD4-positive T cells are Th-1 helper T cells (see Chapter 58).

Circulating antibodies also form, but they play no role in resistance and are not used for diagnostic purposes. Patients deficient in cellular immunity, such as patients with acquired immunodeficiency syndrome (AIDS), are at much higher risk for disseminated, life-threatening tuberculosis. Mutations in the interferon-γ receptor gene are another cause of defective cellular immunity that predisposes to severe tuberculosis. This emphasizes the importance of activation of macrophages by interferon-γ in the host defense against *M. tuberculosis*.

Prior infection can be detected by a positive tuberculin skin test result, which is due to a delayed hypersensitivity reaction. **PPD** is used as the antigen in the tuberculin skin test. The intermediate-strength preparation of PPD, which contains five

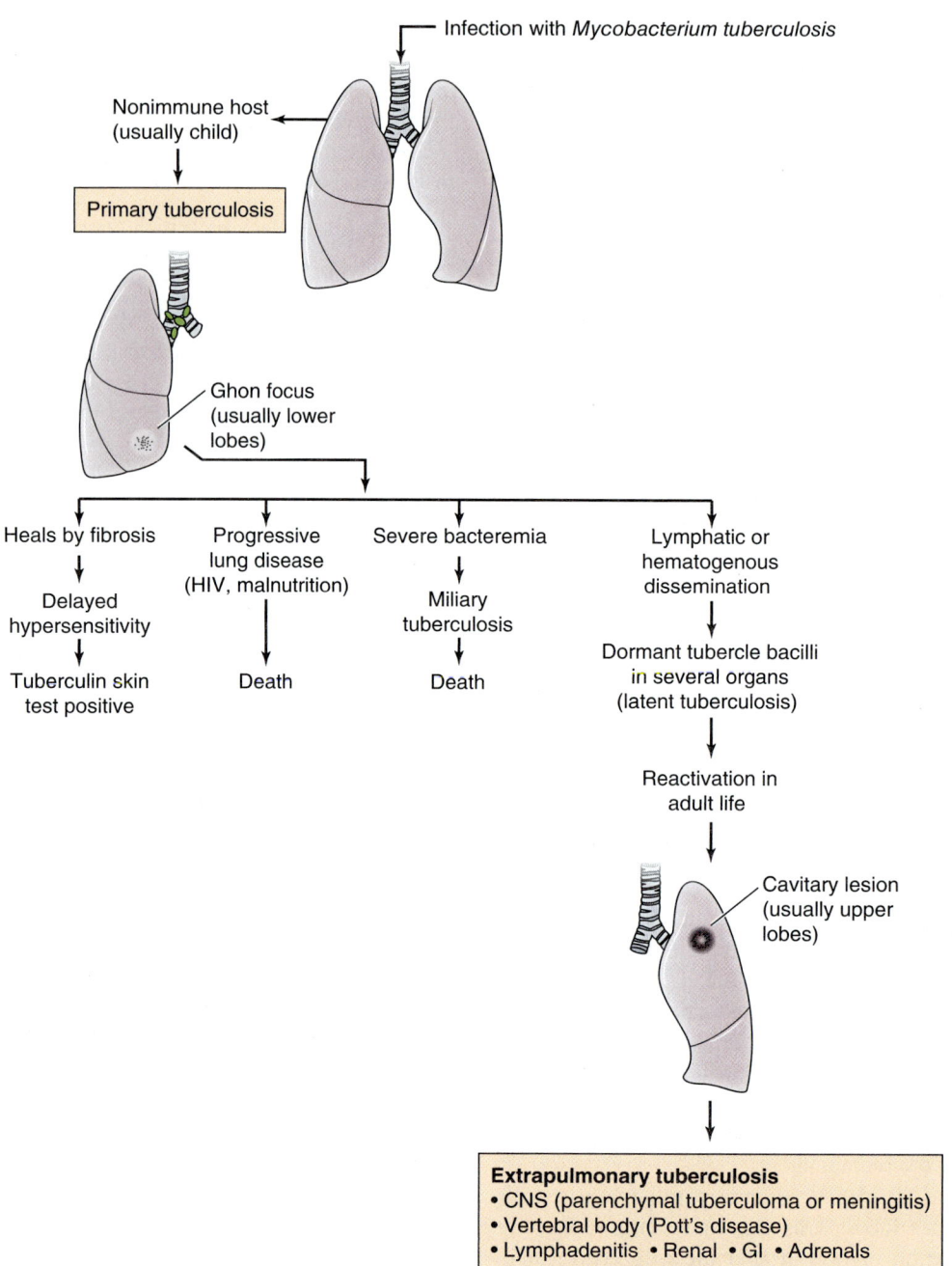

Infection with *Mycobacterium tuberculosis*

Nonimmune host
(usually child)

Primary tuberculosis

Ghon focus
(usually lower
lobes)

Heals by fibrosis → Delayed hypersensitivity → Tuberculin skin test positive

Progressive lung disease (HIV, malnutrition) → Death

Severe bacteremia → Miliary tuberculosis → Death

Lymphatic or hematogenous dissemination → Dormant tubercle bacilli in several organs (latent tuberculosis) → Reactivation in adult life

Cavitary lesion
(usually upper
lobes)

Extrapulmonary tuberculosis
• CNS (parenchymal tuberculoma or meningitis)
• Vertebral body (Pott's disease)
• Lymphadenitis • Renal • GI • Adrenals

FIGURE 21–2 Pathogenesis by *Mycobacterium tuberculosis*. CNS = central nervous system; GI = gastrointestinal; HIV = human immunodeficiency virus. (Reproduced with permission from Le T, Bhushan V, Sochat M: *First Aid for the USMLE Step 1*, 25th ed. New York, NY: McGraw Hill; 2015.)

tuberculin units, is usually used. The skin test is evaluated by measuring the diameter of the **induration** surrounding the skin test site (Figure 21–3). Note that induration (thickening), not simply erythema (reddening), must be observed.

The diameter required to judge the test as positive varies depending on the status of the individual being tested. Induration of 15 mm or more is positive in a person who has no known risk factors. Induration of 10 mm or more is positive in a person with high-risk factors, such as a homeless person, an intravenous drug user, or a nursing home resident. Induration of 5 mm

or more is positive in a person who has deficient cell-mediated immunity (e.g., AIDS patients) or has been in close contact with a person with active tuberculosis.

A positive skin test result indicates previous infection by the organism but not necessarily active disease. The tuberculin test becomes positive 4 to 6 weeks after infection. Immunization with bacillus Calmette-Guérin (BCG) vaccine (see page 182) can cause a positive test, but the reactions are usually only 5 to 10 mm and tend to decrease with time. People with PPD reactions of 15 mm or more are assumed to be infected with

FIGURE 21–3 Tuberculin skin test. Purified protein derivative (PPD) was injected intradermally, and 48 hours later, the diameter of induration was measured with a caliper. (Reproduced with permission from Talaro KP: *Foundations in Microbiology.* 8th ed. New York, NY: McGraw Hill; 2011.)

M. tuberculosis even if they have received the BCG vaccine. A positive skin test reverts to negative in about 5% to 10% of people. Reversion to negative is more common in the United States now than many years ago because now a person is less likely to be exposed to the organism and therefore less likely to receive a boost to the immune system.

The skin test itself does *not* induce a positive response in a person who has not been exposed to the organism. It can, however, "boost" a weak or negative response in a person who has been exposed to produce a positive reaction. The clinical implications of this "booster effect" are beyond the scope of this book.

Tuberculin reactivity is mediated by the cellular arm of the immune system; it can be transferred by CD4-positive T cells but not by serum. Infection with measles virus can suppress cell-mediated immunity, resulting in a loss of tuberculin skin test reactivity and, in some instances, reactivation of dormant organisms and clinical disease.

Clinical Findings

The clinical findings are varied, and many organs can be involved, but the lungs are the main site of infection. Constitutional symptoms such as fever, fatigue, night sweats, and weight loss are common. Anemia of chronic infection can occur.

In **pulmonary tuberculosis**, the main findings are cough and hemoptysis. The chest X-ray findings in reactivation tuberculosis of the lung include an infiltrate in the upper lobe with or without a cavity.

Scrofula is mycobacterial cervical lymphadenitis that presents as swollen, nontender lymph nodes, usually unilaterally. *M. tuberculosis* causes most cases of scrofula, but nontuberculous mycobacteria (NTM), such as *Mycobacterium scrofulaceum*, can also cause scrofula. Lymphadenitis is the most common extrapulmonary manifestation of tuberculosis. Patients infected with

human immunodeficiency virus (HIV) are more likely to have multifocal lymphadenitis than those not infected with HIV.

Erythema nodosum, characterized by tender nodules along the extensor surfaces of the tibia and ulna, is a manifestation of primary infection seen in patients who are controlling the infection with a potent cell-mediated response (Figure 21–4). **Miliary tuberculosis** is characterized by multiple disseminated lesions that resemble millet seeds. **Tuberculous meningitis** and **tuberculous osteomyelitis**, especially vertebral osteomyelitis (Pott's disease), are important disseminated forms.

Gastrointestinal tuberculosis is characterized by abdominal pain and diarrhea accompanied by more generalized symptoms of fever and weight loss. Intestinal obstruction or hemorrhage may occur. The ileocecal region is the site most often involved. Tuberculosis of the gastrointestinal tract can be caused by either *M. tuberculosis* when it is swallowed after being coughed up from a lung lesion or by *M. bovis* when it is ingested in unpasteurized milk products. **Oropharyngeal tuberculosis** typically presents as a painless ulcer accompanied by local adenopathy.

In **renal tuberculosis**, dysuria, hematuria, and flank pain occur. "Sterile pyuria" is a characteristic finding. The urine contains white blood cells, but cultures for the common urinary tract bacterial pathogens show no growth. However, mycobacterial cultures are often positive.

Note that most (approximately 90%) infections with *M. tuberculosis* are asymptomatic. Asymptomatic infections, also

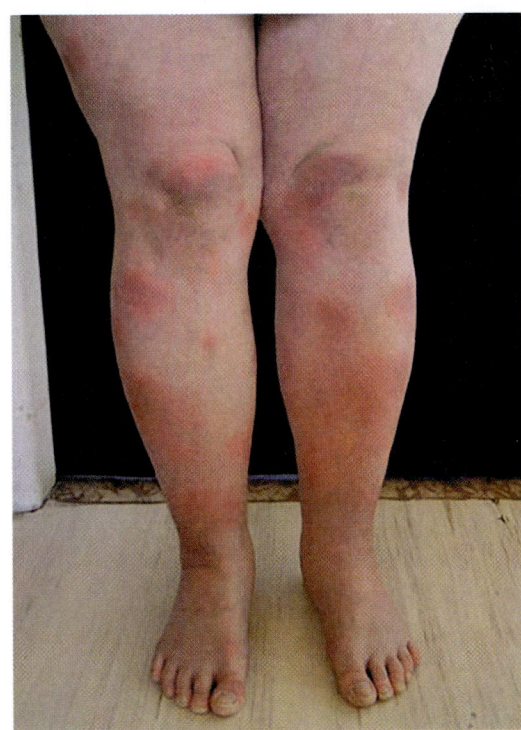

FIGURE 21–4 Erythema nodosum. Note erythematous nodules over the anterior surface of the tibia bilaterally. (Used with permission from Dr. Hanus Rozsypal.)

known as **latent infections**, can reactivate and cause symptomatic tuberculosis. **The most important predisposing factor to overt disease is a host's reduced cell-mediated immune (CMI) response.** For example, AIDS patients have a very high rate of reactivation of prior asymptomatic infection and of rapid progression of the disease. In these patients, untreated disease caused by *M. tuberculosis* has a 50% mortality rate.

Furthermore, administration of **infliximab** (Remicade), a monoclonal antibody that neutralizes tumor necrosis factor (TNF), has activated latent tuberculosis in some patients. The explanation for this is that TNF activates helper T cells to make gamma-interferon that activates macrophages to kill *M. tuberculosis*. So if TNF is neutralized by infliximab then gamma-interferon is not made and killing of *M. tuberculosis* by macrophages is reduced. Remicade is used in the treatment of rheumatoid arthritis (see Chapter 66). Diabetics are also predisposed to reactivation and progression of disease.

In some patients with AIDS who are infected with *M. tuberculosis*, treating the patient with highly active antiretroviral therapy (HAART) causes an exacerbation of symptoms. This phenomenon is called **immune reconstitution inflammatory syndrome (IRIS)**. The explanation of the exacerbation of symptoms is that HAART increases the number of CD4 cells, which increases the inflammatory response. To prevent this, patients should be treated for the underlying infection before starting HAART.

Asymptomatic adults who are at high risk of having a latent infection should be screened using either the PPD skin test or the interferon-γ release assay (IGRA) test. Examples of adults at high risk are those who have lived in countries with increased prevalence of tuberculosis and those who are homeless. If found to be positive, these patients should be treated for latent infection. Screening tests and the treatment of latent infections are described later in this chapter.

Laboratory Diagnosis

Acid-fast staining of sputum or other specimens is the usual initial test (see Figure 21–1). Either the Kinyoun version of the acid-fast stain or the older Ziehl-Neelsen version can be used. The acid-fast stain has low sensitivity, as evidenced by the finding that approximately 50% of "smear-negative" samples are "culture-positive." For rapid screening purposes, auramine stain, which can be visualized by fluorescence microscopy, is used.

In addition to performing an acid-fast stain, the specimen should be cultured. After digestion of the specimen by treatment with NaOH and concentration by centrifugation, the material is cultured on special media, such as Löwenstein-Jensen agar or Middlebrook agar, for up to 8 weeks. It will *not* grow on a blood agar plate.

In liquid BACTEC medium, radioactive metabolites are present, and growth can be detected by the production of radioactive carbon dioxide in about 2 weeks. A liquid medium is preferred for isolation because the organism grows more rapidly and reliably than it does on agar. If growth in the culture

occurs, the organism can be identified by biochemical tests. For example, *M. tuberculosis* produces **niacin**, whereas almost no other mycobacteria do. It also produces catalase.

Nucleic acid amplification tests (NAATs) can be used to detect the presence of *M. tuberculosis* directly in clinical specimens such as sputum. NAATs are available that detect either the ribosomal RNA or the DNA of the organism. These tests are highly specific, but their sensitivity varies. In sputum specimens that are acid-fast stain positive, the sensitivity is high, but in "smear-negative" sputums, the sensitivity is significantly lower. These tests are quite useful in deciding whether to initiate therapy prior to obtaining the culture results.

Because drug resistance, especially to isoniazid (see later), is a problem, susceptibility tests should be performed. However, the organism grows very slowly, and susceptibility tests usually take several weeks, which is too long to guide the initial choice of drugs. To address this problem, molecular tests are available that detect mutations in the chromosomal genes that encode either the catalase gene, which mediates resistance to isoniazid, or the RNA polymerase gene, which mediates resistance to rifampin. Whole genome sequencing can detect genotypic changes that determine resistance to isoniazid, rifampin, ethambutol, and pyrazinamide.

A urine test for active tuberculosis is available. The test detects the lipoarabinomannan antigen with high sensitivity and specificity. Its main drawback is that it does not provide drug susceptibility, so additional testing is required.

The **luciferase assay**, which can detect drug-resistant organisms in a few days, is also used. Luciferase is an enzyme isolated from fireflies that produces flashes of light in the presence of adenosine triphosphate (ATP). If the organism isolated from the patient is resistant, it will not be damaged by the drug (i.e., it will make a normal amount of ATP), and the luciferase will produce the normal amount of light. If the organism is sensitive to the drug, less ATP will be made and less light produced.

There are two approaches to the diagnosis of latent infections. One is the PPD skin test as described in the "Immunity & Hypersensitivity" section earlier in this chapter. Because there are problems both in the interpretation of the PPD test and with the person returning for the skin test to be read, a quantifiable laboratory-based test is valuable.

This laboratory test is an interferon-γ release assay (**IGRA**), and there are two versions available: QuantiFERON-TB Gold (QFT-G) and T-SPOT.TB. In the IGRA assay, blood cells from the patient are exposed to antigens from *M. tuberculosis*, and the amount of interferon-γ released from the cells is measured. The sensitivity and specificity of the IGRA are as good as those of the PPD skin test. Because the antigens used in the test are specific for *M. tuberculosis* and are not present in BCG, the test is not influenced by whether a person has been previously immunized with the BCG vaccine.

Note that the IGRA and PPD tests are positive in both latent disease and in active tuberculosis, so any person with a positive test must be evaluated for the presence of active disease by obtaining a chest X-ray and a sputum sample.

Treatment & Resistance

Multidrug therapy is used to prevent the emergence of drug-resistant mutants during the long (6- to 9-month) duration of treatment. (Organisms that become resistant to one drug will be inhibited by the other.) **Isoniazid** (INH), a bactericidal drug, is the mainstay of treatment. Treatment for most patients with drug-susceptible pulmonary tuberculosis is with four drugs (rifampin, isoniazid, pyrazinamide, and ethambutol; often abbreviated RIPE) for 2 months and two drugs (isoniazid and rifampin) for 4 months. In patients who are immunocompromised (e.g., AIDS patients), who have disseminated disease, or who are likely to have INH-resistant organisms, a fourth drug, ethambutol, is added, and all four drugs are given for 9 to 12 months. A shorter, 4 month course of a combination of moxifloxacin and rifapentine is also useful in the treatment of tuberculosis caused by drug-susceptible strains.

Although therapy is usually given for months, the patient's sputum becomes **noninfectious within 2 to 3 weeks**. The necessity for protracted therapy is attributed to (1) the intracellular location of the organism; (2) caseous material, which blocks penetration by the drug; (3) the slow growth of the organism; and (4) metabolically inactive "persisters" within the lesion. Because metabolically inactive organisms may not be killed by antitubercular drugs, treatment may not eradicate the infection, and reactivation of the disease may occur in the future.

Lymphadenitis, including cervical lymphadenitis (scrofula) caused by *M. tuberculosis*, should be treated with the drug regimens described earlier for disseminated disease. Scrofula caused by *M. scrofulaceum* can be treated by surgical excision of the single cervical lymph node, but alternative approaches exist. A complete discussion of these is beyond the scope of this book.

Treatment of **latent (asymptomatic) infections** consists of INH taken for 6 to 9 months or INH plus rifapentine for 3 months. A course of rifampin for 4 months is also effective. This approach is most often used in asymptomatic patients whose PPD skin test or IGRA test recently converted to positive. The risk of symptomatic infection is greatest within the first 2 years after infection, so INH is particularly indicated for these "recent converters." INH is also used in children exposed to patients with symptomatic tuberculosis. Patients who receive INH should be evaluated for drug-induced hepatitis, especially those over the age of 35 years. Rifampin can be used in those exposed to INH-resistant strains. A combination of rifampin and pyrazinamide should not be used because it can cause severe liver injury.

Resistance to INH and other antituberculosis drugs is being seen with increasing frequency in the United States, especially in immigrants from Southeast Asia and Latin America. Strains of *M. tuberculosis* **resistant to multiple drugs** (MDR strains) have emerged, primarily in AIDS patients. The most common pattern is resistance to both INH and rifampin, but some isolates are resistant to three or more drugs. The treatment of MDR organisms usually involves the use of five drugs (Table 21–4). The precise recommendations depend on the resistance pattern of the isolate and are beyond the scope of this book.

Previous treatment for tuberculosis predisposes to the selection of these MDR organisms. **Noncompliance** (i.e., the failure

TABLE 21-4 Choice of Five Drugs for Treatment of MDR-Tuberculosis

Process	Drugs
Include these drugs	Levofloxacin or moxifloxacin Bedaquiline Linezolid
Choose one or both of these drugs	Clofazimine Cycloserine
If a regimen cannot be designed with the above oral drugs, consider using one of these injectable drugs	Amikacin Streptomycin
If oral drugs are preferred over injectable, then add one or more of these drugs	Delamanid Pyrazinamide Ethambutol

of patients to complete the full course of therapy) is a major factor in allowing the resistant organisms to survive. One approach to the problem of noncompliance is **directly observed therapy (DOT)**, in which healthcare workers observe the patient taking the medication.

The strains of *M. tuberculosis* resistant to INH, rifampin, a fluoroquinolone, and at least one additional drug are called **extensively drug-resistant (XDR)** strains. XDR strains emerged in 2005 among HIV-infected patients in South Africa.

Note that *M. tuberculosis* produces β-lactamase, rendering the organism resistant to many penicillins and cephalosporins. Trials using amoxicillin-clavulanate to treat active tuberculosis were unsuccessful.

Prevention

The most effective measure to prevent tuberculosis disease is to treat latent (asymptomatic) infections as described in the Treatment section. The IGRA test is the most reliable method for detecting latent infections, although the PPD skin test is also used.

The incidence of tuberculosis began to decrease markedly even before the advent of drug therapy in the 1940s. This is attributed to better housing and nutrition, which have improved host resistance. At present, prevention of the spread of the organism depends largely on the prompt identification and adequate treatment of patients who are coughing up the organism. The use of masks and other respiratory isolation procedures to prevent spread to medical personnel is also important. Contact tracing of individuals exposed to patients with active pulmonary disease who are coughing should be done.

An important component of prevention is the use of the PPD skin test to detect recent converters and to institute treatment for latent infections as described earlier. Groups that should be screened with the PPD skin test include people with HIV infection, close contacts of patients with active tuberculosis, low-income populations, alcoholics and intravenous drug users, prison inmates, and foreign-born individuals from countries with a high incidence of tuberculosis.

Because there are some problems associated with PPD skin tests, such as the measurement and the interpretation of results and the inconvenience of the patient having to return for the skin test to be read, a laboratory test to detect latent infections was developed. This IGRA test measures the amount of interferon-γ released from the patient's lymphocytes after exposure to antigens from *M. tuberculosis* in cell culture. The test requires only a single blood specimen and determines the amount of interferon-γ by an enzyme-linked immunosorbent assay (ELISA) test.

BCG vaccine can be used to induce partial resistance to tuberculosis. The vaccine contains a strain of live, attenuated *M. bovis* called BCG. The vaccine is effective in preventing the appearance of tuberculosis as a clinical disease, especially in children, although it does *not* prevent infection by *M. tuberculosis*. However, a major problem with the vaccine is its variable effectiveness, which can range from 0% to 70%. It is used primarily in areas of the world where the incidence of the disease is high. It is *not* usually used in the United States because of its variable effectiveness and because the incidence of the disease is low enough that it is not cost-effective.

The skin test reactivity induced by the vaccine given to children wanes with time, and the interpretation of the skin test reaction in adults is not altered by the vaccine. For example, skin test reactions of 10 mm or more should not be attributed to the vaccine unless it was administered recently. In the United States, use of the vaccine is limited to young children who are in close contact with individuals with active tuberculosis and to military personnel. BCG vaccine should not be given to immunocompromised people because the live BCG organisms can cause disseminated disease.

BCG vaccine is also used to treat bladder cancer. The vaccine is instilled into the bladder and serves to nonspecifically stimulate cell-mediated immunity, which can inhibit the growth of the carcinoma cells.

As an alternative to BCG vaccine, a trial vaccine containing two recombinant *M. tuberculosis* proteins as immunogen was shown to be effective in preventing active disease.

Pasteurization of milk and destruction of infected cattle are important in preventing intestinal tuberculosis.

ATYPICAL MYCOBACTERIA

Several species of mycobacteria are characterized as atypical because they differ in certain aspects from the prototype, *M. tuberculosis*. For example, atypical mycobacteria are widespread in the **environment** and are not pathogenic for guinea pigs,

whereas *M. tuberculosis* is found only in humans and is highly pathogenic for guinea pigs. The atypical mycobacteria are sometimes called mycobacteria other than tuberculosis (MOTT) or non-tuberculous mycobacteria (NTM).

The atypical mycobacteria are classified into four groups according to their rate of growth and whether they produce pigment under certain conditions (Table 21–5). The atypical mycobacteria in groups I, II, and III grow slowly, at a rate similar to that of *M. tuberculosis*, whereas those in group IV are "rapid growers," producing colonies in fewer than 7 days. Group I organisms produce a yellow-orange–pigmented colony only when exposed to light (**photochromogens**), whereas group II organisms produce the pigment chiefly in the dark (**scotochromogens**). Group III mycobacteria produce little or no yellow-orange pigment, irrespective of the presence or absence of light (**nonchromogens**). A summary of the clinical features of the important atypical mycobacteria is presented in Table 21–6.

Group I (Photochromogens)

M. kansasii causes lung disease clinically resembling tuberculosis. Because it is antigenically similar to *M. tuberculosis*, patients are frequently tuberculin skin test positive. Its habitat in the environment is unknown, but infections by this organism are localized to the midwestern states and Texas. It is susceptible to the standard antituberculosis drugs.

Mycobacterium marinum causes "**swimming pool granuloma**," also known as "fish tank granuloma." These granulomatous, ulcerating lesions occur in the skin at the site of abrasions incurred at swimming pools and aquariums. The natural habitat of the organism is both fresh and salt water. Treatment with a tetracycline such as minocycline is effective.

Group II (Scotochromogens)

M. scrofulaceum causes scrofula, a granulomatous cervical adenitis, usually in children. (*M. tuberculosis* also causes scrofula.) The organism enters through the oropharynx and infects the draining lymph nodes. Its natural habitat is environmental water sources, but it has also been isolated as a saprophyte from the human respiratory tract. Scrofula caused by *M. scrofulaceum* can often be cured by surgical excision of the affected lymph nodes.

Group III (Nonchromogens)

Mycobacterium avium-intracellulare complex (MAI, MAC) is composed of two species, *M. avium* and *M. intracellulare*, that are very difficult to distinguish from each other by standard

TABLE 21–5 Runyon's Classification of Atypical Mycobacteria

Group	Growth Rate	Pigment Formation		Typical Species
		Light	Dark	
I	Slow	+	–	M. kansasii, M. marinum
II	Slow	+	+	M. scrofulaceum
III	Slow	–	–	M. avium-intracellulare complex
IV	Rapid	–	–	M. fortuitum-chelonae complex

TABLE 21-6 Clinical Features of Atypical Mycobacteria

Primarily Lung Disease	Primarily Lymphadenitis	Primarily Skin and Soft Tissue Disease
M. kansasi M. avium-intracellulare complex (MAC)	M. scrofulaceum	M. marinum M. abscessus M ulcerans M fortuitum-cheloni complex

laboratory tests. They cause pulmonary disease clinically indistinguishable from tuberculosis, primarily in immunocompromised patients such as those with AIDS who have CD4 cell counts of less than 200/μL. MAI is the most common bacterial cause of disease in AIDS patients. The organisms are widespread in the environment, including water and soil, particularly in the southeastern United States. MAI is much less communicable from person to person than is *M. tuberculosis*. They are highly resistant to antituberculosis drugs, and as many as six drugs in combination are frequently required for adequate treatment. Current drugs of choice are azithromycin or clarithromycin plus one or more of the following: ethambutol, rifabutin, or ciprofloxacin. Azithromycin is currently recommended for preventing disease in AIDS patients.

Mycobacterium ulcerans causes necrotizing skin lesions that progress to ulcerated lesions (Buruli ulcer). It is a common pathogen in tropical, rural, wetland areas. Transmission is related to skin trauma plus exposure to contaminated water. Children are often affected. The organism grows preferentially at 30°C rather than 37°C. Treatment involves at least 8 weeks of a combination of rifampin and clarithromycin. Surgical debridement of the lesions may be useful.

Group IV (Rapidly Growing Mycobacteria)

Mycobacterium fortuitum-chelonae complex is composed of two similar species, *M. fortuitum* and *M. chelonae*. They are saprophytes, found chiefly in soil and water, and rarely cause human disease. Infections occur chiefly in two populations: (1) immunocompromised patients and (2) individuals with prosthetic hip joints and indwelling catheters. Skin and soft tissue infections occur at the site of puncture wounds (e.g., at tattoo sites). They are often resistant to antituberculosis therapy, and therapy with multiple drugs in combination plus surgical excision may be required for effective treatment. Current drugs of choice are amikacin plus doxycycline.

Mycobacterium abscessus is another rapidly growing mycobacteria acquired from the environment. It causes chronic lung infections, especially in cystic fibrosis patients, as well as infections of the skin, bone, and joints. It is highly antibiotic-resistant. Current drugs of choice are amikacin plus imipenem or cefoxitin plus clarithromycin.

Mycobacterium smegmatis is a rapidly growing mycobacterium that is not associated with human disease. It is part of the normal flora of smegma, the material that collects under the foreskin of the penis.

MYCOBACTERIUM LEPRAE

Disease

This organism causes leprosy (Hansen's disease).

Important Properties

M. leprae has **not been grown** in the laboratory, either on artificial media or in cell culture. It can be grown in experimental animals, such as mice and armadillos. Humans are the natural hosts, although the armadillo appears to be a reservoir for human infection in the Mississippi Delta region where these animals are common. In view of this, leprosy can be thought of as a zoonotic disease, at least in certain southern states, such as Louisiana and Texas.

The optimal temperature for growth (30°C) is lower than body temperature; therefore, *M. leprae* grows preferentially in the skin and superficial nerves. It grows very slowly, with a doubling time of 14 days. This makes it the slowest-growing human bacterial pathogen. One consequence of this is that antibiotic therapy must be continued for a long time, usually several years.

Lepromatous leprosy in humans is also caused by *Mycobacterium lepromatosis*, a bacterium found primarily in Mexico and the Caribbean region. This organism also causes lepromatous skin lesions in red squirrels in the British Isles, indicating the probable zoonotic origin of this bacterium.

Transmission

Infection is acquired by **prolonged contact with patients** with lepromatous leprosy, who discharge *M. leprae* in large numbers in nasal secretions and from skin lesions. In the United States, leprosy occurs primarily in Texas, Louisiana, California, and Hawaii. Most cases are found in immigrants from Mexico, the Philippines, Southeast Asia, and India. The disease occurs worldwide, with most cases in the tropical areas of Asia and Africa. The armadillo is unlikely to be an important reservoir because it is not found in many areas of the world where leprosy is endemic.

Pathogenesis

The organism replicates intracellularly, typically within skin histiocytes, endothelial cells, and the Schwann cells of nerves. The nerve damage in leprosy is the result of two processes: damage caused by infection with the bacterium and damage caused by CMI attack on the nerves.

There are two distinct forms of leprosy—**tuberculoid** and **lepromatous**—with several intermediate forms between the two extremes (Table 21–7).

TABLE 21–7 Comparison of Tuberculoid and Lepromatous Leprosy

Feature	Tuberculoid Leprosy	Lepromatous Leprosy
Type of lesion	One or few lesions with little tissue destruction	Many lesions with marked tissue destruction
Number of acid-fast bacilli	Few	Many
Likelihood of transmitting leprosy	Low	High
Cell-mediated response to *M. leprae*	Present	Reduced or absent
Lepromin skin test	Positive	Negative

(1) In tuberculoid (also known as **paucibacillary**) leprosy, the CMI response to the organism limits its growth, very few acid-fast bacilli are seen, and granulomas containing giant cells form. The nerve damage seems likely to be caused by cell-mediated immunity as there are few organisms and the CMI response is strong.

The CMI response consists primarily of CD4-positive cells and a Th-1 profile of cytokines, namely, interferon-γ, interleukin-2, and interleukin-12. It is the CMI response that causes the nerve damage seen in tuberculoid leprosy.

The lepromin skin test result is positive. The lepromin skin test is similar to the tuberculin test (see earlier). An extract of *M. leprae* is injected intradermally, and induration is observed 48 hours later in those in whom a CMI response against the organism exists.

(2) In lepromatous (also known as **multibacillary**) leprosy, the cell-mediated response to the organism is poor, the skin and mucous membrane lesions contain large numbers of organisms, foamy histiocytes rather than granulomas are found, and the lepromin skin test result is negative. The nerve damage seems likely to be caused by bacterial infection as there are many organisms and the CMI response is poor.

There is evidence that people with lepromatous leprosy produce interferon-β (antiviral interferon) in response to *M. leprae* infection, whereas people with tuberculoid leprosy produce interferon-γ. Interferon-β inhibits the synthesis of interferon-γ, thereby reducing the CMI response needed to contain the infection.

Note that in lepromatous leprosy, only the cell-mediated response to *M. leprae* is defective (i.e., the patient is anergic to *M. leprae*). The cell-mediated response to other organisms is unaffected, and the humoral response to *M. leprae* is intact. However, these antibodies are not protective. The T-cell response consists primarily of Th-2 cells.

Clinical Findings

The incubation period averages several years, and the onset of the disease is gradual. In tuberculoid leprosy, hypopigmented macular or plaquelike skin lesions, thickened superficial nerves, and significant anesthesia of the skin lesions occur (Figure 21–5).

In lepromatous leprosy, multiple nodular skin lesions occur, resulting in the typical **leonine** (lionlike) **facies** (Figure 21–6). After the onset of therapy, patients with lepromatous leprosy often develop **erythema nodosum leprosum** (ENL), which is interpreted as a sign that cell-mediated immunity is being restored. ENL is characterized by painful nodules, especially along the extensor surfaces of the tibia and ulna, neuritis, and uveitis.

The disfiguring appearance of the disease results from several factors: (1) the skin anesthesia results in burns and other traumas, which often become infected; (2) resorption of bone leads to loss of features such as the nose and fingertips; and (3) infiltration of the skin and nerves leads to thickening and folding of the skin. In most patients with a single skin lesion,

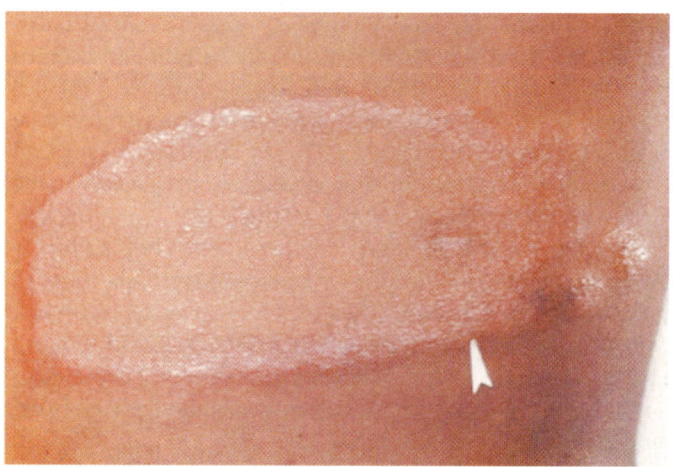

FIGURE 21–5 Tuberculoid leprosy. The tuberculoid form is characterized by a single, flat, hypopigmented lesion that has lost sensation. (Reproduced with permission from Longo DL, Fauci AS, Kasper DL, et al: *Harrison's Principles of Internal Medicine*, 18th ed. New York, NY: McGraw Hill; 2012.)

the disease resolves spontaneously. Patients with forms of the disease intermediate between tuberculoid and lepromatous can progress to either extreme.

Patients with untreated lepromatous leprosy living in the Caribbean region may experience diffuse purpuric and ulcerative lesions called **Lucio's phenomenon**. These lesions are

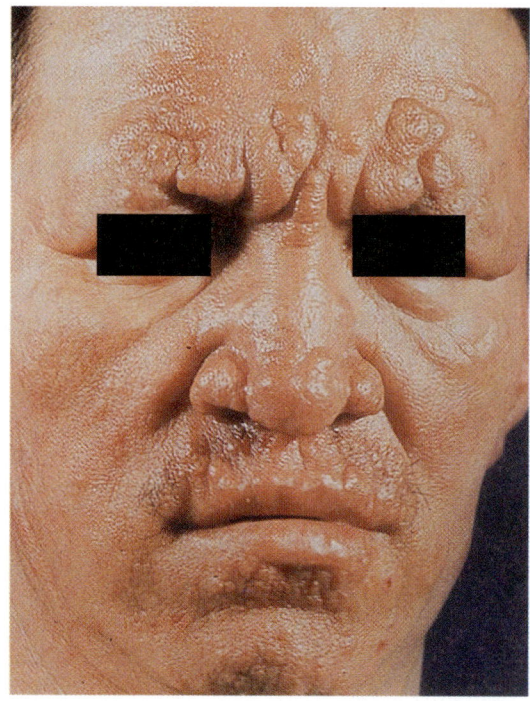

FIGURE 21–6 Lepromatous leprosy. The lepromatous form is characterized by multiple, raised lesions, often with the appearance of leonine facies (the face resembles a lion with a prominent brow). (Used with permission from Robert H. Gelber, MD.)

attributed to immune-complex formation. Secondary infection of the lesions by pyogenic bacteria can cause sepsis and death.

Laboratory Diagnosis

In lepromatous leprosy, the bacilli are easily demonstrated by performing an acid-fast stain of skin lesions or nasal scrapings. Lipid-laden macrophages called "foam cells" containing many acid-fast bacilli are seen in the skin. In the tuberculoid form, very few organisms are seen, and the appearance of typical granulomas is sufficient for diagnosis. Cultures are negative because the organism does not grow on artificial media.

A serologic test for IgM against phenolic glycolipid-1 is useful in the diagnosis of lepromatous leprosy but is not useful in the diagnosis of tuberculoid leprosy. The diagnosis of lepromatous leprosy can be confirmed by using the polymerase chain reaction (PCR) test on a skin sample. False-positive results in the nonspecific serologic tests for syphilis, such as the Venereal Disease Research Laboratory (VDRL) and rapid plasma reagin (RPR) tests, occur frequently in patients with lepromatous leprosy.

Treatment & Prevention

The mainstay of therapy is **dapsone** (diaminodiphenylsulfone), but because sufficient resistance to the drug has emerged, combination therapy is now recommended. For tuberculoid (paucibacillary) leprosy, dapsone and rifampin are given for 6 to 12 months, whereas for lepromatous (multibacillary) leprosy, a combination of dapsone, rifampin, and clofazimine is given for 12 to 24 months. A combination of ofloxacin plus clarithromycin is an alternative regimen. Thalidomide is the treatment of choice for severe ENL reactions.

Prevention involves isolation of all lepromatous patients, coupled with chemoprophylaxis with dapsone for exposed children. There is no vaccine.

SELF-ASSESSMENT QUESTIONS

1. Your patient is a 25-year-old homeless man who complains of a cough for the past month. The cough is now productive of several tablespoons of blood-streaked sputum per day. The sputum is not foul-smelling. He has lost 10 pounds but says that he doesn't eat regularly. On physical exam, temperature is 38°C, and coarse rales are heard in the apex of the left lung. An acid-fast stain of the sputum reveals acid-fast rods. Culture of the sputum shows no growth at 7 days, but buff-colored colonies are visible at 21 days. Of the following organisms, which one is most likely to be the cause of this infection?

    (A) *Mycobacterium fortuitum-chelonae*
    (B) *Mycobacterium leprae*
    (C) *Mycobacterium marinum*
    (D) *Mycobacterium tuberculosis*

2. Which one of the following regimens is optimal initial treatment for the patient in Question 1?

    (A) Isoniazid for 9 months
    (B) Isoniazid and gentamicin for 2 weeks
    (C) Isoniazid and rifampin for 4 months
    (D) Isoniazid, rifampin, ethambutol, and pyrazinamide for 2 months

3. Your patient is a 70-year-old man with progressive weakness in both legs that began about a week ago. He reports back pain and fever for the past month. Magnetic resonance imaging (MRI) of the spine revealed destruction of the seventh thoracic vertebra and a paravertebral mass. Surgical decompression and debridement were performed. Histologic examination of the mass revealed caseating granulomas, and Langhans' giant cells were observed in the granulomas. Gram stain revealed no organisms, but an acid-fast stain showed red rods. Culture showed no growth at 7 days, but growth was seen at 28 days. Of the following, which one is the most likely cause?

    (A) *Mycobacterium fortuitum-chelonae*
    (B) *Mycobacterium leprae*
    (C) *Mycobacterium marinum*
    (D) *Mycobacterium tuberculosis*

4. Your patient is a 30-year-old woman who is infected with HIV and has a low CD4 count. She now has the findings of pulmonary tuberculosis, but you are concerned that she may be infected with *Mycobacterium avium-intracellulare* (MAI). Regarding MAI, which one of the following is most accurate?

    (A) Disseminated disease caused by MAI is typically the result of decreased antibody production, whereas disseminated disease caused by *M. tuberculosis* is typically caused by reduced cell-mediated immunity.
    (B) Immigrants from Southeast Asia are more likely to be infected with MAI than with *M. tuberculosis*.
    (C) In the clinical laboratory, MAI forms colonies in 7 days, whereas *M. tuberculosis* colonies typically require at least 21 days of incubation for colonies to appear.
    (D) MAI is typically susceptible to a drug regimen of isoniazid and rifampin, whereas *M. tuberculosis* is often resistant.
    (E) The natural habitat of MAI is the environment, whereas the natural habitat of *M. tuberculosis* is humans.

5. Regarding the patient in Question 4, if MAI was shown to be the cause of her symptoms, which one of the following is the best choice of antibiotics to prescribe?

    (A) Amikacin and doxycycline
    (B) Clarithromycin, ethambutol, and rifabutin
    (C) Dapsone, rifampin, and clofazimine
    (D) Isoniazid and gentamicin
    (E) Isoniazid, rifampin, ethambutol, and pyrazinamide

6. Your patient is a 20-year-old man with a single, slowly expanding, nonpainful scaly lesion on his chest for the past 2 months. The lesion is nonpruritic, and he has lost sensation at the site of the lesion. He is otherwise well. He is a recent immigrant from Central America. An acid-fast stain of a scraping of the lesion is positive. Which one of the following diseases is he most likely to have?

    (A) Cutaneous tuberculosis
    (B) Fish tank granuloma
    (C) Lepromatous leprosy
    (D) Scrofula
    (E) Tuberculoid leprosy

ANSWERS

(1) **(D)**
(2) **(D)**
(3) **(D)**
(4) **(E)**
(5) **(B)**
(6) **(E)**

SUMMARIES OF ORGANISMS

Brief summaries of the organisms described in this chapter begin on page 675. Please consult these summaries for a rapid review of the essential material.

PRACTICE QUESTIONS: USMLE & COURSE EXAMINATIONS

Questions on the topics discussed in this chapter can be found in the Clinical Bacteriology section of Part XIII: USMLE (National Board) Practice Questions starting on page 737. Also see Part XIV: USMLE (National Board) Practice Examination starting on page 775.

CHAPTER

22

Actinomycetes

CHAPTER CONTENTS

INTRODUCTION

Actinomycetes are a family of bacteria that form **long, branching filaments** that resemble the hyphae of fungi (Figure 22–1). They are gram positive, but some (such as *Nocardia asteroides*) are also weakly acid-fast rods (Figure 22–2) (Table 22–1).

Additional information regarding the clinical aspects of infections caused by the organisms in this chapter is provided in Part IX, entitled Infectious Diseases beginning on page 603.

ACTINOMYCES ISRAELII

Disease

Actinomyces israelii causes actinomycosis.

FIGURE 22–1 *Nocardia asteroides*. Gram stain. Arrow points to area of filaments of gram-positive rods. (Used with permission from Dr. Thomas F. Sellers, Public Health Image Library, Centers for Disease Control and Prevention.)

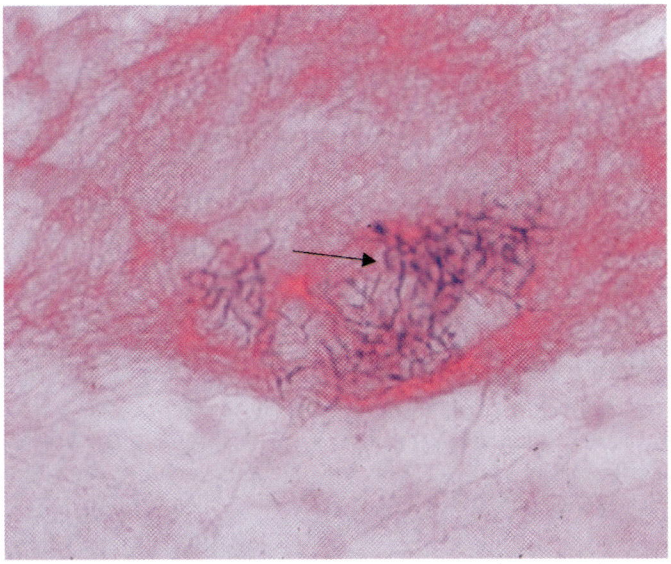

Important Properties & Pathogenesis

A. israelii is an **anaerobe** that forms part of the **normal flora of the oral cavity**. After local trauma such as a broken jaw or dental extraction, it may invade tissues, forming filaments surrounded by areas of inflammation.

Clinical Findings

The typical lesion of actinomycosis appears as a hard, nontender swelling that develops slowly and eventually drains pus through **sinus tracts** (Figure 22–3). Hard, yellow granules (**sulfur granules**) composed of a mass of filaments are formed in pus.

In about 50% of cases, the initial lesion involves the face and neck; in the rest, the chest or abdomen is the site. Pelvic actinomycosis can occur in women who have retained an intrauterine

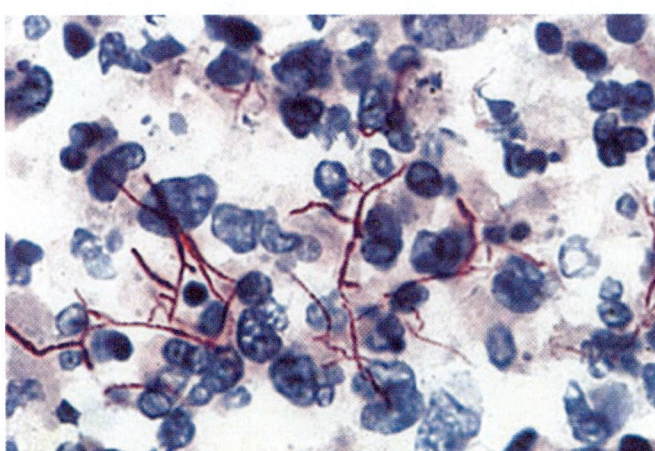

FIGURE 22–2 *Nocardia asteroides*. Acid-fast stain. Note red filamentous rods that are weakly acid-fast. The term "weakly" refers to the use of a weak/(low) concentration of acid used to decolorize the bacteria. (Used with permission from American Registry of Pathology. Travis WD, Colby TV, Koss MN, et al: Non-Neoplastic Disorders of the Lower Respiratory Tract, Atlas of Nontumor Pathology. Fascicle 2. American Registry of Pathology, 2002.)

TABLE 22-1 **Actinomycetes**

Species	Disease	Habitat	Growth in Media	Diagnosis	Treatment
Actinomyces israelii	Actinomycosis (abscess with draining sinus tract and "sulfur granules" in pus)	Oral cavity	Strictly anaerobic	Gram-positive branching filamentous rods; culture (anaerobic)	Penicillin G
Nocardia asteroides	Nocardiosis (abscess in lung and brain, especially in immunodeficient patients, pneumonia)	Environment	Aerobic	Gram-positive branching filamentous rods; often acid-fast; culture (aerobic)	Trimethoprim-sulfamethoxazole

device for a long period of time. *A. israelii* and *Arachnia* species are the most common causes of actinomycosis in humans. The disease is not communicable.

Laboratory Diagnosis

Diagnosis in the laboratory is made by (1) seeing gram-positive branching rods, especially in the presence of sulfur granules and (2) seeing growth when pus or tissue specimens are cultured under anaerobic conditions. Organisms can be identified by immunofluorescence. Note that in contrast to *N. asteroides* (see later), *Actinomyces* is not acid-fast. There are no serologic tests.

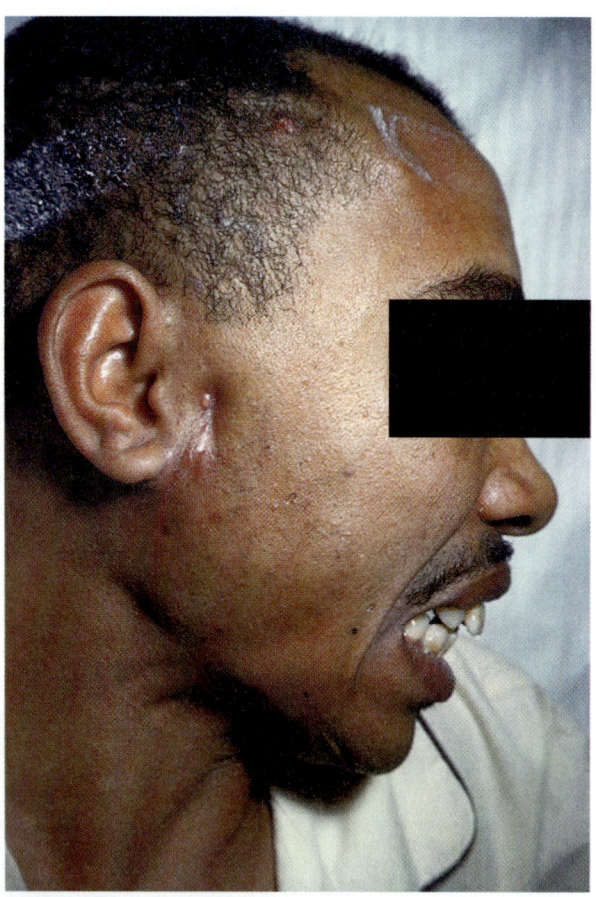

FIGURE 22-3 Actinomycosis. Note inflamed lesion with small sinus tract opening anterior to right ear. Yellowish "sulfur granule" can be seen at the opening. (Used with permission from Dr. Thomas F. Sellers, Public Health Image Library, Centers for Disease Control and Prevention.)

Treatment & Prevention

Treatment consists of prolonged administration of penicillin G, coupled with surgical drainage. There is no significant resistance to penicillin G. No vaccine or prophylactic drug is available.

NOCARDIA ASTEROIDES

Disease

N. asteroides causes nocardiosis.

Important Properties & Pathogenesis

Nocardia species are **aerobes** and are found in the environment, particularly in the **soil**. In immunocompromised individuals, they can produce lung infection and may disseminate. In tissues, *Nocardia* species are thin, branching filaments that are gram-positive on Gram stain.

Many isolates of *N. asteroides* are **weakly acid-fast** (i.e., the staining process uses a weaker solution of hydrochloric acid to decolorize than that used in the stain for mycobacteria). (see Figure 22-2) If the regular-strength acid is used, *N. asteroides* will decolorize. *N. asteroides* is weakly acid-fast because it contains a layer of mycolic acid in its cell wall, similar to, but less than, the mycobacteria.

Clinical Findings

N. asteroides typically causes pneumonia, lung abscess with cavity formation, lung nodules, or empyema. From the lung, the organism can spread to various organs, notably the brain, where it causes brain abscess. Disease occurs most often in immunocompromised individuals, especially those with reduced cell-mediated immunity.

Nocardia brasiliensis, a different species of *Nocardia*, causes skin infections in the southern regions of the United States and mycetoma, usually in tropical regions.

Laboratory Diagnosis

Diagnosis in the laboratory involves (1) seeing branching rods or filaments that are gram-positive (see Figure 22-1) or weakly acid-fast in an acid-fast stain and (2) seeing aerobic growth on bacteriologic media in a few days.

Treatment & Prevention

The drug of choice is trimethoprim-sulfamethoxazole. For the treatment of severe infections, a second drug, such as amikacin, imipenem, or ceftriaxone can be added, depending on results of susceptibility tests. A prolonged course of treatment, for example 3 to 6 months, is often required. Surgical drainage may also be needed. Occasional drug resistance occurs. No vaccine or prophylactic drug is available.

SELF-ASSESSMENT QUESTIONS

1. Your patient is a 75-year-old woman with fever and a painful nodule on her forearm. She also has a nonproductive cough that she says is worse than her usual smoking-related cough. She is taking high-dose corticosteroids (prednisone) for an autoimmune disease. Chest X-ray reveals a nodular lesion in the right upper lobe. A biopsy of the nodule on her arm was obtained. Gram stain of the specimen showed filaments of gram-positive rods. The rods were also weakly acid-fast. Regarding the causative organism, which one of the following is most accurate?

   (A) Culture of the organism should be done under anaerobic conditions.
   (B) The natural habitat of the organism is the soil.
   (C) It produces an exotoxin that inhibits protein synthesis by ADP-ribosylation.
   (D) Sulfur granules are often seen in the skin lesion.
   (E) The vaccine against this organism contains the capsular polysaccharide as the immunogen.

2. Your patient is a 20-year-old man who was in a fist fight in a bar about 3 weeks ago. He took a punch that broke his left second molar. He now has a 3-cm inflamed area on the skin overlying the broken tooth that is draining pus. A Gram stain of the pus reveals gram-positive filamentous rods. The rods did not appear red in the acid-fast stain. Regarding the causative organism, which one of the following is most accurate?

   (A) Infections caused by this organism occur primarily in the Ohio and Mississippi River Valley area.
   (B) The natural habitat of the organism is the soil.
   (C) This organism is resistant to both penicillins and aminoglycosides.
   (D) Sulfur granules are often seen in the pus located at the orifice of the sinus tract in the skin lesion.
   (E) The vaccine against this organism contains a toxoid as the immunogen.

ANSWERS

(1) **(B)**
(2) **(D)**

SUMMARIES OF ORGANISMS

Brief summaries of the organisms described in this chapter begin on page 675. Please consult these summaries for a rapid review of the essential material.

PRACTICE QUESTIONS: USMLE & COURSE EXAMINATIONS

Questions on the topics discussed in this chapter can be found in the Clinical Bacteriology section of Part XIII: USMLE (National Board) Practice Questions starting on page 737. Also see Part XIV: USMLE (National Board) Practice Examination starting on page 775.

Mycoplasmas

INTRODUCTION

Mycoplasmas are a group of very small, **wall-less** organisms, of which *Mycoplasma pneumoniae* is the major pathogen.

Additional information regarding the clinical aspects of infections caused by the organisms in this chapter is provided in Part IX, entitled Infectious Diseases beginning on page 603.

MYCOPLASMA PNEUMONIAE

Disease

M. pneumoniae causes "atypical" pneumonia.

Important Properties

Mycoplasmas are the **smallest free-living organisms**; many are as small as 0.3 μm in diameter. Their most striking feature is the **absence of a cell wall. As a result, Mycoplasmas stain poorly with Gram stain, and antibiotics that inhibit cell wall (peptidoglycan) synthesis (e.g., penicillins and cephalosporins) are ineffective**.

Their outer surface is a flexible cell membrane; hence, these organisms can assume a variety of shapes. Mycoplasmas have the only bacterial membrane that contains **cholesterol**, a sterol usually found in eukaryotic cell membranes.

Mycoplasmas can be grown in the laboratory on artificial media, but they have complex nutritional requirements, including several lipids. They grow slowly and require at least 1 week to form a visible colony. The colony frequently has a characteristic "fried-egg" shape, with a raised center and a thinner outer edge.

Pathogenesis & Epidemiology

M. pneumoniae, a pathogen **only for humans**, is transmitted by **respiratory droplets**. In the lungs, the organism is rod-shaped, with a tapered tip that contains specific proteins that serve as

the point of attachment to the respiratory epithelium. The respiratory mucosa is not invaded, but ciliary motion is inhibited and necrosis of the epithelium occurs. The mechanism by which *M. pneumoniae* causes inflammation is uncertain. It does produce hydrogen peroxide, which contributes to the damage to the respiratory tract cells.

M. pneumoniae has only one serotype and is antigenically distinct from other species of *Mycoplasma*. Immunity is incomplete, and second episodes of disease can occur. During *M. pneumoniae* infection, autoantibodies are produced against red cells (**cold agglutinins**) and brain, lung, and liver cells. These antibodies are involved in some of the extrapulmonary manifestations of infection.

M. pneumoniae infections occur worldwide, with an increased incidence in the winter. This organism is the **most common cause of pneumonia in people between the ages of 5 to 15 years**. It is responsible for outbreaks in groups with close contacts such as families, military personnel, and college students. It is estimated that only 10% of infected individuals actually get pneumonia. *Mycoplasma* pneumonia accounts for about 5% to 10% of all community-acquired pneumonia.

Clinical Findings

Mycoplasma pneumonia is the most common type of atypical pneumonia. It was formerly called **primary atypical pneumonia**. (Atypical pneumonia is also caused by *Legionella pneumophila* [Legionnaires' disease], *Chlamydia pneumoniae*, *Chlamydia psittaci* [psittacosis], *Coxiella burnetii* [Q fever], and viruses such as such as influenza virus and adenovirus. The term *atypical* means that a causative bacterium cannot be isolated on routine media in the diagnostic laboratory or that the disease does not resemble pneumococcal pneumonia.)

The onset of *Mycoplasma* pneumonia is gradual, usually beginning with a nonproductive cough, sore throat, or earache. Small amounts of whitish, nonbloody sputum are produced.

Constitutional symptoms of fever, headache, malaise, and myalgias are pronounced. The paucity of findings on chest examination is in marked contrast to the prominence of the infiltrates seen on the patient's chest X-ray. The disease resolves spontaneously in 10 to 14 days. In addition to pneumonia, *M. pneumoniae* also causes bronchitis.

The extrapulmonary manifestations include Stevens-Johnson syndrome, erythema multiforme, mucositis, Raynaud's phenomenon, cardiac arrhythmias, arthralgias, hemolytic anemia, and neurologic manifestations such as encephalitis and Guillain-Barré syndrome.

Laboratory Diagnosis

Diagnosis is usually *not* made by culturing sputum samples; it takes at least 1 week for colonies to appear on special media. Culture on regular media reveals only normal flora.

Currently, a polymerase chain reaction (PCR) assay that detects *M. pneumoniae*-specific nucleic acids in sputum or in respiratory secretions is the best diagnostic procedure.

Serologic testing for the presence of antibodies in the patient's serum may also be useful. A cold-agglutinin titer of 1:128 or higher is indicative of recent infection. Cold agglutinins are IgM autoantibodies against type O red blood cells that agglutinate these cells at 4°C but not at 37°C. However, only half of patients with *Mycoplasma* pneumonia will be positive for cold agglutinins. The test is nonspecific; false-positive results occur in influenza virus and adenovirus infections. The diagnosis of *M. pneumoniae* infection can be confirmed by a fourfold or greater rise in specific antibody titer in either a complement fixation or an enzyme-linked immunosorbent assay (ELISA) test.

Treatment & Prevention

The treatment of choice is either a macrolide, such as erythromycin or azithromycin, or a tetracycline, such as doxycycline. The fluoroquinolone levofloxacin is also effective. These drugs can shorten the duration of symptoms, although, as mentioned earlier, the disease resolves spontaneously. Penicillins and cephalosporins are **inactive** because the organism has no cell wall.

There is no vaccine or other specific preventive measure.

OTHER MYCOPLASMAS

Mycoplasma hominis has been implicated as an infrequent cause of pelvic inflammatory disease.

Mycoplasma genitalium causes urethritis, predominantly in men. It is estimated to cause approximately 20% of nongonococcal urethritis (NGU). Infections in women are typically asymptomatic but cervicitis may occur. The clinical features resemble those of NGU caused by *Chlamydia trachomatis*.

A nucleic acid-based laboratory test is available. Doxycycline is the drug of choice. Moxifloxacin is effective in eradicating persistent infections.

Ureaplasma urealyticum may cause approximately 20% of cases of NGU. Ureaplasmas can be distinguished from mycoplasmas by their ability to produce the enzyme urease, which degrades urea to ammonia and carbon dioxide. Azithromycin can be used to treat symptomatic infections.

SELF-ASSESSMENT QUESTIONS

1. *Mycoplasma pneumoniae* is an important cause of atypical pneumonia. Regarding this organism, which one of the following is the most accurate?
   (A) Amoxicillin is the drug of choice for pneumonia caused by this organism.
   (B) Antibody in a patient's serum will agglutinate human red blood cells at 4°C, but not at 37°C.
   (C) Gram stain of the sputum reveals small gram-negative rods.
   (D) It is an obligate intracellular parasite that can only grow within human cells in the clinical laboratory.
   (E) People with cystic fibrosis are predisposed to pneumonia caused by this organism.
2. Which one of the following is the drug of choice for atypical pneumonia caused by *M. pneumoniae*?
   (A) Amoxicillin
   (B) Azithromycin
   (C) Ceftriaxone
   (D) Gentamicin
   (E) Vancomycin

ANSWERS

(1) **(B)**
(2) **(B)**

SUMMARIES OF ORGANISMS

Brief summaries of the organisms described in this chapter begin on page 675. Please consult these summaries for a rapid review of the essential material.

PRACTICE QUESTIONS: USMLE & COURSE EXAMINATIONS

Questions on the topics discussed in this chapter can be found in the Clinical Bacteriology section of Part XIII: USMLE (National Board) Practice Questions starting on page 737. Also see Part XIV: USMLE (National Board) Practice Examination starting on page 775.

Spirochetes

INTRODUCTION

Three genera of spirochetes cause human infection: (1) *Treponema*, which causes syphilis and the nonvenereal treponematoses; (2) *Borrelia*, which causes Lyme disease and relapsing fever; and (3) *Leptospira*, which causes leptospirosis (Table 24–1).

Spirochetes are thin-walled, **flexible, spiral rods** (Figure 24–1). They are motile through the undulation of axial filaments that lie under the outer sheath. Treponemes and leptospirae are so thin that they are seen only by dark-field microscopy, silver impregnation, or immunofluorescence. Borreliae are larger, accept Giemsa and other blood stains, and can be seen in the standard light microscope.

Additional information regarding the clinical aspects of infections caused by the organisms in this chapter is provided in Part IX, entitled Infectious Diseases beginning on page 603.

TABLE 24–1 **Spirochetes of Medical Importance**

Species	Disease	Mode of Transmission	Diagnosis	Morphology	Growth in Bacteriologic Media	Treatment
Treponema pallidum	Syphilis	Intimate (sexual) contact; across the placenta	Microscopy; sero-logic tests	Thin, tight, spirals, seen by dark-field illumination, silver stain, or immuno-fluorescent stain	–	Penicillin G
Borrelia burgdorferi	Lyme disease	Tick bite	Clinical observa-tions; microscopy	Large, loosely coiled; stain with Giemsa stain	+	Doxycycline or amoxicillin for early stage; penicillin G for late stage
Borrelia recurrentis	Relapsing fever	Louse bite	Clinical observa-tions; microscopy	Large, loosely coiled; stain with Giemsa stain	+	Tetracycline
Leptospira interrogans	Leptospirosis	Food or drink contaminated by urine of infected animals (rats, dogs, pigs, cows)	Serologic tests	Thin, tight spirals, seen by dark-field illumination	+	Penicillin G

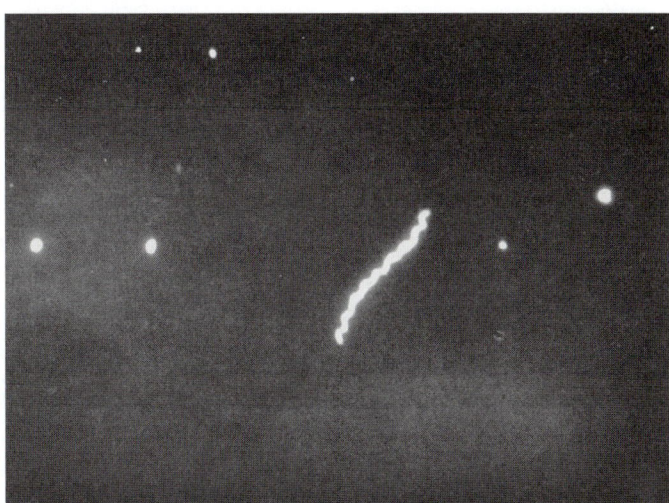

FIGURE 24–1 *Treponema pallidum*—dark-field microscopy. The coiled form of this spirochete is in the center of the field. (Used with permission from Dr. Schwartz, Centers for Disease Control and Prevention.)

TREPONEMA

1. *Treponema pallidum*

Disease

Treponema pallidum causes syphilis.

Important Properties

T. pallidum has **not been grown** on bacteriologic media or in cell culture. Nonpathogenic treponemes, which are part of the normal flora of human mucous membranes, can be cultured.

T. pallidum grows **very slowly**. The medical importance of that fact is that antibiotics must be present at an effective level for several weeks to kill the organisms and cure the disease (see "Treatment" section later). For example, benzathine penicillin is the form of penicillin used to treat primary and secondary syphilis because the penicillin is released very slowly from this depot preparation, and bactericidal concentrations are present for weeks after administration of the antibiotic.

The antigens of *T. pallidum* induce specific antibodies, which can be detected by immunofluorescence or hemagglutination tests in the clinical laboratory. They also induce nonspecific antibodies (**reagin**),[1] which can be detected by the flocculation of lipids (cardiolipin) extracted from normal mammalian tissues (e.g., beef heart). Both specific antitreponemal antibody and nonspecific reagin are used in the serologic diagnosis of syphilis.

Transmission & Epidemiology

T. pallidum is transmitted from spirochete-containing lesions of skin or mucous membranes (e.g., genitalia, mouth, and anus)

---

[1]Syphilitic reagin (IgM and IgG) should not be confused with the reagin (IgE) antibody involved in allergy.

of an infected person to other persons by **intimate contact**. It can also be transmitted from pregnant women to their fetuses. Rarely, blood for transfusions collected during early syphilis is also infectious. *T. pallidum* is a human organism only. There is no animal reservoir.

Syphilis occurs worldwide, and its incidence is increasing. It is one of the leading notifiable diseases in the United States. Many cases are believed to go unreported, which limits public health efforts. There has been a marked increase in the incidence of syphilis in men who have sex with men in recent years.

Pathogenesis & Clinical Findings

T. pallidum produces no important toxins or enzymes. The organism often infects the endothelium of small blood vessels, causing endarteritis. This occurs during all stages of syphilis but is particularly important in the pathogenesis of the brain and cardiovascular lesions seen in tertiary syphilis.

In **primary** syphilis, the spirochetes multiply at the site of inoculation, and a local, nontender ulcer (**chancre**) usually forms in 2 to 10 weeks (Figure 24–2). The ulcer heals spontaneously, but spirochetes spread widely via the bloodstream (bacteremia) to many organs.

One to three months later, the lesions of **secondary syphilis** may occur. These often appear as a maculopapular rash, notably on the **palms** and **soles** (Figure 24–3), or as moist papules on skin and mucous membranes (mucous patches). Moist lesions on the genitals are called **condylomata lata** (Figure 24–4). These lesions are rich in spirochetes and are highly infectious, but they also heal spontaneously. Patchy alopecia may also occur.

Constitutional symptoms of secondary syphilis include low-grade fever, malaise, anorexia, weight loss, headache, myalgias, and generalized lymphadenopathy. Pharyngitis, meningitis, nephritis, and hepatitis may also occur. In some individuals, the symptoms

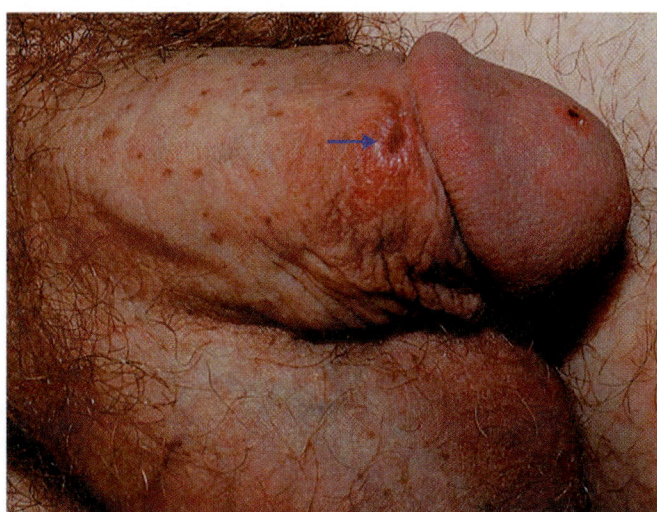

FIGURE 24–2 Chancre of primary syphilis. Note the shallow ulcer with a rolled edge (blue arrow) that is typical of a syphilitic chancre. (Reproduced with permission from Wolff K, Johnson R: *Fitzpatrick's Color Atlas & Synopsis of Clinical Dermatology*, 6th ed. New York, NY: McGraw Hill; 2009.)

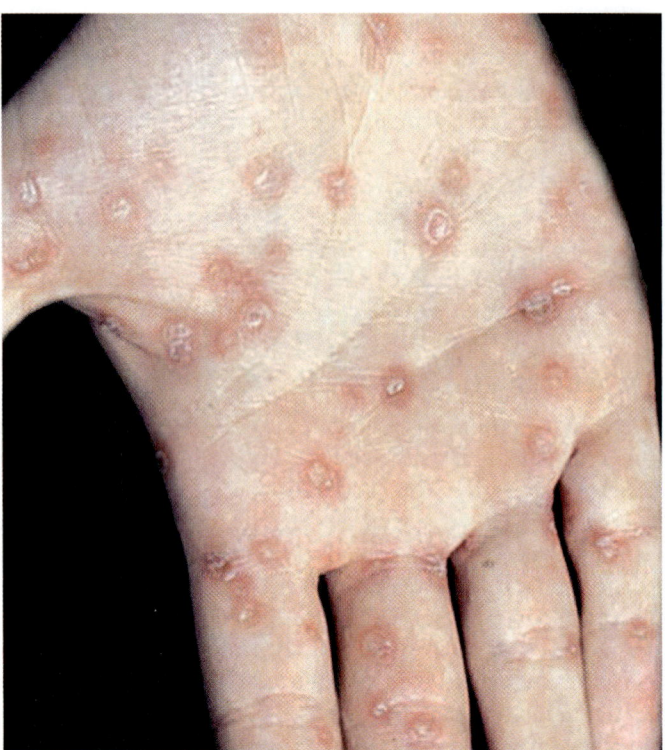

FIGURE 24–3 Palmar lesions of secondary syphilis. Note the papulosquamous lesions on the right palm. Palmar lesions are typically bilateral. (Reproduced with permission from Wolff K, Johnson R: *Fitzpatrick's Color Atlas & Synopsis of Clinical Dermatology*, 6th ed. New York, NY: McGraw Hill; 2009.)

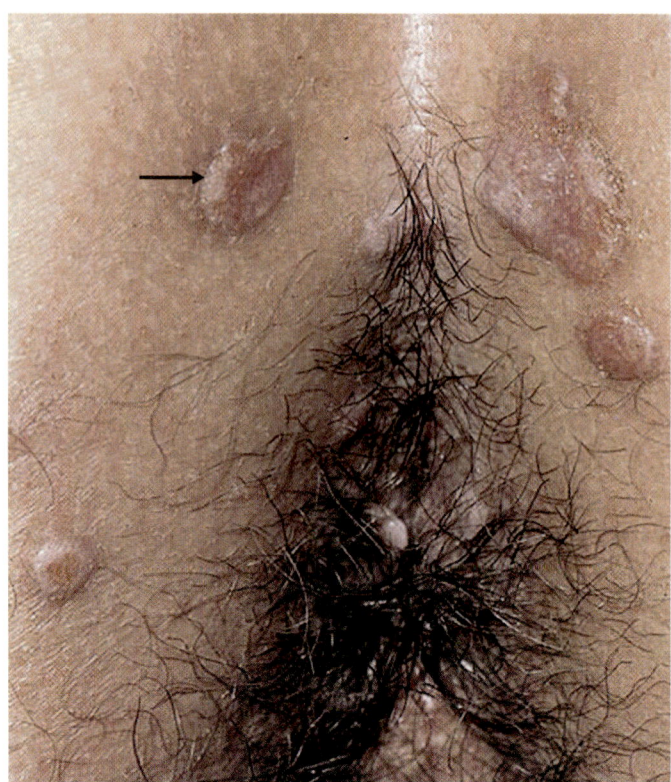

FIGURE 24–4 Condylomata lata of secondary syphilis. Note the flat, moist perianal lesions (black arrow). (Reproduced with permission from Wolff K, Johnson R: *Fitzpatrick's Color Atlas & Synopsis of Clinical Dermatology*, 6th ed. New York, NY: McGraw Hill; 2009.)

of the primary and secondary stages may not occur, and yet the disease may progress.

About one-third of these early (primary and secondary) syphilis cases will "cure" themselves without treatment. Another third remain **latent** (i.e., no lesions appear, but positive serologic tests indicate continuing infection). The latent period can be divided into **early** and **late** stages. In the early latent period, which can last for 1 or 2 years after the secondary stage, the symptoms of secondary syphilis can reappear and patients can infect others. In the late latent period, which can last for many years, no symptoms occur and patients are not infectious.

In the remaining one-third of people, the disease progresses to the **tertiary** stage. Tertiary syphilis may show granulomas (gummas), especially of skin and bones; central nervous system involvement, also known as neurosyphilis (e.g., tabes, paresis); or cardiovascular lesions (e.g., aortitis, aneurysm of the ascending aorta). In tertiary lesions, treponemes are rarely seen.

T. pallidum also causes **congenital syphilis**. The organism is transmitted across the placenta, typically after the third month of pregnancy, and fetal infection can occur. In the infected neonates, skin and bone lesions, such as Hutchinson's teeth, mulberry molars, saber shins, saddle nose, rhagades, snuffles, and frontal bossing, are common. Other findings, such as hepatosplenomegaly, interstitial keratitis, and eighth nerve deafness, also occur. Fetal infection can also result in stillbirth. All pregnant women should be tested for syphilis at least twice during pregnancy.

Immunity to syphilis is incomplete. Antibodies to the organism are produced but do not stop the progression of the disease. Patients with early syphilis who have been treated can contract syphilis again. Patients with late syphilis are relatively resistant to reinfection.

Laboratory Diagnosis

There are three important approaches.

Microscopy

Spirochetes are demonstrated in the lesions of primary or secondary syphilis, such as chancres or condylomata lata, by **dark-field** microscopy or by direct fluorescent antibody (DFA) test. They are *not* seen on a Gram-stained smear because they are too thin to be visualized by light microscopy. In biopsy specimens, such as those obtained from the gummas seen in tertiary syphilis, histologic stains such as silver stain or fluorescent antibody can be used.

Nonspecific Serologic Tests

These tests involve the use of **nontreponemal** antigens. Extracts of normal mammalian tissues (e.g., **cardiolipin** from beef heart) react with antibodies in serum samples from patients with syphilis. These antibodies, which are a mixture of IgG and IgM, are called "reagin" antibodies (see earlier). Flocculation tests (e.g., Venereal Disease Research Laboratory [VDRL] and rapid plasma reagin [RPR] tests) detect the presence of these antibodies. These tests are positive in most cases of primary syphilis and are almost always positive in secondary syphilis. The titer of these nonspecific antibodies **decreases with effective treatment**, in contrast to the specific antibodies, which are positive for life (see later).

False-positive reactions occur in infections such as leprosy, hepatitis B, and infectious mononucleosis and in various autoimmune diseases, such as systemic lupus and antiphospholipid syndrome. Therefore, positive results have to be confirmed by specific tests (see later). Results of nonspecific tests usually **become negative after treatment** and should be used to determine the response to treatment.

These tests can also be falsely negative as a result of the **prozone phenomenon**. In the prozone phenomenon, the titer of antibody is too high (antibody excess), and no flocculation will occur. On dilution of the serum, however, the test result becomes positive (see Chapter 64). These tests are inexpensive and easy to perform and therefore are used as a method of screening the population for infection. The nonspecific tests and the specific tests (see later) are described in more detail in Chapter 9.

The laboratory diagnosis of congenital syphilis is based on the finding that the infant has a higher titer of antibody in the VDRL test than the mother. Furthermore, if a positive VDRL test result in the infant is a false-positive one because maternal antibody has crossed the placenta, the titer will decline with time. If the infant is truly infected, the titer will remain high. However, irrespective of the VDRL test results, any infant whose mother has syphilis should be treated.

Specific Serologic Tests

These tests involve the use of treponemal antigens and therefore are more specific than those described earlier. In these tests, *T. pallidum* reacts in immunofluorescence (FTA-ABS)[2] or hemagglutination (TPHA, MHA-TP)[3] assays with specific treponemal antibodies in the patient's serum. The *Treponema pallidum* particle agglutination test (TPPA) is also used.

These antibodies arise within 2 to 3 weeks of infection; therefore, the test results are positive in most patients with primary syphilis. These **tests remain positive for life** after effective treatment and *cannot* be used to determine the response to treatment or reinfection. They are more expensive and more difficult to perform than the nonspecific tests and therefore are not used as screening procedures.

Treatment

Penicillin G is effective in the treatment of all stages of syphilis. A single injection of benzathine penicillin G (2.4 million units)

can eradicate *T. pallidum* and cure early (primary and secondary) syphilis. Note that **benzathine penicillin** is used because the penicillin is released very slowly from this depot preparation. *T. pallidum* grows very slowly, which requires that the penicillin be present in bactericidal concentration for weeks. If the patient is allergic to penicillin, doxycycline can be used but must be given for prolonged periods.

Tertiary syphilis manifesting with gummas or with cardiovascular findings can also be treated with benzathine penicillin G, but three doses are recommended, However, in *neurosyphilis*, high doses of *aqueous penicillin G* are administered because benzathine penicillin penetrates poorly into the central nervous system. No resistance to penicillin has been observed in *T. pallidum*. However, strains resistant to azithromycin have emerged.

Pregnant women with syphilis should be treated promptly with the type of penicillin used for the stage of their disease. Neonates with a positive serologic test should also be treated. Although it is possible that the positive test is caused by maternal antibody rather than infection of the neonate, it is prudent to treat without waiting several months to determine whether the titer of antibody declines.

More than half of patients with secondary syphilis who are treated with penicillin experience fever, chills, myalgias, and other influenzalike symptoms a few hours after receiving the antibiotic. This response, called the **Jarisch-Herxheimer reaction**, is attributed to the lysis of the treponemes and the release of endotoxin-like substances. Patients should be alerted to this possibility, advised that it may last for up to 24 hours, and told that symptomatic relief can be obtained with aspirin. The Jarisch-Herxheimer reaction also occurs after treatment of other spirochetal diseases such as Lyme disease, leptospirosis, and relapsing fever. Tumor necrosis factor (TNF) is an important mediator of this reaction because passive immunization with antibody against TNF can prevent its symptoms.

Prevention

Prevention depends on early diagnosis and adequate treatment, use of condoms, administration of antibiotic after suspected exposure, and serologic follow-up of infected individuals and their contacts. To prevent congenital syphilis, all pregnant women should be screened by using a treponemal test such as FTA-ABS.

The presence of any sexually transmitted disease makes testing for syphilis mandatory, because several different infections are often transmitted simultaneously. There is no vaccine against syphilis.

2. Nonvenereal Treponematoses

These are infections caused by spirochetes that are virtually indistinguishable from those caused by *T. pallidum*. They are endemic in populations and are transmitted by direct contact. All these infections result in positive (nontreponemal and treponemal) results on serologic tests for syphilis. None of these spirochetes have been grown on bacteriologic media. The diseases include bejel in Africa, yaws (caused by *T. pallidum*

---

[2]FTA-ABS is the fluorescent treponemal antibody-absorbed test. The patient's serum is absorbed with nonpathogenic treponemes to remove cross-reacting antibodies prior to reacting with *T. pallidum*.

[3]TPHA is the *T. pallidum* hemagglutination assay. MHA-TP is a hemagglutination assay done in a microtiter plate.

subspecies *pertenue*) in many humid tropical countries, and pinta (caused by *Treponema carateum*) in Central and South America. All can be cured by penicillin.

BORRELIA

Borrelia species are irregular, loosely coiled spirochetes that stain readily with Giemsa and other stains. They can be cultured in bacteriologic media containing serum or tissue extracts. They are transmitted from animal resevoirs to humans by **arthropods**. They cause two major diseases, Lyme disease and relapsing fever.

1. Borrelia burgdorferi

Disease

Borrelia burgdorferi causes Lyme disease (named after a town in Connecticut). Lyme disease is also known as Lyme borreliosis. Lyme disease is the **most common tick-borne disease in the United States**. It is also the **most common vector-borne disease in the United States**. Approximately 20,000 cases each year are reported to the Centers for Disease Control and Prevention, and that number is thought to be significantly less than the actual number. Another species of *Borrelia*, *Borrelia mayonii*, causes Lyme disease in the upper midwestern states of the United States.

Important Properties

B. burgdorferi is a flexible, motile spirochete that can be visualized by dark-field microscopy and by Giemsa and silver stains. It can be grown in certain bacteriologic media, but routine cultures obtained from patients (e.g., blood, spinal fluid) are typically negative. In contrast, culture of the organism from the tick vector is usually positive.

Transmission & Epidemiology

B. burgdorferi is transmitted by tick bite (Figures 24–5 through 24–7). The tick *Ixodes scapularis* is the vector on the East Coast and in the Midwest; *Ixodes pacificus* is involved on the West Coast. The organism is found in a much higher percentage of *I. scapularis* (35%–50%) than *I. pacificus* (approximately 2%) ticks. This explains the lower incidence of disease on the West Coast. The main reservoir of the organism consists of small mammals, especially the white-footed mouse, upon which the nymph form of the tick feeds.

Large mammals, especially deer, are an obligatory host in the tick's life cycle but are not an important reservoir of the organism.

The nymphal stage of the tick transmits the disease more often than the adult and larval stages do. Nymphs feed primarily in the summer, which accounts for the high incidence of disease during the months of May to September.

The tick must feed for 24 to 48 hours to transmit an infectious dose. This implies that inspecting the skin after being exposed can prevent the disease. However, the nymphs are quite small and can easily be missed. There is no human-to-human spread.

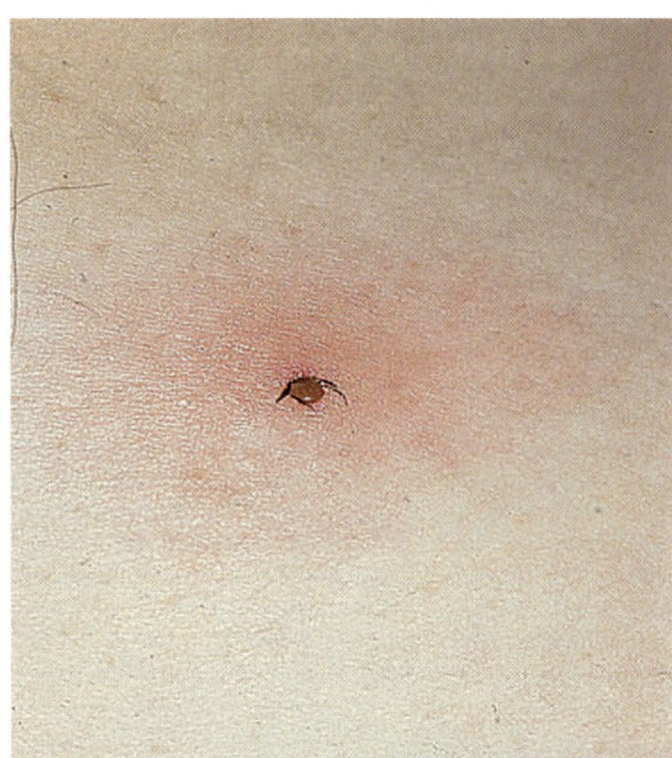

FIGURE 24–5 *Ixodes* tick. Nymph form of tick with head buried in skin surrounded by an erythematous macular rash. (Reproduced with permission from Wolff K, Johnson R: *Fitzpatrick's Color Atlas & Synopsis of Clinical Dermatology*, 6th ed. New York, NY: McGraw Hill; 2009.)

The disease occurs worldwide. In the United States, three regions are primarily affected: the states along the North Atlantic seaboard, the northern midwestern states (e.g., Wisconsin), and the West Coast, especially California. Approximately 80% of the reported cases occurred in four states: New York, Connecticut, Pennsylvania, and New Jersey.

FIGURE 24–6 *Ixodes scapularis*—"blacklegged" tick. Engorged female tick after feeding. (Used with permission from Dr. Gary Alpert, Centers for Disease Control and Prevention.)

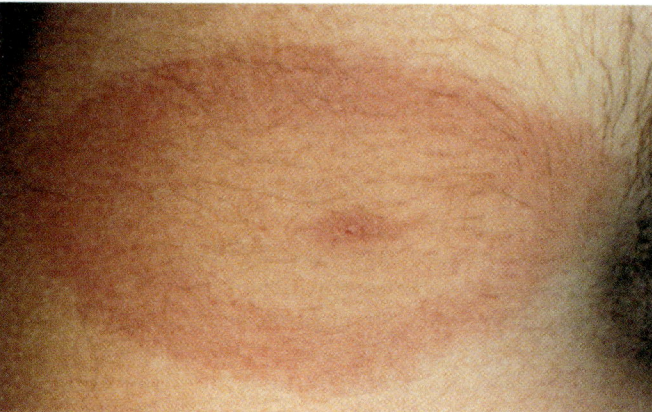

FIGURE 24–8 Erythema chronicum migrans rash of Lyme disease. Note oval-shaped expanding erythematous macular "bull's eye" rash of primary Lyme disease. (Used with permission from Vijay K. Sikand, MD.)

FIGURE 24–7 *Ixodes* tick on a blade of grass questing for a host, such as a deer or human. (Used with permission from Dr. Amanda Loftis, Dr. Will Reeves, and Dr. Chris Paddock, Centers for Disease Control and Prevention.)

Lyme disease is the most common vector-borne disease in the United States. The major bacterial diseases transmitted by ticks in the United States are Lyme disease, Rocky Mountain spotted fever, ehrlichiosis, anaplasmosis, relapsing fever, and tularemia. *I. scapularis* ticks transmit three diseases: two bacterial diseases, Lyme disease and human granulocytic ehrlichiosis, and the protozoan disease, babesiosis. Coinfection with *B. burgdorferi* and *Babesia* occurs, especially in endemic areas such as Massachusetts and other northeastern states.

Pathogenesis

Pathogenesis is associated with spread of the organism from the bite site through the surrounding skin followed by dissemination via the blood (bacteremia) to various organs, especially the heart, joints, and central nervous system. No exotoxins, enzymes, or other important virulence factors have been identified.

Note that the organism must adapt to two markedly different hosts, the tick and the mammal (either mice or humans). It does so by changing its outer surface protein (OSP). These OSPs vary antigenically within humans.

Multiple episodes of Lyme disease are due to reinfection, rather than relapse caused by reactivation of the organism. There is no evidence for a latent stage of *B. burgdorferi*.

Clinical Findings

The clinical findings have been divided into three stages; however, this is a progressive disease, and the stages are not discrete. In stage 1 (early localized stage), the most common finding is **erythema chronicum migrans** (also called **erythema migrans**), an expanding, erythematous, macular rash that often has a "target" or "bull's eye" appearance (Figure 24–8).

The rash appears between 3 and 30 days after the tick bite. Both the tick bite and the rash are painless and nonpruritic. The rash expands over the course of days to weeks and resolves spontaneously in a few weeks.

The rash may be accompanied by nonspecific "flulike" symptoms such as fever, chills, fatigue, myalgia, and headache. Secondary skin lesions frequently occur. Arthralgias, but not arthritis, are another common finding in this early stage. In approximately 25% of cases of Lyme disease, no rash is seen, either because the rash did not occur or because it occurred in an area of the body that is not easily visualized.

In stage 2 (early disseminated stage), which occurs weeks to months later, cardiac and neurologic involvement predominates. Myocarditis, accompanied by various forms of heart block, occurs. Acute (aseptic) meningitis and cranial neuropathies, such as facial nerve palsy (Bell's palsy), are prominent during this stage. Bilateral facial nerve palsy is highly suggestive of Lyme disease. Peripheral neuropathies also occur.

A latent phase lasting weeks to months typically ensues. In stage 3 (late disseminated stage), arthritis, usually of the large joints (e.g., knees), is a characteristic finding. Lyme arthritis is thought to be autoimmune in origin. Encephalopathy also occurs in stage 3.

Some patients treated for Lyme infection continue to have prolonged subjective symptoms of fatigue, joint pains, or mental status changes after objective findings have disappeared. This is called "chronic Lyme disease" or "posttreatment Lyme disease" by some, but there is controversy about whether this exists. No confirmed microbiologic evidence for *B. burgdorferi* infection has been detected in those patients, and prolonged antibiotic therapy does not relieve the symptoms.

Laboratory Diagnosis

Although the organism can be grown in the laboratory, cultures are rarely positive and hence are usually not performed. The diagnosis is typically made serologically by detecting either IgM antibody or a rising titer of IgG antibody with an enzyme-linked immunosorbent assay (ELISA). IgM is typically detectable 2 weeks after infection and peaks at 3 to 6 weeks. Serologic tests done before 2 weeks are likely to yield negative results. Thirty days after infection, tests for IgG are more reliable.

Unfortunately, there are problems with the specificity and sensitivity of these tests because of the presence of cross-reacting antibodies against spirochetes in the normal flora. A positive test result should be confirmed with a Western blot (immunoblot) analysis. The presence of IgM and IgG in the patient's serum can be assayed in the immunoblot assay. In addition, patients treated early in the disease may not develop detectable antibodies. The FDA has approved a more specific ELISA test that can be used as the second confirmatory test and can replace the Western Blot test.

A polymerase chain reaction (PCR) test that detects the organism's DNA is also available. The most appropriate specimens for the PCR test are skin biopsy, joint fluid, or spinal fluid rather than blood or urine. This test is specific but not very sensitive so a negative result does not rule out disease.

Treatment & Prevention

The treatment of choice for stage 1 disease or other mild manifestations is either doxycycline or amoxicillin. Amoxicillin should be used in pregnant women and young children, as doxycycline is contraindicated. For more severe forms or late-stage disease, ceftriaxone is recommended. There is no significant antibiotic resistance.

Prevention involves wearing protective clothing and using insect repellents. Examining the skin carefully for ticks is also very important because the tick must feed for 24 to 48 hours to transmit an infective dose.

Should prophylactic antibiotics be given to people who have been bitten by a tick? The decision depends on two main factors: the percentage of infected ticks in the area and the length of time the tick has fed on the person. If the percentage of infected ticks is high and the length of time is more than 48 hours, it may be cost-effective to prescribe doxycycline prophylactically. Any person bitten by a tick should be advised to watch carefully for a rash or flulike symptoms for the next 3 weeks.

A vaccine containing recombinant outer surface protein A of *B. burgdorferi* as the immunogen was available but has been withdrawn.

2. Borrelia recurrentis & Borrelia hermsii

Borrelia recurrentis, *Borrelia hermsii*, and several other borreliae cause relapsing fever. During infection, the **antigens** of these organisms **undergo variation**. As antibodies develop against one antigen, variants emerge and produce relapses of the illness. This can be repeated 3 to 10 times.

B. recurrentis is transmitted from person to person by the **human body louse**. Humans are the only hosts. *B. hermsii* and many other *Borrelia* species are transmitted to humans by soft **ticks** (*Ornithodoros*). Rodents and other small animals are the main reservoirs. These species of *Borrelia* are passed transovarially in the ticks, a phenomenon that plays an important role in maintaining the organism in nature.

During infection, the arthropod bite introduces spirochetes, which then multiply in many tissues, producing fever, chills, headaches, and multiple-organ dysfunction. Each attack is terminated as antibodies arise.

Diagnosis is usually made by seeing the large spirochetes in stained smears of peripheral blood. They can be cultured in special media. Serologic tests are rarely useful. Tetracycline may be beneficial early in the illness and may prevent relapses. Avoidance of arthropod vectors is the best means of prevention.

3. Borrelia miyamotoi

Borrelia miyamotoi causes a relapsing feverlike syndrome. It was discovered in 1995 in Japan but causes disease worldwide, including the United States. It is transmitted by *Ixodes* ticks. Clinically, the disease begins with an influenzalike syndrome (fever, headache, and myalgia) accompanied by hepatitis and thrombocytopenia. Relapsing episodes occur. The manifestations can resemble anaplasmosis (see Chapter 26) that is also transmitted by *Ixodes* ticks. There is no rash, unlike Lyme disease.

The diagnosis is typically made serologically by detecting IgM antibody or by PCR assay testing for the gene encoding the Glp Q protein that is specific for *B. miyamotoi*. Doxycycline and ceftriaxone are effective treatment choices. There is no vaccine. Wearing clothing impregnated with Permethrin can reduce the risk of tick bites.

LEPTOSPIRA

Leptospiras are tightly coiled spirochetes with hooked ends. They stain poorly with dyes and so are not seen by light microscopy, but they are seen by dark-field microscopy. They grow in bacteriologic media containing serum.

Leptospira interrogans is the cause of leptospirosis. Leptospirosis is common in tropical countries, especially in the rainy season, but is rare in the United States. *L. interrogans* is divided into serogroups that occur in different animals and geographic locations. Each serogroup is subdivided into serovars by the response to agglutination tests.

Leptospiras infect various animals, including **rats** and other rodents, domestic livestock, and household pets. In the United States, dogs are the most important reservoir. Animals excrete leptospiras in **urine**, which contaminates water and soil.

Swimming in contaminated water or consuming contaminated food or drink can result in human infection. Outbreaks have occurred among participants in triathlons and adventure tours involving swimming in contaminated waters. Miners, farmers, and people who work in sewers are at high risk. In the United States, the urban poor have a high rate of infection as determined by the presence of antibodies. Person-to-person transmission is rare.

Human infection results when leptospiras are ingested or pass through mucous membranes or skin. They circulate in the blood and multiply in various organs, producing fever and dysfunction of the liver (jaundice), kidneys (uremia), lungs (hemorrhage), and central nervous system (aseptic meningitis). The illness is typically **biphasic**, with fever, chills, intense headache, and conjunctival suffusion (diffuse reddening of the conjunctivae) appearing early in the disease, followed by a short period of resolution of these symptoms as the organisms are cleared from the blood. The second, "immune," phase is most often characterized by the findings of aseptic meningitis and, in severe cases, liver damage (jaundice) and impaired kidney function. Serovar-specific immunity develops with infection.

Weil's disease is a severe form of acute leptospirosis. It is characterized by jaundice, kidney failure, and bleeding into the skin and GI tract. Acute respiratory distress syndrome (ARDS) including pulmonary hemorrhage can also occur.

Diagnosis is based on history of possible exposure, suggestive clinical signs, and a marked rise in IgM antibody titers. Occasionally, leptospiras are isolated from blood and urine cultures.

The treatment of choice is penicillin G. There is no significant antibiotic resistance. Prevention primarily involves avoiding contact with the contaminated environment. Doxycycline is effective in preventing the disease in exposed persons.

OTHER SPIROCHETES

Anaerobic saprophytic spirochetes are prominent in the normal flora of the human mouth. These spirochetes participate in mixed anaerobic infections, such as infected human bites and stasis ulcers.

SELF-ASSESSMENT QUESTIONS

1. Your patient is a 65-year-old man with gradually increasing confusion and unsteadiness while walking. A lumbar puncture revealed clear spinal fluid, a normal glucose, and an elevated protein. There were 96 cells/µL, of which 86% were lymphocytes. Gram stain of the cerebrospinal fluid (CSF) was negative. Magnetic resonance imaging (MRI) of the brain was normal. A sample of CSF reacted with beef heart cardiolipin at a titer of 1/1024. Regarding the causative organism of his infection, which one of the following is most accurate?

   (A) It is transmitted by tick bite.
   (B) Resistance to penicillin G is common, so ceftriaxone should be used.
   (C) It has never been grown on bacteriologic media in the clinical laboratory.
   (D) It is unlikely to be eradicated because beef cattle are a major reservoir for the organism.
   (E) A confirmatory test for this organism utilizes an agglutination reaction with the capsular polysaccharide of the organism.

2. Your patient is a 20-year-old man with an erythematous, macular, nonpainful rash on the right arm for the past 4 days. The rash is approximately 10 cm in diameter. He also has a fever to 100°F and a mild headache. He reports hiking on several weekends recently in New York State. You suspect the rash is erythema migrans and that he has Lyme disease. Which one of the following is the best approach to confirm your clinical diagnosis?

   (A) Detect IgM antibodies in an ELISA assay
   (B) Determine the titer in a VDRL test
   (C) Gram stain and culture on blood agar incubated aerobically
   (D) Gram stain and culture on blood agar incubated anaerobically
   (E) Grow on human cells in cell culture and identify with fluorescent antibody

3. Assume the patient in Question 2 does have Lyme disease. Which one of the following antibiotics is the most appropriate to treat his infection?

   (A) Azithromycin or trimethoprim-sulfamethoxazole
   (B) Doxycycline or amoxicillin
   (C) Gentamicin or amikacin
   (D) Metronidazole or clindamycin
   (E) Penicillin G or levofloxacin

4. Regarding syphilis, which one of the following is most accurate?

   (A) The characteristic lesion of primary syphilis is a painful vesicle on the genitals.
   (B) In secondary syphilis, the number of organisms is low, so the chance of transmitting the disease to others is low.
   (C) In secondary syphilis, both the rapid plasma reagin (RPR) and the fluorescent treponemal antibody-absorbed (FTA-ABS) tests are usually positive.
   (D) The antibody titer in the FTA-ABS test typically declines when the patient has been treated adequately.
   (E) In congenital syphilis, no antibody is formed against *Treponema pallidum* because the fetus is tolerant to the organism.

5. Regarding *Borrelia burgdorferi* and Lyme disease, which one of the following is most accurate?

   (A) *Borrelia burgdorferi* infects a larger percentage of the rodent reservoir in western states, such as California, than in northeastern states, such as New York.
   (B) Pathogenesis of Lyme disease is based on the production of an exotoxin that induces interleukin-2 production by T-helper cells.
   (C) The vaccine against Lyme disease contains the capsular polysaccharide of all four serotypes as the immunogen.
   (D) Close family members of those infected with *Borrelia burgdorferi* should be given ciprofloxacin.
   (E) *Borrelia burgdorferi* is transmitted to humans by the bite of ticks of the genus *Ixodes*.

6. Benzathine penicillin G is used to treat primary and secondary syphilis rather than procaine penicillin G. Which one of the following is the best reason for this choice?

   (A) Patients allergic to procaine penicillin G are not allergic to benzathine penicillin G.
   (B) Benzathine penicillin G has a higher minimal inhibitory concentration than procaine penicillin G.
   (C) Benzathine penicillin G penetrates the central nervous system to a greater degree than procaine penicillin G.
   (D) Benzathine penicillin G is a depot preparation that provides a long-lasting, high level of drug that kills the slow-growing *T. pallidum*.

ANSWERS

(1) **(C)**
(2) **(A)**
(3) **(B)**
(4) **(C)**
(5) **(E)**
(6) **(D)**

SUMMARIES OF ORGANISMS

Brief summaries of the organisms described in this chapter begin on page 675. Please consult these summaries for a rapid review of the essential material.

PRACTICE QUESTIONS: USMLE & COURSE EXAMINATIONS

Questions on the topics discussed in this chapter can be found in the Clinical Bacteriology section of Part XIII: USMLE (National Board) Practice Questions starting on page 737. Also see Part XIV: USMLE (National Board) Practice Examination starting on page 775.

Chlamydiae

INTRODUCTION

Chlamydiae are obligate intracellular bacteria (i.e., they can grow *only* within cells). They are the agents of common sexually transmitted diseases, such as urethritis and cervicitis, as well as other infections, such as pneumonia, psittacosis, trachoma, and lymphogranuloma venereum.

Additional information regarding the clinical aspects of infections caused by the organisms in this chapter is provided in Part IX, entitled Infectious Diseases beginning on page 603.

Diseases

Chlamydia trachomatis causes eye (conjunctivitis, trachoma), respiratory (pneumonia), and genital tract (urethritis, lymphogranuloma venereum) infections. *C. trachomatis* is the **most common bacterial cause of sexually transmitted disease** in the United States. (Human papilloma virus infection is the **most common sexually transmitted infection** overall in the United States.)

Infection with *C. trachomatis* is also associated with Reiter's syndrome, an autoimmune disease. Approximately 40% of non-gonococcal urethritis is caused by *C. trachomatis*.

Chlamydia pneumoniae causes atypical pneumonia. *Chlamydia psittaci* causes psittacosis, also a disease characterized mainly by pneumonia (Table 25–1).

C. pneumoniae and *C. psittaci* are sufficiently different molecularly from *C. trachomatis* that they have been reclassified into a new genus called *Chlamydophila*. Taxonomically, they are now *Chlamydophila pneumoniae* and *Chlamydophila psittaci*. However, medically, they are often still called *Chlamydia pneumoniae* and *Chlamydia psittaci*.

Important Properties

Chlamydiae are **obligate intracellular** bacteria. They lack the ability to produce sufficient energy to grow independently and therefore can grow only inside host cells. They have a rigid cell wall but do not have a typical peptidoglycan layer.

TABLE 25–1 Chlamydiae of Medical Importance

Species	Disease	Natural Hosts	Mode of Transmission to Humans	Number of Immunologic Types	Diagnosis	Treatment
Chlamydia trachomatis	Urethritis, pneumonia, conjunctivitis, lymphogranuloma venereum, trachoma	Humans	Sexual contact; perinatal transmission	More than 15	Inclusions in epithelial cells seen with Giemsa stain or by immuno-fluorescence; Nucleic acid amplification test (NAAT)	Doxycycline, erythromycin
Chlamydia pneumoniae	Atypical pneumonia	Humans	Respiratory droplets	1	Serologic test	Doxycycline
Chlamydia psittaci	Psittacosis (pneumonia)	Birds	Inhalation of dried bird feces	1	Serologic test NAAT	Doxycycline

Their cell walls resemble those of gram-negative bacteria but lack muramic acid.

Chlamydiae have a replicative cycle different from that of all other bacteria. The cycle begins when the extracellular, metabolically inert, "sporelike" **elementary body** enters the cell and reorganizes into a larger, metabolically active **reticulate body** (Figure 25–1). The latter undergoes repeated cycles of binary fission to form progeny reticulate bodies, which then develop into elementary bodies, which are released from the cell. Within cells, the site of replication appears as an inclusion body in the cytoplasm, which can be stained and visualized microscopically (Figure 25–2). These inclusions are useful in the identification of these organisms in the clinical laboratory.

All chlamydiae share a group-specific lipopolysaccharide antigen, which is detected by complement fixation tests. They also possess species-specific and serovar-specific antigens (proteins), which are detected by immunofluorescence. *C. psittaci* and *C. pneumoniae* each have one serovar, whereas *C. trachomatis* has at least 15 serovars.

Transmission & Epidemiology

C. trachomatis infects **only humans** and is usually transmitted by close personal contact (e.g., **sexually** or by **passage through the birth canal**). Individuals with **asymptomatic genital tract infections** are an important reservoir of infection for others. In trachoma, *C. trachomatis* is transmitted by finger-to-eye or fomite-to-eye contact.

C. pneumoniae infects only humans and is transmitted from person to person by aerosol. *C. psittaci* infects **birds**

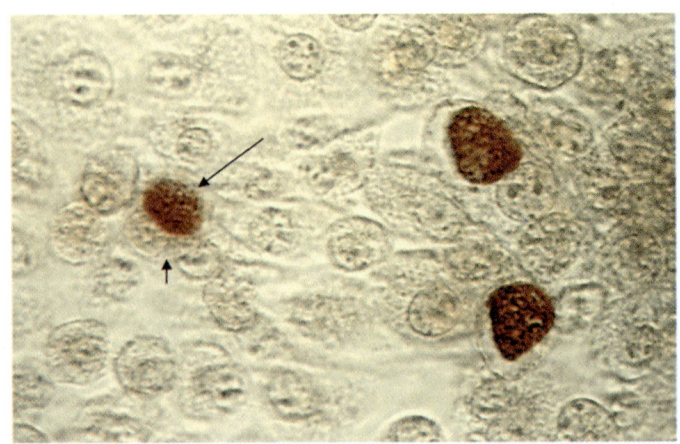

FIGURE 25–2 *Chlamydia trachomatis*—light microscopy of cell culture. Long arrow points to cytoplasmic inclusion body of *C. trachomatis*; short arrow points to nucleus of cell. (Used with permission from Dr. E. Arum and Dr. N. Jacobs, Public Health Image Library, Centers for Disease Control and Prevention.)

(e.g., parrots, pigeons, and poultry, and many mammals including humans). Humans are infected primarily by inhaling organisms in airborne dry bird feces.

Sexually transmitted disease caused by *C. trachomatis* occurs worldwide, but trachoma is most frequently found in developing countries in dry, hot regions such as northern Africa. Trachoma is a leading cause of blindness in those countries.

Patients with a sexually transmitted disease are **coinfected** with both *C. trachomatis* and *Neisseria gonorrheae* in approximately 10% to 30% of cases.

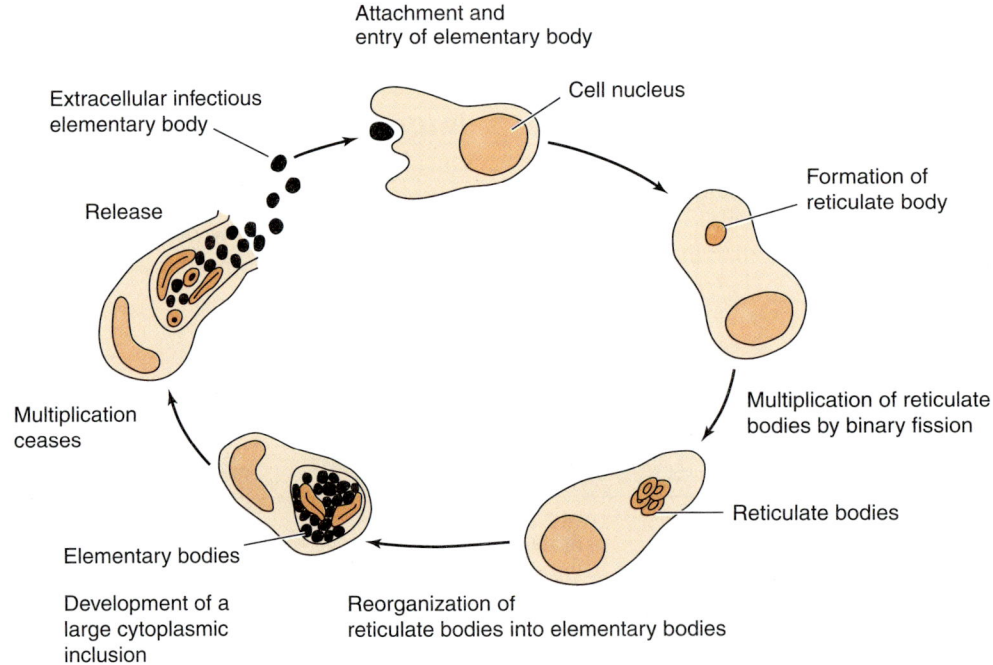

FIGURE 25–1 Life cycle of *Chlamydia*. The extracellular, inert elementary body enters an epithelial cell and changes into a reticulate body that divides many times by binary fission. The progeny reticulate bodies change into elementary bodies and are released from the epithelial cell. The cytoplasmic inclusion body, which is characteristic of chlamydial infections, consists of many progeny reticulate and elementary bodies. (Reproduced with permission from Ryan K: *Sherris Medical Microbiology*, 3rd ed. New York, NY: McGraw Hill; 1994.)

Pathogenesis & Clinical Findings

Chlamydiae infect primarily **epithelial cells** of the mucous membranes or the lungs. They rarely cause invasive, disseminated infections.

CHLAMYDIA TRACHOMATIS

C. trachomatis has more than 15 serovars (A–L). Serovars A, B, and C cause **trachoma**, a chronic conjunctivitis endemic in Africa and Asia. Trachoma may recur over many years and may lead to blindness but causes no systemic illness.

Serovars D–K cause **genital tract infections**. In men, *C. trachomatis* is the most common cause of nongonococcal urethritis (often abbreviated NGU), which is characterized by dysuria and a watery, nonpurulent urethral discharge (Figure 25–3). The discharge may be slight, detectable only by staining of underwear overnight. This infection may progress to epididymitis, prostatitis, or proctitis.

In women, cervicitis accompanied by a vaginal discharge develops and may progress to salpingitis and pelvic inflammatory disease (PID). Women with PID may also have a rare perihepatitis characterized by right upper quadrant pain. This is called Fitz-Hugh-Curtis syndrome and is characterized by "violin-string" adhesions. This syndrome is also caused by *Neisseria gonorrhoeae*.

Repeated episodes of salpingitis or PID can result in infertility or ectopic pregnancy. Urethritis accompanied by dysuria also occurs.

Asymptomatic genital tract infections are very common in both men and women.

Infants born to infected mothers often develop mucopurulent conjunctivitis (neonatal **inclusion conjunctivitis**) 7 to 12 days after delivery, and some develop chlamydial pneumonia 2 to 12 weeks after birth. Chlamydial conjunctivitis also occurs in adults as a result of the transfer of organisms from the genitals to the eye.

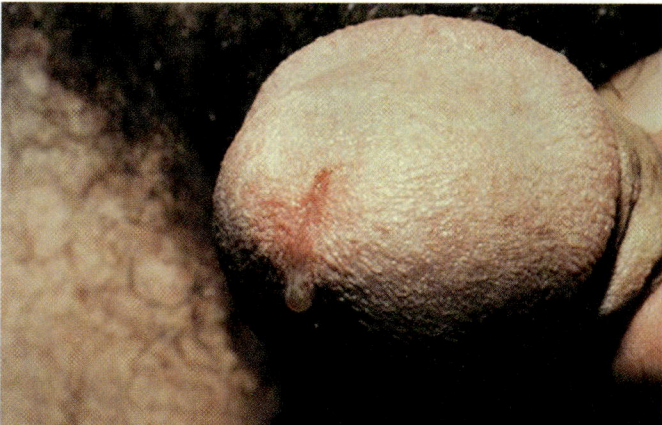

FIGURE 25–3 Nongonococcal urethritis. Note watery, nonpurulent discharge caused by *Chlamydia trachomatis*. The urethral discharge caused by *Neisseria gonorrheae* is more mucoid and purulent. (Reproduced with permission from Seattle STD/HIV Prevention Training Center.)

Patients with genital tract infections caused by *C. trachomatis* have a high incidence of **reactive arthritis** and **Reiter's syndrome**, which is characterized by urethritis, arthritis, and uveitis. These are autoimmune diseases caused by antibodies formed against *C. trachomatis* cross-reacting with antigens on the cells of the urethra, joints, and uveal tract (see Chapter 66).

C. trachomatis L1–L3 serovars cause **lymphogranuloma venereum**, a sexually transmitted disease with lesions on genitalia and in lymph nodes. The primary lesion consists of a small papule in the genital area that heals spontaneously. Days to weeks later, a second stage consisting of tender local lymphadenopathy and systemic symptoms (fever, headache, myalgia) typically occur. Diagnosis is typically made by using a **nucleic acid amplification test** (NAAT). Doxycycline is the drug of choice.

Infection by *C. trachomatis* leads to formation of antibodies and cell-mediated reactions but not to resistance to reinfection or elimination of organisms.

CHLAMYDIA PNEUMONIAE

C. pneumoniae causes upper and lower respiratory tract infections, especially bronchitis and pneumonia, in young adults. Most infections are mild or asymptomatic. The clinical picture resembles other atypical pneumonias, especially that caused by *Mycoplasma pneumoniae*. It is unclear whether *C. pneumoniae* causes upper respiratory infections such as sinusitis and otitis media.

CHLAMYDIA PSITTACI

C. psittaci infects the lungs primarily. The infection may be asymptomatic (detected only by a rising antibody titer) or may produce high fever and pneumonia. Human psittacosis is not generally communicable from human to human.

Although most infections are mild, some are quite severe and involve organs other than the lung. The respiratory infection typically manifests with fever, cough, dyspnea, myalgias, and headache. The most common extrapulmonary organs involved are the liver (hepatomegaly, jaundice), heart (myocarditis, pericarditis), and nervous system (hearing loss, transverse myelitis, and encephalitis).

Laboratory Diagnosis

Chlamydiae form **cytoplasmic inclusions**, which can be seen with special stains (e.g., Giemsa stain) or by immunofluorescence (see Figure 25–2). In general, the Gram stain is *not* useful as the organisms are too small to visualize within the cytoplasm. However, a Gram stain of a urethral discharge that shows neutrophils but no gram-negative diplococci resembling *N. gonorrheae* is presumptive evidence for infection by *C. trachomatis*.

NAATs using the patient's urine are widely used to diagnose chlamydial sexually transmitted disease. Tests not involving culture, such as NAAT, are now more commonly used than culture-based tests (see later).

In exudates, the organism can be identified within epithelial cells by fluorescent antibody staining or hybridization with a

DNA probe. Chlamydial antigens can also be detected in exudates or urine by enzyme-linked immunosorbent assay (ELISA).

Chlamydiae can be grown in cell cultures treated with cycloheximide, which inhibits host cell but not chlamydial protein synthesis, thereby enhancing chlamydial replication. In culture, *C. trachomatis* forms inclusions containing glycogen, whereas *C. psittaci* and *C. pneumoniae* form inclusions that do not contain glycogen. The glycogen-filled inclusions are visualized by staining with iodine. Exudates from the eyes, respiratory tract, or genital tract yield positive cultures in about half of cases.

Serologic tests are used to diagnose infections by *C. psittaci* and *C. pneumoniae* but are rarely helpful in diagnosing disease caused by *C. trachomatis* because the frequency of infection is so high that many people already have antibodies.

Treatment

All chlamydiae are susceptible to tetracyclines, such as doxycycline, and macrolides, such as erythromycin and azithromycin. The drugs of choice for *C. trachomatis* sexually transmitted diseases is single dose azithromycin or doxycycline for seven days. Because the rate of coinfection with gonococci and *C. trachomatis* is high, any patient with a diagnosis of gonorrhea should also be treated for *C. trachomatis* with either doxycycline or azithromycin. Sex partners should be offered treatment.

The drug of choice for neonatal inclusion conjunctivitis and pneumonia caused by *C. trachomatis* is oral erythromycin. The drug of choice for *C. psittaci* and *C. pneumoniae* infections and for lymphogranuloma venereum is doxycycline.

Prevention

There is no vaccine against any chlamydial disease. The best preventive measure against *C. trachomatis* sexually transmitted diseases is to limit transmission by safe sex practices and prompt treatment of both the patient and the sexual partners, including persons who are asymptomatic. Sexual contacts should be traced, and those who had contact within 60 days should be treated. Screening of sexually active, asymptomatic young women and treatment of those who are positive is cost-effective because it may prevent PID and ectopic pregnancy.

Several types of sexually transmitted diseases are often present simultaneously. Thus, diagnosis of one requires a search for other causative agents.

Oral erythromycin given to newborn infants of infected mothers can prevent inclusion conjunctivitis and pneumonitis caused by *C. trachomatis*. Note that erythromycin ointment used to prevent neonatal gonococcal conjunctivitis is much less effective against neonatal chlamydial conjunctivitis. Oral erythromycin should be used. It is thought that the best way to prevent *neonatal* chlamydial conjunctivitis is to diagnose and treat the *mother* during pregnancy.

Psittacosis in humans is controlled by restricting the importation of psittacine birds, treating or destroying sick birds, and adding tetracycline to bird feed. Domestic flocks of turkeys and ducks are tested for the presence of *C. psittaci*.

SELF-ASSESSMENT QUESTIONS

1. Your patient is a 20-year-old man with a urethral discharge. Gram stain of the pus reveals many neutrophils but no bacteria. You suspect this infection may be caused by *Chlamydia trachomatis*. Which one of the following is the laboratory result that best supports your clinical diagnosis?

   (A) Gram stain of the pus reveals small gram-positive rods.
   (B) The organism produces β-hemolytic colonies on blood agar plates when incubated aerobically.
   (C) The organism produces α-hemolytic colonies on blood agar plates when incubated anaerobically.
   (D) Fluorescent antibody staining demonstrates cytoplasmic inclusions in epithelial cells in the exudate.
   (E) There is a fourfold or greater rise in antibody titer against *C. trachomatis*.

2. Regarding chlamydiae, which one of the following is the **most** accurate?

   (A) Lifelong immunity usually follows an episode of disease caused by these organisms.
   (B) The reservoir host for the three species of chlamydiae that cause human infection is humans.
   (C) Their life cycle consists of elementary bodies outside of cells and reticulate bodies within cells.
   (D) They can only replicate within cells because they lack the ribosomes to synthesize their proteins.
   (E) The vaccine against *C. pneumoniae* contains the capsular polysaccharide as the immunogen conjugated to a carrier protein.

3. Which one of the following is the drug of choice for sexually transmitted disease (urethritis, cervicitis) caused by *C. trachomatis*?

   (A) Ampicillin
   (B) Azithromycin
   (C) Ciprofloxacin
   (D) Metronidazole
   (E) Rifampin

ANSWERS

(1) **(D)**
(2) **(C)**
(3) **(B)**

SUMMARIES OF ORGANISMS

Brief summaries of the organisms described in this chapter begin on page 675. Please consult these summaries for a rapid review of the essential material.

PRACTICE QUESTIONS: USMLE & COURSE EXAMINATIONS

Questions on the topics discussed in this chapter can be found in the Clinical Bacteriology section of Part XIII: USMLE (National Board) Practice Questions starting on page 737. Also see Part XIV: USMLE (National Board) Practice Examination starting on page 775.

Rickettsiae

INTRODUCTION

Rickettsiae are obligate intracellular bacteria; that is, they can grow *only* within cells. They are the agents of several important diseases, namely typhus, spotted fevers such as Rocky Mountain spotted fever, Q fever, ehrlichiosis, and anaplasmosis. Other less important rickettsial diseases such as endemic and scrub typhus occur primarily in developing countries.

Outbreaks of endemic typhus occur among the homeless in large cities in the United States. This disease is caused by *Rickettsia typhi*. The organism is transmitted by fleas with rats as the reservoir. Rickettsial pox, caused by *Rickettsia akari*, is a rare disease found in certain densely populated cities in the United States.

Additional information regarding the clinical aspects of infections caused by the organisms in this chapter is provided in Part IX, entitled Infectious Diseases beginning on page 603.

RICKETTSIA RICKETTSII & RICKETTSIA PROWAZEKII

Diseases

Rickettsia rickettsii causes Rocky Mountain spotted fever, a life-threatening disease that occurs primarily in the Southeastern states, for example, North Carolina of the United States. *Rickettsia prowazekii* causes epidemic typhus, also a life-threatening disease that occurs mainly in crowded, unsanitary living conditions during wartime.

Important Properties

Rickettsiae are very short rods that are barely visible in the light microscope. Structurally, their cell wall resembles that of gram-negative rods, but they stain poorly with the standard Gram stain.

Rickettsiae are **obligate intracellular parasites**, because they are unable to produce sufficient energy to replicate extracellularly. Therefore, rickettsiae must be grown in cell culture, embryonated eggs, or experimental animals. Rickettsiae divide by binary fission within the host cell, in contrast to chlamydiae, which are also obligate intracellular parasites but replicate by a distinctive intracellular cycle (see Chapter 25).

Several rickettsiae, such as *R. rickettsii*, *R. prowazekii*, and *Rickettsia tsutsugamushi* (renamed *Orientia tsutsugamushi*), possess antigens that cross-react with antigens of the OX strains of *Proteus vulgaris*. The **Weil-Felix** test, which detects antirickettsial antibodies in a patient's serum by agglutination of the *Proteus* organisms, is based on this cross-reaction.

Transmission

The most striking aspect of the life cycle of the rickettsiae is that they are maintained in nature in certain arthropods such as ticks, lice, fleas, and mites and, with one exception, are transmitted to humans by the **bite of the arthropod**. The rickettsiae circulate widely in the bloodstream (bacteremia), infecting primarily the endothelium of the blood vessel walls.

The exception to arthropod transmission is *Coxiella burnetii*, the cause of Q fever, which is transmitted by aerosol and inhaled into the lungs (see later). Virtually all rickettsial diseases are zoonoses (i.e., they have an animal reservoir), with the prominent exception of **epidemic typhus, which occurs only in humans**. It occurs only in humans because the causative organism, *R. prowazekii*, is transmitted by the human body louse. A summary of the vectors and reservoirs for selected rickettsial diseases is presented in Table 26–1.

The incidence of the disease depends on the geographic distribution of the arthropod vector and on the risk of exposure, which is enhanced by such things as poor hygienic conditions and camping in wooded areas. These factors are discussed later with the individual diseases.

TABLE 26–1 Summary of Selected Rickettsial Diseases

Disease	Organism	Vector	Mammalian Reservoir	Important in the United States
Spotted fevers				
Rocky Mountain spotted fever	*R. rickettsii*	Ticks	Dogs, rodents	Yes (especially in southeastern states such as North Carolina)
Rickettsial pox	*R. akari*	Mites	Mice	No
Typhus group				
Epidemic	*R. prowazekii*	Lice	Humans	No
Endemic	*R. typhi*	Fleas	Rodents	No
Scrub	*R. tsutsugamushi*	Mites	Rodents	No
Others				
Q fever	*C. burnetii*	None	Cattle, sheep, goats	Yes
Anaplasmosis	*A. phagocytophilum*	Ticks	Dogs, rodents	Yes
Ehrlichiosis	*E. chaffeensis*	Ticks	Dogs	Yes

Pathogenesis

The typical lesion caused by these rickettsiae is a **vasculitis**, particularly in the endothelial lining of the vessel wall, where the organism is found. Damage to the vessels of the skin results in the characteristic rash and in edema and hemorrhage caused by increased capillary permeability. Vasculitis of the vessels in the brain leads to the prominent headache.

The basis for pathogenesis by these organisms is unclear. There is some evidence that endotoxin is involved, which is in accord with the nature of some of the lesions such as fever, petechiae, and thrombocytopenia, but its role has not been confirmed. No exotoxins or cytolytic enzymes have been found.

Clinical Findings & Epidemiology

Rocky Mountain Spotted Fever

This disease is characterized by the acute onset of nonspecific symptoms (e.g., fever, severe headache, myalgias, and prostration). The typical rash, which appears 2 to 6 days later, begins with macules that frequently progress to petechiae (Figure 26–1). The rash usually appears first on the hands and feet and then moves inward to the trunk. In addition to headache, other profound central nervous system changes such as delirium and coma can occur. Disseminated intravascular coagulation, edema, and circulatory collapse may ensue in severe cases. The diagnosis must be made on clinical grounds and therapy started promptly, because the laboratory diagnosis is delayed until a rise in antibody titer can be observed.

The name of the disease is misleading because it occurs primarily along the **East Coast** of the United States (in the southeastern states of Virginia, North Carolina, and Georgia), where the dog tick, *Dermacentor variabilis*, is located. The name "Rocky Mountain spotted fever" is derived from the region in which the disease was first found. In the western United States, it is transmitted by the wood tick, *Dermacentor andersoni*.

The **tick** is an important reservoir of *R. rickettsii* as well as the vector; the organism is passed by the transovarian route from tick to tick, and a lifetime infection results. Certain mammals, such as dogs and rodents, are also reservoirs of the organism. Humans are accidental hosts and are not required for the perpetuation of the organism in nature; there is no person-to-person transmission. Most cases occur in children during spring and early summer, when the ticks are active. Rocky Mountain spotted fever accounts for 95% of the rickettsial disease in the United States; there are about 1000 cases per year. It can be fatal if untreated, but if it is diagnosed and treated, a prompt cure results.

Typhus

There are several forms of typhus, namely, louse-borne epidemic typhus caused by *R. prowazekii*, flea-borne endemic typhus caused by *R. typhi*, chigger-borne scrub typhus caused by *O. tsutsugamushi*, and several other quite rare forms. Cases of flea-borne endemic typhus, also called murine typhus, occur in small numbers in the southern regions of California and Texas. The following description is limited to epidemic typhus, the most important of the typhus group of diseases.

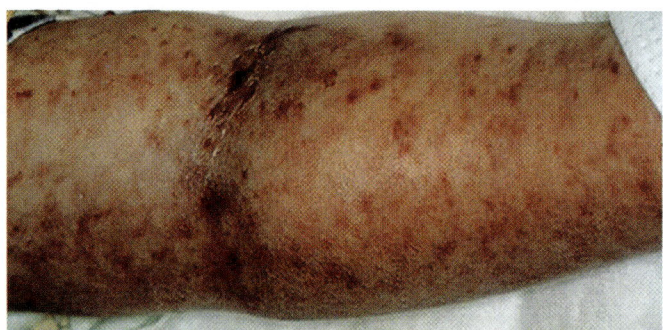

FIGURE 26–1 Rocky Mountain spotted fever. Note widespread petechial rash. (Reproduced with permission from Chapman AS, Bakken JS, Folk SM, et al. Diagnosis and management of tickborne rickettsial diseases: Rocky Mountain spotted fever, ehrlichioses, and anaplasmosis—United States: a practical guide for physicians and other health-care and public health professionals. MMWR Recomm Rep. 2006 Mar 31;55[RR-4]:1–27.)

Typhus begins with the sudden onset of chills, fever, headache, and other influenzalike symptoms approximately 1 to 3 weeks after the louse bite occurs. Between the fifth and ninth days after the onset of symptoms, a maculopapular rash begins on the trunk and spreads peripherally. The rash becomes petechial and spreads over the entire body but spares the face, palms, and soles. Signs of severe meningoencephalitis, including delirium and coma, begin with the rash and continue into the second and third weeks. In untreated cases, death occurs from peripheral vascular collapse or from bacterial pneumonia.

Epidemic typhus is transmitted from person to person by the **human body louse**, *Pediculus*. When a bacteremic patient is bitten, the organism is ingested by the louse and multiplies in the gut epithelium. It is excreted in the feces of the louse during the act of biting the next person and autoinoculated by the person while scratching the bite. The infected louse dies after a few weeks, and there is no louse-to-louse transmission; therefore, human infection is an obligatory stage in the cycle. Epidemic typhus is associated with wars and poverty; at present it is found in developing countries in Africa and South America but not in the United States.

A recurrent form of epidemic typhus is called Brill-Zinsser disease. The signs and symptoms are similar to those of epidemic typhus but are less severe, of shorter duration, and rarely fatal. Recurrences can appear as long as 50 years later and can be precipitated by another intercurrent disease. In the United States, the disease is seen in older people who had epidemic typhus during World War II in Europe. Brill-Zinsser disease is epidemiologically interesting; persistently infected patients can serve as a source of the organism should a louse bite occur.

Laboratory Diagnosis

Laboratory diagnosis of rickettsial diseases is based on serologic analysis rather than isolation of the organism. Although rickettsiae can be grown in cell culture or embryonated eggs, this is a hazardous procedure that is not available in the standard clinical laboratory. A PCR-based assay to detect *R. rickettsii* DNA is available.

Of the serologic tests, the indirect immunofluorescence and enzyme-linked immunosorbent assay (ELISA) tests are most often used. The Weil-Felix test is of historic interest but is no longer performed because its specificity and sensitivity are too low. The basis of the Weil-Felix test is described later.

A fourfold or greater rise in titer between the acute and convalescent serum samples is the most common way the laboratory diagnosis is made. This is usually a retrospective diagnosis, because the convalescent sample is obtained 2 weeks after the acute sample. If the clinical picture is typical, a single acute-phase titer of 1:128 or greater is accepted as presumptive evidence. If the test is available, a diagnosis can be made during the acute phase of the disease by immunofluorescence assay on tissue obtained from the site of the petechial rash.

The Weil-Felix test is based on the cross-reaction of an antigen present in many rickettsiae with the O antigen polysaccharide found in *P. vulgaris* OX-2, OX-19, and OX-K. The test measures the presence of antirickettsial antibodies in the patient's serum by their ability to agglutinate *Proteus* bacteria. The specific rickettsial organism can be identified by the agglutination observed with one or another of these three different strains of *P. vulgaris*. However, as mentioned, this test is no longer used in the United States.

Treatment & Prevention

The treatment of choice for all rickettsial diseases is doxycycline. Prevention of many of these diseases is based on reducing exposure to the arthropod vector by wearing protective clothing and using insect repellent. Frequent examination of the skin for ticks is important in preventing Rocky Mountain spotted fever; the tick must be attached for several hours to transmit the disease. There is no vaccine against Rocky Mountain spotted fever. Prophylactic antibiotics are not recommended in the asymptomatic person bitten by a tick.

Prevention of typhus is based on personal hygiene and "delousing" with DDT. A typhus vaccine containing formalin-killed *R. prowazekii* organisms is effective and useful in the military during wartime but is not available to civilians in the United States.

COXIELLA BURNETII

Disease

C. burnetii causes Q fever. Q stands for "Query"; the cause of this disease was a question mark (i.e., was unknown) when the disease was first described in Australia in 1937.

Important Properties

C. burnetii has a sporelike stage that is highly resistant to drying, which enhances its ability to cause infection. It also has a very low ID_{50}, estimated to be approximately one organism.

C. burnetii exists in two phases that differ in their antigenicity and their virulence: phase I organisms are isolated from the patient, are virulent, and synthesize certain surface antigens, whereas phase II organisms are produced by repeated passage in culture, are nonvirulent, and have lost the ability to synthesize certain surface antigens. The clinical importance of phase variation is that patients with chronic Q fever have a much higher antibody titer to phase I antigens than those with acute Q fever.

Transmission

C. burnetii, the cause of Q fever, is transmitted by aerosol and inhaled into the lungs. Q fever is the one rickettsial disease that is *not* transmitted to humans by the bite of an arthropod. The important reservoirs for human infection are cattle, sheep, and goats. *C. burnetii* causes an inapparent infection in these reservoir hosts and is found in high concentrations in the urine, feces, placental tissue, and amniotic fluid of the animals.

It is transmitted to humans by inhalation of aerosols of these materials.

Clinical Findings & Epidemiology

Unlike other rickettsial diseases, the main organ involved in Q fever is the lungs. It begins suddenly with fever, severe headache, cough, and other influenzalike symptoms. This is all that occurs in many patients, but pneumonia ensues in about half. Hepatitis is frequent enough that the combination of pneumonia and hepatitis should suggest Q fever. A rash is rare, unlike in most of the other rickettsial diseases. In general, Q fever is an acute disease, and recovery is expected even in the absence of antibiotic therapy. Rarely, chronic Q fever characterized by life-threatening endocarditis occurs.

The disease occurs worldwide, chiefly in individuals whose occupations expose them to livestock, such as shepherds, abattoir employees, and farm workers. Ingestion of cow's milk is usually responsible for subclinical infections rather than disease in humans. Pasteurization of milk kills the organism.

Laboratory Diagnosis

Serologic tests, such as the indirect immunofluorescence assay, are used rather than isolation of the organism. *C. burnetii* can be grown in cell culture or embryonated eggs, but this is a hazardous procedure that is not available in the standard clinical laboratory. A PCR-based assay to detect *C. burnetii* DNA is available.

Treatment & Prevention

The treatment of choice is doxycycline. Regarding prevention, persons at high risk of contracting Q fever, such as veterinarians, shepherds, abattoir workers, and laboratory personnel exposed to *C. burnetii*, should receive the vaccine that consists of the killed organism. Pasteurization of milk kills *C. burnetii*.

EHRLICHIA CHAFFEENSIS

Ehrlichia chaffeensis is a member of the *Rickettsia* family and causes human monocytic ehrlichiosis (HME). This disease resembles Rocky Mountain spotted fever, except that the typical rash usually does not occur. High fever, severe headache, and myalgias are prominent symptoms. The organism is endemic in dogs and is transmitted to humans by ticks, especially the dog tick, *Dermacentor*, and the Lone Star tick, *Amblyomma*. Ticks of the genus *Ixodes* are also vectors.

E. chaffeensis primarily infects mononuclear leukocytes and forms characteristic **morulae** in the cytoplasm. (A morula is an inclusion body that resembles a mulberry. It consists of many *E. chaffeensis* cells (Figure 26-2).) Lymphopenia, thrombocytopenia, and elevated liver enzyme values are seen. In the United States, the disease occurs primarily in the southern states, especially Arkansas. The diagnosis is usually made serologically by detecting a rise in antibody titer or by using a PCR-based assay to detect the DNA of the organism. Doxycycline is the treatment of choice.

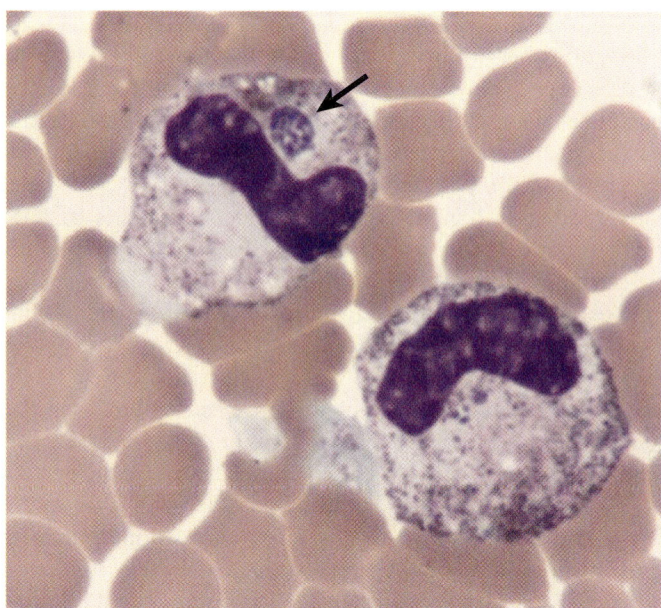

FIGURE 26-2 Morula in cytoplasm of leukocyte infected with *Anaplasma phagocytophilum*. Arrow points to "mulberry-shaped" inclusion body. A morula produced by infection with *Ehrlichia chafeensis* has a similar appearance. (Reproduced with permission from McKean SC, Ross JJ, Dressler DD, et al. Principles and Practice of Hospital Medicine, 2nd ed. New York, NY: McGraw Hill; 2017.)

ANAPLASMA PHAGOCYTOPHILUM

Anaplasma phagocytophilum is a member of the *Rickettsia* family that causes human granulocytic anaplasmosis (HGA). Disease is endemic in northeastern and north central states (e.g., Connecticut and Wisconsin). Distribution is similar to that of Lyme disease. *Ixodes* ticks are the main vectors. Rodents and dogs are important reservoirs.

In HGA, granulocytes rather than mononuclear cells are infected, but the disease is clinically indistinguishable from that caused by *E. chaffeensis* (see earlier). The organism forms an inclusion body called a **morula** in the cytoplasm of infected cells. The morula, which is shaped like a mulberry, is indistinguishable from that formed by *Ehrlichia* (Figure 26-2). The diagnosis is made serologically by detecting a rise in antibody titer or by using a PCR-based assay to detect the DNA of the organism. Doxycycline is the treatment of choice. This organism was formerly known as *Ehrlichia equi*, and the disease it caused was formerly known as human granulocytic ehrlichiosis (HGE).

SELF-ASSESSMENT QUESTIONS

1. Your patient is a 40-year-old woman with the sudden onset of fever to 40°C, severe headache, and petechial rash over most of her body including the palms. Blood cultures are negative. Unfortunately, despite antibiotics and other support, she dies the day after presentation. An autopsy is performed, and immunohistochemical tests on her brain tissue reveal an infection by *Rickettsia rickettsii*. Of the following, which one is the **most** accurate?

(A) It is likely she lives in Colorado and was bitten by a tick.

(B) It is likely she lives in Colorado and was bitten by a mosquito.

(C) It is likely she lives in Virginia and was bitten by a tick.

(D) It is likely she lives in Virginia and was bitten by a flea.

(E) It is likely she lives in Connecticut and was bitten by a mosquito.

2. Regarding Q fever, which one of the following is **most** accurate?

(A) The causative organism is transmitted by tick bite.

(B) The natural habitat of the causative agent is the white-footed mouse.

(C) The diagnosis is made primarily by Gram stain and culture on chocolate agar.

(D) Occupations that predispose people to Q fever include veterinarians and abattoir workers.

(E) Patients with Q fever often have a petechial rash involving the palms.

3. Regarding *Ehrlichia chaffeensis*, which one of the following is most accurate?

(A) It is transmitted primarily by mosquito bite.

(B) It forms β-hemolytic colonies on blood agar.

(C) Its most common clinical presentation is acute meningitis.

(D) It is endemic on the islands off the coast of Massachusetts (e.g., Nantucket).

(E) It forms an inclusion body called a morula in the cytoplasm of infected cell.

ANSWERS

(1) **(C)**

(2) **(D)**

(3) **(E)**

SUMMARIES OF ORGANISMS

Brief summaries of the organisms described in this chapter begin on page 675. Please consult these summaries for a rapid review of the essential material.

PRACTICE QUESTIONS: USMLE & COURSE EXAMINATIONS

Questions on the topics discussed in this chapter can be found in the Clinical Bacteriology section of Part XIII: USMLE (National Board) Practice Questions starting on page 737. Also see Part XIV: USMLE (National Board) Practice Examination starting on page 775.

Minor Bacterial Pathogens

BACTERIA OF MINOR MEDICAL IMPORTANCE

The bacterial pathogens of lesser medical importance are briefly described in this chapter. Experts may differ on their choice of which organisms to put in this category. Nevertheless, separating the minor from the major pathogens should allow the reader to focus on the more important pathogens while providing at least some information about the less important ones.

These organisms are presented in alphabetical order. Table 27–1 lists the organisms according to their appearance on Gram stain.

Additional information regarding the clinical aspects of infections caused by the organisms in this chapter is provided in Part IX, entitled Infectious Diseases beginning on page 603.

TABLE 27–1 Minor Bacterial Pathogens

Type of Bacterium	Genus or Species
Gram-positive cocci	*Abiotrophia, Micrococcus, Peptococcus, Peptostreptococcus, Sarcina, Streptococcus suis*
Gram-positive rods	*Arachnia, Arcanobacterium, Bifidobacterium, Erysipelothrix, Eubacterium, Lactobacillus, Mobiluncus, Propionibacterium, Rhodococcus*
Gram-negative cocci	*Veillonella*
Gram-negative rods	*Achromobacter, Actinobacillus (Aggregatibacter), Aeromonas, Alcaligenes, Arizona, Bartonella quintana* and *B. bacilliformis, Bradyrhizobium, Branhamella, Burkholderia pseudomallei, Calymmatobacterium, Capnocytophaga, Cardiobacterium, Chromobacterium, Chryseobacterium, Citrobacter, Corynebacterium jeikeium, Corynebacterium minutissimum, Edwardsiella, Eikenella, Erwinia,* HACEK group, *Haemophilus aegyptius, Haemophilus ducreyi, Hafnia, Kingella, Moraxella, Plesiomonas, Porphyromonas, Shewanella, Sphingomonas, Spirillum, Streptobacillus, Yersinia enterocolitica, Yersinia pseudotuberculosis*
Rickettsia	*Wolbachia*
Unclassified	*Tropheryma*

Abiotrophia

Abiotrophia species were formerly known as nutritionally deficient streptococci. They are members of the normal flora of the mouth and can cause subacute bacterial endocarditis.

Achromobacter

Achromobacter species are gram-negative coccobacillary rods found chiefly in water supplies. They are opportunistic pathogens and are involved in sepsis, pneumonia, and urinary tract infections.

Actinobacillus (Aggregatibacter)

Actinobacillus species are gram-negative coccobacillary rods. *Actinobacillus actinomycetemcomitans* is found as part of the normal flora in the upper respiratory tract. It is a rare opportunistic pathogen, causing endocarditis on damaged heart valves and sepsis. *A. actinomycetemcomitans* has been renamed *Aggregatibacter actinomycetemcomitans*, but the former genus name *Actinobacillus* is often used.

Aeromonas

Aeromonas species are gram-negative rods found in water, soil, food, and animal and human feces. *Aeromonas hydrophila* causes wound infections, diarrhea, and sepsis, especially in immunocompromised patients.

Alcaligenes

Alcaligenes species are gram-negative coccobacillary rods found in soil and water and are associated with water-containing materials such as respirators in hospitals. *Alcaligenes faecalis* is an opportunistic pathogen, causing sepsis and pneumonia.

Arachnia

Arachnia species are anaerobic gram-positive rods that form long, branching filaments similar to those of *Actinomyces*. They are found primarily in the mouth (associated with dental plaque) and in the tonsillar crypts. *Arachnia propionica*, the major species, causes abscesses similar to those of *Actinomyces israelii*, including the presence of "sulfur granules" in the lesions.

Arcanobacterium

Arcanobacterium haemolyticum is a club-shaped, gram-positive rod that closely resembles corynebacteria. It is a rare cause of pharyngitis and chronic skin ulcers. The pharyngitis can be accompanied by a rash resembling the rash of scarlet fever.

Arizona

Arizona species are gram-negative rods in the family Enterobacteriaceae; they ferment lactose slowly. *Arizona hinshawii* is found in the feces of chickens and other domestic animals and causes diseases similar to those caused by *Salmonella*, such as enterocolitis and enteric fevers. The organism is usually transmitted by contaminated food (e.g., dried eggs).

Bartonella quintana & Bartonella bacilliformis

Bartonella quintana is the cause of trench fever and also is implicated as the cause of some cases of bacillary angiomatosis. Trench fever is transmitted by body lice, and humans are the reservoir for the organism. *Bartonella bacilliformis* causes two rare diseases: Oroya fever and verruga peruana, both of which are stages of Carrión's disease. The disease occurs only in certain areas of the Andes Mountains, and an animal reservoir is suspected.

Bifidobacterium

Bifidobacterium eriksonii is a gram-positive, filamentous, anaerobic rod found as part of the normal flora in the mouth and gastrointestinal tract. It occurs in mixed anaerobic infections.

Bradyrhizobium

Bradyrhizobium enterica is a gram-negative rod that is thought to be the cause of cord colitis. Cord colitis manifests as nonbloody diarrhea in patients who have received an allogeneic hematopoietic stem-cell transplant of umbilical cord cells. It was identified by using DNA sequencing and polymerase chain reaction (PCR) assays on infected tissue from biopsies of the colon.

Bradyrhizobium species are common soil bacteria that fix nitrogen in leguminous plants. *B. enterica* is the first member of the genus to be identified as a human opportunistic pathogen.

Branhamella

Branhamella catarrhalis has been renamed *Moraxella catarrhalis* (see *Moraxella,* later).

Burkholderia pseudomallei

Burkholderia pseudomallei (formerly known as *Pseudomonas pseudomallei*) is a gram-negative rod that causes melioidosis, a rare disease found primarily in Southeast Asia. The organism is found in soil and is transmitted most often when soil contaminates skin abrasions. This disease has been seen in the United States because infections acquired by members of the armed forces during the Vietnam War have reactivated many years later. The acute disease is characterized by high fever and bloody, purulent sputum. Untreated cases can proceed to sepsis and death. In the chronic form, the disease can appear as pneumonia or lung abscess or may resemble tuberculosis. Diagnosis is made by culturing the organism from blood or sputum. The treatment of choice is ceftazidime, which is administered for several weeks.

Calymmatobacterium

Calymmatobacterium granulomatis is a gram-negative rod that causes granuloma inguinale (also known as donovanosis), a sexually transmitted disease characterized by genital ulceration and soft tissue and bone destruction. It is rare in the United States but endemic in many developing countries. The diagnosis is made by visualizing the stained organisms (Donovan bodies) within large macrophages from the lesion. Either doxycycline or azithromycin is an effective treatment for this disease. *C. granulomatis* is also known as *Klebsiella granulomatis.*

Capnocytophaga

Capnocytophaga gingivalis is a gram-negative fusiform rod that is associated with periodontal disease, but it can also be an opportunistic pathogen, causing sepsis and mucositis in immunocompromised patients. *Capnocytophaga canimorsus* is a member of the oral flora of dogs and causes infections following dog bites. It can also cause sepsis in immunocompromised patients, especially those without a spleen and those who abuse alcohol.

Cardiobacterium

Cardiobacterium hominis is a gram-negative pleomorphic rod. It is a member of the normal flora of the human colon, but it can be an opportunistic pathogen, causing mainly endocarditis.

Chromobacterium

Chromobacterium violaceum is a gram-negative rod that produces a violet pigment. It is found in soil and water and can cause wound infections, especially in subtropical parts of the world.

Chryseobacterium

Chryseobacterium species are gram-negative rods found in soil and water. *Chryseobacterium meningosepticum,* the major pathogen in this genus, is an opportunistic pathogen, causing meningitis and sepsis, especially in premature infants. In adults, it causes outbreaks of nosocomial pneumonia, especially in intubated patients. It is resistant to most antibiotics but is noteworthy as the only gram-negative bacterium that is susceptible to vancomycin. The genus *Chryseobacterium* was formerly called *Flavobacterium.*

Citrobacter

Citrobacter species are gram-negative rods (members of the Enterobacteriaceae) related to *Salmonella* and *Arizona.* They occur in the environment and in the human colon and can cause sepsis in immunocompromised patients.

Corynebacterium jeikeium

Corynebacterium jeikeium is a small gram-positive rod primarily found on the skin of hospitalized patients. It causes sepsis in immunocompromised patients, most often those who are neutropenic. Infections are often associated with indwelling catheters and prosthetic heart valves. The drug of choice is vancomycin. Hospital-acquired strains are resistant to many other antibiotics.

Corynebacterium minutissimum

Corynebacterium minutissimum is a small gram-positive rod that causes erythrasma. Erythrasma is characterized by pruritic, scaly, brownish macules on the skin of the genital region. The diagnosis is usually made by visualizing a coral-red fluorescence with a Wood's lamp rather than by culturing the organism. The drug of choice is oral erythromycin.

Edwardsiella

Edwardsiella species are gram-negative rods (members of the Enterobacteriaceae) resembling *Salmonella.* They can cause enterocolitis, sepsis, and wound infections.

Eikenella

Eikenella corrodens is a gram-negative rod that is a member of the normal flora in the human mouth. It causes skin and bone infections associated with **human bites** and "clenched fist" injuries. It also causes sepsis and soft tissue infections of the head and neck, especially in immunocompromised patients and in drug abusers who lick needles prior to injection. *E. corrodens* is also called *Bacteroides ureolyticus.*

Erwinia

Erwinia species are gram-negative rods (members of the Enterobacteriaceae) found in soil and water and are rarely involved in human disease.

Erysipelothrix

Erysipelothrix rhusiopathiae is a gram-positive rod that causes erysipeloid, a skin infection that resembles erysipelas (caused by streptococci). Erysipeloid usually occurs on the hands of persons who handle meat and fish.

Eubacterium

Eubacterium species are gram-positive, anaerobic, non–spore-forming rods that are present in large numbers as part of the normal flora of the human colon. They rarely cause human disease.

HACEK Group

This is a group of small gram-negative rods that have in common the following: slow growth in culture, the requirement for high CO_2 levels to grow in culture, and the ability to cause endocarditis. They are members of the human oropharyngeal flora and can enter the bloodstream from that site. The name "HACEK" is an acronym of the first letters of the genera of the following bacteria: *Haemophilus aphrophilus* and *Haemophilus paraphrophilus, A. (Aggregatibacter) actinomycetemcomitans, C. hominis, E. corrodens,* and *Kingella kingae.*

Haemophilus aegyptius

Haemophilus aegyptius (Koch-Weeks bacillus) is a small gram-negative rod that is an important cause of conjunctivitis in children. Certain strains of *H. aegyptius* cause Brazilian purpuric fever, a life-threatening childhood infection characterized by purpura and shock. This organism is also known as *Haemophilus influenzae* biogroup *aegyptius.*

Haemophilus ducreyi

This small gram-negative rod causes the sexually transmitted disease **chancroid** (soft chancre), which is common in tropical countries but uncommon in the United States. The disease begins with penile lesions, which are painful; nonindurated (soft) ulcers; and local lymphadenitis (bubo). The diagnosis is made by isolating *H. ducreyi* from the ulcer or from pus aspirated from a lymph node. The organism requires heated (chocolate) blood agar supplemented with X factor (heme) but, unlike *H. influenzae,* does not require V factor (NAD). Chancroid can be treated with erythromycin, azithromycin, or ceftriaxone. Because many strains of *H. ducreyi* produce a plasmid-encoded penicillinase, penicillins cannot be used.

Hafnia

Hafnia species are gram-negative rods (members of the Enterobacteriaceae) found in soil and water and are rare opportunistic pathogens.

Kingella

K. kingae is a gram-negative rod in the normal flora of the human oropharynx. It is a rare cause of opportunistic infection and endocarditis.

Lactobacillus

Lactobacilli are gram-positive non–spore-forming rods found as members of the normal flora in the mouth, colon, and female genital tract. In the mouth, they may play a role in the production of dental caries. In the vagina, they are the main source of lactic acid, which keeps the pH low. Lactobacilli are rare causes of opportunistic infection.

Micrococcus

Micrococci are gram-positive cocci that are part of the normal flora of the skin. They are rare human pathogens.

Mobiluncus

Mobiluncus species are anaerobic gram-positive, curved rods that often stain gram-variable. They are associated with **bacterial vaginosis** in women. *Gardnerella* (see above), a facultative rod, is often found in this disease as well.

Moraxella

Moraxella species are gram-negative coccobacillary rods that resemble the diplococcal appearance of the Neisseriae. *Moraxella catarrhalis* is the major pathogen in this genus. It causes otitis media and sinusitis, primarily in children, as well as bronchitis and pneumonia in older people with chronic obstructive pulmonary disease. It is found only in humans and is transmitted by respiratory aerosol. Trimethoprim-sulfamethoxazole or amoxicillin-clavulanate can be used to treat these infections. Most clinical isolates produce β-lactamase. *Moraxella nonliquefaciens* is one of the two common causes of blepharitis (infection of the eyelid); *Staphylococcus aureus* is the other. The usual treatment is local application of antibiotic ointment, such as erythromycin.

Peptococcus

Peptococci are anaerobic gram-positive cocci, resembling staphylococci, found as members of the normal flora of the mouth and colon. They are also isolated from abscesses of various organs, usually from mixed anaerobic infections.

Peptostreptococcus

Peptostreptococci are anaerobic gram-positive cocci found as members of the normal flora of the mouth and colon. They are also isolated from abscesses of various organs, usually from mixed anaerobic infections.

Plesiomonas

Plesiomonas shigelloides is a gram-negative rod associated with water sources. It causes self-limited gastroenteritis, primarily in tropical areas, and can cause invasive disease in immunocompromised individuals.

Porphyromonas

Porphyromonas gingivalis and *Porphyromonas endodontalis* are anaerobic gram-negative rods found in the mouth.

They cause periodontal infections, such as gingivitis and dental abscesses.

Propionibacterium

Propionibacteria are pleomorphic, anaerobic, gram-positive rods found on the skin and in the gastrointestinal tract. *Propionibacterium acnes* is part of the normal flora of the skin and can cause catheter and shunt infections. It is involved in mixed infections associated with cat and dog bites and in head and neck abscesses.

P. acnes is also involved in the pathogenesis of acne, a condition that affects more than 85% of teenagers. The pathogenesis of acne involves impaction of the sebaceous gland followed by inflammation caused by the presence of *P. acnes*. The pustules of acne are composed of sebum, inflammatory cells such as neutrophils and lymphocytes, and the organism. Antibiotics, such as erythromycin, administered either topically or orally, are effective, especially when coupled with other agents such as benzoyl peroxide or retinoids. *P. acnes* has been renamed *Cutibacterium acnes*.

Rhodococcus

Rhodococcus equi is a gram-positive bacterium whose shape varies from a coccus to a club-shaped rod. It is a rare cause of pneumonia and cavitary lung disease in patients whose cell-mediated immunity is compromised. The diagnosis is made by isolating the organism on laboratory agar and observing salmon-pink colonies that do not ferment most carbohydrates. It may appear acid-fast and, if so, can be confused with *Mycobacterium tuberculosis*. The treatment of choice is a combination of rifampin and erythromycin. (*Rhodococcus equi* used to be called *Corynebacterium equi*.)

Sarcina

Sarcina species are anaerobic gram-positive cocci grouped in clusters of four or eight. They are minor members of the normal flora of the colon and are rarely pathogens.

Shewanella

Shewanella haliotis is a gram-negative rod that causes food-borne gastro-intestinal infections such as appendicitis and abdominal abscesses. It is acquired by ingestion of raw or undercooked marine fish or shellfish. Disease occurs primarily in Asian countries. Treatment with piperacillin-tazobactam is often used. *Shewanella* species also cause skin and soft tissue infections when traumatized skin is exposed to sea water.

Sphingomonas

Sphingomonas paucimobilis and *Sphingomonas koreensis* are aerobic, non-fermenting gram-negative rods that cause hospital-acquired infections, such as catheter-associated sepsis. They are relatively rare causes of human infections. They are often found associated with water sources especially hospital plumbing. *S. paucimobilis* was formerly known as *Pseudomonas paucimobilis*.

Spirillum

Spirillum minor is a gram-negative, spiral-shaped rod that causes rat-bite fever ("sodoku"). *S. minor* causes rat-bite fever more commonly in Asian countries than in the United States. The disease is characterized by a reddish-brown rash spreading from the bite, accompanied by fever and local lymphadenopathy. The diagnosis is made by a combination of microscopy and animal inoculation.

Streptobacillus

Streptobacillus moniliformis is a gram-negative rod that causes another type of rat-bite fever (see *Spirillum*, preceding paragraph). *S. moniliformis* is a more common cause of rat-bite fever in the United States than is *S. minor*.

Streptococcus suis

In August 2005, it was reported that *Streptococcus suis* caused the death of 37 farmers in China. The illness is characterized by the sudden onset of hemorrhagic shock. This species is known to cause disease in pigs but only rarely in people prior to this outbreak. Spread of the bacteria from the index case to others has not occurred.

Tropheryma

Tropheryma whipplei is the cause of Whipple's disease, a rare disease characterized by prolonged weight loss, diarrhea, and polyarthritis. Without antibiotic treatment, it is ultimately fatal. Infiltrates of "foamy" macrophages in affected tissue, especially in the small intestine, are commonly seen. The reservoir of the organism, its mode of transmission, and pathogenesis are unknown.

The nature of this organism was unknown for many years. In 1992, it was identified as an actinomycete when ribosomal RNA taken from bacilli seen in duodenal lesions was compared with ribosomal RNA of other bacteria. *Tropheryma* is an intracellular organism that has been grown in human cell culture, but that procedure is not used to diagnose the disease. Laboratory diagnosis is typically made by periodic acid-Schiff (PAS) staining of biopsy specimens of the small bowel in which inclusions are seen in the macrophages. PAS staining, however, is nonspecific, and PCR assays, which are more specific, are used to confirm the diagnosis. First-line treatment typically involves 2 weeks of ceftriaxone, followed by at least 1 year of trimethoprim-sulfamethoxazole.

Veillonella

Veillonella parvula is an anaerobic gram-negative diplococcus that is part of the normal flora of the mouth, colon, and vagina. It is a rare opportunistic pathogen that causes abscesses of the sinuses, tonsils, and brain, usually in mixed anaerobic infections.

Wolbachia

Wolbachia species are *Rickettsia*-like bacteria found intra-cellularly within filarial nematodes such as *Wuchereria* and

Onchocerca (see Chapter 56). *Wolbachia* release endotoxin-like molecules that are thought to play a role in the pathogenesis of *Wuchereria* and *Onchocerca* infections. Treatment of patients with *Wuchereria* and *Onchocerca* infections with doxycycline to kill *Wolbachia* results in a significant decrease in the number of filarial worms in the patient.

Wolbachia themselves are not known to cause human disease but do infect many species of insects worldwide. Experimental infection of mosquitoes with *Wolbachia* reduces the transmission of dengue virus. Also, it is interesting that the neurotoxin of black widow spiders is encoded by WO virus, a bacteriophage that infects *Wolbachia* bacteria.

Yersinia enterocolitica & *Yersinia pseudotuberculosis*

Yersinia enterocolitica and *Yersinia pseudotuberculosis* are gram-negative, oval rods that are larger than *Yersinia pestis*. The virulence factors produced by *Y. pestis* are not made by these species. These organisms are transmitted to humans by contamination of food with the excreta of domestic animals such as dogs, cats, and cattle. *Yersinia* infections are relatively infrequent in the United States, but the number of documented cases has increased during the past few years, perhaps as a result of improved laboratory procedures.

Y. enterocolitica causes enterocolitis that is clinically indistinguishable from that caused by *Salmonella* or *Shigella*. Both *Y. enterocolitica* and *Y. pseudotuberculosis* can cause **mesenteric adenitis** that clinically resembles acute appendicitis. Mesenteric adenitis is the main finding in appendectomies in which a normal appendix is found. Rarely, these organisms are involved in bacteremia or abscesses of the liver or spleen, mainly in persons with underlying disease.

Yersinia infection is associated with two autoimmune diseases: reactive arthritis and Reiter's syndrome. Other enteric pathogens such as *Salmonella*, *Shigella*, and *Campylobacter* also trigger these diseases. Reactive arthritis and Reiter's syndrome are described further in Chapter 66.

Y. enterocolitica is usually isolated from stool specimens and forms a lactose-negative colony on MacConkey's agar. It grows better at 25°C than at 37°C; most biochemical test results are positive at 25°C and negative at 37°C. Incubation of a stool sample at 4°C for 1 week, a technique called *cold enrichment*, increases the frequency of recovery of the organism. *Y. enterocolitica* can be distinguished from *Y. pseudotuberculosis* by biochemical reactions.

The laboratory is usually not involved in the diagnosis of *Y. pseudotuberculosis*; cultures are rarely performed in cases of mesenteric adenitis, and the organism is rarely recovered from stool specimens. Serologic tests are not available in most hospital clinical laboratories.

Enterocolitis and mesenteric adenitis caused by the organisms do not require treatment. In cases of bacteremia or abscess, either trimethoprim-sulfamethoxazole or ciprofloxacin is usually effective. There are no preventive measures except to guard against contamination of food by the excreta of domestic animals.

SELF-ASSESSMENT QUESTIONS

1. Regarding *Haemophilus ducreyi*, which one of the following is most accurate?
   (A) It requires both X and V factors to grow on MacConkey's agar.
   (B) Gram stain of exudate from the lesion shows large gram-positive rods.
   (C) Penicillin G is the drug of choice to treat infections caused by *H. ducreyi*.
   (D) It causes chancroid, which is characterized by a painful ulcer on the genitals.

2. Regarding *Yersinia enterocolitica*, which one of the following is most accurate?
   (A) It causes mesenteric adenitis, which can mimic appendicitis.
   (B) It is a gram-negative diplococcus found primarily within neutrophils.
   (C) It is the most common cause of enterocolitis in the United States.
   (D) Its natural habitat is the human oropharynx, and there is no animal reservoir.

ANSWERS

(1) **(D)**
(2) **(A)**

SUMMARIES OF ORGANISMS

Brief summaries of the organisms described in this chapter begin on page 675. Please consult these summaries for a rapid review of the essential material.

PRACTICE QUESTIONS: USMLE & COURSE EXAMINATIONS

Questions on the topics discussed in this chapter can be found in the Clinical Bacteriology section of Part XIII: USMLE (National Board) Practice Questions starting on page 737. Also see Part XIV: USMLE (National Board) Practice Examination starting on page 775.

PART III   BASIC VIROLOGY

The other infectious agents described in this book, namely bacteria, fungi, protozoa, and worms, are either single cells or composed of many cells. Cells are capable of independent replication, can synthesize their own energy and proteins, and can be seen in the light microscope. In contrast, viruses are not cells; they are not capable of independent replication, can synthesize neither their own energy nor their own proteins, and are too small to be seen in the light microscope.

Viruses are characterized by the following features:

(1) Viruses are particles composed of an internal core containing *either* DNA *or* RNA (but not both) covered by a protective protein coat. Some viruses have an outer lipoprotein membrane, called an envelope, external to the coat. Viruses do not have a nucleus, cytoplasm, mitochondria, or ribosomes. Cells, both prokaryotic and eukaryotic, have *both* DNA and RNA. Eukaryotic cells, such as fungal, protozoal, and human cells, have a nucleus, cytoplasm, mitochondria, and ribosomes. Prokaryotic cells, such as bacteria, are not divided into nucleus and cytoplasm and do not have mitochondria but do have ribosomes; therefore, they can synthesize their own proteins.

(2) Viruses must reproduce (replicate) within cells, because they cannot generate energy or synthesize proteins. Because they can reproduce only within cells, viruses are **obligate intracellular parasites**. (The only bacteria that are obligate intracellular parasites are Chlamydiae and Rickettsiae. They cannot synthesize sufficient energy to replicate independently.)

(3) Viruses replicate in a manner different from that of cells (i.e., viruses do not undergo binary fission or mitosis). One virus can replicate to produce hundreds of progeny viruses, whereas one cell divides to produce only two daughter cells.

Table III–1 compares some of the attributes of viruses and cells.

TABLE III–1 Comparison of Viruses and Cells

Property	Viruses	Cells
Type of nucleic acid	DNA or RNA but not both	DNA and RNA
Proteins	Few	Many
Lipoprotein membrane	Envelope present in some viruses	Cell membrane present in all cells
Ribosomes	Absent[1]	Present
Mitochondria	Absent	Present in eukaryotic cells but not in prokaryotic cells
Enzymes	None or few	Many
Multiplication by binary fission or mitosis	No	Yes

[1]Arenaviruses have a few nonfunctional ribosomes.

Structure

SIZE & SHAPE OF VIRUSES

Viruses range from 20 to 300 nm in diameter; this corresponds roughly to a range of sizes from that of the largest protein to that of the smallest cell (see Figure 2–2). Their shapes are frequently referred to in colloquial terms (e.g., spheres, rods, bullets, or bricks), but in reality, they are complex structures of precise geometric symmetry (see later). The shape of virus particles is determined by the arrangement of the **repeating subunits** that form the protein coat (**capsid**) of the virus. The shapes and sizes of some important viruses are depicted in Figure 28–1.

VIRAL NUCLEIC ACIDS

The anatomy of two representative types of virus particles is shown in Figure 28–2. The viral nucleic acid (genome) is located internally and can be either single- or double-stranded DNA or single- or double-stranded RNA.[1]

Only viruses have genetic material (a genome) composed of single-stranded DNA or of single-stranded or double-stranded RNA. The nucleic acid can be either linear or circular. The DNA is always a single molecule; the RNA can exist either as a single molecule or in several pieces. For example, both influenza virus and rotavirus have a segmented RNA genome. A single-stranded RNA genome can have a positive-polarity or a negative polarity (also called positive-sense and negative-sense). Positive-polarity RNA genomes have the same base sequence as the viral m-RNA whereas negative-polarity genomes have a base sequence that is the base-paired opposite of the m-RNA (see Chapter 29 for more information).

Almost all viruses contain only a single copy of their genome (i.e., they are haploid). The exception is the retrovirus family, whose members have two copies of their RNA genome (i.e., they are diploid).

VIRAL CAPSID & SYMMETRY

The nucleic acid is surrounded by a protein coat called a **capsid** made up of subunits called capsomers. Each capsomer, consisting of one or several proteins, can be seen in the electron microscope as a spherical particle, sometimes with a central hole.

The structure composed of the nucleic acid genome and the capsid proteins is called the **nucleocapsid**. The arrangement of capsomers gives the virus structure its geometric symmetry. Viral nucleocapsids have two forms of symmetry: (1) **icosahedral**, in which the capsomers are arranged in 20 triangles that form a symmetric figure (an icosahedron) with the approximate outline of a sphere and (2) **helical**, in which the capsomers are arranged in a hollow coil that appears rodshaped. The helix can be either rigid or flexible. All human viruses that have a helical nucleocapsid are enclosed by an outer membrane called an **envelope** (i.e., there are no naked helical viruses). Viruses that have an icosahedral nucleocapsid can be either enveloped or naked (see Figure 28–2).

The advantage of building the virus particle from identical protein subunits is twofold: (1) it reduces the need for genetic information and (2) it promotes self-assembly (i.e., no enzyme or energy is required). In fact, functional virus particles have been assembled in the test tube by combining the purified nucleic acid with the purified proteins in the absence of cells, energy source, and enzymes.

---

[1]The nature of the nucleic acid of each virus is listed in Tables 31–1 and 31–2.

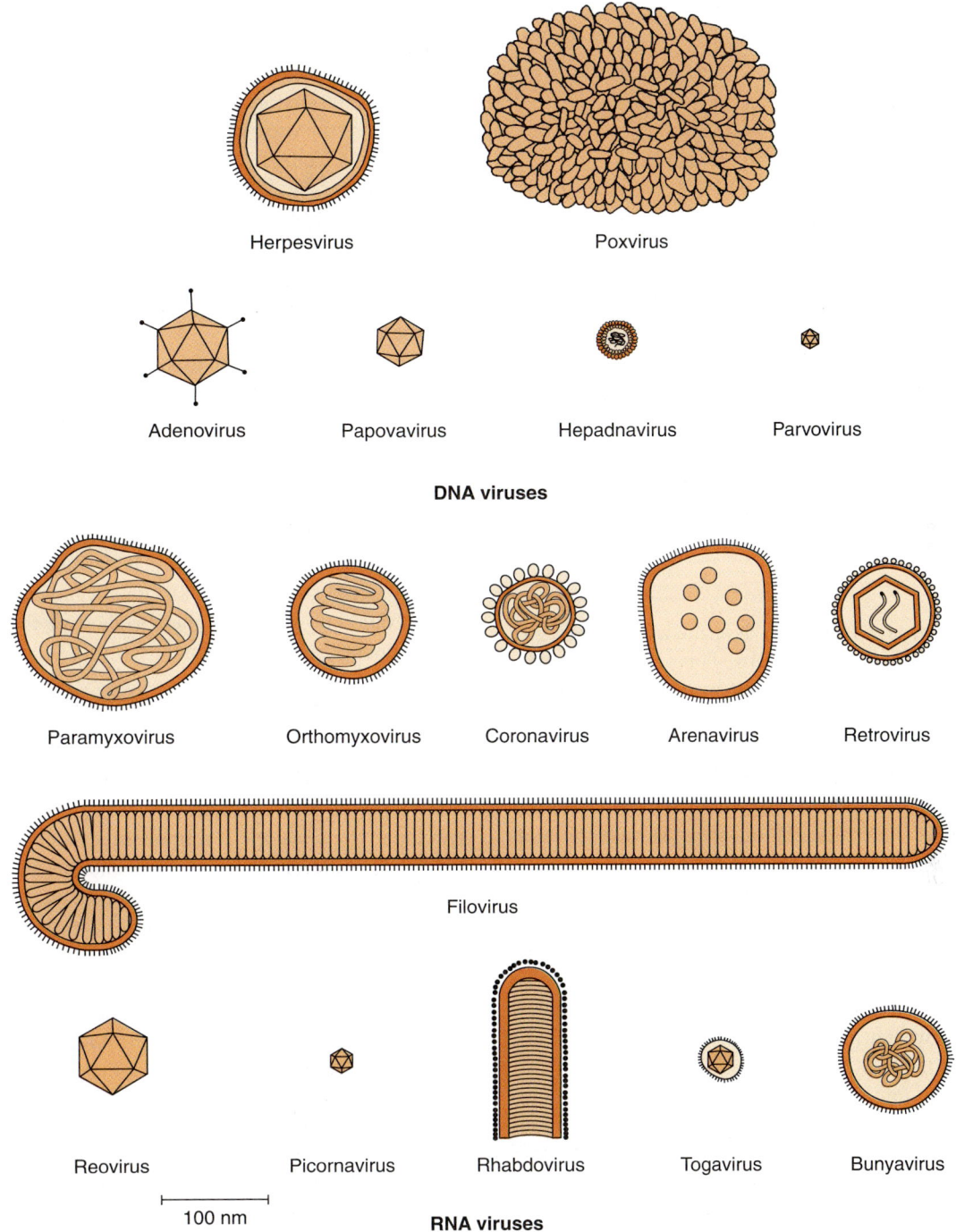

FIGURE 28-1 Shapes and sizes of medically important viruses. (Reproduced with permission from Fenner F, White DO: *Medical Virology,* 4th ed. Philadelphia, PA: Academic Press/Elsevier; 1994.)

VIRAL PROTEINS

Viral proteins serve several important functions. The capsid proteins **protect the genome** DNA or RNA from degradation by nucleases. The proteins on the surface of the virus **mediate the attachment** of the virus to specific receptors on the host cell surface. This interaction of the viral proteins with the cell receptor is the major determinant of **species and organ specificity**.

Outer viral proteins are also **important antigens** that induce neutralizing antibody and activate cytotoxic T cells to kill virus-infected cells. These outer viral proteins not only induce antibodies but are also the target of antibodies (i.e., antibodies bind to these viral proteins and prevent ["neutralize"] the virus from entering the cell and replicating). The outer proteins induce these immune responses following both the natural infection and the immunization (see later).

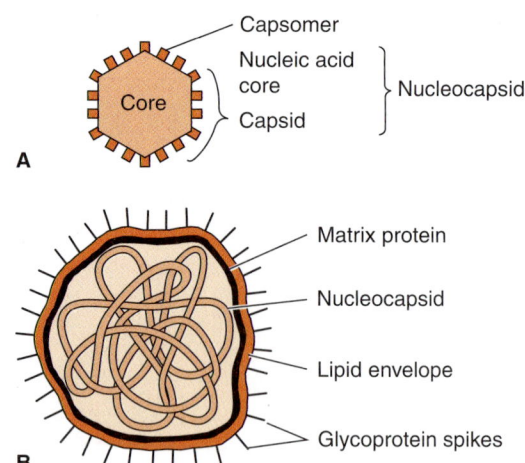

FIGURE 28–2 Cross-section of two types of virus particles. **A:** Nonenveloped virus with an icosahedral nucleocapsid. **B:** Enveloped virus with a helical nucleocapsid. (Reproduced with permission from Brooks GF, Butel JS, Ornston LN: *Jawetz, Melnick & Adelberg's Medical Microbiology*, 20th ed. New York, NY: McGraw Hill; 1995.)

The term "**serotype**" is used to describe a subcategory of a virus based on its surface antigens. For example, measles virus has one serotype, polioviruses have three serotypes, and rhinoviruses have over 100 serotypes. This is because all measles viruses have only one antigenic determinant on their surface protein that induces neutralizing antibody capable of preventing infection. In contrast, polioviruses have three different antigenic determinants on their surface proteins (i.e., poliovirus type 1 has one kind of antigenic determinant, poliovirus type 2 has a different antigenic determinant, and poliovirus type 3 has a different antigenic determinant from types 1 and 2); hence, polioviruses have three serotypes. Some viruses have multiple serotypes that change rapidly, some as rapidly as days, weeks, or months. Examples of these include influenza virus, human immunodeficiency virus (HIV), and hepatitis C virus (HCV).

There are three important medical implications of a virus having multiple serotypes. First is that a person can be immune (have antibodies) to poliovirus type 1 and still get the disease, poliomyelitis, caused by poliovirus types 2 or 3. Second is that the polio vaccine must contain all three serotypes in order to be completely protective. The third implication is that it is difficult to produce an effective vaccine if the antigenicity of the virus is constantly changing. For example, the virus in the influenza vaccine changes every year or two because new antigens appear. Further, the rapid change in the antigens of HIV and HCV has played an important role in the failure to produce a vaccine against these viruses.

Some of the internal viral proteins are structural (e.g., the capsid proteins of the enveloped viruses), whereas others are enzymes (e.g., the polymerases that synthesize the viral mRNA). The internal viral proteins vary depending on the virus. Some viruses have a DNA or RNA polymerase attached to the genome; others do not. If a virus has an envelope, then a matrix protein that mediates the interaction between the capsid proteins and the envelope proteins is present.

Some viruses produce proteins that act as "superantigens," similar in their action to the superantigens produced by bacteria, such as the toxic shock syndrome toxin of *Staphylococcus aureus* (see Chapters 15 and 60). Viruses known to produce superantigens include two members of the herpesvirus family, namely, Epstein–Barr virus and cytomegalovirus, and the retrovirus mouse mammary tumor virus. The current hypothesis offered to explain why these viruses produce a superantigen is that activation of CD4-positive T cells is required for replication of these viruses to occur.

Members of the herpesvirus family, such as herpes simplex virus and cytomegalovirus, have a structure called the **tegument**, located between the nucleocapsid and the envelope. The tegument contains a variety of regulatory proteins include transcription and translation factors that control either viral or cellular processes. For example, a tegument protein called viral protein-16 (VP-16) is a transcription factor that initiates immediate–early m-RNA and protein synthesis which is required for viral replication to proceed (see Chapter 37). Other tegument proteins inhibit the activation of interferon, thereby allowing the virus to evade this component of innate immunity.

VIRAL ENVELOPE

In addition to the capsid and internal proteins, there are two other types of proteins, both of which are associated with the envelope. The **envelope** is a **lipoprotein** membrane composed of lipid derived from the host cell membrane and protein that is virus-specific. Furthermore, there are frequently glycoproteins in the form of spikelike projections on the surface, which attach to host cell receptors during the entry of the virus into the cell.

The viral envelope is acquired as the virus exits from the cell in a process called "budding" (see Chapter 29). The envelope of most viruses is derived from the cell's outer membrane, with the notable exception of herpesviruses that derive their envelope from the cell's nuclear membrane.

In general, the presence of an envelope confers **instability** on the virus. Enveloped viruses are *more* sensitive to heat, drying, detergents, and lipid solvents such as alcohol and ether than are nonenveloped (nucleocapsid) viruses, which are composed only of nucleic acid and capsid proteins.

An interesting clinical correlate of this observation is that virtually all viruses that are transmitted by the fecal–oral route (those that have to survive in the environment) do *not* have an envelope; that is, they are naked nucleocapsid viruses. These include viruses such as hepatitis A virus, poliovirus, Coxsackie virus, echovirus, norovirus, and rotavirus. In contrast, enveloped viruses are most often transmitted by direct contact, such as by blood or by sexual transmission. Examples of these include HIV, herpes simplex virus type 2, and hepatitis B and C viruses. Other enveloped viruses are transmitted directly by insect bite (e.g., yellow fever virus and West Nile virus) or by animal bite (e.g., rabies virus).

Many other enveloped viruses are transmitted from person to person in respiratory aerosol droplets, such as influenza virus, measles virus, rubella virus, respiratory syncytial virus, and varicella-zoster virus. If the droplets do not infect directly, they can dry out in the environment, and these enveloped viruses are rapidly inactivated. Note that rhinoviruses, which are transmitted by respiratory droplets, are naked nucleocapsid viruses and can survive in the environment for significant periods. Therefore, they can also be transmitted by hands that make contact with the virus on contaminated surfaces.

As described earlier in this chapter, the surface proteins of the virus, whether they are the capsid proteins or the envelope glycoproteins, are the principal **antigens** against which the host mounts its immune response to viruses. They are also the determinants of type specificity (often called the **serotype**). There is often little cross-protection between different serotypes. Viruses that have multiple serotypes (i.e., have antigenic variants) have an enhanced ability to evade our host defenses because antibody against one serotype will not protect against another serotype.

ATYPICAL VIRUS-LIKE AGENTS

There are four exceptions to the typical virus as described earlier:

(1) **Defective** viruses are composed of viral nucleic acid and proteins but cannot replicate without a "helper" virus, which provides the missing function. Defective viruses usually have a mutation or a deletion of part of their genetic material. During the growth of most human viruses, many more defective than infectious virus particles are produced. The ratio of defective to infectious particles can be as high as 100:1. Because these defective particles can interfere with the growth of the infectious particles, it has been hypothesized that the defective viruses may aid in recovery from an infection by limiting the ability of the infectious particles to grow.

(2) **Pseudovirions** contain host cell DNA instead of viral DNA within the capsid. They are formed during infection with certain viruses when the host cell DNA is fragmented and pieces of it are incorporated within the capsid protein. Pseudovirions can infect cells, but they do not replicate.

(3) **Viroids** consist solely of a single molecule of circular RNA without a protein coat or envelope. There is extensive homology between bases in the viroid RNA, leading to large double-stranded regions. The RNA is quite small (molecular weight 1×10^5) and apparently does not code for any protein. Nevertheless, viroids replicate, but the mechanism is unclear. They cause several plant diseases but are not implicated in any human disease.

(4) **Prions** are infectious particles that are composed **solely of protein** (i.e., they contain no detectable nucleic acid). They are implicated as the cause of certain "slow" diseases called **transmissible spongiform encephalopathies**, which include such diseases as Creutzfeldt-Jakob disease in humans and scrapie in sheep (see Chapter 44). Because neither DNA nor RNA has been detected in prions, they are clearly different from viruses (Table 28–1). Furthermore, electron microscopy

TABLE 28–1 Comparison of Prions and Conventional Viruses

Feature	Prions	Conventional Viruses
Particle contains nucleic acid	No	Yes
Particle contains protein	Yes, encoded by cellular genes	Yes, encoded by viral genes
Inactivated rapidly by ultraviolet light or heat	No	Yes
Appearance in electron microscope	Filamentous rods (amyloid-like)	Icosahedral or helical symmetry
Infection induces antibody	No	Yes
Infection induces inflammation	No	Yes

reveals filaments rather than virus particles. Prions are much **more resistant** to inactivation by ultraviolet light and heat than are viruses. They are remarkably resistant to formaldehyde and nucleases. However, they are inactivated by hypochlorite, NaOH, and autoclaving. Hypochlorite is used to sterilize surgical instruments and other medical supplies that cannot be autoclaved.

Prions are composed of a single glycoprotein with a molecular weight of 27,000 to 30,000. With scrapie prions as the model, it was found that this protein is encoded by a single **cellular** gene. This gene is found in equal numbers in the cells of both infected and uninfected animals. Furthermore, the amount of prion protein mRNA is the same in uninfected as in infected cells. In view of these findings, **posttranslational** modifications of the prion protein are hypothesized to be the important distinction between the protein found in infected and uninfected cells.

There is evidence that a change in the conformation from the normal alpha-helical form (known as PrP^C, or prion protein cellular) to the abnormal beta-pleated sheet form (known as PrP^{SC}, or prion protein scrapie) is the important modification. The abnormal form then recruits additional normal forms to change their configuration, and the number of abnormal pathogenic particles increases. Although prions are composed only of proteins, specific cellular RNAs enhance the conversion of the normal alpha-helical form to the pathologic beta-pleated sheet form.

Evidence that recruitment is an essential step comes from "knockout" mice in which the gene for the prion protein is nonfunctional and no prion protein is made. These mice do not get scrapie despite the injection of the pathogenic scrapie prion protein.

The function of the normal prion protein is uncertain but it appears that it is one of the signal transduction proteins in neurons. There is some evidence that its function is to regulate the N-methyl-D-aspartate receptor on the postsynaptic terminal by binding copper ions.

Knockout mice in which the gene encoding the prion protein is inactive appear normal. The prion protein in normal cells is protease-sensitive, whereas the prion protein in infected cells is protease-resistant, probably because of the change in conformation.

The observation that the prion protein is the product of a normal cellular gene may explain why **no immune response** is formed against this protein (i.e., tolerance occurs). Similarly, there is **no inflammatory response** in infected brain tissue.

A vacuolated (**spongiform**) appearance is found, without inflammatory cells. Prion proteins in infected brain tissue form rod-shaped particles that are morphologically and histochemically indistinguishable from **amyloid**, a substance found in the brain tissue of individuals with various central nervous system diseases (as well as diseases of other organs).

PEARLS

Virus Size & Structure

- Viruses range in size from that of large proteins (~20 nm) to that of the smallest cells (~300 nm). Most viruses appear as spheres or rods in the electron microscope.
- Viruses contain **either DNA or RNA, but not both**.
- All viruses have a **protein coat called a capsid** that covers the genome. The capsid is composed of repeating subunits called capsomers. In some viruses, the capsid is the outer surface, but in other viruses, the capsid is covered with a lipoprotein **envelope** that becomes the outer surface. The structure composed of the nucleic acid genome and the capsid proteins is called the **nucleocapsid**.
- The repeating subunits of the capsid give the virus a symmetric appearance that is useful for classification purposes. Some viral nucleocapsids have **spherical (icosahedral) symmetry**, whereas others have **helical symmetry**.
- All human viruses that have a helical nucleocapsid are enveloped (i.e., there are no naked helical viruses that infect humans). Viruses that have an icosahedral nucleocapsid can be either enveloped or naked.

Viral Nucleic Acids

- The genome of some viruses is **DNA**, whereas the genome of others is **RNA**. These DNA and RNA genomes can be either **single-stranded** or **double-stranded**.
- Some RNA viruses, such as influenza virus and rotavirus, have a **segmented genome** (i.e., the genome is in several pieces).
- All viruses have one copy of their genome (haploid) except retroviruses, which have two copies (diploid).

Viral Proteins

- Viral surface proteins mediate **attachment to host cell receptors**. This interaction **determines the host specificity and organ specificity** of the virus.
- The surface proteins are the **targets of antibody** (i.e., antibody bound to these surface proteins prevents the virus from attaching to the cell receptor). This "neutralizes" (inhibits) viral replication.
- Viruses also have internal proteins, some of which are **DNA or RNA polymerases**.

- The **matrix protein** mediates the interaction between the viral nucleocapsid proteins and the envelope proteins.
- Some viruses produce **antigenic variants** of their surface proteins that allow the viruses to evade our host defenses. Antibody against one antigenic variant (**serotype**) will not neutralize a different serotype. Some viruses have one serotype; others have multiple serotypes.

Viral Envelope

- The viral **envelope** consists of a membrane that contains lipid derived from the host cell and proteins encoded by the virus. Typically, the envelope is acquired as the virus exits from the cell in a process called **budding**.
- Viruses with an envelope are less stable (i.e., they are more easily inactivated) than naked viruses (those without an envelope). In general, enveloped viruses are transmitted by direct contact via blood and body fluids, whereas naked viruses can survive longer in the environment and can be transmitted by indirect means such as the fecal–oral route.

Prions

- **Prions** are infectious particles composed **entirely of protein**. They have **no DNA or RNA**.
- They cause diseases such as Creutzfeldt-Jakob disease and kuru in humans and mad cow disease and scrapie in animals. These diseases are called **transmissible spongiform encephalopathies**. The term **spongiform** refers to the spongelike appearance of the brain seen in these diseases. The holes of the sponge are vacuoles resulting from dead neurons.
- Prion proteins are **encoded by a cellular gene**. When these proteins are in the **normal, alpha-helix configuration, they are nonpathogenic**, but when their configuration changes to a **beta-pleated sheet, they aggregate into filaments, which disrupts neuronal function and results in the symptoms of disease**.
- Prions are **highly resistant to inactivation by ultraviolet light, heat**, and other inactivating agents. As a result, they have been inadvertently transmitted by human growth hormone and neurosurgical instruments.
- **Because prions are normal human proteins, they do not elicit an inflammatory response or an antibody response** in humans.

SELF-ASSESSMENT QUESTIONS

1. The proteins on the external surface of viruses serve several important functions. Regarding these proteins, which one of the following statements is most accurate?

   (A) They are the antigens against which neutralizing antibodies are formed.

   (B) They are the polymerases that synthesize viral messenger RNA.

   (C) They are the proteases that degrade cellular proteins leading to cell death.

   (D) They are the proteins that regulate viral transcription.

   (E) Change in conformation of these proteins can result in prion-mediated diseases such as Creutzfeldt-Jakob disease.

2. If a virus has an envelope, it is more easily inactivated by lipid solvents and detergents than viruses that do not have an envelope. Which one of the following viruses is the most sensitive to inactivation by lipid solvents and detergents?

   (A) Coxsackie virus

   (B) Hepatitis A virus

   (C) Herpes simplex virus

   (D) Poliovirus

   (E) Rotavirus

3. Regarding the tegument, which one of the following is most accurate?

   (A) It uncoats the virion within the phagocytic vesicle.

   (B) It mediates the binding of the virion to the cell surface.

   (C) It guides the viral core from the cytoplasm to the nucleus.

   (D) It is the site at which new virions bud from the surface of the infected cell.

   (E) It is the location of proteins in the virion that act as viral transcription factors.

ANSWERS

(1) **(A)**

(2) **(C)**

(3) **(E)**

PRACTICE QUESTIONS: USMLE & COURSE EXAMINATIONS

Questions on the topics discussed in this chapter can be found in the Basic Virology section of Part XIII: USMLE (National Board) Practice Questions starting on page 744. Also see Part XIV: USMLE (National Board) Practice Examination starting on page 775.

Replication

INTRODUCTION

The viral replication cycle is described in this chapter in two different ways. The first approach is a growth curve, which shows the amount of virus produced at different times after infection. The second is a stepwise description of the specific events within the cell during virus growth.

VIRAL GROWTH CURVE

The growth curve depicted in Figure 29–1 shows that when one **virion** (one virus particle) infects a cell, it can replicate in approximately 10 hours to produce hundreds of virions within that cell. This remarkable amplification explains how viruses spread rapidly from cell to cell. Note that the time required for the growth cycle varies; it is minutes for some bacterial viruses and hours for some human viruses.

The first event shown in Figure 29–1 is quite striking: the virus disappears, as represented by the solid line dropping to the x axis. Although the virus particle, as such, is no longer present, the viral nucleic acid continues to function and begins to accumulate within the cell, as indicated by the dotted line. The time during which no virus is found inside the cell is known as the **eclipse period**. The eclipse period ends with the appearance of virus (solid line). The **latent period**, in contrast, is defined as the time from the onset of infection to the appearance of virus extracellularly. Note that infection begins with one virus particle and ends with several hundred virus particles having been produced; this type of reproduction is unique to viruses.

Alterations of cell morphology accompanied by marked derangement of cell function begin toward the end of the latent period. This **cytopathic effect** (CPE) culminates in the lysis and death of cells. CPE can be seen in the light microscope and, when

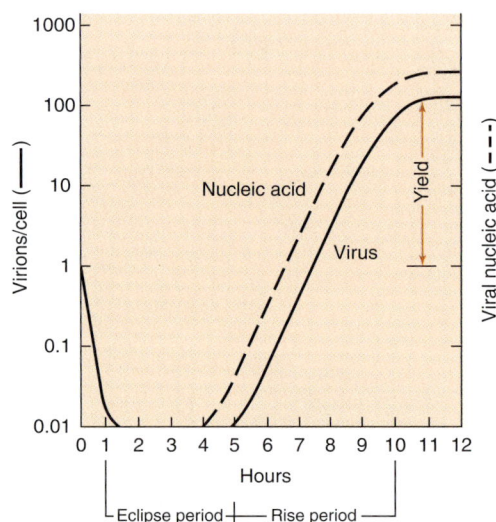

FIGURE 29–1 Viral growth curve. The figure shows that one infectious virus particle (virion) entering a cell at the time of infection results in more than 100 infectious virions 10 hours later, a remarkable increase. Note the eclipse period during which no infectious virus is detectable within the infected cells. In this growth curve, the amount of infecting virus is 1 virion/cell (i.e., 1 infectious unit/cell). (Reproduced with permission from Brooks GF, Butel JS, Ornston LN: *Jawetz, Melnick & Adelberg's Medical Microbiology*, 20th ed. New York, NY: McGraw Hill; 1995.)

observed, is an important initial step in the laboratory diagnosis of viral infection. Not all viruses cause CPE; some can replicate while causing little morphologic or functional change in the cell.

SPECIFIC EVENTS DURING THE GROWTH CYCLE

An overview of the events is described in Table 29–1 and presented in diagrammatic fashion in Figure 29–2. The infecting parental virus particle attaches to the cell membrane and then penetrates the host cell. The viral genome is "uncoated" by removing the capsid proteins, and the genome is free to function. Early mRNA and proteins are synthesized; the **early proteins are enzymes** used to replicate the viral genome. Late mRNA and proteins are then synthesized. These **late proteins are the structural, capsid proteins**. The progeny virions are assembled from the replicated genetic material, and newly made capsid proteins and are then released from the cell.

Another, more general way to describe the growth cycle is as follows: (1) early events (i.e., **attachment, penetration,** and **uncoating**); (2) middle events (i.e., **viral messenger RNA synthesis, viral protein synthesis and processing,** and **viral genome replication**); and (3) late events (i.e., **viral assembly** and **release**). With this sequence in mind, each stage will be described in more detail.

Attachment, Penetration, & Uncoating

The proteins on the surface of the virion attach to specific receptor proteins on the cell surface through weak, noncovalent bonding. The **specificity of attachment determines the host range** of the virus. Some viruses have a narrow range, whereas others have quite a broad range. For example, poliovirus can

TABLE 29–1   Stages of the Viral Growth Cycle

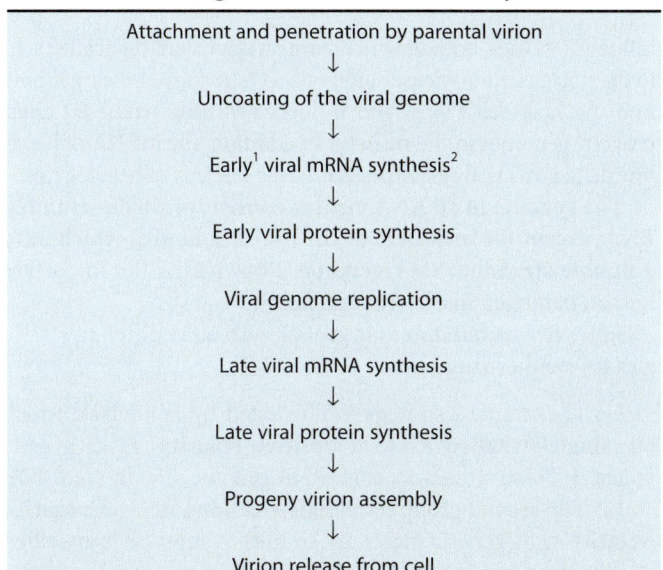

Attachment and penetration by parental virion
↓
Uncoating of the viral genome
↓
Early[1] viral mRNA synthesis[2]
↓
Early viral protein synthesis
↓
Viral genome replication
↓
Late viral mRNA synthesis
↓
Late viral protein synthesis
↓
Progeny virion assembly
↓
Virion release from cell

[1]*Early* is defined as the period before genome replication. Not all viruses exhibit a distinction between early and late functions. In general, early proteins are enzymes, whereas late proteins are structural components of the virus.

[2]In some cases, the viral genome is functionally equivalent to mRNA; thus early mRNA need not be synthesized.

enter the cells of only humans and other primates, whereas rabies virus can enter all mammalian cells. The **organ specificity** of viruses is governed by receptor interaction as well.

The receptors for viruses on the cell surface are proteins that have other functions in the life of the cell. Probably the best known is the CD4 protein that serves as one of the receptors for human immunodeficiency virus (HIV) but whose normal function is the binding of class 2 major histocompatibility complex (MHC) proteins involved in the activation of helper T cells. A few other examples will serve to illustrate the point: rabies virus binds to the acetylcholine receptor, Epstein–Barr virus binds to a complement receptor, herpes simplex virus (HSV) type 1 binds to the fibroblast growth factor receptor, and vaccinia virus binds to the receptor for epidermal growth factor.

Enveloped viruses undergo another process of attachment called **fusion in which the envelope of the virion fuses with the outer membrane of the cell.** For example, cell surface proteases are required to cleave the hemagglutinin of influenza virus and the spike protein of coronaviruses to reveal a fusion protein that mediates entry of these viruses into the host cell. The clinical importance of fusion is illustrated by the antiviral drug, enfuvirtide, which blocks HIV from entering the cell by inhibiting the fusion process.

The virus particle penetrates by being engulfed in a vesicle, within which the process of uncoating begins. A low pH within the vesicle favors uncoating. Rupture of the vesicle or fusion of the outer layer of virus with the vesicle membrane deposits the inner core of the virus into the cytoplasm.

The term "uncoating" refers to the process by which most proteins of the incoming virus are removed and the incoming genome is free to function. Note that one protein that is *not* removed in the uncoating process is the polymerase that some viruses require to continue the replication cycle. Important examples of these polymerases are the reverse transcriptase and integrase of HIV and the transcriptase of viruses with a negative-polarity RNA genome such as influenza virus, rabies virus, and measles virus.

It is appropriate at this point to describe the phenomenon of **infectious nucleic acid** because it provides a transition between the concepts of host specificity described earlier and early genome functioning, which is discussed later. Note that we are discussing whether the purified genome is infectious. All viruses are "infectious" in a person or in cell culture, but not all *purified genomes* are infectious.

Infectious nucleic acid is purified viral DNA or RNA (without any protein) that can carry out the entire viral growth cycle and result in the production of complete virus particles. This is interesting from three points of view:

(1) The observation that purified nucleic acid is infectious is the definitive proof that nucleic acid, not protein, is the genetic material.

(2) Infectious nucleic acid can bypass the host range specificity provided by the viral protein–cell receptor interaction. For example, although intact poliovirus can grow only in primate cells, purified poliovirus RNA can enter nonprimate cells, go through its usual growth cycle, and produce normal poliovirus. The poliovirus produced in the nonprimate cells can infect only primate cells because it now has its capsid proteins. These observations

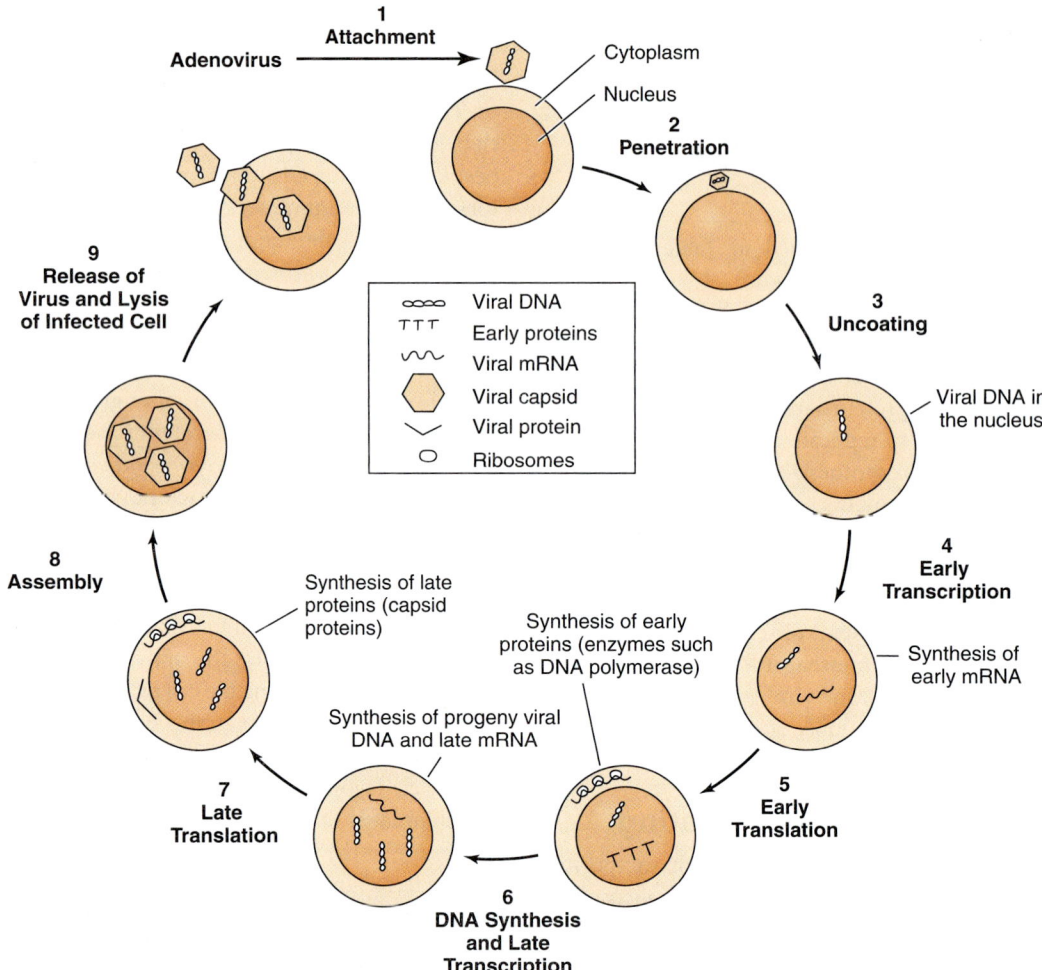

FIGURE 29–2 Viral growth cycle. The growth cycle of adenovirus, a nonenveloped DNA virus, is shown. (Reproduced with permission from Jawetz E, Melnick JL, Adelberg EA: *Review of Medical Microbiology,* 16th ed. New York, NY: McGraw Hill; 1984.)

indicate that the internal functions of the nonprimate cells are capable of supporting viral growth once entry has occurred.

(3) Only certain viruses yield infectious nucleic acid. The reason for this is discussed later. Note that all viruses are infectious, but not all *purified* viral DNAs or RNAs (genomes) are infectious.

Viral Messenger RNA Synthesis

The first step in viral gene expression is **mRNA synthesis**. It is at this point that viruses follow different pathways depending on the nature of their nucleic acid and the part of the cell in which they replicate (Figure 29–3).

DNA viruses, with one exception, **replicate in the nucleus** and use the host cell DNA-dependent RNA polymerase to synthesize their mRNA. The poxviruses are the exception because they replicate in the cytoplasm, where they do not have access to the host cell RNA polymerase. They therefore carry their own polymerase within the virus particle. **The genome of all DNA viruses consists of double-stranded DNA, except for the parvoviruses, which have a single-stranded DNA genome** (Table 29–2).

Most RNA viruses undergo their entire replicative cycle in the cytoplasm. The two principal exceptions are retroviruses and

influenza viruses, both of which have an important replicative step in the nucleus. Retroviruses integrate a DNA copy of their genome into the host cell DNA, and influenza viruses synthesize their progeny genomes in the nucleus. In addition, the mRNA of hepatitis delta virus is also synthesized in the nucleus of hepatocytes.

The genome of all RNA viruses consists of single-stranded RNA, except for members of the reovirus family, which have a double-stranded RNA genome. Rotavirus is the important human pathogen in the reovirus family.

RNA viruses fall into four groups with quite different strategies for synthesizing mRNA (Table 29–3).

(1) The simplest strategy is illustrated by poliovirus, which has **single-stranded RNA** of **positive polarity**[1] as its genetic material. These viruses use their RNA genome directly as mRNA.

(2) The second group has **single-stranded RNA of negative polarity** as its genetic material. An mRNA must be transcribed

---

[1]Positive polarity is defined as an RNA with the same base sequence as the mRNA. RNA with negative polarity has a base sequence that is complementary to the mRNA. For example, if the mRNA sequence is an A-C-U-G, an RNA with negative polarity would be U-G-A-C and an RNA with positive polarity would be A-C-U-G.

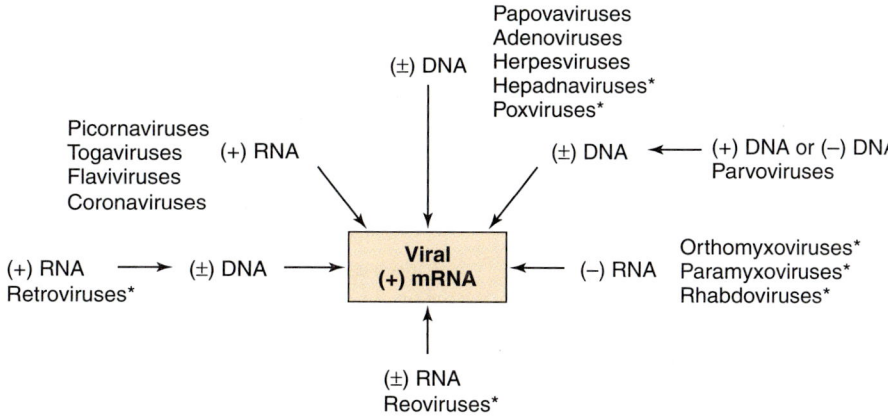

Legend: (+) = Strand with same polarity as mRNA   (±) = Double-stranded
(−) = Strand complementary to mRNA   * = These viruses contain a polymerase in the virion

FIGURE 29–3 Synthesis of viral mRNA by medically important viruses. The following information starts at the top of the figure and moves clockwise: Viruses with a double-stranded DNA genome (e.g., papovaviruses such as human papillomavirus) use host cell RNA polymerase to synthesize viral mRNA. Note that hepadnaviruses (e.g., hepatitis B virus) contain a virion DNA polymerase that synthesizes the missing portion of the DNA genome, but the viral mRNA is synthesized by host cell RNA polymerase. Parvoviruses use host cell DNA polymerase to synthesize viral double-stranded DNA and host cell RNA polymerase to synthesize viral mRNA. Viruses with a single-stranded, negative-polarity RNA genome (e.g., orthomyxoviruses such as influenza virus) use a virion RNA polymerase to synthesize viral mRNA. Viruses with a double-stranded RNA genome (e.g., reoviruses) use a virion RNA polymerase to synthesize viral mRNA. Some viruses with a single-stranded, positive-polarity RNA genome (e.g., retroviruses) use a virion DNA polymerase to synthesize a DNA copy of the RNA genome but a host cell RNA polymerase to synthesize the viral mRNA. Some viruses with a single-stranded, positive-polarity RNA genome (e.g., picornaviruses) use the virion genome RNA itself as their mRNA. (Reproduced with permission from Ryan K: *Sherris Medical Microbiology*, 3rd ed. New York, NY: McGraw Hill; 1994.)

TABLE 29–2 Important Features of DNA Viruses

DNA Genome	Location of Replication	Virion Polymerase	Infectivity of Genome	Prototype Human Virus
Single strand	Nucleus	No[1,2]	Yes	Parvovirus B19
Double strand				
Circular	Nucleus	No[1]	Yes	Papillomavirus
Circular; partially single strand	Nucleus	Yes[3]	No	Hepatitis B virus
Linear	Nucleus	No[1]	Yes	Herpesvirus, adenovirus
Linear	Cytoplasm	Yes	No	Smallpox virus, vaccinia virus

[1]mRNA is synthesized by host cell RNA polymerase in the nucleus.

[2]Single-stranded genome DNA is converted to double-stranded DNA by host cell polymerase. A virus-encoded DNA polymerase then synthesizes progeny DNA.

[3]Hepatitis B virus uses a virion-encoded RNA-dependent DNA polymerase to synthesize its progeny DNA with full-length mRNA as the template. This enzyme is a type of "reverse transcriptase" but functions at a different stage in the replicative cycle than does the reverse transcriptase of retroviruses.

Note: All DNA viruses encode their own DNA polymerase that replicates the genome. They do not use the host cell DNA polymerase (with the minor exception of the parvoviruses as mentioned earlier).

TABLE 29–3 Important Features of RNA Viruses

RNA Genome	Polarity	Virion Polymerase	Source of mRNA	Infective Genome	Prototype Human Virus
Single strand, nonsegmented	+	No	Genome	Yes	Poliovirus, SARS-CoV-2[4]
Single strand					
Nonsegmented	−	Yes	Transcription	No	Measles virus, rabies virus
Segmented	−	Yes	Transcription	No	Influenza virus
Double strand, segmented	±	Yes	Transcription	No	Rotavirus
Single strand, diploid	+	Yes[1]	Transcription[2]	No[3]	HTLV, HIV[4]

[1]Retroviruses contain an RNA-dependent DNA polymerase.

[2]mRNA transcribed from DNA intermediate.

[3]Although the retroviral genome RNA is not infectious, the DNA intermediate is.

[4]HTLV = human T-cell leukemia virus; HIV = human immunodeficiency virus. SARS-CoV-2 = severe acute respiratory syndrome-coronavirus-2

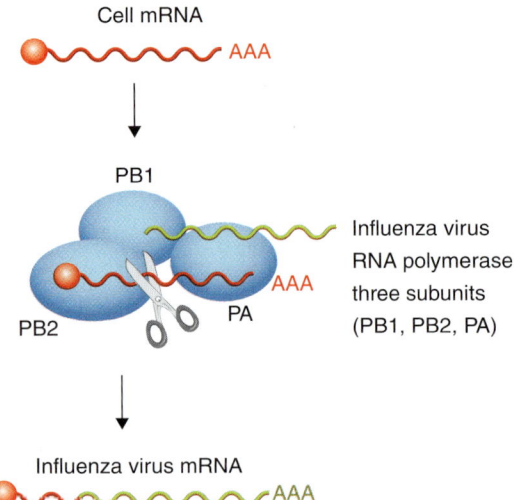

Cell mRNA

PB1

Influenza virus RNA polymerase three subunits (PB1, PB2, PA)

PB2   PA

Influenza virus mRNA

FIGURE 29–4 Cap snatching by influenza virus. **Top**: Cellular mRNA (red wavy line) is shown with its 5′ methylated cap (red ball) and poly A tail. **Center**: Influenza virus RNA polymerase is shown with its three subunits PB1, PB2, and PA. The PB2 subunit of the viral polymerase complex binds the 5′ cap and the adjacent 10 to 15 nucleotides. The PA subunit is a ribonuclease that cleaves the cap (scissor) from cellular mRNA. The PB1 subunit initiates synthesis of viral mRNA (green wavy line) and the cap and adjacent nucleotides are transferred to the viral mRNA. **Bottom**: The viral mRNA with its 5′ cap can now be translated into viral protein by cellular ribosomes. (Modified with permission from Boivin S, Cusack S, Ruigrok RW, Hart DJ. Influenza A virus polymerase: structural insights into replication and host adaptation mechanisms, J Biol Chem 2010 Sep 10; 285(37):28411-28417.)

by using the negative strand as a template. Because the cell does not have an RNA polymerase capable of using RNA as a template, the virus carries its own virus-encoded **RNA-dependent RNA polymerase**. There are two subcategories of negative-polarity RNA viruses: those that have a single piece of RNA (e.g., measles virus [a paramyxovirus] or rabies virus [a rhabdovirus]) and those that have multiple pieces of RNA (e.g., influenza virus [a myxovirus]).

Influenza virus adds a methylated guanosine "cap" to its messenger RNAs. The cap is obtained from cellular mRNAs in a process called **cap snatching** (see Figure 29–4). An influenza virus-encoded ribonuclease mediates this reaction. The cap allows efficient translation of viral mRNAs in the cytoplasm of the infected cell. Baloxavir inhibits the viral ribonuclease and is approved for use in the treatment of influenza (see Chapter 35).

Certain viruses, such as arenaviruses and some bunyaviruses, have a segmented RNA genome, most of which is

negative stranded, but there are some positive-strand regions as well. RNA segments that contain both positive-polarity and negative-polarity regions are called **ambisense.**

(3) The third group has **double-stranded RNA** as its genetic material. Because the cell has no enzyme capable of transcribing this RNA into mRNA, the virus carries its own polymerase. Note that plus strand in double-stranded RNA cannot be used as mRNA because it is hydrogen-bonded to the negative strand. Rotavirus, an important cause of diarrhea in children, has 11 segments of double-stranded RNA.

(4) The fourth group, exemplified by retroviruses, has single-stranded RNA of positive polarity that is transcribed into double-stranded DNA by the RNA-dependent DNA polymerase (**reverse transcriptase**) carried by the virus. This DNA copy is then transcribed into viral mRNA by the regular host cell RNA polymerase (polymerase II). Retroviruses are the only family of viruses that are **diploid** (i.e., that have two copies of their genome RNA).

These differences explain why some viruses yield infectious nucleic acid and others do not. **Viruses that do not require a polymerase in the virion can produce infectious DNA or RNA.** By contrast, viruses such as the poxviruses, the negative-stranded RNA viruses, the double-stranded RNA viruses, and the retroviruses, which require a virion polymerase, cannot yield infectious nucleic acid. Several additional features of viral mRNA are described in the "Viral messenger RNA" box.

Note that two families of viruses utilize a reverse transcriptase (an RNA-dependent DNA polymerase) during their replicative cycle, but the purpose of the enzyme during the cycle is different. As described in Table 29–4, **retroviruses, such as HIV, use their genome RNA as the template to synthesize a DNA copy early in the replicative cycle.** However, **hepadnaviruses, such as hepatitis B virus (HBV), use RNA as the template to produce their DNA genome late in the replicative cycle.** The clinical importance of this is that some antiviral drugs such as lamivudine are effective against infections caused by both HIV and HBV because they inhibit the reverse transcriptase of both viruses.

Note that the DNA polymerase of HBV has **both DNA-dependent and RNA-dependent activity that function at different stages of the replicative cycle.** The DNA-dependent DNA polymerase activity in the virion synthesizes the missing section of the genome and produces a complete covalent circular DNA shortly after entering the cell. The RNA-dependent DNA polymerase activity uses a full-length RNA copy of the DNA genome as the template to synthesize the progeny DNA genomes late in the replicative cycle. See Chapter 41 for additional information on HBV replication.

TABLE 29–4 Comparison of Reverse Transcriptase Activity of HIV (Retroviruses) and HBV (Hepadnaviruses)

Type of Virus	RNA Template for Reverse Transcriptase	DNA Product of Reverse Transcriptase	Phase of Replication When Reverse Transcriptase Is Active
HIV (retrovirus)	Genome	Not genome	Early
HBV (hepadnavirus)	Not genome	Genome	Late

HBV = hepatitis B virus; HIV = human immunodeficiency virus.

TABLE 29-5   Origin of the Genes That Encode the Polymerases That Synthesize the Viral Genome

Type of Polymerase	Polymerase Encoded By	Medically Important Viruses
DNA	Cell	Parvovirus B19, human papillomavirus
DNA	Virus	Herpesviruses (HSV, VZV, CMV, EBV), adenovirus, hepatitis B virus, smallpox virus
RNA	Cell	HIV, HTLV, HDV
RNA	Virus	Poliovirus, HAV, HCV, influenza virus, measles virus, respiratory syncytial virus, rabies virus, rubella virus, rotavirus, Ebola virus, arenavirus, hantavirus, SARS-CoV-2

CMV = cytomegalovirus; EBV = Epstein–Barr virus; HAV = hepatitis A virus; HCV = hepatitis C virus; HDV= hepatitis D virus; HIV = human immunodeficiency virus; HSV = herpes simplex virus; HTLV = human T-cell leukemia virus; SARS-CoV-2 = severe acute respiratory syndrome-coronavirus-2; VZV = varicella zoster virus.

Viral Protein Synthesis

Once the viral mRNA of either DNA or RNA viruses is synthesized, it is translated by host cell ribosomes into viral proteins, some of which are **early proteins** (i.e., **enzymes** required for replication of the viral genome) and others of which are **late proteins** (i.e., **structural proteins**) of the progeny viruses. (The term *early* is defined as occurring before the replication of the genome, and *late* is defined as occurring after genome replication.) The most important of the early proteins for many RNA viruses is the polymerase that will synthesize many copies of viral genetic material for the progeny virus particles.

No matter how a virus makes its mRNA, most viruses make a virus-encoded polymerase (a **replicase**) that replicates the genome (i.e., that makes many copies of the parental genome that will become the genome of the progeny virions). Table 29–5 describes which viruses encode their own replicase and which viruses use host cell polymerases to replicate their genome.

Viral Protein Processing

Some viral mRNAs are translated into **precursor polypeptides that must be cleaved by proteases** to produce the functional structural proteins (Figure 29–5 and Table 29–6), whereas other

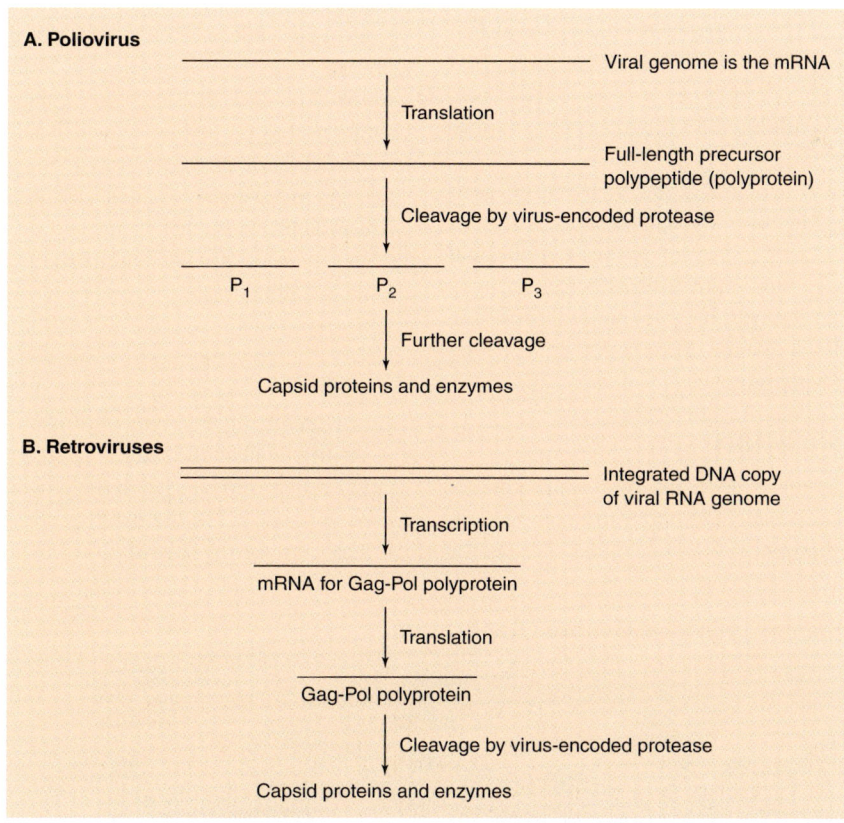

FIGURE 29-5 Synthesis of viral precursor polypeptides. **A:** Poliovirus mRNA is translated into a full-length precursor polypeptide, which is cleaved by the virus-encoded protease into the functional viral proteins. **B:** Retroviral mRNAs are translated into precursor polypeptides, which are then cleaved by the virus-encoded protease into the functional viral proteins. The cleavage of the Gag-Pol precursor polyprotein by the virion protease occurs in the immature virion after it has budded out from the cell membrane. The cleavage produces the capsid protein (p24), the matrix protein (p17), and enzymes such as the reverse transcriptase and the integrase. The cleavage of the Env polyprotein is carried out by a cellular protease, not by the virion protease. Inhibitors of the virion protease are effective drugs against human immunodeficiency virus.

TABLE 29-6 **Virus-Encoded Proteases of Medically Important Viruses**

Virus Family	Nature of Polyprotein	Site of Proteolytic Cleavage	Medically Important Viruses
Picornavirus	Single polypeptide formed by translation of entire genome RNA	Cytoplasm	Poliovirus, rhinovirus, hepatitis A virus, coxsackievirus
Flavivirus	Single polypeptide formed by translation of entire genome RNA	Cytoplasm	Hepatitis C virus, yellow fever virus, dengue virus
Togavirus	More than one polypeptide formed by translation of subgenomic mRNAs	Cytoplasm	Eastern and Western equine encephalitis viruses, rubella virus
Coronaviruses	Two polyproteins formed by translation of two-thirds of genome RNA	Cytoplasm	SARS-CoV-2
Retroviruses	More than one polypeptide formed by translation of subgenomic mRNAs	Budding virion	Human immunodeficiency virus, human T-cell leukemia virus

viral mRNAs are translated directly into structural proteins. A striking example of the former occurs during the replication of picornaviruses (e.g., poliovirus, rhinovirus, and hepatitis A virus), in which the genome RNA, acting as mRNA, is translated into a **single polypeptide**, which is then cleaved by a virus-coded protease into various proteins. This protease is one of the proteins in the single polypeptide, an interesting example of a protease acting on its own polypeptide.

Another important family of viruses in which precursor polypeptides are synthesized is the retrovirus family. For example, the *gag* and *pol* genes of HIV are translated into precursor polypeptides, which are then cleaved by a virus-encoded protease. It is this protease that is inhibited by the drugs classified as **protease inhibitors**.

Coronaviruses, such as SARS-CoV-2 and flaviviruses, such as hepatitis C virus and yellow fever virus, also synthesize precursor polypeptides that must be cleaved to form functional proteins by a virus-encoded protease. In contrast, other viruses, such as influenza virus and rotavirus, have segmented genomes, and each segment encodes a specific functional polypeptide rather than a precursor polypeptide.

Viral Genome Replication

Replication of the viral genome is governed by the principle of **complementarity**, which requires that a strand with a complementary base sequence be synthesized; this strand then serves as the template for the synthesis of the actual viral genome. The following examples from Table 29-7 should make this clear: (1) poliovirus makes a negative-strand intermediate, which is the template for the positive-strand genome; (2) influenza, measles, and rabies viruses make a positive-strand intermediate, which is the template for the negative-strand genome; (3) rotavirus makes a positive strand that acts both as mRNA and as the template for the negative strand in the double-stranded genome RNA; (4) retroviruses use the negative strand of the DNA intermediate to make positive-strand progeny genome RNA; (5) hepatitis B virus uses its mRNA as a template to make progeny double-stranded DNA; and (6) the other double-stranded DNA viruses replicate their DNA by the same semiconservative process by which cell DNA is synthesized.

Synthesis of progeny virus DNA by herpesviruses involves a virus-encoded enzyme called a **terminase**. The genome of herpesviruses is linear double-stranded DNA within the virion but that DNA becomes circular within the cell. Circular DNA replicates using a "rolling circle" process resulting in a concatenate (multiple copies of progeny genome DNAs joined head to tail). Terminase cleaves the concatenate to form the linear genome used in the progeny virions (see Figure 29-6). It also guides the genome into the virion during assembly of progeny virions (see next section).

As the replication of the viral genome proceeds, the structural capsid proteins to be used in the progeny virus particles are synthesized. In some cases, the newly replicated viral genomes can serve as templates for the late mRNA to make these capsid proteins.

TABLE 29-7 **Complementarity in Viral Genome Replication**

Prototype Virus	Parental Genome[1]	Intermediate Form	Progeny Genome
Poliovirus, Coronavirus	+ ssRNA	− ssRNA	+ ssRNA
Influenza virus, measles virus, rabies virus	− ssRNA	+ ssRNA	− ssRNA
Rotavirus	dsRNA	+ ssRNA	dsRNA
Retrovirus	+ ssRNA	dsDNA	+ ssRNA
Parvovirus B19	ssDNA	dsDNA	ssDNA
Hepatitis B virus	dsDNA	+ ssRNA	dsDNA
Papovavirus, adenovirus, herpesvirus, poxvirus	dsDNA	dsDNA	dsDNA

[1]Code: ss = single-stranded; ds = double-stranded; + = positive polarity; − = negative polarity.

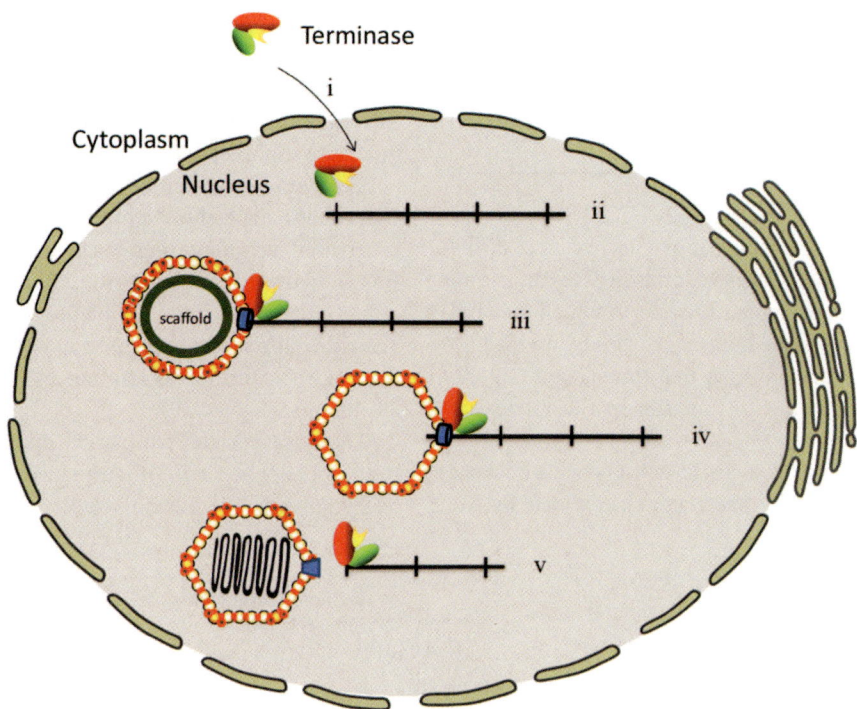

FIGURE 29–6 Action of terminase of herpesviruses. In step (i), terminase synthesized by ribosomes in the cytoplasm enters the nucleus. In step (ii), terminase is about to bind to the progeny viral DNA concatenate (shown as three full-length genomes as black lines). In step (iii), the terminase-DNA complex binds to the portal protein of the newly made empty capsid. In step (iv), terminase transports the progeny viral DNA into the empty capsid. In step (v), terminase cleaves the concatenate and a full-length progeny DNA genome (wavy black line) resides within the capsid, now called a nucleocapsid. (Reproduced with permission from Linlin Y, Yang Q, Wang M, et al: Terminase large subunit provides a new drug target for herpesvirus treatment. Viruses 2019;11[3]:219.)

Assembly of Progeny Virions

The progeny particles are assembled by packaging the viral nucleic acid within the capsid proteins. Certain viruses can be assembled in the test tube by using only purified RNA and purified protein. This indicates that the specificity of the interaction resides within the RNA and protein and that the action of enzymes and expenditure of energy are not required.

In the assembly of cytomegalovirus (CMV) in the nucleus of the cell, the viral DNA enters the interior of the intact capsid after the capsid has formed. In this process, the capsomers first aggregate to form a hollow capsid shell. The genome DNA is then threaded into the interior of the capsid through a "portal protein" located at an apex of the capsid. The virion-encoded **terminase** interacts with the portal protein to guide the progeny DNA into the capsid (see previous section and Figure 29–6). The drug letermovir (see Chapters 35 and 37) inhibits the terminase and thereby inhibits CMV replication.

Release of Progeny Virions

Virus particles are released from the cell by either of two processes. One is rupture of the cell membrane and release of the mature particles; this usually occurs with nonenveloped viruses. The other, which occurs with enveloped viruses, is release of viruses by **budding** through the outer cell membrane (Figure 29–7). (An exception is the **herpesvirus** family, whose members acquire their

envelopes from the **nuclear membrane** rather than from the outer cell membrane.) The budding process begins when virus-specific proteins enter the cell membrane at specific sites. The viral nucleocapsid then interacts with the specific membrane site mediated by **the matrix protein**. The cell membrane evaginates at that site, and an enveloped particle buds off from the membrane. Budding frequently does not damage the cell, and in certain instances, the

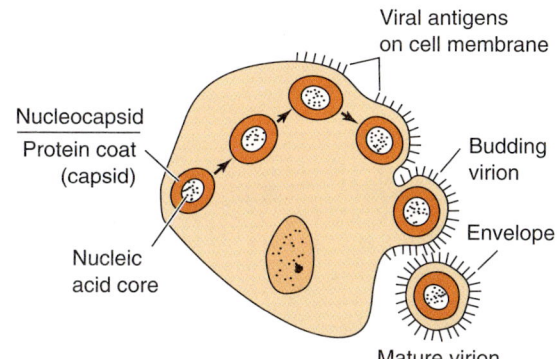

FIGURE 29–7 Budding. Most enveloped viruses derive their lipoprotein envelope from the cell membrane. The matrix protein mediates the interaction between the viral nucleocapsid and the viral envelope. (Reproduced with permission from Mims CA: *The Pathogenesis of Infectious Disease*, 3rd ed. Philadelphia, PA: Academic Press/Elsevier; 1987.)

cell survives while producing large numbers of budding virus particles.

LYSOGENY

Overview of Lysogeny

The typical replicative cycle described above occurs most of the time when viruses infect cells. However, some viruses can use an alternative pathway, called the **lysogenic cycle**, in which the viral DNA becomes integrated into the host cell chromosome and no progeny virus particles are produced at that time (Figure 29–8). The viral nucleic acid continues to function in the integrated state in a variety of ways.

Lysogeny is important from a medical point of view because several exotoxins synthesized by bacteria are encoded by the genes of the integrated bacteriophage (**prophage**). Important examples include **diphtheria toxin, botulinum toxin, cholera toxin**, and **erythrogenic toxin**. This means that a bacterium, e.g., *Corynebacterium diphtheriae* that is not lysogenized will not cause diphtheria.

Lysogenic conversion is the term applied to the new properties that a bacterium acquires as a result of expression of the **integrated prophage genes** (Figure 29–9). Lysogenic conversion is mediated by the transduction of bacterial genes from the donor bacterium to the recipient bacterium by bacteriophages. **Transduction** is the term used to describe the transfer of genes from one bacterium to another by viruses (Figures 29–9 and 29–10 and see page 20).

The lysogenic or "temperate" cycle is described for lambda bacteriophage, because it is the best-understood model system (see Figure 29–10). Several aspects of infections by retroviruses,

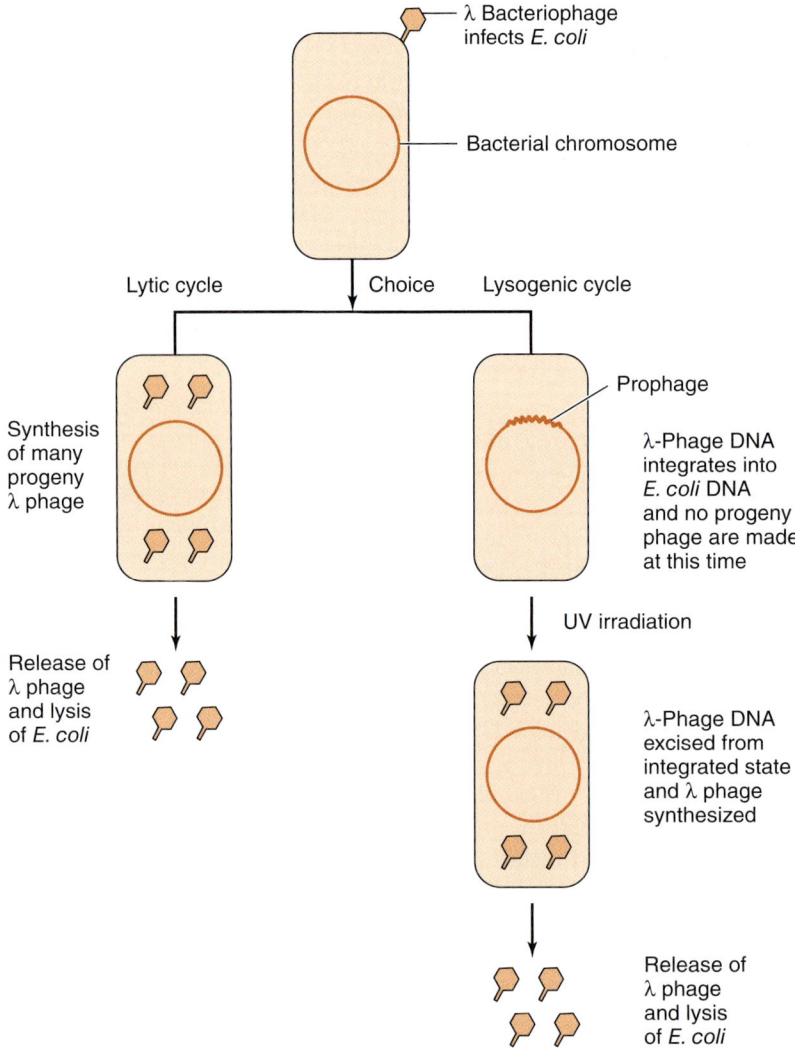

FIGURE 29–8 Comparison of the lytic and lysogenic cycles of bacteriophage (phage) replication. In the lytic cycle, replication of the phage is completed without interruption. In the lysogenic cycle, replication of the phage is interrupted, and the phage DNA integrates into the bacterial DNA. The integrated DNA is called a prophage and can remain in the integrated state for long periods. If the bacteria are exposed to certain activators such as ultraviolet (UV) light, the prophage DNA is excised from the bacterial DNA and the phage enters the lytic cycle, which ends with the production of progeny phage.

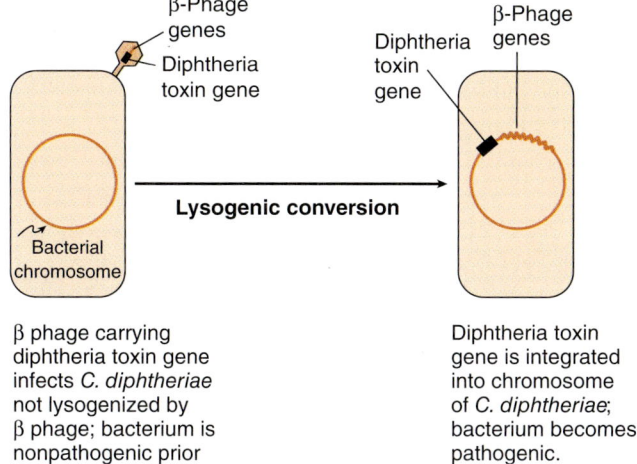

β phage carrying diphtheria toxin gene infects *C. diphtheriae* not lysogenized by β phage; bacterium is nonpathogenic prior to infection by β phage.

Diphtheria toxin gene is integrated into chromosome of *C. diphtheriae*; bacterium becomes pathogenic.

FIGURE 29–9 Lysogenic conversion. In the left-hand panel, transduction of the diphtheria toxin gene by beta bacteriophage results in lysogenic conversion of the nonlysogenized, nonpathogenic *Corynebacterium diphtheriae*. In the right-hand panel, the recipient lysogenized bacterium can now produce diphtheria toxin and can cause the disease diphtheria. Note that no progeny phages are made within the lysogenized bacterium because the diphtheria toxin gene has replaced some of the β-phage genes required for replication. The β phage therefore cannot replicate. The lysogenized bacterium is not killed by the phage and can multiply, produce diphtheria toxin, and cause disease.

such as HIV, and herpesviruses are similar to the events in the lysogenic cycle of lambda phage.

Infection by lambda phage in *Escherichia coli* begins with injection of the linear, double-stranded DNA genome through the phage tail into the cell. The linear DNA becomes a circle as the single-stranded regions on the ends pair their complementary bases. A ligating enzyme makes a covalent bond in each strand to close the circle. Circularization is important because it is the circular form that integrates into host cell DNA.

The choice between the pathway leading to lysogeny and that leading to full replication is made as early protein synthesis begins. Simply put, the choice depends on the balance between two proteins, the **repressor** produced by the *c*-I gene and the **antagonizer of the repressor** produced by the *cro* gene (Figure 29–11). If the repressor predominates, transcription of other early genes is shut off and lysogeny ensues. Transcription is inhibited by binding of the repressor to the two operator sites that control early protein synthesis. If the *cro* gene product prevents the synthesis of sufficient repressor, replication and lysis of the cell result.

The next important step in the lysogenic cycle is the **integration** of the viral DNA into the cell DNA. This occurs by the matching of a specific attachment site on the lambda DNA to a homologous site on the *E. coli* DNA and the integration (breakage and rejoining) of the two DNAs mediated by a phage-encoded recombination enzyme. The integrated viral DNA is called a **prophage**. Most lysogenic phages integrate at one or a few specific sites, but some, such as the Mu (or mutator) phage, can integrate their DNA at many sites, and other phages, such as the P1 phage, never actually integrate but remain in a "temperate" state extrachromosomally, similar to a plasmid.

Because the integrated viral DNA is replicated along with the cell DNA, each daughter cell inherits a copy. However, the prophage is not permanently integrated. It can be induced to resume its replicative cycle by the action of ultraviolet (UV) light and certain chemicals that damage DNA. UV light induces the synthesis of a protease, which cleaves the repressor. Early genes then function, including the genes coding for the enzymes that excise the prophage from the cell DNA. The virus then completes its replicative cycle, leading to the production of progeny virus and lysis of the cell.

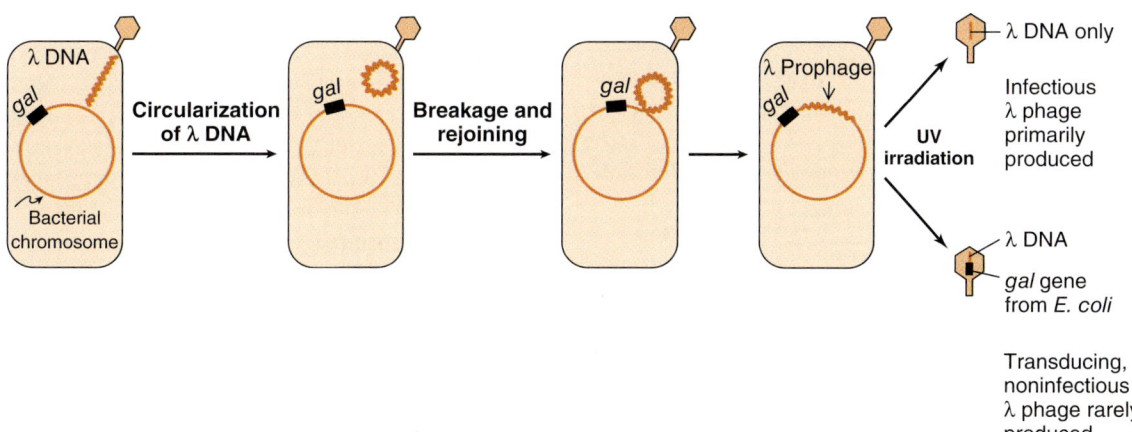

FIGURE 29–10 Lysogeny. The linear lambda (λ) phage DNA is injected into the bacterium, circularizes, and then integrates into the bacterial DNA. When integrated, the phage DNA is called a prophage. When the prophage is induced to enter the replicative cycle, aberrant excision of the phage DNA can occur (i.e., part of the phage DNA and part of the bacterial DNA including the adjacent *gal* gene are excised). The *gal* gene can now be transduced to another bacterium. Transduction is also described in Figure 4–4. (Reproduced with permission from Jawetz E, Melnick JL, Adelberg EA: *Review of Medical Microbiology*, 17th ed. New York, NY: McGraw Hill; 1986.)

VIRAL MESSENGER RNA

There are four interesting aspects of viral mRNA and its expression in eukaryotic cells. (1) Viral mRNAs have three attributes in common with cellular mRNAs: on the 5′ end there is a methylated GTP "cap," which is linked by an "inverted" (3′-to-5′) bond instead of the usual 5′-to-3′ bond; on the 3′ end there is a tail of 100 to 200 adenosine residues [poly(A)]; and the mRNA is generated by splicing from a larger transcript of the genome. In fact, these three modifications were first observed in studies on viral mRNAs and then extended to cellular mRNAs. (2) Some viruses use their genetic material to the fullest extent by making more than one type of mRNA from the same piece of DNA by "shifting the reading frame." This is done by starting transcription one or two bases downstream from the original initiation site. (3) With some DNA viruses, there is temporal control over the region of the genome that is transcribed into mRNA. During the beginning stages of the growth cycle, before DNA replication begins, only the early region of the genome is transcribed, and therefore, only certain early proteins are made. One of the early proteins is a repressor of the late genes; this prevents transcription until the appropriate time. (4) Four different processes are used to generate the monocistronic mRNAs that will code for a single protein from the polycistronic viral genome:

(1) Individual mRNAs are transcribed by starting at many specific initiation points along the genome, which is the same mechanism used by eukaryotic cells and by herpesviruses, adenoviruses, and the DNA and RNA tumor viruses.

(2) In the reoviruses and influenza viruses, the genome is segmented into multiple pieces, each of which codes for a single mRNA.

(3) In polioviruses, the entire RNA genome is translated into one long polypeptide, which is then cleaved into specific proteins by a protease encoded by the virus.

(4) In coronaviruses, the 5′ end of the genome is translated into two large polypeptides which are cleaved by a virus-encoded protease. A frame-shift mechanism is used to generate the second polypeptide. The 3′ end of the genome RNA is transcribed into a set of nested m-RNAs. Each of the nested m-RNAs is translated into one viral protein by a mechanism that translates only the 5′ end of the m-RNA.

During replication, what determines whether an RNA becomes a messenger RNA or a genome RNA? In human immunodeficiency virus (HIV), this control is a function of the number of guanosines in the 5′ cap. If the RNA has one guanosine it is selected as a genome but if it has 2 or 3 guanosines it is translated as m-RNA.

Relationship of Lysogeny in Bacteria to Latency in Human Cells

Members of the herpesvirus family, such as HSV, varicella-zoster virus, CMV, and Epstein–Barr virus, exhibit latency—the phenomenon in which no or very little virus is produced after the initial infection but, at some later time, reactivation and full virus replication occur.

The parallel to lysogeny with bacteriophage is clear. But note that in lysogeny, the phage DNA is integrated into the bacterial DNA whereas in herpesvirus latency the viral DNA is *not*

integrated into human DNA. Only retroviruses, such as HIV, integrate a DNA copy of the RNA genome into human DNA as an obligatory step in viral replication. Retroviruses encode an integrase to perform this function.

What is known about how the herpesviruses initiate and maintain the latent state? In HSV latency, a tegument protein called viral protein-16 (VP-16) plays an important role. VP-16 is a transcription factor that initiates immediate–early mRNA and protein synthesis that is required for viral replication to proceed (see Chapter 37). When HSV replicates in epithelial cells, VP-16 performs its required function. But when the progeny HSV

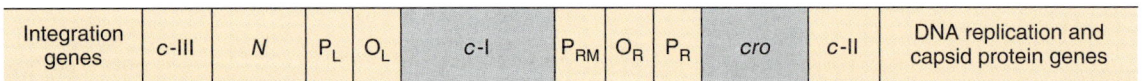

Integration genes	c-III	N	P_L	O_L	c-I	P_RM	O_R	P_R	cro	c-II	DNA replication and capsid protein genes

FIGURE 29–11 Control of lysogeny. Shortly after infection, transcription of the *N* and *cro* genes begins. The *N* protein is an antiterminator that allows transcription of c-II and c-III and the genes to the right of c-II and to the left of c-III. The c-II protein enhances the production of the c-I repressor protein. c-I has two important functions: (1) it inhibits transcription at P_RO_R and P_LO_L, thereby preventing phage replication, and (2) it is a positive regulator of its own synthesis by binding to P_{RM}. The crucial decision point in lysogeny is the binding of either c-I repressor or the *cro* protein to the O_R site. If c-I repressor occupies O_R, lysogeny ensues; if *cro* protein occupies O_R, viral replication occurs. c-I = repressor gene; c-II and c-III = genes that influence the production of c-I; *cro* = gene that antagonizes the c-I repressor; *N* = antiterminator gene; P_LO_L = left promoter and operator; P_RO_R = right promoter and operator; P_{RM} = promoter for repressor maintenance.

infects adjacent neurons, VP-16 does not function. The HSV DNA migrates to the nucleus of the neuron but VP-16 is inactive so replication cannot occur and the virus remains in a latent state.

Furthermore, shortly after HSV infects neurons, a set of **latency-associated transcripts** (LATS) are synthesized. These are noncoding, regulatory RNAs that suppress viral replication. The precise mechanism by which they do so is unclear.

Reactivation of viral replication at a later time occurs when the genes encoding LATS are excised.

CMV employs different mechanisms. The CMV genome encodes microRNAs that inhibit the translation of mRNAs required for viral replication. Also, the CMV genome encodes both a protein and an RNA that inhibit apoptosis in infected cells. This allows the infected cell to survive.

PEARLS

Viral Growth Curve

- One virion infects a cell and hundreds of progeny virions are produced within hours. This is a remarkable amplification and explains the rapid spread of virus from cell to cell.
- The **eclipse period** is the time when no virus particles are detected within the infected cell. It occurs soon after the cell is infected.
- **Cytopathic effect** (CPE) is the term used to describe the damage, both morphologic and functional, inflicted on the cell by the virus. In the clinical laboratory, the presence of a virus in the patient's specimen is often detected by seeing a CPE in cell culture.

Viral Growth Cycle

- **Attachment:** The interaction of proteins on the surface of the virus with specific receptor proteins on the surface of the cell is one of the main determinants of both the **species specificity** and the **organ specificity** of the virus.
- **Infectious nucleic acid** is viral genome DNA or RNA, purified free of all proteins, that can undergo the entire replicative cycle within a cell and produce infectious progeny viruses. Infectious nucleic acid, because it has no associated protein, can enter and replicate within cells that the intact virion cannot.
- **Polarity of viral genome RNA:** Genome RNA that has the same base sequence as the mRNA is, by definition, positive-polarity RNA. Most positive-polarity genomes are translated into viral proteins without the need for a polymerase in the virion. The exception is the retroviruses, which use reverse transcriptase in the virion to transcribe the genome RNA into DNA. Genome RNA that has a base sequence complementary to mRNA has, by definition, negative polarity. A virus with a negative-polarity RNA genome must have an RNA polymerase in the virion to synthesize its mRNA.
- **Viral gene expression:** All viruses require virus-specific messenger RNA to synthesize virus-specific proteins.
- **RNA viruses:** Some RNA viruses, such as poliovirus, have a positive-polarity RNA genome that serves as the mRNA (i.e., the genome is the mRNA). Other viruses, such as influenza virus, have a negative-polarity RNA genome and have an RNA polymerase in the virion that synthesizes the viral mRNA. Rotavirus has a double-stranded RNA genome and has an RNA polymerase in the virion that synthesizes the viral mRNA. Retroviruses, such as HIV, have a positive-polarity RNA genome and have a DNA polymerase in the virion that synthesizes a DNA copy of the RNA genome. This DNA is the template used by the host cell RNA polymerase to synthesize the viral mRNA.
- **DNA viruses:** Most DNA viruses, such as herpesviruses, adenoviruses, and papillomaviruses, have a double-stranded DNA genome and use the host cell RNA polymerase to synthesize the viral mRNA. Poxviruses have a double-stranded DNA genome but have an RNA polymerase in the virion that synthesizes the viral mRNA. Poxviruses have an RNA polymerase in the virion because they replicate in the cytoplasm and do not have access to the host cell RNA polymerase in the nucleus.
- **Viral replication:** All DNA viruses replicate in the nucleus, except poxviruses, which replicate in the cytoplasm. All RNA viruses replicate in the cytoplasm, except retroviruses, influenza virus, and hepatitis D virus, which require an intranuclear step in their replication. Many viruses encode a replicase, which is a DNA or RNA polymerase that synthesizes the many copies of the progeny viral genomes.
- **Viral genome:** The genome of all DNA viruses is double-stranded except for that of parvoviruses, which is single-stranded. The genome of all RNA viruses is single-stranded except for that of reoviruses (e.g., rotavirus), which is double-stranded.
- **Viral proteins:** Early proteins are typically enzymes used in the synthesis of viral nucleic acids, whereas late proteins are typically structural proteins of the progeny viruses. Some viruses, such as poliovirus and retroviruses, translate their mRNA into precursor polyproteins, which must be cleaved by proteases to produce functional proteins.
- **Assembly and release:** All enveloped viruses acquire their envelope by budding through the external cell membrane as they exit the cell, except herpesviruses, which acquire their envelope by budding through the nuclear membrane. The matrix protein mediates the interaction of the nucleocapsid with the envelope.
- **Lysogeny** is the process by which viral DNA becomes integrated into host cell DNA, replication stops, and no progeny virus is made. Later, if DNA is damaged by, for example, UV light, viral DNA is excised from the host cell DNA, and progeny viruses are made. The integrated viral DNA is called a **prophage**. Bacterial cells carrying a prophage can acquire new traits, such as the ability to produce exotoxins such as diphtheria toxin. **Transduction** is the process by which viruses carry genes from one cell to another. **Lysogenic conversion** is the term used to indicate that the cell has acquired a new trait as a result of the integrated prophage.

SELF-ASSESSMENT QUESTIONS

1. Many viruses are highly specific regarding the type of cells they infect. Of the following, which one is the most important determinant of this specificity?
   (A) The matrix protein
   (B) The polymerase in the virion
   (C) The protease protein
   (D) The surface glycoprotein
   (E) The viral mRNA

2. Your summer research project is to study the viruses that cause upper respiratory tract infections. You have isolated a virus from a patient's throat and find that its genome is RNA. Furthermore, you find that the genome is the complement of viral mRNA within the infected cell. Of the following, which is the most appropriate conclusion you could draw?
   (A) The genome RNA is infectious.
   (B) The genome RNA is segmented.
   (C) The virion contains a polymerase.
   (D) The virion has a lipoprotein envelope.
   (E) A single-stranded DNA is synthesized during replication.

3. The purified genome of certain viruses can enter a cell and elicit the production of progeny viruses (i.e., the genome is infectious). Regarding these viruses, which one of the following statements is most accurate?
   (A) Their genome RNA has positive polarity.
   (B) Their genome RNA is double-stranded.
   (C) They have a polymerase in the virion.
   (D) They have a segmented genome.
   (E) They require tegument proteins in order to be infectious.

4. Regarding viral replication, which one of the following is most accurate?
   (A) The cytopathic effect typically occurs during the eclipse period.
   (B) The early proteins are typically enzymes, whereas the late proteins are typically capsid proteins.
   (C) The assembly of a nonenveloped virus typically occurs as the virion buds from the cell membrane.
   (D) Influenza viruses synthesize their mRNA using host cell–encoded RNA-dependent RNA polymerase.
   (E) Retroviruses (e.g., HIV) synthesize their mRNA using an enzyme in the virion called reverse transcriptase.

5. Regarding the viral growth cycle, which one of the following is most accurate?
   (A) During the lysogenic phase, the typical result is the production of hundreds of progeny virions.
   (B) Hepatitis B virus has an RNA polymerase in the virion that is required to synthesize messenger RNA from the positive strand of the viral DNA.
   (C) Herpesviruses have an RNA-dependent DNA polymerase in the virion.
   (D) Lysogenic conversion is the process by which bacteria acquire new genes due to transduction by a lysogenic bacteriophage.
   (E) Smallpox virus translates its genome into a single polypeptide, which is then cleaved into structural and nonstructural proteins.

6. Which one of the following choices names two viruses that both translate their messenger RNA into precursor polypeptides that must be cleaved by virion-encoded proteases?
   (A) Herpes simplex virus and human papillomavirus
   (B) Human immunodeficiency virus and poliovirus
   (C) Influenza virus and measles virus
   (D) Rabies virus and hepatitis B virus
   (E) Rotavirus and parvovirus

ANSWERS

(1) **(D)**
(2) **(C)**
(3) **(A)**
(4) **(B)**
(5) **(D)**
(6) **(B)**

PRACTICE QUESTIONS: USMLE & COURSE EXAMINATIONS

Questions on the topics discussed in this chapter can be found in the Basic Virology section of Part XIII: USMLE (National Board) Practice Questions starting on page 744. Also see Part XIV: USMLE (National Board) Practice Examination starting on page 775.

Genetics & Gene Therapy

INTRODUCTION

The study of viral genetics falls into two general areas: (1) mutations and their effect on replication and pathogenesis and (2) the interaction of two genetically distinct viruses that infect the same cell. In addition, viruses serve as **vectors** in gene therapy and in recombinant vaccines, two areas that hold great promise for the treatment of genetic diseases and the prevention of infectious diseases.

MUTATIONS

Mutations in viral DNA and RNA occur by the same processes of base substitution, deletion, and frameshift as those described for bacteria in Chapter 4. Probably the most important practical use of mutations is in the production of **vaccines containing live, attenuated virus**. These attenuated mutants have lost their pathogenicity but have retained their antigenicity; therefore, they induce immunity without causing disease.

There are two other kinds of mutants of interest. The first are **antigenic variants** such as those that occur frequently with influenza viruses, which have an altered surface protein and are therefore no longer inhibited by a person's preexisting antibody. The variant can thus cause disease, whereas the original strain cannot. Human immunodeficiency virus and hepatitis C virus also produce many antigenic variants. These viruses have an **error-prone polymerase** that causes the mutations. The second are **drug-resistant mutants**, which are insensitive to an antiviral drug because the target of the drug, usually a viral enzyme, has been modified.

Conditional lethal mutations are extremely valuable in determining the function of viral genes. These mutations function normally under permissive conditions but fail to replicate or to express the mutant gene under restrictive conditions. For example, **temperature-sensitive** conditional lethal mutants express their phenotype normally at a low (permissive) temperature, but at a higher (restrictive) temperature, the mutant gene product is inactive. To give a specific example, temperature-sensitive mutants of Rous sarcoma virus can transform cells to malignancy at the permissive temperature of 37°C. When the transformed cells are grown at the restrictive temperature of 41°C, their phenotype reverts to normal appearance and behavior. The malignant phenotype is regained when the permissive temperature is restored.

Note that temperature-sensitive mutants have now entered clinical practice. Temperature-sensitive mutants of influenza virus are now being used to make a vaccine. The mutant virus will grow in the cooler, upper airways where it does not cause symptoms but will induce antibodies. Note that it will *not* grow in the warmer, lower respiratory tract, so will not cause influenza and pneumonia (see Table 30–1).

TABLE 30–1 Temperature-Sensitive Mutants of Influenza Virus Used in Vaccine

Type of Influenza Virus	Growth in Nose (32°C)	Growth in Lungs (37°C)
Normal infectious virus	Yes	Yes
Temperature-sensitive mutant used in vaccine	Yes	No

INTERACTIONS BETWEEN VIRUSES

When two genetically distinct viruses infect a cell, four different phenomena can ensue.

(1) **Recombination** is the exchange of genes between two chromosomes that is based on crossing over within regions of significant base sequence homology. Recombination can be readily demonstrated for viruses with double-stranded DNA as the genetic material and has been used to determine their genetic map. However, recombination by RNA viruses occurs at a very low frequency, if at all.

(2) **Reassortment** is the term used when viruses with segmented genomes, such as influenza virus, exchange segments (Figure 30–1). This usually results in a much higher frequency of gene exchange than does recombination. Reassortment of influenza virus RNA segments encoding the viral hemagglutinin and neuraminidase is involved in the major antigenic changes in the virus that are the basis for recurrent influenza epidemics.

(3) **Complementation** can occur when either one or both of the two viruses that infect the cell have a mutation that results in a nonfunctional protein (Figure 30–2). The nonmutated virus "complements" the mutated one by making a functional protein that serves for both viruses. Complementation is an important method by which a helper virus permits replication of a defective virus. One

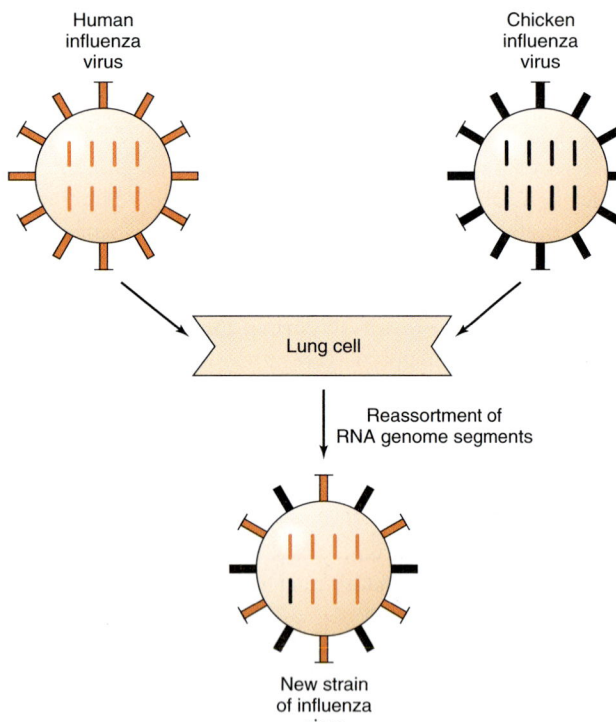

FIGURE 30–1 Reassortment in influenza virus. A human strain of influenza virus containing the gene encoding one antigenic type of hemagglutinin (colored orange) infects the same lung cell as a chicken strain of influenza virus containing the gene encoding a different antigenic type of hemagglutinin (colored black). Reassortment of the genome RNA segments that encode the hemagglutinin occurs, and a new strain of influenza virus is produced containing the chicken type of hemagglutinin (colored black).

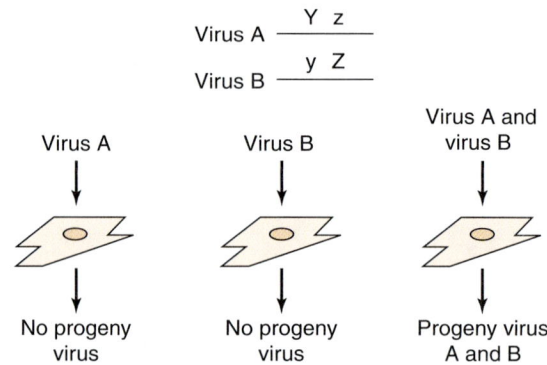

FIGURE 30–2 Complementation. If *either* virus A *or* virus B infects a cell, no virus is produced because each has a mutated gene. If *both* virus A and virus B infect a cell, the protein product of gene Y of virus A will complement virus B, the protein product of gene Z of virus B will complement virus A, and progeny of both virus A and virus B will be produced. Note that no recombination has occurred and that the virus A progeny will contain the mutated z gene and the virus B progeny will contain the mutant y gene. y, z = mutated, nonfunctional genes Y, Z = functional genes.

clinically important example of complementation is hepatitis B virus providing its surface antigen to hepatitis delta virus, which is defective in its ability to produce its own outer protein.

(4) In **phenotypic mixing**, the genome of virus type A can be coated with the surface proteins of virus type B (Figure 30–3). This phenotypically mixed virus can infect cells as determined by its type B protein coat. However, the progeny virus from this infection has a type A coat; it is encoded solely by its type A genetic material.

An interesting example of phenotypic mixing is that of **pseudotypes**, which consist of the nucleocapsid of one virus and the envelope of another. Pseudotypes composed of the genome of vesicular stomatitis virus (a rhabdovirus) and the spike protein of SARS-CoV-2 (a coronavirus) are currently being used to study the immune response to SARS-CoV-2.

GENE THERAPY & RECOMBINANT VACCINES

Viruses are being used as genetic vectors in two novel ways: (1) to deliver new, functional genes to patients with genetic diseases (gene therapy) and (2) to produce new viral vaccines that contain recombinant viruses carrying the genes of several different viruses, thereby inducing immunity to several diseases with one immunization.

Gene Therapy

Retroviruses are currently being used as vectors of the gene encoding adenine deaminase (ADA) in patients with immunodeficiencies resulting from a defective ADA gene. Retroviruses are excellent vectors because a DNA copy of their RNA genome is stably integrated into the host cell DNA and the integrated genes are expressed efficiently. Retroviral vectors are constructed by

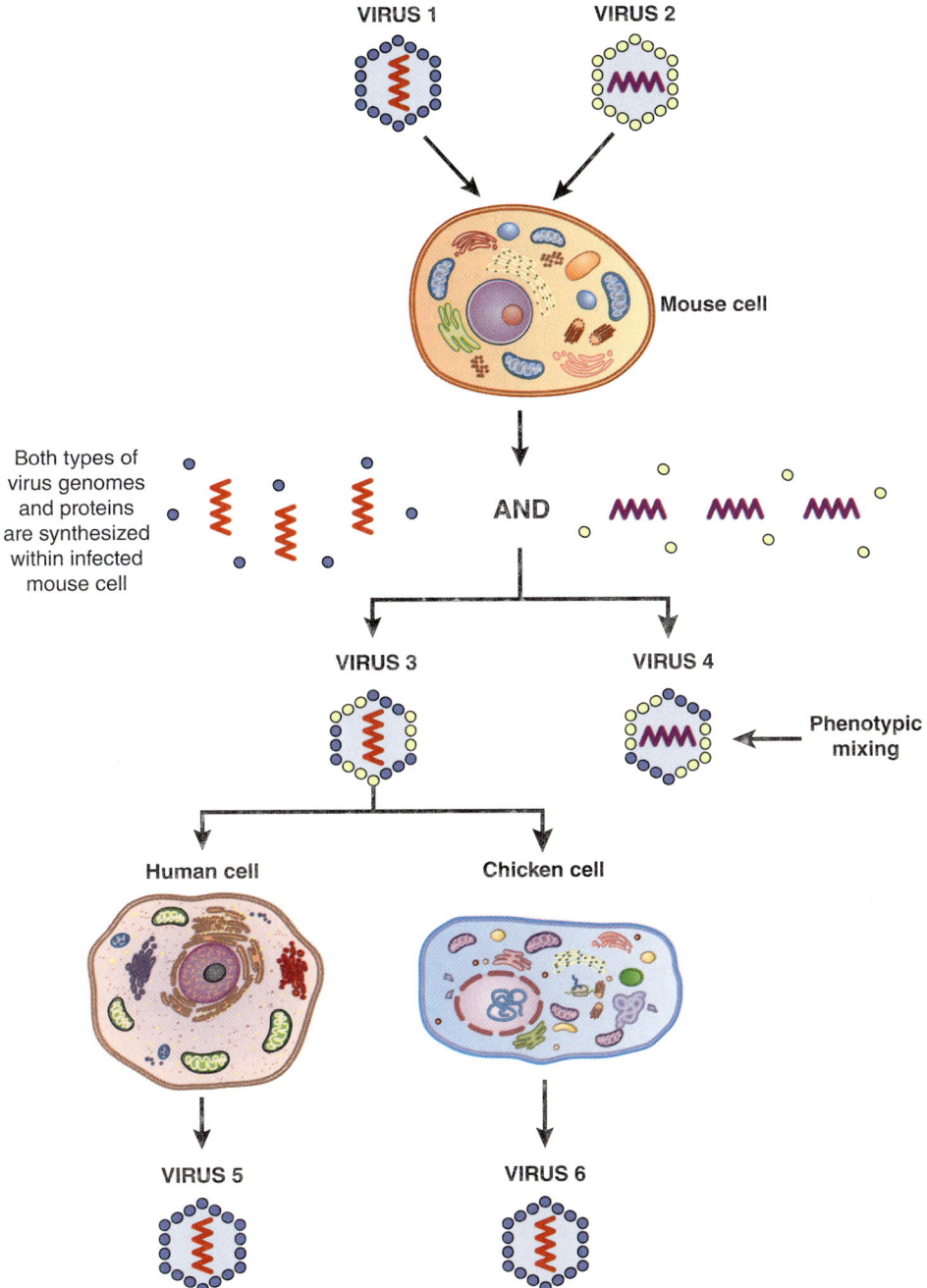

FIGURE 30–3 Phenotypic mixing. Initially, Virus 1 (blue capsid proteins and vertical genome) and Virus 2 (yellow capsid proteins and horizontal genome) infect the same mouse cell. Assume that Virus 1 can infect human cells but not chicken cells (a property determined by the blue surface proteins) and that Virus 2 can infect chicken cells but not human cells (a property determined by the yellow surface proteins). However, both Virus 1 and Virus 2 can infect a mouse cell. Within the mouse cell, both genomes are replicated and both blue and yellow capsid proteins are synthesized. As shown, some of the progeny viruses (Viruses 3 and 4) exhibit *phenotypic mixing* because they have both the blue and the yellow surface proteins and therefore can infect both chicken cells and human cells. Note that in the next round of infection, when progeny Virus 3 infects either human cells or chicken cells, the progeny of that infection (Viruses 5 and 6) is determined by the vertical genome and will be identical to Virus 1 with only blue capsid proteins and a vertical genome. Similarly (but not shown), when progeny Virus 4 infects either human cells or chicken cells, the progeny of that infection is determined by the horizontal genome and will be identical to Virus 2.

removing the genes encoding several viral proteins from the virus and replacing them with the human gene of interest (e.g., the ADA gene). Virus particles containing the human gene are produced within "helper cells" that contain the deleted viral genes and therefore can supply, by complementation, the missing viral proteins necessary for the virus to replicate. The retroviruses produced by the helper cells can infect the patient's cells and introduce the human gene into the cells, but the viruses cannot replicate because they lack several viral genes. This inability of these viruses to replicate is an important advantage in human gene therapy.

Recombinant (Vectored) Vaccines

Recombinant (vectored) vaccines contain viruses that have been genetically engineered to carry the genes of other viruses. To construct the recombinant virus, any gene of the vector virus that is not essential for viral replication is deleted, and the gene from the other virus that encodes the antigen that elicits neutralizing antibody is inserted. For example, the gene for the surface antigen of Ebola virus has been inserted into vesicular stomatitis virus and is expressed in infected cells. The Ebola vaccine containing the vectored virus as the immunogen is currently in clinical use. Another example is one of the rotavirus vaccines in which the gene encoding the capsid protein of the human rotavirus is inserted into a bovine rotavirus that is nonpathogenic in humans.

PEARLS

- Mutations in the viral genome can produce antigenic variants and drug-resistant variants. Mutations can also produce **attenuated** (weakened) variants that cannot cause disease but retain their antigenicity and are useful in vaccines.

- Temperature-sensitive mutants can replicate at a low (permissive) temperature but not at a high (restrictive) temperature. Temperature-sensitive mutants of influenza virus are used in one of the vaccines against this disease.

- **Reassortment** (exchange) of segments of the genome RNA of influenza virus is important in the pathogenesis of the worldwide epidemics caused by this virus.

- **Complementation** occurs when one virus produces a protein that can be used by another virus. A medically important example is hepatitis D virus, which uses the surface antigen of hepatitis B virus as its outer coat protein.

- **Phenotypic mixing** occurs when two different viruses infect the same cell and progeny viruses contain proteins of both parental viruses. This can endow the progeny viruses with the ability to infect cells of species that ordinarily parental virus could not.

SELF-ASSESSMENT QUESTIONS

1. In the lab, a virologist was studying the properties of HIV. She infected the same cell with both HIV and rabies virus. (HIV can infect only human CD4-positive cells, whereas rabies virus can infect both human cells and dog cells.) Some of the progeny virions were able to infect dog cells, within which she found HIV-specific RNA. Which one of the following is the term used to describe these results?

   (A) Complementation
   (B) Phenotypic mixing
   (C) Reassortment
   (D) Recombination

2. You have isolated two mutants of poliovirus, one mutated at gene X and the other mutated at gene Y. If you infect cells with each one alone, no virus is produced. If you infect a single cell with both mutants, which one of the following statements is most accurate?

   (A) If complementation between the mutant gene products occurs, both X and Y progeny viruses will be made.
   (B) If phenotypic mixing occurs, then both X and Y progeny viruses will be made.
   (C) If the genome is transcribed into DNA, then both X and Y viruses will be made.
   (D) Because reassortment of the genome segments occurs at high frequency, both X and Y progeny viruses will be made.

ANSWERS

(1) **(B)**
(2) **(A)**

PRACTICE QUESTIONS: USMLE & COURSE EXAMINATIONS

Questions on the topics discussed in this chapter can be found in the Basic Virology section of Part XIII: USMLE (National Board) Practice Questions starting on page 744. Also see Part XIV: USMLE (National Board) Practice Examination starting on page 775.

Classification of Medically Important Viruses

PRINCIPLES OF CLASSIFICATION

The classification of viruses is based on chemical and morphologic criteria. The two major components of the virus used in classification are (1) the nucleic acid (its molecular weight and structure) and (2) the capsid (its size and symmetry and whether it is enveloped). A classification scheme based on these factors is presented in Tables 31–1 and 31–2 for DNA and RNA viruses, respectively. This scheme was simplified from the complete classification to emphasize organisms of medical importance. Only the virus families are listed; subfamilies are described in the chapter on the specific virus.

TABLE 31–1 Classification of DNA Viruses

Virus Family	Envelope Present	Capsid Symmetry	Virion Size (nm)	DNA MW (×10⁶)	DNA Structure[1]	Medically Important Viruses
Parvovirus	No	Icosahedral	22	2	ss linear	B19 virus
Polyomavirus	No	Icosahedral	45	3	ds circular, supercoiled	JC virus, BK virus
Papillomavirus	No	Icosahedral	55	5	ds circular, supercoiled	Human papillomavirus
Adenovirus	No	Icosahedral	75	23	ds linear	Adenovirus
Hepadnavirus	Yes	Icosahedral	42	1.5	ds incomplete circular	Hepatitis B virus
Herpesvirus	Yes	Icosahedral	100[2]	100–150	ds linear	Herpes simplex virus, varicella-zoster virus, cytomegalovirus, Epstein–Barr virus
Poxvirus	Yes	Complex	250 × 400	125–185	ds linear	Smallpox virus, molluscum contagiosum virus

[1]ss = single-stranded; ds = double-stranded.

[2]The herpesvirus nucleocapsid is 100 nm, but the envelope varies in size. The entire virus can be as large as 200 nm in diameter.

TABLE 31–2 Classification of RNA Viruses

Virus Family	Envelope Present	Capsid Symmetry	Particle Size (nm)	RNA MW ($\times 10^6$)	RNA Structure[1]	Medically Important Viruses
Picornavirus	No	Icosahedral	28	2.5	ss linear, nonsegmented, positive polarity	Poliovirus, rhinovirus, hepatitis A virus
Hepevirus	No	Icosahedral	30	2.5	ss linear, nonsegmented, positive polarity	Hepatitis E virus
Calicivirus	No	Icosahedral	38	2.7	ss linear, nonsegmented, positive polarity	Norovirus
Reovirus	No	Icosahedral	75	15	ds linear, 10 or 11 segments	Rotavirus
Flavivirus	Yes	Icosahedral	45	4	ss linear, nonsegmented, positive polarity	Yellow fever virus, dengue virus, West Nile virus, hepatitis C virus, Zika virus
Togavirus	Yes	Icosahedral	60	4	ss linear, nonsegmented, positive polarity	Rubella virus
Retrovirus	Yes	Icosahedral	100	7[2]	ss linear, 2 identical strands (diploid), positive polarity	HIV, human T-cell leukemia virus
Orthomyxovirus	Yes	Helical	80–120	4	ss linear, 8 segments, negative polarity	Influenza virus
Paramyxovirus	Yes	Helical	150	6	ss linear, nonsegmented, negative polarity	Measles virus, mumps virus, respiratory syncytial virus
Rhabdovirus	Yes	Helical	75 × 180	4	ss linear, nonsegmented, negative polarity	Rabies virus
Filovirus	Yes	Helical	80[3]	4	ss linear, nonsegmented, negative polarity	Ebola virus, Marburg virus
Coronavirus	Yes	Helical	100	10	ss linear, nonsegmented, positive polarity	SARS-CoV-2
Arenavirus	Yes	Helical	80–130	5	ss circular, 2 segments with cohesive ends, negative polarity	Lassa fever virus, lymphocytic choriomeningitis virus
Bunyavirus	Yes	Helical	100	5	ss circular, 3 segments with cohesive ends, negative polarity	California encephalitis virus, hantavirus
Deltavirus	Yes	Uncertain[4]	37	0.5	ss circular, closed circle, negative polarity	Hepatitis delta virus

[1]ss = single-stranded; ds = double-stranded.

[2]Retrovirus genome RNA contains two identical molecules, each with a molecular weight (MW) of 3.5×10^6.

[3]Particles are 80 nm wide but can be thousands of nanometers long.

[4]The nucleocapsid appears spherical but its symmetry is unknown.

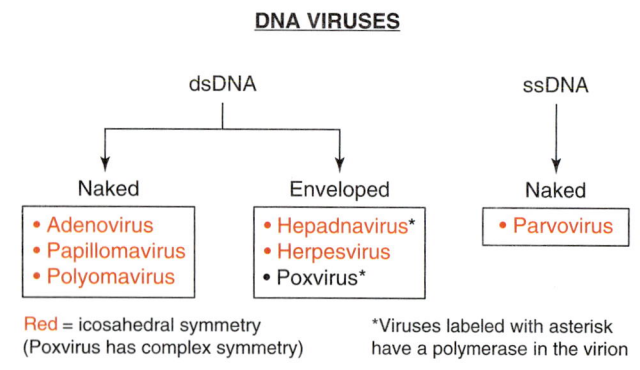

FIGURE 31–1 Classification scheme for DNA viruses.

Figures 31–1 and 31–2 show a classification outline for DNA viruses and RNA viruses, respectively, based on the type of genome, the nature of the nucleocapsid, and whether an envelope is present.

DNA VIRUSES

The families of DNA viruses are described in Table 31–1. The four **naked** (i.e., nonenveloped) icosahedral virus families—the parvoviruses, polyomaviruses, papillomaviruses, and adenoviruses—are presented in order of increasing particle size, as are the three **enveloped** families. The hepadnavirus family, which includes hepatitis B virus (HBV), and the herpesviruses

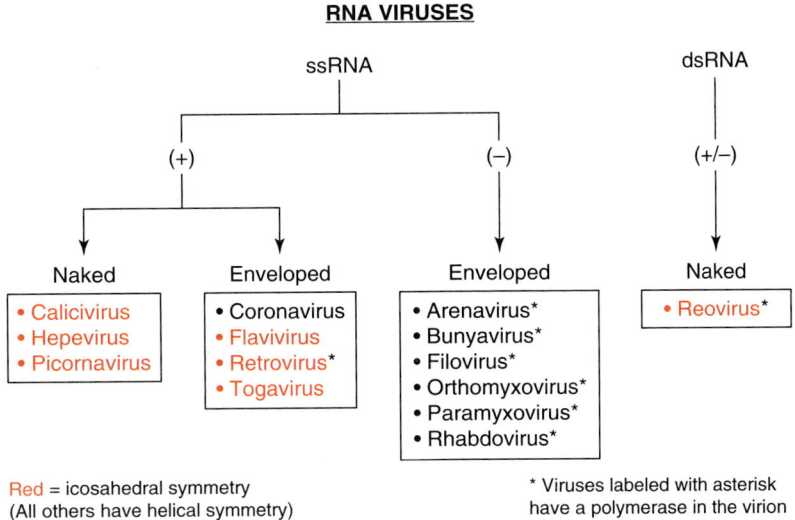

RNA VIRUSES

FIGURE 31–2 Classification scheme for RNA viruses.

are enveloped icosahedral viruses. The largest viruses, the poxviruses, have a complex internal symmetry.

Parvoviruses

These are very small (22 nm in diameter), naked icosahedral viruses with single-stranded linear DNA. There are two types of parvoviruses: defective and nondefective. The defective parvoviruses (e.g., adeno-associated virus) require a helper virus for replication. The DNA of defective parvoviruses is unusual because plus-strand DNA and minus-strand DNA are carried in separate particles. The nondefective parvoviruses are best illustrated by parvovirus B19 virus, which has a minus-strand DNA genome. This virus causes hydrops fetalis, aplastic anemia in sickle cell anemia patients, and erythema infectiosum (fifth disease, slapped cheeks syndrome)—an innocuous childhood disease characterized by an erythematous "slapped cheeks" rash.

Polyomaviruses

These are naked icosahedral viruses (45 nm in diameter) with double-stranded circular supercoiled DNA. Two human polyomaviruses are JC virus, isolated from patients with progressive multifocal leukoencephalopathy, and BK virus, isolated from the urine of immunosuppressed kidney transplant patients. Polyomavirus and simian vacuolating virus 40 (SV40 virus) are polyomaviruses of mice and monkeys, respectively, which induce malignant tumors in a variety of species.

Papillomaviruses

Papillomaviruses are naked icosahedral viruses (55 nm in diameter) with double-stranded supercoiled DNA. The human pathogen in the family is human papillomavirus (HPV). It causes papillomas (warts) of many body sites, and certain strains cause carcinoma of the cervix. Many animal species are infected by papillomaviruses, but those viruses are species-specific and do not infect humans.

Adenoviruses

These are naked icosahedral viruses (75 nm in diameter) with double-stranded linear DNA. They cause pharyngitis, upper and lower respiratory tract disease, and a variety of other less common infections. There are at least 40 antigenic types, some of which cause sarcomas in animals but no tumors in humans.

Hepadnaviruses

These are double-shelled viruses (42 nm in diameter) with an icosahedral capsid covered by an envelope. The DNA is a double-stranded circle that is unusual because the complete strand is not a covalently closed circle and the other strand is missing approximately 25% of its length. HBV is the human pathogen in this family.

Herpesviruses

These are enveloped viruses (100 nm in diameter) with an icosahedral nucleocapsid and double-stranded linear DNA. They are noted for causing latent infections. The five important human pathogens are herpes simplex virus types 1 and 2, varicella-zoster virus, cytomegalovirus, and Epstein–Barr virus (the cause of infectious mononucleosis).

Poxviruses

These are the largest viruses, with a bricklike shape, an envelope with an unusual appearance, and a complex capsid symmetry. They are named for the skin lesions, or "pocks," that they cause. Smallpox virus and molluscum contagiosum virus are the two important members.

RNA VIRUSES

The 14 families of RNA viruses are described in Table 31–2. The three **naked icosahedral** virus families are listed first and are followed by the three **enveloped icosahedral** viruses.

The remaining eight families are **enveloped helical** viruses; the first five have single-stranded linear RNA as their genome, whereas the last three have single-stranded circular RNA.

Picornaviruses

These are the smallest (28 nm in diameter) RNA viruses. They have single-stranded, linear, nonsegmented, positive-polarity RNA within a naked icosahedral capsid. The name "picorna" is derived from *pico* (small), *RNA*-containing. There are two groups of human pathogens: (1) enteroviruses, such as poliovirus, Coxsackie virus, echovirus, and hepatitis A virus and (2) rhinoviruses.

Hepeviruses

These are naked viruses (30 nm in diameter) with an icosahedral nucleocapsid. They have single-stranded, linear, nonsegmented, positive-polarity RNA. The main human pathogen is hepatitis E virus.

Caliciviruses

These are naked viruses (38 nm in diameter) with an icosahedral capsid. They have single-stranded, linear, nonsegmented, positive-polarity RNA. The main human pathogen is norovirus.

Reoviruses

These are naked viruses (75 nm in diameter) with two icosahedral capsid coats. They have 10 or 11 segments of double-stranded linear RNA. The name is an acronym of *respiratory enteric orphan*, because they were originally found in the respiratory and enteric tracts and were not associated with any human disease. The main human pathogen is rotavirus, which causes diarrhea, mainly in infants. The rotavirus genome has 11 segments of double-stranded RNA.

Flaviviruses

These are enveloped viruses with an icosahedral capsid and single-stranded, linear, nonsegmented, positive-polarity RNA. The flaviviruses include hepatitis C virus, yellow fever virus, dengue virus, West Nile virus, Zika virus, and St. Louis and Japanese encephalitis viruses.

Togaviruses

These are enveloped viruses with an icosahedral capsid and single-stranded, linear, nonsegmented, positive-polarity RNA. There are two major groups of human pathogens: the alphaviruses and rubiviruses. The alphavirus group includes eastern and western encephalitis viruses; the rubivirus group consists only of rubella virus.

Retroviruses

These are enveloped viruses with an icosahedral capsid and two identical strands (said to be "diploid") of single-stranded, linear,

positive-polarity RNA. The term *retro* pertains to the reverse transcription of the RNA genome into DNA. A virus-encoded RNA-dependent DNA polymerase (reverse transcriptase) performs this function. Only retroviruses, integrate a DNA copy of the RNA genome into cellular DNA as an obligatory step in viral replication. Retroviruses encode an integrase that performs this function.

A large part of the retroviral genome consists of three structural genes, *gag*, *pol*, and *env*, in that order. Each gene is transcribed into a single mRNA that is translated into a polyprotein. The gag and pol precursor proteins are cleaved into functional proteins by a virion-encoded protease. For example, the POL polyprotein is cleaved to form reverse transcriptase, integrase, and RNAse H by the protease that is also part of that polyproteins.

There are two medically important groups: (1) the oncovirus group, which contains the sarcoma and leukemia viruses (e.g., human T-cell leukemia virus [HTLV]) and (2) the lentivirus ("slow virus") group, which includes human immunodeficiency virus (HIV) and certain animal pathogens (e.g., visna virus).

A third group, spumaviruses, is not medically important and is described in Chapter 46. Spumaviruses are very unusual because although they are retroviruses, they have double-stranded *DNA* in the virion. They have a retroviral sequence of genes (namely *gag, pol, env*) and use reverse transcriptase to synthesize virion DNA at the end of the replicative cycle within the nascent progeny virions. Also, the virion DNA integrates into cellular DNA during viral replication.

Orthomyxoviruses

These viruses (myxoviruses) are enveloped, with a helical nucleocapsid and eight segments of linear, single-stranded, negative-polarity RNA. The term *myxo* refers to the affinity of these viruses for mucins, and *ortho* is added to distinguish them from the paramyxoviruses. Influenza virus is the main human pathogen.

Paramyxoviruses

These are enveloped viruses with a helical nucleocapsid and single-stranded, linear, nonsegmented, negative-polarity RNA. The important human pathogens are measles, mumps, parainfluenza, and respiratory syncytial viruses.

Rhabdoviruses

These are bullet-shaped enveloped viruses with a helical nucleocapsid and a single-stranded, linear, nonsegmented, negative-polarity RNA. The term *rhabdo* refers to the bullet shape. Rabies virus is the only important human pathogen.

Filoviruses

These are enveloped viruses with a helical nucleocapsid and single-stranded, linear, nonsegmented, negative-polarity RNA. They are highly pleomorphic, long filaments that are 80 nm in diameter but can be thousands of nanometers long. The term *filo* means "thread" and refers to the long filaments. The two human pathogens are Ebola virus and Marburg virus.

Coronaviruses

These are enveloped viruses with a helical nucleocapsid and a single-stranded, linear, nonsegmented, positive-polarity RNA. The term *corona* refers to the prominent halo of spikes protruding from the envelope. Coronaviruses cause respiratory tract infections, such as the common cold, severe acute respiratory syndrome (SARS), and coronavirus infectious disease-19 (COVID-19) in humans. SARS-CoV-2 is the virus that caused the world-wide pandemic of COVID-19 in the years 2019 to 2021.

Arenaviruses

These are enveloped viruses with a helical nucleocapsid and a single-stranded, circular, negative-polarity RNA in two segments. (A part of both segments is positive-polarity RNA, and the term *ambisense RNA* is used to describe this unusual genome.) The term *arena* means "sand" and refers to granules on the virion surface that are nonfunctional ribosomes. Two human pathogens are lymphocytic choriomeningitis virus and Lassa fever virus.

Bunyaviruses

These are enveloped viruses with a helical nucleocapsid and a single-stranded, circular, negative-polarity RNA in three segments. Some bunyaviruses contain ambisense RNA in their genome (see Arenaviruses). The term *bunya* refers to the prototype, Bunyamwera virus, which is named after the place in Africa where it was isolated. These viruses cause encephalitis and various fevers such as Korean hemorrhagic fever and Crimean-Congo hemorrhagic fever. Hantaviruses, such as Sin Nombre virus (see Chapter 46), are members of this family.

Deltavirus

Hepatitis deltavirus (HDV) is the only member of this genus. It is an enveloped virus with an RNA genome that is a single-stranded, negative-polarity, covalently closed circle. The symmetry of the nucleocapsid is uncertain. It is a defective virus because it cannot replicate unless HBV is present within the same cell. HBV is required because it encodes hepatitis B surface antigen (HBsAg), which serves as the outer protein coat of HDV. The RNA genome of HDV encodes only one protein, the internal core protein called delta antigen.

PEARLS

- The classification of viruses is based primarily on the nature of the genome and whether the virus has an envelope.
- Poxviruses, herpesviruses, and hepadnaviruses are DNA viruses with an envelope, whereas adenoviruses, polyomaviruses, papillomaviruses, and parvoviruses are DNA viruses without an envelope (i.e., they are naked nucleocapsid viruses). Parvoviruses have single-stranded DNA, whereas all the other families of DNA viruses have double-stranded DNA. The DNA of hepadnaviruses (hepatitis B virus) is mostly double-stranded but has a single-stranded region.
- Picornaviruses, hepeviruses, caliciviruses, and reoviruses are RNA viruses without an envelope, whereas all the other families of RNA viruses have an envelope. Reoviruses have double-stranded RNA; all the other families of RNA viruses have single-stranded RNA. Reoviruses and influenza viruses have segmented RNA; all the other families of RNA viruses have nonsegmented RNA. Picornaviruses, hepeviruses, caliciviruses, flaviviruses, togaviruses, retroviruses, and coronaviruses have positive-polarity RNA, whereas all the other families have negative-polarity RNA.

PRACTICE QUESTIONS: USMLE & COURSE EXAMINATIONS

Questions on the topics discussed in this chapter can be found in the Basic Virology section of Part XIII: USMLE (National Board) Practice Questions starting on page 744. Also see Part XIV: USMLE (National Board) Practice Examination starting on page 775.

Pathogenesis

INTRODUCTION

The ability of viruses to cause disease can be viewed on two distinct levels: (1) the changes that occur within individual cells and (2) the process that takes place in the infected patient.

THE INFECTED CELL

There are four main effects of virus infection on the cell: (1) death, (2) fusion of cells to form multinucleated cells, (3) malignant transformation, and (4) no apparent morphologic or functional change.

Death of the cell is probably due to inhibition of macromolecular synthesis. Inhibition of host cell protein synthesis frequently occurs first and is probably the most important effect. It is important to note that synthesis of **cellular** proteins is inhibited but **viral** protein synthesis still occurs. For example, poliovirus inactivates an initiation factor (IF) required for cellular mRNA to be translated into cellular proteins, but poliovirus mRNA has a special ribosome-initiating site that allows it to bypass the IF so that viral proteins can be synthesized.

Apoptosis is also involved in cell death caused by viruses. Apoptosis is mediated by caspases, a family of cysteine proteases. The activation of caspases by cytochrome C released from mitochondria damaged by viral infection is an important mechanism.

Infected cells frequently contain **inclusion bodies**, which are discrete areas containing viral proteins or viral particles. They have a characteristic intranuclear or intracytoplasmic location and appearance depending on the virus. One of the best examples of inclusion bodies that can assist in clinical diagnosis is that of **Negri bodies**, which are eosinophilic cytoplasmic inclusions

found in rabies virus-infected brain neurons. Another important example is the **owl's eye inclusion** seen in the nucleus of cytomegalovirus (CMV)-infected cells. Electron micrographs of inclusion bodies can also aid in the diagnosis when virus particles of typical morphology are visualized.

Fusion of virus-infected cells produces **multinucleated** giant cells, which characteristically form after infection with **herpesviruses** and **paramyxoviruses**. Fusion occurs as a result of cell membrane changes, which are probably caused by the insertion of viral proteins into the membrane. The clinical diagnosis of herpesvirus skin infections is aided by the finding of multinucleated giant cells with intranuclear inclusions in skin scrapings.

A hallmark of viral infection of the cell is the **cytopathic effect** (CPE). This change in the appearance of the infected cell usually begins with a rounding and darkening of the cell and culminates in either lysis (disintegration) or giant cell formation. In the clinical laboratory, detection of virus in a patient's specimen is frequently based on the appearance of CPE in cell culture. In addition, CPE is the basis for the plaque assay, an important method for quantifying the amount of virus in a sample.

Infection with certain viruses causes **malignant transformation**, which is characterized by unrestrained growth, prolonged survival, and morphologic changes such as focal areas of rounded, piled-up cells. These changes are described in more detail in Chapter 43.

Infection of the cell accompanied by virus production can occur **without** any morphologic or gross functional changes. This observation highlights the wide variations in the nature of the interaction between the virus and the cell, ranging from rapid destruction of the cell to a benign relationship in which the cell survives and multiplies despite the replication of the virus.

THE INFECTED PATIENT

Pathogenesis in the infected patient involves (1) transmission of the virus and its entry into the host; (2) replication of the virus and damage to cells; (3) spread of the virus to other cells and organs; (4) the innate and adaptive immune responses, both as a host defense and as a contributing cause of certain diseases (see Chapter 33); and (5) persistence of the virus in some instances.

The stages of a typical viral infection are the same as those described for a bacterial infection in Chapter 7, namely, an **incubation period** during which the patient is asymptomatic, a **prodromal period** during which nonspecific symptoms occur, a **specific-illness period** during which the characteristic symptoms and signs occur, and a **recovery period** during which the illness wanes and the patient regains good health. In some patients, the infection persists and a chronic carrier state or a latent infection occurs (see later).

Transmission & Portal of Entry

Viruses are transmitted to the individual by many different routes, and their portals of entry are varied (Table 32–1). For example, person-to-person spread occurs by transfer of respiratory secretions, saliva, blood, or semen and by fecal contamination of water or food. The transfer of blood, either by transfusion or by sharing needles during intravenous drug use, can transmit various viruses (and bacteria). The screening of donated blood for human immunodeficiency virus (HIV), human T-cell

TABLE 32–1 Main Portal of Entry of Important Viral Pathogens

Portal of Entry	Virus	Disease
Respiratory tract[1]	Influenza virus	Influenza
	Rhinovirus	Common cold
	Coronavirus	Common cold, pneumonia (COVID-19)
	Respiratory syncytial virus	Bronchiolitis
	Epstein–Barr virus	Infectious mononucleosis
	Varicella zoster virus	Chickenpox
	Herpes simplex virus type 1	Herpes labialis
	Cytomegalovirus	Mononucleosis syndrome
	Measles virus	Measles
	Mumps virus	Mumps
	Rubella virus	Rubella
	Hantavirus	Pneumonia
	Adenovirus	Pneumonia
	Parvovirus B19	Slapped cheeks syndrome
Gastrointestinal tract[2]	Hepatitis A virus	Hepatitis A
	Poliovirus	Poliomyelitis
	Rotavirus	Diarrhea
	Norovirus	Diarrhea
Skin	Rabies virus[3]	Rabies
	Yellow fever virus[4]	Yellow fever
	Dengue virus[4]	Dengue
	Zika virus[4]	Zika fever
	Human papillomavirus	Papillomas (warts)
Genital tract	Herpes simplex virus type 2	Genital herpes & neonatal herpes
	Human papillomavirus	Papillomas (warts)
	Hepatitis B virus	Hepatitis B
	Zika virus	Zika fever
	Human immunodeficiency virus	Acquired immunodeficiency syndrome (AIDS)
Blood	Hepatitis B virus	Hepatitis B
	Hepatitis C virus	Hepatitis C
	Hepatitis D virus	Hepatitis D
	Human T-cell lymphotropic virus	Leukemia
	Human immunodeficiency virus	AIDS
	Cytomegalovirus	Mononucleosis syndrome or pneumonia
Transplacental	Cytomegalovirus	Congenital abnormalities
	Rubella	Congenital abnormalities
	Zika	Congenital abnormalities
	Parvovirus B19	Hydrops fetalis

[1]Transmission of these viruses is typically by respiratory aerosols or saliva.

[2]Transmission of these viruses is typically by the fecal–oral route in contaminated food or water.

[3]Transmission of these viruses is typically by the bite of an infected mammal.

[4]Transmission of these viruses is typically by the bite of an infected mosquito.

TABLE 32–2 Viruses That Commonly Cause Fetal and Neonatal Infections

Type of Transmission	Virus
Transplacental[1]	Cytomegalovirus Parvovirus B19 virus Rubella virus Zika virus
Perinatal (at time of birth)[2]	Hepatitis B virus Hepatitis C virus Herpes simplex virus type 2 Human immunodeficiency virus[3] Human papillomavirus
Breast-feeding	Cytomegalovirus Human T-cell lymphotropic virus

[1]Note that there are important bacteria, namely, *Treponema pallidum* and *Listeria monocytogenes*, and an important protozoan, namely, *Toxoplasma gondii*, that are also transmitted transplacentally.

[2]Note that there are important bacteria, namely, *Neisseria gonorrhoeae, Chlamydia trachomatis*, and group B *Streptococcus* that are also transmitted at the time of birth.

[3]Human immunodeficiency virus is also transmitted transplacentally and in breast milk.

lymphotropic virus, hepatitis B virus, hepatitis C virus (HCV), and West Nile virus (as well as *Treponema pallidum*) has greatly reduced the risk of infection by these pathogens.

Transmission of viruses by the respiratory route can occur either by **droplets** or **aerosols**. Droplets are **large particles** of respiratory secretions that fall quickly by gravity. The uninfected recipient person needs to be close to the infected person to be infected. Aerosols are **small particles** of respiratory secretions that do *not* fall quickly. They remain airborne for minutes to hours and can travel some distance via air currents. Another distinction is that large droplets are typically deposited in the upper airways whereas small aerosols can be inhaled into the lower respiratory tract where they can cause pneumonia.

Aerosol transmission underlies the public health protocols for masks, social distancing, and ventilation that are being employed during the coronavirus pandemic of 2019. Note that the division of particles into large and small is arbitrary and that particles of intermediate size with various characteristics also occur.

Transmission can also occur between mother and offspring in utero across the placenta, at the time of delivery (perinatal), or during breast-feeding (Table 32–2). Transmission between mother and offspring is called **vertical transmission**. Person-to-person transmission that is not from mother to offspring is called **horizontal transmission**.

Animal-to-human transmission can take place either directly from the bite of a reservoir host as in rabies or indirectly through the bite of an insect vector, such as a mosquito, which transfers the virus from an animal reservoir to the person. The zoonotic diseases caused by viruses are described in Table 32–3. Note that in addition to transmission from an animal reservoir, both Zika virus and Ebola virus can be transmitted from person to person via semen.

Specificity of Interaction of Viruses With Cells and Organs

The specificity of the interaction of viruses with cells is primarily determined by the **binding of surface proteins on the virus to receptors on the human cell surface**. One of the best examples of this interaction is the binding of the gp120 on the envelope of HIV to the CD4 protein (receptor) on the surface of CD4-postive T lymphocyte.

The type of cell infected determines, in large part, the clinical features of the disease. For example, rabies virus causes encephalitis because it infects neurons by attaching to the acetylcholine receptor on neurons and hepatitis B virus causes hepatitis because it attaches to a taurocholate transporter on hepatocytes.

TABLE 32–3 Medically Important Viruses That Have an Animal Reservoir

Virus	Animal Reservoir	Mode of Transmission	Disease
Rabies virus	In the United States, skunks, raccoons, and bats; in developing countries, dogs	Usually bite of infected animal; also aerosol of bat saliva	Rabies
Ebola virus	Fruit bats are suspected	Unknown	Ebola hemorrhagic fever
Hantavirus[1]	Deer mice	Aerosol of dried excreta	Hantavirus pulmonary syndrome (pneumonia)
Yellow fever virus	Monkeys	Bite of *Aedes* mosquito	Yellow fever
Dengue virus	Monkeys	Bite of *Aedes* mosquito	Dengue fever
Zika virus	Monkeys	Bite of *Aedes* mosquito	Zika fever
Encephalitis viruses[2]	Wild birds (e.g., sparrows)	Bite of various mosquitoes	Encephalitis
SARS[3] coronavirus SARS-CoV-2	Bats, civet cat Bats, (?pangolin)	Aerosols, droplets Aerosols, droplets	SARS COVID-19
Avian influenza virus (H5N1)	Chickens and other fowl	Aerosol droplets, guano	Influenza

[1]Sin Nombre virus is the most important hantavirus in the United States.

[2]Important encephalitis viruses in the United States include eastern and western equine encephalitis viruses, West Nile virus, and St. Louis encephalitis virus.

[3]SARS = severe acute respiratory syndrome.

Note that these receptors on the cell surface serve various normal functions in the life of the cell. The virus has adapted to these proteins in order to enter the cell and replicate.

Localized or Disseminated Infections

Most viral infections are either **localized** to the portal of entry or spread **systemically** through the body. The best example of the localized infection is the common cold caused by rhinoviruses, which involves only the upper respiratory tract. Influenza is localized primarily to the upper and lower respiratory tracts. Respiratory viruses have a short incubation period because they replicate directly in the mucosa, but systemic infections such as poliomyelitis and measles have a long incubation period because viremia and secondary sites of replication are required.

One of the best-understood systemic viral infections is paralytic poliomyelitis (Figure 32–1). After poliovirus is ingested, it infects and multiplies within the cells of the small intestine and then spreads to the mesenteric lymph nodes, where it multiplies again. It then enters the bloodstream and is transmitted to certain internal organs, where it multiplies again. The virus reenters the bloodstream and is transmitted to the central nervous system, where damage to the anterior horn cells occurs, resulting in the characteristic muscle paralysis. It is during this obligatory viremia that circulating immunoglobulin (Ig) G antibodies induced by the polio vaccine can prevent the virus from infecting the central nervous system. Viral replication in the gastrointestinal tract results in the presence of poliovirus in the feces, thus perpetuating its transmission to others.

Some viral infections spread systemically, not via the bloodstream, but rather by retrograde axonal flow within neurons. Four important human pathogens do this: rabies virus, herpes simplex type 1, herpes simplex type 2, and varicella-zoster virus. As an example, rabies virus is introduced into the body at the site of an animal bite. The virus infects a local sensory neuron and ascends into the central nervous system by retrograde axonal flow, where it causes encephalitis.

Pathogenesis & Immunopathogenesis

The signs and symptoms of most viral diseases undoubtedly are the result of cell killing by virus-induced inhibition of macromolecular synthesis. Death of the virus-infected cells results in a loss of function and in the symptoms of disease. For example, when poliovirus kills motor neurons, paralysis of the muscles innervated by those neurons results. Also, the hemorrhages caused by Ebola virus are due to the damage to the vascular endothelial cells caused by the envelope glycoprotein of the virus.

There are other diseases in which **cell killing by immunologic attack** plays an important role in pathogenesis. Both **cytotoxic T cells and antibodies play a role in immunopathogenesis**.

(1) **Cytotoxic T cells are involved in the pathogenesis of hepatitis caused by hepatitis A, B, and C viruses**. These viruses do not cause a CPE, and the damage to the hepatocytes is the result of the recognition of viral antigens on the hepatocyte surface by

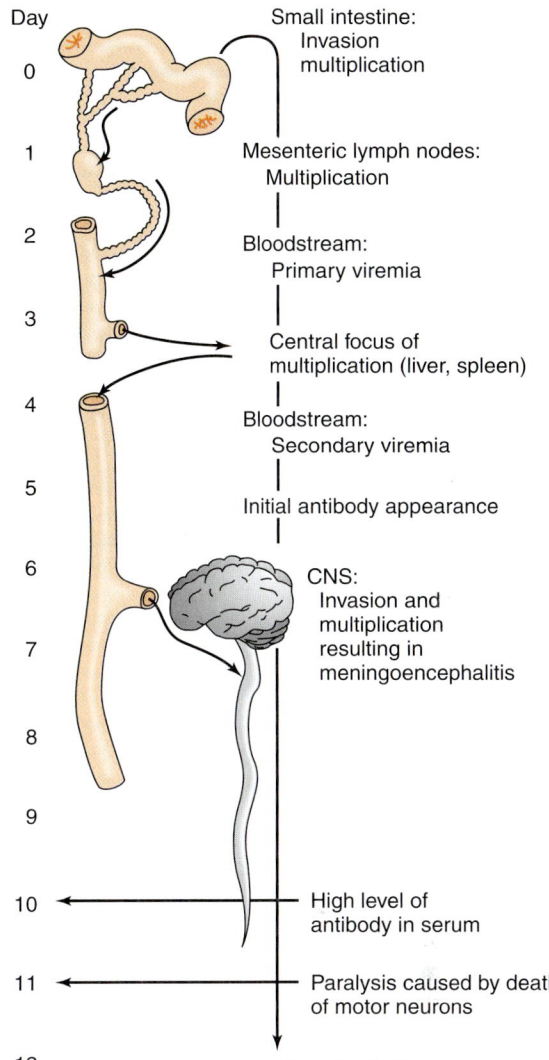

FIGURE 32–1 Systemic viral infection by poliovirus, resulting in paralytic poliomyelitis. CNS = central nervous system. (Reproduced with permission from Brooks GF, Butel JS, Ornston LN: *Jawetz, Melnick & Adelberg's Medical Microbiology*, 20th ed. New York, NY: McGraw Hill; 1995.)

cytotoxic T cells. The rash of measles is similarly caused by these cells attacking the infected vascular endothelium in the skin.

(2) **Immune-mediated pathogenesis also occurs when virus–antibody–complement complexes form and are deposited in various tissues**. This occurs in hepatitis B virus infection, in which immune complexes play a role in producing the arthritis characteristic of the early stage of hepatitis B. Immune complexes also cause the arthritis seen in parvovirus B19 and rubella virus infections. The pathogenesis of pneumonia caused by respiratory syncytial virus in infants is attributed to immune complexes formed by maternal IgG and viral antigens.

Virulence

Strains of viruses differ greatly in their ability to cause disease. For example, there are strains of poliovirus that have mutated

TABLE 32–4 Important Mechanisms by Which Viruses Evade Host Defenses

Host Defense Affected	Mechanism of Evasion	Virus That Employs the Mechanism
Cytotoxic T cells	Reduces MHC class I proteins, thereby decreasing killing by cytotoxic T cells	HIV, HSV, CMV, adenovirus
Helper (Th-1) T cells	Blocks IL-12, which reduces formation of Th-1 cells, thereby decreasing cell-mediated immunity	Measles virus
Interferon	Blocks synthesis of interferon by virus-infected cells	EBV, HCV, SARS-CoV-2
Interferon	Blocks synthesis of kinase that phosphorylates initiation factor-2	HIV, influenza, and HSV
Interferon	NS1 protein blocks action of protein kinase and ribonuclease that inhibit viral replication	Influenza
Interferon	Blocks action of protein kinase	Hepatitis C virus
Interleukins	Encodes receptors for immune mediators; receptors are secreted by infected cells, binds mediators, and inactivates them	Vaccinia virus encodes receptor for IL-1
Chemokines	Encodes chemokine receptor; this blocks action of chemokine, thereby inhibiting migration of inflammatory cells to site of infection	Vaccinia virus, CMV
Complement	Encodes protein that binds to complement protein C3b; this blocks opsonizing action of C3b as well as its ability to participate in forming the membrane attack complex	HSV
Antibody	Glycans, especially oligomannose, coat envelope proteins decreasing antibody synthesis and binding	HIV, SARS-CoV-2, HCV

CMV = cytomegalovirus; EBV = Epstein–Barr virus; HIV = human immunodeficiency virus; HSV = herpes simplex virus; HCV= hepatitis C virus; IL = interleukin; MHC = major histocompatibility complex; SARS-CoV-2 = severe acute respiratory syndrome-coronavirus-2.

sufficiently such that they have lost the ability to cause polio in immunocompetent individuals (i.e., they are **attenuated**). These strains are used in vaccines. The viral genes that control the virulence of the virus are poorly characterized, and the process of virulence is poorly understood. The role of evasion of host defenses in viral virulence is discussed below.

Evasion of Host Defenses

Viruses have several ways by which they evade our host defenses (Table 32–4). These processes are often called **immune evasion**. Three very important processes are (1) **synthesis of receptors for immune mediators**, (2) **reduction of expression of class I major histocompatibility complex (MHC) proteins**, and (3) **inhibition of interferon synthesis and action** (see Chapter 33).

Some viruses encode the receptors for various mediators of immunity such as interleukin-1 (IL-1) and tumor necrosis factor (TNF). For example, vaccinia virus encodes a protein that binds to IL-1, and fibroma virus encodes a protein that binds to TNF. CMV encodes a chemokine receptor that binds to several chemokines. When released from virus-infected cells, these proteins bind to the immune mediators and block their ability to interact with receptors on their intended targets, our immune cells that mediate host defenses against the viral infection. By reducing our host defenses, the virulence of the virus is enhanced. These virus-encoded proteins that block host immune mediators are often called **cytokine decoys**.

In addition, some viruses (e.g., HIV and herpesviruses, such as herpes simplex virus [HSV] and CMV) can reduce the expression of class I MHC proteins, thereby reducing the ability of cytotoxic T cells to kill the virus-infected cells, and other viruses (e.g., HSV) inhibit complement.

The N (nucleocapsid) protein of SARS-CoV-2 blocks the synthesis of interferon. Chronic infection caused by HCV is based in large part, by the ability of the virus to block the synthesis and the action of interferon.

Several viruses (HIV, Epstein–Barr virus, and adenovirus) synthesize RNAs that block the phosphorylation of an initiation factor (eIF-2), which reduces the ability of interferon to block viral replication (see Chapter 33). The NS1 protein of influenza virus blocks the action of interferon by inhibiting the protein kinase and ribonuclease that mediate interferon's antiviral action.

CMV encodes a microRNA that binds to the mRNA of a cell surface ligand for natural killer cells. Binding of the microRNA prevents synthesis of the ligand, which prevents killing of the CMV-infected cells by the natural killer cells. Measles virus blocks synthesis of IL-12, thereby reducing an effective Th-1 response. Ebola virus synthesizes two proteins; one protein blocks the induction of interferon, whereas the other blocks its action. Collectively, these viral virulence factors are called **virokines**.

Some viruses (HIV, SARS-CoV-2, and HCV) have glycans, especially oligomannose, attached to their envelope proteins. These glycans contribute to immune evasion ("glycan shield") by masking the antigenic sites of the protein thereby reducing both the induction of antibody synthesis and the ability of antibody to neutralize infectivity of the virus.

Another important way by which viruses evade our host defenses is by producing **multiple antigenic types** (also known as multiple serotypes) (Table 32–5). The clinical importance of a virus having multiple serotypes is that a patient can be infected with one serotype, recover, and have antibodies that protect from infection by that serotype in the future; however, that person can be infected by another serotype of that virus.

The classic example of a virus with multiple serotypes is rhinovirus, which has more than 100 serotypes. This is the

TABLE 32–5 Serotypes of Some Medically Important Viruses

Viruses With Multiple, Rapidly Changing Serotypes	Viruses With Multiple Serotypes That Do Not Change	Viruses With a Single Stable Serotype for Which There Is a Vaccine	Viruses With a Single Stable Serotype for Which There Is No Vaccine
Influenza virus	Poliovirus	Measles	Herpes simplex virus, types 1 and 2
Human immunodeficiency virus	Rhinovirus	Mumps	Cytomegalovirus
Hepatitis C virus	Adenovirus	Rubella	Epstein–Barr virus
	Human papillomavirus	Varicella-zoster virus	Parvovirus B19
	Rotavirus	Hepatitis A virus	
	Norovirus	Hepatitis B virus	
	Respiratory syncytial virus	Rabies virus	
		Yellow fever virus	
		Smallpox virus	

reason why the "common cold" caused by rhinoviruses is so common. Influenza virus also has multiple serotypes, and the severe worldwide epidemics of influenza are attributed to the emergence of new antigenic types. HIV and HCV have multiple serotypes, which contribute to the difficulty in obtaining a vaccine against these viruses. Note that only some viruses have multiple serotypes. Many important human pathogens (e.g., measles virus, rubella virus, varicella-zoster virus, and rabies virus) have only one serotype, and some have only a few serotypes (e.g., poliovirus has three serotypes).

Persistent Viral Infections

In most viral infections, the virus does not remain in the body for a significant period after clinical recovery. However, in certain instances, the virus persists for long periods either intact or in the form of the genome. The mechanisms that may play a role in the persistence of viruses include (1) integration of a DNA provirus into host cell DNA, as occurs with retroviruses; (2) immune tolerance, because neutralizing antibodies are not formed against the virus; (3) formation of virus–antibody complexes, which remain infectious; (4) location of the virus within an immunologically sheltered "sanctuary" (e.g., the brain); (5) rapid antigenic variation of the virus; (6) spread of the virus from cell to cell without an extracellular phase, so that virus is not exposed to antibody; and (7) immunosuppression, as in acquired immunodeficiency syndrome (AIDS).

There are three types of persistent viral infections of clinical importance. They are distinguished primarily by whether virus is usually produced by the infected cells and by the timing of the appearance both of the virus and of the symptoms of disease.

Chronic-Carrier Infections

Some patients who have been infected with certain viruses continue to produce significant amounts of the virus for long periods. This **carrier state** can follow an asymptomatic infection as well as the actual disease and can itself either be asymptomatic or result in chronic illness. Important clinical examples are chronic hepatitis, which occurs in hepatitis B and HCV carriers,

and neonatal rubella virus and CMV infections, in which carriers can produce virus for years.

Latent Infections

In these infections, best illustrated by the herpesvirus group, the patient recovers from the initial infection and virus production stops. Subsequently, the symptoms may **recur**, accompanied by the production of virus. In HSV infections, the virus enters the latent state in the cells of the sensory ganglia. Latent HSV infection is caused by the loss of function of tegument protein VP16 in the neuron. VP16 is required for viral replication. When it doesn't function, viral DNA remains as an episome in the nucleus of the neuron. See Chapter 37 for more information.

HSV type 1, which causes infections primarily of the eyes and face, is latent in the trigeminal ganglion, whereas HSV type 2, which causes infections primarily of the genitals, is latent in the lumbar and sacral ganglia. Varicella-zoster virus, another member of the herpesvirus family, causes varicella (chickenpox) as its initial manifestation and then remains latent, primarily in the trigeminal or thoracic ganglion cells. It can recur in the form of the painful vesicles of zoster (shingles), usually on the face or trunk.

Slow Virus Infections

The term *slow* refers to the **prolonged period** between the initial infection and the onset of disease, which is usually measured in years. In instances in which the cause has been identified, the virus has been shown to have a normal, not prolonged, growth cycle. It is not, therefore, that virus growth is slow; rather, the incubation period and the progression of the disease are prolonged. Two of these diseases are caused by conventional viruses, namely, subacute sclerosing panencephalitis, which follows several years after measles virus infections, and progressive multifocal leukoencephalopathy (PML), which is caused by JC virus, a papovavirus. PML occurs primarily in patients who have lymphomas or are immunosuppressed. Other slow infections in humans (e.g., Creutzfeldt-Jakob disease and kuru) are caused by unconventional agents called **prions** (see Chapter 28). Slow virus infections are described in Chapter 44.

PEARLS

The Infected Cell

- Death of infected cells is probably caused by inhibition of cellular protein synthesis. Translation of viral mRNA into viral proteins preempts the ribosomes, preventing synthesis of cellular proteins.
- **Inclusion bodies** are aggregates of virions in specific locations in the cell that are useful for laboratory diagnosis. Two important examples are **Negri bodies** in the cytoplasm of rabies virus-infected cells and **owl's eye inclusions** in the nucleus of cytomegalovirus-infected cells.
- Multinucleated giant cells form when cells are infected with certain viruses, notably herpesviruses and paramyxoviruses such as respiratory syncytial virus.
- **Cytopathic effect** (CPE) is a visual or functional change in infected cells typically associated with the death of cells.
- Malignant transformation occurs when cells are infected with oncogenic viruses. Transformed cells are capable of unrestrained growth.
- Some virus-infected cells appear visually and functionally normal, yet are producing large numbers of progeny viruses.

The Infected Patient

- Viral infection in the person typically has four stages: incubation period, prodromal period, specific-illness period, and recovery period.
- The main portals of entry are the respiratory, gastrointestinal, and genital tracts, but through the skin, across the placenta, and via blood are important as well.
- Transmission from mother to offspring is called **vertical transmission**; all other modes of transmission (e.g., fecal–oral, respiratory aerosol, insect bite) are **horizontal transmission**. Transmission can be from human to human or from animal to human.
- Most serious viral infection are systemic (i.e., the virus travels from the portal of entry via the blood to various organs). However, some are localized to the portal of entry, such as the common cold, which involves only the upper respiratory tract.
- The specificity of the interaction of viruses with cells is primarily determined by the **binding of surface proteins on the virus to receptors on the human cell surface**.

Pathogenesis

- The symptoms of viral diseases are usually caused by **death of the infected cells and a consequent loss of function**. For example, poliovirus kills neurons, resulting in paralysis.
- **Immunopathogenesis** is the process by which the symptoms of viral diseases are caused by the immune system rather than by the killing of cells directly by the virus. One type of immunopathogenesis is the **killing of virus-infected cell by the attack of cytotoxic T cells** that recognize viral antigens on the cell surface. Damage to the liver caused by hepatitis viruses occurs by this mechanism. Another is the **formation of virus–antibody complexes that are deposited in tissues**. Arthritis associated with parvovirus B19 or rubella virus infection occurs by this mechanism.
- Virulence of viruses differs markedly from one virus to another and among different strains of the same virus. The genetic basis for these differences is not well understood. Strains with weakened (attenuated) virulence are often used in vaccines.
- Viruses can evade host defenses by producing **multiple antigens**, thereby avoiding inactivation by antibodies, and by **reducing the synthesis of class I MHC proteins**, thereby decreasing the ability of a cell to present viral antigens and blunting the ability of cytotoxic T cells to kill the virus-infected cells. Many viruses **inhibit the synthesis of interferon**. Viruses also produce receptors for immune mediators, such as interleukin (IL)-1 and tumor necrosis factor (TNF), thereby preventing the ability of these mediators to activate antiviral processes.

Persistent Viral Infections

- **Carrier state** refers to people who produce virus for long periods of time and can serve as a source of infection for others. The carrier state that is frequently associated with hepatitis C virus infection is a medically important example.
- **Latent infections** are those infections that are not producing virus at the present time but can be reactivated at a subsequent time. The latent infections that are frequently associated with herpes simplex virus infection are a medically important example.
- **Slow virus infections** refer to the diseases with a long incubation period, often measured in years. Some, such as progressive multifocal leukoencephalopathy, are caused by viruses, whereas others, such as Creutzfeldt-Jakob disease, are caused by prions. The brain is often the main site of these diseases.

SELF-ASSESSMENT QUESTIONS

1. Viruses can cause changes in individual cells that are visible in the light microscope after suitable staining. Which one of the following is most characteristic of the changes seen in rabies virus–infected cells?

   (A) Inclusion bodies in the cytoplasm of macrophages
   (B) Inclusion bodies in the cytoplasm of neurons
   (C) Inclusion bodies in the nucleus of neurons
   (D) Multinucleated giant cells composed of neurons
   (E) Multinucleated giant cells composed of macrophages

2. Many viruses use the upper respiratory tract (mouth, nasopharynx) as their important portal of entry. One feature of the portal of entry is that it is the site where the virus first infects and replicates. Which one of the following viruses is most likely to enter via the upper respiratory tract?

   (A) Dengue virus
   (B) Epstein–Barr virus
   (C) Hepatitis A virus
   (D) Hepatitis B virus
   (E) Rotavirus

3. The term *vertical transmission* refers to:

   (A) transmission by insect vector from reservoir to patient.
   (B) transmission from a sex worker to a client.
   (C) transmission from mother to child.
   (D) transmission from one child to another at school.
   (E) transmission from person to person within a family.

4. Some viruses are known for their ability to cause perinatal infections. Which one of the following viruses is most likely to cause perinatal infections?

   (A) Cytomegalovirus
   (B) Epstein–Barr virus
   (C) JC virus
   (D) Norovirus
   (E) Poliovirus

5. Which one of the following viruses that causes human disease has an animal reservoir?

   (A) Cytomegalovirus
   (B) Hepatitis C virus
   (C) Smallpox virus
   (D) Varicella-zoster virus
   (E) Yellow fever virus

6. Which one of the following best describes the mechanism by which immunopathogenesis occurs?

   (A) Ability of antibodies to block pathogenesis by viruses
   (B) Ability of cytotoxic T cells to block pathogenesis by viruses
   (C) Ability of neutrophils to block pathogenesis by viruses
   (D) Ability of cytotoxic T cells to cause pathogenesis by viruses
   (E) Ability of eosinophils to cause pathogenesis by viruses

ANSWERS

(1) **(B)**
(2) **(B)**
(3) **(C)**
(4) **(A)**
(5) **(E)**
(6) **(D)**

PRACTICE QUESTIONS: USMLE & COURSE EXAMINATIONS

Questions on the topics discussed in this chapter can be found in the Basic Virology section of Part XIII: USMLE (National Board) Practice Questions starting on page 744. Also see Part XIV: USMLE (National Board) Practice Examination starting on page 775.

Host Defenses

INTRODUCTION

Host defenses against viruses fall into two major categories: (1) **nonspecific**, of which the most important are interferons and natural killer (NK) cells and (2) **specific**, including both antibody and cell-mediated immunity. Interferons are an early, first-line defense, whereas humoral immunity and cell-mediated immunity are effective only later because it takes several days to induce the humoral and cell-mediated arms of the immune response.

A description of how viruses evade our host defenses appears in Chapter 32.

NONSPECIFIC DEFENSES

Evidence for the importance of interferons includes (1) people who have a defective interferon response are prone to frequent and severe viral infections; (2) those who have an auto-immune response to interferon, i.e., produce antibody to it are also predisposed to severe viral infections; (3) several viruses, e.g., SARS, coronavirus-2, and influenza virus synthesize proteins that inhibit either the synthesis or action of interferon.

1. Alpha & Beta Interferons

Alpha and beta interferons are a group of proteins produced by human cells after viral infection (or after exposure to other inducers). They inhibit the growth of viruses by **blocking the synthesis of viral proteins**. They do so by two main mechanisms: One is a ribonuclease that degrades mRNA, and the other is a protein kinase that inhibits protein synthesis.

Interferons are divided into three types based on the cell of origin, namely, leukocyte, fibroblast, and lymphocyte. They are also known as alpha, beta, and gamma interferons, respectively. Alpha and beta interferons, collectively known as type I interferon, are induced by viruses, whereas gamma (T cell, immune) interferon, known as type II interferon, is induced by antigens and is one of the effectors of cell-mediated immunity (see Chapter 58). The following discussion of alpha and beta interferons focuses on the induction and action of their antiviral effect (Figure 33–1).

Lambda (λ) interferon, known as type III interferon, is active against intestinal viruses, especially rotavirus and norovirus. It reduces the long-term persistence of virus in intestinal mucosal cells. The role of lambda interferon in human disease is uncertain and will not be discussed further.

Induction of Alpha & Beta Interferons

The strong inducers of these interferons are **viruses** and **double-stranded RNAs**. Induction is not specific for a particular virus; many DNA and RNA viruses are competent inducers, although they differ in effectiveness. The finding that double-stranded RNA, but not single-stranded RNA or DNA, is a good inducer has led to the conclusion that a double-stranded RNA is synthesized as part of the replicative cycle of all inducing viruses. The double-stranded RNA poly (rI-rC) is one of the strongest inducers and was under consideration as an antiviral agent, but toxic side effects prevented its clinical use. The weak inducers of microbiologic interest include a variety of intracellular bacteria and protozoa, as well as certain bacterial substances such as endotoxin.

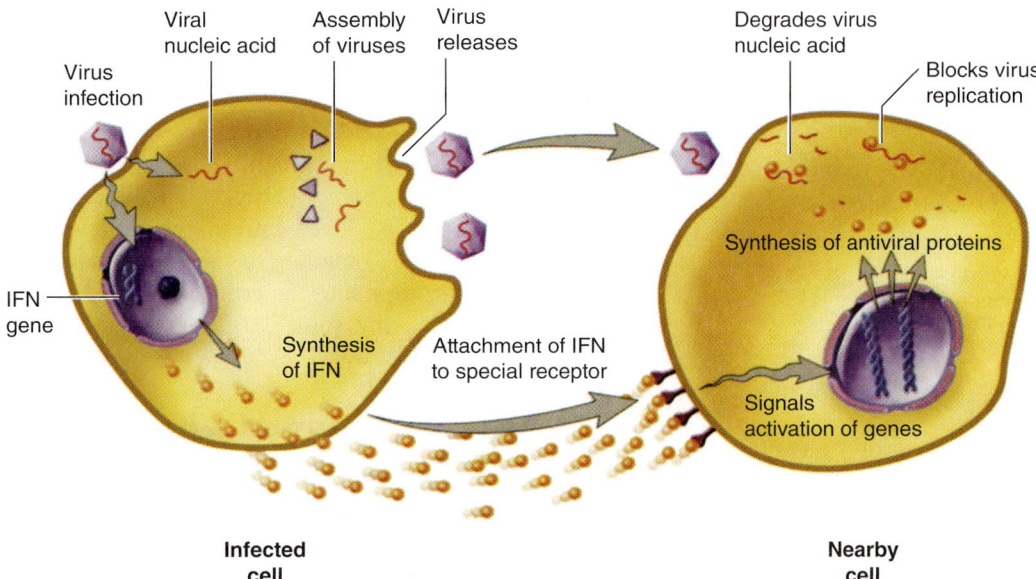

FIGURE 33–1 Induction and action of interferon. **Left side:** Virus infection induces the synthesis of interferon, which then leaves the infected cell. **Right side:** Interferon binds to the surface receptor of an uninfected cell and induces the synthesis of three new cell-encoded enzymes (antiviral proteins). A new virion enters the cell, but viral replication is inhibited by the interferon-induced antiviral proteins. One of these antiviral proteins is a ribonuclease that degrades mRNA and another is a protein kinase that phosphorylates an initiation factor that inhibits protein synthesis. (Reproduced with permission from Willey J, Sherwood L, Woolverton CJ: *Microbiology.* 7th ed. New York, NY: McGraw Hill; 2007.)

This extensive list of inducers makes it clear that **induction** of these interferons is **not specific**. Similarly, their inhibitory **action** is **not specific** for any particular virus. However, they are typically **specific** in regard to the **host species** in which they act (i.e., interferons produced by human cells are active in human cells but are much less effective in cells of other species). It is clear, therefore, that other animals cannot be used as a source of interferons for human therapy. Rather, the genes for human interferons have been cloned, and interferon for medical use is now produced by genetic engineering techniques.

Action of Alpha & Beta Interferons

Interferons inhibit the intracellular replication of a **wide variety** of DNA and RNA viruses but have little effect on the metabolism of normal cells. The selectivity arises from the presence of double-stranded RNA in virus-infected cells, which is not present in uninfected cells.

Interferons have **no effect** on extracellular virus particles. Interferons act by binding to a receptor on the cell surface that signals the cell to synthesize three proteins, thereby inducing the **antiviral state** (Figure 33–2). These three proteins are inactive precursor proteins until they are activated by double-stranded RNA synthesized during viral replication. As a result, these proteins are active in virus-infected cells but not in uninfected cells.

The three cellular proteins are (1) a **2,5-oligo A synthetase** that synthesizes an adenine trinucleotide (2,5-oligo A), (2) a **ribonuclease** that is activated by 2,5-oligo A and degrades viral and cellular mRNAs, and (3) a **protein kinase** that

phosphorylates an initiation factor (eIF-2) for protein synthesis, thereby inactivating it. The end result is that both viral and cellular protein synthesis is inhibited and the infected cell dies. No virus is produced by that cell, and the spread of the virus is reduced.

Because interferons are produced within a few hours of the initiation of viral replication, they act in the early phase of viral diseases to limit the spread of virus. In contrast, antibody begins to appear in the blood several days after infection.

Interferon is an effective part of our innate immune response against viruses. Nevertheless, viruses have various mechanisms for overcoming its antiviral effects. For example, the NS-1 protein of influenza virus inhibits the protein kinase and the ribonuclease that mediate the antiviral effect of interferon. Further, the nucleocapsid (N) protein of SARS-CoV-2 inhibits the synthesis of interferon. Additional information regarding the evasion of our host defenses is described in Chapter 32.

Alpha interferon has been approved for use in patients with condyloma acuminatum and chronic active hepatitis caused by hepatitis B and C viruses. Beta interferon is used in the treatment of multiple sclerosis. Gamma interferon reduces recurrent infections in patients with chronic granulomatous disease (see Chapter 68). Interferons are also used clinically in patients with cancers such as Kaposi's sarcoma and hairy cell leukemia.

2. Natural Killer Cells

NK cells are an important part of the innate defenses against virus-infected cells. They are called "natural" killer cells because they are active without the necessity of being exposed

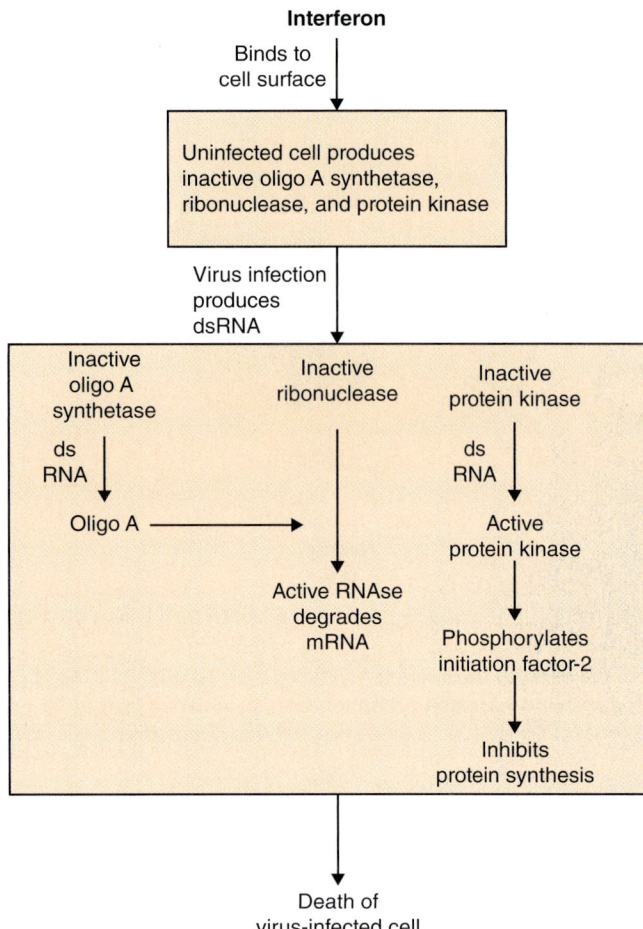

FIGURE 33–2 Interferon induces an antiviral state within an uninfected cell. Interferon binds to the surface of the uninfected cell and induces three proteins that remain inactive until a virus infects the cell. These proteins are oligo A synthetase, ribonuclease, and protein kinase. When a virus infects that cell, a double-stranded RNA (dsRNA) is synthesized as part of the viral replicative cycle. The dsRNA activates oligo A synthetase, which synthesizes oligo A that then activates the ribonuclease to degrade viral (and cell) mRNA. The dsRNA also activates the protein kinase that phosphorylates initiation factor-2 (eIF-2). This inhibits both viral and cell protein synthesis. The cell dies without producing progeny virus, thereby limiting the spread of infection.

to the virus previously and because they are not specific for any virus. NK cells are a type of T lymphocyte but do not have an antigen receptor. They recognize virus-infected cells by the absence of class I major histocompatibility complex (MHC) proteins on the surface of the virus-infected cell. They kill virus-infected cells by secreting perforins and granzymes, which cause apoptosis of the infected cells. (See Chapter 58 for more information.)

3. Phagocytosis

Macrophages, particularly fixed macrophages of the reticulo-endothelial system and alveolar macrophages, are the important cell types in limiting virus infection by phagocytosis.

In contrast, polymorphonuclear leukocytes are the predominant cellular defense in bacterial infections.

4. α-Defensins

α-Defensins are a family of positively charged peptides with antiviral activity. (They also have antibacterial activity; see Chapter 8.) They interfere with human immunodeficiency virus (HIV) binding to the CXCR4 receptor and block entry of the virus into the cell. The production of α-defensins may explain why some HIV-infected individuals are long-term "nonprogressors."

5. Apolipoprotein B RNA-Editing Enzyme (APOBEC3G)

APOBEC3G is an important member of the innate host defenses against retroviral infection, especially against HIV. APOBEC3G is an enzyme that causes hypermutation in retroviral DNA by deaminating cytosines in both mRNA and retroviral DNA, thereby inactivating these molecules and reducing infectivity. HIV defends itself against this innate host defense by producing Vif (viral infectivity protein), which counteracts APOBEC3G, thereby preventing hypermutation from occurring.

6. Fever

Elevated body temperature may play a role in host defenses, but its importance is uncertain. Fever may act in two ways: (1) The higher body temperature may directly inactivate the virus particles, particularly enveloped viruses, which are more heat-sensitive than nonenveloped viruses and (2) replication of some viruses is reduced at higher temperatures; therefore, fever may inhibit replication.

7. Mucociliary Clearance

The mucociliary clearance mechanism of the respiratory tract may protect the host. Its damage (e.g., from smoking) results in an increased frequency of viral respiratory tract infections, especially influenza.

8. Circumcision

There is evidence that circumcision prevents infection by three sexually transmitted viruses: HIV, human papillomavirus (HPV), and herpes simplex virus type 2 (HSV-2).

9. Factors That Modify Host Defenses

Several factors influence host defenses in a nonspecific or multifactorial way:

(1) Age is a significant variable in the outcome of viral infections. In general, infections are more severe in neonates and in the elderly than in older children and young adults. For example, influenza is typically more severe in older people than

in younger adults, and HSV infections are more severe in neonates than in adults.

(2) Increased corticosteroid levels predispose to more severe infections with some viruses, such as varicella-zoster virus; the use of topical cortisone in herpetic keratitis can exacerbate eye damage. It is not clear how these effects are mediated, because corticosteroids can cause a variety of pertinent effects, namely, lysis of lymphocytes, decreased recruitment of monocytes, inhibition of interferon production, and stabilization of lysosomes.

(3) Malnutrition leads to more severe viral infections (e.g., there is a much higher death rate from measles in developing countries than in developed ones). Poor nutrition causes decreased immunoglobulin production and phagocyte activity as well as reduced skin and mucous membrane integrity. Vitamin A deficiency is a well-known predisposing factor to severe measles disease.

SPECIFIC DEFENSES

There is evidence for natural resistance to some viruses in certain species, which is probably based on the absence of receptors on the cells of the resistant species. For example, some people are resistant to HIV infection because they lack one of the chemokine receptors that mediate entry of the virus into the cell. However, by far, the most important type of defense is **acquired immunity**, either actively acquired by exposure to the virus or passively acquired by the transfer of immune serum. Active immunity can be elicited by contracting the actual disease, by having an inapparent infection, or by being immunized.

1. Active Immunity

Active immunity, in the form of both antibodies and cytotoxic T cells, is very important in the prevention of viral diseases. The first exposure to a virus, whether it causes an inapparent infection or symptomatic disease, stimulates the production of antibodies and the activation of cytotoxic T cells. The role that antibodies and cytotoxic T cells play in the recovery from this first infection is uncertain and may vary from virus to virus, but it is clear that they play an essential role in protecting against disease when exposed to the same virus at some time in the future.

The duration of protection varies; disseminated viral infections such as measles and mumps confer lifelong immunity against recurrences, but localized infections such as the common cold usually impart only a brief immunity of several months. Immunoglubulin (Ig) A confers protection against viruses that enter through the respiratory and gastrointestinal mucosa, and IgM and IgG protect against viruses that enter or are spread through the blood.

The lifelong protection against systemic viral infections such as the childhood diseases measles, mumps, rubella, and chickenpox (varicella) is a function of the anamnestic (secondary) response of IgG. For certain respiratory viruses such as parainfluenza and respiratory syncytial viruses, the IgA titer in respiratory secretions correlates with protection, whereas the IgG titer does not. Unfortunately, protection by IgA against most respiratory tract viruses usually lasts less than 5 years.

The role of active immunity in recovery from a viral infection is uncertain. Because recovery usually precedes the appearance of detectable humoral antibody, Igs may not be important. Also, children with agammaglobulinemia recover from measles infections normally and can be immunized against measles successfully, indicating that cell-mediated immunity plays an important role. This is supported by the observation that children with congenital T-cell deficiency are vulnerable to severe infections with measles virus and herpesviruses. T cells are important in recovery from many but not all viral illnesses.

The protection offered by active immunity can be affected by the phenomenon of **original antigenic sin**. This term refers to the observation that when a person is exposed to a virus that cross-reacts with another virus to which that individual was previously exposed, more antibody may be produced against the original virus than against the current one. It appears that the immunologic memory cells can respond to the original antigenic exposure to a greater extent than to the subsequent one. This was observed in people with antibodies to the A_1 type of influenza virus, who, when exposed to the A_2 type, produced large amounts of antibody to A_1 but very little antibody to the A_2 virus. It is also the underlying cause of severe hemorrhagic dengue fever (see Chapter 42). This phenomenon has two practical consequences as well: (1) attempts to vaccinate people against the different influenza virus strains may be less effective than expected and (2) epidemiologic studies based on measurement of antibody titers may yield misleading results.

How does antibody inhibit viruses? There are two main mechanisms. The first is **neutralization** of the infectivity of the virus by antibody binding to the proteins on the outer surface of the virus. This binding has two effects: (1) It can prevent the interaction of the virus with cell receptors and (2) it can cross-link the viral proteins and stabilize the virus so that uncoating does not occur. As a result, the virus cannot replicate.

Furthermore, antibody-coated virus is more rapidly phagocytized than normal virus, a process similar to the opsonizing effect of antibody on bacteria. Antibody does not degrade the virus particle; fully infectious virus can be recovered by dissociating the virus–antibody complex. Incomplete, also called "blocking," antibody can interfere with neutralization and form immune complexes, which are important in the pathogenesis of certain diseases. Some viruses, such as herpesviruses, can spread from cell to cell across intercellular bridges, eluding the neutralizing effect of antibody.

Antibodies that interfere with the adherence (adsorption and penetration) of viruses to cell surfaces are called neutralizing antibodies. Note that neutralizing antibody is directed against the surface proteins of the virus, typically the proteins involved with the interaction of the virus with receptors on the surface of the host cell. Antibodies formed against internal components of the virus (e.g., the core antigen of hepatitis B virus) do not neutralize the infectivity of the virus.

The second main mechanism is the **lysis of virus-infected cells** in the presence of antibody and complement. Antibody binds to new virus-specific antigens on the cell surface and then binds complement, which enzymatically degrades the cell membrane. Because the cell is killed before the full yield of virus is produced, the spread of virus is significantly reduced.

Lysis of virus-infected cells is also caused by **cytotoxic T lymphocytes**. These CD8-positive T cells recognize viral antigen only when it is presented in association with class I MHC proteins (see Chapter 58). They kill virus-infected cells by three methods: (1) by releasing **perforins**, which make holes in the cell membrane of the infected cells; (2) by releasing proteolytic enzymes called **granzymes** into the infected cell, which degrade the cell contents; and (3) by activating the **FAS protein**, which causes programmed cell death (**apoptosis**).

Not all virus infections induce antibodies. **Tolerance** to viral antigens can occur when the virus infection develops in a fetus or newborn infant. The model system in which tolerance has been demonstrated is lymphocytic choriomeningitis (LCM) infection in mice. If LCM virus is inoculated into a newborn mouse, the virus replicates widely, but no antibodies are formed during the lifetime of the animal. The virus is recognized as "self," because it was present at the time of maturation of the immune system. If LCM virus is given to an adult mouse, antibodies are formed normally. There is no example of total tolerance to a virus in humans; even in congenital rubella syndrome, in which the virus infects the fetus, some antibody against rubella virus is made. However, virus production and shedding can go on for months or years.

Suppression of the cell-mediated response can occur during infection by certain viruses. The best-known example is the loss of tuberculin skin test reactivity during measles infection. Infection by cytomegalovirus or HIV can also cause suppression. Some viruses can "downregulate" (reduce) the amount of class I and class II MHC protein made by cells, which may be a mechanism by which these viruses suppress cell-mediated immunity.

2. Passive Immunity

Transfer of human serum containing the appropriate antibodies provides prompt short-term immunity for individuals exposed to certain viruses. The term *passive* refers to the administration of **preformed antibodies**. Two types of immune globulin preparations are used for this purpose. One has a high titer of antibody against a specific virus, and the other is a pooled sample from plasma donors that contains a heterogeneous mixture of antibodies with lower titers. The immune globulins are prepared by alcohol fractionation, which removes any viruses in the serum. The three most frequently used high-titer preparations are used after exposure to hepatitis B, rabies, and varicella-zoster viruses. Low-titer immune globulin is used mainly to prevent hepatitis A in people traveling to areas where this infection is hyperendemic.

Two specialized examples of passive immunity include the transfer of IgG from mother to fetus across the placenta and the transfer of IgA from mother to newborn in colostrum.

FIGURE 33–3 Herd immunity. Immunization of the nine people (tan color) can protect the one unimmunized person (red color) by interrupting transmission. Immunization levels of 90% are generally regarded as sufficient to protect the unimmunized individual.

3. Herd Immunity

"Herd immunity" (also known as "community immunity") is the protection of an individual from infection by virtue of the other members of the population (the herd) being **incapable of transmitting the virus** to that individual (Figure 33–3). Herd immunity can be achieved by immunizing a population with a vaccine that interrupts transmission, such as the live, attenuated polio vaccine, but not with a vaccine that does not interrupt transmission, such as the killed polio vaccine (even though it protects the immunized individual against disease). Note that herd immunity occurs with the live polio vaccine primarily because it induces secretory IgA in the gut, which inhibits infection by virulent virus, thereby preventing its transmission to others. In addition, the live virus in the vaccine can replicate in the immunized person and spread to other members of the population, thereby increasing the number of people protected. However, the important feature as far as herd immunity is concerned is the induction of IgA, which prevents transmission.

Herd immunity can be achieved by natural infection as well as vaccines. For example, if a viral disease, such as measles, occurred in approximately 90% of a group, and if those who recovered from the disease had sufficient immunity to prevent them from becoming infected and serving as a source of virus for others, then the remaining 10% of the group are protected by herd immunity.

PEARLS

Interferons

- **Viruses and double-stranded RNA are the most potent inducers** of interferons. Many viruses induce interferons, and many viruses are inhibited by interferons (i.e., neither the induction of interferons nor its action is specific).
- Interferons act by binding to a receptor on the cell surface that signals the cell to synthesize ribonuclease, protein kinase, and oligo A synthetase in an inactive form. Double-stranded RNA made by the infecting virus activates these proteins. Interferons do not enter the cell and have no effect on extracellular viruses.
- Interferons inhibit virus replication by blocking protein synthesis, primarily by **degrading mRNA and by inactivating elongation factor-2**.
- Alpha and beta interferons have a stronger antiviral action than gamma interferon. The latter acts primarily as an interleukin that activates macrophages.

Other Nonspecific Defenses

- Natural killer (NK) cells are lymphocytes that **destroy cells infected by many different viruses (i.e., they are nonspecific)**. NK cells do not have an antigen receptor on their surface, unlike T and B lymphocytes. Rather, NK cells **recognize and destroy cells that do not display class I MHC proteins on the surface**. They kill cells by the same mechanisms as do cytotoxic T cells (i.e., by secreting perforins and granzymes).

- Phagocytosis by macrophages and the clearance of mucus by the cilia of the respiratory tract are also important defenses. Damage to these defenses predisposes to viral infection.
- Increased corticosteroid levels suppress various host defenses and predispose to severe viral infections, especially disseminated herpesvirus infections. Malnutrition predisposes to severe measles infections in developing countries. The very young and the very old have more severe viral infections.

Specific Defenses

- **Active immunity to viral infection** is mediated by **both antibodies and cytotoxic T cells**. It can be elicited either by exposure to the virus or by immunization with a viral vaccine.
- **Passive immunity consists of antibodies preformed in another person or animal.**
- The **duration of active immunity is much longer than that of passive immunity**. Active immunity is measured in years, whereas passive immunity lasts a few weeks to a few months.
- **Passive immunity is effective immediately, whereas it takes active immunity 7 to 10 days in the primary response** (or 3–5 days in the secondary response) to stimulate detectable amounts of antibody.
- **Herd immunity** is the protection of an individual that results from immunity in many other members of the population (the herd) that interrupts transmission of the virus to the individual. Herd immunity can be achieved either by immunization or by natural infection of a sufficiently high percentage of the population.

SELF-ASSESSMENT QUESTIONS

1. Regarding the mode of action of interferon, which one of the following is the most accurate?
   - (A) It acts by inhibiting the virion protease.
   - (B) It acts by inhibiting the virion polymerase.
   - (C) It acts by inducing a ribonuclease that degrades viral mRNA.
   - (D) It acts by binding to the extracellular virion, thereby preventing entry into the cell.
   - (E) It acts against viruses with a DNA genome but not against viruses that have RNA as their genome.

2. Regarding immunologic aspects of viral diseases, which one of the following is most accurate?
   - (A) Antibodies protect against some viral diseases by inhibiting the synthesis of mRNA by the RNA polymerase in the virion.
   - (B) IgG plays a major role in neutralizing virus infectivity during the primary infection.
   - (C) IgA exerts an antiviral effect by preventing virus from infecting the mucosal cells of the respiratory and gastrointestinal tracts.

   - (D) IgE can prevent viral infection by activating complement, which leads to the production of the membrane attack complex.
   - (E) Interleukin-2 is important in protecting uninfected cells from viral infection by inhibiting the release of virus from infected cells.

ANSWERS

(1) **(C)**
(2) **(C)**

PRACTICE QUESTIONS: USMLE & COURSE EXAMINATIONS

Questions on the topics discussed in this chapter can be found in the Basic Virology section of Part XIII: USMLE (National Board) Practice Questions starting on page 744. Also see Part XIV: USMLE (National Board) Practice Examination starting on page 775.

Laboratory Diagnosis

INTRODUCTION

There are five approaches to the diagnosis of viral diseases by the use of clinical specimens: (1) identification of the virus in cell culture, (2) microscopic identification directly in the specimen, (3) serologic procedures to detect a rise in antibody titer or the presence of immunoglobulin (Ig) M antibody, (4) detection of viral antigens in blood or body fluids, and (5) detection of viral nucleic acids in blood or the patient's cells.

The sensitivity and speed of several types of laboratory tests used to diagnose viral infections are described in Table 34–1.

IDENTIFICATION IN CELL CULTURE

The growth of viruses requires cell cultures because viruses replicate only in living cells, not on cell-free media the way most bacteria can. Because many viruses are inactivated at room temperature, it is important to inoculate the specimen into the cell culture as soon as possible; brief transport or storage at 4°C is acceptable.

Virus growth in cell culture frequently produces a characteristic **cytopathic effect** (CPE) that can provide a **presumptive identification**. CPE is a change in the appearance of the virus-infected cells. This change can be in such features as size, shape, and the fusion of cells to form multinucleated giant cells (syncytia). CPE is usually a manifestation of virus-infected cells that are dying or dead. The time taken for the CPE to appear and the type of cell in which the virus produces the CPE are important clues in the presumptive identification.

If the virus does not produce a CPE, its presence can be detected by several other techniques:

(1) **Hemadsorption** (i.e., attachment of erythrocytes to the surface of virus-infected cells). This technique is limited to viruses with a hemagglutinin protein on their envelope, such as mumps, parainfluenza, and influenza viruses.

(2) **Interference** with the formation of a CPE by a second virus. For example, rubella virus, which does not cause a CPE, can be detected by interference with the formation of a CPE by certain enteroviruses, such as echovirus or Coxsackie virus.

TABLE 34–1 Comparison of Sensitivity and Speed of Viral Lab Tests

	Viral Culture: Detection of Cytopathic Effect	Polymerase Chain Reaction (PCR): Detection of Viral DNA or RNA	Direct Fluorescent Antibody (DFA): Detection of Viral Antigens in Cells	Rapid Antigen Test: Detection of Influenza Proteins
Sensitivity	High (3+)	High (3+)	Intermediate (2+)	Least (1+)
Relative speed when results become available	Slowest	Fast	Fast	Fastest

(3) A decrease in acid production by infected, dying cells. This can be detected visually by a color change in the phenol red (a pH indicator) in the culture medium. The indicator remains red (alkaline) in the presence of virus-infected cells but turns yellow in the presence of metabolizing normal cells as a result of the acid produced. This technique can be used to detect certain enteroviruses.

A **definitive identification** of the virus grown in cell culture is made by using known antibody in one of several tests. Complement fixation, hemagglutination inhibition, and neutralization of the CPE are the most frequently used tests. Other procedures such as fluorescent antibody, radioimmunoassay, enzyme-linked immunosorbent assay (ELISA), and immunoelectron microscopy are also used in special instances. A brief description of these tests follows. They are described in more detail in the section on immunology.

Complement Fixation

If the antigen (the unknown virus in the culture fluid) and the known antibody are homologous, complement will be fixed (bound) to the antigen–antibody complex. This makes it unavailable to lyse the "indicator" system, which is composed of sensitized red blood cells.

Hemagglutination Inhibition

If the virus and antibody are homologous, the virus is blocked from attaching to the erythrocytes and no hemagglutination occurs. Only viruses that agglutinate red blood cells can be identified by this method.

Neutralization

If the virus and antibody are homologous, the antibody bound to the surface of the virus blocks its entry into the cell. This neutralizes viral infectivity because it prevents viral replication and subsequent CPE formation or animal infection.

Fluorescent Antibody Assay

If the virus-infected cells and the fluorescein-tagged antibody are homologous, the typical apple-green color of fluorescein is seen in the cells by ultraviolet (UV) microscopy.

Radioimmunoassay

If the virus and the antibody are homologous, there is less antibody remaining to bind to the known radiolabeled virus.

Enzyme-Linked Immunosorbent Assay

In the ELISA test to identify a virus, known antibody is bound to a surface. If the virus is present in the patient's specimen, it will bind to the antibody. A sample of the antibody linked to an enzyme is added, which will attach to the bound virus. The substrate of the enzyme is added, and the amount of the bound enzyme is determined.

Immunoelectron Microscopy

If the antibody is homologous to the virus, aggregates of virus–antibody complexes are seen in the electron microscope.

MICROSCOPIC IDENTIFICATION

Viruses can be detected and identified by direct microscopic examination of clinical specimens such as biopsy material or skin lesions. Three different procedures can be used. (1) Light microscopy can reveal characteristic inclusion bodies or multinucleated giant cells. The Tzanck smear, which shows herpesvirus-induced multinucleated giant cells in vesicular skin lesions, is a good example. (2) UV microscopy is used for fluorescent antibody staining of the virus in infected cells. (3) Electron microscopy detects virus particles, which can be characterized by their size and morphology.

SEROLOGIC PROCEDURES

A rise in the titer[1] of antibody to the virus can be used to diagnose current infection. **Seroconversion** is the term used to describe the finding of antibody to a virus (or any microbe) in a patient's serum when the patient previously had no antibody. Stated another way, the patient's serum has converted from antibody-negative to antibody-positive.

A serum sample is obtained as soon as a viral etiology is suspected (**acute-phase**), and a second sample is obtained **10 to 14 days later** (**convalescent-phase**). If the antibody titer in the convalescent-phase serum sample is at least **fourfold higher** than the titer in the acute-phase serum sample, the patient is considered to be infected. For example, if the titer in the acute-phase serum sample is 1/4 and the titer in the convalescent-phase serum sample is 1/16 or greater, the patient has had a significant rise in antibody titer and has been recently infected. If, however, the titer in the convalescent-phase serum sample is 1/8, this is not a significant rise and should not be interpreted as a sign of recent infection.

It is important to realize that an antibody titer on a single sample does not distinguish between a previous infection and a current one. The antibody titer can be determined by many of the immunologic tests mentioned previously. These serologic diagnoses are usually made retrospectively because the disease has frequently run its course by the time the results are obtained.

In certain viral diseases, the presence of IgM antibody is used to diagnose current infection. For example, the presence of IgM antibody to core antigen indicates infection by hepatitis B virus.

DETECTION OF VIRAL ANTIGENS

Viral antigens can be detected in the patient's blood or body fluids by various tests, but most often by an ELISA. Tests for the p24 antigen of human immunodeficiency virus (HIV) and

---

[1]Titer is a measure of the concentration of antibodies in the patient's serum. It is defined as the highest dilution of serum that gives a positive reaction in the test. See Chapter 64 for a discussion of titer and various serologic tests.

the surface antigen of hepatitis B virus are common examples of this approach.

DETECTION OF VIRAL NUCLEIC ACIDS

Viral nucleic acids (i.e., either the viral genome or viral mRNA) can be detected in the patient's blood or tissues with complementary DNA or RNA (cDNA or cRNA) as a probe. If only small amounts of viral nucleic acids are present in the patient, the polymerase chain reaction (PCR) can be used to amplify the viral nucleic acids. Assays for the RNA of HIV and hepatitis C virus and the DNA of hepatitis B virus in the patient's blood (**viral load**) are commonly used to monitor the course of the disease and to evaluate the patient's prognosis.

In serious respiratory virus infections, the laboratory diagnosis can be done by using PCR-based assays on respiratory tract secretions. A panel of PCR assays is used to diagnose infections caused by viruses such as influenza virus, parainfluenza virus, respiratory syncytial virus, rhinovirus, human metapneumovirus, and adenovirus.

PEARLS

Identification in Cell Culture

- The presence of a virus in a patient's specimen can be detected by seeing a "cytopathic effect" (CPE) in cell culture. CPE is not specific (i.e., many viruses cause it). A specific identification of the virus usually involves an antibody-based test such as fluorescent antibody, complement fixation, or enzyme-linked immunosorbent assay (ELISA).

Microscopic Identification

- **Inclusion bodies**, formed by aggregates of many virus particles, can be seen in either the nucleus or cytoplasm of infected cells. They are not specific. Two important examples are the nuclear inclusions formed by certain herpesviruses and the cytoplasmic inclusions formed by rabies virus (Negri bodies).
- **Multinucleated giant cells** are formed by several viruses, notably certain herpesviruses, respiratory syncytial virus, and measles virus.
- Fluorescent antibody staining of cells obtained from the patient or of cells infected in culture can provide a rapid, specific diagnosis.
- Electron microscopy is not often used in clinical diagnosis but is useful in the diagnosis of certain viruses such as Ebola virus, which has a characteristic appearance and is dangerous to grow in culture.

Serologic Procedures

- The **presence of IgM** can be used to **diagnose current infection**.
- **The presence of IgG cannot be used to diagnose current infection** because the antibody may be due to an infection in the past. As a result, an acute and convalescent serum sample should be analyzed. An antibody titer that is fourfold or greater in the convalescent serum sample compared with the acute sample can be used to make a diagnosis.

Detection of Viral Antigens & Nucleic Acids

- The **presence of viral proteins**, such as p24 of HIV and hepatitis B surface antigen, is commonly used in diagnosis.
- The **presence of viral DNA or RNA** is increasingly becoming the "gold standard" in viral diagnosis. Labeled probes are highly specific, and results are rapidly obtained. Small amounts of viral nucleic acids can be amplified using reverse transcriptase to produce amounts detectable by the probes. An important example is the **viral load assay of HIV RNA**.

SELF-ASSESSMENT QUESTIONS

1. Regarding the diagnosis of viral infections in the clinical laboratory, which one of the following provides the **MOST** specific diagnosis?
   - **(A)** Cytopathic effect produced by a virus that replicates on human foreskin cells.
   - **(B)** Cytoplasmic inclusion bodies produced by a virus that replicates in the cytoplasm.
   - **(C)** Multinucleated giant cells produced by a virus that replicates in human skin cells.
   - **(D)** Neutralization of infectivity using antibody against the viral surface protein.
   - **(E)** Intranuclear inclusion bodies produced by a virus that replicates in the nucleus.

2. Seeing multinucleated giant cells in a Tzanck smear can be used to make a presumptive diagnosis of infection by which one of the following viruses?
   - **(A)** Epstein–Barr virus
   - **(B)** Herpes simplex virus
   - **(C)** Human papillomavirus
   - **(D)** Parvovirus B19
   - **(E)** Rubella virus

ANSWERS

(1) **(D)**
(2) **(B)**

PRACTICE QUESTIONS: USMLE & COURSE EXAMINATIONS

Questions on the topics discussed in this chapter can be found in the Basic Virology section of Part XIII: USMLE (National Board) Practice Questions starting on page 744. Also see Part XIV: USMLE (National Board) Practice Examination starting on page 775.

Antiviral Drugs

PRINCIPLES OF ANTIVIRAL THERAPY

Compared with the number of drugs available to treat bacterial infections, the number of antiviral drugs is **relatively small**. The major reason for this difference is the **difficulty in obtaining selective toxicity** against viruses; their replication is intimately involved with the normal synthetic processes of the cell. Despite the difficulty, several virus-specific replication steps have been identified that are the site of action of effective antiviral drugs (Table 35–1). Table 35–2 describes the mode of action of antiviral drugs that block early events in viral replication, and Table 35–3 describes the mode of action of antiviral drugs that block viral nucleic acid synthesis. Figure 35–1 shows the replication of a model virus and the site of action of drugs used to treat various viral infections. Figure 35–2 shows the replication of human immunodeficiency virus (HIV) and the site of action of drugs used to treat HIV infection.

Another limitation of antiviral drugs is that they are relatively ineffective because many cycles of viral replication occur during the incubation period when the patient is well. By the time the patient has a recognizable systemic viral disease, the virus has spread throughout the body and it is too late to interdict it. Furthermore, some viruses (e.g., herpesviruses) become latent within cells, and no current antiviral drug can eradicate them.

Another limiting factor is the emergence of drug-resistant viral mutants. For example, when drug-resistant mutants of HIV emerge, it requires that drug regimens be changed. Also, treatment of HIV infection uses multiple drugs, often from different classes, so that if mutants resistant to one drug emerge, another drug will still be effective.

At the end of this chapter is a discussion of oncolytic viruses. These are viruses that preferentially kill cancer cells. This topic is not one that, strictly speaking, belongs in a chapter on drugs used to treat viral infections. However, it seemed to fit better here than in other chapters.

INHIBITION OF EARLY EVENTS

Amantadine (α-adamantanamine, Symmetrel) is a three-ring compound (Figure 35–3) that blocks the replication of influenza A virus. It prevents replication by **inhibiting uncoating of the virus** by blocking the "ion channel" activity of the matrix protein (M2 protein) in the virion. Absorption and penetration occur normally, but transcription by the virion RNA polymerase does not because uncoating cannot occur. This drug specifically inhibits influenza A virus; influenza B and C viruses are not affected.

Despite its efficacy in preventing influenza, it is not recommended for use in the United States because the vaccine

TABLE 35-1 Stage of Viral Replication Inhibited by Antiviral Drugs

Stage of Viral Replication Inhibited	Effective Antiviral Drugs
Early events (entry or uncoating of the virus)	Amantadine, rimantadine, enfuvirtide, maraviroc, fostemsavir, palivizumab
Nucleic acid synthesis by herpesviruses	Acyclovir, ganciclovir, valacyclovir, valganciclovir, penciclovir, famciclovir, cidofovir, trifluridine, foscarnet
Nucleic acid synthesis by human immunodeficiency virus (HIV)	Zidovudine, lamivudine, emtricitabine, didanosine, stavudine, abacavir, tenofovir, nevirapine, delavirdine, efavirenz, etravirine, rilpivirine, doravirine
Nucleic acid synthesis by hepatitis B virus (HBV)	Adefovir, entecavir, lamivudine, telbivudine, tenofovir
Nucleic acid synthesis by hepatitis C virus (HCV)	RNA polymerase inhibitors: sofosbuvir, dasabuvir NS5A inhibitors: ledipasvir, ombitasvir
Nucleic acid synthesis by SARS-CoV-2 coronavirus	Remdesivir
Nucleic acid synthesis by other viruses	Ribavirin
Cap-snatching to produce viral mRNA by influenza virus	Baloxavir
Integrase that integrates HIV DNA into cellular DNA	Raltegravir, elvitegravir, dolutegravir
Cleavage of precursor DNA by cytomegalovirus	Letermovir
Cleavage of precursor polypeptides	Protease inhibitors of HIV: saquinavir, indinavir, ritonavir, nelfinavir, amprenavir, atazanavir, darunavir, lopinavir, tipranavir Protease inhibitors of hepatitis C virus: boceprevir, simeprevir, telaprevir, paritaprevir
Protein synthesis directed by viral mRNA	Interferon, fomivirsen
Release of influenza virus from infected cell	Oseltamivir, zanamivir

TABLE 35-2 Antiviral Drugs That Block Early Events

Antiviral Drug	Mode of Action	Virus Inhibited
Amantadine, rimantadine	Inhibits uncoating by blocking M2 matrix protein	Influenza virus
Enfuvirtide	Inhibits fusion by binding to gp41 of human immunodeficiency virus (HIV)	HIV
Maraviroc	Inhibits attachment to cell surface receptor CCR-5	HIV
Fostemsavir	Inhibits attachment by binding to gp120 on envelope of HIV	HIV
Ibalizumab	Monoclonal antibody against CD4 protein that blocks binding of HIV to CCR-5 and CXCR-4 coreceptors	HIV
Palivizumab	Monoclonal antibody that blocks binding of viral fusion protein to receptor on respiratory mucosal cell	Respiratory syncytial virus

TABLE 35-3 Antiviral Drugs That Block Viral Nucleic Acid Synthesis

Mode of Action	Antiviral Drugs
Inhibition of DNA polymerase of herpesviruses	1. Nucleoside inhibitors: acyclovir, ganciclovir, valacyclovir, valganciclovir, penciclovir, famciclovir, cidofovir, trifluridine 2. Nonnucleoside inhibitors: foscarnet
Inhibition of reverse transcriptase of human immunodeficiency virus (HIV)	1. Nucleoside inhibitors: zidovudine, lamivudine, emtricitabine, didanosine, stavudine, abacavir, tenofovir alafenamide, tenofovir disoproxil fumarate 2. Nonnucleoside inhibitors: nevirapine, efavirenz, etravirine, rilpivirine, doravirine, delavirdine
Inhibition of reverse transcriptase of hepatitis B virus	Adefovir, entecavir, lamivudine, telbivudine
Inhibition of RNA polymerase of hepatitis C virus	Sofosbuvir, dasabuvir
Inhibition of NS5A protein of hepatitis C virus	Ledipasvir, ombitasvir
Inhibition of Nucleic acid synthesis by SARS-CoV-2 coronavirus	Remdesivir
Inhibition of nucleic acid synthesis by other viruses	Ribavirin

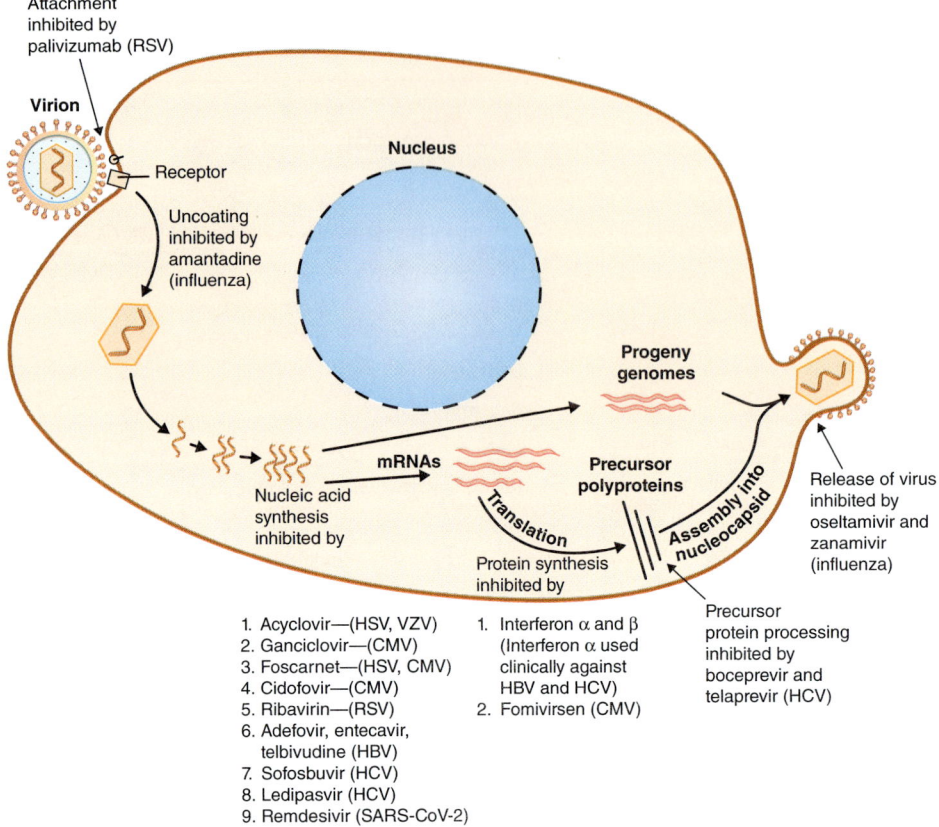

FIGURE 35–1 Replicative cycle of a model virus showing the site of action of drugs used to treat various viral infections. Note that drugs used to treat HIV infections are shown in Figure 35-2. CMV = cytomegalovirus; HBV = hepatitis B virus; HCV = hepatitis C virus; HSV = herpes simplex virus; RSV = respiratory syncytial virus; VZV = varicella-zoster virus.

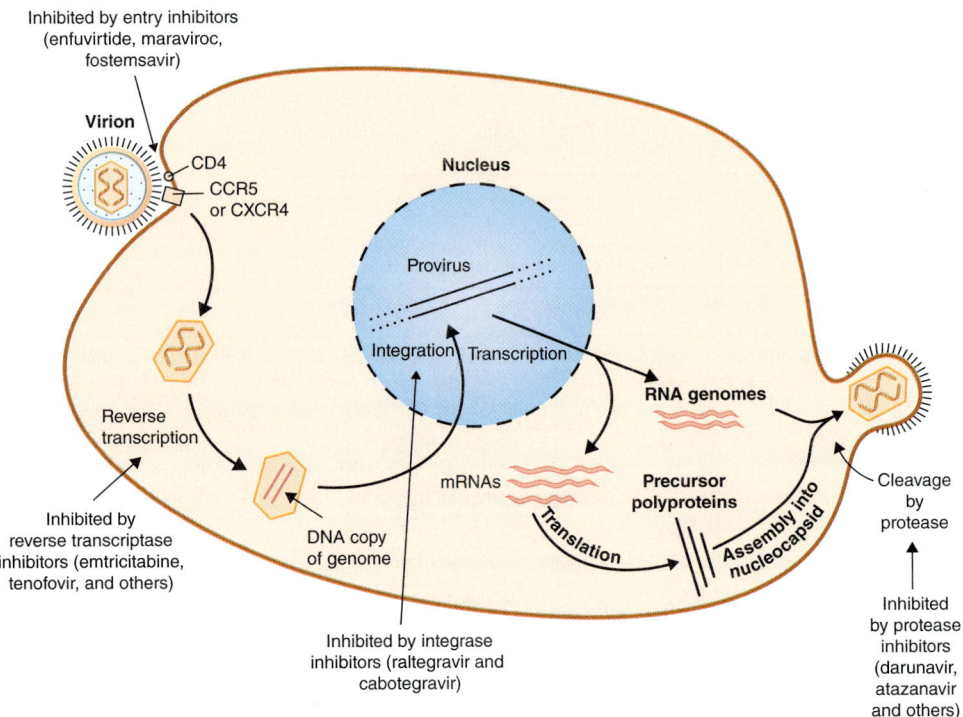

FIGURE 35–2 Replicative cycle of human immunodeficiency virus (HIV) showing the site of action of drugs used to treat HIV infection.

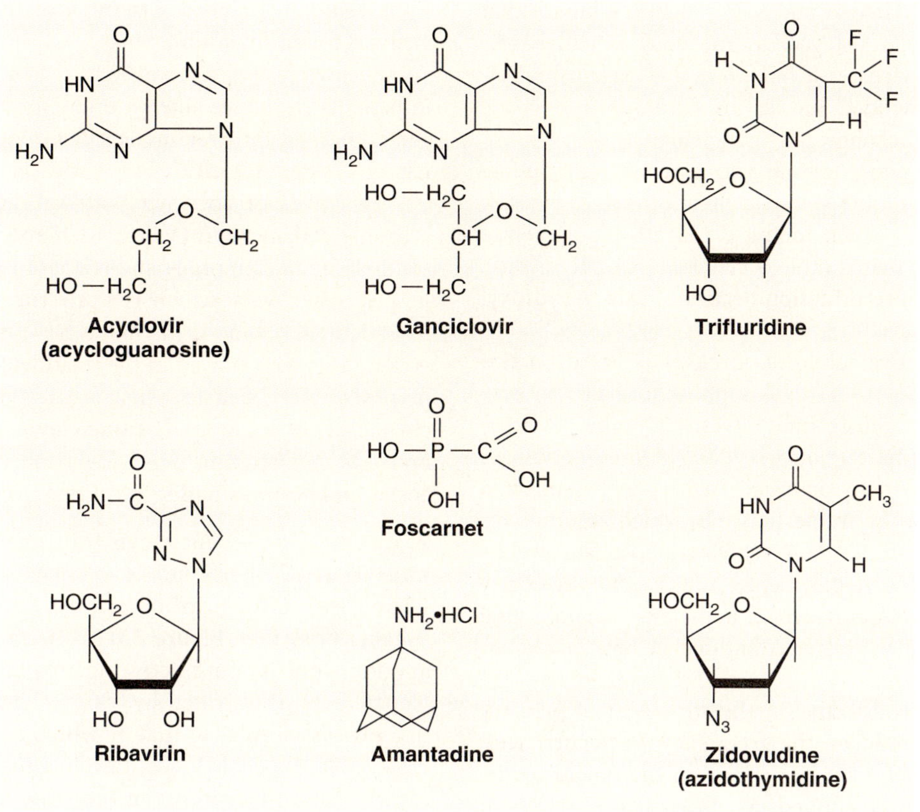

FIGURE 35–3 Structures of some medically important antiviral drugs.

is preferred for the high-risk population. Furthermore, most strains of influenza virus have become resistant to amantadine. The main side effects of amantadine are central nervous system alterations such as dizziness, ataxia, and insomnia. **Rimantadine** (Flumadine) is a derivative of amantadine and has the same mode of action but fewer side effects.

Enfuvirtide (Fuzeon) is a synthetic peptide that binds to gp41 on the surface of HIV, thereby blocking the entry of the virus into the cell. It is the first of a new class of anti-HIV drugs known as "fusion inhibitors" (i.e., they prevent the fusion of the viral envelope with the cell membrane).

Maraviroc (Selzentry) blocks the binding of HIV to CCR5—an important coreceptor for those strains of HIV that use CCR5 for entry into the cell. The drug binds to CCR5 and blocks the interaction of gp120, an HIV envelope protein, to CCR5 on the cell surface.

Fostemsavir (Rukobia) is an attachment inhibitor of HIV infection. It binds to the gp120 protein on the surface of the HIV envelope preventing attachment. It is a prodrug that is converted to the active drug, temsavir.

Ibalizumab (Trogarzo) is a monoclonal antibody against CD4 protein that blocks entry of HIV. It is a "post-attachment" inhibitor, which means it blocks HIV from binding to CCR5 or CXCR4 coreceptors after HIV has bound to the CD4 protein.

Palivizumab (Synagis) is a monoclonal antibody directed against the fusion protein of respiratory syncytial virus (RSV). Palivizumab neutralizes RSV by binding to the fusion protein on the surface of RSV, thereby preventing the virus from binding to receptors on the surface of respiratory tract mucosal cells. It is used to prevent bronchiolitis and pneumonia in premature or immunocompromised infants.

INHIBITION OF VIRAL NUCLEIC ACID SYNTHESIS

Inhibitors of Herpesviruses

Nucleoside Inhibitors

These drugs are nucleoside analogues that inhibit the DNA polymerase of one or more members of the herpesvirus family. For example, acyclovir inhibits the DNA polymerase herpes simplex virus types 1 and 2 (HSV-1 and -2) and varicella-zoster virus (VZV), but not cytomegalovirus (CMV).

1. Acyclovir—Acyclovir (acycloguanosine, Zovirax) is a guanosine analogue that has a three-carbon fragment in place of the normal sugar, ribose, which has five carbons (see Figure 35–3). The term *acyclo* refers to the fact that the three-carbon fragment does not have a sugar ring structure (*a* = without, *cyclo* = ring).

Acyclovir is active primarily against HSV-1 and -2 and VZV. It is relatively nontoxic because it is activated preferentially within virus-infected cells. This is due to the **virus-encoded thymidine kinase**, which phosphorylates acyclovir much more effectively than does the cellular thymidine kinase.

Because only HSV-1, HSV-2, and VZV encode a kinase that efficiently phosphorylates acyclovir, the drug is active primarily against these viruses. It has no activity against CMV because the CMV-encoded kinase does not phosphorylate acyclovir. Rather, it phosphorylates ganciclovir.

Once acyclovir is phosphorylated to acyclovir monophosphate by the viral thymidine kinase, cellular kinases synthesize acyclovir triphosphate, which inhibits viral DNA polymerase much more effectively than it inhibits cellular DNA polymerase. Acyclovir causes **chain termination** because it lacks a hydroxyl group in the 3′ position.

To recap, the selective action of acyclovir is based on two features of the drug. (1) Acyclovir is phosphorylated to acyclovir monophosphate much more effectively by herpesvirus-encoded thymidine kinase than by cellular thymidine kinase. It is therefore preferentially activated in herpesvirus-infected cells and much less so in uninfected cells, which accounts for its relatively few side effects. (2) Acyclovir triphosphate inhibits herpesvirus-encoded DNA polymerase much more effectively than it does cellular DNA polymerase. It, therefore, inhibits viral DNA synthesis to a much greater extent than cellular DNA synthesis (Figure 35–4).

Topical acyclovir is effective in the treatment of primary genital herpes and reduces the frequency of recurrences. However, it has **no effect on latency** or on the rate of recurrences after treatment is stopped. Acyclovir is the treatment of choice for HSV-1 encephalitis and is effective in preventing systemic infection by HSV-1 or VZV in immunocompromised patients.

Acyclovir-resistant mutants have been isolated from HSV-1- and VZV-infected patients. Resistance is most often due to mutations in the gene encoding the viral thymidine kinase.

This results in reduced activity of or the total absence of the virus-encoded thymidine kinase.

Acyclovir is well tolerated and causes few side effects—even in patients who have taken it orally for many years to suppress genital herpes. Intravenous acyclovir may cause renal or central nervous system toxicity.

Derivatives of acyclovir with various properties are now available. **Valacyclovir** (Valtrex) achieves a high plasma concentration when taken orally and is used in herpes genitalis and in herpes zoster. Valacyclovir is a prodrug composed of acyclovir to which the amino acid, valine has been added. **Penciclovir** cream (Denavir) is used in the treatment of recurrent orolabial herpes simplex. Penciclovir is a diacetyl derivative of an acycloguanosine compound. **Famciclovir** (Famvir), when taken orally, is converted to penciclovir and is used to treat herpes zoster and herpes simplex infections.

2. Ganciclovir—Ganciclovir (dihydroxypropoxymethylguanine, DHPG, Cytovene) is a nucleoside analogue of guanosine with a four-carbon fragment in place of the normal sugar, ribose (see Figure 35–3). It is structurally similar to acyclovir but is more active against CMV than is acyclovir. Ganciclovir is activated by a CMV-encoded phosphokinase in a process similar to that by which acyclovir is activated by HSV. Isolates of CMV resistant to ganciclovir have emerged, mostly due to mutations in the *UL97* gene that encodes the phosphokinase.

Ganciclovir is effective in the treatment of retinitis caused by CMV in patients with acquired immunodeficiency syndrome (AIDS) and is useful in other disseminated infections, such as colitis and esophagitis, caused by this virus. The main side effects of ganciclovir are leukopenia and thrombocytopenia as a

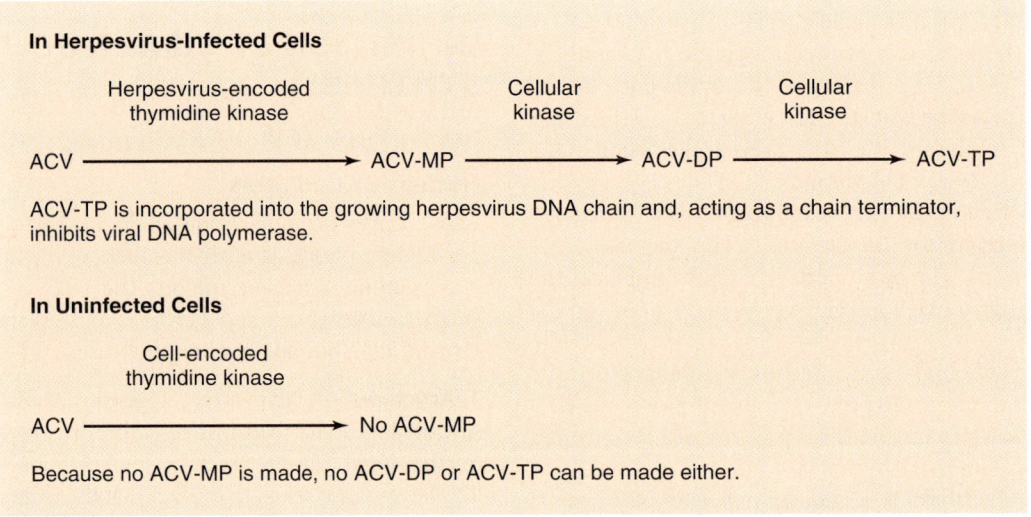

In Herpesvirus-Infected Cells

Herpesvirus-encoded   Cellular   Cellular
thymidine kinase   kinase   kinase

ACV ──────────→ ACV-MP ──────────→ ACV-DP ──────────→ ACV-TP

ACV-TP is incorporated into the growing herpesvirus DNA chain and, acting as a chain terminator, inhibits viral DNA polymerase.

In Uninfected Cells

Cell-encoded
thymidine kinase

ACV ──────────→ No ACV-MP

Because no ACV-MP is made, no ACV-DP or ACV-TP can be made either.

FIGURE 35–4 Acyclovir (ACV) is phosphorylated to ACV-MP very effectively by herpesvirus-encoded thymidine kinase but very poorly by cell-encoded thymidine kinase. The thymidine kinases encoded by herpes simplex virus (HSV)-1, HSV-2, and varicella-zoster virus (VZV) are particularly active on ACV; the thymidine kinases encoded by cytomegalovirus and Epstein–Barr virus are not. This accounts for the selective action of ACV in cells infected by HSV-1, HSV-2, and VZV. The fact that ACV-TP is not made in uninfected cells explains why ACV has so few side effects (i.e., why DNA synthesis is not inhibited in uninfected cells). ACV-DP = ACV diphosphate; ACV-MP = ACV monophosphate; ACV-TP = ACV triphosphate.

result of bone marrow suppression. **Valganciclovir**, which can be taken orally, is also effective against CMV retinitis. Valganciclovir is a prodrug composed of ganciclovir to which the amino acid, valine has been added.

3. Cidofovir—Cidofovir (hydroxyphosphonylmethoxypropylcytosine, HPMPC, Vistide) is an analogue of cytosine that lacks a ribose ring. Cidofovir does **not** have to be phosphorylated and therefore is not dependent on the action of a virus-encoded phosphokinase. It is useful in the treatment of retinitis caused by CMV and in severe human papillomavirus infections. It may be useful in the treatment of severe molluscum contagiosum in immunocompromised patients. Kidney damage is the main side effect.

4. Trifluridine—(trifluorothymidine, Viroptic) is a nucleoside analogue in which the methyl group of thymidine contains three fluorine atoms instead of three hydrogen atoms (see Figure 35–3). The drug is phosphorylated to the triphosphate by cellular kinases and incorporated into DNA. Because it has a high frequency of mismatched pairing to adenine, it causes the formation of faulty progeny DNA and mRNA. However, because it is incorporated into normal cell DNA as well as viral DNA, it is too toxic to be used systemically. It is one of the drugs of choice for the topical treatment of keratoconjunctivitis caused by HSV.

Nonnucleoside Inhibitors

Nonnucleoside inhibitors inhibit the DNA polymerase of herpesviruses by mechanisms distinct from the nucleoside analogues described previously. Foscarnet is the only approved drug in this class at this time.

1. Foscarnet—Foscarnet (trisodium phosphonoformate, Foscavir), unlike the previous drugs, which are nucleoside analogues, is a pyrophosphate analogue (see Figure 35–3). It binds to DNA polymerase at the pyrophosphate cleavage site and prevents removal of the phosphates from nucleoside triphosphates (dNTP). This inhibits the addition of the next dNTP and, as a consequence, the extension of the DNA strand.

Foscarnet inhibits the DNA polymerases of all herpesviruses, especially HSV and CMV. Unlike acyclovir, it does not require activation by thymidine kinase. It is useful in the treatment of retinitis caused by CMV, but ganciclovir is the drug of first choice for this disease. Foscarnet is also used to treat patients infected with acyclovir-resistant mutants of HSV-1 and VZV.

Inhibitors of Human Immunodeficiency Virus

Nucleoside Inhibitors

The selective toxicity of zidovudine, lamivudine, emtricitabine, didanosine, zalcitabine, stavudine, abacavir, and tenofovir is based on their ability to **inhibit DNA synthesis by the reverse transcriptase** of HIV to a much greater extent than they inhibit DNA synthesis by the DNA polymerase in human cells.

These drugs are collectively called nucleoside reverse transcriptase inhibitors (NRTIs). The effect of these drugs on the replication of HIV is depicted in Figure 35–2.

The following four drugs are currently being used in antiretroviral regimens

1. Lamivudine—Lamivudine (dideoxythiacytidine, Epivir, 3TC) is a nucleoside analogue that causes chain termination during DNA synthesis by the reverse transcriptase of HIV. When used in combination with AZT, it is very effective both in reducing the viral load and in elevating the CD4 cell count. Lamivudine is also used in the treatment of chronic hepatitis B because it inhibits the reverse transcriptase of hepatitis B virus (HBV). It is one of the best tolerated of the nucleoside inhibitors, but adverse effects such as neutropenia, pancreatitis, and peripheral neuropathy do occur.

2. Emtricitabine—Emtricitabine (Emtriva), a derivative of lamivudine, is also useful and well tolerated. A combination of emtricitabine and tenofovir (Truvada) can be used for preexposure prophylaxis for men who have sex with men as well as for postexposure prophylaxis.

3. Abacavir—Abacavir (Ziagen) is a nucleoside analogue of guanosine that causes chain termination during DNA synthesis. It is available through the "expanded access" program to those who have failed currently available drug regimens. Abacavir is used in combination with either a protease inhibitor, typically darunavir and ritonavir, or zidovudine plus lamivudine. The main adverse effects are liver damage and severe hypersensitivity reactions. Patients who have an HLA-B1701 allele are more likely to have a severe hypersensitivity reaction, such as fever, rash, or respiratory problems, to abacavir. Patients should be tested for this gene before being prescribed abacavir. If patients develop hypersensitivity symptoms, abacavir should be immediately and permanently discontinued.

4. Tenofovir—Tenofovir (Viread) is an acyclic phosphonate that is an analogue of adenosine monophosphate. It is a reverse transcriptase inhibitor that acts by chain termination. There are two versions: tenofovir alafenamide (TAF) and tenofovir disoproxil fumarate (TDF). TAF has less bone and kidney toxicity whereas TDF is associated with lower lipid levels.

Tenofovir is approved for use in patients who have developed resistance to other reverse transcriptase inhibitors and in those who are starting treatment for the first time. It should be used in combination with other anti-HIV drugs. The main adverse effects are liver damage, lactic acidosis, and renal failure.

The following three drugs are rarely used at present due to their adverse effects.

1. Zidovudine—Zidovudine (azidothymidine, Retrovir, AZT) is a nucleoside analogue that causes **chain termination** during DNA synthesis; it has an azido group in place of the hydroxyl group on the ribose (see Figure 35–3). It is particularly effective against DNA synthesis by the reverse transcriptase of HIV and inhibits the growth of the virus in cell culture. The main adverse effects of zidovudine are bone marrow suppression and myopathy.

2. Didanosine—Didanosine (dideoxyinosine, ddI, Videx) is a nucleoside analogue that causes chain termination during DNA synthesis because it is missing hydroxyl groups on the ribose. The administered drug ddI is metabolized to ddATP, which is the active compound. It is effective against DNA synthesis by the reverse transcriptase of HIV and is used to treat patients with AIDS who are intolerant of or resistant to zidovudine. The main adverse effects of didanosine are pancreatitis and peripheral neuropathy.

3. Stavudine—Stavudine (didehydrodideoxythymidine, d4T, Zerit) is a nucleoside analogue that causes chain termination during DNA synthesis. It inhibits DNA synthesis by the reverse transcriptase of HIV and is used to treat patients with advanced AIDS who are intolerant of or resistant to other approved therapies. The main adverse effect is peripheral neuropathy.

Nonnucleoside Inhibitors

Unlike the drugs described earlier, the drugs in this group are not nucleoside analogues and do not cause chain termination. The nonnucleoside reverse transcriptase inhibitors (NNRTIs) act by binding near the active site of the reverse transcriptase and inducing a conformational change in the enzyme that inhibits its ability to synthesize viral DNA. NNRTIs should not be used as monotherapy because resistant mutants emerge rapidly. Strains of HIV resistant to one NNRTI are usually resistant to others as well. NNRTIs are typically used in combination with one or two nucleoside analogues.

1. Nevirapine—Nevirapine (Viramune) is usually used in combination with zidovudine and didanosine. There is no cross-resistance with the nucleoside inhibitors of reverse transcriptase described previously. The main side effect of nevirapine is a severe skin rash (Stevens-Johnson syndrome).

2. Efavirenz—Efavirenz (Sustiva) is effective in combination with zidovudine plus lamivudine. The most common side effects are referable to the central nervous system, such as dizziness, insomnia, and headaches.

3. Etravirine—Etravirine (Intelence) is a second-generation NNRTI useful in treatment-experienced patients who have significant viremia. It is most effective when given in combination with two protease inhibitors, darunavir and ritonavir. The most common adverse effect is a rash, and Stevens-Johnson syndrome has occurred, albeit rarely.

4. Rilpivirine—Rilpivirine (Edurant) is a second-generation NNRTI useful in treatment-naïve adult patients. It is most effective when used in combination with either tenofovir or emtricitabine. The most common adverse effects are depression and insomnia.

5. Doravirine—(Pifeltro) is a pyridinone inhibitor of reverse transcriptase that was approved by the Food and Drug Administration (FDA) in 2018. It is used in combination with other antiretroviral agents. The most common adverse effects are nausea and diarrhea.

6. Delavirdine—Delavirdine (Rescriptor) is effective in combination with either zidovudine or zidovudine plus didanosine. The main side effect of delavirdine is a skin rash. Currently, it is rarely used.

Inhibitors of Hepatitis B Virus

Adefovir

Adefovir (Hepsera) is a nucleotide analogue of adenosine monophosphate that inhibits the DNA polymerase (reverse transcriptase) of HBV. It is used for the treatment of chronic active hepatitis caused by this virus.

Entecavir

Entecavir (Baraclude) is a guanosine analogue that inhibits the DNA polymerase (reverse transcriptase) of HBV. It has no activity against the DNA polymerase (reverse transcriptase) of HIV. It is approved for the treatment of adults with chronic HBV infection.

Lamivudine

Lamivudine is described in the section "Inhibitors of Retroviruses."

Telbivudine

Telbivudine (Tyzeka) is a thymidine analogue that inhibits the DNA polymerase (reverse transcriptase) of HBV but has no effect on the reverse transcriptase of HIV. It is useful in the treatment of chronic HBV infection.

Tenofovir

Tenofovir is described in the section "Inhibitors of Retroviruses."

Inhibitors of Hepatitis C Virus

1. RNA polymerase inhibitors

Sofosbuvir

Sofosbuvir (Sovaldi) is a uridine analogue that inhibits the RNA polymerase of hepatitis C virus (HCV). It acts as a chain-terminating drug. It is useful in the treatment of chronic HCV infection caused by genotypes 1, 2, 3, and 4.

Dasabuvir

Dasabuvir is a nonnucleoside inhibitor of the RNA polymerase of HCV. It is available in combination with ombitasvir (an NS5A inhibitor), paritaprevir (a protease inhibitor), and ritonavir (a booster of protease inhibitor [PI] activity). This four-drug combination is called Viekira.

2. NS5A inhibitors

Ledipasvir

Ledipasvir is an inhibitor of NS5A, an RNA-binding protein required for the activity of the RNA polymerase of HCV.

Ledipasvir is available in combination with sofosbuvir (Harvoni) and is useful in the treatment of chronic HCV infection caused by genotype 1.

Ombitasvir

Ombitasvir is another NS5A inhibitor that is available in combination with dasabuvir (a polymerase inhibitor), paritaprevir (a protease inhibitor), and ritonavir (a booster of PI activity). This four-drug combination is called Viekira.

Daclatasvir, Elbasvir, Velpatasvir, Pibrentasvir

These drugs also inhibit the action of the NS5A protein.

Inhibitor of SARS-CoV-2 Coronavirus

Remdesivir

Remdesivir is a nucleoside analogue of adenosine that inhibits RNA synthesis by the polymerase of SARS-CoV-2 coronavirus. It is a chain-terminating drug. It is used in the treatment of severe COVID-19 disease.

Inhibitors of Other Viruses

Ribavirin

Ribavirin (Virazole) is a nucleoside analogue in which a triazole-carboxamide moiety is substituted in place of the normal purine precursor aminoimidazole-carboxamide (see Figure 35–3). The drug inhibits the synthesis of guanine nucleotides, which are essential for both DNA and RNA viruses. It also inhibits the 5′ capping of viral mRNA. Ribavirin aerosol is used clinically to treat pneumonitis caused by RSV in infants and to treat severe influenza B infections.

Baloxavir

Baloxavir marboxil (Xofluza) inhibits the **cap-snatching** endonuclease (ribonuclease) required for synthesis of influenza virus mRNA (see Figure 29–4). Briefly, cap-snatching is the process by which the 5′-methyl guanosine cap on cell mRNA is excised (snatched) from cell mRNA and used as a primer by viral polymerase to synthesize influenza viral mRNA. This process is described in the "Summary of Replicative Cycle" section in the Influenza section of Chapter 38. The drug is effective against a wide range of influenza viruses.

INHIBITION OF INTEGRASE

Raltegravir (Isentress) is an integrase inhibitor (i.e., it blocks the HIV-encoded integrase that mediates the integration of the newly synthesized viral DNA into host cell DNA). Several additional integrase inhibitors are available: **dolutegravir** (Tivicay), **elvitegravir** (available as either Stribild or Genvoya in fixed combination with other drugs), and **bictegravir** (available as Biktarvy in fixed combination with emtricitabine and tenofovir). **Cabotgravir** is a long-acting, injectable integrase inhibitor. The collective abbreviation INSTI is often used for these drugs. INSTI stands for INtegrase Strand Transfer Inhibitor.

INHIBITION OF CLEAVAGE OF PRECURSOR NUCLEIC ACIDS

Replication of CMV progeny DNA involves production of a chain of DNA (called a concatenate) that is cleaved into single pieces of progeny DNA by a virion-encoded enzyme called a terminase (see Figure 29-6). The drug letermovir (Prevymis) inhibits the terminase and thereby inhibits CMV replication. Note that letermovir does not inhibit the terminase of HSV or VZV.

INHIBITION OF CLEAVAGE OF PRECURSOR POLYPEPTIDES (PROTEASE INHIBITORS)

Inhibitors of Human Immunodeficiency Virus

Members of the PI class of drugs currently being used include **ritonavir** (Norvir), **lopinavir/ritonavir** (Kaletra), **atazanavir** (Reyataz), **tipranavir** (Aptivus), **amprenavir** (Agenerase), and **darunavir** (Prezista). Note that four other PIs: **indinavir** (Crixivan), **nelfinavir** (Viracept), **saquinavir** (Invirase, Fortovase), and **fosamprenavir** (Lexiva), are rarely used at this time.

These PI drugs inhibit the protease encoded by HIV (Figure 35–5). The protease cleaves the *gag* and *pol* precursor polypeptides to produce several nucleocapsid proteins (e.g., p24) and enzymatic proteins (e.g., reverse transcriptase) required for viral replication. These inhibitors contain peptide bonds that bind to the active site of the viral protease, thereby preventing the protease from cleaving the viral precursor proteins. These drugs inhibit production of infectious virions but do not affect the proviral DNA and therefore do not cure the infection.

Monotherapy with PIs should not be used because resistant mutants emerge rapidly. These drugs typically are prescribed in combination with reverse transcriptase inhibitors, such as zidovudine and lamivudine. Ritonavir is typically used in combination with another PI, as in the commonly used combination lopinavir/ritonavir (Kaletra). Ritonavir inhibits the cytochrome

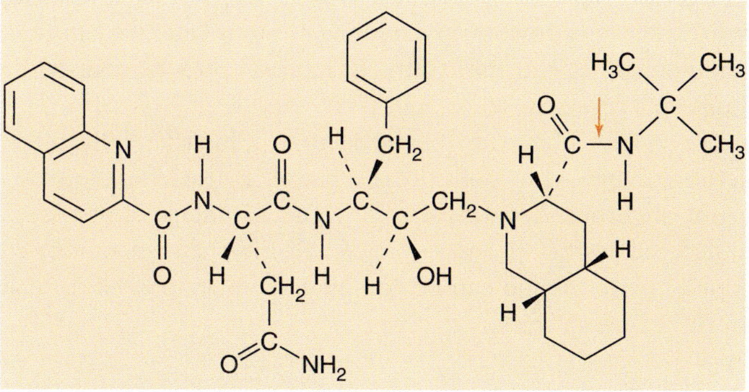

FIGURE 35–5 Structure of the protease inhibitor saquinavir. Note the presence of several peptide bonds, which interact with the active site of the protease. A *red arrow* indicates one of the peptide bonds.

P450 enzymes that metabolize the other PI, which effectively raises the concentration of the other drug (e.g., lopinavir in the Kaletra combination). Ritonavir "boosts" lopinavir is the way to remember it.

Another drug that inhibits cytochrome P450 enzymes is **cobicistat**. It is particularly effective in enhancing the antiviral effect of elvitegravir, an integrase inhibitor. It is also useful in enhancing the effect of two PIs, namely darunavir and atazanavir.

The side effects of PIs include nausea, diarrhea, and abnormal fat accumulation in the back of the neck that can result in a "buffalo hump" appearance. These abnormal fat deposits can be disfiguring and cause patients to stop taking the drug. The fat deposits are a type of lipodystrophy; the metabolic process by which this occurs is unknown. Indinavir can cause kidney stones; thus, extra water should be consumed to reduce the likelihood of stone formation.

Inhibitors of Hepatitis C Virus

Boceprevir (Victrelis), **simeprevir** (Olysio), **telaprevir** (Incivek), and **paritaprevir** (component of Viekira) are PIs that block a serine protease required for the replication of HCV. They are approved for the treatment of chronic hepatitis C caused by HCV (genotype 1) in combination with peginterferon and ribavirin. The most important adverse effect of these drugs is anemia. **Paritaprevir** is an inhibitor of HCV protease that is available in combination with ombitasvir (an NS5A inhibitor), dasabuvir (a polymerase inhibitor), and ritonavir (a booster of PI activity). This four-drug combination is called Viekira.

INHIBITION OF VIRAL PROTEIN SYNTHESIS

Interferon

The mode of action of interferon is described in Chapter 33. Recombinant alpha interferon is effective in the treatment of some patients with chronic hepatitis B and chronic hepatitis C infections. Note that the use of alpha interferon and pegylated alpha interferon (see next paragraph) for chronic HCV infection has been significantly reduced due to the availability of newer less toxic drug regimens. Alpha interferon also causes regression of condylomata acuminata lesions caused by human papillomavirus and the lesions of Kaposi's sarcoma caused by human herpesvirus-8.

Pegylated interferon (peginterferon), which is alpha interferon conjugated to polyethylene glycol, is used for the treatment of chronic hepatitis B and C. The advantage of pegylated interferon is that it has a longer half-life than unconjugated alpha interferon and can be administered once a week instead of three times a week.

Fomivirsen

Fomivirsen (Vitravene) is an antisense DNA that blocks the replication of CMV. Antisense DNA is a single-stranded DNA whose base sequence is the complement of the viral mRNA. Antisense DNA binds to the mRNA within the infected cell and prevents it from being translated into viral protein. Fomivirsen is approved for the intraocular treatment of CMV retinitis. It is the first and, at present, the only antisense molecule to be approved for the treatment of human disease.

INHIBITION OF RELEASE OF VIRUS

Oseltamivir (Tamiflu), zanamivir (Relenza), and peramivir (Rapivab) inhibit the neuraminidase of influenza virus. This enzyme is located on the surface of influenza virus and is required for the release of the virus from infected cells. Inhibition of release of influenza virus limits the infection by reducing the spread of virus from one cell to another. These drugs are effective against both influenza A and B viruses, in contrast to amantadine, which is effective only against influenza A virus. These drugs are effective against strains of influenza virus resistant to amantadine.

CHEMOPROPHYLAXIS

In most instances, the antiviral agents described in this chapter are used to *treat* infectious diseases. However, there are times when they are used to *prevent* diseases from occurring—a process called **chemoprophylaxis**. Table 35–4 describes the drugs used for this purpose and the situations in which they are used. For more information, see the chapters on the individual viruses.

ONCOLYTIC VIRUSES

Oncolytic viruses are viruses that infect and kill cancer cells. Note that these viruses are designed to selectively kill cancer cells and not damage normal cells.

One of the best examples of an oncolytic virus is a genetically modified strain of the live, attenuated poliovirus used in the oral (Sabin) vaccine. It is in clinical trials to treat patients with glioblastoma multiforme. The virus binds to CD155, a protein that is present in large amounts on the glioblastoma cell surface. This provides the specificity of the virus for cancer cells. The oncolytic poliovirus is infused directly into the glioblastoma. Some patients treated with this virus have survived for several years compared to untreated patients who live approximately one year.

In 2021, a report described the use of a disabled strain of HSV-1 to kill glioma cells in several patients. The virus was inoculated directly into the brain tumor and caused death of the glioma cells both by lysis and by immune-mediated killing.

These results lend support to the possibility of using oncolytic viruses to treat cancer. Other viruses, such as adenovirus, reovirus, measles virus, and vaccinia virus also show some promise as an oncolytic virus.

TABLE 35–4   Chemoprophylactic Use of Drugs Described in This Chapter

Drug	Use	Number of Chapter for Additional Information
Acyclovir	Prevention of disseminated HSV or VZV disease in immunocompromised patients	37
Ganciclovir	Prevention of disseminated CMV disease in immunocompromised patients, especially retinitis in AIDS patients	37
Letermovir	Prevention of disseminated CMV disease in patients receiving hematopoietic stem cell transplants	37
Oseltamivir	Prevention of influenza during outbreaks caused by influenza A and B viruses	39
Amantadine	Prevention of influenza during outbreaks caused by influenza A virus (Not used in the United States)	39
Palivizumab	Prevention of bronchiolitis and pneumonia caused by respiratory syncytial virus in infants	39
Zidovudine or nevirapine	Prevention of HIV infection of neonate	45
Tenofovir, emtricitabine, and raltegravir	Prevention of HIV infection in needle-stick injuries	45
Tenofovir, plus emtricitabine	Preexposure prophylaxis of HIV infection in high-risk individuals	45

AIDS = acquired immunodeficiency syndrome; CMV = cytomegalovirus; HIV = human immunodeficiency virus; HSV = herpes simplex virus; VZV = varicella-zoster virus.

PEARLS

- **Selective toxicity** is the ability of a drug to inhibit viral replication without significantly damaging the host cell. It is difficult to achieve a high degree of selective toxicity with antiviral drugs because the virus can only replicate within cells and uses many cellular functions during replication.

Inhibitors of Early Events

- Amantadine inhibits the uncoating of influenza A virus by blocking "ion channel" activity of the viral matrix protein (M2 protein). The drug has no effect on influenza B or C viruses.
- Maraviroc binds to the CCR5 receptor on the cell thereby preventing the binding of the gp120 of human immunodeficiency virus (HIV).
- Fostemsavir binds to gp120 in the envelope of HIV thereby preventing the attachment of HIV to the cell.
- Enfuvirtide is a "fusion inhibitor." It inhibits the fusion of HIV with the cell membrane by binding to gp41, an envelope protein of HIV.
- Ibalizumab is a monoclonal antibody against CD4 protein that blocks entry of HIV.
- Palivizumab is a monoclonal antibody against the fusion protein of respiratory syncytial birus.

Inhibitors of Herpesviruses: Nucleoside Inhibitors

- **Acyclovir inhibits the DNA polymerase** of herpes simplex virus (HSV) type 1, HSV-2, and varicella-zoster virus (VZV). **Acyclovir must be activated within the infected cell by a virus-encoded thymidine kinase** that phosphorylates the drug. Acyclovir is not phosphorylated in uninfected cells, and

cellular DNA synthesis is not inhibited. Selective toxicity is high, and there are very few adverse effects.

- **Acyclovir is a chain-terminating drug** because it lacks a hydroxyl group in the 3' position. It does not have a ribose ring (i.e., it is *acyclo*, meaning without a ring). The absence of this hydroxyl group means the next nucleoside triphosphate cannot be added and the replicating DNA chain is terminated.
- **Acyclovir inhibits viral replication but has no effect on the latency** of HSV-1, HSV-2, and VZV.
- Ganciclovir action is very similar to that of acyclovir, but it is effective against cytomegalovirus (CMV), whereas acyclovir is not.

Inhibitors of Herpesviruses: Nonnucleoside Inhibitors

- **Foscarnet inhibits the DNA polymerase** of all herpesviruses but is clinically useful against HSV and CMV. It is a **pyrophosphate analogue** that inhibits the cleavage of pyrophosphate from the nucleoside triphosphate that has been added to the growing DNA chain.

Inhibitors of Retroviruses: Nucleoside Inhibitors (NRTIs)

- **Zidovudine inhibits the DNA polymerase** (reverse transcriptase) **of HIV**. It is a **chain-terminating drug** because it has an azide group in place of the hydroxyl group in the 3' position. Unlike acyclovir, it does not require a viral-encoded kinase to be phosphorylated. Cellular kinases phosphorylate the drug, so it is active in uninfected cells, and significant adverse effects can occur.

- Other drugs with the same mode of action include lamivudine, didanosine, emtricitabine, stavudine, abacavir, and tenofovir.

Inhibitors of Retroviruses: Nonnucleoside Inhibitors (NNRTIs)

- **Nevirapine**, delavirdine, efavirenz, etravirine, doravirine, and rilpivirine **inhibit the DNA polymerase (reverse transcriptase) of HIV** but are not nucleoside analogues.

Inhibitors of Hepatitis B Virus

- **Adefovir**, entecavir, lamivudine, and telbivudine inhibit the DNA polymerase of hepatitis B virus (HBV). These drugs are useful in the treatment of chronic HBV infection.

Inhibitors of Hepatitis C Virus

- **Sofosbuvir inhibits RNA polymerase** and is useful in the treatment of chronic hepatitis C virus (HCV) infection.
- **Ledipasvir** inhibits NS5A protein of HCV, thereby inhibiting synthesis of the progeny RNA genomes of HCV.

Inhibitor of SARS-CoV-2 Coronavirus

Remdesivir inhibits the RNA polymerase of SARS-CoV-2.

Inhibitors of Other Viruses

- Ribavirin is a guanosine analogue that can inhibit nucleic acid synthesis of several viruses. It is useful in severe respiratory syncytial virus infections.
- Baloxavir inhibits the **"cap-snatching"** endonuclease (ribonuclease) required for synthesis of influenza virus mRNA. It is useful in the treatment of influenza.

Integrase Inhibitors

- **Raltegravir, elvitegravir, and dolutegravir inhibit the integrase** encoded by HIV, which blocks the integration of HIV DNA into host cell DNA.

Inhibitor of Precursor DNA Cleavage

- **Letermovir inhibits the terminase encoded by** CMV which cleaves precursor DNA concatenates **into progeny genome DNA**.

Protease Inhibitors

- **Indinavir and other similar drugs inhibit the virus-encoded protease of HIV**. Inhibition of the protease prevents cleavage of precursor polypeptides, which prevents formation of the structural proteins of the virus. Synthesis of infectious virus is inhibited, but the viral DNA integrated into the host cell DNA is unaffected.
- **Boceprevir, simeprevir, telaprevir, and paritaprevir inhibit the protease** of HCV.

Inhibitors of Viral Protein Synthesis

- **Interferons inhibit virus replication** by **degrading mRNA and blocking protein synthesis**. (See Chapter 33 for more information.) Pegylated alpha interferon is used in the treatment of chronic hepatitis B and acute hepatitis C.
- Fomivirsen is an antisense DNA that binds to the mRNA of CMV, which prevents the mRNA from being translated into viral proteins.

Inhibitors of Release of Virus

- **Zanamivir and oseltamivir inhibit the neuraminidase of both influenza A and B viruses. This inhibits the release of progeny virus**, which reduces the spread of virus to neighboring cells.

SELF-ASSESSMENT QUESTIONS

1. Regarding the mode of action of antiviral drugs, which one of the following is the **MOST** accurate?
    (A) Amantadine inhibits the virus-encoded DNA polymerase that is required to synthesize viral progeny DNA.
    (B) Lamivudine inhibits the cell-encoded RNA polymerase that is required to synthesize viral genome.
    (C) Raltegravir inhibits the translation of viral mRNA into viral proteins.
    (D) Ritonavir inhibits the virus-encoded protease that is required to cleave viral precursor polypeptides into functional proteins.
    (E) Zidovudine inhibits the virus-encoded RNA polymerase that is required to synthesize viral mRNA.
2. Which one of the following best describes the action of oseltamivir (Tamiflu)?
    (A) Inhibits reverse transcriptase
    (B) Inhibits the RNA-dependent RNA polymerase in the virion

    (C) Inhibits the DNA-dependent RNA polymerase in the infected cell
    (D) Inhibits viral protein synthesis by binding to the 60S ribosomal subunit
    (E) Inhibits the neuraminidase required for release of virus from the infected cell
3. Which one of the following is a well-described adverse effect of the protease inhibitors used in the treatment of HIV infection?
    (A) Bone marrow suppression
    (B) Central nervous system disturbances
    (C) Drug-induced hepatitis
    (D) Lipodystrophy
    (E) Peripheral neuropathy
4. Regarding acyclovir, which one of the following is the **MOST** accurate?
    (A) Bone marrow suppression is a significant adverse effect.
    (B) It terminates the latent state of both herpes simplex virus type 1 and type 2.

(C) It inhibits the virus-encoded thymidine kinase that is required to synthesize viral DNA.

(D) Resistance to acyclovir is primarily caused by proton pumps that export the drug from the cell.

(E) It is a chain-terminating drug because it does not have a complete ribose ring and therefore lacks a hydroxyl group in the correct position.

5. Regarding baloxavir, which one of the following is the **MOST** accurate?

(A) It is effective against the reverse transcriptase of influenza virus.

(B) It prevents the action of the influenza virus-encoded protease so functional viral proteins are not synthesized.

(C) It inhibits the influenza virus-encoded endonuclease that transfers methylated guanosine from cell mRNA to viral mRNA, a process called cap-snatching.

(D) It inhibits the integrase required for integration of the influenza genome into cell DNA.

(E) It inhibits release of influenza virus from the infected cell.

6. Regarding the action of ledipasvir against hepatitis C virus (HCV), which one of the following is the **MOST** accurate?

(A) It inhibits the NS5A protein required for activity of the HCV viral RNA polymerase.

(B) It prevents the action of the HCV virus-encoded protease so functional viral proteins are not synthesized.

(C) It inhibits the virus-encoded thymidine kinase required to activate the viral RNA polymerase.

(D) It inhibits the RNA-dependent DNA polymerase required for HCV replication.

(E) It inhibits entry of HCV into the hepatocyte by binding to the surface receptor.

ANSWERS

(1) **(D)**
(2) **(E)**
(3) **(D)**
(4) **(E)**
(5) **(C)**
(6) **(A)**

PRACTICE QUESTIONS: USMLE & COURSE EXAMINATIONS

Questions on the topics discussed in this chapter can be found in the Basic Virology section of Part XIII: USMLE (National Board) Practice Questions starting on page 744. Also see Part XIV: USMLE (National Board) Practice Examination starting on page 775.

36

Viral Vaccines

INTRODUCTION

Vaccines are the best approach to prevent viral diseases. Prevention of viral diseases can be achieved by the use of vaccines that induce active immunity or by the administration of preformed antibody that provides passive immunity.

ACTIVE IMMUNITY

There are three main types of vaccines that induce **active** immunity: those that contain **live virus** whose pathogenicity has been **attenuated**, those that contain **killed virus**, and those that contain purified viral proteins (**subunits**). An *attenuated* virus is one that is unable to cause disease, but retains its antigenicity and can induce protection. Nucleic acid (mRNA) vaccines, such as those used being used against SARS-CoV-2, are also important and hold great promise for widespread use in the future.

Some vaccines, such as the hepatitis B vaccine, contain purified viral proteins and are often called **subunit** vaccines. The features of subunit vaccines resemble those of killed vaccines because no viral replication occurs in these vaccines. The attributes of live and killed vaccines are listed in Table 36–1.

In general, live vaccines are preferred to vaccines containing killed virus because their protection is **greater** and **longer-lasting**. With live vaccines, the virus multiplies in the host, producing a prolonged antigenic stimulus, and immunoglobulin A (IgA) and IgG are elicited when the vaccine is administered by the natural route of infection (e.g., when polio vaccine is given orally). Killed vaccines, which are usually given intramuscularly, do not stimulate a major IgA response. Killed vaccines typically do not stimulate a cytotoxic T-cell response because the virus in the vaccine does not replicate. In the absence of replication, no viral epitopes are presented in association with class I major histocompatibility complex (MHC) proteins, and the cytotoxic T-cell response is not activated (see Chapter 58). Although live vaccines stimulate a long-lasting response, booster doses are now recommended with measles and polio vaccines.

One unique form of a live, attenuated viral vaccine is the influenza vaccine that contains a **temperature-sensitive** mutant of the virus as the immunogen. The temperature-sensitive mutant will replicate in the cooler air passages of the nose, where

TABLE 36–1 **Characteristics of Live and Killed Viral Vaccines**

Characteristic	Live Vaccine	Killed Vaccine
Duration of immunity	Longer	Shorter
Effectiveness of protection	Greater	Lower
Immunoglobulin (Ig) produced	IgA[1] and IgG	IgG
Cell-mediated immunity produced	Yes	Weakly or none
Interruption of transmission of virulent virus	More effective	Less effective
Reversion to virulence	Possible	No
Stability at room temperature	Low	High
Excretion of vaccine virus and transmission to nonimmune contacts	Possible	No
Can cause disease in immuno-compromised patients	Yes	No

[1]If the vaccine is given by the natural route.

it induces IgA-based immunity, whereas it will not replicate in the warmer lung tissue and therefore will not cause disease.

There are three concerns about the use of live vaccines:

(1) They are composed of attenuated viral mutants, which can **revert to virulence** either during vaccine production or in the immunized person. Reversion to virulence during production can be detected by quality control testing, but there is no test to predict whether reversion will occur in the immunized individual. Of the commonly used live vaccines, only polio vaccine has had problems regarding revertants; measles, mumps, rubella, and varicella vaccines have not.

Even if the virus in the live vaccine does not revert, it can still cause disease because, although attenuated (weakened), it can still be pathogenic in a host with reduced immunity. For this reason, live viral vaccines should *not* be given to immunocompromised people or to pregnant women because the fetus may become infected.

(2) The live vaccine can be **excreted** by the immunized person. This is a double-edged sword. It is advantageous if the spread of the virus successfully immunizes others, as occurs with the live polio vaccine. However, it could be a problem if, for example, a virulent poliovirus revertant spreads to a susceptible person. Rare cases of paralytic polio occur in the United States each year by this route of infection.

(3) A second virus could **contaminate** the vaccine if it was present in the cell cultures used to prepare the vaccine. This concern exists for both live and killed vaccines, although, clearly, the live vaccine presents a greater problem, because the process that inactivates the virus in the killed vaccine could inactivate the contaminant as well. It is interesting, therefore, that the most striking incidence of contamination of a vaccine occurred with the *killed* polio vaccine. In 1960, it was reported that live simian vacuolating virus 40 (SV40 virus), an inapparent "passenger" virus in monkey kidney cells, had contaminated some lots of polio vaccine and was resistant to the formaldehyde used to inactivate the poliovirus. There was great concern when it was found that SV40 virus causes sarcomas in a variety of rodents. Fortunately, it has not caused cancer in the individuals inoculated with the contaminated polio vaccine.

Certain viral vaccines, namely, influenza, measles, mumps, and yellow fever vaccines, are grown in chick embryos. These vaccines should *not* be given to those who have had an **anaphylactic reaction to eggs**. People with allergies to chicken feathers can be immunized.

In addition to the disadvantages of the killed vaccines already mentioned—namely, that they induce a **shorter duration** of protection, are **less protective**, and **induce fewer IgA antibodies**—there is the potential problem that the inactivation process might be inadequate. Although this is rare, it happened in the early days of the manufacture of the killed polio vaccine. However, killed vaccines do have two advantages: They **cannot revert to virulence**, and they are **more heat-stable**. Therefore, they can be used more easily in tropical climates.

Most viral vaccines are usually given before a known exposure (i.e., they are administered **preexposure**). However, there

are two vaccines, the vaccines against rabies and hepatitis B, which are also effective when given **postexposure** because the incubation period of these diseases is long enough that the vaccine-induced immunity can prevent the disease. Thus, the rabies vaccine is most often used in people after they have received a bite from a potentially rabid animal, and the hepatitis B vaccine is used in people who have sustained a needle-stick injury.

The prospect for the future is that some of the disadvantages of current vaccines will be bypassed by the use of purified viral antigens produced from genes cloned in either bacteria or yeasts by recombinant DNA techniques. The advantages of antigens produced by the cloning process are that they contain no viral nucleic acid and so cannot replicate or revert to virulence, they have no contaminating viruses from cell culture, and they can be produced in large amounts. A disadvantage of these cloned vaccines is that they are unlikely to stimulate a cytotoxic T-cell response because no viral replication occurs.

Another prospect for the future is the use of "DNA vaccines." These vaccines contain purified DNA encoding the appropriate viral proteins genetically engineered into a viral vector or plasmid. Immunization with this composite DNA elicits both antibody and cytotoxic T cells and protects against disease in experimental animals.

Certain live viral vaccines, such as the vaccines containing vaccinia virus, adenovirus, and poliovirus, are being used experimentally to immunize against other viruses such as HIV. This is done by splicing the HIV gene into the live viral genome and then infecting the experimental animal with the constructed virus. The advantage of this procedure is that a cytotoxic T-cell response is elicited (because the virus is replicating), whereas if the purified antigen alone were used to immunize the animal, an antibody response but not a cytotoxic T-cell response would be elicited.

The viral vaccines currently in use are described in Table 36–2. The vaccines, both viral and bacterial, recommended for children from 0 to 6 years of age are listed in Table 36–3. Table 36–4 lists some important viral (and bacterial) vaccines recommended for travelers. Advice regarding vaccines for travelers can be found at the website for the Centers for Disease Control and Prevention: www.cdc.gov/travel.

TYPES OF VIRAL VACCINES

Killed Vaccines

The vaccines against hepatitis A virus and rabies virus are killed vaccines. Also, one version of the poliovirus vaccine is a killed vaccine. Purified virus is inactivated with formaldehyde (or other inactivating chemicals). The ability of the virus to cause disease is inactivated but the ability to activate the immune response is retained.

Live, Attenuated Vaccines

The vaccines against measles virus, mumps virus, and rubella virus are live, attenuated vaccines. One version of the varicella vaccine, one version of the polio vaccine, and one version of the influenza vaccine are live, attenuated vaccines as well.

TABLE 36–2 Current Viral Vaccines

A. Vaccines in Common Use

Vaccine	Live Virus, Killed Virus, Subunit or Nucleic Acid of Virus	Comment
Measles	Live	Usually given in combination with mumps and rubella viruses (MMR vaccine)
Mumps	Live	
Rubella	Live	
Varicella (chickenpox)	Live	There are two vaccines that contain live varicella-zoster virus: one that prevents varicella (Varivax) and another that prevents zoster (Zostavax).
Zoster (shingles)	Live and subunit	There is one subunit vaccine (Shingrix) that contains recombinant VZV envelope glycoprotein that prevents zoster
Polio	Live and killed	Only the killed vaccine is recommended for routine immunizations in the United States.
Influenza	Live, killed, and subunit	The live vaccine contains a temperature-sensitive mutant of influenza virus. The viral hemagglutinin is the main antigen in the killed vaccine. The subunit vaccine contains hemagglutinin made by inserting its gene into insect cells via a baculovirus vector
Hepatitis A	Killed	
Hepatitis B	Subunit	Recombinant vaccine contains hepatitis B virus surface antigen only
Rotavirus	Live	There are two live vaccines: a pentavalent reassortant vaccine and a monovalent attenuated vaccine.
Human papillomavirus	Subunit	Contains the surface protein of the 9 most important types of HPV
SARS-CoV-2 coronavirus	Messenger RNA for spike protein enclosed in lipid nanoparticle.	Emergency use authorization approved in the United States in December 2020
	DNA for spike protein in a non-replicating human adenovirus	Emergency use authorization approved in the United States in February 20201

B. Vaccines Used in Special Situations

Vaccine	Live Virus, Killed Virus, Subunit or Nucleic Acid of Virus	Comment
Rabies	Killed	Used in those bitten by wild animal and those with possible occupational exposure, e.g., veterinarians
Yellow fever	Live	Used when traveling in endemic areas
Japanese encephalitis	Killed	Used when traveling in endemic areas
Dengue	Live recombinant vaccine with a yellow fever vaccine virus backbone containing genes for the envelope and premembrane proteins of dengue virus.	Used in those who have antibodies against dengue virus only in countries where the disease is common.
Ebola	Live recombinant vaccine with a vesicular stomatitis virus backbone containing gene encoding Ebola surface glycoprotein.	Used in outbreaks in endemic areas
Smallpox	Live vaccinia virus	Used in military personnel and certain medical personnel such as "first responders" and emergency room staff.
Adenovirus	Live	Used in military personnel

The attenuated virus in the measles, mumps, rubella, varicella, and polio vaccines is weakened in its pathogenicity as a result of mutations in the virulence genes of the virus. The live, attenuated influenza vaccine contains a temperature-sensitive mutant of the virus. The virus in this influenza vaccine is attenuated because, although it grows well in the nose at 32°C, it will not grow in the lungs at 37°C, that is, it is temperature-sensitive (see Table 36-5).

Vectored Vaccines

The term "vector" typically refers to an insect that carries a pathogen from person to another. For example, mosquitoes are the vector for the *Plasmodium* protozoan that causes malaria. In this context, a vector is a nonpathogenic virus that carries the gene encoding the antigen that will induce protective antibodies in the recipient of the vaccine.

TABLE 36–3 Vaccines Recommended for Children Aged 0 to 6 Years[1,2]

Bacterial Vaccines	Viral Vaccines[3]
Diphtheria toxoid, tetanus toxoid, acellular pertussis (DTaP)	Hepatitis A
Haemophilus influenzae type b (Hib)	Hepatitis B
Meningococcal	Influenza
Pneumococcal	Measles, mumps, rubella (MMR)
	Poliovirus, inactivated
	Rotavirus
	Varicella (varivax)

[1]Vaccines are listed in alphabetical order.

[2]A complete description of the vaccine schedule is available on the Centers for Disease Control and Prevention website, www.cdc.gov.

[3]Human papillomavirus vaccine (Gardasil 9) is recommended for boys and girls aged 9 to 26 years.

Four vectored vaccines are in current use. One is the Ebola vaccine in which the nonpathogen vesicular stomatitis virus (VSV) is engineered to carry the gene for the surface glycoprotein of Ebola virus. After immunization, VSV replicates in the person and in the process produces the surface glycoprotein that induces protective antibodies.

Another vectored vaccine is the rotavirus vaccine. In this vaccine, a bovine rotavirus that is nonpathogenic for humans carries the gene for the human rotavirus surface protein. The vaccine is given orally and replication of the vaccine virus in the intestinal tract produces protective IgA antibody to the human rotavirus surface protein.

The third is the dengue virus vaccine that contains the genes for the dengue virus envelope and premembrane proteins inserted into the live attenuated yellow fever virus which is used used in the yellow fever vaccine.

The fourth is the SARS-CoV-2 virus vaccine that contains the gene for the spike protein of the coronavirus inserted into the DNA of human adenovirus serotype 26.

Subunit Vaccines

Subunit vaccines contain the purified antigenic protein as the immunogen. There are four subunit vaccines. Three of them are commonly used, namely the vaccines against hepatitis B virus (HBV), human papillomavirus (HPV), and the Shingrix version of the zoster vaccine. Less commonly used is the influenza subunit vaccine that contains hemagglutinin (HA) made by inserting the gene encoding HA into insect cells via a baculovirus vector.

TABLE 36–4 Some Important Vaccines Recommended for Travelers

Bacterial Vaccines	Viral Vaccines
Meningococcal	Hepatitis A
Cholera	Hepatitis B
Typhoid	Polio
	Yellow fever
	Rabies
	Japanese encephalitis

TABLE 36–5 Temperature-Sensitive Mutants of Influenza Virus Used in Vaccine

Type of Influenza Virus	Growth in Nose (32°C)	Growth in Lungs (37°C)
Normal infectious virus	Yes	Yes
Temperature-sensitive mutant used in vaccine	Yes	No

The viral subunit proteins used in the vaccines are made by recombinant DNA techniques. The genes encoding the surface protein antigens of HBV, HPV, and varicella-zoster virus (VZV) are inserted into yeast or hamster ovary cells which produce large amounts of the immunogenic protein used in the vaccine.

DNA and RNA Vaccines

In these vaccines, purified DNA and RNA that encode the viral protein antigens are injected into the person. Messenger RNA is translated and the resulting viral antigens stimulate the immune system to produce protective antibodies. The two vaccines against SARS-CoV-2 coronavirus that were approved for emergency use in December 2020 contain the messenger RNA for the spike protein enclosed in a lipid nanoparticle.

There is a DNA vaccine against West Nile virus (WNV) for use in horses. It consists of a DNA plasmid that contains the genes for the premembrane and envelope proteins. The two genes in the RNA genome of WNV are reverse transcribed into DNA that is integrated into the plasmid DNA. Injection of the plasmid elicits protective antibodies to WNV.

PASSIVE IMMUNITY

Passive immunity is provided by the administration of preformed antibody in preparations called immune globulins. **Passive–active** immunity is induced by giving both immune globulins to provide immediate protection and a vaccine to provide long-term protection. This approach is described in the sections on rabies and hepatitis B.

The following preparations are available:

(1) **Rabies** immune globulin (RIG) is used in the prevention of rabies in people who may have been exposed to the virus. It is administered by injecting as much RIG as possible into the tissue at the bite site, and the remainder is given intramuscularly. The preparation contains a high titer of antibody made by hyperimmunizing human volunteers with rabies vaccine. RIG is obtained from humans to avoid hypersensitivity reactions. In addition to RIG, the vaccine containing killed rabies virus made in human diploid cells should be given. RIG and the vaccine should be given at different sites. This is an example of passive–active immunization.

(2) **Hepatitis B** immune globulin (HBIG) is used in the prevention of hepatitis B in people who may have been exposed to the virus either by needle-stick or as a neonate born of a

mother who is a carrier of HBV. The preparation contains a high titer of antibody to HBV and is obtained from humans to avoid hypersensitivity reactions. HBIG is often used in conjunction with hepatitis B vaccine, an example of passive–active immunization.

(3) **Varicella-zoster** immune globulin (VZIG) is used in the prevention of disseminated zoster in people who may have been exposed to the virus and who are immunocompromised. The preparation contains a high titer of antibody to VZV and is obtained from humans to avoid hypersensitivity reactions.

(4) **Vaccinia** immune globulins (VIGs) can be used to treat some of the complications of the smallpox vaccination.

(5) Immune globulins are useful in the prevention (or mitigation) of **hepatitis A** or **measles** in people who may have been exposed to these viruses. For example, hepatitis A immune globulins are commonly used prior to traveling to areas of the world where hepatitis A virus is endemic. Immune globulins contain pooled serum obtained from a large number of human volunteers who have not been hyperimmunized. Their effectiveness is based on antibody being present in many members of the pool.

HERD IMMUNITY

Herd immunity (also known as community immunity) occurs when a sufficiently large percentage of the population (the "herd") is immunized so that an unimmunized individual is protected (see Chapter 33). For herd immunity to occur, the vaccine must prevent transmission of the virus as well as prevent disease. For example, the live, attenuated polio vaccine can provide good herd immunity because it induces intestinal IgA, which prevents poliovirus from replicating in the gastrointestinal tract and being transmitted to others. However, the killed polio vaccine does not induce herd immunity because secretory

IgA is not produced, and immunized individuals (although protected from poliomyelitis) can still serve as a source of poliovirus for others.

WANING IMMUNITY TO VACCINES

The duration of protection provided by vaccines can decline with time, that is, it can wane. It also varies greatly from one vaccine to another. Evidence that waning immunity is a medical problem is based on the observation that some cases of vaccine-preventable diseases occur in those who have been immunized. For example, in the 2019 outbreak of measles in the United States, about 3% of adults who contracted measles received the measles, mumps, rubella (MMR) vaccine.

Some vaccines provide adequate protection for months and others for many years. For example, the protection induced by the live mumps vaccine wanes rapidly during the first few years but the protection induced by the live smallpox vaccine lasts for at least a decade.

One suggested explanation for waning immunity is that the ability of memory cells to survive and to retain function declines with time. One explanation offered for this decline is that the response of memory cells may depend upon the precise nature of the antigen used in the vaccine. Determining the criteria required to induce long-lived immunity will be important in developing more effective vaccines.

ADJUVANTS FOR VIRAL VACCINES

Adjuvants are substances added to vaccines to enhance the ability of the antigen to induce an immune response. For example, lipid A (endotoxin) from *Salmonella* is added to the zoster vaccine to enhance the immunogenicity of the surface glycoprotein of VZV. Table 36–6 describes the adjuvants used

TABLE 36-6   Adjuvants in Some Important Viral Vaccines

Disease Prevented	Name of Vaccine	Immunogen in Vaccine	Adjuvant in Vaccine
Hepatitis A	Havrix	Surface (capsid) protein	Aluminum salt
Hepatitis B	Heplisav-B	Hepatitis B surface antigen	CpG deoxynucleotide (22-mer)
Hepatitis B	Recombivax HB	Hepatitis B surface antigen	Aluminum salt
Influenza (seasonal)	Fluad	Hemagglutinin of 2 Influenza A and 2 Influenza B viruses	MF59 (squalene-based emulsion)
Papilloma and cervical cancer	Gardasil 9	Surface (capsid) protein of 9 types of human papillomavirus	Aluminum salt
Poliomyelitis (Also diphtheria, tetanus, and pertussis)	Quadracel	Surface (capsid) protein of inactivated poliovirus	Aluminum salt
Zoster (shingles)	Shingrix	Surface glycoprotein E of varicella-zoster virus	AS01 (Lipid A from *Salmonella*, i.e., endotoxin) plus saponin to form liposomes

in several vaccines currently in use. Additional information on adjuvants is presented in Chapter 57 entitled "Overview of Immunity."

The mechanism of action of adjuvants varies. Some, such as lipid A and CpG deoxynucleotide, stimulate Toll-receptors on the surface of antigen-presenting cells to enhance the production of interleukins, whereas others such as squalene emulsions prolong the availability of the antigen. Aluminum salts, such as alum, were thought to act primarily by prolonging the exposure to the antigen. However, current evidence shows that alum can activate the inflammasome and increase production of interleukin-1, a proinflammatory cytokine.

PEARLS

Active Immunity

- Active immunity is most often elicited by vaccines containing killed viruses, purified protein subunits, or live, attenuated (weakened) viruses. The vaccine against SARS-CoV-2 coronavirus contains the messenger RNA for the spike protein enclosed in a lipid nanoparticle.

- In general, **live viral vaccines are preferable to killed vaccines** for three reasons: (1) they induce a higher titer of antibody and hence longer-lasting protection; (2) they induce a broader range of antibody (e.g., both IgA and IgG, not just IgG); and (3) they activate cytotoxic T cells, which kill virus-infected cells.

- There are some potential **problems with live viral vaccines, the most important of which is reversion to virulence.** Transmission of the vaccine virus to others who may be immunocompromised is another concern. Also there may be a second, unwanted virus in the vaccine that was present in the cells used to make the vaccine virus. This second virus may cause adverse effects.

- **Live viral vaccines should not be given to immunocompromised individuals or to pregnant women.**

- Vaccines grown in chick embryos, especially influenza vaccine, should not be given to those who have had an anaphylactic reaction to eggs.

Passive Immunity

- **Passive immunity is immunity acquired by an individual by the transfer of preformed antibodies** made in either other humans or in animals. These antibody preparations are often called **immune globulins**. Passive immunity also occurs naturally when IgG is transferred from the mother to the fetus across the placenta and when IgA is transferred from the mother to the newborn in colostrum.

- **The main advantage of passive immunity** is that it **provides immediate protection**. The main disadvantage is that it does not provide long-term protection (i.e., it is active only for a few weeks to a few months).

- Immune globulin preparations against rabies virus, hepatitis A virus, hepatitis B virus, and varicella-zoster virus are effective.

- **Passive–active immunity consists of administering both immune globulins and a viral vaccine.** This provides both immediate as well as long-term protection. For example, protection against rabies in an unimmunized person who has been bitten by a potentially rabid animal consists of both rabies immune globulins and the rabies vaccine.

Herd Immunity

- **Herd immunity** is the protection of an individual that results from immunity in many other members of the population (the "herd") that interrupts transmission of the virus to the individual. Herd immunity can be achieved either by active immunization or by natural infection of a sufficiently high percentage of the population. Herd immunity is unlikely to be achieved by passive immunity because, although antibodies can protect the individual against spread of virus through the bloodstream, they are unlikely to prevent viral replication at the portal of entry and consequent transmission to others.

SELF-ASSESSMENT QUESTIONS

1. Regarding viral vaccines, which one of the following is the **MOST** accurate?
   (A) Killed vaccines induce a longer-lasting response than do live, attenuated vaccines.
   (B) Killed vaccines are no longer used in this country because they do not induce secretory IgA.
   (C) Killed vaccines induce a broader range of immune responses than do live, attenuated vaccines.
   (D) Killed vaccines are safer to give to immunocompromised patients than are live, attenuated vaccines.

2. Individuals who have had an anaphylactic reaction to egg proteins should **NOT** receive which one of the following vaccines?
   (A) Hepatitis A vaccine
   (B) Hepatitis B vaccine
   (C) Influenza vaccine
   (D) Polio vaccine
   (E) Rabies vaccine

3. Induction of passive–active immunity is useful in the prevention of which one of the following sets of two viral diseases?
   (A) Hepatitis A and dengue
   (B) Hepatitis B and rabies
   (C) Influenza and varicella
   (D) Mumps and yellow fever
   (E) Rubella and measles

4. Protection of the unimmunized individual based on immunization of a sufficient number of other members of the population is a description of which one of the following?

(A) Active immunity
(B) Herd immunity
(C) Passive immunity
(D) Passive–active immunity
(E) Postexposure immunity

ANSWERS

(1) **(D)**
(2) **(C)**
(3) **(B)**
(4) **(B)**

PRACTICE QUESTIONS: USMLE & COURSE EXAMINATIONS

Questions on the topics discussed in this chapter can be found in the Basic Virology section of Part XIII: USMLE (National Board) Practice Questions starting on page 744. Also see Part XIV: USMLE (National Board) Practice Examination starting on page 775.

PART IV CLINICAL VIROLOGY

For medical purposes, viruses are most usefully described in terms of either their main site of infection, their mode of transmission, or the type of lesions and diseases they cause. Chapters 37 to 45 describe the clinically important viruses organized in this medically relevant manner. Several less prominent viruses are described in Chapter 46.

In this brief introduction, the clinically important viral pathogens are categorized into groups according to their main structural characteristics (i.e., DNA enveloped viruses, DNA nonenveloped[1] viruses, RNA enveloped viruses, and RNA nonenveloped viruses) (Table IV–1).

DNA ENVELOPED VIRUSES

Herpesviruses

These viruses are noted for their ability to cause latent infections. This family includes (1) herpes simplex virus types 1 and 2, which cause painful vesicles on the face and genitals, respectively; (2) varicella-zoster virus, which causes varicella (chickenpox) typically in children and, when it recurs, zoster (shingles); (3) cytomegalovirus, an important cause of congenital malformations; (4) Epstein–Barr virus, which causes infectious mononucleosis; and (5) human herpesvirus 8, which causes Kaposi's sarcoma. (See Chapter 37.)

Hepatitis B Virus

This virus is one of the important causes of viral hepatitis. In contrast to hepatitis A virus (an RNA nucleocapsid virus), hepatitis B virus causes a more severe form of hepatitis, results more frequently in a chronic carrier state, and is implicated in the induction of hepatocellular carcinoma, the most common cancer worldwide. (See Chapter 41.)

Poxviruses

Poxviruses are the largest and most complex of the viruses. The disease smallpox has been eradicated by effective use of the vaccine.

[1]Nonenveloped viruses are also called naked nucleocapsid viruses.

Molluscum contagiosum virus is the only poxvirus that causes human disease in the United States at this time. (See Chapter 37.)

DNA NONENVELOPED VIRUSES

Adenoviruses

These viruses are best known for causing upper and lower respiratory tract infections, including pharyngitis and pneumonia. (See Chapter 38.)

Papillomaviruses

These viruses cause papillomas on the skin and mucous membranes of many areas of the body. Some types are implicated as a cause of cancer (e.g., carcinoma of the cervix). (See Chapter 37.)

Parvovirus B19

This virus causes "slapped cheeks" syndrome, hydrops fetalis, and severe anemia, especially in those with hereditary anemias such as sickle cell anemia. (See Chapter 39.)

TABLE IV–1 Major Viral Pathogens

Structure	Viruses
DNA enveloped viruses	Herpesviruses (herpes simplex virus types 1 and 2, varicella-zoster virus, cytomegalovirus, Epstein–Barr virus, human herpesvirus 8), hepatitis B virus, smallpox virus
DNA nucleocapsid viruses	Adenovirus, papillomaviruses, parvovirus B19
RNA enveloped viruses	Influenza virus, parainfluenza virus, respiratory syncytial virus, coronavirus, measles virus, mumps virus, rubella virus, rabies virus, human T-cell lymphotropic virus, human immunodeficiency virus, hepatitis C virus, West Nile virus, dengue virus, Zika virus, yellow fever virus, Ebola virus
RNA nucleocapsid viruses	Enteroviruses (poliovirus, coxsackievirus, echovirus, hepatitis A virus), rhinovirus, rotavirus, noroviruses, hepatitis E virus

RNA ENVELOPED VIRUSES

Respiratory Viruses

(1) Influenza A and B viruses. Influenza A virus is the major cause of recurrent epidemics of influenza. (See Chapter 38.)

(2) Parainfluenza viruses. These viruses are the leading cause of croup in young children and an important cause of common colds in adults. (See Chapter 38.)

(3) Respiratory syncytial virus. This virus is the leading cause of bronchiolitis and pneumonia in infants. (See Chapter 38.)

(4) Human metapneumovirus. This virus causes significant upper and lower respiratory tract disease. (See Chapter 38.)

(5) Coronaviruses. These viruses cause both the common cold and pneumonia. SARS-CoV-2 is the coronavirus that caused the pandemic in 2019–2021. (See Chapter 38.)

Measles, Mumps, and Rubella Viruses

These viruses cause well-known childhood diseases and are the viral components of the MMR vaccine. Widespread use of the vaccine has markedly reduced the incidence of these diseases in the United States. These viruses are well known for the complications associated with the diseases they cause (e.g., rubella virus infection in a pregnant woman can cause congenital malformations). (See Chapter 39.)

Rabies Virus

This virus causes almost invariably fatal encephalitis following the bite of a rabid animal. In the United States, wild animals such as skunks, foxes, raccoons, and bats are the major sources, but human infection is rare. (See Chapter 39.)

Hepatitis C Virus

This virus causes hepatitis C, the most prevalent form of viral hepatitis in the United States. It causes a very high rate of chronic carriers and predisposes to chronic hepatitis and hepatic carcinoma. (See Chapter 41.)

Human T-Cell Lymphotropic Virus

This virus causes T-cell leukemia in humans. It also causes an autoimmune disease called tropical spastic paraparesis. (See Chapter 43.)

Human Immunodeficiency Virus

Human immunodeficiency virus (HIV) causes acquired immunodeficiency syndrome (AIDS). (See Chapter 45.)

West Nile Virus, Dengue Virus, Zika Virus, and Yellow Fever Virus

These viruses are arboviruses (*arthropod-borne* viruses) transmitted by mosquitoes from animal reservoirs to humans. (See Chapter 42.)

Ebola Virus

This virus causes Ebola hemorrhagic fever. (See Chapter 42.)

RNA NONENVELOPED VIRUSES

Enteroviruses

These viruses infect the enteric tract and are transmitted by the fecal–oral route. Poliovirus rarely causes disease in the United States because of the vaccine but remains an important cause of aseptic meningitis and paralysis in developing countries. Of more importance in the United States are Coxsackie viruses, which cause aseptic meningitis, myocarditis, and pleurodynia; and echoviruses, which cause aseptic meningitis. (See Chapter 40.)

Rhinoviruses

These viruses are the most common cause of the common cold. They have a large number of antigenic types, which may account for their ability to cause disease so frequently. (See Chapter 38.)

Rotaviruses

These viruses possess an unusual genome composed of double-stranded RNA in 11 segments. Rotaviruses are an important cause of viral gastroenteritis in young children. (See Chapter 40.)

Hepatitis A Virus

This virus is an important cause of hepatitis. It is an enterovirus but is described in this book in conjunction with hepatitis B virus. It is structurally different from hepatitis B virus, which is a DNA enveloped virus. Furthermore, it is epidemiologically distinct (i.e., it primarily affects children, is transmitted by the fecal–oral route, and rarely causes a prolonged carrier state). (See Chapter 41.)

Noroviruses

Noroviruses are a common cause of gastroenteritis, especially in adults. They are a well-known cause of outbreaks of vomiting and diarrhea in hospitals, in nursing homes, and on cruise ships (See Chapter 40).

Hepeviruses

The main human pathogen in the hepevirus family is hepatitis E virus (HEV). It causes hepatitis acquired by fecal–oral transmission similar to hepatitis A virus. HEV is a nonenveloped virus with a positive-polarity single-stranded RNA genome. (See Chapter 41.)

OTHER CATEGORIES

Chapter 42 describes the large and varied group of arboviruses, which have the common feature of being transmitted by an arthropod. Chapter 43 discusses tumor viruses, and Chapter 44 covers the "slow" viruses that cause degenerative central nervous system diseases primarily. Chapter 45 describes HIV, the cause of AIDS. The less common viral pathogens are described in Chapter 46.

Herpesviruses, Poxviruses, & Human Papilloma Virus

INTRODUCTION

Most of the viruses in this chapter cause skin lesions as their primary clinical manifestation. As described in Table 37–1, herpes simplex viruses (HSV) 1 and 2 and varicella-zoster virus (VZV) cause vesicles. Human herpesvirus-6 causes roseola infantum characterized by a pink macular or maculopapular rash on the trunk. Human herpesvirus 8 causes Kaposi's sarcoma (KS) characterized by purple macular or nodular lesions.

TABLE 37–1 Features of Skin Lesions of Herpesviruses, Poxviruses, and Human Papillomavirus

Name of Virus	Typical Skin Lesion
Herpes simplex virus type 1	Vesicle
Herpes simplex virus type 2	Vesicle
Varicella-zoster virus	Vesicle
Cytomegalovirus	None
Epstein–Barr virus	None
Human herpesvirus 8 (Kaposi's sarcoma virus)	Flat or nodular purple lesion
Smallpox virus	Pustule
Molluscum contagiosum virus	Fleshy papule with umbilicated center
Human papillomavirus	Papule with rough, irregular surface and spiny or cauliflower-like projections (papilloma, wart)

Smallpox virus causes pustules, but the virus has been eradicated, so these lesions are not seen in medical practice today. Molluscum contagiosum virus (MCV), a member of the poxvirus family, causes fleshy papules on the skin. Human papillomavirus (HPV) causes papillomas (warts) on skin and mucous membranes of organs such as the cervix and larynx. Of the viruses described in this chapter, only two of the herpesviruses, cytomegalovirus (CMV) and Epstein–Barr virus (EBV), do not cause skin lesions.

All of the viruses in this chapter have DNA as their genome (Table 37–2). The herpesviruses and poxviruses have linear double-stranded DNA, whereas HPV has circular double-stranded DNA.

TABLE 37–2 Properties of Herpesviruses, Poxviruses, and Human Papillomavirus

Property	Herpesviruses	Poxviruses	Human Papillomavirus
Virus family	Herpesviruses	Poxviruses	Papillomaviruses
Genome	Double-stranded DNA; linear	Double-stranded DNA; linear	Double-stranded DNA; circular
Virion DNA polymerase	No	No	No
Virion RNA polymerase	No	Yes	No
Nucleocapsid	Icosahedral	Complex	Icosahedral
Envelope	Yes	Yes	No

Herpesviruses and HPV replicate in the nucleus of infected cells, whereas poxviruses replicate in the cytoplasm.

Additional information regarding the clinical aspects of infections caused by the viruses in this chapter is provided in Part IX, entitled Infectious Diseases beginning on page 603.

HERPESVIRUSES

OVERVIEW

The herpesvirus family contains seven important human pathogens: HSV types 1 and 2, VZV, CMV, EBV, human herpesvirus 6 (HHV-6), and human herpesvirus 8 (also known as Kaposi's sarcoma–associated herpesvirus [KSHV]).

All herpesviruses are structurally similar. Each has an **icosahedral** core surrounded by a lipoprotein **envelope** (Figure 37–1). The genome is linear double-stranded DNA. The virion does not contain a polymerase. They are large (120–200 nm in diameter), second in size only to poxviruses.

Herpesviruses replicate in the nucleus, form intranuclear inclusions, and are the only viruses that obtain their envelope by budding from the nuclear membrane. The virions of herpesviruses possess a **tegument** located between the nucleocapsid and the envelope. This structure contains regulatory proteins, such as transcription and translation factors, which play a role in viral replication.

Herpesviruses are noted for their ability to cause life-long **latent infections**. In these infections, the acute disease is followed by an asymptomatic period during which the virus remains in a quiescent (latent) state. When the patient is exposed to an inciting agent or immunosuppression occurs, reactivation of virus replication and disease can occur.[1] With some herpesviruses (e.g., HSV), the symptoms of the subsequent episodes are similar to those of the initial one; however, with others (e.g., VZV), they are different (Table 37–3).

Some information is available regarding the mechanism by which HSV and CMV initiate and maintain the latent state. In HSV latency, a tegument protein called viral protein-16 (VP-16) plays an important role. VP-16 is a transcription factor that initiates immediate–early mRNA and protein synthesis. Immediate early (IE) proteins are required for viral replication to proceed. When HSV replicates in epithelial cells, VP-16 performs its required function. But when the progeny HSV infects adjacent neurons, VP-16 does not function. The HSV DNA migrates to the nucleus of the neuron but VP-16 does not initiate immediate–early mRNA synthesis, so replication cannot occur and the viral DNA remains in a latent state. Note that the viral DNA is in the nucleus but is not integrated into cellular DNA.

Furthermore, shortly after HSV infects sensory neurons, a set of "**latency-associated transcripts**" (LATS) are synthesized. These noncoding, regulatory RNAs suppress viral replication by inhibiting early gene expression. HSV DNA persists in the nucleus of infected cells. The process by which latency is terminated, viral replication is activated, and infectious HSV is produced is unclear, but various triggers such as sunlight, fever, and stress are known.

Similarly, CMV establishes latency by producing micro RNAs that inhibit the translation of mRNAs required for viral replication. Also, the CMV genome encodes a protein and an RNA that have the ability to inhibit apoptosis in infected cells. Inhibition of apoptosis allows the infected cell to survive.

Three of the herpesviruses, HSV types 1 and 2 and VZV, cause a **vesicular rash**, both in primary infections and in reactivations. Primary infections are usually more severe than reactivations. The other two herpesviruses, CMV and EBV, do not cause a vesicular rash.

Four herpesviruses, namely HSV types 1 and 2, VZV, and CMV, induce the formation of **multinucleated giant cells**, which can be seen microscopically in the lesions. The importance of giant cells is best illustrated by the Tzanck smear, which reveals multinucleated giant cells in a smear taken from the painful vesicles of the genitals caused by HSV type 2 (Figure 37–2).

The herpesvirus family can be subdivided into three categories based on the type of cell most often infected and the site of latency. The alpha herpesviruses, consisting of HSV types 1 and 2 and VZV, infect epithelial cells primarily and cause latent infection in neurons. The beta herpesviruses, consisting of CMVs and HHV-6, infect and become latent in a variety of tissues. The gamma herpesviruses, consisting of EBV and human herpesvirus 8 (HHV-8, Kaposi's sarcoma–associated virus), infect and become latent primarily in lymphoid cells. Table 37–4 describes some important clinical features of the common herpesviruses.

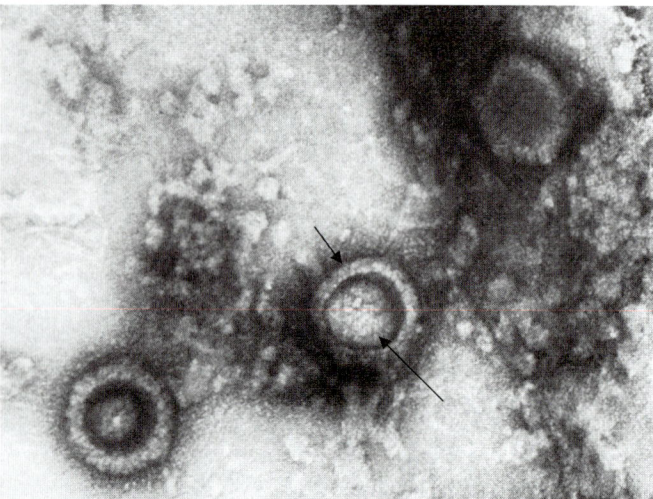

FIGURE 37–1 Herpes simplex virus (HSV). Electron micrograph. Three HSV virions are visible. Short arrow points to the envelope of an HSV virion. Long arrow points to the nucleocapsid of the virion. The dark area between the inner nucleocapsid and the outer envelope is the tegument. (Used with permission from Dr. John Hierholzer, Public Health Image Library, Centers for Disease Control and Prevention.)

TABLE 37–3 Important Features of Common Herpesvirus Infections

Virus	Primary Infection	Usual Site of Latency	Recurrent Infection	Route of Transmission
HSV-1	Gingivostomatitis[1]	Cranial sensory ganglia	Herpes labialis,[2,3] encephalitis, keratitis	Via respiratory secretions and saliva
HSV-2	Herpes genitalis, perinatal disseminated disease	Lumbar or sacral sensory ganglia	Herpes genitalis[2,3]	Sexual contact, perinatal infection
VZV	Varicella	Cranial or thoracic sensory ganglia	Zoster[2]	Via respiratory secretions
EBV	Infectious mononucleosis[1]	B lymphocytes	Asymptomatic shedding[3,4]	Via respiratory secretions and saliva
CMV	Congenital infection (in utero), mononucleosis[1]	Monocytes	Asymptomatic shedding[2]	Intrauterine infection, transfusions, sexual contact, via secretions (e.g., saliva and urine)
HHV-8[5]	Uncertain[6]	Uncertain	Kaposi's sarcoma	Sexual or organ transplantation

CMV = cytomegalovirus; EBV = Epstein–Barr virus; HHV-8 = human herpesvirus 8; HSV = herpes simplex virus; VZV = varicella-zoster virus.

[1]Primary infection is often asymptomatic.

[2]In immunocompromised patients, dissemination of virus can cause life-threatening disease.

[3]Asymptomatic shedding also occurs.

[4]Latent EBV infection predisposes to B-cell lymphomas.

[5]Also known as Kaposi's sarcoma–associated herpesvirus.

[6]A mononucleosis-like syndrome has been described. Kaposi's sarcoma itself also can result from a primary infection.

Certain herpesviruses are associated with or cause cancer in humans (e.g., EBV is associated with Burkitt's lymphoma and nasopharyngeal carcinoma, and HHV-8 causes KS). Several herpesviruses cause cancer in animals (e.g., leukemia in monkeys and lymphomatosis in chickens) (see Chapter 43).

HERPES SIMPLEX VIRUSES (HSV)

HSV type 1 (HSV-1) and type 2 (HSV-2) are distinguished by two main criteria: antigenicity and location of lesions. Lesions caused by HSV-1 are, in general, above the waist, whereas those caused by HSV-2 are below the waist. Table 37–5 describes some

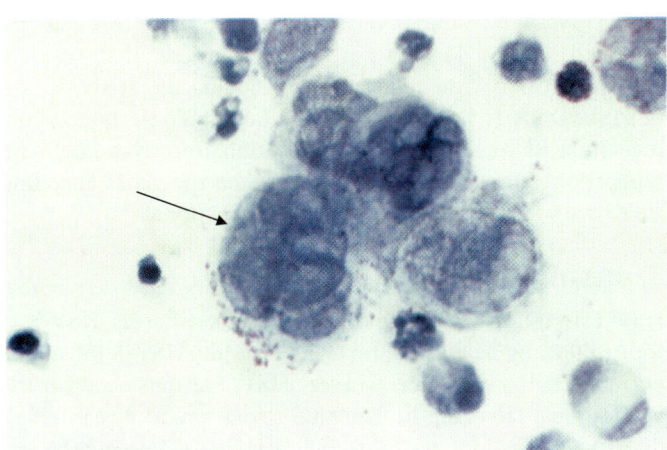

FIGURE 37–2 Herpes simplex virus type 2. Multinucleated giant cells in Tzanck smear. Arrow points to a multinucleated giant cell with approximately eight nuclei. (Used with permission from Dr. Joe Miller, Public Health Image Library, Centers for Disease Control and Prevention.)

important differences between the diseases caused by HSV-1 and HSV-2.

Diseases

HSV-1 causes acute gingivostomatitis, recurrent herpes labialis (cold sores), keratoconjunctivitis (keratitis), and encephalitis, primarily in adults. HSV-2 causes herpes genitalis (genital herpes), neonatal encephalitis and other forms of neonatal herpes, and aseptic meningitis. Infection by HSV-1 or HSV-2 is a common cause of erythema multiforme.

Important Properties

HSV-1 and HSV-2 are structurally and morphologically indistinguishable. They can, however, be differentiated by the restriction endonuclease patterns of their genome DNA and by type-specific monoclonal antisera against glycoprotein G. Humans are the natural hosts of both HSV-1 and HSV-2.

Summary of Replicative Cycle

The cycle begins when HSV-1 binds first to heparan sulfate on the cell surface and then to a second receptor, nectin. Following fusion of the viral envelope with the cell membrane, the nucleocapsid and the tegument proteins are released into the cytoplasm. The viral nucleocapsid is transported to the nucleus, where it docks to a nuclear pore and the genome DNA enters the nucleus along with tegument protein VP-16. The linear genome DNA now becomes circular. VP-16 interacts with cellular transcription factors to activate transcription of viral IE genes by host cell RNA polymerase. IE mRNA is translated into IE proteins that are transported back to the nucleus where they activate the synthesis of early viral proteins such as the **DNA**

TABLE 37–4 Clinical Features of Herpesviruses

Virus	Giant Cells Produced	Fetal or Neonatal Disease Important	Important Laboratory Diagnostic Technique	Antiviral Therapy Commonly Used	Vaccine Available
HSV-1	Yes	No	PCR (NAAT), culture, Tzanck smear (multinucleated giant cells)	Acyclovir[1]	No
HSV-2	Yes	Yes	PCR (NAAT), culture, Tzanck smear (multinucleated giant cells)	Acyclovir	No
VZV	Yes	No	PCR (NAAT), culture, Tzanck smear (multinucleated giant cells)	Acyclovir[2]	Yes
CMV	Yes	Yes	PCR (NAAT), culture, "owl's eye" intranuclear inclusions, pp65 antigen in leukocytes	Ganciclovir[3]	No
EBV	No	No	Heterophile antibody (Monospot test), IgM antibody to viral capsid antigen (VCA), atypical lymphocytes	None	No
HHV-8	No	No	Biopsy of lesions, PCR (NAAT)	Alpha interferon	No

CMV = cytomegalovirus; EBV = Epstein-Barr virus; HHV-8 = human herpes virus 8; HSV = herpes simplex virus; NAAT= nucleic acid amplification test; PCR = polymerase chain reaction; VZV = varicella-zoster virus.

[1]Not used in recurrent herpes labialis.

[2]Not used in varicella in immunocompetent children.

[3]Used in CMV retinitis and other severe forms of disease.

polymerase that replicates the genome, and **thymidine kinase**. These two proteins are important because they are involved in the action of **acyclovir**, which is the most important drug effective against HSV.

Note that early protein synthesis by HSV can be subdivided into two categories: *immediate early* and *early*. IE proteins are those whose mRNA synthesis is activated by a protein brought in by the incoming parental virion (i.e., no new viral protein synthesis is required for the production of the five IE proteins). The early proteins, on the other hand, do require the synthesis of new viral regulatory proteins to activate the transcription of their mRNAs.

TABLE 37–5 Comparison of Diseases Caused by HSV-1 and HSV-2

Site	Disease Caused by HSV-1	Disease Caused by HSV-2
Skin	Vesicular lesions above the waist	Vesicular lesions below the waist (especially genitals)
Mouth	Gingivostomatitis	Rare
Eye	Keratoconjunctivitis	Rare
Central nervous system	Encephalitis (temporal lobe)	Meningitis
Neonate	Rare[1]	Skin lesions, encephalitis, and disseminated infection[2]
Dissemination to viscera in immunocompromised patients	Yes	Rare

HSV = herpes simplex virus.

[1]Infection acquired after birth from HSV-1–infected person.

[2]Infection acquired during passage through birth canal.

The viral DNA polymerase replicates the circular genome DNA using a "rolling circle" process. This produces a chain of multiple copies of HSV progeny DNA consisting of head-to-tail linked genomes called a concatenate. The concatenate is cleaved into single pieces of progeny DNA by a virion-encoded enzyme called a **terminase**. The terminase then interacts with the portal protein to guide the progeny DNA into the newly formed capsid (see Figure 29–6).

When progeny viral DNA is made, early protein synthesis is shut off and late protein synthesis begins. These late, structural proteins are transported to the nucleus, where assembly of the nucleocapsid occurs. The nucleocapsid obtains its envelope by budding through the nuclear membrane and the mature progeny virion exits the cell via tubules or vacuoles that communicate with the exterior.

In latently infected cells, such as HSV-infected neurons, circular HSV DNA resides in the nucleus and is not integrated into cellular DNA. VP-16 does not initiate synthesis of IE m-RNA and proteins so viral replication cannot occur. Transcription of HSV DNA is limited to several **LATS**. These noncoding, regulatory RNAs suppress viral replication. Reactivation of viral replication can occur at a later time when the genes encoding LATS are silenced and VP-16 becomes functional.

Transmission & Epidemiology

HSV-1 is transmitted primarily in **saliva**, whereas HSV-2 is transmitted by **sexual contact**. As a result, HSV-1 infections occur mainly on the face, whereas HSV-2 lesions occur in the genital area. However, oral–genital sexual practices can result in HSV-1 infections of the genitals and HSV-2 lesions in the oral cavity (this occurs in about 10%–20% of cases). Although transmission occurs most often when active lesions are present, **asymptomatic shedding** of both HSV-1 and HSV-2 does occur and plays an important role in transmission.

The number of HSV-2 infections has markedly increased in recent years, whereas that of HSV-1 infections has not. Roughly 80% of people in the United States are infected with HSV-1, and 40% have recurrent herpes labialis. Most primary infections by HSV-1 occur in childhood, as evidenced by the early appearance of antibody. In contrast, antibody to HSV-2 does not appear until the age of sexual activity.

Pathogenesis & Immunity

HSV replicates in the skin or mucous membrane at the initial site of infection. Progeny virus infects adjacent neurons and the nucleocapsid migrates up the axon by retrograde axonal flow to the nucleus of the neuron. The viral DNA enters the nucleus, becomes a closed circular DNA and remains as an episome in the nucleus. The viral DNA does *not* integrate into human DNA.

HSV becomes **latent in sensory ganglion cells**. In general, HSV-1 becomes latent in the **trigeminal ganglia**, whereas HSV-2 becomes latent in the **lumbar** and **sacral ganglia**.

The virus can be reactivated from the latent state by a variety of inducers (e.g., sunlight, hormonal changes, trauma, stress, and fever), at which time it migrates down the neuron and replicates in the skin, causing lesions.

The typical skin lesion caused by both HSV-1 and HSV-2 is a **vesicle** that contains serous fluid filled with virus particles and cell debris. When the vesicle ruptures, virus is liberated and can be transmitted to other individuals. **Multinucleated giant cells** are typically found at the base of herpesvirus lesions.

The skin lesions progress from erythema to papules, then vesicles, followed by ulcers, and ending with crusts. Prodromal itching or tingling can occur. Frequent recurrences are more often seen in HSV-2 infections than in HSV-1 infections.

Immunity is type-specific, but some cross-protection exists. For example, patients with preexisting antibody to HSV-1 frequently have asymptomatic infections with HSV-2. However, immunity is incomplete, and both reinfection and reactivation occur in the presence of circulating IgG. **Cell-mediated immunity** (CMI) is important in limiting herpesviruses, because reduced CMI often results in reactivation, spread, and severe disease.

Clinical Findings

1. HSV-1

HSV-1 causes several forms of primary and recurrent disease:

(1) **Gingivostomatitis** occurs primarily in children and is characterized by fever, irritability, and vesicular lesions in the mouth. The primary disease is more severe and lasts longer than recurrences. The lesions heal spontaneously in 2 to 3 weeks. Many children have asymptomatic primary infections.

(2) **Orolabial herpes** (herpes labialis, fever blisters, or cold sores) is the milder, recurrent form and is characterized by crops of vesicles, usually at the mucocutaneous junction of the lips or nose (Figure 37–3). Recurrences frequently reappear at the same site.

(3) **Keratoconjunctivitis** is characterized by corneal ulcers and lesions of the conjunctival epithelium. Recurrences can lead to scarring and blindness.

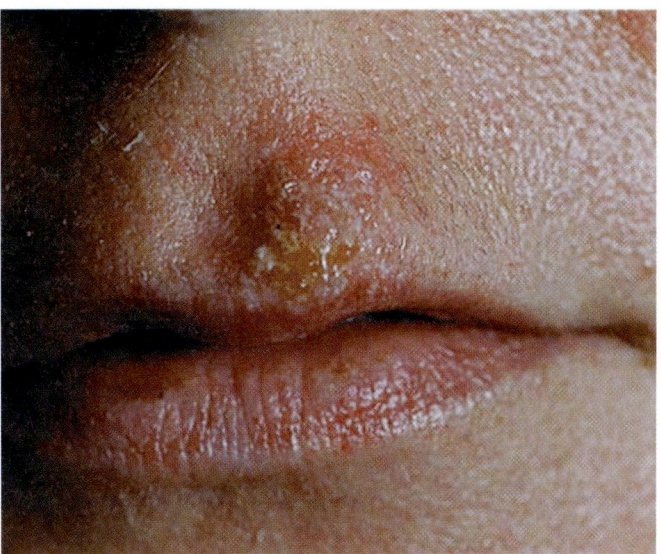

FIGURE 37–3 Herpes labialis. Note vesicles on upper lip adjacent to the vermillion border of the lip caused by herpes simplex virus type 1. (Used with permission from Jack Resneck, Sr., MD.)

(4) **Encephalitis** caused by HSV-1 is characterized by a necrotic lesion in one temporal lobe. Fever, headache, vomiting, seizures, and altered mental status are typical clinical features. The onset may be acute or protracted over several days. The disease occurs as a result of either a primary infection or a recurrence. Magnetic resonance imaging often reveals the lesion. Examination of the spinal fluid typically shows a moderate increase of lymphocytes, a moderate elevation in the amount of protein, and a normal amount of glucose. HSV-1 encephalitis has a high mortality rate and causes severe neurologic sequelae in those who survive.

(5) **Herpetic whitlow** is a pustular lesion of the skin of the finger or hand. It can occur in medical personnel as a result of contact with patient's lesions.

(6) **Herpes gladiatorum**, as the name implies, occurs in wrestlers and others who have close body contact. It is caused primarily by HSV-1 and is characterized by vesicular lesions on the head, neck, and trunk.

(7) **Eczema herpeticum** (Kaposi's varicelliform eruption) is an infection of the skin of a patient with atopic dermatitis. Vesicular lesions are seen at the site of the atopic dermatitis (eczema). Most cases occur in children.

(8) **Disseminated infections**, such as esophagitis and pneumonia, occur in immunocompromised patients with depressed T-cell function.

(9) Some cases of **genital herpes** are caused by HSV-1 as a result of oral–genital contact.

2. HSV-2

HSV-2 causes several diseases, both primary and recurrent:

(1) **Genital herpes** (herpes genitalis) is characterized by painful vesicular lesions of the male and female genitals and anal area (Figure 37–4). The lesions are more severe and

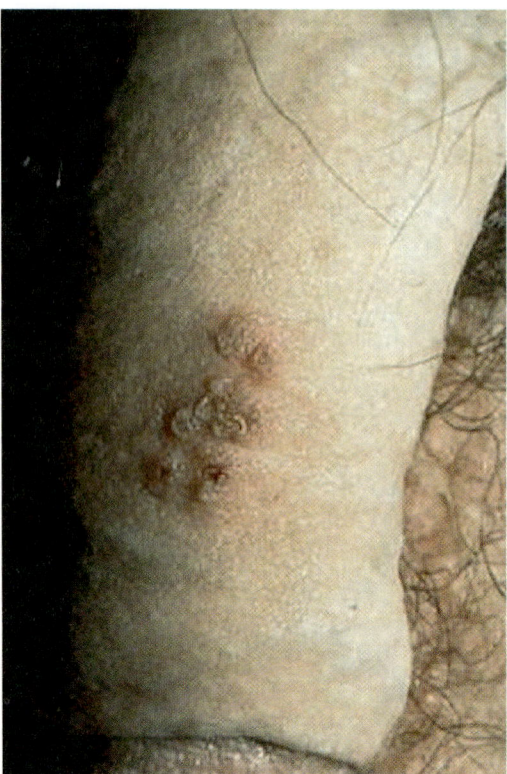

FIGURE 37–4 Herpes genitalis. Note vesicles on shaft of penis caused by herpes simplex virus type 2. (Used with permission from Jack Resneck, Sr., MD.)

protracted in primary disease than in recurrences. Primary infections are associated with fever and inguinal adenopathy.

Many infections are asymptomatic (i.e., many people have antibody to HSV-2 but have no history of disease). Asymptomatic infections occur in both men (in the prostate or urethra) and women (in the cervix). Shedding of virus from those with an asymptomatic infection occurs sporadically and can be a source of infection of other individuals.

Approximately 80% to 90% of herpes genitalis cases are caused by HSV-2. The remainder is caused by HSV-1 as a result of oral–genital contact. The clinical importance of this is that suppressive chemoprophylaxis for HSV-2 lesions should be considered because lesions caused by HSV-2 are more likely to recur than lesions caused by HSV-1. Approximately 70% of patients will have a recurrence within 1 year after their primary HSV-2 infection. The typical patient with genital herpes has four to five recurrences each year, whereas the typical patient with HSV-1 infection has only one recurrence.

(2) **Neonatal herpes** originates chiefly from contact with vesicular lesions within the birth canal. In some cases, although there are no visible lesions, HSV-2 is shed into the birth canal (asymptomatic shedding) and can infect the child during birth. Women who acquire a primary HSV-2 infection in the third trimester of pregnancy have the highest rate of giving birth to a neonate who will manifest symptomatic neonatal herpes.

Neonatal herpes varies from severe disease (e.g., disseminated lesions or encephalitis) to milder local lesions (skin, eye, mouth) to asymptomatic infection. Neonatal disease may be prevented by performing cesarean section on women with either active lesions or positive viral cultures. Both HSV-1 and HSV-2 can cause severe neonatal infections that are acquired after birth from carriers handling the child. Despite their association with neonatal infections, neither HSV-1 nor HSV-2 causes congenital abnormalities to any significant degree.

Serious neonatal infection is more likely to occur when the mother is experiencing a primary herpes infection than a recurrent infection for two reasons: (1) the amount of virus produced during a primary infection is greater than during a secondary infection and (2) mothers who have been previously infected can pass IgG across the placenta, which can protect the neonate from serious disseminated infection.

(3) Aseptic meningitis caused by HSV-2 is usually a mild, self-limited disease with few sequelae.

Both HSV-1 and HSV-2 infections are associated with erythema multiforme. The rash of erythema multiforme appears as a central red area surrounded by a ring of normal skin outside of which is a red ring ("target" or "bull's eye" lesion). The lesions are typically macular or papular and occur symmetrically on the trunk, hands, and feet. The rash is thought to be an immune-mediated reaction to the presence of HSV antigens. Acyclovir is useful in preventing recurrent episodes of erythema multiforme, probably by reducing the amount of HSV antigens. Many drugs, especially sulfonamides among the antimicrobial drugs, commonly cause erythema multiforme. Other prominent infectious causes include *Mycoplasma pneumoniae* and viruses such as hepatitis B virus and hepatitis C virus.

Erythema multiforme major, also known as **Stevens-Johnson syndrome**, is characterized by fever, erosive oral lesions, and extensive desquamating skin lesions. *M. pneumoniae* infection is the most common infectious cause of Stevens-Johnson syndrome.

Laboratory Diagnosis

The polymerase chain reaction (PCR, NAAT) assay and viral culture are typically used for diagnosis. The PCR (NAAT) assay for viral DNA is more rapid and sensitive than viral culture and is able to promptly distinguish infection by HSV-1 from HSV-2.

Regarding viral culture, the typical cytopathic effect occurs in 1 to 3 days, after which the virus is identified by fluorescent antibody staining of the infected cells or by detecting virus-specific glycoproteins in enzyme-linked immunosorbent assays (ELISAs). HSV-1 can be distinguished from HSV-2 by using monoclonal antibody against glycoprotein G often in an ELISA test.

A rapid presumptive diagnosis can be made from skin lesions by using the **Tzanck smear**, in which cells from the base of the vesicle are stained with Giemsa stain. The presence of multinucleated giant cells suggests herpesvirus infection (see Figure 37–2). In addition, vesicular fluid from skin lesions can be used to detect herpesvirus-infected cells by using a direct fluorescent antibody (DFA) test.

If herpes encephalitis is suspected, a rapid diagnosis can be made by detecting HSV DNA in the spinal fluid by using a

TABLE 37–6 Use of Acyclovir, Valacyclovir, Famciclovir, Ganciclovir, and Valganciclovir to Treat HSV, VZV, and CMV Infections

Type of Infection	Drugs Used	Route of Administration
Orolabial herpes	Acyclovir, valacyclovir, famciclovir[1]	Oral
Genital herpes	Acyclovir, valacyclovir, famciclovir[1]	Oral
HSV encephalitis	Acyclovir	IV
HSV in neonate	Acyclovir	IV
HSV in immunocompromised	Acyclovir	IV
HSV keratoconjunctivitis	Ganciclovir,[2] trifluridine	Topical
VZV-Varicella (chicken pox)	None or acyclovir or valacyclovir	Oral
VZV-Zoster	Acyclovir, valacyclovir, famciclovir	Oral
VZV in immunocompromised	Acyclovir	IV
CMV retinitis, colitis, esophagitis	Ganciclovir	IV
CMV prevention in transplant recipients	Valganciclovir	Oral

[1]These drugs are also useful to prevent recurrences.

[2]Topical acyclovir is not available in the United States.

PCR assay. The diagnosis of neonatal herpes infection typically involves the use of viral cultures or PCR assay.

Serologic tests such as the neutralization test can be used in the diagnosis of primary infections because a significant rise in antibody titer is readily observed. However, they are of no use in the diagnosis of recurrent infections because many adults already have circulating antibodies, and recurrences rarely cause a rise in antibody titer.

Treatment

Acyclovir (acycloguanosine, Zovirax) is the drug of choice for orolabial herpes, encephalitis, and systemic disease caused by HSV-1 (Table 37–6). It is also useful for the treatment of primary and recurrent genital herpes. It **shortens the duration** of the lesions and **reduces the extent of shedding** of the virus but does *not* cure the latent state.

Acyclovir is also used to treat neonatal infections caused by HSV-2. Treatment of meningitis caused by HSV-2 in adults typically does not require antiviral drugs but acyclovir is used in severe cases. Treatment of genital herpes with acyclovir in a pregnant woman appears to be safe for both mother and developing fetus. Mutants of HSV-1 resistant to acyclovir have been isolated from patients; foscarnet can be used in these cases.

For HSV-1 eye infections, other nucleoside analogues (e.g., trifluridine [Viroptic]) are used topically. Oral acyclovir is also used for HSV keratitis. Penciclovir (a derivative of acyclovir) or

docosanol (a long-chain saturated alcohol) can be used to treat recurrences of orolabial HSV-1 infections in immunocompetent adults. Oral valacyclovir (Valtrex) and famciclovir (Famvir) are used in the treatment of genital herpes and in the suppression of recurrences.

Note that no drug treatment of the primary infection prevents establishment of the latent state. Drugs also do **not eradicate the latent state**, but prophylactic, long-term administration of acyclovir, valacyclovir, or famciclovir can suppress clinical recurrences.

Prevention

Valacyclovir (Valtrex) and famciclovir (Famvir) are used in the suppression of recurrent lesions, especially in those with frequent recurrences caused by HSV-2. Some studies have shown that valacyclovir is more effective than famciclovir in the suppression of recurrences of genital lesions. Suppressive chemoprophylaxis also reduces shedding of the virus and, as a result, transmission to others. Prevention also involves avoiding contact with the vesicular lesion or ulcer by using a condom. Cesarean section is recommended for women who are at term and who have genital lesions or positive viral cultures. Circumcision reduces the risk of infection by HSV-2. There is no vaccine against HSV-1 or HSV-2.

VARICELLA-ZOSTER VIRUS (VZV)

Disease

Varicella (chickenpox) is the primary disease; zoster (shingles) is the recurrent form.

Important Properties

VZV is structurally and morphologically similar to other herpesviruses but is antigenically different. It has a single serotype. The same virus causes both varicella and zoster. Humans are the natural hosts.

Summary of Replicative Cycle

The cycle is similar to that of HSV (see page 287).

Transmission & Epidemiology

The virus is transmitted by **respiratory droplets** and by direct contact with the lesions. Varicella is a highly contagious disease of childhood; more than 90% of people in the United States have antibody by age 10 years. Varicella occurs worldwide. Prior to 2001, there were more cases of chickenpox than any other notifiable disease, but the widespread use of the vaccine has significantly reduced the number of cases.

There is infectious VZV in zoster vesicles. This virus can be transmitted, usually by direct contact, to children and can cause varicella. The appearance of either varicella or zoster in a hospital is a major infection control problem because the virus can be transmitted to immunocompromised patients and causes life-threatening disseminated infection.

Pathogenesis & Immunity

VZV infects the mucosa of the upper respiratory tract, and then spreads via the blood to the skin, where the typical **vesicular rash** occurs. **Multinucleated giant cells** with intranuclear inclusions are seen in the base of the lesions. The virus infects sensory neurons and is carried by retrograde axonal flow into the cells of the **dorsal root ganglia**, where the virus becomes **latent**.

In latently infected cells, VZV DNA is located in the nucleus and is not integrated into cellular DNA. Later in life, frequently at times of reduced CMI or local trauma, the virus is activated and causes the vesicular skin lesions and **nerve pain** of zoster.

Immunity following varicella is lifelong. A person gets varicella only once, but zoster can occur despite this immunity to varicella. Zoster usually occurs only once. The frequency of zoster increases with advancing age, perhaps as a consequence of waning immunity.

Clinical Findings

Varicella

After an incubation period of 14 to 21 days, brief prodromal symptoms of fever and malaise occur. A papulovesicular rash then appears in crops on the trunk and spreads to the head and extremities (Figure 37–5). The rash evolves from papules to vesicles, pustules, and, finally, crusts. Itching (pruritus) is a prominent symptom, especially when vesicles are present. Varicella is mild in children but more severe in adults. Varicella pneumonia and encephalitis are the major rare complications, occurring more often in adults. **Reye's syndrome**, characterized by encephalopathy and liver degeneration, is associated with VZV and influenza B virus infection, especially in children given aspirin. Its pathogenesis is unknown.

Zoster

The occurrence of painful vesicles along the course of a sensory nerve of the head or trunk is the usual picture (Figure 37–6). The pain can last for weeks, and **postherpetic neuralgia (PNH)** can be debilitating. In immunocompromised patients, especially in stem-cell transplant patients, life-threatening disseminated infections such as pneumonia can occur.

Laboratory Diagnosis

Although most diagnoses of varicella and zoster are made clinically, laboratory tests are available. PCR tests to detect viral DNA and DFA tests on skin lesion specimens are often used.

A presumptive diagnosis can be made by using the Tzanck smear. Multinucleated giant cells are seen in VZV as well as HSV lesions (see Figure 37–2). The definitive diagnosis can also be made by isolation of the virus in cell culture and identification with specific antiserum. A rise in antibody titer can be used to diagnose varicella but is less useful in the diagnosis of zoster.

Treatment

No antiviral therapy is necessary for chickenpox or zoster in immunocompetent children, although oral acyclovir started within 24 hours of the onset of the rash decreases the severity of the symptoms (see Table 37–6). Immunocompetent adults with

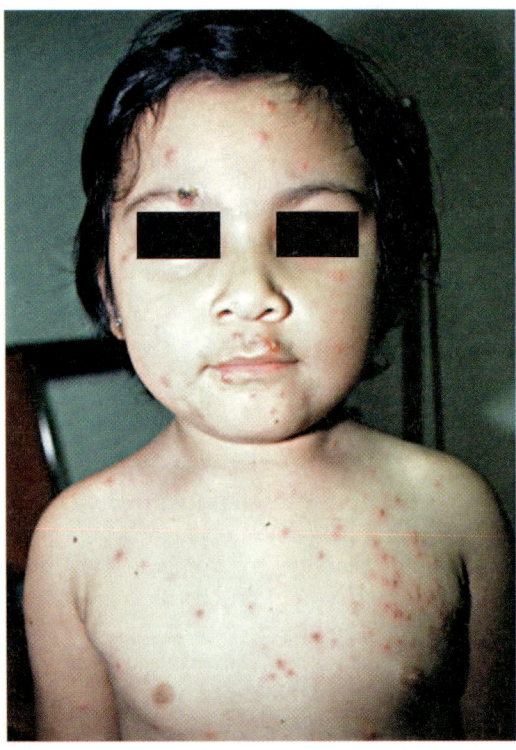

FIGURE 37–5 Varicella (chickenpox). Note vesicles on an erythematous base caused by varicella-zoster virus. (Reproduced with permission from Usatine RP, Smith MA, Mayeaux EJ Jr, et al: *The Color Atlas of Family Medicine.* New York, NY: McGraw Hill; 2009. Photo contributor: Richard P. Usatine, MD.)

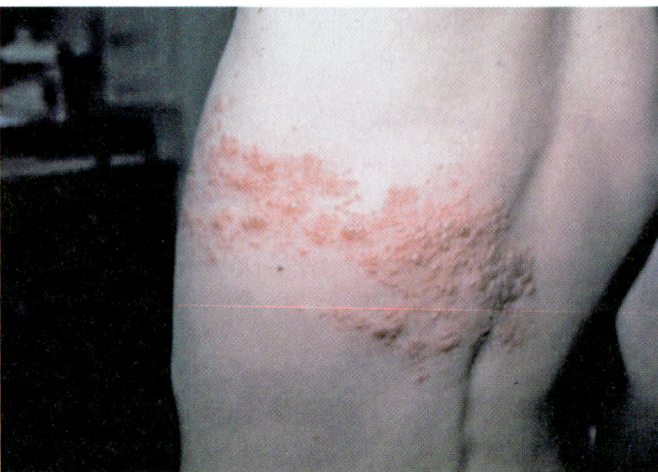

FIGURE 37–6 Zoster (shingles). Note vesicles along the dermatome of a thoracic nerve caused by varicella-zoster virus. (Reproduced with permission from Usatine RP, Smith MA, Mayeaux EJ Jr, et al: *The Color Atlas of Family Medicine.* New York, NY: McGraw Hill; 2009. Photo contributor: Richard P. Usatine, MD.)

TABLE 37–7 Vaccines Against Varicella-Zoster Virus

Name of Vaccine	Immunogen in Vaccine	Purpose of Vaccine	Comment
Varivax	Live, attenuated virus	Prevent varicella in children	Do not give to immunocompromised
Shingrix	Recombinant envelope glycoprotein	Prevent zoster in adults	Can be used in immunocompromised

either moderate or severe cases of chickenpox or zoster often are treated with **acyclovir** because it can reduce the duration and severity of symptoms and reduce the shedding of virus. Immunocompromised children and adults with severe chickenpox, zoster, or disseminated disease should be treated with intravenous acyclovir. Disease caused by acyclovir-resistant strains of VZV can be treated with foscarnet.

Two drugs similar to acyclovir, famciclovir (Famvir), and valacyclovir (Valtrex) can be used in patients with zoster to accelerate healing of the lesions, but none of these drugs can cure the latent state. Whether use of these drugs prevents PNH is uncertain. Treatment of PNH includes drugs that ameliorate neuropathic pain, such as gabapentin and lidocaine patches.

Prevention

There are two vaccines against VZV: one designed to prevent varicella, called Varivax, and one designed to prevent zoster, called Shingrix (Table 37–7). Varivax contains **live, attenuated VZV,** whereas Shingrix is a recombinant vaccine containing the VZV envelope glycoprotein (glycoprotein E) as the immunogen. The zoster vaccine is effective in preventing the symptoms of zoster, but does not eradicate the latent state of VZV. Note that Zostavax, a live attenuated vaccine for the prevention of zoster, has been withdrawn from the market.

The varicella vaccine is recommended for individuals older than 12 months of age, whereas the zoster vaccine is recommended for adults of any age. Because Varivax vaccine contains live virus, it should not be given to immunocompromised people or pregnant women.

Acyclovir is useful in preventing varicella and disseminated zoster in immunocompromised people exposed to the virus. **Varicella-zoster immune globulin** (VZIG), which contains a high titer of antibody to the virus, is also used for such prophylaxis.

CYTOMEGALOVIRUS (CMV)

Diseases

CMV causes cytomegalic inclusion disease (especially congenital abnormalities) in neonates. It is the **most common cause of congenital abnormalities** in the United States. CMV is a very important cause of pneumonia and other diseases in immunocompromised patients such as recipients of bone marrow (stem cell) and solid organ transplants. It also causes heterophil-negative mononucleosis in immunocompetent individuals. CMV infection is very common, ranging from approximately 70% in the United States to almost 100% in parts of Africa.

Important Properties

CMV is structurally and morphologically similar to other herpesviruses but is antigenically different. It has a single serotype. Although CMV has a single serotype (single antigenic type), it has many genotypes. The genotypic changes occur in regions of the viral DNA that do not affect the antigenicity of the surface protein. The large DNA genome is divided into a unique long (UL) and a unique short (US) region

Humans are the natural hosts; animal CMV strains do not infect humans. Giant cells are formed, hence the name *cytomegalo*.

Summary of Replicative Cycle

The cycle is similar to that of HSV (see page 287). One unique feature of CMV replication is that some of its "immediate early proteins" are translated from mRNAs brought into the infected cell by the parental virion rather than being translated from mRNAs synthesized in the newly infected cell.

Like other herpesviruses, CMV encodes a DNA polymerase that is the site of action of ganciclovir and other effective drugs. CMV also encodes a phosphokinase (the product of the UL97 gene) that activates ganciclovir. CMV also encodes proteins that prevent HLA class-1 proteins from reaching the cell surface thereby preventing cytotoxic T cells from killing CMV-infected cells.

Replication of CMV progeny DNA involves production of a chain of DNA (called a concatenate) that is cleaved into single pieces of progeny DNA by a virion-encoded enzyme called a terminase. The drug letermovir (Prevymis) inhibits the terminase and thereby inhibits CMV replication.

Transmission & Epidemiology

CMV is transmitted by a **variety of modes**. Early in life, it is transmitted across the placenta, within the birth canal, and quite commonly in breast milk. In young children, its most common mode of transmission is via saliva. Later in life, it is transmitted sexually; it is present in both semen and cervical secretions. It can also be transmitted during blood transfusions and organ transplants. CMV infection occurs worldwide, and more than 80% of adults have antibody against this virus.

Pathogenesis & Immunity

Infection of the fetus can cause **cytomegalic inclusion disease**, characterized by multinucleated giant cells with prominent intranuclear inclusions. Many organs are affected, and widespread congenital abnormalities result. Infection of the fetus occurs mainly when a **primary infection** occurs in the pregnant woman (i.e., when she has no antibodies that will neutralize the virus before it can infect the fetus). The fetus usually will not be infected if the pregnant woman has antibodies against the virus. Congenital abnormalities are **more common when a fetus is infected during the first trimester** than later in gestation, because the first trimester is when development of organs occurs and the death of any precursor cells can result in congenital defects.

Infections of children and adults are usually asymptomatic, except in immunocompromised individuals. CMV enters a **latent** state primarily in monocytes and can be reactivated when CMI is decreased. CMV can also persist in kidneys for years. Reactivation of CMV from the latent state in cervical cells can result in infection of the newborn during passage through the birth canal.

CMV has a specific mechanism of "immune evasion" that allows it to maintain the latent state for long periods. In CMV-infected cells, assembly of the major histocompatibility complex (MHC) class I–viral peptide complex is unstable, so viral antigens are not displayed on the cell surface and killing by cytotoxic T cells does not occur. In addition, CMV encodes several microRNAs, one of which binds to and prevents the translation of the cell's mRNA for the class I MHC protein. This prevents viral proteins from being displayed on the infected cell surface, and killing by cytotoxic T cells does not occur.

CMV also encodes a protein that functions as a chemokine receptor. When this protein is released from CMV-infected cells, the protein binds to chemokines thereby preventing the chemokines from serving as a signal for host immune cells to migrate to the site of CMV infection.

CMV infection causes an immunosuppressive effect by inhibiting T cells. Host defenses against CMV infection include both circulating antibody and CMI. Cellular immunity is more important, because its suppression can lead to systemic disease.

Clinical Findings

Most CMV infections are asymptomatic but it is an important cause of congenital abnormalities and serious systemic diseases in immunocompromised individuals.

Approximately 20% of infants infected with CMV during gestation show clinically apparent manifestations of cytomegalic inclusion disease such as microcephaly, seizures, deafness, jaundice, and purpura. The purpuric lesions are often called "blueberry muffin" lesions. Hepatosplenomegaly is very common. Cytomegalic inclusion disease is one of the leading causes of mental retardation in the United States. Infected infants can continue to excrete CMV, especially in the urine, for several years.

In immunocompetent adults, CMV can cause **heterophil-negative mononucleosis**, which is characterized by fever, lethargy, and the presence of abnormal lymphocytes in peripheral blood smears.

In immunocompromised patients, systemic CMV infections, especially pneumonitis, esophagitis, and hepatitis, occur in a high proportion of those individuals (e.g., those with renal and stem-cell transplants). In patients with AIDS, CMV commonly infects the intestinal tract and causes intractable colitis with diarrhea. CMV also causes retinitis in AIDS patients, which can lead to blindness. Anemia and thrombocytopenia often occur.

Laboratory Diagnosis

PCR-based assays for CMV DNA or viral m-RNA in tissue or body fluids, such as spinal fluid and amniotic fluid, are often used. The PCR assays are highly sensitive, specific, and provide results rapidly.

Another approach involves culturing the virus in special tubes called **shell vials** coupled with the use of immunofluorescent antibody to detect the immediate early antigen (EA) of CMV. This shell vial culture technique can make a diagnosis as early as 72 hours. The advantage of culturing is that the virus obtained in the culture can then be used to determine the drug susceptibility to ganciclovir.

Other diagnostic methods include fluorescent antibody and histologic staining of inclusion bodies in giant cells in urine and in tissue. The inclusion bodies are intranuclear and have an oval **owl's eye** shape (Figure 37–7). Serologic tests that detect IgM antibody in the patient's serum are also useful to determine recent infection.

CMV antigenemia can be measured by detecting pp65 within blood leukocytes using an immunofluorescence assay. pp65 is a protein located in the nucleocapsid of CMV and can be identified within infected leukocytes using fluorescein-labeled monoclonal antibody specific for pp65.

Treatment

Ganciclovir (Cytovene) is moderately effective in the treatment of CMV retinitis and pneumonia in patients with AIDS (see Table 37–6). Valganciclovir, which can be taken orally, is also effective against CMV retinitis. CMV strains resistant to ganciclovir and valganciclovir have emerged, mostly due to mutations in the gene (UL97) that encodes the phosphokinase. Drug susceptibility testing can be done.

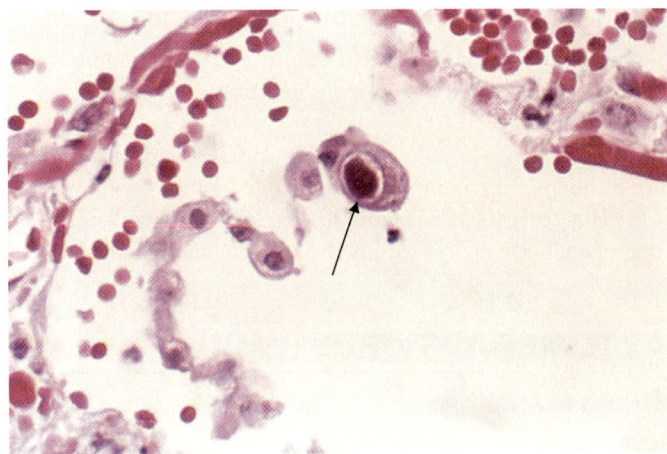

FIGURE 37–7 Cytomegalovirus. Owl's eye inclusion body. Arrow points to an "owl's eye" inclusion body in the nucleus of an infected cell. (Used with permission from Dr. Edwin Ewing, Jr., Public Health Image Library, Centers for Disease Control and Prevention.)

Foscarnet (Foscavir) is useful to treat disease in patients infected with ganciclovir-resistant CMV but it causes more side effects. Unlike HSV and VZV, CMV is largely resistant to acyclovir. Cidofovir (Vistide) is also useful in the treatment of CMV retinitis. Fomivirsen (Vitravene) is an antisense DNA approved for the intraocular treatment of CMV retinitis.

Prevention

There is no vaccine. Ganciclovir can suppress progressive retinitis in AIDS patients. It is also useful to prevent disease in transplant patients. Letermovir, an inhibitor of the terminase of CMV, is used to prevent recurrences in patients who have received a hematopoietic stem cell transplant.

Infants with cytomegalic inclusion disease who are shedding virus in their urine should be kept isolated from other infants. Blood for transfusion to newborns should be CMV antibody-negative. If possible, only organs from CMV antibody-negative donors should be transplanted to antibody-negative recipients. A high-titer immune globulin preparation (CytoGam) is used to prevent disseminated CMV infections in organ transplant patients.

EPSTEIN–BARR VIRUS (EBV)

Diseases

EBV causes infectious mononucleosis. It is associated with Burkitt's lymphoma, other B-cell lymphomas, and nasopharyngeal carcinoma. EBV also causes hairy leukoplakia and post-transplant lymphoproliferative disorder (PTLD) (lymphoma). EBV is the most common cause of hemophagocytic lymphohistiocytosis (HLH).

Important Properties

EBV is structurally and morphologically similar to other herpesviruses but is antigenically different. The most important antigen is the **viral capsid antigen** (VCA), because it is used most often in diagnostic tests. EA, which is produced prior to viral DNA synthesis, and Epstein–Barr nuclear antigen (EBNA), which is located in the nucleus bound to chromosomes, are sometimes diagnostically helpful as well. Two other antigens, lymphocyte-determined membrane antigen and viral membrane antigen, have also been detected. Neutralizing activity is directed against the viral membrane antigen.

Humans are the natural hosts. EBV infects mainly lymphoid cells, primarily **B lymphocytes**. EBV also infects the epithelial cells of the pharynx, resulting in the prominent sore throat. In latently infected cells, EBV DNA is in the nucleus and is not integrated into cellular DNA. Some, but not all, genes are transcribed, and only a subset of those is translated into protein.

Summary of Replicative Cycle

The cycle is similar to that of HSV (see page 287). EBV enters B lymphocytes at the site of the receptor for the C3 component of complement.

Transmission & Epidemiology

EBV is transmitted primarily by the exchange of **saliva** (e.g., during kissing). The saliva of people with a reactivation of a latent infection as well as people with an active infection can serve as a source of the virus. In contrast to CMV, blood transmission of EBV is very rare.

EBV infection is one of the most common infections worldwide; more than 90% of adults in the United States have antibody. Infection in the first few years of life is usually asymptomatic. Early infection tends to occur in individuals in lower socioeconomic groups. The frequency of clinically apparent infectious mononucleosis, however, is highest in those who are exposed to the virus later in life (e.g., college students).

Pathogenesis & Immunity

The infection first occurs in the oropharynx and then spreads to the blood, where it infects B lymphocytes. Cytotoxic T lymphocytes react against the infected B cells. The T cells are the "atypical lymphs" seen in the blood smear. EBV remains **latent within B lymphocytes**.

The immune response to EBV infection consists first of IgM antibody to the VCA. IgG antibody to the VCA follows and persists for life. The IgM response is therefore useful for diagnosing acute infection, whereas the IgG response is best for revealing prior infection. Lifetime immunity against second episodes of infectious mononucleosis is based on antibody to the viral membrane antigen.

In addition to the EBV-specific antibodies, nonspecific **heterophil antibodies** are found. The term *heterophil* refers to antibodies that are detected by tests using antigens different from the antigens that induced them. The heterophil antibodies formed in infectious mononucleosis agglutinate sheep or horse red blood cells in the laboratory. (Cross-reacting Forssman antibodies in human serum are removed by adsorption with guinea pig kidney extract prior to agglutination.) Note that these antibodies do not react with any component of EBV. It seems likely that EBV infection modifies a cell membrane constituent such that it becomes antigenic and induces the heterophil antibody. Heterophil antibodies usually disappear within 6 months after recovery. These antibodies are not specific for EBV infection and are also seen in individuals with hepatitis B and serum sickness.

Clinical Findings

Infectious mononucleosis is characterized primarily by fever, sore throat, lymphadenopathy, and splenomegaly. Anorexia and lethargy are prominent. Hepatitis is frequent; encephalitis occurs in some patients. Spontaneous recovery usually occurs in 2 to 3 weeks. Splenic rupture, associated with contact sports such as football, is a feared but rare complication of the splenomegaly.

In addition to the common form of infectious mononucleosis described in the previous paragraph, EBV causes a severe, often fatal, progressive form of infectious mononucleosis that occurs in children with an inherited immunodeficiency called

X-linked lymphoproliferative syndrome. The mutated gene encodes a signal transduction protein required for both T-cell and natural killer-cell function. The mortality rate is 75% by age 10. Bone marrow or cord blood transplants may cure the underlying immunodeficiency.

EBV also causes **hairy leukoplakia**—a whitish, nonmalignant lesion with an irregular "hairy" surface on the lateral side of the tongue (Figure 37–8). It occurs in immunocompromised individuals, especially AIDS patients.

EBV infection is associated with several cancers, namely **Burkitt's lymphoma**, some forms of Hodgkin's lymphoma (especially diffuse large B-cell lymphoma), and **nasopharyngeal carcinoma**. The word *associated* refers to the observation that EBV infection is the initiating event that causes the cells to divide, but that event itself does not cause a malignancy. It requires additional steps for malignant transformation to occur. Reduced CMI predisposes to the uncontrolled growth of the EBV-infected cells.

Another EBV-associated disease is **PTLD**, also known as post-transplant lymphoma. The most common form of PTLD is a B-cell lymphoma. PTLD occurs following both bone marrow transplants and solid organ transplants. The main predisposing factor to PTLD is the immunosuppression required to prevent rejection of the graft. The lymphoma will regress if the degree of immunosuppression is reduced.

EBV is also the most common virus associated with HLH. It is a rare disease that occurs primarily in young children. It manifests with a variety of symptoms including fever, hepatosplenomegaly, and cytopenias. Pathogenesis involves prolonged activation of CD-8 positive cytotoxic T cells that leads to cellular and organ damage. It is a life-threatening disease that requires immunosuppression and, eventually a hematopoietic stem cell transplant.

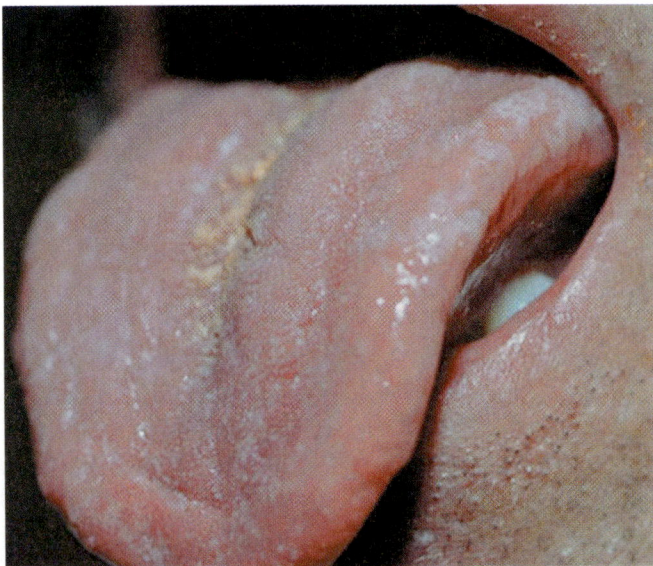

FIGURE 37–8 Hairy leukoplakia. Note whitish plaques on lateral aspect of tongue caused by Epstein–Barr virus. (Reproduced with permission from Wolff K, Johnson R: *Fitzpatrick's Color Atlas & Synopsis of Clinical Dermatology*, 6th ed. New York, NY: McGraw Hill; 2009.)

Laboratory Diagnosis

The diagnosis of infectious mononucleosis in the clinical laboratory is based primarily on serologic tests. The test for IgM antibody to VCA is the preferred test to diagnose acute infection. The Mono Spot test that detects the heterophile antibody is also widely used but false-positive and false-negative results may occur. Note that the heterophile antibody is a nonspecific antibody and occurs in other diseases. More information on these antibodies can be found in the section on Pathogenesis and Immunity above.

There are two types of serologic tests:

(1) The **Monospot test** is an agglutination test used to detect the heterophil antibody. In this test, the patient's serum is mixed with horse red blood cells. If heterophile antibodies are present, the horse red cells are agglutinated. The CDC no longer recommends this test for general use as it can produce false-positive and false-negative results.

(2) The **EBV-specific antibody tests** are commonly used. The IgM VCA antibody response can be used to detect early illness; the IgG VCA antibody response can be used to detect prior infection. In certain instances, antibodies to EA and EBNA can be useful diagnostically.

In addition to serologic tests, the diagnosis of infectious mononucleosis is aided by examination of the patient's blood. An absolute lymphocytosis occurs, and as many as 30% of abnormal lymphocytes are seen on a stained smear. These **atypical lymphs** are enlarged, have an expanded nucleus, and have an abundant, often vacuolated cytoplasm (Figure 37–9). They are cytotoxic T cells that are reacting against the EBV-infected B cells.

In difficult-to-diagnose EBV infections or in diseases such as PTLD, PCR-based assays can be used to detect EBV DNA in the patient's blood.

Although EBV can be isolated from clinical samples such as saliva by morphologic transformation of cord blood lymphocytes, it is a technically difficult procedure and is not readily available. No virus is synthesized in the cord lymphocytes; its presence is detected by fluorescent antibody staining of the nuclear antigen.

Treatment & Prevention

No antiviral therapy is necessary for uncomplicated infectious mononucleosis. Acyclovir has little activity against EBV, but administration of high doses may be useful in life-threatening EBV infections. There is no EBV vaccine.

Association with Cancer

EBV infection is associated with cancers of lymphoid origin: **Burkitt's lymphoma** in African children, other B-cell lymphomas, nasopharyngeal carcinoma in the Chinese population, and thymic carcinoma in the United States. The initial evidence of an association of EBV infection with Burkitt's lymphoma was the production of EBV by the lymphoma cells in culture. In fact, this was how EBV was discovered by Epstein and Barr in 1964. Additional evidence includes the finding of EBV DNA and EBNA

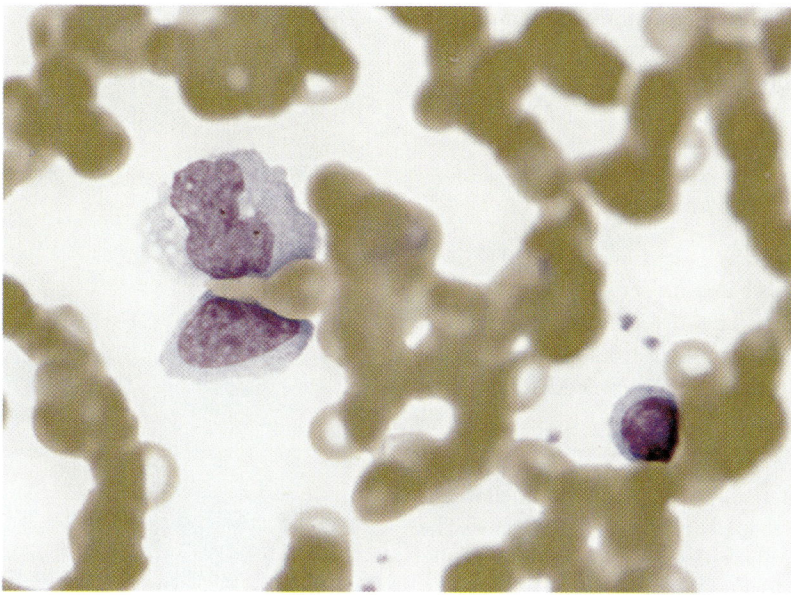

FIGURE 37–9 Atypical lymphocytes in infectious mononucleosis. Note two atypical lymphocytes, each with an enlarged nucleus and abundant cytoplasm on the left side. The lymphocyte on the right side appears normal. (Reproduced with permission from Fauci AS, Braunwald E, Kasper DL et al: *Harrison's Principles of Internal Medicine*, 17th ed. New York, NY: McGraw Hill; 2008.)

in the cells of nasopharyngeal and thymic carcinomas. EBV can induce malignant transformation in B lymphocytes in vitro.

In Burkitt's lymphoma, oncogenesis is a function of the **translocation of the *c-myc* oncogene** to a site adjacent to an immunoglobulin gene promoter. This enhances synthesis of the c-myc protein, a potent oncoprotein. The c-myc protein is a transcriptional regulator that enhances the synthesis of kinases that activate the cell cycle.

HUMAN HERPESVIRUS 6

HHV-6 is the cause of roseola infantum (also known as exanthem subitum). Roseola is a common disease in young children and is described more fully in Chapter 39 on Important Childhood Viruses.

HHV-6 is lymphotropic and infects both T and B cells. It remains latent within these cells but can reactivate in immunocompromised patients and cause pneumonia, encephalitis, or hepatitis. It is one of the most common causes of encephalitis in patients who have received a stem-cell transplant. Many virologic and clinical features of HHV-6 are similar to those of CMV, another member of the herpesvirus family.

HUMAN HERPESVIRUS 8 (KAPOSI'S SARCOMA–ASSOCIATED HERPESVIRUS)

Diseases

Human herpesvirus 8 (HHV-8), also known as KSHV, causes KS, KS is one of the most common cancers in people infected with HIV. HHV-8 also causes primary effusion lymphoma and

multimeric Castleman's disease. It can cause Inflammatory Cytokine Syndrome in those coinfected with HIV.

Important Properties

As a member of the herpesvirus family, HHV-8 is an enveloped virus with a linear, double-stranded DNA genome. Within the latently-infected malignant KS cells, the genome is often circular and is episomal, that is, not integrated into cellular DNA.

Based on DNA analysis, HHV-8 resembles the lymphotropic herpesviruses (e.g., EBV) more than it does the neurotropic herpesviruses, such as HSV and VZV. The electron micrographic appearance of HHV-8 is indistinguishable from other herpesviruses.

The HHV-8 genome contains genes "pirated" from the human genome. These include viral homologues of the cellular genes for cyclin D, interleukin-6, and an interferon-regulatory protein that inhibits synthesis of interferon.

There are 6 main sub-types of HHV-8 (A–F) based on variations in the DNA sequence of ORF-1 (Open Reading Frame-1). These are found in different geographical regions.

Summary of Replicative Cycle

HHV-8 undergoes both a lytic cycle and a latent cycle. The lytic cycle produces progeny virus that then can infect others. However, the latent cycle is more important as that is involved with malignant transformation. The main site of latent infection is B lymphocytes. Endothelial cells also undergo latent infection.

Three HHV-8 genes are expressed in latent infections, the most important of which is the gene encoding the latency-associated nuclear antigen (LANA). LANA inhibits transcription of the p53 gene, an important tumor suppressor. Also expressed is the viral gene encoding cyclin D that overrides the control of cell-cycle growth. The third gene encodes a protein that inhibits Fas death pathway allowing the latently-infected cell to survive attack by cytotoxic T cells.

Transmission & Epidemiology

Transmission of HHV-8 occurs primarily via saliva and by sexual intercourse. It is also transmitted in transplanted organs such as kidneys and is the cause of transplantation-associated KS.

The prevalence of the virus based on serologic assays varies greatly from country to country. The current estimate of HHV-8 infection in the general population ranges from about 3% in the United States and England to about 50% in East Africa.

Pathogenesis & Immunity

KS in AIDS patients is a malignancy of vascular endothelial cells that contains many spindle-shaped cells and erythrocytes. The spindle cells form microvascular channels containing red cells. The spindle cells are infected with HHV-8.

HHV-8 causes malignant transformation by a mechanism similar to that of other DNA viruses (e.g., HPV), namely, inactivation of a tumor suppressor gene. A protein encoded by HHV-8 called LANA inactivates the p53 tumor suppressor proteins, resulting in malignant transformation of endothelial cells.

Note that HHV-8 is necessary but not sufficient to cause KS. Most infections with HHV-8 are asymptomatic and remain latent. The latent infections progress to KS when immunosuppression occurs.

Two proteins, MIR-1 and MIR-2 (MIR stands for modulator of immune recognition) encoded by HHV-8 act to down-regulate the expression of MHC class-1 proteins on the surface of the infected cell. This limits the killing of infected cells by cytotoxic T cells.

Clinical Findings

The lesions of KS are reddish to dark purple, flat to nodular, and often appear at multiple sites such as the skin, oral cavity, and soles of the feet (Figure 37–10). Internally, lesions occur commonly in the gastrointestinal tract and the lungs. The extravasated red cells give the lesions their purplish color.

HHV-8 also infects B cells, inducing them to proliferate and produce a type of lymphoma called primary effusion lymphoma. Multicentric Castleman's disease (MCD) is also a B cell proliferative disease. In MCD, the HHV-8 infected B cells secrete cytokines that attract large numbers of uninfected B cells.

Kaposi-associated Inflammatory Cytokine Syndrome (KISC) occurs in those who are infected with both HHV-8 and HIV. High fever, cachexia, and cytopenias are seen but without prominent lymphadenopathy. A high level of the proinflammatory cytokine, interleukin-6 is found.

Laboratory Diagnosis

Laboratory diagnosis of KS is often made by biopsy of the skin lesions. HHV-8 DNA and RNA are present in most spindle cells

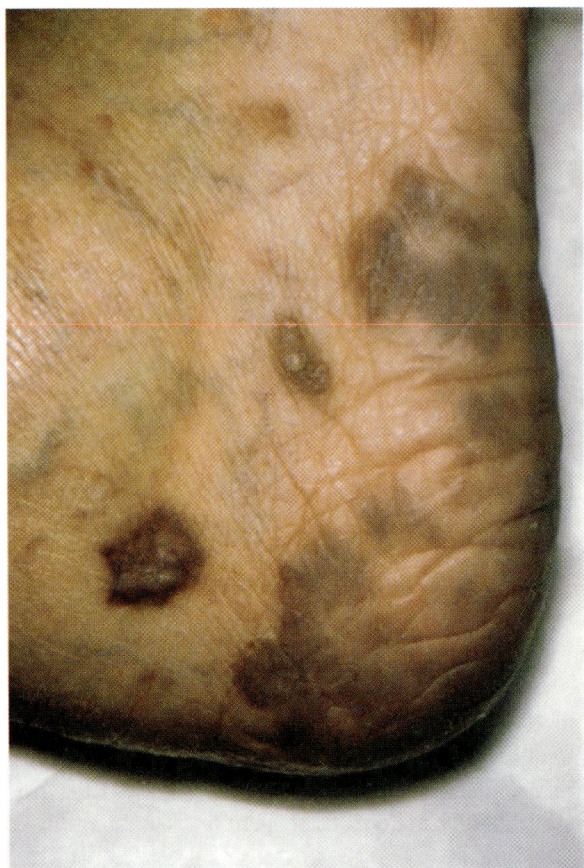

FIGURE 37–10 Kaposi's sarcoma. Note several dark purple lesions on the foot caused by human herpesvirus 8 (Kaposi's sarcoma–associated virus). (Reproduced with permission from Public Health Image Library, Centers for Disease Control and Prevention.)

and can be detected by PCR assay. Virus is not grown in culture. Serologic tests for the presence of antibodies are available to determine whether a person has been infected.

Treatment

The type of treatment depends on the site and number of the lesions. Surgical excision or intralesional vinblastine can be used for localized lesions. Radiation therapy is used for more extensive lesions. Systemic chemotherapy with pegylated liposomal doxorubicin or liposomal daunorubicin is used for advanced KS. In early HIV-associated KS, antiretroviral drugs (ART) can be effective treatment. Note that antiherpesvirus drugs, such as acyclovir, foscarnet, and cidofovir, are not effective treatment for KS.

Prevention

ART also can prevent KS as evidenced by a decreased incidence of KS in HIV-infected patients treated with an effective ART drug regimen. Ganciclovir and foscarnet, but not acyclovir, appear to reduce the development of KS in HIV-infected individuals. There is no vaccine against HHV-8.

POXVIRUSES

The poxvirus family includes three viruses of medical importance: smallpox virus, vaccinia virus, and MCV. Poxviruses are the largest and most complex viruses.

SMALLPOX VIRUS

Disease

Smallpox virus, also called variola virus, is the agent of smallpox, the only human disease that has been eradicated from the face of the Earth. **Eradication** was achieved by the widespread use of the smallpox vaccine. There is concern regarding the use of smallpox virus as an agent of bioterrorism. Poxviruses of animal origin, such as cowpox and monkey pox, are described in Chapter 46. (Note that rinderpest, a disease primarily of cattle, has also been eradicated by using the vaccine against rinderpest virus [RPV]. RPV is a paramyxovirus related to measles virus.)

Important Properties

Poxviruses are brick-shaped particles containing linear double-stranded DNA, a disk-shaped core within a double membrane, and a lipoprotein envelope (see Table 37–2). The virion contains a DNA-dependent RNA polymerase. This enzyme is required because the virus replicates in the cytoplasm and does not have access to the cellular RNA polymerase, which is located in the nucleus.

Smallpox virus has a single, stable serotype, which is the key to the success of the vaccine. If the antigenicity varied as it does in influenza virus, eradication would not have succeeded. Smallpox virus infects only humans; there is no animal reservoir.

Summary of Replicative Cycle

The following description of the replicative cycle is based on studies with vaccinia virus, as it is much less likely to cause human disease than smallpox virus. After penetration of the cell and uncoating, the virion DNA-dependent RNA polymerase synthesizes early mRNA, which is translated into early, nonstructural proteins, mainly enzymes required for subsequent steps in viral replication. The viral DNA then is replicated, after which late, structural proteins are synthesized that will form the progeny virions. The virions are assembled and acquire their envelopes by budding from the cell membrane as they are released from the cell. Note that all steps in replication occur in the cytoplasm, which is unusual for a DNA virus.

Transmission & Epidemiology

Smallpox virus is transmitted via respiratory aerosol or by direct contact with virus either in the skin lesions or on fomites such as bedding.

Prior to the 1960s, smallpox was widespread throughout large areas of Africa, Asia, and South America, and millions of people were affected. In 1967, the World Health Organization embarked on a vaccination campaign that led to the eradication of smallpox. The last naturally occurring case was in Somalia in 1977.

Pathogenesis & Immunity

Smallpox begins when the virus infects the upper respiratory tract and local lymph nodes and then enters the blood (primary viremia). Internal organs are infected; then the virus reenters the blood (secondary viremia) and spreads to the skin. These events occur during the incubation period, when the patient is still well. The rash is the result of virus replication in the skin, followed by damage caused by cytotoxic T cells attacking virus-infected cells.

Immunity following smallpox disease is lifelong; immunity following vaccination lasts about 10 years.

Clinical Findings

After an incubation period of 7 to 14 days, there is a sudden onset of prodromal symptoms such as fever and malaise. This is followed by the rash, which is worse on the face and extremities than on the trunk (i.e., it has a centrifugal distribution). The rash evolves through stages from macules to papules, vesicles, pustules, and, finally, crusts in 2 to 3 weeks.

Laboratory Diagnosis

In the past when the disease occurred, the diagnosis was made either by growing the virus in cell culture or chick embryos or by detecting viral antigens in vesicular fluid by immunofluorescence.

Treatment

Although smallpox disease has been eradicated, smallpox virus still exists and therefore, it is possible that an outbreak may occur. A drug, tecovirimat has been developed to treat disease should that happen. Tecovirimat inhibits p37, a viral protein required for formation and egress of progeny virions.

Prevention

The disease was eradicated by global use of the **vaccine**, which contains live, attenuated **vaccinia virus**. The success of the vaccine is dependent on five critical factors: (1) smallpox virus has a single, stable serotype; (2) there is no animal reservoir, and humans are the only hosts; (3) the antibody response is prompt, and therefore exposed persons can be protected; (4) the disease is easily recognized clinically, and therefore exposed persons can be immunized promptly; and (5) there is no carrier state or subclinical infection.

The vaccine is inoculated intradermally, where virus replication occurs. The formation of a vesicle is indicative of a "take" (success).

Although the vaccine was relatively safe, it became apparent in the 1970s that the incidence of side effects such as encephalitis, generalized vaccinia, and vaccinia gangrenosum exceeded the incidence of smallpox. Routine vaccination of civilians was discontinued, and it is no longer a prerequisite for international travel. Military personnel are still vaccinated.

In response to the possibility of a bioterrorism attack using smallpox virus, the U.S. federal government has instituted a program to vaccinate "first responders" so that they can give emergency medical care without fear of contracting the disease. To protect the unimmunized general population, the concept of "ring vaccination" will be used. This is based on the knowledge that **an exposed individual can be immunized as long as 4 days after exposure and be protected**. Therefore, if an attack occurs, people known to be exposed will be immunized as well as the direct contacts of those people and then the contacts of the contacts, in an expanding ring. Several military personnel and civilians have experienced myocarditis following vaccination, and as of this writing, caution has been urged regarding expanding this program to the general population.

Vaccinia immune globulins (VIG), containing high-titer antibodies against vaccinia virus, can be used to treat most of the complications of vaccination. In the past, methisazone was used to treat the complications of vaccination and could be useful again. Rifampin inhibits viral DNA-dependent RNA polymerase but was not used clinically against smallpox.

MOLLUSCUM CONTAGIOSUM VIRUS

MCV is a member of the poxvirus family but is quite distinct from smallpox and vaccinia viruses. The lesion of molluscum contagiosum is a small (2–5 mm), flesh-colored papule on the skin or mucous membrane that is painless, nonpruritic, and not inflamed (Figure 37–11). The lesions have a characteristic cup-shaped (umbilicated) crater with a white core. The lesion is composed of hyperplastic epithelial cells within which a cytoplasmic inclusion body can be seen. The inclusion body contains progeny MCV.

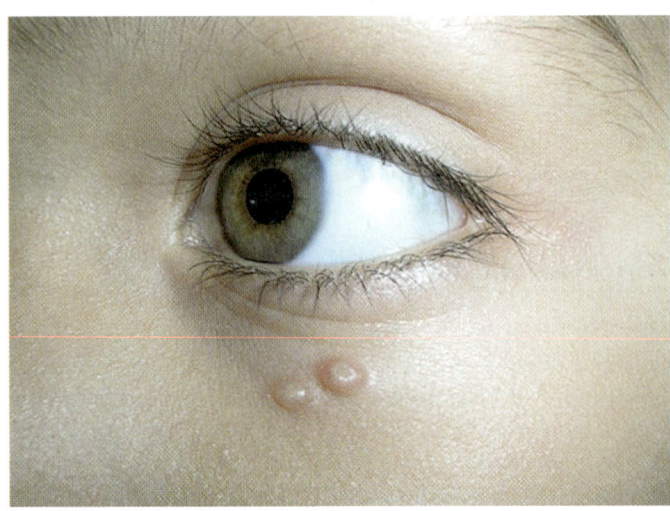

FIGURE 37–11 Molluscum contagiosum. Note two fleshy papular lesions under the eye caused by molluscum contagiosum virus, a member of the poxvirus family. (Reproduced with permission from Usatine RP, Smith MA, Mayeaux EJ Jr, et al: *The Color Atlas of Family Medicine.* New York, NY: McGraw Hill; 2009.)

Note that these lesions are different from warts, which are caused by HPV, a member of the papovavirus family (see next section).

MCV is transmitted by close personal contact, including sexually. The disease is quite common in children, in whom lesions often occur around the eyes and on the trunk. Adults often have lesions in the genital area. The lesions can be large and numerous in patients with reduced cellular immunity, such as AIDS patients. In immunocompetent patients, the lesions are self-limited but may last for months.

The diagnosis is typically made clinically; the virus is not isolated in the clinical laboratory, and antibody titers are not helpful. Removal of the lesions by curettage or with liquid nitrogen is often effective. There is no established antiviral therapy, but cidofovir may be useful in the treatment of the extensive lesions that occur in immunocompromised patients. In AIDS patients, antiretroviral therapy may restore sufficient immunity to cause the lesions to resolve. There is no vaccine.

HUMAN PAPILLOMAVIRUS

Diseases

HPV causes papillomas, which are benign tumors of squamous cells (e.g., warts on the skin). Some HPV types, especially types 16 and 18, cause **carcinoma of the cervix, penis, and anus**. HPV infection is the **most common sexually transmitted infection** in the United States.

Important Properties

Papillomaviruses are nonenveloped viruses with double-stranded circular DNA and an icosahedral nucleocapsid (see Table 37–2). The HPV genome has seven early genes (*E1–E7*) and two late genes (*L1* and *L2*). The early genes encode proteins involved in the synthesis of viral mRNA and in the replication of the progeny DNA genomes, and the late genes encode the structural proteins of the progeny virions.

Two of the early genes, **E6** and **E7**, are implicated in carcinogenesis. They encode proteins that are involved in the inactivation of proteins encoded by tumor suppressor genes in human cells (e.g., the *p53* gene and the retinoblastoma [*RB*] gene, respectively). For example, the E6 protein recruits ubiquitin ligase that ubiquinates p53 leading to its degradation within proteasomes. Inactivation of the p53 and RB proteins is an important step in the process by which a normal cell becomes a cancer cell.

There are at least 100 types of papillomaviruses, classified primarily on the basis of DNA restriction fragment analysis. There is a pronounced **predilection of certain types to infect certain tissues**. For example, skin warts are caused primarily by HPV-1 through HPV-4, whereas genital warts are usually caused by HPV-6 and HPV-11. Approximately 30 types of HPV infect the genital tract.

Summary of Replicative Cycle

After attachment and uncoating, the genome DNA moves to the nucleus. Messenger RNA is synthesized by *host cell* RNA polymerase with early viral protein E2 acting as a transcriptional activator. Early viral protein E1 acts as a helicase that separates the DNA strands of the incoming viral genome. This allows the *host cell* DNA polymerase to synthesize the progeny DNA genomes. The initial progeny genomes are maintained as episomes in the nucleus. Most of the synthesis of progeny viral DNA occurs in conjunction with cellular DNA synthesis during S phase.

Late mRNAs encode both the major structural protein (L1) and the minor structural protein (L2). L1 protein comprises the capsid of HPV virions. L1 has the ability to self-assemble into capsids in vitro, and it is this form that is the immunogen in the HPV vaccine. L2 protein aids in the packaging of genome DNA into the progeny virions as well as in uncoating the genome when they infect the next cell.

In human tissues, infectious virus particles are found in the terminally differentiated squamous cells rather than in the basal cells (Figure 37–12A). Note that HPV initially infects the cells of the basal layer in the skin, but no virus is produced by the basal cells. Rather, infectious virions are produced by squamous cells on the surface, which enhances the likelihood that efficient transmission will occur.

In malignant cells, viral DNA is integrated into host cell DNA in the vicinity of cellular proto-oncogenes, and *E6* and *E7* are overexpressed (Figure 37–12B). However, in latently infected, nonmalignant cells, the viral DNA is episomal, and *E6* and *E7* are not overexpressed. This difference occurs because another early gene, *E2*, controls *E6* and *E7* expression. The *E2* gene is functional when the viral DNA is episomal but is inactivated when it is integrated.

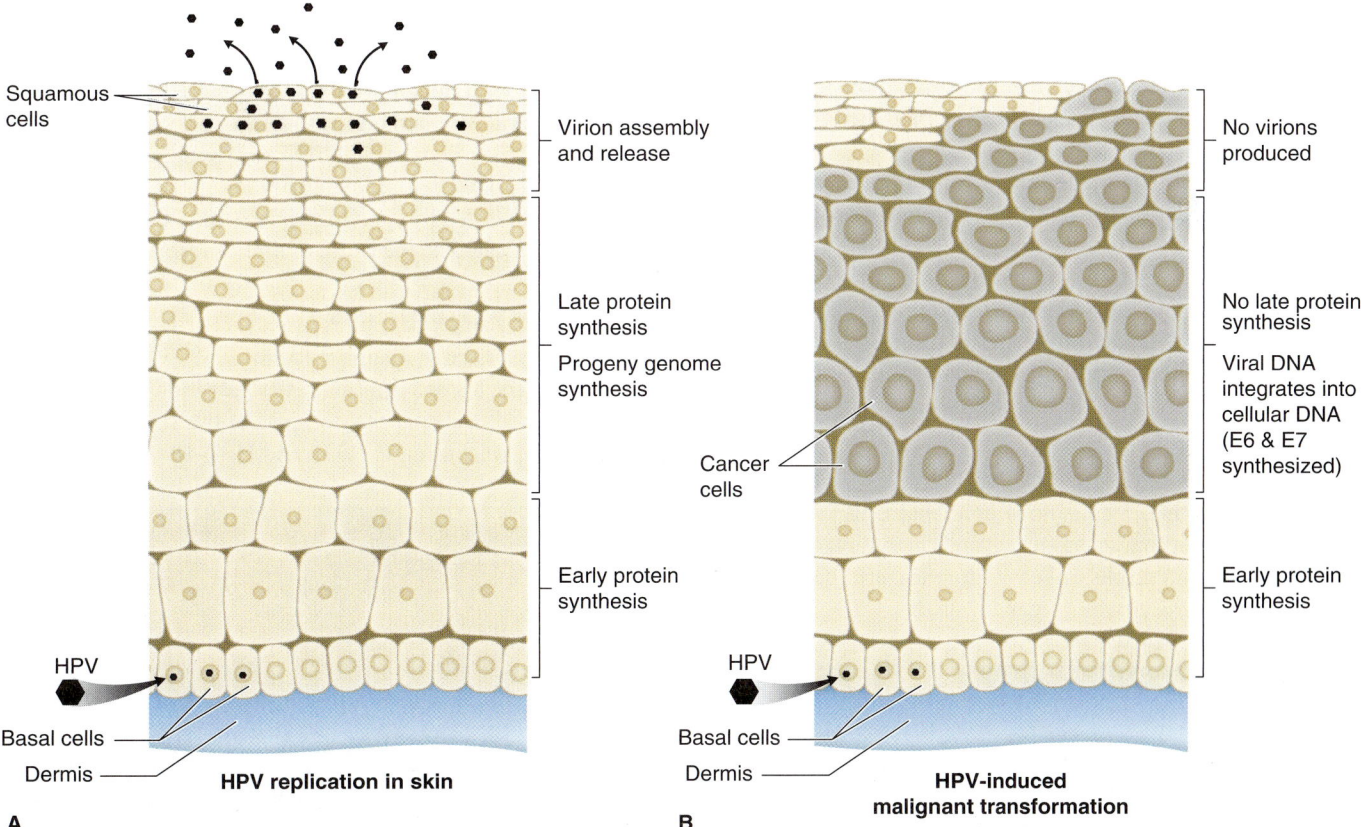

FIGURE 37–12 A. Replication of human papillomavirus (HPV) in the skin. HPV initiates replication in the basal cells of the skin at site of abrasion to skin surface. Small black dots in nucleus of three cells represent viral genome DNA. Early protein synthesis occurs followed by progeny genome synthesis. Late proteins are then produced and progeny virions are released from squamous cells on the surface of the skin. Large black dots at top of figure represent progeny virions. **B.** Malignant transformation by HPV in the skin. HPV initiates replication in the basal cells of the skin. Early protein synthesis occurs. Viral DNA integrates into cell DNA and large amounts of viral E6 and E7 proteins are produced. E6 and E7 proteins inactivate tumor suppressor proteins p53 and RB and the cell becomes malignant. No late viral proteins and no progeny virions are produced.

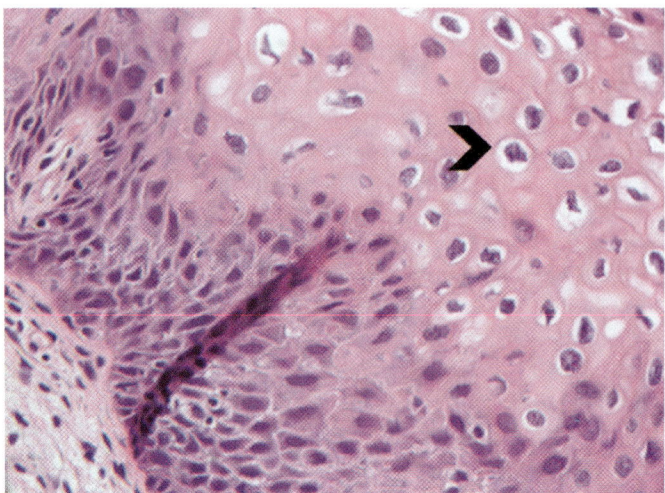

FIGURE 37–13 Koilocytes. The black arrowhead points to a koilocyte, seen here in a biopsy specimen of cervical intraepithelial neoplasia caused by human papillomavirus. Koilocytes have a small condensed nucleus and a large perinuclear cytoplasmic vacuole. 400× magnification. (Reproduced with permission from Kemp WL, Burns DK, Brown TG: *Pathology: The Big Picture.* New York, NY: McGraw Hill; 2008.)

Transmission & Epidemiology

Papillomaviruses are transmitted primarily by skin-to-skin contact, including genital contact. Micro-abrasions in the skin allow access to the basal epithelial cells where infection begins (see Figure 37–12A).

Genital warts are among the **most common sexually transmitted diseases.** Skin warts are more common in children and young adults and tend to regress in older adults. HPV transmitted from an infected mother to the neonate during childbirth causes warts in the mouth and in the respiratory tract, especially on the larynx, of the infant. Many species of animals are infected with their own types of papillomaviruses, but these viruses are not an important source of human infection.

Pathogenesis & Immunity

Papillomaviruses infect squamous epithelial cells and induce within those cells a characteristic perinuclear cytoplasmic vacuole. These vacuolated cells, called **koilocytes,** are the hallmark of infection by these viruses (Figure 37–13).

Most warts are benign and do not progress to malignancy. However, HPV infection is associated with carcinoma of the uterine cervix and penis. The proteins encoded by viral genes *E6* and *E7* mediate the inactivation of the proteins encoded by the *p53* and *RB* tumor suppressor genes and thereby contribute to oncogenesis by these viruses. The E6 and E7 proteins of HPV-16 are more effective than the E6 and E7 proteins of HPV types not implicated in carcinomas—a finding that explains why HPV-16 causes carcinomas more frequently than the other types of HPV.

Both CMI and antibody are induced by viral infection and are involved in the spontaneous regression of warts. Immunosuppressed patients (e.g., patients with AIDS) have more extensive warts, and women infected with HIV have a very high rate of carcinoma of the cervix.

Clinical Findings

Papillomas of various organs are the predominant finding. These papillomas are caused by specific HPV types. For example, skin and plantar warts (Figure 37–14) are caused primarily by HPV-1 through HPV-4, whereas genital warts (**condylomata acuminata**) (Figure 37–15) are caused primarily by HPV-6 and HPV-11. HPV-6 and HPV-11 also cause respiratory tract papillomas, especially laryngeal papillomas, in young children.

Carcinomas of the uterine cervix, the penis, and the anus, as well as premalignant lesions called **intraepithelial neoplasia**, are associated with infection by HPV-16 and HPV-18. The premalignant lesions are named for the organ affected (e.g., cervical intraepithelial neoplasia [CIN] and penile intraepithelial neoplasia [PIN]). Occult premalignant lesions of the cervix and penis can be revealed by applying acetic acid to the tissue. HPV-16 is also implicated as the cause of oral cancers.

Laboratory Diagnosis

Infections are usually diagnosed clinically. The presence of koilocytes in the lesions indicates HPV infection. A PCR-based test can be used to detect the presence of the DNA of 14 high-risk genotypes, including HPV-16 and HPV-18.

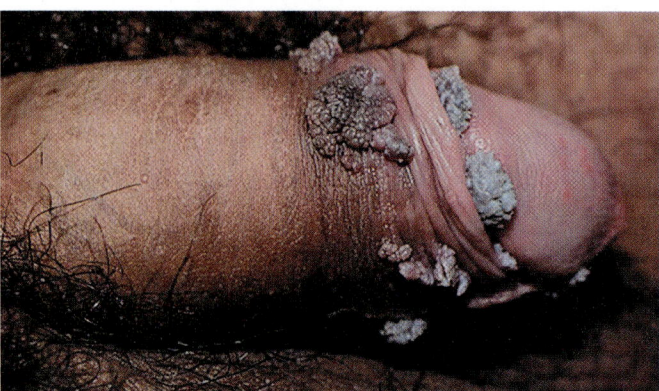

FIGURE 37–15 Papillomas (warts) on penis (condylomata acuminata). Note dry, raised verrucous lesions caused by human papillomavirus. (Reproduced with permission from Wolff K, Johnson R: *Fitzpatrick's Color Atlas & Synopsis of Clinical Dermatology*, 6th ed. New York, NY: McGraw Hill; 2009.)

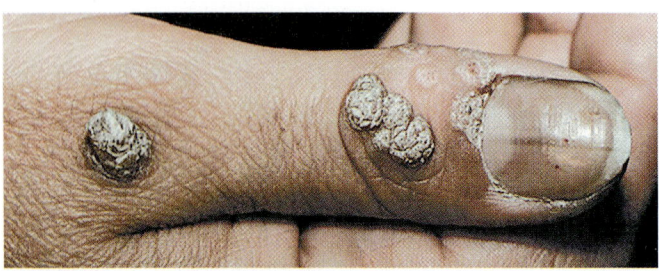

FIGURE 37–14 Papillomas (warts) on finger. Note dry, raised verrucous lesions caused by human papillomavirus. (Reproduced with permission from Wolff K, Johnson R: *Fitzpatrick's Color Atlas & Synopsis of Clinical Dermatology*, 6th ed. New York, NY: McGraw Hill; 2009.)

Diagnostic tests based on detection of antibodies in a patient's serum or on isolation of the virus from a patient's tissue are not used.

Treatment

Topical imiquimod or podophyllin is used for the treatment of genital warts. Liquid nitrogen is commonly used for skin warts. Plantar warts can be removed surgically or treated with salicylic acid topically. Cidofovir is useful in the treatment of severe HPV infections, especially in immunocompromised patients.

Prevention

The HPV vaccine is very effective in preventing carcinoma of the cervix, anal carcinoma, and genital warts. Note that HPV vaccines have no effect on existing papillomas.

Gardasil 9 is the only HPV vaccine available in the United States as of 2020. It is a recombinant subunit vaccine against nine types of HPV. It contains the major (L1) capsid protein of types 6 and 11, which cause genital warts, and types 16 and 18, which are the two most common causes of cervical, penile, and anal carcinoma. It also contains the L1 capsid protein of five more types (31, 33, 45, 52, and 58) that are less common causes of these cancers. It is recommended for both males and females, between the ages of 9 and 26 years.

The role of cesarean section in preventing transmission of HPV from a mother with genital warts to her newborn is uncertain. Circumcision reduces the risk of infection by HPV.

SELF-ASSESSMENT QUESTIONS

1. Your patient is a 30-year-old man who has frequent episodes of herpes labialis. He asks you to tell him something about herpes simplex virus type 1 (HSV-1). Which one of the following would be the most accurate statement to make?

    (A) Acyclovir can eradicate the latent state of HSV-1 but not HSV-2.

    (B) The main site of latency by HSV-1 is the neurons in the sensory ganglia of the face.

    (C) HSV-1 is an enveloped virus that has a DNA genome and a DNA polymerase in the virion.

    (D) The lesions of primary HSV-1 infections are less extensive and less severe than the lesions of recurrent HSV-1 infections.

    (E) The laboratory diagnosis of HSV-1 infections typically involves the detection of a greater than fourfold rise in antibody titer against the virus.

2. Your patient is a woman who is due to give birth next week. She asks you about the risk of her baby becoming infected with herpes simplex virus type 2 (HSV-2). Which one of the following is the most accurate response?

    (A) HSV-2 is a significant cause of congenital abnormalities.

    (B) The risk is higher if the mother has visible lesions than if she does not.

    (C) The risk is higher if the mother has IgG antibody to HSV-2 than if she has IgM antibody.

    (D) The risk is higher if the delivery occurs by cesarean section than if the delivery is performed vaginally.

    (E) The risk is higher if the mother is having an episode of recurrent disease caused by HSV-2 than if it were a primary episode.

3. Regarding varicella-zoster virus (VZV), which one of the following is most accurate?

    (A) High-dose acyclovir can eliminate the latent state caused by VZV.

    (B) The principal site of latency of VZV is in the nucleus of motor neurons.

    (C) Domestic animals, such as pigs and chickens, are the main reservoir for VZV.

    (D) The vaccine against varicella contains all three serotypes of formalin-killed VZV as the immunogen.

    (E) When zoster occurs in an immunocompromised patient, acyclovir should be given to prevent disseminated infection.

4. Regarding cytomegalovirus (CMV), which one of the following is most accurate?

    (A) CMV is usually acquired by the fecal–oral route in adults.

    (B) Neonates born from infected mothers should be given the subunit vaccine.

    (C) Reactivation of CMV in sensory ganglion cells leads to painful vesicles along nerves.

    (D) Lamivudine should be used to treat CMV infections in immunocompromised patients.

    (E) CMV infection of a fetus during the first trimester results in more congenital abnormalities than infection in the third trimester.

5. Regarding Epstein–Barr virus (EBV) and infectious mononucleosis, which one of the following is most accurate?

    (A) EBV enters the latent state primarily in CD4-positive helper T cells.

    (B) Approximately 10% of people in the United States have been exposed to EBV.

    (C) People with infectious mononucleosis produce antibodies that agglutinate sheep red cells.

    (D) The atypical lymphs in the blood of people with infectious mononucleosis are EBV-infected T helper cells.

    (E) Patients with deficient cell-mediated immunity should receive passive–active immunization against EBV.

6. Naturally occurring smallpox disease has been eradicated from the face of the Earth. Eradication was achieved by the use of the vaccine. Regarding this vaccine, which one of the following is the most accurate?

    (A) The vaccine should be given in conjunction with preformed antibody to the virus.

    (B) Administration of the vaccine 1 day after exposure to the virus does not protect against disease.

    (C) The vaccine contains killed smallpox virus so the virus in the vaccine does not cause adverse effects.

    (D) Smallpox virus has a single stable serotype, so new formulations of the vaccine do not have to be made each year.

    (E) Because domestic animals such as cows are the main reservoir for smallpox virus, the vaccine must interrupt transmission from these sources.

7. Your patient is a 35-year-old man who had a grand-mal seizure this morning. Magnetic resonance imaging revealed a lesion in the temporal lobe. A brain biopsy showed multinucleated giant cells with intranuclear inclusion bodies. Which one of the following is the most likely cause of this disease?

   (A) Cytomegalovirus
   (B) Epstein–Barr virus
   (C) Herpes simplex virus type 1
   (D) Human herpesvirus 8
   (E) Varicella-zoster virus

8. Regarding the patient in Question 7, which one of the following is the best choice of drug to treat his infection?

   (A) Acyclovir
   (B) Lamivudine
   (C) Oseltamivir
   (D) Ritonavir
   (E) Zidovudine

9. Your patient is a 22-year-old woman with several episodes of bloody diarrhea. She is HIV antibody positive with a CD4 count of 50. Stool cultures for *Shigella*, *Salmonella*, and *Campylobacter* were negative. An assay for *Clostridium difficile* toxin was negative. Colonoscopy revealed many ulcerated lesions. Biopsy revealed cells with "owl's eye" inclusions in the nucleus. Which one of the following is the most likely cause of this disease?

   (A) Cytomegalovirus
   (B) Epstein–Barr virus
   (C) Herpes simplex virus type 1
   (D) Human herpesvirus 8
   (E) Varicella-zoster virus

10. Regarding the patient in Question 9, which one of the following is the best choice of drug to treat her infection?

    (A) Amantadine
    (B) Enfuvirtide
    (C) Ganciclovir
    (D) Nevirapine
    (E) Ribavirin

11. Regarding human papillomavirus (HPV), which one of the following statements is most accurate?

    (A) There is no vaccine available against HPV.
    (B) Acyclovir is effective in preventing lesions caused by HPV but does not cure the latent state.
    (C) Antigen–antibody complexes play an important role in the pathogenesis of warts caused by HPV.
    (D) The early proteins of HPV play a more important role in malignant transformation than the late proteins.
    (E) The diagnosis of HPV infection is usually made by detecting cytoplasmic inclusions within giant cells in the lesions.

12. A 24-year-old woman is seen by her gynecologist for a routine Pap smear. The smear shows cervical intraepithelial neoplasia grade 3 (CIN 3). You decide to examine her long-term male sexual partner. Which one of the following is the most likely finding?

    (A) Condylomata lata
    (B) Condylomata acuminata
    (C) Penile intraepithelial neoplasia associated with HPV-6
    (D) Penile intraepithelial neoplasia associated with HPV-16

ANSWERS

(1) **(B)**
(2) **(B)**
(3) **(E)**
(4) **(E)**
(5) **(C)**
(6) **(D)**
(7) **(C)**
(8) **(A)**
(9) **(A)**
(10) **(C)**
(11) **(D)**
(12) **(D)**

SUMMARIES OF ORGANISMS

Brief summaries of the organisms described in this chapter begin on page 691. Please consult these summaries for a rapid review of the essential material.

PRACTICE QUESTIONS: USMLE & COURSE EXAMINATIONS

Questions on the topics discussed in this chapter can be found in the Clinical Virology section of Part XIII: USMLE (National Board) Practice Questions starting on page 747. Also see Part XIV: USMLE (National Board) Practice Examination starting on page 775.

Respiratory Viruses

INTRODUCTION

The viruses described in this chapter are the "professional" respiratory tract viruses whose primary clinical manifestations are in the upper and/or lower respiratory tract (Table 38–1).

Many other viruses, such as measles virus, mumps virus, rubella virus, and varicella-zoster virus, initially infect the respiratory tract, but their characteristic clinical findings are seen elsewhere. These viruses are described in other chapters.

TABLE 38–1 Clinical Features of Respiratory Viruses

Virus	Important Disease	Number of Serotypes	Causes Worldwide Epidemics (Pandemics)	Main Clinical Findings	Vaccine Available	Treatment
Influenza virus	Influenza	Many	Yes	Sudden-onset headache, shaking chill, sore throat, cough, and myalgias	Yes	Oseltamivir, zanamivir, baloxavir
Parainfluenza virus	Croup	Four	No	Barking cough	No	None
Respiratory syncytial virus	Bronchiolitis in infants	Two	No	Cough, dyspnea, retractions, wheezing	No	Ribavirin
Human metapneumovirus	Common cold, bronchiolitis, pneumonia	Two	No	Various (coryza, wheezing, cough)	No	None
Coronaviruses, especially SARS-CoV-2	Common cold, SARS,[1] MERS,[2] COVID-19[3]	Seven (4 common cold serotypes; 3 pneumonia serotypes)	Yes	Various (coryza, cough, severe pneumonia)	Yes, for SARS-CoV-2	Remdesivir; Emergency Use Authorization for several drugs and monoclonal antibodies (see Text)
Rhinovirus	Common cold	Many	No	Runny nose (coryza), sneezing, no fever	No	None
Adenovirus	Pharyngitis, pneumonia, conjunctivitis	Many	No	Sore throat, cough, pneumonia, "pink eye"	Yes[4]	None

[1]SARS is severe acute respiratory syndrome. Also SARS-CoV-2.

[2]MERS is Middle East respiratory syndrome.

[3]COVID-19 is Coronavirus Disease-19.

[4]For military only.

Almost all of the respiratory tract viruses have RNA as their genome; one has DNA. Most are enveloped viruses, whereas two, rhinovirus and adenovirus, are nonenveloped. In addition, the enveloped respiratory viruses belong to several different virus families, namely, orthomyxoviruses, paramyxoviruses, and coronaviruses. So they are quite varied in their structure and replication. The feature that unites all of these viruses is their ability to infect the mucosal cells of the respiratory tract and cause significant symptoms there.

In serious respiratory virus infections, a laboratory diagnosis can be made by using polymerase chain reaction (PCR)-based assays on respiratory tract secretions. A panel of PCR assays is used to diagnose infections caused by viruses such as influenza virus, parainfluenza virus, respiratory syncytial virus (RSV), rhinovirus, human metapneumovirus (HMPV), and adenovirus.

Additional information regarding the clinical aspects of infections caused by the viruses in this chapter is provided in Part IX, entitled Infectious Diseases beginning on page 603.

INFLUENZA VIRUS

Influenza virus is an important human pathogen because it causes both outbreaks of influenza that sicken and kill thousands of people each year as well as infrequent but devastating worldwide epidemics (pandemics). The disease influenza affects the upper and lower respiratory tract. It causes symptoms in the pharynx, larynx, trachea, and bronchi. In some case it causes pneumonia as well.

Influenza virus is the only member of the orthomyxovirus family. The orthomyxoviruses differ from the paramyxoviruses primarily in that the former have a segmented RNA genome (usually eight pieces), whereas the RNA genome of the latter consists of a single piece. The term *myxo* refers to the observation that these viruses interact with mucins (glycoproteins on the surface of cells). Table 38–2 shows a comparison of the

properties of influenza A virus with several other viruses that infect the respiratory tract.

Most cases of influenza in humans are caused by H1N1 and H3N2 strains of influenza A virus. However, in 1997, an outbreak of human influenza (avian influenza, bird flu) caused by an H5N1 strain of influenza A virus began. This outbreak and subsequent outbreaks are described on page 311. In 2009, there was an outbreak of human influenza caused by H1N1 influenza A virus of swine origin (swine-origin influenza virus, S-OIV). This outbreak and the subsequent pandemic are described on page 312. In 2013, an outbreak of human influenza caused by another bird-related strain (H7N9) of influenza virus occurred.

1. Human Influenza Virus

Disease

Influenza A virus causes worldwide epidemics (pandemics) of influenza, influenza B virus causes major outbreaks of influenza, and influenza C virus causes mild respiratory tract infections but does not cause outbreaks of influenza. Pandemics occur when a variant of influenza A virus that contains a new hemagglutinin against which people do not have preexisting antibodies is introduced into the human population.

The pandemics caused by influenza A virus occur infrequently (the last one was in 1968), but major outbreaks caused by this virus occur virtually every year in many countries. In most years, influenza is the most common cause of respiratory tract infections that result in physician visits and hospitalizations in the United States.

In the 1918 influenza pandemic, more Americans died than in World War I, World War II, the Korean War, and the Vietnam War combined. There is molecular evidence that the 1918 pandemic was caused by an H1N1 strain containing genes of avian origin. Influenza B virus does not cause pandemics, and the major outbreaks caused by this virus do not occur as often as those caused by influenza A virus. It is estimated that

TABLE 38–2   Properties of Respiratory Viruses

Property	Influenza Virus	Parainfluenza Virus, Respiratory Syncytial Virus, and Human Metapneumovirus	Coronavirus	Rhinovirus	Adenovirus
Virus family	Orthomyxovirus	Paramyxovirus	Coronavirus	Picornavirus	Adenovirus
Genome	Segmented single-stranded RNA; negative polarity	Nonsegmented single-stranded RNA; negative polarity	Nonsegmented single-stranded RNA; positive polarity	Nonsegmented single-stranded RNA; positive polarity	Double-stranded DNA
RNA polymerase in virion	Yes	Yes	No	No	No
Capsid	Helical	Helical	Helical	Icosahedral	Icosahedral
Envelope	Yes	Yes	Yes	No	No
Fusion protein on surface	No	Yes	No	No	No
Giant cell formation	No	Yes	No	No	No

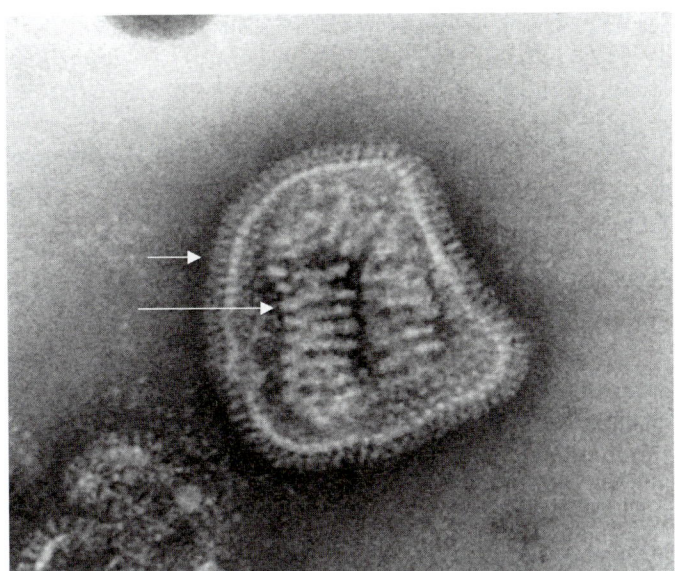

FIGURE 38–1 Influenza virus. Electron micrograph. Long arrow points to the helical nucleocapsid of influenza virus. The nucleocapsid contains the segmented, negative-polarity genome RNA. Short arrow points to the spikes on the virion envelope. The spikes are the hemagglutinin and neuraminidase proteins. (Used with permission from Dr. Erskine Palmer and Dr. M. Martin, Public Health Image Library, Centers for Disease Control and Prevention.)

approximately 36,000 people die of influenza each year in the United States.

Important Properties

Influenza virus is composed of a **segmented** single-stranded RNA genome, a helical nucleocapsid, and an outer lipoprotein envelope (Figure 38–1). The genome RNA of influenza A and influenza B viruses consists of 8 segments. The virion contains RNA-dependent **RNA polymerase**, which transcribes the **negative-polarity** genome into mRNA.

The envelope is covered with two different types of spikes, a **hemagglutinin** and a **neuraminidase**. Influenza A virus has 16 antigenically distinct types of hemagglutinin and 9 antigenically distinct types of neuraminidase. As discussed later, some of these types cause disease in humans, but most of the types typically cause disease in other animal species such as birds, horses, and pigs.

The function of the hemagglutinin is to bind to the cell surface receptor (neuraminic acid, sialic acid) to initiate infection of the cell. In the clinical laboratory, the hemagglutinin agglutinates red blood cells, which is the basis of a diagnostic test called the hemagglutination inhibition test. The hemagglutinin is also the target of neutralizing antibody (i.e., antibody against the hemagglutinin inhibits infection of the cell).

The neuraminidase cleaves neuraminic acid (sialic acid) to release progeny virus from the infected cell. The hemagglutinin functions at the beginning of infection, whereas the neuraminidase functions at the end. Neuraminidase also degrades the protective layer of mucus in the respiratory tract. This enhances

the ability of the virus to gain access to the respiratory epithelial cells.

Influenza viruses, especially influenza A virus, show **changes in the antigenicity** of their hemagglutinin and neuraminidase proteins; this property contributes to their capacity to cause devastating **worldwide epidemics (pandemics)**. There are two types of antigenic changes: (1) **antigenic shift**, which is a major change based on the reassortment of segments of the genome RNA and (2) **antigenic drift**, which is a minor change based on mutations in the genome RNA. Note that in reassortment, entire segments of RNA are exchanged, each one of which codes for a single protein (e.g., the hemagglutinin) (Figure 38–2).

Influenza A virus has two matrix proteins: The M1 matrix protein is located between the internal nucleoprotein and the envelope and provides structural integrity. The **M2 matrix protein forms an ion channel** between the interior of the virus and the external milieu. This ion channel plays an essential role in the **uncoating of the virion** after it enters the cell. It transports protons into the virion causing the disruption of the envelope, which frees the nucleocapsid containing the genome RNA, allowing it to migrate to the nucleus.

Influenza viruses have both **group-specific** and **type-specific** antigens.

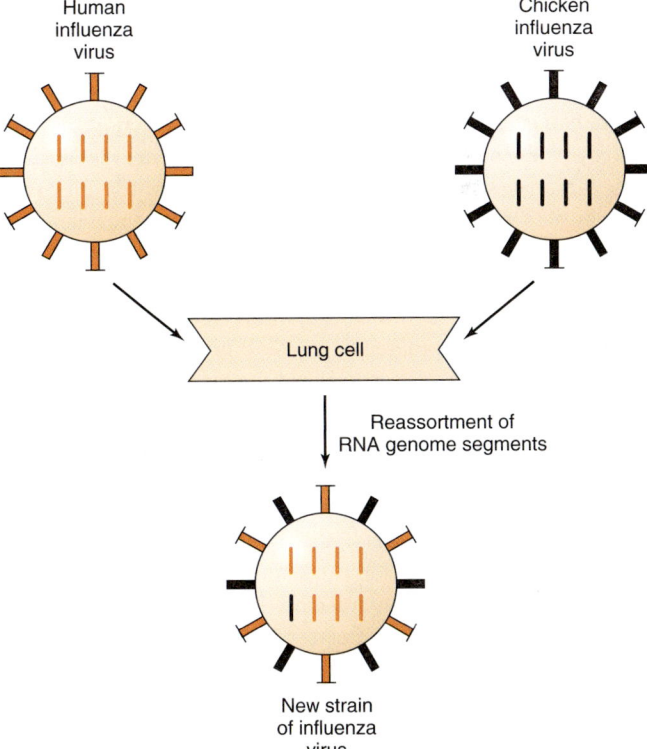

FIGURE 38–2 Antigenic shift in influenza virus. A human strain of influenza virus containing the gene encoding one antigenic type of hemagglutinin (colored orange) infects the same lung cell as a chicken strain of influenza virus containing the gene encoding a different antigenic type of hemagglutinin (colored black). Reassortment of the genome RNA segments that encode the hemagglutinin occurs, and a new strain of influenza virus is produced containing the chicken type of hemagglutinin (colored black).

(1) The internal ribonucleoprotein in the nucleocapsid is the group-specific antigen that distinguishes influenza A, B, and C viruses.

(2) The hemagglutinin and the neuraminidase are the type-specific antigens located on the surface. Antibody against the hemagglutinin neutralizes the infectivity of the virus (and prevents disease), whereas antibody against the group-specific antigen (which is located internally) does not. Antibody against the neuraminidase does not neutralize infectivity but does reduce disease by decreasing the amount of virus released from the infected cell and thus reducing spread of the virus to adjacent cells.

An important determinant of the virulence of this virus is a nonstructural protein called NS-1 encoded by the genome RNA of influenza virus. NS-1 has several functions, but the one pertinent to virulence is its ability to inhibit the synthesis of interferon and to block the action of interferon by inhibiting the protein kinase and ribonuclease that mediate interferon's antiviral action. As a result, innate defenses are reduced and viral virulence is correspondingly enhanced.

Many species of animals (e.g., aquatic birds, chickens, swine, and horses) have their own influenza A viruses. These **animal viruses are the source of the RNA segments** that encode the antigenic shift variants that cause epidemics among humans. For example, if an avian and a human influenza A virus infect the same cell (e.g., in a farmer's respiratory tract), reassortment could occur and a new variant of the human A virus, bearing the avian virus hemagglutinin, may appear (see Figure 38–1).

There is evidence that aquatic birds (waterfowl) are a common source of these new genes and that the reassortment event leading to new human strains occurs in pigs. In other words, pigs may serve as the "mixing bowl" within which the human, avian, and swine viruses reassort. There are 16 types of hemagglutinin (H1–H16) and 9 types of neuraminidase (N1–N9) found in waterfowl. In humans, three types of hemagglutinin (H1, H2, and H3) and two types of neuraminidase (N1 and N2) predominate.

Because influenza B virus is only a human virus, there is no animal source of new RNA segments. Influenza B virus therefore does not undergo antigenic shifts. It does, however, undergo enough antigenic drift that the current strain must be included in the new version of the influenza vaccine produced each year. Influenza B virus has two antigenically distinct subtypes, the Victoria lineage and the Yamagata lineage. Both are present in the quadrivalent vaccine. Influenza B virus has no antigens in common with influenza A virus.

A/Philippines/82 (H3N2) illustrates the nomenclature of influenza viruses. "A" refers to the group antigen. Next are the location and year the virus was isolated. H3N2 is the designation of the hemagglutinin (H) and neuraminidase (N) types. The H1N1 and H3N2 strains of influenza A virus are the most common at this time and are the strains included in the current vaccine. The H2N2 strain caused a pandemic in 1957.

Summary of Replicative Cycle

The virus adsorbs to the cell when the viral hemagglutinin interacts with sialic acid receptors on the cell surface. The hemagglutinin is cleaved by cellular proteases to reveal a fusion protein that mediates fusion with the cell membrane. (Note similarity to the coronavirus entry process described later in this chapter.)

The virus then enters the cell in vesicles and uncoats within an endosome. Uncoating is facilitated by the low pH within the endosome. Protons pass through the ion channel formed by the M2 protein into the interior of the virion. This disrupts the virion envelope and frees the nucleocapsid to enter the cytoplasm and then migrate to the nucleus where the genome RNA is transcribed into mRNA.

One molecule of the virion RNA polymerase is attached to each of the eight genome negative polarity RNA segments. The polyemrase transcribes the eight genome segments into eight mRNAs in the nucleus. Synthesis of the eight mRNAs occurs in the nucleus because a methylated guanosine "cap" is required. The cap (methyl-guanosine plus approximately 10 adjacent nucleotides) is obtained from cellular mRNAs in a process called **cap snatching**. (See Figure 29–4.) The cap serves as the primer for the viral polymerase to synthesize viral mRNA. The capped viral mRNAs then move to the cytoplasm, where they are translated into viral proteins.

In addition, also in the nucleus, full-length positive-strand viral RNA transcripts (lacking methyl-guanosine cap and polyadenosine tail) serve as the template for the synthesis of the negative-strand RNA genomes for the progeny virions.

The polymerase of influenza virus consists of three subunits: the PA subunit is the endonucleases (ribonuclease) that mediates cap-snatching, the PB1 subunit is an RNA polymerase that synthesizes viral mRNAs and progeny genome RNAs, and the PB2 subunit binds capped cellular mRNAs prior to the transfer of the cap by PA that initiates synthesis of viral mRNA (see Figure 29–4). Each of the three subunits is encoded by a separate gene segment and one polymerase heterotrimer (PA, PB1, and PB2) is attached to each gene segment.

Two newly synthesized proteins, NP protein and matrix protein, bind to the progeny RNA genome segments in the nucleus, and that complex is transported to the cytoplasm. The helical ribonucleoprotein assembles in the cytoplasm, matrix protein mediates the interaction of the nucleocapsid with the envelope, and the virion is released from the cell by budding from the outer cell membrane at the site where the hemagglutinin and neuraminidase are located. The neuraminidase releases the virus by cleaving neuraminic acid on the cell surface at the site of the budding progeny virions. Influenza virus, hepatitis delta virus, and retroviruses are the **only RNA viruses** that have an important stage of their replication occurring in the **nucleus**.

Transmission & Epidemiology

The virus is transmitted by **airborne respiratory droplets**. The ability of influenza A virus to cause epidemics is dependent on antigenic changes in the hemagglutinin and neuraminidase. As mentioned previously, influenza A virus undergoes both major antigenic shifts as well as minor antigenic drifts. Antigenic shift variants appear infrequently, whereas drift variants appear virtually every year.

Although the emphasis is placed on the striking ability of the virus to cause pandemics, it should be noted that influenza A virus causes up to half a million deaths worldwide annually, 90% of which occur in older adults.

Influenza occurs primarily in the Winter months of December to February in the Northern Hemisphere, when influenza and bacterial pneumonia secondary to influenza cause a significant number of deaths, especially in older people. In the Southern Hemisphere (e.g., in Australia and New Zealand), influenza occurs primarily in the Winter months of June through August. In the tropics, influenza occurs year round with little seasonal variation.

Pathogenesis & Immunity

Examination of pathologic slides of tissue from patients with influenza reveals inflammation in the pharynx, larynx, trachea, and bronchi. Influenza, therefore, can be described as a pharyngo-laryngo-tracheo-bronchitis. Pneumonia, which involves the alveoli, may also occur.

After the virus has been inhaled, the neuraminidase degrades the protective mucus layer, allowing the virus to gain access to the cells of the upper and lower respiratory tract. The infection is limited primarily to this area because the proteases that cleave the hemagglutinin are located in the respiratory tract.

Despite systemic symptoms, viremia rarely occurs. The systemic symptoms, such as severe myalgias, are due to cytokines circulating in the blood. These cytokines include interleukin-6 (IL-6), interleukin-8 (IL-8), and tumor necrosis factor (TNF). There is necrosis of the superficial layers of the respiratory epithelium. Influenza virus pneumonia, which can complicate influenza, is interstitial in location.

Immunity depends mainly on secretory IgA in the respiratory tract. IgG is also produced but is less protective. Cytotoxic T cells also play a protective role.

Clinical Findings

After an incubation period of 24 to 48 hours, fever, myalgias, headache, sore throat, and cough develop suddenly. Severe myalgias (muscle pains) coupled with respiratory tract symptoms are typical of influenza. Vomiting and diarrhea are rare. The symptoms usually resolve spontaneously in 4 to 7 days, but influenzal or bacterial pneumonia may complicate the course. One of the well-known complications of influenza is pneumonia caused by either *Staphylococcus aureus* or *Streptococcus pneumoniae*. Patients at high risk for severe influenza include children less than 5 years of age, adults over 65 years of age, pregnant women, obese individuals, residents of chronic care facilities, and immunosuppressed persons.

Reye's syndrome, characterized by encephalopathy and liver degeneration, is a rare, life-threatening complication in children following some viral infections, particularly influenza B and chickenpox. Aspirin given to reduce fever in viral infections has been implicated in the pathogenesis of Reye's syndrome.

Laboratory Diagnosis

Although most diagnoses of influenza are made on clinical grounds, laboratory tests are available. PCR-based tests that detect *influenza virus RNA* in respiratory specimens are commonly used in hospitals. Nucleic acid amplification tests (NAAT) have high specificity and acceptably specificity. These tests will diagnose infections caused by influenza A (both H3 and H1) and influenza B virus.

The test most commonly used in the doctor's office is an enzyme-linked immunosorbent assay (ELISA) for viral *antigen* in respiratory secretions such as nasal or throat washings, nasal or throat swabs, or sputum. Several rapid ELISA tests are available. Two tests (FLU OIA and QuickVue Influenza Test) are based on detection of viral antigen using monoclonal antibodies, and a third test (ZstatFlu) is based on detection of viral neuraminidase using a substrate of the enzyme that changes color when cleaved by neuraminidase. The rationale for using the rapid tests is that treatment with the neuraminidase inhibitors should be instituted within 48 hours of the onset of symptoms. The sensitivity of these antigen-based tests is low so false negatives occur.

Influenza can also be diagnosed by the detection of antibodies in the patient's serum. A rise in antibody titer of at least fourfold in paired serum samples taken early in the illness and 10 days later is sufficient for diagnosis. Either the hemagglutination inhibition or complement fixation (CF) test can be used to assay the antibody titer. Because the second sample is taken 10 days later, this approach is used to make a retrospective diagnosis, often for epidemiologic purposes.

Other tests such as direct fluorescent antibody on respiratory specimens and virus isolation in cell culture can also be used.

Treatment

There are four drugs available for the treatment of influenza. Three drugs are neuraminidase inhibitors, namely **oseltamivir, zanamivir, and peramivir**. The fourth drug, **baloxavir** inhibits the endonuclease (ribonuclease) required for the synthesis of viral mRNA. It inhibits a process called "cap-snatching." (Table 38–3 and the paragraph below.) Amantadine is no longer used because many influenza A isolates are resistant.

Oseltamivir (Tamiflu) taken orally and **zanamivir** (Relenza) inhaled into the nose are the two most commonly used drugs for the treatment of influenza. A third drug, peramivir (Rapivab), is administered intravenously. They are members of a class of drugs called **neuraminidase inhibitors**, which act by inhibiting the release of virus from infected cells. This limits the extent of the infection by reducing the spread of virus from one cell to another. These drugs are effective against both influenza A and B viruses.

Resistance to Tamiflu occurs but currently is not clinically significant. Some isolates of H1N1 influenza virus are resistant to Tamiflu. However, H3N2 strains are still susceptible to Tamiflu. Both H1N1 and H3N2 strains remained susceptible to Relenza. All influenza B strains are susceptible to Tamiflu and Relenza.

TABLE 38-3 Drugs for Influenza

Name of Drug	Currently Recommended	Spectrum of Action	Mode of Administration	Mode of Action
Oseltamivir (Tamiflu)	Yes	Both influenza A and B	Oral	Inhibition of neuraminidase thereby blocking release of progeny virus
Zanamivir (Relenza)	Yes	Both influenza A and B	Nasal spray	Inhibition of neuraminidase thereby blocking release of progeny virus
Peramivir (Rapivab)	Yes	Both influenza A and B	Intravenous	Inhibition of neuraminidase thereby blocking release of progeny virus
Baloxavir (Xofluza)	Yes	Both influenza A and B	Oral	Inhibition of cap-snatching ribonuclease thereby blocking viral mRNA synthesis
Amantadine (Symmetrel) Rimantadine (Flumadine)	No[1]	Influenza A only	Oral	Inhibition of M2 ion channel thereby blocking uncoating

[1]Not recommended because too many isolates are resistant.

Tamiflu pills are administered orally, whereas Relenza is delivered by inhaling the powder directly into the respiratory tract. Clinical studies showed they reduce the duration of symptoms by 1 to 2 days. They also reduce the amount of virus produced and therefore reduce the chance of spread to others. These drugs are most effective when taken *within 48 hours* of the onset of symptoms. In 2015, some concern regarding the efficacy of Tamiflu and Relenza arose. Additional studies are needed to resolve this issue.

In 2018, the FDA approved a new drug, **baloxavir marboxil**, (Xofluza) for the treatment of influenza. The drug inhibits the "cap-snatching" ribonuclease required for the synthesis of influenza virus mRNA (see Figure 29–4). Only one oral dose is required and the drug should be given within the first 48 hours of symptoms for maximum effectiveness. It is effective against a wide range of influenza virus strains including oseltamivir-resistant strains. It is effective against both influenza A and influenza B viruses.

Amantadine (Symmetrel) is approved for both the treatment and prevention of influenza A. However, 90% of the H3N2 strains in the United States are resistant to amantadine (and rimantadine, see later), and so *these drugs are no longer recommended*. These drugs block the M2 ion channel, thereby inhibiting uncoating. Resistance is caused primarily by mutations in the gene for the M2 protein.

Note that amantadine is effective only against influenza A, not against influenza B. **Rimantadine** (Flumadine), a derivative of amantadine, can also be used for treatment and prevention of influenza A and has fewer side effects than amantadine. It should be emphasized that the vaccine is preferred over these drugs in the prevention of influenza.

Prevention

The main mode of prevention is the **vaccine**, which contains both influenza A and B viruses. The vaccine is *usually reformulated each year* to contain the current antigenic strains. A major research goal is to formulate a **universal vaccine** that will be effective against all antigenic variants. This would eliminate the need to reformulate the vaccine each year.

The vaccine contains the latest drift mutants that are selected in the Spring of the year for the vaccine to be administered in the Fall and Winter of that year. It takes approximately 6 months to prepare and safety test the new vaccine. This process is designed to induce immunity to the influenza virus strain most likely to cause disease that Winter. One problem with this 6-month process is that both the virus in the community and the virus in the vaccine being prepared continue to drift in the interim and in some years can result in significantly reduced vaccine efficacy.

There are two main types of influenza vaccines available in the United States, **a killed vaccine and a live, attenuated vaccine** (Table 38–4). Both a trivalent vaccine containing recent isolates of two A strains (H1N1 and H3N2) and one B strain and a quadrivalent vaccine containing two A strains and two B strains are available.

The vaccine most often used is the killed vaccine made in chicken eggs. The virus is inactivated with formaldehyde and then treated with a lipid solvent that disaggregates the virions. Note that the **hemagglutinin is the most important antigen** because it elicits neutralizing antibody. This vaccine is typically administered intramuscularly. A high-dose killed vaccine that contains four times as much hemagglutinin as the standard vaccine is recommended for those over 65 years of age. A killed influenza vaccine that can be administered intradermally is also available.

The other vaccine is a live, attenuated vaccine containing temperature-sensitive mutants of influenza A and B viruses. These temperature-sensitive mutants can replicate in the cooler (33°C) nasal mucosa where they induce IgA, but not in the warmer (37°C) lower respiratory tract. The live virus in the vaccine therefore immunizes but does not cause disease. There is no evidence of reversion to virulence.

This vaccine is administered by spraying into the nose ("nasal mist"). The live vaccine is recommended for children, whereas the inactivated vaccine is recommended for adults. The live vaccine should not be given to pregnant women or to immunocompromised individuals.

Note that in 2018, a new guideline stated that egg-allergic patients can be vaccinated safely with any flu vaccine with no special precaution, regardless of egg-reaction history. This changes the previous recommendation which was that anyone who has a significant allergy to chicken egg proteins (e.g., anaphylaxis) should *not* receive these vaccines. Note also that a killed influenza vaccine (Flucelvax) made in calf kidney cell culture is available for patients allergic to egg protein (see Table 38–4). This vaccine has two advantages: it contains no chicken proteins and it has a short turnaround time, so the latest drift mutant can be used.

In addition, a recombinant vaccine (Flublok) is available. This vaccine is made by inserting the gene encoding the viral hemagglutinin into an insect virus (baculovirus) that is propagated in insect cell culture. The insect cells produce the hemagglutinin of influenza virus. Flublok contains purified hemagglutinin as the immunogen without chicken egg proteins. This vaccine also has a short turnaround time and can be given to those with egg allergy.

The killed vaccine induces serum IgG and IgA as well as secretory IgA. However, protection wanes rapidly and typically lasts only 6 months. Yearly boosters are recommended and should be given shortly before the flu season (e.g., in October). These boosters also provide an opportunity to immunize against the latest antigenic changes.

The vaccine should be given to all persons 6 months and older who do not have a contraindication to receive the vaccine. It is especially important that people with chronic diseases, particularly respiratory and cardiovascular conditions, receive the vaccine. It should also be given to healthcare personnel who are likely to transmit the virus to those at high risk. Immunization of pregnant women with the killed vaccine is recommended as that decreases the risk of influenza in the newborn. Transplacental IgG protects the newborn during the first 6 months when the child is not capable of responding vigorously to the vaccine itself.

One side effect of the influenza vaccine used in the 1970s containing the swine influenza strain that caused influenza in humans was an increased risk of Guillain-Barré syndrome, which is characterized by an ascending paralysis. Analysis of the side effects of the influenza vaccines in use during the last 10 years has shown no increased risk of Guillain-Barré syndrome.

In addition to the vaccine, influenza can be prevented by using oseltamivir (Tamiflu), or baloxavir (Xofluza) which are described in the treatment section earlier. These drugs are particularly useful in elderly people who have not been immunized and who may have been exposed. Note that these drugs should not be thought of as a substitute for the vaccine. Immunization is the most reliable mode of prevention.

2. Avian Influenza Virus Infection in Humans

H5N1 Influenza Virus

In 1997, the H5N1 strain of influenza A virus that causes **avian influenza**, primarily in chickens, caused an aggressive form of human influenza with high mortality in Hong Kong. In the Winter of 2003 to 2004, an outbreak of avian influenza caused by H5N1 strain killed thousands of chickens in several Asian countries. Millions of chickens were killed in an effort to stop the spread of the disease. Four hundred eight human cases of H5N1 influenza occurred between 2003 and February 2009, resulting in 254 deaths (a mortality rate of 62%). Note that these 408 people were infected directly from chickens. Both the respiratory secretions and the chicken guano contain infectious virus.

The spread of the H5N1 strain from person to person occurs rarely but remains a major concern because it could increase dramatically if reassortment with the human-adapted strains occurs. In 2005, the H5N1 virus spread from Asia to Siberia and into Eastern Europe, where it killed thousands of birds but has not caused human disease. As of this writing (January 2021), there have been no cases of human influenza caused by an H5N1 virus in the United States. However, there have been two cases of human influenza caused by an H7N2 strain of avian influenza virus.

The ability of the H5N1 strain to infect chickens (and other birds) more effectively than humans is due to the presence of a certain type of viral receptor throughout the mucosa of the chicken respiratory tract. In contrast, humans have this type of receptor only in the alveoli, not in the upper respiratory tract. This explains why humans are rarely infected with the H5N1 strain. However, when the exposure is intense, the virus is able to reach the alveoli and causes severe pneumonia.

The virulence of the H5N1 strain is significantly greater than the H1N1 and H3N2 strains that have been causing disease in humans for many years. This is attributed to two features of the H5N1 strain, namely, relative resistance to interferon and increased induction of cytokines, especially TNF. The increase in cytokines is thought to mediate the pathogenesis of the pneumonia and acute respiratory distress syndrome (ARDS) seen in H5N1 infection.

The H5N1 strain is sensitive to the neuraminidase inhibitors, oseltamivir (Tamiflu) and zanamivir (Relenza), but not to amantadine and rimantadine. Tamiflu is the drug of choice for both treatment and prevention. A vaccine against the H5N1 strain of influenza A virus is available.

TABLE 38–4 Types of Influenza Vaccines

Vaccine Made in Chicken Eggs	Vaccine Not Made in Chicken Eggs
1. Contains inactivated virus (killed vaccine)	1. Virus grown in calf kidney cell culture then inactivated
2. Contains live, attenuated temperature-sensitive mutant virus (live vaccine)	2. Recombinant insect virus containing influenza virus hemagglutinin (HA) gene grown in insect cells; purified HA is the antigen

Note that other H5 subtypes, such as H5N8, also cause both human and avian disease. These subtypes have caused widespread outbreaks of influenza in both wild and domestic birds.

H7N9 Influenza Virus

In 2013, an outbreak of influenza caused by an H7N9 strain of influenza virus A in humans occurred. Prior to this time, the H7N9 strain affected only birds, especially chickens. Annual outbreaks have occurred, up to and including 2017. A total of 1258 infections in humans have been documented, and 41% of these patients died. Cases have occurred primarily in China and Taiwan. There has been no sustained person-to-person spread.

All of the genes of this virus are of avian origin. It acquired its H7 gene from ducks and its N9 gene from wild birds, and all the other genes are from an influenza strain that infects bramblings, a bird common in Asia and Europe. This H7N9 strain is susceptible to the neuraminidase inhibitors, oseltamivir and zanamivir. Candidate vaccines are being developed, but none are available as of this writing.

3. Swine Influenza Virus Infection in Humans

In April 2009, a novel swine-origin strain of influenza A (H1N1) virus (S-OIV) caused an outbreak of human influenza, which appeared first in Mexico, then in the United States, followed by spread to 208 countries by December 2009. The Centers for Disease Control and Prevention (CDC) uses the name "novel influenza A (H1N1)" for this virus.

As of December 2009, millions of cases have occurred worldwide. There have been so many cases that most countries have stopped documenting the number of cases. Worldwide there have been 9596 deaths, of which 1445 have occurred in the United States. On June 11, 2009, the World Health Organization (WHO) declared a level 6 pandemic (the highest level alert). By August 2010, the number of cases had declined significantly and the pandemic warning was rescinded. As of this writing in July 2021, no cases have been reported in the United States.

The disease affected primarily young people (60% of cases were 18 years old or younger). Symptoms were in general mild, with the few fatalities occurring in medically compromised patients. There was no outbreak of swine influenza in pigs prior to this human outbreak. Eating pork does not transmit the virus.

S-OIV is a quadruple reassortant: The hemagglutinin, nucleoprotein, and nonstructural protein genes are of North American swine origin, the neuraminidase and matrix protein genes are of Eurasian swine origin, the genes that encode two subunits of the polymerase are of North American avian origin, and the gene that encodes the third subunit of the polymerase is of human H3N2 origin.

A triple reassortant strain circulated in North American swine for several years prior to 2009 but caused human influenza only rarely. In the triple reassortant strain, all five of the genes that are not polymerase genes are of North American swine origin and the polymerase genes have the same origin as the quadruple reassortant. This strain does not have genes of Eurasian swine origin.

The key point is that most people worldwide do not have protective antibodies against the swine hemagglutinin of S-OIV even though they may have antibodies against the seasonal strain of H1N1 virus acquired either by immunization or by exposure to the virus itself. Note also that S-OIV spreads readily from human to human in contrast to the avian H5N1 strain, which does not.

A PCR test for the diagnosis of S-OIV infection is available. S-OIV is sensitive to oseltamivir and zanamivir but resistant to amantadine and rimantadine. Both an inactivated and a live, attenuated vaccine against S-OIV became widely available in November 2009.

PARAINFLUENZA VIRUS

Diseases

Parainfluenza virus causes croup (acute laryngotracheobronchitis), laryngitis, bronchiolitis, and pneumonia in children and a disease resembling the common cold in adults.

Important Properties

The genome RNA and nucleocapsid are those of a typical paramyxovirus (see Table 38–2). The surface spikes consist of hemagglutinin (H), neuraminidase (N), and fusion (F) proteins. The fusion protein mediates the formation of multinucleated giant cells. The H and N proteins are on the same spike; the F protein is on a separate spike. Both humans and animals are infected by parainfluenza viruses, but the animal strains do not infect humans. There are four types, which are distinguished by antigenicity, cytopathic effect, and pathogenicity (see later). Antibody to either the H or the F protein neutralizes infectivity.

Summary of Replicative Cycle

After adsorption to the cell surface via its hemagglutinin, the virus penetrates and uncoats and the virion RNA polymerase transcribes the negative-strand genome into mRNA. Multiple mRNAs are synthesized, each of which is translated into the specific viral proteins; no polyprotein analogous to that synthesized by poliovirus is made. The helical nucleocapsid is assembled, the matrix protein mediates the interaction with the envelope, and the virus is released by budding from the cell membrane.

Transmission & Epidemiology

These viruses are transmitted via **respiratory droplets**. They cause disease worldwide, primarily in the Winter months.

Pathogenesis & Immunity

These viruses cause upper and lower respiratory tract disease without viremia. A large proportion of infections are subclinical. Parainfluenza viruses 1 and 2 are **major causes of croup**.

Parainfluenza virus 3 is the most common parainfluenza virus isolated from children with lower respiratory tract infection in the United States. Parainfluenza virus 4 rarely causes disease, except for the common cold.

Clinical Findings

Parainfluenza viruses are best known as the main cause of croup in children younger than 5 years of age. Croup is characterized by a harsh, barking cough, and hoarseness. In addition to croup, these viruses cause a variety of respiratory diseases such as the common cold, pharyngitis, laryngitis, otitis media, bronchitis, and pneumonia.

Laboratory Diagnosis

Most infections are diagnosed clinically. A laboratory diagnosis can be made by detecting parainfluenza virus RNA in respiratory tract specimens by using a PCR-based assay. The diagnosis can also be made by isolating the virus in cell culture, by detecting viral antigens using fluorescent antibody, or by observing a fourfold or greater rise in antibody titer.

Treatment & Prevention

There is neither antiviral therapy nor a vaccine available.

RESPIRATORY SYNCYTIAL VIRUS

Diseases

RSV is the most important cause of pneumonia and bronchiolitis in infants. It is also an important cause of otitis media in children and of pneumonia in the elderly and in patients with chronic cardiopulmonary diseases.

Important Properties

The genome RNA and nucleocapsid are those of a typical paramyxovirus (see Table 38–2). Its surface spikes are **fusion proteins**, not hemagglutinins or neuraminidases. The fusion protein causes cells to fuse, forming **multinucleated giant cells (syncytia)**, which give rise to the name of the virus (Figure 38–3).

Humans are the natural hosts of RSV. RSV has two serotypes, designated subgroup A and subgroup B. Antibody against the fusion protein neutralizes infectivity.

Summary of Replicative Cycle

Replication is similar to that of parainfluenza virus (see page 312).

Transmission & Epidemiology

Transmission occurs via **respiratory droplets** and by direct contact of contaminated hands with the nose or mouth. RSV causes **outbreaks** of respiratory infections every Winter, in contrast to many other "cold" viruses, which reenter the community every few years. It occurs worldwide, and virtually everyone has been infected by the age of 3 years. RSV also causes outbreaks of

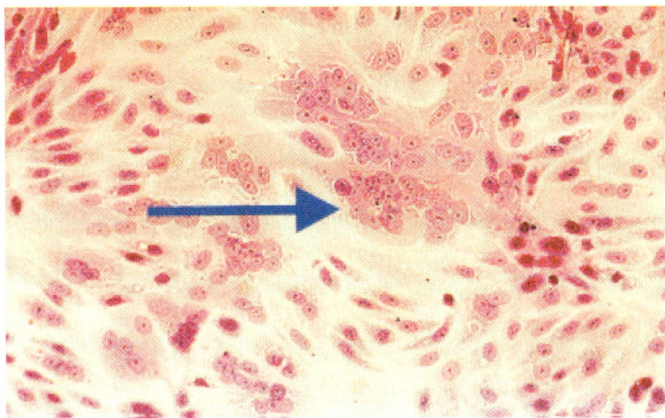

FIGURE 38–3 Multinucleated giant cells of respiratory syncytial virus (RSV). The blue arrow points to a multinucleated giant cell produced by RSV infection of fibroblasts in cell culture. (Used with permission from Dr. L. Stannard. Copyright University of Cape Town, 2016.)

respiratory infections in **hospitalized infants**; these outbreaks can be controlled by handwashing and use of gloves, which interrupt transmission by hospital personnel.

Pathogenesis & Immunity

RSV infection in **infants is more severe** and more often involves the lower respiratory tract than in older children and adults. The infection is localized to the respiratory tract; viremia does not occur.

The severe disease in infants may have an **immunopathogenic** mechanism. Maternal antibody passed to the infant may react with the virus, form immune complexes, and damage the respiratory tract cells. Trials with a killed vaccine resulted in more severe disease, an unexpected finding that supports such a mechanism.

Most individuals have multiple infections caused by RSV, indicating that immunity is incomplete. The reason for this is unknown, but it is not due to antigenic variation of the virus. IgA respiratory antibody reduces the frequency of RSV infection as a person ages.

Clinical Findings

In infants, RSV is an important cause of lower respiratory tract diseases such as bronchiolitis and pneumonia. RSV is also an important cause of otitis media in young children. In older children and young, healthy adults, RSV causes respiratory tract infections such as the common cold and bronchitis. However, in the elderly (people older than 65 years of age) and in adults with chronic cardiopulmonary diseases, RSV causes severe lower respiratory tract disease, including pneumonia.

Laboratory Diagnosis

A laboratory diagnosis can be made by detecting the RNA of RSV in respiratory tract specimens using a PCR-based assay. An enzyme immunoassay ("rapid antigen test") that detects the presence of RSV antigens in respiratory secretions is also commonly used.

The presence of the virus can be detected by immunofluorescence on smears of respiratory epithelium or by isolation in cell culture. The cytopathic effect in cell culture is characterized by the formation of multinucleated giant cells. A fourfold or greater rise in antibody titer is also diagnostic.

Treatment

Aerosolized ribavirin (Virazole) is recommended for severely ill hospitalized infants, but there is uncertainty regarding its effectiveness. Ribavirin is not recommended for general use. A combination of ribavirin and hyperimmune globulins against RSV may be more effective.

Prevention

There is no vaccine. Previous attempts to protect with a killed vaccine resulted in an increase in severity of symptoms. Passive immunization with a monoclonal antibody directed against the fusion protein of RSV (palivizumab, Synagis) can be used for prophylaxis in premature or immunocompromised infants. Hyperimmune globulins (RespiGam) are also available for prophylaxis in these infants and in children with chronic lung disease. Nosocomial outbreaks can be limited by handwashing and use of gloves.

HUMAN METAPNEUMOVIRUS

HMPV was first reported in 2001 as a cause of severe bronchiolitis and pneumonia in young children in the Netherlands. It is a member of the paramyxovirus family and, as such, is an enveloped virus with a single-stranded, nonsegmented, negative-polarity RNA genome. One of the spikes on its surface is a fusion protein that causes multinucleated giant cells in infected respiratory tract tissue. The fusion protein mediates attachment to the cell, and antibody against the fusion protein prevents infection. HMPV has two genotypes and several subtypes.

It is similar to RSV (also a paramyxovirus) in the range of respiratory tract disease it causes, namely, mild upper respiratory infections to bronchiolitis to severe pneumonia. Fever, coryza, wheezing, and cough are the most common symptoms. Serologic studies showed that most children have been infected by 5 years of age. Immunity is incomplete, and reinfection occurs despite development of an antibody response.

Laboratory diagnosis typically involves detection of viral RNA in respiratory tract samples by using a PCR assay. Treatment is supportive. There is no effective antiviral drug and no vaccine.

CORONAVIRUS

Diseases

Coronaviruses are an important cause of the common cold, probably second only to rhinoviruses in frequency. In 2002, a new disease, an atypical pneumonia called severe acute

respiratory syndrome (SARS), emerged. In 2012, another severe pneumonia called Middle East respiratory syndrome (MERS) emerged in that area of the world. These pneumonias are caused by SARS coronavirus (SARS-CoV) and MERS coronavirus (MERS-CoV), respectively.

In December 2019, an outbreak of pneumonia in Wuhan, China caused by a new coronavirus was reported. This virus, now named SARS-CoV-2, is causing a world-wide pandemic. This virus is discussed separately in a section below entitled "Coronavirus Outbreak and Global Pandemic in 2019 to 2021."

Important Properties

Coronavirus has a nonsegmented, single-stranded, positive-polarity RNA genome (see Table 38–2). It is an enveloped virus with a helical nucleocapsid. There is no polymerase in the virion. In the electron microscope, prominent club-shaped spikes in the form of a corona (halo) can be seen (Figure 38–4). A cross-sectional model of a coronavirus is shown in Figure 38–5. Note the spike proteins on the surface and the coiled RNA genome in the interior of the virion.

There are seven serotypes of human coronaviruses, four of which cause upper respiratory tract infections, such as the common cold. The other three cause lower respiratory tract infection, namely pneumonia. The coronaviruses that cause pneumonia are SARS-CoV, SARS-CoV-2, and MERS-CoV. The antigenicity of the viral spike protein of these three viruses is different from each other. The antigenicity of the common cold strains is relatively stable. However, antigenic variants of SARS-CoV-2 have appeared during the pandemic and are causing significant disease.

The receptor for the SARS-CoV and SARS-CoV-2 on the surface of human cells is angiotensin-converting enzyme-2 (ACE-2) protein. The other coronaviruses use different cell surface peptidases as their receptor.

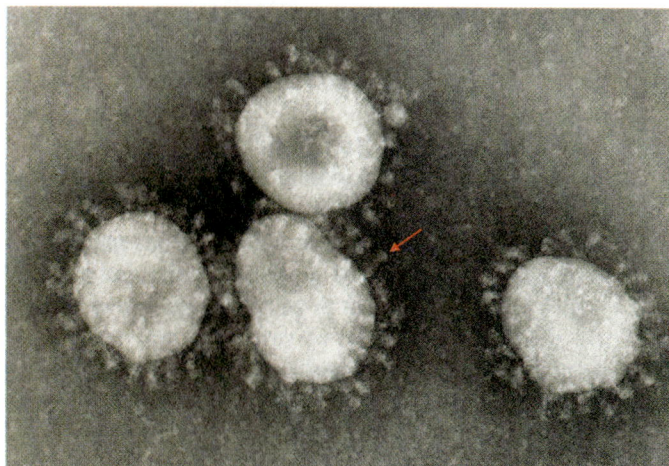

FIGURE 38–4 Electron micrograph of coronavirus. Note crown (corona) of spikes protruding from virion envelope. Arrow points to one of the spikes. (Used with permission from Dr. Fred Murphy and Dr. Sylvia Whitfield, Public Health Image Library, Centers for Disease Control and Prevention.)

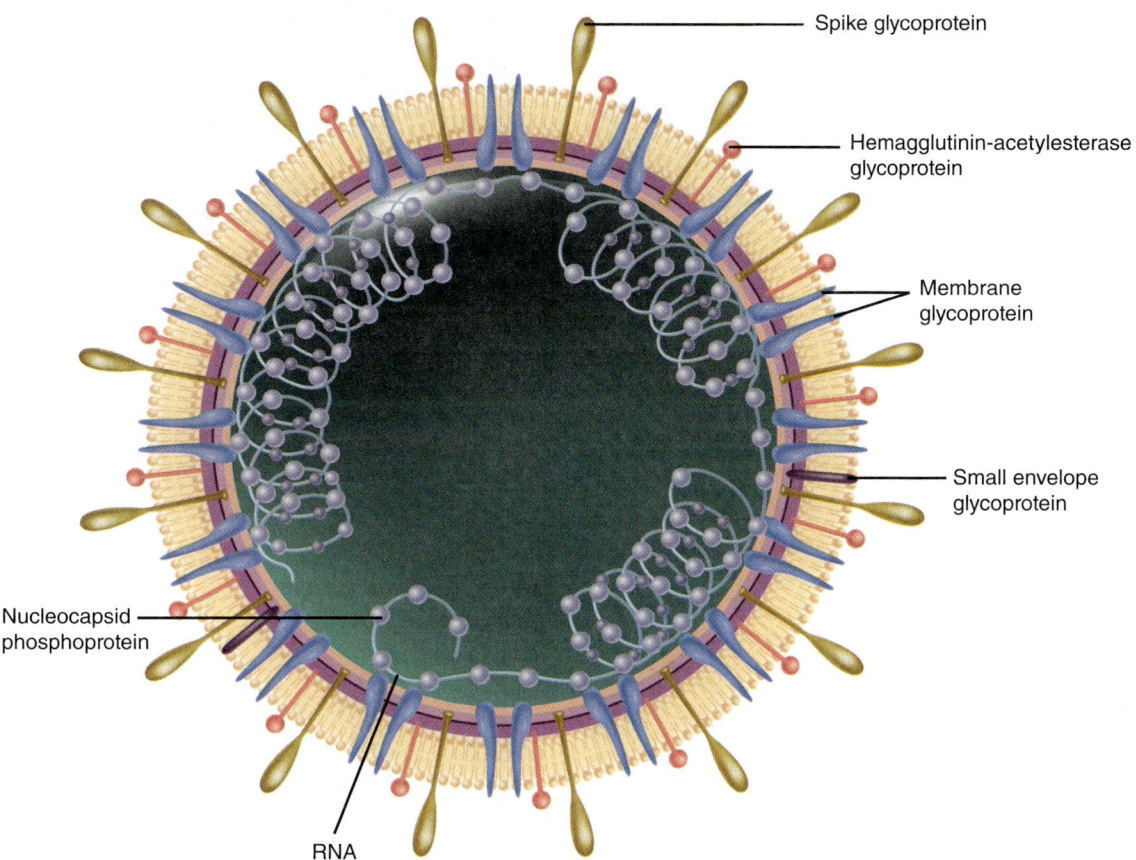

Spike glycoprotein

Hemagglutinin-acetylesterase
glycoprotein

Membrane
glycoprotein

Small envelope
glycoprotein

Nucleocapsid
phosphoprotein

RNA

FIGURE 38–5 Cross-section of a coronavirus. Note the club-shaped spike proteins on the surface that appear as a halo or corona in the electron microscope. Spike proteins attach to receptors on the surface of respiratory tract mucosa. Other surface proteins are shown as well. The genome RNA is depicted as the coiled strand within the virion. (Reproduced with permission from Ryan K, Ahmad N, Alspaugh JA, et al: *Sherris Medical Microbiology*, 7th ed. New York, NY: McGraw Hill; 2018.)

Summary of Replicative Cycle

The replicative cycle begins when the spike protein on the surface of the virion binds to the receptor, often the ACE-2 protein, on the cell surface. A cell surface protease cleaves the spike protein to reveal a fusion protein that mediates entry into the cytoplasm, where the virion is uncoated. The positive-strand genome is translated into two large polypeptides, which are cleaved by two virus-encoded proteases into functional viral proteins. One of these proteins is the RNA polymerase that synthesizes both the progeny genome and the mRNAs that are translated into the structural proteins of the progeny virions. The mRNAs form a set of "nested" RNAs that are a characteristic feature of coronaviruses. The virus is assembled and obtains its envelope from the endoplasmic reticulum, not from the plasma membrane. Replication occurs in the cytoplasm.

Transmission & Epidemiology

Coronaviruses are transmitted primarily by the respiratory route via sneezing and coughing. Transmission via contact of hands with contaminated surfaces also occurs. Infection by the common cold coronaviruses occurs worldwide and early in life, as evidenced by finding antibody in more than half of children.

Outbreaks occur primarily in the Winter on a 2-to 3-year cycle. This seasonality is less dramatic than that of influenza virus.

SARS-CoV originated in China in November 2002 and spread rapidly to other countries. As of this writing, there have been 8300 cases and 785 deaths—a fatality rate of approximately 9%. Human-to-human transmission occurs, and some patients with SARS are known to be "super-spreaders." Early in the outbreak, many hospital personnel were affected, but instituting strict respiratory infection control procedures greatly reduced the spread within hospitals.

There are many animal coronaviruses in both domestic and wild animals. They are suspected of being the source of SARS-CoV. The horseshoe bat appears to be the natural reservoir for SARS-CoV, with the civet cat serving as an intermediate host.

In 2012 to 2013, a new human coronavirus caused an outbreak of serious, often fatal pneumonia in Saudi Arabia and other countries in that region. The disease is called MERS, and the virus is called MERS-CoV. As of 2019, approximately 2400 cases of MERS have been reported with a mortality rate of 35%.

The closest relative of MERS-CoV is a bat coronavirus, and bats are thought to be a reservoir. Close contact with camels appears to be the mode of transmission to humans. The risk of person-to-person transmission is low but has occurred in hospitals with inadequate infection control.

Pathogenesis & Immunity

Infection by the "common cold" coronaviruses typically is limited to the mucosal cells of the respiratory tract. Approximately 50% of infections are asymptomatic. Immunity following infection appears to be brief, up to 2 years, and reinfection can occur.

Pneumonia caused by SARS-CoV is characterized by diffuse edema in the alveoli resulting in hypoxia. Infection by SARS-CoV-2 involves not only the lung but other organs as well. See the separate section on SARS-CoV-2 later.

Clinical Findings

The common cold caused by coronavirus is characterized by coryza (rhinorrhea, runny nose), scratchy sore throat, and low-grade fever. This illness typically lasts several days and has no long-term sequelae. Coronaviruses also cause bronchitis.

SARS is a severe atypical pneumonia characterized by a fever of at least 38°C, nonproductive cough, dyspnea, and hypoxia. Chills, rigors, malaise, and headache commonly occur, but sore throat and rhinorrhea are uncommon. Chest X-ray reveals interstitial "ground-glass" infiltrates that do not cavitate. Leukopenia and thrombocytopenia are seen. The incubation period for SARS ranges from 2 to 10 days, with a mean of 5 days. The clinical findings of MERS are similar to those of SARS.

Laboratory Diagnosis

If SARS or MERS is suspected, PCR-based tests can be used to detect coronavirus RNA in respiratory tract specimens. Antibody-based tests to detect a rise in antibody titer can be used for epidemiologic purposes.

Treatment & Prevention

There is no antiviral therapy available for SARS-CoV, MERS-CoV, or any of the common cold strains. Some antiviral drugs are effective against SARS-CoV-2 and are discussed in the next section. There is no vaccine available against any common cold coronavirus, SARS-CoV, or MERS-CoV.

Coronavirus Outbreak and Global Pandemic in 2019

Introduction to the Pandemic

In December 2019, an outbreak of pneumonia in Wuhan, China caused by a new, novel Coronavirus occurred. This virus is named SARS-CoV-2 and the disease is named COVID-19. (COVID stands for Coronavirus Disease and 19 stands for the year 2019.) **WHO declared it a global pandemic on March 11, 2020**.

As of this writing on September 7, 2021, the virus has caused approximately 221 million cases, and more than 4.5 million deaths world-wide. The fatality rate is approximately 2%. In the United States, approximately 39 million cases and 643 thousand deaths have been reported. The fatality rate in the United States is approximately 1.6%. World-wide, approximately 45.3 billion vaccine doses have been administered.

As of June 2021, the number of cases and deaths from COVID-19 declined significantly, mostly as a result of widespread immunization. However, in July and August 2021, a fourth wave of infections occurred. This increase in infections is the result of three factors, **waning immunity** induced by the vaccine, the **delta variant** that has increased transmissibility, and **relaxation of public health measures** such as masking.

The fundamental reason why this virus caused a global pandemic is that it is a new (novel) strain of coronavirus to which the human population had no preexisting immunity. The virus emerged from the animal population with spike protein antigens on its surface to which no one had antibodies.

Origin of the Novel Coronavirus

Based on genome RNA sequencing results, SARS-CoV-2 closely resembles a coronavirus of bats and that animal is likely to be the natural reservoir. A second, intermediate reservoir, the pangolin, may be involved. Sequences of the pangolin CoV spike protein are found in the SARS-CoV-2 spike protein.

In May 2021, the original hypothesis that SARS-COV-2 spread from an animal to people at "wet animal" market in Wuhan, China was questioned. An alternative possibility that workers at the Virus Laboratory in Wuhan China may have been accidently infected and unintentionally spread the virus to others was raised.

Sequencing of the spike protein gene during the course of this pandemic revealed several mutations that are causing disease in many areas. For example, a mutation designated D614G in the gene encoding the spike protein favors an "open" configuration of the receptor-binding domain on the spike protein thereby increasing the ability of the virus to infect the cell. The mutant D614G virus (also called G strain) was, at the time, the most common virus isolated from COVID-19 patients world-wide.

In December 2020, a new mutant strain designated B.1.1.7 (now known as Alpha variant) was reported to be spreading widely. This variant has 17 mutations, 8 of which are in the gene for the spike protein. This variant is more transmissible and may be more virulent than the existing strain. The current vaccines elicit antibodies that appear to be protective against this new variant but booster immunizations specific for the variants may be necessary.

In January 2021, additional variants appeared in South Africa, Brazil, India, and in the United States. Initial studies showed that these variants have increased transmission, and may be more virulent, that is, cause death more frequently. In addition to the B.1.1.7 variant mentioned above, the B.1.351 (Beta) variant that emerged in South Africa, the P.1 (Gamma) variant that emerged in Japan and Brazil, and the B.1.617.2 (Delta) variant that emerged in India, are all circulating in the United States. The Delta variant is now the dominant strain in the United States.

Recent studies have shown that two doses of the existing vaccines will provide reasonable protection against these new variants.

These four variants, B.1.1.7 (Alpha), B.1.351 (Beta), P.1 (Gamma), and B.1.617.2 (Delta) have been designated "Variants of Concern" by WHO and CDC.

Attachment and Entry of Virus into Cell

The main receptor for SARS-CoV-2 is the **ACE-2 protein** on the surface of respiratory tract epithelium. Binding of the spike protein on the surface of the virus to the ACE-2 receptor is the first step in the entry of the virus into the cell. After binding, a **cell surface protease, transmembrane protease serine-2 (TMPRSS-2)**, cleaves the spike protein to reveal a fusion protein sub-unit that mediates fusion of the virus with the cell membrane and entry of the virion into the cytoplasm.

Antibody to the virus elicited either by natural infection or by a vaccine neutralizes the virus by preventing the binding of the spike protein to the ACE-2 receptor. The relatively low number of cases of COVID-19 in children is attributed to the low number of the ACE-2 receptor displayed on their cells.

Another receptor on the cell surface for SARS-CoV-2 is neuropilin-1 (NRP-1). The spike protein of the virus binds to NRP-1 and the virus can enter the cell. Antibody to NRP-1 prevents infection of the cell suggesting that it could be an additional target for drugs or vaccines.

Transmission

The primary mode of transmission is inhalation of respiratory droplets generated by coughing, sneezing, or talking. Respiratory aerosols also play important role in transmission of this virus. Note that aerosols are smaller than droplets so stay in the air longer and can be distributed over a distance of more than 6 feet by air currents.

Shedding of virus by an infected patient typically begins 2 to 3 days before symptom onset and lasts for about 7 days. Shedding of virus prior to the appearance of symptoms explains the well-recognized phenomenon of *asymptomatic transmission*.

Transmission by hand contact with surfaces containing virus also occurs. Fingers transport the virus on the surface to the recipient's eyes, nose, or mouth.

The majority of infections are acquired by transmission from *asymptomatic* carriers, probably by droplets/aerosols generated by talking, singing, or shouting. This indicates that a large amount of virus is present in the upper respiratory tract. People with asymptomatic infections are thought to be capable of transmitting the virus during a period that begins several days after the time of infection and lasts for about a week. A rough approximation is, therefore, about 10 days after the time of infection.

Vertical transmission from an infected mother to the fetus is very rare. Transmission from an infected parent to a neonate is uncommon but can result in symptomatic COVID-19 disease in the neonate.

"Super-spreader" events at which large numbers of people have been infected have been observed. These typically involve large numbers of people gathering, often indoors, and not wearing masks or observing the 6 foot social distance. It is thought

TABLE 38–5 Clinical Features of COVID-19

Organ Affected	Clinical Manifestation
Lung	Pneumonia, ARDS with "cytokine storm"
Heart	Myocarditis
Nervous system/Brain	Encephalopathy, anosmia, dysgeusia
Gastrointestinal tract	Nausea, vomiting, diarrhea
Kidney	Renal failure
Blood vessels	Thrombi, emboli

that "super-spreading" is caused not by an individual producing exceptionally large amounts of infectious virus but rather by the combined effect of the environmental factors conducive to spread as mentioned in the previous sentence.

Nursing homes, prisons, and homeless shelters have high rates of infection due to the crowded living conditions. Nursing homes in particular have many serious infections because the population is elderly.

Coronavirus infections exhibit less seasonality that do influenza virus infections. The increase in SARS-CoV-2 infections in the summer of 2020 indicates that this coronavirus is not exhibiting a drop-off in infections during the warmer months the way influenza does. Nevertheless, a world-wide study performed in January to March 2020 indicated a correlation of outbreaks of COVID-19 within a narrow band of latitude with low temperature and low humidity.

Clinical Findings

The incubation period ranges from 2 to 14 days with a mean of 5 days. A quarantine period of 10 to 14 days has, therefore, been instituted.

The main clinical findings are fever, dry cough, and shortness of breath. Sore throat may occur but is not a prominent feature. Systemic symptoms such as fatigue, shaking chills, headache, and myalgia also may occur.

In addition to the respiratory tract, other organs such as the heart, kidney, brain, and gastrointestinal tract can be affected (Table 38–5). A severe myocarditis with symptoms resembling a myocardial infarction has occurred in some patients. Loss of ability to smell (anosmia) and abnormal ability to taste (dysgeusia) are the initial symptoms in some patients. Anosmia and dysgeusia are important diagnostic features of COVID-19. Encephalopathy has also been observed. Nausea, vomiting, and diarrhea have occurred in some patients. Blood clots occur in some patients resulting in thrombosis or emboli leading to an increased risk of stroke.

Many of these findings are caused not by the virus directly but by the **cytokine release syndrome**, also known as **cytokine storm**. Viral infection triggers an overproduction of cytokines such as interferon-gamma, TNF, IL-6, bradykinin, and other pro-inflammatory cytokines (see Pathophysiology section below).

Other clinical findings include inflammation of the toes ("COVID toes") and a Kawasaki disease-like syndrome that has occurred in children. The latter has been given the name "pediatric multisystem inflammatory syndrome." It is also known as multisystem inflammatory syndrome in children (MIS-C). Another important clinical finding is the observation that superimposed bacterial infections in the respiratory tract or blood occur in many patients.

Lymphopenia is common. Serum C-reactive protein and lactate dehydrogenase are elevated. Elevated D-dimers in the plasma occur in some patients resulting from clotting abnormalities. Respiratory tract specimens are typically negative for other respiratory viruses, including influenza virus.

Chest X-ray typically reveals bilateral opacities, often with a "ground-glass" appearance. No pleural effusions are seen. Mechanical ventilation is often required in patients with severe disease.

Death is due to hypoxemic respiratory failure. Cytokine storm contributes to death in some patients. A commonly listed cause of death is ARDS.

Overall hospital mortality rates range from 15% to 20% but up to 40% for ICU admissions. Mortality rates differ greatly by age. The death rate per 1000 COVID-19 cases is 1.1 for those in the 18 to 29 year age group but 210.5 for those 75 to 84 years old and 304.9 for those 85 years or older.

In patients who have recovered from COVID-19 disease, the duration of antibody-mediated immunity appears to be at least 6 months. IgG antibody to spike protein was detectable for more than 6 months as were memory B cells. However, CD-4 positive T cells and CD-8 positive T cells declined with a half-life of 3 to 5 months.

Some patients who have recovered from the initial severe symptoms, continue to have symptoms, such as prolonged cough, shortness of breath, chest pain, joint pain, fatigue, dizziness, and confusion for several months. These after-effects cause significant limitations on the quality of life. This is occurring in many who did not have prior preexisting conditions. Some also continue to have positive PCR tests indicating that virus is still present. WHO has named this syndrome "Long COVID." These patients are also referred to as "long haulers." These long-term symptoms are also called "post-acute sequelae of SARS-CoV-2" abbreviated PASC.

Approximately 50% of infections are asymptomatic and many others have only mild respiratory tract disease. Older adults (especially those over 70 years of age), people with compromised immunity, diabetics, and those with chronic heart, kidney, or respiratory tract disease are more likely to have serious COVID-19 disease. Obese people, specifically those with a body mass index of 30 or greater, are at high risk of serious disease. People with a history of smoking or vaping have a significant risk of severe COVID-19 disease.

In general, the symptoms of COVID-19 in children are less severe and death occurs much less often. This is attributed to two observations: that there are fewer ACE-2 receptors in children than adults and that children mount a stronger innate immune response than do older adults.

In August 2020, reports of reinfection began to appear. For example, a patient with documented COVID-19 infection in April 2020, recovered, and tested negative. In August, although asymptomatic, he tested positive again. Genome analysis revealed he was infected with two different clades of the virus.

Men have more COVID-19 infections than women. This is attributed to the up-regulation by **testosterone** of TMPRSS-2, a protease on the cell surface involved with entry of the virus into the cell.

Pathology

Microscopic examination of the lung tissue of deceased patients showed marked alveolar damage, edema, and infiltration of T lymphocytes. Severe endothelial damage was observed. Widespread thrombosis of capillaries was seen in many organs. Microangiopathy and angiogenesis were also observed.

Pathophysiology

The respiratory manifestations are likely to have two pathogenetic mechanisms: one is the killing of alveolar cells by the virus. Accumulated cell debris blocks diffusion of oxygen into the capillaries resulting in hypoxia.

The other is killing of the endothelial cells of the capillaries lining the alveoli. This triggers blood clots and an immune-mediated "cytokine storm" (also known as cytokine release syndrome) resulting in further damage to the alveolar membrane and ARDS. High levels of proinflammatory cytokines such as IL-1, IL-6, bradykinin, and TNF are found. Damage to endothelial cells and the resulting blood clots are the cause of many of the extra-respiratory manifestations of COVID-19 disease.

SARS-CoV-2 inhibits interferon synthesis, an important part of the innate immune response. The N (nucleocapsid) protein blocks the action of the RIG receptor that detects dsRNA in the cytoplasm thereby inhibiting interferon synthesis.

Laboratory Diagnosis

There are three types of laboratory tests. Two of these detect the presence of the virus, either by PCR test detecting the viral RNA or by enzyme immunoassay detecting a viral antigen. The third type of test detects antibodies to the virus in a person's serum.

Laboratory diagnosis for the presence of *viral RNA* is made by PCR testing on respiratory tract specimens, such as nasopharyngeal swabs. Although very sensitive and specific, PCR test can be falsely negative if the specimen is taken too soon after infection. The test is most sensitive at 3 days after symptom onset. A rapid isothermal amplification assay for use in the field has been developed. This test does not require cycles of heating and cooling.

Rapid laboratory tests for *viral antigen*, typically the spike protein, are also available. These enzyme immunoassay tests are not as sensitive as PCR tests so false-negative results can occur. A negative result does not rule out infection. The advantage of this type of test is that they are rapid, inexpensive, and can be done by the user at home, much like a pregnancy test. In August 2020, the FDA approved a "point-of-care" antigen test that detects the nucleocapsid antigen of SARS-CoV-2 using a nasal swab specimen.

Tests to detect IgM and IgG *antibodies to the virus* are also available. IgM antibodies can be detected by 5 days after infection but IgG antibodies are first detected 14 days after symptom onset. Note that these tests indicate an antibody response but they do *not* indicate that the virus is present at the time of the test. A positive test indicates that the person has been infected but not that the person can transmit the virus to others.

Antibody tests have greater than 95% specificity. Tests that detect total (IgM and IgG) antibody are the most sensitive. There is a question about which antigen, the nucleocapsid (NC) protein or the spike protein (or both) should be used in the tests. NC protein is most abundant so testing for that protein would be most sensitive. But tests that detect antibody to the spike protein are most likely to detect neutralizing antibody.

Another question is how long antibody elicited by infection, as distinct from immunization, will protect from reinfection. In July 2020, a study reported that patients with mild disease showed a rapid decline in antibody titer during the 3 months following recovery. Note that the duration of detectable antibody in patients from the SARS outbreak in 2003 lasted about 2 to 3 years.

PCR tests for viral RNA are likely to be positive during the first week of symptomatic infection but antibody tests are likely to be positive only after 3 weeks of symptomatic infection. It is recommended that people wait at least 3 weeks after symptoms appear to have antibody testing done.

PCR tests are currently reported as either positive or negative, not as the viral load as is done with HIV infection. There is a proposal to report the Cycle Threshold (CT) as an indirect measure of the amount of virus present in the patient's specimen. The PCR test undergoes repeated cycles of heating and cooling hence the term "Cycle" is used. If no viral RNA is detected after 40 cycles, the test is reported as negative. Note that if the specimen turns positive at a *low* CT then there is a *large* amount of virus present. Knowing this would be helpful in evaluating how likely the patient was to transmit this virus to others.

Treatment

Supportive care including supplemental oxygen should be instituted. Some hospitalized patients need respiratory support provided by mechanical ventilators. Prone positioning helps patients to breathe.

Therapeutic treatment modalities fall into four categories. There are drugs directed against viral replication, antibodies directed against viral replication, drugs directed against cytokine storm and antibodies directed against cytokine storm (Table 38–6).

On October 22, 2020, the FDA approved the use of **remdesivir** in hospitalized patients. Clinical trials showed that remdesivir shortened the hospital stay of seriously ill patients by 4 days and decreased mortality from 11% to 7%.

Remdesivir is an adenosine analogue that inhibits viral RNA-dependent RNA polymerase. It is a "chain-terminating" drug. **Favipiravir**, another RNA polymerase inhibitor, is also

TABLE 38–6 Important Treatment Modalities Used in COVID-19

Therapeutic Purpose	Drug	Antibody
Inhibit viral replication	Remdesivir (EUA)	1. Cocktail of two monoclonal antibodies (EUA); either casirivimab and imdevimab or bamlanimumab and etesevimab 2. Bamlanimumab (EUA) 3. Convalescent plasma (EUA)
Inhibit cytokine release syndrome (Cytokine Storm)	1. Dexamethasone 2. Baricitinib and remdesivir (EUA)	Monoclonal antibody against IL-6 receptor (e.g., tocilizumab) (EUA)

EUA = Emergency Use Authorization.

being tested. Hydroxychloroquine, the antimalarial drug, is not effective.

In addition to the viral polymerase, a second drug target is the virus-encoded protease. Unfortunately, a trial of the protease inhibitor combination lopinavir/ritonavir reported no clinical improvement. A trial of the protease inhibitor, atazanavir, is in progress.

In August 2020, the use of convalescent plasma received Emergency Use Authorization (EUA) from the FDA. Note that plasma is being used, not hyperimmune globulins.

Humanized monoclonal antibodies can be used for either treatment or prevention. Monoclonal antibodies avoid the risk of using human plasma which may transmit viruses in the donor plasma. On November 9, 2020, the FDA approved emergency authorization for bamlanivimab, an IgG monoclonal antibody directed against the spike protein of SARS-CoV-2. It should be used to treat those who have tested positive, have mild-to moderate disease, and are at risk of developing severe disease. It is not for hospitalized patients with severe disease.

On November 21, 2020, the FDA granted EUA for a combination of two monoclonal antibodies (casirivimab and imdevimab) for the treatment of mild to moderate COVID-19. These two antibodies bind to different areas on the Receptor Binding Domain of the spike protein. In February 2021, the FDA approved a EUA for the use of a second monoclonal antibody combination containing bamlanimumab and etesevimab. These monoclonal antibodies target different sections of the receptor-binding domain of the spike protein.

Nanobodies are single domain antibodies composed of the variable region of the heavy chain of an immunoglobulin. Their small size makes them very stable, an attribute that allows delivery via aerosol into the respiratory tract. Nanobodies directed against the CoV spike protein (termed Aeronabs) are being considered for either treatment of early disease or for prophylaxis.

A trial of intranasal interferon-beta showed treated patients were significantly more likely to recover than patients who did not receive the drug.

Regarding the treatment of cytokine storm, both **dexamethasone**, a corticosteroid, and **tocilizumab**, a monoclonal antibody against the IL-6 receptor, are approved for use. One recommended regimen consists of dexamethasone with or without remdesivir plus either tocilizumab or baricitinib, a Janus kinase inhibitor (see next paragraph).

Baricitinib and **ruxolitinib** (Janus kinase [JAK] inhibitors), are also being evaluated as an inhibitor of cytokine release. A combination of **baricitinib** and **remdesivir** reduced time to recovery in hospitalized patients in a recent clinical trial. On November 19, 2020, the FDA granted EUA for the combination of **baricitinib** and **remdesivir**.

Antioxidant compounds, such as glutathione (GSH) and N-acetyl **cysteine** inhibit NF kappa-B, a proinflammatory transcription factor. A trial using GSH to treat ARDS has shown some efficacy. Mesenchymal stem cell therapy is also undergoing clinical trial to reduce the effects of cytokine storm.

Prevention

There are three vaccines that have received an EUA as of February 27, 2021. On December 11, 2020, the FDA approved the Pfizer mRNA vaccine and 1 week later an EUA for the Moderna mRNA vaccine was granted. Both vaccines contain the mRNA for the coronavirus spike protein enclosed in a lipid nanoparticle. On August 23, 2021, the FDA gave full approval to the Pfizer vaccine.

On February 27, 2021, the FDA approved an EUA for the Johnson & Johnson vectored vaccine. This vaccine contains a replication-deficient human adenovirus into which the DNA encoding the spike protein of SARS-CoV-2 has been inserted. Table 38–7 describes the leading vaccines that have either received a EUA or are in advanced clinical trials.

Note that these vaccines are *monovalent* and stimulate an immune response to the spike protein of the original strain of SARS-CoV-2. Because of waning immunity and the surge of delta variant cases, booster doses for the Pfizer, Moderna, and Johnson & Johnson vaccines have been approved.

As of August 2021, a third dose (booster dose) of the two mRNA vaccines is being administered to immunocompromised people. Because of waning immunity and the surge of delta variant cases, booster doses may be administered to the general population in the Fall of 2021.

Despite a vaccine effectiveness of 80% to 90% in preventing severe COVID disease, people immunized with two doses nevertheless have contracted COVID-19 disease. In most cases, this "breakthrough" disease was mild, although in some, it was severe and a few patients died.

An unusual adverse effect of these mRNA vaccines is a rash at or near the site of infection called "COVID-Arm." It appears to be a delayed hypersensitivity response to some component of the vaccine. Onset of the rash occurred approximately 8 days after inoculation and the rash resolved without sequelae. Most patients did not experience a rash upon receiving the second dose but a few did.

TABLE 38–7 Important Types of Vaccines against COVID-19

Type of Vaccine	Immunogen in Vaccine	Comment
RNA	Messenger RNA for spike protein	Enclosed in lipid nanoparticles. Pfizer/BioNTech and Moderna vaccines have EUA approved by FDA .
Protein subunit	Recombinant spike protein	Saponin-based adjuvant added.
Vector	Human adenovirus (serotype 26) containing DNA encoding spike protein	Adenovirus is engineered to be nonreplicating. Johnson & Johnson vaccine has EUA approved by FDA.
Vector	Chimpanzee adenovirus containing DNA encoding spike protein	Chimpanzee adenovirus is non-replicating and non-pathogenic in humans. Oxford/Astra-Zenica vaccine approved by WHO.

Monoclonal antibodies can be used for prevention as well as treatment. These antibody preparations can be used either preexposure in people working in high-risk situations or postexposure in people known to be exposed to the virus. These antibodies can provide immediate protection that can last for weeks to months.

Prevention centers on public health and hygiene measures to interrupt transmission. Wearing a mask and social (physical) distancing at least 6 feet apart are the most important measures. Other basic measures include frequent hand washing, practicing cough and sneeze etiquette, avoiding touching eyes, nose, and mouth, avoiding people with respiratory symptoms, and staying at home if you are sick.

Those infected should be isolated. Isolation should be maintained for at least 10 days after onset of symptoms. Those who have been exposed but who are not known to be infected should be quarantined for 10 to 14 days.

Contact tracing to identify those who have been exposed is another important tool to interrupt the chain of transmission. Personal protective equipment (PPE) for medical personnel is essential to interrupt transmission in hospitals.

In the year 2020, many schools and colleges were closed or classes presented remotely. Large gatherings such as sporting events and concerts were cancelled or postponed. Many states issued "Stay at Home" orders, allowing only essential services to remain open.

The safe transition to the "New Normal" will depend on the presence of antibodies in the individual acquired either by the infection or by the vaccine. If enough people develop immunity, then **herd immunity** may protect those who are still susceptible. The percentage of immune people required to develop effective herd immunity against SARS-CoV-2 is unknown at this time, but is estimated to be 60% to 70%.

Whether this virus becomes endemic and recurs at regular intervals or, like the SARS-CoV-2003 coronavirus no longer causes disease in humans, remains to be seen.

RHINOVIRUS

Disease

This virus is the main cause of the common cold.

Important Properties

Rhinovirus has a nonsegmented, single-stranded, positive-polarity RNA genome. It is a nonenveloped virus with an icosahedral nucleocapsid. There is no polymerase within the virion (see Table 38–2).

There are **more than 100 serologic types**, which explains why the common cold is so common. They **replicate better at 33°C** than at 37°C, which explains why they affect primarily the nose and conjunctiva rather than the lower respiratory tract. Because they are **acid-labile**, they are killed by gastric acid when swallowed. This explains why they do not infect the gastrointestinal tract, unlike the enteroviruses. The host range is limited to humans and chimpanzees.

Summary of Replicative Cycle

Rhinovirus replication begins with the attachment of the infecting virion to a cell surface receptor called intercellular adhesion molecule-1 (ICAM-1). The virion enters the cytoplasm, and the capsid proteins are then removed (uncoated). After uncoating, the genome RNA functions as mRNA and is translated into **one large polypeptide**. This polypeptide is cleaved by a virus-encoded protease to form both the capsid proteins of the progeny virions and several noncapsid proteins, including the RNA polymerase that synthesizes the progeny RNA genomes. Replication of the genome occurs by synthesis of a complementary negative strand, which then serves as the template for the positive strands. Some of these positive strands function as mRNA to make more viral proteins, and the remainder become progeny virion genome RNA. Assembly of the progeny virions occurs by coating of the genome RNA with capsid proteins. Progeny virions accumulate in the cell cytoplasm and are released upon death of the cell.

Transmission & Epidemiology

There are **two modes** of transmission for these viruses. In the past, it was accepted that they were transmitted directly from person to person via aerosols of respiratory droplets. However, now it appears that an indirect mode, in which respiratory droplets are deposited on the hands or on a surface such as a table and then transported by fingers to the nose or eyes, is also important.

The common cold is reputed to be the most common human infection, although data are difficult to obtain because it is not a well-defined or notifiable disease. Millions of days of work and school are lost each year as a result of "colds."

Rhinoviruses occur worldwide, causing disease particularly in the Fall and Winter. The reason for this seasonal variation is unclear. Low temperatures per se do not predispose to the common cold, but the crowding that occurs at schools, for example, may enhance transmission during Fall and Winter. The frequency of colds is high in childhood and tapers off during adulthood, presumably because of the acquisition of immunity.

A few serotypes of rhinoviruses are prevalent during one season, then are replaced by other serotypes during the following season. It appears that the population builds up immunity to the prevalent serotypes but remains susceptible to the others.

Pathogenesis & Immunity

The portal of entry is the upper respiratory tract, and the infection is limited to that region. Rhinoviruses rarely cause lower respiratory tract disease, probably because they grow poorly at 37°C.

Immunity is serotype-specific and is a function of nasal secretory IgA rather than humoral antibody.

Clinical Findings

After an incubation period of 2 to 4 days, sneezing, nasal discharge, sore throat, cough, and headache are common. A chilly sensation may occur, but there are few other systemic symptoms. The illness lasts about 1 week. Note that other viruses such as coronaviruses, adenoviruses, influenza C virus, and Coxsackie viruses also cause the common cold syndrome.

Laboratory Diagnosis

A laboratory diagnosis can be made by detecting the RNA of rhinoviruses in respiratory tract specimens using a PCR-based assay. Serologic tests are not done as there are too many serotypes.

Treatment & Prevention

No specific antiviral therapy is available. Vaccines appear impractical because of the large number of serotypes. Paper tissues impregnated with a combination of citric acid (which inactivates rhinoviruses) and sodium lauryl sulfate (a detergent that inactivates enveloped viruses such as influenza virus and RSV) limit transmission when used to remove viruses from fingers contaminated with respiratory secretions. High doses of vitamin C have little ability to prevent rhinovirus-induced colds. Lozenges containing zinc gluconate are available for the treatment of the common cold, but their efficacy remains uncertain. Use of intranasal zinc solution may result in anosmia (loss of ability to smell).

ADENOVIRUS

Diseases

Adenovirus causes a variety of upper and lower respiratory tract diseases such as pharyngitis, conjunctivitis ("pink eye"), the common cold, and pneumonia. Keratoconjunctivitis, hemorrhagic cystitis, and gastroenteritis also occur. Some adenoviruses cause sarcomas in rodents.

Important Properties

Adenoviruses are **nonenveloped** viruses with double-stranded linear DNA and an **icosahedral** nucleocapsid (see Table 38–2). They are the only viruses with a **fiber** protruding from each of the 12 vertices of the capsid. The fiber is the organ of attachment and is a hemagglutinin. When purified free of virions, the fiber is toxic to human cells.

There are 41 known antigenic types; the fiber protein is the main type-specific antigen. All adenoviruses have a common group-specific antigen located on the hexon protein.

Certain serotypes of human adenoviruses (especially 12, 18, and 31) cause **sarcomas** at the site of injection in laboratory rodents such as newborn hamsters. There is no evidence that adenoviruses cause tumors in humans.

Summary of Replicative Cycle

After attachment to the cell surface via its fiber, the virus penetrates and uncoats, and the viral DNA moves to the nucleus. Host cell DNA-dependent RNA polymerase transcribes the early genes, and splicing enzymes remove the RNA representing the introns, resulting in functional mRNA. (Note that introns and exons, which are common in eukaryotic DNA, were first described for adenovirus DNA.) Early mRNA is translated into nonstructural proteins in the cytoplasm. Progeny viral DNA genomes are synthesized by a virion-encoded DNA polymerase in the nucleus. After viral DNA replication, late mRNA is transcribed and then translated into structural virion proteins. Viral assembly occurs in the nucleus, and the virus is released by lysis of the cell, not by budding.

Transmission & Epidemiology

Adenoviruses are transmitted by several mechanisms: **aerosol** droplet, **fecal–oral** route, and **direct inoculation** of conjunctivas by tonometers or fingers. The fecal–oral route is the most common mode of transmission among young children and their families. Many species of animals are infected by strains of adenovirus, but these strains are not pathogenic for humans.

Adenovirus infections are endemic worldwide, but outbreaks occur among military recruits, apparently as a result of the close living conditions that facilitate transmission. Certain serotypes are associated with specific syndromes (e.g., types 3, 4, 7, and 21 cause respiratory disease, especially in military recruits; types 8 and 19 cause epidemic keratoconjunctivitis; types 11 and 21 cause hemorrhagic cystitis; and types 40 and 41 cause infantile gastroenteritis).

Pathogenesis & Immunity

Adenoviruses infect the mucosal epithelium of several organs (e.g., the **respiratory tract** [both upper and lower], the **gastrointestinal tract**, and the **conjunctivas**). Immunity based on neutralizing antibody is type-specific and lifelong.

In addition to acute infection leading to death of the cells, adenoviruses cause a latent infection, particularly in the adenoidal and tonsillar tissues of the throat. In fact, these viruses were named for the adenoids, from which they were first isolated in 1953.

Clinical Findings

In the upper respiratory tract, adenoviruses cause infections such as pharyngitis, pharyngoconjunctival fever, and acute respiratory disease, characterized by fever, sore throat, coryza (runny nose), and conjunctivitis. In the lower respiratory tract, they cause bronchitis and atypical pneumonia. Hematuria and dysuria are prominent in hemorrhagic cystitis. Gastroenteritis with nonbloody diarrhea occurs mainly in children younger than 2 years of age. Most adenovirus infections resolve spontaneously. Approximately half of all adenovirus infections are asymptomatic.

Laboratory Diagnosis

The most common method of laboratory diagnosis is a PCR-based assay that detects the DNA of adenoviruses in respiratory tract specimens. In addition, adenoviruses can be isolated in cell culture and detected by fluorescent antibody techniques. The detection of a fourfold or greater rise in antibody titer to adenoviruses can also be used.

Treatment & Prevention

There is no antiviral therapy. Regarding prevention, three live, nonattenuated vaccines against serotypes 4, 7, and 21 are available but are used only by the military. Each of the three vaccines is monovalent (i.e., each contains only one serotype). The vaccines are delivered in an enteric-coated capsule, which protects the live virus from inactivation by stomach acid. The virus infects the gastrointestinal tract, where it causes an asymptomatic infection and induces immunity to respiratory disease. This vaccine is not available for civilian use.

Epidemic keratoconjunctivitis is an iatrogenic disease, preventable by strict asepsis and hand washing by healthcare personnel who examine eyes.

SELF-ASSESSMENT QUESTIONS

1. Regarding influenza virus, which one of the following statements is most accurate?
   (A) The virion contains an RNA-dependent DNA polymerase.
   (B) Its surface proteins, hemagglutinin and neuraminidase, have multiple serologic types.
   (C) The protein that undergoes antigenic variation most often is the internal ribonucleoprotein.
   (D) Antigenic drift involves major changes in antigenicity that result from reassortment of the segments of its RNA genome.
   (E) The neuraminidase on the virion surface mediates the interaction of the virus with the receptors on the respiratory tract epithelium.

2. Regarding influenza virus and the disease influenza, which one of the following statements is most accurate?

   (A) Both the killed and the live, attenuated vaccines induce life-long immunity.

   (B) Influenza A virus causes more severe disease and more widespread epidemics than does influenza B virus.

   (C) The genome of influenza A virus has eight segments, but the genome of influenza B virus is in one piece.

   (D) The classification of influenza viruses into A, B, and C viruses is based on antigenic differences in their hemagglutinin.

   (E) Chronic carriers (i.e., patients from whom influenza virus is isolated at least 6 months after the acute disease) are an important source of human infection.

   (F) This virus has only one antigenic type, and lifelong immunity occurs in patients who have had measles.

3. Regarding respiratory syncytial virus (RSV), which one of the following statements is most accurate?

   (A) RSV is an important cause of bronchiolitis in infants.

   (B) RSV causes tumors in newborn animals but not in humans.

   (C) The RSV vaccine is recommended for all children prior to entering school.

   (D) Amantadine should be given to elderly nursing home residents to prevent outbreaks of disease caused by RSV.

   (E) RSV forms intranuclear inclusion bodies within neutrophils that are important in diagnosis by the clinical laboratory.

4. Your patient is a 75-year-old woman with fever, chills, and myalgias that began yesterday. It is January and an outbreak of influenza is occurring in the retirement community in which she lives. A rapid test for influenza antigen is positive. Which one of the following is the best choice of drug to treat the infection?

   (A) Acyclovir

   (B) Amantadine

   (C) Interferon

   (D) Oseltamivir

   (E) Ribavirin

5. Regarding rhinoviruses, which one of the following is most accurate?

   (A) Rhinoviruses are an important cause of viral meningitis and myocarditis.

   (B) The rhinovirus vaccine is recommended for all children over 2 years of age.

   (C) Rhinoviruses have many serologic types, so a person can have many infections caused by these viruses.

   (D) Rhinoviruses are not inactivated by stomach acid, so they infect the upper gastrointestinal tract and are one of the causes of viral diarrhea.

   (E) An important feature of the laboratory diagnosis of rhinoviruses is finding cytopathic effect in cell culture consisting of multinucleated giant cells.

6. Regarding adenoviruses, which one of the following statements is most accurate?

   (A) Acyclovir is the drug of choice for life-threatening infections.

   (B) They cause pharyngitis, pneumonia, and conjunctivitis ("pink eye").

   (C) They are often transmitted across the placenta and cause hydrocephalus in the fetus.

   (D) The adenovirus vaccine is recommended for all children prior to entering first grade.

   (E) Laboratory diagnosis depends on seeing multinucleated giant cells on biopsy as the virus has not been grown in cell culture.

7. Regarding SARS coronavirus-2 (SARS-CoV-2), which one of the following statements is most accurate?

   (A) It is a nonenveloped virus with a negative-polarity RNA genome.

   (B) SARS-Cov-2 is transmitted primarily by the fecal-oral route

   (C) Genetic variants of SARS-CoV-2 rarely occur.

   (D) The main receptor for the spike protein of SARS CoV-2 is the ACE-2 protein on the cell surface.

   (E) Wild birds, primarily sparrows, are the main reservoir for SARS CoV-2.

8. Regarding Coronavirus Disease-19 (COVID-19) caused by SARS-CoV-2, which one of the following statements is most accurate?

   (A) Symptoms of COVID-19 are limited to the lower respiratory tract.

   (B) Asymptomatic infections rarely occur.

   (C) The laboratory diagnosis can be made by either detection of viral RNA or detection of antibodies to the spike protein.

   (D) Oseltamivir, a neuraminidase inhibitor, is used to treat severe COVID-19 disease.

   (E) A vaccine that protects against COVID-19 contains the messenger RNA encoding the SARS-CoV-2 spike protein.

ANSWERS

(1) **(B)**

(2) **(B)**

(3) **(A)**

(4) **(D)**

(5) **(C)**

(6) **(B)**

(7) **(D)**

(8) **(E)**

SUMMARIES OF ORGANISMS

Brief summaries of the organisms described in this chapter begin on page 691. Please consult these summaries for a rapid review of the essential material.

PRACTICE QUESTIONS: USMLE & COURSE EXAMINATIONS

Questions on the topics discussed in this chapter can be found in the Clinical Virology section of Part XIII: USMLE (National Board) Practice Questions starting on page 747. Also see Part XIV: USMLE (National Board) Practice Examination starting on page 775.

Important Childhood Viruses

INTRODUCTION

The viruses that cause measles, mumps, rubella, roseola, and slapped cheek syndrome are typically thought of as childhood diseases, although they can cause disease in adults as well. Measles, mumps, and rubella viruses are united as components of the widely used, very successful measles, mumps, and rubella (MMR) vaccine. Note that measles and rubella are characterized by a rash, whereas mumps is not. The prominent feature of mumps is parotid gland swelling. Slapped cheek syndrome, as the name implies, is characterized by a rash on the face and is caused by parvovirus B19. Roseola infantum is a childhood disease characterized by high fever and a rash. It is caused by human herpesvirus-6 (HHV-6). Coxsackievirus which causes hand, foot, and mouth disease in children is discussed in Chapter 40.

Additional information regarding the clinical aspects of infections caused by the viruses in this chapter is provided in Part IX, entitled Infectious Diseases beginning on page 603.

MEASLES VIRUS

Disease

This virus causes measles, a disease characterized by a maculopapular rash. It occurs primarily in childhood.

Important Properties

The genome of measles virus consists of single-stranded, nonsegmented RNA with a negative polarity (Table 39–1). It is an enveloped virus with a helical nucleocapsid. The virus has a single serotype. Humans are the natural host.

Summary of Replicative Cycle

After adsorption to the cell surface via its hemagglutinin, the virus penetrates and uncoats and the virion RNA polymerase transcribes the negative-strand genome into mRNA.

TABLE 39–1 Properties of Important Childhood Viruses

Property	Measles Virus	Mumps Virus	Rubella Virus	Parvovirus B19	Human Herpesvirus-6
Virus family	Paramyxoviruses	Paramyxoviruses	Togavirus	Parvovirus	Herpesvirus
Genome	Single-stranded RNA; negative polarity	Single-stranded RNA; negative polarity	Single-stranded RNA; positive polarity	Single-stranded DNA	Double-stranded DNA
Virion RNA polymerase	Yes	Yes	No	No	No
Nucleocapsid	Helical	Helical	Icosahedral	Icosahedral	Icosahedral
Envelope	Yes	Yes	Yes	No	Yes
Number of serotypes	One	One	One	One	Two

Multiple mRNAs are synthesized, each of which is translated into the specific viral proteins; no polyprotein analogous to that synthesized by poliovirus is made. The helical nucleocapsid is assembled, the matrix protein mediates the interaction with the envelope, and the virus is released by budding from the cell membrane.

Transmission & Epidemiology

Measles virus is transmitted via **respiratory droplets** produced by coughing and sneezing both during the prodromal period and for a few days after the rash appears. Measles occurs worldwide, usually in outbreaks every 2 to 3 years, when the number of susceptible children reaches a high level. The World Health Organization (WHO) estimates there are 7 million cases of measles each year worldwide.

In the year 2000, the Centers for Disease Control and Prevention (CDC) declared that measles is eliminated from the United States. Elimination meant that sustained transmission within the United States no longer occurred. However, cases acquired abroad (imported cases) followed by small outbreaks continue to occur. In 2016, measles was declared eradicated from the Western Hemisphere.

The attack rate is one of the highest of viral diseases; most children contract the clinical disease on exposure. When this virus is introduced into a population that has not experienced measles, such as the inhabitants of the Hawaiian Islands in the 1800s, devastating epidemics occur.

In malnourished children, especially those in developing countries, measles is a much more serious disease than in well-nourished children. **Vitamin A deficiency** is especially important in this regard, and supplementation of this vitamin greatly reduces the severity of measles. Patients with deficient cell-mediated immunity (e.g., acquired immunodeficiency syndrome [AIDS] patients) have a severe, life-threatening disease when they contract measles.

Pathogenesis & Immunity

After infecting the cells lining the upper respiratory tract, the virus enters the blood and infects reticuloendothelial cells, where it replicates again. It then spreads via the blood to the skin. The **rash** is caused primarily by cytotoxic T cells attacking the measles virus-infected vascular endothelial cells in the skin. Antibody-mediated vasculitis may also play a role. Shortly after the rash appears, the virus can no longer be recovered and the patient can no longer spread the virus to others. **Multinucleated giant cells**, which form as a result of the fusion protein in the spikes, are characteristic of the lesions.

Lifelong immunity occurs in individuals who have had the disease. The hemagglutinin on the surface of the virion is the antigen against which neutralizing antibody is directed. Although IgG antibody may play a role in neutralizing the virus during the viremic stage, cell-mediated immunity is more important. The importance of cell-mediated immunity is illustrated by the fact that agammaglobulinemic children have

a normal course of disease, are subsequently immune, and are protected by immunization. Maternal antibody passes the placenta, and infants are protected during the first 6 months of life.

Infection with measles virus can **transiently depress cell-mediated immunity** against other intracellular microorganisms, such as *Mycobacterium tuberculosis*, leading to a loss of purified protein derivative (PPD) skin test reactivity, reactivation of dormant organisms, and clinical disease. The proposed mechanism for this unusual finding is that when measles virus binds to its receptor (called CD46) on the surface of human macrophages, the production of interleukin-12 (IL-12), which is necessary for cell-mediated immunity to occur, is suppressed.

Clinical Findings

After an incubation period of 10 to 14 days, a prodromal phase characterized by fever, and the three C's (conjunctivitis, coryza, and coughing) occurs. **Koplik's spots** appear several days before the rash and are virtually diagnostic. They are bright red lesions with a white, central dot that are located on the buccal mucosa (Figure 39–1). A few days later, a maculopapular rash appears on the face and proceeds gradually down the body to the lower extremities, including the palms and soles (Figure 39–2 and Table 39–2). The rash develops a brownish hue several days later.

The complications of measles can be quite severe. Encephalitis occurs at a rate of 1 per 1000 cases of measles. The mortality rate of encephalitis is 10%, and there are permanent sequelae, such as deafness and mental retardation, in 40% of cases. In addition, both primary measles (giant cell) pneumonia and secondary bacterial pneumonia occur. Bacterial otitis media are quite common. Subacute sclerosing panencephalitis (SSPE) is a rare, fatal disease of the central nervous system that occurs several years after measles (see Chapter 44).

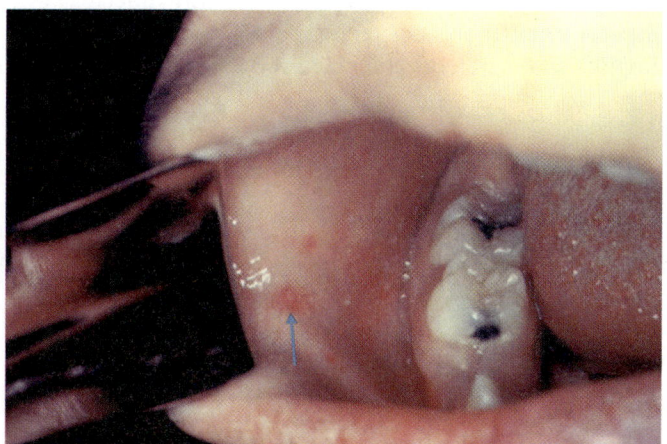

FIGURE 39–1 Koplik's spots of measles. Blue arrow points to one of several erythematous lesions with a white center on the buccal mucosa. (Reproduced with permission from the Centers for Disease Control and Prevention.)

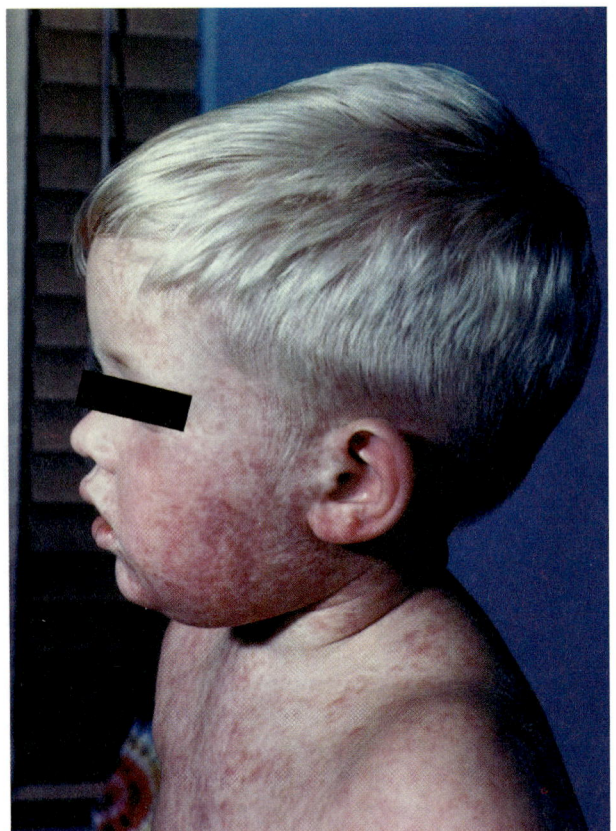

FIGURE 39–2 Measles. Note splotchy "morbilliform" macular-papular rash. (Reproduced with permission from Public Health Image Library, Centers for Disease Control and Prevention.)

Measles in a pregnant woman leads to an increased risk of stillbirth rather than congenital abnormalities. Measles virus infection of the fetus is more severe than rubella virus infection, so the former typically causes fetal death, whereas the latter causes congenital abnormalities.

Atypical measles occurs in some people who were given the killed vaccine and were subsequently infected with measles virus. It is characterized by an atypical rash without Koplik's spots. Because the killed vaccine has not been used for many years, atypical measles occurs only in adults and is infrequent.

Laboratory Diagnosis

Most diagnoses are made on clinical grounds, but a polymerase chain reaction (PCR) assay for measles virus RNA can be performed in cases that are difficult to diagnose. Detection of IgM antibody to measles virus or a greater than fourfold rise in antibody titer to measles virus can also be used.

Treatment & Prevention

There is no antiviral therapy available. Prevention rests on immunization with the **live, attenuated vaccine**. The vaccine is effective and causes few side effects. Two doses are recommended. The first dose should be given between 12 and 15 months of age and the second dose at 4 to 6 years of age. It is usually given in combination with rubella and mumps vaccines (MMR vaccine).

The vaccine should not be given to children prior to **12 months of age because maternal antibody in the child can neutralize the virus** and reduce the immune response. Because immunity can wane, a **booster dose** at 4 to 6 years is recommended. The vaccine contains live virus, so it should not be given to immunocompromised persons or pregnant women.

The vaccine has decreased the number of cases of measles greatly in the United States; in recent years, there are typically less than 1000 cases in the entire country. However, outbreaks still occur among unimmunized individuals (e.g., children in inner cities and in developing countries).

Immune globulin can be used to modify the disease if given to unimmunized individuals early in the incubation period. This is especially necessary if the unimmunized individuals are immunocompromised.

MUMPS VIRUS

Disease

This virus causes mumps, a disease characterized by salivary gland swelling. It occurs primarily in childhood.

TABLE 39–2 Clinical Features of Important Childhood Viruses

Virus	Disease	Rash Is a Prominent Feature	Causes Congenital Malformations	Infection Causes Lifelong Immunity to Disease	Vaccine Available	Treatment
Measles virus	Measles	Yes	No	Yes	Yes	No antiviral drug
Mumps virus	Mumps	No	No	Yes	Yes	No antiviral drug
Rubella virus	Rubella	Yes	Yes	Yes	Yes	No antiviral drug
Parvovirus B19	Slapped cheeks syndrome; hydrops fetalis	Yes	Yes	Yes	No	No antiviral drug
Human herpesvirus-6	Roseola	Yes	No	No. Reactivation of latent infection occurs in immunocompromised	No	No antiviral drug

Important Properties

The genome of mumps virus consists of single-stranded, non-segmented RNA with a negative polarity (see Table 39–1). It is an enveloped virus with a helical nucleocapsid. The virus has a single serotype. Humans are the natural host.

Summary of Replicative Cycle

Replication is similar to that of measles virus (see page 324).

Transmission & Epidemiology

Mumps virus is transmitted via respiratory droplets. Mumps occurs worldwide, with a peak incidence in the winter. About 30% of children have a subclinical (inapparent) infection, which confers immunity. In recent years, there are typically less than 1000 cases in the United States—a finding attributed to the widespread use of the vaccine. However, in 2006, 2016, and 2017, a resurgence of mumps occurred, with more than 6000 cases being recorded despite a high (87%) coverage rate for the vaccine.

Pathogenesis & Immunity

The virus infects the upper respiratory tract and then spreads through the blood to infect the salivary glands, especially the parotid gland, testes, ovaries, pancreas, and, in some cases, meninges. Alternatively, the virus may ascend from the buccal mucosa up Stensen's duct to the parotid gland.

Lifelong immunity occurs in persons who have had the disease. There is a popular misconception that unilateral mumps can be followed by mumps on the other side. Mumps occurs only once; subsequent cases of parotitis can be caused by other viruses such as parainfluenza viruses, by bacteria, and by duct stones. Maternal antibody passes the placenta and provides protection during the first 6 months of life.

Clinical Findings

After an incubation period of 18 to 21 days, a prodromal stage of fever, malaise, and anorexia is followed by tender swelling of the salivary glands, either unilateral or bilateral (Figure 39–3). There is a characteristic increase in parotid gland pain when drinking citrus juices. The disease is typically benign and resolves spontaneously within 1 week. A rash does *not* occur in mumps (see Table 39–2).

Two complications are of significance. One is orchitis in postpubertal males, which, if bilateral, can result in sterility. Postpubertal males have a fibrous tunica albuginea, which resists expansion, thereby causing pressure necrosis of the spermatocytes. Unilateral orchitis, although quite painful, does not lead to sterility. The other complication is meningitis, which is usually benign, self-limited, and without sequelae. Mumps virus, Coxsackie virus, and echovirus are the three most frequent causes of viral (aseptic) meningitis. The widespread use of the vaccine in the United States has led to a marked decrease in the incidence of mumps meningitis.

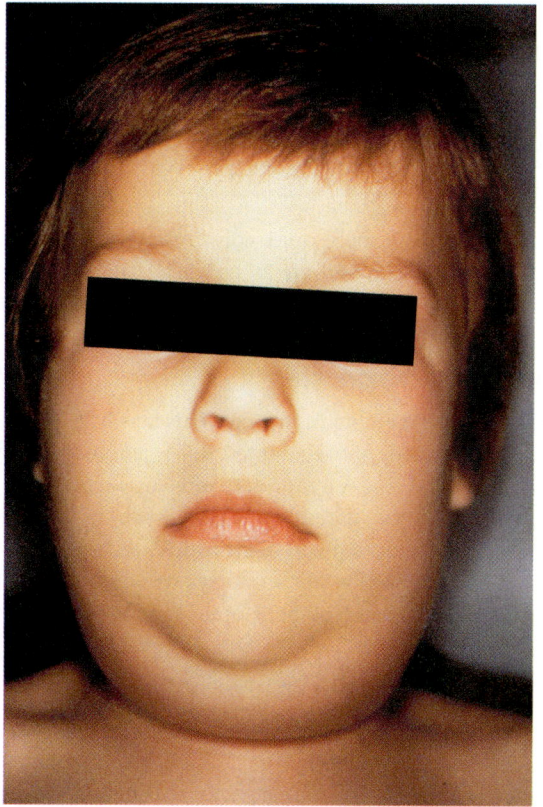

FIGURE 39–3 Mumps. Note bilateral swelling of neck due to inflammation of salivary glands. Note also absence of a rash as mumps is not a rash disease, unlike measles and rubella. (Used with permission from Dr. Patricia Smith and Dr. Barbara Rice, Public Health Image Library, Centers for Disease Control and Prevention.)

Laboratory Diagnosis

Most cases of mumps are diagnosed based on clinical features, but a PCR assay for mumps virus RNA is available. Serologic tests, which detect IgM antibody to mumps virus or a greater than fourfold rise in antibody titer to mumps virus, can also be used.

Treatment & Prevention

There is no antiviral therapy for mumps. Prevention consists of immunization with the **live, attenuated vaccine**. The vaccine is effective and long-lasting (at least 10 years) and causes few side effects. Two immunizations are recommended, one at 15 months and a booster dose at 4 to 6 years, usually in combination with measles and rubella vaccines. Because it is a live vaccine, it should not be given to immunocompromised persons or pregnant women. Immunoglobulin is not useful for preventing or mitigating mumps orchitis.

In recent years, for example, in 2016 and 2017, outbreaks of mumps occurred in the United States, especially on college campuses. Some of these occurred in individuals who had received two doses of the vaccine. In many individuals, more than 10 years had elapsed since their last MMR immunization, indicating that waning immunity may play a role.

A third mumps immunization can be given to limit transmission in outbreaks.

RUBELLA VIRUS

Diseases

This virus causes rubella and congenital rubella syndrome. Congenital rubella syndrome is characterized by **congenital malformations**.

Important Properties

Rubella virus is a member of the togavirus family. It is composed of a single-stranded, nonsegmented RNA genome, an **icosahedral** nucleocapsid, and a lipoprotein **envelope** (see Table 39–1). However, unlike the paramyxoviruses, such as measles and mumps viruses, it has a **positive-strand** RNA and therefore has no virion polymerase. Its surface spikes contain hemagglutinin. The virus has a single serotype. Humans are the natural host.

Summary of Replicative Cycle

After the virion enters the cell and uncoats, the plus-strand RNA genome is translated into several nonstructural and structural proteins. One of the nonstructural rubella proteins is an RNA-dependent RNA polymerase, which replicates the genome first by making a minus-strand template and then, from that, plus-strand progeny. Both replication and assembly occur in the cytoplasm, and the envelope is acquired from the outer membrane as the virion exits the cell.

Transmission & Epidemiology

The virus is transmitted via **respiratory droplets** and from mother to fetus **transplacentally**. The disease occurs worldwide. In areas where the vaccine is not used, epidemics occur every 6 to 9 years.

In 2005, the CDC declared rubella is eliminated from the United States, and in 2015, rubella was declared eliminated from the Western Hemisphere. The few cases that occur in the United States are acquired outside and imported into this country. Elimination was made possible by the widespread use of the vaccine. As a result, cytomegalovirus is a much more common cause of congenital malformations in the United States than is rubella virus.

Pathogenesis & Immunity

Initial replication of the virus occurs in the nasopharynx and local lymph nodes. From there, it spreads via the blood to the internal organs and skin. The origin of the rash is unclear; it may be due to antigen/antibody-mediated vasculitis.

Natural infection leads to **lifelong immunity**. Second cases of rubella do not occur; similar rashes are caused by other viruses, such as Coxsackie viruses and echoviruses. Antibody crosses the placenta and protects the newborn.

Clinical Findings

Rubella

Rubella is a milder, shorter disease than measles. After an incubation period of 14 to 21 days, a brief prodromal period with fever and malaise is followed by a maculopapular rash, which starts on the face and progresses downward to involve the extremities (Figure 39–4 and Table 39–2). Posterior auricular lymphadenopathy is characteristic. The rash typically lasts 3 days. When rubella occurs in adults, especially women, polyarthritis caused by immune complexes often occurs.

Congenital Rubella Syndrome

The significance of rubella virus is not as a cause of mild childhood disease but as a **teratogen**. When a nonimmune pregnant woman is **infected during the first trimester**, especially the first month, significant congenital malformations can occur as a result of maternal viremia and fetal infection (see Table 39–2). The increased rate of abnormalities during the early weeks of pregnancy is attributed to the very sensitive organ development that occurs at that time. The malformations are widespread and involve primarily the heart (e.g., patent ductus arteriosus), the eyes (e.g., cataracts), and the brain (e.g., deafness and mental retardation).

In addition, some children infected in utero can **continue to excrete** rubella virus for months after birth, which is a significant public health hazard because the virus can be transmitted to pregnant women. Some congenital shedders are asymptomatic and without malformations and hence can be diagnosed

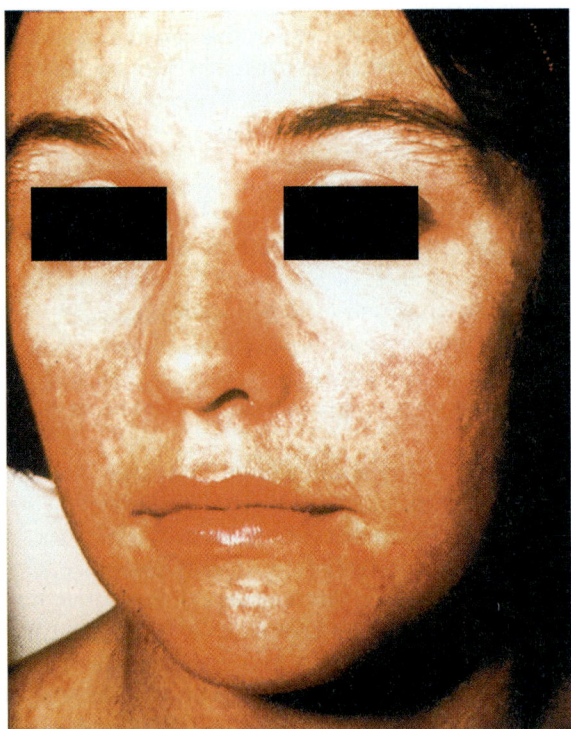

FIGURE 39–4 Rubella. Note fine, almost confluent macular-papular rash. (Used with permission from Stephen E. Gellis, MD.)

only if the virus is isolated. Congenitally infected infants also have significant IgM titers and persistent IgG titers long after maternal antibody has disappeared.

Laboratory Diagnosis

A laboratory diagnosis of rubella virus infection in either adult or newborn specimens or in amniotic fluid can be made by detecting the RNA of the virus by using a PCR-based assay.

The diagnosis can also be made by observing a fourfold or greater rise in antibody titer between acute-phase and convalescent-phase sera or by observing the presence of IgM antibody in a single acute-phase serum sample.

In a pregnant woman, the presence of **IgM antibody indicates recent infection**, whereas a 1:8 or greater titer of IgG antibody indicates immunity and consequent protection of the fetus. If recent infection has occurred, an **amniocentesis** can reveal whether there is rubella virus in the amniotic fluid, which indicates definite fetal infection.

Treatment & Prevention

There is no antiviral therapy. Prevention involves immunization with the **live, attenuated vaccine**. The vaccine is effective and long-lasting (at least 10 years) and causes few side effects, except for transient arthralgias in some women. It is given subcutaneously to children at 15 months of age (usually in combination with measles and mumps vaccine) and to unimmunized young adult women if they are not pregnant and will use contraception for the next 3 months. There is no evidence that the vaccine virus causes malformations. Because it is a live vaccine, it should not be given to immunocompromised patients or to pregnant women.

The vaccine has caused a significant reduction in the incidence of both rubella and congenital rubella syndrome. It induces some respiratory IgA, thereby interrupting the spread of virulent virus by nasal carriage.

Immune serum globulins (ISG) can be given to pregnant women in the first trimester who have been exposed to a known case of rubella and for whom termination of the pregnancy is not an option. The main problems with giving ISG are that there are instances in which it fails to prevent fetal infection and that it may confuse the interpretation of serologic tests. If termination of the pregnancy is an option, it is recommended to attempt to determine whether the mother and fetus have been infected as described in the preceding "Laboratory Diagnosis" section.

To protect pregnant women from exposure to rubella virus, many hospitals require their personnel to demonstrate immunity, either by serologic testing or by proof of immunization.

PARVOVIRUS B19

Diseases

Parvovirus B19 causes erythema infectiosum (slapped cheek syndrome, fifth disease), aplastic anemia (especially in patients with sickle cell anemia), and fetal infections, including hydrops fetalis.

Important Properties

Parvovirus B19 is a very small (22nm) nonenveloped virus with a **single-stranded DNA genome** (see Table 39–1). The genome is negative-strand DNA, but there is no virion polymerase. The capsid has icosahedral symmetry. There is one serotype and three genotypes.

Summary of Replicative Cycle

After adsorption to its host cell receptor (blood group P antigen), the virion penetrates and moves to the nucleus, where replication occurs. The single-stranded genome DNA has "hairpin" loops at both of its ends that provide double-stranded areas for the cellular DNA polymerase to initiate the synthesis of the progeny genomes. The viral mRNA is synthesized by cellular RNA polymerase from the double-stranded DNA intermediate. The progeny virions are assembled in the nucleus. B19 virus replicates only when a cell is in S phase, which explains why the virus replicates in nucleated red cell precursors but not in nonnucleated mature red cells.

Transmission & Epidemiology

B19 virus is transmitted primarily by the respiratory route; transplacental transmission also occurs. Blood donated for transfusions also can transmit the virus. B19 virus infection occurs worldwide, and about half the people in the United States older than 18 years of age have antibodies to the virus. Humans are the natural reservoir of B19 virus. Animal parvoviruses, for example canine parvovirus, do not cause human infection.

Pathogenesis & Immunity

B19 virus infects primarily two types of cells: **red blood cell precursors** (erythroblasts) in the bone marrow, which accounts for the aplastic anemia, and endothelial cells in the blood vessels, which accounts, in part, for the rash associated with erythema infectiosum. Immune complexes composed of virus and IgM or IgG also contribute to the pathogenesis of the rash and to the arthritis that is seen in some adults infected with B19 virus. Infection provides lifelong immunity against reinfection.

Hydrops fetalis manifests as massive edema of the fetus. This is secondary to congestive heart failure precipitated by severe anemia caused by the death of parvovirus B19-infected erythroblasts in the fetus.

Clinical Findings

There are five important clinical presentations.

Erythema Infectiosum (Slapped Cheek Syndrome, Fifth Disease)

This is a mild disease, primarily of childhood, characterized by a bright red rash that is most prominent on the cheeks (Figure 39–5), accompanied by low-grade fever, runny nose (coryza), and sore

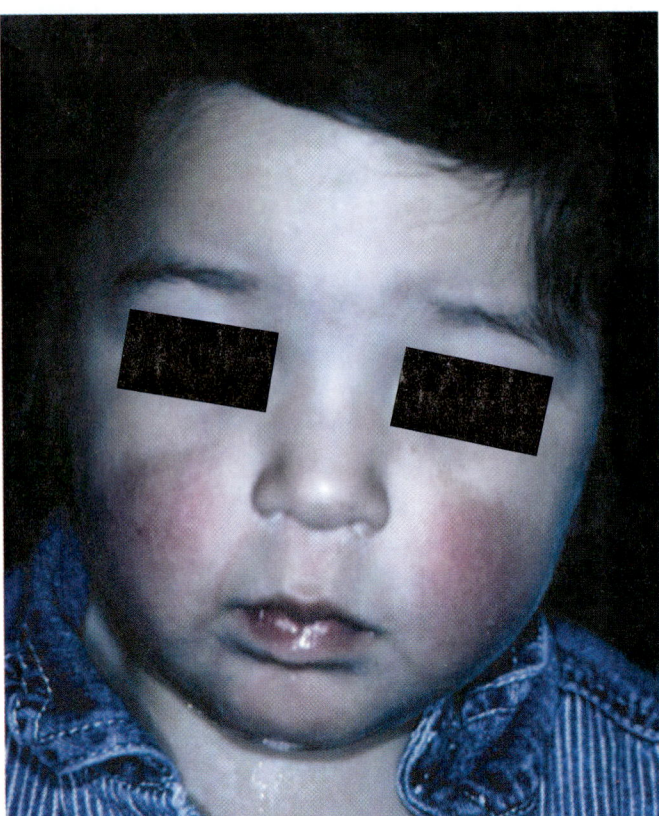

FIGURE 39–5 Slapped cheek syndrome. Note erythematous macular rash on cheeks bilaterally caused by parvovirus B19. (Reproduced with permission from Usatine RP, Smith MA, Mayeaux EJ Jr, et al: *The Color Atlas of Family Medicine.* New York, NY: McGraw Hill; 2009. Photo contributor: Richard P. Usatine, MD.)

throat. A "lacy," less intense, erythematous rash appears on the body. The symptoms resolve in about 1 week.

The disease in children is also called fifth disease. The four other macular or maculopapular rash diseases of childhood are measles, rubella, scarlet fever, and roseola.

Aplastic Anemia

Children with chronic anemia, such as sickle cell anemia, thalassemia, and spherocytosis, can have transient but severe aplastic anemia (aplastic crisis) when infected with B19 virus. People with normal red blood cells do not have clinically apparent anemia, although their red blood cell precursors are infected.

Fetal Infections

If a woman is infected with B19 virus during the first or second trimester of pregnancy, the virus may cross the placenta and infect the fetus. Infection during the first trimester is associated with fetal death, whereas infection during the second trimester leads to **hydrops fetalis** (see Table 39–2). Third-trimester infections do not result in important clinical findings. B19 virus is not a common cause of congenital abnormalities, probably because the fetus dies when infected early in pregnancy.

Arthritis

Parvovirus B19 infection in adults, especially women, can cause arthritis mainly involving the small joints of the hands and feet bilaterally. It resembles rheumatoid arthritis. Other viral infections that cause an immune complex-related arthritis include hepatitis B and rubella.

Chronic B19 Infection

People with immunodeficiencies, especially HIV-infected, chemotherapy, or transplant patients, can have chronic anemia, leukopenia, or thrombocytopenia as a result of chronic B19 infection.

Laboratory Diagnosis

Fifth disease and aplastic anemia are usually diagnosed by detecting IgM antibodies to parvovirus B19. Fetal infection can be determined by PCR analysis of amniotic fluid to detect the DNA of parvovirus B19. In immunocompromised patients, antibodies may not be detectable; therefore, viral DNA in the blood can be assayed by PCR methods.

Treatment & Prevention

There is no specific treatment of B19 infection. Pooled immune globulins may have a beneficial effect on chronic B19 infection in patients with immunodeficiencies. There is no vaccine or chemoprophylaxis.

HUMAN HERPESVIRUS-6

HHV-6 is the cause of **roseola** infantum (also known as exanthem subitum). Roseola is a common disease in young children characterized by the sudden onset of a high fever that lasts for a few days. When the fever abates, a transient macular or maculopapular erythematous rash on the face and trunk is often seen (Figure 39–6). Most cases occur in children who are 6 months to 3 years of age. In immunocompetent children, recovery without sequelae typically occurs. HHV-6 is found worldwide, and up to 80% of people are seropositive.

HHV-6 is lymphotropic and infects both T and B cells. It remains latent within these cells but can reactivate in immunocompromised patients and cause pneumonia, encephalitis, or hepatitis. It is one of the most common causes of encephalitis in patients who have received a stem-cell transplant. Many virologic and clinical features of HHV-6 are similar to those of cytomegalovirus, another member of the herpesvirus family. There are two serotypes of HHV-6 called HHV-6A and HHV-6B. HHV-6B is the major cause of roseola.

The laboratory diagnosis can be made by either PCR assay to detect viral DNA or by finding a fourfold or greater rise in antibody titer in the convalescent serum specimen compared to the titer in the acute specimen. There is no antiviral drug needed for roseola in immunocompetent children. Ganciclovir can be used in immunocompromised patients with serious disease. There is no vaccine.

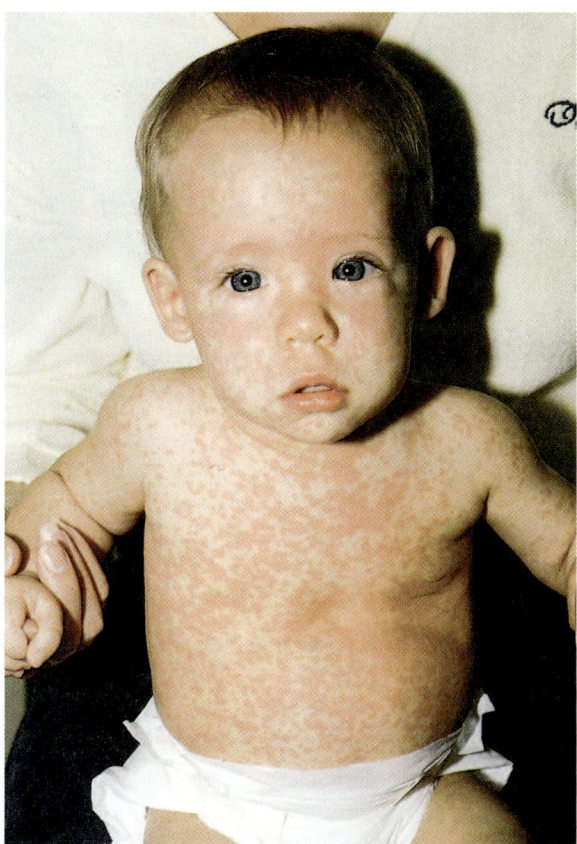

FIGURE 39–6 Roseola. Note erythematous macular rash on face and trunk caused by human herpesvirus-8. (Reproduced with permission from Knoop KJ, Stack LB, Storrow AB, et al: *The Atlas of Emergency Medicine*, 4th ed. New York, NY: McGraw Hill; 2016. Photo contributor Raymond C. Baker, MD.)

SELF-ASSESSMENT QUESTIONS

1. Regarding measles virus and the disease measles, which one of the following statements is most accurate?

   (A) The measles vaccine contains killed virus as the immunogen.

   (B) One of the main sequelae of measles is autoimmune glomerulonephritis and kidney failure.

   (C) Measles is unlikely to be eradicated because there is a significant animal reservoir for this virus.

   (D) Fecal–oral transmission during the diaper stage is the main mode of acquisition of measles virus.

   (E) This virus has only one antigenic type, and lifelong immunity occurs in patients who have had measles.

2. Regarding rubella virus, which one of the following statements is most accurate?

   (A) Systemic infection with rubella virus often causes severe liver damage resulting in cirrhosis.

   (B) If a pregnant woman is infected during the first trimester, significant fetal abnormalities typically result.

   (C) The main source of virus is adults who have recovered from the disease but are chronic carriers of the virus.

   (D) Immunization of both male and female healthcare workers with the formalin-inactivated vaccine is recommended.

   (E) The significant changes in the antigenicity of this virus are attributed to reassortment of the segments of its genome.

3. Regarding parvovirus B19, which one of the following statements is most accurate?

   (A) A vaccine is available that contains killed virus as the immunogen.

   (B) Patients infected by parvovirus B19 can be diagnosed in the laboratory using the cold agglutinin test.

   (C) Parvovirus B19 causes a severe anemia because it preferentially infects erythrocyte precursors such as erythroblasts.

   (D) It commonly infects neutrophils, resulting in an immunodeficiency that predisposes to infections by pyogenic bacteria.

   (E) Parvoviruses have a double-stranded DNA genome but require a DNA polymerase in the virion because they replicate in the cytoplasm.

4. Regarding human herpesvirus-6 (HHV-6), which one of the following statements is the most accurate?

   (A) It is a common cause of rubella.

   (B) It causes a latent infection in lymphocytes.

   (C) It replicates in the cytoplasm of infected cells.

   (D) The drug of choice to treat severe infections is acyclovir.

   (E) The vaccine against HHV-6 is a live, attenuated vaccine.

ANSWERS

(1) **(E)**
(2) **(B)**
(3) **(C)**
(4) **(B)**

SUMMARIES OF ORGANISMS

Brief summaries of the organisms described in this chapter begin on page 691. Please consult these summaries for a rapid review of the essential material.

PRACTICE QUESTIONS: USMLE & COURSE EXAMINATIONS

Questions on the topics discussed in this chapter can be found in the Clinical Virology section of Part XIII: USMLE (National Board) Practice Questions starting on page 747. Also see Part XIV: USMLE (National Board) Practice Examination starting on page 775.

Viruses That Infect
the Enteric Tract

INTRODUCTION

The viruses described in this chapter are transmitted by the fecal–oral route and enter the body via the enteric tract. Some, such as norovirus and rotavirus, cause diarrheal disease, whereas others, such as poliovirus, Coxsackie virus, echovirus, and enterovirus (EV) 68, 70, and 71 cause disease primarily outside the enteric tract. Polio, Coxsackie, and echoviruses are well-known causes of central nervous system disease, such as meningitis and encephalitis. Coxsackie virus also causes hand, foot, and mouth disease and myocarditis. EV D68 is suspected of causing acute flaccid myelitis.

Poliovirus, Coxsackie virus, echovirus, and EV 68, 70, and 71 are members of a group of viruses called enteroviruses within the picornavirus family. The term *enterovirus* refers to the enteric tract as an important site of viral replication and to the feces as a common source of infection and a common specimen from which these viruses are isolated. Note, however, that

Coxsackie virus, echovirus, and EV 68, 70, and 71 also replicate and cause disease symptoms in the upper respiratory tract and in the central nervous system.

All of the viruses described in this chapter are naked nucleocapsid viruses (i.e., they do not have an envelope). Viruses *without* an envelope are *more* stable in the environment, a feature that allows them to survive outside the body and to be transmitted by the fecal–oral route.

Note that other viruses also infect via the enteric tract such as hepatitis A virus and hepatitis E virus. These are discussed in Chapter 41 with the other hepatitis viruses. Note also that two viruses, Astrovirus and Sapporo virus, cause diarrhea. Because these are less common, they are described in Chapter 46, "Minor Viral Pathogens."

Additional information regarding the clinical aspects of infections caused by the viruses in this chapter is provided in Part IX, "Infectious Diseases" beginning on page 603.

VIRUSES THAT PRIMARILY CAUSE DIARRHEA

NOROVIRUS

Disease

Norovirus is one of the most common causes of viral gastroenteritis in adults both in the United States and worldwide. Norovirus is also the most common cause of viral gastroenteritis in children in the United States because the rotavirus vaccine has lowered the incidence of disease caused by that virus. Norwalk

virus is an important norovirus and is named for an outbreak of gastroenteritis in a school in Norwalk, Ohio, in 1969.

Important Properties

Norovirus is a member of the calicivirus family (*Caliciviridae*). It is a nonenveloped virus with an icosahedral nucleocapsid and has a nonsegmented, single-stranded, positive-polarity RNA genome (Table 40–1). There is no polymerase within the virion.

TABLE 40–1   Properties of Viruses Commonly Infecting the Intestinal Tract

Property	Norovirus	Rotavirus	Poliovirus	Coxsackie Virus	Echovirus
Virus family	Caliciviruses	Reoviruses	Picornavirus	Picornavirus	Picornavirus
Genome	Single-stranded RNA; positive polarity	Double-stranded RNA; 11 segments	Single-stranded RNA; positive polarity	Single-stranded RNA; positive polarity	Single-stranded RNA; positive polarity
Virion RNA polymerase	No	Yes	No	No	No
Nucleocapsid	Icosahedral	Icosahedral	Icosahedral	Icosahedral	Icosahedral
Envelope	No	No	No	No	No
Number of serotypes	Two or more	At least six	Three	Many	Many

In the electron microscope, 10 prominent spikes and 32 cup-shaped depressions can be seen. There are two or more serotypes; the exact number is uncertain. Six genogroups have been identified. Most human infections are caused by members of genogroup II.

Summary of Replicative Cycle

It is difficult to cultivate norovirus in cell culture, so the replicative cycle is incomplete. It is known that the virus binds to blood group antigens on the surface of the intestinal mucosa. Replication occurs in the cytoplasm. Most of the positive-strand RNA genome is translated into a large precursor polypeptide that is cleaved by the virion-encoded protease into several nonstructural proteins including the RNA polymerase and helicase. A second open reading frame encodes capsid protein one and another encodes capsid protein 2. Assembly of progeny virions occurs in the cytoplasm.

Transmission & Epidemiology

Norovirus is transmitted by the fecal–oral route, often involving the ingestion of contaminated seafood or water. Outbreaks typically occur in group settings such as cruise ships (especially in the Caribbean region), schools, camps, hospitals, and nursing homes. Person-to-person transmission also occurs, especially in group settings. There are many animal caliciviruses, but there is no evidence that they cause human infection.

Infection is enhanced by several features of the virus: low infectious dose, excretion of virus in the stool both before the onset of symptoms and for several weeks after recovery, and resistance to inactivation by chlorination and to drying in the environment. It is thought to remain infectious for several days in water, uncooked food, and on environmental surfaces such as door handles.

Pathogenesis & Immunity

Norovirus infection is typically limited to the mucosal cells of the intestinal tract. Watery diarrhea without red cells or white cells occurs. The pathogenesis of the diarrhea is likely to be based on two factors: (1) blunting of the villi resulting in reduced surface area and consequent decreased fluid uptake by enterocytes and (2) damage to the enterocytes resulting in increased fluid secretion in the gut lumen.

Many asymptomatic infections occur, as determined by the detection of antibodies. Immunity following infection appears to be brief, and reinfection can occur. New strains appear every 2 to 4 years and cause widespread infections.

Clinical Findings

Disease is characterized by sudden onset of vomiting and diarrhea accompanied by low-grade fever and abdominal cramping (Table 40–2). Neither the emesis nor the stool contains blood. The illness typically lasts 2 to 3 days, is self-limited, and there are no long-term sequelae, except in some immunocompromised patients in whom chronic gastroenteritis can occur. In some outbreaks, certain patients manifest signs of central nervous system involvement such as headache, meningismus, photophobia, and obtundation.

The incubation period is 1 to 2 days and virus is shed in the stool for 7 to 10 days after resolution of symptoms. Approximately 30% of norovirus infections are asymptomatic.

TABLE 40–2   Clinical Features of Viruses Commonly Infecting the Intestinal Tract

Virus	Disease	Main Clinical Findings	Vaccine Available	Antiviral Therapy
Norovirus	Gastroenteritis	Watery diarrhea	No	No
Rotavirus	Gastroenteritis	Watery diarrhea, especially in infants	Yes	No
Poliovirus	Poliomyelitis	Paralysis due to death of motor neurons	Yes	No
Coxsackie virus	1. Hand, foot, and mouth disease 2. Meningitis 3. Myocarditis	1. Vesicular lesions on hands and feet and in mouth 2. Fever, headache, and stiff neck 3. Congestive heart failure	No	No
Echovirus	Meningitis	Fever, headache, and stiff neck	No	No

Laboratory Diagnosis

A polymerase chain reaction (PCR)-based test on the stool or vomitus is performed when a specific diagnosis is required. However, the diagnosis is often a clinical one.

Treatment & Prevention

There is no antiviral therapy available. Dehydration and electrolyte imbalance caused by the vomiting and diarrhea may require oral rehydration or intravenous fluids. Regarding prevention, there is no vaccine available. Personal hygiene, such as handwashing, and public health measures, such as proper sewage disposal and disinfection of contaminated surfaces, are useful.

ROTAVIRUS

Disease

Rotavirus is a common cause of viral gastroenteritis, especially in young children.

Important Properties

Rotavirus has a **segmented, double-stranded RNA genome** surrounded by a double-layered icosahedral capsid without an envelope (see Table 40–1). The rotavirus genome has 11 segments. The virion contains an **RNA-dependent RNA polymerase**. A virion polymerase is required because human cells do not have an RNA polymerase that can synthesize mRNA from a double-stranded RNA template.

Many domestic animals are infected with their own strains of rotaviruses, but these are not a source of human disease. There are at least six serotypes of human rotavirus. The outer surface protein (also known as the viral hemagglutinin) is the type-specific antigen and elicits protective antibody.

Summary of Replicative Cycle

Rotavirus attaches to the cell surface at the site of the β-adrenergic receptor. After entry of the virion into the cell, the RNA-dependent RNA polymerase synthesizes mRNA from each of the 11 segments within the cytoplasm. The 11 mRNAs are translated into the corresponding number of structural and nonstructural proteins. One of these, an RNA polymerase, synthesizes minus strands that will become part of the genome of the progeny virus. Capsid proteins form an incomplete capsid around the minus strands, and then the plus strands of the progeny genome segments are synthesized. The virus is released from the cytoplasm by lysis of the cell, not by budding.

Transmission & Epidemiology

Rotavirus is transmitted by the **fecal–oral** route. Infection occurs worldwide, and by age 6 years, most children have antibodies to at least one serotype.

Pathogenesis & Immunity

Rotavirus replicates in the mucosal cells of the small intestine, resulting in the excess secretion of fluids and electrolytes into the bowel lumen. The consequent loss of salt, glucose, and water leads to diarrhea. No inflammation occurs, and the diarrhea is nonbloody. Blunting of the villi occurs which decreases the surface area for uptake of water from the gastrointestinal (GI) tract. In addition, a viral nonstructural protein, NSP4, acts as a pore-forming toxin causing release of calcium from the cell membranes both in the infected cell and adjacent cells, resulting in diarrhea.

Immunity to rotavirus infection is unclear. It is likely that intestinal immunoglobulin (Ig) A directed against specific serotypes protects against reinfection and that colostrum IgA protects newborns up to the age of 6 months.

Clinical Findings

Rotavirus infection is characterized by nausea, vomiting, and watery, nonbloody diarrhea (see Table 40–2). **Gastroenteritis is most serious in young children**, in whom dehydration and electrolyte imbalance are a major concern. Adults usually have minor symptoms.

Laboratory Diagnosis

Although the diagnosis of most cases of viral gastroenteritis does not involve the laboratory, a diagnosis can be made by **detection of rotavirus antigen in the stool** by using an enzyme-linked immunosorbent assay (ELISA). A PCR-based assay for rotavirus RNA in the stool is the most sensitive method of diagnosis. The diagnosis can also be made by detecting a fourfold or greater rise in antibody titer.

Treatment

There is no antiviral therapy available. Dehydration and electrolyte imbalance caused by the vomiting and diarrhea may require oral rehydration or intravenous fluids.

Prevention

There are two rotavirus vaccines available. Both contain live virus and are given orally. One is a live, attenuated vaccine (Rotarix), which contains the single most common rotavirus serotype (G1) causing disease in the United States. The other is a live reassortant vaccine (RotaTeq), which contains five rotavirus strains (G1, G2, G3, G4, and G9). An increased risk of intussusception has been reported with both vaccines. Patients with a history of intussusception should not receive either vaccine.

The five rotaviruses in the RotaTeq vaccine are reassortants in which the gene for the outer surface protein of a human rotavirus is inserted into a bovine strain of rotavirus. (Recall that rotavirus has a segmented genome.) The bovine strain is nonpathogenic for humans, but the human outer surface protein in the vaccine virus elicits protective (IgA) immunity in the GI tract.

A previously approved vaccine (Rotashield) was withdrawn when a high rate of intussusception occurred in vaccine recipients. Hygienic measures such as proper sewage disposal, disinfection of surfaces, and handwashing are helpful.

ENTERIC ADENOVIRUSES

Enteric adenoviruses type 40 and 41 are the second leading cause of diarrhea in young children worldwide, after rotavirus. The diarrheal stool is watery and nonbloody, and is accompanied by nausea, vomiting, and abdominal cramps. It typically occurs in the summer months. The symptoms resolve spontaneously and there is no specific antiviral drug or vaccine.

VIRUSES THAT CAUSE DISEASE PRIMARILY OUTSIDE THE GI TRACT

POLIOVIRUS

Disease

This virus causes poliomyelitis, the most prominent symptom of which is paralysis.

Important Properties

Poliovirus has a nonsegmented, single-stranded, positive-polarity RNA genome (see Table 40–1). It is a nonenveloped virus with an icosahedral nucleocapsid. There is no polymerase within the virion. There are three serotypes.

The host range is limited to primates (i.e., humans and nonhuman **primates** such as apes and monkeys). This limitation is due to the binding of the viral capsid protein to a receptor found only on primate cell membranes. However, note that purified viral RNA (without the capsid protein) can enter and replicate in many nonprimate cells—the RNA can bypass the cell membrane receptor (i.e., it is "infectious RNA").

There are **three serologic (antigenic) types** based on different antigenic determinants on the outer capsid proteins. Because there is little cross-reaction, protection from disease requires the presence of antibody against each of the three types.

Summary of Replicative Cycle

The virion interacts with specific cell receptors on the cell membrane and then enters the cell. The capsid proteins are then removed. After uncoating, the genome RNA functions as mRNA and is translated into **one very large polypeptide** called noncapsid viral protein 00. This polypeptide is cleaved by a **virus-encoded protease** in multiple steps to form both the capsid proteins of the progeny virions and several noncapsid proteins, including the RNA polymerase that synthesizes the progeny RNA genomes. Replication of the genome occurs by synthesis of a complementary negative strand, which then serves as the template for the positive strands. Some of these positive strands function as mRNA to make more viral proteins, and the remainder become progeny virion genome RNA. Assembly of the progeny virions occurs by coating of the genome RNA with capsid proteins. Virions accumulate in the cell cytoplasm and are released upon death of the cell. They do not bud from the cell membrane.

Transmission & Epidemiology

Poliovirus is transmitted by the **fecal–oral** route. It replicates in the oropharynx and intestinal tract. Humans are the only natural hosts.

As a result of the success of the vaccine, poliomyelitis caused by naturally occurring "wild-type" virus has been **eradicated** from the United States and, indeed, **from the entire Western Hemisphere**. The rare cases in the United States occur mainly in (1) people exposed to virulent revertants of the attenuated virus in the live vaccine and (2) unimmunized people exposed to wild-type poliovirus while traveling abroad. Before the vaccine was available, epidemics occurred in the summer and fall.

The World Health Organization (WHO) set the eradication of paralytic polio by 2005 as a goal. Unfortunately, this goal was not achieved. In 1988, there were 388,000 cases of paralytic polio worldwide, whereas in 2005, there were fewer than 2000. Despite this remarkable decrease, paralytic polio continues to occur. As of 2017, there were less than 100 cases worldwide and they occurred in only three countries—Afghanistan, Pakistan, and Nigeria—so progress toward eradication continues to occur. In 2020, the WHO declared the entire continent of Africa, including Nigeria, to be "polio-free." This leaves Afghanistan and Pakistan as the only two countries where disease caused by "wild-type poliovirus" occurs. Thus far, smallpox is the only human infectious disease that has been eradicated, a consequence of the worldwide use of the smallpox vaccine.

Pathogenesis & Immunity

After replicating in the oropharynx and small intestine, especially in lymphoid tissue, the virus spreads through the bloodstream to the central nervous system. It can also spread retrograde along nerve axons.

In the central nervous system, poliovirus preferentially replicates in the **motor neurons** located in the **anterior horn** of the spinal cord. Death of these cells results in paralysis of the muscles innervated by those neurons. Paralysis is not due to virus infection of muscle cells. The virus also affects the brainstem, leading to "bulbar" poliomyelitis (with respiratory paralysis), but rarely damages the cerebral cortex.

In infected individuals, the immune response consists of both intestinal IgA and humoral IgG to the specific serotype. Infection provides lifelong type-specific immunity.

Clinical Findings

The range of responses to poliovirus infection includes (1) inapparent, asymptomatic infection; (2) abortive poliomyelitis; (3) nonparalytic poliomyelitis; and (4) paralytic poliomyelitis (see Table 40–2). Asymptomatic infection is quite common. Roughly 1% of infections are clinically apparent. The incubation period is usually 10 to 14 days.

The most common clinical form is abortive poliomyelitis, which is a mild, febrile illness characterized by headache, sore throat, nausea, and vomiting. Most patients recover spontaneously. Nonparalytic poliomyelitis manifests as aseptic meningitis with fever, headache, and a stiff neck. This also usually resolves spontaneously. In paralytic poliomyelitis, flaccid paralysis is the predominant finding, but brainstem involvement can lead to life-threatening respiratory paralysis. Painful muscle spasms also occur. The motor nerve damage is permanent, but some recovery of muscle function occurs as other nerve cells take over. In paralytic polio, both the meninges and the brain parenchyma (meningoencephalitis) are often involved. If the spinal cord is also involved, the term *meningomyeloencephalitis* is often used.

A **postpolio syndrome** that occurs many years after the acute illness has been described. Marked deterioration of the residual function of the affected muscles occurs many years after the acute phase. The cause of this deterioration is unknown.

No permanent carrier state occurs following infection by poliovirus, but virus excretion in the feces can occur for several months.

Laboratory Diagnosis

The diagnosis is made either by isolation of the virus or by a rise in antibody titer. Virus can be recovered from the throat, stool, or spinal fluid by inoculation of cell cultures. The virus causes a cytopathic effect (CPE) and can be identified by neutralization of the CPE with specific antisera. A PCR-based assay for poliovirus RNA is also available.

Treatment

There is no antiviral therapy. Treatment is limited to symptomatic relief and respiratory support, if needed. Physiotherapy for the affected muscles is important.

Prevention

Poliomyelitis can be prevented by both the **killed** vaccine (Salk vaccine, inactivated vaccine, IPV) and the **live, attenuated** vaccine (Sabin vaccine, oral vaccine, OPV) (Table 40–3). Both vaccines induce humoral antibodies, which neutralize virus entering the blood and hence prevent central nervous system infection and disease. Both the killed and the live vaccines contain all three serotypes. At present, the **inactivated vaccine** is preferred for reasons that are described later.

The current version of the inactivated vaccine is called **enhanced polio vaccine**, or **eIPV**. It contains all three serotypes. It has a higher seroconversion rate and induces a higher titer of antibody than the previous IPV. eIPV also induces some mucosal immunity IgA, making it capable of interrupting transmission, but the amount of secretory IgA induced by eIPV is significantly less than the amount induced by OPV. OPV is therefore preferred for eradication efforts. The only version of polio vaccine currently produced and used in the United States is eIPV.

In the past, the live vaccine was preferred in the United States for two main reasons: (1) It interrupts fecal–oral transmission

TABLE 40–3 Important Features of Poliovirus Vaccines

Attribute	Killed (Salk)	Live (Sabin)
Prevents disease	Yes	Yes
Interrupts transmission	No	Yes
Induces humoral IgG	Yes	Yes
Induces intestinal IgA	No	Yes
Affords secondary protection by spread to others	No	Yes
Interferes with replication of virulent virus in gut	No	Yes
Reverts to virulence	No	Yes (rarely)
Coinfection with other enteroviruses may impair immunization	No	Yes
Can cause disease in the immunocompromised	No	Yes
Route of administration	Injection	Oral
Requires refrigeration	No	Yes
Duration of immunity	Shorter	Longer

by inducing secretory IgA in the GI tract. (2) It is given orally and so is more readily accepted than the killed vaccine, which must be injected.

The live vaccine has four disadvantages: (1) Rarely, **reversion** of the attenuated virus to virulence will occur, and disease may ensue (especially for the type 3 virus). (2) It can cause disease in immunodeficient persons and therefore should not be given to them. (3) Infection of the GI tract by other enteroviruses can limit replication of the vaccine virus and reduce protection. (4) It must be kept refrigerated to prevent heat inactivation of the live virus.

Outbreaks of paralytic polio caused by vaccine-derived poliovirus (VDPV) continue to occur, especially in areas where there are large numbers of unimmunized people. These VDPV strains have lost their attenuation by acquiring genes from wild-type enteroviruses by recombination and by back mutation in the case of type 2. Outbreaks of VDPV-associated paralytic polio have been contained by campaigns to immunize people in the affected area with the oral (Sabin) vaccine that interrupts fecal–oral transmission.

The duration of immunity is thought to be longer with the live than with the killed vaccine, but a booster dose is recommended with both.

The currently approved vaccine schedule in the United States consists of four doses of inactivated vaccine administered at 2 months, 4 months, 6 to 18 months, and upon entry to school at 4 to 6 years. One booster (lifetime) is recommended for adults who travel to endemic areas. The use of the inactivated vaccine should prevent approximately 10 cases per year of vaccine-associated paralytic polio.

In 2016, WHO decided to use only the bivalent (containing types 1 and 3) inactivated vaccine worldwide. The decision to stop using the oral vaccine was based on the unacceptable number of cases of paralytic polio caused by the vaccine strains.

The decision to omit type 2 from the vaccine was based on the finding that type 2 had not caused disease since 1999. However, cases of polio caused by VDPV type 2 continue to occur, albeit rarely. In 2020, a new vaccine containing genetically engineered type 2 poliovirus was tested in clinical trials. It was engineered to prevent back mutations to virulence and to reduce recombination with other enteroviruses.

In the past, some lots of poliovirus vaccines were contaminated with a papovavirus, SV40 virus, which causes sarcomas in rodents. SV40 virus was a "passenger" virus in the monkey kidney cells used to grow the poliovirus for the vaccine. Fortunately, no increase in cancer occurred in persons inoculated with the SV40 virus-containing polio vaccine. However, there is some evidence that SV40 DNA can be found in certain human cancers such as non-Hodgkin's lymphoma; the role of SV40 as a cause of cancer in persons immunized with early versions of the polio vaccine is unresolved. At present, cell cultures used for vaccine purposes are carefully screened to exclude the presence of adventitious viruses.

Passive immunization with immune serum globulin is available for protection of unimmunized individuals known to have been exposed. Passive immunization of newborns as a result of passage of maternal IgG antibodies across the placenta also occurs.

Quarantine of patients with disease is not effective, because fecal excretion of the virus occurs in infected individuals prior to the onset of symptoms and in those who remain asymptomatic.

COXSACKIE VIRUS

Coxsackie virus is named for the town of Coxsackie, New York, where they were first isolated.

Diseases

Coxsackie viruses cause a variety of diseases. Group A viruses cause, for example, herpangina, acute hemorrhagic conjunctivitis, and hand, foot, and mouth disease, whereas group B viruses cause pleurodynia, myocarditis, and pericarditis. Both types cause non-specific upper respiratory tract disease (common cold), febrile rashes, and aseptic meningitis. Coxsackie viruses and echoviruses (see next section) together cause approximately 90% of cases of viral (aseptic) meningitis. Both types also cause acute flaccid paralysis resembling paralytic poliomyelitis (Table 40–4).

TABLE 40–4 Clinical Features of Coxsackie Viruses

Type of Coxsackie Virus	Diseases Caused
Type A	Hand, foot, and mouth disease; herpangina; hemorrhagic conjunctivitis
Type B	Myocarditis, pericarditis, pleurodynia, myositis
Type A and type B	Aseptic meningitis, common cold, pharyngitis, febrile rash, acute flaccid paralysis

Important Properties

The size and structure of the virion and the nature of the genome RNA are similar to those of poliovirus (see Table 40–1). The classification of Coxsackie viruses into group A or B is based on pathogenicity in mice. Group A viruses cause widespread myositis and flaccid paralysis, which is rapidly fatal, whereas group B viruses cause generalized, less severe lesions of the heart, pancreas, and central nervous system and focal myositis. At least 24 serotypes of Coxsackie virus A and 6 serotypes of Coxsackie virus B are recognized.

Summary of Replicative Cycle

Replication is similar to that of poliovirus.

Transmission & Epidemiology

Coxsackie viruses are transmitted primarily by the **fecal–oral** route, but respiratory **aerosols** also play a role. They replicate in the oropharynx and the intestinal tract. Humans are the only natural hosts. Coxsackie virus infections occur worldwide, primarily in the summer and fall.

Pathogenesis & Immunity

Group A viruses have a predilection for skin and mucous membranes, whereas group B viruses cause disease in various organs such as the heart, pleura, pancreas, and liver. Both group A and B viruses can affect the meninges and the motor neurons (anterior horn cells) to cause paralysis. From their original site of replication in the oropharynx and GI tract, they disseminate via the bloodstream.

Immunity following infection is provided by type-specific IgG antibody.

Clinical Findings
Group A-Specific Diseases

Herpangina is characterized by fever, sore throat, and tender vesicles in the oropharynx. **Hand, foot, and mouth disease** (see Table 40–2) is characterized by a vesicular rash on the hands and feet and ulcerations in the mouth, mainly in children. Coxsackie A24 virus causes acute hemorrhagic conjunctivitis.

Group B-Specific Diseases

Pleurodynia (Bornholm disease, epidemic myalgia, "devil's grip") is characterized by fever and severe pleuritic-type chest pain. Note that pleurodynia is pain due to an infection of the intercostal muscles (myositis), not of the pleura.

Myocarditis and pericarditis are characterized by fever, chest pain, and signs of congestive failure. Dilated cardiomyopathy with global hypokinesia of the myocardium is a feared sequela that often requires cardiac transplantation to sustain life. **Diabetes** in mice can be caused by pancreatic damage as a result of infection with Coxsackie virus B4. This virus is suspected to have a similar role in juvenile diabetes in humans.

Diseases Caused by Both Groups

Both groups of viruses can cause **aseptic meningitis**, mild paresis, and acute flaccid paralysis similar to poliomyelitis. Upper respiratory infections, pharyngitis, and minor febrile illnesses with or without rash can occur also.

Laboratory Diagnosis

A PCR-based test for Coxsackie virus RNA in the spinal fluid is useful for making a prompt diagnosis of viral meningitis because culture techniques typically take days to obtain a result. The diagnosis can also be made by isolating the virus in cell culture or suckling mice or by observing a rise in titer of neutralizing antibodies.

Treatment & Prevention

There is neither antiviral drug therapy nor a vaccine available against these viruses. No passive immunization is recommended.

ECHOVIRUS AND PARECHOVIRUS

The prefix ECHO is an acronym for *enteric cytopathic human orphan*. Although called "orphans" because they were not initially associated with any disease, they are now known to cause a variety of diseases such as aseptic meningitis, upper respiratory tract infection, febrile illness with and without rash, infantile hepatitis, and hemorrhagic conjunctivitis.

The structure of echoviruses is similar to that of other enteroviruses (see Table 40–1). More than 30 serotypes have been isolated. In contrast to Coxsackie viruses, they are not pathogenic for mice. Unlike polioviruses, they do not cause disease in monkeys. They are transmitted by the **fecal–oral** route and occur worldwide. Pathogenesis is similar to that of the other enteroviruses.

Based on genome sequencing, echovirus types 22 and 23 have been reclassified as **parechoviruses**. They cause respiratory, GI, and central nervous system disease in young children.

Along with Coxsackie viruses, echoviruses are one of the **leading causes of aseptic (viral) meningitis**. The diagnosis is typically made by PCR assay that detects echovirus RNA. Serologic tests are of little value because there are a large number of serotypes and no common antigen. There is no antiviral therapy or vaccine available.

OTHER ENTEROVIRUSES

In view of the difficulty in classifying many enteroviruses, all new isolates have been given a simple numerical designation since 1969.

Enterovirus 68 (EV68 and EV-D68) is a common cause of respiratory tract disease that ranges from a mild common cold to pneumonia and respiratory failure. It is also implicated as a cause of **acute flaccid myelitis** ("polio-like") in children (see Chapter 72). A PCR test is available. There is no antiviral therapy and no vaccine.

Enterovirus 70 is the main cause of acute hemorrhagic conjunctivitis, characterized by petechial hemorrhages on the bulbar conjunctivas. Complete recovery usually occurs, and there is no therapy.

Enterovirus 71 is one of the leading causes of viral central nervous system disease, including meningitis, encephalitis, and paralysis. It also causes diarrhea, pulmonary hemorrhages, hand, foot, and mouth disease, and herpangina. Enterovirus 72 is hepatitis A virus, which is described in Chapter 41.

SELF-ASSESSMENT QUESTIONS

1. Regarding poliovirus and the disease poliomyelitis, which one of the following is most accurate?
   - (A) Poliovirus is transmitted primarily by the fecal–oral route.
   - (B) New antigenic variants arise by coinfection with animal strains of poliovirus.
   - (C) Paralytic poliomyelitis is the most common manifestation of poliovirus infection.
   - (D) Poliovirus has single-stranded RNA as its genome and a polymerase in the virion that synthesizes its mRNA.
   - (E) The current vaccine recommendation is to give the live, attenuated vaccine for the first three immunizations to prevent the child from acting as a reservoir, followed by boosters using the killed vaccine.

2. A 70-year-old retired carpenter has signed up with a volunteer organization to build houses in a developing country where polio is still endemic. He plans to be there about 9 months. He thinks he has never been immunized against polio. Which one of the following is the most appropriate thing to do?
   - (A) Give immune serum globulins (ISG).
   - (B) Give the killed vaccine containing only type 3.
   - (C) Give the killed vaccine containing types 1, 2, and 3.
   - (D) Give the live vaccine containing only type 3.
   - (E) Give the live vaccine containing types 1, 2, and 3.

3. Regarding norovirus, which one of the following is most accurate?
   - (A) The diarrhea is caused by an exotoxin that increases cyclic adenosine monophosphate (cAMP).
   - (B) There are no neutrophils or red cells in the stool.
   - (C) Ritonavir, a protease inhibitor, is the drug of choice for chronic diarrhea caused by norovirus.
   - (D) Ingestion of undercooked hamburger is a common mode of acquisition of norovirus as cattle are a major reservoir of the virus.
   - (E) The diagnosis of norovirus-induced diarrhea is typically made by the detection of a fourfold or greater rise in antibody titer to the virus.

4. Regarding rotavirus, which one of the following is most accurate?
   - (A) Rotavirus is a major cause of nosocomial diarrhea in intensive care units.
   - (B) The vaccine against rotavirus contains live, attenuated virus as the immunogen.
   - (C) Rotavirus has a nonsegmented, single-stranded RNA genome, and there is no polymerase in the virion.
   - (D) The diagnosis of rotavirus diarrhea is typically made by the detection of a fourfold or greater rise in antibody titer to the virus.
   - (E) Diarrhea caused by rotavirus is due to a viral protein that increases the release of IgA from many submucosal B lymphocytes.

ANSWERS

(1) **(A)**
(2) **(C)**
(3) **(B)**
(4) **(B)**

SUMMARIES OF ORGANISMS

Brief summaries of the organisms described in this chapter begin on page 691. Please consult these summaries for a rapid review of the essential material.

PRACTICE QUESTIONS: USMLE & COURSE EXAMINATIONS

Questions on the topics discussed in this chapter can be found in the Clinical Virology section of Part XIII: USMLE (National Board) Practice Questions starting on page 747. Also see Part XIV: USMLE (National Board) Practice Examination starting on page 775.

Hepatitis Viruses

INTRODUCTION

Many viruses cause hepatitis. Of these, five medically important viruses are commonly described as "hepatitis viruses" because their main site of infection is the liver. These five are hepatitis A virus (HAV), hepatitis B virus (HBV), hepatitis C virus (HCV), hepatitis D virus (HDV, delta virus), and hepatitis E virus (HEV) (Tables 41–1 and 41–2). Other viruses, such as Epstein–Barr virus (the cause of infectious mononucleosis), cytomegalovirus, and yellow fever virus, infect the liver but also infect other sites in the body and therefore are not exclusively hepatitis viruses. They are discussed elsewhere.

Additional information regarding the clinical aspects of infections caused by the viruses in this chapter is provided in Part IX, titled Infectious Diseases beginning on page 603.

Note that these viruses belong to different viral families; some are DNA viruses, whereas others are RNA viruses, and some are enveloped, whereas others are nonenveloped. They are united by their ability to infect hepatocytes because they have proteins on their surface that attach to receptors on the surface of hepatocytes.

TABLE 41–1 Glossary of Hepatitis Viruses and Their Serologic Markers

Abbreviation	Name and Description
HAV	Hepatitis A virus, a picornavirus (nonenveloped RNA virus)
IgM HAV Ab	IgM antibody to HAV; best test to detect acute hepatitis A
HBV	Hepatitis B virus, a hepadnavirus (enveloped, partially double-stranded DNA virus); also known as Dane particle
HBsAg	Antigen found on surface of HBV, also found on noninfectious particles in patient's blood; positive during acute disease; continued presence indicates carrier state
HBsAb	Antibody to HBsAg; provides immunity to hepatitis B
HBcAg	Antigen associated with core of HBV
HBcAb	Antibody to HBcAg; positive during window phase; IgM HBcAb is an indicator of recent disease
HBeAg	A second, different antigenic determinant in the HBV core; important indicator of transmissibility
HBeAb	Antibody to e antigen; indicates low transmissibility
Non-A, non-B	Hepatitis viruses that are neither HAV nor HBV
HCV	Hepatitis C virus, a flavivirus (enveloped RNA virus); one of the non-A, non-B viruses
HDV	Hepatitis D virus, small RNA virus with HBsAg envelope; defective virus that replicates only in HBV-infected cells
HEV	Hepatitis E virus, a hepevirus (nonenveloped RNA virus); one of the non-A, non-B viruses

TABLE 41–2 Important Properties of Hepatitis Viruses

Virus	Genome	Replication Defective	DNA Polymerase in Virion	HBsAg in Envelope	Virus Family
HAV	ssRNA	No	No	No	Picornavirus
HBV	dsDNA[1]	No	Yes	Yes	Hepadnavirus
HCV	ssRNA	No	No	No	Flavivirus
HDV	ssRNA[2]	Yes	No	Yes	Deltavirus
HEV	ssRNA	No	No	No	Calicivirus

ds = double stranded; ss = single stranded.

[1]Interrupted, circular dsDNA.

[2]Circular, negative-stranded ssRNA.

Note also that these viruses are all noncytotoxic (i.e., they do not kill hepatocytes directly). The death of hepatocytes is mediated by cytotoxic T cells directed against viral antigen displayed on the surface of the hepatocyte in association with class I major histocompatibility complex (MHC) proteins.

HEPATITIS A VIRUS (HAV)

Disease

HAV causes hepatitis A.

Important Properties

HAV is a typical **enterovirus** classified in the picornavirus family. It has a single-stranded RNA genome and a nonenveloped icosahedral nucleocapsid and replicates in the cytoplasm of the cell. It is also known as enterovirus 72. It has one serotype, and there is no antigenic relationship to HBV or other hepatitis viruses.

Summary of Replicative Cycle

HAV has a replicative cycle similar to that of other enteroviruses (the replicative cycle of poliovirus is discussed in Chapter 40).

Transmission & Epidemiology

HAV is transmitted by the **fecal–oral** route. Humans are the reservoir for HAV. Virus appears in the feces roughly 2 weeks before the appearance of symptoms, so quarantine of patients is ineffective. **Children are the most frequently infected** group, and outbreaks occur in special living situations such as summer camps and boarding schools and among the homeless. Common-source outbreaks arise from fecally contaminated water or food such as oysters grown in polluted water and eaten raw. Unlike HBV, HAV is **rarely transmitted via the blood**, because the level of viremia is low and chronic infection does not occur. About 50% to 75% of adults in the United States have been infected, as evidenced immunoglobulin (Ig) G antibody.

Pathogenesis & Immunity

The virus replicates in the gastrointestinal tract, spreads to the liver via the portal vein, and infects hepatocytes.

Attack by cytotoxic T cells causes the damage to the hepatocytes. The infection is subsequently cleared, the damage is repaired, and no chronic infection ensues. Hepatitis caused by the different viruses cannot be distinguished pathologically.

The immune response consists initially of IgM antibody, which is detectable at the time jaundice appears. It is therefore important in the laboratory diagnosis of hepatitis A. The appearance of IgM is followed 1 to 3 weeks later by the production of IgG antibody, which provides lifelong protection.

Clinical Findings

The clinical manifestations of acute hepatitis are virtually the same, regardless of which hepatitis virus is the cause (Table 41–3). Fever, anorexia, nausea, vomiting, and jaundice are typical. Dark urine, pale feces, and elevated transaminase levels are seen. Most cases resolve spontaneously in 2 to 4 weeks.

Hepatitis A has a short incubation period (3–4 weeks) in contrast to that of hepatitis B, which is 10 to 12 weeks. Most HAV infections in children are asymptomatic whereas HAV infections in adults are often symptomatic. Asymptomatic HAV infections are detected by the presence of IgG antibody. No chronic hepatitis or chronic carrier state occurs, and there is no predisposition to hepatocellular carcinoma (HCC).

Laboratory Diagnosis

The detection of **IgM antibody** is the most important test. A fourfold rise in IgG antibody titer can also be used. Isolation of the virus in cell culture is possible but not available in the clinical laboratory.

Treatment & Prevention

No antiviral therapy is indicated for acute hepatitis A. Prevention of hepatitis A is achieved by **active immunization** with a vaccine containing inactivated HAV. The virus in the vaccine is grown in human cell culture and inactivated with formalin. Two doses, an initial dose followed by a booster 6 to 12 months later, should be given. No subsequent booster dose is recommended.

The vaccine is recommended for travelers to developing countries, for children ages 2 to 18 years, and for men who have sex with men. If an unimmunized person must travel to

TABLE 41–3   Clinical Features of Hepatitis Viruses

Virus	Mode of Transmission	Chronic Carriers	Laboratory Test Usually Used for Diagnosis	Vaccine Available	Immune Globulins Useful
HAV	Fecal–oral	No	IgM HAV	Yes	Yes
HBV	Blood, sexual, at birth	Yes	HBsAg, HBsAb, IgM HBcAb	Yes	Yes
HCV	Blood, sexual[1]	Yes	HCV Ab	No	No
HDV	Blood, sexual[1]	Yes	Ab to delta Ag	No	No
HEV	Fecal–oral	No	None	No	No

Ab = antibody; Ag = antigen.

[1]Sexual transmission seems likely but is poorly documented.

an endemic area within 4 weeks, then passive immunization (see later) should be given to provide immediate protection and the vaccine given to provide long-term protection. This is an example of **passive–active immunization**.

Because many adults have antibodies to HAV, it may be cost-effective to determine whether antibodies are present before giving the vaccine. The vaccine is also effective in postexposure prophylaxis if given within 2 weeks of exposure. A combination vaccine that immunizes against both HAV and HBV called Twinrix is available. Twinrix contains the same immunogens as the individual HAV and HBV vaccines.

Passive immunization with immune serum globulin prior to infection or within 14 days after exposure can prevent or mitigate the disease. Observation of proper hygiene (e.g., sewage disposal and handwashing after bowel movements) is of prime importance.

HEPATITIS B VIRUS (HBV)

Disease

HBV causes hepatitis B.

Important Properties

HBV is a member of the hepadnavirus family. It is a 42-nm **enveloped** virion,[1] with an icosahedral nucleocapsid core containing a **partially double-stranded circular** DNA genome (Figure 41–1 and Table 41–2).

The envelope contains a protein called the **surface antigen** (HBsAg), which is important for laboratory diagnosis and immunization.[2]

Within the core is a **DNA polymerase**. The genome contains four genes (four open reading frames) that encode five proteins; namely, the S gene encodes the surface antigen, the C gene encodes the core antigen and the e antigen, the P gene encodes the polymerase, and the X gene encodes the X protein (HBx).

HBx is an activator of viral RNA transcription and may be involved in oncogenesis because it can inactivate the p53 tumor suppressor protein (see Chapter 43). The DNA polymerase has both RNA-dependent (reverse transcriptase) and DNA-dependent activity.

Electron microscopy of a patient's serum reveals three different types of particles: a few 42-nm virions and many 22-nm **spheres** and long **filaments** 22-nm wide, which are composed of surface antigen (Figure 41–2). HBV is the only human virus that produces these spheres and filaments in such large numbers in the patient's blood. The ratio of filaments and small spheres to virions is 1000:1.

In addition to HBsAg, there are two other important antigens both located in the core of the virus: the **core antigen** (HBcAg) and the **e antigen** (HBeAg). The core antigen, as the name implies, is located on the nucleocapsid protein that forms the core of the virion. The e antigen is produced by proteolytic cleavage of the core protein during passage through the endoplasmic reticulum. The e antigen is soluble and is released from infected cells into the blood. The e antigen is an important indicator of **transmissibility**.

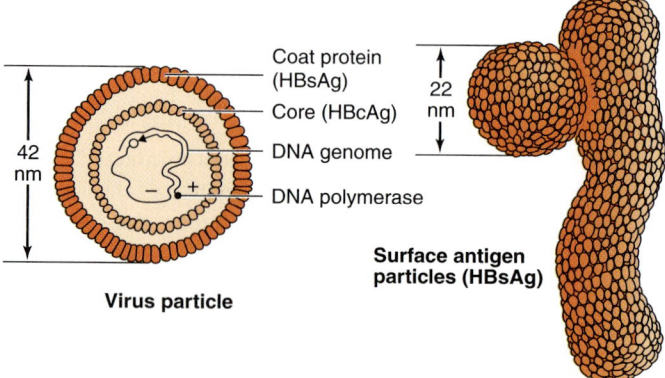

FIGURE 41–1 Hepatitis B virus (HBV). **Left:** Cross-section of the HBV virion. **Right:** The 22-nm spheres and filaments composed only of hepatitis B surface antigen. Because there is no viral DNA in the spheres and filaments, they are not infectious. (Reproduced with permission from Ryan K: *Sherris Medical Microbiology*, 3rd ed. New York, NY: McGraw Hill; 1994.)

Labels in figure: Coat protein (HBsAg); Core (HBcAg); DNA genome; DNA polymerase; 42 nm; 22 nm; **Surface antigen particles (HBsAg)**; **Virus particle**

---

[1]Also known as a Dane particle (named for the scientist who first published electron micrographs of the virion).

[2]HBsAg was known as Australia antigen because it was first found in the serum of an Australian aborigine.

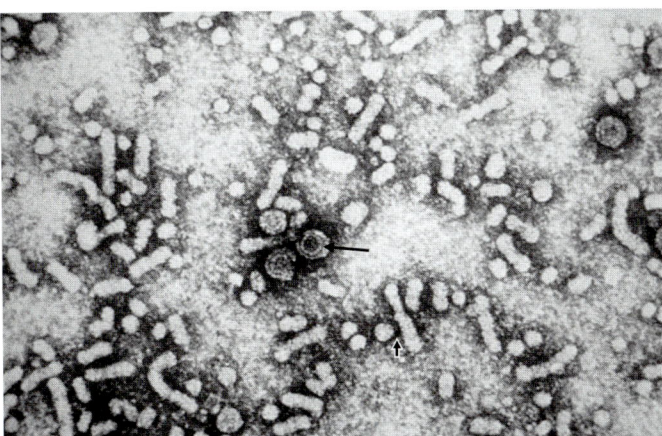

FIGURE 41–2 Hepatitis B virus. Electron micrograph. Long arrow points to a typical virion of hepatitis B virus. Short arrow points to a small sphere (just left of arrowhead) and a long rod (just right of arrowhead), both composed only of hepatitis B surface antigen. (Reproduced with permission from Public Health Image Library, Centers for Disease Control and Prevention.)

For vaccine purposes, HBV has one serotype based on HBsAg. However, for epidemiologic purposes, there are four serologic subtypes of HBsAg based on a group-specific antigen, "a," and two sets of mutually exclusive epitopes, d or y and w or r. This leads to four serotypes—adw, adr, ayw, and ayr—which are useful in epidemiologic studies because they are concentrated in certain geographic areas.

The specificity of HBV for liver cells is based on two properties: virus-specific receptors located on the hepatocyte cell membrane (facilitate entry) and transcription factors found only in the hepatocyte that enhance viral mRNA synthesis (act postentry).

Humans are the only natural hosts of HBV. There is no animal reservoir.

Summary of Replicative Cycle

The replicative cycle of HBV is depicted in Figure 41–3. The infecting virion attaches to the surface of the hepatocyte via its receptor, the sodium taurocholate cotransporting polypeptide. After entry of the virion into the cell and its uncoating, the nucleocapsid moves to the nucleus via microtubules. The nucleocapsid binds to a nuclear pore and the DNA genome is released into the nucleus. In the nucleus, the host cell DNA polymerase synthesizes the missing portion of DNA, and a double-stranded circular DNA is formed (CCC DNA). The minus strand of CCC DNA serves as a template for mRNA synthesis by cellular RNA polymerase. After the individual mRNAs are made and translated into viral proteins, a full-length positive-strand RNA is made, which is the template for the minus strand of the progeny DNA. The minus strand then serves as the template for the plus strand of the genome DNA.

Note that the synthesis of the progeny DNA genome is catalyzed by the virus-encoded RNA-dependent DNA polymerase (**reverse transcriptase**). Synthesis of the genome takes place within the newly assembled virion nucleocapsid core in the cytoplasm. The completed nucleocapsid acquires its envelope containing HBsAg in the endoplasmic reticulum. Progeny virions are released from the cell by budding through the cell membrane.

Hepadnaviruses are the *only* viruses that produce **genome DNA** by reverse transcription with viral RNA as the template. (Note that this type of RNA-dependent DNA synthesis is similar to but different from the process in retroviruses, in which the genome RNA is transcribed into a DNA intermediate.)

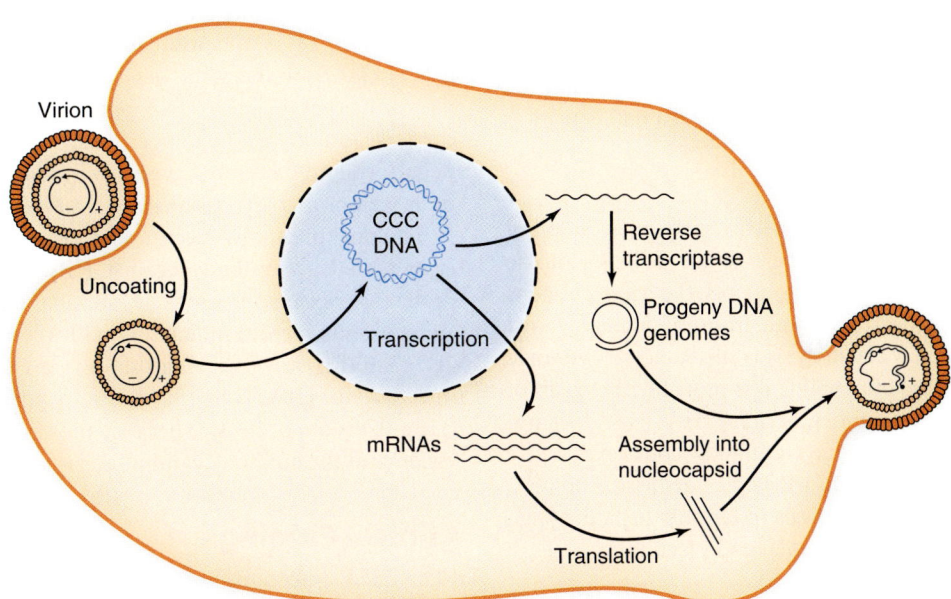

FIGURE 41–3 Replication cycle of hepatitis B virus. Note virus-encoded reverse transcriptase synthesizes the progeny DNA genomes using viral RNA as the template. CCC DNA is covalently closed circular DNA in the nucleus (blue circle).

In chronic HBV infection, a carrier state occurs and progeny HBV continues to be made. In this carrier state, most of the circular HBV DNA is found free in the nucleus as an episome. A small amount of HBV DNA is integrated into host cell DNA. How episomal HBV DNA is maintained in the carrier state for many years is unclear.

Transmission & Epidemiology

The three main modes of transmission are via blood, during sexual intercourse, and perinatally from mother to newborn. The observation that needle-stick injuries can transmit the virus indicates that only very small amounts of blood are necessary. HBV infection is especially prevalent in those who use intravenous drugs. Screening of blood for the presence of HBsAg has greatly decreased the number of transfusion-associated cases of hepatitis B.[3]

However, because blood transfusion is a modern procedure, there must be another natural route of transmission. HBV is found in semen and vaginal fluids, so it is likely that **sexual** transmission is important. Transmission from **mother to child during birth** is another important natural route. Transplacental transmission, if it occurs, is rare. There is no evidence that transmission of HBV occurs during breast-feeding.

Note that enveloped viruses, such as HBV, are more sensitive to environmental damage, for example, heat and dryness, than nonenveloped viruses and hence are more efficiently transmitted by intimate contact (e.g., sexual contact). Nonenveloped viruses, such as HAV, are quite stable and are transmitted well via the environment (e.g., fecal–oral transmission).

Hepatitis B is found worldwide but is particularly prevalent in Asia. Globally, more than 300 million people are chronically infected with HBV, and about 75% of them are Asian. There is a high incidence of **HCC** in many Asian countries—a finding that indicates that HBV is likely to be a human tumor virus (see Chapter 43). Immunization against HBV has significantly reduced the incidence of HCC in children. It appears that the HBV vaccine is the **first vaccine to prevent a human cancer**.

Pathogenesis & Immunity

After entering the blood, the virus infects hepatocytes, and viral antigens are displayed on the surface of the cells. Cytotoxic T cells mediate an immune attack against the viral antigens and inflammation and necrosis occur. **Immune attack** against viral antigens on infected hepatocytes is mediated by cytotoxic T cells. The pathogenesis of hepatitis B is probably the result of this cell-mediated immune injury, because HBV itself does not cause a cytopathic effect. Antigen–antibody complexes cause some of the early symptoms (e.g., arthralgias, arthritis, and urticaria) and some of the complications in chronic hepatitis (e.g., glomerulonephritis, cryoglobulinemia, and vasculitis).

About 5% of adult patients with HBV infection become chronic carriers. In contrast, 90% of infected newborns become chronic carriers (see later). A chronic carrier is someone who has **HBsAg persisting in their blood for 6 months or longer**. The chronic carrier state is attributed to a persistent infection of the hepatocytes, which results in the prolonged presence of HBV and HBsAg in the blood. The main determinant of whether a person clears the infection or becomes a chronic carrier is the adequacy of the cytotoxic T-cell response. HBV DNA exists primarily as an episome in the nucleus of persistently infected cells; a small number of copies of HBV DNA are integrated into cell DNA.

A high rate of **HCC occurs in chronic carriers**. The *HBx* gene may be an oncogene because the HBx protein inactivates the p53 tumor suppressor protein (see Chapter 43). In addition, HCC may be the result of persistent cellular regeneration that attempts to replace the dead hepatocytes. Alternatively, malignant transformation could be the result of insertional mutagenesis, which could occur when the HBV DNA integrates into the hepatocyte DNA. Integration of the HBV DNA could activate a cellular oncogene, leading to a loss of growth control. Almost all HCC cells have HBV DNA integrated into the cell DNA. Note that although viral DNA is integrated into cell DNA in most HCC cells, integration of viral DNA is *not* a required step in HBV replication. (See section on viral replication above.)

Chronic carriage is more likely to occur when infection occurs in a newborn than in an adult, probably because a newborn's cytotoxic T cells are less competent than those of an adult. **Approximately 90% of infected neonates become chronic carriers**. Chronic carriage resulting from neonatal infection is associated with a high risk of HCC.

Some chronic carriers make e antigen (they are said to be **e antigen positive**) and therefore have a high probability of making infectious virions and being able to transmit the disease. The **e antigen is the indicator of transmissibility** because it is encoded by the same gene that encodes the core protein indicating that the HBV DNA genome is present. Some chronic carriers do not make e antigen (they are said to be **e antigen negative**), and therefore have a low probability of making infectious virions and are less likely to transmit the disease.

Lifelong immunity occurs after the natural infection and is mediated by antibody against HBsAg (HBsAb). HBsAb is protective because it binds to surface antigen on the virion and prevents it from interacting with receptors on the hepatocyte. Another way of saying this is that HBsAb neutralizes the infectivity of HBV. Note that antibody against the core antigen (HBcAb) is *not* protective because the core antigen is inside the virion and the antibody cannot interact with it.

Clinical Findings

Many HBV infections are asymptomatic and are detected only by the presence of antibody to HBsAg. The mean incubation period for hepatitis B is 10 to 12 weeks, which is much longer than that of hepatitis A (3–4 weeks). The clinical appearance of acute hepatitis B is similar to that of hepatitis A. However, with

---

[3]In the United States, donated blood is screened for HBsAg and antibodies to HBcAg, HCV, HIV-1, HIV-2, and HTLV-1. Two other tests are also performed: a VDRL test for syphilis and a transaminase assay, which, if elevated, indicates liver damage and is a surrogate marker of viral infection.

hepatitis B, symptoms tend to be more severe, and life-threatening hepatitis can occur. Most chronic carriers are asymptomatic, but some have chronic active hepatitis, which can lead to cirrhosis and death.

Chronic hepatitis B can be managed but it cannot be cured. HBV infection persists for the lifetime of the individual, not only in those who have chronic hepatitis but also in those who are asymptomatic and have antibody to HB surface antigen. The clinical importance of this is that patients who are immunosuppressed later in life can manifest a reactivation of disease, including active hepatitis and liver failure. Patients at high risk for this reactivation include those who are HBsAg positive and who are treated with anti-CD20 drugs, such as rituximab, or have a hematopoietic stem cell transplant. Patients in this high-risk category should be treated prophylactically with either tenofovir or entecavir to prevent this severe complication (see Treatment section later).

In addition to liver-related findings, extrahepatic manifestations of HBV infection occur. In acute infection, serum sickness-like symptoms, such as fever, rash, and arthralgias, can occur. In chronic HBV infection, neuropathies, glomerulonephritis, and polyarteritis nodosa (a vasculitis of small- and medium-sized arteries) may occur. Autoantibodies, such as cryoglobulins and rheumatoid factor, may be detected.

Patients coinfected with both HBV and human immunodeficiency virus (HIV) may have increased hepatic damage if HIV is treated prior to treating HBV. This occurs because the **immune reconstitution** that results when HIV is treated successfully leads to increased damage to the hepatocytes by the restored, competent cytotoxic T cells. For this reason, it is suggested that HBV be treated prior to treating HIV.

Laboratory Diagnosis

The two most important serologic tests for the diagnosis of early hepatitis B are the tests for **HBsAg** and for **IgM antibody to the core antigen**. Both appear in the serum early in the disease. The hallmark test for acute hepatitis B is IgM antibody to core antigen. The test for HBsAg is also positive but that occurs in chronic hepatitis B also so is not a specific marker for acute infection.

HBsAg appears during the incubation period and is detectable in most patients during the prodrome and acute disease (Figure 41–4). It falls to undetectable levels during convalescence in most cases; its **prolonged presence** (at least 6 months) indicates the carrier state and the risk of chronic hepatitis and hepatic carcinoma.

As described in Table 41–4, HBsAb is not detectable in the chronic carrier state. Note that HBsAb is, in fact, being made but is not detectable in the laboratory tests because it is bound to the large amount of HBsAg present in the blood. HBsAb is also being made during the acute disease but is similarly undetectable because it is bound in antigen–antibody complexes.

Note that there is a period of several weeks when HBsAg has disappeared but HBsAb is not yet detectable. This is the **window phase**. At this time, the HBcAb is always positive and

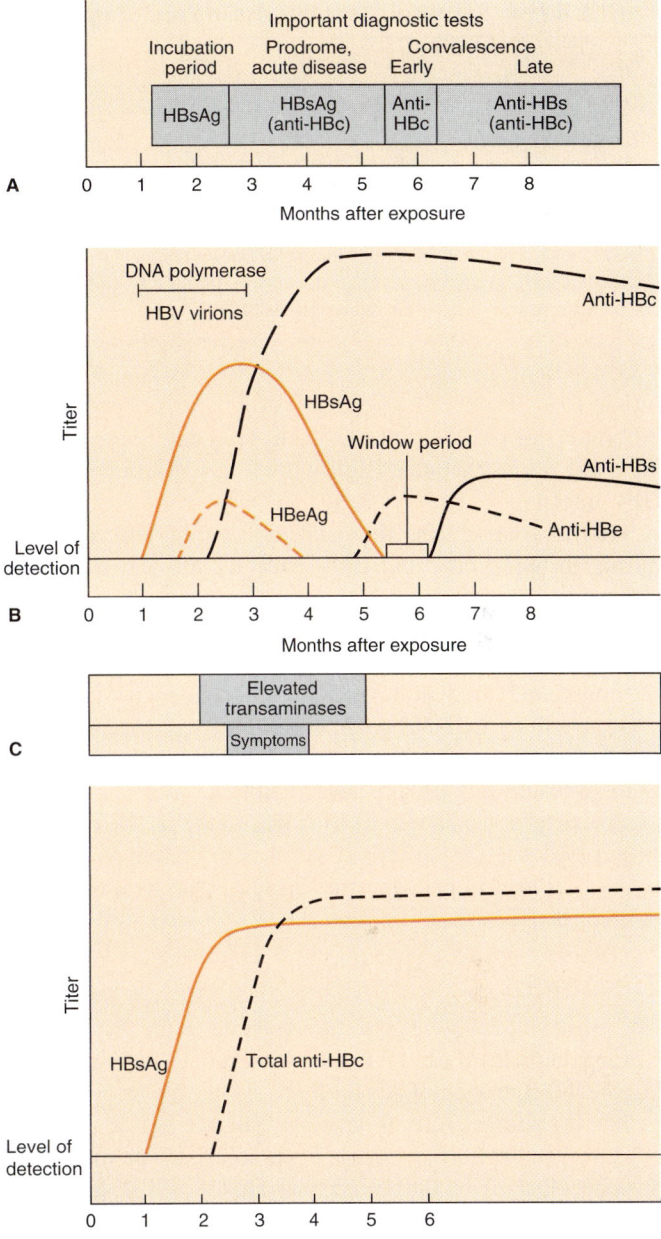

FIGURE 41–4 A: Important diagnostic tests during various stages of hepatitis B. **B:** Serologic findings in a patient with acute hepatitis B. **C:** Duration of increased liver enzyme activity and of symptoms in a patient with acute hepatitis B. **D:** Serologic findings in a patient with chronic hepatitis B. anti-HBc = hepatitis B core antibody; anti-HBe = hepatitis B e antibody; anti-HBs = hepatitis B surface antibody; HBeAg = hepatitis B e antigen; HBsAg = hepatitis B surface antigen; HBV = hepatitis B virus. (Adapted with permission from Lennette EH: *Manual of Clinical Microbiology*. 4th ed. Washington, DC: ASM Press; 1985.)

can be used to make the diagnosis. HBcAb is present in those with acute infection and chronic infection, as well as in those who have recovered from acute infection. Therefore, it cannot be used to distinguish between acute and chronic infection. The IgM form of HBcAb is present during acute infection and disappears approximately 6 months after infection. The test for

TABLE 41–4 Serologic Test Results in Four Stages of HBV Infection

Test	Acute Disease	Window Phase	Complete Recovery	Chronic Carrier State[1]
HBsAg	Positive	Negative	Negative	Positive
HBsAb	Negative	Negative	Positive	Negative[2]
HBcAb	Positive[3]	Positive	Positive	Positive

Note: People immunized with HBV vaccine have HBsAb but not HBcAb because the immunogen in the vaccine is purified HBsAg.

[1]Chronic carriers who are HBeAg positive are highly likely to transmit HBV, whereas those who are HBeAb positive are less likely to transmit HBV.

[2]Chronic carriers have negative antibody tests, but HBsAb is being made by these individuals. It is undetected in the tests because it is bound to the large amount of HBsAg present in the plasma. They are not tolerant to HBsAg.

[3]IgM is found in the acute stage; IgG is found in subsequent stages.

HBcAg is not readily available. Table 41–4 describes the serologic test results that characterize the four important stages of HBV infection.

HBeAg arises during the incubation period and is present during the prodrome and early acute disease and in certain chronic carriers. Its presence in chronic carriers indicates a **high likelihood of transmissibility**, and, conversely, the absence of HBeAg indicates a low likelihood of transmission. In addition, the finding of HBeAb indicates a lower likelihood, but transmission can still occur. DNA polymerase activity is detectable during the incubation period and early in the disease, but the assay is not available in most clinical laboratories.

The detection of viral DNA (**viral load**) in the serum is strong evidence that infectious virions are present. Reduction of the viral load in patients with chronic hepatitis B is used to monitor the success of drug therapy.

Treatment

No antiviral therapy is typically used in acute hepatitis B. Chronic hepatitis B can be treated with various antiviral drugs but the infection cannot be cured.

For chronic hepatitis B, entecavir (Baraclude) or tenofovir (Viread, Vemlidy) are the drugs of choice. They are nucleoside analogues that inhibit the reverse transcriptase of HBV. Interferon in the form of peginterferon alfa-2a (Pegasys) may also be used. Other nucleoside analogues such as lamivudine (Epivir-HBV), adefovir (Hepsera), and telbivudine (Tyzeka) are used less frequently. A combination of tenofovir and emtricitabine (Emtriva) is also used. Resistance to several of these antiretroviral drugs has emerged during long-term treatment of chronic HBV.

These drugs reduce hepatic inflammation, reduce the risk of cirrhosis and hepatocellular carcinoma, and lower the viral load of HBV in patients with chronic active hepatitis. Note, however, that neither interferon nor the nucleoside analogues cure the HBV infection. In most patients, when the drug is stopped, HBV replication resumes.

Patients coinfected with HBV and HIV should be prescribed highly active antiretroviral therapy (HAART) with caution because recovery of cell-mediated immunity can result in an exacerbation of hepatitis (immune reconstitution syndrome, IRIS). Consideration should be given to treat the HBV infection prior to starting HAART.

Prevention

The two main modes of prevention involve the use of either the **vaccine** or **hyperimmune globulin** or both.

(1) **The subunit vaccine (e.g., Recombivax, Engerix-B, or Heplisav-B) contains HBsAg** produced in yeasts by recombinant DNA techniques. The vaccine is highly effective in preventing hepatitis B and has few side effects. The seroconversion rate is approximately 95% in healthy adults. Recombivax and Engerix-B are given in a three-dose regimen whereas Heplisav-B is given as a two-dose regimen.

It is indicated for people who are frequently exposed to blood or blood products, such as certain healthcare personnel (e.g., medical students, surgeons, and dentists), patients receiving multiple transfusions or dialysis, patients with frequent sexually transmitted disease, and users of illicit intravenous drugs. Travelers who plan a long stay in areas of endemic infection, such as many countries in Asia and Africa, should receive the vaccine. The U.S. Public Health Service recommends that all newborns and adolescents receive the vaccine.

At present, booster doses after the initial two or three-dose regimen are not recommended. However, if antibody titers have declined in immunized patients who are at high risk, such as dialysis patients, then a booster dose should be considered.

Widespread immunization with the HBV vaccine has significantly reduced the incidence of HCC in children. A vaccine called Twinrix that contains both HBsAg and inactivated HAV provides protection against both hepatitis B and hepatitis A.

(2) **Hepatitis B immune globulin (HBIG) contains a high titer of HBsAb**. It is used to provide immediate, passive protection to individuals known to be exposed to HBsAg-positive blood (e.g., after an accidental needle-stick injury).

Precise recommendations for the use of the vaccine and HBIG are beyond the scope of this book. However, the recommendation regarding one common concern of medical students, the needle-stick injury from a patient with HBsAg-positive blood, is that both the vaccine and HBIG be given (at separate sites). This is true even if the patient's blood is HBeAb positive.

Both the vaccine and HBIG should also be given to a newborn whose mother is HBsAg positive. This regimen is very effective in reducing the infection rate of newborns whose mothers are chronic carriers. The regimen of vaccine plus

HBIG in those with needle-stick injuries and in neonates is a good example of **passive–active** immunization, in which both immediate protection and long-term protection are provided.

The effectiveness of cesarean section to reduce HBV infection of neonates is uncertain. It is currently not recommended. Breast-feeding of immunized neonates by mothers who are chronic carriers entails little risk of infection of the neonate.

All blood for transfusion should be screened for HBsAg. No one with a history of hepatitis (of any type) should donate blood, because other viruses, such as HCV, may be present. Screening of high-risk populations to detect chronic carriers using serologic testing should be done because identification and treatment of carriers will reduce transmission.

Note that preexposure prophylaxis using Truvada (tenofovir plus emtricitabine) to prevent HIV infection also prevents HBV infection. These two drugs inhibit the reverse transcriptase of both HBV and HIV.

NON-A, NON-B HEPATITIS VIRUSES

The term "non-A, non-B hepatitis" was coined to describe the cases of hepatitis for which existing serologic tests had ruled out all known viral causes. The term is not often used because the main cause of non-A, non-B hepatitis, namely, HCV, has been identified. In addition, HDV and HEV have been described. Cross-protection experiments indicate additional hepatitis viruses exist.

HEPATITIS C VIRUS (HCV)

Disease

HCV causes hepatitis C.

Important Properties

HCV is a member of the flavivirus family. It is an enveloped virion containing a genome of single-stranded, nonsegmented, positive-polarity RNA. It has no polymerase in the virion.

HCV has at least six genotypes and multiple subgenotypes based on differences in the genes that encode one of its two envelope glycoproteins. This genetic variation results in a "hypervariable" region in the envelope glycoprotein. The genetic variability is due to the high mutation rate in the envelope gene coupled with the absence of a proofreading function in the virion-encoded RNA polymerase. As a result, multiple subspecies (quasispecies) often occur in the blood of an infected individual at the same time. Genotype 1 causes approximately 75% of infections in the United States.

More than 50% of HCV infections result in **chronic infection**, a rate much higher than that of HBV. Chronic HCV infection predisposes to **HCC**. The high rate of chronic infection is attributed to the ability of the HCV protease to inactivate a signal protein involved in inducing interferon in hepatocytes.

Summary of Replicative Cycle

HCV replicates in the cytoplasm and translates its genome RNA into large polyproteins, from which functional viral proteins are cleaved by a virion-encoded protease. This protease is the target of potent anti-HCV therapy (see Treatment section). In addition, HCV genome RNA encodes a protein called NS5A that cooperates with the RNA polymerase of the virus to synthesize progeny genome RNAs. The NS5A protein also functions in the assembly of progeny virions. The NS5A protein is the target of potent anti-HCV therapy (see Treatment section).

The replication of HCV in the liver is enhanced by a liver-specific micro-RNA called miR-122. This micro-RNA acts by increasing the synthesis of HCV mRNA. (Micro-RNAs are known to enhance cellular mRNA synthesis in many tissues.) A clinical trial of an antisense nucleotide called miravirsen that bound to and blocked the activity of miR-122 showed prolonged reduction in HCV RNA levels in infected patients.

Transmission & Epidemiology

Humans are the reservoir for HCV. It is transmitted primarily via **blood**. At present, injection drug use accounts for almost all new HCV infections. Transmission from mother to child during birth is another important mode of transmission. Transmission via blood transfusion rarely occurs because donated blood containing antibody to HCV is discarded. Transmission via needle-stick injury occurs, but the risk is lower than for HBV. Sexual transmission is uncommon, and there is no evidence for transmission across the placenta or during breast-feeding.

HCV is the **most prevalent blood-borne pathogen** in the United States. (In the nationally reported incidence data, HCV ranks below HIV and HBV as a blood-borne pathogen, but it is estimated that HCV is more prevalent.) Approximately 4 million people in the United States (1%–2% of the population) are chronically infected with HCV. Unlike yellow fever virus, another flavivirus that infects the liver and is transmitted by mosquitoes, there is no evidence for an insect vector for HCV. Worldwide, it is estimated that 180 million people are infected with HCV.

Many infections are **asymptomatic**, so screening of high-risk individuals for HCV antibody should be done. In addition, screening of those who were born between 1945 and 1965 should be done because they have a high rate of infection.

In the United States, about 1% of blood donors have antibody to HCV. People who share needles when taking intravenous drugs are very commonly infected. Commercially prepared Ig preparations are generally very safe, but several instances of the transmission of HCV have occurred.

Pathogenesis & Immunity

HCV infects hepatocytes primarily, but there is no evidence for a virus-induced cytopathic effect on the liver cells. Rather, death of the hepatocytes is probably caused by immune attack by cytotoxic T cells. **HCV infection strongly predisposes to HCC**, but there is no evidence for an oncogene in the viral genome or for insertion of a copy of the viral genome into the DNA of the cancer cells.

Alcoholism greatly enhances the rate of HCC in HCV-infected individuals. This supports the idea that the cancer is

TABLE 41–5 Laboratory Test Results at Different Stages of HCV Infection

Diagnostic Test	Acute HCV Infection	Chronic HCV Infection	Recovered from HCV Infection
Antibody to HCV	Positive in 6–24 weeks. Negative early in infection	Positive	Positive
Viral load (HCV RNA in serum)	Detectable within 1–2 weeks	Detectable	Undetectable
Transaminase (alanine amino-transferase, ALT)	Elevated	Typically elevated but fluctuates to near normal	May be normal but may be positive and fluctuate

caused by prolonged liver damage and the consequent rapid growth rate of hepatocytes as the cells attempt to regenerate rather than by a direct oncogenic effect of HCV. Added support for this idea is the observation that patients with cirrhosis of any origin, not just alcoholic cirrhosis, have an increased risk of HCC.

Antibodies against HCV are made, but **approximately 75% of patients are chronically infected** and continue to produce virus for at least 1 year. (**Note that the rate of chronic carriage of HCV is much higher than the rate of chronic carriage of HBV.**) Chronic infection is likely to be encouraged by the suppression of interferon synthesis by the protease of HCV that cleaves a signal transduction protein required to activate interferon synthesis. Chronic active hepatitis and cirrhosis occur in approximately 10% of patients with chronic HCV infection.

Clinical Findings

The acute infection is often asymptomatic. If symptoms such as malaise, nausea, and right upper quadrant pain do occur, they are milder than with infection by the other hepatitis viruses. Fever, anorexia, nausea, vomiting, and jaundice are common. Dark urine, pale feces, and elevated transaminase levels are seen.

Hepatitis C resembles hepatitis B as far as the ensuing chronic liver disease, cirrhosis, and the predisposition to HCC are concerned. Note that a chronic carrier state occurs *much more often* with HCV infection than with HBV.

Liver biopsy is often done in patients with chronic infection to evaluate the extent of liver damage and to guide treatment decisions. Many infections with HCV, including both acute and chronic infections, are asymptomatic and are detected only by the presence of antibody. The mean incubation period is 8 weeks. Cirrhosis resulting from chronic HCV infection is the most common indication for liver transplantation.

HCV infection also leads to significant extrahepatic autoimmune reactions, including vasculitis, arthralgias, purpura, and membranoproliferative glomerulonephritis. Hepatitis C infection also markedly increases the risk of B-cell non-Hodgkin's lymphoma.

HCV is the main cause of essential mixed cryoglobulinemia. Cryoglobulins are defined by their ability to precipitate at cold temperature (cryo = cold). The cryoprecipitates are immune complexes composed of HCV antigens and antibodies. Peripheral neuropathy is one of the most common complications of chronic HCV infections. Damage to the nerves is thought to be caused by cryoprecipitates.

Reinfection can occur after a patient has a documented undetectable viral load following drug therapy. This occurs primarily in intravenous drug users who are frequently exposed to HCV. Reinfection in the face of existing antibody is thought to be caused by the high rate of antigenic variants combined with an diminishing T cell response. High risk patients should be screened every 6 to 12 months with a PCR assay to detect viral RNA to determine whether reinfection has occurred.

Laboratory Diagnosis

HCV infection is diagnosed by detecting antibodies to HCV in an enzyme-linked immunosorbent assay (ELISA) (Table 41–5). The antigen in the assay is a recombinant protein formed from three immunologically stable HCV proteins and does not include the highly variable envelope proteins. The test does not distinguish between IgM and IgG and does not distinguish between an acute, chronic, and resolved infection.

If the result of ELISA antibody test is positive, a PCR-based test that detects the presence of viral RNA (**viral load**) in the serum should be performed to determine whether active disease exists. Reduction of the viral load in patients with hepatitis C is used to monitor the success of drug therapy. Isolation of the virus from patient specimens is not done. A chronic infection is characterized by elevated transaminase levels, a positive ELISA antibody test, and **detectable viral RNA for at least 6 months**.

Treatment

Treatment of **acute** hepatitis C with peginterferon alfa significantly decreases the number of patients who become chronic carriers. An oral regimen of ledipasvir and sofosbuvir is now used for acute hepatitis C infection by genotype-1 HCV.

The treatment of choice for **chronic** hepatitis C is a combination of drugs from three classes: RNA polymerase inhibitors, NS5A inhibitors, and protease inhibitors (Figure 41–5 and Table 41–6). These drugs are administered orally, which is an improvement over the drugs in previous regimens that often included pegylated interferon alfa, which is administered parenterally and has significant adverse effects.

Currently available combinations are described in Table 41–7. Note that these drug combinations are more effective against certain genotypes of HCV, although some drug combinations are effective against all six of the major genotypes. These various drug combinations offer the prospect of a "cure" for chronic hepatitis C.

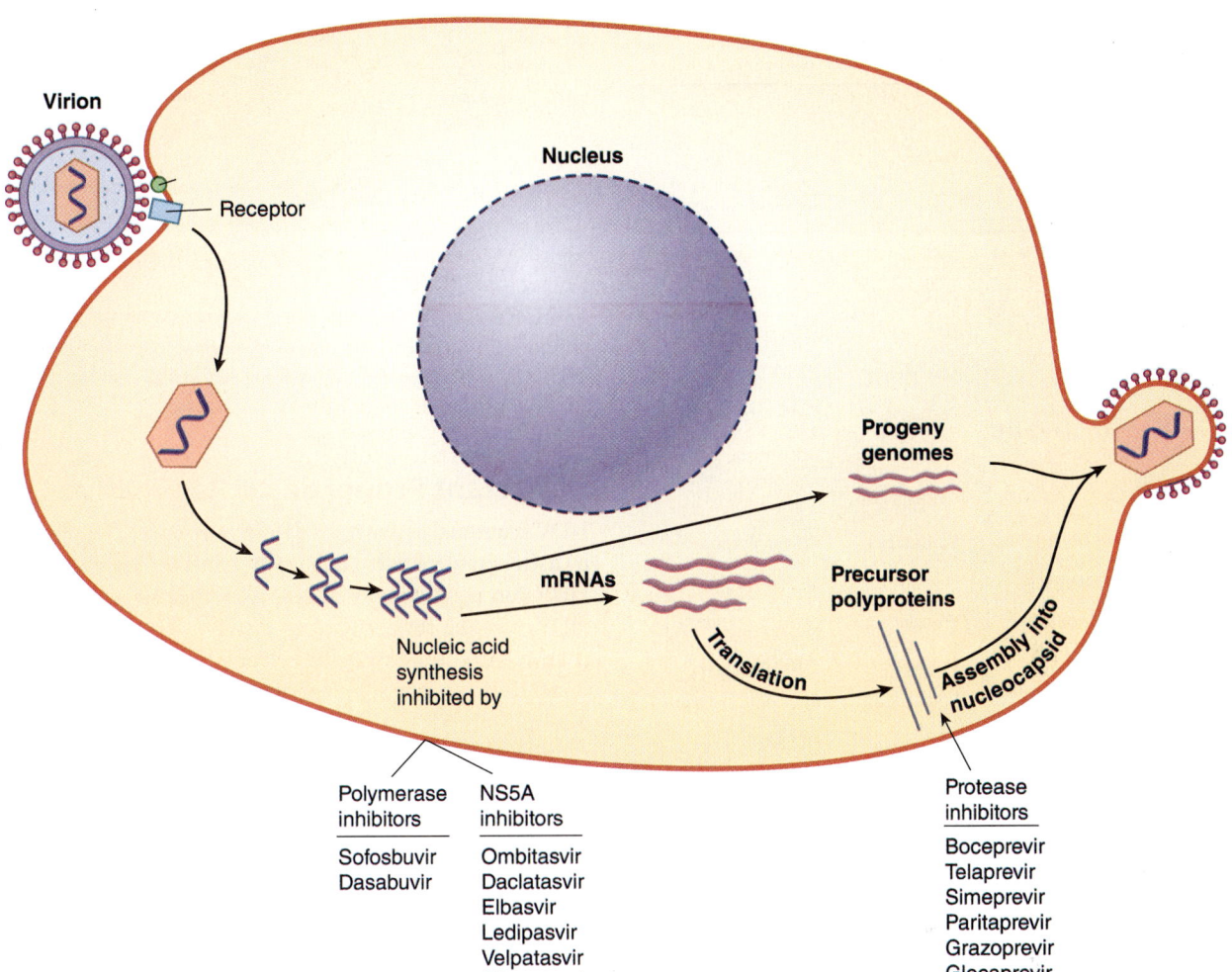

FIGURE 41–5 Site of action of drugs used in treatment of chronic hepatitis C virus infection.

TABLE 41–6 Oral Drugs for Chronic HCV Infection

Class of Drug	Name of Drug	Mechanism of Action
RNA poly-merase inhibitor	Sofosbuvir	Inhibits synthesis of genome RNA; nucleoside (uridine) analogue; chain-terminating drug
RNA poly-merase inhibitor	Dasabuvir	Inhibits synthesis of genome RNA; nonnucleoside inhibitor
NS5A inhibitor	Ledipasvir Ombitasvir Daclatasvir Elbasvir Velpatasvir Pibrentasvir	Inhibits synthesis of genome RNA; blocks action of NS5A protein, which is a cofactor required for RNA polymerase activity
Protease inhibitor	Boceprevir Simeprevir Telaprevir Paritaprevir Grazoprevir Glecaprevir Voxilaprevir	Inhibits cleavage of precursor polypeptide; blocks production of functional HCV structural and nonstructural proteins

One important adverse effect of these drugs used in the treatment of chronic HCV infection is the reactivation of HBV infection. The mechanism of reactivation of HBV is unknown.

Prevention

There is no vaccine, and hyperimmune globulins are not available. Pooled immune serum globulins are not useful for postexposure prophylaxis. There is no effective regimen for prophylaxis following needle-stick injury; only monitoring is recommended.

Blood found to contain antibody is discarded—a procedure that has prevented virtually all cases of transfusion-acquired HCV infection since 1994, when screening began. Screening of individuals born in the United States between 1945 and 1965 for HCV antibody is recommended because they have a high rate of infection. Treatment of those who are antibody-positive should reduce transmission.

Patients with chronic HCV infection should be advised to reduce or eliminate their consumption of alcoholic beverages to reduce the risk of HCC and cirrhosis. Patients with chronic HCV infection and cirrhosis should be monitored with

TABLE 41–7   Drug Combinations Effective for Treatment of Chronic HCV Infection

HCV Genotype Against Which Treatment Is Effective	Name of Drug	Name of HCV Inhibitor	Site of Action of HCV Inhibitor
1	Sovaldi plus	Sofosbuvir	RNA polymerase
	Olysio	Simeprevir	Protease
1	Viekira Pak	Dasabuvir	RNA polymerase
		Ombitasvir	NS5A
		Paritaprevir	Protease
		Ritonavir	Booster of protease inhibitor
1 or 3	Sovaldi plus	Sofosbuvir	RNA polymerase
	Daklinza	Daclatasvir	NS5A
1 or 4	Zepatier	Elbasvir	NS5A
		Grazoprevir	Protease
1, 4, 5, or 6	Harvoni	Sofosbuvir	RNA polymerase
		Ledipasvir	NS5A
4	Technivie	Ombitasvir	NS5A
		Paritaprevir	Protease
		Ritonavir	Booster of protease inhibitor
All 6	Epclusa	Sofosbuvir	RNA polymerase
		Velpatasvir	NS5A
All 6	Mavyret	Pibrentasvir	NS5A
		Glecaprevir	Protease
All 6	Vosevi	Sofosbuvir	RNA polymerase
		Velpatasvir	NS5A
		Voxilaprevir	Protease

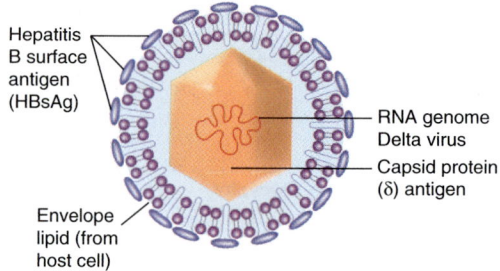

FIGURE 41–6 Hepatitis D virus. Note that hepatitis B surface antigen (HBsAg) forms the outer envelope and the genome consists of circular RNA. (Reproduced with permission from Ryan KJ, Ray C: *Sherris Medical Microbiology*, 6th ed. New York, NY: McGraw Hill; 2014.)

Important Properties & Replicative Cycle

HDV is unusual in that it is a **defective** virus (i.e., it cannot replicate by itself because it does not have the genes for its envelope protein). HDV can replicate only in cells also infected with HBV because HDV uses the surface antigen of HBV (HBsAg) as its envelope protein. HBV is therefore the helper virus for HDV (Figure 41–6).

HDV is an enveloped virus with an RNA genome that is a single-stranded, negative-polarity, covalently closed circle. The RNA genome of HDV is very small and encodes only one protein, the internal core protein called **delta antigen**. HDV genome RNA has no sequence homology to HBV genome DNA. HDV has no virion polymerase; the genome RNA is replicated and transcribed by the host cell RNA polymerase. HDV genome RNA is a "ribozyme" (i.e., it has the ability to self-cleave and self-ligate—properties that are employed during replication of the genome). HDV replicates in the nucleus, but the specifics of the replicative cycle are complex and beyond the scope of this book.

HDV has one serotype because HBsAg has only one serotype. There is no evidence for the existence of an animal reservoir for HDV.

Transmission & Epidemiology

HDV is transmitted by the same means as is HBV (i.e., sexually, by blood, and perinatally). In the United States, the incidence of HDV infections is low; most infections occur in intravenous drug users who share needles. HDV infections occur worldwide, with a similar distribution to that of HBV infections.

Pathogenesis & Immunity

It seems likely that the pathogenesis of hepatitis caused by HDV and HBV is the same (i.e., the virus-infected hepatocytes are damaged by cytotoxic T cells). There is some evidence that delta antigen is cytopathic for hepatocytes.

IgG antibody against delta antigen is not detected for long periods after infection; it is therefore uncertain whether long-term immunity to HDV exists.

Clinical Findings

Because HDV can replicate only in cells also infected with HBV, hepatitis delta can occur only in a person infected with HBV.

alpha-fetoprotein tests and liver sonograms to detect carcinoma at an early stage. Patients with liver failure due to HCV infection can receive a liver transplant, but infection of the graft with HCV typically occurs.

Patients coinfected with HCV and HIV should be prescribed HAART with caution because recovery of cell-mediated immunity can result in an exacerbation of hepatitis (IRIS). Consideration should be given to treat the HCV infection prior to starting HAART.

HEPATITIS D VIRUS (HDV, DELTA VIRUS)

Disease

HDV causes hepatitis D (hepatitis delta).

A person can either be infected with both HDV and HBV at the same time (i.e., be "coinfected") or be previously infected with HBV and then "superinfected" with HDV.

Hepatitis in patients coinfected with HDV and HBV is more severe than in those infected with HBV alone, but the incidence of chronic hepatitis is about the same in patients infected with HBV alone. However, hepatitis in chronic carriers of HBV who become superinfected with HDV is much more severe, and the incidence of fulminant, life-threatening hepatitis, chronic hepatitis, and liver failure is significantly higher.

Laboratory Diagnosis

The diagnosis of HDV infection in the laboratory is made by detecting either delta antigen or IgM antibody to delta antigen in the patient's serum. Tests for HDV RNA in the blood are also available.

Treatment & Prevention

Peginterferon alfa can mitigate some of the effects of the chronic hepatitis caused by HDV but does not eradicate the chronic carrier state. There is no specific antiviral therapy against HDV. There is no vaccine against HDV, but a person immunized against HBV will not be infected by HDV because HDV cannot replicate unless HBV infection also occurs.

HEPATITIS E VIRUS (HEV)

HEV is a major cause of hepatitis transmitted by the fecal–oral route. It is thought to be more common than HAV in many developing countries. It is a common cause of waterborne epidemics of hepatitis in Asia, Africa, India, and Mexico but is uncommon in the United States. HEV is a nonenveloped, single-stranded RNA virus classified as a member of the hepevirus family.

Clinically, the disease resembles hepatitis A, with the exception of a high mortality rate in pregnant women. Most cases resolve without sequelae. Chronic infection resulting in chronic hepatitis and cirrhosis, but not HCC, occurs in immunocompromised individuals such as HIV-infected patients, those who are receiving cancer chemotherapy, and patients who are receiving immunosuppressive drugs to prevent rejection of solid organ transplants.

The diagnosis is typically made by detecting IgM antibody to HEV. A PCR assay that detects HEV RNA in patient specimens is available. There is no antiviral drug available for acute infection in immunocompetent patients. In immunocompromised patients, ribavirin cleared HEV viremia in solid organ transplant recipients. There is no vaccine.

HEPATITIS G VIRUS (HGV)

In 1996, hepatitis G virus (HGV) was isolated from patients with posttransfusion hepatitis. HGV is a member of the flavivirus family, as is HCV. However, unlike HCV, which is clearly the cause of both acute hepatitis and chronic active hepatitis and predisposes to HCC, HGV has not been documented to cause any of these clinical findings. The role of HGV in the causation of liver disease has yet to be established, but it can cause a chronic infection lasting for decades. Approximately 60% to 70% of those infected clear the virus and develop antibodies.

HGV is transmitted via sexual intercourse and blood. It is carried in the blood of millions of people worldwide. In the United States, it is found in the blood of approximately 2% of random blood donors, 15% of those infected with HCV, and 35% of those infected with HIV. Patients coinfected with HIV and HGV have a lower mortality rate and have less HIV in their blood than those infected with HIV alone. It is hypothesized that HGV may interfere with the replication of HIV. (HGV is also known as GB virus C, human hepegivirus, and pegivirus H.)

SELF-ASSESSMENT QUESTIONS

1. An outbreak of jaundice occurs in several young children who attend the same day care center. If the outbreak was caused by a virus, which one of the following is the most likely cause?

    (A) Hepatitis A virus
    (B) Hepatitis B virus
    (C) Hepatitis C virus
    (D) Hepatitis D virus

2. Regarding hepatitis A virus (HAV), which one of the following statements is most accurate?

    (A) The HAV vaccine contains live, attenuated virus as the immunogen.
    (B) The screening of blood for transfusion has greatly reduced the spread of this virus.
    (C) The diagnosis is typically made by serologic tests rather than by culturing the virus.
    (D) Multiple episodes of hepatitis A are common because it has three serotypes.
    (E) It has a segmented, negative-polarity, single-stranded RNA genome and an RNA polymerase in the virion.

3. A woman who is hepatitis B surface antigen (HBsAg) positive and hepatitis B surface antibody (HBsAb) negative has just given birth. Which one of the following is the most appropriate thing to do for the newborn?

    (A) Nothing. The child is protected against hepatitis B.
    (B) Immunize with the vaccine containing HBsAg (HBV vaccine).
    (C) Give hepatitis B hyperimmune globulin (HBIG).
    (D) Give both the HBV vaccine and HBIG.

4. Regarding hepatitis B virus (HBV) and the disease hepatitis B, which one of the following is most accurate?

    (A) The most reliable indicator that a person can transmit HBV is the presence of HBsAg in the blood.
    (B) HBV has a circular, partially double-stranded DNA as its genome and has a DNA polymerase in the virion.
    (C) Healthcare personnel who sustain a needle-stick injury while taking blood from a person with hepatitis B should receive acyclovir.
    (D) HBV infection induces antibody to HBcAg (core antigen), which protects the person from a second attack by the same strain of HBV.
    (E) A person in the "window period" can be diagnosed as having been infected by HBV if antibody to the surface antigen (HBsAg) is present.

5. Regarding hepatitis C virus (HCV), which one of the following is most accurate?

(A) Chronic infection with HCV predisposes to hepatocellular carcinoma.

(B) HCV is a defective virus that requires concurrent hepatitis B virus (HBV) infection in order to replicate.

(C) Chronic infection with HCV occurs less frequently than chronic infection with HBV.

(D) The killed vaccine against HCV is poorly immunogenic, so booster doses must be given at least every 5 years.

(E) Proper sewage disposal has significantly decreased the incidence of hepatitis C.

6. Regarding hepatitis D virus (HDV), which one of the following is most accurate?

(A) Alpha interferon can eradicate the latent state established by HDV.

(B) Immunization against hepatitis B virus (HBV) will reduce the incidence of hepatitis caused by HDV.

(C) HDV has DNA as its genome and an RNA-dependent DNA polymerase in the virion.

(D) The laboratory diagnosis of HDV infection is made by growing HDV in cells coinfected with HBV.

(E) Many HDV infections occur in young children in the diaper stage in day care centers because the virus is transmitted primarily by the fecal–oral route.

7. Your patient is a 35-year-old man who complains that the whites of his eyes have turned yellow. After taking a history and doing a physical, you order serologic tests to determine whether he has viral hepatitis. On the basis of the results, you tell him that he has a mild form of hepatitis that does not cause long-term damage to the liver. Your conclusion is based on a positive result on which one of the following tests?

(A) Antibody to hepatitis C virus

(B) Hepatitis B surface antigen

(C) Hepatitis delta antigen

(D) IgM antibody to hepatitis A virus

8. Your patient is a 20-year-old woman with chronic hepatitis B that was diagnosed by detecting hepatitis B antigen in her blood more than 6 months after her acute infection. Which one of the following is the best choice of drug to treat her chronic hepatitis B?

(A) Acyclovir

(B) Foscarnet

(C) Entecavir

(D) Ritonavir

(E) Zidovudine

9. Your patient is a 27-year-old man with a history of intravenous drug use who now is diagnosed with chronic hepatitis C. He is HIV antibody negative. Which one of the following is the best choice of drugs to treat his chronic hepatitis C?

(A) Acyclovir and foscarnet

(B) Ganciclovir and enfuvirtide

(C) Sofosbuvir and velpatasvir

(D) Zidovudine and lamivudine

(E) Tenofovir and simeprevir

10. Treatment with a drug whose mode of action is inhibition of reverse transcriptase is used in which one of the following diseases?

(A) Acute hepatitis A

(B) Chronic hepatitis A

(C) Acute hepatitis B

(D) Chronic hepatitis B

(E) Acute hepatitis C

(F) Chronic hepatitis C

11. Treatment with a drug whose mode of action is inhibition of NS5A protein is used in which one of the following diseases?

(A) Acute hepatitis A

(B) Chronic hepatitis A

(C) Acute hepatitis B

(D) Chronic hepatitis B

(E) Acute hepatitis C

(F) Chronic hepatitis C

12. Treatment with a drug whose mode of action is inhibition of a virus-encoded protease is used in which one of the following diseases?

(A) Acute hepatitis A

(B) Chronic hepatitis A

(C) Acute hepatitis B

(D) Chronic hepatitis B

(E) Acute hepatitis C

(F) Chronic hepatitis C

ANSWERS

(1) **(A)**

(2) **(C)**

(3) **(D)**

(4) **(B)**

(5) **(A)**

(6) **(B)**

(7) **(D)**

(8) **(C)**

(9) **(C)**

(10) **(D)**

(11) **(F)**

(12) **(F)**

SUMMARIES OF ORGANISMS

Brief summaries of the organisms described in this chapter begin on page 691. Please consult these summaries for a rapid review of the essential material.

PRACTICE QUESTIONS: USMLE & COURSE EXAMINATIONS

Questions on the topics discussed in this chapter can be found in the Clinical Virology section of Part XIII: USMLE (National Board) Practice Questions starting on page 747. Also see Part XIV: USMLE (National Board) Practice Examination starting on page 775.

Arboviruses, Rabies Virus, & Ebola Virus

The viruses described in this chapter are united by having an animal reservoir. This indicates that these viruses can replicate within both the cells of the host animal and within human cells. Most viruses that cause human disease are limited to replicating in human cells as the attachment proteins on the viral surface interact only with receptors on the surface of human cells.

The animal reservoir for arboviruses is quite varied. Birds are the most common, but monkeys and rodents serve as the reservoir for some arboviruses. Many mammals serve as a reservoir for rabies virus. In the United States, bats, skunks, and raccoons are common reservoirs, whereas worldwide, dogs are the most common. The animal reservoir for Ebola virus appears likely to be fruit bats.

Several uncommon viruses with an animal reservoir, such as hantavirus, Lassa fever virus, and Marburg virus are discussed in Chapter 46.

Additional information regarding the clinical aspects of infections caused by the viruses in this chapter is provided in Part IX, titled Infectious Diseases beginning on page 603.

ARBOVIRUSES

Arbovirus is an acronym for *ar*thropod-*bo*rne virus. This highlights the fact that these viruses are transmitted by **arthropods**, primarily mosquitoes and ticks. It is a collective name for a large group of diverse viruses, more than 600 at last count. In general,

they are named either for the diseases they cause (e.g., yellow fever virus [YFV]) or for the place where they were first isolated (e.g., St. Louis encephalitis [SLE] virus).

As an aside, another group of viruses called **roboviruses** can be mentioned here for comparison. The term *robo* refers to the fact that these viruses are *ro*dent-*bo*rne (i.e., they are transmitted directly from rodents to humans *without* an arthropod vector). Transmission occurs when dried rodent excrement is inhaled into the human lung, as when sweeping the floor of a cabin. Two roboviruses cause a respiratory distress syndrome that is often fatal: Sin Nombre virus (a hantavirus) and Whitewater Arroyo virus (an arenavirus). These viruses are described in Chapter 46.

Important Properties of Arboviruses

Most arboviruses are classified into three families,[1] namely, togaviruses, flaviviruses, and bunyaviruses (Table 42–1).

(1) Togaviruses[2] are characterized by an icosahedral nucleocapsid surrounded by an envelope and a single-stranded,

---

[1]A few arboviruses belong to two other families. For example, Colorado tick virus is a reovirus; Kern Canyon virus and vesicular stomatitis virus are rhabdoviruses.

[2]*Toga* means cloak.

TABLE 42–1   Classification of Major Arboviruses

Family	Genus	Viruses of Medical Interest in the Americas
Togavirus	Alphavirus[1]	Eastern equine encephalitis virus, western equine encephalitis virus, chikungunya virus
Flavivirus	Flavivirus[2]	St. Louis encephalitis virus, yellow fever virus, dengue virus, West Nile virus, Zika virus
Bunyavirus	Bunyavirus[3]	California encephalitis virus
Reovirus	Orbivirus	Colorado tick fever virus

[1]Alphaviruses of other regions include Chikungunya, Mayaro, O'Nyong-Nyong, Ross River, and Semliki Forest viruses.

[2]Flaviviruses of other regions include Japanese encephalitis, Kyasanur Forest, Murray Valley encephalitis, Omsk hemorrhagic fever, Powassan encephalitis viruses, and West Nile viruses.

[3]Bunyaviruses of other regions include the Bunyamwera complex of viruses and Oropouche virus.

positive-polarity RNA genome. They are 70 nm in diameter, in contrast to the flaviviruses, which are 40 to 50 nm in diameter (see later). Togaviruses are divided into two families, alphaviruses and rubiviruses. Only alphaviruses are considered here. The only rubivirus that causes disease in humans is rubella virus, which was discussed in Chapter 39.

(2) Flaviviruses[3] are similar to togaviruses in that they also have an icosahedral nucleocapsid surrounded by an envelope and a single-stranded, positive-polarity RNA genome, but the flaviviruses are only 40 to 50 nm in diameter, whereas the togaviruses have a diameter of 70 nm.

(3) Bunyaviruses[4] have a helical nucleocapsid surrounded by an envelope and a genome consisting of three circular segments of single-stranded negative-polarity RNA. The three segments are called large (L), medium (M), and small (S).

Transmission of Arboviruses

The life cycle of the arboviruses is based on the ability of these viruses to multiply in *both* the vertebrate host and the bloodsucking vector (Figure 42–1). For effective transmission to occur, the virus must be present in the bloodstream of the vertebrate host (viremia) in sufficiently high titer to be taken up in the small volume of blood ingested during an insect bite. After ingestion, the virus replicates in the gut of the arthropod and then spreads to other organs, including the salivary glands. Only the female of the species serves as the vector of the virus, because only she requires a blood meal in order for progeny to be produced. An obligatory length of time, called the **extrinsic incubation period**,[5] must pass before the virus has replicated

---

[3]*Flavi* means yellow, as in yellow fever.

[4]"Bunya" is short for Bunyamwera-the town in Africa where the prototype virus was located.

[5]The intrinsic incubation period is the interval between the time of the bite and the appearance of symptoms in the human host.

sufficiently for the saliva of the vector to contain enough virus to transmit an infectious dose. For most viruses, the extrinsic incubation period ranges from 7 to 14 days.

In addition to transmission through vertebrates, some arboviruses are transmitted by vertical "transovarian" passage from the mother tick to her offspring. Vertical transmission has important survival value for the virus if a vertebrate host is unavailable.

Humans are involved in the transmission cycle of arboviruses in two different ways. Usually, humans are **dead-end hosts** because the concentration of virus in human blood is too low and the duration of viremia is too brief for the next bite to transmit the virus. However, in some diseases (e.g., yellow fever and dengue), humans have a high-level viremia and act as reservoirs of the virus.

Infection by arboviruses usually does not result in disease either in the arthropod vector or in the vertebrate animal that serves as the natural host. Disease occurs primarily when the virus infects dead-end hosts. For example, YFV cycles harmlessly among the jungle monkeys in South America, but when the virus infects a human, the disease yellow fever can occur.

Clinical Findings & Epidemiology of Arboviruses

Most human arboviral infections are asymptomatic. Of those infections that are symptomatic, most are acute febrile illnesses. A minority of infections cause neuroinvasive disease, such as encephalitis and meningitis. The diseases caused by arboviruses range in severity from mild to rapidly fatal.

The clinical picture usually fits one of three categories: (1) **encephalitis;** (2) **hemorrhagic fever;** or (3) **fever with systemic symptoms** such as myalgias, arthralgias, and nonhemorrhagic rash. The pathogenesis of these diseases involves not only the cytocidal effect of the virus but also, in some, a prominent immunopathologic component. After recovery from the disease, immunity is usually lifelong.

The arboviral diseases occur primarily in the **tropics** but are also found in temperate zones such as the United States and as far north as Alaska and Siberia. They have a tendency to cause sudden outbreaks of disease, generally at the interface between human communities and jungle or forest areas.

IMPORTANT ARBOVIRUSES THAT CAUSE DISEASE IN THE UNITED STATES

West Nile Virus

West Nile virus (WNV) is the most common cause of neuroinvasive (encephalitis, meningitis) arboviral disease in the United States. WNV caused an outbreak of encephalitis in New York City and environs in July, August, and September 1999. This is the first time WNV caused disease in the United States. In this outbreak, there were 27 confirmed cases and 23 probable

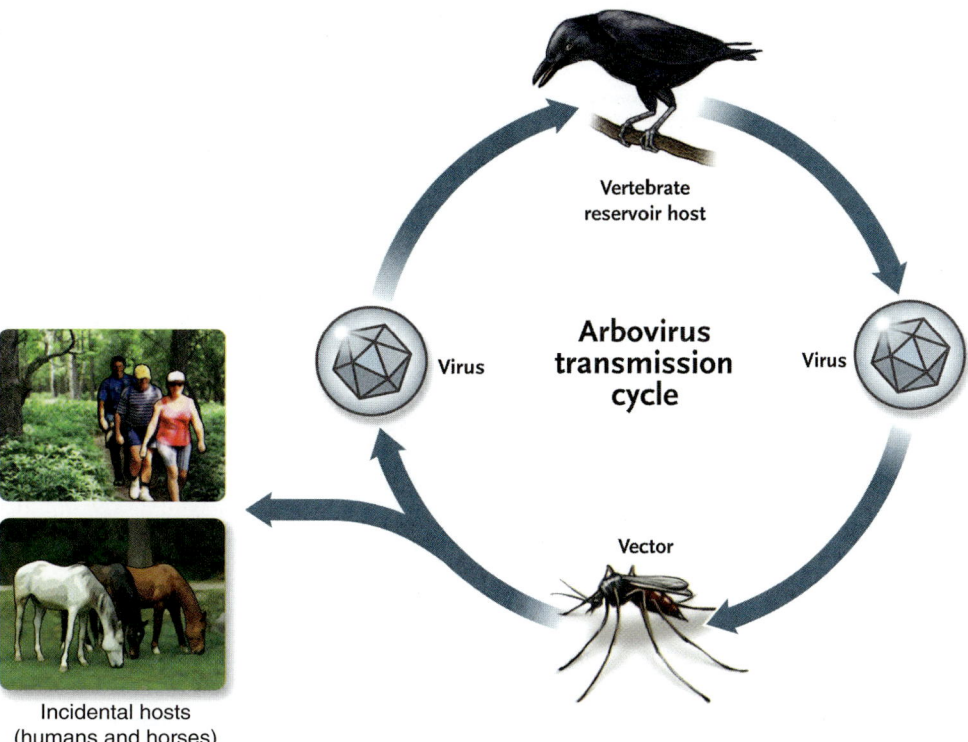

FIGURE 42–1   Arbovirus transmission cycle. Arboviruses typically cycle between the vertebrate reservoir host, often a bird, and the vector, often a mosquito. The infected vector can also bite other hosts, such as humans and horses, which are "dead-end" hosts because their viremia is too low to provide the vector with an infectious dose. (Reproduced with permission from Centers for Disease Control and Prevention.)

cases, including 5 deaths. Many birds, especially crows, died as well. No human cases occurred after area-wide spraying of mosquito-control compounds and the onset of cooler weather. It is not known how WNV entered the United States, but either an infected traveler or an infected mosquito brought by an airplane seems likely to be involved.

In the ensuing years, outbreaks of West Nile encephalitis involving hundreds to thousands of people occurred as the virus progressed westward in the United States. For example, in 2012, there were 3142 reported cases and 134 deaths. Each year, WNV causes the highest number of deaths due to mosquito-borne encephalitis in the United States.

WNV is a flavivirus that is classified in the same antigenic group as SLE virus. It is endemic in Africa but has caused encephalitis in areas of Europe and Asia as well. Wild birds are the main reservoir of this virus, which is transmitted by mosquitoes, especially *Culex* species. Humans are dead-end hosts. Transmission of the virus via solid organ transplants has also occurred.

The most important clinical picture is encephalitis with or without signs of meningitis, typically in a person over 60 years of age. Encephalitis occurs in about 1% of infections, fever and headache without encephalitis occur in about 20% (called West Nile fever), and roughly 80% of infections are asymptomatic.

The laboratory diagnosis is typically made by detecting immunoglobulin (Ig) M antibodies in spinal fluid or blood. Polymerase chain reaction (PCR)-based assays are also available.

No antiviral therapy or vaccine for humans is available. In an attempt to prevent blood-borne transmission, blood banks screen donated blood for the presence of WNV using nucleic acid probes specific for the virus.

Eastern Equine Encephalitis Virus

Of the encephalitis viruses listed in Table 42–2, eastern equine encephalitis (EEE) virus causes the **most severe** disease and is associated with the highest fatality rate (approximately 50%). In its natural habitat, the virus is transmitted primarily by the swamp **mosquito,** *Culiseta,* among the small wild birds of the Atlantic and Gulf Coast states. Species of *Aedes* mosquitoes are suspected of carrying the virus from its **wild bird reservoir** to the principal **dead-end hosts, horses,** and **humans.** The number of cases of human encephalitis caused by EEE virus in the United States usually ranges from zero to four per year, but outbreaks involving hundreds of cases also occur. Subclinical infections greatly exceed the number of overt cases.

The encephalitis is characterized by the sudden onset of severe headache, nausea, vomiting, and fever. Changes in mental status, such as confusion and stupor, ensue. A rapidly progressive downhill course with nuchal rigidity, seizures, and coma occurs. If the patient survives, the central nervous system sequelae are usually severe. Immunity following the infection is lifelong.

The diagnosis is made by either isolating the virus or demonstrating a rise in antibody titer. Clinicians should have a high

TABLE 42–2 Epidemiology of Important Arbovirus Diseases in the United States

Disease[1]	Vector	Animal Reservoir	Geographic Distribution	Approximate Incidence Per Year[2]
West Nile encephalitis	Mosquito	Wild birds	Endemic in Africa; Widespread in the United States	700–1000
EEE	Mosquito	Wild birds[3]	Atlantic and Gulf states	0–4
WEE	Mosquito	Wild birds[3]	West of Mississippi	5–20[4]
SLE	Mosquito	Wild birds	Widespread in southern, central, and western states	10–30[4]
CE	Mosquito	Small mammals	Northcentral states	40–80
CTF	Tick	Small mammals	Rocky Mountains	100–300

CE = California encephalitis; CTF = Colorado tick fever; EEE = eastern equine encephalitis virus; SLE = St. Louis encephalitis; WEE = western equine encephalitis virus.

[1]Venezuelan equine encephalitis virus causes disease in the United States too rarely to be included.

[2]Human cases.

[3]Horses are dead-end hosts, not reservoirs.

[4]Hundreds of cases during an outbreak.

index of suspicion in the summer months in the appropriate geographic areas. The disease does not occur in the winter because mosquitoes are not active. It is not known how the virus survives the winter—in birds, mosquitoes, or perhaps some other animal.

No antiviral therapy is available. A killed vaccine is available to protect horses but not humans. The disease is too rare for production of a human vaccine to be economically feasible.

Western Equine Encephalitis Virus

Western equine encephalitis (WEE) virus causes disease more frequently than does EEE virus, but the illness is less severe. Inapparent infections outnumber the apparent by at least 100:1. The number of cases in the United States usually ranges between 5 and 20 per year, and the fatality rate is roughly 2%.

The virus is transmitted primarily by *Culex* **mosquitoes** among the **wild bird** population of the western states, especially in areas with irrigated farmland.

The clinical picture of WEE virus infection is similar but less severe than that caused by EEE virus. Sequelae are less common. The diagnosis is made by isolating the virus or observing a rise in antibody titer. There is no antiviral therapy. There is a killed vaccine for horses but not for humans.

St. Louis Encephalitis Virus

SLE virus causes disease over a wider geographic area than do EEE and WEE viruses. It is found in the southern, central, and western states and causes 10 to 30 cases of encephalitis per year in the United States.

The virus is transmitted by several species of *Culex* **mosquitoes** that vary depending on location. Again, small **wild birds**, especially English sparrows, are the reservoir, and humans are dead-end hosts. Although EEE and WEE viruses are predominantly rural, SLE virus occurs in **urban areas** because these mosquitoes prefer to breed in stagnant wastewater.

SLE virus causes moderately severe encephalitis with a fatality rate that approaches 10%. Most infections are inapparent. Sequelae are uncommon.

The diagnosis is usually made serologically because the virus is difficult to isolate. No antiviral therapy or vaccine is available.

California Encephalitis Virus

California encephalitis (CE) virus was first isolated from mosquitoes in California in 1952, but its name is something of a misnomer because most human disease occurs in the north-central states. The strain of CE virus that causes encephalitis most frequently is called La Crosse for the city in Wisconsin where it was isolated. CE virus is the only one of the four major encephalitis viruses in the United States that is a member of the **bunyavirus** family.

La Crosse virus is transmitted by the **mosquito** *Aedes triseriatus* among forest **rodents**. The virus is passed transovarially in mosquitoes and thus survives the winter when mosquitoes are not active. The clinical picture can be mild, resembling enteroviral meningitis, or severe, resembling herpes encephalitis. Death rarely occurs. Diagnosis is usually made serologically rather than by isolation of the virus. No antiviral therapy or vaccine is available.

Colorado Tick Fever Virus

Of the five diseases described in Table 42–2, Colorado tick fever (CTF) is the most easily distinguished from the others, both biologically and clinically. CTF virus is a **reovirus** transmitted by the wood **tick** *Dermacentor andersoni* among the small **rodents** (e.g., chipmunks and squirrels) of the Rocky Mountains. There are approximately 100 to 300 cases per year in the United States.

The disease occurs primarily in people hiking or camping in the Rocky Mountains and is characterized by fever, headache, retro-orbital pain, and severe myalgia. The diagnosis is made either by isolating the virus from the blood or by detecting a rise in antibody titer. No antiviral therapy or vaccine is available. Prevention involves wearing protective clothing and inspecting the skin for ticks.

IMPORTANT ARBOVIRUSES THAT CAUSE DISEASE PRIMARILY OUTSIDE THE UNITED STATES

Although yellow fever and dengue are not endemic in the United States, extensive travel by Americans to tropical areas means that imported cases occur. It is reasonable, therefore, that physicians in the United States be acquainted with these two diseases.

YFV, dengue virus, and Zika virus (ZIKV) are members of the flavivirus family whereas chikungunya virus is a member of the togavirus family. Table 42–3 describes the epidemiology of the important arboviral diseases that occur primarily outside the United States. Japanese encephalitis virus, also a flavivirus and an important cause of epidemic encephalitis in Asia, is described in Chapter 46.

Note that all four viruses described in this section: YFV, dengue virus, chikungunya virus, and ZIKV **are transmitted by** *Aedes* **mosquitoes.**

Yellow Fever Virus

This virus is a flavivirus with one serotype. Most people infected with YFV are asymptomatic but about 15% manifest severe, often fatal disease. Yellow fever occurs primarily in the tropical regions of Africa and South America.

As the name implies, yellow fever is characterized by jaundice and fever. It is a severe, life-threatening disease that begins with the sudden onset of fever, headache, myalgias, and photophobia. After this prodrome, the symptoms progress to involve the liver, kidneys, and heart. Prostration and shock occur, accompanied by upper gastrointestinal tract hemorrhage with hematemesis ("black vomit").

In the epidemiology of yellow fever, **two distinct cycles** exist in nature, with different reservoirs and vectors.

(1) Jungle yellow fever is a disease of **monkeys** in tropical Africa and South America; it is transmitted primarily by the treetop **mosquitoes** of the *Haemagogus* species. Monkeys are the permanent reservoir, whereas humans are accidental hosts. Humans (e.g., tree cutters) are infected when they enter the jungle occupationally.

(2) In contrast, urban yellow fever is a disease of **humans** that is transmitted by the **mosquito** *Aedes aegypti*, which breeds in stagnant water. In the urban form of the disease, humans are the reservoir. For effective transmission to occur, the virus must replicate in the mosquito during the 12- to 14-day extrinsic incubation period. After the infected mosquito bites the person, the intrinsic incubation period is 3 to 6 days.

Diagnosis in the laboratory can be made either by PCR assay or by detecting either IgM or a rise in IgG antibody titer. No antiviral therapy is available, and the mortality rate is high. If the patient recovers, no chronic infection ensues and lifelong immunity is conferred.

Prevention of yellow fever involves mosquito control and immunization with the **vaccine** containing live, attenuated YFV. Travelers to and residents of endemic areas should be immunized. Protection lasts up to 10 years, and boosters are required every 10 years for travelers entering certain countries. Epidemics still occur in parts of tropical Africa and South America. Because it is a live vaccine, it should not be given to immunocompromised people or to pregnant women.

Dengue Virus

Although dengue fever is **not endemic** in the United States, some tourists to the Caribbean and other tropical areas return with this disease. In recent years, there were 100 to 200 cases per year in the United States, mostly in the southern and eastern states. No indigenous transmission occurred within the United States. It is estimated that about 20 million people are infected with dengue virus each year worldwide. **Dengue is the most common mosquito-borne viral disease in the world.** Humans are the main reservoir of the virus but forest monkeys may also be involved in maintaining the virus in Nature.

TABLE 42–3 Epidemiology of Important Arboviral Diseases Outside the United States

Disease	Vector	Animal Reservoir	Geographic Distribution	Vaccine Available
Yellow fever				Yes
1. Urban	*Aedes* mosquito	Humans	Tropical Africa and South America	
2. Jungle	*Haemagogus* mosquito	Monkeys	Tropical Africa and South America	
Dengue	*Aedes* mosquito	Humans; probably monkeys also	Tropical areas, especially Caribbean	Yes
Chikungunya virus	*Aedes* mosquito	Humans	Tropical areas, especially Caribbean	No
Zika virus	*Aedes* mosquito	Humans and nonhuman primates	Tropical areas of Central and South America	No

There are four serotypes of dengue virus. Infection with one type induces immunity against that serotype but not against the other three serotypes. In fact, subsequent infection with another serotype can cause a **strong cross-reacting (heterotypic) antibody response**, meaning that a strong antibody response is elicited against the *first* serotype by infection with the second serotype. This has severe consequences as it causes dengue hemorrhagic fever (see later).

Classic dengue fever (**breakbone fever**) begins suddenly with an influenza-like syndrome consisting of fever, malaise, retro-orbital pain, and headache. Severe pains in muscles (myalgia) and joints (arthralgia, "breakbone") occur. Enlarged lymph nodes, facial flushing, a maculopapular rash, and leukopenia are common. After a week or so, the symptoms regress, but weakness may persist. Although unpleasant, this typical form of dengue is rarely fatal and has few sequelae.

In contrast, **dengue hemorrhagic fever** is a much more severe disease, with a fatality rate that approaches 10%. The initial picture is the same as classic dengue, but then shock and hemorrhage, especially into the gastrointestinal tract and skin, develop. Dengue hemorrhagic fever occurs particularly in southern Asia, whereas the classic form is found in tropical areas worldwide.

Hemorrhagic shock syndrome is due to the production of large amounts of **cross-reacting antibody** at the time of a second dengue infection. The pathogenesis is as follows: The patient recovers from classic dengue caused by one of the four serotypes, and antibody against that serotype is produced. When the patient is infected with another serotype of dengue virus, an anamnestic, heterotypic response occurs, and large amounts of cross-reacting antibody to the first serotype are produced.

There are two hypotheses about what happens next. One is that immune complexes comprising virus and antibody are formed that activate complement, causing increased vascular permeability and thrombocytopenia. The other is that the antibodies increase the entry of virus into monocytes and macrophages, with the consequent liberation of a large amount of cytokines. In either scenario, shock and hemorrhage result.

Dengue virus is transmitted by the *A. aegypti* **mosquito**, which is also the vector of YFV. Humans are the reservoir for dengue virus, but a jungle cycle involving monkeys as the reservoir and other *Aedes* species as vectors is suspected.

The diagnosis can be made in the laboratory either by isolation of the virus in cell culture or by serologic tests that demonstrate the presence of IgM antibody or a fourfold or greater rise in antibody titer in acute and convalescent sera. A PCR assay that detects virus in the blood is also available.

No antiviral therapy for dengue is available. Outbreaks are controlled by using insecticides and draining stagnant water that serves as the breeding place for the mosquitoes. Personal protection includes using mosquito repellent and wearing clothing that covers the entire body.

A vaccine against dengue (Dengvaxia) is available for use primarily in countries where the disease is common. It was approved for use in the United States in 2019. It is a live, attenuated tetravalent recombinant vaccine. It is composed of a backbone of yellow fever vaccine virus containing the genes encoding the envelope and premembrane proteins of four serotypes of dengue virus.

The vaccine should be used only in those who have a laboratory documented *prior* infection by dengue virus. This is because the vaccine acts as a first infection that could predispose to dengue hemorrhagic fever if the vaccinee is subsequently infected with a different serotype. (See explanation earlier in this section.)

Another approach to prevention is the experimental infection of mosquitoes with *Wolbachia*, which reduces the transmission of dengue virus. *Wolbachia* are *Rickettsia*-like bacteria that reside permanently within infected cells (see Chapter 27).

Zika Virus

This is a flavivirus that causes Zika fever, an illness similar to dengue characterized by fever, arthralgia, myalgia, pruritic maculopapular rash, and nonpurulent conjunctivitis. Approximately 80% of infections are asymptomatic. The typical illness lasts a few days to a week. Most symptomatic adults recover without sequelae. However, infection predisposes to Guillain-Barré syndrome. Unlike dengue, hemorrhagic shock does not occur.

The most important aspect of ZIKV infection is vertical transmission from mother to fetus across the placenta. Infection in pregnant women can cause **serious fetal abnormalities, including microcephaly**. The risk of microcephaly is greatest when the mother is infected in the first trimester. Other fetal abnormalities include visual defects, hearing loss, and cerebral calcifications. In addition to the brain, other organs can be affected, and the term "Zika congenital syndrome" is used to describe the various effects on the fetus. Fetal death also occurs. ZIKV is the only arbovirus documented to cause microcephaly.

The vector is the *Aedes* mosquito, and the vertebrate hosts are humans and nonhuman primates, especially monkeys. Almost all infections are transmitted by mosquitoes, but semen can contain the virus and sexual transmission occurs. There is no confirmed report of transmission by blood transfusion or breast milk. ZIKV has a single serotype.

Disease caused by ZIKV was very rare until 2007 when an outbreak occurred on Yap, an island in Micronesia. In 2015, ZIKV caused outbreaks in the Caribbean and South America. By March 2016, ZIKV had spread to 33 countries in the Americas, causing disease in thousands of people.

The diagnosis is established either by finding viral RNA in the serum or urine using a PCR assay or by finding IgM antibody in the serum using an ELISA test. Interpretation of the antibody test is complicated by cross-reacting antibody in people infected with other flaviviruses, such as dengue virus, or in those who have received the yellow fever vaccine.

Treatment of ZIKV infection in adults is symptomatic. There is no effective antiviral drug and no vaccine against ZIKV.

The main goal of prevention is to eliminate infection in pregnant women. As there is no vaccine, prevention involves

TABLE 42–4 Public Health Measures to Prevent Zika Virus Transmission

Type of Prevention	Specific Measures
Environmental measures	1. Use screens on doors and windows 2. Remove standing water around home 3. Sleep under mosquito net 4. Use insecticide spray around home
Personal protection	1. Use mosquito repellant 2. Wear long sleeves and long pants 3. Do not travel to endemic area if pregnant
Prevent sexual transmission	Use condoms or abstain from sexual activity during pregnancy

TABLE 42–5 Properties of Rabies Virus and Ebola Virus

Property	Rabies Virus	Ebola Virus
Virus family	Rhabdoviruses	Filoviruses
Genome	Single-stranded RNA; negative polarity	Single-stranded RNA; negative polarity
Virion RNA polymerase	Yes	Yes
Nucleocapsid	Helical	Helical
Envelope	Yes	Yes
Number of serotypes	One	The five strains are cross-reactive

various environmental and personal measures, such as protective clothing, mosquito repellent, bed nets, and screened windows (Table 42–4). Pregnant women should not travel to areas where outbreaks are occurring. A man who has been diagnosed with ZIKV infection should use condoms or refrain from having sex for at least 6 months after symptoms began. A man who has traveled to an area where ZIKV is endemic and who is asymptomatic should refrain from having sex for at least 2 months after return from endemic area.

Chikungunya Virus

This virus causes chikungunya fever characterized by the sudden onset of high fever and joint pains, especially of the wrists and ankles. Joint involvement is bilateral and symmetric. Severe arthritis, especially of the hands, can last for months. A macular or maculopapular rash over much of the body is common. Encephalitis may occur.

Outbreaks involving millions of people in India, Africa, and the islands in the Indian Ocean have occurred in the years from 2004 to 2006. In 2013 to 2014, this virus moved to the Western Hemisphere, causing outbreaks involving thousands of people on many Caribbean islands and in the state of Florida.

Chikungunya virus is an RNA enveloped virus and is a member of the togavirus family. It has a single-stranded, positive-polarity RNA genome. It is transmitted by species of *Aedes* mosquitoes, both *A. aegypti* and *Aedes albopictus*. The latter mosquito is found in the United States, so the potential for outbreaks exists. Humans are the most important reservoir, but infection of nonhuman primates is thought to sustain the virus in nonpopulated areas.

Individuals returning to the United States from areas where outbreaks have occurred have been diagnosed with chikungunya fever. Laboratory diagnosis involves detecting the virus in blood either by PCR assay for viral RNA or by enzyme-linked immunosorbent assay (ELISA) for IgM antibody. There is no antiviral therapy, and no vaccine is available.

RABIES VIRUS

Disease

This virus causes rabies, a disease characterized by encephalitis.

Important Properties

Rabies virus is the only medically important member of the rhabdovirus family. It has a nonsegmented **single-stranded RNA** enclosed within a **bullet-shaped capsid** surrounded by a lipoprotein envelope (Table 42–5). Because the genome RNA has **negative polarity**, the virion contains an RNA-dependent **RNA polymerase**. Rabies virus has a single serotype. The antigenicity resides in the envelope glycoprotein spikes.

Rabies virus has a **broad host range:** It can infect all mammals, but only certain mammals are important sources of infection for humans (see later).

Summary of Replicative Cycle

Rabies virus attaches to the **acetylcholine receptor** on the cell surface. After entry into the cell, the virion RNA polymerase synthesizes five mRNAs that code for viral proteins. After replication of the genome viral RNA by a virus-encoded RNA polymerase, progeny RNA is assembled with virion proteins to form the nucleocapsid, and the envelope is acquired as the virion buds through the cell membrane.

Transmission & Epidemiology

The virus is transmitted by the **bite** of a rabid animal that manifests aggressive, biting behavior induced by the viral encephalitis. The virus is in the saliva of the rabid animal. In the United States, transmission is usually from the bite of **wild animals** such as skunks, raccoons, and bats; dogs and cats are frequently immunized and therefore are rarely sources of human infection. In recent years, **bats** have been the source of most cases of human rabies in the United States. Rodents and rabbits do not transmit rabies. In developing countries, unimmunized dogs are the most common reservoir.

Human rabies has also occurred in the United States in people who have not been bitten, so-called "nonbite" exposures. The most important example of this type of transmission is exposure to aerosols of bat secretions containing rabies virus. Another rare example is transmission in transplants of corneas taken from patients who died of undiagnosed rabies.

In the United States, fewer than 10 cases of rabies occur each year (mostly imported), whereas in developing countries, there are hundreds of cases, mostly due to rabid dogs. In 2007, the United States was declared "canine-rabies free"—the result of the widespread immunization of dogs. Worldwide, approximately 50,000 people die of rabies each year.

The country of origin and the reservoir host of a strain of rabies virus can often be identified by determining the base sequence of the genome RNA. For example, a person developed clinical rabies in the United States, but sequencing of the genome RNA revealed that the virus was the Mexican strain. It was later discovered that the man had been bitten by a dog while in Mexico several months earlier.

Pathogenesis & Immunity

The virus multiplies locally at the bite site, infects the sensory neurons, and **moves by axonal transport to the central nervous system**. During its transport within the nerve, the virus is sheltered from the immune system and little, if any, immune response occurs. The virus multiplies in the central nervous system and then travels down the peripheral nerves to the salivary glands and other organs. From the salivary glands, it enters the saliva to be transmitted by the bite. There is no viremic stage.

Within the central nervous system, **encephalitis** develops, with the death of neurons and demyelination. Infected neurons contain an eosinophilic cytoplasmic inclusion called a **Negri body**, which is important in laboratory diagnosis of rabies (Figure 42–2). Because so few individuals have survived rabies, there is no information regarding immunity to disease upon being bitten again.

Clinical Findings

The incubation period varies, according to the location of the bite, from as short as 2 weeks to 16 weeks or longer. It is shorter when bites are sustained on the head rather than on the leg, because the virus has a shorter distance to travel to reach the central nervous system (Table 42–6).

Clinically, the patient exhibits a prodrome of nonspecific symptoms such as fever, anorexia, and changes in sensation

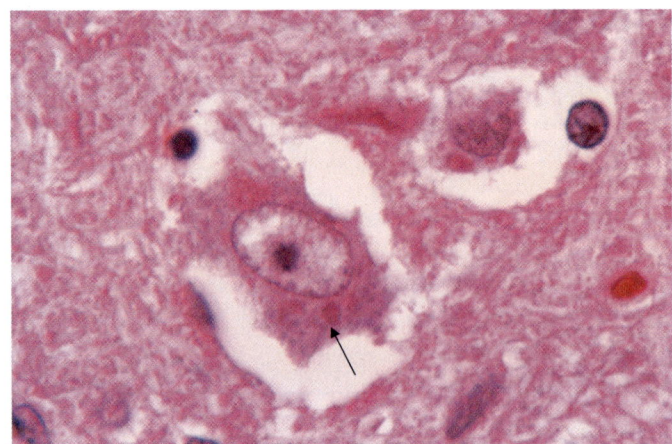

FIGURE 42–2 Rabies virus—Negri body. Arrow points to a "Negri body," an inclusion body in the cytoplasm of an infected neuron. (Reproduced with permission from Public Health Image Library, Centers for Disease Control and Prevention.)

at the bite site called paresthesias. After the prodrome, rabies encephalitis manifests as either of two forms: **furious** (encephalitic) or **dumb** (paralytic). The furious form occurs in about 80% of cases. In the furious form, agitation, delirium, seizures, and **hydrophobia** occur. Hydrophobia is an aversion to swallowing water because of painful spasm of the pharyngeal muscles. In contrast, in the dumb form, these symptoms do not occur. Rather, the spinal cord is primarily involved, and an ascending paralysis occurs. Death almost invariably occurs following both forms, but with the advent of life support systems, a few individuals have survived.

Laboratory Diagnosis

Rapid diagnosis of rabies infection in the *animal* is usually made by examination of brain tissue by using either PCR assay, fluorescent antibody to rabies virus, or histologic staining of Negri bodies in the cytoplasm of hippocampal neurons (see Figure 42–2). The virus can be isolated from the animal brain by growth in cell culture, but this takes too long to be useful in the decision of whether to give the vaccine.

TABLE 42–6   Clinical Features of Rabies Virus and Ebola Virus

Virus	Disease	Animal Reservoir	Main Clinical Findings	Vaccine Available	Treatment
Rabies virus	Rabies (encephalitis)	1. In the United States: Bats, skunks, raccoons 2. In developing countries: Dogs	Two forms: 1. Furious (delirium, seizures) 2. Dumb (paralysis)	Yes (Killed vaccine)	No antiviral drug
Ebola virus	Ebola hemorrhagic fever	Fruit bats are suspected	Bleeding into skin, gastrointestinal tract, and brain (headache); DIC; thrombocytopenia	Yes (Live, recombinant vaccine)	No antiviral drug

DIC = disseminated intravascular coagulation.

Rabies in *humans* can be diagnosed by PCR assay; by fluorescent antibody staining of a biopsy specimen, usually taken from the skin of the neck at the hairline; by isolation of the virus from sources such as saliva, spinal fluid, and brain tissue; or by a rise in titer of antibody to the virus. Negri bodies can be demonstrated in corneal scrapings and in autopsy specimens of the brain.

Treatment & Prevention

There is no antiviral therapy for a patient with rabies. Only supportive treatment is available. Prevention of rabies in the United States is provided by the **rabies vaccine** that contains inactivated virus grown in human diploid cells. (Vaccine grown in monkey lung cells or chick embryo cells is also available.)

In other countries, the duck embryo vaccine or various nerve tissue vaccines are available as well. Duck embryo vaccine has low immunogenicity, and the nerve tissue vaccines can cause an allergic encephalomyelitis as a result of a cross-reaction with human myelin. For these reasons, the human diploid cell vaccine (HDCV) is preferred.

There are two approaches to prevention of rabies in humans: **preexposure** and **postexposure immunization**. Preexposure immunization with rabies vaccine should be given to individuals in high-risk groups, such as veterinarians, zookeepers, and travelers to areas of hyperendemic infection (e.g., Peace Corps members). Preexposure immunization consists of three doses given on days 0, 7, and 21 or 28. Booster doses are given as needed to maintain an antibody titer of 1:5.

The rabies vaccine is also used routinely after exposure (i.e., after the person has been exposed to the virus via animal bite). The long incubation period of the disease allows the virus in the vaccine sufficient time to induce protective immunity.

Postexposure immunization involves the use of both the **vaccine and human rabies immune globulin** (RIG, obtained from hyperimmunized persons) plus immediate cleaning of the wound. This is an example of passive–active immunization. Tetanus immunization should also be considered.

The decision to give postexposure immunization depends on a variety of factors, such as (1) the type of animal (all wild animal attacks demand immunization); (2) whether an attack by a domestic animal was provoked, whether the animal was immunized adequately, and whether the animal is available to be observed; and (3) whether rabies is endemic in the area. The advice of local public health officials should be sought. Hospital personnel exposed to a patient with rabies need not be immunized unless a significant exposure has occurred (e.g., a traumatic wound to the healthcare worker).

If the decision is to immunize, both HDCV and RIG are recommended. Five doses of HDCV are given (on days 0, 3, 7, 14, and 28), but RIG is given only once with the first dose of HDCV (at a different site). HDCV and RIG are given at different sites to prevent neutralization of the virus in the vaccine by the antibody in the RIG. As much as possible of the RIG is given into the bite site, and the remainder is given intramuscularly. If the animal has been captured, it should be observed for 10 days and euthanized if symptoms develop. The brain of the animal should be examined by immunofluorescence.

EBOLA VIRUS

Disease

Ebola virus causes **Ebola hemorrhagic fever** (EHF). The virus is named for the river in Zaire that was the site of the first known outbreak of EHF in 1976. A devastating epidemic of EHF occurred in several West African countries, especially Liberia, Sierra Leone, and Guinea, in 2014 and 2015.

Important Properties

Ebola virus is a member of the filovirus family. Filoviruses are long filamentous (filo = thread) enveloped viruses. They are the longest viruses, often measuring thousands of nanometers (Figure 42–3). Ebola virus has a single-stranded, nonsegmented, negative-polarity RNA genome (see Table 42–5). There is an RNA-dependent RNA polymerase in the virion. The nucleocapsid has helical symmetry. Ebola virus has five strains (see below), and the serotypes cross-react.

Ebola virus is one of the most virulent human viruses and is cultured only under the highest biosafety containment (BSL-4). It can be inactivated by lipid solvents and bleach (hypochlorite).

There are five strains: Ebola-Zaire is the most pathogenic for humans and Ebola-Reston is pathogenic for monkeys but not for humans. Ebola-Sudan is also highly pathogenic. The degree of pathogenicity of Ebola-Ivory Coast (Tai Forest) and Ebola-Bundibugyo for humans is uncertain because the number of cases is relatively small. The Zaire, Sudan, Ivory Coast, and Bundibugyo types are found in Africa, whereas the Reston type originated in the Philippines.

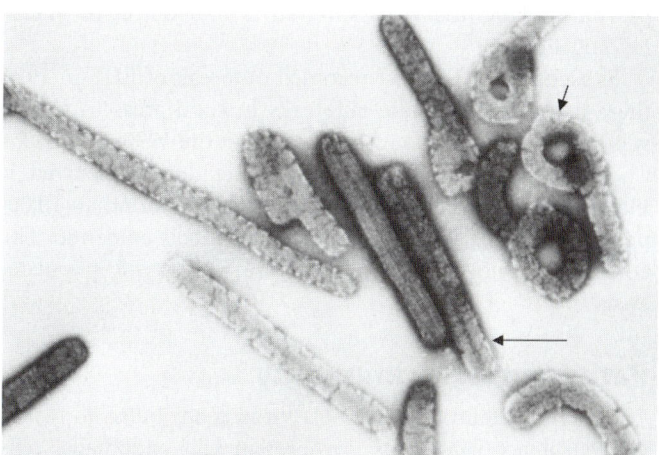

FIGURE 42–3 Ebola virus. Electron micrograph. Long arrow points to a typical virion of Ebola virus. Short arrow points to the "shepherd's crook" appearance of some Ebola virions. (Used with permission from Dr. Erskine Palmer and Dr. Russell Regnery, Public Health Image Library, Centers for Disease Control and Prevention.)

Summary of Replicative Cycle

The general outline of its replication is similar to that of other negative-stranded RNA enveloped viruses. After the virion envelope glycoproteins bind to the surface of the human cell, the nucleocapsid enters the cytoplasm where the virion RNA polymerase transcribes the seven genes into individual messenger RNAs (mRNAs). The mRNAs are translated into structural and nonstructural proteins. The negative-strand progeny genome is synthesized by the virus-encoded RNA polymerase using a plus-strand template. The newly synthesized nucleocapsid proteins surround both the progeny genome and the virion RNA polymerase. The matrix protein then mediates the interaction of the nucleocapsid protein with the outer cell membrane at the site of the progeny envelope proteins. The progeny virus then buds from the surface of the infected cell.

Transmission & Epidemiology

The natural reservoir of Ebola virus is thought to be fruit bats. The mode of transmission from the reservoir host to humans is unknown.

Transmission from human to human occurs via blood and body fluids. Hospital personnel without adequate protection are especially at risk. Many cases arise by secondary transmission from contact with the patient's blood or secretions (e.g., in hospital staff). Reuse of needles and syringes has been implicated in the spread in some hospitals in resource-poor countries. There is no evidence for human disease occurring via airborne transmission or by casual personal contact.

There is evidence of Ebola virus persisting in the semen of survivors of the disease. Transmission in one outbreak is attributed to the semen of a survivor 500 days after his infection. In 2021, it was found that the strain of Ebola virus involved in the outbreak in Guinea, West Africa, has virtually the same genome sequence as the strain that caused an outbreak 5 years previously. How the virus survived is the subject of intense investigation.

Subsequent to the first recorded outbreak of EHF in 1976, there have been sporadic outbreaks in rural areas in various sub-Saharan African countries, mostly in the 1990s and 2000s. Then in 2014 and 2015, the largest epidemic occurred in Liberia, Sierra Leone, and Guinea in which more than 10,000 people died. The fatality rate was 60% in this epidemic. This epidemic included cases in both rural and highly populated areas.

Pathogenesis & Immunity

The high mortality rate of Ebola virus is attributed to several viral virulence factors: Its glycoprotein kills endothelial cells, resulting in hemorrhage, and two other proteins inhibit the induction and action of interferon. Lymphocytes, macrophages, and dendritic cells are killed. As a result, the antibody response is often ineffective in preventing disease. Hepatocytes are also killed, leading to liver failure.

Clinical Findings

The incubation period is typically 5 to 7 days but may be up to 21 days. EHF begins with a constellation of symptoms, some of which are fever, headache, sore throat, myalgia, arthralgia, epigastric pain, vomiting, and diarrhea (see Table 42–6). Later, bleeding into the skin and gastrointestinal tract occurs, followed by shock and disseminated intravascular coagulation leading to multiorgan failure. The hemorrhages are the result of both severe thrombocytopenia and death of endothelial cells. Marked lymphopenia occurs. The mortality rate associated with this virus can be up to 90%.

In some patients who recover from EHF, a post-Ebola syndrome (PES) occurs. The findings in PES include eye pain, blurred vision, cataracts, hearing loss, headache, joint pain, fatigue, and insomnia. In one patient with uveitis, infectious Ebola virus was recovered from fluid aspirated from the interior of his eye several months after recovery.

Laboratory Diagnosis

Diagnosis is most often made by detecting viral RNA using a PCR assay. Tests detecting viral antigens in serum using an ELISA, or detecting IgM antibody in the serum are also used. (Extreme care must be taken when handling specimens in the laboratory.) The virus can be cultured in monkey cells in BSL-4 containment facility. Electron microscopy may reveal the long rod shape of a filovirus, implicating either Ebola virus or Marburg virus.

Treatment

No approved antiviral therapy is available. Supportive therapy including intravenous fluids and electrolytes is useful. Treatment with immune serum globulins containing antibody to Ebola virus has had variable results. Experimental monoclonal antibodies (ZMapp and MAb114) and experimental drugs (remdesivir and favipiravir) are being used in outbreaks but their effectiveness is uncertain.

Prevention

Prevention centers on limiting secondary spread by proper handling of patient's secretions and blood and by the wearing of personal protective equipment (PPE). Quarantine of individuals thought to be exposed for 21 days is also important.

Also available for prevention is a live recombinant vaccine (rZEBOV) containing a backbone of vesicular stomatitis virus into which the gene for the Ebola virus surface glycoprotein has been inserted. It is used both pre- and postexposure in outbreaks of EHF.

Two other vaccines are also being used in some countries. One is a recombinant adenovirus containing the gene for the Ebola surface glycoprotein and the other is a recombinant vaccinia virus containing the genes for proteins of three strains of Ebola virus and one strain of Marburg virus.

SELF-ASSESSMENT QUESTIONS

1. An outbreak of dengue hemorrhagic fever (DHF) recently occurred in two Central American countries. Regarding dengue and DHF, which one of the following is the most accurate?

   (A) Humans are dead-end hosts for dengue virus.

   (B) DHF occurs primarily in individuals who are deficient in the late-acting complement components.

   (C) Dengue virus is transmitted by *Aedes* mosquitoes, and monkeys are an important natural reservoir.

   (D) The vaccine containing live, attenuated dengue virus is recommended for those living or traveling in endemic areas.

   (E) DHF occurs more often in people infected for the first time than when they are reinfected because antibody protects against reinfection.

2. Yellow fever still exists in many tropical areas of the globe. Which one of the following is the best reason yellow fever still exists?

   (A) Sewage disposal is inadequate in many areas.

   (B) Both humans and monkeys are reservoirs for yellow fever virus.

   (C) The virus has mutated, so the existing vaccine is no longer effective.

   (D) The vaccine has been withdrawn because it was found to have unacceptable side effects.

   (E) The people in developing countries cannot afford to take amantadine when they enter endemic areas.

3. Regarding West Nile virus (WNV), which one of the following is the most accurate?

   (A) Rodents are the main reservoir for WNV.

   (B) WNV does not cause disease in the United States.

   (C) WNV is transmitted primarily by *Ixodes* ticks.

   (D) Most infections are asymptomatic, but the elderly are at risk for encephalitis.

   (E) The live, attenuated vaccine should be administered to elderly adults in endemic areas.

4. An outbreak of febrile disease involving severe joint pain and an erythematous macular rash has occurred on several Caribbean islands. Infection with chikungunya virus is suspected. Which one of the following is correct regarding this virus?

   (A) Its genome is composed of double-stranded DNA.

   (B) It is transmitted by *Aedes* mosquitoes.

   (C) Wild birds are the most important reservoir.

   (D) Laboratory diagnosis involves electron microscopy to observe the very long filamentous shape of the virus.

   (E) The killed vaccine should be administered to travelers to endemic regions.

5. Regarding rabies virus and the disease rabies, which one of the following statements is most accurate?

   (A) Finding intranuclear inclusion bodies within macrophages is presumptive evidence of rabies virus infection.

   (B) Lamivudine is used to treat rabies because it inhibits the RNA-dependent DNA polymerase in the virion.

   (C) In the United States, skunks and bats are more likely to transmit rabies virus to people than are dogs and cats.

   (D) The incubation period of the disease is usually 2 to 4 days, leading to the rapid progression of the encephalitis and death.

   (E) After the animal bite, rabies virus enters the bloodstream, replicates in internal organs such as the liver, and then reaches the central nervous system during the secondary viremia.

6. A woman was hiking in an isolated area when a skunk appeared and bit her on the leg. She now presents to your emergency room about an hour after the bite. Which one of the following is the most appropriate action to take?

   (A) Give rabies vaccine and hyperimmune globulin immediately.

   (B) Reassure her that rabies is not a problem because skunks do not carry rabies.

   (C) Quarantine the animal for 10 days and only treat her if signs of rabies appear in the animal.

   (D) Test the patient's serum for antibodies now and in 10 days to see if there is a rise in antibody titer before treating her.

   (E) Administer ribavirin intravenously.

7. Regarding Ebola virus, which one of the following is most accurate?

   (A) Skunks and raccoons are the main natural reservoirs for Ebola virus.

   (B) In endemic areas, most people are latently infected with Ebola virus.

   (C) People known to be exposed to Ebola virus should be given ganciclovir to prevent disease.

   (D) Ebola hemorrhagic fever occurs primarily in people with deficient cell-mediated immunity.

   (E) The appearance of Ebola virus in the electron microscope is that of a long thread, which often has a curved end.

ANSWERS

(1) **(C)**
(2) **(B)**
(3) **(D)**
(4) **(B)**
(5) **(C)**
(6) **(A)**
(7) **(E)**

SUMMARIES OF ORGANISMS

Brief summaries of the organisms described in this chapter begin on page 691. Please consult these summaries for a rapid review of the essential material.

PRACTICE QUESTIONS: USMLE & COURSE EXAMINATIONS

Questions on the topics discussed in this chapter can be found in the Clinical Virology section of Part XIII: USMLE (National Board) Practice Questions starting on page 747. Also see Part XIV: USMLE (National Board) Practice Examination starting on page 775.

Tumor Viruses

INTRODUCTION

Viruses can cause benign or malignant tumors in many species of animals (e.g., frogs, fishes, birds, and mammals). Despite the common occurrence of tumor viruses in animals, only a few viruses are associated with **human** tumors, and evidence that they are truly the causative agents exists for very few.

Tumor viruses have no characteristic size, shape, or chemical composition. Some are large, and some are small; some are enveloped while others are nonenveloped; some have DNA as their genetic material, and others have RNA. The factor that unites all of them is their common ability to cause cancer.

Tumor viruses are at the forefront of cancer research for two main reasons:

(1) They are more rapid, reliable, and efficient tumor producers than either chemicals or radiation. For example, many of these viruses can cause tumors in all susceptible animals in 1 or 2 weeks and can produce malignant transformation in cultured cells in just a few days.

(2) They have a small number of genes compared with a human cell (only three, four, or five for many retroviruses), and hence their role in the production of cancer can be readily analyzed and understood. To date, the genomes of many tumor viruses have been sequenced and the number of genes and their functions have been determined; all of this has provided important information.

MALIGNANT TRANSFORMATION OF CELLS

The term *malignant transformation* refers to changes in the growth properties, shape, and other features of the tumor cell (Table 43–1). Malignant transformation can be induced by tumor viruses not only in animals but also in cultured cells. In culture, the following changes occur when cells become malignantly transformed.

Altered Morphology

Malignant cells lose their characteristic differentiated shape and appear rounded and more refractile when seen in a microscope. The rounding is due to the disaggregation of actin filaments, and the reduced adherence of the cell to the surface of the culture dish is the result of changes in the surface charge of the cell.

TABLE 43–1 Features of Malignant Transformation

Feature	Description
Altered morphology	Loss of differentiated shape Rounded as a result of disaggregation of actin filaments and decreased adhesion to surface More refractile
Altered growth control	Loss of contact inhibition of growth Loss of contact inhibition of movement Reduced requirement for serum growth factors Increased ability to be cloned from a single cell Increased ability to grow in suspension Increased ability to continue growing ("immortalization")
Altered cellular properties	Induction of DNA synthesis Chromosomal changes Appearance of new antigens Increased agglutination by lectins
Altered biochemical properties	Reduced level of cyclic AMP Enhanced secretion of plasminogen activator Increased anaerobic glycolysis Loss of fibronectin Changes in glycoproteins and glycolipids

Altered Growth Control

Malignant cells grow in a disorganized, piled-up pattern in contrast to normal cells, which have an organized, flat appearance. The term applied to this change in growth pattern in malignant cells is **loss of contact inhibition**. Contact inhibition is a property of normal cells that refers to their ability to stop their growth and movement upon contact with another cell. Malignant cells lose this ability and consequently move on top of one another, continue to grow to large numbers, and form a random array of cells.

Malignant cells are easily cloned (i.e., they can grow into a colony of cells starting with a single cell), whereas normal cells cannot do this effectively. Infection of a cell by a tumor virus "immortalizes" that cell by enabling it to continue growing long past the time when its normal counterpart would have died. Normal cells in culture have a lifetime of about 50 generations, but malignantly transformed cells grow indefinitely.

Altered Cellular Properties

(1) DNA synthesis is induced. If cells resting in the G_1 phase are infected with a tumor virus, they will promptly enter the S phase (i.e., synthesize DNA and go on to divide).

(2) The karyotype becomes altered (i.e., there are changes in the number and shape of the chromosomes as a result of deletions, duplications, and translocations).

(3) New antigens, different from those in normal cells, appear. These new antigens can be either virus-encoded proteins, preexisting cellular proteins that have been modified, or previously repressed cellular proteins that are now being synthesized.

Altered Biochemical Properties

(1) Levels of cyclic adenosine monophosphate (AMP) are reduced in malignant cells. Addition of cyclic AMP will cause malignant cells to revert to the appearance and growth properties of normal cells.

(2) Increased anaerobic glycolysis leads to increased lactic acid production (Warburg effect). The mechanism for this change is unknown.

ROLE OF TUMOR VIRUSES IN MALIGNANT TRANSFORMATION

Malignant transformation is a permanent change in the behavior of the cell. One important question regarding the role of the virus in malignant transformation is must the viral genetic material be present and functioning at all times, or can it alter some cell component and not be required subsequently? The answer to this question was obtained by using a temperature-sensitive mutant of Rous sarcoma virus. This mutant has an altered transforming gene that is functional at the low, permissive temperature (35°C) but not at the high, restrictive temperature (39°C). When chicken cells were infected at 35°C they transformed as expected, but when incubated at 39°C, they regained their normal morphology and behavior within a few hours. Days or weeks later, when these cells were returned to 35°C, they recovered their transformed phenotype. Thus, continued production of some functional virus-encoded protein is required for the maintenance of the transformed state.

PROVIRUSES & ONCOGENES

The two major concepts of the way viral tumorigenesis occurs are expressed in the terms **provirus** and **oncogene**. These contrasting ideas address the fundamental question of the source of the genes for malignancy.

(1) In the provirus model, the genes enter the cell at the time of infection carried by the tumor virus.

(2) In the oncogene model, the genes for malignancy are already present in all cells of the body by virtue of being present in the initial sperm and egg. These oncogenes encode proteins that encourage cell growth (e.g., fibroblast growth factor). In the oncogene model, carcinogens such as chemicals, radiation, and tumor viruses activate cellular oncogenes to overproduce these growth factors. This initiates inappropriate cell growth and malignant transformation (Figure 43–1).

Both proviruses and oncogenes can play a role in malignant transformation. Evidence for the provirus mode consists of finding copies of viral DNA integrated into cell DNA only in

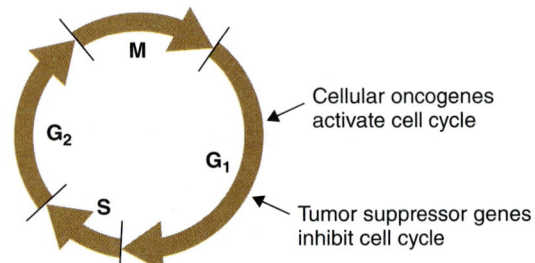

FIGURE 43–1 Effect of cellular oncogenes and tumor suppressor genes on the cell cycle. The oncoproteins encoded by cellular oncogenes activate the cell cycle by allowing passage from the G_1 phase into the S phase. The proteins encoded by tumor suppressor genes, notably p53 and RB, inhibit the cell cycle in the G_1 phase. Inactivation of these proteins activates the cell cycle by allowing passage from the G_1 phase into the S phase. G_1 = gap 1; G_2 = gap 2; M = mitosis; S = synthesis of DNA. (Reproduced with permission from Murray RK, Bender D, Botham KM, et al: *Harper's Illustrated Biochemistry*, 29th ed. New York, NY: McGraw Hill; 2012.)

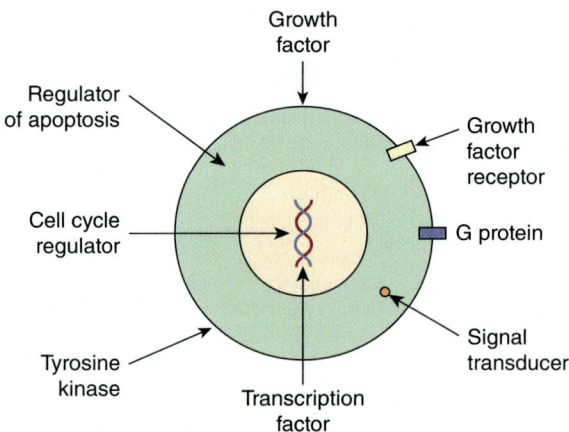

FIGURE 43–2 Functions of cellular oncoproteins. Cellular oncogenes encode proteins with a variety of functions that are shown in the figure. These oncoproteins activate the cell cycle and cause cell to grow in an unregulated manner. (Reproduced with permission from Rodwell WV, Bender D, Botham KM, et al: *Harper's Illustrated Biochemistry*, 31st ed. New York, NY: McGraw Hill; 2018.)

cells that have been infected with the tumor virus. The corresponding uninfected cells have no copies of the viral DNA.

1. Role of Cellular Oncogenes in Tumorigenesis

The first direct evidence that oncogenes exist in normal cells was based on results of experiments in which a DNA copy of the *src* gene of the chicken retrovirus Rous sarcoma virus was used as a probe. DNA in normal chicken embryonic cells hybridized to the probe, indicating that the cells contain a gene homologous to the viral gene. It is hypothesized that the **cellular oncogenes** may be the precursors of viral oncogenes.

Proto-oncogenes are the normal precursors of the cellular oncogenes. Proto-oncogenes encode normal cellular proteins and are under regulatory control. Cellular oncogenes have acquired mutations that cause them to escape regulatory control and overproduce altered proteins. Figure 43–2 shows the functions of important oncoproteins encoded by the cellular oncogenes.

Although cellular oncogenes and viral oncogenes are similar, they are not identical. They differ in base sequence at various points; and cellular oncogenes have exons and introns, whereas viral oncogenes do not. It seems likely that viral oncogenes were acquired by incorporation of cellular oncogenes into retroviruses lacking these genes. Retroviruses can be thought of as **transducing agents**, carrying oncogenes from one cell to another. (See Chapter 4 for a discussion of transduction.)

Since this initial observation, **more than 20 cellular oncogenes** have been identified by using either the Rous sarcoma virus DNA probe or probes made from other viral oncogenes. Table 43–2 describes the function of several important cellular oncogenes and their relationship to various human cancers. Many cells contain several different cellular oncogenes. In addition, the same cellular oncogenes have been found in species as diverse as fruit flies, rodents, and humans. Such conservation

through evolution suggests a normal physiologic function for these genes. Some are known to be expressed during normal embryonic development.

A marked **diversity** of viral oncogene function has been found. Some oncogenes, such as the *src* gene, encode a **protein kinase** that specifically phosphorylates the amino acid tyrosine, in contrast to the commonly found protein kinase of cells, which preferentially phosphorylates serine. There is evidence that the kinase phosphorylates signal transduction factors that activate synthesis of cyclins. This drives the cell into S phase and subsequent mitosis.

Other oncogenes have a base sequence almost identical to that of the gene for certain cellular **growth factors** (e.g., epidermal growth factor). Several proteins encoded by oncogenes have

TABLE 43–2 Examples of Cellular Oncogenes Involved in Human Cancer

Cellular Oncogene	Function of Oncogene	Important Human Cancer
abl	Signaling tyrosine kinase	Chronic myelogenous leukemia
erb B-2 (*her/neu*)	Receptor tyrosine kinase	Carcinoma of breast and ovary; neuroblastoma
ras	G protein	Carcinoma of colon, lung, and thyroid; melanoma
myc	Transcription factor	Burkitt's lymphoma; carcinoma of breast and ovary
jun/fos	Transcription regulator	Carcinoma of breast and lung
src	Signaling tyrosine kinase	Carcinoma of colon
pi3k	Signaling lipid kinase	Carcinoma of colon

their effect at the cell membrane (e.g., the *ras* oncogene encodes a G protein), whereas some act in the nucleus by binding to DNA (e.g., the *myc* oncogene encodes a transcription factor). These observations indicate that growth control is a multistep process and that carcinogenesis can be induced by affecting one or more of several steps.

On the basis of the known categories of oncogenes, the following model of growth control can be constructed. After a **growth factor** binds to its **receptor** on the cell membrane, membrane-associated **G proteins** and **tyrosine kinases** are activated. These, in turn, interact with **cytoplasmic proteins** or produce **second messengers**, which are transported to the nucleus and interact with nuclear factors. DNA synthesis is activated, and cell division occurs. Overproduction or inappropriate expression of any of the preceding factors **in boldface type** can result in malignant transformation.

Note that not all tumor viruses of the retrovirus family contain *onc* genes. How do these viruses cause malignant transformation? It appears that the DNA copy of the viral RNA integrates near a cellular oncogene, causing a marked increase in its expression. This process is called **insertional mutagenesis**. **Overexpression** of the cellular oncogene may play a key role in malignant transformation by these viruses.

Although it has been demonstrated that viral oncogenes can cause malignant transformation, it has not been directly shown that cellular oncogenes can do so. However, as described in Table 43–3, the following evidence suggests that they do:

(1) DNA-containing cellular oncogenes isolated from certain tumor cells can transform normal cells in culture. When the base sequence of these "transforming" cellular oncogenes

was analyzed, it was found to have a **single base change** from the normal cellular oncogene (i.e., it had **mutated**). In several tumor cell isolates, the altered sites in the gene are the same.

(2) In certain tumors, characteristic **translocations** of chromosomal segments can be seen. In Burkitt's lymphoma cells, a translocation occurs that moves a cellular oncogene (c-*myc*) from its normal site on chromosome 8 to a new site adjacent to an immunoglobulin heavy chain gene on chromosome 14. This shift enhances expression of the c-*myc* gene.

(3) In chronic myelogenous leukemia (CML) cells, a truncated chromosome called a **Philadelphia chromosome** is seen. This chromosome has a characteristic translocation that results in the overexpression of the *bcr-abl* oncogene that encodes a tyrosine kinase. Increased kinase activity increases the rate of cell division and inhibits DNA repair resulting in leukemia. Drugs that inhibit this kinase, such as imatinib (Gleevec), induce a prolonged remission and are well tolerated.

(4) Some tumors have multiple copies of the cellular oncogenes, either on the same chromosome or on multiple tiny chromosomes. The **amplification** of these genes results in overexpression of their mRNA and proteins.

(5) **Insertion** of the DNA copy of the retroviral RNA (proviral DNA) near a cellular oncogene stimulates expression of the c-*onc* gene.

(6) Certain cellular oncogenes isolated from normal cells can cause malignant transformation if they have been modified to be **overexpressed** within the recipient cell.

In summary, two different mechanisms—**mutation** and **increased expression**—appear to be able to activate the quiescent "proto-oncogene" into a functioning oncogene capable of transforming a cell. Cellular oncogenes provide a rationale for carcinogenesis by chemicals and radiation (e.g., a chemical carcinogen might act by enhancing the expression of a cellular oncogene). Furthermore, DNA isolated from cells treated with a chemical carcinogen can malignantly transform other normal cells. The resulting tumor cells contain cellular oncogenes from the chemically treated cells, and these genes are expressed with high efficiency.

2. Role of Cellular Tumor Suppressor Genes in Tumorigenesis

There is another mechanism of carcinogenesis involving cellular genes, namely, mutation of a **tumor suppressor** gene (see Figure 43–1). A well-documented example is the retinoblastoma susceptibility gene, which normally acts as a suppressor of retinoblastoma formation. When both alleles of this **antioncogene** are mutated (made nonfunctional), retinoblastoma occurs. Human papillomavirus (HPV) and SV40 virus produce a protein that inactivates the protein encoded by the retinoblastoma gene.

HPV also produces a protein that inactivates the protein encoded by the *p53* gene, another tumor suppressor gene in human cells. The *p53* gene encodes a transcription factor that activates the synthesis of a second protein, which blocks the

TABLE 43–3 Evidence that Cellular Oncogenes (c-*onc*) Can Cause Tumors

Evidence	Description
Mutation of c-*onc* gene	DNA isolated from tumor cells can transform normal cells. This DNA has a c-*onc* gene with a mutation consisting of a single base change.
Translocation of c-*onc* gene	Movement of c-*onc* gene to a new site on a different chromosome results in malignancy accompanied by increased expression of the gene.
Amplification of c-*onc* gene	The number of copies of c-*onc* gene is increased, resulting in enhanced expression of their mRNA and proteins.
Insertion of retrovirus near c-*onc* gene	Proviral DNA inserts near c-*onc* gene, which alters its expression and causes tumors.
Overexpression of c-*onc* gene by modification in the laboratory	Addition of an active promoter site enhances expression of the c-*onc* gene, and malignant transformation occurs.

TABLE 43–4 Examples of Tumor Suppressor Genes Involved in Human Cancer

Tumor Suppressor Gene	Important Human Cancer
Rb	Retinoblastoma; carcinoma of breast, bladder, and lung
p53	Carcinoma of breast, colon, and lung; astrocytoma
WT1	Wilms' tumor of kidney
DCC	Carcinoma of colon
BRCA	Breast cancer
PTEN	Prostate cancer

cyclin-dependent kinases required for cell division to occur. The p53 protein also promotes apoptosis of cells that have sustained DNA damage or contain activated cellular oncogenes. Apoptosis-induced death of these cells has a "tumor-suppressive" effect by killing those cells destined to become cancerous.

Inactivation of tumor suppressor genes appears likely to be an important general mechanism of viral oncogenesis. Tumor suppressor genes are involved in the formation of other cancers as well (e.g., breast and colon carcinomas and various sarcomas). For example, in many colon carcinomas, two genes are inactivated, the *p53* gene and the *DCC* (*d*eleted in *c*olon *c*arcinoma) gene. Another tumor suppressor gene, the *BRCA* gene, encodes a protein involved in DNA repair. Mutation of the *BRCA* gene predisposes to breast cancer because there is a failure of DNA repair.

Table 43–4 lists several important tumor suppressor genes and their relationship to various human cancers. **More than half of human cancers have a mutated *p53* gene in the DNA of malignant cells**.

3. Role of Cellular Micro-RNA Genes in Tumorigenesis

Micro-RNA genes do not encode proteins but rather exert their regulatory effect by being transcribed into micro-RNA that can bind to sequences in mRNA and prevent that mRNA from being translated into proteins. For example, there are micro-RNAs that bind to ("silence") mRNA transcribed from a tumor suppressor gene. As a result, the tumor suppressor protein is not synthesized, which enhances the likelihood of tumorigenesis.

OUTCOME OF TUMOR VIRUS INFECTION

The outcome of tumor virus infection is dependent on the virus and the type of cell. Some tumor viruses go through their entire replicative cycle with the production of progeny virus, whereas others undergo an interrupted cycle, analogous to lysogeny, in which the **proviral DNA is integrated** into cellular DNA and limited expression of proviral genes occurs. Therefore, malignant transformation does not require that progeny virus be

produced. Rather, all that is required is the expression of one or, at most, a few viral genes. Note, however, that some tumor viruses transform by inserting their proviral DNA in a manner that activates a cellular oncogene.

In most cases, the DNA tumor viruses such as the papovaviruses transform only cells in which they do not replicate. These cells are called "nonpermissive" because they do not permit viral replication. Cells of the species from which the DNA tumor virus was initially isolated are "permissive" (i.e., the virus replicates and usually kills the cells, and no tumors are formed). For example, SV40 virus replicates in the cells of the African green monkey (its species of origin) and causes a cytopathic effect but no tumors. However, in rodent cells, the virus does not replicate, expresses only its early genes, and causes malignant transformation. In the "nonproductive" transformed cell, the viral DNA is integrated into the host chromosome and remains there through subsequent cell divisions. The underlying concept applicable to both DNA and RNA tumor viruses is that **only viral gene expression**, not replication of the viral genome or production of progeny virus, is required for transformation.

The essential step required for a DNA tumor virus (e.g., SV40 virus) to cause malignant transformation is expression of the **early genes** of the virus (Table 43–5). (The early genes are those expressed prior to the replication of the viral genetic material.) These required early genes produce a set of early proteins called **T antigens**.[1]

The large T antigen, which is both necessary and sufficient to induce transformation, binds to SV40 virus DNA at the site of initiation of viral DNA synthesis. This is compatible with the finding that the large T antigen is required for the initiation of cellular DNA synthesis in the virus-infected cell. Biochemically, large T antigen has protein kinase and adenosine triphosphate (ATPase) activity. Almost all of the large T antigen is located in the cell nucleus, but some of it is in the outer cell membrane. In that location, it can be detected as a transplantation antigen called **tumor-specific transplantation antigen** (TSTA). TSTA is the antigen that induces the immune response against the transplantation of virally transformed cells. Relatively little is known about the SV40 virus small T antigen, except that if it is not synthesized, the efficiency of transformation decreases. In polyomavirus-infected cells, the middle T antigen plays the same role as the SV40 virus large T antigen.

In RNA tumor virus-infected cells, this required gene has one of several different functions, depending on the retrovirus. The oncogene of Rous sarcoma virus and several other viruses codes for a protein kinase that phosphorylates tyrosine. Some viruses have a gene for a factor that regulates cell growth (e.g., epidermal growth factor or platelet-derived growth factor), and

---

[1]In SV40 virus-infected cells, two T antigens, large (molecular weight [MW] 100,000) and small (MW 17,000), are produced, whereas in polyomavirus-infected cells, three T antigens, large (MW 90,000), middle (MW 60,000), and small (MW 22,000), are made. Other tumor viruses such as adenoviruses also induce T antigens, which are immunologically distinct from those of the two papovaviruses.

TABLE 43–5 Viral Oncogenes

Characteristic	DNA Virus	RNA Virus
Prototype virus	SV40 virus	Rous sarcoma virus
Name of gene	Early-region A gene	*src* gene
Name of protein	T antigen	Src protein
Function of protein	Protein kinase, ATPase activity, binding to DNA, and stimulation of DNA synthesis	Protein kinase that phosphorylates tyrosine[1]
Location of protein	Primarily nuclear, but some in plasma membrane	Plasma membrane
Required for viral replication	Yes	No
Required for cell transformation	Yes	Yes
Gene has cellular homologue	No	Yes

[1]Some retroviruses have *onc* genes that code for other proteins such as platelet-derived growth factor and epidermal growth factor.

still others have a gene that codes for a protein that binds to DNA. The conclusion is that normal growth control is a multistep process that can be affected at any one of several levels. The addition of a viral oncogene perturbs the growth control process, and a tumor cell results.

The viral genetic material remains stably integrated in host cell DNA by a process similar to lysogeny. In the lysogenic cycle, bacteriophage DNA becomes stably integrated into the bacterial genome. The linear DNA genome of the temperate phage, lambda, forms a double-stranded circle within the infected cell and then covalently integrates into bacterial DNA (Table 43–6). A repressor is synthesized that prevents transcription of most of the other lambda genes.

Similarly, the double-stranded circular DNA of the DNA tumor virus covalently integrates into eukaryotic-cell DNA, and only early genes are transcribed. With RNA tumor viruses (retroviruses), the single-stranded linear RNA genome is transcribed into a double-stranded linear DNA that integrates into cellular DNA. In summary, despite the differences in their genomes and in the nature of the host cells, these viruses go through the common pathway of a double-stranded DNA intermediate followed by covalent integration into cellular DNA and subsequent expression of certain genes.

Just as a lysogenic bacteriophage can be induced to enter the replicative cycle by ultraviolet radiation and certain chemicals, tumor viruses can be induced by several mechanisms. Induction is one of the approaches used to determine whether tumor viruses are present in human cancer cells (e.g., human T-cell lymphotropic virus was discovered by inducing the virus from leukemic cells with iododeoxyuridine).

Three techniques have been used to induce tumor viruses to replicate in the transformed cells:

(1) The most frequently used method is the addition of nucleoside analogues (e.g., iododeoxyuridine). The mechanism of induction by these analogues is uncertain.

(2) The second method involves fusion with "helper" cells (i.e., the transformed, nonpermissive cell is fused with a permissive cell) in which the virus undergoes a normal replicative cycle. Within the heterokaryon (a cell with two or more nuclei that is formed by the fusion of two different cell types), the tumor virus is induced and infectious virus is produced. The mechanism of induction is unknown.

(3) In the third method, helper viruses provide a missing function to complement the integrated tumor virus. Infection with the helper virus results in the production of both the integrated tumor virus and the helper virus.

The process of rescuing tumor viruses from cells revealed the existence of **endogenous viruses**. Treatment of *normal, uninfected* embryonic cells with nucleoside analogues resulted in the production of retroviruses. Retroviral DNA is integrated within the chromosomal DNA of all cells and serves as the template for viral replication. This proviral DNA probably arose by retrovirus infection of the germ cells of some prehistoric ancestor.

Endogenous retroviruses, which have been rescued from the cells of many species (including humans), differ depending on the species of origin. Endogenous viruses are xenotropic (*xeno* means foreign; *tropism* means to be attracted to; i.e., they infect cells of other species more efficiently than they infect the cells of the species of origin). Entry of the endogenous virus into the cell of origin is limited as a result of defective viral envelope–cell receptor interaction. Although they are retroviruses, most endogenous viruses are not tumor viruses (i.e., only a few cause leukemia).

TRANSMISSION OF TUMOR VIRUSES

Tumor virus transmission in experimental animals can occur by two processes, vertical and horizontal. **Vertical transmission** indicates movement of the virus from mother to newborn offspring, whereas **horizontal transmission** describes the passage of virus between animals that do not have a mother–offspring relationship. Vertical transmission occurs by three methods: (1) the viral genetic material is in the sperm or the egg; (2) the virus is passed across the placenta; and (3) the virus is transmitted in the breast milk.

When vertical transmission occurs, exposure to the virus early in life can result in tolerance to viral antigens and, as a consequence, the immune system will not eliminate the virus. Large amounts of virus are produced, and a high frequency of cancer occurs. In contrast, when horizontal transmission occurs, the immunocompetent animal produces antibody against the virus, and the frequency of cancer is low. If an immunocompetent animal is experimentally made immunodeficient, the frequency of cancer increases greatly.

Horizontal transmission probably does not occur in humans; those in close contact with cancer patients (e.g., family members and medical personnel) do not have an increased frequency

TABLE 43–6 Lysogeny as a Model for the Integration of Tumor Viruses

Type of Virus	Name	Genome	Integration	Limited Transcription of Viral Genes
Temperate phage	Lambda phage	Linear dsDNA	+	+
DNA tumor virus	SV40 virus	Circular dsDNA	+	+
RNA tumor virus	Rous sarcoma virus	Linear ssRNA	+	+[1]

ds = double-stranded; ss = single-stranded.

[1]Limited transcription in some cells or under certain conditions but full transcription with viral replication in others.

of cancer. There have been "outbreaks" of leukemia in several children at the same school, but these have been interpreted statistically to be random, rare events that happen to coincide.

HUMAN TUMOR VIRUSES

There are seven known human tumor viruses (Table 43–7). Two are RNA viruses, namely, human T-cell lymphotropic virus and hepatitis C virus (HCV). The other five are DNA viruses, namely, HPV, Epstein–Barr virus (EBV), human herpesvirus 8 (HHV-8) (Kaposi's sarcoma [KS] virus), hepatitis B virus (HBV), and Merkel cell polyomavirus (MCPV).

1. RNA Tumor Viruses

Human T-Cell Leukemia Virus

There are two important human retroviruses: human T-cell leukemia virus (HTLV), which is described here, and human immunodeficiency virus (HIV), which is described in Chapter 45.

Disease

HTLV-1 causes two distinctly different diseases: a cancer called adult T-cell leukemia/lymphoma (ATL) and a neurologic disease called HTLV-associated myelopathy (HAM) (also known as tropical spastic paraparesis or chronic progressive myelopathy). HTLV-2 also appears to cause these diseases, but the association is less clearly documented. (All information in this section refers to HTLV-1 unless otherwise stated.)

Important Properties

HTLV and HIV are the two medically important members of the retrovirus family. Both are enveloped viruses with reverse transcriptase in the virion and two copies of a single-stranded, positive-polarity RNA genome. However, HTLV does not kill T cells, whereas HIV does. In fact, HTLV does just the opposite; it causes malignant transformation that "immortalizes" the infected T cells and allows them to proliferate in an uncontrolled manner.

The genes in the HTLV genome whose functions have been clearly identified are the three structural genes common to all retroviruses, namely, *gag*, *pol*, and *env*, plus two regulatory genes, *tax* and *rex*. In general, HTLV genes and proteins are similar to those of HIV in size and function, but the genes differ in base sequence, and therefore, the proteins differ in amino acid sequence (and antigenicity). For example, p24 is the major nucleocapsid protein in both HTLV and HIV, but they differ antigenically. The virions of both HTLV and HIV contain a reverse transcriptase, integrase, and protease. The envelope proteins of HTLV are gp46 and gp21, whereas those of HIV are gp120 and gp41.

TABLE 43–7 Varieties of Tumor Viruses

Genome Nucleic Acid	Virus Family	Human Tumor Viruses	Animal Tumor Viruses
1. RNA	Retrovirus Flavivirus	Human T-cell leukemia virus Hepatitis C virus	Sarcoma, leukemia, and carcinoma viruses in many avian and mammalian species
2. DNA	Papillomavirus	Human papillomavirus	Papillomaviruses of many mammals
	Herpesvirus	Epstein–Barr virus; human herpesvirus 8 (Kaposi's sarcoma-associated virus)	Herpesvirus saimiri causes lymphomas in monkeys; Marek's disease virus of chickens
	Hepadnavirus	Hepatitis B virus	Hepatitis viruses of ducks and squirrels
	Polyomavirus	Merkel cell polyomavirus	Polyomavirus and SV40 virus cause various cancers in rodents
	Adenovirus		Human adenovirus serotypes 12, 18, and 31 cause sarcomas in rodents
	Poxvirus		Myxoma-fibroma virus; Yaba monkey tumor virus

The proteins encoded by the *tax* and *rex* genes play the same functional roles as those encoded by the HIV regulatory genes, *tat* and *rev*. The Tax protein is a transcriptional activator, and the Rex protein governs the processing of viral mRNA and its export from the nucleus to the cytoplasm.

The *tax* gene is an oncogene and the Tax protein is required for malignant transformation of T cells. The Tax protein activates the synthesis of interleukin-2 (IL-2; which is T-cell growth factor) and of the IL-2 receptor. IL-2 promotes rapid T-cell growth and eventually malignant transformation of the T cell.

The stability of the genes of HTLV is much greater than that of HIV. As a consequence, HTLV does not show the high degree of variability of the antigenicity of the envelope proteins that occurs in HIV.

Summary of Replicative Cycle

The replication of HTLV is thought to follow a typical retroviral cycle, but specific information has been difficult to obtain because the virus grows poorly in cell culture. HTLV primarily infects CD4-positive T lymphocytes. The cellular receptor for the virus is unknown. Within the cytoplasm, reverse transcriptase synthesizes a DNA copy of the genome, which migrates to the nucleus and integrates into cell DNA. Viral mRNA is made by host cell RNA polymerase, and transcription is upregulated by Tax protein, as mentioned earlier. The Rex protein controls the synthesis of the *gag/pol* mRNA, the *env* mRNA, and their subsequent transport to the cytoplasm, where they are translated into structural viral proteins. Full-length RNA destined to become progeny genome RNA is also synthesized and transported to the cytoplasm. The virion nucleocapsid is assembled in the cytoplasm, and budding occurs at the outer cell membrane. Cleavage of precursor polypeptides into functional structural proteins is mediated by the virus-encoded protease.

Transmission & Epidemiology

HTLV is transmitted primarily by intravenous drug use, sexual contact, or breast-feeding. Transplacental transmission has been rarely documented. Transmission by blood transfusion has greatly decreased in the United States with the advent of screening donated blood for antibodies to HTLV and discarding those that are positive. Transmission by processed blood products, such as immune serum globulins, has not occurred. Transmission is thought to occur primarily by the transfer of infected cells rather than free, extracellular virus. For example, whole blood, but not plasma, is a major source, and infected lymphocytes in semen are the main source of sexually transmitted virus.

HTLV infection is endemic in certain geographic areas, namely, the Caribbean region including southern Florida, eastern South America, western Africa, and southern Japan. The rate of seropositive adults is as high as 20% in some of these areas, but infection can occur anywhere because infected individuals migrate from these areas of endemic infection. At least half the people in the United States who are infected with HTLV are infected with HTLV-2, usually acquired via intravenous drug use.

Pathogenesis & Immunity

HTLV causes two distinct diseases, each with a different type of pathogenesis. One disease is ATL in which HTLV infection of CD4-positive T lymphocytes induces malignant transformation. As described earlier, HTLV-encoded Tax protein enhances synthesis of IL-2 (T-cell growth factor) and IL-2 receptor, which initiates the uncontrolled growth characteristic of a cancer cell. All the malignant T cells contain the same integrated proviral DNA, indicating that the malignancy is monoclonal (i.e., it arose from a single HTLV-infected cell). HTLV remains latent within the malignant T cells (i.e., HTLV is typically not produced by the malignant cells).

The other disease is HAM, also known as tropical spastic paraparesis or chronic progressive myelopathy. HAM is a demyelinating disease of the brain and spinal cord, especially of the motor neurons in the spinal cord. HAM is caused either by an autoimmune cross-reaction in which the immune response against HTLV damages the neurons or by cytotoxic T cells that kill HTLV-infected neurons.

Clinical Findings

ATL is characterized by lymphadenopathy, hepatosplenomegaly, lytic bone lesions, and skin lesions. These features are caused by proliferating T cells infiltrating these organs. In the blood, the malignant T cells have a distinct "flower-shaped" nucleus. Hypercalcemia due to increased osteoclast activity within the bone lesions is seen. Patients with ATL often have reduced cell-mediated immunity, and opportunistic infections with fungi and viruses are common.

The clinical features of HAM include gait disturbance, weakness of the lower limbs, and low back pain. Loss of bowel and bladder control may occur. Loss of motor function is much greater than sensory loss. T cells with a "flower-shaped" nucleus can be found in the spinal fluid. Magnetic resonance imaging of the brain shows nonspecific findings. Progression of symptoms occurs slowly over a period of years. HAM occurs primarily in women of middle age. The disease resembles multiple sclerosis except that HAM does not exhibit the remissions characteristic of multiple sclerosis.

Both ATL and HAM are relatively rare diseases. The vast majority of people infected with HTLV develop asymptomatic infections, usually detected by the presence of antibody. Only a small subset of those infected develops either ATL or HAM.

Laboratory Diagnosis

Infection with HTLV is determined by detecting antibodies against the virus in the patient's serum using the enzyme-linked immunosorbent assay (ELISA) test. The Western blot assay is used to confirm a positive ELISA result. Polymerase chain reaction (PCR) assay can detect the presence of HTLV RNA or DNA within infected cells. The laboratory tests used to screen donated blood contain only HTLV-1 antigens, but because there is cross-reactivity between HTLV-1 and HTLV-2, the presence of antibodies against both viruses is usually detected. However, some HTLV-2 antibodies are missed in these routine screening tests. Isolation of HTLV in cell culture from the patient's specimens is not done.

ATL is diagnosed by finding malignant T cells in the lesions. The diagnosis of HAM is supported by the presence of HTLV antibody in the spinal fluid or finding HTLV nucleic acids in cells in the spinal fluid.

Treatment & Prevention

There is no specific antiviral treatment for HTLV infection, and no antiviral drug will cure latent infections by HTLV. ATL is treated with anticancer chemotherapy regimens. Antiviral drugs have not been effective in the treatment of HAM. Corticosteroids and danazol have produced improvement in some patients.

There is no vaccine against HTLV. Preventive measures include screening donated blood for the presence of antibodies, using condoms to prevent sexual transmission, and encouraging women with HTLV antibodies to refrain from breast-feeding.

Hepatitis C Virus

Chronic infection with HCV, like HBV, also predisposes to hepatocellular carcinoma. HCV is an RNA virus that has no oncogene and forms no DNA intermediate during replication. It does cause chronic hepatitis, which seems likely to be the main predisposing event. (Additional information regarding HCV can be found in Chapter 41.)

2. DNA Tumor Viruses

Human Papillomavirus

HPV is one of the two viruses definitely known to cause tumors in humans. Papillomas (warts) are benign but can progress to form carcinomas, especially in an immunocompromised person. HPV primarily infects keratinizing or mucosal squamous epithelium. (Additional information regarding HPV can be found in Chapter 38.)

Papillomaviruses are DNA nucleocapsid viruses with double-stranded, circular, supercoiled DNA and an icosahedral nucleocapsid. Carcinogenesis by HPV involves two proteins encoded by HPV genes *E6* and *E7* that interfere with the activity of the proteins encoded by two tumor suppressor genes, *p53* and *Rb* (retinoblastoma), found in normal cells. In cancer cells, the viral DNA is integrated into the cellular DNA, and the E6 and E7 proteins are produced.

There are at least 100 different types of HPV, many of which cause distinct clinical entities. For example, HPV-1 through HPV-4 cause plantar warts on the soles of the feet, whereas HPV-6 and HPV-11 cause anogenital warts (condylomata acuminata) and laryngeal papillomas. Certain types of HPV, especially types 16 and 18, are implicated as the cause of carcinoma of the cervix, penis, and anus.

Epstein–Barr Virus

EBV is a herpesvirus that was isolated from the cells of an East African individual with **Burkitt's lymphoma**. EBV, the cause of infectious mononucleosis, transforms B lymphocytes in culture

and causes lymphomas in marmoset monkeys. It is also associated with **nasopharyngeal carcinoma**, a tumor that occurs primarily in China, and with thymic carcinoma and B-cell lymphoma in the United States. However, cells from Burkitt's lymphoma patients in the United States show no evidence of EBV infection. (Additional information regarding EBV can be found in Chapter 37.)

Cells isolated from East African individuals with Burkitt's lymphoma contain EBV DNA and EBV nuclear antigen. Only a small fraction of the many copies of EBV DNA is integrated; most viral DNA is in the form of closed circles in the cytoplasm.

The difficulty in proving that EBV is a human tumor virus is that infection by the virus is widespread but the tumor is rare. The current hypothesis is that EBV infection induces B cells to proliferate, thus increasing the likelihood that a second event (e.g., activation of a cellular oncogene) will occur. In Burkitt's lymphoma cells, a cellular oncogene, c-*myc*, which is normally located on chromosome 8, is **translocated** to chromosome 14 at the site of immunoglobulin heavy chain genes. This translocation brings the c-*myc* gene in juxtaposition to an active promoter, and large amounts of c-*myc* RNA are synthesized. It is known that the c-*myc* oncogene encodes a transcription factor, but the role of this factor in oncogenesis is uncertain.

Human Herpesvirus 8

HHV-8, also known as Kaposi's sarcoma-associated herpesvirus (KSHV), causes Kaposi's sarcoma (KS). KS is a malignancy of vascular endothelial cells that contains many spindle-shaped cells and erythrocytes. It is the most common cancer in patients with acquired immunodeficiency syndrome (AIDS). KSHV is transmitted both sexually and by saliva. A protein encoded by KSHV called latency-associated nuclear antigen (LANA) inactivates RB and p53 tumor suppressor proteins, which causes malignant transformation of the endothelial cells. (Additional information regarding HHV-8 can be found in Chapter 37.)

Hepatitis B Virus

HBV infection is significantly more common in patients with primary hepatocellular carcinoma (**hepatoma**) than in control subjects. This relationship is striking in areas of Africa and Asia, where the incidence of both HBV infection and hepatoma is high. Chronic HBV infection commonly causes cirrhosis of the liver; these two events are the main predisposing factors to hepatoma. Part of the HBV genome is integrated into cellular DNA in malignant cells. However, no HBV gene has been definitely implicated in oncogenesis. The integration of HBV DNA may cause insertional mutagenesis, resulting in the activation of a cellular oncogene. In addition, the HBx protein may play a role because it inhibits the p53 tumor suppressor protein. (Additional information regarding HBV can be found in Chapter 41.)

Merkel Cell Polyomavirus

MCPV causes a carcinoma of Merkel cells in the skin. (Merkel cells are neuroreceptors for pressure and touch.) The carcinoma occurs most often on skin exposed to the sun such as the face and neck. Immunosuppressed individuals and the elderly are predisposed to this cancer.

Members of the polyomavirus family are small, nonenveloped, double-stranded DNA viruses known to cause cancer in animals (see later section on animal tumor viruses). Infection with MCPV is common, as indicated by the presence of antibody to the virus in many healthy blood donors. The mode of transmission is uncertain.

In carcinoma cells, the DNA of MCPV is integrated into cell DNA. The gene for the large T antigen is mutated so the virus cannot replicate but the T antigen continues to be synthesized. The T antigen causes the cell to become malignant by inhibiting tumor suppressor proteins such as p53 and RB. Because MCPV does not replicate in the carcinoma cells, patients are not infectious to others.

Diagnosis is made by pathologic analysis of surgical specimens. There is no virus-based laboratory test clinically available. There is no antiviral drug or vaccine available. Prevention involves reducing sun exposure, use of sunscreen, and frequent skin examinations to detect the cancer before it metastasizes.

VACCINES AGAINST CANCER

There are two vaccines designed to prevent human cancer: the HBV vaccine and the HPV vaccine. The widespread use of the HBV vaccine in Asia has significantly reduced the incidence of hepatocellular carcinoma. The vaccine against HPV, the cause of carcinoma of the cervix, was approved for use in the United States in 2006.

DO ANIMAL TUMOR VIRUSES CAUSE CANCER IN HUMANS?

There is no evidence that animal tumor viruses cause tumors in humans. In fact, the only available information suggests that they do not, because (1) people who were inoculated with poliovirus vaccine contaminated with SV40 virus have no greater incidence of cancers than do uninoculated controls, (2) soldiers inoculated with yellow fever vaccine contaminated with avian leukemia virus do not have a high incidence of tumors, and (3) members of families whose cats have died of leukemia caused by feline leukemia virus show no increase in the occurrence of leukemia over control families. Note, however, that some human tumor cells, namely, non-Hodgkin's lymphoma, contain SV40 DNA, but the relationship of that DNA to malignant transformation is uncertain.

ANIMAL TUMOR VIRUSES

1. RNA Tumor Viruses

RNA tumor viruses have been isolated from a large number of species, namely, snakes, birds, and mammals, including nonhuman primates. The important RNA tumor viruses are listed in

Table 43–7. They are important because of their ubiquity, their ability to cause tumors in the host of origin, their small number of genes, and the relationship of their genes to cellular oncogenes (see page 366).

These viruses belong to the retrovirus family (the prefix *retro* means reverse), so named because a **reverse transcriptase** is located in the virion. This enzyme transcribes the genome RNA into double-stranded proviral DNA and is essential to their replication. The viral genome consists of two identical molecules of positive-strand RNA. Each molecule has a molecular weight of approximately 2×10^6 (these are the only viruses that are diploid [i.e., have two copies of their genome in the virion]). The two molecules are hydrogen-bonded together by complementary bases located near the 5′ end of both RNA molecules. Also bound near the 5s′ end of each RNA is a transfer RNA (tRNA) that serves as the primer[2] for the transcription of the RNA into DNA.

The icosahedral capsid is surrounded by an envelope with glycoprotein spikes. Some internal capsid proteins are group-specific antigens, which are common to retroviruses within a species. There are three important morphologic types of retroviruses, labeled B, C, and D, depending primarily on the location of the capsid or core. Most of the retroviruses are C-type particles, but mouse mammary tumor virus is a B-type particle, and HIV, the cause of AIDS, is a D-type particle.

The gene sequence of the RNA of a typical avian sarcoma virus is *gag*, *pol*, *env*, and *src*. The nontransforming retroviruses have three genes; they are missing *src*. The *gag* region codes for the group-specific antigens, the *pol* gene codes for the reverse transcriptase, the *env* gene codes for the two envelope spike proteins, and the *src* gene codes for the protein kinase. There is evidence that the kinase phosphorylates signal transduction factors that activate synthesis of cyclins that drive the cell into S phase and subsequent mitosis.

The sequences at the 5′ and 3′ ends function in the integration of the proviral DNA and in the transcription of mRNA from the integrated proviral DNA by host cell RNA polymerase II. At each end is a sequence[3] called an LTR that is composed of several regions, one of which, near the 5′ end, is the binding site for the primer tRNA.

After infection of the cell by a retrovirus, the following events occur: Using the genome RNA as the template, the reverse transcriptase (RNA-dependent DNA polymerase) synthesizes double-stranded proviral DNA. The DNA then integrates into cellular DNA. Integration of the proviral DNA is an obligatory step, but there is no specific site of integration. Insertion of the viral LTR can enhance the transcription of adjacent host cell genes. If this host gene is a cellular oncogene, malignant transformation may result. This explains how retroviruses without viral oncogenes can cause transformation.

---

[2]The purpose of the primer tRNA is to act as the point of attachment for the first deoxynucleotide at the start of DNA synthesis. The primers are normal-cell tRNAs that are characteristic for each retrovirus.

[3]The length of the sequence varies from 250 to 1200 bases, depending on the virus.

2. DNA Tumor Viruses

Papovaviruses

The two best-characterized oncogenic papovaviruses are **polyomavirus** and **SV40 virus**. Polyomavirus (*poly* means many; *oma* means tumor) causes a wide variety of histologically different tumors when inoculated into newborn rodents. Its natural host is the mouse. SV40 virus, which was isolated from normal rhesus monkey kidney cells, causes sarcomas in newborn hamsters.

Polyomavirus and SV40 virus share many chemical and biologic features (e.g., double-stranded, circular, supercoiled DNA of molecular weight 3×10^6 and a 45-nm icosahedral nucleocapsid). However, the sequence of their DNA and the antigenicity of their proteins are quite distinct. Both undergo a lytic (permissive) cycle in the cells of their natural hosts, with the production of progeny virus. However, when they infect the cells of a heterologous species, the nonpermissive cycle ensues, no virus is produced, and the cell is malignantly transformed.

In the transformed cell, the viral DNA integrates into the cell DNA, and only early proteins are synthesized. Some of these proteins (e.g., the T antigens described on page 368) are required for induction and maintenance of the transformed state.

JC virus, a human papovavirus, is the cause of progressive multifocal leukoencephalopathy (see Chapter 44). It also causes brain tumors in monkeys and hamsters. There is no evidence that it causes human cancer.

Adenoviruses

Some human adenoviruses, especially serotypes 12, 18, and 31, induce sarcomas in newborn hamsters and transform rodent cells in culture. There is no evidence that these viruses cause tumors in humans, and no adenoviral DNA has been detected in the DNA of any human tumor cells.

Adenoviruses undergo both a permissive cycle in some cells and a nonpermissive, transforming cycle in others. The linear genome DNA circularizes within the infected cell, but—in contrast to the papovaviruses, whose entire genome integrates—only a small region (10%) of the adenovirus genome does so; yet transformation still occurs. This region codes for several proteins, one of which is the T (tumor) antigen. Adenovirus T antigen is required for transformation and is antigenically distinct from the polyomavirus and SV40 virus T antigens.

Herpesviruses

Several animal herpesviruses are known to cause tumors. Four species of herpesviruses cause **lymphomas** in nonhuman primates. Herpesviruses saimiri and ateles induce T-cell lymphomas in New World monkeys, and herpesviruses pan and papio transform B lymphocytes in chimpanzees and baboons, respectively.

A herpesvirus of chickens causes Marek's disease, a contagious, rapidly fatal neurolymphomatosis. Immunization of chickens with a live, attenuated vaccine has resulted in a considerable decrease in the number of cases. A herpesvirus is implicated as the cause of kidney carcinomas in frogs.

Poxviruses

Two poxviruses cause tumors in animals; these are the fibroma–myxoma virus, which causes fibromas or myxomas in rabbits and other animals, and Yaba monkey tumor virus, which causes benign histiocytomas in animals and human volunteers. Little is known about either of these viruses.

SELF-ASSESSMENT QUESTIONS

1. Regarding viruses that play a role in human carcinogenesis, which one of the following statements is the most accurate?
   (A) Epstein–Barr virus is implicated as the cause of nasopharyngeal carcinoma primarily in Asia, where it is transmitted by mosquitoes in rural areas.
   (B) Evidence for hepatitis C virus (HCV) as a cause of hepatocellular carcinoma includes finding a DNA copy of the HCV genome integrated into the DNA of hepatocytes.
   (C) Hepatitis B virus is implicated as the cause of hepatocellular carcinoma because countries with a high incidence of chronic hepatitis B also have a high incidence of hepatocellular carcinoma.
   (D) Human T-cell leukemia virus is a retrovirus that was found to be associated with leukemia in Japan but is not found in the United States.

2. Regarding the oncogenes of DNA tumor viruses, which one of the following is most accurate?
   (A) They encode protein kinases that phosphorylate p53 protein.
   (B) They interact with cellular proto-oncogenes and activate them.
   (C) They encode cellular growth factors that activate S-phase DNA synthesis.
   (D) They encode proteins that bind to the proteins encoded by tumor suppressor genes.

3. Regarding the main mechanism by which oncogenic retroviruses cause malignant transformation, which one of the following is most accurate?
   (A) They cause point mutations in cellular regulatory genes.
   (B) They carry the genes for proteins that act as cellular growth factors.
   (C) They synthesize a protein that inhibits the action of the cellular p53 protein.
   (D) They encode a recombinase that causes translocation of certain chromosomes.
   (E) They encode a DNA polymerase that increases the rate of cellular DNA synthesis.

4. Human T-cell leukemia virus (HTLV) causes T-cell leukemia in adults. Regarding this virus, which one of the following statements is most accurate?
   (A) HTLV is transmitted primarily by the fecal–oral route.
   (B) Oseltamivir cures the latent state established by HTLV within T cells.
   (C) The genome of HTLV consists of double-stranded RNA; therefore, there is no polymerase in the virion.
   (D) Infection by HTLV is diagnosed in the clinical laboratory by observing cytoplasmic inclusion bodies.
   (E) Oncogenesis by HTLV is related to a viral transcription factor that activates the production of interleukin-2 and its receptor.

ANSWERS

(1) **(C)**
(2) **(D)**
(3) **(B)**
(4) **(E)**

SUMMARIES OF ORGANISMS

Brief summaries of the organisms described in this chapter begin on page 691. Please consult these summaries for a rapid review of the essential material.

PRACTICE QUESTIONS: USMLE & COURSE EXAMINATIONS

Questions on the topics discussed in this chapter can be found in the Clinical Virology section of Part XIII: USMLE (National Board) Practice Questions starting on page 747. Also see Part XIV: USMLE (National Board) Practice Examination starting on page 775.

Slow Viruses & Prions

INTRODUCTION

"Slow" infectious diseases are caused by a heterogeneous group of agents containing both conventional viruses and unconventional agents that are not viruses (e.g., prions). **Prions** are **protein-containing particles** with **no detectable nucleic acid** that are highly resistant to inactivation by heat, formaldehyde, and ultraviolet light at doses that will inactivate viruses. Note that prions are resistant to the temperatures usually employed in cooking, a fact that may be important in their suspected ability to be transmitted by food (see variant Creutzfeldt-Jakob disease [vCJD] later). Prions are, however, inactivated by protein- and lipid-disrupting agents such as phenol, ether, NaOH, and hypochlorite (see Chapter 28).

The prion protein is encoded by a normal cellular gene and is thought to function in a signal transduction pathway in neurons.

There is some evidence that the function of the normal cellular prion protein is to regulate the N-methyl-D-aspartate receptor on the postsynaptic terminal by binding copper ions.

The normal prion protein (known as PrP^C, or prion protein cellular) has a significant amount of alpha-helical conformation. When the alpha-helical conformation changes to a beta-pleated sheet (known as PrP^{SC}, or prion protein scrapie), these abnormal forms aggregate into filaments, which disrupt neuron function and cause cell death. Prions, therefore, "reproduce" by the abnormal beta-pleated sheet form recruiting normal alpha-helical forms to change their conformation. Note that the normal alpha-helical form and the abnormal beta-pleated sheet form have the same amino acid sequence. It is only their conformation that differs. A specific cellular RNA enhances this conformational change. Prions are described in more detail in Chapter 28.

Pathogenic prion proteins can be thought of conceptually as **misfolded proteins**. These misfolded proteins not only cause CJD in humans and "mad cow" disease in cattle but are suspected of being involved in the pathogenesis of other important diseases of the central nervous system, such as Alzheimer's disease and Parkinson's disease.

In humans, the "slow" agents cause **central nervous system** diseases characterized by a long incubation period, a gradual onset, and a progressive, invariably fatal course. There is no antimicrobial therapy for these diseases. Note that the term *slow* refers to the disease, not to the rate of replication of those viruses that cause these slow diseases. The replication rate of these viruses is similar to that of most other viruses.

The human prion-mediated diseases (e.g., kuru and CJD) are called **transmissible spongiform encephalopathies (TSE)**. The term *spongiform* refers to the spongy, Swiss cheese-like holes seen in the brain parenchyma that are caused by the death of the neurons (Figure 44–1). No virus particles are seen in the brain of people with these diseases.

The term *encephalopathy* refers to a pathologic process in the brain without signs of inflammation. In contrast, *encephalitis* refers to an inflammatory brain process in which either neutrophils or lymphocytes are present. In TSEs, there are no inflammatory changes in the brain.

The transmissibility of the agent of kuru and CJD ("prions") was initially established by inoculation of material from the

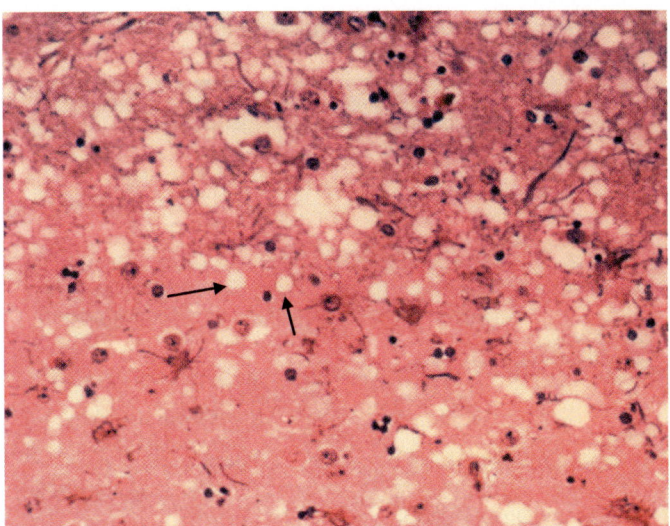

FIGURE 44–1 Prion-mediated spongiform encephalopathy (mad cow disease). Two arrows point to the spongiform appearance (Swiss cheese-like holes) in the brain of a cow with mad cow disease. The brain of a patient with Creutzfeldt-Jakob disease has a similar appearance. (Used with permission from Dr. Al Jenny, Public Health Image Library, Centers for Disease Control and Prevention.)

brains of infected patients into the brains of primates followed by serial transfer to the brains of other primates.

Note, however, that kuru, variant CJD, and bovine spongiform encephalopathy (BSE, mad cow disease) are acquired by ingestion. In this route, the prion protein must survive digestion in the intestinal tract and then penetrates the gut mucosa. The prion protein is then amplified within follicle dendritic cells in lymphatic tissue, such as Peyer's patches. Prions then spread to the spleen, carried by migrating dendritic cells. From the spleen, prions spread to the central nervous system probably via the sympathetic nerves.

It is also possible that prions reach the brain within lymphocytes, as there is a documented case of CJD that was acquired by transfused blood. In addition, CJD has been transmitted **iatrogenically** (i.e., in a medical context, via corneal transplants, dura mater grafts, implanted brain electrodes, and growth hormone extracts made from human pituitary glands).

There is evidence that quinacrine and other acridine analogues inhibit the formation of the pathologic PrP^{SC} form in cell culture. These drugs are currently being tested in animal models for their ability to treat or prevent prion diseases.

Prion-caused diseases can be classified into three categories: some are clearly **transmissible (infectious)**, such as kuru; some

are clearly **hereditary (genetic)**, such as fatal familial insomnia; and others are **sporadic** (neither infectious nor hereditary), such as most cases of CJD. The sporadic cases seem likely to be due to spontaneous somatic mutations in the affected individual.

SLOW DISEASES CAUSED BY CONVENTIONAL VIRUSES

Progressive Multifocal Leukoencephalopathy

Progressive multifocal leukoencephalopathy (PML) is a fatal demyelinating disease of the white matter (*leuko* meaning white) and involves multiple areas of the brain (i.e., it is multifocal). Note that it is not an encephalitis because there is no inflammation in the brain.

The clinical picture includes visual field defects, mental status changes, and weakness. The disease rapidly progresses to blindness, dementia, and coma, and most patients die within 6 months. It occurs primarily in individuals with compromised cell-mediated immunity, especially patients with acquired immunodeficiency syndrome (AIDS) and those who are receiving cancer chemotherapy and immunosuppressive drugs following organ transplantation. Some patients undergoing treatment for multiple sclerosis with the monoclonal antibody natalizumab develop PML, and others receiving mycophenolate to prevent transplant rejection have also developed PML. Table 44–1 describes some important features of slow viral diseases in humans caused by conventional viruses.

PML is caused by JC virus, a member of the polyomavirus family. Polyomaviruses are nonenveloped viruses with a circular, double-stranded DNA genome. JC virus infects and kills oligodendroglia, causing demyelination.

Antibodies to JC virus are found in approximately 75% of normal human sera, indicating that infection is widespread. Disease occurs when latent JC virus is activated in an immunocompromised patient. The virus persists in the kidney and is excreted in the urine. The diagnosis is typically made by polymerase chain reaction (PCR) assay of a brain biopsy specimen or spinal fluid. There is no effective antiviral treatment and no vaccine.

Subacute Sclerosing Panencephalitis

Subacute sclerosing panencephalitis (SSPE) is a slowly progressive disease characterized by inflammatory lesions in many areas of the brain. It is a rare disease of **children** who were infected by **measles virus** several years earlier. Unlike PML, immunosuppression is *not*

TABLE 44–1 Important Features of Slow Viral Diseases Caused by Conventional Viruses

Disease	Virus	Virus Family	Important Characteristics
Progressive multifocal leukoencephalopathy	JC virus	Papovavirus	Infection widespread; disease only in immunocompromised
Subacute sclerosing panencephalitis	Measles virus	Paramyxovirus	Disease in young children with defective virus in brain

a predisposing factor. SSPE begins with mild changes in personality and ends with dementia and death.

SSPE is a persistent infection by a variant of measles virus that cannot complete its replication. SSPE has virtually disappeared in the United States since the onset of widespread immunization with measles vaccine.

SLOW DISEASES CAUSED BY PRIONS

There are five human TSEs caused by prions: kuru, CJD, variant CJD, Gerstmann-Sträussler-Scheinker (GSS) syndrome, and fatal familial insomnia. Table 44–2 describes some important features of slow viral diseases in humans caused by prions.

Kuru

This fatal disease is characterized by progressive tremors and ataxia but not dementia. It occurs *only* among the **Fore tribes in New Guinea**. It was transmitted during a ritual in which the skulls of the dead were opened and the brains eaten. There are two ways the disease could have been acquired: either by eating the brains or via cuts in the skin that occurred during the preparation of the brains at which time brain tissue was introduced into the body. Since the ritual practice was stopped, kuru has almost disappeared. The agents of kuru and CJD (see next) have been transmitted serially in primates.

Creutzfeldt-Jakob Disease

Pathologic examination of the brains of patients with CJD and kuru reveals a spongiform (sponge or Swiss cheese) appearance similar to that associated with scrapie in sheep (see later). The spongiform changes are the result of neuronal vacuolation and neuronal loss rather than demyelination. No inflammatory cells are seen in the brains. Prions have been found in the brains of CJD patients.

In contrast to kuru, CJD is **found sporadically worldwide** and affects both sexes. The incidence of CJD is approximately 1 case per 1 million population, and there is no increased risk associated with dietary habits, occupation, or animal exposure. Vegetarians and meat eaters have the same rate. The rate of CJD is the same in countries whose animals have scrapie (see later) and those whose animals do not. There is no evidence for person-to-person or transplacental transmission.

There is no increased risk for medical caregivers; therefore, gowns and masks are unnecessary. The standard precautions for obtaining infectious specimens should be observed. It has been transmitted **iatrogenically** (e.g., in a corneal transplant, via intracerebral electrodes, in hormones extracted from human pituitaries, and in grafts of cadaveric dura mater). There is only one confirmed case of CJD being transmitted by blood transfusion, and intravenous drug use does not increase the risk. Proper sterilization of CJD agent-contaminated material consists of either autoclaving or treating with sodium hypochlorite.

The main clinical findings of CJD are dementia (including behavioral changes, memory loss, and confusion) and myoclonic jerking. Additional findings include ataxia, aphasia,

TABLE 44–2 Important Features of Slow Viral Diseases Caused by Prions

Disease	Pathogenesis	Important Feature
Kuru	Transmissible/infectious	Caused by ingesting or handling brain tissue; occurred in New Guinea tribes people
Creutzfeldt-Jakob disease	1. Transmissible/infectious	Iatrogenic transmission by corneal transplant, brain electrodes, and growth hormone
	2. Hereditary/genetic	Mutation in germ cells
	3. Sporadic	No relationship to any known cause; possible new mutation in somatic cells; most common form
Variant Creutzfeldt-Jakob disease	Transmissible/infectious	Probably acquired by eating meat or nervous tissue from animals with mad cow disease
Gerstmann-Sträussler-Scheinker syndrome	Hereditary/genetic	Mutation in germ cells
Fatal familial insomnia	Hereditary/genetic	Mutation in germ cells

visual loss, and hemiparesis. The symptoms typically appear gradually and progress inexorably. In the terminal stage, the patient becomes mute and akinetic and then comatose. About 80% of those affected die within 1 year. Most cases occur in people who are 50 to 70 years of age.

A presumptive diagnosis of CJD can be made pathologically by detecting spongiform changes in a brain biopsy specimen. Neuronal loss and gliosis are seen. Amyloid plaques are also seen in some cases of CJD. In variant CJD, "florid" plaques composed of flowerlike amyloid plaques surrounded by a halo of vacuoles are seen. Brain imaging and the electroencephalogram may show characteristic changes. There is no evidence of inflammation (i.e., no neutrophils or lymphocytes are seen). The blood count and routine spinal fluid test results are normal. The finding of a normal brain protein called 14-3-3 in the spinal fluid supports the diagnosis.

The specific diagnosis of CJD can be made on the patient's cerebrospinal fluid (CSF) by using a "quaking" assay that detects PrP^{SC}. In this assay, recombinant PrP^{C} is added to the patient's CSF and the mixture is shaken vigorously (quaking). If any PrP^{SC} is present in the CSF, the PrP^{C} will be converted to PrP^{SC}. The newly formed PrP^{SC} is detected by binding of the fluorescent dye, thioflavin T.

The diagnosis of CJD can also be made by immunohistochemistry in which labeled antiprion antibodies are used to stain the patient's brain specimen. Because we do not make antibodies to prion proteins, there are no serologic diagnostic tests. No antibodies are made in humans because humans are tolerant to our prion proteins. (The antibodies used in the immunohistochemical lab tests are made in other animals in which the human prions are immunogenic.) Unlike viruses, prions cannot be grown in culture, so there are no culture-based diagnostic tests.

Tonsillar tissue obtained from patients with variant CJD was positive for prion protein using monoclonal antibody-based assays. The use of tonsillar or other similar lymphoid tissue may obviate the need for a brain biopsy. Pathologic prion proteins have also been detected in the olfactory epithelium of patients with CJD.

There is no treatment for CJD, and there is no drug or vaccine available for prevention.

Although most cases of CJD are sporadic, about 10% are hereditary. The hereditary (familial) form is inherited as an autosomal dominant trait. In these patients, 12 different point mutations and several insertion mutations in the prion protein gene have been found. One of these, a point mutation in codon 102, is the same mutation found in patients with **GSS syndrome**—another slow central nervous system disease of humans. The main clinical features of GSS syndrome are cerebellar ataxia and spastic paraparesis. The hereditary forms of these diseases may be prevented by the detection of carriers and genetic counseling.

The origin of these spongiform encephalopathies is three-fold: **infectious, hereditary**, and **sporadic**. The infectious forms are kuru and probably variant CJD (see next section). Transmission of the infectious agent was documented by serial passage of brain material from a person with CJD to chimpanzees. The hereditary form is best illustrated by GSS syndrome (see preceding paragraph) and by a disease called fatal familial insomnia. The term *sporadic* refers to the appearance of the disease in the absence of either an infectious or a hereditary cause.

Fatal familial insomnia is a very rare disease characterized by progressive insomnia, dysautonomia (dysfunction of the autonomic nervous system) resulting in various symptoms, dementia, and death. A specific mutation in the prion protein is found in patients with this disease.

Variant Creutzfeldt-Jakob Disease

In 1996, several cases of CJD occurred in Great Britain due to ingestion of beef. These cases are a new variant of CJD (vCJD, also called nvCJD) because they occurred in much younger people than usual and had certain clinical and pathologic findings different from those found in the typical form of the disease. None of those affected had consumed cattle or sheep brains, but brain material may have been admixed into processed meats such as sausages.

Only people whose native prion protein is homozygous for methionine at amino acid 129 contract vCJD. People whose native prion protein is homozygous for valine at amino acid 129 or who are heterozygotic do not contract vCJD. These findings indicate that prion proteins with methionine are more easily folded into the pathologic beta-pleated sheet form.

The prions isolated from the "variant CJD" cases in humans chemically resemble the prions isolated from mad cow disease more than they resemble other prions, which is evidence to support the hypothesis that variant CJD originated by eating beef. There is no evidence that eating lamb is associated with variant CJD.

As of November 2020, vCJD has been diagnosed in 232 people, 178 of whom have lived in the United Kingdom. Three cases of vCJD have occurred in the United States; two of them are thought to have acquired it in the United Kingdom. All cases of vCJD have occurred in individuals who lived or traveled in a country where BSE has been detected.

It is unknown how many people harbor the pathogenic prion in a latent (asymptomatic) form. The possibility that there may be people who are asymptomatic carriers of the vCJD prion and who could be a source for infection of others (e.g., via blood transfusions) has led blood banks in the United States to eliminate from the donor pool people who have lived in Great Britain for more than 6 months.

SLOW DISEASES OF ANIMALS

The slow transmissible diseases of animals are important models for human diseases. Scrapie and visna are diseases of sheep, and BSE (mad cow disease) is a disease of cattle that appears to have arisen from the ingestion of sheep tissue by the cattle. Chronic wasting disease (CWD) occurs in deer and elk. Visna is caused by a virus, whereas the other three are prion-mediated diseases.

Scrapie

Scrapie is a disease of sheep, characterized by tremors, ataxia, and itching, in which the sheep scrape off their wool against fence posts. It has an incubation period of many months. Spongiform degeneration without inflammation is seen in the brain tissue of affected animals. It has been transmitted to mice and other animals via a brain extract that contained no recognizable virus particles. Studies of mice revealed that the infectivity is associated with a 27,000-molecular-weight protein known as a prion (see Chapter 28).

Bovine Spongiform Encephalopathy

BSE is also known as mad cow disease. The cattle become aggressive, ataxic, and eventually die. Cattle acquire BSE by eating feed supplemented with organs (e.g., brains) obtained from sheep infected with scrapie prions. (It is also possible that BSE arose in cattle by a mutation in the gene encoding the prion protein.)

BSE is endemic in Great Britain. Supplementation of feed with sheep organs was banned in Great Britain in 1988, and thousands of cattle were destroyed, two measures that have led to a marked decline in the number of new cases of BSE. BSE has been found in cattle in other European countries such as France, Germany, Italy, and Spain, and there is significant concern in those countries that variant CJD may emerge in humans. Two cases of BSE in cattle in the United States have been reported.

Chronic Wasting Disease

CWD of deer, moose, and elk is a prion-mediated disease that exists in the United States. Because vCJD is strongly suspected to be transmitted by ingesting meat, there is concern regarding the consequences of eating deer and elk meat (venison). In 2002, it was reported that neurodegenerative diseases occurred in three men who ate venison in the 1990s. One of these diseases was confirmed as CJD. Whether there is a causal relationship is unclear, and surveillance continues. This concern was heightened in 2006 when prions were detected in the muscle of deer with CWD but not in the muscle of normal deer.

As of 2020, CWD in deer, moose, and elk has occurred in 24 states but there is no evidence for transmission to humans. CWD is fatal to the animals and there is no treatment or vaccine.

Visna

Visna is a disease of sheep that is characterized by pneumonia and demyelinating lesions in the brain. It is caused by visna virus, a member of the lentivirus subgroup of retroviruses. As such, it has a single-stranded, diploid RNA genome and an RNA-dependent DNA polymerase in the virion. It is thought that integration of the DNA provirus into the host cell DNA may be important in the persistence of the virus within the host and, consequently, in its long incubation period and prolonged, progressive course.

SELF-ASSESSMENT QUESTIONS

1. Regarding "slow viruses" and their diseases, which one of the following is the most accurate?

    (A) The viruses that cause slow diseases, such as progressive multifocal leukoencephalopathy (PML), have a slow rate of replication that accounts for the long latent period and slow progression of the disease.

    (B) PML is caused by a virus that causes widespread inapparent infections early in life but causes the disease PML primarily in people with reduced cell-mediated immunity.

    (C) Creutzfeldt-Jakob disease (CJD) is caused by CJ virus, a retrovirus that integrates a DNA copy of its genome into the DNA of brain neurons.

    (D) CJD occurs primarily in immunocompromised people, but infection with the virus that causes CJD is common, as evidenced by the presence of antibodies.

2. Regarding prions, which one of the following is the most accurate?

    (A) The genome of prions consists of a negative-polarity RNA that has a defective polymerase gene.

    (B) Prion proteins are characterized by having changes in conformation from the alpha-helical form to the beta-pleated sheet form.

    (C) Prions are very sensitive to ultraviolet (UV) light, which is why UV light is used in hospital operating rooms to prevent their transmission.

    (D) The main host defense against prions consists of an inflammatory response composed primarily of macrophages and CD4-positive T cells.

3. Regarding progressive multifocal leukoencephalopathy (PML), which one of the following is the most accurate?

    (A) It is caused by a defective mutant of measles virus.

    (B) The virus remains latent in hepatocytes for many years.

    (C) Lesions occur in several areas of the brain, resulting in diverse symptoms.

    (D) Acyclovir is the drug of choice for patients in the early stages of PML.

    (E) It is characterized by an inflammatory reaction in the brain containing many neutrophils.

4. Regarding prion-mediated diseases, which one of the following is the most accurate?

    (A) Prion-mediated diseases are characterized by vacuoles in the brain called "spongiform changes."

    (B) Variant Creutzfeldt-Jakob disease is a disease of cattle caused by the ingestion of sheep brain mixed into cattle feed.

    (C) Kuru is a prion-mediated disease for which the diagnosis can be confirmed in the laboratory by a fourfold or greater rise in antibody titer.

    (D) In Creutzfeldt-Jakob disease, only neurons latently infected by JC virus produce the prion filaments that disrupt neuronal function.

    (E) Creutzfeldt-Jakob disease occurs primarily in children under the age of 2 years because they cannot mount an adequate immune response to the prion protein.

ANSWERS

(1) **(B)**
(2) **(B)**
(3) **(C)**
(4) **(A)**

SUMMARIES OF ORGANISMS

Brief summaries of the organisms described in this chapter begin on page 691. Please consult these summaries for a rapid review of the essential material.

PRACTICE QUESTIONS: USMLE & COURSE EXAMINATIONS

Questions on the topics discussed in this chapter can be found in the Clinical Virology section of Part XIII: USMLE (National Board) Practice Questions starting on page 747. Also see Part XIV: USMLE (National Board) Practice Examination starting on page 775.

Human Immunodeficiency Virus

Disease

Human immunodeficiency virus (HIV) is the cause of acquired immunodeficiency syndrome (AIDS).

Both HIV-1 and HIV-2 cause AIDS, but HIV-1 is found worldwide, whereas HIV-2 is found primarily in West Africa. This chapter refers to HIV-1 unless otherwise noted.

Important Properties

HIV is one of the two important human T-cell lymphotropic retroviruses (human T-cell leukemia virus is the other). HIV preferentially infects and **kills helper (CD4) T lymphocytes**, resulting in the loss of cell-mediated immunity and a high probability that the host will develop **opportunistic infections**. Other cells (e.g., macrophages and monocytes) that have CD4 proteins on their surfaces can be infected also.

HIV is classified within the lentivirus subgroup of retroviruses. The virion has a cylinder-shaped (type D) core surrounded by an envelope containing virus-specific glycoproteins (gp120 and gp41) (Figures 45–1 and 45–2). The genome of HIV consists of two identical molecules of single-stranded, positive-polarity RNA and is said to be **diploid**. (Note that this is *not* double-stranded RNA, which consists of one positive strand and one negative strand.)

The HIV genome is the most complex of the known retroviruses (Figure 45–3). In addition to the three typical retroviral genes *gag, pol,* and *env*, which encode the structural proteins, the genome RNA has six regulatory genes (Table 45–1). Two of these regulatory genes, *tat* and *rev*, are required for replication, and the other four, *nef, vif, vpr,* and *vpu*, are not required for replication and are termed "accessory" genes.

The *gag* gene encodes the internal "core" proteins, the most important of which is the p24 protein. It is important medically as it is the antigen in the initial serologic test that determines whether the patient has antibody to HIV (i.e., has been infected with HIV). (See "Laboratory Diagnosis" section in this chapter.)

The *pol* gene encodes several proteins, including the virion "reverse transcriptase," which synthesizes DNA by using the

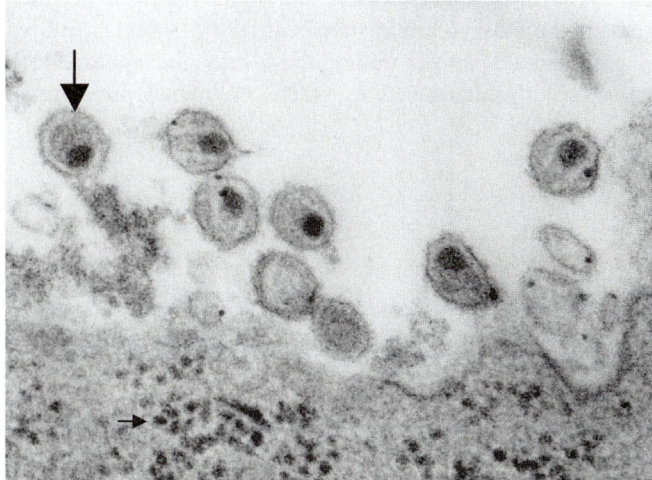

FIGURE 45–1 Human immunodeficiency virus (HIV). Electron micrograph. Large arrow points to a mature virion of HIV that has just been released from the infected lymphocyte at the bottom of the figure. Small arrow (in bottom left of image) points to several nascent virions in the cytoplasm just prior to budding from the cell membrane. (Used with permission from Dr. A. Harrison, Dr. P. Feirino, and Dr. E. Palmer, Public Health Image Library, Centers for Disease Control and Prevention.)

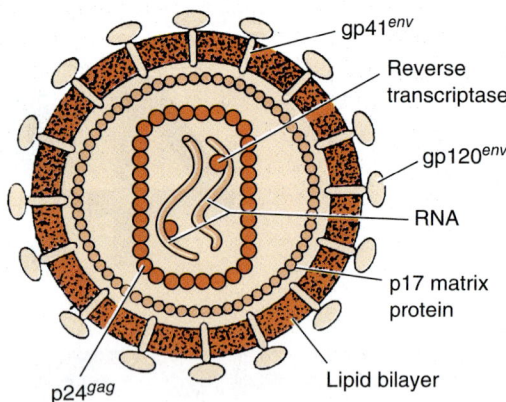

FIGURE 45–2 Cross-section of human immunodeficiency virus (HIV). In the interior, two molecules of viral RNA are shown associated with reverse transcriptase. Surrounding those structures is a rectangular nucleocapsid composed of p24 proteins. Note that the viral protease and integrase are also located within the nucleocapsid (in addition to the reverse transcriptase), but, for lack of space, are not shown in the figure. On the exterior are the two envelope proteins, gp120 and gp41, which are embedded in the lipid bilayer derived from the cell membrane. (Reproduced with permission from Green WC: The molecular biology of human immunodeficiency virus type 1 infection. N Engl J Med. 1991 Jan 31;324[5]:308-317.)

genome RNA as a template, an integrase that integrates the viral DNA into the cellular DNA, and a protease that cleaves the various viral precursor proteins. The *env* gene encodes gp160, a precursor glycoprotein that is cleaved to form the two envelope (surface) glycoproteins, gp120 and gp41.

Differences in the base sequence of the gp120 gene are used to subdivide HIV into subtypes called **clades**. Different clades are found in different areas of the world. For example, the B clade is the most common subtype in North America. Subtype B preferentially infects mononuclear cells and appears to be passed readily during anal sex, whereas subtype E preferentially

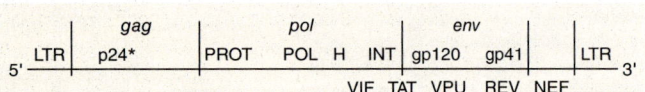

FIGURE 45–3 The genome of human immunodeficiency virus (HIV). Above the line are the three genes for the main structural proteins: (1) *gag* encodes the structural proteins of the capsid (e.g., p24); (2) *pol* encodes proteins that have four enzymatic activities: protease (PROT), polymerase that functions as a reverse transcriptase (POL), RNase H (H), and integrase (INT); and (3) *env* encodes the two envelope glycoproteins, gp120 and gp41. Below the line are five regulatory proteins: viral infectivity factor (VIF), transactivating protein (TAT), viral protein U (VPU), regulator of expression of virion protein (REV), and negative regulatory factor (NEF). At both ends are long terminal repeats (LTR), which are transcription initiation sites. Within the 5′ LTR is the binding site for the TAT protein, called the transactivation response element (TAR). TAT enhances the initiation and elongation of viral mRNA transcription. (*p24 and other smaller proteins such as p17 and p7 are encoded by the *gag* gene.)

TABLE 45–1 Genes and Proteins of Human Immunodeficiency Virus

Gene	Proteins Encoded by Gene	Function of Proteins
I. Structural genes found in all retroviruses		
gag	p24, p7	Nucleocapsid
	p17	Matrix
pol	Reverse transcriptase[1]	Transcribes RNA genome into DNA
	Protease	Cleaves precursor polypeptides
	Integrase	Integrates viral DNA into host cell DNA
env	gp120	Attachment to CD4 protein
	gp41	Fusion with host cell
II. Regulatory genes found in human immunodeficiency virus that are required for replication		
tat	Tat	Activation of transcription of viral genes
rev	Rev	Transport of late mRNAs from nucleus to cytoplasm
III. Regulatory genes found in human immunodeficiency virus that are not required for replication (accessory genes)		
nef	Nef	Decreases CD4 proteins and class I MHC proteins on surface of infected cells; induces death of uninfected cytotoxic T cells; important for pathogenesis by SIV[2]
vif	Vif	Enhances infectivity by inhibiting the action of APOBEC3G (an enzyme that causes hypermutation in retroviral DNA)
vpr	Vpr	Transports viral core from cytoplasm into nucleus in nondividing cells
vpu	Vpu	Enhances virion release from cell

MHC = major histocompatibility complex.

[1]Reverse transcriptase also contains ribonuclease H activity, which degrades the genome RNA to allow the second strand of DNA to be made.

[2]Mutants of the *nef* gene of simian immunodeficiency virus (SIV) do not cause acquired immunodeficiency syndrome in monkeys.

infects female genital tract cells and appears to be passed readily during vaginal sex. Note that these clades (e.g., B and E) are subtypes of group M (Major), by far the most common group of HIV-1 worldwide.

Three enzymes are located within the nucleocapsid of the virion: **reverse transcriptase, integrase**, and **protease** (see Figure 45–2).

Reverse transcriptase is the RNA-dependent DNA polymerase that is the source of the family name retroviruses. This enzyme transcribes the RNA genome into the proviral DNA. Reverse transcriptase is a bifunctional enzyme; it also has ribonuclease H activity. Ribonuclease H degrades RNA when it is in the form of an RNA–DNA hybrid molecule. The degradation of the viral RNA genome is an essential step in the synthesis of the double-stranded proviral DNA. Integrase, another important enzyme within the virion, mediates the integration of the proviral DNA into the host cell DNA. The viral protease cleaves the precursor polyproteins into functional viral polypeptides.

One essential regulatory gene is the **tat** (transactivation of transcription)[1] gene, which encodes a protein that enhances viral (and perhaps cellular) gene transcription.

The Tat protein and another HIV-encoded regulatory protein called Nef repress the synthesis of class I major histocompatibility complex (MHC) proteins, thereby reducing the ability of cytotoxic T cells to kill HIV-infected cells. The other essential regulatory gene, *rev*, controls the passage of late mRNA from the nucleus into the cytoplasm. The function of the four accessory genes is described in Table 45–1.

The accessory protein Vif (*viral infectivity*) enhances HIV infectivity by inhibiting the action of APOBEC3G, an enzyme that causes hypermutation in retroviral DNA. APOBEC3G is "apolipoprotein B RNA-editing enzyme" that deaminates cytosines in both mRNA and retroviral DNA, thereby inactivating these molecules and reducing infectivity. APOBEC3G is considered to be an important member of the innate host defenses against retroviral infection. HIV defends itself against this innate host defense by producing Vif, which counteracts APOBEC3G, thereby preventing hypermutation from occurring.

There are several important antigens of HIV:

(1) gp120 and gp41 are the **type-specific envelope glycoproteins**. gp120 protrudes from the surface and interacts with the CD4 receptor (and a second protein, a chemokine receptor) on the cell surface. gp41 is embedded in the envelope and mediates the fusion of the viral envelope with the cell membrane at the time of infection. The gene that encodes gp120 mutates rapidly, resulting in many **antigenic variants**. The most immunogenic region of gp120 is called the V3 loop; it is one of the sites that varies antigenically to a significant degree. Antibody against gp120 neutralizes the infectivity of HIV, but the rapid appearance of gp120 variants is one of the main reasons why there is no effective vaccine. The high mutation rate appears to be due to lack of an editing function in the reverse transcriptase.

(2) The group-specific antigen, p24, is located in the nucleocapsid core and is not known to vary. Antibodies against p24 do not neutralize HIV infectivity but serve as important serologic markers of infection.

The natural host range of HIV is limited to humans, although certain primates can be infected in the laboratory. HIV is **not an endogenous virus** of humans (i.e., no HIV sequences are found in normal human cell DNA). The origin of HIV and how it entered the human population remains uncertain. There is evidence that chimpanzees living in West Africa were the source of HIV-1. If chimpanzees are the source of HIV in humans, it would be a good example of a virus "jumping the species barrier."

In addition to HIV-1, two other similar retroviruses are worthy of comment:

(1) Human immunodeficiency virus type 2 (HIV-2) was isolated from AIDS patients in West Africa in 1986. The proteins of HIV-2 are only about 40% identical to those of the original HIV isolates. HIV-2 remains localized primarily to West Africa and is much less transmissible than HIV-1. HIV-2 is closely related to a simian immunodeficiency virus (SIV) from a species of monkey called sooty mangabey. Accidental infection of a person with SIVsmm is thought to be the origin of HIV-2.

(2) SIVs have been isolated from various nonhuman primates such as monkeys and chimpanzees (SIVcpz). SIVs cause persistent infection in these species and are thought to be the source of HIV in humans via accidental inoculation with blood of the nonprimate. For example, the genome RNA of HIV-1 is closely related to that SIVcpz. Unlike HIV in humans, SIV infection in nonhuman primates is often asymptomatic. However, an AIDS-like illness caused by SIVcpz does occur in some chimpanzees.

Summary of Replicative Cycle

In general, the replication of HIV follows the typical retroviral cycle (Figure 45–4 and Table 45-2). The initial step in the entry of HIV into the cell is the binding of the virion gp120 envelope protein to the CD4 protein on the cell surface. The virion gp120 protein then interacts with a second protein on the cell surface, one of the **chemokine receptors** (see next paragraph). Next, the virion gp41 protein mediates fusion of the viral envelope with the cell membrane, and the virion core containing the nucleocapsid, RNA genome, and reverse transcriptase enters the cytoplasm.

Chemokine receptors, such as CXCR4 and CCR5 proteins, are required for the entry of HIV into CD4-positive cells. The T cell-tropic strains of HIV bind to CXCR4, whereas the macrophage-tropic strains bind to CCR5. Mutations in the gene encoding CCR5 endow the individual with protection from infection with HIV. People who are homozygotes are completely resistant to infection, and heterozygotes progress to disease more slowly. Approximately 1% of people of Western European ancestry have homozygous mutations in this gene, and about 10% to 15% are heterozygotes. One of the best-characterized mutations is the delta-32 mutation, in which 32 base pairs are deleted from the *CCR5* gene.

In the cytoplasm, **reverse transcriptase** transcribes the genome RNA into double-stranded DNA, which migrates to the nucleus, where it integrates into the host cell DNA. The viral DNA can integrate at different sites in the host cell DNA, and multiple copies of viral DNA can integrate. Integration is mediated by a virus-encoded endonuclease (**integrase**). Viral mRNA is transcribed from the integrated (proviral) DNA by host cell RNA polymerase (augmented by virus-encoded Tat protein) and translated into several large polyproteins. The Gag and Pol polyproteins are cleaved by the **viral protease**, whereas the Env polyprotein is cleaved by a cellular protease.

The Gag polyprotein is cleaved to form the main core protein (p24), the matrix protein (p17), and several smaller proteins. The Pol polyprotein is cleaved to form the reverse transcriptase, integrase, and protease. The immature virion containing the precursor polyproteins forms in the cytoplasm, and cleavage

---

[1]*Transactivation* refers to activation of transcription of genes distant from the gene (i.e., other genes on the same proviral DNA or on cellular DNA). One site of action of the Tat protein is the long terminal repeat at the 5α end of the viral genome.

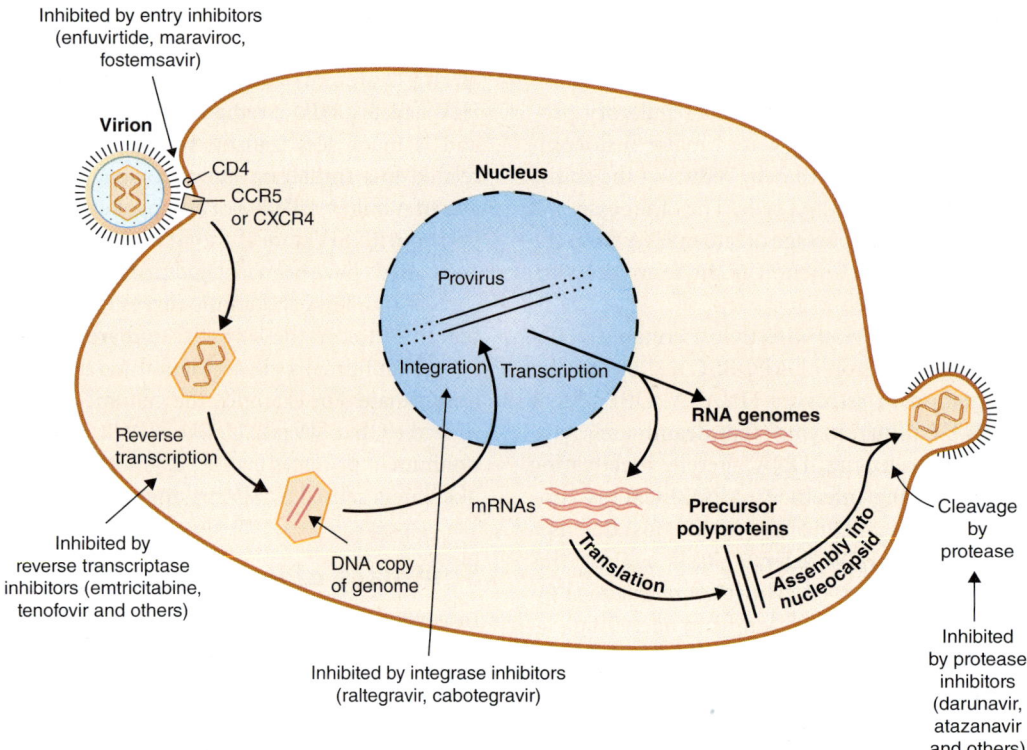

FIGURE 45–4 Replicative cycle of human immunodeficiency virus (HIV) showing the sites of action of the important drugs used to treat HIV infection. The mode of action of the reverse transcriptase inhibitors, the entry inhibitors, the integrase inhibitor, and the PIs is described in Chapter 35. On the right side of the figure, "cleavage by protease" describes the process by which the virus-encoded protease cleaves the Gag-Pol polyprotein into functional viral proteins as the virion buds from the cell membrane. These newly formed functional proteins are transported by the mature virion to the next cell and function within that newly infected cell. The viral reverse transcriptase and integrase are two such proteins. (Reproduced with permission from Ryan K: *Sherris Medical Microbiology*, 3rd ed. New York, NY: McGraw Hill; 1994.)

TABLE 45-2 Important Steps in HIV Replication

Steps in Replication	Process
Attachment to CD4-positive cell	Viral gp120 binds to CD4 protein on cell surface
Co-receptor binding	Viral gp120 binds to chemokine receptor, e.g., CCR-5
Fusion of viral envelope with cell membrane	Viral gp 41 is a transmembrane protein involved in fusion of viral envelope with cell membrane
Entry of nucleocapsid into cytoplasm and reverse transcription	Virus-encoded reverse transcriptase transcribes genome RNA into ds DNA in cytoplasm
Movement of DNA into nucleus and integration into cellular DNA	Virus-encoded integrase mediates integration of viral DNA into cellular DNA
Transcription of viral DNA into mRNA and genome RNA	Cellular RNA polymerase synthesizes polygenic viral mRNAs and full-length genome RNA. Virus-encoded Tat protein enhances transcription by cellular RNA polymerase.
Translation of viral mRNAs into polyproteins	Viral mRNAs are exported to cytoplasm and translated into Gag-Pol polyprotein and Env polyprotein, gp160. Virus-encoded Rev protein enhances export of viral mRNAs to cytoplasm.
Assembly of genome RNA and Gag-Pol polyprotein into progeny nucleocapsids	
Progeny nucleocapsids acquire envelope Gp 120 and gp41	Cleavage of gp160 into gp120 and gp41 by furin, a cellular protease
Progeny virions exit the cell by budding from cell membrane.	
After budding from cell, Gag-Pol polyprotein is cleaved into functional proteins	Virus-encoded protease cleaves Gag-Pol polyprotein into p24, reverse transcriptase, and integrase to form mature infectious virion.

by the viral protease occurs during or soon after the immature, noninfectious virion buds from the cell membrane. It is this cleavage process that results in the mature, infectious virion.

In addition to viral mRNA, the cellular RNA polymerase synthesizes the genome RNAs. These genome RNAs have one guanosine in their 5′ cap which allows them to be distinguished from mRNAs which have two or three guanosines in the cap.

Note that HIV replication is dependent on cell proteins as well as viral proteins. First, there are the cell proteins required during the early events, namely, CD4, and the chemokine receptors, CCR5 and CXCR4. Cell proteins, such as actin and tubulin, are involved with the movement of viral DNA into the nucleus. The cell protein cyclin T1 and the viral protein Tat are part of the complex that transcribes viral mRNA. Cell proteins are also involved in the budding process by which the virus exits the cell.

Transmission & Epidemiology

Transmission of HIV occurs primarily by sexual contact and by transfer of infected blood. Perinatal transmission from infected mother to neonate also occurs, either across the placenta, at birth, or via breast milk. It is estimated that more than 50% of neonatal infections occur at the time of delivery and that the remainder is split roughly equally between transplacental transmission and transmission via breast-feeding. There is no evidence for airborne, waterborne, or insect transmission of HIV.

Infection occurs by the transfer of either HIV-infected cells or free HIV (i.e., HIV that is not cell-associated). Although small amounts of virus have been found in other fluids (e.g., saliva and tears), there is no evidence that they play a role in infection. In general, transmission of HIV follows the pattern of hepatitis B virus (HBV), except that HIV infection is much less efficiently transferred (i.e., the dose of HIV required to cause infection is much higher than that of HBV). People with sexually transmitted diseases, especially those with ulcerative lesions such as syphilis, chancroid, and herpes genitalis, have a significantly higher risk of acquiring HIV. Uncircumcised males have a higher risk of acquiring HIV than do circumcised males.

Transmission of HIV is significantly reduced by drug treatment because it lowers the amount of virus in the blood and other body fluids. Early diagnosis and treatment are so important that current Guidelines recommend that all sexually active adults be tested at least once for HIV and that those at high risk for infection be tested annually.

Transmission of HIV via blood transfusion has been greatly reduced by screening donated blood for the presence of antibody to HIV. However, there is a "window" period early in infection when the blood of an infected person can contain HIV but antibodies are not detectable. Blood banks now test for the presence of p24 antigen in an effort to detect blood that contains HIV.

The Centers for Disease Control and Prevention (CDC) estimates that at the end of 2018, there were approximately 1.2 million people infected with HIV living in the United States. The transmission rate has declined markedly, primarily due to increased prevention efforts and improved treatments for HIV; the latter reduces the number of people with high titers of HIV. CDC estimates that approximately 38,000 new infections occur each year. CDC also estimates that 15% of those who are infected with HIV do not know it because they have not been tested.

As of 2019, it is estimated that approximately 38 million people worldwide are infected, two-thirds of whom live in sub-Saharan Africa. Three regions, Africa, Asia, and Latin America, have the highest rates of new infections.

As a result of public health measures such as condom use, plus the use of antiretroviral drugs that lower the viral load, the number of new HIV cases fell by 36% worldwide between 2000 and 2019. Similarly, the number of HIV-related deaths fell by 38% during the same period.

In the United States and Europe during the 1980s, HIV infection and AIDS occurred primarily in men who have sex with men (especially those with multiple partners), intravenous drug users, and hemophiliacs. Heterosexual transmission was rare in these regions in the 1980s but is now rising significantly. Heterosexual transmission is the predominant mode of infection in African countries.

Very few healthcare personnel have been infected despite continuing exposure and needle-stick injuries, supporting the view that the infectious dose of HIV is high. The risk of being infected after percutaneous exposure to HIV-infected blood is estimated to be about 0.3%. The transmission of HIV from healthcare personnel to patients is exceedingly rare.

Pathogenesis & Immunity

HIV infects helper T cells (CD4-positive cells) and kills them, resulting in **suppression of cell-mediated immunity**. This predisposes the host to various opportunistic infections and certain cancers such as Kaposi's sarcoma and lymphoma. HIV does not directly cause these tumors because HIV genes are not found in these cancer cells. The initial infection of the genital tract occurs in dendritic cells that line the mucosa (Langerhans' cells), after which the local CD4-positive helper T cells become infected. HIV is first found in the blood 4 to 11 days after infection.

HIV infection also targets a subset of CD4-positive cells called **Th17 cells**. These cells are an important mediator of **mucosal immunity**, especially in the gastrointestinal tract. Many mucosal Th17 cells are killed early in HIV infection. Th17 cells produce interleukin-17 (IL-17), which attracts neutrophils to the site of bacterial infection. The loss of Th17 cells predisposes HIV-infected individuals to bloodstream infections by bacteria in the normal flora of the colon, such as *Escherichia coli*.

HIV also infects brain monocytes and macrophages, producing multinucleated giant cells and significant central nervous system symptoms. The fusion of HIV-infected cells in the brain and elsewhere mediated by gp41 is one of the main pathologic findings. The cells recruited into the syncytia ultimately die. The death of HIV-infected cells is also the result of immunologic attack by cytotoxic CD8 lymphocytes. Effectiveness of the cytotoxic T cells may be limited by the ability of the viral Tat and Nef proteins to reduce class I MHC protein synthesis (see later).

The main mechanism by which HIV kills CD-4 positive T cells is **activation of caspases that cause apoptosis**. Note that HIV-infected cells can also kill adjacent "bystander" cells by apoptosis induced by death ligands such as FAS ligand.

Persistent noncytopathic infection of T lymphocytes also occurs. Persistently infected cells continue to produce HIV, which may help sustain the infection in vivo. Lymphoid tissue (e.g., lymph nodes) is the main site of ongoing HIV infection.

In addition, **a true latent infection can occur** in which no HIV is produced. This occurs in resting CD4-positive memory T cells within which an integrated HIV genome is found. The latent period can last for months to years, but if the resting cell is activated, HIV can be produced. HIV replication depends on host cell transcription factors made in activated, but not resting, CD4-positive cells.

A person infected with HIV is considered to be **infected for life**. This seems likely to be the result of integration of viral DNA into the DNA of infected cells. Although the use of powerful antiviral drugs (see "Treatment" section later) can significantly reduce the amount of HIV being produced, the silent, latent infection in CD4-positive memory T cells can be activated and serve as a continuing source of virus.

Elite controllers are a rare group of HIV-infected people (less than 1% of those infected) who have no detectable HIV in their blood. Their CD4 counts are normal without using antiretroviral drugs. The ability to be an elite controller does not depend on gender, race, or mode of acquisition of the virus. Although the mechanism is unclear, there is evidence that certain HLA alleles are protective and that an inhibitor of the cyclin-dependent kinase known as p21 plays an important role.

In addition, there is a group of HIV-infected individuals who have lived for many years without opportunistic infections and without a reduction in the number of their helper T (CD4) cells. The strain of HIV isolated from these individuals has mutations in the *nef* gene, indicating the importance of this gene in pathogenesis. The Nef protein decreases class I MHC protein synthesis, and the inability of the mutant virus to produce functional Nef protein allows the cytotoxic T cells to retain their activity.

Another explanation why some HIV-infected individuals are long-term "nonprogressors" may lie in their ability to produce large amounts of α-defensins. α-Defensins are a family of positively charged peptides with antibacterial activity that also have antiviral activity. They interfere with HIV binding to the CXCR4 receptor and block entry of the virus into the cell.

In addition to the detrimental effects on T cells, abnormalities of B cells occur. Polyclonal activation of B cells is seen, with resultant high immunoglobulin levels. Autoimmune diseases, such as thrombocytopenia, occur.

The main immune response to HIV infection consists of cytotoxic CD8-positive lymphocytes. These cells respond to the initial infection and control it for many years. Mutants of HIV, especially in the *env* gene encoding gp120, arise, but new clones of cytotoxic T cells proliferate and control the mutant strain. It is the ultimate failure of these cytotoxic T cells that results in the clinical picture of AIDS. Cytotoxic T cells lose their effectiveness because so many CD4 helper T cells have died; thus, the supply of lymphokines, such as interleukin-2 (IL-2), required to activate the cytotoxic T cells is no longer sufficient.

There is evidence that "escape" mutants of HIV are able to proliferate unchecked because the patient has no clone of cytotoxic T cells capable of responding to the mutant strain. Furthermore, mutations in any of the genes encoding class I MHC proteins result in a more rapid progression to clinical AIDS. The mutant class I MHC proteins cannot present HIV epitopes, which results in cytotoxic T cells being incapable of recognizing and destroying HIV-infected cells.

Antibodies against various HIV proteins, such as p24, gp120, and gp41, are produced, but they neutralize the virus poorly in vivo and appear to have little effect on the course of the disease.

HIV has three main mechanisms by which it evades the immune system: (1) integration of viral DNA into host cell DNA, resulting in a persistent infection; (2) a high rate of mutation of the *env* gene; and (3) the production of the Tat and Nef proteins that downregulate class I MHC proteins required for cytotoxic T cells to recognize and kill HIV-infected cells. The ability of HIV to infect and kill CD4-positive helper T cells further enhances its capacity to avoid destruction by the immune system.

Clinical Findings

The clinical picture of HIV infection can be divided into three stages: an early, acute stage; a middle, latent stage; and a late, immunodeficiency stage (Figure 45–5). In the **acute stage**, which

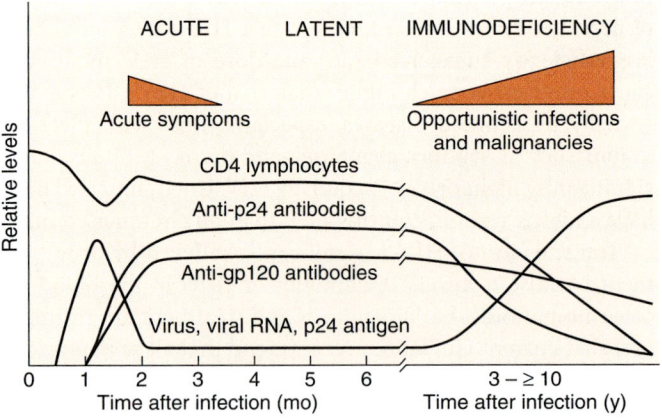

FIGURE 45–5 Time course of human immunodeficiency virus (HIV) infection. The three main stages of HIV infection—acute, latent, and immunodeficiency—are shown in conjunction with several important laboratory findings. Note that the levels of virus and viral RNA (viral load) are high early in the infection, become low for several years, and then rise during the immunodeficiency stage. The level of CD4 lymphocytes remains more or less normal for many years but then falls. This results in the immunodeficiency stage, which is characterized by opportunistic infections and malignancies. Not shown in the figure is the marked loss of the Th-17 subset of CD4-positive T cells early in the infection. (Adapted with permission from Weiss RA: How does HIV cause AIDS? Science. 1993 May 28;260[5112]:1273–1279.)

usually begins 2 to 4 weeks after infection, a mononucleosis-like picture of fever, lethargy, sore throat, and generalized lymphadenopathy occurs. A maculopapular rash on the trunk, arms, and legs (but sparing the palms and soles) is also seen. Leukopenia occurs, but the number of CD4 cells is usually normal. A high-level viremia typically occurs, and the infection is readily transmissible during this acute stage. This acute stage typically resolves spontaneously in about 2 weeks. Resolution of the acute stage is usually accompanied by a lower level of viremia and a rise in the number of CD8-positive (cytotoxic) T cells directed against HIV.

Antibodies to HIV typically appear 10 to 14 days after infection, and most patients will have seroconverted by 3 to 4 weeks after infection. Note that the inability to detect antibodies prior to that time can result in "false-negative" serologic tests (i.e., the person is infected, but antibodies are not detectable at the time of the test). This has important implications because HIV can be transmitted to others during this period. If the antibody test is negative but HIV infection is still suspected, then a polymerase chain reaction (PCR)-based assay for viral RNA in the plasma should be done.

Of those who become seropositive during the acute infection, approximately 87% are symptomatic (i.e., about 13% experience an asymptomatic initial infection).

After the initial viremia, a viral **set point** occurs, which can differ from one person to another. The set point represents the amount of virus produced (i.e., the **viral load**) and tends to remain "set," or constant, for years. The higher the set point at the end of the initial infection, the more likely the individual is to progress to symptomatic AIDS. It is estimated that an infected person can produce up to 10 billion new virions each day. This viral load can be estimated by using an assay for viral RNA in the patient's plasma. (The assay detects the RNA in free virions in the plasma, not cell-associated virions.)

The amount of viral RNA serves to guide treatment decisions and the prognosis. For example, if a drug regimen fails to reduce the viral load, the drugs should be changed. As far as the prognosis is concerned, a patient with more than 10,000 copies of viral RNA/mL of plasma is significantly more likely to progress to AIDS than a patient with fewer than 10,000 copies.

The number of CD4-positive T cells is another important measure that guides the management of infected patients. It is used to determine whether a patient needs chemoprophylaxis against opportunistic organisms, to determine whether a patient needs anti-HIV therapy, and to determine the response to this therapy. The lower limit of CD4 count considered as normal is 500 cells/μL. People with this level or higher are usually asymptomatic. The frequency and severity of opportunistic infections significantly increase when the CD4 counts fall below 200/μL. A CD4 count of 200/μL or below is an AIDS-defining condition.

In the **middle stage** of HIV infection, a long latent period, measured in years, usually ensues. In untreated patients, the latent period typically lasts for 7 to 11 years. The patient is asymptomatic during this period. Although the patient is asymptomatic and viremia is low or absent, a large amount of HIV is being produced by lymph node cells but remains sequestered within the lymph nodes. This indicates that during this period of clinical latency, the virus itself does not enter a latent state.

A syndrome called AIDS-related complex (ARC) can occur during the latent period. The most frequent manifestations are persistent fevers, fatigue, weight loss, and lymphadenopathy. ARC often progresses to AIDS.

The **late stage** of HIV infection is AIDS, manifested by a decline in the number of CD4 cells to below 200/μL and an increase in the frequency and severity of opportunistic infections. Table 45–3 describes some of the common opportunistic

TABLE 45–3   Common Opportunistic Infections in AIDS Patients

Site of Infection	Disease or Symptom	Causative Organism
Lung	1. Pneumonia 2. Tuberculosis	*Pneumocystis jiroveci*, cytomegalovirus *Mycobacterium tuberculosis*
Mouth	1. Thrush 2. Hairy leukoplakia 3. Ulcers	*Candida albicans* Epstein–Barr virus Herpes simplex virus-1, *Histoplasma capsulatum*
Esophagus	1. Thrush 2. Esophagitis	*C. albicans* Cytomegalovirus, herpes simplex virus-1
Intestinal tract	Diarrhea	*Salmonella* species, *Shigella* species, cytomegalovirus, *Cryptosporidium parvum*, *Giardia lamblia*
Central nervous system	1. Meningitis 2. Brain abscess 3. Progressive multifocal leukoencephalopathy	*Cryptococcus neoformans* *Toxoplasma gondii* JC virus
Eye	Retinitis	Cytomegalovirus
Skin	1. Kaposi's sarcoma 2. Zoster 3. Subcutaneous nodules	Human herpesvirus 8 Varicella-zoster virus *C. neoformans*
Reticuloendothelial system	Lymphadenopathy or splenomegaly	*Mycobacterium avium* complex, Epstein–Barr virus

infections and their causative organisms seen in HIV-infected patients during the late, immunocompromised stage of the infection.

The two most characteristic manifestations of AIDS are *Pneumocystis* pneumonia and Kaposi's sarcoma. However, many other opportunistic infections occur with some frequency. These include viral infections such as disseminated herpes simplex, herpes zoster, and cytomegalovirus infections and progressive multifocal leukoencephalopathy; fungal infections such as thrush (caused by *Candida albicans*), cryptococcal meningitis, and disseminated histoplasmosis; protozoal infections such as toxoplasmosis and cryptosporidiosis; and disseminated bacterial infections such as those caused by *Mycobacterium avium-intracellulare* and *Mycobacterium tuberculosis*. Many AIDS patients have severe neurologic problems (e.g., dementia and neuropathy), which can be caused by either HIV infection of the brain or by many of these opportunistic organisms.

Two cancers are very common in AIDS patients, namely non-Hodgkin's B-cell lymphoma caused by Epstein–Barr virus and Kaposi's sarcoma caused by human herpesvirus-8 (Kaposi's sarcoma-associated herpesvirus). Cancers caused by human papillomavirus, namely anal, cervical, oral, pharyngeal, penile, and vulvar also occur commonly in AIDS patients.

Laboratory Diagnosis

The diagnosis of early HIV-1 and HIV-2 infection is made by using immunoassays that detect antibody to HIV-1 and HIV-2, and p24 antigen in serum. This combination ("Combo") test is useful for the diagnosis of early infections because p24 antigen is typically detectable earlier in infection than antibody (Figure 45–6).

Specimens positive in the Combo test proceed to an antibody test to distinguish HIV-1 from HIV-2 and to a PCR-based test (nucleic acid amplification test, NAT) to detect viral RNA.

OraQuick is a rapid, screening immunoassay for HIV antibody that uses an oral swab sample in an enzyme-linked immunosorbent assay (ELISA)-type test that can be done at home. Because there are some false-positive tests with oral specimens and dried blood spots, a Western blot (immunoblot) test is performed on positive specimens. Figure 64–9 depicts a Western blot (immunoblot) test used to diagnose HIV infection.

After HIV infection has been established, the amount of viral RNA in the plasma (i.e., the viral load) can also be determined using PCR-based assays. The amount of viral RNA is used to guide treatment decisions and to predict the risk of progression to AIDS.

Other laboratory tests that are important in the management of an HIV-infected person include CD4 cell counts and tests for drug resistance of the strain of HIV infecting the patient. Drug resistance tests are described at the end of the "Treatment" section in this chapter. HIV can be grown in culture from clinical specimens, but this procedure is available only at a few medical centers.

Treatment

The treatment of HIV infection has resulted in a remarkable reduction in mortality and improvement in the quality of life of infected individuals. The two specific goals of treatment are (1) to restore immunologic function by increasing the CD4 count, which reduces opportunistic infections and certain malignancies and (2) to reduce viral load, which reduces the chance of

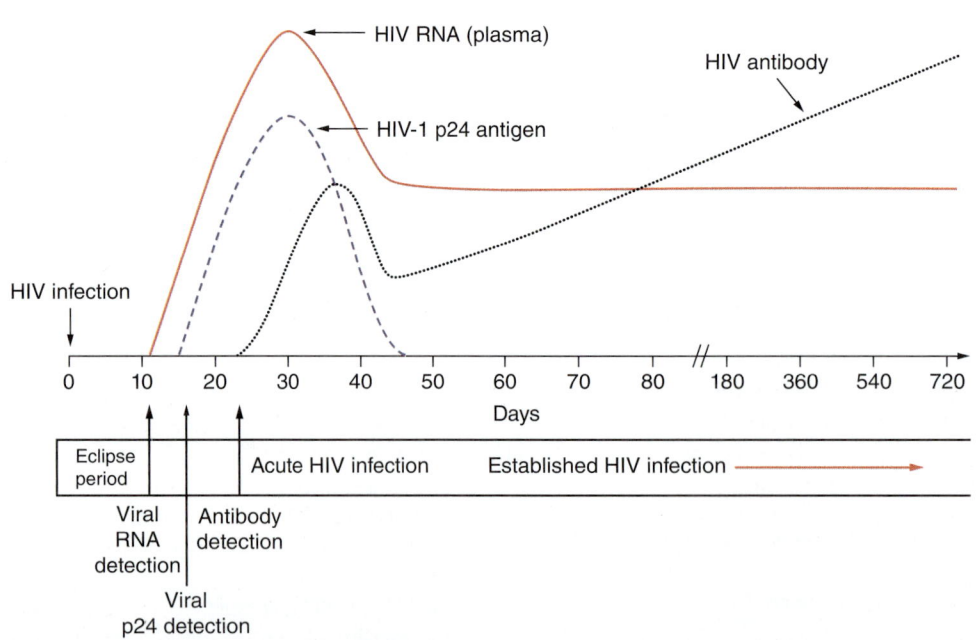

FIGURE 45–6 Time of appearance of human immunodeficiency virus (HIV) RNA, HIV p24 antigen, and HIV antibody. Note that HIV RNA in the plasma is the earliest laboratory finding followed by p24 antigen. Antibody to HIV is the last to become detectable. (Modified with permission from Laboratory Testing for the Diagnosis of HIV Infection. Centers for Disease Control and Prevention. June 27, 2014.)

transmission to others. There is evidence that starting drug therapy as soon as possible after making the diagnosis of HIV infection is the best way to achieve these goals. The recommended approach is to begin treatment within one week of diagnosis. If suppression of viral load is achieved, patients can expect a near normal life span.

Unfortunately, no drug regimen results in a "cure" (i.e., eradicates the virus from the body), but long-term suppression can be achieved. However, if drugs are stopped, the virus resumes active replication, and large amounts of infectious virus reappear.

Transmission of HIV is significantly reduced by drug treatment because it lowers the amount of virus in the blood and other body fluids. Although drugs do not cure the infection, they can lower the amount of virus so that the frequency of transmission is very low. Adherence to drug treatment regimens is important both for the health of the infected patient as well as to protect uninfected individuals. When the viral load is undetectable, the likelihood of transmissions is very low leading to the expression "undetectable equals untransmittable."

Treatment of HIV infection typically involves multiple antiretroviral drugs. The use of a single drug (monotherapy) for treatment is not done because of the high rate of mutation to drug resistance.

The choice of drugs is complex and depends on several factors (e.g., whether it is an initial infection or an established infection, the number of CD4 cells, the viral load, the resistance pattern of the virus, and whether the patient is pregnant or is coinfected with HBV or hepatitis C virus [HCV]). Table 45–4 describes the mechanism of action of the drugs and their main adverse effects. The number of drugs and the various determining factors mentioned previously make describing all the treatments beyond the scope of this book. The reader is advised to consult the Department of Health and Human Services Antiretroviral Therapy Guidelines or other reliable sources, such as the *Medical Letter*.

The seven-drug combinations recommended in 2019 for the treatment of newly infected adults are listed in Table 45–5. In brief, most of them consist of an integrase inhibitor plus two nucleoside reverse transcriptase inhibitors. Four of the five contain tenofovir alafenamide and emtricitabine. One regimen contains only two drugs, dolutegravir (integrase inhibitor) and lamivudine (nucleoside reverse transcriptase inhibitor). Another regimen containing only two drugs is dolutegravir (integrase inhibitor) and rilpivirine (non-nucleoside reverse transcriptase inhibitor).

These combinations are known as **highly active antiretroviral therapy (HAART)**. HAART is very effective in prolonging life, improving quality of life, and reducing viral load but does not cure the chronic HIV infection (i.e., replication of HIV within CD4-positive cells continues indefinitely). Discontinuation of HAART almost always results in viremia (a return of the viral load to its pretreatment set point) and a fall in the CD4 count.

In January 2021, the Food and Drug Administration (FDA) approved an injectable treatment consisting of a combination of cabotegravir and rilpivirine (Cabenuva). Two monthly doses are given, each drug is injected separately. The treatment suppressed HIV to undetectable levels for 2 years.

Nucleoside/Nucleotide Reverse Transcriptase Inhibitors (NRTIs)

Table 45–4 describes six nucleoside reverse transcriptase inhibitors (abacavir, didanosine, emtricitabine, lamivudine, stavudine, and zidovudine [ZVD]) and a single nucleotide reverse transcriptase inhibitor (tenofovir). Tenofovir is supplied as tenofovir alafenamide (TAF) and as tenofovir disoproxil fumarate (TDF). They are prodrugs of the active drug, tenofovir.

These drugs are characterized by *not* having a 3α hydroxyl group on the ribose ring and therefore are **chain-terminating drugs**. They inhibit HIV replication by interfering with proviral DNA synthesis by reverse transcriptase. They cannot cure an infected cell of an already integrated copy of proviral DNA. Additional information on these "nucleoside analogue" drugs and the other antiretroviral drugs can be found in Chapter 35. Note that zalcitabine (Hivid), a nucleoside/nucleotide reverse transcriptase inhibitor (NRTI) analogue of cytosine, is no longer available.

Two main problems limit the use of NTRIs: the emergence of resistance and adverse effects. The main adverse effects are described in Table 45–4. For example, the long-term use of ZDV is limited by suppression of the bone marrow leading to anemia and neutropenia. This hematotoxicity is due to the inhibition of the mitochondrial DNA polymerase. Nevertheless, ZDV is used in postexposure prophylaxis (PEP) and to prevent vertical transmission from mother to fetus. Lamivudine and its analogue emtricitabine have the same mechanism of action as ZDV but are better tolerated, and emtricitabine is a common component of HAART. Abacavir is also used in one version of HAART. Patients who have an HLA-B1701 allele are more likely to have a severe hypersensitivity reaction to abacavir. Patients should be tested for this gene before being prescribed abacavir.

Nonnucleoside Reverse Transcriptase Inhibitors

Table 45–4 describes five nonnucleoside reverse transcriptase inhibitors (delavirdine, efavirenz, etravirine, nevirapine, and rilpivirine) that are effective against HIV. Unlike the NRTIs, these drugs are not base analogues. Efavirenz is a component of some HAART regimens, especially a single pill containing efavirenz, tenofovir, and emtricitabine (Atripla). Nevirapine is often used to prevent vertical transmission of HIV from mother to fetus. Both nevirapine and efavirenz can cause skin rashes and Stevens-Johnson syndrome. Rilpivirine is available as Odefsey, a fixed drug combination containing emtricitabine and tenofovir.

Protease Inhibitors

Table 45–4 describes the currently available protease inhibitors (PIs). PIs, when combined with nucleoside analogues, are very effective in inhibiting viral replication and increasing CD4 cell counts. Darunavir and atazanavir are used in some HAART regimens. In general, PIs should be given in combination with "booster drugs," either ritonavir or cobicistat.

Lopinavir and ritonavir are given in combination because ritonavir inhibits the degradation of lopinavir by cytochrome

TABLE 45-4 Drugs Used for the Treatment of HIV Infection

Class of Drug	Name of Drug	Main Adverse Effect (AE) or Comment
Reverse transcriptase inhibitors		
Nucleosides (NRTI)	Abacavir (ABC) (Ziagen)	AE: severe multiorgan hypersensitivity reaction, especially in patients with HLA-B5701
	Emtricitabine (FTC) (Emtriva)	A derivative of lamivudine; well tolerated
	Lamivudine (3TC) (Epivir)	Well tolerated. Also used to treat hepatitis B virus infection
	Didanosine (ddI) (Videx)	Not often used AE: pancreatitis and peripheral neuropathy
	Stavudine (d4T) (Zerit)	Not often used. AE: peripheral neuropathy, lipoatrophy, lactic acidosis with hepatic steatosis, pancreatitis
	Zidovudine (AZT, ZDV) (Retrovir)	Not routinely used. AE: bone marrow suppression (anemia, neutropenia)
Nucleotides	Tenofovir disoproxil fumarate (TDF) (Viread)	AE: Bone loss and renal toxicity occurs
	Tenofovir alafenamide (TAF) (Vemlidy)	AE: Less bone and renal toxicity than TDF. TAF is used in fixed combination with other drugs. These combinations include Genvoya, Odefsey, Biktarvy, Descovy, and Symtuza.
Nonnucleosides (NNRTI)	Delavirdine (Rescriptor)	AE: rash; avoid in pregnancy; rarely used
	Efavirenz (Sustiva)	AE: CNS changes and rash; possibly teratogenic so avoid in pregnancy
	Rilpivirine (Edurant)	AE: rash
	Etravirine (Intelence)	AE: Stevens-Johnson syndrome; hepatotoxicity
	Nevirapine (Viramune)	AE: depression, insomnia
Protease inhibitors[1]	Amprenavir (Agenerase)	AE: rash, hemolytic anemia
	Atazanavir (Reyataz)	AE: hyperbilirubinemia, prolonged PR interval
	Darunavir (Prezista)	AE: hepatotoxicity
	Fosamprenavir (Lexiva)	A prodrug of amprenavir; metabolized by phosphatases in gut epithelium to amprenavir. AE: rash
	Indinavir (Crixivan)	AE: crystalluria and nephrolithiasis due to poor solubility. Infrequently used
	Lopinavir/ritonavir (Kaletra)	Ritonavir inhibits CYP3A metabolism of lopinavir, thereby increasing the effective concentration of lopinavir
	Ritonavir (Norvir)	See lopinavir/ritonavir, saquinavir, and tipranavir
	Saquinavir (Invirase and Fortovase)	Invirase must be taken with ritonavir; Fortovase can be taken without ritonavir. Infrequently used
	Tipranavir (Aptivus)	AE: if taken with ritonavir (Norvir), severe liver disease may occur
Entry inhibitors		
Fusion inhibitor	Enfuvirtide (Fuzeon)	Binds to viral gp41 and blocks fusion of virus with cell membrane AE: injection site reactions
Coreceptor antagonist	Maraviroc (Selzentry)	Blocks binding of viral gp120 to CCR5 coreceptor on cell membrane; effective against CCR5-tropic viruses but not against CXCR4-tropic viruses
Attachment inhibitor	Fostemsavir (Rukobia)	AE: hepatotoxicity, especially in those infected with HBV and HCV Binds to gp120 on HIV envelope. AE: liver damage in those infected with HBV or HCV
Post-attachment inhibitor	Ibalizumab (Trogarzo)	Monoclonal antibody against CD4 protein that blocks binding to coreceptors, CCR-5 and CXCR-4.
Integrase inhibitor	Raltegravir (Isentress)	Inhibits integration of proviral DNA into cellular DNA AE: nausea, diarrhea, rash
	Elvitegravir (Stribild or Genvoya)	Available in combination with cobicistat, tenofovir, and emtricitabine AE: diarrhea
	Dolutegravir (Tivicay)	AE: insomnia, headache Available in combination with emtricitabine and tenofovir alafenamide
	Bictegravir (Biktarvy)	AE: severe exacerbation of hepatitis B may occur in those coinfected with HBV and HIV and who discontinue Biktarvy.
	Cabotgravir	Long term, injectable drug. Packaged in nanoparticles

CNS = central nervous system.

[1]All protease inhibitors cause lipodystrophy ("buffalo hump") and central obesity as an adverse effect. Nausea and diarrhea are also quite common. Hepatotoxicity occurs, especially in those infected with hepatitis B or C virus.

TABLE 45-5 Drugs Combinations Recommended for the Treatment of Newly HIV-Infected Adults

Integrase Inhibitor	Nucleoside Reverse Transcriptase Inhibitor
Dolutegravir	Tenofovir alafenamide plus emtricitabine
Bictegravir	Tenofovir alafenamide plus emtricitabine
Raltegravir	Tenofovir alafenamide plus emtricitabine
Elvitegravir/cobicistat	Tenofovir alafenamide plus emtricitabine
Dolutegravir	Lamivudine
Dolutegravir	Rilpivirine
Dolutegravir	Abacavir plus lamivudine[1]

[1]Only for patients who are HLA-B*5701 negative.

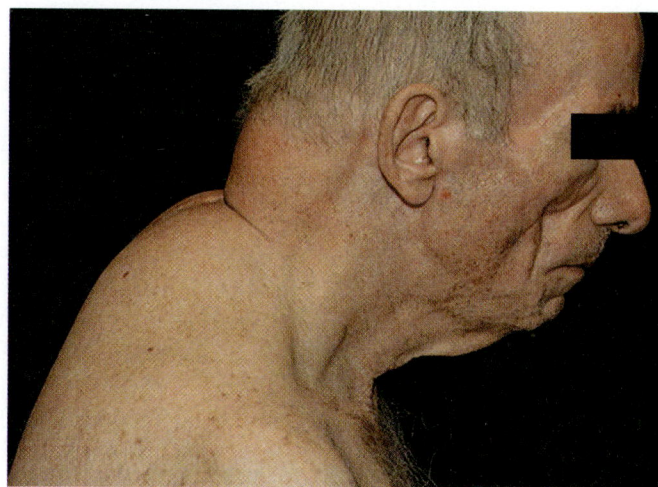

FIGURE 45-7 Lipodystrophy. Note enlarged fat pad on back of neck. This is known as a "buffalo hump" and is an adverse effect of the protease inhibitor class of antiretroviral drugs. (Reproduced with permission from Wolff K, Johnson R: *Fitzpatrick's Color Atlas & Synopsis of Clinical Dermatology*, 6th ed. New York, NY: McGraw Hill; 2009.)

P450 enzymes, thereby increasing the concentration of lopinavir. A briefer way of saying that is **ritonavir boosts lopinavir**.

Another drug that inhibits cytochrome P450 enzymes is **cobicistat**. It is particularly effective in enhancing the antiviral effect of elvitegravir, an integrase inhibitor. It is used in two four-drug, fixed-dose combinations marketed as Stribild and Genvoya. Both drugs contain elvitegravir, cobicistat, emtricitabine, and a tenofovir derivative. Cobicistat is also useful in enhancing the effect of PIs. The drug Prezcobix contains darunavir plus cobicistat and the drug Evotaz contains atazanavir and cobicistat.

Mutants of HIV resistant to PIs can be a significant clinical problem. Resistance to one PI often conveys resistance to all; however, the combination of two PIs, namely, ritonavir and lopinavir (Kaletra), is effective against both mutant and nonmutant strains of HIV. Also, darunavir is effective against many strains of HIV that are resistant to other PIs. Mutants of HIV resistant to both PIs and reverse transcriptase inhibitors have been recovered from patients.

A major side effect of PIs is abnormal fat deposition in specific areas of the body, such as the back of the neck (Figure 45-7). The fat deposits in the back of the neck are said to give the person a **buffalo hump** appearance. These abnormal fat deposits are a type of lipodystrophy; the metabolic process by which this occurs is unknown. Saquinavir and indinavir are infrequently used because of toxicity and nelfinavir is no longer recommended.

Entry Inhibitors

Table 45-4 describes four entry inhibitors, enfuvirtide, maraviroc, fostemsavir, and ibalizumab. Enfuvirtide (Fuzeon) is the first of a new class of anti-HIV drugs known as **fusion inhibitors** (i.e., they prevent the fusion of the viral envelope with the cell membrane). Enfuvirtide is a synthetic peptide that binds to gp41 on the viral envelope, thereby blocking the entry of HIV into the cell. It must be administered by injection and is quite expensive.

Maraviroc (Selzentry) also prevents the entry of HIV into cells. It **blocks the binding of the gp120** envelope protein of HIV to CCR-5, which is an important coreceptor on the cell

surface. Before prescribing maraviroc, a laboratory test (Trofile assay) should be performed to ensure that the tropism of the patient's strain of HIV is CCR5. Maraviroc should be used in combination with other antiretroviral drugs in patients infected with CCR5-tropic strains of HIV and in treatment-experienced adults infected with an HIV strain that is resistant to other antiretroviral drugs.

Fostemsavir is an attachment inhibitor that acts **by binding to the gp120 protein** on the surface of the envelope of HIV. It is a prodrug of temsavir, the active drug. It is approved for use in treatment-experienced patients with difficult-to-treat HIV infection. Severe liver damage may occur in patients also infected with either HBV or HCV.

Ibalizumab (Trogarzo) is a monoclonal antibody against CD4 protein that blocks entry of HIV. It is a "post-attachment" inhibitor, which means it blocks HIV from binding to CCR-5 or CXCR4 coreceptors after HIV has bound to the CD4 protein. It is approved for patients infected with multidrug-resistant HIV.

Integrase Inhibitors

These drugs **inhibit the HIV-encoded integrase** thereby preventing the proviral DNA from integrating into cellular DNA (see Table 45-4). They are commonly used in HAART regimens Four orally administered integrase inhibitors are available: raltegravir (Isentress), dolutegravir (Tivicay), elvitegravir (available as either Stribild or Genvoya in fixed combination with other drugs), and bictegravir (available as Biktarvy in fixed combination with emtricitabine and tenofovir). Dolutegravir is available in fixed combination with lamivudine (Dovato) and with rilpivirine (Juluca). The collective abbreviation INSTI is often used for these drugs. INSTI stands for INtegrase Strand Transfer Inhibitor.

Cabotegravir is an injectable integrase inhibitor with a long half-life that is effective in preventing HIV infection for up to 2 months. The long half-life is achieved by packaging the drug in nanoparticles. The structure of cabotegravir is similar to that of dolutegravir. Monthly injections of cabotegravir and rilpivirine provide long-term suppression of HIV.

Resistance to Antiretroviral Drugs

Drug-resistant mutants of HIV have emerged that significantly affect the ability of both reverse transcriptase inhibitors and PIs to sustain their clinical efficacy. Approximately 10% of newly infected patients are infected with a strain of HIV resistant to at least one antiretroviral drug. Laboratory tests to detect mutant strains include both genotypic and phenotypic analysis. Genotyping reveals the presence of specific mutations in either the reverse transcriptase (*RT*) or protease (*PR*) genes. Phenotyping determines the ability of the virus to grow in cell culture in the presence of the drug. One method of phenotyping recovers the *RT* and *PR* genes from the patient's virus and splices them into a test strain of HIV, which is then used to infect cells in culture. Another laboratory test can determine the tropism of the patient's isolate (i.e., whether it uses CCR5 as its coreceptor). If so, then maraviroc can be used for treatment.

Immune Reconstitution Inflammatory Syndrome

Immune reconstitution inflammatory syndrome (IRIS) may occur in HIV-infected patients who are treated with a HAART regimen and who are coinfected with other microbes such as HBV, HCV, *M. tuberculosis*, *M. avium* complex, *Cryptococcus neoformans*, and *Toxoplasma gondii*. In this syndrome, an exacerbation of clinical symptoms occurs because the antiretroviral drugs enhance the ability to mount an inflammatory response. HIV-infected patients with a low CD4 count have a reduced capacity to produce inflammation, but HAART restores the inflammatory response, and as a result, symptoms become more pronounced. To avoid IRIS, the coinfection should be treated prior to instituting HAART whenever possible.

Prevention

No vaccine is available. Multiple trials of a variety of experimental vaccines have failed to induce protective antibodies, protective cytotoxic T cells, or mucosal immunity. Prevention consists of taking measures to avoid exposure to the virus (e.g., using condoms, not sharing needles, and discarding donated blood that is contaminated with HIV).

Postexposure prophylaxis (PEP), such as that given after a needle-stick injury or a high-risk nonoccupational exposure, employs three drugs: the preferred regimen consists of the combination of tenofovir (TDF) and emtricitabine (given as Truvada) plus dolutegravir. Other alternative drug regimens are available. PEP should be given as soon as possible after exposure and continued for 28 days.

Preexposure prophylaxis (PrEP) using Truvada is indicated for individuals at high risk of infection, such as men who have sex with men. The FDA has also approved the combination of emtricitabine plus tenofovir alafenamide (Descovy) for PrEP. Trials using injectable cabotegravir prevented HIV infection for

TABLE 45–6 Drugs Used for the Prevention of Opportunistic Infections in AIDS Patients

Name of Drug	Infection Prevented
Trimethoprim-sulfamethoxazole	1. *Pneumocystis* pneumonia 2. Toxoplasmosis
Fluconazole	Cryptococcal meningitis
Clotrimazole	Thrush caused by *Candida albicans*
Ganciclovir	Retinitis caused by cytomegalovirus
Azithromycin	*Mycobacterium avium* complex (MAC) infection

as long as 2 months. This promises to be a useful form of PrEP as it bypasses the need to take a daily Truvada pill.

Two steps can be taken to reduce the number of cases of HIV infection in children: antiretroviral therapy should be given to HIV-infected mothers and neonates, and HIV-infected mothers should not breast-feed. The choice of antiretroviral drugs is dependent on several factors, so current guidelines should be consulted. In addition, the risk of neonatal HIV infection is lower if delivery is accomplished by cesarean section rather than by vaginal delivery. Circumcision reduces HIV infection.

Several drugs are commonly taken by patients in the advanced stages of AIDS to prevent certain opportunistic infections (Table 45–6). Some examples are trimethoprim-sulfamethoxazole to prevent *Pneumocystis* pneumonia, fluconazole to prevent recurrences of cryptococcal meningitis, ganciclovir to prevent recurrences of retinitis caused by cytomegalovirus, and oral preparations of antifungal drugs, such as clotrimazole, to prevent thrush caused by *C. albicans*.

SELF-ASSESSMENT QUESTIONS

1. Regarding the structure and replication of human immunodeficiency virus (HIV), which one of the following is most accurate?

   (A) Viral mRNA is the template for the synthesis of the genome RNA.

   (B) During entry of HIV into the cell, the viral p24 protein interacts with the CD4 protein on the cell surface.

   (C) HIV contains an integrase within the virion that integrates copies of the viral genome into the progeny virions.

   (D) HIV has an enzyme in the virion that synthesizes double-stranded DNA using the single-stranded genome RNA as the template.

   (E) The HIV genome encodes a protease that cleaves cellular ribosomal proteins, resulting in the inhibition of cell-specific protein synthesis.

2. Regarding clinical aspects of human immunodeficiency virus (HIV), which one of the following is most accurate?

   (A) During the primary infection with HIV, *Pneumocystis* pneumonia commonly occurs.

   (B) During the long asymptomatic period that can last for years, no HIV is synthesized.

   (C) During the period when many opportunistic infections occur, HIV usually cannot be detected in the blood.

(D) The antibody response to a primary HIV infection usually is detected within 7 to 10 days after infection.

(E) People with a high level of viral RNA in their plasma are more likely to have symptomatic AIDS (i.e., opportunistic infections) than those with low levels.

3. Regarding the laboratory diagnosis of human immunodeficiency virus (HIV), which one of the following is most accurate?

(A) The initial screening of blood for antibodies to HIV is done by the complement fixation test.

(B) Viral load is the term used to describe the amount of infectious virus produced by the patient's CD4-positive T lymphocytes in cell culture.

(C) After infection with HIV, antibodies to the virus can be detected before the polymerase chain reaction (PCR) test can detect nucleic acids specific to HIV.

(D) Because false-positive results occur in the screening test for HIV, a confirmatory test called the Western blot assay should be performed for those with a positive result on the screening test.

4. Regarding the mode of action of drugs used in the treatment of human immunodeficiency virus (HIV) infection, which one of the following is most accurate?

(A) Maraviroc acts by inhibiting the reverse transcriptase in the virion.

(B) Raltegravir inhibits the integration of HIV DNA into host cell DNA.

(C) Zidovudine is a nucleoside analog that inhibits messenger RNA (mRNA) synthesis of HIV.

(D) Ritonavir acts by binding to the Tat protein, which prevents budding and release of the HIV virion.

(E) Lamivudine is a "chain-terminating" drug because it inhibits the growing polypeptide chain by causing misreading of the viral mRNA.

5. Regarding the adverse effects of drugs used in the treatment of human immunodeficiency virus (HIV) infection, which one of the following is most likely to cause bone marrow suppression?

(A) Lamivudine
(B) Lopinavir
(C) Nevirapine
(D) Maraviroc
(E) Zidovudine

6. Regarding the adverse effects of drugs used in the treatment of human immunodeficiency virus (HIV) infection, which one of the following is most likely to cause lipodystrophy (i.e., abnormal fat deposits)?

(A) Lamivudine
(B) Lopinavir
(C) Nevirapine
(D) Maraviroc
(E) Zidovudine

7. Regarding the adverse effects of drugs used in the treatment of human immunodeficiency virus (HIV) infection, which one of the following is most likely to cause Stevens-Johnson syndrome?

(A) Lamivudine
(B) Lopinavir
(C) Nevirapine
(D) Maraviroc
(E) Zidovudine

8. Which of the following modes of transmission of human immunodeficiency virus (HIV) occurs significantly **MORE** often than the others?

(A) Direct skin contact
(B) During childbirth
(C) Fecal–oral route
(D) Respiratory aerosols

9. Your patient is a 25-year-old man who was just found to be infected with HIV based on a positive enzyme-linked immunosorbent assay (ELISA) and a positive Western blot test. His CD4 count is 125, and his viral load is 7000. He has not received any antiretroviral medications. Which one of the following is the best regimen to treat his infection?

(A) Acyclovir, foscarnet, and ribavirin
(B) Enfuvirtide, raltegravir, and maraviroc
(C) Lamivudine, ribavirin, and ritonavir/lopinavir
(D) Emtricitabine, tenofovir, efavirenz, and atazanavir plus ritonavir

ANSWERS

(1) **(D)**
(2) **(E)**
(3) **(D)**
(4) **(B)**
(5) **(E)**
(6) **(B)**
(7) **(C)**
(8) **(B)**
(9) **(D)**

SUMMARIES OF ORGANISMS

Brief summaries of the organisms described in this chapter begin on page 691. Please consult these summaries for a rapid review of the essential material.

PRACTICE QUESTIONS: USMLE & COURSE EXAMINATIONS

Questions on the topics discussed in this chapter can be found in the Clinical Virology section of Part XIII: USMLE (National Board) Practice Questions starting on page 747. Also see Part XIV: USMLE (National Board) Practice Examination starting on page 775.

46

Minor Viral Pathogens

VIRUSES OF MINOR MEDICAL IMPORTANCE

These viruses are presented in alphabetical order. They are listed in Table 46–1 in terms of their nucleic acid and the presence of an envelope.

ASTROVIRUS

Astrovirus is a nonenveloped RNA virus similar in size to poliovirus. It has a characteristic five- or six-pointed star appearance in the electron microscope. It causes watery diarrhea, especially in children. Most adults have antibodies against astrovirus, suggesting that infection occurs commonly. No antiviral drugs or preventive measures are available.

BK VIRUS

BK virus is a member of the polyomavirus family. Polyomaviruses are nonenveloped viruses with a circular, double-stranded DNA genome. BK virus and JC virus (see Chapter 44) are the two polyomaviruses that infect humans.

BK virus infection is widespread as determined by the presence of antibody and is typically acquired in childhood, and infection is not associated with any disease at that time. It does, however, cause nephropathy and graft loss in immunosuppressed renal transplant patients. Asymptomatic shedding of BK virus in the urine of immunocompromised patients and pregnant women in the third trimester occurs. There is no antiviral therapy effective against BK virus.

BORNA VIRUS

Borna virus is an enveloped virus with a nonsegmented, single-strand, negative-polarity RNA genome. It has the smallest genome of any virus with this type of RNA and is the only virus of this type to replicate in the nucleus of the infected cell. DNA sequences homologous to the Borna virus genome are integrated into human cellular DNA. It is a neurotropic virus known to infect regions of the brain such as the hippocampus.

TABLE 46-1 Minor Viral Pathogens

Characteristics	Representative Viruses
DNA-enveloped viruses	Herpes B virus, poxviruses of animal origin (cowpox virus, monkeypox virus)
DNA-nonenveloped viruses	BK virus, human bocavirus
RNA-enveloped viruses	Borna virus, Cache Valley virus, Crimean-Congo hemorrhagic virus, hantaviruses, Heartland virus, Hendra virus, Jamestown Canyon virus, Japanese encephalitis virus, Lassa fever virus, Lujo virus, lymphocytic choriomeningitis virus, Nipah virus, Powassan virus, spumaviruses, Tacaribe complex of viruses (e.g., Junin and Machupo viruses), Whitewater Arroyo virus
RNA-nonenveloped viruses	Astrovirus, Sapporo virus

Borna is the name of a town in Germany where the virus caused a disease in horses in 1885. It is primarily a zoonotic virus causing disease in domestic animals, such as cattle, sheep, dogs, and cats. In humans, Borna virus often causes fatal encephalitis. In addition, there is evidence that it is associated with human psychiatric diseases characterized by abnormal behavior, such as bipolar disorder.

CACHE VALLEY VIRUS

This virus was first isolated in Utah in 1956 but is found throughout the Western Hemisphere. It is a bunyavirus transmitted by *Aedes, Anopheles*, or *Culiseta* mosquitoes from domestic livestock to people. It is a rare cause of encephalitis in humans. There is no treatment or vaccine for Cache Valley virus infections.

CRIMEAN-CONGO HEMORRHAGIC VIRUS

Crimean-Congo hemorrhagic virus (CCHV) causes Crimean-Congo hemorrhagic fever (CCHF), which is characterized by fever, hemorrhage into the skin (ecchymoses) and gastrointestinal tract, and severe liver necrosis. Death often occurs due to shock and multiorgan failure. A broad spectrum of disease occurs, from mild flulike symptoms to severe hemorrhagic fever. There is no antiviral drug therapy and no vaccine.

CCHV is an enveloped, negative-polarity RNA virus that is a member of the bunyavirus family. It is transmitted most often by the bite of ticks of the genus *Hyalomma*. Nosocomial transmission also occurs.

HANTAVIRUS

Hantaviruses are members of the bunyavirus family. The prototype virus is Hantaan virus, the cause of Korean hemorrhagic fever (KHF). KHF is characterized by headache, petechial hemorrhages, shock, and renal failure. It occurs in Asia and Europe but not in North America and has a mortality rate of about 10%. Hantaviruses are part of a heterogeneous group of viruses called **roboviruses**, which stands for "*rodent-borne*" viruses. Roboviruses are transmitted from rodents directly (without an arthropod vector), whereas arboviruses are "*arthropod-borne*."

In 1993, an outbreak of a new disease, characterized by influenzalike symptoms followed rapidly by acute respiratory failure, occurred in the western United States, centered in New Mexico and Arizona. This disease, now called **hantavirus pulmonary syndrome**, is caused by a hantavirus (Sin Nombre virus) endemic in deer mice (*Peromyscus*) and is acquired by inhalation of aerosols of the rodent's urine and feces. It is not transmitted from person to person. Very few people have antibody to the virus, indicating that asymptomatic infections are not common.

The diagnosis is made by detecting viral RNA in lung tissue with the polymerase chain reaction (PCR) assay, by performing immunohistochemistry on lung tissue, or by detecting IgM antibody in serum. The mortality rate of hantavirus pulmonary syndrome is very high, approximately 35%. Between 1993 and December 2009, a total of 534 cases of hantavirus pulmonary syndrome have been reported in the United States. Most cases occurred in the states west of the Mississippi, particularly in New Mexico, Arizona, California, and Colorado, in that order.

There is no effective drug; ribavirin has been used but appears to be ineffective. There is no vaccine for any hantavirus.

HEARTLAND VIRUS

This virus was first recognized as a human pathogen in 2012, when it caused fever, thrombocytopenia, and leukopenia in two men in the state of Missouri. It is a member of the bunyavirus family. It is transmitted by the bite of the Lone Star tick, *Amblyomma*. There is no antiviral treatment or vaccine for this virus.

HENDRA VIRUS

This virus was first recognized as a human pathogen in 1994, when it caused severe respiratory disease in Hendra, Australia. It is a paramyxovirus resembling measles virus and was previously called equine morbillivirus. The human infections were acquired by contact with infected horses, but fruit bats appear to be the natural reservoir. There is no treatment or vaccine for Hendra virus infections.

HERPES B VIRUS

This virus (monkey B virus or herpesvirus simiae) causes a rare, often fatal encephalitis in persons in close contact with monkeys or their tissues (e.g., zookeepers or cell culture technicians). The virus causes a latent infection in monkeys that is similar to herpes simplex virus (HSV)-1 infection in humans.

Herpes B virus and HSV-1 cross-react antigenically, but antibody to HSV-1 does not protect from herpes B encephalitis.

The presence of HSV-1 antibody can, however, confuse serologic diagnosis by making the interpretation of a rise in antibody titer difficult. The diagnosis can therefore be made only by recovering the virus. Acyclovir may be beneficial. Prevention consists of using protective clothing and masks to prevent exposure to the virus. Immune globulin-containing antibody to herpes B virus should be given after a monkey bite.

HUMAN BOCAVIRUS

Human bocavirus (HBoV) is a parvovirus isolated from young children with respiratory tract infections. Antibody to HBoV is found in most adults worldwide. A description of this virus was first reported in 2005, and its precise role in respiratory tract disease has yet to be defined.

JAMESTOWN CANYON VIRUS

Jamestown Canyon virus (JCV) is a member of the bunyavirus family that causes encephalitis. It is transmitted by mosquito bite, most commonly by *Aedes* species. JCV circulates widely among deer in North America, but human disease is rare. In the United States, cases are primarily in the northeastern and midwestern states. There is no antiviral treatment or vaccine for JCV infections.

JAPANESE ENCEPHALITIS VIRUS

This virus is the most common cause of **epidemic encephalitis** in rural areas of Asia. The disease is characterized by fever, headache, nuchal rigidity, altered states of consciousness, tremors, incoordination, and convulsions. The mortality rate is high, and neurologic sequelae are severe and can be detected in most survivors. The disease occurs throughout Asia but is most prevalent in Southeast Asia. The rare cases seen in the United States have occurred in travelers returning from that continent. American military personnel in Asia have been affected.

Japanese encephalitis virus is a member of the flavivirus family. It is transmitted to humans by certain species of *Culex* mosquitoes endemic to Asian rice fields. There are two main reservoir hosts—birds and pigs. The diagnosis can be made by isolating the virus, by detecting IgM antibody in serum or spinal fluid, or by staining brain tissue with fluorescent antibody. There is no antiviral therapy. Prevention consists of an inactivated vaccine and personal protection against mosquito bites. Immunization is recommended for individuals living in areas of endemic infection for several months or longer.

LASSA FEVER VIRUS

Lassa fever virus was first seen in 1969 in the Nigerian town of that name. It causes a severe, **hemorrhagic fever** characterized by multiorgan involvement. The disease begins slowly with fever, headache, vomiting, and diarrhea and progresses to involve the lungs, heart, kidneys, and brain. A petechial rash and gastrointestinal tract hemorrhage ensue, followed by death from vascular collapse. The fatality rate is approximately 20%.

Lassa fever virus is a member of the arenavirus family, which includes other infrequent human pathogens such as lymphocytic choriomeningitis virus and certain members of the Tacaribe group. Arenaviruses ("arena" means sand) are united by their unusual appearance in the electron microscope. Their most striking feature is the "sandlike" particles on their surface, which are ribosomes. The function, if any, of these ribosomes is unknown. Arenaviruses are enveloped viruses with surface spikes, a helical nucleocapsid, and single-stranded RNA with negative polarity.

The natural host for Lassa fever virus is the small rodent *Mastomys*, which undergoes a chronic, lifelong infection. The virus is transmitted to humans by contamination of food or water with animal urine. Secondary transmission among hospital personnel also occurs. Asymptomatic infection is widespread in areas of endemic infection.

The diagnosis is made either by isolating the virus or by detecting a rise in antibody titer. Ribavirin reduces the mortality rate if given early, and hyperimmune serum, obtained from persons who have recovered from the disease, has been beneficial in some cases. No vaccine is available, and prevention centers on proper infection control practices and rodent control.

LUJO VIRUS

Lujo virus is an arenavirus that causes a hemorrhagic fever similar to Lassa fever. This virus emerged in Zambia in 2008 and caused an outbreak in which four of the five infected patients died. The one survivor was treated with ribavirin. The identification of this virus was made by sequencing the viral RNA from the liver and serum of patients. The animal reservoir and mode of transmission are unknown, but other arenaviruses are transmitted by rodent excreta.

LYMPHOCYTIC CHORIOMENINGITIS VIRUS

Lymphocytic choriomeningitis virus is a member of the arenavirus family. It is a rare cause of aseptic meningitis and cannot be distinguished clinically from the more frequent viral causes (e.g., echovirus, Coxsackie virus, or mumps virus). The usual picture consists of fever, headache, vomiting, stiff neck, and changes in mental status. Spinal fluid shows an increased number of cells, mostly lymphocytes, with an elevated protein level and a normal or low sugar level.

The virus is endemic in the mouse population, in which chronic infection occurs. Animals infected transplacentally become healthy lifelong carriers. The virus is transmitted to humans via food or water contaminated by mouse urine or feces. There is no human-to-human spread (i.e., humans are accidental dead-end hosts), although transmission of the virus via solid organ transplants has occurred. In 2005, seven of eight transplant recipients who became infected died. Diagnosis is made by isolating the virus from the spinal fluid or by detecting an increase in antibody titer. No antiviral therapy or vaccine is available.

This disease is the prototype used to illustrate **immunopathogenesis**. If immunocompetent adult mice are inoculated, meningitis and death ensue. If, however, newborn mice or X-irradiated immunodeficient adults are inoculated, no meningitis occurs despite extensive viral replication. If sensitized T cells are transplanted to the immunodeficient adults, meningitis and death occur. The immunodeficient adult mice, who are apparently well, slowly develop glomerulonephritis. It appears that the mice are partially tolerant to the virus in that their cell-mediated immunity is inactive, but sufficient antibody is produced to cause immune complex disease.

MARBURG VIRUS

Marburg virus and Ebola virus are similar in that they both cause **hemorrhagic fever** and are members of the filovirus family; however, they are antigenically distinct. Marburg virus was first recognized as a cause of human disease in 1967 in Marburg, Germany. The common feature of the infected individuals was their exposure to African green monkeys that had recently arrived from Uganda. As with Ebola virus, the natural reservoir of Marburg virus is unknown, although bats are suspected. Ebola virus is described in Chapter 42.

Marburg hemorrhagic fever begins with a constellation of symptoms some of which are fever, headache, sore throat, myalgia, arthralgia, epigastric pain, vomiting, and diarrhea. Later, bleeding into the skin and gastrointestinal tract occurs, followed by shock and disseminated intravascular coagulation leading to multiorgan failure. The hemorrhages are the result of both severe thrombocytopenia and death of endothelial cells. Marked lymphopenia occurs. The mortality rate associated with this virus can be up to 90%.

In 2005, an outbreak of hemorrhagic fever caused by Marburg virus killed hundreds of people in Angola. No cases of disease caused by Marburg virus occurred in the United States prior to 2008. However, in that year, a U.S. traveler became ill after visiting a cave in Uganda inhabited by fruit bats. He returned to the United States, where he was diagnosed with Marburg hemorrhagic fever. He recovered without sequelae.

The diagnosis is made by PCR assay or detecting a rise in IgM antibody titer. No antiviral therapy or vaccine is available. As with Ebola virus, secondary cases among medical personnel have occurred; therefore, stringent infection control practices must be instituted to prevent nosocomial spread.

NIPAH VIRUS

Nipah virus is a paramyxovirus that causes encephalitis, primarily in the South Asian countries of Bangladesh, Malaysia, and Singapore. The fatality rate is approximately 70%. The natural reservoir appears to be fruit bats. People who have contact with pigs are particularly at risk for encephalitis, and some human-to-human transmission occurs. In general, paramyxoviruses are transmitted by saliva or sputum and that is likely to be the natural mode of transmission. Exposure to body fluids also transmits the virus. There is no treatment or vaccine for Nipah virus infections.

POWASSAN VIRUS

Powassan virus is a flavivirus that causes severe encephalitis with significant sequelae. It is transmitted by *Ixodes* ticks, and rodents are the reservoir. It is the only flavivirus transmitted by ticks.

It is named for the town of Powassan, Ontario, Canada, where one of the first cases occurred. Most cases in the United States occur in Minnesota and Wisconsin. There are typically 0 to 10 cases in the United States each year. The diagnosis can be made by PCR or serologic tests. There is no antiviral drug or vaccine.

POXVIRUSES OF ANIMAL ORIGIN

Four poxviruses cause disease in animals and also cause poxlike lesions in humans on rare occasions. They are transmitted by contact with the infected animals, usually in an occupational setting.

Cowpox virus causes vesicular lesions on the udders of cows and can cause similar lesions on the skin of persons who milk cows. Pseudocowpox virus causes a similar picture but is antigenically distinct. Orf virus is the cause of contagious pustular dermatitis in sheep and of vesicular lesions on the hands of sheepshearers.

Monkeypox virus is different from the other three; it causes a human disease that resembles smallpox. It occurs almost exclusively in Central Africa. In 2003, an outbreak of monkeypox occurred in Wisconsin, Illinois, and Indiana. In this outbreak, the source of the virus was animals imported from Africa. It appears that the virus from the imported animals infected local prairie dogs, which then were the source of the human infection. None of those affected died. In Africa, monkeypox has a death rate of between 1% and 10%, in contrast to 50% for smallpox. There is no effective antiviral treatment. The vaccine against smallpox appears to have some protective effect against monkeypox.

Any new case of smallpox-like disease must be precisely diagnosed to ensure that it is not due to smallpox virus. There has not been a case of smallpox in the world since 1977,[1] and smallpox immunization has been allowed to lapse.

For these reasons, it is important to ensure that new cases of smallpox-like disease are due to monkeypox virus. Monkeypox virus can be distinguished from smallpox virus in the laboratory both antigenically and by the distinctive lesions it causes on the chorioallantoic membrane of chicken eggs.

---

[1]With the exception of two laboratory-acquired cases in 1978.

SAPPORO VIRUS

Sapporo virus is a calicivirus that causes acute gastroenteritis. The main symptoms are vomiting and diarrhea, especially in young children. It is transmitted by ingestion of food or water contaminated with human feces. There is no antiviral drug or vaccine. It is named after Sapporo, Japan, the city where it was first isolated. It is a member of the sapovirus group of viruses which are nonenveloped, single-stranded, positive polarity RNA viruses.

SPUMAVIRUS

Spumaviruses are members of the foamy virus family of retroviruses. They cause a foamy appearance in cultured cells *in vitro*. Spumaviruses are unusual in that they have DNA as their genome, whereas typical retroviruses have RNA. However, they do have the typical retroviral sequence of genes, namely *gag*, *pol*, and *env*. The genome DNA is synthesized by virus-encoded reverse transcriptase using viral RNA as the template.

They can present a problem in the production of viral vaccines if they contaminate the cell cultures used to make the vaccine. There are no known human diseases caused by spumaviruses.

TACARIBE COMPLEX OF VIRUSES

The Tacaribe complex contains several human pathogens, all of which cause hemorrhagic fever.

The best known are Sabia virus in Brazil, Junin virus in Argentina, and Machupo virus in Bolivia. Hemorrhagic fevers, as the name implies, are characterized by fever and bleeding into the gastrointestinal tract, skin, and other organs. The bleeding is due to thrombocytopenia. Death occurs in up to 20% of cases, and outbreaks can involve thousands of people. Agricultural workers are particularly at risk.

Similar to other arenaviruses such as Lassa fever virus and lymphocytic choriomeningitis virus, these viruses are endemic in the rodent population and are transmitted to humans by accidental contamination of food and water by rodent excreta. The diagnosis can be made either by isolating the virus or by detecting a rise in antibody titer. In a laboratory-acquired Sabia virus infection, ribavirin was an effective treatment. No vaccine is available.

WHITEWATER ARROYO VIRUS

This virus is the cause of a hemorrhagic fever/acute respiratory distress syndrome in the western part of the United States. It is a member of the arenavirus family, as is Lassa fever virus, a cause of hemorrhagic fever in Africa (see earlier in this chapter).

Wood rats are the reservoir of this virus, and it is transmitted by inhalation of dried rat excrement. This mode of transmission is the same as that of the hantavirus, Sin Nombre virus (see earlier in this chapter). There is no established antiviral therapy, and there is no vaccine.

SELF-ASSESSMENT QUESTIONS

1. Regarding Sin Nombre virus (a hantavirus), which one of the following is most accurate?
    (A) Its main clinical manifestation is encephalitis.
    (B) The main reservoir is domestic animals such as pigs.
    (C) Infection is acquired by inhalation of dried mouse feces and urine.
    (D) Oseltamivir is an effective prophylactic drug if given within 48 hours of exposure.
    (E) Immunization of children at the age of 15 months with the killed vaccine has greatly reduced the incidence of disease.
2. Regarding Japanese encephalitis virus (JEV), which one of the following is most accurate?
    (A) The principal reservoir of JEV is bats.
    (B) It is transmitted by the bite of the dog tick, *Dermacentor*.
    (C) Acyclovir is the drug of choice for encephalitis caused by JEV.
    (D) The killed vaccine should be given to those living in an endemic area.
    (E) JEV is a nonenveloped virus with a circular double-stranded RNA genome.

ANSWERS

(1) **(C)**
(2) **(D)**

SUMMARIES OF ORGANISMS

Brief summaries of the organisms described in this chapter begin on page 691. Please consult these summaries for a rapid review of the essential material.

PRACTICE QUESTIONS: USMLE & COURSE EXAMINATIONS

Questions on the topics discussed in this chapter can be found in the Clinical Virology section of Part XIII: USMLE (National Board) Practice Questions starting on page 747. Also see Part XIV: USMLE (National Board) Practice Examination starting on page 775.

Basic Mycology

STRUCTURE & GROWTH

Because fungi (yeasts and molds) are **eukaryotic** organisms, whereas bacteria are prokaryotic, they differ in several fundamental respects (Table 47–1). Two fungal cell structures are important medically:

(1) The fungal cell wall consists primarily of chitin (not peptidoglycan as in bacteria); thus, fungi are insensitive to certain antibiotics, such as penicillins and cephalosporins, that inhibit peptidoglycan synthesis. Chitin is a polysaccharide composed of long chains of *N*-acetylglucosamine. The fungal cell wall contains other polysaccharides as well, the most important of which is β-glucan, a long polymer of D-glucose. The medical importance of β-glucan is that it is the site of action of the antifungal drug caspofungin.

(2) The fungal cell membrane contains ergosterol, in contrast to the human cell membrane, which contains cholesterol. The selective action of amphotericin B and azole drugs, such as fluconazole and ketoconazole, on fungi is based on this difference in membrane sterols.

There are two types of fungi: yeasts and molds. **Yeasts** grow as **single cells** that reproduce by asexual budding. **Molds** grow as **long filaments** (**hyphae**) and form a mat (**mycelium**). Some hyphae form transverse walls (**septate hyphae**), whereas others do not (**nonseptate hyphae**). Nonseptate hyphae are multinucleated (coenocytic). The growth of hyphae occurs by extension of the tip of the hypha, not by cell division all along the filament.

Several medically important fungi are thermally **dimorphic** (i.e., they form different structures at different temperatures). They exist as molds in the environment at ambient temperature and as yeasts (or other structures such as the spherules of *Coccidioides*) in human tissues at body temperature.

Most fungi are obligate aerobes; some are facultative anaerobes; but none are obligate anaerobes. All fungi require a preformed organic source of carbon—hence their frequent association with decaying matter. The natural habitat of most fungi is, therefore, the **environment**. An important exception is *Candida albicans*, which is part of the normal human flora.

Some fungi reproduce sexually by mating and forming sexual spores (e.g., **zygospores, ascospores**, and **basidiospores**). Zygospores are single large spores with thick walls; ascospores are formed in a sac called ascus; and basidiospores are formed externally on the tip of a pedestal called a basidium. The classification of these fungi is based on their sexual spores.

TABLE 47–1   **Comparison of Fungi and Bacteria**

Feature	Fungi	Bacteria
Diameter	Approximately 4 μm (*Candida*)	Approximately 1 μm (*Staphylococcus*)
Nucleus	Eukaryotic	Prokaryotic
Cytoplasm	Mitochondria and endoplasmic reticulum present	Mitochondria and endoplasmic reticulum is absent
Cell membrane	Sterols present	Sterols absent (except *Mycoplasma*)
Cell wall content	Chitin	Peptidoglycan
Spores	Sexual and asexual spores for reproduction	Endospores for survival, not for reproduction
Thermal dimorphism	Yes (some)	No
Metabolism	Require organic carbon; no obligate anaerobes	Many do not require organic carbon; many obligate anaerobes

Fungi that do not form sexual spores are termed "imperfect" and are classified as **fungi imperfecti**.

Most fungi of medical interest propagate asexually by forming **conidia** (asexual spores) from the sides or ends of specialized structures (Figure 47–1). The shape, color, and arrangement of conidia aid in the identification of fungi. Some important conidia are (1) **arthrospores**,[1] which arise by fragmentation of the ends of hyphae and are the mode of transmission of *Coccidioides immitis*; (2) **chlamydospores**, which are rounded, thick-walled, and quite resistant (the terminal chlamydospores of *C. albicans* aid in its identification); (3) **blastospores**, which are formed by the budding process by which yeasts reproduce asexually (some yeasts, e.g., *C. albicans*, can form multiple buds that do not detach, thus producing sausagelike chains called **pseudohyphae**, which can be used for identification); and (4) **sporangiospores**, which are formed within a sac (sporangium) on a stalk by molds such as *Rhizopus* and *Mucor*.

Although this book focuses on the fungi that are human pathogens, it should be remembered that fungi are used in the production of important foods (e.g., bread, cheese, wine, and beer). Fungi are also responsible for the spoilage of certain foods. Because molds can grow in a drier, more acidic, and higher osmotic pressure environment than bacteria, they tend to be involved in the spoilage of fruits, grains, vegetables, and jams.

PATHOGENESIS

The response to infection with many fungi is the formation of **granulomas**. Granulomas are produced in the major systemic fungal diseases (e.g., coccidioidomycosis, histoplasmosis, and blastomycosis, as well as several others). The cell-mediated immune response is involved in granuloma formation. Acute suppuration, characterized by the presence of neutrophils in the exudate, also occurs in certain fungal diseases such as aspergillosis and sporotrichosis. Fungi do not have endotoxin in their cell walls and do not produce bacterial-type exotoxins.

Activation of the cell-mediated immune system results in a **delayed hypersensitivity skin test** response to certain fungal antigens injected intradermally. A positive skin test indicates exposure to the fungal antigen. It does *not* imply current infection because the exposure may have occurred in the past. A negative skin test makes the diagnosis unlikely unless the patient is immunocompromised. Because most people carry *Candida* as part of the normal flora, skin testing with *Candida* antigens can be used to determine whether cell-mediated immunity is normal.

The transmission and geographic locations of some important fungi are described in Table 47–2.

Intact skin is an effective host defense against certain fungi (e.g., *Candida*, dermatophytes), but if the skin is damaged, organisms can become established. Fatty acids in the skin inhibit dermatophyte growth, and hormone-associated skin changes at

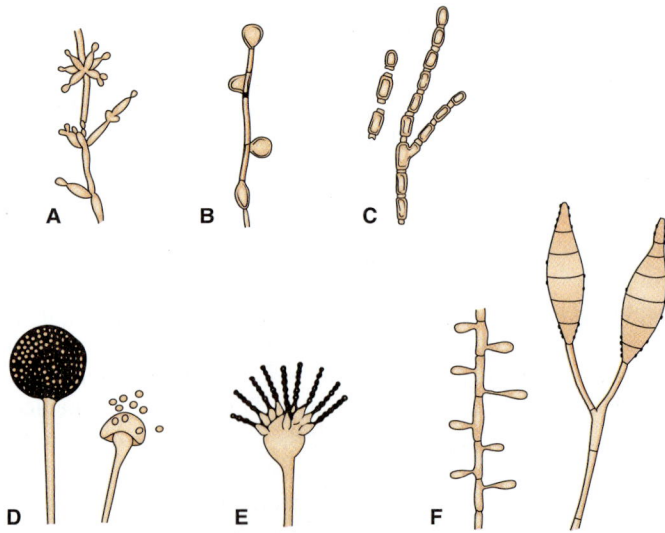

FIGURE 47–1   Asexual spores. **A:** Blastoconidia and pseudohyphae (*Candida*). **B:** Chlamydospores (*Candida*). **C:** Arthrospores (*Coccidioides*). **D:** Sporangia and sporangiospores (*Mucor*). **E:** Microconidia (*Aspergillus*). **F:** Microconidia and macroconidia (*Microsporum*). (Reproduced with permission from Conant NF: *Manual of Clinical Mycology*, 3rd ed. Philadelphia, PA: Saunders/Elsevier; 1971.)

---

[1]The term spores can be replaced with conidia (e.g., arthroconidia).

TABLE 47–2 Transmission and Geographic Location of Some Important Fungi

Genus	Habitat	Form of Organism Transmitted	Portal of Entry	Endemic Geographic Location
Coccidioides	Soil	Arthrospores	Inhalation into lungs	Southwestern United States and Latin America
Histoplasma	Soil (associated with bird feces)	Microconidia	Inhalation into lungs	Mississippi and Ohio River valleys in the United States; many other countries
Blastomyces	Soil	Microconidia	Inhalation into lungs	States east of Mississippi River in the United States; Africa
Paracoccidioides	Soil	Uncertain	Inhalation into lungs	Latin America
Cryptococcus	Soil (associated with pigeon feces)	Yeast	Inhalation into lungs	Worldwide
Aspergillus	Soil and vegetation	Conidia	Inhalation into lungs	Worldwide
Candida	Human body	Yeast	Normal flora of skin, mouth, gastrointestinal tract, and vagina	Worldwide

puberty limit ringworm of the scalp caused by *Trichophyton*. The normal flora of the skin and mucous membranes suppress fungi. When the normal flora is inhibited (e.g., by antibiotics), overgrowth of fungi such as *C. albicans* can occur.

Innate immunity to fungi is initiated by dectin, a receptor on the surface of macrophages, neutrophils, and dendritic cells. Dectin recognizes β-glucan on the surface of yeast cells.

In the respiratory tract, the important host defenses are the mucous membranes of the nasopharynx, which trap inhaled fungal spores, and alveolar macrophages. Circulating IgG and IgM are produced in response to fungal infection, but their role in protection from disease is uncertain. The cell-mediated immune response is protective; its suppression can lead to reactivation and dissemination of asymptomatic fungal infections and to disease caused by opportunistic fungi.

FUNGAL TOXINS & ALLERGIES

In addition to mycotic infections, there are two other kinds of fungal disease: (1) **mycotoxicoses**, caused by ingested toxins, and (2) **allergies** to fungal spores. The best-known mycotoxicosis occurs after eating *Amanita* mushrooms. These fungi produce five toxins, two of which—amanitin and phalloidin—are among the most potent hepatotoxins. The toxicity of amanitin is based on its ability to inhibit cellular RNA polymerase, which prevents mRNA synthesis. Another mycotoxicosis, ergotism, is caused by the mold *Claviceps purpurea*, which infects grains and produces alkaloids (e.g., ergotamine and lysergic acid diethylamide [LSD]) that cause pronounced vascular and neurologic effects.

Other ingested toxins, **aflatoxins**, are coumarin derivatives produced by *Aspergillus flavus* that cause liver damage and tumors in animals and are suspected of causing hepatic carcinoma in humans. Aflatoxins are ingested with spoiled grains and peanuts and are metabolized by the liver to the epoxide, a potent carcinogen. Aflatoxin B1 induces a mutation in the *p53*

tumor suppressor gene, leading to a loss of p53 protein and a consequent loss of growth control in the hepatocyte.

Allergies to fungal spores, particularly those of *Aspergillus*, are manifested primarily by an asthmatic reaction (rapid bronchoconstriction mediated by IgE), eosinophilia, and a "wheal and flare" skin test reaction. These clinical findings are caused by an immediate hypersensitivity response to the fungal spores.

LABORATORY DIAGNOSIS

There are four commonly used approaches to the laboratory diagnosis of fungal diseases: (1) direct microscopic examination, (2) culture of the organism, (3) polymerase chain reaction (PCR) tests, and (4) serologic tests. In addition, a fifth type of test called mass spectrometry (matrix-assisted laser desorption ionization-time of flight [MALDI-TOF]) assay is used to detect the proteins of *Candida* species.

Direct microscopic examination of clinical specimens such as sputum, lung biopsy material, and skin scrapings depends on finding characteristic asexual spores, hyphae, or yeasts in the light microscope. The specimen is either treated with 10% potassium hydroxide (KOH) to dissolve tissue material, leaving the alkali-resistant fungi intact, or stained with special fungal stains. Some examples of diagnostically important findings made by direct examination are (1) the spherules of *C. immitis* and (2) the wide capsule of *Cryptococcus neoformans* seen in India ink preparations of spinal fluid. Calcofluor white is a fluorescent dye that binds to fungal cell walls and is useful in the identification of fungi in tissue specimens. Methenamine silver stain is also useful in the microscopic diagnosis of fungi in tissue.

Fungi are frequently cultured on Sabouraud's agar, which facilitates the appearance of the slow-growing fungi by inhibiting the growth of bacteria in the specimen. Inhibition of bacterial growth is due to the low pH of the medium and to the penicillin, streptomycin, and cycloheximide that are frequently added.

TABLE 47–3 Mechanism of Action and Adverse Effects of Antifungal Drugs

Usage	Name of Drug	Mechanism of Action	Important Adverse Reactions
Systemic use (intravenous, oral)	Amphotericin B	Binds to ergosterol and disrupts fungal cell membranes	Renal toxicity, fever, and chills; monitor kidney function; use test dose; liposomal preparation reduces toxicity
	Azoles such as fluconazole, ketoconazole, itraconazole, voriconazole, posaconazole	Inhibits ergosterol synthesis	Ketoconazole inhibits human cytochrome P450; this decreases synthesis of gonadal steroids resulting in gynecomastia; voriconazole causes periostitis
	Echinocandins such as caspofungin, micafungin	Inhibits synthesis of D-glucan, a component of fungal cell wall	Well-tolerated
	Flucytosine (FC)	Inhibits DNA synthesis; FC converted to fluorouracil, which inhibits thymidine synthetase	Bone marrow toxicity
	Griseofulvin	Disrupts mitotic spindle by binding to tubulin	Liver toxicity
Topical use (skin only); too toxic for systemic use	Azoles such as clotrimazole, miconazole	Inhibits ergosterol synthesis	Well-tolerated on skin
	Terbinafine	Inhibits ergosterol synthesis	Well-tolerated on skin
	Tolnaftate	Inhibits ergosterol synthesis	Well-tolerated on skin
	Nystatin	Binds to ergosterol and disrupts fungal cell membranes	Well-tolerated on skin

The appearance of the mycelium and the nature of the asexual spores are frequently sufficient to identify the organism.

PCR-based tests using DNA probes can identify colonies growing in culture at an earlier stage of growth than can tests based on visual detection of the colonies. As a result, the diagnosis can be made more rapidly. At present, PCR assays are available for *Coccidioides, Histoplasma, Blastomyces,* and *Cryptococcus.*

Tests for the presence of antibodies in the patient's serum or spinal fluid are useful in diagnosing systemic mycoses but less so in diagnosing other fungal infections. As is the case for bacterial and viral serologic testing, a significant rise in the antibody titer must be observed to confirm a diagnosis. The complement fixation test is most frequently used in suspected cases of coccidioidomycosis, histoplasmosis, and blastomycosis. In cryptococcal meningitis, the presence of the polysaccharide capsular antigens of *C. neoformans* in the spinal fluid can be detected by the latex agglutination test. This test is often called the cryptococcal antigen test, abbreviated CRAG.

ANTIFUNGAL THERAPY

The drugs used to treat bacterial diseases have no effect on fungal diseases. For example, penicillins and aminoglycosides inhibit the growth of many bacteria but do not affect the growth of fungi. This difference is explained by the presence of certain structures in bacteria (e.g., peptidoglycan and 70S ribosomes) that are absent in fungi.

The most effective antifungal drugs, amphotericin B and the various azoles, exploit the presence of **ergosterol** in fungal cell membranes that is not found in bacterial or human cell membranes. Amphotericin B (Fungizone) disrupts fungal cell membranes at the site of ergosterol, and azole drugs inhibit the synthesis of ergosterol, which is an essential component

of fungal membranes. Another antifungal drug, caspofungin (Cancidas), inhibits the synthesis of β-glucan, which is found in fungal cell walls but not in bacterial cell walls. Human cells do not have a cell wall.

The mechanism of action of these drugs is described below. Table 47–3 summarizes the mechanism of action and the important adverse effects of the major antifungal drugs. Two well-known adverse effects are (1) the fever and chills caused by amphotericin B that has been ameliorated by using liposomal amphotericin B and (2) the gynecomastia caused by ketoconazole. Figure 47–2 depicts a typical fungal cell showing the site of action of important antifungal drugs.

MECHANISM OF ACTION OF ANTIFUNGAL DRUGS

Inhibition of Fungal Cell Wall Synthesis

Echinocandins, such as caspofungin (Cancidas) and micafungin (Mycamine), are lipopeptides that block fungal cell wall synthesis by inhibiting β-glucan synthase, the enzyme that synthesizes β-glucan (see Figure 47–2). β-Glucan is a polysaccharide composed of long chains of D-glucose, which is an essential component of certain medically important fungal pathogens.

Caspofungin inhibits the growth of *Aspergillus* and *Candida* but not *Cryptococcus* or *Mucor.* Caspofungin is used for the treatment of disseminated candidiasis and for the treatment of invasive aspergillosis that does not respond to amphotericin B. Micafungin is approved for the treatment of esophageal candidiasis and the prophylaxis of invasive *Candida* infections in bone marrow transplant patients. Anidulafungin is approved for the treatment of esophageal candidiasis and other serious *Candida* infections.

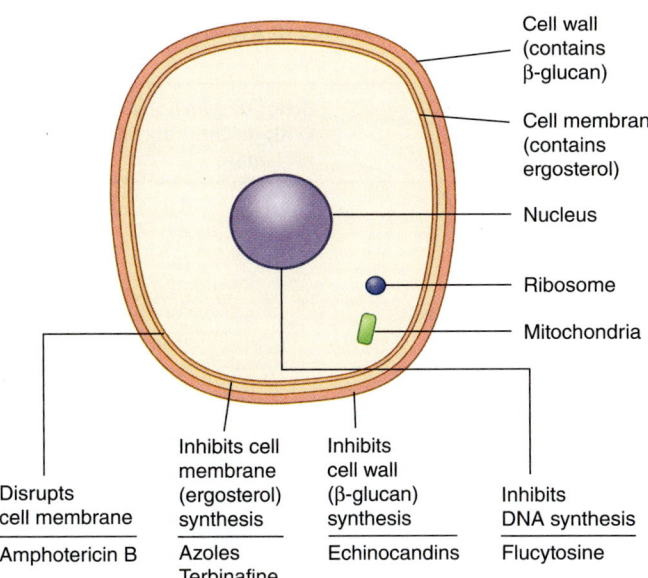

Cell wall
(contains
β-glucan)

Cell membrane
(contains
ergosterol)

Nucleus

Ribosome

Mitochondria

Disrupts cell membrane	Inhibits cell membrane (ergosterol) synthesis	Inhibits cell wall (β-glucan) synthesis	Inhibits DNA synthesis
Amphotericin B	Azoles Terbinafine	Echinocandins	Flucytosine

FIGURE 47–2 Model of typical fungal cell showing sites of action of important antifungal drugs.

Alteration of Fungal Cell Membranes

Amphotericin B & Nystatin

Amphotericin B, the most important antifungal drug, is used in the treatment of a variety of disseminated fungal diseases. It is a polyene with a series of seven unsaturated double bonds in its macrolide ring structure (*poly* means many, and *-ene* is a suffix indicating the presence of double bonds; Figure 47–3). It disrupts the cell membrane of fungi because of its affinity for **ergosterol**, a component of fungal membranes but not of bacterial or human cell membranes. Fungi resistant to amphotericin B have been recovered from patient specimens, but are uncommon.

Amphotericin B has significant renal toxicity; measurement of serum creatinine levels is used to monitor the dose. Nephrotoxicity is significantly reduced when the drug is administered in liposomes, but liposomal amphotericin B is expensive. Fever, chills, nausea, and vomiting are common side effects.

Nystatin is another polyene antifungal agent, which, because of its toxicity, is used topically for infections caused by the yeast *Candida*.

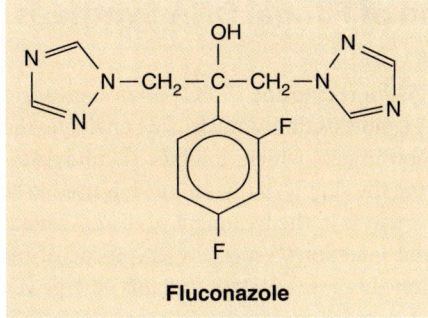

FIGURE 47–4 Fluconazole.

Azoles & Terbinafine

Azoles are antifungal drugs that act by **inhibiting ergosterol synthesis**. They block cytochrome P450-dependent demethylation of lanosterol, the precursor of ergosterol. Fluconazole, ketoconazole, voriconazole, posaconazole, itraconazole, and isavuconazonium (a prodrug of isavuconazole) are used to treat systemic fungal diseases. Clotrimazole and miconazole are used only topically because they are too toxic to be given systemically. The two nitrogen-containing azole rings of fluconazole can be seen in Figure 47–4.

Ketoconazole is useful in the treatment of blastomycosis, chronic mucocutaneous candidiasis, coccidioidomycosis, and skin infections caused by dermatophytes. Fluconazole is useful in the treatment of candidal and cryptococcal infections. Itraconazole is used to treat histoplasmosis, blastomycosis, and sporotrichosis.

Voriconazole is used in the treatment of the three infections mentioned in the previous sentence as well as *Candida* infections, including infections caused by fluconazole-resistant *Candida* species.

Posaconazole is used for the treatment of oropharyngeal candidiasis and mucormycosis, and in the prevention of *Candida* and *Aspergillus* infections in immunocompromised individuals. Isavuconazonium is used to treat invasive aspergillosis. Miconazole and clotrimazole, two other imidazoles, are useful for topical therapy of *Candida* infections and dermatophytoses. Fungi resistant to the azole drugs are an increasing problem.

Terbinafine blocks ergosterol synthesis by inhibiting squalene epoxidase. It is used in the treatment of dermatophyte infections of the skin, fingernails, and toenails.

FIGURE 47–3 Amphotericin B.

Inhibition of Fungal DNA Synthesis

Flucytosine

Flucytosine (5-fluorocytosine, 5-FC) is an antifungal drug that inhibits DNA synthesis. It is a nucleoside analogue that is metabolized to fluorouracil, which inhibits thymidylate synthetase, thereby limiting the supply of thymidine. It is used in combination with amphotericin B in the treatment of disseminated cryptococcal or candidal infections, especially cryptococcal meningitis. It is not used alone because resistant mutants emerge very rapidly.

MECHANISM OF RESISTANCE TO ANTIFUNGAL DRUGS

Resistance to antifungal drugs is becoming clinically important (Table 47–4). Certain species of *Candida*, especially *Candida glabrata*, are resistant to fluconazole and to voriconazole. *Candida auris* is highly drug resistant. *Cryptococcus* species, especially *Cryptococcus gattii*, are increasingly resistant to fluconazole. *Aspergillus fumigatus* is showing increased resistance to azole drugs such as itraconazole and voriconazole. *Scedosporium apiospermum*, the asexual form of *Pseudallescheria boydii*, is resistant to virtually all antifungal drugs.

One of the most common mechanisms of drug resistance in fungi is export of the drug by efflux pump transporters located in the fungal cell membrane. Alteration of the drug target caused by mutations in fungal DNA is another important mechanism. Also, the presence of fungi within biofilms impedes access of the drug to the fungal cell. Widespread use of antifungal drugs, both in medical practice and in agricultural use of fungicides, has selected for drug-resistant fungi.

TABLE 47–4   Drug Resistance in Medically Important Fungi

Name of Fungus	Drugs to Which the Fungus Exhibits Clinically Significant Resistance
1. Yeasts	
Candida krusei	Fluconazole
Candida glabrata	Fluconazole, voriconazole, echinocandins
Candida auris	Fluconazole, amphotericin B, echinocandins
Cryptococcus species, especially *C. gattii*	Echinocandins
2. Molds	
Aspergillus fumigatus	Azole drugs
Aspergillus terreus	Amphotericin B
Scedosporium apiospermum	Often resistant to all antifungal drugs

Additional Antifungal Drug Mechanisms

Griseofulvin is an antifungal drug that is useful in the treatment of hair and nail infections caused by dermatophytes. It binds to tubulin in microtubules and may act by preventing formation of the mitotic spindle.

Pentamidine is active against fungi and protozoa. It is widely used to prevent or treat pneumonia caused by *Pneumocystis jiroveci*. It inhibits DNA synthesis by an unknown mechanism.

PEARLS

Structure & Growth

- Fungi are eukaryotic organisms that exist in two basic forms: **yeasts and molds**. Yeasts are single cells, whereas molds consist of long filaments of cells called hyphae. Yeasts reproduce by **budding**, a process in which the daughter cells are unequal in size, whereas molds reproduce by cell division (daughter cells are equal in size).

- Some fungi are **dimorphic** (i.e., they can exist either as yeasts or molds, depending on the temperature). At room temperature (e.g., 25°C), dimorphic fungi are molds, whereas at body temperature they are yeasts (or some other form such as a spherule).

- The fungal cell wall is made of **chitin**; the bacterial cell wall is made of peptidoglycan. Therefore, antibiotics that inhibit peptidoglycan synthesis such as penicillins, cephalosporins, and vancomycin are not effective against fungi.

- The fungal cell membrane contains **ergosterol**, whereas the bacterial cell membrane does not contain ergosterol. Therefore, antibiotics that inhibit ergosterol synthesis (e.g., the azole drugs) are not effective against bacteria. Similarly, amphotericin B that binds to fungal cell membranes at the site of ergosterol is not effective against bacteria.

Pathogenesis

- Infection with certain systemic fungi, such as *Histoplasma* and *Coccidioides*, elicits a **granulomatous host defense response** (composed of macrophages and helper T cells). Infection with other fungi, notably *Aspergillus*, *Mucor*, and *Sporothrix*, elicits a **pyogenic response** (composed of neutrophils).

- Infection with the systemic fungi, such as *Histoplasma* and *Coccidioides*, can be detected by using **skin tests**. An antigen extracted from the organism injected intradermally elicits a **delayed hypersensitivity reaction**, manifested as **induration** (thickening) of the skin. Note that a positive skin test only indicates that infection has occurred, but it is not known whether that infection occurred in the past or at the present time. Therefore, a positive skin test does not indicate that the disease the patient has now is caused by that organism. Note also that a false-negative skin test can occur in patients with reduced cell-mediated immunity, such as those with a low CD4 count.

To determine whether the patient can mount a delayed hypersensitivity response, a control skin test with a common antigen, such as *Candida albicans*, can be used.

- Reduced cell-mediated immunity predisposes to disseminated disease caused by the systemic fungi, such as *Histoplasma* and *Coccidioides*, whereas a reduced number of neutrophils predispose to disseminated disease caused by fungi such as *Aspergillus* and *Mucor*.

Fungal Toxins & Allergies

- Ingestion of *Amanita* mushrooms causes **liver necrosis** due to the presence of two fungal toxins, amanitin and phalloidin. **Amanitin** inhibits the RNA polymerase that synthesizes cellular mRNA.
- Ingestion of peanuts and grains contaminated with *Aspergillus flavus* causes **liver cancer** due to the presence of **aflatoxin**. Aflatoxin epoxide induces a mutation in the *p53* gene that results in a loss of the p53 tumor suppressor protein.
- Inhalation of the spores of *Aspergillus fumigatus* can cause **allergic bronchopulmonary aspergillosis**. This is an IgE-mediated immediate hypersensitivity response.

Laboratory Diagnosis

- Microscopic examination of a **KOH preparation** can reveal the presence of fungal structures. The purpose of the KOH is to dissolve the human cells, allowing visualization of the fungi.

- **Sabouraud's agar** is often used to grow fungi because its low pH inhibits the growth of bacteria, allowing the slower-growing fungi to emerge.
- PCR tests can be used to identify fungi growing in culture at a much earlier stage (i.e., when the colony size is much smaller).
- Tests for the presence of fungal antigens and for the presence of antibodies to fungal antigens are often used. Two commonly used tests are those for cryptococcal antigen in spinal fluid and for *Coccidioides* antibodies in the patient's serum.

Antifungal Therapy

- The selective toxicity of amphotericin B and the azole group of drugs is based on the presence of **ergosterol** in fungal cell membranes, in contrast to the cholesterol found in human cell membranes and the absence of sterols in bacterial cell membranes.
- Amphotericin B binds to fungal cell membranes at the site of ergosterol and disrupts the integrity of the membranes.
- Azole drugs, such as itraconazole, fluconazole, and ketoconazole, inhibit the synthesis of ergosterol.
- The selective toxicity of echinocandins, such as caspofungin, is based on the presence of a cell wall in fungi, whereas human cells do not have a cell wall. Echinocandins inhibit the synthesis of **D-glucan**, which is a component of the fungal cell wall.

SELF-ASSESSMENT QUESTIONS

1. Regarding the structure and reproduction of fungi, which one of the following is most accurate?

   (A) Peptidoglycan is an important component of the cell wall of fungi.

   (B) Molds are fungi that grow as single cells and reproduce by budding.

   (C) Some fungi are dimorphic (i.e., they are yeasts at room temperature and molds at body temperature).

   (D) The fungal cell membrane contains ergosterol, whereas the human cell membrane contains cholesterol.

   (E) As most fungi are anaerobic, they should be cultured under anaerobic conditions in the clinical laboratory.

2. Regarding fungal pathogenesis, which one of the following is most accurate?

   (A) Ingestion of *Amanita* mushrooms typically causes kidney failure.

   (B) The host response to infection by the systemic fungi, such as *Histoplasma* and *Coccidioides*, consists of granulomas formation.

   (C) The fever seen in systemic fungal infections is caused by endotoxin-induced release of interlukin-1.

   (D) Ingestion of aflatoxin produced by *Aspergillus flavus* can cause adenocarcinoma of the colon.

   (E) A positive result in the skin test to fungal antigens, such as coccidioidin, is caused by an immediate hypersensitivity reaction.

3. Regarding the mode of action of antifungal drugs, which one of the following is most accurate?

   (A) Azole drugs, such as fluconazole, act by inhibiting ergosterol synthesis.

   (B) Amphotericin B acts by inhibiting fungal protein syntheses at the 40S ribosomal subunit.

   (C) Terbinafine acts by inhibiting fungal DNA synthesis but has no effect on DNA synthesis in human cells.

   (D) Echinocandins, such as caspofungin, act by inhibiting messenger RNA synthesis in yeasts but not in molds.

4. The selective toxicity of amphotericin B is based on the presence in fungi of which one of the following?

   (A) 30S ribosomal subunit

   (B) Dihydrofolate reductase

   (C) DNA gyrase

   (D) Ergosterol

   (E) Mycolic acid

ANSWERS

(1) **(D)**
(2) **(B)**
(3) **(A)**
(4) **(D)**

PRACTICE QUESTIONS: USMLE & COURSE EXAMINATIONS

Questions on the topics discussed in this chapter can be found in the Mycology section of Part XIII: USMLE (National Board) Practice Questions starting on page 752. Also see Part XIV: USMLE (National Board) Practice Examination starting on page 775.

Cutaneous & Subcutaneous Mycoses

48

CHAPTER CONTENTS

INTRODUCTION

Medical mycoses can be divided into four categories: (1) **cutaneous**, (2) **subcutaneous**, (3) **systemic**, and (4) **opportunistic**. Some features of the important fungal diseases are described in Table 48–1. Cutaneous and subcutaneous mycoses are discussed in this chapter, and important features of the causative organisms are described in Table 48–2. The systemic and opportunistic mycoses are discussed in Chapters 49 and 50, respectively.

Additional information regarding the clinical aspects of infections caused by the fungi in this chapter is provided in Part IX, entitled Infectious Diseases beginning on page 603.

CUTANEOUS MYCOSES

Dermatophytoses

Dermatophytoses are caused by fungi (**dermatophytes**) that infect only superficial keratinized structures (skin, hair, and nails), not deeper tissues. The most important dermatophytes are classified into three genera: *Trichophyton, Epidermophyton,* and *Microsporum.* All of these dermatophytes are molds. They are *not* dimorphic.

The dermatophytes are spread from person to person by direct contact. *Microsporum* is also spread from animals such as dogs and cats. This indicates that to prevent reinfection by *Microsporum,* the animal must also be treated.

TABLE 48–1 Features of Important Fungal Diseases

Type	Anatomic Location	Representative Disease	Genus of Causative Organism(s)	Seriousness of Illness[1]
Cutaneous	Dead layer of skin	Tinea versicolor	*Malassezia*	1+
	Epidermis, hair, nails	Dermatophytosis (ringworm)	*Microsporum, Trichophyton, Epidermophyton*	2+
Subcutaneous	Subcutis	Sporotrichosis	*Sporothrix*	2+
		Mycetoma	Several genera	2+
Systemic	Internal organs	Coccidioidomycosis	*Coccidioides*	4+
		Histoplasmosis	*Histoplasma*	4+
		Blastomycosis	*Blastomyces*	4+
		Paracoccidioidomycosis	*Paracoccidioides*	4+
Opportunistic	Internal organs	Cryptococcosis	*Cryptococcus*	4+
		Candidiasis	*Candida*	2+ to 4+
		Aspergillosis	*Aspergillus*	4+
		Mucormycosis	*Mucor, Rhizopus*	4+

[1]1+ = not serious, treatment may or may not be given; 2+ = moderately serious, treatment often given; 4+ = serious, treatment given especially in disseminated disease.

TABLE 48–2 Important Features of Skin and Subcutaneous Fungal Diseases

Genus	Forms in Tissue Seen by Microscopy	Mode of Transmission	Important Clinical Findings	Laboratory Diagnosis
Trichophyton, Epidermophyton	Hyphae	Human to human	Tinea capitis, tinea pedis, etc., ("ringworm");—ring of inflammatory, pruritic vesicles with a healing center	Potassium hydroxide (KOH) prep shows septate hyphae culture on Sabouraud's agar
Microsporum	Hyphae	Animal to human as well as human to human	Tinea capitis, tinea pedis, etc., ("ringworm");—ring of inflammatory, pruritic vesicles with a healing center	KOH prep shows septate hyphae culture on Sabouraud's agar
Malassezia	Hyphae and yeasts	Human to human	Scaly plaques on trunk; often hypopigmented; often nonpruritic	KOH prep shows mixture of hyphae and yeasts
Sporothrix	Yeasts	Penetrating lesion in garden implants fungal spores, e.g., rose thorn	Pustule or ulcer on hands often with nodules on arms	KOH prep shows cigar-shaped yeasts culture at 20°C shows hyphae with daisy-like conidia

Dermatophytoses (tinea, ringworm) are chronic infections often located in the warm, humid areas of the body (e.g., athlete's foot and jock itch).[1] Typical ringworm lesions have an inflamed circular border containing papules and vesicles surrounding a clear area of relatively normal skin. The lesions are typically pruritic. Broken hairs and damaged nails are often seen.

The disease is typically named for the affected body part (i.e., tinea capitis [head], tinea corporis [body], tinea cruris [groin], and tinea pedis [foot]) (Figure 48–1). Tinea unguium, also called onychomycosis, is a disease of the nails, especially toe nails. The nails become thickened, broken, and discolored.

Trichophyton tonsurans is the most common cause of outbreaks of **tinea capitis** in children and is the main cause of endothrix (inside the hair) infections. *Trichophyton rubrum* is also a very common cause of tinea capitis. *Trichophyton schoenleinii* is the cause of **favus**, a form of tinea capitis in which crusts are seen on the scalp. *Trichophyton* species also cause an inflammatory pustular lesion on the scalp called a **kerion**. The marked inflammation is caused by an intense T-cell–mediated reaction to the presence of the fungus.

In some infected persons, hypersensitivity causes **dermatophytid ("id")** reactions (e.g., vesicles on the fingers). Id lesions are a response to circulating fungal antigens; the lesions do not contain hyphae. Patients with tinea infections show positive skin tests with fungal extracts (e.g., trichophyton).

Scrapings of skin or nail placed in 10% potassium hydroxide (KOH) on a glass slide show septate hyphae under microscopy. Cultures on Sabouraud's agar at room temperature develop typical hyphae and conidia. Tinea capitis lesions caused by *Microsporum* species can be detected by seeing fluorescence when the lesions are exposed to ultraviolet light from a Wood's lamp.

Treatment involves local antifungal creams, such as terbinafine (Lamisil), undecylenic acid (Desenex), miconazole (Micatin), or tolnaftate (Tinactin). Oral griseofulvin (Fulvicin) or oral itraconazole (Sporanox) can also be used. Tinea unguium can be treated with efinaconazole solution applied topically to the nails. Prevention centers on keeping skin dry and cool.

Tinea Versicolor

Tinea versicolor (pityriasis versicolor), a superficial skin infection of cosmetic importance only, is caused by *Malassezia* species. The lesions are usually noticed as hypopigmented areas, especially on tanned skin in the summer. There may be slight scaling or itching, but usually the infection is asymptomatic. It occurs more frequently in hot, humid weather. The lesions contain both budding yeast cells and hyphae. Diagnosis is usually made by observing this mixture in KOH preparations of skin scrapings. Culture is not usually done. The treatment of choice is topical miconazole, but the lesions have a tendency to recur. Oral antifungal drugs, such as fluconazole or itraconazole, can be used to treat recurrences.

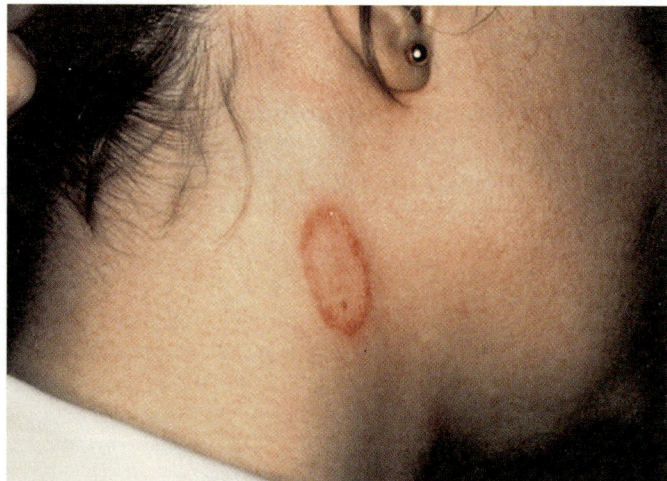

FIGURE 48–1   Tinea corporis (ringworm). Note oval, ring-shaped inflamed lesion with central clearing. Caused by dermatophytes such as *Epidermophyton*, *Trichophyton*, and *Microsporum*. (Reproduced with permission from Fauci AS, Braunwald E, Kasper DL et al: *Harrison's Principles of Internal Medicine*, 17th ed. New York, NY: McGraw Hill; 2008.)

---

[1]These infections are also known as tinea pedis and tinea cruris, respectively.

Tinea Nigra

Tinea nigra is an infection of the keratinized layers of the skin. It appears as a brownish spot caused by the melanin-like pigment in the hyphae. The causative organism, *Cladosporium werneckii*, is found in the soil and transmitted during injury. In the United States, the disease is seen in the southern states. Diagnosis is made by microscopic examination and culture of skin scrapings. The infection is treated with a topical keratolytic agent (e.g., salicylic acid).

SUBCUTANEOUS MYCOSES

These are caused by fungi that grow in soil and on vegetation and are introduced into subcutaneous tissue through **trauma**.

Sporotrichosis

Sporothrix schenckii is a **dimorphic** fungus. The mold form lives on plants, and the yeast form occurs in human tissue. When spores of the mold are introduced into the skin, typically by a thorn, it causes a local pustule or ulcer with nodules along the draining lymphatics (Figure 48–2). The lesions are typically painless, and there is little systemic illness. Untreated lesions may wax and wane for years. In human immunodeficiency virus (HIV)-infected patients with low CD4 counts, disseminated sporotrichosis can occur. Sporotrichosis occurs most often in **gardeners, especially those who prune roses**, because they may be stuck by a rose thorn.

In the clinical laboratory, round or cigar-shaped budding yeasts are seen in tissue specimens. In culture at room temperature, hyphae occurs bearing oval conidia in clusters at the tip of slender conidiophores (resembling a daisy). The drug of choice for skin lesions is itraconazole (Sporanox). It can be prevented by protecting skin when touching plants, moss, and wood.

Chromomycosis

This is a slowly progressive granulomatous infection that is caused by several soil fungi (*Fonsecaea, Phialophora, Cladosporium*, etc.) when introduced into the skin through trauma. These fungi are collectively called **dematiaceous** fungi, so named because their conidia or hyphae are dark-colored, either

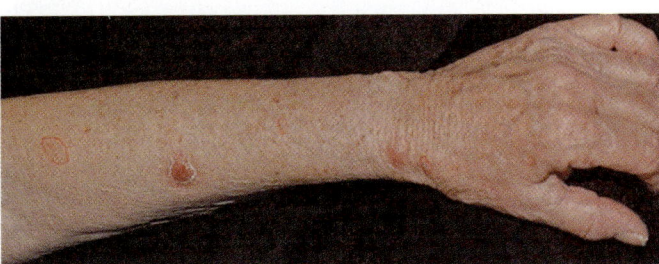

FIGURE 48–2 Sporotrichosis. Note papular lesions on left hand and forearm. Caused by *Sporothrix schenckii*. (Reproduced with permission from Wolff K, Johnson R: *Fitzpatrick's Color Atlas & Synopsis of Clinical Dermatology*, 6th ed. New York, NY: McGraw Hill; 2009.)

gray or black. Wartlike lesions with crusting abscesses extend along the lymphatics. The disease occurs mainly in the tropics and is found on bare feet and legs. In the clinical laboratory, dark brown, round fungal cells are seen in leukocytes or giant cells. The disease is treated with oral flucytosine or thiabendazole, plus local surgery.

Mycetoma

Soil fungi (*Petriellidium, Madurella*) enter through wounds on the feet, hands, or back and cause abscesses, with pus discharged through sinuses. The pus contains compact colored granules. Actinomycetes such as *Nocardia* can cause similar lesions (actinomycotic mycetoma). Sulfonamides may help the actinomycotic form. There is no effective drug against the fungal form; surgical excision is recommended.

SELF-ASSESSMENT QUESTIONS

1. Regarding ringworm and the dermatophytes, which one of the following is most accurate?

    (A) The dermatophytes are molds and are not thermally dimorphic.

    (B) The drug of choice for the treatment of ringworm lesions is amphotericin B.

    (C) The purpose of the KOH prep is to observe fungal antigens within infected cells.

    (D) The dermatophytid reaction refers to the necrotic area typically seen in the center of ringworm lesions.

    (E) The principal reservoir of dermatophytes in the genus *Trichophyton* is domestic animals such as dogs and cats.

2. Regarding sporotrichosis and *Sporothrix schenckii*, which one of the following is most accurate?

    (A) The main reservoir of *Sporothrix* is dog feces.

    (B) Laboratory diagnosis involves seeing a nonseptate mold in an aspirate of the lesion.

    (C) *Sporothrix* is often acquired by penetrating wounds sustained while gardening.

    (D) The treatment of choice for sporotrichosis is surgical removal of the lesion because there is no effective drug.

    (E) Disease occurs primarily in patients who are deficient in the late-acting complement components.

3. Your patient is a 65-year-old woman with a 2-cm ulcerated lesion on the palm of her hand that has been gradually getting bigger during the past month. The lesion is only slightly tender and is not red, hot, or painful. A careful history reveals that she was making holly wreaths for use at Christmas. (Holly leaves have sharp points.) She is afebrile and otherwise well. An aspirate of the lesion was obtained. Which one of the following would best support a diagnosis of sporotrichosis?

    (A) A culture on blood agar at 25°C revealed white, beta-hemolytic colonies.

    (B) A methenamine silver stain examined in the light microscope revealed budding yeasts.

    (C) A KOH preparation examined in the light microscope revealed septate hyphae.

    (D) A culture on Sabouraud's agar at 37°C revealed a brownish mycelium with green spores.

    (E) An unstained sample examined in the dark field microscope revealed nonseptate hyphae.

4. Your patient is a 10-year-old boy with tinea pedis (athlete's foot). Which one of the following is the best choice of drug to treat his infection?

   (A) Amphotericin B
   (B) Caspofungin
   (C) Flucytosine
   (D) Terbinafine

ANSWERS

(1) **(A)**
(2) **(C)**
(3) **(B)**
(4) **(D)**

SUMMARIES OF ORGANISMS

Brief summaries of the organisms described in this chapter begin on page 703. Please consult these summaries for a rapid review of the essential material.

PRACTICE QUESTIONS: USMLE & COURSE EXAMINATIONS

Questions on the topics discussed in this chapter can be found in the Mycology section of Part XIII: USMLE (National Board) Practice Questions starting on page 752. Also see Part XIV: USMLE (National Board) Practice Examination starting on page 775.

Systemic Mycoses

INTRODUCTION

These infections result from **inhalation** of the spores of **dimorphic** fungi that have their **mold** forms in the **soil**. Within the **lungs**, the spores differentiate into **yeasts** or other specialized forms, such as spherules.

Most lung infections are asymptomatic and self-limited. However, in some persons, disseminated disease develops in which the organisms grow in other organs, cause destructive lesions, and may result in death. Infected persons do *not* communicate these diseases to others.

Important features of the systemic fungal diseases are described in Table 49–1. Systemic fungi are also called endemic fungi because they are endemic (localized) to certain geographic areas.

Additional information regarding the clinical aspects of infections caused by the fungi in this chapter is provided in Part IX, entitled Infectious Diseases beginning on page 603.

COCCIDIOIDES

Disease

Coccidioides immitis and *Coccidioides posadasii* cause coccidioidomycosis. The clinical manifestations of disease caused by these two species are the same, but the geographical distribution differs. For simplicity, the original species name, *C. immitis*, will be used most often in this chapter.

TABLE 49–1 Important Features of Systemic Fungal Diseases

Genus	Form in Tissue Seen by Microscopy	Geographic Location	Important Clinical Findings	Laboratory Diagnosis
Coccidioides	Spherule	Southwestern United States and Latin America	Valley fever in immunocompetent; dissemination to bone and meninges in immunocompromised, pregnant women, African Americans, and Filipinos	Culture at 20°C grows mold with arthrospores; serologic test for IgM and IgG
Histoplasma	Yeasts within macrophages	Ohio and Mississippi River valleys; worldwide; associated with bird and bat guano	Cavitary lung lesions; granulomas in liver and spleen; pancytopenia and tongue ulcer in immunocompromised	Culture at 20°C grows mold with tuberculate macroconidia; serologic test for IgM and IgG; urinary antigen
Blastomyces	Yeasts with single broad-based bud	Central and southeastern United States; Africa	Ulcerated lesions of the skin	Culture at 20°C grows mold
Paracoccidioides	Yeasts with multiple buds	Latin America, especially Brazil	Ulcerated lesions of the face and mouth	Culture at 20°C grows mold; serologic test for IgM and IgG

Properties

Coccidioides species are **dimorphic** fungi that exist as a **mold** in soil and as a **spherule** in tissue (Figure 49–1). *C. immitis* and *C. posadasii* are distinguished by genotyping but not by routine diagnostic tests in the clinical laboratory.

Transmission & Epidemiology

The fungus is **endemic** in the soil of arid regions of the **southwestern United States** and **Latin America**. People who live in Central and Southern California, Arizona, New Mexico, Western Texas, and Northern Mexico, a geographic region called the Lower Sonoran Life Zone, are often infected. The organism is also found in the soil in areas of Central and South America. *C. immitis* is found in California, whereas *C. posadasi* is found in other southwestern states and in Latin America.

In soil, it forms hyphae with alternating **arthrospores** and empty cells (Figure 49–2). Arthrospores are very light and are carried by the wind. They can be **inhaled** and infect the lungs.

Pathogenesis

In the lungs, arthrospores form **spherules** that are large (30 mm in diameter), have a thick, doubly refractive wall, and are filled with **endospores** (Figure 49–3). Upon rupture of the wall, endospores are released and differentiate to form new spherules. The organism can spread within a person by direct extension or via the bloodstream. Granulomatous lesions can occur in virtually any organ but are found primarily in bones and the central nervous system (meningitis).

Dissemination from the lungs to other organs occurs in people who have a defect in cell-mediated immunity. Most people who are infected by *C. immitis* develop a cell-mediated (delayed hypersensitivity) immune response that restricts the growth of the organism. One way to determine whether a person has produced adequate cell-mediated immunity to the

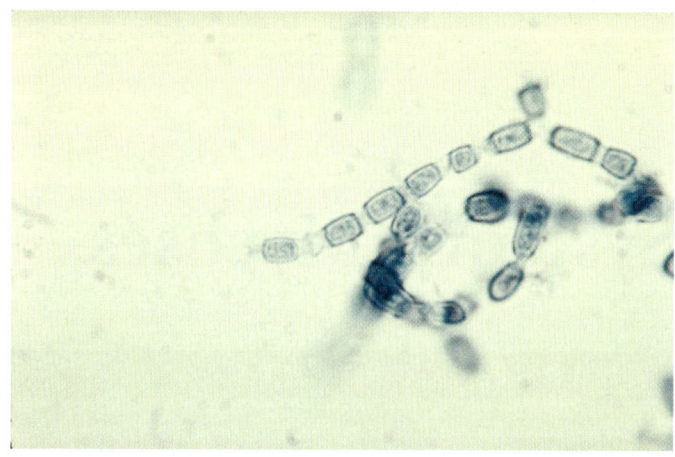

FIGURE 49–2 *Coccidioides immitis.* Arthrospores. Barrel-shaped, rectangular arthrospores appear blue with lactophenol-cotton blue stain. Arthrospores are also called arthroconidia. (Used with permission from Dr. Hardin, Public Health Image Library, Centers for Disease Control and Prevention.)

organism is to do a skin test (see later). In general, a person who has a positive skin test reaction has developed sufficient immunity to prevent disseminated disease from occurring. If, at a later time, a person's cellular immunity is suppressed by drugs or disease, disseminated disease can occur.

Clinical Findings

Infection of the lungs is often asymptomatic and is evident only by a positive skin test and the presence of antibodies. Some infected persons have an influenza-like illness with fever and cough, resembling a community-acquired pneumonia. About 50% have changes in the lungs (infiltrates, adenopathy, or effusions) as seen on chest X-ray, and 10% develop erythema nodosum (EN) (see later) or arthralgias. This syndrome is called

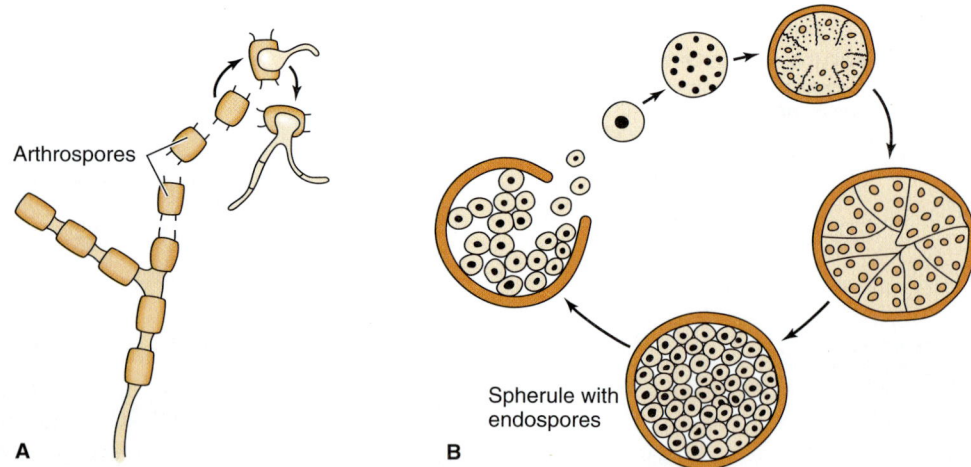

FIGURE 49–1 Stages of *Coccidioides immitis.* **A:** Arthrospores form at the ends of hyphae in the soil. They germinate in the soil to form new hyphae. If inhaled, the arthrospores differentiate into spherules. **B:** Endospores form within spherules in tissue. When spherules rupture, endospores disseminate and form new spherules. (Reproduced with permission from Brooks GF, Butel JS, Ornston LN: *Jawetz, Melnick & Adelberg's Medical Microbiology,* 20th ed. New York, NY: McGraw Hill; 1995.)

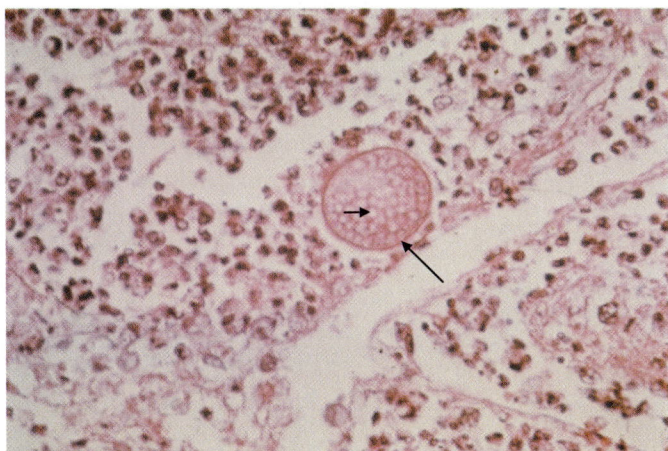

FIGURE 49–3 *Coccidioides immitis*. Spherule. Long arrow points to a spherule in lung tissue. Spherules are large thick-walled structures containing many endospores. Short arrow points to an endospore. (Used with permission from Dr. Georg L, Public Health Image Library, Centers for Disease Control and Prevention.)

"valley fever" (in the San Joaquin Valley of California) or "desert rheumatism" (in Arizona); it tends to subside spontaneously.

Disseminated disease can occur in almost any organ; the meninges (meningitis), bone (osteomyelitis), and skin (nodules) are important sites. The overall incidence of dissemination in persons infected with *C. immitis* is 1%, although the incidence in Filipinos and African Americans is 10 times higher. Women in the third trimester of pregnancy also have a markedly increased incidence of dissemination. Patients with compromised cell-mediated immunity such as those with acquired immunodeficiency syndrome (AIDS), cancer chemotherapy, or drugs to prevent transplant rejection also have an increased risk of disseminated disease. Infected patients in these high-risk categories are often treated even though they do *not* have signs of disseminated disease.

EN manifests as red, tender nodules ("desert bumps") on extensor surfaces such as the skin over the tibia and ulna. It is a delayed (cell-mediated) hypersensitivity response to fungal antigens and thus is an indicator of a good prognosis. There are no organisms in these lesions; they are not a sign of disseminated disease. EN is not specific for coccidioidomycosis; it occurs in other granulomatous diseases (e.g., histoplasmosis, tuberculosis, and leprosy).

In infected persons, **skin tests** with fungal extracts (coccidioidin or spherulin) cause at least a 5-mm induration 48 hours after injection (delayed hypersensitivity reaction). Skin tests become positive within 2 to 4 weeks of infection and remain so for years but are often negative (anergy) in patients with disseminated disease.

Laboratory Diagnosis

In tissue specimens, spherules are seen microscopically. The presence of spherules is pathognomonic for *Coccidioides* infection.

Cultures on Sabouraud's agar incubated at 25°C show septate hyphae with arthrospores (see Figure 49–2). To confirm that the fungus recovered from the patient is *Coccidioides*, a probe for the DNA of the organism can be used. (*Caution*: Cultures

are highly infectious; precautions against inhaling arthrospores must be taken.)

Serologic testing is a common procedure for diagnosing *Coccidioides* infection. IgM antibodies are detected by a tube precipitin test, and their presence indicates an acute infection. IgG antibodies are detected by a complement fixation (CF) test, and their presence indicates a long-standing infection or a disseminated infection. A titer of 1/16 or greater is common in coccidioidal meningitis. Enzyme-linked immunosorbent assays (ELISA) are also used to detect IgM and IgG antibodies to *Coccidioides*.

In immunosuppressed patients, serological tests may not be useful. In those patients, tests for *Coccidioides* antigen in serum and urine can be done. A polymerase chain reaction (PCR) assay that detects nucleic acids of *Coccidioides* is also available.

Treatment & Prevention

No treatment is needed in asymptomatic or mild primary infection. Fluconazole or itraconazole is used for persisting lung lesions or mild disseminated disease. Severe disseminated disease including bone lesions should be treated with amphotericin B. If meningitis occurs, fluconazole is the drug of choice. Intrathecal amphotericin B may be required and may induce remission, but long-term results are often poor.

There is no vaccine. Prevention involves avoiding travel to endemic areas. Patients with *Coccidioides* infection who are significantly immunocompromised (e.g., organ transplant patients) should receive fluconazole. Patients who have recovered from coccidioidal meningitis should receive long-term suppressive therapy with fluconazole.

HISTOPLASMA

Disease

Histoplasma capsulatum causes histoplasmosis.

Properties

H. capsulatum is a **dimorphic** fungus that exists as a **mold** in soil and as a **yeast** in tissue. The mold forms two types of asexual spores (Figure 49–4): (1) **tuberculate macroconidia**, with typical thick walls and fingerlike projections that are important in laboratory identification and (2) **microconidia**, which are smaller, thin, smooth-walled spores that, if inhaled, transmit the infection.

Transmission & Epidemiology

This fungus occurs in many parts of the world. In the United States, it is **endemic** in central and eastern states, especially in the **Ohio and Mississippi River valleys**. It grows in soil, particularly if the soil is heavily contaminated with **bird droppings**, especially from starlings. Although the birds are not infected, bats can be infected and can excrete the organism in their guano. In areas of endemic infection, excavation of the

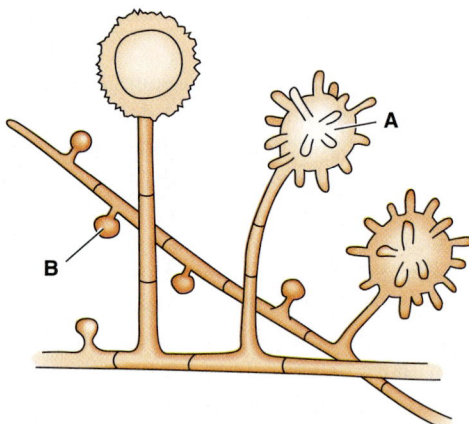

FIGURE 49–4 Asexual spores of *Histoplasma capsulatum.* **A:** Tuberculate macroconidia. **B:** Microconidia. (Reproduced with permission from Brooks GF, Jawetz E: *Medical Microbiology,* 19th ed. New York, NY: McGraw Hill; 1991.)

soil during construction or exploration of bat-infested caves has resulted in a significant number of infected individuals.

In several tropical African countries, histoplasmosis is caused by *Histoplasma duboisii.* The clinical picture is different from that caused by *H. capsulatum.* A description of the differences between African histoplasmosis and that seen in the United States is beyond the scope of this book.

Pathogenesis & Clinical Findings

Inhaled spores are engulfed by **macrophages** and develop into yeast forms. In tissues, *H. capsulatum* occurs as an **oval budding yeast inside macrophages** (Figures 49–5 and 49–6). The yeasts survive within the phagolysosome of the macrophage by producing alkaline substances which raise the pH and thereby inactivate the degradative enzymes of the phagolysosome.

The organisms spread widely throughout the body, especially to the liver and spleen, but most infections remain asymptomatic, and the small granulomatous foci heal by calcification. Acute symptomatic infection typically manifests with fever, headache, chills, cough, and chest pain. With intense exposure (e.g., in a chicken house or bat-infested cave), pneumonia and cavitary lung lesions may become clinically manifest. Chronic histoplasmosis is characterized by fever, dyspnea, and productive cough in a patient with underlying lung disease. Apical infiltrates and cavities may be seen on chest radiographs.

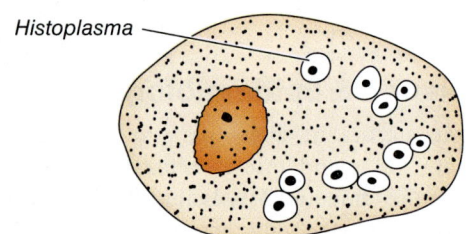

Histoplasma

FIGURE 49–5 *Histoplasma capsulatum.* Yeasts are located within the macrophage. (Reproduced with permission from Brooks GF, Jawetz E: *Medical Microbiology,* 19th ed. New York, NY: McGraw Hill; 1991.)

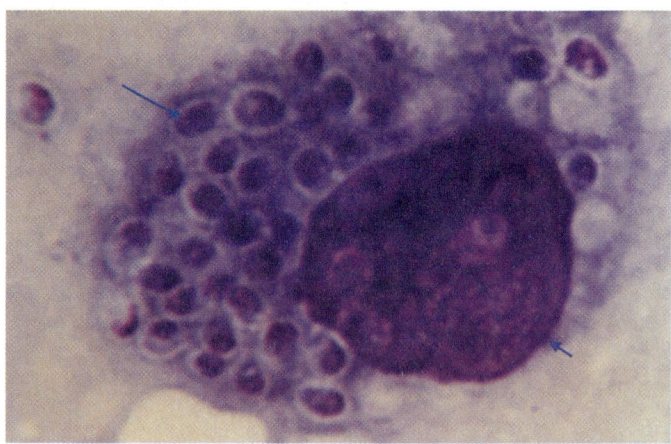

FIGURE 49–6 *Histoplasma capsulatum.* Yeasts within a macrophage. Long blue arrow points to one of many yeasts within the cytoplasm of a macrophage. Short blue arrow points to nucleus of the macrophage. Giemsa stain. (Used with permission from Dr. McClellan JT, Public Health Image Library, Centers for Disease Control and Prevention.)

These clinical features resemble tuberculosis and the two must be distinguished because the treatment is different.

Severe disseminated histoplasmosis develops in a small minority of infected persons, especially infants and individuals with reduced cell-mediated immunity, such as patients with AIDS. In AIDS patients, pancytopenia and ulcerated lesions on the tongue are typical of disseminated histoplasmosis. In immunocompetent people, EN can occur (see description of EN in earlier section on *Coccidioides*). EN is a sign that cell-mediated immunity is active and the organism will probably be contained.

A skin test using histoplasmin (a mycelial extract) becomes positive (i.e., shows at least 5 mm of induration) within 2 to 3 weeks after infection and remains positive for many years. However, because there are many false-positive reactions (due to cross-reactivity) and many false-negative reactions (in disseminated disease), the skin test is not useful for diagnosis. Furthermore, the skin test can stimulate an antibody response and confuse the serologic tests. The skin test is useful for epidemiologic studies, and up to 90% of individuals have positive results in areas of endemic infection.

Laboratory Diagnosis

In tissue biopsy specimens or bone marrow aspirates, oval **yeast cells within macrophages** are seen microscopically (see Figure 49–6). Cultures on Sabouraud's agar show hyphae with tuberculate macroconidia when grown at low temperature (e.g., 25°C) and yeasts when grown at 37°C. Tests that detect a *Histoplasma* polysaccharide antigen by ELISA and *Histoplasma* RNA with DNA probes are also useful. In immunocompromised patients with disseminated disease, tests for *Histoplasma* antigen in the urine are especially useful because antibody tests may be negative.

Two serologic tests are useful for diagnosis: CF and immunodiffusion (ID). An antibody titer of 1:32 in the CF test with yeast phase antigens is considered to be diagnostic. However, cross-reactions with other fungi, especially, *Blastomyces*, occur. CF titers fall when the disease becomes inactive and rise in disseminated disease. The ID test detects precipitating antibodies (precipitins) by forming two bands, M and H, in an agar-gel diffusion assay. The ID test is more specific but less sensitive than the CF test.

Treatment & Prevention

No therapy is needed in asymptomatic or mild primary infections. With progressive lung lesions, oral itraconazole is effective. In disseminated disease, parenteral itraconazole (or amphotericin B) is the treatment of choice. Liposomal amphotericin B should be used in patients with preexisting kidney damage. In meningitis, fluconazole is often used because it penetrates the spinal fluid well. Oral itraconazole is used for chronic suppression in patients with AIDS. There are no means of prevention except avoiding exposure in areas of endemic infection.

BLASTOMYCES

Disease

Blastomyces dermatitidis causes blastomycosis, also known as North American blastomycosis.

Properties

B. dermatitidis is a **dimorphic** fungus that exists as a mold in soil and as a yeast in tissue. The yeast is round with a doubly refractive wall and a single **broad-based bud** (Figures 49–7 and 49–8). Note that this organism forms a broad-based bud, whereas *Cryptococcus neoformans* is a yeast that forms a narrow-based bud.

Transmission & Epidemiology

This fungus is **endemic** primarily in eastern North America, especially in the region bordering the Ohio, Mississippi, and St. Lawrence rivers, and the Great Lakes region. Less commonly, blastomycosis has also occurred in Central and South America,

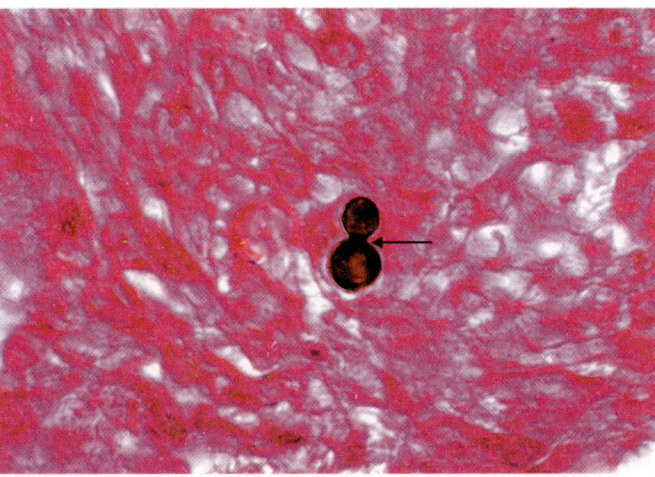

FIGURE 49–8 *Blastomyces dermatitidis.* Broad-based budding yeast. Arrow points to the broad base of the budding yeast. (Used with permission from Dr. Ajello L, Public Health Image Library, Centers for Disease Control and Prevention.)

Africa, and the Middle East. It grows in moist soil rich in organic material, forming hyphae with small pear-shaped conidia. Inhalation of the conidia causes human infection.

Pathogenesis & Clinical Findings

Infection occurs mainly via the respiratory tract. Asymptomatic or mild cases are rarely recognized. Dissemination may result in ulcerated granulomas of skin, bone, or other sites.

Laboratory Diagnosis

In tissue biopsy specimens, thick-walled yeast cells with single broad-based buds are seen microscopically (see Figure 49–8). Hyphae with small pear-shaped conidia are visible on culture. The skin test lacks specificity and has little value. Serologic tests have little value. A PCR assay that detects nucleic acids of *Blastomyces* is available.

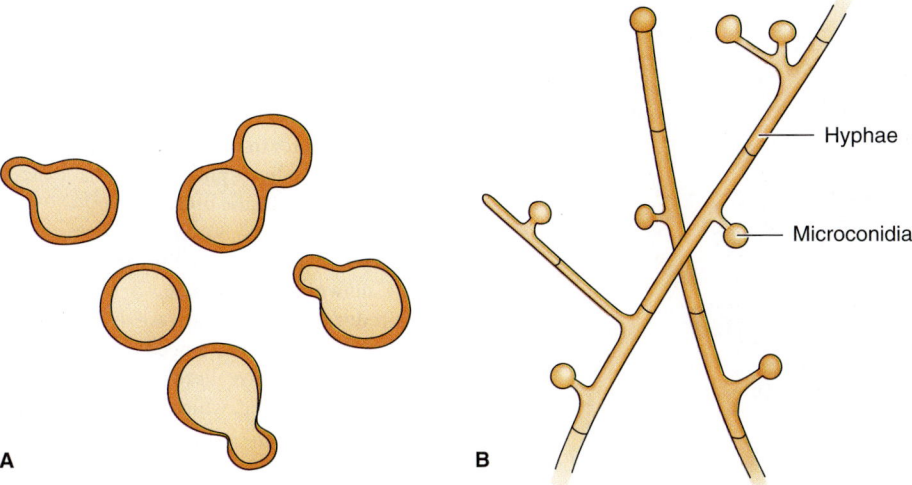

A

B

FIGURE 49–7 *Blastomyces dermatitidis.* **A:** Yeast with a broad-based bud at 37°C. **B:** Mold with microconidia at 20°C. (Reproduced with permission from Brooks GF, Jawetz E: *Medical Microbiology,* 19th ed. New York, NY: McGraw Hill; 1991.)

FIGURE 49–9 *Paracoccidioides brasiliensis.* Note the multiple buds of the yeast form of *Paracoccidioides*, in contrast to the single bud of *Blastomyces*.

Treatment & Prevention

Itraconazole is the drug of choice for most patients, but amphotericin B should be used to treat severe disease. Surgical excision may be helpful. There are no means of prevention.

PARACOCCIDIOIDES

Disease

Paracoccidioides brasiliensis causes paracoccidioidomycosis, also known as South American blastomycosis.

Properties

P. brasiliensis is a **dimorphic** fungus that exists as a mold in soil and as a yeast in tissue. The yeast is thick-walled with **multiple buds**, in contrast to *B. dermatitidis*, which has a single bud (Figures 49–9 and 49–10).

Transmission & Epidemiology

This fungus grows in the soil and is endemic in rural Latin America. Disease occurs only in that region.

Pathogenesis & Clinical Findings

The spores are **inhaled**, and early lesions occur in the lungs. Asymptomatic infection is common. Alternatively, oral mucous membrane lesions and lymph node enlargement can occur. Disseminated disease in many organs may develop.

FIGURE 49–10 *Paracoccidioides.* Yeasts with multiple buds resembling a "ship captain's wheel." Methenamine silver stain. (Used with permission from Dr. Lucille Georg, Public Health Image Library, Centers for Disease Control and Prevention.)

Laboratory Diagnosis

In pus or tissues, yeast cells with multiple buds resembling a "ship captain's wheel" are seen microscopically. A specimen cultured for 2 to 4 weeks may grow typical organisms. Skin tests are rarely helpful. Serologic testing shows that when significant antibody titers (by ID or CF) are found, active disease is present.

Treatment & Prevention

The drug of choice is itraconazole taken orally for several months. There are no means of prevention.

SELF-ASSESSMENT QUESTIONS

1. Regarding coccidioidomycosis and *Coccidioides immitis*, which one of the following is most accurate?
   (A) *C. immitis* is a mold in the soil and a yeast in the body.
   (B) The diagnosis of acute coccidioidomycosis can be made by detecting IgM antibodies in the patient's serum.
   (C) Travelers to the Philippines are at high risk of acquiring the disease.
   (D) The nodules of erythema nodosum are a typical finding in disseminated coccidioidomycosis.
   (E) Infection typically occurs when arthrospores enter the skin (e.g., through a wound caused by a rose thorn).

2. Regarding histoplasmosis and *Histoplasma capsulatum*, which one of the following is most accurate?
   (A) In tissue biopsies, *H. capsulatum* is found as a yeast within macrophages.
   (B) The laboratory diagnosis is made by seeing germ tubes when incubated at 37°C.
   (C) Histoplasmosis occurs primarily in the tropical areas of Central and South America.
   (D) To prevent disease, people who live in endemic areas should receive the vaccine containing histoplasmin.
   (E) Most infections are acquired by ingesting food accidentally contaminated with fungal spores from the soil.

3. Regarding *Blastomyces dermatitidis*, which one of the following is most accurate?
   (A) It forms a mycelium in culture at 37°C in the clinical lab.
   (B) Humoral immunity is the main host defense against this organism.
   (C) It causes a dermatophytid ("id") reaction when it disseminates to the skin.
   (D) The most important virulence factor of this organism is endotoxin in its cell wall.
   (E) It is a dimorphic fungus that exists as a mold in the soil and a yeast in the body.

4. Your patient is a 30-year-old woman who is in her third trimester of pregnancy, is of Filipino origin, and lives in the Central Valley of California. She complains of severe low back pain of several weeks in duration. An X-ray reveals a lesion in the fourth lumbar vertebra. Material from a needle biopsy of the lesion is examined by a pathologist who calls to tell you the patient has coccidioidomycosis. Of the following, which one did the pathologist see in the biopsy?
   (A) Nonseptate hyphae
   (B) Septate hyphae
   (C) Spherules containing endospores
   (D) Yeasts with a single bud
   (E) Yeasts with multiple buds

5. Your patient is a 30-year-old man who is human immunodeficiency virus (HIV) antibody positive with a CD4 count of 100. He has an ulcerated lesion on his tongue, and biopsy of the lesion reveals yeasts within macrophages. A diagnosis of disseminated histoplasmosis is made. Which one of the following is the best choice of drug to treat his disseminated histoplasmosis?

 (A) Amphotericin B
 (B) Caspofungin
 (C) Clotrimazole
 (D) Flucytosine
 (E) Terbinafine

ANSWERS

(1) **(B)**
(2) **(A)**
(3) **(E)**
(4) **(C)**
(5) **(A)**

SUMMARIES OF ORGANISMS

Brief summaries of the organisms described in this chapter begin on page 703. Please consult these summaries for a rapid review of the essential material.

PRACTICE QUESTIONS: USMLE & COURSE EXAMINATIONS

Questions on the topics discussed in this chapter can be found in the Mycology section of Part XIII: USMLE (National Board) Practice Questions starting on page 752. Also see Part XIV: USMLE (National Board) Practice Examination starting on page 775.

Opportunistic Mycoses

INTRODUCTION

Opportunistic fungi fail to induce disease in most immunocompetent persons but can do so in those with **impaired host defenses**. There are five genera of medically important fungi: *Candida, Cryptococcus, Aspergillus, Mucor,* and *Rhizopus*. Important features of the opportunistic fungal diseases are described in Table 50–1.

Additional information regarding the clinical aspects of infections caused by the fungi in this chapter is provided in Part IX, entitled Infectious Diseases beginning on page 603.

CANDIDA

Diseases

Candida albicans, the most important species of *Candida*, causes thrush, vaginitis, esophagitis, diaper rash, and chronic mucocutaneous candidiasis (CMC). It also causes disseminated infections such as right-sided endocarditis (especially in intravenous drug users), bloodstream infections (candidemia), and endophthalmitis. Infections related to indwelling intravenous, central lines, and urinary catheters are also important.

TABLE 50–1 Important Features of Opportunistic Fungal Diseases

Genus	Form in Tissue Seen by Microscopy	Geographic Location	Important Clinical Findings	Laboratory Diagnosis
Candida	Yeast forms pseudohyphae (also hyphae)	Worldwide	Thrush in mouth and vagina; endocarditis in intravenous drug users	Gram-positive; culture grows yeast colonies; *Candida albicans* forms germ tubes; polymerase chain reaction (PCR) assay; MALDI-TOF assay
Cryptococcus	Yeast with large capsule	Worldwide	Meningitis	India ink stain shows yeast with large capsule; culture grows very mucoid colonies; PCR assay
Aspergillus	Mold with septate hyphae	Worldwide	Fungus ball in lung; wound and burn infections; indwelling catheter infections; sinusitis	Culture grows mold with green spores; conidia in radiating chains
Mucor and Rhizopus	Mold with nonseptate hyphae	Worldwide	Necrotic lesion formed when mold invades blood vessels; predisposing factors are diabetic ketoacidosis, renal acidosis, and cancer	Culture grows mold with black spores; conidia enclosed in a sac called a sporangium

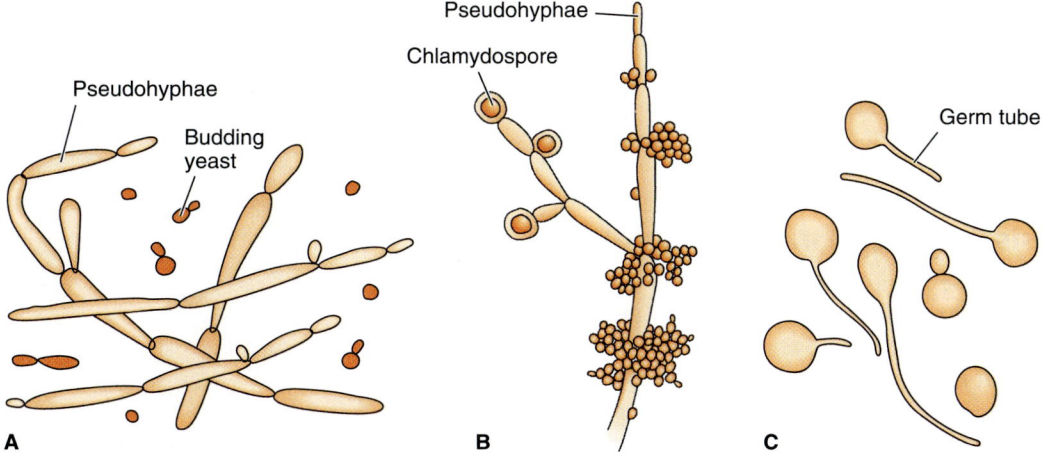

FIGURE 50-1 *Candida albicans.* **A:** Budding yeasts and pseudohyphae in tissues or exudate. **B:** Pseudohyphae and chlamydospores in culture at 20°C. **C:** Germ tubes at 37°C. (Reproduced with permission from Brooks GF, Butel JS, Ornston LN: *Jawetz, Melnick & Adelberg's Medical Microbiology,* 20th ed. New York, NY: McGraw Hill; 1995.)

Candida glabrata is the second most common cause of disseminated candidal infections and is more drug resistant than *C. albicans. Candida auris* causes serious bloodstream infections and is highly antibiotic resistant.

Properties

C. albicans is an **oval yeast with a single bud** (Figures 50–1 and 50–2). It is part of the **normal flora** of mucous membranes of the upper respiratory, gastrointestinal, and female genital tracts. In tissues, it appears most often as yeasts or as **pseudohyphae** (Figures 50–1 and 50–3). Pseudohyphae are elongated yeasts that visually resemble hyphae but are not true hyphae. True hyphae are also formed when *C. albicans* invades tissues. *C. albicans* forms germ tubes whereas most other candidal species do not.

Carbohydrate fermentation reactions can be used to differentiate it from other species (e.g., *Candida tropicalis, Candida parapsilosis, Candida krusei,* and *C. glabrata*) that cause human infections.

Candida dubliniensis is closely related to *C. albicans.* It also causes opportunistic infections in immunocompromised patients, especially acquired immunodeficiency syndrome (AIDS) patients. Both species form chlamydospores, but *C. albicans* grows at 42°C, whereas *C. dubliniensis* does not.

C. auris does not form germ tubes and is very difficult to distinguish from other candidal species by routine laboratory tests.

Transmission

As a member of the normal flora, *C. albicans* is already present on the skin and mucous membranes. In addition to the skin, *C. albicans* is found throughout the gastrointestinal tract (especially the mouth and esophagus) and in the vagina.

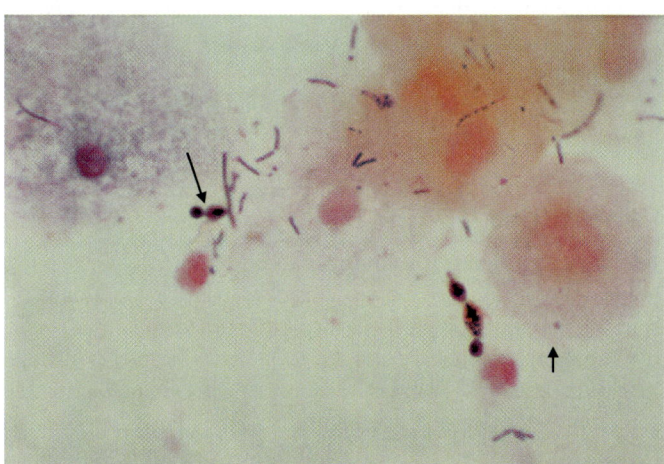

FIGURE 50-2 *Candida albicans.* Yeast. Long arrow points to a budding yeast. Short arrow points to the outer membrane of a vaginal epithelial cell. In this Gram-stained specimen, various bacteria that are part of the normal flora of the vagina can be seen. (Used with permission from Dr. Brown S, Public Health Image Library, Centers for Disease Control and Prevention.)

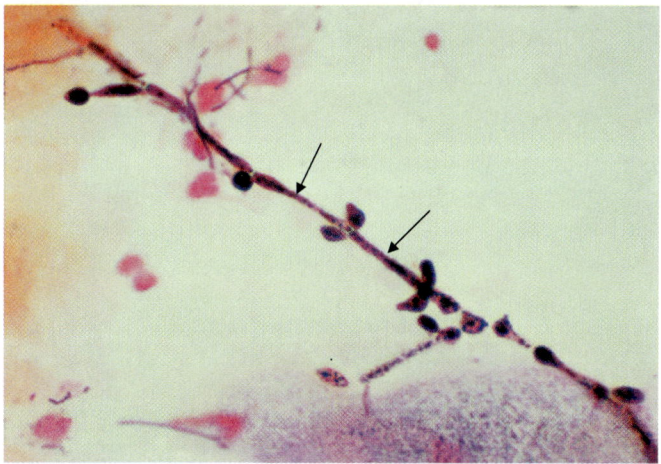

FIGURE 50-3 *Candida albicans.* Pseudohyphae. Two arrows point to pseudohyphae of *Candida albicans.* (Used with permission from Dr. Brown S, Public Health Image Library, Centers for Disease Control and Prevention.)

Thrush in the newborn is the result of passage through a birth canal heavily colonized by the organism. The presence of *C. albicans* on the skin predisposes to infections involving instruments that penetrate the skin, such as needles (intravenous drug use) and indwelling catheters. It is often found in the urine of patients with indwelling urinary (Foley) catheters.

Pathogenesis & Clinical Findings

The first line of defense against *Candida* infections is **intact skin and mucous membranes**. The second line is cell-mediated immunity, especially **Th-1 cells** producing gamma-interferon that activates efficient killing by macrophages. **Neutrophils** are also important as evidenced by the finding that neutropenia predisposes to disseminated *Candida* infections.

When local or systemic host defenses are impaired, disease may result. Overgrowth of *C. albicans* in the mouth produces white patches called **thrush** (Figure 50–4). (Note that thrush is a *pseudomembrane*, a term that is defined in Chapter 7). Vaginitis with itching and discharge is favored by high pH, diabetes, or **use of antibiotics**. Antibiotics suppress the normal flora *Lactobacillus*, which keep the pH low. As a result, the pH rises, which favors the growth of *Candida*.

Skin invasion occurs in warm, moist areas, which become red and weeping. Fingers and nails become involved when repeatedly immersed in water; persons employed as dishwashers in restaurants are commonly affected. Thickening or loss of the nail can occur. Diaper rash in infants occurs when wet diapers are not changed promptly (Figure 50–5).

In immunosuppressed individuals, *Candida* may disseminate to many organs or cause CMC. CMC is a prolonged infection of the skin, oral and genital mucosa, and nails that occurs in individuals deficient in T-cell immunity. Patients with mutations in the gene encoding interleukin-17 (IL-17) and the receptor for IL-17 are predisposed to CMC. After organ transplantation,

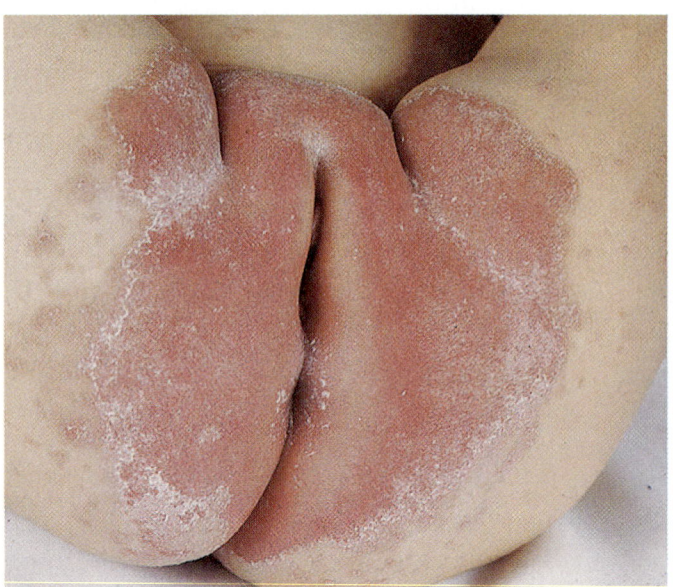

FIGURE 50–5 *Candida albicans*. Diaper rash. Note extensive area of inflammation in perineal region. (Reproduced with permission from Wolff K, Johnson R: *Fitzpatrick's Color Atlas & Synopsis of Clinical Dermatology*, 6th ed. New York, NY: McGraw Hill; 2009.)

patients receiving immunosuppressive drugs to prevent rejection are predisposed to invasive *Candida* infections.

Intravenous drug abuse, indwelling intravenous catheters, and hyperalimentation also predispose to disseminated candidiasis, especially right-sided endocarditis and endophthalmitis (infection within the eye). **Biofilm** formation on catheters or on prosthetic devices is an important predisposing factor to *Candida* central line infections.

Candida esophagitis, often accompanied by involvement of the stomach and small intestine, is seen in patients with leukemia and lymphoma. Subcutaneous nodules are often seen in neutropenic patients with disseminated disease. *C. albicans* is the most common species to cause disseminated disease in these patients, but *C. tropicalis*, *C. auris*, and *C. parapsilosis* are important pathogens also. *C auris* often causes disease in an intensive care setting.

Laboratory Diagnosis

In exudates or tissues, budding yeasts and pseudohyphae appear gram-positive and can be visualized by using calcofluor-white staining. In culture, typical yeast colonies are formed that resemble large staphylococcal colonies. *C. albicans* forms **germ tubes** in serum at 37°C, whereas most other species of pathogenic *Candida* species do not (see Figure 50–1). **Chlamydospores** are typically formed by *C. albicans* but not by most other species of *Candida*. Note that *C. dubliniensis* also forms chlamydospores but will not grow at 42°C, whereas *C. albicans* will. Serologic testing is rarely helpful.

Molecular methods are also useful for the diagnosis of *Candida* infections. Two current methods are (1) polymerase chain reaction (PCR)-based assays that detect the DNA encoding

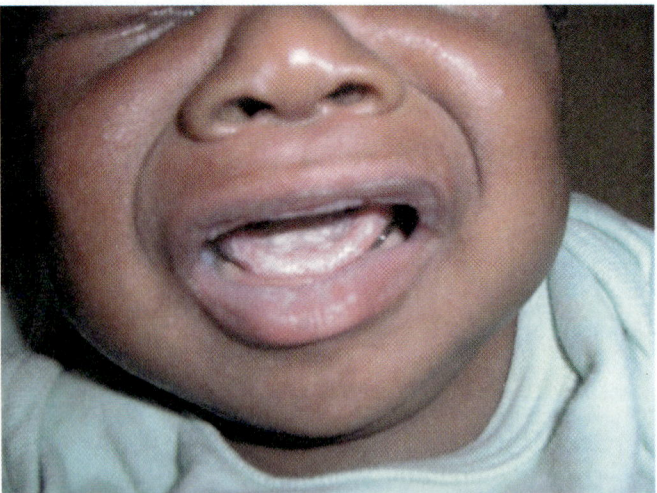

FIGURE 50–4 *Candida albicans*. Thrush in mouth. Note whitish plaques on tongue. (Reproduced with permission from Usatine RP, Smith MA, Mayeaux EJ Jr, et al: *The Color Atlas of Family Medicine*. New York, NY: McGraw Hill; 2009. Photo contributor: Richard P. Usatine, MD.)

the ribosomal RNA of *Candida* species and (2) mass spectrometry (matrix-assisted laser desorption ionization-time of flight [MALDI-TOF]) assays that detect the proteins of *Candida* species. *C. auris* is often difficult to distinguish from other candidal species, so PCR and genome sequencing techniques are being used.

Treatment & Prevention

The drug of choice for most candidal infections is fluconazole, including oropharyngeal or esophageal thrush. Itraconazole and voriconazole are also effective. An echinocandin, such as caspofungin or micafungin, can also be used for esophageal candidiasis.

Treatment of skin infections consists of topical antifungal drugs (e.g., clotrimazole or nystatin). *Candida* vaginitis is treated either with topical (intravaginal) azole drugs, such as clotrimazole or miconazole, or with oral fluconazole. CMC can be controlled by fluconazole or itraconazole. Treatment of disseminated candidiasis consists of either fluconazole or an echinocandin such as caspofungin.

Strains of *C. albicans* resistant to azole drugs have emerged in patients with AIDS receiving long-term prophylaxis with fluconazole. Most isolates of *C. glabrata* are resistant to fluconazole and voriconazole. An echinocandin such as caspofungin or amphotericin B should be used. *C. auris* is often multidrug resistant, including resistance to fluconazole, other azoles, and amphotericin B. Most strains of *C. auris* are susceptible to echinocandins but some are resistant to that class of drugs as well. Rare stains are pan-resistant, that is, they are resistant to all antifungal drugs.

Certain candidal infections (e.g., thrush) can be prevented by oral clotrimazole troches, buccal miconazole tablets, or nystatin "swish and swallow." Fluconazole is useful in preventing candidal infections in high-risk patients, such as those undergoing bone marrow transplantation and premature infants. Micafungin can also be used.

Treatment of candidal infections with antifungal drugs should be supplemented by reduction of predisposing factors. There is no vaccine against any *Candida* species.

CRYPTOCOCCUS

Disease

Cryptococcus neoformans causes cryptococcosis, especially cryptococcal meningitis. Cryptococcosis is the most common, life-threatening, invasive fungal disease worldwide. It is especially important in AIDS patients. Another species, *Cryptococcus gattii*, causes human disease less frequently than *C. neoformans*.

Properties

C. neoformans is an **oval, budding yeast** surrounded by a **wide polysaccharide capsule** (Figures 50–6 and 50–7). It is not dimorphic. Note that this organism forms a narrow-based bud, whereas the yeast form of *Blastomyces dermatitidis* forms a broad-based bud.

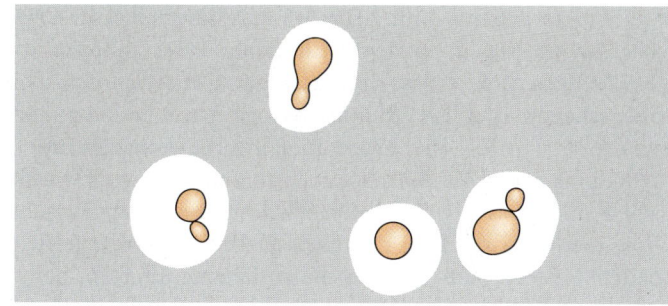

FIGURE 50–6 *Cryptococcus neoformans*. India ink preparation shows budding yeasts with a wide capsule. India ink forms a dark background; it does not stain the yeast itself. (Reproduced with permission from Brooks GF, Butel JS, Ornston LN: *Jawetz, Melnick & Adelberg's Medical Microbiology*, 20th ed. New York, NY: McGraw Hill; 1995.)

Transmission

C. neoformans occurs widely in nature and grows abundantly in **soil containing bird (especially pigeon) droppings**. The birds are not infected. Human infection results from **inhalation** of the organism. There is no human-to-human transmission. *C. gattii* is associated with eucalyptus trees, most often in the northwestern states of the United States. It is also found in subtropical and tropical areas of many countries.

Pathogenesis & Clinical Findings

Lung infection is often asymptomatic or may produce pneumonia. Disease caused by *C. neoformans* occurs mainly in patients with reduced cell-mediated immunity, especially AIDS patients, in whom the organism disseminates to the central nervous system (meningitis) and other organs. Subcutaneous nodules are often seen in disseminated disease. Note, however, that roughly half the patients with cryptococcal meningitis fail to show evidence of immunosuppression.

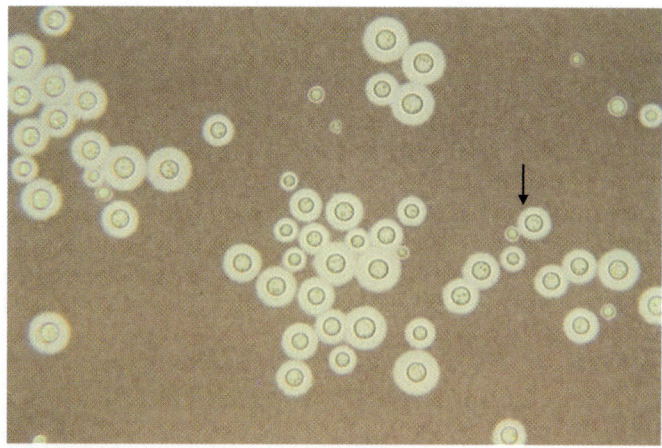

FIGURE 50–7 *Cryptococcus neoformans*. India ink preparation. Arrow points to a budding yeast of *C. neoformans*. Note the thick, translucent polysaccharide capsule outlined by the dark India ink particles. (Used with permission from Dr. Haley L, Public Health Image Library, Centers for Disease Control and Prevention.)

In some patients with AIDS who are infected with *Cryptococcus*, treating the patient with highly active antiretroviral therapy (HAART) causes an exacerbation of symptoms. This phenomenon is called immune reconstitution inflammatory syndrome (IRIS). The explanation of the exacerbation of symptoms is that HAART increases the number of CD4 cells, which increases the inflammatory response. Some patients have died as a result of cryptococcal IRIS. To prevent IRIS, patients should be treated for the underlying infection before starting HAART.

C. gattii causes human disease less frequently but is more capable of causing disease in an immunocompetent person than *C. neoformans*. *C. gattii* is more likely to cause cryptococcomas (granulomas), especially in the brain, than *C. neoformans*.

Laboratory Diagnosis

In spinal fluid mixed with **India ink**, the yeast cell is seen microscopically surrounded by a wide, unstained capsule. Appearance of the organism in Gram stain is unreliable, but stains such as **periodic acid–Schiff (PAS stain), methenamine silver**, and **mucicarmine** will allow the organism to be visualized (Figure 50–8). The organism can be cultured from spinal fluid and other specimens. The colonies are highly mucoid—a reflection of the large amount of capsular polysaccharide produced by the organism.

Serologic tests can be done for both antibody and antigen. In infected spinal fluid, **capsular antigen** occurs in high titer and can be detected by the **latex particle agglutination test**. This test is called the cryptococcal antigen test, often abbreviated as "CRAG." PCR-based assays that detect the ribosomal DNA of *Cryptococcus* are also useful.

Distinguishing between *C. neoformans* and *C. gattii* in the laboratory requires specialized media not generally available, so many *C. gattii* infections may go undiagnosed.

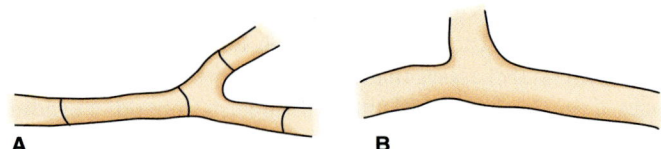

FIGURE 50–9 *Aspergillus* and *Mucor* in tissue. **A:** *Aspergillus* has septate hyphae with V-shaped branching. **B:** *Mucor* has nonseptate hyphae with right-angle branching.

Treatment & Prevention

Combined treatment with amphotericin B and flucytosine is used in meningitis and other disseminated disease. Liposomal amphotericin B should be used in patients with preexisting kidney damage. There are no specific means of prevention. Fluconazole is used in AIDS patients for long-term suppression of cryptococcal meningitis. *C. gattii* is less responsive to antifungal drugs than is *C. neoformans*.

ASPERGILLUS

Disease

Aspergillus species, especially *Aspergillus fumigatus*, cause infections of the skin, eyes, ears, and other organs; "fungus ball" in the lungs; and allergic bronchopulmonary aspergillosis (ABPA).

Properties

Aspergillus species exist **only as molds**; they are not dimorphic. They have **septate hyphae** that form V-shaped (dichotomous) branches (Figures 50–9 and 50–10). The walls are more or less parallel, in contrast to *Mucor* and *Rhizopus* walls, which are irregular (Figures 50–9 and 50–11). The conidia of *Aspergillus* form radiating chains, in contrast to those of *Mucor* and *Rhizopus*, which are enclosed within a sporangium (Figure 50–12).

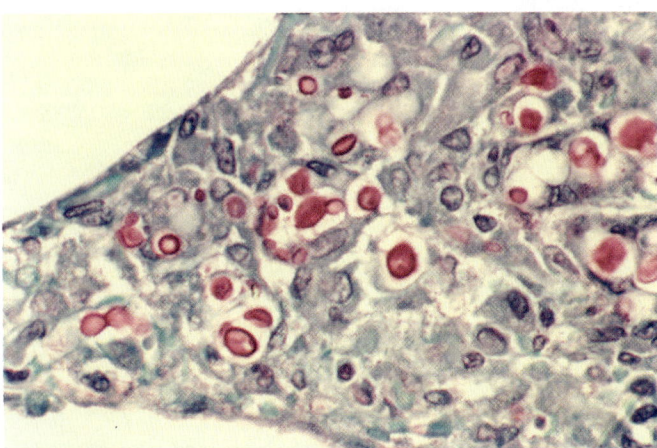

FIGURE 50–8 *Cryptococcus neoformans.* Mucicarmine stain. Note many red, oval yeasts of *C. neoformans* in lung tissue of patient with AIDS. (Used with permission from Dr. Edwin P. Ewing, Jr, Public Health Image Library, Centers for Disease Control and Prevention.)

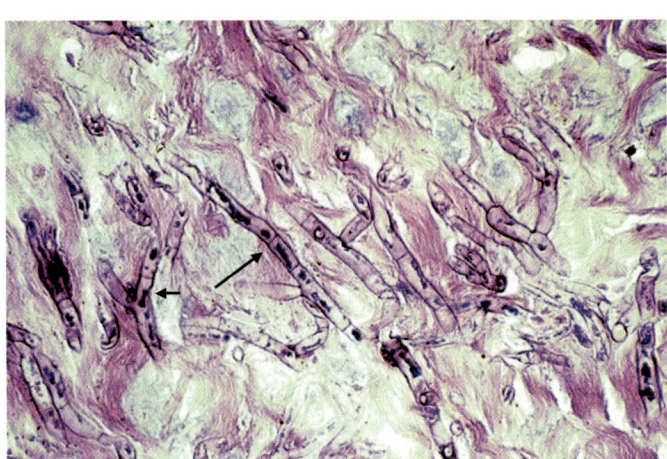

FIGURE 50–10 *Aspergillus fumigatus.* Septate hyphae. Long arrow points to the septate hyphae of *Aspergillus*. Note the straight parallel cell walls of this mold. Short arrow points to the typical low-angle, Y-shaped branching. (Used with permission from Prof. Henry Sanchez, University of California, San Francisco School of Medicine.)

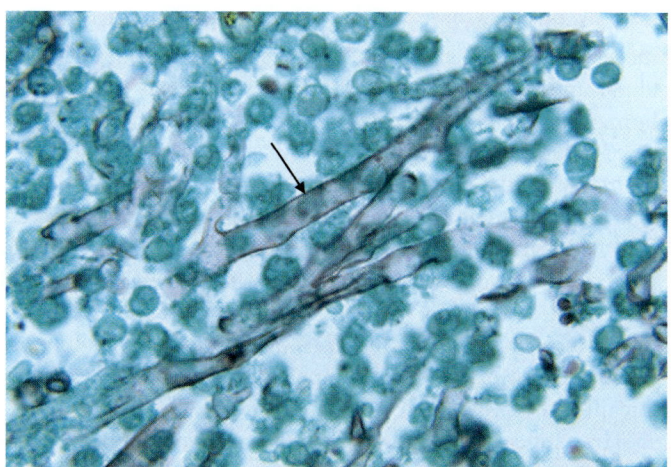

FIGURE 50–11 *Mucor* species. Nonseptate hyphae. Arrow points to irregular-shaped, nonseptate hyphae of *Mucor*. (Used with permission from Dr. Ajello L, Public Health Image Library, Centers for Disease Control and Prevention.)

Transmission

These molds are widely distributed in nature. They grow on decaying vegetation, producing chains of conidia. Transmission is by **airborne conidia**.

Pathogenesis & Clinical Findings

A. fumigatus can colonize and later invade abraded skin, wounds, burns, the cornea, the external ear, or paranasal sinuses. It is the most common cause of fungal sinusitis. In immunocompromised persons, especially those with neutropenia, it can invade the lungs, producing hemoptysis, and the brain, causing an abscess. Neutropenic patients are also predisposed to intravenous catheter infections caused by this organism.

Aspergilli are well-known for their ability to grow in cavities within the lungs, especially cavities caused by tuberculosis. Within the cavities, they produce an aspergilloma (**fungus ball**), which can be seen on chest X-ray as a radiopaque structure that changes its position when the patient is moved from an erect to a supine position.

ABPA is a hypersensitivity reaction to the presence of *Aspergillus* in the bronchi. Patients with ABPA have asthmatic symptoms and a high IgE titer against *Aspergillus* antigens, and they expectorate brownish bronchial plugs containing hyphae. Asthma caused by the inhalation of airborne conidia, especially in certain occupational settings, also occurs. *Aspergillus flavus* growing on cereals or nuts produces aflatoxins that may be carcinogenic or acutely toxic.

Laboratory Diagnosis

Biopsy specimens show **septate, branching hyphae** invading tissue (see Figure 50–10). Cultures show colonies with characteristic radiating chains of conidia (see Figure 50–12). However, positive cultures do not prove disease because colonization is common. In persons with invasive aspergillosis, there may be high titers of galactomannan antigen in serum. Patients with ABPA have high levels of IgE specific for *Aspergillus* antigens and prominent eosinophilia. IgG precipitins are also present.

Treatment & Prevention

Voriconazole is the drug of choice for invasive aspergillosis. Liposomal amphotericin B, posaconazole, caspofungin, and isavuconazonium are alternative drugs. A fungus ball growing in a sinus or in a pulmonary cavity can be surgically removed. Patients with ABPA can be treated with corticosteroids and antifungal agents, such as itraconazole. There are no specific means of prevention.

MUCOR & RHIZOPUS

Mucormycosis (zygomycosis, phycomycosis) is a disease caused by saprophytic **molds** (e.g., *Mucor*, *Rhizopus*, and *Absidia*) found widely in the environment. They are not dimorphic. These organisms are transmitted by airborne asexual spores and invade tissues of patients with reduced host defenses. They proliferate in the walls of blood vessels, particularly of the paranasal sinuses, lungs, or gut, and cause infarction and necrosis of tissue distal to the blocked vessel (Figure 50–13).

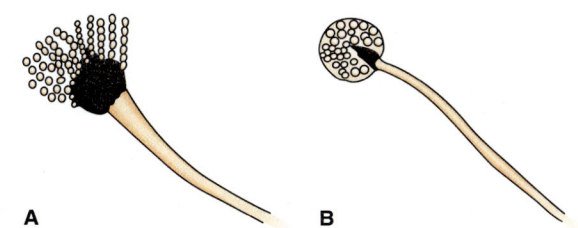

FIGURE 50–12 *Aspergillus* and *Mucor* in culture. **A:** *Aspergillus* spores form in radiating columns. **B:** *Mucor* spores are contained within a sporangium.

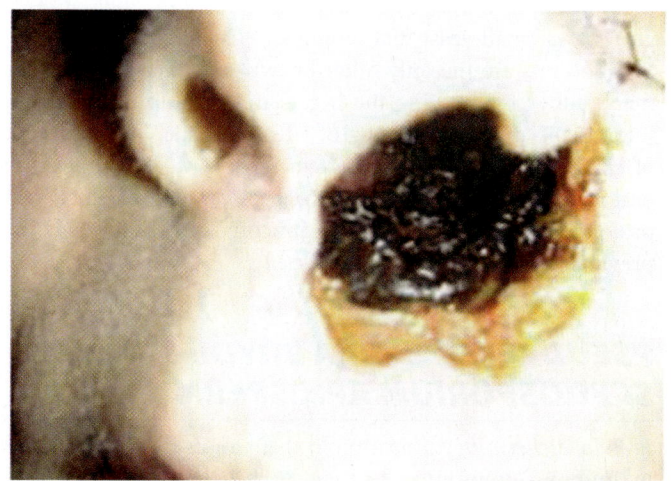

FIGURE 50–13 *Mucor* species. Mucormycosis. Note necrotic area involving the nose and face in a patient with acute lymphocytic leukemia. (Reproduced with permission from Lichtman MA, Shafer MS, Felgar RE, et al: *Lichtman's Atlas of Hematology*. New York, NY: McGraw Hill; 2007.)

Patients with **diabetic ketoacidosis**, burns, bone marrow transplants, or leukemia are particularly susceptible. Diabetic patients are particularly susceptible to **rhinocerebral mucormycosis**, in which mold spores in the sinuses germinate to form hyphae that invade blood vessels that supply the brain. One species, *Rhizopus oryzae*, causes about 60% of cases of mucormycosis.

In biopsy specimens, organisms are seen microscopically as **nonseptate hyphae** with broad, irregular walls and branches that form more or less at right angles (see Figures 50–9 and 50–11). Cultures show colonies with spores contained within a sporangium (see Figure 50–12). These organisms are difficult to culture because they are a single, very long cell, and damage to any part of the cell can limit its ability to grow.

If diagnosis is made early, treatment of the underlying disorder, plus administration of amphotericin B and surgical removal of necrotic infected tissue, has resulted in some remissions and cures. Liposomal amphotericin B should be used in patients with preexisting kidney damage. Posaconazole can also be used to treat mucormycosis. Posaconazole is also used for prophylaxis against *Mucor* infections in severely immunocompromised patients.

PNEUMOCYSTIS

Pneumocystis jiroveci is classified as a yeast on the basis of molecular analysis, but it has many characteristics of a protozoan. Some regard it as an "unclassified" organism. A summary of the important clinical information is presented here, and a more detailed description is presented in Chapter 52 with the blood and tissue protozoa. In 2002, taxonomists renamed the human species of *Pneumocystis* as *P. jiroveci* and recommended that *Pneumocystis carinii* be used only to describe the rat species of *Pneumocystis*.

Pneumocystis is acquired by inhalation of airborne organisms into the lungs. An inflammatory exudate composed primarily of plasma cells occurs, oxygen exchange is reduced, and dyspnea occurs. A reduced number of CD4-positive T lymphocytes, such as occurs in AIDS, predisposes to *Pneumocystis* pneumonia. Most immunocompetent people have asymptomatic infections.

The clinical findings of *Pneumocystis* pneumonia include fever, nonproductive cough, and dyspnea. Rales are heard bilaterally, and the chest X-ray shows a "ground-glass" pattern. The mortality rate of untreated *Pneumocystis* pneumonia is approximately 100%.

The diagnosis is typically made by finding the cysts of *Pneumocystis* in bronchial lavage specimens. Fluorescent antibody stains or tissue stains, such as methenamine silver or Giemsa, are used to identify the organism. PCR-based tests are also used. Serologic tests are not useful.

The drug of choice for *Pneumocystis* pneumonia is trimethoprim-sulfamethoxazole. Trimethoprim-sulfamethoxazole or aerosolized pentamidine can be used for prophylaxis in patients with CD4 counts below 200.

FUNGI OF MINOR IMPORTANCE

TALAROMYCES MARNEFFEI

Talaromyces marneffei is a dimorphic fungus that causes tuberculosis-like disease in AIDS patients, particularly in Southeast Asian countries such as Thailand. It grows as a mold that produces a rose-colored pigment at 25°C but at 37°C grows as a small yeast that resembles *Histoplasma capsulatum*. Bamboo rats are the only other known hosts. The diagnosis is made either by growing the organism in culture or by using fluorescent antibody staining of affected tissue. The treatment of choice consists of amphotericin B for 2 weeks followed by oral itraconazole for 10 weeks. Relapses can be prevented with prolonged administration of oral itraconazole. *T. marneffei* was previously named *Penicillium marneffei*.

PSEUDALLESCHERIA BOYDII & SCEDOSPORIUM APIOSPERMUM

Pseudallescheria boydii is a mold that causes disease primarily in immunocompromised patients. The clinical findings and the microscopic appearance of the septate hyphae in tissue closely resemble those of *Aspergillus*. In culture, the appearance of the conidia (pear-shaped) and the color of the mycelium (brownish-gray) of *P. boydii* are different from those of *Aspergillus*. The drug of choice is either ketoconazole or itraconazole because the response to amphotericin B is poor. Debridement of necrotic tissue is important as well.

Scedosporium apiospermum is the asexual form of *P. boydii*. *Scedosporium* primarily causes disease in immunocompromised patients but also causes mycetoma in immunocompetent individuals. In immunocompromised patients, *Scedosporium* causes angioinvasive disease, especially pneumonia and disseminated abscesses. A microbiologic diagnosis is made by seeing septate hyphae in tissue and growing colorless mold colonies on fungal media. *Scedosporium* is resistant to all currently used antifungal drugs. Mortality rates in immunocompromised patients with disseminated disease range from 85% to 100%.

FUSARIUM SOLANI

Fusarium solani is a mold that causes disease primarily in neutropenic patients. Fever and skin lesions are the most common clinical features. The organism is similar to *Aspergillus* in that it is a mold with septate hyphae that tends to invade blood vessels. Blood cultures are often positive in disseminated disease. In culture, banana-shaped conidia are seen. Liposomal amphoteric

B is the drug of choice. Indwelling catheters should be removed or replaced. In 2006, an outbreak of *Fusarium* keratitis (infection of the cornea) occurred in people who used a certain contact lens solution.

MICROSPORIDIA

Microsporidia are a group of opportunistic organisms that were formerly classified as protozoa but are now considered to be fungi. They are characterized by obligate intracellular replication and spore formation. As the name implies, the spores are quite small, approximately 1 to 3 mm, about the size of *Escherichia coli*. One unique feature of these spores is a "polar tube," which is coiled within the spore and extrudes to attach to the human cells upon infection. The protoplasm of the spore then enters the human cell via the polar tube.

Enterocytozoon bieneusi and *Encephalitozoon intestinalis* are two important microsporidial species that cause severe, persistent, watery diarrhea in AIDS patients. The organisms are transmitted from human to human by the fecal–oral route. Microsporidia are also implicated in infections of the central nervous system, the genitourinary tract, and the eye. It is uncertain whether an animal reservoir exists. Diagnosis is made by visualization of spores in stool samples or intestinal biopsy samples. The treatment of choice is albendazole.

SELF-ASSESSMENT QUESTIONS

1. Regarding *Candida albicans*, which one of the following is most accurate?
   (A) The diagnosis of disseminated candidiasis is typically made by detecting IgM antibodies.
   (B) It exists as a yeast on mucosal surfaces but forms pseudohyphae when it invades tissue.
   (C) Antibody-mediated immunity is a more important host defense than cell-mediated immunity.
   (D) A positive skin test can be used to confirm the diagnosis of skin infection caused by *C. albicans*.
   (E) In the clinical laboratory, it is diagnosed by isolating a mold with nonseptate hyphae when cultures are grown at room temperature.

2. Regarding *Cryptococcus neoformans*, which one of the following is most accurate?
   (A) It is a dimorphic fungus, growing as a mold in the soil and a yeast in the body.
   (B) It is acquired primarily by ingestion of food contaminated with pigeon guano.
   (C) Dark field microscopy is typically used to visualize the organism in spinal fluid.
   (D) Pathogenesis involves an exotoxin that acts as a superantigen recruiting lymphocytes into the spinal fluid.
   (E) Laboratory diagnosis of cryptococcal meningitis can be achieved by detecting the capsular polysaccharide of the organism in the spinal fluid.

3. Regarding *Aspergillus fumigatus* and aspergillosis, which one of the following is most accurate?
   (A) The natural habitat of *A. fumigatus* is the hair follicles of the human skin.
   (B) In the clinical laboratory, cultures of *A. fumigatus* incubated at 37°C form yeast colonies.
   (C) The India ink stain is typically used to visualize *A. fumigatus* in the clinical laboratory.
   (D) *A. fumigatus* causes "fungus balls" in patients with lung cavities caused by tuberculosis.
   (E) The main predisposing factor to allergic bronchopulmonary aspergillosis is neutropenia.

4. Regarding *Mucor* species, which one of the following is most accurate?
   (A) Infection is acquired by the ingestion of food contaminated by spores of the organism.
   (B) Diabetic ketoacidosis is a major predisposing factor for invasive mucormycosis.
   (C) *Mucor* species have septate hyphae in contrast to *Aspergillus* species, which have nonseptate hyphae.
   (D) In biopsy specimens obtained from patients with invasive disease, *Mucor* species appear as pseudohyphae.
   (E) Skin tests using mucoroidin as the immunogen are used to determine whether the patient has been infected with *Mucor* species.

5. Your patient is a 20-year-old woman who is human immunodeficiency virus (HIV) antibody positive with a CD4 count of 50. She has recovered from cryptococcal meningitis. Which one of the following is the best choice of drug to use as long-term prophylaxis to prevent another episode of cryptococcal meningitis?
   (A) Amphotericin B
   (B) Caspofungin
   (C) Fluconazole
   (D) Flucytosine
   (E) Terbinafine

6. Your patient is a 1-month-old infant with whitish lesions in the mouth that are diagnosed as oropharyngeal candidiasis (thrush). Which one of the following is the best choice of drug to treat this infection?
   (A) Amphotericin B
   (B) Caspofungin
   (C) Fluconazole
   (D) Flucytosine
   (E) Terbinafine

7. Your patient is a 50-year-old woman with leukemia who is neutropenic from her cancer chemotherapy. She now has disseminated aspergillosis that does not respond to amphotericin B. Which one of the following is the best choice of drug to treat this infection?
   (A) Amphotericin B
   (B) Caspofungin
   (C) Fluconazole
   (D) Flucytosine
   (E) Terbinafine

ANSWERS

(1) **(B)**
(2) **(E)**
(3) **(D)**
(4) **(B)**
(5) **(C)**
(6) **(C)**
(7) **(B)**

SUMMARIES OF ORGANISMS

Brief summaries of the organisms described in this chapter begin on page 703. Please consult these summaries for a rapid review of the essential material.

PRACTICE QUESTIONS: USMLE & COURSE EXAMINATIONS

Questions on the topics discussed in this chapter can be found in the Mycology section of Part XIII: USMLE (National Board) Practice Questions starting on page 752. Also see Part XIV: USMLE (National Board) Practice Examination starting on page 775.

PART VI   PARASITOLOGY

Parasites occur in two distinct forms: single-celled **protozoa** and multicellular metazoa called **helminths** or worms. For medical purposes, protozoa are classified according to their most important site of infection, namely, the **intestinal** protozoa such as *Giardia*, the **urogenital** protozoa such as *Trichomonas*, the **blood** protozoa such as *Plasmodium* (the cause of malaria), and **tissue** protozoa such as *Toxoplasma*. This book discusses the protozoa according to these categories. In some contexts, the protozoa are classified into four groups: Sarcodina (amebas), Sporozoa (sporozoans), Mastigophora (flagellates), and Ciliata (ciliates).

Metazoa are subdivided into two phyla: the Platyhelminthes (**flatworms**) and the Nemathelminthes (**roundworms**, nematodes). The phylum Platyhelminthes contains two medically important classes: Cestoda (**tapeworms**) and Trematoda (**flukes**). This classification is shown in Figure VI–1. Examples of medically important flatworms include *Taenia solium*, the tapeworm that causes cysticercosis, and *Schistosoma mansoni*, the fluke that causes schistosomiasis. Medically important roundworms (nematodes) include the pinworm (*Enterobius*), the hookworms (*Ancylostoma* and *Necator*), the threadworm (*Strongyloides*; the cause of strongyloidiasis), and *Trichinella* (the cause of trichinosis).

Understanding the life cycle and pathogenesis of protozoa and helminths requires an explanation of certain terms. Many protozoa have a life cycle consisting of a **trophozoite**, which is the motile, feeding, reproducing form surrounded by a flexible cell membrane, and a **cyst**, which is the nonmotile, nonmetabolizing, nonreproducing form surrounded by a thick wall. The cyst form survives well in the environment and so is often involved in transmission. Certain protozoa, such as *Leishmania* and *Trypanosoma*, have flagellated forms called **promastigotes or trypomastigotes** and nonflagellated forms called **amastigotes**.

Transmission of the intestinal protozoa typically occurs by ingestion of cysts, whereas transmission of the blood and tissue protozoa usually occurs via insect vectors such as the mosquito in the case of *Plasmodium* (malaria), the reduvid bug in the case

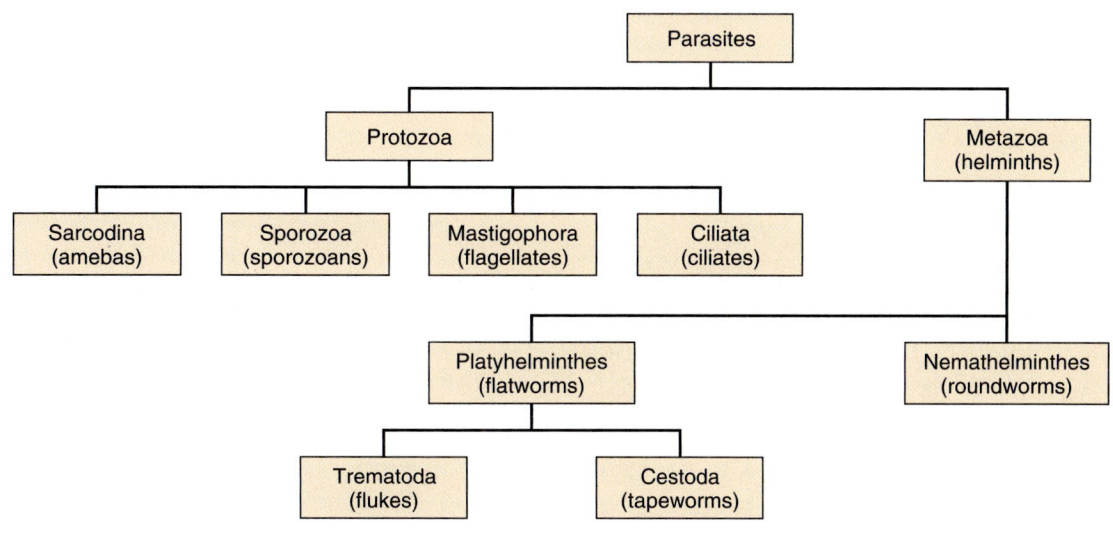

FIGURE VI–1 Relationships of the medically important parasites.

of *Trypanosoma cruzi* (Chagas' disease), the tsetse fly in the case of *Trypanosoma brucei* (sleeping sickness), and the sandfly in the case of *Leishmania donovani* (visceral leishmaniasis or kala-azar). The main exception to this is *Toxoplasma* which is a tissue protozoan that is transmitted primarily by ingestion of cysts in cat feces and across the placenta from mother to fetus.

Prevention of these diseases involves **interrupting the chain of transmission**, in particular via proper sewage disposal and water purification in the case of the intestinal protozoa and insect control in the case of the blood protozoa.

Many helminths have a life cycle that progresses from **egg to larva to adult**. The egg contains an embryo that, upon hatching, differentiates into a larval form, which then matures into the adult form that produces the eggs.

There are special terms applied to the host of certain parasites as they proceed through their life cycle. A **definitive host** is one in which the sexual cycle occurs or the adult is present, and the **intermediate host** is one in which the asexual cycle occurs or the larva is present. In some helminthic infections, humans are **dead-end hosts**—that is, the larval form in the human is not transmitted to other humans or animals. Humans are dead-end hosts for *Taenia solium* (cysticercosis), *Echinococcus* (hydatid cyst disease), and *Trichinella* (trichinosis).

Transmission occurs by three main modes: ingestion of eggs (*T. solium, Enterobius, Ascaris*), penetration of the skin by larvae (*Schistosoma*, hookworms, *Strongyloides*), or insect bite (*Wuchereria, Onchocerca, Dracunculus*). As with the protozoa, prevention of these diseases involves interrupting the chain of transmission, in particular via proper sewage disposal, water purification, insect control, and personal protection such as insect repellent and bed nets. In addition, avoiding bathing or swimming in certain freshwater sources (*Schistosoma*) and wearing shoes (hookworms and *Strongyloides*) prevent infection by these organisms.

Regarding the laboratory diagnosis of helminthic infections, examination of the stool for **ova and parasites (O&P)** is often done. The term **ova** refers to the eggs, and the term **parasites** refer to the larval or adult forms.

Eosinophilia is associated with several helminth infections, especially when roundworm larvae migrate through tissue. High eosinophil counts are seen in infections caused by the following roundworms: *Ascaris, Strongyloides, Trichinella, Toxocara,* and the hookworms, *Necator* and *Ancylostoma*. Infections with the flatworm (fluke) *Schistosoma* also elicit eosinophilia. Eosinophils are an important component of the host defense against these parasites. Immunoglobulin (Ig) E is also elevated in these infections. In addition, cell-mediated immunity (CMI) is important in some helminth infections. For example, in *Strongyloides* infection, reduced CMI (as a result of, for example, acquired immunodeficiency syndrome [AIDS] high-dose corticosteroids) may result in disseminated strongyloidiasis, a life-threatening complication.

Additional information regarding the clinical aspects of infections caused by the protozoa and helminths in chapters 51 to 56 is provided in Part IX, entitled Infectious Diseases beginning on page 603.

Intestinal & Urogenital—Protozoa

INTRODUCTION

In this book, the major protozoan pathogens are grouped according to the location in the body where they most frequently cause diseases. The intestinal and urogenital protozoa are described in this chapter, and the blood and tissue protozoa are described in Chapter 52.

(1) Within the intestinal tract, three organisms—the ameba, *Entamoeba histolytica*; the flagellate, *Giardia lamblia*; and the sporozoan, *Cryptosporidium hominis*—are the most important.

(2) In the urogenital tract, the flagellate *Trichomonas vaginalis* is the important pathogen.

(3) The blood and tissue protozoa are a varied group consisting of the flagellates *Trypanosoma* and *Leishmania* and the sporozoans *Plasmodium* and *Toxoplasma*. The important opportunistic lung pathogen *Pneumocystis* will be discussed in this group, although there is molecular evidence that it should be classified as a fungus.

The major and minor pathogenic protozoa are listed in Table 51–1.

Although immigrants and Americans returning from abroad can present to physicians in the United States with any parasitic disease, certain parasites are much more likely to occur outside the United States. The features of the medically important protozoa, including their occurrence in the United States, are described in Table 51–2.

The medically important stages in the life cycle of the intestinal protozoa are described in Table 51–3.

INTESTINAL PROTOZOA

ENTAMOEBA

Diseases

E. histolytica causes amebic dysentery and liver abscess.

Important Properties

The life cycle of *E. histolytica* is shown in Figure 51–1. The life cycle has two stages: the motile **ameba (trophozoite)** and the nonmotile **cyst** (Figures 51–2A and B, 51–3, and 51–4). The trophozoite is found within the intestinal and extraintestinal lesions and in diarrheal stools. The cyst predominates in nondiarrheal stools. These cysts are not highly resistant and are readily killed by boiling but not by chlorination of water supplies. They are removed by filtration of water.

The cyst has **four nuclei**, an important diagnostic criterion. Upon excystation in the intestinal tract, an ameba with four nuclei emerges and then divides to form eight trophozoites. The mature trophozoite has a single nucleus with an even lining of peripheral chromatin and a prominent central nucleolus (karyosome).

Antibodies are formed against trophozoite antigens in invasive amebiasis, but they are not protective; previous infection does not prevent reinfection. The antibodies are useful, however, for serologic diagnosis.

TABLE 51–1 Major and Minor Pathogenic Protozoa

Type and Location	Species	Disease
Major protozoa		
Intestinal tract	Entamoeba histolytica	Amebiasis
	Giardia lamblia	Giardiasis
	Cryptosporidium hominis	Cryptosporidiosis
Urogenital tract	Trichomonas vaginalis	Trichomoniasis
Blood and tissue	Plasmodium species	Malaria
	Toxoplasma gondii	Toxoplasmosis
	Pneumocystis jiroveci	Pneumonia
	Trypanosoma species	Trypanosomiasis
	T. cruzi	Chagas' disease
	T. gambiense[1]	Sleeping sickness
	T. rhodesiense[1]	Sleeping sickness
	Leishmania species	Leishmaniasis
	L. donovani	Kala-azar
	L. tropica	Cutaneous leishmaniasis[2]
	L. mexicana	Cutaneous leishmaniasis[2]
	L. braziliensis	Mucocutaneous leishmaniasis
Minor protozoa		
Intestinal tract	Balantidium coli	Dysentery
	Isospora belli	Isosporiasis
	Enterocytozoon bieneusi	Microsporidiosis
	Septata intestinalis	Microsporidiosis
	Cyclospora cayetanensis	Cyclosporiasis
Blood and tissue	Naegleria species	Meningitis
	Acanthamoeba species	Meningitis
	Babesia microti	Babesiosis

[1]Also known as *T. brucei gambiense* and *T. brucei rhodesiense, respectively.*

[2]*L. tropica* and *L. mexicana* cause Old World and New World cutaneous leishmaniasis, respectively.

Pathogenesis & Epidemiology

The organism is acquired by ingestion of cysts that are transmitted primarily by the **fecal–oral** route in contaminated food and water. Anal–oral transmission (e.g., among male homosexuals) also occurs. There is **no animal reservoir**. The ingested cysts differentiate into trophozoites in the ileum but tend to colonize the cecum and colon.

The trophozoites invade the colonic epithelium and secrete enzymes that cause localized necrosis. Little inflammation occurs at the site. As the lesion reaches the muscularis layer, a typical **flask-shaped ulcer** forms that can undermine and destroy large areas of the intestinal epithelium (Figure 51–5). Progression into the submucosa leads to invasion of the portal circulation by the trophozoites. By far, the most frequent site of systemic disease is the **liver**, where abscesses containing trophozoites form.

Infection by *E. histolytica* is found worldwide but occurs most frequently in tropical countries, especially in areas with poor sanitation. About 1% to 2% of people in the United States are affected. Infection is common in men who have sex with men.

Clinical Findings

Acute intestinal amebiasis presents as **dysentery** (i.e., bloody, mucus-containing diarrhea) accompanied by lower abdominal discomfort, flatulence, and tenesmus. Chronic amebiasis with low-grade symptoms such as occasional diarrhea, weight loss, and fatigue also occurs. Roughly 90% of those infected have asymptomatic infections, but they may be carriers, whose feces contain cysts that can be transmitted to others. In some patients, a granulomatous lesion called an **ameboma** may form in the cecal or rectosigmoid areas of the colon. These lesions can resemble an adenocarcinoma of the colon and must be distinguished from them.

Amebic abscess of the liver is characterized by right-upper-quadrant pain, weight loss, fever, and a tender, enlarged liver. Right-lobe abscesses can penetrate the diaphragm and cause lung disease. Most cases of amebic liver abscess occur in patients who have not had overt intestinal amebiasis. Aspiration of the liver abscess yields brownish-yellow pus with the appearance and consistency of **anchovy paste**.

Laboratory Diagnosis

Diagnosis of intestinal amebiasis rests on finding either trophozoites in diarrheal stools or cysts in formed stools (see Figures 51–3 and 51–4). Diarrheal stools should be examined within 1 hour of collection to see the ameboid motility of the trophozoite. Trophozoites characteristically contain ingested red blood cells. The most common error is to mistake fecal leukocytes for trophozoites. Because cysts are passed intermittently, at least three specimens should be examined. The O&P test is insensitive, and false negatives commonly occur. Also, about half of the patients with extraintestinal amebiasis have negative stool examinations.

E. histolytica can be distinguished from other amebas by two major criteria: (1) The first is the nature of the **nucleus** of the trophozoite. The *E. histolytica* nucleus has a small central nucleolus and fine chromatin granules along the border of the nuclear membrane. The nuclei of other amebas are quite different. (2) The second is **cyst size and number of its nuclei**. Mature cysts of *E. histolytica* are smaller than those of *Entamoeba coli* and contain four nuclei, whereas *E. coli* cysts have eight nuclei.

The trophozoites of *Entamoeba dispar*, a nonpathogenic species of *Entamoeba*, are morphologically indistinguishable from those of *E. histolytica*; therefore, a person who has trophozoites in the stool is only treated if symptoms warrant it. Two tests are highly specific for *E. histolytica* in the stool: one detects *E. histolytica* antigen, and the other detects nucleic acids of the organism in a polymerase chain reaction (PCR)-based assay.

A complete examination for cysts includes a wet mount in saline, an iodine-stained wet mount, and a fixed,

TABLE 51–2 Features of Medically Important Protozoa

Organism	Mode of Transmission	Occurrence in the United States	Diagnosis	Treatment
I. Intestinal and urogenital protozoa				
Entamoeba	Ingestion of cysts in food	Yes	Trophozoites or cysts in stool; serology	Metronidazole or tinidazole
Giardia	Ingestion of cysts in food	Yes	Trophozoites or cysts in stools	Metronidazole or tinidazole
Cryptosporidium	Ingestion of cysts in food	Yes	Cysts on acid-fast stain	Nitazoxanide in immunocompetent; paromomycin may be useful in AIDS patient
Trichomonas	Sexual	Yes	Trophozoites in wet mount	Metronidazole or tinidazole
II. Blood and tissue protozoa				
Trypanosoma				
T. cruzi	Reduviid bug	Rare	Blood smear, bone marrow, xenodiagnosis	Nifurtimox
T. gambiense, T. rhodesiense	Tsetse fly	No	Blood smear	Suramin[1]
Leishmania				
L. donovani	Sandfly	No	Bone marrow, spleen, or lymph node	Stibogluconate
L. tropica, L. mexicana, L. braziliensis	Sandfly	No	Fluid from lesion	Stibogluconate
Plasmodium				
P. vivax, P. ovale, P. malariae	*Anopheles* mosquito	Rare	Blood smear	Chloroquine if sensitive; also primaquine for *P. vivax* and *P. ovale*
P. falciparum	*Anopheles* mosquito	No	Blood smear	If uncomplicated, use Coartem or Malarone. If complicated, use artesunate.
Toxoplasma	Ingestion of cysts in raw meat; contact with cat feces	Yes	Serology; microscopic examination of tissue; mouse inoculation	Sulfadiazine and pyrimethamine for congenital disease and immunocompromised patients
Pneumocystis	Inhalation	Yes	Lung biopsy or lavage	Trimethoprim-sulfamethoxazole; also pentamidine or atovaquone

[1]Melarsoprol is used if the central nervous system is involved.

trichrome-stained preparation, each of which brings out different aspects of cyst morphology. These preparations are also helpful in distinguishing amebic from bacillary dysentery. In the latter, many inflammatory cells such as polymorphonuclear leukocytes are seen, whereas in amebic dysentery, they are not.

Serologic testing is useful for the diagnosis of invasive amebiasis. The indirect hemagglutination test is usually positive in patients with invasive disease but is frequently negative in asymptomatic individuals who are passing cysts.

Treatment & Prevention

The treatment of choice for symptomatic intestinal amebiasis or hepatic abscesses is metronidazole (Flagyl) or tinidazole. Hepatic abscesses need not be drained. Asymptomatic cyst carriers should be treated with iodoquinol or paromomycin.

Prevention involves avoiding fecal contamination of food and water and observing good personal hygiene such as handwashing. Purification of municipal water supplies is usually effective, but outbreaks of amebiasis in city dwellers still occur

TABLE 51–3 Medically Important Stages in Life Cycle of Intestinal Protozoa

Organism	Insect Vector	Stage that Infects Humans	Stage(s) in Humans Most Associated With Disease	Important Stage(s) Outside of Humans
Entamoeba	None	Cyst	Trophozoites cause bloody diarrhea and liver abscess	Cyst
Giardia	None	Cyst	Trophozoites cause watery diarrhea	Cyst
Cryptosporidium	None	Cyst	Trophozoites cause watery diarrhea	Cyst
Trichomonas	None	Trophozoite	Trophozoites cause vaginal discharge	None

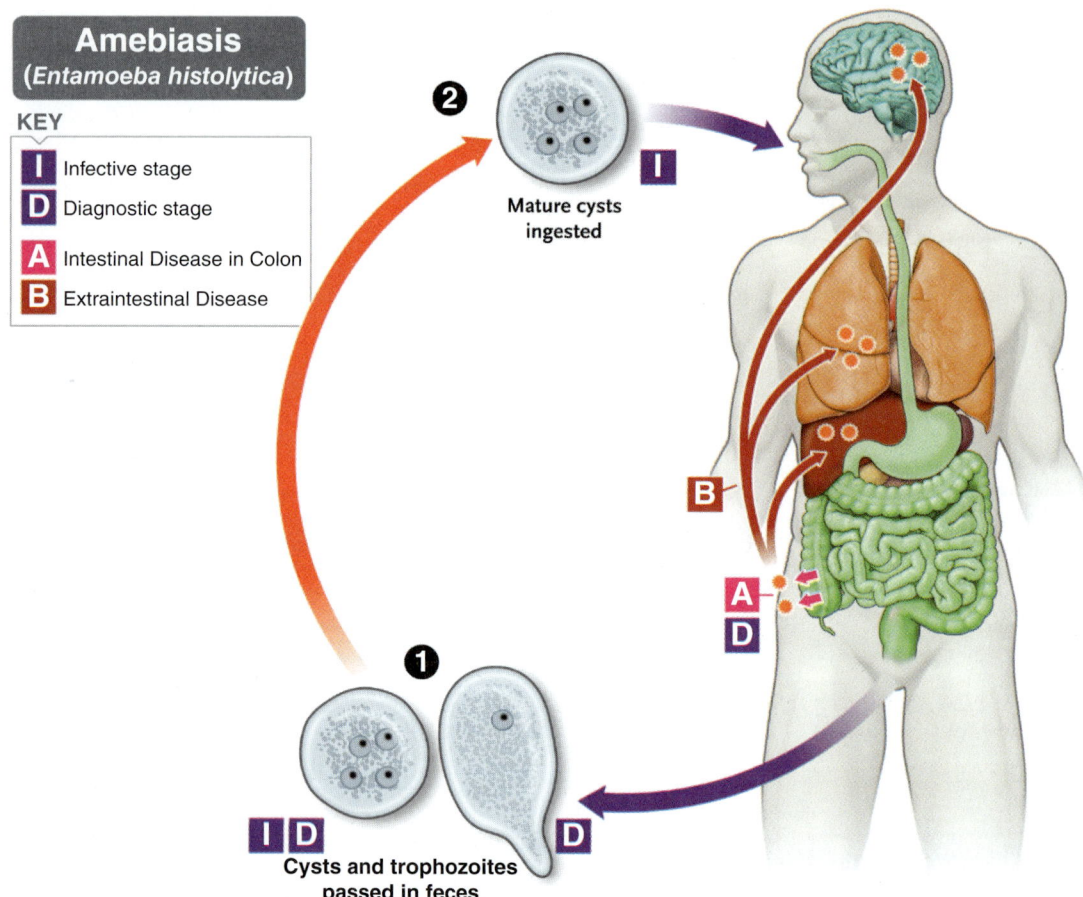

FIGURE 51–1 *Entamoeba histolytica*. Life cycle. Top blue arrow shows cysts being ingested. Within the intestine, the cyst produces trophozoites that cause amebic dysentery in the colon and can spread to the liver (most often), lung, and brain (Boxes A and B). Bottom blue arrow shows cysts and trophozoites being passed in the stool and entering the environment. Red arrow indicates survival of cysts in the environment. (Reproduced with permission from Centers for Disease Control and Prevention.)

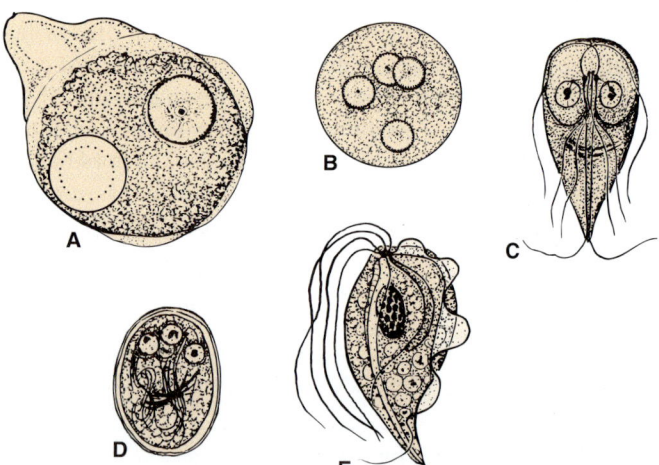

FIGURE 51–2 **A:** *Entamoeba histolytica* trophozoite with one ingested red blood cell and one nucleus (circle with inner dotted line represents a red blood cell). **B:** *E. histolytica* cyst with four nuclei. **C:** *Giardia lamblia* trophozoite. **D:** *G. lamblia* cyst. **E:** *Trichomonas vaginalis* trophozoite (1200×).

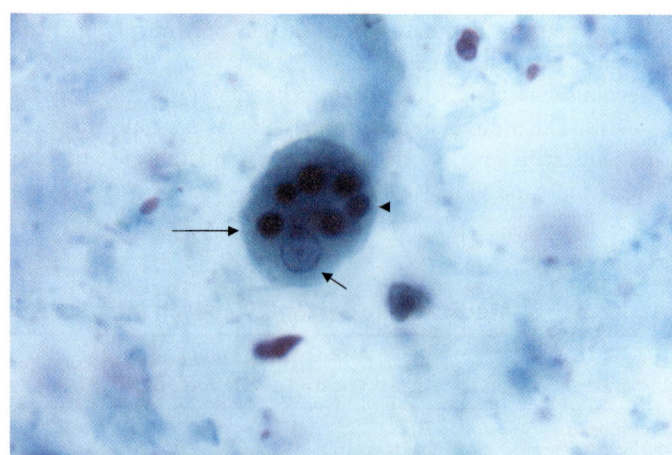

FIGURE 51–3 *Entamoeba histolytica*. Trophozoite. Long arrow points to trophozoite of *E. histolytica*. Short arrow points to the nucleus of the trophozoite. Arrowhead points to one of the six ingested red blood cells. (Reproduced with permission from Centers for Disease Control and Prevention.)

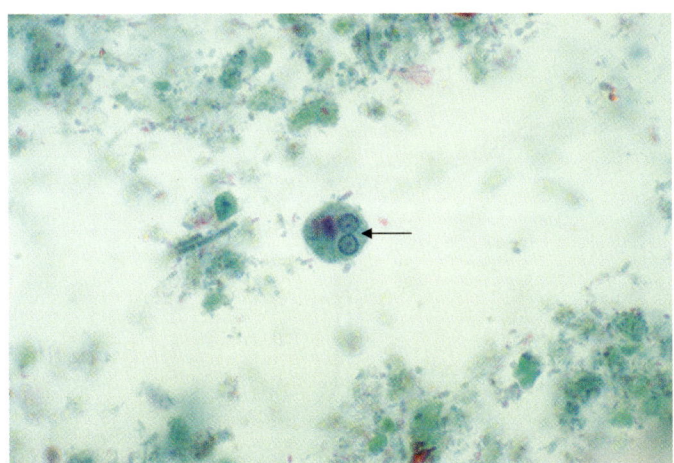

FIGURE 51–4 *Entamoeba histolytica*. Cyst. Arrow points to a cyst of *E. histolytica*. Two of the four nuclei are visible just to the left of the head of the arrow. (Reproduced with permission from Centers for Disease Control and Prevention.)

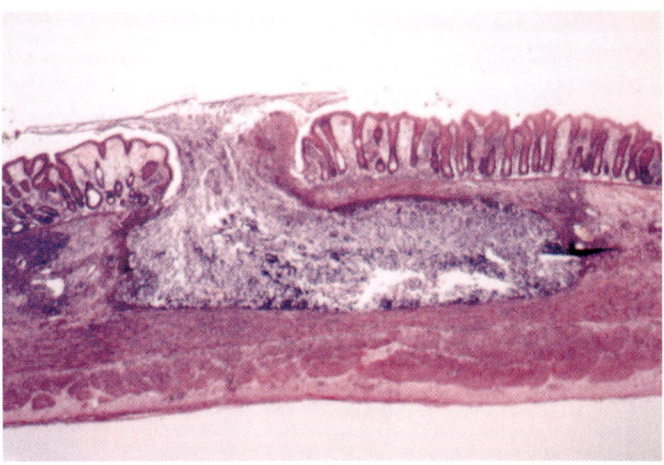

FIGURE 51–5 *Entamoeba histolytica*. Flask-shaped ulcer forms in colonic mucosa resulting in bloody diarrhea. (Reproduced with permission from Centers for Disease Control and Prevention.)

when contamination is heavy. The use of "night soil" (human feces) for fertilization of crops should be prohibited. In areas of endemic infection, vegetables should be cooked.

GIARDIA

Disease

G. lamblia causes giardiasis. (*G. lamblia* is also known as *G. duodenalis* and *G. intestinalis*)

Important Properties

The life cycle of *G. lamblia* is shown in Figure 51–6. The life cycle consists of two stages: the **trophozoite** (Figures 51–2C and 51–7) and the **cyst** (Figures 51–2D and 51–8). The trophozoite is pear-shaped with two nuclei, four pairs of flagella, and a suction disk with which it attaches to the intestinal wall. The oval cyst is thick-walled with four nuclei and several internal fibers. Each cyst gives rise to two trophozoites during excystation in the intestinal tract.

Pathogenesis & Epidemiology

Transmission occurs by ingestion of cysts in **fecally contaminated** food and water. Excystation takes place in the duodenum, where the trophozoite attaches to the gut wall but does *not* invade the mucosa and does not enter the bloodstream. The trophozoite causes inflammation of the duodenal mucosa, leading to **malabsorption** of protein and fat.

The organism is found worldwide; about 5% of stool specimens in the United States contain *Giardia* cysts. Approximately half of those infected are asymptomatic carriers who continue to excrete the cysts for years. IgA deficiency greatly predisposes to symptomatic infection.

In addition to being endemic, giardiasis occurs in outbreaks related to contaminated water supplies. Chlorination does not kill the cysts, but filtration removes them. Hikers who drink untreated stream water are frequently infected. Many species of mammals as well as humans act as the reservoirs. They pass cysts in the stool, which then contaminates water sources. Giardiasis is common in male homosexuals as a result of oral–anal contact. The incidence is high among children in day care centers and among patients in mental hospitals.

Clinical Findings

Watery (nonbloody), foul-smelling diarrhea is accompanied by nausea, anorexia, flatulence, and abdominal cramps persisting for weeks or months. There is no fever.

Laboratory Diagnosis

Diagnosis is made by finding trophozoites or cysts or both in diarrheal stools (see Figures 51–7 and 51–8). In formed stools (e.g., in asymptomatic carriers), only cysts are seen. An enzyme-linked immunosorbent assay (ELISA) test that detects *Giardia* antigen in the stool is also very useful. Tests for antibody in the serum are not routinely available. A PCR-based test that detects *Giardia* nucleic acid in the stool is available.

If those tests are negative and symptoms persist, the **string test**, which consists of swallowing a weighted piece of string until it reaches the duodenum, may be useful. The trophozoites adhere to the string and can be visualized after withdrawal of the string.

Treatment & Prevention

The treatment of choice is either tinidazole (Tindamax) or metronidazole (Flagyl). Tinidazole is better tolerated. Prevention involves drinking boiled, filtered, or iodine-treated water in endemic areas and while hiking. No prophylactic drug or vaccine is available.

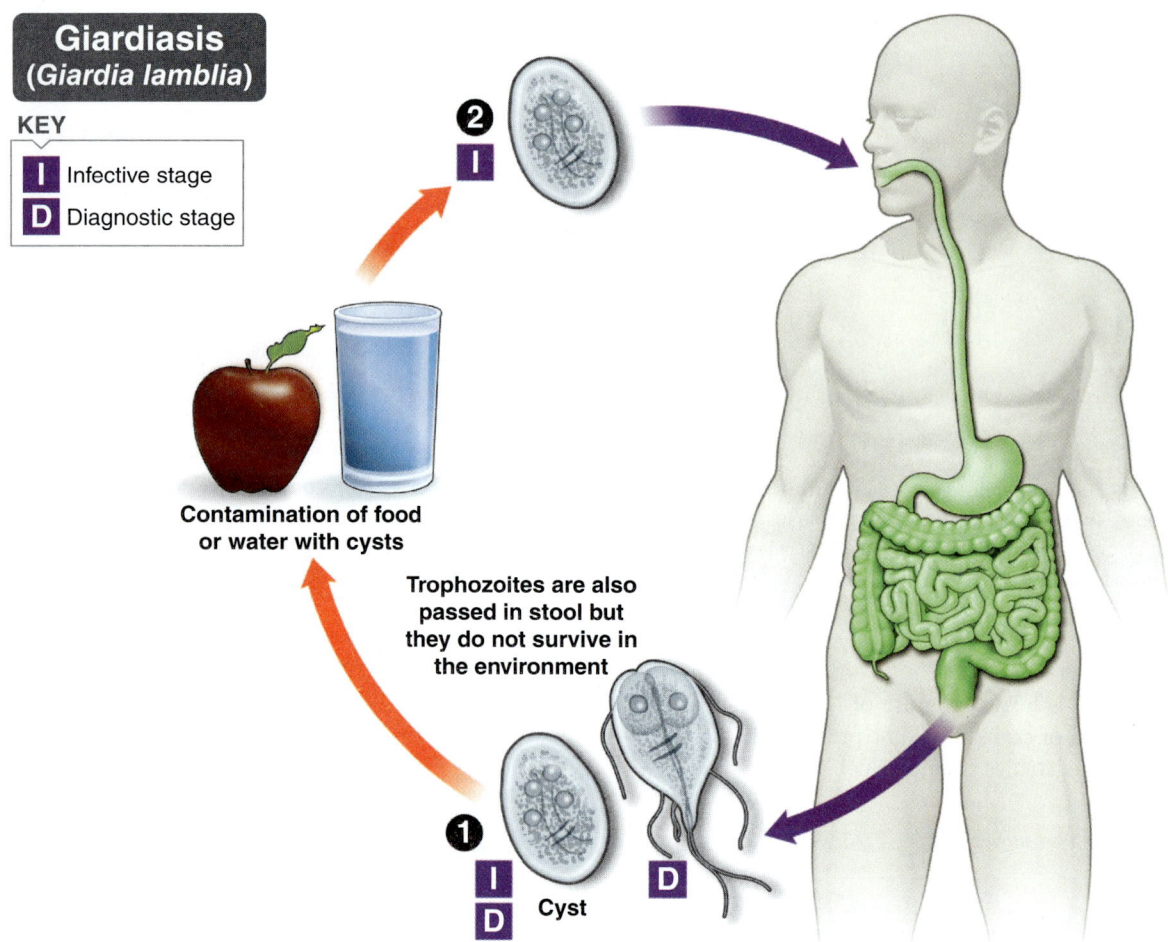

FIGURE 51–6 *Giardia lamblia*. Life cycle. Top blue arrow shows cysts being ingested. Within the intestine, the cyst produces trophozoites that cause diarrhea. Bottom blue arrow shows cysts and trophozoites being passed in the stool and entering the environment. Red arrow indicates survival of cysts in the environment. (Used with permission from Dr. Alexander J. da Silva and Melanie Moser, Centers for Disease Control and Prevention.)

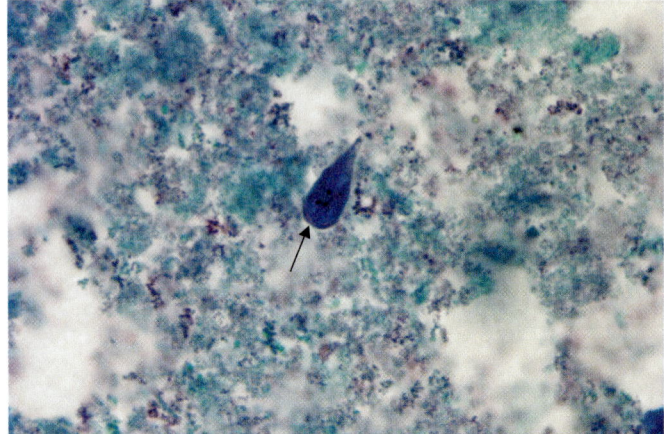

FIGURE 51–7 *Giardia lamblia*. Trophozoite. Arrow points to a pear-shaped trophozoite of *G. lamblia*. (Used with permission from Dr. M. Mosher, Centers for Disease Control and Prevention.)

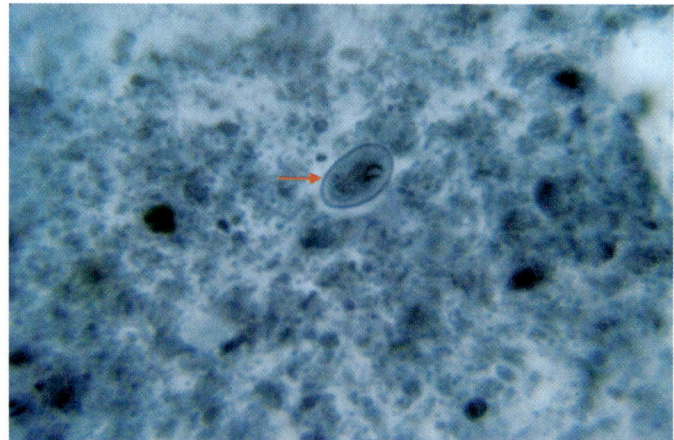

FIGURE 51–8 *Giardia lamblia*. Cyst. Arrow points to an oval cyst of *G. lamblia*. (Used with permission from Dr. George Healy, Centers for Disease Control and Prevention.)

CRYPTOSPORIDIUM

Disease

C. hominis causes cryptosporidiosis, the main symptom of which is diarrhea. The diarrhea is most severe in **immunocompromised** patients (e.g., those with AIDS). *Cryptosporidium parvum* is the former name that is no longer used.

Important Properties

The life cycle of *C. hominis* is shown in Figure 51–9. Some aspects of the life cycle remain uncertain, but the following stages have been identified. Oocysts release sporozoites, which form trophozoites. Several stages ensue, involving the formation of schizonts and merozoites. Eventually microgametes and macrogametes form; these unite to produce a zygote, which differentiates into an oocyst. This cycle has several features in common with other sporozoa (e.g., *Isospora*). Taxonomically, *Cryptosporidium* is in the subclass Coccidia.

Pathogenesis & Epidemiology

The organism is acquired by **fecal–oral** transmission of oocysts from either human sources (primarily) or from animal sources, for example, cattle (occasionally). The oocysts excyst in the small intestine, where the trophozoites (and other forms) attach to the gut wall. Invasion does not occur. The jejunum is the site most heavily infested. The pathogenesis of the diarrhea is uncertain; no toxin has been identified.

Cryptosporidia cause diarrhea worldwide. Large outbreaks of diarrhea caused by cryptosporidia in several cities in the United States are attributed to inadequate purification of drinking water. Other outbreaks are related to swimming in fecally contaminated pools and lakes. The cysts are highly resistant to chlorination but are killed by pasteurization and can be removed by filtration.

Clinical Findings

The disease in immunocompromised patients presents primarily as a watery, nonbloody **diarrhea** causing large fluid loss.

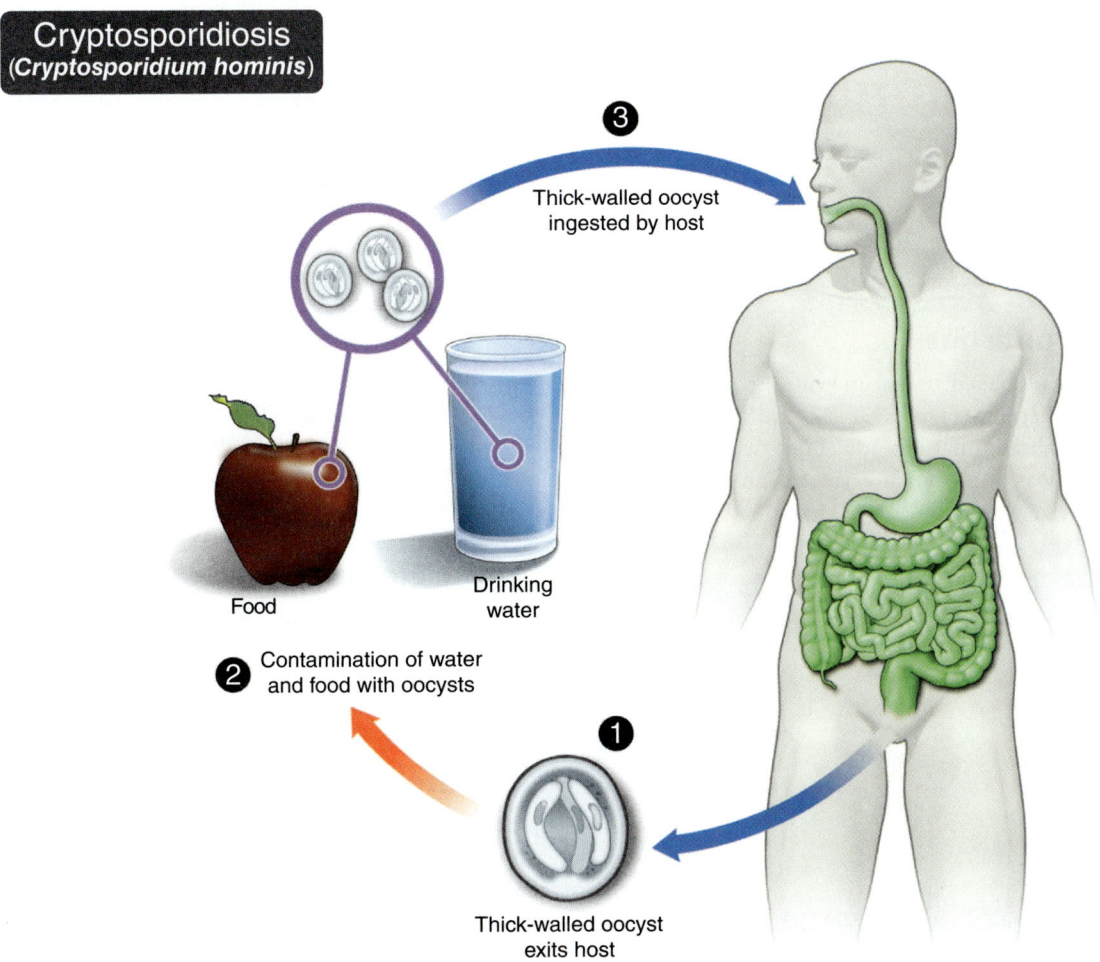

FIGURE 51–9 *Cryptosporidium hominis.* Life cycle. Top blue arrow shows cysts being ingested. Within the intestine, the oocyst produces trophozoites that cause diarrhea. Bottom blue arrow shows cysts being passed in the stool and entering the environment. Red arrow indicates survival of cysts in the environment. (Used with permission from Dr. Alexander J. da Silva and Melanie Moser, Centers for Disease Control and Prevention.)

Symptoms persist for long periods in immunocompromised patients, whereas they are self-limited in immunocompetent patients. Although immunocompromised patients usually do not die of cryptosporidiosis, the fluid loss and malnutrition are severely debilitating.

Laboratory Diagnosis

Diagnosis is made by finding oocysts in fecal smears when using a modified Kinyoun acid-fast stain (Figure 51–10). A test for *Cryptosporidium* antigen in the stool is also useful. A PCR-based test that detects *Cryptosporidium* nucleic acid in the stool is available.

Treatment & Prevention

Nitazoxanide is the drug of choice for patients not infected with human immunodeficiency virus (HIV). There is no effective drug therapy for severely immunocompromised patients, but paromomycin may be useful in reducing diarrhea.

There is no vaccine or other specific means of prevention. Purification of the water supply, including filtration to remove

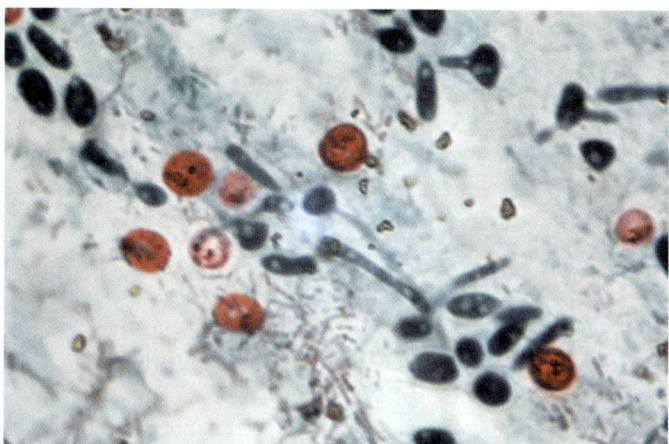

FIGURE 51–10 *Cryptosporidium hominis.* Cysts. Acid-fast stain of cysts in stool. Cysts appear red on a blue background. (Reproduced with permission from Ma P, Soave R: Three-step stool examination for cryptosporidiosis in 10 homosexual men with protracted watery diarrhea. J Infect Dis. 1983 May;147[5]:824–828.)

the cysts, which are resistant to the chlorine used for disinfection, can prevent cryptosporidiosis.

UROGENITAL PROTOZOA

TRICHOMONAS

Disease

T. vaginalis causes trichomoniasis.

Important Properties

T. vaginalis is a pear-shaped organism with a central nucleus and four anterior flagella (Figures 51–2E and 51–11). It has an undulating membrane that extends about two-thirds of its length. It exists **only as a trophozoite**; there is no cyst form.

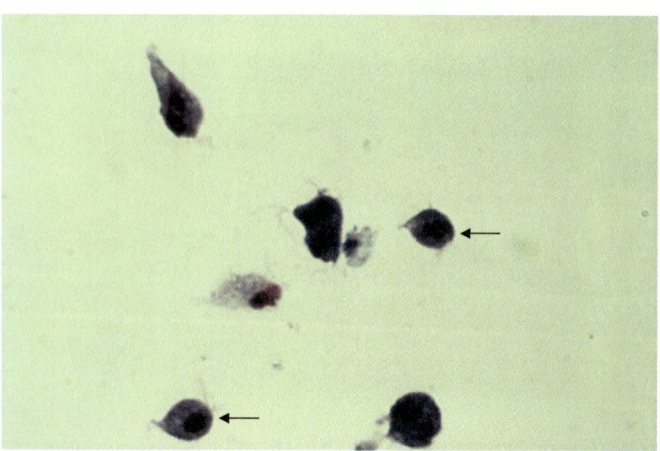

FIGURE 51–11 *Trichomonas vaginalis.* Trophozoite. Arrows point to two trophozoites. (Reproduced with permission from Centers for Disease Control and Prevention.)

Pathogenesis & Epidemiology

The organism is transmitted by sexual contact, and hence there is no need for a durable cyst form. The primary locations of the organism are the vagina and the prostate. It is found only in humans; there is no animal reservoir.

Trichomoniasis is one of the most common infections worldwide. Roughly 25% to 50% of women in the United States harbor the organism. The frequency of symptomatic disease is highest among sexually active women in their thirties and lowest among postmenopausal women. Asymptomatic infections are common in both men and women.

Clinical Findings

In women, a watery, foul-smelling, greenish vaginal discharge accompanied by itching and burning occurs. Infection in men is usually asymptomatic, but about 10% of infected men have urethritis.

Laboratory Diagnosis

In a wet mount of vaginal discharge, the pear-shaped trophozoites have a typical jerky motion (see Figure 51–11). Nucleic acid amplification tests (NAATs) are often used because they are highly specific and sensitive. There is no serologic test.

Treatment & Prevention

The treatment of choice is either tinidazole (Tindamax) or metronidazole (Flagyl) for both partners to prevent reinfection. Tinidazole is better tolerated. Maintenance of the low pH of the vagina is helpful. Condoms limit transmission. No prophylactic drug or vaccine is available.

SELF-ASSESSMENT QUESTIONS

1. Regarding *Entamoeba histolytica*, which one of the following is most accurate?

   (A) *E. histolytica* causes "flask-shaped" ulcerations in the colon mucosa.

   (B) Domestic animals such as dogs and cats are the main reservoir of *E. histolytica*.

   (C) In the microscope, *E. histolytica* is recognized by having two sets of paired flagella.

   (D) *E. histolytica* infections are limited to the intestinal mucosa and do not spread to other organs.

   (E) The infection is typically acquired by the ingestion of the trophozoite in contaminated food and water.

2. Regarding *Giardia lamblia*, which one of the following is most accurate?

   (A) The drug of choice for giardiasis is chloroquine.

   (B) In giardiasis, ova and parasite (O&P) analysis of the stool reveals sporozoites in the feces.

   (C) *Giardia lamblia* produces an enterotoxin that increases cyclic AMP within the enterocyte, resulting in diarrhea.

   (D) *Giardia lamblia* infection is acquired by ingestion of food or water contaminated with human feces only (i.e., there is no animal reservoir for this organism).

   (E) Infection by *G. lamblia* occurs principally in the small intestine, frequently resulting in the malabsorption of fat and foul-smelling, frothy, fat-containing stools.

3. Regarding *Cryptosporidium hominis*, which one of the following is most accurate?

   (A) Humans are the only reservoir for *C. hominis*.

   (B) Microscopic examination of the diarrheal stool reveals both red cells and white cells.

   (C) Laboratory diagnosis involves seeing cysts of the organism in an acid-fast stain of the stool.

   (D) *C. hominis* is typically acquired by the ingestion of trophozoites in contaminated food or water.

   (E) In immunocompromised patients, such as AIDS patients with a very low CD4 count, disseminated disease occurs that typically involves the brain and meninges.

4. Regarding *Trichomonas vaginalis*, which one of the following is most accurate?

   (A) The drug of choice for trichomoniasis is metronidazole.

   (B) Domestic animals, such as dogs and cats, are the principal reservoir of the organism.

   (C) *T. vaginalis* is typically acquired by contact with the cysts of the organism during sexual intercourse.

   (D) Laboratory diagnosis typically involves the detection of a greater than fourfold rise in the titer of IgA antibody.

   (E) The asymptomatic male sex partner of a woman with *T. vaginalis* infection should not be treated because asymptomatic men are rarely the source of the organism.

5. Your patient is a 30-year-old woman who returned from traveling in Eastern Europe 1 week ago. While on the trip, she experienced anorexia, nausea but no vomiting, and abdominal bloating. For the past 2 days, she has had explosive watery diarrhea. An examination of her stool revealed pear-shaped, flagellated, motile organisms. Of the following, which one is the most likely cause of this infection?

   (A) *Cryptosporidium hominis*

   (B) *Entamoeba histolytica*

   (C) *Giardia lamblia*

   (D) *Trichomonas vaginalis*

6. Regarding the patient in Question 5, which one of the following is the best antibiotic to treat the infection?

   (A) Chloroquine

   (B) Metronidazole

   (C) Nifurtimox

   (D) Praziquantel

   (E) Stibogluconate

7. Your patient is a 30-year-old Peace Corps volunteer who has recently returned from Central America. She now has fever and right-upper-quadrant pain. She reports that she had bloody diarrhea 2 months ago. A computed tomography scan reveals a radiolucent area in the liver that is interpreted to be an abscess. Aspiration of material from the abscess was performed. Microscopic examination revealed motile, nonflagellated trophozoites with ameboid movement. Of the following, which one is the most likely cause of this infection?

   (A) *Cryptosporidium hominis*

   (B) *Entamoeba histolytica*

   (C) *Giardia lamblia*

   (D) *Trichomonas vaginalis*

8. Your patient is a 30-year-old man with persistent watery diarrhea for 2 weeks. He is HIV antibody positive with a CD4 count of 10. Routine stool culture revealed no bacterial pathogen. Ova and parasite analysis revealed cysts that stained red in an acid-fast stain. Of the following, which one is the most likely cause of this infection?

   (A) *Cryptosporidium hominis*

   (B) *Entamoeba histolytica*

   (C) *Giardia lamblia*

   (D) *Trichomonas vaginalis*

ANSWERS

(1) **(A)**

(2) **(E)**

(3) **(C)**

(4) **(A)**

(5) **(C)**

(6) **(B)**

(7) **(B)**

(8) **(A)**

SUMMARIES OF ORGANISMS

Brief summaries of the organisms described in this chapter begin on page 706. Please consult these summaries for a rapid review of the essential material.

PRACTICE QUESTIONS: USMLE & COURSE EXAMINATIONS

Questions on the topics discussed in this chapter can be found in the Parasitology section of Part XIII: USMLE (National Board) Practice Questions starting on page 755. Also see Part XIV: USMLE (National Board) Practice Examination starting on page 775.

Blood & Tissue Protozoa

INTRODUCTION

The medically important organisms in this category of protozoa consist of the sporozoans *Plasmodium* and *Toxoplasma* and the flagellates *Trypanosoma* and *Leishmania*. *Pneumocystis* is discussed in this book as a protozoan because it is considered as such from a medical point of view. However, molecular data indicate that it is related to yeasts such as *Saccharomyces cerevisiae*. Table 51–2 summarizes several important features of these blood and tissue protozoa.

The medically important stages in the life cycle of the blood and tissue protozoa are described in Table 52–1.

TABLE 52–1 Medically Important Stages in Life Cycle of Blood and Tissue Protozoa

Organism	Insect Vector	Stage That Infects Humans	Stage(s) in Humans Most Associated With Disease	Important Stage(s) Outside of Humans
Plasmodium	Female mosquito (*Anopheles*)	Sporozoite in mosquito saliva	Trophozoites and merozoites in red blood cells	Mosquito ingests gametocytes → fuse to form zygote → ookinete → sporozoites
Toxoplasma	None	Tissue cyst (pseudo-cysts) in undercooked meat or oocyst in cat feces	Rapidly multiplying trophozoites (tachyzoites) within various cell types; tachyzoites can pass placenta and infect fetus; slowly multiplying trophozoites (bradyzoites) in tissue cysts	Cat ingests tissue cysts containing bradyzoites → gametes → ookinete → oocysts in feces
Pneumocystis	None	Uncertain; probably cyst	Cysts	None known
Trypanosoma cruzi	Reduviid bug (*Triatoma*)	Trypomastigote in bug feces	Amastigotes in cardiac muscle and neurons	Bug ingests trypomastigote in human blood → epimastigote → trypomastigote
Trypanosoma gambiense and *Trypanosoma rhodesiense*	Tsetse fly (*Glossina*)	Trypomastigote in fly saliva	Trypomastigotes in blood and brain	Fly ingests trypomastigote in human blood → epimastigote → trypomastigote
Leishmania donovani	Sandfly (*Phlebotomus* and *Lutzomyia*)	Promastigotes in fly saliva	Amastigotes in macrophages in spleen, liver, and bone marrow	Fly ingests macrophages containing amastigotes → promastigotes
Leishmania tropica and others	Sandfly (*Phlebotomus* and *Lutzomyia*)	Promastigotes in fly saliva	Amastigotes in macrophages in skin	Fly ingests macrophages containing amastigotes → promastigotes

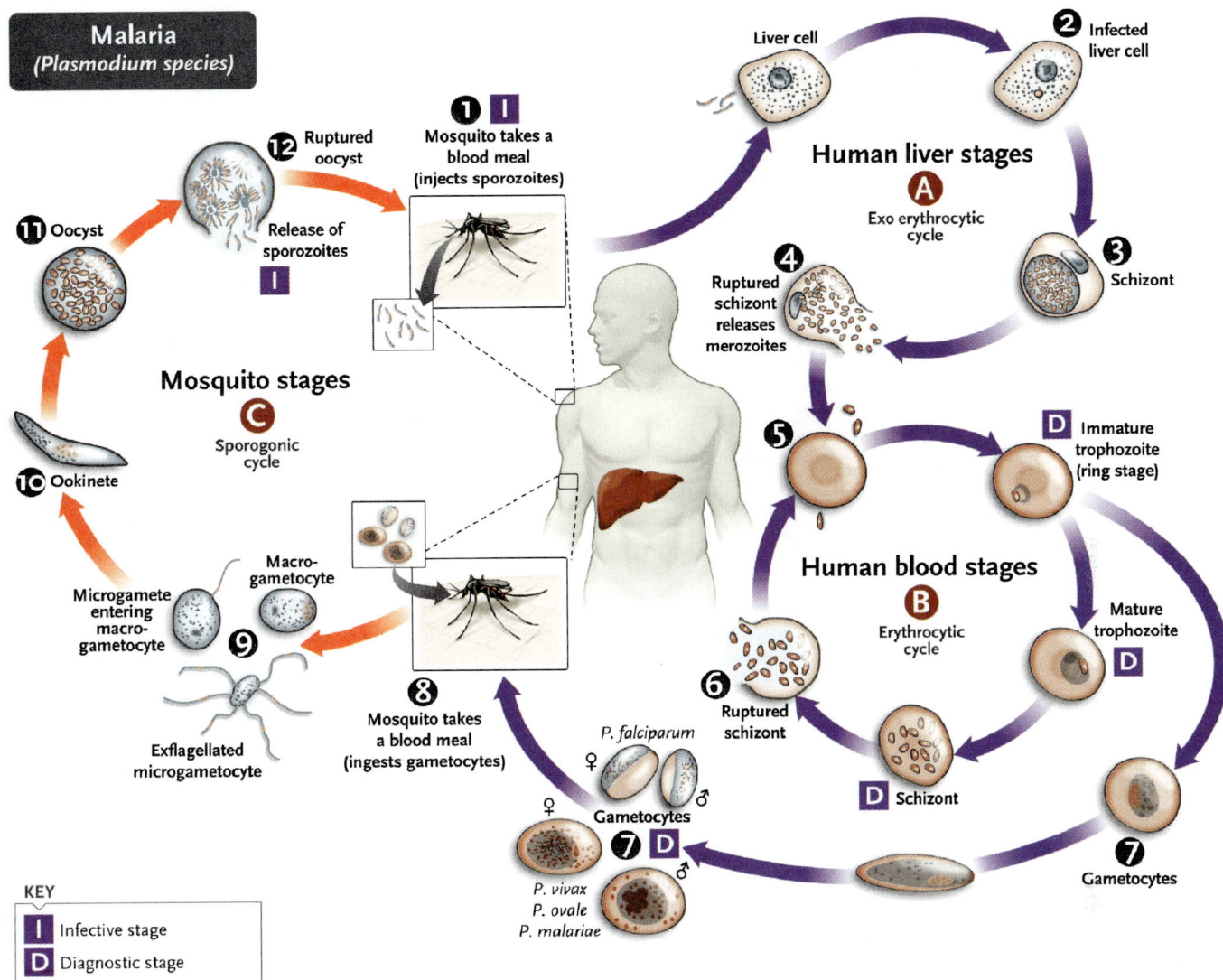

FIGURE 52–1 *Plasmodium* species. Life cycle. Right side of figure describes the stages within the human (blue arrows). Cycle A (top right) is the exoerythrocyte stage that occurs in the liver. Note that only *P. vivax* and *P. ovale* produce hypnozoites (a latent form) in liver cells (not shown). Cycle B (bottom right) is the erythrocyte stage that occurs in the red blood cell. Note that at step 6 in the cycle, merozoites released from the ruptured schizonts then infect other red blood cells. The synchronized release of merozoites causes the periodic fever and chills characteristic of malaria. Left side of figure describes the stages within the mosquito (red arrows). Humans are infected at step 1 when mosquito injects sporozoites. Mosquito is infected at step 8 when mosquito ingests gametocytes in human blood. (Used with permission from Dr. Alexander J. da Silva and Melanie Moser, Centers for Disease Control and Prevention.)

PLASMODIUM

Disease

Malaria is caused primarily by four plasmodia: *Plasmodium vivax*, *Plasmodium ovale*, *Plasmodium malariae*, and *Plasmodium falciparum*. *P. vivax* and *P. falciparum* are more common causes of malaria than are *P. ovale* and *P. malariae*. *P. vivax* is most widely distributed and *P. falciparum* causes the most serious disease. A fifth species, *Plasmodium knowlesi*, is found in Southeast Asia.

Worldwide, malaria is one of the most common infectious diseases and one of the leading causes of death.

Important Properties

The life cycle of *Plasmodium* species is shown in Figure 52–1. The vector and definitive host for plasmodia is the **female Anopheles** mosquito (only the female takes a blood meal). There are two phases in the life cycle: the sexual cycle, which occurs primarily in mosquitoes, and the asexual cycle, which occurs in humans, the intermediate hosts.[1]

---

[1]The sexual cycle is initiated in humans with the formation of gametocytes within red blood cells (gametogony) and completed in mosquitoes with the fusion of the male and female gametes, oocyst formation, and production of many sporozoites (sporogony).

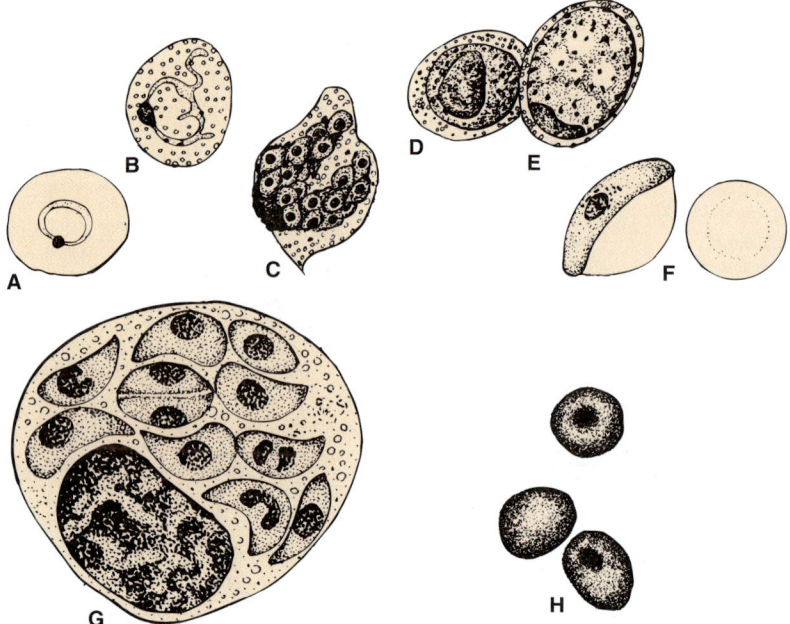

FIGURE 52–2 **A:** *Plasmodium vivax* signet-ring trophozoite within a red blood cell. **B:** *P. vivax* ameboid trophozoite within a red blood cell, showing Schüffner's dots. **C:** *P. vivax* mature schizont with merozoites inside. **D:** *P. vivax* microgametocyte. **E:** *P. vivax* macrogametocyte. **F:** *Plasmodium falciparum* "banana-shaped" gametocyte with attached red cell ghost. **G:** *Toxoplasma gondii* trophozoites within macrophage. **H:** *Pneumocystis jiroveci* cysts. (A–G, 1200×; H, 800×.)

The sexual cycle is called **sporogony** because sporozoites are produced (sporogonic cycle is labeled C in Figure 52–1), and the asexual cycle is called **schizogony** because schizonts are made.

The life cycle in humans begins with the introduction of sporozoites into the blood from the saliva of the biting mosquito. The sporozoites are taken up by hepatocytes within 30 minutes. This "exoerythrocytic" phase (labeled A in Figure 52–1) consists of cell multiplication and differentiation into **merozoites**. *P. vivax* and *P. ovale* produce a latent form (**hypnozoite**) in the liver; this form is the cause of relapses seen with vivax and ovale malaria.

Merozoites are released from the liver cells and infect red blood cells. During the erythrocytic phase (labeled B in Figure 52–1), the organism differentiates into a ring-shaped trophozoite (Figures 52–2A and B and 52–3). The ring form grows into an ameboid form and then differentiates into a schizont filled with merozoites (Figure 52–2C). After release, the merozoites infect other erythrocytes (step 6 in Figure 52–1). This cycle in the red blood cell repeats at regular intervals typical for each species. The periodic release of merozoites causes the typical recurrent symptoms of chills, fever, and sweats seen in malaria patients.

The sexual cycle begins in the human red blood cells when some merozoites develop into male and others into female gametocytes (Figures 52–2D to F and 52–4, and step 7 in Figure 52–1). The gametocyte-containing red blood cells are ingested by the female *Anopheles* mosquito and, within her gut, produce a female macrogamete and eight spermlike male microgametes. After fertilization, the diploid zygote differentiates into a motile

ookinete that burrows into the gut wall, where it grows into an oocyst within which many haploid sporozoites are produced. The sporozoites are released and migrate to the salivary glands, ready to complete the cycle when the mosquito takes her next blood meal.

A very important feature of *P. falciparum* is **chloroquine resistance**. Chloroquine-resistant strains now predominate in

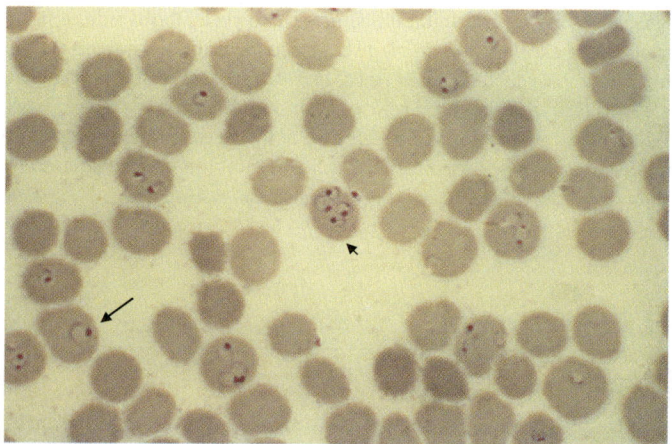

FIGURE 52–3 *Plasmodium falciparum*. Ring-shaped trophozoite. Long arrow points to a red blood cell containing a ring-shaped trophozoite. Arrowhead points to a red blood cell containing four ring-shaped trophozoites. Note the very high percentage of red cells containing ring forms. This high-level parasitemia is more often seen in *Plasmodium falciparum* infection than in infection by the other plasmodia. (Used with permission from Dr. Glenn S., Public Health Image Library, Centers for Disease Control and Prevention.)

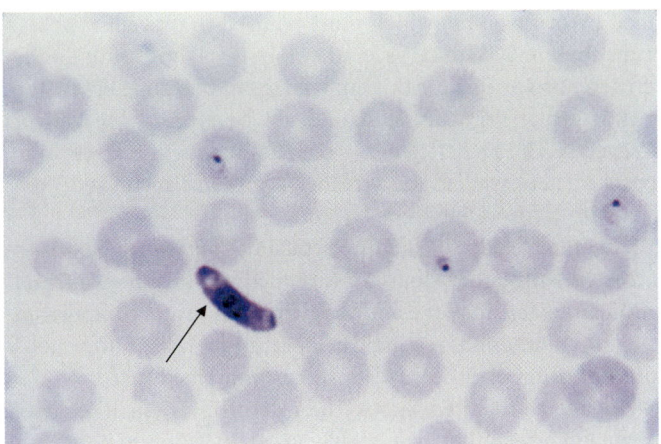

FIGURE 52–4 *Plasmodium falciparum*. Gametocyte. Arrow points to a "banana-shaped" gametocyte of *P. falciparum*. (Used with permission from Dr. Glenn S., Public Health Image Library, Centers for Disease Control and Prevention.)

most areas of the world where malaria is endemic. Chloroquine resistance is mediated by a mutation in the gene encoding the chloroquine transporter in the cell membrane of the organism.

Pathogenesis & Epidemiology

Most of the pathologic findings of malaria result from the **destruction of red blood cells**. Red cells are destroyed both by the release of the merozoites and by the action of the spleen to first sequester the infected red cells and then to lyse them. The enlarged spleen characteristic of malaria is due to congestion of sinusoids with erythrocytes, coupled with hyperplasia of lymphocytes and macrophages.

Malaria caused by *P. falciparum* is **more severe** than that caused by other plasmodia. It is characterized by infection of far more red cells than the other malarial species and by occlusion of the capillaries with aggregates of parasitized red cells. This leads to life-threatening hemorrhage and necrosis, particularly in the brain (cerebral malaria). Furthermore, extensive hemolysis and kidney damage occur, with resulting hemoglobinuria. The dark color of the patient's urine has given rise to the term "blackwater fever." The hemoglobinuria can lead to acute renal failure.

The timing of the fever cycle is 72 hours for *P. malariae* and 48 hours for the other plasmodia. Disease caused by *P. malariae* is called quartan malaria because it recurs every fourth day, whereas malaria caused by the other plasmodia is called tertian malaria because it recurs every third day. Tertian malaria is subdivided into malignant malaria, caused by *P. falciparum*, and benign malaria, caused by *P. vivax* and *P. ovale*.

P. falciparum causes a high level of parasitemia because it can infect red cells of all ages. In contrast, *P. vivax* infects only reticulocytes and *P. malariae* infects only mature red cells; therefore, they produce much lower levels of parasites in the blood. Individuals with sickle cell trait (heterozygotes) are protected against malaria because their red cells have too little ATPase activity and cannot produce sufficient energy to

support the growth of the parasite. People with homozygous sickle cell anemia are also protected but rarely live long enough to obtain much benefit.

The receptor for *P. vivax* is the Duffy blood group antigen. People who are homozygous recessive for the genes that encode this protein are resistant to infection by *P. vivax*. More than 90% of black West Africans and many of their American descendants do not produce the Duffy antigen and are thereby resistant to vivax malaria.

People with glucose-6-phosphate dehydrogenase (G6PD) deficiency are also protected against the severe effects of falciparum malaria. G6PD deficiency is an X-linked hemoglobinopathy found in high frequency in tropical areas where malaria is endemic. Both male and female carriers of the mutated gene are protected against malaria.

Malaria is transmitted primarily by mosquito bites, but transmission across the placenta, in blood transfusions, and by intravenous drug use also occurs.

Partial immunity based on humoral antibodies that block merozoites from invading the red cells occurs in infected individuals. A low level of parasitemia and low-grade symptoms result; this condition is known as **premunition**. In contrast, a nonimmune person, such as a first-time traveler to an area where falciparum malaria is endemic, is at risk of severe, life-threatening disease.

More than 200 million people worldwide have malaria, and more than 1 million die of it each year, making it the most common lethal infectious disease. It occurs primarily in tropical and subtropical areas, especially in Asia, Africa, and Central and South America. Malaria in the United States is seen in Americans who travel to areas of endemic infection without adequate chemoprophylaxis and in immigrants from areas of endemic infection. It is not endemic in the United States. Certain regions in Southeast Asia, South America, and East Africa are particularly affected by chloroquine-resistant strains of *P. falciparum*. People who have lived or traveled in areas where malaria occurs should seek medical attention for febrile illnesses up to 3 years after leaving the malarious area.

Clinical Findings

Malaria presents with abrupt onset of fever and chills, accompanied by headache, myalgias, and arthralgias, about 2 weeks after the mosquito bite. Fever may be continuous early in the disease; the typical periodic cycle does not develop for several days after onset. The fever spike, which can reach 41°C, is frequently accompanied by shaking chills, nausea, vomiting, and abdominal pain. The fever is followed by drenching sweats. Patients usually feel well between the febrile episodes. Splenomegaly is seen in most patients, and hepatomegaly occurs in roughly one-third. Anemia is prominent.

Untreated malaria caused by *P. falciparum* is potentially life-threatening as a result of extensive brain (cerebral malaria) and kidney (blackwater fever) damage. Malaria caused by the other three plasmodia is usually self-limited, with a low mortality rate. However, relapses of *P. vivax* and *P. ovale* malaria can occur up to several years after the initial illness as a result of hypnozoites latent in the liver.

Laboratory Diagnosis

Diagnosis rests on microscopic examination of blood, using both **thick** and **thin** Giemsa-stained smears. The thick smear is used to screen for the presence of organisms, and the thin smear is used for species identification. It is important to identify the species because the treatment of different species can differ. Ring-shaped trophozoites can be seen within infected red blood cells (see Figure 52–3). The gametocytes of *P. falciparum* are **crescent-shaped** ("banana-shaped"), whereas those of the other plasmodia are spherical (Figure 52–2F). If more than 5% of red blood cells are parasitized, the diagnosis is usually *P. falciparum* malaria.

Plasmodium species typically produce **hemozoin** pigment in infected red blood cells, whereas *Babesia* species (see Chapter 53) do not. Plasmodia metabolize heme in the red cells to produce hemozoin. Also found within *P. vivax*– and *P. ovale*–infected red cells are **Schüffner's dots**. These are intracytoplasmic granules that stain red using the Romanowsky stain.

If blood smears do not reveal the diagnosis, then a polymerase chain reaction (PCR)-based test for *Plasmodium* nucleic acids or an enzyme-linked immunosorbent assay (ELISA) test for a protein specific for *P. falciparum* can be useful.

Treatment

The treatment of malaria is complicated, and the details are beyond the scope of this book. Table 52–2 presents the drugs commonly used in the United States. The main criteria used for choosing specific drugs are the severity of the disease and whether the organism is resistant to chloroquine. Chloroquine resistance is determined by the geographical location where the infection was acquired rather than by laboratory testing.

Chloroquine is the drug of choice for treatment of uncomplicated malaria caused by non-falciparum species in areas without chloroquine resistance. Chloroquine kills the merozoites, thereby reducing the parasitemia, but does not affect the hypnozoites of *P. vivax* and *P. ovale* in the liver. These are

killed by primaquine, which must be used to prevent relapses. Primaquine may induce severe hemolysis in those with G6PD deficiency, so testing for this enzyme should be done before the drug is given. Primaquine should not be given if the patient is severely G6PD deficient. If primaquine is not given, one approach is to wait to see whether symptoms recur and then treat with chloroquine.

Uncomplicated, chloroquine-resistant *P. falciparum* infection is treated with either Coartem (artemether plus lumefantrine) or Malarone (atovaquone and proguanil). In severe complicated cases of chloroquine-resistant falciparum malaria, intravenous administration of either artesunate or quinidine is used.

Outside the United States, the artemisinins, such as artesunate or artemether, are widely used in combination with other antimalarial drugs. The artemisinins are inexpensive and have few side effects. However, *P. falciparum* has developed resistance to artemisinins in mainland Southeast Asia (e.g., Vietnam, Cambodia, Myanmar, and Thailand) and in Africa (e.g., Equatorial Guinea). It is recommended to add primaquine to artemisinin-based treatment regimens in areas where resistance has developed.

Prevention

Chemoprophylaxis of malaria for travelers to areas where chloroquine-resistant *P. falciparum* is endemic consists of mefloquine or doxycycline. A combination of atovaquone and proguanil (Malarone), in a fixed dose, can also be used. Drugs used to prevent malaria are described in Table 52-3.

Chloroquine should be used in areas where *P. falciparum* is sensitive to that drug. Travelers to areas where the other three plasmodia are found should take chloroquine starting 2 weeks before arrival in the endemic area and continuing for 4 weeks after leaving the endemic area. This should be followed by a 2-week course of primaquine if exposure was high. Primaquine will kill the hypnozoites of *P. vivax* and *P. ovale*.

Tafenoquine is a long-acting analog of primaquine and is approved for once-weekly prophylaxis against all species and stages of Plasmodia. It is contraindicated in persons with G6PD deficiency.

TABLE 52–2 Drugs Commonly Used for the Treatment of Malaria in the United States

Species	Drug(s)	Comments
Chloroquine-sensitive *Plasmodium falciparum* and *Plasmodium malariae*	Chloroquine	Oral
Chloroquine-sensitive *Plasmodium vivax* and *Plasmodium ovale*	Chloroquine plus primaquine	Oral Do not use primaquine if G6PD deficient
Chloroquine-resistant *P. falciparum*; uncomplicated infection	Coartem (artemether and lumefantrine) or Malarone (atovaquone and proguanil)	Oral
Chloroquine-resistant *P. falciparum*; severe complicated infection	Artesunate[1] or quinidine[2]	Intravenous

G6PD = glucose-6-phosphate dehydrogenase.

[1]Available in the United States through the Centers for Disease Control and Prevention.

[2]Intravenous quinidine is no longer available in the United States.

TABLE 52-3 Drugs Commonly Used for Prevention of Malaria[1]

Drug	Duration of Travel	Geographic Location	Comment
Chloroquine	Start 1 week before; stop 4 weeks after leaving zone[2]	Use only in area where *Plasmodium* is chloroquine sensitive	
Atovaquone/proguanil (Malarone)	Start 1–2 days before; stop 1 week after leaving zone	Useful in area where *Plasmodium* is either chloroquine sensitive or resistant	
Doxycycline	Start at least 2 weeks before; stop 4 weeks after leaving zone	Useful in area where *Plasmodium* is either chloroquine sensitive or resistan	1. Can also prevent rickettsial infections 2. Do not use in children less than 8 years old
Mefloquine	Start 1 week before; stop 4 weeks after leaving	Useful in area where *Plasmodium* is chloroquine resistant	Mefloquine-resistance reported in some areas in Southeast Asia
Primaquine	Start 1–2 days before; stop 1 week after leaving Zone	1. Used to prevent relapse of *P. vivax* and *P. ovale* caused by hypnozoites in liver 2. Used in areas where a high rate of *P. vivax* infection occurs	Requires testing for G6PD deficiency
Tafenoquine	Start 3 days before; stop 1 week after leaving zone	1. Used to prevent relapse of *P. vivax* and *P. ovale* caused by hypnozoites in liver 2. Useful in area where *Plasmodium* is either chloroquine resistant or mefloquine resistant	1. Long-acting analog of primaquine 2. Requires testing for G6PD deficiency

[1] Consult CDC Guidelines to determine the countries with chloroquine-sensitive and resistant *Plasmodium*.

[2] Zone means geographic area where risk of contracting malaria exists.

Other preventive measures include the use of insecticide-treated mosquito netting, window screens, protective clothing, and insect repellents. The mosquitoes feed from dusk to dawn, so protection is particularly important during the night. Communal preventive measures are directed against reducing the mosquito population. Many insecticide sprays, such as DDT, are no longer effective because the mosquitoes have developed resistance. Drainage of stagnant water in swamps and ditches reduces the breeding areas. There is no vaccine.

TOXOPLASMA

Disease

Toxoplasma gondii causes toxoplasmosis, including congenital toxoplasmosis.

Important Properties

The life cycle of *T. gondii* is shown in Figure 52–5. The definitive host is the **domestic cat** and other felines; humans and other mammals are intermediate hosts. Infection of humans begins with the **ingestion of cysts** in undercooked meat or from accidental contact with cysts in cat feces. In the small intestine, the cysts rupture and release forms that invade the gut wall, where they are ingested by macrophages and differentiate into rapidly multiplying trophozoites (**tachyzoites**), which kill the cells and infect other cells (Figures 52–2G and 52–6). Cell-mediated immunity usually limits the spread of tachyzoites, and the parasites enter host cells in the brain, muscle, and other tissues, where they develop into cysts in which the parasites multiply slowly. These forms are called **bradyzoites**. These tissue cysts are both an important diagnostic feature and a source of organisms when the tissue cyst breaks in an immunocompromised patient.

The cycle within the cat begins with the ingestion of cysts in raw meat (e.g., mice). Bradyzoites are released from the cysts in the small intestine, infect the mucosal cells, and differentiate into male and female gametocytes, whose gametes fuse to form oocysts that are excreted in cat feces. The cycle is completed when soil contaminated with cat feces is accidentally ingested. Human infection usually occurs from eating undercooked meat (e.g., lamb and pork) from animals that grazed in soil contaminated with infected cat feces.

Pathogenesis & Epidemiology

T. gondii is usually acquired by **ingestion** of cysts in uncooked meat or in food accidentally contaminated by cat feces.

Transplacental transmission from an infected mother to the fetus occurs also. Human-to-human transmission, other than transplacental transmission, does not occur. After infection of the intestinal epithelium, the organisms spread to other organs, especially the brain, lungs, liver, and eyes. Progression of the infection is usually limited by a competent immune system. **Cell-mediated immunity** plays the major role, but circulating antibody enhances killing of the organism. Most initial

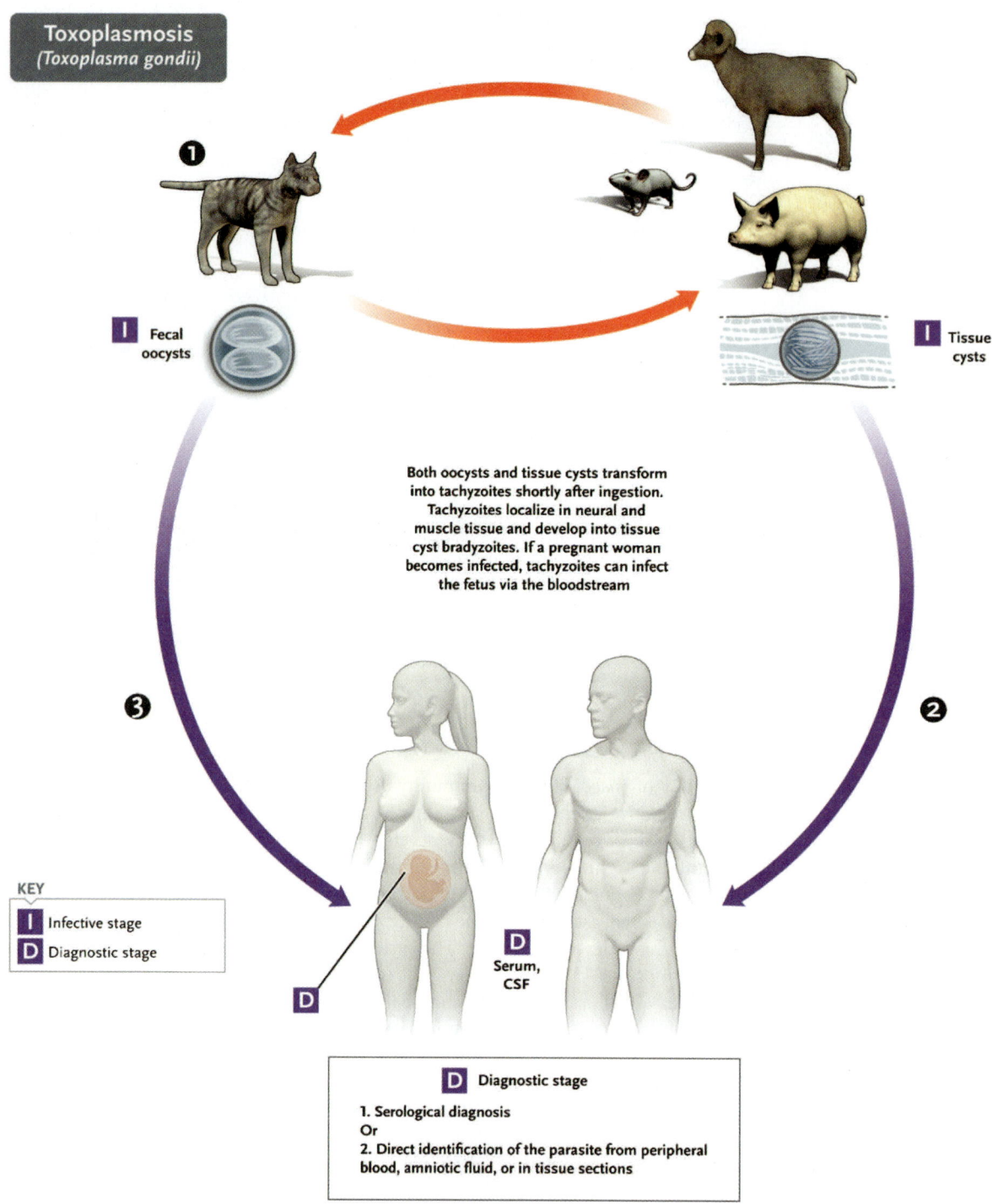

FIGURE 52–5 *Toxoplasma gondii*. Life cycle. Top red arrows show the natural life cycle as *T. gondii* circulates between cats (#1), which excrete oocysts in the feces that are eaten by mice, but also by domestic animals such as pigs and sheep. Cysts form in tissue such as muscle and brain. The natural cycle is completed when cats eat mice. Humans are accidental hosts. They can be infected by the ingestion of under-cooked pork and lamb (blue arrow #2) containing tissue cysts in muscle or by ingestion of food contaminated with cat feces containing oocysts (blue arrow #3). (Used with permission from Dr. Alexander J. da Silva and Melanie Moser, Centers for Disease Control and Prevention.)

infections are asymptomatic. When contained, the organisms persist as cysts within tissues. There is no inflammation, and the individual remains well unless immunosuppression allows activation of organisms in the cysts.

Congenital infection of the fetus occurs *only* when the mother is infected during pregnancy. If she is infected before the pregnancy, the organism will be in the cyst form and there will be no trophozoites to pass through the placenta. The mother who is

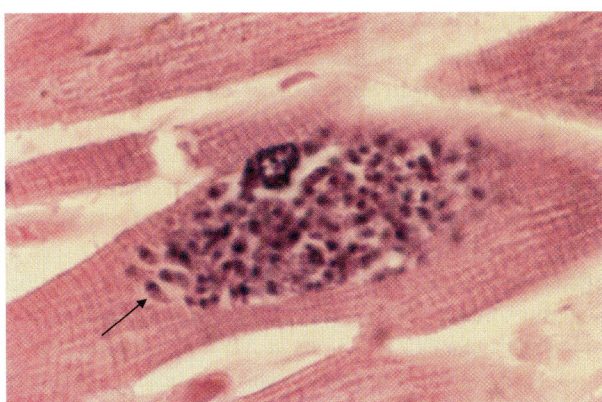

FIGURE 52–6 *Toxoplasma gondii.* Tachyzoite. Arrow points to a tachyzoite of *T. gondii* in cardiac muscle. (Used with permission from Dr. E. Ewing, Jr., Public Health Image Library, Centers for Disease Control and Prevention.)

reinfected during pregnancy but who has immunity from a previous infection will not transmit the organism to her child. Roughly one-third of mothers infected during pregnancy give birth to infected infants, but only 10% of these infants are symptomatic.

Infection by *T. gondii* occurs worldwide. Serologic surveys reveal that in the United States antibodies are found in 5% to 50% of people in various regions. Infection is usually sporadic, but outbreaks associated with ingestion of raw meat or contaminated water occur. Approximately 1% of domestic cats in the United States shed *Toxoplasma* cysts.

Clinical Findings

Most primary infections in immunocompetent adults are asymptomatic, but some resemble infectious mononucleosis, except that the heterophil antibody test is negative. Congenital infection can result in abortion, stillbirth, or neonatal disease with encephalitis, **chorioretinitis**, and hepatosplenomegaly. Fever, jaundice, and **intracranial calcifications** are also seen. Most infected newborns are asymptomatic, but chorioretinitis or mental retardation will develop in some children months or years later. Congenital infection with *Toxoplasma* is one of the leading causes of blindness in children.

In patients with reduced cell-mediated immunity (e.g., patients with acquired immunodeficiency syndrome [AIDS]), life-threatening disseminated disease, primarily encephalitis, occurs. A rim-enhancing lesion representing a brain abscess is often seen on brain imaging (Figure 52–7).

Laboratory Diagnosis

For the diagnosis of acute and congenital infections, an immunofluorescence assay for **IgM antibody** is used. IgM is used to diagnose congenital infection, because IgG can be maternal in origin. Tests of IgG antibody can be used to diagnose acute infections if a significant rise in antibody titer in paired sera is observed.

Microscopic examination of Giemsa-stained preparations shows crescent-shaped trophozoites during acute infections.

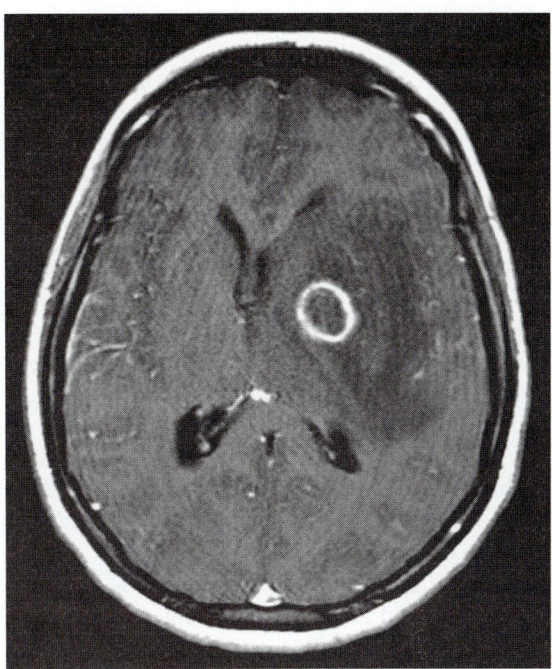

FIGURE 52–7 *Toxoplasma gondii.* A rim-enhancing lesion in the brain of an AIDS patient caused by *Toxoplasma.* (Reproduced with permission from Ropper AH, Samuels M, Klein J: *Adams and Victor's Principles of Neurology*, 11th ed. New York, NY: McGraw Hill; 2019.)

Cysts may be seen in the tissue. The organism can be grown in cell culture. Inoculation into mice can confirm the diagnosis.

Treatment

Congenital toxoplasmosis, whether symptomatic or asymptomatic, should be treated with a combination of sulfadiazine and pyrimethamine. These drugs also constitute the treatment of choice for disseminated disease in immunocompromised patients. Acute toxoplasmosis in an immunocompetent individual is usually self-limited, but any patient with chorioretinitis should be treated.

Prevention

The most effective means of preventing toxoplasmosis is to cook meat thoroughly to kill the cysts. Pregnant women should be especially careful to avoid undercooked meat and contact with cat feces. They should refrain from emptying cat litter boxes. Cats should not be fed raw meat. Trimethoprim-sulfamethoxazole is used to prevent *Toxoplasma* encephalitis in patients infected with human immunodeficiency virus (HIV).

PNEUMOCYSTIS

Disease

Pneumocystis jiroveci is an important cause of pneumonia in immunocompromised individuals. In 2002, taxonomists renamed the human species of *Pneumocystis* as *P. jiroveci* and recommended that *Pneumocystis carinii* be used only to describe the rat species of *Pneumocystis*.

Important Properties

The classification and life cycle of *Pneumocystis* are unclear. Many aspects of its biochemistry indicate that it is a yeast, but it also has several attributes of a protozoan. An analysis of rRNA sequences published in 1988 indicates that *Pneumocystis* should be classified as a **fungus** related to yeasts such as *S. cerevisiae*. Subsequent analysis of mitochondrial DNA and of various enzymes supports the idea that it is a fungus. However, it does not have ergosterol in its membranes as do the fungi. It has cholesterol.

Medically, it is still thought of as a protozoan. In tissue, it appears as a cyst that resembles the cysts of protozoa (Figures 52–2H and 52–8). The findings that it does not grow on fungal media and that antifungal drugs are ineffective have delayed acceptance of its classification as a fungus.

Pneumocystis species are found in domestic animals such as horses and sheep and in a variety of rodents, but it is thought that these animals are not a reservoir for human infection. Each mammalian species is thought to have its own species of *Pneumocystis*.

Pneumocystis species have a major surface glycoprotein that exhibits significant antigenic variation in a manner similar to that of *Trypanosoma brucei*. *Pneumocystis* species have multiple genes encoding these surface proteins, but only one is expressed at a time. This process of programmed rearrangements was first observed in *T. brucei*.

Pathogenesis & Epidemiology

Transmission occurs by **inhalation**, and infection is predominantly in the lungs. The presence of cysts in the alveoli induces an inflammatory response consisting primarily of plasma cells, resulting in a frothy exudate that blocks oxygen exchange. (The presence of plasma cells has led to the name "plasma cell pneumonia.") The organism does not invade the lung tissue.

Pneumonia occurs when host defenses (e.g., the number of CD4-positive [helper] T cells) are reduced. This accounts for the

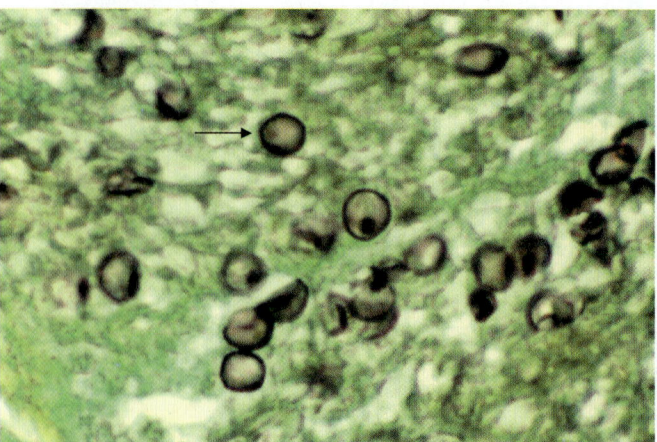

FIGURE 52–8 *Pneumocystis jiroveci.* Arrow points to a cyst of *P. jiroveci* in lung tissue. (Used with permission from Dr. E. Ewing, Jr., Public Health Image Library, Centers for Disease Control and Prevention.)

prominence of *Pneumocystis* pneumonia in patients with AIDS and in premature or debilitated infants. Hospital outbreaks do not occur, and patients with *Pneumocystis* pneumonia are not isolated.

P. jiroveci is distributed worldwide. It is estimated that 70% of people have been infected. Most 5-year-old children in the United States have antibodies to this organism. Asymptomatic infection is therefore quite common. Prior to the advent of immunosuppressive therapy, *Pneumocystis* pneumonia was rarely seen in the United States. Its incidence has paralleled the increase in immunosuppression and the rise in the number of AIDS cases.

Most *Pneumocystis* infections in AIDS patients are new rather than a reactivation of a prior latent infection. This conclusion is based on the finding that *Pneumocystis* recovered from AIDS patients shows resistance to drugs that the patients have not taken.

Clinical Findings

The sudden onset of fever, nonproductive cough, dyspnea, and tachypnea is typical of *Pneumocystis* pneumonia. Bilateral rales and rhonchi are heard, and the chest X-ray shows a diffuse interstitial pneumonia with "ground glass" infiltrates bilaterally. In infants, the disease usually has a more gradual onset. Extrapulmonary *Pneumocystis* infections occur in the late stages of AIDS and affect primarily the liver, spleen, lymph nodes, and bone marrow. The mortality rate of untreated *Pneumocystis* pneumonia approaches 100%.

Laboratory Diagnosis

Diagnosis is made by finding the typical cysts by microscopic examination of lung tissue or fluids obtained by bronchoscopy, bronchial lavage, or open lung biopsy (see Figure 52–7). Sputum is usually less suitable. The cysts can be visualized with methenamine silver, Giemsa, or other tissue stains. Fluorescent antibody staining is also commonly used for diagnosis. PCR-based tests using respiratory tract specimens are also useful. The organism stains poorly with Gram stain. There is no serologic test, and the organism has not been grown in culture.

Treatment & Prevention

The treatment of choice is a combination of trimethoprim and sulfamethoxazole (TMP-SMX) (Bactrim, Septra). Pentamidine and atovaquone are alternative drugs. Regarding prevention, TMP-SMX should be used as chemoprophylaxis in patients whose CD4 counts are below 200. If the patient has had a severe reaction to TMP-SMX, atovaquone can be used.

TRYPANOSOMA

The genus *Trypanosoma* includes three major pathogens: *Trypanosoma cruzi*, *Trypanosoma gambiense*, and *Trypanosoma rhodesiense*.[2]

---

[2]Taxonomically, the last two organisms are morphologically identical species called T. brucei gambiense and T. brucei rhodesiense, but the shortened names are used here.

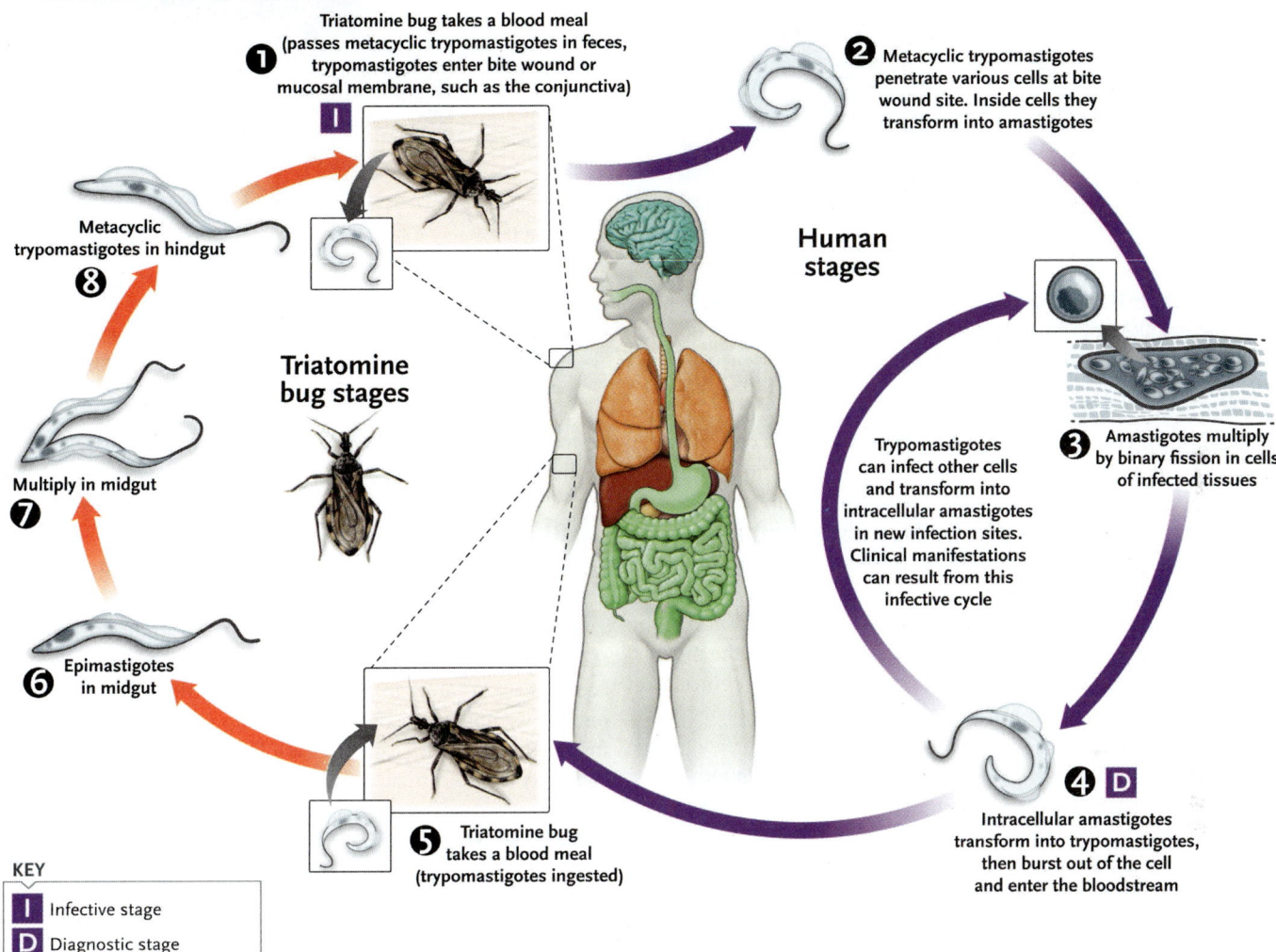

Chagas' Disease
(American Trypanosomiasis)
(Trypanosoma cruzi)

Triatomine bug takes a blood meal
① (passes metacyclic trypomastigotes in feces, trypomastigotes enter bite wound or mucosal membrane, such as the conjunctiva)

② Metacyclic trypomastigotes penetrate various cells at bite wound site. Inside cells they transform into amastigotes

Human stages

Metacyclic trypomastigotes in hindgut ⑧

Triatomine bug stages

Trypomastigotes can infect other cells and transform into intracellular amastigotes in new infection sites. Clinical manifestations can result from this infective cycle

③ Amastigotes multiply by binary fission in cells of infected tissues

Multiply in midgut ⑦

Epimastigotes in midgut ⑥

⑤ Triatomine bug takes a blood meal (trypomastigotes ingested)

④ Intracellular amastigotes transform into trypomastigotes, then burst out of the cell and enter the bloodstream

KEY
I Infective stage
D Diagnostic stage

FIGURE 52–9 *Trypanosoma cruzi.* Life cycle. Right side of figure describes the stages within the human (blue arrows). Humans are infected at step 1 when triatomine (reduviid) bug bites human and defecates at bite site. Trypomastigotes in feces enter bite wound. Amastigotes form within cells, especially heart muscle and neural tissue. Reduviid bug is infected at step 5 when it ingests trypomastigotes in human blood. Left side of figure describes the stages within the reduviid bug (red arrows). (Used with permission from Dr. Alexander J. da Silva and Melanie Moser, Centers for Disease Control and Prevention.)

1. Trypanosoma cruzi

Disease

T. cruzi is the cause of Chagas' disease (American trypanosomiasis).

Important Properties

The life cycle of *T. cruzi* is shown in Figure 52–9. The life cycle involves the **reduviid bug** (*Triatoma*, cone-nose or kissing bug) as the vector and both humans and animals as reservoir hosts. The animal reservoirs include domestic cats and dogs and wild species such as the armadillo, raccoon, and rat. The cycle in the reduviid bug begins with ingestion of trypomastigotes in the blood of the reservoir host. In the insect gut, they multiply and differentiate first into epimastigotes and then into trypomastigotes. When the bug bites again, the site is contaminated with feces containing trypomastigotes, which enter the blood of the person (or other reservoir) and form nonflagellated amastigotes within host cells. Many cells can be affected, but myocardial, glial, and reticuloendothelial cells are the most frequent sites. To complete the cycle, amastigotes differentiate into trypomastigotes, which enter the blood and are taken up by the reduviid bug (Figures 52–10A to C and 52–11).

Pathogenesis & Epidemiology

Chagas' disease occurs primarily in rural Central and South America. Acute Chagas' disease occurs rarely in the United

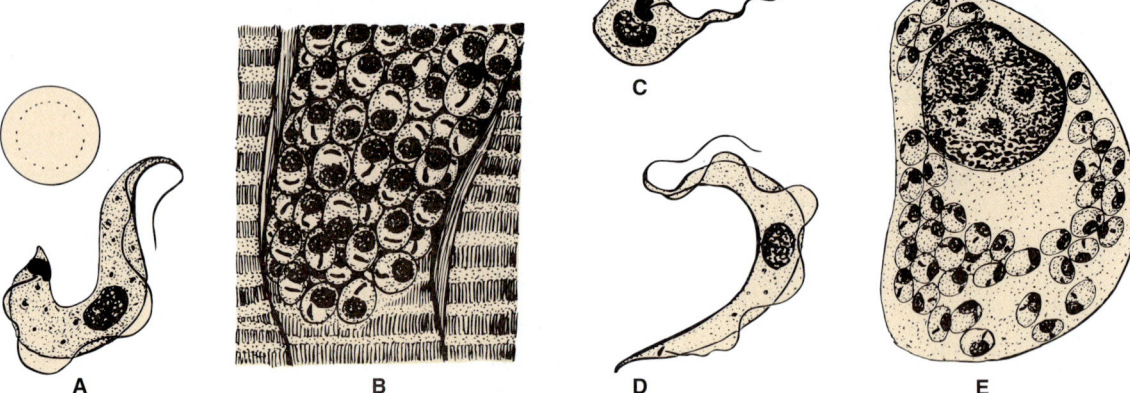

FIGURE 52–10 **A:** *Trypanosoma cruzi* trypomastigote found in human blood (1200×). **B:** *T. cruzi* amastigotes found in cardiac muscle (850×). **C:** *T. cruzi* epimastigote found in reduviid bug (1200×). **D:** *Trypanosoma brucei gambiense* or *rhodesiense* trypomastigote found in human blood (1200×). **E:** *Leishmania donovani* amastigotes within splenic macrophages (1000×). (Circle with inner dotted line represents a red blood cell.)

States, but the chronic form causing myocarditis and congestive heart failure is seen with increasing frequency in immigrants from Latin America. The disease is seen primarily in rural areas because the reduviid bug lives in the walls of rural huts and feeds at night. It bites preferentially around the mouth or eyes, hence the name "kissing bug."

The amastigotes can kill cells and cause inflammation, consisting mainly of mononuclear cells. **Cardiac muscle** is the most frequently and severely affected tissue. In addition, neuronal damage leads to cardiac arrhythmias and loss of tone in the colon (**megacolon**) and esophagus (**megaesophagus**). During the acute phase, there are both trypomastigotes in the blood and amastigotes intracellularly in the tissues. In the chronic phase, the organism persists in the amastigote form.

Chagas' disease has occurred in the United States in recipients of either blood transfusions or organ transplants from infected donors. The organism can also be transmitted congenitally from an infected mother to the fetus across the placenta.

Clinical Findings

The acute phase of Chagas' disease consists of facial edema and a nodule (chagoma) near the bite, coupled with fever, lymphadenopathy, and hepatosplenomegaly. A bite around the eye can result in unilateral palpebral swelling called Romaña's sign. The acute phase resolves in about 2 months. Most individuals then remain asymptomatic, but some progress to the chronic form with myocarditis and megacolon. Death from chronic Chagas' disease is usually due to cardiac arrhythmias or congestive heart failure.

Laboratory Diagnosis

Acute disease is diagnosed by demonstrating the presence of trypomastigotes in thick or thin films of the patient's blood. Both stained and wet preparations should be examined, the latter for motile organisms. Because the trypomastigotes are not numerous in the blood, other diagnostic methods may be required, namely, (1) a stained preparation of a bone marrow aspirate or muscle biopsy specimen (which may reveal amastigotes); (2) culture of the organism on special medium; and (3) **xenodiagnosis**, which consists of allowing an uninfected, laboratory-raised reduviid bug to feed on the patient and, after several weeks, examining the intestinal contents of the bug for the organism.

Serologic tests can be helpful also. The indirect fluorescent antibody test is the earliest to become positive. Indirect hemagglutination and complement fixation tests are also available. Diagnosis of chronic disease is difficult because there are few trypomastigotes in the blood. Xenodiagnosis and serologic tests are used.

Treatment & Prevention

The drug of choice for the acute phase is nifurtimox, which kills trypomastigotes in the blood but is much less effective against amastigotes in tissue. Benznidazole is an alternative drug. There is no effective drug against the chronic form.

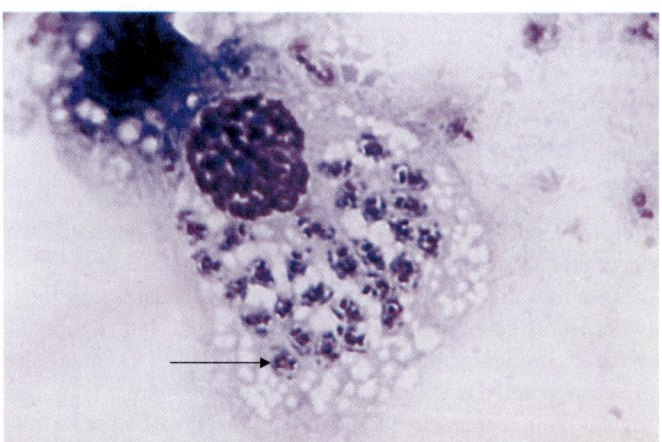

FIGURE 52–11 *Trypanosoma cruzi.* Amastigotes. Arrow points to an amastigote (nonflagellated form) in cytoplasm. (Used with permission from Dr. Sulzer A. J., Public Health Image Library, Centers for Disease Control and Prevention.)

Prevention involves protection from the reduviid bite, improved housing, and insect control. No prophylactic drug or vaccine is available. Blood for transfusion is tested for the presence of antibodies to *T. cruzi*. Blood containing antibodies should not be used.

2. *Trypanosoma gambiense* & *Trypanosoma rhodesiense*

Disease

These organisms cause sleeping sickness (African trypanosomiasis). They are also known as *T. brucei gambiense* and *T. brucei rhodesiense*.

Important Properties

The life cycle of *T. brucei* is shown in Figure 52–12. The morphology and life cycle of the two species are similar. The vector for both is the **tsetse fly**, *Glossina*, but different species of fly are involved for each. Humans are the reservoir for *T. gambiense*, whereas *T. rhodesiense* has reservoirs in both domestic animals (especially cattle) and wild animals (e.g., antelopes).

The 3-week life cycle in the tsetse fly begins with ingestion of trypomastigotes in a blood meal from the reservoir host. They multiply in the insect gut and then migrate to the salivary glands, where they transform into epimastigotes, multiply further, and then form metacyclic trypomastigotes, which are transmitted by the tsetse fly bite. The organisms in the saliva are

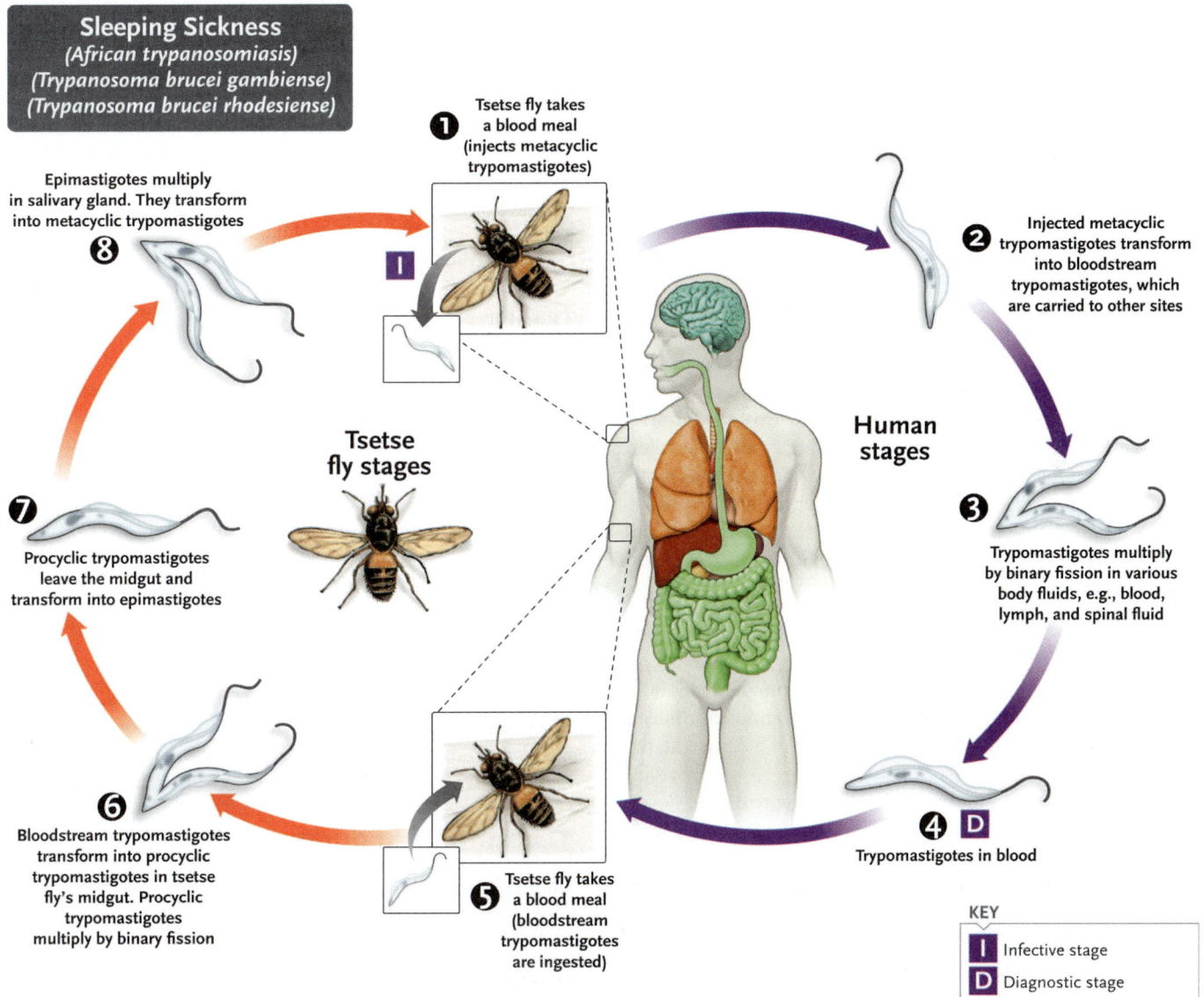

FIGURE 52–12 *Trypanosoma brucei*. Life cycle. Right side of figure describes the stages within the human (blue arrows). Humans are infected at step 1 when the tsetse fly bites human and injects trypomastigotes into bloodstream. Tsetse fly is infected at step 5 when it ingests trypomastigotes in human blood. Left side of figure describes the stages within the tsetse fly (red arrows). (Used with permission from Dr. Alexander J. da Silva and Melanie Moser, Centers for Disease Control and Prevention.)

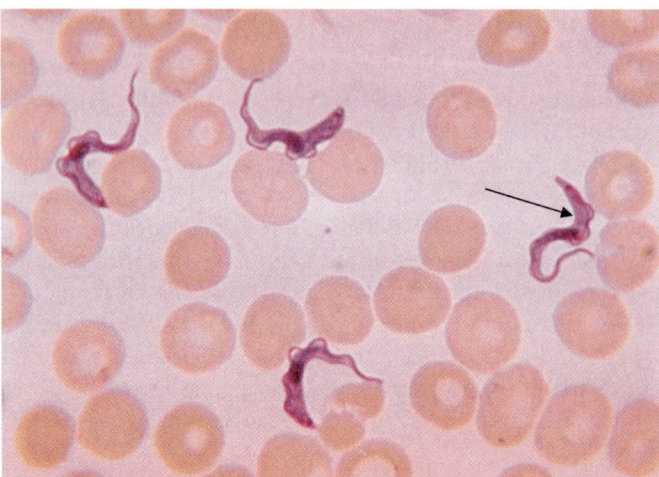

FIGURE 52–13 *Trypanosoma brucei.* Trypomastigotes. Arrow points to a trypomastigote (the flagellated form) in the blood. (Used with permission from Dr. Schultz M., Public Health Image Library, Centers for Disease Control and Prevention.)

injected into the skin, where they enter the bloodstream, differentiate into blood-form trypomastigotes, and multiply, thereby completing the cycle (Figures 52–10D and 52–13). Note that these species are rarely found as amastigotes in tissue, in contrast to *T. cruzi* and *Leishmania* species, in which amastigotes are commonly found.

These trypanosomes exhibit remarkable **antigenic variation** of their surface glycoproteins, with hundreds of antigenic types found. One antigenic type will coat the surface of the parasites for approximately 10 days, followed by other types in sequence in the new progeny. This variation is due to sequential movement of the glycoprotein genes to a preferential location on the chromosome, where only that specific gene is transcribed into mRNA. These antigenic variations allow the organism to continually evade the host immune response.

Pathogenesis & Epidemiology

The trypomastigotes spread from the skin through the blood to the lymph nodes and the brain. The typical somnolence (**sleeping sickness**) progresses to coma as a result of a demyelinating encephalitis.

In the acute form, a cyclical fever spike (approximately every 2 weeks) occurs that is related to antigenic variation. As antibody-mediated agglutination and lysis of the trypomastigotes occur, the fever subsides. However, a few antigenic variants survive, multiply, and cause a new fever spike. This cycle repeats itself over a long period. The lytic antibody is directed against the surface glycoprotein.

The disease is endemic in sub-Saharan Africa, the natural habitat of the tsetse fly. Both sexes of fly take blood meals and can transmit the disease. The fly is infectious throughout its 2- to 3-month lifetime. *T. gambiense* is the species that causes the disease along watercourses in West Africa, whereas *T. rhodesiense* is found in the arid regions of East Africa. Both species are found in central Africa.

Clinical Findings

Although both species cause sleeping sickness, the progress of the disease differs. *T. gambiense*–induced disease runs a low-grade chronic course over a few years, whereas *T. rhodesiense* causes a more acute, rapidly progressive disease that, if untreated, is usually fatal within several months.

The initial lesion is an indurated skin ulcer ("trypanosomal chancre") at the site of the fly bite. After the organisms enter the blood, intermittent weekly fever and lymphadenopathy develop. Enlargement of the posterior cervical lymph nodes (Winterbottom's sign) is commonly seen. The encephalitis is characterized initially by headache, insomnia, and mood changes, followed by muscle tremors, slurred speech, and apathy that progress to somnolence and coma. Untreated disease is usually fatal as a result of pneumonia.

Laboratory Diagnosis

During the early stages, microscopic examination of the blood (either wet films or thick or thin smears) reveals trypomastigotes (see Figure 52–13). An aspirate of the chancre or enlarged lymph node can also demonstrate the parasites. The presence of trypanosomes in the spinal fluid, coupled with an elevated protein level and pleocytosis, indicates that the patient has entered the late, encephalitic stage. Serologic tests, especially the ELISA for IgM antibody, can be helpful.

Treatment & Prevention

Treatment must be initiated before the development of encephalitis, because suramin, the most effective drug, does not pass the blood–brain barrier well. Suramin can cure the infection if given early. Pentamidine is an alternative drug. If central nervous system symptoms are present, suramin (to clear the parasitemia) followed by melarsoprol should be given.

The most important preventive measure is protection against the fly bite, using netting and protective clothing. Clearing the forest around villages and using insecticides are helpful measures. No vaccine is available.

LEISHMANIA

The genus *Leishmania* includes four major pathogens: *Leishmania donovani*, *Leishmania tropica*, *Leishmania mexicana*, and *Leishmania braziliensis*.

1. Leishmania donovani

Disease

L. donovani is the cause of kala-azar (visceral leishmaniasis).

Important Properties

The life cycle of *L. donovani* is shown in Figure 52–14. The life cycle involves the **sandfly**[3] as the vector and a variety of mammals such as dogs, foxes, and rodents as reservoirs.

---

[3]Phlebotomus species in the Old World; Lutzomyia species in South America.

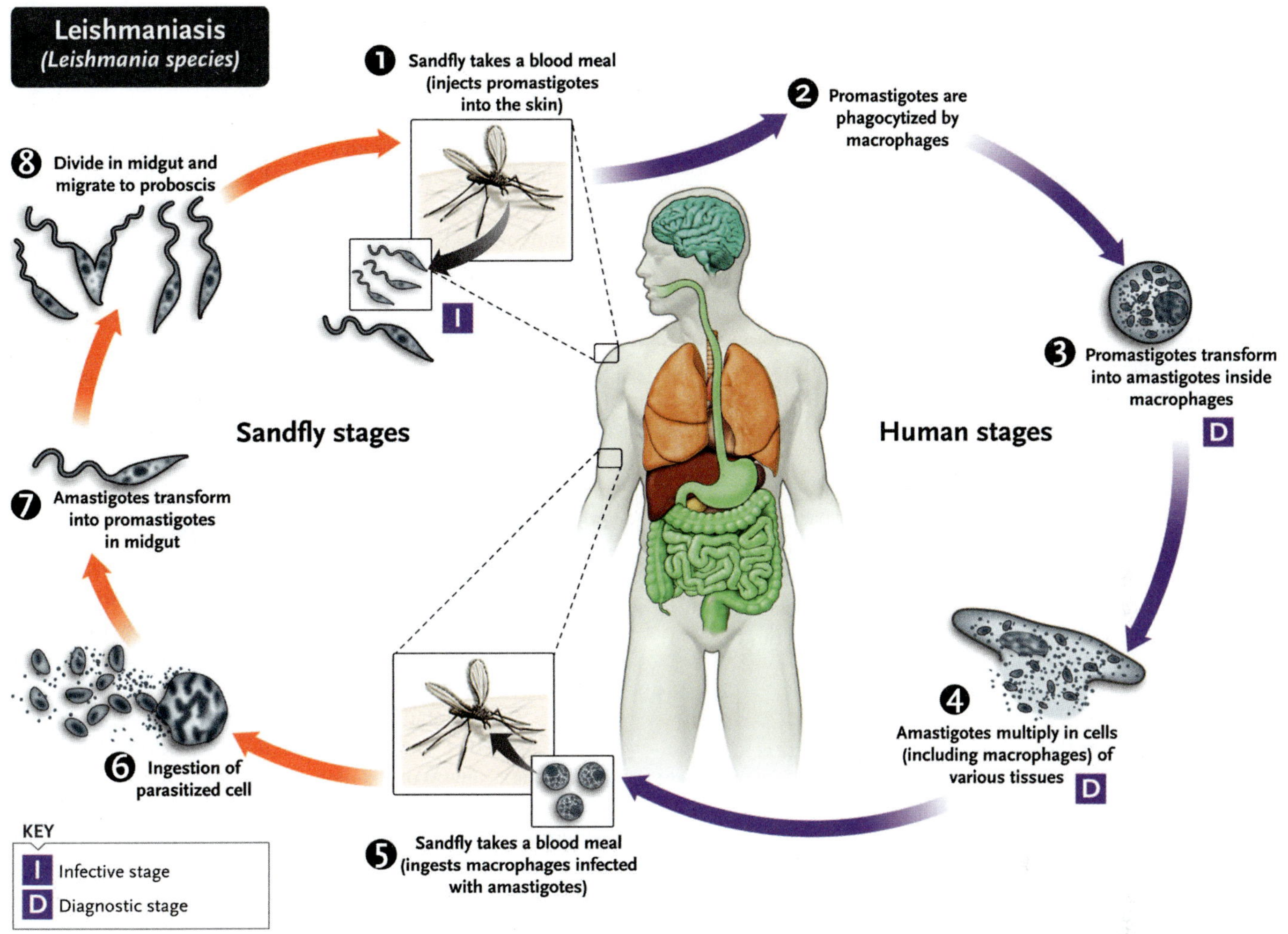

Leishmaniasis
(Leishmania species)

Sandfly stages

❽ Divide in midgut and migrate to proboscis

❶ Sandfly takes a blood meal (injects promastigotes into the skin)

I

❼ Amastigotes transform into promastigotes in midgut

❻ Ingestion of parasitized cell

❺ Sandfly takes a blood meal (ingests macrophages infected with amastigotes)

KEY

I Infective stage
D Diagnostic stage

Human stages

❷ Promastigotes are phagocytized by macrophages

❸ Promastigotes transform into amastigotes inside macrophages **D**

❹ Amastigotes multiply in cells (including macrophages) of various tissues **D**

FIGURE 52–14 *Leishmania donovani.* Life cycle. Right side of figure describes the stages within the human (blue arrows). Humans are infected at step 1 when the sandfly bites human and injects promastigotes. Sandfly is infected at step 5 when it ingests macrophages containing amastigotes in human blood. Left side of figure describes the stages within the sandfly (red arrows). (Used with permission from Dr. Alexander J. da Silva and Melanie Moser, Centers for Disease Control and Prevention.)

Only female flies are vectors because only they take blood meals (a requirement for egg maturation). When the sandfly sucks blood from an infected host, it ingests **macrophages containing amastigotes** (Figures 52–10E and 52–15).[4]

After dissolution of the macrophages, the freed amastigotes differentiate into promastigotes in the gut. They multiply and then migrate to the pharynx and proboscis, where they can be transmitted during the next bite. The cycle in the sandfly takes approximately 10 days.

Shortly after an infected sandfly bites a human, the promastigotes are engulfed by macrophages, where they transform into amastigotes (Figure 52–9E). Amastigotes can remain in the cytoplasm of macrophages because they can prevent fusion of the vacuole with lysosomes.

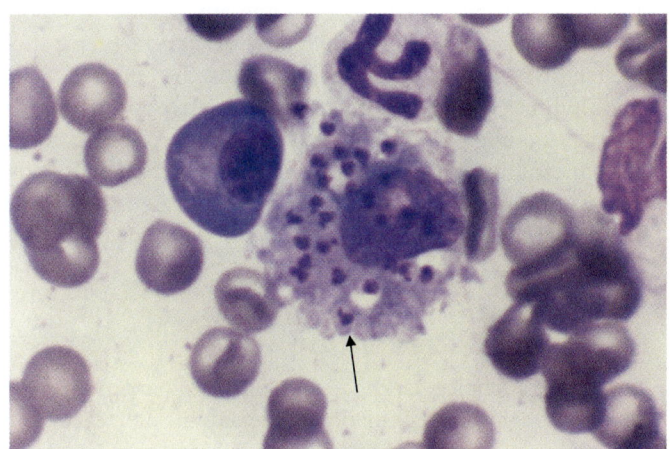

FIGURE 52–15 *Leishmania donovani.* Amastigotes. Arrow points to an amastigote (nonflagellated form) in cytoplasm of bone marrow cell. (Used with permission from Dr. Francis Chandler, Public Health Image Library, Centers for Disease Control and Prevention.)

[4] Amastigotes are nonflagellated, in contrast to promastigotes, which have a flagellum with a characteristic anterior kinetoplast.

The infected cells die and release progeny amastigotes that infect other macrophages and reticuloendothelial cells. The cycle is completed when the fly ingests macrophages containing the amastigotes.

Pathogenesis & Epidemiology

In visceral leishmaniasis, the organs of the **reticuloendothelial** system (liver, spleen, and bone marrow) are the most severely affected. Reduced bone marrow activity, coupled with cellular destruction in the spleen, results in anemia, leukopenia, and thrombocytopenia. This leads to secondary infections and a tendency to bleed. The striking **enlargement of the spleen** is due to a combination of proliferating macrophages and sequestered blood cells. The marked increase in IgG is neither specific nor protective.

Kala-azar occurs in three distinct epidemiologic patterns. In one area, which includes the Mediterranean basin, the Middle East, southern Russia, and parts of China, the reservoir hosts are primarily dogs and foxes. In sub-Saharan Africa, rats and small carnivores (e.g., civets) are the main reservoirs. A third pattern is seen in India and neighboring countries (and Kenya), in which humans appear to be the only reservoir.

Clinical Findings

Symptoms begin with intermittent fever, weakness, and weight loss. Massive enlargement of the spleen is characteristic. Hyperpigmentation of the skin is seen in light-skinned patients (kala-azar means **black sickness**). The course of the disease runs for months to years. Initially, patients feel reasonably well despite persistent fever. As anemia, leukopenia, and thrombocytopenia become more profound, weakness, infection, and gastrointestinal bleeding occur. Untreated severe disease is nearly always fatal as a result of secondary infection.

Laboratory Diagnosis

Diagnosis is usually made by detecting amastigotes in a bone marrow, spleen, or lymph node biopsy or "touch" preparation (see Figure 52–15). The organisms can also be cultured. Serologic (indirect immunofluorescence) tests are positive in most patients. Although not diagnostic, a very high concentration of IgG is indicative of infection. A skin test using a crude homogenate of promastigotes (leishmanin) as the antigen is available. The skin test is negative during active disease but positive in patients who have recovered.

Treatment & Prevention

The drug of choice is either liposomal amphotericin B or sodium stibogluconate. With proper therapy, the mortality rate is reduced to almost 5%. Recovery results in permanent immunity.

Prevention involves protection from sandfly bites (use of netting, protective clothing, and insect repellents) and insecticide spraying.

2. *Leishmania tropica*, *Leishmania mexicana*, & *Leishmania braziliensis*

Disease

L. tropica and *L. mexicana* both cause cutaneous leishmaniasis; the former organism is found in the Old World, whereas the latter is found only in the Americas. *L. braziliensis* causes mucocutaneous leishmaniasis, which occurs only in Central and South America.

Important Properties

Sandflies are the vectors for these three organisms, as they are for *L. donovani*, and forest rodents are their main reservoirs. The life cycle of these parasites is essentially the same as that of *L. donovani*.

Pathogenesis & Epidemiology

The lesions are confined to the skin in cutaneous leishmaniasis and to the mucous membranes, cartilage, and skin in mucocutaneous leishmaniasis. A granulomatous response occurs, and a necrotic ulcer forms at the bite site. The lesions tend to become superinfected with bacteria.

Old World cutaneous leishmaniasis (Oriental sore, Delhi boil), caused by *L. tropica*, is endemic in the Middle East, Africa, and India. New World cutaneous leishmaniasis (chicle ulcer, bay sore), caused by *L. mexicana*, is found in Central and South America. Mucocutaneous leishmaniasis (espundia), caused by *L. braziliensis*, occurs mostly in Brazil and Central America, primarily in forestry and construction workers.

Clinical Findings

The initial lesion of cutaneous leishmaniasis is a red papule at the bite site, usually on an exposed extremity. This enlarges slowly to form multiple satellite nodules that coalesce and ulcerate (Figure 52–16). There is usually a single lesion that heals

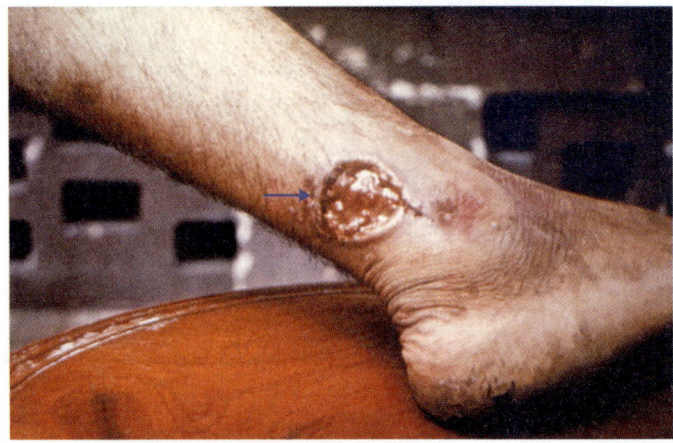

FIGURE 52–16 Cutaneous leishmaniasis. Blue arrow points to shallow ulcer near right ankle. (Used with permission from Dr. Mae Melvin, Public Health Image Library, Centers for Disease Control and Prevention.)

spontaneously in patients with a competent immune system. However, in certain individuals, if cell-mediated immunity does not develop, the lesions can spread to involve large areas of skin and contain enormous numbers of organisms.

Mucocutaneous leishmaniasis begins with a papule at the bite site, but then metastatic lesions form, usually at the mucocutaneous junction of the nose and mouth. Disfiguring granulomatous, ulcerating lesions destroy nasal cartilage but not adjacent bone. These lesions heal slowly, if at all. Death can occur from secondary infection.

Laboratory Diagnosis

Diagnosis is usually made microscopically by demonstrating the presence of **amastigotes** in a smear taken from the skin lesion. The leishmanin skin test becomes positive when the skin ulcer appears and can be used to diagnose cases outside the area of endemic infection.

Treatment & Prevention

The drug of choice is sodium stibogluconate, but the results are frequently unsatisfactory. Prevention involves protection from sandfly bites by using netting, window screens, protective clothing, and insect repellents.

SELF-ASSESSMENT QUESTIONS

1. Regarding *Plasmodium* species, which one of the following is most accurate?

   (A) These organisms are transmitted by the bite of female *Anopheles* mosquitoes.

   (B) The bite of the vector injects merozoites into the bloodstream that then infect red blood cells.

   (C) Both male and female gametocytes are formed in the vector and are injected into the person at the time of the bite.

   (D) Hypnozoites are produced by *Plasmodium falciparum* and can cause relapses of malaria after the acute phase is over.

   (E) Malaria caused by *Plasmodium vivax* is characterized by a cerebral malaria and blackwater fever more often than malaria caused by the other three species.

2. Regarding drugs used to treat or prevent malaria, which one of the following is most accurate?

   (A) The combination of atovaquone and proguanil is useful for the treatment of acute malaria but not for prevention.

   (B) Chloroquine is the drug of choice in malaria caused by *Plasmodium falciparum* because resistance to the drug is rare.

   (C) Mefloquine is useful for the prevention of chloroquine-sensitive *Plasmodium falciparum* but not for chloroquine-resistant strains.

   (D) Artemisinin derivatives, such as artesunate and artemether, are effective in the treatment of multiple-drug resistant *Plasmodium falciparum*.

   (E) Primaquine is useful in the treatment of infections caused by *Plasmodium falciparum* because it kills the hypnozoites residing in the liver.

3. Regarding *Toxoplasma gondii*, which one of the following is most accurate?

   (A) One way to prevent this infection is to advise pregnant women not to drink unpasteurized milk.

   (B) The form of *Toxoplasma* found in the tissue cysts in humans is the rapidly dividing tachyzoite.

   (C) The most important definitive host (the host in which the sexual cycle occurs) for *Toxoplasma* is the domestic cat.

   (D) Infection in people with reduced cell-mediated immunity, such as AIDS patients, is characterized by persistent watery (nonbloody) diarrhea.

   (E) If your patient is a pregnant woman who has IgM antibody to *Toxoplasma* in her blood, then you can tell her that it is unlikely that her fetus is at risk for infection.

4. Regarding *Pneumocystis jiroveci*, which one of the following is most accurate?

   (A) The treatment of choice is a combination of penicillin G and an aminoglycoside.

   (B) Finding oval cysts in bronchial lavage fluid supports a diagnosis of *Pneumocystis* pneumonia.

   (C) Large domestic animals such as cows and sheep are an important reservoir of human infection with this organism.

   (D) Patients with a CD4 count below 200 should receive the vaccine containing the surface glycoprotein as the immunogen.

   (E) Transmission occurs by the ingestion of food contaminated with the organism, after which it enters the bloodstream and is transported to the lung.

5. Regarding *Trypanosoma cruzi*, which one of the following is most accurate?

   (A) Humans are the main reservoir of *T. cruzi*.

   (B) The drug of choice for the acute phase of Chagas' disease is chloroquine.

   (C) The vector for *T. cruzi*, the cause of Chagas' disease, is the reduviid (cone-nosed) bug.

   (D) Seeing trypomastigotes in a muscle biopsy supports the diagnosis of Chagas' disease.

   (E) The main site of disease caused by *T. cruzi* is skeletal muscle, resulting in severe muscle pain.

6. Regarding leishmaniasis, which one of the following is most accurate?

   (A) Mefloquine is effective in preventing disease caused by *Leishmania donovani*.

   (B) Large domestic animals such as cattle are the principal reservoir of *L. donovani*.

   (C) Both visceral leishmaniasis and cutaneous leishmaniasis are transmitted by the bite of sandflies.

   (D) Marked enlargement of the heart on chest X-ray is a typical finding of visceral leishmaniasis.

   (E) Pathologists examining a specimen for the presence of *L. donovani* should look primarily at eosinophils in the peripheral blood.

7. Your patient is a 20-year-old man who, while playing soccer, experienced palpitations and dizziness and then fainted. An electrocardiogram showed right bundle branch block. Holter monitoring showed multiple runs of ventricular tachycardia. A ventricular myocardial biopsy was performed. Microscopic examination revealed a lymphocytic inflammatory process surrounding areas containing amastigotes. The patient was born

and raised in rural El Salvador and came to this country 2 years ago. Of the following, which one is the most likely cause?

(A) *Leishmania donovani*
(B) *Plasmodium falciparum*
(C) *Toxoplasma gondii*
(D) *Trypanosoma brucei*
(E) *Trypanosoma cruzi*

8. Your patient is a 25-year-old man with fever and weight loss for the past 3 weeks. He is a soldier in the U.S. Army who recently returned from a tour of duty in the Middle East. Physical exam was noncontributory. Laboratory tests revealed anemia and leukopenia. Multiple blood cultures for bacteria and fungi were negative, as was a test for the p24 antigen of HIV. Computed tomography (CT) scan of the abdomen revealed splenomegaly. A bone marrow biopsy was performed, and a stained sample revealed amastigotes within mononuclear cells. Of the following, which one is the most likely cause?

(A) *Leishmania donovani*
(B) *Plasmodium falciparum*
(C) *Toxoplasma gondii*
(D) *Trypanosoma brucei*
(E) *Trypanosoma cruzi*

9. Your patient is a 55-year-old man with fever and increasing fatigue during the past week. Today, he was so weak he "could barely stand up." He had been working in Cameroon and Chad for 2 months and returned 2 weeks ago. On examination, he was febrile to 40°C, hypotensive, and tachycardic. Pertinent lab work revealed anemia and thrombocytopenia. Blood smear revealed ring-shaped trophozoites within red blood cells. Of the following, which one is the most likely cause?

(A) *Leishmania donovani*
(B) *Plasmodium falciparum*
(C) *Toxoplasma gondii*
(D) *Trypanosoma brucei*
(E) *Trypanosoma cruzi*

10. Your patient is a 35-year-old woman who has just had a seizure. A CT scan shows a ring-enhancing lesion in her brain. History reveals that she is an intravenous drug user and is HIV antibody positive with a CD4 count of 30. Serologic tests confirm that the patient is infected with *Toxoplasma gondii*. Which one of the following is the best choice of drug to treat her cerebral toxoplasmosis?

(A) Artemether
(B) Atovaquone
(C) Mefloquine
(D) Metronidazole
(E) Pyrimethamine and sulfadiazine

11. Regarding the patient in Question 10, she was treated and recovered without sequelae. Antiretroviral therapy was instituted. As long as her CD4 count remains below 100, she should receive chemoprophylaxis to prevent recurrent disease caused by *Toxoplasma gondii*. Which one of the following is the best chemoprophylactic drug?

(A) Artesunate
(B) Metronidazole
(C) Pentamidine
(D) Primaquine
(E) Trimethoprim-sulfamethoxazole

ANSWERS

(1) **(A)**
(2) **(D)**
(3) **(C)**
(4) **(B)**
(5) **(C)**
(6) **(C)**
(7) **(E)**
(8) **(A)**
(9) **(B)**
(10) **(E)**
(11) **(E)**

SUMMARIES OF ORGANISMS

Brief summaries of the organisms described in this chapter begin on page 706. Please consult these summaries for a rapid review of the essential material.

PRACTICE QUESTIONS: USMLE & COURSE EXAMINATIONS

Questions on the topics discussed in this chapter can be found in the Parasitology section of Part XIII: USMLE (National Board) Practice Questions starting on page 755. Also see Part XIV: USMLE (National Board) Practice Examination starting on page 775.

53

Minor Protozoan Pathogens

The medically important stages in the life cycle of certain minor protozoa are described in Table 53–1.

ACANTHAMOEBA & NAEGLERIA

Acanthamoeba castellanii and *Naegleria fowleri* are free-living **amebas** that cause **meningoencephalitis**. The organisms are found in warm freshwater lakes and in soil. Their life cycle involves trophozoite and cyst stages. Cysts are quite resistant and are not killed by chlorination.

Naegleria trophozoites usually enter the body through mucous membranes while an individual is **swimming**. They can penetrate the nasal mucosa and cribriform plate to produce a purulent meningitis and encephalitis that are usually rapidly fatal (Figure 53–1). *Acanthamoeba* is carried into the skin or eyes during trauma. *Acanthamoeba* infections occur primarily in immunocompromised individuals, whereas *Naegleria* infections occur in otherwise healthy persons, usually children. In the United States, these rare infections occur mainly in the southern states and California.

Diagnosis is made by finding amebas in the spinal fluid. The prognosis is poor even in treated cases. Amphotericin B may be effective in *Naegleria* infections. Pentamidine, ketoconazole, or flucytosine may be effective in *Acanthamoeba* infections.

Acanthamoeba also causes **keratitis**—an inflammation of the cornea that occurs primarily in those who wear contact lenses. With increasing use of contact lenses, keratitis has become the most common disease associated with *Acanthamoeba* infection. The amebas have been recovered from contact lenses, lens cases, and lens disinfectant solutions. Tap water contaminated with amebas is the source of infection for lens users.

BABESIA

Babesia microti causes babesiosis—a zoonosis acquired chiefly in the coastal areas and islands off the northeastern coast of the United States (e.g., Nantucket Island). The sporozoan organism is endemic in rodents and is transmitted by the bite of the **tick** *Ixodes dammini* (renamed *Ixodes scapularis*), the same species of tick that transmits *Borrelia burgdorferi*, the agent of Lyme disease. *Babesia* infects red blood cells, causing them to lyse, but unlike plasmodia, it has no exoerythrocytic phase. Asplenic patients and patients being treated with rituximab are affected more severely.

TABLE 53–1 Medically Important Stages in Life Cycle of Certain Minor Protozoa

Organism	Insect Vector	Stage that Infects Humans	Stage(s) in Humans Most Associated with Disease	Important Stage(s) Outside of Humans
Acanthamoeba and *Naegleria*	None	Trophozoite	Trophozoites in meninges	Cyst
Babesia	Tick (*Ixodes*)	Sporozoite in tick saliva	Trophozoites and merozoites in red blood cells	None

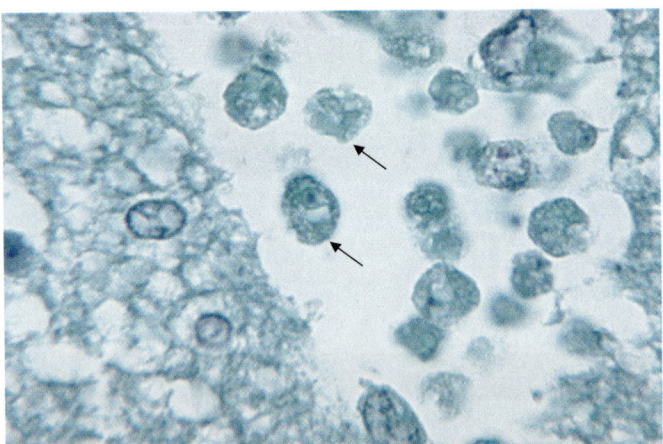

FIGURE 53–1 *Naegleria fowleri.* Trophozoite. Arrows point to two ameba-shaped trophozoites in brain tissue. (Reproduced with permission from Public Health Image Library, Centers for Disease Control and Prevention.)

The influenza-like symptoms begin gradually and may last for several weeks. Hepatosplenomegaly and anemia occur. Diagnosis is made by seeing intraerythrocytic ring-shaped parasites on Giemsa-stained blood smears. The intraerythrocytic ring-shaped trophozoites are often in tetrads in the form of a **Maltese cross** (Figure 53–2). Unlike the case with plasmodia, there is no pigment in the erythrocytes. The treatment of choice for mild to moderate disease is a combination of atovaquone and azithromycin. Patients with severe disease should receive a combination of quinidine and clindamycin. Exchange transfusion should also be considered in patients with severe disease. Prevention involves protection from tick bites and, if a person is bitten, prompt removal of the tick.

BALANTIDIUM

Balantidium coli is the **only ciliated protozoan** that causes human disease (i.e., **diarrhea**). It is found worldwide but only infrequently in the United States. Domestic animals, especially

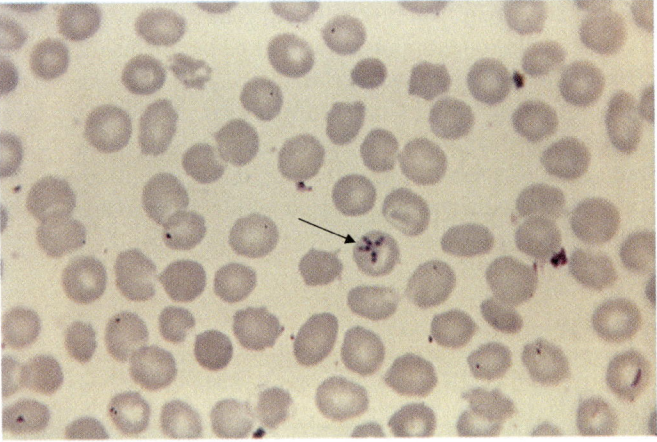

FIGURE 53–2 *Babesia microti.* Trophozoites in tetrads. Arrow points to a red blood cell containing four trophozoites in a tetrad resembling a Maltese cross. (Used with permission from Dr. Glenn S., Public Health Image Library, Centers for Disease Control and Prevention.)

pigs, are the main reservoir for the organism, and humans are infected after ingesting the cysts in food or water contaminated with animal or human feces. The trophozoites excyst in the small intestine, travel to the colon, and, by burrowing into the wall, cause an ulcer similar to that of *Entamoeba histolytica*. However, unlike the case with *E. histolytica*, extraintestinal lesions do not occur.

Most infected individuals are asymptomatic; diarrhea rarely occurs. Diagnosis is made by finding large ciliated trophozoites or large cysts with a characteristic V-shaped nucleus in the stool. There are no serologic tests. The treatment of choice is tetracycline. Prevention consists of avoiding contamination of food and water by domestic animal feces.

CYCLOSPORA

Cyclospora cayetanensis is an intestinal protozoan that causes watery diarrhea in both immunocompetent and immunocompromised individuals. It is classified as a member of the Coccidia.[1]

The organism is acquired by fecal–oral transmission, especially via contaminated food and water supplies. Several outbreaks in the United States were caused by the ingestion of imported fresh produce, such as basil and raspberries. There is no evidence for an animal reservoir.

The diarrhea can be prolonged and relapsing, especially in immunocompromised patients. Infection occurs worldwide. The diagnosis is made microscopically by observing the spherical oocysts in a modified acid-fast stain of a stool sample. A PCR-based test that detects *Cyclospora* nucleic acid in the stool is available. There are no serologic tests. The treatment of choice is trimethoprim-sulfamethoxazole.

ISOSPORA

Isospora belli is an intestinal protozoan that causes **diarrhea**, especially in **immunocompromised patients** (e.g., those with acquired immunodeficiency syndrome [AIDS]). Its life cycle parallels that of other members of the Coccidia. The organism is acquired by fecal–oral transmission of oocysts from either human or animal sources. The oocysts excyst in the upper small intestine and invade the mucosa, causing destruction of the brush border.

The disease in immunocompromised patients presents as a chronic, profuse, watery diarrhea. The pathogenesis of the diarrhea is unknown. Diagnosis is made by finding the typical oocysts in fecal specimens. Serologic tests are not available. The treatment of choice is trimethoprim-sulfamethoxazole.

---

[1]Coccidia is a subclass of Sporozoa.

SELF-ASSESSMENT QUESTIONS

1. Regarding *Acanthamoeba* and *Naegleria* species, which one of the following is most accurate?

   (A) They are free-living amebas that live in warm freshwater.

   (B) *Naegleria* is a well-recognized cause of otitis media, primarily in children.

   (C) The drug of choice for infections caused by these organisms is chloroquine.

   (D) Their main clinical presentation is pneumonia acquired when water is aspirated into the lung.

2. Regarding *Babesia microti*, which one of the following is most accurate?

   (A) It infects macrophages, causing them to lyse.

   (B) Doxycycline is the drug of choice for babesiosis.

   (C) It is transmitted by the bite of *Culex* mosquitoes.

   (D) Seeing sporozoites within red cells supports the diagnosis of babesiosis.

   (E) *B. microti* causes disease primarily in the northeastern region of the United States.

3. Your patient is a 10-year-old girl with a fever and a severe headache for the past 2 days. Pertinent history includes swimming in a pond near their home in rural California in August. On exam, nuchal rigidity was observed and a lumbar puncture was performed. The spinal fluid white blood cell count was 12,200 with 80% neutrophils. Microscopic examination of a wet mount of spinal fluid revealed motile trophozoites. Of the following, which one is the most likely cause?

   (A) *Babesia microti*

   (B) *Cryptosporidium parvum*

   (C) *Naegleria fowleri*

   (D) *Toxoplasma gondii*

   (E) *Trypanosoma cruzi*

4. Your patient is a 50-year-old man with a fever and shaking chills who had been vacationing 2 weeks ago on one of the islands off the coast of Massachusetts. Microscopic examination of a blood smear reveals ring-shaped trophozoites in tetrads within red blood cells. Of the following, which one is the most likely cause?

   (A) *Babesia microti*

   (B) *Cryptosporidium parvum*

   (C) *Naegleria fowleri*

   (D) *Toxoplasma gondii*

   (E) *Trypanosoma cruzi*

ANSWERS

(1) **(A)**

(2) **(E)**

(3) **(C)**

(4) **(A)**

SUMMARIES OF ORGANISMS

Brief summaries of the organisms described in this chapter begin on page 706. Please consult these summaries for a rapid review of the essential material.

PRACTICE QUESTIONS: USMLE & COURSE EXAMINATIONS

Questions on the topics discussed in this chapter can be found in the Parasitology section of Part XIII: USMLE (National Board) Practice Questions starting on page 755. Also see Part XIV: USMLE (National Board) Practice Examination starting on page 775.

Cestodes

INTRODUCTION

Platyhelminthes (*platy* means flat; *helminth* means worm) are divided into two classes: Cestoda (tapeworms) and Trematoda (flukes). The trematodes are described in Chapter 55.

Tapeworms consist of two main parts: a rounded head called a **scolex** and a flat body consisting of multiple segments. Each segment is called a **proglottid**. The scolex has specialized means of attaching to the intestinal wall, namely, suckers, hooks, or sucking grooves. The worm grows by adding new proglottids from its germinal center next to the scolex. The oldest proglottids at the distal end are gravid and produce many eggs, which are excreted in the feces and transmitted to various intermediate hosts such as cattle, pigs, and fish.

Humans usually acquire the infection when undercooked meat or fish containing the larvae is ingested. However, in two important human diseases, cysticercosis and hydatid disease, it is the eggs that are ingested and the resulting larvae cause the disease.

There are four medically important cestodes: *Taenia solium*, *Taenia saginata*, *Diphyllobothrium latum*, and *Echinococcus granulosus*. Their features are summarized in Table 54–1, and the medically important stages in the life cycle of these organisms are described in Table 54–2. Three cestodes of lesser importance, *Echinococcus multilocularis*, *Hymenolepis nana*, and *Dipylidium caninum*, are described at the end of this chapter.

TABLE 54–1 Features of Medically Important Cestodes (Tapeworms)

Cestode	Mode of Transmission	Intermediate Host(s)	Main Sites Affected in Human Body	Diagnosis	Treatment
Taenia solium	1. Ingest larvae in undercooked pork	Pigs	Intestine	Proglottids in stool	Praziquantel
	2. Ingest eggs in food or water contaminated with human feces		Brain and eyes (cysticerci)	Biopsy, computed tomography (CT) scan	Praziquantel, albendazole, or surgical removal of cysticerci
Taenia saginata	Ingest larvae in undercooked beef	Cattle	Intestine	Proglottids in stool	Praziquantel
Diphyllobothrium latum	Ingest larvae in undercooked fish	Copepods and fish	Intestine	Operculated eggs in stool	Praziquantel
Echinococcus granulosus	Ingest eggs in food contaminated with dog feces	Sheep	Liver, lungs, and brain (hydatid cysts)	Biopsy, CT scan, serology	Albendazole or surgical removal of cyst

TABLE 54–2 Medically Important Stages in Life Cycle of Cestodes (Tapeworms)

Organism	Insect Vector	Stage that Infects Humans	Stage(s) in Humans Most Associated with Disease	Important Stage(s) Outside of Humans
Taenia solium	None	1. Larvae in undercooked pork	Adult tapeworm in intestine	Larvae in muscle of pig
		2. Eggs in food or water contaminated with human feces	Cysticercus, especially in brain	None
Taenia saginata	None	Larvae in undercooked beef	Adult tapeworm in intestine	Larvae in muscle of pig
Diphyllobothrium latum	None	Larvae in undercooked fish	Adult tapeworm in intestine can cause vitamin B_{12} deficiency	Larvae in muscle of freshwater fish
Echinococcus granulosus	None	Eggs in food or water contaminated with dog feces	Hydatid cysts, especially in liver and lung	Adult tapeworm in dog intestine produces eggs

TAENIA

There are two important human pathogens in the genus *Taenia*: *T. solium* (the pork tapeworm) and *T. saginata* (the beef tapeworm).

1. Taenia solium

Disease

The adult form of *T. solium* causes taeniasis. *T. solium* larvae cause cysticercosis.

Important Properties

The life cycle of *T. solium* is shown in Figure 54–1. *T. solium* can be identified by its scolex, which has **four suckers and a circle of hooks**, and by its gravid proglottids, which have 5 to 10 primary uterine branches (Figures 54–2A, B and 54–3). The eggs appear the same microscopically as those of *T. saginata* and *Echinococcus* species (Figure 54–4A).

In taeniasis, the adult tapeworm is located in the human intestine (see Figure 54–1). This occurs when humans are infected by eating raw or undercooked **pork** containing the larvae, called **cysticerci**. (A cysticercus consists of a pea-sized fluid-filled bladder with an invaginated scolex.) In the small intestine, the larvae attach to the gut wall and take about 3 months to grow into adult worms measuring up to 5 m. The gravid terminal proglottids containing many eggs detach daily, are passed in the feces, and are accidentally eaten by pigs. Note that pigs are infected by the worm eggs; therefore, it is the larvae (cysticerci) that are found in the pig. A six-hooked embryo (oncosphere) emerges from each egg in the pig's intestine. The embryos burrow into a blood vessel and are carried to skeletal muscle. They develop into cysticerci in the muscle, where they remain until eaten by a human. Humans are the definitive hosts, and pigs are the intermediate hosts.

In cysticercosis, a more dangerous sequence occurs when a person **ingests the worm eggs** in food or water that has been contaminated with human feces (Figure 54–5). Note that in cysticercosis, humans are infected by eggs excreted in human feces, *not* by ingesting undercooked pork. Also, pigs do not have the adult worm in their intestine, so they are not the source of the eggs that cause human cysticercosis. The eggs hatch in the small intestine, and the oncospheres burrow through the wall into a blood vessel. They can disseminate to many organs, especially the eyes, skin, and brain, where they encyst to form cysticerci (Figure 54–6). Each cysticercus contains a larva.

Pathogenesis & Epidemiology

The adult tapeworm attached to the intestinal wall causes little damage. The cysticerci, on the other hand, can become very large, especially in the **brain**, where they manifest as a **space-occupying lesion** (see Figure 54–6). Living cysticerci do not cause inflammation, but when they die, they can release substances that provoke an inflammatory response. Eventually, the cysticerci calcify.

The epidemiology of taeniasis and cysticercosis is related to the access of pigs to human feces and to consumption of raw or undercooked pork. The disease occurs worldwide but is endemic in areas of Asia, South America, and Eastern Europe. Most cases in the United States are imported.

Clinical Findings

Most patients with adult tapeworms are asymptomatic, but anorexia and diarrhea can occur. Some may notice proglottids in the stools. Cysticercosis in the brain causes headache, vomiting, and seizures. Cysticercosis in the eyes can appear as uveitis or retinitis, or the larvae can be visualized floating in the vitreous. Subcutaneous nodules containing cysticerci commonly occur. Cysts also are commonly found in skeletal muscle.

Laboratory Diagnosis

Identification of *T. solium* consists of finding gravid proglottids with 5 to 10 primary uterine branches in the stools. In contrast, *T. saginata* proglottids have 15 to 20 primary uterine branches. Eggs are found in the stools less often than are proglottids. Diagnosis of cysticercosis depends on demonstrating the presence of the cyst in tissue, usually by computed tomography (CT) scan. Serologic tests (e.g., enzyme-linked immunoelectrotransfer blot assay) that detect antibodies to *T. solium* antigens are available through CDC&P, but they may be negative in neurocysticercosis.

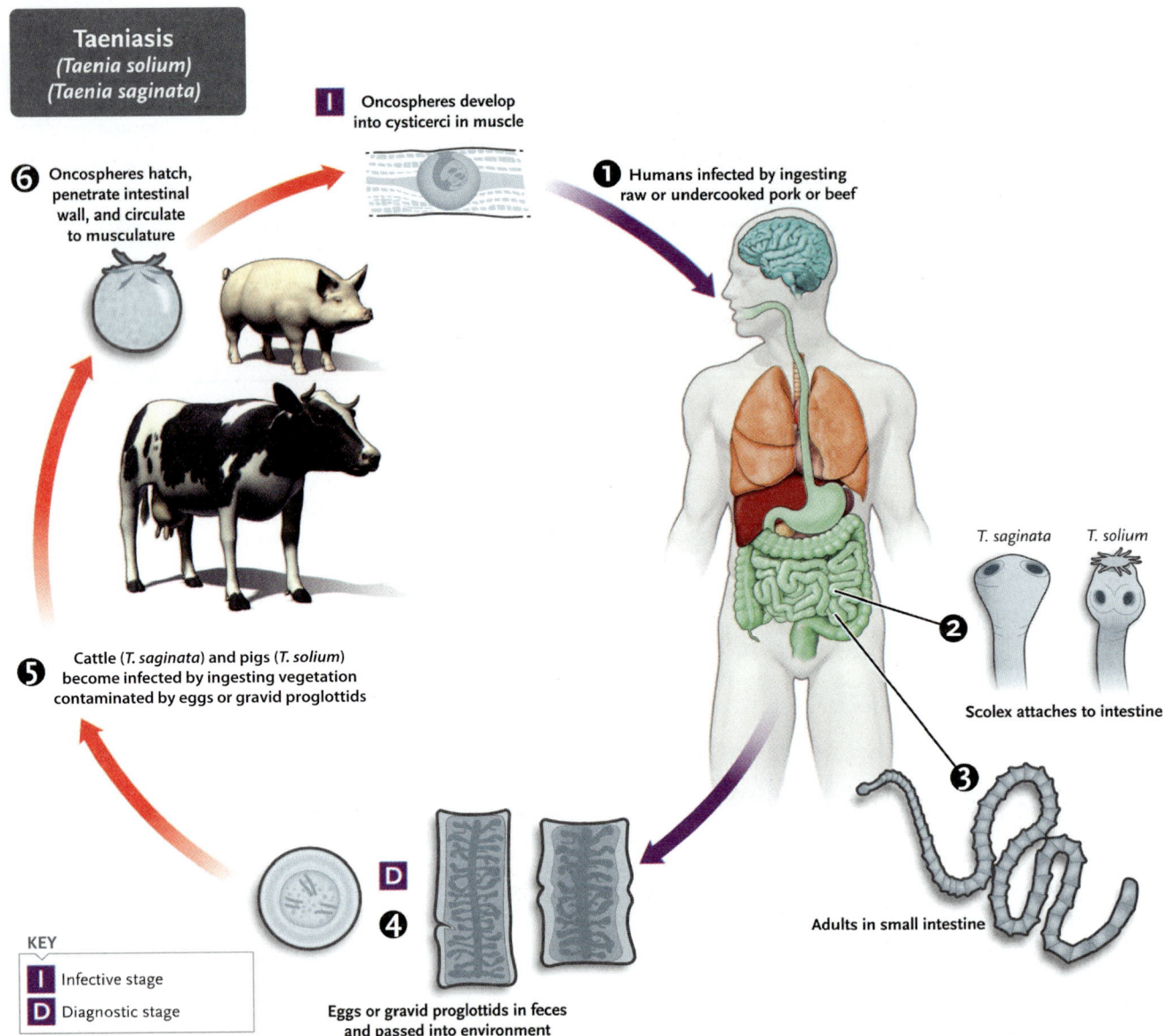

Taeniasis
(Taenia solium)
(Taenia saginata)

I Oncospheres develop into cysticerci in muscle

6 Oncospheres hatch, penetrate intestinal wall, and circulate to musculature

1 Humans infected by ingesting raw or undercooked pork or beef

5 Cattle (*T. saginata*) and pigs (*T. solium*) become infected by ingesting vegetation contaminated by eggs or gravid proglottids

T. saginata   *T. solium*

2

Scolex attaches to intestine

3

Adults in small intestine

D

4

KEY

I Infective stage

D Diagnostic stage

Eggs or gravid proglottids in feces and passed into environment

FIGURE 54–1 *Taenia solium* and *Taenia saginata*. Life cycle. Right side of figure describes the stages within the human (blue arrows). Humans are infected at step 1 when they ingest undercooked pork (*T. solium*) or beef (*T. saginata*) containing cysticerci (larval stage). Adult tapeworms form in intestine and lay eggs. Pigs and cattle are infected when they ingest either the eggs or proglottids in human stool. Left side of figure describes the stages within the pigs and cattle (red arrows). (Reproduced with permission from Public Health Image Library, Centers for Disease Control and Prevention.)

Treatment

The treatment of choice for the intestinal worms is praziquantel. The treatment for cysticercosis is either albendazole alone or in combination with praziquantel.

Prevention

Prevention of taeniasis involves cooking pork adequately and disposing waste properly so that pigs cannot ingest human feces. Prevention of cysticercosis consists of treatment of patients to prevent autoinfection plus observation of proper hygiene, including handwashing, to prevent contamination of food with the eggs.

2. Taenia saginata

Disease

T. saginata causes taeniasis. *T. saginata* larvae do not cause cysticercosis.

Important Properties

T. saginata has a scolex with four suckers but, in contrast to *T. solium*, **no hooklets**. Its gravid proglottids have 15 to 25 primary uterine branches, in contrast to *T. solium* proglottids, which have 5 to 10 (see Figure 54–2C and D). The eggs are morphologically indistinguishable from those of *T. solium*.

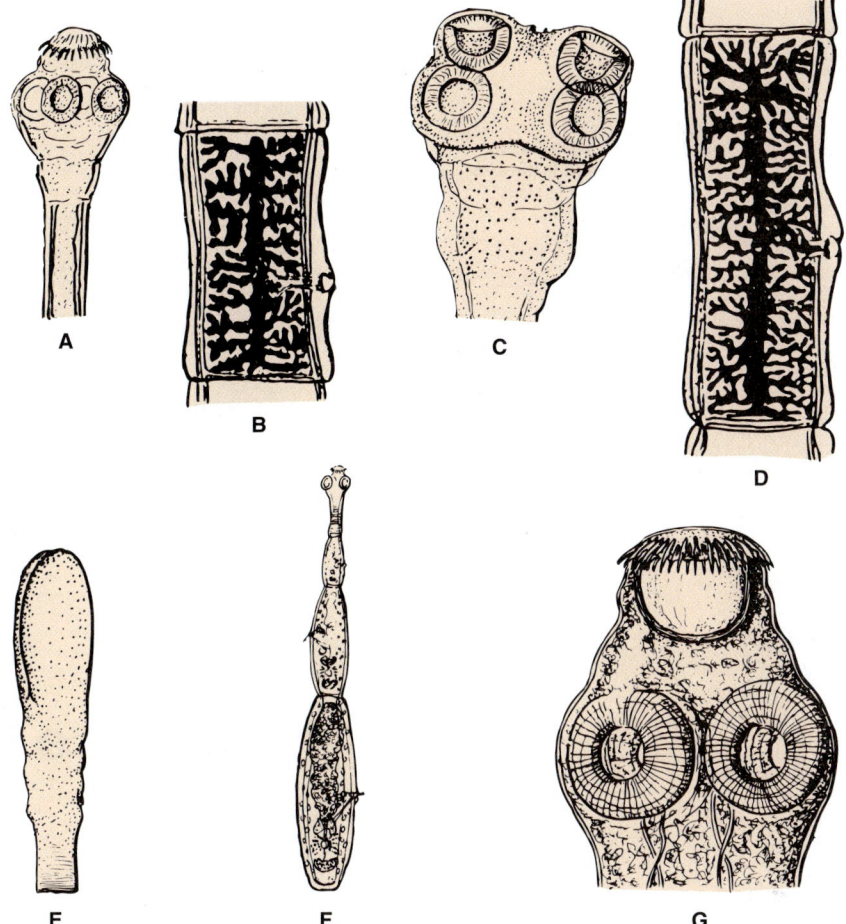

FIGURE 54–2 **A:** *Taenia solium* scolex with suckers and hooks (10×). **B:** *Taenia solium* gravid proglottid. This has fewer uterine branches than does the proglottid of *Taenia saginata* (see panel D) (2×). **C:** *T. saginata* scolex with suckers (10×). **D:** *T. saginata* gravid proglottid (2×). **E:** *Diphyllobothrium latum* scolex with sucking grooves (7×). **F:** Entire adult worm of *Echinococcus granulosus* (7×). **G:** *E. granulosus* adult scolex (70×).

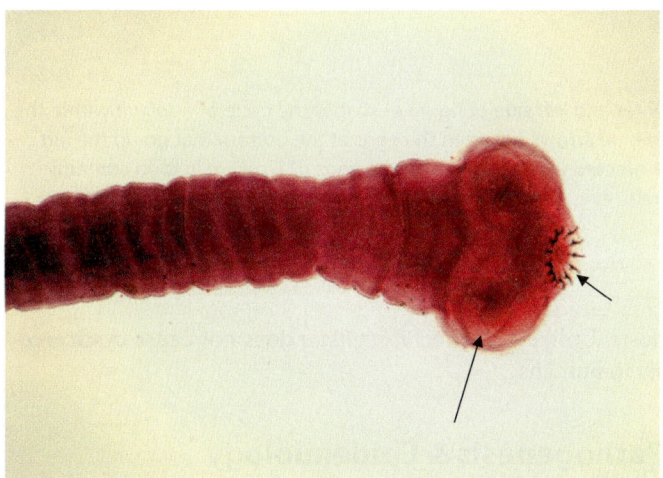

FIGURE 54–3 *Taenia solium*. Scolex and several proglottids. Long arrow points to one of the four suckers on the scolex of *T. solium*. Short arrow points to the circle of hooklets. Proglottids can be seen extending from the scolex toward the left side of the image. (Used with permission from Dr. Melvin M., Public Health Image Library, Centers for Disease Control and Prevention.)

FIGURE 54–4 **A:** *Taenia solium* egg containing oncosphere embryo. Four hooklets are visible. *Taenia saginata* and *Echinococcus granulosus* eggs are very similar to the *T. solium* egg but do not have hooklets. **B:** *Diphyllobothrium latum* egg with an operculum on the top (300×).

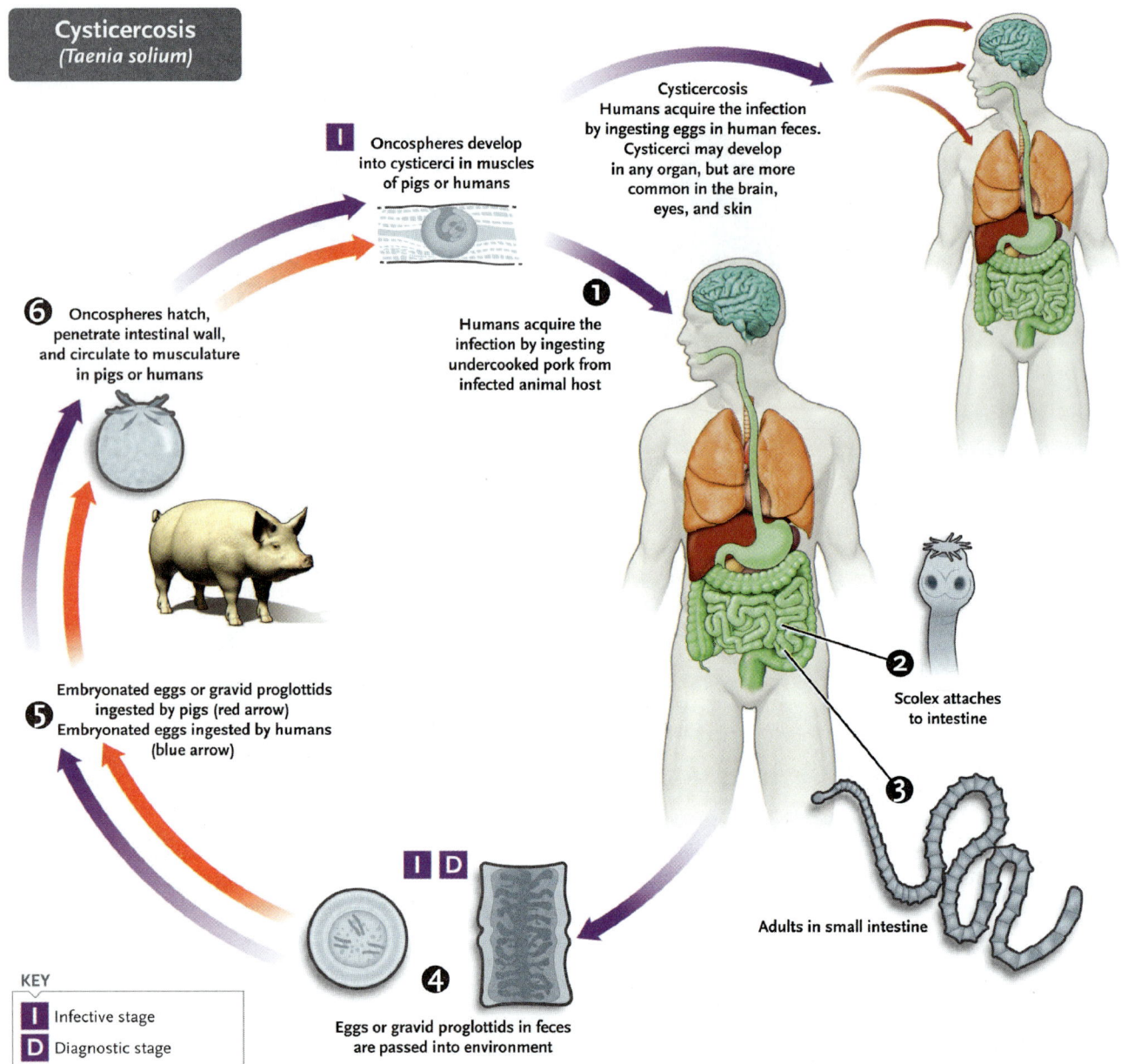

FIGURE 54–5 *Taenia solium.* Life cycle including cysticercosis stage. Center and left side of figure describes the cycle of *T. solium* within the human and the pig similar to Figure 54–1. Note, however, that there are now blue arrows between the eggs at the bottom that go up the left side of the figure to the person at the top right. In cysticercosis, humans are infected when they ingest the eggs of *T. solium* in food contaminated with human feces. The eggs differentiate into cysticerci primarily in brain, eyes, and skin. (Used with permission from Dr. Alexander J. da Silva and Melanie Moser, Centers for Disease Control and Prevention.)

The life cycle of *T. saginata* is shown in Figure 54–1. Humans are infected by eating raw or undercooked **beef** containing larvae (cysticerci). In the small intestine, the larvae attach to the gut wall and take about 3 months to grow into adult worms measuring up to 10 m (Figure 54–7). The gravid proglottids detach, are passed in the feces, and are eaten by cattle. The embryos (**oncospheres**) emerge from the eggs in the cow's intestine and burrow into a blood vessel, where they are carried to skeletal muscle. In the muscle, they develop into cysticerci. The cycle is completed when the cysticerci are ingested. Humans are the definitive hosts and cattle are the intermediate

hosts. Unlike *T. solium, T. saginata* **does not cause cysticercosis** in humans.

Pathogenesis & Epidemiology

Little damage results from the presence of the adult worm in the small intestine. The epidemiology of taeniasis caused by *T. saginata* is related to the access of cattle to human feces and to the consumption of raw or undercooked beef. The disease occurs worldwide but is endemic in Africa, South America, and Eastern Europe. In the United States, most cases are imported.

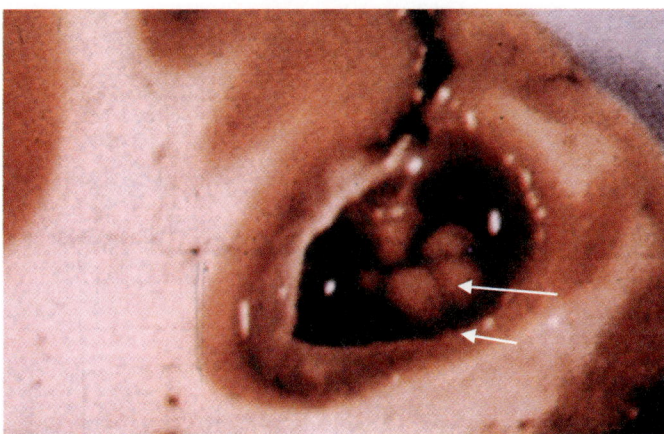

FIGURE 54–6 Cysticercus of *Taenia solium* in brain. Long arrow points to a larva of *T. solium*. Short arrow points to the wall of the cysticercus (sac) that surrounds the larva. (Used with permission from Rhodes B. Holliman, Ph.D., Professor Emeritus, Virginia Tech.)

Clinical Findings

Most patients with adult tapeworms are asymptomatic, but malaise and mild cramps can occur. In some, proglottids appear in the stools and may even protrude from the anus. The proglottids are motile and may cause pruritus ani as they move on the skin adjacent to the anus.

Laboratory Diagnosis

Identification of *T. saginata* consists of finding gravid proglottids with 15 to 20 uterine branches in the stools. Eggs are found in the stools less often than are the proglottids.

Treatment & Prevention

The treatment of choice is praziquantel. Prevention involves cooking beef adequately and disposing waste properly so that cattle cannot consume human feces.

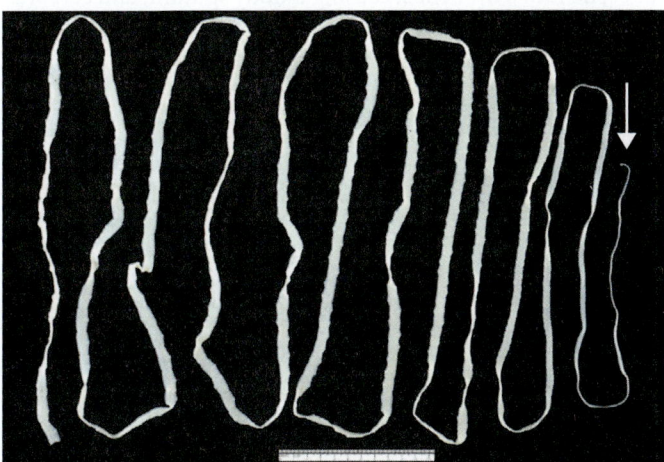

FIGURE 54–7 *Taenia saginata*. Adult tapeworm. Note the tiny scolex on the right side of the image and the gravid proglottids on the left side of the image. White arrow points to the scolex. Ruler is 12 inches long. (Reproduced with permission from Public Health Image Library, Centers for Disease Control and Prevention.)

DIPHYLLOBOTHRIUM

Disease

D. latum, the fish tapeworm, causes diphyllobothriasis.

Important Properties

In contrast to the other cestodes, which have suckers, the scolex of *D. latum* has two elongated **sucking grooves** by which the worm attaches to the intestinal wall (see Figure 54–2E). The scolex has no hooks, unlike *T. solium* and *Echinococcus*. The proglottids are wider than they are long, and the gravid uterus is in the form of a rosette. Unlike other tapeworm eggs, which are round, *D. latum* eggs are oval and have a lid-like opening (operculum) at one end (see Figure 54–4B). *D. latum* is the longest of the tapeworms, measuring up to 13 m.

Humans are infected by ingesting raw or undercooked **fish** containing larvae (called plerocercoid or sparganum larvae). In the small intestine, the larvae attach to the gut wall and develop into adult worms. Gravid proglottids release fertilized eggs through a genital pore, and the eggs are then passed in the stools. The immature eggs must be deposited in freshwater for the life cycle to continue. The embryos emerge from the eggs and are eaten by tiny copepod crustacea (first intermediate hosts). There, the embryos differentiate and form procercoid larvae in the body cavity. When the copepod is eaten by freshwater fish (e.g., pike, trout, and perch), the larvae differentiate into plerocercoids in the muscle of the fish (second intermediate host). The cycle is completed when raw or undercooked fish is eaten by humans (definitive hosts).

Pathogenesis & Epidemiology

Infection by *D. latum* causes little damage in the small intestine. In some individuals, megaloblastic anemia occurs as a result of vitamin B_{12} deficiency caused by preferential uptake of the vitamin by the worm.

The epidemiology of *D. latum* infection is related to the ingestion of raw or inadequately cooked fish and to contamination of bodies of freshwater with human feces. The disease is found worldwide but is endemic in areas where eating raw fish is the custom, such as Scandinavia, northern Russia, Japan, Canada, and certain north-central states of the United States.

Clinical Findings

Most patients are asymptomatic, but abdominal discomfort and diarrhea can occur.

Laboratory Diagnosis

Diagnosis depends on finding the typical eggs (i.e., oval, yellow-brown eggs with an operculum at one end) in the stools. There is no serologic test.

Treatment & Prevention

The treatment of choice is praziquantel. Prevention involves adequate cooking of fish and proper disposal of human feces.

ECHINOCOCCUS

Disease

E. granulosus (dog tapeworm) causes echinococcosis. The larva of *E. granulosus* causes unilocular hydatid cyst disease. Multilocular hydatid disease is caused by *E. multilocularis*, which is a minor pathogen and is discussed later.

Important Properties

E. granulosus is composed of a scolex and only three proglottids, making it **one of the smallest tapeworms** (see Figures 54–2F and G and Figure 54–8). The scolex has a circle of hooks and four suckers similar to *T. solium*. The larval form is called a **protoscolex** and is found within the hydatid cyst (Figure 55–9).

Dogs are the most important definitive hosts. The intermediate hosts are usually **sheep**. Humans are almost always dead-end intermediate hosts.

The life cycle of *E. granulosus* is shown in Figure 54–10. In the typical life cycle, worms in the dog's intestine liberate thousands of eggs, which are ingested by sheep (or humans) (see Figure 54–4). The oncosphere embryos emerge in the small intestine and migrate primarily to the liver but also to the lungs, bones, and brain. The embryos develop into large fluid-filled **hydatid cysts**, the inner germinal layer of which generates many protoscoleces (larval form) (see Figure 54–9) within "brood capsules." The life cycle is completed when the entrails (e.g., liver containing hydatid cysts) of slaughtered sheep are eaten by dogs.

Pathogenesis & Epidemiology

E. granulosus usually forms one large fluid-filled cyst (unilocular) that contains thousands of individual protoscoleces as well as many daughter cysts within the large cyst. Individual protoscoleces lying at the bottom of the large cyst are called "hydatid

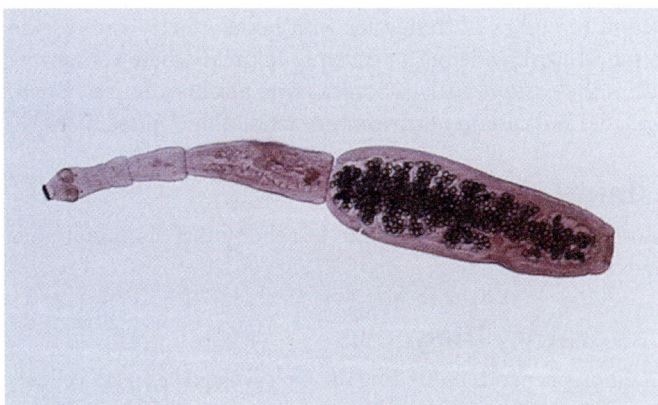

FIGURE 54–8 *Echinococcus granulosus.* Adult worm showing the scolex with hooks and suckers and three proglottids. The terminal proglottid shows many uterine branches with eggs. (Used with permission from Dr. Peter Schantz, Centers for Disease Control and Prevention.)

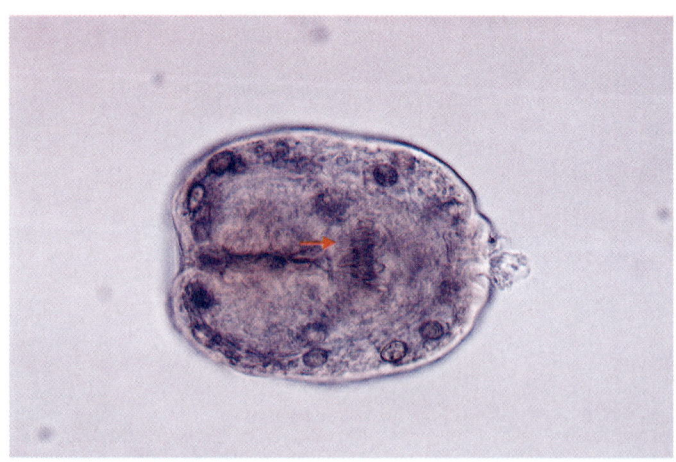

FIGURE 54–9 *Echinococcus granulosus.* Protoscolex of *Echinococcus.* Protoscolices are the "hydatid sand" within the hydatid cyst. Red arrow points to area where several hooklets can be seen. (Used with permission from Dr. Moore L., Jr., Centers for Disease Control and Prevention.)

sand." The cyst acts as a space-occupying lesion, putting pressure on adjacent tissue. The outer layer of the cyst is thick, fibrous tissue produced by the host. The cyst fluid contains parasite antigens, which can sensitize the host. Later, if the cyst ruptures spontaneously or during trauma or surgical removal, life-threatening **anaphylactic shock** can occur. Rupture of a cyst can also spread protoscoleces widely.

The disease is found primarily in shepherds living in the Mediterranean region, the Middle East, and Australia. In the United States, the western states report the largest number of cases.

Clinical Findings

Many individuals with hydatid cysts are asymptomatic, but **liver cysts** may cause hepatic dysfunction. Cysts in the lungs can erode into a bronchus, causing bloody sputum, and cerebral cysts can cause headache and focal neurologic signs. Rupture of the cyst can cause fatal anaphylactic shock.

Laboratory Diagnosis

Diagnosis is based either on microscopic examination demonstrating the presence of brood capsules containing multiple protoscoleces or on serologic tests (e.g., the indirect hemagglutination test).

Treatment & Prevention

Treatment involves albendazole with or without surgical removal of the cyst. Extreme care must be exercised to prevent release of the protoscoleces during surgery. A protoscolicidal agent (e.g., hypertonic saline) should be injected into the cyst to kill the organisms and prevent accidental dissemination. Prevention of human disease involves not feeding the entrails of slaughtered sheep to dogs.

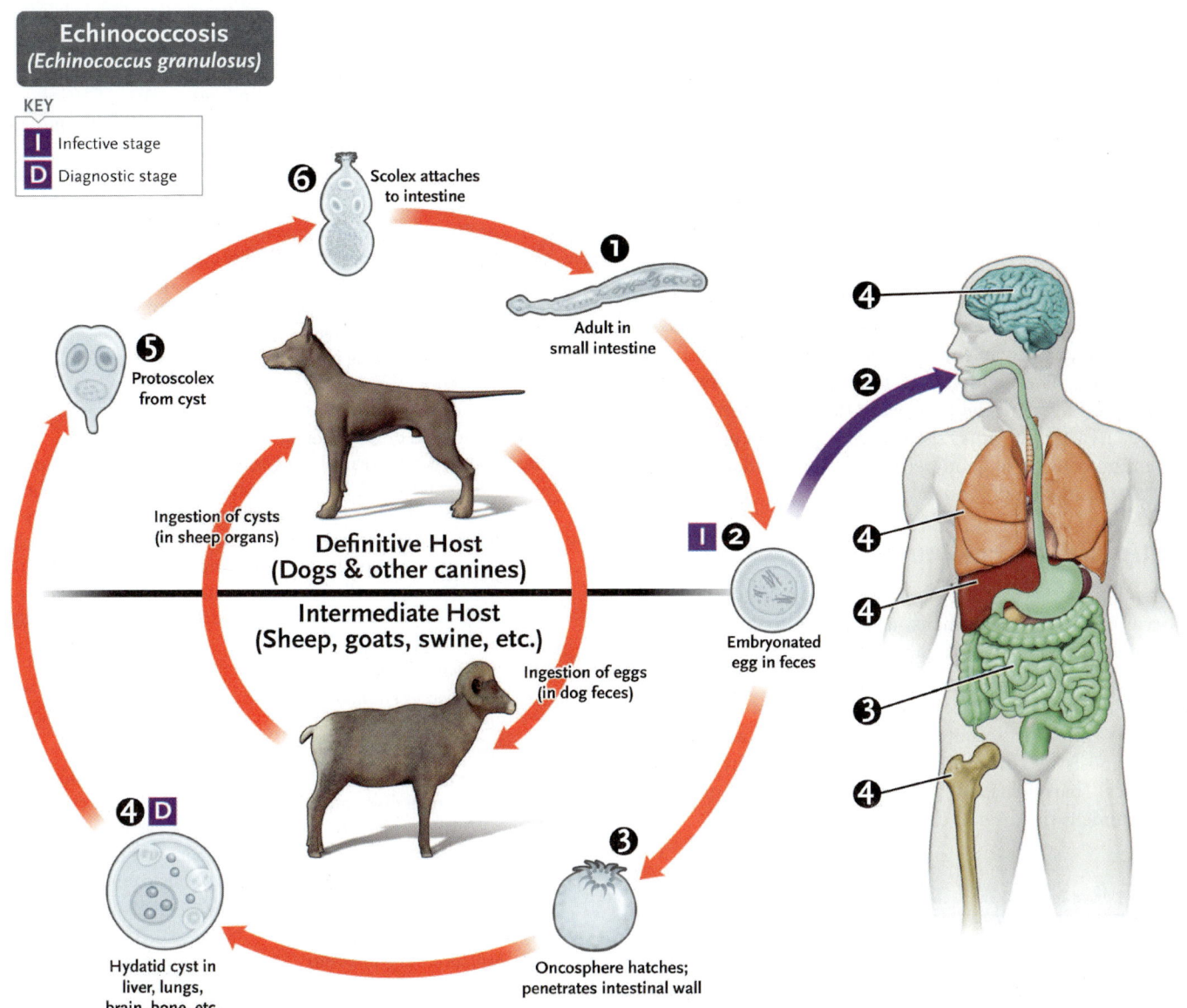

FIGURE 54–10 *Echinococcus granulosus*. Life cycle. Center and left side of figure describes the natural cycle of *E. granulosus* within dogs (top half) and sheep (bottom half). Dogs are the definitive hosts and contain the adult tapeworm in the intestines. Sheep is an important intermediate host and ingest the eggs in dog feces. Hydatid cysts containing larvae form in the sheep. Humans are accidental intermediate "dead-end" hosts when they ingest food contaminated with dog feces containing the eggs (#2 at blue arrow at right). Eggs hatch oncosphere embryos in human intestine (#3 in human figure). Hydatid cysts containing protoscolices (larvae) form primarily in the liver, lung, brain, and bone (#4 in human figure). (Used with permission from Dr. Alexander J. da Silva and Melanie Moser, Centers for Disease Control and Prevention.)

CESTODES OF MINOR IMPORTANCE

1. Echinococcus multilocularis

Many of the features of this organism are the same as those of *E. granulosus*, but the definitive hosts are mainly foxes, and the intermediate hosts are various rodents. Humans are infected by accidental ingestion of food contaminated with fox feces. The disease occurs primarily in hunters and trappers and is endemic in northern Europe, Siberia, and the western provinces of Canada. In the United States, it occurs in North and South Dakota, Minnesota, and Alaska.

Within the human liver, the larvae form multiloculated cysts with few protoscoleces. No outer fibrous capsule forms, so the cysts continue to proliferate, producing a honey-comb effect of hundreds of small vesicles. The clinical picture usually involves jaundice and weight loss. The prognosis is poor. Albendazole treatment may be successful in some cases. Surgical removal may be feasible.

2. Hymenolepis nana

H. nana (dwarf tapeworm) is the **most frequently** found tapeworm in the United States. It is only 3 to 5 cm long and is different from other tapeworms because its eggs are **directly infectious** for humans (i.e., ingested eggs can develop into adult worms without an intermediate host). Within the duodenum, the eggs hatch and differentiate into cysticercoid larvae and then into adult worms. Gravid proglottids detach, disintegrate, and release fertilized eggs. The eggs either pass in the stool or can reinfect the small intestine (autoinfection). In contrast to infection by other tapeworms, where only one adult worm is present, many *H. nana* worms (sometimes hundreds) are found.

Infection causes little damage, and most patients are asymptomatic. The organism is found worldwide, commonly in the tropics. In the United States, it is most prevalent in the southeastern states, usually in children. Diagnosis is based on finding eggs in stools. The characteristic feature of *H. nana* eggs is the 8 to 10 polar filaments lying between the membrane of the six-hooked larva and the outer shell. The treatment is praziquantel. Prevention consists of good personal hygiene and avoidance of fecal contamination of food and water.

3. Dipylidium caninum

D. caninum is the most common tapeworm of dogs and cats. It occasionally infects humans, usually young children, while playing with their pets. Human infection occurs when dog or cat fleas carrying cysticerci are ingested. The cysticerci develop into adult tapeworms in the small intestine. Most human infections are asymptomatic, but diarrhea and pruritus ani can occur. The diagnosis in animals and humans is made by observing the typical "barrel-shaped" proglottids in the stool or diapers. Niclosamide is the drug of choice.

SELF-ASSESSMENT QUESTIONS

1. Regarding *Taenia solium*, which one of the following is most accurate?
    (A) The scolex of *T. solium* has four suckers and a circle of hooklets.
    (B) The drug of choice for the adult worm in humans is metronidazole.
    (C) The cysticercus of *T. solium* contains the mature eggs of the organism.
    (D) In the laboratory, identification of adult worms is based on finding the typical scolex in the stool.
    (E) Ingestion of the terminal proglottids of *T. solium* by pigs results in mature tapeworms in the pig's intestine.
2. Cysticercosis is most likely to be acquired by:
    (A) drinking water contaminated with feces of an infected pig.
    (B) drinking water contaminated with feces of an infected cow.
    (C) drinking water contaminated with feces of an infected human.
    (D) ingestion of undercooked pork from an infected pig.
    (E) ingestion of undercooked beef from an infected cow.
3. Regarding *Diphyllobothrium latum*, which one of the following is most accurate?
    (A) Cattle are the most important intermediate hosts.
    (B) Megaloblastic anemia may occur as a result of vitamin B_{12} deficiency.

    (C) The laboratory diagnosis depends on finding a scolex with hooklets in the stool.
    (D) Infection is acquired by the ingestion of eggs in food or water contaminated with human feces.
    (E) Larvae migrate from the gastrointestinal tract via the portal circulation to the liver, where abscesses can occur.
4. Regarding *Echinococcus granulosus*, which one of the following is most accurate?
    (A) The drug of choice for *E. granulosus* infection is metronidazole.
    (B) Dogs are a required part of the life cycle of the causative organism.
    (C) *E. granulosus* is one of the longest tapeworms, sometimes measuring 10 ft in length.
    (D) *E. granulosus* larvae typically migrate to skeletal muscle, where they cause an abscess.
    (E) The main mode of transmission to humans is ingestion of eggs in food or water contaminated with human feces.
5. Your patient is a 15-year-old girl with a 2-week history of headache and vomiting and a 3-day history of confusion and incoherent speech. She was born in Ecuador but moved to this country 5 years ago. Magnetic resonance imaging (MRI) of the brain reveals multiple lesions bilaterally. The following day, she has a seizure and dies. On autopsy, the brain lesions consist of a cyst-like sac containing a larva. Of the following, which one is the most likely cause?
    (A) *Diphyllobothrium latum*
    (B) *Echinococcus granulosus*
    (C) *Taenia saginata*
    (D) *Taenia solium*
6. Your patient is a 40-year-old man with occasional mild right upper abdominal discomfort but is otherwise well. On examination, his liver is enlarged. An MRI reveals a cystic mass in the liver. On questioning, he says that he was born and raised in rural Argentina on a sheep ranch and came to this country 10 years ago. Of the following, which one is the most likely cause?
    (A) *Diphyllobothrium latum*
    (B) *Echinococcus granulosus*
    (C) *Taenia saginata*
    (D) *Taenia solium*
7. Your patient is a 20-year-old woman who is a recent immigrant from Central America. On routine exam, a stool ova and parasite test reveal eggs resembling those of *Taenia solium*. Which one of the following is the best choice of drug to treat this patient?
    (A) Ivermectin
    (B) Pentamidine
    (C) Praziquantel
    (D) Pyrimethamine and sulfadiazine
    (E) Stibogluconate

ANSWERS

(1) **(A)**
(2) **(C)**
(3) **(B)**
(4) **(B)**
(5) **(D)**
(6) **(B)**
(7) **(C)**

SUMMARIES OF ORGANISMS

Brief summaries of the organisms described in this chapter begin on page 706. Please consult these summaries for a rapid review of the essential material.

PRACTICE QUESTIONS: USMLE & COURSE EXAMINATIONS

Questions on the topics discussed in this chapter can be found in the Parasitology section of Part XIII: USMLE (National Board) Practice Questions starting on page 755. Also see Part XIV: USMLE (National Board) Practice Examination starting on page 775.

Trematodes

INTRODUCTION

Trematoda (flukes) and Cestoda (tapeworms) are the two large classes of parasites in the phylum Platyhelminthes. The most important trematodes are *Schistosoma* species (blood flukes), *Clonorchis sinensis* (liver fluke), and *Paragonimus westermani* (lung fluke). Schistosomes have by far the greatest impact in terms of the number of people infected, morbidity, and mortality.

Features of the medically important trematodes are summarized in Table 55–1, and the medically important stages in the life cycle of these organisms are described in Table 55–2. Three trematodes of lesser importance, such as *Fasciola hepatica*,

Fasciolopsis buski, and *Heterophyes heterophyes*, are described at the end of this chapter.

The life cycle of the medically important trematodes involves a sexual cycle in humans (definitive host) and asexual reproduction in **freshwater snails** (intermediate hosts) (Figure 55–1). Transmission to humans takes place either via penetration of the skin by the free-swimming **cercariae** of the schistosomes (Figures 55–2D and 55–3) or via ingestion of cysts in undercooked (raw) fish or crabs in *Clonorchis* and *Paragonimus* infection, respectively.

Trematodes that cause human disease are not endemic in the United States. However, immigrants from tropical areas, especially Southeast Asia, are frequently infected.

TABLE 55–1 Features of Medically Important Trematodes (Flukes)

Trematode	Mode of Transmission	Main Sites Affected	Intermediate Host(s)	Diagnostic Features of Eggs	Endemic Area(s)	Treatment
Schistosoma mansoni	Penetrate skin	Veins of colon	Snail	Large lateral spine	Africa, Latin America (Caribbean)	Praziquantel
Schistosoma japonicum	Penetrate skin	Veins of small intestine, liver	Snail	Small lateral spine	Asia	Praziquantel
Schistosoma haematobium	Penetrate skin	Veins of urinary bladder	Snail	Large terminal spine	Africa, Middle East	Praziquantel
Clonorchis sinensis	Ingested with raw fish	Liver	Snail and fish	Operculated	Asia	Praziquantel
Paragonimus westermani	Ingested with raw crab	Lung	Snail and crab	Operculated	Asia, India	Praziquantel

TABLE 55–2 Medically Important Stages in Life Cycle of Trematodes (Flukes)

Organism	Insect Vector	Stage that Infects Humans	Stage(s) in Humans Most Associated with Disease	Important Stage(s) Outside of Humans
Schistosoma mansoni, Schistosoma haematobium, Schistosoma japonicum	None	Cercariae penetrate skin	Adult flukes living in mesenteric or bladder veins lay eggs that cause granulomas	Miracidium (ciliated larvae) infect snails → cercariae infect humans
Clonorchis	None	Larvae in undercooked fish	Adult flukes live in biliary ducts	Eggs ingested by snails → cercariae infect fish
Paragonimus	None	Larvae in undercooked crab	Adult flukes live in lung	Eggs ingested by snails → cercariae infect crab

SCHISTOSOMA

Disease

Schistosoma species cause schistosomiasis. *Schistosoma mansoni* and *Schistosoma japonicum* affect the **gastrointestinal tract**,[1] whereas *Schistosoma haematobium* affects the **urinary tract**.

Important Properties

The life cycle of *Schistosoma* species is shown in Figure 55–1. In contrast to the other trematodes, which are hermaphrodites, adult schistosomes exist as **separate sexes** but live attached to each other. The female resides in a groove in the male, the gynecophoric canal ("schist"), where he continuously fertilizes her eggs (see Figure 55–2A). The three species can be distinguished by the appearance of their eggs in the microscope: *S. mansoni* eggs have a **prominent lateral spine**, whereas *S. japonicum* eggs have a very small lateral spine and *S. haematobium* eggs have a terminal spine (Figures 55–4A and B, 55–5, and 55–6). *S. mansoni* and *S. japonicum* adults live in the **mesenteric veins**, whereas *S. haematobium* lives in the veins draining the urinary bladder. Schistosomes are therefore known as **blood flukes**.

Humans are infected when the free-swimming, fork-tailed **cercariae** penetrate the skin (see Figures 55–2D and 55–3). They differentiate to larvae (schistosomula), enter the blood, and are carried via the veins into the arterial circulation. Those that enter the superior mesenteric artery pass into the portal circulation and reach the liver, where they mature into adult flukes. *S. mansoni* and *S. japonicum* adults migrate against the portal flow to reside in the mesenteric venules. *S. haematobium* adults reach the bladder veins through the venous plexus between the rectum and the bladder.

In their definitive venous site, the female lays fertilized eggs, which penetrate the vascular endothelium and enter the gut or bladder lumen, respectively. The eggs are excreted in the stools or urine and must enter freshwater where they release ciliated, swimming larvae called **miracidia**. The miracidia then penetrate **snails** and undergo further development and multiplication to produce many cercariae. (The three schistosomes use different species of snails as intermediate hosts.) Cercariae leave the snails, enter freshwater, and complete the cycle by penetrating human skin.

Pathogenesis & Epidemiology

Most of the pathologic findings are caused by the presence of eggs in the liver, spleen, or wall of the gut or bladder. Eggs in the liver induce granulomas, which lead to fibrosis, hepatomegaly, and portal hypertension. The granulomas are formed in response to antigens secreted by the eggs. Hepatocytes are usually undamaged, and liver function tests remain normal. Portal hypertension leads to **splenomegaly**.

S. mansoni eggs damage the wall of the distal colon (inferior mesenteric venules), whereas *S. japonicum* eggs damage the walls of both the small and large intestines (superior and inferior mesenteric venules). The damage is due both to digestion of tissue by proteolytic enzymes produced by the egg and to the host inflammatory response that forms granulomas in the venules. The eggs of *S. haematobium* in the wall of the bladder induce granulomas and fibrosis, which can lead to **carcinoma of the bladder**.

Schistosomes have evolved a remarkable process for **evading the host defenses**. There is evidence that their surface becomes coated with host antigens, thereby limiting the ability of the immune system to recognize them as foreign.

The epidemiology of schistosomiasis depends on the presence of the specific freshwater snails that serve as intermediate hosts. *S. mansoni* is found in Africa and Latin America (including Puerto Rico), whereas *S. haematobium* is found in Africa and the Middle East. *S. japonicum* is found only in Asia and is the only one for which domestic animals (e.g., water buffalo and pigs) act as important reservoirs. More than 150 million people in the tropical areas of Africa, Asia, and Latin America are affected.

Clinical Findings

Most patients are asymptomatic, but chronic infections may become symptomatic. The acute stage, which begins shortly after cercarial penetration, consists of itching and dermatitis

---

[1]As does Schistosoma mekongi.

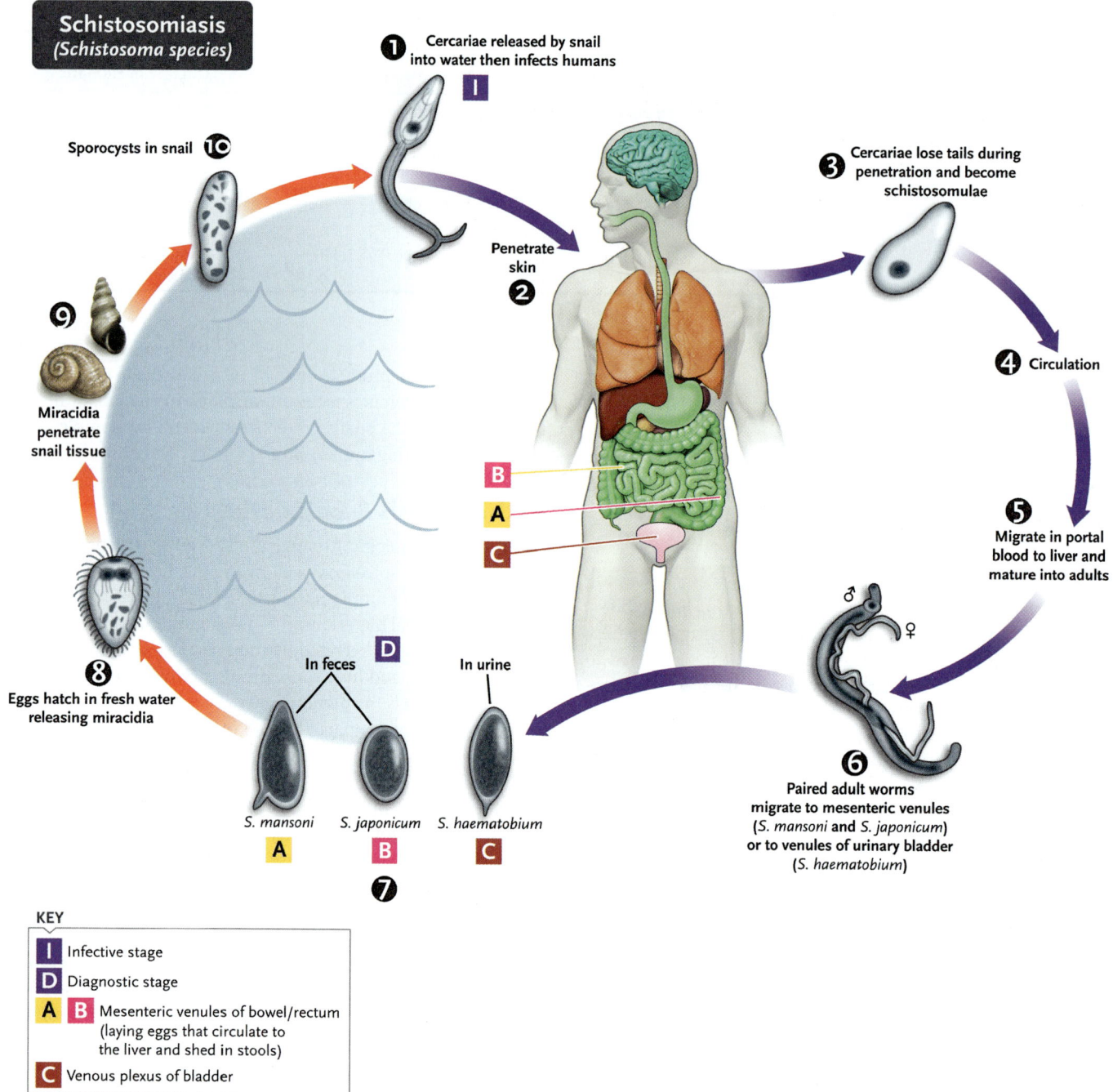

FIGURE 55–1 *Schistosoma* species. Life cycle. Right side of figure describes the stages within the human (blue arrows). Humans are infected at step 2 when free-swimming cercariae penetrate human skin. Cercariae differentiate into adult worms (two sexes) that migrate to the mesenteric veins (*Schistosoma mansoni* and *Schistosoma japonicum*) or the venous plexus of the urinary bladder (*Schistosoma haematobium*). The adult worms lay eggs, which appear in the stool (*S. mansoni* and *S. japonicum*) or the urine (*S. haematobium*). The eggs pass into freshwater, where the miracidia stage infects snails, which produce cercariae. Left side of figure describes the stages in freshwater and in the snail (red arrows). (Used with permission from Dr. Alexander J. da Silva and Melanie Moser, Centers for Disease Control and Prevention.)

followed 2 to 3 weeks later by fever, chills, diarrhea, lymphadenopathy, and hepatosplenomegaly. Eosinophilia is seen in response to the migrating larvae. This stage usually resolves spontaneously.

The chronic stage can cause significant morbidity and mortality. In patients with *S. mansoni* or *S. japonicum* infection,

gastrointestinal hemorrhage, hepatomegaly, and massive splenomegaly can develop. The most common cause of death is exsanguination from ruptured esophageal varices. Patients infected with *S. haematobium* have hematuria as their chief early complaint. Superimposed bacterial urinary tract infections occur frequently.

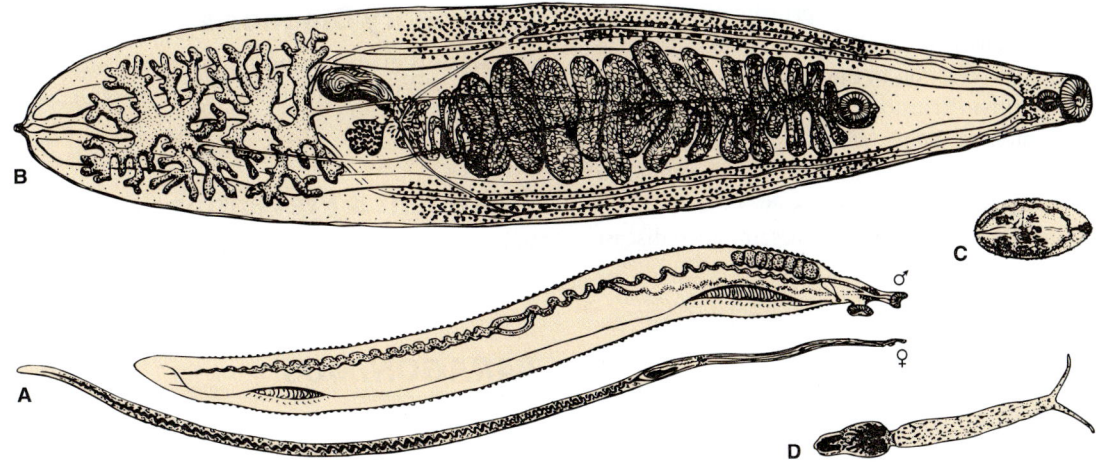

FIGURE 55–2 **A:** Male and female *Schistosoma mansoni* adults. The female lives in the male's schist (shown as a ventral opening) (6×). **B:** *Clonorchis sinensis* adult (6×). **C:** *Paragonimus westermani* adult (0.6×). **D:** *Schistosoma mansoni* cercaria (300×).

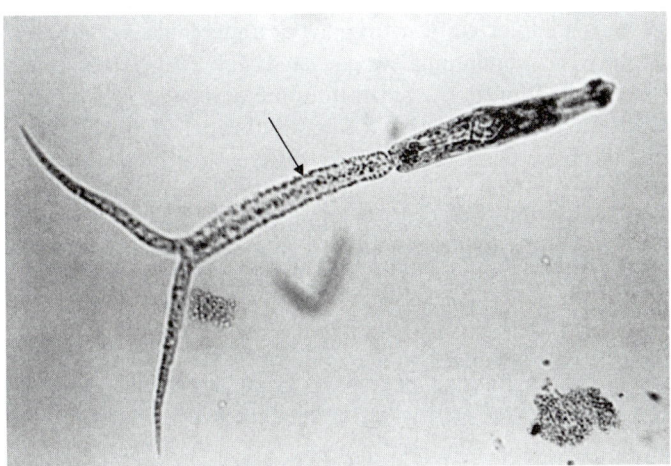

FIGURE 55–3 *Schistosoma*. Cercaria. Arrow points to a cercaria of *Schistosoma*. Note the typical forked tail on the left side of the image. (Reproduced with permission from the Minnesota Department of Health, R.N. Barr Library; Librarians Rethlefson M. and Jones M.; Prof. Wiley W., Public Health Image Library, Centers for Disease Control and Prevention.)

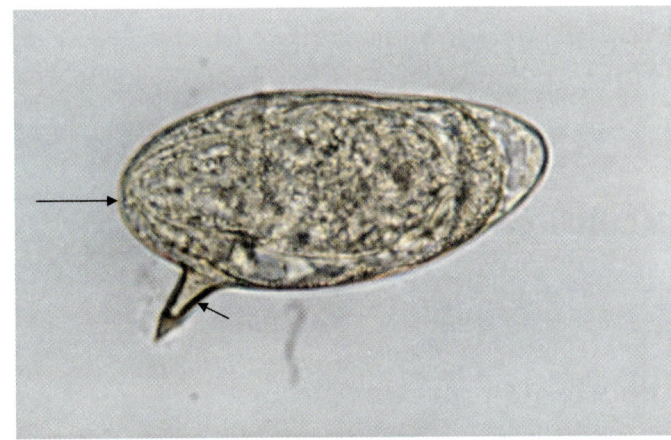

FIGURE 55–5 *Schistosoma mansoni*. Egg. Long arrow points to an egg of *Schistosoma mansoni*. Short arrow points to its large lateral spine. (Reproduced with permission from Public Health Image Library, Centers for Disease Control and Prevention.)

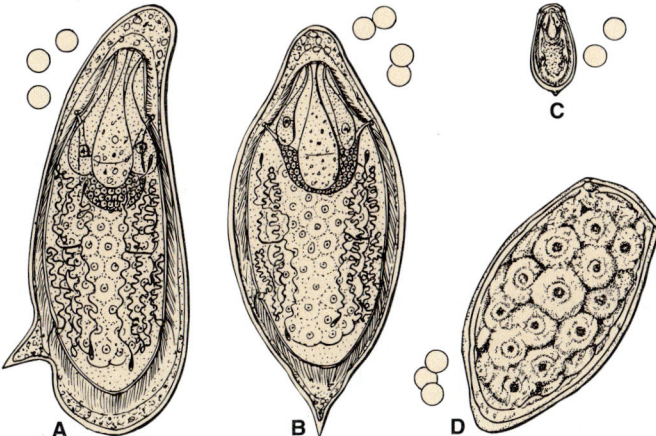

FIGURE 55–4 **A:** *Schistosoma mansoni* egg with lateral spine. **B:** *Schistosoma haematobium* egg with terminal spine. **C:** *Clonorchis sinensis* egg with operculum. **D:** *Paragonimus westermani* egg with operculum (300×). (Circles represent red blood cells.)

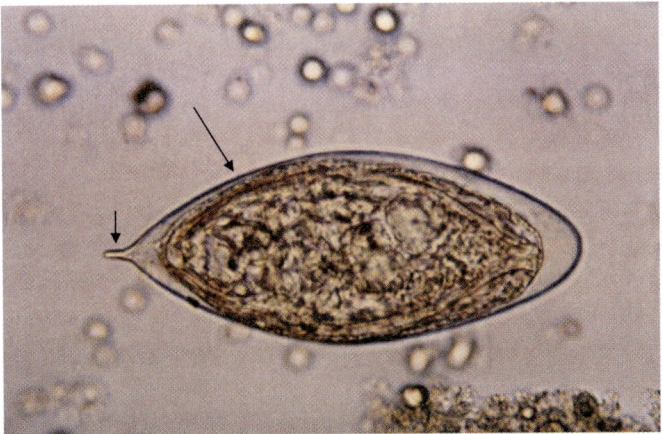

FIGURE 55–6 *Schistosoma haematobium*. Egg. Long arrow points to an egg of *Schistosoma haematobium*. Short arrow points to its terminal spine. (Reproduced with permission from Public Health Image Library, Centers for Disease Control and Prevention.)

"Swimmer's itch," which consists of pruritic papules, is a frequent problem in many lakes in the United States. The papules are an immunologic reaction to the presence in the skin of the cercariae of nonhuman schistosomes. The pruritic papules appear within minutes to hours after exposure, indicating that this is an immediate (immunoglobulin [Ig] E–mediated) hypersensitivity. These nonhuman schistosomes are incapable of replicating in humans and do not cause disseminated disease.

Laboratory Diagnosis

Diagnosis depends on finding the characteristic ova in the feces or urine. The large lateral spine of *S. mansoni* and the rudimentary spine of *S. japonicum* are typical, as is the large terminal spine of *S. haematobium* (see Figures 55–4A and B, 55–5, and 55–6). Serologic tests are not useful. Moderate eosinophilia occurs.

Treatment & Prevention

Praziquantel is the treatment of choice for all three species. Prevention involves proper disposal of human waste and eradication of the snail host when possible. Swimming in areas of endemic infection should be avoided.

CLONORCHIS

Disease

C. sinensis causes clonorchiasis (Asian liver fluke infection).

Important Properties

Humans are infected by eating raw or undercooked **fish** containing the encysted larvae (metacercariae). After excystation in the duodenum, immature flukes enter the **biliary ducts** and differentiate into adults (see Figure 55–2B). The hermaphroditic adults produce eggs, which are excreted in the feces (see Figure 55–4C). Upon reaching freshwater, the eggs are ingested by snails, which are the first intermediate hosts. The eggs hatch within the gut and differentiate first into larvae (rediae) and then into many free-swimming cercariae. Cercariae encyst under the scales of certain freshwater fish (second intermediate hosts), which are then eaten by humans.

Pathogenesis & Epidemiology

In some infections, the inflammatory response can cause hyperplasia and fibrosis of the biliary tract, but often there are no lesions. Clonorchiasis is endemic in China, Japan, Korea, and Indochina, where it affects about 20 million people. It is seen in the United States among immigrants from these areas.

Clinical Findings

Most infections are asymptomatic. In patients with a heavy worm burden, upper abdominal pain, anorexia, hepatomegaly, and eosinophilia can occur.

Laboratory Diagnosis

Diagnosis is made by finding the typical small, brownish, operculated eggs in the stool (see Figure 55–4C). Serologic tests are not useful.

Treatment & Prevention

Praziquantel is an effective drug. Prevention centers on adequate cooking of fish and proper disposal of human waste.

PARAGONIMUS

Disease

P. westermani, the lung fluke, causes paragonimiasis.

Important Properties

Humans are infected by eating raw or undercooked **crab meat** (or crayfish) containing the encysted larvae (metacercariae). After excystation in the small intestine, immature flukes penetrate the intestinal wall and migrate through the diaphragm into the **lung** parenchyma. They differentiate into hermaphroditic adults (see Figure 55–2C) and produce eggs that enter the bronchioles and are coughed up or swallowed (see Figure 55–4D). Eggs in either sputum or feces that reach freshwater hatch into miracidia, which enter snails (first intermediate hosts). There, they differentiate first into larvae (rediae) and then into many free-swimming cercariae. The cercariae infect and encyst in freshwater crabs (second intermediate hosts). The cycle is completed when undercooked infected crabs are eaten by humans.

Pathogenesis & Epidemiology

Within the lung, the worms exist in a fibrous capsule that communicates with a bronchiole. Secondary bacterial infection frequently occurs, resulting in bloody sputum. Paragonimiasis is endemic in Asia and India. In the United States, it occurs in immigrants from these areas.

Clinical Findings

The main symptom is a chronic cough with bloody sputum. Dyspnea, pleuritic chest pain, and recurrent attacks of bacterial pneumonia occur. The disease can resemble tuberculosis.

Laboratory Diagnosis

Diagnosis is made by finding the typical operculated eggs in sputum or feces (see Figure 55–4D). Serologic tests are not useful.

Treatment & Prevention

Praziquantel is the treatment of choice. Cooking crabs properly is the best method of prevention.

TREMATODES OF MINOR IMPORTANCE

Fasciola

F. hepatica, the sheep liver fluke, causes disease primarily in sheep and other domestic animals in Latin America, Africa, Europe, and China. Humans are infected by **eating watercress** (or other aquatic plants) contaminated by larvae (metacercariae) that excyst in the duodenum, penetrate the gut wall, and reach the liver, where they mature into adults. Hermaphroditic adults in the bile ducts produce eggs, which are excreted in the feces. The eggs hatch in freshwater, and miracidia enter the snails. Miracidia develop into cercariae, which then encyst on aquatic vegetation. Sheep and humans eat the plants, thus completing the life cycle.

Symptoms are due primarily to the presence of the adult worm in the biliary tract. In early infection, right-upper-quadrant pain, fever, and hepatomegaly can occur, but most infections are asymptomatic. Months or years later, obstructive jaundice can occur. Halzoun is a painful pharyngitis caused by the presence of adult flukes on the posterior pharyngeal wall. The adult flukes are acquired by eating raw sheep liver.

Diagnosis is made by identification of eggs in the feces. There is no serologic test. The drug of choice is triclabendazole. Adult flukes in the pharynx and larynx can be removed surgically. Prevention involves not eating wild aquatic vegetables or raw sheep liver.

Fasciolopsis

F. buski is an intestinal parasite of humans and pigs that is endemic to Asia and India. Humans are infected by **eating aquatic vegetation** that carries the cysts. After excysting in the small intestine, the parasites attach to the mucosa and differentiate into adults. Eggs are passed in the feces; on reaching freshwater, they differentiate into miracidia. The ciliated miracidia penetrate snails and, after several stages, develop into cercariae that encyst on aquatic vegetation. The cycle is completed when plants carrying the cysts are eaten.

Pathologic findings are due to damage of the intestinal mucosa by the adult fluke. Most infections are asymptomatic, but ulceration, abscess formation, and hemorrhage can occur. Diagnosis is based on finding typical eggs in the feces. Praziquantel is the treatment of choice. Prevention consists of proper disposal of human sewage.

Heterophyes

H. heterophyes is an intestinal parasite of people living in Africa, the Middle East, and Asia who are infected by **eating raw fish** containing cysts. Larvae excyst in the small intestine, attach to the mucosa, and develop into adults. Eggs are passed in the feces and, on reaching brackish water, are ingested by snails. After several developmental stages, cercariae are produced that encyst under the scales of certain fish. The cycle is completed when fish carrying the infectious cysts are eaten.

Pathologic findings are due to inflammation of the intestinal epithelium as a result of the presence of the adult flukes. Most infections are asymptomatic, but abdominal pain and non-bloody diarrhea can occur. Diagnosis is based on finding the typical eggs in the feces. Praziquantel is the treatment of choice. Prevention consists of proper disposal of human sewage.

SELF-ASSESSMENT QUESTIONS

1. Regarding schistosomes, which one of the following statements is the most accurate?
   (A) The visual appearance of male and female schistosomes is the same.
   (B) Humans are infected by schistosomes when cercariae penetrate the skin.
   (C) Infection of freshwater fish is a required part of the life cycle of schistosomes.
   (D) The pathology of schistosomiasis is principally caused by the cercariae entering hepatocytes and killing them.
   (E) Infection by nonhuman schistosomes can cause meningitis in people who swim in certain lakes in the United States.

2. Regarding *Schistosoma mansoni*, which one of the following statements is the most accurate?
   (A) The main site of *S. mansoni* in the human body is the mesenteric venules.
   (B) Schistosomiasis caused by *S. mansoni* has been eradicated from the Western hemisphere.
   (C) The laboratory diagnosis of *S. mansoni* depends on seeing eggs with a terminal spine in the stool.
   (D) Adult schistosomes are passed in the stool, and it is obligatory that they be ingested by freshwater snails to continue the life cycle.
   (E) Swimmer's itch occurs when *S. mansoni* eggs spread from the liver to the skin, where they induce a histamine-mediated immediate (type 1) hypersensitivity reaction.

3. Which one of the following is the drug of choice for infections with *Schistosoma mansoni* and *Schistosoma haematobium*?
   (A) Albendazole
   (B) Metronidazole
   (C) Nifurtimox
   (D) Praziquantel
   (E) Stibogluconate

4. Your patient is a 30-year-old man with low-grade perineal pain for several weeks who had an episode of painful ejaculation and postcoital hematuria yesterday. He is in a long-standing monogamous relationship. He has traveled extensively throughout the world during the past 10 years. Urinalysis and urine culture were negative. Cytologic examination of cells in the urine revealed no tumor cells. Cystoscopy revealed several polypoid lesions, and a biopsy of a lesion was taken. The tissue was examined in the light microscope, and eggs with a terminal spine were seen. Of the following, which one is the **MOST** likely cause?
   (A) *Clonorchis sinensis*
   (B) *Paragonimus westermani*
   (C) *Schistosoma haematobium*
   (D) *Schistosoma japonicum*
   (E) *Schistosoma mansoni*

ANSWERS

(1) **(B)**
(2) **(A)**
(3) **(D)**
(4) **(C)**

SUMMARIES OF ORGANISMS

Brief summaries of the organisms described in this chapter begin on page 706. Please consult these summaries for a rapid review of the essential material.

PRACTICE QUESTIONS: USMLE & COURSE EXAMINATIONS

Questions on the topics discussed in this chapter can be found in the Parasitology section of Part XIII: USMLE (National Board) Practice Questions starting on page 755. Also see Part XIV: USMLE (National Board) Practice Examination starting on page 775.

Nematodes

INTRODUCTION

Nematodes (also known as Nemathelminthes) are roundworms with a cylindrical body and a complete digestive tract, including a mouth and an anus. The body is covered with a noncellular, highly resistant coating called a cuticle. Nematodes have separate sexes; the female is usually larger than the male. The male typically has a coiled tail.

The medically important nematodes can be divided into two categories according to their primary location in the body, namely, **intestinal** and **tissue** nematodes.

(1) The intestinal nematodes include *Enterobius* (pinworm), *Trichuris* (whipworm), *Ascaris* (giant roundworm), *Necator* and *Ancylostoma* (the two hookworms), *Strongyloides* (small roundworm), and *Trichinella*. *Enterobius*, *Trichuris*, and *Ascaris* are transmitted by ingestion of eggs; the others are transmitted as larvae. There are two larval forms: the first- and second-stage (**rhabditiform**) larvae are noninfectious, feeding forms; the third-stage (**filariform**) larvae are the infectious, nonfeeding forms. As adults, these nematodes live within the human body, except for *Strongyloides*, which can also exist in the soil.

(2) The important tissue nematodes *Wuchereria*, *Onchocerca*, and *Loa* are called the "filarial worms," because they produce motile embryos called **microfilariae** in blood and tissue fluids. These organisms are transmitted from person to person by bloodsucking mosquitoes or flies. A fourth species is the guinea worm, *Dracunculus*, whose larvae inhabit tiny crustaceans (copepods) and are ingested in drinking water.

The nematodes described above cause disease as a result of the presence of adult worms within the body. In addition, several species cannot mature to adults in human tissue, but their larvae can cause disease. The most serious of these diseases is visceral larva migrans, caused primarily by the larvae of the dog ascarid, *Toxocara canis*. Cutaneous larva migrans, caused mainly by the larvae of the dog and cat hookworm, *Ancylostoma caninum*, is less serious. A third disease, anisakiasis, is caused by the ingestion of *Anisakis* larvae in raw seafood.

In some nematode infections, the larvae pass through the lung and cause symptoms of pneumonitis, namely *Ascaris*, *Strongyloides*, hookworms (*Ancylostoma* and *Necator*), and *Toxocara*. *Ascaris* and *Toxocara* are acquired by ingestion of eggs whereas *Strongyloides* and the hookworms are acquired by penetration of the skin by filariform larvae. *Ascaris* is the most common cause of eosinophilic pneumonia.

In infections caused by certain nematodes that migrate through tissue (e.g., *Strongyloides*, *Trichinella*, *Ascaris*, and the two hookworms *Ancylostoma* and *Necator*), a striking increase in the number of eosinophils (**eosinophilia**) occurs. Eosinophils do not ingest the organisms; rather, they attach to the surface of the parasite via immunoglobulin (Ig) E and secrete cytotoxic enzymes contained within their eosinophilic granules. Host defenses against helminths are stimulated by interleukins

TABLE 56–1 Features of Medically Important Nematodes

Primary Location	Species	Common Name or Disease	Mode of Transmission	Endemic Areas	Diagnosis	Treatment
Intestines	Enterobius	Pinworm	Ingestion of eggs	Worldwide	Eggs on skin	Albendazole, mebendazole, or pyrantel pamoate
	Trichuris	Whipworm	Ingestion of eggs	Worldwide, especially tropics	Eggs in stools	Albendazole
	Ascaris	Ascariasis	Ingestion of eggs	Worldwide, especially tropics	Eggs in stools	Albendazole, mebendazole, or ivermectin
	Ancylostoma and Necator	Hookworm	Penetration of skin by larvae	Worldwide, especially tropics (Ancylostoma), the United States (Necator)	Eggs in stools	Albendazole, mebendazole, or pyrantel pamoate
	Strongyloides	Strongyloidiasis	Penetration of skin by larvae; also autoinfection	Tropics primarily	Larvae in stools	Ivermectin
	Trichinella	Trichinosis	Ingestion of larvae in undercooked meat	Worldwide	Larvae encysted in muscle; serology	Albendazole plus prednisone against larvae; Mebendazole against adult worm
	Anisakis	Anisakiasis	Ingestion of larvae in undercooked seafood	Japan, the United States, the Netherlands	Clinical	No drug available
Tissue	Wuchereria	Filariasis	Mosquito bite	Tropics primarily	Blood smear	Diethylcarbamazine
	Onchocerca	Onchocerciasis (river blindness)	Blackfly bite	Africa, Central America	Skin biopsy	Ivermectin
	Loa	Loiasis	Deer fly bite	Tropical Africa	Blood smear	Diethylcarbamazine
	Dracunculus	Guinea worm	Ingestion of copepods in water	Tropical Africa and Asia	Clinical	Gradual extraction of worm
	Toxocara larvae	Visceral larva migrans	Ingestion of eggs	Worldwide	Clinical and serologic	Albendazole or mebendazole
	Ancylostoma larvae	Cutaneous larva migrans	Penetration of skin by larvae	Worldwide	Clinical	Albendazole or ivermectin

(ILs) synthesized by the Th-2 subset of helper T cells (e.g., the production of IgE is increased by IL-4, and the number of eosinophils is increased by IL-5) (see Chapter 58). Cysteine proteases produced by the worms to facilitate their migration through tissue are the stimuli for IL-5 production.

Features of the medically important nematodes are summarized in Table 56–1. The medically important stages in the life cycle of the intestinal nematodes are described in Table 56–2, and those of the tissue nematodes are described in Table 56–3.

INTESTINAL NEMATODES

ENTEROBIUS

Disease

Enterobius vermicularis causes pinworm infection (enterobiasis).

Important Properties

The life cycle of *E. vermicularis* is shown in Figure 56–1. Infection occurs **only in humans**; there is no animal reservoir or vector. The infection is acquired by ingesting the worm eggs. The eggs hatch in the small intestine, where the larvae differentiate into adults and migrate to the colon. The adult male and female worms live in the colon, where mating occurs (Figure 56–2A). At night, the female migrates from the anus and releases thousands of fertilized eggs on the perianal skin and into the environment. Within 6 hours, the eggs develop into embryonated eggs (Figures 56–3A and 56–4) and become infectious. Reinfection can occur if they are carried to the mouth by fingers after scratching the itching skin.

Pathogenesis & Clinical Findings

Perianal pruritus is the most prominent symptom. Pruritus is thought to be an allergic reaction to the proteins of either the adult female or the eggs. Scratching predisposes to secondary bacterial infection.

TABLE 56-2 Medically Important Stages in Life Cycle of Intestinal Nematodes (Roundworms)

Organism	Insect Vector	Stage that Infects Humans	Stage(s) in Humans Most Associated with Disease	Important Stage(s) Outside of Humans
Enterobius	None	Eggs	Female worm migrates out anus and lays eggs on perianal skin, causing itching	None
Trichuris	None	Eggs	Adult worms in colon may cause rectal prolapse	Eggs survive in environment
Ascaris	None	Eggs	Larvae migrate to lung, causing pneumonia	Eggs survive in environment
Ancylostoma and *Necator*	None	Filariform larvae enter skin	Adult worms in small intestine cause blood loss (anemia)	Egg → rhabditiform larvae → filariform larvae
Strongyloides	None	Filariform larvae enter skin	Larvae disseminate to various tissues in immunocompromised (autoinfection)	Egg → rhabditiform larvae → filariform larvae; also "free living" cycle in soil
Trichinella	None	Larvae in meat ingested	Larvae encyst in muscle causing myalgia	Larvae in muscle of pig, bear, and other animals
Anisakis	None	Larvae in fish ingested	Larvae in submucosa of gastrointestinal tract	Larvae in muscle of fish

Epidemiology

Enterobius is found worldwide and is the **most common** helminth in the United States. Children younger than 12 years are the most commonly affected group.

Laboratory Diagnosis

The eggs are recovered from perianal skin by using the **Scotch tape** technique and can be observed microscopically (see Figure 56–4).

Unlike those of other intestinal nematodes, these **eggs are not found in the stools**. The small, whitish adult worms can be found in the stools or near the anus of diapered children. No serologic tests are available.

Treatment & Prevention

The drug of choice is albendazole, mebendazole, or pyrantel pamoate. These drugs kill the adult worms in the colon but not

the eggs, so retreatment in 2 weeks is suggested. Reinfection is very common. Household members should also be treated.

There are no specific means of prevention, but washing hands when preparing food and washing bed sheets, towels, diapers, and clothing to remove eggs are helpful.

TRICHURIS

Disease

Trichuris trichiura causes whipworm infection (trichuriasis).

Important Properties

Humans are **infected** by ingesting worm eggs in food or water contaminated with human feces (see Figures 56–3B and 56–5). The eggs hatch in the small intestine, where the larvae differentiate into immature adults. These immature adults migrate to the colon, where they mature, mate, and produce thousands of fertilized eggs daily, which are passed in the feces. Eggs deposited

TABLE 56-3 Medically Important Stages in Life Cycle of Tissue Nematodes (Roundworms)

Organism	Insect Vector	Stage that Infects Humans	Stage(s) in Humans Most Associated with Disease	Important Stage(s) Outside of Humans
Wuchereria	Mosquito	Larvae	Adult worms in lymphatics (elephantiasis)	Mosquito ingests microfilariae in human blood → larvae
Onchocerca	Blackfly	Larvae	Adult worms in skin; microfilariae in eye (blindness)	Blackfly ingests microfilariae in human skin → larvae
Loa	Deer fly (mango fly)	Larvae	Adult worms in tissue (skin, conjunctivae)	Deer fly ingests microfilariae → larvae
Dracunculus	None	Larvae in copepods are swallowed in drinking water	Female worms cause skin blister; see head of worm	Copepods ingest larvae
Toxocara canis	None	Eggs in dog feces	Larvae in internal organs	Adult worms in dog intestine → eggs
Ancylostoma caninum	None	Filariform larvae penetrate skin	Larvae in subcutaneous tissue	Adult worms in dog intestine → eggs → larvae

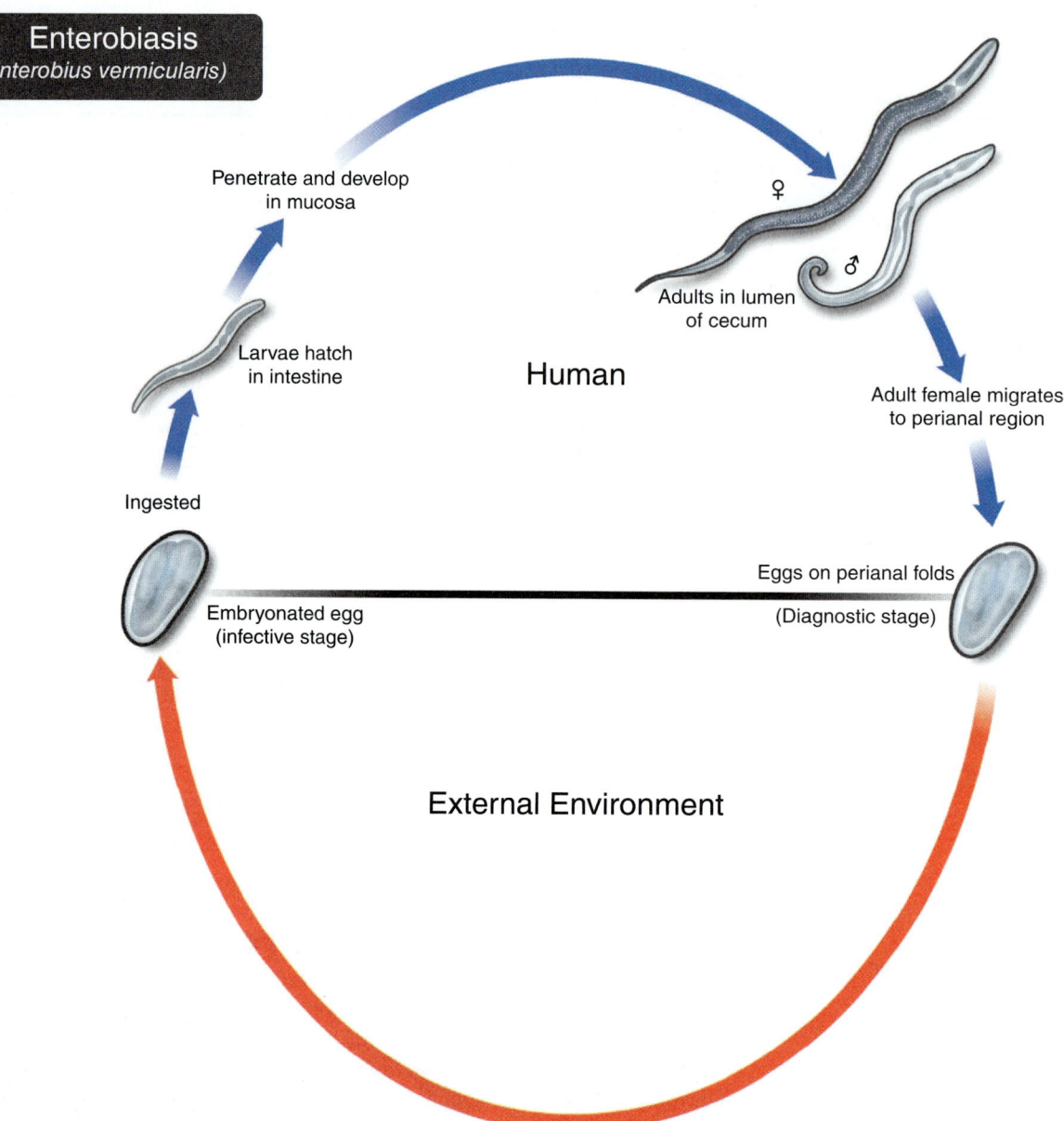

Enterobiasis
(Enterobius vermicularis)

Penetrate and develop
in mucosa

♀

♂

Adults in lumen
of cecum

Larvae hatch
in intestine

Human

Adult female migrates
to perianal region

Ingested

Eggs on perianal folds

Embryonated egg
(infective stage)

(Diagnostic stage)

External Environment

FIGURE 56–1 *Enterobius vermicularis.* Life cycle. **Top:** Blue arrow at top left shows eggs being ingested. Adult pinworms form in colon. Female migrates out anus and lays eggs on perianal skin. **Bottom:** Red arrow indicates survival of eggs in the environment. (Reproduced with permission from Public Health Image Library, Centers for Disease Control and Prevention.)

in warm, moist soil form embryos. When the embryonated eggs are ingested, the cycle is completed. Figure 56–2B illustrates the characteristic "whiplike" appearance of the adult worm.

Pathogenesis & Clinical Findings

Although adult *Trichuris* worms burrow their hair-like anterior ends into the intestinal mucosa, they do not cause significant anemia, unlike the hookworms. *Trichuris* may cause diarrhea, but most infections are asymptomatic.

   Trichuris may also cause **rectal prolapse** in children with heavy infection. Prolapse results from increased peristalsis that occurs in an effort to expel the worms. The whitish worms may be seen on the prolapsed mucosa.

Epidemiology

Whipworm infection occurs worldwide, especially in the tropics; more than 500 million people are affected. In the United States, it occurs mainly in the southern states.

Laboratory Diagnosis

Diagnosis is based on finding the typical eggs (i.e., barrel-shaped [lemon-shaped] with a plug at each end) in the stool (see Figures 56–3B and 56–5).

Treatment & Prevention

Albendazole is the drug of choice. Proper disposal of feces prevents transmission.

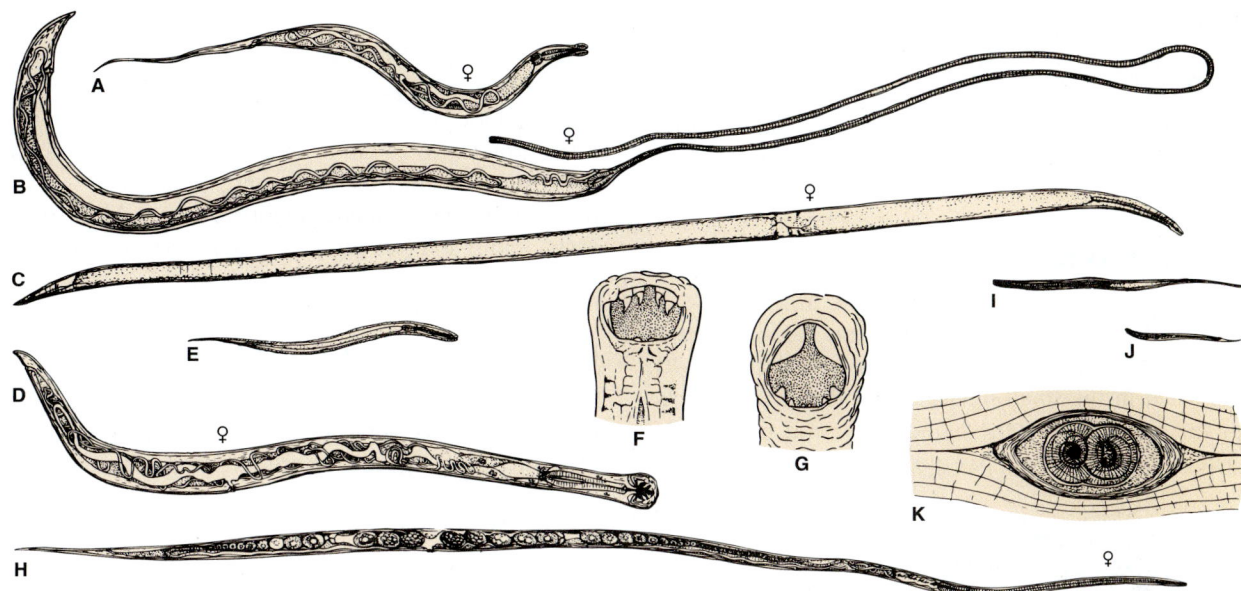

FIGURE 56–2 **A:** *Enterobius vermicularis* female adult (6×). **B:** *Trichuris trichiura* female adult. Note the thin anterior (whiplike) end (6×). **C:** *Ascaris lumbricoides* female adult (0.6×). **D:** *Ancylostoma duodenale* female adult (6×). **E:** *Ancylostoma duodenale* filariform larva (60×). **F:** *Ancylostoma duodenale* head with teeth (25×). **G:** *Necator americanus* head with cutting plates (25×). **H:** *Strongyloides stercoralis* female adult (60×). **I:** *Strongyloides stercoralis* filariform larva (60×). **J:** *Strongyloides stercoralis* rhabditiform larva (60×). **K:** *Trichinella spiralis* cyst containing two larvae in muscle (60×).

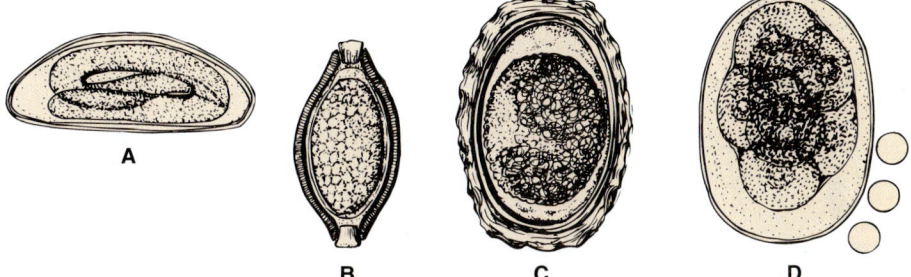

FIGURE 56–3 **A:** *Enterobius vermicularis* egg. **B:** *Trichuris trichiura* egg. **C:** *Ascaris lumbricoides* egg. **D:** *Ancylostoma duodenale* or *Necator americanus* egg (300×). (Circles represent red blood cells.)

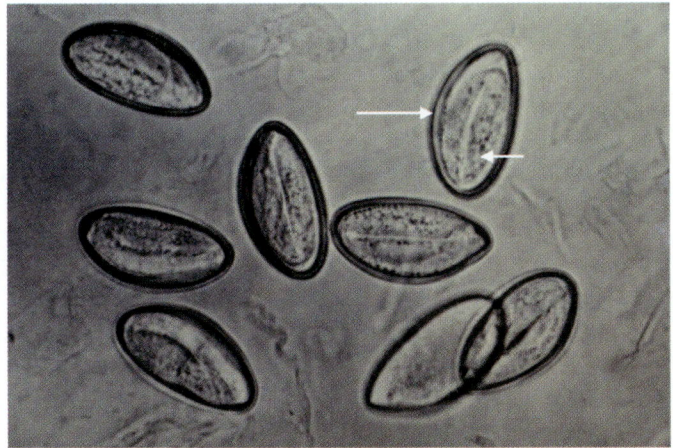

FIGURE 56–4 *Enterobius vermicularis*. Eggs. Long arrow points to an egg of the pinworm, *E. vermicularis* recovered on "Scotch tape." Short arrow points to the embryo inside the egg. (Reproduced with permission from Public Health Image Library, Centers for Disease Control and Prevention.)

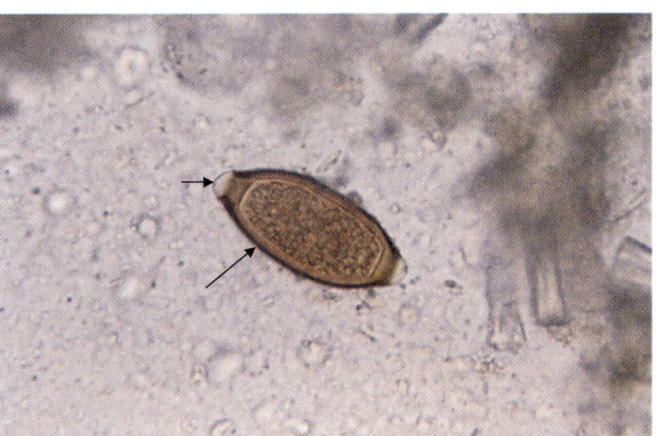

FIGURE 56–5 *Trichuris trichiura*. Egg. Long arrow points to an egg of *T. trichiura*. Short arrow points to one of the two "plugs" on each end of the egg. (Used with permission from Dr. Melvin M., Public Health Image Library, Centers for Disease Control and Prevention.)

ASCARIS

Disease

Ascaris lumbricoides causes ascariasis. *Ascaris* larvae migrating through the lung can cause eosinophilic pneumonia (Loeffler's syndrome).

Important Properties

The life cycle of *A. lumbricoides* is shown in Figure 56–6. Humans are infected by **ingesting worm eggs** in food or water contaminated with human feces (Figures 56–3C and 56–7). The eggs hatch in the small intestine, and the larvae migrate through the gut wall into the bloodstream and then to the lungs. They enter the alveoli, pass up the bronchi and trachea,

and are swallowed. Within the small intestine, they become adults (Figures 56–2C and 56–8). They live in the lumen, do not attach to the wall, and derive their sustenance from ingested food. The adults are the **largest intestinal nematodes**, often growing to 25 cm or more. *A. lumbricoides* is known as the "giant roundworm." Thousands of eggs are laid daily, are passed in the feces, and differentiate into embryonated eggs in warm, moist soil (see Figure 56–3C). Ingestion of the embryonated eggs completes the cycle.

Pathogenesis & Clinical Findings

The major damage occurs during larval migration rather than from the presence of the adult worm in the intestine. The principal sites of tissue reaction are the **lungs**, where inflammation

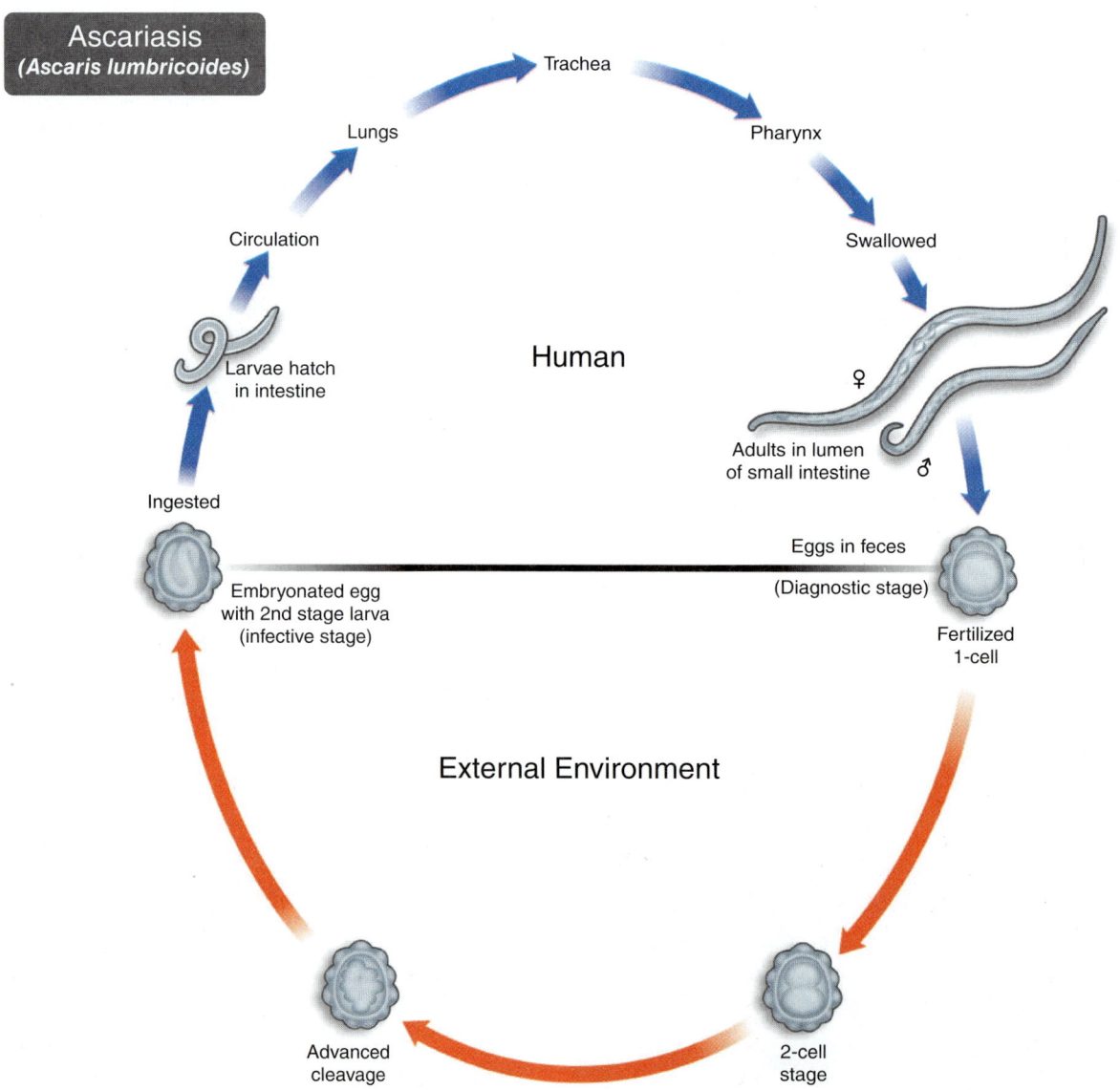

FIGURE 56–6 *Ascaris lumbricoides.* Life cycle. **Top:** Blue arrow at top left shows eggs being ingested. Larvae emerge in the intestinal tract, enter the bloodstream, and migrate to the lungs. They then enter the alveoli, ascend into the bronchi and trachea, migrate to the pharynx, and are swallowed. Adult *Ascaris* worms form in small intestine. Eggs pass in human feces. **Bottom:** Red arrow indicates maturation of eggs in the soil. (Reproduced with permission from Public Health Image Library, Centers for Disease Control and Prevention.)

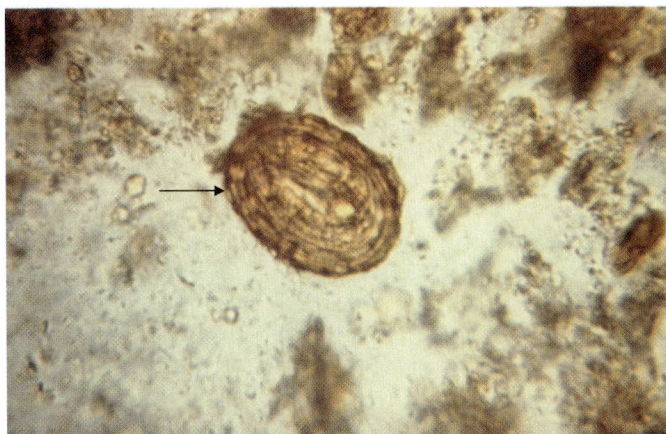

FIGURE 56–7 *Ascaris lumbricoides*. Egg. Arrow points to an egg of *Ascaris*. Note the typical "scalloped" edge of the *Ascaris* egg. (Reproduced with permission from Public Health Image Library, Centers for Disease Control and Prevention.)

with an **eosinophilic exudate** occurs in response to larval antigens. Because the adults derive their nourishment from ingested food, a heavy worm burden may contribute to malnutrition, especially in children in developing countries.

FIGURE 56–8 *Ascaris lumbricoides*. Adult worms. (Used with permission from Dr. Henry Bishop, Public Health Image Library, Centers for Disease Control and Prevention.)

Most infections are asymptomatic. **Ascaris pneumonia** with fever, cough, and eosinophilia can occur with a heavy larval burden. Abdominal pain and even obstruction can result from the presence of adult worms in the intestine.

Epidemiology

Ascaris infection is very common, especially in the tropics; hundreds of millions of people are infected. In the United States, most cases occur in the southern states.

Laboratory Diagnosis

Diagnosis is usually made microscopically by detecting eggs in the stools. The egg is oval with an irregular surface (see Figures 56–3C and 56–7). Occasionally, the patient sees adult worms in the stools.

Treatment & Prevention

Albendazole, mebendazole, and ivermectin are effective. Proper disposal of feces can prevent ascariasis.

ANCYLOSTOMA & NECATOR

Disease

Ancylostoma duodenale (Old World hookworm) and *Necator americanus* (New World hookworm) cause hookworm infection.

Important Properties

The life cycle of the hookworms is shown in Figure 56–9. Humans are infected when **filariform larvae in moist soil penetrate the skin**, usually of the feet or legs (Figures 56–2E and 56–10). They are carried by the blood to the lungs, migrate into the alveoli and up the bronchi and trachea, and then are swallowed. They develop into adults in the small intestine, attaching to the wall with either cutting plates (*Necator*) or teeth (*Ancylostoma*) (Figures 56–2D, F, and G, and 56–11). They feed on blood from the capillaries of the intestinal villi. Thousands of eggs per day are passed in the feces (Figures 56–3D and 56–12). Eggs develop first into noninfectious, feeding (rhabditiform) larvae and then into third-stage, infectious, nonfeeding (filariform) larvae (see Figure 56–2E), which penetrate the skin to complete the cycle.

Pathogenesis & Clinical Findings

The major damage is due to the **loss of blood** at the site of attachment in the small intestine. Up to 0.1 to 0.3 mL per worm can be lost per day. Blood is consumed by the worm and oozes from the site in response to an anticoagulant made by the worm. Weakness and pallor accompany the microcytic anemia caused by blood loss. These symptoms occur in patients whose nutrition cannot compensate for the blood loss. "Ground itch," a pruritic papule or vesicle, can occur at the site of entry of the larvae into the skin. The human hookworms also cause cutaneous

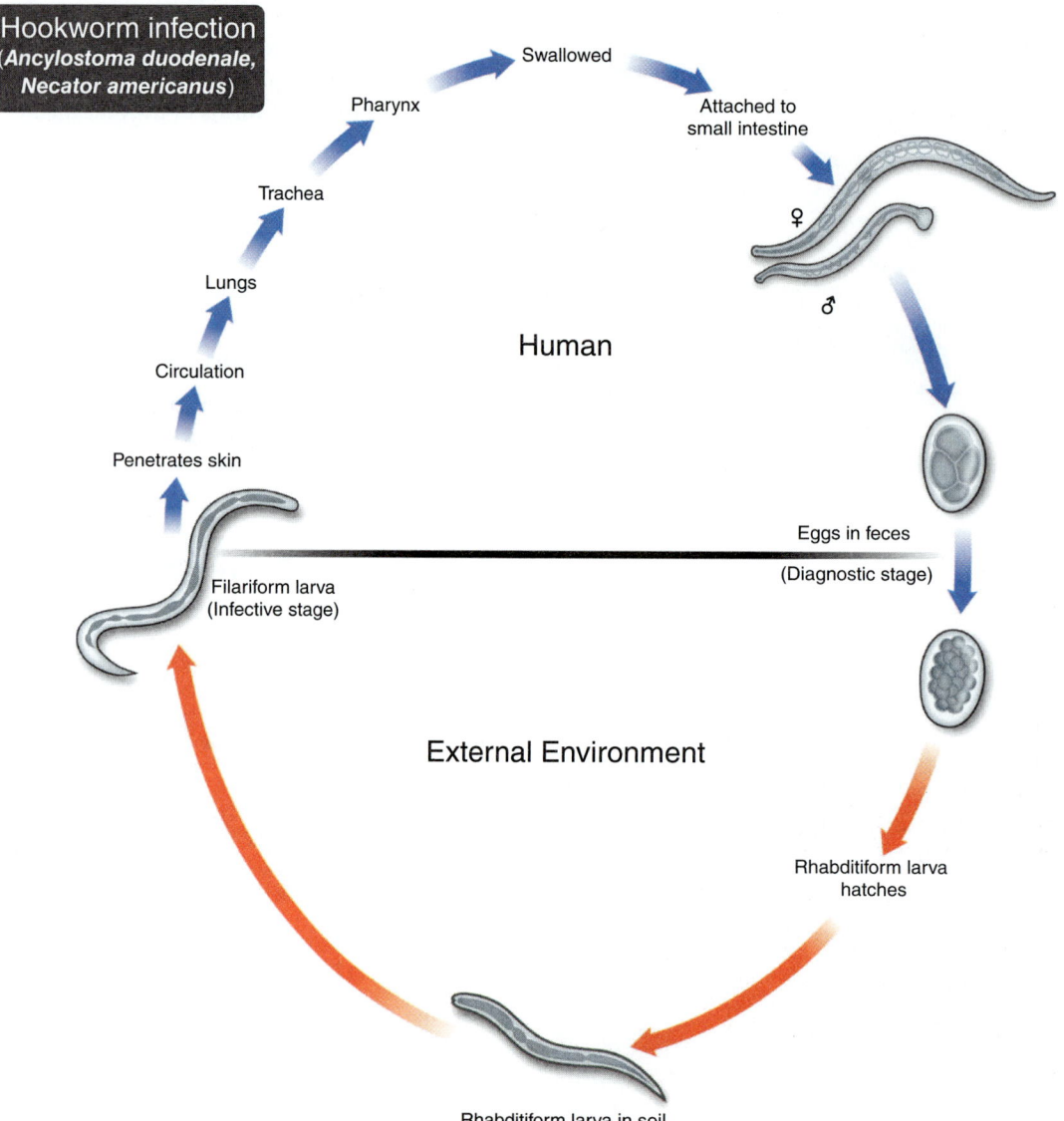

FIGURE 56–9 Hookworms (*Necator* and *Ancylostoma*). Life cycle. **Top:** Blue arrow on left shows filariform larvae penetrating skin. Larvae migrate through lung and may cause pneumonia. Adult hookworms attach to intestinal mucosa and cause bleeding and anemia. Eggs pass in human feces. **Bottom:** Red arrow indicates maturation of eggs in the soil to form rhabditiform larvae and then infective filariform larvae. (Reproduced with permission from Public Health Image Library, Centers for Disease Control and Prevention.)

larva migrans. Pneumonia with eosinophilia can be seen during larval migration through the lungs.

Epidemiology

Hookworm is found worldwide, especially in tropical areas. In the United States, *Necator* is endemic in the rural southern states. Walking barefooted on soil predisposes to infection. An important public health measure was requiring children to wear shoes to school.

Laboratory Diagnosis

Diagnosis is made microscopically by observing the eggs in the stools (see Figures 56–3D and 56–12). Occult blood in the stools is frequent. Eosinophilia is typical.

Treatment & Prevention

The drug of choice is albendazole, mebendazole, or pyrantel pamoate. Disposing of sewage properly and wearing shoes are effective means of prevention.

STRONGYLOIDES

Disease

Strongyloides stercoralis causes strongyloidiasis.

Important Properties

The life cycle of *S. stercoralis* is shown in Figure 56–13. *S. stercoralis* has **two distinct life cycles**, one within the human body and the other free-living in the soil. The life cycle in the human

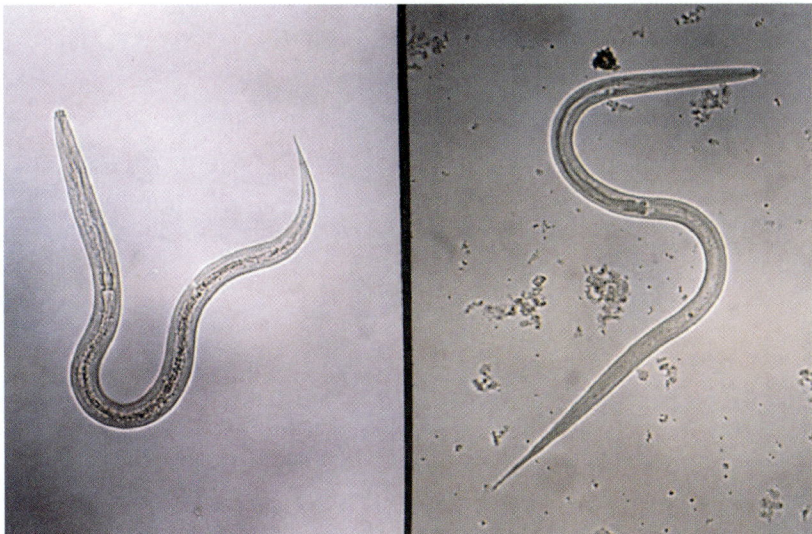

FIGURE 56–10 *Necator* and *Strongyloides*. Filariform larvae. Filariform larva of *Necator* on the left and *Strongyloides* on the right. Filariform larva is the infective form that penetrates the skin. (Reproduced with permission from Public Health Image Library, Centers for Disease Control and Prevention.)

body begins with the **penetration of the skin**, usually of the feet, by **infectious (filariform) larvae** (see Figures 56–2I and 56–10) and their migration to the lungs. They enter the alveoli, pass up the bronchi and trachea, and then are swallowed. In the small intestine, the larvae molt into adults (see Figure 56–H) that enter the mucosa and produce eggs.

The eggs usually hatch within the mucosa, forming rhabditiform larvae (see Figure 56–2J) that are passed in the feces. Some larvae molt to form filariform larvae, which penetrate the intestinal wall directly without leaving the host and migrate to the lungs (**autoinfection**). Filariform larvae can also exit the anus and reinfect through the perianal skin. In immunocompetent patients, this is an infrequent, clinically unimportant event.

However, in immunocompromised patients (e.g., those who have acquired immunodeficiency syndrome [AIDS] or are taking high-dose corticosteroids or tumor necrosis factor [TNF] inhibitors) or patients who are severely malnourished, autoinfection can lead to **massive reinfection (hyperinfection)**, with larvae passing to many organs and with severe, sometimes fatal consequences. Reinfection can also occur in those infected with human T-cell lymphotropic virus (HTLV) because their ability to mount a protective T-cell response is diminished.

If larvae are passed in the feces and enter warm, moist soil, they molt through successive stages to form adult male and female worms. After mating, the entire life cycle of egg, larva, and adult can occur in the soil. After several free-living

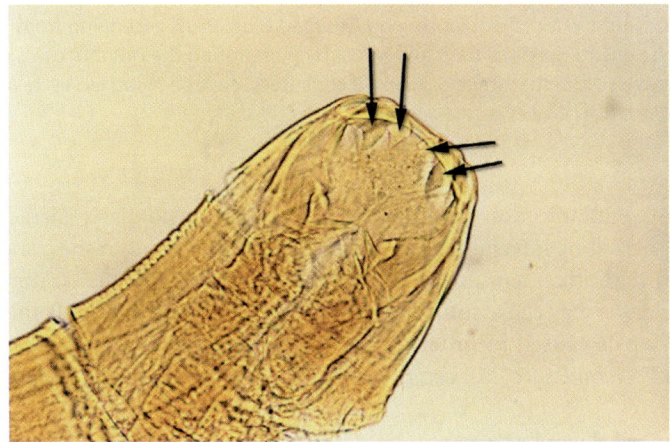

FIGURE 56–11 *Ancylostoma duodenale*. Head of the adult hookworm. Arrows point to the four cutting teeth in the mouth of *Ancylostoma*. (Used with permission from Dr. Melvin M., Public Health Image Library, Centers for Disease Control and Prevention)

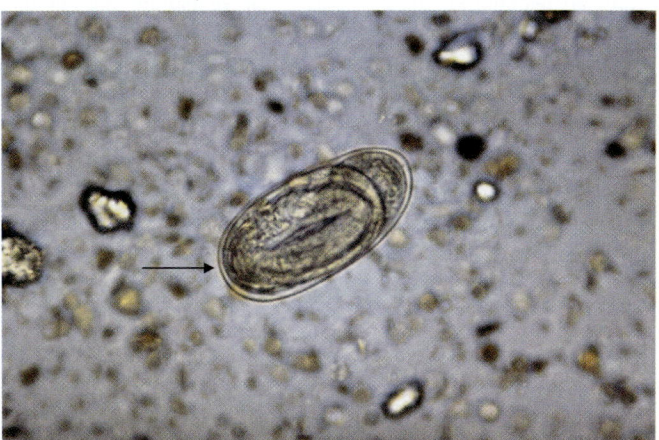

FIGURE 56–12 *Necator* and *Ancylostoma* (hookworms). Egg. Arrow points to an egg of a hookworm. The eggs of *Necator* and *Ancylostoma* are indistinguishable. Note the embryo coiled up inside. (Reproduced with permission from Public Health Image Library, Centers for Disease Control and Prevention.)

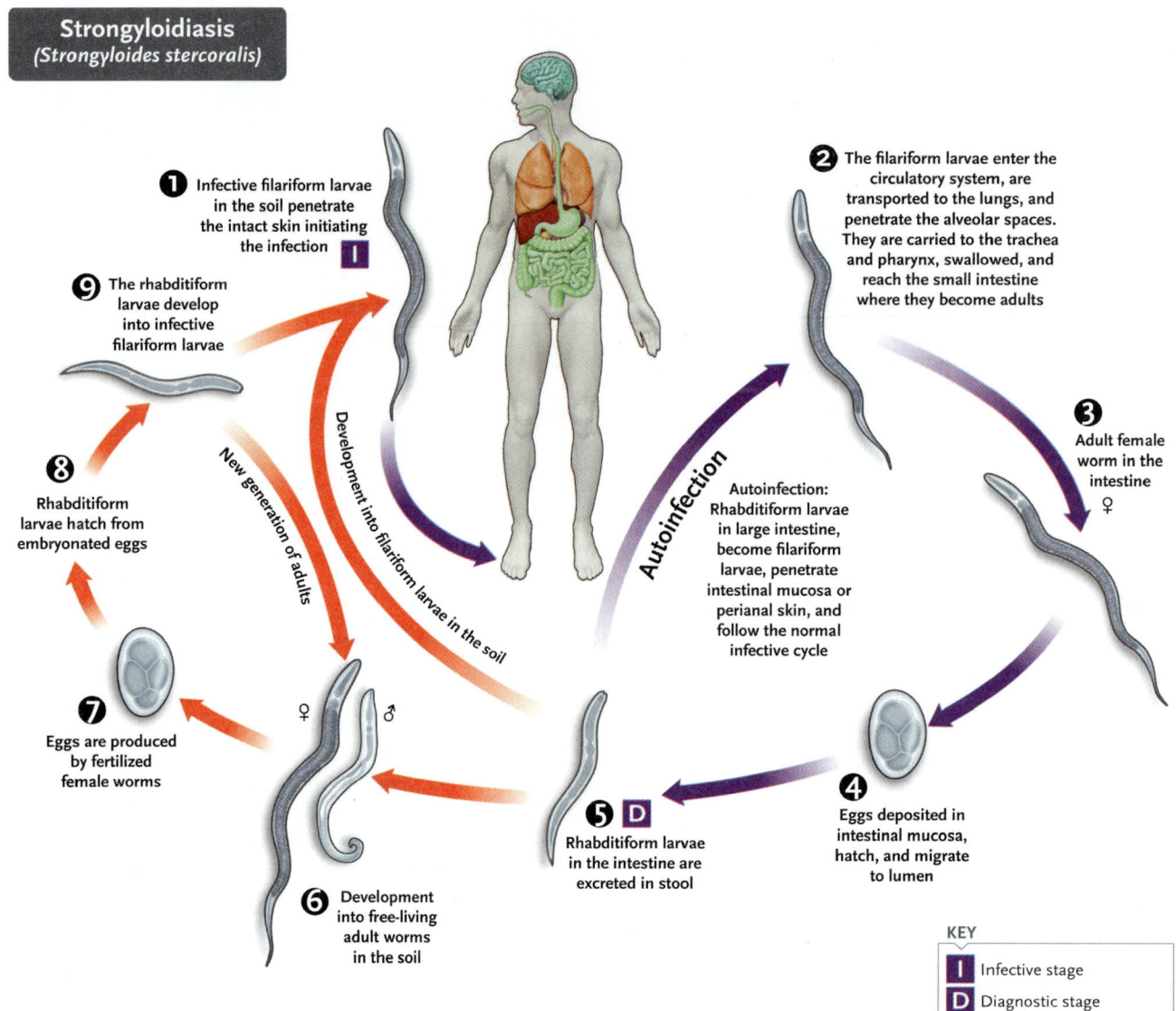

Strongyloidiasis
(Strongyloides stercoralis)

1 Infective filariform larvae in the soil penetrate the intact skin initiating the infection **I**

9 The rhabditiform larvae develop into infective filariform larvae

8 Rhabditiform larvae hatch from embryonated eggs

7 Eggs are produced by fertilized female worms

6 Development into free-living adult worms in the soil

New generation of adults

Development into filariform larvae in the soil

5 D Rhabditiform larvae in the intestine are excreted in stool

Autoinfection

Autoinfection: Rhabditiform larvae in large intestine, become filariform larvae, penetrate intestinal mucosa or perianal skin, and follow the normal infective cycle

2 The filariform larvae enter the circulatory system, are transported to the lungs, and penetrate the alveolar spaces. They are carried to the trachea and pharynx, swallowed, and reach the small intestine where they become adults

3 Adult female worm in the intestine ♀

4 Eggs deposited in intestinal mucosa, hatch, and migrate to lumen

KEY
I Infective stage
D Diagnostic stage

FIGURE 56–13 *Strongyloides stercoralis.* Life cycle. Center and right side of figure describe the stages within the human (blue arrows). Filariform larvae penetrate the skin (step 1). Larvae migrate through lung and may cause pneumonia. Adult *Strongyloides* worms form in small intestine. Eggs hatch in intestinal mucosa, and rhabditiform larvae are excreted in human feces, not worm eggs. Curved blue arrow ascending from step 5 describes the autoinfection cycle in which filariform larvae form in the gastrointestinal tract and infect by penetrating the gut mucosa or perianal skin. Left side of figure describes the maturation in the soil (red arrows). Note that steps 6, 7, and 8 constitute the free-living life cycle in the soil. (Used with permission from Dr. Alexander J. da Silva and Melanie Moser, Centers for Disease Control and Prevention.)

cycles, filariform larvae are formed. When they contact skin, they penetrate and again initiate the parasitic cycle within humans.

Pathogenesis & Clinical Findings

Most patients are asymptomatic, especially those with a low worm burden. Adult female worms in the wall of the small intestine can cause inflammation, resulting in watery diarrhea. Larvae in the lungs can produce a pneumonitis similar to that caused by *Ascaris*. Pruritus (ground itch) can occur at the site of larval penetration of the skin, as with hookworm. *S. stercoralis* also causes cutaneous larva migrans.

Autoinfection can result in chronic strongyloidiasis characterized by intermittent abdominal pain, fluctuating rashes, and intermittent eosinophilia. In hyperinfection, the penetrating larvae may cause sufficient damage to the intestinal mucosa that **sepsis caused by enteric bacteria**, such as *Escherichia coli* and *Bacteroides fragilis*, can occur.

Epidemiology

Strongyloidiasis occurs primarily in the tropics, especially in Southeast Asia. Its geographic pattern is similar to that of hookworm because the same type of soil is required. In the United States, *Strongyloides* is endemic in the southeastern states.

Laboratory Diagnosis

Diagnosis depends on finding **larvae**, rather than eggs, in the stool (see Figure 56–10). As with many nematode infections in which larvae migrate through tissue, **eosinophilia can be striking**. Serologic tests are useful when the larvae are not visualized. An enzyme immunoassay that detects antibody to larval antigens is available through the Centers for Disease Control and Prevention (CDC) in Atlanta.

Treatment & Prevention

Ivermectin is the drug of choice. Albendazole is an alternative drug.

Prevention involves disposing of sewage properly and wearing shoes. To prevent *Strongyloides* hyperinfection in patients scheduled to receive immunosuppressive drugs (e.g., corticosteroids, TNF inhibitors) and who have lived in an area of *Strongyloides* endemicity, serologic tests to determine whether antibodies to *Strongyloides* are present should be performed. If antibodies are found, the patient should be treated with ivermectin before immunosuppression is undertaken, if possible.

TRICHINELLA

Disease

Trichinella spiralis causes trichinosis. *T. spiralis* is also called the trichina worm.

Important Properties

The life cycle of *T. spiralis* is shown in Figure 56–14. Any mammal can be infected, but **pigs** are the most important reservoirs of human disease in the United States (except in Alaska, where bears constitute the main reservoir). Humans are infected by **eating raw** or **undercooked meat** containing larvae encysted in the muscle (see Figure 56–2K). The larvae excyst and mature into adults within the mucosa of the small intestine. Eggs hatch within the adult females, and larvae are released and distributed via the bloodstream to many organs; however, they develop only in **striated muscle cells**. Within these **nurse cells**, they encyst within a fibrous capsule and can remain viable for several years but eventually calcify (Figure 56–15).

The parasite is maintained in nature by cycles within reservoir hosts, primarily swine and rats. Humans are **end-stage hosts**, because the infected flesh is not consumed by other animals.

Pathogenesis & Clinical Findings

A few days after eating undercooked meat, usually pork, the patient experiences diarrhea followed 1 to 2 weeks later by **fever, muscle pain, periorbital edema**, and **eosinophilia**. Subconjunctival hemorrhages are an important diagnostic criterion. Signs of cardiac and central nervous system disease are frequent, because the larvae migrate to these tissues as well. Death, which is rare, is usually due to congestive heart failure or respiratory paralysis.

Epidemiology

Trichinosis occurs worldwide, especially in Eastern Europe and West Africa. In the United States, it is related to eating home-prepared sausage, usually on farms where the pigs are fed uncooked garbage. Bear and seal meat are also sources. In many countries, the disease occurs primarily in hunters who eat undercooked wild game.

Laboratory Diagnosis

Muscle biopsy reveals **larvae within striated muscle** (see Figures 56–2K and 56–15). Serologic tests, especially the bentonite flocculation test, become positive 3 weeks after infection.

Treatment & Prevention

There is no effective treatment for trichinosis when the larvae have infected the muscle, but for patients with severe symptoms, steroids plus albendazole can be useful. Mebendazole is effective against the adult intestinal worms early in infection. The disease can be prevented by properly cooking pork and by feeding pigs only cooked garbage.

TISSUE NEMATODES

WUCHERERIA

Disease

Wuchereria bancrofti causes filariasis. Elephantiasis is a striking feature of this disease. Tropical pulmonary eosinophilia (TPE) is an immediate hypersensitivity reaction to *W. bancrofti* in the lung. *Brugia malayi* causes filariasis and TPE in Malaysia.

Important Properties

The life cycle of *W. bancrofti* is shown in Figure 56–16. Humans are infected when the **female mosquito** (especially *Anopheles* and *Culex* species) deposits infective larvae on the skin while biting. The larvae penetrate the skin, enter a lymph node, and, after 1 year, mature to adults that produce **microfilariae** (Figure 56–17A and 56–18). These circulate in the blood, chiefly at night, and are ingested by biting mosquitoes. Within the mosquito, the microfilariae produce infective larvae that are transferred with the next bite. Humans are the only definitive hosts.

Pathogenesis & Clinical Findings

Adult worms in the lymph nodes cause inflammation that eventually obstructs the lymphatic vessels, causing edema. Massive

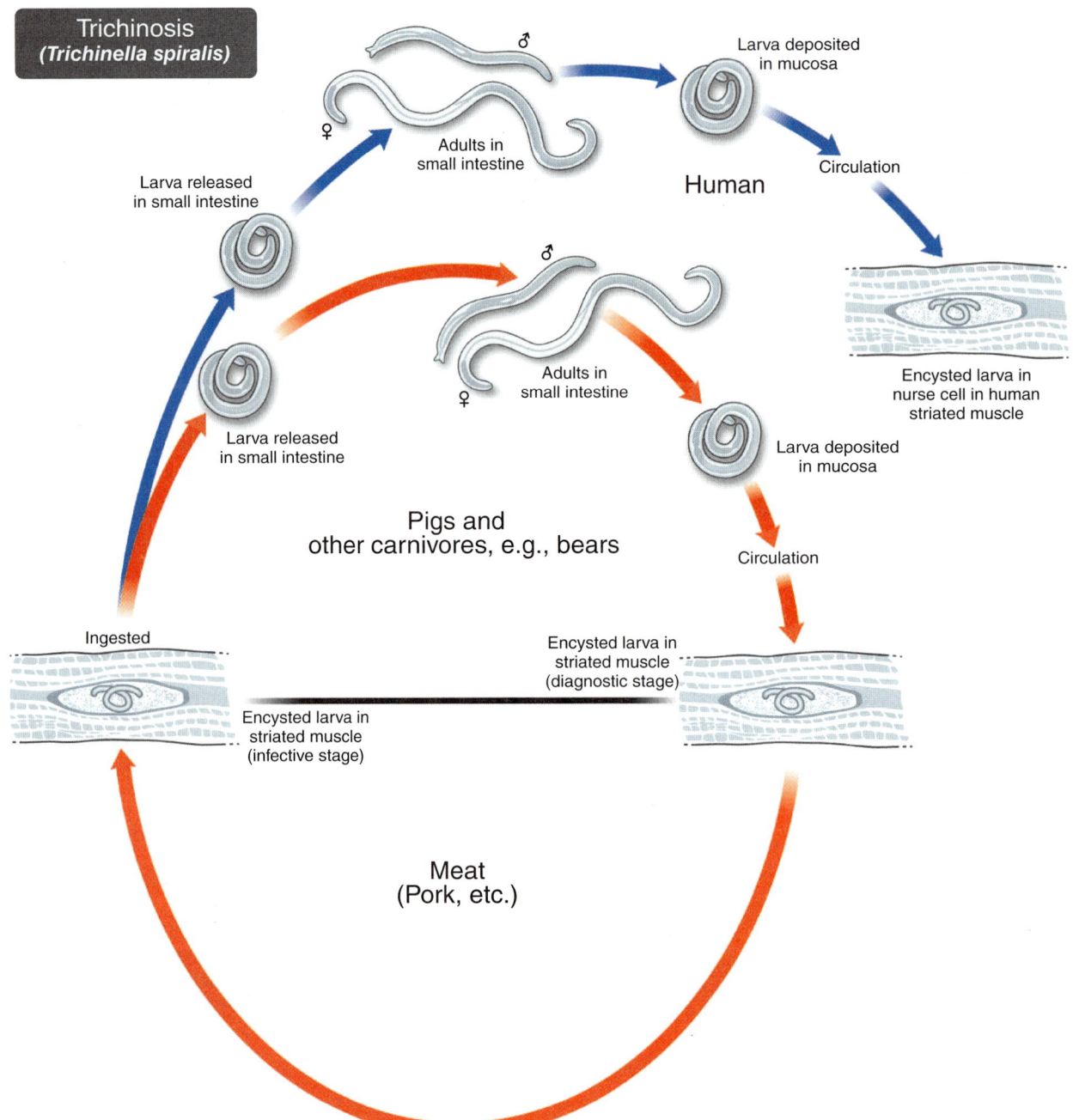

FIGURE 56–14 *Trichinella spiralis.* Life cycle. **Top:** Blue arrow on left shows ingestion of meat (muscle) containing encysted *Trichinella* larva. Adult worms form in intestine and produce larvae that enter bloodstream and encyst in human muscle. **Bottom:** Red circular arrow describes the natural cycle in which *Trichinella* circulates between pigs and various carnivores such as bears. (Reproduced with permission from Public Health Image Library, Centers for Disease Control and Prevention.)

edema of the legs is called **elephantiasis** (Figure 56–19). Note that microfilariae do *not* cause symptoms.

Early infections are asymptomatic. Later, fever, lymphangitis, and cellulitis develop. Gradually, the obstruction leads to edema and fibrosis of the legs and genitalia, especially the scrotum. Elephantiasis occurs mainly in patients who have been repeatedly infected over a long period. Tourists, who typically are infected only once, do *not* get elephantiasis.

Wolbachia species are *Rickettsia*-like bacteria found intracellularly within filarial nematodes such as *Wuchereria* and *Onchocerca*. *Wolbachia* release endotoxin-like molecules that are thought to play a role in the pathogenesis of *Wuchereria* and *Onchocerca* infections. Evidence for this includes the use of doxycycline, which kills the *Wolbachia*, resulting in a reduction in the number of microfilaria and in the inflammatory response to the nematode infection.

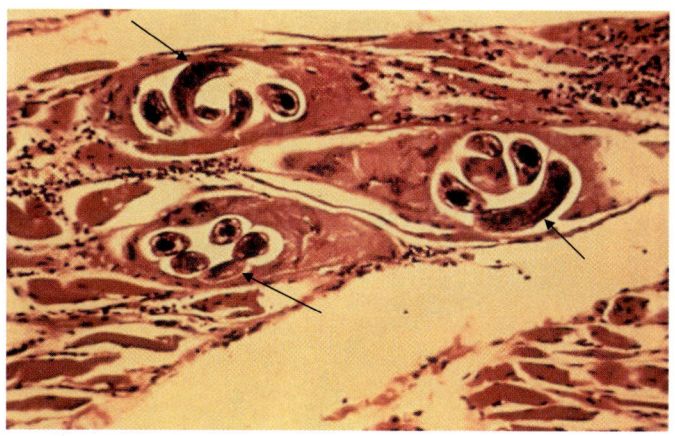

FIGURE 56–15 *Trichinella spiralis.* Larvae in skeletal muscle. Three arrows point to *Trichinella* larvae within "nurse cells" in skeletal muscle. (Reproduced with permission from Public Health Image Library, Centers for Disease Control and Prevention.)

TPE is characterized by coughing and wheezing, especially at night. These symptoms are caused by microfilariae in the lung that elicit an immediate hypersensitivity reaction characterized by a high IgE concentration and eosinophilia.

Epidemiology

This disease occurs in the tropical areas of Africa, Asia, and Latin America. The species of mosquito that acts as the vector varies from area to area. Altogether, 200 to 300 million people are infected.

Laboratory Diagnosis

Thick blood smears taken from the patient at night reveal the microfilariae (see Figure 56–18). Serologic tests are not useful.

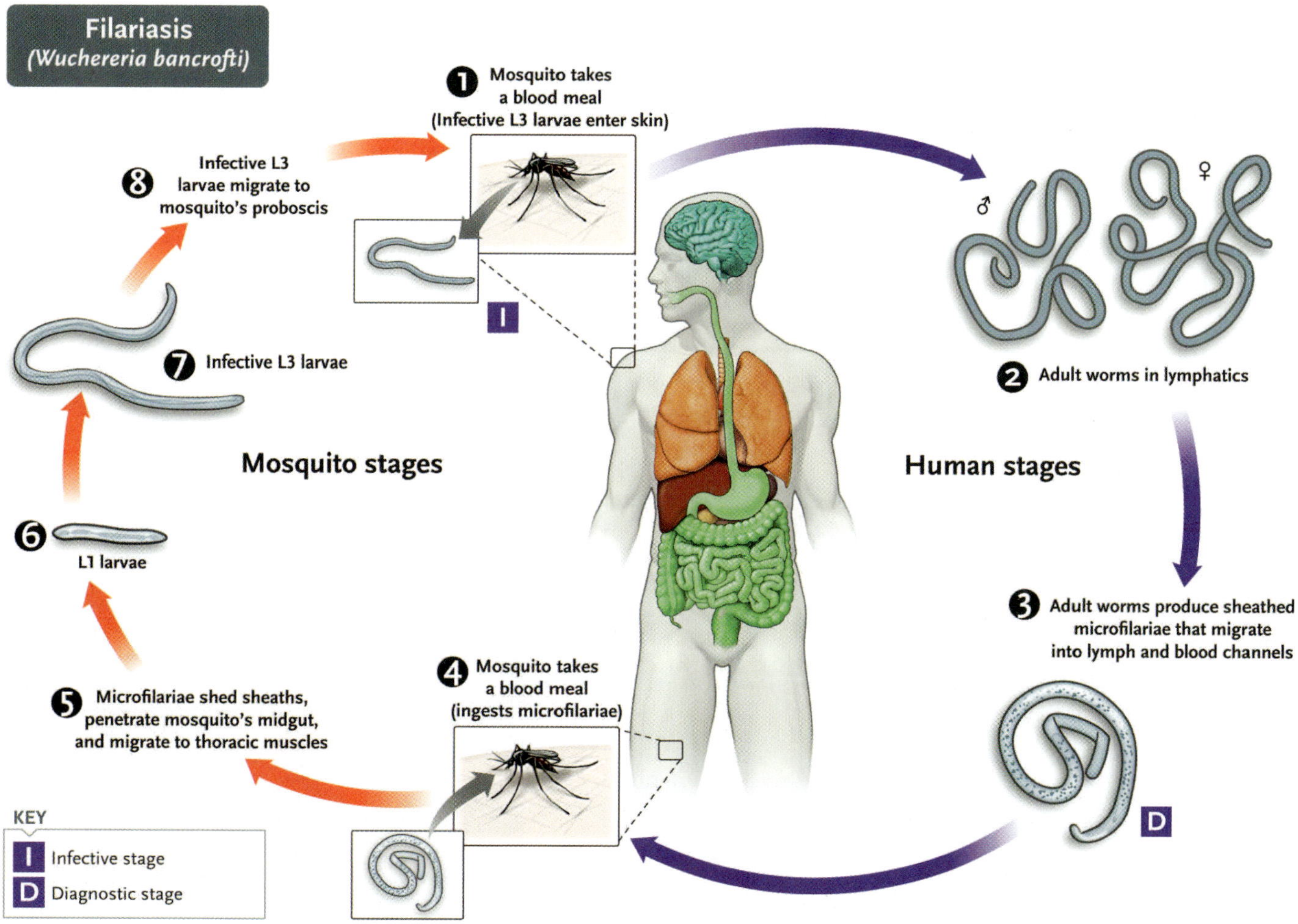

Filariasis
(Wuchereria bancrofti)

1 Mosquito takes a blood meal
(Infective L3 larvae enter skin)

8 Infective L3 larvae migrate to mosquito's proboscis

7 Infective L3 larvae

Mosquito stages

6 L1 larvae

5 Microfilariae shed sheaths, penetrate mosquito's midgut, and migrate to thoracic muscles

4 Mosquito takes a blood meal
(ingests microfilariae)

KEY

I Infective stage
D Diagnostic stage

♂ ♀

2 Adult worms in lymphatics

Human stages

3 Adult worms produce sheathed microfilariae that migrate into lymph and blood channels

FIGURE 56–16 *Wuchereria bancrofti.* Life cycle. Right side of figure describes the stages within the human (blue arrows). Humans are infected at step 1 when mosquito bites human and larvae enter bloodstream. Adult *Wuchereria* worms are formed in lymphatics. Mosquito is infected at step 4 when it ingests microfilariae in human blood. Left side of figure describes the stages within the mosquito (red arrows). (Used with permission from Dr. Alexander J. da Silva and Melanie Moser, Centers for Disease Control and Prevention.)

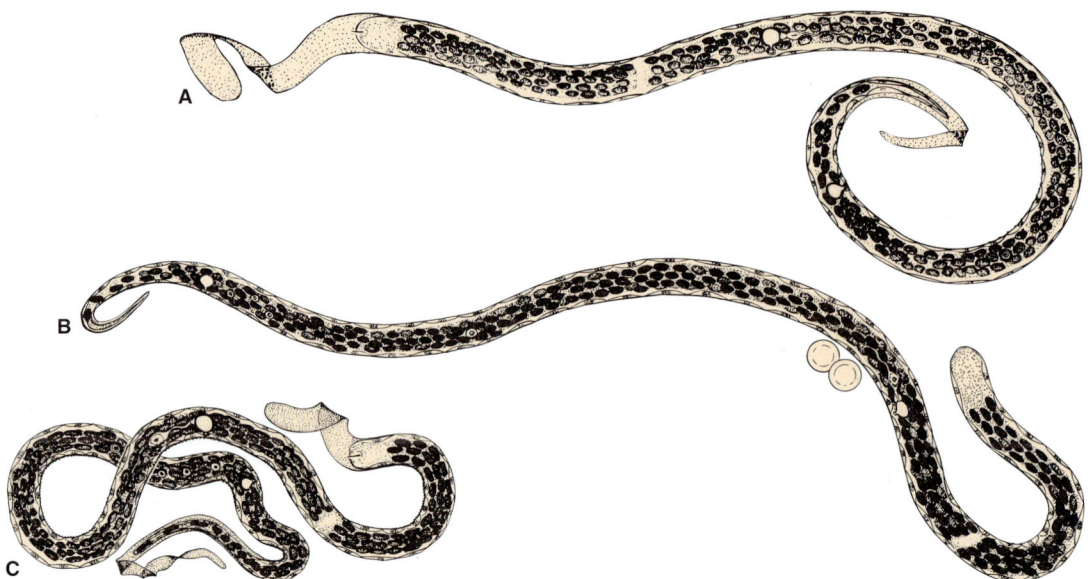

FIGURE 56–17 **A:** *Wuchereria bancrofti* microfilaria in blood. Note that the pointed tail is free of nuclei (225–300 × 8–10 μm). **B:** *Onchocerca volvulus* microfilaria in skin (rare) in blood (300–350 × 5–9 μm). **C:** *Loa loa* microfilaria in blood. Note that the pointed tail contains nuclei (250–300 × 6–9 μm). (Circles represent red blood cells.)

Treatment & Prevention

Diethylcarbamazine is effective only against microfilariae; no drug therapy for adult worms is available. Treatment of patients with *Wuchereria* (and *Onchocerca*) infections with doxycycline to kill *Wolbachia* results in a significant decrease in the number of microfilariae in the patient.

Prevention involves mosquito control with insecticides and the use of protective clothing, mosquito netting, and repellents.

ONCHOCERCA

Disease

Onchocerca volvulus causes onchocerciasis.

Important Properties

Humans are infected when the **female blackfly**, *Simulium*, deposits infective larvae while biting. The larvae enter the wound and migrate into the subcutaneous tissue, where they differentiate into adults, usually within **dermal nodules**. The female produces microfilariae (see Figure 56–17B) that are ingested when another blackfly bites. The microfilariae develop

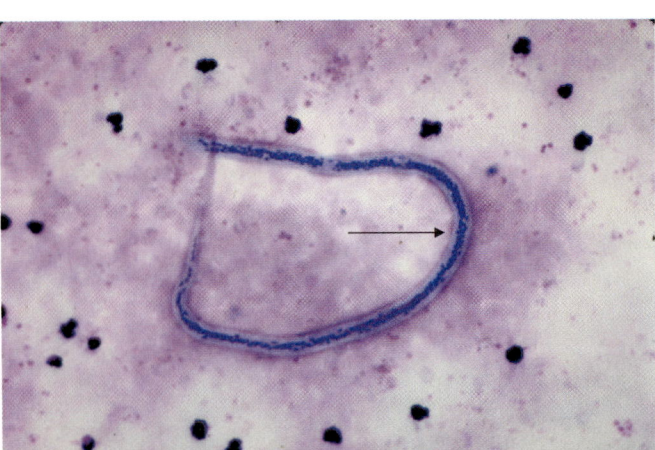

FIGURE 56–18 *Wuchereria bancrofti*. Filarial worm in blood. Arrow points to filarial worm in blood smear. (Used with permission from Dr. Melvin M., Public Health Image Library, Centers for Disease Control and Prevention.)

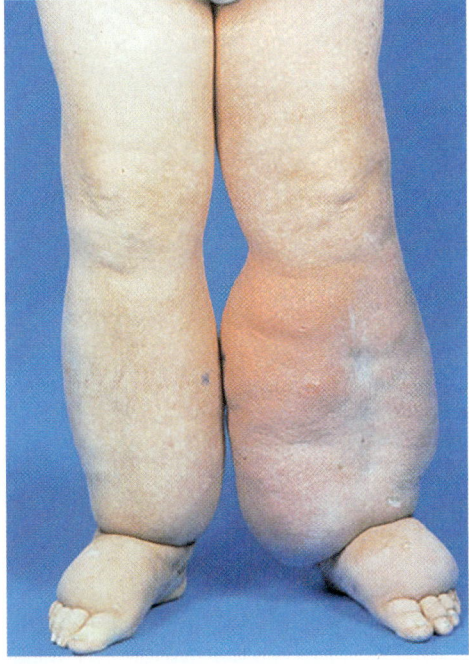

FIGURE 56–19 *Wuchereria bancrofti*. Elephantiasis. Note massive swelling of legs bilaterally. (Used with permission from Jay S. Keystone, MD, FRCPC.)

into infective larvae in the fly to complete the cycle. Humans are the only definitive hosts.

Pathogenesis & Clinical Findings

Inflammation occurs in subcutaneous tissue, and pruritic papules and nodules form in response to the adult worm proteins. Microfilariae migrate through subcutaneous tissue, ultimately concentrating in the eyes. There they cause lesions that can lead to blindness. Loss of subcutaneous elastic fibers leads to wrinkled skin, which is called "hanging groin" when it occurs in the inguinal region. Thickening, scaling, and dryness of the skin accompanied by severe itching are the manifestations of a dermatitis often called "lizard skin."

The role of *Wolbachia* in the pathogenesis of onchocerciasis has been discussed earlier in "*Wuchereria*."

Epidemiology

Millions of people are affected in Africa and Central America. The disease is a major cause of blindness. It is called **river blindness**, because the blackflies develop in rivers and people who live along those rivers are affected. Infection rates are often greater than 80% in areas of endemic infection.

Laboratory Diagnosis

Biopsy of the affected skin reveals microfilariae (see Figure 56–17B). Examination of the blood for microfilariae is not useful because they do not circulate in the blood. Eosinophilia is common. Serologic tests are not helpful.

Treatment & Prevention

Ivermectin is effective against microfilariae but not adults. Suramin kills adult worms but is quite toxic and is used particularly in those with eye disease. Skin nodules can be removed surgically, but new nodules can develop; therefore, a surgical cure is unlikely in areas of endemic infection.

Prevention involves control of the blackfly with insecticides. Ivermectin prevents the disease.

LOA

Disease

Loa loa causes loiasis.

Important Properties

Humans are infected by the bite of the **deer fly** (mango fly), *Chrysops*, which deposits infective larvae on the skin. The larvae enter the bite wound, wander in the body, and develop into adults. The females release microfilariae (see Figure 56–17C) that enter the blood, particularly during the day. The microfilariae are taken up by the fly during a blood meal and differentiate into infective larvae, which continue the cycle when the fly bites the next person.

Pathogenesis & Clinical Findings

There is no inflammatory response to the microfilariae or adults, but a hypersensitivity reaction causes transient, localized, non-erythematous, subcutaneous edema (Calabar swellings). The most dramatic finding is an adult worm **crawling across the conjunctiva** of the eye, a harmless but disconcerting event.

Epidemiology

The disease is found only in tropical Central and West Africa, the habitat of the vector *Chrysops*.

Laboratory Diagnosis

Diagnosis is made by visualization of the microfilariae in a blood smear (see Figure 56–17C). There are no useful serologic tests.

Treatment & Prevention

Diethylcarbamazine eliminates the microfilariae and may kill the adults. Worms in the eyes may require surgical excision. Control of the fly by insecticides can prevent the disease.

DRACUNCULUS

Disease

Dracunculus medinensis (guinea fire worm) causes dracunculiasis. This disease is on the verge of being eradicated (see Epidemiology section below).

Important Properties

Humans are infected when tiny **crustaceans** (copepods) containing infective larvae are **swallowed in drinking water**. The larvae are released in the small intestine and migrate into the body, where they develop into adults. Meter-long adult females cause the skin to ulcerate and then release motile larvae into freshwater. Copepods eat the larvae, which molt to form infective larvae. The cycle is completed when these are ingested in the water.

Pathogenesis & Clinical Findings

The adult female produces a substance that causes inflammation, blistering, and ulceration of the skin, usually of the lower extremities. The inflamed papule **burns and itches**, and the ulcer can become secondarily infected. Diagnosis is usually made clinically by finding **the worm in the skin ulcer**.

Epidemiology

The global eradication campaign sponsored by the World Health Organization (WHO) to provide clean drinking water has greatly reduced the number of cases. During the year 2015, only 22 new cases were detected worldwide. The cases occurred in four African countries (Chad, Ethiopia, Mali, and South Sudan). Prior to the campaign, the disease occurred over large areas of tropical Africa, the Middle East, and India, where tens of millions of people were infected.

Laboratory Diagnosis

The laboratory usually does not play a role in diagnosis.

Treatment & Prevention

The time-honored treatment consists of gradually extracting the worm by winding it up on a stick over a period of days. Prevention consists of filtering or boiling drinking water.

NEMATODES WHOSE LARVAE CAUSE DISEASE

TOXOCARA

Disease

T. canis is the major cause of visceral larva migrans. *Toxocara cati* and several other related nematodes also cause this disease.

Important Properties

The definitive host for *T. canis* is the dog. The adult *T. canis* female in the dog intestine produces eggs that are passed in the feces into the soil. Humans ingest soil containing the eggs, which hatch into larvae in the small intestine. The larvae migrate to many organs, especially the liver, brain, and eyes. The larvae eventually are encapsulated and die. The life cycle is not completed in humans; humans are therefore accidental, dead-end hosts.

Pathogenesis & Clinical Findings

Pathology is related to the granulomas that form around the dead larvae as a result of a delayed hypersensitivity response to larval proteins. The most serious clinical finding is blindness associated with retinal involvement. Fever, hepatomegaly, and eosinophilia are common. A pruritic urticarial rash may occur.

Epidemiology

Young children are primarily affected, because they are likely to ingest soil containing the eggs. *T. canis* is a common parasite of dogs in the United States.

Laboratory Diagnosis

Serologic tests are commonly used, but the definitive diagnosis depends on visualizing the larvae in tissue. The presence of hyper-gammaglobulinemia and eosinophilia supports the diagnosis.

Treatment & Prevention

The treatment of choice is either albendazole or mebendazole, but there is no proven effective treatment. Many patients recover without treatment. Regarding prevention, dogs should be dewormed, and children should be prevented from eating soil.

ANCYLOSTOMA

Cutaneous larva migrans is caused by the filariform larvae of *A. caninum* (dog hookworm) and *Ancylostoma braziliense* (cat hookworm), as well as other nematodes. The organism cannot complete its life cycle in humans. The larvae penetrate the skin and **migrate through subcutaneous tissue**, causing an inflammatory response. The lesions ("creeping eruption") are extremely pruritic (Figure 56–20).

The larvae are typically confined to the epidermis as they lack the collagenase necessary to break through the basement membrane. Most infections are localized in the lower leg as it is the common site of larval penetration. The eruption appears to migrate as the larvae move up to a few centimeters daily.

The disease occurs primarily in the southern United States, in children and construction workers who are exposed to infected soil. The diagnosis is made clinically; the laboratory is of little value. Albendazole or ivermectin is usually effective.

ANGIOSTRONGYLUS

The larvae of the rat lung nematode *Angiostrongylus cantonensis* cause eosinophilic meningitis (i.e., a meningitis characterized by many eosinophils in the spinal fluid and in the blood). Usually at least 10% of the white cells are eosinophils. The larvae are typically ingested in undercooked seafood, such as crabs, prawns, and snails. Infection by this organism most often occurs in Asian countries. The diagnosis is made primarily on clinical grounds, but occasionally, the laboratory will find a larva in the spinal fluid. There is no treatment. Most patients recover spontaneously without major sequelae.

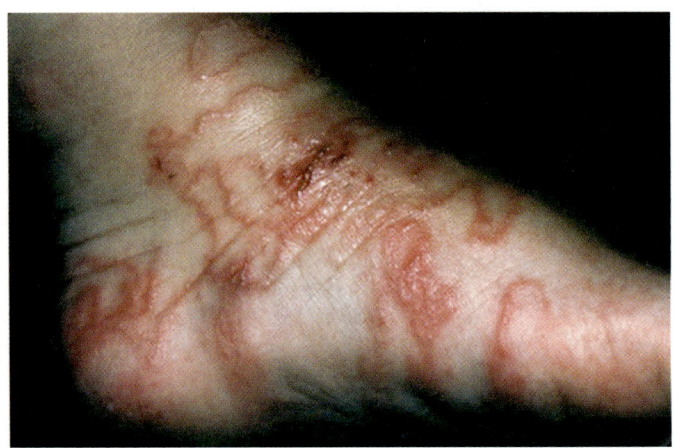

FIGURE 56–20 *Ancylostoma caninum.* Cutaneous larva migrans. Note serpiginous rash on foot. (Reproduced with permission from Usatine RP: A rash on the feet and buttocks. West J Med. 1999 Jun;170[6]:334–335.)

Eosinophilic meningitis is also caused by the larvae of two additional nematodes. *Gnathostoma spinigerum*, an intestinal nematode of cats and dogs, is acquired by eating undercooked fish, and *Baylisascaris procyonis*, a raccoon roundworm, is acquired by accidentally ingesting raccoon feces. These organisms cause more severe disease than *Angiostrongylus*, and fatalities occur. Albendazole may be effective against *Gnathostoma*, but there is no treatment for *Baylisascaris*.

ANISAKIS

Anisakiasis is caused by the larvae of the nematode, *Anisakis simplex*. The larvae are **ingested in raw seafood** and can penetrate the submucosa of the stomach or intestine. The adult worms live in the intestines of marine mammals such as whales, dolphins, and seals. The eggs produced by the adults are eaten by crustaceans, which are then eaten by marine fish such as salmon, mackerel, and herring. Gastroenteritis, abdominal pain, eosinophilia, and occult blood in the stool typically occur. Acute infection can resemble appendicitis, and chronic infection can resemble gastrointestinal cancer.

Most cases in the United States have been traced to eating sushi and sashimi (especially salmon and red snapper) in Japanese restaurants. The diagnosis is typically made endoscopically or on laparotomy. Microbiologic and serologic tests are not helpful in the diagnosis. There are no effective drugs. Surgical removal may be necessary. Prevention consists of cooking seafood adequately or freezing it for 24 hours before eating.

Another member of the Anisakid family of nematodes is *Pseudoterranova decipiens*, whose larvae cause a noninvasive form of anisakiasis. The larvae are acquired by eating undercooked fish and cause vomiting and abdominal pain. The diagnosis is made by finding the larvae in the intestinal tract or in the vomitus. There is no drug treatment. The larvae can be removed during endoscopy.

SELF-ASSESSMENT QUESTIONS

1. You are a volunteer with Doctors Without Borders in sub-Saharan Africa. In certain villages, you detect anemia in a significant number of children. This is most likely due to infection with which one of the following?
   (A) *Ancylostoma duodenale*
   (B) *Ascaris lumbricoides*
   (C) *Enterobius vermicularis*
   (D) *Trichinella spiralis*
   (E) *Wuchereria bancrofti*

2. In the same villages as described in Question 1, you observe that some people are eating unwashed raw vegetables. Which one of the following organisms is most likely to cause infection in these people?
   (A) *Ancylostoma duodenale*
   (B) *Ascaris lumbricoides*
   (C) *Enterobius vermicularis*
   (D) *Trichinella spiralis*
   (E) *Wuchereria bancrofti*

3. Which one of the following nematodes is transmitted by a filariform larva penetrating the skin?
   (A) *Anisakis simplex*
   (B) *Onchocerca volvulus*
   (C) *Strongyloides stercoralis*
   (D) *Toxocara canis*
   (E) *Trichuris trichiura*

4. One of the most important public health measures in the United States in the twentieth century was recommending that children in rural areas wear shoes. This effort was designed to prevent infection through the feet with which one of the following organisms?
   (A) *Ascaris lumbricoides*
   (B) *Enterobius vermicularis*
   (C) *Necator americanus*
   (D) *Onchocerca volvulus*
   (E) *Trichuris trichiura*

5. The larvae of certain nematodes migrate through the lung and cause pneumonitis characterized by cough or wheezing. Infection by which one of the following nematodes is most likely to cause this clinical picture?
   (A) *Anisakis simplex*
   (B) *Ascaris lumbricoides*
   (C) *Enterobius vermicularis*
   (D) *Trichinella spiralis*
   (E) *Trichuris trichiura*

6. Of the following drugs, which one is the **MOST** effective in nematode infections?
   (A) Albendazole
   (B) Chloroquine
   (C) Praziquantel
   (D) Primaquine
   (E) Stibogluconate

7. Your patient is a 60-year-old man with abdominal pain, vomiting, and weight loss for the past 2 months. He has a history of asthma that requires 20 mg of prednisone daily to control. He lived most of his life in Cuba, moved to Spain 10 years ago, and has lived in this country for 1 year. Abdominal exam is normal, and radiographic studies are unrevealing. His white blood cell count is 10,900 with 16% eosinophils. Examination of the stool reveals rhabditiform larvae. Of the following, which organism is the **MOST** likely cause?
   (A) *Ascaris lumbricoides*
   (B) *Onchocerca volvulus*
   (C) *Strongyloides stercoralis*
   (D) *Toxocara canis*
   (E) *Trichinella spiralis*

8. Regarding the patient in Question 7, which one of the following is the best drug to treat the infection?
   (A) Ivermectin
   (B) Metronidazole
   (C) Nifurtimox
   (D) Pentamidine
   (E) Praziquantel

9. Your patient is a 40-year-old man with fever, myalgia, and facial swelling. White blood cell count was 14,400 with 24% eosinophils. Additional history reveals that he shot a bear in Canada and ate some of it about 6 weeks ago. He emphasized that he likes his meat rarely. A muscle biopsy was performed, and a

hematoxylin and eosin (H&E) stain of the tissue showed coiled larvae within skeletal muscle. Of the following, which one is the most likely cause?

(A) *Ancylostoma caninum*
(B) *Anisakis simplex*
(C) *Necator americanus*
(D) *Trichinella spiralis*
(E) *Wuchereria bancrofti*

10. Your patient is a 35-year-old woman with severe upper abdominal pain for the past hour. There is no nausea, vomiting, or diarrhea. You suspect she may have cholecystitis, pancreatitis, or a perforated viscus but first ask her if she has ingested raw fish recently. She says yes and tells you that she had sashimi the night before last. Endoscopy reveals a larva in the gastric mucosa. Of the following, which one is the most likely cause?

(A) *Ancylostoma caninum*
(B) *Ancylostoma duodenale*
(C) *Anisakis simplex*
(D) *Toxocara canis*
(E) *Trichuris trichiura*

11. Your patient is a 5-year-old boy who complains of perianal itching, especially at night. A "Scotch tape" preparation reveals the eggs of *Enterobius* in the microscope. Which one of the following is the best drug to treat his pinworm infection?

(A) Ivermectin
(B) Mebendazole
(C) Pentamidine
(D) Praziquantel
(E) Pyrimethamine and sulfadiazine

ANSWERS

(1) **(A)**
(2) **(B)**
(3) **(C)**
(4) **(C)**
(5) **(B)**
(6) **(A)**
(7) **(C)**
(8) **(A)**
(9) **(D)**
(10) **(C)**
(11) **(B)**

SUMMARIES OF ORGANISMS

Brief summaries of the organisms described in this chapter begin on page 706. Please consult these summaries for a rapid review of the essential material.

PRACTICE QUESTIONS: USMLE & COURSE EXAMINATIONS

Questions on the topics discussed in this chapter can be found in the Parasitology section of Part XIII: USMLE (National Board) Practice Questions starting on page 755. Also see Part XIV: USMLE (National Board) Practice Examination starting on page 775.

CHAPTER

57

Overview of Immunity

FUNCTION OF THE IMMUNE SYSTEM

The main function of the immune system is to **prevent or limit infections** due to viruses, bacteria, fungi, protozoa, and worms. The first line of defense against microorganisms is the barrier formed by **intact skin and mucous membranes**. If microorganisms breach this defense, then a second line of defense can rapidly detect foreign material and destroy harmful agents. These components of the immune system are active even before an infectious exposure, and therefore, this arm of host defense is called **innate immunity** (Table 57–1). Innate immunity works immediately upon the first encounter with a microorganism. The innate arm is nonspecific in that it can recognize **patterns** many microorganisms share (described in more detail in Chapter 58). For example, a neutrophil can sense, ingest, and destroy many different kinds of bacteria by exploiting features common among bacterial cells.

Some microbes can mutate to resist the tactics of innate immunity. For these microbes, there is a more targeted defense that is specific for individual infectious agents, which is provided by the **adaptive (acquired)** arm of the immune system (often considered the third line of defense). The adaptive arm

TABLE 57–1 Important Features of Innate and Adaptive Immunity

	Specificity	Time from Exposure to Action	Has Memory	Examples of Cell-Mediated Immunity	Examples of Humoral Proteins
Innate	Nonspecific	Rapid—acts within minutes	No	Natural killer cells, macrophages	Complement
Adaptive	Highly specific	Slow—requires several days before becoming effective	Yes	Helper T cells, cytotoxic T cells	Antibodies (produced by B cells)

takes days to become fully functional, but once engaged, it remembers an infectious agent and responds more quickly to repeat encounters. For example, after receiving the first dose of the pneumococcal vaccine, it takes 7 to 10 days to produce protective levels of antibodies, but when you get a booster, this takes only 2 to 3 days. Table 57–1 provides a summary of the features of innate and adaptive immunity.

The immune system has a **cell-mediated arm** (orchestrated by **T lymphocytes**) and a **humoral arm** (circulating factors, such as **antibodies** and **complement proteins**). The combined effects of immune cells (e.g., T cells, macrophages) and proteins (e.g., antibodies, complement) produce **inflammation** (see Chapter 8). This chapter will introduce the innate and adaptive immune system, and subsequent chapters will discuss how they cooperate during normal immune responses and how their failure can cause disease.

INNATE & ADAPTIVE IMMUNITY

1. Innate Immunity

At the time of birth, you already have a powerful arsenal of immune defenses at work. These defenses exist, fully encoded in your genes, **prior to exposure** to any microbes, and because of this, they are called **innate**. Innate immunity is **nonspecific** and includes barriers (e.g., skin and mucous membranes), certain cells (e.g., **macrophages** and **natural killer [NK] cells**), and certain proteins (e.g., **complement**) (Table 57–2). In addition to host defense, another important function of innate immunity is to heal damaged tissue and clear away dead cells in a way that does not harm the body. Thus, **innate immunity can function independently of adaptive immunity**, although innate immunity is often amplified by the adaptive immune arm. In addition, **innate immunity has no memory**, whereas adaptive immunity is characterized by long-term memory.

Innate host defenses perform two major functions: **killing invaders** and **activating adaptive immunity**. Some components of the innate arm, such as neutrophils, only kill microbes, whereas others, such as **macrophages** and **dendritic cells**, kill microbes and also communicate with T cells (described below). To do this, innate immunity must first **recognize** molecular patterns common among microbial families through **pattern recognition receptors**. Once they recognize a microbe, the phagocytic cells of the innate immune system, including **macrophages**, attempt to ingest and kill it. (The process of phagocytosis and killing of the ingested microbe within the phagocyte is described in Chapters 8 and 58.)

These and other innate effector cells release **cytokines** (proteins that immune cells use to communicate) and **chemokines** (proteins that recruit other cells to join in) and other **inflammatory signals**. The phagocytic cells also break down ingested microbes and display pieces of the microbial proteins on their surface to alert cells of the adaptive immune system, specifically T cells. These peptide fragments are called **antigens**, and the processing of microbial products into peptides for T-cell activation is called **antigen presentation**. Another innate cell, the **NK cell**, kills virus-infected cells or malignant host cells.

Although innate immunity is often successful in eliminating microbes, it clearly is not enough, as children with severe combined immunodeficiency disease (SCID), who have intact innate immunity but no adaptive immunity, suffer from repeated, life-threatening infections (see Chapter 68).

2. Adaptive (Acquired) Immunity

In contrast to innate immunity, adaptive immunity occurs only **after exposure** to an agent, is **specific** for the agent, and **improves upon repeated exposure**. It is mediated by **B lymphocytes** (or B cells, so-called because their development mainly occurs in the *bone* marrow) and by **T lymphocytes** (or

TABLE 57–2 Important Components of Innate Immunity

Factor	Mode of Action
I. Factors that limit entry of microorganisms into the body	
Keratin layer of intact skin	Acts as mechanical barrier
Lysozyme in tears and other secretions	Degrades peptidoglycan in bacteria cell wall
Respiratory cilia	Elevate mucus-containing trapped organisms
Low pH in stomach and vagina; fatty acids in skin	Retard growth of microbes
Surface phagocytes (e.g., alveolar macrophages)	Ingest and destroy microbes
Defensins (cationic peptides)	Create pores in microbial membrane
Normal flora of throat, colon, and vagina	Occupy receptors, which prevents colonization by pathogens
II. Factors that limit growth of microorganisms within the body	
Natural killer cells	Kill virus-infected cells
Neutrophils	Ingest and destroy microbes
Macrophages and dendritic cells	Ingest and destroy microbes and present antigen to helper T cells
Interferons	Inhibit viral replication
Complement	C3b is an opsonin; membrane attack complex creates holes in bacterial membranes
Transferrin and lactoferrin	Sequester iron required for bacterial growth
Fever	Elevated temperature retards bacterial growth
Inflammatory response	Limits spread of microbes
APOBEC3G (apolipoprotein B RNA-editing enzyme)	Causes hypermutation in retroviral DNA and mRNA

T cells, so-called because their development mainly occurs in the *t*hymus). Unlike innate immune cells, T cells and B cells recognize **antigens**, rather than universal microbial "patterns."

B and T lymphocytes share three important features: (1) they exhibit remarkable **diversity** (i.e., collectively they can respond to millions of different antigens); (2) they have a long **memory** (i.e., they can respond many years after the initial exposure); and (3) they exhibit exquisite **specificity** (i.e., their actions are specifically directed against the antigen that initiated the response).

Some of the major functions of T cells and B cells are shown in Table 57–3, including examples where they are protective or cause disease. Figure 57–1 shows how the components of adaptive immunity enhance the activity of components of innate immunity.

T cells can be further divided based on their function and based on molecules on the cell surface called **cluster of differentiation** (or **CD**). These proteins are important for the function of these cells and are used to distinguish them. **CD8** marks the cells that are called **cytotoxic T lymphocytes (CTLs)**, whereas **CD4** marks the cells that are called **T helper (Th) cells**. As described earlier, innate phagocytic cells ingest microbes and present microbial antigens. The adaptive arm is typically activated **only after** the innate arm has interacted with the microbe. Figure 57–2 is a summary of how phagocytic cells interact with helper T cells through the **major histocompatibility complex (MHC) proteins**. The role of innate immune cells as effector cells and antigen-presenting cells is described in Chapter 58, and the various types of T cells are described in Chapter 60.

The main function of **cytotoxic (CD8-positive) T cells** is to recognize and kill any cell that has foreign ("nonself") proteins on its surface. Cells might contain foreign proteins because the cells have been infected or because they are cancer cells that form new nonself proteins. **Helper (CD4-positive) T cells** instruct B cells to make antibody and enhance the activity of innate cells, such as macrophages.

B cells can proliferate and differentiate into **plasma cells** that secrete large amounts of highly specific **antibodies** (also called **immunoglobulins [Ig]**). Antibodies have a variety of functions such as **neutralizing toxins and viruses** and **opsonizing**

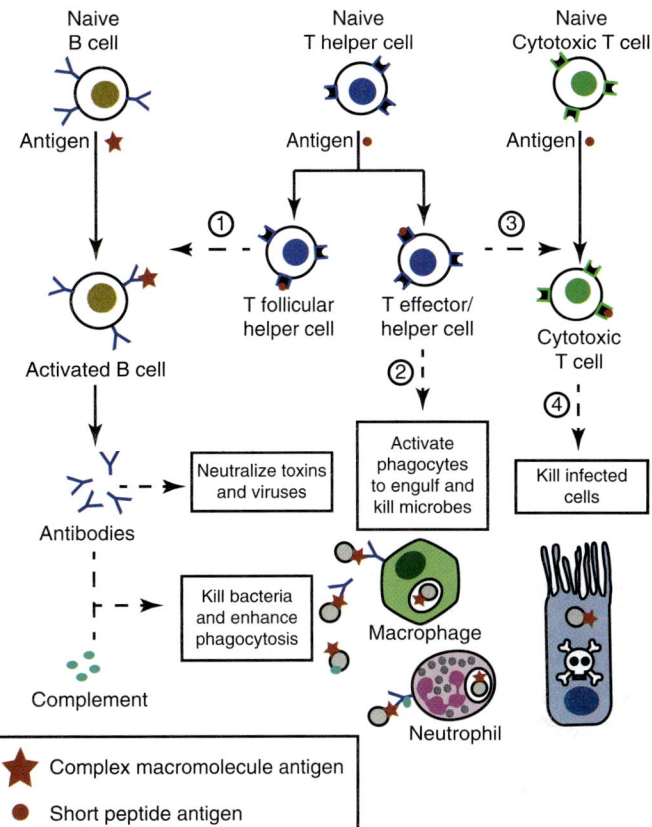

FIGURE 57–1 Introduction to the interactions and functions of the major components of the immune system. **Left:** Antibody-mediated immunity. Combined with complement, antibodies assist neutrophils and other cells in our defense against extracellular, encapsulated, pyogenic bacteria such as staphylococci and streptococci (see Chapters 8 and 63 for information about complement). Antibodies also neutralize toxins, such as tetanus toxin, as well as viruses. Antibodies recognize complex structures of many types of large molecules, represented by the red stars. **Right:** T-cell–mediated immunity. There are four distinct components. (1) CD4-positive T follicular helper (Tfh) cells help B cells to make antibody in the lymphoid tissue follicles. (2) CD4-positive T helper/effector (Th) cells activate macrophages to provide defense against intracellular bacteria and fungi. (3) CD4-positive Th cells produce interleukins that support the proliferation and survival of other T cells. (4) CD8-positive CTLs are an important defense against intracellular bacteria and viruses and act by destroying infected cells. T cells only recognize short peptide chains, represented by the red circles. In the figure, these four processes are indicated by arrows accompanied by circled numbers.

microbes (see Chapter 61). B cells that have been activated with an antigen can also become **memory B cells**, which respond more rapidly to a rechallenge.

ACTIVE & PASSIVE IMMUNITY

Active immunity is a host immune response induced after **contact** with foreign antigens (e.g., microorganisms). This contact may occur through an infection or an immunization with microbial toxins or antigens. In all these instances, the

TABLE 57–3 Major Functions of T Cells and B Cells

B-Cell (Antibody) Functions	T-Cell Functions
Host defense against infection (opsonize bacteria, neutralize toxins and viruses)	Host defense against infection (especially *Mycobacterium tuberculosis*, fungi, and virus-infected cells) Tumor rejection Coordination and regulation of adaptive immune response (helper T cells)
Allergy/hypersensitivity (e.g., hay fever, anaphylactic shock)	Allergy/hypersensitivity (e.g., poison oak)
Autoimmunity	Autoimmunity Transplant graft rejection

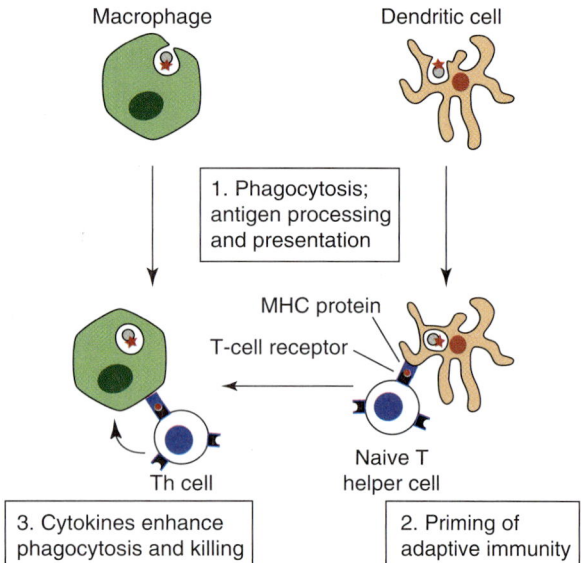

Macrophage   Dendritic cell

1. Phagocytosis; antigen processing and presentation

MHC protein

T-cell receptor

Th cell   Naive T helper cell

3. Cytokines enhance phagocytosis and killing

2. Priming of adaptive immunity

FIGURE 57–2 Macrophages and dendritic cells participate in both the innate arm and the adaptive arm of immune responses. (1) These cells are considered part of the innate arm because they phagocytize many types of microbes and also produce cytokines that cause inflammation. (2) Dendritic cells are also essential for the adaptive arm because they present antigen in association with major histocompatibility complex (MHC) proteins to activate naïve CD4-positive helper T cells. (Not shown: dendritic cells also present antigen complexed with MHC proteins to activate CD8-positive cytotoxic T cells.) (3) T helper/effector (Th) cells migrate to peripheral tissues and, upon contacting macrophages displaying the same antigen, release inflammatory cytokines that help the macrophages kill pathogens and repair damaged barriers.

host actively responds by making antibodies and activated T lymphocytes (i.e., adaptive immunity).

The main advantage of active immunity is that resistance is **long-term** (Table 57–4). Its major disadvantage is its **slow onset**, especially the primary response (see Chapter 61).

Passive immunity is given to a person in the form of immune components that were **preformed** in another person or animal. Hospitals have supplies of antibody against toxins produced by bacteria that cause, for example, tetanus and botulism. Administering these to a patient transfers large amounts of **antitoxin** that is immediately available to neutralize the toxins. Likewise, preformed antibodies to rabies and other viruses can be injected to neutralize viral multiplication. Other forms of passive immunity are IgG passed from mother to fetus during

TABLE 57–4 Characteristics of Active and Passive Immunity

	Mediators	Advantages	Disadvantages
Active immunity	Antibody and T cells	Long duration (years)	Slow onset
Passive immunity	Antibody only	Immediate availability	Short duration (months)

pregnancy and IgA passed from mother to newborn during breast-feeding. Passive immunity can even occur between species, as when snake-bite victims are given the antibody-rich serum from an animal (usually horse or sheep) that was previously inoculated with the venom so that the serum contains high levels of specific antivenom antibodies.

The advantage of passive immunization is the **prompt availability** of large amounts of antibody. However, in bypassing an active immune response, passive immunization does not confer T cell or B cell memory. Therefore, because antibodies only last a few weeks, the main disadvantage of passive immunization is its **short durability**. Another disadvantage is the risk of **hypersensitivity reactions**, if serum from animals is used (see section on serum sickness in Chapter 65).

In **passive–active immunity**, a patient gets both preformed antibodies to provide immediate protection and a vaccine to provide long-term protection. These preparations are given at different sites in the body to prevent the antibodies from neutralizing the vaccine. This approach is used to prevent tetanus (see Chapters 12 and 17), rabies (see Chapters 36 and 42), and hepatitis B (see Chapters 36 and 41).

IMMUNOGENS

An **immunogen** is any molecule that induces an immune response. As described earlier, **antigens** are immunogens that react with the highly specific receptors on T cells or B cells.

1. Antigens

At this point, you might ask why certain molecules are **immunogenic**. The features that determine immunogenicity are as follows.

Foreignness

In general, only molecules recognized as "nonself" or foreign are immunogenic (i.e., we are tolerant to our own molecules) (see Chapter 66).

Molecular Size

The most potent immunogens are large proteins (i.e., molecular weights above 100,000 g/mol), whereas mid-sized molecules (i.e., molecular weight below 10,000 g/mol) are weakly immunogenic, and very small ones (e.g., amino acids) are nonimmunogenic.

Chemical–Structural Complexity

Some chemical complexity is required for immunogenicity. For example, peptide "homopolymers" that contain a single type of amino acid are less immunogenic than peptides containing diverse amino acids.

Antigenic Determinants (Epitopes)

Epitopes are the chemical features on an antigen that physically bind to antibody or T-cell receptors. Most antigens have more than one epitope (i.e., they are multivalent).

Dosage, Route, and Timing of Antigen Exposure

Vaccine research aims to optimize these factors to find regimens with the best balance of highest immunogenicity, fewest side effects, and most convenient administration.

Host Genetics

The genetic makeup of each individual (especially the genes that form the MHC, described above and in Chapter 62) can determine that individual's response to a particular immunogen.

2. Haptens

In contrast to an antigen, a **hapten** is a molecule that is not immunogenic by itself but can react with specific antibody. Haptens can be small molecules, nucleic acids, lipids, or drugs (e.g., penicillins), but because they are not peptides, they **cannot activate helper T cells**. Haptens stimulate a primary adaptive response only when covalently bound to a "carrier" protein (Figure 57–3). In this process, the hapten interacts with the B-cell receptor of a naïve B cell and the entire hapten–carrier protein complex is internalized. The B cell processes this complex and presents a peptide from the carrier protein in association with its MHC protein to helper T cells, and a nearby helper T cell that recognizes that peptide then provides the help that stimulates the B cells to produce antibody to the hapten. This is how **conjugate vaccines** work; a weak immunogen is "conjugated" to a strong peptide antigen such that T cells (recognizing the peptide) can help B cells (recognizing the weaker immunogen) to produce protective antibody. These T cell–B cell interactions are covered in more detail in Chapter 61.

Two additional ideas are needed to understand how haptens interact with our immune system. The first is that **many haptens bind to our normal proteins** and modify these proteins. Some examples of haptens that do this are drugs (e.g., penicillin) and poison oak oil. The hapten–protein combination now becomes immunogenic (i.e., the hapten modifies the protein sufficiently such that when the hapten–peptide combination is presented by the MHC protein, it is recognized as foreign).

The second idea is that a nonimmunogenic hapten can active cells if **many hapten molecules bound to a carrier protein bind and cluster antibodies together**. The best example of this occurs in mast cells, which are innate cells that become activated when a large number of antibodies are gathered together on the cell surface, a process called **receptor cross-linking**. When many molecules of a hapten bind to a host protein, they can cause cross-linking of many penicillin-specific IgE molecules on the mast cell surface. This activates the mast cell, which releases the mediators that cause hives (mast cells in the skin), bronchoconstriction (mast cells in the lungs), and anaphylaxis (mast cells in the systemic vasculature).

3. Adjuvants

Adjuvants enhance the immune response to an immunogen, but they do so without binding to antibody or to the immunogen.

Adjuvants can act by causing slow release of immunogen, thereby prolonging the stimulus; they can enhance uptake of immunogen by antigen-presenting cells; they can speed up the migration of antigen-presenting cells into the lymphoid tissues; and they can induce costimulatory molecules ("second signals," described in Chapter 60). Another way adjuvants can work is by binding Toll-like receptors (see Chapter 58) on the surface of macrophages and B cells, which results in cytokine production that enhances T cell and B cell responses to the immunogen.

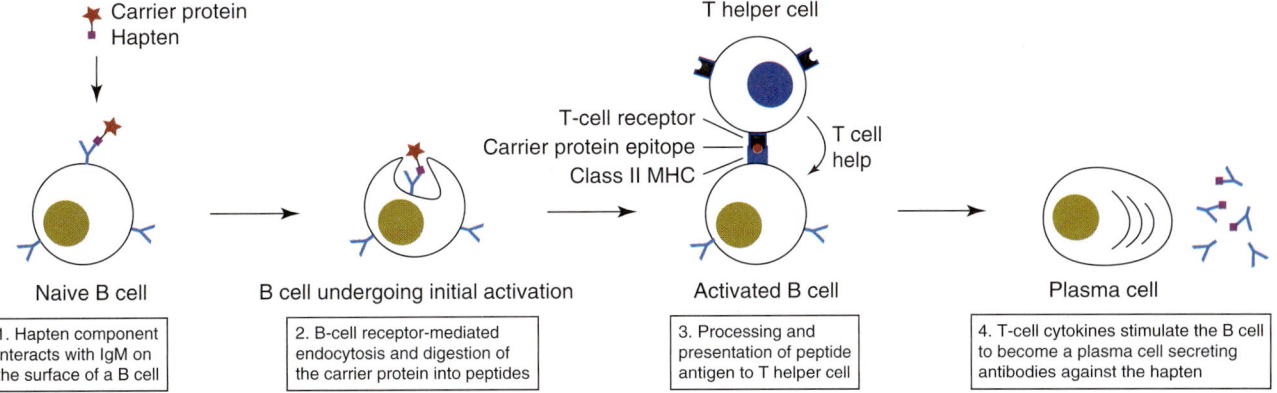

FIGURE 57–3 Hapten–carrier conjugate induces antibody against the hapten. A hapten bound to a carrier protein can induce antibody to a hapten by the mechanism depicted in the figure. (1) A hapten can bind the surface immunoglobulin receptor on the B cell specific for the hapten. (2) The hapten–carrier conjugate is taken up by the B cell, which processes the carrier protein into peptides. (3) But a hapten alone cannot induce antibody, because only peptides (not haptens) can be loaded onto major histocompatibility complex (MHC) proteins to present to CD4-positive T helper cells. (4) T-cell recognition of carrier protein epitope by the T-cell receptor prompts production of T helper cytokines that are necessary to stimulate the B cell to differentiate. Once stimulated, the B-cell clone matures into a plasma cell that secretes antibody against the hapten.

AGE & THE IMMUNE RESPONSE

Immunity is less than optimal at both ends of life (i.e., in the newborn and the elderly). In newborns, natural barriers, such as the intestine, are not fully developed until 3 to 4 weeks of life, and innate cells, such as phagocytes, are less sensitive to cytokines and chemokines. Newborns have more circulating lymphocytes than adults, but the newborn's lymphocytes are individually less effective.

IgG and IgA production begins after birth and only reaches protective levels at around 1 year. As a consequence, until around 6 months, most of the circulating IgG is, in fact, maternal-derived antibody that crossed the placenta before birth, and the mucosal surface of the gastrointestinal tract is similarly protected by maternal IgA that is secreted into breast milk.

The precise reason why newborns have reduced immunity is unknown, but the same phenomenon is observed in all mammals, suggesting that this state confers an evolutionary survival benefit during the fetus-to-newborn transition. One possible explanation is that the sudden move from the relatively sterile uterus to the outside world comes with an enormous increase in the amount of foreign material the newborn encounters, most of which is harmless (e.g., commensal microbes). The newborn immune system needs to take a "tolerant" stance, because responding to all of these as pathogenic invaders would result in an overwhelming and inappropriate inflammatory response that would cause collateral damage to the newborn's own tissues.

The neonatal window leaves infants highly susceptible to infections, and vaccines must be timed carefully; they should be given as early as possible so that the infant gets timely protection, but if given too soon, then the infant cannot mount an effective response. For example, the pneumococcal vaccine containing unconjugated polysaccharides does not induce protective immunity when given prior to 18 months of age (see Chapter 61). But the pneumococcal vaccine containing the polysaccharides conjugated to a carrier protein is effective when given as early as 2 months of age.

Because adaptive immunity provides long-term memory, you might expect that immunity gets increasingly stronger with age, and this is true up to a point. However, as we become elderly, immunity declines. The thymus, which is the source of all new T cells (see Chapter 59), begins to atrophy during puberty, and by the time we reach age 60, we mostly rely on memory T cells for immunity because our ability to generate T cells that recognize new antigens is greatly decreased. B cells similarly trend toward a more experienced and "exhausted" state later in life. As a result, the immune responses to certain vaccines and infections are blunted.

As in the very young, the elderly experience a somewhat increased frequency and severity of infections, such as influenza. In addition, the elderly can develop "reactivation" of a latent infection, caused by, for example, *Mycobacterium tuberculosis* or varicella-zoster virus, which was previously held in check by their "young" immune system.

This phenomenon, known as *immunosenescence*, might explain why older age groups are so hard-hit by COVID-19. And there is another troubling implication: vaccines, which activate the immune system to fight off invaders, often perform poorly in older people. The best strategy for quelling the pandemic might fail in exactly the group that needs it most.

SELF-ASSESSMENT QUESTIONS

1. Which one of the following is an attribute of the innate, rather than the adaptive (acquired), arm of our host defenses?

    (A) Is highly specific in its response to individual bacterial species

    (B) Responds to viruses and fungi, but not bacteria

    (C) Exhibits memory following exposure to bacteria

    (D) Is part of our host defense against bacteria but not against fungi

    (E) Is as effective the first time it is exposed to bacteria as it is subsequent times

2. Regarding haptens, which one of the following is the most accurate?

    (A) A hapten is the antigen-binding site of an immunoglobulin.

    (B) A hapten cannot induce antibody by itself but can do so when covalently bound to a carrier protein.

    (C) A hapten can bind to the antigen receptors of CD4-positive T cells without being processed by antigen-presenting cells.

    (D) A hapten is defined by its ability to bind to the smaller of the two polypeptides that comprise the class I MHC proteins.

3. Certain components of our immune system are characterized by two attributes: being able (1) to respond specifically to microbes and (2) to exhibit memory of having responded to a particular microbe previously. Which one of the following has BOTH specificity and memory?

    (A) B cells

    (B) Natural killer cells

    (C) Dendritic cells

    (D) Macrophages

    (E) Neutrophils

4. Your patient says that she must travel on business 3 days from now to a country where hepatitis A is endemic. She just read in the newspaper that there are two types of protection against this disease: one is a vaccine that contains killed hepatitis A virus, and the other is a serum globulin preparation that contains antibodies to the virus. She asks which you would recommend and for what reason?

    (A) The vaccine containing killed hepatitis A virus is best because it induces the most antibody.

    (B) The vaccine containing killed hepatitis A virus is best because it provides the most long-lived immunity.

    (C) The serum globulin preparation containing antibodies against the virus is best because it provides immunity in the shortest time.

    (D) The serum globulin preparation containing antibodies against the virus is best because it provides the most long-lived immunity.

ANSWERS

(1) **(E)**
(2) **(B)**
(3) **(A)**
(4) **(C)**

PRACTICE QUESTIONS: USMLE & COURSE EXAMINATIONS

Questions on the topics discussed in this chapter can be found in the Immunology section of Part XIII: USMLE (National Board) Practice Questions starting on page 757. Also see Part XIV: USMLE (National Board) Practice Examinations starting on page 775.

Innate Immunity

The **innate immune system** is composed of physical barriers, cells, and circulating factors that are always active and ready to repel microbes. They form a boundary between you and the viruses, bacteria, and fungi that live on and inside you. Innate immune cells also clean up debris and dying cells. Advantages of the innate immune system are that it **responds rapidly** and **targets molecular patterns** widely shared among microbes.

THE BARRIER

An extremely important, but often overlooked, component of host defense is **the barrier formed by skin and mucous membranes**. The epithelia covering our skin, respiratory, gastrointestinal, and genitourinary tracts provide the first line of defense against invaders. The central mission of the barrier is to separate you from the outside world, allowing some microbes to survive in niches along the surface, but preventing any from gaining a foothold to cause disease.

Mechanical Barrier

The outermost layer of the barrier is the epithelial cells, connected to one another by tight junctions. The skin's **epidermis** is covered by keratinized squamous cells that are continuously

shed ("desquamated"). Similarly, nonskin barriers are called **mucous membranes** because they are coated by **mucins**, a sticky mixture of glycoproteins produced by secretory cells of the respiratory, gastrointestinal, and genitourinary tracts. Cells of the mucous membranes also rapidly divide, slough off, and are replaced continuously. Keratinized epidermis, cell sloughing, sweat secretion, and mucins all prevent microbes from attaching and invading.

The ciliated cells lining the respiratory tract, the gastrointestinal tract's peristalsis, and the continuous flow of urine from the kidney through the bladder and urethra ensure that microbes cannot attach and invade these sites. Failure of any of these mechanisms is a common predisposing factor to infection by bacteria or fungi that otherwise colonize us harmlessly. In patients with severe skin burns, pulmonary ciliary cell disorders, or bowel or urinary obstruction, infection is the key cause of morbidity and mortality.

Chemical Barrier

Epithelial cells also produce a number of chemicals and proteins that inhibit microbes from growing or attaching. The skin and the stomach excrete concentrated **hydrochloric acid** that kills bacteria. **Lysozyme** is an enzyme in saliva and tears that makes

TABLE 58–1 Components of the Barrier

Anatomic Site	Mechanical	Chemical	Biological
Skin	Keratinized squamous epidermis cells	Fatty acids; defensins	Skin flora
Gastrointestinal tract	Mucins; peristalsis; normal shedding of epithelial cells	Gastric acid; digestive enzymes; defensins; lysozyme; iron-binding proteins	Gut flora; IgA
Genitourinary tract	Urine flow	Low pH	Vaginal flora; IgA
Respiratory tract	Airflow; ciliated airway cells; coughing	Surfactant proteins	Nose, mouth, and pharyngeal flora; IgA

Ig = immunoglobulin.

holes in bacterial cell walls by breaking linkages in their peptidoglycan molecules. In addition, **antimicrobial peptides**, such as **defensins**, are produced throughout the skin and mucous membranes.

Defensins are highly positively charged peptides, primarily produced in the gastrointestinal and lower respiratory tracts, that create pores in lipid membranes of bacteria, fungi, and even some viruses. Neutrophils and Paneth cells in the intestinal crypts contain one type of defensin (α-defensins), which may have antiviral activity, whereas the respiratory tract produces different defensins called β-defensins, which are antibacterial. **Surfactants** are lipoproteins produced in the lung alveoli that bind to the surface of microbes, which can facilitate their phagocytosis or undermine their cell membranes.

Biological Barrier

Some microbes have evolved ways to counteract chemical barriers, granting access and forming "ecological microenvironments" throughout our bodies. The epithelial barrier harbors many harmless (commensal) microbes, collectively called the **microbiome**, that inhabit distinct niches. These include bacteria, fungi, protozoa, and even *Demodex* mites that inhabit our hair follicles! Each of these species competes for nutrients, evolving strategies to coexist and defend their niche. When we give patients antibiotics for an infection, we are also killing some of the commensals and disrupting the normal microbiome. This might predispose patients to infectious diseases, including *Clostridium difficile* colitis, and inflammatory diseases, such as inflammatory bowel disease.

Additional biological barriers are **immunoglobulin (Ig) A** and **IgG**, two classes of antibody (see Chapters 59 and 61). IgA secreted from our mucosal surfaces binds bacteria, viruses, and toxins, preventing them from attaching to the epithelial layer. We will discuss IgA and the other classes of antibody further in Chapters 59 and 61. Table 58–1 summarizes the major components of barrier defenses.

In addition to IgG, **complement** is another biological barrier present in the blood. It is a system of approximately 20 proteins that are activated either by binding to the surface of microbes or by the binding of antigen to IgM or IgG antibody. The main actions of complement are (1) lysis of bacteria, especially those of the genus *Neisseria*, by the membrane-attack complex,

(2) opsonization of bacteria by the C3b component of complement, and (3) attraction of neutrophils to the site of infection by C5a, a chemokine component of complement. Complement is discussed in more detail in Chapter 63.

Some infectious diseases are caused by pathogens that have evolved ways to bypass barrier defenses. For example, herpesviruses and respiratory viruses attach and invade via specific receptors on the mucosal epithelial cells. Vector-borne pathogens, such as arboviruses and malaria, have evolved to move between hosts inside insects and bypass the barrier when we are bitten.

In addition, most of the bacteria and fungi that commonly live in the environment or in our intestines are harmless at that site, but they can cause life-threatening infections if a breakdown of the barrier allows them entry. Some examples of these include respiratory infections in people with cystic fibrosis, who have defects in mucus clearance, or infections of the bloodstream in people who require intravenous catheters.

PHAGOCYTES AND OTHER MYELOID CELLS

Innate immune cells are the second line of defense, and the most important of these cells are **phagocytes**. The name "phagocyte" derives from Greek, emphasizing these cells' ability to "eat" foreign material and debris. (Innate immune phagocytes also play a role in repairing damage, even in the absence of a foreign infection.)

There are several key phagocytic cells: **tissue macrophages** are the first cells to encounter invaders; **neutrophils** then migrate into damaged tissue and contribute to inflammation; and **dendritic cells** carry away microbial material to the lymph nodes or spleen, where they activate the adaptive immune response. A fourth cell, called a **monocyte**, can be recruited into inflamed tissue to take on the role played by macrophages or dendritic cells. Phagocytes belong to the family of immune cells called **myeloid** cells because they originate from bone marrow "myeloid" progenitor stem cells (Figure 58–1).

Origin of the Myeloid Lineage Cells

Throughout life, **stem cells** in the bone marrow continuously produce daughter cells that give rise to all red and white blood

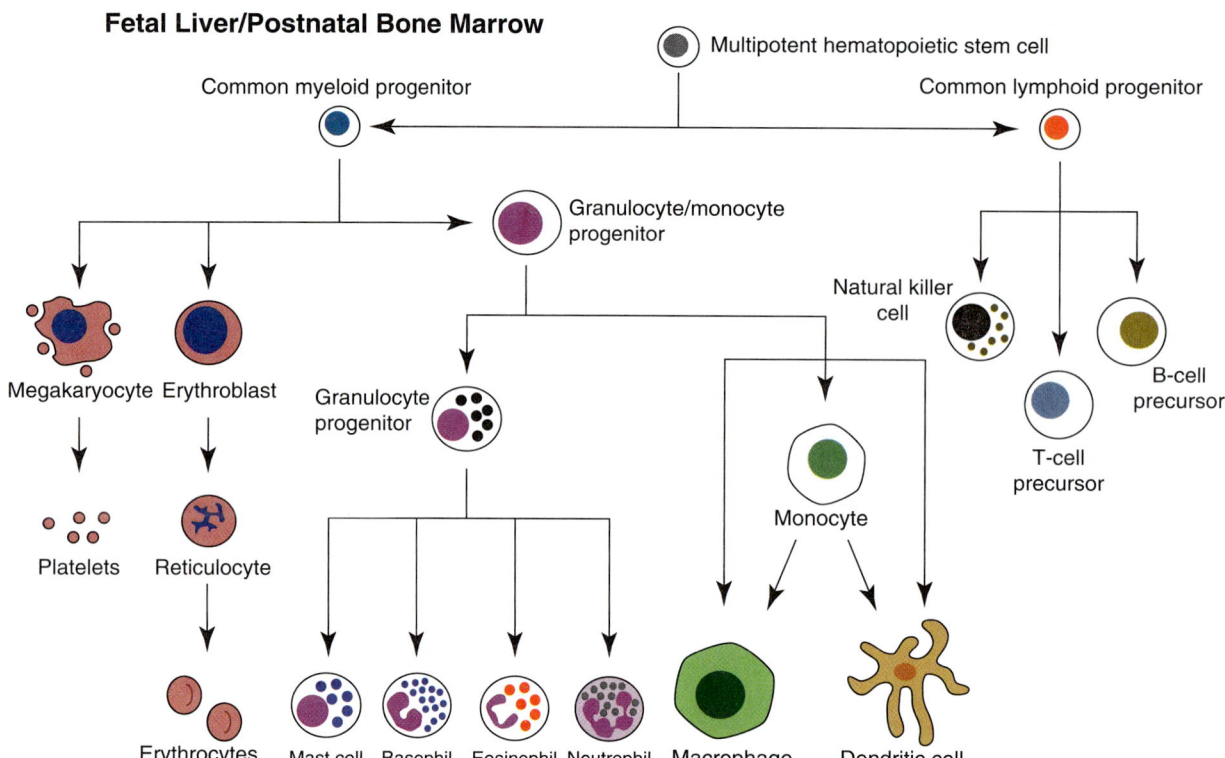

FIGURE 58–1   Origin of hematopoietic cells. Stem cells in the bone marrow (or fetal liver) are the precursors of all blood cells. Stem cells differentiate into myeloid or lymphoid progenitor cells. The myeloid cells are the source of platelets, erythrocytes, granulocytes, macrophages, and dendritic cells. Monocytes are a special type of myeloid cell that can differentiate into macrophages or dendritic cells when needed. The lymphoid cells are the source of T lymphocytes, B lymphocytes, and natural killer (NK) cells. The development of lymphocytes is covered in detail in Chapter 59.

cells. Immune cells are often called white blood cells, or **leukocytes**, and include **lymphoid cells** and **myeloid cells**. In contrast to T lymphocytes and B lymphocytes, which differentiate from lymphoid stem cells, most cells of the innate immune system arise from myeloid precursors. (There are exceptions, such as **natural killer [NK] cells**, which are covered later in this chapter).

PATTERN RECOGNITION RECEPTORS OF INNATE IMMUNE CELLS

The first step of an immune response is that innate immune cells **recognize** foreign material. To identify what is foreign, innate immune cells have receptors, called **pattern recognition receptors (PRRs)**, that recognize a molecular pattern, called a **pathogen-associated molecular pattern (PAMP)**, that is common among many microbes but—very importantly—*is not present in human cells and is difficult for those microbes to alter through mutation*. Some examples of PAMPs include carbohydrates, lipids, or nucleic acids (e.g., double-stranded RNA). With the ability to recognize PAMPs, innate immune cells do not need a highly specific receptor for each individual microbe strain but can still broadly distinguish what is foreign from what is "self." Table 58–2 lists examples of the four major classes of PRRs.

There are two classes of PRRs (Toll-like receptors [TLRs] and C-type lectin receptors [CLRs]) that *recognize microbes that are outside of cells or within the cells' vesicles*. Two other classes of receptors in the cytoplasm of cells (NOD-like receptors [NLRS] and RIG-I helicase-like receptors [RLRs]) *recognize microbes that have invaded the cell's cytoplasm*. Mutations in the genes encoding these PRRs result in a failure to recognize pathogens and predispose to severe bacterial, viral, and fungal infections.

PAMP recognition results in a rapid innate immune response, triggered by a particular microbe in a particular location. The type of adaptive immune response that may result after innate immunity is activated differs depending on the type of microbe and where it is found. Part of the function of innate immunity is to ensure that an appropriate attack is called in against that type of microbe in that location. Therefore, the **immune response is shaped by the PRRs activated during the initial encounter with innate immunity**.

For example, antibody-mediated responses are most effective against extracellular (especially encapsulated) bacteria, whereas T-cell–mediated responses are required against intracellular microbes, such as viruses or *Mycobacterium tuberculosis*. The innate "first responders" produce the signals that dictate the immune response, and these signals depend on which PRRs are activated by the foreign organism.

TABLE 58–2 Pattern Recognition Receptors

Location	Receptor Class	Examples	Activating Microbial Ligands
Extracellular	Toll-like receptors (TLRs)	TLR2	Peptidoglycan (gram-positive bacteria)
		TLR4	Lipopolysaccharide (bacterial endotoxin)
		TLR9	Bacterial DNA
	C-type lectin receptors (CLRs)	MBL	Mannose (viral, fungal, and bacterial carbohydrate)
		Dectin-1	Glucans (fungal cell wall)
Intracellular	NOD-like receptors (NLRs)	NOD2	Peptidoglycan (gram-positive bacteria)
		NLRP3	*Staphylococcus* α toxin
	RIG-I helicase-like receptors (RLRs)	RIG-I	Double-stranded RNA (viral)

MBL = mannan-binding lectin; NOD = nucleotide-binding oligomerization domain; RIG-I = retinoic acid-inducible gene I.

Some important examples of PRRs are described in the following sections.

Toll-Like Receptors

TLRs are a family of 10 receptors found on the surface of many cells, including epithelial cells and immune cells. Each of the ten TLRs recognizes a core microbial building block (e.g., endotoxin or peptidoglycan), and all of them signal through transcription factors to release proinflammatory **cytokines** and **cell surface molecules** (see Chapter 60) that enhance adaptive immune activation. Endotoxin is a lipopolysaccharide (LPS) found on the surface of most gram-negative bacteria (but not on human cells). When released from the bacterial surface, LPS combines with LPS-binding protein, a normal component of plasma, that transfers LPS to a receptor on the surface of macrophages called CD14. LPS then stimulates a PRR called **Toll-like receptor 4** (TLR4), which transmits a signal to the nucleus of the cell. Note that a different TLR, **TLR2**, signals the presence of peptidoglycan from gram-positive bacteria, which has a different molecular pattern but produces the same innate cell activation. Excessive macrophage TLR activation may contribute to systemic inflammation and septic shock.

C-Type Lectin Receptors

A PRR called **mannan-binding lectin (MBL)** is one of the **CLR** family and can be found both in the plasma and attached to the surface of dendritic cells and macrophages. Many bacteria and yeasts have mannan (a polymer of the sugar mannose) on their surface that is not present on human cells. MBL binds mannose on the surface of the microbes and then activates complement (see Chapter 63), resulting in death of the microbe. MBL also enhances phagocytosis (acts as an **opsonin**) via receptors to which it binds on the surface of phagocytes, such as macrophages. MBL is a normal serum protein whose concentration in the plasma is greatly increased during the **acute-phase response** (see later). A different CLR, called **dectin-1**, recognizes beta-glucan in the cell wall of fungi such as *Candida albicans*.

NOD-Like Receptors

NLRs recognize part of the peptidoglycan cell wall of bacteria (NOD stands for nucleotide-binding oligomerization domain).

These receptors are located within the cytoplasm of human cells (e.g., macrophages, dendritic cells, and epithelial cells); hence, they are important in the innate response to intracellular bacteria such as *Listeria*.

RIG-I Helicase-Like Receptors

Finally, **RLRs** recognize microbial nucleic acids in the cytoplasm of infected cells (RIG-I stands for retinoic acid–inducible gene I). For example, certain viruses synthesize double-stranded RNA during replication that is recognized by RLRs. Activation of these RLRs results in the synthesis of alpha- and beta-interferons that promote antiviral immune responses (see below).

EFFECTOR MECHANISMS OF INNATE IMMUNE CELLS

Once innate recognition occurs, **proinflammatory signals** are produced that can:

(1) Activate cells to kill the invader and recruit other immune cells to the area

(2) Block the infection from causing disease beyond the local site of inflammation

(3) Repair the damaged barrier.

The cells that exert these functions are antigen-presenting cells (APCs), granulocytes, and innate lymphocytes called NK cells.

Antigen-Presenting Cells: Macrophages & Monocytes

All nucleated cells express a protein called **class I major histocompatibility complex (MHC)**, and cells present this protein on their surface in complex with peptides from cytosol for recognition by **cytotoxic T cells** (see Chapter 60). There is a different protein called **class II MHC**, that is presented on the cell surface in complex with peptides from endosomes, or vesicles, to be recognized by **helper T cells**. Class II MHC is only found on specialized APCs.

The most abundant APCs are myeloid cells called **macrophages**. They derive from precursors in the yolk sac and liver during fetal development, or from bone marrow in adults. Some examples include the microglial cells in the brain, the alveolar macrophages in the lung, and the Kupffer cells in the liver. In addition to tissue-resident macrophages, there are other cells, called **monocytes**, which are short-lived myeloid cells that patrol the body throughout life, reacting to inflammation by rapidly entering inflamed tissue and **differentiating** into macrophages or dendritic cells on demand.

Tissue-resident macrophages and **monocyte-derived macrophages** have three main functions: phagocytosis, antigen presentation, and cytokine production (Figure 58–2 and Table 58–3).

(1) **Phagocytosis.** Macrophages, neutrophils, and dendritic cells ingest bacteria, viruses, and other foreign particles. They are activated to do this when their PRRs recognize foreign molecular patterns (see Table 58–2). Phagocytes also have two other important types of receptors: one type for C3b, part of the **complement system** that binds to microbes making them easier to ingest (see Chapter 63), and another type for **IgG** that similarly enhance the uptake of Ig-bound microbes. (Factors such complement and immunoglobulins that bind to microbes and enhance phagocytosis are called **opsonins**.)

After ingestion, the phagosome containing the microbe fuses with a lysosome. The microbe is killed within this **phagolysosome** by proteases and reactive oxygen and reactive nitrogen

TABLE 58–3 **Important Features of Macrophages**

Function	Mechanisms
Phagocytosis	Ingestion and killing of microbes in phagolysosomes. Killing caused by reactive oxygen intermediates such as superoxides, reactive nitrogen intermediates such as nitric oxide, and lysosomal enzymes such as proteases, nucleases, and lysozyme.
Antigen presentation	Presentation of short peptide antigens in association with class II MHC proteins to helper T cells. Co-stimulatory signals are also required (see Chapter 60).
Cytokine production	Synthesis and release of cytokines, such as IL-1, IL-6, IL-8, and TNF.

IL = interleukin; MHC = major histocompatibility complex; TNF = tumor necrosis factor.

radicals (generated by **NADPH oxidase** and nitric oxide synthase, respectively). This reaction is called the **oxidative burst**, and it is a critical innate immune mechanism for killing many microorganisms.

Genetic defects in NADPH oxidase cause **chronic granulomatous disease (CGD)**, a condition in which phagocytes are unable to generate an oxidative burst. This causes severe infections as the macrophages and neutrophils, unable to kill the

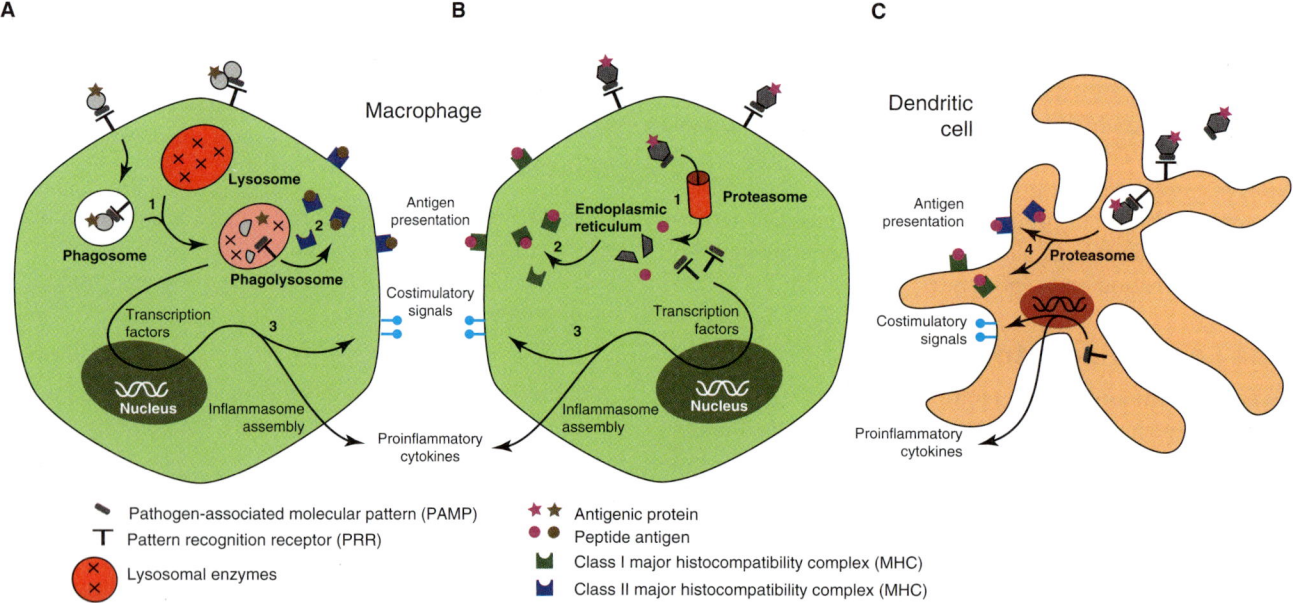

FIGURE 58–2 The functions of phagocytic antigen-presenting cells. **A:** (1) Microbes such as bacteria, viruses, or fungi are sensed by surface pattern recognition receptors (PRRs) and/or receptors for antibody (opsonins), facilitating phagocytosis by macrophages. The receptors are engulfed together with the microbe into the phagosome within the cell. Once there, the phagosome and lysosome are fused, exposing the microbe to degradative enzymes and free radicals. (2) The microbe is killed and its proteins are cleaved into short peptides, which are then complexed with class II major histocompatibility complex (MHC) proteins for presentation on the cell surface. (3) At the same time, killing of the microbe exposes more ligands for PRRs, which leads to transcription of inflammatory cytokine genes, activation of the inflammasome, and expression of costimulatory signals and cytokines that drive further inflammation. **B:** (1) Intracellular (cytosolic) microbes are degraded by the proteasome, releasing (1) antigens for loading onto class I MHC proteins and (2) ligands for PRRs that activate the macrophage to express costimulatory signals and inflammatory cytokines. **C:** Dendritic cells perform these functions in much the same fashion, and in addition, some dendritic cells can (4) complex endosomal antigens with both class I and class II MHC proteins. (For simplification, some aspects of the pathway are omitted.)

microbes they have ingested, resort to forming large granulomas to wall off the microbe (see Chapter 68). Note that the ingestion and killing of microbes is further enhanced by adaptive immunity: **antibodies**, especially IgG, can act as opsonins (see Chapter 61), and phagocytes' killing mechanisms are enhanced by cytokines, especially gamma interferon (IFN-γ), from activated T cells (i.e., **T-cell–mediated immunity**; see Chapter 61).

(2) **Antigen presentation.** After foreign material is ingested and degraded, fragments of antigen are presented on the macrophage cell surface in conjunction with **class II MHC proteins** (for interaction with **helper T cells**). The antigen fragments presented with MHC are short peptides. (See Table 58–3 and Chapters 60, 61, and 62 for more details about T-cell interactions with class I MHC and class II MHC proteins.) APCs also provide surface "co-stimulatory" signals that tell the T cell that the peptide came from a foreign source.

(3) **Cytokine production.** Macrophages produce cytokines, such as interleukin (IL)-1, IL-6, and tumor necrosis factor-α (TNF-α), that further enhance inflammation. In addition, macrophages produce IL-8, a **chemokine** that attracts other cells to the site of infection.

The macrophage's phagocytic ability, antigen presentation, and cytokine production are greatly enhanced when a process called **macrophage activation** occurs. Macrophages are activated by microbial PAMPs (such as bacterial endotoxin, peptidoglycan, or DNA) that interact with TLRs and other macrophage **PRRs**, as described above, and signal the cell to increase its expression of co-stimulatory molecules and its production of cytokines, including TNF-α.

Macrophages are also activated by cytokines. For example, the cytokine **IFN-γ**, produced by T cells and NK cells, enhances antigen presentation by increasing the synthesis of class II MHC proteins. IFN-γ also increases the microbial killing activity of macrophages by increasing the synthesis of NADPH oxidase.

Antigen-Presenting Cells: Dendritic Cells

Dendritic cells are the **main inducers of the adaptive immune response**, thus serving as a bridge between innate and adaptive immunity. They are called "dendritic" because their long, narrow branches make them very efficient at making contact with foreign material (déndron is Greek for "tree").

Dendritic cells are primarily located in barrier tissues, including the skin and the mucosal barriers. Similar to macrophages, dendritic cells ingest foreign material, process it into peptide antigens, and present the antigens with MHC proteins to cytotoxic T cells (through class I MHC proteins) and helper T cells (through class II MHC proteins).

But two very important features of dendritic cells distinguish them from macrophages. First is their ability to collect antigens and then migrate from these barrier locations, through the draining lymphatic vessels, and into local lymph nodes. To do this, the dendritic cell displays the C-C chemokine receptor 7, or **CCR7**, on its cell surface when it recognizes it has engulfed a pathogen (i.e., its PRRs are activated). The dendritic cells use the CCR7 receptor to sense and migrate toward chemokines

that are produced by cells in lymphoid tissue. Once there, the dendritic cell presents antigen complexed with MHC proteins to "naïve" T cells in the **T-cell zone**. Thus, dendritic cells are chiefly responsible for priming naïve T cells to become activated during the initiation of an immune response (see Chapter 60), whereas macrophages only interact with already-activated T cells in the peripheral inflamed tissue.

The second special feature of dendritic cells is that some of them can present **endosomal** antigens on class I MHC. As described earlier, all nucleated cells express **cytosolic** peptides on class I MHC. Usually, these peptides are innocuous "self" antigens and do not elicit an immune response, but if a cell is infected by a virus, those viral peptides will be presented in complex with class I MHC for recognition by cytotoxic T cells.

A particular subset of dendritic cells is able to phagocytize viral particles into their endosomes and present them on **both** class I MHC and class II MHC, bypassing the step in which the cell becomes infected with the virus. This process is called **cross-presentation**, and it allows dendritic cells to prime naïve cytotoxic T cells to recognize tissue-tropic viruses, such as hepatitis B virus, without the dendritic cell itself actually being infected (Figure 58–3). The process by which certain dendritic cells are capable of cross-presentation is not fully known.

Granulocytes: Neutrophils

Neutrophils are the most abundant immune cell in the blood. They are phagocytic myeloid cells in a subgroup called **granulocytes**, named for their cytoplasmic **granules** visible with Wright stain. Neutrophils are a very important component of our innate host defenses, and severe bacterial and fungal infections occur if they are too few in number (neutropenia) or are deficient in function (as in some of the immune disorders discussed in Chapter 68).

Neutrophil granules stain a pale pink (neutral) color with Wright stain, in contrast to eosinophils and basophils, whose granules stain red and blue, respectively. (The differences in the staining color of the various types of granulocytes are due to differences in the charge of their various granules' contents.) The pink granules are lysosomes, which contain a variety of degradative enzymes that are important in the microbicidal action of these cells. The process of phagocytosis and killing by neutrophils is described in detail in Chapter 8. Like macrophages, neutrophils have surface receptors for IgG, making it easier for them to phagocytize opsonized microbes. Neutrophils also have receptors for C3b allowing them to opsonize bacteria before class switching to IgG occurs.

Neutrophils can be thought of as a "two-edged" sword. The positive edge of the sword is their powerful microbicidal activity, but the negative edge is the tissue damage caused by the release of degradative enzymes. For example, the neutrophil gelatinase-associated lipocalin (**NGAL**, also known as **lipocalin-2**) is a protease that is also a urine biomarker of acute kidney injury, which can occur during acute poststreptococcal glomerulonephritis. In this disease, **immune complexes** composed of antibody, streptococcal antigens, and complement attach to the

	Characteristics	Key Functions	Interaction with Adaptive Immunity
Macrophage	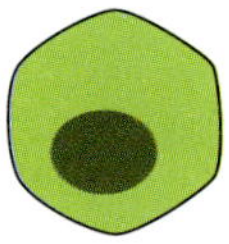 Large phagocytes, located in all tissues, class II MHC expression	Engulfs and kills many classes of microbes, removal of debris, tissue repair	Possess surface IgG receptors that facilitate phagocytosis (opsonization), activated by IFN-γ, TNF-α from T cells. Professional APC expressing class II MHC
Dendritic cell	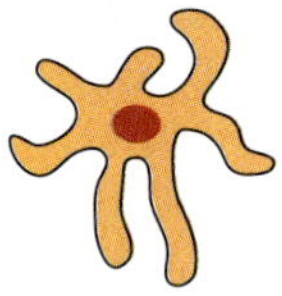 Sentinel cells with long branches; reside in epithelial barriers and secondary lymphoid organs, class II MHC expression	Antigen uptake and presentation including cross-presentation	Professional APC expressing class II MHC, responsible for priming of naive T cells
Neutrophil	 Most common leukocyte in blood, first responder in inflamed or necrotic tissue	Engulfs and kills bacteria and fungi, digests cellular debris	Attracted into tissues by chemokines, which are increased by T cell-derived IL-17
Eosinophil	 Eosinophili granules contain major basic protein; recruited into inflamed tissue by eotaxin	Granule proteins are toxic to cells; involved in asthma and allergic diseases, and protective against invasive helminth infections	Surface IgE receptors; maturation and survival supported by IL-5 from T cells
Basophil	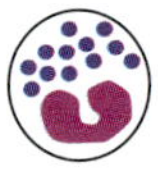 Present at low frequency in the blood	Release histamine, proteases, chemokines, and cytokines; contribute to allergic disease and anaphylaxis	IgE receptors hold IgE molecules that survey for antigen
Mast cell	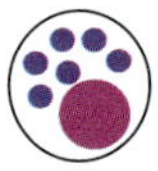 Distributed throughout the tissues around the vasculature		

FIGURE 58–3 Key features of myeloid-derived innate immune cells. The figure lists some of the distinguishing characteristics and functions of myeloid cells, as well as their interactions with adaptive immunity (T cells and antibody). Note that only macrophages and dendritic cells are "professional" antigen-presenting cells (APCs) and that dendritic cells are primarily responsible for initial activation of the T-cell response. See text for abbreviations.

glomerular membrane. Neutrophils that are attracted into the glomeruli and activated by the immune complexes, release their enzymes causing kidney damage.

Granulocytes: Eosinophils

Eosinophils are white blood cells with cytoplasmic **granules** that appear red when stained with Wright stain. The red color is caused by the negatively charged eosin dye binding to the positively charged **major basic protein** in the granules. The eosinophil count is elevated in two medically important types of diseases: **parasitic diseases**, especially those caused by tissue-invading nematodes and trematodes (see Chapters 55 and 56, respectively) and **hypersensitivity diseases**, such as asthma and serum sickness (see Chapter 65). Diseases caused by protozoa are typically *not* characterized by eosinophilia.

It seems likely that eosinophils defend against migratory parasites, such as *Strongyloides* and *Trichinella*. The larvae of these parasites become coated with IgE, and eosinophils, which have receptors for IgE, can then attach to the surface of larvae

and discharge the contents of their eosinophilic granules, damaging the cuticle of the larvae. The granules of the eosinophils also contain **leukotrienes** and **peroxidases**, which can damage tissue and cause inflammation. Eosinophil major basic protein can damage respiratory epithelium and contributes to the pathogenesis of asthma.

Another function of eosinophils may be to reduce inflammation. The granules of eosinophils contain **histaminase**, an enzyme that degrades histamine, which is an important mediator of immediate hypersensitivity (allergic) reactions. The growth and differentiation of eosinophils are stimulated by the cytokine **IL-5**, and **eotaxin** is a chemokine (see below) that attracts eosinophils from the blood into tissues.

Granulocytes: Basophils & Mast Cells

Basophils are white blood cells with cytoplasmic granules that appear blue when stained with Wright stain. The blue color is caused by the positively charged methylene blue dye binding to several negatively charged molecules in the granules. Basophils circulate in the bloodstream, whereas mast cells are fixed in tissue, especially under the skin and in the mucosa of the respiratory and gastrointestinal tracts.

Basophils and mast cells have receptors on the cell surface for the Fc portion of the heavy chain of IgE. When adjacent IgE molecules are cross-linked by antigen, the cells release **preformed inflammatory mediators** from their granules. Some examples of these mediators are **histamine**, proteolytic **enzymes**, and **proteoglycans** such as heparin. They also release **newly generated** eicosanoids, such as **prostaglandins** and **leukotrienes**. These cause inflammation and, when produced in large amounts, cause a wide range of immediate hypersensitivity reactions: the mildest form is **urticaria** (hives), while the most severe form is **systemic anaphylaxis**.

Natural Killer Cells

NK cells play two important roles in immunity:

(1) They kill virus-infected cells and tumor cells
(2) They produce IFN-γ that activates macrophages to kill ingested bacteria (see Chapter 60).

NK cells are called "natural" because, unlike adaptive cells, they do not use antigen receptors to recognize their target cells or microbes and they do not exhibit an enhanced response (i.e., memory) upon repeat challenge.

Rather, NK cells target cells to be killed by detecting **other features of cell dysfunction**, for example, the *lack* of class I MHC proteins on the cell surface. This detection process is effective because many cells lose their ability to synthesize class I MHC proteins after they have been infected by a virus.

NK cells kill virus-infected cells and tumor cells by secreting cytotoxins (**perforins** and **granzymes**) that induce apoptosis. They can do this without antibody, but antibody (IgG) enhances their effectiveness, a process called **antibody-dependent cellular cytotoxicity** (ADCC) (see Chapter 61). IL-12 produced by macrophages and interferons alpha and beta produced by

virus-infected cells are potent activators of NK cells. Approximately 5% to 10% of peripheral lymphocytes are NK cells. Humans who lack functional NK cells are predisposed to severe infections with herpesviruses and human papillomavirus, as well as various cancers.

INFLAMMATORY MEDIATORS

Local inflammation at the site of an infection causes the four classic symptoms of **pain, redness, warmth**, and **swelling**. These symptoms reflect the immune system's efforts to recruit leukocytes to the area and limit the infection from spreading. Innate signals of tissue damage, including **lipid mediators** (i.e., prostaglandins and leukotrienes), **histamine**, activated **complement**, and components of **coagulation cascade**, cause redness, warmth, and swelling through vasodilation and vascular leak. **Nitric oxide** (NO) is made by macrophages and neutrophils and causes vasodilation. **Bradykinin** also causes vasodilation, vascular leak, and pain. It is released from macrophages and is a potent vasodilator.

Note that these mediators are released within seconds to minutes and are **nonspecific** signals of tissue damage, regardless of the injury. Once inflammation starts, it can be greatly amplified by the presence of microbial products that stimulate PRRs.

In addition to these mediators, **cytokines** and **chemokines** of the innate immune system are engaged to recruit and activate leukocytes. Leukocytes usually patrol the bloodstream and lymphatics, but they can migrate from blood vessels into tissue in a process called **extravasation** (Figure 58–4), which involves the three key steps of **rolling, adhesion**, and **migration**, and the combined action of **selectins, integrins**, and **chemokines**:

(1) First, the endothelial cells that line the capillaries produce "sticky" **selectins** on their surface. Leukocytes have the ligands that bind these selectins, causing them to slow down and **roll** along the capillary wall.

(2) Next, higher levels of local chemokines cause the leukocytes to produce activated **integrins** on their surface, which **adhere** to the **cell adhesion molecules** on the endothelial cells. This stops the leukocyte rolling.

(3) Finally, the leukocytes squeeze and spread their cell bodies, passing between endothelial cells, and **transmigrate** by pulling themselves along the cell adhesion molecules out of the capillary and into the surrounding tissue.

Cytokines

The epithelial cells and the leukocytes of the innate immune system work together to detect and fend off foreign invaders. When necessary, the also recruit and activate adaptive immune cells. This network of cell–cell communication tunes up or tunes down inflammation. **Cytokines** are the language of the immune system, and immune cells use cytokines to communicate with other cells. Inflammation initiates a **cascade** of cytokines, including some that later turn off inflammation when it is no longer needed.

TABLE 58–4 Important Features of Natural Killer (NK) Cells

I. Nature of NK Cells
- Large granular lymphocytes
- Lack T-cell receptor, CD3 proteins, and surface IgM and IgD
- Thymus not required for development
- Normal numbers in severe combined immunodeficiency disease (SCID) patients
- Activity not enhanced by prior exposure
- Have no memory

II. Function of NK Cells
- Recognize virus-infected cells by detecting lack of class I MHC proteins on the surface of the infected cells
- Kill virus-infected cells and cancer cells using perforin and granzyme
- Killing is nonspecific and is not dependent on foreign antigen presentation by class I or II MHC proteins
- Produce gamma interferon that activates macrophages to kill ingested bacteria

Ig = immunoglobulin; MHC = major histocompatibility complex.

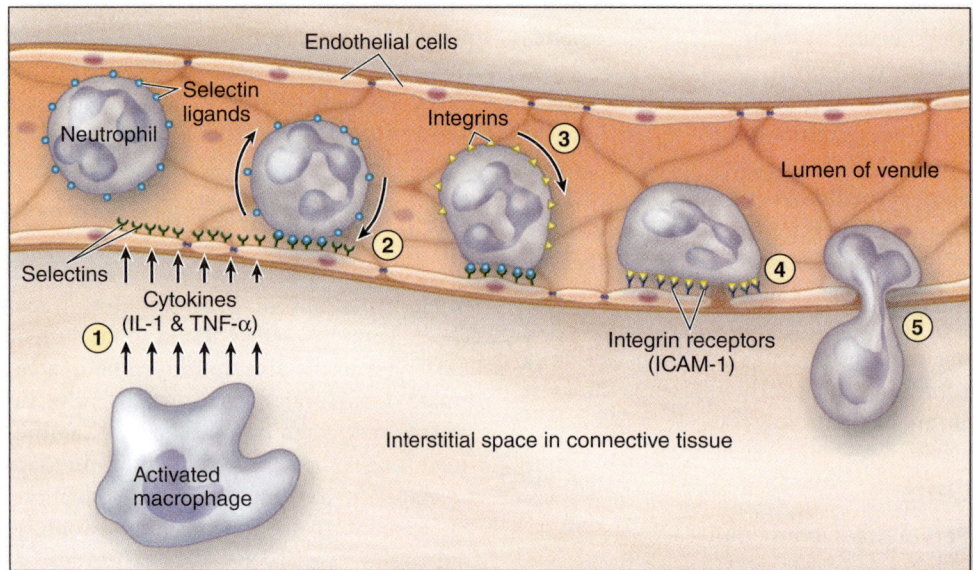

FIGURE 58–4 The five steps of leukocyte extravasation. (1) Capillary endothelial cells increase their surface selectins in response to inflammatory mediators such as tumor necrosis factor (TNF) and interleukin-1 (IL-1). (2) Leukocytes (e.g., neutrophils) patrolling the bloodstream slow down and *roll* along the luminal surface of the endothelial cells when their selectin ligands bind to the selectins. (3) Local inflammatory chemokines bind to the leukocyte's chemokine receptors, triggering a conformational change in their surface integrins from a state that is low affinity to high affinity. (4) The high-affinity integrins interact with endothelial cell adhesion molecules, causing the leukocyte to stop rolling and *adhere* along the surface of the vessel wall. (5) Using cell adhesion molecules, the leukocytes change shape and *transmigrate* between the gaps between endothelial cells and into the inflamed tissue, following the chemokine gradient. (Reproduced with permission from Mescher AL: *Junqueira's Basic Histology: Text and Atlas*, 15th ed. New York, NY: McGraw Hill; 2018.)

Historically, cytokines were named based on a function that was originally discovered for them, such as TNF or "granulocyte colony-stimulating factor" (G-CSF), but more recently, they have mostly been given the name "interleukin," with a number assignment corresponding to the order in which they were discovered (Table 58–5). Over time, new functions have been discovered for many cytokines, and their names have become less reflective of their function. Note that medical therapies that target many of these cytokines are now in use to either *boost* immunity (through agonism/activation) or *limit* excessive inflammation (through antagonism/blockade).

(1) **TNF-α** is a proinflammatory cytokine produced primarily by macrophages. At low concentrations, it increases the synthesis of adhesion molecules by endothelial cells, which allows neutrophils to adhere to blood vessel walls at the site of infection. It also activates the respiratory burst within neutrophils, thereby enhancing the killing power of these phagocytes. TNF-α is also an **endogenous pyrogen**, a cytokine that causes fever.

At high concentrations, TNF-α mediates **endotoxin-induced** septic shock by inducing fever and causing hypotension through vasodilation and increased capillary leakiness. (The action of endotoxin is described in Chapters 7 and 79.)

TNF-α is also known as **cachectin** because it inhibits lipoprotein lipase in adipose tissue, thereby reducing the utilization of fatty acids. This results in cachexia, or wasting. TNF-α, as its name implies, causes the **death and necrosis of certain tumors**

in experimental animals. It may do this by promoting intravascular coagulation that causes infarction of the tumor tissue. Note the similarity of this intravascular coagulation with the

TABLE 58–5 Innate/Acute Immune Cytokines

Cytokine	Cell Source	Important Cellular Targets and Cytokine Function
Interleukin-1	Macrophages, DCs (others)	Endothelial cells: Increased selectins Hypothalamus: fever Liver: synthesis of acute-phase proteins
Interleukin-6	Macrophages, endothelial cells	Liver: Synthesis of acute-phase proteins B cells: proliferation of antibody-producing cells
Tumor necrosis factor-α	Macrophages, NK cells	Endothelial cells: Increased selectins Hypothalamus: fever Muscle/fat: catabolism (cachexia) Neutrophils and macrophages: Activation
Interferon-α; interferon-β	Macrophages, DCs, fibroblasts, and epithelial cells	All cells: Innate defense against viruses through inhibition of protein synthesis and induction of ribonuclease that degrades mRNA NK cells: Activation of killing function

DC = dendritic cell; NK = natural killer.

disseminated intravascular coagulation (DIC) of septic shock, both of which are caused by TNF-α.

(2) IL-1 is a proinflammatory cytokine that is stored in an inactive form, called **pro-IL-1**, ready for use, inside macrophages and epithelial cells. How does it become active IL-1? As described earlier, NLRs are PRRs that sense microbial products inside the cell's cytosol. When activated, these NLRs can assemble into a multiprotein complex, called the **inflammasome**, which starts a chain reaction whereby inactive proteins are processed into their active forms. The result is that pro-IL-1 is converted to IL-1 and released from the cell. IL-1's function is primarily to increase the "stickiness" of endothelial cells in the blood vessels and increase the production of chemokines (discussed below), which in turn increases recruitment of more inflammatory cells. Like TNF-α, IL-1 is also an endogenous pyrogen that causes fever.

(3) IL-6 is released by macrophages and mast cells and probably also by nonimmune cells such as muscle and fat cells. Its primary function is to signal the liver to increase production of **acute-phase proteins**, which enter the circulation and cause fever and cachexia. The acute-phase response is described in more detail below. IL-6 also triggers the release of new neutrophils from the bone marrow, which is detected as an elevated white blood cell count, or **leukocytosis**.

(4) Two other important cytokines that stimulate leukocyte migration out of the bone marrow are **granulocyte colony-stimulating factor** (G-CSF, or CSF1) and **granulocyte-macrophage colony-stimulating factor** (GM-CSF, or CSF2). G-CSF and GM-CSF are made by various cells and enhance the development of neutrophils (in the case of G-CSF) or of all granulocytes and monocytes (in the case of GM-CSF) from bone marrow stem cells. Both of these cytokines are used therapeutically to boost leukocyte numbers and prevent infections in patients who have received cancer chemotherapy and/or stem cell transplantation.

(5) **Interferons** are glycoproteins that were originally named because they **interfere** with virus replication, but in fact, they are innate cytokines that have a variety of effects on cells. The **type I interferons** (also called **alpha interferon** or **IFN-α**, made by leukocytes, and **beta interferon** or **IFN-β**, made by nonhematopoietic cells) are induced when cells detect they contain a virus.

Both IFN-α and IFN-β signal nearby cells to make degradative enzymes that will inactivate the virus when it infects those cells. The nearby cells can prevent the virus from replicating and thereby prevent the spread of virus from cell to cell. They do this through key enzymes that:

(1) Degrade viral messenger RNA

(2) Block translation of new proteins

(3) Initiate apoptosis pathways so that the cell dies before its machinery can be used to help the virus spread.

Type I interferons can also increase the expression of class I and class II MHC proteins in virus-infected cells, making the presence of viral infection more easily recognizable by other immune cells (see Chapter 33.)

Type II interferon (also called **gamma interferon or IFN-γ**) is primarily produced by activated T cells and NK cells. Its name comes from the fact that it is structurally related to the other interferons, but unlike those cytokines, the main function of IFN-γ is to activate macrophages (discussed in Chapter 60) rather than to provide innate antiviral activity.

In 2021, it was reported that gamma interferon induced the synthesis of apolipoprotein L3 in nonimmune cells such as fibroblasts and epithelial cells. Apolipoprotein L3 was shown to disrupt the cell membrane of Salmonella bacteria demonstrating it has bactericidal activity. This indicates that gamma interferon has antimicrobial activity beyond its ability to enhance killing within macrophages.

Chemokines

Chemokines are cytokines that attract leukocytes and help them migrate to where they are needed. The term *chemokine* is a contraction of **chemo**tactic and cyto**kine**. Chemokines are produced by various cells; in infected areas, endothelial cells and macrophages produce chemokines to recruit more inflammatory cells (see Figure 58–4). Other chemokines are produced in lymph nodes and spleen, where their gradients guide immune cells toward areas where they encounter and communicate with other cells.

Approximately 50 chemokines have been identified; they are small polypeptides ranging in size from 68 to 120 amino acids. They are often classified based on their structure: alpha-chemokines have two adjacent cysteines separated by another amino acid (Cys-X-Cys), whereas the beta-chemokines have two adjacent cysteines (Cys-Cys) (Table 58–6).

The alpha-chemokines generally attract neutrophils, monocytes, dendritic cells, and NK cells. **IL-8** and **eotaxins** are important members of this group. The beta-chemokines attract macrophages and monocytes and are produced by activated T cells. **RANTES** and **CCL2** (also called MCAF or MCP-1) are important beta-chemokines. A few chemokines have unusual structures that do not fall into these groups, including the **lymphotoxins** (C chemokines) and **fractalkine** (the only CX3C chemokine). **C5a**, the cleavage product of complement component

TABLE 58–6 Chemokines of Medical Importance

Class	Chemistry	Attracts	Produced by	Examples
Alpha	C-X-C	Neutrophils	Activated mononuclear cells	Interleukin-8, eotaxins
Beta	C-C	Monocytes	Activated T cells	RANTES,[1] MCAF[2]

[1]RANTES is an abbreviation for *regulated upon activation, normal T expressed and secreted*.

[2]MCAF is an abbreviation for *macrophage chemoattractant and activating factor*.

C5, is also a powerful chemoattractant, although it is not structurally related to the other chemokines (see Chapter 63).

There are specific receptors for chemokines on the surface of leukocytes. As described earlier, interaction of the chemokine with its receptor results in changes in cell surface proteins that allow the cell to **adhere to** and **migrate between** endothelial cells to the site of infection (see Figure 58–4). Chemotactic factors for neutrophils, basophils, and eosinophils selectively attract each of these different cell types. For example, **IL-8** and complement component **C5a** are strong attractants for neutrophils. **C3a** is a weak chemoattractant for mast cells and eosinophils.

THE ACUTE-PHASE RESPONSE

In contrast to the **local** effects of inflammation described earlier, the **acute-phase response** is a rapid, **systemic** increase in various plasma proteins in response to innate inflammation. Together, these cause systemic symptoms of fever, malaise, elevated heart rate, and loss of appetite that we commonly associate with infection. As described earlier, macrophages and other cells that are triggered through their PRRs assemble **inflammasomes**, leading to IL-1 release. IL-6 and TNF-α are also made by macrophages in response to IL-1 and PRR stimulation.

IL-1, IL-6, and TNF-α are **proinflammatory cytokines**, meaning that they enhance the inflammatory response in various ways (see Table 58–5). They signal to the hypothalamus to change the body's thermostat, causing fever. They also signal to the liver hepatocytes to increase production of C-reactive protein, MBL, proteins of the complement cascade (covered in Chapter 63), ferritin, and other acute-phase proteins. **Mast cells** can directly sense microbial patterns through PRRs or through receptors for antibodies (i.e., IgE receptors), or can be stimulated by IL-1 to release IL-6, leukotrienes, vasoactive signals, and other proinflammatory mediators.

Some of the acute-phase proteins are anticoagulants that improve blood flow to inflamed tissues. Other acute-phase proteins, such as ferritin, sequester iron, which bacteria need for replication, or bind to the surface of microbes and activate complement. For example, as mentioned earlier, MBL binds to mannan (mannose) on the surface of many bacteria, fungi, and protozoa, and C-reactive protein similarly binds to other components of the microbial cell walls. Upon binding, these host proteins can enhance complement activation (see Chapter 63) and phagocytosis (i.e., act as opsonins). Finally, many acute-phase proteins increase the migration of new neutrophils and other leukocytes from the bone marrow and enhance their homing, phagocytic, and microbicidal functions.

TRAINED INNATE IMMUNITY

The concept of trained innate immunity is based on the observation that immunization with certain vaccines provides protection not only against the target disease but against other infectious diseases as well. For example, several epidemiological studies have shown that administration of the BCG vaccine to

prevent tuberculosis in children reduces *both* the incidence of symptomatic pulmonary tuberculosis and also the incidence of viral respiratory tract infections and neonatal sepsis.

This protection, which is not pathogen-specific, is likely mediated by enhanced activity of innate immune cells, such as macrophages and NK cells. It has been demonstrated using animal models that the innate immune cells participating in an immune response can undergo epigenetic changes that enhance the cells' transcription of genes encoding proinflammatory cytokines, chemokines, and pattern-recognition molecules. This makes the cells able to respond more quickly to the next challenge. Trained immunity can last for months to years but should not be confused with immune memory, which is a feature of adaptive immunity.

SELF-ASSESSMENT QUESTIONS

1. Which one of the following is the most accurate statement?
   (A) Loss of the epithelial barrier predisposes to fungal infections but not bacterial or viral infections.
   (B) The main function of mast cells and eosinophils is to engulf microbes and debris.
   (C) Eosinophils and natural killer cells are both innate cells of the myeloid lineage.
   (D) Dendritic cells are the primary cell responsible for initiating an adaptive immune response.

2. Local swelling, redness, warmth, and pain associated with inflammation are primarily due to which one of the following?
   (A) Pathogen-associated molecular patterns (PAMPs) recognized by neurons and endothelial cells.
   (B) Release of preformed mediators such as leukotrienes and histamine and activated complement.
   (C) Cytokines such as tumor necrosis factor-alpha released by T cells.
   (D) IgE bound to the surface of eosinophils.

3. Which one of the following is an accurate statement about antigen-presenting cells (APCs)?
   (A) Monocytes can enter inflamed tissue and differentiate into macrophages and dendritic cells.
   (B) Dendritic cells turn off CCR7 upon recognition of pathogen-associated molecular patterns.
   (C) Macrophages are "professional" APCs that express class II major histocompatibility complex (MHC) but not class I MHC.
   (D) Cytosolic antigens are degraded by the proteasome and displayed in complex with class II MHC.

4. Which one of the following is NOT a primary function of phagocytes?
   (A) Engulfing and killing invading microbes
   (B) Expression of proinflammatory cytokines and chemokines
   (C) Attacking cells with perforins and granzymes
   (D) Production of free oxidative radicals
   (E) Presentation of antigen peptides in complex with MHC to T cells

5. Which one of the following describes the immune signals responsible for fever?
   (A) Histamine and proteases produced by mast cells
   (B) Interleukin-1 (IL-1) and tumor necrosis factor (TNF) produced by macrophages

(C) Type I interferons produced by virus-infected cells

(D) Type II interferon and perforin produced by natural killer (NK) cells

6. The pathogenesis of chronic granulomatous disease is BEST described as which one of the following?

   (A) A defect of chemokine signaling causing impaired granulocyte exit from the bone marrow

   (B) A defect in complement receptors causing impaired granulocyte activation

   (C) A defect in integrin signaling causing impaired granulocyte migration into inflamed tissue

   (D) A defect in oxidative burst causing impaired ability of granulocytes to kill microbes

7. Regarding chemokines, which one of the following is the most accurate?

   (A) Chemokines penetrate the membranes of target cells during attack by cytotoxic T cells.

   (B) Chemokines bind to the T-cell receptor outside of the antigen-binding site and activate many T cells.

   (C) Chemokines attract neutrophils to the site of bacterial infection, thereby playing a role in the inflammatory response.

   (D) Chemokines induce gene switching in B cells, which increases the amount of IgE synthesized, thereby predisposing to allergies.

ANSWERS

(1) **(D)**

(2) **(B)**

(3) **(A)**

(4) **(C)**

(5) **(B)**

(6) **(D)**

(7) **(C)**

PRACTICE QUESTIONS: USMLE & COURSE EXAMINATIONS

Questions on the topics discussed in this chapter can be found in the Immunology section of Part XIII: USMLE (National Board) Practice Questions starting on page 757. Also see Part XIV: USMLE (National Board) Practice Examination starting on page 775.

Adaptive Immunity: Lymphocyte Antigen Receptors

Some microbes have evolved ways to subvert or evade the **innate immune system**. The next line of host defense is the **adaptive immune system**, which is composed of **lymphocytes** (also called lymphoid cells) and their secreted factors (see Table 57–1).

A critical property of adaptive immunity is that the immune response is **specifically tailored** against different microbes. This is achieved by first generating an enormous number of diverse lymphocytes, each with a unique antigen specificity. Before they see their antigen, these lymphocytes are called **naïve** (Figure 59–1). How these cells function is closely linked to how they develop from stem cells, so in order to understand how lymphocytes work, it is first necessary to review lymphocyte development.

ORIGIN OF LYMPHOID CELLS

As described in Chapter 58, all white and red blood cells originate from stem cells (see Figure 58–1). The **common lymphoid progenitor** is a stem cell that gives rise to lymphocytes of the adaptive immune system, including **B cells** and **T cells**. The common lymphoid progenitor is also the source of innate lymphocytes, such as **natural killer (NK) cells**. The process by which common lymphoid progenitors develop into lymphocytes depends on cytokines, and mutations in cytokine signaling can cause severe combined immunodeficiency, a complete absence of mature lymphocytes (see Chapter 68). Table 59–1 compares various important features of B cells and T cells. These features will be discussed in detail in this and later chapters.

LYMPHOCYTE RECEPTOR DIVERSITY

All vertebrates produce enormously diverse pools of antigen receptors; in humans, this pool is estimated to comprise 100 million different specificities, protecting us from millions of potential pathogens. How do we accomplish this with a genome that only contains approximately 20,000 genes? The solution is that during their development, T and B lymphocytes do something extremely unconventional. They switch on a program of **DNA rearrangement**, cutting their DNA, removing pieces, and shuffling other pieces, to form entirely new coding sequences in their antigen receptor genes. The two most important enzymes in this process are **recombinases**, called *RAG-1* and *RAG-2* (recombination-activating genes). The *RAG* genes have been found in our first vertebrate ancestors from 500 million years ago.

This DNA rearrangement is absolutely necessary for proper functioning of our adaptive immune system. Mutations in these *RAG* genes halt the development of lymphocytes and result in severe combined immunodeficiency (see Chapter 68). However, DNA rearrangement is also risky. The RAG proteins are supposed to work only in specific locations (i.e., in the immunoglobulin and T-cell receptor gene loci, as discussed below), but errors do occur. If the recombinase makes a rearrangement with the wrong gene, it could kill the cell, or worse, it could cause it to divide uncontrollably, resulting in a leukemia or lymphoma.

First, we will discuss the development of B cells, which detect antigen using **immunoglobulins (Igs)**, and then we will discuss T cells, which detect antigen using **T-cell receptors (TCR)**.

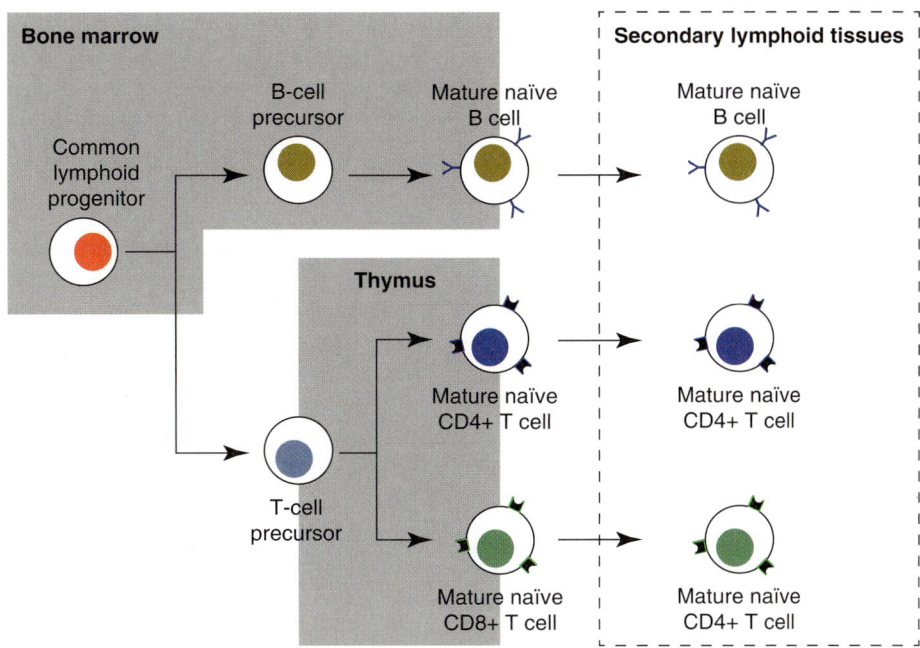

FIGURE 59-1 Development of naïve lymphocytes. Common lymphoid progenitors give rise to B-cell precursors, which develop into mature B lymphocytes in the bone marrow, and T-cell precursors, which leave the bone marrow and complete their development into mature CD4-positive and CD8-positive T cells in the thymus. Mature naïve lymphocytes migrate throughout the secondary lymphoid tissue surveying for antigen. CD = cluster of differentiation.

B CELLS

B cells perform two important functions: (1) they differentiate into plasma cells that produce **antibodies** (also called **Igs**) and (2) they can become long-lived **memory B cells** that can rapidly respond to a reinfection.

*The Ig on the B-cell surface is its antigen receptor (**B-cell receptor** or **BCR**) and the ability of a B-cell precursor to make*

TABLE 59-1 Comparison of T Cells and B Cells

Feature	T Cells	B Cells
Antigen receptors on surface	Yes	Yes
Antigen receptor recognizes only processed peptides in association with MHC protein	Yes	No
Antigen receptor recognizes whole, unprocessed proteins and has no requirement for presentation by MHC protein	No	Yes
IgM on surface	No	Yes
CD3 proteins on surface	Yes	No
Clonal expansion after contact with specific antigen	Yes	Yes
Immunoglobulin synthesis	No	Yes
Regulator of antibody synthesis	Yes	No
IL-2, IL-4, IL-5, and gamma interferon synthesis	Yes	No
Effector of cell-mediated immunity	Yes	No
Maturation in thymus	Yes	No
Maturation in bone marrow	No	Yes

IgM = immunoglobulin M; IL = interleukin; MHC = major histocompatibility complex.

*this antigen receptor determines whether it is allowed to develop into a **mature** B cell.*

By the time of birth, B-cell precursors exist in the bone marrow. Unlike T cells, B cells do not require the thymus for maturation. The maturation of B cells has two phases: the first is the **antigen-independent phase**, which consists of stem cells, pre-B cells, and B cells, and it is during this phase that the B cell recombines its Ig genes to make a unique antigen receptor. For pre-B cells to differentiate into B cells, a functional Ig must be present on the cell surface. A protein called **Bruton's tyrosine kinase (BTK)** detects this Ig and signals to the cell to continue to divide and differentiate. A mutation in the gene encoding this protein causes **X-linked agammaglobulinemia**, a condition in which cells cannot progress to the pre-B cell stage and no antibodies are made (see Chapter 68).

During the second phase, which is the **antigen-dependent phase**, mature B cells with functional antigen receptors interact with antigens. This phase will be covered in more detail in Chapter 61.

The Ig, or **BCR**, of a mature B cell is an IgM molecule with an additional region at the end of its heavy chain that tethers it to the B-cell surface. Approximately 10^9 B cells are produced each day, but only a small fraction of these make it out of the bone marrow. In this chapter, we will explore the structure of BCRs, and in Chapter 61, we will describe B-cell activation and how antibodies work.

Antibody Structure

Antibodies are glycoproteins made up of **light** (L) and **heavy** (H) polypeptide chains. The terms *light* and *heavy* refer to molecular weight; light chains have a molecular weight of about 25 kDa, whereas heavy chains have a molecular weight of about 50 kDa. The simplest antibody molecule has a Y shape (Figure 59–2) and consists of four polypeptide chains: two **identical** H chains and two **identical** L chains. In other words, even though you received a copy of the H and L chain genes from each of your parents, each B cell ultimately synthesizes only one of the H chain genes and one of the L chain genes to form an antibody, and therefore, all of the subsequent antibodies from that B cell and its progeny use the same H and L chains. Table 59–2 is a summary of the properties of the human lymphocyte antigen receptors.

One end of the Y is composed of two identical pieces that bind the antigen, and therefore, this is called the **antigen-binding fragment** (or **Fab**). The Fab includes the **variable** region of the L chain (V_L) and the **variable** region of the H chain (V_H), as well as the **constant** region of the L chain (C_L) and the first **constant** region of the H chains (C_H1). The portions of the L and H chains that actually bind the antigen are only 5 to 10 amino acids long, each composed of three *extremely* variable (**hypervariable**) amino acid sequences. Antigen–antibody binding involves electrostatic and van der Waals' forces and hydrogen and hydrophobic bonds rather than covalent bonds. The remarkable specificity of antibodies is due to these hypervariable regions.

The other end of the Y is a single stalk, where the H chains come together, and it is made of the remaining three or four **constant** regions of each of the H chains (C_H2, etc.). This is called the **constant** or **crystallizable** fragment (or **Fc**). You might think that the Fab is the most important part of the antibody because it binds the antigen, but the Fc is also very important because it is needed to attach the antibody to host cells (e.g., via Fc receptors) or to complement (at the C_H2 domain). The Fc is also the region that is used to fuse IgM and IgA together into larger "multimers." It is also necessary for transport of IgA across epithelial barriers and transport of IgG from mother to fetus through the placenta.

There are five classes of antibodies: IgM, IgD, IgG, IgE, and IgA. Each class has structural differences that make it unique. For example, IgG and IgA have three C_H domains, whereas IgM and IgE have four. The structural differences between the antibody classes translate into important functional differences. Mature naïve B cells start out making only IgM and IgD but later "switch" to making the other classes. We will discuss the different antibody functions and how the B cells class-switch in Chapter 61.

L chains can be of two types, κ (**kappa**) or λ (**lambda**), which differ in their constant regions. Either type can pair with

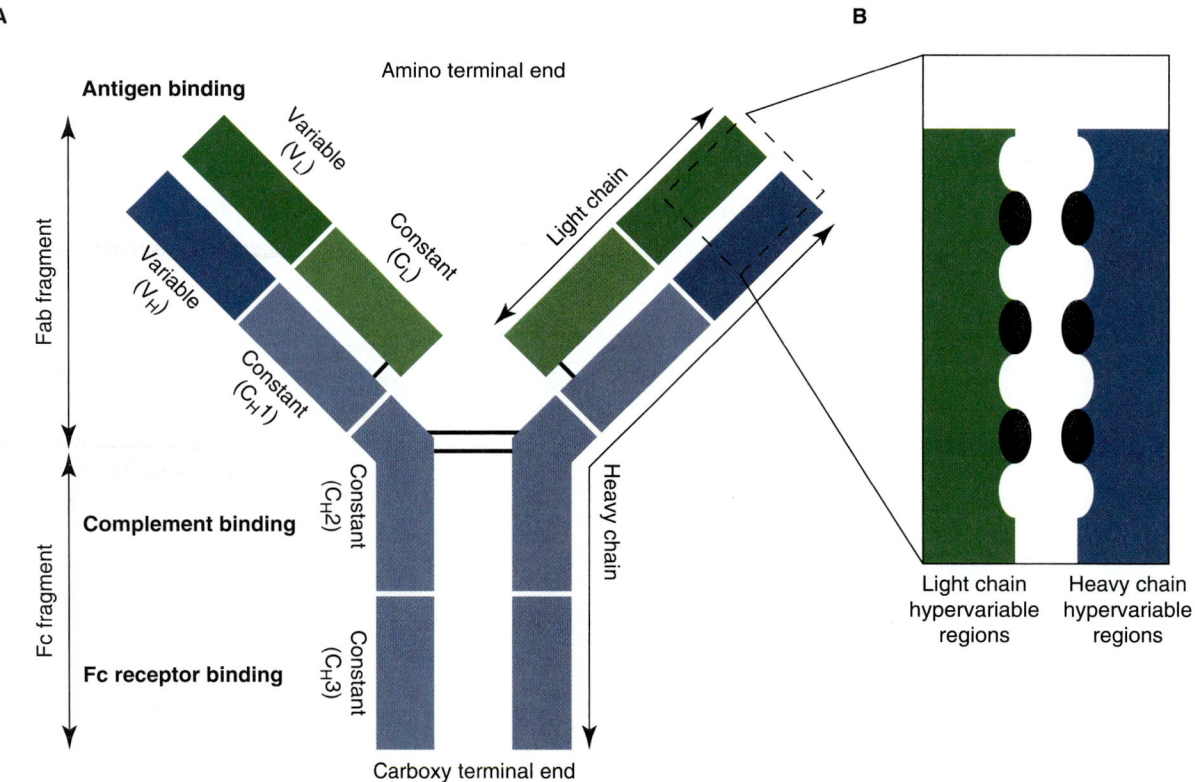

FIGURE 59–2 Structure of immunoglobulin G (IgG). **A:** The Y-shaped IgG molecule consists of two light chains and two heavy chains. Each light chain consists of a variable region (dark green) and a constant region (light green). Each heavy chain consists of a variable region (dark blue) and a constant region (light blue) that is divided into three domains: C_H1, C_H2, and C_H3. The C_H2 domain contains the complement-binding site, and the C_H3 domain is the site of attachment of IgG to receptors on neutrophils and macrophages. The antigen-binding site is formed by the variable regions of both the light and heavy chains. **B:** The specificity of the antigen-binding site is a function of the amino acid sequence of the hypervariable regions, shown in magnified view. (Adapted with permission from Brooks GF, Butel JS, Ornston LN: *Jawetz, Melnick & Adelberg's Medical Microbiology*, 20th ed. New York, NY: McGraw Hill; 1995.)

TABLE 59–2 Properties of Lymphocyte Antigen Receptors

Cell Type	Types of Chains	Types of Antigens Recognized
B cells	Heavy	Macromolecules, including large proteins, carbohydrates, lipids, nucleic acids
	Light (κ)	
	Light (λ)	
αβ T cells	α	Peptides, complexed with class I or II MHC (Rare exceptions are iNKT cells, which recognize glycolipids, complexed with CD1d; and MAIT cells, which recognize metabolites of mucosa-associated bacteria, complexed with MR1)
	β	
γδ T cells	γ	Possibly small-molecule metabolites of mycobacteria and plasmodia; mechanism of antigen presentation unknown
	δ	

iNKT = invariant natural killer T cell; MAIT cell = mucosal associated invariant T cell; MHC = major histocompatibility complex.

H chains in all classes of Igs (IgG, IgM, etc.), but once a B cell chooses to use κ or λ, it shuts off the other L chain gene, so that all of the immunoglobulin from any one B cell contains only one type of L chain. In humans, the ratio of Igs containing κ chains to those containing λ chains is approximately 2:1, and a value dramatically different can be a sign of a monoclonal Ig-producing malignancy such as multiple myeloma.

H chains are distinct for each of the five Ig **classes** and are designated γ (**gamma**), α (**alpha**), μ (**mu**), ε (**epsilon**), and δ (**delta**). The V_H hypervariable region of the H chain joins with V_L in binding antigen; the opposite regions of the V_H chains form the Fc fragment, which determines which class the antibody is and, therefore, what its biologic activities will be (see Chapter 61).

Antibody Genes

As described earlier, each antibody is composed of four Ig chains (two light chains and two heavy chains). There are two light chain gene clusters, one encoding kappa light chain (κL), on human chromosome 2, and one encoding lambda light chain (λL), on chromosome 22. All heavy chain genes (μH, δH, γH, εH, and αH) are clustered together on chromosome 14. The heavy chains and light chains are assembled after recombining gene segments within their respective gene clusters, a process that is directed by **recombinase enzymes** (**RAG1** and **RAG2**). A schematic diagram of gene recombination is shown in Figure 59–3.

First, the V_H and V_L genes are recombined. Each cluster contains dozens of different V gene segments widely separated from the D (diversity, seen only in H chains), J (joining), and C gene segments. The V_H region of each heavy chain is encoded by three gene segments (V + D + J). In the synthesis of a heavy chain, one V region (out of ~45) is translocated to lie close to one D segment (out of ~23), one J segment (out of 6), and one C segment.

The V_H/C_H combination is transcribed together on an RNA molecule and spliced to produce an mRNA that codes for the complete heavy chain, encoded by a single V, D, and J segment attached to a C segment. Why are IgM and IgD the first antibodies that are produced? The newly assembled V + D + J gene segments are closest to the Cμ and Cδ genes! In Chapter 61, we will describe how class switching leads to IgG, IgE, and IgA, which are further downstream in the heavy chain locus.

The V_L region of each L chain is encoded by two gene segments (V + J). In the assembly of an L chain, the same process occurs except that there are slightly fewer possible V segments (~30–35 in kappa and lambda), and neither of the L chains have D segments. Also, the kappa chain gene has a single Cκ, whereas the lambda chain gene has four Cλ segments, one already associated with each J segment. The L chain comes from a similar translocation in which a single V and J are brought close together and then transcribed and translated with the appropriate C segment. Note that the DNA of the unused V, D, and J genes is discarded; once a particular B cell has recombined its light and heavy chains, it is committed to making antibody with only one specificity.

The H and L chains are synthesized as separate peptides and then folded and assembled in the cytoplasm by means of disulfide bonds to form H2L2 units. Finally, an oligosaccharide is added to the constant region of the heavy chain, and the BCR molecule is transported to the cell surface.

Clonal Selection

Note that the genetic recombination outlined earlier can lead to an enormous number of possible combinations. There are approximately 10^{11} possible heavy chain–light chain combinations! Antibody **diversity** depends on (1) multiple gene segments, (2) their rearrangement into different sequences, (3) the combining of different L and H chains in the assembly of Ig molecules, and (4) mutations. A fifth mechanism called junctional diversity applies primarily to the antibody heavy chain. Junctional diversity occurs by the addition of new nucleotides at the splice junctions between the V-D and D-J gene segments. The resulting antibodies have the potential to recognize the three-dimensional structure of a wide range of proteins, carbohydrates, nucleic acids, and lipids.

Despite the enormous potential diversity, the actual specificities represented among the pool of circulating B cells that we each have is somewhat smaller (about 10^6). Each immunologically responsive B cell bears copies of a single BCR on its surface (initially composed of its VDJ + Cμ or Cδ, paired with a VJ + Cκ or VJ + Cλ chain) that can react with one antigen (or closely related group of antigens). Even after that B cell divides, all of its progenies, or **clones**, will continue to make antibodies with the same antigen specificity.

There are two steps by which B-cell precursors are "auditioned" and **selected** to become mature B cells. (This process is

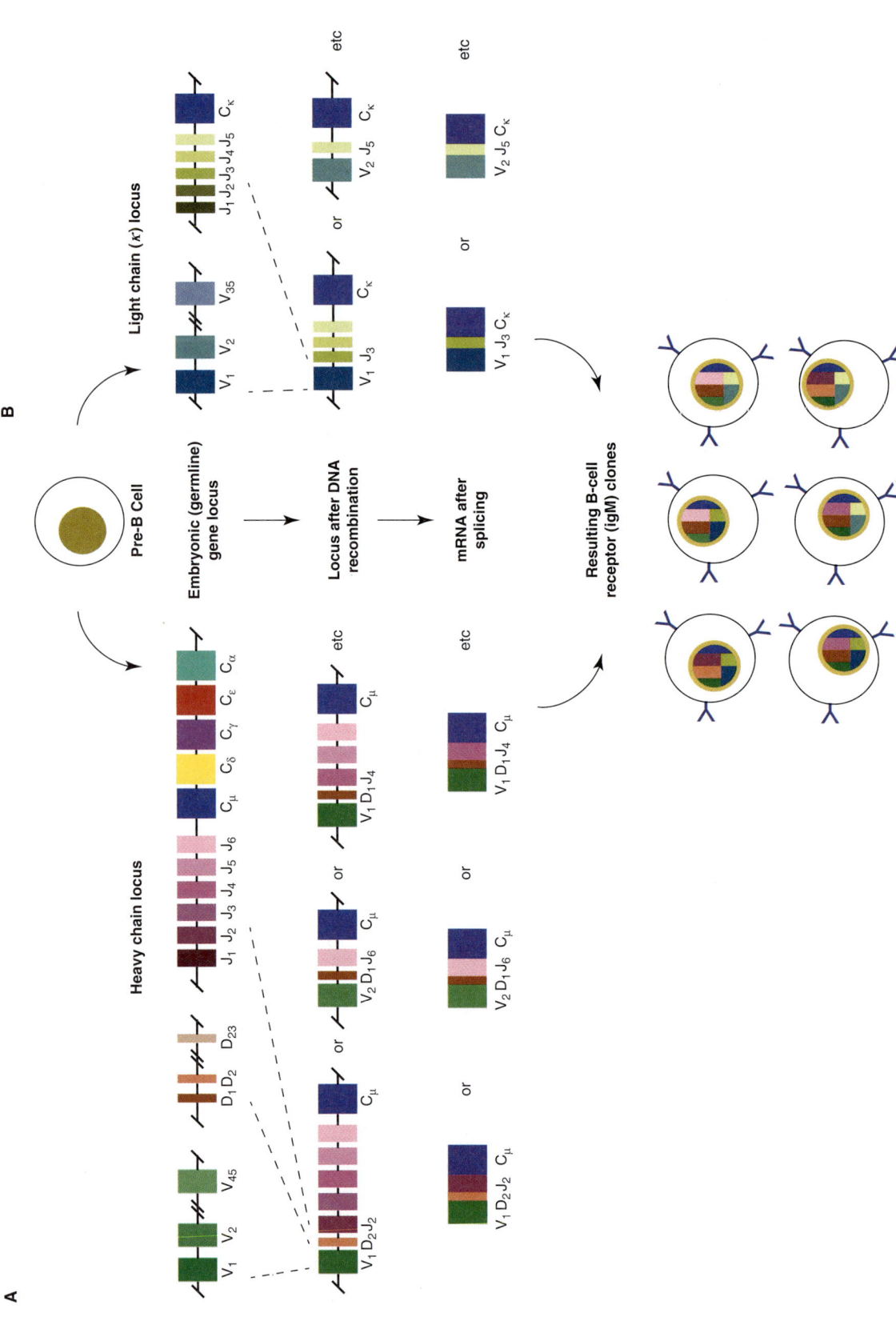

FIGURE 59-3 Producing diverse immunoglobulin (Ig) M molecules by light chain (κ) and heavy chain gene rearrangement. The pre-B cell (top) has no Ig on its surface. **A:** The heavy chain antigen-binding site is formed after recombinase activating gene (RAG) proteins make double-strand DNA breaks and one of the V_H segments, one of the D_H segments, and one of the J_H segments are chosen at random to be joined together. The heavy chain C segment determines the Ig class (i.e., isotype). (Only three VDJC examples are shown of the many possible combinations.) **B:** The κ light chain is shown. Light chain genes do not have D_H segments; the κ antigen-binding site is formed by randomly joining one of the V_H segments and one of the J_H segments. There is only one C segment for the κ gene. (Only two VJC examples are shown of the many possible combinations.) In both heavy chain and light chain gene recombination, the unused intervening DNA is discarded. After transcription and splicing, one possible heavy chain and one possible light chain mRNA are translated to produce a single species of IgM molecule. C = constant segments; D = diversity segments; J = joining segments; V = variable segments.

similar to that of T-cell clonal selection, described later in this chapter and in Figure 59–6, but selection of B cells occurs in the bone marrow rather than in the thymus.) The first step of B-cell clonal selection is called **positive selection**. Pre-B cells lack surface BCR. If a B-cell precursor fails to rearrange its immunoglobulin gene segments and generate a functional BCR, it dies before it reaches the mature B-cell stage. This is called positive selection because only those cells that **do generate a BCR** are allowed to survive and mature. For example, mutations in the genes encoding the **recombinase enzymes** (see above) result in a failure to generate antigen receptors and therefore a deficiency of lymphocytes (**severe combined immunodeficiency**).

Similarly, a mutation in the gene on the X chromosome that encodes **BTK**, which is important for transmitting the BCR signal from the cell surface, results in the disease **X-linked agammaglobulinemia**, in which B cells and antibodies are absent. These patients are more susceptible to bacterial infections in their sinuses, lungs, and gastrointestinal tract because they lack the antibodies that usually protect these barrier surfaces (see Chapter 68).

Pre-B cells that do successfully generate surface IgM pass through positive selection and progress to become B cells. At this stage, their IgM BCRs immediately encounter self-antigens. Remember that, whereas TCRs can only bind peptides complexed with major histocompatibility complex (MHC) proteins, the BCR can potentially bind to **any circulating proteins, lipids, carbohydrates, or nucleic acids**. However, because this phase of development occurs in the bone marrow rather than in the peripheral tissues or secondary lymphoid organs, **all of the antigens that the B cell could encounter at this stage are self-antigens**. During this phase, called **negative selection**, if the BCR strongly binds a self-antigen, this indicates high potential for *autoreactivity*.

A cell with this BCR will be removed from the pool of mature B-cell clones, although it has one chance to escape this fate by a process called **receptor editing**: an alternate V_L combination using an unused light chain allele can replace the previous allele, creating a new IgM receptor. But if this receptor is also autoreactive, the B cells are either killed by **apoptosis** or rendered **anergic** (their production of surface IgM is turned off and they become insensitive to activation). It is estimated that 25% to 50% of circulating B cells have undergone receptor editing. This phase is called **negative selection** because it ensures that only B cells that do **not** strongly bind self-antigens are allowed to leave the bone marrow and, therefore, will be **self-tolerant**.

Class Switching

Initially, all B cells that exit the bone marrow carry IgM specific for antigen. At this stage, they may be considered **mature**, because they have a functional BCR, but **naïve**, because they have not yet encountered their cognate antigen. Later, in a process called **class switching**, further gene rearrangement enables new antibodies that use the same V_H but different C_H chains. (In Chapter 61, we will describe how activation of B cells causes this class switching and the function of the different Ig classes.) A B cell that has class switched from IgM can never go back.

Allelic Exclusion

A single B cell has one maternal and one paternal copy of the L chain genes (both κ and λ) and the H chain gene. As described earlier, B cells that recognize self-antigens during clonal selection can attempt "receptor editing," swapping in the alternate allele, to escape apoptosis or anergy. But once they have succeeded in exiting the bone marrow as a mature B cell, the alleles that gave them the successful BCR are fixed and the others are silenced. This is called **allelic exclusion**. We all have a diverse mixture of B-cell clones expressing different combinations of paternal and maternal genes. The precise mechanism of how the alternate alleles are turned off is unknown.

T CELLS

Like B cells, T cells derive from common lymphoid progenitors. But unlike B cells, T-cell development includes a step in which the precursors migrate through a specialized organ called the **thymus**, which is why they are abbreviated "T" cells. It is during passage through the thymus that each T-cell precursor begins to express a **unique antigen receptor (TCR)**. The thymus is required for normal development of T cells, as patients with a congenital disease called **DiGeorge syndrome**, who are born without a thymus, are T-cell deficient and die at an early age of infection if they are not treated (see Chapter 68). Upon exiting the thymus, the mature T cells are called **naïve** because they have not yet seen foreign antigens.

T-Cell Receptor Structure

Earlier, we described BCRs as having two identical light chains (lambda or kappa) and two identical heavy chains (mu, delta, gamma, alpha, or epsilon). With rare exceptions, which we will discuss later in the chapter, the TCR is composed of a single α (**alpha**) chain and a single β (**beta**) chain (Figure 59–4 and Table 59–2). Each chain includes a **variable** region, which includes the **hypervariable** region that binds to the peptide–MHC complex, and a **constant** region, which attaches the α chain and β chain to each other.

The α chain and β chain are fixed to the cell membrane by a transmembrane domain and a short cytoplasmic tail. The tail binds to a molecule called CD3ζ (**CD3-zeta**). Although it is not actually part of the antigen receptor, **all T cells have CD3 proteins in association with TCR**. The purpose of CD3 is to transmit the TCR peptide recognition signal from the surface to the inside of the cell. This is achieved through intracellular tyrosine kinases that are bound to CD3 and phosphorylate downstream second messengers.

T cells are named by markers we can detect on their cell surface, called **cluster of differentiation (CD)** markers: helper T cells are CD4-positive (CD4+), whereas cytotoxic T cells are

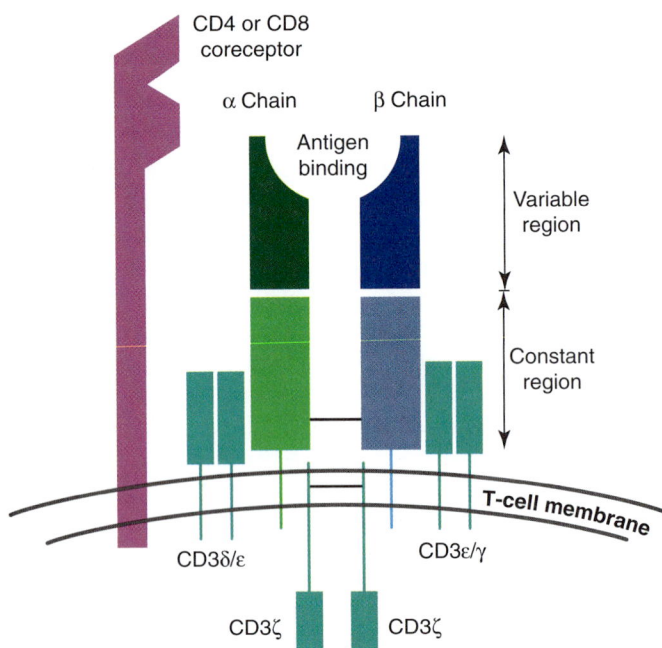

FIGURE 59–4 Schematic of T-cell receptor (TCR) structure. The TCR is composed of one alpha (α) and one beta (β) chain, each of which has a variable and a constant region. Upon TCR stimulation, the TCR associates with delta (δ), epsilon (ε), gamma (γ), and zeta (ζ) subunits of the CD3 molecule that transmits the signal into the cell through tyrosine kinases. CD4 and CD8 are coreceptors that participate in TCR signaling by binding to the peptide-bound major histocompatibility complex (MHC) molecule (not shown).

CD8-positive (CD8+). The CD4 and CD8 proteins are **coreceptors** for the TCR; they sit in the T-cell membrane and bind to nonpolymorphic regions on MHC (class II and class I, respectively). The cytoplasmic domains of CD4 and CD8 amplify the TCR signal transmission, also through a cytoplasmic tyrosine kinase (see Chapter 60).

T-Cell Receptor Genes

Each T-cell precursor creates a different, highly specific TCR. Rearrangement of **variable, diversity**, and **joining** gene segments, analogous to those that encode the B-cell Ig receptor, accounts for the remarkable ability of T cells to recognize millions of different antigens.

Construction of the TCR starts with recombination of the β chain, located on human chromosome 7. Like the heavy chain genes of immunoglobulins, The TCR is composed of V, D, and J segments that are selected and joined at random through a process that requires the **recombinase enzymes (RAG1** and **RAG2)**. One of approximately 48 Vβ segments, one of two Dβ segments, and one of approximately 13 Jβ segments are randomly selected and translocated next to one another, followed by translocation close to one of two Cβ segments. The new V + D + J + Cβ gene is then transcribed, and its mRNA is spliced and translated into a functional TCR β chain (Figure 59–5).

A similar process occurs to recombine the TCR α chain locus, although, like the light chain genes of BCRs, the α chain

gene of the TCR has no "D" regions. The α chain is therefore constructed by translocating one of approximately 45 Vα segments (selected at random) adjacent to one of approximately 50 Jα segments (also selected at random), a process that also requires RAG recombination. (Unlike the BCR light chain genes, which occur at two separate κ and λ loci, there is only one TCR α chain gene locus, on human chromosome 14.) Next, the randomly assembled V + J segments are translocated close to a Cα region, and after splicing the resulting mRNA, the TCR α chain protein is synthesized.

As occurs in B cells during V(D)J recombination, when the RAG proteins recombine the DNA of the TCR α and β chain gene loci, the unused pieces of DNA are permanently excised, with their ends joined together to form rings. These rings of DNA are called **T-cell receptor excision circles (or TRECs)** (see Figure 59-6). TRECs are easily detected using a polymerase chain reaction (PCR) assay that amplifies them in the blood of people with *normal* T-cell development.

In fact, many public health departments in the United States now use this assay in all newborns as a screening test for T-cell deficiency; because over 70% of T cells generate TRECs as a byproduct of their TCR, the *absence* of these TRECs prompts further testing to find out why a baby's T cells are failing to reach this stage. This test is highly cost-effective because detecting T-cell deficiency at the time of birth allows doctors to anticipate immunodeficiency and prepare for a stem cell transplant *before* the baby has severe infectious complications (see Chapter 68).

Positive and Negative Thymic Selection

Exactly analogous to B-cell development, each T-cell precursor in the thymus has a unique TCR, and, even after it divides, its progeny, or **clones**, will carry an identical TCR. Because of the large number of Vα and Vβ segments, including those that come from the paternal and maternal copies of the α and β chain genes, TCR recombination can theoretically generate 10^{13} different receptor combinations! (But we only have around 10^7 T cells in our bodies.) Before they can exit the thymus, T-cell precursors are subjected to **thymic selection**, a rigorous two-step "audition," analogous to that described earlier for B cells, that generates functional T cells ready to identify and respond to foreign (but not "self") antigen (Figure 59–7).

(1) First, T-cell precursors in the **thymus cortex** migrate past specialized thymus cells bearing MHC proteins and presenting "self"-peptides. The precursors, which start *without* CD4 or CD8 (i.e., **double-negative**), begin to express *both* CD4 and CD8 (becoming **double-positive**) and recombine their TCR genes to generate unique TCRs. If these double-positive cells bind to the MHC proteins they encounter, they are given a necessary survival signal through their newly formed TCR, leading them to divide into a population of **clones** (see Figure 59–6A). This is called **positive selection**, because only the T-cell precursors that **do bind** to MHC are chosen to survive. (The self-peptides presented at this stage serve as a proxy for the diversity of peptides that the mature T cells will encounter when they leave the thymus.)

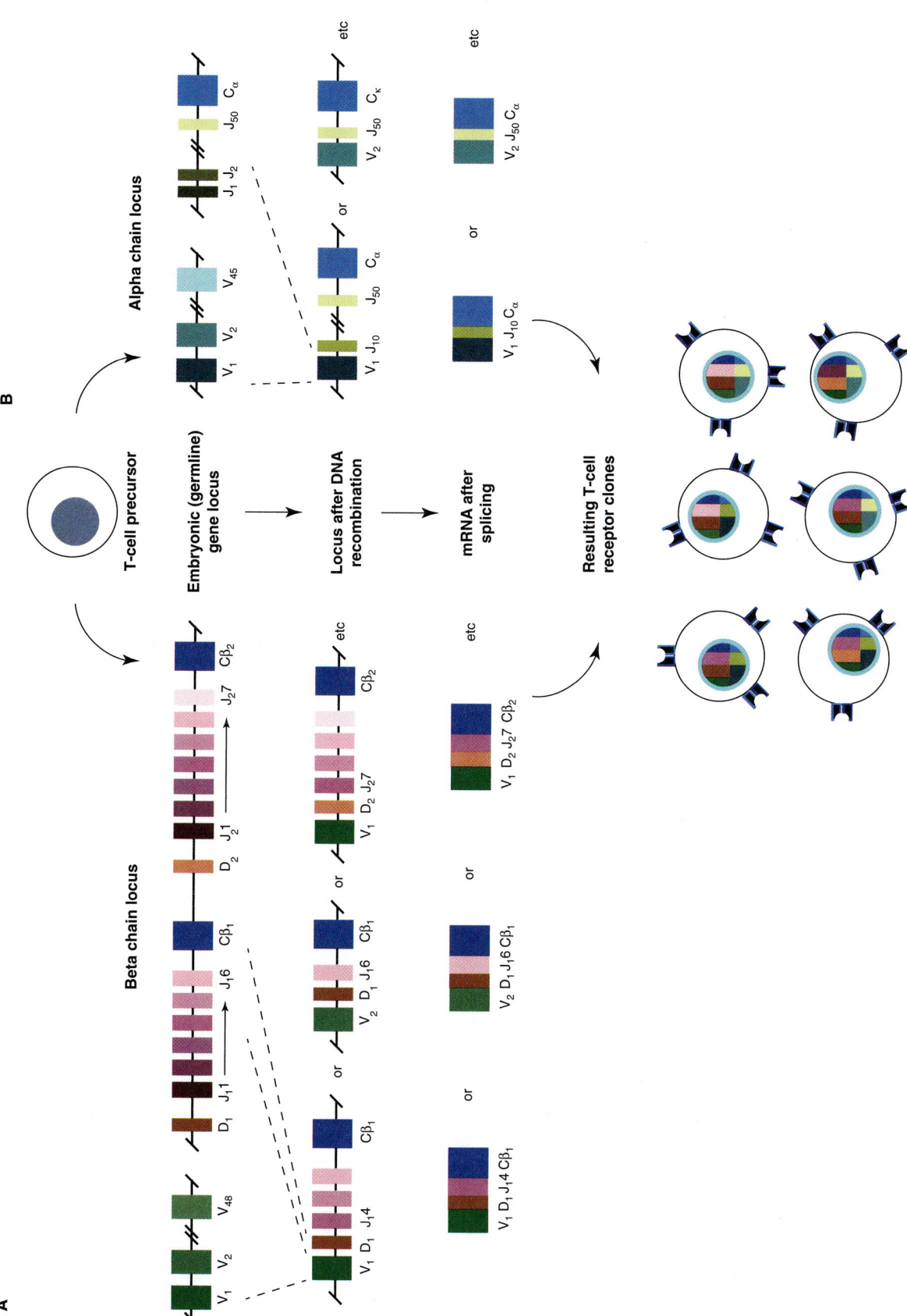

FIGURE 59–5 Producing diverse T-cell receptors (TCRs) by alpha (α) and beta (β) chain gene rearrangement. The T-cell precursor (top) has no TCR. **A:** The beta chain is formed after RAG proteins make double-strand DNA breaks and one of the Vβ segments, one of the Dβ segments, and one of the Jβ segments are chosen at random to be joined together with a corresponding Cβ segment. (Only three VDJC examples are shown of the many possible combinations.) **B:** The alpha chain does not have D segments; its antigen-binding site is formed by randomly choosing one of the Vα and one of the Jα segments and translocating them close to the Cα segment. (Only two VJC examples are shown of the many possible combinations.) The resulting gene transcript is spliced to encode one TCR β chain and one TCR α chain per cell. Note that the excised pieces of DNA form T-cell receptor excision circles (TRECs) that are discarded from the genome. C = constant segments; D = diversity segments; J = joining segments; V = variable segments.

519

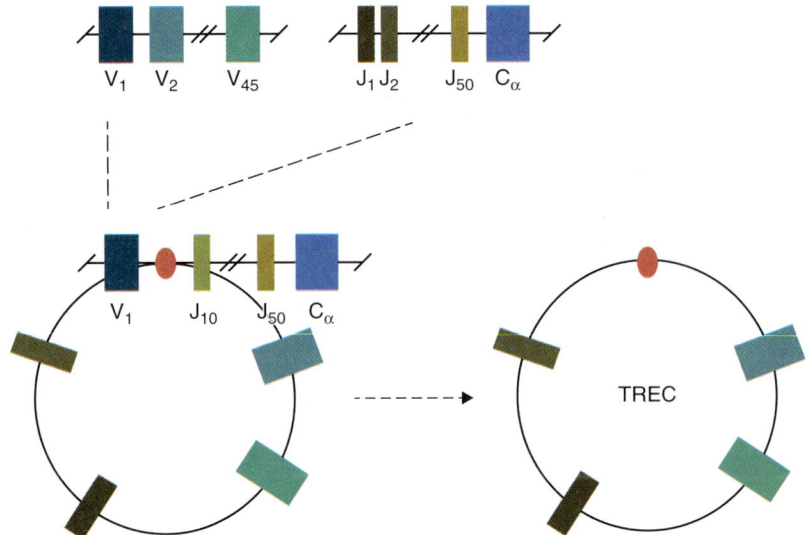

FIGURE 59-6 During recombination of the TCR alpha chain locus, a random V segment is brought into proximity to a random J segment. To make the double-strand break, the intervening genome forms a circular piece of DNA, which is discarded as a T-cell receptor excision circle (TREC). Note that the specific segments included in the TRECs are not labeled because they will vary from cell to cell, but certain constant regions exist with all TRECs that can be detected by PCR.

Cells that fail this step can make further attempts; because all cells have maternal and paternal alleles of their TCR genes, a T cell stuck at this stage can continue to mutate and re-audition TCR genes to make a more suitable MHC-binding receptor. However, most double-positive cells do not survive positive selection. In addition, the type of MHC that is bound by the TCR during positive selection will determine which type of single-positive T cell will develop; for example, if the cell binds strongly to **class II MHC**, then it turns off CD8 expression and remains (single) **CD4-positive**, and the opposite is true if the cell binds class I MHC. (This is sometimes referred to as "the rule of eight"

because CD4-positive cells bind to class II MHC [$4 \times 2 = 8$], and CD8-positive cells bind to class I MHC [$8 \times 1 = 8$]).

Why is this important? For a T cell to function, it is essential that its TCR can interact strongly with an appropriate MHC molecule, and positive selection guarantees that the T cells that eventually exit the thymus have functional MHC-binding TCRs.

(2) There is a second selection process that occurs as double-positive cells that have survived positive selection begin to move to the **thymus medulla**. As described earlier, these cells are continuing to contact thymus cells that display self-antigens

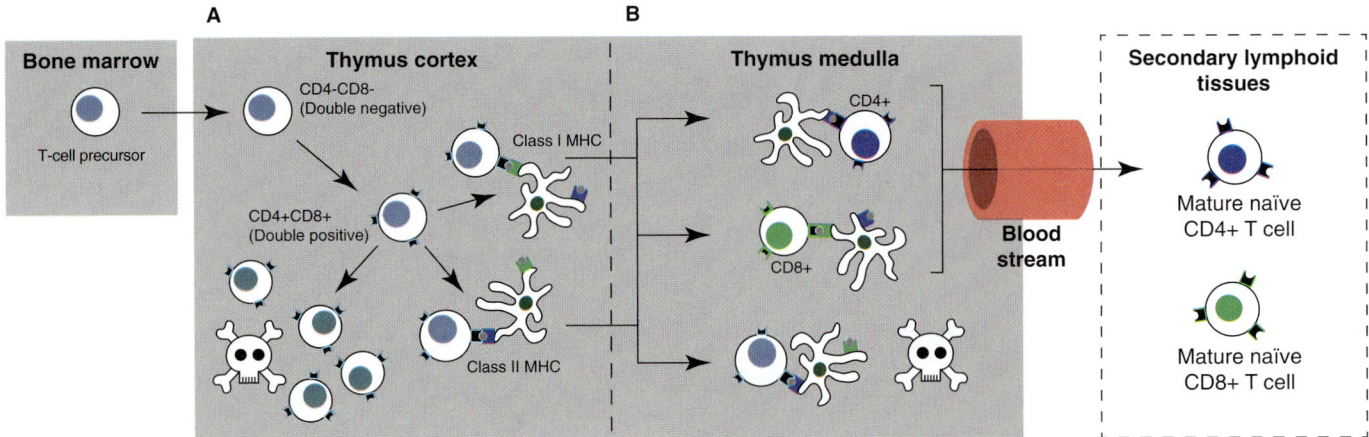

FIGURE 59–7 Thymic selection generates mature naïve T cells. **A: Positive Selection.** T-cell precursors arrive in the thymus cortex lacking CD4 and CD8 (double-negative) and lacking a T-cell receptor (TCR). They then become CD4+CD8+ (double-positive) while rearranging their TCR genes. Next, double-positive cells meet specialized thymus cells expressing a wide range of self-peptides complexed with class I and class II major histocompatibility complex (MHC). Only a few cells are able to make functional contacts with these MHC complexes and will receive survival signals (positive selection). **B: Negative Selection.** The surviving cells move into the thymus medulla, where those that make strong contacts with self-antigens are killed by apoptosis (negative selection). The remaining cells that weakly bind self-antigens survive. These cells exit the thymus as mature naïve T cells and migrate throughout the circulation and secondary lymphoid tissues surveying for antigen.

complexed with class I MHC or class II MHC. Any T cell that binds *too* strongly to these self-presenting cells is deleted by a process of programmed cell death, or **apoptosis** (see Figure 59–6B). This is called **negative selection**, because only the T cells that **do not strongly bind** to self-peptides are allowed to survive. For negative selection to be efficient, the thymic cells must display a wide repertoire of self-peptides. A transcription factor called the **autoimmune regulator** (AIRE) directs this array of self-peptides in the thymus.

Why is this important? The removal of self-reactive cells ensures that the naïve T cells that eventually exit the thymus are not specific for self-antigens, and this **self-tolerance** is one of the key ways the immune system discriminates between what is self and what is foreign.

Thymic selection uses a tightly controlled signal threshold that ensures TCR binding to self MHC is strong (positive selection) and TCR binding to self-antigens is weak (negative selection). Stated another way, if TCR binding to self MHC is **too weak**, the cell is deleted, and if TCR binding to self-antigens is **too strong**, the cell is deleted. The end result is a mature T cell that binds well to self MHC and not well to self-antigen. Mutations in genes that control TCR signaling or in the gene encoding AIRE can cause **autoimmune disease** due to defective thymic selection (see Chapter 66).

INNATE-LIKE T CELLS

Approximately 95% of the body's T cells are CD4-positive or CD8-positive cells that carry αβ TCRs, as described earlier. These cells have a highly **diverse TCR repertoire**, capable of responding to a wide range of potential infectious agents. A few other T cells develop in an unusual manner: although they still pass through the thymus, they have a highly **restricted TCR repertoire**, capable of responding quickly but to a narrow range of antigens. Therefore, because these cells respond more rapidly and are less diverse than other T cells, they are often called "innate-like" (see Table 59–2).

One type of innate-like T cell is the **NK-T cell**. As their name implies, NK-T cells share many features with innate NK cells, including surface receptors and markers that are important for NK cell function (see Chapter 58). But do not confuse the two cell types! NK-T cells are **not** innate cells; they have an αβ T-cell receptor and require the thymus for their development.

The best-described NK-T cell, called the "invariant" NK-T (iNKT) cell, uses a highly limited set of V, D, and J gene segments to create the α and β receptor chains of its TCR (see Table 59–2). Instead of recognizing peptides complexed with MHC, the TCRs of iNKT cells recognize lipids and glycolipids complexed with an alternate antigen presentation molecule called CD1d. The precise role of NK-T cells is unknown, but they may be important in host defense against organisms that contain certain lipids or in responding to situations of host tissue stress in which lipids are released from damaged cells.

Another unusual T cell is the **mucosal-associated invariant T (MAIT) cell**. Like NK-T cells, MAIT cells develop in the thymus and use a limited set of α and β gene segments in their TCRs, but they are restricted to a different antigen presentation molecule called MR1 (see Table 59–2). Like class I MHC, MR1 is expressed on a large variety of cells. However, rather than only presenting peptides, MR1 activates MAIT cells with a wide range of other types of antigens. Not all of the ligands bound by MR1 that activate MAIT cell have been identified, but at least some of them are small-molecule metabolites produced by bacteria. This may explain why MAIT cells populate peripheral barrier surfaces, such as the lung, intestine, and liver, where bacterial products are often encountered.

Finally, the **gamma-delta** (γδ) **T cell** is perhaps the most unusual innate-like T cell in that it does **not** have an αβ TCR. Instead, for reasons that are not well understood, at the time when thymic T-cell precursors begin to recombine the α and β chain genes, γδ T cells instead recombine the γ and δ chain genes (located on human chromosomes 7 and 14, respectively). The γ chain gene is composed of V, D, and J segments (similar to the β chain), and the δ chain gene is composed of V and J segments (similar to the α chain). These alternate TCR chains then combine with CD3ζ on the cell surface.

Little is known about the antigens recognized by γδ TCRs, but some of them may be lipids or microbial products. Also, γδ T cells may respond to these antigens *in the absence of the usual antigen-presenting molecules*, such as MHC. Like MAIT cells, γδ T cells primarily reside in barrier tissues, such as skin and gut. In the blood of healthy individuals, γδ T cells are rare (<5% of lymphocytes), and although they increase in number in response to infections, such as tuberculosis, *Listeria*, and malaria, their precise role in host defense is unclear.

SELF-ASSESSMENT QUESTIONS

1. Regarding the genes that encode antibodies, which one of the following is most accurate?

   (A) Hypervariable regions are encoded by the genes of both the light and heavy chains.

   (B) The genes for the light and heavy chains are linked on the same chromosome adjacent to the human leukocyte antigen (HLA) locus.

   (C) During the production of immunoglobulin (Ig) G, the light and heavy chains acquire the same antigen-binding sites by translocation of the same variable genes.

   (D) The gene for the constant region of the gamma heavy chain is first in the sequence of heavy chain genes, and that is why IgG is made in greatest amounts.

2. Regarding events that occur in the thymus during the maturation of T cells, which one of the following is the most accurate?

   (A) T cells bearing antigen receptors that recognize self-antigens are deleted, a process known as "negative selection."

   (B) "Positive selection" ensures that CD4-positive T cells and CD8-positive T cells recognize antigen presented by class I major histocompatibility complex (MHC) proteins and class II MHC proteins, respectively.

   (C) T cells bearing antigen receptors that recognize antigen in association with foreign MHC proteins survive, a process known as "positive selection."

(D) Most mature T cells have both CD4 and CD8 proteins on their surface that ensures their ability to react with antigen presented by either MHC class I or MHC class II proteins.

3. Which one of the following is a mechanism used by B and T lymphocytes to recognize a diverse range of microbes?

 (A) A receptor encoded from birth in the germline

 (B) A receptor that recognizes molecular motifs common among many different microbes

 (C) A process of clonal selection that eliminates self-reactive cells

 (D) A process of DNA recombination that generates clones with unique antigen receptors

 (E) A process that shuts off alternate alleles

4. Which one of the following lists the components of mRNA that might be found in a mature naïve B cell in the secondary lymphoid tissue?

 (A) mRNA containing V, D, J, and Cμ segments; mRNA containing V, J, and Cκ segments; and mRNA containing V, J, and Cλ segments

 (B) mRNA containing V, D, J, and Cμ segments and mRNA containing V, J, and Cκ segments

 (C) mRNA containing V, D, J, and Cγ segments and mRNA containing V, D, J, and Cκ segments

 (D) mRNA containing V, J, and Cγ segments and mRNA containing V, D, J, and Cλ segments

 (E) mRNA containing V, D, J, and Cγ segments; mRNA containing V, J, and Cκ segments; and mRNA containing V, J, and Cλ segments

 (F) mRNA containing V, J, and Cκ segments and mRNA containing V, J, and Cλ segments

5. You are seeing a child with a suspected immunodeficiency, and on testing, you find that he has absent B cells and undetectable antibody levels but slightly elevated numbers of T cells and NK cells. What component of lymphocyte development is most likely defective?

 (A) Abnormal assembly of the lambda and kappa light chains

 (B) Abnormal function of the recombinase enzymes encoded by recombinase activating genes

 (C) Abnormal development of the thymus

 (D) Abnormal survival and differentiation of common lymphoid progenitor cells

 (E) Abnormal expression of MHC proteins

6. In many states, newborn children are screened for immunodeficiency by testing for:

 (A) Mutations in Bruton's tyrosine kinase (BTK) gene

 (B) T-cell excision circles

 (C) Concentration of IgG in blood of newborn

 (D) Mutations in recombinase activating gene (RAG)

ANSWERS

(1) **(A)**
(2) **(A)**
(3) **(D)**
(4) **(B)**
(5) **(A)**
(6) **(B)**

PRACTICE QUESTIONS: USMLE & COURSE EXAMINATIONS

Questions on the topics discussed in this chapter can be found in the Immunology section of Part XIII: USMLE (National Board) Practice Questions starting on page 757. Also see Part XIV: USMLE (National Board) Practice Examination starting on page 775.

Adaptive Immunity: T-Cell–Mediated Immunity

60

Innate immunity (see Chapter 58) and antibodies (see Chapter 61) are important mechanisms for preventing infections from taking hold, but in many infectious diseases, it is primarily the T cells that orchestrate resistance and recovery. Furthermore, T cells are important for cancer surveillance, and they are responsible for most autoimmune diseases and rejection of organ transplants. The strongest evidence for the importance of T cells comes from the increase in infections and cancers that occurs when T-cell function is reduced by immunosuppressive drugs, by acquired diseases such as human immunodeficiency virus (HIV), or in congenital (primary) immunodeficiency syndromes.

The constituents of the T-cell–mediated immune system include several cell types:

(1) **Macrophages** and **dendritic cells (DCs)**, which phagocytize microbes and present antigens to T cells (see Chapter 58)

(2) **Effector/helper CD4-positive T cells**, which use *antigen receptors* to recognize antigen and make cytokines that enhance or suppress immune functions

(3) **Cytotoxic CD8-positive T cells**, which use *antigen receptors* to detect and kill infected cells

(4) **Natural killer (NK) cells**, which detect and kill infected cells using *innate receptors.*

The major defining feature of cell-mediated immunity, covered in detail in this chapter, is that it is critically dependent on **cytokines** produced by these cells. Although the interactions

between various cells are complex, the result is relatively simple: **opportunistic microbes** only cause disease when T-cell–mediated immunity is compromised.

ACTIVATION OF T CELLS

As discussed in previous chapters, lymphocyte precursors develop into mature **B** cells and **T** cells in the bone marrow and thymus, respectively, and these are therefore called **primary lymphoid organs** (see Chapter 59). The result is an enormous diversity of adaptive immune cell "clones," and each clone has a unique and specific antigen receptor, which is either a B-cell receptor (**BCR**) or a T-cell receptor (**TCR**). At this stage, a lymphocyte is considered **mature** because it has a functional antigen receptor, but **naïve**, because it has not yet encountered an antigen that can strongly bind to its TCR or BCR. Note that only a *few lymphocyte clones* might be specific for any given antigen.

How do lymphocyte clones survey our entire barrier, bloodstream, and organs for microbial antigens? **Secondary lymphoid organs** concentrate and filter antigenic material so that immune cells can sample it and remove it if necessary. After lymphocytes complete their maturation (see Chapter 59), they exit to circulate through the secondary lymphoid organs via **blood** and **lymphatic** vessels (Figure 60–1).

Lymphatics are a specialized circulatory system parallel to the blood system, with one-way valves that keep the lymph

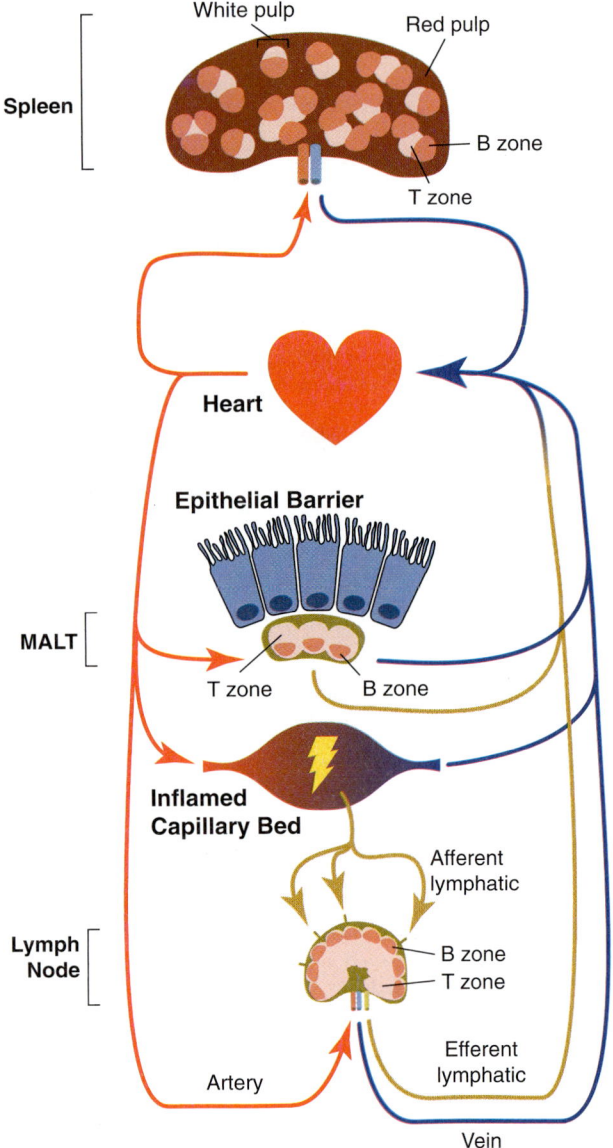

FIGURE 60–1 Schematic of the circulation through secondary lymphoid organs. (1) The spleen has arterial (red) and venous (blue) circulation. The "red pulp" filters and clears the blood of damaged red blood cells and circulating microbes, and the "white pulp" contains T and B lymphocytes that survey the blood for antigens. (2) The mucosa-associated lymphoid tissue (MALT) is in direct contact with—and shares the blood supply of—epithelial barrier surfaces. Epithelial cells shuttle microbes and antigens into the MALT, where immune cells ingest and kill microbes and survey for antigens. (3) Lymph nodes have arterial and venous blood supply but also receive lymphatic circulation from all of the body's tissues by way of *afferent* lymphatic vessels. Afferent lymphatics run alongside blood capillary beds and carry free antigens or antigen-presenting cells away from inflamed, infected tissues. (4) Cells that exit the MALT and the draining lymph nodes do so primarily via *efferent* lymphatic vessels, but a few may also leave through veins. Lymphatics move the filtered lymph fluid by means of one-way valves into progressively larger vessels that eventually join the venous circulation.

circulating in one direction. The lymphatic vessels drain all of your body's tissues, filtering foreign material through **draining lymph nodes**. The **initial phase of T-cell activation** occurs in the secondary lymphoid organs when the antigen receptors of T cells and/or B cells recognize antigens. The initial activation, or **priming**, of naïve T cells depends on antigen-presenting cells (APCs), which are usually **DCs**.

Lymphoid Tissue Architecture

Secondary lymphoid organs can be divided into zones, and chemokines direct cell migration to the appropriate zones. The **follicle** is a zone mostly made up of B cells, with an adjacent or surrounding **T-cell zone**. DCs that pick up antigens first encounter naïve T cells in the T-cell zone.

DCs process foreign proteins that are either extracellular (which they take up into vesicles) or cytoplasmic. They break these proteins down into **small peptides** and then load them onto major histocompatibility complex (MHC) proteins. The peptide–MHC complex is transported to the surface of the DC for presentation to nearby T cells.

How do DCs and T cells meet one another in the T-cell zone? T cells circulate freely through the bloodstream and lymphatics, using the **chemokine receptor CCR7** to migrate toward chemokines produced by structural fibroblast cells in the T-cell zones of the secondary lymphoid tissue (Figure 60–2). After a DC ingests a microbe, the accompanying pathogen-associated molecular patterns (PAMPs) stimulate the DC to express the same chemokine receptor, CCR7. Therefore, the chemokine gradient attracts both DCs and T cells to the T-cell zone.

Note that some DCs and macrophages wait in the outer layers of the secondary lymphoid tissue to ingest freely circulating antigens. This is particularly important in the secondary lymphoid organs that receive antigen directly. These include the **spleen**, which filters the blood, so that damaged or infected red blood cells and bloodborne pathogens can be cleared and processed for antigen presentation. This also occurs in the **mucosa-associated lymphoid tissues** (MALT), which receive antigen directly from the mucosal barrier. In **lymph nodes**, there is an additional route of entry, namely the **afferent lymphatics** that drain all tissues. Antigens can either travel through these lymphatics to be ingested and processed when they arrive in a draining lymph node, or else the DCs in the peripheral tissue can take up the antigens and carry them through the afferent lymphatics to the T-cell zone using CCR7.

T-Cell Receptor Signaling

T cells recognize *only* polypeptide antigens, in the form of short peptide chains. The specific polypeptide that binds to a TCR is called its **cognate antigen** (or **cognate peptide**). Furthermore, TCRs recognize their cognate antigens only when presented in association with MHC proteins. (As described in Chapter 59, there are rare innate-like T cells that recognize antigens presented by nonclassical MHC antigen presentation proteins, but these are minor exceptions!)

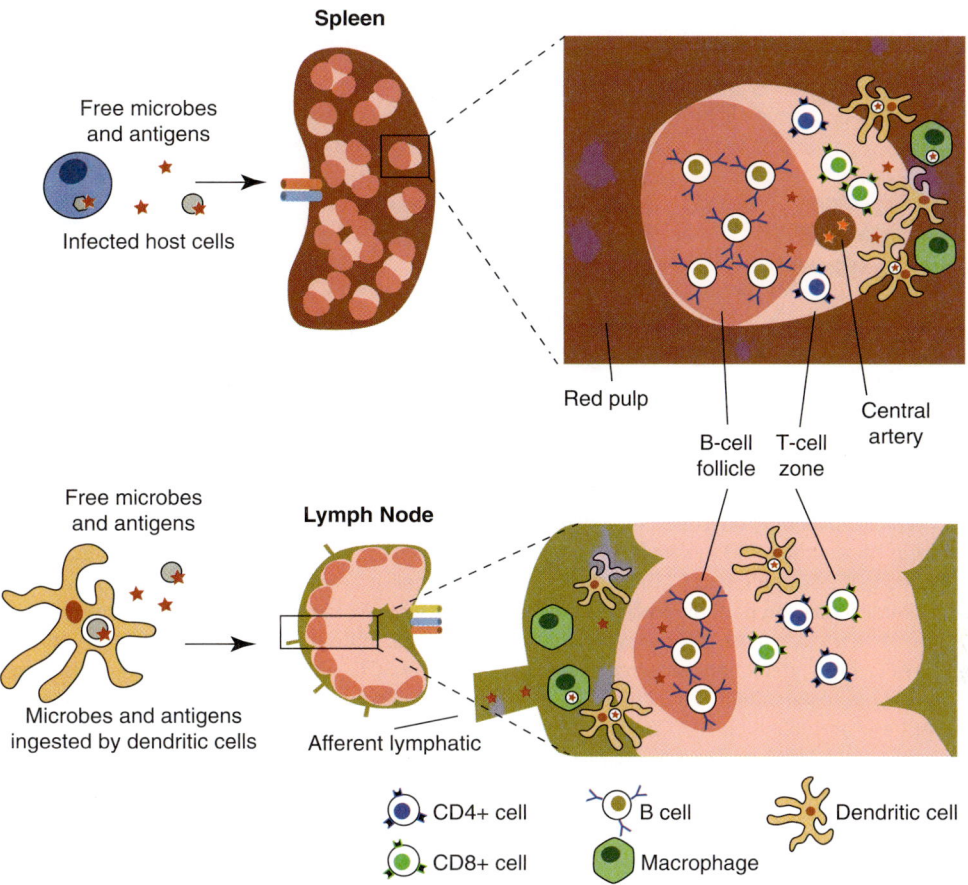

FIGURE 60–2 Antigens are delivered to naïve T cells in secondary lymphoid tissues. **Top:** The spleen filters the bloodstream. Circulating microbes (gray circle) or their antigens, (red stars), or infected host cells (large blue circle) arrive via central arteries. The antigens are either detected and engulfed by dendritic cells and macrophages or they drift into the B-cell follicle where they can be detected by B cells with surface immunoglobulin (Ig; B-cell receptor). **Bottom:** Lymph nodes drain the tissue capillary beds through afferent lymphatics. Microbes or their antigens can be engulfed by dendritic cells in the nonlymphoid tissue and transported to the lymph node via afferent lymphatics. Alternatively, free antigens can be carried through the afferent lymphatics to be deposited in lymph nodes. The antigens are either detected and engulfed by antigen-presenting cells or drift into the B-cell follicle.

Remember the "rule of eight": CD4-positive T cells recognize antigen in association with class II MHC proteins ($4 \times 2 = 8$), whereas CD8-positive T cells recognize antigen in association with class I MHC proteins ($8 \times 1 = 8$). This is called **MHC restriction** because each type of T cell is "restricted" to recognize antigen *only* when presented by the appropriate class of MHC protein. As described in Chapter 59, MHC restriction is a feature of thymic *positive selection* and is mediated by specific binding sites on the TCR as well as on the CD4 and CD8 proteins that bind to specific regions on the MHC proteins. Furthermore, the specific class I and class II genes you inherit from your parents ensure that your T cells are selected to only recognize antigen presented by your APCs (see Chapter 62).

When the TCR interacts with the peptide–MHC complex, the CD4 or CD8 protein on the surface of the T cell also interacts with the class II or class I MHC protein on the APC. This binding is reinforced by other protein interactions (e.g.,

lymphocyte function-associated antigen 1 [LFA-1] binding to intracellular adhesion molecule 1 [ICAM-1]), stabilizing the contact between the T cell and the APC. The initial activation of naïve T cells is called **priming**, and it occurs when the TCR recognizes a peptide–MHC complex presented by a DC in the T-cell zone of a secondary lymphoid organ. A series of cell–cell interactions provide **two signals** that prime the naïve T cell (Figure 60–3).

Signal 1 is the interaction of the TCR with its *cognate peptide* complexed with the MHC protein. When the peptide–MHC protein complex on the DC is strongly bound by the TCR, a signal is initiated by the **CD3** protein complex that leads to an influx of calcium into the cell. (The structure of the TCR is presented in Chapter 59.) Calcium activates calcineurin, which increases production of interleukin-2 (**IL-2**) and the **high-affinity IL-2 receptor**. (Calcineurin function is blocked by cyclosporine, one of the most effective drugs used to prevent rejection of organ transplants [see Chapter 62].)

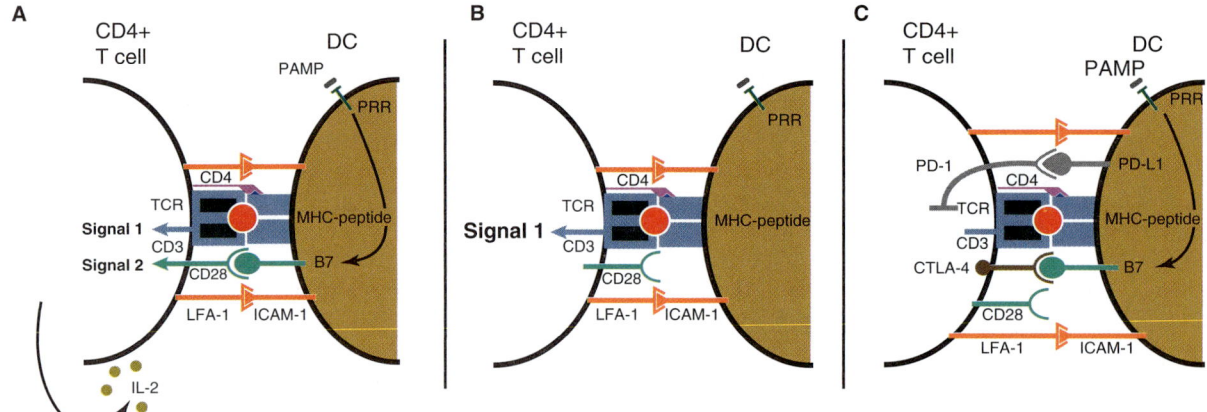

FIGURE 60–3 The signals and adhesion molecules required for initial T-cell receptor (TCR) priming. **A: Activation and Proliferation.** Signal 1 is initiated when a CD4-positive T cell's TCR (black and blue box) recognizes its *cognate antigen* (red circle) presented by class II major histocompatibility complex (MHC; blue box) on a dendritic cell (DC). The CD4 molecule (purple) acts as a co-receptor, stabilizing the TCR–MHC interaction. Signal 1 proceeds through the CD3 molecule (blue arrow), which increases LFA-1/ICAM-1 (orange) interactions, ensuring strong adhesion. Signal 2 is provided by CD28 binding to B7 (both turquoise). An antigen-presenting cell (APC) that was stimulated by pathogen-associated molecular patterns (PAMPs) through pattern recognition receptors (PRRs; green) can provide signal 2, for example, by increased B7 levels, leading to T-cell production of interleukin (IL)-2 and *clonal proliferation of the T cell*. **B: Anergy.** Without PAMP–PRR activation, B7 levels are low, so there is no signal 2. This APC–T cell interaction leads to *anergy*. **C: Suppression.** CTLA-4 on the T-cell surface is an inhibitory "checkpoint" that outcompetes CD28 because it has higher affinity for B7 molecules, but unlike CD28 does not provide signal 2. PD-1 is another inhibitor "checkpoint" that recognizes PD-L1 and acts to inhibit TCR/CD3 signaling. Note that the same processes occur in CD8-positive T-cell activation except that class I MHC presents the peptide antigen. CTLA-4 = cytotoxic T lymphocyte antigen-4; ICAM-1 = intracellular adhesion molecule 1; LFA-1 = lymphocyte function-associated antigen 1.

Co-stimulation (Signal 2)

The process of activating naïve T cells does not occur as a simple "on–off" switch. As mentioned earlier, two signals are required in the initial activation of naïve T cells, and the interactions that provide **signal 2** are called **co-stimulation**. The best example of co-stimulation is the B7 protein on a DC that interacts with the CD28 protein on the T cell (see Figure 60–3A). Resting APCs express low levels of B7 proteins, but they increase these levels upon stimulation of their pattern recognition receptors by microbial products, such as PAMPs, or adjuvants, which are nonantigenic ingredients in vaccines (see Chapter 58).

The requirement for co-stimulation is important because it prevents inadvertent T-cell activation by benign antigens.

THERAPEUTIC APPLICATIONS OF SIGNALING CHECKPOINTS

Once the T cell has received an activating signal through TCR–peptide–MHC and B7/CD28 co-stimulation, inhibitory mechanisms, called **checkpoints**, prevent unrestrained T-cell activation. A different protein called cytotoxic T lymphocyte antigen-4 (**CTLA-4**) appears on the T-cell surface, and because it has higher affinity for B7 proteins, it competes with and displaces CD28. CTLA-4 thus acts as a "brake" on T-cell activation. **Boosting** T-cell responses by blocking CTLA-4 has been a successful strategy in cancer immunotherapy, and alternatively, **suppressing** T-cell responses by administering a CTLA-4 "mimic" has been a successful strategy in treating organ transplant rejection and autoimmune disease.

In addition to CTLA-4, there is another inhibitory "checkpoint" protein on the surface of T cells called **PD-1** (**Programmed cell Death-1**). When PD-1 interacts with its ligands, **PDL-1** or **PDL-2** (also B7 family members), that are on the surface of APCs, a signal is transmitted into the T cell that blocks the phosphorylation cascade initiated by TCR/CD3 and CD28. Many tumor cells also express high levels of PDL-1. In these tumors, monoclonal antibodies that block PD-1/PDL-1 interactions enhance T-cell activation and therefore have been quite effective as anticancer immunotherapy.

Unsurprisingly, the immune-boosting benefit of CTLA-4– and PD-1–blocking drugs can occasionally cause adverse effects due to autoimmunity. Manipulating the strength of T-cell activation in this way could enable future breakthroughs in vaccine development and cancer therapy.

For example, if the TCR recognizes a cognate antigen on an APC but the co-stimulatory signal is **absent**, the T cell adopts a state of unresponsiveness called **anergy** (see Figure 60–3B). As described earlier, foreign antigens from pathogens generally contain PAMPs that stimulate the pattern recognition receptors of APCs, whereas "self"-antigens do not. Co-stimulation (signal 2) tells the T cell that its cognate antigen is being presented in an inflammatory context.

Note that Signal 1 and Signal 2 are required for activation of **both** CD4-positive and CD8-positive T cells. However, a unique characteristic of CD8-positive cells is that they require additional "help" in the form of cytokines from CD4-positive cells to become fully functional effector cells (see Figure 57–1). As discussed below, CD8-positive cells can be *extremely* lethal to host cells. Because they recognize peptide presented by class I MHC proteins, which are expressed by all nucleated cells, the additional requirement that CD8-positive cells get "help" is added insurance that the CD8-positive cells are not activated inadvertently. (The central role of CD4-positive cells in orchestrating so many different components of immune responses, as described below, explains why HIV-associated deficiency of CD4-positive cells predisposes to such a variety of severe opportunistic infections.)

T-CELL EFFECTOR FUNCTIONS

The end result of TCR stimulation is the activation of the T cell to produce various cytokines (e.g., IL-2), as well as to express the high-affinity IL-2 receptor. IL-2, also known as

T-cell growth factor, stimulates the T cells to multiply, resulting in **clonal proliferation** of a population of antigen-specific T cells. As they proliferate, different progeny cells of this clonal population take on one of a number of essential functions. Some of these cells remain in the secondary lymphoid organ, while others exit via the blood or efferent lymphatics and migrate to inflamed tissues. Figure 60–4 is an overview of the **priming** of naïve CD4-positive and CD8-positive T cells in a lymph node draining a site of infection.

T-cell functions can be divided into four main categories: CD4-positive cells become:

(1) **Effector/helper (Teff or Th) cells**, which leave the lymphoid organ and coordinate immune responses in inflamed tissue

(2) **Follicular helper (Tfh) cells**, which move into the B-cell follicle of the lymphoid organ and **help** the B cells

(3) **Regulatory T (Treg) cells**, which suppress inflammation. Finally, CD8-positive cells become

(4) **Cytotoxic T cells** (or cytotoxic T lymphocytes, usually abbreviated **CTL**), which kill virus-infected cells and tumor cells.

Figure 60–5 is an overview of the "help" provided by the various subsets of CD4-positive effector/helper (Th-1, Th-2, and Th-17) and follicular helper T (Tfh) cells.

Effector/Helper T Cells

CD4-positive T lymphocytes perform a variety of functions that **help an immune response by enhancing the functions of**

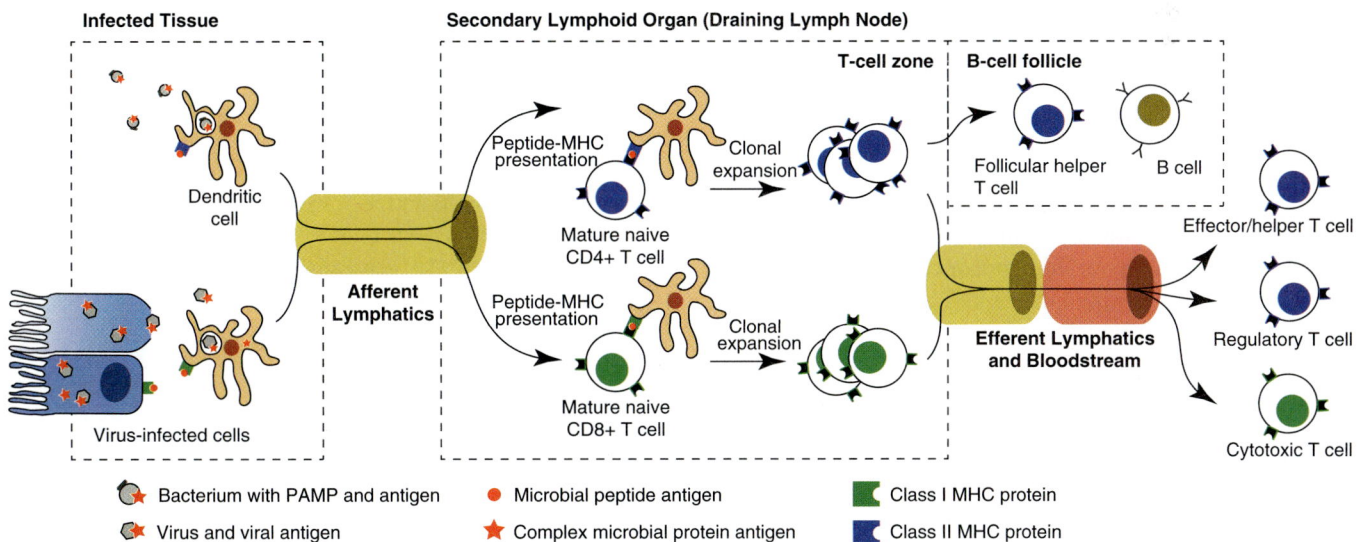

FIGURE 60–4 Overview of the initial priming and differentiation of CD4-positive and CD8-positive T cells in a secondary lymphoid organ. Dendritic cells (DC) from an infected tissue take up antigen and migrate through afferent lymphatics to the T-cell zone of a draining lymph node. Endosomal antigens are presented on class II major histocompatibility complex (MHC) proteins to CD4-positive T cells (blue), and cytosolic (e.g., viral) antigens are presented on class I MHC proteins to CD8-positive T cells (green). Naïve T cells' T-cell receptors (TCRs) recognize their *cognate* peptide antigens presented by the corresponding MHC proteins and become activated. As the T cells proliferate, they differentiate: CD4-positive cells become follicular helper (Tfh) cells, effector/helper T (Th) cells, or regulatory (Treg) cells, whereas CD8-positive cells become cytotoxic T cells. Depending on their function, differentiated T cells either stay in the lymph nodes or migrate through the lymphatics or bloodstream to sites of inflammation. PAMP = pathogen-associated molecular pattern.

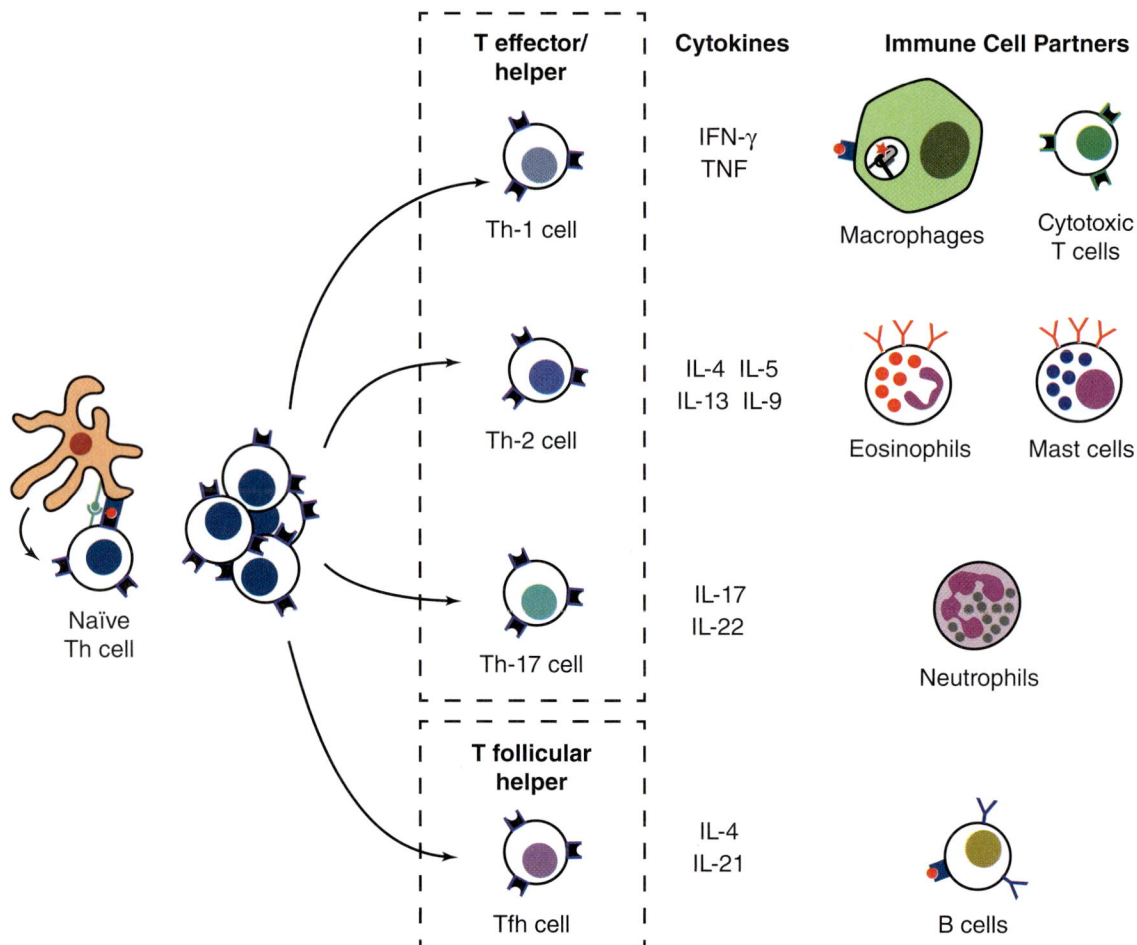

FIGURE 60–5 Subsets of CD4-positive helper T cells. After activation by a dendritic cell, in the presence of particular cytokines, a naïve CD4-positive T cell divides and differentiates into effector/helper (Th-1, Th-2 or Th-17) or follicular helper (Tfh) subsets. Each type of cell produces cytokines that "help" by activating other immune cell partners.

other cells. T effector/helper (Th) cells leave the lymph node, migrate to inflamed tissues in the body, and produce cytokines. Different infectious pathogens must be handled by the immune system in different ways. In order to provide a targeted immune defense against a specific organism, Th cells must produce the **appropriate cytokines** for the **appropriate organisms**. This programming happens when the Th cell is initially primed (i.e., when the naïve T cell Signal 1 and Signal 2 from a nearby DC), and additional signals that drive it to one of several specialist **Th subsets**. Table 60–1 lists the major innate cytokines that influence Th subset differentiation. The original naïve cell clone has the potential to become any of the subsets, but as the progeny cells divide epigenetic modifications reinforce the subset identity through **signature transcription factors** and **cytokines**.

Most of our understanding of Th cell subsets comes from studies in which Th cell clones can be transferred among genetically identical mice. Through new rounds of infectious and inflammatory stimuli, these Th cells and their progeny continue to have the same signature cytokines of their original subset. For example, all effector/helper Th cell subsets express the gene *PRDM1* (encoding the protein BLIMP-1) and have the ability to

make IL-2, migrate from the secondary lymphoid organ to the site of infection, and express further cytokines upon re-stimulation of their TCR. Table 60–2 lists the major cytokines produced by the various Th cell subsets and their main target cells.

TABLE 60–1 Innate Cytokines that Influence Th Cell Subset Differentiation

Cytokine	Cell Source	Key Biologic Effects
IL-1	Macrophages, dendritic cells	Th-17 differentiation
IL-4	Mast cells, basophils, T cells	Th-2 differentiation
IL-6	Macrophages, endothelial cells, T cells	Th-17 differentiation and activation
IL-12	Macrophages, Dendritic cells	Th-1 differentiation; in NK cells and CD8+ T cells, increases cytotoxic activity and IFN-γ production
IL-23	Macrophages, Dendritic cells	Th-17 differentiation and activation

IFN = interferon; IL = interleukin; NK = natural killer; Th = T helper.

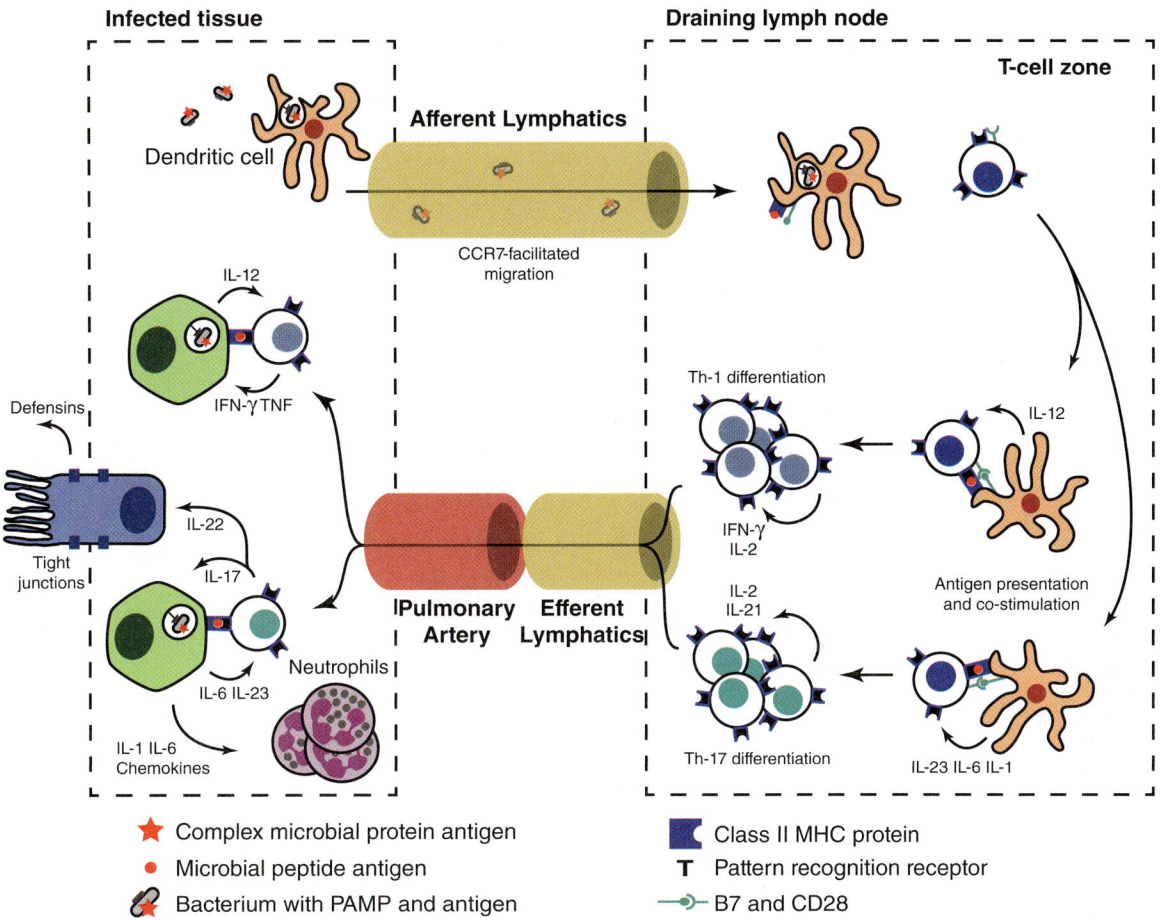

FIGURE 60–6 The Th-1 immune response protects against phagocytized bacteria during infection. Free antigen circulates through the lymph into the T-cell zone of the draining lymph node, and dendritic cells (DC) that have phagocytized bacteria migrate to the T-cell zone by increasing their surface expression of the chemokine receptor CCR7. DCs process the complex protein antigens of endosomal bacteria (red stars) and present them as short peptides complexed with the class II major histocompatibility complex (MHC) proteins to CD4-positive T cells, along with co-stimulatory B7 (turquoise). During this process, if DCs produce interleukin (IL)-12, this promotes the activated antigen-specific Th cells to differentiate into Th-1 cells as they make IL-2 and undergo clonal proliferation. Alternatively, if the DCs produce IL-23, IL-1, and IL-6, this promotes the activated Th cells to differentiate into Th-17 cells. These cells make IL-2 and undergo clonal proliferation.

(1) **Th-1 cells** are primarily responsible for "classical" activation of macrophages, leading to enhanced phagocytosis, phagolysosome free radical production, and granuloma formation (Figure 60–6 and Table 60–2).

In certain infections, the DCs produce **IL-12** at the time of Th cell activation, leading to the differentiation of these CD4-positive cells into **Th-1 cells**. Th-1 cells express the signature transcription factor *TBX21* and produce the cytokine **gamma interferon (IFN-γ)**. The activated Th-1 cell clones then move out of the lymphoid tissue, into the arterial circulation, and into the inflamed tissue by means of inflammation-induced extravasation (see Figure 58–4). There, they come in contact with **macrophages**, and after recognizing the *same peptide presented with class II MHC* by these macrophages, the Th-1 cells make more IFN-γ as well as **tumor necrosis factor (TNF)**. These cytokines activate the macrophages to be more effective killers of **phagocytized intracellular organisms** and help the macrophages form large granulomas to wall off microbes that are hard to kill.

Th-1 cells and macrophages play a role in host defense against many bacteria, fungi, and viruses, as well as against tumors, but individuals with deficiencies in IL-12 or IFN-γ are *particularly* susceptible to **mycobacterial infections**, such as **tuberculosis**. In addition to their role in controlling phagocytized pathogens, overactive Th-1 cells are associated with autoimmune and inflammatory diseases, including Crohn's disease, psoriasis, and rheumatoid arthritis.

(2) The **Th-17 cell** subset is closely related to the Th-1 cell subset but is generated by high levels of **IL-1, IL-6, and IL-23** at the time of initial activation by DCs (see Figure 60–6 and Table 60–2). Th-17 cells express the signature transcription factors *RORC* and *STAT3*, which are reinforced by autocrine signaling from the cytokine IL-21. Th-17 cells also produce the cytokines **IL-17** (the source of their name), which stimulates phagocytes and mucosal epithelial cells to increase the production of IL-1, IL-6, and neutrophil-attracting chemokines. Th-17 cells also make **IL-22**, which stimulates mucosal epithelial cells to increase the production of antimicrobial defensins and tight

TABLE 60–2 **Important Effector Cytokines Produced by T Cells**

Cytokine	Key Cell Source	Key Target Cells and Biologic Effects
IL-2	All T cells	T and NK cells: proliferation Treg cells: differentiation, survival, and function
IL-4	CD4+ T cells (Th-2 and Tfh), primarily in the secondary lymphoid tissue	B cells: isotype switching to IgE T cells: reinforce Th-2 cell differentiation Other cells: same effects as IL-13 below
IL-5	CD4+ T cells (Th-2)	Eosinophils: survival, maturation, and mobilization from bone marrow
IL-10	CD4+ T regulatory cells (Treg)	Myeloid cells: suppression of co-stimulatory signals and proinflammatory cytokines
IL-13	CD4+ T cells (Th-2), primarily in the nonlymphoid tissue	Epithelial cells: mucus Eosinophils: chemokines Smooth muscle cells: hypercontractility Macrophages: "alternative" activation and collagen deposition
IL-17	CD4+ T cells (Th-17)	Macrophages: increased chemokine and cytokine production (e.g., IL-1, IL-6, TNF) Epithelial cells: increased IL-1 Neutrophils: chemokines, antimicrobial peptides
IL-21	CD4+ T cells (Th-17 and Tfh)	B cells: isotype switching to all Ig subclasses and affinity maturation Tfh cells: reinforces Tfh cell differentiation Th-17 cells: reinforces Th-17 differentiation
IL-22	CD4+ T cells (Th-17)	Epithelial cells: increased production of defensins and tight junction adhesion proteins
IFN-γ (i.e., type II interferon)	T cells (Th-1, CD8+ cells, Tfh), NK cells (and others)	B cells: isotype switching to opsonizing and complement-fixing IgG subclasses; increased MHC class I and class II expression T cells: reinforces Th-1 cell differentiation Macrophages: "classical" activation, enhanced phagocytosis, and intracellular killing
TNF	T cells (Th-1, Th-17)	Endothelial cells: inflammatory chemokines Macrophages: activation Neutrophils: activation

IFN = interferon; Ig = immunoglobulin; IL = interleukin; NK = natural killer; Tfh = follicular helper T cell; Th = T helper cell; TNF = tumor necrosis factor; Treg = regulatory T cell.

junction proteins. Together, the cytokines from Th-17 cells and the neutrophils they recruit **defend the barrier tissues against bacterial and fungal infections**.

Patients with genetic mutations causing IL-17 deficiency have particular susceptibility to mucocutaneous infections from the yeast *Candida albicans*. In addition, loss of Th-17 cells in HIV disease is associated with chronic translocation of small numbers of bacteria from the intestinal lumen across the bowel wall and into the portal circulation. Like Th-1 cells, overactive Th-17 cells are associated with autoimmune and inflammatory diseases.

(3) The **Th-2 cell** subset is most commonly associated with infection by certain helminth worms, such as *Schistosoma* and *Strongyloides*, which have a tissue-invasive stage of their life cycle (Figure 60–7 and Table 60–2). The signature Th-2 transcription factor is *GATA3*, and the signature Th-2 cytokines are **IL-4** and **IL-13**, two cytokines that share the same receptor and therefore have similar effects. These cytokines increase the production of mucus by goblet cells at barrier surfaces, cause smooth muscle cells to be hypercontractile, and cause "alternative" activation of macrophages, leading to collagen deposition often seen in wound healing. IL-4 also signals in autocrine fashion to reinforce the Th-2 transcriptional program (analogous to IL-21 for Th-17 cells). Th-2 cells also make **IL-5**, which is the specific factor that recruits and maintains eosinophils, and **IL-9**, which activates mast cells.

When dysregulated, Th-2 cells cause allergic disease, such as atopic dermatitis, allergic asthma, and eosinophilic gastrointestinal disease. As shown in Figure 60–7, another important part of the Th2 immune response is IgE, which mast cells and eosinophils use to detect antigen. Tfh cells are likely the main source of IL-4 that helps B cells mature to become IgE-producing plasma cells.

Follicular Helper T Cells

Tfh cells differentiate from naïve CD4-positive T cells like the other activated T cells, but rather than disperse from the lymphoid tissue to other sites, they migrate into the B-cell follicles (see Figure 60–7 and Table 60–2). **The positioning of Tfh cells within the lymphoid tissue is absolutely critical to their function**, as it dictates which cells they will encounter. They find the follicle by downregulating the chemokine receptor CCR7 and upregulating the chemokine receptor CXCR5, which senses chemokines produced by the stromal cells of the follicle. The earliest signals that start the Tfh cell transcriptional program are unknown, but as they proliferate and migrate to the follicle, Tfh cells begin to express the transcription factor *BCL6* and suppress expression of *PRDM1* (the gene encoding BLIMP-1). In addition, Tfh cells produce **IL-21**, which, in an autocrine fashion, increases their own BCL-6 and CXCR5 levels.

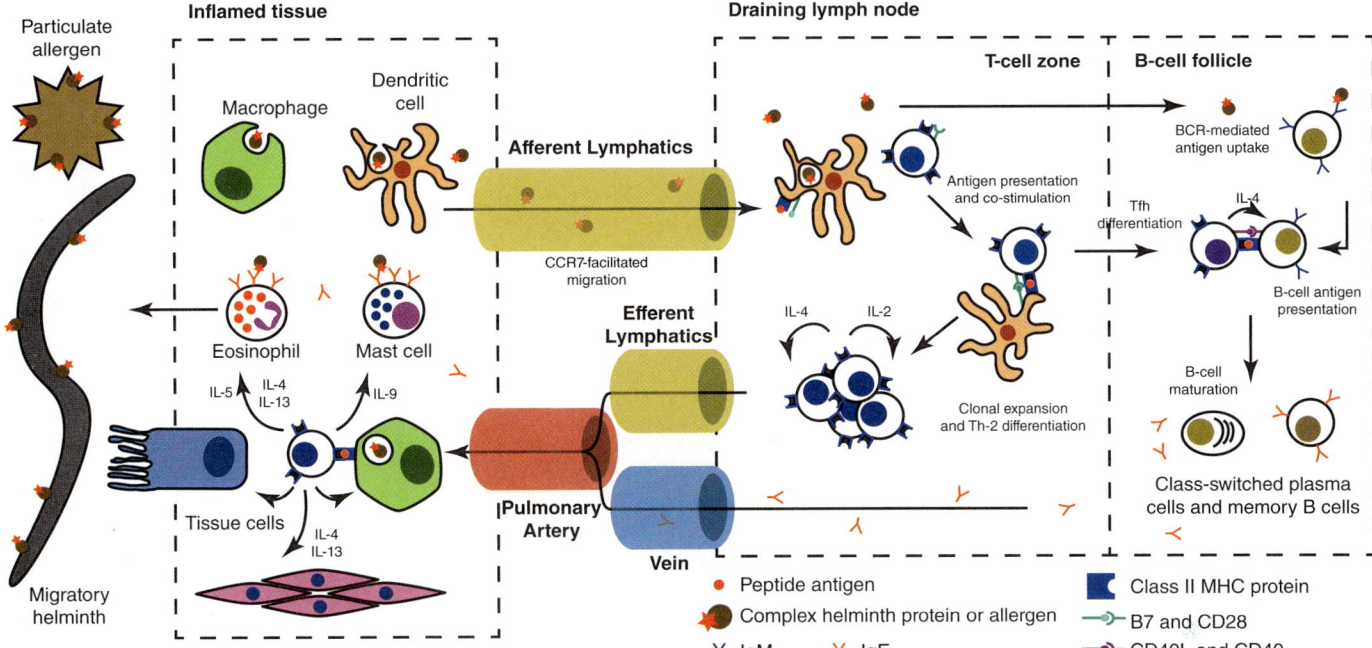

FIGURE 60-7 The Th-2 immune response reacts to migratory helminths and allergens. In this example, free antigen circulates from the inflamed tissue through the lymph into the T-cell zone of the draining lymph node. Dendritic cells (DCs) that have phagocytized bacteria also migrate to the T-cell zone by increasing their surface expression of the chemokine receptor CCR7. Some free antigens also circulate to the B-cell follicle, where they are taken up by the B cells through B-cell receptor (BCR)-mediated endocytosis. DCs and B cells process the complex protein antigens of endosomal bacteria (red stars) and present them as short peptides complexed with the class II major histocompatibility complex (MHC) proteins to CD4-positive T cells (blue). The DCs also produce co-stimulatory B7 (turquoise). A few of the naïve CD4-positive clones differentiate into T follicular helper cells, which migrate to the B-cell follicle and provide help for naïve B cells (yellow) in the form of CD40-ligand (CD40L) and interleukin (IL)-4.

Tfh cells help B cells primarily through the production of cytokines such as IL-4 and IL-21 and also by expression of CD40 ligand (**CD40L**) on the Tfh cell surface, which interacts with **CD40** on the surface of the B cell. (For more details on how Tfh cells promote B-cell activities, see Chapter 61.) Because of their central role in B-cell function, Tfh cells are particularly important for antibody responses, including responses to vaccines. A mutation in the gene encoding CD40L causes **hyper-IgM syndrome**, in which B cells are unable to "class switch" from IgM to the more mature immunoglobulin isotypes, as discussed in Chapter 68. Inappropriate Tfh activation can lead to self-reactive antibodies in autoimmune disease.

Regulatory T Cells

Normal immune responses can become pathologic if they are unrestrained. This can cause tissue damage, either from excessive inflammation at the site of an infection or from inappropriate activation of self-reactive (i.e., autoimmune) adaptive cells. The subset of CD4-positive cells called suppressor or **regulatory T cells** (**Tregs**) is responsible for limiting immune responses and maintaining **tolerance** of **self-antigens** and harmless **commensal antigens**. Some Tregs are programmed to be suppressive cells in the thymus, whereas others are programmed at the time of priming in secondary lymphoid organs.

The Treg cell subset expresses the signature transcription factor *FOXP3*. Patients with mutations in this gene have absent Tregs, enlarged secondary lymphoid organs, and severe autoimmune disease in numerous tissues. There are multiple mechanisms by which Tregs probably suppress immunity, although the most well-described mechanisms involve **inhibiting Th and CTL activation** (see Chapter 66). There is significant interest in targeting these pathways because *boosting* Treg function could be beneficial for transplant rejection and autoimmune disease, and *inhibiting* Treg function could be beneficial for cancer immunotherapy and chronic infections.

Cytotoxic T Cells

CD8-positive cytotoxic (or cytolytic) T lymphocytes (CTLs) are particularly effective at killing virus-infected cells. Mature naïve CD8-positive T cells arise in the thymus and recognize non–self-peptide antigens complexed with class I MHC proteins (see Chapter 59). Because **all** nucleated cells express class I MHC, **all** nucleated cells bearing the cognate peptide for that CTL are a potential APC. To prevent inadvertent activation of self-reactive CTLs, there is an additional requirement that, during their activation, CTLs receive IL-2 produced by nearby CD4-positive T cells also undergoing activation.

For example, if a virus (e.g., influenza virus) infects and lyses a respiratory epithelial cell, virus particles (virions) are

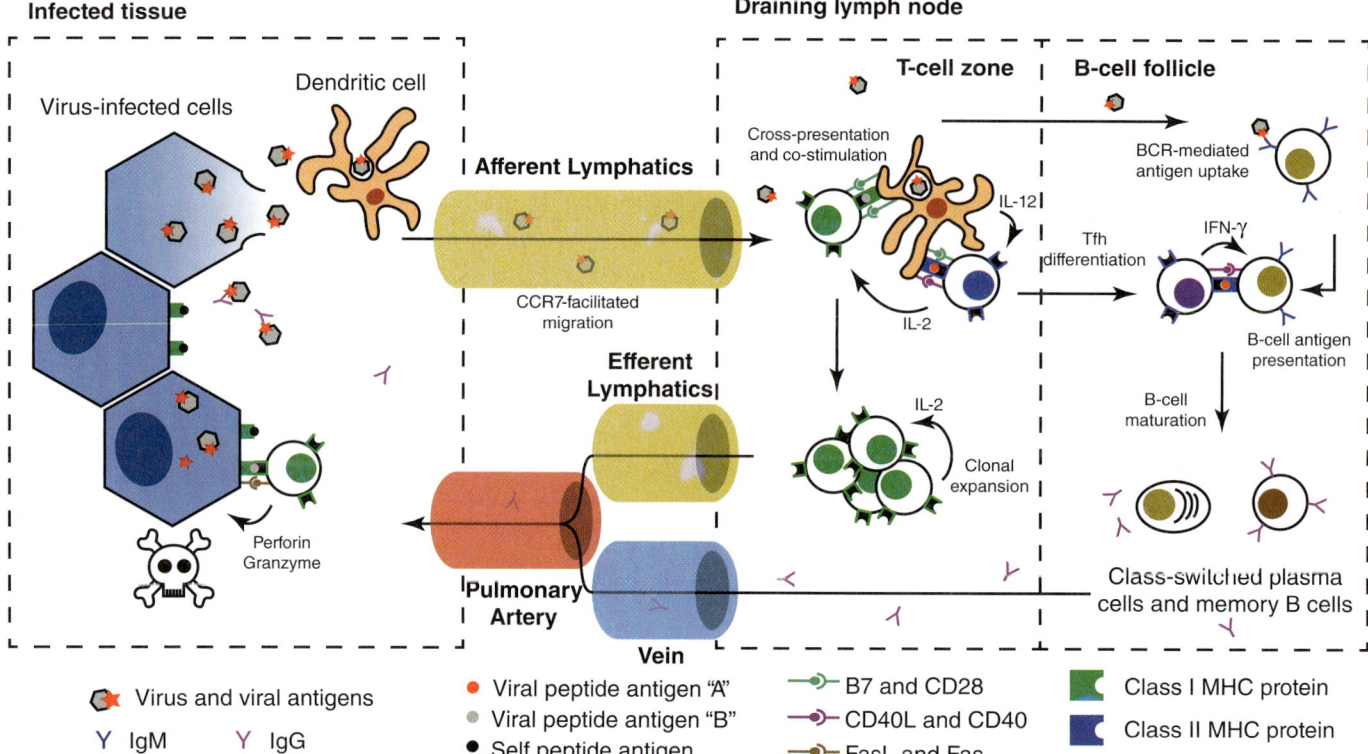

FIGURE 60–8 Cytotoxic T lymphocyte (CTL) activation and killing of virus-infected cells. In this example, virus-infected cells release free virus, which circulates from the inflamed tissue through the lymph into the T-cell zone of the draining lymph node. Dendritic cells (DCs) that engulf virus also migrate to the T-cell zone by increasing their surface expression of the chemokine receptor CCR7. Some free virus particles also circulate to the B-cell follicle, where they are taken up by the B cells through B-cell receptor (BCR)-mediated endocytosis. DCs and B cells process the complex viral protein antigens (red stars) and present short peptide "A" (red circle) complexed with class II major histocompatibility complex (MHC) proteins to CD4-positive T cells (blue). DCs also cross-present peptide "B" (gray) with class I MHC proteins to CD8-positive T cells (green).

released to be phagocytized by DCs (Figure 60–8). The DCs transport these particles into the secondary lymphoid tissues, and viral peptide antigens appear on the surface of the DC in association with MHC proteins. Viral "peptide A" (red circle) is presented with class II MHC and recognized by the TCR of a **CD4-positive T cell**. In addition, through the process of "cross-presentation" (see Chapter 58), viral "peptide B" (gray circle) is presented with class I MHC and recognized by the TCR of a **CD8-positive T cell** (cell with green nucleus).

The CD4-positive cells help the DC–CD8-positive cell interaction in two ways: (1) When activated, the "A"-specific CD4-positive T cell expresses high surface levels of **CD40L**, which interacts with **CD40** on the surface of the DC. This signals to the DC to further increase its expression of co-stimulatory molecules, ensuring activation of the "B"-specific CD8-positive cell. (Note that this mechanism of CD40L "licensing" of DCs is similar to the CD40L that Tfh cells provide to B cells, mentioned earlier and described in Chapter 61.) (2) The "A"-specific CD4-positive T cell also secretes IL-2, which directly signals the "B"-specific CD8-positive T-cell clone to proliferate. These new CTLs are *virus-specific* killers, able to recognize and kill any cell that displays viral peptide "B" on its surface.

The main function of CTLs is to secrete **perforins** and **proteases** into infected cells. Perforins form a channel through

the membrane, which allows the cell contents to leak out. It also allows the proteases to enter the cell cytosol and degrade cellular proteins. One of these proteases, **granzyme B**, cleaves procaspases into their active form, initiating apoptosis. Another mechanism by which CTLs kill target cells is the **Fas-Fas ligand (FasL)** interaction. Fas is a protein displayed on the surface of many cells. FasL is induced on the surface of the cytotoxic cell when its TCR recognizes its cognate antigen on the surface of a target cell. When Fas and FasL interact, the caspases that initiate apoptosis in the target cell are activated. After killing the virus-infected cell, the CTL itself is not damaged and can continue to kill other cells infected with the same virus.

Like Th-1 cells, CTLs express the transcription factor *TBX21* and produce the cytokine IFN-γ. CTLs are especially important as an immune defense against viruses and some intracellular bacteria, such as *Listeria monocytogenes*. This is because these **intracellular pathogens** reside within host cells and use the cell machinery to divide and spread. These pathogens spend little time outside of host cells, meaning they are not susceptible to antibody and complement, so the best way to defeat them is for CTLs to kill the host cells, allowing phagocytes to engulf the remains. (Cytotoxic T cells have no effect on free virus, only on virus-infected cells.) In some cases, the killing effect of CTLs is actually pathogenic:

the severe liver damage caused by hepatitis viruses is *not* the result of viral cytotoxicity, but rather the result of a robust CTL response that kills virus-infected hepatocytes.

CTLs are also important in the surveillance of the body for cancer: when malignant cells accumulate somatic mutations, they begin to generate novel (non-"self") proteins, and CTLs that recognize and are activated by DCs presenting these peptide "neoantigens" can infiltrate the tumor and kill the malignant cells expressing those proteins. Transplanted cells from an allograft can similarly be recognized as non-"self" based on the presence of different human leukocyte antigen (HLA) polymorphisms and are therefore targets of CTLs (see Chapter 62).

Memory T Cells

Each of the above T-cell types can contribute to the pool of **memory T cells** that patrol the body and respond rapidly to reinfection. Memory T cells, as the name implies, endow our host defenses with the ability to respond rapidly and vigorously for many years after the initial exposure to a microbe or other foreign material. The **primary immune response** occurs after the initial exposure to the antigen, when the naïve T cells are first primed. The specific T-cell clones primed during the first encounter **proliferate** to large numbers, outnumbering many of the other T-cell clones in the circulation. For example, it is estimated that during infectious mononucleosis caused by Epstein–Barr virus (EBV), 40% of all circulating CD8-positive T cells are specific for EBV lytic phase proteins. After the infection has resolved, the T cell pool **contracts** as many die by apoptosis, and the remaining few T cells persist as **memory cells**.

Memory cells live for many years and reproduce over many cell generations. On subsequent exposure to the antigen, these few T-cell clones rapidly proliferate again as part of a **secondary immune response**, generating many more specific T cells. This secondary response to a specific antigen is stronger and faster because: (1) the starting pool of memory cells is greater than the starting pool of that clone during the primary response, so it takes less time to reexpand; (2) compared with naïve cells, memory cells have a lower threshold of activation, meaning smaller amounts of antigen and co-stimulation are required; and (3) activated memory cells produce greater amounts of cytokines than do naïve T cells.

EFFECT OF SUPERANTIGENS ON T CELLS

Certain proteins, particularly staphylococcal enterotoxins and toxic shock syndrome toxin, act as "superantigens" (Figure 60–9). They do this by binding to the MHC proteins and the TCRs on the surfaces of adjacent APCs and T cells, respectively, forcing the signaling molecules together. As a result, the T cell receives a strong TCR signal *regardless of the peptide that is displayed in complex with the MHC molecule.* Superantigens are "super" not because they activate each individual T cell more strongly, but rather because they activate a vastly larger number of the available T cells, in many cases bypassing the need for co-stimulation. For

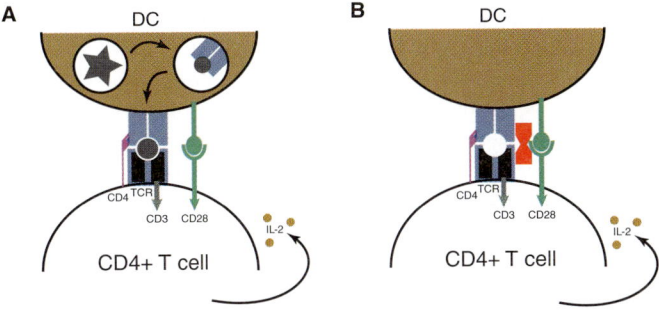

FIGURE 60–9 Activation of helper T cells by superantigen. **A: Activation by an Antigen that is not a Superantigen.** An antigen (gray star) is processed by a dendritic cell (DC) and presented to a CD4-positive T cell in association with a class II major histocompatibility complex (MHC) protein along with co-stimulation. (Note that additional cell–cell adhesion proteins are not shown.) Only the T cells with T-cell receptors (TCRs) specific for that antigen will be activated. **B: Activation by a Superantigen.** The CD4-positive T cell is activated by the binding of *unprocessed* superantigen (red dumbbell) to the Vβ portion of the TCR *outside* of its antigen-specific site. Because it bypasses the antigen-binding site, superantigen can activate many more helper T cells. Note that the activation signal "per cell" is the same, represented in this case by the cells producing the same amount of interleukin (IL)-2.

example, staphylococcal toxic shock syndrome toxin-1 (TSST-1) binds class II MHC proteins directly to the variable portion of the β chain of the TCR, specifically Vβ2. This causes unrestrained activation of any CD4-positive T cells that use this Vβ in their TCR, regardless of that TCR's antigen specificity and regardless of the peptide complexed with the MHC protein.

Because a large percentage of human T cells use Vβ2 (up to 30%), if all of these T cells are activated, it causes massive amounts of IL-2 released from the T cells and IL-1 and TNF from macrophages. These cytokines account for many of the findings seen in toxin-mediated staphylococcal diseases, such as toxic shock syndrome. Certain viral proteins (e.g., those of mouse mammary tumor virus [a retrovirus]) also possess superantigen activity. (Although not all superantigens bind Vβ2, they all cause pathology by activating an excessive number of T cells irrespective of those cells' TCR specificities.)

T CELL EXHAUSTION

What happens when an infection becomes chronic and effector T cells are continuously stimulated by cognate antigen? Chronic, unrestrained T cell activation could cause immune-related damage to the host. To prevent this, the T cells adopt a dormant state, **T cell exhaustion**, in which they express more checkpoint molecules, such as PD-1, and respond poorly to antigen. Exhausted T cells have been found in the setting of chronic viral infection, such as with HIV, and in certain cancers, and it is believed that in certain settings, T cell exhaustion limits what could be a beneficial immune response to chronic diseases. Therapies that reverse T cell exhaustion could therefore be way to restore the immune system's ability to recognize and fight these diseases.

TESTS FOR EVALUATION OF CELL-MEDIATED IMMUNITY

Evaluation of "immunocompetence" often depends on demonstrating intact T-cell–mediated immune responses to commonly present "harmless" antigens (i.e., a T-cell–mediated hypersensitivity reaction) or, more specifically, on laboratory assessments of T-cell **numbers** and **function**.

Enumeration of T Cells & Subpopulations

The number of each type of immune cell can be precisely counted by use of a **flow cytometer** (see Chapter 64). Counting cells with specific markers permit the enumeration of **total** T cells and the **percentage** that are CD4-positive, CD8-positive, regulatory, etc. The normal number of CD4-positive cells in adults is between 500 and 1500 cells/µL, whereas in patients with advanced HIV/AIDS, this number drops to less than 200 cells/µL. (In Chapter 59, we described the polymerase chain reaction [PCR] assay for T-cell receptor excision circles [TRECs], which is an inexpensive newborn screening test to identify T-cell deficiencies.)

In Vivo Tests for T-Cell Competence (Skin Tests)

Skin Tests for Preexisting (Memory) T-Cell–Mediated Hypersensitivity

Most normal people respond with inflammatory reactions to skin test antigens of *Candida* and other benign environmental antigens because of immune memory of past exposure to these antigens. After pricking the skin with a small quantity of these protein antigens, normal memory T-cell responses take 2 to 3 days to develop, causing an area of skin induration and redness. Absence of this immune response suggests impairment of T-cell–mediated immunity. This is also the principle behind skin testing for latent tuberculosis infection.

Skin Tests for Newly Developed T-Cell–Mediated Hypersensitivity

Most normal persons readily develop reactivity to simple chemicals (e.g., dinitrochlorobenzene [DNCB]) applied to their skin in lipid solvents. When the same chemical is applied to the same area 7 to 14 days later, the host's newly primed T cells generate a skin reaction. Immunocompromised persons with inadequate development of T-cell–mediated responses fail to generate a reaction on DNCB rechallenge.

In Vitro Tests for T-Cell Proliferation & Function

As described earlier, flow cytometry can be used to count specific cell populations from a pool of the patient's cells. It can also be used to identify **cell proliferation** by determining the percentage of cells that incorporate alkyne-modified nucleotides added to a cell culture. Incorporation of these nucleotides only occurs in dividing cells.

Similarly, flow cytometry can be used to assay cell activation by determining the level of certain surface markers or measuring the production of certain cytokines. These tests can be used in conjunction with a variety of stimuli to test either *antigen-specific* T-cell responses, as in the case of **IFN-γ release assays (IGRAs)** used to diagnose latent tuberculosis, or *nonspecific* T-cell responses, as in the case of **mitogens** such as phytohemagglutinin or concanavalin A, which are plant extracts that bypass the TCR to stimulate T cells.

Finally, CTL cytotoxic function can be assayed by culturing CD8-positive cells with MHC-matched cells displaying foreign (e.g., viral) peptides. Some fraction of the CD8-positive cells should recognize these foreign peptides, increase their expression of activation markers and cell-killing effector proteins, and cause widespread death of the target cells in the culture.

SELF-ASSESSMENT QUESTIONS

1. T-cell–mediated immunity is the main host defense against which one of the following organisms?
   (A) *Escherichia coli*
   (B) *Mycobacterium leprae*
   (C) *Pseudomonas aeruginosa*
   (D) *Staphylococcus aureus*
   (E) *Streptococcus pneumoniae*
2. You would like to target one of the cells involved in a certain autoimmune disease, described as a CD3-positive CD4-positive cell. Which one of the following is the most accurate about this cell's function?
   (A) Produces IgG
   (B) Produces interleukin-2
   (C) Kills virus-infected cells
   (D) Presents antigen in association with class II major histocompatibility complex (MHC) proteins
   (E) Presents antigen in association with class I MHC proteins
3. Which one of the following sets of cells is primarily responsible for presenting antigen to helper T cells?
   (A) B cells and dendritic cells
   (B) B cells and cytotoxic T cells
   (C) Macrophages and eosinophils
   (D) Neutrophils and cytotoxic T cells
   (E) Neutrophils and plasma cells
4. In addition to antigen presentation in association with class I MHC proteins, activation of a CD8-positive T cell requires which one of the following?
   (A) High levels of co-stimulation and interleukin-2 produced by CD4-positive T cells
   (B) High levels of co-stimulation and gamma interferon produced by macrophages
   (C) Presentation of antigen with class II MHC proteins and interleukin-1 produced by macrophages
   (D) Presentation of antigen with class II MHC proteins and interleukin-4 produced by CD4-positive T cells

5. Regarding Th-1, Th-2, and Th-17 cells, which one of the following is the most accurate?

   (A) Th-17 cells produce interleukin-17, which stimulates the production of Th-2 cells.

   (B) The production of Th-1 cells is enhanced by interleukin-4, whereas the production of Th-2 cells is enhanced by interleukin-2.

   (C) Th-2 cells synthesize gamma interferon, which is important in controlling infections caused by *Staphylococcus aureus* and other pyogenic bacteria.

   (D) Th-1 cells control infections caused by *Mycobacterium tuberculosis*.

6. Regarding interleukins, which one of the following is the most accurate?

   (A) IL-2 is made by B cells and increases class switching from IgM to IgG.

   (B) IL-4 is made by cytotoxic T cells and mediates the killing of virus-infected cells.

   (C) IL-21 is made by Th-17 cells and enhances the differentiation of cells that defend epithelial barrier surfaces.

   (D) IL-12 is made by Tfh cells and increases the generation of cytotoxic T cells.

7. Your patient is a 20-year-old woman who experienced the sudden onset of fever, vomiting, myalgias, and diarrhea. This was followed by hypotension and a sunburn-like rash over most of her body. You make a presumptive diagnosis of toxic shock syndrome. Which one of the following is the most accurate description of the pathogenesis of this disease?

   (A) It is caused by the release of large amounts of histamine from basophils.

   (B) It is caused by an insufficient amount of inhibitor of the C1 component of complement.

   (C) It is caused by a superantigen that induces an overproduction of cytokines from T cells.

   (D) It is caused by a delayed hypersensitivity response to procainamide, which she was taking for her atrial fibrillation.

   (E) It is caused by a gene mutation causing excessive signal transduction through the T-cell receptor.

ANSWERS

(1) **(B)**
(2) **(B)**
(3) **(A)**
(4) **(A)**
(5) **(D)**
(6) **(C)**
(7) **(C)**

PRACTICE QUESTIONS: USMLE & COURSE EXAMINATIONS

Questions on the topics discussed in this chapter can be found in the Immunology section of Part XIII: USMLE (National Board) Practice Questions starting on page 757. Also see Part XIV: USMLE (National Board) Practice Examination starting on page 775.

Adaptive Immunity: B Cells & Antibodies

B cells perform two important functions: (1) they differentiate into plasma cells that produce antibodies and (2) they differentiate into long-lasting memory cells that respond robustly and rapidly to reinfection.

Antibodies are the principal defense used by the immune system to *prevent* infection because, by binding to the microbes' surfaces, they can **inhibit them from attaching to target cells** and/or **help innate killing mechanisms**. Antibodies can also **inhibit toxins** such as those made by tetanus and diphtheria. Vaccines work by raising protective, or **neutralizing**, antibodies.

Advances in cell biology have allowed the generation of large quantities of engineered **monoclonal antibodies**. The ability of these antibodies to strongly bind a specific antigen with very limited "cross-reactive" binding of other antigens is the basis for many common diagnostic tests and an increasing array of therapies for various diseases (see Monoclonal Antibodies section later in this chapter).

B-CELL MATURATION

As described in Chapter 59, B cells come from stem cells called **common lymphoid progenitors**, which give rise to all lymphocytes. Each mature B cell represents a **clone**, a group of cells arising from a precursor, all of which have the same heavy chain and light chain rearrangements to form the same **B-cell receptor** (BCR). Because the BCR—and secreted antibodies—from the clone and its progeny all have the same antigen specificity, these antibodies are called **monoclonal**. Figure 61–1 depicts an overview of the phases of B-cell maturation.

B-CELL ACTIVATION

B cells constitute about 30% of circulating lymphocytes. In lymph nodes, they are located in **follicles**; in the spleen, they are located in the **white pulp**. They are also found in gut-associated lymphoid tissue (GALT) such as Peyer's patches. They express the chemokine receptor **CXCR5**, which guides them toward chemokines produced in a region called the **B-cell follicle**. B cells reside in the follicles and survey the lymph and bloodstream for antigens.

After binding an antigen, B cells are stimulated to proliferate and "class switch." Like T cells, B cells generally require two signals to become activated. **Signal 1** is the binding of antigen to the BCR (Figure 61–2). Binding of multiple BCRs leads to *cross-linking* in which the BCRs are clustered together, increasing the signals sent into the cell. The more BCRs that are cross-linked by antigen, the stronger the signal will be. As we will discuss below, **signal 2** can come from a variety of sources. What the various types of signal 2 have in common is that they are *inflammatory*, meaning they only accompany foreign antigens that represent a real threat to the host.

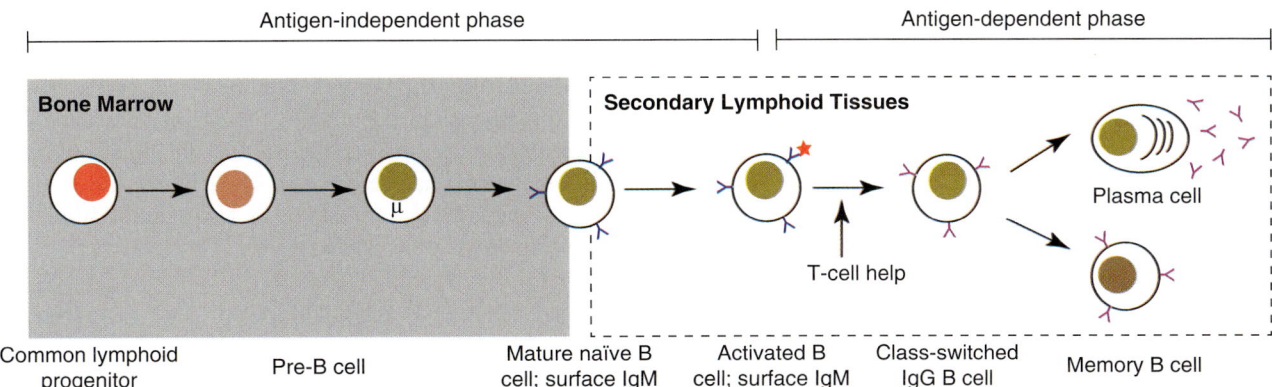

FIGURE 61–1 Maturation of B cells. B cells arise from lymphoid progenitor stem cells and differentiate into pre-B cells expressing μ heavy chains in the cytoplasm and then into mature B cells expressing monomer IgM on the surface. This occurs independent of antigen. Activation of B cells, class switching, and differentiation into memory B cells and plasma cells occurs after exposure to antigen (red star) and is enhanced by T-cell help. μ = mu heavy chains in cytoplasm; Y = IgM (blue) or IgG (purple). (Adapted with permission from Stites DP, Terr A: *Basic & Clinical Immunology*, 7th ed. New York, NY: McGraw Hill; 1991.)

Naïve B cells are short lived; without signal 2, they fail to achieve activation and are either **deleted** by apoptosis or become **anergic**, a state of nonresponsiveness. The requirement for inflammatory input from signal 2 at the time of activation is a safeguard so that B cells are not inadvertently activated by harmless antigens. Upon activation, B cells can either differentiate into long-lived **plasma cells**, which secrete more antibody, or into long-lived **memory B cells**, which wait in the follicles of the secondary lymphoid organs to respond to a reinfection.

T-Cell–Independent Activation

In some circumstances, B cells can be activated by a strong signal 1 and signal 2 and *do not need* T-cell help (see Figure 61–2A). Antigens that activate B cells without T-cell help are usually large multivalent molecules such as the chains of repeating sugars that make up bacterial capsular polysaccharide. The repeated subunits act as a **multivalent antigen** that cross-links many IgM antigen receptors on the B cell and sends a strong activating signal 1 into the B cell. Other macromolecules, such as lipids,

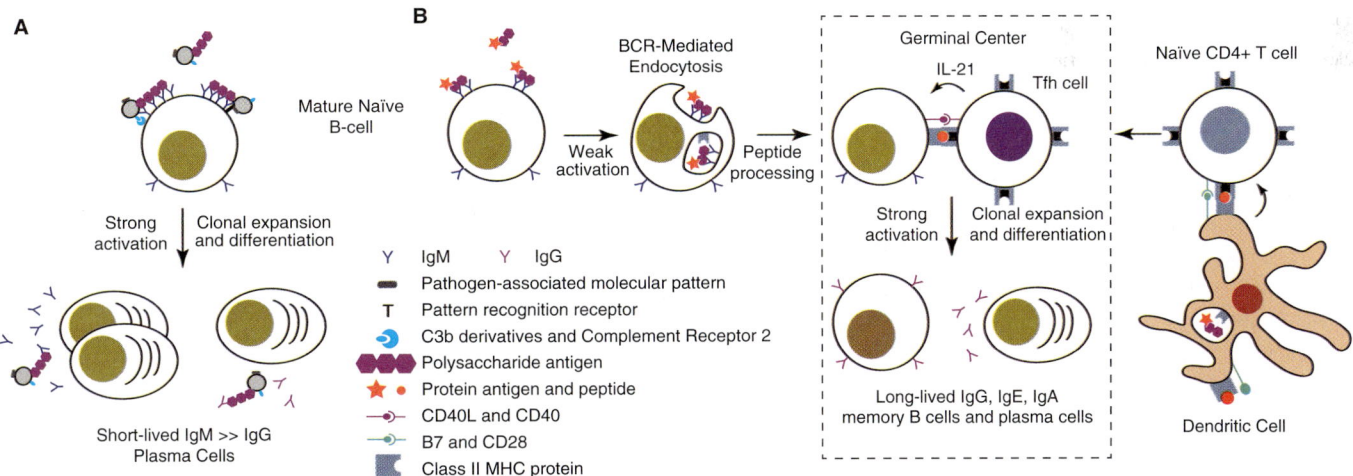

FIGURE 61–2 Overview of B-cell activation. **A: T-cell–independent Response.** T-cell–independent antigens are large multivalent structures, often polysaccharides (purple) of a bacterial cell (gray circles), that cross-link many IgM receptors to achieve a strong activation signal 1. Signal 2 might include complement C3b derivatives (light blue) bound to the bacterial cell (shown on upper left of cell), or pathogen-associated molecular patterns (gray bars, shown binding to pattern recognition receptor on upper right of cell). Without T-cell help, these responses are short-lived and dominated by IgM plasma cells, although some IgG is also generated. Note that the IgM receptors initially recognize and bind the polysaccharide, so all resulting antibodies will be specific for that polysaccharide. **B: T-cell–dependent Response.** A T-cell–dependent antigen must contain at least some protein component. The B-cell receptor (BCR) binds a specific part of the antigen, it is endocytosed by the B cell, and the peptides are processed and complexed with class II major histocompatibility complex (MHC) proteins. During the germinal center reaction, B cells compete to present antigen to peptide-specific T follicular helper (Tfh) cells that had been previously activated by dendritic cells presenting the same peptide fragment of the antigen. In the germinal center, the B-cell clones that receive more CD40 ligand (CD40L) and cytokines are able to proliferate, class switch, and become long-lived memory B cells and plasma cells.

DNA, and RNA, can also provide signal 1 to the B cell if these antigens are recognized by its surface BCR.

During **T-cell–independent activation**, the B cell's signal 2 can come from various **innate** (T-cell–independent) inflammatory sources:

(1) B cells have **pattern recognition receptors**, which recognize pathogen-associated molecular patterns (PAMPs)

(2) B cells have the **complement receptor CR2**, which recognize cleavage products of C3b released during complement activation (see Chapter 63)

(3) Vaccine **adjuvants** can activate a B cell without requiring T-cell help (see Chapter 57).

These events typically occur *outside* the B-cell follicle, and the plasma cells generated by T-cell–independent activation are relatively short-lived.

The T-cell–independent response is significant because it is the main response to bacterial capsular polysaccharides, which are not proteins and therefore not recognized by T cells. For example, the pneumococcal polysaccharide vaccine contains the surface polysaccharides of the most common serotypes of *Streptococcus pneumoniae* along with an adjuvant but no carrier protein. Together, the polysaccharide (signal 1) and adjuvant (signal 2) strongly activate B cells. However, because the vaccine does not contain peptides, activation of B cells by these polysaccharides is considered to be T-cell–independent.

T-Cell–Dependent Activation

The previous example illustrates an important concept for vaccine design, but, in general, antibodies generated independently of T-cell help are short-lived and are less specific for their antigens compared with antibodies generated with T-cell help. The strongest and most specific antibody response requires the participation of **dendritic cells** (DCs) and **T cells**. To describe T-cell–dependent activation of B cells, we first need the activation of naïve T cells (see Figure 61–2B, right side). As described in Chapter 60, CD4-positive T cells are activated by DCs presenting a foreign peptide complexed with class II major histocompatibility complex (MHC) proteins, along with co-stimulation.

Consider a T cell activated by a peptide. As it undergoes clonal proliferation, some of the offspring will differentiate into **T follicular helper (Tfh) cells** (see Chapter 60). Upon activation, a Tfh cell **turns off CCR7**, which held it in the T cell zone, and **turns on CXCR5**, pulling it into the B-cell follicle.

While this happens, antigen fragments of the same foreign entity circulate into the B-cell follicles of the secondary lymphoid tissue and interact directly with the antigen receptors (which are membrane-bound IgM molecules) of **naïve B cells**. The recognized epitopes of these circulating antigens might be lipid, polysaccharide, or nucleic acid components, but some component of the antigen may also contain the same peptide. The B cell then uses its BCR to take up the antigen into endosomes, and the antigen is processed. **This B cell can now function as an antigen-presenting cell**. The antigen is processed and its peptide components are complexed with class II MHC molecules and presented on the B cell's surface to interact with the Tfh cells at the border of the T-cell zone. Note that the Tfh cell recognizes the peptide, but the antibody from that B cell will recognize whatever lipid, polysaccharide, nucleic acid, or protein that initially bound to its BCR; the antibody is *not* determined by the peptide in the DC-T cell interaction.

Class Switching & Affinity Maturation

If a Tfh cell recognizes the antigen peptide presented by the B cell's class II MHC molecules, the Tfh cell provides two key signals back to the B cell: first, **CD40 ligand (CD40L)** molecules on the Tfh cell bind to **CD40** on the B cell; and second, the Tfh cells produce the cytokine **interleukin (IL)-21**. Together, these signals have three important effects on the B cells:

(1) **Rapid proliferation**

(2) **Class switching**, changing from using the Cμ segment to using one of the other heavy chain C_H segments (Cγ, Cε, or Cα) (see Figure 61–3)

(3) **Somatic hypermutation**

Genetic deficiency of the gene encoding CD40L causes an immunodeficiency called **hyper-IgM syndrome**. Patients with this disease have very high immunoglobulin (Ig) M levels and very little IgG, IgA, and IgE because their B cells are unable to receive T-cell help and therefore are unable to proliferate and "class switch." Hyper-IgM syndrome is characterized by severe bacterial infections (see Chapter 68). As a B-cell clone divides, switches its class, and hypermutates, the newly formed cluster of cells is called a **germinal center**.

Both B-cell class switching and somatic hypermutation are directed by the enzyme **activation-induced cytidine deaminase** (AID). For class switching, AID makes double-strand breaks in the DNA of the C_H locus of the heavy chain, removing the intervening DNA between the VDJ region and either Cγ, Cε, or Cα (see Figure 61–3). This causes *irreversible* switching of that IgM-positive B cell to instead express surface IgG, IgE, or IgA. The decision of whether to switch to IgG, IgE, or IgA is made based on the cytokine signals that the B cell receives:

(1) **IL-21 plus gamma interferon (IFN-γ) → IgG.** This makes sense because IFN-γ is the cytokine associated with macrophage activation, and it is the same cytokine that generates the antibody most associated with opsonization and phagocytosis.

(2) **IL-21 plus IL-4 → IgE.** This makes sense because IL-4 is one of the main cytokines associated with Th-2 immunity, and it is the same cytokine that generates the antibody most associated with mast cell, basophil, and eosinophil activity. Patients with allergic diseases caused by excess IgE often have excess IL-4.

(3) **IL-21 plus various "mucosal" cytokines → IgA.** This makes sense because the cytokines in mucosal barriers induce antibodies that are secreted across mucosal surfaces. (A deficiency in the gene encoding the receptor for some these

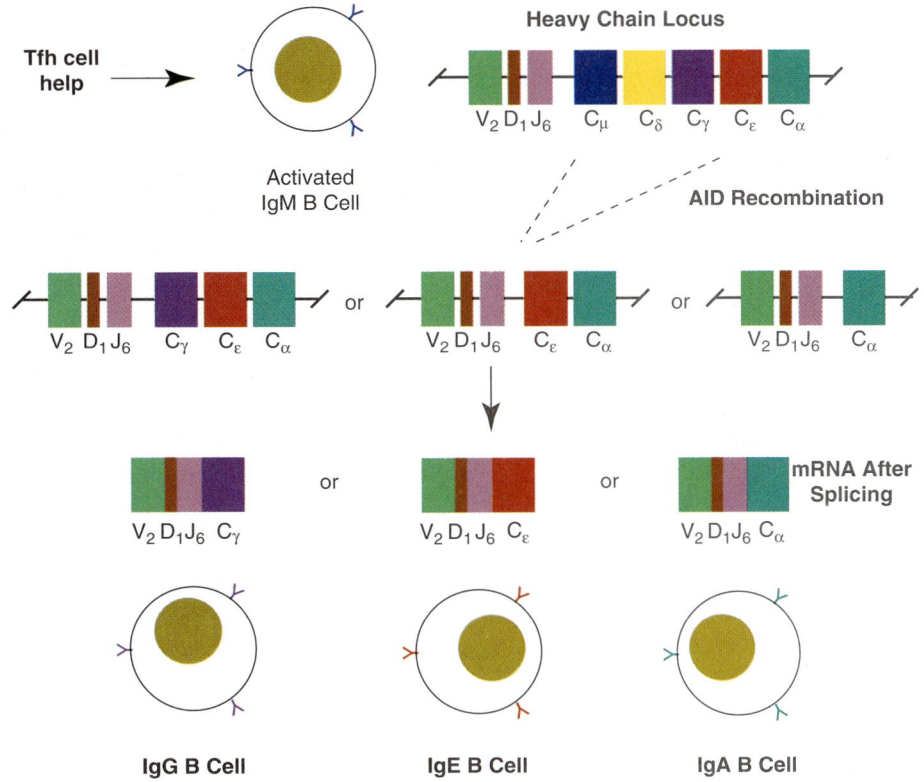

FIGURE 61-3   Class switching. T-cell help induces activation-induced cytidine deaminase (AID)-driven class switching. Activated IgM-positive B cells receive help from T follicular helper (Tfh) cells, including CD40 ligand (CD40L) and interleukin (IL)-21. This causes AID to create double-strand DNA breaks in the heavy chain locus that remove Cμ and Cδ and bring the VDJ region adjacent to one of the other C regions, either γ, ε, or α. After RNA splicing, the B cell begins to express IgG, IgE, or IgA instead of IgM.

cytokines causes **IgA deficiency**, which can present with serious sinopulmonary and gastrointestinal infections.)

Recall that the variable region of an antibody is responsible for binding to antigen, and because the variable region is not affected by *class switching*, the resulting IgG, IgE, or IgA antibodies should have the same antigen specificities. However, AID does something else. It also makes **nucleotide substitutions** in the gene regions that encode the V_H and V_L chains. This results in the exchange of new amino acids into the antigen-binding *hypervariable* region, **massively increasing the potential diversity of the B-cell pool**.

With successive cell division and new mutations, the enlarging pool of B cells continue to **compete** for the circulating antigens that are present in the follicle; those B cells with higher affinity will be more likely to bind and take up the antigens and therefore more likely to present the correct peptides to CD40L-positive Tfh cells, whereas B cells with lower affinity immunoglobulins will be outcompeted, will not receive the Tfh cell's survival signals, and therefore will die. This process is called **affinity maturation**, and over multiple rounds of cell division, mutation, competition, and selection, a pool of highly specific B-cell clones *evolves* from the initial germinal center (Figure 61–4). Many germinal centers in many secondary lymphoid organs are engaged with each infection, ensuring a broad **polyclonal** antibody response.

T-cell "help," in the form of IL-21 and CD40L, is also the main stimulus that drives B cells to differentiate into long-lived **plasma cells**. It might seem that this requirement of T-cell help to make plasma cells is unnecessarily cumbersome, but remember that the T cells have been carefully selected in the thymus *not* to see "self" peptides, and therefore, the involvement of T cells in B-cell activation is an additional safeguard against autoimmunity. Compared with T-cell–independent activation, the presence of Tfh cells generates higher titers of IgG, IgA, and IgE antibodies, longer-lived plasma cells, and a stronger response upon reinfection.

The concept of T-cell help for B cells was used in making an improved pneumococcal "conjugate" vaccine. The polysaccharides from common serotypes of *S. pneumoniae* were **conjugated to a highly immunogenic protein**. The vaccine is taken up by DCs, which process the *protein component* to be recognized by the T cells that become Tfh cells. In contrast, the B cells that are activated by the vaccine recognize the *polysaccharide component*, but once they bind the polysaccharide, they also take up and process the conjugated protein. Like the DCs, the B cells process this protein component and present the peptides to the newly activated Tfh cells. In this way, T-cell help is recruited to the follicle, generating high titers of antibody specific for the polysaccharide (see Figure 61–2).

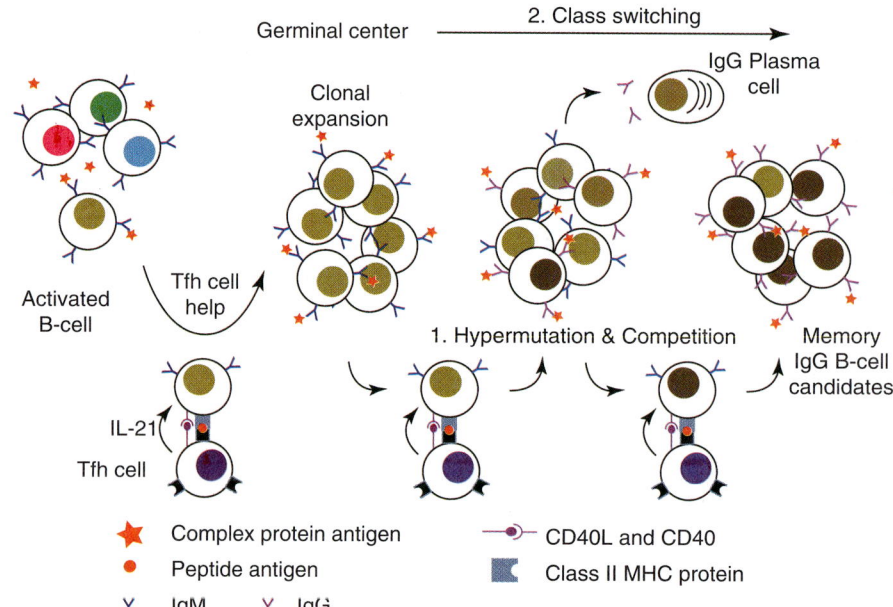

FIGURE 61–4 **Germinal center reaction.** B cells compete for antigen to receive T-cell help in the germinal center reaction. Naïve IgM-positive B cells survey for antigens in the B-cell follicle. Those that bind antigen (light brown) are selectively activated to engulf and process the antigen and present the peptides to T follicular helper (Tfh) cells. If a Tfh cell recognizes the peptide, it provides help (CD40L and interleukin [IL]-21), and B-cell clonal expansion initiates the germinal center reaction: (1) Repeated rounds of activation-induced cytidine deaminase (AID)-driven somatic hypermutation in the clones alter the specificity of the surface IgM for the antigen. Clones that out-compete their neighbors for antigens in the follicle (darker brown nuclei) enable them more interactions with Tfh cells, leading to progressive affinity maturation. (2) Tfh cytokines induce AID-driven class switching. (Note: In this case, gamma interferon [IFN-γ] signaled the B cells to class switch to IgG. If the Tfh cells provided IL-4, the B cells would class switch to IgE.) The successful clones either become long-lived plasma cells that leave the follicle or have the potential to become circulating memory B cells expressing IgG.

THE PRIMARY RESPONSE

The **primary response** occurs the *first* time that an antigen is encountered. This usually involves T-cell–dependent activation of B cells, as described earlier (see Figure 61–2). Most of the B cells activated from an initial exposure undergo class switching and affinity maturation and differentiate into **plasma cells**. Plasma cells secrete thousands of antibody molecules per second for a life span that lasts from a few days to months.

The first antibodies are detectable in the serum after around **7 to 10 days** in a primary response but can be longer depending on the nature and dose of the antigen and the route it takes to the secondary lymphoid organ (e.g., bloodstream or draining lymphatics). Antibody levels continue to rise for several weeks and then decline. As shown in Figure 61–5, the **first** antibodies to appear in the primary response are IgM, followed by IgG, IgE, or IgA, as more class-switched plasma cells are generated. IgM levels decline earlier than IgG levels because these plasma cells have shorter life spans.

THE SECONDARY RESPONSE

A small fraction of activated B cells become **memory B cells**, which can remain quiescent in the B-cell follicles but are activated rapidly upon reexposure of their surface BCR to antigen.

When there is a second encounter with the same or closely related antigen, months or years after the primary response, the secondary response is both more **rapid** (the lag period is typically only **3 to 5 days**) and generates **higher** levels of antibody than did the primary response (see Figure 61–5). The memory B cells, which underwent some degree of affinity maturation during the primary response, now proliferate to form a new germinal center and repeat the process of affinity maturation before generating new plasma cells.

This means that, with each exposure, antibodies tend to bind antigen with **higher affinity** because the memory B cells are subjected to further rounds of affinity maturation (see Figure 61–4). During the secondary response, more long-lived IgG plasma cells are generated, meaning secondary IgG levels are *higher* and *tend to persist longer* (see Figure 61–5). This concept is medically important because the protection from the first dose of a vaccine can be "boosted" with repeat doses.

RESPONSE TO MULTIPLE ANTIGENS ADMINISTERED SIMULTANEOUSLY

When two or more antigens are encountered at the same time, the host reacts by producing antibodies to all of them. Combined immunization is widely used and shown to be just as effective as single immunization (e.g., the diphtheria,

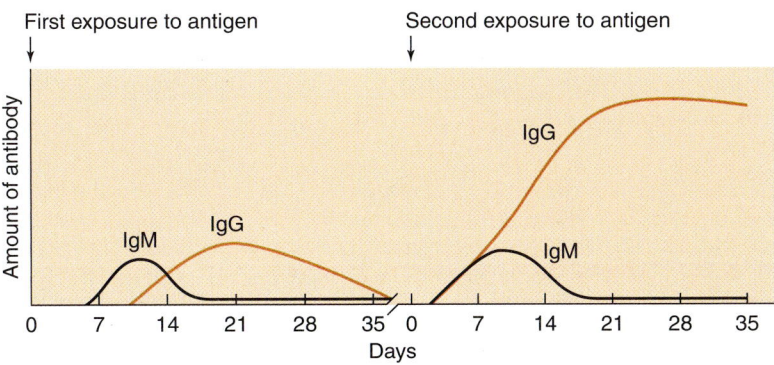

FIGURE 61–5 Antibody synthesis in the primary and secondary responses. In the primary response, immunoglobulin (Ig) M is the first type of antibody to appear. In the secondary response, IgG and IgM appear but IgG shows a more rapid rise and a higher final concentration than in the primary response.

tetanus, pertussis [DTP] vaccine and the measles, mumps, rubella [MMR] vaccine).

EFFECTOR FUNCTIONS OF ANTIBODIES

The primary function of antibodies is to protect against infectious agents or their products (Table 61–1). Antibodies provide protection by:

(1) **Activating complement** to lyse cell membranes and drive inflammation (see Chapter 63)

(2) **Opsonize** bacteria, with or without complement

(3) Stimulate immune cells' Fc receptors to kill a target cell, also called **antibody-dependent cellular cytotoxicity (ADCC)**

(4) Bind and **neutralize** toxins and viruses.

TABLE 61–1 Important Functions of Immunoglobulins

Immunoglobulin	Major Functions
IgG	Main antibody in the secondary response. Opsonizes bacteria, making them easier to phagocytize. Fixes complement, which enhances bacterial killing. Neutralizes bacterial toxins and viruses. Crosses the placenta.
IgA	Secretory IgA prevents attachment of bacteria and viruses to mucous membranes. Does not fix complement.
IgM	Produced in the primary response to an antigen. Fixes complement. Does not cross the placenta. Antigen receptor on the surface of B cells.
IgD	Uncertain. Found on the surface of many B cells as well as in serum.
IgE	Mediates immediate hypersensitivity by causing release of contents from the granules of mast cells and basophils upon exposure to antigen (allergen). Defends against worm infections by causing release of enzymes from eosinophils. Does not fix complement. Important host defense against tissue-invasive helminth infections.

Note that this last mechanism—the binding of an antibody's Fab to its target—is independent of the Fc region. This mechanism is the basis for many *therapeutic blocking antibodies*. Some examples include antibodies that inhibit cytokines, chemokines, or other pathogenic proteins.

Opsonization is the process by which antibodies make microbes more easily ingested by phagocytic cells. This occurs by either of two reactions: (1) the Fc portion of IgG interacts with its receptors on the phagocyte or (2) IgG or IgM activates complement to yield C3b, which interacts with its receptors on the surface of the phagocyte.

Table 61–2 is a summary of the properties of the various classes of immunoglobulins.

ISOTYPES & ALLOTYPES

Two antibodies with the same antigen specificity might nevertheless have important differences:

(1) **Isotype** antibodies are defined by their Fc regions. For example, the different antibody **classes** (IgM, IgD, IgG, IgA, and IgE) are different isotypes; the constant regions of their H chains (μ, δ, γ, α, and ε) are different.

(2) **Allotype** antibodies might be of the same isotype but have additional features that **vary among individuals**. They vary because the genes that encode the light and heavy chains are polymorphic, and individuals can have different alleles. For example, the γ heavy chain contains an allotype called Gm, which varies by one or two amino acids between individuals.

PROPERTIES OF ANTIBODY ISOTYPES (CLASSES)

IgG

Each IgG molecule consists of two L chains and two H chains linked by disulfide bonds (molecular formula H2L2) (see Figure 61–6C). Because it has two identical antigen-binding sites, it is said to be **divalent**. There are four **subclasses**, IgG1 to IgG4, based on differences in the H chains and on the

TABLE 61–2 Properties of Human Immunoglobulins

Property	IgG	IgA	IgM	IgD	IgE
Percentage of total immunoglobulin in serum (approximate)	75	15	9	0.2	0.004
Serum concentration (mg/dL) (approximate)	1000	200	120	3	0.05
Sedimentation coefficient	7S	7S or 11S[1]	19S	7S	8S
Molecular weight (×1000)	150	170 or 400[1]	900	180	190
Structure	Monomer	Monomer or dimer	Monomer or pentamer	Monomer	Monomer
H chain symbol	γ	α	μ	δ	ε
Complement fixation	++	−	++	−	−
Transplacental passage	++	−	−	−	−
Mediation of allergic responses	−	−	−	−	++
Found in secretions	−	++	−	−	−
Opsonization	++	−	−[2]	−	−
Antigen receptor on B cell	−	−	++	?	−
Polymeric form contains J chain	−	++	++	−	−

[1]The 11S form is found in secretions (e.g., saliva, milk, and tears) and fluids of the respiratory, intestinal, and genital tracts.

[2]IgM opsonizes indirectly by activating complement. This produces C3b, which is an opsonin.

number and location of disulfide bonds. IgG1 makes up most (65%) of the total IgG. IgG2 is directed against polysaccharide antigens and is an important host defense against encapsulated bacteria.

IgG is the predominant antibody in the **secondary response** and constitutes an important defense against bacteria and viruses (see Table 60–1). IgG is the only antibody to **cross the placenta**; only its Fc portion binds to receptors on the surface of placental cells (see Table 61–2). This receptor, called **FcRn**, transports maternal IgG across the placenta into the fetal blood. IgG is therefore the **most abundant immunoglobulin in newborns**. This is an example of passive immunity because the IgG is made by the mother, not by the fetus (see Chapter 57). Another important attribute of IgG is that it is one of the two

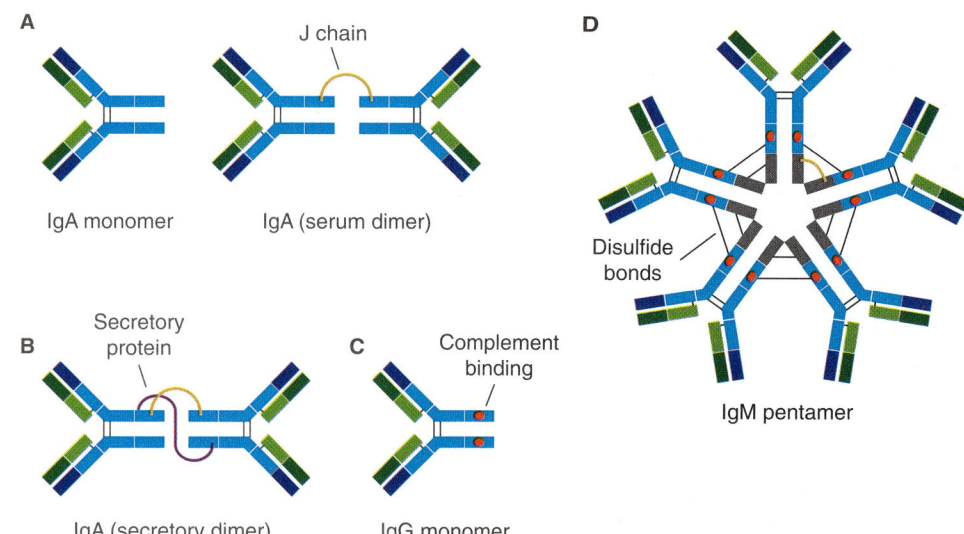

FIGURE 61–6 Structure of immunoglobulin (Ig) A and IgM, showing light chains in green and heavy chains in blue. **A:** Serum IgA monomer and dimer, linked by a J chain (orange). **B:** Secretory IgA dimer, linked by a J chain with an additional secretory protein (purple). **C:** IgG monomer, with complement-binding domains (red ovals). **D:** Pentamer of IgM molecules, which have a fourth C_H domain (gray), complement-binding domains, and a single J chain linking two adjacent molecules. Disulfide between the C_H domains maintains the pentamer structure. (Note that the IgM molecules have a fourth C_H domain.) (Adapted with permission from Stites D, Terr A, Parslow T: *Basic & Clinical Immunology*, 8th ed. New York, NY: McGraw Hill; 1994.)

immunoglobulins that can activate complement; IgM is the other (see Chapter 63).

IgG is the immunoglobulin that **opsonizes**. It can opsonize (i.e., enhance phagocytosis) because there are receptors for the γH chain, called Fcγ receptors, on the surface of phagocytes. These Fcγ receptors are also found on natural killer (NK) cells, which are responsible for **ADCC**. If a target cell's membrane antigens are bound by the Fab portion of an IgG antibody (e.g., if the cell is infected by a virus), then the Fc portion of these antibodies can activate the surface Fcγ receptors of the NK cell. This triggers the NK cell to release its cytotoxic mediators, including **perforins** and **proteases**, killing the target cell.

IgG has various sugars attached to the heavy chains, especially in the C_H2 domain. The medical importance of these sugars is that they determine whether IgG will have a proinflammatory or antiinflammatory effect. For example, if the IgG molecule has a terminal *N*-acetyl glucosamine, it is proinflammatory because it will bind to mannose-binding ligand and activate complement (see Chapter 63 and Figure 63–1). In contrast, if the IgG has a sialic acid side chain, then it will not bind and becomes antiinflammatory. Thus, IgG proteins specific for a single antigen that are of the same clone (i.e., made by the same plasma cell) can, at various times, possess different properties depending on these sugar modifications.

IgA

IgA is the main immunoglobulin in **secretions** such as colostrum, saliva, tears, and respiratory, intestinal, and genital tract secretions. It prevents attachment of microorganisms (e.g., bacteria and viruses) to mucous membranes. Secreted IgA consists of two H2L2 units plus one molecule each of J (joining) chain and secretory component (Figure 61–6A and B). (The J chain is only found in IgA and IgM, which are the only immunoglobulins that exist as multimers. The J chain helps form the disulfide bonds that bind multiple heavy chains into a multimer.) The secretory component is a polypeptide synthesized by epithelial cells that provides for IgA passage to the mucosal surface. It also protects IgA from being degraded by proteases in the intestinal tract. In serum, some IgA exists as monomeric H2L2.

IgM

IgM is the main immunoglobulin produced early in the **primary response**. It is present as a monomer on the surface of virtually all B cells as the BCR. In serum, IgM is a **pentamer** composed of five H2L2 units plus one molecule of J (joining) chain (see Figure 61–6D). IgM cannot bind to Fcγ receptors to facilitate opsonization or ADCC. However, IgM does bind to complement, and the resulting C3b is an opsonin (see Chapter 63). Because the IgM pentamer has 10 antigen-binding sites, it has the **highest avidity** of the immunoglobins and is the **most efficient** in agglutination, complement activation, and other antibody reactions. It can be produced by the fetus in certain infections.

IgD

This immunoglobulin has no known antibody function but may function as an antigen receptor. It is present on the surface of many B lymphocytes and in small amounts in serum.

IgE

IgE is medically important for two reasons: (1) it mediates immediate (anaphylactic) hypersensitivity (see Chapter 65) and (2) it participates in host defenses against certain parasites (e.g., helminths [worms]) (see Chapter 56). The Fc region of IgE binds to **Fcε receptors** on the surface of mast cells and basophils. Bound IgE then becomes a receptor for antigen (allergen). When the antigen-binding sites of adjacent IgEs are cross-linked by allergens, several mediators are released by the cells, and immediate (anaphylactic) hypersensitivity reactions occur (see Figure 65–1). Although IgE is present in **trace** amounts in normal serum persons with allergic reactivity have greatly increased amounts, and IgE may appear in external secretions. IgE does not bind to complement and does not cross the placenta.

IgE may provide host defense against certain important helminth (worm) infections, such as *Strongyloides*, *Trichinella*, *Ascaris*, and the hookworms *Necator* and *Ancylostoma*. The larvae of these worms migrate through tissue causing the increased serum IgE level seen in these infections. Because worms are too large to be ingested by phagocytes, they are killed by eosinophils that release worm-destroying enzymes. IgE specific for worm proteins binds to **Fcε receptors** on eosinophils, triggering the ADCC response with release of major basic protein from the eosinophil's granules.

ANTIBODIES IN THE FETUS

In general, the fetus and the newborn have an underdeveloped immune system that responds weakly to infections and vaccines. Antibodies in the fetus are primarily IgG acquired by transfer of maternal IgG across the placenta. This is why it is important to confirm the mother's vaccine history to ensure the newborn will be protected.

After birth, it generally takes months until newborn infants can make IgG (and other isotypes, such as IgM and IgA), so most vaccines are delayed by several months after birth to ensure the newborn's immune system has developed enough to respond. During this time, maternal IgG gradually declines, and protection from maternal IgG is lost by 3 to 6 months. The risk of infections begins to increase over this time, which is why the first set of vaccines is usually recommended by 2 months.

ANTIBODIES IN THE DIAGNOSIS OF DISEASES

Immunoglobulins themselves can also be detected with **anti-immunoglobulin antibodies**. This is used to diagnose patients with an infection. In other words, rather than trying to detect a particular microbe, it is often easier to detect the antibody specific for that microbe using an anti-immunoglobulin antibody. This is the basis for many tests for infections, including human immunodeficiency virus (HIV) tests. It can also be used to detect self-reactive antibodies, which cause autoimmune diseases such as myasthenia gravis. In addition, antibodies can also be used to treat various diseases as discussed in the section entitled "Creating Therapeutic Antibodies" below.

MONOCLONAL ANTIBODIES

The antibodies you make against a single vaccination or infection are usually made by many **different clones** of B cells (i.e., they heterogeneous, or **polyclonal**). Antibodies that arise from a **single clone of cells** are homogeneous, or **monoclonal**. Often, a plasma cell malignancy, such as **multiple myeloma**, can be diagnosed because a single proliferating clone might produce abnormally high levels of a monoclonal immunoglobulin, usually IgG.

The remarkable **specificity** of an antibody's hypervariable region for its target molecule has made monoclonal antibodies an invaluable resource in laboratory research and clinical applications. In the 1970s, the first method of generating virtually unlimited quantities of monoclonal antibodies in the laboratory was described (Figure 61–7). **Hybridoma cells**, formed by the fusion of two different cells, can be created by (1) isolating **spleen cells** from an animal (e.g., a mouse) previously immunized with an antigen of interest and (2) mixing these cells in a culture dish with **mouse myeloma cells** (which grow indefinitely in culture but do not make antibodies) so that the two cell types fuse. The newly formed *hybridoma* cells produce antibodies against the antigen of interest. More recent

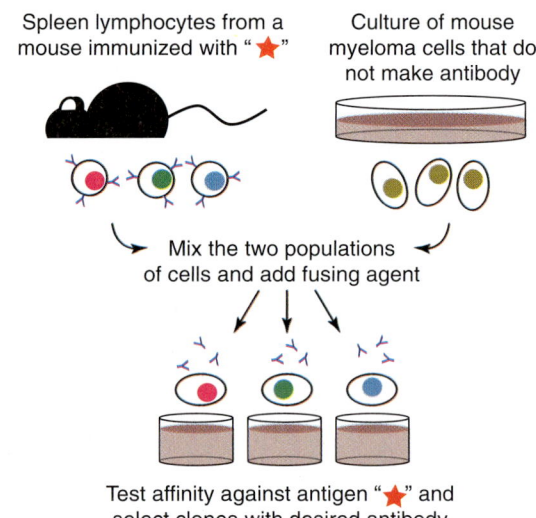

Spleen lymphocytes from a mouse immunized with "★"

Culture of mouse myeloma cells that do not make antibody

Mix the two populations of cells and add fusing agent

Test affinity against antigen "★" and select clones with desired antibody

FIGURE 61–7 Overview of the approach to generating monoclonal hybridomas. A mouse is immunized with an antigen of interest (red star). Spleen cells are isolated and mixed with mouse myeloma cells, which can grow indefinitely in culture but do not make antibody. A "fusing agent" is added to encourage the two cell types to combine. The resulting antibody-producing *hybridoma clones* are sorted into culture wells, one cell to each well. Their antibodies are screened against the antigen of interest, and clones that make high-affinity antibodies are selected to be cultured to make antibody indefinitely.

advances have eased production of therapeutic antibodies (see box "Creating Therapeutic Antibodies") that are now used in a variety of clinical situations, including prevention of infectious disease.

CREATING THERAPEUTIC ANTIBODIES

Hybridoma antibodies were the first monoclonal antibodies to be generated but have significant limitations as therapeutics because different animal species have different heavy chains. This means that many mouse antibodies, for example, bind poorly to human Fc receptors and, therefore, cannot initiate ADCC. These antibodies are also immunogenic if given repeatedly, resulting in hypersensitivity reactions.

This problem was solved with new techniques to cut and fuse pieces of DNA, which allowed the creation of **chimeric** monoclonal antibodies in which the DNA encoding the mouse spleen cell *variable* regions is fused to DNA encoding the human *constant* regions (Figure 61–8). These antibodies are about 65% human. The advantage of the mouse variable region is that it is easy to obtain mouse spleen cells that make

antibodies against, for example, a human or viral protein injected into the mouse. The names of chimeric antibodies end with the suffix *–ximab*, such as infliximab (antitumor necrosis factor [TNF]) and rituximab (anti-CD20).

The variable portion of a chimeric antibody is nonhuman, meaning it can still be immunogenic in a human patient. The next generation of **humanized** monoclonal antibodies is generated by replacing *all* of the mouse DNA with the equivalent human DNA sequence *except* the small part encoding the hypervariable (antigen-binding) region (see Figure 61–8). The resulting antibodies are about 95% human, further reducing the chances of an immune reaction. The names of humanized antibodies end with the suffix *–zumab*, such as omalizumab (anti-IgE) and pembrolizumab (anti-PD-1).

Fully human monoclonal antibodies are the next generation of therapeutics that have even greater affinity and virtually eliminate the risk of hypersensitivity (see Figure 61–8). The entire human heavy and light chain gene loci can be fully expressed, either by cultured cells or by mutant strains of mice in which the mouse immunoglobulin genes have been replaced. This results in an enormous potential diversity of 100% human antibodies that can be tested against antigens of interest. (These antibodies might not reflect the range and effector function of human antibodies generated during a "real-world" infection, and efforts are under way to address this by isolating B cells from patients after, for example, a viral infection, and *immortalizing* these cells in order to generate large amounts of their antibodies.) The names of fully human antibodies end with the suffix *–umab*, such as adalimumab (anti-TNF) and ipilimumab (anti-CTLA-4).

Further advances in antibody engineering have given rise to therapeutic proteins that only consist of a Fab region, for example ranibizumab (anti-VEGF), antibodies that only have a heavy chain, for example, caplacizumab (anti-vWF), "fusion" antibodies in which the Fab has been replaced by a different protein, such as a "decoy receptor," for example etanercept (another anti-TNF), and "bispecific" antibodies, in which the two arms come from different clones and each bind a different antigen, such as catumaxomab (anti-EpCAM/CD3). Each of these approaches offer certain advantages over traditional antibodies, and as engineering methods continue to evolve, antibody technology will be applied to a broader range of diseases.

TESTS FOR EVALUATION OF B CELLS AND ANTIBODIES

Evaluation of B cell function consists primarily of measuring the amount of each of the three important immunoglobulins (i.e., IgG, IgM, and IgA) in the patient's serum. To evaluate patients suspected of having an immunodeficiency, it may be necessary to count B-cell numbers by flow cytometry, as described in Chapter 60 for T cells. It is also possible to test for B-cell function by comparing antibody titers before vs. after immunization (e.g., with a *S. pneumoniae* vaccine). The absence of a normal rise in IgM and IgG after immunization indicates a defect, either an intrinsic defect in the B cells themselves or an extrinsic defect that, for example, inhibits T cells' capacity to provide help to activate the B cells.

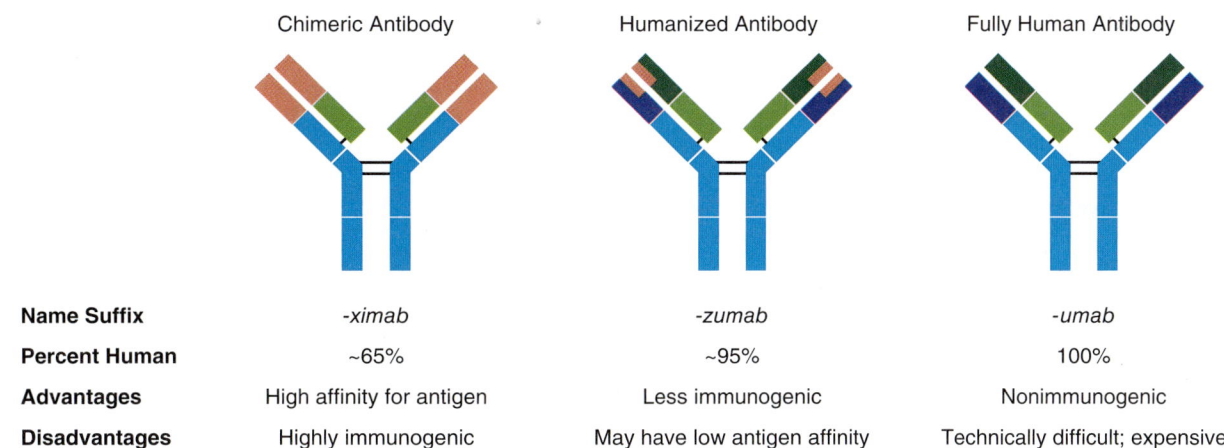

	Chimeric Antibody	Humanized Antibody	Fully Human Antibody
Name Suffix	-ximab	-zumab	-umab
Percent Human	~65%	~95%	100%
Advantages	High affinity for antigen	Less immunogenic	Nonimmunogenic
Disadvantages	Highly immunogenic	May have low antigen affinity	Technically difficult; expensive

FIGURE 61–8 Summary of therapeutic monoclonal antibodies. Chimeric antibodies are generated by fusing DNA encoding mouse variable regions (orange) to DNA encoding human constant regions (green light chain and blue heavy chain). Compared with mouse antibodies, these antibodies (-ximab) have much greater effector function potential because their human Fc fragments bind optimally to human Fc receptors. However, these antibodies are highly immunogenic due to the residual mouse components. Humanized antibodies are generated by replacing all *but* the DNA encoding the hypervariable (antigen-binding) region (orange) with human immunoglobulin gene DNA (green and blue). These antibodies (-zumab) are less immunogenic but may require additional mutation to improve the antigen affinity. Fully human antibodies (-umab) are generated by screening a "phage library" of randomly generated human antigen-binding sites or by immunizing transgenic mice that carry the human immunoglobulin gene loci in place of the mouse genes. These antibodies are the least immunogenic but carry significant technical barriers and cost. Therapeutics that make use of antibody properties but are engineered to have alternative structures include Fab- and Heavy Chain-Only, which can have one or more antigen-binding domains, fusion proteins, and "bispecific" antibodies in which each chain binds a different antigen.

SELF-ASSESSMENT QUESTIONS

1. It's time to play "Who am I?" I am the first class of antibody to appear, so my presence indicates an active infection rather than an infection that occurred in the past. I can fix complement, which is an important defense against many bacterial infections. I am found in plasma as a pentamer.

   (A) IgA
   (B) IgD
   (C) IgE
   (D) IgG
   (E) IgM

2. Regarding IgG, which one of the following statements is the most accurate?

   (A) Each IgG molecule has one antigen-binding site.
   (B) It is the most important antigen receptor on the surface of neutrophils.
   (C) During the primary response, it is made in larger amounts than is IgM.
   (D) The ability of IgG to fix complement resides on the constant region of the light chain.
   (E) It is the only one of the five immunoglobulins that is transferred from mother to fetus in utero.

3. If a person had a mutation in the gene encoding J (joining) chains, which of the following classes of antibodies could NOT be produced?

   (A) IgA and IgM
   (B) IgA and IgG
   (C) IgG and IgE
   (D) IgD and IgE
   (E) IgM and IgE

4. Regarding the function of the different classes of antibodies, which one of the following statements is the most accurate?

   (A) IgE blocks the binding of viruses to the gut mucosa.
   (B) IgA acts as an antigen receptor on the surface of B cells.
   (C) IgD is our most important defense against worm parasites, such as hookworms.
   (D) IgG can activate the alternative pathway of complement, resulting in the production of C3a that degrades the bacterial cell wall.
   (E) There are receptors for the heavy chain of IgG on the surface of neutrophils that mediate a host defense process called opsonization.

5. Regarding the genes that encode antibodies, which one of the following statements is most accurate?

   (A) Hypervariable regions are encoded by the genes of both the light and heavy chains.
   (B) The genes for the light and heavy chains are linked on the same chromosome adjacent to the human leukocyte antigen (HLA) locus.
   (C) During the production of IgG, the light and the heavy chains acquire the same antigen-binding sites by translocation of the same variable genes.
   (D) The gene for the constant region of the gamma heavy chain is first in the sequence of heavy chain genes, and that is why IgG is made in greatest amounts.

6. Regarding T-cell–independent activation of B cells, which one of the following statements is true?

   (A) "Conjugate" vaccines are most effective in this pathway because the antigen is *conjugated* to an ingredient that provides a signal 2.

   (B) Polysaccharide vaccines make use of adjuvants in part to bypass the requirement for T-cell help in generating antibody.
   (C) This pathway is best engaged by short antigens that cross-link relatively few immunoglobulin receptors.
   (D) **Engagement of the** complement system acts to diminish B cell activation.
   (E) Antigen presentation by dendritic cells is a key step in this pathway.

7. Regarding T-cell–dependent activation of B cells, which one of the following statements is true?

   (A) B cell competition for T-cell survival factors usually occurs outside of the B-cell follicle.
   (B) In this pathway, B cells perform as professional antigen-presenting cells that present antigen to activate naïve T cells.
   (C) Chemokines and their receptors are involved in bringing the T cells and dendritic cells together, but they do not have a role in T cell–B cell interactions.
   (D) IL-21 is a key cytokine made by T cells that induces B-cell class switching.
   (E) The AID enzyme directs V(D)J recombination, somatic hypermutation, and class switching.

8. A new monoclonal antibody is being tested in a trial for a type of leukemia. The antibody is directed against an antigen found on the leukemia cells. Which one of the following is NOT a possible mechanism of action of this antibody?

   (A) The antibody binds to the surface of a target cell, and its Fc region is detected by NK cells, which then kill the target cell.
   (B) The antibody binds to the surface of the target cell, and its Fc region recruits complement proteins, which are then activated to kill the cell directly.
   (C) The antibody is captured by Fc receptors on the target cell, and binding of antigen to the Fab region recruits phagocytes to kill the target cell.
   (D) The antibody binds to the surface of a target cell, and this binding blocks the cell's ability to receive essential survival signals, causing it to undergo apoptosis.
   (E) The antibody binds to the surface of a target cell, and its Fc region recruits complement proteins, which are then detected by phagocytes that engulf and kill the cell.

ANSWERS

(1) **(E)**
(2) **(E)**
(3) **(A)**
(4) **(E)**
(5) **(A)**
(6) **(B)**
(7) **(D)**
(8) **(C)**

PRACTICE QUESTIONS: USMLE & COURSE EXAMINATIONS

Questions on the topics discussed in this chapter can be found in the Immunology section of Part XIII: USMLE (National Board) Practice Questions starting on page 757. Also see Part XIV: USMLE (National Board) Practice Examination starting on page 775.

Major Histocompatibility Complex & Transplantation

In transplantation, an organ or tissue from one person is "grafted" to another person. A major barrier to the success of these life-saving procedures is the immune system, which attacks any cells it sees as foreign. Graft survival is largely determined by the donor's and recipient's **major histocompatibility complex (MHC)** proteins, which present antigens to T cells. In humans, these proteins are encoded by the **human leukocyte antigen (HLA)** genes. (Note that we will use MHC and HLA interchangeably.) Three of these genes (HLA-A, HLA-B, and HLA-C) code for the class I MHC proteins. Several HLA-D loci determine the class II MHC proteins (i.e., DP, DQ, and DR) (Figure 62–1). The features of class I and class II MHC proteins are compared in Table 62–1. If the HLA proteins on the donor's cells differ from those on the recipient's cells, then an immune response occurs in the recipient.

Each person has two HLA **haplotypes** (i.e., two sets of these genes—one on the paternal and the other on the maternal chromosome 6). These genes are highly **polymorphic** (i.e., there are many alleles of the class I and class II genes). For example, as of 2020, there are at least 20,000 HLA Class I alleles and 7700 HLA Class II alleles, and more are being discovered. However, an individual inherits only a single allele at each locus from each parent. Expression of these genes is **codominant** (i.e., the proteins encoded by *both* the paternal and maternal genes are produced).

The class I MHC proteins consist of one HLA-encoded polypeptide (an "alpha" chain), so each individual can have one set encoded by paternal genes and one set encoded by maternal genes. However, the class II MHC proteins consist of two HLA-encoded polypeptides (an "alpha" chain and a "beta" chain), so each individual can have as many as four of each class II MHC proteins because the maternal- and paternal-encoded peptides can mix and match.

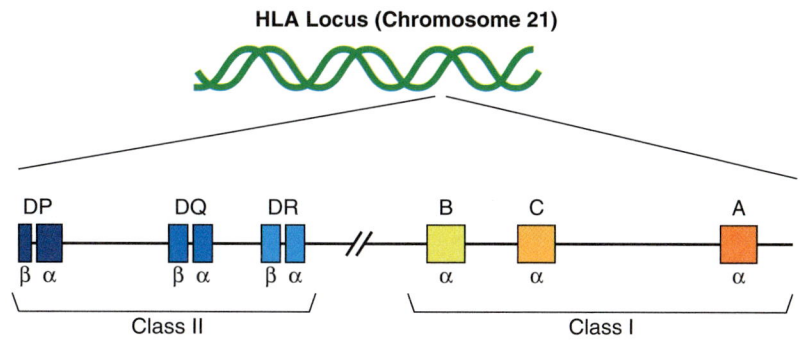

FIGURE 62–1 The human leukocyte antigen (HLA)–gene complex. A, B, and C are class I loci, and each gene encodes an alpha chain that pairs with the same β2-microglobulin. DP, DQ, and DR are class II loci, and each gene encodes alpha and beta chains that pair with each other.

TABLE 62–1 Comparison of Class I and Class II MHC Proteins

Feature	Class I MHC Proteins	Class II MHC Proteins
Present antigen to CD4-positive cells	No	Yes
Present antigen to CD8-positive cells	Yes	No
Found on surface of all nucleated cells	Yes	No
Found on surface of "professional" antigen-presenting cells, such as dendritic cells, macrophages, and B cells	Yes[1]	Yes
Encoded by genes in the HLA locus	Yes	Yes
Expression of genes is codominant	Yes	Yes
Multiple alleles at each gene locus	Yes	Yes
Composed of two peptides encoded in the HLA locus	No	Yes
Composed of one peptide encoded in the HLA locus and a β_2-microglobulin	Yes	No

[1]Note that class I MHC proteins are found on the surface of all nucleated cells, including those that have class II MHC proteins on their surface. Mature red blood cells are nonnucleated; therefore, they do *not* synthesize class I MHC-proteins.

HLA gene **polymorphism** causes us each to have somewhat different proteins on the surface of our cells. Each HLA protein can bind certain peptides better than others. This likely explains our HLA diversity: our ability as a population to recognize diverse infectious agents is an evolutionary advantage, ensuring that some individuals would more likely survive an epidemic. But these protein differences can be recognized as nonself when introduced to another person's immune system. The MHC is called **major** because these HLA differences are often responsible for acute rejection of a transplant, and these are the genes most crucial for matching donors and recipients. Another context where MHC genes and proteins are important is in autoimmune diseases, many of which occur more frequently in people who carry certain MHC genes (see Chapter 66).

In addition to the major antigens encoded by the HLA genes, there is a large number of **minor** antigens encoded by genes at sites other than the HLA locus. These minor antigens are various normal body proteins that have one or more amino acid differences from one person to another (i.e., they are "allelic variants"). Like the HLA proteins, these proteins can therefore be recognized as nonself by another person's immune system. Minor antigens induce a weak immune response, which might result in slow rejection of a transplant or rapid rejection if several minor antigens have a cumulative effect. Predicting rejection on the basis of minor antigens is difficult, so donors and recipients are not routinely tested for specific minor histocompatibility antigens. In view of these differences in minor antigens, all recipients routinely receive immunosuppressive drugs, even if their *major* histocompatibility loci are well-matched.

MHC PROTEINS

Class I MHC Proteins

These are found on the **surface of virtually all nucleated cells**. The complete class I protein is composed of a heavy chain (called the alpha chain) bound to a β_2-microglobulin. The alpha heavy chain is **highly polymorphic** and is similar to an immunoglobulin molecule; it has *hypervariable* regions that bind and present short peptides to T cells. The **polymorphism** of these molecules is important in the **recognition of self** and **nonself**. Stated another way, if these molecules were more similar from individual to individual, our ability to accept foreign grafts would be improved but our species might be more susceptible to certain infections. The heavy chain also has a constant region where the **CD8** protein of a cytotoxic T cell binds.

Class II MHC Proteins

These are found only on the surface of **antigen-presenting cells** (**APCs**), such as dendritic cells, macrophages, and B cells (see Chapter 60). They are composed of two **highly polymorphic** chains (called alpha and beta). Like class I proteins, the class II proteins have hypervariable regions that present the short peptides to T cells and provide much of the polymorphism. Unlike class I proteins, which have only one chain encoded by the MHC locus (pairing with β_2-microglobulin), *both* the alpha and beta chains of the class II proteins are encoded by the MHC locus. The two peptides also have a constant region where the **CD4** protein of a helper T cell binds.

BIOLOGIC IMPORTANCE OF MHC

The ability of T cells to recognize antigen depends on association of the antigen with either class I or class II proteins (see Chapter 60). For example, CD8-positive cytotoxic T cells only respond to antigen *in association with class I MHC proteins*. Thus, a cytotoxic T cell that is activated to kill a virus-infected cell will *only* kill a cell infected with the same virus and presenting antigen with the appropriate class I proteins. (This was determined by mixing cytotoxic T cells from individual "A," bearing one set of class I MHC proteins, with virus-infected cells bearing a set of class I MHC proteins from individual "B." Because of the class I MHC mismatch, no killing of the virus-infected "B" cells occurred.) The requirement that antigen recognition occurs in association with a "self" MHC protein is called **MHC restriction** and is a result of **positive thymic selection** (see Chapter 59).

TRANSPLANTATION

The likelihood that a transplanted organ, or *graft*, is accepted by the recipient's immune system depends on the genetic similarity between the recipient and the donor. On one end of the spectrum, an **autograft** (transfer of an individual's own tissue to another site in the body) is always permanently accepted (i.e., it always "takes"). A **syngeneic graft** is a transfer of tissue between genetically identical individuals (i.e., identical twins) and almost always "takes." On the other end of the spectrum, a **xenograft**, a transfer of tissue between different species, is the least likely to succeed except under certain unusual circumstances.

An **allograft** is a graft between genetically different members of the same species (e.g., from one human to another). Allografts are usually rejected unless the recipient is given immunosuppressive drugs. The rapidity of the rejection will vary depending on the degree of difference between the donor and the recipient at the MHC loci.

Solid Organ Rejection

Even with perfect HLA matching, the presence of minor antigens means that immunosuppression is required after a transplant to prevent **allograft rejection**. As HLA mismatching increases, more immunosuppression is needed. In **acute allograft rejection**, vascularization of the graft is normal initially, but, in 11 to 14 days, there is marked reduction in blood flow and immune cells infiltrate the graft, with eventual necrosis. A **T-cell–mediated reaction is the main cause of acute rejection** of many types of grafts, but antibodies may contribute to the rejection of certain transplants.

A graft that survives an acute allograft reaction can undergo **chronic rejection**. This causes gradual loss of graft function and can occur months to years after engraftment. The main pathologic finding in chronic rejection is atherosclerosis of the vascular endothelium. The immunologic stimulus that causes chronic rejection is complex and multifactorial and can occur even in HLA-matched donor–recipient pairs due to the presence of *minor* histocompatibility antigens. The adverse effects of long-term use of immunosuppressive drugs may also play a role in chronic rejection. Chronic rejection generally does not respond to treatment, and it carries a poor prognosis.

In addition to acute and chronic rejection, a third type called **hyperacute rejection** can occur. Hyperacute rejection typically occurs within minutes of a solid organ transplant graft and is due to the reaction of **preformed** anti-ABO antibodies in the recipient with ABO antigens on the surface of the endothelium of the graft. To prevent this severe rejection reaction, the ABO blood group of donors and recipients must be matched, and a cross-matching test must be done (see Chapter 64).

Depending on the type of graft and the type of rejection, *mismatching* of the HLA-A, HLA-B, and HLA-DR alleles is the most predictive factor in solid organ transplant rejection. The strength of the recipient's T-cell response to alloantigens encoded by these donor MHC alleles can be explained by the observation that there are two immune pathways by which the immune response is triggered (Figure 62–2).

These pathways are summarized as follows and are differentiated by whether the *sensitizing APC* is **donor-derived** or **recipient-derived**:

(1) In the **direct pathway** of allograft recognition, there must be donor–recipient HLA mismatch. In this pathway, the *donor's* APCs contained within the grafted organ migrate to a nearby secondary lymphoid tissue and present peptides in association with their class I and class II MHC proteins. The mere presence of the donor HLA protein that is *presenting* the peptide is enough to make the peptide–MHC complex **appear to be nonself** to the recipient's T cells, *regardless of the peptide*. Unlike the conventional activation of T cells by cognate peptides complexed with MHC (see Chapter 60), "direct" recognition of these nonself HLA proteins triggers a **polyclonal activation of a much larger percentage of recipient T-cell clones**, by some estimates up to 10% of the recipient T cells. This is likely because the diversity of peptides complexed to the donor MHC proteins can trigger a similarly diverse array of T-cell clones.

If the nonself HLA proteins are class I, they will **activate CD8-positive T cells** to become cytotoxic T lymphocytes (CTLs), which infiltrate the graft and kill the graft cells because they express the same class I proteins. "Direct" recognition of class II HLA proteins can also trigger **activation of the recipient's CD4-positive T cells**, which can provide the cytokine "help" that enhances CD8-positive T-cell activation, as described in Chapter 60.

(2) In the **indirect pathway** of allograft recognition, the *recipient's* APCs present the donor's proteins. The donor's proteins that are shed by damaged cells of the graft are taken up by recipient dendritic cells, processed, and presented to T cells as "foreign" proteins in a draining lymph node. (If there is HLA mismatch, the donor HLA proteins are often the antigens responsible for this pathway because HLA proteins are highly polymorphic and immunogenic. But even without HLA mismatch, there are minor antigens that can bring about the indirect pathway of rejection.) This results in activation of CD4-positive helper T cells (see Chapter 60). The newly activated T helper cells can (a) migrate back to the graft and **activate macrophages** and (b) **recruit neutrophils**, or (c) migrate to the B-cell follicle and **induce antibodies** against the graft cells.

Compared with the *direct pathway*, the *indirect pathway* takes longer because the recipient dendritic cells have to enter the graft, take up nonself proteins, and migrate to the draining lymph node to activate the adaptive immune response. As time passes, the donor APCs in the graft are replaced by recipient APCs, and the risk of *direct recognition* being a mechanism for rejection declines. Therefore, whereas both the *direct* and *indirect* pathways can contribute to **acute rejection**, the *indirect* pathway is primarily responsible for **chronic rejection**.

In all rejection scenarios, the activation of T cells is accompanied by inflammatory stimuli, such as pathogen-associated molecular patterns (PAMPs) and damage-associated molecular patterns (DAMPs). These are necessary to induce the costimulatory signals, such as B7 molecules, on the APCs in order to fully activate the T cells (see Chapter 60). This is clinically relevant because limiting inflammation and tissue damage at the time of transplantation significantly limits the likelihood of graft rejection and improves outcomes. This explains why, given a choice, surgeons prefer to transplant organs from live donors whenever possible.

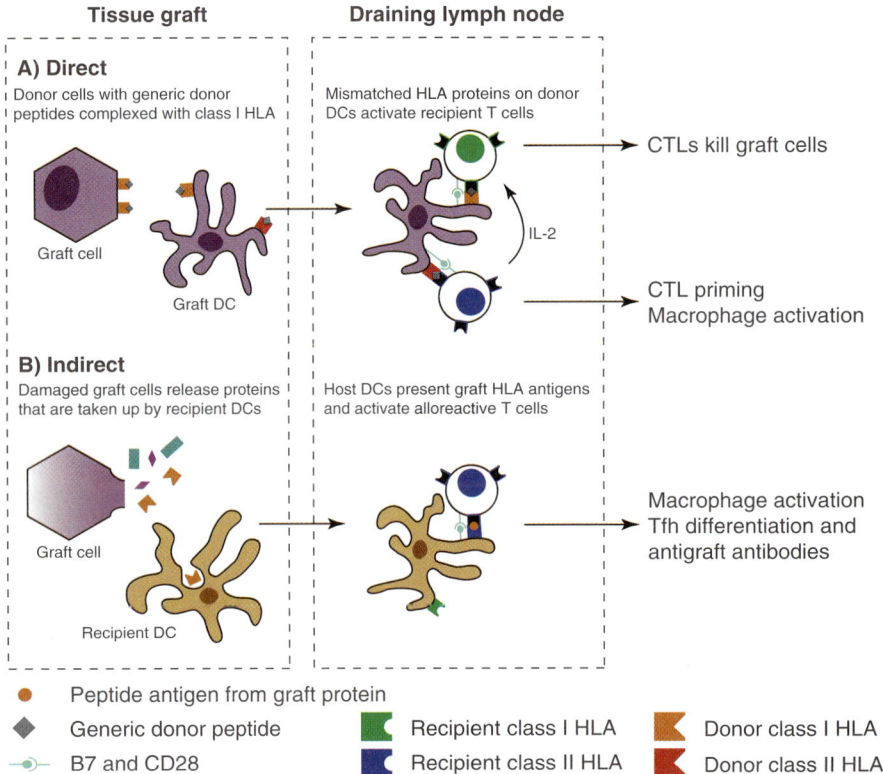

Tissue graft | **Draining lymph node**

A) Direct
Donor cells with generic donor peptides complexed with class I HLA

Graft cell

Graft DC

Mismatched HLA proteins on donor DCs activate recipient T cells

IL-2

→ CTLs kill graft cells

→ CTL priming
Macrophage activation

B) Indirect
Damaged graft cells release proteins that are taken up by recipient DCs

Graft cell

Recipient DC

Host DCs present graft HLA antigens and activate alloreactive T cells

→ Macrophage activation
Tfh differentiation and antigraft antibodies

● Peptide antigen from graft protein
◆ Generic donor peptide
—●— B7 and CD28

▌ Recipient class I HLA   ▌ Donor class I HLA
▌ Recipient class II HLA   ▌ Donor class II HLA

FIGURE 62–2   The direct and indirect pathways of solid organ rejection. **A: Direct Pathway.** In the **direct pathway,** *donor* dendritic cells (DCs) migrate from the graft to the draining lymph node and interact with recipient CD4-positive **T helper** cells (Th cells, blue) and CD8-positive **cytotoxic T** cells (CTLs, green). Because of the human leukocyte antigen (HLA) mismatch (green ≠ orange and red ≠ blue), the generic peptide presented by the DC is irrelevant! The Th cells help activate the CTLs (see Chapter 60), which go into the graft and kill donor cells that express the mismatched class I HLA protein (orange). The Th cells also migrate into the graft and interact with mismatched class II HLA proteins (red), releasing cytokines that activate macrophages to enhance inflammation. **B: Indirect Pathway.** In the **indirect pathway,** *recipient* DCs take up proteins released by damaged graft cells. (This could be any mechanism of damage, including that caused by the direct rejection pathway described above.) The DCs process the antigens into peptides and carry them to the draining lymph node where they interact with recipient Th cells (blue). The Th cells can differentiate into T follicular helper (Tfh) cells that activate B cells to make antibody (see Chapter 61), or they can migrate into the graft and activate recipient macrophages to enhance inflammation. (If there is HLA mismatch, then CTLs might be activated, but they will not be able to kill donor cells.) Note that both pathways require co-stimulatory signals, such as B7 interacting with CD28, and therefore can only occur in inflammatory settings.

Hematopoietic Stem Cell Transplants

Malignancies of the hematologic system, particularly leukemia, are often treated with transplant of **hematopoietic stem cells**. The principle of this approach is to use aggressive chemotherapy to ablate all of the patient's hematopoietic cells, which includes most of the malignant cells, and then replace them with healthy stem cells that can repopulate the hematopoietic system (see Figure 58–1). The stem cells can be obtained from *leukopheresis of peripheral blood* or from a sample of *banked umbilical cord blood*.

Unlike most solid organ transplants, transplanted hematopoietic stem cells can be *autologous* (from the patient's own stem cell pool) or *allogeneic* (from a donor). Autologous cell transplants are safer and avoid the need to find a matched donor, so you might think this would be the preferred approach in all cases. However, the main advantage to using allogeneic cells is that once these cells engraft, the T cells will actually

attack any surviving malignant cells. This **graft-versus-malignancy** effect can occur with HLA-matched or -unmatched stem cells because of the minor antigens recognized by the donor cells. Without this effect, autologous transplants have higher relapse rates.

Graft-Versus-Host Reaction

While it is advantageous to transplant cells that attack the remaining malignant cells, an unfortunate adverse effect in autologous transplants is that the transplanted cells may *attack healthy host cells.* This **graft-versus-host (GVH)** reaction develops in 30% to 70% of recipients, depending on the type of donor cells and other factors. The reaction occurs because grafted immunocompetent T cells proliferate in the immunocompromised host and reject host cells with "foreign" proteins, resulting in severe organ dysfunction. The *donor's* cytotoxic T cells play a major role in destroying the *recipient's* cells. These reactions tend to occur in the skin and gastrointestinal system, causing severe rash, oral ulcers, diarrhea, and hepatitis.

There are three requirements for a GVH reaction to occur:

(1) The graft must contain immunocompetent T cells
(2) The recipient must be immunocompromised
(3) The recipient must express antigens (e.g., MHC proteins) foreign to the donor (i.e., the donor T cells recognize the recipient cells as foreign).

Risk of GVH reactions can be reduced by depleting the donor cell pool of T cells before the transplant, but this also reduces the graft-versus-malignancy effect and therefore increases relapse rates. Once it occurs, patients with GVH disease are treated with immunosuppressive agents, often for their entire lives. Immunosuppression increases the risk of disease relapse (by limiting graft-versus-malignancy effect) and also increases the risk of developing other malignancies as well as opportunistic infections.

HLA Typing in the Laboratory

Prior to transplantation, laboratory tests, commonly called **HLA typing** or **tissue typing**, are performed to match the donor and the recipient. The most important alleles to match are HLA-A, HLA-B, HLA-C, HLA-DR, and HLA-DQ, and a donor–recipient pair in which all 10 of the maternal and paternal alleles of these five genes match is called a "10/10 match." In the past, serologic assays were used to determine the Class 1 and Class II MHC proteins of the donor and recipient. However, serologic assays have now been largely replaced by **DNA sequencing** using polymerase chain reaction (PCR) amplification.

In addition to the tests used for matching, preformed cytotoxic antibodies in the recipient's serum reactive against the graft are detected by observing the lysis of donor lymphocytes by the recipient's serum. This is called **cross-matching** and is done to prevent hyperacute rejections from occurring. In solid organ transplants, the donor and recipient are also matched for the compatibility of their ABO blood groups (see Chapter 64).

Among siblings, there is a 25% chance for both haplotypes to be shared, a 50% chance for one haplotype to be shared, and a 25% chance for no haplotypes to be shared. For example, if the father is haplotype AB, the mother is CD, and the recipient child is AC, there is a 25% chance for a sibling to be AC (i.e., a two-haplotype match), a 50% chance for a sibling to be either BC or AD (i.e., a one-haplotype match), and a 25% chance for a sibling to be BD (i.e., a zero-haplotype match).

The Fetus Is an Allograft That Is Not Rejected

A fetus has MHC genes inherited from the father that are foreign to the mother, yet allograft rejection of the fetus does not occur. The reason that the mother's immune system does not reject the fetus as foreign is not fully understood. The mother can form antibodies against paternal MHC proteins; therefore, the reason is not that the mother is not exposed to fetal antigens.

Some possible explanations are (1) that the placenta does not allow maternal T cells to enter the fetus and (2) the maternal T cells within the placenta are biased toward a T-regulatory subset, which promotes tolerance of fetal antigens (see Chapter 60).

EFFECT OF IMMUNOSUPPRESSION ON GRAFT REJECTION

To reduce the rejection of transplanted cells or to treat GVH disease, immunosuppressive measures are generally required (Table 62–2 and Figure 62–3). These fall under the categories of corticosteroids (prednisone), DNA synthesis inhibitors (azathioprine, methotrexate, mycophenolate), calcineurin inhibitors (cyclosporine and tacrolimus), mammalian target of rapamycin (mTOR) inhibitors (sirolimus), signaling blockade (belatacept, basiliximab, etc.), and cell-depleting antibodies (antithymocyte globulin [ATG]).

Corticosteroids bind to glucocorticoid receptors that result in altered gene transcription in a variety of cell types. In immune cells, steroids act by inhibiting synthesis of leukotrienes, prostaglandins, and cytokines (e.g., IL-2) and by inducing apoptosis of rapidly dividing T cells. Corticosteroids inhibit cytokine production by blocking transcription factors, such as nuclear factor-κB and AP-1, which prevents the mRNA for these cytokines from being synthesized. Glucocorticoid receptors are found in nearly every cell in the body, causing widespread off-target changes in gene transcription. Therefore, the major disadvantage of corticosteroids that limits their chronic use is that they have numerous adverse endocrine, neuropsychiatric, metabolic, and cardiovascular side effects.

Azathioprine (which is converted to 6-mercaptopurine in the body), methotrexate, and mycophenolate mofetil inhibit different aspects of the nucleotide synthesis and metabolism, **shutting down DNA synthesis** and thereby blocking T-cell proliferation. Methotrexate may also have a variety of other effects on cell signaling through inhibition of signaling enzymes.

Cyclosporine prevents the activation of T lymphocytes by inhibiting the synthesis of interleukin (IL)-2 and IL-2 receptor. It does so by **inhibiting calcineurin**—a phosphatase enzyme that is activated by calcium flux following binding of the T-cell receptor, the first step in the cascade that ultimately leads to transcription of the genes encoding IL-2 and the IL-2 receptor. Tacrolimus binds a different protein (FKBP1A) but has a similar effect on calcineurin. Thus, cyclosporine and tacrolimus inhibit one of the earliest steps in the **first phase** of T-cell activation.

Sirolimus inhibits signal transduction through **mTOR**, which is primarily involved in signal transduction *downstream* of IL-2. Therefore, sirolimus inhibits later steps in the **second phase** of T-cell activation, in a pathway different from that of cyclosporine and tacrolimus. These drugs are more selective than steroids and therefore have fewer toxicities.

Belatacept is a fusion protein consisting of cytotoxic T lymphocyte antigen-4 (CTLA-4) fused to the Fc fragment of human IgG. CTLA-4 competes with CD28 for binding to B7 proteins, but does so with higher affinity, thereby **blocking**

TABLE 62–2 Immunosuppressive Therapies Used in Transplantation

Category	Example(s)	Mechanism of Action
Corticosteroids	Prednisone Methylprednisolone	Bind to glucocorticoid receptor, causing many changes in gene transcription. Suppress production of inflammatory mediators (including leukotrienes and prostaglandins) and cytokines (including IL-1 and TNF). Induce apoptosis in T cells and B cells.
DNA synthesis inhibitors	Azathioprine Methotrexate Mycophenolate	Inhibit purine synthesis or purine metabolism, preventing DNA synthesis, which leads to apoptosis in rapidly dividing lymphocytes. Azathioprine is converted to 6-mercaptopurine. Methotrexate may also directly impair T-cell and B-cell function through its effects on various signaling enzymes.
Calcineurin inhibitors	Cyclosporine Tacrolimus	Cyclosporine binds to cyclophilin; tacrolimus binds to FKBP1A. Inhibition of calcineurin blocks calcium-dependent dephosphorylation (activation) of NFAT, a transcription factor required for the **first phase** of T-cell activation. Reduces synthesis of IL-2 and IL-2 receptor.
mTOR inhibitors	Sirolimus/rapamycin Everolimus	Binds to FKBP1A. Inhibits mTOR, which is involved in the **second phase** of T-cell activation. Blocks IL-2 signal transduction and clonal proliferation.
Signaling blockade	Belatacept Abatacept	These are CTLA-4 proteins fused with an Fc region of an immunoglobulin. They act as "mimics" that bind B7 molecules on APCs with high affinity, competing with CD28 and preventing them from providing co-stimulatory signals to T cells.
	Basiliximab Daclizumab	Monoclonal mouse/human mixture antibodies that bind and block the IL-2 receptor on T cells, inhibiting proliferation.
	Muromonab (OKT3)	Monoclonal mouse antibody that binds and blocks CD3, inhibiting T-cell receptor signaling.
Cell-depleting antibodies	Antithymocyte globulin	Polyclonal horse or rabbit antibodies prepared by immunizing animals with human T cells. Antibodies bind to multiple targets (CD3, CD4, CD8, etc.) and cause complement-mediated cell lysis; binding also inhibits signaling to the T cells, suppressing their survival and activation.

APC = antigen-presenting cell; IL = interleukin; CTLA = cytotoxic T-cell associated protein; mTOR = mammalian target of rapamycin; NFAT = nuclear factor of activated T cells; TNF = tumor necrosis factor.

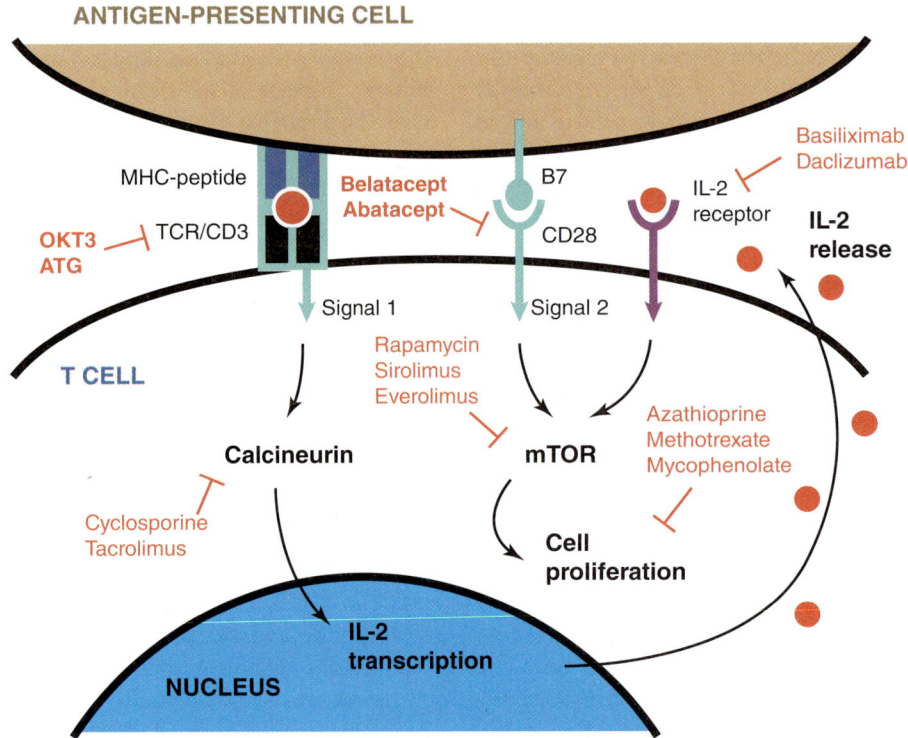

FIGURE 62–3 Mechanism of action of important immunosuppressive drugs, depicted on a schematic of an interaction in which an antigen-presenting cell activates a T cell. Corticosteroids have varied effects and are not described in the figure. ATG = antithymocyte globulin.

co-stimulation of T cells and preventing graft rejection. Muromonab (OKT3) was the first approved monoclonal antibody. It is a mouse antibody against CD3 that **blocks signal transduction through the T-cell receptor**. Basiliximab is a chimeric monoclonal antibody that **blocks the IL-2 receptor**, preventing T-cell proliferation.

ATG is a polyclonal cocktail of horse (Atgam) or rabbit (Thymoglobulin) antibodies against human thymocytes. ATG contains antibodies against many lymphocyte antigens (e.g., CD3, CD4, CD8, and others). After binding to their targets on the surfaces of T cells, these antibodies **kill T cells** through complement-mediated lysis of the cell (among other potential mechanisms). As a consequence, ATG has a broader immunosuppressive effect than do the more targeted monoclonal antibodies described in the previous paragraph.

Unfortunately, immunosuppression greatly enhances the recipient's susceptibility to opportunistic infections and neoplasms. For example, some patients undergoing treatment for multiple sclerosis with the monoclonal antibody natalizumab developed progressive multifocal leukoencephalopathy (see Chapter 44 for a description of this viral disease). The incidence of cancer is increased as much as 100-fold in transplant recipients who have been immunosuppressed for a long time. Common cancers in these patients include squamous cell carcinoma of the skin, adenocarcinoma of the colon and the lung, and lymphoma.

SELF-ASSESSMENT QUESTIONS

1. Regarding solid organ transplantation, which one of the following is the most accurate?
   (A) An allograft is a graft that transfers tissue or an organ from one member of a species to a member of another species.
   (B) The mother or father of the patient is typically the best donor of a graft because they are two-haplotype matches.
   (C) The ABO blood groups of the donor and recipient do not have to be matched because they do not play a role in allograft rejection.
   (D) Even when a donor and a recipient are matched at both the class I and class II MHC loci, rejection can occur and the recipient should be given immunosuppressive drugs.
   (E) If the same donor is the source of tissue for two grafts to a recipient and the second graft is performed 1 month after the first graft is rejected, then the second graft will not be rejected.
2. Regarding the MHC proteins and the genes that encode them, which one of the following is the most accurate?
   (A) The genes encoding class II MHC proteins are highly polymorphic, whereas the genes encoding class I MHC proteins are not.
   (B) The genes encoding class I MHC proteins are located on a different chromosome from the genes encoding class II MHC proteins.

   (C) The genes are codominant, and each person expresses class I and class II MHC genes inherited from both mother and father.
   (D) Class II MHC proteins are found on the surface of all cells, whereas class I MHC proteins are found only on the surface of phagocytes.
3. Regarding the graft-versus-host reaction, which one of the following is the most accurate?
   (A) It occurs primarily when a kidney is transplanted.
   (B) It is caused primarily by mature T cells in the graft.
   (C) It occurs primarily when ABO blood groups are matched.
   (D) It occurs primarily when the donor is immunocompromised.
   (E) It occurs primarily when the haplotypes of the donor and recipient are matched.
4. Listed below are transplants between individuals with various genotypes and the outcome of these transplants. The genotypes are designated A and B for simplicity. A person who is AA or BB is homozygous, whereas a person who is AB is heterozygous. Regarding outcomes X and Y, which one of the following is the most accurate?

Genotype of Donor	Genotype of Recipient	Outcome of Transplant
AA	AA	Accepted
BB	BB	Accepted
AA	BB	Rejected
BB	AA	Rejected
AB	AA	X
AA	AB	Y

(A) X is accepted, and Y is accepted.
(B) X is accepted, and Y is rejected.
(C) X is rejected, and Y is accepted.
(D) X is rejected, and Y is rejected.

ANSWERS

(1) **(D)**
(2) **(C)**
(3) **(B)**
(4) **(C)**

PRACTICE QUESTIONS: USMLE & COURSE EXAMINATIONS

Questions on the topics discussed in this chapter can be found in the Immunology section of Part XIII: USMLE (National Board) Practice Questions starting on page 757. Also see Part XIV: USMLE (National Board) Practice Examination starting on page 775.

Complement

The complement system consists of approximately 20 proteins that are present in normal human (and other animal) serum. The term *complement* refers to the ability of these proteins to complement (i.e., augment) the effects of other components of the immune system (e.g., antibody). Complement is an important component of our innate host defenses.

There are three main effects of complement: (1) **lysis** of cells such as bacteria, allografts, and tumor cells; (2) **generation of mediators** that participate in inflammation and attract neutrophils; and (3) **opsonization** (i.e., enhancement of phagocytosis). Complement proteins are synthesized mainly by the liver.

ACTIVATION OF COMPLEMENT

Several complement components are proenzymes that must be cleaved to form active enzymes. Activation of the complement system can be initiated either by antigen–antibody complexes or by a variety of nonimmunologic molecules (e.g., endotoxin).

Sequential activation of complement components (Figure 63–1) occurs via one of three pathways: the *classical* pathway, the *lectin* pathway, and the *alternative* pathway (see later). Of these pathways, the **lectin and the alternative pathways are more important the first time** we are infected by a microorganism because the antibody required to trigger the classical pathway is not present.

All three pathways lead to the production of **C3b**. The presence of C3b on the surface of a microbe marks it as foreign and targets it for destruction. C3b has three important functions:

(1) Formation of the C5 convertase, leading to the membrane attack complex (MAC)

(2) Opsonizing bacteria for phagocytes with receptors for C3b on their surface

(3) Releasing derivatives that bind to a receptor on B cells that provides "signal 2" for T-cell–independent B-cell activation (see Chapter 61).

Classical Pathway

In the **classical** pathway, complement proteins first become **fixed (bound)** to an antigen–antibody complex (see Figure 63–1). Only IgM and IgG can fix complement proteins, because only the Fc regions of the γ and μ heavy chains have a C1 binding site. Complement fixation is the gathering together of bound proteins, starting a chain reaction of proteases.

In the classical pathway, C1[1] binds and is cleaved to form an active protease, which cleaves C2 and C4 to form a C4b,2b complex. The latter is *C3 convertase*, which cleaves C3 molecules into two fragments, C3a and C3b. C3a, an **anaphylatoxin**, is discussed later. C3b forms a complex with C4b,2b, producing a new enzyme, *C5 convertase* (C4b,2b,3b), which cleaves C5 to form C5a and C5b. C5a is also an anaphylatoxin and a chemotactic factor (see later). C5b binds to C6 and C7 to form a complex that interacts with C8 and C9 to produce the **membrane attack complex** (C5b,6,7,8,9).

The membrane attack complex forms a pore in whichever cell membrane was the original site of complement fixation. The pore causes leakage of water and electrolytes, leading to cytolysis. Note that for each complement protein, the "b" fragment

---

[1] C1 is composed of three proteins, C1q, C1r, and C1s. C1q is an aggregate of 18 polypeptides that binds to the Fc portion of IgG and IgM. It is multivalent and can cross-link several immunoglobulin molecules. Calcium is required for the activation of C1.

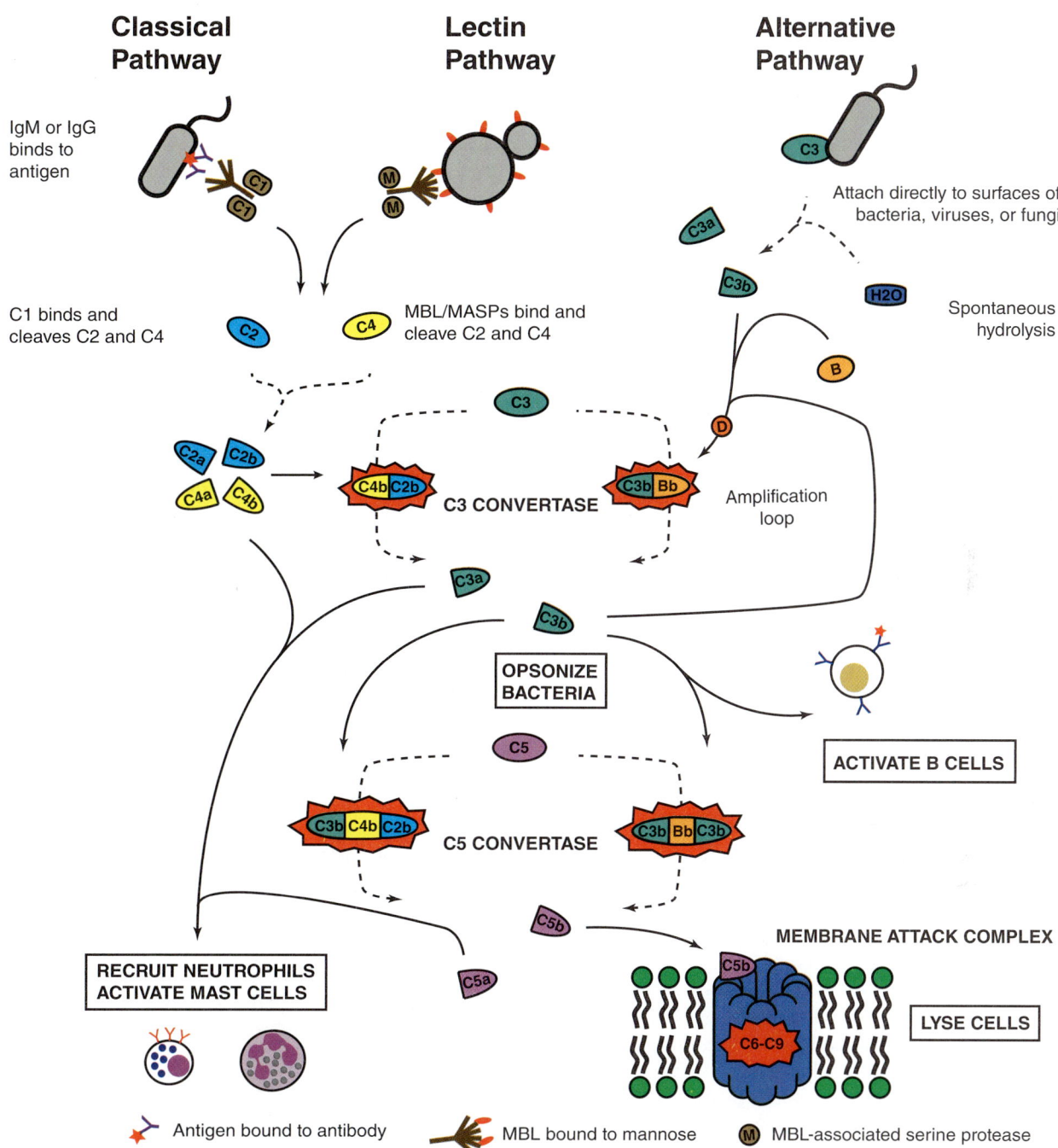

FIGURE 63–1 The classical and alternative pathways of the complement system indicate that proteolytic cleavage of the molecule at the tip of the arrow has occurred; a line over a complex indicates that it is enzymatically active. Note that all small fragments are labeled "a," and all large fragments are labeled "b." Hence, the C3 convertase is depicted as C4b,2b. Note that proteases associated with the mannan-binding lectin cleave C4 as well as C2.

continues in the main pathway, fixed to the target, whereas the "a" fragment is released.

Lectin Pathway

In the **lectin** pathway, mannan-binding lectin (MBL) (also known as mannose-binding protein) binds to the surface of microbes bearing **mannan** (a polymer of the sugar, mannose). This activates proteases associated with MBL that cleave C2 and

C4 components of complement, converging with the classical pathway (see Figure 63–1). Note that this process bypasses the antibody-requiring step and so is protective early in infection before antibody is formed.

Alternative Pathway

In the **alternative** pathway, many unrelated cell surface substances (e.g., bacterial lipopolysaccharides [endotoxin], fungal cell walls, and viral envelopes) can initiate the process by binding $C3(H_2O)$ and factor B. This complex is cleaved by a protease, factor D, to produce C3b,Bb. This acts as a C3 convertase, similar to C4b,2b, that generates more C3b. Like the lectin pathway, the alternative pathway is antibody independent and therefore is protective before antibody is formed.

REGULATION OF THE COMPLEMENT SYSTEM

Unrestrained complement activation can cause tissue injury and even systemic anaphylaxis, and complement is therefore closely regulated in several ways:

(1) In the classical pathway, complement activation requires **antibody**, but complement is not activated by free circulating IgM and IgG despite all being present in the blood. The complement-binding site on the heavy chain of IgM and IgG is unavailable to the C1 component of complement if antigen is not bound to these antibodies. When antigen binds the antibody, a conformational shift occurs allowing the C1 component to bind and initiate the cascade.

(2) **C1 inhibitor** (also called C1 esterase inhibitor) regulates the classical pathway by constantly inactivating the protease activity of C1. This means that initiation of the classical pathway can proceed only if sufficient C1 is fixed to overcome the threshold set by C1 inhibitor. (See below for discussion of clinical syndrome of C1 inhibitor deficiency.)

(3) The alternative pathway is regulated by factor H binding to C3b and cleavage of this complex by factor I. This reduces the amount of available C5 convertase (C3b,Bb). Like the C1 inhibitor, factors H and I create a threshold that requires sufficient C3b for the alternative pathway to proceed. Attachment of C3b to cell membranes and stabilization by properdin protect C3b from degradation by factors H and I.

(4) Human cells have **decay-accelerating factor** (DAF, CD55) and **CD59** on their surface. DAF acts by binding to C3b and C4b and limiting the formation of C3 convertase and C5 convertase. CD59 binds and inhibits C9 polymerization. Both DAF and CD59 therefore prevent the formation of the MAC and protect the cells from lysis.

BIOLOGIC EFFECTS OF COMPLEMENT

Opsonization

Microbes, such as bacteria and viruses, are phagocytized much better in the presence of C3b because there are C3b receptors on the surface of many phagocytes.

Chemotaxis

C5a and the C5,6,7 complex attract neutrophils, which migrate especially well toward C5a. C5a also enhances the adhesiveness of neutrophils to the endothelium.

Anaphylatoxin

C3a, C4a, and C5a cause degranulation of mast cells that releases active mediators, such as histamine. These increase vascular permeability and smooth muscle contraction, especially contraction of the bronchioles, leading to bronchospasm. Anaphylatoxins can also bind directly to smooth muscle cells of the bronchioles and cause bronchospasm. C5a is the most potent of these anaphylatoxins. Anaphylaxis caused by these complement components is less common than anaphylaxis caused by type I (IgE-mediated) hypersensitivity (see Chapter 65).

Cytolysis

Insertion of the MAC into the cell membrane forms a "pore" in the membrane. This disruption allows entry of water and electrolytes, killing many types of cells, including erythrocytes, bacteria, and tumor cells. Gram-negative bacteria, especially *Neisseria* species, are very susceptible to MAC.

Enhancement of Antibody Production

The binding of C3b derivatives to its receptors on the surface of activated B cells (complement receptor 2 [CR2]) provides the signal 2, which greatly enhances antibody production compared with that by B cells that are activated by antigen alone (see Chapter 61). The clinical importance of this is that patients who are deficient in C3b produce significantly less antibody than do those with normal amounts of C3b and are therefore susceptible to repeated pyogenic infections.

CLINICAL ASPECTS OF COMPLEMENT

(1) Inherited (or acquired) deficiency of some complement components, especially **C5–C8 deficiency**, enhances susceptibility to *Neisseria* bacteremia and other infections that are particularly sensitive to killing by the MAC. **Deficiency of C3 leads to severe, recurrent pyogenic sinus and respiratory tract infections**.

(2) **Deficiency of C1 inhibitor results in angioedema**. When the amount of inhibitor is reduced, the activation threshold for C1 is also reduced. C1 inhibitor also inhibits the activity of several coagulation and vasoactive mediators such as Factor XI, Factor XII, plasmin, and kallikrein. The result of a reduced amount of C1 inhibitor is capillary permeability and edema in several organs. C1 inhibitor deficiency can occur as an acquired disease or, rarely, as a genetic disease called **hereditary angioedema**.

(3) **Deficiency of the GPI (glycosylphosphatidylinositol) anchor** for DAF and CD59 in the cell membrane reduces the levels of CD59, a protein that blocks MAC formation. Red blood cells are particularly sensitive to MAC lysis, an effect that may be enhanced in the setting of low blood oxygen and pH. Reduced levels of DAF and CD59 result in increased hemolysis by MAC, causing **paroxysmal nocturnal hemoglobinuria (PNH)**. Clinically, PNH is characterized by episodes of brownish urine (hemoglobinuria), which reflects hemolysis, particularly early in the morning. The complement-mediated

hemolysis may occur more at night because of normal variation in oxygen and pH during sleep.

(4) In **transfusion mismatches** (e.g., when type A blood is given by mistake to a person who has type B blood), antibody to the A antigen in the recipient binds to A antigen on the donor red cells, complement is activated, and large amounts of anaphylatoxins and MACs are generated. The anaphylatoxins cause shock, and the MACs cause red cell hemolysis.

(5) In certain autoimmune or inflammatory diseases, such as acute glomerulonephritis and systemic lupus erythematosus, **immune complexes** formed by antigens and antibodies can bind (fix) complement, attracting neutrophils that damage tissue. Because complement proteins are bound and consumed, the levels of available complement proteins (e.g., C3 and C4) in the blood are low, a feature that can be used to diagnose these diseases.

(6) Patients with **severe liver disease** (e.g., cirrhosis or chronic hepatitis B), who have lost significant liver function and therefore cannot synthesize sufficient complement proteins, are predisposed to infections caused by pyogenic bacteria.

SELF-ASSESSMENT QUESTIONS

1. Regarding the complement pathway, which one of the following is the most accurate?
   - (A) C3 convertase protects normal cells from lysis by complement.
   - (B) C3a is a decay-accelerating factor and causes the rapid decay and death of bacteria.
   - (C) In general, gram-positive bacteria are more likely to be killed by complement than gram-negative bacteria.
   - (D) The membrane attack complex is formed as a result of activation of the classical pathway but not by activation of the alternative pathway.
   - (E) The first time a person is exposed to a microorganism, the alternative pathway of complement is more likely to be activated than the classical pathway.

2. Of the following complement components, which one is the most important opsonin?
   - (A) C1
   - (B) C3a
   - (C) C3b
   - (D) C5a
   - (E) C5b

3. Of the following complement components, which one is the most potent in attracting neutrophils to the site of infection (i.e., acting as a chemokine)?
   - (A) C1
   - (B) C2

   - (C) C3b
   - (D) C5a
   - (E) Mannan-binding lectin

4. Of the following, which one is the most important function of the complex formed by complement components C5b,6,7,8,9?
   - (A) To enhance antibody production
   - (B) To inhibit immune complex formation
   - (C) To opsonize viruses
   - (D) To perforate bacterial cell membranes
   - (E) To release histamine from mast cells

5. A deficiency of which one of the following complement components predisposes to bacteremia caused by members of the genus *Neisseria*?
   - (A) C1
   - (B) C3b
   - (C) C5a
   - (D) C5b
   - (E) C5b,6,7,8,9

6. Your patient is a 20-year-old woman who complains of swellings on her arms and legs and a feeling of fullness in her throat that makes it difficult to breath. The swellings are not red, hot, or tender. You suspect she may have angioedema caused by a complement abnormality. Of the following, which one is the most likely explanation?
   - (A) She has too little C1 inhibitor.
   - (B) She has too little C3b.
   - (C) She has too little factor B.
   - (D) She has too much C5a.
   - (E) She has too much C9.

ANSWERS

(1) **(E)**
(2) **(C)**
(3) **(D)**
(4) **(D)**
(5) **(E)**
(6) **(A)**

PRACTICE QUESTIONS: USMLE & COURSE EXAMINATIONS

Questions on the topics discussed in this chapter can be found in the Immunology section of Part XIII: USMLE (National Board) Practice Questions starting on page 757. Also see Part XIV: USMLE (National Board) Practice Examination starting on page 775.

Antigen–Antibody Reactions in the Laboratory

64

CHAPTER CONTENTS

Reactions of antigens and antibodies are highly specific. Because of the high specificity, reactions between antigens and antibodies are suitable for identifying one by using the other. This is the basis of **serologic** tests. Table 64–1 describes the medical importance of serologic (antibody-based) tests. Their major uses are in the diagnosis of infectious diseases, in the diagnosis of autoimmune diseases, and in the typing of blood and tissues prior to transplantation.

The results of many immunologic tests are expressed as a **titer**, which is defined as the highest *dilution* (or, in other words, the

smallest *concentration*) of the specimen (e.g., serum) that still gives a positive reaction in the test. Note that a patient's serum with an antibody titer of, for example, 1/64 contains **more** antibodies (i.e., is a higher titer) than a serum with a titer of, for example, 1/4.

Cross-reaction can occasionally occur between related antigens and antibodies, limiting the usefulness of certain tests. **Monoclonal antibodies** excel in the identification of antigens because they consist of a single antigen-binding specificity (i.e., they lack cross-reacting antibodies found in polyclonal antibody preparations). Chapter 61 discusses the generation of specific

TABLE 64–1 Major Uses of Serologic (Antibody-Based) Tests

I. Diagnosis of infectious diseases
- When the organism cannot be cultured (e.g., syphilis and hepatitis A, B, and C).
- When the organism is too dangerous to culture (e.g., rickettsial diseases).
- When culture techniques are not readily available (e.g., HIV, EBV).
- When the organism takes too long to grow (e.g., *Mycoplasma*).
  One problem with this approach is that it takes time for antibodies to form (e.g., 7–10 days in the primary response). For this reason, acute and convalescent serum samples are taken, and a fourfold or greater rise in antibody titer is required to make a diagnosis. By this time, the patient has often recovered and the diagnosis becomes a retrospective one. If a test is available that can detect IgM antibody in the patient's serum, it can be used to make a diagnosis of current infection. In certain infectious diseases, an arbitrary IgG antibody titer of sufficient magnitude is used to make a diagnosis.

II. Diagnosis of autoimmune diseases
- Antibodies against various normal body components are used (e.g., antibody against DNA in systemic lupus erythematosus, antibody against human IgG [rheumatoid factor] in rheumatoid arthritis).

III. Determination of blood type and HLA type
- Known antibodies are used to determine ABO and Rh blood types.
- Known antibodies are used to determine class I and class II HLA proteins prior to transplantation, although DNA sequencing is also being used.

EBV = Epstein-Barr virus; HIV = human immunodeficiency virus; HLA = human leukocyte antigen.

antibodies, including monoclonal antibodies used for diagnostic purposes.

TYPES OF DIAGNOSTIC TESTS

Many types of diagnostic tests are performed in the immunology laboratory. Most of these tests are designed to determine the presence of either antigen or antibody. To do this, one of the components, either antigen or antibody, is of a known quantity or concentration and the other is unknown. For example, with a known antigen such as influenza virus, a test can determine whether antibody to the virus is present in the patient's serum. Alternatively, with a known antibody, such as antibody to herpes simplex virus, a test can determine whether viral antigens are present in cells taken from the patient's lesions.

Agglutination

In this test, the antigen is **particulate** (e.g., bacteria and red blood cells) or is an inert particle (latex beads) coated with an antigen. Antibody, because it is divalent or multivalent, cross-links the antigenically multivalent particles and forms a latticework, and clumping (agglutination) can be seen. When red blood cells are used as the particulate antigen, the reaction is called hemagglutination. One commonly used hemagglutination test can determine a person's ABO blood group (Figure 64–1; see the section on blood groups at the end of this chapter and Figure 64–12).

Precipitation (Precipitin)

In this test, the antigen is **in solution**. The antibody cross-links antigen molecules in variable proportions, and aggregates (precipitates) form. In the **zone of equivalence**, optimal proportions of antigen and antibody combine; the maximal amount of precipitates forms, and the supernatant contains neither an excess

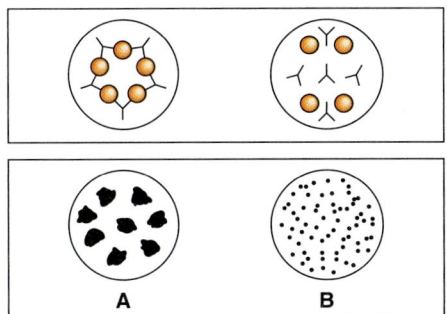

FIGURE 64–1 Agglutination test to determine ABO blood type. On the slide at the bottom of the figure, a drop of the patient's blood was mixed with antiserum against either type A (**left**) or type B (**right**) blood cells. Agglutination (clumping) has occurred in the drop on the left containing the type A antiserum but not in the drop containing the type B antiserum, indicating that the patient is type A (i.e., has A antigen on the red cells). The slide at the top shows that the red cells (circles) are cross-linked by the antibodies ("Y" shapes) in the drop on the left but not in the drop on the right. If agglutination had occurred in the right side as well, it would indicate that the patient was producing B antigen as well as A and was type AB.

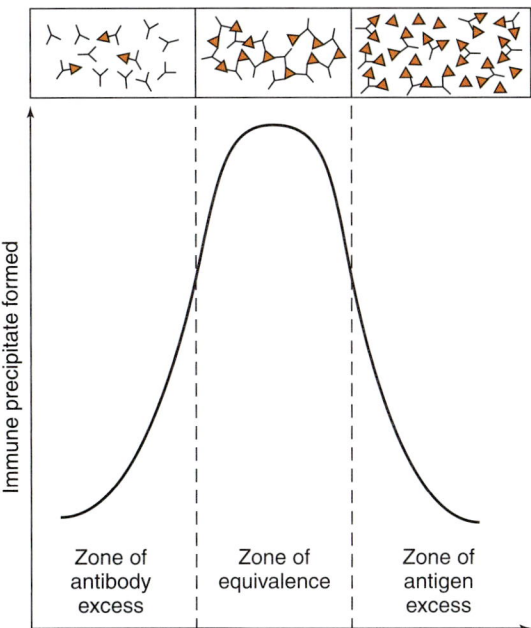

FIGURE 64–2 Precipitin curve. In the presence of a constant amount of antibody, the amount of immune precipitate formed is plotted as a function of increasing amounts of antigen. In the top part of the figure, the binding of antigen (▲) and antibody (Y) in the three zones is depicted. In the zones of antibody excess and antigen excess, a lattice is not formed and precipitation does not occur, whereas in the equivalence zone, a lattice forms and precipitation is maximal. (Reproduced with permission from Stites DP, Terr A, Parslow T: *Basic & Clinical Immunology*, 9th ed. New York, NY: McGraw Hill; 1997.)

of antibody nor an excess of antigen (Figure 64–2). In the **zone of antibody excess**, there is too much antibody for efficient lattice formation, and precipitation is less than maximal.[1] In the **zone of antigen excess**, all antibody has combined, but precipitation is reduced because many antigen–antibody complexes are too small to precipitate (i.e., they are "soluble"). Precipitation in solution can be used to determine the levels of immunoglobulins (IgM, IgG, etc.) in a patient's plasma. Precipitation can also be visualized in a gel matrix. Antigens and antibodies are allowed to mix by passive diffusion or by applying an electric current to spatially organize antigens by charge and size.

Enzyme-Linked Immunosorbent Assay (ELISA)

This method can be used for the quantitation of either antigens or antibodies in patient specimens. It is based on covalently linking an enzyme to a known antigen or antibody, reacting the enzyme-linked material with the patient's specimen, and then assaying for enzyme activity by adding the substrate of the enzyme (Figure 64–3). The method is highly sensitive and can

---

[1]The term "prozone" refers to the failure of a precipitate or flocculate to form because too much antibody is present. For example, a false-negative serologic test for syphilis (VDRL) is occasionally reported because the antibody titer is too high. Dilution of the serum yields a positive result.

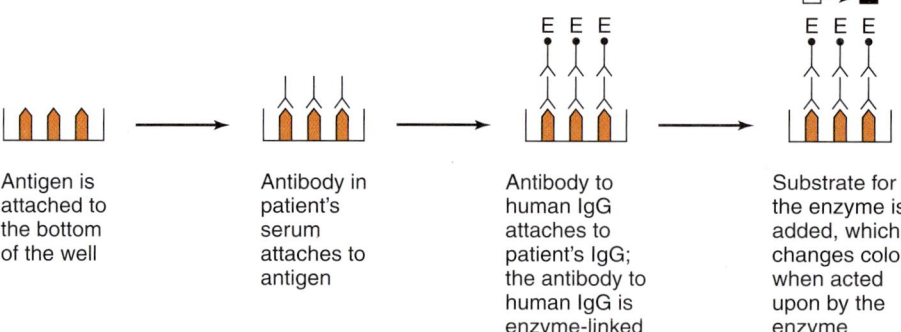

FIGURE 64–3 Enzyme-linked immunosorbent assay (ELISA). The term enzyme-linked refers to the covalent binding (linking) of an enzyme to antibody to human IgG. If the patient has antibodies to the microbial or viral antigen, those antibodies will bind to the microbial or viral antigens. The antibody to human IgG linked to the enzyme will then bind to the patient's antibodies. Then when the substrate of the enzyme is added, the substrate changes color, indicating that the patient's serum contained antibodies.

be performed in a multi-well tray, allowing many tests to be run at the same time.

For measurement of antibody, known antigens are fixed to a surface (e.g., the bottom of small wells on a plastic plate), incubated with dilutions of the patient's serum, washed, and then reincubated with antibody to human IgG labeled with an enzyme (e.g., horseradish peroxidase). Enzyme activity is measured by adding the substrate for the enzyme and estimating the color reaction in a spectrophotometer. The amount of antibody bound is proportional to the enzyme activity. The titer of antibody in the patient's serum is the highest dilution of serum that gives a positive color reaction.

Immunofluorescence (Fluorescent Antibody)

Fluorescent dyes (e.g., fluorescein and rhodamine) can be covalently attached to antibody molecules and made visible by exposing the sample to light of the correct excitation spectrum in a fluorescence microscope. Such "labeled" antibody can be used to identify antigens (e.g., on the surface of bacteria such as streptococci and treponemes, in cells in histologic section, or in other specimens) (Figure 64–4). The immunofluorescence

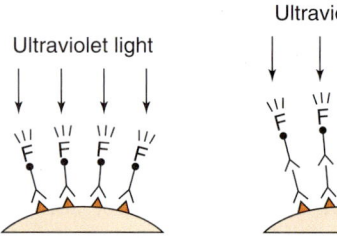
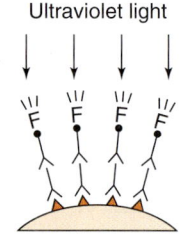

A. Direct fluorescent-antibody test

B. Indirect fluorescent-antibody test

FIGURE 64–4 Fluorescent antibody test. **A:** In the direct fluorescent antibody test, the fluorescent dye is attached directly to the antibody that is interacting with the antigen (dark triangles) on the surface of the cell. **B:** In the indirect fluorescent antibody test, the fluorescent dye is attached to antibody made against human IgG.

reaction is **direct** when known labeled antibody interacts directly with unknown antigen and **indirect** when a two-stage process is used. For example, in the *indirect* assay, known antigen is attached to a slide, the patient's serum (unlabeled) is added, and the preparation is washed; if the patient's serum contains antibody against the antigen, it will remain fixed to it on the slide and can be detected on addition of a fluorescent dye-labeled antibody against human IgG and examination by fluorescent microscopy. The indirect test is often more sensitive than direct immunofluorescence, because more labeled antibody adheres per antigenic site, amplifying the signal.

Complement Fixation

The complement system consists of 20 or more plasma proteins that interact with one another and with cell membranes in a sequential cascade (see Chapter 63). Antigen–antibody complexes are among the triggers that initiate the cascade (e.g., in the classic pathway), and complement fixation can therefore be used to detect antibody or antigen if the other is known (Figure 64–5).

First, the patient's serum is heated to inactivate any complement activity. Next, antigen or antibody (whichever is the "known" quantity in the reaction) is mixed with the serum containing the "unknown" ingredient. For example, to determine whether a patient's serum contains antibodies to a certain antigen, a measured amount of that antigen is added. In addition, a measured amount of complement is added. If the antigen and antibody match, they will combine and use up ("fix") the complement. Finally, an indicator system, consisting of "sensitized" red blood cells (i.e., red blood cells plus anti-red blood cell antibody), is added last.

If the antibody matches the antigen, complement is fixed and is unavailable to attach and lyse the sensitized red blood cells. The red blood cells will remain **unhemolyzed** (i.e., the test is **positive**) because the patient's serum has antibodies to that antigen and all the complement is used up in the first step. If the antibody *does not* match the antigen in the first step, complement remains free to attach to the sensitized red blood cells, and they are **lysed** (i.e., the test is **negative**). The result is expressed

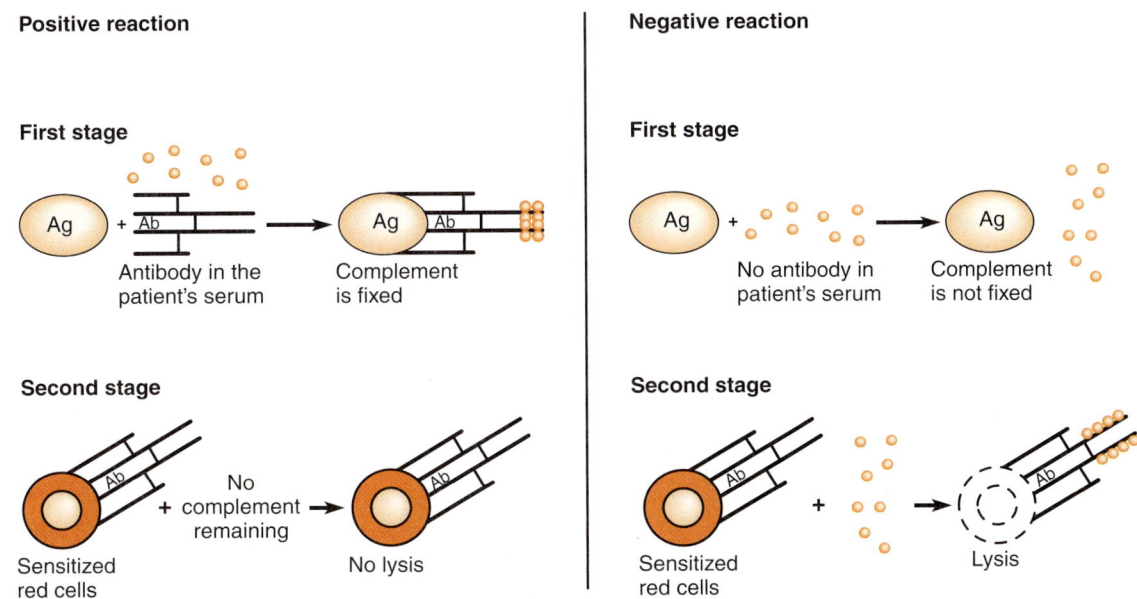

FIGURE 64–5 Complement fixation. **Left:** Positive reaction (i.e., the patient's serum contains antibody). If a known antigen is mixed with the patient's serum containing antibody against that antigen, then complement (solid circles) will be fixed. Because no complement is left over, the sensitized red cells are *not* lysed. **Right:** Negative reaction. If a known antigen is mixed with the patient's serum that does *not* contain antibody against that antigen, complement (solid circles) is *not* fixed. Complement is left over and the sensitized red cells are lysed. Ab = antibody; Ag = antigen.

as the highest dilution (lowest concentration) of serum that gives positive results.

Neutralization Tests

These use the ability of antibodies to block the effect of toxins or the infectivity of viruses. They can be used in cell culture or in host animals. For example, a patient's serum specimen is added to a culture of cells along with a known quantity of virus that kills the cells. If there is a concentration of serum at which the cells are protected from viral killing, this determines the patient's titer of protective (neutralizing) antibodies. Alternatively, this assay can be used to detect unknown quantities of virus by adding measured quantities of antibodies specific for various viruses. If an antibody added to the aliquots blocks whatever virus is in the culture, this identifies the virus in the culture.

Immune Complexes

Immune complexes in tissue sections can be stained with fluorescent complement, which will bind to the Fc portion of IgM and IgG (see Chapters 61 and 63). These can be detected using fluorescent microscopy. Immune complexes in serum can be detected by binding to C1q or by attachment to certain cells in culture.

Antiglobulin (Coombs) Test

Some patients with certain diseases (e.g., hemolytic disease of the newborn [Rh incompatibility] and drug-related hemolytic anemias) become sensitized against red blood cell antigens but do not exhibit overt symptoms of disease. In these patients, antibodies against the red cells are formed and bind to the red cell surface but do not cause hemolysis. These cell-bound antibodies can be detected by the *direct* antiglobulin (Coombs) test, in which antiserum against human immunoglobulin is used to agglutinate the patient's red cells. In some cases, antibody against the red cells is not bound to the red cells but is in the serum, and the *indirect* antiglobulin test for antibodies in the patient's serum should be performed. In the indirect Coombs test, the patient's serum is mixed with normal red cells, and antiserum to human immunoglobulins is added. If antibodies are present in the patient's serum, agglutination occurs.

Western Blot (Immunoprecipitation)

In this test, protein antigens are separated by size in a gel in an electric field. The antigens are then transferred from the gel (i.e., blotted) onto filter paper, and antibodies are then added. Antigen–antibody binding can be detected by direct or indirect methods, using radioactivity or an enzyme-substrate reaction (similar to ELISA). As with the other test methods, either known antigens or antibodies could be used to identify unknown antibody or antigen, respectively. Figure 64–6 illustrates a Western blot test for the presence of HIV antibodies in the patient's serum. Note that the Western blot test to detect HIV antibodies is no longer being used clinically but it is an excellent example of how the test is carried out. Western blot tests are used in the diagnosis of Lyme disease.

Flow Cytometry & Fluorescence-Activated Cell Sorting

This test is commonly used to count the number of various types of immune cells in a sample of blood, bone marrow, or lymphoid tissue (Figure 64–7). For example, it is used

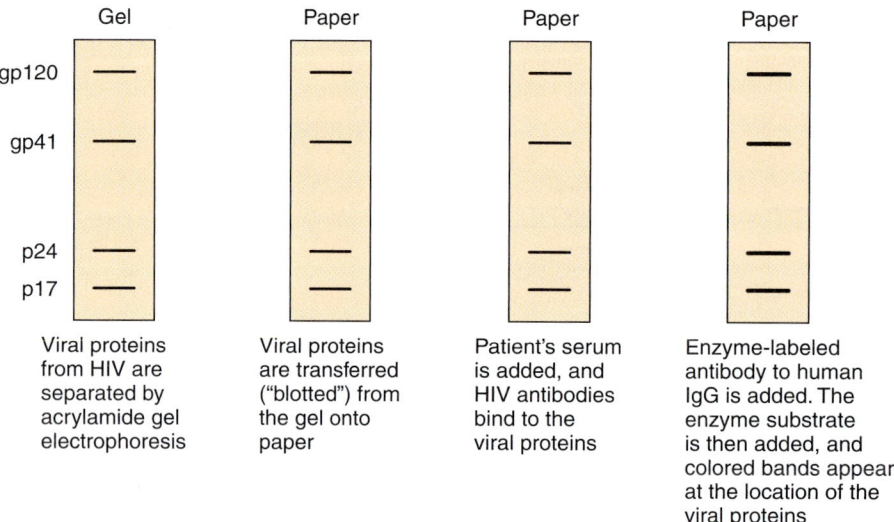

FIGURE 64–6 Western blot (immunoblot test). In this test, microbial or viral proteins are separated on an acrylamide gel and then transferred (blotted) onto paper. The patient's serum then interacts with the separated proteins. If antibodies are present in the patient's serum, they bind to the proteins. The patient's antibodies are then detected by using labeled antibody to human IgG.

in HIV-infected patients to determine the number of CD4-positive T cells. In this test, the patient's cells are mixed with fluorescently tagged monoclonal antibodies specific to different proteins on immune cells of interest (e.g., CD4 protein, if the number of helper T cells is to be determined). The antibodies have a fluorescent tag, such as fluorescein or rhodamine, which

is excited by a specific wavelength of light. The flow cytometer instrument passes the cells one-by-one through a laser beam of the appropriate wavelength of light. If the antibodies are bound to the cell, the fluorescent tag emits a signal that is detected by the instrument, and the number of cells and their fluorescence intensity are recorded (see Figure 64–7B).

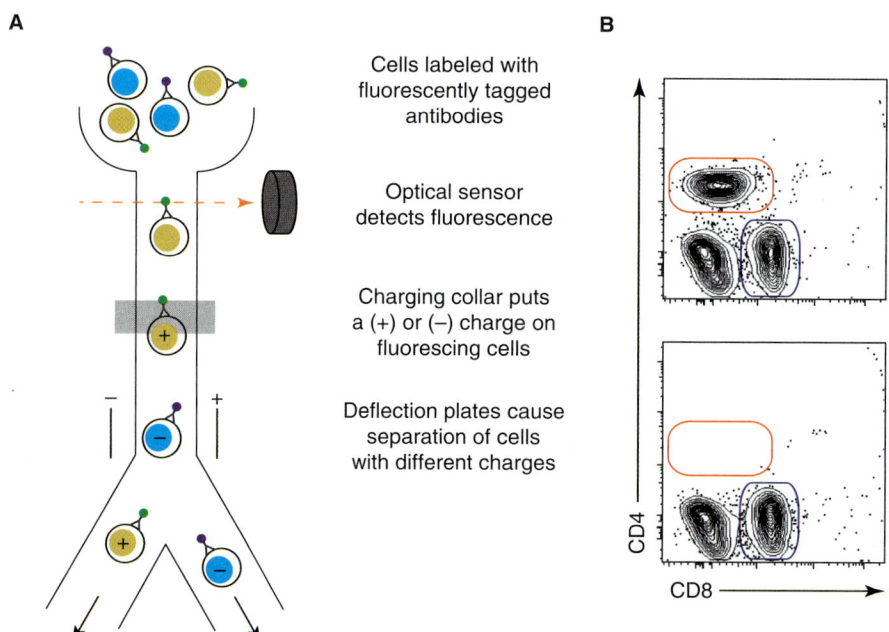

FIGURE 64–7 Flow cytometry. **A:** At the top of the figure, two types of cells interact with monoclonal antibodies labeled with fluorescent dyes. The cells are passed one-by-one through a tube. As the cell passes down the tube, laser light of a specific wavelength causes the dyes to fluoresce and a sensor counts the cells. Depending on the detected fluorescence, farther down the tube, an electrical charge is put on each cell. The fluorescence allows the cells to be counted, and the charge allows the cells to be deflected into a test tube and subjected to additional analysis. **B:** Example of flow cytometry of cells isolated from a lymph node, showing cells that are CD4 positive (red oval) and CD8 positive (blue oval). Cells in the bottom left of the plot have neither CD4 nor CD8 on their surface. The top graph depicts a normal sample, and the bottom graph depicts nearly complete depletion of CD4-positive cells.

A more sophisticated instrument called a fluorescence-activated cell sorter (FACS) does one additional step. A cell sorter isolates each cell within an individual fluid droplet before it passes through the laser beam. Cells that are bound by the fluorescently tagged antibodies are detected and then quickly *sorted* away from other cells in the sample by shunting the droplets into separate sample tubes (see Figure 64–7A).

ANTIGEN–ANTIBODY REACTIONS INVOLVING RED BLOOD CELL ANTIGENS

Many different blood group systems exist in humans. Each system consists of a gene locus specifying antigens on the erythrocyte surface. The two most important blood groupings, ABO and Rh, are described next.

The ABO Blood Groups & Transfusion Reactions

All human erythrocytes contain alloantigens (i.e., antigens that vary among individual members of a species) of the ABO group. A person's ABO blood group is a very important determinant of the success of both blood transfusions and organ transplants.

The A and B alleles of the ABO gene encode enzymes that add specific sugars to the end of a polysaccharide chain on the surface of many cells, including red cells (Figure 64–8). The alleles are codominant, so people who inherit both are type AB, whereas people who inherit neither allele are type O. People

TABLE 64–2 ABO Blood Groups

Group	Antigen on Red Cell	Antibody in Plasma
A	A	Anti-B
B	B	Anti-A
AB	A and B	No anti-A or anti-B
O	No A or B	Anti-A and anti-B

who are either homozygous AA or heterozygous AO are type A (have an extra *N*-acetylgalactosamine on the red blood cell H antigen), whereas people who are either homozygous BB or heterozygous BO are type B (have an extra galactose on the red blood cell H antigen). Thus, the A and B antigens are carbohydrates that only differ by a single sugar! Despite this small difference, A and B antigens are different enough that antibodies that bind one antigen do not "cross-react" with the other.

There are four combinations of the A and B antigens, called A, B, AB, and O (Table 64–2). A person's blood group is determined by mixing the person's blood with antiserum against A antigen on one area on a slide and with antiserum against B antigen on another area (Figure 64–9). If agglutination occurs only with A antiserum, the blood group is A; if it occurs only with B antiserum, the blood group is B; if it occurs with both A and B antisera, the blood group is AB; and if it occurs with neither A nor B antisera, the blood group is O. In the United States, the approximate percentage of each blood group is: type O: 45%, type A: 40%, type B: 11%, and type AB: 4%.

People who are group O have antibodies that bind group A and group B antigens. How does this happen? These antibodies are formed against bacterial polysaccharides, and they happen to cross-react with A or B polysaccharides. Anti-A and anti-B antibodies are formed through T-cell–independent B-cell activation and are therefore primarily of the IgM class (see Chapter 61). They are first detectable at 3 to 6 months of age.

Importantly, group A individuals (with A antigens) have anti-B antibodies but lack anti-A antibodies, and group B individuals (with B antigens) similarly have anti-A antibodies but lack anti-B antibodies. How does this happen? During the development of B-cell precursors in the bone marrow, *negative selection* causes any precursor clones with antigen receptors that strongly recognize "self" antigens to be deleted by apoptosis (see Chapter 59). The result is that any potential B-cell clones that make anti-A immunoglobulins in a person with blood group A are removed from the eventual pool of mature B cells. Therefore, individuals will always be **tolerant** to their own blood group antigens, and the antigens and their corresponding antibodies *do not* coexist in the same person's blood.

Transfusion reactions occur when **incompatible** donor red blood cells are transfused (e.g., if group A blood was transfused into a group B person, i.e., who has anti-A antibodies). The anti-A antibodies bind to the donor red cells forming red cell–antibody complexes. This activates complement (see Chapter 63), and a cascading reaction of **anaphylactic shock** occurs due to large amounts of C3a and C5a (anaphylatoxins) and hemolysis caused by C5, C6, C7, C8, and C9 (membrane attack complex) (Figure 64–10).

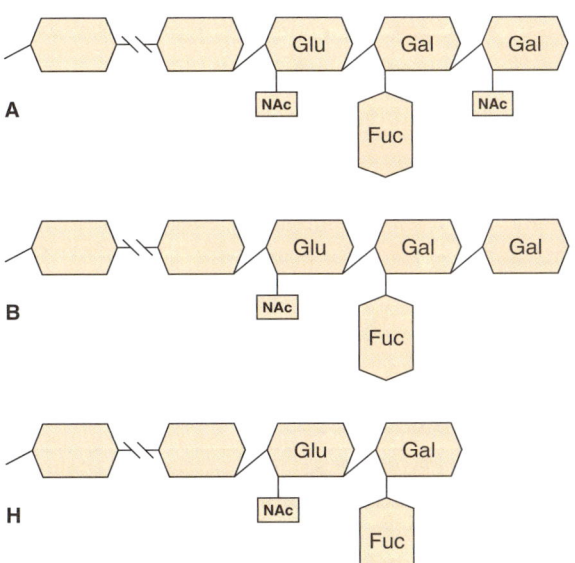

FIGURE 64–8 ABO blood groups. Structures of the terminal sugars that determine ABO blood groups are shown. Blood group O cells have H antigen on their surface; blood group A cells have *N*-acetylgalactosamine added to the end of the H antigen; and blood group B cells have galactosamine added to the end of the H antigen. (Reproduced with permission from Stites DP, Stobo JD, Wells JV: *Basic & Clinical Immunology*, 6th ed. New York, NY: McGraw Hill; 1987.)

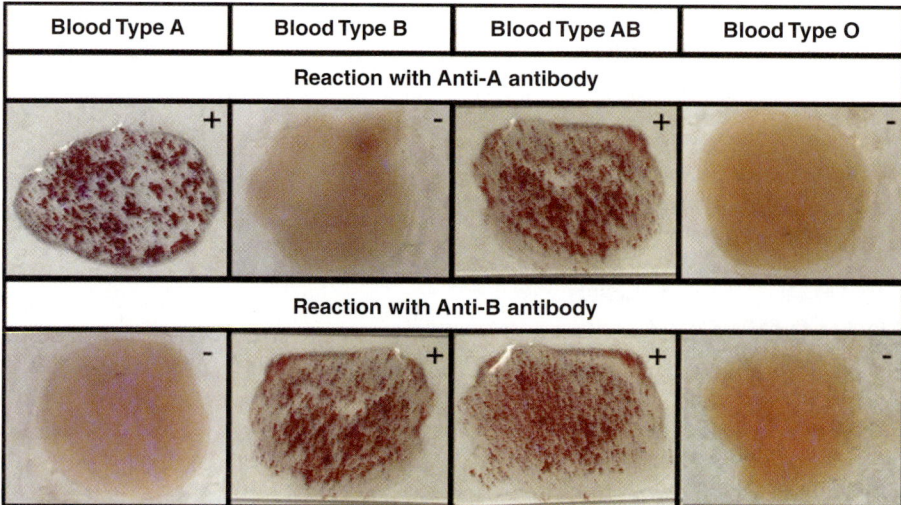

Blood Type A	Blood Type B	Blood Type AB	Blood Type O
Reaction with Anti-A antibody			
+	−	+	−
Reaction with Anti-B antibody			
−	+	+	−

FIGURE 64–9 Determination of ABO blood type. In this agglutination test, the donor's red blood cells are mixed with antiserum against Type A and Type B. In the **left** column, the red cells have agglutinated with Type A antiserum but not Type B antiserum so the donor is Type A. In the right column, no agglutination with either antiserum is observed, so the donor is Type O. The reactions in the middle two columns are self-explanatory. (Used with permission from Professor Elizabeth Joyce, University of California, San Francisco School of Medicine.)

To avoid antigen–antibody reactions that would result in transfusion reactions, all blood for transfusions must be carefully **matched** (i.e., erythrocytes are typed for their surface antigens by specific sera). As shown in Table 64–2, persons with

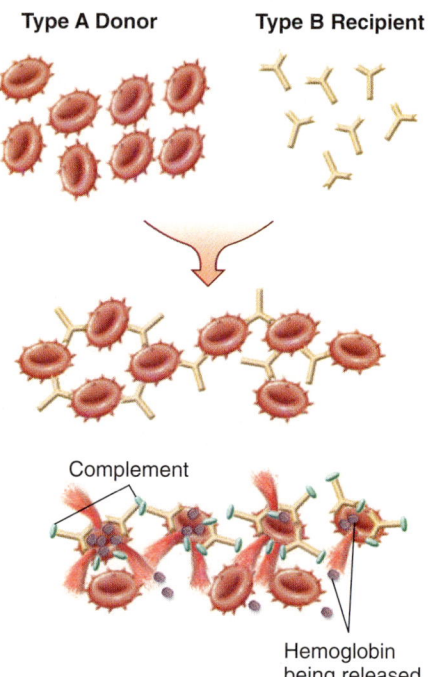

FIGURE 64–10 Transfusion reaction. **Top panel:** Red blood cells bearing A antigen are transfused into a person who is type B and therefore has antibodies to A antigen. **Middle panel:** Anti-A antibodies bind to A antigen on the red cells causing agglutination of red cells that can block movement of blood through capillaries causing anoxia to tissue. **Bottom panel:** Complement is activated by the antigen–antibody complexes, and the membrane attack complex lyses the red cells, causing hemolysis and anemia. (Reproduced with permission from Cowan MK, Talero KP: *Microbiology: A Systems Approach*, 2nd ed. New York, NY: McGraw Hill; 2009.)

group O blood have no A or B *antigens* on their red cells and so are **universal donors** (i.e., they can give blood to people in all four groups) (Table 64–3). Note that type O serum does have A and B *antibodies*. Therefore, when type O blood is given to a person with type A, B, or AB blood, you might expect that the antibodies in the type O serum fraction would cause a reaction. However, a clinically detectable reaction does not actually occur because the blood used for transfusions is usually **packed with red blood cells**, and not **whole blood**. Packed red blood cell transfusions contain *extremely* small amounts of donor antibody, and whatever small amount is present is rapidly diluted below a significant level. Those with group AB blood have neither A nor B antibody and thus are **universal recipients**.

ABO blood group differences can lead to neonatal jaundice and anemia, but the effects on the fetus are usually less severe than those seen in Rh incompatibility (see next section). As described earlier, anti-A and anti-B antibodies are *usually* of the **IgM** class and therefore do not cross the placenta (see Chapter 61). However, when a mother and father have different blood groups, there are rare occasions when the mother's adaptive immune system can

TABLE 64–3 Compatibility of Blood Transfusions Between ABO Blood Groups[1]

Donor	Recipient			
	O	A	B	AB
O	Yes	Yes	Yes	Yes
A (AA or AO)	No	Yes	No	Yes
B (BB or BO)	No	No	Yes	Yes
AB	No	No	No	Yes

[1]Yes indicates that a blood transfusion from a donor with that blood group to a recipient with that blood group is compatible (i.e., no hemolysis will occur). No indicates that the transfusion is incompatible and that hemolysis of the donor's cells will occur.

become *sensitized* to the father's blood group. When this occurs, **IgG** antibodies are generated against the A and/or B antigens absent from the mother. These IgG antibodies *can* pass through the placenta and, if the fetus has the father's blood group, can cause lysis of fetal red cells.

Rh Blood Type & Hemolytic Disease of the Newborn

About 85% of humans have erythrocytes that express the Rh(D) antigen on their surface. They are said to be Rh-positive. The remaining 15% lack the Rh(D) protein.

The Rh status of parents is clinically important because a specific combination can result in **hemolytic disease of the newborn (erythroblastosis fetalis)**. When an **Rh-negative woman** has an **Rh-positive fetus** (the D gene being inherited from the father), the Rh(D) antigen on the fetal red blood cells will *sensitize* the mother's adaptive immune response, leading to development of anti-Rh(D) IgG antibodies (Table 64–4). This sensitization occurs most often during delivery of the first Rh(D)-positive child, when Rh(D) erythrocytes of the fetus leak into the maternal circulation (Figure 64–11).

If the mother does form anti-Rh(D) antibodies in this way, subsequent Rh(D) pregnancies are at risk of hemolytic disease of the newborn (erythroblastosis fetalis). This disease results from the passage of maternal IgG anti-Rh(D) antibodies through the placenta to the fetus, with subsequent lysis of the fetal erythrocytes. The direct antiglobulin (Coombs) test is typically positive (see earlier description of the Coombs test).

The problem can be prevented if the mother's adaptive immune system is not allowed to be sensitized to red cells

carrying Rh(D) antigens. This is achieved by administration of **high-titer Rh(D) immune globulins (Rho-Gam)** to an Rh(D) mother at 28 weeks of gestation and immediately upon the delivery of any Rh(D) child. These antibodies promptly attach to Rh(D) erythrocytes in the maternal circulation and prevent their acting as sensitizing antigen. (Note that Rho-Gam is an IgG antibody and will cross the placenta when given at 28 weeks. But the amount given is calculated to be just enough to neutralize Rh(D) in the maternal circulation, i.e., not enough to bind and cause significant fetal hemolysis.). This prophylaxis is widely practiced and effective.

TABLE 64–4 **Rh Status and Hemolytic Disease of the Newborn**

Rh Status			
Father	Mother	Child	Hemolysis[1]
+	+	+ or −	No
+	−	+	No (1st child) Yes (2nd child and subsequent children)
+	−	−	No
−	+	+ or −	No
−	−	−	No

[1]No indicates that hemolysis of the newborn's red cells will not occur and that hemolytic disease will therefore not occur. Yes indicates that hemolysis of the newborn's red cells is likely to occur and that symptoms of hemolytic disease will therefore probably occur.

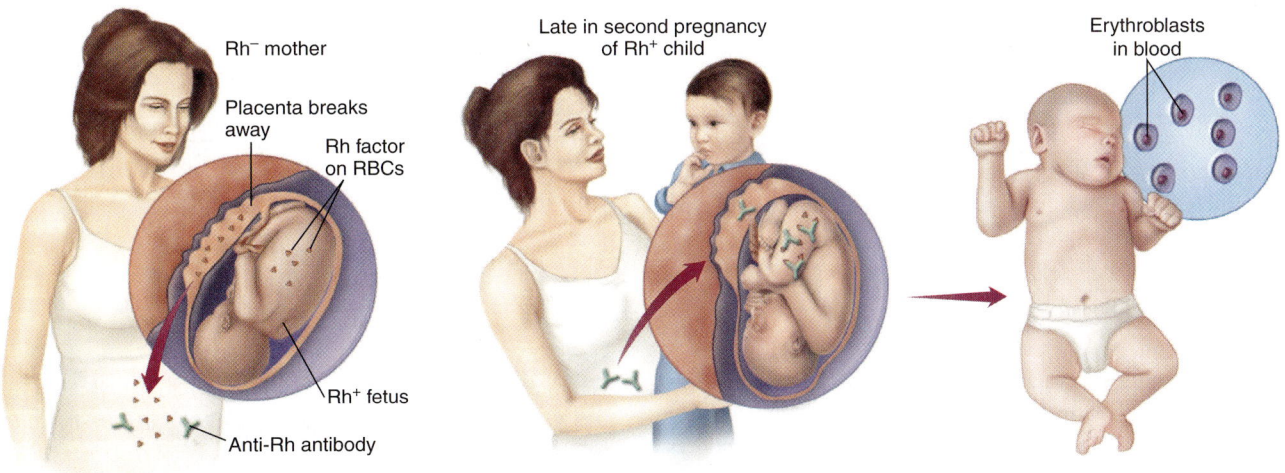

FIGURE 64–11 Hemolytic disease of the newborn (erythroblastosis fetalis). **Left panel:** Fetal red cells (RBCs) bearing the Rh antigen enter the mother's blood when the placenta separates during the birth of the first Rh-positive child. IgG antibodies to Rh antigen are then produced by the mother. **Center panel:** During a second pregnancy with an Rh-positive fetus, IgG antibodies pass from the mother into the fetus via the placenta. The antibodies bind to the fetal red cells, complement is activated, and the membrane attack complex lyses the fetal red cells. **Right panel:** Anemia and jaundice occur in the fetus/newborn. As a result of the anemia, large numbers of erythroblasts are produced by the bone marrow and are seen in the blood of the newborn. (Reproduced with permission from Cowan MK, Talero KP: *Microbiology: A Systems Approach*, 2nd ed. New York, NY: McGraw Hill; 2009.)

SELF-ASSESSMENT QUESTIONS

1. Which one of the following laboratory tests would be the best to determine the number of CD4-positive cells in the blood of a patient infected with HIV?
   (A) Agglutination
   (B) Complement fixation
   (C) Enzyme-linked immunosorbent assay (ELISA)
   (D) Flow cytometry
   (E) Immunoelectrophoresis

2. You have just received a lab report that says your patient is positive for IgM antibody to *Borrelia burgdorferi* in an enzyme-linked immunosorbent assay (ELISA). This supports your clinical impression that the patient has Lyme disease. Which one of the following best describes how the ELISA was performed? (For brevity, the wash steps have been left out.)
   (A) The patient's serum was reacted with antibody to human mu heavy chain. Then *Borrelia* antigens labeled with an enzyme were added. Then the enzyme substrate was added, and a color change was observed.
   (B) The patient's serum was reacted with *Borrelia* antigens. Then antibody to human mu heavy chain labeled with an enzyme was added. Then the enzyme substrate was added, and a color change was observed.
   (C) *Borrelia* antigens were reacted with antibody to human mu heavy chain. Then the patient's serum labeled with an enzyme was added. Then the enzyme substrate was added, and a color change was observed.
   (D) *Borrelia* antigens were reacted with antibody to human mu heavy chain labeled with an enzyme. Then the patient's serum was added. Then the enzyme substrate was added, and a color change was observed.

3. Regarding ABO blood groups, which one of the following is the most accurate?
   (A) People who are blood group O have the O antigen on the surface of their red cells.
   (B) The A and B blood group antigens are located on the surface of red cells but not on the surface of other cells.
   (C) The differences between the A and B blood group antigens are dependent on the presence of different D-amino acids on the cell surface.
   (D) People who are blood group O do not have antibodies to A and B blood group antigens and thus can be given both type A and type B blood.
   (E) The genes that determine ABO blood groups are codominant, so a person who is blood group AB is expressing both genes that encode the enzymes that synthesize the A and the B blood group antigens.

4. Regarding hemolytic disease of the newborn (erythroblastosis fetalis), which one of the following is the most accurate?
   (A) Maternal red cells are the source of the antigen that induces the antibody.
   (B) It typically occurs when the father is Rh-positive and the mother is Rh-negative.
   (C) Maternal IgM anti-Rh antibody enters the fetus and causes damage to the fetal red cells.
   (D) Symptomatic disease is more likely to occur in the first child than in the subsequent children.
   (E) Administration of Rh antigen to the newborn can prevent symptomatic disease if given early enough.

5. You think your patient has secondary syphilis, and you order a VDRL serologic test. The lab reports that the test is negative. If this is a false-negative result due to the "prozone" phenomenon, which one of the following is the most likely explanation?
   (A) The patient's serum has too much antibody, and the reaction is in the zone of antibody excess.
   (B) The patient's serum has too much antigen, and the reaction is in the zone of antigen-excess phase.
   (C) The patient's serum has too little antibody, and the reaction is in the zone of antibody-deficient phase.
   (D) The patient's serum has too little antigen, and the reaction is in the zone of antigen-deficient phase.
   (E) The patient's serum has an amount of antibody that puts it in the zone of equivalence.

6. As part of a murder investigation, the blood group of the victim was determined by analyzing the antibodies in her serum. (Unfortunately, the red cells of the victim were lost by the crime squad, so they had to use her serum.) In this test, red cells known to be either O, A, B, or AB were mixed with her serum and agglutination observed. Based on the results in the following table, what is the blood group of the victim?

Red Blood Cells Used	Agglutination Seen With Victim's Serum
O	No
A	Yes
B	Yes
AB	Yes

   (A) Type O
   (B) Type A
   (C) Type B
   (D) Type AB
   (E) A laboratory error has occurred, and the test should be repeated

ANSWERS

(1) **(D)**
(2) **(B)**
(3) **(E)**
(4) **(B)**
(5) **(A)**
(6) **(A)**

PRACTICE QUESTIONS: USMLE & COURSE EXAMINATIONS

Questions on the topics discussed in this chapter can be found in the Immunology section of Part XIII: USMLE (National Board) Practice Questions starting on page 757. Also see Part XIV: USMLE (National Board) Practice Examination starting on page 775.

Hypersensitivity (Allergy)

Hypersensitivity reactions are exaggerated or inappropriate immune responses to benign antigens. It is the immune response, not the antigens, that is harmful to the host. Usually, hypersensitivity reactions occur in response to *external*, or "non-self," antigens (covered in this chapter), whereas autoimmune reactions (see Chapter 66) occur in response to *internal*, or "self," antigens.

Hypersensitivity reactions are antigen-specific, meaning that the first contact with the antigen sensitizes the immune system (i.e., *primes* the adaptive immune system), and subsequent contacts elicit the hypersensitive (allergic) response. Within an individual, these subsequent antigen exposures elicit similar clinical manifestations, although the severity of the hypersensitivity reactions may increase with time.

Hypersensitivity reactions can be subdivided into four main types. Types I, II, and III are **antibody-mediated**, whereas type IV is **cell–mediated** (Table 65–1). The immunologic reactions are summarized in Table 65–1. The clinical manifestations of the hypersensitivity reactions are described in Table 65–2.

TYPE I: IMMEDIATE (ANAPHYLACTIC) HYPERSENSITIVITY

Sometimes the term *allergy* is used broadly referring to any food or drug intolerance, but a more accurate **definition of allergy** includes only the immediate immunoglobulin (Ig) E-mediated reactions discussed in this section.

TABLE 65–1 Immunologic Aspects of Hypersensitivity Reactions

Type	Sensitization	Immunologic Reaction
I (Immediate, anaphylactic) IgE-mediated	Antigen (allergen) induces IgE antibody that binds to mast cells and basophils.	When exposed to the allergen again, the allergen cross-links the bound IgE on those cells. This causes degranulation and release of mediators (e.g., histamine).
II (Cytotoxic) IgG-mediated	Antigens on a cell surface elicit Tfh and B-cell activation.	Antibody binding to cell membrane antigens leads to complement-mediated lysis of the cells (e.g., transfusion or Rh reactions) or autoimmune hemolytic anemia.
III (Immune complex) Multiple types of antibodies	Soluble antigens elicit Tfh and B-cell activation.	Antigen–antibody immune complexes are deposited in tissues and neutrophils are attracted to the site. They release lysosomal enzymes, causing tissue damage. In the case of IgM or IgG, complement is also involved.
IV (Delayed) T-cell–mediated	CD4 and/or CD8 T cells sensitized by protein antigens.	Memory T cells release cytokines upon second contact with the same antigen. The lymphokines induce inflammation and activate macrophages, which, in turn, release various inflammatory mediators.

Tfh = T follicular helper cell.

TABLE 65–2   **Clinical Manifestations of Hypersensitivity Reactions**

Type	Typical Time of Onset	Clinical Manifestation or Disease
I (Immediate, anaphylactic)	Minutes	Systemic anaphylaxis, urticaria (hives), asthma, hay fever, allergic rhinitis, allergic conjunctivitis, atopic dermatitis (eczema), angioedema; allergies to foreign substances such as food (e.g., nuts, shellfish, eggs), pollen, penicillin, bee venom, or latex gloves
II (Cytotoxic)	Hours to days	Hemolytic anemia, neutropenia, thrombocytopenia, ABO transfusion reactions, Rh incompatibility (erythroblastosis fetalis, hemolytic disease of the newborn), rheumatic fever, Goodpasture's syndrome
III (Immune complex)	2–3 weeks	Systemic lupus erythematosus, rheumatoid arthritis, poststreptococcal glomerulonephritis, IgA nephropathy, serum sickness, hypersensitivity pneumonitis (e.g., farmer's lung)
IV (Delayed)	2–3 days	Contact dermatitis, poison oak/ivy, tuberculin skin test reaction, drug rash, Stevens-Johnson syndrome, toxic epidermal necrolysis, erythema multiforme

An immediate hypersensitivity reaction occurs when an antigen (allergen) binds to IgE on the surface of mast cells, resulting in the release of several mediators (see list of mediators that follows) (Figure 65–1). The process begins with sensitization to the antigen. **IgE antibody** is first formed through the combined activation of T follicular helper (Tfh) cells and class-switched B cells (see Chapters 60 and 61). The IgE binds firmly by its Fc portion to receptors on the surface of basophils and mast cells. Now, upon exposure to the antigen, many cell-surface IgE receptors can become *cross-linked*, activating the cells to release pharmacologically active mediators. Once the immune system has been sensitized by antigen and IgE is formed, mast cell *degranulation* can occur within minutes of antigen reexposure (**immediate phase**). Cyclic nucleotides and calcium play essential roles in release of the mediators.[1] Symptoms such as edema and erythema ("wheal and flare") and itching appear rapidly because these mediators (e.g., histamine) are preformed so releasing them does not require new mRNA or protein synthesis (see Chapter 58 and below).

The **late phase** of IgE-mediated inflammation occurs approximately 6 hours after exposure to the antigen and is due to mediators (e.g., leukotrienes) that are newly synthesized *after* the cells degranulate. These mediators cause an influx of inflammatory cells, such as neutrophils and eosinophils, and symptoms such as erythema and induration occur. For example, eosinophils play a major role in the late-phase reaction in asthma.

Note that the allergens involved in hypersensitivity reactions are substances in the environment, such as pollens, animal danders, foods (nuts, shellfish), and various drugs, to which most people do *not* exhibit clinical symptoms. However, some individuals respond to those substances by producing large amounts of IgE and, as a result, manifest various allergic symptoms. It is unknown why certain antigens produce allergies in certain people. One theory is that proteases, such as those found in fungal allergens, pollens, and dust mite feces, can cleave fibrinogen and other proteins that initiate the response.

The increased IgE is the result of high levels of interleukin (IL)-4 produced by Tfh cells, which signals newly activated B cells to class switch from IgM to IgE, and IL-21, which induces the B cells to mature into antibody-producing plasma cells. Nonallergic individuals might instead respond to the same antigen by producing IgG, which does not cause the release of mediators from mast cells and basophils. (There are no receptors for IgG on those cells.) Why some people are allergic (IgE-biased) and others are nonallergic (IgG-biased) is an area of active immunology research.

Systemic Anaphylaxis

Type I hypersensitivity can appear in various forms (e.g., urticaria [also known as hives], eczema, rhinitis and conjunctivitis [also known as hay fever], and asthma). Which clinical manifestation occurs depends in large part on the route of entry of the allergen and on the location of the mast cells bearing the IgE specific for the allergen. For example, some individuals sensitized to pollens in the air might have hay fever or asthma, whereas those sensitized to allergens in food might get swelling and itching of the lips, tongue, and throat.

The most severe form of type I hypersensitivity is **systemic anaphylaxis**, in which severe bronchoconstriction and hypotension (shock) can be life-threatening. Unlike rhinitis or urticaria, which occur locally, anaphylaxis occurs when large amounts of mediators are suddenly released systemically as a result of a massive dose of antigen abruptly combining with IgE on many mast cells. A sense of doom and dizziness can occur. Other symptoms include wheezing due to bronchoconstriction, hoarseness due to laryngeal edema, pruritus, and urticaria. Tachycardia, arrhythmia, cyanosis, and cardiac arrest can occur.

The most common causes of anaphylaxis are foods such as peanuts and shellfish, bee venom, and drugs such as penicillin. It is the proteins in peanuts, shellfish, and bee venom that cross-link adjacent IgEs and trigger the release of histamine and other

---

[1]An increase in cyclic guanosine monophosphate (GMP) within these cells increases mediator release, whereas an increase in cyclic adenosine monophosphate (AMP) decreases the release. Therefore, drugs that increase intracellular cyclic AMP, such as epinephrine, are used to treat type I reactions. Epinephrine also has sympathomimetic activity, which is useful in treating type I reactions.

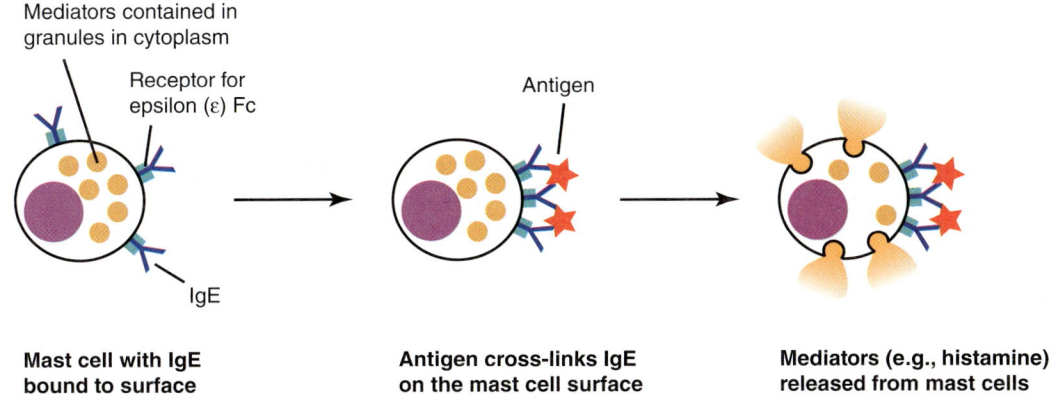

FIGURE 65–1 Immediate (anaphylactic) hypersensitivity. Antigen cross-links IgE on surface of mast cell. Mediators, e.g., histamine, are released from granules in the cytoplasm of the mast cell.

mediators from mast cells and basophils. Drugs such as penicillin are haptens that need to bind to human proteins to cross-link adjacent IgEs (see Chapter 57).

Of particular interest to medical personnel are type I hypersensitivity reactions to the wearing of latex rubber gloves, which include urticaria, asthma, and even systemic anaphylaxis. Table 65–3 summarizes some of the important clinical aspects of immediate hypersensitivities.

One interesting food allergy occurs in some people who have been bitten by a tick. They experience symptoms, such as hives, shortness of breath, or vomiting when they eat red meat (beef, pork, and lamb). This syndrome is sometimes called "alpha-gal" because it is likely caused by injection of a digalactoside in tick saliva during the bite. IgE antibodies to the digalactoside form in the affected person. These antibodies trigger the type 1 hypersensitivity reaction when the person eats red meat containing the digalactoside. Symptom formation is unusual in that it occurs 4 to 6 hours after ingestion of the red meat whereas most food allergies occur within minutes of ingesting the allergen.

No single mediator accounts for all the manifestations of type I hypersensitivity reactions. Some important mediators and their effects are as follows:

(1) **Histamine** is stored in granules of tissue mast cells and basophils in a preformed state. It causes vasodilation, increased capillary permeability, and smooth muscle contraction. Clinically, disorders such as allergic rhinitis (hay fever), urticaria, and angioedema can occur. Histamine also contributes to bronchospasm. Antihistamine drugs block histamine receptor sites and can be relatively effective in allergic rhinitis but not in asthma (see later).

(2) **Slow-reacting substance of anaphylaxis (SRS-A)** consists of several **leukotrienes**, which do not exist in a preformed state but are produced during anaphylactic reactions. This accounts for the slow onset of the effect of SRS-A. Leukotrienes are formed from arachidonic acid by the lipoxygenase pathway and cause increased vascular permeability and smooth muscle contraction. They are the principal mediators in the bronchoconstriction of asthma and are not influenced by antihistamines.

TABLE 65–3 Important Clinical Aspects of Immediate Hypersensitivities

Main Organ Affected	Disease	Main Symptoms	Typical Allergens	Route of Acquisition
Lung	Asthma	Wheezing, dyspnea, tachypnea	Pollens, house dust (feces of dust mite), animal danders, many occupational airborne allergens	Inhalation
Nose and eyes	Rhinitis, conjunctivitis, "hay fever"	Runny nose, redness and itching of eyes	Pollens	Contact with mucous membranes
Skin	1. Eczema (atopic dermatitis) 2. Urticaria (hives)	Pruritic, vesicular lesions Pruritic, bullous lesions	Uncertain 1. Various foods 2. Drugs	Uncertain Ingestion Various
Intestinal tract	Eosinophilic esophagitis	Vomiting, dysphagia, esophageal "rings"	Various foods	Ingestion
Systemic	Anaphylaxis	Shock, hypotension, wheezing, pruritis, urticaria, asphyxia, cardiac arrest	1. Insect venom (e.g., bee venom) 2. Drugs (e.g., penicillin) 3. Foods (e.g., peanuts)	Sting Various Ingestion

(3) **Eosinophil chemotactic factor of anaphylaxis (ECF-A)** is a tetrapeptide that exists preformed in mast cell granules. When released during anaphylaxis, it attracts eosinophils that are prominent in immediate allergic reactions. Eosinophils release histaminase and arylsulfatase, which degrade two important mediators, histamine and SRS-A, respectively, and therefore may reduce the severity of the type I response. On the other hand, eosinophils also release cytotoxic proteins that damage tissue, which, in the lung and intestine, can contribute to organ remodeling and predispose to further reactions upon subsequent allergen exposure.

(4) **Prostaglandins and thromboxanes** are related to leukotrienes. They are derived from arachidonic acid via the cyclooxygenase pathway. Prostaglandins cause dilation and increased permeability of capillaries and bronchoconstriction. Thromboxanes aggregate platelets.

(5) **Platelet-activating factor (PAF)** is a phospholipid produced by mast cells that can cause bronchoconstriction, hypotension, and vascular permeability.

The aforementioned mediators are active only for a few minutes after release; they are enzymatically inactivated and resynthesized slowly. Manifestations of anaphylaxis vary because mediators are released at different rates in different amounts, and tissues vary in their sensitivity to them.

(6) **Cytokines** released by T cells recruit innate cells and amplify their function. As mentioned earlier, Tfh cell-derived IL-4 is responsible for inducing class switching in B cells to generate the IgE that causes mast cell and basophil degranulation. Th-2 cells also make IL-4, IL-13 (which is closely related to IL-4 and shares an IL-4 receptor), IL-5, and IL-9 (see Chapter 60). (All four of the genes encoding these cytokines are in close proximity in a locus on chromosome 5.)

These cytokines work together in tissues to prolong and enhance allergic hypersensitivity. For example, in allergic airway disease (asthma), the airway hyperactivity and eosinophil recruitment are caused by IL-13. Macrophages can be "alternatively activated" by IL-13 to promote fibrosis and remodeling associated with healing injured tissues. The Th-2 cells also make IL-5, which prolongs eosinophil survival and activity in tissues, and IL-9, which activates mast cells to produce additional cytokines. Monoclonal antibodies against IgE, IL-13, and IL-5, as well as small-molecule inhibitors of the IL-4/IL-13 receptor, have all shown benefits in treating patients with severe allergic asthma.

In contrast to anaphylactic reactions, which are IgE-mediated, **anaphylactoid** reactions, which appear clinically similar to anaphylactic ones, are not IgE-mediated. In anaphylactoid reactions, the inciting agents, usually drugs or iodinated contrast media, directly induce the mast cells and basophils to release their mediators without the involvement of IgE.

Atopy

Atopic disorders, such as hay fever, asthma, eczema, and urticaria, are immediate hypersensitivity reactions with both an **environmental trigger** and a strong **familial predisposition**. Processes that play a role in atopy include:

(1) Dysfunction of the **barrier** leading to enhanced antigen exposure

(2) Increased **allergen uptake and presentation** by dendritic cells

(3) Dysregulation of **T cells** (e.g., increased production of IL-4, Th-2 differentiation, and IgE)

(4) More sensitive **response to allergic mediators**.

Atopy is associated with **elevated IgE levels**, and target tissues often contain large numbers of **Th-2 cells**, which play a major role in the pathogenesis of atopic reactions.

Several genes associated with atopy have been identified. Mutations in the gene encoding the alpha chain of the IL-4 receptor strongly predispose to atopy. These mutations enhance the effectiveness of IL-4, resulting in an increased amount of IgE synthesis by B cells. Other genes identified include the gene for IL-4 itself, the gene for the receptor for the IgE heavy chain, and several class II major histocompatibility complex (MHC) genes.

It is estimated that up to 40% of people in the United States have experienced an atopic disorder at some time in their lives. The incidence of allergic diseases, such as asthma, has increased markedly in the developed countries of North America and Europe since the middle of the twentieth century. The developing world has not shown a similar increase. This suggests that, in addition to a genetic disposition, there is also a strong environmental component.

One theory that has been proposed to explain this observation is called the "hygiene hypothesis," which states that **people who are exposed to certain microbes early in childhood development are protected from atopic diseases**. These microbes, which might include viruses, commensal bacteria or fungi, protozoa, and even parasitic worms, are agents that co-evolved with us, helping shape our immune system, and that are still highly prevalent in the developing world. The theory argues that eradicating them from the developed world has caused widespread **immune dysregulation**, leading to an increase in allergic and autoimmune diseases. Improved understanding of the human microbiome and work using animal models have provided some support for this theory, although it is debated which specific microbes might be responsible.

Atopic episodes are triggered by allergen exposure. The allergens are antigens that are typically found in the environment (e.g., pollens released by plants and dust mite feces often found in bedding and carpet) or in foods (e.g., shellfish, eggs, and nuts). Many sufferers will exhibit strong immediate-type reactions at the site of skin pricks containing the offending allergen, and this form of skin testing with a variety of antigens is a rapid way to identify which allergens are responsible for disease flares. A supportive test is measuring the levels of IgE specific for potentially offending allergens.

Drug Hypersensitivity

Drugs, particularly antimicrobial agents such as penicillin, are now among the most common causes of hypersensitivity reactions. Usually, it is not the intact drug that induces antibody formation but rather a metabolic product of the drug that acts as a hapten (see Chapter 57). Type I hypersensitivity occurs if the resulting IgE antibody reacts with the hapten.[2] When reexposed to the drug, the person may exhibit a drug rash, fever, or local or systemic anaphylaxis of variable severity.

Desensitization

Acute desensitization involves the administration of very small amounts of antigen at 15-minute intervals. Antigen–IgE complexes form on a small scale, and not enough mediator is released to cross-link IgE receptors on mast cells systemically. This permits the administration of a drug or foreign protein to a hypersensitive person. However, once the drug is discontinued and cleared from the body, the hypersensitive state returns because the levels of specific IgE will remain elevated.

Chronic desensitization involves the long-term weekly administration of the antigen to which the person is hypersensitive. This stimulates the production of IgA- and IgG-blocking antibodies, which can prevent the binding of the antigen to the specific IgE on the mast cell surface, thus preventing a reaction. It also induces regulatory T cells to produce IL-10, which reduces the synthesis of IgE.

Treatment & Prevention

Anaphylaxis is a life-threatening emergency due to the combination of airway compromise and hypotensive (distributive) shock. Treatment of **anaphylactic reactions** includes (1) drugs to counteract the action of mediators, (2) ensuring a protected airway, and (3) support of respiratory and cardiac function. Epinephrine, antihistamines, corticosteroids, or cromolyn sodium, either singly or in combination, should be given. Cromolyn sodium prevents release of mediators (e.g., histamine) from mast cell granules. Prevention relies on identification of the allergen by a skin test and avoidance of that allergen.

There are several approaches to the **treatment of asthma**. Inhaled β-adrenergic bronchodilators, such as albuterol, are commonly used. Corticosteroids, such as prednisone, are also effective but carry significant toxicity if used chronically. A monoclonal anti-IgE antibody (omalizumab) and monoclonal antibodies that block IL-4, IL-13, and IL-5 signaling might be indicated for patients with severe asthma whose symptoms are not controlled by corticosteroids. For the prevention of asthma, leukotriene receptor inhibitors, such as montelukast (Singulair), and cromolyn sodium are effective.

The treatment of allergic **rhinitis** typically involves antihistamines along with nasal decongestants. For allergic **conjunctivitis**, eye drops containing antihistamines or vasoconstrictors are effective. **Atopic skin disease** is generally treated with topical corticosteroids. In all types of allergic disease, avoidance of the inciting allergens, such as pollens, is helpful in prophylaxis.

Prevention of peanut allergy can be achieved by exposure to small amounts of peanut allergens during the first years of life in children with a strong family history of atopy. Palforzia is a drug approved for the prevention of life-threatening anaphylaxis in individuals who are hypersensitive to peanuts. It is a powder manufactured from defatted peanut flour and is taken orally with food. The mechanism of action is uncertain.

TYPE II: CYTOTOXIC HYPERSENSITIVITY

Cytotoxic hypersensitivity occurs when antibody directed at **antigens of the cell membrane** activates complement (Figure 65–2). This generates a membrane attack complex (see Chapter 63) or activates antibody-dependent cellular cytotoxicity (ADCC) (see Chapter 61), leading to cell death. The antibody (IgG or IgM) attaches to the antigen via its Fab region and its Fc region "fixes" complement. As a result, there is complement-mediated lysis, as in hemolytic anemias, ABO transfusion reactions, or Rh hemolytic disease. In addition to causing lysis, complement cleavage products attract phagocytes and natural killer (NK) cells to the site, and binding of the IgG Fc to Fc-gamma receptors (FcγR) on these innate immune cells activates them to release of enzymes that damage the target cell membranes.

Drugs (e.g., penicillins, phenacetin, quinidine) can attach to surface proteins on red blood cells and initiate antibody formation. Such autoimmune antibodies (IgG) then interact with the red blood cell surface and result in hemolysis. The direct antiglobulin (Coombs) test is typically positive (see Chapter 64).

Other drugs (e.g., quinine) can attach to platelets and induce autoantibodies that lyse the platelets, producing thrombocytopenia and, as a consequence, a bleeding tendency. Others (e.g., hydralazine) may modify host tissue and induce the production of autoantibodies directed at cell DNA. As a result, disease manifestations resembling those of systemic lupus erythematosus (SLE) occur.

Certain infections (e.g., *Mycoplasma pneumoniae* infection) can induce antibodies that cross-react with red cell antigens, resulting in hemolytic anemia. In rheumatic fever, antibodies against the group A streptococci cross-react with cardiac tissue. In Goodpasture's syndrome (see Chapter 66), antibody to basement membranes of the kidneys and lungs bind to those structures and activate complement. Severe damage to the membranes is caused by proteases released from leukocytes attracted to the site by complement component C5a (see Chapter 63).

TYPE III: IMMUNE COMPLEX HYPERSENSITIVITY

Immune complex hypersensitivity occurs when **antigen–antibody complexes** induce an inflammatory response in tissues (Figure 65–3). The key distinction between type III reactions and the process described earlier for the type II reactions is the type and location of antigens: in type II hypersensitivity, the antigens are either bound to or an integral part of, *cellular membranes*, whereas in type III hypersensitivity, the antigens are *freely circulating* (i.e., soluble). Each antibody

---

[2]Some drugs are involved in cytotoxic hypersensitivity reactions (type II) and in serum sickness (type III).

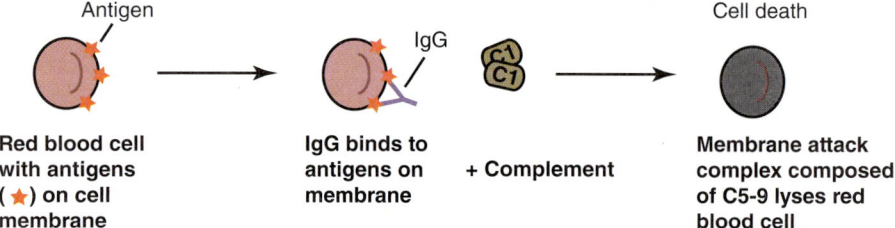

FIGURE 65–2 Cytotoxic hypersensitivity. IgG antibody binds to antigen on the cell surface. Complement is activated and the membrane attack complex lyses the cell.

has two Fab binding sites, and most antigens are multivalent, meaning they can be bound by more than one antibody. As a result, soluble antigens and antibodies can form large web-like *complexes* similar to the precipitation reactions described in Chapter 64.

Normally, immune complexes are promptly removed by the *reticuloendothelial system*, a network of endothelial cells and phagocytes in the liver and other organs. But occasionally, they persist and are **deposited in tissues**, causing disease. In persistent microbial or viral infections, immune complexes may be deposited, for example in the kidney. In autoimmune disorders, "self" antigens may elicit antibodies that deposit in organs as complexes, especially in joints (arthritis), kidneys (nephritis), or blood vessels (vasculitis).

Wherever immune complexes composed of IgM or IgG are deposited, there is also the potential to activate the complement system. Polymorphonuclear cells are attracted to the site, and inflammation and tissue injury occur. Two typical type III hypersensitivity reactions are the Arthus reaction and serum sickness.

Arthus Reaction

The Arthus reaction is **localized inflammation** caused by immune complex deposition. In 1903 Dr. Arthus first showed that animals with high levels of IgG antibody to an antigen (i.e., after repeated antigen exposure) develop intense edema and

hemorrhage, reaching a peak in 3 to 6 hours after that antigen is injected subcutaneously or intradermally. Antigen, antibody, and complement are deposited in vessel walls; neutrophil infiltration and intravascular clumping of platelets then occur. These reactions can lead to vascular occlusion and necrosis. Because this requires that IgG must form large complexes, deposit in the capillaries, and activate the complement system, the Arthus reaction requires much more IgG than, for example, an IgE-mediated type I hypersensitivity reaction where IgE is already bound to mast cells that can degranulate immediately.

A clinical manifestation of the Arthus reaction is hypersensitivity pneumonitis (allergic alveolitis) associated with the inhalation of thermophilic actinomycetes ("farmer's lung") growing in plant material such as hay. There are many other occupation-related examples of hypersensitivity pneumonitis, such as "cheese-worker's lung," "woodworker's lung," and "wheat-miller's lung." Most of these are caused by the inhalation of some microorganism, either bacterium or fungus, growing on the starting material. An Arthus reaction can also occur at the site of tetanus immunizations if they are given at the same site with too short an interval between immunizations. (The minimum interval is usually 5 years.)

Serum Sickness

In contrast to the Arthus reaction, which is localized inflammation, serum sickness is a **systemic inflammatory response**

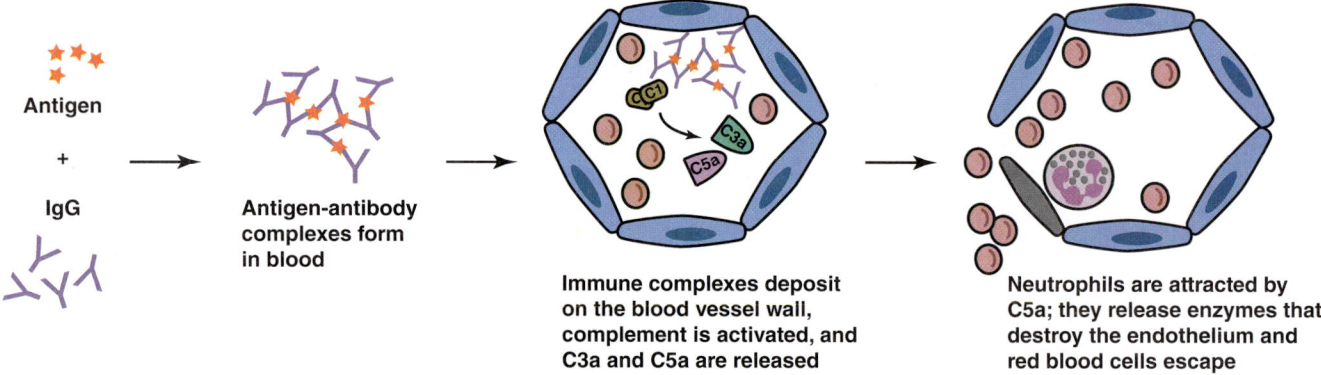

FIGURE 65–3 Immune complex hypersensitivity. Antigen-antibody complexes form in the blood and are deposited on the blood vessel wall. Complement is activated and C5a attracts neutrophils. Neutrophils release enzymes that degrade endothelium and red cells escape into tissue.

to the presence of immune complexes deposited in many areas of the body. The injection of foreign serum (i.e., serum from another animal such as a horse), a monoclonal antibody that contains mouse components (see Chapter 51), or even other nonantibody medications can result in adaptive immune system activation and sensitization. (When this occurs, it is often in the setting of preexisting inflammation, such as underlying cancer, infection, or autoimmune disease, when co-stimulatory signals will be highly expressed, as discussed in Chapter 60.) When the foreign substance has a long half-life or is administered repeatedly, the simultaneous presence of antigen and antibody leads to the formation of immune complexes, which may circulate or be deposited at various sites.

Typical serum sickness results in fever, urticaria, arthralgia, lymphadenopathy, splenomegaly, and eosinophilia a few days to 2 weeks after injection of the foreign serum or drug. Although it takes several days for symptoms to appear, serum sickness is classified as an immediate reaction because symptoms occur promptly after immune complexes form. Symptoms improve as the immune system removes the antigen and subside when the antigen is eliminated. A maculopapular drug-induced rash to penicillins, such as ampicillin, is quite common. Use of antithymocyte globulin, which is extracted from rabbits or horses, to provide immunosuppression in transplant patients may cause serum sickness (see Chapter 62). Note also that diphtheria antitoxin and snake antivenom made in horses are known to cause serum sickness.

Other Diseases with Immune Complex Deposition

Many clinical disorders associated with immune complexes have been described, although the antigen that initiates the disease is often in doubt. Several representative examples are SLE, IgA nephropathy, rheumatoid arthritis, and various types of vasculitis. These are described in Chapter 66. Note that many immune complex diseases primarily affect the kidney because the small capillaries of the glomeruli are particularly susceptible to complex deposition. Poststreptococcal glomerulonephritis, described below, is an excellent example. The exceptions are

rheumatoid arthritis, which does not affect the kidney, and vasculitis, which can affect many different organs. Not all immune complex diseases involve the complement system, as only IgM and IgG can "fix" complement.

Poststreptococcal Glomerulonephritis

Acute poststreptococcal glomerulonephritis is an immune complex disease that follows several weeks after a group A β-hemolytic streptococcal infection, particularly of the skin, and often with "nephritogenic" serotypes of *Streptococcus pyogenes*. Typically, the complement level is lower than the normal range, suggesting an antigen–antibody reaction causing consumption of complement proteins through their cleavage and activation. Lumpy deposits of immunoglobulin and C3 are seen along glomerular basement membranes by immunofluorescence, suggesting the presence of antigen–antibody complexes. These streptococcal antigen–antibody complexes, after being deposited on glomeruli, fix complement and attract neutrophils, which start the inflammatory process (see Chapters 61 and 63).

Similar lesions with "lumpy" glomerular deposits containing immunoglobulin and C3 occur in infective endocarditis, serum sickness, and certain viral infections (e.g., hepatitis B and dengue hemorrhagic fever).

TYPE IV: DELAYED (CELL-MEDIATED) HYPERSENSITIVITY

Delayed hypersensitivity is caused by **T lymphocytes, not antibody** (Figure 65–4). It can be transferred by sensitized T cells, *not* by serum. The response is "delayed" (i.e., it starts hours or days after contact with the antigen and often lasts for days).

In certain contact hypersensitivities, such as poison oak, the pruritic, vesicular skin rash is caused by CD8-positive cytotoxic T cells that attack skin cells that display the plant oil as a foreign antigen. In the tuberculin skin test, the indurated skin rash is caused by *memory* CD4-positive T cells and macrophages that reside near the injection site. Table 65–4 describes some of the important clinical aspects of delayed hypersensitivities.

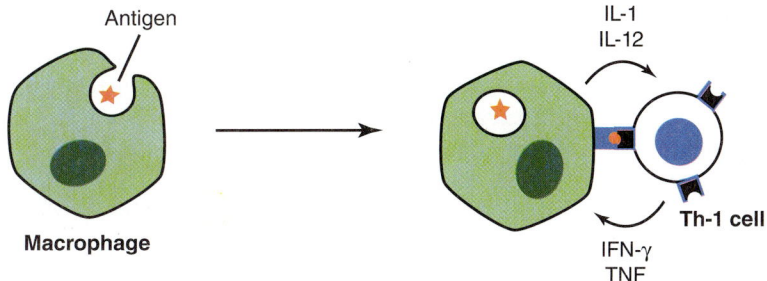

FIGURE 65–4 Delayed (cell-mediated) hypersensitivity. The macrophage ingests the antigen, processes it, and presents an epitope on its surface in association with class II major histocompatibility complex (MHC) protein. The helper T (Th-1) cell is activated and produces gamma interferon, which activates macrophages. These two types of cells mediate delayed hypersensitivity. IFN = interferon; IL = interleukin; TNF = tumor necrosis factor.

TABLE 65–4 Important Clinical Aspects of Delayed Hypersensitivities

Main Immune Cells Involved	Important Disease or Skin Test	Pathologic or Clinical Feature	Common Inducing Agents
CD4 (helper) T cells and macrophages	1. Tuberculosis, coccidioidomycosis	Granuloma	Constituents of bacterium or fungus
	2. Tuberculin or coccidioidin (or spherulin) skin tests	Induration	PPD (purified protein derivative) or coccidioidin (or spherulin)
CD8 (cytotoxic) T cells	1. Contact dermatitis	Pruritic, vesicular rash	Oil of poison oak or poison ivy, topical drugs, soaps, heavy metals (in jewelry)
	2. Erythema multiforme, Stevens-Johnson syndrome, toxic epidermal necrolysis	Target lesion	Herpes simplex virus-1, *Mycoplasma pneumoniae*, and sulfonamides

Contact Hypersensitivity

This manifestation of cell-mediated hypersensitivity occurs after sensitization with simple chemicals (e.g., nickel, formaldehyde), plant materials (e.g., urushiol in poison ivy and poison oak), topically applied drugs (e.g., sulfonamides, neomycin), some cosmetics, soaps, and other substances. Neomycin in topical antibacterial ointment is a very common cause.

In all cases, the small molecules acting as **haptens** enter the skin, attach to body proteins, and modify those proteins enough to "break tolerance." For example, normal skin proteins, to which the T cells tolerate as "self" due to **negative thymic selection** (see Chapter 59), after binding to nickel ions, can be altered enough that they are recognized as foreign. The skin proteins are taken up, processed, and presented to CD4-positive T cells by dendritic cells. The T cells differentiate into Th-1 and Th-17 cells, and upon later skin contact with nickel, these Th cells cause inflammation when they recognize the nickel-bound peptides presented by antigen-presenting cells in the nickel-exposed skin. The sensitized person develops **contact dermatitis** characterized by erythema, itching, vesicles, eczema, or necrosis of skin within 12 to 48 hours. More severe contact dermatitis, such as that seen in poison ivy or poison oak exposure, is the result of activation of both CD4-positive and CD8-positive T cells. Patch testing on a small area of skin can sometimes identify the offending antigen by stimulating memory T cells. Avoidance of the material will prevent recurrences.

Tuberculin-Type Hypersensitivity

Delayed hypersensitivity to antigens of microorganisms occurs in many infectious diseases and has been used as an aid in diagnosis. It is typified by the tuberculin reaction. The basis for this test is that patients with a latent *Mycobacterium tuberculosis* infection will generate a **memory T-cell** response that is dispersed throughout the body, including in the skin. The test is a semiquantitative way to detect the presence of these antigen-specific skin-resident T cells.

When a patient **previously exposed** to *M. tuberculosis* is injected with a small amount of tuberculin (purified protein derivative [PPD]) intradermally, there is little reaction in the first few hours. Gradually, however, induration and redness develop and reach a peak in 48 to 72 hours. (A positive skin test indicates that the person **has been infected** with the agent, but it does *not* confirm the presence of current disease, unless the person recently had a negative result. This switching from a negative result to a positive result is called "PPD conversion.") Measuring the diameter of the induration gives an estimate of the degree of antigen-specific T-cell memory present in the patient. This procedure is often called the tuberculin skin test.

Uninfected persons occasionally have a *false-positive* test result due to cross-reactivity between the antigens contained in the PPD inoculum and antigens present in other mycobacteria species. In addition, infected persons might have a *false-negative* result caused by overwhelming infection, in which memory T cells cannot be activated by low doses of antigen; disorders that suppress T-cell function, such as human immunodeficiency virus (HIV) infection and end-stage renal or liver disease; or immunosuppressing medications.

A positive skin test response assists in diagnosis and helps in decisions regarding whether or not to treat patients with a **latent infection**. It is essential to know a patient's PPD status prior to starting an immunosuppressive medication that might suppress T-cell or macrophage function. These patients must first be treated with antimycobacterial agents for a period of time to prevent reactivation of the latent infection.

The value of a skin test is that it is a well-validated assessment of memory T-cell **function**. Skin tests have been used for a number of fungal, protozoal, and helminth infections that can be difficult to diagnose, but with more sophisticated antibody, antigen, and molecular (nucleic acid) tests, only the PPD test for latent tuberculosis is currently in common use.

Erythema Multiforme, Stevens-Johnson Syndrome, & Toxic Epidermal Necrolysis

Erythema multiforme (EM), Stevens-Johnson syndrome (SJS), and toxic epidermal necrolysis (TEN) are related skin diseases caused primarily by **cytotoxic T-cell attack on skin cells** (keratinocytes). These diseases exist on a spectrum, with EM being the mildest and TEN being the most severe. The most common triggers are herpes simplex virus-1, *M. pneumoniae*, and a

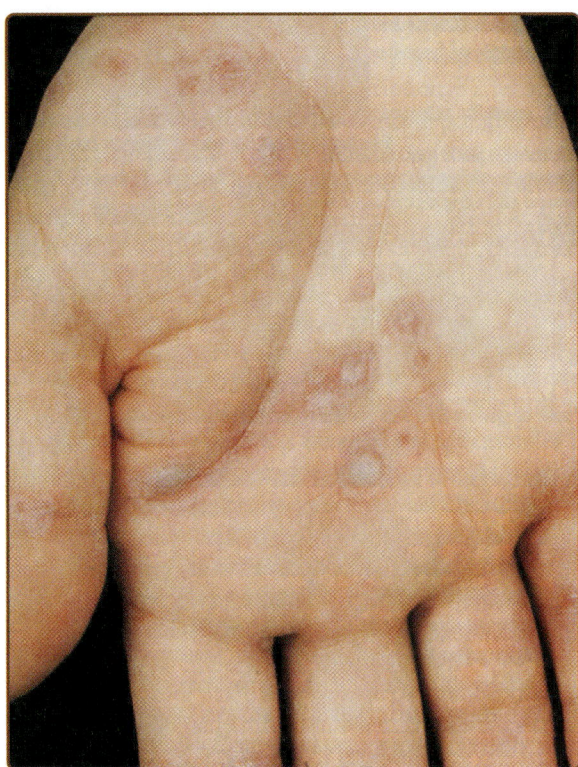

FIGURE 65–5 Erythema multiforme. Target lesions on palm. (Reproduced with permission from Kang S, Amagai M, Bruckner AL, et al: *Fitzpatrick's Dermatology in General Medicine*, 9th ed. New York, NY: McGraw Hill; 2019.)

variety of drugs, including sulfonamides and penicillins. Several human leukocyte antigen (HLA) alleles predispose to these diseases, especially HLA-DQ3 and HLA-B12.

The clinical manifestations of these diseases are characterized by a continuum of symptoms that differ in severity and anatomic location. **EM minor** is characterized by relatively few, localized target lesions on the skin, often involving the extremities (Figure 65–5), with minimal involvement of mucous membranes. The lesions begin to heal in 7 days but may recur. In contrast, **EM major** has more extensive lesions on the skin and involves the mucous membranes, often of the mouth and conjunctivae.

SJS has more extensive blistering lesions, often on the face and trunk, with significant lesions on the mucous membranes. In SJS, 3% to 10% of the body surface is involved. In **TEN**, more than 30% of the body surface is involved. TEN is a life-threatening disease, and treatment in a burn unit is recommended.

SELF-ASSESSMENT QUESTIONS

1. Your patient has episodes of eye tearing, "blood-shot" eyes, and runny nose, which you think may be due to an allergy to some plant pollen. You refer the patient to an allergist, who performs skin tests with various allergens. Within minutes, a wheal-and-flare reaction is seen on the patient's back at the site where several pollens were injected. What is the most likely sequence of events that produced the wheal-and-flare reaction?

(A) Allergen binds to IgE on the surface of B cells, and IL-4 is released.

(B) Allergen binds to IgE on the surface of mast cells, and histamine is released.

(C) Allergen binds to IgE in the plasma, which activates complement to produce C3b.

(D) Allergen binds to IgE in the plasma, and the allergen-IgE complex binds to the surface of macrophages and IL-1 is released.

2. One important test to determine whether your patient has been exposed to *Mycobacterium tuberculosis*, the organism that causes tuberculosis, is to do a purified protein derivative (PPD) skin test. In this test, PPD extracted from the organism is injected intradermally. Of the following, which one is most likely to occur at the site of a positive PPD?

(A) Cytotoxic T cells kill target cells at the site.

(B) Macrophages and CD4-positive T cells infiltrate the site.

(C) Histamine and leukotrienes are liberated from mast cells at the site.

(D) Immune complexes consisting of PPD and IgG are deposited at the site.

3. Your patient is a 77-year-old man with enterococcal endocarditis who was treated with penicillin G and gentamicin. Five days later, fever and a diffuse maculopapular rash developed. There is no urticaria, hypotension, or respiratory compromise. Urinalysis revealed proteinuria and granular casts. You suspect he may have serum sickness. Which one of the following immunopathogenic mechanisms is most likely to be the cause?

(A) One of the drugs formed immune complexes with IgG.

(B) One of the drugs activated CD4-positive T cells and macrophages.

(C) One of the drugs activated the alternative pathway of complement.

(D) One of the drugs cross-linked IgE on the mast cells and caused the release of histamine.

4. Of the following diseases, which one is most likely to be caused by a delayed hypersensitivity reaction?

(A) Autoimmune hemolytic anemia

(B) Contact dermatitis, such as poison oak

(C) Hemolytic disease of the newborn

(D) Poststreptococcal glomerulonephritis

(E) Systemic lupus erythematosus

5. Atopic individuals (i.e., those with a hereditary predisposition to immediate hypersensitivity reactions) produce an increased amount of IgE. Of the following, which one is the most likely explanation for the increased production of IgE?

(A) Large amounts of IL-1 are produced by dendritic cells.

(B) Large amounts of IL-2 are produced by macrophages.

(C) Large amounts of IL-4 are produced by Th-2 cells.

(D) Large amounts of gamma interferon are produced by Th-1 cells.

(E) Large amounts of C3a are produced by the alternative pathway of complement.

6. Of the following four types of hypersensitivity reactions, which one causes the hemolysis that occurs in hemolytic disease of the newborn (erythroblastosis fetalis)?

(A) Type I–immediate hypersensitivity

(B) Type II–cytotoxic hypersensitivity

(C) Type III–immune complex hypersensitivity

(D) Type IV–delayed hypersensitivity

ANSWERS

(1) **(B)**
(2) **(B)**
(3) **(A)**
(4) **(B)**
(5) **(C)**
(6) **(B)**

PRACTICE QUESTIONS: USMLE & COURSE EXAMINATIONS

Questions on the topics discussed in this chapter can be found in the Immunology section of Part XIII: USMLE (National Board) Practice Questions starting on page 757. Also see Part XIV: USMLE (National Board) Practice Examination starting on page 775.

Tolerance & Autoimmune Disease

TOLERANCE

Immune tolerance is the **lack of responsiveness to a specific antigen** that could otherwise elicit an immune response. The best example of antigen tolerance is a host's normal absence of response for "self" antigens, while those same antigens might be considered "foreign" if transplanted to a different host. Because it applies to responses to *antigens*, tolerance is a feature of **adaptive immunity**, although certain antigen-presenting cells can have a *tolerogenic effect* on T cells. In this chapter, we will discuss how immune tolerance to "self" develops and what happens when that tolerance is broken, namely, autoimmune diseases develop.

Whether an antigen will induce *tolerance* rather than *sensitization* is largely determined by:

(1) The immunologic **maturity** of the immune system. In general, antigens that are present during early development *do not stimulate* an immunologic response (i.e., we are tolerant to those antigens). On the other hand, antigens that are not present during the process of immune maturation (i.e., that are encountered first when the body is more immunologically mature) are considered "foreign" and usually elicit an immunologic response.

(2) The **structure** of the antigen. For example, simple molecules (small proteins) are more likely to induce tolerance than complex molecules (polysaccharides).

(3) The antigen's potential to **cross-react** with other immunogenic antigens. T-cell and B-cell receptors are highly specific, but occasionally, they can confuse one antigen with another. When this happens, an appropriate response against a foreign antigen can inappropriately begin to target self-antigens, causing host tissue damage.

(4) The presence of **proinflammatory signals**, such as those induced by pathogen-associated molecular patterns (PAMPs) (see Chapter 58), or **anti-inflammatory treatment**, such as the immunosuppressive drugs described below and in Chapter 62.

(5) The **duration** of antigen exposure. Tolerance is maintained best if the antigen to which the immune system is tolerant continues to be present.

T-Cell Tolerance

Although both B cells and T cells participate in tolerance, it is **T-cell tolerance** that plays the primary role. The main process by which T lymphocytes acquire the ability to distinguish self from nonself occurs in the thymus (see Chapter 59). Tolerance to self-antigens acquired within the thymus is called **central tolerance**. This process, which includes **positive and negative clonal selection**, eliminates T cells ("negative selection") that react strongly to antigens encountered in the thymus.

Tolerance acquired *outside* the thymus is called **peripheral tolerance**. Peripheral tolerance is necessary because some *self-antigens* are not encountered in the thymus, and therefore, some potentially self-reactive T cells are not eliminated by negative selection. In addition, there are *foreign antigens*, such as those from commensal organisms on the barriers of our skin and gastrointestinal tract, which are harmless and should therefore elicit tolerance. A sensitization response to these antigens could be pathogenic (see Chapter 65).

There are several mechanisms involved in peripheral tolerance:

(1) T cells activated *in the absence of co-stimulation* become **anergic** (a state of nonresponsiveness)

(2) Activation of naïve T cells can be **suppressed** by nearby regulatory T cells, resulting in anergy or apoptosis

(3) Naïve T cells themselves can undergo **differentiation into T regulatory cells (Tregs)** at the time of their activation (Figure 66–1).

There are a number of proposed mechanisms by which Tregs suppress other T cells (see Figure 66–1C). For example, Tregs have high baseline levels of the **high-affinity interleukin (IL)-2 receptor**, meaning they act as a competitive "sink" that reduces the amount of IL-2 available to other T cells. Also, Tregs have high levels of the protein **cytotoxic T lymphocyte antigen-4 (CTLA-4)**, which they use to block or remove the co-stimulatory "signal 2" that B7 provides (recall that CTLA-4 has higher affinity for B7 than CD28, as discussed in Chapter 59). Tregs also secrete cytokines such as **IL-10** and **transforming growth factor-beta (TGF-β)** that may be immunosuppressive in certain

contexts. These signals and others can induce T cells to become anergic or induce them to undergo apoptosis.

B-Cell Tolerance

B-cell clones are also tolerant to self-antigens due to **negative selection**, which occurs primarily in the bone marrow (see Chapter 59). However, tolerance in B cells is less complete than in T cells, an observation supported by the finding that most autoimmune diseases are accompanied by self-reactive **autoantibodies**.

B-cell precursors bearing an antigen receptor that strongly binds a self-protein are subject to clonal deletion (apoptosis) or anergy. However, the clones can escape this fate by **receptor editing** (see Chapter 59). In this process, a new, different light

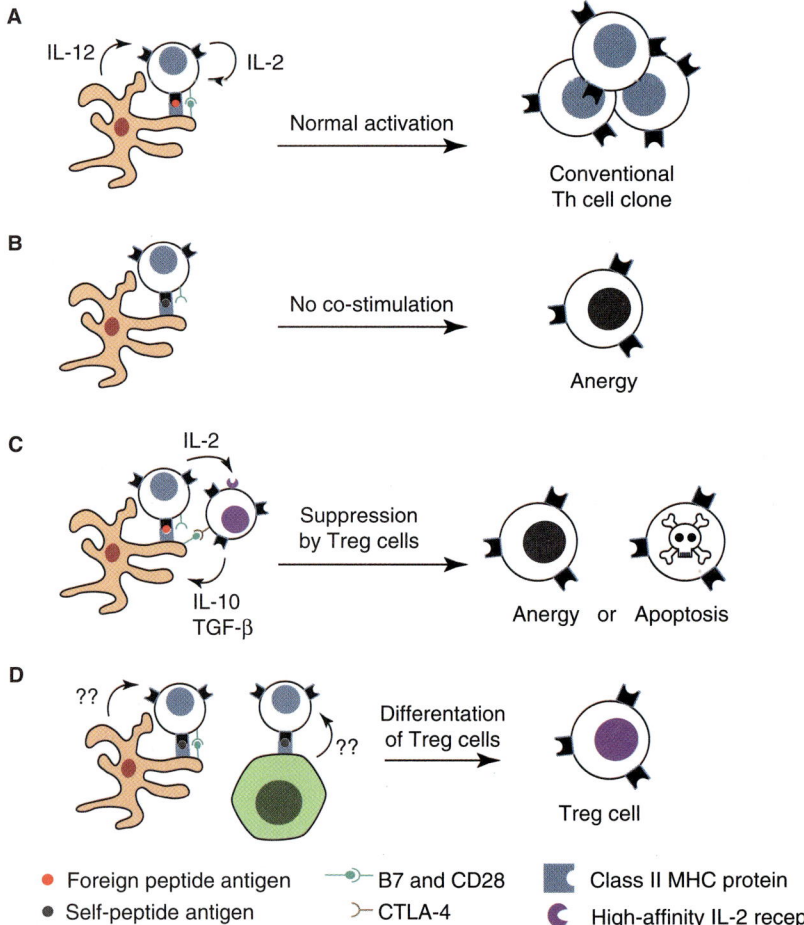

FIGURE 66–1 Pathways to peripheral T-cell tolerance. **A: Normal activation.** During "conventional" T-cell activation, B7 proteins on the dendritic cell (DC) are induced by inflammatory signals, such as pathogen-associated molecular patterns (PAMPs) that accompany uptake and presentation of foreign antigens (red circle). The B7 proteins interact with CD28 (turquoise) to provide co-stimulation, and interleukin (IL)-2 drives clonal proliferation of Th cells. **B: No co-stimulation.** In the absence of co-stimulation, which often occurs in presentation of self-antigens (gray circle), T cells become anergic. **C: Suppression by regulatory T (Treg) cells.** Treg cells suppress by at least three mechanisms: (1) the high-affinity IL-2 receptor (purple) competes for IL-2, depriving the conventional T cell; (2) CTLA-4 competes for B7, preventing co-stimulation of the conventional T cell; and (3) Treg cytokines (e.g., IL-10 and transforming growth factor [TGF]-β) and other factors suppress T-cell activation or induce apoptosis. **D: Differentiation of Treg cells.** Unknown signals provided to CD4-positive T cells at the time of antigen presentation (by a DC or by other antigen-presenting cells) can modify them to differentiate into Treg cells. Note that the same pathways shown in A, B, and C also apply to CD8-positive cytotoxic T cells. CTLA-4 = cytotoxic T lymphocyte antigen-4; MHC = major histocompatibility complex.

chain is produced that changes the specificity of the receptor so that it no longer recognizes a self-protein. This allows the B-cell precursor a chance to exit the bone marrow as a mature B-cell clone. It is estimated that as many as 25% to 50% of B cells that were initially self-reactive undergo receptor editing. T cells do *not* undergo receptor editing.

AUTOIMMUNE DISEASES

In certain circumstances, tolerance may be lost and immune reactions to host antigens may develop, resulting in autoimmune diseases. The most important step in the loss of tolerance is the **activation of self-reactive (autoreactive) CD4-positive T cells.** These self-reactive T cells can differentiate into effector/helper T cells (e.g., Th-1, Th-2, and Th-17 cells) that cause **inappropriate inflammation in tissues.** They can also differentiate into follicular helper T cells (Tfh cells), which provide **inappropriate help to autoantibody-producing B cells.** As described in Table 66–1, most autoimmune diseases are associated with autoantibodies. In some cases, the disease is directly mediated by the antibodies, and in other cases, the disease is the result of immune complex deposition or T-cell–derived cytokines, and the antibodies may only be a marker of loss of tolerance.

Genetic Factors

Many autoimmune diseases run in families, suggesting a **genetic predisposition** to these disorders. There is a strong association of some diseases with certain human leukocyte antigen (HLA) types, especially the class II genes. For example, certain alleles of the *HLA-DR* gene increase the risk of having rheumatoid arthritis, and individuals with these alleles may also have more severe disease. This allele and other alleles of *HLA-DR* and *HLA-DQ* increase the risk of having autoimmune (type 1) diabetes. Ankylosing spondylitis is 100 times more likely to occur in people who carry *HLA-B27* than in those who do not carry that gene. These associations underscore the importance of T-cell antigen recognition in the development of autoimmunity.

Why are certain HLA alleles associated with autoimmune diseases? One theory is that those alleles encode proteins that present autoantigens with greater efficiency. Another theory is that autoreactive T cells escape negative selection because they bind poorly to those class I or class II major histocompatibility complex (MHC) proteins in the thymus. It should be noted, however, that whether a person develops an autoimmune disease or not is multifactorial because some people with HLA genes known to predispose to certain autoimmune diseases do *not* develop the disease (e.g., many people carrying the *HLA-DR* risk alleles do *not* develop rheumatoid arthritis).

Hormonal Factors

Approximately **90% of all autoimmune diseases occur in women.** In general, class II MHC-related diseases (e.g., rheumatoid arthritis, Graves' disease [hyperthyroidism], and systemic lupus erythematosus [SLE]) occur more commonly in women, whereas class I MHC-related diseases (e.g., ankylosing spondylitis and reactive arthritis) occur more commonly in

TABLE 66–1 Important Autoimmune Diseases

Type of Immune Response	Autoimmune Disease	Main Target of the Immune Response
Antibody to receptors	Myasthenia gravis	Acetylcholine receptor
	Graves' disease	TSH receptor
	Insulin-resistant diabetes	Insulin receptor
	Lambert-Eaton myasthenia	Calcium channel receptor
Antibody to other cell components	Systemic lupus erythematosus[1]	dsDNA, histones
	Rheumatoid arthritis[1]	Joint tissue
	Rheumatic fever	Heart and joint tissue
	Hemolytic anemia	RBC membrane
	Idiopathic thrombocytopenic purpura	Platelet membranes
	Goodpasture's syndrome	Basement membrane of kidney and lung
	Pernicious anemia	Intrinsic factor and parietal cells
	Hashimoto's thyroiditis[1]	Thyroglobulin
	Insulin-dependent diabetes mellitus[1]	Islet cells
	Addison's disease	Adrenal cortex
	Acute glomerulonephritis	Glomerular basement membrane
	Polyarteritis nodosa	Small- and medium-sized arteries
	Guillain-Barré syndrome	Myelin protein
	Granulomatosis with polyangiitis	Cytoplasmic enzymes of neutrophils
	Pemphigus	Desmoglein in tight junctions of skin
	IgA nephropathy	Glomerulus
T-cell mediated	Autoimmune encephalomyelitis and multiple sclerosis	Reaction to myelin protein causes demyelination of brain neurons
	Celiac disease	Enterocytes

RBC = red blood cell; TSH = thyroid-stimulating hormone.

[1]These diseases involve a significant cell-mediated response in addition to an antibody-mediated response.

men. Although the explanation for this unequal gender ratio is unclear, there is some evidence from animal models that estrogen can alter the B-cell repertoire and enhance the formation of antibody to DNA. Clinically, the observation that SLE either appears or exacerbates during pregnancy (or immediately postpartum) supports the idea that hormones play an important role in predisposing women to autoimmune diseases.

Environmental Factors

There are several environmental triggers for autoimmune diseases, most of which are either bacteria or viruses. In some cases, there is a known **molecular mimic** in the infectious agent that induces an immune response that cross-reacts with self-proteins (see Molecular Mimicry below). However, in many cases, the causal link between a particular infection and the autoimmune disease is unexplained (see Table 66–2 for examples). It is speculative at this time, but members of the bowel flora are thought to play a role in the genesis, or the maintenance, of inflammatory bowel diseases, such as Crohn's disease and ulcerative colitis. Other environmental triggers include certain drugs such as procainamide or hydralazine, which cause SLE.

In summary, the current theory is that most autoimmune diseases occur in people (1) with a genetic predisposition that is determined by their MHC genes and (2) who are exposed to an environmental agent that triggers a cross-reacting immune response against some component of normal tissue. Furthermore, because autoimmune diseases increase in number with advancing age, another possible factor is a decline in the

number of regulatory T cells, which allows any surviving autoreactive T cells to proliferate and cause disease.

Mechanisms of Autoimmune Diseases

The following mechanisms for autoimmunity have been proposed.

Molecular Mimicry

Various bacteria and viruses are implicated as the source of cross-reacting antigens that trigger the activation of autoreactive T cells or B cells. For example, reactive arthritis occurs following infections with *Shigella* or *Chlamydia*, and Guillain-Barré syndrome occurs following infections with *Campylobacter*. The concept of **molecular mimicry** is that the environmental trigger resembles (mimics) a component of the body sufficiently that an immune attack is directed against the cross-reacting body component. One of the best-characterized examples of molecular mimicry is the relationship between the M protein of *Streptococcus pyogenes* and the myosin of cardiac muscle. Antibodies against certain M proteins cross-react with cardiac myosin, leading to the heart damage seen in rheumatic fever. In some cases, there are identical amino acid sequences in certain foreign and human proteins. For example, there is an identical six–amino acid sequence in the hepatitis B viral polymerase and the human myelin basic protein.

Alteration of Normal Proteins

Drugs can bind to normal proteins and make them immunogenic. Procainamide-induced SLE is an example of this mechanism.

Release of Sequestered Antigens

Certain tissues (e.g., the testes, central nervous system, and the lens and uveal tract of the eye) are kept isolated so that their antigens are **less exposed** to the immune system. These are known as **immunologically privileged** sites. When antigens from these organs do enter the circulation accidentally (e.g., after damage), they can elicit immune responses, producing aspermatogenesis, encephalitis, or endophthalmitis, respectively. Sperm, in particular, must be in a sequestered, immunologically privileged site, because they develop after immunologic maturity has been reached.

Intracellular antigens, such as DNA, histones, and mitochondrial enzymes, are also normally sequestered from the immune system. However, cellular damage from bacterial or viral infection, radiation, and chemicals may cause the release of these sequestered antigens, which then elicit an immune response. Once autoantibodies are formed, new cells can be targeted and damaged, and ongoing release of sequestered antigens results in the formation of immune complexes and the symptoms of the autoimmune disease. Sunlight is known to exacerbate the skin rash in patients with SLE. It is thought that ultraviolet (UV) radiation damages cells, which releases the normally sequestered DNA and histones that are the major antigens in this disease.

TABLE 66–2 Microbial Infections Associated with Autoimmune Diseases

Microbe	Autoimmune Disease
Bacteria	
Streptococcus pyogenes	Rheumatic fever
Campylobacter jejuni	Guillain-Barré syndrome
Escherichia coli	Primary biliary cirrhosis
Chlamydia trachomatis	Reactive arthritis
Shigella species	Reactive arthritis
Yersinia enterocolitica	Reactive arthritis
Borrelia burgdorferi	Lyme arthritis
Viruses	
Hepatitis B virus[1]	Multiple sclerosis
Hepatitis C virus	Mixed cryoglobulinemia
Measles virus	Allergic encephalitis
Coxsackie virus B3[2]	Myocarditis
Coxsackie virus B4[3]	Type 1 diabetes mellitus
Cytomegalovirus	Scleroderma
Human T-cell leukemia virus (HTLV)	HTLV-associated myelopathy

[1]Other viruses, such as Epstein–Barr virus, human herpes virus-6, influenza A virus, and measles virus, are also implicated as the possible cause of multiple sclerosis. No virus has definitely been shown to be the environmental trigger at this time.

[2]Coxsackie virus infects and kills cardiac myocytes, causing the acute symptoms, but the late phase is caused by the attack of cytotoxic T cells on the myocytes.

[3]Causes diabetes mellitus in mice, but it is uncertain whether it is a cause in humans.

Epitope Spreading

Epitope spreading is the term used to describe the new exposure of sequestered autoantigens as a result of damage to cells caused by viral infection. These newly exposed autoantigens stimulate autoreactive T cells. In an animal model, a multiple sclerosis-like disease was caused by infection with an encephalomyelitis virus. Note that the self-reactive T cells were directed against cellular antigens rather than the antigens of the virus.

Failure of Regulatory T Cells

As described earlier, Tregs are CD4-positive cells that suppress the inflammatory effects of other T cells. Rare Mendelian deficiencies, such as mutations of the gene *AIRE* or the gene *FOXP3* (required for Treg differentiation), cause autoimmunity through deficiency of Tregs. Deficiency in *AIRE* causes an autoimmune disease called *autoimmune polyendocrinopathy*. Deficiency of *FOXP3* causes *immunodysregulation polyendocrinopathy X-linked* (IPEX) syndrome. Similarly, rare mutations in genes encoding proteins involved in T-cell receptor signaling can cause both immunodeficiency *and* autoimmunity due to complex effects on thymic selection.

Specific Auotimmune Diseases

Table 66–1 lists some important autoimmune diseases according to the type of immune response causing the disease and the target affected by the autoimmune response. Some examples of autoimmune disease are described in more detail next.

Diseases Involving Primarily One Type of Cell or Organ

(1) **Multiple sclerosis**—In this disease, autoreactive T cells and activated macrophages cause demyelination of the white matter of the brain. The trigger that stimulates the autoreactive T cells is thought to be a viral infection in persons who are genetically susceptible (e.g., those with the HLA-DRB1 allele). The clinical findings in multiple sclerosis typically wax and wane and affect both sensory and motor functions. Magnetic resonance imaging (MRI) of the brain reveals plaques in the white matter. **Oligoclonal bands of IgG** are found in the spinal fluid of most patients. B cells probably play a central role in the pathogenesis, through the production of harmful antibodies and/or through interactions with pathogenic T cells in the lymphoid follicles. Immunosuppressive drugs, including anti-CD20 monoclonal antibodies (such as ocrelizumab) are used to prevent flares or suppress them when they do occur.

(2) **Chronic thyroiditis**—Hashimoto's thyroiditis is an autoimmune disease in which antibodies are formed against **thyroglobulin** and **thyroid peroxidase**. These antibodies may provoke an inflammatory process that leads to fibrosis of the gland. There is also evidence of Th-1 cell and cytotoxic T-cell activation, and the combined effect of these cells causes inflammation and thyroid cell death. Treatment entails thyroid hormone replacement.

(3) **Anemias, thrombocytopenias, and granulocytopenias**—Various forms of these disorders have been attributed to the attachment of autoantibodies to cell surfaces and subsequent cell destruction.

Immune thrombocytopenic purpura is caused by **antibodies directed against platelets**. Platelets coated with antibody are either destroyed in the spleen or lysed by the membrane attack complex of complement.

Several drugs, acting as haptens, bind to the platelet membrane and form a "neoantigen" that induces the cytotoxic antibody that results in **platelet** destruction. Penicillins, cephalothin, tetracyclines, sulfonamides, isoniazid, and rifampin, as well as drugs that are not antimicrobials, can have this effect. **Autoimmune hemolytic anemia** caused by penicillins and cephalosporins is due to the same mechanism, that is, autoantibodies directed against erythrocytes attach and result in cell destruction. Treatment of these conditions generally involves immunosuppression.

Pernicious anemia is not hemolytic; it is caused by antibodies to **intrinsic factor**, a protein secreted by parietal cells of the stomach that facilitates the absorption of vitamin B_{12}. Treatment involves vitamin B_{12} replacement.

(4) **Type I diabetes mellitus**—In this disease, **autoreactive T cells destroy the insulin-producing islet cells** of the pancreas. A main antigen against which the T-cell attack is directed is the islet cell enzyme, glutamic acid decarboxylase. There may also be a role for **autoantibodies targeting islet cell antigens**, including insulin itself. The mainstay of treatment is insulin replacement, but immune-based therapies that aim to restore immune tolerance of islet cells are under investigation.

(5) **Insulin-resistant diabetes, myasthenia gravis, and hyperthyroidism (Graves' disease)**—In these diseases, antibodies to receptors play a pathogenic role. In some patients with insulin-resistant diabetes, **antibodies to insulin receptors** have been demonstrated that interfere with insulin binding. In **myasthenia gravis**, which is characterized by severe muscular weakness, **antibodies to acetylcholine receptors** *block* neuromuscular junction signaling. Muscular weakness also occurs in Lambert-Eaton syndrome, in which antibodies form against the proteins in calcium channels. Some patients with **Graves' disease** have circulating **antibodies to thyrotropin receptors**, which, when they bind to the receptors, resemble thyrotropin in activity and *stimulate* the thyroid to produce more thyroxine.

(6) **Guillain-Barré syndrome**—This disease is the most common cause of acute paralysis in the United States. It follows a variety of infectious diseases such as **viral illnesses** (e.g., upper respiratory tract infections, human immunodeficiency virus [HIV] infection, and mononucleosis caused by Epstein–Barr virus and cytomegalovirus) and infection with *Campylobacter jejuni*. Infection with *C. jejuni*, which causes enteritis and diarrhea, is considered to be the most common antecedent to Guillain-Barré syndrome. **Antibodies against membrane gangliosides** are formed, complement is activated, and the membrane attack complex destroys the myelin sheath, resulting in a demyelinating polyneuropathy. (Unlike multiple sclerosis, this neuropathy occurs in peripheral nerves.) The main symptoms are those of a rapidly progressing ascending paralysis. The treatment involves either intravenous immunoglobulins or plasmapheresis, which replaces the patient's plasma, removing

the harmful antibodies; glucocorticoids have *not* been an effective treatment.

(7) **Pemphigus**—Pemphigus is a skin disease characterized by bullae (blisters). It is caused by **autoantibodies against desmoglein**, a protein in the desmosomes that forms the tight junctions between epithelial cells in the skin. When the tight junctions are disrupted, fluid fills the spaces between cells and forms the bullae. One form of pemphigus, pemphigus foliaceus, is endemic in rural areas of South America, which lends support to the idea that infection with an endemic pathogen is the environmental trigger for this disease. Treatment entails immunosuppression, either with topical or systemic glucocorticoids.

(8) **Celiac disease**—Celiac disease (also known as celiac sprue and gluten enteropathy) is characterized by diarrhea, painful abdominal distention, fatty stools, and failure to thrive. Symptoms are induced by ingestion of gliadin, a protein found primarily in wheat, barley, and rye grains. Most patients have **antibodies to tissue transglutaminase**, and these are often used to aid in making the diagnosis. These autoantibodies may have a role in the disease, but the destruction of enterocytes, which cause villous atrophy, inflammation, and malabsorption, are primarily caused by **cytotoxic T cells that react to the protein antigen gliadin**. Patients who carry certain alleles of *HLA-DQ* are predisposed to celiac disease. A gluten-free diet typically results in marked improvement.

(9) **Inflammatory bowel disease (Crohn's disease and ulcerative colitis)**—These diseases are characterized by diarrhea, often bloody, and crampy lower abdominal pain. These symptoms arise from chronic inflammation, primarily in the **ileum** (in Crohn's disease) and in the **rectosigmoid colon** (in ulcerative colitis). It is thought that the chronic inflammation is caused by an **abnormal immune response to certain members of the normal intestinal flora**. Dysregulated Th-1 and Th-17 cells are involved in the pathogenesis of these diseases. Corticosteroids, DNA synthesis inhibitors (see Table 62–2), and inhibitors of tumor necrosis factor (TNF), IL-12, and IL-23 have been effective.

(10) **IgA nephropathy**—This disease is one of the most common types of glomerulonephritis and is characterized primarily by hematuria, but proteinuria and progression to end-stage renal disease can occur. **The glomeruli are lined with immune complexes containing IgA**. Some patients are asymptomatic, some have mild symptoms, and others progress rapidly to kidney failure. Symptoms are temporally related to viral infections, especially pharyngitis, but no specific virus has been identified. No treatment regimen is clearly effective.

(11) **Psoriasis**—Psoriasis is a chronic autoimmune skin disease characterized by raised erythematous plaques with silvery scales, often on the extensor surfaces of the arms and legs (i.e., elbows, shins, and knees). Skin lesions are the most common manifestation, but psoriatic arthritis also occurs. The inflammatory infiltrate in the skin lesions consists of dendritic cells, macrophages, and T cells. (In contrast, atopic dermatitis, covered in Chapter 65, is often characterized by eosinophilic infiltrate.) There is a strong genetic component to psoriasis susceptibility, with individuals carrying the HLA-Cw6 allele of class I MHC

being particularly predisposed. The environmental trigger is unknown. Topical corticosteroids and UV phototherapy with psoralen are two common treatment modalities, and systemic immunosuppression may be required for severe disease and/or psoriatic arthritis.

Diseases Involving Multiple Organs (Systemic Diseases)

(1) **Systemic lupus erythematosus**—SLE is a chronic inflammatory autoimmune disease that affects the skin of the face, the joints, and the kidneys. Antibodies are formed against DNA and other components of the nucleus of cells. **Antibodies against double-stranded DNA** are the hallmark of SLE. These antibodies form immune complexes that activate complement. Complement activation produces C5a, which attracts neutrophils that release enzymes, thereby damaging tissues (see Chapter 63). **Lumpy glomerular deposits can be seen that are similar to those caused by other types of glomerulonephritis**.

Most of the clinical findings are caused by immune complexes that activate complement and, as a consequence, damage tissues. For example, the characteristic rash on the cheeks is the result of a vasculitis caused by immune complex deposition. The arthritis and glomerulonephritis commonly seen in SLE are also caused by immune complexes. The immune complexes found on the glomerulus contain antibodies (IgG, IgM, or IgA) and the C3 component of complement. However, the anemia, leukopenia, and thrombocytopenia are caused by cytotoxic antibodies rather than immune complexes. The diagnosis is supported by detecting **antinuclear antibodies (ANAs)** with fluorescent antibody tests and **antidouble-stranded DNA antibodies** with enzyme-linked immunosorbent assay (ELISA). Antibodies to several other nuclear components are also detected, as is **a reduced level of complement**.

SLE primarily affects women between the ages of 20 and 60 years. Individuals with *HLA-DR2* or *HLA-DR3* genes are predisposed. The agent that induces these autoantibodies in most patients is unknown. However, two drugs, procainamide and hydralazine, are known to cause SLE. Treatment of SLE varies depending on the severity of the disease and the organs affected.

(2) **Rheumatoid arthritis**—RA is a systemic disease involving not only the joints but other organs as well, most often the lung and pericardium. Serum and synovial fluid of patients often contain **rheumatoid factor**, which is an antibody (usually IgM but occasionally IgG, IgD, IgA, or IgE) whose Fab recognizes and binds to the Fc fragment of normal human IgG. Rheumatoid factor is associated with RA but is not specific for it.

Deposits of immune complexes (containing the normal IgG and rheumatoid factor) on synovial membranes and in blood vessels activate complement and attract polymorphonuclear cells, causing inflammation. The main clinical finding is inflammation of the **proximal interphalangeal** and **metacarpophalangeal** joints of the hands, the small joints of the feet, and the cervical spine, knees, and shoulders. Within the inflamed joints, the synovial membrane is infiltrated with T cells, plasma cells, and macrophages, and the synovial fluid contains high levels of macrophage-produced inflammatory cytokines such as TNF, IL-1, and IL-6.

RA affects primarily women between the ages of 30 and 50 years. People with *HLA-DR4* genes are predisposed to RA. The agent that induces rheumatoid factor is unknown. In addition to the joints, RA can affect the lung and pericardium, although unlike many of the immune complex-related disorders, RA is an exception in that it does *not* involve the kidney. The diagnosis is supported by finding high titers of rheumatoid factor and low titers of complement in serum, especially during periods when the disease is most active. Detection of antibody to citrullinated peptide in the serum also supports the diagnosis.

Treatment of RA typically involves immunosuppressive drugs (especially methotrexate or corticosteroids, see Table 62–2). Inhibitors of TNF, IL-1, and IL-6 also have proven helpful.

(3) Vasculitis—Inflammation of the walls of blood vessels, which can include large, medium, and small arteries and veins, is called *vasculitis*. A number of multisystem autoimmune diseases manifest with vasculitis caused by immune complexes: polyarteritis nodosa, Henoch-Schönlein purpura (IgA vasculitis), cryoglobulin-related vasculitis, and the vasculitis that occurs in SLE. An important example of a cryoglobulin-related vasculitis occurs in hepatitis C virus infection.

One of the more common examples is **granulomatosis with polyangiitis** (formerly called Wegener's granulomatosis). The main pathologic finding in this disease is necrotizing granulomatous vasculitis that primarily affects the upper and lower respiratory tracts and the kidneys. Common clinical findings include sinusitis, otitis media, cough, sputum production, and arthritis. Glomerulonephritis is one of the main features of this disease. The diagnosis is supported by finding **antineutrophil cytoplasmic antibodies** (ANCAs) in the patient's serum. Immunosuppressive therapy with corticosteroids (see Table 62–2) is effective for treating disease flares.

In contrast, some diseases, such as **giant cell arteritis** (GCA), are caused by T cells infiltrating the arterial wall. The most common form of GCA is temporal arteritis that involves the temporal artery.

The symptoms and signs of vasculitis vary depending on the organ affected. Nonspecific findings include fever, weight loss, arthralgia, myalgia, and abdominal pain. Some findings often associated with vasculitis are palpable purpura and mononeuritis multiplex, which often manifests as foot or wrist drop. As with the other immune complex-related diseases, vasculitis often involves the kidney glomeruli.

(4) Reactive arthritis—Reactive arthritis is an acute inflammation of the joints that follows 1 to 3 weeks after various bacterial infections. However, importantly, the infectious agents themselves are *not* cultured from the joint fluid. The inflammation is caused either by cross-reactive immune responses to self-antigens or by immune complexes with foreign antigens that deposit in the joints. Reactive arthritis is associated with **enteric infections** caused by *Shigella, Campylobacter, Salmonella,* and *Yersinia* and with **urethritis** caused by *Chlamydia trachomatis.*

The arthritis is usually oligoarticular and asymmetric. The bacterial infection precedes the arthritis by a few weeks. Men are more commonly affected, and those who carry the HLA-B27 allele are at higher risk. Antibiotics directed against the organism have no effect. Anti-inflammatory agents are typically used. Reactive arthritis often presents as part of a triad of **arthritis, conjunctivitis,** and **urethritis**. The pathogenesis of the disease is unclear, but immune complexes may play a role.

(5) Goodpasture's syndrome—In this syndrome, autoantibodies are formed against the collagen in basement membranes of the kidneys and lungs. Goodpasture's syndrome (GS) affects primarily young men, and those carrying particular *HLA-DR2* alleles are at risk for this disease. The agent that induces these autoantibodies is unknown, but GS often follows a viral infection.

The main clinical findings are hematuria, proteinuria, and pulmonary hemorrhage. The clinical findings are caused by cytotoxic antibodies that activate complement. As a consequence, C5a is produced, neutrophils are attracted to the site, and enzymes are released by the neutrophils that damage the kidney and lung tissue. The diagnosis of GS is supported by detecting **antibody and complement bound to glomerular basement membranes**. Because this is a rapidly progressive, often fatal disease, treatment, including plasma exchange to remove the antibodies, and the use of immunosuppressive drugs, must be instituted promptly.

(6) Other collagen vascular diseases—Other multisystem autoimmune diseases include ankylosing spondylitis (which, like reactive arthritis, is common in people carrying the HLA-B27 allele), polymyositis and dermatomyositis, scleroderma, polyarteritis nodosa, and Sjögren's syndrome.

Treatment of Autoimmune Diseases

The conceptual basis for the treatment of autoimmune diseases is to reduce the patient's immune response or inflammatory response sufficiently to eliminate the symptoms. Immunosuppressive therapy must be given cautiously because of the risk of opportunistic infections. Long-term immunosuppression requires concurrent treatment with antimicrobials to prevent opportunistic infections.

Many of the drugs used for autoimmunity are also used to treat acute transplant rejection, and these were covered in Chapter 62 (see Table 62–2). These include corticosteroids, including prednisone, and antimetabolites, such as methotrexate and azathioprine, that inhibit DNA synthesis in the immune cells. Other drug classes described in Chapter 62 include calcineurin inhibitors and molecules that block cytokines and activation signals, such as B7-CD28. Table 66–3 describes some of the many therapies that are FDA-approved for autoimmune diseases.

Nonsteroidal anti-inflammatory drugs are used for certain autoimmune and inflammatory diseases. These act by inhibiting cyclooxygenase (COX) enzymes, blocking the production of inflammatory mediators, particularly prostaglandins.

Other approaches to therapy include antibody to TNF and soluble receptor for TNF that acts as a decoy. Both infliximab and adalimumab (antibody to TNF) as well as etanercept (TNF receptor) have been shown to ameliorate the joint inflammation of RA and the skin lesions of psoriasis. However, these anti-TNF therapies increase the risk of infections, such as activating latent tuberculosis, serious infections caused by *Legionella*

TABLE 66–3 Drugs Used in the Treatment of Autoimmune Diseases

Class of Drugs	Important Example of Drug	Mechanism of Action	Comment/Adverse Effect
Corticosteroids	Prednisone Methylprednisolone	1. Inhibits synthesis of inflammatory mediators such as IL-1 and TNF 2. Inhibits phospholipase, resulting in decreased arachidonic acid and prostaglandins 3. Induces apoptosis in lymphocytes	Used to control acute flares Long-term therapy limited by off-target toxicity on bones, endocrine systems
Nonsteroidal anti-inflammatory drugs (NSAIDs)	1. Aspirin 2. Ibuprofen, naproxen 3. Celecoxib	1. Inhibits cyclooxygenases (COX-1 and COX-2); this decreases prostaglandins 2. Same as aspirin 3. Inhibits only COX-2 so only inflammatory mediators are decreased	1. GI tract bleeding; Reye's syndrome 2. GI tract bleeding 3. Less GI tract bleeding than the NSAIDs described above
DNA synthesis inhibitors	Methotrexate	Inhibits DHFR, which decreases folate synthesis; this stops cell division of lymphocytes Suppresses T-cell and B-cell function	Bone marrow suppression; teratogenic
TNF inhibitors	1. Etanercept 2. Infliximab, • Adalimuab • Certolizumab • Golimumab	1. TNF receptor fused with an Fc region of an immunoglobulin; neutralizes TNF 2. Antibodies against TNF	Risk of activation of tuberculosis
Inhibitors of Th-1 and Th-17 pathway	1. Brodalumab 2. Ixekizumab • Secukinumab 3. Ustekinumab	1. Antibody against the IL-17 receptor 2. Antibodies against IL-17 3. Antibody against IL-12 and IL-23	Different agents used for psoriasis, rheumatoid arthritis, ankylosing spondylitis, and Crohn's disease
Inhibitors of leukocyte migration	1. Natalizumab 2. Vedolizumab	1. Inhibits *all* α4 integrins 2. Inhibits α4β7 integrins (gut selective)	Progressive multifocal leukoencephalopathy Theoretical increased risk of GI infection
B-cell inhibitors	Rituximab Ocrelizumab	Both bind to CD20 on surface of B cells, resulting in apoptosis of B cells	PML and reactivation of HBV

CTLA = cytotoxic T-lymphocyte associated protein; DHFR = dihydrofolate reductase; GI = gastrointestinal; HBV = hepatitis B virus; IL = interleukin; PML= progressive multifocal leukoencephalopathy; TNF = tumor necrosis factor.

and *Listeria*, and skin and soft tissue infections caused by pyogenic bacteria. These drugs increase the risk of activating latent fungal infections such as histoplasmosis as well.

Monoclonal antibodies against the cytokine IL-17 and the IL-17 receptor block Th-17 cell function. In addition, antibodies that block Th-1 and Th-17 cell activation by neutralizing IL-12 and IL-23. Various antibodies from this family are approved for psoriasis, RA, ankylosing spondylitis, and Crohn's disease.

Rituximab and ocrelizumab are monoclonal antibodies against CD20, a protein located on the surface of B cells but not plasma cells. These antibodies cause the death of B cells either by complement-mediated killing (complement-dependent cytotoxicity), by the attack of natural killer cells (antibody-dependent cell-mediated cytotoxicity), or by directly inducing cell death (apoptosis).

Certain antibody-mediated autoimmune diseases, such as Guillain-Barré syndrome and myasthenia gravis, can be treated either with plasmapheresis, which removes autoimmune antibodies, or with high doses of IgG pooled from healthy donors. The mechanism of how high-dose intravenous IgG (IVIG) suppresses inflammation is not completely understood. One hypothesis is that it binds to Fc receptors on the surface of neutrophils, monocytes, and macrophages and blocks the attachment of the inflammatory immune complexes that activate these cells. Another hypothesis is that excess IgG saturates the FcRn receptor on the surface of vascular endothelial cells, which accelerates the catabolism of IgG, thereby reducing the level of autoimmune antibodies. A third hypothesis is that the IVIG preferentially binds *inhibitory* Fc receptors, which counteracts the immune activation.

SELF-ASSESSMENT QUESTIONS

1. Regarding immunologic tolerance, which one of the following is the most accurate?

   (A) Clonal deletion occurs with T cells but not with B cells.

   (B) Tolerance to certain self-antigens occurs by negative selection of immature T cells in the thymus.

   (C) The presence of B7 on the surface of the antigen-presenting cell is one of the essential steps required to establish tolerance.

   (D) Tolerance is easier to establish in adults than in newborns because more self-reactive T cells have undergone apoptosis in adults than in newborns.

   (E) Once tolerance is established to an antigen, it is permanent (i.e., that individual cannot react against that antigen even though the antigen is no longer present).

2. Antibodies against normal components of the body typically occur in autoimmune diseases. In which one of the following sets of two diseases do antibodies against DNA occur in one disease and antibodies against IgG occur in the other disease?

   (A) Myasthenia gravis and systemic lupus erythematosus
   (B) Pernicious anemia and rheumatic fever
   (C) Rheumatic fever and myasthenia gravis
   (D) Rheumatoid arthritis and pernicious anemia
   (E) Systemic lupus erythematosus and rheumatoid arthritis

3. Regarding the pathogenesis of autoimmune diseases, which one of the following is the most accurate?

   (A) In reactive arthritis, neuropathy occurs following viral respiratory tract infections.
   (B) In myasthenia gravis, antibodies are formed against acetylcholine at the neuromuscular junction.
   (C) In Goodpasture's syndrome, antibodies are formed against the synovial membrane in the large weight-bearing joints.
   (D) In autoimmune hemolytic anemia, the red cells are destroyed by tumor necrosis factor produced by activated macrophages.
   (E) In Graves' disease, antibodies bind to the receptor for thyroid-stimulating hormone, which stimulates the thyroid to produce excess thyroxine.

4. Your patient is a 25-year-old woman with a fever, a malar facial rash, alopecia, and ulcerations on two fingertips. Urinalysis shows proteinuria. You suspect she has systemic lupus erythematosus. Which one of the following is the most likely explanation for her proteinuria?

   (A) Cytotoxic T cells attack the glomerular basement membrane.
   (B) An IgE-mediated response releases histamine and leukotrienes that damage the tubules.
   (C) A delayed hypersensitivity response consisting of macrophages and CD4-positive T cells damages the glomeruli.
   (D) Immune complexes are trapped by glomeruli and activate complement, then C5a attracts neutrophils that damage the glomeruli.

ANSWERS

(1) **(B)**
(2) **(E)**
(3) **(E)**
(4) **(D)**

PRACTICE QUESTIONS: USMLE & COURSE EXAMINATIONS

Questions on the topics discussed in this chapter can be found in the Immunology section of Part XIII: USMLE (National Board) Practice Questions starting on page 757. Also see Part XIV: USMLE (National Board) Practice Examination starting on page 775.

Tumor Immunity

One of the key functions of the immune system is to patrol the body for cancer cells in order to kill these cells before they become malignant tumors. In this chapter, we will discuss (1) how the immune system is able to recognize cancer cells; (2) how cancer cells evolve to hide from and suppress the immune system; and (3) how immunotherapy re-invigorates the immune system's killing mechanisms.

TUMOR-ASSOCIATED ANTIGENS

Malignant cancer cells are genetically similar to healthy cells. They generally display the same major human leukocyte antigen (HLA) and minor histocompatibility proteins as other nucleated "self" cells, and lymphocytes with antigen receptors that could recognize these proteins are largely removed during negative selection (see Chapter 66). Nevertheless, animals carrying a malignant tumor *can* develop an immune response to that tumor and cause its **regression**. In the course of neoplastic transformation, **new antigens (neoantigens)**, called **tumor-associated antigens (TAAs)**, develop at the cell surface, and the host recognizes such cells as "nonself." An immune response then kills the offending cells, preventing the formation of a malignant cancer. A tumor with a higher "mutational burden" (i.e., more somatic mutations that differ from the germline genetic code of the individual) is more likely to have immunogenic TAAs.

TAAs can either be highly specific (i.e., cells of one tumor will have different TAAs from the cells of another tumor) or be shared by different tumors, even tumors that occur in different hosts. For example, virus-induced tumors (such as those due to **human papillomavirus**) tend to have TAAs that *cross-react* with one another if induced by the same virus strain. This feature has spurred interest in developing antitumor vaccines that use the shared, virus-induced TAAs found in all individuals who have that virus-induced tumor.

CARCINOEMBRYONIC ANTIGEN & ALPHA-FETOPROTEIN

Some human tumors contain antigens that normally occur in fetal but not in adult human cells.

(1) **Carcinoembryonic antigen** circulates at elevated levels in the serum of many patients with carcinoma of the colon, pancreas, breast, or liver. It is found in fetal gut, liver, and pancreas and in very small amounts in normal sera. Detection of this antigen can be helpful in the diagnosis of such tumors, and if the level declines after surgery, it suggests that the tumor is not spreading. Conversely, a rise in the level of carcinoembryonic antigen in patients with resected carcinoma of the colon suggests recurrence or spread of the tumor.

(2) **Alpha-fetoprotein** is present at elevated levels in the sera of patients with hepatocellular cancer and is used as a marker for this disease. It is produced by fetal liver and is found in small amounts in some normal sera. However, it is nonspecific; it occurs in several other malignant and nonmalignant diseases.

Monoclonal antibodies directed against new surface antigens on malignant cells (e.g., B-cell lymphomas) can be useful in diagnosis. Monoclonal antibodies coupled to toxins, such as diphtheria toxin or ricin, a product of the *Ricinus* plant, can kill tumor cells in vitro and someday may be useful for cancer therapy.

MECHANISM OF TUMOR IMMUNITY

The immune response that attacks nonself tumor cells is directed by T cells. Such immune responses probably act as a **surveillance** system to detect and eliminate newly arising clones of neoplastic cells. The cells that infiltrate tumors include **natural killer (NK) cells**, which can kill cells directly or can react to cells bound by antibody (antibody-dependent cellular cytotoxicity); **CD8-positive cytotoxic T cells**; and activated **macrophages**, which rely on antigen-specific **Th-1 cells** and cytokines for activation.

In general, the immune response against one or a few tumor cells is effective. However, because tumor cells both proliferate and mutate rapidly, selection pressure favors those cells that can escape immune surveillance by evading or suppressing the anti-tumor immune response. Tumor cells do this through a number of mechanisms, including (1) **decreased expression of major histocompatibility complex (MHC) class I–complexed TAAs**; (2) **release of soluble factors**, such as the enzyme indoleamine 2,3-dioxygenase (IDO), that promote the activity of immunosuppressive leukocytes, including T regulatory (Treg) cells; (3) **expression of cell surface molecules**, such as programmed cell death-ligand 1 (PD-L1) and cytotoxic T lymphocyte-associated antigen-4 (CTLA-4; see Chapter 60), that inhibit the function of cytotoxic T cells and NK cells; and (4) pushing T cells into a dormant state called **T cell exhaustion** through chronic antigen stimulation (see Chapter 60).

Tumor antigens can stimulate the development of specific antibodies as well. Some of these antibodies are **cytotoxic**, but others, called **blocking antibodies**, can actually *enhance* tumor growth. This might be due to blocking recognition of tumor antigens by the host or due to binding of the Fc regions to inhibitory Fc receptors.

CANCER IMMUNOTHERAPY

Therapies that boost the antitumor immune response, known as *immunotherapies*, are clinically effective, and the field is evolving rapidly. One of the first cancer immunotherapies still in use is the **BCG vaccine** (bacillus Calmette-Guérin, a bovine mycobacterium). It was observed that injecting BCG into the lesions of skin melanoma and bladder cancer activates and recruits immune cells into the tumor, leading to tumor regression.

Recent approaches to immunotherapy have sought to enhance the numbers of **tumor-infiltrating lymphocytes (TILs)**. The basis for this approach is the observation that some cancers are infiltrated by lymphocytes (NK cells and cytotoxic T cells) that appear to be destroying the cancer cells. Treatment with **interleukin-2** was shown to improve survival rates in certain melanomas and renal cell cancers, although interleukin-2 treatment is limited by the toxicity of a systemic inflammatory response.

Monoclonal antibodies directed against CTLA-4 and PD-1 (see Chapter 60) are effective in enhancing the immune response against cancer cells. CTLA-4 and PD-1 on T cells inhibit the co-stimulatory signal, and antibody against these proteins blocks this inhibitory effect. This removal of an inhibition, or **checkpoint**, enhances the immune response against the tumor, and therefore, this strategy is called **checkpoint blockade immunotherapy**. Not surprisingly, checkpoint blockade therapies are limited by the fact that some patients develop autoimmune diseases as a consequence of off-target immune activation.

Another approach to cancer immunotherapy involves directly applying cells as a therapeutic. In one example, TILs are recovered from a surgically removed cancer, grown in cell culture until a large number of cells are obtained, activated with interleukin-2, and returned to the patient in the expectation that they will "home in" specifically on the cancer cells and kill them.

Finally, another successful approach has produced sustained remission in patients with certain leukemias and lymphomas. These remissions are induced by infusion of **chimeric antigen-receptor modified T (CAR-T) cells** that target proteins on the surface of the leukemic B cells. When the CAR-T cells recognize these proteins, for example, CD19 that is found on malignant B cells, they release cytokines, perforin, and granzymes, killing the cells much as a TIL recognizes and kills virus-infected cells (see Chapter 60).

These "cell-based" therapies are expensive because they require manipulating each individual patient's cells and then reinfusing them as an autologous transplant. In addition, CAR-T cells can cause systemic inflammatory responses due to excessive cytokine release, and because healthy B cells also have CD19, patients must take pooled intravenous immunoglobulin (IVIG) replacement to counteract the persistent B-cell deficiency that occurs. The first commercial CAR-T cell therapy was approved by the US Food and Drug Administration for B-cell leukemia in 2017.

SELF-ASSESSMENT QUESTIONS

1. Regarding tumor immunity, which one of the following is the most accurate?

   (A) Both cytotoxic T cells and cytotoxic antibodies attack human cancer cells.

   (B) An elevated level of alpha-fetoprotein is a marker for carcinoma of the lung.

   (C) A declining level of carcinoembryonic antigen (CEA) is an indication that the patient's colon cancer has recurred.

   (D) Cancer cells induced by chemicals have new antigens on the surface, but cancer cells induced by viruses do not.

   (E) Natural killer (NK) cells do not participate in the cell-mediated response to cancer cells because they do not have an antigen-specific receptor on their surface.

2. Regarding cancer immunotherapy, which one of the following statements is most accurate?

   (A) Cytotoxic T cells directed against CD-28 protein can kill many B-cell leukemia cells.

   (B) Tumor-infiltrating lymphocytes inhibit interleukin-2 production, thereby preventing T-cell leukemias.

   (C) Checkpoint blockade immunotherapy involves the inhibition of class 2 MHC presentation by antigen-presenting cells.

   (D) Monoclonal antibodies against CTLA-4 and PD-1 are effective in enhancing the immune response against some cancers.

   (E) The antitumor effect of chimeric antigen-receptor modified T (CAR-T) cells is based on their ability to activate CTLA-4 and PD-1 proteins.

ANSWERS

(1) **(A)**
(2) **(D)**

PRACTICE QUESTIONS: USMLE & COURSE EXAMINATIONS

Questions on the topics discussed in this chapter can be found in the Immunology section of Part XIII: USMLE (National Board) Practice Questions starting on page 757. Also see Part XIV: USMLE (National Board) Practice Examination starting on page 775.

Immunodeficiency

Immunodeficiency can occur in any of the four major components of the immune system: (1) B cells (antibody), (2) T cells, (3) complement, and (4) phagocytes. In a patient with a history of infections that are **unusually frequent**, **unusually severe**, or caused by **unusual organisms**, the pattern of these infections can indicate which component(s) of the immune system might be defective. Most immunodeficiencies are *acquired*, and these are frequently caused by immunosuppressive medications or diseases that suppress immunity, such as HIV/AIDS. Although they are less common, *congenital* immunodeficiencies (Table 68–1) are important to understand because (1) the patterns of infections that are seen teach us how various immune components are supposed to function normally, and (2) recent technological advances have allowed us to better diagnose and treat these diseases, preventing the infectious complications.

CONGENITAL IMMUNODEFICIENCIES

T-Cell Deficiencies and Combined T- and B-Cell Deficiencies

Congenital T-cell deficiencies tend to be the most severe and easily recognized immunodeficiency. Because T cells are central to so many aspects of immune responses, including **antiviral immunity** and the activation of **macrophages**, their absence results in a broad range of unusual **opportunistic** infections (i.e., viruses, bacteria, fungi, and protozoa that are rarely seen in healthy hosts). In addition, because T cells are required for maturation of **B cells and antibody**, T-cell deficiencies result in some degree of B-cell deficiency. These diseases often present

within the first 6 to 12 months of life when maternal antibody is waning. Children with severe T-cell deficiency rarely live past the age of 2 years without a hematopoietic stem cell transplant, and these transplants are only possible if the diagnosis is made *before* the onset of severe infectious complications. Newborn screening for T-cell receptor excision circles (TRECs) has greatly increased the early detection of these diseases (see Chapter 59).

An isolated T-lymphocyte deficiency is a life-threatening condition, but even more devastating are deficiencies caused by defects in *all* lymphocyte development, which causes **severe combined immunodeficiency (SCID)**. In this disease, recurrent infections caused by bacteria, viruses, fungi, and protozoa occur in early infancy (3 months of age) because **both B cells and T cells are defective**. Immunoglobulin levels are very low, and tonsils and lymph nodes are absent. Note that innate immunity is not directly affected, but without adaptive immunity, the innate immune system is unable to clear infections.

Pneumocystis pneumonia is the most common presenting infection in infants with combined T- and B-cell deficiency. Infections caused by *Candida albicans* and viruses such as varicella-zoster virus, cytomegalovirus, and respiratory syncytial virus are common and often fatal.

(1) **Wiskott-Aldrich syndrome**—Recurrent **pyogenic infections, eczema**, and **low platelets** characterize this syndrome. The symptoms typically appear during the first year of life. It is an X-linked disease caused by mutations in the *WASp* gene, leading to a defect in actin filament assembly that is important for T cells to respond to antigen presentation *and* for B cells to be activated by signals from the B-cell receptor. These patients

TABLE 68–1 Important Congenital Immunodeficiencies

Deficient Component and Name of Disease	Specific Deficiency	Molecular Defect	Clinical Features
Combined B and T cell			
Severe combined immunodeficiency (SCID)	Deficiency of both B-cell and T-cell function	Various mutations: Defective interleukin (IL)-2 receptor, defective recombinases, defective kinases, absence of class II MHC proteins, or ADA or PNP deficiency	Bacterial, viral, fungal, and protozoal infections
T cell			
Thymic aplasia (DiGeorge's syndrome)	Absence of T cells and suppressed antibody responses	Defective development of pharyngeal pouches, associated with chromosome 22 deletions	Viral, fungal, and protozoal infections; tetany caused by hypoparathyroidism
Chronic mucocutaneous candidiasis	Deficient T-cell response to *Candida*	IL-17 and IL-17 receptor deficiencies have been described	Skin and mucous membrane infections with *Candida*
B cell			
X-linked (Bruton's agammaglobulinemia)	Absence of B cells; very low immunoglobulin (Ig) levels	Mutant tyrosine kinase	Recurrent bacterial infections, especially of respiratory tract, caused by pyogenic bacteria such as pneumococci
Selective IgA	Very low IgA levels	Failure of heavy-chain gene switching	Recurrent infections, especially of the sinuses and lung, caused by pyogenic bacteria
Complement			
C3b	Insufficient C3	Unknown	Pyogenic infections, especially with *Staphylococcus aureus*
C6,7,8	Insufficient C6,7,8	Unknown	*Neisseria* infections
Phagocytes			
Chronic granulomatous disease	Defective bactericidal activity because no oxidative burst	Deficient NADPH oxidase activity	Pyogenic infections, especially with *S. aureus* and *Aspergillus*

ADA = adenosine deaminase; MHC = major histocompatibility complex; NADPH = nicotinamide adenine dinucleotide phosphate hydrogen; PNP = purine nucleoside phosphorylase.

are therefore unable to mount an IgM response to the capsular polysaccharides of bacteria, such as pneumococci.

(2) **Ataxia-telangiectasia**—In this disease, **ataxia** (staggering) and **telangiectasia** (enlarged small blood vessels of the conjunctivas and skin) occur. About two-thirds of patients have lymphopenia and low immunoglobulins, particularly IgA, which results in **recurrent pyogenic upper respiratory infections** that appear by 2 years of age. It is an autosomal recessive disease caused by **mutations in the genes that encode DNA repair enzymes**. In addition to the lymphopenia, these patients frequently develop leukemia, lymphoma, or other cancers.

(3) **Thymic aplasia (DiGeorge's syndrome)**—In this disease, both **the thymus and the parathyroids fail to develop properly** as a result of a defect in the third and fourth pharyngeal pouches. (In **complete** DiGeorge syndrome, the thymus is completely absent, but this is quite rare. More often, the thymus is malformed or small.) A common presenting symptom is **tetany due to hypocalcemia** caused by hypoparathyroidism. Aortic arch malformations and cleft palate can also be seen in these patients. It is usually caused by a spontaneous deletion in chromosome 22 during early development, although rare cases are inherited genetically.

Severe viral, fungal, or protozoal infections occur in affected infants early in life as a result of absence of the thymus, where T-cell precursors recombine their receptors and mature to become

T cells (see Chapter 59). Pneumonia caused by *Pneumocystis jiroveci* and thrush caused by *C. albicans* are two common infections in these patients. Antibody production may be decreased or normal. If decreased, severe pyogenic bacterial infections can occur.

This disease is unusual in that it is *not* caused by a defect in the bone marrow precursor cells. The T-cell precursors themselves are normal, and so performing a **thymus transplant** into these patients can largely correct the defect by allowing the patient's own T-cell precursors to mature.

(4) **Cytokine signaling defects**—Patients with defects in specific cytokines or their receptors have increased susceptibility to specific organisms. For example, **chronic mucocutaneous candidiasis** is an infection of the skin and mucous membranes with *C. albicans*, which, in immunocompetent individuals, is a nonpathogenic member of the normal flora. In rare families with heritable mucocutaneous candidiasis, the overall T-cell and B-cell levels and functions are normal except that there is deficiency **specifically** of interleukin (IL)-17 or the IL-17 receptor.

In contrast, defects in IL-12, gamma interferon (IFN-γ), or the receptors for these cytokines result in recurrent or severe infections with **mycobacteria** and *Salmonella* species. IL-12 normally helps to differentiate naïve CD4-positive T cells into

Th-1 cells, which make the IFN-γ that is required to activate the macrophages that limit these infections (see Chapter 60). A common presentation of IL-12 or IFN-γ deficiency is a child with disseminated infection with bacillus Calmette-Guérin (BCG), the attenuated *Mycobacterium* strain in the BCG vaccine that is given in many countries to prevent severe tuberculosis disease.

Deficiency in the gene encoding the transcription factor **STAT3** causes defects in IL-23, IL-6, and IL-21 signaling. STAT3 is essential for the differentiation of Th-17 cells but is also important in B-cell class switching and neutrophil activation. This leads to **high levels of IgE** and impaired neutrophil migration to barrier surfaces, causing recurrent **staphylococcal skin infections**. These patients can also have **eosinophilia, eczema**, and **skeletal defects**. The autosomal dominant form is referred to as **Job's syndrome** or **hyper-IgE syndrome**.

(5) **X-linked SCID** is caused by a **defect in one protein chain of the IL-2 receptor**, a chain that is also a part of the receptors for several other cytokines, such as IL-7. These cytokines are important for lymphocyte development. Because immunity is so profoundly depressed, children with SCID must be strictly isolated from potentially harmful microorganisms. Live, attenuated viral vaccines should *not* be given. Hematopoietic stem cell transplantation may restore immunity, and because infants with SCID do not reject allografts, these transplants require minimal immunosuppressive drugs to engraft.

The X-linked form is the most common, but autosomal forms of SCID can also occur due to mutations in other genes, including:

(6) The gene encoding **a tyrosine kinase called ZAP-70** that plays a role in signal transduction in T cells.

(7) The gene encoding a different kinase called **Janus kinase 3**, which transmits activation and survival signals from cell surface receptors.

(8) The *RAG-1* or *RAG-2* genes that encode the **recombinase enzymes** that catalyze the recombination of the DNA required to generate the T-cell antigen receptor and the IgM that acts as the B-cell antigen receptor.

(9) **Adenosine deaminase (ADA)** and **purine nucleoside phosphorylase (PNP)**, enzymes that recycle nucleotides for DNA synthesis, which reduces the ability of B-cell and T-cell precursors to divide and survive in the bone marrow. In patients with ADA deficiency, enzyme replacement therapy can boost lymphocyte numbers and reduce the number and severity of infections. Patients with ADA deficiency have benefited from gene therapy, placing a new functional enzyme in their hematopoietic stem cells.

(10) The **class I or class II major histocompatibility complex (MHC) proteins**, causing inability to display antigens to T cells (also called **bare lymphocyte syndrome**).

B-Cell Deficiencies

Congenital deficiencies in the number or function of B cells cause **low or absent antibody levels**. Like the T-cell and combined deficiencies, patients with these deficiencies are protected from infections by maternal antibody until the age of 6 to 12 months, at which point they begin to have recurrent infections. However, unlike the "opportunistic" infections described earlier, **recurrent bacterial infections** and **impaired responses to vaccines** are usually the presenting finding in patients with low antibody levels. Their infections are often in the oropharynx and respiratory tract, including sinusitis, otitis, and pneumonia, as these are the sites protected by antibody. But antibody deficiencies also predispose patients to certain viral infections, infections of the gastrointestinal tract, and bacteremia from encapsulated organisms.

(1) **X-linked hypogammaglobulinemia (Bruton's agammaglobulinemia)**—Boys with this disease have low levels of all immunoglobulins (IgG, IgA, IgM, IgD, and IgE) and a virtual absence of B cells; female carriers are immunologically normal. Pre-B cells are present, but they fail to differentiate into mature B cells. This failure is caused by a mutation in the gene encoding a tyrosine kinase that is an important signal transduction protein. Clinically, **recurrent pyogenic bacterial infections** (e.g., otitis media, sinusitis, and pneumonia caused by *Streptococcus pneumoniae* and *Haemophilus influenzae*) occur in infants at about 6 months of age, when maternal antibody is no longer present in sufficient amount to be protective.

Treatment with pooled gamma globulin reduces the number of infections. The pooled gamma globulin preparation contains purified IgG from more than a thousand donors to ensure a broad range of protective antibodies. It is called "intravenous immunoglobulins" (IVIGs) and is administered monthly.

(2) **IgA deficiency**—This is the most common antibody class deficiency; IgG and IgM deficiencies are rarer. Patients with a deficiency of IgA can have **recurrent sinus and lung infections**. However, with time, the infections become less and less frequent, and some individuals with IgA deficiency do not have frequent infections. This is likely because their IgG and IgM levels confer protection that compensates for the loss of IgA. The cause of IgA deficiency may be a failure of heavy chain gene switching. Patients with a deficiency of IgA should **not** be treated with IVIG preparations because the IgA in the infusion can be immunogenic in these patients, that is, they can form antibodies against the foreign IgA.

Patients with selective IgM deficiency or deficiency of one or more of the IgG subclasses also have recurrent sinopulmonary infections caused by pyogenic bacteria such as *S. pneumoniae, H. influenzae*, or *Staphylococcus aureus*.

(3) **Hyper-IgM syndrome**—In this syndrome, severe, recurrent pyogenic bacterial infections resembling those seen in X-linked hypogammaglobulinemia begin early in life. Patients have a high concentration of IgM but very little IgG, IgA, and IgE. But unlike the X-linked agammaglobulinemia patients, these patients have **normal numbers of B cells**. Instead, the deficiency is in the gene encoding CD40 ligand (CD40L), which CD4-positive T cells express on their surface to bind and activate other cells that express CD40. CD40L is one of the main components of **T-cell help** (see Chapter 60). T follicular helper (Tfh) cells from these patients lack CD40L, and their failure to properly interact with B-cell CD40 results in an inability of the

B cell to switch from the production of IgM to the other classes of antibodies. Treatment with IVIG results in fewer infections.

(4) **Common variable immunodeficiency (CVID)**—This term actually comprises a heterogeneous group of diseases, most of which are idiopathic. Like other antibody deficiencies, patients with CVID present with low IgG levels and recurrent infections caused by pyogenic bacteria (e.g., sinusitis and pneumonia caused by pyogenic bacteria such as *S. pneumoniae* and *H. influenzae*). However, the infections tend to be milder and may even begin to occur later in life, between the ages of 15 and 35 years. (This has led to speculation that some forms of idiopathic CVID are actually acquired rather than congenital.) Like hyper-IgM syndrome (described earlier), the known causes of CVID are often functional defects in **helper T cells** rather than in B cells. The number of B cells is usually normal, but their ability to synthesize IgG (and other immunoglobulins) is greatly reduced. Regular IVIG treatment reduces the number of infections.

Complement Deficiencies

The complement system is an important initiator of many inflammatory processes. It plays a key role in complement-dependent cytotoxicity (CDC) and in immune complex deposition (type III hypersensitivity) reactions seen in many autoimmune and inflammatory disorders.

Chapter 63 describes several conditions in which complement is *overactivated*, including **hereditary angioedema**, caused by C1 inhibitor deficiency, and **paroxysmal nocturnal hemoglobinuria**, caused by a deficiency of the glycosylphosphatidylinositol (GPI) anchor for CD59 protein in the cell membrane. CD59 protects the cell membrane from damage by the membrane attack complex (MAC) of complement. A reduced amount of CD59 results in increased hemolysis by MAC.

In host defense, many components of the complement cascade play overlapping roles with other immune components (e.g., antibody). This means that, although diseases of inherited complement deficiencies do exist, the infectious complications in these patients are relatively rare.

Patients with deficiencies in C1, C3, or C5 or the components recruited later in the cascade, C6, C7, C8, or C9 have an increased susceptibility to bacterial infections. Patients with C3 deficiency are particularly susceptible to sepsis with pyogenic bacteria such as *S. aureus*. Those with reduced levels of the components that form the **MAC** (see Chapter 63) are especially prone to bacteremia with *Neisseria meningitidis* or *Neisseria gonorrhoeae*.

Phagocyte Deficiencies

(1) **Chronic granulomatous disease (CGD)**—Patients with this disease are susceptible to opportunistic infections with certain bacteria and fungi (e.g., *S. aureus*); enteric gram-negative rods, especially *Serratia* and *Burkholderia*; and *Aspergillus fumigatus*. CGD is due to a **defect in the intracellular microbicidal activity of phagocytes** as a result of a lack of **NADPH oxidase** activity (or similar enzymes). Much less hydrogen peroxide and superoxides are produced (i.e., no oxidative burst occurs), and the organisms, although ingested, are not killed as efficiently. Even with this deficiency, a phagocyte can use hydrogen peroxide produced by the microbe itself to generate toxic hypochlorite, but CGD patients have a particular susceptibility to infections with *catalase-positive bacteria*, such as staphylococci, because the microbial catalase further degrades what little peroxide there is to water and oxygen. (Infections with catalase-negative bacteria, such as streptococci, and viral, mycobacterial, and protozoal infections are of less concern in CGD patients than infections caused by catalase-positive bacteria and fungi.) B-cell and T-cell functions are usually normal. In 60% to 80% of cases, this is an X-linked disease that appears by the age of 2 years. In the remaining patients, the disease is autosomal.

In the laboratory, diagnosis can be confirmed by the **nitro blue tetrazolium (NBT)** dye reduction test or by the **dichlorofluorescein (DCF)** test. In the NBT test, normal neutrophils will turn the dye a blue color, whereas the neutrophils of a patient with CGD fail to produce the blue color. In the DCF test, cells that oxidize the DCF are detected by flow cytometry.

Prompt, aggressive treatment of infection with the appropriate antibiotics is important. Chemoprophylaxis using antibacterials can reduce the number of infections. Gamma interferon treatment significantly reduces the frequency of recurrent infections, probably because it increases phagocytosis by macrophages.

The name *chronic granulomatous disease* arises from the widespread granulomas seen in these patients, even in the absence of clinically apparent infection. These granulomas can become large enough to cause obstruction of the stomach, esophagus, or bladder. The cause of these granulomas is unknown.

(2) **Chédiak-Higashi syndrome**—In this autosomal recessive disease, recurrent pyogenic infections, caused primarily by **staphylococci** and **streptococci**, occur. This is due to the failure of the **lysosomes** of neutrophils to fuse with **phagosomes**. The degradative enzymes in the lysosomes are, therefore, not available to kill the ingested organisms. Large granular inclusions composed of abnormal lysosomes are seen. In addition, the neutrophils do not function correctly during chemotaxis as a result of faulty microtubules. The mutant gene in this disease encodes a cytoplasmic protein involved in protein transport. Peroxide and superoxide formation is normal, as are B-cell and T-cell functions. Treatment involves antimicrobial drugs to prevent infections. There is no therapy for the phagocyte defect.

(3) **Leukocyte adhesion deficiency syndrome**—Patients with this syndrome have severe pyogenic infections early in life because they have **defective adhesion (LFA-1) proteins** on the surface of their phagocytes. This is an autosomal recessive disease in which there is a mutation in the gene encoding the β chain of an **integrin** that mediates adhesion. As a result, neutrophils adhere poorly to endothelial cell surfaces and *cannot exit the blood circulation* (see Figure 58–4). Accordingly, although they are immunosuppressed, these patients often have extremely high numbers of leukocytes in the blood.

(4) **Cyclic neutropenia**—In this autosomal dominant disease, patients have a very low neutrophil count (<200/μL) for

3 to 6 days of a 21-day cycle. During the neutropenic stage, patients are susceptible to life-threatening bacterial infections, but when neutrophil counts are normal, patients are not susceptible. Mutations in the gene encoding neutrophil elastase have been identified in these patients, but it is unclear how these contribute to the cyclic nature of the disease. It is hypothesized that irregular production of granulocyte colony-stimulating factor may play a role in the cyclic aspect of the disease.

Pattern-Recognition Receptor Deficiency

Mutations in the genes encoding the pattern recognition receptors (PRRs) on the surface of and within the cells of the innate immune system result in susceptibility to severe infections (Table 68–2). For more information on PRRs, see Chapter 58.

(1) **Receptors on the surface of innate immune cells**—Deficiency of Toll-like receptor-5 (TLR-5) results in a failure to recognize flagellin on bacteria and a marked susceptibility to *Legionella* infections. This deficiency is quite common. Deficiency of mannose-binding lectin (MBL) is also common. It results in a failure to activate complement through the lectin pathway (see Chapter 63). However, there are redundant pathways of complement activation, and therefore MBL deficiency typically causes no clinical disease.

(2) **Receptors within innate immune cells**—NOD receptors in the cytoplasm recognize the peptidoglycan of gram-positive and gram-negative bacteria. Various mutations of NOD-2 have been associated with Crohn's disease, presumably resulting from defects in gut barrier immunity and small amounts of bacteria able to invade the intestinal wall. RIG helicase receptors recognize viral double-stranded RNAs synthesized during replication in the cytoplasm. Deficiency of these receptors results in a reduced interferon response to various viruses (i.e., influenza virus).

TABLE 68–2 Important Pattern Recognition Receptor Deficiencies of Innate Immune Cells

Deficient Receptor	Molecular Defect	Clinical Significance
Receptor on surface of cell		
TLR-5	Failure to recognize flagellin on bacteria	Increased *Legionella* infections
MBL	Failure to activate complement	Unknown
Receptor in cytoplasm of cell		
NOD	Failure to recognize peptidoglycan of bacteria	Defective gut immunity and predisposition to Crohn's disease
RIG helicase	Failure to recognize viral double-stranded RNA	Reduced interferon response and susceptibility to various viruses such as influenza virus

MBL = mannose-binding lectin; NOD = nucleotide-binding oligomerization domain; TLR-5 = Toll-like receptor-5.

ACQUIRED IMMUNODEFICIENCIES

Previous chapters have addressed some of the immunosuppressive medications, such as corticosteroids, that can cause **acquired immunodeficiency** (see Tables 62–2 and 66–3). Cushing's syndrome, which can be caused by excessive exposure to *exogenous* or *endogenous* corticosteroids, is associated with immunosuppression. As discussed in Chapters 57 and 66, aging and pregnancy are associated with relatively depressed immunity. Finally, certain cancers, particularly the malignancies of hematopoietic cells (e.g., leukemia, lymphoma, and myeloma), can cause immunosuppression when excessive proliferation of the malignant cells "crowds out" the leukocyte progenitor cells in the bone marrow.

Acquired T-Cell Deficiencies

(1) **Acquired immunodeficiency syndrome**—Patients with **acquired immunodeficiency syndrome (AIDS)** present with opportunistic infections caused by certain bacteria, viruses, fungi, and protozoa (e.g., *Mycobacterium avium-intracellulare*, herpesviruses, *C. albicans, P. jiroveci,* and *Cryptosporidium*). This is due to **reduced helper T-cell numbers** caused by infection with the retrovirus **human immunodeficiency virus** (HIV; see Chapter 45). This virus specifically infects and kills cells bearing the **CD4 surface protein**. The response to specific immunizations is poor; this is attributed to the loss of helper T-cell activity. Patients with AIDS also have a high incidence of cancers such as lymphomas, which may be the result of a failure of immune surveillance. See Chapter 45 for information on treatment and prevention.

(2) **Measles**—Patients with measles have a transient suppression of delayed hypersensitivity as manifested by a loss of reactivity to purified protein derivative (PPD), which is used in the skin test for latent tuberculosis. Tuberculosis can actually reactivate in these patients. Despite the depressed T-cell function in this disease, immunoglobulins are normal.

Acquired B-Cell Deficiencies

(1) **Malnutrition**—Severe malnutrition can reduce the supply of amino acids and thereby reduce the synthesis of IgG. This predisposes to infection by pyogenic bacteria.

(2) **Asplenia**—Patients occasionally need to have their spleen removed through a surgical splenectomy, either because of a traumatic splenic rupture, a malignancy, or an autoimmune disease such as immune thrombocytopenic purpura. Even patients with spleens can develop "functional asplenia," as occurs in patients with sickle cell disease who suffer numerous splenic infarcts. Without a functioning spleen, antibody responses to *new* antigens are generally suppressed, although antibodies to *previously encountered* antigens are generally intact. This is because the long-lived plasma cells responsible for circulating antibodies usually leave the spleen and reside in the bone marrow. Asplenic patients are predisposed to pneumococcal sepsis and severe babesiosis caused by *Babesia microti*.

Acquired Complement Deficiencies

(1) **Liver failure**—Liver failure caused by alcoholic cirrhosis or by chronic hepatitis B or hepatitis C can reduce the synthesis of complement proteins by the liver to a level that allows severe pyogenic infections to occur.

(2) **Malnutrition**—Severe malnutrition can reduce the supply of amino acids and thereby reduce the synthesis of complement proteins by the liver. This predisposes to infection by pyogenic bacteria.

Acquired Phagocyte Deficiencies

(1) **Neutropenia**—Patients with neutropenia (low circulating neutrophils) present with severe infections caused by pyogenic bacteria, such as *S. aureus* and *S. pneumoniae* and enteric gram-negative rods. This occurs because neutrophils are a major component of the immune barrier in the skin, oropharynx, and gastrointestinal tract. Neutrophil counts below 500/μL predispose to these infections. Common causes of neutropenia include cytotoxic drugs, such as those used in cancer chemotherapy; leukemia, in which the bone marrow is "crowded out" by leukemic cells; and autoimmune destruction of the neutrophils. Ciprofloxacin is used to try to prevent infections in neutropenic patients.

SELF-ASSESSMENT QUESTIONS

1. Your patient is a 2-year-old boy who has had several episodes of pustules and lymphadenitis caused by *Staphylococcus aureus*. His immunoglobulin and complement levels are normal. A nitro blue tetrazolium test reveals defective cells. Which one of the following cells is the most likely to be defective?

    (A) CD4-positive T lymphocytes
    (B) CD8-positive T lymphocytes
    (C) Eosinophils
    (D) Natural killer cells
    (E) Neutrophils

2. Your patient is a 25-year-old woman who has had several serious episodes of bacterial pneumonia in the past 5 months. She has not had frequent or unusual infections prior to the onset of these pneumonias. Which one of the following is the most likely immunodeficiency that predisposes her to these infections?

    (A) She is likely to have a defect in her cytotoxic T cells.
    (B) She is likely to have a reduced level of immunoglobulins.
    (C) She is likely to have a mutation in the gene that encodes the C3a portion of complement.
    (D) She is likely to have a mutation in one of the genes that encode the class I MHC proteins.

3. Regarding Bruton's agammaglobulinemia, which one of the following is most accurate?

    (A) There is very little IgG produced, but IgM and IgA levels are normal.
    (B) Viral infections are more common than pyogenic bacterial infections.
    (C) The number of B cells is normal, but they cannot differentiate into plasma cells.

    (D) There is a mutation in the gene for tyrosine kinase, an important enzyme in the signal transduction pathway in B cells.

4. Which one of the following is the most accurate description of the defect in chronic granulomatous disease?

    (A) There is an inability to produce an oxidative burst.
    (B) There is a failure to produce sufficient interleukin-2.
    (C) There is a deficiency of a late-acting complement component.
    (D) There is a mutant protein kinase in a signal transduction pathway.
    (E) There is a mutation in the gene that encodes a class II MHC protein.

5. Which one of the following immunodeficiencies is most likely to predispose to both pyogenic bacterial infections and viral infections in a young child?

    (A) Bruton's agammaglobulinemia
    (B) Chronic granulomatous disease
    (C) DiGeorge's syndrome
    (D) Job's syndrome
    (E) Severe combined immunodeficiency disease

6. Regarding immunodeficiency diseases, which one of the following is most accurate?

    (A) Patients who have a deficiency of IgA have a high incidence of pyogenic infections of the sinuses and lungs.
    (B) Common variable hypogammaglobulinemia typically occurs in boys under the age of 6 months and results from a virtual absence of B cells.
    (C) In Wiskott-Aldrich syndrome, the combination of antibody deficiency and complement deficiency leads to disseminated viral and fungal infections.
    (D) Patients with DiGeorge's syndrome (congenital thymic aplasia) have a reduced number of both T cells and B cells and have severe infections caused by pyogenic bacteria.
    (E) Patients who cannot produce one or more of the late-acting complement components, such as C6, C7, C8, or C9, have episodes of angioedema, including laryngeal edema that can be fatal.

ANSWERS

(1) **(E)**
(2) **(B)**
(3) **(D)**
(4) **(A)**
(5) **(E)**
(6) **(A)**

PRACTICE QUESTIONS: USMLE & COURSE EXAMINATIONS

Questions on the topics discussed in this chapter can be found in the Immunology section of Part XIII: USMLE (National Board) Practice Questions starting on page 757. Also see Part XIV: USMLE (National Board) Practice Examination starting on page 775.

Ectoparasites That Cause Human Disease

INTRODUCTION

Ectoparasites are organisms that are found either on the skin or only in the superficial layers of the skin. *Ecto* is a prefix meaning "outer." Virtually all ectoparasites are arthropods; that is, they are invertebrates with a chitinous exoskeleton.

The ectoparasites that cause human disease fall into two main categories: insects (six-legged arthropods) and arachnids (eight-legged arthropods). The ectoparasites discussed in this chapter include insects such as lice, flies, and bedbugs and arachnids such as mites, ticks, and spiders.

Many arthropods are vectors that transmit the organisms that cause important infectious diseases. A well-known example is the *Ixodes* tick that transmits *Borrelia burgdorferi*, the cause of Lyme disease. Table XII–3 describes the medically important vectors. However, in this chapter, the arthropods are discussed not as vectors but as the cause of the disease itself. Table 69–1 summarizes the common features of diseases caused by the medically important ectoparasites that are described in this

chapter. Ectoparasites of minor medical importance are briefly described at the end of the chapter.

INSECTS

1. Lice

Disease

Pediculosis is caused by two species of lice: *Pediculus humanus* and *Phthirus pubis*. *P. humanus* has two subspecies: *P. humanus capitus* (head louse), which primarily affects the scalp, and *P. humanus corporis* (body louse), which primarily affects the trunk. *P. pubis* (pubic louse) primarily affects the genital area, but the axilla and eyebrows can be involved as well.

Note that the body louse is the vector for several human pathogens, notably *Rickettsia prowazekii*, the cause of epidemic typhus, whereas the head louse and the pubic louse are not vectors of human disease.

TABLE 69–1   Important Ectoparasites That Cause Human Disease

	Name of Organism	Common Features of Disease
Insects		
1. Lice	*Pediculus humanus* (head or body louse)	Pruritus of scalp or trunk; nits seen on hair shaft
	Phthirus pubis (pubic louse)	Pruritus in pubic area; nits seen on hair shaft
2. Flies	*Dermatobia hominis* (botfly)	Pruritic, painful, and erythematous nodule; larva may be seen emerging from nodule
3. Bedbugs	*Cimex lectularius* (common bedbug)	Pruritic, erythematous wheal
Arachnids		
1. Mites	*Sarcoptes scabiei* (itch mite)	Pruritic, erythematous papules, and linear tracks
2. Ticks	*Dermacentor* species	Ascending paralysis
3. Spiders	*Latrodectus mactans* (black widow spider)	Severe pain and muscle spasms
	Loxosceles reclusa (brown recluse spider)	Necrotic ulcer

Important Properties

Lice are easily visible, being roughly 2 to 4 mm long. They have six legs armed with claws by which they attach to the hair and skin (Figure 69–1). *Pediculus* has an elongated body, whereas *Phthirus* has a short body and resembles a crab, and hence its nickname, the crab louse (Figure 69–2). People infected with *Phthirus* are said to have "crabs."

Nits are the eggs of the louse and are typically found attached to the hair shaft (Figure 69–3). They are white and can be seen with the naked eye. Nits of the body louse are often attached to the fibers of clothing.

Transmission

Head lice are transmitted primarily by fomites such as hats, combs, and towels. These are especially common in school children. Body lice live primarily on clothing and are transmitted either by clothing or by personal contact. Body lice leave the clothing when they require a blood meal. Pubic lice are transmitted primarily by sexual contact.

Widespread infestations of body lice occur when personal hygiene is poor (e.g., during wartime or in crowded refugee camps).

Pathogenesis

Adult lice feed on blood and, in the process, inject saliva into the skin, which induces a hypersensitivity reaction and, as a consequence, pruritus.

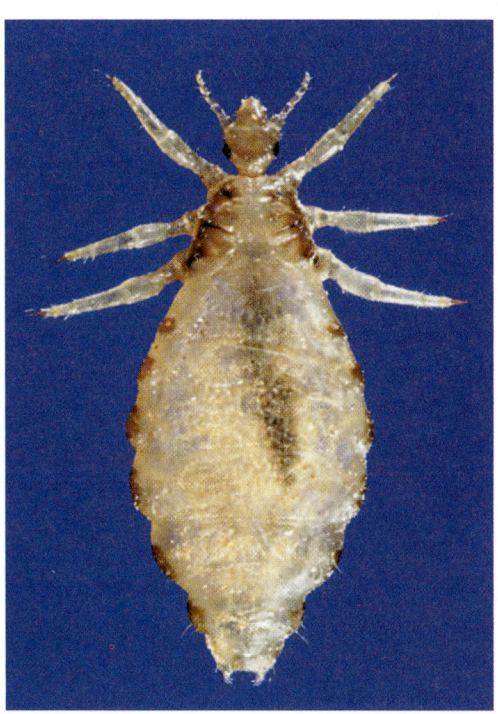

FIGURE 69–1 *Pediculus corporis*. Body louse. Note the elongated abdomen of *Pediculus corporis*. In contrast, the pubic louse has a short "crab-like" abdomen. (Used with permission from Dr. Collins F., Public Health Image Library, Centers for Disease Control and Prevention.)

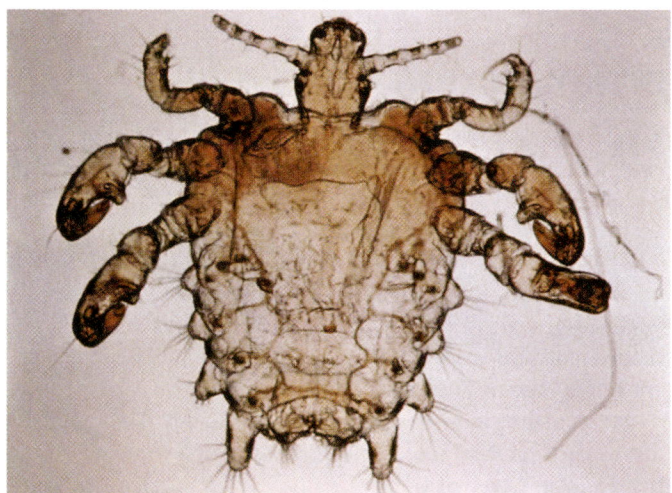

FIGURE 69–2 *Phthirus pubis*. Pubic louse. Note the short, rounded "crab-like" abdomen of *Phthirus pubis*. In contrast, the body louse has an elongated abdomen. (Reproduced with permission from Public Health Image Library, Centers for Disease Control and Prevention. Original content provider is the World Health Organization.)

FIGURE 69-3 *Pediculus capitis.* Egg case (nit). Arrow points to an egg case (also known as a nit) attached to hair shaft. Note embryo within egg. (Used with permission from Dr. Juranek D., Public Health Image Library, Centers for Disease Control and Prevention.)

Clinical Findings

Pruritus is the main symptom. Excoriations may result from scratching, and secondary bacterial infections may occur. In pediculosis capitis, the adult lice are often difficult to see, but the nits are easily visualized. In pediculosis corporis, the adult lice are primarily in the clothing rather than on the body. In pediculosis pubis, the adult lice and nits can be seen attached to the pubic hair.

Laboratory Diagnosis

The laboratory is not involved in diagnosis. Nits fluoresce under ultraviolet light of a Wood's lamp, which can be used to screen the hair of large numbers of people.

Treatment

Permethrin (Nix, RID) is the treatment of choice, as it is both pediculicidal and ovicidal. However, resistance to permethrin is increasing. Ivermectin (Sklice) is also effective, and resistance has not been reported. Nits are removed using a fine-toothed (nit) comb. Patients with body lice often do not need to be treated, but the clothing should be either discarded or treated.

Prevention

Children should not share articles of clothing. Many schools have a policy that children cannot attend school until they are nit-free, but the need for this exclusionary approach is under review. The personal items of affected individuals, such as towels, combs, hairbrushes, clothing, and bedding, should be treated. Sexual partners of those infested with pubic lice should be treated and tested for other sexually transmitted diseases.

2. Flies

Disease

Myiasis is caused by the larva of many species of flies, but the one best known is the botfly, *Dermatobia hominis.* Fly larvae are also known as maggots. Note that maggots are occasionally used to debride nonhealing wounds, but these maggots do not cause myiasis.

Important Properties

The flies that cause myiasis are found worldwide and infest many animals as well as humans. Human infestation occurs most often in tropical areas. *Dermatobia* is common in Central and South America.

Transmission

The precise route of transmission varies depending on the species of fly. In one scenario, the adult fly deposits its egg in a wound and the egg hatches to produce the larva. In another, the fly deposits the egg in the nostrils, in the conjunctiva, or on the lips. In yet another, the fly deposits the egg on unbroken skin and the larva invades the skin.

Dermatobia is especially interesting in that it deposits its egg on a mosquito. When the mosquito bites a human, the warmth of the skin induces the egg to hatch and the larva enters the skin at the site of the mosquito bite.

Pathogenesis

The presence of the larva in tissue induces an inflammatory response.

Clinical Findings

The characteristic lesion is a painful, erythematous papule resembling a furuncle (Figure 69–4). The lesion may also be pruritic. The larva can often be seen within a central pore. Some patients report a sense of movement within the lesion. A history of travel to tropical regions is commonly elicited. Cutaneous myiasis is the most common form, but ocular, intestinal, genitourinary, and cerebral forms occur.

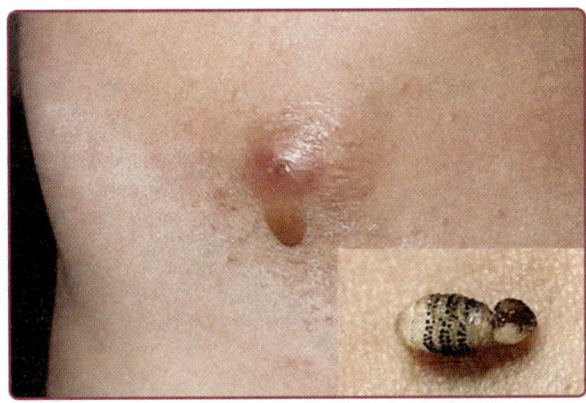

FIGURE 69-4 *Dermatobia hominis.* Myiasis. Botfly larva emerging from the center of erythematous nodular skin lesion. Insert shows intact larva. (Reproduced with permission from Kang S, Amagai M, Bruckner AL, et al: *Fitzpatrick's Dermatology in General Medicine*, 9th ed. New York, NY: McGraw Hill; 2019.)

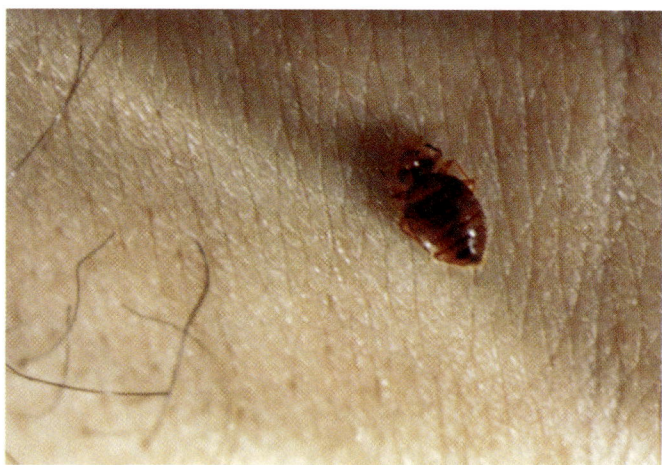

FIGURE 69–5 *Cimex lenticularis*. Bedbug. Bedbug in the process of ingesting blood from skin. They are wingless, reddish-brown, and about 5 mm long. (Reproduced with permission from Public Health Image Library, Centers for Disease Control and Prevention.)

Laboratory Diagnosis

The laboratory is not involved in diagnosis except when identification of the larva is needed.

Treatment

Surgical removal of the larva is the most common mode of treatment. If the larva is visible, manual extraction can be performed. If the larva is not visible, the central pore can be covered with petroleum jelly, thus causing anoxia in the larva. This induces the larva to migrate to the surface.

Prevention

Prevention involves limiting exposure to flies, especially in tropical areas. General measures, such as wearing clothing that covers the extremities, mosquito netting, and insect repellant, are recommended.

3. Bedbugs

Cimex lectularius is the most common bedbug found in the United States. It has an oval, brownish body and is about 5 mm long (Figure 69–5). Bedbugs reside in mattresses and in the crevices of wooden beds. At night, they emerge to take a blood meal from sleeping humans. The main symptom of a bedbug bite is a pruritic wheal caused by a histamine-related hypersensitivity reaction to proteins in the bug saliva (Figure 69–6). Some individuals show little reaction. The bite of a bedbug is not known to transmit any human disease. Calamine lotion can be used to relieve the itching. Malathion or lindane can be used to treat mattresses and beds.

ARACHNIDS

1. Mites

Disease

Scabies is caused by the "itch" mite, *Sarcoptes scabiei*.

FIGURE 69–6 Bedbug bites. Several urticarial wheals surrounded by erythema on the patient's back. (Reproduced with permission from Kang S, Amagai M, Bruckner AL, et al: *Fitzpatrick's Dermatology in General Medicine*, 9th ed. New York, NY: McGraw Hill; 2019.)

Important Properties

The adult female *Sarcoptes* mite is approximately 0.4 mm in length, with a rounded body and eight short legs (Figure 69–7). It is found worldwide, and it is estimated that several hundred million people are affected around the globe.

Transmission

It is transmitted by personal contact or by fomites such as clothing, especially under unhygienic conditions (e.g., in the homeless and during wartime). It is not a vector for other human pathogens.

Pathogenesis

The pruritic lesions result from a delayed hypersensitivity reaction to the antigenic proteins of the eggs or feces of the mite. The mite is located within the stratum corneum of the epidermis.

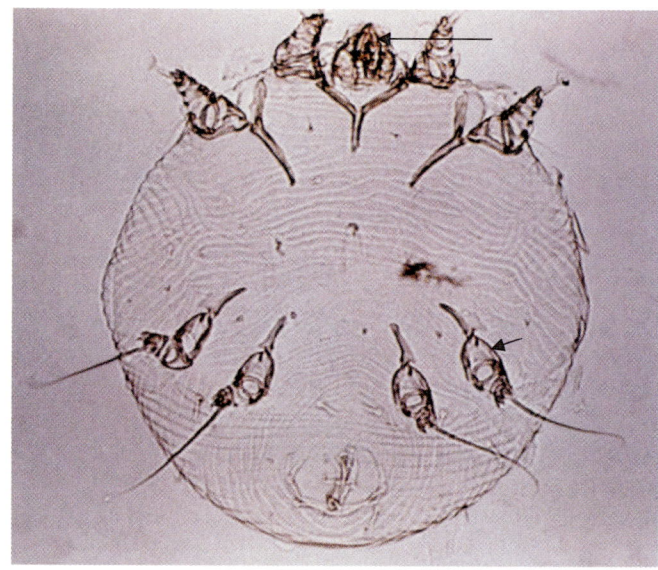

FIGURE 69–7 *Sarcoptes scabiei*. "Itch" mite. Long arrow points to the mouth. Short arrow points to one of the eight legs. This is a ventral view. (Reproduced with permission from Public Health Image Library, Centers for Disease Control and Prevention; donated by the World Health Organization, Geneva, Switzerland.)

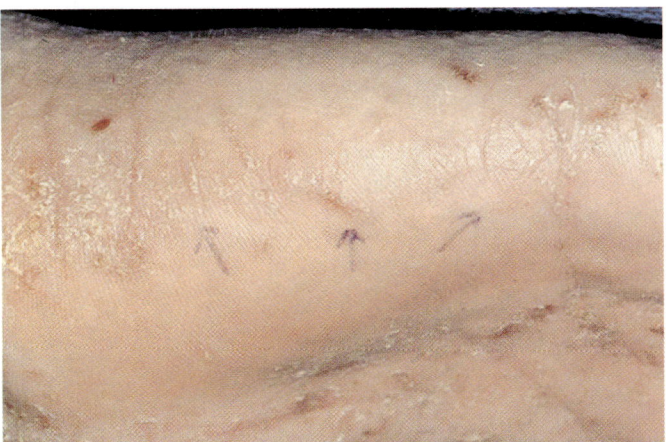

FIGURE 69–8 *Sarcoptes scabiei.* Lesions. Note three arrows that point to linear track-like lesions on the hand. (Reproduced with permission from Wolff K, Johnson R: *Fitzpatrick's Color Atlas & Synopsis of Clinical Dermatology,* 6th ed. New York, NY: McGraw Hill; 2009.)

Clinical Findings

The typical lesions in immunocompetent people are either tracks or papules that are very pruritic (Figure 69–8). The most common sites are the hands, wrists, axillary folds, and genitals. Areas of the body where clothing is tight, such as along the belt line, are often involved. The itching is typically worse at night.

In immunocompromised individuals, an extensive crusted dermatitis (Norwegian scabies) can occur. These patients may be infested with thousands of mites. Excoriations may become infected with *Staphylococcus aureus* or *Streptococcus pyogenes,* resulting in pyoderma.

Laboratory Diagnosis

Microscopic examination of skin scrapings reveals the mites, their eggs, or fecal pellets.

Treatment

Permethrin (Elimite) is the drug of choice. Ivermectin is also effective. Topical steroids are used to relieve the itching.

Prevention

Prevention involves treatment of close contacts of the patient and treating or discarding fomites such as clothing and towels. Mass administration of ivermectin reduced the prevalence of scabies.

2. Ticks

Disease

Tick paralysis is caused by many species of ticks, the most common of which in the United States are *Dermacentor* species. Ticks are vectors for several human diseases, including Lyme disease and Rocky Mountain spotted fever, but in this chapter, we discuss only tick paralysis, which is caused by a toxin produced by the tick itself.

Important Properties and Transmission

Female ticks require a blood meal for maturation of their eggs, and hence it is the female that causes tick paralysis as well as serves as the vector of diseases. Ticks are commonly found in grassy woodland areas and are attracted by carbon dioxide and warmth from humans. A tick attaches to human skin by means of its proboscis.

Dermacentor andersoni, the wood tick, is more common in the western United States, whereas *Dermacentor variabilis,* the dog tick, is more common in the eastern states (Figure 69–9). Both species can cause tick paralysis. Presently in the United States, there are no cases of paralysis caused by *Ixodes* ticks; however, such cases are reported in other countries, especially Australia.

Pathogenesis

Paralysis is mediated by a neurotoxin that blocks acetylcholine release at the neuromuscular junction—an action similar to that of botulinum toxin. The toxin is made in the salivary gland of the tick. The tick must remain attached for at least 4 days prior to the onset of symptoms.

Clinical Findings

An ascending paralysis resembling Guillain-Barré syndrome occurs. Ataxia is an early presenting symptom. The paralysis is symmetrical and can ascend from the legs to the head within several hours. Respiratory failure and death can occur. Recovery typically occurs within 24 hours of removal of the tick.

The tick is often found at the hairline at the back of the neck or near the ear. Children younger than 8 years are most often affected.

FIGURE 69–9 *Dermacentor* tick. This tick causes tick paralysis and is the vector that transmits *Rickettsia rickettsiae,* the cause of Rocky Mountain spotted fever. (Used with permission from Dr. Christopher Paddock, Public Health Image Library, Centers for Disease Control and Prevention.)

Laboratory Diagnosis

The laboratory is not involved in diagnosis.

Treatment

Treatment involves removal of the tick.

Prevention

Tick bites can be prevented by application of insect repellant and wearing clothes that cover the extremities. Searching for and removing ticks promptly is an important preventive measure.

3. Spiders

Two species of spiders cause most of the significant disease in the United States, namely, the black widow spider (*Latrodectus mactans*) and the brown recluse spider (*Loxosceles reclusa*). The black widow spider is about 1 cm in length with a characteristic orange-red hourglass on its ventral surface (Figure 69–10). The brown recluse spider is also about 1 cm in length, but has a characteristic violin-shaped pattern on its dorsal surface (Figure 69–11). It is also called the "fiddleback" spider.

Neurotoxic Disease

The bite of the black widow spider causes neurologic symptoms primarily. Within an hour after the bite, pain and numbness spread from the site. Severe pain and spasms in the extremities and abdominal pain occur. Fever, chills, sweats, vomiting, and other constitutional symptoms can occur. In contrast to the bite of the brown recluse spider, tissue necrosis does not occur. Most patients recover in several days, but some, mainly children, die. Antiserum, if available, to the venom of the black widow should be given in severe cases. The antiserum is made in horses, so testing for hypersensitivity to horse serum should be performed.

FIGURE 69–11   *Loxosceles reclusa* (brown recluse spider). Note "violin" shape on dorsal surface of thorax. (Used with permission from Dr. Andrew J. Brooks, Public Health Image Library, Centers for Disease Control and Prevention.)

It is interesting that the neurotoxin is encoded by WO virus, a bacteriophage that infects *Wolbachia* bacteria. *Wolbachia* bacteria infect many insects, including black widow spiders.

Dermonecrotic Disease

The bite of the brown recluse spider causes tissue necrosis symptoms primarily. The necrosis is due to proteolytic enzymes in the venom. Pain and pruritus at the site of the bite occur early, followed by vesicles and then hemorrhagic bullae (Figure 69–12). The lesion ulcerates, becomes necrotic, and may not heal for weeks to months. Skin grafting may be required. Antiserum to the venom of the brown recluse spider is not available in the United States.

ECTOPARASITES OF MINOR MEDICAL IMPORTANCE

1. Demodex

Demodex mites are also known as hair follicle or eyelash mites. They cause folliculitis, especially on the eyelashes (blepharitis) and on the face. They block the follicles, causing an inflammatory response and loss of eyelashes. Dry eyes and chalazions can occur. They are implicated as a cause of rosacea-like lesions on the face.

These mites are very small. As many as 25 mites have been found in one hair follicle. The diagnosis is made by observing the mite with a slit-lamp biomicroscope. Treatment involves careful debridement of the affected areas plus application of tea tree oil ointment.

2. Trombicula

Trombicula mites are also known as harvest mites or chiggers. The bite of the larva causes papules accompanied by intense

FIGURE 69–10   *Latrodectus mactans* (black widow spider). Note red "hourglass" on ventral surface. (Used with permission from Dr. Paula Smith, Public Health Image Library, Centers for Disease Control and Prevention.)

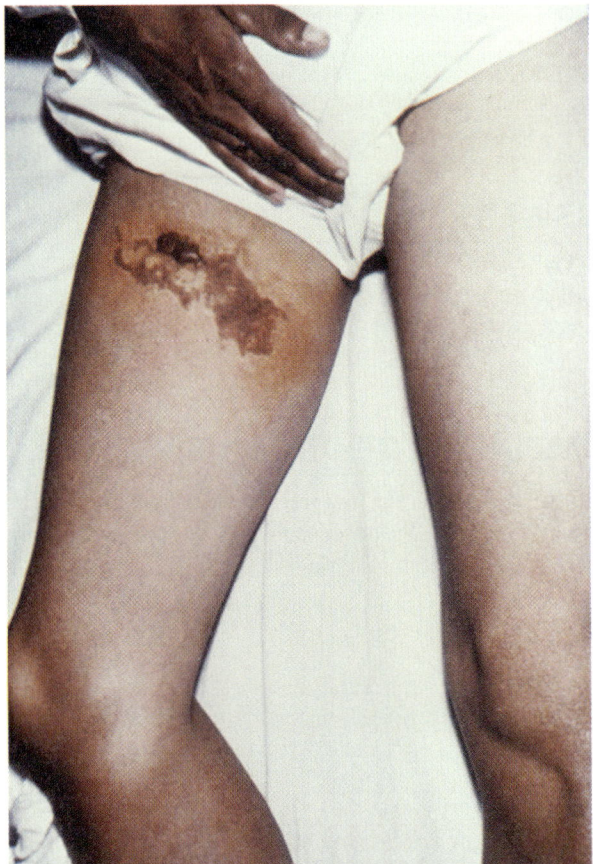

FIGURE 69–12 Recluse spider bite. Note hemorrhagic bullae surrounded by irregular areas of necrosis on right thigh. (Used with permission from Dr. Parsons M.A. and donated by Dr. Rosenfeld G., Head Hospital Vital, of Department of Physiopathology, Brazil; Public Health Image Library, Centers for Disease Control and Prevention.)

itching, most often where tight-fitting clothing meets the skin. The pruritic papules result from an allergic response to proteins in the saliva, which are injected into the skin at the time of the bite. The larvae are not bloodsucking but obtain nutrients from dissolved skin cells. They are found in vegetation in hot, humid regions, such as southeastern states of the United States. Treatment involves the use of oral antihistamines or topical steroids to control itching. The chiggers can be killed with permethrin.

3. Dermatophagoides

Dermatophagoides mites are also known as house dust mites. They feed on exfoliated human skin cells. They do not cause disease in humans directly, but proteins in their feces are powerful allergens for some people. Small particles in house dust can become airborne, be inhaled, and induce asthma and atopic dermatitis.

4. Tunga

The flea, *Tunga penetrans*, causes tungiasis. The small female flea penetrates the skin, usually of the feet, and lays eggs there. An erythematous papule develops at the site that evolves into a

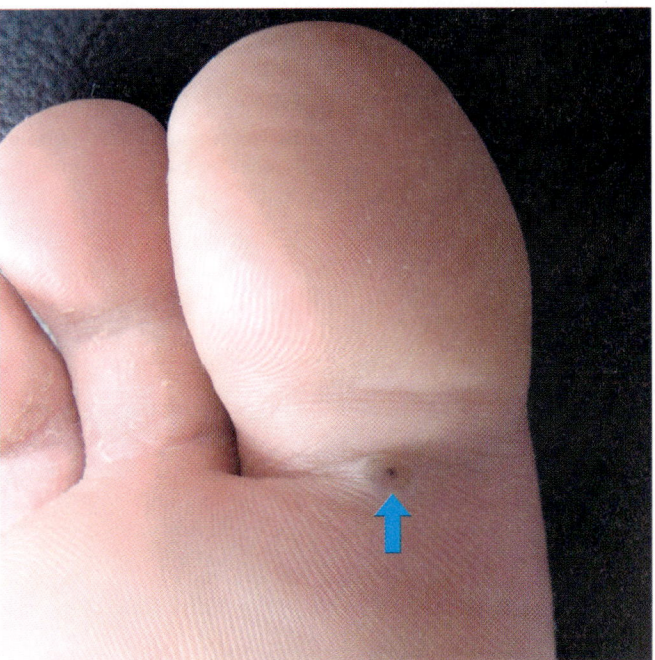

FIGURE 69–13 Tungiasis. Blue arrow points to lesion on sole of foot. The dark center of the lesion is the *Tunga* flea. (Reproduced with permission from Knoop KJ, Stack LB, Storrow AB, et al: *The Atlas of Emergency Medicine*, 4th ed. New York, NY: McGraw Hill; 2016.)

whitish nodule with a dark center (the flea) (Figure 69–13). The lesion causes itching at the outset, then pain.

The fleas are commonly found in the sand of tropical beaches and walking in bare feet is a risk factor. It is widely found in the tropical areas of Latin America, Africa, and Asia. Tungiasis is a disease typically seen in returned travelers.

No drug is effective. Treatment involves removing the flea with a needle. If not removed, the flea will die in 4 to 6 weeks and the lesion will resolve. Cases of tetanus are associated with tungiasis, so a tetanus booster is recommended.

There are many local names for the *Tunga* flea, such as chigoe flea, jigger flea, and bicho de pie (foot bug). Note that, despite the similarity in name, these fleas are not chiggers. Chiggers are mites in the genus *Trombicula* (see above).

SELF-ASSESSMENT QUESTIONS

1. Your patient is a homeless person with several papules on his hands that are very pruritic. One lesion is a linear track. You suspect the patient may have scabies. Which one of the following is most likely to be seen?

   (A) Nits are seen attached to hair.

   (B) Visual inspection reveals a larva in the lesions.

   (C) The nymph form of a tick is seen in the lesions.

   (D) Examination of a skin scraping in the microscope reveals a mite.

2. Regarding the patient in Question 1, which one of the following is the best drug to treat the infection?

   (A) Albendazole

   (B) Ivermectin

(C) Permethrin

(D) Praziquantel

(E) Primaquine

3. Your patient has recently returned from a trip to Central America that included a 2-week trek in the tropical rainforest. She now has a raised erythematous lesion on her leg that is quite painful. A 7-day course of cephalexin has had no effect. Which one of the following is the most likely cause?

(A) *Carex lenticularis*

(B) *Dermatobia hominis*

(C) *Latrodectus mactans*

(D) *Pediculus humanus*

(E) *Phthirus pubis*

4. Regarding pediculosis, which one of the following is most accurate?

(A) Nits are the eggs of the louse and are typically found attached to the hair shaft.

(B) Praziquantel is the drug of choice for pediculosis caused by both *Pediculus* and *Phthirus*.

(C) To visualize the organism, a skin sample should be examined using the 10× objective in a light microscope.

(D) The lesions caused by the body louse are pruritic, but the lesions caused by the pubic louse form a painful necrotic black eschar.

ANSWERS

(1) **(D)**

(2) **(C)**

(3) **(B)**

(4) **(A)**

SUMMARIES OF ORGANISMS

Brief summaries of the organisms described in this chapter begin on page 713. Please consult these summaries for a rapid review of the essential material.

Bone & Joint Infections

INTRODUCTION

Bone and joint infections are serious infections because destruction of bone or cartilage can lead to significant disability. Osteomyelitis and infectious arthritis are caused primarily by bacteria or fungi. In these diseases, the organisms directly infect the bone and joint. In contrast, immune complex arthritis, reactive arthritis, and rheumatic fever are caused by immune reactions to either bacteria or viruses, and organisms are not found in the joints.

The clinical diagnosis of infectious arthritis often involves an analysis of joint fluid. Radiologic studies of joints and bone contribute important information. Microbiologic diagnosis of osteomyelitis and infectious arthritis is typically made by culturing either a bone specimen or joint fluid. Antimicrobial therapy is typically given for long periods (i.e., weeks to months).

OSTEOMYELITIS

Definition

Osteomyelitis is an infection of the bone. The term *osteo* refers to bone, and *myelo* refers to the bone marrow. Osteomyelitis is classified as either acute or chronic.

Pathophysiology

The most common mode by which organisms reach the bone is by hematogenous spread (i.e., either bacteremia or fungemia) from a distant site. Acute bacterial osteomyelitis often arises from a pyogenic skin infection such as a boil, but many sources are undetected. Mycobacterial and fungal osteomyelitis often arise from the initial site of infection in the lung.

In children, hematogenous spread tends to result in osteomyelitis located at the end of long bones (at the metaphyses) that are richly endowed with blood vessels. In adults, hematogenous spread results most commonly in vertebral osteomyelitis and discitis, not osteomyelitis of the long bones.

Osteomyelitis also occurs by direct extension from an infected contiguous site such as a skin or soft tissue infection. It can also occur following trauma that results in an open fracture and direct contamination of the bone.

Chronic osteomyelitis tends to occur in the lower extremity, especially in diabetics who often have vascular insufficiency. They are predisposed to skin and soft tissue infections that extend into the bone.

Clinical Manifestations

The most characteristic clinical manifestations are bone pain and localized tenderness at the site of infection. Most patients

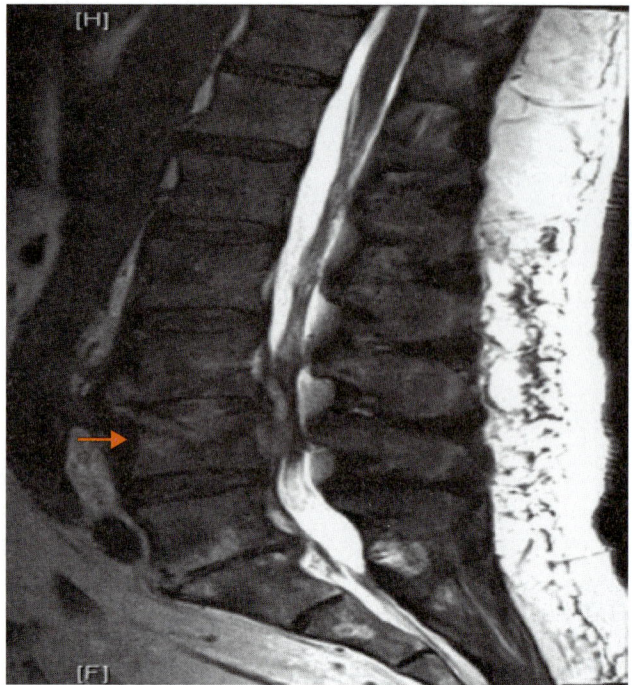

FIGURE 70–1   Vertebral osteomyelitis. Arrow indicates site of vertebral lesion. (Reproduced with permission from McKean SC, Ross JJ, Dressler DD, et al: *Principles and Practice of Hospital Medicine.* New York, NY: McGraw Hill; 2012.)

TABLE 70–1   Organisms Causing Osteomyelitis with Various Predisposing Factors

Predisposing Factor	Common Organisms
Neonates	*Streptococcus agalactiae* (group B *Streptococcus*)
Children and adults	*Staphylococcus aureus*
Prosthetic joints	*Staphylococcus epidermidis*
Adults with vertebral osteomyelitis	*S. aureus, Mycobacterium tuberculosis*
Intravenous drug users	*S. aureus, Pseudomonas aeruginosa, Serratia marcescens, Candida albicans*
Skin infection in diabetic patient	*S. aureus,* anaerobes
Puncture wounds of foot	*P. aeruginosa*
Cat bite	*Pasteurella multocida*
Sickle cell anemia	*Salmonella* species
Exposure in endemic area	*Coccidioides immitis, Histoplasma capsulatum*

also have constitutional symptoms such as fever, night sweats, and fatigue. Limited range of motion of an affected extremity is seen. In vertebral osteomyelitis, the lumbar region is affected more often than the cervical or thoracic regions (Figure 70–1).

In acute osteomyelitis, the symptoms occur abruptly and progress rapidly, whereas in chronic osteomyelitis, the course is more indolent. In chronic osteomyelitis, necrosis of the bone occurs, and a sequestrum (an avascular piece of infected bone)

can form at the site of the lesion (Figure 70–2). Relapses tend to occur in chronic osteomyelitis more than in acute osteomyelitis, and surgical debridement, especially to remove sequestra, is important to minimize the risk of relapse.

Pathogens

The most common bacterial cause of acute osteomyelitis in both children and adults is *Staphylococcus aureus* (Table 70–1). However, vertebral osteomyelitis in adults may be caused by *Mycobacterium tuberculosis* (Pott's disease). Osteomyelitis in patients with hip or knee prostheses is likely to be caused by *Staphylococcus epidermidis* or other skin flora, such as *Propionibacterium acnes.*

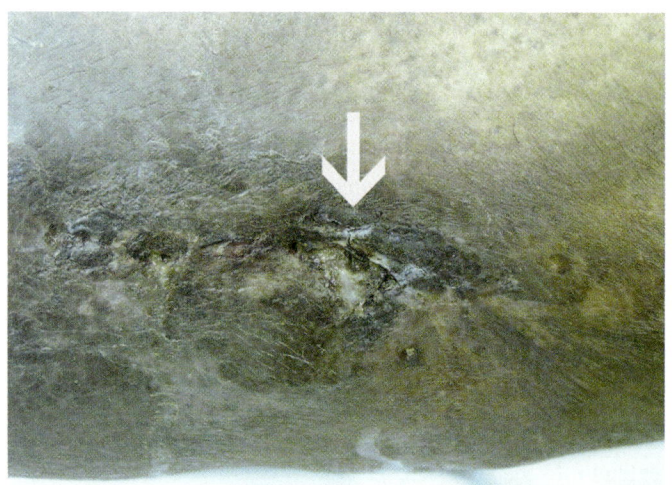

A

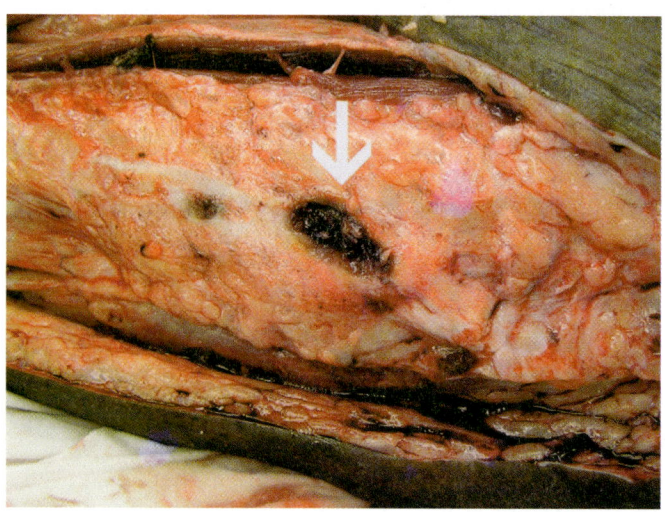

B

FIGURE 70–2   Chronic osteomyelitis. **A.** White arrow points to draining fistula at the site of chronic osteomyelitis. **B.** White arrow points to necrotic bone caused by chronic osteomyelitis. (Reproduced with permission from Kemp WL, Burns DK, Brown TG: *Pathology: The Big Picture.* New York, NY: McGraw Hill; 2008.)

Diabetics often have arterial insufficiency in the lower extremities leading to chronic infections and skin ulcers. These infections can extend into the adjacent bone causing osteomyelitis. *S. aureus* and skin anaerobes such as *P. acnes* are often involved.

When osteomyelitis occurs in an intravenous drug user, it is most often caused by *S. aureus*; however, gram-negative rods, such as *Pseudomonas* and *Serratia*, and yeasts, such as *Candida* species, are also important causes. Osteomyelitis following a puncture wound of the foot through a sneaker is often caused by *Pseudomonas aeruginosa*, and osteomyelitis associated with a cat bite is likely to be caused by *Pasteurella multocida*. Patients with sickle cell anemia are predisposed to osteomyelitis caused by *Salmonella* species.

Fungal osteomyelitis is most often caused by either *Coccidioides immitis* or *Histoplasma capsulatum*. Living in areas where these fungi are endemic is an important predisposing factor. Viruses, protozoa, and helminths do not cause osteomyelitis.

Diagnosis

A microbiologic diagnosis of acute osteomyelitis is most consistently made by culture of a specimen of the bone lesion. Blood cultures are positive in approximately half of the cases.

The typical radiologic finding in acute osteomyelitis is a defect in the bone accompanied by periosteal elevation (Figure 70–3). Early in the disease, X-rays and even computed tomography (CT) scans may be negative. Magnetic resonance imaging (MRI) scans are the most sensitive radiologic tests for diagnosis of osteomyelitis.

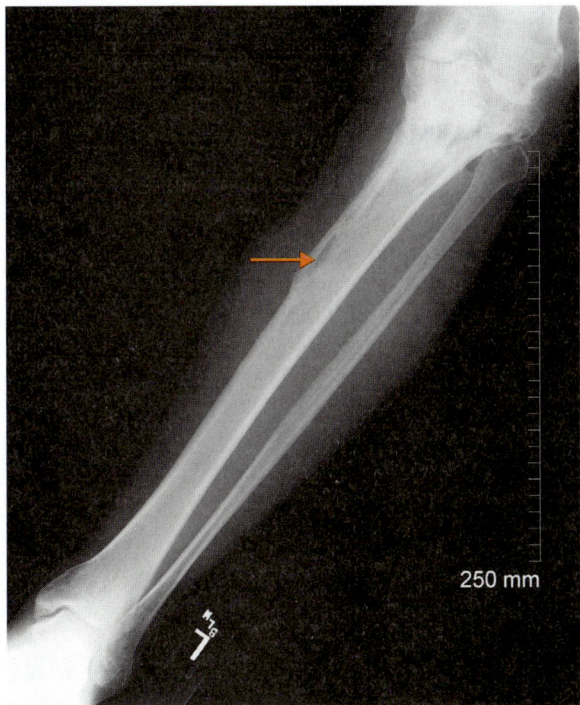

250 mm

FIGURE 70–3 Periosteal elevation (arrow) in acute osteomyelitis of the tibia. (Reproduced with permission from Longo DL, Fauci AS, Kasper DL, et al: *Harrison's Principles of Internal Medicine*, 18th ed. New York, NY: McGraw Hill; 2012.)

Treatment

Empiric therapy for acute osteomyelitis should include drugs that are bactericidal, penetrate well into bone, and include coverage for *S. aureus*. Vancomycin, nafcillin, or cephalexin administered parenterally can be used. Vancomycin is often used until the culture results and the sensitivity of the organism are known. If methicillin-resistant *S. aureus* (MRSA) is the cause then vancomycin, daptomycin, or linezolid can be used. If gram-negative rods are the cause, then ceftriaxone, ceftazidime, or cefepime can be used. The duration of therapy ranges from 3 to 6 weeks or longer. Surgical debridement of chronic osteomyelitis lesions is often necessary.

Prevention

There is no vaccine effective against the common causes of osteomyelitis, and chemoprophylaxis is typically not employed. Generally speaking, prophylactic antibiotics are not recommended prior to dental procedures to prevent prosthetic joint infection. Proper foot care in diabetics can prevent osteomyelitis.

INFECTIOUS (SEPTIC) ARTHRITIS

Definition

Infectious (septic) arthritis is an infection of the joints. The terms **infectious** and **septic** are used to distinguish these infections from immune-mediated arthritis, such as rheumatoid arthritis. Bacteria, especially *S. aureus*, cause the vast majority of cases of infectious (septic) arthritis. Monoarticular involvement of a large weight-bearing joint, such as the hip or knee, is the most common presentation.

Synovial Fluid Analysis

Analysis of synovial fluid aspirated from a swollen joint plays an important role in the diagnosis of arthritis. Table 70–2 shows the findings in the fluid aspirated from an infected joint compared to normal synovial fluid. Synovial fluid from an infected joint may appear cloudy, has at least 20,000 neutrophils/μL, and has a low glucose concentration. Analysis of the fluid from the joints of those with rheumatoid arthritis and those who have a traumatic injury to the joint is included for comparison.

Pathophysiology

Organisms typically reach the joint via the bloodstream from a skin site. Less frequently, organisms enter the joints through penetrating trauma, medical procedures such as arthroscopy, or a contiguous osteomyelitis.

Patients with long-standing rheumatoid arthritis and those with prosthetic hips and knees are predisposed to infectious arthritis.

Clinical Manifestations

The acute onset of an inflamed joint, typically a large weight-bearing joint such as the hip or knee, is the typical manifestation

TABLE 70–2 Synovial Fluid Findings in Arthritis

Disease	Appearance	Cell Number (per µL)	Glucose (Fluid/Blood Ratio)
Normal	Clear	<200 neutrophils	Approximately 1.0
Infectious (septic)	Cloudy	>20,000 neutrophils	<0.25
Rheumatoid arthritis	Opalescent	2000–20,000 neutrophils	~0.5–0.8
Trauma	Clear	200–2000 neutrophils	~1.0

(Figure 70–4). Fever is often present. On physical examination, the affected joint is red, warm, and swollen, and a joint effusion is typically present. Reluctance to use a joint, especially in a child, may be a sign of infectious arthritis.

Pathogens

The most common cause of infectious arthritis overall is *S. aureus* (Table 70–3). Streptococci, such as *Streptococcus pyogenes* and *Streptococcus pneumoniae*, also cause infectious arthritis. In young sexually active adults, *Neisseria gonorrhoeae* is the most common cause. Patients with a prosthetic hip or knee joint are predisposed to infectious arthritis caused by *S. epidermidis*. *S. aureus* and *P. aeruginosa* are the most common causes in intravenous drug users.

Borrelia burgdorferi, the cause of Lyme disease, should also be mentioned as the cause of inflamed joints that resemble those seen in infectious arthritis. However, in Lyme disease, the arthritis is immune-mediated, and organisms are not recovered from the affected joints.

Diagnosis

Visualization of the organisms in the Gram stain of joint fluid is used to guide empiric therapy. A microbiologic diagnosis of infectious arthritis is typically made by culture of a specimen of the joint fluid. Blood cultures are positive in less than 30% of cases.

The typical radiologic finding in infectious arthritis is soft tissue swelling. Evidence of joint destruction can be seen if the infection progresses.

Treatment

Untreated infectious arthritis can lead to joint destruction and loss of mobility, so prompt antibiotic treatment is required for optimal recovery. Empiric therapy for infectious arthritis should include drugs such as vancomycin, nafcillin, or cefazolin that are bactericidal against *S. aureus*. Vancomycin, daptomycin, or linezolid should be used to treat MRSA and methicillin-resistant *S. epidermidis* (MRSE).

Ceftriaxone should be used if there is evidence that *N. gonorrhoeae* is the cause. Removal of joint fluid via arthrocentesis and/or surgical drainage is an important adjunct to antibiotics.

Prevention

There is no vaccine effective against the common causes of infectious arthritis, and chemoprophylaxis is typically not employed.

VIRAL (IMMUNE COMPLEX) ARTHRITIS

Viral arthritis is often called immune complex arthritis because the virus does not infect the joint but, rather, the virus forms immune complexes with antiviral antibody that is deposited in joints and elicits an inflammatory response.

The clinical features of viral arthritis consist of either arthralgia (painful joints but without visible inflammation) or frank arthritis in which inflammation is apparent. Most cases of viral

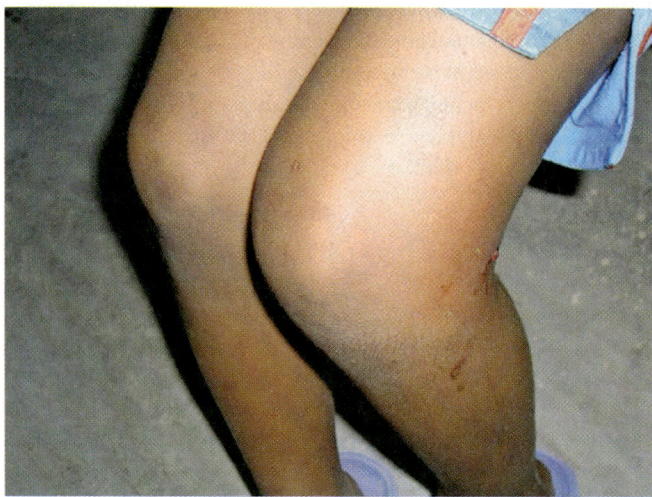

FIGURE 70–4 Septic arthritis of knee. Note swollen and inflamed left knee. (Reproduced with permission from Usatine RP, Smith MA, Mayeaux EJ, et al: *The Color Atlas and Synopsis of Family Medicine*, 3rd ed. New York, NY: McGraw Hill; 2019. Photo contributor: Richard P. Usatine, MD.)

TABLE 70–3 Organisms Causing Infectious Arthritis with Various Predisposing Factors

Predisposing Factor	Common Organisms
Neonates	*Streptococcus agalactiae* (group B Streptococcus)
Children and adults	*Staphylococcus aureus*
Sexually active adults	*Neisseria gonorrhoeae*
Prosthetic hip and knee joints	*S. aureus, Staphylococcus epidermidis*
Intravenous drug users	*S. aureus* and *Pseudomonas aeruginosa*

arthritis are of short duration and resolve spontaneously, but chronic arthritis may occur. The small joints of the hands are most often affected, but large joints can also be involved.

Viral arthritis occurs during the course of infection by several viruses. Rubella virus, either from the natural infection or from the immunization, is a well-recognized cause. Parvovirus B19 is an important cause in that the lesions resemble those of rheumatoid arthritis. The joint lesions of chronic hepatitis C also resemble rheumatoid arthritis. Arthralgia and arthritis occur in the prodromal period of hepatitis B infection. Several arboviruses also cause severe arthralgia, the most common of which is dengue virus. There is no antiviral treatment for viral arthritis.

REACTIVE ARTHRITIS

Reactive arthritis is the term used to describe arthritis that occurs following infection by several bacteria that infect the gastrointestinal or genitourinary tract. The bacteria do not infect the joints. Rather, the arthritis is a result of the immune response to the bacterial infection (see Chapter 66). People who are *HLA-B27* positive are predisposed to reactive arthritis. The bacteria commonly associated with this arthritis are *Campylobacter*, *Shigella*, *Salmonella*, *Yersinia*, and *Chlamydia*.

The main clinical manifestation is an asymmetric arthritis of the knee or ankle accompanied by fever. It typically resolves within a few days or weeks, but chronic arthritis may occur. Recurrences are common. Culture of synovial fluid is negative. Reactive arthritis accompanied by conjunctivitis and urethritis is called Reiter's syndrome. Nonsteroidal anti-inflammatory drugs are considered first-line therapy. Antibiotics have no effect on reactive arthritis.

RHEUMATIC FEVER

Rheumatic fever is an immune-mediated, poststreptococcal disease that affects the joints, heart, brain, and skin. It follows pharyngitis caused by *S. pyogenes* (group A *Streptococcus*) (see Chapter 15). It typically occurs in children ages 5 to 15 years. It is rare in the United States today probably because streptococcal pharyngitis is treated promptly.

TABLE 70–4 Jones Guidelines for the Diagnosis of Acute Rheumatic Fever

Major Manifestations	Minor Manifestations
Polyarthritis	Fever
Carditis	Arthralgia
Chorea	Prolonged P-R interval
Erythema marginatum	Elevated erythrocyte sedimentation rate
Subcutaneous nodules	Elevated C-reactive protein

Rheumatic fever typically begins with a migratory polyarthritis involving the large joints approximately 2 to 3 weeks after the pharyngitis. Carditis often occurs and is the main, life-threatening component of rheumatic fever. The carditis is a pancarditis (i.e., endocarditis, myocarditis, and pericarditis occur, often resulting in congestive heart failure). The mitral valve is most frequently involved. Chorea consisting of involuntary athetoid movements also occurs but is a rare manifestation. Skin involvement consists of erythema marginatum and subcutaneous nodules.

There is no diagnostic test for rheumatic fever. Table 70–4 shows the Jones criteria that are used as a guideline to establish the diagnosis. Two major manifestations or one major plus two minor manifestations suggest the diagnosis. In addition, laboratory evidence of prior infection by *S. pyogenes* is needed. This consists of either (1) a positive throat culture or positive rapid streptococcal antigen test or (2) a rising anti-streptolysin O antibody titer.

The drug of choice is aspirin to reduce the inflammation. Antibiotics such as penicillin G have no effect on the course of the disease but can be given to reduce carriage of streptococci in the pharynx.

Prevention of rheumatic fever involves prompt diagnosis and treatment of strep throat with penicillin G or oral penicillin V. In patients with residual heart disease, prevention of additional damage to heart valves by preventing subsequent episodes of streptococcal pharyngitis is very important. This is achieved by monthly administration of benzathine penicillin G, a depot preparation. This should continue until the patient is at least 20 years old or for 10 years after the last attack.

Cardiac Infections

INTRODUCTION

Cardiac infections are severe, life-threatening infections in many cases. The heart valves (endocardium), myocardium, and pericardium can all be infected. In addition, infection of cardiac devices (pacemakers, defibrillators) is becoming more frequently diagnosed with their increase in use. Diagnosis of cardiac infection can be challenging and usually requires a combination of microbiologic testing and cardiac imaging. Treatment often requires antimicrobial therapy but may also require surgical management for cure.

DIAGNOSTIC TESTING FOR CARDIAC INFECTIONS

Electrocardiogram

An electrocardiogram (ECG) measures electrical activity in the heart using noninvasive monitoring with leads attached to the skin. Cardiac infections can cause disease-specific ECG changes, which can assist in diagnosis.

Echocardiogram

Echocardiography uses Doppler ultrasound to visualize structures and flow of blood through the heart. The test is very helpful in diagnosing most types of cardiac infections. There are two types of echocardiograms, a transthoracic echocardiogram (TTE), where the probe is placed on the chest wall, and a transesophageal echocardiogram (TEE), where the probe is inserted into the esophagus. The TEE often produces higher-quality images, particularly of aortic and mitral valves, since the TEE probe is closer to the heart itself.

ENDOCARDITIS

Definition

Endocarditis is an infection of the valves of the heart.

Pathophysiology

Infection of the heart valves is thought to result from the colonization of damaged valvular endothelium by circulating pathogens. Endothelial damage may result from turbulent blood flow around the valve (congenital or rheumatic heart disease), direct injury from foreign bodies (e.g., intravenous catheters), or repeated intravenous injections of particles in persons who inject drugs. Deposition of platelets and fibrin forms a thrombus at the site of the damaged endothelium. This is called **nonbacterial thrombotic endocarditis** (NBTE).

Organisms enter the bloodstream most often through the mouth (i.e., dental disease or dental procedures) or skin (i.e., trauma, injection drug use). Adhesion of bacteria to the damaged endothelium is enhanced by their ability to produce a glycocalyx.

Once the infection has begun, a combination of organisms and thrombus organize to form a **vegetation** (Figure 71–1).

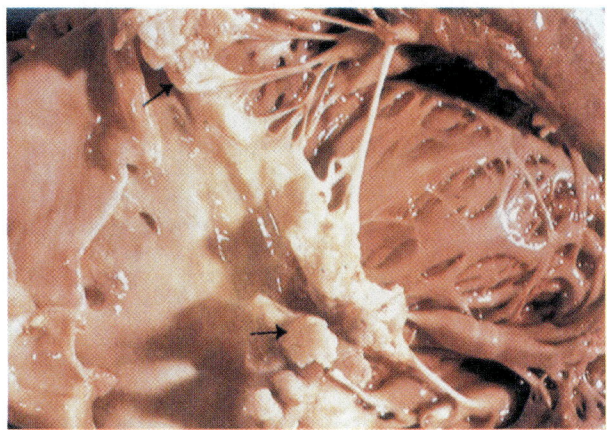

FIGURE 71–1 Endocarditis. Note vegetations on mitral valve. Black arrows point to vegetations. (Reproduced with permission from Longo DL, Fauci AS, Kasper DL, et al: *Harrison's Principles of Internal Medicine*, 18th ed. New York, NY: McGraw Hill; 2012.)

Destruction of the valve occurs at different rates depending on the virulence of the organism. As the valve is destroyed, symptoms of valvular regurgitation can develop. Organisms can spread to surrounding myocardium, resulting in abscess formation and destruction of the electrical conduction system.

As the vegetation on the valve enlarges, fragments can spread via the bloodstream (emboli), resulting in catastrophic effects, such as cerebrovascular accidents (CVAs) and metastatic infections to other organs. Prolonged infection as seen in subacute endocarditis can result in antigen–antibody complex formation. Deposition of these complexes can result in other clinical manifestations, as described in the next section. Artificial materials within the heart, such as prosthetic heart valves, pacemakers, and defibrillators, serve as potential sites for infection.

In summary, the steps in the pathogenesis of endocarditis are as follows:

(1) Formation of NBTE
(2) Transient bacteremia
(3) Adherence of bacteria
(4) Proliferation of bacteria within the vegetation

Clinical Manifestations

The clinical manifestations of infective endocarditis can include any of the following listed below. Depending on the virulence of the infecting pathogen, the time course of illness may be days (acute endocarditis; caused by, e.g., *Staphylococcus aureus*) or weeks to months (subacute endocarditis; caused by, e.g., viridans group streptococci).

- Constitutional symptoms: fever (>80% cases), chills, night sweats, anorexia
- Consequences of destruction of heart valves and associated structures: new murmur, heart failure, atrioventricular (AV) block (PR prolongation seen on ECG; Figure 71–2)
- Embolic phenomena:

  - Left-sided endocarditis: CVAs or brain abscess (Figure 71–3) (new focal neurologic deficits), splenic or renal infarcts (abdominal or flank pain), and emboli to other sites manifesting as splinter hemorrhages (Figure 71–4), Janeway lesions (Figure 71–5), retinal hemorrhages (Figure 71–6), and conjunctival hemorrhages (Figure 71–7).
  - Right-sided endocarditis: septic pulmonary emboli (cough, shortness of breath, chest pain, hemoptysis).
  - Antigen–antibody deposition from uncontrolled infection: Osler's nodes (Figure 71–8), Roth's spots (Figure 71–9), glomerulonephritis (hematuria), and/or arthritis.

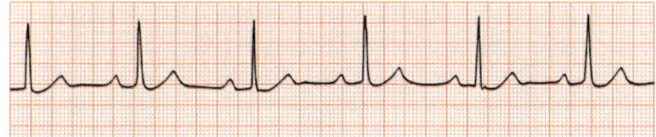

FIGURE 71–2 Atrioventricular block with sinus bradycardia. (Reproduced with permission from McKean SC, Ross JJ, Dressler DD, et al: *Principles and Practice of Hospital Medicine*. New York, NY: McGraw Hill; 2012.)

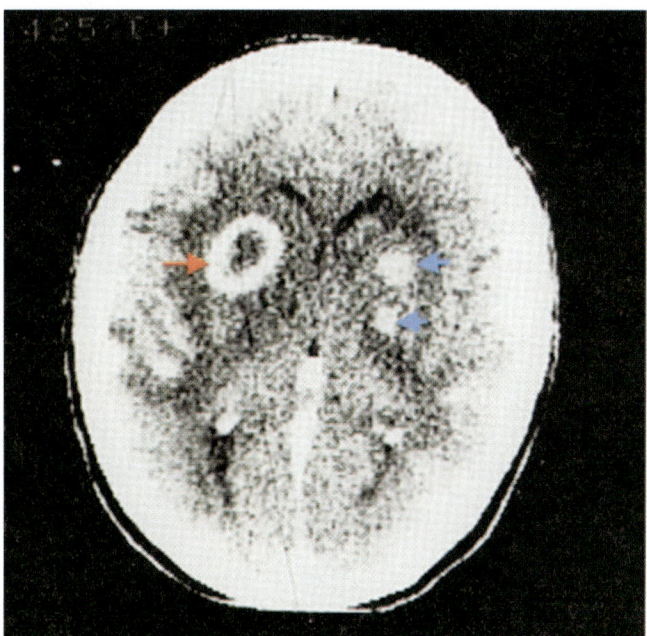

FIGURE 71–3 Brain abscess. Red arrow points to a characteristic ring-enhancing lesion. The blue arrows point to two additional abscesses. (Reproduced with permission from Ropper AH, Samuels MA: *Adams and Victor's Principles of Neurology*, 9th ed. New York, NY: McGraw Hill; 2009.)

Pathogens

Bacteria are the most common causes of endocarditis, but occasionally fungi such as *Candida* are involved as well. The modern classification of pathogens causing endocarditis is divided into

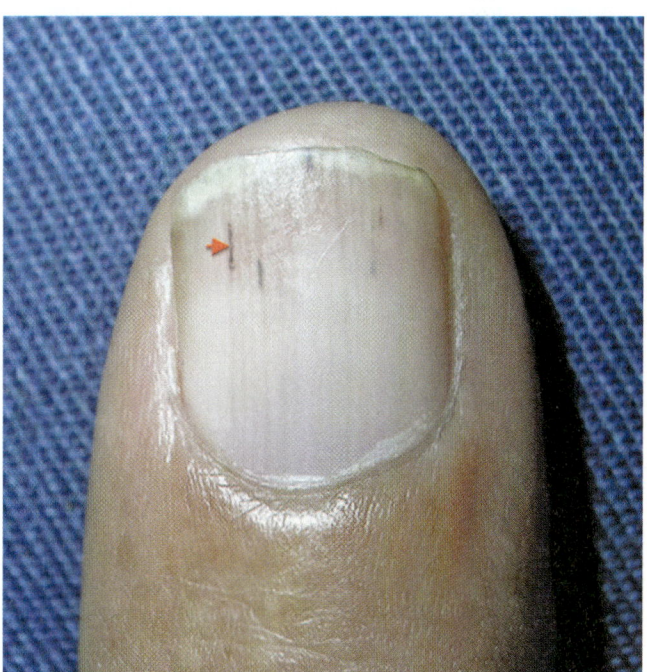

FIGURE 71–4 Splinter hemorrhage. Red arrow points to a splinter hemorrhage under the fingernail. (Reproduced with permission from Usatine RP, Smith MA, Mayeaux EJ Jr, et al: *The Color Atlas of Family Medicine*. New York, NY: McGraw Hill; 2009.)

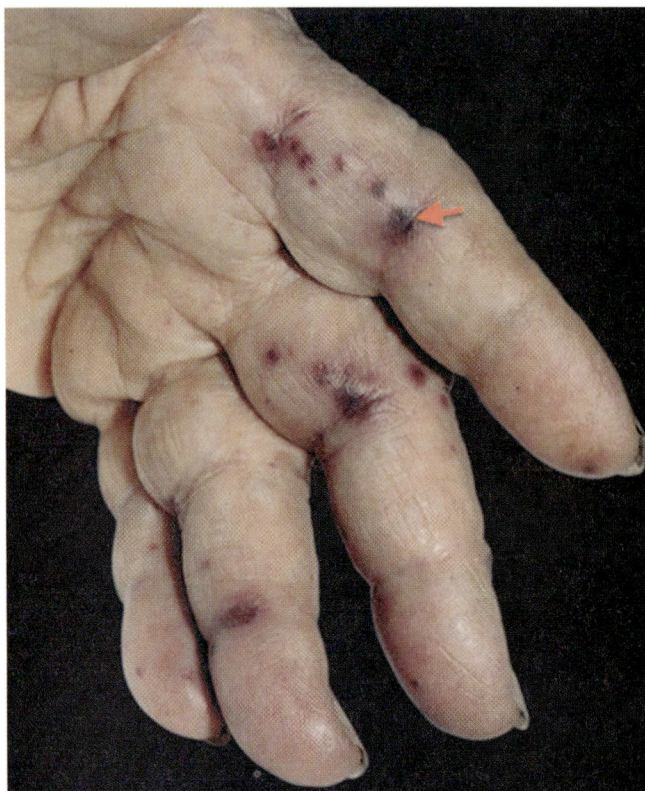

FIGURE 71–5 Janeway lesions. Red arrow points to a Janeway lesion. (Reproduced with permission from Wolff K, Johnson R, Saavedra A: *Fitzpatrick's Color Atlas & Synopsis of Clinical Dermatology*, 7th ed. New York, NY: McGraw Hill; 2013.)

native valve versus prosthetic valve, with subclassifications within each group (Table 71–1). *S. aureus* is most common cause of native-valve infective endocarditis followed by viridans group Streptococci and *Enterococcus* species.

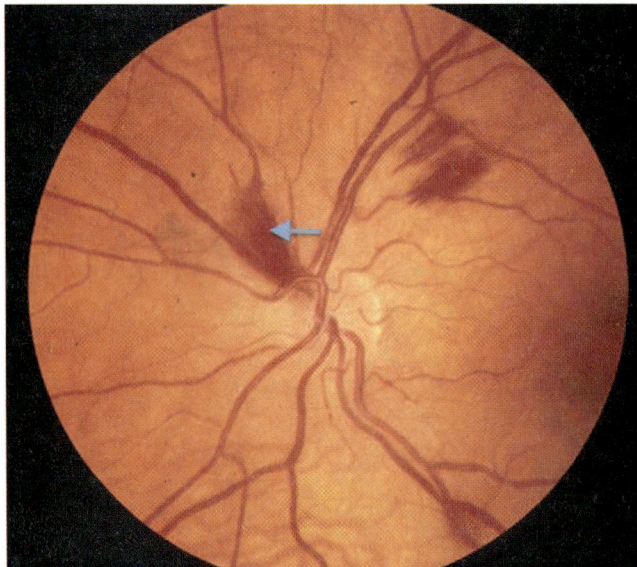

FIGURE 71–6 Retinal hemorrhages. Blue arrow points to a retinal hemorrhage. (Reproduced with permission from Usatine RP, Smith MA, Mayeaux EJ Jr, et al: *The Color Atlas of Family Medicine*. New York, NY: McGraw Hill; 2009. Photo contributor: Paul D. Comeau, MD.)

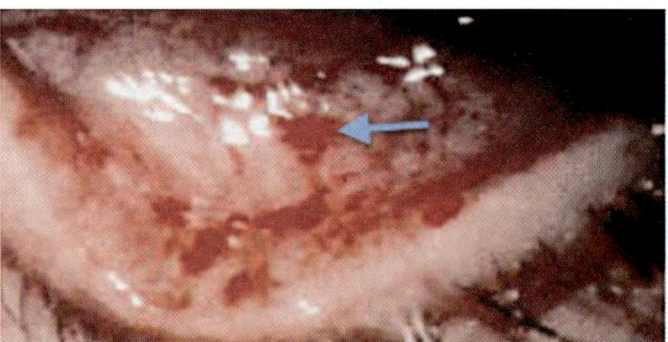

FIGURE 71–7 Conjunctival hemorrhages. Blue arrow points to one of several conjunctival hemorrhages. (Reproduced with permission from Fuster V, Walsh RA, Harrington RA: *Hurst's The Heart*, 13th ed. New York, NY: McGraw Hill; 2011.)

In patients who have prosthetic valves, pacemakers, or defibrillators in place, coagulase-negative staphylococci such as *Staphylococcus epidermidis* and *S. aureus* are the most common pathogens. Other less common pathogens that grow relatively well in routine culture media include the β-hemolytic streptococci, *Streptococcus pneumoniae*, HACEK organisms

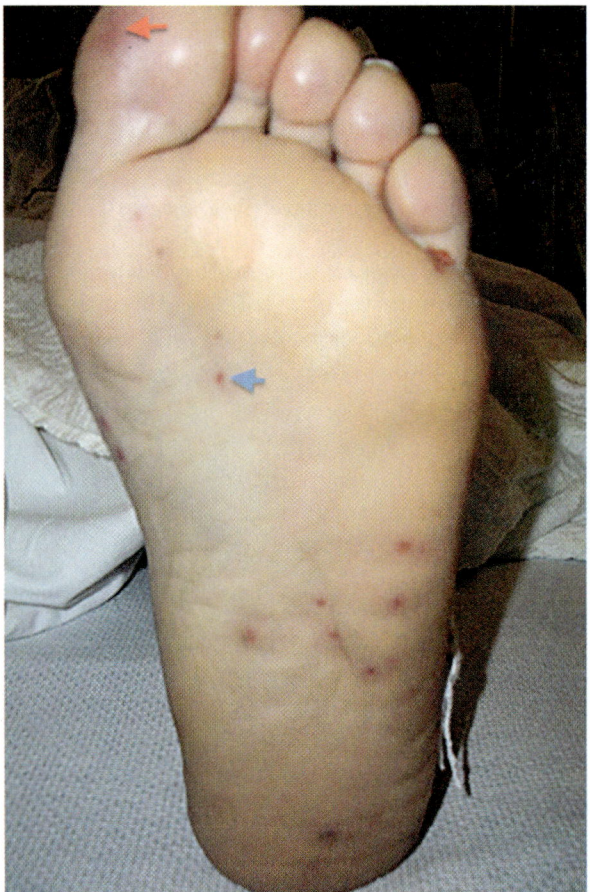

FIGURE 71–8 Osler's node in pulp of big toe. Red arrow points to an Osler's node. Note also Janeway lesions on sole of foot. Blue arrow points to a Janeway lesion. (Reproduced with permission from Usatine RP, Smith MA, Mayeaux EJ Jr, et al: *The Color Atlas of Family Medicine*. New York, NY: McGraw Hill; 2009. Photo contributor: David A. Kasper DO, MBA.)

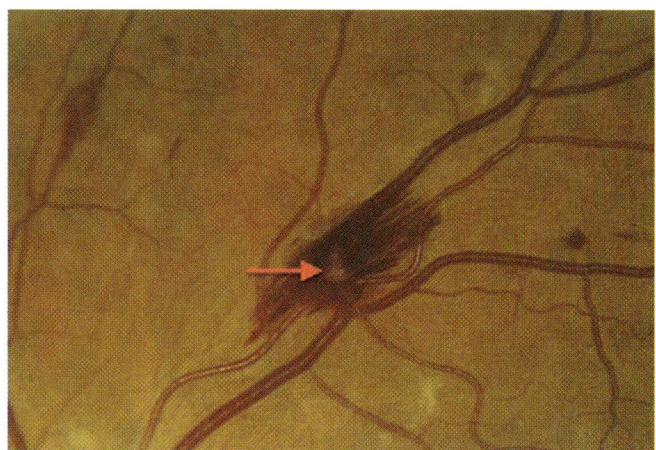

FIGURE 71–9 Roth's spots. Note the central white spots characteristic of Roth's spots (red arrow). (Reproduced with permission from Usatine RP, Smith MA, Mayeaux EJ Jr, et al: *The Color Atlas of Family Medicine*. New York, NY: McGraw Hill; 2009. Photo contributor: Paul D. Comeau, MD.)

(*Haemophilus aphrophilus*, *Aggregatibacter* species, *Cardiobacterium hominis*, *Eikenella corrodens*, and *Kingella kingae*), and *Candida* species.

Pathogens that do not grow in routine blood culture media and require specialized testing for diagnosis include *Bartonella* species, *Coxiella burnetii*, *Brucella* species, and *Tropheryma whipplei*. These are cited as pathogens that cause "culture-negative" endocarditis. The most frequent cause of "culture-negative endocarditis" is the use of antimicrobials prior to obtaining blood cultures.

Diagnosis

A definitive diagnosis of endocarditis requires direct pathologic examination and microbiologic analysis of the heart valve. Because in most cases the heart valve tissue is not available for evaluation, most clinicians use a combination of blood cultures

TABLE 71–1 Etiology of Endocarditis by Category

Category	Pathogen
Native valve	
Community onset	Viridans group streptococci, *Staphylococcus aureus*, *Streptococcus bovis*, *Enterococcus* species
Health care associated	*S. aureus*, *Enterococcus* species, *Staphylococcus epidermidis*
Injection drug use	*S. aureus*, gram-negative rods such as *Pseudomonas*, *Candida* species
Prosthetic valve	
Early	*S. epidermidis*, *S. aureus*
Late	*S. aureus*, viridans group streptococci, *Enterococcus* species, *S. epidermidis*
Pacemaker or defibrillator	*S. epidermidis*, *S. aureus*
Culture negative	Prior antibiotics, *Bartonella* species, *Coxiella burnetii*, *Brucella* species, *Tropheryma whipplei*

TABLE 71–2 Modified Duke Criteria for the Diagnosis of Infective Endocarditis

Definite infective endocarditis
- Pathologic criteria:
  - Microorganism demonstrated by culture or histology in a vegetation or in a vegetation that has embolized or in an intracardiac abscess OR
  - Pathologic lesions, vegetation or intracardiac abscess, confirmed by histology showing active endocarditis
- Clinical criteria:
  - Two major criteria OR one major criterion and three minor criteria OR five minor criteria

Major criteria
- Positive blood cultures of typical organism for infective endocarditis from two separate blood cultures or persistently positive blood culture or single culture positive for culture or serology consistent with *Coxiella burnetii* infection
- Positive echocardiogram for infective endocarditis
- New valvular regurgitation

Minor criteria
- Predisposing heart condition or intravenous drug use
- Fever
- Vascular phenomena (arterial emboli, septic pulmonary infarcts, mycotic aneurysm, etc.)
- Immunologic phenomena (Osler's nodes, Roth's spots, glomerulonephritis, etc.)
- Microbiologic evidence not meeting major criteria

and echocardiographic findings to make the diagnosis of infective endocarditis. The Modified Duke Criteria are the most frequently used criteria for making the diagnosis of endocarditis (Table 71–2) and help guide clinicians to make an accurate diagnosis.

Infecting pathogens are most commonly recovered through blood cultures. To maximize sensitivity of the test, it is recommended to obtain three sets of blood cultures over at least an hour. Whenever possible, blood cultures should be obtained prior to administering antibiotics. In some rare cases of endocarditis due to organisms that do not grow easily in blood culture media (*Bartonella* species), serology can be used to help make the diagnosis.

Evaluation of valves for infection is best accomplished through echocardiography. TTE has reduced sensitivity when compared with a TEE to assess for vegetations and myocardial abscesses but is a less invasive test. Not only can echocardiogram identify new vegetations on valves, which are evidence of infection, but it can also assess the degree of valvular damage and complications such as perivalvular abscesses (Figure 71–10). ECG can be used to detect damage to the conducting system. The most common finding is PR prolongation in patients with aortic valve endocarditis and associated perivalvular abscess (see Figure 71–2).

Treatment

Without treatment, endocarditis is always fatal, so prompt effective therapy is essential. The treatment for endocarditis always includes antimicrobial therapy, and in some cases, surgical removal of the infected valve is indicated as well. Empiric therapy

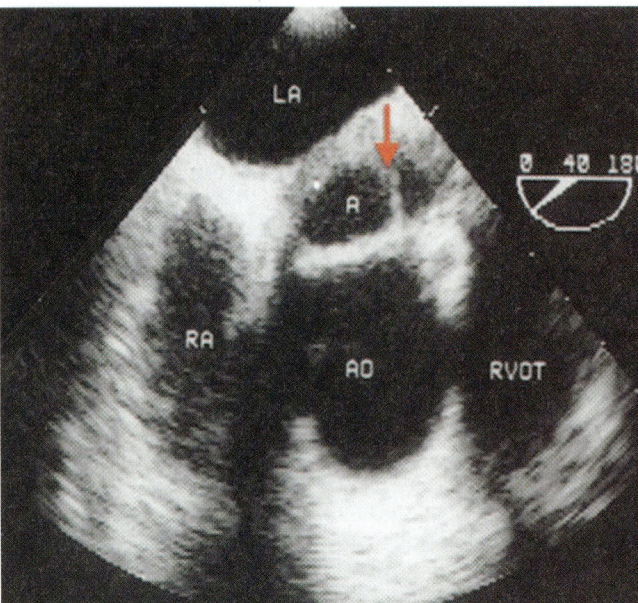

FIGURE 71–10 Transesophageal echocardiography in endocarditis. Segmented abscess cavity (labeled A) between the left atrium (labeled LA) and the aortic root (labeled AO). Red arrow indicates the wall that segments the abscess. RA = right atrium; RVOT = right ventricular outflow tract. (Reproduced with permission from Fuster V, Walsh RA, Harrington RA: *Hurst's The Heart*, 13th ed. New York, NY: McGraw Hill; 2011.)

for endocarditis is recommended in cases where the patient has hemodynamic instability, severe disease, evidence of embolic disease, or large vegetations.

Empiric antimicrobial coverage should be active against methicillin-resistant *S. aureus*, viridans group streptococci, enterococci, and HACEK organisms. A common empiric regimen is vancomycin plus ceftriaxone. Specific antimicrobial therapy should be instituted when the results of blood cultures and antibiotic susceptibility tests are known. Antimicrobial therapy for endocarditis is usually required for 4 to 6 weeks.

Surgical therapy is either indicated or should be strongly considered in patients with severe congestive heart failure, perivalvular abscesses, infections refractory to medical management, and embolic events with large vegetations.

Prevention

In patients with prior endocarditis, a prosthetic heart valve, or select types of congenital heart disease, antibiotic prophylaxis is recommended prior to certain procedures. Guidelines support giving antibiotics, such as amoxicillin, to these high-risk patients at the time of invasive dental procedures (not for routine cleanings), surgery involving respiratory mucosa, or surgery involving infected tissues.

MYOCARDITIS

Definition

Myocarditis is infection of the heart muscle.

Pathophysiology

Infection of the myocardium most frequently occurs following hematogenous spread of virus or other pathogen to the heart muscle, although direct spread from adjacent structures can occur. Infection and inflammation of myocardium may result in cardiac dysfunction, leading to heart failure.

Clinical Manifestations

Patients with myocarditis present with signs and symptoms of heart failure. Depending on the pathogen, the pace of disease progression may be over days or weeks. Patients may have signs and symptoms of a systemic infection as well (fever, constitutional symptoms). Those with associated pericarditis often have chest pain.

Pathogens

Viral pathogens are thought to be the predominant cause of infectious myocarditis, although many cases are idiopathic. Coxsackie viruses are the most common cause, although cytomegalovirus, Epstein–Barr virus, parvovirus B19, and influenza virus have been implicated. Other pathogens include *Trypanosoma cruzi*, the agent of Chagas' disease, and *Trichinella spiralis*.

Diagnosis

A definitive diagnosis requires cardiac muscle biopsy revealing myocardial inflammation and necrosis. However, most cases are presumptively diagnosed in a patient presenting with heart failure, who has (often global) cardiac dysfunction on echocardiogram and elevated cardiac enzymes. The ECG may be abnormal and may show ST changes mimicking an acute myocardial infarction.

Treatment

There is no known treatment for most causes of myocarditis, and supportive care is most often given. Patients may ultimately require heart transplant.

Prevention

There is no known mechanism to prevent myocarditis.

PERICARDITIS

Definition

Pericarditis refers to inflammation of the pericardium, which can be due to infection, autoimmune diseases, trauma, or malignancy.

Pathophysiology

Pathogens reach the pericardium by either hematogenous spread through the blood or direct spread from adjacent intrathoracic structures or, rarely, directly from infected myocardium. Inflammation of the pericardium can result in the formation of pericardial effusion. Pericardial effusions can result in cardiac tamponade. Inflammation can also result in a constrictive physiology. Certain infections causing pericarditis may also be associated with a concomitant myocarditis (see previous "Myocarditis" section).

Clinical Manifestations

Chest pain is the most common manifestation of pericarditis. Pain often worsens with inspiration or coughing. Sitting up and leaning forward often improve the pain associated with pericarditis. Patients may have fever and constitutional symptoms. On exam, a **friction rub** (often consisting of three phases) may be heard when performing auscultation of the heart. This exam finding is very specific for pericarditis. Severe infection may result in cardiac tamponade or constrictive cardiac physiology. These patients present with acute or subacute/chronic onset of symptoms of heart failure, respectively.

Pathogens

Viruses, bacteria, mycobacteria, and fungi have all been reported to cause pericarditis. Among viral infections, Coxsackie virus and echovirus are most common, although human immunodeficiency virus and cytomegalovirus can cause pericarditis as well. Among bacteria, *S. aureus* and *S. pneumoniae* are most common. *Mycobacterium tuberculosis* is one of the most common infectious causes of pericarditis worldwide. Clinical presentation is often subacute and may result in a constrictive pattern. Several fungi such as *Histoplasma capsulatum* and *Coccidioides immitis* can cause pericarditis, which clinically presents similarly to tuberculous pericarditis.

Diagnosis

Culture of pericardial fluid or pericardial tissue may reveal causative bacteria. Viruses are rarely isolated. Additional diagnostic tests that can help make the diagnosis include ECG that reveals changes in the PR and ST segments. If a significant pericardial effusion is present, the ECG may have reduced amplitude in all leads. An echocardiogram and/or cardiac magnetic resonance imaging will often reveal a pericardial effusion and/or pericardial thickening. In addition, chest X-ray may show an enlarged cardiac silhouette (Figure 71–11), and cardiac enzymes can be

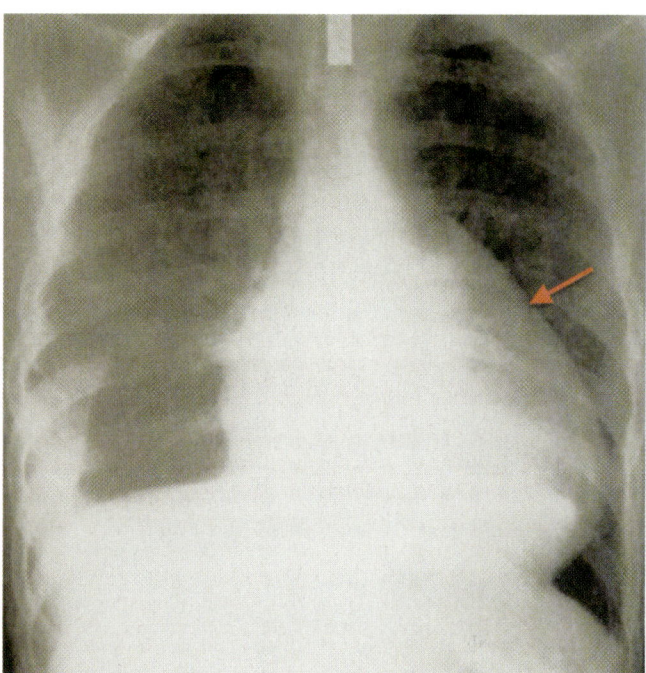

FIGURE 71–11 Chest X-ray of a patient with pericardial effusion. Red arrow indicates left border of dilated pericardial sac containing effusion fluid. (Reproduced with permission from Crawford MH, DiMarco JP: *Cardiology.* London: Mosby; 2001.)

elevated. Recovery of a pathogen often requires a pericardiocentesis or pericardial biopsy.

Treatment

Treatment for infectious pericarditis is dependent on the pathogen. Most viral etiologies are treated with symptomatic management and supportive care, whereas bacterial, mycobacterial, and fungal infections will require directed antimicrobial therapy. In patients with constrictive pericarditis and tamponade, pericardiocentesis can be life-saving. Untreated bacterial pericarditis is rapidly fatal.

Prevention

Immunization against *S. pneumoniae* may be effective. Treatment of early or latent stages of infections (e.g., tuberculosis) may prevent development of pericarditis in some cases.

CHAPTER

72

Central Nervous System Infections

CHAPTER CONTENTS

INTRODUCTION

Central nervous system (CNS) infections are often life-threatening and can have severe sequelae. These infections cause inflammation and edema within the unyielding cranium, resulting in damage to brain tissue and loss of function. The most common causes of CNS infections are bacteria and viruses, but fungi, protozoa, and helminths also cause these infections.

In addition to the history and physical examination, clinical diagnosis of CNS infections requires a spinal fluid analysis combined with neuroimaging using either magnetic resonance imaging (MRI) or computed tomography (CT) scan. Microbiologic diagnosis of bacterial infections frequently is made using Gram stain and culture of spinal fluid and blood. Polymerase chain reaction (PCR) assays and serologic tests are also useful. Antimicrobial therapy requires that the antibiotics be bactericidal and that they penetrate the blood–brain barrier. Some CNS infections, such as a brain abscess, often require surgical drainage.

CEREBROSPINAL FLUID ANALYSIS

Examination of cerebrospinal fluid (CSF) is critical in making the diagnosis of CNS infections. CSF is obtained by performing a lumbar puncture at the L3–L4 interspace. During the process, the CSF pressure is measured and fluid obtained for analysis of cells (both number and cell type, i.e., neutrophils or lymphocytes), protein, and glucose. The results of CSF analysis in acute bacterial meningitis, acute viral meningitis, and subacute meningitis are described in Table 72–1.

Although CSF analysis is a very important step in the diagnosis of many CNS infections, a lumbar puncture should *not* be performed if there are signs of increased intracranial pressure, such as papilledema or focal neurologic signs, because herniation of the brainstem and death may occur. A CT scan should be performed prior to the lumbar puncture to determine whether a mass lesion, such as a brain abscess or cancer, is present. If a mass lesion is seen, a lumbar puncture should not be performed.

METAGENOMIC NEXT-GENERATION SEQUENCING

A causative organism is *not* identified in more than 50% of cases of meningitis and encephalitis. To address this problem, metagenomic next-generation sequencing (mNGS) techniques are being applied.

In the mNGS process, all of the DNA and RNA in the patient's specimen is sequenced and known cellular sequences

TABLE 72–1 Spinal Fluid Findings in Acute and Subacute Meningitis

Etiology	Pressure (mm H$_2$O)	Cells (µL)	Proteins (mg/100 cc)	Glucose (CSF/Blood)
Normal	<200	0–5 Lymphs, 0 Polys	<45	>0.6
Acute bacterial	Increased	200–5000; mostly (>90%) Polys	>100	<0.6
Acute viral	Slight increase	100–700 Lymphs	Slight increase	Normal
Subacute/chronic (TB, fungus)	Increased	25–500 Lymphs	>100	<0.6

CSF = cerebrospinal fluid; Lymphs = lymphocytes; Polys = polymorphonuclear leukocytes (neutrophils); TB = tuberculosis.

are disregarded. The sequences that remain are then compared to those in a database that contains the sequences of the known pathogens, including bacteria, viruses, fungi, protozoa, and helminths.

This procedure obviates the need to have specific probes for all the different causes of infectious diseases in humans. It also provides a mechanism for identifying microbes that do not grow using standard laboratory methods. However, one drawback is that it does not determine whether the organism is alive or dead, only that it is present.

MENINGITIS

Definition

Meningitis is an infection of the meninges, the membranes that line the brain and spinal cord (Figure 72–1). Meningitis can be categorized as acute, subacute, or chronic depending on the speed of onset of the initial presentation and the rate of progression of the illness. Acute meningitis is caused by either pyogenic bacteria, such as *Streptococcus pneumoniae* and

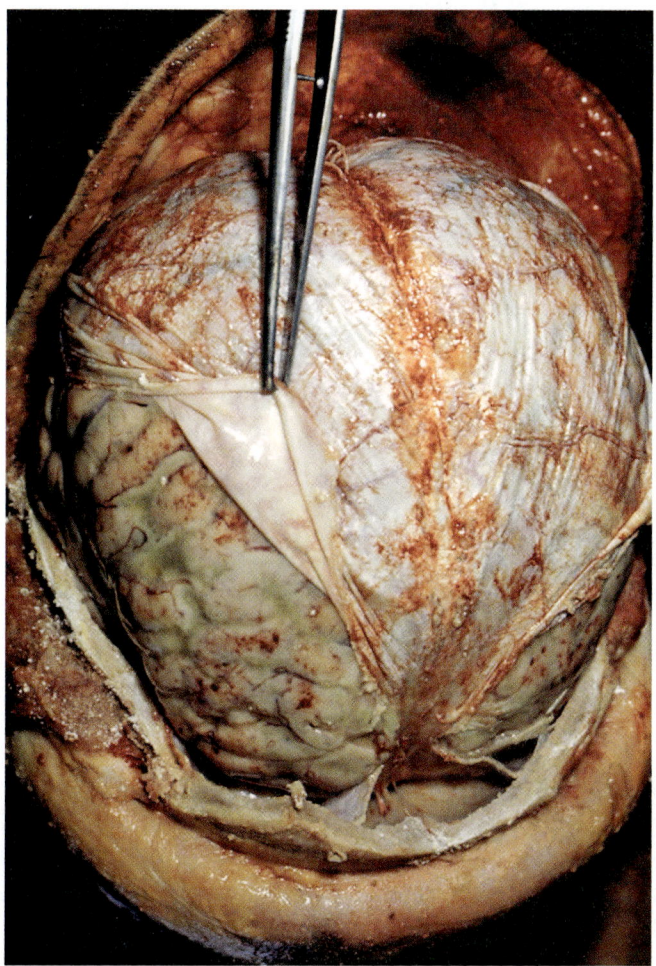

FIGURE 72–1 Purulent meningitis. Note film of greenish pus in the subarachnoid space covering the brain. The dura is reflected back and held by forceps. (Reproduced with permission from Centers for Disease Control and Prevention.)

Neisseria meningitidis, or viruses, such as Coxsackie virus and herpes simplex virus type 2 (HSV-2). Viral meningitis is often called aseptic meningitis because routine cultures for bacterial pathogens are negative. Subacute meningitis is caused by *Mycobacterium tuberculosis* and fungi, such as *Cryptococcus*. The causative organisms are often found in the spinal fluid located in the subarachnoid space.

Pathophysiology

Hematogenous spread (i.e., bacteremia or viremia) is the most common route by which organisms reach the meninges. Direct spread via adjacent infections, such as otitis media and sinusitis; via neurosurgery, such as a shunt to relieve hydrocephalus; or via trauma, such as a fracture of the cribriform plate, occurs less frequently. The importance of hematogenous spread is emphasized by the success of the conjugate vaccines against *S. pneumoniae*, *N. meningitidis*, and *Haemophilus influenzae* type B that induce circulating IgG antibodies that neutralize the bacteria in the blood.

Acute bacterial meningitis begins with nasopharygeal colonization followed by local invasion, entry into the bloodstream, and invasion of the meninges (Figure 72–2). This is followed by an inflammatory response that causes many of the clinical manifestations, especially the edema resulting in increased intracranial pressure leading to headache. Cerebral vasculitis and infarction can also occur.

The **blood–brain barrier** retards the entry of microbes and drugs from the capillaries into the brain. It is formed by strong tight junctions between the endothelial cells. Microbes, such as bacteria and viruses, have difficulty passing through uninflamed endothelium but inflammation weakens the tight junctions and allows passage of microbes into the brain. Similarly, drugs, for example, penicillins, pass more easily through inflamed capillaries than through uninflamed vasculature. In addition, lipophilic drugs pass through the blood–brain barrier more easily than do charged hydrophilic drugs. Molecules required for normal brain function, such as glucose, have specific transporters that allow passage into the brain.

Clinical Manifestations

Early symptoms include the **classic triad of fever, headache, and stiff neck** (nuchal rigidity). Altered mental status also commonly occurs. If untreated, meningitis may progress to vomiting, seizures, photophobia, and focal neurologic deficits. Different pathogens can present with different rates of clinical progression, from acute onset and rapid progression (hours to days) to subacute or chronic onset and slow progression (days to weeks). *N. meningitidis* infection can be associated with disseminated disease (meningococcemia) and result in petechial rash and ultimately purpura fulminans (Figure 72–3).

Pathogens

Bacterial Pathogens Causing Acute Meningitis

The most common bacterial cause of acute meningitis overall is *S. pneumoniae*. However, *Streptococcus agalactiae* (group B

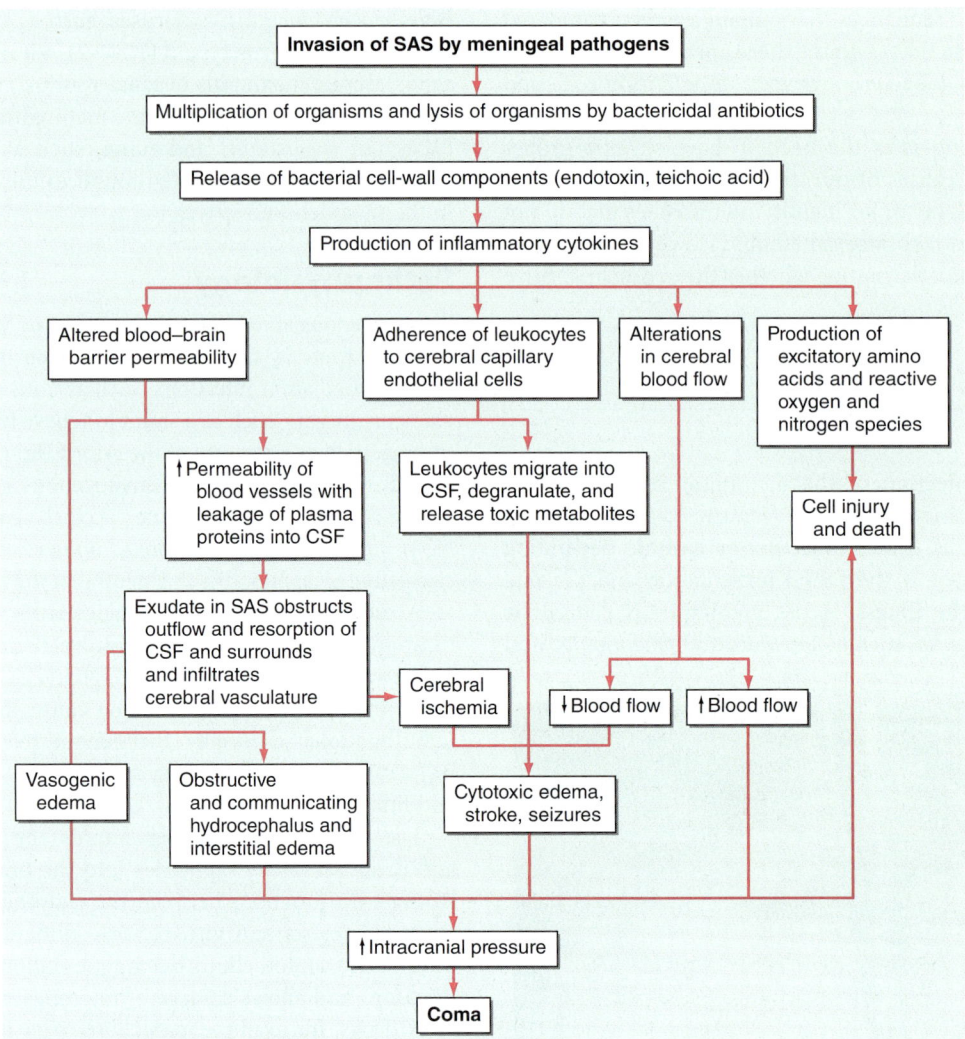

FIGURE 72-2 Pathogenesis of bacterial meningitis. CSF = cerebrospinal fluid; SAS = subarachnoid space. (Reproduced with permission from Longo DL, Fauci AS, Kasper DL, et al: *Harrison's Principles of Internal Medicine*, 18th ed. New York, NY: McGraw Hill; 2012.)

Streptococcus) predominates in neonates, and *N. meningitidis* is common in teenagers and young adults (Table 72–2).

H. influenzae type B used to be an important cause in young children, but the widespread use of the conjugate polysaccharide vaccine has greatly decreased its incidence. *Listeria monocytogenes* is reasonably common in the very young, the very old, and immunocompromised patients. Less common pathogens include *Borrelia burgdorferi* (Lyme disease) and *Treponema pallidum* (syphilis).

Viral Pathogens Causing Acute Meningitis

The most common viral causes of acute meningitis are enteroviruses such as Coxsackie virus, echovirus, and enterovirus 71. Enteroviral meningitis occurs primarily in young children, and the peak incidence is in the summer and fall seasons.

HSV-2 is also a common cause of meningitis. Note that HSV-2 typically causes meningitis, whereas herpes simplex virus type 1 (HSV-1) causes encephalitis. Primary genital infections with HSV-2 are more likely to result in meningitis than

recurrent HSV-2 infections. Primary and reactivation varicella zoster virus (VZV) infection can also be associated with meningitis.

Although arboviruses typically cause encephalitis, arboviruses such as West Nile virus (WNV) and St. Louis encephalitis virus can also cause meningitis. Mumps virus used to be a common cause of meningitis, but widespread use of the mumps vaccine has greatly reduced its incidence.

Pathogens Causing Subacute and Chronic Meningitis

The most common causes of subacute and chronic meningitis are *M. tuberculosis* and fungi such as *Cryptococcus*, *Coccidioides*, and *Histoplasma*. Cryptococcal meningitis occurs most commonly in immunocompromised patients, such as those with acquired immunodeficiency syndrome (AIDS), but can cause subacute and chronic meningitis in immunocompetent patients as well. Other causes of chronic meningitis include *T. pallidum* and *B. burgdorferi*. Human immunodeficiency virus (HIV) is the most common viral cause of chronic meningitis.

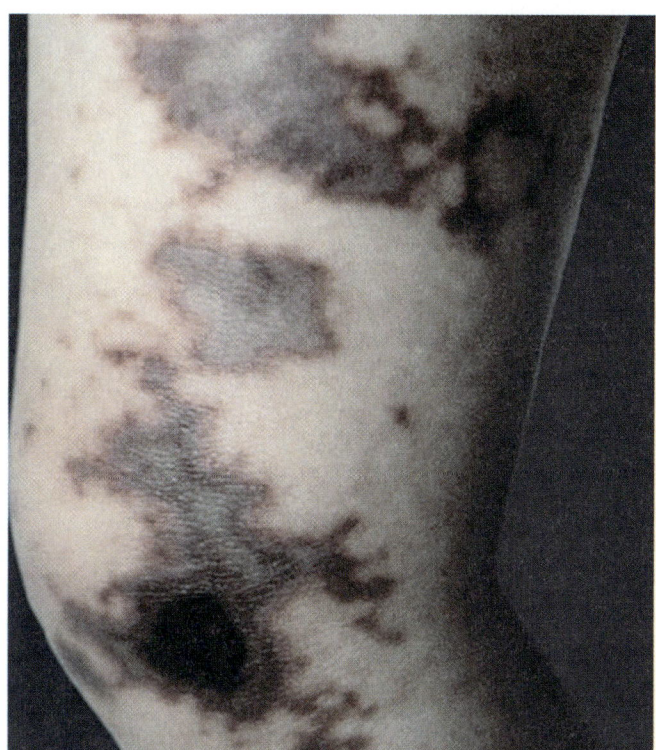

FIGURE 72–3 Purpura fulminans caused by *Neisseria meningitidis*. (Reproduced with permission from Wolff K, Johnson R: *Fitzpatrick's Color Atlas & Synopsis of Clinical Dermatology*, 6th ed. New York, NY: McGraw Hill; 2009.)

Diagnosis

A microbiologic diagnosis of acute bacterial meningitis is typically made by Gram stain and culture of CSF. However, PCR-based tests are being increasingly used because they yield results rapidly and with great accuracy. For example, a PCR-based panel is now available that tests for the presence in spinal fluid of six common bacteria, seven common viruses, and the yeast *Cryptococcus* with a turnaround time of 1 hour.

Analysis of spinal fluid can distinguish between acute bacterial meningitis and viral meningitis (see Table 72–1). While they both tend to have elevated white blood cells (WBCs) and protein in CSF, bacterial infections tend to be neutrophil predominant, whereas viral infections are lymphocyte predominant. Bacterial infections are associated with low glucose concentrations in CSF, whereas viral infections have normal glucose levels.

Subacute and chronic meningitis tend to be lymphocyte predominant with very high protein levels and low glucose. Viral infections are often diagnosed by using PCR assay for viral DNA or RNA in CSF or by serologic tests for specific antibody. Gram stain and bacteriologic cultures of CSF are negative in viral meningitis. Fungal infections can be diagnosed by culture or by serologic tests. In the case of *Cryptococcus*, the India ink test and the cryptococcal antigen test are also useful.

Treatment

Empiric therapy for acute bacterial meningitis must include drugs with excellent penetration into the CSF (able to pass the

TABLE 72–2 Organisms Causing Meningitis With Various Predisposing Factors

Predisposing Factor	Common Organisms
Neonate	*Streptococcus agalactiae* (group B *Streptococcus*), *Escherichia coli*, *Listeria monocytogenes*
Young children	*Streptococcus pneumoniae, Neisseria meningitidis, Haemophilus influenzae*; also enteroviruses
Teenagers and young adults	*S. pneumoniae, N. meningitidis*; also herpes simplex virus type 2
Older adults	*S. pneumoniae, N. meningitidis, L. monocytogenes*
Immunocompromised	*L. monocytogenes*, aerobic gram-negative rods (*Pseudomonas* and *Klebsiella*)
Pregnant women	*L. monocytogenes*
Cerebrospinal fluid leak; splenectomy; sickle cell anemia	*S. pneumoniae*
Deficiency of late-acting complement components; military recruits	*N. meningitidis*
After neurosurgery	*Staphylococcus aureus*
Ventriculoperitoneal shunt	*Staphylococcus epidermidis*
Immunocompromised (HIV/AIDS)	*Cryptococcus neoformans*
Living in or traveling in Central Valley of California (Sonoran life zone)	*Coccidioides immitis*
Swimming/diving in fresh water	*Naegleria fowleri*
Mosquito bite	West Nile virus; other arboviruses
Tick bite	*Borrelia burgdorferi*
Sexually transmitted disease (secondary syphilis)	*Treponema pallidum*

AIDS = acquired immunodeficiency syndrome; HIV = human immunodeficiency virus.

blood–brain barrier), that are bactericidal, and that are active against the most common pathogens. In older children and adults, ceftriaxone or cefotaxime plus vancomycin is a common empiric regimen. Vancomycin is added to cover for penicillin- or cephalosporin-resistant pneumococci. Ampicillin should be added if *Listeria* is a likely cause.

Empiric therapy for neonatal bacterial meningitis includes ampicillin plus either ceftriaxone or cefotaxime, with or without gentamicin. Acyclovir is used for the treatment of HSV and VZV infection.

Prevention

Prevention strategies include both immunization and chemoprophylaxis. Vaccines are effective in preventing bacterial meningitis, caused by *S. pneumoniae, N. meningitidis*, and *H. influenzae* type B. The immunogen in the conjugate vaccines is the capsular polysaccharide of the organism.

The current conjugate pneumococcal vaccine (Prevnar 13) protects against the 13 most common serotypes. The current conjugate *H. influenzae* vaccine protects only against the type B serotype.

The current conjugate meningococcal vaccine protects against four common serotypes (A, C, Y, and W-135). Note, however, it does not contain the type B polysaccharide. The vaccine against type B meningococcus contains factor H binding protein (fHbp) as the immunogen. A second vaccine against type B meningococci containing four surface proteins (fHbp, NadA, NHBA, and PorA) is also available.

Chemoprophylaxis against *S. agalactiae* (group B *Streptococcus*) is aimed at reducing vaginal carriage in the mother. If vaginal or rectal cultures are positive at 35 to 37 weeks of gestation, then ampicillin should be given. Chemoprophylaxis is also used to reduce nasopharyngeal carriage of *N. meningitidis* and *H. influenzae* type B. Close contacts of patients with meningitis caused by these organisms should receive either ciprofloxacin for *Neisseria* or rifampin for *Haemophilus*.

ENCEPHALITIS

Definition

Encephalitis is an infection of the brain parenchyma predominantly caused by viruses. Sometimes both the brain and the meninges are involved, a condition called meningoencephalitis.

Pathophysiology

The mode of acquisition of the viruses that cause encephalitis varies (Table 72–3). Neonates acquire HSV-2 during passage through the birth canal. HSV-2 then reaches the brain by hematogenous spread. Mothers with visible vesicular lesions are much more likely to have newborns with serious HSV-2 infections than mothers who are asymptomatic shedders of HSV-2 because the amount of virus present is significantly greater in the former.

In contrast, HSV-1 probably reaches the temporal lobe by travel within neurons following activation of latent infection in the trigeminal ganglion (Figure 72–4). Rabies virus also reaches the brain by axonal travel from the site of the animal bite.

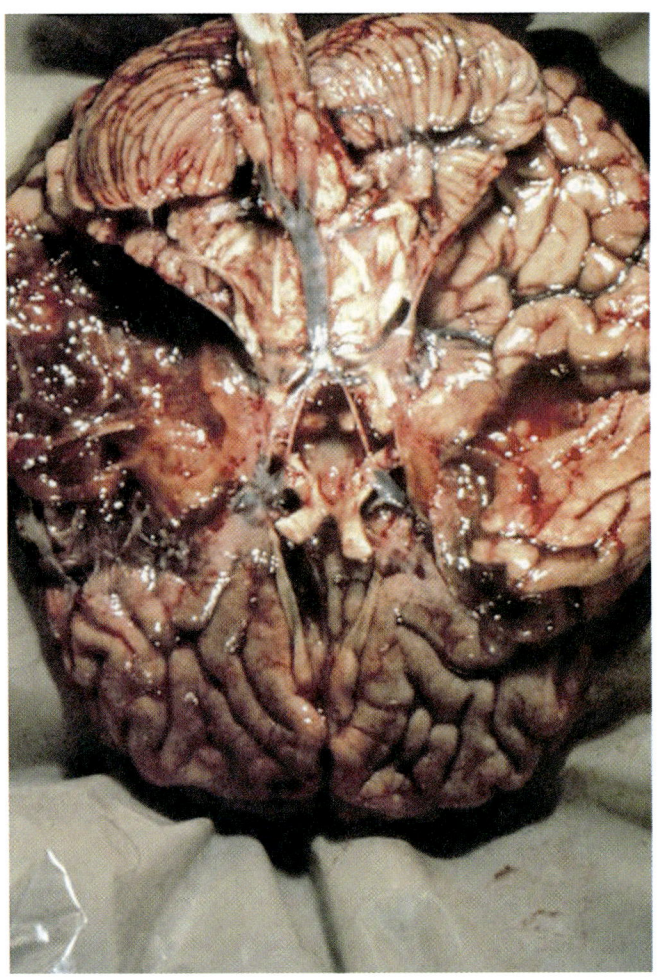

FIGURE 72–4 Encephalitis caused by herpes simplex virus-1. Note destruction of temporal lobe on left side of image. (Used with permission from Dr. John Mills, Monash University, Melbourne, Australia, and Dr. Kim Erlich, University of California School of Medicine, San Francisco, CA.)

Arboviruses, such as WNV, are acquired primarily by mosquito bite and then travel to the brain via the bloodstream. The incidence of arboviral encephalitis peaks in the summer and early fall because that is when mosquitoes are most active.

VZV can cause encephalitis during the primary infection (varicella is also known as chickenpox) or during the

TABLE 72–3   Viruses Commonly Causing Encephalitis with Various Predisposing Factors

Predisposing Factor	Common Viruses	Comment
Neonate	HSV-2	Acquired at the time of birth.
Child over the age of 1 year and adult	HSV-1	Primarily affects temporal lobe. Probably reach the brain by traveling within neurons following activation of latent infection in trigeminal ganglion.
Animal bite (e.g., dog, cat, bat, skunk, raccoon)	Rabies	In the United States, dogs and cats are uncommon reservoirs. Bats are the most common reservoir; raccoons are common reservoirs east of the Mississippi.
Mosquito bite	West Nile virus, Eastern and Western equine encephalitis viruses, St. Louis encephalitis virus, Zika virus	West Nile virus is the most common arboviral infection in the United States.

reactivation infection (zoster is also known as shingles). VZV also causes a postinfectious encephalomyelitis involving the brain and spinal cord after resolution of the primary infection. Cytomegalovirus (CMV) causes encephalitis primarily in immunocompromised individuals such as AIDS patients and those receiving drugs to prevent transplant rejection. Encephalitis caused by Epstein–Barr virus (EBV) is a rare complication of infectious mononucleosis.

Postinfection encephalitis typically follows an infection or an immunization by several weeks. It is a demyelinating disease caused by an immune attack on neurons, primarily those of the white matter.

Note that the lesions in encephalitis are inflammatory (contain WBCs, especially lymphocytes), whereas the lesions of an encephalopathy show degenerating neurons but no inflammation and do not contain WBCs. Encephalopathy is discussed later in a separate section.

Clinical Manifestations

The most characteristic clinical manifestations of encephalitis include fever, headache, and altered mental status, as well as seizures and focal neurologic deficits.

Rabies encephalitis has two clinical manifestations. Most cases of rabies (80%) present with hyperactivity, agitation, delirium, hydrophobia, and seizures (called furious rabies). The other 20% of cases have paralytic symptoms in which an ascending paralysis without hyperactivity is the predominant feature (called dumb rabies). Coma and death are the final common pathway in both forms.

Pathogens

Viruses are the main cause of encephalitis; however, the cause of at least half of the cases of encephalitis is unknown. Approximately 15% are caused by HSV-1. Encephalitis caused by HSV-1 and HSV-2 is very important because HSV-1 and HSV-2 are the most common causes for which antiviral drugs are available, namely acyclovir. About 5% are caused by arboviruses such as WNV. Rabies virus is a rare cause in the United States but occurs more frequently in countries where immunization of dogs is not a common practice. VZV, CMV, and EBV also cause encephalitis.

WNV is the most common arboviral cause of encephalitis in the United States. Most WNV infections (80%) are asymptomatic. Most of the remaining 20% develop an acute febrile "flulike" illness. Less than 1% develop CNS disease, of which half have encephalitis. Other arboviruses that cause encephalitis with some frequency are St. Louis encephalitis virus, the La Crosse strain of California encephalitis virus, and Eastern and Western equine encephalitis viruses (EEE and WEE, respectively). They are all transmitted by either *Culex* or *Aedes* mosquitoes.

Poliovirus causes meningoencephalitis in a few countries as of 2018, such as Nigeria, Pakistan, and Afghanistan. It has been eradicated from the Western Hemisphere and most other countries as a result of widespread use of the vaccine. However, disease may occur in these countries when imported by travelers from countries where the virus is endemic.

Postinfection encephalitis follows immunization or infection caused most often by VZV, measles, and influenza.

Diagnosis

In contrast to meningitis, CSF findings in encephalitis are more variable. A mild elevation in CSF lymphocytes (pleocytosis) can be seen along with an elevation of protein and a normal glucose. A normal CSF pattern can also be seen in encephalitis.

PCR-based testing of CSF is commonly used to determine a specific etiology, such as with HSV and VZV. WNV encephalitis is often diagnosed by finding WNV-specific IgM in the spinal fluid.

Rabies can be diagnosed by direct fluorescent antibody staining of a biopsy of skin from the nape of the neck. A PCR assay using CSF, saliva, or tissue is also available. The PCR assay has the advantage of identifying the animal reservoir and the geographic location of the virus because the base sequence of the RNA genome varies in accord with those two features.

Radiographic findings can be useful as well. In particular, in HSV encephalitis, temporal lobe abnormalities are frequently seen.

Treatment

Intravenous acyclovir is the treatment of choice for HSV-1, HSV-2, and VZV encephalitis. Ganciclovir is used for CMV encephalitis. There is no antiviral therapy for arboviral or rabies encephalitis.

Prevention

Prevention of rabies includes both preexposure (before the bite) and postexposure (after the bite) prophylaxis. Preexposure prophylaxis with the killed vaccine should be given to veterinarians and others at risk of exposure. Postexposure prophylaxis consists of both the killed vaccine and the hyperimmune globulins that contain a high titer of anti-rabies virus antibodies. They are inoculated at different sites so the antibodies do not neutralize the virus in the vaccine. This is an important example of passive–active immunization. There is no vaccine for HSV-1, HSV-2, and WNV.

To reduce the transmission of HSV-2 to the neonate, pregnant women with active lesions late in pregnancy should receive acyclovir and should be considered for cesarean section.

ACUTE FLACCID MYELITIS

Definition

Acute flaccid myelitis (AFM) is an infection of the spinal cord, particularly of the motor neurons, which results in paralysis. AFM is also known as acute flaccid paralysis. The term "myelitis" refers to infection/inflammation of the spinal cord.

Pathophysiology

Following viral infection of the upper respiratory tract and/or the gastrointestinal tract, the virus enters the bloodstream and infects the motor neurons in the spinal cord. Death of the motor neurons results in weakness or paralysis.

Outbreaks of AFM began in the United States in 2014 and have continued at 2-year intervals ever since. AFM affects young, healthy children, especially in the summer and fall seasons.

Clinical Manifestations

Patients with AFM frequently report antecedent symptoms consisting of mild fever and/or respiratory tract infection (rhinorrhea and cough) or gastrointestinal tract infection (vomiting and diarrhea). Sudden onset of weakness or paralysis of an arm or leg occurs after 3 to 10 days. In addition, facial nerve palsy can occur along with slurred speech or swallowing difficulty. Respiratory failure is rare but life-threatening. MRI will show characteristic changes in the gray matter of the spinal cord. CSF analysis often shows a pleocytosis (an increase in the number of cells).

Pathogens

Enterovirus D68 and enterovirus A71 are suspected of causing most cases of AFM. Poliovirus has *not* been detected in patients with AFM in the United States. Many cases of AFM have identified no viral pathogen in the spinal fluid.

Diagnosis

Microbiologic diagnosis is made by isolating viruses from the spinal fluid or stool. Nucleic acid amplification tests (NAATs) for viral RNA are also useful.

Treatment

No antiviral drugs are available for treatment. Supportive care is important.

Prevention

No antiviral drugs or vaccines are available. Handwashing and other infection control techniques are useful.

BRAIN ABSCESS

Definition

A brain abscess is a localized, walled-off collection of pus surrounded by a fibrous capsule within the brain parenchyma. Bacteria are the most common cause of brain abscesses, but fungi and protozoa are also involved. Viruses do not cause brain abscess.

Pathophysiology

Brain abscess is a recognized complication of head and neck pyogenic infections, such as sinusitis, otitis media, and dental infections. Sinusitis predisposes to lesions in the frontal lobe, whereas otitis media predisposes to lesions in the temporal lobe.

TABLE 72–4 Organisms Causing Brain Abscess with Various Predisposing Conditions

Predisposing Factor	Common Organisms
Otitis media or sinusitis	Aerobic and anaerobic streptococci, gram-negative anaerobes such as *Bacteroides*, *Prevotella*, and *Fusobacterium*
Dental infection	Same organisms as above plus *Actinomyces*
Trauma or neurosurgery	*Staphylococcus aureus*, *Staphylococcus epidermidis*, aerobic and anaerobic streptococci
Neutropenia	Aerobic gram-negative rods, e.g., Enterobacteriaceae, *Aspergillus*, *Mucor*
HIV infection	*Toxoplasma gondii*, *Listeria*, *Nocardia*, *Mycobacterium*
Endocarditis	*S. aureus*, viridans streptococci

HIV = human immunodeficiency virus.

Hematogenous spread from an infected site, such as with infective endocarditis, also occurs. Table 72–4 correlates various predisposing conditions with the organisms likely to cause brain abscess.

With increasing use of immunosuppressive drugs, indwelling intravenous catheters, and hyperalimentation, fungal brain abscesses have become more common. Immunocompromised patients, especially those with AIDS, also have brain abscesses caused by *Toxoplasma gondii*.

Clinical Manifestations

Headache alone is the most common symptom of brain abscess, and thus, when headache alone occurs early in the disease, the diagnosis of brain abscess can often be missed. As the lesion progresses, patients may develop fever, behavioral changes, focal neurologic deficits, and seizures.

Pathogens

Bacteria

Streptococci, both aerobic and anaerobic, are most commonly isolated from bacterial brain abscesses. They are typically of oropharyngeal origin, such as *Streptococcus anginosus* and viridans group streptococci. They are typically seen in mixed infections with oral anaerobes such as *Prevotella*, *Fusobacterium*, and *Bacteroides*. *Nocardia asteroides* also causes brain abscesses. Monomicrobial infections with *Staphylococcus aureus* are often associated with infective endocarditis.

Fungi

Fungal abscesses occur primarily in immunocompromised patients. *Aspergillus fumigatus* can occur in neutropenic patients, rhinocerebral mucormycosis (caused by *Mucor* and *Rhizopus* species) in diabetic patients with ketoacidosis, and cryptococcal infection in patients with HIV/AIDS. *Candida* species are also involved.

Protozoa

T. gondii is the main protozoal cause of brain abscess. It is an important cause in immunocompromised patients, especially those with AIDS, patients receiving cancer chemotherapy, or patients on immunosuppressive drugs used to enhance transplant survival. *T. gondii* can be transmitted by solid organ transplant, especially heart transplants, as well as by the more common modes of transmission, namely ingestion of raw meat containing cysts or by exposure to cat feces containing oocytes. Transplacental transmission of *T. gondii* can cause intracranial calcifications in the fetus.

Diagnosis

MRI is an important diagnostic modality, often revealing a "ring-enhancing" lesion (Figure 72–5). A microbiologic diagnosis requires obtaining pus from the abscess and performing a culture for aerobic and anaerobic bacteria and fungi. In bacterial brain abscesses, the Gram stain frequently reveals several types of bacteria indicating a mixed infection. Aspiration of pus from the lesion is both diagnostic and therapeutic, having the effect of draining the abscess.

A microbiologic diagnosis of *Toxoplasma* infection is usually made by identifying specific radiographic findings in an at-risk host (e.g., HIV/AIDS) with a positive *Toxoplasma* IgG and a response to specific antiprotozoal therapy. A PCR-based assay for *Toxoplasma* nucleic acid is also available.

Treatment

Empiric antimicrobial therapy for bacterial brain abscesses consists of a third-generation cephalosporin, such as ceftriaxone or cefotaxime, plus metronidazole. The latter is coverage for the

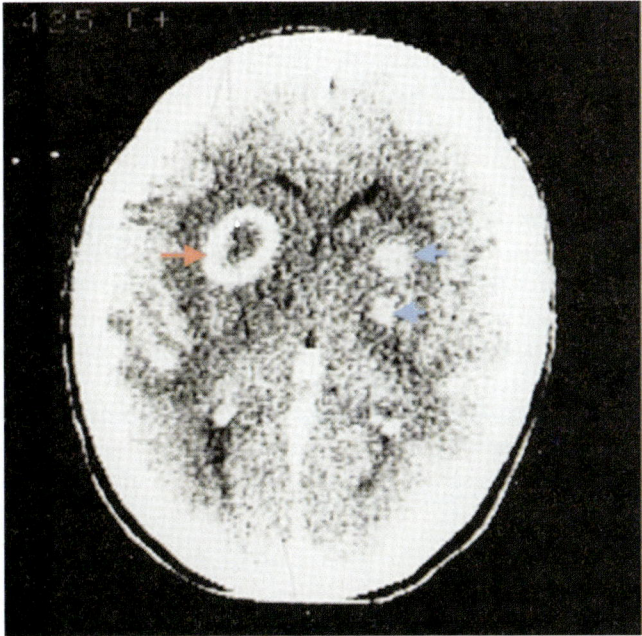

FIGURE 72–5 Brain abscess. Red arrow points to a characteristic ring-enhancing lesion. The blue arrows point to two additional abscesses. (Reproduced with permission from Ropper AH, Samuels MA: *Adams and Victor's Principles of Neurology*, 9th ed. New York, NY: McGraw Hill; 2009.)

anaerobic bacteria. Vancomycin should be added if the patient has undergone a neurosurgical procedure. Treatment of bacterial and fungal brain abscesses may require aspiration of pus from the abscess in addition to antibacterial or antifungal drugs.

Brain abscess caused by *Nocardia* can be treated with trimethoprim-sulfamethoxazole. Brain abscess caused by *Aspergillus* can be treated with voriconazole. Treatment of *Toxoplasma* brain abscess includes a combination of pyrimethamine and sulfadiazine.

Prevention

There are no vaccines to prevent brain abscesses. Early treatment of odontogenic and sinus infections may prevent these complications. Tight control of blood glucose may prevent rhinocerebral mucormycosis in diabetics. Treatment of AIDS patients with antiretroviral therapy may prevent *Toxoplasma* brain abscess, and when the CD4 count is <100 cells/μL, primary prophylaxis with trimethoprim-sulfamethoxazole is recommended in patients who are positive for *Toxoplasma* IgG.

SUBDURAL AND EPIDURAL EMPYEMA

Subdural empyema is a collection of pus on the inner surface of the dura mater, whereas epidural empyema is a collection of pus on the outer surface. They can occur adjacent to the dura of either the brain or spinal cord.

Sinusitis and otitis media are common predisposing factors, and the bacteria causing these empyemas are those that cause sinusitis and otitis media, namely, aerobic and anaerobic streptococci, staphylococci, enteric gram-negative rods such as *Escherichia coli*, and anaerobic gram-negative rods such as *Prevotella*. Mixed infections are common.

The clinical features include fever plus symptoms of increased intracranial pressure, such as headache, vomiting, focal neurologic deficits, and altered mental status. MRI with gadolinium enhancement reveals a mass adjacent to the dura. Microbiologic diagnosis involves aspirating pus from the lesion and performing a Gram stain and culture. Treatment involves surgical drainage of the pus combined with antibiotics appropriate for the bacteria isolated from the aspirated pus.

NEUROCYSTICERCOSIS

Neurocysticercosis (NCC) is a disease caused by the larval form of the pork tapeworm, of *Taenia solium* (pork tapeworm). It forms cysts called cysticerci in the brain that act like mass lesions. NCC is a leading cause of adult-onset seizures in immigrants to the United States from countries in Central and South America.

The most common clinical presentation of NCC is a seizure in an afebrile person. Headache and visual changes may also occur. The diagnosis is typically made by visualizing the cysts on brain imaging. There may be one or multiple cysts.

Serologic testing for antibodies to *T. solium* is used to confirm the diagnosis. A PCR assay for *T. solium* nucleic acid is also available. Peripheral eosinophilia usually is not seen.

Treatment of NCC involves either albendazole if there are one or two cysts or a combination of albendazole plus praziquantel if there are multiple cysts. Corticosteroids should be used to reduce the inflammatory response.

The pathogenesis of NCC involves ingesting the eggs of *T. solium* in food or water contaminated with human feces. The eggs hatch in the small intestine and the resulting larvae enter the bloodstream and migrate to the brain where they form the cysticercus, a cyst-like sac containing the scolex of the worm. Other common sites of cysticercus formation include the skin and bones.

ENCEPHALOPATHY

Encephalopathy refers to altered brain function in the absence of inflammation. In general, patients with encephalopathy do not have fever, headache, seizures, focal neurologic signs, and an increased WBC count in the blood and spinal fluid, whereas patients with encephalitis often do. Common manifestations of encephalopathy include confusion, personality changes, disorientation, aphasia, delirium, and dementia.

There are several infection-related causes of encephalopathy (see later), but most causes are noninfectious (e.g., alcohol, drugs, lead, uremia, or liver failure).

Important infection-related causes of encephalopathy include the following:

- **Progressive multifocal leukoencephalopathy (PML).** PML is caused by JC virus and occurs in immunocompromised patients, notably AIDS patients. Infection with JC virus occurs early in life and remains latent until the immune system is compromised. PML has occurred in multiple sclerosis patients being treated with natalizumab and in transplant recipients being treated with mycophenolate. Microbiologic diagnosis is made by detecting JC virus DNA using PCR assay on brain specimens or spinal fluid. There is no antiviral drug therapy and no vaccine. Additional information can be found in Chapter 44.
- **HIV encephalopathy including AIDS dementia.** Another CNS disease that is seen in HIV-infected individuals is encephalopathy caused by HIV itself. It can vary from mild symptoms such as memory problems and apathy to more serious disease such as profound memory loss and psychosis (AIDS dementia). AIDS dementia is more likely to occur when CD4 counts are below 200/μL and when the viral load in the CSF is high.
- **Creutzfeldt-Jakob disease (CJD) and kuru.** CJD is one of the human transmissible spongiform encephalopathies.

The term "spongiform" refers to the spongy, Swiss cheese-like appearance of the brain of patients with CJD. CJD is caused by prions, a misfolded protein in which the normal alpha-helical configuration has changed to a beta-pleated sheet, thereby altering the function of the protein and leading to death of neurons. Additional information on prions can be found in Chapter 44.

- CJD occurs sporadically worldwide at a rate of about one case per million population. CJD has been transmitted iatrogenically by corneal transplant, intracerebral electrodes, and dura mater grafts. CJD does not have any relationship to the ingestion of any food, unlike variant CJD, which is discussed later.
- The main clinical findings in CJD are dementia and myoclonus. The progression is gradual but inexorable, resulting in coma and death. Definitive diagnosis is made by observing spongiform changes in brain biopsy followed by histochemical staining with anti-prion antibodies. There is no drug treatment for CJD and no vaccine.
- Variant CJD is acquired by the ingestion of prion-containing beef. It is declining as a result of the ban on the addition of animal products to cattle feed.
- Kuru is a spongiform encephalopathy found in the Fore tribe in New Guinea. It is now very rare because the eating rituals that transmitted the agent are no longer practiced.
- **Reye's syndrome.** Reye's syndrome is a postinfectious disease consisting of encephalopathy plus liver failure. It occurs primarily following influenza B and varicella infections in children and is associated with aspirin use. The role of aspirin in pathogenesis is uncertain but a toxic effect on mitochondria has been proposed.
- After the child has recovered from the viral infection, Reye's syndrome begins with prominent vomiting followed by encephalopathic changes such as lethargy and combative behavior progressing to coma and death. Cerebral edema is marked. Fatty degeneration of the liver occurs, and liver enzymes such as transaminases are elevated. Blood ammonia levels are elevated.
- Treatment should be instituted promptly. If coma occurs, treatment is less effective. Antiviral drugs are not effective. Supportive measures such as cooling blanket, ventilator to provide respiratory support, control of intracranial pressure, hemodialysis, and fluid and electrolyte balance are used. Vaccines against varicella and influenza and public health campaigns to reduce aspirin use in febrile children have greatly reduced the incidence of this disease. Acetaminophen should be used to reduce fever in children.

Gastrointestinal Tract Infections

CHAPTER CONTENTS

INTRODUCTION

Infections with a variety of agents can occur in any part of the gastrointestinal (GI) tract from the mouth to the anal canal. Infections can range in severity from self-limited to life-threatening, particularly if infection spreads from the gut to other parts of the body. Infections are typically caused by the ingestion of exogenous pathogens in sufficient quantities to evade host defenses and then cause disease by multiplication, toxin production, or invasion through the GI mucosa to reach the bloodstream and other tissues. In other cases, members of the normal flora of the GI tract can cause disease, particularly when the gut microbiome is disrupted.

ESOPHAGITIS

Definition

Esophagitis is an inflammatory process that can damage the esophagus.

Pathophysiology

Inflammation caused by infection, typically by fungi such as *Candida* or viruses such as herpes simplex virus, causes the symptoms of esophagitis. Most cases occur in immunocompromised patients, especially those with reduced cell-mediated immunity. The extent of damage to the esophagus is typically related to the severity of symptoms.

Clinical Manifestations

Odynophagia (pain on swallowing) and dysphagia (difficulty in swallowing) are the key clinical manifestations of esophagitis.

Pathogens

Candida is the most common cause, particularly among human immunodeficiency virus (HIV)-infected patients and other immunocompromised hosts (Figure 73–1). Less common pathogens include herpesviruses such as cytomegalovirus and herpes simplex virus. Noninfectious causes also occur, such as acid reflux from the stomach and medication-induced disease (e.g., doxycycline).

Diagnosis

Diagnosis may be empiric after a trial of fluconazole results in improvement for presumed *Candida* esophagitis. If an empiric course of fluconazole does not work, then endoscopy for

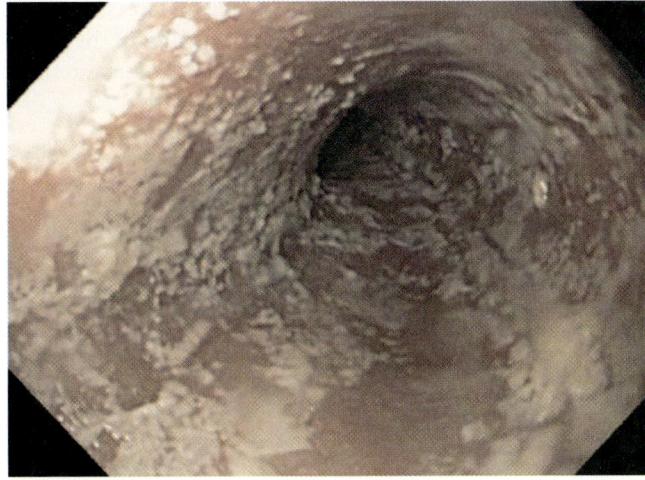

FIGURE 73–1 *Candida* esophagitis. Note the many whitish lesions on the esophageal mucosa seen on endoscopy. (Reproduced with permission from McKean SC, Ross JJ, Dressler DD, et al: *Principles and Practice of Hospital Medicine*. New York, NY: McGraw Hill; 2012.)

visualization and biopsy could be helpful, particularly in immunocompromised hosts. Biopsy samples should be analyzed by using pathologic and microbiologic tests.

Treatment

In a typical patient (e.g., HIV-infected patient) presenting with odynophagia and retrosternal pain, an empiric diagnosis of esophageal candidiasis is made and fluconazole therapy instituted. If there is no effect on symptoms and if *Candida* resistance is not suspected, then further diagnostics as outlined earlier may identify a specific organism that could be targeted for treatment.

Prevention

One option to prevent recurrent esophageal candidiasis is by using fluconazole prophylaxis. However, this is not generally advised given the high risk of selecting for fluconazole-resistant *Candida*. Immune restoration in HIV-infected patients may decrease the incidence of esophageal and oropharyngeal candidiasis.

GASTRITIS

Definition

Gastritis refers to inflammation of the mucosa of the stomach. It may be erosive or nonerosive, depending on histologic and endoscopic findings. A break in the gastric and adjacent duodenal mucosa defines peptic ulcer disease.

Pathophysiology

The mechanism by which one of the main pathogens, *Helicobacter pylori*, causes peptic ulcer disease has been largely elucidated. Following attachment to the gastric mucosa, *H. pylori* causes direct mucosal damage by the combination of ammonia production (from the action of the organism's urease on urea) and the host inflammatory response. The ability of the organism to survive is enhanced by the neutralization of the stomach's acid by the ammonia produced.

Clinical Manifestations

Patients with gastritis typically complain of dyspepsia (epigastric pain, burning), nausea, and vomiting. In the case of peptic ulcer disease, epigastric pain is the primary symptom. Some patients may report alleviation of pain with food, particularly those with duodenal ulcers. GI bleeding is a complication of peptic ulcer disease. Some patients with gastritis may be asymptomatic.

Pathogens

Infectious and noninfectious etiologies are possible. Among infectious causes, *H. pylori* is the most important. Viruses such as cytomegalovirus and fungi such as *Histoplasma capsulatum* and *Mucor* may rarely cause ulcer disease as well, particularly among immunocompromised patients. Following ingestion of raw fish, larvae of *Anisakis* species may become embedded in the gastric mucosa and cause severe abdominal pain. Mycobacteria (tuberculosis and nontuberculosis mycobacteria), *Giardia*, and *Strongyloides* may also cause gastritis. Noninfectious causes such as alcohol and medications (e.g., nonsteroidal anti-inflammatory drugs) are also implicated.

Diagnosis

Upper endoscopy with gastric biopsy is the definitive diagnostic strategy. If abnormal findings are detected, pathologic analysis and further directed testing may be performed. For the most common infectious cause of peptic ulcer disease, *H. pylori*-associated ulcers can be confirmed using a urease test on the biopsy specimen or using noninvasive tests such as the urea breath test or stool antigen test.

Treatment

Treatment is directed at the underlying pathogen, taking the host immune status into consideration. Quadruple drug therapy is recommended for the eradication of *H. pylori*. One regimen consists of a combination of tetracycline, bismuth, metronidazole (or tinidazole), and a proton pump inhibitor (PPI) such as omeprazole. Another regimen consists of amoxicillin, clarithromycin, metronidazole (or tinidazole), and a PPI. A levofloxacin-containing regimen consisting of levofloxacin, nitazoxanide, doxycycline, and omeprazole can also be used.

DIARRHEA (GASTROENTERITIS, ENTEROCOLITIS)

Definition

It is useful to think of diarrhea as acute (lasting <2 weeks) or chronic (persisting >4 weeks). We will focus on acute diarrhea in this chapter because most of the etiologies are infectious in nature. We can further categorize acute diarrhea as noninflammatory (watery, nonbloody) or inflammatory (bloody). Bloody diarrhea is also known as dysentery. For example, bloody diarrhea caused by *Shigella* is often called bacillary dysentery and bloody diarrhea caused by *Entamoeba* is called amebic dysentery. Table 73–1 describes the important features of watery and

TABLE 73–1   Characteristics of Watery Diarrhea Compared to Bloody Diarrhea

Characteristics of Watery Diarrhea	Characteristics of Bloody Diarrhea
No red blood cells or white blood cells in stool, i.e., no inflammation	Typically both red blood cells and white blood cells in stool, i.e., inflammatory response
Typically afebrile	Often febrile
Usually large-volume diarrhea	Usually small-volume diarrhea
Infection typically in small intestine	Infection typically in colon

TABLE 73–2 Important Organisms That Typically Cause Either Watery or Bloody Diarrhea

Organisms That Cause Watery Diarrhea	Organisms That Cause Bloody Diarrhea
Enterotoxigenic *Escherichia coli* (ETEC)	Shiga toxin-producing *Escherichia coli* (STEC)
Vibrio cholerae	*Shigella* species
Staphylococcus aureus	*Salmonella enterica*
Bacillus cereus	*Campylobacter jejuni*
Listeria monocytogenes	*Clostridioides difficile*
Norovirus	*Yersinia enterocolitica*
Rotavirus	*Entamoeba histolytica*
Giardia lamblia	
Cryptosporidium hominis	

bloody diarrhea. Table 73–2 lists the important organisms that cause either watery or bloody diarrhea.

Diarrhea must be calibrated against the patient's normal bowel movements but is usually considered to be greater than three to five bowel movements per day. Most of the infectious agents that cause diarrhea act at the small intestine (where the majority of fluid normally gets absorbed) or the colon.

Acute diarrhea is very common. There are approximately 179 million cases each year in the United States. The most common cause of acute diarrhea in the United States is norovirus. It is an especially prominent cause of *outbreaks* of diarrhea. The second leading cause of outbreaks of diarrhea is *Salmonella*. The most common cause of *fatal* diarrhea is hospital-associated *Clostridioides difficile* (previously known as *Clostridium difficile*).

Pathophysiology

Pathogens or their associated toxins disrupt the normal absorption and secretory processes in the small intestines. Acute diarrhea is usually caused by preformed exotoxins in food or by the infectious agents in the intestinal tract (via either enterotoxin and cytotoxin production or mucosal invasion). Pathogens that produce preformed exotoxins include *Staphylococcus aureus*, *Bacillus cereus*, and *Clostridium perfringens*. Other pathogens that cause noninflammatory acute diarrhea by enterotoxin production include enterotoxigenic *Escherichia coli* (ETEC) and *Vibrio cholerae*. Chapter 7 describes the mechanism of action of these toxins.

Pathogens that cause acute inflammatory diarrhea include *Salmonella*, *Shigella*, *Campylobacter* (via mucosal invasion), Shiga toxin-producing *E. coli* (STEC) such as *E. coli* O157:H7, and *C. difficile* (via cytotoxin production). Antibiotic use predisposes to pseudomembranous colitis caused by *C. difficile*. Chapter 18 provides additional information on these enteric gram-negative rods, and Chapter 17 discusses *C. difficile*.

There are several host factors that predispose to diarrheal illness. Patients taking PPIs are at risk because gastric acid levels are reduced. Recent travel to developing countries and antibiotic treatment are also associated with an increased incidence of diarrhea. Immunosuppressed patients have more frequent and more severe diarrheal illness.

Clinical Manifestations

Table 73–3 describes the clinical presentation caused by important GI tract pathogens. Patients complain of diarrhea accompanied by urgency, abdominal bloating, and cramping. In the case of acute inflammatory diarrhea, there is also blood or pus seen in the stool, and patients can be febrile. If vomiting is a major feature of the clinical presentation, this suggests *S. aureus* food poisoning or viral gastroenteritis. If symptoms begin within 6 hours after ingestion of suspected contaminated food, then preformed toxin of *S. aureus* or *B. cereus* should be suspected. On physical examination, patients may also show signs of dehydration with tachycardia and orthostatic changes in blood pressure.

Children infected with STEC often have bloody diarrhea and may progress to hemolytic–uremic syndrome (HUS). HUS occurs when Shiga toxin produced by STEC enters the bloodstream. The symptoms of HUS include hemolytic anemia, thrombocytopenia, and renal failure. Distorted red blood cells called schistocytes can be seen in blood smears. The use of ciprofloxacin increases the risk of HUS. Ingestion of undercooked hamburger or contaminated produce or contact with animals at petting zoos predisposes to disease caused by STEC.

Pathogens

Most cases of mild, acute watery diarrhea of short duration are caused by viruses. These include norovirus, rotavirus, enteric adenovirus, and less commonly, astrovirus. Outbreaks of norovirus infection commonly occur in closed populations, such as nursing homes, hospitals, cruise ships, and dormitories. Rotavirus is a common cause of diarrhea in children, but the incidence is declining due to increased use of the rotavirus vaccine.

Several bacteria are also important causes of watery, nonbloody diarrhea. *V. cholerae* causes severe, life-threatening watery diarrhea. ETEC is the most common cause of traveler's diarrhea (TD), typically a mild to moderate watery diarrhea. *Listeria monocytogenes* is another bacterial cause.

Most cases of severe, often bloody diarrhea are caused by bacteria. Pathogens such as *Salmonella*, *Shigella*, *Campylobacter*, STEC, and *C. difficile* are implicated in this category. In the United States, *Salmonella* and *Campylobacter* are the most common bacterial causes. Diarrhea caused by these bacteria is typically bloody. In addition to bloody diarrhea, *Yersinia enterocolitica* also causes mesenteric adenitis (see Appendicitis section later).

Protozoa, such as *Giardia*, *Entamoeba histolytica*, *Cryptosporidium*, and *Cyclospora* are less common causes of diarrhea but are suspected in certain scenarios (e.g., in returning travelers or immunocompromised patients). *Giardia* is the most common protozoan cause of diarrhea in the United States. Giardiasis typically occurs in young children in day care, in men who have sex with men, and in hikers who drink untreated ambient water. In HIV-infected patients with very low CD4 counts, *Cryptosporidium* causes prolonged diarrhea and may cause extraintestinal

TABLE 73–3 Clinical Presentation, Diagnosis, and Treatment of Diarrhea Caused by Important Gastrointestinal Tract Pathogens

Pathogen	Clinical Presentation	Diagnosis	Treatment	Comments
1. Acute noninflammatory diarrhea (watery, nonbloody stools; usually no fever)				
A. Bacteria				
Staphylococcus aureus	Vomiting, epigastric pain, diarrhea (mild)	Clinical. Food and stool can be tested for toxin	Supportive care (e.g., fluids, electrolytes)	Usually within 6 hours of consumption of infected food (dairy products, mayonnaise, meat products); recovery in 1–2 days
Bacillus cereus	Vomiting, epigastric pain, diarrhea	Clinical. Food and stool can be tested for toxin	Supportive care (e.g., fluids, electrolytes)	Usually within 6 hours of consumption of infected food (reheated rice)
Enterotoxigenic *Escherichia coli* (ETEC)	Afebrile, watery diarrhea	Clinical. Reference laboratory can perform DNA probe for LT or ST toxins	Ciprofloxacin	"Traveler's diarrhea"
Listeria monocytogenes	Often febrile, vomiting, diarrhea	Suspect *Listeria* when routine stool cultures do not show a pathogen, particularly in an outbreak setting	Supportive care (e.g., fluids, electrolytes)	Acquired by ingestion of unpasteurized soft cheese, deli meats, and raw vegetables. Can grow at refrigerator temperature
Vibrio cholerae	Severe, watery diarrhea with rapid fluid and volume loss. Vomiting in early disease	Clinical. Can be confirmed by stool culture	Supportive care (e.g., aggressive fluid repletion, electrolytes). Antibiotics (e.g., ciprofloxacin) in severe disease	Suspect cholera if watery diarrhea is associated with rapid and severe volume loss or in an outbreak setting
B. Viruses				
Norovirus	Afebrile, vomiting, headaches, diarrhea	Clinical. Stool PCR available	Supportive care (e.g., fluids, electrolytes)	Cruise ship and nursing home outbreaks
Rotavirus	Low-grade fever and vomiting prodrome, then diarrhea	Clinical. Rapid antigen test Stool PCR available	Supportive care (e.g., fluids, electrolytes)	Common in children
C. Protozoa				
Giardia lamblia	Abdominal cramps, flatulence, diarrhea (acute or chronic); stools are fatty, foul-smelling, and may float	Stool ova and parasite analysis may reveal cysts or trophozoites. Stool antigen test increasingly used	Metronidazole or tinidazole	Diarrhea may persist for weeks
Cryptosporidium hominis	Abdominal pain and cramps, watery diarrhea	See cysts in acid-fast stain of stool	Nitazoxanide for severe diarrhea. Antiretroviral therapy to restore immune system in AIDS patients	Cause of large community-wide outbreaks from contaminated water supply; important cause of diarrhea in AIDS patients
2. Acute inflammatory diarrhea (stools can be bloody; can be febrile)				
A. Bacteria				
Shiga toxin–producing *E. coli* (STEC), esp. *E. coli* O157:H7	Bloody diarrhea, abdominal pain, usually afebrile	Stool culture grows *E. coli* that does not ferment sorbitol. Use PCR assay or immunoassay to identify toxin-producing strains	None. Antibiotics may increase risk of hemolytic–uremic syndrome, especially in children	Acquired by ingestion of undercooked ground beef or fruits and vegetables contaminated with cattle manure
Clostridioides difficile	Bloody diarrhea, fever	Stool test for toxin production. Colonoscopy may reveal characteristic yellowish plaques	Oral vancomycin. Fecal microbiota transplantation if recurrent disease.	Traditionally associated with antimicrobial drug use; increasingly, community-acquired cases in patients without traditional risk factors
Shigella	Diarrhea with blood or pus usually; abdominal cramps; can be febrile Syndrome is called bacillary dysentery	Stool culture	Ciprofloxacin	Person-to-person spread can occur; humans are the reservoir; not found in animals

(Continued)

TABLE 73–3 Clinical Presentation, Diagnosis, and Treatment of Diarrhea Caused by Important Gastrointestinal Tract Pathogens (*Continued*)

Pathogen	Clinical Presentation	Diagnosis	Treatment	Comments
Salmonella	Diarrhea can be bloody; low-grade fevers	Stool culture	Ciprofloxacin (if severe illness); supportive care (if mild illness)	Acquired by ingestion of undercooked eggs, unpasteurized dairy products, raw vegetables, or undercooked poultry. Also by exposure to pet snakes and turtles
Campylobacter jejuni	Fever, diarrhea	Stool culture on special medium	Azithromycin or ciprofloxacin	Acquired by ingestion of unpasteurized dairy products or undercooked poultry. Associated with Guillain-Barré syndrome
Yersinia enterocolitica	Fever, diarrhea	Stool culture on special medium	Ciprofloxacin (if severe illness)	Causes mesenteric adenitis that can mimic appendicitis
B. Protozoa *Entamoeba histolytica*	Bloody diarrhea, fever, and abdominal pain. Syndrome is called amebic dysentery	Stool ova and parasite analysis may reveal cysts or trophozoites; serology	Metronidazole or tinidazole to eliminate tissue trophozoites, plus a luminal agent such as paromomycin	Can also cause hepatic abscesses

AIDS = acquired immunodeficiency syndrome; PCR = polymerase chain reaction.

disease involving the biliary and respiratory tracts. *E. histolytica* causes amebic dysentery characterized by bloody diarrhea.

Diagnosis

Diagnosis is generally focused on deciding who and when to test (i.e., determining when a test result may potentially impact the outcome). Because many causes of acute diarrhea are self-limited, this is an important issue. In general, we seek a diagnosis in cases of severe watery diarrhea, in cases of bloody diarrhea, if the patient is febrile, or if the patient is elderly or immunocompromised. Routine stool cultures will identify *Salmonella*, *Shigella*, and *Campylobacter*. If diarrhea is bloody, a special culture (e.g., MacConkey-sorbitol agar) is specifically set up to detect STEC. The basis for the special culture is that STEC strains typically do not ferment sorbitol. The definitive laboratory diagnosis of an STEC strain is made by either polymerase chain reaction (PCR) test or immunoassay for the Shiga toxin. Cultures for enterotoxigenic *E. coli* (ETEC) are not performed in the typical clinical laboratory. Multiplex molecular panels using nucleic acid amplification tests for a variety of bacterial, viral, and protozoal pathogens are increasingly being used.

In addition, if bloody diarrhea is associated with antibiotic use, laboratory tests for the presence of the *C. difficile* toxin in the stool should be done. Colonoscopy may reveal the characteristic yellowish plaques seen in pseudomembranous colitis (Figure 73–2).

Rotavirus infection can be diagnosed by testing the stool for rotaviral antigen or for rotavirus RNA using a PCR assay. A PCR test for norovirus RNA in the stool can be used to diagnose infection by that virus and is especially useful in outbreak situations.

Sending stool samples for analysis of ova and parasites (O&P) is generally not cost-effective, except in immunocompromised patients, patients with a history of recent foreign travel, or when diarrhea is associated with community waterborne outbreaks. Stools for O&P are usually sent on 3 consecutive days given

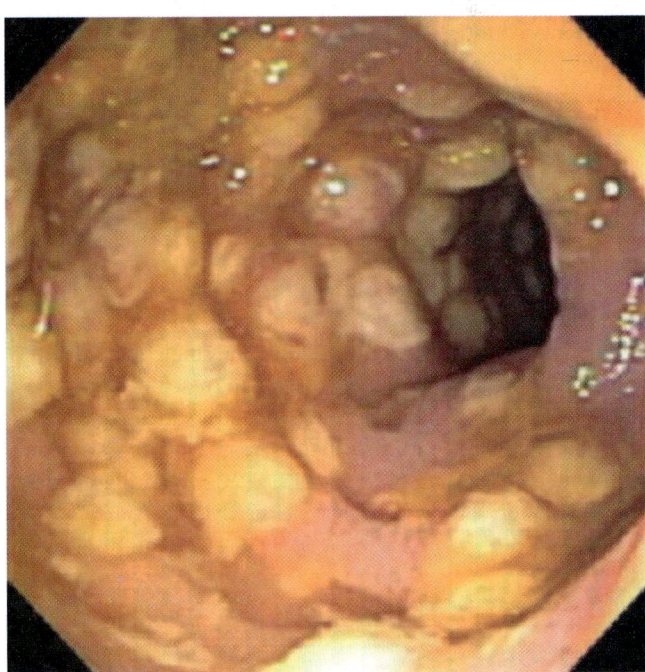

FIGURE 73–2 Pseudomembranous colitis caused by *Clostridioides difficile*. Note yellowish pseudomembranes seen on colonoscopy. (Reproduced with permission from Longo DL, Fauci AS, Kasper DL, et al: *Harrison's Principles of Internal Medicine*, 18th ed. New York, NY: McGraw Hill; 2012.)

that some parasites only intermittently shed eggs or cysts in the stool.

Treatment

The most important treatment modality in diarrhea is hydration. Oral rehydration solution containing water, salt, and sugar has been life-saving in many parts of the world. In general, for most cases of community-acquired diarrhea, empiric treatment with antimicrobials is not beneficial. Mild watery, non-bloody TD is often treated with bismuth subsalicylate (Pepto-Bismol) or loperamide (Imodium). The drug of choice for moderate to severe TD is azithromycin. Ciprofloxacin, rifamycin, or rifaximin can also be used.

The drug of choice for mild or moderate colitis caused by *C. difficile* is oral vancomycin. Severe infections caused by *C. difficile* should also be treated with oral vancomycin. Fidaxomicin is an alternative to vancomycin. For patients with recurrent (3 or more) *C. difficile* infections, fecal microbiota transplant should be offered if available.

Severe cases of shigellosis (bacillary dysentery) should be treated with ciprofloxacin. Severe cases of *Campylobacter* infection should be treated with azithromycin. *Listeria* gastroenteritis does not require antibiotics in immunocompetent, nonpregnant patients. There is no antiviral drug available for either norovirus or rotavirus. *Giardia* infection should be treated with tinidazole.

Probiotics have little value in the treatment of diarrhea.

Prevention

Most of the preventative strategies are directed at travelers to developing countries. They are advised to avoid potentially contaminated water sources as well as fresh fruit and vegetables if not washed in boiled water. For immunocompromised patients and others with severe underlying comorbidities, one current approach is to prescribe the traveler with rifaximin to prevent travelers' diarrhea. Probiotics are generally not recommended to prevent antibiotic-associated colitis caused by *C. difficile*.

There are two rotavirus vaccines available (see Chapter 40). Both contain live virus and are given orally. One is a live, attenuated vaccine (Rotarix), which contains the single most common rotavirus serotype (G1) causing disease in the United States. The other is a live reassortant vaccine (Rotateq), which contains five rotavirus strains. A rare but increased risk of intussusception has been reported with both vaccines. Patients with a history of intussusception should not receive either vaccine.

APPENDICITIS

Definition

Appendicitis is inflammation of the vestigial vermiform appendix. It is one of the most common causes of acute abdomen requiring surgical exploration.

Pathophysiology

Obstruction of the appendix by one of the variety of causes (e.g., fecaliths, infection such as parasites, tumor) leads to an increase in luminal and intramural pressure. Bacterial overgrowth is accompanied by inflammation. If there is necrosis, perforation followed by diffuse peritonitis caused by bacteria of the normal colonic flora (e.g., *E. coli* and *Bacteroides*) may occur.

Clinical Manifestations

Clinical manifestations include abdominal pain (especially periumbilical pain migrating to the right lower quadrant), anorexia, nausea, and vomiting. Low-grade fever and mild leukocytosis may be present. Initial symptoms may be missed because they may be nonspecific (e.g., indigestion). A standard abdominal computed tomography (CT) scan with contrast is often used when appendicitis is suspected.

Pathogens

Early in the course of the disease, the predominant organisms are anaerobic. In late disease, mixed organisms predominate. *E. coli*, *Peptostreptococcus*, *Bacteroides fragilis*, and *Pseudomonas* are commonly isolated. *Yersinia*, *Campylobacter*, and *Salmonella* can cause an acute ileitis and mesenteric adenitis that can mimic appendicitis.

Diagnosis

Clinical manifestations combined with imaging, using either CT scan or ultrasonography, are typically used to make a decision as to whether a patient should be taken to the operating room.

Treatment

Surgery is the definitive treatment for appendicitis, usually in concert with perioperative antibiotics. Laparoscopic appendectomy is preferred to open appendectomy, as the laparoscopic approach has a reduced risk of surgical wound site infections, a reduced risk of bowel obstruction, and a faster recovery time. A course of antibiotics alone (without surgery) is sometimes used, but there is an increased risk of recurrent appendicitis.

Treatment of a perforated appendix includes surgery with drainage coupled with intravenous antibiotics. Ertapenem, eravacycline, piperacillin/tazobactam, or the combination of ceftriaxone plus metronidazole can be used.

DIVERTICULITIS

Definition

Diverticulitis is inflammation of a sac-like protrusion of the colonic wall, usually in the sigmoid colon (Figure 73–3). Perforation of the diverticulum with consequent abscess formation or peritonitis may occur. Diverticulosis describes the presence of one or more diverticula in an asymptomatic person.

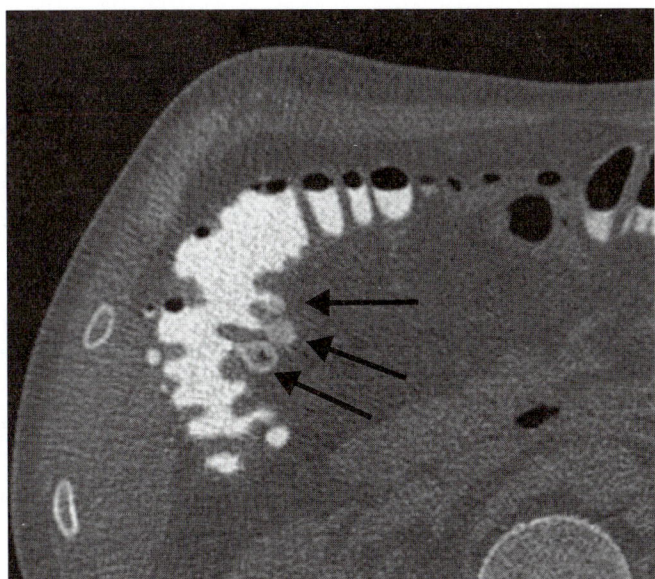

FIGURE 73–3 Diverticulosis. Three black arrows indicate diverticula. A diverticulum is an outpouching from the colonic wall. Diverticulitis is an inflammation of the diverticulum. (Reproduced with permission from Elsayes KM, Oldham SA: *Introduction to Diagnostic Radiology.* New York, NY: McGraw Hill; 2015.)

Pathophysiology

Colonic diverticula may occur following years of a diet deficient in fiber. Additional risk factors include smoking and obesity. There is no evidence that ingestion of seeds or nuts causes diverticulitis.

Clinical Manifestations

There is a range of symptoms depending on the degree of perforation. Patients usually present with dull, aching, left lower quadrant abdominal pain. This is often accompanied by a low-grade fever, leukocytosis, nausea, and vomiting. Diarrhea or constipation may be present. If perforation occurs, patients can present with peritonitis characterized by diffuse abdominal pain and shock. A perforation surrounded by mesentery often results in an abscess.

Pathogens

Mixed bowel flora including anaerobic bacteria, such as *B. fragilis*, and facultative bacteria, such as those in the Enterobacteriaceae family, for example, *E. coli*, are typically involved.

Diagnosis

Abdominal CT will show colonic diverticula and associated wall thickening, fat infiltration, abscesses, and extraluminal air or contrast medium.

Treatment

Oral antibiotics with excellent anaerobic activity (e.g., amoxicillin with clavulanate or a combination of ciprofloxacin plus metronidazole) are used in mild cases. In more serious cases requiring hospitalization, intravenous fluids and antibiotics are given with bowel rest as needed. If there is an associated abdominal abscess or signs of peritonitis, surgical evaluation must be undertaken.

Prevention

Prevention involves increasing the amount of fiber in the diet.

ENTERIC FEVER SUCH AS TYPHOID FEVER

Definition

Enteric fever is a clinical syndrome comprised of constitutional symptoms, such as fever and headache, and nausea, vomiting, and abdominal pain. Although enteric fever can be caused by several *Salmonella* species, "typhoid fever" refers to enteric fever caused by *Salmonella typhi*. *S. typhi* is also known as *Salmonella enterica* serotype Typhi. Typhoid fever is a significant global health problem.

Pathophysiology

Following the consumption of contaminated food, *Salmonella* bacteria enter through the intestinal mucosal epithelium by transcytosis. The microbes then replicate in the macrophages of Peyer patches, mesenteric lymph nodes, and spleen. Bacteremia then occurs with dissemination to lungs, gallbladder, kidneys, or central nervous system.

Humans are the only reservoir for *S. typhi*, so contamination of food or water by human feces should be suspected.

Clinical Manifestations

A prodromal phase is characterized by constitutional symptoms such as malaise, together with abdominal pain, constipation, and headache. Fever increases over the next several days. During the second week of disease, a typical transient rash of pink maculopapular lesions (**rose spots**) may be seen. Splenomegaly occurs more commonly than hepatomegaly, but both may occur. Relative bradycardia and leukopenia are often observed. Diarrhea is uncommon.

The chronic carrier state occurs in approximately 3% of patients with typhoid fever. The organisms typically reside in the gallbladder and are excreted in the stool, serving as a source of infection for others.

Pathogens

S. typhi and other *Salmonella* species, such as *Salmonella paratyphi* A and *S. paratyphi* B, cause typhoid fever.

Diagnosis

A history of travel to endemic areas, together with a compatible clinical presentation, is often used initially. Any fever in a returning traveler should prompt blood cultures and a clinical suspicion for enteric fever. Early in the disease, blood cultures are typically positive and stool cultures are often negative.

Later in the disease and in the carrier state, stool cultures are positive and blood cultures are negative. Stool cultures are positive at this stage because bile from an infected gallbladder carries organisms into the stool.

Treatment

The emergence of drug-resistant *S. typhi* has limited treatment options. Patients with severe disease should be treated with intravenous ceftriaxone. Oral or intravenous ciprofloxacin is an alternative particularly if disease is acquired outside of South Asia. Ciprofloxacin for 4 weeks (if isolate is susceptible) can also be used to eliminate the carrier state. Cholecystectomy should be considered for those chronic carriers who do not respond to antimicrobial therapy.

Prevention

Hygienic measures to protect the food and water supply from human fecal contamination are an important public health intervention. Immunization may not always be effective but can be considered in epidemic outbreaks, for travelers to endemic countries, and for household contacts of typhoid carriers.

Two vaccines against typhoid fever are available in the United States, both providing approximately 50% to 80% protection. The vaccine containing the Vi capsular polysaccharide of *S. typhi* has the advantage of being administered once intramuscularly. The other vaccine contains live attenuated *S. typhi* organisms and is administered orally. It has the advantage of stimulating gut immunity (IgA) thereby interrupting transmission.

Pelvic Infections

INTRODUCTION

Infections in the pelvic organs and surrounding structures compose a heterogeneous group of diseases. They primarily affect sexually active women and men. Many of the pathogens implicated are sexually transmitted, so an important facet of management is partner notification and treatment, as well as patient education regarding safe sexual practices. Among sexually transmitted infections, major syndromes that will be discussed are genital ulcer disease, vaginitis, cervicitis, pelvic inflammatory disease (PID), urethritis, prostatitis, and epididymitis.

Some of the organisms described in this chapter, such as *Treponema pallidum, Neisseria gonorrhoeae, Chlamydia trachomatis*, and herpes simplex virus 2, are transmitted from the mother to the fetus. The acronym TORCHES is used to describe certain important fetal or newborn infections acquired from the mother. There are several versions of this acronym. The following is a commonly used one: T = *Toxoplasma*; O = *o*ther (including parvovirus, human immunodeficiency virus [HIV], and Zika virus); R = *r*ubella; C = *c*ytomegalovirus; HE = *he*rpes simplex virus 2; and S = *s*yphilis. A more complete list of organisms vertically transmitted from mother to child can be found in Part 12, Table XII–10.

GENITAL ULCER DISEASE

Definition

Genital ulcer disease manifests as a breach in the skin or mucosa of the genitalia, usually caused by a sexually transmitted infection. Of these infections, herpes simplex virus type 2 (HSV-2) is the most common etiology in most geographic areas, followed by syphilis and chancroid. The most important noninfectious cause is Behçet's disease.

Pathophysiology

The mechanisms by which ulcers are produced by pathogens are incompletely understood, and there are different mechanisms of injury depending on the pathogen. In chancroid, a cytotoxin secreted by *Haemophilus ducreyi* may be important in epithelial cell injury.

Clinical Manifestations

Although the various lesions may have a characteristic appearance, it is important to note that local epidemiology is an important consideration because lesions may appear in an atypical fashion. The appearance of the ulcer, whether it is painful, and the nature of the associated lymphadenopathy may be clues in the etiology of the ulcer. Figure 74–1 shows several vesicles on the shaft of the penis of a patient with genital herpes. The vesicular lesions are typically painful. The vesicles can then progress to form shallow ulcers. Figure 74–2 shows the chancre of primary syphilis. It is a painless lesion with a shallow base and a firm, rolled edge. Table 74–1 describes the important clinical features of genital ulcer lesions, their diagnostic procedures, and treatment.

Pathogens

Common infectious etiologies of genital ulcer disease include HSV-2 (causing genital herpes) and *T. pallidum* (causing primary syphilis) primarily, as well as *H. ducreyi* (causing chancroid). Less common causes of genital ulcers include *C. trachomatis* serovars L1–3 (causing lymphogranuloma venereum) and *Klebsiella granulomatis* (causing granuloma inguinale, also known as donovanosis).

Diagnosis

A thorough sexual and medical history, followed by the physical examination, are important for diagnosis. Although clinical

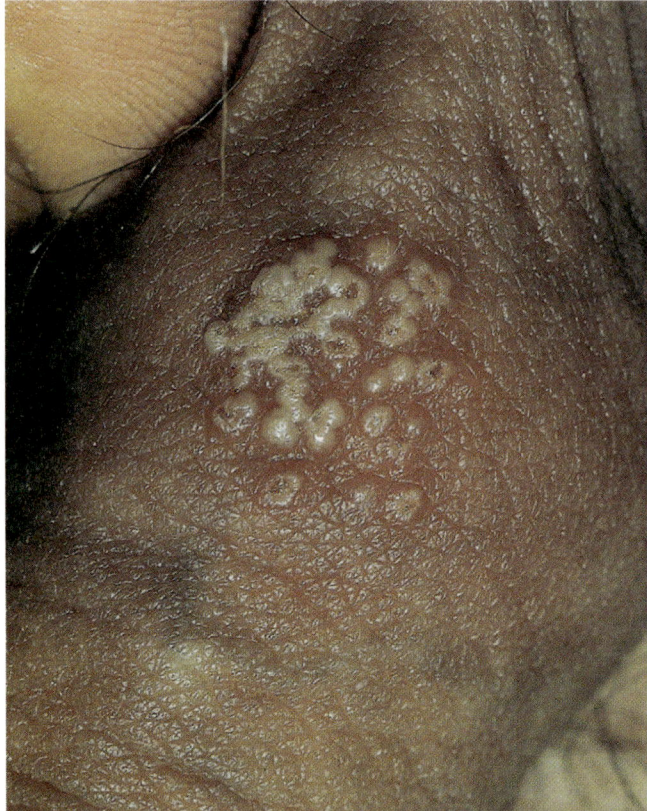

FIGURE 74–1 Genital herpes caused by herpes simplex virus 2. Note group of vesicles on shaft of penis. (Reproduced with permission from Wolff K, Johnson R, Saavedra A: *Fitzpatrick's Color Atlas & Synopsis of Clinical Dermatology*, 7th ed. New York, NY: McGraw Hill; 2013.)

characteristics can be very helpful, there is often overlap in presentation, and there may also be multiple syndromes copresenting. Therefore, diagnostic testing is highly recommended. This typically includes the following tests at the initial visit: direct

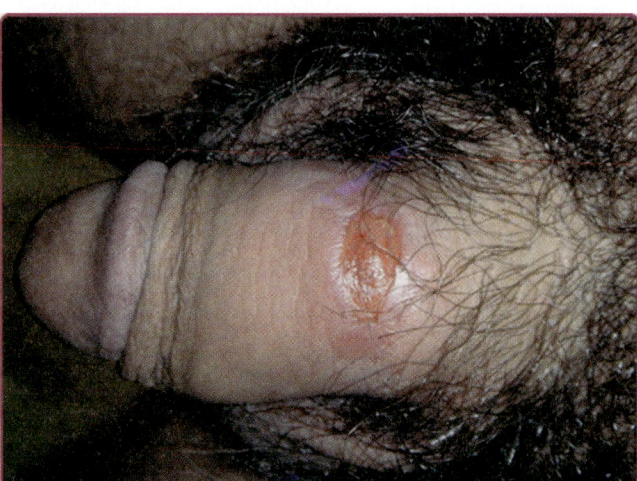

FIGURE 74–2 Chancre of primary syphilis caused by *Treponema pallidum*. Note shallow ulcer with clean base and rolled edge. (Reproduced with permission from Kang S, Amagai M, Bruckner AL, et al: *Fitzpatrick's Dermatology in General Medicine*, 9th ed. New York, NY: McGraw Hill; 2019.)

fluorescent antibody (DFA) test for HSV antigens, viral culture for HSV, or nucleic acid amplification methods for HSV DNA using a specimen taken from the base of the ulcer, and syphilis serologic testing (e.g., rapid plasmin reagin, RPR) using a serum sample. Testing for other sexually transmitted diseases including HIV is also important because there is often cotransmission of multiple pathogens (see Table 74–1).

Treatment

The drug of choice for genital herpes is acyclovir or one of its derivatives, famciclovir or valacyclovir. Primary and secondary syphilis are treated with a long-acting penicillin, benzathine penicillin G. The drug of choice for chancroid is azithromycin, whereas for lymphogranuloma venereum, it is doxycycline (see Table 74–1).

Empiric treatment is often used before diagnostic tests return. As for most sexually transmitted infections, treatment that involves one dose and that is observed is preferred if possible.

Prevention

Consistent use of condoms is an important measure that can prevent genital ulcers. In some cases, primary prevention of HSV infection can be undertaken by treatment of the negative partner in serodiscordant couples with acyclovir or one of its derivatives. Prophylaxis with these drugs can be effective in preventing recurrences of HSV outbreaks in patients who have had frequent occurrences, especially among those who are immunosuppressed. Partner notification and treatment are important prevention strategies as well. There is no vaccine against any of the organisms that cause genital ulcer disease.

VAGINITIS & BACTERIAL VAGINOSIS

Definition

Vaginitis is inflammation of the vagina that can result in discharge, itching, and pain. These symptoms occur primarily in three diseases: candidiasis, trichomoniasis, and bacterial vaginosis (BV). Noninfectious causes include lichen planus and certain medications (e.g., oral contraceptives). Note that a vaginal discharge can occur in both vaginitis and cervicitis. Vaginal symptoms are very common, resulting in millions of visits to healthcare providers each year.

Pathophysiology

The use of antibiotics that inhibit the normal flora of the vagina, especially lactobacilli, predisposes to *Candida* vaginitis. *Candida* is a member of the normal flora of many women. The pathogenesis of BV is uncertain, but it does not appear to be a sexually transmitted disease. BV can be thought of as a dysbiosis of the vaginal microbiome. Lactobacilli are the predominant organisms in the healthy vagina but other bacteria (see Pathogens section later) predominate in the dysbiotic vagina. Trichomoniasis, on the other hand, is a sexually transmitted disease.

TABLE 74-1 Genital Ulcers: Clinical Features, Diagnosis, and Treatment

Syndrome	Pathogen	Appearance of Ulcer	Pain	Adenopathy	Diagnosis	Treatment
Genital herpes	HSV-2 (more common than HSV-1)	Multiple, small vesicles and ulcers with an erythematous base	Painful	Tender lymphadenopathy	Direct fluorescent antibody (DFA) testing and/or viral culture; sample taken from the base of the ulcer; Tzanck smear shows multinucleated giant cells; nucleic acid amplification test (NAAT) may be useful	Acyclovir, famciclovir, valacyclovir
Syphilis	*Treponema pallidum*	Single (usually) indurated ulcer with a clean base; self-resolves and may not be observed	Painless	Painless, regional lymphadenopathy; lymph nodes feel "rubbery"	Serologic screening with nontreponemal test (e.g., rapid plasma reagin [RPR] or Venereal Disease Research Laboratory [VDRL]); treponemal test (e.g., fluorescent treponemal antibody absorption [FTA-ABS]) for confirmation; dark field examination of fluid from the lesion if possible	Benzathine penicillin G
Chancroid	*Haemophilus ducreyi*	Multiple, nonindurated ulcers with a gray or yellow exudate at the base	Very painful	Tender, unilateral inguinal lymphadenopathy; lymph nodes may rupture	Difficult to diagnose; special culture media, if available, or polymerase chain reaction (PCR)	Azithromycin
Lymphogranuloma venereum	*Chlamydia trachomatis* serovars L1–3	Small, shallow ulcers that self-resolve and are not usually observed	Painless	Characteristic appearance of lymph nodes is key feature; may be bilateral, large, and painful; presents with fluctuant "buboes" and sinus tracts	NAAT; serology can also be used	Doxycycline
Granuloma inguinale (donovanosis)	*Klebsiella granulomatis*	Marked, beefy red, vascular ulcer with granulomatous appearance and rolled edges	Painless	Not a major feature; subcutaneous granulomas, "pseudobuboes" may occur	Difficult to diagnose; PCR if available; may also see dark-staining oval organisms on Giemsa staining (Donovan bodies) of biopsy specimen	Doxycycline

Clinical Manifestations

Patients are usually prompted to seek medical attention because of an abnormal vaginal discharge. This may be accompanied by pruritus, pain (including dyspareunia), and symptoms of vaginal irritation. Figure 74–3 depicts the white, "cottage cheese" appearance of vaginal candidiasis. Figure 74–4 shows the "strawberry" cervix of trichomoniasis. There are red, punctate lesions on the cervix, and frothy exudate can be seen at the cervical os. The vaginal discharge in BV is thin and grayish and has an unpleasant odor, often described as "fishy." There is an absence of neutrophils in the discharge. Table 74–2 describes the important clinical features of vaginitis, its diagnostic procedures, and its treatment.

Pathogens

The yeast *Candida albicans* is the most common cause of vaginal candidiasis. The protozoan *Trichomonas vaginalis* is the cause

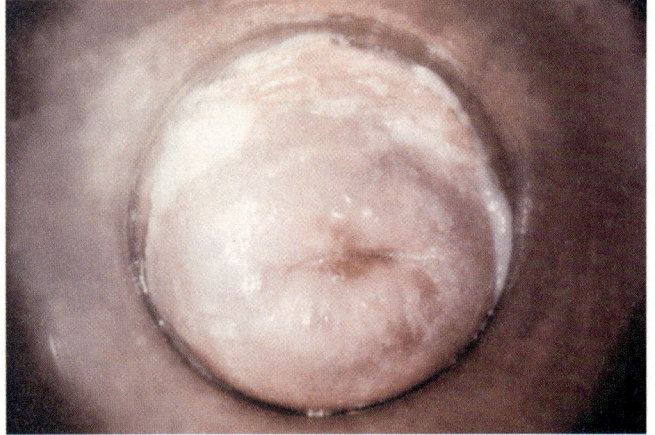

FIGURE 74–3 Vaginal candidiasis caused by *Candida albicans*. Note areas of whitish, "cottage cheese–like" exudate on cervical mucosa. (Reproduced with permission from Centers for Disease Control and Prevention.)

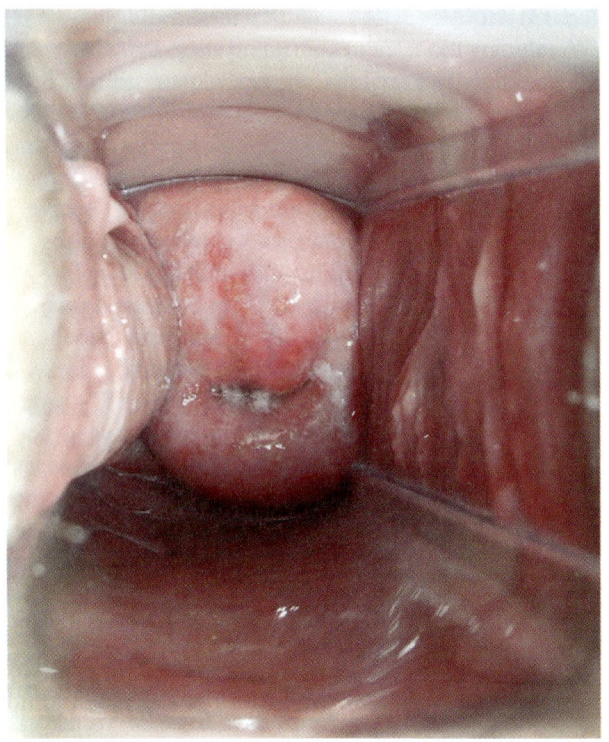

FIGURE 74–4 Trichomoniasis caused by *Trichomonas vaginalis*. Note frothy discharge and punctate "strawberry" lesions on cervix. (Reproduced with permission from Usatine RP, Smith MA, Mayeaux EJ Jr, et al: *The Color Atlas of Family Medicine*. New York, NY: McGraw Hill; 2009. Photo contributor: Richard P. Usatine, MD.)

of trichomoniasis. Overgrowth of bacteria such as *Gardnerella vaginalis* are implicated in BV, but anaerobes such as *Mobiluncus* and *Prevotella* and nonanaerobes such as *Atopobium vaginae*, *Mycoplasma hominis*, and *Ureaplasma* may also be involved.

Diagnosis

A patient's complaint of vaginal discharge should prompt a careful history, including time of last menstrual period, medications, and sexual activity. The physical examination should include a microscopic examination of the vaginal discharge itself on a glass slide using a drop of 0.9% saline solution (to look for motile trichomonads or clue cells), followed by a drop of 10% potassium hydroxide (to look for *Candida*). If microscopy is negative, *Candida* culture and *Trichomonas* nucleic acid amplification test (NAAT) are performed to increase sensitivity.

Trichomonads are shown in Figure 74–5. Figure 74–6 shows clue cells as large, vaginal epithelial cells dotted with bacteria. A Gram stain of clue cells reveals many gram-variable rods on the surface of the epithelial cells. Figure 74–7 shows the appearance of the yeasts and pseudohyphae of *Candida*. Cultures for *Gardnerella* are not done because at least 50% of asymptomatic women carry the organism. See Table 74–2 for additional information.

Treatment

Metronidazole is the drug of choice for both BV and trichomoniasis. Tinidazole is also used because it has fewer adverse effects than metronidazole. Secnidazole can be used to treat BV.

TABLE 74–2   Vaginitis: Clinical Features, Diagnosis, and Treatment

Syndrome	Pathogen(s)	Nature of Vaginal Discharge	pH	Microscopy	Treatment	Other Notes
Normal	*Lactobacillus* species are the predominant normal flora	Could be clear and mucoid, especially at midcycle estrogen surge; during pregnancy, secretions may be thicker and white	<4.5			
Bacterial vaginosis	*Gardnerella vaginalis*; anaerobes such as *Mobiluncus* species are also involved	Malodorous, gray, thin	>4.5	Clue cells are epithelial cells covered with gram-variable bacteria (see Figure 74–6)	Metronidazole (oral or gel), tinidazole (oral) or secnidazole (oral)	An amine-like "fishy" odor occurs after the addition of 10% potassium hydroxide
Vaginal candidiasis	*Candida albicans*	"Cottage cheese"; white and clumpy	<4.5	Yeasts and pseudohyphae seen in KOH prep (10% potassium hydroxide) (see Figure 74–7)	Fluconazole (oral), miconazole (vaginal suppository), butoconazole (vaginal cream) are single-dose options	
Trichomoniasis	*Trichomonas vaginalis*	Malodorous, green-yellow, thin	>4.5	Numerous neutrophils	Metronidazole or tinidazole (oral)	"Strawberry cervix" seen on speculum examination (inflammation and punctate hemorrhage of the cervix) (see Figure 74–4)

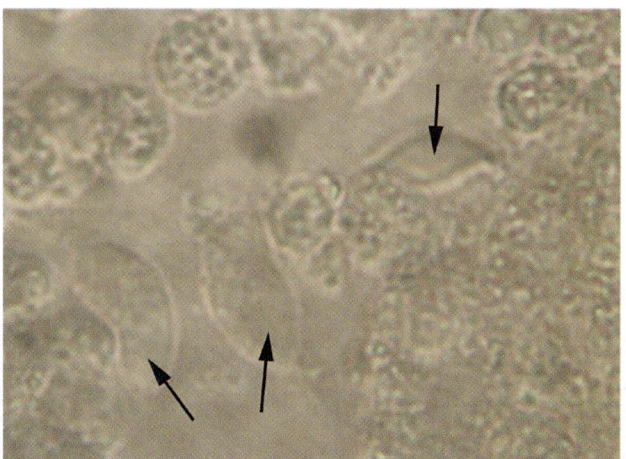

FIGURE 74–5 *Trichomonas vaginalis* in vaginal discharge. Note trichomonads (arrows) mounted in saline and visualized in light microscope. (Reproduced with permission from Usatine RP, Smith MA, Mayeaux EJ Jr, et al: *The Color Atlas of Family Medicine*. New York, NY: McGraw Hill; 2009. Photo contributor: Richard P. Usatine, MD.)

For candidiasis, either oral fluconazole or vaginally administered miconazole or butoconazole is the drug of choice (see Table 74–2). *T. vaginalis* is a sexually transmitted infection, so a one-time treatment regimen of patient and partner is preferred. Following treatment of trichomoniasis, the Centers for Disease Control and

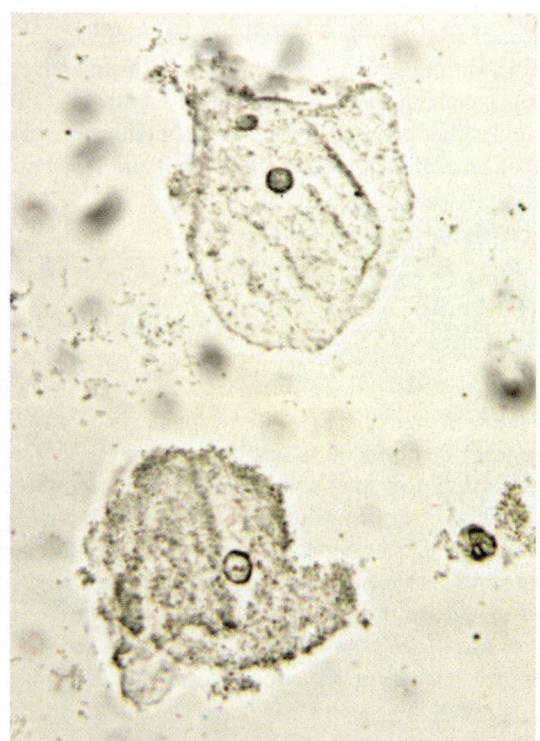

FIGURE 74–6 Clue cells in bacterial vaginosis. Note that the lower epithelial cell is a "clue cell" because its surface is covered with bacteria. The upper epithelial cell is *not* a "clue cell" because its surface has few bacteria. (Reproduced with permission from Usatine RP, Smith MA, Mayeaux EJ Jr, et al: *The Color Atlas of Family Medicine*. New York, NY: McGraw Hill; 2009. Photo contributor: Mayeaux E.J., MD.)

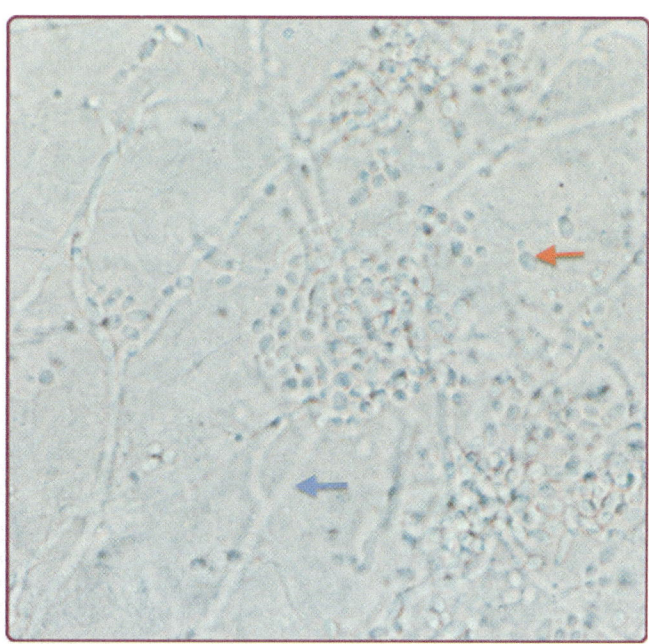

FIGURE 74–7 *Candida* visualized in KOH preparation. Note yeast cells (red arrow) and pseudohyphae (blue arrow). (Reproduced with permission from Kang S, Amagai M, Bruckner AL, et al: *Fitzpatrick's Dermatology in General Medicine*, 9th ed. New York, NY: McGraw Hill; 2019.)

Prevention recommends retesting with NAAT to evaluate for reinfection, regardless of whether the partner was treated.

Prevention

There is no vaccine against any of the organisms that cause vaginitis.

CERVICITIS

Definition

Cervicitis is inflammation of the uterine cervix. Acute cervicitis is usually due to a sexually transmitted infection caused by either *C. trachomatis* or *N. gonorrhoeae* or both. Note that a vaginal discharge can occur in both vaginitis and cervicitis.

Clinical Manifestations

A large proportion of persons with cervicitis are asymptomatic but a vaginal discharge and bleeding between menstrual periods may occur. In many cases, cervicitis is detected on speculum examination (Figure 74–8) and/or following routine screening for *C. trachomatis* and *N. gonorrhoeae*. Individuals who have concomitant urethral infection may have dysuria. On physical examination, increased friability of the cervical tissue after a swab is inserted may be a clue to the diagnosis.

Pathogens

The usual pathogens are *C. trachomatis* serovars D–K and/or *N. gonorrhoeae*. Other less common etiologies include HSV and *T. vaginalis*.

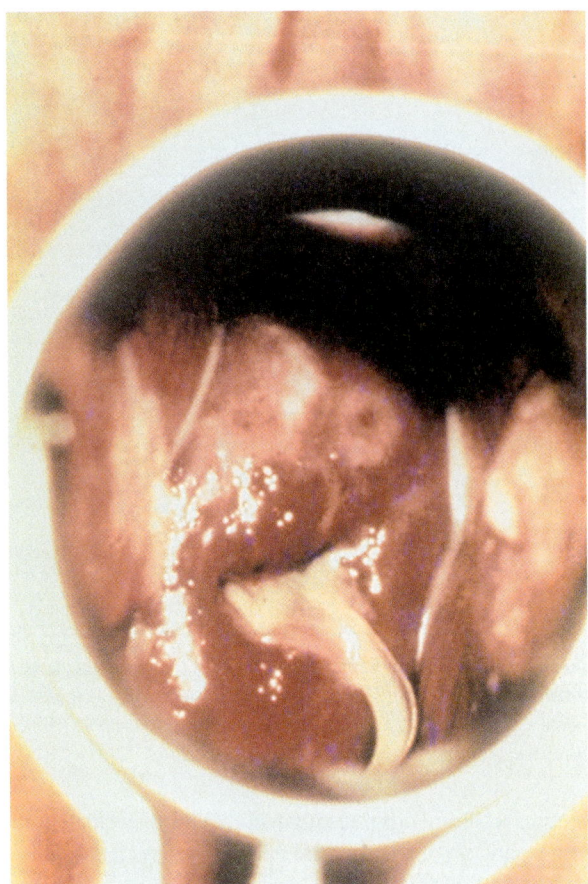

FIGURE 74–8 Cervicitis. Note purulent exudate at cervical os. (Reproduced with permission from Knoop KJ, Stack LB, Storrow AB, et al: *The Atlas of Emergency Medicine*, 3rd ed. New York, NY: McGraw Hill; 2009. Photo contributor: Sue Rist, FNP.)

Diagnosis

A clinical diagnosis may be made based on increased friability of the cervix, with or without mucopurulent discharge. To make a laboratory diagnosis, NAAT for *C. trachomatis* and *N. gonorrhoeae* is routinely performed. If NAAT testing is not available, then Gram stain and culture for gonococci can be used.

Treatment

If there is clinical evidence of cervicitis, empiric treatment for both *C. trachomatis* and *N. gonorrhoeae* (ceftriaxone intramuscularly plus doxycycline orally) is recommended, particularly if follow-up of test results by the patient is uncertain. Sex partners of patients with a confirmed diagnosis should also be notified and treated.

Prevention

Consistent use of condoms is an important measure that can prevent sexually transmitted diseases. Partner notification and treatment are important prevention strategies as well. There is no vaccine against any of the organisms that cause cervicitis.

PELVIC INFLAMMATORY DISEASE

Definition

PID is a polymicrobial infection of upper genital tract structures, namely, the uterus, fallopian tubes, and ovaries. Infection of the uterus is called endometritis and infection of the fallopian tubes is called salpingitis.

Pathophysiology

When the endocervical canal barrier is compromised, vaginal bacteria can ascend into the normally sterile space of the upper genital tract (uterus, fallopian tubes, and ovaries). A sexually transmitted infection affecting the cervix (e.g., *N. gonorrhoeae* and *C. trachomatis*) can initiate the process, permitting the anaerobic bacteria of the vagina to ascend.

Having multiple sex partners increases the risk of PID. Multiple episodes of PID lead to scarring of the fallopian tubes and an increased risk of ectopic pregnancy and sterility. PID is especially common in adolescents and young adults.

Clinical Manifestations

Patients can present with a range of symptoms, from lower back pain to fever, chills, lower abdominal pain, and cervical and adnexal tenderness. The abrupt onset of abdominal pain associated with menses is a common finding in PID. On physical exam, tenderness on motion of the cervix and an abnormal vaginal discharge are important diagnostic signs.

Patients with PID may also have perihepatitis characterized by right upper quadrant pain. This is called Fitz-Hugh-Curtis syndrome and is characterized by "violin-string" adhesions. This syndrome can be caused by both *C. trachomatis* and *N. gonorrhoeae*.

Pathogens

PID is primarily caused by *N. gonorrhoeae* and *C. trachomatis*, together with *Mycoplasma genitalium*, enteric gram-negative rods, and anaerobes.

Diagnosis

Because it is often difficult to diagnose PID precisely (given the nonspecific findings) and because the consequences of not treating PID can be grave, many opt to treat with minimum diagnostic criteria such as uterine, adnexal, or cervical motion tenderness. Fever, the presence of leukocytes on cervical or vaginal discharge, elevated C-reactive protein, and laboratory evidence of cervical infection with *N. gonorrhoeae* or *C. trachomatis*, such as NAATs for these organisms, can increase the specificity of the diagnosis.

Treatment

If symptoms are mild, persons can be treated as outpatients with cefoxitin or ceftriaxone (one dose) plus doxycycline (14 days). Metronidazole is sometimes added to the drug regimen. In the inpatient setting, intravenous therapy is preferred. Cefoxitin or cefotetan with doxycycline and clindamycin plus gentamicin are initial options with oral antibiotics only after 24 hours of improvement of the patient.

Prevention

There is no vaccine against any of the organisms that cause PID.

URETHRITIS

Definition

Urethritis is an inflammation of the urethra. It is usually caused by a sexually transmitted infection, particularly in sexually active persons. Urethritis is often thought of as either gonococcal urethritis or nongonococcal urethritis (NGU). The latter is most often caused by *C. trachomatis*.

An autoimmune cause of urethritis called reactive arthritis (formerly named Reiter's syndrome) is a syndrome that includes urethritis, uveitis, and arthritis. This syndrome is often a sequel to infection by *C. trachomatis* though other bacteria such as *Salmonella, Shigella, Campylobacter, Yersinia, Clostridioides difficile* and *Chlamydia pneumoniae* have been implicated.

Clinical Manifestations

Dysuria is the most common presenting complaint often accompanied by discharge from the urethra (Figure 74–9). Pruritus and burning pain are also common complaints.

Pathogens

N. gonorrhoeae, C. trachomatis, and M. genitalium are the most common organisms implicated. NGU is caused primarily by *C. trachomatis* followed by *M. genitalium* but about half of NGU

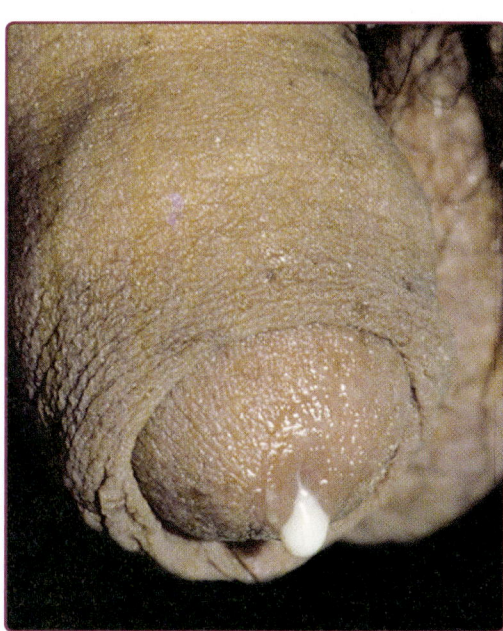

FIGURE 74–9 Urethral discharge in gonorrhea. Note thick purulent urethral discharge. (Reproduced with permission from Kang S, Amagai M, Bruckner AL, et al: *Fitzpatrick's Dermatology in General Medicine*, 9th ed. New York, NY: McGraw Hill; 2019. Photo contributor: Dr. Ted Rosen.)

cases have no specific pathogen identified. Other organisms such as *T. vaginalis*, which are not normally tested, can be involved.

Diagnosis

NAAT for *C. trachomatis* and *N. gonorrhoeae* is routinely performed in many centers. Some centers also perform NAAT testing for *M. genitalium*.

Treatment

If there is clinical evidence of urethritis such as a purulent urethral discharge, empiric treatment for both *N. gonorrhoeae* and *C. trachomatis* (ceftriaxone intramuscularly plus doxycycline orally) is recommended.

Prevention

Consistent use of condoms is an important measure that can prevent sexually transmitted diseases. Partner notification and treatment are important prevention strategies as well. There is no vaccine against any of the organisms that cause urethritis.

EPIDIDYMITIS

Definition

Acute epididymitis is characterized by pain and swelling of the epididymis of less than 6 weeks in duration. Sometimes the testes are involved as well (epididymo-orchitis). Chronic epididymitis typically has similar but more muted findings and is longer than 6 weeks in duration.

Pathophysiology

In sexually active persons, the infection begins as a urethritis that ascends into the epididymis. In older, nonsexually active adults, the infection is related to bladder infections, secondary to prostatic hypertrophy. Predisposing factors include prostate biopsy, urinary tract instrumentation, and immunosuppression.

Clinical Manifestations

Acute epididymitis typically presents with unilateral scrotal pain and tenderness and swelling of the epididymis. On physical exam, an enlarged red scrotum is seen. A hydrocele may be present, and a thickened spermatic cord may be palpated. Torsion of the testicle (noted as sudden testicular pain with an asymmetrical and high-riding testis) must be distinguished from acute epididymitis because torsion is a surgical emergency. Torsion of the testicle typically occurs in young adults several hours after physical activity or trauma.

Chronic epididymitis presents with discomfort in the epididymis, scrotum, or testicle lasting 6 weeks or longer.

Pathogens

Acute epididymitis in sexually active adults is typically caused by *C. trachomatis* or *N. gonorrhoeae*. *Escherichia coli* may be the cause in adults who have sex with adults. *E. coli* is also a common cause in older patients with a concomitant bladder infection.

Chronic epididymitis can be a granulomatous infection caused by *Mycobacterium tuberculosis*. However, often no etiology is identified.

Diagnosis

A microbiologic diagnosis of acute epididymitis in sexually active adults can be made by using Gram stain of urethral secretions and/or NAAT for *C. trachomatis* or *N. gonorrhoeae*. In nonsexually active patients, urinalysis and urine culture should be done.

Scrotal Doppler ultrasonography is the test of choice used to distinguish torsion from epididymitis. Absence of blood flow on the ultrasound indicates torsion of the testicle.

Treatment

Acute epididymitis in sexually active patients should be treated with ceftriaxone and doxycycline to cover the most common causes. In proven *C. trachomatis* or *N. gonorrhoeae* cases, sex partners should also be treated. In nonsexually active patients, oral levofloxacin to cover *E. coli* is appropriate.

Prevention

There is no vaccine against any of the organisms that cause epididymitis.

PROSTATITIS

Definition

Acute bacterial prostatitis is characterized by the presence of irritative voiding symptoms (urinary frequency, hesitancy, feeling of incomplete voiding, dribbling), fever, pyuria, and positive urine cultures. Chronic bacterial prostatitis is characterized by the same voiding symptoms, but fever and pyuria are typically absent. Prostatitis is also discussed in Chapter 78 on urinary tract infections.

Pathophysiology

Bacteria ascend the urethra and then reflux into the prostatic ducts and prostate where infection occurs.

Clinical Manifestations

Patients appear ill in acute prostatitis with fevers, chills, irritative voiding symptoms, and pelvic or perineal pain. Physical examination may reveal a very tender and enlarged prostate. Symptoms in chronic bacterial prostatitis may be more subtle. Patients may present with recurrent urinary tract infections, but only prolonged treatment of prostatitis will result in a cure.

Pathogens

Generally, gram-negative rods that reflect the range of organisms that cause cystitis are involved. These organisms include the Enterobacteriaceae (e.g., *E. coli, Klebsiella,* and *Proteus* species) as well as *Pseudomonas.* In sexually active adults, *N. gonorrhoeae* and *C. trachomatis* can cause prostatitis, especially in association with urethritis and epididymitis.

Diagnosis

A patient with symptoms of prostatitis who has an edematous and tender prostate on examination is considered to have acute bacterial prostatitis. Culture of urine is done to determine the causative organism. Culture of prostatic fluid is not done in acute prostatitis because prostatic massage should not be done during the acute phase. Prostatic massage may be useful in chronic prostatitis.

Treatment

Trimethoprim-sulfamethoxazole or a fluoroquinolone such as ciprofloxacin can be used as empiric therapy until culture results return. These agents exhibit good penetration into the prostate. Therapy is prolonged, usually given for 4 to 6 weeks.

Upper Respiratory Tract Infections

INTRODUCTION

Infections of the upper respiratory tract are a common ambulatory care complaint, resulting in a large proportion of office visits. Although the vast majority of infections are viral and are self-limited, some may require hospitalization, particularly in the pediatric population. Bacterial etiologies of some of the common upper respiratory tract infections may be primary or superinfections of the original viral processes and are amenable to treatment (Table 75–1).

TABLE 75–1 Common Infections of the Upper Respiratory Tract

Infection	Important Pathogens	Treatment
Otitis media	*Streptococcus pneumoniae, Haemophilus influenzae, Moraxella catarrhalis*	Amoxicillin
Acute sinusitis	*S. pneumoniae, H. influenzae, M. catarrhalis*	Amoxicillin if symptoms persist for >10 days
Pharyngitis	*Streptococcus pyogenes* (group A *Streptococcus*), viruses (e.g., adenovirus)	Penicillin or amoxicillin if group A *Streptococcus* diagnosed
Common cold	Rhinovirus, coronavirus (common cold strains), and others	Supportive; zinc may be helpful in reducing duration of symptoms
Croup	Parainfluenza virus	Supportive; corticosteroids and epinephrine if moderate or severe symptoms
Laryngitis	Parainfluenza virus and rhinovirus	Supportive
Epiglottitis	*H. influenzae* type B	Ceftriaxone

OTITIS MEDIA

Definition

Otitis media is an infection of the middle ear caused by either viruses or bacteria. Otitis media can be either acute or chronic. The information in this chapter refers to acute otitis media. Acute otitis media is the second most common diagnosis in children and the most common reason for prescribing antibiotics to a child.

Pathophysiology

Any process that leads to eustachian tube obstruction can result in fluid retention and concomitant infection of the middle ear. The most common predisposing factors are upper respiratory tract infections and seasonal allergic rhinitis. Otitis media is very common in children under the age of 3 years because they have a small opening of the eustachian tube that is easily blocked by the inflammation caused by a viral infection or an allergic response.

Clinical Manifestations

Patients present with ear pain (otalgia) and pressure, often accompanied by an upper respiratory tract infection. In infants, the ear pain may manifest as ear pulling. Patients may also complain of decreased hearing, irritability, poor sleeping, and fever. If the tympanic membrane ruptures, drainage from the ear may occur. On examination, the tympanic membrane is erythematous (Figure 75–1A and B) with a loss of the light reflex and decreased mobility. In some cases, the tympanic membrane may bulge and then rupture.

Pathogens

Both bacteria and viruses cause otitis media. Among bacteria, *Streptococcus pneumoniae* is the most common cause.

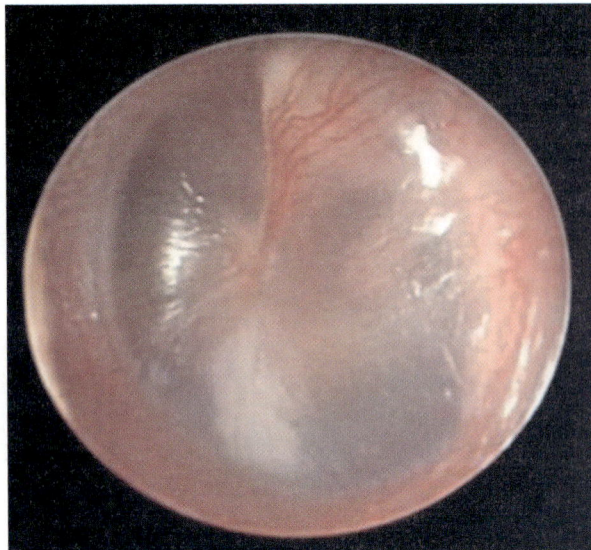 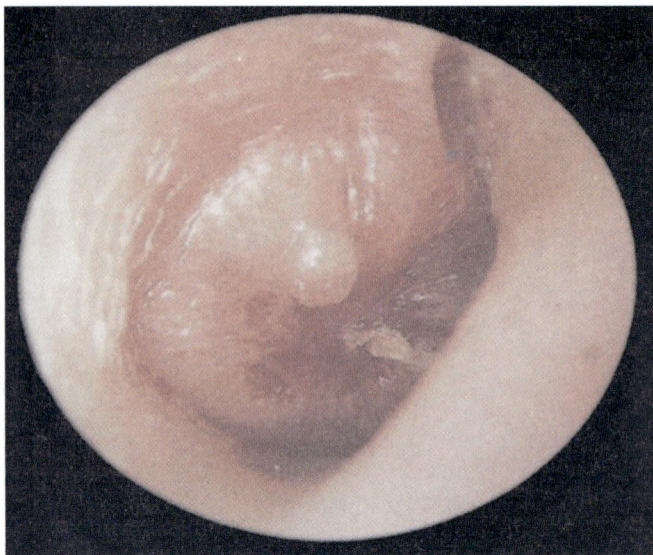

FIGURE 75–1 **A:** Normal tympanic membrane in a 6-year-old child. **B:** Otitis media in a 3-year-old child. Note bulging tympanic membrane and loss of light reflex. (Reproduced with permission from Tintinalli JE Stapczynski J, Ma OJ, et al: *Tintinalli's Emergency Medicine: A Comprehensive Study Guide*, 7th ed. New York, NY: McGraw Hill; 2009. Photo contributor: Dr. Shelagh Cofer, Department of Otolaryngology, Mayo Clinic.)

Haemophilus influenzae (usually non-typeable strains) and *Moraxella catarrhalis* are also common causes. Among viruses, respiratory syncytial virus, coronaviruses (common cold strains), and rhinoviruses are commonly involved.

Diagnosis

Otitis media is usually diagnosed clinically. If the membrane ruptures, a sample of the exudate can be analyzed by Gram stain and culture. If indicated, tympanocentesis can be done to relieve pressure before the drum ruptures and to obtain a specimen for culture.

Treatment

Amoxicillin orally is usually the drug of choice together with nasal decongestants to open the eustachian tube. In cases caused by bacteria resistant to penicillins, amoxicillin-clavulanate (Augmentin) may be used.

Prevention

Recurrent episodes of otitis media can be suppressed by prophylactic antibiotics such as amoxicillin or sulfisoxazole. Ventilating tubes may be inserted as a strategy to prevent recurrent infections. The conjugate pneumococcal vaccine is effective in preventing invasive pneumococcal disease but is less effective in preventing otitis media.

SINUSITIS

Definition

Sinusitis is inflammation of the paranasal sinuses. It can be either acute or chronic. Acute infections are considered those with symptoms lasting less than 4 weeks. The information in this chapter refers to acute sinusitis.

Pathophysiology

Impaired mucociliary clearance caused by viral infection or allergic rhinitis can obstruct the orifice of the sinus. Mucus then accumulates in the sinus cavity. Stasis can lead to bacterial overgrowth and superinfection. Sinusitis frequently involves the maxillary sinus because the ostium of that sinus is located superior to most of the sinus and drainage of mucus has to occur against gravity. Drainage of the other sinuses is aided by gravity.

Clinical Manifestations

Clinical manifestations include purulent nasal discharge, nasal congestion, facial or sinus pain, decreased sense of smell, and fever. Headache and malodorous breath may be present.

Pathogens

Many cases begin with a viral upper respiratory tract infection. Bacterial superinfection can then occur. In the case of acute bacterial sinusitis, common organisms are *S. pneumoniae*, *H. influenzae*, and *M. catarrhalis*, as in the case of acute otitis media. *Staphylococcus aureus* also causes sinusitis but less commonly. In immunocompromised patients and diabetics, sinusitis caused by fungi such as *Aspergillus* or *Mucor* may occur.

Diagnosis

Sinusitis is often diagnosed based on a typical constellation of symptoms and clinical findings. Computed tomography

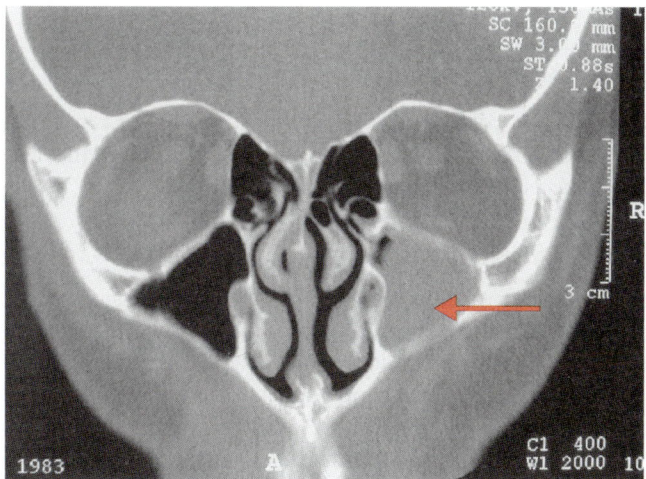

FIGURE 75–2 Sinusitis. Arrow points to opacified maxillary sinus seen in computed tomography scan of head. (Reproduced with permission from Brunicardi FC, Andersen D, Billiar T, et al: *Schwartz's Principles of Surgery*, 8th ed. New York, NY: McGraw Hill; 2004.)

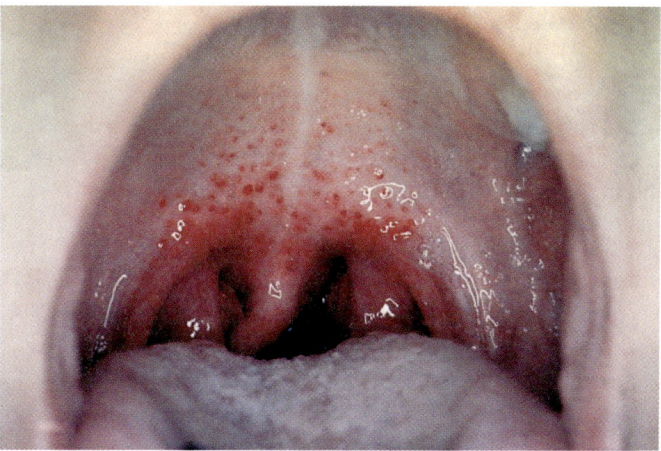

FIGURE 75–3 Pharyngitis caused by *Streptococcus pyogenes*. Note inflamed pharynx, tonsils, and palatal petechiae. (The white circles and curved lines are an artifact of the lighting during photography.) (Used with permission from Dr. Heinz F. Eichenwald, Public Health Image Library, Centers for Disease Control and Prevention.)

scan of the sinuses is a very sensitive modality for indicating inflammatory processes of the sinus. However, in the absence of bony destruction, these are nonspecific findings for diagnosing clinically significant sinusitis requiring antibiotic therapy (Figure 75–2).

Treatment

If symptoms are severe, antibiotics are given in concert with intranasal corticosteroids, as well as nasal decongestants. Amoxicillin is the drug of choice, but if resistance is a concern, then amoxicillin-clavulanate (Augmentin) is used. In mild cases, antibiotics are not normally used unless the symptoms have lasted for longer than 10 to 14 days.

Prevention

There is no convincing evidence that the pneumococcal vaccine and the *H. influenzae* type B vaccine have a significant effect in reducing sinusitis caused by these organisms.

PHARYNGITIS

Definition

Pharyngitis is inflammation of the throat caused primarily by viruses. Approximately 10% of cases of pharyngitis are caused by *Streptococcus pyogenes* (group A *Streptococcus* [GAS]). Streptococcal pharyngitis (strep throat) is important because poststreptococcal immune sequelae, such as rheumatic fever, may occur.

Clinical Manifestations

Patients will complain of sore throat that is worse when swallowing. Low-grade fever may also be present. Typical symptoms associated with an upper respiratory tract infection (rhinorrhea, sinus tenderness, ear pain, cough) may accompany the sore throat. On examination, an inflamed pharynx, tonsils, and palate are typically seen. A grayish exudate is often present on the tonsils. Tender anterior cervical lymphadenopathy may be present. Petechiae on the palate is a diagnostic clue for GAS pharyngitis but may also occur in pharyngitis caused by Epstein–Barr virus (Figure 75–3).

Pathogens

Bacteria

S. pyogenes (GAS) is the most important bacterial cause. Group C and G streptococci also cause pharyngitis but are not antecedents to rheumatic fever. Pharyngitis caused by *Neisseria gonorrhoeae* is likely to be the result of sexual activity and, if it occurs in children, is considered as a sign of child abuse. *Mycoplasma pneumoniae*, *Chlamydia pneumoniae*, and *Arcanobacterium haemolyticum* also cause pharyngitis.

In certain countries where the diphtheria vaccine is not widely used, *Corynebacterium diphtheriae* is a significant cause of pharyngitis, often accompanied by a pseudomembrane. *Fusobacterium necrophorum*, a gram-negative anaerobe, can cause pharyngitis accompanied by septic thrombophlebitis (Lemierre's syndrome). Note that although *S. pneumoniae* and *H. influenzae* colonize the oropharynx, they do not cause pharyngitis.

Viruses

Most cases of pharyngitis are caused by respiratory viruses, such as adenovirus, influenza A and B viruses, parainfluenza virus, rhinovirus, and the common-cold strains of coronavirus. Other viral causes include Coxsackie virus (herpangina), Epstein–Barr virus (infectious mononucleosis), and herpes simplex virus, especially type 1. Human immunodeficiency virus causes an acute retroviral syndrome that includes pharyngitis as one of its components.

Diagnosis

The main strategy in diagnostics is to establish whether there is infection with GAS. This is because GAS is treatable, and timely intervention may prevent complications such as acute rheumatic fever. The Centor criteria are criteria that may be used to aid in the diagnosis of GAS. These criteria include tonsillar exudates, tender anterior cervical adenopathy, fever, and absence of cough. Rapid antigen detection tests for GAS and throat culture are often used to confirm the diagnosis.

It can be difficult to distinguish between a bacterial pharyngitis and a viral pharyngitis when examining the throat. Figure 75–4 shows extensive exudates on the tonsils in pharyngitis caused by Epstein–Barr virus. A throat culture is the most reliable method of determining whether *S. pyogenes* is the cause.

Treatment

If GAS is diagnosed, treatment with penicillin G, penicillin V, or amoxicillin is undertaken. In penicillin-allergic patients, erythromycin or cephalexin can be used.

The overuse of antibiotics, such as penicillin and macrolides, in cases of pharyngitis continues to be a problem. Although only 10% of sore throats are bacterial, data show that 60% of sore throats are treated with antibiotics, resulting in unnecessary expense, adverse effects, and the selection of resistant bacteria.

Prevention

There is a vaccine against *C. diphtheriae* and influenza virus but not against pharyngitis caused by *S. pyogenes* or any other bacterial or viral cause. Long-term carriers of GAS should not be treated because there is no evidence that such treatment prevents spread of the organism to close contacts or the development of complications such as acute rheumatic fever. Note that children who have rheumatic heart disease should receive penicillin orally for many years to prevent infection by *S. pyogenes*, which could cause a flare of their rheumatic heart disease.

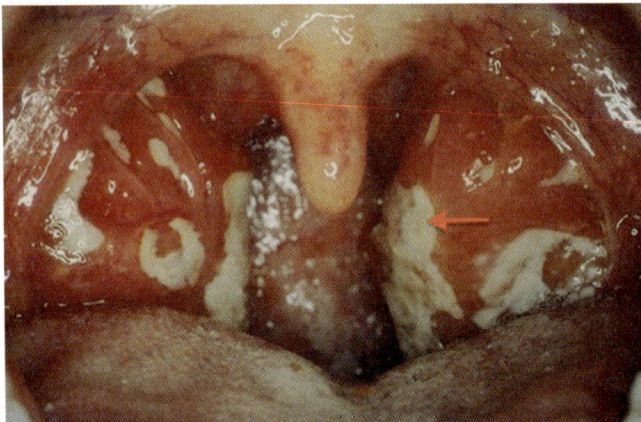

FIGURE 75–4 Pharyngitis caused by Epstein–Barr virus. Note several whitish exudates on tonsils (red arrow). (Reproduced with permission from Kane K, Nambudiri VE, Stratigos AJ: *Color Atlas & Synopsis of Pediatric Dermatology,* 3rd ed. New York, NY: McGraw Hill; 2017.)

COMMON COLD

Definition

The common cold is a viral infection of the upper respiratory tract, including some or all of the following structures: the nose, throat, sinuses, eustachian tubes, trachea, and larynx.

Pathophysiology

The viruses that cause the common cold are transmitted primarily by aerosols generated by sneezing, or by direct contact. Direct contact involves either hand-to-hand contact or hand-to-surface contact. The hand then transfers the virus to the recipient's nose, mouth, or eyes. The nonenveloped viruses such as rhinoviruses and adenoviruses are particularly stable in the environment and are often transmitted by hand-to-surface contact.

The common cold and other respiratory infections such as influenza occur more often in the winter months than in the summer months in both the Northern and Southern hemispheres. The reason for this seasonality is uncertain.

Clinical Manifestations

Clinical manifestations include nasal congestion, decreased sense of smell, rhinorrhea (watery nasal discharge without purulence), and sneezing. Patients also complain of general malaise and sore throat. In some cases, headache may also be reported.

Pathogens

Rhinoviruses (more than 100 serotypes) are the most common etiology (up to 50%). Coronaviruses (common-cold strains), adenoviruses, and enteroviruses such as Coxsackie viruses are other causes. Viruses such as parainfluenza virus and respiratory syncytial virus are also possible causes of the common cold, although they primarily cause other diseases (croup and bronchiolitis, respectively).

Diagnosis

The common cold is usually diagnosed clinically. Erythematous and edematous nasal mucosa is seen on physical examination. Conjunctival and pharyngeal injection may also be seen. (Injection in this context means hyperemia of small blood vessels.)

Treatment

Generally, only symptomatic therapy is offered. It is controversial whether zinc salts may be helpful. Zinc acetate in doses greater than 75 mg/d may reduce the duration of symptoms. Other strategies include oral decongestants and buffered hypertonic saline nasal irrigation. If used for more than a few days, nasal sprays may be associated with rebound congestion after stopping.

There are no antiviral drugs useful against the common cold. Antibacterial drugs should *not* be prescribed for patients with the common cold.

Prevention

Although many vitamins and herbal therapies (e.g., echinacea) have been evaluated, there has been no conclusive evidence that any one therapy is helpful. Vitamin C taken prophylactically may be helpful in a population of cold weather athletes. However, when vitamin C was tested in the general population (rather than athletes), its ability to prevent colds was marginal. Handwashing may prevent the transmission of respiratory viruses. There is no vaccine against any virus that causes the common cold.

CROUP

Definition

Croup is an inflammation of the larynx, trachea, and large bronchi (laryngotracheobronchitis).

Clinical Manifestations

Inspiratory stridor is the key finding, together with a barking cough and a hoarse voice. Symptoms may begin in a subtle fashion with nasal irritation and congestion and then rapidly progress to stridor over a day.

Pathogens

Parainfluenza viruses, especially type 1, are the most common cause. Respiratory syncytial virus and influenza virus account for 1% to 10% of cases.

Diagnosis

The diagnosis is usually made clinically. Plain radiographs may show a "steeple sign" (subglottic tracheal narrowing results in an inverted "V" shape) (Figure 75–5).

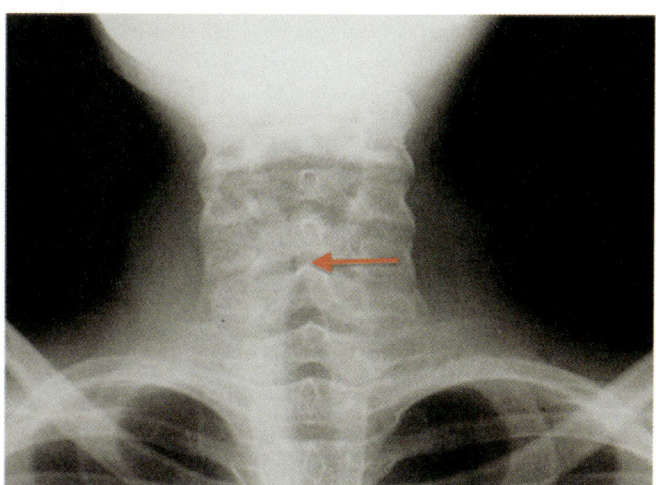

FIGURE 75–5 Croup. This X-ray shows the "steeple sign" of croup in a 12-year-old child. The red arrow points to the upper tip of the steeple. The steeple represents the constriction caused by an inflamed larynx and trachea. (Reproduced with permission from Stone CK, Humphries RL: *Current Diagnosis and Treatment of Emergency Medicine*, 7th ed. New York, NY: McGraw Hill; 2011.)

Treatment

Patients with moderate to severe symptoms may be given corticosteroids such as dexamethasone, with or without epinephrine. There is no antiviral drug therapy.

Prevention

There is no vaccine against parainfluenza virus.

LARYNGITIS

Definition

Laryngitis is inflammation of the vocal folds of the larynx.

Clinical Manifestations

Clinical manifestations include hoarseness and the inability to speak (aphonia). Laryngitis may be accompanied or preceded by an upper respiratory infection.

Pathogens

Parainfluenza viruses and rhinoviruses are the most common causes of laryngitis. Other respiratory viruses such as influenza virus, adenovirus, and coronavirus have been isolated from patients. Bacteria such as *S. pyogenes, M. catarrhalis,* and *H. influenzae* have also been isolated.

Diagnosis

The diagnosis of laryngitis is primarily made clinically.

Treatment

Treatment includes hydration and voice rest. Antibiotics are not needed.

Prevention

There is no vaccine against parainfluenza virus and rhinoviruses. There is no convincing evidence that the influenza virus vaccine and the *H. influenzae* type B vaccine have reduced the number of cases of laryngitis.

EPIGLOTTITIS

Definition

Epiglottitis is an inflammation of the epiglottis.

Clinical Manifestations

Patients present with rapidly worsening sore throat and odynophagia (pain on swallowing) or dysphagia (difficulty in swallowing). Pain may be out of proportion to physical examination findings. Airway obstruction can occur in severe cases. Epiglottitis in young children should be treated as a medical emergency.

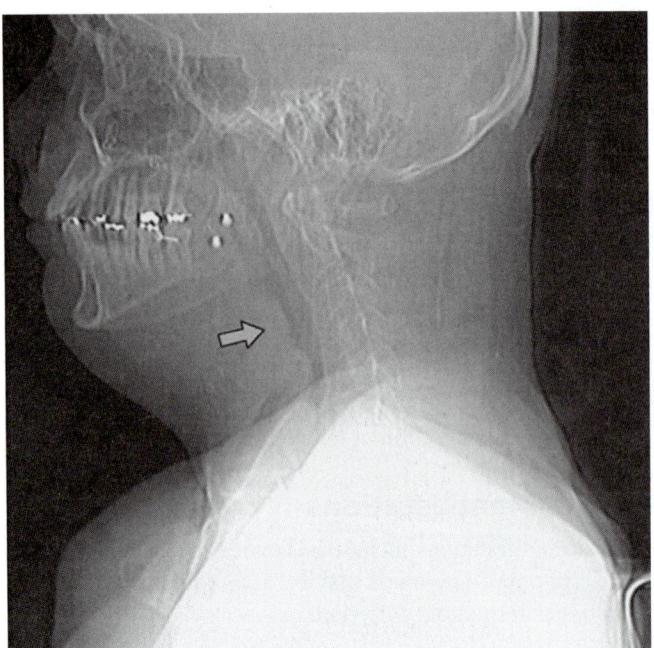

FIGURE 75-6 Epiglottitis. Note enlarged epiglottis (white arrow) in a lateral view of X-ray of neck. (Reproduced with permission from Longo DL, Fauci AS, Kasper DL, et al: *Harrison's Principles of Internal Medicine,* 18th ed. New York, NY: McGraw Hill; 2012.)

Pathogens

H. influenzae type B is, by far, the most common cause, although the widespread use of the vaccine against *H. influenzae* type B has greatly reduced the incidence of epiglottitis. Less common pathogens include other *H. influenzae* types, *S. pneumoniae*, *S. pyogenes*, and *S. aureus*.

Diagnosis

Diagnosis is made by visualization of the epiglottis. If indirect laryngoscopy (done primarily in children) is performed, a swollen and erythematous "cherry-red" epiglottis may be visualized. On lateral plain X-rays, an enlarged epiglottis may be seen as a "thumb" sign (Figure 75–6).

Treatment

Treatment involves intravenous ceftriaxone. Some centers add corticosteroids to reduce inflammation, but its effects are undocumented. An adequate airway must be maintained.

Prevention

Prevention includes immunization against *H. influenzae* type B and *S. pneumoniae*. Rifampin prophylaxis should be given to close household contacts to reduce oropharyngeal carriage.

Lower Respiratory Tract Infections

INTRODUCTION

Lower respiratory tract infections are an important cause of morbidity and mortality worldwide in children and in adults. Community-acquired pneumonia, for example, is the most deadly infectious disease in the United States. This chapter takes an anatomic approach to lower respiratory tract infections, moving from the large bronchi (bronchitis) down to the very small bronchioles (bronchiolitis) and then into the alveoli where pneumonia occurs.

BRONCHITIS

Definition

Bronchitis is an inflammation of the bronchi typically caused by viruses. Acute bronchitis must be distinguished from chronic bronchitis in which patients have a cough for more than 3 months. The information in this chapter refers to acute bronchitis.

Pathophysiology

The coughing so characteristic of bronchitis is an attempt to clear the mucus produced by the inflammatory response to viral infection. Bronchitis occurs more often in the winter months than in the summer. Smoking predisposes to bronchitis (and pneumonia) by damaging the cilia in the bronchi, leading to an inability to clear mucus from the respiratory tract.

Clinical Manifestations

Cough is the most prominent symptom of bronchitis. Initially, bronchitis presents with the symptoms of an upper respiratory infection, namely, nasal congestion, scratchy sore throat, and perhaps a low-grade fever. Physical examination typically reveals expiratory wheezes. However, if cough persists for more than 5 days and pneumonia has been ruled out, acute bronchitis should be suspected. Bronchitis is self-limited and usually resolves in 1 to 2 weeks. However, cough may persist for several more weeks due to airway hyperreactivity.

Pathogens

Respiratory viruses are the most common pathogens (influenza A and B, parainfluenza virus, coronavirus, rhinovirus, respiratory syncytial virus [RSV], and human metapneumovirus). Bacterial pathogens do not play a significant role in acute bronchitis. Acute exacerbations of chronic bronchitis may be caused by *Streptococcus pneumoniae* or *Haemophilus influenzae*.

Diagnosis

The diagnosis is primarily made clinically. Cough, with or without sputum production, which may persist for more than 5 days, is the typical presentation. Patients are usually afebrile but may have a low-grade fever. Sputum cultures are typically not done. In patients with chronic cardiorespiratory disease, a rapid antigen test for influenza virus may be useful because oseltamivir (Tamiflu) can shorten the duration and intensity of symptoms.

Because treatment of both upper respiratory infections and acute bronchitis is largely supportive, these distinctions may have less clinical significance. What may be more important clinically is to distinguish acute bronchitis (usually viral) from pneumonia (mainly bacterial; see section on Pneumonia), which does require antimicrobial therapy. A chest radiograph may be performed to determine whether pneumonia is present.

Treatment

Treatment involves patient education and symptom relief with agents such as over-the-counter cough medications, nonsteroidal anti-inflammatory drugs, and/or a bronchodilator such as albuterol (if wheezing or underlying lung disease is present). If influenza is diagnosed, oseltamivir (Tamiflu) may reduce the length and severity of symptoms. Antibiotics should generally *not* be prescribed. Use of antibiotics for acute bronchitis is a major contributor to antimicrobial misuse worldwide. Acute exacerbations of chronic bronchitis shown to be caused by *S. pneumoniae* or *H. influenzae* by laboratory tests can be treated with amoxicillin-clavulanate.

Prevention

Influenza vaccine can prevent bronchitis and pneumonia caused by influenza A and B viruses. The neuraminidase inhibitor oseltamivir (Tamiflu) should be given to unimmunized individuals with chronic cardiorespiratory disease. Handwashing is recommended to reduce the carriage of respiratory viruses.

BRONCHIOLITIS

Definition

Bronchiolitis is inflammation of the bronchioles—the small airways less than 2 mm in diameter. The focus in this section will be on bronchiolitis among infants and young children where the etiology is primarily infectious.

Pathophysiology

Particularly among children under 2 years of age, viruses can directly damage the epithelial cells of the terminal bronchioles, causing inflammation and obstruction of the small airways. Prematurity is an important predisposing factor.

Clinical Manifestations

Usually children initially have symptoms consistent with an upper respiratory tract infection and then are noticed to have increased respiratory distress. Children under 2 years old in particular may have tachypnea, wheezing, nasal flaring, and chest retractions. In severe cases, hypoxia, apnea, and respiratory failure may ensue. In most cases, recovery occurs in 1 to 2 weeks.

Pathogens

RSV is the most common pathogen. Other etiologies include influenza virus, parainfluenza virus, adenovirus, coronavirus, rhinovirus, and human metapneumovirus. In children, viruses are the main etiology of bronchiolitis. Bacteria are not thought to be involved. In adults, the causes are more varied and range from viruses, to inhaled toxic chemicals in the workplace, to idiopathic causes. Bronchiolitis caused by RSV occurs primarily in the winter months.

Diagnosis

The diagnosis is primarily clinical. Upper respiratory tract infection symptoms followed by lower respiratory tract symptoms and signs (e.g., nasal flaring, wheezing) in a young child during the fall and winter would be very suggestive of bronchiolitis. Chest radiograph typically shows hyperinflation of the lungs. A polymerase chain reaction (PCR) assay that detects the RNA of RSV in respiratory secretions is sometimes used in hospitalized patients. An enzyme immunoassay (EIA) for RSV antigen is also available but molecular assays are more sensitive.

Treatment

Because this is typically a self-limited disease, general supportive measures are adequate in most cases. Patients with moderate or severe respiratory distress will require hospitalization, including supplemental oxygen. Ribavirin, delivered by aerosol into the lungs, is approved for severe disease caused by RSV, but its use is limited to hospitalized infants. Inhaled bronchodilators (e.g., albuterol) may be useful. Antibacterial drugs and systemic glucocorticoids are not recommended.

Prevention

Handwashing to minimize transmission of pathogens is an important strategy. Infection control procedures should be instituted in hospitalized patients to prevent the spread of viruses to others.

Palivizumab is a humanized monoclonal antibody against the RSV F (fusion) envelope protein that may be used in certain populations to decrease the risk of disease progression caused by RSV. These populations include children with bronchopulmonary dysplasia and congenital heart disease and prematurely born infants. An annual influenza vaccine for everyone older than 6 months of age is recommended. There is no viral vaccine against RSV.

PNEUMONIA

Definition

Pneumonia is an inflammation of the lung affecting the alveoli. We consider whether pneumonia is community acquired versus hospital acquired to help us determine the spectrum of potential pathogens that differs based on setting. More importantly, because empiric therapy is often given in pneumonia, therapeutic interventions differ based on the different populations.

The focus in this section will be on community-acquired pneumonia, which is a leading cause of death both in the United States and worldwide. Hospital-acquired pneumonia, also known as nosocomial pneumonia, is pneumonia that occurs 48 hours or more after admission to the hospital and was not present at the time of admission. The term "healthcare-associated" pneumonia is also used.

Pathophysiology

The alveoli of the lungs are continually exposed to microbes from the environment via the upper respiratory tract. Our host defenses usually keep these potential pathogens in check. However, disease can occur when there is a particularly virulent organism, when there is a large burden of organisms inhaled from the environment or aspirated from the oropharynx, or when there is a defect in host immunity.

Predisposing factors to pneumonia include the extremes of age (the very young and very old), chronic obstructive pulmonary disease (COPD) and chronic bronchitis, diabetes mellitus, cystic fibrosis, and congestive heart failure. Injection drug users who overdose, alcoholics, and those with seizure disorders have a high risk of pneumonia because they can aspirate organisms into the lung when unconscious, resulting in aspiration pneumonia. People exposed to water aerosols, especially from air conditioners, are at risk for pneumonia caused by *Legionella*. Hospitalized patients in the intensive care unit are at risk for ventilator-associated pneumonia caused by gram-negative rods such as *Escherichia coli*, *Pseudomonas*, and *Acinetobacter*.

Clinical Manifestations

Symptoms include cough that may be productive of sputum, fever, chills, chest pain, and shortness of breath. "Rusty" (blood-streaked) sputum is a well-known finding in pneumococcal pneumonia. Sputum that has a "currant jelly" appearance can occur in pneumonia caused by *Klebsiella* because the organism is heavily encapsulated. Physical examination findings include tachypnea, rales, and rhonchi. If the lung is consolidated, dullness to percussion may be detected.

Patients who are intubated and who acquire a nosocomial pneumonia may only have fever as a presenting sign, which may be accompanied by increased respiratory secretions or increased oxygen requirements. Pneumonia may be complicated by an infected pleural effusion or a pleural empyema. A pleural empyema is a walled-off collection of pus in the pleural space.

Pathogens

S. pneumoniae is the most common cause of community-acquired pneumonia. Approximately 15% to 30% of isolates of *S. pneumoniae* are resistant to one or more commonly used antibiotics.

Other common bacterial pathogens include *Klebsiella pneumoniae* and *H. influenzae*. Note that it is the nontypeable strains of *H. influenzae* rather than the type B strain that cause pneumonia in elderly patients with COPD. Community-acquired, methicillin-resistant *Staphylococcus aureus* (CA-MRSA) is increasingly found.

Mycoplasma pneumoniae, *Legionella* species, and *Chlamydophila pneumoniae* are other pathogens commonly involved. Infection with *Mycobacterium tuberculosis* can also manifest as a pneumonia. In approximately 30% of adults with community-acquired pneumonia, no pathogen, neither bacteria nor virus, is isolated.

Table 76–1 shows the important causes of community-acquired pneumonia as a function of age. Note that the causes of pneumonia in a neonate are those acquired during passage through the birth canal. The main cause of pneumonia in an infant, *Chlamydia trachomatis*, is also acquired during passage through the birth canal but is a less aggressive pathogen so its onset in delayed. Note that *M. pneumoniae* is the most common cause in young adults.

TABLE 76–1 Important Bacterial and Viral Causes of Community-Acquired Pneumonia by Age (Listed in Order of Frequency)

Age	Bacteria	Viruses
Neonates	Group B streptococci *Escherichia coli*	Respiratory syncytial virus (RSV)
Infants	*Chlamydia trachomatis* *Streptococcus pneumoniae*	RSV Parainfluenza virus
Children	*S. pneumoniae* *Haemophilus influenzae*	RSV Parainfluenza virus
Young adults	*Mycoplasma pneumoniae* *Chlamydophila pneumoniae* *S. pneumoniae*	Various respiratory viruses (e.g., adenovirus)
Older adults	*S. pneumoniae* *H. influenzae* *Legionella pneumophila*	Influenza virus SARS-CoV-2[1]

[1]During the pandemic that began in 2019, SARS-CoV-2, the cause of COVID-19, caused more pneumonia than did influenza virus.

Table 76–2 shows the typical causes of community-acquired pneumonia as a function of various predisposing factors. In certain patient populations, *Pseudomonas aeruginosa* and other gram-negative organisms and *S. aureus* may be important pathogens causing pneumonia. For example, *P. aeruginosa*,

TABLE 76–2 Predisposing Factors Associated with Typical Pathogens Causing Community-Acquired Pneumonia

Predisposing Factors	Typical Pathogens
Alcoholism	*Klebsiella pneumoniae*, oral anaerobes
Bird exposure, especially psittacine birds such as parrots (psittacosis)	*Chlamydophila psittaci*
Chronic obstructive pulmonary disease (COPD), including smoking related	*Haemophilus influenzae*
Cystic fibrosis	*Pseudomonas aeruginosa*
Imported wool, spores in wool (woolsorter's diseases)	*Bacillus anthracis*
Influenza virus infection	*Staphylococcus aureus*
Intubation, postsurgery, and intensive care unit (ICU)	Coliforms,[1] *P. aeruginosa*, *S. aureus*
Mouse droppings exposure, especially in southwestern states	Hantavirus
Obesity	SARS CoV-2
Sheep exposure, especially placental tissue (Q fever)	*Coxiella burnetii*
Travel to or reside in Central Valley of California, Arizona, or New Mexico	*Coccidioides immitis*
Travel to or reside in Ohio or Mississippi river valleys	*Histoplasma capsulatum*
Ventilator associated, especially in ICU	*Acinetobacter* species
Water aerosols, especially from air conditioners	*Legionella pneumophila*

[1]Coliforms such as *Escherichia coli*, *Klebsiella*, *Enterobacter*, *Serratia*, and *Proteus*.

Stenotrophomonas, and *Burkholderia* cause pneumonia in cystic fibrosis patients, and *S. aureus* is a well-recognized cause of pneumonia in patients with influenza. Oral anaerobes are often involved in aspiration pneumonia.

People exposed to certain animals have an increased risk of pneumonia; for example, those exposed to psittacine birds such as parrots are at risk for psittacosis caused by *Chlamydophila psittaci* and those exposed to the placentas of pregnant sheep are at risk for Q fever caused by *Coxiella burnetii*. People exposed to the spores of the anthrax bacillus in sheep wool may get "wool-sorter's disease," a pneumonia caused by *Bacillus anthracis*.

Common pathogens for hospital-acquired pneumonia include gram-negative rods such as *E. coli, K. pneumoniae, P. aeruginosa, Enterobacter* species, *Serratia marcescens, Acinetobacter* species, and gram-positive cocci, especially *S. aureus*.

In most years, the most common viral cause of pneumonia is influenza virus. However, beginning in late 2019, the novel coronavirus, SARS-CoV-2, emerged to cause a world-wide pandemic. The name of the disease is COVID-19 which stands for COronaVIrus Disease-2019. For detailed information on this virus, COVID-19, and the pandemic, please see Chapter 38 on Respiratory Viruses.

Other viral pathogens such as RSV, parainfluenza virus, adenovirus, and human metapneumovirus, also cause pneumonia. In patients with reduced cell-mediated immunity, herpesviruses, such as herpes simplex virus, varicella-zoster virus, and cytomegalovirus, can cause life-threatening pneumonia. In certain geographical areas, such as the rural southwestern part of the United States, outbreaks of pneumonia caused by hantavirus have occurred.

Fungi such as *Coccidioides* and *Histoplasma* also cause pneumonia. *Pneumocystis jiroveci* causes pneumonia, especially in patients with acquired immunodeficiency syndrome (AIDS) with low CD4 counts.

Diagnosis

The "gold standard" for a diagnosis of pneumonia is an infiltrate on a plain chest radiograph (Figure 76–1). Clinical data may help, but ultimately the chest radiograph is the most important diagnostic tool. Sputum analysis for Gram stain and culture and blood cultures may be helpful in the hospitalized patient but are only optional in an outpatient setting because therapy is largely empiric for community-acquired pneumonia. If sputum cultures and blood cultures are indicated, then these specimens should be obtained before antibiotics are started.

In pneumonia caused by one of the encapsulated pyogenic bacteria, such as *S. pneumoniae*, the white blood cell count is frequently elevated and the number of neutrophils is often increased. The urinary antigen test for pneumococcal polysaccharide is also useful.

It is important that sputum (*not saliva*) be sent to the lab for Gram stain and culture. If the specimen contains many neutrophils and few epithelial cells, then the specimen is likely to be sputum and will be analyzed. If, however, the specimen contains many epithelial cells and few neutrophils, then the specimen is saliva and will be rejected by the lab. An "induced sputum"

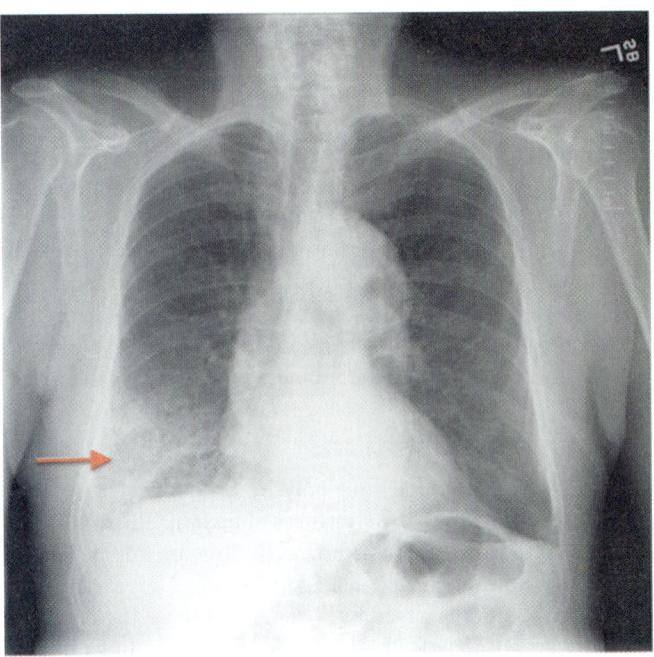

FIGURE 76–1 Lobar pneumonia caused by *Streptococcus pneumoniae*. Arrow points to area of consolidation in right lung. (Reproduced with permission from McKean SC, Ross JJ, Dressler DD, et al: *Principles and Practice of Hospital Medicine*. New York, NY: McGraw Hill; 2012.)

sample produced by using nebulized hypertonic saline increases the probability of obtaining a good sputum sample.

Pneumonia caused by *M. tuberculosis* is diagnosed by acid-fast stain of sputum and culture on mycobacterial medium. A PCR assay done directly on sputum is available also. Pneumonia caused by *Legionella pneumophila* is often diagnosed by urinary antigen or PCR. PCR tests for various respiratory pathogens such as *M. pneumoniae, C. pneumoniae*, influenza virus, SARS CoV-2, and RSV are also useful. The cold agglutinin test is no longer recommended for the diagnosis of pneumonia caused by *Mycoplasma*.

Treatment

Treatment for community-acquired bacterial pneumonia is largely empiric because microbiologic diagnostic strategies are generally insensitive. Outpatients are typically treated with a macrolide such as azithromycin, a tetracycline such as doxycycline, or a respiratory quinolone such as levofloxacin. Amoxicillin-clavulanate with or without a macrolide can also be used. Lefamulin can also be used.

Hospitalized patients with community-acquired pneumonia are often treated with ceftriaxone plus a macrolide or respiratory quinolone monotherapy. If aspiration pneumonia is suspected, metronidazole or clindamycin can be added.

If influenza is occurring in the community, then oseltamivir should be prescribed. Note that influenza can predispose to bacterial pneumonia, especially caused by *S. aureus* and *S. pneumoniae*, so antibiotics may be needed even in a person with established influenza. As of this writing, the main anti-viral drug approved for COVID-19 is remdesivir (see Chapter 38).

Approved treatment for the cytokine storm phase of COVID-19 consists of methylprednisolone and tocilizumab.

Patients with suspected hospital-acquired pneumonia may be given broader spectrum agents such as a carbapenem (e.g., ertapenem), depending on the local epidemiology, given that many hospital-acquired infections are multidrug resistant. A combination of ceftriaxone and azithromycin can also be used. Prompt initiation of antibiotics is important because morbidity and mortality increase after a delay of more than 8 hours. Drainage of an empyema or infected pleural fluid should be performed.

Prevention

The pneumococcal protein-conjugate vaccine is generally recommended for infants and children, and the pneumococcal polysaccharide vaccine for adults at intermediate risk of pneumococcal disease (e.g., cigarette smokers). The pneumococcal polysaccharide and protein-conjugate vaccines are *both* recommended for older and select high-risk adults (e.g., immunocompromised patients). The conjugate vaccine against *H. influenzae* type B is *not* an important source of protection, because it is the nontypeable strains of *H. influenzae*, rather than the type B strain, that are the most common cause of pneumonia. Smoking cessation and treatment of alcohol abuse may also decrease pneumonia risk.

The influenza vaccine is effective in decreasing the likelihood of pneumonia. As of this writing, three vaccines have been approved to prevent COVID-19. Two are m-RNA vaccines and one is a recombinant vaccine containing a human adenovirus carrying the gene for the spike protein of SARS-CoV-2 (see Chapter 38 for more information).

LUNG ABSCESS

Definition

Lung abscess is a necrotic process within the lung parenchyma that frequently results in a cavity with an air-fluid level.

Pathophysiology

Patients may aspirate oropharyngeal bacteria into the lower airways and alveoli. This usually occurs when the patient is in the recumbent position and cannot clear secretions. For example, aspiration can occur when a person is unconscious from drug overdose, excess alcohol intake, or the anesthesia that accompanies surgery. Poor oral hygiene is a common predisposing factor. A pneumonitis may first occur, but this can progress to necrosis in a week or so. A lung abscess caused by *S. aureus* may infect the lung via the bloodstream from a distant site of infection such as right-sided endocarditis in an intravenous drug user.

Clinical Manifestations

Patients present with symptoms typical of pneumonia with fever and productive cough. The sputum is often foul smelling, indicating the presence of anaerobes. These symptoms may be indolent and progress over a period of weeks. Systemic symptoms such as night sweats, fatigue, and weight loss may also be present.

Pathogens

The most common organisms are anaerobes or mixed aerobes and anaerobes that are part of the oral flora. Anaerobes commonly involved include *Peptostreptococcus* species, *Prevotella* species, and *Fusobacterium nucleatum*. Aerobes include *Streptococcus milleri* and *S. aureus*. (Clinicians often use the term "aerobe" rather than facultative to describe bacteria that are not anaerobic.)

Diagnosis

A chest radiograph shows a pulmonary infiltrate with a cavity, often with an air-fluid level (Figure 76–2). An air-fluid level occurs when the abscess erodes a bronchus and some of the pus in the abscess is coughed up and replaced by air. Pleural fluid, if present, and blood cultures may provide microbiologic data, but anaerobes may be difficult to identify.

Treatment

Clindamycin and ampicillin-sulbactam are typical treatment options. Duration of therapy is usually 4 to 6 weeks. Patients who do not respond to antibiotics will require surgical drainage.

Prevention

There is no vaccine against the organisms that cause lung abscess. Preventive measures include good dental hygiene and

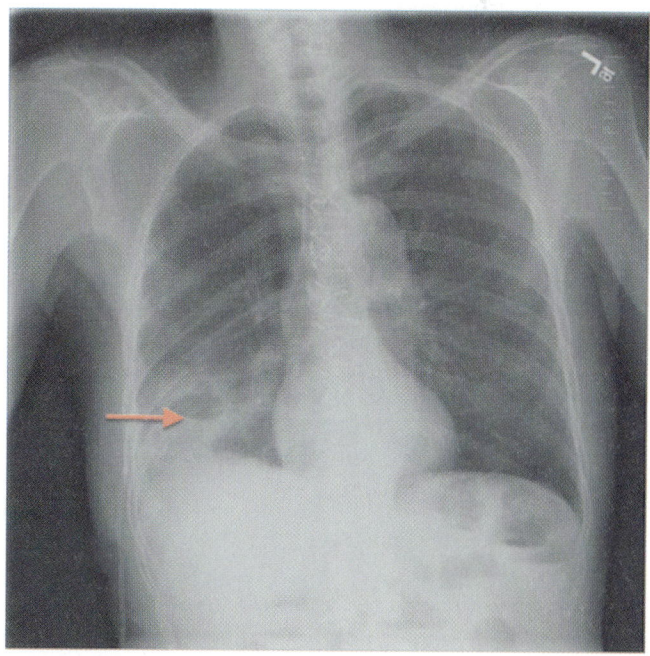

FIGURE 76–2 Lung abscess. Arrow points to air-fluid interface within the abscess. (Reproduced with permission from McKean SC, Ross JJ, Dressler DD, et al: *Principles and Practice of Hospital Medicine.* New York, NY: McGraw Hill; 2012.)

avoidance of unconsciousness caused by drug overdose and alcohol abuse.

LUNG EMPYEMA

Empyema of the lung is a collection of pus in the pleural space. It should be distinguished from a pleural effusion, which is a transudate, whereas an empyema is an exudate containing many neutrophils.

Pneumonia, including aspiration pneumonia, is the most common predisposing factor to empyema. Thoracotomy and trauma account for most of the remaining cases. Diabetes is a common comorbidity associated with empyema.

S. pneumoniae is the most common cause of empyema; however, members of the *Streptococcus anginosus* group are often found. In hospitalized patients, *S. aureus* is a common cause.

Enteric gram-negative rods such as *K. pneumoniae* also cause empyema in hospitalized patients. Anaerobic bacteria such as *Prevotella, Fusobacterium*, and *Peptostreptococcus* are often isolated from empyema fluid associated with aspiration pneumonia.

The clinical findings are those of pneumonia such as fever, cough, and chest pain that do not resolve using appropriate antibiotics. Weight loss and night sweats may occur. On physical exam, decreased breath sounds, dullness to percussion, and splinting on deep breathing may be observed. Ultrasound and chest computed tomography (CT) scan are used to reveal the localized collection of pus.

Microbiologic diagnosis involves aspirating pus from the empyema and performing a Gram stain and culture. Treatment involves surgical drainage of the pus combined with antibiotics appropriate for the bacteria isolated from the aspirated pus.

Infections of the Skin & Skin Structures

INTRODUCTION

Infections of the skin and skin structures are some of the most common infectious diagnoses and result in hundreds of thousands of medical office and emergency room visits each year. These infections often occur following a break in normal skin integrity from either trauma or skin disease (e.g., atopic dermatitis). The vast majority of these infections are caused by *Staphylococcus aureus* and *Streptococcus pyogenes* (Table 77–1). Rarely, infections in patients with burns, diabetes mellitus, or decubitus ulcers may involve gram-negative rods such as *Pseudomonas* or anaerobes. Hematogenous seeding of organisms into the skin can occur but is uncommon. Normal histology of the skin can be seen in Figure 77–1.

IMPETIGO

Definition

Impetigo is an infection of the epidermal layer of skin.

Pathophysiology

There are two modes of acquisition of impetigo, primary infection, which occurs in otherwise normal skin, or secondary impetigo, which occurs following a break in normal skin integrity. Bacteria invade into the epidermal layer and cause local damage. Bullous impetigo occurs when strains of *S. aureus* secrete exfoliative toxin, a protease that degrades desmoglein, resulting in loss of adhesion of the superficial epidermis. This is the same toxin that causes staphylococcal scalded skin syndrome.

Clinical Manifestations

There are three clinical variants of impetigo: (1) classic impetigo, (2) bullous impetigo, and (3) ecthyma. Classic impetigo begins as papules that progress to vesicles surrounded by erythema.

Subsequently, the fluid-filled lesions enlarge and break down to form thick, adherent crusts with a characteristic golden "honey-colored" appearance (Figures 77–2A and 2B). Bullous impetigo is similar to classic impetigo, but bullae form (Figures 77–3A and 3B) via the mechanism described earlier. Ecthyma is an ulcerating form of impetigo where the lesion penetrates through the epidermis into the dermis (Figure 77–4). Some strains of *S. pyogenes* that cause impetigo have been associated with post-streptococcal glomerulonephritis, and providers should be aware of this potential complication. Rheumatic fever is a less common sequel to streptococcal skin infections.

Pathogens

S. aureus and *S. pyogenes* are the two main pathogens that cause impetigo. In neutropenic patients, a clinical syndrome termed ecthyma gangrenosum is due to disseminated *Pseudomonas aeruginosa* infection. Its cutaneous findings are a result of hematogenous seeding of dermal vessels with bacteria, resulting in thrombosis, ischemia, and focal skin necrosis. This is not a superficial skin infection.

Diagnosis

The diagnosis of impetigo is made clinically in most cases. Culture of bullous fluid or pus can be considered when patients do not respond to standard treatment.

Treatment

Antibacterial therapy should be directed against both *S. aureus* and *S. pyogenes*. Topical therapy with mupirocin or retapamulin is preferred when only a few lesions are present. Ozenoxacin cream can also be used. In patients with widespread disease, a systemic antimicrobial is preferred. If concern for methicillin-resistant *S. aureus* (MRSA) exists, clindamycin is recommended; otherwise, cephalexin or dicloxacillin would be appropriate.

TABLE 77–1 **Skin and Skin Structures Infections: Appearance of Lesions, Skin Layers Involved, Common Pathogens, and Treatment Modalities**

Type of Infection	Appearance of Lesion	Description of Lesion	Layer of Skin Involved	Common Pathogens	Treatment
Impetigo		Vesicles with honey-colored crust, often on the face of a child	Epidermis	*Staphylococcus aureus, Streptococcus pyogenes*	Few lesions: topical antibiotics (e.g., mupirocin); numerous lesions: systemic therapy (e.g., cephalexin, clindamycin)
Erysipelas		Erythematous, very painful lesion with sharply demarcated, raised, regular border	Superficial dermis	*S. pyogenes, Streptococcus agalactiae > S. aureus*	Systemic antibiotics (e.g., cephalexin or cefazolin)
Cellulitis		Erythematous, diffuse, flat lesion with irregular border	Deep dermis	*S. pyogenes, S. agalactiae > S. aureus*	Systemic antibiotics (e.g., cephalexin or cefazolin)
Folliculitis		Localized, inflamed papules containing a small amount of pus	Hair follicle	*S. aureus, Pseudomonas aeruginosa* (associated with hot tubs)	Antibiotics often not needed; warm, moist compresses are useful
Skin abscess (also known as a boil, furuncle, carbuncle)		Raised, tender, inflamed nodule with central region of purulence; the area of pus initially is firm but then progresses to fluctuance (becomes movable)	Deep dermis	*S. aureus*	Incision and drainage is mainstay of therapy; antibiotics directed against *S. aureus* in select cases
Necrotizing fasciitis		Very painful area of inflammation with rapid progression to necrosis, bullae, purpura, anesthesia, and systemic toxicity	Fascia and muscle; local blood vessels and nerves also involved	Monomicrobial form: *S. pyogenes, Clostridium perfringens, Vibrio vulnificus* Polymicrobial form: enteric gram-negative rods plus anaerobes	Surgical debridement is critical in addition to broad-spectrum systemic antibiotics

Prevention

Handwashing and covering draining lesions should be used to prevent spread of bacteria.

CELLULITIS/ERYSIPELAS

Definition

Cellulitis and erysipelas are infections of the dermis.

Pathophysiology

These infections occur following a break in normal skin integrity. Cellulitis and erysipelas both involve the dermis, but erysipelas involves the upper dermis and superficial lymphatics, whereas cellulitis involves the deeper dermis and subcutaneous fat.

Clinical Manifestations

Cellulitis and erysipelas both manifest with erythema, swelling, and pain in the affected region plus or minus fever. However,

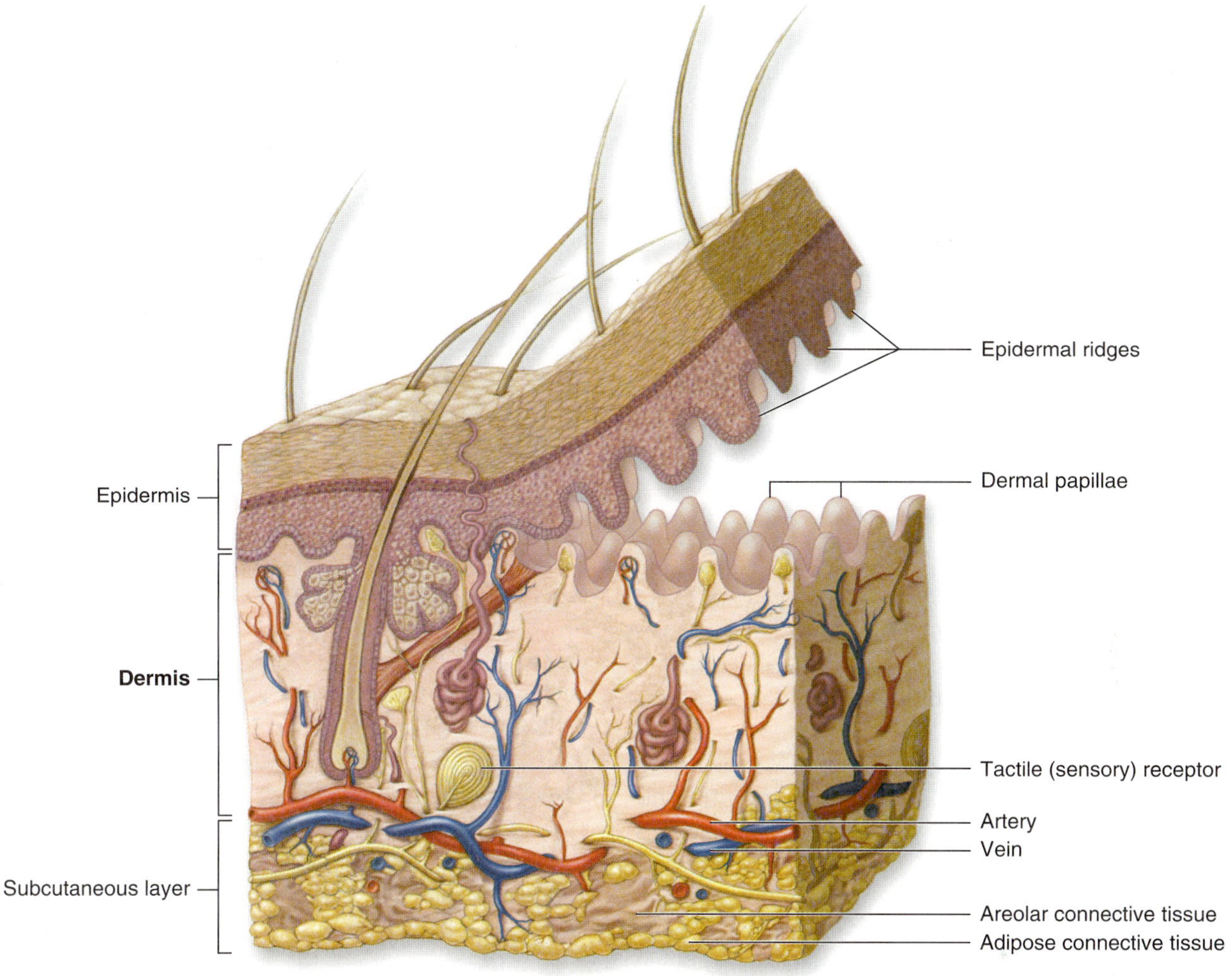

Epidermal ridges

Dermal papillae

Epidermis

Dermis

Tactile (sensory) receptor

Artery

Vein

Subcutaneous layer

Areolar connective tissue

Adipose connective tissue

FIGURE 77-1 Layers of the skin and subcutaneous tissue. Note sebaceous glands (purple coils) and hair follicles with protruding hair. (Reproduced with permission from Mescher AL: *Junqueira's Basic Histology: Text & Atlas*, 13th ed. New York, NY: McGraw Hill; 2013.)

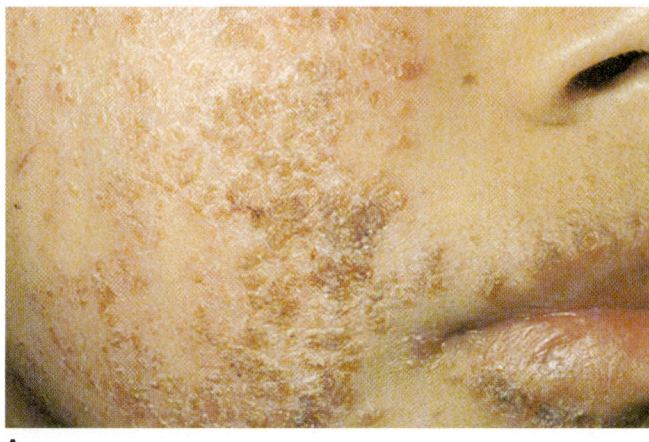

A

FIGURE 77-2A Nonbullous impetigo. Note lesions with a "honey-colored" crust on right side of face. (Reproduced with permission from Kelly AP, Taylor SC, Lom HW, et al: *Taylor and Kelly's Dermatology for Skin of Color*, 2nd ed. New York, NY: McGraw Hill; 2016.)

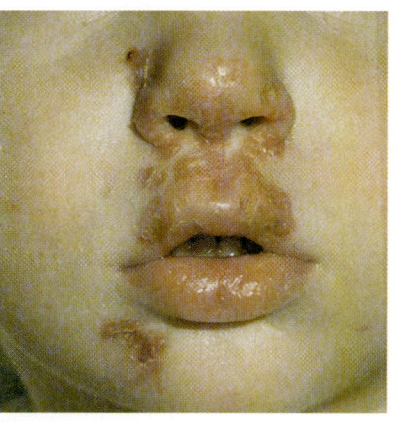

B

FIGURE 77-2B Nonbullous impetigo in young child. Note lesions with a "honey-colored" crust around the nose and mouth. (Reproduced with permission from Wolff K, Johnson R: *Fitzpatrick's Color Atlas & Synopsis of Clinical Dermatology*, 6th ed. New York, NY: McGraw Hill; 2009.)

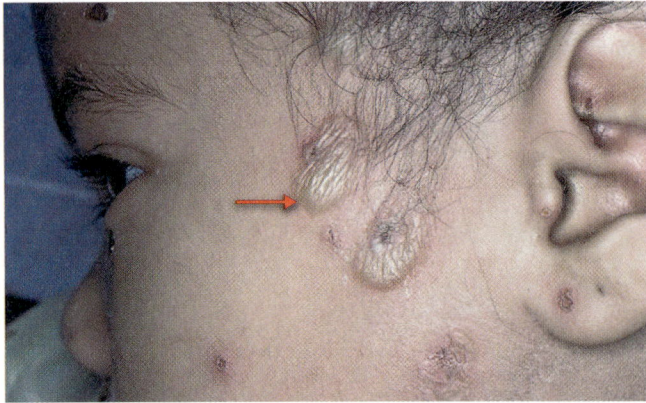

A

FIGURE 77–3A Bullous impetigo. Note bullous lesion (arrow). (Reproduced with permission from Ma OJ, Cline DM, Tintinalli JE, et al: *Emergency Medicine Manual*, 6th ed. New York, NY: McGraw Hill; 2004.)

erysipelas lesions are raised above the level of surrounding skin, and there is a clear line of demarcation between involved and uninvolved tissue (Figure 77–5). In contrast, the lesions of cellulitis are not significantly raised and have an irregular line of demarcation (Figures 77–6A and 6B).

Pathogens

Beta-hemolytic streptococci are the most common pathogens to cause these infections, with *S. pyogenes* and *Streptococcus*

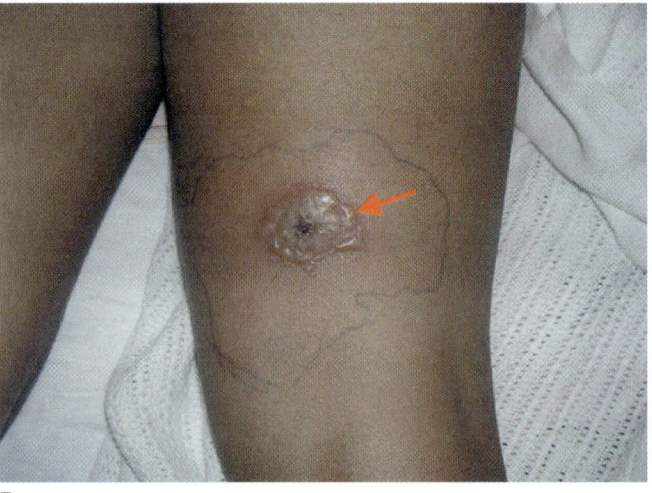

B

FIGURE 77–3B Bullous impetigo. Note bullous lesion (arrow). (Reproduced with permission from Studdiford J, Stonehouse A. Bullous eruption on the posterior thigh, *J Fam Pract* 2005 Dec;54(12):1041-1044.)

agalactiae being the most common species. *S. aureus* is not a common cause of these types of infections. Other less common pathogens are listed in Table 77–2 by exposure risk.

Diagnosis

Diagnosis is made clinically because it is difficult to obtain cultures from the skin in the absence of pus. Sometimes patients will have bacteremia.

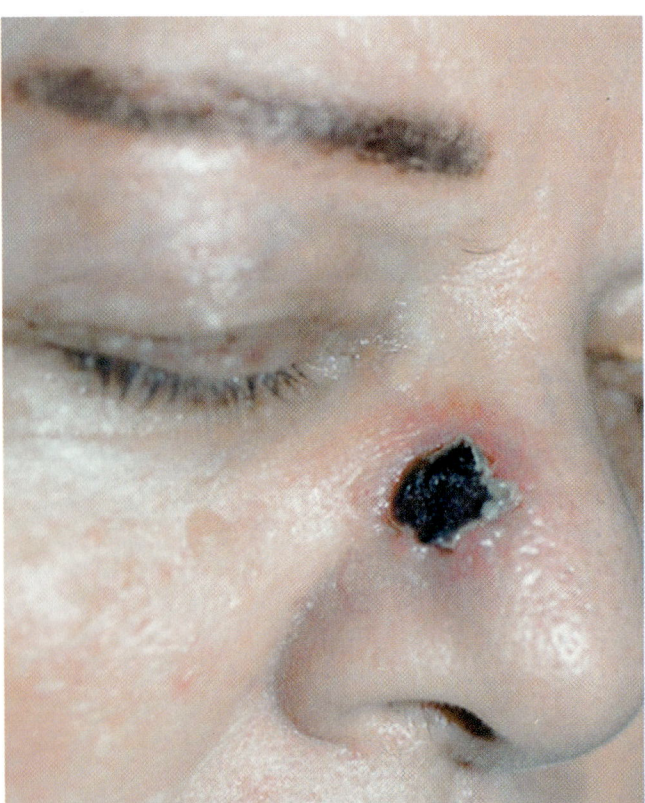

FIGURE 77–4 Ecthyma. Note necrotic lesion on the nose. (Reproduced with permission from Wolff K, Johnson R: *Fitzpatrick's Color Atlas & Synopsis of Clinical Dermatology*, 6th ed. New York, NY: McGraw Hill; 2009.)

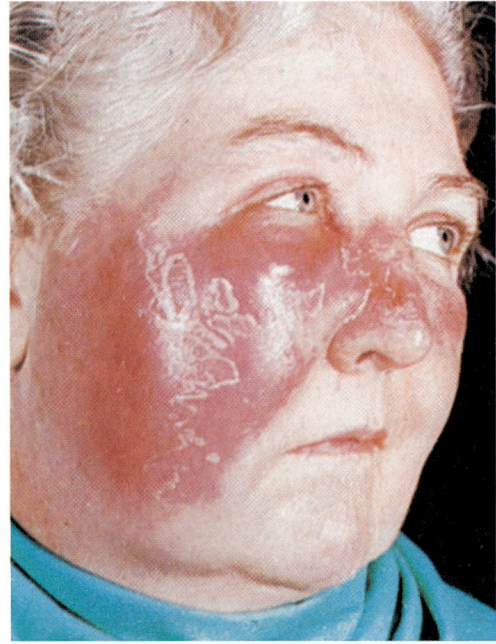

FIGURE 77–5 Erysipelas. Note markedly inflamed lesion with clearly demarcated border on right cheek, across the nose to the left cheek. (Reproduced with permission from Longo DL, Fauci AS, Kasper DL, et al: *Harrison's Principles of Internal Medicine*, 18th ed. New York, NY: McGraw Hill; 2012.)

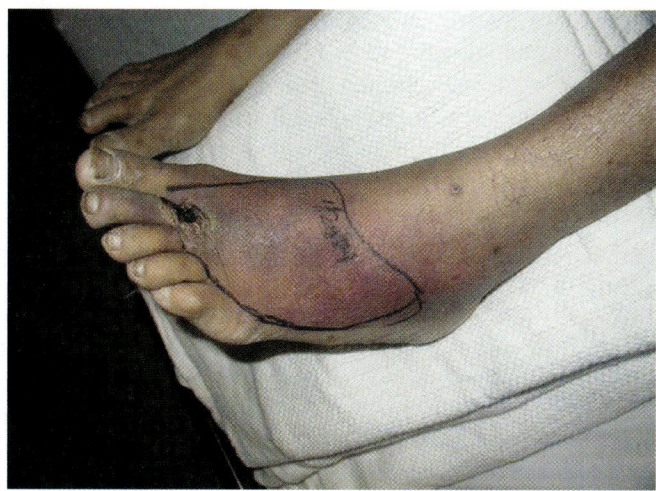

A

FIGURE 77–6A Cellulitis. Note diffuse inflammation on dorsum of left foot. (Reproduced with permission from Usatine RP, Smith MA, Mayeaux EJ, et al: *The Color Atlas and Synopsis of Family Medicine*, 3rd ed. New York, NY: McGraw Hill; 2019. Photo contributor: Richard P. Usatine, MD.)

Treatment

Empiric therapy should be focused on beta-hemolytic streptococci. A systemic oral agent can be used for mild infection (cephalexin, dicloxacillin, or clindamycin), but for severe infection, hospitalization and the administration of intravenous antibiotics (cefazolin or vancomycin) are recommended.

TABLE 77–2   Causes of Skin and Skin Structures Infections and Their Associated Risk Factors

Risk Factor	Pathogen
Animal bite (cats and dogs)	*Pasteurella multocida; Capnocytophaga canimorsus*
Human bite	*Eikenella corrodens*
Contact with fish, crabs	*Erysipelothrix rhusiopathiae*
Exposure to freshwater	*Aeromonas hydrophila*
Exposure to brackish or saltwater	*Vibrio vulnificus*
Exposure to unchlorinated water in hot tub	*Pseudomonas aeruginosa*
Intravenous drug use	*Staphylococcus aureus*, enteric gram-negative rods such as *Serratia* and *Pseudomonas*, *Clostridium botulinum*
Exposure to soil caused by military trauma or vehicle accidents	*Clostridium perfringens*
Surgery	*S. aureus, Streptococcus pyogenes*
Young children	*Haemophilus influenzae* type B
Severe burn wounds	*P. aeruginosa*
Decubitus ulcers and diabetic foot ulcers	*S. aureus*; enteric gram-negative rods, anaerobes (often polymicrobial)

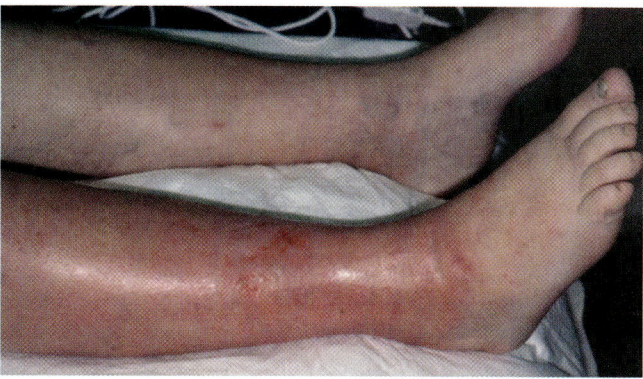

B

FIGURE 77–6B Cellulitis. Note diffuse inflammation on right leg including the skin over the ankle and dorsum of the foot. (Reproduced with permission from Knoop KJ, Stack L, Storrow A, et al: *Atlas of Emergency Medicine*, 3rd ed. New York, NY: McGraw Hill; 2009. Photo contributor: Lawrence B. Stack, MD.)

Prevention

In patients with recurrent cellulitis, a strategy of chronic suppressive antibiotics may effectively prevent subsequent infections.

FOLLICULITIS

Definition

Folliculitis is a superficial infection of the hair follicles.

Pathophysiology

Bacteria and purulent material accumulate in hair follicles in the epidermal layer of the skin.

Clinical Manifestations

Folliculitis presents with pinpoint erythema around individual hair follicles. A small amount of purulence may be seen (Figure 77–7). This can be seen in an isolated body area or throughout the skin.

Pathogens

S. aureus is the most common cause of folliculitis. *P. aeruginosa* is associated with folliculitis following the use of unchlorinated hot tubs. Rarely, *Candida* and certain dermatophytes can cause folliculitis.

Diagnosis

Diagnosis is made clinically, but if purulent material is present, it can be cultured.

Treatment

Folliculitis often resolves on its own, and treatment is not needed. Warm compresses or topical antibiotics can be considered in select cases.

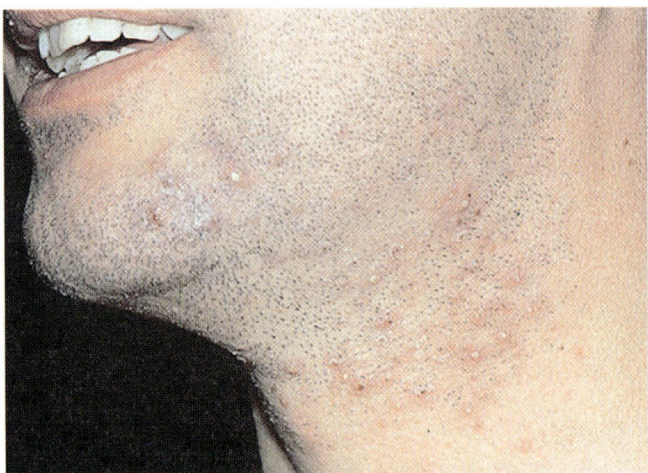

FIGURE 77–7 Folliculitis. Note the multiple, small pustules on the chin and neck. (Reproduced with permission from Wolff K, Goldsmith LA, Katz SI, et al: *Fitzpatrick's Dermatology in General Medicine*, 7th ed. New York, NY: McGraw Hill; 2008.)

Prevention

Handwashing and covering draining lesions should be used to prevent spread of bacteria. Avoiding unchlorinated hot tubs is recommended.

SKIN ABSCESS (FURUNCLE & CARBUNCLE)

Definition

A skin abscess is an infection of the dermis and deeper layers of skin that contains purulent material.

Pathophysiology

Abscesses occur when pathogens enter a break in the skin following trauma or when they spread from infected hair follicles (Figures 77–8A and 8B). When a single follicle is infected and tracks down into the dermis, it is termed a furuncle ("boil"), and when multiple infected hair follicles coalesce, it is termed a carbuncle. Occasionally, an abscess may develop following hematogenous dissemination of an infection.

Clinical Manifestations

A furuncle consists of a central pustule usually surrounded by an area of erythema, warmth, and tenderness with underlying fluctuance. Patients may have multiple furuncles. A carbuncle is a larger, more serious lesion than a furuncle. It is composed of several adjacent furuncles that have coalesced into an inflamed, indurated lesion that typically extends deep into subcutaneous tissue (Figure 77–9). Patients may have signs and symptoms of systemic infection, and this should alert the provider that more severe disease exists.

Pathogens

S. aureus is, by far, the most common cause of skin abscesses (more than 75% of cases). Beta-hemolytic streptococci are also capable of causing these types of infections. Rarely *Mycobacterium tuberculosis*, nontuberculous mycobacteria, and fungi such as *Coccidioides, Candida,* and *Cryptococcus* can cause abscesses.

Diagnosis

Gram stain and culture of purulent material obtained from the abscess allow for diagnosis. Radiographic imaging such as ultrasound or computed tomography (CT) may help further define the size and extent of an abscess.

Treatment

The primary treatment for abscesses is incision and drainage and is highly effective alone. However, a short course (≤7 days) of active antibiotics increased cure rates and can help reduce

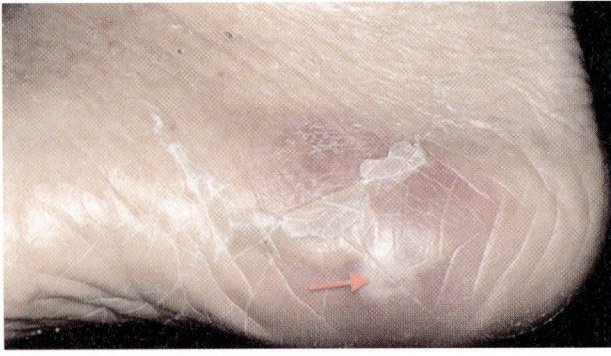

A

FIGURE 77–8A Abscess. Note localized area of inflammation containing a central core of yellowish pus (arrow) on medial aspect of foot. This lesion occurred at the site of a sewing needle injury. (Reproduced with permission from Wolff K, Johnson R: *Fitzpatrick's Color Atlas & Synopsis of Clinical Dermatology*, 6th ed. New York, NY: McGraw Hill; 2009)

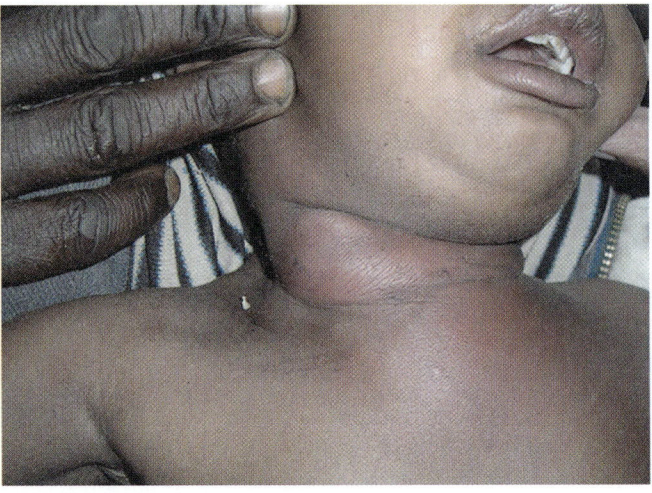

B

FIGURE 77–8B Abscess. Note localized area of inflammation of neck and upper sternum. (Reproduced with permission from Usatine RP, Smith MA, Mayeaux EJ, et al: *The Color Atlas and Synopsis of Family Medicine*, 3rd ed. New York, NY: McGraw Hill; 2019. Photo contributor: Richard P. Usatine, MD.)

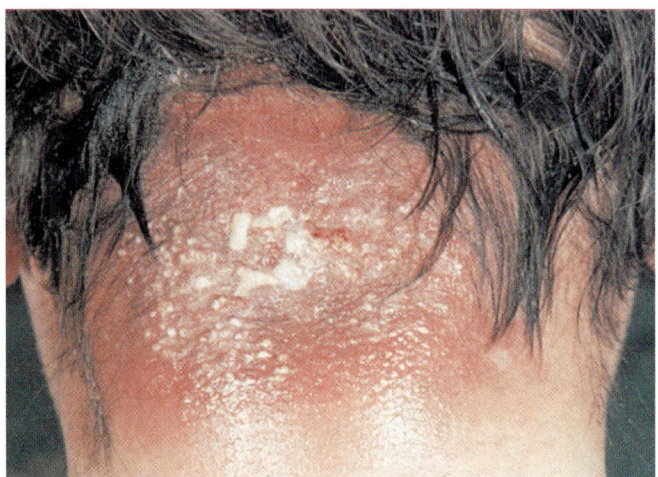

FIGURE 77–9 Carbuncle. Note multiple furuncles that have coalesced to form a large area of inflammatory lesion. (Reproduced with permission from Kang S, Amagai M, Bruckner AL, et al: *Fitzpatrick's Dermatology in General Medicine*, 9th ed. New York, NY: McGraw Hill; 2019.)

the risk of recurrent disease. Antibiotics should also be added when the patient has signs and symptoms of systemic infection, a rapidly progressive or severe infection, infection in a hard-to-drain area of the body, extremes of age, immunocompromised state, or failure to resolve with previous incision and drainage.

When antibiotics are indicated, the patient should be treated with an empiric antibiotic regimen that has activity against MRSA. Such oral antibiotic regimens include clindamycin, trimethoprim-sulfamethoxazole, or doxycycline (Table 77–3). Empiric intravenous regimens include vancomycin or daptomycin. If antibiotic susceptibility testing demonstrates that MSSA is the pathogen, oral regimens can include cephalexin or dicloxacillin, whereas intravenous antibiotics would include nafcillin or cefazolin.

Prevention

Handwashing and covering draining lesions should be used to prevent the spread of bacteria.

TABLE 77–3 Antibiotics Active and Inactive Against Methicillin-Resistant *Staphylococcus aureus* (MRSA)

Antibiotics Active Against MRSA	Antibiotics Not Active Against MRSA
Vancomycin	Penicillins with or without beta-lactamase inhibitor
Clindamycin	Cephalosporins (except ceftaroline)
Daptomycin	Carbapenems
Doxycycline and minocycline	
Trimethoprim-sulfamethoxazole	
Ceftaroline	

NECROTIZING FASCIITIS & MYONECROSIS

Definition

Necrotizing fasciitis is a necrotizing infection of the deep structures of the skin including the underlying fascia. In myonecrosis, the underlying muscle becomes necrotic.

Pathophysiology

A break in the skin caused by trauma or surgery allows for passage of organisms to deeper structures. Infection in the fascial layer results in thrombosis of the vascular supply and adjacent nerve tissue. Destruction of these vital structures manifests as necrosis and anesthesia of the more superficial layers of skin.

Clinical Manifestations

Early symptoms of necrotizing fasciitis are skin erythema, warmth, and tenderness. Patients may have pain out of proportion of the examination findings. These skin changes often spread and progress very quickly and are followed by evidence of skin hypoperfusion, blue-gray coloring, bullae, and anesthesia (Figure 77–10). Crepitus may be felt. Patients often demonstrate signs and symptoms of systemic infection progressing to severe sepsis.

Pathogens

There are two types of necrotizing fasciitis—type I, which is polymicrobial, and type II, which is monomicrobial. Type I infection

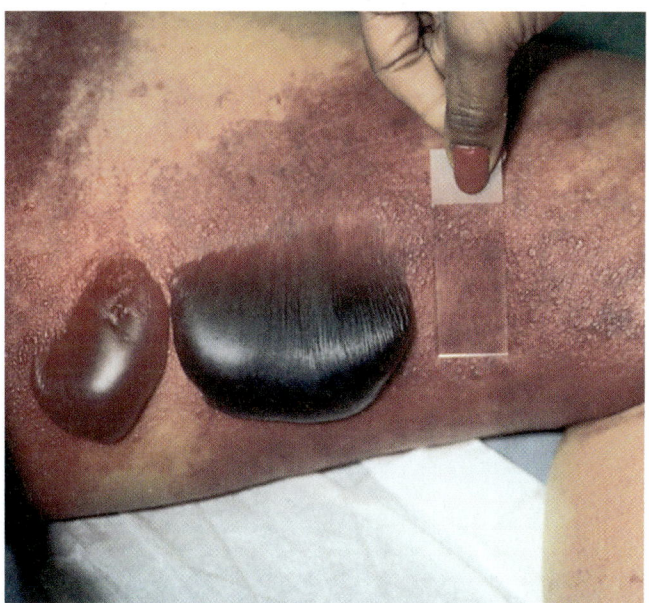

FIGURE 77–10 Necrotizing fasciitis. Note two large hemorrhagic bullae surrounded by dusky red inflamed tissue. (Reproduced with permission from Knoop KJ, Stack L, Storrow A, et al: *Atlas of Emergency Medicine*, 3rd ed. New York, NY: McGraw Hill; 2009. Photo contributor: Lawrence B. Stack, MD.)

is often due to both aerobic and anaerobic bacteria and is more common following intra-abdominal surgery, in diabetics, and in intravenous drug users. It can also be seen in the male perineum when bacteria breach the abdominal wall, a disease called Fournier's gangrene (Figure 77–11). Ludwig's angina, an infection of the sub-mandibular facial space, and Lemierre's syndrome, a thrombophlebitis of the jugular vein, are other examples.

Type II infection is most often due to *S. pyogenes* but can also be caused by *Vibrio vulnificus* following trauma in salt (sea) water, *Aeromonas* species following trauma in freshwater, *Clostridium perfringens* from soil-contaminated wounds caused by motor vehicle/motorcycle accidents or shrapnel (gas gangrene), and community-acquired MRSA.

Diagnosis

Gram stain and culture from debrided tissue can assist in making a microbiologic diagnosis. Radiographic imaging may be useful. Plain films may demonstrate presence of gas in tissues, and a CT scan may reveal enhancement in the fascial plane. Magnetic resonance imaging is the most sensitive approach but is limited in specificity.

Treatment

Necrotizing soft tissue infections are a medical emergency, and treatment requires a combination of surgical debridement of infected tissue and antibiotic therapy. Antibiotic therapy should be directed at *S. pyogenes*, MRSA, and anaerobic and aerobic gram-negative rods. A common empiric regimen would include clindamycin plus vancomycin plus piperacillin-tazobactam. The recommended treatment of gas gangrene caused by clostridia is a combination of penicillin and clindamycin.

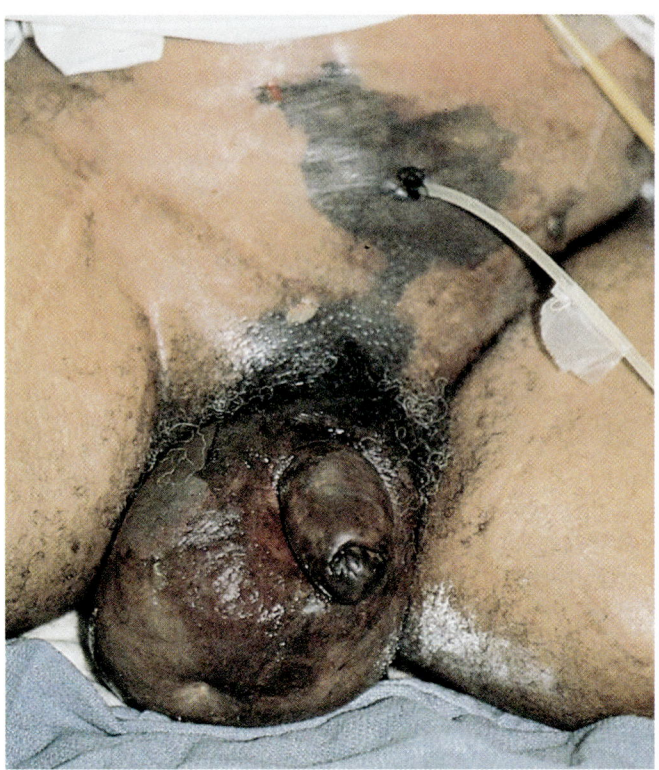

FIGURE 77–11 Fournier's gangrene. Note gangrene of the genitalia and skin of lower abdomen. (Reproduced with permission from Knoop KJ, Stack L, Storrow A, et al: *Atlas of Emergency Medicine*, 3rd ed. New York, NY: McGraw Hill; 2009.)

Prevention

Handwashing and covering draining lesions should be used to prevent the spread of bacteria.

Urinary Tract Infections

INTRODUCTION

Urinary tract infections are a group of common diseases that occur predominantly by ascension of normal enteric flora through the urethra into the bladder. These infections more frequently affect persons with vaginas due to anatomic differences including a shorter urethra. Diagnosis is made by identifying related clinical symptoms in combination with an abnormal urinalysis and growth on urine culture. Antibiotics are often effective therapy, although antibiotic resistance is increasing.

DIAGNOSTIC TESTING FOR URINARY TRACT INFECTIONS

Urine microscopy is the use of a microscope to look at urine. In patients with urinary tract infections, pyuria (elevated white blood cells [WBCs] in urine) are seen, and hematuria (red blood cells in urine), and bacteria are also often observed. The presence of WBC casts indicates pyelonephritis rather than cystitis. A urine sample that has abundant squamous epithelial cells suggests that it is contaminated, and the results of the culture are not reliable.

Urine dipsticks use different chemicals reagents on a strip that is dipped in urine to diagnose urinary tract diseases. Certain dipstick test results are suggestive of infection, namely positive leukocyte esterase, positive nitrite, and positive hemoglobin. The positive nitrite occurs from the conversion of nitrate to nitrite by Enterobacteriaceae.

Urine culture allows identification of the organism causing infection. Urine in the bladder is normally sterile. Because contamination of samples can occur as urine passes through the outer third of the urethra, a numeric threshold of colony-forming units (CFUs) per milliliter has been established to confirm infection. In samples obtained from a midstream void, $\geq 1 \times 10^5$ CFU/mL is consistent with infection. In samples

collected via catheterization, $\geq 1 \times 10^2$ CFU/mL is consistent with infection. Either a voided midstream urine specimen or a specimen obtained by bladder catheterization can be used for urine culture.

CYSTITIS

Definition

Cystitis is an infection of the bladder. The term "cysto" refers to bladder, and "itis" refers to inflammation. Uncomplicated cystitis is defined as cystitis in otherwise healthy women, whereas complicated cystitis is defined as cystitis in all other groups such as men, pregnant women, diabetics, those with anatomic and neurologic problems, and those with recurrent urinary tract infections.

Pathophysiology

Bacteria (rarely fungi) reach the bladder via ascension through the urethra. This is much more common in persons with vaginas due to the short urethra and close approximation of the urethra to the vagina and anus. Preceding infection, the vagina, which is normally colonized by *Lactobacillus* species, will become colonized by enteric organisms such as *Escherichia coli* instead. *E. coli* are able to adhere to the urethral and bladder mucosa via pili. Once bacteria enter the bladder, they are able to reproduce and cause an inflammatory response, resulting in the symptoms of infection.

Medical conditions that cause abnormal emptying of bladder increase risk for urinary tract infections. These include anatomic abnormalities such as cystoceles, neurologic disorders such as spinal cord injuries and multiple sclerosis, and the presence of foreign bodies such as indwelling Foley catheters. In infants less than 3 months of age, uncircumcised children with penises are at higher risk for urinary tract infections than children with vaginas. However, after infancy, children with vaginas are at higher risk for infection than children with penises.

Clinical Manifestations

The most common clinical manifestations of cystitis include dysuria (pain with urination); frequent, low-volume urination; suprapubic tenderness; and gross hematuria. Persons with penises may experience some penile discharge. Most patients with cystitis do not have fever or other systemic symptoms of infection, and when they are present, an upper urinary tract infection (pyelonephritis) should be considered.

Pathogens

E. coli is by far the most common cause of urinary tract infections, especially cystitis. Other enteric gram-negative rods such as *Klebsiella* species and *Proteus* species are regular culprits. *Pseudomonas aeruginosa* can cause urinary tract infection, but this is most common in healthcare-associated infections, patients with anatomic/neurologic abnormalities afflicting their urinary tract, or heavily antibiotic-experienced patients.

Gram-positive pathogens include *Staphylococcus saprophyticus* and *Enterococcus* species. *S. saprophyticus* is common in young persons with vaginas. *Candida* species are found but are rarely a cause of true infection. Viruses such as adenovirus, BK virus, and cytomegalovirus can cause a hemorrhagic cystitis. These viruses almost exclusively cause cystitis in immunocompromised hosts such as those who have undergone stem cell transplants.

Diagnosis

The diagnosis of cystitis requires identifying a combination of pyuria (by seeing WBCs on microscopy or positive leukocyte esterase on urine dipstick) often accompanied by a positive nitrite test and evidence of red blood cells in the urine, *plus* positive urine cultures *plus* clinical symptoms consistent with infection. Either a voided midstream urine specimen or a specimen obtained by bladder catheterization can be used for urine culture.

Treatment

Treatment of cystitis requires antibiotic therapy. Empiric therapy is directed against *E. coli* in cases of uncomplicated cystitis and is accomplished with either trimethoprim-sulfamethoxazole or nitrofurantoin. Empiric therapy for complicated cystitis is usually with a fluoroquinolone (ciprofloxacin or levofloxacin). A combination of meropenem/vaborbactam can also be used. Symptomatic relief of the dysuria can be accomplished using phenazopyridine.

Prevention

There is no known method for primary prevention of cystitis. However, prevention of cystitis in patients with a history of recurrent cystitis may be accomplished with several strategies. These include ways to enhance growth of the normal vaginal flora (*Lactobacillus* species) to prevent colonization with enteric gram-negative rods, such as *E. coli*; intravaginal estrogen in postmenopausal persons with vaginas; and avoidance of spermicide as a form of contraception. In persons with vaginas who frequently experience cystitis following sexual intercourse, postcoital antibiotics can be beneficial.

PYELONEPHRITIS

Definition

Pyelonephritis is an infection of the kidney(s). "Pyelo" refers to the renal pelvis, and "nephritis" means inflammation of the kidney. Uncomplicated pyelonephritis is defined as pyelonephritis in otherwise healthy persons with vaginas, whereas complicated pyelonephritis is pyelonephritis in all other patients.

Pathophysiology

Pyelonephritis may occur either by ascension of bacteria from the urethra to the bladder and then to the kidney(s) or, less commonly, through hematogenous spread from other sites of infection such as endocarditis. Kidney stones predispose to pyelonephritis (Figure 78–1). Urinary tract infections in children can be associated with anatomic abnormalities, and additional workup for diseases such as vesicoureteral reflux should be considered.

Clinical Manifestations

Patients with pyelonephritis typically present with fever, flank pain, nausea, and vomiting. They may or may not have signs and symptoms of lower tract infection (dysuria, frequency, hematuria, suprapubic tenderness).

Pathogens

E. coli is the most common pathogen causing pyelonephritis. Other enteric gram-negative rods such as *Klebsiella* and *Proteus* species are also involved. *P. aeruginosa* can cause pyelonephritis,

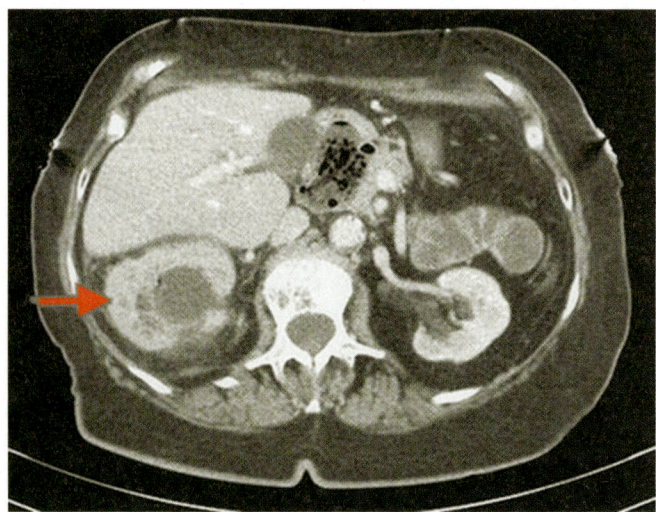

FIGURE 78–1 Pyelonephritis. Red arrow indicates enlarged right kidney caused by a stone at the ureteropelvic junction. (Reproduced with permission from McKean SC, Ross JJ, Dressler DD, et al: *Principles and Practice of Hospital Medicine.* New York, NY: McGraw Hill; 2012.)

but this typically occurs in healthcare-associated infections, patients with anatomic/neurologic abnormalities afflicting their urinary tract, or heavily antibiotic-experienced patients. **Patients with recurrent *Proteus* pyelonephritis should be evaluated for struvite stones.** Infection of the kidney following hematogenous spread of infection can occur with essentially any organism but is seen most commonly with *Staphylococcus aureus*. Hematogenous spread also occurs with *Mycobacterium tuberculosis* and can be seen in disseminated fungal infection as well.

Diagnosis

The most important laboratory test is the urine culture. Urine test findings are similar to those seen in cystitis, but urinary WBC casts can be seen in pyelonephritis (Figure 78–2). Blood WBC counts are frequently elevated, and occasionally blood cultures can be positive. Ultrasound and computed tomography scans can reveal inflammation and can occasionally reveal obstruction or perinephric abscess (see Figure 78–1). Radiographic imaging is not routinely recommended in patients who respond quickly to antibiotics and in whom there is no clinical concern for associated nephrolithiasis or obstruction. Patients with renal tuberculosis may have pyuria in the absence of positive cultures (sterile pyuria) because *M. tuberculosis* does not grow in routine culture media.

Treatment

Antibiotics that are able to obtain high concentrations in the renal parenchyma and have activity against common pathogens are required to treat pyelonephritis. Empiric regimens for community-onset infection include a fluoroquinolone (ciprofloxacin or levofloxacin) or a third-generation cephalosporin such as ceftriaxone. Patients with heavy exposure to prior antibiotics, anatomic abnormalities, or exposure to the healthcare setting should be treated with antibiotics with reliable activity against *Pseudomonas*, such as cefepime, piperacillin, or meropenem. Antibiotic therapy should be narrowed once antibiotic susceptibilities become available.

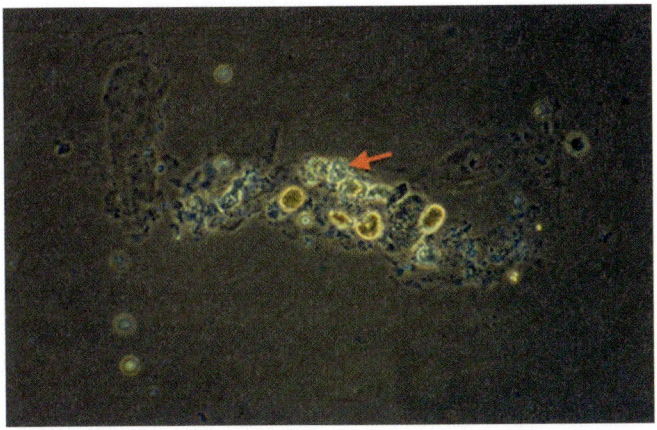

FIGURE 78–2 White blood cell casts. Note cylindrical-shaped casts containing round, refractile white blood cells (arrow). (Used with permission from Agnes B. Fogo.)

Prevention

Patients who have bladder dysfunction that predisposes them to pyelonephritis may require frequent catheterization to allow proper urinary tract drainage. Pregnant women with asymptomatic bacteriuria (see next section) may benefit from antibiotic therapy to prevent pyelonephritis.

ASYMPTOMATIC BACTERIURIA

Definition

Asymptomatic bacteriuria is when bacteria colonize the urinary bladder in the absence of signs or symptoms of upper or lower urinary tract infection. It is defined as the presence of $\geq 1 \times 10^5$ CFU/mL of a single bacterial species on two successive urine cultures in a patient without urinary tract symptoms.

Pathophysiology

Asymptomatic bacteriuria is common in many populations including persons with diabetes, patients with anatomic and neurologic abnormalities of the urinary tract, patients with indwelling Foley catheters, and elderly patients. The bacteria reach the bladder via ascension through the urethra, not from hematogenous dissemination.

Clinical Manifestations

Patients with asymptomatic bacteriuria have no signs or symptoms of upper or lower tract infection.

Pathogens

The same organisms that commonly cause cystitis also cause asymptomatic bacteriuria. Asymptomatic candiduria can occur as well.

Diagnosis

The diagnosis of asymptomatic bacteriuria requires the identification of positive urine cultures. **Patients have pyuria present in over 50% cases of asymptomatic bacteriuria.**

Treatment

Treatment of asymptomatic bacteriuria is indicated in select populations who have been identified to be at risk for subsequent severe infection from presence of bacteriuria. These high-risk groups include (1) pregnant persons, (2) adults scheduled to undergo urinary tract procedures that could cause mucosal bleeding and translocation of bacteria into the blood, and (3) neutropenic patients.

Prevention

Strategies to prevent asymptomatic bacteriuria are not routinely used.

PROSTATITIS

Definition

Prostatitis is inflammation of the prostate, most often caused by bacterial infection. Prostatitis is also discussed in Chapter 74 on Pelvic Infections.

Pathophysiology

Infection most frequently occurs via the urethra then into the prostatic ducts. However, hematogenous seeding of the prostate can occur as well. Microabscesses may develop within the prostate (Figure 78–3).

Clinical Manifestations

Acute prostatitis may present with acute onset of fever, dysuria, urinary frequency, and severe pain with palpation of the prostate. Patients may be very ill and can present with severe sepsis. In contrast, chronic prostatitis presents with more subacute onset of dysuria, frequency, urinary hesitancy, and pelvic discomfort.

Pathogens

In younger patients, *Neisseria gonorrhoeae* and *Chlamydia trachomatis* are the most common causes of prostatitis. However, in older patients, enteric bacteria, such as *E. coli*, are the predominant pathogens. When hematologic seeding occurs, *S. aureus* is a common cause.

Diagnosis

The diagnosis of acute bacterial prostatitis is often confirmed by the finding of an acutely tender prostate on digital rectal exam. Recovery of an organism, when possible, is from urine or blood cultures. Prostatic massage is contraindicated in acute prostatitis. However, in chronic prostatitis, prostate massage following

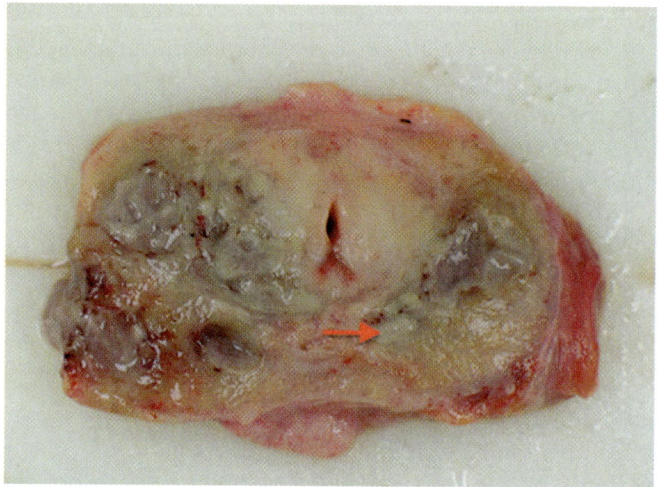

FIGURE 78–3 Prostatitis. Note yellow-green areas of pus (arrow) forming multiple abscesses in prostate gland. (Reproduced with permission from Kemp WL, Burns DK, Brown TG: *Pathology: The Big Picture.* New York, NY: McGraw Hill; 2008.)

collection of prostatic secretion is recommended to obtain a microbiologic diagnosis.

Treatment

Antimicrobial therapy with excellent penetration to the prostatic tissues is recommended for treatment of prostatitis. Fluoroquinolones (e.g., ciprofloxacin or levofloxacin) and trimethoprim-sulfamethoxazole both achieve high levels in the prostate and are good options. Antibiotic susceptibility testing should be used to guide treatment of infecting pathogens.

Prevention

Prompt treatment of acute prostatitis may reduce the risk of development of chronic prostatitis.

Sepsis & Septic Shock

INTRODUCTION

Sepsis is one of the leading causes of morbidity and mortality in the United States. Centers for Disease Control and Prevention (CDC) estimates that more than 1.7 million adults in America develop sepsis each year. Sepsis and septic shock cause approximately 270,000 deaths annually, have fatality rates of 30% to 50% in older patients, and are estimated to cost more than 30 billion dollars each year. The incidence of infections that result in sepsis continues to rise with the increased incidence of **antibiotic-resistant organisms** along with the increased use of **immunosuppressive drugs**, **intravenous and urinary catheters**, and **prosthetic implants**.

DEFINITIONS

Sepsis is defined as life-threatening organ dysfunction caused by a dysregulated host response to infection. Evidence of organ dysfunction includes clinical and laboratory abnormalities of the respiratory system, coagulation, liver, cardiovascular system, nervous system, and kidneys (Table 79–1).

A subset of patients with sepsis can develop **septic shock**, which is defined by profound cellular abnormalities and inadequate organ perfusion. Clinically, septic shock can be identified in patients with sepsis who also have **persistent hypotension** (mean arterial blood pressure below 65 mm Hg) and **elevated serum lactate** despite adequate intravenous fluids.

Bacteremia is an associated term, defined as the presence of bacteria in the bloodstream. Approximately 25% of patients with sepsis have detectable bacteremia. The remaining 75% without bacteremia have organ system infections, most often in the respiratory tract, urinary tract, gall bladder, or intestine. Sepsis without bacteremia may also occur as a result of infection with viruses, fungi, or protozoa.

PATHOPHYSIOLOGY

Sepsis results from the interaction of the infectious agent, usually bacteria, with the host's immune, cardiovascular, neuronal, metabolic, and coagulation systems. Some degree of inflammatory response to infection is normal, but when this response is *dysregulated*, an excess of pro- and anti-inflammatory mediators leads to organ dysfunction.

Sepsis caused by **gram-negative bacteria** is mediated primarily by **endotoxin**, also known as **lipopolysaccharide (LPS)**. The main effects of LPS are caused by its lipid A component. Lipid A combines with LPS-binding protein, and together are bound by Toll receptor 4 (TLR4), a **pattern recognition receptor (PRRs)**

TABLE 79–1 Evidence of Organ Dysfunction in Sepsis

Clinical Evidence	Comment
Hypoxemia	Reduced arterial blood oxygen caused by acute respiratory distress syndrome (ARDS)
Oliguria or elevated creatinine	Evidence of renal failure
Lactic acidosis	Elevated serum lactate caused by end-organ hypoperfusion
Thrombocytopenia	Reduced number of platelets caused by disseminated intravascular coagulation
Altered mental status	Caused by hypoxia to brain and manifested by confusion or obtundation
Elevated liver function tests	Elevated bilirubin and transaminase caused by hepatocyte damage
Global hypokinesis seen on echocardiography	Evidence of myocardial dysfunction

on the surface of macrophages and other innate immune cells, as well as B cells. When macrophage TLR4 is activated, this stimulates the production of interleukin-1 (IL-1), tumor necrosis factor (TNF), and IL-6. These cytokines cause fever, alter the endothelial cells to cause vascular leak, and recruit and activate inflammatory white blood cells. Nitric oxide is also released, causing vasodilation and hypotension, contributing to hypotension (see Chapter 58). Figure 79–1 describes the processes that occur in endotoxin-mediated septic shock. The effects of endotoxin are discussed in more detail in Chapter 7 on Bacterial Pathogenesis.

Endotoxin also activates the coagulation cascade, causing **disseminated intravascular coagulation (DIC)**. Endotoxin initiates DIC by stimulating endothelial cells to produce **tissue factor**. The end result of this cascade is the formation of thrombi (composed of fibrin) in capillaries throughout the body, blocking the flow of blood, and resulting in **anoxia of vital organs**. Petechial hemorrhages and purpuric lesions occur when blood leaks into the tissue spaces at the site of endothelial cells damaged by anoxia (Figure 79–2).

Sepsis caused by gram-positive bacteria is *not* mediated by endotoxin because these bacteria do not contain LPS. Rather, there are surface components such as **peptidoglycan** and **teichoic acid** that stimulate the macrophage, through **PRRs** other than TLR4, to produce the inflammatory mediators mentioned above. Similarly, some fungi, viruses, and protozoa have elements that can trigger macrophages to generate the same effect.

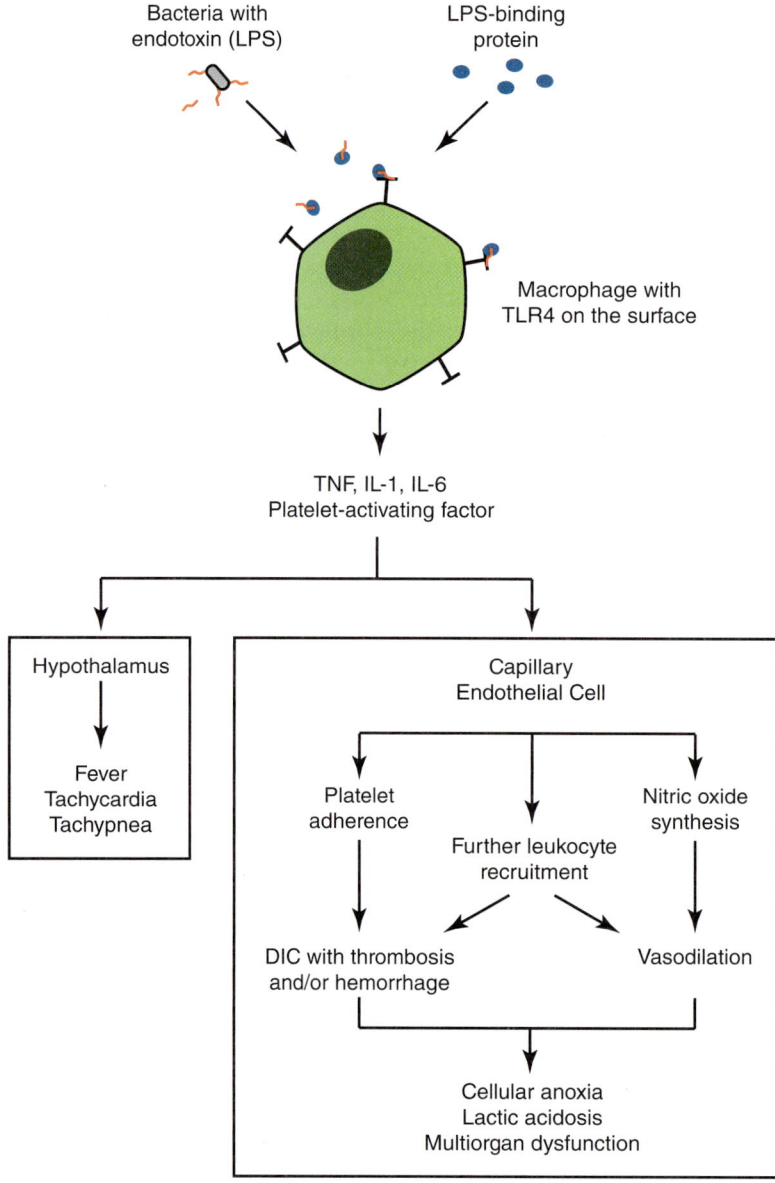

FIGURE 79–1 Processes involved in endotoxin-mediated septic shock. Endotoxin (LPS) from gram-negative bacteria binds to LPS-binding protein, a normal component of blood plasma. This complex binds to Toll-like receptor (TLR)-4 on the surface of macrophages. Activation of TLR-4 induces the synthesis of inflammatory cytokines resulting in the manifestations of sepsis. DIC = disseminated intravascular coagulation; IL = interleukin; LPS = lipopolysaccharide; TNF = tumor necrosis factor.

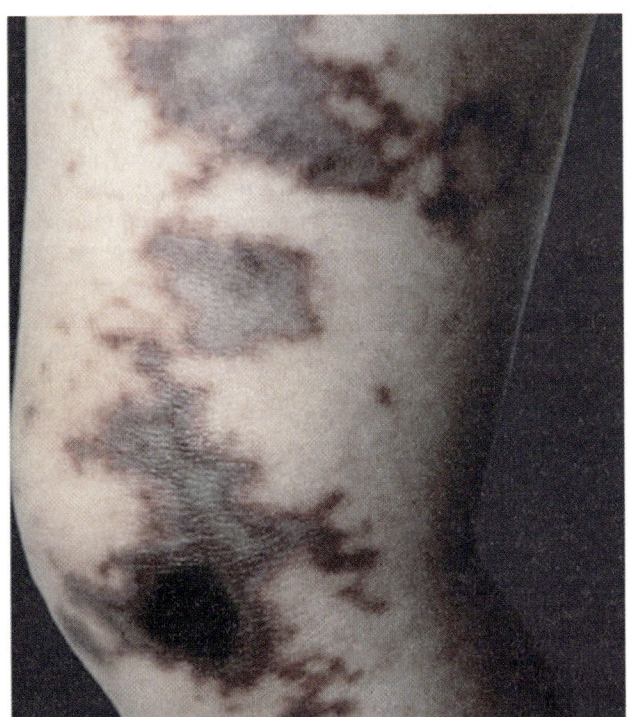

FIGURE 79–2 Disseminated intravascular coagulation (DIC). Note purpuric lesions on leg caused by endotoxin-mediated DIC in a patient with meningococcemia. (Reproduced with permission from Wolff K, Johnson R: *Fitzpatrick's Color Atlas & Synopsis of Clinical Dermatology*, 6th ed. New York, NY: McGraw Hill; 2009.)

Often sepsis is marked by an *elevation* in total blood leukocytes, especially neutrophils, but it can be accompanied by a *reduction* in the number and function of leukocytes, particularly B and T lymphocytes. This limits the adaptive host response, enhances the severity of the infection, and further augments the impact of sepsis. Frequently, there is also an increase in the number of *immature* neutrophils (also called band cells). This is also known as a "shift to the left."

A predisposition to sepsis occurs in the very young (neonatal sepsis), the very old, those with reduced host defenses, and people with chronic diseases such as diabetes, chronic hepatitis, and kidney failure.

Sepsis is also associated with **intravenous drug use**. The organisms most often involved are methicillin-resistant *Staphylococcus aureus* (MRSA), *Streptococcus pyogenes* (Group A streptococci), and the yeast *Candida*.

Neonatal sepsis is the result of the infant's immature immune system coupled with carriage of the bacteria in the female genital tract that are transmitted during the passage through the birth canal. The most common causes are Group B streptococci (*Streptococcus agalactiae*), *Escherichia coli*, and *Listeria monocytogenes*.

CLINICAL MANIFESTATIONS

The clinical manifestations of sepsis often include **fever** and **elevated neutrophils** along with signs of **organ system dysfunction**, including hypoxemia, low platelets, abnormal liver function, hypotension, mental status changes, and kidney injury (see Table 79–1). In septic shock, **hypotension** is severe enough to require vasopressors to maintain the mean arterial pressure at 65 mm Hg. Petechial hemorrhages and purpuric rash can occur as well, which indicate activation of the coagulation cascade. The purpuric rash seen in meningococcemia is called **purpura fulminans** (see Figure 79–2).

PATHOGENS

Sepsis can be caused by a variety of pathogens (Table 79–2). Bacteria cause the vast majority of cases of sepsis and septic shock. However, viruses such as Ebola virus, influenza virus, SARS coronavirus-2, hantavirus, yellow fever virus, and dengue virus (dengue hemorrhagic fever) can also cause the clinical features of sepsis. In addition, yeasts, such as *Candida albicans*, and protozoa, such as *Plasmodium falciparum*, can also cause a sepsis syndrome.

Among the bacteria, Gram-negative rods, such as Enterobacteriaceae (e.g., *E. coli*, *Enterobacter*, *Klebsiella*, *Serratia*, and *Proteus*) and *Pseudomonas* are common causes of endotoxin-mediated sepsis. *Neisseria meningitidis* causes meningococcemia, a common cause of septic shock in young adults.

Gram-positive bacteria, such as *S. aureus* and *Enterococcus faecalis*, are important causes of sepsis. *Streptococcus pneumoniae* is also an important cause of sepsis, especially in asplenic patients. Both *S. aureus* and Group A streptococci (*S. pyogenes*) cause toxic shock syndrome mediated by superantigens (*see* Chapters 7 and 58). Group B streptococci (*S. agalactiae*) are the most common cause of neonatal sepsis. The gram-positive rod, *Listeria*, also causes neonatal sepsis.

TABLE 79–2 Important Infectious Agents Causing Sepsis and Septic Shock

Type of Infectious Agent	Name of Infectious Agent
Bacteria	
1. Gram-positive cocci	*Staphylococcus aureus, Streptococcus pneumoniae, Streptococcus pyogenes, Streptococcus agalactiae, Enterococcus faecalis*
2. Gram-positive rods	*Listeria monocytogenes, Bacillus anthracis*
3. Gram-negative cocci	*Neisseria meningitidis*
4. Gram-negative rods	Enterobacteriaceae (such as *Escherichia coli, Enterobacter, Klebsiella, Serratia,* and *Proteus*), *Pseudomonas, Salmonella typhi, Vibrio vulnificus, Yersinia pestis, Francisella tularensis*
5. Rickettsia	*Rickettsia rickettsiae*
Viruses	Ebola virus, influenza virus, SARS coronavirus-2, hantavirus, yellow fever virus, and dengue virus
Fungi	*Candida albicans*
Protozoa	*Plasmodium falciparum*

Uncommon bacteria that cause sepsis in the United Sates include *Rickettsia rickettsiae* (Rocky Mountain spotted fever), *Salmonella typhi* (typhoid fever), *Francisella tularensis* (tularemia), *Bacillus anthracis* (anthrax), and *Vibrio vulnificus*. *Yersinia pestis* causes plague that can progress to life-threatening septic shock.

DIAGNOSIS

To identify the microbiologic cause of sepsis, **blood cultures** are the mainstay of diagnosis of bloodstream infections. Urinalysis and urine cultures should also be done. Other sites of infection, including skin lesions and sputum, should be cultured. Approximately 25% of septic patients will not have a causative organism identified by culture.

TREATMENT

Sepsis is a life-threatening emergency. It is essential to administer **broad-spectrum bactericidal antibiotics** intravenously to cover the most likely organisms. Antibiotics should be started directly after blood cultures are drawn. The goal is to initiate intravenous antibiotics within the first hour after the diagnosis of sepsis or septic shock is made. The prevalence of antibiotic-resistant organisms should be considered when choosing the antimicrobial drug regimen.

One suggested antibiotic regimen includes vancomycin plus either a third or fourth generation cephalosporin, such as ceftazidime or cefepime, or piperacillin/tazobactam, or a carbapenem such as imipenem. An aminoglycoside or a fluoroquinolone can be added.

In addition to antimicrobial therapy, treatment of sepsis is similar to treatment of other causes of distributive hypotensive shock. **Intravenous fluids**, particularly crystalloid solutions, are used to raise blood volume. If fluids are ineffective, hypotension can also be treated with **vasopressors**, such as norepinephrine. Oxygen supplementation, assisted breathing using a ventilator, and/or dialysis for kidney failure may be necessary.

None of these measures will be successful without **identifying and removing the source of the infection**. In general, a foreign body (e.g., intravenous catheters) that is infected should be removed as soon as is safely possible, and any abscess should be drained and cultured.

PREVENTION

There are vaccines against *N. meningitidis, S. pneumoniae, S. typhi*, influenza virus, and yellow fever virus. There is no vaccine against the enteric gram-negative rods, *Pseudomonas, S. aureus*, or *E. faecalis*.

Preventive antibiotics are given to those who develop immunosuppression and to pregnant women who test positive for group B streptococcus carriage to prevent sepsis in these highly susceptible patients. Prompt removal of unnecessary intravenous and urinary catheters is important because these provide an entry point for bacteria and fungi into the bloodstream. Appropriate hospital infection control practices play an important role in preventing healthcare-associated episodes of sepsis.

Eye Infections

INTRODUCTION

Eye infections are serious infections because loss of vision may occur. Eye infections are named for their anatomical site, that is conjunctivitis affects the conjunctiva, uveitis affects the uveal tract (iris, ciliary body, choroid, and retina), and endophthalmitis affects the vitreous or aqueous humor that occupies most of the space within the eye. Adjacent structures can be involved in which case the compound names are used such as keratoconjunctivitis and chorioretinitis.

Most eye infections are caused by bacteria and viruses, although some fungal and parasitic infections occur as well. Some symptoms associated with eye infections, for example, "a red eye," can be caused by other diseases such as autoimmune uveitis, acute angle glaucoma, and traumatic corneal ulcer.

CONJUNCTIVITIS

Definition

Conjunctivitis is typically caused by an infection of the conjunctiva but allergic and chemical conjunctivitis also occur. It is useful to consider infectious conjunctivitis according to the age of the patient, namely neonatal conjunctivitis (from birth to 4 weeks of age) and adult conjunctivitis (older children to older adults) separately as the mode of acquisition and the types of causative microbes differ.

Ophthalmia neonatorum is another name for neonatal conjunctivitis caused by either *Neisseria gonorrhoeae* or *Chlamydia trachomatis*. The original use of the term "ophthalmia neonatorum" referred to gonococcal conjunctivitis but current common usage includes chlamydial conjunctivitis as well.

Trachoma is the world's leading cause of preventable blindness. It is a conjunctivitis caused by certain strains of

C. trachomatis. Because of its importance, it is discussed separately in the next section.

Pathophysiology

In neonatal conjunctivitis, the bacteria or viruses are acquired during passage through the birth canal. The neonatal infection is acquired from a mother who either has a symptomatic infection or who is an asymptomatic shedder of the organism.

In adult conjunctivitis, the bacteria or viruses are acquired either by direct contact with infected individuals or by fomites (inanimate objects such as towels) that carry the microbe. A common mode of transmission is a finger that contacts infected secretions and transfers the organism to the eye. Other risk factors include contact lens use and swimming in under-chlorinated pools.

Clinical Manifestations

The most characteristic clinical manifestations are a red eye (hyperemia) and a discharge. When the red eye manifests as visibly enlarged blood vessels, the eye is said to be "injected." The discharge in bacterial conjunctivitis, such as gonococcal conjunctivitis, is typically copious and purulent (Figures 80–1 and 80–2), whereas in chlamydial conjunctivitis it is less copious and watery (Figure 80–3). Gonococcal conjunctivitis can be very severe and cause corneal ulceration, scarring, and blindness. Itching, lid edema, and light sensitivity may also occur.

In viral conjunctivitis, the discharge consists a small amount of watery, serous fluid that can cause matting of the eyelashes. Viral conjunctivitis, especially that is caused by adenovirus, is often called "pink eye." (Figure 80–4.) Epidemic keratoconjunctivitis is a severe infection caused by certain strains of adenovirus, typical serotypes 8 and 19. It is highly contagious, causes more severe symptoms, lasts longer, and has significant sequelae, such

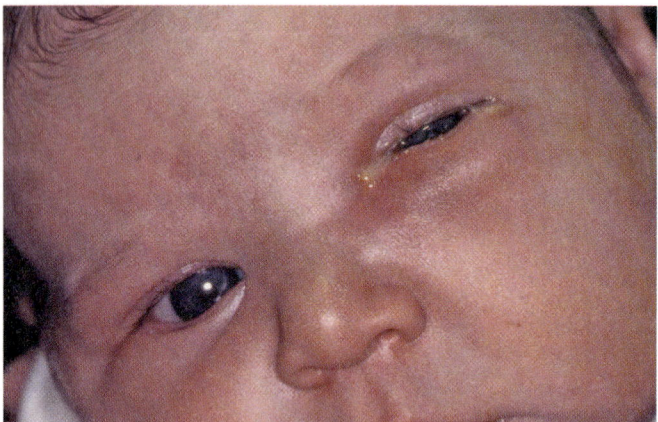

FIGURE 80-1 Gonococcal conjunctivitis in neonate (ophthalmia neonatorum). Note purulent exudate in medial aspect of child's left eye. (Reproduced with permission from Knoop KJ, Stack LB, Storrow AB, et al: *The Atlas of Emergency Medicine*, 4th ed. New York, NY: McGraw Hill; 2016. Photo contributor: David Effron, MD.)

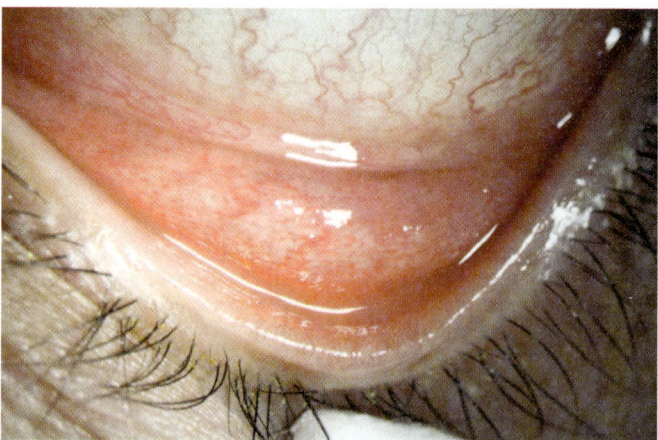

FIGURE 80-3 Chlamydial conjunctivitis. Note hyperemia and watery exudate. (Reproduced with permission from Knoop KJ, Stack LB, Storrow AB, et al: *The Atlas of Emergency Medicine*, 4th ed. New York, NY: McGraw Hill; 2016. Photo contributor: Jeffrey Goshe, MD.)

as scarring and chronic keratitis. Because it is highly contagious, concerns about patients returning to work or to school arise.

Chlamydial conjunctivitis is often called follicular conjunctivitis because follicles are seen on the mucosa of the palpebral conjunctivae. It is also called inclusion conjunctivitis because inclusion bodies are seen in the cytoplasm of infected cells.

Eye pain consists of mild irritation or is absent in conjunctivitis. Visual acuity may be mildly affected. Tearing and "matting together" of the eyelashes are common. Constitutional symptoms such as fever do not occur.

Allergic conjunctivitis presents with many of the same symptoms, that is red eye, tearing, and a serous exudate but itching is more prominent in allergic conjunctivitis.

Pathogens

Neonatal conjunctivitis is commonly caused by two bacteria, *N. gonorrhoeae* and *C. trachomatis*, and by herpes simplex virus type-2 (HSV-2) (Table 80-1). *C. trachomatis* is the most common cause in the United States.

Conjunctivitis in adults is most commonly caused by *Staphylococcus aureus*. *Streptococcus pneumoniae, Haemophilus influenzae,* and *Moraxella* species are common causes in children. *Haemophilus aegypticus* (Koch-Weeks bacillus) is a common cause in tropical areas. *N. gonorrhoeae* and *C. trachomatis* are common causes of conjunctivitis in neonates but also can cause conjunctivitis in sexually active adults.

Adenovirus is the most common viral cause of conjunctivitis but herpes simplex virus type-1 (HSV-1) and varicella–zoster virus (VZV) also cause conjunctivitis. Adenovirus types 8 and 19 are the most common cause of epidemic keratoconjunctivitis. Enterovirus 70 is the main cause of acute hemorrhagic conjunctivitis.

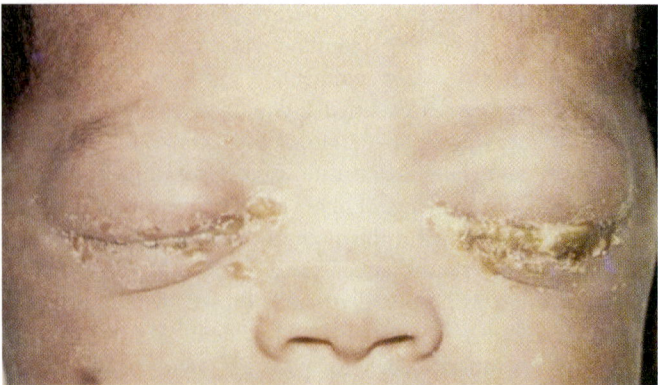

FIGURE 80-2 Gonococcal conjunctivitis in neonate (ophthalmia neonatorum). Note copious exudate in both eyes. (Used with permission from Dr. Pledger J. Public Health Image Library, Centers for Disease Control and Prevention.)

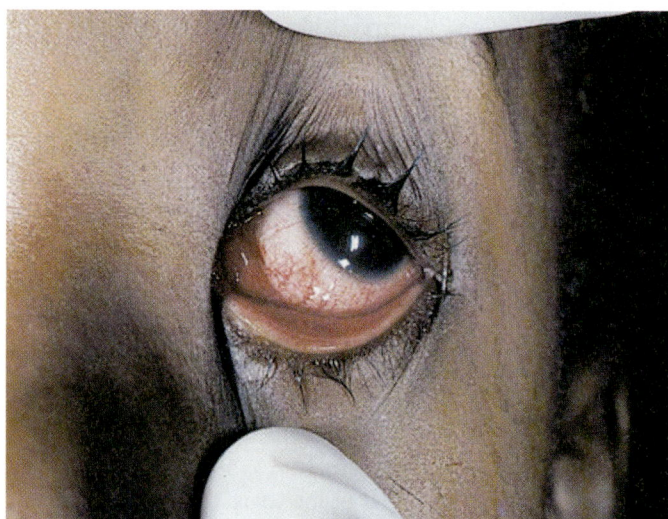

FIGURE 80-4 Viral conjunctivitis caused by adenovirus. Note hyperemia (redness) of both scleral and palpebral conjunctivae and matted eyelashes. (Reproduced with permission from Shah BR, Mahajan P, AModio J, et al: *Atlas of Pediatric Emergency Medicine*, 3rd ed. New York, NY: McGraw Hill; 2019. Photo contributor: Binita R. Shah, MD.)

TABLE 80–1   Important Causes of Conjunctivitis

Age of Patient	Bacterial Causes	Viral Causes	Comment
Neonate (Birth to 4 weeks of age)	*Neisseria gonorrhoeae* *Chlamydia trachomatis*	Herpes simplex-2 virus	Asymptomatic shedding produces fewer organisms and therefore has reduced risk of causing disease in the neonate
Older Children and Adults	*Staphylococcus aureus* *Streptococcus pneumoniae* *Haemophilus influenzae* *Moraxella* species *Neisseria gonorrhoeae* *Chlamydia trachomatis*	Adenovirus Herpes simplex-1 virus Varicella-zoster virus Enterovirus 70	Organisms typically acquired by contact with infected individuals

Conjunctivitis caused by fungi, protozoa, and helminths rarely occurs.

Diagnosis

Most diagnoses are made based on clinical grounds. If serious bacterial conjunctivitis is suspected, microbiologic diagnosis of conjunctivitis can be made by Gram stain and culture of a specimen of the discharge. Polymerase-chain reaction (PCR) assay is the most sensitive and specific test for *C. trachomatis*. Conjunctival scrapings stained with Giemsa stain reveal cytoplasmic inclusion bodies in chlamydial conjunctivitis. Viral conjunctivitis caused by, for example, adenovirus, HSV-1, and HSV-2 can be diagnosed by using PCR assay.

Treatment

Serious conjunctivitis, such as that caused by *N. gonorrhoeae*, should be treated with systemic ceftriaxone in both the neonate and adult. Neonatal chlamydial conjunctivitis should be treated with systemic erythromycin or azithromycin as that can prevent chlamydial pneumonitis as well.

Mild cases of conjunctivitis often resolve without treatment. In mild cases of bacterial conjunctivitis, topical ophthalmic antibiotics such as sulfacetamide, ciprofloxacin, or a polymyxin-trimethoprim combination can be used.

There is no treatment available for adenoviral conjunctivitis. Conjunctivitis caused by HSV-1 can be treated with oral acyclovir or topical ganciclovir. (There is no topical acyclovir available in the United States.) Neonatal conjunctivitis caused by HSV-2 should be treated with intravenous acyclovir.

Prevention

Bacterial neonatal conjunctivitis can be prevented by the application of erythromycin ointment to the eyes of the neonate. Proper hand hygiene can limit spread to others.

TRACHOMA

Trachoma, a conjunctivitis caused by certain strains of *C. trachomatis*, is the world's leading cause of preventable blindness. Repeated episodes cause inversion of the eyelashes that scrape the cornea resulting in scarring and blindness. Millions of people living in 53 countries in Africa, Asia, and Central and South America are at risk. Trachoma is caused by *C. trachomatis* serovars A, A, B, Ba, and C.

Trachoma affects both children and adults living in regions with poor hygiene, poor sanitation, and overcrowding. It is spread by direct contact with infected secretions and by indirect methods such as flies, clothing, and towels.

Clinically, it is a chronic follicular conjunctivitis manifesting with hyperemia (red eye), tearing, exudate and follicles seen visually by eversion of the upper lid (Figure 80–5). Repeated infections cause scarring of the conjunctivae and inward turning of the eyelashes (trichiasis). Scraping of the cornea by the eyelashes results in scarring and blindness (Figure 80–6).

The diagnosis is usually made clinically. PCR tests are available. Conjunctival scrapings stained with Giemsa stain reveal cytoplasmic inclusion bodies. *C. trachomatis* can be grown in cell culture but not on routine bacteriologic media.

The drug of choice is azithromycin orally. Doxycycline is also effective but should not be given to children under 8 years of age. Topical ophthalmic antibiotic ointments/drops of sulfonamides, tetracycline or erythromycin are also effective. Surgical

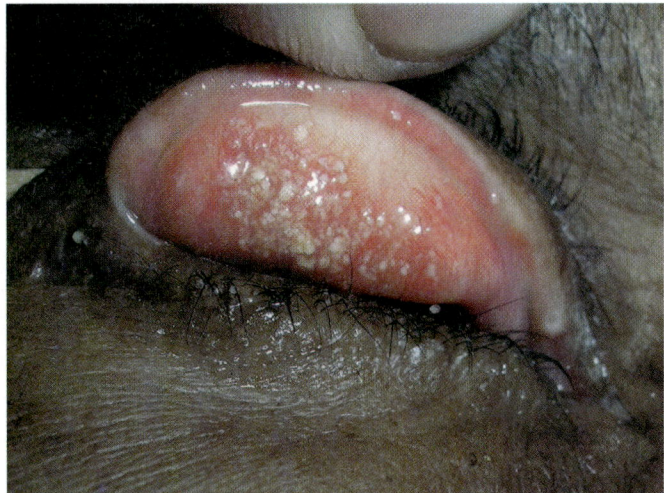

FIGURE 80–5 Trachoma. Note many white follicles on undersurface of upper lid. (Reproduced with permission from Usatine RP, Smith MA, Mayeaux EJ, et al: *The Color Atlas and Synopsis of Family Medicine*, 3rd ed. New York, NY: McGraw Hill; 2019. Photo contributor: Richard P. Usatine, MD.)

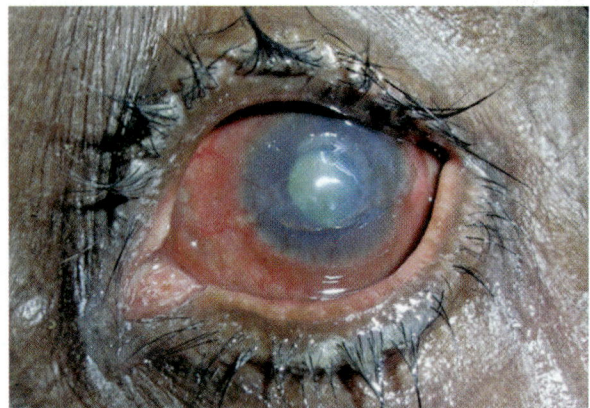

FIGURE 80–6 Trachoma. Note corneal opacity resulting in blindness. Note also marked injection (hyperemia) of the conjunctiva. (Reproduced with permission from Usatine RP, Smith MA, Mayeaux EJ, et al: *The Color Atlas and Synopsis of Family Medicine*, 3rd ed. New York, NY: McGraw Hill; 2019. Photo contributor: Richard P. Usatine, MD.)

correction of trichiasis may be necessary to prevent scarring of the cornea.

Prevention involves improved hygiene. There is no vaccine.

KERATITIS

Definition

Keratitis is an inflammatory lesion of the cornea, most often caused by infection. These infections tend to be serious and can lead to blindness. Some infections involve both the cornea and the conjunctivae and are called keratoconjunctivitis. Bacteria and viruses are the most important causes but fungi and the protozoan, *Acanthamoeba*, are significant causes as well.

Pathophysiology

The main predisposing factor to bacterial keratitis is prolonged contact lens use, especially overnight use. Keratitis related to contact lens use is primarily caused by *Pseudomonas*, but also by *Acanthamoeba*, a protozoan and *Fusarium*, a mold.

Other contributing factors include trauma to the eye, surgical procedures, and contaminated eye solutions. Systemic diseases such as diabetes, Sjögren syndrome, and immunodeficiencies play a role.

Viral keratitis is often related to reactivation of latent infection by HSV-1 and VZV. Keratitis caused by adenovirus is highly contagious and outbreaks occur in schools, workplaces, and doctor's offices.

Keratitis may cause blindness because scarring or perforation of the cornea may occur. Worldwide, trachoma-related keratitis is one of the leading causes of blindness.

Clinical Manifestations

The main manifestations are pain, redness, photophobia, and discharge. Slit-lamp examination shows damage to the corneal epithelium and a cloudy anterior chamber may be seen. A corneal

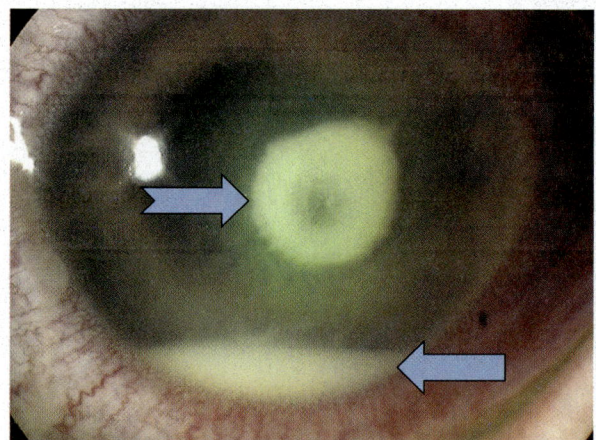

FIGURE 80–7 *Pseudomonas* keratitis causing corneal ulcer and hypopyon. Top notched arrow points to a corneal ulcer. Bottom arrow points to a hypopyon. (Reproduced with permission from Riordan-Eva P, Augsburger JJ: *Vaughan & Asbury's General Ophthalmology*, 19th ed. New York, NY: McGraw Hill; 2017.)

ulcer and a hypopyon (pus at the bottom of the anterior chamber) may occur (Figure 80–7).

The most characteristic lesion of keratitis caused by HSV and VZV is the dendritic ulcer (Figure 80–8).

Pathogens

The principal bacteria that cause keratitis are *Pseudomonas aeruginosa, S. pneumoniae,* and *C. aureus.* Worldwide, *C. trachomatis,* causing trachoma is a very common cause of keratitis. In trachoma, the keratitis (inflammation of the cornea) is due more to eyelid abrasion than to actual infection of the cornea. Furthermore, the damaged cornea can be secondarily infected with other bacteria. Other less common causes are listed in Table 80–2.

The principal viruses that cause keratitis are HSV-1, VZV, and adenovirus. The main fungi that cause keratitis are *Candida*

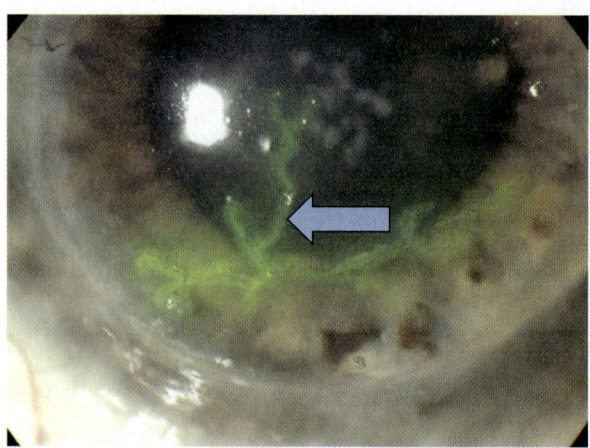

FIGURE 80–8 Dendritic ulcer of cornea caused by herpes simplex virus type-1. Tip of arrow touches one branch of the dendritic ulcer stained green by fluorescein. (Reproduced with permission from Riordan-Eva P, Augsburger JJ: *Vaughan & Asbury's General Ophthalmology*, 19th ed. New York, NY: McGraw Hill; 2017.)

TABLE 80-2 **Important Causes of Keratitis**

	Bacteria	Viruses	Fungi	Protozoa
Important or common causes	*Pseudomonas aeruginosa* *Streptococcus pneumoniae* *Staphylococcus aureus*	Herpes simplex-1 virus Varicella-zoster virus Adenovirus	*Candida albicans* *Aspergilus fumigatus* *Fusarium solani*	*Acanthamoeba*
Important, common cause worldwide	*Chlamydia trachomatis*			
Less important or less common causes	*Staphylococcus epidermidis* *Nocardia asteroides* *Mycobacterium chelonae* *Viridans streptococci*	Herpes simplex-2 virus		

albicans, Aspergillus fumigatus, and Fusarium solani. The protozoan, *Acanthamoeba*, is an important cause of keratitis associated with improperly cleaned contact lens, especially in areas of the world where sterile eye solutions are not readily available.

Diagnosis

Visualization of the organisms in the Gram stain of corneal scraping is used to guide empiric therapy. Giemsa stain can identify the cytoplasmic inclusion bodies of *C. trachomatis*. A microbiologic diagnosis of keratitis is typically made by culture of the specimen or by PCR assay. PCR assay is useful to identify viruses, fungi, and *Acanthamoeba*. MALDI-TOF assay can also be used to identify various microbes, if available.

Treatment

Immediate empiric antibiotic therapy should be instituted to treat bacterial keratitis. Pseudomonas keratitis in a contact-lens user can be treated with a topical fluoroquinolone.

Herpes simplex keratitis can be treated with either topical acyclovir or ganciclovir or trifluridine, or by oral acyclovir. Keratitis caused by VZV should be treated with oral or intravenous acyclovir. Topical glucocorticoids can reduce scarring but carry significant risks so are used only in special circumstances. Fungal keratitis can be treated with topical amphotericin B or itraconazole. *Acanthamoeba* keratitis can be treated with topical or oral voriconazole.

Prevention

The shingles vaccine (Shingrix) can reduce reactivation of VZV. There is a vaccine against *S. pneumoniae* but whether it reduces pneumococcal keratitis is uncertain. There is no vaccine effective against the other causes of keratitis. Proper contact lens cleaning and wearing practices are important. Wearing safety glasses to prevent injury to the cornea is also important.

UVEITIS

Definition

Uveitis is an inflammation of the uvea, caused either by infection or autoimmune disease. Uvea is a collective term comprising the iris, the ciliary body, and the choroid. Infection of the choroid often involves the retina in which case, the term chorioretinitis is used. Autoimmune uveitis is more common than infectious uveitis.

Pathophysiology

Uveitis is often classified as anterior (involving the iris or ciliary body) or posterior (involving the choroid or retina). The anterior form can involve both iris and ciliary body in which case, it is called iridocyclitis. When the posterior form involves both choroid and retina, it is called chorioretinitis.

A minority (10%–20%) of all uveitis is caused by infection. The remainder is split roughly equally between autoimmune uveitis and idiopathic. Infectious uveitis is often caused by a systemic infection that reaches the eye via the blood. Neutrophils are the predominant type of leukocyte in acute uveitis.

Anterior uveitis causes approximately 90% of uveitis, the remainder is either posterior or panuveitis. Anterior uveitis is most often caused by an autoimmune reaction in patients who are HLA-B27-postive.

Clinical Manifestations

Patients with anterior uveitis typically present with eye pain, hyperemia, and decreased vision. Slit lamp examination shows cells in the anterior chamber. The vitreous has few cells and the retina is normal.

In posterior uveitis, there is reduced vision but no pain. Fundoscopic examination shows lesions in the choroid or retina. In ocular toxoplasmosis, there will be cells in the vitreous and a creamy, yellow chorioretinal lesion can be seen. In CMV retinitis, typically there are no cells in the vitreous and white retinal lesions and hemorrhages can be seen.

Acute retinal necrosis (ARN) is a medical emergency. It presents with eye pain and decreased vision. Areas of retinal necrosis with sharp borders are seen. These predispose to retinal detachment.

Pathogens

Most cases of infectious anterior uveitis are caused by a reactivation of HSV-1 latent infection. VZV and CMV can also cause anterior uveitis.

Posterior uveitis (retinitis) is most often caused by *Toxoplasma* in immunocompetent individuals. CMV is a common cause in immunocompromised patients, such as those with

AIDS or organ transplant recipients. Most cases of ARN are caused by VZV. HSV-1 also causes ARN, especially in patients with HSV-1 encephalitis.

Treponema pallidum also causes uveitis which may be anterior, posterior, or panuveitis. Uveitis can occur in both congenital and acquired syphilis. The most common ocular finding in secondary syphilis is iritis. *Mycobacterium tuberculosis* and *Toxocara canis* also cause uveitis.

Diagnosis

The diagnosis is often made clinically. The anterior chamber can be sampled and the aqueous fluid analyzed by PCR assay. Sampling of the vitreous is more difficult and is done only in specific circumstances. In suspected syphilitic uveitis, a VDRL and a treponemal-specific test should be done as older patients may have a false-negative VDRL.

Treatment

Treatment of anterior uveitis caused by HSV is oral acyclovir plus topical corticosteroids. Treatment of ARN caused by VZV or HSV is high-dose intravenous acyclovir. Intravitreal injection may be required. Treatment of posterior uveitis caused by CMV is intravenous ganciclovir. Treatment of posterior uveitis caused by *Toxoplasma* is oral pyrimethamine (plus folinic acid) and sulfadiazine to which systemic corticosteroids should be added.

Prevention

Oral acyclovir is often effective in preventing recurrences of HSV uveitis. There is no vaccine effective against the common causes of uveitis.

ENDOPHTHALMITIS

Definition

Endophthalmitis is an infection of the interior of the eye, involving the aqueous humor or vitreous humor or both. It is typically caused by either bacteria or fungi.

Pathophysiology

Endophthalmitis arises either exogenously following trauma or eye surgery, or endogenously via the bloodstream from a systemic infection, for example, endocarditis. Cataract surgery or injections into the eye are important exogenous causes.

Clinical Manifestations

The main manifestations are an acute onset of eye pain and reduced vision. A hypopyon, a collection of pus (white blood cells) in the aqueous humor, is frequently seen.

Pathogens

The most common cause of post-cataract surgical endophthalmitis and injection endophthalmitis is *Staphylococcus*

epidermidis. The most common bacterial causes of external post-traumatic endophthalmitis are *S. epidermidis* and *Bacillus cereus*. Two molds, *Aspergillus* and *Fusarium*, are also important causes of external post-traumatic endophthalmitis. *Candida* endophthalmitis usually is the result of bloodstream spread from a systemic infection.

Diagnosis

Gram stain and culture is used to identify the bacterial cause in aspirates of the aqueous humor or vitreous humor. Similarly, fungal stains and culture are used to identify the fungi. MALDI-TOF assays can also be used. Blood cultures can be used to identify the organism in those cases involving spread from a systemic infection.

Treatment

Intravitreal antibiotics are used to treat endophthalmitis. Antibiotics delivered systemically are used to treat infections that spread via the bloodstream to the eye (metastatic infections). Vitrectomy may be required.

Prevention

There is no vaccine effective against the common causes of endophthalmitis.

INFECTIONS OF THE EYELIDS AND LACRIMAL SAC

Hordeolum (Stye)

A hordeolum is a localized infection of a sebaceous gland of the eyelid. It is commonly known as a stye. Patients typically present with focal pain, swelling, and erythema of the eyelid near the eyelashes. (Figure 80–9.) A painful pustule may be seen. The most common cause is *S. aureus*. Treatment consists of warm compresses and topical bacitracin or erythromycin ointment.

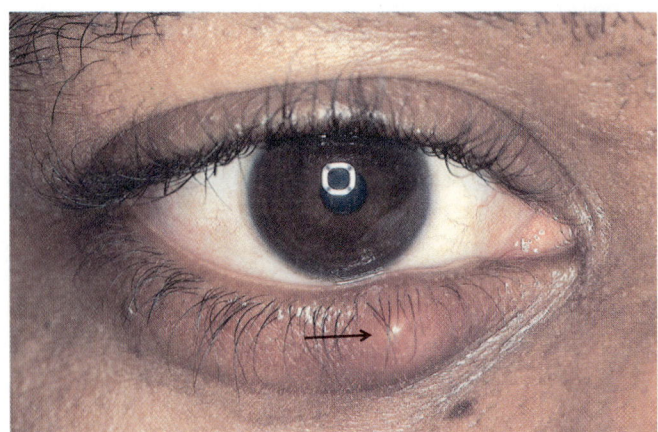

FIGURE 80–9 Hordeolum. Arrow points to focal inflammation on lower eyelid. (Reproduced with permission from Knoop KJ, Stack LB, Storrow AB, et al: *The Atlas of Emergency Medicine*, 4th ed. New York, NY: McGraw Hill; 2016. Photo contributor: Frank Birinyi, MD.)

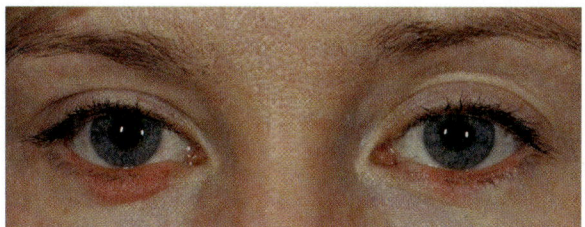

FIGURE 80–10 Blepharitis. Diffuse inflammation of eyelid margin bilaterally. (Reproduced with permission from Riordan-Eva P, Augsburger JJ: *Vaughan & Asbury's General Ophthalmology*, 19th ed. New York, NY: McGraw Hill; 2017.)

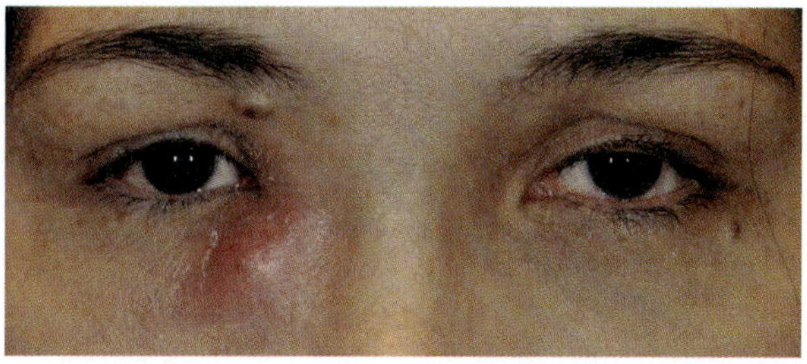

FIGURE 80–11 Dacryocystitis. Localized inflammation below nasal corner of right eye bilaterally. (Reproduced with permission from Riordan-Eva P, Augsburger JJ: *Vaughan & Asbury's General Ophthalmology*, 19th ed. New York, NY: McGraw Hill; 2017.)

Chalazion

A chalazion resembles a stye in appearance, but is *not* an infection. It is a response to blocked secretions in a meibomian gland. It is a painless, small pea-sized nodule in the tarsus part of the eyelid. Warm compresses are usually sufficient treatment but incision and drainage to remove the inspissated material may be necessary. Injection of corticosteroids can also be useful.

Blepharitis

Blepharitis is a diffuse inflammation of eyelid margin. It is quite common. It is due to changes in the meibomian gland, including inflammation and plugging of the gland orifice. Infection with *S. aureus* can occur. Patients typically present with erythema, burning, and itching of the lid margin. (Figure 80–10.) Scales are often seen clinging to the eyelashes. It can become chronic and remitting. It is often associated with seborrheic dermatitis or rosacea. Treatment consists of eyelid washing with baby shampoo or commercial lid scrub products to remove scales and topical bacitracin ointment.

Dacryocystitis

Dacryocystitis is an infection of the lacrimal sac. It occurs because the lacrimal duct becomes obstructed and tears pool in the sac which is proximal to the duct. As a consequence of the obstruction, infection occurs. *S. aureus* is the most common cause but *Streptococcus* species and *Escherichia coli* can also be involved. Patients typically present with pain, erythema, tearing, and swelling near the nasal corner of the eye. (Figure 80–11.) Treatment consists of systemic antibiotics to treat the causative organism. Incision and drainage of a lacrimal sac abscess may be required.

PART X BRIEF SUMMARIES OF MEDICALLY IMPORTANT ORGANISMS

SUMMARIES OF MEDICALLY IMPORTANT BACTERIA

GRAM-POSITIVE COCCI (CHAPTER 15)

Staphylococcus aureus

Diseases—Abscesses of many organs, skin and soft tissue infections, endocarditis, osteomyelitis, septic arthritis, and impetigo. Also hospital-acquired pneumonia, surgical wound infections, and sepsis. Also exotoxin-mediated diseases such as gastroenteritis (food poisoning), toxic shock syndrome, and scalded skin syndrome. It is one of the most common causes of human infections.

Methicillin-resistant *Staphylococcus aureus* (MRSA) is the most common cause of skin abscesses in the United States. MRSA is also an important cause of pneumonia, necrotizing fasciitis, and sepsis in immunocompetent patients. *S. aureus*, especially MRSA, is the most common cause of infections in users of intravenous drugs.

Characteristics—Gram-positive cocci in clusters. Coagulase-positive. Catalase-positive. Most isolates produce β-lactamase. Some isolates have an altered penicillin-binding protein making it resistant to methicillin and nafcillin. These are called MRSA strains.

Habitat and Transmission—Main habitat is human nose; also found on human skin. Transmission is via hands.

Pathogenesis—Abscess containing pus is the most common lesion. These pyogenic lesions are often in the skin but can occur in any organ if *S. aureus* enters the bloodstream.

Three exotoxins can also be made if the specific strain contains the genes that encode the toxin. Toxic shock syndrome toxin is a superantigen and causes toxic shock syndrome by stimulating many helper T cells to release large amounts of lymphokines, especially interleukin (IL)-2. Enterotoxin, which causes food poisoning, is also a superantigen. Food poisoning has a short incubation period because it is preformed in food. Scalded skin syndrome toxin is a protease that cleaves desmoglein in tight junctions in the skin.

Protein A is an important virulence factor because it binds to the heavy chain of IgG. This reduces phagocytosis because the gamma heavy chain cannot bind to its receptor on the surface of neutrophils and macrophages. Predisposing factors to infection include breaks in the skin, foreign bodies such as sutures, neutrophil levels below 500/μL, intravenous drug use (predisposes to right-sided endocarditis), and tampon use (predisposes to toxic shock syndrome).

Laboratory Diagnosis—Gram-stained smear and culture. Yellow or gold colonies on blood agar; colonies often β-hemolytic. *S. aureus* is coagulase-positive; *Staphylococcus epidermidis* is coagulase-negative. Serologic tests not useful.

Treatment—Penicillin G for sensitive isolates; β-lactamase–resistant penicillins such as nafcillin for resistant isolates; vancomycin for isolates resistant to nafcillin. About 85% are resistant to penicillin G. Plasmid-encoded β-lactamase mediates most resistance. Resistance to methicillin and nafcillin is caused by changes in penicillin-binding proteins. Vancomycin-resistant strains have emerged.

Prevention—Cefazolin is used to prevent surgical wound infections. No vaccine is available. Handwashing reduces transmission.

Staphylococcus epidermidis

Diseases—Endocarditis on prosthetic heart valves, prosthetic hip infection, intravascular catheter infection, cerebrospinal fluid shunt infection, neonatal sepsis.

Characteristics—Gram-positive cocci in clusters. Coagulase-negative. Catalase-positive.

Habitat and Transmission—Normal flora of the human skin and mucous membranes. It is probably the patient's own strains that cause infection, but transmission from person to person via hands may occur.

Pathogenesis—Glycocalyx-producing strains adhere well to foreign bodies such as prosthetic implants and catheters. It is a low-virulence organism that causes disease primarily in immunocompromised patients and in those with implants. It is a major cause of hospital-acquired infections. Unlike *S. aureus*, no exotoxins have been identified.

Laboratory Diagnosis—Gram-stained smear and culture. Whitish, nonhemolytic colonies on blood agar. It is coagulase-negative. *S. epidermidis* is sensitive to novobiocin, whereas the other coagulase-negative *Staphylococcus, Staphylococcus saprophyticus*, is resistant. Serologic tests are not useful.

Treatment—Vancomycin plus either rifampin or an aminoglycoside. It produces β-lactamases and is resistant to many antibiotics.

Prevention—There is no drug or vaccine.

Staphylococcus saprophyticus

Gram-positive cocci in clusters. Coagulase-negative. Resistant to novobiocin in contrast to *S. epidermidis*, which is sensitive. Causes community-acquired urinary tract infections (UTIs) in young women (but *Escherichia coli* is a much more common cause).

Streptococcus pyogenes (Group A Streptococcus)

Diseases—Suppurative (pus-producing) diseases (e.g., pharyngitis and cellulitis); toxigenic diseases, for example necrotizing fasciitis ("flesh-eating" streptococci) and scarlet fever; immunologic (antibody-mediated) diseases (e.g., rheumatic fever and acute glomerulonephritis).

Characteristics—Gram-positive cocci in chains. β-Hemolytic colonies. Catalase-negative. Bacitracin-sensitive. β-Hemolytic streptococci are subdivided into group A, B, etc., by differences in the antigenicity of their cell wall carbohydrate.

Habitat and Transmission—Habitat is the human throat and skin. Transmission is via respiratory droplets.

Pathogenesis—For suppurative infections, hyaluronidase ("spreading factor") mediates subcutaneous spread seen in cellulitis. For toxigenic infections, erythrogenic toxin and pyrogenic exotoxin A act as superantigens and cause scarlet fever and streptococcal toxic shock syndrome, respectively. Exotoxin B (a protease) causes necrotizing fasciitis. M protein that forms the pilus impedes phagocytosis.

For nonsuppurative (immunologic) diseases, rheumatic fever is caused by immunologic cross-reaction between bacterial antigen and human heart and joint tissue (i.e., antibody against streptococcal M protein reacts with myosin in cardiac muscle), and acute glomerulonephritis is caused by immune complexes formed between streptococcal antigens and antibody to those antigens. The immune complexes are trapped by glomeruli,

complement is activated, neutrophils are attracted to the site by C5a, and proteases produced by neutrophils damage glomeruli.

Laboratory Diagnosis—The diagnosis of suppurative infections (e.g., cellulitis) differs from immunologic diseases (e.g., rheumatic fever). For suppurative infections, use Gram-stained smear and culture. β-Hemolytic colonies on blood agar. (Hemolysis due to streptolysins O and S.) If isolate is sensitive to bacitracin, it is identified as *S. pyogenes*. Rapid enzyme-linked immunosorbent assay (ELISA) tests for group A streptococcal antigens in throat swabs are available. Assay for antibody in patient's serum is not done for suppurative infections.

If rheumatic fever is suspected, patient's antistreptolysin O (ASO) antibody titer is tested to determine whether previous exposure to *S. pyogenes* has occurred. If acute glomerulonephritis is suspected, antibody to streptococcal DNase B is used as evidence of a previous skin infection by *S. pyogenes*.

Treatment—Penicillin G (no significant resistance). Oral penicillin V or amoxicillin is often used.

Prevention—Penicillin is used in patients with rheumatic fever to prevent recurrent *S. pyogenes* pharyngitis. This prevents additional damage to heart valves. There is no vaccine.

Streptococcus agalactiae (Group B Streptococcus)

Diseases—Neonatal meningitis and sepsis.

Characteristics—Gram-positive cocci in chains. β-Hemolytic colonies. Catalase-negative. Bacitracin-resistant. β-Hemolytic streptococci are subdivided into group A, B, etc., by differences in the antigenicity of their cell wall carbohydrate.

Habitat and Transmission—Main habitat is the human vagina. Transmission occurs during birth.

Pathogenesis—Pyogenic organism. No exotoxins identified. Predisposing factors to neonatal infection include rupture of membranes more than 18 hours before delivery, labor prior to 37 weeks (infant is premature), absence of maternal antibody, and heavy colonization of the genital tract by the organism.

Laboratory Diagnosis—Gram-stained smear and culture. β-Hemolytic (narrow zone) colonies on blood agar that are resistant to bacitracin. Organisms hydrolyze hippurate and are CAMP test-positive.

Treatment—Penicillin G.

Prevention—No vaccine. Ampicillin should be given to mothers if prolonged rupture of membranes occurs, if mother has a fever, or if the neonate is premature.

Enterococcus faecalis

Diseases—Urinary tract and biliary tract infections are most frequent. Endocarditis is rare but life-threatening.

Characteristics—Gram-positive cocci in chains. Catalase-negative.

Habitat and Transmission—Habitat is the human colon; urethra and female genital tract can be colonized. May enter bloodstream during gastrointestinal (GI) or genitourinary tract procedures. May infect other sites (e.g., endocarditis).

Pathogenesis—Pyogenic organism. No exotoxins identified.

Laboratory Diagnosis—Gram-stained smear and culture. α- or β-hemolytic or nonhemolytic colonies on blood agar. Grows in 6.5% NaCl and hydrolyzes esculin in the presence of 40% bile. Serologic tests not useful.

Treatment—Penicillin or vancomycin plus an aminoglycoside such as gentamicin is bactericidal. Organism is resistant to either drug given individually, but given together, they have a synergistic effect. Aminoglycoside alone is ineffective because it cannot penetrate. Penicillin or vancomycin weakens the cell wall, allowing the aminoglycoside to penetrate. Vancomycin-resistant enterococci (VRE) are important causes of nosocomial (hospital-acquired) infections. Linezolid can be used to treat VRE.

Prevention—Penicillin and gentamicin should be given to patients with damaged heart valves prior to intestinal or urinary tract procedures. No vaccine is available.

Streptococcus pneumoniae (Pneumococcus)

Diseases—The most common diseases are pneumonia and meningitis in adults and otitis media and sinusitis in children.

Characteristics—Gram-positive "lancet-shaped" cocci in pairs (diplococci) or short chains. α-Hemolytic colonies. Catalase-negative. Growth is inhibited by optochin in contrast to viridans streptococci, which are resistant. Colonies are bile-soluble. Prominent polysaccharide capsule. Eighty-five serotypes based on antigenicity of polysaccharide capsule. One of the three classical encapsulated pyogenic bacteria (*Neisseria meningitidis* and *Haemophilus influenzae* are the other two).

Habitat and Transmission—Habitat is the human upper respiratory tract. Transmission is via respiratory droplets.

Pathogenesis—Induces pyogenic inflammatory response. No known exotoxins. Polysaccharide capsule retards phagocytosis. Antipolysaccharide antibody opsonizes the organism and provides type-specific immunity. IgA protease degrades secretory IgA on respiratory mucosa, allowing colonization. Viral respiratory infection predisposes to pneumococcal pneumonia by damaging mucociliary elevator; splenectomy predisposes to sepsis. Skull fracture with spinal fluid leakage from nose predisposes to meningitis.

Laboratory Diagnosis—Gram-stained smear and culture. α-Hemolytic colonies on blood agar. Growth inhibited by bile and optochin. Quellung reaction occurs (swelling of capsule with type-specific antiserum). Serologic tests for antibody not useful. Tests for capsular antigen in spinal fluid and C polysaccharide in urine can be diagnostic.

Treatment—Penicillin G. Low-level and high-level resistance to penicillin is caused by alterations in penicillin-binding proteins. No β-lactamase is made.

Prevention—Two vaccines are available. The one used in adults contains capsular polysaccharide of the 23 serotypes that cause bacteremia most frequently. The other, which is used primarily in children under the age of 2 years, contains capsular polysaccharide of 13 serotypes coupled to carrier protein (diphtheria toxoid). Oral penicillin is used in immunocompromised children.

Viridans Group Streptococci (e.g., Streptococcus sanguis, Streptococcus mutans)

Diseases—Endocarditis is the most important disease. Also brain abscess, especially in mixed infections with mouth anaerobes. *S. mutans* implicated in dental caries.

Characteristics—Gram-positive cocci in chains. α-Hemolytic colonies. Catalase-negative. Growth is resistant to optochin in contrast to pneumococci, which are inhibited. Colonies are not dissolved by bile.

Habitat and Transmission—Habitat is the human oropharynx. Organism enters bloodstream during dental procedures.

Pathogenesis—Low virulence organism. Bacteremia from dental procedures spreads organism to damaged heart valves. Organism is protected from host defenses within vegetations. No known toxins. Glycocalyx composed of polysaccharide enhances adhesion to heart valves.

Laboratory Diagnosis—Gram-stained smear and culture. α-Hemolytic colonies on blood agar. Growth not inhibited by bile or optochin, in contrast to pneumococci. Viridans streptococci are classified into species by using various biochemical tests. Serologic tests not useful.

Treatment—Penicillin G with or without an aminoglycoside.

Prevention—Penicillin to prevent endocarditis in patients with damaged or prosthetic heart valves who undergo dental procedures.

GRAM-NEGATIVE COCCI (CHAPTER 16)

Neisseria meningitidis (Meningococcus)

Diseases—Meningitis and meningococcemia.

Characteristics—Gram-negative "kidney-bean" diplococci. Oxidase-positive. Large polysaccharide capsule. One of the three classic encapsulated pyogenic bacteria (*S. pneumoniae* and *H. influenzae* are the other two).

Habitat and Transmission—Habitat is the human upper respiratory tract; transmission is via respiratory droplets.

Pathogenesis—After colonizing the upper respiratory tract, the organism reaches the meninges via the bloodstream. Endotoxin in cell wall causes symptoms of septic shock seen in meningococcemia. No known exotoxins; IgA protease produced. Polysaccharide capsule is antiphagocytic and is the main virulence factor of this aggressive pathogen. Deficiency in late complement components predisposes to recurrent meningococcal infections.

Laboratory Diagnosis—Gram-stained smear and culture. Oxidase-positive colonies on chocolate agar. Ferments maltose in contrast to gonococci, which do not. Serologic tests not useful.

Treatment—Penicillin G (no significant resistance).

Prevention—The vaccines against groups A, C, Y, and W-135 meningococci contain the polysaccharide capsule as the immunogen. The vaccine against group B meningococci contains factor H-binding protein as the immunogen. The polysaccharide vaccine exists in two forms: the conjugate vaccine contains the polysaccharides coupled to a carrier protein such as diphtheria toxoid, and the nonconjugate vaccine contains only the polysaccharides. Rifampin or ciprofloxacin is given to close contacts to decrease oropharyngeal carriage.

Neisseria gonorrhoeae (Gonococcus)

Disease—Gonorrhea. Also neonatal conjunctivitis and pelvic inflammatory disease.

Characteristics—Gram-negative "kidney-bean" diplococci. Oxidase-positive. Insignificant capsule.

Habitat and Transmission—Habitat is the human genital tract. Transmission in adults is by sexual contact. Transmission to neonates is during birth.

Pathogenesis—Organism invades mucous membranes and causes inflammation. Endotoxin present but weaker than that of meningococcus, so less severe disease when bacteremia occurs. No exotoxins identified. IgA protease and pili are virulence factors.

Laboratory Diagnosis—Gram-stained smear and culture. Organism visible intracellularly within neutrophils in urethral exudate. Oxidase-positive colonies on Thayer-Martin medium. Gonococci do not ferment maltose, whereas meningococci do. Serologic tests not useful. Nucleic acid amplification tests (NAATs) are used as a screening test in urogenital infections.

Treatment—Ceftriaxone for uncomplicated cases. Doxycycline or azithromycin added for urethritis caused by coinfection with *Chlamydia trachomatis*. High-level resistance to penicillin is caused by plasmid-encoded penicillinase.

Low-level resistance to penicillin is caused by reduced permeability and altered binding proteins.

Prevention—No drug or vaccine. Condoms offer protection. Trace contacts and treat to interrupt transmission. Treat eyes of newborns with erythromycin ointment or silver nitrate to prevent conjunctivitis.

GRAM-POSITIVE RODS (CHAPTER 17)

Bacillus anthracis

Disease—Anthrax.

Characteristics—Aerobic, gram-positive, spore-forming rods. Capsule composed of poly-D-glutamate. *B. anthracis* is the only medically important organism that has a capsule composed of amino acids rather than polysaccharides.

Habitat and Transmission—Habitat is soil. Transmission is by contact with infected animals or inhalation of spores from animal hair and wool.

Pathogenesis—Anthrax toxin consists of three proteins: edema factor, which is an adenylate cyclase; lethal factor, which kills cells by inhibiting a signal transduction protein involved in cell division; and protective antigen, which mediates the entry of the other two components into the cell. The capsule is antiphagocytic.

Laboratory Diagnosis—Gram-stained smear plus aerobic culture on blood agar. *B. anthracis* is nonmotile, in contrast to other *Bacillus* species. Rise in antibody titer in indirect hemagglutination test is diagnostic.

Treatment—Penicillin G (no significant resistance).

Prevention—Vaccine consisting of protective antigen is given to individuals in high-risk occupations.

Bacillus cereus

Disease—Gastroenteritis (Food poisoning).

Characteristics—Aerobic, gram-positive, spore-forming rod.

Habitat and Transmission—Habitat is grains, such as rice. Spores survive boiling during preparation of rice, then germinate when rice is held at warm temperature.

Pathogenesis—Two enterotoxins are produced: one acts like cholera toxin (i.e., cyclic adenosine monophosphate [AMP] is increased within enterocytes); the other acts like staphylococcal enterotoxin (i.e., it is a superantigen).

Laboratory Diagnosis—Not done.

Treatment—Symptomatic only.

Prevention—No vaccine.

Clostridium tetani

Disease—Tetanus.

Characteristics—Anaerobic, gram-positive, spore-forming rods. Spore is at one end ("terminal spore"), so organism looks like a tennis racket.

Habitat and Transmission—Habitat is the soil. Organism enters through traumatic breaks in the skin.

Pathogenesis—Spores germinate under anaerobic conditions in the wound. Organism produces exotoxin, which blocks release of inhibitory neurotransmitters (glycine and γ-aminobutyric acid [GABA]) from spinal neurons. Excitatory neurons are unopposed, and extreme muscle spasm (tetanus, spastic paralysis) results. "Lock-jaw" and "risus sardonicus" are two examples of the muscle spasms. Tetanus toxin (tetanospasmin) is a protease that cleaves proteins involved in the release of the inhibitory neurotransmitters.

Laboratory Diagnosis—Primarily a clinical diagnosis. Organism is rarely isolated. Serologic tests not useful.

Treatment—Hyperimmune human globulin to neutralize toxin. Also penicillin G and spasmolytic drugs (e.g., Valium). No significant resistance to penicillin.

Prevention—Toxoid vaccine (toxoid is formaldehyde-treated toxin). Usually given to children in combination with diphtheria toxoid and acellular pertussis vaccine (DTaP). If patient is injured and has not been immunized, give hyperimmune globulin plus toxoid (passive–active immunization). Debride wound. Give tetanus toxoid booster every 10 years.

Clostridium botulinum

Disease—Botulism.

Characteristics—Anaerobic, gram-positive, spore-forming rods.

Habitat and Transmission—Habitat is the soil. Organism and botulinum toxin transmitted in improperly preserved food.

Pathogenesis—Botulinum toxin is a protease that cleaves proteins involved in the release of acetylcholine at the myoneural junction, causing flaccid paralysis. Failure to sterilize food during preservation allows spores to survive. Spores germinate in anaerobic environment and produce toxin. The toxin is heat-labile; therefore, foods eaten without proper cooking are usually implicated.

Laboratory Diagnosis—Presence of toxin in patient's serum or stool or in food. Detection of toxin involves either antitoxin in serologic tests or production of the disease in mice. Serologic tests for antibody in the patient are not useful.

Treatment—Antitoxin to types A, B, and E made in horses. Respiratory support may be required.

Prevention—Observing proper food preservation techniques, cooking all home-canned food, and discarding bulging cans.

Clostridium perfringens

Diseases—Gas gangrene (necrotizing fasciitis, myonecrosis) and food poisoning.

Characteristics—Anaerobic, gram-positive, spore-forming rods.

Habitat and Transmission—Habitat is soil and human colon. Myonecrosis results from contamination of wound with soil or feces. Food poisoning is transmitted by ingestion of contaminated food.

Pathogenesis—Gas gangrene in wounds is caused by germination of spores under anaerobic conditions and the production of several cytotoxic factors, especially alpha toxin, a lecithinase that cleaves cell membranes. Gas in tissue (CO_2 and H_2) is produced by organism's anaerobic metabolism. Food poisoning is caused by production of enterotoxin within the gut. Enterotoxin acts as a superantigen, similar to that of *S. aureus*.

Laboratory Diagnosis—Gram-stained smear plus anaerobic culture. Spores not usually seen in clinical specimens; the organism is growing, and nutrients are not restricted. Production of lecithinase is detected on egg yolk agar and identified by enzyme inhibition with specific antiserum. Serologic tests not useful.

Treatment—Penicillin G plus debridement of the wound in gas gangrene (no significant resistance to penicillin). Only symptomatic treatment needed in food poisoning.

Prevention—Extensive debridement of the wound plus administration of penicillin decreases probability of gas gangrene. There is no vaccine.

Clostridium difficile

Disease—Pseudomembranous colitis.

Characteristics—Anaerobic, gram-positive, spore-forming rods.

Habitat and Transmission—Habitat is the human colon. Transmission is fecal–oral.

Pathogenesis—Antibiotics suppress normal flora of colon, allowing *C. difficile* to overgrow and produce large amounts of exotoxins. Exotoxins A and B inhibit GTPases, causing inhibition of signal transduction and depolymerization of actin filaments. This leads to apoptosis and death of enterocytes. The pseudomembranes seen in the colon are the visual result of the death of enterocytes.

Laboratory Diagnosis—Exotoxin in the stool is typically detected by using known antibody to the toxin in an ELISA test or by polymerase chain reaction (PCR) assay. Exotoxin in stool can also be detected by cytopathic effect on cultured cells. Identified by neutralization of cytopathic effect with known antibody. A screening test that detects the glutamine dehydrogenase of the organism in stool is available.

Treatment—Oral vancomycin or fidaxomicin can be used. If life-threatening, use vancomycin plus metronidazole.

Prevention—No vaccine or drug is available.

Corynebacterium diphtheriae

Disease—Diphtheria.

Characteristics—Club-shaped gram-positive rods arranged in V or L shape. Granules stain metachromatically. Aerobic, non–spore-forming organism.

Habitat and Transmission—Habitat is the human throat. Transmission is via respiratory droplets.

Pathogenesis—Organism secretes an exotoxin that inhibits protein synthesis by adding ADP-ribose to elongation factor-2 (EF-2). Toxin has two components: subunit A, which has the ADP-ribosylating activity, and subunit B, which binds the toxin to cell surface receptors. Pseudomembrane in throat caused by death of mucosal epithelial cells.

Laboratory Diagnosis—Gram-stained smear and culture. Black colonies on tellurite plate. Document toxin production with precipitin test or by disease produced in laboratory animals. Serologic tests not useful.

Treatment—Antitoxin made in horses neutralizes the toxin. Penicillin G kills the organism. No significant resistance to penicillin.

Prevention—Toxoid vaccine (toxoid is formaldehyde-treated toxin), usually given to children in combination with tetanus toxoid and acellular pertussis vaccine (DTaP).

Listeria monocytogenes

Diseases—Meningitis and sepsis in newborns and immunocompromised adults. Gastroenteritis.

Characteristics—Small gram-positive rods. Aerobic, non–spore-forming organism.

Habitat and Transmission—Organism colonizes the GI and female genital tracts; in nature, it is widespread in animals, plants, and soil. Transmission is across the placenta or by contact during delivery. Outbreaks of sepsis in neonates and gastroenteritis in the general population are related to ingestion of unpasteurized milk products (e.g., cheese).

Pathogenesis—Listeriolysin is an exotoxin that degrades cell membranes. Reduced cell-mediated immunity and immunologic immaturity as in neonates predispose to disease.

Intracellular pathogen that moves from cell to cell via "actin rockets."

Laboratory Diagnosis—Gram-stained smear and culture. Small, β-hemolytic colonies on blood agar. Tumbling motility. Serologic tests not useful.

Treatment—Ampicillin with or without gentamicin.

Prevention—Pregnant women and immunocompromised patients should not ingest unpasteurized milk products or raw vegetables. Trimethoprim-sulfamethoxazole given to immunocompromised patients to prevent *Pneumocystis* pneumonia can also prevent listeriosis. No vaccine is available.

Gardnerella vaginalis

Facultative gram-variable rod. Involved in bacterial vaginosis, along with *Mobiluncus* species, which are anaerobic. See "clue cells," which are vaginal epithelial cells covered with *G. vaginalis* cells. Positive "whiff" test found in bacterial vaginosis.

GRAM-NEGATIVE RODS RELATED TO THE ENTERIC TRACT (CHAPTER 18)

Escherichia coli

Diseases—Urinary tract infection (UTI), sepsis, neonatal meningitis, and "traveler's diarrhea" are the most common. Also hemolytic-uremic syndrome.

Characteristics—Facultative gram-negative rods; ferment lactose.

Habitat and Transmission—Habitat is the human colon; it colonizes the vagina and urethra. From the urethra, it ascends and causes UTI. Acquired during birth in neonatal meningitis and by the fecal–oral route in diarrhea.

Pathogenesis—Endotoxin in cell wall causes septic shock. Two enterotoxins are produced by enterotoxigenic *E. coli* (ETEC) strains. The heat-labile toxin (LT) stimulates adenylate cyclase by ADP-ribosylation. Increased cyclic AMP causes outflow of chloride ions and water, resulting in diarrhea. The heat-stable toxin (ST) causes diarrhea, perhaps by stimulating guanylate cyclase. Virulence factors include pili for attachment to mucosal surfaces and a capsule that impedes phagocytosis. Shiga toxin (verotoxin) is an enterotoxin produced by *E. coli* strains (STEC) with the O157:H7 serotype. It causes bloody diarrhea and hemolytic-uremic syndrome associated with eating undercooked meat. Shiga toxin (verotoxin) inhibits protein synthesis by removing adenine from the 28S rRNA of human ribosomes.

Predisposing factors to UTI in women include the proximity of the anus to the vagina and urethra, as well as a short urethra. This leads to colonization of the urethra and vagina by the fecal flora. Abnormalities (e.g., strictures, valves, and stones) predispose as well. Indwelling urinary catheters and intravenous lines predispose to UTI and sepsis, respectively. Colonization of the vagina leads to neonatal meningitis acquired during birth. The main virulence factor for neonatal meningitis is the K1 capsular polysaccharide.

Laboratory Diagnosis—Gram-stained smear and culture. Lactose-fermenting colonies on eosin–methylene blue (EMB) or MacConkey's agar. Green sheen on EMB agar. Triple sugar iron (TSI) agar shows acid slant and acid butt with gas but no H_2S. Differentiate from other lactose-positive organisms by biochemical reactions. For epidemiologic studies, type organism by O and H antigens by using known antisera. Serologic tests for antibodies in patient's serum not useful.

Treatment—Ampicillin or sulfonamides for UTIs. Third-generation cephalosporins for meningitis and sepsis. Rehydration is effective in traveler's diarrhea; trimethoprim-sulfamethoxazole may shorten duration of symptoms. Antibiotic resistance mediated by plasmid-encoded enzymes (e.g., β-lactamase and aminoglycoside-modifying enzymes).

Prevention—Prevention of UTI involves limiting the frequency and duration of urinary catheterization. Prevention of sepsis involves promptly removing or switching sites of intravenous lines. Traveler's diarrhea is prevented by eating only cooked food and drinking boiled water in certain countries. Prophylactic doxycycline or Pepto-Bismol may prevent traveler's diarrhea. There is no vaccine that prevents any of the diseases caused by *E. coli*.

Salmonella typhi

Disease—Typhoid fever.

Characteristics—Facultative gram-negative rods. Non–lactose-fermenting. Produces H_2S.

Habitat and Transmission—Habitat is the human colon only, in contrast to other salmonellae, which are found in the colon of animals as well. Transmission is by the fecal–oral route.

Pathogenesis—Infects the cells of the reticuloendothelial system, especially in the liver and spleen. Endotoxin in cell wall causes fever. Capsule (Vi antigen) is a virulence factor. No exotoxins known. Decreased stomach acid resulting from ingestion of antacids or gastrectomy predisposes to *Salmonella* infections. Chronic carrier state established in gallbladder. Organism excreted in bile results in fecal–oral spread to others.

Laboratory Diagnosis—Gram-stained smear and culture. Non–lactose-fermenting colonies on EMB or MacConkey's agar. TSI agar shows alkaline slant and acid butt, with no gas and a small amount of H_2S. Biochemical and serologic reactions used to identify species. Identity can be determined by using known antisera against O, H, and Vi antigens in agglutination test. Widal test detects agglutinating antibodies to O and H antigens in patient's serum, but its use is limited.

Treatment—Most effective drug is ceftriaxone. Ampicillin and trimethoprim-sulfamethoxazole can be used in patients who are not severely ill. Resistance to chloramphenicol and ampicillin is mediated by plasmid-encoded acetylating enzymes and β-lactamase, respectively.

Prevention—Public health measures (e.g., sewage disposal, chlorination of the water supply, stool cultures for food handlers, and handwashing prior to food handling). Two vaccines are in common use; one vaccine contains purified Vi polysaccharide capsule as the immunogen and the other contains live, attenuated *S. typhi* as the immunogen.

Salmonella enterica (often called Salmonella enteritidis)

Diseases—Enterocolitis. Sepsis with metastatic abscesses occasionally.

Characteristics—Facultative gram-negative rods. Non–lactose-fermenting. Produces H_2S. Motile, in contrast to *Shigella*. More than 1500 serotypes.

Habitat and Transmission—Habitat is the enteric tract of humans and animals (e.g., poultry and domestic livestock). Transmission is by the fecal–oral route.

Pathogenesis—Invades the mucosa of the small and large intestines. Can enter blood, causing sepsis. Infectious dose is at least 100,000 organisms, much greater than the infectious dose of *Shigella*. Infectious dose is high because organism is inactivated by stomach acid. Endotoxin in cell wall; no exotoxin. Predisposing factors include lowered stomach acidity from either antacids or gastrectomy. Sickle cell anemia predisposes to *Salmonella* osteomyelitis.

Laboratory Diagnosis—Gram-stained smear and culture. Non–lactose-fermenting colonies on EMB or MacConkey's agar. TSI agar shows alkaline slant and acid butt, with gas and H_2S. Biochemical and serologic reactions used to identify species. Can identify the organism by using known antisera in agglutination assay. Widal test detects antibodies in patient's serum to the O and H antigens of the organism but is not widely used.

Treatment—Antibiotics usually not recommended for uncomplicated enterocolitis. Ceftriaxone or other drugs are used for sepsis, depending on sensitivity tests. Resistance to ampicillin and chloramphenicol is mediated by plasmid-encoded β-lactamases and acetylating enzymes, respectively.

Prevention—Public health measures (e.g., sewage disposal, chlorination of the water supply, stool cultures for food handlers, and handwashing prior to food handling). Do not eat raw eggs or meat. No vaccine is available.

Shigella species (e.g., Shigella dysenteriae, Shigella sonnei)

Disease—Enterocolitis (dysentery).

Characteristics—Facultative gram-negative rods. Non–lactose-fermenting. Nonmotile, in contrast to *Salmonella*.

Habitat and Transmission—Habitat is the human colon only; unlike *Salmonella enterica*, there are no animal carriers for *Shigella*. Transmission is by the fecal–oral route.

Pathogenesis—Invades the mucosa of the ileum and colon but does not penetrate farther; therefore, sepsis is rare. Endotoxin in cell wall. Infectious dose is much lower (1–10 organisms) than that of *Salmonella*. The infectious dose of *Shigella* is low because it is resistant to stomach acid. Children in mental institutions and day care centers experience outbreaks of shigellosis. No chronic carrier state.

Laboratory Diagnosis—Gram-stained smear and culture. Non–lactose-fermenting colonies on EMB or MacConkey's agar. TSI agar shows an alkaline slant with an acid butt and no gas or H_2S. Identified by biochemical reactions or by serology with anti-O antibody in agglutination test. Serologic tests for antibodies in the patient's serum are not done.

Treatment—In most cases, fluid and electrolyte replacement only. In severe cases, ciprofloxacin. Resistance is mediated by plasmid-encoded enzymes (e.g., β-lactamase, which degrades ampicillin, and a mutant pteroate synthetase, which reduces sensitivity to sulfonamides).

Prevention—Public health measures (e.g., sewage disposal, chlorination of the water supply, stool cultures for food handlers, and handwashing prior to food handling). Prophylactic drugs not used. No vaccine is available.

Vibrio cholerae

Disease—Cholera.

Characteristics—Comma-shaped gram-negative rods. Oxidase-positive, which distinguishes them from Enterobacteriaceae.

Habitat and Transmission—Habitat is the human colon and shellfish. Transmission is by the fecal–oral route.

Pathogenesis—Massive, watery diarrhea caused by enterotoxin that activates adenylate cyclase by adding ADP-ribose to the stimulatory G protein. Increase in cyclic AMP activates cyclic AMP-dependent kinase that phosphorylates a membrane ion channel. This causes an outflow of chloride ions and water.

Toxin has two components: subunit A, which has the ADP-ribosylating activity; and subunit B, which binds the toxin to cell surface receptors. Organism produces mucinase, which enhances attachment to the intestinal mucosa. Role of endotoxin is unclear. Infectious dose is high (>10^7 organisms). Carrier state rare.

Laboratory Diagnosis—Gram-stained smear and culture. (During epidemics, cultures not necessary.) Agglutination of the isolate with known antisera confirms the identification.

Treatment—Treatment of choice is fluid and electrolyte replacement. Tetracycline is not necessary but shortens duration and reduces carriage.

Prevention—Public health measures (e.g., sewage disposal, chlorination of the water supply, stool cultures for food handlers, and handwashing prior to food handling). For travelers to endemic areas, oral vaccine containing live, attenuated bacteria is available in United States. Tetracycline used for close contacts.

Vibrio parahaemolyticus

Comma-shaped gram-negative rod found in warm sea water. Causes watery diarrhea. Acquired by eating contaminated raw seafood. Outbreaks have occurred on cruise ships in Caribbean. Diarrhea is mediated by enterotoxin similar to cholera toxin.

Vibrio vulnificus

Comma-shaped gram-negative rod found in warm sea water. Causes cellulitis and life-threatening sepsis with hemorrhagic bullae. Acquired either by trauma to skin, especially in shellfish handlers, or by ingestion of raw shellfish, especially in patients who are immunocompromised or have reduced complement caused by liver damage.

Campylobacter jejuni

Disease—Enterocolitis.

Characteristics—Comma-shaped gram-negative rods. Microaerophilic. Grows well at 42°C.

Habitat and Transmission—Habitat is human and animal feces. Transmission is by the fecal–oral route.

Pathogenesis—Invades mucosa of the colon but does not penetrate; therefore, sepsis rarely occurs. No enterotoxin known.

Laboratory Diagnosis—Gram-stained smear plus culture on special agar (e.g., Skirrow's agar) at 42°C in high-CO_2, low-O_2 atmosphere. Serologic tests not useful.

Treatment—Usually symptomatic treatment only; erythromycin for severe disease.

Prevention—Public health measures (e.g., sewage disposal, chlorination of the water supply, stool cultures for food handlers, and handwashing prior to food handling). No preventive vaccine or drug is available.

Helicobacter pylori

Diseases—Gastritis and peptic ulcer. Risk factor for gastric carcinoma.

Characteristics—Curved gram-negative rod.

Habitat and Transmission—Habitat is the human stomach. Transmission is by ingestion.

Pathogenesis—Organisms synthesize urease, which produces ammonia that damages gastric mucosa. Ammonia also neutralizes acid pH in stomach, which allows the organism to live in gastric mucosa.

Laboratory Diagnosis—Gram stain and culture. Urease-positive. Serologic tests for antibody and the "urea breath" test are useful.

Treatment—Quadruple drug therapy because resistance has emerged. One regimen is amoxicillin, clarithromycin, metronidazole, and a proton pump inhibitor such as omeprazole.

Prevention—No vaccine or drug is available.

Klebsiella pneumoniae

Diseases—Pneumonia, UTI, and sepsis.

Characteristics—Facultative gram-negative rods with large polysaccharide capsule.

Habitat and Transmission—Habitat is the human upper respiratory and enteric tracts. Organism is transmitted to the lungs by aspiration from upper respiratory tract and by inhalation of respiratory droplets. It is transmitted to the urinary tract by ascending spread of fecal flora.

Pathogenesis—Endotoxin causes fever and shock associated with sepsis. No exotoxin known. Organism has large capsule, which impedes phagocytosis. Chronic pulmonary disease predisposes to pneumonia; catheterization predisposes to UTI.

Laboratory Diagnosis—Gram-stained smear and culture. Characteristic mucoid colonies are a consequence of the organism's abundant polysaccharide capsule. Lactose-fermenting colonies on MacConkey's agar. Differentiated from *Enterobacter* and *Serratia* by biochemical reactions.

Treatment—Cephalosporins alone or with aminoglycosides, but antibiotic sensitivity testing must be done. Resistance is mediated by plasmid-encoded enzymes, especially β-lactamase.

Prevention—No vaccine or drug is available. Urinary and intravenous catheters should be removed promptly.

Enterobacter cloacae

Enteric gram-negative rod similar to *K. pneumoniae*. Causes hospital-acquired pneumonia, UTI, and sepsis. Highly antibiotic-resistant.

Serratia marcescens

Enteric gram-negative rod similar to *K. pneumoniae*. Causes hospital-acquired pneumonia, UTI, and sepsis. Red-pigmented colonies. Highly antibiotic-resistant.

Proteus species (e.g., *Proteus vulgaris, Proteus mirabilis*)

Diseases—UTI and sepsis.

Characteristics—Facultative gram-negative rods. Non–lactose-fermenting. Highly motile. Produce urease, as do *Morganella* and *Providencia* species (see later). Antigens of OX strains of *P. vulgaris* cross-react with many rickettsiae.

Habitat and Transmission—Habitat is the human colon and the environment (soil and water). Transmission to urinary tract is by ascending spread of fecal flora.

Pathogenesis—Endotoxin causes fever and shock associated with sepsis. No exotoxins known. Urease is a virulence factor because it degrades urea to produce ammonia, which raises the pH. This leads to "struvite" stones, which can obstruct urine flow, damage urinary epithelium, and serve as a nidus for recurrent infection by trapping bacteria within the stone. Organism is highly motile, which may facilitate entry into the bladder. Predisposing factors are colonization of the vagina, urinary catheters, and abnormalities of the urinary tract such as strictures, valves, and stones.

Laboratory Diagnosis—Gram-stained smear and culture. "Swarming" (spreading) effect over blood agar plate as a consequence of the organism's active motility. Non–lactose-fermenting colonies on EMB or MacConkey's agar. TSI agar shows an alkaline slant and acid butt with H_2S. Organism produces urease, whereas *Salmonella*, which can appear similar on TSI agar, does not. Serologic tests not useful. *P. mirabilis* is indole-negative, whereas *P. vulgaris, M. morganii,* and *Providencia* species are indole-positive.

Treatment—Trimethoprim-sulfamethoxazole or ampicillin is often used for uncomplicated UTIs, but a third-generation cephalosporin should be used for serious infections. The indole-negative species *P. mirabilis* is more likely to be sensitive to antibiotics such as ampicillin than are the indole-positive species. Antibiotic sensitivities should be tested. Resistance is mediated by plasmid-encoded enzymes.

Prevention—No vaccine or drug is available. Prompt removal of urinary catheters helps prevent UTIs.

Morganella morganii

Enteric gram-negative rod similar to *Proteus* species. Causes UTIs and sepsis. Highly motile and produces urease. Indole-positive and more resistant to antibiotics than *P. mirabilis*.

Providencia rettgeri

Enteric gram-negative rod similar to *Proteus* species. Causes UTIs and sepsis. Highly motile and produces urease. Indole-positive and more resistant to antibiotics than *P. mirabilis*.

Pseudomonas aeruginosa

Diseases—Wound infection, UTI, pneumonia, and sepsis. One of the most important causes of nosocomial infections, especially in burn patients and those with cystic fibrosis. Causes endocarditis in intravenous drug users.

Characteristics—Aerobic gram-negative rods. Non–lactose-fermenting. Pyocyanin (blue-green) pigment produced. Oxidase-positive, which distinguishes it from members of the Enterobacteriaceae family.

Habitat and Transmission—Habitat is environmental water sources (e.g., in hospital respirators and humidifiers). Also inhabits the skin, upper respiratory tract, and colon of about 10% of people. Transmission is via water aerosols, aspiration, and fecal contamination.

Pathogenesis—Endotoxin is responsible for fever and shock associated with sepsis. Produces exotoxin A, which acts like diphtheria toxin (inactivates EF-2). Pili and capsule are virulence factors that mediate attachment and inhibit phagocytosis, respectively. Glycocalyx-producing strains predominate in chronic infections in cystic fibrosis patients. Strains with type III secretion systems are more virulent than those without. Severe burns and neutropenia are important predisposing factors.

Laboratory Diagnosis—Gram-stained smear and culture. Non–lactose-fermenting colonies on EMB or MacConkey's agar. TSI agar shows an alkaline slant and an alkaline butt because the sugars are not fermented. Oxidase-positive. Serologic tests not useful.

Treatment—Antibiotics must be chosen on the basis of antibiotic sensitivities because resistance is common. Antipseudomonal penicillin and aminoglycoside are often used. Resistance is mediated by a variety of plasmid-encoded enzymes (e.g., β-lactamases and acetylating enzymes).

Prevention—Disinfection of water-related equipment in the hospital, handwashing, and prompt removal of urinary and intravenous catheters. There is no vaccine.

Burkholderia cepacia

Gram-negative rod resembling *P. aeruginosa*. Important cause of chronic infections in patients with cystic fibrosis. Formerly called *Pseudomonas cepacia*.

Stenotrophomonas maltophilia

Gram-negative rod resembling *P. aeruginosa*. Important cause of chronic infections in patients with cystic fibrosis. Formerly called *Pseudomonas maltophilia*.

Bacteroides fragilis

Diseases—Sepsis, peritonitis, and abdominal abscess.

Characteristics—Anaerobic, gram-negative rods.

Habitat and Transmission—Habitat is the human colon, where it is the predominant anaerobe. Transmission occurs by spread from the colon to the blood or peritoneum.

Pathogenesis—Lipopolysaccharide in cell wall is chemically different from and less potent than typical endotoxin. No exotoxins known. Capsule is antiphagocytic and promotes abscess formation. Predisposing factors to infection include bowel surgery and penetrating abdominal wounds.

Laboratory Diagnosis—Gram-stained smear plus anaerobic culture. Identification based on biochemical reactions and gas chromatography. Serologic tests not useful.

Treatment—Metronidazole, clindamycin, and cefoxitin are all effective. Abscesses should be surgically drained. Resistance to penicillin G, some cephalosporins, and aminoglycosides is common. Plasmid-encoded β-lactamase mediates resistance to penicillin.

Prevention—In bowel surgery, perioperative cefoxitin can reduce the frequency of postoperative infections. No vaccine is available.

Prevotella melaninogenica

Anaerobic gram-negative rod resembling *B. fragilis*. Member of normal flora found primarily above the diaphragm (e.g., mouth) in contrast to *B. fragilis*, which is found below (e.g., colon). Often involved in brain and lung abscesses. Formerly called *Bacteroides melaninogenicus*.

Fusobacterium nucleatum

Anaerobic gram-negative rod with pointed ends. Member of the normal human flora in mouth, colon, and female genital tract. Causes brain, lung, abdominal, and pelvic abscesses, typically in combination with other anaerobes and facultative bacteria.

GRAM-NEGATIVE RODS RELATED TO THE RESPIRATORY TRACT (CHAPTER 19)

Haemophilus influenzae

Diseases—Sinusitis, otitis media, and pneumonia are common. Epiglottitis is uncommon, but *H. influenzae* is the most important cause. *H. influenzae* used to be a leading cause of meningitis, but the vaccine has greatly reduced the number of cases.

Characteristics—Small gram-negative (coccobacillary) rods. Requires factors X (hemin) and V (NAD) for growth. Of the six capsular polysaccharide types, type b causes 95% of invasive disease. Type b capsule is polyribitol phosphate.

Habitat and Transmission—Habitat is the upper respiratory tract. Transmission is via respiratory droplets.

Pathogenesis—Polysaccharide capsule is the most important determinant of virulence. Unencapsulated ("untypeable") strains cause mucosal infections but not invasive infections. IgA protease is produced. Most cases of meningitis occur in children younger than 2 years of age, because maternal antibody has waned and the immune response of the child to capsular polysaccharides can be inadequate. No exotoxins identified.

Laboratory Diagnosis—Gram-stained smear plus culture on chocolate agar. Growth requires both factors X and V. Determine serotype by using antiserum in various tests (e.g., latex agglutination). Capsular antigen can be detected in serum or cerebrospinal fluid. Serologic test for antibodies in patient's serum not useful.

Treatment—Ceftriaxone is the treatment of choice for meningitis. Approximately 25% of strains produce β-lactamase.

Prevention—Vaccine containing the type b capsular polysaccharide conjugated to diphtheria toxoid or other protein is given between 2 and 18 months of age. Rifampin can prevent meningitis in close contacts.

Bordetella pertussis

Disease—Whooping cough (pertussis).

Characteristics—Small gram-negative rods.

Habitat and Transmission—Habitat is the human respiratory tract. Transmission is via respiratory droplets.

Pathogenesis—Pertussis toxin stimulates adenylate cyclase by adding ADP-ribose onto the inhibitory G protein. Toxin has two components: subunit A, which has the ADP-ribosylating activity, and subunit B, which binds the toxin to cell surface receptors. Pertussis toxin causes lymphocytosis in the blood by inhibiting chemokine receptors. Inhibition of these receptors prevents lymphocytes from entering tissue, resulting in large numbers being retained in the blood. Inhibition of chemokine receptors occurs because pertussis toxin ADP-ribosylates the inhibitory G protein, which prevents signal transduction within the cell. In addition, extracellular adenylate cyclase is produced, which can inhibit killing by phagocytes. Tracheal cytotoxin damages ciliated epithelium of respiratory tract.

Laboratory Diagnosis—Gram-stained smear plus culture on Bordet-Gengou agar. Identified by biochemical reactions and slide agglutination with known antisera. PCR tests, if available, are both sensitive and specific. Serologic tests for antibody in patient's serum not useful.

Treatment—Azithromycin.

Prevention—The acellular vaccine containing pertussis toxoid and four other purified proteins is recommended rather than the killed vaccine, which contains whole organisms. Usually given to children in combination with diphtheria and tetanus toxoids (DTaP). Azithromycin is useful in unimmunized people who are known to be exposed.

Legionella pneumophila

Disease—Legionnaires' disease ("atypical" pneumonia).

Characteristics—Gram-negative rods, but stain poorly with standard Gram stain. Require increased iron and cysteine for growth in culture. Sixteen serogroups; most cases caused by serogroup 1.

Habitat and Transmission—Habitat is environmental water sources. Transmission is via aerosol from the water source. Person-to-person transmission does not occur.

Pathogenesis—Aside from endotoxin, no toxins, enzymes, or virulence factors are known. Predisposing factors include being older than 55 years, smoking, and having a high alcohol intake. Immunosuppressed patients (e.g., renal transplant recipients) are highly susceptible. The organism replicates intracellularly; therefore, cell-mediated immunity is an important host defense. Smoking damages alveolar macrophages, which explains why it predisposes to pneumonia.

Laboratory Diagnosis—Microscopy with silver impregnation stain or fluorescent antibody. Culture on charcoal yeast extract agar containing increased amounts of iron and cysteine. Urinary antigen provides rapid diagnosis for serogroup 1 bacteria only. Diagnosis can be made serologically by detecting rise in antibody titer in patient's serum.

Treatment—Azithromycin or erythromycin. Rifampin can be added in severe cases.

Prevention—No vaccine or prophylactic drug is available.

Acinetobacter baumannii

Acinetobacter baumannii is a small coccobacillary gram-negative rod that causes pneumonia and UTIs. Disease occurs primarily in immunocompromised hospitalized patients associated with respiratory therapy equipment (ventilator-associated pneumonia) and indwelling catheters. It is highly antibiotic-resistant. Imipenem and colistin are effective in susceptible strains. There is no vaccine.

GRAM-NEGATIVE RODS RELATED TO ANIMAL SOURCES (ZOONOTIC ORGANISMS) (CHAPTER 20)

Brucella species (e.g., Brucella abortus, Brucella suis, Brucella melitensis)

Disease—Brucellosis (undulant fever).

Characteristics—Small gram-negative rods.

Habitat and Transmission—Reservoir is domestic livestock. Transmission is via unpasteurized milk and cheese or direct contact with the infected animal.

Pathogenesis—Organisms localize in reticuloendothelial cells, especially the liver and spleen. Able to survive and replicate intracellularly. No exotoxins. Predisposing factors include consuming unpasteurized dairy products and working in an abattoir.

Laboratory Diagnosis—Gram-stained smear plus culture on blood agar plate. Identified by biochemical reactions and by agglutination with known antiserum. Diagnosis may be made serologically by detecting antibodies in patient's serum.

Treatment—Tetracycline plus rifampin.

Prevention—Pasteurize milk; vaccinate cattle. No human vaccine is available.

Francisella tularensis

Disease—Tularemia.

Characteristics—Small gram-negative rods.

Habitat and Transmission—Reservoir is many species of wild animals, especially rabbits, deer, and rodents. Transmission is by ticks (e.g., *Dermacentor*), aerosols, contact, and ingestion.

Pathogenesis—Organisms localize in reticuloendothelial cells. No exotoxins.

Laboratory Diagnosis—Culture is rarely done because special media are required and there is a high risk of infection of laboratory personnel. Diagnosis is usually made by serologic tests that detect antibodies in patient's serum.

Treatment—Streptomycin.

Prevention—Live, attenuated vaccine for persons in high-risk occupations. Protect against tick bites.

Pasteurella multocida

Disease—Wound infection (e.g., cellulitis).

Characteristics—Small gram-negative rods.

Habitat and Transmission—Reservoir is the mouth of many animals, especially cats and dogs. Transmission is by animal bites.

Pathogenesis—Spreads rapidly in skin and subcutaneous tissue. No exotoxins.

Laboratory Diagnosis—Gram-stained smear and culture.

Treatment—Penicillin G.

Prevention—Ampicillin should be given to individuals with cat bites. There is no vaccine.

Yersinia pestis

Disease—Bubonic and pneumonic plague.

Characteristics—Small gram-negative rods with bipolar ("safety pin") staining. One of the most virulent organisms (i.e., very low ID_{50}).

Habitat and Transmission—Reservoir is wild rodents (e.g., rats, prairie dogs, and squirrels). Transmission is by flea bite.

Pathogenesis—Virulence factors include endotoxin, an exotoxin, two antigens (V and W), and an envelope (capsular) antigen that protects against phagocytosis. V and W proteins allow organism to grow within cells. V antigen suppresses synthesis of gamma-interferon and tumor necrosis factor thereby reducing host immune response. Bubo is a swollen inflamed lymph node, usually located in the region of the flea bite.

Laboratory Diagnosis—Gram-stained smear. Other stains (e.g., Wayson's) show typical "safety-pin" appearance more clearly. Cultures are hazardous and should be done only in specially equipped laboratories. Organism is identified by immunofluorescence. Diagnosis can be made by serologic tests that detect antibody in patient's serum.

Treatment—Streptomycin either alone or in combination with doxycycline. Strict quarantine for 72 hours.

Prevention—Control rodent population and avoid contact with dead rodents. Killed vaccine is available for persons in high-risk occupations. Close contacts should be given tetracycline.

Bartonella henselae

Diseases—Cat-scratch disease (CSD) and bacillary angiomatosis (BA).

Characteristics—Small gram-negative rod.

Habitat and Transmission—Reservoir is the cat's mouth and transmitted by scratch or bite.

Pathogenesis—Low virulence organism. CSD is self-limited in immunocompetent individuals, but BA occurs in immunocompromised individuals.

Laboratory Diagnosis—Diagnosis of CSD usually made by serologic tests. Biopsy of BA lesion shows pleomorphic rods using Warthin-Starry stain.

Treatment—None for CSD. Doxycycline or erythromycin for BA.

Prevention—No vaccine.

MYCOBACTERIA (CHAPTER 21)

Mycobacterium tuberculosis

Disease—Tuberculosis.

Characteristics—Aerobic, acid-fast rods. High lipid content of cell wall, which prevents dyes used in Gram stain from staining organism. Lipids include mycolic acids and wax D. Grows very slowly, which requires that drugs be present for long periods (months). Produces catalase, which is required to activate isoniazid to the active drug.

Habitat and Transmission—Habitat is the human lungs. Transmission is via respiratory droplets produced by coughing.

Pathogenesis—Granulomas and caseation mediated by cellular immunity (i.e., macrophages and CD4-positive T cells [delayed hypersensitivity]). Cord factor (trehalose mycolate) correlates with virulence. No exotoxins or endotoxin. Suppression of cell-mediated immunity increases risk of reactivation and dissemination.

Laboratory Diagnosis—Acid-fast rods seen with Ziehl-Neelsen (or Kinyoun) stain. Slow-growing (3–6 weeks) colony on Löwenstein-Jensen medium. Organisms produce niacin and are catalase-positive. Serologic tests for antibody in patient's serum not useful.

Skin Test—Purified protein derivative (PPD) skin test is positive if induration measuring 10 mm or more appears 48 hours after inoculation. Induration is caused by a delayed hypersensitivity response. Positive skin test indicates that the person has been infected but not necessarily that the person has the disease tuberculosis.

Treatment—Long-term therapy (6–9 months) with three drugs: isoniazid, rifampin, and pyrazinamide. A fourth drug, ethambutol, is used in severe cases (e.g., meningitis), in immunocompromised patients (e.g., those with acquired immunodeficiency syndrome [AIDS]), and where the chance of isoniazid-resistant organisms is high, as in Southeast Asians. Most patients become noninfectious within 2 weeks of adequate therapy. Treatment of latent (asymptomatic) infections consists of isoniazid taken for 6 to 9 months or isoniazid plus rifapentine for 3 months. Multidrug-resistant (MDR) strains have emerged and require other drug combinations.

Prevention—Bacillus Calmette-Guérin (BCG) vaccine containing live, attenuated *Mycobacterium bovis* organisms may prevent or limit extent of disease but does not prevent infection with *M. tuberculosis*. Vaccine used rarely in the United States but widely used in parts of Europe and Asia.

Atypical Mycobacteria

These Mycobacteria are called atypical because they differ from *M. tuberculosis* in various ways. The most important difference is that the atypicals are found in the environment, whereas *M. tuberculosis* is found only in humans. The atypicals are also called "Mycobacteria other than *M. tuberculosis*," or MOTTS.

The atypicals are subdivided into slow growers and rapid growers based on whether they form colonies in more than or less than 7 days. (Pigment production by the slow growers need not concern us here.)

The following are important slow growers:

(1) *Mycobacterium avium-intracellulare* complex (MAC) causes tuberculosis-like disease, especially in immunocompromised patients, such as those with AIDS. It is highly antibiotic-resistant.

(2) *Mycobacterium kansasii* also causes tuberculosis-like disease but is less antibiotic resistant than MAC.

(3) *Mycobacterium marinum* causes "swimming pool granuloma" or "fish tank granuloma," which is a skin lesion at the site of an abrasion acquired in a swimming pool or an aquarium.

(4) *Mycobacterium scrofulaceum* causes scrofula, which manifests as swollen, nontender cervical lymph nodes (cervical adenitis).

(5) *Mycobacterium ulcerans* causes Buruli ulcer, a necrotizing skin lesion that progresses to an ulcer. It is common in tropical rural wetland areas.

The important rapid grower is *Mycobacterium fortuitum-chelonae* complex, which causes infections of prosthetic joints and indwelling catheters. It also causes skin and soft tissue infections at the site of puncture wounds. The organisms are usually resistant to most antituberculosis drugs.

Mycobacterium leprae

Disease—Leprosy.

Characteristics—Aerobic, acid-fast rods. Cannot be cultured in vitro. Optimal growth at less than body temperature, so lesions are on cooler parts of the body, such as skin, nose, and superficial nerves.

Habitat and Transmission—Humans are the main reservoir. Also found in armadillos. Most important mode of transmission is nasal secretions of patients with the lepromatous form. Patients with the lepromatous form are more likely to transmit than those with the tuberculoid form because they have much higher numbers of organisms than those with tuberculoid leprosy. Prolonged exposure is usually necessary.

Pathogenesis—Lesions usually occur in the cooler parts of the body (e.g., skin and peripheral nerves). In tuberculoid leprosy, destructive lesions are due to the cell-mediated response to the organism. Damage to fingers is due to burns and other trauma because nerve damage causes loss of sensation. In lepromatous leprosy, the cell-mediated response to *M. leprae* is lost, and large numbers of organisms appear in the lesions and blood. No toxins or virulence factors are known.

Laboratory Diagnosis—Acid-fast rods are abundant in lepromatous leprosy, but few are found in the tuberculoid form. Cultures and serologic tests not done. Lepromin skin test is positive in the tuberculoid but not in the lepromatous form. A serologic test for IgM against phenolic glycolipid-1 is useful in the diagnosis of lepromatous leprosy.

Treatment—Dapsone plus rifampin for the tuberculoid form. Clofazimine is added to that regimen for the lepromatous form or if the organism is resistant to dapsone. Treatment is for at least 2 years.

Prevention—Dapsone for close family contacts. No vaccine is available.

ACTINOMYCETES (CHAPTER 22)

Actinomyces israelii

Disease—Actinomycosis (abscesses with draining sinus tracts).

Characteristics—Anaerobic, gram-positive filamentous, branching rods.

Habitat and Transmission—Habitat is human mouth, especially anaerobic crevices around the teeth. Transmission into tissues occurs during dental disease or trauma. Organism also aspirated into lungs, causing thoracic actinomycosis. Retained intrauterine device (IUD) predisposes to pelvic actinomycosis.

Pathogenesis—No toxins or virulence factors known. Organism forms sinus tracts that open onto skin and contain "sulfur granules," which are mats of intertwined filaments of bacteria.

Laboratory Diagnosis—Gram-stained smear plus anaerobic culture on blood agar plate. "Sulfur granules" visible in the pus. No serologic tests.

Treatment—Penicillin G and surgical drainage.

Prevention—No vaccine or drug is available.

Nocardia asteroides

Disease—Nocardiosis (especially lung and brain abscesses).

Characteristics—Aerobic, gram-positive filamentous, branching rods. Weakly acid-fast.

Habitat and Transmission—Habitat is the soil. Transmission is via airborne particles, which are inhaled into the lungs.

Pathogenesis—No toxins or virulence factors known. Immunosuppression and cancer predispose to infection.

Laboratory Diagnosis—Gram-stained smear and modified Ziehl-Neelsen stain. Aerobic culture on blood agar plate. No serologic tests.

Treatment—Sulfonamides.

Prevention—No vaccine or drug is available.

MYCOPLASMAS (CHAPTER 23)

Mycoplasma pneumoniae

Disease—"Atypical" pneumonia.

Characteristics—Smallest free-living organisms. Not seen on Gram-stained smear because they have no cell wall, so dyes are not retained. Penicillins and cephalosporins are not effective because there is no cell wall (peptidoglycan). The only bacteria with cholesterol in cell membrane. Can be cultured in vitro.

Habitat and Transmission—Habitat is the human respiratory tract. Transmission is via respiratory droplets.

Pathogenesis—No exotoxins produced. No endotoxin because there is no cell wall. Produces hydrogen peroxide, which can damage the respiratory tract.

Laboratory Diagnosis—Gram stain not useful. Can be cultured on special bacteriologic media but takes at least 10 days to grow, which is too long to be clinically useful. Positive cold–agglutinin test is presumptive evidence. Complement fixation test for antibodies to *Mycoplasma pneumoniae* is more specific.

Treatment—Azithromycin or doxycycline.

Prevention—No vaccine or drug is available.

SPIROCHETES (CHAPTER 24)

Treponema pallidum

Disease—Syphilis.

Characteristics—Spirochetes. Not seen on Gram-stained smear because organism is too thin. Not cultured in vitro.

Habitat and Transmission—Habitat is the human genital tract. Transmission is by sexual contact and from mother to fetus across the placenta.

Pathogenesis—Organism multiplies at site of inoculation and then spreads widely via the bloodstream. Many features of syphilis are attributed to blood vessel involvement causing vasculitis. Primary (chancre) and secondary lesions heal spontaneously. Tertiary lesions consist of gummas (granulomas in bone, muscle, and skin), aortitis, or central nervous system inflammation. No toxins or virulence factors known.

Laboratory Diagnosis—Seen by dark-field microscopy or immunofluorescence. Serologic tests important: VDRL and RPR are nontreponemal (nonspecific) tests used for screening; FTA-ABS is the most widely used specific test for *Treponema pallidum*. Antigen in VDRL and RPR is beef heart cardiolipin; antigen in FTA-ABS is killed *T. pallidum*. VDRL declines with treatment, whereas FTA-ABS remains positive for life.

Treatment—Penicillin is effective in the treatment of all stages of syphilis. In primary and secondary syphilis, use benzathine penicillin G (a depot preparation) because *T. pallidum* grows slowly, so drug must be present for a long time. There is no resistance.

Prevention—Benzathine penicillin given to contacts. No vaccine is available.

Borrelia burgdorferi

Disease—Lyme disease.

Characteristics—Spirochetes. Gram stain not useful. Can be cultured in vitro but not usually done.

Habitat and Transmission—The main reservoir is the white-footed mouse. Transmitted by the bite of ixodid ticks, especially in three areas in the United States: Northeast (e.g., Connecticut), Midwest (e.g., Wisconsin), and West Coast (e.g., California). Eighty percent of cases are in the northeastern states of Connecticut, New York, and New Jersey. Very small nymph stage of ixodid tick (deer tick) is the most common vector. Tick must feed on person for at least 24 hours to deliver an infectious dose of *B. burgdorferi*.

Pathogenesis—Organism invades skin, causing a rash called erythema migrans. It then spreads via the bloodstream to involve primarily the heart, joints, and central nervous system. No toxins or virulence factors identified.

Laboratory Diagnosis—Diagnosis usually made serologically (i.e., by detecting IgM antibody). Confirm positive serologic test with Western blot assay.

Treatment—Doxycycline for early stages; penicillin G for late stages.

Prevention—No vaccine available. Avoid tick bite. Can give doxycycline or amoxicillin to people who are bitten by a tick in endemic areas.

Leptospira interrogans

Disease—Leptospirosis.

Characteristics—Spirochetes that can be seen on dark-field microscopy but not light microscopy. Can be cultured in vitro.

Habitat and Transmission—Habitat is wild and domestic animals. Transmission is via animal urine. In the United States, transmission is chiefly via dog, livestock, and rat urine.

Pathogenesis—Two phases: an initial bacteremic phase and a subsequent immunopathologic phase with meningitis. No toxins or virulence factors known.

Laboratory Diagnosis—Dark-field microscopy and culture in vitro are available but not usually done. Diagnosis usually made by serologic testing for antibodies in patient's serum.

Treatment—Penicillin G. There is no significant antibiotic resistance.

Prevention—Doxycycline effective for short-term exposure. Vaccination of domestic livestock and pets. Rat control.

Borrelia recurrentis

Causes relapsing fever. Transmitted by human body louse. Organism well known for its rapid antigenic changes, which account for the relapsing nature of disease. Antigenic changes are due to programmed rearrangements of bacterial DNA encoding surface proteins.

CHLAMYDIAE (CHAPTER 25)

Chlamydia trachomatis

Diseases—Nongonococcal urethritis, cervicitis, inclusion conjunctivitis, lymphogranuloma venereum, and trachoma. Also pneumonia in infants.

Characteristics—Obligate intracellular parasites. Not seen on Gram-stained smear. Exists as inactive elementary body extracellularly and as metabolically active, dividing reticulate body intracellularly.

Habitat and Transmission—Habitat is the human genital tract and eyes. Transmission is by sexual contact and during passage of neonate through birth canal. Transmission in trachoma is chiefly by hand-to-eye contact.

Pathogenesis—No toxins or virulence factors known.

Laboratory Diagnosis—Nucleic acid amplification test (NAAT) using the patient's urine is used to diagnose chlamydial sexually transmitted disease. Gram stain of urethral exudates that show neutrophils but no gram-negative diplococci (gonococci) is presumptive evidence for chlamydial infection. Cytoplasmic inclusions seen on Giemsa-stained or fluorescent antibody–stained smear of exudate. Organism grows in cell culture and embryonated eggs, but these are not often used.

Treatment—A tetracycline (e.g., doxycycline) or a macrolide (e.g., azithromycin).

Prevention—Erythromycin effective in infected mother to prevent neonatal disease. No vaccine is available.

Chlamydia pneumoniae

Disease—Atypical pneumonia.

Characteristics—Same as *C. trachomatis*.

Habitat and Transmission—Habitat is human respiratory tract. Transmission is by respiratory aerosol.

Pathogenesis—No toxins or virulence factors known.

Laboratory Diagnosis—NAAT and serologic tests for antibody in patient's serum.

Treatment—A tetracycline, such as doxycycline.

Prevention—No vaccine or drug is available.

Chlamydia psittaci

Disease—Psittacosis.

Characteristics—Same as *C. trachomatis*.

Habitat and Transmission—Habitat is birds, both psittacine and others. Transmission is via aerosol of dried bird feces.

Pathogenesis—No toxins or virulence factors known.

Laboratory Diagnosis—NAAT or testing for antibodies in patient's serum.

Treatment—Tetracycline.

Prevention—No vaccine or drug is available.

RICKETTSIAE (CHAPTER 26)

Rickettsia rickettsii

Disease—Rocky Mountain spotted fever.

Characteristics—Obligate intracellular parasites. Not seen well on Gram-stained smear. Antigens cross-react with OX strains of *P. vulgaris* (Weil-Felix reaction).

Habitat and Transmission—*Dermacentor* (dog) ticks are both the vector and the main reservoir. Transmission is via tick bite. Dogs and rodents can be reservoirs as well.

Pathogenesis—Organism invades endothelial lining of capillaries, causing vasculitis. No toxins or virulence factors identified.

Laboratory Diagnosis—Diagnosis made by detecting antibody in serologic tests such as the ELISA test. Weil-Felix test is no longer used. Stain and culture rarely done.

Treatment—Doxycycline.

Prevention—Protective clothing and prompt removal of ticks. Tetracycline effective in exposed persons. No vaccine is available.

Rickettsia prowazekii

Disease—Typhus.

Characteristics—Same as *R. rickettsii*.

Habitat and Transmission—Humans are the reservoir, and transmission is via the bite of the human body louse.

Pathogenesis—No toxins or virulence factors known.

Laboratory Diagnosis—Serologic tests for antibody in patient's serum.

Treatment—Doxycycline.

Prevention—A killed vaccine is used in the military but is not available for civilian use.

Coxiella burnetii

Disease—Q fever.

Characteristics—Obligate intracellular parasites. Not seen well on Gram-stained smear.

Habitat and Transmission—Habitat is domestic livestock. Transmission is by inhalation of aerosols of urine, feces, amniotic fluid, or placental tissue. The only rickettsia not transmitted to humans by an arthropod.

Pathogenesis—No toxins or virulence factors known.

Laboratory Diagnosis—Diagnosis usually made by serologic tests. Weil-Felix test is negative. Stain and culture rarely done.

Treatment—Doxycycline.

Prevention—Killed vaccine for persons in high-risk occupations. No drug is available.

Anaplasma phagocytophilum

Member of *Rickettsia* family. Causes human granulocytic anaplasmosis. Transmitted from reservoir (rodents, dogs) to humans by ticks, especially *Ixodes*, the deer tick. Endemic in northeastern and northcentral states (e.g., Connecticut and Wisconsin). Forms morulae in cytoplasm of granulocytes. (A *morula* is a "mulberry-shaped" inclusion body composed of many *A. phagocytophilum* cells.) Doxycycline is the drug of choice. There is no vaccine.

Ehrlichia chaffeensis

Member of *Rickettsia* family. Causes human monocytic ehrlichiosis. Transmitted from dog reservoir to humans by ticks, especially *Dermacentor*, the dog tick. Endemic in southern states (e.g., Arkansas). Forms morulae in cytoplasm of monocytes. (A *morula* is a "mulberry-shaped" inclusion body composed of many *E. chaffeensis* cells.) Doxycycline is the drug of choice. There is no vaccine.

MINOR BACTERIAL PATHOGENS (CHAPTER 27)

Only the most important of the minor bacterial pathogens are summarized in this section.

Eikenella corrodens

Gram-negative rod that is a member of the normal flora in the human mouth. It causes skin and bone infections associated with human bites and "clenched fist" injuries.

Haemophilus ducreyi

Small gram-negative rod. Causes chancroid. Sexually transmitted disease with painful ulcer on genitals (in contrast to syphilis, which is painless). To grow in culture, it requires factor X (heme) but not factor V (in contrast to *H. influenzae*, which requires both).

Moraxella catarrhalis

Small coccobacillary gram-negative rod that resembles the cocci of the genus *Neisseria*. Causes otitis media and sinusitis primarily in children. Also causes bronchitis and pneumonia, primarily in older people with chronic obstructive pulmonary disease. It is found only in humans and is transmitted by respiratory aerosol.

Yersinia enterocolitica

Gram-negative rods. Causes enterocolitis similar to that caused by *Shigella* and *Salmonella*. Also causes mesenteric adenitis, which can mimic appendicitis. Found in domestic animals and transmitted to humans by fecal contamination of food.

SUMMARIES OF MEDICALLY IMPORTANT VIRUSES

HERPESVIRUSES, POXVIRUSES, AND HUMAN PAPILLOMAVIRUS (CHAPTER 37)

Herpes Simplex Virus Type 1

Diseases—Herpes labialis (fever blisters or cold sores), keratitis, encephalitis.

Characteristics—Enveloped virus with icosahedral nucleocapsid and linear double-stranded DNA. No virion polymerase. One serotype; cross-reaction with herpes simplex virus (HSV) type 2 occurs. HSV-1 can be distinguished from HSV-2 by using monoclonal antibody against glycoprotein G. No herpes group-specific antigen.

Transmission—By saliva or direct contact with virus from the vesicle.

Pathogenesis—Initial vesicular lesions occur in the mouth or on the face. The virus then travels up the axon and becomes latent in sensory (trigeminal) ganglia. Recurrences occur in skin innervated by affected sensory nerve and are induced by fever, sunlight, stress, etc. Dissemination to internal organs occurs in patients with depressed cell-mediated immunity with life-threatening consequences. HSV-1 encephalitis often affects the temporal lobe.

Laboratory Diagnosis—Both NAAT assay and viral culture are used for diagnosis. Virus causes cytopathic effect (CPE) in cell culture. If CPE is observed, HSV is identified by antibody neutralization or fluorescent antibody test. Tzanck smear of cells from the base of the vesicle reveals multinucleated giant cells with intranuclear inclusions. These giant cells are not specific for HSV-1; they are seen in the vesicular lesions caused by HSV-2 and varicella-zoster virus as well. A rise in antibody titer can be used to diagnose a primary infection but not recurrences. HSV encephalitis can be diagnosed using a NAAT assay to detect HSV-1 DNA in spinal fluid.

Treatment—Acyclovir for encephalitis and disseminated disease. Acyclovir has no effect on the latent state of the virus. Trifluridine or acyclovir for keratitis. Primary infections and localized recurrences are self-limited.

Prevention—Recurrences can be prevented by avoiding the specific inciting agent such as intense sunlight. Acyclovir, valacyclovir, or famciclovir is used to reduce recurrences. No vaccine is available.

Herpes Simplex Virus Type 2

Diseases—Herpes genitalis, aseptic meningitis, and neonatal infection.

Characteristics—Enveloped virus with icosahedral nucleocapsid and linear double-stranded DNA. No virion polymerase. One serotype; cross-reaction with HSV-1 occurs. HSV-2 can be distinguished from HSV-1 by using monoclonal antibody against glycoprotein G. No herpes group–specific antigen.

Transmission—Sexual contact in adults and during passage through the birth canal in neonates.

Pathogenesis—Initial vesicular lesions occur on genitals. The virus then travels up the axon and becomes latent in sensory (lumbar or sacral) ganglion cells. Recurrences are less severe than the primary infection. HSV-2 infections in neonate can be life-threatening because neonates have reduced cell-mediated immunity. Asymptomatic shedding of HSV-2 in the female genital tract is an important contributing factor to neonatal infections.

Laboratory Diagnosis—Both NAAT assay and viral culture are used for diagnosis. Virus causes CPE in cell culture. Identify by antibody neutralization or fluorescent antibody test. Tzanck smear reveals multinucleated giant cells but is not specific for HSV-2. A rise in antibody titer can be used to diagnose a primary infection but not recurrences.

Treatment—Acyclovir is useful in the treatment of primary and recurrent genital infections as well as neonatal infections. It has no effect on the latent state.

Prevention—Primary disease can be prevented by protection from exposure to vesicular lesions. Recurrences can be reduced by the long-term use of oral acyclovir, valacyclovir, or famciclovir. Neonatal infection can be prevented by delivering the child by cesarean section if the mother has visible vesicular lesions in the birth canal. There is no vaccine.

Varicella-Zoster Virus

Diseases—Varicella (chickenpox) in children and zoster (shingles) in adults.

Characteristics—Enveloped virus with icosahedral nucleocapsid and linear double-stranded DNA. No virion polymerase. One serotype.

Transmission—Varicella is transmitted primarily by respiratory droplets. Zoster is not transmitted; it is caused by a reactivation of latent virus.

Pathogenesis—Initial infection is in the oropharynx. It spreads via blood to the internal organs such as the liver and then to the skin. After the acute episode of varicella, the virus remains latent in the sensory ganglia and can reactivate to cause zoster years later, especially in older and immunocompromised individuals.

Laboratory Diagnosis—PCR assays and direct fluorescent antibody on specimens containing vesicular fluid can be used. Virus causes CPE in cell culture and can be identified by fluorescent antibody test. Multinucleated giant cells seen in smears from the base of the vesicle. Intranuclear inclusions seen in infected cells. A fourfold or greater rise in antibody titer in convalescent-phase serum is diagnostic.

Treatment—No antiviral therapy is indicated for varicella or zoster in the immunocompetent patient. In the immunocompromised patient, acyclovir can prevent dissemination.

Prevention—The varicella vaccine and one version of the zoster vaccine contain live, attenuated varicella-zoster virus. Another version of the zoster vaccine contains recombinant envelope glycoprotein as the immunogen. Immunocompromised patients exposed to the virus should receive passive immunization with varicella-zoster immune globulin (VZIG) and acyclovir to prevent disseminated disease.

Cytomegalovirus

Diseases—Most common cause of congenital abnormalities in the United States. Cytomegalic inclusion body disease in infants. Mononucleosis in transfusion recipients. Pneumonia and hepatitis in immunocompromised patients. Retinitis and enteritis, especially in AIDS patients.

Characteristics—Enveloped virus with icosahedral nucleocapsid and linear double-stranded DNA. No virion polymerase. One serotype.

Transmission—Virus is found in many human body fluids, including blood, saliva, semen, cervical mucus, breast milk, and urine. It is transmitted via these fluids, across the placenta, or by organ transplantation.

Pathogenesis—Initial infection usually in the oropharynx. In fetal infections, the virus spreads to many organs (e.g., central nervous system and kidneys). In adults, lymphocytes are frequently involved. A latent state occurs in monocytes. Disseminated infection in immunocompromised patients can result from either a primary infection or reactivation of a latent infection.

Laboratory Diagnosis—Both PCR assay and viral culture are used for diagnosis. The virus causes CPE in cell culture and can be identified by fluorescent antibody test. "Owl's eye" nuclear inclusions are seen. A fourfold or greater rise in antibody titer in convalescent-phase serum is diagnostic.

Treatment—Ganciclovir is used to treat pneumonia and retinitis. Acyclovir is ineffective.

Prevention—No vaccine is available. Ganciclovir suppresses retinitis. Letermovir, an inhibitor of the terminase of CMV, is used to prevent CMV infection in patients who have received a hematopoietic stem cell transplant. Do not transfuse cytomegalovirus antibody-positive blood into newborns or antibody-negative immunocompromised patients.

Epstein–Barr Virus

Disease—Infectious mononucleosis. EBV is also associated with Burkitt's lymphoma in East African children.

Characteristics—Enveloped virus with icosahedral nucleocapsid and linear double-stranded DNA. No virion polymerase. One serotype.

Transmission—Virus found in human oropharynx and B lymphocytes. It is transmitted primarily by saliva.

Pathogenesis—In infectious mononucleosis, infection begins in the pharyngeal epithelium, spreads to the cervical lymph nodes, and then travels via the blood to the liver and spleen. EBV establishes latency in B lymphocytes. In Burkitt's lymphoma, oncogenesis is a function of the translocation of the *c-myc* oncogene to a site adjacent to an immunoglobulin gene promoter. This enhances synthesis of the c-myc protein, a potent oncoprotein.

Laboratory Diagnosis— A significant rise in EBV-specific antibody to viral capsid antigen is diagnostic. PCR assays are available. Heterophil antibody is typically positive (Monospot test) but this test is no longer recommended because false-positive and false-negative results occur. Heterophil antibody agglutinates sheep or horse red blood cells. In infectious mononucleosis, lymphocytosis, including atypical lymphocytes, occurs. The virus is rarely isolated.

Treatment—No effective drug is available for infectious mononucleosis.

Prevention—There is no drug or vaccine.

Human Herpesvirus 8

Causes Kaposi's sarcoma (KS), especially in AIDS patients. Transmitted sexually. Diagnosis made by pathologic examination of lesion biopsy. See spindle cells and extravasated red blood cells. Purple color of lesions due to collections of venous blood. Human herpesvirus 8 (HHV-8) DNA can be detected within spindle cells by PCR assay.

Treatment consists of surgical excision, radiation, chemotherapy, or immunomodulatory drugs, such as alpha interferon. In early human immunodeficiency virus (HIV)-associated KS, highly active antiretroviral drugs (HAART) can be effective. Note that antiherpesvirus drugs, such as acyclovir are not effective. HAART also can prevent KS. There is no vaccine against HHV-8.

Smallpox Virus

Disease—Smallpox. The disease smallpox has been eradicated by use of the vaccine. The last known case was in 1977 in Somalia.

Characteristics—Poxviruses are the largest viruses. Enveloped virus with linear double-stranded DNA. DNA-dependent RNA polymerase in virion. One serologic type.

Transmission—By respiratory droplets or direct contact with the virus from skin lesions.

Pathogenesis—The virus infects the mucosal cells of the upper respiratory tract, then spreads to the local lymph nodes and by viremia to the liver and spleen and later the skin. Skin lesions progress in the following order: macule, papule, vesicle, pustule, crust.

Laboratory Diagnosis—Virus identified by CPE in cell culture or "pocks" on chorioallantoic membrane. Electron microscopy reveals typical particles; cytoplasmic inclusions seen in light microscopy. Viral antigens in the vesicle fluid can be detected by precipitin tests. A fourfold or greater rise in antibody titer in the convalescent-phase serum is diagnostic.

Treatment—None.

Prevention—Vaccine contains live, attenuated vaccinia virus. Vaccine is no longer used except by the military, because the disease has been eradicated.

Molluscum Contagiosum Virus

Causes molluscum contagiosum. See pinkish, papular skin lesions with an umbilicated center. Lesions usually on the face, especially around the eyes. Transmitted by direct contact. Diagnosis made clinically; laboratory is not involved. There is no established antiviral therapy and no vaccine. Cidofovir may be useful in the treatment of the extensive lesions that occur in immunocompromised patients.

Human Papillomavirus (HPV)

Diseases—Papillomas (warts); condylomata acuminata (genital warts); associated with carcinoma of the cervix and penis.

Characteristics—Nonenveloped virus with icosahedral nucleocapsid and circular double-stranded DNA. No virion polymerase. There are at least 60 types, which are determined by DNA sequence not by antigenicity. Many types infect the epithelium and cause papillomas at specific body sites.

Transmission—Direct contact of skin or genital lesions.

Pathogenesis—Two early viral genes, *E6* and *E7*, encode proteins that inhibit the activity of proteins encoded by tumor suppressor genes (e.g., the *p53* gene and the retinoblastoma gene, respectively).

Laboratory Diagnosis—Diagnosis is made clinically by finding koilocytes in the lesions. PCR assays that detect HPV DNA are available. Virus isolation and serologic tests are not done.

Treatment—Treatment varies according to the site of lesions: liquid nitrogen is used for skin lesions, podophyllin for genital lesions, and salicylic acid for plantar lesions.

Prevention—Three vaccines are available: one contains the capsid proteins of nine HPV types (6, 11, 16, and 18 and five others) and another contains only the capsid proteins of the four most common HPV types (6, 11, 16, and 18) that cause cancer. The third vaccine contains the capsid proteins of two types (16 and 18) that are the most common cause of cervical cancer.

RESPIRATORY VIRUSES (CHAPTER 38)

Influenza Virus

Disease—Influenza. Influenza A virus is the main cause of worldwide epidemics (pandemics) of influenza.

Characteristics—Enveloped virus with a helical nucleocapsid and segmented, single-stranded RNA of negative polarity. RNA polymerase in virion. The two major antigens are the hemagglutinin (HA) and the neuraminidase (NA) on separate surface spikes. Antigenic *shift* in these proteins as a result of reassortment of RNA segments accounts for the epidemics of influenza caused by influenza A virus. Influenza A viruses of animals are the source of the new RNA segments. Antigenic *drift* due to mutations also contributes. The virus has many serotypes because of these antigenic shifts and drifts. The antigenicity of the internal nucleocapsid protein determines whether the virus is an A, B, or C influenza virus.

Transmission—Respiratory droplets from human to human. H5N1 strains transmitted from birds to humans.

Pathogenesis—Infection is limited primarily to the epithelium of the respiratory tract.

Laboratory Diagnosis—A PCR assay that detects viral RNA in respiratory tract specimens is used for diagnosis. A rapid ELISA test to detect influenza viral antigen in respiratory secretions is also used. Virus grows in cell culture and embryonated eggs and can be detected by hemadsorption or hemagglutination. It is identified by hemagglutination inhibition or complement fixation. A fourfold or greater antibody titer rise in convalescent-phase serum is diagnostic.

Treatment—The NA inhibitor, oseltamivir (Tamiflu), is the drug of choice. Zanamivir, another NA inhibitor, is also available. Baloxavir (Xofluza) inhibits the "cap-snatching" ribonuclease required for the synthesis of influenza virus mRNA. Amantadine and rimantadine are no longer used due to widespread resistance.

Prevention—Two main types of vaccines are available: (1) a killed (subunit) vaccine containing purified HA and NA and (2) a vaccine containing a live, temperature-sensitive mutant of influenza virus. The virus in the live vaccine replicates in cool nasal passages, where it induces secretory IgA, but not in warm lower respiratory tract. Both vaccines contain the strains of influenza A and B virus currently causing disease. The killed vaccine is not a good immunogen and must be given annually.

The vaccine against "standard" influenza contains either two A strains (H1N1 and H3N2) and one B strain (trivalent) or those two A strains and two B strains (quadrivalent). Most of these vaccines are made in eggs, so anyone who has had a severe anaphylactic response to egg proteins should not receive the egg-derived vaccine. In view of this, two vaccines not made in eggs are now available. One vaccine is made in calf kidney cells and the other is made in insect cells. Oseltamivir (Tamiflu) can be used for prophylaxis in unimmunized people who have been exposed.

Parainfluenza Virus

Diseases—Bronchiolitis in infants, croup in young children, and the common cold in adults.

Characteristics—Enveloped virus with helical nucleocapsid and one piece of single-stranded, negative-polarity RNA. RNA polymerase in virion. Unlike influenza viruses, the antigenicity of its HA and NA is stable. There are four serotypes.

Transmission—Respiratory droplets.

Pathogenesis—Infection and death of respiratory epithelium without systemic spread of the virus.

Laboratory Diagnosis— A PCR assay that detects viral RNA in respiratory tract specimens is used for diagnosis. Isolation of the virus in cell culture is detected by hemadsorption. Immunofluorescence is used for identification. A fourfold or greater rise in antibody titer can also be used for diagnosis.

Treatment—None.

Prevention—No vaccine or drug is available.

Respiratory Syncytial Virus

Diseases—Most important cause of bronchiolitis and pneumonia in infants. Also causes otitis media in older children.

Characteristics—Enveloped virus with a helical nucleocapsid and one piece of single-stranded, negative-polarity RNA. RNA polymerase in virion. Unlike other paramyxoviruses, it has only a fusion protein in its surface spikes. It has no HA. It has two serotypes.

Transmission—Respiratory droplets.

Pathogenesis—Infection involves primarily the lower respiratory tract in infants without systemic spread. Immune response probably contributes to pathogenesis. Multinucleated giant cells caused by the viral fusion protein are a hallmark.

Laboratory Diagnosis—A PCR assay that detects viral RNA in respiratory tract specimens is used for diagnosis. Enzyme immunoassay (rapid antigen test) that detects respiratory syncytial virus (RSV) antigens in respiratory secretions is also used. If grown in cell culture, see multinucleated giant cells. Immunofluorescence is used for identification. Serology is not useful for diagnosis in infants.

Treatment—Aerosolized ribavirin for very sick infants.

Prevention—Passive immunization with palivizumab (monoclonal antibody) or immune globulins in infants who have been exposed is effective. Handwashing and the use of gloves may prevent nosocomial outbreaks in the newborn nursery.

Human Metapneumovirus

Diseases—Common cold, bronchiolitis, and pneumonia.

Characteristics—Enveloped virus with helical nucleocapsid and one piece of single-stranded, negative-polarity RNA. RNA polymerase in virion. A fusion protein on the envelope produces multinucleated giant cells in respiratory tract. There are two serotypes.

Transmission—Respiratory droplets.

Pathogenesis—Infection and death of respiratory epithelium without systemic spread of the virus.

Laboratory Diagnosis—Detect viral RNA in respiratory tract specimens by PCR assay.

Treatment—None.

Prevention—No vaccine or drug is available.

Coronavirus

Disease—Common cold, SARS (severe acute respiratory syndrome), MERS (Middle-east respiratory syndrome), and

COVID-19 (coronavirus infectious disease). Pandemic caused by SARS-CoV-2 began in 2019. Delta variant of SARS CoV-2 is causing widespread disease in 2021.

Characteristics—Enveloped virus with helical nucleocapsid and single-stranded, positive-polarity RNA. No virion polymerase. There are seven serotypes: four common cold serotypes and three pneumonia serotypes.

Transmission—Primarily respiratory droplets and aerosols. Animal coronaviruses appear to be the source of the three pneumonia serotypes, for example bat coronavirus is the source of SARS-CoV-2.

Pathogenesis—Infection is typically limited to the mucosal cells of the respiratory tract in common cold strains. However, SARS-CoV-2 causes symptoms in many organs. Entry into cells is mediated by spike protein of SARS-CoV-2 binding to ACE-2 receptor on respiratory tract epithelial cells.

At least 50% of SARS-CoV-2 infections are asymptomatic but can be source of infection for others. Immunity to common cold strains is brief and reinfection occurs. Duration of immunity to SARS-CoV-2, both to natural infection and vaccine is unknown.

Laboratory Diagnosis— A PCR assay that detects viral RNA in respiratory tract specimens is used for diagnosis. Antibody-based tests are also available.

Treatment—Remdesivir for severe SARS-CoV-2 infections. Dexamethasone and tocilizumab for cytokine release syndrome (cytokine storm). No drug available for common cold strains, SARS, and MERS.

Prevention—Three vaccines are available for COVID-19. Two are mRNA vaccines (mRNA for the spike protein) and one is a vectored vaccine containing a nonpathogenic human adenovirus with the gene for the spike protein of SARS-CoV-2 inserted into the adenovirus genome DNA.

Rhinoviruses

Disease—Common cold.

Characteristics—Naked nucleocapsid viruses with single-stranded, positive-polarity RNA. No virion polymerase. There are more than 100 serotypes, which explains why the common cold is so common. Rhinoviruses are destroyed by stomach acid and therefore do not replicate in the GI tract, in contrast to other picornaviruses such as poliovirus, Coxsackie virus, and echovirus, which are resistant to stomach acid.

Transmission—Aerosol droplets and hand-to-nose contact.

Pathogenesis—Infection is limited to the mucosa of the upper respiratory tract and conjunctiva. The virus replicates best at the low temperatures of the nose and less well at 37°C, which explains its failure to infect the lower respiratory tract.

Laboratory Diagnosis—Laboratory tests are rarely used clinically. A PCR assay that detects viral RNA in respiratory tract specimens can be used for diagnosis. Serologic tests are not useful.

Treatment—No antiviral therapy is available.

Prevention—No vaccine is available because there are too many serotypes.

Adenovirus

Diseases—Upper and lower tract respiratory disease, especially pharyngitis and pneumonia. Also conjunctivitis (pink-eye). Enteric strains cause diarrhea. Some strains cause sarcomas in certain animals but not humans.

Characteristics—Nonenveloped virus with icosahedral nucleocapsid and linear double-stranded DNA. No virion polymerase. There are 41 serotypes, some associated with specific diseases.

Transmission—Respiratory droplet primarily; iatrogenic transmission in eye disease; fecal–oral transmission with enteric strains.

Pathogenesis—Virus preferentially infects epithelium of respiratory tract and eyes. After acute infection, persistent, low-grade virus production without symptoms can occur in the pharynx.

Laboratory Diagnosis—A PCR assay that detects viral DNA in respiratory tract specimens is used for diagnosis. Virus causes CPE in cell culture and can be identified by fluorescent antibody. Antibody titer rise in convalescent-phase serum is diagnostic.

Treatment—None.

Prevention—Live vaccine against types 3, 4, and 7 is used in the military to prevent pneumonia.

IMPORTANT CHILDHOOD VIRUSES (CHAPTER 39)

Measles Virus

Disease—Measles. Subacute sclerosing panencephalitis is a rare late complication.

Characteristics—Enveloped virus with a helical nucleocapsid and one piece of single-stranded, negative-polarity RNA. RNA polymerase in virion. It has a single serotype.

Transmission—Respiratory droplets.

Pathogenesis—Initial site of infection is the upper respiratory tract. Virus spreads to local lymph nodes and then via the blood to other organs, including the skin. Giant cell pneumonia and encephalitis can occur. The maculopapular rash is due to cell-mediated immune attack by cytotoxic T cells on virus-infected vascular endothelial cells in the skin.

Laboratory Diagnosis—A PCR assay that detects viral RNA can be used for diagnosis. The virus is rarely isolated. Serologic tests are used if necessary.

Treatment—No antiviral therapy is available.

Prevention—Vaccine contains live, attenuated virus. Usually given in combination with mumps and rubella vaccines.

Mumps Virus

Disease—Mumps. Sterility due to bilateral orchitis is a rare complication.

Characteristics—Enveloped virus with a helical nucleocapsid and one piece of single-stranded, negative-polarity RNA. RNA polymerase in virion. It has a single serotype.

Transmission—Respiratory droplets.

Pathogenesis—The initial site of infection is the upper respiratory tract. The virus spreads to local lymph nodes and then via the bloodstream to other organs, especially the parotid glands, testes, ovaries, meninges, and pancreas.

Laboratory Diagnosis—A PCR assay that detects viral RNA can be used for diagnosis. The virus can be isolated in cell culture and detected by hemadsorption. Diagnosis can also be made serologically.

Treatment—No antiviral therapy is available.

Prevention—Vaccine contains live, attenuated virus. Usually given in combination with measles and rubella vaccines.

Rubella Virus

Disease—Rubella. Congenital rubella syndrome is characterized by congenital malformations, especially affecting the cardiovascular and central nervous systems, and by prolonged virus excretion. The incidence of congenital rubella has been greatly reduced by the widespread use of the vaccine.

Characteristics—Enveloped virus with an icosahedral nucleocapsid and one piece of single-stranded positive-polarity RNA. No polymerase in virion. It has a single serotype.

Transmission—Respiratory droplets and across the placenta from mother to fetus.

Pathogenesis—The initial site of infection is the nasopharynx, from which it spreads to local lymph nodes. It then disseminates to the skin via the bloodstream. The rash is attributed to both viral replication and immune injury. During maternal infection, the virus replicates in the placenta and then spreads to fetal tissue. If infection occurs during the

first trimester, a high frequency of congenital malformations occurs. Maternal antibody protects against fetal infection.

Laboratory Diagnosis—A PCR assay that detects viral RNA can be used for diagnosis. To determine whether an adult woman is immune, a single serum specimen to detect IgG antibody in the hemagglutination inhibition test is used. To detect whether recent infection has occurred, either a single serum specimen for IgM antibody or a set of acute- and convalescent-phase sera for IgG antibody can be used.

Treatment—No antiviral therapy is available.

Prevention—Vaccine contains live, attenuated virus. Usually given in combination with measles and mumps vaccine.

Parvovirus B19

Diseases—Slapped cheek syndrome (erythema infectiosum), aplastic anemia, arthritis, and hydrops fetalis.

Characteristics—Nonenveloped virus with icosahedral symmetry and single-stranded DNA genome. Virion contains no polymerase. There is one serotype.

Transmission—Respiratory droplets and transplacental.

Pathogenesis—Virus preferentially infects erythroblasts, causing aplastic anemia in patients with hereditary anemias; immune complexes cause rash and arthritis. Virus can infect fetus and cause severe anemia, leading to congestive heart failure and edema (hydrops fetalis). Maternal antibody protects fetus from infection.

Laboratory Diagnosis—A PCR assay that detects viral DNA in amniotic fluid or in blood can be used for diagnosis.

Treatment—None.

Prevention—There is no drug or vaccine.

Human Herpesvirus-6

Human herpesvirus-6 (HHV-6) causes roseola, a disease characterized by high fever and a maculopapular rash primarily on the trunk. HHV-6 is lymphotropic and is latent within both T and B lymphocytes. It can reactivate in immunocompromised patients, causing a systemic disease such as pneumonia, encephalitis, or hepatitis. Laboratory diagnosis is by PCR assay for viral DNA. There is no antiviral drug and no vaccine.

VIRUSES THAT INFECT THE ENTERIC TRACT (CHAPTER 40)

Norovirus

Disease—Gastroenteritis (watery diarrhea).

Characteristics—Nonenveloped virus with icosahedral nucleocapsid and one piece of single-stranded, positive-polarity

RNA. No virion polymerase. Many serotypes; exact number is uncertain.

Transmission—Fecal–oral route.

Pathogenesis—Infection is typically limited to the mucosal cells of the intestinal tract. Many infections are asymptomatic. Immunity is brief and reinfection occurs.

Laboratory Diagnosis—The diagnosis is primarily a clinical one. A PCR-based test that detects viral RNA in stool or vomitus is available.

Treatment—No antiviral drugs available. Treat diarrhea with fluid and electrolytes.

Prevention—No vaccine or drug available. Handwashing and disinfection of surfaces are helpful.

Rotavirus

Disease—Rotavirus causes gastroenteritis (diarrhea), especially in young children.

Characteristics—Naked double-layered capsid with 11 segments of double-stranded RNA. RNA polymerase in virion. Rotavirus is resistant to stomach acid and hence can reach the small intestine. There are at least six serotypes.

Transmission—Rotavirus is transmitted by the fecal–oral route.

Pathogenesis—Rotavirus infection is limited to the GI tract, especially the small intestine.

Laboratory Diagnosis—A PCR-based test that detects viral RNA in stool is available. Detection of rotavirus in the stool by ELISA is also done. Isolation of the virus from clinical specimens is not done.

Treatment—No antiviral drug is available.

Prevention—There are two rotavirus vaccines. One is a live attenuated vaccine that contains the single most common rotavirus serotype (G1), and the other is a live reassortant vaccine that contains five rotavirus strains.

Poliovirus

Diseases—Paralytic poliomyelitis and aseptic meningitis. Poliomyelitis has been eradicated in the Western Hemisphere and in many other countries.

Characteristics—Naked nucleocapsid virus with single-stranded, positive-polarity RNA. Genome RNA acts as mRNA and is translated into one large polypeptide, which is cleaved by virus-encoded protease to form functional viral proteins. No virion polymerase. There are three serotypes.

Transmission—Fecal–oral route. Humans are the natural reservoir.

Pathogenesis—The virus replicates in the pharynx and the GI tract. It can spread to the local lymph nodes and then through the bloodstream to the central nervous system. Most infections are asymptomatic or very mild. Aseptic meningitis is more frequent than paralytic polio. Paralysis is the result of death of motor neurons, especially anterior horn cells in the spinal cord. Pathogenesis of postpolio syndrome is unknown.

Laboratory Diagnosis—Recovery of the virus from spinal fluid indicates infection of the central nervous system. Isolation of the virus from stools indicates infection but not necessarily disease. It can be found in the GI tract of asymptomatic carriers. The virus can be detected in cell culture by CPE and identified by neutralization with type-specific antiserum. A significant rise in antibody titer in convalescent-phase serum is also diagnostic. A PCR-based test that detects viral RNA in stool is available.

Treatment—No antiviral therapy is available.

Prevention—Disease can be prevented by both the inactivated (Salk) vaccine and the live, attenuated (Sabin) vaccine; both induce humoral antibody that neutralizes the virus in the bloodstream. However, only the oral vaccine induces intestinal IgA, which interrupts the chain of transmission by preventing GI tract infection. For that reason and because it induces immunity of longer duration and is orally administered rather than injected, the Sabin vaccine has been the preferred vaccine for many years. However, there have been a few vaccine-associated cases of paralytic polio caused by poliovirus in the vaccine that reverted to virulence. In view of this, the current recommendation in the United States is to use the killed vaccine.

Coxsackie Viruses

Diseases—Aseptic meningitis, herpangina, pleurodynia, myocarditis, pericarditis, and hand, foot, and mouth disease are the most important diseases. Also Coxsackie virus B4 may cause juvenile diabetes, as it will do so in mice.

Characteristics—Naked nucleocapsid virus with single-stranded, positive-polarity RNA. No virion polymerase. Group A and B viruses are defined by their different pathogenicity in mice. There are multiple serotypes in each group.

Transmission—Fecal–oral route.

Pathogenesis—The initial site of infection is the oropharynx, but the main site is the GI tract. The virus spreads through the bloodstream to various organs.

Laboratory Diagnosis—A PCR-based test that detects viral RNA in spinal fluid or blood is available. The virus can be detected by CPE in cell culture and identified by neutralization. A significant rise in antibody titer in convalescent-phase serum is diagnostic.

Treatment—No antiviral therapy is available.

Prevention—No vaccine is available.

HEPATITIS VIRUSES (CHAPTER 41)

Hepatitis A Virus

Disease—Hepatitis A.

Characteristics—Naked nucleocapsid virus with a single-stranded, positive-polarity RNA. No virion polymerase. Virus has a single serotype.

Transmission—Fecal–oral route. In contrast to hepatitis B virus (HBV) and hepatitis C virus (HCV), blood-borne transmission of hepatitis A virus (HAV) is uncommon because viremia is brief and of low titer.

Pathogenesis—The virus replicates in the GI tract and then spreads to the liver during a brief viremic period. The virus is not cytopathic for the hepatocyte. Hepatocellular injury is caused by immune attack by cytotoxic T cells.

Laboratory Diagnosis—The most useful test to diagnose acute infection is IgM antibody. Isolation of the virus from clinical specimens is not done.

Treatment—No antiviral drug is available.

Prevention—Vaccine contains killed virus. Administration of immune globulin during the incubation period can mitigate the disease.

Hepatitis B Virus

Diseases—Hepatitis B; implicated as a cause of hepatocellular carcinoma.

Characteristics—Enveloped virus with incomplete circular double-stranded DNA (i.e., one strand has about one-third missing and the other strand is "nicked" [not covalently bonded]). DNA polymerase in virion. HBV-encoded DNA polymerase acts as a reverse transcriptase by using viral mRNA as the template for the synthesis of progeny genome DNA. There are three important antigens: the surface antigen, the core antigen, and the e antigen. Another protein, HBx, inactivates p53 tumor suppressor protein, a process involved in causing hepatocellular carcinoma. In the patient's serum, long rods and spherical forms composed solely of HBsAg predominate. HBV has one serotype based on the surface antigen.

Transmission—Transmitted by blood, during birth, and by sexual intercourse.

Pathogenesis—Hepatocellular injury due to immune attack by cytotoxic (CD8) T cells. Chronic carrier state occurs in 5% of adult infections but in 90% of neonatal infections because neonates have poor cytotoxic T-cell activity. Chronic carrier state can lead to chronic hepatitis, cirrhosis, and hepatocellular carcinoma. Hepatocellular carcinoma may be related to the integration of part of the viral DNA into hepatocyte DNA and subsequent synthesis of HBx protein. Antigen–antibody complexes cause arthritis, rash, and glomerulonephritis.

Laboratory Diagnosis—HBV has not been grown in cell culture. Three serologic tests are commonly used: surface antigen (HBsAg), surface antibody (HBsAb), and core antibody (HBcAb). Detection of HBsAg for more than 6 months indicates a chronic carrier state. The presence of e antigen indicates a chronic carrier who is making infectious virus. The presence of e antigen is an important indicator of transmissibility. An HBV-infected person who has neither detectable HBs antigen nor HBs antibody is said to be in the "window" phase. Diagnosis of this patient is made by detecting HB core antibody. See Chapter 41 for a discussion of the results of these tests.

Treatment—No treatment is given for acute hepatitis B. For chronic hepatitis B, a reverse transcriptase inhibitor, such as tenofovir or entecavir, can reduce the inflammation associated with chronic hepatitis B but does not cure the carrier state. A combination of tenofovir and emtricitabine is also effective.

Prevention—There are three main approaches: (1) vaccine that contains HBsAg as the immunogen; (2) hyperimmune serum globulins obtained from donors with high titers of HBsAb; and (3) education of chronic carriers regarding precautions. Passive–active immunization using both vaccine and immune globulins can prevent infection in neonates and those with needle-stick injuries.

Hepatitis C Virus

Disease—Hepatitis C; associated with hepatocellular carcinoma. HCV is the most prevalent bloodborne pathogen in the United States.

Characteristics—Enveloped virus with one piece of single-stranded, positive-polarity RNA. No polymerase in virion. HCV has six genotypes.

Transmission—Most transmission is perinatal or via blood. Sexual transmission is less common.

Pathogenesis—Hepatocellular injury caused by cytotoxic T cells. HCV replication itself does not kill cells (i.e., does not cause a cytopathic effect). More than 50% of infections result in the chronic carrier state. The chronic carrier state predisposes to chronic hepatitis and to hepatocellular carcinoma.

Laboratory Diagnosis—Serologic testing detects antibody to HCV. A PCR-based assay for "viral load" can be used to evaluate whether active infection is present.

Treatment—Treatment of acute hepatitis C with pegylated interferon alpha significantly reduce the number of patients who become chronic carriers. Treat chronic hepatitis C with a combination of drugs from three classes: an RNA polymerase inhibitor such as sofosbuvir, an NS5A inhibitor such as ledipasvir, and a protease inhibitor such as paritaprevir. Choice of drugs depends on genotype of HCV.

Prevention—Posttransfusion hepatitis can be prevented by discarding donated blood if antibody to HCV is detected. There is no vaccine, and hyperimmune globulins are not available.

Hepatitis D Virus

Disease—Hepatitis D (hepatitis delta).

Characteristics—Defective virus that uses hepatitis B surface antigen as its protein coat. Hepatitis D virus (HDV) can replicate only in cells already infected with HBV (i.e., HBV is a helper virus for HDV). Genome is one piece of single-stranded, negative-polarity, circular RNA. No polymerase in virion. HDV has one serotype (because HBV has only one serotype).

Transmission—Transmitted by blood, sexually, and from mother to child.

Pathogenesis—Hepatocellular injury probably caused by cytotoxic T cells. Chronic hepatitis and chronic carrier state occur.

Laboratory Diagnosis—Serologic testing detects either delta antigen or antibody to delta antigen.

Treatment—Pegylated alpha interferon mitigates symptoms but does not eradicate the carrier state.

Prevention—Prevention of HBV infection by using the HBV vaccine and the HBV hyperimmune globulins will prevent HDV infection also.

Hepatitis E Virus

Causes outbreaks of hepatitis, primarily in developing countries. Similar to HAV in the following ways: transmitted by fecal–oral route, no chronic carrier state, no cirrhosis, and no hepatocellular carcinoma. No antiviral therapy and no vaccine.

ARBOVIRUSES, RABIES VIRUS, & EBOLA VIRUS (CHAPTER 42)

All arboviruses are transmitted by arthropods (*arthropod-borne*) such as mosquitoes and ticks from the wild animal reservoir to humans.

West Nile Virus

Disease—Encephalitis. Most infections are asymptomatic.

Characteristics—Enveloped virus with icosahedral nucleocapsid and single-stranded, positive-polarity RNA. No polymerase in virion.

Transmission—Bite of *Culex* mosquito. Wild birds are reservoir. Humans are dead-end hosts.

Pathogenesis—Virus transmitted via blood from bite site to brain.

Laboratory Diagnosis—Virus isolation from blood, spinal fluid, or brain. Also antibodies in patient's serum. PCR test is available.

Treatment—No antiviral treatment.

Prevention—No vaccine or drug is available. Blood for transfusion is screened for antibodies.

Eastern Equine Encephalitis Virus

Member of the togavirus family. Causes encephalitis along the East Coast of the United States. Encephalitis is severe but uncommon. Transmitted to humans (and horses) by mosquitoes from small wild birds, such as sparrows. Humans and horses are "dead-end" hosts because viremia is low. There is no antiviral therapy and no vaccine for humans.

Western Equine Encephalitis Virus, St. Louis Encephalitis Virus, and California Encephalitis Virus

The transmission of these encephalitis viruses is similar (i.e., they are transmitted to humans by mosquitoes from small wild birds). However, they differ in details (i.e., they belong to different virus families and cause disease in different geographic areas). Please consult Chapter 42 in the text for specific information.

Yellow Fever Virus

Member of the flavivirus family. Causes yellow fever in the tropical areas of Africa and South America. "Jungle" yellow fever is transmitted from monkeys to humans by mosquitoes. "Urban" yellow fever is transmitted from human to human by *Aedes* mosquitoes (i.e., humans are the reservoir in the urban form). Humans are not a "dead-end" host because viremia is high. There is no antiviral therapy. There is a live, attenuated vaccine for humans.

Dengue Virus

Member of the flavivirus family. Causes dengue fever in the Caribbean region and other tropical areas. Dengue is the most common insect-borne viral disease in the world. Transmitted by *Aedes* mosquitoes from one human to another. A monkey reservoir is suspected. Second episodes may result in dengue hemorrhagic fever, a life-threatening complication. There is no antiviral therapy.

A vaccine against dengue (Dengvaxia) is available for use primarily in countries where the disease is common. It is a live, attenuated tetravalent recombinant vaccine. It is composed of a backbone of yellow fever vaccine virus containing the genes encoding the envelope and premembrane proteins of 4 sero-types of dengue virus.

Chikungunya Virus

Member of the togavirus family. Causes chikungunya fever in the Caribbean region and other tropical areas. Transmitted by *Aedes* mosquitoes from one human to another. There is no antiviral therapy and no vaccine.

Zika Virus

Member of the flavivirus family. Causes Zika fever and Zika congenital syndrome primarily in the tropical regions of Central and South America. Causes microcephaly and other fetal abnormalities. It is the *only* arbovirus to cause fetal abnormalities. Infection also predisposes to Guillain-Barré syndrome. Transmitted by *Aedes* mosquitoes from one human to another. Also transmitted in semen. There is no antiviral therapy and no vaccine.

Rabies Virus

Disease—Rabies is an encephalitis.

Characteristics—Bullet-shaped enveloped virus with a helical nucleocapsid and one piece of single-stranded, negative-polarity RNA. RNA polymerase in virion. The virus has a single serotype.

Transmission—Main reservoir is wild animals such as skunks, raccoons, and bats. Transmission to humans is usually by animal bite, but the virus is also transmitted by aerosols of bat saliva. In the United States, dogs are infrequently involved because canine immunization is so common, but in developing countries, they are often involved.

Pathogenesis—Viral receptor is the acetylcholine receptor. Replication of virus occurs at the site of the bite, followed by axonal transport up the nerve to the central nervous system. After replicating in the brain, the virus migrates peripherally to the salivary glands, where it enters the saliva. When the animal is in the agitated state as a result of encephalitis, virus in the saliva can be transmitted via a bite.

Laboratory Diagnosis—PCR assay can provide rapid diagnosis in both humans and animals. Tissue can be stained with fluorescent antibody or with various dyes to detect cytoplasmic inclusions called Negri bodies. The virus can be grown in cell culture, but the process takes too long to be useful in determining whether a person should receive the vaccine. Serologic testing is useful only to make the diagnosis in the clinically ill patient. Antibody does not form quickly enough to help in the decision whether or not to immunize the person who has been bitten. Serologic testing is also used to evaluate the antibody response to the vaccine given before exposure to those in high-risk occupations.

Treatment—No antiviral therapy is available.

Prevention—Preexposure prevention of rabies consists of the vaccine only. Postexposure prevention consists of (1) washing the wound; (2) giving rabies immune globulins (passive immunization), mostly into the wound; and (3) giving the inactivated vaccine (active immunization) made in human cell culture. The decision to give the immune serum and the vaccine depends on the circumstances. Prevention of rabies in dogs and cats by using a killed vaccine has reduced human rabies significantly.

Ebola Virus

Member of the Filovirus family. Causes Ebola hemorrhagic fever, which has a very high mortality rate. Bats are suspected to be the animal reservoir. The mode of transmission of the virus from bats to humans is unknown. Human-to-human transmission, especially in hospital setting, is by blood and other body fluids. Diagnosis is usually a clinical one, but PCR tests are available. In addition, detection of IgM in serum and detection of viral antigen in serum are also used. In electron microscope, seen as long "thread-like" viruses. Culturing the virus is very dangerous and should be done only in special laboratories.

No approved antiviral therapy is available. Experimental monoclonal antibodies (ZMapp and MAb114) and experimental drugs (remdesivir and favipiravir) are being tested but their effectiveness is uncertain.

A recombinant vaccine (rZEBOV) containing a backbone of vesicular stomatitis virus plus the gene encoding the Ebola virus surface glycoprotein is being used in African countries.

TUMOR VIRUSES (CHAPTER 43)

1. Human Cancer Viruses With RNA Genome

Human T-Cell Lymphotropic Virus

Disease—Adult T-cell leukemia/lymphoma and human T-cell lymphotropic virus (HTLV)-associated myelopathy (also known as tropical spastic paraparesis or chronic progressive myelopathy).

Characteristics—HTLV is a member of the retrovirus family. It causes malignant transformation of CD4-positive T cells (in contrast to HIV, which kills those cells). HTLV has three structural genes common to all retroviruses, namely, *gag*, *pol*, and *env*, plus two regulatory genes, *tax* and *rex*. The Tax protein is required for malignant transformation. It activates the synthesis of IL-2 (which is T-cell growth factor) and of the IL-2 receptor. IL-2 promotes rapid T-cell growth, which predisposes to malignant transformation.

Transmission—HTLV is transmitted primarily by intravenous drug use, sex, and breast-feeding. Transmission by donated blood has greatly decreased in the United States because donated blood that has antibodies to HTLV is discarded. HTLV infection is endemic in certain geographic areas, namely, the Caribbean region including southern Florida, eastern South America, western Africa, and southern Japan.

Pathogenesis—HTLV induces malignant transformation of CD4-positive T lymphocytes by activating IL-2 synthesis as described previously. It also causes HTLV-associated myelopathy (HAM), which is a demyelinating disease of the brain and spinal cord caused either by an autoimmune cross-reaction in which the immune response against HTLV damages the neurons or by cytotoxic T cells that kill HTLV-infected neurons.

Laboratory Diagnosis—Anti-HTLV antibodies can be detected in the patient's serum using the ELISA test. Western blot assay is used to confirm a positive ELISA result. PCR assay can detect the presence of HTLV RNA or DNA within infected cells.

Treatment and Prevention—No specific antiviral treatment for HTLV infection, and no antiviral drug will cure latent infections by HTLV. No vaccine against HTLV. Preventive measures include discarding donated blood if anti-HTLV antibodies are present, using condoms to prevent sexual transmission, and encouraging women with HTLV antibodies to refrain from breast-feeding.

Hepatitis C Virus

HCV causes hepatocellular carcinoma in chronic carriers of HCV. The mechanism of oncogenesis by HCV is unclear. It appears to be a consequence of the rapid cell division that occurs in an effort to replace the killed hepatocytes. An oncogene has not been identified in the HCV genome. For further information, see summary of HCV in Chapter 41.

2. Human Cancer Viruses With DNA Genome

Human Papillomavirus

Human papillomavirus (HPV) primarily causes carcinoma of the cervix, penis, and anus. Oncogenesis is a function of the *E6* and *E7* genes of HPV. The E6 and E7 proteins inactivate the tumor suppressor proteins, p53 and RB, respectively. For further information, see summary of HPV in Chapter 37.

Epstein–Barr Virus

Epstein–Barr virus (EBV) primarily causes Burkitt's lymphoma and nasopharyngeal carcinoma. Oncogenesis is a function of the translocation of the *c-myc* oncogene to a site adjacent to an immunoglobulin gene promoter. This enhances synthesis of the c-myc protein, a potent oncoprotein. For further information, see summary of EBV in Chapter 37.

Human Herpesvirus 8

HHV-8 causes KS. Oncogenesis is primarily a function of an early protein analogous to the E7 protein of HPV that inactivates the tumor suppressor protein RB. For further information, see summary of HHV-8 in Chapter 37.

Hepatitis B Virus

HBV causes hepatocellular carcinoma in chronic carriers of HBV. Oncogenesis is primarily a function of the Hbx protein that inactivates the tumor suppressor protein p53. For further information, see summary of HBV in Chapter 41.

Merkel Cell Polyomavirus

Merkel cell polyomavirus (MCPV) causes carcinoma of Merkel cells in the skin, often on sun-exposed areas such as the face and neck. MCPV is a nonenveloped virus with a double-stranded DNA genome. The T antigen protein of MCPV inhibits tumor suppressor proteins, p53 and RB. Infection with MCPV is widespread, but the carcinoma is rare. The carcinoma cells do not produce virus, so transmission from patients with the carcinoma to others does not occur. Diagnosis is made by microscopic analysis of surgical specimens. There is no virus-based assay. There is no antiviral drug or vaccine.

SLOW VIRUSES & PRIONS (CHAPTER 44)

JC Virus

Member of the papovavirus family. Causes progressive multifocal leukoencephalopathy (PML). Infection with JC virus is widespread, but PML occurs only in immunocompromised patients, such as those with AIDS. Invariably fatal. No antiviral therapy and no vaccine.

Prions

Diseases—Creutzfeldt-Jakob disease (CJD), variant CJD, and kuru. These are transmissible spongiform encephalopathies. There is a hereditary form of CJD called Gerstmann-Sträussler-Scheinker (GSS) syndrome.

Characteristics—Prions are composed of protein only. They have no detectable nucleic acid and are highly resistant to ultraviolet (UV) light, formaldehyde, and heat. They are encoded by a cellular gene. The pathogenic form increases in amount by inducing conformational change in normal form. Normal conformation is alpha helix; abnormal is beta-pleated sheet. In GSS syndrome, a mutation occurs that enhances the probability of the conformational change to the beta-pleated sheet form.

Transmission—In most cases of CJD, mode of transmission is unknown. CJD has been transmitted by pituitary extracts, brain electrodes, and corneal transplants. Kuru was transmitted by ingestion or inoculation of human brain tissue. Variant CJD probably is transmitted by ingestion of cow brain tissue in undercooked food.

Pathogenesis—Aggregation of prion filaments within neurons occurs and vacuoles within neurons cause spongiform changes in brain. No inflammation or immune response occurs.

Laboratory Diagnosis—Brain biopsy shows spongiform changes. No serologic tests are useful. Prions cannot be grown in culture.

Treatment—None.

Prevention—There is no drug or vaccine.

HUMAN IMMUNODEFICIENCY VIRUS (CHAPTER 45)

Disease—AIDS.

Characteristics—Enveloped virus with two copies (diploid) of a single-stranded, positive-polarity RNA genome. RNA-dependent DNA polymerase (reverse transcriptase) makes a DNA copy of the genome, which integrates into host cell DNA. Precursor polypeptides must be cleaved by virus-encoded protease to produce functional viral proteins. The tat gene encodes a protein that activates viral transcription. Antigenicity of the gp120 protein changes rapidly; therefore, there are many serotypes.

Transmission—Transfer of body fluids (e.g., blood and semen). Also transplacental and perinatal transmission.

Pathogenesis—Two receptors are required for HIV to enter cells. One receptor is CD4 protein found primarily on helper T cells. HIV infects and kills helper T cells, which predisposes to opportunistic infections. Other cells bearing CD4 proteins on the surface (e.g., astrocytes) are also infected. The other receptor for HIV is a chemokine receptor such as CCR5. The NEF protein is an important virulence factor. It reduces class I MHC protein synthesis, thereby reducing the ability of cytotoxic T cells to kill HIV-infected cells. Cytotoxic T cells are the main host defense against HIV.

Laboratory Diagnosis—HIV can be isolated from blood or semen, but this procedure is not routinely available. Diagnosis of early infection is made by immunoassays that detect antibody to HIV and to p24 antigen. If positive, then determine whether viral RNA is present in the blood using PCR test. In established infections, determine the "viral load" (i.e., the amount of HIV RNA in the plasma) using PCR-based assays. A high viral load predicts a more rapid progression to AIDS than a low viral load.

Treatment—HAART consists of several drugs combined into various regimens. For the treatment of newly infected adults, the combination of an integrase inhibitor, for example dolutegravir, plus two nucleoside inhibitors, tenofovir and emtricitabine, is often used. Clinical improvement occurs, but the virus persists for a lifetime, that is, the patient is *not* cured by drug therapy.

Nucleoside analogues, such as zidovudine, lamivudine, emtricitabine, tenofovir, and others inhibit HIV replication by inhibiting reverse transcriptase. Nonnucleoside inhibitors of reverse transcriptase, such as efavirenz, nevirapine, and others, are also used. Protease inhibitors (e.g., indinavir, ritonavir, and others) prevent cleavage of precursor polypeptides. Integrase inhibitors, such as raltegravir, dolutegravir, and elvitegravir, block the integration of HIV DNA into host cell DNA by inhibiting the integrase of HIV. Enfuvirtide, a "fusion inhibitor" that blocks entry of HIV, and maraviroc, which inhibits binding of the gp120 envelope protein of HIV to the cell coreceptor CCR-5, are also useful. Treatment of the opportunistic infection depends on the organism.

Prevention—There is no vaccine. Screening of blood prior to transfusion for the presence of antibody. "Safe sex," including the use of condoms. Preexposure prophylaxis (PrEP) using Truvada is indicated for individuals at high risk of infection. Injectable cabotegravir provides long-term prevention. Post-exposure prophylaxis (PEP), such as that given after a needle-stick injury or a high-risk nonoccupational exposure, employs three drugs: the preferred regimen consists of the combination of tenofovir and emtricitabine (given as Truvada) plus raltegravir. To reduce the number of cases of HIV infection in children: antiretroviral therapy should be given to HIV-infected mothers and neonates, and HIV-infected mothers should not breast-feed.

MINOR VIRAL PATHOGENS (CHAPTER 46)

Only the most important of the minor viral pathogens are summarized in this section.

Hantavirus (Sin Nombre Virus)

Member of the bunyavirus family. Causes hantavirus pulmonary syndrome. Sin Nombre virus (SNV) is a robovirus (i.e., it is *rodent-borne*). Deer mice are the reservoir, and the virus is acquired by inhalation of dried urine and feces. Diagnosis is made by detecting viral RNA in lung tissue or by serologic tests. No antiviral therapy and no vaccine.

Japanese Encephalitis Virus

Member of the flavivirus family. Causes outbreaks of encephalitis in Asian countries. Transmitted to humans by mosquitoes from the reservoir hosts, birds, and pigs. No antiviral therapy. An inactivated vaccine is available.

SUMMARIES OF MEDICALLY IMPORTANT FUNGI

FUNGI CAUSING CUTANEOUS & SUBCUTANEOUS MYCOSES (CHAPTER 48)

Dermatophytes (e.g., *Trichophyton, Microsporum, Epidermophyton species*)

Disease—Dermatophytoses (e.g., tinea capitis, tinea cruris, and tinea pedis).

Characteristics—These fungi are molds that use keratin as a nutritional source. Not dimorphic. Habitat of most dermatophytes that cause human disease is human skin, with the exception of *Microsporum canis*, which infects dogs and cats also.

Transmission—Direct contact with skin scales.

Pathogenesis—These fungi grow only in the superficial keratinized layer of the skin. They do not invade underlying tissue. The lesions are due to the inflammatory response to the fungi. Frequency of infection is enhanced by moisture and warmth (e.g., inside shoes). An important host defense is provided by the fatty acids produced by sebaceous glands. The "id" reaction is a hypersensitivity response in one skin location (e.g., fingers) to the presence of the organism in another (e.g., feet).

Laboratory Diagnosis—Skin scales should be examined microscopically in a KOH preparation for the presence of hyphae. The organism is identified by the appearance of its mycelium and its asexual spores on Sabouraud's agar. Serologic tests are not useful.

Skin Test—Trichophyton antigen can be used to determine the competence of a patient's cell-mediated immunity. Not used for diagnosis of tinea.

Treatment—Topical agents, such as miconazole, clotrimazole, or tolnaftate, are used. Undecylenic acid is effective against tinea pedis. Griseofulvin is the treatment of choice for tinea unguium and tinea capitis.

Prevention—Skin should be kept dry and cool.

Sporothrix schenckii

Disease—Sporotrichosis.

Characteristics—Thermally dimorphic. Mold in the soil, yeast in the body at 37°C. Habitat is soil or vegetation.

Transmission—Mold spores enter skin in puncture wounds caused by rose thorns and other sharp objects in the garden.

Pathogenesis—Local abscess or ulcer with nodules in draining lymphatics.

Laboratory Diagnosis—Cigar-shaped budding yeasts visible in pus. Culture on Sabouraud's agar shows typical morphology.

Skin Test—None.

Treatment—Itraconazole.

Prevention—Skin should be protected when gardening.

FUNGI CAUSING SYSTEMIC MYCOSES (CHAPTER 49)

Histoplasma capsulatum

Disease—Histoplasmosis.

Characteristics—Thermally dimorphic (i.e., a yeast at body temperature and a mold in the soil at ambient temperature). The mold grows preferentially in soil enriched with bird droppings. Endemic in Ohio and Mississippi River Valley areas.

Transmission—Inhalation of airborne asexual spores (microconidia).

Pathogenesis—Microconidia enter the lung and differentiate into yeast cells. The yeast cells are ingested by alveolar macrophages and multiply within them. An immune response is mounted, and granulomas form. Most infections are contained at this level, but suppression of cell-mediated immunity can lead to disseminated disease.

Laboratory Diagnosis—Sputum or tissue can be examined microscopically and cultured on Sabouraud's agar. Yeasts visible within macrophages. The presence of tuberculate chlamydospores in culture at 25°C is diagnostic. A rise in antibody titer is useful for diagnosis, but cross-reaction with other fungi (e.g., *Coccidioides*) occurs.

Skin Test—Histoplasmin, a mycelial extract, is the antigen. Useful for epidemiologic purposes to determine the incidence of infection. A positive result indicates only that infection has occurred; it cannot be used to diagnose active disease. Because skin testing can induce antibodies, serologic tests must be done first.

Treatment—Amphotericin B or itraconazole for disseminated disease; itraconazole for pulmonary disease.

Prevention—No vaccine is available. Itraconazole can be used for chronic suppression in AIDS patients.

Coccidioides immitis

Disease—Coccidioidomycosis.

Characteristics—Thermally dimorphic. At 37°C in the body, it forms spherules containing endospores. At 25°C, either in the soil or on agar in the laboratory, it grows as a mold. The cells at the tip of the hyphae differentiate into asexual spores (arthrospores). Natural habitat is the soil of arid regions (e.g., San Joaquin Valley in California and parts of Arizona and New Mexico).

Transmission—Inhalation of airborne arthrospores.

Pathogenesis—Arthrospores differentiate into spherules in the lungs. Spherules rupture, releasing endospores that form new spherules, thereby disseminating the infection within the body. A cell-mediated immune response contains the infection in most people, but those who have reduced cell-mediated immunity are at high risk for disseminated disease.

Laboratory Diagnosis—Sputum or tissue should be examined microscopically for spherules and cultured on Sabouraud's agar. A rise in IgM (using precipitin test) antibodies indicates recent infection. A rising titer of IgG antibodies (using complement-fixation test) indicates dissemination; a decreasing titer indicates a response to therapy.

Skin Test—Either coccidioidin, a mycelial extract, or spherulin, an extract of spherules, is the antigen. Useful in determining whether the patient has been infected. A positive test indicates prior infection but not necessarily active disease.

Treatment—Amphotericin B or itraconazole for disseminated disease; ketoconazole for limited pulmonary disease.

Prevention—No vaccine or prophylactic drug is available.

Blastomyces dermatitidis

Disease—Blastomycosis.

Characteristics—Thermally dimorphic. Mold in the soil, yeast in the body at 37°C. The yeast form has a single, broad-based bud and a thick, refractile wall. Natural habitat is rich soil (e.g., near beaver dams), especially in the upper midwestern region of the United States.

Transmission—Inhalation of airborne spores (conidia).

Pathogenesis—Inhaled conidia differentiate into yeasts, which initially cause abscesses followed by formation of granulomas. Dissemination is rare, but when it occurs, skin and bone are most commonly involved.

Laboratory Diagnosis—Sputum or skin lesions examined microscopically for yeasts with a broad-based bud. Culture on Sabouraud's agar also. Serologic tests are not useful.

Skin Test—Little value.

Treatment—Itraconazole is the drug of choice.

Prevention—No vaccine or prophylactic drug is available.

Paracoccidioides brasiliensis

Disease—Paracoccidioidomycosis.

Characteristics—Thermally dimorphic. Mold in the soil, yeast in the body at 37°C. The yeast form has multiple buds (resembles the steering wheel of a ship).

Transmission—Inhalation of airborne conidia.

Pathogenesis—Inhaled conidia differentiate to the yeast form in lungs. Can disseminate to many organs.

Laboratory Diagnosis—Yeasts with multiple buds visible in pus or tissues. Culture on Sabouraud's agar shows typical morphology.

Skin Test—Not useful.

Treatment—Itraconazole.

Prevention—No vaccine or prophylactic drug is available.

FUNGI CAUSING OPPORTUNISTIC MYCOSES (CHAPTER 50)

Candida albicans

Diseases—Thrush, disseminated candidiasis, and chronic mucocutaneous candidiasis.

Characteristics—*C. albicans* is a yeast when part of the normal flora of mucous membranes but forms pseudohyphae and hyphae when it invades tissue. The yeast form produces germ tubes when incubated in serum at 37°C. Not thermally dimorphic.

Transmission—Part of the normal flora of skin, mucous membranes, and GI tract. No person-to-person transmission.

Pathogenesis—Opportunistic pathogen. Predisposing factors include reduced cell-mediated immunity, altered skin and mucous membrane, suppression of normal flora by antibiotics, and presence of foreign bodies. Thrush is most common in infants, immunosuppressed patients, and persons receiving antibiotic therapy. Skin lesions occur frequently on moisture-damaged skin. Disseminated infections, such as endocarditis and endophthalmitis, occur in immunosuppressed patients and intravenous drug users. Chronic mucocutaneous candidiasis occurs in children with a T-cell defect in immunity to *Candida*.

Laboratory Diagnosis—Microscopic examination of tissue reveals yeasts and pseudohyphae. If only yeasts are found, colonization is suggested. The yeast is gram-positive. Forms colonies of yeasts on Sabouraud's agar. Germ tube formation and production of chlamydospores distinguish *C. albicans* from virtually all other species of *Candida*. Serologic tests not useful.

Skin Test—Used to determine competency of cell-mediated immunity rather than to diagnose candidal disease.

Treatment—Skin infections can be treated with topical antifungal agents such as nystatin or clotrimazole. Oral thrush is treated with fluconazole. Esophageal thrush can be treated with fluconazole or caspofungin. Vaginitis can be treated with either intravaginal clotrimazole or oral fluconazole. Disseminated disease can be treated with either amphotericin B, fluconazole or caspofungin. Chronic mucocutaneous candidiasis can be controlled with fluconazole.

Prevention—Predisposing factors should be reduced or eliminated. Oral thrush can be prevented by using clotrimazole troches or nystatin "swish and swallow." Fluconazole is used to prevent disseminated infection in immunocompromised patients. There is no vaccine.

Cryptococcus neoformans

Disease—Cryptococcosis, especially cryptococcal meningitis.

Characteristics—Heavily encapsulated yeast. Not dimorphic. Habitat is soil, especially where enriched by pigeon droppings.

Transmission—Inhalation of airborne yeast cells.

Pathogenesis—Organisms cause influenza-like syndrome or pneumonia. They spread via the bloodstream to the meninges. Reduced cell-mediated immunity predisposes to severe disease, but some cases of cryptococcal meningitis occur in immunocompetent people who inhale a large dose of organisms.

Laboratory Diagnosis—Visualization of the encapsulated yeast in India ink preparations of spinal fluid. Culture of sputum or spinal fluid on Sabouraud's agar produces colonies of yeasts. Cryptococcal antigen test (CRAG) is a latex agglutination test that detects polysaccharide capsular antigen in spinal fluid.

Skin Test—Not available.

Treatment—Amphotericin B plus flucytosine for meningitis.

Prevention—Cryptococcal meningitis can be prevented in AIDS patients by using oral fluconazole. There is no vaccine.

Aspergillus fumigatus

Diseases—Invasive aspergillosis is the major disease. Allergic bronchopulmonary aspergillosis and aspergilloma (fungus ball) are important also.

Characteristics—Mold with septate hyphae that branch at a V-shaped angle (low-angle branching). Not dimorphic. Habitat is the soil.

Transmission—Inhalation of airborne spores (conidia).

Pathogenesis—Opportunistic pathogen. In immunocompromised patients, invasive disease occurs. The organism invades blood vessels, causing thrombosis and infarction. A person with a lung cavity (e.g., from tuberculosis) may develop a "fungal ball" (aspergilloma). An allergic (hypersensitive) person (e.g., one with asthma) is predisposed to allergic bronchopulmonary aspergillosis mediated by IgE antibody.

Laboratory Diagnosis—Septate hyphae invading tissue are visible microscopically. Invasion distinguishes disease from

colonization. Forms characteristic mycelium when cultured on Sabouraud's agar. See chains of conidia radiating from a central stalk. Serologic tests detect IgG precipitins in patients with aspergillomas and IgE antibodies in patients with allergic bronchopulmonary aspergillosis.

Skin Test—None available.

Treatment—Amphotericin B or voriconazole for invasive aspergillosis. Some lesions (e.g., fungus balls) can be surgically removed. Corticosteroids plus itraconazole are recommended for allergic bronchopulmonary aspergillosis.

Prevention—No vaccine or prophylactic drug is available.

Mucor & Rhizopus species

Disease—Mucormycosis.

Characteristics—Molds with nonseptate hyphae that typically branch at a 90-degree angle (wide-angle branching). Not dimorphic. Habitat is the soil.

Transmission—Inhalation of airborne spores.

Pathogenesis—Opportunistic pathogens. They cause disease primarily in ketoacidotic diabetic and leukemic patients. The sinuses and surrounding tissue are typically involved. Hyphae invade the mucosa and progress into underlying tissue and vessels, leading to necrosis and infarction.

Laboratory Diagnosis—Microscopic examination of tissue for the presence of nonseptate hyphae that branch at wide angles. Forms mycelium when cultured on Sabouraud's agar showing spores contained within a sac called a sporangium. Serologic tests are not available.

Skin Test—None.

Treatment—Amphotericin B and surgical debridement.

Prevention—No vaccine or prophylactic drug is available. Control of underlying disease (e.g., diabetes) tends to prevent mucormycosis.

Pneumocystis jirovecii

Although there is molecular evidence that *P. jirovecii* is a fungus, it is described in these brief summaries in the section on protozoa that cause blood and tissue infections (see Chapter 52).

Microsporidia

Group of spore-forming, obligate intracellular fungi. Two important species are *Enterocytozoon bieneusi* and *Septata intestinalis*. Cause diarrhea, especially in immunocompromised (e.g., AIDS) patients. Acquired by fecal–oral transmission from human sources. Diagnosis is made by finding spores within cells in feces or intestinal biopsy specimens. Treat with albendazole. No specific means of prevention.

SUMMARIES OF MEDICALLY IMPORTANT PARASITES

PROTOZOA CAUSING INTESTINAL & UROGENITAL INFECTIONS (CHAPTER 51)

Entamoeba histolytica

Diseases—Amebic dysentery and liver abscess.

Characteristics—Intestinal protozoan. Motile ameba (trophozoite); forms cysts with four nuclei. Life cycle: Humans ingest cysts, which form trophozoites in small intestine. Trophozoites pass to the colon and multiply. Cysts form in the colon, which then pass in the feces.

Transmission and Epidemiology—Fecal–oral transmission of cysts. Human reservoir. Occurs worldwide, especially in tropics.

Pathogenesis—Trophozoites invade colon epithelium and produce flask-shaped ulcer. Can spread to liver and cause amebic abscess.

Laboratory Diagnosis—Trophozoites or cysts visible in stool. Serologic testing (indirect hemagglutination test) positive with invasive (e.g., liver) disease.

Treatment—Metronidazole or tinidazole for symptomatic disease. Iodoquinol or paromomycin for asymptomatic cyst carriers.

Prevention—Proper disposal of human waste. Water purification. Handwashing.

Giardia lamblia

Disease—Giardiasis, especially diarrhea.

Characteristics—Intestinal protozoan. Pear-shaped, flagellated trophozoite, forms cyst with four nuclei. Life cycle: Humans ingest cysts, which form trophozoites in duodenum. Trophozoites form cysts that are passed in feces.

Transmission and Epidemiology—Fecal–oral transmission of cysts. Human and animal reservoir. Occurs worldwide.

Pathogenesis—Trophozoites attach to wall but do not invade. They interfere with absorption of fat and protein.

Laboratory Diagnosis—Trophozoites or cysts visible in stool. String test used if necessary.

Treatment—Metronidazole.

Prevention—Water purification. Handwashing.

Cryptosporidium hominis

Disease—Cryptosporidiosis, especially diarrhea.

Characteristics—Intestinal protozoan. Life cycle: Oocysts release sporozoites; they form trophozoites. After schizonts and merozoites form, microgametes and macrogametes are produced; they unite to form a zygote and then an oocyst.

Transmission and Epidemiology—Fecal–oral transmission of cysts. Human and animal reservoir. Occurs worldwide.

Pathogenesis—Trophozoites attach to wall of small intestine but do not invade.

Laboratory Diagnosis—Oocysts visible in stool with acid-fast stain.

Treatment—No effective therapy; however, paromomycin may reduce symptoms.

Prevention—None.

Trichomonas vaginalis

Disease—Trichomoniasis.

Characteristics—Urogenital protozoan. Pear-shaped, flagellated trophozoites. No cysts or other forms.

Transmission and Epidemiology—Transmitted sexually. Human reservoir. Occurs worldwide.

Pathogenesis—Trophozoites attach to the wall of vagina and cause inflammation and discharge.

Laboratory Diagnosis—Trophozoites visible in secretions.

Treatment—Metronidazole for both sexual partners.

Prevention—Condoms limit transmission.

PROTOZOA CAUSING BLOOD & TISSUE INFECTIONS (CHAPTER 52)

Plasmodium species (Plasmodium vivax, Plasmodium ovale, Plasmodium malariae, & Plasmodium falciparum)

Disease—Malaria.

Characteristics—Protozoan that infects red blood cells and tissue (e.g., liver, kidney, and brain). Life cycle: Sexual cycle consists of gametogony (production of gametes) in humans and sporogony (production of sporozoites) in mosquitoes; asexual cycle (schizogony) occurs in humans. Sporozoites in saliva of female *Anopheles* mosquito enter the human bloodstream and rapidly invade hepatocytes (exoerythrocytic phase). There they multiply and form merozoites (*P. vivax* and *P. ovale* also form hypnozoites, a latent form). Merozoites leave the hepatocytes and infect red cells (erythrocytic phase). There they form schizonts that release more merozoites, which infect other red cells in a synchronous pattern (3 days for *P. malariae*; 2 days for the others). Some merozoites become male and female gametocytes, which, when ingested by female *Anopheles*, release male and female gametes. These unite to produce a zygote, which forms an oocyst containing many sporozoites. These are released and migrate to salivary glands.

Transmission and Epidemiology—Transmitted by female *Anopheles* mosquitoes. Occurs primarily in the tropical areas of Asia, Africa, and Latin America.

Pathogenesis—Merozoites destroy red cells, resulting in anemia. Cyclic fever pattern is due to periodic release of merozoites. *P. falciparum* can infect red cells of all ages and cause aggregates of red cells that occlude capillaries. This can cause tissue anoxia, especially in the brain (cerebral malaria) and the kidney (blackwater fever). Hypnozoites can cause relapses.

Laboratory Diagnosis—Organisms visible in blood smear. Thick smear is used to detect the presence of organism and thin smear to speciate.

Treatment—Chloroquine if sensitive. For chloroquine-resistant *P. falciparum*, use Coartem (artemether plus lumefantrine) or Malarone (atovaquone and proguanil). Primaquine for hypnozoites of *P. vivax* and *P. ovale*. In severe cases, use parenteral artesunate or quinidine.

Prevention—Chloroquine in areas where organisms are sensitive. For those in areas with a high risk of chloroquine resistance, Malarone, mefloquine, or doxycycline. Primaquine to prevent relapses of *P. vivax* or *P. ovale*. Tafenoquine, a long-acting analog of primaquine, can also be used. Protection from bites. Control mosquitoes by using insecticides and by draining water from breeding areas.

Toxoplasma gondii

Disease—Toxoplasmosis, including congenital toxoplasmosis.

Characteristics—Tissue protozoan. Life cycle: Cysts in cat feces or in meat are ingested by humans and differentiate in the gut into forms that invade the gut wall. They infect macrophages and form trophozoites (tachyzoites) that multiply rapidly, kill cells, and infect other cells. Cysts containing bradyzoites form later. Cat ingests cysts in raw meat, and bradyzoites excyst, multiply, and form male and female gametocytes. These fuse to form oocysts in cat gut, which are excreted in cat feces.

Transmission and Epidemiology—Transmitted by ingestion of cysts in raw meat and in food contaminated with cat feces. Also by passage of trophozoites transplacentally from mother to fetus. Infection of fetus occurs only when mother is infected during pregnancy and when she is infected for the first time (i.e., she has no protective antibody). Cat is definitive host; humans and other mammals are intermediate hosts. Occurs worldwide.

Pathogenesis—Trophozoites infect many organs, especially brain, eyes, and liver. Cysts persist in tissue, enlarge, and cause symptoms. Severe disease in patients with deficient cell-mediated immunity (e.g., encephalitis in AIDS patients).

Laboratory Diagnosis—Serologic tests for IgM and IgG antibodies are usually used. Trophozoites or cysts visible in tissue.

Treatment—Sulfadiazine plus pyrimethamine for congenital or disseminated disease.

Prevention—Meat should be cooked. Pregnant women should not handle cats, cat litter boxes, or raw meat. Trimethoprim-sulfamethoxazole is used to prevent *Toxoplasma* encephalitis in HIV-infected patients.

Pneumocystis jirovecii

Disease—Pneumonia.

Characteristics—Respiratory pathogen. Reclassified in 1988 as a yeast based on molecular evidence but medically has several attributes of a protozoan. Life cycle: uncertain.

Transmission and Epidemiology—Transmitted by inhalation. Humans are reservoir. Occurs worldwide. Most infections asymptomatic.

Pathogenesis—Organisms in alveoli cause inflammation. Immunosuppression predisposes to disease.

Laboratory Diagnosis—Organisms visible in silver stain of lung tissue or lavage fluid.

Treatment—Trimethoprim-sulfamethoxazole is drug of choice. Pentamidine is an alternative drug.

Prevention—Trimethoprim-sulfamethoxazole or aerosolized pentamidine in immunosuppressed individuals.

Trypanosoma cruzi

Disease—Chagas' disease.

Characteristics—Blood and tissue protozoan. Life cycle: Trypomastigotes in blood of reservoir host are ingested by reduviid bug and form epimastigotes and then trypomastigotes in the gut. When the bug bites, it defecates and feces containing trypomastigotes contaminate the wound. Organisms enter the blood and form amastigotes within cells; these then become trypomastigotes.

Transmission and Epidemiology—Transmitted by reduviid bugs. Humans and many animals are reservoirs. Occurs in rural Latin America.

Pathogenesis—Amastigotes kill cells, especially cardiac muscle, leading to myocarditis. Also neuronal damage, leading to megacolon and megaesophagus.

Laboratory Diagnosis—Trypomastigotes visible in blood, but bone marrow biopsy, culture in vitro, xenodiagnosis, or serologic tests may be required.

Treatment—Nifurtimox or benznidazole for acute disease. No effective drug for chronic disease.

Prevention—Protection from bite. Insect control. Blood for transfusion should not be used if antibodies to *T. cruzi* are present.

Trypanosoma gambiense & Trypanosoma rhodesiense

Disease—Sleeping sickness (African trypanosomiasis).

Characteristics—Blood and tissue protozoan. Life cycle: Trypomastigotes in blood of human or animal reservoir are ingested by tsetse fly. They differentiate in the gut to form epimastigotes and then metacyclic trypomastigotes in salivary glands. When fly bites, trypomastigotes enter the blood. Repeated variation of surface antigen occurs, which allows the organism to evade the immune response.

Transmission and Epidemiology—Transmitted by tsetse flies. *T. gambiense* has a human reservoir and occurs primarily in west Africa. *T. rhodesiense* has an animal reservoir (especially wild antelope) and occurs primarily in east Africa.

Pathogenesis—Trypomastigotes infect brain, causing encephalitis.

Laboratory Diagnosis—Trypomastigotes visible in blood in early stages and in cerebrospinal fluid in late stages. Serologic tests useful.

Treatment—Suramin in early disease. Suramin plus melarsoprol if central nervous system symptoms exist.

Prevention—Protection from bite. Insect control.

Leishmania donovani

Disease—Kala-azar (visceral leishmaniasis).

Characteristics—Blood and tissue protozoan. Life cycle: Human macrophages containing amastigotes are ingested by sandfly. Amastigotes differentiate in fly gut to promastigotes, which migrate to pharynx. When sandfly bites a human, promastigotes enter blood macrophages and form amastigotes. These can infect other reticuloendothelial cells, especially in spleen and liver.

Transmission and Epidemiology—Transmitted by sandflies (*Phlebotomus* or *Lutzomyia*). Animal reservoir (chiefly dogs,

small carnivores, and rodents) in Africa, Middle East, and parts of China. Human reservoir in India.

Pathogenesis—Amastigotes kill reticuloendothelial cells, especially in liver, spleen, and bone marrow.

Laboratory Diagnosis—Amastigotes visible in bone marrow smear. Serologic tests useful. Skin test indicates prior infection.

Treatment—Sodium stibogluconate.

Prevention—Protection from bite. Insect control.

Leishmania tropica, Leishmania mexicana, and Leishmania braziliensis

L. tropica and *L. mexicana* cause cutaneous leishmaniasis; *L. braziliensis* causes mucocutaneous leishmaniasis. *L. tropica* occurs primarily in the Middle East, Asia, and India, whereas *L. mexicana* and *L. braziliensis* occur in Central and South America. All are transmitted by sandflies. Forest rodents are the main reservoir. Diagnosis is made by observing amastigotes in smear of skin lesion. Treatment is sodium stibogluconate. No specific means of prevention.

MINOR PROTOZOAN PATHOGENS (CHAPTER 53)

Acanthamoeba castellanii

Ameba that causes meningoencephalitis. Also causes keratitis in contact lens wearers. Life cycle includes trophozoite and cyst stages. Found in freshwater lakes and soil. Transmitted via trauma to skin or eyes. Disease occurs primarily in immunocompromised patients. Diagnosis made by finding ameba in spinal fluid. Treatment with pentamidine, ketoconazole, or flucytosine may be effective. No specific means of prevention.

Naegleria fowleri

Ameba that causes meningoencephalitis. Found in freshwater lakes and soil. Life cycle includes trophozoite and cyst stages. Transmitted while swimming or diving in contaminated lake. Disease occurs primarily in healthy individuals. Diagnosis made by finding ameba in spinal fluid. Treatment with amphotericin B may be effective. No specific means of prevention.

Babesia microti

Sporozoan that causes babesiosis. Endemic in rodents along the northeast coast of the United States. Transmitted by *Ixodes* ticks to humans. Infects red blood cells, causing them to lyse, and anemia results. Asplenic patients have severe disease. Diagnosis is made by observing organism in "Maltese cross" tetrad pattern in red blood cells. Treat with combination of atovaquone and azithromycin for mild disease. Use a combination of quinine and clindamycin for serious disease. No specific means of prevention.

Balantidium coli

Only ciliated protozoan to cause human disease. Causes diarrhea. Acquired by fecal–oral transmission from domestic animals, especially pigs. Diagnosis is made by finding trophozoites or cysts in feces. Treat with tetracycline. No specific means of prevention.

Cyclospora cayetanensis

Coccidian protozoan. Causes diarrhea, especially in immunocompromised (e.g., AIDS) patients. Acquired by fecal–oral transmission. No evidence for animal reservoir. Diagnosis is made by finding oocytes in acid-fast stain of feces. Treat with trimethoprim-sulfamethoxazole. No specific means of prevention.

Isospora belli

Coccidian protozoan. Causes diarrhea, especially in immunocompromised (e.g., AIDS) patients. Acquired by fecal–oral transmission from either human or animal sources. Diagnosis is made by finding oocytes in acid-fast stain of feces. Treat with trimethoprim-sulfamethoxazole. No specific means of prevention.

CESTODES (CHAPTER 54)

Diphyllobothrium latum

Disease—Diphyllobothriasis.

Characteristics—Cestode (fish tapeworm). Scolex has two elongated sucking grooves; no circular suckers or hooks. Gravid uterus forms a rosette. Oval eggs have an operculum at one end. Life cycle: Humans ingest undercooked fish containing sparganum larvae. Larvae attach to gut wall and become adults containing gravid proglottids. Eggs are passed in feces. In freshwater, eggs hatch and the embryos are eaten by copepods. When these are eaten by freshwater fish, larvae form in the fish muscle.

Transmission and Epidemiology—Transmitted by eating raw or undercooked freshwater fish. Humans are definitive hosts; copepods are the first and fish the second intermediate hosts, respectively. Occurs worldwide but endemic in Scandinavia, Japan, and northcentral United States.

Pathogenesis—Tapeworm in gut causes little damage.

Laboratory Diagnosis—Eggs visible in stool.

Treatment—Praziquantel.

Prevention—Adequate cooking of fish. Proper disposal of human waste.

Echinococcus granulosus

Disease—Hydatid cyst disease.

Characteristics—Cestode (dog tapeworm). Scolex has four suckers and a double circle of hooks. Adult worm has only three proglottids. Life cycle: Dogs are infected when they ingest the entrails of sheep (e.g., liver) containing hydatid cysts. The adult worms develop in the gut, and eggs are passed in the feces. Eggs are ingested by sheep (and humans) and hatch hexacanth larvae in the gut that migrate in the blood to various organs, especially the liver and brain. Larvae form large, unilocular hydatid cysts containing many protoscoleces and daughter cysts.

Transmission and Epidemiology—Transmitted by ingestion of eggs in food contaminated with dog feces. Dogs are main definitive hosts; sheep are intermediate hosts; humans are dead-end hosts. Endemic in sheep-raising areas (e.g., Mediterranean, Middle East, some western states of the United States).

Pathogenesis—Hydatid cyst is a space-occupying lesion. Also, if cyst ruptures, antigens in fluid can cause anaphylaxis.

Laboratory Diagnosis—Serologic tests (e.g., indirect hemagglutination). Pathologic examination of excised cyst.

Treatment—Albendazole or surgical removal of cyst.

Prevention—Sheep entrails should not be fed to dogs.

Taenia saginata

Disease—Taeniasis.

Characteristics—Cestode (beef tapeworm). Scolex has four suckers but no hooks. Gravid proglottids have 15–20 uterine branches. Life cycle: Humans ingest undercooked beef containing cysticerci. Larvae attach to gut wall and become adult worms with gravid proglottids. Terminal proglottids detach, pass in feces, and are eaten by cattle. In the gut, oncosphere embryos hatch, burrow into blood vessels, and migrate to skeletal muscles, where they develop into cysticerci.

Transmission and Epidemiology—Transmitted by eating raw or undercooked beef. Humans are definitive hosts; cattle are intermediate hosts. Occurs worldwide but endemic in areas of Asia, Latin America, and Eastern Europe.

Pathogenesis—Tapeworm in gut causes little damage. In contrast to *Taenia solium*, cysticercosis does not occur.

Laboratory Diagnosis—Gravid proglottids visible in stool. Eggs seen less frequently.

Treatment—Praziquantel.

Prevention—Adequate cooking of beef. Proper disposal of human waste.

Taenia solium

Diseases—Taeniasis and cysticercosis.

Characteristics—Cestode (pork tapeworm). Scolex has four suckers and a circle of hooks. Gravid proglottids have 5–10 uterine branches. Life cycle: Humans ingest undercooked pork containing cysticerci. Larvae attach to gut wall and develop into adult worms with gravid proglottids. Terminal proglottids detach, pass in feces, and are eaten by pigs. In gut, oncosphere (hexacanth) embryos burrow into blood vessels and migrate to skeletal muscle, where they develop into cysticerci. If humans eat *T. solium* eggs in food contaminated with human feces, the oncospheres burrow into blood vessels and disseminate to organs (e.g., brain, eyes), where they encyst to form cysticerci.

Transmission and Epidemiology—Taeniasis acquired by eating raw or undercooked pork. Cysticercosis acquired only by ingesting eggs in fecally contaminated food or water. Humans are definitive hosts; pigs or humans are intermediate hosts. Occurs worldwide but endemic in areas of Asia, Latin America, and southern Europe.

Pathogenesis—Tapeworm in gut causes little damage. Cysticerci can expand and cause symptoms of mass lesions, especially in brain.

Laboratory Diagnosis—Gravid proglottids visible in stool. Eggs seen less frequently.

Treatment—Praziquantel for intestinal worms and for cerebral cysticercosis.

Prevention—Adequate cooking of pork. Proper disposal of human waste.

Hymenolepis nana

H. nana infection is the most common tapeworm in the United States. Infection is usually asymptomatic. It is endemic in the southeastern states, mostly in children. It is called the dwarf tapeworm because of its small size. It is also different from other tapeworms because the eggs are directly infectious for humans without the need for an intermediate animal host. Diagnosis is made by finding eggs in feces. Treat with praziquantel. No specific means of prevention.

TREMATODES (CHAPTER 55)

Schistosoma (Schistosoma mansoni, Schistosoma japonicum, & Schistosoma haematobium)

Disease—Schistosomiasis.

Characteristics—Trematode (blood fluke). Adults exist as two sexes but are attached to each other. Eggs are distinguished by spines: *S. mansoni* has large lateral spine; *S. japonicum* has small lateral spine; *S. haematobium* has terminal spine. Life cycle: Humans are infected by cercariae penetrating skin. Cercariae form larvae that penetrate blood vessels and are carried to the liver, where they become adults. The adult flukes migrate retrograde in the portal vein to reach the mesenteric venules (*S. mansoni* and *S. japonicum*) or urinary bladder venules (*S. haematobium*). Eggs penetrate the gut or bladder wall, are excreted, and hatch in freshwater. The ciliated larvae (miracidia) penetrate snails and multiply through generations to produce many free-swimming cercariae.

Transmission and Epidemiology—Transmitted by penetration of skin by cercariae. Humans are definitive hosts; snails are intermediate hosts. Endemic in tropical areas: *S. mansoni* in Africa and Latin America, *S. haematobium* in Africa and Middle East, *S. japonicum* in Asia.

Pathogenesis—Eggs in tissue induce inflammation, granulomas, fibrosis, and obstruction, especially in liver and spleen. *S. mansoni* damages the colon (inferior mesenteric venules), *S. japonicum* damages the small intestine (superior mesenteric venules), and *S. haematobium* damages the bladder. Bladder damage predisposes to carcinoma.

Laboratory Diagnosis—Eggs visible in feces or urine. Eosinophilia occurs.

Treatment—Praziquantel.

Prevention—Proper disposal of human waste. Swimming in endemic areas should be avoided.

Clonorchis sinensis

Disease—Clonorchiasis.

Characteristics—Trematode (liver fluke). Life cycle: Humans ingest undercooked fish containing encysted larvae (metacercariae). In duodenum, immature flukes enter biliary duct, become adults, and release eggs that are passed in feces. Eggs are eaten by snails; the eggs hatch and form miracidia. These multiply through generations (rediae) and then produce many free-swimming cercariae, which encyst under scales of fish and are eaten by humans.

Transmission and Epidemiology—Transmitted by eating raw or undercooked freshwater fish. Humans are definitive hosts; snails and fish are first and second intermediate hosts, respectively. Endemic in Asia.

Pathogenesis—Inflammation of biliary tract.

Laboratory Diagnosis—Eggs visible in feces.

Treatment—Praziquantel.

Prevention—Adequate cooking of fish. Proper disposal of human waste.

Paragonimus westermani

Disease—Paragonimiasis.

Characteristics—Trematode (lung fluke). Life cycle: Humans ingest undercooked freshwater crab meat containing encysted larvae (metacercariae). In gut, immature flukes enter peritoneal cavity, burrow through diaphragm into lung parenchyma, and become adults. Eggs enter bronchioles and are coughed up or swallowed. In freshwater, eggs hatch, releasing miracidia that enter snails, multiply through generations (rediae), and then form many cercariae that infect and encyst in crabs.

Transmission and Epidemiology—Transmitted by eating raw or undercooked crab meat. Humans are definitive hosts; snails and crabs are first and second intermediate hosts, respectively. Endemic in Asia and India.

Pathogenesis—Inflammation and secondary bacterial infection of lung.

Laboratory Diagnosis—Eggs visible in sputum or feces.

Treatment—Praziquantel.

Prevention—Adequate cooking of crabs. Proper disposal of human waste.

NEMATODES (CHAPTER 56)

1. Intestinal Infection

Ancylostoma duodenale & Necator americanus

Disease—Hookworm.

Characteristics—Intestinal nematode. Life cycle: Filariform larvae penetrate skin, enter the blood, and migrate to the lungs. They enter alveoli, pass up the trachea, then are swallowed. They become adults in small intestine and attach to walls via teeth (*Ancylostoma*) or cutting plates (*Necator*). Eggs are passed in feces and form noninfectious rhabditiform larvae and then infectious filariform larvae in soil.

Transmission and Epidemiology—Filariform larvae in soil penetrate skin of feet. Humans are the only hosts. Endemic in the tropics.

Pathogenesis—Anemia due to blood loss from GI tract.

Laboratory Diagnosis—Eggs visible in feces. Eosinophilia occurs.

Treatment—Mebendazole or pyrantel pamoate.

Prevention—Use of footwear. Proper disposal of human waste.

Ascaris lumbricoides

Disease—Ascariasis. Also eosinophilic pneumonia (Loeffler's syndrome) when larvae pass through lung.

Characteristics—Intestinal nematode. Life cycle: Humans ingest eggs, which form larvae in gut. Larvae migrate through the blood to the lungs, where they enter the alveoli, pass up the trachea, and are swallowed. In the gut, they become adults and lay eggs that are passed in the feces. They embryonate (i.e., become infective) in soil.

Transmission and Epidemiology—Transmitted by food contaminated with soil containing eggs. Humans are the only hosts. Endemic in the tropics.

Pathogenesis—Larvae in lung can cause pneumonia. Heavy worm burden can cause intestinal obstruction or malnutrition.

Laboratory Diagnosis—Eggs visible in feces. Eosinophilia occurs.

Treatment—Mebendazole or pyrantel pamoate.

Prevention—Proper disposal of human waste.

Enterobius vermicularis

Disease—Pinworm infection.

Characteristics—Intestinal nematode. Life cycle: Humans ingest eggs, which develop into adults in gut. At night, females migrate from the anus and lay many eggs on skin and in environment. Embryo within egg becomes an infective larva within 4 to 6 hours. Reinfection is common.

Transmission and Epidemiology—Transmitted by ingesting eggs. Humans are the only hosts. Occurs worldwide.

Pathogenesis—Worms and eggs cause perianal pruritus.

Laboratory Diagnosis—Eggs visible by "Scotch tape" technique. Adult worms found in diapers.

Treatment—Mebendazole or pyrantel pamoate.

Prevention—None.

Strongyloides stercoralis

Disease—Strongyloidiasis.

Characteristics—Intestinal nematode. Life cycle: Filariform larvae penetrate skin, enter the blood, and migrate to the lungs. They move into alveoli and up the trachea and are swallowed. They become adults and enter the mucosa, where females produce eggs that hatch in the colon into noninfectious, rhabditiform larvae that are usually passed in feces. Occasionally, rhabditiform larvae molt in the gut to form infectious, filariform larvae that can enter the blood and migrate to the lung (autoinfection). The noninfectious larvae passed in feces form infectious filariform larvae in the soil. These larvae can either penetrate the skin or form adults. Adult worms in soil can undergo several entire life cycles there. This free-living cycle can be interrupted when filariform larvae contact the skin.

Transmission and Epidemiology—Filariform larvae in soil penetrate skin. Endemic in the tropics.

Pathogenesis—Little effect in immunocompetent persons. In immunocompromised persons, massive superinfection can occur, accompanied by secondary bacterial infections.

Laboratory Diagnosis—Larvae visible in stool. Eosinophilia occurs.

Treatment—Ivermectin is the drug of choice. Thiabendazole is an alternative.

Prevention—Proper disposal of human waste.

Trichinella spiralis

Disease—Trichinosis.

Characteristics—Intestinal nematode that encysts in tissue. Life cycle: Humans ingest undercooked meat containing encysted larvae, which mature into adults in small intestine. Female worms release larvae that enter blood and migrate to skeletal muscle or brain, where they encyst.

Transmission and Epidemiology—Transmitted by ingestion of raw or undercooked meat, usually pork. Reservoir hosts are primarily pigs and rats. Humans are dead-end hosts. Occurs worldwide but endemic in Eastern Europe and west Africa.

Pathogenesis—Larvae encyst within striated muscle cells called "nurse cells," causing inflammation of muscle.

Laboratory Diagnosis—Encysted larvae visible in muscle biopsy. Eosinophilia occurs. Serologic tests positive.

Treatment—Thiabendazole effective early against adult worms. For severe symptoms, steroids plus mebendazole can be tried.

Prevention—Adequate cooking of pork.

Trichuris trichiura

Disease—Whipworm infection.

Characteristics—Intestinal nematode. Life cycle: Humans ingest eggs, which develop into adults in gut. Eggs are passed in feces into soil, where they embryonate (i.e., become infectious).

Transmission and Epidemiology—Transmitted by food or water contaminated with soil containing eggs. Humans are the only hosts. Occurs worldwide, especially in the tropics.

Pathogenesis—Worm in gut usually causes little damage.

Laboratory Diagnosis—Eggs visible in feces.

Treatment—Mebendazole.

Prevention—Proper disposal of human waste.

2. Tissue Infection

Dracunculus medinensis

Disease—Dracunculiasis.

Characteristics—Tissue nematode. Life cycle: Humans ingest copepods containing infective larvae in drinking water. Larvae are released in gut, migrate to body cavity, mature, and mate. Fertilized female migrates to subcutaneous tissue and forms a papule, which ulcerates. Motile larvae are released into water, where they are eaten by copepods and form infective larvae.

Transmission and Epidemiology—Transmitted by copepods in drinking water. Humans are major definitive hosts. Many domestic animals are reservoir hosts. Endemic in tropical Africa, Middle East, and India.

Pathogenesis—Adult worms in skin cause inflammation and ulceration.

Laboratory Diagnosis—Not useful.

Treatment—Thiabendazole or metronidazole. Extraction of worm from skin ulcer.

Prevention—Purification of drinking water.

Loa loa

Disease—Loiasis.

Characteristics—Tissue nematode. Life cycle: Bite of deer fly (mango fly) deposits infective larvae, which crawl into the skin and develop into adults that migrate subcutaneously. Females produce microfilariae, which enter the blood. These are ingested by deer flies, in which the infective larvae are formed.

Transmission and Epidemiology—Transmitted by deer flies. Humans are the only definitive hosts. No animal reservoir. Endemic in central and west Africa.

Pathogenesis—Hypersensitivity to adult worms causes "swelling" in skin. Adult worm seen crawling across conjunctivas.

Laboratory Diagnosis—Microfilariae visible on blood smear.

Treatment—Diethylcarbamazine.

Prevention—Deer fly control.

Onchocerca volvulus

Disease—Onchocerciasis (river blindness).

Characteristics—Tissue nematodes. Life cycle: Bite of female blackfly deposits larvae in subcutaneous tissue, where they mature into adult worms within skin nodules. Females produce microfilariae, which migrate in interstitial fluids and are ingested by blackflies, in which the infective larvae are formed.

Transmission and Epidemiology—Transmitted by female blackflies. Humans are the only definitive hosts. No animal reservoir. Endemic along rivers of tropical Africa and Central America.

Pathogenesis—Microfilariae in eye ultimately can cause blindness ("river blindness"). Adult worms induce inflammatory nodules in skin. See scaly dermatitis called "lizard skin." Also loss of subcutaneous tissue called "hanging groin."

Laboratory Diagnosis—Microfilariae visible in skin biopsy, not in blood.

Treatment—Ivermectin affects microfilariae, not adult worms. Suramin for adult worms.

Prevention—Blackfly control and ivermectin.

Wuchereria bancrofti

Disease—Filariasis.

Characteristics—Tissue nematodes. Life cycle: Bite of female mosquito deposits infective larvae that penetrate bite wound, form adults, and produce microfilariae. These circulate in the blood, chiefly at night, and are ingested by mosquitoes, in which the infective larvae are formed.

Transmission and Epidemiology—Transmitted by female mosquitoes of several genera, especially *Anopheles* and *Culex*, depending on geography. Humans are the only definitive hosts. Endemic in many tropical areas.

Pathogenesis—Adult worms cause inflammation that blocks lymphatic vessels (elephantiasis). Chronic, repeated infection required for symptoms to occur.

Laboratory Diagnosis—Microfilariae visible on blood smear.

Treatment—Diethylcarbamazine affects microfilariae. No treatment for adult worms.

Prevention—Mosquito control.

3. Nematodes Whose Larvae Cause Disease

Toxocara canis

Disease—Visceral larva migrans.

Characteristics—Nematode larvae cause disease. Life cycle in humans: *Toxocara* eggs are passed in dog feces and ingested by humans. They hatch into larvae in small intestine; larvae enter the blood and migrate to organs, especially liver, brain, and eyes, where they are trapped and die.

Transmission and Epidemiology—Transmitted by ingestion of eggs in food or water contaminated with dog feces. Dogs are definitive hosts. Humans are dead-end hosts.

Pathogenesis—Granulomas form around dead larvae. Granulomas in the retina can cause blindness.

Laboratory Diagnosis—Larvae visible in tissue. Serologic tests useful.

Treatment—Albendazole or mebendazole.

Prevention—Dogs should be dewormed.

Ancylostoma caninum & Ancylostoma braziliense

The filariform larvae of *A. caninum* (dog hookworm) and *A. braziliense* (cat hookworm) cause cutaneous larva migrans. The larvae in the soil burrow through the skin, then migrate within the subcutaneous tissue, causing a pruritic rash called "creeping eruption." These organisms cannot complete their life cycle in humans. The diagnosis is made clinically. Thiabendazole is effective.

Anisakis simplex

The larvae of *A. simplex* cause anisakiasis. They are ingested in raw seafood, such as sashimi and sushi, and migrate into the submucosa of the intestinal tract. Acute infection resembles appendicitis. Diagnosis is not dependent on the clinical laboratory. There is no effective drug therapy. Larvae can be removed when visualized during gastroscopy. Prevention consists of not eating raw fish.

SUMMARIES OF MEDICALLY IMPORTANT ECTOPARASITES

ECTOPARASITES THAT CAUSE HUMAN DISEASE (CHAPTER 69)

1. Lice

Pediculus humanus & Phthirus pubis

Disease—Pediculosis.

Characteristics—Lice are easily visible. *P. humanus* has an elongated body, whereas *P. pubis* has a short body resembling a crab. Nits are the eggs of the louse, often attached to the hair shaft or clothing.

Transmission—Hair and body lice are transmitted from human to human by contact, especially fomites such as hats and combs. Pubic lice are transmitted by sexual contact.

Pathogenesis—Itching is caused by a hypersensitivity response to saliva of the louse.

Laboratory Diagnosis—Not involved.

Treatment—Permethrin. Ivermectin is also effective. Nits are removed from hair with a comb.

Prevention—Personal items should be treated or discarded.

2. Flies
Dermatobia hominis
Disease—Myiasis.

Characteristics—Fly larvae (maggots) cause the disease, not the adult flies.

Transmission—*Dermatobia* deposits its egg on a mosquito, and when the mosquito bites, the eggs are then deposited on the skin. The warmth of the skin causes the egg to hatch, and the larva enters the skin at the site of the mosquito bite.

Pathogenesis—Larva induces an inflammatory response.

Laboratory Diagnosis—Not involved.

Treatment—Surgical removal of larva.

Prevention—Limit exposure to flies and mosquitoes.

3. Mites
Sarcoptes scabiei
Disease—Scabies.

Characteristics—Round body with eight short legs. Too small to be seen with naked eye.

Transmission—Person-to-person contact or fomites such as clothing.

Pathogenesis—Itching is caused by a hypersensitivity response to feces of the mite.

Laboratory Diagnosis—Microscopic examination reveals mites and their feces.

Treatment—Permethrin.

Prevention—Treat contacts and discard fomites.

4. Ticks
Dermacentor Species
Disease—Tick paralysis.

Characteristics—Certain species of ticks produce a neurotoxin.

Transmission—Ticks reside in grassy areas and attach to human skin.

Pathogenesis—Female tick requires a blood meal and toxin enters in tick saliva at bite site. Neurotoxin blocks release of acetylcholine at neuromuscular junction. Similar action as botulinum toxin.

Laboratory Diagnosis—Not involved.

Treatment—Removal of tick results in prompt reversal of paralysis.

Prevention—Remove ticks; wear protective clothing.

5. Spiders
Latrodectus mactans (Black Widow Spider)
Disease—Spider bite.

Characteristics—Black widow spiders have an orange-red hourglass on their ventral surface.

Pathogenesis—Neurotoxin causes pain in extremities and abdomen. Numbness, fever, and vomiting also occur.

Laboratory Diagnosis—Not involved.

Treatment—Antivenom should be given in severe cases.

Loxosceles reclusa (Brown Recluse Spider)
Disease—Spider bite.

Characteristics—Brown recluse spiders have a violin-shaped pattern on their dorsal surface.

Pathogenesis—Dermotoxin is a protease that causes painful necrotic lesions.

Laboratory Diagnosis—Not involved.

Treatment—Antivenom is not available in the United States.

PART XI   CLINICAL CASES

These brief clinical case vignettes are typical presentations of common infectious diseases. Learning the most likely causative organisms of these classic cases will help you answer the USMLE questions and improve your diagnostic skills. These cases are presented in random order similar to the way they are on the USMLE. The important features of the case are written in **boldface**.

CASE 1

A 22-year-old woman has a severe sore throat. Findings on physical examination include an inflamed throat, swollen cervical lymph nodes, and an enlarged spleen. **Her heterophile agglutinin test (Monospot test) is positive**.
Diagnosis: Infectious mononucleosis is caused by Epstein–Barr virus. Other viruses and bacteria, especially *Streptococcus pyogenes*, can cause pharyngitis and cervical lymphadenopathy, but an enlarged spleen and a positive Monospot test make infectious mononucleosis the most likely diagnosis. See page 295 for additional information.

CASE 2

A 5-year-old boy with diabetic ketoacidosis has ptosis of his right eyelid, periorbital swelling, and a black, necrotic skin lesion under his eye. Biopsy of the skin lesion shows **nonseptate hyphae with wide-angle branching**.
Diagnosis: Mucormycosis caused by *Mucor* or *Rhizopus* species. Diabetic ketoacidosis and renal acidosis predispose to mucormycosis. Fungal spores are inhaled into the sinuses, resulting in lesions on the face. See page 423 for additional information.

CASE 3

A 40-year-old man complains of watery, foul-smelling diarrhea and flatulence for the past 2 weeks. He drank untreated water on a camping trip about a month ago. See **pear-shaped flagellated trophozoites** in stool.
Diagnosis: Giardiasis caused by *Giardia lamblia*. Of the protozoa that are common causes of diarrhea, *Giardia* and *Cryptosporidium* cause watery diarrhea, whereas *Entamoeba* causes bloody diarrhea. See page 433 for additional information on *Giardia*, page 435 for additional information on *Cryptosporidium*, and page 429 for additional information on *Entamoeba*.

CASE 4

A 35-year-old man who is human immunodeficiency virus (HIV) antibody positive has had a persistent headache and a low-grade fever (temperature, 100°F) for the past 2 weeks. See **budding yeasts with a wide capsule in India ink preparation** of spinal fluid.
Diagnosis: Meningitis caused by *Cryptococcus neoformans*. The latex agglutination test that detects the capsular polysaccharide antigen of *Cryptococcus* in the spinal fluid is a more sensitive and specific test than is the test with India ink. See page 421 for additional information. If **acid-fast rods** are seen in spinal fluid, think *Mycobacterium tuberculosis*. See page 175 for additional information.

CASE 5

A 12-year-old boy has a painful arm that he thought he had injured while pitching in a Little League baseball game. The pain has gotten worse over a 2-week period, and he now has a temperature of 100°F. X-ray of the humerus reveals raised periosteum. Aspirate of lesion reveals **gram-positive cocci in clusters**.

Diagnosis: Osteomyelitis caused by *Staphylococcus aureus*. This organism is the **most common cause of osteomyelitis in children**. Osteomyelitis in prosthetic joints is often caused by *Staphylococcus epidermidis*. See page 104 for additional information on staphylococci.

CASE 6

A 50-year-old woman receiving chemotherapy via a subclavian catheter for acute leukemia has the sudden onset of blindness in her right eye. Her total white blood cell (WBC) count is 120/μL. Blood cultures grew **budding yeasts that formed germ tubes**.

Diagnosis: Endophthalmitis (infection inside the eye) caused by *Candida albicans*. A catheter-related infection gave rise to an embolus containing the organism, which traveled through the bloodstream to reach the eye. *C. albicans* is a member of the normal flora of the skin and enters through a break in the skin at the catheter site. See page 418 for additional information. If the blood culture grew colonies of gram-positive cocci in clusters that were coagulase-negative, think *S. epidermidis*, another member of the skin flora that is also a common cause of catheter-associated infections. See page 110 for additional information.

CASE 7

A 60-year-old man has had a nonproductive cough and fever (temperature, 101°F) for 1 week. He received a kidney transplant 6 weeks ago and has had one episode of rejection that required increased prednisone. There was no response to erythromycin, indicating that *Legionella* and *Mycoplasma* are unlikely causes. See **owl's-eye inclusion bodies within the nucleus** of infected cells in bronchoalveolar lavage fluid.

Diagnosis: Pneumonia caused by cytomegalovirus (CMV). These intranuclear inclusions are typical findings in CMV infections. Immunosuppression predisposes to disseminated CMV infections. See page 293 for additional information.

CASE 8

A 45-year-old woman complains that her right arm has become increasingly weak during the past few days. This morning, she had a generalized seizure. She recently finished a course of cancer chemotherapy. Magnetic resonance imaging (MRI) of the brain reveals a lesion resembling an abscess. Brain biopsy shows **gram-positive rods in long filaments**. Organism is **weakly acid-fast**.

Diagnosis: Brain abscess caused by Nocardia asteroides. *N. asteroides* initially infects the lung, where it may or may not cause symptoms in immunocompetent people. Dissemination to the brain is common in immunocompromised patients. See page 188 for additional information.

CASE 9

A 20-year-old man has a severe headache and vomiting that began yesterday. He is now confused. On examination, his temperature is 39°C and his neck is stiff. Spinal fluid reveals no bacteria on Gram stain, 25 lymphs, normal protein, and normal glucose. Culture of the spinal fluid on blood agar shows no bacterial colonies.

Diagnosis: Viral meningitis, which is most often caused by Coxsackie virus. Can isolate the virus from spinal fluid. See page 337 for additional information.

CASE 10

A 60-year-old man with a history of tuberculosis now has a cough productive of bloody sputum. Chest X-ray reveals a round opaque mass within a cavity in his left upper lobe. Culture of the sputum grew an organism with **septate hyphae that had straight, parallel walls**. The hyphae exhibited **low-angle branching**.

Diagnosis: Fungus ball caused by *Aspergillus fumigatus*. Fungal spores are inhaled into the lung, where they grow within a preexisting cavity caused by infection with *M. tuberculosis*. See page 422 for additional information.

CASE 11

A 3-month-old girl has watery, nonbloody diarrhea. Stool culture reveals only normal enteric flora.

Diagnosis: Think rotavirus, the most common cause of diarrhea in infants. The enzyme-linked immunosorbent assay (ELISA) test for rotavirus antigen in the stool is positive, which confirms the diagnosis. See page 334 for additional information.

CASE 12

A 30-year-old woman has a painless ulcer on her tongue. She is HIV antibody positive and has a CD4 count of 25. Her serum is nonreactive in the VDRL test. Biopsy of the lesion revealed **yeasts within macrophages**.

Diagnosis: Disseminated histoplasmosis caused by *Histoplasma capsulatum.* Patients with a low CD4 count have severely reduced cell-mediated immunity, which predisposes to disseminated disease caused by this dimorphic fungus. A negative VDRL test indicates the ulcer was not caused by *Treponema pallidum.* See page 413 for additional information on *Histoplasma.*

CASE 13

A 20-year-old man has a swollen, red, hot, tender ankle, accompanied by a temperature of 100°F for the past 2 days. There is no history of trauma. See **gram-negative diplococci** in joint fluid aspirate. Organism is **oxidase-positive**.

Diagnosis: Arthritis caused by *Neisseria gonorrhoeae,* the most common cause of infectious arthritis in sexually active adults. Sugar fermentation tests were used to identify the organism as *N. gonorrhoeae.* See page 127 for additional information.

CASE 14

A 40-year-old woman has blurred vision and slurred speech. She is afebrile. She is famous in her neighborhood for her home-canned vegetables and fruits.

Diagnosis: Botulism caused by *Clostridium botulinum.* Botulinum toxin causes a descending paralysis that starts with the cranial nerves, typically appearing initially as diplopia. The toxin is a **protease that cleaves the proteins involved in the release of acetylcholine** at the neuromuscular junction. Treat with antiserum immediately. **Confirm diagnosis with mouse protection test** or ELISA test using a sample of food suspected of containing the toxin. See page 134 for additional information. Wound botulism occurs in heroin users (e.g., users of black tar heroin), especially in those who "skin pop." Bacterial spores in the heroin germinate in the anaerobic conditions in necrotic skin tissue.

CASE 15

A neonate was born with a small head (microcephaly), jaundice, and hepatosplenomegaly. Urine contained **multinucleated giant cells with intranuclear inclusions**.

Diagnosis: Cytomegalovirus (CMV) infection acquired in utero. CMV is the **leading cause of congenital abnormalities**. For fetal infection to occur, the mother must be infected for the first time during pregnancy. She therefore would have no preexisting antibodies to neutralize the virus prior to its infecting the placenta and the fetus. See page 293 for additional information.

CASE 16

A 14-year-old girl has a rapidly spreading, painful, erythematous rash on her leg. The rash is warm and tender, and her temperature is 38°C. **Gram-positive cocci in chains** were seen in an aspirate from the lesion. Culture of the aspirate on blood agar grew colonies surrounded by **clear (beta) hemolysis**. Growth of the organism was **inhibited by bacitracin.**

Diagnosis: Cellulitis caused by *S. pyogenes.* The rapid spread of cellulitis caused by *S. pyogenes* is due to hyaluronidase (spreading factor) that degrades hyaluronic acid in subcutaneous tissue. **Acute glomerulonephritis (AGN)** can follow skin infections caused by *S. pyogenes.* AGN is an immunologic disease caused by **antigen–antibody complexes**. See page 111 for additional information.

CASE 17

A 4-year-old boy wakes up at night because his anal area is itching. See **worm eggs in Scotch tape preparation**.

Diagnosis: Pinworm infection (enterobiasis) caused by *Enterobius vermicularis.* Pinworm infection is the most common helminth disease in the United States. See page 476 for additional information.

CASE 18

A 25-year-old woman has a painful, inflamed swollen hand. She was bitten by a cat about 8 hours ago. See **small gram-negative rods** in the exudate from lesion.

Diagnosis: Cellulitis caused by *Pasteurella multocida*. Organism is **normal flora in cat's mouth**. See page 172 for additional information.

CASE 19

A 7-year-old girl has bloody diarrhea and fever (temperature, 38°C) but no nausea or vomiting. Only lactose-fermenting colonies are seen on EMB agar.

Diagnosis: Think either *Campylobacter jejuni* or enterohemorrhagic strains of *Escherichia coli* (*E. coli* O157:H7). If *Campylobacter* is the cause, see colonies on *Campylobacter* agar containing **curved gram-negative rods**, and the colonies on EMB agar are likely to be nonpathogenic *E. coli*. If *E. coli* O157:H7 is the cause, the organism in the lactose-fermenting colonies on EMB agar is **unable to ferment sorbitol**. The absence of non–lactose-fermenting colonies indicates that *Shigella* and *Salmonella* are not the cause. See page 154 for additional information on *Campylobacter* and page 146 for additional information on *E. coli* O157:H7.

CASE 20

A 15-year-old girl has had a nonproductive cough and temperature of 100°F for the past 5 days. The symptoms came on gradually. Lung examination shows few scattered rales. Chest X-ray shows patchy infiltrate in left lower lobe but no consolidation. **Cold agglutinin test is positive**.

Diagnosis: Atypical pneumonia caused by *Mycoplasma pneumoniae*. This organism is the most common cause of atypical pneumonia in teenagers and young adults. In the cold agglutinin test, antibodies in the patient's serum agglutinate human red blood cells in the cold (4°C). These antibodies do not react with *Mycoplasma*. If sputum is available, a PCR test can confirm *Mycoplasma* infection. See page 190 for additional information.

CASE 21

A 45-year-old man sustained a skull fracture in an automobile accident. The following day, he noted clear fluid dripping from his nose, but he did not notify the hospital personnel. The following day, he spiked a fever to 39°C and complained of a severe headache. Nuchal rigidity was found on physical examination. Spinal fluid analysis revealed a WBC count of 5200/μL, 90% of which were neutrophils. Gram stain showed gram-positive diplococci.

Diagnosis: Meningitis caused by *Streptococcus pneumoniae*. Patients with a **fracture of the cribriform plate who leak spinal fluid into the nose** are predisposed to meningitis by this organism. Pneumococci can colonize the nasal mucosa and enter the subarachnoid space through the fractured cribriform plate. See page 118 for additional information.

CASE 22

A 7-year-old girl was well until about 3 weeks ago, when she began complaining of being "tired all the time." On examination, her temperature is 38°C, and there is tenderness below the right knee. Hemoglobin: 10.2; WBC: 9600 with increased neutrophils. A sickle cell prep shows a moderate sickling tendency. **Gram-negative rods** grew in the blood culture.

Diagnosis: Osteomyelitis caused by *Salmonella* species. Sickle cell anemia predisposes to osteomyelitis caused by *Salmonella* species. The abnormally shaped sickle cells are trapped in the small capillaries of the bone and cause microinfarcts. These microinfarcts enhance the likelihood of infection by *Salmonella*. See page 149 for additional information.

CASE 23

A 3-month-old boy has a persistent cough and severe wheezing for the past 2 days. On physical examination, his temperature is 39°C and coarse rhonchi are heard bilaterally. Chest X-ray shows interstitial infiltrates bilaterally. Diagnosis was made by **ELISA that detected viral antigen in nasal washings**.

Diagnosis: Think pneumonia caused by respiratory syncytial virus (RSV), RSV is the most common cause of pneumonia and bronchiolitis in infants. RSV causes **giant cells (syncytia)** that can be seen in respiratory secretions and in cell culture. See page 313 for additional information.

CASE 24

A 34-year-old man was in his usual state of health until last night, when he felt feverish, had a shaking chill, and became short of breath at rest. Temperature 39°C, blood pressure 110/60, pulse 104, respirations 18. Scattered rales were heard in both bases. A new murmur consistent with tricuspid insufficiency was heard. Needle tracks were seen on both forearms. **Gram-positive cocci in clusters** grew in blood culture.

Diagnosis: Acute endocarditis caused by *S. aureus.* This organism is the most common cause of acute endocarditis in intravenous drug users. The valves on the right side of the heart are often involved. See page 104 for additional information.

CASE 25

A 2-week-old infant was well on discharge from the hospital 10 days ago and remained so until last night, when he appeared drowsy and flushed. His skin felt hot to the touch. On physical examination, the infant was very difficult to arouse, but there were no other positive findings. His temperature was 40°C. Blood culture grew **gram-positive cocci in chains.** A narrow zone of **clear (beta) hemolysis** was seen around the colonies. **Hippurate hydrolysis** test was positive.

Diagnosis: Neonatal sepsis caused by *Streptococcus agalactiae* (group B streptococci). Group B streptococci are the most common cause of neonatal sepsis. Think *E. coli* if gram-negative rods are seen or *Listeria monocytogenes* if gram-positive rods are seen. See page 115 for additional information on group B streptococci, page 146 for additional information on *E. coli*, and page 138 for additional information on *L. monocytogenes*.

CASE 26

A 70-year-old woman had a hip replacement because of severe degenerative joint disease. She did well until a year later, when a fall resulted in a fracture of the femur and the prosthesis had to be replaced. Three weeks later, bloody fluid began draining from the wound site. The patient was afebrile, and the physical examination was otherwise unremarkable. Two days later, because of increasing drainage, the wound was debrided and pus was obtained. Gram stain of the pus was negative, but an **acid-fast stain revealed red rods**.

Diagnosis: Prosthetic joint infection caused by *Mycobacterium fortuitum-chelonae* complex. Think *S. epidermidis* if gram-positive cocci in clusters are seen. See page 183 for additional information on *M. fortuitum-chelonae* complex and page 110 for additional information on *S. epidermidis*.

CASE 27

An 80-year-old man complains of a painful rash on his left forehead. The rash is vesicular and only on that side. He is being treated with chemotherapy for leukemia. Smear of material from the base of the vesicle reveals **multinucleated giant cells with intranuclear inclusions**.

Diagnosis: Herpes zoster (shingles) caused by varicella-zoster virus. The rash of zoster follows the dermatome of the neuron that was latently infected. Herpes simplex virus type 1 can cause a similar picture. These viruses can be distinguished using fluorescent antibody assay. See page 291 for additional information.

CASE 28

A 55-year-old woman has an inflamed ulcer on her right hand and several tender nodules on the inner aspect of her right arm. She is an avid gardener and especially enjoys pruning her roses. Biopsy of the lesion reveals budding yeasts.

Diagnosis: Sporotrichosis caused by *Sporothrix schenckii.* The organism is a mold in the soil and a yeast in the body (i.e., it is **dimorphic**). Infection occurs when spores produced by the mold form are introduced into the skin by a penetrating injury. See page 409 for additional information.

CASE 29

A 15-year-old boy sustained a broken tooth in a fist fight several weeks ago. He now has an inflamed area on the skin over the broken tooth, in the center of which is a draining sinus tract. Gram stain of the drainage fluid reveals **filamentous gram-positive rods**.

Diagnosis: Actinomycosis caused by *Actinomyces israelii.* See **sulfur granules** in the sinus tract. These granules are particles composed of interwoven filaments of bacteria. See page 187 for additional information.

CASE 30

A 24-year-old woman experienced the sudden onset of high fever, myalgias, vomiting, and diarrhea. Her vital signs were as follows: temperature 40°C, blood pressure 70/30, pulse 140, respirations 30. A sunburn-like rash appeared over most of her body. Blood cultures and stool cultures are negative. She is recovering from a surgical procedure on her maxillary sinus, and the bleeding was being staunched with nasal tampons. Gram-positive cocci in clusters were seen in blood adherent to the nasal tampon. **Diagnosis: Toxic shock syndrome caused by *S. aureus.*** Toxic shock syndrome toxin is a **superantigen that stimulates the release of large amounts of cytokines from many helper T cells**. See page 107 for additional information.

CASE 31

An 8-year-old girl has a pruritic rash on her chest. Lesions are round or oval with an inflamed border and central clearing. The lesions contain both papules and vesicles. See **hyphae in KOH prep** of scrapings from the lesion.
Diagnosis: Tinea corporis (ringworm) caused by one of the dermatophytes, especially species of *Microsporum*, *Trichophyton*, or *Epidermophyton.* Dermatophytes use **keratin** as a nutrient source, so lesions are limited to the skin. See page 407 for additional information.

CASE 32

A 25-year-old woman has a papular rash on her trunk, arms, and palms. She says the rash does not itch. Vaginal examination reveals two flat, moist, slightly raised lesions on the labia. Material from a labial lesion examined in a **dark-field microscope revealed spirochetes**.
Diagnosis: Secondary syphilis caused by *T. pallidum.* The rash on the palms coupled with the vaginal lesions (condylomata lata) is compatible with secondary syphilis. **Serologic tests, such as the nonspecific test (VDRL) and the specific test (FTA-ABS), were positive**. See page 193 for additional information.

CASE 33

A 5-year-old girl complains of an earache for the past 2 days. On examination, she has a temperature of 39°C, the right external canal contained dried blood, the drum was perforated, and a small amount of purulent fluid was seen. Gram stain of the pus revealed **gram-positive diplococci**. Colonies formed **green (alpha) hemolysis on blood agar**. Growth was inhibited by **optochin**.
Diagnosis: Otitis media caused by *S. pneumoniae.* Think *Haemophilus influenzae* if small gram-negative rods are seen. These organisms colonize the oropharynx and enter the middle ear via the eustachian tube. See page 118 for additional information on *S. pneumoniae* and page 163 for additional information on *H. influenzae*.

CASE 34

A 25-year-old woman was well until the sudden onset of high fever (temperature, 40°C) accompanied by several purple skin lesions (ecchymoses, purpura). The lesions are scattered over the body, are irregularly shaped, and are not raised. Her blood pressure is 60/10, and her pulse rate is 140. Blood culture grew **gram-negative diplococci**.
Diagnosis: Meningococcemia caused by *Neisseria meningitidis.* The endotoxin (lipopolysaccharide, or LPS) of the organism triggers release of interleukin-1, tumor necrosis factor, and nitric oxide from macrophages. These cause the high fever and low blood pressure. The purpuric lesions are a manifestation of **disseminated intravascular coagulation (DIC)**. Endotoxin activates the coagulation cascade, causing DIC. **Lipid A** is the toxic part of LPS. See page 124 for additional information.

CASE 35

A 40-year-old woman was well until 2 days ago, when she experienced the sudden onset of fever, shaking chills, and profuse sweating. Today, she also complains of headache and abdominal pain but no nausea, vomiting, or diarrhea. She does not have a stiff neck, rash, or altered mental status. Travel history reveals she returned from an extended trip to several countries in central Africa 1 week ago. Blood smear reveals **ring-shaped trophozoites within red blood cells**.
Diagnosis: Malaria caused by *Plasmodium* species. If **banana-shaped gametocytes** are seen in the blood smear, think *Plasmodium falciparum*. *P. falciparum* is the species that causes the life-threatening complications of malaria, such as cerebral malaria. The fever and chills experienced by the patient coincide with the release of merozoites from infected red blood cells and occur in either a tertian or quartan pattern. See page 439 for additional information.

CASE 36

A 35-year-old man is seen in the emergency room (ER) complaining of severe headache and vomiting that began last night. His temperature is 40°C. While in the ER, he is increasingly combative and has a grand mal seizure. He is "foaming at the mouth" and cannot drink any liquids. Analysis of his spinal fluid reveals no abnormality, and no organisms are seen in the Gram stain. Two days later, despite supportive measures, he dies. Pathologic examination of the brain reveals **eosinophilic inclusion bodies in the cytoplasm of neurons**.

Diagnosis: Rabies (an encephalitis) caused by rabies virus. The inclusions are **Negri bodies**. Diagnosis can be confirmed by using fluorescent antibody assays. The patient was a farm worker who was **bitten by a bat** about a month prior to the onset of symptoms. Note the long incubation period, which can be as long as 6 months. People bitten by a bat (or any wild animal) should receive rabies immunization consisting of the inactivated vaccine plus rabies immune globulins (passive–active immunization). See page 359 for additional information.

CASE 37

A 70-year-old man was admitted to the hospital after suffering extensive third-degree burns. Three days later, he spiked a fever, and there was pus on the dressing that had a **blue–green color**. Gram stain of the pus revealed **gram-negative rods**.

Diagnosis: Wound (burn) infection caused by *Pseudomonas aeruginosa*. The blue-green color is caused by **pyocyanin**, a pigment produced by the organism. See page 157 for additional information.

CASE 38

A 65-year-old woman reports that she has had several episodes of confusion and memory loss during the past few weeks. On examination, she is afebrile but has a staggering gait, and myoclonus can be elicited. Over the next several months, her condition deteriorates and death ensues. On autopsy, microscopic examination of the brain reveals **many vacuoles** but no viral inclusion bodies.

Diagnosis: Creutzfeldt-Jakob disease (CJD) caused by prions. CJD is a **spongiform encephalopathy**. The vacuoles give the brain a sponge-like appearance. See page 378 for additional information.

CASE 39

A 20-year-old man complains of several episodes of blood in his urine. He has no dysuria or urethral discharge. He is not sexually active. He is a college student but was born and raised in Egypt. Physical examination reveals no penile lesions. Urinalysis shows many red cells, no white cells, and several large **eggs with terminal spines**.

Diagnosis: Schistosomiasis caused by *Schistosoma haematobium*. Schistosome eggs in venules of the bladder damage the bladder epithelium and cause bleeding. The eggs are excreted in the urine. See page 469 for additional information.

CASE 40

A 35-year-old man complains of night sweats, chills, and fatigue at varying intervals during the past 2 months. These episodes began while he was traveling in Latin America. When questioned, he says that cheeses, especially the unpasteurized varieties, are some of his favorite foods. On examination, his temperature is 39°C, and his liver and spleen are palpable. His hematocrit is 30%, and his WBC count is 5000. Blood culture grew **small gram-negative rods**.

Diagnosis: Brucellosis caused by *Brucella* species. Domestic animals such as cows and goats are the main reservoir for *Brucella*, and it is often transmitted in **unpasteurized dairy products**. This patient could also have typhoid fever caused by *Salmonella typhi*, but *S. typhi* is only a human pathogen (i.e., there is no animal reservoir). See page 169 for additional information on *Brucella* species and page 149 for additional information on *S. typhi*.

CASE 41

A 6-year-old girl has a rash on her face that appeared yesterday. The rash is **erythematous and located over the malar eminences** bilaterally. The rash is macular; there are no papules, vesicles, or pustules. A few days prior to the appearance of the rash, she had a runny nose and anorexia.

Diagnosis: Slapped cheek syndrome caused by parvovirus B19. This virus also causes **aplastic anemia because it preferentially infects and kills erythroblasts**. It also **infects the fetus, causing hydrops fetalis**, and causes an immune complex–mediated **arthritis**, especially in adult women. See page 329 for additional information.

CASE 42

A 20-year-old man fell off his motorcycle and suffered a compound fracture of the femur. The fracture was surgically reduced and the wound debrided. Forty-eight hours later, he spiked a fever (temperature, 40°C), and the wound area became necrotic. Crepitus was felt, and a foul-smelling odor was perceived originating from the wound. Marked anemia and a WBC count of 22,800 were found. Gram stain of the exudate showed **large gram-positive rods**. Colonies grew on blood agar incubated **anaerobically** but not aerobically.

Diagnosis: Gas gangrene (myonecrosis) caused by *Clostridium perfringens.* The main virulence factor produced by this organism is an **exotoxin that is a lecithinase**. It causes necrosis of tissue and lysis of red blood cells (causing hemolytic anemia). The spores of the organism are in the soil and enter the wound site. A foul-smelling exudate is characteristic of infections caused by anaerobic bacteria. See page 134 for additional information.

CASE 43

A 30-year-old woman complains of a burning feeling in her mouth and pain on swallowing. Sexual history reveals she is a commercial sex worker and has had unprotected vaginal, oral, and anal intercourse with multiple partners. On examination, whitish lesions are seen on the tongue, palate, and pharynx. No vesicles are seen. The test for HIV antibody is positive, and her CD4 count is 65. Gram stain of material from the lesions reveals **budding yeasts and pseudohyphae.**

Diagnosis: Thrush caused by *C. albicans.* This organism forms pseudohyphae when it invades tissue. The absence of vesicles indicates that her symptoms are not caused by herpes simplex virus type 2. See page 418 for additional information.

CASE 44

You're a physician at a refugee camp in sub-Saharan Africa, when an outbreak of diarrhea occurs. Massive amounts of watery stool, without blood, are produced by the patients. **Curved gram-negative rods** are seen in a Gram stain of the stool.

Diagnosis: Cholera caused by *Vibrio cholerae.* There are three genera of curved gram-negative rods: ***Vibrio, Campylobacter,*** and ***Helicobacter.*** *V. cholerae* causes watery, nonbloody diarrhea, whereas *C. jejuni* typically causes bloody diarrhea. *Helicobacter pylori* causes gastritis and peptic ulcer, not diarrhea. Enterotoxigenic *E. coli* causes watery diarrhea by producing an exotoxin that has the same mode of action as does the exotoxin produced by *V. cholerae.* However, *E. coli* is a straight gram-negative rod, not a curved one. If an outbreak of bloody diarrhea had occurred in the refugee camp, then *Shigella dysenteriae* would be the most likely cause. See the following pages for additional information: *Vibrio,* page 152; *Campylobacter,* page 154; *Helicobacter,* page 155; *Escherichia,* page 146; and *Shigella,* page 151.

CASE 45

A 40-year-old man with low-grade fever and night sweats for the past 4 weeks now has increasing fatigue and shortness of breath. He says he has difficulty climbing the one flight of stairs to his apartment. Pertinent past history includes rheumatic fever when he was 15 years old and the extraction of two wisdom teeth about 3 weeks before his symptoms began. No chemoprophylaxis was given at the time of the extractions. There is no history of intravenous drug use. His temperature is 38.5°C, and a loud holosystolic murmur can be heard over the precordium. His spleen is palpable. He is anemic, and his WBC count is 13,500. Blood cultures grow **gram-positive cocci in chains that produce green (alpha) hemolysis on blood agar**. Growth is **not inhibited by optochin**.

Diagnosis: Subacute bacterial endocarditis caused by one of the viridans group streptococci, such as *Streptococcus sanguinis.* The laboratory findings are also compatible with *Enterococcus faecalis,* but the history of dental surgery makes the viridans group streptococci more likely to be the cause. Endocarditis caused by *E. faecalis* is associated with gastrointestinal or genitourinary tract surgery. See page 116 for additional information on both viridans group streptococci and *E. faecalis.*

CASE 46

A 60-year-old woman is asymptomatic but has a lung nodule seen on chest X-ray. Pertinent past history includes her cigarette smoking (2 packs per day for 40 years) and her occupation as an archaeologist, digging primarily in Arizona and New Mexico. Because of concern that the nodule may be malignant, it was surgically removed. Pathologic examination revealed **large (25 μm) round structures with thick walls and many round spores inside**. No malignant cells were seen.

Diagnosis: Coccidioidomycosis caused by *Coccidioides immitis.* These structures are **spherules**, which are pathognomonic for this disease. The mold form of the organism is found in the soil of the southwestern United States, and the organism is acquired by inhalation of arthrospores produced by the mold. The inhaled arthrospores form spherules in the lung. *C. immitis* is dimorphic and forms spherules at 37°C. See page 411 for additional information.

CASE 47

A 20-year-old woman in her 30th week of pregnancy had an ultrasound examination that revealed a growth-retarded fetus with a large head (indicating hydrocephalus) and calcifications within the brain. Umbilical blood was cultured, and **crescent-shaped trophozoites** were grown.

Diagnosis: Toxoplasmosis caused by *Toxoplasma gondii*. Detection of IgM antibody in the Sabin-Feldman dye test can also be used to make a diagnosis. The main reservoir is domestic cats. Domestic farm animals, such as cattle, acquire the organism by accidentally eating cat feces. Pregnant women should **not be exposed to cat litter or eat undercooked meat**. See page 443 for additional information.

CASE 48

A 10-day-old neonate has several vesicles on the scalp and around the eyes. The child is otherwise well, afebrile, and feeding normally. A Giemsa-stained smear of material from the base of a vesicle revealed **multinucleated giant cells with intranuclear inclusions**.

Diagnosis: Neonatal infection caused by herpes simplex virus type 2. Infection is acquired during passage through the birth canal. Life-threatening encephalitis and disseminated infection of the neonate also occur. See page 289 for additional information.

CASE 49

A 40-year-old woman has just had a grand mal seizure. There is a history of headaches for the past week and one episode of vertigo but no previous seizures. She is afebrile. She is a native of Honduras but has lived in the United States for the past 5 years. MRI reveals a mass in the parietal lobe. The mass **is a cyst-like sac with what appears to be a larva inside**.

Diagnosis: Cysticercosis caused by the larva of *Taenia solium*. Infection is acquired by ingesting the tapeworm eggs, *not* by ingesting undercooked pork. This clinical picture can also be caused by a brain abscess, a granuloma such as a tuberculoma, or a brain tumor. See page 459 for additional information.

CASE 50

A 1-week-old neonate has a yellowish exudate in the corners of both eyes. The child is otherwise well, afebrile, and feeding normally. Gram stain of the exudate reveals no gram-negative diplococci. A Giemsa-stained smear of the exudate reveals **a large cytoplasmic inclusion**.

Diagnosis: Conjunctivitis caused by *Chlamydia trachomatis*. Confirm the diagnosis with direct fluorescent antibody test. Infection is acquired during passage through the birth canal. The inclusion contains large numbers of the **intracellular replicating forms called reticulate bodies**. See page 203 for additional information.

PART XII   PEARLS FOR THE USMLE

Many questions on the USMLE can be answered by knowing the meaning of the epidemiologic information provided in the case description. In order to do this, the student should know the reservoir of the organism, its mode of transmission, and the meaning of factors such as travel, occupation, and exposure to pets, farm animals, or wild animals. Knowledge of the microbes that typically cause disease in individuals with specific immunodeficiencies will also be helpful.

In addition to being useful for the USMLE, this information will prove valuable to make the diagnosis of infectious diseases on the wards and in your clinical practice.

The "Pearls" are presented in tables entitled:

TABLE XII-1   Farm Animals and Household Pets as Reservoirs of Medically Important Organisms

Animal	Mode of Transmission	Important Organisms	Disease
Cattle/cows	1. Ingestion of meat[1]	1. *Escherichia coli* O157	Enterocolitis and hemolytic-uremic syndrome
		2. *Salmonella enterica*	Enterocolitis
		3. Prions	Variant Creutzfeldt-Jakob disease
		4. *Taenia saginata*	Taeniasis (intestinal tapeworm)
		5. *Toxoplasma gondii*	Toxoplasmosis
	2. Ingestion of milk products[2]	1. *Listeria monocytogenes*	Neonatal sepsis
		2. *Brucella* species	Brucellosis
		3. *Mycobacterium bovis*	Intestinal tuberculosis
	3. Contact with animal hides	*Bacillus anthracis*	Anthrax
Sheep	Inhalation of amniotic fluid	*Coxiella burnetii*	Q fever
Goats	Ingestion of milk products[2]	*Brucella* species	Brucellosis
Pigs	Ingestion of meat[1]	1. *Taenia solium*	Taeniasis (intestinal tapeworm)[3]
		2. *Trichinella spiralis*	Trichinosis
Poultry (chickens; turkeys)	Ingestion of meat or eggs[1]	1. *S. enterica*	Enterocolitis
		2. *Campylobacter jejuni*	Enterocolitis
Dogs	1. Ingestion of dog feces	1. *Echinococcus granulosus*	Echinococcosis
		2. *Toxocara canis*	Visceral larva migrans
	2. Ingestion of dog urine	*Leptospira interrogans*	Leptospirosis
	3. Dog bite	1. Rabies virus	Rabies
		2. *Capnocytophaga canimorsus*	Sepsis
	4. Direct contact	*Microsporum canis*	Tinea corporis
Cats	1. Ingestion of cat feces	*T. gondii*	Toxoplasmosis
	2. Cat bite/scratch	1. *Pasteurella multocida*	Cellulitis
		2. *Bartonella henselae*	Cat-scratch disease; bacillary angiomatosis
		3. Rabies virus	Rabies

[1]Raw or undercooked.

[2]Unpasteurized.

[3]Ingestion of eggs in human feces, not ingestion of pork, results in cysticercosis.

TABLE XII-2   Wild Animals as Reservoirs of Medically Important Organisms

Animal	Mode of Transmission	Important Organisms	Disease
Rats	1. Flea bite	*Yersinia pestis*	Plague
	2. Ingestion of urine	*Leptospira interrogans*	Leptospirosis
Mice	1. Tick bite	*Borrelia burgdorferi*	Lyme disease
	2. Inhale aerosol of droppings	Hantavirus	Hantavirus
			Pulmonary syndrome
Bats, skunks, raccoons, and foxes	Bite	Rabies virus	Rabies
Rabbits	Contact	*Francisella tularensis*	Tularemia

(Continued)

TABLE XII–2 Wild Animals as Reservoirs of Medically Important Organisms (*Continued*)

Animal	Mode of Transmission	Important Organisms	Disease
Bats	Inhale aerosol	Coronavirus—SARS-1&2	Pneumonia
Camels	Inhale aerosol	Coronavirus—MERS	Pneumonia
Monkeys	Mosquito bite	Yellow fever virus	Yellow fever
Birds			
1. Psittacine birds (e.g., parrots)	Inhale aerosol	*Chlamydia psittaci*	Psittacosis
2. Chickens	Inhale aerosol	Influenza virus	Influenza
3. Pigeons	Inhale aerosol	*Cryptococcus neoformans*	Meningitis, pneumonia
4. Starlings	Inhale spores	*Histoplasma capsulatum*	Histoplasmosis
5. Sparrows	Mosquito bite	Encephalitis viruses (e.g., West Nile virus)	Encephalitis
Snakes and turtles	Fecal–oral	*Salmonella enterica*	Enterocolitis
Beaver	Fecal–oral	*Giardia lamblia*	Giardiasis
Fish	Ingestion of fish[1]	*Anisakis simplex*	Anisakiasis
		Diphyllobothrium latum	Diphyllobothriasis

MERS = Middle East respiratory syndrome; SARS = severe acute respiratory syndrome.

[1]Raw or undercooked.

TABLE XII–3 Insects as Vectors of Medically Important Organisms

Insects	Important Organisms	Reservoir	Disease
Ticks			
1. *Ixodes* (deer tick)	1. *Borrelia burgdorferi*	Mice	Lyme disease
	2. *Babesia microti*	Mice	Babesiosis
2. Dermacentor (dog tick)	1. *Rickettsia rickettsii*	Rodents, dogs	Rocky Mountain spotted fever
	2. *Ehrlichia chaffeensis*	Dogs	Ehrlichiosis
	3. *Anaplasma phagocytophilum*	Rodents, dogs	Anaplasmosis
Lice			
Human body louse	*Rickettsia prowazekii*	Humans	Typhus
Mosquitoes			
1. *Anopheles*	*Plasmodium falciparum, P. vivax, P. ovale, P. malariae*	Humans	Malaria
2. *Aedes*	Yellow fever virus	Humans and monkeys	Yellow fever
3. *Aedes*	Dengue virus	Humans	Dengue
4. *Culex*	Encephalitis viruses, such as West Nile virus	Birds	Encephalitis
5. *Anopheles and Culex*	*Wuchereria bancrofti*	Humans	Filariasis, especially elephantiasis
Fleas			
Rat flea	*Yersinia pestis*	Rats	Plague
Flies			
1. Sandfly	*Leishmania donovani*	Various animals	Leishmaniasis
2. Tsetse fly	*Trypanosoma brucei*	Humans and various animals	Sleeping sickness
3. Blackfly	*Onchocerca volvulus*	Humans	Onchocerciasis
Bugs			
Reduviid bug	*Trypanosoma cruzi*	Various animals	Chagas disease

TABLE XII-4 Environmental Sources of Medically Important Organisms

Environmental Source	Important Organisms	Mode of Transmission	Disease
Water	1. *Legionella pneumophila*	Inhale aerosol	Pneumonia
	2. *Pseudomonas aeruginosa*	Inhale aerosol or direct contact	Pneumonia, burn, and wound infections
	3. *Mycobacterium marinum*	Skin abrasion	Swimming pool granulomas
	4. Vibrio cholerae	Ingestion	Cholera
	5. *Vibrio vulnificus*	Skin abrasion, ingestion of seafood	Cellulitis, sepsis
	6. *Schistosoma mansoni, S. haematobium*	Cercariae enter skin	Schistosomiasis
	7. *Naegleria fowleri*	Ameba enter nose while swimming	Meningoencephalitis
	8. *Giardia lamblia*	Ingestion	Diarrhea
	9. *Cryptosporidium hominis*	Ingestion	Diarrhea
Soil	1. *Clostridium tetani*	Spores in soil enter wound	Tetanus
	2. *Clostridium botulinum*	Spores in soil contaminate food that is improperly canned	Botulism
	3. *Clostridium perfringens*	Spores in soil enter wound	Gas gangrene
	4. *Bacillus anthracis*	Spores in soil enter wound	Anthrax
	5. Atypical mycobacteria (e.g., *Mycobacterium avium-intracellulare*)	Inhale aerosol	Tuberculosis-like disease
	6. *Nocardia asteroides*	Inhale aerosol	Nocardiosis
	7. *Cryptococcus neoformans*	Inhale yeast in aerosol of soil contaminated by pigeon guano	Meningitis, pneumonia
	8. *Histoplasma capsulatum*	Inhale spores in aerosol of soil contaminated by starling guano	Histoplasmosis
	9. *Coccidioides immitis*	Inhale spores in aerosol of soil dust	Coccidioidomycosis
	10. *Sporothrix schenckii*	Spores on plants (rose thorn) enter skin wound	Sporotrichosis
	11. *Ancylostoma duodenale* and *Necator americanus*	Filariform larvae enter skin	Hookworm, especially anemia
	12. *Strongyloides stercoralis*	Filariform larvae enter skin	Strongyloidiasis
	13. *Ancylostoma caninum*	Filariform larvae enter skin	Cutaneous larva migrans

TABLE XII-5 Main Geographical Location of Medically Important Organisms

Main Geographical Location	Important Organism	Disease
Within the United States		
1. South central states (e.g., North Carolina and Virginia)	*Rickettsia rickettsii*	Rocky Mountain spotted fever
2. Northeastern states (e.g., Connecticut, New York, and New Jersey)	*Borrelia burgdorferi*	Lyme disease
3. Midwestern states in the Ohio and Mississippi River valleys (e.g., Missouri, Indiana, and Illinois)	*Histoplasma capsulatum*	Histoplasmosis
4. Southwestern states (e.g., California and Arizona)	*Coccidioides immitis*	Coccidioidomycosis
Outside the United States		
1. Tropical areas of Africa, Asia, and South America	*Plasmodium* species	Malaria
2. Central America	*Trypanosoma cruzi*	Chagas' disease
3. Caribbean Islands, South America, and Africa	Dengue virus	Dengue fever

(Continued)

TABLE XII–5 Main Geographical Location of Medically Important Organisms (*Continued*)

Main Geographical Location	Important Organism	Disease
4. West Africa	Ebola virus	Ebola hemorrhagic fever
5. Tropical areas of Africa and South America	Yellow fever virus	Yellow fever
6. Sub-Saharan Africa	*Neisseria meningitidis* (Group A)	Meningococcal meningitis
7. Central Africa	*Trypanosoma brucei*	African sleeping sickness
8. Middle East, Africa, and India	*Leishmania donovani*	Visceral leishmaniasis (kala-azar)
9. Middle East, Africa, and India	*Leishmania tropica*	Cutaneous leishmaniasis
10. Central and South America	*Leishmania brasiliensis*	Mucocutaneous leishmaniasis
	Zika virus	Zika fever

TABLE XII–6 Occupations and Avocations That Increase Exposure to Medically Important Organisms

Occupation/Avocation	Predisposing Factor	Important Organism	Disease
Hiking/camping	Tick exposure	*Borrelia burgdorferi*	Lyme disease
Rancher/farm worker	Skin wound contaminated with soil	*Bacillus anthracis*	Anthrax
Sewer worker	Exposure to rat urine	*Leptospira interrogans*	Leptospirosis
Cave explorer (spelunker) in bat-infested caves	Exposure to aerosol of bat saliva	Rabies virus	Rabies
Cave explorer (spelunker) or construction worker	Exposure to aerosol of bat guano	*Histoplasma capsulatum*	Histoplasmosis
Archaeologist or construction worker digging in soil	Exposure to soil containing spores	*Coccidioides immitis*	Coccidioidomycosis
Pigeon fancier	Exposure to soil containing bird guano	*Cryptococcus neoformans*	Cryptococcosis
Psittacine birds as pets	Exposure to aerosol of bird guano	*Chlamydia psittaci*	Psittacosis
Bear hunter in Alaska	Ingestion of bear meat	*Trichinella spiralis*	Trichinosis
Aquarium personnel/swimming pool	Abrasion of skin	*Mycobacterium marinum*	Swimming pool granuloma
Hot tub user	Inadequate chlorination of water	*Pseudomonas aeruginosa*	Folliculitis

TABLE XII–7 Hospital-Related Events That Predispose to Infection by Medically Important Organisms

Hospital-Related Event	Important Organism	Disease
Surgery	*Staphylococcus aureus*	Wound infection
Urinary catheter	1. *Escherichia coli* primarily, but also other enteric gram-negative rods (e.g., *Proteus, Serratia,* and *Pseudomonas*)	Urinary tract infection
	2. *Enterococcus faecalis*	Urinary tract infection
Intravenous catheter	*Staphylococcus epidermidis, Candida albicans*	Catheter-related infection, bacteremia
Prosthetic device (e.g., hip or heart valve)	1. *S. epidermidis*	Osteomyelitis or endocarditis
	2. *Mycobacterium fortuitum-chelonae*	Osteomyelitis
Respiratory therapy	*Pseudomonas aeruginosa, Acinetobacter baumannii*	Pneumonia
Burn therapy	*P. aeruginosa*	Wound infection
Intracerebral electrodes	Prion	Creutzfeldt-Jakob disease
Needlestick	1. HBV, HCV	Hepatitis B or C
	2. HIV	AIDS
Nursery for premature infants	Respiratory syncytial virus	Bronchiolitis or pneumonia

AIDS = acquired immunodeficiency syndrome; HBV = hepatitis B virus; HCV = hepatitis C virus; HIV = human immunodeficiency virus.

TABLE XII-8 Organisms That Commonly Cause Disease in Patients With Immunodeficiencies or Reduced Host Defenses

Immunodeficiency or Reduced Host Defense	Organisms
Reduced antibodies (e.g., agammaglobulinemia and IgA deficiency)	Encapsulated bacteria (e.g., *Streptococcus pneumoniae*, *Haemophilus influenzae* type b)
Reduced phagocytosis (e.g., chronic granulomatous disease, cancer chemotherapy [neutropenia])	*Staphylococcus aureus*, *Pseudomonas aeruginosa*, *Aspergillus fumigatus*
Reduced complement	
1. C3b	*S. pneumoniae*, *H. influenzae* type b, *S. aureus*
2. C6,7,8,9 (membrane attack complex)	*Neisseria meningitidis*
Reduced cell-mediated immunity	*Candida albicans*, *Pneumocystis jirovecii*
1. Thymic aplasia (DiGeorge's syndrome)	Intracellular bacteria (e.g., *Mycobacterium tuberculosis*, MAI, *Listeria*, *Salmonella*)
2. HIV infection (AIDS), corticosteroids	Opportunistic fungi (e.g., *Candida*, *Cryptococcus*)
	Herpesviruses (e.g., herpes simplex virus, varicella-zoster virus, cytomegalovirus)
	Protozoa (e.g., *Toxoplasma*, *Cryptosporidium*)
	Pneumocystis
Disrupted epithelial surface (e.g., burns)	*P. aeruginosa*
Splenectomy	*S. pneumoniae*, *Babesia microti*
Diabetes mellitus	*S. aureus*, *Mucor* species, *P. aeruginosa*

AIDS = acquired immunodeficiency syndrome; HIV = human immunodeficiency virus; IgA = immunoglobulin A; MAI = *Mycobacterium avium-intracellulare* complex.

TABLE XII-9 Important Factors That Predispose to Infections by Specific Organisms

Predisposing Factor	Organism	Disease	Pathogenetic Mechanism
Cystic fibrosis	*Pseudomonas aeruginosa*	Pneumonia	Tenacious mucus traps bacteria in airways
Sickle cell anemia	*Salmonella enterica*	Osteomyelitis	Abnormally shaped red cells block blood vessels in bone and trap bacteria
	Streptococcus pneumoniae	Sepsis	Abnormally shaped red cells block blood vessels in spleen causing infarction of spleen
Intravenous drug use	*Staphylococcus aureus*	Right-sided endocarditis	Skin flora enter venous blood at site of needle
Antibiotic use	*Clostridium difficile*	Pseudomembranous colitis	Antibiotics suppress enteric normal flora, allowing *C. difficile* to grow
Aortic aneurysm	*S. enterica*[1]	Vascular graft infection	Uncertain
Tampon use (either vaginal or nasal tampon)	*S. aureus*	Toxic shock syndrome	Tampon blocks flow of blood, allowing *S. aureus* to grow and produce toxin
Dental surgery	Viridans group streptococci	Endocarditis	These bacteria are normal flora in the mouth and enter the blood at the site of the surgical wound
Prosthetic heart valve	*Staphylococcus epidermidis*	Endocarditis	Skin flora enter bloodstream at site of catheter or skin wound
Prosthetic joint	*S. epidermidis*	Osteomyelitis	Skin flora enter bloodstream at site of catheter or skin wound
Motorcycle accident	*Clostridium perfringens*	Gas gangrene (myonecrosis)	Spores in soil enter wound site
Contact lenses	*P. aeruginosa*, *Acanthamoeba castellani*	Keratitis	Abrasions caused by lenses provide entry site for organisms

[1]Especially *S. enterica* serotype Choleraesuis and serotype Dublin.

TABLE XII–10 Maternal Infections That Pose Significant Risk to the Fetus or Neonate[1]

Microbe	Transplacental or Perinatal Transmission to Fetus	Comment
A. Virus		
Cytomegalovirus	Transplacental	The leading cause of congenital abnormalities
Parvovirus B-19	Transplacental	Important cause of congenital abnormalities, including hydrops fetalis
Rubella virus	Transplacental	Vaccine has greatly reduced the incidence of fetal infection
Human immunodeficiency virus	Perinatal	Most are perinatal but transplacental and via breast milk also occurs
Hepatitis B virus (HBV)	Perinatal	Neonatal HBV infection greatly increases the risk of chronic carrier state
Hepatitis C virus (HCV)	Perinatal	Neonatal HCV infection greatly increases the risk of chronic carrier state
Herpes simplex type 2 virus	Perinatal	Important cause of encephalitis
Zika virus	Transplacental	Causes microcephaly; the only arbovirus that causes congenital abnormalities
B. Bacteria		
Treponema pallidum	Transplacental	Causes congenital syphilis
Neisseria gonorrhoeae	Perinatal	Important cause of conjunctivitis (ophthalmia neonatorum)
Chlamydia trachomatis	Perinatal	Important cause of conjunctivitis and pneumonia
Streptococcus agalactiae (group B *Streptococcus*)	Perinatal	Important cause of meningitis and sepsis
Escherichia coli	Perinatal	Important cause of meningitis and sepsis
Listeria monocytogenes	Perinatal	Important cause of meningitis and sepsis
C. Yeast		
Candida albicans	Perinatal	Causes thrush of the oropharynx
D. Protozoan		
Toxoplasma gondii	Transplacental	Important cause of congenital abnormalities, especially of eye and brain

[1]The acronym TORCHES is used to describe certain important fetal or neonatal infections acquired from the mother. There are several versions of this acronym. The following is a commonly used one: T = *Toxoplasma*; O = Other (including parvovirus B19, human immunodeficiency virus, and Zika virus); R = Rubella; C = Cytomegalovirus; HE = HErpes simplex virus 2; and S = Syphilis.

TABLE XII–11 Important Skin Lesions Caused by Microorganisms

Name or Type of Lesion	Causative Organism	Description of Lesion	Comment
A. Single or localized lesions			
Black eschar of anthrax	*Bacillus anthracis*	Crust over a necrotic ulcer	Caused by lethal toxin of *B. anthracis*
Carbuncle	*Staphylococcus aureus*	Group of furuncles (see below), often on neck	Poor personal hygiene predisposes
Cellulitis	*Streptococcus pyogenes*	Red, hot, tender, rapidly spreading, irregular shape	Hyaluronidase is "spreading factor"
Chancre of primary syphilis	*Treponema pallidum*	Painless, moist, shallow ulcer	Dark-field microscopy shows motile spirochetes
Cutaneous larva migrans	*Ancylostoma caninum*	Pruritic track, often on foot	Larva of dog hookworm migrates in skin
Ecthyma gangrenosum	Most often *Pseudomonas aeruginosa*	Necrotic ulcer with black eschar	Neutropenia predisposes
Erysipelas	*S. pyogenes*	Raised, red, tender, with defined border	Rapid progression (minutes to hours); diabetes predisposes
Erythema chronicum migrans (ECM) of Lyme disease	*Borrelia burgdorferi*		Lesion is at site of tick bite
Furuncle (boil, folliculitis)	1. *S. aureus*	Expanding erythematous macule[1]	1. Contains neutrophils and gram-positive cocci
	2. *P. aeruginosa*	Small pustule[1] at hair follicle	2. Causes "hot tub" folliculitis

(Continued)

TABLE XII–11 Important Skin Lesions Caused by Microorganisms (*Continued*)

Name or Type of Lesion	Causative Organism	Description of Lesion	Comment
Impetigo	*S. pyogenes* and *S. aureus*	Vesicles[1] with honey-colored crust	*S. pyogenes* skin infections predispose to acute glomerulonephritis
Malignant otitis externa	*P. aeruginosa*	Necrotic lesion on pinna of ear	Diabetes predisposes
Papilloma (warts)	Human papillomavirus (HPV)	Raised, dry, noninflamed papules[1]	Benign tumors except HPV 16 and 18 cause carcinoma of cervix
Ringworm	*Trichophyton, Epidermophyton, Microsporum*	Oval, inflamed, pruritic border with central clearing	See hyphae in KOH prep
Scabies	*Sarcoptes scabiei*	Pruritic track or papule[1]	*S. scabiei* is called the "itch mite"
Slapped cheeks syndrome	Parvovirus B19	Erythematous, macular, nontender rash on cheeks	
Zoster (shingles)	Varicella-zoster virus (VZV)	Painful, vesicles[1] along sensory nerve	Reactivation of latent VZV infection
B. Multiple or disseminated lesions			
Disseminated gonococcal infection (DGI)	*Neisseria gonorrhoeae*	Scattered pustules and inflamed tendons, especially of wrists and fingers (tenosynovitis)	Immunologic response to circulating antigen; no organisms in lesion
Erythema nodosum	Systemic fungi (e.g., *Coccidioides*) and mycobacteria (e.g., *Mycobacterium tuberculosis* and *Mycobacterium leprae*)	Erythematous, tender nodules on skin over tibia or ulna	See Koplik's spots on buccal mucosa; rash caused by cytotoxic T-cell attack on virus-infected cells
Hand, foot, and mouth disease	Coxsackie virus	Vesicles in those locations	A sign of disseminated intravascular coagulation (DIC) that occurs in sepsis; can enlarge to form purpuric (ecchymotic) lesions
Measles	Measles virus	Maculopapular splotchy (morbilliform) rash, especially on head and trunk	*Rickettsia* infect and kill vascular endothelium, resulting in hemorrhage into skin
Petechial hemorrhage	Many bacteria, (e.g., *Neisseria meningitidis*) and viruses (e.g., Ebola virus)	Small area of bleeding into the skin	Latent infection in lymphocytes
Rocky Mountain spotted fever	*Rickettsia rickettsiae*	Petechial hemorrhages including on palms and soles	Milder disease than measles
Roseola	Human herpesvirus 6	Maculopapular rash on face and trunk	Protease that cleaves desmoglein causes desquamation
Rubella	Rubella virus	Maculopapular, nonconfluent rash on face and trunk	Caused by strains of *S. pyogenes* that produces erythrogenic toxin that is a superantigen
Scalded skin syndrome	*S. aureus*	Desquamation over large area of body	Sign of emboli from vegetation on heart valve
Scarlet fever	*S. pyogenes*	Diffuse, macular, red (scarlet) rash; also strawberry tongue and circumoral pallor	Toxic shock syndrome toxin (TSST) is a superantigen
Secondary syphilis	*T. pallidum*	Maculopapular rash on trunk, palms, and soles	
Splinter hemorrhage	Viridans streptococci, *S. aureus* and other causes of endocarditis	Linear, black "splinters" under nails	
Toxic shock syndrome	*S. aureus*	Macular "sunburn-like" rash that desquamates later	
Varicella (chickenpox)	VZV	Pruritic vesicles on face and trunk	

[1]Description of certain important skin lesions: Macule is a flat, erythematous lesion. Papule is a raised, erythematous lesion with no visible fluid inside; resembles a mosquito bite. Vesicle is a raised, erythematous lesion with yellowish fluid (resembling plasma) inside; approximately the same size as a papule. Pustule is a raised, erythematous lesion with cloudy fluid (pus) inside; typically larger than a papule or vesicle.

These practice questions are presented in the format used by the United States Medical Licensing Examination (USMLE) Step 1. Note that in the computerized version of the USMLE, all questions are of the "ONE-BEST-ANSWER" type. There are no questions of the "EXCEPT" or "LEAST ACCURATE" type in which you are asked to determine the one wrong answer. Nevertheless, for studying purposes, the EXCEPT or LEAST ACCURATE type of questions are excellent learning tools because they provide you with several correct statements and only one incorrect statement rather than several incorrect ones. In view of this learning advantage, many practice questions in Part XIII of this book are of the EXCEPT or LEAST ACCURATE type. However, in Part XIV, the questions in the USMLE Practice Examination are presented in the ONE-BEST-ANSWER format, and no EXCEPT type questions are used.

After the questions regarding the specific content areas (i.e., bacteriology, virology, mycology, parasitology, and immunology), there are two additional sections, one containing questions in an extended matching format and the other containing questions based on infectious disease cases. The questions in the computerized version of the USMLE have 4 to 10 answer choices. Although the format of the questions in the extended matching section of this book is different from the format used in the USMLE, the questions in this section are designed to be a highly time-effective way of transmitting the important information.

BASIC BACTERIOLOGY

DIRECTIONS (Questions 1–39): Select the ONE lettered answer that is BEST in each question.

1. Each of the following statements concerning the surface structures of bacteria is correct EXCEPT:

   (A) Pili mediate the interaction of bacteria with mucosal epithelium.
   (B) Polysaccharide capsules retard phagocytosis.
   (C) Both gram-negative rods and cocci have lipopolysaccharide ("endotoxin") in their cell wall.
   (D) Bacterial flagella are nonantigenic in humans because they closely resemble human flagella in chemical composition.

2. Each of the following statements concerning peptidoglycan is correct EXCEPT:

   (A) It has a backbone composed of alternating units of muramic acid and acetylglucosamine.
   (B) Cross-links between the tetrapeptides involve D-alanine.
   (C) It is thinner in gram-positive than in gram-negative cells.
   (D) It can be degraded by lysozyme.

3. Each of the following statements concerning bacterial spores is correct EXCEPT:

   (A) Their survival ability is based on their enhanced metabolic activity.
   (B) They are formed by gram-positive rods.
   (C) They can be killed by being heated in an autoclave to 121°C for 15 minutes.
   (D) They are formed primarily when nutrients are limited.

4. Which one of the statements is the MOST accurate comparison of human, bacterial, and fungal cells?

   (A) Human cells undergo mitosis, whereas neither bacteria nor fungi do.
   (B) Human and fungal cells have a similar cell wall, in contrast to bacteria, whose cell wall contains peptidoglycan.
   (C) Human and bacterial cells have plasmids, whereas fungal cells do not.
   (D) Human and fungal cells have similar ribosomes, whereas bacterial ribosomes are significantly different.

5. Which statement is MOST accurate regarding the drug depicted in the diagram?

(A) It inhibits DNA synthesis.

(B) It is bacteriostatic.

(C) It binds to 30S ribosomes.

(D) It prevents formation of folic acid.

6. Each of the following statements regarding the selective action of antibiotics on bacteria is correct EXCEPT:

(A) Chloramphenicol affects the large subunit of the bacterial ribosome, which is different from the large subunit of the human ribosome.

(B) Isoniazid affects the DNA polymerase of bacteria but not that of human cells.

(C) Sulfonamides affect folic acid synthesis in bacteria, a pathway that does not occur in human cells.

(D) Penicillins affect bacteria rather than human cells because bacteria have a cell wall, whereas human cells do not.

7. Each of the following statements concerning endotoxins is correct EXCEPT:

(A) They are less toxic (i.e., less active on a weight basis) than exotoxins.

(B) They are more stable on heating than exotoxins.

(C) They bind to specific cell receptors, whereas exotoxins do not.

(D) They are part of the bacterial cell wall, whereas exotoxins are not.

8. The MAIN host defense against bacterial exotoxins is:

(A) Activated macrophages secreting proteases

(B) IgG and IgM antibodies

(C) Helper T cells

(D) Modulation of host cell receptors in response to the toxin

9. Which one of the following processes involves a sex pilus?

(A) Transduction of a chromosomal gene

(B) Transposition of a mobile genetic element

(C) Integration of a temperate bacteriophage

(D) Conjugation resulting in transfer of an R (resistance) factor

10. Each of the following statements concerning the normal flora is correct EXCEPT:

(A) The most common organism found on the skin is *Staphylococcus epidermidis*.

(B) *Escherichia coli* is a prominent member of the normal flora of the throat.

(C) The major site where *Bacteroides fragilis* is found is the colon.

(D) One of the most common sites where *Staphylococcus aureus* is found is the nose.

11. Each of the following statements concerning the mechanism of action of antimicrobial drugs is correct EXCEPT:

(A) Vancomycin acts by inhibiting peptidoglycan synthesis.

(B) Quinolones, such as ciprofloxacin, act by inhibiting the DNA gyrase of bacteria.

(C) Erythromycin is a bactericidal drug that disrupts cell membranes by a detergent-like action.

(D) Aminoglycosides such as streptomycin are bactericidal drugs that inhibit protein synthesis.

12. Each of the following statements concerning the resistance of bacteria to antimicrobial drugs is correct EXCEPT:

(A) Resistance to chloramphenicol is known to be due to an enzyme that acetylates the drug.

(B) Resistance to penicillin is known to be due to reduced affinity of transpeptidases.

(C) Resistance to penicillin is known to be due to cleavage by β-lactamase.

(D) Resistance to tetracycline is known to be due to an enzyme that hydrolyzes the ester linkage.

13. Of the following choices, the MOST important function of antibody in host defenses against bacteria is:

(A) Activation of lysozyme that degrades the cell wall

(B) Acceleration of proteolysis of exotoxins

(C) Facilitation of phagocytosis

(D) Inhibition of bacterial protein synthesis

14. Which of the following events is MOST likely to be due to bacterial conjugation?

(A) A strain of *Corynebacterium diphtheriae* produces a toxin encoded by a prophage.

(B) A strain of *Pseudomonas aeruginosa* produces β-lactamase encoded by a plasmid similar to a plasmid of another gram-negative organism.

(C) An encapsulated strain of *Streptococcus pneumoniae* acquires the gene for capsule formation from an extract of DNA from another encapsulated strain.

(D) A gene encoding resistance to gentamicin in the *Escherichia coli* chromosome appears in the genome of a bacteriophage that has infected *E. coli*.

15. Which one of the following BEST describes the mode of action of endotoxin?

(A) Degrades lecithin in cell membranes

(B) Inactivates elongation factor-2

(C) Blocks release of acetylcholine

(D) Causes the release of tumor necrosis factor

16. The identification of bacteria by serologic tests is based on the presence of specific antigens. Which one of the following bacterial components is LEAST likely to contain useful antigens?

(A) Capsule

(B) Flagella

(C) Exotoxins

(D) Ribosomes

17. Each of the following statements concerning bacterial spores is correct EXCEPT:

(A) Spores are formed under adverse environmental conditions such as the absence of a carbon source.

(B) Spores are resistant to boiling.

(C) Spores are metabolically inactive and contain dipicolinic acid, a calcium chelator.

(D) Spores are formed primarily by organisms of the genus *Neisseria*.

18. Each of the following statements concerning the mechanism of action of antibacterial drugs is correct EXCEPT:

(A) Cephalosporins are bactericidal drugs that inhibit the transpeptidase reaction and prevent cell wall synthesis.

(B) Tetracyclines are bacteriostatic drugs that inhibit protein synthesis by blocking tRNA binding.

(C) Aminoglycosides are bacteriostatic drugs that inhibit protein synthesis by activating ribonuclease, which degrades mRNA.

(D) Erythromycin is a bacteriostatic drug that inhibits protein synthesis by blocking translocation of the polypeptide.

19. Each of the following is a typical property of obligate anaerobes EXCEPT:

    (A) They generate energy by using the cytochrome system.
    (B) They grow best in the absence of air.
    (C) They lack superoxide dismutase.
    (D) They lack catalase.

20. Each of the following statements concerning the Gram stain is correct EXCEPT:

    (A) *Escherichia coli* stains pink because it has a thin peptidoglycan layer.
    (B) *Streptococcus pyogenes* stains blue because it has a thick peptidoglycan layer.
    (C) *Mycobacterium tuberculosis* stains blue because it has a thick lipid layer.
    (D) *Mycoplasma pneumoniae* is not visible in the Gram stain because it does not have a cell wall.

21. Each of the following statements concerning the killing of bacteria is correct EXCEPT:

    (A) Lysozyme in tears can hydrolyze bacterial cell walls.
    (B) Silver nitrate can inactivate bacterial enzymes.
    (C) Detergents can disrupt bacterial cell membranes.
    (D) Ultraviolet light can degrade bacterial capsules.

22. In the Gram stain, the decolorization of gram-negative bacteria by acetone-alcohol is MOST closely related to:

    (A) Proteins encoded by F plasmids
    (B) Lipids in the outer cell membrane
    (C) 70S ribosomes
    (D) Branched polysaccharides in the capsule

23. Chemical modification of benzylpenicillin (penicillin G) has resulted in several beneficial changes in the clinical use of this drug. Which one of the following is NOT one of those beneficial changes?

    (A) Lowered frequency of anaphylaxis
    (B) Increased activity against gram-negative rods
    (C) Increased resistance to stomach acid
    (D) Reduced cleavage by penicillinase

24. Each of the following statements concerning resistance to antibiotics is correct EXCEPT:

    (A) Resistance to aminoglycosides can be due to phosphorylating enzymes encoded by R plasmids.
    (B) Resistance to sulfonamides can be due to enzymes that hydrolyze the five-membered ring structure.
    (C) Resistance to penicillins can be due to alterations in binding proteins in the cell membrane.
    (D) Resistance to cephalosporins can be due to cleavage of the β-lactam ring.

25. The effects of endotoxin include each of the following EXCEPT:

    (A) Opsonization
    (B) Fever
    (C) Activation of the coagulation cascade
    (D) Hypotension

26. Bacterial surface structures that show antigenic diversity include each of the following EXCEPT:

    (A) Pili
    (B) Capsules
    (C) Flagella
    (D) Peptidoglycan

27. The effects of antibody on bacteria include each of the following EXCEPT:

    (A) Lysis of gram-negative bacteria in conjunction with complement
    (B) Augmentation of phagocytosis
    (C) Increase in the frequency of lysogeny
    (D) Inhibition of adherence of bacteria to mucosal surfaces

28. Each of the following statements concerning exotoxins is correct EXCEPT:

    (A) When treated chemically, some exotoxins lose their toxicity and can be used as immunogens in vaccines.
    (B) Some exotoxins are capable of causing disease in purified form, free of any bacteria.
    (C) Some exotoxins act in the gastrointestinal tract to cause diarrhea.
    (D) Some exotoxins contain lipopolysaccharides as the toxic component.

29. Each of the following statements concerning bacterial and human cells is correct EXCEPT:

    (A) Bacteria are prokaryotic (i.e., they have one molecule of DNA, are haploid, and have no nuclear membrane), whereas human cells are eukaryotic (i.e., they have multiple chromosomes, are diploid, and have a nuclear membrane).
    (B) Bacteria derive their energy by oxidative phosphorylation within mitochondria in a manner similar to human cells.
    (C) Bacterial and human ribosomes are of different sizes and chemical compositions.
    (D) Bacterial cells possess peptidoglycan, whereas human cells do not.

30. Each of the following statements concerning penicillin is correct EXCEPT:

    (A) An intact β-lactam ring of penicillin is required for its activity.
    (B) The structure of penicillin resembles that of a dipeptide of alanine, which is a component of peptidoglycan.
    (C) Penicillin is a bacteriostatic drug because autolytic enzymes are not activated.
    (D) Penicillin inhibits transpeptidases, which are required for cross-linking peptidoglycan.

31. Each of the following statements concerning the mechanisms of resistance to antimicrobial drugs is correct EXCEPT:

    (A) R factors are plasmids that carry the genes for enzymes that modify one or more drugs.
    (B) Resistance to some drugs is due to a chromosomal mutation that alters the receptor for the drug.
    (C) Resistance to some drugs is due to transposon genes that code for enzymes that inactivate the drugs.
    (D) Resistance genes are rarely transferred by conjugation.

32. Each of the following statements concerning endotoxins is correct EXCEPT:

    (A) The toxicity of endotoxins is due to the lipid portion of the molecule.
    (B) Endotoxins are found in most gram-positive bacteria.
    (C) Endotoxins are located outside of the cell wall peptidoglycan.
    (D) The antigenicity of somatic (O) antigen is due to repeating oligosaccharides.

33. Each of the following statements concerning exotoxins is correct EXCEPT:

(A) Exotoxins are polypeptides.

(B) Exotoxins are more easily inactivated by heat than are endotoxins.

(C) Exotoxins are less toxic than the same amount of endotoxins.

(D) Exotoxins can be converted to toxoids.

34. Each of the following statements concerning the killing of bacteria is correct EXCEPT:

(A) A 70% solution of ethanol kills more effectively than absolute (100%) ethanol.

(B) An autoclave uses steam under pressure to reach the killing temperature of 121°C.

(C) The pasteurization of milk kills pathogens but allows many organisms and spores to survive.

(D) Iodine kills by causing the formation of thymine dimers in bacterial DNA.

35. Each of the following statements concerning the drug depicted in the diagram is correct EXCEPT:

(A) The drug is bacteriostatic.

(B) The drug inhibits cell wall synthesis.

(C) The drug is made by a fungus.

(D) The portion of the molecule required for activity is labeled B.

36. Each of the following statements concerning the normal flora is correct EXCEPT:

(A) The normal flora of the colon consists predominantly of anaerobic bacteria.

(B) The presence of the normal flora prevents certain pathogens from colonizing the upper respiratory tract.

(C) Fungi (e.g., yeasts) are not members of the normal flora.

(D) Organisms of the normal flora are permanent residents of the body surfaces.

37. Each of the following statements concerning the structure and chemical composition of bacteria is correct EXCEPT:

(A) Some gram-positive cocci contain teichoic acid external to the peptidoglycan.

(B) Some gram-positive rods produce spores that are resistant to boiling.

(C) Some gram-negative rods contain lipid A in their outer cell membrane.

(D) Some mycoplasmas contain pentaglycine in their peptidoglycan.

38. Each of the following statements concerning the normal flora is correct EXCEPT:

(A) *Streptococcus mutans* is found in the mouth and contributes to the formation of dental caries.

(B) The predominant organisms in the alveoli are viridans streptococci.

(C) *Bacteroides fragilis* is found in greater numbers than *Escherichia coli* in the colon.

(D) *Candida albicans* is part of the normal flora of both men and women.

39. Each of the following statements concerning cholera toxin is correct EXCEPT:

(A) Cholera toxin inhibits elongation factor-2 in the mucosal epithelium.

(B) Binding of cholera toxin to the mucosal epithelium occurs via interaction of the B subunit of the toxin with a ganglioside in the cell membrane.

(C) Cholera toxin acts by adding ADP-ribose to a G protein.

(D) Cholera toxin activates the enzyme adenylate cyclase in the enterocyte.

Answers (Questions 1–39)

1. (D)	9. (D)	17. (D)	25. (A)	33. (C)
2. (C)	10. (B)	18. (C)	26. (D)	34. (D)
3. (A)	11. (C)	19. (A)	27. (C)	35. (A)
4. (D)	12. (D)	20. (C)	28. (D)	36. (C)
5. (C)	13. (C)	21. (D)	29. (B)	37. (D)
6. (B)	14. (B)	22. (B)	30. (C)	38. (B)
7. (C)	15. (D)	23. (A)	31. (D)	39. (A)
8. (B)	16. (D)	24. (B)	32. (B)	

DIRECTIONS (Questions 40–51): Select the ONE lettered option that is MOST closely associated with the numbered items. Each lettered option may be selected once, more than once, or not at all.

Questions 40–43

(A) Penicillins

(B) Aminoglycosides

(C) Chloramphenicol

(D) Rifampin

(E) Sulfonamides

40. Inhibit(s) bacterial RNA polymerase

41. Inhibit(s) cross-linking of peptidoglycan

42. Inhibit(s) protein synthesis by binding to the 30S ribosomal subunit

43. Inhibit(s) folic acid synthesis

Questions 44–46

(A) Transduction

(B) Conjugation

(C) DNA transformation

(D) Transposition

44. During an outbreak of gastrointestinal disease caused by an *Escherichia coli* strain sensitive to ampicillin, tetracycline, and chloramphenicol, a stool sample from one patient yields *E. coli* with the same serotype resistant to the three antibiotics.

45. A mutant cell line lacking a functional thymidine kinase gene was exposed to a preparation of DNA from normal cells; under appropriate growth conditions, a colony of cells was isolated that makes thymidine kinase.

46. A retrovirus without an oncogene does not induce leukemia in mice; after repeated passages through mice, viruses recovered from a tumor were highly oncogenic and contained a new gene.

Questions 47–51

(A) Diphtheria toxin

(B) Tetanus toxin

(C) Botulinum toxin

(D) Toxic shock syndrome toxin

(E) Cholera toxin

47. Causes paralysis by blocking release of acetylcholine
48. Inhibits protein synthesis by blocking elongation factor-2
49. Stimulates T cells to produce cytokines
50. Stimulates the production of cyclic AMP by adding ADP-ribose to a G protein
51. Inhibits the release of inhibitory neurotransmitters causing muscle spasms

Answers (Questions 40–51)

40. **(D)**	43. **(E)**	46. **(A)**	49. **(D)**
41. **(A)**	44. **(B)**	47. **(C)**	50. **(E)**
42. **(B)**	45. **(C)**	48. **(A)**	51. **(B)**

CLINICAL BACTERIOLOGY

DIRECTIONS (Questions 52–136): Select the ONE lettered answer that is BEST in each question.

52. An outbreak of sepsis caused by *Staphylococcus aureus* has occurred in the newborn nursery. You are called upon to investigate. According to your knowledge of the normal flora, what is the MOST likely source of the organism?

    (A) Colon
    (B) Nose
    (C) Throat
    (D) Vagina

53. Each of the statements about the classification of streptococci is correct EXCEPT:

    (A) Pneumococci (*Streptococcus pneumoniae*) are α-hemolytic and can be serotyped on the basis of their polysaccharide capsules.
    (B) Enterococci are group D streptococci and can be classified by their ability to grow in 6.5% sodium chloride.
    (C) Although pneumococci and the viridans streptococci are α-hemolytic, they can be differentiated by the bile solubility test and their susceptibility to optochin.
    (D) Viridans streptococci are identified by Lancefield grouping, which is based on the C carbohydrate in the cell wall.

54. Each of the following agents is a recognized cause of diarrhea EXCEPT:

    (A) *Clostridium perfringens*
    (B) *Enterococcus faecalis*
    (C) *Escherichia coli*
    (D) *Vibrio cholerae*

55. Each of the following organisms is an important cause of urinary tract infections EXCEPT:

    (A) *Escherichia coli*
    (B) *Proteus mirabilis*
    (C) *Klebsiella pneumoniae*
    (D) *Bacteroides fragilis*

56. Your patient is a 30-year-old woman with nonbloody diarrhea for the past 14 hours. Which one of the following organisms is LEAST likely to cause this illness?

    (A) *Clostridium difficile*
    (B) *Streptococcus pyogenes*
    (C) *Shigella dysenteriae*
    (D) *Salmonella enteritidis*

57. Each of the following statements concerning *Mycobacterium tuberculosis* is correct EXCEPT:

    (A) After being stained with carbol fuchsin, *M. tuberculosis* resists decolorization with acid alcohol.

    (B) *M. tuberculosis* has a large amount of mycolic acid in its cell wall.
    (C) *M. tuberculosis* appears as a red rod in Gram-stained specimens.
    (D) *M. tuberculosis* appears as a red rod in acid-fast–stained specimens.

58. A 50-year-old homeless alcoholic has a fever and is coughing up 1 cup of green, foul-smelling sputum per day. You suspect that he may have a lung abscess. Which one of the following pairs of organisms is MOST likely to be the cause?

    (A) *Listeria monocytogenes* and *Legionella pneumophila*
    (B) *Nocardia asteroides* and *Mycoplasma pneumoniae*
    (C) *Fusobacterium nucleatum* and *Peptostreptococcus intermedius*
    (D) *Clostridium perfringens* and *Chlamydia psittaci*

59. Which one of the following diseases is BEST diagnosed by a serologic (antibody-based) test?

    (A) Q fever
    (B) Pulmonary tuberculosis
    (C) Gonorrhea
    (D) Actinomycosis

60. Your patient has subacute bacterial endocarditis caused by a member of the viridans group of streptococci. Which one of the following sites is MOST likely to be the source of the organism?

    (A) Skin
    (B) Colon
    (C) Oropharynx
    (D) Urethra

61. A culture of skin lesions from a patient with pyoderma (impetigo) shows numerous colonies surrounded by a zone of β-hemolysis on a blood agar plate. A Gram-stained smear shows gram-positive cocci. If you found the catalase test to be negative, which one of the following organisms would you MOST probably have isolated?

    (A) *Streptococcus pyogenes*
    (B) *Staphylococcus aureus*
    (C) *Staphylococcus epidermidis*
    (D) *Streptococcus pneumoniae*

62. The coagulase test, in which the bacteria cause plasma to clot, is used to distinguish:

    (A) *Streptococcus pyogenes* from *Enterococcus faecalis*
    (B) *S. pyogenes* from *Staphylococcus aureus*
    (C) *S. aureus* from *Staphylococcus epidermidis*
    (D) *S. epidermidis* from *Neisseria meningitidis*

63. Which one of the following is a virulence factor for *Staphylococcus aureus*?

    (A) A heat-labile toxin that inhibits glycine release at the internuncial neuron
    (B) An oxygen-labile hemolysin
    (C) Resistance to novobiocin
    (D) Protein A that binds to the Fc portion of IgG

64. Which one of the following host defense mechanisms is the MOST important for preventing dysentery caused by *Salmonella*?

    (A) Gastric acid
    (B) Salivary enzymes
    (C) Normal flora of the mouth
    (D) Alpha interferon

65. The MOST important protective function of the antibody stimulated by tetanus immunization is:

    (A) To opsonize the pathogen (*Clostridium tetani*)
    (B) To prevent growth of the pathogen
    (C) To prevent adherence of the pathogen
    (D) To neutralize the toxin of the pathogen

66. Five hours after eating reheated rice at a restaurant, a 24-year-old woman and her husband both developed nausea, vomiting, and diarrhea. Which one of the following organisms is the MOST likely to be involved?

    (A) *Clostridium perfringens*
    (B) Enterotoxigenic *Escherichia coli*
    (C) *Bacillus cereus*
    (D) *Salmonella typhi*

67. Which one of the following bacteria has the LOWEST 50% infectious dose (ID_{50})?

    (A) *Shigella sonnei*
    (B) *Vibrio cholerae*
    (C) *Salmonella typhi*
    (D) *Campylobacter jejuni*

68. For which one of the following enteric illnesses is a chronic carrier state MOST likely to develop?

    (A) *Campylobacter* enterocolitis
    (B) *Shigella* enterocolitis
    (C) Cholera
    (D) Typhoid fever

69. Which one of the following zoonotic illnesses has NO arthropod vector?

    (A) Plague
    (B) Lyme disease
    (C) Brucellosis
    (D) Epidemic typhus

70. Which one of the following organisms principally infects vascular endothelial cells and as a result often causes a petechial rash?

    (A) *Salmonella typhi*
    (B) *Rickettsia rickettsii*
    (C) *Haemophilus influenzae*
    (D) *Coxiella burnetii*

71. Which one of the following statements MOST accurately depicts the ability of the organism to be cultured in the laboratory?

    (A) *Treponema pallidum* from a chancre can be grown on a special artificial medium supplemented with cholesterol.
    (B) *Mycobacterium leprae* can be grown in the armadillo and the mouse footpad but not on any artificial media.

    (C) *Mycobacterium tuberculosis* can be grown on enriched artificial media and produces visible colonies in 48 to 96 hours.
    (D) Atypical mycobacteria are found widely in soil and water but cannot be cultured on artificial media in the laboratory.

72. Each of the following statements concerning chlamydiae is correct EXCEPT:

    (A) Chlamydiae are strict intracellular parasites because they cannot synthesize sufficient adenosine triphosphate (ATP).
    (B) Chlamydiae possess both DNA and RNA and are bounded by a cell wall.
    (C) *Chlamydia trachomatis* has multiple serotypes that can cause different diseases.
    (D) Most chlamydiae are transmitted by arthropods.

73. For which one of the following bacterial vaccines are adverse effects an important concern?

    (A) The vaccine containing pneumococcal polysaccharide
    (B) The vaccine containing killed *Bordetella pertussis*
    (C) The vaccine containing tetanus toxoid
    (D) The vaccine containing diphtheria toxoid

74. Each of the following statements concerning *Staphylococcus aureus* is correct EXCEPT:

    (A) Gram-positive cocci in grapelike clusters are seen on Gram-stained smear.
    (B) The coagulase test is positive.
    (C) Treatment should include a β-lactamase-resistant penicillin.
    (D) Endotoxin is an important pathogenetic factor.

75. Your patient is a 70-year-old man who underwent bowel surgery for colon cancer 3 days ago. He now has a fever and abdominal pain. You are concerned that he may have peritonitis. Which one of the following pairs of organisms is MOST likely to be the cause?

    (A) *Bacteroides fragilis* and *Klebsiella pneumoniae*
    (B) *Bordetella pertussis* and *Salmonella enteritidis*
    (C) *Actinomyces israelii* and *Campylobacter jejuni*
    (D) *Clostridium botulinum* and *Shigella dysenteriae*

76. A 65-year-old man develops dysuria and hematuria. A Gram stain of a urine sample shows gram-negative rods. Culture of the urine on EMB agar reveals lactose-negative colonies without evidence of swarming motility. Which one of the following organisms is MOST likely to be the cause of his urinary tract infection?

    (A) *Enterococcus faecalis*
    (B) *Pseudomonas aeruginosa*
    (C) *Proteus vulgaris*
    (D) *Escherichia coli*

77. A 25-year-old man complains of a urethral discharge. You perform a Gram stain on a specimen of the discharge and see neutrophils but no bacteria. Of the organisms listed, the one MOST likely to cause the discharge is:

    (A) *Treponema pallidum*
    (B) *Chlamydia trachomatis*
    (C) *Candida albicans*
    (D) *Coxiella burnetii*

78. Two hours after a delicious Thanksgiving dinner of barley soup, roast turkey, stuffing, sweet potato, green beans, cranberry

sauce, and pumpkin pie topped with whipped cream, the Smith family of four experience vomiting and diarrhea. Which one of the following organisms is MOST likely to cause these symptoms?

(A) *Shigella flexneri*

(B) *Campylobacter jejuni*

(C) *Staphylococcus aureus*

(D) *Salmonella enteritidis*

79. Your patient has a brain abscess that was detected 1 month after a dental extraction. Which one of the following organisms is MOST likely to be involved?

(A) Anaerobic streptococci

(B) *Mycobacterium smegmatis*

(C) *Lactobacillus acidophilus*

(D) *Mycoplasma pneumoniae*

80. The MOST important contribution of the capsule of *Streptococcus pneumoniae* to virulence is:

(A) To prevent dehydration of the organisms on mucosal surfaces

(B) To retard phagocytosis by polymorphonuclear leukocytes

(C) To inhibit polymorphonuclear leukocyte chemotaxis

(D) To accelerate tissue invasion by its collagenase-like activity

81. The MOST important way the host counteracts the function of the pneumococcal polysaccharide capsule is via:

(A) T lymphocytes sensitized to polysaccharide antigens

(B) Polysaccharide-degrading enzymes

(C) Anticapsular antibody

(D) Activated macrophages

82. The pathogenesis of which one of the following organisms is MOST likely to involve invasion of the intestinal mucosa?

(A) *Vibrio cholerae*

(B) *Shigella sonnei*

(C) Enterotoxigenic *Escherichia coli*

(D) *Clostridium botulinum*

83. Which one of the following organisms that infects the gastrointestinal tract is the MOST frequent cause of bacteremia?

(A) *Shigella flexneri*

(B) *Campylobacter jejuni*

(C) *Vibrio cholerae*

(D) *Salmonella typhi*

84. A 30-year-old woman with systemic lupus erythematosus is found to have a positive serologic test for syphilis (VDRL test). She denies having had sexual contact with a partner who had symptoms of a venereal disease. The next best step would be to:

(A) Reassure her that the test is a false-positive reaction related to her autoimmune disorder

(B) Trace her sexual contacts for serologic testing

(C) Treat her with penicillin

(D) Perform a fluorescent treponemal antibody-absorbed (FTA-ABS) test on a specimen of her serum

85. Each of the following statements concerning *Treponema* is correct EXCEPT:

(A) *Treponema pallidum* produces an exotoxin that stimulates adenylate cyclase.

(B) *T. pallidum* cannot be grown on conventional laboratory media.

(C) Treponemes are members of the normal flora of the human oropharynx.

(D) Patients infected with *T. pallidum* produce antibodies that react with beef heart cardiolipin.

86. Each of the following statements concerning clostridia is correct EXCEPT:

(A) Pathogenic clostridia are found both in the soil and in the normal flora of the colon.

(B) Antibiotic-associated (pseudomembranous) colitis is due to a toxin produced by *Clostridium difficile.*

(C) Anaerobic conditions at the wound site are not required to cause tetanus, because spores will form in the presence of oxygen.

(D) Botulism, which is caused by ingesting preformed toxin, can be prevented by boiling food prior to eating.

87. Each of the following statements concerning *Bacteroides fragilis* is correct EXCEPT:

(A) *B. fragilis* is a gram-negative rod that is part of the normal flora of the colon.

(B) *B. fragilis* forms endospores, which allow it to survive in the soil.

(C) The capsule of *B. fragilis* is an important virulence factor.

(D) *B. fragilis* infections are characterized by foul-smelling pus.

88. Each of the following statements concerning staphylococci is correct EXCEPT:

(A) *Staphylococcus aureus* is differentiated from *Staphylococcus epidermidis* by the production of coagulase.

(B) *S. aureus* infections are often associated with abscess formation.

(C) The majority of clinical isolates of *S. aureus* produce penicillinase; therefore, penicillin G should not be used for antibiotic therapy for *S. aureus* infections.

(D) Scalded skin syndrome caused by *S. aureus* is due to enzymatic degradation of epidermal desmosomes by catalase.

89. Acute glomerulonephritis is a nonsuppurative complication that follows infection by which one of the following organisms?

(A) *Enterococcus faecalis*

(B) *Streptococcus pyogenes*

(C) *Streptococcus pneumoniae*

(D) *Streptococcus agalactiae*

90. Each of the following statements concerning gram-negative rods is correct EXCEPT:

(A) *Escherichia coli* is part of the normal flora of the colon; therefore, it does not cause diarrhea.

(B) *E. coli* ferments lactose, whereas the enteric pathogens *Shigella* and *Salmonella* do not.

(C) *Klebsiella pneumoniae*, although a cause of pneumonia, is part of the normal flora of the colon.

(D) *Proteus* species are highly motile organisms that are found in the human colon and cause urinary tract infections.

91. A 70-year-old man is found to have a hard mass in his prostate, which is suspected to be a carcinoma. Twenty-four hours after surgical removal of the mass, he develops fever to 39°C and has several shaking chills. Of the organisms listed, which one is LEAST likely to be involved?

(A) *Escherichia coli*

(B) *Enterococcus faecalis*

(C) *Klebsiella pneumoniae*

(D) *Legionella pneumophila*

92. Five days ago, a 65-year-old woman with a lower urinary tract infection began taking ampicillin. She now has a fever and severe diarrhea. Of the organisms listed, which one is MOST likely to be the cause of the diarrhea?

    (A) *Clostridium difficile*
    (B) *Bacteroides fragilis*
    (C) *Proteus mirabilis*
    (D) *Bordetella pertussis*

93. The pathogenesis of which one of the following diseases does NOT involve an exotoxin?

    (A) Scarlet fever
    (B) Typhoid fever
    (C) Toxic shock syndrome
    (D) Botulism

94. Regarding the effect of benzylpenicillin (penicillin G) on bacteria, which one of the following organisms is LEAST likely to be resistant?

    (A) *Staphylococcus aureus*
    (B) *Enterococcus faecalis*
    (C) *Streptococcus pyogenes*
    (D) *Neisseria gonorrhoeae*

95. Which one of the following organisms is MOST likely to be the cause of pneumonia in an immunocompetent young adult?

    (A) *Nocardia asteroides*
    (B) *Serratia marcescens*
    (C) *Mycoplasma pneumoniae*
    (D) *Legionella pneumophila*

96. Each of the following statements concerning chlamydial genital tract infections is correct EXCEPT:

    (A) Infection can be diagnosed by finding antichlamydial antibody in a serum specimen.
    (B) Infection can persist after administration of penicillin.
    (C) Symptomatic infections can be associated with urethral or cervical discharge containing many polymorphonuclear leukocytes.
    (D) There is no vaccine against these infections.

97. Which one of the following illnesses is NOT a zoonosis?

    (A) Typhoid fever
    (B) Q fever
    (C) Tularemia
    (D) Rocky Mountain spotted fever

98. Which one of the following is NOT a characteristic of the *Staphylococcus* associated with toxic shock syndrome?

    (A) Release of a superantigen
    (B) Coagulase production
    (C) Appearance of the organism in grapelike clusters on Gram-stained smear
    (D) Catalase-negative reaction

99. Which one of the following is NOT an important characteristic of either *Neisseria gonorrhoeae* or *Neisseria meningitidis*?

    (A) Polysaccharide capsule
    (B) IgA protease
    (C) M protein
    (D) Pili

100. Which one of the following is NOT an important characteristic of *Streptococcus pyogenes*?

    (A) Protein A
    (B) M protein

(C) β-Hemolysin
(D) Polysaccharide group-specific substance

101. Each of the following is a feature of the Lancefield group B streptococci (*Streptococcus agalactiae*) EXCEPT:

    (A) Common cause of pyoderma (impetigo)
    (B) Vaginal carriage in 5% to 25% of normal women of child-bearing age
    (C) Neonatal sepsis and meningitis
    (D) β-Hemolysis

102. Three organisms, *Streptococcus pneumoniae*, *Neisseria meningitidis*, and *Haemophilus influenzae*, cause the vast majority of cases of bacterial meningitis. What is the MOST important pathogenic component they share?

    (A) Protein A
    (B) Capsule
    (C) Endotoxin
    (D) β-Lactamase

103. Diarrhea caused by which one of the following agents is characterized by the presence of fecal leukocytes?

    (A) *Campylobacter jejuni*
    (B) Rotavirus
    (C) *Clostridium perfringens*
    (D) Enterotoxigenic *Escherichia coli*

104. Each of the following statements concerning *Chlamydia trachomatis* is correct EXCEPT:

    (A) It is an important cause of nongonococcal urethritis.
    (B) It is the cause of lymphogranuloma venereum.
    (C) It is an important cause of subacute bacterial endocarditis.
    (D) It is an important cause of conjunctivitis.

105. Each of the following statements concerning *Actinomyces* and *Nocardia* is correct EXCEPT:

    (A) *Actinomyces israelii* is an anaerobic rod found as part of the normal flora in the mouth.
    (B) Both *Actinomyces* and *Nocardia* are branching, filamentous rods.
    (C) *Nocardia asteroides* causes infections primarily in immunocompromised patients.
    (D) Infections are usually diagnosed by detecting a significant rise in antibody titer.

106. Which one of the following types of organisms is NOT an obligate intracellular parasite and therefore can replicate on bacteriologic media?

    (A) *Chlamydia*
    (B) *Mycoplasma*
    (C) Adenovirus
    (D) *Rickettsia*

107. Tissue-degrading enzymes play an important role in the pathogenesis of several bacteria. Which one of the following is NOT involved in tissue or cell damage?

    (A) Lecithinase of *Clostridium perfringens*
    (B) Hyaluronidase of *Streptococcus pyogenes*
    (C) M protein of *Streptococcus pneumoniae*
    (D) Leukocidin of *Staphylococcus aureus*

108. The soil is the natural habitat for certain microorganisms of medical importance. Which one of the following is LEAST likely to reside there?

    (A) *Clostridium tetani*
    (B) *Mycobacterium avium-intracellulare*

    (C) *Bacillus anthracis*

    (D) *Chlamydia trachomatis*

109. Which one of the following organisms is the MOST common bacterial cause of pharyngitis?

    (A) *Staphylococcus aureus*

    (B) *Streptococcus pneumoniae*

    (C) *Streptococcus pyogenes*

    (D) *Neisseria meningitidis*

110. Several pathogens are transmitted either during gestation or at birth. Which one of the following is LEAST likely to be transmitted at these times?

    (A) *Haemophilus influenzae*

    (B) *Treponema pallidum*

    (C) *Neisseria gonorrhoeae*

    (D) *Chlamydia trachomatis*

111. Each of the following statements concerning exotoxins is correct EXCEPT:

    (A) Some strains of *Escherichia coli* produce an enterotoxin that causes diarrhea.

    (B) Cholera toxin acts by stimulating adenylate cyclase.

    (C) Diphtheria is caused by an exotoxin that inhibits protein synthesis by inactivating an elongation factor.

    (D) Botulism is caused by a toxin that hydrolyzes lecithin (lecithinase), thereby destroying nerve cells.

112. Each of the following statements concerning the VDRL test for syphilis is correct EXCEPT:

    (A) The antigen is composed of inactivated *Treponema pallidum*.

    (B) The test is usually positive in secondary syphilis.

    (C) False-positive results are more frequent than with the fluorescent treponemal antibody-absorbed (FTA-ABS) test.

    (D) The antibody titer declines with appropriate therapy.

113. Each of the following statements concerning the fluorescent treponemal antibody-absorbed (FTA-ABS) test for syphilis is correct EXCEPT:

    (A) The test is specific for *Treponema pallidum*.

    (B) The patient's serum is absorbed with saprophytic treponemes.

    (C) Once positive, the test remains so despite appropriate therapy.

    (D) The test is rarely positive in primary syphilis.

114. Each of the following statements concerning *Corynebacterium diphtheriae* is correct EXCEPT:

    (A) *C. diphtheriae* is a gram-positive rod that does not form spores.

    (B) Toxin production is dependent on the organism's being lysogenized by a bacteriophage.

    (C) Diphtheria toxoid should not be given to children younger than 3 years because the incidence of complications is too high.

    (D) Antitoxin should be used to treat patients with diphtheria.

115. Each of the following statements concerning certain gram-negative rods is correct EXCEPT:

    (A) *Pseudomonas aeruginosa* causes wound infections that are characterized by blue-green pus as a result of pyocyanin production.

    (B) In unimmunized individuals, invasive disease caused by *Haemophilus influenzae* is most often due to strains possessing a type b polysaccharide capsule.

    (C) *Legionella pneumophila* infection is acquired by inhalation of aerosols from environmental water sources.

    (D) Whooping cough, which is caused by *Bordetella pertussis*, is on the rise because changing antigenicity of the organism has made the vaccine relatively ineffective.

116. Each of the following statements concerning enterotoxins is correct EXCEPT:

    (A) Enterotoxins typically cause bloody diarrhea with leukocytes in the stool.

    (B) *Staphylococcus aureus* produces an enterotoxin that causes vomiting and diarrhea.

    (C) *Vibrio cholerae* causes cholera by producing an enterotoxin that increases adenylate cyclase activity within the enterocyte.

    (D) *Escherichia coli* enterotoxin mediates ADP-ribosylation of a G protein.

117. Each of the following statements concerning plague is correct EXCEPT:

    (A) Plague is caused by a gram-negative rod that can be cultured on blood agar.

    (B) Plague is transmitted from the animal reservoir to humans by flea bite.

    (C) The main reservoirs in nature are small rodents.

    (D) Plague is of concern in many underdeveloped countries but has not occurred in the United States since 1968.

118. Which one of the following statements concerning *Brucella* species, the organisms that cause brucellosis, is CORRECT?

    (A) They are transmitted primarily by tick bite.

    (B) The principal reservoirs of these bacteria are small rodents.

    (C) They commonly infect reticuloendothelial cells in the liver, spleen, and bone marrow.

    (D) They are obligate intracellular parasites that are usually identified by growth in human cell culture.

119. Each of the following statements concerning epidemic typhus is correct EXCEPT:

    (A) The disease is characterized by a rash.

    (B) The Weil-Felix test can aid in diagnosis of the disease.

    (C) The disease is caused by a *Rickettsia*.

    (D) The causative organism is transmitted from rodents to humans by a tick.

120. Which one of the following organisms causes diarrhea by producing an enterotoxin that increases adenylate cyclase activity within enterocytes?

    (A) *Escherichia coli*

    (B) *Bacteroides fragilis*

    (C) *Staphylococcus aureus*

    (D) *Enterococcus faecalis*

121. Each of the following statements concerning Rocky Mountain spotted fever is correct EXCEPT:

    (A) The causative organism forms β-hemolytic colonies on blood agar.

    (B) Headache, fever, and rash are characteristic features of the disease.

    (C) The disease occurs primarily in states east of the Mississippi river.

    (D) The disease is caused by a *Rickettsia*.

122. Each of the following statements concerning *Clostridium perfringens* is correct EXCEPT:

(A) It causes gas gangrene.

(B) It causes food poisoning.

(C) It produces an exotoxin that degrades lecithin and causes necrosis and hemolysis.

(D) It is a gram-negative rod that does not ferment lactose.

123. Each of the following statements concerning *Clostridium tetani* is correct EXCEPT:

(A) It is a gram-positive, spore-forming rod.

(B) Pathogenesis is due to the production of an exotoxin that blocks inhibitory neurotransmitters.

(C) Laboratory diagnosis is based on seeing beta-hemolytic colonies on a blood agar plate in the presence of room air.

(D) Its natural habitat is primarily the soil.

124. Each of the following statements concerning spirochetes is correct EXCEPT:

(A) Species of *Treponema* are part of the normal flora of the mouth.

(B) Species of *Borrelia* cause a tick-borne disease called relapsing fever.

(C) The species of *Leptospira* that cause leptospirosis grow primarily in humans and are usually transmitted by human-to-human contact.

(D) Species of *Treponema* cause syphilis and yaws.

125. Each of the following statements concerning gonorrhea is correct EXCEPT:

(A) Infection in men is more frequently symptomatic than in women.

(B) A presumptive diagnosis can be made by finding gram-negative kidney bean-shaped diplococci within neutrophils in a urethral discharge.

(C) The definitive diagnosis can be made by detecting antibodies to *Neisseria gonorrhoeae* in the patient's serum.

(D) Gonococcal conjunctivitis of the newborn rarely occurs in the United States because silver nitrate or erythromycin is commonly used as prophylaxis.

126. Each of the following statements concerning *Mycobacterium tuberculosis* is correct EXCEPT:

(A) Some strains of *M. tuberculosis* isolated from patients exhibit multiple drug resistance (i.e., they are resistant to both isoniazid and rifampin).

(B) *M. tuberculosis* contains a small amount of lipid in its cell wall and therefore stains poorly with the Gram stain.

(C) *M. tuberculosis* grows slowly, often requiring 3 to 6 weeks before colonies appear.

(D) The antigen in the tuberculin skin test is a protein extracted from the organism.

127. Which one of the following statements concerning immunization against diseases caused by clostridia is CORRECT?

(A) Antitoxin against tetanus protects against botulism as well, because the two toxins share antigenic sites.

(B) Vaccines containing alpha toxin (lecithinase) are effective in protecting against gas gangrene.

(C) The toxoid vaccine against *Clostridium difficile* infection should be administered to immunocompromised patients.

(D) Immunization with tetanus toxoid induces effective protection against tetanus toxin.

128. Each of the following statements concerning *Neisseria meningitidis* and *N. gonorrhoeae* is correct EXCEPT:

(A) They are gram-negative diplococci.

(B) They produce IgA protease as a virulence factor.

(C) They are oxidase-positive.

(D) In the laboratory, they form beta-hemolytic colonies on Thayer-Martin medium.

129. Which one of the following statements concerning *Legionella pneumophila* is CORRECT?

(A) It is part of the normal flora of the colon.

(B) It cannot be grown on laboratory media.

(C) It does not have a cell wall.

(D) It causes atypical pneumonia, especially in those with reduced cell-mediated immunity.

130. Each of the following statements concerning wound infections caused by *Clostridium perfringens* is correct EXCEPT:

(A) An exotoxin plays a role in pathogenesis.

(B) Gram-positive rods are found in the exudate.

(C) The organism grows only in human cell culture.

(D) Anaerobic culture of the wound site should be ordered.

131. Each of the following statements concerning infection with *Chlamydia psittaci* is correct EXCEPT:

(A) *C. psittaci* can be isolated by growth in cell culture and will not grow in blood agar.

(B) The organism appears purple in Gram-stained smears of sputum.

(C) The infection is more readily diagnosed by serologic tests than by isolation of the organism.

(D) The infection is more commonly acquired from a nonhuman source than from another human.

132. Ticks are vectors for the transmission of each of the following diseases EXCEPT:

(A) Rocky Mountain spotted fever

(B) Epidemic typhus

(C) Tularemia

(D) Lyme disease

133. Each of the following statements concerning pneumonia caused by *Mycoplasma pneumoniae* is correct EXCEPT:

(A) Pneumonia caused by *M. pneumoniae* is associated with a rise in the titer of cold agglutinins.

(B) Pneumonia caused by *M. pneumoniae* occurs primarily in immunocompetent individuals.

(C) Pneumonia caused by *M. pneumoniae* is an "atypical" pneumonia.

(D) *M. pneumoniae* cannot be cultured in vitro because it has no cell wall.

134. Each of the following statements concerning *Neisseria meningitidis* is correct EXCEPT:

(A) It is an oxidase-positive, gram-negative diplococcus.

(B) It contains endotoxin in its cell wall.

(C) It produces an exotoxin that stimulates adenylate cyclase.

(D) It has a polysaccharide capsule that is antiphagocytic.

135. Each of the following statements concerning Q fever is correct EXCEPT:

(A) Rash is a prominent feature.

(B) It is transmitted by respiratory aerosol.

(C) Farm animals are an important reservoir.

(D) It is caused by *Coxiella burnetii*.

136. Each of the following statements concerning *Mycobacterium leprae* is correct EXCEPT:

(A) In lepromatous leprosy, large numbers of organisms are usually seen in acid-fast–stained smears.

(B) The organism will grow on bacteriologic media in 3 to 6 weeks.

(C) Prolonged therapy (9 months or longer) is required to prevent recurrence.

(D) Loss of sensation due to nerve damage is often seen in leprosy.

Answers (Questions 52–136)

52. (B)	69. (C)	86. (C)	103. (A)	120. (A)
53. (D)	70. (B)	87. (B)	104. (C)	121. (A)
54. (B)	71. (B)	88. (D)	105. (D)	122. (D)
55. (D)	72. (D)	89. (B)	106. (B)	123. (C)
56. (B)	73. (B)	90. (A)	107. (C)	124. (C)
57. (C)	74. (D)	91. (D)	108. (D)	125. (C)
58. (C)	75. (A)	92. (A)	109. (C)	126. (B)
59. (A)	76. (B)	93. (B)	110. (A)	127. (D)
60. (C)	77. (B)	94. (C)	111. (D)	128. (D)
61. (A)	78. (C)	95. (C)	112. (A)	129. (D)
62. (C)	79. (A)	96. (A)	113. (D)	130. (C)
63. (D)	80. (B)	97. (A)	114. (C)	131. (B)
64. (A)	81. (C)	98. (D)	115. (D)	132. (B)
65. (D)	82. (B)	99. (C)	116. (A)	133. (D)
66. (C)	83. (D)	100. (A)	117. (D)	134. (C)
67. (A)	84. (D)	101. (A)	118. (C)	135. (A)
68. (D)	85. (A)	102. (B)	119. (D)	136. (B)

DIRECTIONS (Questions 137–158): Select the ONE lettered option that is MOST closely associated with the numbered items. Each lettered option may be selected once, more than once, or not at all.

Questions 137–140

(A) *Mycobacterium avium-intracellulare*
(B) *Treponema pallidum*
(C) *Rickettsia prowazekii*
(D) *Mycoplasma pneumoniae*

137. Is an obligate intracellular parasite
138. Its natural habitat is the soil
139. Has no cell wall
140. Is an acid-fast rod

Questions 141–143

(A) *Borrelia burgdorferi*
(B) *Helicobacter pylori*
(C) *Pasteurella multocida*
(D) *Brucella melitensis*

141. Peptic ulcer in a 45-year-old salesman
142. Cellulitis of the hand following a cat bite
143. Expanding, bull's eye–shaped red rash in a 6-year-old boy after a camping trip

Questions 144–147

(A) *Corynebacterium diphtheriae*
(B) *Listeria monocytogenes*
(C) *Bacillus anthracis*
(D) *Clostridium botulinum*

144. Causes both skin lesions and a severe pneumonia
145. Causes flaccid paralysis
146. Causes a pseudomembrane in the throat, which can cause respiratory tract obstruction
147. Causes meningitis in neonates and the immunosuppressed

Questions 148–150

(A) *Escherichia coli*
(B) *Klebsiella pneumoniae*
(C) *Salmonella enteritidis*
(D) *Proteus mirabilis*

148. Is frequently implicated in nosocomial infections, is an important cause of community-acquired pneumonia in adults, and has a thick, mucoid capsule
149. Is the most common cause of urinary tract infections
150. It causes urinary tract infections, produces urease, and exhibits swarming motility

Questions 151–154

(A) *Staphylococcus aureus*
(B) *Streptococcus pyogenes*
(C) *Enterococcus faecalis*
(D) *Streptococcus pneumoniae*

151. Grows in 6.5% sodium chloride
152. Is bile soluble
153. Produces enterotoxin
154. Is associated with rheumatic fever

Questions 155–158

(A) *Bacteroides fragilis*
(B) *Haemophilus influenzae*
(C) *Pseudomonas aeruginosa*
(D) *Chlamydia pneumoniae*

155. Coccobacillary gram-negative rod that causes meningitis in young children
156. Oxidase-positive gram-negative rod that is an important cause of wound and burn infections
157. Causes atypical pneumonia in immunocompetent adults
158. Anaerobic gram-negative rod that is an important cause of peritonitis

Answers (Questions 137–158)

137. (C)	143. (A)	149. (A)	155. (B)
138. (A)	144. (C)	150. (D)	156. (C)
139. (D)	145. (D)	151. (C)	157. (D)
140. (A)	146. (A)	152. (D)	158. (A)
141. (B)	147. (B)	153. (A)	
142. (C)	148. (B)	154. (B)	

BASIC VIROLOGY

DIRECTIONS (Questions 159–192): Select the ONE lettered answer that is BEST in each question.

159. Viruses enter cells by adsorbing to specific sites on the outer membrane of cells. Each of the following statements regarding this event is correct EXCEPT:

    (A) The interaction determines the specific target organs for infection.
    (B) The interaction determines whether the purified genome of a virus is infectious.
    (C) The interaction can be prevented by neutralizing antibody.
    (D) If the sites are occupied, interference with virus infection occurs.

160. Many viruses mature by budding through the outer membrane of the host cell. Each of the following statements regarding these viruses is correct EXCEPT:

    (A) Some of these viruses cause multinucleated giant cell formation.
    (B) Some new viral antigens appear on the surface of the host cell.
    (C) Some of these viruses contain host cell lipids.
    (D) Some of these viruses do not have an envelope.

161. Biochemical analysis of a virus reveals the genome to be composed of eight unequally sized pieces of single-stranded RNA, each of which is complementary to viral mRNA in infected cells. Which one of the following statements is UNLIKELY to be correct?

    (A) Different proteins are encoded by each segment of the viral genome.
    (B) The virus particle contains a virus-encoded enzyme that can copy the genome into its complement.
    (C) Purified RNA extracted from the virus particle is infectious.
    (D) The virus can acquire new antigens via reassortment of its RNA segments.

162. Latency is an outcome particularly characteristic of which one of the following virus groups?

    (A) Polioviruses
    (B) Herpesviruses
    (C) Rhinoviruses
    (D) Influenza viruses

163. Each of the following statements concerning viral serotypes is correct EXCEPT:

    (A) In naked nucleocapsid viruses, the serotype is usually determined by the outer capsid proteins.
    (B) In enveloped viruses, the serotype is usually determined by the outer envelope proteins, especially the spike proteins.
    (C) Some viruses have multiple serotypes.
    (D) Some viruses have an RNA polymerase that determines the serotype.

164. The ability of a virus to produce disease can result from a variety of mechanisms. Which one of the following mechanisms is LEAST likely?

    (A) Cytopathic effect in infected cells
    (B) Malignant transformation of infected cells
    (C) Immune response to virus-induced antigens on the surface of infected cells
    (D) Production of an exotoxin that activates adenylate cyclase

165. Which one of the following forms of immunity to viruses would be LEAST likely to be lifelong?

    (A) Passive immunity
    (B) Passive–active immunity
    (C) Active immunity
    (D) Cell-mediated immunity

166. Which one of the following statements concerning alpha, beta, and gamma interferons is LEAST accurate?

    (A) Interferons inhibit a broad range of viruses, not just the virus that induced the interferon.
    (B) Interferons are synthesized only by virus-infected cells.
    (C) Interferons induce the synthesis of a protein kinase that phosphorylates an elongation factor, thereby inactivating protein synthesis.
    (D) Interferons induce the synthesis of a ribonuclease that degrades viral mRNA.

167. You have isolated a virus from the stool of a patient with diarrhea and shown that its genome is composed of multiple pieces of double-stranded RNA. Which one of the following is LEAST LIKELY to be true?

    (A) Each piece of RNA encodes a different protein.
    (B) The virus encodes an RNA-directed RNA polymerase.
    (C) The virion contains an RNA polymerase.
    (D) The genome integrates into the host chromosome.

168. A temperate bacteriophage has been induced from a new pathogenic strain of *Escherichia coli* that produces a toxin. Which one of the following is the MOST convincing way to show that the phage encodes the toxin?

    (A) Carry out conjugation of the pathogenic strain with a nonpathogenic strain.
    (B) Infect an experimental animal with the phage.
    (C) Lysogenize a nonpathogenic strain with the phage.
    (D) Look for transposable elements in the phage DNA.

169. Each of the following statements concerning retroviruses is correct EXCEPT:

    (A) The virion carries an RNA-directed DNA polymerase encoded by the viral genome.
    (B) The viral genome consists of three segments of double-stranded RNA.
    (C) The virion is enveloped and enters cells via an interaction with specific receptors on the host cell.
    (D) During infection, the virus synthesizes a DNA copy of its RNA, and this DNA becomes covalently integrated into host cell DNA.

170. A stock of virus particles has been found by electron microscopy to contain 10^8 particles/mL, but a plaque assay reveals only 10^5 plaque-forming units/mL. The BEST interpretation of these results is that:

    (A) Only one particle in 1000 is infectious.
    (B) A nonpermissive cell line was used for the plaque assay.
    (C) Several kinds of viruses were present in the stock.
    (D) The virus is a temperature-sensitive mutant.

171. Reasonable mechanisms for viral persistence in infected individuals include all of the following EXCEPT:

    (A) Generation of defective-interfering particles
    (B) Virus-mediated inhibition of host DNA synthesis

(C) Integration of a provirus into the genome of the host

(D) Host tolerance to viral antigens

172. Each of the following statements concerning viral surface proteins is correct EXCEPT:

(A) They elicit antibody that neutralizes infectivity of the virus.

(B) They determine the species specificity of the virus–cell interaction.

(C) They participate in active transport of nutrients across the viral envelope membrane.

(D) They protect the genetic material against nucleases.

173. Each of the following statements concerning viral vaccines is correct EXCEPT:

(A) In live, attenuated vaccines, the virus has lost its ability to cause disease but has retained its ability to induce neutralizing antibody.

(B) In live, attenuated vaccines, the possibility of reversion to virulence is of concern.

(C) With inactivated vaccines, IgA mucosal immunity is usually induced.

(D) With inactivated vaccines, protective immunity is due mainly to the production of IgG.

174. The major barrier to the control of rhinovirus upper respiratory infections by immunization is:

(A) The poor local and systemic immune response to these viruses

(B) The large number of serotypes of the rhinoviruses

(C) The side effects of the vaccine

(D) The inability to grow the viruses in cell culture

175. The feature of the influenza virus genome that contributes MOST to the antigenic shift exhibited by the virus is:

(A) A high G + C content, which augments binding to euchromatin

(B) Inverted repeat regions, which create "sticky ends"

(C) Segmented nucleic acid

(D) Unique methylated bases

176. Which one of the following is the BEST explanation for the selective action of acyclovir (acycloguanosine) in herpes simplex virus (HSV)-infected cells?

(A) Acyclovir binds specifically to viral receptors only on the surface of the HSV-infected cell.

(B) Acyclovir is phosphorylated by a virus-encoded phosphokinase only within HSV-infected cells.

(C) Acyclovir selectively inhibits the RNA polymerase in the HSV virion.

(D) Acyclovir specifically blocks the matrix protein of HSV, thereby preventing release of progeny HSV.

177. Each of the following statements concerning interferon is correct EXCEPT:

(A) Interferon inhibits the growth of both DNA and RNA viruses.

(B) Interferon is induced by double-stranded RNA.

(C) Interferon made by cells of one species acts more effectively in the cells of that species than in the cells of other species.

(D) Interferon acts by preventing viruses from entering the cell.

178. Each of the following statements concerning the viruses that infect humans is correct EXCEPT:

(A) Only viruses with a negative polarity RNA genome have a polymerase in the virion.

(B) The purified nucleic acid of some viruses is infectious, but at a lower efficiency than the intact virions.

(C) Some viruses contain lipoprotein envelopes derived from the plasma membrane of the host cell.

(D) The nucleic acid of some viruses is single-stranded DNA and that of others is double-stranded RNA.

179. Which one of the following statements about virion structure and assembly is CORRECT?

(A) Most viruses acquire surface glycoproteins by budding through the nuclear membrane.

(B) Helical nucleocapsids are found primarily in DNA viruses.

(C) The symmetry of virus particles prevents inclusion of any nonstructural proteins, such as enzymes.

(D) Enveloped viruses use a matrix protein to mediate interactions between viral glycoproteins in the plasma membrane and structural proteins in the nucleocapsid.

180. Each of the following statements concerning viruses is correct EXCEPT:

(A) Viruses can reproduce only within cells.

(B) The proteins on the surface of the virus mediate the entry of the virus into host cells.

(C) Neutralizing antibody is directed against proteins on the surface of the virus.

(D) Viruses replicate by binary fission.

181. Viruses are obligate intracellular parasites. Each of the following statements concerning this fact is correct EXCEPT:

(A) Viruses cannot generate energy outside of cells.

(B) Viruses cannot synthesize proteins outside of cells.

(C) Viruses must degrade host cell DNA in order to obtain nucleotides.

(D) Enveloped viruses require host cell membranes to obtain their envelopes.

182. Each of the following statements concerning lysogeny is correct EXCEPT:

(A) Viral genes replicate independently of bacterial genes.

(B) Viral genes responsible for lysis are repressed.

(C) Viral DNA is integrated into bacterial DNA.

(D) Some lysogenic bacteriophages encode toxins that cause human disease.

183. Each of the following viruses possesses an outer envelope of lipoprotein EXCEPT:

(A) Varicella-zoster virus

(B) Human papillomavirus

(C) Influenza virus

(D) Human immunodeficiency virus

184. Which one of the following viruses possesses a genome of single-stranded RNA that is infectious when purified?

(A) Influenza virus

(B) Rotavirus

(C) Measles virus

(D) Poliovirus

185. Each of the following viruses possesses an RNA polymerase in the virion EXCEPT:

(A) Hepatitis A virus

(B) Smallpox virus

(C) Mumps virus

(D) Rotavirus

186. Each of the following viruses possesses a DNA polymerase in the virion EXCEPT:

    (A) Human immunodeficiency virus
    (B) Human T-cell lymphotropic virus
    (C) Epstein–Barr virus
    (D) Hepatitis B virus

187. Each of the following viruses possesses double-stranded nucleic acid as its genome EXCEPT:

    (A) Coxsackie virus
    (B) Herpes simplex virus
    (C) Rotavirus
    (D) Adenovirus

188. Which one of the following viruses performs cap-snatching during the synthesis of its mRNA?

    (A) Herpes simplex virus
    (B) Influenza virus
    (C) Hepatitis A virus
    (D) Hepatitis B virus
    (E) Zika virus

189. Each of the following statements about both measles virus and rubella virus is correct EXCEPT:

    (A) They are RNA enveloped viruses.
    (B) Their virions contain an RNA polymerase.
    (C) They have a single antigenic type.
    (D) They are transmitted by respiratory aerosol.

190. Each of the following statements about both influenza virus and rabies virus is correct EXCEPT:

    (A) They are enveloped RNA viruses.
    (B) Their virions contain an RNA polymerase.
    (C) A killed vaccine is available for both viruses.
    (D) They each have a single antigenic type.

191. Each of the following statements about both poliovirus and rhinoviruses is correct EXCEPT:

    (A) They are nonenveloped RNA viruses.
    (B) They have multiple antigenic types.
    (C) Their virions contain an RNA polymerase.
    (D) They do not integrate their genome into host cell DNA.

192. Each of the following statements about human immunodeficiency virus (HIV) is correct EXCEPT:

    (A) HIV is an enveloped RNA virus.
    (B) The virion contains an RNA-dependent DNA polymerase.
    (C) A DNA copy of the HIV genome integrates into host cell DNA.
    (D) Acyclovir inhibits HIV replication.

Answers (Questions 159–192)

159. **(B)**	166. **(B)**	173. **(C)**	180. **(D)**	187. **(A)**
160. **(D)**	167. **(D)**	174. **(B)**	181. **(C)**	188. **(B)**
161. **(C)**	168. **(C)**	175. **(C)**	182. **(A)**	189. **(B)**
162. **(B)**	169. **(B)**	176. **(B)**	183. **(B)**	190. **(D)**
163. **(D)**	170. **(A)**	177. **(D)**	184. **(D)**	191. **(C)**
164. **(D)**	171. **(B)**	178. **(A)**	185. **(A)**	192. **(D)**
165. **(A)**	172. **(C)**	179. **(D)**	186. **(C)**	

DIRECTIONS (Questions 193–211): Select the one lettered option that is MOST CLOSELY associated with the numbered items. Each lettered option may be selected once, more than once, or not at all.

Questions 193–196

    (A) DNA enveloped virus
    (B) DNA nonenveloped virus
    (C) RNA enveloped virus
    (D) RNA nonenveloped virus

193. Herpes simplex virus
194. Human T-cell lymphotropic virus
195. Human papillomavirus
196. Rotavirus

Questions 197–201

    (A) Attachment and penetration of virion
    (B) Viral mRNA synthesis
    (C) Viral protein synthesis
    (D) Viral genome DNA synthesis
    (E) Assembly and release of progeny virus

197. Main site of action of acyclovir
198. Main site of action of maraviroc
199. Function of virion polymerase of influenza virus
200. Main site of action of antiviral antibody
201. Step at which budding occurs

Questions 202–206

    (A) Poliovirus
    (B) Epstein–Barr virus
    (C) Prions
    (D) Hepatitis B virus
    (E) Respiratory syncytial virus

202. Part of the genome DNA is synthesized by the virion polymerase.
203. The translation product of viral mRNA is a polyprotein that is cleaved to form virion structural proteins.
204. It is remarkably resistant to ultraviolet light.
205. It causes latent infection of B cells.
206. An envelope protein induces the formation of giant cells.

Questions 207–211

    (A) Hepatitis A virus
    (B) Hepatitis B virus
    (C) Hepatitis C virus
    (D) Hepatitis D virus

207. Enveloped DNA virus that is transmitted by blood
208. Enveloped RNA virus that has the surface antigen of another virus
209. Enveloped RNA virus that is the most common cause of non-A, non-B hepatitis
210. Nonenveloped RNA virus that is transmitted by the fecal–oral route
211. Purified surface protein of this virus is the immunogen in a vaccine

Answers (Questions 193–211)

193. **(A)**	197. **(D)**	201. **(E)**	205. **(B)**	209. **(C)**
194. **(C)**	198. **(A)**	202. **(D)**	206. **(E)**	210. **(A)**
195. **(B)**	199. **(B)**	203. **(A)**	207. **(B)**	211. **(B)**
196. **(D)**	200. **(A)**	204. **(C)**	208. **(D)**	

CLINICAL VIROLOGY

DIRECTIONS (Questions 212–275): Select the ONE lettered answer that is BEST in each question.

212. Which one of the following outcomes is MOST common following a primary herpes simplex virus infection?

    (A) Complete eradication of virus and virus-infected cells
    (B) Persistent asymptomatic viremia
    (C) Establishment of latent infection
    (D) Persistent cytopathic effect in infected cells

213. Each of the following pathogens is likely to establish chronic or latent infection EXCEPT:

    (A) Cytomegalovirus
    (B) Hepatitis A virus
    (C) Hepatitis B virus
    (D) Herpes simplex virus

214. Each of the following statements regarding poliovirus and its vaccine is correct EXCEPT:

    (A) Poliovirus is transmitted by the fecal–oral route.
    (B) Pathogenesis by poliovirus primarily involves the death of sensory neurons.
    (C) The live, attenuated vaccine contains all three serotypes of poliovirus.
    (D) An unimmunized adult traveling to countries where there is a known risk of being infected with poliovirus should receive the inactivated vaccine.

215. Which one of the following strategies is MOST likely to induce lasting intestinal mucosal immunity to poliovirus?

    (A) Parenteral (intramuscular) immunization with inactivated vaccine
    (B) Oral administration of poliovirus immune globulin
    (C) Parenteral immunization with live vaccine
    (D) Oral immunization with live vaccine

216. Each of the following clinical syndromes is associated with infection by picornaviruses EXCEPT:

    (A) Myocarditis/pericarditis
    (B) Hepatitis
    (C) Mononucleosis
    (D) Meningitis

217. Each of the following statements concerning rubella vaccine is correct EXCEPT:

    (A) The vaccine prevents reinfection, thereby limiting the spread of virulent virus.
    (B) The immunogen in the vaccine is killed rubella virus.
    (C) The vaccine induces antibodies that prevent systemic dissemination of the virus by neutralizing it during the viremic stage.
    (D) The incidence of both childhood rubella and congenital rubella syndrome has decreased significantly since the advent of the vaccine.

218. Each of the following statements concerning the rabies vaccine for use in humans is correct EXCEPT:

    (A) The vaccine contains live, attenuated rabies virus.
    (B) If your patient is bitten by a wild animal (e.g., a skunk), the rabies vaccine should be given.
    (C) When the vaccine is used for postexposure prophylaxis, rabies immune globulin should also be given.
    (D) The virus in the vaccine is grown in human cell cultures, thus decreasing the risk of allergic encephalomyelitis.

219. Each of the following statements concerning influenza is correct EXCEPT:

    (A) World-wide epidemics (pandemics) of the disease are caused more commonly by influenza A viruses than by influenza B and C viruses.
    (B) Likely sources of new antigens for influenza A viruses are the viruses that cause influenza in animals.
    (C) Major antigenic changes (shifts) of viral surface proteins are seen primarily in influenza A viruses rather than in influenza B and C viruses.
    (D) The antigenic changes that occur with antigenic drift are due to reassortment of the multiple pieces of the influenza virus genome.

220. Each of the following statements concerning the prevention and treatment of influenza is correct EXCEPT:

    (A) The inactivated influenza vaccine contains H_1N_1 virus, whereas the live, attenuated influenza vaccine contains H_3N_2 virus.
    (B) The vaccine is recommended to be given each year because the antigenicity of the virus drifts.
    (C) Oseltamivir (Tamiflu) is effective against both influenza A and influenza B viruses.
    (D) The main antigen in the vaccine that induces protective antibody is the hemagglutinin.

221. A 6-month-old child develops a persistent cough and a fever. Physical examination and chest X-ray suggest pneumonia. Which one of the following organisms is LEAST likely to cause this infection?

    (A) Respiratory syncytial virus
    (B) Adenovirus
    (C) Parainfluenza virus
    (D) Rotavirus

222. A 45-year-old man was attacked by a bobcat and bitten repeatedly about the face and neck. The animal was shot by a companion and brought back to the public health authorities. Once you decide to immunize against rabies virus, how would you proceed?

    (A) Use hyperimmune serum only
    (B) Use active immunization only
    (C) Use hyperimmune serum and active immunization
    (D) Use active immunization and follow this with hyperimmune serum if adequate antibody titers are not obtained in the patient's serum

223. Each of the following statements concerning mumps is correct EXCEPT:

    (A) Mumps virus is a paramyxovirus and hence has a single-stranded RNA genome.
    (B) Meningitis is a recognized complication of mumps.
    (C) Mumps orchitis in children prior to puberty often causes sterility.
    (D) During mumps, the virus spreads through the bloodstream (viremia) to various internal organs.

224. Each of the following statements concerning respiratory syncytial virus (RSV) is correct EXCEPT:

    (A) RSV has a single-stranded RNA genome.
    (B) RSV induces the formation of multinucleated giant cells.
    (C) RSV causes pneumonia primarily in children.
    (D) RSV infections can be effectively treated with acyclovir.

225. The principal reservoir for the antigenic shift variants of influenza virus appears to be:

(A) People in isolated communities such as the Arctic

(B) Animals, specifically pigs, horses, and fowl

(C) Soil, especially in the tropics

(D) Sewage

226. The role of an infectious agent in the pathogenesis of kuru was BEST demonstrated by which one of the following observations?

(A) A 16-fold rise in antibody titer to the agent was observed.

(B) The viral genome was isolated from infected neurons.

(C) Electron micrographs of the brains of infected individuals demonstrated intracellular structures resembling paramyxovirus nucleocapsids.

(D) The disease was serially transmitted to experimental animals.

227. A 64-year-old man with chronic lymphatic leukemia develops progressive deterioration of mental and neuromuscular function. At autopsy the brain shows enlarged oligodendrocytes whose nuclei contain naked, icosahedral virus particles. The MOST likely diagnosis is:

(A) Herpes encephalitis

(B) Creutzfeldt-Jakob disease

(C) Subacute sclerosing panencephalitis

(D) Progressive multifocal leukoencephalopathy

(E) Rabies

228. A 20-year-old man, who for many years had received daily injections of growth hormone prepared from human pituitary glands, develops ataxia, slurred speech, and dementia. At autopsy the brain shows widespread neuronal degeneration, a spongy appearance due to many vacuoles between the cells, no inflammation, and no evidence of virus particles. The MOST likely diagnosis is:

(A) Herpes encephalitis

(B) Creutzfeldt-Jakob disease

(C) Subacute sclerosing panencephalitis

(D) Progressive multifocal leukoencephalopathy

(E) Rabies

229. A 24-year-old woman has had fever and a sore throat for the past week. Moderately severe pharyngitis and bilateral cervical lymphadenopathy are seen on physical examination. Which one of the following viruses is LEAST likely to cause this picture?

(A) Norovirus

(B) Adenovirus

(C) Coxsackie virus

(D) Epstein–Barr virus

230. Scrapie and kuru possess all of the following characteristics EXCEPT:

(A) A histologic picture of spongiform encephalopathy

(B) Transmissibility to animals associated with a long incubation period

(C) Slowly progressive deterioration of brain function

(D) Prominent intranuclear inclusion bodies in oligodendrocytes

231. Each of the following statements concerning subacute sclerosing panencephalitis is correct EXCEPT:

(A) Immunosuppression is a frequent predisposing factor.

(B) Aggregates of helical nucleocapsids are found in infected cells.

(C) High titers of measles antibody are found in cerebrospinal fluid.

(D) Gradual progressive deterioration of brain function occurs.

232. The slow virus disease that MOST clearly has immunosuppression as an important factor in its pathogenesis is:

(A) Progressive multifocal leukoencephalopathy

(B) Subacute sclerosing panencephalitis

(C) Creutzfeldt-Jakob disease

(D) Kuru

233. You think your patient may be in the "window period" of hepatitis B virus (HBV) infection because his blood tests for HBs antigen and anti-HBs antibody are negative. Which one of the following additional tests is MOST useful to establish that he has been infected with HBV and is in the "window period"?

(A) HBe antigen

(B) Anti-HBc antibody

(C) Anti-HBe antibody

(D) Delta antigen

234. Which one of the following is the MOST reasonable explanation for the ability of hepatitis B virus to cause chronic infection?

(A) Infection does not elicit the production of antibody.

(B) The liver is an "immunologically sheltered" site.

(C) Viral DNA can persist within the host cell.

(D) Many humans are immunologically tolerant to HBs antigen.

235. The routine screening of transfused blood has greatly reduced the problem of posttransfusion hepatitis. For which one of the following viruses has screening eliminated a large number of cases of posttransfusion hepatitis?

(A) Hepatitis A virus

(B) Hepatitis C virus

(C) Cytomegalovirus

(D) Epstein–Barr virus

236. A 35-year-old man addicted to intravenous drugs has been a carrier of HBs antigen for 10 years. He suddenly develops acute fulminant hepatitis and dies within 10 days. Which one of the following laboratory tests would contribute MOST to a diagnosis?

(A) Anti-HBs antibody

(B) HBe antigen

(C) Anti-HBc antibody

(D) Anti-delta virus antibody

237. Which one of the following is the BEST evidence on which to base a decisive diagnosis of acute mumps disease?

(A) A positive skin test

(B) A fourfold rise in antibody titer to mumps antigen

(C) A history of exposure to a child with mumps

(D) Orchitis in young adult male

238. Varicella-zoster virus and herpes simplex virus share many characteristics. Which one of the following characteristics is LEAST accurate?

(A) Primary infection rarely produces symptoms

(B) Persistence of latent virus occurs after recovery from acute disease

(C) Cause a vesicular rash

(D) Virions possess a linear, double-stranded DNA genome

239. Herpes simplex virus and cytomegalovirus share many features. Which one of the following features is LEAST likely to be shared?

(A) Important cause of morbidity and mortality in the newborn

(B) Congenital abnormalities due to transplacental passage

(C) Important cause of serious disease in immunosuppressed individuals

(D) Mild or inapparent infection

240. The eradication of smallpox was facilitated by several features of the virus. Which one of the following contributed LEAST to eradication?

    (A) It has one antigenic type.
    (B) Inapparent infection is rare.
    (C) Administration of live vaccine reliably induces immunity.
    (D) It multiplies in the cytoplasm of infected cells.

241. Which one of the following statements concerning infectious mononucleosis is the MOST accurate?

    (A) Multinucleated giant cells are found in the skin lesions.
    (B) Infected T lymphocytes are abundant in peripheral blood.
    (C) Isolation of virus is necessary to confirm the diagnosis.
    (D) Infectious mononucleosis is transmitted by virus in saliva.

242. Which one of the following statements about genital herpes is LEAST accurate?

    (A) Acyclovir reduces the number of recurrent disease episodes by eradicating latently infected cells.
    (B) Genital herpes can be transmitted in the absence of apparent lesions.
    (C) Multinucleated giant cells with intranuclear inclusions are found in the lesions.
    (D) Initial disease episodes are generally more severe than recurrent episodes.

243. There are several influenza vaccines administered in the United States. Regarding these vaccines, which one of the following statements is LEAST accurate?

    (A) One of the vaccines contains purified peptide subunits of neuraminidase produced in yeast.
    (B) One of the vaccines is an inactivated vaccine consisting of formalin-treated influenza virions.
    (C) One of the vaccines contains a temperature-sensitive mutant of influenza virus that replicates in the nose but not in the lungs.
    (D) Influenza vaccines contain influenza A and B strains but not C strains.

244. Which of the following is the MOST common lower respiratory pathogen in infants?

    (A) Respiratory syncytial virus
    (B) Adenovirus
    (C) Rhinovirus
    (D) Coxsackie virus

245. Which of the following conditions is LEAST likely to be caused by adenoviruses?

    (A) Conjunctivitis
    (B) Pneumonia
    (C) Pharyngitis
    (D) Glomerulonephritis

246. Regarding the serologic diagnosis of infectious mononucleosis, which one of the following is CORRECT?

    (A) A heterophil antibody is formed that reacts with a capsid protein of Epstein–Barr virus.
    (B) A heterophil antibody is formed that agglutinates sheep or horse red blood cells.
    (C) A heterophil antigen occurs that cross-reacts with *Proteus* OX19 strains.
    (D) A heterophil antigen occurs following infection with cytomegalovirus.

247. Herpes simplex virus type 1 (HSV-1) is distinct from HSV-2 in several different ways. Which one of the following is the LEAST accurate statement?

    (A) HSV-1 causes lesions above the umbilicus more frequently than HSV-2 does.
    (B) Infection by HSV-1 is not associated with any tumors in humans.
    (C) Antiserum to HSV-1 neutralizes HSV-1 much more effectively than HSV-2.
    (D) HSV-1 causes frequent recurrences, whereas HSV-2 infection rarely recurs.

248. Which one of the following statements about the *src* gene and src protein of Rous sarcoma virus is INCORRECT?

    (A) The src protein inactivates a protein encoded by *p53*, a tumor suppressor gene.
    (B) The src protein is a protein kinase that preferentially phosphorylates tyrosine in cellular proteins.
    (C) The continued presence of src protein is required to maintain neoplastic transformation of infected cells.
    (D) The viral *src* gene is derived from a cellular gene found in many vertebrate species.

249. Each of the following statements supports the idea that cellular protooncogenes participate in human carcinogenesis EXCEPT:

    (A) The *c-abl* gene is rearranged on the Philadelphia chromosome in myeloid leukemias and encodes a protein with increased tyrosine kinase activity.
    (B) The *N-myc* gene is amplified as much as 100-fold in many advanced cases of neuroblastoma.
    (C) The receptor for platelet-derived growth factor is a transmembrane protein that exhibits tyrosine kinase activity.
    (D) The *c-Ha-ras* gene is mutated at specific codons in several types of human cancer.

250. Each of the following statements concerning human immunodeficiency virus (HIV) is correct EXCEPT:

    (A) Screening tests for antibodies are useful to prevent transmission of HIV through transfused blood.
    (B) The opportunistic infections seen in AIDS are primarily the result of a loss of cell-mediated immunity.
    (C) Zidovudine (azidothymidine) inhibits the RNA-dependent DNA polymerase.
    (D) The presence of circulating antibodies that neutralize HIV is evidence that an individual is protected against HIV-induced disease.

251. Which one of the following statements concerning viral meningitis and viral encephalitis is CORRECT?

    (A) Herpes simplex virus type 2 is the leading cause of viral meningitis.
    (B) Herpes simplex virus type 1 is an important cause of viral encephalitis.
    (C) The spinal fluid protein is usually decreased in viral meningitis.
    (D) The diagnosis of viral meningitis can be made by using the India ink stain on a sample of spinal fluid.

252. Each of the following statements is correct EXCEPT:
    (A) Coxsackie viruses are enteroviruses and can replicate in both the respiratory and gastrointestinal tracts.
    (B) Influenza viruses have multiple serotypes based on hemagglutinin and neuraminidase proteins located on the envelope surface.
    (C) Flaviviruses are RNA enveloped viruses that replicate in animals as well as humans.
    (D) Adenoviruses are RNA enveloped viruses that are an important cause of sexually transmitted disease.

253. Which one of the following statements concerning the prevention of viral disease is CORRECT?
    (A) Adenovirus vaccine contains purified penton fibers and is usually given to children in conjunction with polio vaccine.
    (B) Coxsackie virus vaccine contains live virus that induces IgA, which prevents reinfection by homologous serotypes.
    (C) Flavivirus immunization consists of hyperimmune serum plus a vaccine consisting of subunits containing the surface glycoprotein.
    (D) One of the influenza virus vaccines contains killed virus that induces neutralizing antibody directed against the hemagglutinin.

254. Each of the following statements concerning hepatitis C virus (HCV) and hepatitis D virus (HDV) is correct EXCEPT:
    (A) HCV is an RNA virus that causes posttransfusion hepatitis.
    (B) HDV is a defective virus that can replicate only in a cell that is also infected with hepatitis B virus.
    (C) HDV is transmitted primarily by the fecal–oral route.
    (D) People infected with HCV commonly become chronic carriers of HCV and are predisposed to hepatocellular carcinoma.

255. Each of the following statements concerning measles virus is correct EXCEPT:
    (A) Measles virus is an enveloped virus with a single-stranded RNA genome.
    (B) One of the important complications of measles is encephalitis.
    (C) The initial site of measles virus replication is the upper respiratory tract, from which it spreads via the blood to the skin.
    (D) Latent infection by measles virus can be explained by the integration of provirus into the host cell DNA.

256. Each of the following statements concerning measles vaccine is correct EXCEPT:
    (A) The vaccine contains live, attenuated virus.
    (B) The vaccine should not be given at the same time as the mumps vaccine because the immune system cannot respond to two viral antigens given at the same time.
    (C) Virus in the vaccine contains only one serotype.
    (D) The vaccine should not be given prior to 12 months of age because maternal antibodies can prevent an immune response.

257. Each of the following statements concerning rubella is correct EXCEPT:
    (A) Congenital abnormalities occur primarily when a pregnant woman is infected during the first trimester.
    (B) Women who say that they have never had rubella can, nevertheless, have neutralizing antibody in their serum.
    (C) In a 6-year-old child, rubella is a mild, self-limited disease with few complications.
    (D) Acyclovir is effective in the treatment of congenital rubella syndrome.

258. Each of the following statements concerning rabies and rabies virus is correct EXCEPT:
    (A) The virus has a lipoprotein envelope and single-stranded RNA as its genome.
    (B) The virus has a single antigenic type (serotype).
    (C) In the United States, dogs are the most common reservoir.
    (D) The incubation period is usually long (several weeks) rather than short (several days).

259. Each of the following statements concerning arboviruses is correct EXCEPT:
    (A) The pathogenesis of dengue hemorrhagic shock syndrome is associated with the heterotypic anamnestic response.
    (B) Wild birds are the reservoir for encephalitis viruses but not for yellow fever virus.
    (C) Ticks are the main mode of transmission for both encephalitis viruses and yellow fever virus.
    (D) There is a live, attenuated vaccine that effectively prevents yellow fever.

260. Each of the following statements concerning rhinoviruses is correct EXCEPT:
    (A) Rhinoviruses are picornaviruses (i.e., small, nonenveloped viruses with an RNA genome).
    (B) Rhinoviruses are an important cause of lower respiratory tract infections, especially in patients with chronic obstructive pulmonary disease.
    (C) Rhinoviruses do not infect the gastrointestinal tract because they are inactivated by the acid pH in the stomach.
    (D) There is no vaccine against rhinoviruses because they have too many antigenic types.

261. Each of the following statements concerning herpes simplex virus type 2 (HSV-2) is correct EXCEPT:
    (A) Primary infection with HSV-2 does not confer immunity to primary infection with HSV-1.
    (B) HSV-2 causes vesicular lesions, typically in the genital area.
    (C) HSV-2 can cause alterations of the cell membrane, leading to cell fusion and the formation of multinucleated giant cells.
    (D) Recurrent disease episodes due to reactivation of latent HSV-2 are usually more severe than the primary episode.

262. Each of the following statements concerning Epstein–Barr virus is correct EXCEPT:
    (A) Many infections are mild or inapparent.
    (B) The earlier in life primary infection is acquired, the more likely the typical picture of infectious mononucleosis will be manifest.
    (C) Latently infected B lymphocytes regularly persist following an acute episode of infection.
    (D) Infection confers immunity against second episodes of infectious mononucleosis.

263. Each of the following statements regarding rotaviruses is correct EXCEPT:
    (A) The rotavirus vaccine contains recombinant RNA polymerase as the immunogen.
    (B) Rotaviruses are a leading cause of diarrhea in young children.

(C) Rotaviruses are transmitted primarily by the fecal–oral route.

(D) Rotaviruses have a double-stranded, segmented RNA genome.

264. Each of the following statements concerning the antigenicity of influenza A virus is correct EXCEPT:

(A) Antigenic shifts, which represent major changes in antigenicity, occur infrequently and are due to the reassortment of segments of the viral RNA genome.

(B) Antigenic shifts affect both the hemagglutinin and the neuraminidase.

(C) The worldwide epidemics caused by influenza A virus are due to antigenic shifts.

(D) The protein involved in antigenic drift is primarily the internal ribonucleoprotein.

265. Each of the following statements concerning adenoviruses is correct EXCEPT:

(A) Adenoviruses are composed of a double-stranded DNA genome and a capsid without an envelope.

(B) Adenoviruses cause both sore throat and pneumonia.

(C) Adenoviruses have only one serologic type.

(D) Adenoviruses are implicated as a cause of tumors in animals but not humans.

266. Each of the following statements concerning the prevention of viral respiratory tract disease is correct EXCEPT:

(A) To prevent disease caused by adenoviruses, a live enteric-coated vaccine that causes asymptomatic enteric infection is used in the military.

(B) To prevent disease caused by influenza A virus, an inactivated vaccine is available for the civilian population.

(C) There is no vaccine available against respiratory syncytial virus.

(D) To prevent disease caused by rhinoviruses, a vaccine containing purified capsid proteins is used.

267. Each of the following statements concerning herpesvirus latency is correct EXCEPT:

(A) Exogenous stimuli can cause reactivation of herpesvirus replication in latently infected cells.

(B) During latency, antiviral antibody is not demonstrable in the sera of infected individuals.

(C) Reactivation of latent herpesviruses is more common in patients with impaired cell-mediated immunity than in immunocompetent patients.

(D) Herpesvirus genome DNA persists in latently infected cells.

268. Each of the following statements concerning rhinoviruses is correct EXCEPT:

(A) Rhinoviruses are the most common cause of the common cold.

(B) Rhinoviruses grow better at 33°C than at 37°C; hence, they tend to cause disease in the upper respiratory tract rather than the lower respiratory tract.

(C) Rhinoviruses are members of the picornaviruses family and hence resemble poliovirus in their structure and replication.

(D) The immunity provided by the rhinovirus vaccine is excellent because there is only one serotype.

269. Which one of the following statements concerning poliovirus infection is CORRECT?

(A) Congenital infection of the fetus is an important complication.

(B) The virus replicates extensively in the gastrointestinal tract.

(C) A skin test is available to determine prior exposure to the virus.

(D) Oseltamivir is an effective preventive agent.

270. Each of the following statements concerning yellow fever is correct EXCEPT:

(A) Yellow fever virus is transmitted by the *Aedes aegypti* mosquito in the urban form of yellow fever.

(B) Infection by yellow fever virus causes significant damage to hepatocytes.

(C) Nonhuman primates in the jungle are a major reservoir of yellow fever virus.

(D) Acyclovir is an effective treatment for yellow fever.

271. Which one of the following statements concerning mumps is CORRECT?

(A) Although the salivary glands are the most obvious sites of infection, the testes, ovaries, and pancreas can be involved as well.

(B) Because there is no vaccine against mumps, passive immunization is the only means of preventing the disease.

(C) The diagnosis of mumps is made on clinical grounds because the virus cannot be grown in cell culture and serologic tests are inaccurate.

(D) Second episodes of mumps can occur because there are two serotypes of the virus and protection is type-specific.

272. Many of the oncogenic retroviruses carry oncogenes closely related to normal cellular genes, called proto-oncogenes. Which one of the following statements concerning proto-oncogenes is INCORRECT?

(A) Several proto-oncogenes have been found in mutant form in human cancers that lack evidence for viral etiology.

(B) Several viral oncogenes and their progenitor proto-oncogenes encode protein kinases specific for tyrosine.

(C) Some proto-oncogenes encode cellular growth factors and receptors for growth factors.

(D) Proto-oncogenes are closely related to transposons found in bacteria.

273. Each of the following statements concerning human immunodeficiency virus is correct EXCEPT:

(A) The CD4 protein on the T-cell surface is one of the receptors for the virus.

(B) There is appreciable antigenic diversity in the envelope glycoprotein of the virus.

(C) One of the viral genes codes for a protein that augments the activity of the viral transcriptional promoter.

(D) A major problem with testing for antibody to the virus is its cross-reactivity with human T-cell leukemia virus type I.

274. Each of the following statements concerning human immunodeficiency virus (HIV) is correct EXCEPT:

(A) Patients infected with HIV typically form antibodies against both the envelope glycoproteins (gp120 and gp41) and the internal group-specific antigen (p24).

(B) HIV probably arose as an endogenous virus of humans because HIV proviral DNA is found in the DNA of certain normal human cells.

(C) Transmission of HIV occurs primarily by the transfer of blood or semen in adults, and neonates can be infected at the time of delivery.

(D) The diagnosis of early HIV infection is made by testing the patient's serum for antibodies against HIV and for p24 antigen.

275. Each of the following statements concerning hepatitis A virus (HAV) is correct EXCEPT:

    (A) The hepatitis A vaccine contains inactivated HAV as the immunogen.
    (B) HAV commonly causes asymptomatic infection in children.
    (C) The diagnosis of hepatitis A is usually made by isolating HAV in cell culture.
    (D) Gamma globulin is used to prevent hepatitis A in exposed persons.

Answers (Questions 212–275)

212. (C)	225. (B)	238. (A)	251. (B)	264. (D)
213. (B)	226. (D)	239. (B)	252. (D)	265. (C)
214. (B)	227. (D)	240. (D)	253. (D)	266. (D)
215. (D)	228. (B)	241. (D)	254. (C)	267. (B)
216. (C)	229. (A)	242. (A)	255. (D)	268. (D)
217. (B)	230. (D)	243. (A)	256. (B)	269. (B)
218. (A)	231. (A)	244. (A)	257. (D)	270. (D)
219. (D)	232. (A)	245. (D)	258. (C)	271. (A)
220. (A)	233. (B)	246. (B)	259. (C)	272. (D)
221. (D)	234. (C)	247. (D)	260. (B)	273. (D)
222. (C)	235. (B)	248. (A)	261. (D)	274. (B)
223. (C)	236. (D)	249. (C)	262. (B)	275. (C)
224. (D)	237. (B)	250. (D)	263. (A)	

DIRECTIONS (Questions 276–294): Select the ONE lettered option that is MOST closely associated with the numbered items. Each lettered option may be selected once, more than once, or not at all.

Questions 276–279

    (A) Yellow fever virus
    (B) Rabies virus
    (C) Rotavirus
    (D) Rubella virus
    (E) Rhinovirus

276. Causes diarrhea
277. Causes jaundice
278. Causes congenital abnormalities
279. Causes encephalitis

Questions 280–284

    (A) Bronchiolitis
    (B) Meningitis
    (C) Pharyngitis
    (D) Shingles
    (E) Subacute sclerosing panencephalitis

280. Adenovirus causes
281. Measles virus causes
282. Respiratory syncytial virus causes
283. Coxsackie virus causes
284. Varicella-zoster virus causes

Questions 285–289

    (A) Adenovirus
    (B) Parainfluenza virus
    (C) Rhinovirus
    (D) Coxsackie virus
    (E) Epstein–Barr virus

285. Causes myocarditis and pleurodynia
286. Grows better at 33°C than 37°C
287. Causes tumors in laboratory rodents
288. Causes croup in young children
289. Causes infectious mononucleosis

Questions 290–294

    (A) Hepatitis C virus
    (B) Cytomegalovirus
    (C) Human papillomavirus
    (D) Dengue virus
    (E) St. Louis encephalitis virus

290. It causes carcinoma of the cervix.
291. Wild birds are an important reservoir.
292. It is an important cause of pneumonia in immunocompromised patients.
293. Donated blood containing antibody to this RNA virus should not be used for transfusion.
294. It causes a hemorrhagic fever that can be life-threatening.

Answers (Questions 276–294)

276. (C)	280. (C)	284. (D)	288. (B)	292. (B)
277. (A)	281. (E)	285. (D)	289. (E)	293. (A)
278. (D)	282. (A)	286. (C)	290. (C)	294. (D)
279. (B)	283. (B)	287. (A)	291. (E)	

MYCOLOGY

DIRECTIONS (Questions 295–317): Select the ONE lettered answer that is BEST in each question.

295. Which one of the following fungi is MOST likely to be found within reticuloendothelial cells, such as macrophages?

    (A) *Histoplasma capsulatum*
    (B) *Candida albicans*
    (C) *Cryptococcus neoformans*
    (D) *Sporothrix schenckii*

296. Your patient is a woman with a vaginal discharge. You suspect, on clinical grounds, that it may be due to *Candida albicans*.

Which one of the following statements is LEAST accurate or appropriate?

    (A) A Gram stain of the discharge should reveal budding yeasts.
    (B) Culture of the discharge on Sabouraud's agar should produce a white mycelium with aerial conidia.
    (C) The clinical laboratory can use germ tube formation to identify the isolate as *C. albicans*.
    (D) Antibiotics predispose to *Candida* vaginitis by killing the normal flora lactobacilli that keep the vaginal pH low.

297. You have made a clinical diagnosis of meningitis in a 50-year-old immunocompromised woman. A latex agglutination test on the spinal fluid for capsular polysaccharide antigen is positive. Of the following organisms, which one is the MOST likely cause?

    (A) *Histoplasma capsulatum*
    (B) *Cryptococcus neoformans*
    (C) *Aspergillus fumigatus*
    (D) *Candida albicans*

298. Fungi often colonize lesions due to other causes. Which one of the following is LEAST likely to be present as a colonizer?

    (A) *Aspergillus*
    (B) *Mucor*
    (C) *Sporothrix*
    (D) *Candida*

299. Your patient complains of an "itching rash" on her abdomen. On examination, you find that the lesions are red, circular, with a vesiculated border and a healing central area. You suspect tinea corporis. Of the following choices, the MOST appropriate laboratory procedure to make the diagnosis is a:

    (A) Potassium hydroxide mount of skin scrapings
    (B) Giemsa stain for multinucleated giant cells
    (C) Fluorescent antibody stain of the vesicle fluid
    (D) Fourfold rise in antibody titer against the organism

300. Each of the following statements concerning *Cryptococcus neoformans* is correct EXCEPT:

    (A) Its natural habitat is the soil, especially associated with pigeon feces.
    (B) Pathogenesis is related primarily to the production of exotoxin A.
    (C) Budding yeasts are found in the lesions.
    (D) The initial site of infection is usually the lung.

301. A woman who pricked her finger while pruning some rose bushes develops a local pustule that progresses to an ulcer. Several nodules then develop along the local lymphatic drainage. The MOST likely agent is:

    (A) *Cryptococcus neoformans*
    (B) *Candida albicans*
    (C) *Sporothrix schenckii*
    (D) *Aspergillus fumigatus*

302. Several fungi are associated with disease in immunocompromised patients. Which one of the following is the LEAST frequently associated?

    (A) *Cryptococcus neoformans*
    (B) *Aspergillus fumigatus*
    (C) *Malassezia furfur*
    (D) *Mucor* species

303. Fungal cells that reproduce by budding are seen in the infected tissues of patients with:

    (A) Candidiasis, cryptococcosis, and sporotrichosis
    (B) Mycetoma, candidiasis, and mucormycosis
    (C) Tinea corporis, tinea unguium, and tinea versicolor
    (D) Sporotrichosis, mycetoma, and aspergillosis

304. Infection by a dermatophyte is MOST often associated with:

    (A) Intravenous drug abuse
    (B) Inhalation of the organism from contaminated bird feces
    (C) Adherence of the organism to perspiration-moist skin
    (D) Fecal–oral transmission

305. Aspergillosis is recognized in tissue by the presence of:

    (A) Budding cells
    (B) Septate hyphae
    (C) Metachromatic granules
    (D) Pseudohyphae

306. Which one of the following is NOT a characteristic of histoplasmosis?

    (A) Person-to-person transmission
    (B) Specific geographic distribution
    (C) Yeasts in the tissue
    (D) Mycelial phase in the soil

307. Each of the following statements concerning mucormycosis is correct EXCEPT:

    (A) The fungi that cause mucormycosis are transmitted by airborne asexual spores.
    (B) Tissue sections from a patient with mucormycosis show budding yeasts.
    (C) Hyphae typically invade blood vessels and cause necrosis of tissue.
    (D) Ketoacidosis in diabetic patients is a predisposing factor to mucormycosis.

308. Each of the following statements concerning fungi is correct EXCEPT:

    (A) Yeasts are fungi that reproduce by budding.
    (B) Molds are fungi that have elongated filaments called hyphae.
    (C) Thermally dimorphic fungi often exist as yeasts at 37°C and as molds at 25°C.
    (D) Both yeasts and molds have a cell wall made of peptidoglycan.

309. Each of the following statements concerning yeasts is correct EXCEPT:

    (A) Yeasts have chitin in their cell walls and ergosterol in their cell membranes.
    (B) Yeasts form ascospores when they invade tissue.
    (C) Yeasts have eukaryotic nuclei and contain mitochondria in their cytoplasm.
    (D) Yeasts produce neither endotoxin nor exotoxins.

310. Each of the following statements concerning the mode of action of antifungal drugs is correct EXCEPT:

    (A) Fluconazole inhibits the synthesis of ergosterol, a component of the fungal cell membrane.
    (B) Amphotericin B disrupts fungal cell membranes at the site of ergosterol.
    (C) Flucytosine inhibits the synthesis of chitin, a component of the fungal cell wall.
    (D) Caspofungin inhibits the synthesis of beta-glucan, a component of the fungal cell wall.

311. You suspect that your patient's disease may be caused by *Cryptococcus neoformans*. Which one of the following findings would be MOST useful in establishing the diagnosis?

    (A) A positive heterophil agglutination test for the presence of antigen
    (B) A history of recent travel in the Mississippi River valley area
    (C) The finding of encapsulated budding cells in spinal fluid
    (D) Recovery of an acid-fast organism from the patient's sputum

312. Each of the following statements concerning *Candida albicans* is correct EXCEPT:

(A) *C. albicans* is a budding yeast that forms pseudohyphae when it invades tissue.

(B) *C. albicans* is transmitted primarily by respiratory aerosol.

(C) *C. albicans* causes thrush.

(D) Impaired cell-mediated immunity is an important predisposing factor to disease.

313. Each of the following statements concerning *Coccidioides immitis* is correct EXCEPT:

(A) The mycelial phase of the organism grows primarily in the soil, which is its natural habitat.

(B) In the body, spherules containing endospores are formed.

(C) A rising titer of complement-fixing antibody indicates disseminated disease.

(D) Most infections are symptomatic and require treatment with amphotericin B.

314. Each of the following statements concerning *Histoplasma capsulatum* is correct EXCEPT:

(A) The natural habitat of *H. capsulatum* is the soil, where it grows as a mold.

(B) *H. capsulatum* is transmitted by airborne conidia, and its initial site of infection is the lung.

(C) Within the body, *H. capsulatum* grows primarily intracellularly within macrophages.

(D) Passive immunity in the form of high titer antibodies should be given to those known to be exposed.

315. Each of the following statements concerning infection caused by *Coccidioides immitis* is correct EXCEPT:

(A) *C. immitis* is a dimorphic fungus.

(B) *C. immitis* is acquired by inhalation of arthrospores.

(C) More than 50% of clinical isolates are resistant to amphotericin B.

(D) Infection occurs primarily in the southwestern states and California.

316. Each of the following statements concerning *Blastomyces dermatitidis* is correct EXCEPT:

(A) *B. dermatitidis* grows as a mold in the soil in North America.

(B) *B. dermatitidis* is a dimorphic fungus that forms yeast cells in tissue.

(C) *B. dermatitidis* infection is commonly diagnosed by serologic tests because it does not grow in culture.

(D) *B. dermatitidis* causes granulomatous skin lesions.

317. *Aspergillus fumigatus* can be involved in a variety of clinical conditions. Which one of the following is LEAST likely to occur?

(A) Tissue invasion in immunocompromised host

(B) Allergy following inhalation of airborne particles of the fungus

(C) Colonization of tuberculous cavities in the lung

(D) Thrush

Answers (Questions 295–317)

295. (A)	301. (C)	307. (B)	313. (D)
296. (B)	302. (C)	308. (D)	314. (D)
297. (B)	303. (A)	309. (B)	315. (C)
298. (C)	304. (C)	310. (C)	316. (C)
299. (A)	305. (B)	311. (C)	317. (D)
300. (B)	306. (A)	312. (B)	

DIRECTIONS (Questions 318–325): Select the ONE lettered option that is MOST closely associated with the numbered items. Each lettered option may be selected once, more than once, or not at all.

Questions 318–321

(A) *Histoplasma capsulatum*

(B) *Candida albicans*

(C) *Aspergillus fumigatus*

(D) *Sporothrix schenckii*

318. A budding yeast that is a member of the normal flora of the vagina

319. A dimorphic organism that is transmitted by trauma to the skin

320. A dimorphic fungus that typically is acquired by inhalation of asexual spores

321. A mold that causes pneumonia in immunocompromised patients

Questions 322–325

(A) *Coccidioides immitis*

(B) *Rhizopus nigricans*

(C) *Blastomyces dermatitidis*

(D) *Cryptococcus neoformans*

322. A yeast acquired by inhalation that causes meningitis primarily in immunocompromised patients

323. A mold that invades blood vessels primarily in patients with diabetic ketoacidosis

324. A dimorphic fungus that is acquired by inhalation by people living in certain areas of the southwestern states in the United States

325. A dimorphic fungus that causes granulomatous skin lesions in people living in many areas of North America

Answers (Questions 318–325)

318. (B)	320. (A)	322. (D)	324. (A)
319. (D)	321. (C)	323. (B)	325. (C)

PARASITOLOGY

DIRECTIONS (Questions 326–352): Select the ONE lettered answer that is BEST in each question.

326. Children at day care centers in the United States have a high rate of infection with which one of the following?

(A) *Ascaris lumbricoides*
(B) *Entamoeba histolytica*
(C) *Enterobius vermicularis*
(D) *Necator americanus*

327. The main anatomic location of *Schistosoma mansoni* adult worms is:

(A) Lung alveoli
(B) Intestinal venules
(C) Renal tubules
(D) Bone marrow

328. In malaria, the form of plasmodia that is transmitted from mosquito to human is the:

(A) Sporozoite
(B) Gametocyte
(C) Merozoite
(D) Hypnozoite

329. Which one of the following protozoa primarily infects macrophages?

(A) *Plasmodium vivax*
(B) *Leishmania donovani*
(C) *Trypanosoma cruzi*
(D) *Trichomonas vaginalis*

330. Each of the following parasites has an intermediate host as part of its life cycle EXCEPT:

(A) *Trichomonas vaginalis*
(B) *Taenia solium*
(C) *Echinococcus granulosus*
(D) *Toxoplasma gondii*

331. Each of the following parasites passes through the lung during human infection EXCEPT:

(A) *Strongyloides stercoralis*
(B) *Necator americanus*
(C) *Enterobius vermicularis*
(D) *Ascaris lumbricoides*

332. Each of the following parasites is transmitted by flies EXCEPT:

(A) *Schistosoma mansoni*
(B) *Onchocerca volvulus*
(C) *Trypanosoma gambiense*
(D) *Loa loa*

333. Each of the following parasites is transmitted by mosquitoes EXCEPT:

(A) *Leishmania donovani*
(B) *Wuchereria bancrofti*
(C) *Plasmodium vivax*
(D) *Plasmodium falciparum*

334. Pigs or dogs are the source of human infection by each of the following parasites EXCEPT:

(A) *Echinococcus granulosus*
(B) *Taenia solium*
(C) *Ascaris lumbricoides*
(D) *Trichinella spiralis*

335. Each of the following parasites is transmitted by eating inadequately cooked fish or seafood EXCEPT:

(A) *Diphyllobothrium latum*
(B) *Ancylostoma duodenale*
(C) *Paragonimus westermani*
(D) *Clonorchis sinensis*

336. Laboratory diagnosis of a patient with a suspected liver abscess due to *Entamoeba histolytica* should include:

(A) Stool examination and indirect hemagglutination test
(B) Stool examination and blood smear
(C) Indirect hemagglutination test and skin test
(D) Xenodiagnosis and string test

337. Each of the following statements concerning *Toxoplasma gondii* is correct EXCEPT:

(A) *T. gondii* can be transmitted across the placenta to the fetus.
(B) *T. gondii* can be transmitted by cat feces.
(C) *T. gondii* can cause encephalitis in immunocompromised patients.
(D) *T. gondii* can be diagnosed by finding trophozoites in the stool.

338. Each of the following statements concerning *Giardia lamblia* is correct EXCEPT:

(A) *G. lamblia* has both a trophozoite and a cyst stage in its life cycle.
(B) *G. lamblia* is transmitted by the fecal–oral route from both human and animal sources.
(C) *G. lamblia* causes hemolytic anemia.
(D) *G. lamblia* can be diagnosed by the string test in which a weighted string is swallowed and passes into the upper GI tract.

339. Each of the following statements concerning malaria is correct EXCEPT:

(A) The female *Anopheles* mosquito is the vector.
(B) Early in infection, sporozoites enter hepatocytes.
(C) Release of merozoites from red blood cells causes periodic fever and chills.
(D) The principal site of gametocyte formation is the human gastrointestinal tract.

340. Each of the following statements concerning *Trichomonas vaginalis* is correct EXCEPT:

(A) *T. vaginalis* is transmitted sexually.
(B) *T. vaginalis* can be diagnosed by visualizing the trophozoite.
(C) *T. vaginalis* can be treated effectively with metronidazole.
(D) *T. vaginalis* causes bloody diarrhea.

341. Which one of the following agents can be used to prevent malaria?

(A) Mebendazole
(B) Chloroquine
(C) Inactivated vaccine
(D) Praziquantel

342. Each of the following statements concerning *Pneumocystis carinii* is correct EXCEPT:

   (A) *P. carinii* infections primarily involve the respiratory tract.

   (B) *P. carinii* can be diagnosed by seeing cysts in tissue.

   (C) *P. carinii* infections are symptomatic primarily in immunocompromised patients.

   (D) *P. carinii* symptomatic infections can be prevented by administering penicillin orally.

343. Each of the following statements concerning *Trypanosoma cruzi* is correct EXCEPT:

   (A) *T. cruzi* is transmitted by the reduviid bug.

   (B) *T. cruzi* occurs primarily in tropical Africa.

   (C) *T. cruzi* can be diagnosed by seeing amastigotes in a bone marrow aspirate.

   (D) *T. cruzi* typically affects heart muscle, leading to cardiac failure.

344. Each of the following statements concerning sleeping sickness is correct EXCEPT:

   (A) Sleeping sickness is caused by a trypanosome.

   (B) Sleeping sickness is transmitted by tsetse flies.

   (C) Sleeping sickness can be diagnosed by finding eggs in the stool.

   (D) Sleeping sickness occurs primarily in tropical Africa.

345. Each of the following statements concerning kala-azar is correct EXCEPT:

   (A) Kala-azar is caused by *Leishmania donovani*.

   (B) Kala-azar is transmitted by the bite of sandflies.

   (C) Kala-azar occurs primarily in rural Latin America.

   (D) Kala-azar can be diagnosed by finding amastigotes in bone marrow.

346. Each of the following statements concerning *Diphyllobothrium latum* is correct EXCEPT:

   (A) *D. latum* is transmitted by undercooked fish.

   (B) *D. latum* has operculated eggs.

   (C) *D. latum* causes a megaloblastic anemia due to vitamin B_{12} deficiency.

   (D) *D. latum* is a tapeworm that has a scolex with a circle of hooks.

347. Each of the following statements concerning hydatid cyst disease is correct EXCEPT:

   (A) The disease is caused by *Echinococcus granulosus*.

   (B) The cysts occur primarily in the liver.

   (C) The disease is caused by a parasite whose adult form lives in dogs' intestines.

   (D) The disease occurs primarily in tropical Africa.

348. Each of the following statements concerning *Schistosoma haematobium* is correct EXCEPT:

   (A) *S. haematobium* is acquired by humans when cercariae penetrate the skin.

   (B) Snails are intermediate hosts of *S. haematobium*.

   (C) *S. haematobium* eggs have no spine.

   (D) *S. haematobium* infection predisposes to bladder carcinoma.

349. Each of the following statements concerning hookworm infection is correct EXCEPT:

   (A) Hookworm infection can cause anemia.

   (B) Hookworm infection is acquired by humans when filariform larvae penetrate the skin.

   (C) Hookworm infection is caused by *Necator americanus*.

   (D) Hookworm infection can be diagnosed by finding the trophozoite in the stool.

350. Each of the following statements concerning *Ascaris lumbricoides* is correct EXCEPT:

   (A) *A. lumbricoides* is one of the largest nematodes.

   (B) *A. lumbricoides* is transmitted by ingestion of eggs.

   (C) Both dogs and cats are intermediate hosts of *A. lumbricoides*.

   (D) *A. lumbricoides* can cause pneumonia.

351. Each of the following statements concerning *Strongyloides stercoralis* is correct EXCEPT:

   (A) *S. stercoralis* is acquired by ingestion of eggs.

   (B) *S. stercoralis* undergoes a free-living life cycle in soil.

   (C) Migrating larvae of *S. stercoralis* induce a marked eosinophilia.

   (D) *S. stercoralis* produces filariform larvae.

352. Each of the following statements concerning trichinosis is correct EXCEPT:

   (A) Trichinosis is acquired by eating undercooked pork.

   (B) Trichinosis is caused by a protozoan that has both a trophozoite and a cyst stage in its life cycle.

   (C) Trichinosis can be diagnosed by seeing cysts in muscle biopsy specimens.

   (D) Eosinophilia is a prominent finding.

Answers (Questions 326–352)

326. **(C)**	332. **(A)**	338. **(C)**	344. **(C)**	350. **(C)**
327. **(B)**	333. **(A)**	339. **(D)**	345. **(C)**	351. **(A)**
328. **(A)**	334. **(C)**	340. **(D)**	346. **(D)**	352. **(B)**
329. **(B)**	335. **(B)**	341. **(B)**	347. **(D)**	
330. **(A)**	336. **(A)**	342. **(D)**	348. **(C)**	
331. **(C)**	337. **(D)**	343. **(B)**	349. **(D)**	

DIRECTIONS (Questions 353–386): Select the ONE lettered option that is MOST closely associated with the numbered items. Each lettered option may be selected once, more than once, or not at all.

Questions 353–360

   (A) *Dracunculus medinensis*

   (B) *Loa loa*

   (C) *Onchocerca volvulus*

   (D) *Wuchereria bancrofti*

   (E) *Toxocara canis*

353. Causes river blindness

354. Transmitted by mosquito

355. Acquired by drinking contaminated water

356. Treated by extracting worm from skin ulcer

357. Transmitted by the bite of a deer fly or mango fly

358. Causes visceral larva migrans

359. Causes filariasis

360. Acquired by ingestion of worm eggs

Questions 361–372

   (A) *Giardia lamblia*

   (B) *Plasmodium vivax*

   (C) *Taenia saginata*

   (D) *Clonorchis sinensis*

   (E) *Enterobius vermicularis*

361. A trematode (fluke) acquired by eating undercooked fish

362. A cestode (tapeworm) acquired by eating undercooked beef

363. A nematode (roundworm) transmitted primarily from child to child

364. A protozoan transmitted by mosquito

365. A protozoan transmitted by the fecal–oral route

366. Primarily affects the biliary ducts

367. Causes diarrhea as the most prominent symptom

368. Causes perianal itching as the most prominent symptom

369. Causes fever, chills, and anemia

370. Can be treated with metronidazole

371. Can be treated with mebendazole or pyrantel pamoate

372. Can be treated with chloroquine and primaquine

Questions 373–386

(A) *Entamoeba histolytica*

(B) *Plasmodium falciparum*

(C) *Taenia solium*

(D) *Paragonimus westermani*

(E) *Strongyloides stercoralis*

373. A cestode (tapeworm) acquired by eating undercooked pork

374. A nematode (roundworm) acquired when filariform larvae penetrate the skin

375. A protozoan transmitted by the fecal–oral route

376. A trematode (fluke) acquired by eating undercooked crab meat

377. A protozoan that infects red blood cells

378. Laboratory diagnosis based on finding eggs in sputum

379. Causes cysticercosis in humans

380. Chloroquine-resistant strains occur

381. Autoinfection within humans, especially in immunocompromised patients

382. Causes blackwater fever

383. Causes bloody diarrhea and liver abscesses

384. Produces "banana-shaped" gametocytes

385. Produces cysts with four nuclei

386. Has a scolex with suckers and a circle of hooks

Answers (Questions 353–386)

353. **(C)**	362. **(C)**	371. **(E)**	380. **(B)**
354. **(D)**	363. **(E)**	372. **(B)**	381. **(E)**
355. **(A)**	364. **(B)**	373. **(C)**	382. **(B)**
356. **(A)**	365. **(A)**	374. **(E)**	383. **(A)**
357. **(B)**	366. **(D)**	375. **(A)**	384. **(B)**
358. **(E)**	367. **(A)**	376. **(D)**	385. **(A)**
359. **(D)**	368. **(E)**	377. **(B)**	386. **(C)**
360. **(E)**	369. **(B)**	378. **(D)**	
361. **(D)**	370. **(A)**	379. **(C)**	

IMMUNOLOGY

DIRECTIONS (Questions 387–474): Select the ONE lettered answer that is BEST in each question.

387. Which category of hypersensitivity BEST describes hemolytic disease of the newborn caused by Rh incompatibility?

(A) Atopic or anaphylactic

(B) Cytotoxic

(C) Immune complex

(D) Delayed

388. The principal difference between cytotoxic (type II) and immune complex (type III) hypersensitivity is:

(A) The class (isotype) of antibody

(B) Whether the antibody reacts with the antigen on the cell or reacts with antigen before it interacts with the cell

(C) The participation of complement

(D) The participation of T cells

389. A child stung by a bee experiences respiratory distress within minutes and lapses into unconsciousness. This reaction is probably mediated by:

(A) IgE antibody

(B) IgG antibody

(C) Sensitized T cells

(D) Complement

(E) IgM antibody

390. A patient with rheumatic fever develops a sore throat from which β-hemolytic streptococci are cultured. The patient is started on treatment with penicillin, and the sore throat resolves within several days. However, 7 days after initiation of penicillin therapy, the patient develops a fever of 103°F, a generalized rash, and proteinuria. This MOST probably resulted from:

(A) Recurrence of the rheumatic fever

(B) A different infectious disease

(C) An IgE response to penicillin

(D) An IgG-IgM response to penicillin

(E) A delayed hypersensitivity reaction to penicillin

391. A kidney biopsy specimen taken from a patient with acute glomerulonephritis and stained with fluorescein-conjugated anti-human IgG antibody would probably show:

(A) No fluorescence

(B) Uniform fluorescence of the glomerular basement membrane

(C) Patchy, irregular fluorescence of the glomerular basement membrane

(D) Fluorescent B cells

(E) Fluorescent macrophages

392. A patient with severe asthma gets no relief from antihistamines. The symptoms are MOST likely to be caused by:

(A) Interleukin-2

(B) Slow-reacting substance A (leukotrienes)

(C) Serotonin

(D) Bradykinin

393. Hypersensitivity to penicillin and hypersensitivity to poison oak are both:

(A) Mediated by IgE antibody

(B) Mediated by IgG and IgM antibody

(C) Initiated by haptens

(D) Initiated by Th-2 cells

394. A recipient of a 10/10 HLA-matched kidney from a relative still needs immunosuppression to prevent graft rejection because:

(A) Graft-versus-host disease is a problem
(B) Class II MHC antigens will not be matched
(C) Minor histocompatibility antigens will not be matched
(D) Complement components will not be matched

395. Bone marrow transplantation in immunocompromised patients presents which major problem?

(A) Potentially lethal graft-versus-host disease
(B) High risk of T-cell leukemia
(C) Inability to use a live donor
(D) Delayed hypersensitivity

396. What is the role of class II MHC proteins on donor cells in graft rejection?

(A) They are the receptors for interleukin-2, which is produced by macrophages when they attack the donor cells.
(B) They are recognized by helper T cells, which then activate cytotoxic T cells to kill the donor cells.
(C) They induce the production of blocking antibodies that protect the graft.
(D) They induce IgE, which mediates graft rejection.

397. Grafts between genetically identical individuals (i.e., identical twins):

(A) Are rejected slowly as a result of minor histocompatibility antigens
(B) Are subject to hyperacute rejection
(C) Are not rejected, even without immunosuppression
(D) Are not rejected if a kidney is grafted, but skin grafts are rejected

398. Penicillin is a hapten in both humans and mice. To explore the hapten–carrier relationship, a mouse was injected with penicillin covalently bound to bovine serum albumin and, at the same time, with egg albumin to which no penicillin was bound. Of the following, which one will induce a secondary response to penicillin when injected into the mouse 1 month later? (An explanation of the answer to this question is given on page 765.)

(A) Penicillin
(B) Penicillin bound to egg albumin
(C) Egg albumin
(D) Bovine serum albumin

399. AIDS is caused by a human retrovirus that kills:

(A) B lymphocytes
(B) Lymphocyte stem cells
(C) CD4-positive T lymphocytes
(D) CD8-positive T lymphocytes

400. Which one of the following mechanisms is a feature of tumor cells that successfully avoid killing by the adaptive immune system?

(A) Increased expression of class I MHC proteins
(B) Increased expression of programmed cell death ligand-1 (PD-L1)
(C) Increased expression of gamma interferon
(D) Increased expression of neoantigens

401. Polyomavirus (a DNA virus) causes tumors in mice lacking a thymus, but not in normal mice. The BEST interpretation is that:

(A) Macrophages are required to reject polyomavirus-induced tumors.
(B) Natural killer cells can reject polyomavirus-induced tumors without help from T lymphocytes.

(C) T lymphocytes play an important role in the rejection of polyomavirus-induced tumors.
(D) B lymphocytes play no role in rejection of polyomavirus-induced tumors.

402. C3 is cleaved to form C3a and C3b by C3 convertase. C3b is involved in all of the following EXCEPT:

(A) Increasing vascular permeability
(B) Promoting phagocytosis
(C) Forming alternative-pathway C3 convertase
(D) Forming C5 convertase

403. After binding to its specific antigen, a B lymphocyte may switch its:

(A) Immunoglobulin light chain isotype
(B) Immunoglobulin heavy chain class
(C) Variable region of the immunoglobulin heavy chain
(D) Constant region of the immunoglobulin light chain

404. Diversity is an important feature of the immune system. Which one of the following statements about it is INCORRECT?

(A) Humans can make antibodies with about 10^{11} different $V_H \times V_L$ combinations.
(B) A single cell can synthesize IgM antibody and then switch to IgA antibody.
(C) The hematopoietic stem cell carries the genetic potential to create all possible immunoglobulin proteins.
(D) A single B lymphocyte can produce antibodies of many different specificities, but a plasma cell is monospecific.

405. C3a and C5a can cause:

(A) Bacterial lysis
(B) Increased vascular permeability
(C) Phagocytosis of IgE-coated bacteria
(D) Aggregation of C4 and C2

406. Neutrophils are attracted to an infected area by:

(A) IgM
(B) C1
(C) C5a
(D) C8

407. Complement fixation refers to:

(A) The ingestion of C3b-coated bacteria by macrophages
(B) The destruction of complement in serum by heating at 56°C for 30 minutes
(C) The binding of complement components by antigen–antibody complexes
(D) The interaction of C3b with mast cells

408. The classic complement pathway is initiated by interaction of C1 with:

(A) Antigen
(B) Factor B
(C) Antigen–IgG complexes
(D) Bacterial lipopolysaccharides

409. Patients with severely reduced C3 levels tend to have:

(A) Increased numbers of severe viral infections
(B) Increased numbers of severe bacterial infections
(C) Low gamma globulin levels
(D) Frequent episodes of hemolytic anemia

410. Individuals with a genetic deficiency of C6 have:

(A) Decreased resistance to viral infections
(B) Increased hypersensitivity reactions
(C) Increased frequency of cancer
(D) Increased frequency of *Neisseria* bacteremia

411. Natural killer cells are:

(A) B cells that can kill without complement

(B) Cytotoxic T cells

(C) Increased by immunization

(D) Able to kill virus-infected cells without prior sensitization

412. A positive tuberculin skin test (a delayed hypersensitivity reaction) indicates that:

(A) A humoral immune response has occurred

(B) A cell-mediated immune response has occurred

(C) Both the T- and B-cell systems are functional

(D) Only the B-cell system is functional

413. Reaction to poison ivy or poison oak is:

(A) An IgG-mediated response

(B) An IgE-mediated response

(C) A T-cell–mediated response

(D) An Arthus reaction

414. A child disturbs a wasp nest, is stung repeatedly, and goes into shock within minutes, manifesting respiratory failure and vascular collapse. This is MOST likely to be due to:

(A) Systemic anaphylaxis

(B) Serum sickness

(C) An Arthus reaction

(D) Cytotoxic hypersensitivity

415. "Isotype switching" of immunoglobulin classes by B cells involves:

(A) Simultaneous insertion of V_H genes adjacent to each C_H gene

(B) Successive insertion of a V_H gene adjacent to different C_H genes

(C) Activation of homologous genes on chromosome 6

(D) Switching of light chain types (kappa and lambda)

416. Which one of the following pairs of genes is linked on a single chromosome?

(A) V gene for lambda chain and C gene for kappa chain

(B) C gene for gamma chain and C gene for kappa chain

(C) V gene for lambda chain and V gene for heavy chain

(D) C gene for gamma chain and C gene for alpha chain

417. Antigen-binding determinants are located within:

(A) Hypervariable regions of heavy and light chains

(B) Constant regions of light chains

(C) Constant regions of heavy chains

(D) The hinge region

418. A primary immune response in an adult human requires approximately how much time to produce detectable antibody levels in the blood?

(A) 12 hours

(B) 3 days

(C) 1 week

(D) 3 weeks

419. The membrane IgM and IgD on the surface of an individual B cell:

(A) Have identical heavy chains but different light chains

(B) Are identical except for their C_H regions

(C) Are identical except for their V_H regions

(D) Have different V_H and V_L regions

420. During the maturation of a B lymphocyte, the first immunoglobulin heavy chain synthesized is the:

(A) Mu chain

(B) Gamma chain

(C) Epsilon chain

(D) Alpha chain

421. In the immune response to a hapten–protein conjugate, in order to get anti-hapten antibodies, it is essential that:

(A) The hapten be recognized by helper T cells

(B) The protein be recognized by helper T cells

(C) The protein be recognized by B cells

(D) The hapten be recognized by suppressor T cells

422. Which one of the following is NOT an example of a "second signal" that leads to proliferation and activation of B cells?

(A) CD40 ligand (CD40L) that binds B-cell CD40

(B) Degradation components of C3 that bind B-cell complement receptor 2

(C) Pathogen-associated molecular patterns (PAMPS) that bind B-cell TLRs

(D) Antigen binding to the B-cell surface immunoglobulin

423. Which one of the following sequences is appropriate for testing a patient for antibody against the AIDS virus with the ELISA procedure? (The assay is carried out in a plastic plate with an incubation and a wash step after each addition except the final one.)

(A) Patient's serum/enzyme substrate/HIV antigen/enzyme-labeled antibody against HIV

(B) HIV antigen/patient's serum/enzyme-labeled antibody against human gamma globulin/enzyme substrate

(C) Enzyme-labeled antibody against human gamma globulin/patient's serum/HIV antigen/enzyme substrate

(D) Enzyme-labeled antibody against HIV/HIV antigen/patient's serum/enzyme substrate

424. The BEST method to demonstrate IgG on the glomerular basement membrane in a kidney tissue section is the:

(A) Precipitin test

(B) Complement fixation test

(C) Agglutination test

(D) Indirect fluorescent-antibody test

425. A woman had a high fever, hypotension, and a diffuse macular rash. When all cultures showed no bacterial growth, a diagnosis of toxic shock syndrome was made. Regarding the mechanism by which the toxin causes this disease, which one of the following is LEAST accurate?

(A) The toxin is not processed within the macrophage.

(B) The toxin binds to both the class II MHC protein and the T-cell receptor.

(C) The toxin activates many CD4-positive T cells, and large amounts of interleukins are released.

(D) The toxin has an A-B subunit structure—the B subunit binds to a receptor, and the A subunit enters the cells and activates them.

426. A patient with a central nervous system disorder is maintained on the drug methyldopa. Hemolytic anemia develops, which resolves shortly after the drug is withdrawn. This is MOST probably an example of:

(A) Atopic hypersensitivity

(B) Cytotoxic hypersensitivity

(C) Immune-complex hypersensitivity

(D) Cell-mediated hypersensitivity

427. Which one of the following substances is NOT a major cytokine released by activated helper T cells?

(A) Alpha interferon

(B) Gamma interferon

(C) Interleukin-2

(D) Interleukin-4

428. A delayed hypersensitivity reaction is characterized by:

(A) Edema without a cellular infiltrate

(B) An infiltrate composed of neutrophils

(C) An infiltrate composed of helper T cells and macrophages

(D) An infiltrate composed of eosinophils

429. Two dissimilar inbred strains of mice, A and B, are crossed to yield an F_1 hybrid strain, AB. If a large dose of spleen cells from an adult A mouse is injected into an adult AB mouse, which one of the following is MOST likely to occur? (An explanation of the answer to this question is given on page 765.)

(A) The spleen cells will be destroyed.

(B) The spleen cells will survive and will have no effect in the recipient.

(C) The spleen cells will induce a graft-versus-host reaction in the recipient.

(D) The spleen cells will survive and induce tolerance of strain A grafts in the recipient.

430. This question is based on the same strains of mice described in the previous question. If adult AB spleen cells are injected into a newborn B mouse, which one of the following is MOST likely to occur? (An explanation of the answer to this question is given on page 765.)

(A) The spleen cells will be destroyed.

(B) The spleen cells will survive without any effect on the recipient.

(C) The spleen cells will induce a graft-versus-host reaction in the recipient.

(D) The spleen cells will survive and induce tolerance of strain A grafts in the recipient.

431. The minor histocompatibility antigens on cells:

(A) Are detected by reaction with antibodies and complement

(B) Are controlled by several genes in the major histocompatibility complex

(C) Are unimportant in human transplantation

(D) Induce reactions that can cumulatively lead to a strong rejection response

432. Which one of the following is NOT true of class I MHC antigens?

(A) They can be assayed by a cytotoxic test that uses antibody and complement.

(B) One of their two polypeptide chains is a beta-2 microglobulin.

(C) They are encoded by at least three gene loci in the major histocompatibility complex.

(D) They are found mainly on B cells, macrophages, and activated T cells.

433. An antigen found in relatively high concentration in the plasma of normal fetuses and a high proportion of patients with progressive carcinoma of the colon is:

(A) Viral antigen

(B) Carcinoembryonic antigen

(C) α-Fetoprotein

(D) Heterophil antigen

434. An antibody directed against the hypervariable regions of a human IgG antibody would react with:

(A) The Fc part of the IgG

(B) An IgM antibody produced by the same plasma cell that produced the IgG

(C) All human kappa chains

(D) All human gamma chains

435. Which one of the following is NOT true of the gene segments that combine to make up a heavy chain gene?

(A) Many V region segments are available.

(B) Several J segments and several D segments are available.

(C) V, D, and J segments combine to encode the antigen-binding site.

(D) A V segment and a J segment are preselected by an antigen to make up the variable-region portion of the gene.

436. When immune complexes from the serum are deposited on glomerular basement membrane, damage to the membrane is caused mainly by:

(A) Gamma interferon

(B) Phagocytosis

(C) Cytotoxic T cells

(D) Enzymes released by polymorphonuclear cells

437. If an individual was genetically unable to make J chains, which immunoglobulin(s) would be affected?

(A) IgG

(B) IgM

(C) IgA

(D) IgG and IgM

(E) IgM and IgA

438. The antigen-binding site on antibodies is formed primarily by:

(A) The constant regions of H and L chains

(B) The hypervariable regions of H and L chains

(C) The hypervariable regions of H chains

(D) The variable regions of H chains

(E) The variable regions of L chains

439. The class of immunoglobulin present in highest concentration in the blood of a human newborn is

(A) IgG

(B) IgM

(C) IgA

(D) IgD

(E) IgE

440. Individuals of blood group type AB:

(A) Are Rh(D)-negative

(B) Are "universal recipients" of transfusions

(C) Have circulating anti-A and anti-B antibodies

(D) Have the same haplotype

441. Cytotoxic T cells induced by infection with virus A will kill target cells:

(A) From the same host infected with any virus

(B) Infected by virus A and identical at class I MHC loci of the cytotoxic T cells

(C) Infected by virus A and identical at class II MHC loci of the cytotoxic T cells

(D) Infected with a different virus and identical at class I MHC loci of the cytotoxic cells

(E) Infected with a different virus and identical at class II MHC loci of the cytotoxic cells

442. Antigen-presenting cells that activate helper T cells must express which one of the following on their surfaces?

(A) IgE

(B) Gamma interferon

(C) Class I MHC antigens

(D) Class II MHC antigens

443. Which one of the following does NOT contain C3b?

(A) Classic-pathway C5 convertase

(B) Alternative-pathway C5 convertase

(C) Classic-pathway C3 convertase

(D) Alternative-pathway C3 convertase

444. Which one of the following is NOT true regarding the alternative complement pathway?

(A) It can be triggered by infectious agents in absence of antibody.

(B) It does not require C1, C2, or C4.

(C) It cannot be initiated unless C3b fragments are already present.

(D) It has the same terminal sequence of events as the classic pathway.

445. In setting up a complement fixation test to detect antibody in the patient's serum, the reactants should be added in what sequence? (Ag = antigen; C = complement; EA = antibody-coated indicator erythrocytes.)

(A) Ag + EA + C/wait/ + patient's serum

(B) C + patient's serum + EA/wait/ + Ag

(C) Ag + patient's serum + EA/wait/ + C

(D) Ag + patient's serum + C/wait/ + EA

446. An effective new therapy is approved for autoimmune diseases in which Th-17 cells are overactivated. Which one of the following interleukins (ILs) is most likely to be inhibited by this new therapy?

(A) IL-10

(B) IL-12

(C) IL-13

(D) IL-23

447. Complement lyses cells by:

(A) Enzymatic digestion of the cell membrane

(B) Activation of adenylate cyclase

(C) Insertion of complement proteins into the cell membrane

(D) Inhibition of elongation factor-2

448. Graft and tumor rejection are mediated primarily by:

(A) Non–complement-fixing antibodies

(B) Phagocytic cells

(C) Mast cells

(D) Cytotoxic T cells

449. Which one of the following properties of antibodies is NOT dependent on the structure of the heavy chain constant region?

(A) Ability to cross the placenta

(B) Isotype (class)

(C) Ability to fix complement

(D) Affinity for antigen

450. In which one of the following situations would a graft-versus-host reaction be MOST likely to occur? (Mouse strains A and B are highly inbred; AB is an F$_1$ hybrid between strain A and strain B.) (An explanation of the answer to this question is given on page 765.)

(A) Newborn strain A spleen cells injected into a strain B adult

(B) X-irradiated adult strain A spleen cells injected into a strain B adult

(C) Adult strain A spleen cells injected into an x-irradiated strain AB adult

(D) Adult strain AB spleen cells injected into a strain A newborn

451. A researcher takes cells from patient A, cultures them in a lab, and infects them with a virus. CD8-positive T cells from patient B are then added to the virus-infected cell culture. After several days of culture, no cytokine production or cytotoxicity is seen. Of the following, which one is the BEST explanation?

(A) Patient B was previously exposed to this virus.

(B) Patient A and patient B do not share HLA alleles.

(C) The virus caused an increase in expression of B7 proteins on patient A's cells.

(D) The cultured, virus-infected cells from patient A were not dendritic cells or B cells.

452. A patient skin-tested with purified protein derivative (PPD) to determine previous exposure to *Mycobacterium tuberculosis* develops induration at the skin test site 48 hours later. Histologically, the reaction site would MOST probably show:

(A) Eosinophils

(B) Neutrophils

(C) Helper T cells and macrophages

(D) B cells

453. Hemolytic disease of the newborn caused by Rh blood group incompatibility requires maternal antibody to enter the fetal bloodstream. Therefore, the mediator of this disease is:

(A) IgE antibody

(B) IgG antibody

(C) IgM antibody

(D) IgA antibody

454. An Rh-negative woman married to a heterozygous Rh-positive man has three children. The probability that all three of their children are Rh-positive is:

(A) 1:2

(B) 1:4

(C) 1:8

(D) Zero

455. Which one of the following statements BEST explains the relationship between inflammation of the heart (carditis) and infection with group A β-hemolytic streptococci?

(A) Streptococcal antigens induce antibodies cross-reactive with heart tissue.

(B) Streptococci are polyclonal activators of B cells.

(C) Streptococcal antigens bind to IgE on the surface of heart tissue, and histamine is released.

(D) Streptococci are ingested by neutrophils that release proteases that damage heart tissue.

456. Your patient became ill 10 days ago with a viral disease. Laboratory examination reveals that the patient's antibodies against this virus have a high ratio of IgM to IgG. What is your conclusion?

(A) It is unlikely that the patient has encountered this organism previously.

(B) The patient is predisposed to IgE-mediated hypersensitivity reactions.

(C) The information given is irrelevant to previous antigen exposure.

(D) It is likely that the patient has an autoimmune disease.

457. If you measure the ability of cytotoxic T cells from an HLA-B27 person to kill virus X–infected target cells, which one of the following statements is CORRECT?

(A) Any virus X–infected target cell will be killed.

(B) Only virus X–infected cells of HLA-B27 type will be killed.

(C) Any HLA-B27 cell will be killed.

(D) No HLA-B27 cell will be killed.

458. You have a patient who makes autoantibodies against his own red blood cells, leading to hemolysis. Which one of the following mechanisms is MOST likely to explain the hemolysis?

(A) Perforins from cytotoxic T cells lyse the red cells.

(B) Neutrophils release proteases that lyse the red cells.

(C) Interleukin-2 binds to its receptor on the red cells, which results in lysis of the red cells.

(D) Complement is activated, and membrane attack complexes lyse the red cells.

459. Your patient is a child who has no detectable T or B cells. This immunodeficiency is most probably the result of a defect in:

(A) The thymus

(B) The membrane attack complex of complement

(C) T cell–B cell interaction

(D) Stem cells originating in the bone marrow

460. The role of dendritic cells during an antibody response is to:

(A) Make antibody

(B) Lyse virus-infected target cells

(C) Activate cytotoxic T cells

(D) Process antigen and present it

461. The structural basis of blood group A and B antigen specificity is:

(A) A single terminal sugar residue

(B) A single terminal amino acid

(C) Multiple differences in the carbohydrate portion

(D) Multiple differences in the protein portion

462. Complement can enhance phagocytosis because of the presence on macrophages and neutrophils of receptors for:

(A) Factor D

(B) C3b

(C) C6

(D) C9

463. The main advantage of passive immunization over active immunization is that:

(A) It can be administered orally.

(B) It provides antibody more rapidly.

(C) Antibody persists for a longer period.

(D) It contains primarily IgM.

464. On January 15, a patient developed an illness suggestive of influenza, which lasted 1 week. On February 20, she had a similar illness. She had no influenza immunization during this period. Her hemagglutination inhibition titer to influenza A virus was 10 on January 18, 40 on January 30, and 320 on February 20. Which one of the following is the MOST appropriate interpretation?

(A) The patient was ill with influenza A on January 15.

(B) The patient was ill with influenza A on February 20.

(C) The patient was not infected with influenza virus.

(D) The patient has an autoimmune disease.

465. An individual who is heterozygous for Gm allotypes contains two allelic forms of IgG in serum, but individual lymphocytes produce only one of the two forms. This phenomenon, known as "allelic exclusion," is consistent with:

(A) A rearrangement of a heavy chain gene on only one chromosome in a lymphocyte

(B) Rearrangements of heavy chain genes on both chromosomes in a lymphocyte

(C) A rearrangement of a light chain gene on only one chromosome in a lymphocyte

(D) Rearrangements of light chain genes on both chromosomes in a lymphocyte

466. Each of the following statements concerning class I MHC proteins is correct EXCEPT:

(A) They are cell surface proteins on virtually all cells.

(B) They are recognition elements for cytotoxic T cells.

(C) They are codominantly expressed.

(D) They are important in the skin test response to *Mycobacterium tuberculosis*.

467. Which one of the following is the BEST method of reducing the effect of graft-versus-host disease in a bone marrow recipient?

(A) Matching the complement components of donor and recipient

(B) Administering alpha interferon

(C) Removing mature T cells from the graft

(D) Removing pre-B cells from the graft

468. Regarding Th-1 and Th-2 cells, which one of the following is LEAST accurate?

(A) Th-1 cells produce gamma interferon and promote cell-mediated immunity.

(B) Th-2 cells produce interleukin-4 and interleukin-5 and promote allergy.

(C) Both Th-1 and Th-2 cells have both CD3 and CD4 proteins on their outer cell membrane.

(D) Before naive Th cells differentiate into Th-1 or Th-2 cells, they are double-positives (i.e., they produce both gamma interferon and interleukin-4).

469. Each of the following statements concerning the variable regions of heavy chains and the variable regions of light chains in a given antibody molecule is correct EXCEPT:

(A) They have the same amino acid sequence.

(B) They define the specificity for antigen.

(C) They are encoded on different chromosomes.

(D) They contain the hypervariable regions.

470. Each of the following statements concerning class II MHC proteins is correct EXCEPT:

(A) They are found on the surface of both B and T cells.

(B) They have a high degree of polymorphism.

(C) They are involved in the presentation of antigen by macrophages.

(D) They have a binding site for CD4 proteins.

471. Which one of the following statements concerning immunoglobulin allotypes is CORRECT?

(A) Allotypes are found only on heavy chains.

(B) Allotypes are determined by class I MHC genes.

(C) Allotypes are confined to the variable regions.

(D) Allotypes are due to genetic polymorphism within a species.

472. Each of the following statements concerning immunologic tolerance is correct EXCEPT:

(A) Tolerance is not antigen-specific (i.e., paralysis of the immune cells results in a failure to produce a response against many antigens).

(B) Tolerance is more easily induced in T cells than in B cells.

(C) Tolerance is more easily induced in neonates than in adults.

(D) Tolerance is more easily induced by simple molecules than by complex ones.

473. Each of the following statements concerning a hybridoma cell is correct EXCEPT:

(A) The spleen cell component provides the ability to form antibody.

(B) The myeloma cell component provides the ability to grow indefinitely.

(C) The antibody produced by a hybridoma cell is IgM, because heavy chain switching does not occur.

(D) The antibody produced by a hybridoma cell is homogeneous (i.e., it is directed against a single epitope).

474. Each of the following statements concerning haptens is correct EXCEPT:

(A) A hapten can combine with (bind to) an antibody.

(B) A hapten cannot induce an antibody by itself; rather, it must be bound to a carrier protein to be able to induce antibody.

(C) In both penicillin-induced anaphylaxis and poison ivy, the allergens are haptens.

(D) Haptens must be processed by CD8$^+$ cells to become immunogenic.

Answers (Questions 387–474)

387. (B)	405. (B)	423. (B)	441. (B)	459. (D)
388. (B)	406. (C)	424. (D)	442. (D)	460. (D)
389. (A)	407. (C)	425. (D)	443. (C)	461. (A)
390. (D)	408. (C)	426. (B)	444. (C)	462. (B)
391. (C)	409. (B)	427. (A)	445. (D)	463. (B)
392. (B)	410. (D)	428. (C)	446. (D)	464. (A)
393. (C)	411. (D)	429. (C)	447. (C)	465. (A)
394. (C)	412. (B)	430. (D)	448. (D)	466. (D)
395. (A)	413. (C)	431. (D)	449. (D)	467. (C)
396. (B)	414. (A)	432. (D)	450. (C)	468. (D)
397. (C)	415. (B)	433. (B)	451. (B)	469. (A)
398. (D)	416. (D)	434. (B)	452. (C)	470. (A)
399. (C)	417. (A)	435. (D)	453. (B)	471. (D)
400. (B)	418. (C)	436. (D)	454. (C)	472. (A)
401. (C)	419. (B)	437. (E)	455. (A)	473. (C)
402. (A)	420. (A)	438. (B)	456. (A)	474. (D)
403. (B)	421. (B)	439. (A)	457. (B)	
404. (D)	422. (D)	440. (B)	458. (D)	

DIRECTIONS (Questions 475–535): Select the ONE lettered option that is MOST closely associated with the numbered items. Each lettered option may be selected once, more than once, or not at all.

Questions 475–480

(A) T cells

(B) B cells

(C) Macrophages

(D) B cells and macrophages

(E) T cells, B cells, and macrophages

475. Major source of interleukin-1

476. Acted on by interleukin-1

477. Major source of interleukin-2

478. Express class I MHC proteins

479. Express class II MHC proteins

480. Express surface immunoglobulin

Questions 481–484

(A) Primary antibody response

(B) Secondary antibody response

481. Appears more quickly and persists longer

482. Relatively richer in IgG

483. Relatively richer in IgM

484. Typically takes 7 to 10 days for antibody to appear

Questions 485–488

(A) Blood group A

(B) Blood group O

(C) Blood groups A and O

(D) Blood group AB

485. People with this type have circulating anti-A antibodies

486. People with this type have circulating anti-B antibodies

487. People with this type are called "universal donors"

488. People with this type are called "universal recipients"

Questions 489–494

(A) Variable region of light chain

(B) Variable region of heavy chain

(C) Variable regions of light and heavy chains

(D) Constant region of heavy chain

(E) Constant regions of light and heavy chains

489. Determines immunoglobulin class

490. Determines allotypes

491. Determines isotypes

492. Binding of IgG to macrophages

493. Activation of complement by IgG

494. Antigen-binding site

Questions 495–498

In the following diagram of an IgM pentamer, identify the labeled structures from the following list. (An explanation of the answer to this question is given on page 765.)

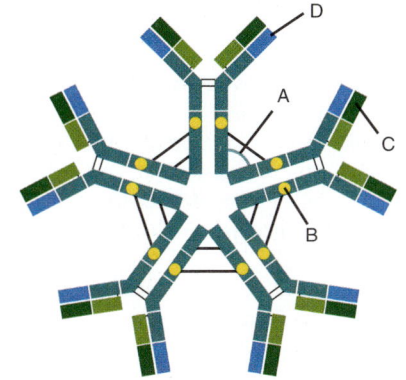

495. Composed of recombined V, D, and J segments

496. Consists of kappa chains or lambda chains, but not both

497. Required for antibody dimers and pentamers

498. Binds to complement

Questions 499–501

(A) Immediate hypersensitivity

(B) Cytotoxic hypersensitivity

(C) Immune-complex hypersensitivity

(D) Delayed hypersensitivity

499. Irregular deposition of IgG along glomerular basement membrane

500. Involves mast cells and basophils

501. Involves macrophages and helper T cells

Questions 502–505

(A) IgM

(B) IgG

(C) IgA

(D) IgE

502. Crosses the placenta

503. Can contain a polypeptide chain not synthesized by a B lymphocyte

504. Found in highest concentration in the milk of lactating women

505. Binds firmly to mast cells and triggers anaphylaxis

Questions 506–509

(A) Agglutination

(B) Precipitin test

(C) Immunofluorescence

(D) Enzyme immunoassay

506. Concentration of IgG in serum

507. Surface IgM on cells in a bone marrow smear

508. Growth hormone in serum

509. Type A blood group antigen on erythrocytes

Questions 510–513

(A) IgA

(B) IgE

(C) IgG

(D) IgM

510. Present in highest concentration in serum

511. Present in highest concentration in secretions

512. Present in lowest concentration in serum

513. Contains 10 heavy and 10 light chains

Questions 514–517

Match each of the following characteristics with one family of pattern recognition receptors listed below:

(A) Intracellular receptors that activate the inflammasome.

(B) Primary receptor for extracellular and endosomal molecular patterns, including lipopolysaccharide.

(C) Activates type I interferon production in response to intracellular viral RNA.

(D) This family includes members that bind to mannose, participating in the mannose pathway of complement activation.

514. Toll-like receptors (TLRs)

515. Nod-like receptors (NLRs)

516. C-type lectin receptors (CLRs)

517. RIG-I helicase-like receptors (RLRs)

Questions 518–521

(A) Class I MHC proteins

(B) Class II MHC proteins

518. Involved in the presentation of antigen to CD4-positive cells

519. Involved in the presentation of antigen to CD8-positive cells

520. Involved in antibody responses to T-dependent antigens

521. Involved in target cell recognition by cytotoxic T cells

Questions 522–525

(A) Fab fragment of IgG

(B) Fc fragment of IgG

522. Contains an antigen-combining site

523. Contains hypervariable regions

524. Contains a complement-binding site

525. Is crystallizable

Questions 526–530

(A) Severe combined immunodeficiency disease (SCID)

(B) X-linked hypogammaglobulinemia

(C) Thymic aplasia

(D) Chronic granulomatous disease

(E) Hereditary angioedema

526. Caused by a defect in the ability of neutrophils to kill microorganisms

527. Caused by a development defect that results in a profound loss of T cells

528. Caused by a deficiency in an inhibitor of the C1 component of complement

529. Caused by a marked deficiency of B cells

530. Caused by a virtual absence of both B and T cells

Questions 531–535

(A) Systemic lupus erythematosus

(B) Rheumatoid arthritis

(C) Rheumatic fever

(D) Graves' disease

(E) Myasthenia gravis

531. Associated with antibody to the thyroid-stimulating hormone (TSH) receptor

532. Associated with antibody to IgG

533. Associated with antibody to the acetylcholine receptor

534. Associated with antibody to DNA

535. Associated with antibody to *Streptococcus pyogenes*

Answers (Questions 475–535)

475. **(C)**	488. **(D)**	501. **(D)**	514. **(B)**	527. **(C)**
476. **(A)**	489. **(D)**	502. **(B)**	515. **(A)**	528. **(E)**
477. **(A)**	490. **(E)**	503. **(C)**	516. **(D)**	529. **(B)**
478. **(E)**	491. **(D)**	504. **(C)**	517. **(C)**	530. **(A)**
479. **(D)**	492. **(D)**	505. **(D)**	518. **(B)**	531. **(D)**
480. **(B)**	493. **(D)**	506. **(D)**	519. **(A)**	532. **(B)**
481. **(B)**	494. **(C)**	507. **(C)**	520. **(B)**	533. **(E)**
482. **(B)**	495. **(D)**	508. **(D)**	521. **(A)**	534. **(A)**
483. **(A)**	496. **(C)**	509. **(A)**	522. **(A)**	535. **(C)**
484. **(A)**	497. **(A)**	510. **(C)**	523. **(A)**	
485. **(B)**	498. **(B)**	511. **(A)**	524. **(B)**	
486. **(C)**	499. **(C)**	512. **(B)**	525. **(B)**	
487. **(B)**	500. **(A)**	513. **(D)**	526. **(D)**	

Explanation of question 398: Bovine serum albumin is the correct answer because it activates helper T cells that are required to provide the interleukins needed for a secondary response. Penicillin alone is incorrect because it is a hapten and cannot activate helper T cells. Choices B and C refer to egg albumin, which can activate helper T cells but not the ones that were activated by the initial stimulus that contained penicillin bound to bovine serum albumin.

Explanation of question 429: Spleen cells from the adult donor A will recognize the B antigen on the recipient's cells as foreign. Spleen cells from the adult donor will contain mature CD4 and CD8 cells that will attack the recipient cells, causing a graft-versus-host reaction; therefore, answer C is correct. Because the recipient is tolerant to antigen A, the donor A spleen cells will not be destroyed; therefore, answer A is incorrect. Answer B is incorrect because although the donor cells will survive, they will have an effect on the recipient. Answer D is incorrect because the recipient is already tolerant to antigen A.

Explanation of question 430: Because the donor AB spleen cells will not see any foreign antigen in the recipient, no graft-versus-host reaction will occur; therefore, answer C is incorrect. The immune cells of the newborn mouse do not have the capability to kill the donor cells; therefore, answer A is incorrect. Answer D is more correct than answer B because the donor cells will survive and induce tolerance to antigen A in the newborn recipient.

Explanation of question 450: Graft-versus-host (GVH) reaction is most likely to occur when the recipient is immunocompromised and functioning donor T cells recognize a foreign antigen in the recipient. Answer C is correct because the recipient is X-irradiated (immunocompromised) and the donor A T cells recognize the B antigen in the recipient as foreign. In answers A and B, the recipient is not immunocompromised and in answer D, the newborn recipient is relatively immunocompromised but the donor T cells are tolerant to the A antigen in the recipient so no GVH reaction occurs.

Explanation of questions 495–498: The diagram depicts labels pointing to (A) the J chain, which is required for antibody multimers, (B) the complement-binding domain of the IgM Fc region, (C) the variable region of the light chain, which is composed of V and J segments and is either a kappa or lambda protein, and (D) the variable region of the heavy chain, which is composed of V, D, and J segments.

EXTENDED MATCHING QUESTIONS

DIRECTIONS (Questions 536–593): Each set of matching questions in this section consists of a list of lettered options followed by several numbered items. For each numbered item, select the ONE lettered option that is MOST closely associated with it. Each lettered option may be selected once, more than once, or not at all.

(A) Capsule
(B) Periplasmic space
(C) Peptidoglycan
(D) Lipid A
(E) 30S ribosomal subunit
(F) G protein
(G) Pilus
(H) ADP-ribosylating enzyme
(I) Flagellum
(J) Transposon

536. Is the site of action of lysozyme
537. Mediates adherence of bacteria to mucous membranes
538. Is the toxic component of endotoxin

(A) Skin
(B) Colon
(C) Nose
(D) Stomach
(E) Vagina
(F) Mouth
(G) Outer third of urethra
(H) Gingival crevice
(I) Pharynx

539. Anatomic location where *Bacteroides fragilis* is most commonly found
540. Anatomic location where *Actinomyces israelii* is most commonly found

(A) Toxic shock syndrome toxin
(B) Tetanus toxin
(C) Diphtheria toxin
(D) Cholera toxin
(E) Coagulase
(F) Botulinum toxin
(G) Alpha toxin of *Clostridium perfringens*
(H) M protein
(I) Endotoxin
(J) Verotoxin

541. Blocks release of acetylcholine
542. Its lipid component causes fever and shock by inducing tumor necrosis factor (TNF)
543. Causes fever and shock by binding to the T-cell receptor
544. Inhibits protein synthesis by ADP-ribosylation of elongation factor-2
545. Increases cyclic AMP by ADP-ribosylation of a G protein

(A) Ampicillin
(B) Nafcillin
(C) Clindamycin
(D) Gentamicin
(E) Tetracycline
(F) Amphotericin B
(G) Ciprofloxacin
(H) Rifampin
(I) Sulfonamide
(J) Erythromycin

546. Inhibits protein synthesis by blocking formation of the initiation complex so that no polysomes form
547. Inhibits DNA gyrase
548. Inhibits folic acid synthesis; analogue of para-aminobenzoic acid
549. Inhibits peptidoglycan synthesis; resistant to β-lactamase

550. Inhibits RNA polymerase
- **(A)** *Streptococcus pneumoniae*
- **(B)** *Streptococcus pyogenes*
- **(C)** *Haemophilus influenzae*
- **(D)** *Salmonella typhi*
- **(E)** *Staphylococcus aureus*
- **(F)** *Enterococcus faecalis*
- **(G)** *Clostridium tetani*
- **(H)** *Bordetella pertussis*
- **(I)** *Escherichia coli*
- **(J)** *Streptococcus agalactiae*
- **(K)** *Staphylococcus epidermidis*
- **(L)** *Streptococcus mutans*

551. The vaccine contains a single serotype of a capsular polysaccharide coupled to a protein carrier

552. Immunogen in the vaccine is a toxoid

553. Causes acute glomerulonephritis; is β-hemolytic

554. Causes urinary tract infections; grows in 6.5% NaCl

555. Causes neonatal meningitis; is bacitracin-resistant

556. Causes meningitis in adults; is α-hemolytic and optochin-sensitive

557. Causes food poisoning; is coagulase-positive
- **(A)** *Escherichia coli*
- **(B)** *Shigella sonnei*
- **(C)** *Salmonella typhi*
- **(D)** *Salmonella enteritidis*
- **(E)** *Proteus mirabilis*
- **(F)** *Pseudomonas aeruginosa*
- **(G)** *Vibrio cholerae*
- **(H)** *Campylobacter jejuni*
- **(I)** *Helicobacter pylori*
- **(J)** *Bacteroides fragilis*

558. Causes gastritis and peptic ulcer; produces urease

559. Causes bloody diarrhea; does not ferment lactose and does not produce H_2S

560. Causes peritonitis; is an obligate anaerobe

561. Causes wound infections with blue-green pus; is oxidase-positive

562. Comma-shaped rod; causes high-volume watery diarrhea
- **(A)** *Legionella pneumophila*
- **(B)** *Yersinia pestis*
- **(C)** *Haemophilus influenzae*
- **(D)** *Corynebacterium diphtheriae*
- **(E)** *Pasteurella multocida*
- **(F)** *Bordetella pertussis*
- **(G)** *Brucella melitensis*
- **(H)** *Listeria monocytogenes*
- **(I)** *Clostridium perfringens*
- **(J)** *Neisseria gonorrhoeae*

563. Gram-positive spore-forming rod that causes myonecrosis

564. Gram-negative rod that is transmitted by cat bite

565. Gram-negative rod that causes cough and lymphocytosis
- **(A)** *Mycobacterium tuberculosis*
- **(B)** *Borrelia burgdorferi*
- **(C)** *Nocardia asteroides*
- **(D)** *Treponema pallidum*
- **(E)** *Coxiella burnetii*
- **(F)** *Mycoplasma pneumoniae*
- **(G)** *Mycobacterium leprae*
- **(H)** *Chlamydia trachomatis*

- **(I)** *Rickettsia rickettsii*
- **(J)** *Leptospira interrogans*

566. Spirochete that does not have an animal reservoir

567. Obligate intracellular parasite that forms elementary bodies

568. Respiratory pathogen without a cell wall
- **(A)** Influenza virus
- **(B)** Adenovirus
- **(C)** Hepatitis A virus
- **(D)** Hepatitis B virus
- **(E)** Herpes simplex virus
- **(F)** Measles virus
- **(G)** Human immunodeficiency virus
- **(H)** Rabies virus
- **(I)** Rotavirus

569. Nonenveloped virus with a genome composed of single-stranded, positive-polarity RNA

570. Enveloped virus with a genome composed of two identical strands of positive-polarity RNA

571. Enveloped virus with a genome composed of double-stranded DNA and has a DNA polymerase in the virion

572. Enveloped virus with a genome composed of segmented, negative-polarity, single-stranded RNA

573. Nonenveloped virus with a genome composed of segmented double-stranded RNA
- **(A)** Herpes simplex virus type 1
- **(B)** Rabies virus
- **(C)** Varicella-zoster virus
- **(D)** Measles virus
- **(E)** Epstein–Barr virus
- **(F)** Influenza virus
- **(G)** Rubella virus
- **(H)** Herpes simplex virus type 2
- **(I)** Mumps virus
- **(J)** Cytomegalovirus
- **(K)** Parainfluenza virus
- **(L)** Respiratory syncytial virus

574. Leading cause of congenital malformations; no vaccine available

575. Causes a painful vesicular rash along the course of a thoracic nerve

576. Causes encephalitis; killed vaccine available

577. Causes pharyngitis, lymphadenopathy, and a positive heterophil test

578. Causes retinitis and pneumonia in patients deficient in helper T cells

579. Causes encephalitis, especially in the temporal lobe

580. Causes pneumonia primarily in infants; induces giant cells

581. Causes orchitis that can result in sterility
- **(A)** Human papillomavirus
- **(B)** Hepatitis A virus
- **(C)** Rotavirus
- **(D)** Adenovirus
- **(E)** Hepatitis delta virus (HDV)
- **(F)** Parvovirus B19
- **(G)** Human immunodeficiency virus
- **(H)** Hepatitis B virus
- **(I)** Sin Nombre virus (Hantavirus)
- **(J)** Human T-cell lymphotropic virus
- **(K)** Prion
- **(L)** Hepatitis C virus

582. Most important cause of diarrhea in infants
583. A vaccine containing purified viral protein is available
584. Defective virus with an RNA genome
    (A) *Coccidioides immitis*
    (B) *Cryptococcus neoformans*
    (C) *Blastomyces dermatitidis*
    (D) *Sporothrix schenckii*
    (E) *Aspergillus fumigatus*
    (F) *Candida albicans*
    (G) *Histoplasma capsulatum*
    (H) *Mucor* species
    (I) *Microsporum canis*
585. Dimorphic fungus that enters the body through puncture wounds in the skin
586. Nonseptate mold that invades tissue, especially in acidotic patients
587. Yeast that forms pseudohyphae when it invades tissue
    (A) *Giardia lamblia*
    (B) *Plasmodium vivax*
    (C) *Leishmania donovani*
    (D) *Entamoeba histolytica*
    (E) *Toxoplasma gondii*
    (F) *Trypanosoma cruzi*
    (G) *Pneumocystis carinii*
    (H) *Plasmodium falciparum*
    (I) *Naegleria fowleri*
    (J) *Trichomonas vaginalis*
588. Acquired while swimming; causes meningitis
589. Transmitted by reduviid bug and invades cardiac muscle
590. Amastigotes found within macrophages
    (A) *Echinococcus granulosus*
    (B) *Clonorchis sinensis*

(C) *Strongyloides stercoralis*
(D) *Taenia solium*
(E) *Necator americanus*
(F) *Enterobius vermicularis*
(G) *Schistosoma haematobium*
(H) *Wuchereria bancrofti*
(I) *Trichinella spiralis*
(J) *Taenia saginata*

591. Infection predisposes to bladder carcinoma
592. Ingestion of eggs can cause cysticercosis
593. Acquired by penetration of feet by larvae; causes anemia

Answers (Questions 536–593)

536. (C)	548. (I)	560. (J)	572. (A)	584. (E)
537. (G)	549. (B)	561. (F)	573. (I)	585. (D)
538. (D)	550. (H)	562. (G)	574. (J)	586. (H)
539. (B)	551. (C)	563. (I)	575. (C)	587. (F)
540. (H)	552. (G)	564. (E)	576. (B)	588. (I)
541. (F)	553. (B)	565. (F)	577. (E)	589. (F)
542. (I)	554. (F)	566. (D)	578. (J)	590. (C)
543. (A)	555. (J)	567. (H)	579. (A)	591. (G)
544. (C)	556. (A)	568. (F)	580. (L)	592. (D)
545. (D)	557. (E)	569. (C)	581. (I)	593. (E)
546. (D)	558. (I)	570. (G)	582. (C)	
547. (G)	559. (B)	571. (D)	583. (H)	

CLINICAL CASE QUESTIONS

DIRECTIONS (Questions 594–654): Select the ONE lettered answer that is BEST in each question.

CASE 1. Your patient is a 20-year-old woman with the sudden onset of fever to 104°F and a severe headache. Physical examination reveals nuchal rigidity. You suspect meningitis and do a spinal tap. Gram stain of the spinal fluid reveals many neutrophils and many gram-negative diplococci.

594. Of the following bacteria, which one is MOST likely to be the cause?
    (A) *Haemophilus influenzae*
    (B) *Neisseria meningitidis*
    (C) *Streptococcus pneumoniae*
    (D) *Pseudomonas aeruginosa*
595. Additional history reveals that she has had several serious infections with this organism previously. On the basis of this, which one of the following is the MOST likely predisposing factor?
    (A) She is HIV antibody positive.
    (B) She is deficient in CD8-positive T cells.

(C) She is deficient in one of the late-acting complement components.
(D) She is deficient in antigen presentation by her macrophages.

CASE 2. Your patient is a 70-year-old man with a long history of smoking who now has a fever and a cough productive of greenish sputum. You suspect pneumonia, and a chest X-ray confirms your suspicion.

596. If a Gram stain of the sputum reveals very small gram-negative rods and there is no growth on a blood agar but colonies do grow on chocolate agar supplemented with NAD and heme, which one of the following bacteria is the MOST likely cause?
    (A) *Chlamydia pneumoniae*
    (B) *Legionella pneumophila*
    (C) *Mycoplasma pneumoniae*
    (D) *Haemophilus influenzae*

CASE 3. Your patient is a 50-year-old woman who returned yesterday from a vacation in Peru, where there is an epidemic of cholera. She now has multiple episodes of diarrhea.

597. Of the following, which one is MOST compatible with cholera?

   (A) Watery diarrhea without blood, no polys in the stool, and growth of curved gram-negative rods in the blood culture
   (B) Watery diarrhea without blood, no polys in the stool, and no organisms in the blood culture
   (C) Bloody diarrhea, polys in the stool, and growth of curved gram-negative rods in the blood culture
   (D) Bloody diarrhea, polys in the stool, and no organisms in the blood culture

CASE 4. Your patient is a 55-year-old man who is coughing up greenish blood-streaked sputum. For the past 2 weeks, he has had fever and night sweats. He thinks he has lost about 10 lb. On physical examination, there are crackles in the apex of the right lung, and a chest X-ray shows a cavity in that location.

598. Of the following, which one is the LEAST likely finding?

   (A) Gram stain of the sputum shows no predominant organism.
   (B) Culture of the sputum on blood agar shows no predominant organism.
   (C) Culture of the sputum on Löwenstein-Jensen medium shows tan colonies after incubation for 4 weeks.
   (D) Rapid plasma reagin test reveals the causative organism.

CASE 5. Your patient is a 5-year-old girl with bloody diarrhea and no vomiting. There is no history of travel outside of the United States. Stool culture grows both lactose-positive and lactose-negative colonies on EMB agar.

599. Of the following organisms, which one is MOST likely to be the cause?

   (A) *Shigella sonnei*
   (B) *Salmonella typhi*
   (C) *Campylobacter jejuni*
   (D) *Helicobacter pylori*

CASE 6. Your patient is a 25-year-old woman with acute onset of pain in her left lower quadrant. On pelvic examination, there is a cervical exudate and tenderness in the left adnexa. You conclude that she has pelvic inflammatory disease (PID) and order laboratory tests.

600. Of the following, which one is the LEAST informative laboratory result?

   (A) Gram stain of the cervical exudate shows gram-negative diplococci within polys.
   (B) Culture of the cervical exudate on Thayer-Martin agar shows oxidase-positive colonies.
   (C) Fluorescent-antibody test shows cytoplasmic inclusions.
   (D) Complement fixation test shows a rise in antibody titer.

CASE 7. Your patient is a 22-year-old man with fever, fatigue, and a new diastolic murmur. You suspect endocarditis and do a blood culture.

601. Which of the following statements is LEAST accurate?

   (A) If he had dental surgery recently, one of the most likely organisms to grow would be a viridans group streptococcus.
   (B) If he is an intravenous drug user, one of the most likely organisms to grow would be *Candida albicans.*
   (C) If he had colon surgery recently, one of the most likely organisms to grow would be *Enterococcus faecalis.*
   (D) If he has a prosthetic aortic valve, one of the most likely organisms to grow would be *Streptococcus agalactiae.*

In fact, none of the above organisms grew in the blood culture. What did grow was a gram-positive coccus arranged in clusters. When subcultured on blood agar, the colonies were surrounded by a zone of clear hemolysis, and a coagulase test was positive.

602. In view of this, which one of the following is MOST accurate?

   (A) He is probably an intravenous drug user.
   (B) He probably lives on a farm and has had contact with pregnant sheep.
   (C) He probably has a common sexually transmitted disease.
   (D) He probably has been camping and was bitten by a tick.

CASE 8. Your patient is a 70-year-old woman who had a hysterectomy for carcinoma of the uterus 3 days ago. She has an indwelling urinary catheter in place and now has a fever of 39°C, and the urine in the collection bottle is cloudy. A Gram stain of the urine specimen shows many neutrophils and gram-positive cocci in chains. You also do a urine culture.

603. Which one of the following is the MOST likely set of findings on the urine culture?

   (A) β-Hemolytic colonies that are bacitracin-sensitive
   (B) α-Hemolytic colonies that are optochin-sensitive
   (C) Nonhemolytic colonies that grow in 6.5% sodium chloride
   (D) Nonhemolytic colonies that grow only anaerobically

CASE 9. Your patient is a 27-year-old woman who was treated with oral ampicillin for cellulitis caused by *Streptococcus pyogenes*. Several days later, she developed bloody diarrhea. You suspect that she may have pseudomembranous colitis.

604. Regarding the causative organism of pseudomembranous colitis, which one of the following is the MOST accurate?

   (A) It is an anaerobic gram-positive rod that produces exotoxins.
   (B) It is a comma-shaped gram-negative rod that grows best at 41°C.
   (C) It is an obligate intracellular parasite that grows in cell culture but not on blood agar.
   (D) It is a yeast that forms germ tubes when incubated in human serum at 37°C.

CASE 10. Your patient is a 10-year-old girl who has had pain in her left arm for the past 5 days. On physical examination, her temperature is 38°C, and there is tenderness of the humerus near her deltoid. On X-ray of the humerus, an area of raised periosteum and erosion of bone is seen. You do a blood culture.

605. Which one of the following is the MOST likely set of findings?

   (A) Gram-negative rods that grow on EMB agar, forming purple colonies and a green sheen
   (B) Gram-positive cocci that grow on blood agar, causing a clear zone of hemolysis and are coagulase-positive
   (C) Gram-positive rods that grow only anaerobically and form a double zone of hemolysis on blood agar
   (D) Gram-negative diplococci that grow on blood agar, are oxidase-positive, and ferment maltose

CASE 11. Your patient is a 30-year-old man who is HIV antibody positive and has a history of *Pneumocystis* pneumonia 2 years ago. He now has an ulcerating lesion on the side of his tongue. A Giemsa stain of the biopsy specimen reveals budding yeasts within macrophages. A culture of the specimen grows an organism that is a budding yeast at 37°C but produces hyphae at 25°C.

606. Of the following, which one is the MOST likely organism to cause this infection?

(A) *Coccidioides immitis*
(B) *Aspergillus fumigatus*
(C) *Histoplasma capsulatum*
(D) *Cryptococcus neoformans*

CASE 12. Your patient is a 10-year-old boy who is receiving chemotherapy for acute leukemia. He develops fever, headache, and a stiff neck, and you make a presumptive diagnosis of meningitis and do a lumbar puncture. A Gram stain reveals a small gram-positive rod, and culture of the spinal fluid grows a β-hemolytic colony on blood agar.

607. Regarding this organism, which one of the following is MOST accurate?

(A) It has more than 100 serologic types.
(B) It produces an exotoxin that inhibits elongation factor-2.
(C) It is commonly acquired by eating unpasteurized dairy products.
(D) There is a toxoid vaccine available against this organism.

CASE 13. Ms. Jones calls to say that she, her husband, and their child have had nausea and vomiting for the past hour or so. Also, they have had some nonbloody diarrhea. You ask when their last meal together was, and she says they had a picnic lunch in the park about 3 hours ago. They have no fever.

608. Which one of the following is the MOST likely finding?

(A) Gram stain of the leftover food would show many gram-positive cocci in clusters.
(B) Gram stain of the stool would show many gram-negative diplococci.
(C) KOH prep of the leftover food would show many budding yeasts.
(D) Acid-fast stain of the stool would show many acid-fast rods.

CASE 14. Your patient is a 9-year-old boy who was sent home from school because his teacher thought he was acting strangely. This morning, he had a seizure and was rushed to the hospital. On physical examination, his temperature is 40°C and he has no nuchal rigidity. A computed tomography (CT) scan is normal. A lumbar puncture is done, and the spinal fluid protein and glucose are normal. A Gram stain of the spinal fluid reveals no organisms and no polys. He is treated with various antibiotics but becomes comatose and dies 2 days later. The blood culture and spinal fluid culture grow no bacteria or fungi. On autopsy of the brain, eosinophilic inclusion bodies are seen in the cytoplasm of neurons.

609. Of the following, which one is the MOST likely cause?

(A) Prions
(B) JC virus
(C) Rabies virus
(D) Herpes simplex virus type 1

CASE 15. Your patient is a 20-year-old man who was in a fistfight and suffered a broken jaw and lost two teeth. Several weeks later, he developed an abscess at the site of the trauma that drained to the surface of the skin, and yellowish granules were seen in the pus.

610. Regarding this disease, which one of the following is MOST accurate?

(A) The causative organism is a gram-positive rod that forms long filaments.
(B) The causative organism is a comma-shaped gram-negative rod that produces an exotoxin which increases cyclic AMP.

(C) The causative organism cannot be seen in the Gram stain but can be seen in an acid-fast stain.
(D) A combination of gram-negative cocci and spirochetes cause this disease.

CASE 16. Your patient is a 25-year-old man who is HIV antibody positive and has a CD4 count of 120 cells (normal, 1000–1500). He has had a mild headache for the past week and vomited once yesterday. On physical examination, he has a temperature of 38°C and mild nuchal rigidity but no papilledema. The rest of the physical examination is negative.

611. Of the following, which one is the MOST likely to be found on examination of the spinal fluid?

(A) Lymphs and gram-positive cocci resembling *Streptococcus pneumoniae*
(B) Lymphs and budding yeasts resembling *Cryptococcus neoformans*
(C) Polys and anaerobic gram-negative rods resembling *Bacteroides fragilis*
(D) Polys and septate hyphae resembling *Aspergillus fumigatus*

CASE 17. Your patient is a 25-year-old woman with a sore throat since yesterday. On physical examination, her throat is red, but no exudate is seen. Two enlarged, tender cervical lymph nodes are palpable. Her temperature is 101°F. A throat culture reveals no β-hemolytic colonies. After receiving this result, you do another physical examination, which reveals an enlarged spleen. A heterophil antibody test finds that sheep red blood cells are agglutinated by the patient's serum.

612. Which one of the following is the MOST likely cause of this disease?

(A) *Streptococcus pyogenes*
(B) *Corynebacterium diphtheriae*
(C) Epstein–Barr virus
(D) Influenza virus

CASE 18. Your patient is a 15-year-old boy with migratory polyarthritis, fever, and a new, loud cardiac murmur. You make a clinical diagnosis of rheumatic fever.

613. Which one of the following laboratory results is MOST likely to be found in this patient?

(A) A blood culture is positive for *Streptococcus pyogenes* at this time.
(B) A throat culture is positive for *Streptococcus pyogenes* at this time.
(C) A Gram stain of the joint fluid shows gram-positive cocci in chains at this time.
(D) An anti-streptolysin O assay is positive at this time.

614. Which one of the following modes of pathogenesis is MOST compatible with a diagnosis of rheumatic fever?

(A) Bacteria attach to joint and heart tissue via pili, invade, and cause inflammation.
(B) Bacteria secrete exotoxins that circulate via the blood to the joints and heart.
(C) Bacterial antigens induce antibodies that cross-react with joint and heart tissue.
(D) Bacterial endotoxin induces interleukin-1 and tumor necrosis factor, which cause inflammation in joint and heart tissue.

615. Which one of the following approaches is MOST likely to prevent endocarditis in patients with rheumatic fever?

(A) They should take the streptococcal polysaccharide vaccine.

(B) They should take penicillin if they have dental surgery.

(C) They should take the toxoid vaccine every 5 years.

(D) They should take rifampin if they have abdominal surgery.

CASE 19. Your patient is a 10-year-old girl who has leukemia and is receiving chemotherapy through an indwelling venous catheter. She now has a fever of 39°C but is otherwise asymptomatic. You do a blood culture, and the laboratory reports growth of *Staphylococcus epidermidis*.

616. Which one of the following results is LEAST likely to be found by the clinical laboratory?

(A) Gram-positive cocci in clusters were seen on Gram stain of the blood culture.

(B) Subculture of the blood culture onto blood agar revealed nonhemolytic colonies.

(C) A coagulase test on the colonies was negative.

(D) A catalase test on the colonies was negative.

CASE 20. Your patient is a 25-year-old woman with several purpuric lesions indicative of bleeding into the skin. Her vital signs are as follows: temperature, 38°C; blood pressure, 70/40; pulse, 140; respiratory rate, 24. You think she has septic shock and do a blood culture.

617. Which one of the following organisms is LEAST likely to be the cause of her septic shock?

(A) *Corynebacterium diphtheriae*

(B) *Neisseria meningitidis*

(C) *Clostridium perfringens*

(D) *Escherichia coli*

618. Of the following mechanisms, which one is LEAST likely to be involved with the pathogenesis of her septic shock?

(A) Increased amount of interleukin-1

(B) Activation of the alternate pathway of complement

(C) Increased amount of tumor necrosis factor

(D) Increased amount of antigen–antibody complexes

CASE 21. Your patient is a 55-year-old man with severe cellulitis of the right leg, high fever, and a teeth-chattering chill. He is a fisherman who was working on his boat in the waters off the Texas coast yesterday.

619. Which one of the following organisms is MOST likely to be the cause of his disease?

(A) *Yersinia pestis*

(B) *Vibrio vulnificus*

(C) *Pasteurella multocida*

(D) *Brucella melitensis*

CASE 22. Your patient is a 30-year-old woman with facial nerve paralysis. She also has fever and headache but does not have a stiff neck. On physical examination, she has a circular, erythematous, macular rash on the back of her thigh. You suspect that she has Lyme disease.

620. Of the following, which one is the MOST appropriate test to order to confirm a diagnosis of Lyme disease?

(A) Blood culture to grow the organism

(B) Stain for inclusion bodies within cells involved in the rash

(C) Test for serum antibody against the organism

(D) Dark-field microscopy

CASE 23. Your patient is a 60-year-old man with confusion for 2 months. He has no history of fever or stiff neck. On physical examination, he was ataxic and his coordination was abnormal. A diagnosis of tertiary syphilis was made by the laboratory.

621. Of the following tests, which one is the MOST appropriate to make a diagnosis of tertiary syphilis?

(A) Spinal fluid culture to grow the organism

(B) Stain for inclusion bodies in the lymphocytes in the spinal fluid

(C) Test for antibody in the spinal fluid that reacts with cardiolipin

(D) ELISA for the antigen in the spinal fluid

CASE 24. Your patient is a 65-year-old man who had an adenocarcinoma of the pancreas that was surgically removed. Several blood transfusions were given, and he did well until 2 weeks later, when fever, vomiting, and diarrhea began. Blood and stool cultures were negative, and the tests for *Clostridium difficile* and hepatitis B surface antigen were negative. A liver biopsy revealed intranuclear inclusion bodies.

622. Of the following, which one is the MOST likely cause?

(A) Adenovirus

(B) Cytomegalovirus

(C) Hepatitis A virus

(D) Rotavirus

CASE 25. Your patient is a 3-year-old girl with fever and pain in her right ear. On physical examination, the drum is found to be perforated, and a bloody exudate is seen. A Gram stain of the exudate reveals gram-positive diplococci.

623. Of the following, which one is the MOST likely cause?

(A) *Streptococcus pyogenes*

(B) *Staphylococcus aureus*

(C) *Corynebacterium diphtheriae*

(D) *Streptococcus pneumoniae*

CASE 26. Your patient is a 70-year-old man with a fever of 40°C and a very painful cellulitis of the right buttock. The skin appears necrotic, and there are several fluid-filled bullae. Crepitus can be felt, indicating gas in the tissue. A Gram stain of the exudate reveals large gram-positive rods.

624. Of the following, which one is the MOST likely cause?

(A) *Clostridium perfringens*

(B) *Bacillus anthracis*

(C) *Corynebacterium diphtheriae*

(D) *Actinomyces israelii*

CASE 27. Your patient is a 45-year-old woman with a cadaveric renal transplant that is being rejected despite immunosuppressive therapy. She is now in renal failure with a blood pH of 7.32. This morning, she awoke with a pain near her right eye. On physical examination, her temperature is 38°C, and the skin near her eye is necrotic. A biopsy specimen of the lesion contains nonseptate hyphae invading the blood vessels.

625. Of the following, which one is the MOST likely cause?

(A) *Histoplasma capsulatum*

(B) *Aspergillus fumigatus*

(C) *Cryptococcus neoformans*

(D) *Mucor* species

CASE 28. Your patient is a 35-year-old man who is HIV antibody positive and has a CD4 count of 85 cells. He recently had a seizure, and a magnetic resonance imaging (MRI) scan indicates a lesion in the temporal lobe. A brain biopsy specimen reveals multinucleated giant cells with intranuclear inclusions.

626. Of the following, which one is the MOST likely cause?

   (A) Herpes simplex virus type 1
   (B) Parvovirus B19
   (C) Coxsackie virus
   (D) Western equine encephalitis virus

CASE 29. Your patient is a 40-year-old woman who had a severe attack of diarrhea that began yesterday on the airplane while she was returning from a vacation in the Middle East. She had multiple episodes of watery, nonbloody diarrhea but no vomiting. She is now asymptomatic. A stool culture reveals only lactose-fermenting colonies on EMB agar.

627. Of the following, which one is the MOST likely cause?

   (A) *Shigella sonnei*
   (B) *Helicobacter pylori*
   (C) *Escherichia coli*
   (D) *Pseudomonas aeruginosa*

CASE 30. Your patient is a 20-year-old man with a sore throat for the past 3 days. On physical examination, his temperature is 38°C, the pharynx is red, and several tender submaxillary nodes are palpable.

628. Of the following, which one is the MOST likely organism to cause this infection?

   (A) *Streptococcus agalactiae* (group B *Streptococcus*)
   (B) *Streptococcus sanguis* (a viridans group *Streptococcus*)
   (C) Parvovirus B19
   (D) Epstein–Barr virus

You do a throat culture, and many small, translucent colonies that are β-hemolytic grow on blood agar. Gram stain of one of these colonies reveals gram-positive cocci in chains.

629. Of the following, which one is the MOST likely organism to cause this infection?

   (A) *Streptococcus pneumoniae*
   (B) *Streptococcus pyogenes*
   (C) *Streptococcus agalactiae* (group B *Streptococcus*)
   (D) *Peptostreptococcus* species

CASE 31. Your patient is a 55-year-old woman with a lymphoma who is receiving chemotherapy via intravenous catheter. She suddenly develops fever, shaking chills, and hypotension.

630. Of the following, which one is the LEAST likely organism to cause this infection?

   (A) *Streptococcus pneumoniae*
   (B) *Klebsiella pneumoniae*
   (C) *Mycoplasma pneumoniae*
   (D) *Proteus mirabilis*

631. If a blood culture drawn from the patient described in case 31 grows a gram-negative rod, which one of the following is the LEAST likely organism to cause this infection?

   (A) *Bordetella pertussis*
   (B) *Escherichia coli*
   (C) *Pseudomonas aeruginosa*
   (D) *Serratia marcescens*

632. Of the following virulence factors, which one is the MOST likely to cause the fever and hypotension seen in the patient described in case 31?

   (A) Pilus
   (B) Capsule
   (C) Lecithinase
   (D) Lipopolysaccharide

CASE 32. Your patient is a 30-year-old woman who was part of a tour group visiting a Central American country. The day before leaving, several members of the group developed fever, abdominal cramps, and bloody diarrhea.

633. Of the following, which one is the LEAST likely organism to cause this infection?

   (A) *Shigella dysenteriae*
   (B) *Salmonella enteritidis*
   (C) *Vibrio cholerae*
   (D) *Campylobacter jejuni*

A stool culture reveals no lactose-negative colonies on the EMB agar.

634. Which one of the following is the MOST likely organism to cause this infection?

   (A) *Shigella dysenteriae*
   (B) *Salmonella enteritidis*
   (C) *Vibrio cholerae*
   (D) *Campylobacter jejuni*

CASE 33. Your patient is a 78-year-old man who had an episode of acute urinary retention and had to be catheterized. He then underwent cystoscopy to determine the cause of the retention. Two days later, he developed fever and suprapubic pain. Urinalysis revealed 50 white blood cells and 10 red blood cells per high-power field. Culture of the urine revealed a thin film of bacterial growth over the entire blood agar plate, and the urease test was positive.

635. Which one of the following is the MOST likely organism to cause this infection?

   (A) *Escherichia coli*
   (B) *Proteus mirabilis*
   (C) *Enterococcus faecalis*
   (D) *Moraxella catarrhalis*

CASE 34. Your patient is a 40-year-old man with a depigmented lesion on his chest that appeared about a month ago. The skin of the lesion is thickened and has lost sensation. He has lived most of his life in rural Louisiana.

636. Of the following tests, which one is the MOST appropriate to do to reveal the cause of this disease?

   (A) Perform a biopsy of the lesion and do an acid-fast stain
   (B) Culture on Sabouraud's agar and look for germ tubes
   (C) Culture on blood agar anaerobically and do a Gram stain
   (D) Obtain serum for a Weil-Felix agglutination test

CASE 35. Your patient is a 28-year-old man with third-degree burns over a large area of his back and left leg. This morning, he spiked a fever to 40°C and had two teeth-chattering chills. A blood culture grows a gram-negative rod that is oxidase-positive and produces a blue-green pigment.

637. Of the following, which one is the MOST likely organism to cause this infection?

    (A) *Prevotella melaninogenica*

    (B) *Pseudomonas aeruginosa*

    (C) *Proteus mirabilis*

    (D) *Haemophilus influenzae*

CASE 36. Your patient is a 32-year-old moving-van driver who lives in St. Louis. He arrived in San Francisco about 10 days ago after picking up furniture in Little Rock, Dallas, Albuquerque, and Phoenix. He now has a persistent cough and fever to 101°F, and he feels poorly. On physical examination, crackles are heard in the left lower lobe, and chest X-ray reveals an infiltrate in that area.

638. Of the following, which one is the LEAST accurate statement?

    (A) He probably has spherules containing endospores in his lung.

    (B) If dissemination to the bone occurs, this indicates a failure of his cell-mediated immunity.

    (C) He probably acquired this disease by inhaling arthrospores.

    (D) The causative organism of this disease exists as a yeast in the soil.

CASE 37. Your patient is a 25-year-old man with an ulcerated lesion on his penis that is not painful. You suspect that it may be a chancre.

639. Which one of the following tests is the MOST appropriate to do with the material from the lesion?

    (A) Dark-field microscopy

    (B) Gram stain

    (C) Acid-fast stain

    (D) Culture on Thayer-Martin agar

640. Which one of the following tests is the MOST appropriate to do with the patient's blood?

    (A) Culture on blood agar

    (B) Assay for antibodies that react with cardiolipin

    (C) Assay for neutralizing antibody in human cell culture

    (D) Heterophil antibody test

CASE 38. Your patient is a 6-year-old boy with papular and pustular skin lesions on his face. A serous, "honey-colored" fluid exudes from the lesions. You suspect impetigo. A Gram stain of the pus reveals many neutrophils and gram-positive cocci in chains.

641. If you cultured the pus on blood agar, which one of the following would you be MOST likely to see?

    (A) Small β-hemolytic colonies containing bacteria that are bacitracin-sensitive

    (B) Small α-hemolytic colonies containing bacteria that are resistant to optochin

    (C) Large nonhemolytic colonies containing bacteria that are oxidase-positive

    (D) Small nonhemolytic colonies containing bacteria that grow in 6.5% NaCl

CASE 39. Your patient is a 66-year-old woman being treated with chemotherapy for lymphoma. She develops fever to 38°C and a nonproductive cough. A chest X-ray reveals an infiltrate. You treat her empirically with an appropriate antibiotic. The following day, several vesicles appear on her chest.

642. Which one of the following viruses is the MOST likely cause of her disease?

    (A) Measles virus

    (B) Respiratory syncytial virus

    (C) Varicella-zoster virus

    (D) Rubella virus

CASE 40. Your patient is a 40-year-old woman with systemic lupus erythematosus who is being treated with high-dose prednisone during a flare of her disease. She develops a fever to 38°C and a cough productive of a small amount of greenish sputum. On physical examination, you hear coarse breath sounds in the left lower lobe. Chest X-ray reveals an infiltrate in that region. Gram stain of the sputum reveals long filaments of gram-positive rods.

643. Which one of the following organisms is the MOST likely cause of this disease?

    (A) *Mycobacterium kansasii*

    (B) *Listeria monocytogenes*

    (C) *Nocardia asteroides*

    (D) *Mycoplasma pneumoniae*

CASE 41. Your patient is a 10-year-old girl with acute leukemia who responded well to her first round of chemotherapy but not to the most recent one. In view of this, she had a bone marrow transplant and is on an immunosuppressive regimen. She is markedly granulocytopenic. Ten days after the transplant, she spikes a fever and coughs up bloody, purulent sputum. Chest X-ray shows pneumonia. A wet mount of the sputum shows septate hyphae with dichotomous (Y-shape) branching.

644. Which one of the following organisms is the MOST likely cause of this disease?

    (A) *Histoplasma capsulatum*

    (B) *Aspergillus fumigatus*

    (C) *Rhizopus nigricans*

    (D) *Candida albicans*

CASE 42. Your patient is a 30-year-old man with acute onset of fever to 40°C and a swollen, very tender right femoral node. His blood pressure is 90/50, and his pulse is 110. As you examine him, he has a teeth-chattering shaking chill. He returned from a camping trip in the Southern California desert 2 days ago.

645. Regarding this disease, which one of the following is MOST accurate?

    (A) An aspirate of the node will reveal a small gram-negative rod with bipolar staining (appears like a "safety pin").

    (B) The organism was probably acquired by eating food contaminated with rodent excrement.

    (C) The aspirate of the node should be cultured on Löwenstein-Jensen agar and an acid-fast stain performed.

    (D) The organism causes disease primarily in people with impaired cell-mediated immunity.

CASE 43. Your patient is a 62-year-old woman with a history of carcinoma of the sigmoid colon that was removed 5 days ago. The surgery was complicated by the escape of bowel contents into the peritoneal cavity. She now has fever and pain in the perineum and left buttock. On physical examination, her temperature is 39°C, and myonecrosis with a foul-smelling discharge is found. A Gram stain of the exudate reveals gram-negative rods.

646. Of the following, which one is the MOST likely organism to cause this infection?

    (A) *Helicobacter pylori*

    (B) *Bacteroides fragilis*

(C) *Salmonella typhi*
(D) *Vibrio parahaemolyticus*

CASE 44. Your patient is an 18-year-old woman with a swollen left ankle. Two days ago, when the ankle began to swell, she thought she had twisted it playing soccer. However, today she has a fever to 38°C, and the ankle has become noticeably more swollen, warm, and red. Her other joints are asymptomatic. You aspirate fluid from the joint.

647. Using the joint fluid, which one of the following procedures is MOST likely to provide diagnostic information?

   (A) Acid-fast stain and culture on Löwenstein-Jensen medium
   (B) Gram stain and culture on chocolate agar
   (C) Dark field microscopy and the VDRL test
   (D) India ink stain and culture on Sabouraud's agar

CASE 45. Your patient is a 6-year-old boy with a history of several episodes of pneumonia. A sweat test revealed an increased amount of chloride, indicating that he has cystic fibrosis. He now has a fever and is coughing up a thick, greenish sputum. A Gram stain of the sputum reveals gram-negative rods.

648. Of the following, which one is the MOST likely organism to cause this infection?

   (A) *Pseudomonas aeruginosa*
   (B) *Haemophilus influenzae*
   (C) *Legionella pneumophila*
   (D) *Bordetella pertussis*

CASE 46. Your patient is a 7-year-old boy with fever, two episodes of vomiting, and a severe headache that began this morning. He has no diarrhea. On physical examination, his temperature is 39°C, and nuchal rigidity is found. Examination of the spinal fluid revealed a white cell count of 800, of which 90% were lymphs, and a normal concentration of both protein and glucose. A Gram stain of the spinal fluid revealed no bacteria.

649. Of the following, which one is the MOST likely to cause this infection?

   (A) *Chlamydia trachomatis*
   (B) *Mycobacterium avium-intracellulare*
   (C) Coxsackie virus
   (D) Adenovirus

CASE 47. Your patient is a 22-year-old man who has been on a low-budget trip to India, where he ate many of the local foods. He has had a low-grade fever, anorexia, and mild abdominal pain for about a month. You suspect that he may have typhoid fever.

650. If he does have typhoid fever, which one of the following is the LEAST likely laboratory finding?

   (A) Culture of the blood reveals gram-negative rods.
   (B) Culture of the stool grows lactose-negative colonies in EMB agar.
   (C) His serum contains antibodies that agglutinate *Salmonella typhi*.
   (D) His serum contains antibodies that cause a positive Weil-Felix reaction.

CASE 48. Your patient is a 30-year-old man who is HIV antibody positive and has had two episodes of *Pneumocystis pneumonia*. He now complains of pain in his mouth and difficulty swallowing. On physical examination, you find several whitish plaques on his oropharyngeal mucosa.

651. Regarding the most likely causative organism, which one of the following statements is MOST accurate?

   (A) It is a filamentous gram-positive rod that is part of the normal flora in the mouth.
   (B) It is an anaerobic gram-negative rod that is part of the normal flora in the colon.
   (C) It is a yeast that forms pseudohyphae when it invades tissue.
   (D) It is a spirochete that grows only in cell culture.

CASE 49. Your patient is a 20-year-old woman with a rash that began this morning. She has been feeling feverish and anorexic for the past few days. On physical examination, there is a papular rash bilaterally over the chest, abdomen, and upper extremities including the hands. There are no vesicles and no petechiae. Cervical and axillary lymph nodes were palpable. Her temperature was 38°C. White blood count was 9000 with a normal differential.

652. Of the following organisms, which one is the MOST likely cause of her disease?

   (A) *Histoplasma capsulatum*
   (B) *Coxiella burnetii*
   (C) *Neisseria meningitidis*
   (D) *Treponema pallidum*

CASE 50. Your patient is a 10-year-old boy who fell, abraded the skin of his thigh, and developed cellulitis (i.e., the skin was red, hot, and tender). Several days later, the infection was treated with a topical antibiotic ointment, and the cellulitis gradually healed. However, 2 weeks later, he told his mother that his urine was cloudy and reddish, and she noted that his face was swollen. You suspect acute glomerulonephritis.

653. Regarding the causative organism, what is the MOST likely appearance of a Gram stain of the exudate from the skin infection?

   (A) Gram-positive cocci in grape-like clusters
   (B) Gram-positive cocci in chains
   (C) Gram-positive diplococci
   (D) Gram-negative diplococci

654. What is the pathogenesis of the cloudy urine and facial swelling?

   (A) Toxin-mediated
   (B) Direct invasion by the bacteria
   (C) Immune complex-mediated
   (D) Cell-mediated immunity (delayed hypersensitivity)

Answers (Questions 594–654)

594. **(B)**	607. **(C)**	620. **(C)**	633. **(C)**	646. **(B)**
595. **(C)**	608. **(A)**	621. **(C)**	634. **(D)**	647. **(B)**
596. **(D)**	609. **(C)**	622. **(B)**	635. **(B)**	648. **(A)**
597. **(B)**	610. **(A)**	623. **(D)**	636. **(A)**	649. **(C)**
598. **(D)**	611. **(B)**	624. **(A)**	637. **(B)**	650. **(D)**
599. **(A)**	612. **(C)**	625. **(D)**	638. **(D)**	651. **(C)**
600. **(D)**	613. **(D)**	626. **(A)**	639. **(A)**	652. **(D)**
601. **(D)**	614. **(C)**	627. **(C)**	640. **(B)**	653. **(B)**
602. **(A)**	615. **(B)**	628. **(D)**	641. **(A)**	654. **(C)**
603. **(C)**	616. **(D)**	629. **(B)**	642. **(C)**	
604. **(A)**	617. **(A)**	630. **(C)**	643. **(C)**	
605. **(B)**	618. **(D)**	631. **(A)**	644. **(B)**	
606. **(C)**	619. **(B)**	632. **(D)**	645. **(A)**	

PART XIV

USMLE (NATIONAL BOARD) PRACTICE EXAMINATION

This practice examination consists of two blocks, each containing 40 microbiology and immunology questions. You should be able to complete each block in 50 minutes. The proportion of the questions devoted to bacteriology, virology, mycology, parasitology, and immunology is approximately that of the USMLE. As in the USMLE, the questions are randomly assorted (i.e., they are not grouped according to subject matter).

All of the questions have between 4 and 10 answer choices. Each question has a single "BEST" answer; there are no "EXCEPT" type questions. The answer choices are listed either in alphabetical order or in order of the length of the answer. The answer key is located at the end of each block.

QUESTIONS

BLOCK ONE

Directions (Questions 1–40)—Select the ONE lettered answer that is BEST in each question.

1. A 9-year-old girl was playing soccer when she began to limp. She has a pain in her leg and points to her upper thigh when asked where it hurts. Her temperature is 101°F. X-ray of the femur reveals that the periosteum is eroded. You order a blood culture. Which one of the following would be the MOST likely blood culture findings?

   (A) Gram-negative rods that grow on EMB agar, forming purple colonies and a green sheen
   (B) Gram-positive cocci that grow on blood agar, causing a clear zone of hemolysis, and are coagulase-positive
   (C) Gram-positive rods that grow only anaerobically and form a double zone of hemolysis on blood agar
   (D) Gram-negative diplococci that grow on chocolate agar, are oxidase-positive, and ferment maltose
   (E) Gram-positive cocci that grow on blood agar, causing a green zone of hemolysis, and are not inhibited by optochin and bile

2. Your summer research project is to study the viruses that cause upper respiratory tract infections. You have isolated a virus from a patient's throat and find that its genome is RNA. Furthermore, you find that the genome is the complement of viral mRNA within the infected cell. Of the following, which one is the MOST appropriate conclusion you could draw?

   (A) The virion contains a polymerase.
   (B) The purified genome RNA is infectious.
   (C) The genome RNA is segmented.
   (D) A single-stranded DNA is synthesized during replication.
   (E) The genome RNA encodes a precursor polypeptide that must be cleaved by a protease.

3. A 25-year-old man has a history of four episodes of boils in the last year. Boils are abscesses caused by *Staphylococcus aureus*. Which one of the following is MOST likely to be the underlying immunologic factor that predisposes him to multiple episodes of boils?

   (A) A deficient amount of the C8 component of complement in his plasma
   (B) An inability of his macrophages to present antigen in association with class I MHC proteins
   (C) A failure to release granzymes from his cytotoxic T cells
   (D) An insufficient amount of IgG in his plasma

4. You are reading an article that says that otitis media is commonly caused by nonencapsulated strains of *Haemophilus influenzae*. You are surprised that nonencapsulated strains can cause this disease. Which one of the following BEST explains why your surprise is justified?

   (A) Nonencapsulated strains would not have endotoxin.
   (B) Nonencapsulated strains cannot secrete exotoxin A.
   (C) Nonencapsulated strains should be easily phagocytized.
   (D) Nonencapsulated strains should be rapidly killed by ultraviolet light.
   (E) Nonencapsulated strains should be susceptible to killing by cytotoxic T cells.

5. A 35-year-old man is HIV antibody positive and has a CD4 count of 50/mL (normal, 1000–1500). He has had a fever of 101°F for a few weeks and "feels tired all the time." He has no other symptoms, and findings on physical examination are normal. Complete blood cell count, urinalysis, and chest X-ray are normal. Blood, stool, and urine cultures show no growth. A bone marrow biopsy reveals granulomas, and a culture grows an organism that is a budding yeast at 37°C but produces hyphae at 25°C. Of the following, which one is the MOST likely cause?

   (A) *Aspergillus fumigatus*
   (B) *Cryptococcus neoformans*
   (C) *Mucor* species
   (D) *Histoplasma capsulatum*
   (E) *Coccidioides immitis*

6. A 70-year-old woman has sustained third-degree burns over a significant area of her body. Despite appropriate burn care in the hospital, she spiked a fever to 39°C, and the nurse reports blue-green pus on the dressing covering the burned area. Gram stain of the pus reveals gram-negative rods, and antibiotic sensitivity tests show resistance to most antibiotics. Which one of the following organisms is MOST likely to cause this disease?

   (A) *Nocardia asteroides*
   (B) *Vibrio vulnificus*
   (C) *Bacteroides fragilis*
   (D) *Haemophilus influenzae*
   (E) *Pseudomonas aeruginosa*

7. A 20-year-old woman has had several episodes of high fever, shaking chills, and a severe headache. She has a hematocrit of 30%. She has recently returned from Africa, where she was a Peace Corps volunteer. Which one of the following is MOST likely to be seen in the blood smear sample from this patient?

   (A) Acid-fast rods
   (B) Banana-shaped gametocytes
   (C) Nonseptate hyphae
   (D) Spherules
   (E) Tachyzoites

8. Certain microorganisms, such as the protozoan *Trypanosoma* and the bacterium *Neisseria gonorrhoeae,* can change their surface antigens quite frequently. This allows the organisms to evade our host defenses. Which one of the following BEST explains how this frequent change in antigenicity occurs?

   (A) It is due to the transposition of existing genes into an active expression site.
   (B) It is due to the acquisition of new fertility plasmids by transduction.
   (C) It is due to conjugation, during which the recipient obtains new chromosomal genes.
   (D) It is due to new mutations that occur at "hot spots" in the genome.

9. A 60-year-old woman had an adenocarcinoma of the colon that was surgically removed. Several blood transfusions were given, and she did well until 3 weeks after surgery, when fever, vomiting, and diarrhea began. Blood and stool cultures were negative for bacteria, and the tests for *Clostridium difficile* and hepatitis B surface antigen were negative. A liver biopsy revealed intranuclear inclusion bodies. Which one of the following is the MOST likely cause?

   (A) Cytomegalovirus
   (B) Dengue virus
   (C) Hepatitis A virus
   (D) Rotavirus
   (E) Yellow fever virus

10. Which one of the immunoglobulins BEST fits the following description: It is found in plasma as a dimer with a J chain. As it passes through mucosal cells, it acquires a secretory piece that protects it from degradation by proteases.

    (A) IgM
    (B) IgG
    (C) IgA
    (D) IgD
    (E) IgE

11. *Mycobacterium tuberculosis* (MTB) and *Mycobacterium avium-complex* (MAC) are important causes of disease, especially in immunocompromised patients. (MAC is also known as *Mycobacterium avium-intracellulare.*) Regarding MTB and MAC, which one of the following statements is the MOST accurate?

    (A) Cell-mediated immunity is the most important host defense mechanism against MTB, whereas antibody-mediated immunity is the most important host defense mechanism against MAC.
    (B) In the clinical laboratory, MAC can be distinguished from MTB by the fact that MAC forms colonies in 7 days, whereas MTB does not.
    (C) Multidrug-resistant strains of MAC are much less common than multidrug-resistant strains of MTB.
    (D) MAC is found in the environment and is not transmitted from person to person, whereas MTB is found in humans and is transmitted from person to person.

12. In the laboratory, a virologist was studying the properties of mutant viruses. When she infected cells with mutant virus #1, no progeny viruses were produced. When she infected cells with mutant virus #2, no progeny viruses were produced. But when she infected cells with both mutant virus #1 and mutant virus #2, progeny viruses of both virus #1 and virus #2 were produced. Which one of the following is the term that BEST describes this phenomenon?

    (A) Phenotypic mixing
    (B) Complementation
    (C) Reassortment
    (D) Recombination

13. Your patient has been treated for endocarditis with penicillin G for the past 2 weeks. She now has a fever and maculopapular erythematous rash over her chest and abdomen. A urinalysis shows significant protein in the urine. If the fever, rash, and proteinuria are immunologic in origin, which one of the following is MOST likely to be involved?

    (A) IgG and complement
    (B) IgE and histamine
    (C) IL-2 and cytotoxic T cells
    (D) Gamma interferon and macrophages

14. Endotoxin is an important underlying cause of septic shock and death, especially in hospitalized patients. Regarding endotoxin, which one of the following is the MOST accurate?

    (A) It acts by phosphorylating the G stimulating protein.
    (B) It is a polypeptide with an A-B subunit configuration.
    (C) It induces the synthesis of tumor necrosis factor.
    (D) It is found primarily in gram-positive rods.
    (E) It can be treated with formaldehyde to form an effective toxoid vaccine.

15. A 12-year-old girl had a seizure this morning and was rushed to the hospital. On examination, her temperature was 40°C, and she had no nuchal rigidity. Computed tomography (CT) scan revealed no abnormality. A spinal tap was done, and the protein and glucose were normal. Gram stain of the spinal fluid showed no organisms and no polys. She was treated with various antibiotics but became comatose and died 2 days later. The routine blood culture and spinal fluid culture grew no organism. On autopsy of the brain, eosinophilic inclusion bodies were seen in the cytoplasm of neurons. Of the following, which one is the MOST likely cause?

    (A) Prions
    (B) JC virus
    (C) Rabies virus
    (D) Parvovirus B19
    (E) Herpes simplex virus type 1

16. A 70-year-old woman presents with rapid onset of fever to 39°C and a cough productive of greenish sputum. She is not hospitalized and not immunocompromised. A chest X-ray reveals a left lower lobe infiltrate. Of the following, which set of findings describes the MOST likely causative organism found in the sputum culture?

    (A) Gram-positive diplococci that form an α-hemolytic colony
    (B) Gram-negative diplococci that form an oxidase-positive colony
    (C) Gram-positive rods that form a β-hemolytic colony
    (D) Gram-negative rods that form an oxidase-positive colony
    (E) Gram-negative cocci that grow only anaerobically

17. Regarding the function of chemokines in host defenses, which one of the following is the MOST accurate?

    (A) Chemokines bind to the T-cell receptor outside of the antigen-binding site and activate many T cells.
    (B) Chemokines induce gene switching in B cells that increases the amount of IgE synthesized, thereby predisposing to allergies.
    (C) Chemokines penetrate the membranes of target cells during attack by cytotoxic T cells.
    (D) Chemokines attract neutrophils to the site of bacterial infection, thereby playing a role in the inflammatory response.

18. Which one of the following answer choices consists of bacteria, BOTH of which produce exotoxins that act by ADP-ribosylation?

    (A) *Salmonella typhi* and *Vibrio cholerae*
    (B) *Vibrio cholerae* and *Corynebacterium diphtheriae*
    (C) *Salmonella typhi* and *Clostridium perfringens*
    (D) *Corynebacterium diphtheriae* and *Staphylococcus aureus*
    (E) *Clostridium perfringens* and *Streptococcus pyogenes*

19. Regarding hepatitis C virus (HCV) and hepatitis D virus (HDV), which one of the following is MOST accurate?

    (A) HCV is transmitted by blood, but HDV is not.
    (B) More than half of HCV infections result in a chronic carrier state.
    (C) There is an effective vaccine against HCV but not against HDV.
    (D) Both HCV and HDV are defective RNA viruses and require concurrent HBV infection to replicate.

20. Which one of the following is MOST likely to induce an IgM antibody response without the participation of helper T cells?

    (A) Bacterial capsular polysaccharide
    (B) Toxic shock syndrome toxin
    (C) Penicillin–bovine serum albumin (BSA) complex
    (D) Tetanus toxoid

21. A 25-year-old pregnant woman in the third trimester comes to the emergency room saying that about 12 hours ago she began to feel feverish and weak. On examination, she has a temperature of 40°C but no other pertinent findings. A blood culture grows small gram-positive rods that cause β-hemolysis on a blood agar plate incubated in room air. Which one of the following bacteria is the MOST likely cause?

    (A) *Clostridium perfringens*
    (B) *Streptococcus pyogenes*
    (C) *Bacillus cereus*
    (D) *Listeria monocytogenes*
    (E) *Brucella abortus*

22. Regarding the mode of action of antiviral drugs, which one of the following is MOST accurate?

    (A) Oseltamivir inhibits influenza A virus by inhibiting the RNA polymerase carried by the virion.
    (B) Ganciclovir inhibits varicella-zoster virus by inhibiting the RNA polymerase carried by the virion.
    (C) Acyclovir action is greater in herpesvirus-infected cells than in uninfected cells because herpesvirus-infected cells contain an enzyme that phosphorylates acyclovir very efficiently.
    (D) Emtricitabine inhibits human immunodeficiency virus (HIV) by inhibiting viral mRNA synthesis more efficiently than cellular mRNA synthesis.
    (E) Tenofovir blocks HIV replication by inhibiting the protease required for the envelope protein gp120 to bind to the CD8 protein on the surface of the T cell.

23. Which one of the following diseases is MOST likely to be caused by a delayed hypersensitivity reaction?

    (A) Serum sickness
    (B) Poststreptococcal glomerulonephritis
    (C) Systemic lupus erythematosus
    (D) Hemolytic disease of the newborn
    (E) Contact dermatitis

24. Members of the genus *Mycobacterium* stain better with the acid-fast stain than with the Gram stain. Which one of the following is the BEST explanation for this finding?

    (A) They lack a cell wall; therefore, they cannot adsorb the crystal violet.
    (B) They have a very thin cell wall that does not retain the crystal violet.
    (C) They have a thick polysaccharide capsule that prevents entry of the iodine solution.
    (D) They have a large amount of lipid in their cell wall that prevents entry of the crystal violet.

25. A 50-year-old man with a cadaveric renal transplant is rejecting the transplant despite immunosuppressive drugs. He is now in renal failure with a blood pH of 7.31. Yesterday, he developed a pain near his left eye that has become progressively more severe. On examination, his temperature is 37.5°C, and the skin near his eye is swollen and necrotic. Microscopic examination of a biopsy of the lesion reveals nonseptate hyphae with right-angle branching. Which one of the following organisms is the MOST likely cause?

    (A) *Candida albicans*
    (B) *Coccidioides immitis*
    (C) *Cryptococcus neoformans*
    (D) *Histoplasma capsulatum*
    (E) *Mucor* species

26. A 60-year-old woman had surgery for ovarian carcinoma 4 days ago and has an indwelling urinary catheter in place. She now spikes a fever to 39°C and has cloudy urine in the collection bottle. Gram stain of the urine shows many polys and gram-positive cocci in chains. Which one of the following would be the MOST likely finding in the urine culture?

    (A) α-Hemolytic colonies on the blood agar plate that are optochin-sensitive
    (B) β-Hemolytic colonies on the blood agar plate that are bacitracin-sensitive
    (C) β-Hemolytic colonies on the blood agar plate that hydrolyze hippurate
    (D) Nonhemolytic colonies on the blood agar plate that grow in 6.5% sodium chloride

27. Your patient is a 40-year-old man with a history of confusion for the past 2 days and a grand mal seizure that occurred this morning. He is HIV antibody positive and has a CD4 count of 100/mL. On examination, his temperature is 37.5°C, and the findings of the remainder of the examination are within normal limits. Magnetic resonance imaging (MRI) reveals several "ring-enhancing" cavitary brain lesions. He has not traveled outside of the United States, is employed as the manager of a supermarket, is a strict vegetarian, and has several household pets, namely, a dog, a cat, a parrot, and a turtle. Which one of the following organisms is the MOST likely cause?

    (A) *Toxocara canis*
    (B) *Toxoplasma gondii*
    (C) *Taenia saginata*
    (D) *Trichinella spiralis*
    (E) *Trypanosoma cruzi*

28. The emergence of antibiotic-resistant bacteria, especially in enteric gram-negative rods, is an extremely important phenomenon. The acquisition of resistance most commonly occurs by a process that involves a sex pilus and the subsequent transfer of plasmids carrying one or more transposons. Which one of the following is the name that BEST describes this process?

    (A) Conjugation
    (B) Combination
    (C) Transformation
    (D) Transduction
    (E) Translocation

29. Regarding the diagnosis, treatment, and prevention of HIV, which one of the following is the MOST accurate?

    (A) The drug zidovudine (AZT) is a "chain terminating" drug; that is, it inhibits the growing polypeptide chain by causing misreading of the viral mRNA.
    (B) The drug lamivudine (3TC) acts by binding to the integrase, which prevents integration of the viral DNA into cellular DNA.
    (C) In the laboratory test for early HIV infection, the patient's serum is tested for both antibody to HIV and for the p24 protein of HIV.
    (D) A major limitation to our ability to produce a vaccine against HIV is that there are many serologic types of the viral p24 protein.

30. Regarding haptens, which one of the following statements is the MOST accurate?

    (A) They are typically polypeptides that are resistant to proteolytic cleavage within the antigen-presenting cell.
    (B) They bind to class II MHC proteins but not to class I MHC proteins.
    (C) They cannot induce antibodies unless they are bound to a carrier protein.
    (D) They activate complement by binding to the Fc part of the heavy chain of IgG.

31. Your patient is a 20-year-old man with a urethral discharge. Gram stain of the pus reveals many neutrophils but no bacteria. Which one of the following organisms is the MOST likely cause?

    (A) *Treponema pallidum*
    (B) *Haemophilus ducreyi*
    (C) *Mycobacterium marinum*
    (D) *Candida albicans*
    (E) *Chlamydia trachomatis*

32. Regarding host defenses against viruses, which one of the following is MOST accurate?

    (A) IgA exerts its main antiviral effect by enhancing the cytopathic effect of natural killer cells—a process called antibody-dependent cellular cytotoxicity.
    (B) IgG plays a major role in neutralizing virus infectivity during the primary infection.
    (C) Complexes of virus and IgE are the cause of the inflammatory arthritis seen in several viral infections, such as hepatitis B and rubella.
    (D) Alpha and beta interferons exert their antiviral action by inducing a ribonuclease that degrades viral mRNA and a protein kinase that inactivates protein synthesis.
    (E) Alpha and beta interferons exert their antiviral effect against viruses with RNA genomes but not against those with DNA genomes.

33. Allergic rhinitis is characterized by sneezing, rhinorrhea, nasal congestion, and itching of the eyes and nose. Persons with allergic rhinitis have "X" that binds to high-affinity receptors on "Y." On reexposure to antigen, the "Y" of patients with allergic rhinitis degranulate, releasing "Z" and other mediators. Which one of the following sets BEST describes X, Y, and Z?

    (A) X is IgE, Y is macrophages, and Z is tumor necrosis factor.
    (B) X is IgE, Y is basophils, and Z is histamine.
    (C) X is IgG, Y is eosinophils, and Z is histamine.
    (D) X is IgG, Y is neutrophils, and Z is tumor necrosis factor.
    (E) X is IgA, Y is eosinophils, and Z is interleukin-5.

34. An outbreak of postsurgical wound infections caused by *Staphylococcus aureus* has occurred. The infection control team was asked to determine whether the organism could be carried by one of the operating room personnel. Using your knowledge of normal flora, which one of the following body sites is the MOST likely location for this organism?

    (A) Colon
    (B) Gingival crevice
    (C) Nose
    (D) Throat
    (E) Vagina

35. A 35-year-old man who is HIV antibody positive and has a CD4 count of 30 says, "I can't remember the simplest things." You are concerned about dementia. An MRI indicates several widely scattered lesions in the brain. Over the next 4 months, he develops visual field defects, becomes paralyzed, and dies. Autopsy reveals that many neurons of the brain have lost myelin and contain intranuclear inclusions. Electron microscopy reveals the inclusions contain nonenveloped viruses. Which one of the following viruses is the MOST likely cause?

    (A) Adenovirus
    (B) Cytomegalovirus
    (C) Herpes simplex virus
    (D) JC virus
    (E) Coxsackie virus

36. A 75-year-old man with substernal chest pain was found to have angina pectoris caused by syphilitic aortitis that affected his coronary arteries. Of the following, which one is the MOST likely way that the diagnosis of syphilis was made?

    (A) Blood culture
    (B) Culture on Thayer-Martin medium (chocolate agar with antibiotics)

(C) Detecting antibodies to cardiolipin in his blood

(D) Detecting treponemal antigen in his blood

(E) Western blot assay

37. A 22-year-old woman has an erythematous rash on the malar eminences of her face that gets worse when she is out in the sun. She has lost about 10 lb and feels tired much of the time. She took her temperature a few times, and it was 99°F. Physical examination was normal except for the rash. Laboratory tests revealed a hemoglobin of 11 and a white blood cell count of 5500. Urinalysis showed albumin in the urine but no red cells, white cells, or bacteria. Which one of the following is the MOST likely laboratory finding in this disease?

(A) Decreased number of helper (CD4-positive) T cells

(B) High level of antibodies to double-stranded DNA

(C) Increased number of cytotoxic (CD8-positive) T cells

(D) Low level of C1 inhibitor

(E) Low microbicidal activity of neutrophils

38. Regarding antimicrobial drugs that act by inhibiting nucleic acid synthesis in bacteria, which one of the following is the MOST accurate?

(A) Quinolones, such as ciprofloxacin, inhibit the RNA polymerase in bacteria by acting as nucleic acid analogues.

(B) Rifampin inhibits the RNA polymerase in bacteria by binding to the enzyme and inhibiting messenger RNA synthesis.

(C) Sulfonamides inhibit the DNA polymerase in bacteria by causing chain termination of the elongating strand.

(D) Trimethoprim inhibits the DNA polymerase in bacteria by preventing the unwinding of double-stranded DNA.

39. Regarding parvovirus B19, which one of the following is the MOST accurate?

(A) Parvovirus B19 has a double-stranded DNA genome but requires a DNA polymerase in the virion because it replicates in the cytoplasm.

(B) Parvovirus B19 is transmitted primarily by sexual intercourse.

(C) Parvovirus B19 causes severe anemia because it preferentially infects erythrocyte precursors.

(D) Patients infected by parvovirus B19 can be diagnosed in the laboratory using the cold agglutinin test.

(E) Patients with disseminated disease caused by parvovirus B19 should be treated with acyclovir.

40. Which one of the following laboratory tests would be the BEST in order to determine the number of CD4-positive cells in a patient infected with HIV?

(A) Agglutination

(B) Enzyme-linked immunosorbent assay (ELISA)

(C) Flow cytometry

(D) Immunoelectrophoresis

(E) Ouchterlony gel assay

ANSWERS TO BLOCK ONE

1. **(B)**	11. **(D)**	21. **(D)**	31. **(E)**
2. **(A)**	12. **(B)**	22. **(C)**	32. **(D)**
3. **(D)**	13. **(A)**	23. **(E)**	33. **(B)**
4. **(C)**	14. **(C)**	24. **(D)**	34. **(C)**
5. **(D)**	15. **(C)**	25. **(E)**	35. **(D)**
6. **(E)**	16. **(A)**	26. **(D)**	36. **(C)**
7. **(B)**	17. **(D)**	27. **(B)**	37. **(B)**
8. **(A)**	18. **(B)**	28. **(A)**	38. **(B)**
9. **(A)**	19. **(B)**	29. **(C)**	39. **(C)**
10. **(C)**	20. **(A)**	30. **(C)**	40. **(C)**

BLOCK TWO

1. A 4-year-old girl has papular and pustular lesions on her face. The lesions are exuding a honey-colored serous fluid. You make a clinical diagnosis of impetigo. A Gram stain of the exudate reveals gram-positive cocci in chains, and a culture reveals β-hemolytic colonies on blood agar. For which one of the following sequelae is she MOST at risk?

(A) Bloody diarrhea

(B) Blurred vision

(C) Paralysis of the facial nerve (Bell's palsy)

(D) Red blood cells and albumin in her urine

(E) Rusty-colored sputum

2. The purified genome of certain RNA viruses can enter a cell and elicit the production of progeny viruses (i.e., the genome is infectious). Regarding these viruses, which one of the following statements is MOST accurate?

(A) They have a segmented genome.

(B) They have a polymerase in the virion.

(C) Their genome RNA is double-stranded.

(D) They encode a protease that cleaves a precursor polypeptide.

(E) Their genome RNA has the same base sequence as mRNA.

3. A 77-year-old man with enterococcal endocarditis needed to be treated with penicillin G but had a history of a severe penicillin reaction. He was therefore skin tested using penicilloyl-polylysine as the antigen. Which one of the following is MOST likely to occur in a positive skin test?

(A) The antigen forms immune complexes with IgG.

(B) The antigen activates CD4-positive T cells and macrophages.

(C) The antigen activates the alternative pathway of complement.

(D) The antigen activates CD8-positive T cells by binding to class I MHC proteins.

(E) The antigen cross-links IgE on the mast cells and causes the release of histamine.

4. Regarding the Gram stain, which one of the following is the MOST accurate?

    (A) After adding crystal violet and Gram's iodine, both gram-positive bacteria and gram-negative bacteria will appear blue.

    (B) If you forget to stain with the red dye (safranin or basic fuchsin), both gram-positive bacteria and gram-negative bacteria will appear blue.

    (C) If you forget to heat-fix, both gram-positive bacteria and gram-negative bacteria will appear blue.

    (D) One reason why bacteria have a different color in this stain is because the gram-positive bacteria have lipid in their membrane, whereas gram-negative bacteria do not.

5. A 35-year-old man with a CD4 count of 50 presents with a skin nodule on his chest. The nodule is about 3 cm in diameter and is not red, hot, or tender. He says it has been slowly growing bigger for the past 3 weeks. You biopsy the nodule, and the pathologist calls to say that the patient has disseminated cryptococcosis. Which one of the following is the BEST description of what the pathologist saw in the biopsy specimen?

    (A) Spherules

    (B) Nonseptate hyphae

    (C) Germ tubes

    (D) Budding yeasts with a thick capsule

    (E) Septate hyphae with low-angle branching

6. A 22-year-old woman complains of a persistent nonproductive cough and a fever of 101°F that came on slowly over the last 4 days. Physical examination reveals some rales in the left lung base. A patchy infiltrate is seen on chest X-ray. She works as a secretary in a law office and has not traveled recently. She is not immunocompromised and has not been hospitalized recently. Which one of the following organisms is the MOST likely cause of her disease?

    (A) *Aspergillus fumigatus*

    (B) *Schistosoma mansoni*

    (C) *Mycoplasma pneumoniae*

    (D) Zika virus

    (E) *Toxoplasma gondii*

7. The mother of a 4-year-old child notes that her child is sleeping poorly and scratching his anal area. You suspect the child may have pinworms. Which one of the following is the BEST method to make that diagnosis?

    (A) Examine the stool for the presence of cysts

    (B) Examine the stool for the presence of trophozoites

    (C) Examine a blood smear for the presence of microfilaria

    (D) Determine the titer of IgE antibody against the organism

    (E) Examine transparent adhesive tape for the presence of eggs

8. Regarding bacterial spores, which one of the following is the MOST accurate?

    (A) One spore germinates to form one bacterium.

    (B) They are produced primarily within human red blood cells.

    (C) They are killed by boiling at sea level but not at high altitude.

    (D) They are produced by anaerobes only in the presence of oxygen.

    (E) They contain endotoxin, which accounts for their ability to cause disease.

9. A 22-year-old woman had fever of 100°F and anorexia for the past 2 days, and this morning she appears jaundiced. On examination, her liver is enlarged and tender. She has a total bilirubin of 5 mg/dL (normal, <1) and elevated transaminases. She received the complete course of the hepatitis B vaccine 2 years ago but has not had the hepatitis A vaccine. The results of her hepatitis serologies are as follows: HAV-IgM negative, HAV-IgG positive, HBsAg negative, HBsAb positive, HBcAb negative, HCV-Ab positive. Of the following, which one is the MOST accurate?

    (A) She probably has hepatitis A now, probably has not been infected with hepatitis B virus (HBV), and probably had hepatitis C in the past.

    (B) She probably has hepatitis A now, probably has been infected with HBV in the past, and probably had hepatitis C in the past.

    (C) She has been infected with hepatitis A virus (HAV) in the past, probably has not been infected with HBV, and probably has hepatitis C now.

    (D) She has been infected with HAV in the past, probably has hepatitis B now, and probably had hepatitis C in the past.

10. Regarding the function of the different classes of antibodies, which one of the following statements is the MOST accurate?

    (A) IgA acts as an antigen receptor on the surface of B cells.

    (B) IgG activates the alternative pathway of complement, resulting in the production of C3a that degrades the bacterial cell wall.

    (C) IgG binds to the bacterial surface and makes the bacteria more easily ingested by phagocytes.

    (D) IgM defends against worm parasites, such as hookworms.

    (E) IgE blocks the binding of viruses to the gut mucosa.

11. A 6-year-old boy fell and sustained a deep wound from a rusty nail that penetrated his thigh. His mother removed the nail and cleaned the wound with soap and water. The next morning, he had a temperature of 102°F, and his thigh was very painful and swollen. In the emergency room, crepitus (gas in the tissue) was noted. A Gram stain of exudate from the wound area revealed large gram-positive rods. Which one of the following is the MOST likely cause?

    (A) *Actinomyces israelii*

    (B) *Clostridium perfringens*

    (C) *Clostridium tetani*

    (D) *Listeria monocytogenes*

    (E) *Mycobacterium fortuitum-chelonae* complex

    (F) *Nocardia asteroides*

    (G) *Pseudomonas aeruginosa*

12. The two most common types of viral vaccines are killed vaccines and live, attenuated vaccines. Regarding these vaccines, which one of the following statements is the MOST accurate?

    (A) Killed vaccines induce a longer-lasting response than do live, attenuated vaccines.

    (B) Killed vaccines are no longer used in this country because they do not induce secretory IgA.

    (C) Killed vaccines induce a broader range of immune responses than do live, attenuated vaccines.

    (D) Killed vaccines are safer to give to immunocompromised patients than are live, attenuated vaccines.

13. Regarding immediate (type I) and immune complex (type III) hypersensitivities, which one of the following is the MOST accurate?

    (A) IgE is involved in both immediate and immune complex hypersensitivities.

    (B) Complement is involved in both immediate and immune complex hypersensitivities.

    (C) Less antigen is typically needed to trigger an immediate reaction than an immune complex reaction.

    (D) Neutrophils play a more important role in immediate reactions than in immune complex reactions.

14. Disease caused by which one of the following bacteria can be prevented by a toxoid vaccine?

    (A) *Actinomyces israelii*
    (B) *Bacteroides fragilis*
    (C) *Borrelia burgdorferi*
    (D) *Corynebacterium diphtheriae*
    (E) *Haemophilus influenzae*
    (F) *Listeria monocytogenes*
    (G) *Neisseria meningitidis*
    (H) *Salmonella typhi*
    (I) *Streptococcus pneumoniae*
    (J) *Yersinia pestis*

15. A 50-year-old woman has had a gradual onset of headaches that have become increasingly more severe during the past 3 weeks. On examination, she is confused regarding time, place, and person, and she is febrile to 39°C. Her spinal fluid reveals a normal glucose, normal protein, and 17 cells, all of which were lymphocytes. Gram stain of the spinal fluid shows no organism. An MRI reveals a 2-cm radiolucent lesion in the temporal lobe. A biopsy of the brain lesion was performed. A Giemsa stain of the tissue shows multinucleated giant cells with intranuclear inclusion bodies. Which one of the following is the MOST likely causative organism?

    (A) Adenovirus
    (B) Coxsackie virus
    (C) Cytomegalovirus
    (D) Herpes simplex virus type 1
    (E) Influenza virus
    (F) Measles virus
    (G) Parvovirus B19
    (H) Poliovirus
    (I) Prion
    (J) Rabies virus

16. An 80-year-old man had a carcinoma of the colon removed 3 days ago. He was doing well until this morning, when he spiked a fever to 39°C and complained of severe abdominal pain. Examination revealed a "board-like" abdomen indicative of peritonitis. He was taken to the operating room, where it was discovered that his anastomosis had broken down and bowel contents had spilled into the peritoneal cavity. A foul-smelling exudate was observed. A Gram stain of the peritoneal exudate revealed many gram-negative rods. Which one of the following sets of bacteria is the MOST likely cause of this infection?

    (A) *Escherichia coli* and *Brucella melitensis*
    (B) *Enterobacter cloacae* and *Salmonella enteritidis*
    (C) *Fusobacterium nucleatum* and *Bacteroides fragilis*
    (D) *Haemophilus influenzae* and *Actinomyces israelii*
    (E) *Shigella dysenteriae* and *Serratia marcescens*

17. Regarding the primary and secondary antibody responses, which one of the following statements is MOST accurate?

    (A) The IgM made in the primary response is made primarily by memory B cells.
    (B) The lag phase is shorter in the response than in the secondary response.
    (C) In the primary response, memory B cells are produced, but memory T cells are not.
    (D) Antigen must be processed and presented in the primary response but not in the secondary response.
    (E) The amount of IgG made in the secondary response is greater than the amount made in the primary response.

18. A 70-year-old man who is receiving chemotherapy for leukemia develops a fever to 40°C and has two episodes of teeth-chattering chills, and his blood pressure drops to 80/20 mmHg. Of the following factors, which one is MOST likely to be the cause of his fever, chills, and hypotension?

    (A) Coagulase
    (B) Dipicolinic acid
    (C) Glycocalyx
    (D) Lipid A
    (E) Mycolic acid
    (F) Pili
    (G) Polysaccharide capsule

19. A 22-year-old woman presents with "the worst sore throat I've ever had." She also complains of fatigue and anorexia. She is not immunocompromised and has not been hospitalized recently. On examination, she is febrile to 38°C, the pharynx is inflamed, and there are a few tender cervical nodes bilaterally. There are no white lesions on the tongue or pharynx. A throat culture grows α-hemolytic colonies on blood agar that are optochin-resistant. Of the following, which one is the MOST likely cause?

    (A) *Candida albicans*
    (B) Epstein–Barr virus
    (C) Parvovirus B19
    (D) *Pneumocystis carinii*
    (E) Poliovirus
    (F) *Serratia marcescens*
    (G) *Streptococcus mutans*
    (H) *Streptococcus pneumoniae*
    (I) *Streptococcus pyogenes*
    (J) *Strongyloides stercoralis*

20. Regarding the complement pathway, which one of the following is MOST accurate?

    (A) C5a mediates chemotaxis and attracts neutrophils to the site of infection.
    (B) C5b plays an important role in the opsonization of gram-negative bacteria.
    (C) C3a is a decay-accelerating factor that causes the rapid decay and death of bacteria.
    (D) C1 binds to the surface of gram-positive bacteria, which initiates the classic pathway.
    (E) The membrane attack complex is produced in the classic pathway but not in the alternative pathway.

21. A 65-year-old woman had symptoms of dementia. An MRI revealed significant cortical atrophy. It was determined that her intraventricular pressure was very high, and a ventriculoperitoneal shunt (from the brain, tunneling under the skin into the peritoneal cavity) was placed to relieve the pressure. Three weeks later, she developed a fever to 38°C, malaise, and anorexia but no other symptoms. Of the following, which one BEST describes the MOST likely organism causing her current symptoms?

    (A) A gram-positive coccus that does not clot plasma
    (B) A curved gram-negative rod that produces urease
    (C) An acid-fast rod that does not grow on bacteriologic media
    (D) An obligate intracellular parasite that forms a cytoplasmic inclusion body
    (E) A spirochete that induces an antibody that agglutinates a lipid from a cow's heart

22. Two mutants of poliovirus, one mutated at gene X and the other mutated at gene Y, have been isolated. If a cell is infected with each mutant alone, no virus is produced. If a cell is infected with both mutants, which one of the following is MOST likely to occur?

    (A) Complementation between the mutant gene products may occur, and, if so, both X and Y progeny viruses will be made.
    (B) Phenotypic mixing may occur, and, if so, both X and Y progeny viruses will be made.
    (C) Reassortment of the genome segments may occur, and, if so, both X and Y progeny viruses will be made.
    (D) The genome may be transcribed into DNA, and, if so, both X and Y viruses will be made.

23. A 40-year-old woman has a history of chronic inflammation of the small joints of the hands bilaterally. You suspect rheumatoid arthritis. Which one of the following statements is the MOST accurate regarding the pathogenesis of this disease?

    (A) It is caused by sensitized CD4-positive T lymphocytes and macrophages invading the joints.
    (B) It is caused by antibody against human IgG-forming immune complexes within the joints.
    (C) It is caused by the release of mediators from mast cells when environmental agents cross-link adjacent IgEs within the joints.
    (D) It is caused by superantigens inducing the release of large amounts of lymphokines from helper T cells within the joints.

24. Listed below are five bacteria paired with a mode of transmission. Which one of the pairings is MOST accurate?

    (A) *Borrelia burgdorferi*—mosquito bite
    (B) *Coxiella burnetii*—bat guano
    (C) *Haemophilus influenzae*—penetrating wound contaminated with soil
    (D) *Rickettsia rickettsii*—contaminated food
    (E) *Yersinia pestis*—flea bite

25. A 70-year-old man with leukemia initially responded to chemotherapy but now is refractory. He therefore underwent a bone marrow transplant and is now receiving large doses of cyclosporine A and prednisone. Three weeks after the transplant, he became febrile to 39°C and began coughing up purulent sputum. A chest X-ray revealed pneumonia. A Gram stain of the sputum did not reveal a predominant organism, but a KOH prep of the sputum revealed septate hyphae with parallel walls and low-angle branching. Of the following organisms, which one is MOST likely to be the cause of this pneumonia?

    (A) *Aspergillus fumigatus*
    (B) *Candida albicans*
    (C) *Coccidioides immitis*
    (D) *Cryptococcus neoformans*
    (E) *Rhizopus nigricans*

26. Your patient is a 20-year-old woman with severe diarrhea that began yesterday. She has just returned from a 3-week trip to Peru, where she ate some raw shellfish at the farewell party. She now has watery diarrhea, perhaps 20 bowel movements a day, and is feeling quite weak and dizzy. Her stool is guaiac-negative, a test that determines whether there is blood in the stool. A Gram stain of the stool reveals curved gram-negative rods. Of the following organisms, which one is MOST likely to be the cause of her diarrhea?

    (A) *Bacteroides fragilis*
    (B) *Campylobacter jejuni*
    (C) *Entamoeba histolytica*

    (D) *Helicobacter pylori*
    (E) *Shigella dysenteriae*
    (F) *Vibrio cholerae*
    (G) *Yersinia enterocolitica*

27. A 50-year-old man has had low-grade, persistent headaches for several months. In the past few days, nausea, vomiting, and blurred vision have occurred. An MRI reveals several cystlike lesions in the brain parenchyma. The patient lived for many years on one of the small Caribbean islands. On the basis of a positive serologic test, a diagnosis of neurocysticercosis was made. Which one of the following is the MOST likely mode by which this disease was acquired?

    (A) Sandfly bite
    (B) Mosquito bite
    (C) Sexual intercourse
    (D) Ingestion of the larvae of the organism in raw fish
    (E) Ingestion of the eggs of the organism in contaminated food
    (F) Penetration of the skin by the organism while walking bare-footed
    (G) Penetration of the skin by the organism while bathing in freshwater

28. A 30-year-old woman with a previous history of rheumatic fever now has a fever for the past 2 weeks. Physical examination reveals a new heart murmur. You suspect endocarditis and do a blood culture, which grows a viridans group *Streptococcus* later identified as *Streptococcus sanguinis*. Of the following body sites, which one is the MOST likely source of this organism?

    (A) Colon
    (B) Mouth
    (C) Skin
    (D) Stomach
    (E) Vagina

29. Regarding poliovirus, which one of the following is MOST accurate?

    (A) Poliovirus remains latent within sensory ganglia, and reactivation occurs primarily in immunocompromised patients.
    (B) When the live, attenuated virus in the oral vaccine replicates, revertant mutants can occur that can cause paralytic polio.
    (C) The widespread use of the killed vaccine in the countries of North and South America has led to the virtual elimination of paralytic polio in those areas.
    (D) The current recommendation is to give the live, attenuated vaccine for the first three immunizations to prevent the child from acting as a reservoir, followed by boosters using the killed vaccine.

30. Regarding ABO and Rh blood types, which one of the following is the MOST accurate?

    (A) People with type O are called universal recipients because they have antibodies against H substance but not against A and B antigens.
    (B) If the father is Rh-positive and the mother is Rh-negative, hemolytic disease of the newborn only occurs when the child is Rh-negative.
    (C) People who are Rh-negative usually have antibodies to the Rh antigen because they are exposed to cross-reacting antigen located on bacteria in the colon.
    (D) If type A blood is transfused into a person with type B blood, complement will be activated, and the membrane attack complex will cause lysis of the type A red cells.

31. A 25-year-old man was in a motorcycle accident 3 days ago, in which he sustained severe head trauma. He has had spinal fluid leaking from his nose since the accident and now develops a severe headache. His temperature is 39°C, and on examination, you find nuchal rigidity. You do a lumbar puncture and find that the spinal fluid is cloudy and contains 5000 WBC/mL, 90% of which are polys. Of the following, which one is the MOST likely result observed in the laboratory analysis of the spinal fluid?

    (A) Gram-negative rods that grew only on Thayer-Martin medium
    (B) A motile spirochete that formed beta-hemolytic colonies on blood agar
    (C) Gram-positive cocci that formed alpha-hemolytic colonies on blood agar
    (D) Gram-positive rods that grew only on chocolate agar supplemented with X and V factors
    (E) No organism was seen using Gram stain, but tissue stains revealed cytoplasmic inclusion bodies

32. Regarding prions and prion-caused diseases, which one of the following is MOST accurate?

    (A) Prions are highly resistant to both ultraviolet light and to boiling but are inactivated by hypochlorite.
    (B) Prions are protein-containing particles surrounded by a lipoprotein envelope with a DNA polymerase in the envelope.
    (C) The diagnosis of prion-caused diseases such as Creutzfeldt-Jakob disease is typically made by observing cytopathic effect in cell culture.
    (D) Creutzfeldt-Jakob disease occurs primarily in children younger than the age of 2 years because they cannot mount an adequate immune response to the prion protein.

33. A 2-year-old boy has had several infections of the sinuses and lungs and is being evaluated to determine whether he has chronic granulomatous disease. Regarding this disease, which one of the following is the MOST accurate?

    (A) There is a deficiency in NADPH oxidase activity.
    (B) The defect is primarily in antigen-presenting cells such as macrophages.
    (C) *Pneumocystis jiroveci* infections are common in patients with this disease.
    (D) The diagnosis is primarily made by ELISA, in which antibody against the affected cell component is detected.

34. Regarding Chlamydiae, which one of the following is MOST accurate?

    (A) They are gram-positive rods that do not form spores.
    (B) They exhibit swarming motility on a blood agar plate.
    (C) Their life cycle consists of a metabolically inactive particle in the extracellular phase.
    (D) They can replicate only within cells because they lack the ability to produce certain essential mRNAs.
    (E) They replicate in the nucleus of infected cells, where they form inclusions that are useful diagnostically.

35. Regarding human papillomavirus (HPV), which one of the following is MOST accurate?

    (A) Blood and blood products are an important mode of transmission of HPV.
    (B) HPV is an enveloped virus with a genome composed of double-stranded RNA.
    (C) Oseltamivir is a chain-terminating drug that inhibits HPV replication by blocking DNA synthesis.
    (D) HPV induces the formation of koilocytes in the skin that are an important diagnostic feature of HPV infection.
    (E) The P2 capsid protein of HPV activates the *c-src* oncogene in human cells, which is the process by which HPV predisposes to malignancy.

36. Regarding Lyme disease, which one of the following is MOST accurate?

    (A) The causative organism is a small gram-positive rod.
    (B) Mice are the main reservoir of the causative organism.
    (C) The Lyme disease vaccine contains toxoid as the immunogen.
    (D) Fleas are the principal mode of transmission of the causative organism.
    (E) The diagnosis in the clinical laboratory is typically made by culturing the organism on chocolate agar.

37. Regarding Bruton's agammaglobulinemia, which one of the following is the MOST accurate?

    (A) VDJ gene switching does not occur.
    (B) There is very little IgG, but IgM and IgA levels are normal.
    (C) The number of B cells is normal, but they cannot differentiate into plasma cells.
    (D) There is a defect in a tyrosine kinase, one of the enzymes in the signal transduction pathway.
    (E) Viral infections are more common in patients with this disease than are pyogenic bacterial infections.

38. A 20-year-old woman presents with a history of vaginal discharge for the past 3 days. On pelvic examination, you see a mucopurulent exudate at the cervical os, and there is tenderness on palpation of the right fallopian tube. You do a Gram stain and culture on the cervical discharge. The culture is done on Thayer-Martin medium, which is a chocolate agar that contains antibiotics that inhibit the growth of normal flora. Of the following, which findings are the MOST likely to be found?

    (A) A Gram stain reveals many neutrophils and spirochetes, and culture on Thayer-Martin medium reveals no colonies.
    (B) A Gram stain reveals many neutrophils and gram-variable rods, and culture on Thayer-Martin medium reveals β-hemolytic colonies.
    (C) A Gram stain reveals many neutrophils and gram-negative diplococci, and culture on Thayer-Martin medium reveals oxidase-positive colonies.
    (D) A Gram stain reveals many neutrophils but no gram-negative diplococci are seen, and culture on Thayer-Martin medium reveals coagulase-positive colonies.

39. Regarding human immunodeficiency virus (HIV), which one of the following is MOST accurate?

    (A) The term *viral load* refers to the concentration of HIV RNA in the patient's blood.
    (B) Both emtricitabine and tenofovir block HIV replication by inhibiting cleavage of the precursor polypeptide by the virion-encoded protease.
    (C) The antigenicity of the GAG protein of HIV is highly variable, which is a significant impediment to the development of a vaccine against HIV.
    (D) Highly active antiretroviral therapy (HAART) consists of a "backbone" of zidovudine and lamivudine plus acyclovir.

40. Regarding Th-1 and Th-2 cells, which one of the following is MOST accurate?

(A) Th-1 cells produce gamma interferon and promote cell-mediated immunity.

(B) Th-2 cells produce interleukin-17, which inhibits the formation of Th-1 cells.

(C) Both Th-1 and Th-2 cells have class II MHC proteins on their outer cell membrane.

(D) Before they differentiate into Th-1 or Th-2 cells, naïve Th cells are double-positives (i.e., they produce both gamma interferon and interleukin-4).

ANSWERS TO BLOCK TWO

1. (D)	11. (B)	21. (A)	31. (C)
2. (E)	12. (D)	22. (A)	32. (A)
3. (E)	13. (C)	23. (B)	33. (A)
4. (A)	14. (D)	24. (E)	34. (C)
5. (D)	15. (D)	25. (A)	35. (D)
6. (C)	16. (C)	26. (F)	36. (B)
7. (E)	17. (E)	27. (E)	37. (D)
8. (A)	18. (D)	28. (B)	38. (C)
9. (C)	19. (B)	29. (B)	39. (A)
10. (C)	20. (A)	30. (D)	40. (A)

Index

Note: Page numbers followed by *f* and *t* indicate figures and tables, respectively; those followed by *b* and *n* indicate boxes and notes, respectively; those followed by *s* indicate summaries; and those in **boldface** indicate main discussions.